French-English Military Techincal Dictionary

A FRENCH-ENGLISH MILITARY TECHNICAL DICTIONARY

By
Cornélis De Witt Willcox

Colonel, United States Army

Professor of Modern Languages,
United States Military Academy

GOVERNMENT REPRINTS PRESS
Washington, D.C.

© Ross & Perry, Inc. 2001 All rights reserved.

No claim to U.S. government work contained throughout this book.

Protected under the Berne Convention. Published 2001

Printed in The United States of America
Ross & Perry, Inc. Publishers
717 Second St., N.E., Suite 200
Washington, D.C. 20002
Telephone (202) 675-8300
Facsimile (202) 675-8400
info@RossPerry.com

SAN 253-8555

Government Reprints Press Edition 2001

Government Reprints Press is an Imprint of Ross & Perry, Inc.

Library of Congress Control Number: 2001092400

http://www.GPOreprints.com

ISBN 1-931641-17-X

♾ The paper used in this publication meets the requirements for permanence established by the American National Standard for Information Sciences "Permanence of Paper for Printed Library Materials" (ANSI Z39.48-1984).

All rights reserved. No copyrighted part of this publication may be reproduced, stored in a retrieval system, or transmitted, in any form or by any means, electronic, photocopying, recording, or otherwise, without the prior written permission of the publisher.

PREFACE TO THE 1917 EDITION.

The following pages contain the Dictionary as originally published by the War Department, with a supplement containing the more important recent military and technical terms.

It had been my intention, whenever time should serve, to incorporate these terms, as well as thousands of others, in the main work, so as to make a new book of it; but such a book would have called for at least two years' steady labor. When, therefore, The Adjutant General asked me if I could furnish any additions to the original text, the only possible course was to collect them into a supplement.

C. De W. W.

West Point, N. Y.,
    July 26, 1917.

PREFACE.

In spite of the events of 1870–71, perhaps because of them, the French language still holds its own as one whose military literature is of deep interest to the student of military progress. Apart from professional works of purely French origin, the stake of the French people in their army is so great that, sooner or later, every technical advance, every military achievement, whether afloat or ashore, is laid under contribution, if not to furnish its share of possible improvement to the French service, at least to satisfy the French people of the excellence of that service. And it follows that with French alone an officer can keep abreast of his profession almost as well as though he had at his command all the other foreign languages whose military literatures are important in a professional point of view.

In some respects the preparation of a French-English military technical dictionary is a task of great difficulty. This arises partly from the inability of the English language in all cases to supply the exact technical equivalent of a given French term, but more particularly from the extremely detailed nature of technical French nomenclature. Where we in English are frequently satisfied to apply the class name to each of the objects or things of one and the same class, the French will almost invariably modify this class name by some adjunct limiting or defining its application. Lexicographically, the result is to force the substitution of a description for a name. It is pertinent to remark, however, that this substitution is unavoidable in those cases, relatively numerous, in which the thing to be defined is peculiar to the French. It is hoped that these, as well as other difficulties of which it is unnecessary to speak here, may be borne in mind by those who may use this book; for while all care has been taken to make it as accurate as possible, still it would be vain to hope that all errors, whether of omission or of commission, have been excluded. The author takes this occasion to say that he will be greatly obliged to officers and others if they will bring to his attention any mistakes that may fall under their notice.

The author desires gratefully to thank the various friends that have helped him. Chief among these stands Col. Peter S. Michie, of the Military Academy. Maj. William A. Simpson, of the Adjutant-General's Department; Dr. Charles E. Munroe, of the Corcoran Scientific School; Capt. John R. Williams and Lieut. I. N. Lewis, of the Artillery; Capt. J. E. Kuhn, of the Engineers; Lieut. E. B. Cassatt, of the Cavalry, and Mr. Laurence V. Benét, of the Hotchkiss Ordnance Company, Limited, have laid him under obligations that are cheerfully acknowledged. Mr. Wilfred Stevens, of the Adjutant-General's Office, has given substantial aid in preparing the manuscript for the press and in reading the proof.

C. DE W. W.

WAR DEPARTMENT, ADJUTANT-GENERAL'S OFFICE,
May, 1899.

AUTHORITIES.

An asterisk (*) denotes a text-book of the French Artillery and Engineer School at Fontainebleau. These texts are not published, being lithographed simply for the use of the school.

A dagger (†) indicates that the work is an official publication of the French Government.

  Administration de l'armée, Lassalle, 1892.
* Administration militaire, Masselin, 1893.
  Aide-mémoire d'artillerie (ponts militaires, reconnaissances), 1883.
  Aide-mémoire de marine, Durassier et Valentino, 1892.
  Aperçus sur la tactique de demain, Coumès, 1892.
* Application de la fortification au terrain, Corbin, 1890.
* Application de l'électricité, Prérart, 1892.
  Argot and Slang, Barrère, London, 1889.
  Les armes à feu portatives, 1894.
  Les armes à feu portatives, Bornecque.
* Armes portatives, 1885.
  L'Artillerie à l'exposition de 1889, 1890.
  Artillerie moderne, Hennebert, about 1891.
* Attaque et défense des places, Sandier, 1893.
  Balistique expérimentale, Vallier, 1894.
* Bouches à feu, Lechaussée, 1894.
  Les cahiers du capitaine Coignet, 1889.
  Canet: pamphlets, etc., relating to Canet guns.
  Canon, torpilles et cuirasse, Croneau, about 1893.
  Capitulations, Thoumas.
  Catalogue de la "Brown & Sharpe Co." (French), 1893.
  Le cheval, Cuyer et Alix, 1886.
* Construction des batteries, Richard, 1893.
* Construction en bois et fer, Goetschy, 1885.
* Contremines, guerre de mines, Sandier, 1893.
  Cours des écoles de tir, 1886–87.
  Cours d'hippologie pratique, Saumur, 1893–94.
  Cours de topographie, Lehagre, Vols. I and III.
  Cours spécial sur le matériel de côte, Delaissey, 1890.
  Cours spécial à l'usage des sous-officiers d'artillerie, 1888.
  Défense des côtes d'Europe, Didelot, 1894.
  La défense des états et la fortification à la fin du XIXème siècle, Brialmont, 1895.
  Dictionary of French and English Military Terms, Barrère, London, 1896.
  Dictionnaire de la marine, Dabovich, Pola, 1887.
  Dictionnaire militaire (Berger Levrault), as far as published.
  Dictionnaire technologique, Bergman, Wiesbaden, 1887.
† Droits et usages de la guerre, Masselin, 1893.
  La dynamite de guerre, Dumas-Guilin, 1887.
  Équipages de ponts militaires, Haillot.

AUTHORITIES.

Équitation diagonale dans le mouvement en avant, 1892.
Les expériences de Bucarest, J. v. Schütz, 1886.
Explosifs modernes, Chalon, 1889.
*Fabrication des armes portatives, Walter, 1892.
*Fabrication des assemblages métalliques et des affûts, Couhard, 1889.
*Fabrication des bouches à feu; opérations métallurgiques, Couhard, 1885.
*Fabrication des projectiles, Couhard, 1891.
*Fabrication des substances explosives, munitions et artifices, Gages, 1893.
Hotchkiss: pamphlets, etc., published by the Hotchkiss Company.
Influence du tir plongeant, Brialmont, 1888.
*Instruction pratique sur le lever de reconnaissance, Goulier, ed. by Jardinet, 1893.
*Instruction pratique sur le lever d'ensemble, Romieux, Bosson, 1890.
*Instruction pratique sur le lever de position, Bosson, 1891.
*Instruction pratique sur le lever expédié, Jardinet, 1892.
*Instruction practique sur le lever de fortification, Romieux, 1893.
†Instruction sur l'emploi des agrès dont doivent être pourvus les batteries de 80 de campagne appelées à manoeuvrer en pays de Montagne, 1894.
†Instruction sur l'armement, les munitions, les champs de tir et le matériel de l'infanterie, 1891.
†The same, revised edition, 1895.
†Instruction sur le service de l'infanterie en campagne, 1893.
†Instruction pratique sur le service de la cavalerie en campagne, 1893.
†Instruction sur l'emploi de l'artillerie dans le combat, 1893.
†Instruction sur le remplacement des munitions en campagne, 1893.
*Instruments et procédés topographiques, Guillemard, Bosson, ed. by Jardinet, 1891.
Littré Dictionnaire de la langue française.
*Machines motrices (moteurs hydrauliques et moteurs thermiques), Paloque, 1891–92.
*Machines opératrices (wood, metal, and machine tools) de Beauchamp, 1886–1889–1891.
*Magasins à poudre et à dynamite; Stands, Goetschy, 1884.
†Maniement et emploi de la lance, 1891.
†Manuel d'escrime, 1889.
Manuel de fortification, Plessix et Legrand-Girarde, 1890.
†Manuel de gymnastique, 1888.
Maxim-Nordenfeldt: pamphlets relating to their guns.
Mémoires du Général Baron de Marbot.
*Métallurgie du fer, 1884.
*Métallurgie du fer: Travail des métaux, 1893.
*Méthodes de lever, Romieux, ed. by Jardinet, 1893.
*Mines, Simoutre, 1889.
Mines sous-marines, torpilles et torpédos, Dupont, 1895.
Naval and Military Dictionary, Burn, London, 1852.
Notices sur de nouveaux appareils balistiques, Sebert, 1881.
*Notions sommaires sur le matériel d'artillerie, 1893.
Nouveau dictionnaire militaire, Baudoin, 1892.
Nouveau manuel de fortification, anon., 1895.
Nouveau matériel naval (except purely naval parts), Ledieu et Cadiat, 1889, 1890.
†Observations sur l'emploi des troupes de cavalerie appelées à opérer avec des détachement de toutes armes, 1893.
*Organisation des affûts, Gautier, 1893.
*Organisation de l'artillerie, Gautier, 1893.
*Organisation des services de l'arrière, Delanne, 1885.
*Ponts militaires, Collet, 1887.

AUTHORITIES. IX

Ponts militaires et passages des rivières, Meurdra.
* Procédés de reproduction des dessins, cartes, etc.; emploi de la photographie, Prérart, 1893.
* Projectiles et fusées, Lachaussée, 1893.
  Les régions fortifiées, Brialmont, 1890.
† Règlement sur le service des canons de 155 long sur affûts de tourelle et de casemate, 1893.
† Règlement sur le service du canon de 120 court, 1896.
† Règlement sur le service intérieur de l'artillerie et du train des équipages, 1895.
† Règlement sur le service intérieur de la cavalerie, 1893.
† Règlement sur le service intérieur de l'infanterie, n. d.
* Règlement sur le service intérieur de l'infanterie (revised edition), 1895.
† Règlement sur le service dans les places de guerre, etc., 1896.
† Règlement sur le service des armées en campagne, 1892.
† Règlement sur le service des batteries de montagne, 1894.
† Règlement sur les manœuvres des batteries attelées, 1896.
† Règlement sur le service des canons de 80 et de 90, 1893.
† Règlement sur le service des canons de 80 et de 90, 1894.
† Règlement sur les exercices de la cavalerie, 1892.
† Règlement sur les exercices et les manœuvres de l'infanterie, 1893.
† Règlement sur l'instruction du tir, 1893.
† Règlement sur l'instruction du tir (revised edition), 1895.
† Règlement sur le service du train des équipages, 1895.
† Règlement sur le service des canons de 95 mm., mle. 1888, montés sur affût de campagne, 1896.
† Règlements sur le service des bouches à feu de siège et de place.
    Vol. I. Service des bouches à feu, 1894.
    Vol. II. Manœuvres de force, 1893.
    Vol. III. Matériel, 1891.
    Vol. IV. Renseignements spéciaux destinés aux officiers et aux sous-officiers, 1893.
† Règlement provisoire sur le service des bouches à feu de côte (marine), 1897.
† Règlement sur le service des bouches à feu de côte.
    Vol. I. Service des bouches à feu, 1895.
    Vol. II. Manœuvres de force et mouvements de matériel, 1898.
    Vol. III. Description du matériel, 1898.
* Résistance des matériaux, Paloque, 1894.
  Reuleaux, Le Constructeur (translation), 2d edition, Paris, 1881.
  Revue d'artillerie, from October, 1886, to January 1, 1898.
* Rôle et emploi de la fortification dans la défense des états, Corbin, 1890.
* Sapes, Mourral, 1891.
  Science et guerre (télégraphie optique; crytographie; l'éclairage électrique à la guerre; poste par pigeons), 1888.
* Service de l'artillerie dans la défense des places, Hermary, 1885.
* Service de l'artillerie dans les siéges, Hermary, 1887.
* Service de l'artillerie en campagne, Richard, 1892.
  Stratégie, Maillard.
* Tactique de combat, Roulin, 1892.
* Tactique générale, Roulin, 1892.
* Tactique de marche et de stationnement, Roulin, 1893.
  Topographie militaire, Bertrand.
  Traité de topographie, Pelletan.
  Virmaître, Dictionnaire d'argot fin-de-siècle, Paris, 1894.

SERVICE AND OTHER ABBREVIATIONS.

[Taken partly from Stavenhagen, Petit Dictionaire Militaire Français-Allemand, and the Almanach de Gotha.]

A.

A..............allongement; artillerie (on the *fleuron* of a bridle); altitude (on the réglette de lecture); Altesse.
Aduc., adsse...archiduc, archiduchesse.
A.-M..........aide-mémoire.
$A\frac{m}{n}$..........(of naval powders). A represents proportions and mode of fabrication; m, thickness of presscake; n, width of strips into which it is cut.
Amb...........ambassadeur.
Amb. C. (Div.)..ambulance de corps (divisionnaire).
Æ..............arrière (naval, and also in gun construction).
AR............Alger (on cartridge bags).
Ar. G..........arrière garde.
Art.............artillerie.
Art. C..........artillerie de corps.
Art. 1. Div....artillerie de la première division.
Art. T..........artillerie territoriale.
$AS\frac{m}{n}$..........powder manufactured at Sevran Livry. A represents dosage and mode of fabrication; m, thickness of galette; and n, maximum dimension of the grain.
AT............artillerie territoriale.
Atm............atmosphère.
Att.............attaché.
AV.............avant (naval, and gun construction).
Av. G..........avant garde.
Av. P..........avant poste.

B.

B..............symbol of smokeless powders (Vieille); (on base of projectiles) obus à balles; Bouchet (powder works); breveté d'é-tat-major.
Bat............bataillon.
Batt...........batterie.
BC............symbol of smokeless powder for field guns (C=campagne).
BF............symbol of small-arm smokeless powder (F=fusil).
Bie............batterie.
BNn...........symbol of smokeless powders (Vieille; N=nitratée).
BO............boulets ogivaux, on F. M. Art. sights; Bulletin officiel du ministère de la guerre.
B. O., P. R....bulletin officiel, etc., partie règlementaire.
B. O., P. S....bulletin officiel, etc., partie supplémentaire.
B^{on}, B^{onne}......Baron, Baronne.
Bon............bataillon.
Br.............breveté d'état-major; brigade.
Brig............brigade.
BSP...........symbol of smokeless powders for siege and fortress guns (SP=siège et place).

C.

C..............campagne; cube; (of guns) court; centime; commandeur (de la légion d'honneur).
Cn.............symbol of black powders (C=campagne).
C. A...........corps d'armée.
C. a. d.........c'est à dire.
Cal............calibre.
Cap............capitaine.
Cav............cavalerie.
Cav. T..........cavalerie territoriale.
C. d. f., Ch. d. f..chemin de fer.
Ceaux..........(for) faisceaux, in commands only.
C. G...........consul général.
Ch. d'aff........chargé d'affaires.
Chamb..........chambellan.
Chev...........chevalier.
Cie.............compagnie.
Cm.............centimètre.
Cm²............centimètre(s) carré(s).
Cm³............centimètre(s) cube(s).
Col............colonel.
Comm..........commandeur, commandant.
Com^t...........commandant.
C^{on} D^é.........canon déclassé.
Cslt............consulat.
Cslt G..........consulat-général.
Cte,* Ctesse....comte, comtesse.

D.

D..............docteur en médicine (after the name of a pharmacist, in the Annuaire de l'armée).
Dcm...........décimètre.
Dcm²..........décimètre(s) carré(s).
Dcm³..........décimètre(s) cube(s).
Dép...........département.
Dét^t...........détachement.
Div............division.

*Written also Cte, and similarly of some other titles.

SERVICE AND OTHER ABBREVIATIONS.

DP............(on base of projectile) obus à double paroi.
DPB..........(on base of projectile) obus à double paroi et à balles.
Dr............docteur (en droit, ès sciences, etc.).
Dsse..........duchesse.

E.

E.............épreuve (surmounted by a crown, and stamped on gun barrels after firing test).
ECP..........École centrale de Pyrotechnie.
E. e..........envoyé extraordinaire.
E. e. et M. pl..envoyé extraordinaire et ministre plénipotentiaire.
E. M..........état-major.
Esc...........escadron; —— 2 Div., escadron divisionnaire de la 2e division.
Ette..........(for) baïonnette (in commands only).
Ex............(on sights) exercice, indicates that corresponding graduations are for drill projectiles.

F.

F.............franc (coin and unit).
F à DE........fusée à double effêt.
Fn............symbol of smokeless powder (F= fusil).
F. é. m........force électro-motrice.
Fl. G..........flanc garde.
F. M..........feld maréchal.
FP............fusée percutante.
F. Z. M........feldzeugmestre.

G.

G.............génie.
Gal...........général.
GB............grosses balles (boîte à mitraille).
GC............grand-croix (de la légion d'honneur).
Gén...........général; génie.
Gén. T........génie territorial.
GM............génie mobilier (stamped on barrack furniture, etc., for which the engineers are responsible).
GO............grand-officier (de la légion d'honneur).
Gr............groupe; gramme.
GV............grande vitesse (on handwheel of turrets operated by steam).

H.

H.............hauteur (on réglette de lecture).
H. C..........hors cadre.
HS............hors service (cf. I. C., U. S. A.).

I.

Illustr.........illustrissime.
Inf............infanterie.
Inf. T.........Infanterie territoriale.
IR............initials stamped on hospital furnishings, articles, etc.

J.

J. M..........Journal militaire.

K.

K.............kilogramme(s).
Kg............kilogramme(s).
Kilo..........kilogramme(s).
Km...........kilomètre(s).
Km^2.........kilomètre(s) carré(s).

L.

L.............long (of guns).
LC............lumière centrale (on cartridge bags).
Lég...........légation.
L. F. M.......lieutenant feld maréchal.
Lieut..........lieutenant.
Lieut. Col.....lieutenant-colonel.
LL. AA........Leurs Altesses.
LL. EE........Leurs Éminences.
L. M. R.......ligne de moindre résistance.
LN............Lyon (on cartridge bags).
Lt............lieutenant.

M.

M.............montagne (of guns), mètre, monsieur, modèle, modifié; e. g., M. 1886=model of 1886; but, 1886 M.=of 1886, modified.
MA............manufacture d'armes.
Mad...........madame.
Maj...........major.
Mc............mètre(s) cube(s).
MC$_n$..........(of powders) symbol indicating that powder has been trituré n minutes (M=meule), and is intended for guns (C=canon).
M^d, M^{de}........marchand, marchande.
Me............maître (lawyer, notary).
Mgr...........monseigneur.
Mil............millimètre(s).
Min...........ministre.
Mlle...........mademoiselle.
MM...........messieurs.
Mm...........millimètre(s).
Mm^2.........millimètre(s) carré(s).
Mm^3.........millimètre(s) cube(s).
MmC..........millimètres court; e. g., pièce de 155^{mm}C.
MmL..........millimètres long; e. g., pièce de 155^{mm}L.
Mme..........madame.
Mmq..........millimètre(s) carré(s).
Mle...........modèle.
Mod...........modèle.
M. pl..........ministre plénipotentiaire.
Mq............mètre(s) carré(s).
Mqs(e)........marquis, marquise.
Mr............monsieur (rare).
M. R..........ministre résident.

N.

N.............navire (on réglette de lecture).
NC............non classé (of men that have not qualified at target practice).
N. D..........Notre Dame.
N^t............négociant.

SERVICE AND OTHER ABBREVIATIONS. XIII

N. S. J. C......Notre Seigneur Jésus-Christ.
Ngt............négociant.

O.

O...............officier (de la légion d'honneur).
Off..............officier, officiel.
On.............(for) canon (in commands only).

P.

PA.............symbol of black prismatic powder.
P. art..........parc d'artillerie.
PBn............symbol of poudres brunes.
PBn S..........poudre brune de siège.
PC.............poudre composition.
P. C...........portion centrale.
PD.............pièce droite (in a turret).
P G............pièce gauche (in a turret).
P. I............point initial.
P. L. M........Paris-Lyon-Méditerranée (railway).
P. O...........par ordre.
Po..............pouce (inch).
P. P...........portion principale.
Pr..............prince.
Prov...........province.
PS.............pur sang (of a horse).
Psse...........princesse.
PV.............petite vitesse (on handwheel of turrets operated by steam).

Q.

Q..............carré, e. g., mmq. (v. s. v. m.).
Q. G...........quartier général.

R.

R..............(on fuses) Rubin (name of inventor).
Róg............régiment.
Résid..........résidence; résident.
R. F...........République française.

S.

S..............siège; saint.
S. A. I. (R.)....Son Altesse Impériale Royale).
S. C...........service en campagne.
S. É...........Son Éminence.
S. Exc.........Son Excellence.
S. G...........Sa Grandeur.
S. G. D. G.....Sans garantie du gouvernement.
SL.............Sevran Livery (on cartridge bags).
S. Lt..........sous-lieutanant.

SM.............siège et montague.
S. M...........Sa Majesté; station-magasin.
S. M. A........section de munitions d'artillerie.
S. M. I........section de munitions d'infanterie.
S. P...........Le Saint Père.
SP.............siège et place (guns).
SnPn...........symbol for powders (siege and fortress guns).
SS.............saints.
S. S...........Sa Sainteté.
Sr.............sieur (in legal papers).
St., Ste........saint, sainte.

T.

T..............territorial, transformé.
TE.............train des équipages.
T. E. G........(station) tête d'étapes de guerre.
T. E. R........tête d'étapes de route.
Tr.............transformé.
T. R...........tir rapide.
Tr. Comb......train de combat.
Tr. Rég........train régimentaire.

V.

V. C...........vice-consul.
V. C. G........vice consul général.
Vl.............vitesse du vent dans la direction de la ligne de tir.
Vr.............vitesse totale du vent.
Vt.............vitesse du vent dans rection perpendiculaire à la ligne de tir.
Vcte...........vicomte.
Vte., Vtesse....vicomte, vicomtesse.

W.

W..............Wetteren (powder).
W$\frac{m}{n}$............Wetteren powder; m = thickness ness of grain; n = length of side of base.

X.

X..............(not an abbreviation) stamp on an arme de théorie, q. v.

Z.

Z..............(not an abbreviation) série Z. (Scabbards i s s u e d without bayonets, e. g., to infirmiers, belong to série Z.)

LEGION OF HONOR.

G C ✤ *Grand-Croix.*
G O ✤ *Grand-Officier.*
C ✤ *Commandeur.*
O ✤ *Officier.*
✤ *Chevalier.*

Ⓜ *Médaille militaire.*

ABBREVIATIONS USED ON THE FRENCH GENERAL STAFF MAPS.

[Taken mostly from Stavenhagen, Petit Dictionnaire Militaire Français-Allemand.]

Abb^e........Abbaye.	Emb^{re}.......Embarcadère.	Pap^{ie}........Papeterie.
Aig^{le}.........Aiguille.	Emb^{ure}......Embouchure.	PF..........Préfecture.
Aq^{uc}.........Aqueduc.	E. Min.......Eau minérale.	P^{ge}..........Passage.
Arb..........Arbre.	Étab^{nt}......Établissement.	Ph..........Phare.
Aub^{ge}.......Auberge.	Étg..........Étang.	Pl^{au}.........Plateau.
B............Bois.	Ét^{le}.........Étoile.	P^{on}..........Pavillon.
Bat^{ie}........Batterie.	Fab^e........Fabrique.	Poud^{ie}......Poudrerie.
B^{che}........Bouche.	F^{bg}..........Faubourg.	P^t...........Petit; Pont; Port
B^{de}..........Borde.	F^e...........Ferme.	P^{te}..........Pointe; Porte.
B^{ide}........Bastide.	F^{ge}..........Forge.	P^{te} de D.....Poste de Douane.
B^{ie}..........Bergerie.	Fl...........Fleuve.	Q^r...........Quartier.
B^{in}..........Barin; Bassin.	F^{me}..........Ferme.	R............Rivière; Rue.
B^{on}..........Buisson; Buron.	F^{ne}, Font^{ne}..Fontaine.	R^{au}..........Radeau.
B^{que}........Baraque.	Fr^{ie}..........Fonderie.	Red^e........Redoute.
B^{re}..........Barrière.	F^t...........Forêt; Fort.	R^{er}..........Rocher.
Briq^{ie}......Briqueterie.	G^d...........Grand.	Retr^{nt}......Retranchement.
C............Cap.	G^{ge}..........Gorge.	R^{ise}..........Remise.
Cab^{et}......Cabaret.	Gl^{er}..........Glacier.	Riv..........Rivière.
Ca^l..........Canal; Cortal.	Hab^t.......Habert.	Roub^{ne}......Roubine.
Carr^e........Carrière.	H^{au}..........Hameau.	R^{te}..........Route.
Carref^r......Carrefour.	H^{ts}..........hauts.	Sa^l..........Signal.
Cay^r........Cayolar.	I............Île.	Sal..........Saline.
Ch^{au}........Château.	J^{ée}..........Jetée.	Salp^{ie}........Salpêtrerie.
Ch^{ee}........Chaussée.	J^{se}..........Jasse.	Sap^{re}........Sapinière.
Ch^{et}........Châlet.	K............Ker.	Sc^{ie}..........Scierie.
Ch^{lle}........Chapelle.	L............Lac.	Sém..........Sémaphore.
Ch^{ne}........Chaîne.	Lag..........Lagune.	Sig^l..........Signal.
Ch^{née}......Cheminée.	L^{de}..........Lande.	Som^t........Sommet.
Cim^{re}......Cimetière.	Loc^{re}......Locature.	SP...........Sous-préfecture.
Cit^{lle}........Citadelle.	M............Mas.	St., St^n, St^{on}..Station.
C^{ne}..........Cabane.	Mal^{re}......Maladrerie.	Télég^e......Télégraphe.
Colomb^r....Colombier.	Manuf^{re}....Manufacture.	T^{ie}..........Tuilerie.
Couv^t......Couvent.	Métr^{ie}......Métairie.	T^{nt}..........Torrent.
Cr^x..........Croix.	Mg^{ne}........Montagne.	T^r...........Tour.
C^{se}..........Cense.	M^{in}..........Moulin.	U^{se}..........Usine.
CT...........Canton.	M^{on}..........Maison.	Vac^{ie}........Vacherie.
Dép^{le}......Départementale.	M^s..........Marais.	Vé^e, Valé^e....Vallée.
Dig..........Digue.	M^t...........Mont.	V^{er}..........Verger.
D^{ne}..........Douane.	Nat^{le}......Nationale.	V^{on}, Val^{on}...Vallon.
Dom^e......Domaine.	N. D.......Notre Dame.	Vr^{ie}..........Verrerie.
Éc^{ie}........Écurie.	O^y..........Orry.	V^x..........Vieux.
Éc^{se}........Écluse.	P............Parc à bestiaux; Pic.	
Ég^{se}........Église.		

LIST OF ABBREVIATIONS USED IN THE FOLLOWING PAGES.

a..........adjective.
adm........administration.
adv........adverb.
a. p........armor-piercing.
art........artillery.
artif......fireworks, pyrotechny.
ball.......ballistics.
carp.......carpentry.
cav........cavalry.
chem.......chemistry.
c. o.......commanding officer.
cons.......construction, art of building.
cord.......cordage.
elec.......electricity (and magnetism).
eng........engineering, engineers.
esp........especially.
expl.......explosives.
f..........feminine noun.
fam........familiar.
farr.......farriery.
fenc.......fencing.
fig........figuratively.
F. m. art..French marine artillery.
fond.......founding.
fort.......fortification; field fortification.
Fr.........French.
Fr. a......French army.

gym........gymnastics.
harn.......harness.
hipp.......hippology.
hydr.......hydrography.
inf........infantry.
in gen.....in general.
inst.......instruments.
lit........literally.
m..........masculine noun.
mach.......machinery.
man........manège, equitation, riding in general.
mas........masonry.
med........medicine, medical.
met........metallurgy.
mil........military.
min........mining.
m. l.......muzzle loader.
mus........music.
nav........naval.
n. c. o....noncommissioned officer.
n. e.......no equivalent.
obs........obsolete.
opt........optics.
pl.........plural.
p. p.......past participle.
phot.......photography.
pont.......bridges (especially military bridges).

powd.......gunpowder and its manufacture.
prep.......preposition.
q.v.,qq.v..which see.
r. f.......rapid fire.
R. A.......Royal Artillery.
R. M. A....Royal Military Academy.
r. r.......railroads.
s. b.......smoothbore gun.
s. c.......sea coast.
s. v.......sub verbo.
sig........signaling.
sm. a......small arms.
surv.......surveying.
tech.......technical, technically.
teleg......telegraphy.
top........topography.
torp.......torpedoes.
t. p.......infantry target practice.
unif.......uniform.
U. S. A....United States Army.
U.S.M.A....United States Military Academy.
U. S. N....United States Navy.
v..........see.
v. a.......active verb.
v. n.......neuter verb.
v. r.......reflexible verb.

XV

A.

abaissement, m., sinking, depression, subsidence; letting, putting, or throwing down; vertical distance of a point below another; fall (of level, of pressure, etc.); (*ball.*) negative jump;
— *de l'horizon,* dip of the horizon;
— *du terre-plein,* (*fort.*) lowering of terreplein, so as to keep it 2.5^m below the plane of defilade;
— *du tir,* (*ball.*) for equal ranges, the dropping of points of impact below the normal position, on account of variations of temperature, heating of barrel, etc.;
— *de la trajectoire,* (*ball.*) vertical distance between any point of the trajectory and the line of fire produced.

abaisser, v. a., to lower, let fall; to bring down, decrease, abate;
— *un intervalle,* (*mil.*) to reduce an interval.

abandon, m., abandonment, departure; cession (of rights, etc.);
à l' —, at random, adrift;
— *de son poste,* (*mil.*) quitting of one's post (of a sentry, before being relieved); (*in gen.*) desertion of one's post of duty.

abandonner, v. a., to abandon, forsake, give up, let go, desert, quit; (*man.*) to set a horse at full speed;
— *le combat,* (*mil.*) to quit the fight, give up the contest, battle;
s' — *sur la selle* (or simply *s'* —), (*man.*) to lounge in the saddle.

abaquarrage, m., the act or operation of forming an *abaque* (graphic table).

abaquarrer, v. a., to form or make an *abaque* (graphic table).

abaque, m., abacus, counting frame; any graphic table; (*art.*) range-finding attachment of a s. c. artillery telescope;
— *graphique de tir,* (*ball.*) graphic table of fire, or range table;
— *de marche,* (*mil.*) march diagram;
— *des probabilités,* (*ball.*) graphic table of probabilities (of fire);
— *des trajectoires,* (*ball.*) graphic table of fire, or range table.

abatage, m., cutting down, felling (of trees); destruction, demolition (as by high explosives); lowering; bearing down, as on a lever; mechanical power, or purchase; holding-up hammer, dolly; throwing down (as of coal, in a mine); (*mil.*) destruction of diseased horses;
— *d'arbres,* (*mil.*) felling of trees for defensive purposes;
— *d'armes,* (*mil.*) the bringing of the rifle down to position of load, etc.;
— *de chevaux,* (*mil.*) destruction of diseased horses;
— *du chien,* (*sm. a.*) letting down of hammer or of cocking-piece;
commission d' —, (*mil.*) board of officers to pass on diseased horses;
— *de la crosse,* (*art.*) lever made of trail of gun; *faire un,* —, to form a purchase under a body; *mandrin d'* —, holding-up hammer, dolly.

abatant, m., shutter of a skylight.

abatis, m., (*mil.*) abatis;
— *actif,* defensible abatis;
— *artificiel,* abatis of trees, etc., transported to desired position;
— *défensif,* defensible abatis;
— *de franc-étable,* houses, walls, or trees thrown down to form a sort of abatis;

abatis *naturel,* abatis of trees left *in situ;*
— *passif,* abatis that serves as a mere obstacle;
— *de transport,* abatis of branches, etc., dragged or transported into position.

abat-jour, m., reflector; shade (of a lamp); skylight.

abattoir, m., slaughter-house.

abattre, v. a., to throw, throw down; to overthrow, demolish, cast down; to cut down (trees); to bear down on a lever; (*min.*) to throw down, bring down, as coal in a mine; v. r., (*hipp.*) to break down, to fall down;
— *l'arme,* (*mil.*) to bring the piece down to position of load, of charge bayonets, etc.;
— *du bois,* to fell trees;
— *un cheval,* (*mil.*) to throw, cast, destroy, a horse;
— *l'eau d'un cheval,* (*hipp.*) to rub a horse dry;
— *la muraille,* (*farr.*) to pare down a horse's hoof;
— *du pied,* (*farr.*) v. — *la muraille;*
— *un rivet,* to clinch a rivet;
— *les tentes,* (*mil.*) to strike tents.

abattu, p. p., down, throw down;
à l' —, (*sm. a.*) said of hammer when down; of bolt when shot forward and lever turned down; (*mech.*) said of a spring when free from pressure.

abat-vent, m., penthouse.

abélite, f., (*expl.*) nitroglycerine.

aberration, f., aberration;
— *chromatique,* chromatic aberration;
— *de sphéricité,* spherical aberration.

abois, m. pl., *aux* —, (of game) at bay; (*mil.*) (of a garrison) reduced to extremities.

abonnataire, m., contractor, who agrees to perform service within a given time and for given pay.

abonnement, m., subscription; contract, agreement; allowance, idemnification; (*esp. Fr. adm.*) contract for given time and given pay.

abord, m., landing place; approach (of a bridge);
—*s,* (*mil.*) ground or terrain immediately in front of, or near, a position, or within close range of;
en —, (*nav.*) close to ship's side (of a gun so mounted, as distinguished from turret mounts, etc.).

abordable, a., easy of access, accessible.

abordage, m., (*mil.*) contact with enemy (in an attack, or as result of an attack); (*nav.*) collision; boarding (of a ship).

abordé, m., (*nav.*) ship boarded, run down, or rammed.

aborder, v. a., to come alongside; to board; to collide with; to reach, attack; to charge; (*mil.*) to attack by assault; to occupy the *abords* (q. v.) of a position; to get close to the enemy;
— *à l'éperon,* (*nav.*) to ram.

abordeur, m., (*nav.*) boarder, member of a boarding party; boarding ship, ramming ship.

abouchement, m., interview; butt-joining, junction (of pipes, etc.).

aboucher, v. a., to bring together, for an interview;
s' — *avec un ennemi,* (*mil.*) to parley with an enemy.

abouement, m., flush-joint.

abourgri, a., stunted; knotty; cross-grained.

aboûment, m., *v. abouement.*

3877°—17——1   (1)

about, m., butt, butt-end; shoulder of a tenon; abutment; beveled piece of timber; (*in joints*) the end of the piece that is assembled to another piece;
joindre en ——, to lengthen a timber by joining on another;
plaque d' ——, shoe;
portée d' ——, the bearing surface, the surface receiving the *about;*
—— *de rive*, ornamental or special tile at end of rafter.

aboutement, m., butt, abutment.

abouter, v. a., to join end to end.

aboutir, v. n., to lead to (as a road to a village); to lead up to (a result); to abut.

aboutissant, a., abutting, bordering on; m., abutment, adjacent or abutting part;
les tenants et les ——*s*, the houses, fields, etc., adjacent.

aboutissement, m., eking-piece.

abraquer, v. a., (*cord.*) to haul or pull (upon a rope);
—— *le mou*, to take up the slack of a rope.

abreuvage, m., watering (of horses).

abreuver, v. a., to water (a horse); to steep, to soak.
—— *un bateau*, to soak a boat, so as to swell the seams.

abreuvoir, m., watering-place (for animals); watering-trough; horse-pond; (of timber), sort of gutter or hollow formed in stem when branches are torn off.

abréviation, f., abbreviation;
—— *de service*, (*mil. sig.*) conventional abbreviation.

abri, m., shelter; (*mil.*) cover;
à l' ——, sheltered, protected, covered;
—— *actif*, defensible shelter or cover;
—— *de bivouac*, any sort of shelter in a bivouac;
—— *blindé*, bombproof;
cantonnement- ——, defensible cantonment;
—— *de champ de tir*, (*t. p.*) marker's shelter;
—— *de chargement*, (*siege*) filling shelter or cover;
—— *en clayonnage*, shelter of wattled work;
—— *défensif*, bombproof cover;
—— *à l'épreuve*, (*fort.*) bombproof, bombproof shelter;
à l' —— *du feu* sheltered from fire; (*in gen.*) fireproof;
—— *en feuillage*, shelter of branches, etc., against wind, rain, etc.;
—— *des marqueurs*, (*t. p.*) marker's shelter;
—— *passif*, mere cover;
—— *-remise*, (*fort.*) proof storehouse for matériel;
—— *sous-traverse*, (*fort*) traverse cover or shelter;
tente- ——, shelter-tent;
—— *de tirailleurs*, shelter-trench, rifle-pit;
tranchée- ——, shelter-trench;
traverse- ——, (*fort.*) traverse and bombproof in one.

abri-caverne, m., (*fort.*) shelter cut out of the rock.

abriter, v. a., to cover, shelter; (*mil.*) to put under cover; —— *les attelages*, —— *les avant-trains*, (*art.*) to cover the teams, the limbers, in action.

abrivé, p. p., under way (of a boat).

abriver, v. n., to give way (boats).

abri-vent, m., (*mil.*) wind guard; cover or shelter against wind.

abri-voûte, m., (*fort.*) any vaulted or arched bombproof.

abrupt, a., (*top.*) rough; having steep slopes or sides.

abruti, m., (*mil. slang*) plodding cadet (Polytechnique); "fly-specker" (U. S. M. A.); "swat" (R. M. A.).

abscisse, f., abscissa.

absence, f., absence;
—— *illégale* (*mil.*) absence without leave;
—— *légale* (*mil.*) legal (justifiable) absence (hospital, capture, captivity);
—— *régulière* (*mil.*) authorized absence (leave);
solde d' ——, (*mil.*) leave pay, half-pay.

absent, a., absent.

absenter, v. r., (*mil.*) to be absent without leave.

absolu, a., peremptory; complete;
congé ——, (*mil.*) discharge from service.

absolution, f., (*mil. law*) release of accused by reviewing authority, provided act committed does not call for punishment.

absorbant, m., (*expl.*) dope.

absorption, f., annual ceremony at Polytechnique, at which upper classmen are entertained by the newcomers (*melons*, q. v.).

abtheilung, m., (*art.*) German name for group of 3 or 4 batteries; brigade; so-called battalion, U. S. artillery.

abus, m., abuse, misuse;
—— *d'autorité*, (*mil.*) exceeding of one's authority.

acacia, m., acacia (wood and tree).

académie, f., academy, school; (*esp.*) riding school;
faire son ——, (*man.*) to learn to ride.

acajou, m., mahogany (wood and tree).

acatène, a., chainless; m., chainless bicycle.

accabler, v. a., to crush, overthrow, overwhelm.

accaparer, v. a., to buy up all the stores, etc., in the market; to "corner" the market.

accastillage, m. (*nav.*) poop and forecastle.

accélérateur, a., accelerating;
pas ——, (*mil.*) quick time.

accélération, f., acceleration;
contre- ——, (*ball.*) retardation.

accélérer, v. a., to accelerate, quicken, hasten.

accélérographe, m., (*ball.*) accelerograph;
—— *à chariot*, slide accelerograph;
—— *à diapason électrique*, electric fork accelerograph.
—— *à diapason mécanique*, mechanical fork accelerograph.

accéléromètre, m., (*ball.*) accelerometer;
—— *à diapason*, fork accelerometer;
—— *à poids*, weight accelerometer;
—— *à ressort*, spring accelerometer;
—— *simple à poids*, simple weight accelerometer.

accepter, v. a., to accept;
—— *le combat*, (*mil.*) to accept the combat, to risk a battle.

accès, m., access, approach.

accessible, a., accessible.

accessoires, m. pl., fittings, mountings; (*sm. a.*) all parts that may be removed from barrel, stock, etc.;
—— *d'armes*, (*sm. a.*) cleaning kit;
—— *de batterie*, (*siege*) supplies for service of a battery for 24 hours;
—— *de bouches à feu*, (*art.*) all removable parts of breech mechanism;
—— *de chargement*, loading outfit.
—— *d'embarquement*, (*mil. r. r.*) ramps, etc., for loading and unloading;
—— *de rechange*, spare parts;
—— *de la selle*, (*harn.*) saddle fittings;
—— *de solde* (*mil. adm.*) additions to regular pay, extra pay, extra-duty pay (U. S. A.);
—— *du tir à la cible*, (*t. p.*) target practice equipments.

accident, m., accident; (*top.*) accident, variation, unevenness (of the ground or terrain).

accidenté, a., (*top.*) uneven, rough, varied, broken, intersected (of the ground or terrain).

acclamper, v. a., to clamp together, to clamp.

acclimater, v. a., to acclimatize.

accolade, f., brace.

accolé, a., (*mil.*) placed side by side (in battle, camp, etc., of companies, battalions, regiments, etc).

accolement, m., (*mil.*) state of being side by side (in battle, or camp, etc., of various units).

accoler, v. a., (*mil.*) to place side by side (of various units, in battle, camp, etc.).

accon, m., flat-bottomed boat, scow, lighter.

accord, m., agreement;
—— *des aides*, (*man.*) combination of the aids.

accore, a., (*top.*) steep, perpendicular (really a naval term); m., prop, stay, shore; (*top.*) contour of a reef, bank.
accorer, v. a., to prop, stay, shore up.
accostage, m., coming alongside (of a boat, lighter, etc.).
ponton d' ——, lighter.
accoster, v. a. and n., to go or come alongside of; to cause to bear, to touch, accurately, (of plates to be riveted together);
s' ——, to come alongside of.
accotement, m., footpath on each side of a road (*chaussée*), berm; (*r. r.*) space between exterior rail and edge of ballast.
accoter, v. a., to chock, scotch, a wheel.
accotoir, m., prop, stay; chock, scotch; (*med.*) hand-rail or frame of a cacolet, of a mule-litter.
accoudement, m., (*mil.*) touch of elbow in ranks.
accouple, f., leash.
accouplement, m., (*mach.*) coupling, connection; (*hipp.*, etc.) copulation; pairing (as of carrier pigeons);
—— *à articulations en croix*, (*mach.*) universal joint, Cardan coupling, Hooke's coupling;
—— *articulé*, (*mach.*) jointed coupling;
—— *à croisillon articulé*, (*mach.*) universal joint;
—— *à débrayage*, (*mach.*) releasing, clutch, coupling;
—— *dirigé*, (*hipp.*, etc.) pairing by selection;
—— *à enveloppe conique*, (*mach.*) cone coupling;
—— *facultatif*, (*hipp.*, etc.) pairing by chance;
—— *fixe*, (*mach.*) rigid coupling;
—— *par manchon*, (*mach.*) muff coupling;
—— *par plateaux*, (*mach.*) plate coupling.
accoupler, v. a. r., (*mach.*) to couple, to connect; (*art.*) to couple horses; (*hipp.*, etc.) to couple, pair.
accréditer, v. a., to accredit; to present one's vouchers and have them accepted.
accrochage, m., coupling, connecting, joining.
accrocher, v. a., to hook; (*nav.*) to grapple an enemy's ship;
—— *un soldat*, (*mil. slang*) to confine a soldier to barracks.
accul, m., cove, creek; blind alley.
acculé, a., (*hipp.*) liable to rear, to throw himself on his haunches.
acculement, m., (*hipp.*) state or condition of a horse that backs, rears, etc.
acculer, v. a., to bring to a stand; (*fam.*) to "corner;" (*mil.*) to drive back an army against an obstacle, as a river, lake, etc.; to push an army, so far into enemy's country that it must either break through or surrender.
accumulateur, m., (*mach.*, etc.) accumulator (air, steam, water, etc.); (*elec.*) storage battery, secondary battery, accumulator; (*fort.*) weight, counterpoise weight (used to raise turrets to firing position);
—— *de travail*, storer-up of energy.
accumulateur-multiplicateur, m., (*mach.*) multiplying-accumulator (e. g., of pump).
accusation, f., (*mil. law*) charge and specification.
accusé, m., (*mil. law*) the accused, prisoner before a court-martial; a., prominent (as a hill, ridge, steeple, etc.);
—— *de réception*, receipt (for a letter, message, parcel, etc.).
accuser, v. a., to accuse; to prefer charges against.
acéré, a., sharp, sharp-edged.
acérer, v. a., to edge, point; to overlay with steel; to steel, to temper.
acétone, f., (*chem.*) acetone.
achat, m., purchase;
—— *à caisse ouverte*, purchase at fixed time, place, and price (open market), by the military administration, of certain supplies;

achat *à commission*, —— *par voie de commission*, purchase by agent on commission allowed by state;
—— *direct*, ordinary purchase, in open market;
—— *par marché*, purchase by contract, or according to conditions published or furnished.
acheminement, m., preliminary step, preparatory measure.
acheminer, v. a., r., to forward, dispatch, send; to proceed, set out, get on;
—— *un cheval*, (*man.*) to teach a young horse to advance in a straight line.
acheteur, m., buyer;
officier ——, (*mil.*) member of horse-purchasing board.
achever, v. a., to finish;
—— *un cheval*, (*man.*) to finish breaking a horse.
achromatique, a., achromatic.
achromatisme, m., achromatism.
achromatopsie, f., color-blindness.
acide, a. and m., (*chem*) acid;
—— *azoteux*, nitrous acid;
—— *azotique*, nitric acid;
—— *fulminique*, fulminic acid;
—— *marin*, hydrochloric acid;
—— *picrique*, (*expl.*) picric acid;
—— *sulfoglycérique*, (*expl.*) glycerol-sulphuricacid;
—— *sulfonitrique*, (*expl.*) mixture of nitric and sulphuric acids in French process of making nitroglycerine.
acier, m., (*met.*) steel;
—— *adouci*, annealed steel;
—— *affiné*, fined steel;
—— *Bessemer*, Bessemer steel;
—— *au bore*, boron steel;
—— *boursouflé*, blister steel;
—— *bronze*, steel bronze;
—— *brûlé*, burnt steel;
—— *brut*, crude, raw steel;
—— *à canon*, (*art.*) gun steel;
—— *à*, —— *de*, *cémentation*, cement steel;
—— *cémenté*, cement steel;
—— *centrifugé*, steel treated by centrifugal process to get rid of blowholes;
—— *au chrome*, —— *de chrome*, —— *chromé*, chrome steel;
—— *commun*, steely iron;
—— *comprimé liquide*, fluid compressed steel;
—— *corroyé*, shear steel, weld steel;
—— *une fois corroyé*, shear steel;
—— *deux fois corroyé*, double shear steel;
—— *coulé*, cast steel;
—— *de Damas*, —— *damassé*, damask steel;
—— *diamant*, tool steel, tungsten steel;
—— *doux*, mild, soft steel;
—— *dur*, —— *durci*, hard steel;
—— *ferreux*, half steel;
—— *fondu*, cast steel, ingot steel; homogeneous iron;
—— *fondu au creuset*, crucible steel;
—— *de fonte*, furnace steel, hearth steel;
—— *de forge*, fined steel, charcoal hearth steel;
—— *des Indes*, —— *indien*, wootz steel;
—— *de l'ingot*, ingot steel;
—— *laminé*, rolled steel;
—— *malléable*, forge steel;
—— *manganèse*, —— *au manganèse*, —— *manganèsé*, manganese steel, Hadfield steel;
—— *marchand*, rolled steel;
—— *moulé*, cast steel;
—— *naturel*, natural steel, hearth steel;
—— *nickel*, —— *au nickel*, —— *nickelé*, nickel steel;
—— *ordinaire*, common steel;
—— *à outils*, tool steel;
—— *poule*, blister steel;
—— *puddlé*, puddled steel;
—— *raffiné*, refined steel;
—— *raffiné à une marque*, shear steel;
—— *raffiné à deux marques*, double shear steel;
—— *raffiné à trois marques*, steel heated and welded three times;

acier *par réaction*, reaction steel (Martin process, "pig and scrap");
—— *recuit*, annealed steel;
—— *à ressort*, spring steel;
—— *au réverbère*, reverberatory furnace steel;
—— *en rubans*, hoop steel;
—— *sans-soufflures*, Terre-Noire steel without blowholes;
—— *sauvage*, wild steel;
—— *soudable*, welding steel;
—— *soudé*, weld steel;
—— *spécial*, special steel;
—— *supérieur*, hard steel;
—— *trempé*, hard, hardened steel;
—— *au tungstène*, —— *de tungstène*, tungsten steel;
—— *vif*, hard steel.

aciération, f., steeling.

aciérer, v. a., to steel, to overlay with steel; (*met.*) to convert iron into steel.

aciéreux, a., steely.

aciérie, f., steel works.

acompte, m., (*adm.*) payment on account, before complete delivery and acceptance of things bought.

a-coup, m., (*man.*) shy (of a horse); jerk, sudden start; jerk or sudden pull on a bridle; (*mil.*) jerk or start in a column or line; sudden and accidental halt on march;
par ——, step by step, by fits and starts.

acoustique, f., acoustics; a., acoustic.

acquit, m., discharge, receipt;
—— *à caution*, permit;
pour ——, "paid," "received payment."

acquittement, m., discharge; payment; (*mil. law*) acquittal of a prisoner.

acquitter, v. a., to acquit.

acte, m., legal paper, document, or instrument; act or deed; action or step or procedure of legal authorities; declaration; register, certificate;
—— *d'accusation*, document setting forth crime or misdemeanor charged;
—— *administratif*, —— *d'administration*, any document proceeding from administrative authorities;
—— *conservatoire*, document safe-guarding rights, etc., of absent person, or of his heirs, or of the state;
—— *de courage*, brilliant deed;
—— *de décès*, certificate of death;
—— *de disparition*, (*mil.*) certificate of disappearance of a soldier;
—— *d'engagement*, (*mil.*) contract of enlistment for young man that has never served before;
—— *de l'état civil*, register or certificate of births, marriages, and deaths;
—— *d'hostilité*, (*mil.*) any act equivalent to a declaration of war, to the violation of a treaty, or to the cessation of an armistice or truce;
—— *judiciaire*, legal document that must be passed on by a court or verified by a judge;
—— *notarié*, document drawn up by or before a notary;
—— *de notoriété*, voucher of identification by two or more witnesses;
—— *s privés de l'état civil*, documents concerning powers of attorney, wills, etc.;
—— *de procédure*, (*mil.*) document reporting a military crime or misdemeanor to the corps commander;
—— *de recours*, (*mil. law*) document informing *conseil de revision* (q. v.) that an appeal will be laid before it;
—— *de rengagement*, (*mil.*) contract of reenlistment by a soldier whose time is up, or who is entering his last year of service;
—— *séditieux*, (*mil.*) any seditious or mutinous act (revolt against laws or regulations is so considered);
—— *de substitution*, (*mil.*) document setting forth that two brothers of same class and same home are authorized to exchange numbers drawn by lot;
—— *de vigueur*, (*mil.*) serious attack, serious effort; desperate, final effort.

actif, a., active; (*mil.*, *fort.*) organized for defensive purposes; (*mil.*) on the active list, with the colors; m., assets;
à l'——, to the credit of;
armée ——*ve*, active army (as opposed to reserve, etc.);
service ——, (*mil.*) active service, with the colors.

action, f., deed, act, action; share (of stock); (*mil.*) engagement, affair; (*law*) the right of prosecution;
en ——, (*art.*) action, (command of execution, n. e. U. S. Art.);
—— *d'un cheval*, (*man.*) action of a horse;
—— *civile*, (*law*) right of individual to bring suit;
—— *à distance*, (*mil.*) action or effect at a distance, as by artillery fire;
à double ——, (*mil.*) with both sides represented (drill);
—— *d'éclat*, (*mil.*) brilliant feat of arms;
—— *éloignée*, (*siege*) operations carried on by artillery fire; operations against outer works of the attack;
entrer en ——, (*mil.*) to begin operations, to get to work;
—— *à force ouverte*, (*mil.*) open attack;
hautes ——*s*, (*man.*) high action, showy action;
hors d'——, (*mach.*) out of gear;
manœuvres à double ——, (*drill*) maneuver in which problem is worked out with both sides represented;
—— *de la police judiciaire*, (*mil. law*) first step in procedure of a court-martial, looking up of evidence of crime or misdemeanor, and delivery of accused to military authorities;
—— *publique*, the right of the state to bring suit;
—— *rapprochée*, (*siege*) operation near the point defended or attacked, close attack;
simple ——, (*drill*) maneuver in which only one side is represented;
—— *de vigueur*, (*mil.*) principal engagement or attack or combat, serious attack;
—— *de vive force*, (*mil.*) open assault of a fortress without preliminary siege operations.

actionnaire, m., shareholder, stockholder.

activer, v. a., to urge on, hasten, quicken; press;
—— *le feu*, —— *les feux*, to stir, urge the fires, so as to make them burn better;
—— *le tir*, (*mil.*) to increase the rapidity, the intensity of fire.

activité, f., activity; (*mil.*) active service, active duty; (*mach.*) operation;
en ——, at work, in operation, in blast (of a furnace); (*mil.*) on the active list, employed, on full duty;
en —— *de service*, (*mil.*) on active service, on full duty, employed;
non-——, (*mil.*) state of being unemployed, of an officer not employed under his commission;
en non-——, (*mil.*) unemployed;
en pleine ——, (*mil.*, *etc.*) employed, on full duty;
position d'——, (*mil.*) status of *activité*.

adapter, v. a., to adjust, fit, adapt.

adent, m., (*carp.*) notch, scarf; indentation for scarfing; dovetail joint;
assemblage à ——, scarfing, joggling;
réunir à ——, to joggle or scarf together;
tailler et assembler en ——, to joggle or scarf together.

adenter, v. a., (*carp.*) to joggle, to scarf together.

adhérence, f., adhesion.

adhérer, v. n., to adhere.

adhésif, a., adhesive.

adhésion, f., adhesion.

adiabaticité, f., adiabaticity.

adiabatique, a., adiabatic.

adiabatisme, m., adiabatism.

adjoint, m., assistant; (*Fr. a.*) assistant of a *chef de service*; a., attached to;
officier d'administration ——, v. s. v. *officier*;
officier —— *à l'armement*, v. *officier d'armement*;

adjoint *au chef de corps*, first lieutenant (*inf.*) detailed to assit commanding officer;
—— *du génie*, military employee of engineer service, takes an oath of service. They are:
—— *principal de 1ère classe*, with rank of major;
—— *de 1ère classe*, rank of captain; —— *de 2ème classe*, rank of lieutenant; —— *de 3ème classe*, rank of sub-lieutenant;
officier —— *à l'habillement*, assistant of *capitaine d'habillement*, q. v.;
—— *à l'intendance militaire*, functionary of the *intendance*, q. v., either a captain or an *officier d'intendance*, q. v.;
—— *au trésorier*, assistant of *capitaine trésorier*, q. v.

adjudant, m., (*Fr. a.*) ranking n. c. o. of a company, troop, or battery (n. e.);
—— *de bataillon*, assisted the *adjudant-major* (almost the same as regimental sergeant major, U. S. A., suppressed in 1893);
capitaine —— *major*, adjutant (almost the same as battalion adjutant);
—— *de casernement*, has charge of lights, bedding, etc., in barracks;
—— *de compagnie*, assists the captain (somewhat like first sergeant, U. S. A.);
—— *élève d'administration*, (*adm.*) n. c. o. graduated from *Ecole d'administration* and graded as *adjudant* in the bureaux of the *administration* until promoted;
—— *d'état-major*, one belonging to the *petit état-major*, q. v.; assists *adjudant-major*;
—— *de la garnison*, officer who assists the *major de la garnison* (q. v.) in a *place de guerre*;
—— *général*, used by Marbot as equivalent to *colonel d'état-major*; sometimes (formerly) used for *aide-de-camp*;
—— *major*, battalion adjutant, adjutant;
—— *de pavillon*, (*nav.*) flag-lieutenant;
—— *de place*, assistant of c. o. in *places de guerre*, suppressed in 1875;
—— *principal*, (*Fr. nav.*) functionary attached to each of the five divisions of the personnel of the fleet;
—— *sous-officier*, highest ranking in n. e. o.
—— *vaguemestre*, adjudant who performs duties of *vaguemestre*, q. v.

adjudantur, m., (*mil.*) the German *adjutantur*.
adjudicataire, m., bidder, lowest or highest bidder; person who gets a contract.
adjudication, f., adjudication; assignment of contract;
—— *précédée d'une séance préparatoire*, award in which admissibility of contractor is made known to him after a meeting of the *commission d'admission* (case requiring contractor of assured responsibility);
—— *simple*, award made in open meeting before competing contractors, whose soundness is taken for granted by the fact of their presence;
—— *par voie d'enchères*, assignment of government property (e. g., condemned horses) to highest bidder;
—— *par voie des soumissions*, assignment of contract to lowest bidder (the government buys).
administrateur, m., administrator; any person having administrative duties; (*Fr. a.*) person or functionary who has administrative duties and reports directly to the minister of war (not a strictly defined term).
administratif, a., administrative, coming from the administration.
administration, f., administration; a general term, including all the methods, means, etc., necessary to carry on a given business;
—— *de l'armée*, (*Fr. a.*) generic term for the whole business of organization, instruction, supply, discipline, mobilization, recruiting, etc., of the army. It includes specifically the *services* of artillery, engineers, *intendance*, "powders and saltpeters," and the medical department. The minister of war is the responsible head. The term is also used to denote the administrative personnel.

administration *centrale de la guerre*, (*Fr. a.*) the ministry of war, including the office (*cabinet*) of the minister, the general staff of the army, the *service intérieur* (q. v.), and the various *directions* (q. v.);
conseil d' ——, v. s. v. *conseil*;
—— *intérieure des corps de troupe et des établissements*, (*Fr. a.*) management and supervision of all questions relating to troops and to establishments.
administrativement, adv., in accordance with administrative regulations.
administrer, v. a., to administer, using legal means and resources.
admissibilité, f., fitness for admission (as, to military schools);
épreuves d' ——, written examination, to test fitness for admission.
admission, f., admission, (e. g., to a school); introduction, admission (of steam into cylinder, etc.); act of admitting;
épreuves d' ——, oral and practical examination, to which are admitted only those who have passed the *épreuves d'admissibilité*.
adonner, v. n., (of a rope) to stretch.
ados, m., slope, talus, embankment.
adossement, m., (*fort.*) slope.
adosser, v. a., r., to prop; to back up, or be backed up, against something; (*mil.*) to have a river, lake, hill, etc., close in one's rear or back.
adouber, v. a., (*nav.*) to mend, repair.
adoucir, v. a., to soften; to remove a sharp edge, to smoothen, make smooth; to file smooth; to cut down, reduce (a slope); to round off (an edge, a crest); to grind (metal, etc.); (*sm. a., art.*) to fine-bore; (*met.*) to anneal.
adoucissage, m., grinding, filing down; reduction; removal of sharp edges, etc.
adoucissement, m., union of two bodies by means of a bevel; (*met.*) annealing, softening.
adoucisseur, m., (*art.*, *sm. a.*) fine-borer, finishing-borer; (*in gen.*) workman who polishes.
adresse, f., address (place); skill, dexterity.
à droite, m., (foot troops, from a halt) an individual right face; (foot troops marching) an individual by the right flank; (of a squad, section, company) right; (*cav.*) (of a trooper) by the right flank, (of a platoon, troop) right; (*art.*) (school of the driver) by the right flank, (platoon, battery) right wheel; (*mountain art.*) (of a mule) an individual by the right flank, (of a section) right wheel.
aduction, f., v. *adution*.
aduire, v. n., to exercise, train, young carrier pigeons to fly about their cote without flying off; to "home."
aduition, f., training of young carrier pigeons to fly about their cote without escaping; homing.
adversaire, m., adversary, enemy.
aérage, m., ventilation;
puits d' ——, air-shaft.
aération, f., ventilation.
aérer, v. a., to air, to ventilate.
aéroduct, m., air-passage in a wall.
aéromètre, m., (*inst.*) aerometer.
aéromoteur, m., (*mach.*) hot-air engine (hot air and vapor of water).
aéronaute, m., aeronaut.
aéronautique, a. and f., aeronautic; aeronautics.
aéronef, f., air-ship.
aéroplane, m., aeroplane, flying machine.
aérostat, m., balloon.
aérostatier, m., aeronaut.
aérostation, f., aerostation.
aérostatique, a. and f., aerostatic; aerostatics.
aérosteur, m., aeronaut.
aérostier, m., aeronaut; (*mil.*) man serving in balloon section.
affaiblir, v. a., to weaken.
affaiblissement, m., weakening; loss of strength, of intensity (as of fire).

affaire, f., affair, business, piece of business; (*mil.*) affair, combat, engagement;
— *d'avant-postes,* (*mil.*) outpost engagement, or skirmish;
— *s indigènes,* — *s arabes,* (*Fr. a.*) all matters relating to administration of the military territory in Algiers.

affaissement, m., settling, subsidence; (*sm. a.*) shortening of bullet, due to pressure of powder gases; (*art.*) depression in cast guns, due to projection in mold.

affaisser, v. a. r., to sink, settle, sag, subside, collapse, give way.

affaler, v. a., to overhaul (a rope, etc.);
— *un palan,* to overhaul a tackle;
s'—, to stumble (of a horse).

affamer, v. a., (*mil.*) to starve, to reduce to starvation (garrison, fort).

affectation, f., (*mil.*) assignment to a particular arm, unit, or duty;
état nominatif d'—, (*Fr. a.*) list of men sent each year from active army to the reserve;
— *spéciale,* special assignment in certain administrative and technical branches of the public service.

affecter, v. a., to assign, to allot; to tell off or detail for a particular duty, etc.

affermage, m., (*adm.*) renting out of slopes and of arable land in or near fortifications.

affermir, v. a., to strengthen; fasten, fix; establish, consolidate.

affermissement, m., strengthening, consolidation.

affichage, m., posting of public notices; (*esp., Fr. a.*) posting of notices dealing with the army, as of lists, drawing of lots, etc.

affiche, f., poster.

afficher, v. a., to post (a notice).

affilage, m., sharpening, setting; act of sharpening, etc.

affiler, v. a., to sharpen, set (a tool, etc.).

affilure, f., operation of pointing (as a nail).

affinage, m., (*met.*) fining, refining;
méthode d'— *anglaise,* puddling.

affiner, v. a., (*met.*) to fine, refine; (of the weather) to clear.

affinerie, f., (*met.*) fining-hearth, finery.

affineur, m., (*met.*) finer (workman).

affleuré, a., flush (said of two adjacent or contiguous solids);
assemblage —, flush-joint.

affleurement, m., outcrop (of a ledge, rock, etc.).

affleurer, v. a., to make flush, level, with each other.

afflouer, v. a., to float (a stranded boat).

affluent, m., (*top.*) affluent, tributary stream; a., tributary, flowing into;
à marée, tide-stream.

affolé, a., (*elec.*) perturbed, disturbed (of magnetic needle).

affolement, m., (*elec.*) perturbation, unsteadiness (of magnetic needle).

affouillable, a., liable to erosion, to undermining (of soil, by water); crumbling, friable.

affouillement, m., caving in; undermining; washout; exfoliation (as by rust); (*art.*) exposing of a foundation, by artillery fire;
— *des balles,* (*sm. a.*) small cavity in bullets of pure lead, due to powder gases;
— *des bouches à feu,* (*art.*) pitting; long and sinuous cavities in bronze guns (obs.);
— *du terrain,* wash, washout, caused by running waters.

affouiller, v. a., to undermine, to wash away; (*art.*) to lay bare a foundation (by artillery fire).

affourager, v. a., to give fodder to animals (in the stable).

affourchement, m., slit-and-tongue joint, tongue joint, split joint.

affourcher, v. a., to join by an open mortise; (*nav.*) to moor.

affranchir, v. a., to free, discharge; release, deliver; let go, exempt; pay, prepay; accomplish; shake off;
— *d'eau,* to pump a vessel free of water;

— *les bouts,* to trim off ends (e. g., to turn the base of a projectile in a lathe).

affranchissement, m., action of freeing, accomplishing, etc.; prepayment (of postage, carriage, etc.).

affrètement, m., chartering, freighting (of a vessel);
— *à la cueillette,* chartering by administration of commercial vessel in desired direction, when government cargo will not fill a whole ship.

affréter, v. a., to charter, to freight.

affréteur, m., charterer.

affronter, v. a., (*mil.*) to face, to attack boldly, to expose one's self.

affût, m., (*s. c. art.*) top-carriage (U. S.), slide-carriage (R. A.); (*r. f. art.*) mount, mounting; carriage (field artillery); frame or bed (as of a saw, etc.);
(Except where otherwise indicated, the following terms relate to artillery.)
— *à aiguille,* platform-carriage;
— *à balancier,* any carriage in which the gun is directly supported by levers (*balancier*) rotating about a horizontal axis;
— *de batterie,* (*nav. art.*) broadside carriage, central-battery carriage;
— *à bêche de crosse,* trail-spade carriage;
— *à bêche d'essieu,* carriage equipped with an axle-spade;
— *de bord,* (*nav. art.*) naval carriage;
— *à bouclier,* any carriage provided with a shield;
— *à brancard,* shaft carriage;
— *de campagne,* field carriage;
— *de caponnière,* caponier mount;
— *de casemate,* casemate carriage;
— *de casemate à pivot-avant,* front-pintle casemate carriage;
— *de cavalerie,* galloping carriage;
— *-chandelier,* — *à chandelier,* (*r. f. art.*) forkpivot mount, pillar-mounting;
— *à châssis,* slide-carriage, châssis-carriage; also, a generic term for this class of carriages, as distinguished from the field-carriage type;
— *à châssis circulaire,* carriage with circular châssis (Canet);
— *à châssis horizontal,* carriage with horizontal châssis;
— *à châssis oscillant,* carriage with rocking or oscillating châssis;
— *à châssis à pivot-avant,* front-pintle carriage;
— *à châssis à pivot central,* center-pintle carriage;
— *à col de cygne,* swan-neck carriage (siege);
— *de côte,* seacoast artillery carriage;
— *à crinoline,* (*r. f. art.*) crinoline mount; elastic cone mount; cage-mount (U. S. N.);
— *à crosse,* trail-carriage; (*r.f. art.*) mount fitted with a stock;
crosse d'—, trail;
— *cuirassé,* armored carriage, mount; French translation of German (Schumann) name for turret or cupola;
— *cuirassé à éclipse,* disappearing armored carriage;
— *de débarquement,* (*nav.*) landing-carriage;
— *à déformation,* generic term for carriages in which recoil is taken up by springs, cylinders, etc.;
— *démontable,* carriage that takes apart, take-down carriage;
— *à dépression,* depression carriage;
— *à éclipse,* disappearing carriage;
— *élastique,* (*r.f. art.*) elastic mount; (*art.*) generic term for carriage in which buffers, cylinders, etc., are used to take up shock of recoil;
— *élevé,* barbette, overbank carriage;
— *d'embarcation,* (*nav.*) boat-carriage;
— *d'embrasure,* embrasure carriage;
— *-embrasure,* (*r. f. art.*) caponier or flanking casemate mount;
— *à embrasure minima,* minimum-port carriage;
— *à embrasure réduite,* minimum-port carriage;
seacoast carriage (R. A.) of limited field of fire;
— *fixe,* stationary carriage, (seacoast, etc.);
— *à flasques,* bracket-trail carriage;

affût *à flèche*, block-trail carriage;
— *à flèche élastique*, telescoping-trail carriage, elastic-trail carriage (Canet);
— *à frein circulaire*, carriage fitted with a circular brake (Canet);
— *à frein hydraulique*, carriage with hydraulic recoil-check;
— *à freins à lames*, compressor-bar carriage;
— *à fusée*, (mil.) rocket-stand; rocket-frame;
— *de gaillard*, (nav.) gun-deck carriage;
— *glissant*, (siege art.) carriage sliding on skids during recoil;
— *à glissement*, (Fr. art.) 220ᵐᵐ mortar carriage (slides on platform during recoil);
grand —, lower carriage of French 120ᵐᵐ short;
— *indépendant*, unarmored carriage;
— *d'infanterie* (r. f. art.) infantry mounting;
— *libre*, generic term for movable (wheeled) carriages;
— *à limonière*, shaft carriage; galloping carriage;
— *à manœuvre électrique*, electrically controlled carriage;
— *à manœuvre hydraulique*, hydraulically controlled carriage;
— *marin*, naval carriage;
de montagne, mountain-gun carriage;
— *à*, — *de*, *mortier*, mortar carriage, mortar bed;
— *d'obusier*, howitzer carriage;
— *omnibus*, (Fr. art.) a carriage so called because it can be adapted to several different guns, field and siege;
— *oscillant*, oscillating, rocking carriage;
petit —, top carriage, French 120ᵐᵐ short;
— *à pivot*, pintle carriage;
— *à pivot antérieur*, — *à pivot avant*, front-pintle carriage;
— *à pivot central*, center-pintle carriage;
— *à pivot élastique*, (r. f. art.) elastic mount (Engström);
— *à pivotement fictif et à tourillonnement autour de la bouche*, generic term for minimum port carriage under metallic cover;
— *de place*, fortress carriage; garrison carriage (England);
— *à plaque tournante*, turntable carriage;
— *de plat-bord*, (nav. r. f.) gunnel-mount, rail-mount;
— *plateforme*, (Fr. a.) carriage of 120ᵐᵐ short;
— *à plateforme tournante*, turntable carriage;
— *de pont*, (nav. r. f.) deck mount;
queue d'—, rear end of top carriage;
— *à rabattement*, disappearing carriage;
— *à rappel automatique*, spring-return carriage;
— *à recul*, recoil carriage;
— *à recul limité*, limited recoil carriage;
— *sans recul*, (r. f. art.) nonrecoil carriage;
— *de rempart*, (r. f. art.) rampart mount; (in gen.) rampart carriage;
— *à ressorts de rappel*, spring-return carriage;
— *rigide*, generic term for carriage in which recoil is not taken up by buffers, cylinders, etc.;
— *roulant*, wheeled carriage; roller carriage; (siege) carriage on wheels, as distinguished from — *glissant*, q. v.;
— *à roulettes*, truck carriage;
— *de siège*, siege carriage;
— *de siège et place*, siege and fortress carriage;
— *à soulèvement*, (Fr. art., obs.) elevator carriage (lets down from rollers before firing);
— *à tampons élastiques*, Engelhardt field-mortar carriage;
— *de teugue*, (nav.) forecastle carriage (swivel and chase guns);
— *de tir*, (em. a.) rifle-rest;
— *à*, — *de*, *tourelle*, turret carriage;
— *de tourelle-barbette*, barbette turret mount;
— *à traineau*, — *traineau*, sledge-carriage;
— *à trépied*, (r. f. art.) tripod mount;
— *à treuil de brague*, Brookwell carriage (obs.);
— *truc*. r. r. gun carriage, platform on rails, from which gun may be fired;
— *à tubes*, (mil.) rocket volley carriage;
— *à vis*, generic term for turret-carriages worked by screw-gearing, as opposed to hydraulically controlled carriages.

affûtage, m., outfit of tools; sharpening, setting; act of sharpening.

affûter, v. a., to sharpen a tool;
machine à —, tool-grinding machine.

affûteuse, f., tool-grinding machine.

agaric, m., punk.

à gauche, m., v. *à droite*, substituting *gauche* (left) for *droite* (right).

âge, m., age;
cheval hors d'—, (hipp.) old horse, horse that has lost the mark;
limite d'—, limits of age of entry into schools; minimum age of entry into certain departments; (mil.) retiring age;
être atteint par la limite d'—, (mil.) to reach retiring age.

agenda, m., memorandum book.

agent, m., agent; (mil.) subordinate employé of pay, post-office, telegraph, and railroad departments;
— *des attelages*, (art.) agent of communication between captain (at guns in position) and limbers and teams under cover (he is the *chef de caisson de la section du centre*);
— *de liaison*, (mil.) officer, n. c. o., or man who acts as agent of communication between various bodies of troops, or of parts of a single body (e.g., *batterie de tir* and *échelon de combat*, qq. v.) (frequently *agent* alone is used, when not an officer);
— *principal des prisons*, (Fr. a.), n. c. o. of the grade of *adjudant*, in charge of interior economy and of lower personnel of a military prison;
— *réducteur*, — *révélateur*, (phot.) developer.

agglomération, f., agglomeration.

aggloméré, m., (elec.) mixture of gas-coke and of bioxide of manganese (in one kind of Leclanché cell).

agir, v. a., to act, work; (mil.) to operate;
— *sur le cordon*, (art.) to pull the lanyard.

agitateur, m., (mach., etc.) agitator; mechanical stirrer; shaker.

agiter, v. a. r., to agitate, shake, stir; to become rough (of the sea, or water); to rise (of the sea).

agrafage, m., clamping, clasping, cramping, etc.;
— *à ressaut*, (art.) locking of hoops by means of shoulders.

agrafe, f., cramp, cramping-iron; lip, clip; (mil.) clasp of a cuirass or breastplate; clasp of a medal; (art.) filling-ring; coupling-pin; locking-lip (of a hoop); locking-shoulder (of a hoop); clip, guide-clip, guide-hook; holding-down clip; catch; longitudinal strain-bar;
— *d'armement*, (art.) cocking-clip (of a fuse);
— *d'artillerie*, hook or clasp to bind various parts together.
— *d'assemblage à vis*, (art.) locking-ring;
— —*guide*, (art.) guide-holding-down clip.
— *et porte*, hook and eye;
— *de sûreté*, (art.) safety-clip, safety-spring catch (of a fuse).

agrafer, v. a., to hook, clasp, clamp; (mach.) to gib; (art.) to lock by an *agrafe*.
s'—, (art.) to lock (gun construction);
— *à ressaut*, (art.) to lock hoops by means of shoulders.

agrandir, v. a., to enlarge.

agrandissement, m., enlargement.

agréer, v. a., to rig, (v. *gréer*).

agrégation, f., (mil.) formation (applied to all *regular* assemblies).

agréments, m. pl., (unif.) trimmings or insignia or ornaments of uniform, of headdress, and of equipment.

agrès, m. pl., (*pont.*) bridge equipment, stores and supplies; (*art.*) all implements for mechanical maneuvers; (*gym.*) appliances, apparatus;
—— *simples* (*art.*) in mechanical maneuvers, any implements used directly, e. g., a lever;
—— *spéciaux*, (*art.*) in mechanical maneuvers, any special apparatus or gear.

agresseur, m., aggressor.

agression, f., aggression; (*mil.*) initiative of the offensive; armed incursion.

aguerri, a., (*mil.*) disciplined; experienced in war; war-hardened; m., trained and disciplined soldier;
mal ——, undisciplined, inexperienced in war.

aguerrir, v. a., (*mil.*) to inure to war, to season (troops).

aguets, m. pl., *être aux* ——, to be on the watch, on the lookout.

agui, m., (*cord.*) girtline, sling.
nœud d'——, standing bowline knot.

aide, m., assistant, coadjutor (*aide* is prefixed to the names of various grades and offices, to denote either a lower rank of the grade or an assistant in the duty indicated); (*surv.*) chainman;
——-*cargot*, (*mil. slang*) canteen assistant;
——-*chaîneur*, (*surv.*) fore-chainman;
——*de cuisine*, (*mil.*) assistant cook, second cook (U. S. A.);
deuxième ——, (*surv.*) fore-chainman
——-*maréchal*, —— -*maréchal ferrant*, under farrier; *premier* ——, (*surv.*) hind-chainman;
——-*servant*, (*art.*) supernumerary cannoneer;
——-*vétérinaire*, assistant veterinary.

aide, f., assistance, help, succor, relief, aid; (*man.*) aid;
accord des ——-*s*, (*man.*) combination of the aids in horsemanship;
·*désaccord des* ——*s*, (*man.*) failure of the aids to combine for desired result;
——*s diagonales*, (*man.*) leg and rein, etc., of opposite sides;
——*s latérales*, (*man.*) leg and rein, etc., of same side.

aide de camp, m., (*mil.*) aid-de-camp.

aide-major, m. (*Fr. a.*) —— *de 1ère classe*, surgeon-lieutenant; —— *de 2ème classe*, surgeon second lieutenant (n. e.).

aigle, m., eagle; f., (*mil.*) eagle (ensign);
porte- ——, (*mil.*) ensign-, standard-bearer.

aigre, a., (*met.*) cold short, brittle.

aigrette, f., (*unif.*) tuft, plume, aigrette;
pot d'——, (*artif.*) firework, so called.

aigrir, v. a., (*met.*) to make or render brittle.

aigu, a., sharp, pointed, shrill.

aiguade, f., fresh water (for ships) (*obs.*); place where fresh water may be obtained (for ships).

aiguayer, v. a., to water, take in water; to wash a horse in a pond or river; to rinse.

aiguillage, m., (*r. r.*) switch;
—— *à l'anglaise*, automatic switch.

aiguille, f., needle; draught-tree (of a carriage); (*top.*) needle, sharp top of a mountain; (*inst.*) index, hand, pointer; (*fond.*) vent-wire; (*r. r.*) switch-, point-, rail; (*sm. a.*) firing-pin; (*in plural*) shear-legs;
—— *aimantée*, (*elec.*) magnetic, compass, needle;
arme à ——, (*sm. a.*) needle gun;
—— *d'artificier*, (*artif.*) rod or pin or drift, for making fireworks;
—— *astatique*, (*elec.*) astatic needle;
—— *de boussole*, compass-needle;
—— *de chemin de fer*, (*r. r.*) switch;
—— *à contrepoids*, (*r. r.*) counterpoise switch;
—— *d'emballage*, packing needle, burlap needle;
fausse ——, (*r. r., mil.*) contrivance consisting of dislodging point rails, so as to cause a run-off;
—— *de ferme*, (*cons.*) roof-post, truss-post;
—— *de fusil* (*sm. a.*) firing-pin;
—— *d'inclinaison*, (*elec.*) dipping needle;
—— *mobile*, (*r. r.*) point-rail; point;
—— *pendante*, (*cons.*) hanging-post (king-, queen-, post);

aiguille, *de raccordement*, (*r. r.*) switch;
—— *à tricoter les côtes*, (*mil. slang*) sword (cf. "frog-sticker," "cheese-knife," "toasting-fork";
—— *trotteuse*, (*inst.*) split-second hand of a stop-watch, etc.).

aiguiller, v. a., (*r. r.*) to work or set a switch.

aiguilletage, m., (*cord.*) lashing, seizing; mousing, frapping, racking;
faire un ——, to frap, seize, etc.;
—— *en portugaise*, rose-lashing.

aiguilleter, v. a., (*cord.*) to lash, seize, mouse, frap, etc.

aiguillette, f., (*cord.*) lashing, seizing; (*unif.*) aiguillette; (*sm. a.*) needle, firing-pin; (*in plural*) shears;
nouer l' ——, (*man.*) to kick out violently with both legs.

aiguilleur, m., (*r. r.*) switchman (U. S.); points-man (England).

aiguillon, m., (*top*) needle, sharp top of a mountain.

aiguillot, m., pintle.

aiguisage, m., sharpening (of a cutting tool); grinding down of swords, etc., to proper form and weight.

aiguiser, v. a., to sharpen, set, whet;
pierre à ——, whetstone.

aiguiserie, f., shop where arms, etc., are ground down to shape and weight.

aiguiseur, m., workman who does the *aiguisage*, q. v.

aile, f., wing; blade (of a propeller); fluke of an anchor); pallet; flange (of angle iron); wing (to direct a flame in a furnace, hence a sort of fire bridge); (*mil.*) wing (of an army); (*fort.*) branch, wing; long face (e. g., of a crown or horn-work); (*mach.*) tooth of a pinion; (*pont.*) widening (of the approach to a bridge);
—— *décisive*, (*mil.*) that wing or part of the army by which the real attack is made;
—— *démonstrative*, (*mil.*) that wing or part by which the front attack or demonstrative attack is made, so as to give time and opportunity to the *aile décisive*;
—— *de fortification*, (*fort.*) long branch on flank of open gorge work;
—— *d'hélice*, blade of a propeller;
—— *de lance*, (*sm. a.*) hilt of a lance;
—— *marchante*, (*drill*) marching flank;
—— *de mouche*, cramp iron;
—— *de pont*, (*pont.*) widening or splay of the approach to a bridge;
à tire d' ——, as fast as possible.

aileron, m., (*torp.*) blade; (*fort.*) shoulder-caponier; simple, single caponier; (sometimes) flank of a small work.

ailette, f., lug, tenon, stud; projecting piece; small wing of a building; (*mach., etc.*) paddle; (*art.*) stud (of a projectile, *obs.*); (*sm. a.*) locking-lug;
—— *directrice*, (*torp.*) guide-fin, guide-vane.

aimant, m., (*elec.*) magnet, loadstone;
armer un ——, to arm a magnet;
—— *artificiel*, artificial loadstone; magnet;
—— *directeur*, compensating, directive, magnet;
—— *en fer à cheval*, horseshoe magnet;
—— *naturel*, natural magnet;
pierre d' ——, loadstone.

aimantation, f., (*elec.*) magnetization;
—— *par la double touche*, magnetization by double touch;
—— *par la simple touche*, magnetization by simple touch;
—— *par la touche séparée*, magnetization by divided touch.

aimanter, v. a., (*elec.*) to magnetize;
—— *à saturation*, to magnetize to saturation;
—— *par la double*, etc., *touche*, v. s. v. *aimantation*.

ain, m., (*unif.*) (abbreviation of *centain*) space or length containing 100 threads (manufacture of cloth).

aîné, m., eldest son.

air, m., air, the atmosphere; blast (in, or of, a furnace); (*man.*) air;
bâche de la pompe à ——, hot-well;

air *bıs,* (*man.*) air in which the horse does not raise or lift his body far from the ground;
bielle de la pompe à ——, pump side-rod;
boîte à ——, air-box;
—— *chaud,* hot blast;
—— *diagonal,* (*man.*) air based on the diagonal efforts of the horse;
échappement de l' ——, escape, air escape;
être en l' ——, (*mil.*) to be too far advanced; to be without proper support; to be in the air;
—— *froid,* cold blast, cold-air blast;
à l' —— *libre,* (*expl.*) in free air;
——*s de manège,* (*man.*) riding-school paces, graces, airs;
matelas d'——, (*steam, etc.*) air-cushion;
mettre la crosse en l' ——, (*mil.*) to surrender;
pompe à ——, air-pump;
—— *près de terre,* (*man.*) v. *air bas;*
—— *relevé,* (*man.*) air in which horse raises his body in the air;
réservoir, seau, de la pompe à ——, air-pump bucket;
tige d'articulation de la pompe à ——, pump-rod and links;
tirer en l'——, (*mil.*) to fire in the air (e. g., to frighten rioters).

airain, m., (*met.*) brass.

aire, f., area, superficies; floor, flooring, floor area, floor space; (*in barracks*) flooring of ground floor when covered with neither wood nor flagstones; any even or smooth plat or flat; hearth, sole (of a furnace); (*nav.*) way, headway;
—— *de champ de tir,* (*ball.*) plane surface suitably marked and prepared to receive projectiles, so that coordinates of points of fall may be obtained;
—— *d'enclume,* anvil face or plate;
—— *de grille,* (*steam*) grate-area;
—— *d'un pont,* roadway;
—— *de vent,* —— *de vent de la rose,* point of the compass; rhumb.

ais, m., board, plank;
—— *de carton,* pasteboard.

aisance, f., ease; (*mil.*) "rear" (*fam.*); room, elbow-room (in ranks);
—— *des coudes,* (*mil.*) space between adjacent men in ranks, room, elbow-room.

aisceau, m., adze, cooper's adze.

aissante, f., (*cons.*) shingle.

aisseau, m., (*cons.*) shingle.

aisselier, m., (*carp.*) brace, strut; angle-brace.

aisselle, f., armpit; throat (of an anchor).

aissette, f., adze.

aissieu, m., v. *essieu.*

aisson, m., grapnel, four-fluked anchor.

ajourné, m., (*Fr. a.*) man who has had an *ajournement* (q. v.).

ajournement, m., postponement; (*Fr. a.*) delay of two years, accorded young men found too short or otherwise physically unfitted for military service, before they are examined again; postponement or delay of joining for annual instruction, of reservists and of territorialists, for one year, for private reasons;
—— *d'appel,* (*mil.*) v. *ajournement* (*Fr. a.*);
—— *de rappel de solde,* (*Fr. a.*) delay of six months in drawing pay accumulated, to be suffered by soldier who does not present required vouchers on rejoining;
—— *de réception d'effets,* (*Fr. a. adm.*) postponement of acceptance, for reexamination, of articles found defective by *commission de réception,* and susceptible of repair.

ajourner, v. a., to put off, to postpone; (*Fr. a.*) to put off entry into service, of young men, reservists, territorialists (v. *ajournement*).

ajoutoir, m., v. *ajutage.*

ajust, m., (*cord.*) bend or knot to fasten ends of two ropes together.

ajuste, m., v. *ajust.*

ajustage, m., fitting (of parts together); performing of final operations; assembling of various parts of machines, arms, etc.); finishing. touching up of a piece, a mechanism, etc.; (*mach.*) adjusting, fitting.

ajusté, a., (*art., sm. a.*) well aimed;
feu ——, well-aimed fire.

ajustement, m., adjusting, fitting, sizing; arrangement, settlement.

ajuster, v. a., to adjust, to fit; (*mach.*) to fit; (*art. sm. a.*) to aim at (with the notion of *care*), to take aim at;
—— *un cheval,* (*man.*) to complete the training of a horse;
—— *cône,* (*mach.*) to taper;
—— *à frottement doux, dur, libre,* to give an easy, tight, free, fit;
—— *à refus,* to fit as tight, as close, as possible;
—— *les rênes,* (*man.*) to get the reins evenly in the hand.

ajusture, f., (*farr.*) fitting of upper face of a horseshoe to the foot;
—— *en bateau,* fitting in which the branches curve in, so that toe, quarter, etc., do not bear on ground;
—— *entolée,* fitting in which the branches are kept from ground by their outer edge;
—— *à éponges renversées,* fitting in which the quarters (*éponges*) are turned down;
—— *de mulet,* fitting with toe in air, and the rest of the shoe flat on the ground.

ajut, m., (*cord.*) v. *ajust.*

ajutage, m., adjutage, tube.

ajutoir, m., adjutage.

alaise, f., v. *alèse.*

alarme, f., alarm; cry of alarm, warning cry; warning;
faire ——, to give the alarm;
fausse ——, false alarm;
place d' ——, (*mil.*) point designated for assembly of troops under arms, when the *générale* is beaten.

albugo, (*hipp.*) albugo.

album, m., illustrated catalogue (of manufactures, etc.).

alcaloïde, m., (*chem.*) alkaloid.

alcool, m., (*chem.*) alcohol.

alcoolique, a., alcoholic.

alcoolomètre, m., alcoholometer.

alène, f., awl, pricker; saddler's awl;
—— *à coudre,* common awl;
—— *plate,* brad awl.

alerte, f., (*mil.*) alarm; turning out the guard;
fausse ——, false alarm.

alésage, m., (*mach., etc.*) broaching, boring, drilling, fine-boring; (*art.*) boring of guns; boring to caliber; (*sm. a.*) fine-boring (of barrel);
banc d' ——, boring bench;
barre d'——, boring bar.

alèse, f., furring; (*sm. a.*) wooden lining of (metal) scabbard.

alésé, a., shouldered; bored; drilled.

aléser, v. a., to bore, to broach, to drill; to fine-bore (a gun); to rebore, to bore up; to turn true;
machine à ——, drill, drilling machine.

aléseuse, f., (*mach.*) drill, drilling machine;
—— *à revolver,* (*mach.*) turret-drill, turret-head drill.

alésoir, m., (*mach.*) broach; reamer; rose-reamer) fine borer; polishing-bit, finishing-bit; boring machine;
—— *conique,* angular reamer;
—— *de lumière,* (*art.*) vent-bit;
—— *rond,* round broach;
tête d' ——, broaching-head, boring-head.

alésure, f., boring-chip.

alezan, a., (*hipp.*) chestnut (color); m., chestnut (horse).
—— *brûlé,* color of roasted coffee;
—— *cerise,* reddish chestnut;
—— *châtain,* chestnut sorrel;
—— *clair,* light chestnut;
—— *cuivré,* chestnut of coppery red reflection;
—— *doré,* chestnut of a golden reflection;
—— *foncé,* brownish chestnut;
—— *lavé,* light chestnut (light sorrel, according to some);
—— *ordinaire,* approaching a cinnamon color;
—— *roux,* red or light bay;
—— *saure,* sorrel (according to some).

alèze — **allumeur**

alèze, f., v. *alèse*.
alfa, m., (*cord.*) Spanish grass.
algarade, f., (*mil.*) feint, sham attack, bold dash, etc., to excite alarm in enemy.
algarde, f., false alarm.
alidade, f., (*inst.*) alidade;
 —— *autoréductrice*, alidade giving distances and differences of level;
 —— *à éclimètre*, clinometer alidade;
 —— *à lunette*, telescopic alidade;
 —— *nivelatrice*, leveling alidade;
 —— *à pinnules*, open-sight alidade;
 —— *plongeante*, telescopic alidade; one fitted with a sighting tube;
 —— *plongeante à lunette*, telescopic alidade proper;
 —— *à télescope*, telescopic alidade;
 —— *à viseur*, sighting-tube alidade.
alignement, m., alignment; (*surv.*) bearing; line of direction; (*cons.*) lining-out; (*mil.*) alignment, dress, dressing; in commands, "dress!"
 —— *des approvisionnements*, (*mil.*) state or condition of keeping supplies and stores up to regulation standard;
 —— *en arrière*, (*mil.*) dressing on a line in rear;
 —— *des comptes*, (*adm.*) condition of accounts so that balance, final statement, may be struck;
 —— *direct*, (*mil.*) dressing forward on a line;
 —— *droit*, (*r. r.*) straight part of a track;
 —— *à droite*, (*gauche*), (*mil.*) right (left) dress;
 être d' ——, to be in line;
 —— *oblique*, (*mil.*) dressing up obliquely;
 perdre l' ——, (*mil.*) to lose the dress, to get out of dress;
 rectifier l' ——, (*mil.*) to rectify the alignment;
 —— *topographique*, (*top.*) condition of having three points in same vertical plane.
aligner, v. a., to line out, up; to set out, mark out; to align (in general); (*art.*) to lay a piece for direction (rare); (*surv.*) to stake out; (*mil.*) to keep up to a certain quantity, as rations, etc.; to bring up to a certain amount; to get on the same line;
 s' ——, (*mil.*) to dress up to a line;
 —— *les hommes à deux jours de vivres*, (*mil.*) to furnish the men with two days' rations;
 —— *les rations pour tant de jours*, (*mil.*) to issue, furnish so many days' rations, to bring amount up to so many days' rations.
aliment, m., food, aliment.
alimentaire, a., feed (as, feed-water); m., (*mach.*) feed mechanism;
 eau ——, (*mach.*) feed-water;
 pompe ——, feed-pump;
 tuyau ——, feed-pipe.
alimentation, f., victualing, rationing, supply; (*mil.*) subsistence; (*mach.*) feed, feeding;
 appareil d' ——, (*mach.*) feed, feed apparatus;
 eau d' ——, feed-water;
 pompe d' ——, feed-pump;
 tuyau d' ——, feed-pipe.
alimenter, v. a., (*mach.*) to feed; (*in gen.*) to supply, keep up; (*mil.*) to subsist.
alizé, m., —— *s*, trade winds;
 vent ——, trade wind.
alkali, m., (*chem.*) alkali, potash.
alkalinité, f., (*chem.*) alkalinity.
allège, f., lighter (boat); (*r. r.*) tender.
allégement, m., relief, easing, lightening, alleviation.
alléger, v. a., to ease, relieve, lessen, alleviate; to unload; (*man.*) to ease (a horse); (*cord.*) to stick out (a rope).
aller, v. n., to go, proceed, repair, depart, sail, move; lie or rest on one's oars; (*mach.*) to work, function;
 —— *l'amble*, (*man.*) to amble;
 —— *en arrière*, to back;
 —— *en avant*, to go forward, ahead;
 —— *à bord*, (*nav.*) to go on board;
 —— *aux cartouches*, (*mil.*) to go for cartridges, on battlefield;
 —— *de conserve*, (*nav.*) to sail in company, under convoy;

aller *en course*, (*nav.*) to go privateering; to cruise;
 —— *en dérive*, (*nav., etc.*) to drift;
 —— *à l'épée*, (*fenc.*) to uncover one's self in a parry;
 faire ——, to set in motion;
 —— *au feu*, (*mil.*) to go into action;
 —— *au fond*, to sink;
 —— *le galop*, (*man.*) to gallop;
 laisser ——, to give way together (boats);
 —— *le pas*, (*man.*) to pace;
 —— *au recul*, (*art.*) to recoil, to go to position of "from battery;"
 —— *le trot*, (*man.*) to trot.
alliage, m., (*met.*) alloy.
alliance, f., alliance, confederacy; (*harn.*) connecting loop or strap (metal or leather);
 acte d' ——, *traité d'* ——, treaty of alliance.
allié, m., (*mil.*) ally.
allier, v. a. r., to combine, unite; ally; blend; (*met.*) to alloy.
allingue, f., pile driven into the bed of a river.
allocation, f., allowance, grant;
 —— *de chauffage*, (*mil. adm.*) fuel allowance;
 —— *de convoi*, (*mil. adm.*) transportation in kind (by team, in certain cases provided by regulations);
 —— *en deniers*, (*mil. adm.*) money allowance, to soldier (v. *haute paye*, *indemnité*, *prime*);
 —— *de frais de route*, (*mil. adm.*) travel allowance;
 —— *en matière*, —— *en nature*, (*mil. adm.*) issues in kind;
 —— *de solde*, (*mil.*) administrative term for *pay*.
allogne, f., (*cord.*) cordage used in construction of flying bridges, and on trestle bridges.
allonge, f., extension-piece; eking-piece; joint (e. g., to lengthen a handle); (*harn.*) lunging rein; (*hipp.*) strain of hip joint; (*mach.*) coupling-rod, sliding-rod;
 —— *de ceinturon*, belt strap.
allongé, a., (*man.*) increased (of gaits);
 trot ——, trot out.
allongement, m., extension; extension, elongation, of a test piece in a testing machine; (*mil.*) elongation of a column (on march); (*cord.*) stretch, lengthening of a rope (under stress); (*art.*) term descriptive of field artillery method of regulating practice.
 —— *du pas*, —— *du trot*, (*man.*) increase in rate of walk, trot.
allonger, v. a., to stretch, lengthen, elongate; to wire-draw; (*man.*) to increase the gait; (*fenc.*) to thrust, lunge, advancing right without moving left foot;
 —— *la courroie*, —— *la ficelle*, (*mil. slang*) to add to a penalty;
 —— *le feu*, (*art.*) to increase the intensity of fire;
 —— *la nage*, to lengthen the stroke (rowing);
 —— *le pas*, (*man.*) to walk out;
 —— *le tir*, (*art.*) to fire over the heads of one's own men (in an assault), to extend the fire from the hostile firing line to the reserves behind it.
allotir, v. a., to group into lots.
allotropique, a., (*chem.*) allotropic.
allouer, v. a., (*adm.*) to verify and approve title to an *allocation*.
alluchon, m., (*mach.*) cog;
 à —— *s*, cogged, toothed.
allumage, m., act of lighting, lighting; (*mil.*, *min.*) touching off a mine chamber.
allumelle f., (*nav.*) bar- or handspike hole of a capstan; bush of a capstan-hole.
allumer, v. a., to light; to fire (a locomotive, engine, furnace, etc.).
allumette, f., match, friction match;
 —— *amorphe*, safety-match;
 —— *de sûreté*, fusee, vesuvian.
allumeur, m., igniter; any burning substance used to touch off a charge; inflammable core of any fuse; touch-paper;
 —— *Bickford*, (*expl.*) igniter composed of tube 25^{mm} long and 4^{mm} or 5^{mm} in diameter;

allumeur *fusant*, (*art.*) "exciter" or primer of a torpedo-shell detonator;
——*s de gaz*, (*mil. slang*) lancers.

allure, f., (*mil.*) gait, rate of march (of troops); (*man.*) gait, pace; (*mach.*) behavior, work; (*met.*) behavior, working (of a furnace).
  I. Behavior of a furnace:
——*chaude*, hot working;
——*crue*, (*lit.* "raw," hence) poor, irregular;
——*extra-chaude*, extra-hot working;
——*extra-froide*, extra-cold working;
——*froide*, cold working;
——*normale*, regular, steady working.
  II. Gait (of a horse, etc.);
——*acquise*, acquired gait;
allonger une ——, to lengthen, increase a gait, rate;
——*artificielle*, artificial gait, air of *manège*;
——*basse*, low gait (action near ground);
——*coulante*, easy, flowing gait;
——*défectueuse*, broken, defective gait (e. g., amble, traquenard, etc.);
——*détendue*, loose gait;
——*enlevée*, lively, quick gait;
éteindre l'——, to reduce speed slowly and gradually;
faire des ——*s*, to put a horse through his paces;
——*haute*, high gait (action well above ground);
——*marchée*, marched (walking) gait;
montrer ses ——*s*, to show, go through his paces (of a horse);
——*naturelle*, natural gait;
——*raccourcie*, short, shortened gait;
——*réglée*, steady, well-regulated gait;
——*régulière*, regular gait;
——*relevée*, high-action gait;
——*sautée*, leaped gait (e. g., gallop);
——*vive*, rapid gait.

alluvion, m., wash, alluvium, sedimentary deposit.

almanach, m., almanac;
——*nautique*, ephemeris, nautical almanac.

alogne, f. v., *allogne*.

aloi, m., alloy; standard (of gold and of silver).

aloyage, m., v. *aloi*.

aloyer, v. a., to give to gold and to silver the standard fixed by law.

alpin, a., alpine, (*mil.*) applied to certain troops in France and in Italy destined and trained for service in the Alps;
chasseur ——, v. *chasseur*.

alquifoux, m., black lead, plumbago.

altérer, v. a., to alter, impair, injure; to excite thirst.

alternateur, m., (*elec.*) alternator.

alternatif, a., alternate, alternating; (*elec.*) alternating; (*mach.*) reciprocating;
courant ——, (*elec.*) alternating current;
mouvement ——, (*mach.*) reciprocating, back-and-forth motion.

altimétrie, f., (*top.*) altimetry, determination of differences of level.

altitude, f., altitude, height; (strictly) height above sea level;
cote d'——, height in figures, reference;
——*du moment*, actual height above sea level at a given moment;
——*moyenne*, height above mean level of the sea;
——*positive*, height above sea level.

alumelle, f., v. *allumelle*.

alumine, f., (*chem.*) alumina.

alumineux, a., (*chem.*) aluminous.

aluminium, m., (*met.*) aluminium.

alun, m., (*chem.*) alum.

alvéole, m., cell, socket; (*sm. a.*) globular mold for bullets; (*expl.*) detonator seat; (*art.*) seat (as for a rotating band);
——*porte-capsule*, (*sm. a.*) primer cavity of a cartridge case.

amande, f., (*sm. a.*) middle part of bow of sword hilt.

amadou, m., tinder.

amaigrir, v. a., to bring down, reduce, shrink;
——*l'argile*, to reduce, thin, clay by a mixture of quartz.

amalgamation, f., (*chem.*) amalgamation.

amalgame, m., (*chem.*) amalgam; (*mil.*) mixture of troops, as of new with veteran; fusion of two regiments, or parts of two or more regiments.

amalgamer, v. a., to amalgamate.

amarinage, m., (*nav.*) manning of a prize.

amariner, v. a., (*nav.*) to man a prize; to inure to the sea.

amarque, f., (*nav.*) buoy, beacon.

amarrage, m., (*cord.*) tying, lashing, seizing, mooring; knot; action of lashing, etc.; (*nav. art.*) securing (a gun);
anneau d'——, (*pont.*) mooring-ring;
bouée d'——, (*nav., etc.*) mooring-buoy;
caisse d'——, (*pont.*) makeshift anchor, case filled with stones;
——*croisé*, racking seizing;
——*par demi-clefs*, two half-hitches;
——*en étrive*, —— *en étrive avec bridure*, throat-seizing;
faire un ——, to lash, to seize;
——*en fouet*, tail-block frapping; rolling-hitch;
——*au milieu*, quarter-seizing;
——*en patte d'oie*, bridle mooring, double cable clinch;
——*plat*, flat seizing;
——*plat avec bridure*, throat seizing on the bight;
——*en portugaise*, shear lashing, racking seizing;
——*près du bout*, end-seizing;
——*en tête d'alouette*, running bowline.

amarre, f., (*cord.*) rope; cable; hawser; lashing; seizing; mooring, mooring-rope; fast, land-fast, shore-fast; painter; (*pont.*) breast-line;
——*d'arrière*, —— *de l'arrière*, stern-fast;
——*d'avant*, —— *de l'avant*, bow-rope, head-fast;
——*de bout*, bow-rope, head-fast, head-cable, hawser with its anchor;
être sur ses ——*s*, to be at her moorings (of a boat or ship);
——*de grue*, cheek of a crane;
pieu d'——, *poteau d'*——, mooring-post;
——*de poupe*, —— *de retenue*, stern-rope, stern-fast;
——*de terre*, shore-fast;
——*de touée*, towline;
——*de travers*, (*pont.*) breast-fast, breast-line.

amarrer, v. a., (*cord.*) to moor, make fast, seize, lash, belay, hitch, nip, fasten with ropes, chains, cords, etc., by any kind of knot; (*nav.*) to secure (as a gun);
——*sur n brins*, to moor a capstan by an n-fold mooring-chain;
s'—— *à terre*, to be moored (of a ship) by shore fastenings;
——*en vache*, (*nav. art.*) to house or secure a gun parallel to the gunwale (obs.).

amas, m., mass, heap, pile.

amasser, v. a., to mass, pile, heap up, store, lay by, etc.

amateur, m., (*mil. slang*) a civilian; officer who gives himself no trouble in his profession.

ambassade, f., embassy; headquarters of an embassy.

ambassadeur, m., ambassador.

ambassadrice, wife of an ambassador.

amble, m., (*man.*) amble, pace;
aller à l'——, to amble, pace;
——*rompu*, broken amble (a pace in which hind foot comes down a little ahead of fore on same side).

ambler, v. n., (*man.*) to amble, pace.

ambleur, m., pacer, pacing horse.

ambulance, f., (*med.*) the personnel and material of the medical service in the field; a field hospital; ambulance;
—— *de brigade de cavalerie*, field hospital accompanying corps cavalry;
—— *de division*, —— *divisionnaire*, the field hospital of a division;
—— *d'évacuation*, field hospital receiving sick and wounded that can stand moving, and sends them to home hospitals;

ambulance   12   **amorce**

ambulance de *gare*, r. r. station hospital; cares for sick and wounded in transit;
— *de quartier général*, ambulance at corps headquarters, (serves as a sort of reserve);
— *de tranchée*, (*siege*) ambulance established in trenches during attacks.

ambulancier, m., (*mil.*) hospital attendant; (also) attendant in civilian *ambulance*.

âme, f., (*art.*, *sm. a.*) bore; (*cord.*) core (of a cable); (*fond.*) core (of a mold); (*fort.*) interior part of a fascine; (*cons.*) web (of a rail, girder, of channel-iron, T-beams, plate-beams, etc.), vertical plate or brace of iron; (*artif.*) core, or filling, of a quick or other match; (*sm. a.*) hollow (cylindrical) of breech-casing (Lebel);
— *de chargement*, (*art.*) loading-tube;
fausse ——, (*art.*) loading-hole of the Krupp fermeture;
fausse —— *de chargement*, (*art.*) loading-tube;
fond de l' ——, (*sm. a.*, *art.*) bottom of the bore;
— *lisse*, (*art.*) smooth bore;
— *rayée*, (*art.*) rifled bore.

amélioration, f., improvement, bettering;
— *de la hausse*, (*art.*) correction of the elevation, in ranging.

améliorer, v. a., to better, correct;
— *la hausse*, (*art.*) to correct the elevation, in ranging.

amelotte, f., v. *allumelle*.

aménagement, m., selection, etc., of trees for cutting and felling; (*mil.*, *Fr. a.*) any labor undertaken by engineers to prepare rooms, etc., to furnish arm-racks, breadboards, etc., for the service of troops.

aménager, v. a., to alter; to regulate the felling of trees; to saw up, cut up a log into lumber.

amende, f., fine; money penalty (of a contractor);
frapper d' ——, to impose a fine.

amener, v. a., to bring up, on; to lead, to take; (*cord.*) to lower, to ease off; (*nav.*) to haul down, strike (colors).
— *l'avant-train*, (*art.*) to limber up, (*esp.*) to limber *rear*.
— *l'avant-train en avant*, (*art.*) to limber front;
— *en douceur*, — *en paquet*, to lower handsomely;
— *pavillon*, — *son pavillon* (*nav.*) to strike one's colors;
— *à retour*, to lower with a turn (around a cleat, etc.);
— *un signal*, to haul down a signal;
— *l'un par l'autre*, to bring into one (as two masts).

amer, m., beacon, sea- or land-mark.

ameublement, m., generic term for furniture in barracks, headquarters, etc.

ameublir, v. a., to loosen earth (as with a pick); to make the ground soft by turning over the surface.

ameulonner, v. a., to stack hay, etc.

amiante, m., fibrous asbestos, amianthus; earth flax; mountain flax.

amiantine, f., (*art.*) cartridge-bag cloth;
toile ——, (*art.*) cartridge-bag cloth.

amidogène, m., (*expl.*) amidogen.

amidon, m., starch;
— *nitré*, (*expl.*) Uchatius white powder.

aminci, m., any thin part of a device, mechanism (i. e., any part that has been reduced in thickness, that is thinner than adjacent parts);
— *circulaire* (or simply *aminci*), (*sm. a.*) circular chamfer in breech end of barrel, corresponding to seat of extractor.

amincir, v. a., to thin, to reduce in thickness.

amincissement, m., thinning, reduction of thickness;
— *tactique*, (*mil.*) term used to describe the gradual loss of depth of formation of troops, due to effectiveness of modern fire.

amiral, m., (*nav.*) admiral; flagship;
contre- ——, rear-admiral;
vaisseau ——, flagship;
vice- ——, vice-admiral.

amirauté, f., (*nav.*) admiralty;
bureau de l' ——, admiralty, admiralty office.

ammoniacal, a., (*chem.*) ammoniacal.

ammoniadynamite, f., (*expl.*) amidogen, ammoniadynamite.

ammoniagélatine, f., (*expl.*) ammoniagelatine.

ammoniaque, f., (*chem.*) ammonia.

amnistie, f., amnesty.

amnistier, v. a., to grant an amnesty.

amoise, f., (*carp.*) tie, brace, binding-piece.

amolette, f., v. *allumelle*.

amont, m., the height from which a river or stream flows, i. e., higher level with respect to mouth or to any point below the one considered;
d' ——, *en* ——, upstream;
ancre d' ——, (*pont.*) upstream anchor;
vent d' ——, land breeze.

amorçage, m., priming, action of priming; starting (of a pump, a dynamo); (*sm. a.*) action of seating a primer in a cartridge case; (*in gen.*) action of fitting a primer (as, e. g., to a mine, a torpedo, etc.); priming charge; primer-holder.

amorce, f., beginning, outline, sketch; (*art.*, *artif.*, *mil.*, *min.*, *expl.*, *sm. a.*) primer, priming; fuse.
I. Fuse; II. Beginning, indication.
—— *à* . . . ., primer to detonate a given explosive; one containing a given ingredient, e. g., — *à dynamite*, to detonate dynamite; — *au fulminate de mercure*, one containing mercury fulminate;
—— *d'âme*, (*mil. min.*) central primer, one in contact with the core (*âme*) of a Bickford or other fuse;
boîte d' ——, (*torp.*, etc.) primer or detonator casing, box, or tube;
bouchon d' ——, (*torp.*) fuse-plug;
branche d' ——, fuse-wire, fuse-lead;
—— *à broche*, (*sm. a.*) pin primer (*obs.*);
sans brûler une ——, (*mil.*) without firing a shot;
câble d' ——, (*torp.*) fuse-cable;
—— *Canouil*, (*art.*) priming for French canon *à balles*:
—— *chimique*, chemical fuse;
—— *cirée*, pellet primer, wax-primer;
—— *composée*, (*art.*) a primer or priming in which fire is communicated to a secondary igniting charge;
couvre- ——, (*sm. a.*) primer cover;
—— *électrique*, (*art.*, etc.) electric primer or fuse;
—— *à étincelles*, (*elec.*) high-tension fuse;
—— *à fil continu*, low-tension, quantity fuse;
—— *à fil discontinu*, high-tension fuse;
—— *à fil métallique*, low-tension, quantity fuse;
—— *à fil de platinum*, low-tension, quantity fuse;
—— *à friction*, (*art.*, etc.) friction primer; friction primer composition;
—— *fulminante*, any fulminating primer;
—— *galvanique*, quantity fuse; wire-bridge fuse, low-tension fuse;
—— *d'induction*, high-tension, induction fuse;
—— *mécanique*, mechanical fuse (percussion, friction, etc.);
—— *des parcs du génie*, French service fuse;
—— *à percussion*, percussion primer;
—— *à percussion centrale*, (*sm. a.*) center-fire primer;
—— *périphérique*, (*sm. a.*) rim-fire primer;
porte- ——, (*art.*) central channel (of a fuse);
—— *pyrotechnique*, (*mil. min.*, etc.) pyrotechnic fuse (like Bickford, etc.);
—— *de quantité*, low-tension, quantity fuse;
—— *simple*, (*art.*) fuse that communicates fire directly to bursting charge;
—— *de tension*, high-tension fuse;
trou d' ——, (*torp.*) fuse hole;
tube d' ——, (*torp.*) fuse-, primer-casing;
—— *voltaïque*, low-tension fuse.
(According to some writers, *amorce* is a fuse to explode gunpowder, while *détonateur* is used for high explosives. But this distinction is not observed.)

II. Beginning, indication, etc.:
amorce *d'araignée*, (*mil.*, *min.*) opening or entrance of a branch gallery;
— *de courbe*, (*top.*, *surv.*) indication, sketching in, of parts of contours;
— *de galerie*, (*mil.*, *min.*) opening of a gallery;
— *de glacis*, (*fort. drawing*) indication of parts near covered way (remainder omitted);
— *de lame à canon*, (*sm. a.*) scarf;
— *de lamette* (*de palonnier*), (*art.*) loop of swingletree cramp;
— *de parallèle*, (*siege*) break, breaking out of parallels;
— *de pont*, shore end or fastening of a ponton bridge;
— *de rameau*, (*mil.*, *min.*) opening or entrance of a branch gallery;
— *d'un système de mines*, (*fort.*) beginning of a system of mines for permanent fort, to be completed when needed.
amorcer, v. a., to start, sketch out, lay down a beginning; to start (a pump, an injector); to scarf; (*sm. a.*) to prime a cartridge case, to seat a primer in its case; (*art.*) to insert friction primer, to prime a gun; (*in gen.*) to prime (a torpedo, etc.); (*mil.*) to decoy, to make a feint; (*elec.*) to start a dynamo;
s'—, (*elec.*) of the magnetic field, to build up (e. g., *la machine s'amorce*).
amorceur, m., primer.
amorçoir, m., wimble, auger; (*mach.*) centerpunch; (*art., etc.*) small box for holding primers.
amorphe, a., amorphous.
amortir, v. a., to deaden, lessen (an effort, shock, velocity, etc.; to settle, take up a debt, obligation, etc.; to damp (a magnetic needle).
amortissement, m., deadening, settling; damping (of a needle); taking up of an obligation; settlement, as of war expenses, etc.;
voie d'—, (*mil.*) (of promotion) absorption, e. g., where *two* vacancies are needed to make *one* promotion, etc.
amour, m., (*sm. a.*) small portion of iron attached to socket of bayonet, to which neck is welded.
amovible, a., removable; held during pleasure.
ampère, m., (*elec.*) ampere.
ampère-heure, m., (*elec.*) ampere-hour.
ampèremètre, m., (*elec.*) ammeter, amperemeter.
ampliation, f., copy, duplicate, of a receipt, of any document;
pour —"true copy" (the chief of staff countersigns "*pour ampliation*," which is thus like "official," U. S. A.).
amplitude, f., (*ball.*) obsolete term for *portée* (range); (*art.*) limit of elevation, of traversing, of a gun;
— *du pointage en hauteur*, limit of elevation;
— *de pointage latéral*, limit of traversing for guncarriage;
— *de pointage vertical*, v. — *en hauteur*, *supra*.
ampoule, f., water-blister; (*met.*) blister (in steel), blowhole; (*elec.*) bulb, globe, of an incandescent light.
ampoulé, a., (*met.*) blistered, of steel.
ampoulette, f., hour-glass; truncated cone of soft wood, serving as fuse-case (obs.).
amputation, f., amputation.
amputé, m., man who has lost a leg or arm by amputation.
amputer, v. a., to amputate.
anallatique, a., (*opt.*) anallatic.
anallatiseur, a., (*opt.*) anallatizing.
anallatisme, m. (*opt.*) anallatism.
analyse, f., analysis; analytical mathematics;
— *chimique*, chemical analysis;
— *spectrale*, spectrum analysis.
analyseur, m., (*opt.*) analyzer (polarized light).

anamorphose, f., (*top.*) anamorphic view or image.
anche, f., reed (of musical instruments).
ancien, a., ancient, old; former, late, ex-; (*mil.*) senior; m., (*mil.*) oldster;
moins — *que*, (*mil.*) junior to;
plus — *que*, (*mil.*) senior to;
le plus —, (*mil.*) the ranking (officer, etc.);
le plus — *dans le grade le plus élevé*, (*mil.*) the ranking officer.
ancienneté, f., (*mil.*) seniority;
à l'—, by seniority;
— *absolue*, lineal rank;
avancement à l'—, *avancer à l'*—, promotion, to be promoted, by seniority;
classement d'—, lineal list;
— *de grade*, — *dans le grade*, relative rank in a grade;
liste d'—, lineal list;
— *minima*, minimum length of service in a grade before promotion to next higher;
— *de service*, length of service, seniority by length of service.
ancrage, m., anchorage, anchoring; cramping down; anchoring, mooring, of a bridge
caisse d'—, (*pont.*) box filled with stones, etc., for anchoring a bridge;
panier d'—, basket anchor, wicker anchor.
ancre, f., anchor; grappling-iron; stay; iron brace; an S;
— *d'amont*, (*pont.*) upstream anchor;
— *articulée*, anchor with movable arms;
— *d'aval*, (*pont.*) downstream anchor;
— *de ballon*, balloon anchor; grapnel;
— *borgne*, one-armed anchor, mooring anchor;
— *de bossoir*, bower;
— *de branchage*, brushwood anchor;
cordage d'—, cable, mooring rope;
— *à demeure*, mooring anchor;
— *de détroit*, stream anchor;
— *engagée*, foul anchor;
— *de flot*, flood anchor;
— *flottante*, drag sail;
grande —, best bower;
— *de jet*, small bower;
— *de jusant*, ebb anchor;
— *de miséricorde*, sheet anchor;
mouiller l'—, to cast anchor.
nœud d'—, cable clinch;
orin d'—, buoy line of anchor;
— *à pattes mobiles*, anchor with movable arms;
repêcher l'—, to weigh anchor;
— *sarpée*, anchor fouled by the flukes;
— *surjalée*, anchor fouled by the stock;
— *surpattée*, anchor fouled by the flukes;
— *de terre*, shore anchor;
— *de touée*, kedge anchor;
— *de veille*, sheet anchor.
ancre-chaîne, f., balloon anchor or grapnel, consisting of several small anchors in series on same rope.
ancrer, v. a. n., to anchor, cast anchor, grapple.
andaillot, m., hank, grommet, ring.
âne, m., ass, donkey; (*tech.*) trestle, horse;
dos d'—, (*top.*) sharp ridge.
ânée, f., ass load.
anémographe, m., anemograph.
anémomètre, anemometer, wind-gauge; (*mach.*) draught-gauge.
anémoscope, m., wind-gauge.
anémométrographe, m., instrument for measuring the direction, force, and velocity of wind.
anéroïde, m., aneroid, aneroid barometer.
anfractuosité, f.; (*top.*) indentation (of a coast).
anglaiser, v. a., (*hipp.*) to dock a horse's tail.
angle, m., angle; corner.
  I. Artillery, ballistics, and small arms;
  II. Fortification; III. Technical and miscellaneous.
  I. Artillery, ballistics, etc.
— *d'abaissement*, (*ball.*) angle of negative jump; negative jump;

angle *d'abaissement au but*, (*ball.*) angle of sight (target below horizontal plane through muzzle); angle made with horizontal plane, by line of sight, by line from muzzle to object;
— *d'arrivée*, (*ball.*) angle of incidence, angle made with surface of ground, by tangent to trajectory at point of impact;
— *du bond*, (*art.*) grazing angle;
— *de chute*, (*ball.*) angle of fall, angle made with horizontal by tangent to trajectory at point of fall; (*esp. sm. a.*) angle of line of sight and tangent at point of fall;
— *de couche*, (*sm. a.*) angle between axis of barrel and line from point of application of recoil to shoulder, drop;
— *de départ*, (*ball.*) angle of departure, of projection;
— *de dépression*, (*ball.*) angle of depression, angle of sight (target below plane of muzzle);
— *de dérivation*, (*ball.*) angle of drift measured from the plane of fire, with the piece as origin;
— *de déviation initiale*, (*ball.*) name proposed for angle of jump;
— *d'écart horizontal*, (*ball.*) horizontal projection of angle of jump; angle of lateral jump, of initial lateral deviation;
— *d'écart initial*, (*ball.*) angle of jump, jump;
— *d'égalité d'actions*, (*torp.*) angle made with horizontal by axis of torpedo (Whitehead) when piston and pendulum neutralize each other;
— *d'élévation du but*, angle of sight (target above plane of muzzle);
— *d'incidence*, (*ball.*) angle under which a projectile, fired in no matter what fashion, strikes no matter what surface; angle of incidence;
— *d'incidence de l'affût*, (*art.*) angle of incidence of the trail, angle made by trail-axis with surface of ground;
— *limite*, (*art.*) limiting angle of fire (depression and elevation);
— *de mire*, (*ball.*) angle between line of sight and axis of gun, sight angle of elevation (called angle of elevation by some English writers, angle of tangent elevation by others; it becomes angle of elevation for American writers only when line of sight is horizontal, as this angle in United States is measured from horizontal to axis of piece);
— *de mire artificiel*, (*ball.*) artificial angle of sight;
— *de mire naturel*, (*ball.*) natural angle of sight;
— *naturel*, angle between line to object and horizontal;
— *ogival*, (*art.*) one-half the vertical angle of tangent cone to vertex of a projectile;
— *de pointage*, (*ball.*) angle of elevation (rare);
— *de plus grande portée*, (*ball.*) angle corresponding to the greatest range;
— *de projection*, (*ball.*) angle of projection, of departure;
— *de recul*, (*art.*) angle of recoil; angle made with horizontal by line from center of trunnions to point of contact of trail with ground;
— *de relèvement*, (*ball.*) angle of jump, jump (frequently *relèvement* simply);
— *du ricochet*, (*art.*) angle under which a projectile bounds after ricochetting;
— *de site*, (*ball.*) angle of elevation or of depression of the target, measured from the horizontal, angle of sight, inclination of line of sight to horizontal (in sm. a. practice this angle may be, and frequently is, measured from horizontal to line connecting muzzle and target);
— *de soulèvement*, (*art.*) angle of lift, limiting inclination of axis of piece, beyond which carriage would tend to overturn in first part of recoil;
— *de tête*, (*ball.*) v. — *d'élévation du but;*

angle *de tir*, (*ball.*) angle between axis of gun and horizontal, quadrant-angle of elevation (or depression); this is the angle of elevation in the United States. In the absence of expressed conditions, *angle de tir* and *angle de mire* are taken as synonymous, they are rigorously equivalent only when line of sight is horizontal;
— *de tir négatif*, angle of depression (quadrant);
— *de tir positif*, angle of elevation (quadrant);
— *de vibration*, (*ball.*) German name of jump;

II. Fortification:
— *du bastion*, bastion-angle;
— *de courtine*, curtain angle (more frequently called — *de flanc*);
— *de défense*, angle of defense;
— *diminué*, diminished angle;
— *d'épaule*, shoulder angle;
— *extérieur*, — *externe*, salient angle;
— *de flanc*, angle of the flank;
— *flanquant*, flanking angle;
— *flanqué*, flanked, salient angle;
— *du fossé*, diminished angle;
— *de la gorge*, gorge angle;
— *mort*, dead angle, dead space;
— *perdu*, dead angle, dead space;
— *de polygone*, angle made by two adjacent sides of polygon to be fortified;
— *rentrant*, reentering, reentrant angle;
— *saillant*, salient angle;
— *sortant*, salient angle;
— *de tenaille*, — *tenaillé*, angle made by intersection of lines of defense;
— *vif*, salient angle;

III. Technical and miscellaneous:
abattre les —*s*, to edge, chamfer;
— *abordable*, angle within which a dirigible balloon may *make* any point;
— *aigu*, acute angle;
— *d'avance*, (*steam*) angular advance, angle of advance;
— *de calage*, (*elec.*) lead of the brushes (of a dynamo); (*steam*) angle between eccentric radius and crank, 90° plus angular advance;
— *de coupe*, (of cutting tools) cutting angle;
— *de déviation*, (*mach.*) leading off angle of a belt);
— *dièdre*, diedral angle;
— *droit*, right angle.
émousser les —*s*, (*mil.*) to break off corners of a square, so as to get fire in all directions;
— *d'enroulement*, (*mach.*) angle of contact (of a belt);
— *d'équerre*, right angle;
faire —, to elbow;
— *de frottement*, angle of friction;
— *d'incidence*, angle of relief (of a cutting tool);
— *moyen*, (*surv.*) angle corrected for spherical excess; corrected angle, so that when added to the two others of a triangle their sum shall be 180°.
— *oblique*, oblique angle; bevel-rule, miter square;
observer un —, to take an angle, to observe an angle;
— *obtus*, obtuse angle;
— *optique*, visual angle;
— *de pente*, (*top.*) angle of slope, gradient;
— *plan*, plane angle;
rapporter un —, to protract an angle;
— *rapporteur*, (*inst.*) protractor;
— *relevé*, observed angle;
relever un —, to take, measure, observe, an angle;
— *en retour*, (*surv.*) back-reading;
— *de rupture*, angle of rupture;
— *sphérique*, spherical angle;
— *de tournant*, locking-angle (of vehicles);
— *de traction*, angle of traction (vehicles);
— *tranchant*, angle between faces (of cutting tools), cutting angle;
— *trièdre*, triedral angle;
— *visuel*, visual angle.

anglet, m., indenture, channel; miter-square.
anguille, f., (*pont.*) small balk or beam between bays of a raft-bridge.
angulaire, a., angular;
— *pierre* ——, (*cons.*) corner stone.
angulé, a., (*fort.*) angular, having angles (e. g., a line of works, etc.).
anguleux, a., v. *angulé*.
anhydre, a., (*chem.*) anhydrous.
anhygrométrique, a., nonhygrometric.
aniline, f., (*chem.*) aniline.
ankylose, f., (*hipp.*) anchylosis;
—— *fausse*, false anchylosis; adherence, dryness, of synovial membrane;
—— *vraie*, true anchylosis, union of adjacent surfaces.
anneau, m., ring; collar; hoop; link; link (of a chain); loop; band; rim; sleeve; (*elec.*) ring-armature; (*art.*) lug or ear of a (spherical) shell; (*harn.*, *sm. a.*) clapper-ring; (*gym.*) rings (in pl.).
I. Artillery; II. Harness, carriages, and miscellaneous.

I. Artillery:
—— *anti-friction*, obturator-spindle washer (*U. S. art.*);
—— *d'attelage d'un affût*, lunette;
—— *de bombe*, ear, lug (spherical shell) (obs.);
—— *de brague*, breeching-loop (*nav. art.*) (obs.);
—— *Broadwell*, Broadwell ring (obturator, Krupp);
—— *de calage*, key-ring;
—— *carré de manœuvre*, trail-handle;
—— *de compression*, grip-ring (Gardner gun);
—— *de crosse*, pointing ring;
—— *élingue*, trunnion-sling, -ring, -bale;
—— *expansif*, obturating ring, obturator;
—— *de levage*, carrying-ring;
—— *de lunette*, lunette, lunette ring;
—— *de manœuvre*, trail-handle;
—— *obturateur*, gas-check, gas-check ring; obturating ring;
—— *plastique*, gas-check pad;
—— *poignée*, breech-handle, block-handle;
—— *de pointage*, pointing ring;
—— *porte-armements*, equipment-ring, lashing-ring;
—— *porte-écouvillon*, sponge-ring;
—— *porte-levier de pointage*, trail handspike ring;
—— *porte-levier portereau*, trail handspike ring (Fr. 80mm);
—— *porte-manche de hachette*, hatchet-ring;
—— *porte-refouloir*, rammer-ring;
—— *de prolonge*, prolonge-ring;
—— *protecteur*, glacis armor;
—— *rebord*, circular flange (of fuses);
—— *de refouloir*, sponge-ring;
—— *de réglage*, time-scale of a fuse; time-setting ring;
—— *ressort*, restraining-ring (of a fuse);
—— *de rotation*, rotating-ring;
—— *de sûreté*, safety ring (fuse);

II. Harness and miscellaneous:
—— *d'amarrage*, (*pont.*) mooring-ring;
—— *d'attelage*, shoulder link;
—— *des attelles*, hames-ring;
—— *de baïonnette*, (*sm. a.*) bayonet ring or clasp;
—— *de bivouac*, (*Fr. cav.*) sort of grommet made with forage cord, as a picketing-ring (the *anneau italien* is a metallic ring, but used for the same purpose);
—— *de bout d'essieu*, linch-hoop, nave-band;
—— *de brélage*, (*pont.*, etc.) lashing-ring; lashing-ring (of a pack saddle); keep-chain ring;
—— *de calotte*, (*sm. a.*) pistol-butt ring;
—— *de campement*, v. —— *de bivouac*;
—— *de caveçon*, lunging ring;
—— *de chaîne de timon*, pole-chain ring;
—— *de collier*, hames-ring;
—— *de corde*, (*cord.*) grommet; slip noose; running bowline knot;
—— *coulant*, sliding-ring, sliding-loop;
—— *à crochet*, ring and hook;
—— *de crochet d'attelage*, ring of any draught hook; shoulder link;

anneau-*dé*, D-ring;
—— *denté*, (*mach.*, etc.) ratchet-circle; pawl-ring;
—— *d'embrelage*, keep-chain ring, (*pont.*) lashing-ring;
—— *à émerillon*, swivel-ring;
—— *d'essieu*, linch-loop;
—— *à fiche*, eyebolt, ringbolt;
—— *de garniture*, (*mach.*) packing-ring, piston-ring;
—— *à happe*, ring and staple;
—— *italien*, (*Fr. cav.*) v. —— *de bivouac;*
—— *de longe*, lunging-ring;
—— *de manœuvre*, lashing-ring;
—— *mobile*, runner, running loop;
—— *oculaire*, (*inst.*) annular diaphragm of the eye tube of a telescope;
—— *de pansage*, grooming-ring- stable-ring, hitching ring (during grooming);
—— *à pattes*, flanged ring, pole-ring;
—— *à patte porte-palonnier*, splinter-bar loop or ring;
—— *à piton*, ringbolt, eyebolt;
—— *plat de palonnier*, flat swingletree ring;
—— *plat de volée*, flat splinter-bar hook or ring;
—— *porte-extracteur*, (*sm. a.*) extractor-sleeve, extractor collar (Spanish Mauser);
—— *porte-fils*, (*inst.*) diaphragm, reticule-holder (of a telescope);
—— *porte-mors*, (*harn.*) bit-ring;
—— *porte-rêne*, (*harn.*) bit-ring, bridle-ring;
—— *porte-sabre*, sabre-ring on near side of saddle;
—— *porte-servante*, pole-prop loop;
—— *-rond*, D-ring;
—— *de sabre*, (*sm. a.*) scabbard-ring;
—— *de suspension*, (*pont.*) trestle-cap ring-bolt;
—— *de volée de bout de timon*, ring or hoop of the master swingletree bar.

année, f., year;
—— *de campagne*, (*Fr. a.*) campaign counted as so much service (generally one year) in addition to time actually spent (n. e.);
—— *de campagne d'embarquement*, (*Fr. a.*) service afloat, counted as an additional half campaign, in peace; as an additional full campaign, in war (service in seacoast works or ports, in war, counts as an additional half campaign);
—— *de campagne hors d'Europe*, (*Fr. a.*) campaign in army out of Europe, counting double in addition to time actually spent (one year and two campaigns; except in Algiers, one year and one campaign);
—— *de campagne de terre*, (*Fr. a.*) campaign in Europe, counts as two years (one year and one campaign);
—— *de grade*, time to be served in a grade before promotion;
—— *de services effectifs*, time actually passed in service, up to a given date.
anneler, v. a., (*hipp.*) to ring a mare.
annelet, m., small ring, annulet.
annexer, v. a., to annex.
annexion, f., annexation.
annuaire, m., any annual publication;
—— *de l'armée*, —— *militaire*, (*mil.*) army register (U. S. A.), army list (British);
passer l'—— *sous le bras*, (*mil. slang*) to be promoted according to seniority.
annuité, f., (*mil. adm.*) deduction of service or of money due for a period of one year;
demi- ——, deduction based on a length of time varying from 15 days to 6½ months.
annulation, f., annulment, act of annulling; lapsing of an unexpended appropriation; canceling, cancellation (as of an enlistment contract);
—— *de jugement*, (*mil. law*) disapproval of the proceedings or findings or sentence, or all three, of a court-martial;
timbre d'——, cancellation stamp.
annuler, v. a., to annul, to cancel; (*mil. law*) to disapprove proceedings, sentence, etc., of a court-martial.
anode, m., (*elec.*) anode.
anonyme, a., joint stock (of commercial, etc., companies).

anse, f., handle, ear; (*art.*) maneuvering handle (on gun); (*top.*) arm or bay (of the sea);
—— *de bombe*, (*art.*) lug, ear, of a (spherical) shell;
—— *de panier*, (lit. "basket handle") curve used in road-making, made up of three arcs of circles;
—— *à vis*, screw eyebolt.

ansette, f., small handle; (*mil.*) kind of clasp into which ribbon of an order is passed.

anspect, m., handspike; windlass handspike; *barre d'*——, handspike.

antenne, f., catch, hook; (*torp.*) striker (e. g., of Whitehead);
—— *à ressort*, (*F. m. art.*) implement for drawing a projectile from a gun, mortar.

antérieur, a., (*hipp.*) fore (as, leg); m., fore leg of a horse.

antestature, f., barricades improvised of gabions, sandbags, fascines, etc.

anthéximètre, m., instrument for testing papers, etc., tissues in general.

anthracite, m., anthracite coal, hard coal.

anticipation, f., anticipation;
—— *de service*, (*Fr. a.*) credit for service allowed certain classes of men (sons of foreigners, pupils of certain government schools, etc.).

antidérapant, a., (of a bicycle tire) nonslipping; m., pebble-tread tire.

anti-fortificateur, m., (*mil.*) opponent of fortifications as a means of defense.

anti-friction, a., antifriction;
métal ——, antifriction metal.

anti-incrustant, a., (*steam*) scale-preventing.

antimoine, m., (*met.*) antimony.

antiparallèle, a. and f., antiparallel, antiparallel line.

antique, m., (*mil. slang*) cadet of the Polytechnique who has completed the regular course of studies.

antiseptique, a., antiseptic.

antréomètre, m., instrument for measuring the size of head-wear.

aperçu, a., (*sig.*) answering signal, (*nav.*) answering pennant.

apériodique, a., (*elec.*) dead-beat, aperiodic.

a-pic, m., (*top.*) bluff.

aplanétique, a., (*opt.*) aplanetic.

aplanir, v. a., to smooth, smooth off, level, bring to a regular surface.

aplati, a., (*fort.*) blunted (of a *tracé*);
lunette ——*e*, blunted lunette;
redan ——, blunted redan.

aplatir, v. a., to flatten;
—— *un ressort à bloc*, to close down a spiral spring completely between two flat surfaces.

aplatissement, m., (*ball.*) flatness of trajectory (rare).

aplomb, m., perpendicularity; (*hipp.*) stand or way of standing of a horse, with respect to the position of his legs;
—— *bon*, —— *régulier*, (*hipp.*) regular distribution of the weight on the legs, such a disposition as is most favorable to either standing or moving;
à l'—— *de*, immediately underneath;
d'——, vertical;
——*s du cheval*, (*hipp.*) v. *supra*;
prendre l'——, to try with a plummet line.

apomécométrie, f., pedantic term for art of measuring distances by pacing.

aponévrose, f., (*hipp.*) aponeurosis.

aposter, v. a., (*mil.*) to post a (small) detachment, either to guard a point or at a point from which a surprise is to be made.

appareil, m., apparatus, device, gear, fixture, arrangement, disposition, contrivance, mechanism; plant; purchase, hold; (*mds.*) bond, stonecutting. *Appareil*, followed by *à* or *de*, etc., is combined with a very great number of words to denote the gear or apparatus for the thing, or consisting of the thing, represented by the dependent word or words.

I. Artillery and military in general;
II. Technical, machinery, and miscellaneous.

I. Artillery and military in general:
appareil, (*in pl.*) implements, equipments;
—— *de bouche*, (*ball.*) muzzle "target," in getting small-arm muzzle velocities;
—— *de campagne*, —— *de campagne à lentilles*, French field visual signaling apparatus;
—— *de charge*, —— *de chargement*, loading implements;
—— *à chargement central*, device in certain Canet turrets by which guns may be loaded from any radial position;
—— *de choc*, buffer;
—— *concutant*, the striker and related parts of a double-action fuse;
—— *à contrepoids*, rolling-table and counterweight, for testing eccentricity of the center of gravity of projectiles;
—— *crusher*, crusher gauge;
—— *de déformation*, generic expression for pressure (crusher) gauges;
—— *de détente*, (*r. f. art., sm. a.*) trigger and related parts;
—— *à écrasement*, generic expression for crusher gauges;
—— *de fermeture*, breech-closing mechanism;
—— *à fracture*, (*med.*) splints, bandages, etc.;
—— *de frein*, recoil-checking device;
—— *fusant*, the time element of a combination fuse;
grand ——, (*Fr. a.*) search light (complete);
—— *à grenade et à obus*, (*Fr. art.*) sort of projectile consisting of a half-barrel full of shells or of grenades, fired from a mortar (known as —— *Moisson*);
—— *de grenage*, (*powd.*) granulating machine;
—— *grimpeur*, creepers, climbing outfit (cavalry pioneers, etc.);
—— *de lancement tournant*, (*torp.*) torpedo carriage;
—— *lance-torpille*, (*torp.*) torpedo-tube, torpedo-launching gear;
—— *à lentille*, French field signaling apparatus;
—— *de mise en batterie*, mechanism for returning a piece to battery;
—— *de mise de feu*, any device, etc., for firing a gun, mine, torpedo, etc.;
—— *Moisson*, v. —— *à grenade et à obus*;
—— *d'obturation*, gas-check, obturator;
—— *optique de campagne*, French field signal apparatus;
—— *à pansement*, (*med.*) portable box containing medical supplies, dressings, etc.;
—— *percutant*, the percussion element of a combination fuse; firing gear (e. g., of r. f. guns);
—— *de pointage*, —— *de pointage en hauteur*, elevating gear;
—— *de pointage Deport*, the Deport depression range-finder;
—— *de position*, French field signaling apparatus;
——*s prothétiques*, generic term for artificial limbs, etc.;
—— *de relèvement*, —— *de relèvement rapide*, gear for bringing a gun quickly into loading position after firing;
—— *Rodman*, Rodman's pressure gauge;
—— *secondaire*, French army search light (complete on one carriage);
—— *de signaleurs*, French field signaling apparatus;
—— *à soulèvement*, (siege and fortress artillery) arrangement or gear (eccentric) for raising a carriage so that it may easily be run into battery;
—— *de sûreté*, safety device;
—— *télescopique*, French field signaling apparatus;
—— *à tige cannelée*, (*Fr. art.*) device (insertion of a grooved rod in a mortar) by which several spherical projectiles may be fired at once from the same piece;
—— *de tir*, (*in gen.*) firing gear;

appareil *de tir à mitraille des mortiers*, (*Fr. art.*) apparatus for using *mitraille* in mortars (v. *supra*, —— *à grenade et à obus*, —— *à tige cannelée*).
 II. Technical, machinery, and miscellaneous:
—— *à air chaud*, (*met.*) hot-blast stove;
—— *alimentaire Giffard*, (*steam*) Giffard injector;
—— *d'alimentation*, (*mach.*, *etc.*) feed-gear;
—— *par assises réglées*, (*mas.*) plain work;
—— *à cadran*, (*teleg.*) dial instrument;
—— *de changement de marche*, (*mach.*) reversing gear;
—— *de changement de voie*, (*r. r.*) switch;
—— *à chargement*, (*met.*) charging apparatus of a blast-furnace, etc.;
—— *des chemins de fer*, apparatus for making falling weight tests;
—— *de choc*, buffer (e. g., of a railroad car);
—— *de cinglage*, (*met.*) shingler;
—— *de commande*, (*mach.*) general term for any operating gear, any controlling gear;
—— *Cowper*, (*met.*) Cowper hot-blast stove;
—— *à croisettes*, (*mas.*) cross-bond;
—— *de débrayage*, (*mach.*) any device for throwing out of gear;
—— *de détente*, (*steam*) expansion gear;
—— *de la distribution*, (*steam*) slide-valve gear;
—— (*dit*) *anglais*, (*mas.*) old English bond;
—— (*dit*) *flamand*, (*mas.*) headers and stretchers, Flemish bond;
—— *dynamique*, —— *dynamométrique*, (*mach.*) dynamometer;
—— *d'éclairage*, illuminating plant; generic term for any lighting apparatus;
—— *électrique*, (*elec.*) electrical machine;
—— *d'embrayage*, (*mach.*) any device for throwing in gear;
—— *à enclanchement*, (*r. r.*, *etc.*) interlocking gear;
—— *en épi*, (*mas.*) herringbone work;
—— *étalon*, (*elec.*) calibrator, standard;
—— *de l'excentrique*, (*mach.*) eccentric gear;
—— *à feu de cave*, sort of smoke-proof dress for entering rooms, etc., full of smoke;
—— *fumivore*, smoke-consuming apparatus;
—— *graisseur*, —— *de graissage*, (*mach.*) lubricator;
—— *de la locomotion*, (*hipp.*) mechanism of a horse's motion;
—— *de marche à contrevapeur*, (*mach.*) reversing gear;
—— *en moellons*, (*mas.*) ashlar-work;
—— *moteur*, (*mach.*) driver, driving engine;
—— *multiplicateur*, (*mach.*) any device for increasing power, speed, etc.;
—— *à percussion*, (*min.*) generic term for jumpers, drilling bars;
—— *de perforation*, (*min.*) borer, earth auger, etc.;
—— *photographique*, (*phot.*) camera;
—— *d'une pierre*, (*mas.*) height of a stone;
—— *à retour rapide*, (*mach.*) quick-return gear, motion;
—— *à rotation*, (*min.*) generic expression for earth-augers;
—— *de sauvetage*, any life-saving device; life-belt; air-supplying device, making possible entrance into places full of smoke, asphyxiating gases, etc.; fire-escape;
—— *de sondage*, boring apparatus (foundations, etc.);
—— *télégraphique*, telegraph key and sounder;
—— *de la transmission*, (*mach.*) transmission gear;
—— *de la voie*, (*r. r.*) switch;
—— *Whitwell*, (*met.*) Whitwell stove.

appareillage, m., (*nav.*) preparation for sailing, getting under way, weighing of anchor; (*hipp.*) matching (of horses, for a pair, team); (*mas.*) bonding, bond; matching (of bricks, stones, etc., in construction).

appareiller, v. a., (*nav.*) to hoist anchor, to get under way; (*mas.*) to bond, to match (stones, bricks, etc.), to give a model for stone-cutting; (*hipp.*) to match (horses, for a pair, team).

appel, m., call; (*mil.*) roll-call, call or signal for roll-call; ruffle (of a drum); call (drum, bugle, trumpet); call of a sentry to the guard; levy; general term for the operations of census-taking, drawing by lot, assignment, etc., for the military service; order to a *classe* to join the colors; muster; (*sm. a.*) click of a gun-lock, or of the sear mechanism; (*fenc.*) appel, beat or stamp of the foot, by way of challenge; (*tech.*) tension, pull; draught (of air); (*law*) appeal to a higher court.
 I. Military; II. Technical and miscellaneous.
 I. Military:
—— *à l'activité*, summons or order to join the colors;
battre l'——, to beat a call;
billet d'——, roll-call list (showing absentees and passmen);
contre-——, check roll-call;
demi-——, v. *demi-appel*;
devancement d'——, joining before date on which a new class is ordered to leave home (French navy and colonial troops);
devancer l'——, v. *devancement d'——*;
—— *échelonné*, summons to service, by individual order, at any time of the year, by corps commander, of certain classes of men, performing special duties; as, reservists of artillery, workmen, etc.;
faire l'——, to call the roll;
feuille d'——, muster roll;
—— *de la journée*, roll-call during the day (in the field);
manquer à l'——, to be absent from roll-call;
—— *du matin*, morning roll-call (half hour after reveille);
—— *de mobilisation*, levy or summons on mobilization;
—— *de pansage*, stable-call;
passer l'——, to pass muster;
recevoir l'——, to receive a roll-call;
rendre l'——, to report the results of a roll-call;
répondre à l'——, to answer to one's name;
retard à l'——, delay in reporting for enlistment (*bureau de recrutement*); delay in joining regiment or corps after having been drawn for service;
—— *du soir*, evening roll-call (half hour after the *retraite*).
 II. Technical and miscellaneous:
—— *d'air*, draught, supply of air;
faire —— *d'air*, to draw a supply of air from (for ventilation);
—— *à la concurrence*, (*adm.*) advertisement for bids;
—— *de cravache*, (*man.*) signal made with riding whip (e. g., tap on boot);
—— *de langue*, (*man.*) cluck (of the tongue);
—— *mixte*, telephone call used in connection either with an electric bell or with a buzzer;
—— *phonique*, telephone call, buzzer;
—— *de la vapeur*, steam-blast.

appelé, m., (*mil.*) member of a class summoned to service.

appeler, v. a., n., to call; (*sm. a.*) to click (of the sear); (*mil.*) to call the roll; to summon into service, to call out a contingent; (*sig.*) to call up (a station); (*tech.*) to draw (of a chimney); to drive (air, ventilating machinery);
—— *à l'activité*, —— *sous les drapeaux*, (*mil.*) to call out for service in the active army, for active service.

appendice, m., neck (of a balloon); projection, stud; (*adm.*) detailed list of baggage of a body of troops changing station, or moving in general;
—— *de prise de vapeur*, (*steam*) interior steam-pipe.

appentis, m., shed, lean-to; (*siege*) protective plate or shield used in trench-work, sap-shield.

application, f., application;
—— *de peine*, (*mil. law*) determination of suitable punishment after finding of guilty.

applique, f., a piece, applied or affixed to another; reenforce.

appliquer, v. a., to apply (as one piece to another, to strengthen or adorn it).

appoint, m., anything over; (as, odd pennies over, odd inches over a certain number of feet, etc., hence frequently) addition, advantage; (*art.*) priming charge of black powder, for a cartridge of brown powder; (*top.*) the difference between total distance to be measured and the distance corresponding to the whole number of times the measure is contained in the distance (similarly of *weights*);
— *l'—— d'un bataillon*, (*mil.*) the addition of a battalion, the having a battalion over and above.

appointage, m., pointing (of branches, stakes, etc.).

appointements, m. pl., pay (especially of officers).

appointer, v. a., (*mil.*) to pay (obs.); to sharpen (a branch, stake, etc.);
— *de corvée*, — *de garde, etc.*, (*mil.*) to put on extra fatigue, extra guard, etc., by way of punishment.

appondure, f., binding-piece (on a raft).

appontement, m., (*nav.*) flying-bridge; mooring buoy.

apport, m., —— *dotal*, (*Fr. a.*) dowry a bride must bring her husband, if in army.

apposer, v. a., to affix (a seal, a signature).

apposition, f., affixing (of a seal, of a signature).

appréciation, f., estimation;
— *des angles*, measurement of angles;
— *des distances*, (*mil.*) estimation of distances;
— *des pentes*, measurement of slopes.

apprécier, v. a., to estimate;
— *les distances*, (*mil.*) to estimate distances;
— *trop court*, (*mil.*) to underestimate (distances);
— *trop long*, (*mil.*) to overestimate (distances);
— *les vitesses*, (*art.*) to estimate speed of vessels.

apprêter, v. a., to prepare, make ready;
— *l'arme*, (*drill*) to come to a "ready."

approche, f., approach; approach to a position;
— *s*, (*pl., siege*) approaches;
— *directe*, (*siege*) double sap, method of advance by double sap;
lignes d'——, (*siege*) trenches;
marche d'——, (*mil.*) v. *marche*;
travaux d'——, (*siege*) trench-work, trenches, approaches;
— *en zigzags*, (*siege*) zigzags.

approfondir, v. a., to deepen (as trenches); (*fig.*) to go profoundly into a subject.

approprié, a., adapted (as, *art.*, a carriage to a gun not its own).

approvisionnement, m., supply, store; (*sm. a.*) filling of magazine; m. pl., (*mil.*) stores, supplies (any and all kinds of supplies, as, —— *d'artillerie*, artillery stores —— *du génie*, etc.);
— *s de campagne de 1ère ligne*, supplies (i. e., subsistence) carried by and accompanying mobilized troops;
— *s de campagne de stations-magasins*, supplies kept in the *stations-magasins*, q. v., and drawn in case of war;
— *s de concentration*, supplies for use during the concentration that follows the *transports stratégiques*;
— *s de corps*, regimental supplies, kept on hand for current service;
— *s de débarquement*, supplies for first needs in zone of concentration;
— *s de l'état*, reserve of supplies for general needs of a mobilization;
— *s de mobilisation*, supplies that are to be drawn on when mobilization is ordered (i. e., — *des 20 jours*, — *des transports stratégiques*, — *de concentration*, qq. v.);
— *s en munitions*, supply of ammunition (*inf.* and *art.*);
— *s régionaux*, 8 days' supplies, kept in each corps region for the troops mobilized therein;

approvisionnement, —— *s de la réserve de guerre*, comprehensive term for all supplies and stores, kept up in view of mobilization;
— *s du service courant*, supplies for current service;
— *s de siège*, supplies stored in *places fortes*, in view of war;
— *s de sûreté*, stores, etc., kept up in a *place forte* during peace, so as to be sure to have them at hand in war;
— *s des transports stratégiques*, supplies organized in the *stations halte-repas*, q. v., to subsist men in the *transports stratégiques*, q. v.;
— *s des vingt jours*, 20 days' supplies for subsisting men and horses during the first 20 days of mobilization.

approvisionner, v. a., to victual, to supply, to furnish stores; (*sm. a.*) to fill the magazine; (*art.*) to supply ammunition to a piece, to a battery in action.

appui, m., buttress; prop; stay; support; rest; sill (of a window, frame building; (*man.*) movement in which shoulders and hindquarters go on parallel tracks; the bearing of the foot on the ground, in the walk;
n'avoir point d'——, (*man.*) to have a tender mouth (horse);
— *de la cuirasse*, (*nav.*) armor-shelf, for belt armor;
donner de l'——, un ——, à un cheval, (*man.*) to let a horse bear on the hand (while urging him forward), by an increased tension of the reins;
à hauteur d'——, breast-high;
— *mobile*, (*pont.*) support or prop for raising or lowering the cap of the Belgian trestle;
pièce à l'——, (*adm.*) voucher;
plaque d'——, (*art., etc.*) wheel-guard plate;
point d'——, (*tech.*) fulcrum, pivot; (*mil.*) any accident of the ground (natural or artificial) that troops may take advantage of in attack or defense;
donner un point d'—— à un cheval, (*man.*) v. *donner d'——*, *supra*;
— *pour les roues*, (*art., etc.*) v. *plaque d'——*;
— *de tir*, (*sm. a.*) firing-rest; aiming-rest (aiming drill).

appuyer, m., (*man.*) passage.

appuyer, v. a., to support, to prop, sustain; to strengthen, protect; to incline to the right (left); to enforce (as, a signal by a cannon shot); (*man.*) to cause a horse to passage, to bear to the right or left; (*adm., etc.*) to support or bear out a statement, account, etc., by vouchers;
— *une armée à* . . . . , (*mil.*) to rest an army on (such and such a natural, etc., support, as a swamp, lake, etc.);
— *la botte*, (*fenc.*) to press the foil after touching the adversary;
— *un cheval*, (*man.*) to passage a horse; to cause a horse to bear on the hand by increasing the tension of the reins;
— *l'éperon*, (*man.*) to spur briskly;
— *au guide*, (*drill*) to touch to the pivot;
— *son pavillon*, (*nav.*) to fire a shot under one's true colors.

appuyeur, m., holding-up hammer.

aptitude, f., fitness;
— *militaire*, fitness for military service;
— *professionnelle*, fitness for certain duties (as, *ouvrier d'artillerie*).

apurement, m., (*adm.*) auditing of accounts.

apurer, v. a., (*adm.*) to audit accounts.

apyrite, f., (*expl.*) apyrite.

aquarelle, f., water-color (sketch).

aqueduc, m., aqueduct, culvert.

araignée, f., spider; the American buggy; (*mil. min.*) system of branches of a mine, each terminating in a mine chamber (obs.);
fils d'——, cross-hairs of a telescope;
pattes d'——, (*mach.*) grease-channels.

arasement, m., act of bringing to the same level, leveling (as, a course of masonry, etc.).

araser, v. a., to make flush or level; to level (as, the top of a wall, an embankment or parapet); to be flush with.
arases, f. pl., leveling-courses of masonry.
arbalète, f., the leader of a spike-team; (*carp.*) hanging-post truss.
arbalèter, v. a., to shore up, prop.
arbalétrier, m., (*cons.*) main rafter of a roof, head-rafter.
arbalatrière, f., gangway of a platform.
arbitrage, m., arbitration; (*mil.*) jurisdiction and judgment of umpires.
arbitral, a., relating to arbitration.
arbitre, m., abritrator; (*mil.*) umpire at manœuvers.
arbitrer, v. a., to arbitrate.
arborer, v. a., (*nav.*) to hoist colors, a flag, a signal;
—— *un pavillon*, to hoist a flag;
—— *le pavillon blanc*, to hoist the white flag, to surrender.
arborescence, f., arborescence; (*met.*) treelike formation or arborescence (defect in steel lacking in homogeneity).
arbre, m., tree; (*mach.*, *etc.*) arbor, shaft, spindle, axis, axle; boring-bar, cutter-bar; stem or post of a crane.
(Except where otherwise indicated, the following terms relate to machinery:)
—— *à ailettes*, axle of cruciform section;
—— *de brin*, tree of a single stem, i. e., straight-stemmed tree;
—— *à cames*, cam-shaft;
—— *de changement de marche*, reversing-shaft;
—— *chargé*, shafting subjected to bending stresses;
—— *de commande*, the particular shaft driving a particular machine; controlling shaft (of a particular machine);
—— *du communicateur*, gearing-shaft;
—— *de couche*, horizontal shaft; main, driving-shaft; crank shaft; the shaft running the length of a machine shop, i. e., power-shaft, distributing-shaft;
—— *de couche à manivelles rapportées*, built-up crank shaft;
—— *à un seul coude*, simple crank-axle;
—— *coudé*, cranked shaft, cranked axle, double crank;
—— *à coudes multiples*, multiple crank-shaft;
—— *creux*, hollow shaft;
—— *debout*, arbor, spindle (vertical);
—— *de distribution*, valve-stem;
—— *à double-joint universel*, spindle fitted with a double ball-and-socket joint;
faux ——, (*fond.*) false core or spindle (casting hollow projectiles);
—— *à feuillage persistant*, evergreen tree;
—— *flexible*, flexible shaft;
—— *de frein*, brake-spindle; (*art.*) recoil-check spindle, compressor-shaft;
—— *de grue*, crane-post;
—— *d'hélice*, (*nav.*) screw-, propeller-shaft;
—— *intermédiaire*, countershaft;
ligne d'——s, line-shafting;
—— *de la machine*, engine-shaft;
—— *à manivelle*, crank-shaft;
—— *de mise en train*, starting-shaft;
—— *moteur*, main shaft, driving-shaft, driving-axle;
—— *premier moteur*, the engine-shaft; any shaft which, transmitting all the power, is fitted with a fly-wheel or sheave;
—— *second moteur*, transmitting shaft, intermediate shaft, has no fly-wheel, and either receives or transmits its power by gearing;
—— *troisième moteur*, a shaft that receives power from belts or light gearing, or which controls oscillatory movements only;
—— *à nervures*, ribbed axle, axle of cruciform section;
—— *à noyau*, (*fond.*) core-spindle, core-bar;
—— *oscillant*, rocking-shaft;
—— *de pointage en direction*, (*art.*) traversing-shaft;
—— *de pointage en hauteur*, (*art.*) elevating shaft;
—— *porte-fraise*, overhanging arm (machine tools);
—— *porte-molettes*, cutter-arbor;

arbre *porte-outil*, tool-spindle, tool-holder;
—— *porte-pièce*, work-spindle;
—— *rainé*, slotted shaft;
—— *de réglage*, (*art.*, *mach.*) adjusting-shaft, regulating-spindle;
—— *de serrage*, (*art.*) compressor-shaft, tightening spindle;
—— *suspendu*, overhead shaft;
—— *du tiroir*, —— *des tiroirs*, eccentric shaft, way shaft;
—— *de tour*, lathe-spindle, mandrel;
—— *de transmission*, secondary transmission-shaft, transmission-shaft; countershaft, distributing-shaft;
—— *vecteur*, main mill-shaft;
—— *en vilebrequin*, crank-shaft;
—— *à vis sans fin*, worm.
arc, m., bow; arch; fork, bow, of a saddle (rare);
—— *en berceau*, round, Roman arch;
—— *denté*, (*mach.*) toothed arc, segmental rack;
—— *denté de pointage*, (*art.*) elevating arc;
—— *doubleau*, recessed, compound arch, facing-arch (as of a row of casemates);
—— *embrassé*, (*mach.*) arc of contact (of a belt on a pulley);
—— *gradué*, (*inst.*) graduated arc, limb;
—— *en plein cintre*, v. —— *en berceau*;
—— *de pointage*, (*art.*) elevating arc;
—— *de roulement*, (*mach.*) rolling arc (of gearing);
—— *stérile*, arc in which there is no circulation (of a ventilator);
—— *voltaïque*, (*elec.*) electric arc (light).
arcade, f., arcade; curved brace; (*harn.*) fork or bow of a saddle, of a pack saddle; (*art.*) curved transom;
—— *dentaire*, (*hipp.*) dental arcade, incisive arcade;
—— *de derrière*, (*harn.*) cantle (of a saddle);
—— *de devant*, (*harn.*) bow, peak (of a saddle);
pointe d' ——, (*harn.*) point or lower end of the saddle-fork;
—— *renversée*, counter arch.
arcanson, m., colophany, black rosin.
arc-boutant, m., buttress, flying-buttress; stay, spur, strut, prop; (*art.*) axletree stay; (*hipp.*) bar, stay (of the foot).
arc-boutement, m., staying; buttressing; (*sm. a.*) jamming of cartridges in a tubular magazine; *point d'* ——, point of support.
arc-bouter, v. a., to brace, prop, support, buttress; (*art.*) to jam; (*sm. a.*) to jam, as cartridges in a tubular magazine.
arc-doubleau, m., v. s. v. *arc*.
arceau, m., small arch, vault;
—— *en décharge*, (*fort.*, *etc.*) relieving-arch.
archal, m., brass; (used only in) *fil d'* ——, brass wire.
arche, f., arch;
largeur d'une ——, span.
archet, m., bow drill;
drille à ——, bow drill.
architecture, f., architecture;
—— *militaire*, (*mil.*) a general term for the art of military construction, forts, bridges, etc. (almost equivalent to military engineering, in respect of constructions).
archives, f. pl., archives, records; record office.
archiviste, m., (*Fr. a.*) clerk (in the staff). (The grades are: —— *principal de 1ère*, *2ème classe*; —— *de 1ère*, *2ème classe*; *3ème classe*: the grades are not to be assimilated.)
arc-niveau, m., (*art.*) pointing-arc; gunner's quadrant; (generally —— *de pointage*).
arçon, m., crank of a centerbit; (*harn.*) saddle-tree;
bande d' ——, side-bar, saddle-bar;
être ferme sur les ——s, (*man.*) to have a sure seat;
perdre les ——s, (*man.*) to lose one's seat, to be thrown from a horse;
pistolet d' ——, (*sm. a.*) horse pistol;
pointe de bande d' ——, (*harn.*) rear horn or point of a side-bar;
se remettre dans les ——s, (*man.*) to recover one's stirrups;
vider les ——s, (*man.*) to be thrown from a horse.

arçonnerie, f., business or art of saddle-framing; generic term for all objects related to same; *atelier d'——*, shop where saddletrees are made (*Fr. a.*, at Saumur).

arcure, f., (*art.*) defect in cast guns, swell or rising in exterior surface; buckling, bending, or warping, as of a gun heated to receive a tube.

ardent, a., burning; high-spirited; of high mettle; *verre ——*, burning-glass.

ardillon, m., tongue of a buckle.

ardoise, f., slate; roofing-slate.

are, m., are, French unit of area=100 square meters.

aréage, m., land-surveying.

arène, f., sand, gravel; arena; cockpit.

aréographie, f., areography.

arête, f., edge, sharp edge (not a cutting edge), angle; (*top.*) crest, ridge; (*fort.*) salient angle of intersection of two faces; (*sm. a.*) ridge or rib on a sword-blade, on a bayonet; (*art. and sm. a.*) edge of the lands; (*unif.*) ridge on breast of cuirass; (*hipp.*) rat-tail, grapes, mangy tumors on a horse's legs; —— *vive, vive ——*, sharp edge (e. g., square edge of a screw head not countersunk, etc.); *à vives ——s*, sharp-edged.

arêtier, m., ridge; (*cons.*) any salient of a roof, due to intersection of faces; hip-rafter (of a roof).

arganeau, m., ringbolt; anchor ring.

argent, m., (*met.*) silver; —— *détonant*, (*expl.*) detonating silver (obs.); —— *fulminant*, (*expl.*) fulminate of silver.

argenter, v. a., to silver, to overlay with silver.

argile, m., clay; —— *réfractaire*, fire-clay.

argileux, a., clayey.

argilo-siliceux, a., argilo-siliceous.

arigot, m., a familiar term for fifes, flutes, flageolets.

arithmographie, f., arithmography.

arithmomètre, m., counting machine; comptometer; computing machine.

armand, m., (*hipp.*) horse-drench; mash.

armateur, m., shipowner; owner of a privateer; privateer.

armature, f., fitting, brace, strengthening piece, mounting; frame, truss; Y of a telescope; (*elec.*) sheathing, armoring (of a cable); armature (of a magnet, of a dynamo); (*fort.*) armor-plating (rare); (*sm. a.*) holder (as for a quick loader); —— *enfer*, shoe of a pile; iron facing (as for a skid, to roll guns on); —— *de pont*, truss-frame of a bridge; *poser une —— à une poutre*, to truss a timber; —— *simple*, (*cons.*) simple truss-frame, kingpost truss; —— *des soupapes*, (*mach.*) valve-gear.

arme, f., arm, weapon; arm of the service; (in pl.) fencing, campaign; (figuratively) efforts made by force of arms, war, fight, troops.
    I. General terms and classes of arms;
    II. Small arms (fire); III. Manual of arms; IV. Miscellaneous.

I. General terms and classes of arms.

—— *d'abordage*, (*nav.*) boarding weapon;
—— *blanche*, side arm (sword, bayonet, lance);
—— *à canon lisse*, smooth bore;
—— *à canon rayé*, rifle;
—— *se chargeant par la bouche*, muzzle-loader;
—— *se chargeant par la culasse*, breechloader;
—— *de chasse*, sporting weapon, shotgun;
—— *de choc*, striking, crushing, weapon (mace);
—— *contondante*, v. —— *de choc*;
—— *courtoise*, tilting, blunted weapon; arm of parade;
—— *s dardelles*, darts, javelins, etc.;
—— *défensive*, any defensive weapon;
—— *d'escrime*, fencing weapon, foil;
—— *d'estoc*, thrusting weapon;
—— *d'estoc et de taille*, cut-and-thrust weapon;

arme à feu, firearm;
—— *à feu portative*, hand firearm;
—— *de guerre*, military weapon;
—— *d'hast*, arm fitted with a shaft (lance);
hautes ——s, v. —— *d'hast*;
—— *d'honneur*, presentation weapon given in France as a reward for gallantry, after 1793, and before the institution of the Legion of Honor;
—— *de jet*, ballistic, projectile weapon;
—— *à lame*, generic term for sword;
—— *de luxe*, private weapon (generally highly ornamented), as opposed to *arme de guerre*;
—— *de main*, any hand arm (exclusive of firearms);
—— *offensive*, any weapon of offense;
—— *à outrance*, deadly weapon (in tilting);
—— *de parade*, arm of parade, v. —— *courtoise*;
—— *à percussion*, any percussion firearm, (*esp.*) the now obsolete (*mil.*) hammer-gun;
—— *de petit calibre*, the modern small-caliber rifle;
—— *petite*, side arm;
—— *pneumatique*, air-gun;
—— *de pointe*, thrusting weapon;
—— *portative*, hand arm;
—— *de précision*, arm of precision (modern firearms);
—— *pyrobalistique*, pedantic term for firearm;
—— *rayée*, rifle;
—— *à rotation*, generic term for Hotchkiss revolving cannon and its similars;
—— *sous-marine aggressive*, any automobile torpedo;
—— *sous-marine défensive*, torpedo, groundmine, mine;
—— *de taille*, cutting weapon;
—— *à tir rapide*, any r. f. firearm;
—— *de trait*, missile weapon;
—— *tranchante*, cutting weapon;
—— *transformée*, converted arm (from one model to another).

II. Small arms (fire):

—— *à aiguille*, needle gun;
—— *d'arçon*, holster pistol, horse pistol;
—— *automotrice*, Maxim automatic gun;
—— *à barillet*, rifle with revolving cylindrical breech (Spitalsky, Werndl);
—— *à bloc*, rifle of the falling-block system (Peabody);
—— *à broche*, arm firing a *cartouche à broche*, q. v.;
—— *à canon mobile*, sporting gun, shotgun;
—— *à chargement multiple*, magazine rifle;
—— *à chargement rapide*, rapid loader, quick loader;
—— *à chargement successif*, single-loader; magazine rifle that may be used as a single-loader;
—— *à charnière*, —— *à clapet*, rifle with hinged breechblock (original Springfield);
—— *à deux coups*, double-barreled shotgun;
—— *à six coups*, six-shooter (revolver);
—— *à un coup*, —— *à un seul coup*, (*mil.*) single-loader; (*sporting*) single-barreled shotgun;
—— *à culasse glissante*, rifle with a sliding breechblock;
—— *à culasse mobile*, an arm with movable breechblock;
—— *à culasse mobile par glissement*, arm with sliding breechblock;
—— *à culasse mobile par rotation*, arm with rotating breechblock;
—— *à culasse tombante*, arm with falling block (Peabody, Martini-Henry);
—— *à culasse tournante*, arm with rotating breechblock;
—— *à extracteur*, arm having or fitted with an extractor;
—— *à fermeture rectiligne*, straight-pull rifle (Mannlicher);
—— *à feu portative*, small arm proper (fire);
—— *à feu de vitesse et à répétition*, the modern magazine rifle;
—— *à longue portée*, long-range rifle;

arme à *magasin*, magazine rifle;
—— à *magasin fixe*, rifle with fixed magazine;
—— à *magasin séparable*, rifle with detachable magazine;
—— à *mouvement rectiligne*, straight-pull rifle (Mannlicher);
—— à *pène*, arm with hinged breechblock (original Springfield);
—— à *percussion*, the old hammer type of gun; any percussion arm;
—— *de petit calibre*, the modern small-caliber rifle;
—— à *platine*, hammer arm, small arm in which rotary motion of hammer is converted into translation of the firing-pin, or is applied directly to the cartridge or charge (includes the obsolete percussion-cap system);
—— *sans platine*, lockless arm (the modern military rifle);
—— *polygonale*, Whitworth rifle (twisted hexagonal prisms);
—— *portative à feu*, small-arm proper (fire);
—— à *répétition*, repeating rifle;
—— *s de réserve*, (*Fr. a.*) rifles kept in store for the reserve of the army;
—— *revolver*, arm in which repetition is effected by a revolving cylinder;
—— à *rotation longitudinale*, arm in which the axis of rotation of the breechblock is parallel to the axis of the barrel;
—— à *rotation rétrograde*, arm in which the breechblock rotates to the rear (Remington);
—— à *rotation transversale*, arm in which the breechblock rotates at right angles to the barrel;
—— *s de service courant*, (*Fr. a.*) arms in hands of active army;
—— *simple*, single-loader;
—— à *système complexe*, any rifle not falling into any one of the distinct classes of rifles;
—— à *tabatière*, arm in which axis of breechblock is on the side and parallel to axis of barrel;
—— *de théorie*, (*Fr. a.*) rifle issued (4 to a company) for training young soldiers in mounting and dismounting; "hack" rifle (always marked X, and never fired);
—— à *tige*, "tige" rifle, having a pin (*tige*) at the bottom of the bore in the axis of the barrel;
—— à *tir rapide*, repeating rifle;
—— à *tiroir*, arm in which breechblock slides up and down at right angles to barrel (Treuille de Beaulieu);
—— à *verrou*, bolt gun (usual type of modern armies).

III. Manual of arms:
arme, (in commands) "arms!";
apprêter l'——, to make ready, to come to a ready;
attaquer l'——, to seize the piece briskly in the execution of the manual;
l'—— *au bras*, (position of) support arms;
l'—— *sous le bras gauche*, arms under the left arm (funerals);
l'—— à *la bretelle*, rifle slung;
l'—— *sur l'épaule droite*, (position of) shoulder arms;
faisceau d'——*s*, stack;
haut l'——, (position of) advance carbine; also command for same;
—— *haute*, piece at a ready (obs.);
maniement d'——*s, de l'*——, manual of arms;
—— *au pied*, (position of) order arms;
port d'——, *port des* ——*s*, (position of) the carry;
porter l'——, *les* ——*s*, to carry arms;
présenter l'——, *les* ——*s*, to present arms;
replacer l'——, to recover arms;
reposé sur l'——, (position of) order arms;
reposer l'——, to order arms;
—— à *volonté*, pieces at will (route step).

IV. Miscellaneous:
appeler sous les ——*s*, to turn out under arms; (*in gen.*) to call to arms;

arme, *assaut d'*——*s*, assault at arms, fencing match;
aux ——*s!* to arms; (*mil.*) turn out the guard! (also) beat of drums to call guards, *piquets*, and posts under arms;
les ——*s basses*, rifles carried low (of troops entering or leaving a trench);
—— à *cheval*, (*mil.*) mounted arm;
cheval d'——, charger;
coffre d'——*s*, arm chest;
compagnon d'——, brother officer; companion in arms;
contrôleur d'——*s*, (*Fr. a.*) v. s. v. *contrôleur*;
déposer les ——*s*, to lay down one's arms, surrender;
à ——*s égales*, on equal terms;
en ——*s*, under arms;
faire des (*les*) ——*s*, to fence;
faire ses premières ——*s*, to make one's first campaign, to see real service for the first time;
fait d'——*s*, brilliant deed of arms;
huissier d'——*s*, sergeant-at-arms;
les ——*s sont journalières*, one can not win all the time (proverb);
les ——*s à la main*, armed;
maître d'——*s*, fencing master;
mettre bas les ——*s*, to lay down one's arms, to surrender;
se mettre sous les ——*s*, to get under arms, to turn out under arms;
pas d'——*s*, passage of arms;
passer l'—— à *gauche*, (*mil. slang*) to die;
passer par les ——*s*, to be shot (military execution);
faire passer par les ——*s*, to execute by shooting (military execution);
—— à *pied*, dismounted arm of the service;
place d'——*s*, stronghold; drill ground, parade ground; alarm post; place of assembly in a cantonment; (*fort.*) place of arms;
port d'——, shooting license; license to bear arms;
porter les ——*s*, to bear arms; to make war;
poser les ——*s*, to lay down one's arms, surrender, make peace;
prendre les ——*s*, to take up arms, go to war;
prendre les —— *contre*, to levy war against;
rendre les ——*s*, to surrender, give up one's arms;
salle d'——*s*, armory; fencing school;
avoir tant d'années de salle d'——*s*, to have taken fencing lessons for so many years;
—— *savante*, —— *scientifique*, —— *spéciale*, technical, scientific arm of the service;
sous les ——*s*, under arms;
sortir en ——*s*, to turn out under arms;
suspension d'——*s*, armistice, suspension of arms, of hostilities;
tirer les ——*s*, to fence;
tireur d'——*s*, fencer.

armé, a., fitted with, fitted out, equipped; (*art.*) supplied with its equipments, etc. (as a turret); fused (of a shell); (*sm. a.*) cocked;
à *l'*—— (*sm. a.*) full-cocked, at the full cock;
—— *automatique*, (*sm. a.*) automatic cocking of modern rifles;
cran de l'——, (*sm. a.*) full-cock notch, cocking notch;
—— *en guerre*, armed; fitted out for active service (as a ship);
à *main* ——*e*, by force of arms, by open force;
—— *de toutes pièces*, completely armed.

armée, f., army;
—— *active*, the active army, as distinguished from the reserve and the territorial army;
—— *en bataille*, army drawn up in array, for attack or defense;
—— à *cheval*, the cavalry preceding advance of the regular field army (modern tactics);
—— *coloniale*, (in France) marine infantry and artillery; colonial troops in general;
corps d'——, army corps (strategic unit);
entrer dans l'——, to go into the army, enter the service;
être à l'——, to be in the army, with the army;
gros de l'——, main body (as distinguished from the advance guard);

armée *métropolitaine*, "home" army, as opposed to colonial troops;
— *de mer*, sea forces, i. e., the navy proper, and the various marine troops, as infantry, artillery, etc.; (sometimes) fleet, navy;
— *d'observation*, army of observation, to watch a point from which danger is threatened (e. g., to watch a relieving army in a siege, a frontier in case of probable intervention of a third party, etc.);
— *d'occupation*, army of occupation;
— *d'opérations*, army of operations, i. e., whose first business is to fight;
— *permanente*, standing army;
— *de secours*, relieving army (to raise a siege, relieve a besieged fortress, or succor another army);
— *de siège*, besieging army; besieging forces proper, siege corps;
— *de terre*, land forces (as distinguished from naval);
— *territoriale*, (France) territorial army, consisting of men that have served ten years in the active army and its reserve.

armement, m., armament, arming (fort, ship, army, etc.); equipping, equipment; fitting out (commercial as well as military); arms; warlike preparations; (*in pl.*, *art.*) equipments; (*art.*) arming (of a fuse);
— *d'une batterie*, arming, supplying a battery with guns;
— *de combat*, (*art.*, *fort.*) guns of an intrenched camp (Brialmont);
— *complémentaire*, (*art.*, *fort.*) in a permanent work, guns added to, complementary of, — *de sûreté*, q. v.;
— *des corps de troupe*, (*Fr. a.*) the supplying of arms, etc., to troops (done by the artillery);
— *cuirassé*, (*art.*) a generic term for guns in armored emplacements;
— *de défense*, (*art.*, *fort.*) the armament of a permanent work, comprising — *de sûreté*, — *complémentaire*, qq. v., and the general reserve of guns; the (increased) armament of an attacked sector of defense (Brialmont);
— *d'interdiction*, (*art.*, *fort.*) in a *fort d'arrêt*, q. v., the guns covering the roads closed or commanded by the fort;
— *de mobilisation*, (*art.*, *fort.*) the armament of permanent works of the first order, installed in place, and intended to offset open assault and to assure the defense until the zone of attack is known;
— *normal* (*art.*, *fort.*) the armament necessary to repel an open assault (Brialmont);
officier d'——, (*mil.*) v. *officier*;
— *d'une place de guerre*, (*art.*, *fort.*) the armament of a fortified position, consisting of — *de défense*, *batteries mobiles*, q. v., and spare guns;
— *de réserve*, (*Fr. a.*) arms, etc., for men who are to join on mobilization, for the reserve;
— *du service courant*, (*Fr. a.*) arms, etc., for the active army;
— *en service*, arms, etc., in the hands of troops;
— *de sûreté*, (*art.*, *fort.*) the armament of a permanent work, destined to prevent a surprise or an open assault.

armer, v. a. r., to arm; to furnish with arms; to make preparations for war; to take up arms; to fit with, to equip with; to build up; to fit out (commercial as well as military); to fit out a vessel (trade or war); to steel (a tool); (*art.*) to mount guns in a battery, in a fort; to arm (a fuse); (*nav.*) to put a ship in commission; to man (a boat); (*sm. a.*) to cock a piece; (*hipp.*) to take the bit in the teeth, to bend down the head until cheeks of bridle rest on chest;
— *les avirons*, (*boats*) to ship the oars;
— *un bâtiment*, (*nav.*) to put a ship in commission;
— *une batterie*, (*art.*) to mount guns in a battery, to furnish a battery with guns;
— *un cabestan*, to man a capstan;
— *un câble*, to sheathe, armor, a cable; to wind a cable with wire;

armer *en course*, (*nav.*) to equip, fit out, for privateering;
— *une dynamo*, (*elec.*) to wind a dynamo;
— *un frein*, (*art.*, *etc.*) to set a brake;
— *une fusée*, (*art.*) to arm a fuse;
— *sur la gâchette*, (*sm. a.*) to move the bolt, etc., violently forward, so as to cause the head of the sear to be struck;
— *en guerre*, (*nav.*) to commission (a ship) for war;
— *un obus*, (*art.*) to fit or insert a fuse in a shell;
— *une prise*, (*nav.*) to man a prize.

armistice, m., armistice, truce;
dénoncer un ——, to close an armistice, to end it;
— *général*, general armistice, affecting all the armies engaged;
— *particulier*, truce affecting only a part of the field of operations.

armoire, f., cupboard, press;
— *à glace*, — *à poils*, (*mil. slang*) soldier's knapsack.

armoire-étagère, f., clothes-press.

armoiries, f. pl., coat-of-arms.

armon, m., (*art.*) hound (or side-rail); futchel (limber); side-rail (caisson);
queue d'——*s*, fork.

armure, f., armor; (*elec.*) pole piece (of a dynamo), armature of a magnet; (*nav.*) armor (of a vessel);
— *du génie*, (*mil.*) armor worn by two leading sappers in the full sap (abandoned in 1878).

armurier, m., armorer; (*Fr. a.*) employé of the artillery who repairs, etc., arms in the regiment (called *chef* ——; there are two classes, ranking next after *adjudant*.

aronde, f., (old name for *hirondelle*, swallow);
queue d'——, (*carp.*) dovetail;
à, *en*, *queue d'* ——, dovetailed.

arpent, m., old (and varying) unit of area.

arpentage, m., (*surv.*) land-surveying (to determine areas).

arpenter, v. a., (*surv.*) to survey (land, for area).

arpenteur, m., land surveyor;
chaîne d'——, (*inst.*) surveyor's chain;
équerre d'——, (*inst.*) surveyor's square (for angles of 45°, 90°).

arqué, p. p., bent; (*esp. hipp.*) bent or curved (of a horse's legs).

arquebusier, m., (*sm. a.*) gunsmith, gunmaker.

arquer, v. a. and n., to arch, curve, camber; to be or become bent, curved, crooked, etc.

arrache-cartouche, m., (*sm. a.*) extractor.

arrache-clou, m., nail-drawer.

arrache-culot, m., (*art.*) cartridge-head extractor, used in *canon à balles*.

arrachement, m., pulling, wrenching; (*art.*) stripping (of a rifling-band); (*met.*) break, tearing, or rupture (in a fracture, especially of fibrous metal); metallic threads noticeable in metal that has been broken with considerable violence.

arrache-pieux, m., pile-drawing engine;
vérin ——, withdrawing-screw;
levier ——, withdrawing-lever.

arracher, v. a., to tear off, out, apart; to rend, wrench; (*art.*, *sm. a.*) to strip (of jacket on bullet, on lead-covered shells, of rifling-band, etc.).

arraisonner, v. a., (*nav.*) to force a (commercial) vessel to answer questions, to question a vessel.

arraser, v. a., v. *araser*.

arrérages, m. pl., arrearage.

arrestation, f., arrest, confinement; custody.

arrêt, m., stop; (*mil.*) check, in a column on the march, of a wagon in a train; (*mach.*, *tech.*) part of a machine or mechanism intended to stop or limit the action or play of some other parts; (hence) stop, catch, lug, rib, etc.; (*mil.*, *etc.*) judgment, decision; (*man.*) stopping or halting (of a horse in motion, by its rider); (*in pl.*, *mil.*), arrest (of officers, disciplinary punishment);
d'——, (*mach.*, *art.*, *etc.*) stop, keep (in many relations);

arrêt, *aux* ――*s*, (*mil.*) in, under, arrest;
―― *de barillet*, (*sm. a.*) cylinder stop;
―― *de brêlage*, (*art.*) lashing stop;
――*-butoir*, (*F. m. art.*) V-shaped lug on traversing head of sight;
camper aux ――*s*, (*mil.*) to clap into arrest;
―― *de cartouche*, (*sm. a.*) cartridge stop;
―― *de chaîne*, chain-nipper, chain-stop;
―― *des coffres à munitions*, (*art.*) keep-plate;
crochet d' ――, (*mach., etc.*) stop;
―― *de la culasse mobile*, (*sm. a.*) bolt-stop;
―― *de défense*, safe-conduct;
demi- ――, (*man.*) slowing or reduction of horse's speed by action of hand on reins;
――*s forcés*, (*mil.*) close arrest;
―― *de fonctionnement*, (*art., sm. a.*) failure of a mechanism to work; jamming;
fort d' ――, (*fort.*) barrier fort; "stop" fort (also known by its French name);
――*s de forteresse*, (*mil.*) arrest carrying with it confinement to a fortress;
gagner les ――*s de*, (*mil.*) to be put in arrest by;
garder les ――*s*, (*mil.*) to stay in arrest, to observe one's arrest;
lever les ――*s*, (*mil.*) to release from arrest;
―― *de la main*, (*man.*) stopping of a horse by action of hand on reins;
―― *de manœuvre*, (*r. f. art.*) drill-stop, drill-hook;
mettre aux ――*s*, (*mil.*) to put in arrest;
pièce d' ――, (*mach., etc.*) lock, locking piece;
plaque d' ――, stop-, keep-plate;
point d' ――, stopping-, sticking-place;
―― *de prise*, (*nav. int. law*) temporary forbidding of commercial vessels in a port to leave their anchorages (by way of reprisal or punishment);
se rendre aux ――*s*, (*mil.*) to consider one's self in arrest;
―― *de répétition*, (*sm. a.*) cut-off, magazine cut-off; (in the Loewe loader) lid-closing device allowing but one cartridge in the piece at a time;
――*s de rigueur*, (*mil.*) close arrest;
robinet d' ――, stop-cock;
rompre les ――*s*, (*mil.*) to commit a breach of arrest, to break one's arrest;
rondelle d' ――, (*mach.*) nut or screw of a safety-valve lever;
rude sur l' ――, (*man.*) hard-mouthed;
――*s simples*, (*mil.*) open arrest (with limits of post, U. S. A.);
―― *de sûreté*, safety stop (machine guns, etc.);
temps d' ――, stoppage, delay; (*mil.*) halt (as, in the *marche d'approche*, etc.);
―― *des trains*, (*r. r.*) stop of a train.
arrêtage, m., stop (i. e., device for stopping).
arrête, m., (*mil.*) cry of a sentry at interior gate of a town to sentry at exterior, to prevent carriages from coming through until way is clear;
―― *là bas*, (*mil.*) cry of sentry at exterior of a gate, to prevent anyone from coming out.
arrêté, m., decision, formal decision, by a court, assembly, administrative authority, etc.;
―― *de compte*, (*adm.*) statement of account; operation of drawing up a statement of account.
arrête-gaz, m., (*art.*) (literal translation of) gas-check.
arrêter, v. a., n., to stop, hold; fasten, fix; arrest; apprehend; to bring to a close; to come to a stop; to throw out of gear, to be thrown out of gear; to post (a ledger); to close an account, a return;
―― *et rendre*, (*man.*) to raise and lower the wrist (in backing).
arrêtoir, m., stop, catch, nib, lug; keep-plate, stop-plate; (*sm. a.*) catch, catch-lever; packet-catch;
―― *de baïonnette*, (*sm. a.*) bayonet-stop;
―― *de chargeur*, (*sm. a.*) magazine catch, packet catch;
―― *de coffre*, (*art.*) stay, keep-plate;
―― *de culasse mobile*, (*sm. a.*) bolt-stop;
―― *de cylindre*, (*sm. a.*) bolt-stop, cylinder stop;
―― *d'écouvillon*, (*art.*) sponge-stop or -catch;

arrêtoir *de levier de pointage*, (*art.*) trail-hand-spike stop or catch;
―― *de madrier*, (*pont.*) chess-stop (in a chess-wagon);
―― *de poutrelle*, (*pont.*) back-stop (of a ponton wagon);
vis- ――, (*sm. a., art., etc.*) stop-screw.
arrière, adv., back, in rear; m., (*mil.*) the rear (of an army); (*nav.*) stern (of a vessel, boat);
en ――, backward;
aller en ――, (*mil., fam.*) to go to the "rear";
coup d' ――, back stroke;
service, services, de l' ――, (*mil.*) a comprehensive term of wide application, including all operations and duties (supply, transportation of sick and wounded, etc.), of whatever nature, between an army or armies, the "front" generally, and the national territory (base).
arriéré, a., behindhand, in arrears, unexecuted; m., (*adm.*) expenses not met or paid for in the fiscal year to which they belong.
arrière-bec, m., (*pont.*) stern, after-peak, of a ponton boat (all that part of the stern the bottom of which curves upward); a downstream cutwater, starling.
arrière-bord, m., rear edge.
arrière-coup, m., (*surv.*) back-sight.
arrière-garde, f., (*mil.*) rear guard;
faire l' ――, to do or be the rear guard;
gros de l' ――, main body of the rear guard;
pointe d' ――, point of the rear guard;
extrême pointe d' ――, extreme point of the rear guard;
tête de l' ――, head of the rear guard.
arrière-main, f., (*hipp.*) hind quarters of a horse (everything behind the rider).
arrière-molaire, f., (*hipp.*) back molar tooth.
arrière-plan, m., background.
arrière-rang, m., (*mil.*) rear rank.
arrière-saison, f., the end of fall and beginning of winter.
arrière-train, m., (*art.*) hind carriage, i. e., the gun and carriage, as opposed to the limber; rear portion, after-body, of any four-wheeled carriage; (*hipp.*) hind quarters of a horse (everything behind the rider); (of a bicycle) rear wheel and related parts.
arrimage, m., packing, stowing; (*art.*) packing of ammunition in chests; (*mil.*) the careful packing, according to regulations, of the knapsack, of the various articles carried by the trooper on his saddle; the operation of hitching the car and accessories to the balloon.
arrimer, v. a., to stow, to pack carefully (ammunition, knapsack, etc.); to attach a car and accessories to a balloon; to pack a train (mountain battery on mules);
―― *un ballon*, to attach the car, etc., to a balloon.
arrimeur, m., packer; balloon-packer (i. e., helps in the inflation, etc., of a balloon).
arrivé, p. p., ―― *par les rangs*, (*mil.*) risen from the ranks.
arrivée, f., arrival; (*steam, etc.*) admission pipe;
rive d' ――, (*pont.*) the farther bank in building or dismantling a ponton bridge.
arrondi, a., rounded; m., swell or rounded part of anything, rounded surface (as, *art.*, between lands and grooves, etc.);
―― *en dos d'âne*, barreled (of a road).
arrondir, v. a., to round, round off;
―― *un cheval*, (*man.*) to break a horse.
arrondissement, m., rounding; (*fort.*) rounding of the counterscarp in front of a salient; rounding of a reentrant angle; (*mil.*) district, military district (not territorial, variable in area);
―― *d'artillerie*, ―― *du génie*, ―― *de l'intendance*,
―― *d'inspection générale*, artillery, engineer, intendance, q. v., inspection districts, respectively.
arrosage, m., irrigation; sprinkling; (*powd.*) moistening of gunpowder in manufacture.
arroser, v. a., to sprinkle; soak; moisten; irrigate;

arroser *ses galons*, (*mil. slang*) of a new-made n. c. o., to treat one's comrades, to wet his chevrons;
—— *les poudres*, (*mil.*) to flood, wet, powder in case of fire or of danger.
arrosoir, m., watering-pot;
coup d'——, (*art.*) a spray of shrapnel bullets, fragments, etc.
ars, m., (*hipp.*) point of junction of arm and trunk of horse, where he is usually bled.
arsenal, m., arsenal;
—— *d'artillerie*, gunshop, gun factory; shop where all military material is made;
—— *de construction*, (*art.*) workshop;
—— *de dépôt*, store, depot;
—— *de marine*, —— *maritime*, —— *de port*, (*nav.*) navy-yard (U. S.), dock yard (England).
arsenic, m., (*chem.*) arsenic.
arséniure, f., (*chem.*) arsenide.
art, m., art;
construction d'——, (*r. r.*) generic term for bridges, culverts, etc., as distinguished from mere earthwork;
—— *de la guerre*, —— *militaire*, art of war;
ouvrage d'——, v. *construction d'*——;
ouvrier d'——, skilled workman, v. *ouvrier*.
artère, f., artery; highway of communication; (*elec.*) feeder.
arthrogoniomètre, m., (*hipp.*) arthrogoniometer.
artichaut, m., spiked fence or gate; (*fort.*) iron spikes.
article, m., (*mach.*, etc.) link, joint.
articulation, f., (*mach.*, etc.) joint, articulation; link; jointing;
—— *circulaire*, rule joint;
—— *à fourche*, forked joint, forked head;
—— *à lamelles*, (*mach.*) compound link;
—— *à lames*, (*mach.*) plate joint intersection;
—— *à nœud de compas*, forked joint;
—— *sphérique*, ball-and-socket joint;
tige d'——, pump rod and links.
articuler, v. a., to joint, unite by means of a joint; to articulate;
—— *une colonne*, (*mil.*) to break up a column into march-units;
s'—— *à rotule*, (*mach.*) to be, or form, a ball-and-socket joint.
artifice, m., artifice, contrivance; strategem; (*artif.*) pyrotechnic composition; (*artif., in pl.*) fireworks;
atelier d'——*s*, laboratory;
—— *de communication du feu*, any pyrotechnic means of communicating fire (to a charge, mine, etc.); priming, train;
—— *de conservation du feu*, any means of preserving fire for use when wanted (*mèche à canon*, q. v., etc.).
—— *d'éclairage*, any illuminating firework (in sieges, to light on a dangerous pass, etc.);
feu d'——, fireworks;
feu d'—— *Bengale*, Bengal light;
—— *pour feux de Bengale*, blue-light composition;
—— *de guerre*, military, service fireworks; (*mil.*) ruse, strategem;
—— *incendiaire*, incendiary fireworks;
—— *d'inflammation*, pyrotechnic means of firing directly (*lance à feu, porte-feu Bickford*, qq. v., etc.);
—— *de mise de feu*, generic term for primers, fuses, detonators, etc.;
—— *de production de feu*, generic term for primers, etc.;
—— *de réjouissance*, fireworks proper, i. e., mere display (not military in character);
—— *de rupture*, any explosive, suitably put up, used for demolition purposes (torpedo, fougasse, petard, etc.);
—— *de signaux*, signal fireworks;
—— *de transmission du feu*, any pyrotechnic means of communicating fire.
artificier, m., any man who makes fireworks; (*art.*) artificer, laboratory man; (*Fr. art.*) artilleryman employed in fireworks laboratory;
chef ——, (*Fr. art.*) master artificer (one to each regiment);

artificier *de la gaine*, (*art.*) artificer responsible for supplying turret guns with ammunition;
—— *du montecharges*, (*art.*) artificer responsible for the supply of ammunition to the turret depot;
sous-chef ——, (*Fr. art.*) battery artificer.
artillé, a., (*nav.*) carrying guns;
vaisseau —— *de n pièces*, ship mounting *n* guns (obs.).
artillerie, f., artillery (arm of the service; personnel of the arm; guns, ordnance; science of gunnery);
—— *de bord*, (*nav.*) naval artillery;
—— *de campagne*, field artillery (horse and mounted);
—— *à cheval*, horse artillery;
combat d'——, v. *duel d'*——;
—— *de corps*, (*mil.*) corps artillery;
—— *de côte*, seacoast artillery;
—— *cuirassée*, (*art., fort.*) a general term for guns under cupolas or other metallic cover;
—— *divisionnaire*, divisional artillery;
duel d'——, (*mil.*) the artillery combat preceding an attack by infantry;
—— *de forteresse*, fortress artillery;
garde d'——, (*Fr. a.*) v. s. v. *garde*;
—— *à grande puissance*, high-power guns;
grosse ——, heavy guns (seacoast, etc.);
—— *légère*, light artillery, field artillery;
lutte d'——, v. *duel d'*——;
—— *de marine*, —— *de la marine*, marine artillery;
—— *de montagne*, mountain artillery;
—— *montée*, field artillery (proper, does not include horse);
ouvrier d'——, v. s. v. *ouvrier*;
—— *de petit calibre*, generic term for r. f. artillery, etc.; secondary battery or artillery;
—— *à pied*, foot artillery (in France, fortress artillery);
—— *à pied montée*, field artillery;
—— *de place*, garrison artillery;
—— *pneumatique*, guns using compressed air (Zalinski);
—— *pontonniers*, in France, 2 regiments of artillery pontoniers used to constitute the personnel of the bridge train; turned over to the engineers in 1894;
—— *de position*, field howitzers, mortars, etc.;
—— *de préparation*, (*mil.*) the artillery that prepares the way for the infantry attack;
—— *de siège*, siege artillery;
—— *de terre*, land, as distinguished from marine, artillery;
—— *volante*, antiquated expression for horse artillery.
artilleur, m., artilleryman (officer or man); gunner; artillerist.
artison, m., moth, woodworm.
artisonné, a., (of wood) worm-eaten.
arzel, m., *cheval* ——, (*hipp.*) horse with off hind foot white or white spotted.
as, m., ace;
—— *de carreau*, (*mil. slang*) knapsack, ribbon of the Legion of Honor;
—— *de pique*, (*fort.*) pointed caponier, shaped somewhat like the ace of spades;
six —— *et* ——, (*powd.*) v. s. v. *six*.
ascenseur, m., lift, elevator; (*art.*) shell-hoist;
—— *écluse*, (*mach.*) canal-boat elevator.
ascension, f., ascension, balloon ascension; (*mach.*) upstroke of a piston;
—— *captive*, ascent of a captive balloon;
—— *libre*, ascent of a free balloon.
asphaline, f., (*expl.*) asphaline.
asphalte, m., asphalt; bitumen.
aspirail, m., (*mach.*, etc.) vent-hole, grating.
aspirant, m., (*nav.*) midshipman;
—— *de 1ère classe*, naval cadet (at sea, U. S. N.), passed midshipman (R. N.);
—— *de 2ème classe*, midshipman (n. e.).
aspirateur, m., exhauster; a., exhausting.
aspiration, f., suction, inspiration; exhaustion;
d'——, (*mach.*, etc.) exhausting;
soupape d'——, suction valve;
tuyau d'——, suction-pipe, exhaust pipe.

aspirer, v. a., (*mach.*, *etc.*) to exhaust; to suck, to suck in, up.

assablement, m., sand bank, shoal.

assabler, v. a., to fill, choke with sand.

assaillant, m., assailant; (*mil.*) the assailant, the attack (storming of a breach, etc.).

assaillir, v. a., (*mil.*) to assault, storm, attack vigorously; to march to the assault.

assainir, v. a., to make healthy, introduce sanitary measures.

assainissement, m., sanitation, sanitary measures.

assaut, m., (*mil.*) assault, storming (of a breach, position, fort, etc.); the final or decisive attack of infantry, in battle; storm (wind and rain);
—— *d'armes*, (*fenc.*) assault at arms, fencing match;
colonne d'——, (*mil.*) storming party;
donner l'——, (*mil.*) to storm, to attack with vigor;
emporter, enlever, d'——, (*mil.*) to carry, take by storm;
—— *d'escrime*, (*fenc.*) fencing match;
faire ——, (*fenc.*) to have a fencing match;
faire, livrer, un ——, (*mil.*) to make an assault, to storm;
monter à l'——, (*mil.*) to storm;
prendre d'——, (*mil.*) to capture, take by storm, by open attack;
prise d'——, (*mil.*) capture by storm, by open assault;
soutenir l'——, (*mil.*) to withstand an assault.

assèchement, m., drainage.

assécher, v. a., to drain.

assemblage, m., assemblage, assembling, collection, bringing together; (*carp.*) scarfing, joint; (*mach.*) shaft-coupling, joint, coupling; (*elec.*) connecting, joining up; (*mas.*) bond.

I. Carpentry; II. Miscellaneous.

I. Carpentry:

—— *par acollement*, overlapping joint (one piece resting on the other);
—— *affleuré*, flush-joint;
—— *en biaisement*, miter-joint;
—— *bout à bout*, butt-joint;
—— *en bout*, joining butt on butt;
—— *carré*, square miter-joint, square joint;
—— *à clef*, keyed joint;
—— *à, en, crémaillère*, joggling;
—— *par croisement*, joint in which the pieces are at right angles;
—— *à dent*, —— *en adent*, dovetail and miter;
—— *par embrèvement*, open mortise and tongue;
—— *à embrèvement longitudinal simple (double)*, simple (double) panel-joint;
—— *par encastrement*, rabbeting;
—— *par endentement*, joggling;
—— *à entailles*, —— *à simple entaille*, cross-grain notch, simple notch-joint;
—— *à enture*, any end joint (pieces end to end), ends notched, etc., together;
—— *avec goujons*, doweling;
—— *à mi-bois*, halving, scarfing;
—— *mobile*, temporary joint (e. g., leg in trestle-cap);
—— *à moise*, joint in which the pieces are caught between *moises*, q. v.;
—— *à mortaise*, mortise and tenon joint;
—— *à, d', en, onglet*, miter-joint;
—— *à oulice*, triangular tenon joint (the supporting piece receiving a triangular tenon from the supported);
—— *à paume*, undercut in end of a timber resting by the end on another at right angles to itself;
—— *à paume et à repos*, undercut and shoulder in end of a timber resting by the end on another timber at right angles to itself;
—— *par pénétration*, any joint in which one piece is bedded in the other;
—— *à quart-bois*, notch cut one-quarter of thickness;
—— *à queue d'aronde (d'hironde)*, dovetail joint, dovetailing;
—— *à queue d'aronde perdue*, dovetail and miter;

assemblage *à queue d'hironde simple*, joint dovetailed on one side only;
—— *à rainure et languette*, tonguing and grooving;
—— *à, en, sifflet*, skew scarf;
—— *à tenon et mortaise*, mortise and tenon;
—— *à tenon et mortaise avec embrèvement*, slit and tongue joint;
—— *à tenon passant*, joint through tenoned and keyed;
tête d'——, connecting point;
—— *à tiers-bois*, notch cut one-third through;
—— *à trait de Jupiter*, skew tabled scarf and key.

II. Miscellaneous:
—— *à baïonnette*, bayonet joint;
—— *à boulons déchargés*, (*mach.*) unloaded bolt connection;
—— *par boulons*, (*mach.*) bolt, pin, screw connection;
—— *à bride*, (*mach.*, *etc.*) flange joint;
—— *à clavette*, (*mach.*) keyed connection;
—— *à clavettes de hauteur*, (*mach.*) edged keying;
—— *à clavettes longitudinales*, (*mach.*) longitudinal keying, or keyed connection;
—— *à clavettes transversales*, (*mach.*) crosskeyed connection;
—— *douille*, socket, socket joint;
—— *à enforcement*, forked joint;
—— *en fer*, iron joint;
—— *à joint plat*, lap-joint (iron);
—— *à manchon*, (*mach.*, *etc.*) box-coupling, sleeve coupling;
—— *à manchon fileté*, (*mach.*, *etc.*) screw sleeve coupling;
—— *à mi-fer*, any joint in which half the thickness is cut away (iron, steel, etc.);
—— *mixte*, (*elec.*) joining up in series and parallel;
—— *à oreille simple*, single flange joint (pieces flanged on one and same side);
—— *à double oreille*, double flange joint (pieces flanged on both sides);
—— *en quantité*, (*elec.*) joining up in parallel, in quantity, in compound circuit;
—— *à rebords*, (*cons.*, *mach.*) flange joint;
—— *à rivets*, (*mach.*) rivet joint, connection;
—— *en série*, (*elec.*) joining up in series;
—— *en tension*, (*elec.*) joining up in series;
—— *de tiges mobiles*, turn-buckle;
—— *à vis*, (*mach.*, *etc.*) screw coupling.

assemblée, f., assembly, convention; (*mil.*) assembly (call), signal to fall in (for guard-mounting);
quartier d'——, (*mil.*) alarm post, assembling point.

assembler, v. a., r., to assemble, meet; to bring together; to collect; (*mil.*) to muster; (*carp.*) to scarf, joint, score, tenon, etc.; (*elec.*) to join up, connect. (For the various adjuncts of this verb, v. *assemblage*, supra.)
—— *à mi-bois*, —— *à mi-fer*, to halve and lap;
—— *en quantité*, (*elec.*) to join up in parallel, in quantity;
—— *en série*, —— *en tension*, (*elec.*) to join up in series.

assener, v. a., to strike hard, home (of a blow).

asseoir, v. a., to set, seat, lay; put, fix; base, ground; (*mil.*) to fix a camp; to fix the position of a besieging army;
s' —— *devant une place*, (*siege*) to lay siege to a place;
—— *un camp*, (*mil.*) to pitch a camp;
—— *un cheval*, (*man.*) to cause a horse to throw his weight well on his haunches; to teach him to gallop with his croup lower than his shoulders;
—— *les fondations*, to sink a foundation;
—— *une pension*, to settle a pension;
—— *une tente*, to pitch a tent.

assermenté, a., sworn, having taken an oath; (*Fr. a.*) sworn, of certain functionaries (*adjoint du génie, garde d'artillerie*, etc.); m., person who has taken the oath (of service, etc.).

assermenter, v. a., to swear in a public functionary.

asservir, v. a., to reduce to subjection; (*mach.*) to have under complete control (e. g., turret-gear, steering-gear), to put a mechanism, etc., under complete control (said of control by a servo-motor).

asservissement, m., subjection; (*mach.*) state of being under complete control (turret-gear, etc., by a servo-motor).

assiégeant, m., (*mil.*) besieger.

assiéger, v. a., (*mil.*) to besiege, invest.

assiette, f., seating, fixing; (*man.*) seat; (*r. r.*) plane or base on which the superstructure rests;
— *d'un camp*, (*mil.*) laying out, disposition of a camp;
— *du casernement*, (*Fr. a.*, *adm.*) evaluation and assignment to troops, of barracks and accessory buildings, and determination of the number of men and of horses they shall house;
déplacement d' ——, (*man.*) unsteadiness of seat;
enlever l' ——, (*man.*) in trotting, to post too high, to rise too high from the saddle;
faire l' —— *d'une torpille*, (*torp.*) to ballast a torpedo (Whitehead, etc.) suitably;
— *de la fortification*, (*fort.*) manner in which a fortification is adapted to the site; adaptation of works to the site;
— *du logement*, (*Fr. a.*, *adm.*) general term for estimate of capacity, use, etc., of public military buildings in garrison towns, or of any buildings that may be occupied or used by troops or filled with military stores;
— *d'une place forte*, (*fort.*) situation and extent of a fortified position, plan and form of; situation of a fortified position from a political or a strategic point of view;
— *d'une ville*, topographical position of a town.

assignation, f., (*mil. law*) summons to a military person to appear as a witness before a court-martial.

assimilation, f., (*mil.*) assimilation of rank);
— *de grade*, —— *de rang*, assimilation of grade, of rank.

assimilé, a., (*mil.*) assimilated; **m.,** (*mil.*) non-combatant ranking as an officer.

assimiler, v. a., (*mil.*) to assimilate.

assise, f., course, layer, bed; base (as of a machine tool);
par ——*s*, (*mas.*) coursed;
maçonnerie par ——*s*, *par* ——*s réglées* (*mas.*), coursed work;
sans ——, (*mas.*) uncoursed.

association, f., *des Dames françaises*, association of French women to help medical service in war by nursing, etc.

associer, v. a., (*elec.*) to join up, connect;
— *en quantité*, to join up in parallel;
— *en série*, —— *en tension*, to join up in series.

assortiments, m. pl., (*art.*) small stores.

assouplir, v. a., to supple, render flexible;
— *un cheval*, (*man.*) to break a horse, give him an easy motion.

assouplissement, m., suppling; (*in pl.*, *mil.*) setting-up exercises, of drill;
exercices d' ——, (*mil.*) setting-up exercises.

assourdir, v. a., to muffle (a drum, an oar).

assujétir, assujettir, v. a., to subdue, bring into subjection; to fix anything in position; to key, wedge, in.

assurance, f., insurance, assurance;
coup d' ——, (*nav.*) shot fired by a ship on display of her colors, to show that she belongs to the nation whose colors are displayed.

assurer, v. a., to assure; to perform, furnish; to secure, make fast (a rope, etc.); (*mil.*) to rectify, align (guides and markers at drill); (*art.*) to lock (as a breech);
s' ——, to slip, to settle, into good bearing (of a chain, rope, etc., on a drum, etc.);
— *la bouche d'un cheval*, (*man.*) to accustom a horse to the bit;
— *un bout*, (*cord.*) to make an end fast;

assurer *un cheval*, (*man.*) to train a horse to steadiness and to regularity in all his movements;
— *garde*, (*fenc.*) to take position of "guard," in bayonet exercise;
— *une muraille*, to prop up a wall;
— *le pavillon*, (*nav.*) to fire a gun under a ship's true colors;
— *le service*, (*mil.*) to carry out a duty, duties; to make it sure that a *service* will be performed.

astatique, a., (*elec.*) astatic.

astic, m., (*mil.*) polisher of wood, for spreading polish over leathern part of equipment.

astiquage, astique, m., (*mil.*) cleaning of equipments, polishing of.

astiquer, v. a., (*mil.*) to polish, render brilliant (equipment, etc.);
— *au blanc d' Espagne*, to pipeclay.

astragale, m., (*hipp.*) astragalus; (*art.*) astragal (*obs.*); (*in gen.*) astragal.

atelier, m., workshop; (*fort.*, *siege*) task; (*r. r.*) working party, gang of workmen; (*in gen.*) a working party; (*mil.*, *siege*, *fort.*) working party; (*Fr. a.*) subdivision of *section de 1ère ligne*, of the telegraph service, i. e. construction party; (also, *Fr. a.*) skilled telegraph squad in each regiment of cavalry;
— *d'ajustage*, (*mach.*, etc.) fitting shop;
— *d'artificiers*, (*art.*) laboratory;
— *d'assemblage*, (*mach.*, etc.) erecting shop,
— *de chargement*, (*fort.*) filling room;
chef d' ——, (*in. gen.*) foremen; (*mil.*) n. c. o. in charge of a subdivision of a working party;
— *civil de confection militaire*, (*Fr. a.*) civilian shop supplying materials for army, under the military administration;
— *de confection*, any manufacturing shop; (*siege*) place where ammunition is prepared (part of *parc aux munitions*, q. v.);
— *de construction*, (*Fr. a.*) artillery workshop;
— *du corps*, (*Fr. a.*) regimental workshop; personnel of regimental workshop;
— *d'entretien de la voie*, (*r. r.*) repairing gang;
— *de fascinage*, (*siege*) place where fascines, etc., are made (attack);
— *de granulation*, (*powd.*) graining room, v. *grenoir*;
— *mécanique*, machine shop (for manufacture of machines);
— *de précision*, shop where accurate work or fitting is done, in which instruments of precision are assembled, adjusted, etc.; (*Fr. art.*) establishment in which instruments of precision are manufactured, adjusted, etc.;
— *régimentaire du génie*, (*Fr. eng.*) engineer workshop;
— *de remplissage*, (*art.*) filling room;
— *de travaux publics*, (*Fr. a.*) penitentiary workshop for men convicted by court-martial of various offenses, as well as for others who have had their punishment changed to this correctional form.

atlas, m., atlas.

atmosphère, f., atmosphere.

atmosphérique, a., atmospheric;
tube ——, *tuyau* ——, air tube.

atomicité, f., (*chem.*) atomicity, valency.

âtre, m., fireplace, hearth (of a forge, etc.).

attache, f., cord, strap, band; thong, tether; brace, rivet, fastening;
— *à bouton*, button-fastening;
— *fixe*, (*hipp.*) the origin of a muscle;
— *mobile*, (*hipp.*) the insertion of a muscle.

attaché, m., attaché;
— *militaire*, —— *naval*, military, naval, attaché.

attachement, m., fastening, attachment; (*top.*) the name of an *état de lieux*. *s*, q. v., of limited area to be cleared; (in pl.) statement or notes of work done, of progress accomplished, said work and progress being impossible of inspection or of verification after the whole construction is completed (building, etc.).

attachement *écrit*, written data;
—— *figuré*, sketches, drawings, etc.;
—— *du mineur*, (*mil. min.*) attaching the miner, attachment of the miner (obs.).
attacher, v. a., to tie, fasten, attach, make fast, bind; (*mil.*) to attach (to a regiment, department, etc.);
—— *le mineur*, (*mil. min.*) to attach the miner (obs.).
attaquable, a., liable to attack, open to attack.
attaquant, m., (*mil.*) the assailant, the attacking body.
attaque, f., attack; (*in gen., mil.*) attack, any offensive operation involving or leading to fighting; onset; (*ball.*) attack (i. e., trial, test) of an armor plate; (*mil. min.*) the offensive, the besiegers; (*mil. min.*) each separate mining operation; (*siege*) attack, i. e., a subdivision of the general attack, having an entirety of its own (e. g., right attack, left attack, etc.); (in pl., frequently) approaches; (*teleg.*) call; (*man.*) the rational use of the spur as an aid, not as a punishment;
—— *abrégée*, (*siege*) irregular attack;
batterie d'——, (*art.*) attacking, fighting, battery; (in proving grounds) the testing battery (for armor-plate trials);
—— *brusquée*, attack of a fort by surprise, bombardment, open assault, etc.; in short, any attack not accompanied by trench work;
canon d'——, (proving ground) the testing gun in armor-plate trials;
contre ——, (*mil.*) counter attack;
corps d'——, attacking party;
—— *décisive*, (*mil.*) the decisive (and therefore the real) attack, as opposed to a demonstration (on flank, generally), hence especially the attack made by the *aile décisive*, q. v.;
—— *démonstrative*, (*mil.*) demonstration, (esp. to-day) front attack;
—— *directe*, (*nav., torp.*) case where ship touches a towed torpedo directly; (*mil.*) front attack;
—— *double*, (*siege*) two-storied descent into the ditch, v. *descente;*
—— *d'emblée*, (*siege*) attack without opening trenches;
—— *en échelons*, (*mil.*) attack in echelon;
—— *en échiquier*, (*mil.*) attack in squares, in checkerboard disposition;
—— *par enlacement*, (*nav. torp.*) case in which the ship touches the towline of a (towed) torpedo and the latter doubles on the ship;
—— *à l'éperon*, (*nav.*) ram attack, ramming;
fausse ——, (*mil.*) feint;
—— *de flanc*, (*mil.*) flank attack, (esp. to-day) the real, the decisive attack;
—— *dans les formes*, (*siege*) regular siege;
—— *en fourrageurs*, (*cav.*) attack as foragers, in extended order.
—— *de front*, (*mil.*) front attack, direct attack, (esp. to-day) demonstrative attack;
—— *par intelligence*, (*mil.*) attack or assault by treachery (by an understanding with traitors in enemy's lines);
—— *irrégulière*, attack of a fort by surprise, assault, etc. (no trenches opened);
—— *en laisse*, (*nav. torp.*) case where (towed) torpedo, at first held in leash, is let go, and strikes the ship in taking position of equilibrium;
—— *en masse*, (*mil.*) attack of a point in the enemy's line; (*in gen.*) general advance on the enemy;
—— *en muraille*, (*cav.*) charge of cavalry in close order, with very small or no intervals;
ordre d'——, (*mil., nav.*) formation for attack;
—— *ouverte*, (*mil.*) open attack;
—— *des places*, general expression for any operations leading to capture of a fortified position;
—— *rapprochée*, (*siege*) attack by means of approaches, trenches, etc., at close quarters;
—— *en règle*, (*siege*) regular, formal siege;
simulacre d'——, —— *simulée*, (*mil.*) sham attack, feint;

attaque *souterraine*, (*siege*) attack by mining;
—— *par surprise*, (*siege*) surprise of a fortified position;
—— *de vive force*, (*siege*) storming, assault;
—— *volante*, (*mil.*) feint, sham attack (rare).
attaquer, v. a., to attack, to assault; (*art.*) to attack an armor plate in proving-ground trials;
s'——, (*expl.*) to operate against;
—— *l'arme*, (*drill*) to seize the piece (small arm) briskly in the execution of the manual of arms;
—— *un cheval*, (*man.*) to spur a horse, to use force to reduce him to obedience, (more specially) to use the spur, etc., as an aid;
—— *l'ennemi à l'éperon*, (*nav.*) to ram the enemy.
atteindre, v. a., n., to reach; v. r., (*hipp.*) to interfere, overreach; (*art., sm. a.*) to hit;
—— *au but*, —— *le but directement*, —— *directement au but*, (*art., sm. a.*) to make a direct hit;
—— *de plein fouet*, (*art., sm. a.*) to make a direct hit (i. e., not a ricochet hit);
—— *au vol*, to hit on the wing (game).
atteint, p. p., —— *par la limite d'âge*, (*mil.*) having reached retiring age.
atteinte, f., blow, stroke, reach; attempt; (*hipp.*) interference, overreach (more strictly, the former); wound caused by interference, overreaching; (*sm. a.*) hit (target practice, also in action);
—— *encornée*, (*hipp.*) interference that causes a detachment, and is followed by suppuration;
—— *simple*, (*hipp.*) interference productive of only a slight contusion;
—— *sourde*, (*hipp.*) interference accompanied by severe pain.
attelage, m., (*mil.*) draught; draught-harness; team (of horses, mules, etc.); (*r. r.*) coupling; (*art.*) team;
—— *à l'allemande*, pole-draught;
—— *à la comtoise*, v. —— *à la française;*
—— *de derrière*, (*art.*) wheel-team, wheelers;
—— *à deux*, (*quatre, etc.*) two- (four-) horse team, etc.;
—— *de devant*, (*art.*) lead team, leaders;
—— *à l'espagnole*, (*mil., train*) draught with sort of yoke or bar over horses' shoulders;
—— *de, en, file*, single draught, tandem;
—— *à deux, trois, etc., files*, double, treble draught, etc.;
—— *à la française*, rear horse in shafts, all the horses of the team being in single file;
—— *de front*, horses hitched up abreast, double draught;
—— *haut-le-pied*, spare team;
—— *à limonière*, v. —— *à la française;*
—— *du milieu*, (*art.*) swing-team;
—— *en selle*, (*mil., art.*) saddle-team;
—— *à timon*, pole draught, double draught.
atteler, v. a., to hitch up, harness, put (the horses) to; (*art.*) to belong to a certain piece (*lit.*, to hitch it), to be or serve with a certain piece;
—— *à, de, n chevaux*, to put in n horses;
—— *en cheville*, to hitch a horse in front of the shaft horse;
—— *en flèche*, to hitch up tandem;
s'—— *à n chevaux*, to be drawn by n horses.
attelle, f., (*harn.*) hame; (*surgery*) splint;
courroie d'——, (*harn.*) hame-strap.
attente, f., expectation, expectancy; (*unif.*) epaulette loop or strap (in pl.);
pierre d'——, (*mas.*) toothing;
position d'——, (*mil.*) position in which enemy, attack, etc., is expected.
attention, f., attention.
atterrage, m., (*nav.*) landing, making land; point where a cable goes ashore (torpedoes, electric cables, etc.). (*Atterrissage* is now generally used instead.)
atterrer, v. a., n., to throw down, overturn, demolish, etc.; (*nav.*) to make land. (As a naval term, *atterrir* is preferred by sailors.)
atterrir, v. n., (*nav.*) to land, to make, approach, land; (of a balloon) to descend to earth, to alight; (of a cable) to come ashore, to leave the water.

atterrissage, m., (*nav.*) sighting of land; landing, alighting (of a balloon); (*torp., elec.*) point where a cable goes ashore.

atterrissement, m., bank (in bend of river); (*mil.*) landing of troops (from boats, flotilla, etc.).

attestation, f., attestation, certificate;
—— *d'aptitude*, (*Fr. a.*) certificate of fitness for service in chosen arm;
—— *de repentir*, (*mil., F. a.*) "certificate of repentance," given to soldiers discharged from the *compagnies de discipline*, provided they have borne themselves well. (They can not obtain a certificate of good conduct.)

attila, m., (*unif.*) hussar-jacket, dolman (not necessarily a cavalry jacket).

attirail, m., gear, plant; implements, apparatus; (*mil.*) equipage, train, luggage, stores; (*esp. art.*) implements for mechanical maneuvers, and other parts of the artillery service;
—— *de guerre*, (*mil.*) generic term for the material necessary to make war.

attiser, v. a., to stir, poke, rake up, a fire; to stoke.

attiseur, m., stoker.

attisoir, attisonnoir, m., poker.

attouchement, m., touch, feeling, feel;
point d'——, point of contact.

attrapage, m., —— *au 1er numéro*, (*mil. slang*) serious duel.

attrape, f., (*cord.*) pendant, tackle guy, relieving tackle, tripping rope or tackle.

attraper, v. a., r., to catch, take, hit, strike; obtain, get; (*hipp.*) to overreach, to interfere.

attraquer, v. n., to come alongside a quay, wharf (ships).

attribut, m., (*mil.*) sign or mark of functions, of arms of the service, or of some honor won or conferred, worn on head gear or uniform; symbol, device.

attribution, f., attribution, allowance; (*mil., etc.*) the ensemble of rights and of duties belonging to each grade, or to each employment.

attroupement, m., mob;
auteur d'——, (*mil.*) soldier (officer or man) regarded as instigator of a riot;
chef d'——, (*mil.*) ringleader of a riot.

aubade, f., (*mil.*) serenade to a c. o., or to the head of the state.

aubage, m., generic term for paddle, etc., of a hydraulic wheel.

aubaine, f., any unexpected advantage (e. g., over enemy).

aube, f., dawn, daybreak; float, float-board; paddle-board (of a side wheel); bucket (of a turbine, water-wheel, etc.); (*harn.*) side-bar of a pack saddle;
—— *mobile*, feathering paddle;
roue à ——*s*, paddle-wheel.

aubère, a., (*hipp.*) composed of white and of red hairs intermingled (coat); m., coat of intermingled red and white hairs, horse of such a color. (According to some, *aubère*=flea-bitten, but that is rather *moucheté*.)

auberisé, a., (*hipp.*) containing white hairs (said of a chestnut or bay coat, flea-bitten, according to some; but see *aubère*).

auberon, m., bolt-staple; catch of a lock.

auberonnière, f., clasp (of a lock), staple-plate.

aubier, m., sap-wood;
double ——, (*lit.,* "double sap") frozen sapwood remaining in a tree (causes decay, etc.).

aubin, m., (*man.*) defective gait, consisting of a gallop with fore, and trot or amble with hind, legs;
aller de l'——, to go at the *aubin;*
—— *de derrière, aubin* in which the hind legs do the galloping.
—— *de devant, aubin* in which the fore legs gallop.

aubiner, v. n., (*man.*) to go at the *aubin*.

audience, f., audience, hearing; (*law*) hearing of a case; (*mil.*) sitting, session, of a court-martial.

auditeur, m., (*mil.*) term used in France to-day for the judge advocate of foreign services;
—— *général*, judge-advocate-general (of services other than French);
—— *supérieur*, deputy judge-advocate-general (not French).

audition, f., (*mil. law.*) hearing of witnesses.

auditoriat, m., (*mil. law.*) judge-advocate-general's department (of services other than French).

auge, f., hod, trough; watering trough; bucket (of a wheel); (*hipp.*) hollow or channel in the under part of a horse's head, between his lower jaws;
—— *-abreuvoir*, horse trough (wood or concrete);
—— *à goudron*, tar bucket;
—— *de roulage*, (*powd.*) under roller-stone, bed;
—— *de trempage*, (*expl.*) dipping-trough (guncotton manufacture).

augée, f., hodful, troughful.

auget, m., small trough, bucket (of a water-wheel); (*sm. a.*) spoon cartridge-carrier, feed plate, platform, cartridge-hopper; (*r. f. art.*) feed-slot, feed-hole; (*cons.*) trough-shaped filling (plaster) in floors between beams; (*art.*) groove of a rocket frame; (*mil. min.*) wooden trough or channel to hold a *saucisson*, q. v.;
—— *de chargement*, (*r. f. art.*) loading-hole;
boîte d'——, (*sm. a.*) carrier box or casing;
—— *pivotant*, (*sm. a.*) rotating cartridge-carrier;
—— *de répétition*, (*sm. a.*) feed-trough, spoon, cartridge-carrier;
roue à ——*s*, overshot wheel;
—— *à tiroir*, (*sm. a.*) sliding platform or cartridge-carrier.

aulne, f., canvas awning.

aulne, m., alder (wood and tree).

aumônerie, f., chaplain's headquarters;
—— *militaire*, body of chaplains.

aumônier, m., chaplain.

aune, m., alder (wood and tree); f., ell (measure).

auraie, f., (*nav.*) mooring stone or block.

aurore, f., dawn, daybreak.

aussière, f., (*cord*) hawser;
en ——, hawser laid;
commis en ——, hawser laid;
—— *à touer*, towline.

austo, m., (*mil. slang*) guard-room, cell.

austral, a., southern.

autarelle, f., rowlock (i. e., thole, pin).

autel, m., altar; (*téch.*) fire-bridge; part of furnace through which flame passes on leaving fire hole;
—— *d'eau*, (*steam*) water-bridge;
grand ——, (*met.*) fire-bridge;
—— *renversé*, (*steam*) water-bridge.

autoclave, m., (*steam*) manhole lid (French boiler).

autocopiste, m., copying, reproducing, machine (letters, etc.).

autodiagrammateur, a., giving a diagram automatically, autodiagrammatic.

auto-excitateur, a., (*elec.*) self-exciting (dynamo, etc.).

auto-excitation, f., (*elec.*) self-excitement.

autographe, m., autograph;
—— *instantané*, improved cyclostyle (copying instrument).

autographie, f., autography; press-copy; transfer (of a drawing, etc., to the lithographic stone, by means of prepared paper).

autographier, v. a., to make a press-copy.

automate, m., automaton; a., automatic.

automaticité, f., general term for automatic qualities (as of an automatic gun, etc.).

automatique, a., automatic.

automètre, m., (*inst.*) autometer.

autonome, a., autonomous; (*mil.*) said of departments depending directly on the minister or secretary of war.

autopsie, f., autopsy.

autoréduction, f., (*top., etc.*) direct determination of length of a line of sight, reduced to the horizon.

auto-réparateur, m., puncture-stop (bicycle).
autorisation, f., authority, consent, license, warrant.
autoriser, v. a., to authorize, commission, empower; to license, warrant, qualify.
autorité, f., authority;
abus d'——, abuse of authority;
les ——*s*, the authorities.
auvent, m., shed., penthouse; (*unif.*) visor of a helmet (obs.).
auxiliaire, a., auxiliary, assisting, helping; m., (in pl.) auxiliary troops.
aval, m., adv., downstream direction;
d' ——, from downstream; (also) down, downstream;
en ——, downstream, down;
vent d' ——, sea breeze.
avalaison, f., flood, torrent (i. e., sudden, due to rain or to melting snow); pile of stones left on the bank by the current.
avalanche, f., avalanche.
avalasse, f., v. *avalaison*.
avalé, p. p., sunken, drooping (of parts of body);
croupe ——*e*, (*hipp.*) croup that is too low.
avalement, m., upsetting, jumping (as a tube, etc.);
—— *de la loupe*, (*met.*) end of the operation of refining the bloom.
avaler, v. n., to descend a river (obs.); v. a., to swallow; to upset, jump (a tube, etc.);
—— *de l'air*, (*hipp.*) to suck wind (*vice*);
—— *une branche*, to lop off a branch close to the trunk;
—— *la loupe*, (*met.*) to finish the operation of refining the bloom.
avaloire, f., (*harn.*) breeching;
bras du bas d' ——, hip strap;
bras du haut d' ——, breech strap proper;
porte- ——, hip or flank strap.
avalure, f., (*hipp.*) soft and bad hoof; accidental depression of wall of hoof; (also) normal descent, regular growth of the hoof.
avance, f., advance, start; distance in advance, lead, ground gained; projection; gain, progress; sum of money lent; payment in advance; (in pl.) first steps, preliminary steps, advances; (*steam, mach., etc.*) lead (as of a slide valve, of a crank, etc.);
—— *à l'admission*, (*steam*) lead, steam-lead, outside lead;
angle d' ——, —— *angulaire*, (*mach., steam*) angular advance (of an eccentric);
avoir de l' ——, to have the start of;
—— *diurne*, daily gain;
—— *à l'échappement*, —— *à l'évacuation*, (*steam*) exhaust lead, inside lead;
gagner de l' ——, to get a start;
—— *à l'introduction*, (*steam*) outside lead, steam lead;
—— *linéaire*, —— *linéaire du tiroir*, (*steam*) linear advance, lead of the slide-valve;
—— *de la manivelle*, (*mach.*) lead of the crank;
—— *à l'ordre*, (*mil.*) challenge from corporal at guardhouse to any person or party approaching (rounds, patrols, etc.); challenge given when two patrols, rounds, bodies of troops, meet; challenge from the chief of any post to party or person approaching. The answer to the challenge is the *mot d'ordre*, q. v., and is itself answered by the *mot de ralliement*, q. v. (nearest equivalent is "advance (or advance one), and give the countersign");
—— *de la phase du courant*, (*elec.*) lead;
prendre de l' —— *sur*, to get a start of;
—— *au ralliement*, (*mil.*) challenge by sentry (not No. 1) to advance and give the *mot de ralliement*, q. v.;
—— *du tiroir*, (*steam*) lead of the slide valve.
avancé, a., (*mil.*) advanced;
ouvrage ——, (*fort.*) advanced work, detached work;
poste ——, (*mil.*) advanced post, picket.

avancée, f., (*fort.*) outlying post, in front of the first barrier (obs.); (used loosely for) advanced work or post (in gen.); advance-guard (rare);
à l' ——, (*mil.*) thrown out, in an advanced position.
avancement, m., (*mil.*) promotion;
—— *à l'ancienneté*, promotion by seniority;
—— *sur l'arme*, linear promotion in one's own arm;
—— *à brevet*, brevet promotion (U. S. A.);
—— *au choix*, promotion by selection;
proposition d' ——, list of names submitted for promotion;
recevoir de l' ——, to be promoted;
tableau d' ——, promotion list or roll;
—— *par unité*, promotion in the unit to which one belongs, as regiment, etc.
avancer, v. a. and **n.**, to advance, come forward, move forward; accelerate, hasten, urge; bring up, forward; to grow (of the moon); (*mil.*) to march forward; to be promoted; (*mach.*) to feed;
—— *à l'ancienneté*, to be promoted lineally, by seniority;
—— *sur l'arme*, to be promoted lineally in one's own arm;
—— *au choix*, to be promoted by selection.
avant, m., head, prow, fore part; (*nav.*) bow of (a ship, boat); prep., before; adv., forward, far, advanced;
en ——, in advance, ahead; (*mil.*) forward (*command and call*);
aller à l' ——, to have headway (boat);
aller de l' ——, to be under headway;
—— *d'aval*, downstream starling of a bridge;
coup d' ——, front stroke, forward stroke;
—— *à l'éperon*, (*nav.*) ram-bow;
gagner l' ——, to get ahead of;
vogue ——, (*boats*) stroke.
Commands in boat drill:
avant, give way;
en ——, give way;
—— *bâbord et scie tribord*, give way port; back starboard;
—— *partout*, give way together;
—— *qui peut*, pull with oars that are shipped;
—— *tribord* (*bâbord*), pull starboard (port).
avantage, m., advantage, benefit; victory; whip-hand; odds; vantage-ground; (*sm. a.*) (of a double-barreled shotgun) the distance of the axis of the stock to the right of the axis of symmetry of the barrels;
prendre l' ——, *son* ——, *pour monter à cheval*, (*man.*) to use an elevation in order to mount more easily;
—— *du vent*, (*nav.*) weather-gauge.
avantageur, m., (*mil.*) avantageur (German army).
avant-bataille, f., (*mil.*) (a synonym of) advanced guard.
avant-bec, m., (*pont.*) head, fore peak of a ponton boat (all that part of the bow the bottom of which curves up); (in some systems of pontons) detachable bow (or stern); upstream cutwater of a bridge.
avant-bord, m., front edge; (*pont.*) upper edge and borders of a ponton boat.
avant-bras, m., (*hipp.*) forearm (or sometimes "arm" simply).
avant-chemin, m., —— *couvert*, (*fort.*) advanced covered way, covered way at foot of glacis of main covered way; sort of covered way of a detached fort.
avant-contrescarpe, f., (*fort.*) advanced counterscarp.
avant-corps, m., fore or projecting part.
avant-coureur, m., forerunner.
avant-creuset, m., (*met.*) fore-heart (of a blast furnace).
avant-cuirasse, f., (*fort.*) glacis armor (of a turret); circular ring of metal protecting the parapet over the vulnerable parts of a cupola or turret; glacis plate.
avant-duc, m., (*pont.*) abutment of a wooden or boat bridge.
avant-fossé, m., (*fort.*) advanced ditch, at foot of glacis of the *enceinte,* outer ditch, second ditch.

avant-foyer, m., (*art.*) fire-guard (of a traveling-forge).
avant-garde, f., (*mil.*) advance guard;
  extrême-pointe de l'——, extreme point of the advance guard;
  faire l'——, to do advance-guard duty, to be the advance guard;
  gros de l'—— main body of the advance guard (reserve of, U. S. A.);
  pointe de l'——, point, leading group of the advance guard;
  tête de l'——, head, van of the advance guard (support, U. S. A.).
avant-glacis, m., (*fort.*) glacis in front of the *avant-chemin couvert;* secondary glacis in front of the principal glacis; sort of advanced covered way.
avant-ligne, f., (*mil.*) line occupied in front of the main position, but not intended to be permanently held, the object being to break the enemy here before he gets to the real position; in the field, a slightly fortified position, in front of the main and real position, and intended to deceive the enemy as to the latter by throwing his reconnaissances off the scent; front or advanced line, in infantry formation for attack; (*siege*) line of outposts of a line of investment; (*siege*) subsidiary works thrown up by the advanced posts of the besiegers in order to establish themselves thoroughly on the ground held by them.
avant-main, f., (*hipp.*) forehand, forequarters;
  avoir de l'——, to have a fine forehand.
avant-mètré, m., (*adm.*) estimate in figures (numbers) of the quantity of work to be done (in cases of submission to *adjudication publique*).
avant-molaire, f., (*hipp.*) front molar tooth.
avant-mur, m., screen wall, outer wall; (*fort.*) masonry wall farthest from main body (*corps*) of a *place*.
avant-parallèle, f., (*siege*) intermediate place of arms, constructed between the batteries of the first position and the first parallel, when works between these are not sufficient to establish the latter.
avant-pied, m., vamp.
avant-port, m., outer port or harbor.
avant-poste, m., (*mil.*) outpost;
  —— *de cavalerie,* cavalry outpost, (*esp.*) one furnished by the *cavalerie d'exploration,* q. v.;
  —— *de combat,* regular outpost, infantry outpost;
  —— *cosaque,* —— *à la cosaque,* —— *irrégulier,* (*cav.*) cossack post (the distinction between regular and irregular outposts has been broken down in France so far as the cavalry is concerned—1894);
  —— *irrégulier,* special outpost, one to meet a special case (no longer exists in the cavalry, v. —— *cosaque*);
  —— *de jour,* first line of outposts, furnished by cavalry;
  ligne d'——s, line of outposts;
  ligne de résistance des ——, line of grand guards, line of resistance;
  —— *mixte,* outpost consisting of cavalry and of infantry;
  —— *régulier,* usual or regulation outpost;
  réseau d'——s, network or system of outposts;
  —— *de sûreté,* outpost (so called as being part of the *service de sûreté,* q. v.).
avant-projet, m., preliminary project, rough estimate.
avant-terrain, m., foreground.
avant-train, m., (*in gen.*) fore body, fore carriage (of a four-wheeled vehicle); (*art.*) limber; (*hipp.*) forehand (of a horse); (of a bicycle) fore wheel and related parts;
  amener l'——, to limber rear;
  amener l'—— *en avant,* to limber front;
  —— *à bras de limonière,* limber with shafts;
  —— *de caisson,* caisson-limber;
  —— *de campagne,* field-gun limber;
  —— *à contre-appui,* limber (e. g., of a siege gun) in which the trail rests on the limber-axle;
  décrocher les ——s, to unlimber;

avant-train, *mettre l'*——, to limber, limber up;
  ôter l'——, to unlimber;
  —— *à sellette,* limber with bolster and pintle;
  —— *de siège,* siege-gun limber;
  —— *à suspension,* limber in which the lunette fits over a pintle, the usual type of field limber;
  —— *à timon,* pole limber;
  —— *de tombereau,* (*Fr. art.*) the tumbril or box limber of the *affût à soulèvement,* q. v.
avarie, f., damage; average; dues (usually a commercial term).
avarié, a., damaged, spoiled.
avarier, v. a., to spoil, damage.
à vau, adv., (used only in) —— *-l'eau,* downstream, with the current;
  —— *-de-route,* (*mil.*) in great disorder, in confusion (of a defeated army or body of troops).
aventure, f., adventure.
aventuré, a., (*mil.*) exposed, in a critical position.
aventurer, v. a., to risk; v. r., to expose one's self to hazard; to run a risk.
aventurier, m., soldier of fortune.
averse, f., sudden shower.
avertir, v. a., to notify, warn;
  —— *un cheval,* (*man.*) to gather a horse; to urge, to encourage a horse.
avertissement, m., warning, notice; (*Fr. mil. law*) notice given to accused, before a court-martial, that if he does not select counsel, counsel will be assigned to him by the president of the court; notice given by the *greffier* (q. v.) of a court, to prisoner found guilty, that he has 24 hours in which to lodge an appeal;
  commandement d'——, (*mil.*) preparatory command.
avertisseur, a., warning, indicating; m., (*elec.*) call-bell, annunciator;
  —— *d'incendie,* fire alarm.
aveu, m., confession;
  —— *extrajudiciaire,* confession made out of court;
  —— *judiciaire,* confession in court.
aveugler, v. a., to blind; to stop up;
  —— *une casemate,* (*mil.*) to make a casemate useless by stopping up the embrasure;
  —— *une voie d'eau,* to stop a leak.
aveuglette, *marcher à l'*——, (*mil.*) to march at random, feeling one's way.
aviateur, m., person fond of, or student of, the art of aerial locomotion by flight.
aviation, f., flying, aerial locomotion (in imitation of the flight of birds).
avir, v. a., to beat or hammer down edges.
avire, m., flying machine.
aviron, m., oar, sweep;
  aller à l'——, to row;
  —— *de l'arrière,* stroke oar;
  assourdir les ——s, to muffle the oars;
  —— *de l'avant,* bow oar;
  —— *du brigadier,* bow oar;
  —— *du chef de nage,* stroke oar;
  coup d'——, stroke of an oar;
  ——s *à couple,* double-banked oars;
  dégarnir, désarmer, les ——s, to unship, take in, oars;
  être à l'——, to be rowing;
  —— *à gouverner,* steering oar;
  mâter les ——s, to toss the oars;
  ——s *à pointe,* single-banked oars;
  rentrer les ——s, to unship, take in, oars.
avis, m., notice, information; warning, caution; advice; (*Fr. a.*) report to minister of war by a technical committee;
  —— *d'adjudication,* (*adm.*) advertisement for bids;
  —— *écrit,* memorandum;
  —— *d'expédition,* (*Fr. a., adm.*) notice of transportation of material by rail, sent and received by the interested transportation officers (*sous-intendant*);
  —— *motivé,* reasons (e. g., as by indorsements);
  —— *de passage,* (*Fr. a., adm.*) notice of movement of troops, sent by corps commander to c. o. of regions, and to prefects of departments, through which troops must pass;

avis *de transport*, (*Fr. a., adm.*) notice of journey of a detachment by rail, sent by c. o. of regiment to the railway station of departure.
aviso, m., (*nav.*) dispatch boat;
— *canonnière* —, gunboat, gun vessel;
— *mortier*, mortar gunboat, mortar boat;
— *torpilleur*, torpedo gunboat.
avisse, f., screw-nail.
avitaillement, m., storing, victualing.
avitailler, v. a., to store, victual, provide stores, supplies.
aviver, v. a., to sharpen, to make clean and smooth (as an edge); to hew timber (square).
avives, f. pl., (*hipp.*) vives; parotid glands.
avocat, m., (*law*) advocate; pleader before a court-martial.
avoine, f., oats; (*mil. slang*) brandy; (*in pl.*) standing oats;
boire son —, (*hipp.*) to swallow oats, etc., without mastication (sign of disease);
— *cariée*, rusty, foxy oats;
— *charançonnée*, weevil-eaten oats;
— *charbonnée*, smutty oats;
— *ergotée*, oats affected by ergot;
folle —, wild oats;
— *germée*, oats that have begun to sprout;
— *moisie*, musty oats;
— *du sac*, one day's supply of oats carried in the battery sack;
— *terreuse*, dirty, badly winnowed oats.
avoir, v. a., to have, own, possess;
— *à la masse*, — *de masse*, — *la masse*, (*Fr. a., adm.*) excess of receipts over expenditures, of the *masse individuelle*, q. v.
avoué, m., (*law*) solicitor, counsel.
axe, m., axis, axle, spindle, arbor; (sometimes) rocking-shaft; shaft;
— *des abscisses*, axis of abscissas;
— *d'arrière*, trailing axle (of a locomotive);
— *d'avant*, leading axle (of a locomotive);

axe *du barillet*, (*sm. a.*) center-pin (of a revolver);
— *boulon d'* —, (*mach.*) a bolt that serves as an axis or spindle;
— *coudé*, (*mach.*) crank, cranked axle;
— *à deux leviers opposés*, (*mach.*) cross axle;
— *d'entraînement*, (*r. f. art.*) hollow spindle communicating motion to the breech-plug (Canet);
— *d'explosion*, (*min.*) line of least resistance;
— -*galet*, (*mach.*) rolling or movable axis (e. g., a pin rolling in a groove and serving as a movable axis);
goupille d' —, (*mach.*) pin that serves as an axis;
— *moteur*, (*mach.*) shaft, driving-shaft, driving-axle;
— *des ordonnées*, axis of ordinates;
— *du régulateur*, (*steam*) governor spindle;
— *en* ⊤, (*mach.*) cross axle;
— *tournant à volonté*, (*mach.*) revolving axle (e. g., of a railway truck);
trou d' —, (*mach.*) hole serving as a center of rotation for a spindle;
— *de vis*, (*mach.*) screw arbor.
axonge, m., hog-grease, melted and strained (constituent of lubricant for care of arms).
ayant droit, m., (*law*) legally entitled person; rightful claimant or owner.
azimut, m., azimuth; (*inst.*) azimuth circle.
azor, m., (*mil. slang*) knapsack;
etre à cheval sur —, to shoulder the knapsack.
azotate, m., (*chem.*) nitrate.
azote, m., (*chem.*) nitrogen.
azoté, a., (*chem.*) nitrogenized;
poudre — *e*, (*expl.*) nitrogen powder, generic term for modern nitrogenized smokeless powders.
azotyle, m., (*chem.*) azotyl.

B.

bâbord, m., (*nav.*) port, port side of a vessel, boat; (in steering) starboard (of the helm).
bac, m., ferry; ferryboat; (*tech.*) vat;
— *oblique*, oblique ferry, trail ferry;
— *à pontons*, ponton raft;
— *à traille*, trail ferry;
— *à vapeur*, steam ferry;
— *volant*, flying ferry.
bacalas, m., cleat, fastening; sort of knee projecting above the rowlocks.
bâche, f., wagon cover (of canvas), cart tilt; pack cover, manta; small boat; (*steam*) cistern or tank of a steam engine;
— *alimentaire*, (*steam, etc.*) feed-water cistern;
— *à eau froide*, cold-water cistern;
— *de gonflement*, inflating-canvas (of a balloon);
— *à huile*, (*met.*) oil-tank;
— *de la pompe à air*, hot well.
bâcher, v. a., to cover a wagon with an awning; to raise a tent.
bachot, m., small boat, generally flat bottomed.
bâclage, m., mooring of boats side by side, for unloading, etc.; closing of a port by chains, cables, etc.
bâcler, v. a., to make fast; to chain, bar up (a window, a door); to close a port by chains, cables, etc.; to moor boats, side by side, for loading and unloading.
baderne, f., mat or paunch.
badigeon, m., color or wash laid over a wall or large surface.
badigeonner, v. a., to wash (a wall, etc.) with color.
badinant, m., spare horse in a team (obs.).
bagage, m., baggage; kit; luggage;
gros — *s*, (*mil.*) heavy baggage, main baggage;
— *menu*, (*mil.*) baggage or effects that can be carried by the men;
plier —, to run away, decamp.

bagne, m., galleys, hulk.
bagnolette, f., tarpaulin (for covering guns, etc.).
baguage, m., (*art.*) operation of screwing on the *bague* or base-ring in De Reffye bronze guns.
bague, f., (*mach., art., etc.*) ring, hoop, strap, collar, sleeve; (*cord.*) grommet; (*art.*) centering-band, rotating-band; base-ring (bronze De Reffye gun); (*sm. a.*) circular impression left in bore of rifle by the neck of a broken cartridge-case;
— *d'appui*, (*in gen.*) ring, collar, that serves as a support; (*art.*) assembling or supporting hoop (e. g., Hotchkiss 155ᵐᵐ;) flange or shoulder against which the winding abuts (Longridge wire gun);
— *d'assemblage*, collar, sleeve;
— *d'assemblage à vis*, screw collar or sleeve;
— *d'attache*, (*art.*) ring in which the wires start (wire gun);
— *de baïonnette*, (*sm. a.*) locking-ring of a bayonet;
— *à boulets rouges*, (*art.*) hot shot bearer (obs.);
— *de centrage*, (*art.*) centering band (Hotchkiss 155ᵐᵐ);
— -*directrice*, (*sm. a.*) guide-ring;
— *à*, — *d'*, *élingue*, (*cord.*) rope sling;
— -*écrou*, bush washer-nut;
— *d'excentrique*, (*mach.*) eccentric strap;
— *fendue*, (*art.*) split ring;
— *de garniture*, (*steam, etc.*) packing-ring, piston ring;
— *de presse-étoupes*, (*steam, etc.*) packing-ring;
— *à ressort*, (*art.*) spring clip (of the De Bange fermeture);
— *à sertir*, — *de sertissage*, (*art.*) band-setter (of a projectile), seating-collar or clamp;
— *à tube*, (*steam*) ferrule, tube-ferrule.
bagué, a., (*sm. a.*) (of a barrel) defective (having a ring or shoulder in the surface of the bore).

bagues, f. pl., baggage, (used only in) *sortir vie et —— sauves,* (*mil.*) to leave a fort, etc., with only such baggage, arms, etc., as one may carry on one's back.

baguette, f., small rod, wand; (*mus.*) drumstick; (*artif.*) drift; rocket-stick; (*sm. a.*) ejector-rod (of a revolver), cleaning-rod, ramrod;
—— *d'artifice,* (*artif.*) drift (for making fuses, rockets, etc.);
avaler ses ——s, (*mil. slang*) to die;
—— *-baïonnette,* (*sm. a.*) rod-bayonet;
—— *de carton,* (*artif.*) pasteboard rocket stick;
—— *de chargement,* (*artif.*) v. —— *d'artifice;*
—— *à charger,* (*artif.*) v. —— *d'artifice;*
—— *de déchargement,* (*sm. a.*) ejector-rod (of a revolver); (*art.*) ejector-rod of a subcaliber tube;
—— *de direction,* —— *directrice,* (*art.*) rocket-stick;
—— *directrice de chargement,* (*art.*) small rod used in making up cartridges of prismatic powder;
—— *-expulseur,* (*sm. a.*) ejector-rod (of a revolver);
—— *de fusée,* —— *de fusée volante,* (*art.*) rocket-stick;
—— *de fusil,* (*sm. a.*) ramrod;
—— *à laver,* —— *-lavoir,* (*art., sm. a.*) cleaning rod;
—— *en massif,* (*artif.*) v. —— *d'artifice:*
—— *à mèche,* (*mach.*) boring-bit;
passer par les ——s, (*mil.*) to run the gantlet;
—— *à refouler,* v. *d'artifice;*
—— *à rouler,* (*artif.*) former;
—— *de tambour,* (*mus.*) drumstick.

bahut, m., chest, trunk;
mur à ——, (*fort.*) low wall on top of scarp, at foot of exterior slope;
—— *spécial,* (*St. Cyr slang*) the military school at St. Cyr.

bahuté, a., (*St. Cyr slang*) smart, soldier-like.
bahuter, v. a., (*St. Cyr slang*) to make a disturbance.

bai, a. and m., (*hipp.*) bay (coat) and bay horse;
—— *acajou,* mahogany bay;
—— *brun,* brown bay (almost black);
—— *cerise,* cherry, red or blood bay;
—— *châtain,* uniform light-brown bay;
—— *clair,* light, bright bay;
—— *doré,* yellow bay (golden reflections);
—— *foncé,* dark bay (shading on brown);
—— *lavé,* light bay;
—— *marron,* chestnut bay;
—— *à miroir,* dappled bay;
—— *mirouetté,* bright dappled bay;
—— *ordinaire,* red bay;
—— *sanguin,* blood bay.

baie, f., bay; opening a hole in a wall; window or door opening; (*nav.*) opening in a ship's deck.

baignade, f., swim, bath (in a river, the sea).
baigner, v. a., to bathe, dip in the water; (of rivers) to flow through, by, or around; (of the sea) to wash.

bail, m., lease.
baille, f., tub;
—— *de combat,* (*art.*) sponge-bucket.
bailler, v. a., to gape, to open (as a joint).
baillet, a., (*hipp.*) sorrel.
baillon, m., gag.
baillonner, v. a., to gag; to wedge (as a door).

bain, m., bath; (*met., chem.*) bath;
—— *de développement,* (*phot.*) developing bath;
—— *fixateur,* (*phot.*) fixing bath;
—— *de fonte,* (*met.*) bath of melted pig iron;
—— *d'huile,* (*chem., etc.*) oil bath;
—— *-marie,* (*chem.*) water bath;
—— *de mortier,* (*mas.*) bed of mortar, thick layer of mortar;
—— *de plomb,* (*met.*) lead bath;
—— *de sable,* (*chem.*) sand bath;
—— *de trempe,* (*met.*) tempering bath;
—— *de vapeur,* (*chem.*) steam bath;
—— *de virage,* (*phot.*) toning bath.

baïonnette, f., (*sm. a.*) bayonet;
la —— en avant, with fixed bayonets;
baguette——, rod bayonet;
—— *à bout,* with fixed bayonets;
—— *au canon,* (*drill*) fix bayonets! trumpet call for same;
charge à la ——, (*mil.*) bayonet charge;
charger à la ——, (*mil.*) to charge with the bayonet;
coude de ——, neck, bend, elbow, of a bayonet;
couteau- ——, knife bayonet;
croiser la ——, (*drill*) to charge bayonets;
—— *à douille,* ordinary socket bayonet;
enlever à la ——, (*mil.*) to carry at the point of the bayonet;
épée- ——, sword-bayonet (sometimes, —— *-épée*);
escrime à la ——, bayonet exercise;
fausse —— d'escrime, fencing bayonet (bayonet exercise);
—— *à fourche,* (*torp.*) spar-torpedo fork (French navy);
—— *au fusil,* with fixed bayonets;
mettre la —— au canon, (*drill, etc.*) to fix bayonets;
poignard- ——, dagger-bayonet;
remettre la ——, (*drill, etc.*) to unfix bayonets; trumpet call for same;
sabre- ——, sword-bayonet (sometimes —— *-sabre*).

baiser, m., (*sm. a.*) v. *amour.*
baiser, v. r., (*cord.*) to be two-block, chock-a-block.
baisse, f., fall (of a river, of the tide).
baisser, v. a. and n., to fall; bow, lower; strike (colors); fall off; depress; decrease; diminish; go down, let down; ebb;
—— *la main à un cheval,* (*man.*) to slack the hand, to give a horse his head;
—— *le pavillon,* (*nav.*) to strike one's colors, haul down the flag.
baissier, m., (*top.*) shoal water.
baisure, f., place where one loaf touches another in baking; kissing crust.
bajoyer, m., side wall of a lock or sluice gate.
bal, m., (*mil. slang*) extra drill.
balafre, f., gash (cutting weapon); scar left by a cutting weapon (especially on the face).
balai, m., broom; (*elec.*) brush, collecting-brush;
—— *de communication,* (*elec.*) commutator brush;
—— *électrique,* (*elec.*) dynamo-brush;
—— *de fils métalliques,* wire brush;
—— *-frotteur,* elec brush (e. g., gramme machine);
—— *de laveur,* mop, swab;
—— *à plumes,* (*mil. slang*) plumes of a shako.
balance, f., balance, scales, pair of scales;
—— *à bascule,* weighing machine;
—— *manométrique,* (*ball.*) manometric pressure balance (gives pressure at a point, at any time, in the gun);
—— *des moments,* (*art.*) sort of weigh-bridge for determining eccentricity of projectiles;
plat de ——, scale-pan;
—— *à ressort,* spring balance;
—— *romaine,* steelyard;
trait de la ——, turn of the scale.
balancement, m., rocking (as of a bridge, a suspension bridge, a locomotive, etc.); unsteadiness (of horses in harness); balancing, poising, swinging; (*torp.*) test and verification of various parts of a torpedo (Whitehead) by means of compressed air;
—— *de l'avant à l'arrière,* overbalancing.
balancer, v. a., n., to balance, poise; counterbalance, counterpoise; to swing, rock; (of horses) to be unsteady in harness; (*fig.*) to hesitate, waver; (*gym.*) to swing.
balancier, m., beam, lever; pendulum; any beam or lever working a counterweight; beam of a balance); outrigger (of a boat); (*steam*) beam (of a beam engine); walking beam; rocking beam or lever; (*mach.*) press, fly-press, stamping press; flyer, fly of a wheel; (*art.*) lever or arm of a disappearing carriage (e. g., Crozier's);
—— *de boussole,* —— *de compas,* gimbals;

balancier *découpoir*, cutting-press (washers, etc.);
—— *à friction*, screw fly press, worked by a friction wheel;
—— *de pompe*, spear of a pump;
—— *sertisseur*, (*mach.*, *art.*) seating press, crimping-in press;
tête de ——, (*mach.*) beam end;
—— *transversal*, cross beam;
—— *à vis*, fly or screw press.
balancine, f., (*cord.*) topping lift;
—— *de gui*, topping lift of a boom.
balant, m., (*cord.*) bight or slack of a rope; *embraquer le* ——, to take in the slack.
balast, m., ballast (usually *ballast*).
balayage, m., sweeping.
balayer, v. a., to sweep, to clear; (*mil.*) to clear a position, an area, of the enemy; (also) to sweep by fire.
balayette, f., (*powd.*) small hair brush.
balayure, f., sweepings;
—— *de poudrerie*, (*powd.*) gunpowder dust, sweepings.
baldaquin, m., baldaquin, canopy with curtains.
balisage, m., setting, planting of buoys.
balise, f., beacon, buoy;
bateau-——, *bâtiment-*——, light-ship;
—— *de direction*, range-buoy;
—— *à signal*, signal beacon.
balisticien, m., expert in ballistics, writer on ballistics; ballistician.
balistique, a. and f., ballistic; ballistics;
—— *expérimentale*, practical, experimental ballistics;
—— *extérieure*, exterior ballistics;
—— *graphique*, graphical ballistics;
—— *intérieure*, interior ballistics;
—— *de pénétration*, ballistics of penetration;
—— *rationnelle*, rational, theoretical ballistics.
balistite, f., (*expl.*) balistite (Italian smokeless powder, also spelled *ballistite*).
baliveau, m., tree left standing when timber is cut, so as to reach full growth.
ballage, m., (*met.*) operation of making *fer ballé*, q. v.
ballast, m., (*r. r.*) ballast; crowbar.
ballastage, m., (*r. r.*) ballasting.
balle, f., ball; (*sm. a.*, *art.*) bullet;
—— *ardente*, (*artif.*) light-ball;
—— *d'artifice*, (*artif.*) fire-ball;
—— *en bouche*, (*mil.*) with the honors of war;
—— *calepinée*, patched bullet;
—— *de calibre*, —— *à calibrer*, proof-bullet;
—— *cannelée*, —— *à cannelures*, cannelured bullet;
cartouche à ——, ball cartridge;
—— *conique*, conical bullet;
cribler de ——*s*, to riddle with bullets;
—— *à culot*, bullet with capsule (*culot*) in rear, to cause expansion (obs.);
—— *doublée*, (in target practice) a bullet hole cut or enlarged by a subsequent bullet (where two or more bullets strike a bullet hole, the *balle* is said to be *triplée*, *quadruplée*);
——, *à éclairer*, (*siege*, etc.) light-ball;
—— *égarée*, stray bullet;
—— *ensabotée*, bullet seated in a pasteboard sabot (obs.);
—— *à enveloppe*, jacketed bullet;
—— *d'essai*, (*t. p.*) sighting shot;
—— *évidée*, bullet with hollowed base;
—— *expansive*, expanding bullet;
faire ——, (of bird shot) to stick together when fired, instead of spreading;
—— *à feu*, (*artif.*) fire-ball;
—— *à feu à carcasse*, (*artif.*) carcass;
—— *à fumée*, (*artif.*) smoke-ball;
—— *de fusil*, rifle bullet;
—— *luisante*, (*artif.*) light-ball;
—— *lumineuse*, v. —— *à feu*;
—— *mise*, (*t. p.*) hit (on target);
—— *à mitraille*, (*art.*) case-shot bullet, canister bullet;
—— *morte*, spent bullet;

balle *multiple*, fragmentation bullet (*canon à balles*);
—— *oblongue*, ovoidal bullet (theoretically the best, but completely abandoned);
—— *perdue*, spent bullet; unaimed shot, shot fired at hazard;
—— *de plein fouet*, (*t. p.*) direct hit;
—— *quadruplée*, v. —— *doublée*;
——*s de remplissage*, (*art.*) shrapnel bullets;
—— *à sabot*, bullet with wooden sabot (obs.);
—— *sphérique*, spherical bullet, (obs., except in gallery practice and for shrapnel and canister);
—— *à téton et à évidement*, bullet in hollowed base of which was fitted a lug or nipple (*téton*) to cause expansion (obs.);
tirer à ——, to fire ball cartridge;
tire- ——, (*sm. a.*) worm;
—— *triplée*, v. —— *doublée*;
—— *tubulaire*, tubular bullet;
—— *vive*, a bullet whose energy is not all spent.
baller, v. a., (*met.*) to make *fer ballé*, q. v.
ballon, m., balloon; big-bellied jar; (*top.*) rounded or dome-like hill or mountain;
—— *d'artifices*, (*artif.*) fire-balloon;
—— *auxiliaire*, (*Fr. balloon park*) second or additional balloon, will take up one man;
—— *captif*, captive balloon;
—— *captif type*, the balloon of the French balloon park;
—— *dirigeable*, dirigible balloon;
—— *d'essai*, pilot balloon;
—— *gazomètre*, (*Fr. balloon park*) a balloon that carries a reserve of gas, does not ascend;
grappin de ——, balloon anchor;
—— *libre*, free balloon;
monter un ——, to be, or go up, in a balloon;
—— *montgolfière*, hot-air balloon;
nacelle de ——, balloon car;
—— *non monté*, balloon having no man in car, balloon without a car;
—— *de signaux*, signal balloon;
tir sur ——, (*art.*) firing at a balloon;
tirer sur ——, (*art.*) to fire at a balloon.
ballonné, a., (*hipp.*) hidebound.
ballonnement, m., swelling.
ballonnet, m., small balloon.
ballot, m., bale.
ballottade, f., (*man.*) ballotade.
ballottement, m., shaking, jolting; (*art.*) balloting of a projectile in a smooth-bore gun; "tumbling," irregularity of a projectile in its flight.
ballotter, v. a., to shake, jolt; to wobble (as an imperfectly seated rivet); (*art.*) to ballot (projectile in smooth-bore gun); to tumble.
balustrade, f., balustrade; any breast-high, open-work fence.
balzan, a., (*hipp.*) *cheval* ——, black or bay with white marks on his feet (used also as a noun).
balzane, f., (*hipp.*) white mark or markings on a horse's leg (balzane is, strictly speaking, "white foot," the white not rising above fetlock);
grande ——, half-stocking, halfway up canon. large white foot;
grande —— *chaussée*, white stocking;
grande —— *haut chaussée*, high stocking;
grande —— *très haut chaussée*, a very high stocking (practically a white leg);
—— *incomplète*, incomplete white foot, not surrounding the member;
petite ——, white marking on coronet and pastern, small white foot.
bambou, m., bamboo.
ban, m., ban; banishment; proclamation; (*mil.*) roll of a drum or trumpet call, giving notice of an official announcement (e. g., investiture of insignia of Legion of Honor, appearance of an officer before the men he is to command);
battre un ——, to give notice by beat of drum;
fermeture du ——, roll of drum following announcement;
ouverture du ——, roll of drum preceding announcement;

3877°—17——3

ban *de réception*, call before presentation of medal or cross of Legion of Honor, or appearance of an officer before his men for the first time.
bananier, m., (*cord.*) manila hemp.
banasse, f., handbarrow.
banc, m., bench; bank; reef, shoal; thwart (of a boat); (*mach.*) frame of a lathe;
—— *d'alésage*, (*sm. a.*) boring-bench;
—— *d'artifices*, (*artif.*) laboratory table;
—— *de brouillard*, fog-bank;
—— *de chargement*, (*art.*) filling-bench;
—— *d'épreuve*, (*sm. a.*) proof-bench;
—— *à étalonner*, gauge-bench, standard bench, for measures of length;
—— *à étrangler*, (*artif.*) choking-frame, choking-press;
—— *à forer*, —— *de forerie*, (*mach.*) boring-bench;
—— *de mouleur*, molder's bench;
—— *de nage*, thwart (of a boat);
pied de ——, (*mil. slang*) sergeant;
—— *à rabuttement*, folding-bench, folding-back bench.
—— *de rameur*, thwart (of a boat);
—— *de rayage*, (*art., sm. a.*) rifling bench;
—— *de sable*, (*top.*) sand bank.
bancal, m., curved saber (Revolution and First Empire); curved saber (popular).
bandage, m., bandage; hoop; tire (of a wheel); surgical bandage;
—— *composé*, a surgical bandage made of several pieces and shaped for a given purpose;
—— *à fracture*, bandage for a broken limb;
—— *herniaire*, truss;
—— *mécanique*, surgical brace or frame; e. g., a truss for hernia;
—— *pneumatique*, pneumatic tire (bicycle, etc.);
—— *de roue*, tire;
—— *simple*, ordinary surgical bandage.
bande, f., band, strap; back-strap (of a shoe); tire (of a wheel); gang, set, crew, band; compression (of a spring); flight (of pigeons); (*steam*) lip of a slide-valve; (*hipp.*) leg-bandage, leg-band, strap; (*mach.*) belt or cord, band (over a pulley); (*sm. a.*) upper part of revolver frame, connecting *console* and *rempart*; (*harn.*) side-bar (of a saddle); (*nav.*) list, roll (to starboard, port); (*unif.*) trouser stripe (gold or silver) worn by officers of the military household of the President and by personal staffs of ministers of war and of navy, and of marshals of France;
—— *d'arçon*, (*harn.*) side-bar, saddletree bar;
—— *de calage*, (*art.*) chocking racer;
—— *de cartouches*, Maxim cartridge belt;
—— *de chargement*, loading belt (Maxim);
—— *de cible*, (*t p.*) horizontal stripe on target, at 6.50ᵐ, 1.10ᵐ, 1.60ᵐ, respectively, above ground, to enable rapid estimate of effect of fire at those heights;
—— *circulaire*, fifth wheel (of a wagon, etc.);
—— *cruciale*, (*hipp.*) cross on withers;
—— *du cylindre*, (*steam*) cylinder face;
—— *diagonale*, (*in gen.*) diagonal brace;
—— *d'écartement*, stay bar or rod; distance-piece;
—— *de l'embouchoir*, (*sm. a.*) upper band;
—— *d'épreuve*, test piece, test specimen;
—— *d'essieu*, axletree band;
—— *d'excentrique*, (*mach.*) eccentric strap;
—— *à fourche*, (*mach., etc.*) forked strap;
—— *de friction*, (*mach., etc.*) friction band or strap;
—— *de frottement*, friction plate (of a wheel, etc.); friction strap;
—— *de frottement du tiroir*, (*steam*) valve-face;
—— *de fusil*, (*sm. a.*) band, slide;
—— *du garrot*, (*harn.*) head plate (of a saddle);
—— *de groupement*, (*t. p.*) probability zone or band (imaginary) on a target;
—— *de jonction*, (*carp., etc.*) butt-strap;
—— *latérale*, (*art.*) side-strap, cheek-strap;
—— *molletière*, (*unif.*) puttee;
—— *à oreilles*, axletree band;
—— *à pansement*, surgical bandage, strip;

bande *de pantalon*, (*unif.*) trouser stripe;
plate- ——, v. s. v. *plate;*
—— *de recouvrement*, (*art.*) garnish-plate, cheek-plate;
—— *de renfort*, (*art., etc.*) any reenforcing plate or strap;
—— *d'un ressort*, set, bending, compression of a spring;
—— *de revolver*, v. supra s. v. *bande;*
—— *de rognon*, (*harn.*) cantle-plate (of a saddle);
—— *de roue*, tire of a wheel;
—— *de sassoire*, sweep-bar band;
—— *de selle*, v. —— *d'arçon;*
sous- ——, (*art.*) trunnion bedplate;
—— *de support*, (*art.*) shaft-brace, of a mountain-gun carriage;
sus- ——, (*art.*) cap-square;
—— *de tiroir*, (*steam*) lap of a slide-valve;
—— *à tourillons*, (*art.*) axle-socket strap or band (of a gin).
bandé, a., (*sm. a., r. f. art.*) cocked; (of a spring) set, compressed; (*mil., etc.*) blind-folded;
au ——, (*sm. a., r.f. art.*) at full cock;
cran du ——, (*sm. a.*) full-cock notch;
position du ——, (*sm. a., r.f. art.*) cocked position, position of full cock.
bandeau, m., bandage, (*esp. mil.*) a bandage over the eyes; (*unif.*) band around a forage cap; (*mas.*) string-course; (*tech.*) iron plate, wider than *bande* and narrower than *plaque;*
—— *de bec*, (*pont.*) bow (or stern) strap of a ponton-boat.
bandelette, f., small strap (esp. of iron, tin, etc.); narrow strap or strip;
contre- ——, small strap (so called because placed over against a *bandelette*).
bander, v. a., to bend, stretch, compress, wind up, set (a spring); to bind, tie; to tine at the edges; (*sm. a., r. f. art.*) to cock (a piece);
se ——, to compress itself (as a spiral spring);
—— *un tambour*, (*mil.*) to tighten the strings of a drum;
—— *les yeux*, to blindfold.
bandereau, m., trumpet-cord, bugle-cord.
banderole, f., banderole, streamer; small flag; bugle-cord; shoulder-belt (of man-harness); pistol-belt; support of cavalry sapper's tool; belt, cross-belt (obs.); sling (of a flag, etc.);
—— *de bricole*, shoulder strap or belt (of man-harness);
coup de ——, (*fenc.*) diagonal cut (from shoulder toward belt).
bandière, f., flag, ensign (obs.);
front de ——, (*mil.*) front of an army; line or ground in front of a camp, or of an army in camp or in bivouac, and on which stacks are made; ground along front toward enemy, when troops are camped in column; color-line of a camp (U. S. A.).
bandoulière, f., shoulder-belt (obs.);
en ——, slung across the shoulder.
banlieue, f., suburbs; (*adm.*) territory in neighborhood of a fort, and depending on it, in an administrative point of view.
bannière, f., banner, ensign, flag, standard (not a military term).
bannir, v. a., to banish, exile.
bannissement, m., banishment, exile.
banquet, m., (*harn.*) lower part of the cheek of a bit; rivet-hole in same.
banquette, f., long bench; footpath (pavement) (rare); stretcher (of a boat); (*art.*) platform of chassis of seacoast carriage; (*fort.*) banquette;
—— *d'artillerie*, (*fort.*) artillery banquette;
—— *d'infanterie*, (*fort.*) infantry banquette;
—— *de pointage*, (*art.*) aiming platform;
talus de ——, (*fort.*) banquette slope.
banquise, f., ice jam, ice pack (in a river); ice pack, in general.
baptême, m., baptism;
—— *de feu*, (*mil.*) first exposure to fire.
baquet, m., bucket, tub, trough;
—— *de propreté*, (*Fr. a.*) latrine tub.

baraque, f., (*mil.*) hut, soldier's hut; any light building taking place of a regular barrack; (*mil. slang*) service stripe;
—— *-gourbi*, (*mil.*) hut of wattles, etc., covered with clay and chopped straw;
—— *en torchis*, (*mil.*) mud hut (chopped straw and clay, etc.).
baraquées, f. pl., (*mil.*) troops lodged in huts.
baraquement, m., (*mil.*) hut camp; action of lodging in huts; lodging or quartering in barracks (synonym of *casernement*).
baraquer, v. a., (*mil.*) to build huts; v. r., to be lodged in huts;
—— *la troupe*, to lodge men in huts.
baraquette, f., sister-block.
baraterie, f., barratry.
barbarou, m., v. *avant-mur*.
barbe, f., beard; burr; m., barb, Barbary horse; f., (*hipp.*) chuck, part of under jaw on which curb-chain rests; (*art.*) fringe around a shot hole in an armor plate;
—— *à poux*, (*mil. slang*) pioneer;
Sainte Barbe, patron saint of artillerymen and of sailors;
royaume de Sainte Barbe, (*art.*) artillery in its widest extension;
tirer en ——, (*art.*) to fire over bank, to fire in barbette.
barbelé, a., barbed, bearded;
cheville ——*-e*, *clou* ——, spike-nail.
barbeler, v. a., to barb.
barbette, f., (*art. fort.*) gun-bank, barbette;
la ——, (*mil. slang*) the engineers, course of military engineering at military schools;
le 2ème, etc., ——, (*mil. slang*) the 2d, etc., regiment of engineers;
batterie à ——, (*art.*) barbette battery;
en ——, in barbette (usually *à* ——);
tir à ——, (*art.*) barbette fire;
tirer à ——, to fire in barbette.
barbillon, m., barb (of a hook); (*hipp.*) small fold in mucous membrane of a horse's mouth, barbles.
barbotage, m., stirring, mixing; (*hipp.*) mixture of bran or of barley with drinking water.
barboteur, m., (*expl.*) mixer (gun-cotton manufacture);
—— *à palettes*, paddle-mixer (gun-cotton manufacture).
barbotin, m., sprocket-wheel or pulley; chain-pulley.
barbouiller, v. a., to smear, besmear.
bard, m., handbarrow.
bardeau, m., (*cons.*) shingle, slate.
bardelle, f., sort of saddle (of canvas, and stuffed).
barder, v. a., to barb a horse; to load upon a handbarrow, to carry off stone, etc., on a handbarrow.
bardot, m., (*hipp.*) offspring of a horse and she-ass.
barème, m., ready reckoner;
—— *graphique*, graphic table or diagram;
—— *de solde*, pay table.
barge, f., barge; levee (along a river).
baril, m., barrel;
—— *ardent*, (*siege*) fire-barrel;
—— *d'artifice*, (*siege*) fire-barrel filled with grenades;
—— *à bourse*, (*art.*) budge-barrel;
—— *à ébarber*, (*powd.*) polishing barrel;
—— *éclairant*, —— *à éclairer*, (*siege*) light-barrel, illuminating barrel;
—— *à feu*, v. —— *ardent*;
—— *foudroyant*, (*siege*) v. —— *d'artifice*;
—— *infernal*, v. —— *ardent*;
—— *à poudre*, powder barrel; (*mil. min.*) powder holder, i. e., a water-tight barrel;
—— *à poudre en zinc*, (*mil. min.*) powder holder (used when water is feared, to contain the charges).
barillet, m., small barrel, keg; (*sm. a.*) cylinder (of a revolver); any cylindrical breechblock (as of Werndl rifle); (*art.*) time-train cone (of a combination fuse);
—— *Accles*, Accles "feed" (Gatling gun);
axe du ——, (*sm. a.*) center-pin (revolver);
—— *de chargement*, (*art.*) shell-hoist scoop or scuttle;

barillet *de fusée*, (*art.*) case or casing of a fuse;
—— *porte-outils*, (*mach.*) turret (of a lathe, etc.);
support de ——, (*sm. a.*) crane of a revolver.
baritel, m., winding-machine;
—— *à chevaux*, horse-whim, whim-gin.
barlong, a., (of quadrilaterals) longer on one side than on the other.
baromètre, m., barometer;
—— *altimétrique*, leveling barometer, giving altitudes by direct reading;
—— *anéroïde*, aneroid barometer;
—— *à cadran*, wheel barometer;
—— *à cuvette*, cup barometer;
—— *à mercure*, mercurial barometer;
—— *orométrique*, mountain barometer;
—— *portatif*, portable barometer;
—— *à siphon*, siphon barometer.
barque, f., small boat, skiff.
barrage, m., dam, weir; barrier, toll-bar;
—— *-déversoir*, weir;
—— *éclusé*, sluice-dam.
barre, f., bar, rod, cross-bar; band; boom; lever; bar (of a river or harbor); bar (of a bit); (*hipp.*) bar (space between molars and incisors, of a horse's mouth); (*elec.*) bus-bar; (*harn.*) hip-strap;
—— *en acier à boudin*, steel bulb bar;
—— *en acier à* U, channel steel;
—— *d'alésage*, (*mach.*) boring-bar;
—— *d'alésoir*, (*mach.*) boring-bar;
—— *d'aller*, (*elec.*) positive bar, bus-bar (of a *tableau de distribution*, q. v.);
—— *d'appui*, hand-rail;
—— *d'appui de ringard*, pricker-bar;
—— *d'attelage*, drag-bar;
—— *à boudin*, bulb-bar;
—— *de cabestan*, capstan bar;
—— *directrice*, (*mach.*) guide-bar, motion bar;
—— *de distribution*, (*elec.*) bus-bar (of a *tableau de distribution*);
—— *d'écartement*, spreader, distance-bar;
—— *s échantillonnées*, generic term for iron in bars of determined shapes;
—— *d'écurie*, swing-bar, stall-bar;
—— *d'excentrique*, (*mach.*) eccentric-rod;
—— *à fileter*, (*mach.*) chasing-bar;
—— *fixe*, (*gym.*) horizontal bar;
—— *de fourneau*, —— *du foyer*, —— *de grille*, fire-bar, grate-bar;
—— *de manège*, (*man.*) leaping bar;
—— *de manœuvre*, handspike, maneuvering lever; (*met.*) porter bar;
—— *à mine*, —— *à mines*, (*min.*) boring bar, blasting bar, jumper;
—— *à mine à rallonges*, (*mil. min.*) extensible, extension jumper or borer, jointed borer (joints assembled by screw-sleeves);
—— *de mire*, (*art.*) sight bar;
—— *de mors*, (*harn.*) bar of a bit;
—— *d'obturation*, (*art.*) throttling bar;
—— *s parallèles*, (*gym.*) parallel bars;
—— *de percussion*, (*min.*) jumper;
—— *de port*, boom;
—— *profilée*, special iron or steel;
—— *de rayage*, (*art.*) rifling-bar;
—— *de recul*, (*art.*) recoil-bar (Schumann);
—— *de réglage*, (*art.*) throttling bar;
—— *de retour*, (*elec.*) negative bar, bus-bar (of a *tableau de distribution*);
—— *de support*, —— *-support*, (*mach.*) overhanging arm (e. g., in a Brown & Sharpe milling machine);
—— *à*, —— *de*, *suspension*, (*gym.*) horizontal bar;
—— *à* T, T-iron, T-steel;
—— *à* T *en acier avec boudin*, T-bulb steel;
—— *à* T *double en acier*, girder steel;
—— *à* T *simple*, T-iron, T-steel;
—— *témoin*, (*met.*) trial bar, tap-bar (cementation process);
toucher ——, to arrive and leave immediately.
barreau, m., bar; bar (of a court); the bar, legal profession; test bar, test specimen;
—— *de choc*, bar for falling test;
—— *d'essai*, test bar, test specimen;
—— *grille*, fire-bar, grate-bar;
—— *de ployage*, bending test-bar;
—— *de traction*, tensile test-bar.

barreaudage, m., barring, grating, bars (of a window, etc.).
barrer, v. a., to bar, stop or close up; to dam a stream; to draw the pen through a word or words.
barrette, f., small bar, rib; test strip or specimen; (in a balloon) the short bar from which the *pinceaux* (q. v.) hang; (*mil.*) bar (of a medal); (*harn.*) bar (of a mule bit); (*sm. a.*) hand (of a revolver); (*mach.*) slide-bar, slide-face, valve face; (*fond.*) core-box stay, spindle centering piece (as in casting of projectiles);
—— *de détente*, (*sm. a.*) hand of a revolver;
—— *d'essai*, test strip or bar.
barricade, f., barricade.
barrière, f., barrier; fence, rail, stile; (*fort.*) heavy wooden door in exits of fortifications; (*man.*) hurdle;
—— *double*, (*fort.*) double-swing barrier (field work);
—— *flottante*, floating barricade, boom;
—— *de péage*, turnpike gate;
—— *à pivot central*, (*fort.*) central pivot barrier (field work);
—— *à un vantail*, —— *à deux vantaux*, (*fort.*) single, double leaf barrier (field work).
barrique, f., hogshead.
barrot, m., (*nav.*) deck beam.
barrotage, m., (*nav.*) deck beaming.
baryte, f., (*chem.*) baryta.
bas, m., stocking; lowest part of anything; lower course of a stream;
—— *d'un bâtiment*, ground sill of a building;
—— *du bois d'un fusil*, (*sm. a.*) butt-end of stock;
—— *de la branche*, (*harn.*) leg or branch of a bit;
—— *de l'eau*, low water, time of low water;
—— *élastique*, (*med.*) elastic stocking (for varicose veins);
—— *de soie*, (*fam.*) irons, fetters;
a., low; overcast (of the weather);
eau —— *se*, low water;
faire main —— *se sur*, to put everybody to the sword;
mer —— *se*, low water;
mettre —— *les armes*, (*mil.*) to surrender;
mettre pavillon ——, (*nav.*) to strike one's colors;
—— *se pression*, (*steam*) low pressure.
basane, f., sheepskin, sheep (leather);
—— *s*, (*unif.*) false boots, leather legs of French cavalry trowsers.
basaner, v. a., to darken (leather);
—— *un pantalon en cuir*, (*unif.*) to reenforce a pair of trowsers with leather.
bascule, f., plyer, lever (as of a bridge, etc.), swipe; swing-gate; weigh-bridge; weighing-machine; (*mach.*) rocking, reciprocating, seesaw motion; (*inst.*) needle-lifter (of a compass); (*sm. a.*) break-off, and related parts, of the breech of a shotgun; (*sm. a.*) false breech;
à ——, balance, balancing, rocking;
branche de ——, hinder end of a plyer;
—— *à contrepoids*, counterpoise lever (of a drawbridge);
flèche de ——, fore part of a plyer;
pont à ——, weigh-bridge, weighing-machine;
—— *de pont-levis*, (*pont.*) swipe;
—— *de serrure*, lever of a lock.
basculer, v. a., to tilt, dip, drop, swing, balance, rock; (*art.*) to "tumble" (of a projectile).
base, f., base, basis; foundation; support, rest; foot, pedestal; (*top.*) base-line; (*mil.*) base; basis or base of formation, at drill; (*man.*) horse's feet considered as forming a line or polygon of support;
—— *accidentelle*, (*mil.*) temporary base of operations;
—— *d'alignement*, basis of alignment (*drill*);
—— *d'approvisionnements*, (*mil.*) base of supplies;
—— *auxiliaire*, (*top.*) secondary base;
—— *brisée*, (*top.*) broken base-line;
de ——, (*drill*) directing (said of a unit on which movement is made);

base *diagonale*, (*man.*) diagonally opposed feet of a horse, considered as points of support in his movements;
—— *en équerre*, (*mil.*) enveloping base;
—— *de formation*, (*drill*) the unit on which, as a base, a movement is made or maneuvor carried out;
à —— *invariable*, said of a cipher in which a symbol retains same meaning throughout;
—— *latérale*, (*man.*) feet on same side of horse, considered as points of support;
—— *d'opérations*, (*mil.*) base of operations; (also, more narrowly) line of points at which an army leaves the railways;
—— *quadrupédale*, (*man.*) the four feet of the horse, as points of support;
—— *secondaire*, (*mil.*) secondary base, intermediate base of operations on the line of operations;
—— *tripédale*, (*man.*) any three feet of the horse, as points of support;
à —— *variable*, said of a cipher, in which the same symbol changes meaning in the same message.
bas-fond, m., (*top.*) low ground, bottom, bottom land; swampy bottom (frequently); shoal, shoal water; deep place in a stream, deep water.
bas-foyer, m., (*met.*) generic term for a low furnace (less than 1ᵐ high), as opposed to *haut-fourneau*.
basique, a., (*chem.*) basic.
bas-jointé, a., (*hipp.*) low-pasterned (of horse and of pastern).
bas-off, m., noncommissioned officer (slang, Polytechnic).
bas-officier, m., noncommissioned officer (obs.).
basque, f., (*unif.*) skirt, tail, of a coat, tunic, etc.
basse, f., (*hydr.*) shoal water, hiding a bank of sand or of rocks; sand bank, or rocks never uncovered, but near surface.
basse-eau, f., ebb, ebb-tide, low water.
basse-enceinte, f., (*fort.*) false bray.
basse-marée, f., ebb, ebb-tide, low water, low tide.
basse-mer, f., low water, low tide, ebb, ebb-tide.
basse-palée, f., (*pont.*) foundation piles.
bassin, m., basin (vessel); scale-pan (of a balance); basin (of a port); geographical basin; so much of a river or canal as lies between two bridges, in a town; (*hipp.*) pelvis;
—— *de chasse*, (*fort.*) flush-basin (in a ditch that may be flooded);
—— *de construction*, dry dock;
—— *à flot*, wet dock;
—— *flottant*, floating dock;
—— *de fuite*, (*fort.*) escape-basin (after a rush of water through the ditch);
—— *hydrographique*, basin drained by a river;
—— *lacustre*, basin whose waters discharge into a lake;
—— *maritime*, maritime basin;
—— *de partage*, reservoir at summit level of a canal;
—— *de radoub*, dry dock, graving dock;
—— *de réception*, basin or area drained by a torrent;
—— *de retenue*, (*fort.*) flushing-basin (for a rush of water through the ditch);
—— *sec*, dry dock.
bassine, f., flat, shallow, metal dish.
bassinet, m., (*art.*, etc.) cup of a fuse; (*sm. a.*) pan of a flintlock (obs.).
couvre- ——, (*sm. a.*) pan-cover (obs.).
basson, m., (*mus.*) bassoon.
bastide, f., (*fort.*) ancient bastion, blockhouse, small fortress (obs.).
bastille, f., (*fort.*) bastille, fortress (obs.); *the Bastille.*
bastillon, m., (*fort.*) small bastille (obs.).
bastin, m., (*cord.*) coir.
basting, m., (of timber) standard form or size.
bastingage, m., (*nav.*) hammock nettings.
bastion, m., (*fort.*) bastion;
—— *camus*, bastion, with one of the faces suppressed;

bastion

bastion *casematé,* casemated bastion;
— *composé,* composite bastion;
— *creux,* hollow bastion;
— *déforme,* irregular bastion;
demi- —, half-bastion;
— *détaché,* detached bastion;
— *double,* double bastion;
— *mixte,* composite bastion;
— *plein,* solid, full bastion;
— *retranché,* retrenched bastion;
— *vide,* hollow bastion.
bastionné, a., *(fort.)* bastioned;
front —, bastioned front;
tracé —, bastioned trace.
bastionner, v. a., *(fort.)* to fortify by means of bastions, to fortify on a bastioned trace.
bastionnet, m., *(fort.)* bastionnet, small caponier;
— *de flanquement,* half-bastion for flanking.
bastonnade, f., bastinado.
bât, m., pack, pack-saddle;
— *d'affût,* *(art.)* carriage-pack (mountain battery);
animal de —, pack animal;
— *de caisse,* *(art.)* ammunition pack (mountain battery);
cheval de —, pack-horse;
étrier de —, pack-saddle stirrup or support, for loads that can not be hooked on;
mulet de —, pack mule;
— *de pièce,* *(art.)* gun pack (mountain battery);
— *porte-caisses,* *(art.)* ammunition-box pack (of a landing gun, French);
— *du train des équipages,* *(mil.)* pack outfit for removal of wounded.
batail, m., bell-clapper.
bataille, f., battle, array, battle array;
en —, in order of battle; in line *(art. and cav. drill)*; into line *(art. and cav. drill,* of movements and in commands); *(unif.)* (of a cocked hat) worn with axis parallel to shoulders;
— *acharnée,* desperate battle;
champ de —, field of battle;
cheval de —, charger;
corps de —, main body of an army (i. e., the fighting element, as opposed to noncombatants);
— *défensive,* defensive battle;
— *défensive-offensive,* defensive in one part of the field and offensive in another;
— *démonstrative,* attack or battle intended to deceive as to real point of attack;
donner —, to give battle;
fort de la —, thick of the fight;
— *imprévue,* unforeseen battle, chance battle;
livrer —, to give battle;
— *navale,* sea fight, naval battle;
— *offensive,* offensive battle;
ordre de —, order of battle, battle array, array or list of troops engaged or forming an army;
— *préméditée,* — *prévue,* battle foreseen, and hence provided for in orders;
— *rangée,* pitched battle;
ranger en —, to draw up in battle order, in battle array;
— *de rencontre,* unforeseen, unpremeditated, unexpected battle.
batailler, v. n., to give battle (obs.); *(fig.)* to dispute tenaciously.
batailleur, m., one fond of battles; *(fig.)* one fond of disputes.
bataillon, m., *(mil.)* battalion *(foot troops)*;
— *-cadre,* skeleton battalion;
— *de canonniers sédentaires,* a battalion of foot artillery so called, kept up at Lille in remembrance of its siege in 1792;
chef de —, major *(inf.)*;
— *-école,* instruction battalion;
école de —, battalion drill, school of the battalion;
— *encadré,* a battalion *framed* in by others, acting in conjunction with others on its right and its left;
— *de guerre,* full battalion, battalion on a war footing;

37

bâtiment

bataillon *d'infanterie légère d'Afrique,* punishment battalion (serving in Africa);
— *d'instruction,* drill battalion;
— *isolé,* detached battalion, operating alone;
— *de marche,* makeshift battalion (1789 and 1870), composed of fragments drawn from different regiments;
— *scolaire,* v. — *-école.*
bâtard, a., (of animals) not full bred; *(in gen.)* said of what can not be put into a definite class.
bâtarde, f., round hand; engrossing;
écriture —, round hand.
batardeau, m., dam, coffer-dam; *(fort.)* batardeau.
bateau, m., boat, skiff; *(pont.)* ponton boat;
— *d'avant-garde,* *(pont.)* advanced-guard ponton boat;
— *de ballon,* balloon-car;
— *-canon,* boat considered merely as a carriage for a gun (e. g., the *Gabriel Charmes* of Admiral Aube); gunboat;
— *charbonnier,* collier;
— *à contrepoids,* *(pont.)* boat with counterpoise, used in construction of trestle bridges (the counterpoise counterbalances the trestle in the operation);
— *délesteur,* lighter;
— *divisible,* *(pont.)* ponton boat that takes apart in two pieces;
— *-éperon,* French name for Admiral Ammen's ram;
— *d'équipage,* *(pont.)* regulation ponton-boat of the bridge train;
— *-feu,* light-ship;
hangar à —, boat shed, boathouse;
— *-hôpital,* hospital ship;
— *de passage,* ferryboat;
— *plat,* flat-bottomed boat;
— *pliable,* *(pont.)* collapsible or folding boat;
— *plongeur,* submarine torpedo boat;
— *ponté* (— *non ponté*), decked (open) boat;
— *-porte,* draw (of a bridge of boats);
— *porte-torpille,* spar torpedo boat;
— *à successifs,* *(pont.)* successive pontons;
— *-torpilleur,* torpedo boat;
— *torpilleur lance-torpilles,* regular torpedo boat;
— *torpilleur porte-torpilles,* spar torpedo boat;
— *traille,* ordinary boat used as a trail in a flying bridge;
— *à vapeur,* steamboat;
— *à vapeur postal,* mail steamer.
batelage, m., boating; transportation by boats.
batelée, f., boat load.
bateler, v. a., to transport by boat.
batelier, m., boatman, waterman;
— *aide-portier,* (Fr. a.) engineer employee, assistant to *portier-consigne,* q. v., in works provided with wet ditches.
batellerie, f., generic term for small boats.
bâter, v. a., to put on a pack saddle.
bat-flanc, m., swing pole, swing bale (between stalls in a stable).
bâti, m., *(mach., etc.)* frame, framework, support; engine frame or bed; stand;
— *américain,* engine bed of the Corliss type;
— *composé,* combined support;
— *à quatre paliers,* fourfold bearings;
— *simple,* simple support.
bâtiment, m., building, pile, structure; *(nav.)* vessel, ship;
— *armé en course,* privateer;
— *armé en guerre,* ship equipped for war;
— *en armement,* vessel in commission;
— *balise,* light-ship;
— *blindé,* armored vessel;
— *charbonnier,* collier;
— *convoyeur,* convoying ship;
— *croiseur,* cruiser;
— *cuirassé,* armored ship;
— *-école,* school-ship;
— *de guerre,* war ship;
— *de haut bord,* vessel of high free-board;

bâtiment *militaire*, (*mil.*) any building used for service purposes by the regular establishment;
— *neutre*, neutral vessel;
— *d'observation*, lookout ship;
— *parlementaire*, cartel vessel;
— *plongeur*, submarine boat;
— *pour la pose des câbles*, cable-ship;
— *à tourelles*, turret ship;
— *de transport pour les troupes*, troop ship;
— *à vapeur*, steamship.

bâton, m., stick, baton, marshal's baton; (*fenc.*) singlestick;
être réduit au — *blanc*, to be ruined, in extremities;
sortir d'une place le — *blanc à la main*, (*mil.*) to depart without arms or baggage;
— *de commandement*, truncheon;
court-——, (*fenc.*) sort of half-pike or staff;
— *d'enseigne*, (*nav.*) flagstaff;
— *de Jacob*, (*inst.*) Jacob's staff;
— *de maréchal*, (*mil.*) marshal's baton;
— *à mèche*, (*art.*) linstock (obs.);
— *de mesure*, (*mus.*) baton of band leader;
— *de pavillon*, (*nav.*) flagstaff;
— *de soufre*, roll of sulphur;
— *à surfaix*, (*hipp.*) sort of surcingle twitch.

bâtonnée, f., — *d'eau*, throw of a pump.

bâtonner, v. a., to beat with a stick; to strike out (a word, a line, etc.).

bâtonnet, m., short stick; (*hipp.*) blood stick; (*art.*) grooved stick to facilitate explosion of bursting charge in shells.

battage, m., beating (as in making *pisé*); driving (of piles); (*met.*) beating, striking; (*powd.*) stamping.

battant, m., leaf (of a door); bell-clapper; fly (of a flag); stile (of a window frame); (*mach.*) movable or oscillating frame (e. g., in a copying lathe, touches the model and moves toward or from work as needed); (*nav.*) sufficiency of free-board, command; (*sm. a.*) swivel;
— *de crosse*, (*sm. a.*) stock-, butt-swivel;
à deux — *s*, folding (of doors);
— *de grenadière*, (*sm. a.*) middle band swivel;
— *meneau*, inner stile (of a window frame);
— *de noix*, outer stile (of a window frame);
— *de pavillon*, fly of a flag;
— *d'une poulie*, pay of a pulley;
— *de sous-garde*, (*sm. a.*) lower band swivel.

battant, a., at work, going; (*nav.*) fit or ready for battle (of a ship);
mener — , (*mil.*) to pursue a defeated enemy close and hard;
pluie — *e*, pelting rain;
porte — *e*, swing-door;
tambour — , (*mil.*) with drums beating, by beat of drum;
sortir tambour — , (*mil.*) to march out with the honors of war;
vaisseau de guerre bien — , (*nav.*) ship in which the guns are well served.

batte, f., rammer, beetle; (*harn.*) peak or bow of a saddle; knee-puff; (*sm. a.*) sputcheon (of a scabbard);
— *d'armes*, — *de la cuvette*, (*sm. a.*) sputcheon;
— *à bourrer*, tamper, earth-rammer;
— *de selle à la hussarde*, (*harn.*) fore peak of a hussar saddle.

battellement, m., (*cons.*) eaves, eaves-course (of a roof).

battement, m., beating, stamping; oscillation; (*art.*) balloting of projectile in s. b. guns, indentation made by balloting in s. b. guns, (*r. r.*) long interval of time between departures of consecutive trains; (*in gen.*) any interval of time necessary to the good working of the line, e. g., interval between arrival of troops at a station and departure; (*fenc.*) beat of point on point;
— *d'épée*, (*fenc.*) act of striking adversary's blade, to remove it from the line of the body;
— *de tambour*, beating of a drum.

batterie, f., (*art., fort.*) battery (unit of organization; the guns themselves, the personnel; work or emplacement); (*nav.*) ship's battery; (*elec.*) battery; (*sm. a., r. f. art.*) lock; (*mil.*) call or signal (drum); drum corps (of a regiment, etc.), (*obs.*); (*powd.*) set of mortars and pestles (obs.).
I. Battery in respect of functions; II. Battery in respect of site or of nature of emplacement; III. Miscellaneous.

I. Battery in respect of functions:

batterie, *d'Afrique*, (*Fr. art.*) battery serving, or for service, in Africa;
— *d'approche*, (*siege*) battery of second artillery position;
— *d'attaque*, attacking battery; testing battery, proving-ground battery;
— *attelée*, horsed battery (field or horse);
— *de bombardement*, battery to deliver high-angle fire, shelling battery (esp. in sieges against broad targets, e. g., towns, magazines, storehouses, etc.); (*in gen.*) any battery that assists in a bombardment; (*s. c. art.*) a battery armed with guns for high-angle fire;
— *de brèche*, (*siege*) breaching battery;
— *de bricole*, oblique battery, one delivering an oblique fire;
— *s par camarade*, batteries uniting their fire on the same target;
— *de campagne*, field battery;
— *de cavalerie*, horse battery;
— *s en chapelet*, cross-fire batteries;
— *à cheval*, horse battery;
— *de circonstance*, (*siege*) battery constructed when and where needed, in the defense of a fortified position (classed with the *réserve générale*, q. v.);
— *circulaire*, (*nav.*) floating battery (circular);
— *de combat*, (*Fr. art.*) subdivision of a field battery, on a war footing (horse or mounted) with respect to marches and combats, consisting of the first 8 *pièces*, q. v., that is, 6 guns, 9 caissons, the battery wagon, and the forge; subdivision of a mountain battery on a war footing, consisting of the 6 guns, and various accessory supplies and the necessary personnel: [see supplement];
— *à la Congrève*, rocket battery;
contre- ——, (*siege*) counter-battery, direct-fire battery, esp. such a battery established in the crowning of the covered way;
— *de corps*, battery forming part of the corps artillery;
— *s croisées*, cross-fire batteries;
— *de débarquement*, (*nav.*) landing battery;
— *de démolition*, (*siege*) demolition battery, whose fire is directed against case-mates, redoubts, and other masonry constructions;
— *à démonter*, (*siege*) battery whose special business it is to dismount the guns on the work fired at, counter-battery;
deuxième ——, (*siege*) extra battery put up when third parallel masked the *première batterie*, q. v. (obs.);
— *directe*, (*siege*) battery whose line of fire is normal to line fired at (synonym of — *à démonter*);
— *divisionnaire*, battery forming part of the divisional artillery;
— *d'écharpe*, — *en écharpe*, (*siege*) battery whose line of fire is oblique to line fired at;
— *de deuxième échelon*, (*siege*) battery somewhat in advance of first artillery position;
— *d'encadrement*, name formerly applied to battery accompanying infantry in its advance on hostile position;
— *d'enfilade*, enfilading battery (esp. of a siege);
— *fixe*, battery fixed in position (as distinguished from — *mobile*), permanent battery (seacoast, etc.);
— *flanquante*, — *de flanquement*, (*fort.*) flanking battery;
— *flottante*, (*nav., s. c.*) floating battery;
— *de France*, (*Fr. art.*) battery serving, or for service, in France;
— *de fusées*, rocket battery;

batterie *de guerre,* field battery on a war footing (composed, in France, of —— *de combat,* q. v., and *train régimentaire,* q. v.);
—— *-guide,* that battery of a group to which as assigned the duty of getting the range (*régler le tir*);
—— *haute,* (*fort.*) any battery situated in the higher parts of a work and having a considerable command; (in seacoast artillery) battery on a high site, for delivering a plunging fire (see under II);
—— *d'investissement,* (*siege*) investing battery, established at outset of siege;
—— *lance-torpilles,* —— *de lancement de torpilles,* —— *de lancement de torpilles,* (seacoast *fort.*) torpedo battery;
—— *lointaine,* (*siege*) battery at 5 or 6 km. from most advanced salients (attack); (sometimes synonymous with —— *de première période*);
—— *de manœuvre,* maneuver battery, (*Fr. art.*) 6 pieces and 3 caissons; battery for drill and instruction;
—— *mobile,* any field battery; (*siege*) battery accompanying mobile troops in the defense of a place, sortie battery (part of *armement de place de guerre*); (*s. c. art.*) auxiliary batteries of field pieces (against landing parties, etc.);
—— *de montagne,* mountain battery;
—— *montée,* mounted battery;
—— *de mortiers,* mortar battery;
—— *de deuxième période,* (*siege*) battery thrown up by the attack in the second period of the siege, and used for the distant artillery attack of the main position, battery of the second artillery position;
—— *de première période,* (*siege*) battery thrown up by the attack at the outset of a siege, battery of the first artillery position, opened before the first parallel is opened;
—— *à pied,* foot battery; field as opposed to horse battery (rare);
—— *à pied montée,* old name for mounted battery;
—— *de plein fouet* (*siege*) direct-fire battery;
—— *plongeante,* (*seacoast fort.*) battery to deliver plunging fire;
—— *de deuxième position* (*siege*) battery of the second artillery position;
—— *de première position,* (*siege*) battery of the first artillery position (according to some writers the term *batterie de 1ère, 2ème position* is limited to batteries armed with b. l. rifles, while *batterie de 1ère, 1ème période* applies to batteries armed with m. l. rifles, now obsolete; but this distinction is fine drawn);
—— *première, première ——,* (*siege*) battery of first artillery position, more often in front of second parallel (obs.);
—— *de préparation de combat,* name formerly applied to battery accompanying infantry in its attack on a hostile position;
—— *de protection,* battery protecting a *center of resistance* in the infantry line of a fortified position; (*siege*) battery of light pieces to repel attacks on the *position de combat* of the besiegers;
—— *de revers,* (*siege*) reverse-fire battery;
—— *en rouage,* (*siege*) enfilade battery when directed against another battery;
—— *de rupture,* battery of armor-piercing guns;
—— *secondaire,* (*nav.*) secondary battery;
—— *seconde,* v. *deuxièmei*
—— *de siège,* (*siege*) siege battery (guns and works);
—— *de sortie,* (*siege*) field battery accompanying a sortie or attack made by *troupes mobiles* (defense), v. —— *mobile;*
—— *de tir,* (*Fr. art.*) that part of a field battery (horse or mounted) that goes into action as a whole from its very beginning, i. e., 6 guns and 3 caissons (2 or 3 horse battery); in a mountain battery, 6 guns and 2 ammunition mules; (*batterie de guerre= batterie de combat+train régimentaire,* and *batterie de combat=batterie de tir+écheion de combat*); [see supplement];
—— *à tir indirect,* (*siege*) indirect-fire battery;

batterie *à tir normal,* (*siege*) battery whose line of fire is normal to line or work fired at;
—— *de tir plongeant,* —— *de tir vertical,* (*siege*) high-angle fire, vertical fire, battery; (sometimes mortar battery, e. g., in. s. c. artillery);

II. Battery in respect of site, etc.:
——*annexe,* (*fort.*) wing battery; emplacement outside of fort, and protected by it, for pieces that can not conveniently be placed in the work itself;
—— *barbette,* —— *à barbette,* barbette battery;
—— *basse,* guns along the lower crests of works; (*seacoast*) water battery, any battery less than 30^m to 35^m above water level (also) any battery that can not use a depression range-finder; (sometimes) ricochet battery; casemated guns for flanking the ditch (bastioned fort);
—— *blindée,* battery furnished with overhead cover, blinded battery;
—— *casematée,* —— *à casemates,* —— *sous casemate,* (*fort.*) casemate battery, casemate;
——*-caverne,* (*fort.*) batteries carved out of rock, or constructed in rocks, for defense of defile, etc., in mountainous countries;
—— *à ciel ouvert,* open battery (no overhead cover);
—— *de circonstance,* (*fort.*) a synonym of —— *détachée,* q. v.;
—— *de côte,* (*fort.*) seacoast battery;
—— *de couronnement,* (*siege*) battery established by the attack in covered way, when crowned;
—— *à,* —— *en crémaillère,* indented battery;
—— *de crête,* (*fort.*) any battery firing over a crest; (*siege*) a battery behind a covering crest, but retaining possibility of direct fire while using indirect aiming;
—— *cuirassée,* armored battery;
—— *détachée,* (*fort.*) detached battery; any battery sufficiently far off to be considered isolated from some main work; (according to some writers) a synonym of —— *de circonstance;*
—— *à distance,* (*siege*) distant battery;
—— *à embrasures,* (*fort., etc.*) embrasure battery;
—— *encaissée,* —— *enfoncée,* —— *enterrée,* (*siege, etc.*) sunken battery, one whose parapet is constituted by the natural and undisturbed ground, whose terreplein is in the trench;
—— *à étages,* tiered battery, with guns in tiers;
—— *exhaussée,* a battery whose terreplein is on a *remblai* or artificial mound;
—— *de fortresse,* (*fort.*) fortress battery;
—— *en gradins,* battery whose guns are on different levels;
—— *haute,* any (fixed) battery on an elevated site; (*Fr. s. c. art.*) any battery having a command of 30^m or more if on Mediterranean coast, of 35^m or more if on Atlantic or Channel; (in gen., s. c. art.) any battery using a depression range-finder;
—— *isolée,* (*s. c. fort.*) battery not connected with or dependent on a *place;*
—— *masquée,* screened, masked battery;
—— *de mobilisation,* battery constructed on mobilization in a *place de guerre;*
—— *de niveau,* battery on level of the ground;
—— *à double parapet,* battery permitting fire in opposite directions;
—— *de place,* fortress battery, fixed or permanent battery;
—— *rapide,* (*siege*) battery thrown up where fire must be opened as rapidly as possible, hence name (can open fire after 12 hours' work);
—— *rasante,* (*s. c. fort.*) battery near water level, low or water battery; (sometimes) ricochet battery;
—— *à redans,* indented battery;
—— *à ressauts,* battery with guns on different levels;

batterie *en sacs à terre,* (*siege*) battery made of sandbags;
—— *simplifiée,* (*siege*) a sunken battery simplified so that it can open fire after 24 hours' work;
—— *au dessus du sol,* (*siege*) v. —— *exhaussée;*
—— *sur le sol,* (*siege*) battery with its terreplein on the natural surface of the ground;
—— *-traditore,* (*fort.*) casemated battery in the gorge or under the flank of the work;
—— *avec* (*sans*) *traverses,* (*siege*) traversed (untraversed) battery (type);
—— *volante,* emergency battery, established or used on occasion, when and where needed;
—— *à vues directes,* (*siege*) battery in which the pieces may be aimed directly at their target;
—— *dérobée aux vues,* —— *hors de vue,* (*siege*) battery completely concealed from the enemy.

III. Miscellaneous:

en ——, (*art.*) in battery (movement and command); mounted (of guns, esp. of position);
—— *d'accumulateurs,* (*elec.*) storage battery;
chariot de ——, (*art.*) battery wagon;
—— *de chaudières,* (*steam*) range, battery of boilers;
—— *à courant constant,* (*elec.*) constant-current cell or battery;
école de ——, (*art.*) school of the battery;
—— *électrique,* (*elec.*) electric battery;
entrer en —— (*art.*) of detachments, to take posts (at guns);
—— *de fusil,* (*sm. a.*) gunlock (*mil.,* obs.);
—— *galvanique,* (*elec.*) galvanic battery;
gardien de ——, (*Fr. art.*) artillery storekeeper (ordnance sergeant, U. S. A.)
hors de ——, (*art.*) from battery;
—— *à percussion,* (*r.f. art.*) percussion lock;
rentrer en ——, (*art.*) to go to "in battery;"
ressort de ——, (*sm. a.*) hammer spring;
retour en ——, (*art.*) return to "in battery;"
sortir de ——, (*art.*) of detachments, to execute "detachment rear," followed by closing on leading detachment;
—— *de 80,* —— *de 90,* etc., name of battery with respect to caliber of guns (80mm, 90mm, etc.).

battiture, f., scale, forge-scale, hammer-scale.

battoir, m., rammer.

battre, v. a., n., to beat, batter, strike; to ram, to drive in (as a pile); to be loose (of parts of a mechanism); (*met.*) to forge, hammer; (*mil.*) to defeat, to worst; to beat or sound a call or signal on the drum; (*art.*) to cover, sweep with fire, to command or sweep (a water area, a pass, etc.); to direct the fire upon a given point;
 I. Artillery and miscellaneous; II. Drum signals.
 I. Artillery and miscellaneous:

à ——, (*art.*) to be fired on, to be attacked by fire;
—— *un ban,* to give public notice by beat of drum;
—— *en brèche,* (*siege*) to breach;
—— *de bricole,* to fire so as to glance off and hit some other part of the work (obs.);
—— *par camarade(s),* (*art.*) to concentrate fire on same target;
—— *la campagne,* (*mil.*) to beat up the country, to scout;
—— *du canon,* to cannonade;
—— *en chasse,* (*nav.*) to fire the bow, the forward guns;
—— *à chaud,* (*met.*) to hammer hot;
contre—— (*art.*) to counter-batter;
—— *la couverte,* (*mil. slang*) to sleep;
—— *à dos,* (*art.*) to take an object by fire in rear;
—— *à dos en revers,* to take an object by fire directly in rear;
—— *d'écharpe,* (*art.*) to fire obliquely (halfway between direct and enfilade);
—— *l'estrade,* (*mil.*) to beat up a country, to scout (cav.);
—— *du flanc,* (*hipp.*) to heave (of a horse);

battre *en flanc,* (*art., etc.*) to take in flank;
—— *à fond,* (*art.*) to scale a gun (obs.);
—— *à froid,* (*met.*) to hammer-harden, to hammer cold;
—— *de front,* (*art.*) to fire normally to a line or position, etc.;
—— *une ligne,* (*tech.*) to mark out, set out, work; to line out;
—— *à la main,* (*hipp.*) to throw the head up and down;
—— *la mer,* (*nav.*) to cruise up and down over the same area, to continue cruising in the same latitude or station;
—— *une passe,* (*fort*) to command a pass;
—— *pavillon,* —— *le pavillon,* (*nav.*) to fly, show (such and such) a flag, to show one's colors;
—— *un pieu,* (*tech.*) to sink a pile;
—— *à plate couture,* (*mil.*) to defeat thoroughly, to rout;
point à ——, (*art., etc.*) target, point to be reached by fire;
—— *en refus,* —— *jusqu'à refus de mouton,* (*tech.*) to drive a pile home till it will sink no farther;
—— *en retraite,* (*mil.*) to retreat, to fall back; to maintain a running fight;
—— *à,* —— *de, revers,* (*art., etc.*) to take in reverse;
—— *à ricochet,* (*art.*) to deliver a ricochet fire;
—— *en rouage,* (*art.*) to take a hostile work in flank;
—— *en ruine,* (*art.*) to batter down;
—— *en salve,* (*mil.*) to fire a salvo;
—— *en sape,* (*siege*) to undermine by fire.

II. Drum signals (those marked by an asterisk are the French regulation):

—— *l'alarme,* to beat an alarm;
—— *l'appel,* to beat a signal for calling the roll;
——* *l'assemblée,* to beat the assembly;
——* *le ban,* v. s. v. *ban;*
——* *la berloque* (more commonly *breloque*), to sound the signal for "dismissed," to break ranks (also mess call, issue, but this is not official);
—— *la chamade,* to sound a parley (obs.);
——* *aux champs,* to beat or sound the President's march; (*in gen.*) to beat or sound off in honor of the chief of the state, of general officers, of troops passing, etc. (v. s. v. *champ;*)
—— *la charge,* (*mil.*) to beat the charge (in action);
——* *la corvée de l'ordinaire,* to beat off mess fatigue;
——* *la corvée de quartier,* to beat police call (barracks, etc.);
——* *le cours préparatoire,* to sound school call;
—— *le dernier,* to beat the assembly (obs.);
——* *la diane,* to beat reveille (camp or quarters; no longer used, though still in the list of calls);
——* *au drapeau,* to sound to the color;
——* *l'extinction des feux,* to beat off "taps;"
—— *la fascine,* to beat a call for working parties (obs.);
——* *aux fourriers de distribution,* to beat "issue;"
—— *la fricassée,* to beat the signal for trooping or lodging the colors, for a battalion to take up or leave its position in line (obs.);
——* *le garde à vous,* to beat "attention;"
——* *la générale,* to beat the long roll;
——* *aux hommes punis,* to beat signal for prisoners, etc., to turn out;
——* *aux malades,* to beat sick call;
—— *la marche,* to beat a march;
—— *la messe,* to beat church call;
——* *à l'ordre,* to beat signal for reporting for orders (somewhat like first sergeant's call, U. S. A.);
——* *le pas accéléré,* (*drill*) to beat quick time;
——* *le pas de charge,* (*drill, etc.*) to beat "double time;"
——* *au piquet,* to beat signal for a *piquet* to turn out;
—— *le premier,* to beat the first call (obs.);

battre *la prière,* to beat church call;
— *le rappel,* to sound, beat, the assembly;
— *le rappel, (aux tambours et aux clairons)* to sound the signal for field music to fall in;
— *le rappel de pied ferme, (tambours et clairons réunis)* to sound signal for men to take their places in ranks (on a march, etc.);
— *la retraite,* to beat off tattoo;
— *le réveil,* to sound reveille;
— *le rigodon,* lively beat, when bullet hits target at target practice;
— *le roulement,* to beat a roll (reveille, taps, etc.);
— *le second,* to beat the asembly (obs.);
— *la soupe,* to sound mess call;
— *le tambour,* — *du tambour,* to beat a drum; *tambour battant, (mil.)* with drums beating;
— *la veillée,* to sound reveille (obs.).

battu, p. p., beaten;
— *tout juste, (fort.)* said of the glacis when superior slope produced passes 0.50ᵐ over the crest and 1ᵐ over the foot of the glacis.

battue, f., *(man.)* beat (i. e., noise) of a horse's hoofs on the ground; the last tread (in the gallop) before taking a jump;
— *de départ,* last tread or beat before taking a jump;
— *diagonale,* diagonal beat;
— *d'enlever,* v. — *de départ.*

bau, m., *(nav.)* deck beam;
— *composé,* trussed beam;
— *à T,* T-beam;
— *à double T,* channel iron, H-iron.

bauchérisme, m., *(man.)* Baucher's method in all its extension.

baudet, m., *(lit.,* donkey) trestle, horse, frame.

baudre, m., *(cord.)* thick stem of hemp.

baudrier, m., cross-belt, shoulder belt;
— *porte sabre,* sabre support (on saddle).

baudruche, f., gold-beater's skin; thin, transparent skin, for carrying and using staff maps.

bauge, f., *(mas.)* pugging-mortar.

baume, m., balsam.

bauxite, f., *(met. and chem.)* bauxite.

bavard, m., *(mil. slang)* punishment leaf in a soldier's book.

bave, f., foam, froth (animals).

bavette, f., upper part of the mouth of a scabbard, turned in on wood or leather (rare).

bavières, f. pl., *(unif.)* cloth ornaments or fittings on light helmets; sort of chin strap.

bavure, f., seam, beard, burr; roughness or ridge caused by repeated shocks or blows; *(art.)* fringe around a shot hole; swell or bulge in a gun; *(sm. a.)* seam of mold on bullet, fash, mold-mark.

bazarder, v. a., *(mil. slang)* to pillage a house, to wreck it.

beaupré, m., *(nav.)* bowsprit.

bec, m., nose, beak, bill; nozzle; prow; stud; projection (esp. a sharp projection); jib (of a crane); catch; jaw (of a clamp, vise);
— *d'amont, (pont.)* upstream cutwater;
— *d'ancre,* bill of an anchor;
— *d'âne,* mortising chisel; crosscut chisel;
— *d'âne double,* mortising chisel with two cutting edges;
— *d'âne simple,* mortising chisel with only one cutting edge;
arrière —, *(pont.)* stern of a ponton boat;
— *d'aval, (pont.)* downstream cutwater;
avant —, bow of a ponton boat;
— *de bateau, (pont.)* end of a boat;
blanc —, greenhorn;
— *Bunsen, (chem.)* Bunsen burner;
— *de cane,* spring bolt (of a lock); miner's pick-ax;
— *de capucine, (sm. a.)* prolongation of lower band in direction of ramrod groove;
— *carcel,* photometric unit (lamp burning 42 grams of purified colza oil per hour, with a flame 40ᵐᵐ long, under specified conditions);
— *de corbeau,* rave-hook, calker's tool;
— *de corbin, (hipp.)* parrot-mouth;
— *à,* — *de, corvin,* v. — *de corbeau;*

bec *de crosse, (sm. a.)* toe of the butt;
— *de gâchette, (sm. a.)* sear-nose;
— *de gaz,* gas-burner;
— *de grue,* arm or jib of a crane;
— *de lampe,* burner of a lamp;
— *de loquet,* nose of a latch;
— *de mâchoire,* edge, jaw, of a clamp or vise;
— *de pile, (pont.)* starling, cutwater;
— *de ponton, (pont.)* bow piece (esp. of Birago system).

bécane, f., popular term for steam engine; also, a "smart" name for the bicycle.

bêche, f., spade; *(art.)* trail-spade;
— *à articulation élastique, (art.)* elastic trail-spade;
— *de crosse, (art.)* trail-spade;
— *élastique, (art.)* elastic trail-spade;
— *d'essieu, (art.)* recoil-checking axle-spade;
pelle- —, (Fr.) infantry intrenching-tool.

bêcher, v. a., to dig, to spade.

bédane, m., mortising-chisel;
ciseau —, reenforced chisel.

bedon, m., metal-boring bit.

bégayer, v. n., *(hipp.)* to throw the head up and down.

bégu, m., *(hipp.)* horse that never razes or loses mark of mouth.

béguité, f., *(hipp.)* quality of never losing the mark;
fausse —, persistence of external dental cavity, when it should have disappeared.

beige, m., wool of natural color, undyed wool.

bélier, m., ram, monkey; battering-ram; *(nav.)* ram;
— *cuirassé, (nav.)* armored ram;
— *hydraulique, (mach.)* hydraulic ram;
— *lance-torpilles, (nav.)* torpedo-ram;
— *torpilleur, (nav.)* torpedo-ram.

bélière, f., *(unif.)* sword-belt sling; *(sm. a.)* chape of a scabbard; clapper ring;
— *antérieure,* short sling (of a sword belt);
grande —, long sling (of a sword belt);
petite —, v. — *antérieure;*
— *postérieure,* v. *grande* —.

belle, f., *(nav.)* waist of a ship;
tir en —, fire at hull, body, of a ship;
tirer en —, to fire at the body, the hull, of a ship.

belleface, a., *(hipp.)* having a white head, or one side of head white.

belligérant, m. and a., *(mil.)* belligerent.

belliqueux, a., warlike, martial.

bellite, f., *(expl.)* bellite.

bénard, m., heavy stone truck.

bénédiction, f., blessing, consecration;
— *de drapeaux, (mil.)* consecration of colors.

bengaline, f., *(expl.)* bengaline.

benne, f., cage, basket (as of an elevator, of a hoist); *(art.)* shot platform or cradle of a shell hoist;
— *d'extraction, (min., etc.)* miner's car.

benzine, f., benzine.

béquille, f., crutch; *(mach.)* link; *(art.)* prop or support (of a carriage), firing support (15ᶜᵐ field mortar, Canet);
— *de sûreté,* sort of prop used as a brake on slopes, 0.60ᵐ siege railway.

berceau, m., cradle; vaulted chamber, barrel vault; *(fond.)* grate furnace, to dry molds; *(art.)* cradle or saddle (for very heavy guns); *(r. f. art.)* cradle;
— *à tourillons, (r. f. art.)* trunnion cradle (e. g., Hotchkiss 65ᵐᵐ, etc.).

bercer, v. r., *(hipp.)* to rock from side to side (horse).

béret, m., *(unif.)* tam-o'-shanter worn by Alpine troops.

berge, f., border of a road, bank of a road; narrow skiff; *(top.)* bank of a river; slope along a valley, side or flank of a valley; steep shore.

berloque, f., *(mil. mus.)* drumbeat in which one stick makes twice as many beats as the other; signal for certain fatigues, and particularly for "dismissed"; trumpet call for the same.

berme, f., *(fort.)* berm.

berne, f., *pavillon en* ——, (*nav.*) at half-staff signal of distress or of mourning.
berry, m., (*slang, Polytechnic*) tunic.
bersaglier, m., (*mil.*) bersagliere, Italian light infantry soldier.
besace, f., wallet, small bag, sack.
besaiguë, f., double-ended carpenter's tool, one end being a mortising and the other a common chisel; twibill.
besicles, f. pl., spectacles.
besogne, f., task;
 belle ——, fine piece of work.
bétail, m., cattle;
 gros ——, horse, mule, ass, ox;
 menu ——, *petit* ——, sheep, goat, hog, etc.;
 —— *sur pied,* cattle on the hoof.
 n. têtes de ——, *n.* head of cattle.
bête, f., beast, animal; old hack, good-for nothing horse;
 —— *épaulée,* (*hipp.*) shoulder-shot horse;
 —— *de somme,* beast of burden, pack animal.
béton, m., beton, concrete;
 —— *aggloméré,* beton aggloméré;
 —— *de ciment,* beton made of cement;
 —— *gras,* beton rich in cement;
 —— *maigre,* beton poor in cement;
 —— *au quart,* mixture of 1 volume of cement to 4 volumes of siliceous pebbles;
 —— *au septième* (*huitième*), mixture of 1 volume of cement to 7 (8) of pebbles.
bétonnage, m., concrete work, concrete foundation; concrete (generic term for).
bétonner, v. a., to concrete, to cover with concrete or breton.
bétonnière, f., concrete mixer.
béveau, m., bevel-rule, miter-rule; bevel;
 en ——, beveled.
biais, m., slope, inclination, bevel; a., skew, oblique;
 de ——, *en* ——, obliquely, sloping, aslant, beveled.
biaisement, m., bevel, slope, slant.
biaiser, v. a., to bevel, slope, slant.
bibliothèque, f., library;
 —— *de garnison,* (*mil.*) post library;
 —— *de régiment,* —— *régimentaire,* (*mil.*) regimental library;
 —— *de troupe,* (*mil.*) soldiers' library.
biche, f., doe;
 pied-de- ——, claw-lever; crowbar; prypole of a gin; (*cord.*) double splice.
bichonner, v. a., —— *coco,* (*mil. slang*) to groom one's horse.
bickford, m., (*artif.*) Bickford fuse (frequently so called).
biconcave, a., biconcave, double concave.
biconique, a., biconic;
 frettage ——, (*art.*) biconic hooping.
biconvexe, a., biconvex, double convex.
bicoq, m., prypole of a gin.
bicoque, f., small town of no importance; (*fort.*) poorly fortified place or post.
bicorne, m., (*unif.*) cocked hat.
bicycle, m., bicycle (esp. though not necessarily, the ordinary);
 grand ——, the (old-fashioned) ordinary.
bicyclette, f., safety bicycle;
 —— *accouplée,* duplex bicycle;
 —— *de course,* racing machine;
 —— *de dames,* woman's bicycle;
 —— *de demi-route,* semiroadster;
 —— *pliante,* folding bicycle;
 —— *de route,* roadster, road machine;
 —— *suspendue,* spring bicycle;
 —— *tandem* tandem (bicycle).
bicycliste, m., bicyclist (*mil.* and *gen.*).
bidet, m., (*hipp.*) pony, nag; courier's pony.
bidon, m., can; (*mil.*) water-bottle, canteen; mess-tin; (*sm. a.*) skelp, bar of steel, from which a blade or barrel is made;
 grand ——, (*mil.*) camping vessel (5.5 liters);
 —— *à huile,* oil-can;
 petit ——, (*mil.*) soldier's canteen proper;
 petit —— *de cavalerie,* (*Fr. cav.*) trooper's canteen.
bief, m., mill-race, race; (*mach.*) water conduit of a hydraulic motor;
 —— *d'amont,* head-race;
 —— *d'aval,* tail-race;
 —— *de canal,* reach of a canal;

bief *de partage,* summit level pond.
bielle, f., (*mach.*, steam, etc.) connecting-rod; strut (in a truss); any arm, link, or rod, in a mechanism, particularly when the action is that of a crank; (*art.*) arm, rocking-arm, lever, of a disappearing carriage; (*sm. a.*) link (of the Borchardt pistol);
 —— *d'accouplement,* coupling rod;
 bride de la ——, connecting-rod strap;
 —— *de connexion,* connecting-rod;
 corps de ——, connecting-rod proper;
 corps de —— *à nervures et à ailettes,* channeled and ribbed connecting-rod;
 —— *directe,* direct-acting connecting-rod;
 —— *directrice,* driving-rod;
 grande ——, main connecting-rod;
 —— *motrice,* (*r. r.*) driving-rod;
 —— *de parallélogramme,* (*steam*) parallel-rod, motion rod;
 —— *pendante du grand piston,* (*steam*) side-rod;
 —— *de la pompe à air,* pump side-rod;
 —— *de pointage,* (*art.*) elevation-arm;
 —— *renversée,* —— *en retour,* back-action connecting-rod;
 tête de ——, connecting-rod end or head; (v. s. v. *tête*);
 —— *du tiroir,* valve-rod;
 —— *de traction,* (*art.*) pole of the *cabestan de carrier,* q. v.
bien, m., good;
 —— *du service,* (*mil.*) good of the service.
bien-fonds, m., real estate, landed property.
bienvenue, f., (*mil.*), (lit., welcome) sort of scot or tax paid by recruit to men of his squad room on first joining.
bière, f., beer; (funeral) bier.
biez, m., v. *bief*.
bifurcation, f., branching, fork (of a road); (*r. r.*) branching.
bifurquer, v. r., to fork, branch (of roads, *r. r.*).
bigornage, m., (*met.*) working or forming on a *bigorne,* on a swinging mandrel (latter *esp.* of gun construction).
bigorne, f., beaked anvil;
 à ——, beaked.
bigorner, v. a., (*met.*) to work, to shape, on a *bigorne,* on a swinging mandrel.
bigue, f., derrick; sheers;
 grandes ——*s,* hoisting sheers;
 —— *de harnais,* awning-framing of a hospital boat;
 —— *à osillation,* oscillating sheers;
 ponton- ——, floating sheers.
billarder, v. n., (*hipp.*) to dish, to paddle.
bille, f., log of wood, for further sawing; (*r. r.*) sleeper, cross-tie; (*mach.*) ball-bearing;
 ——*s de roulement,* ball, spherical bearings.
billebaude, f., confusion, disorder;
 feu à la ——, (*mil.*) independent firing, fire at will (obs.).
billet, m., ticket, railway ticket; note; (*mil.*) billet;
 —— *d'aller et retour,* (*r. r.*) round-trip ticket, return ticket;
 —— *d'appel,* (*mil.*) roll-call list, showing absentees and passmen;
 —— *d'arrêts,* (*mil.*) sealed note received by an officer in arrest, informing him of the reasons for his arrest and of its duration;
 —— *de banque,* bank note;
 —— *collectif,* (*Fr. adm.*) general ticket issued by station master for *bons de chemin de fer,* q. v., and retained by c. o. of detachment under way;
 —— *direct,* (*r. r.*) through ticket;
 —— *d'honneur,* challenge (rare);
 —— *d'hôpital,* (*mil.*) ticket turned over by officer or soldier on presenting himself at a hospital;
 —— *de logement,* (*mil.*) billeting ticket or order, billet;
 —— *à ordre,* check; bill, promissory note;
 —— *de salle,* (*mil.*) order for admission to hospital;
 —— *de santé,* bill of health;
 —— *de service,* (*mil.*) detail (as for guard); cocked hat;
 —— *simple,* (*r. r.*) single ticket;

billet *de sortie*, (*mil.*) order of discharge from hospital.
billot, m., block of wood; short roller; toggle (as of a *longe*, etc.); (*pont.*) rack stick; (*artif.*) fuse-block;
—— *d'appui*, (*art.*) supporting standard or blocks of Canet 15cm mortar;
—— *de brêlage*, (*pont.*) rack stick;
—— *à charger*, (*artif.*) fuse-block;
—— *d'enclume*, anvil-block;
—— *à enrayer*, wheel-drag;
—— *d'épreuve*, (*sm. a.*) proof-block (for sword blades);
—— *de guindage*, (*pont.*) rack stick.
binaire, a., binary; m., (*powd.*) mixture of two ingredients of gunpowder;
—— *salpêtre*, (*powd.*) charcoal and niter;
—— *soufre*, (*powd.*) charcoal and sulphur.
binocle, f., binocle, double eyeglass.
binoculaire, a., binocular;
télescope ——, binocular telescope.
binômes, m., pl., (*slang, Polytechnic*) pair of chums who work together.
bipède, a., two-footed; m., (*man.*) any two feet or legs of the horse;
—— *antérieur*, fore feet;
—— *diagonal*, diagonally opposed feet or legs;
—— *diagonal droit*, off fore and near hind foot;
—— *diagonal gauche*, near fore and off hind foot;
—— *latéral*, feet on same side;
—— *latéral droit* (*gauche*), feet on off (near) side;
—— *postérieur*, hind feet.
bipolaire, a., (*elec.*) bipolar.
biribi, m., (*mil. slang*) term of service in punishment companies.
bisaiguë, f., v. *besaiguë*.
biscaïen, m., (*art.*) grapeshot (obs.); (*sm. a*) long-barreled musket and its bullet.
biscuit, m., biscuit;
—— *de fourrage*, compressed or concentrated forage for horses;
—— *de mer*, sea biscuit.
biscuité, a., *pain* ——, ordinary bread cooked longer than usual, 1 hour and 40 minutes.
bise, f., (*in gen.*) north wind; (more esp.) N.NE. wind.
biseau, m., slope, bevel, chamfer; (*sm. a.*) back chamfer of a sword blade, back edge near point; double-edged part of a bayonet blade;
en ——, beveled, chamfered.
biseauter, v. a., to bevel, to chamfer.
biset, m., sort of carrier pigeon; rock-pigeon; (*fam.*) national guardsman, on duty in plain clothes;
en ——, in plain clothes.
bismuth, m., bismuth.
bisquain, m., sheepskin with the wool on.
bissac, m., (*mil.*) grain bag, (*art.*, and *cav.* serving in Africa, officers); (*inf.*) ammunition bag for supply of line from company wagons.
bistouri, m., bistoury.
bistournage, m., (*hipp.*) castration by torsion.
bistourné, a., twisted, crooked.
cheval ——, (*hipp.*) horse castrated by torsion.
bistourner, v. a., (*hipp.*) to castrate by torsion.
bistre, a., brown.
bitord, m., (*cord.*) spun yarn, marline;
—— *blanc*, untarred spun yarn;
—— *goudronné*, tarred spun yarn.
bitte, f., bitts.
bitter, v. a., (*cord.*) to bitt a cable.
bitton, m., kevel-head, mooring-post.
biture, f., (*cord.*) range of a cable.
bitume, m., bitumen, mineral pitch;
—— *de Judée*, compact bitumen, Judæa bitumen;
—— *solide*, mineral pitch.
bituminer, v. a., to cover with bitumen.
bitumineux, a., bituminous.
bivac, m., v. *bivouac*.
bivaquer, v. n., v. *bivouaquer*.
bivoie, f., fork of a road.

bivouac, m., (*mil.*) bivouac;
abri de ——, v. s. v. *abri;*
anneau de ——, v. s. v. *anneau.*
bivouaquer, v. n., (*mil.*) to bivouac.
blaireau, m., badger; shaving-brush; (*mil. slang*) recruit.
blanc, a., white; (*cord.*) untarred, white;
arme —— *he*, (*mil.*, etc.) any offensive weapon except firearms;
le bâton —— *à la main*, (*mil.*) without arms or baggage;
coupe —— *he, coupe à* —— *estoc, coupe à* —— *être*, complete clearing of woodland, leaving neither trees nor bushes;
eau —— *he*, (*hipp.*) water mixed with bran;
en ——, unpainted; in blank (of documents); (*cord.*) untarred;
épée —— *he*, bare sword;
fer ——, (*met.*) tin;
—— *-seing*, —— *signé*, blank signature, signature in blank;
m., white color; blank (in a form or document); chalk, whiting; (*met.*) white heat; (*t. p.*) bull's eye;
—— *d'argent*, commercial term for the best white lead;
—— *argenté*, (*hipp.*) white giving a pale silvery reflection;
—— *de baleine*, spermaceti;
boire dans son ——, (*hipp.*) said of a horse when the tip of his nose is white down to his mouth;
but en ——, (*ball.*) point-blank;
cartouche à ——, (*sm. a.*) blank cartridge;
—— *de céruse*, white lead;
—— *de chaux*, whitewash;
donner un —— *de chaux*, to whitewash;
donner dans le ——, (*t. p.*) to hit the bull's eye, to make a hit;
—— *d'eau*, (*top.*) pool of water in a swamp; (*fort.*) shallow sheet of water, due to inundation, and used as an accessory defense;
—— *d'Espagne*, (*mil.*) pipeclay;
astiquer au —— *d'Espagne*, (*mil.*) to pipeclay;
—— *de lune*, (*met.*) white heat, welding heat;
—— *mat*, (*hipp.*) white giving a milky reflection;
—— *de plomb*, white lead;
—— *porcelaine*, (*hipp.*) white with the bluish tint of china;
porter au ——, (*met.*) to raise to a white heat;
—— *rosé*, (*hipp.*) white that is pinkish in spots (due to absence of cutaneous pigment);
—— *sale*, (*hipp.*) yellowish white;
—— *soudant*, —— *suant*, (*met.*) white heat, welding heat;
tir à ——, (*art., sm. a.*) fire with blank cartridges;
tirer à ——, (*art., sm. a.*) to fire blank cartridges;
—— *de Troyes*, Spanish white;
—— *de zinc*, oxide of zinc.
blanchet, m., strainer, filter; (*harn.*) reenforce of leather sewed on a second piece of leather to strengthen the latter.
blanchiment, m., blanching; whitening; washing; whitewashing.
blanchir, v. a., to whiten; clean, wash; to dress trim down (as woodwork); to plane, polish, smoothen; to file down (as a piece of rough iron); (*met.*) to pickle (as copper); to refine (as gray pig);
—— *à la chaux*, to whitewash;
—— *le pied*, (*farr.*) to remove charring left after trying the hot shoe.
blanchissage, m., washing (clothes); whitewashing (walls, etc.).
blasé, a., (*hipp.*) indifferent (as a horse to the touch).
blaude, f., frock, groom's frock (provincial).
blé, m., wheat;
grands ——*s*, wheat and rye;
petits ——*s*, oats, barley;
—— *de Turquie*, maize, Indian corn.
bleime, f., (*hipp.*) contusion of the sole (heels and quarters), bleyme;
—— *foulée*, mere bruise of heels or of quarters;

bleime sèche, deep-seated bruise of heels or of quarters;
— *suppurée*, running sore in heels or quarters, leading to separation of parts.
bleimeux, a., (*hipp.*) suffering from bleyme, from contusion of heels or of quarters.
blénomètre, m., spring gauge or tester.
blessé, a., m., wounded; man who has been wounded.
blesser, v. a., to wound;
— *à mort*, to wound mortally;
— *au vif*, to wound to the quick.
blessure, f., wound;
— *mortelle*, fatal wound.
bleu, a., blue; m., blue color; (*mil. slang*) young soldier, recruit (until he has passed the school of the platoon); republican soldier (1793, Vendée);
coup —, (*art., sm. a.*), miss.
bleuir, v. a., to blue (as, *sm. a.*, metallic parts).
blin, m., beetle, rammer; (*nav.*) boom-iron of a spar-torpedo boat.
blindage, m., (*fort.*) blindage, overhead cover (*in gen.*) any cover not a mere parapet and made of various materials, either singly or in combination with one another, blindage; (*fort., nav.*) armor, armor plating;
boulon de —, armor bolt;
— *de ceinture*, (*nav.*) belt armor;
— *de muraille*, (*nav.*) side armor, side plating;
plaque de —, armor plate;
— *de pont*, (*nav.*) deck plating, deck armor.
blinde, f., (*siege, fort.*) piece of wood for a *blindage*; blind; (*siege*) wooden frame to support planks, fascines, in *descente de fossé blindée*, q. v.; (*siege*) blindage frame.
blindé, a., (*fort., nav.*) blinded, armor plated;
abri —, v. s. v. *abri*.
blinder, v. a., (*fort., nav.*) to armor, to plate; to protect, cover.
bliner, v. a., to ram (as, paving stones).
bloc, m., block, lump; (*sm. a., art., r. f. art.*) breech-block; (*mil. slang*) cell;
à —, "home," tight, pushed home (as a cartridge); (*cord.*) two-block, chock-a-block;
— *d'appui*, (*r. f. art.*) resistance block;
— *à bascule*, (*sm. a.*) oscillating block;
— *de calage*, (*art.*) guide-block, of a dynamite-gun cartridge;
— *de compression*, (*mil. min.*) besieger's mine containing a very large quantity of powder (2,000 to 3,000 kilograms);
— *de culasse*, — *de fermeture*, (*art., sm. a.*) breechblock;
— *à fouetter les armes*, (*sm. a.*) testing block for sword blades;
— *de pression*, (*mach.*) pressure block;
— *à rotation*, (*sm. a.*) rotating block;
serrer à —, to screw, set, turn, etc., home.
blocage, m., (*mach.*) clamping, locking; (*mas.*) pebble work, rubble work, rubble.
blocaille, f., (*mas.*) rubble, pebble work.
blochet, m., (*carp.*) small tie-beam; (*art.*) tie-piece, for sleepers of a platform.
blockaus, m., (*mil. slang.*) shako.
blockhaus, m., (*fort.*) blockhouse; (German name for) small caponier; (*nav.*) conning tower;
— *à étage*, two-storied blockhouse.
block-notes, m., memorandum block, pad.
block-système, m., (*r. r.*) block system.
blocus, m., (*siege, nav.*) investment, blockade;
escadre de —, (*nav.*) blockading squadron;
faire le — *de*, to blockade, invest;
forcer le —, (*nav.*) to run, force, the blockade;
forceur de —, (*nav.*) blockade runner;
lever le —, to raise the blockade;
— *pacifique*, pacific blockade, (e. g., by way of coercion);
relever le —, to relieve the blockade;
rupture de —, breaking, breaking up, of a blockade.
bloom, m., (*met.*) the *loupe* after passage under shingling hammer.

bloquable, a., (*mach.*) admitting of locking, clamping.
bloquer, v. a., (*siege, nav.*) to blockade, invest, hem in; (*mach.*) to lock, clamp; (*mil. slang*) to imprison, confine.
blouse, f., frock, smock.
bluette, f., spark; flake or scale of hot iron.
blutage, m., bolting, sifting.
bluteau, m., sieve, bolter; bolting machine or mill.
bluter, v. a., to bolt, to sift.
blutoir, m., v. *bluteau*.
bobine, f., spool, drum, bobbin; coil of wire; drum of wire; (*art.*) slide of a lanyard;
— *de commettage*. (*cord.*) winding-on spool;
— *de dérivation*, (*art.*) shunt-coil;
— *de déroulement*, reel;
— *d'induction*, — *inductrice*, (*elec.*) induction coil, primary coil;
— *induite*, (*elec.*) secondary coil;
— *primaire*, (*elec.*) primary, induction, coil;
— *à réaction*, (*elec.*) choking-coil, kicking-coil, reaction coil, impedance coil;
— *de résistance*, (*elec.*) resistance coil;
— *de Ruhmkorff*, (*elec.*) Ruhmkorff coil;
— *secondaire*, (*elec.*) secondary coil.
bocage, m., grove, clump of trees, coppice; (*met.*) scrap (of iron, steel, etc.).
bocal, m., wide-mouthed jar; (*mus.*) mouth-piece of a trumpet.
bocard, m., (*met.*) ore crusher, stamp; stamping mill.
bocardage, m., (*met.*) ore-crushing, stamping.
bocarder, v. a., (*met.*) to crush ore.
bogue, f., helve-ring (of a forge hammer).
boire, v. a., to drink; to absorb;
— *l'avoine*, (*hipp.*) to swallow oats without mastication;
— *la bride*, (*hipp.*) to take the bit between the teeth;
— *dans son blanc*, (*hipp.*) said of a horse with a white muzzle.
bois, m., wood, timber; (*top.*) wood (forest);
— *d'acajou*, mahogany;
— *d'aubier*, sapwood;
— *blanc*, softwood, generic term for nonresinous softwood;
— *de bourdaine*, black alder (for gunpowder);
— *de brin*, rough-hewn logs (hewn square); small timber;
— *carré*, — *au carré*, squared timber;
— *chablis*, wind-fallen wood;
— *de charpente*, construction timber, building timber;
— *de charronnage*, cartwright's wood (oak, etc.);
— *de chauffage*, firewood;
— *cintré*, bent wood;
— *de construction*, v. — *de charpente*;
— *qui se couronne*, tree that begins to die from age;
— *débité*, dimension timber; timber cut into scantlings;
— *déjeté*, warped timber;
— *desséché*, seasoned timber;
— *doux*, softwood;
— *droit*, straight-grained wood;
— *dur*, hardwood;
— *éclaté*, shaky wood;
— *équarri*, — *d'équarrissage*, squared timber;
faire du —, to get wood, to go for wood;
— *de fascinages*, (*fort.*) fascine stuff;
— *de fer*, ironwood;
— *feuillu*, wood of deciduous trees;
fil de —, grain;
— *de fil*, wood split with the grain;
— *fin*, close-grained, fine wood (for cabinet-maker's use);
— *flâche*, timber that can not be economically squared;
— *de fusil*, (*sm. a.*) stock of a rifle; (also, more specially) forestock;
— *de gaïac*, boxwood, lignum vitæ;
— *gélif*, frost-cracked wood;
— *gras*, soft wood;

bois *en grume*, timber with or without bark, obtained by a first operation of felling, trimming, and cutting into lengths (generally understood as timber with the bark on);
—— *de haute futaie*, forest timber; wood of lofty trees;
—— *de lance*, (*sm. a.*) lance-pole, lance-shaft;
—— *léger*, light wood, wood suitable for gunpowder charcoal;
—— *liant*, tough wood, whose fibers resist separation;
—— *de lit*, bedstead;
—— *de mine*, (*mil. min.*) uprights, sills, and caps;
mort- ——, wood of little value;
—— *mort*, dead wood; brambles, underbrush;
—— *mort sur pied*, tree dead standing;
—— *d'ouvrage*, —— *d'œuvre*, timber;
—— *de palissandre*, violet ebony;
—— *parfait*, heart wood, mature wood, as distinguished from sap;
petits ——, (*cons.*) window-pane framing;
petits —— *en fer*, (*cons.*) iron framing for window panes;
—— *de refend*, split timber;
—— *sur pied*, standing timber;
—— *à polir*, polishing stick;
—— *rebours*, —— *à rebours*, cross-grained wood;
—— *de recette*, timber fit for service;
—— *résineux*, resinous wood;
—— *sur le retour*, tree decaying from age;
—— *de rose*, rosewood;
—— *de selle*, (*harn.*) saddletree;
—— *de service*, timber, lumber;
—— *de taille*, —— *taillis*, underwood, wood of underbrush;
—— *tendre*, soft wood;
—— *tordu*, —— *tors*, —— *tortillard*, twisted timber, wood;
—— *de travail*, lumber;
—— *vert*, —— *vif*, unseasoned, green timber;
—— *à vives arêtes*, full-edged timber.

boisage, m., timber-casing.

boisé, a., (*top.*) wooded; (*cons.*) wainscoted.

boiser, v. a., to line or case with wood; (*cons.*) to wainscot.

boiserie, f., (*cons.*) wainscoting; boarding; joining-work.

boisseau, m., shell of a faucet; (*popular slang*) shako.

boitard, m., (*mach.*) upper journal or bearing of a vertical shaft.

boîte, f., box; case; chest; (*mach.*) bearing (*mil. slang*) prisoner's room in a guardhouse.
I. Artillery, small arms, military mining, etc.; II. Machinery, miscellaneous.

I. Artillery, etc.
—— *d'alimentation*, (*r. f. art.*) feed-box;
—— *d'amorce*, (*mil. min., torp.*) primer-box, priming tube, casing of a primer or fuse;
—— *d'auget*, (*sm. a.*) carrier box or casing;
avoir la ——, (*mil.*) to be foot-sore from marching;
—— *à balles*, (*art.*) case-shot; sort of canister for French s. b. mortars;
—— *à grosses balles*, (*art.*) canister containing large bullets;
—— *à petites balles*, (*art.*) canister containing small bullets;
—— *de boules*, (*mil. min.*) box-trap;
—— *à boulets*, (*art.*) sort of canister used in French s. b. mortars;
—— *à caffûts*, v. —— *à boulets*;
—— *-chargeur*, (*sm. a.*) magazine packet; loader, loading-packet, loading-case; quick loader;
—— *de la cheville-ouvrière*, (*art.*) pintle-socket;
—— *de culasse*, (*sm. a.*) breech-casing, receiver; (*r. f. art.*) breech piece, Hotchkiss revolving cannon;
—— *de distribution*, (*mil. min.*) sort of switch board, or connecting box (for electric wires);
—— *éclairante*, (*artif.*) box containing illuminating composition;
—— *d'écouvillon*, (*art.*) sponge head on which sheepskin or bristles are fastened;
—— *à étoupilles*, (*art.*) friction primer box;
—— *à feu*, (*mil. min.*) box-trap;
—— *à fumée*, (*mil. min.*) box of asphyxiating composition for subterranean warfare;
—— *à gargousses*, (*art.*) pass-box;
—— *à graisse*, (*sm. a.*) grease-box for care of arms;
—— *de hausse*, (*art.*) rear-sight seat, rear-sight box or socket;
—— *d'inflammateur*, (*torp., etc.*) igniter-, primer-box;
—— *d'inflammation*, (*mil. min.*) igniting box (for producing simultaneous explosions);
—— *de levier*, (*art., etc.*) handspike-, lever-socket;
—— *à livrets*, (*mil. min.*) case for *livrets*, q. v.;
—— *à marques*, (*mil.*) box for marking implements;
—— *à marrons*, (*mil.*) box in which the officer making rounds deposits his *marron*, q. v.;
—— *de mire*, (*art.*) sight box;
—— *à mitraille*, (*art.*) case-shot, canister (used in preference to —— *à balles*);
—— *de mitraille*, (*art.*) the case or casing of canister or case-shot;
—— *moyenne*, (*art.*) canister containing bullets of same size as those of —— *à grosses balles*, but fewer of them;
—— *de pilon*, (*powd.*) shoe (of a stamp);
—— *à plaques d'identité*, (*mil.*) box to hold *plaques d'identité*, q. v.;
—— *de pointage*, (*art.*) elevator-pinion box; elevating-screw frame, box; elevating-gear frame;
—— *à poudre*, (*mil. min.*) powder box (used in relatively weak mines of small charge);
—— *à pulvérin*, (*art.*) dredging-box (obs.);
—— *quintuple*, casing of the turret-reversing gear (five valves) on one or two French ships (e. g., *Tonnant*);
—— *à ressort*, (*r. f. art.*) spring (recoil) cylinder; spring buffer;
—— *de ronde*, v. —— *à marrons;*
—— *de secours*, (*mil. min.*) "first-aid" box for asphyxiated miners; life-saving box;
—— *à tampons*, —— *de tampon*, (*art.*) recoil-buffer casing; buffer box;
—— *à tiroir*, (*mil. min.*) trap, box-trap;
—— *à tourillons*, (*art., etc.*) journal-box;

II. Machinery, miscellaneous:
—— *de buttoir*, buffer-box;
—— *de câble*, sleeve or box (for fastening the end of a wire cable to rigid metallic objects);
—— *de choc*, buffer-box;
—— *à clapet*, (*steam*) valve-box, valve casing;
—— *à compas*, compass-box;
—— *cornée*, (*hipp.*) hoof (proper, horny part);
—— *crânienne*, (*hipp.*) brain pan;
dessous de ——, (*r. r., etc.*) lower axle-box;
dessus de ——, (*r. r., etc.*) upper axle-box;
—— *de distribution*, (*steam*) valve-box, steam-chest;
—— *étouffante*, (*steam, etc.*) stuffing-box (rare);
—— *à étoupe(s)*, (*steam, etc.*) stuffing-box;
—— *étrière*, (*cons.*) chair, stirrup;
—— *à ferrer*, (*farr.*) farrier's tool-box;
—— *à feu*, (*steam*) fire-box;
—— *de fonte*, coak (of a block);
—— *à foret*, (*mach.*) drill-stock;
—— *à fumée*, smoke-box (furnaces, etc.);
—— *à garniture*, (*steam*) stuffing-box;
—— *à graisse*, (*mach.*) lubricator; axle-box; grease-cup;
grande ——, —— *de gros bout*, body and box of the nave;
—— *à huile*, (*mach.*) lubricator;
—— *à incendie*, fire-alarm box;
—— *aux lettres*, letter-box;
—— *de levier*, lever socket;
—— *de manœuvre*, (*r. r.*) switch-box;
—— *de mathématiques*, box of mathematical (drawing) instruments;
—— *de mitre*, (*carp.*) mitre-box;

boîte à *noyau*, (*fond.*) core-box;
— à *oreilles*, axletree box, with lugs to prevent turning in nave;
petite ——, —— *de petit bout*, end-box of the nave;
—— *de régulateur*, (*steam*) governor sleeve;
—— *de résistance*, (*elec.*) resistance box;
—— à *ressorts*, buffer-box;
—— *de retenue*, (*mach.*) check-valve box;
—— *de roue*, nave-box; coak;
—— à *soupape*, (*steam, etc.*) valve-chest;
——*-support*, (*harn.*) shaft-loop;
—— à *tiroir*, (*steam*) slide-valve chest;
—— à *vapeur*, (*steam*) valve-box.

boitement, m., limping, halting; (*mach.*) irregular action.

boiter, v. n., to limp, to be lame, to hobble;
—— *de* . . . . , (*hipp.*, etc.) to be lame in . . . . ;
—— *tout bas*, (*hipp.*) to be quite lame.

boiterie, f., lameness;
—— *ancienne*, (*hipp.*) chronic, deep-seated lameness;
—— à *chaud*, lameness that reappears after exercise;
—— à *froid*, lameness that reappears after rest.

boiteux, a., lame, halt.

bombarde, f., (*nav.*) bomb vessel.

bombardement, m., (*art.*, *siege*) bombardment, shelling;
batterie de ——, battery to deliver high-angle fire, shelling battery;
canon de ——, (*art., etc.*) shelling gun.

bombarder, v. a., (*art.*, *siege*) to bombard; to shell.

bombardier, m., (*mil.*) bombardier (obs.).

bombe, f., (*art.*) (spherical) mortar-shell; bomb; shell (of a helmet);
—— *flamboyante*, (*artif.*) incendiary shell, lightball.

bombé, a., arched, curved, bulged; rounded, crowning (of a pulley); (*tech.*) snap-riveted;
chemin ——, barreled road;
poulie ——*e*, crowning pulley.

bombement, m., bulge, bulging; rounding, crowning, of a pulley.

bomber, v. a., n., to cause to bulge; to bulge, swell out; to barrel (a road).

bon, m., check, order for money; (*mil. adm.*) order or requisition receipt for supplies or for service, forming at one and the same time a request, a receipt, and a voucher;
—— *de chauffage*, fuel order;
—— *de chemin de fer*, order for transportation by rail;
—— *de convoi*, order for transportation of troops, etc., by land roads, by wagon (*par convois de terre*, q. v.);
—— *de denrées*, requisition for supplies (rations);
—— *d'effets et d'armes*, requisition for clothes, etc., and arms;
—— *de fourrage*, forage order or requisition;
—— *mensuel*, order of requisition presented by a captain on monthly drawing of effects chargeable to the *masse d'entretien et d'habillement*, q. v.;
—— *partiel*, order presented by an *isolé*, q. v.;
—— *sur la poste*, post-office money order;
—— *de réapprovisionnement*, voucher for stores drawn from *convois administratifs*, during fall maneuvers;
—— *de subsistances*, ration return;
—— *de tabac*, order on canteen for tobacco;
—— *total*, quarterly summary of ——*s particls*;
—— *de vivres*, ration return.

bonbonne, f., (*chem., etc.*) vat; (*in gen.*) sort of demijohn.

bond, m., bound; (*inf.*) rush (used also, but rarely, of artillery); (*art.*) ricochet, graze, of a projectile; in ranging, the passage ("jump") from one elevation to the next one used or tried (evaluated in numbers of turns of elevating-screw);
faire faux ——, (*man.*) of a horse, to start aside;
premier ——, (*art.*) first graze;
——*s successifs*, (*inf.*) successive rushes;

bond(s) *successifs de durée*, (*art.*) in ranging, the passage from one observed duration or interval of burst (time-shell) to the next, and so on;
—— *de 1, 2, ... n tours*, (*art.*) in ranging, the difference of elevation due to 1, 2, ... n turns of the elevating screw.

bonde, f., bunghole; bung.

bon-dieu, m., (*mil. slang*) sword.

bondon, m., bung, plug.

bondonner, v. a., to bung; to tap, broach.

bonhomme, m., bolt, bolt-catch; pin, trip-pin;
—— à *ressort*, spring-bolt;
—— *de sûreté*, (*art.*) safety-bolt.

boni, m., sum left over after a financial operation;
—— *de l'ordinaire*, (*mil. adm.*) excess of receipts over expenses in the mess.

bonnet, m., cap, hat, bonnet;
—— à *cornes*, (*unif.*) cocked hat;
—— *de fourrure*, (*unif.*) hussar busby;
—— *de fusée*, (*art.*) fuse-cap;
gros ——, (*mil. slang, etc.*) officer of high rank;
—— *d'ourson*, (*unif.*) bearskin;
—— à *poil*, (*unif.*) bearskin; busby (hussars');
—— *de police*, (*unif.*) forage-cap; (*mil. slang*) raw recruit;
—— *de prêtre*, (*fort.*) priest-cap, sort of fieldwork, shaped somewhat like a swallowtail.

bonnetage, m., (*artif.*) capping of fuses, rockets, etc.

bonnette, f., (*fort.*) bonnette; (*nav.*) studding-sail;
—— *lardée*, (*nav.*) collision-mat.

borax, m., borax.

bord, m., lip, edge, border; flange, rim; shore, bank; side of a boat or ship; side, edge, of a road;
à ——, (*nav.*) on board;
—— à ——, (*nav.*) yardarm and yardarm, broadside on;
à *contre* ——, (*nav.*) broadside on;
franc ——, (*nav.*) free-board;
de haut ——, (*nav.*) of high free-board;
à *pleins* ——*s*, (of water in rivers) up to the top of the banks;
—— *supérieur*, rim (of a well, shaft, etc.).

bordage, m., (*nav., etc.*) planking; (*pont.*) side planking (of a ponton boat);
—— à *clin*, clinker-work.

bordé, m., planking; sheathing behind backing, in test plates (armor); (*unif*) lace or trimming on a general officer's cocked hat;
—— à *clin*, (*nav.*) clinker-work;
—— *diagonal*, diagonal planking;
—— *intérieur*, —— *sous la cuirasse*, —— *en tôle*, skin plating, skin behind armor plating;
—— *du vent*, (*nav.*) weather side;
—— *de sous le vent*, (*nav.*) lee side.

bordé, a., planked, covered with planking; (*hipp.*) bordered (of horse's coat).

bordée, f., (*nav.*) broadside (both guns and fire); (*art.*) salvo (rare);
donner une ——, (*nav.*) to give a broadside;
feu de ——, (*nav.*) broadside;
lâcher une ——, v. *donner la* ——;
tirer une ——, (*mil. slang*) to escape, run off, away.

border, v. a., to edge, line, border; to plank, sheath;
—— *les avirons*, (*boats*) to ship the oars;
—— *la haie*, (*mil.*) to draw up troops along a road for a funeral or other ceremony;
—— *un parapet*, (*mil.*) to line, man, a parapet.

bordereau, m., (*adm.*) memorandum, register, of issues, payments, etc.; summary of documents, of receipts and expenditures, etc.; abstract (U. S. A.);
—— *d'émission*, daily statement of *mandats de paiement* issued;
—— *énonciatif*, abstract giving detailed account of all the vouchers covered by it;
—— *énumératif*, abstract giving merely class and name of vouchers covered by it;

bordereau *d'envoi*, return voucher, setting forth that goods, etc., sent have been received in good condition;
— *nominatif*, list of names, etc., of recruits forwarded to a regiment by recruiting officer;
— *de prêt*, abstract of payments (on *feuilles de prêt*) made by paymaster during the day (shows total expended).
bordure, f., edge, border;
— *du pavé*, curbstone.
bore, m., boron.
boréal, a., northern.
borgne, a., blind of one eye;
ancre —, single-fluked anchor;
trou —, hole that does not go quite through (as in armor plate);
vis —, screw that does not go through the piece (as in armor plate).
bornage, m., marking or setting off by means of boundary stones.
borne, f., limit boundary; boundary stone; curbstone, spur-post; (*in pl.*) geographical boundaries; (*art.*) support or pillar of (Engelhardt) field mortar carriage; (*elec.*) binding-post, bar of a switch board; (*teleg.*) post of a sending instrument;
— *d'attache*, (*elec.*) binding-post;
— *-fontaine*, water-plug;
— *-limite*, boundary stone;
— *milliaire*, milestone;
— *-presse*, (*elec.*) the connection or attachment to which (on a *tableau de connexion*, q. v.) the wires lead from the dynamo;
— *repère*, (*surv.*) bench mark;
— *signal*, (*surv.*) boundary signal or mark.
borner, v. a., to limit, bound, contract.
bornoyer, v. a., to stake off, to mark off by pickets, stakes, etc.; to verify an edge, a surface, etc., by looking at it with one eye, the other being closed.
bossage, m., (*mach.*) crank-pin boss; (*in gen.*) boss, swell.
bosse, f., boss, bump, swell; (*cord.*) stopper;
en —, in relief;
— *du saut*, (*hipp.*) vertex of the croup.
bosser, v. a., (*cord.*) to stopper, to seize.
bossette, f., (*harn.*) boss of a bit; bridle-stud; (*sm. a.*) boss or swell of a trigger piece (causing double pull off, when there are two).
botte, f., boot; bundle, truss (of hay, fodder, etc.); coil (of wire); (*sm. a.*) lance-, carbine-, etc., bucket; (*mil.*) popular term for horses' meal; (*fenc.*) thrust lunge;
— *arabe*, (*unif.*) boot worn by native officers of *spahis* and of *tirailleurs algériens*;
— *d'asperges*, (*fort.*) fascines for stuffing a gabion;
— *à* —, (*cav.*) boot to boot;
— *de carabine*, (*sm. a.*) carbine-bucket;
— *de cheval*, riding-boot;
demi- —, half-boot;
— *demi-forte*, (*unif.*) sort of jack-boot worn by the *garde républicaine*;
donner une —, (*fenc.*) to thrust, lunge;
— *à l'écuyère*, riding-boot;
— *d'étendard*, (*mil.*) color-bucket;
— *d'étrier*, (*sm. a.*) stirrup-bucket for lance;
fausses —*s*, (*unif.*) spatterdashes, worn by French cavalry;
— *-s fortes*, jack-boots;
— *de fourrages*, truss, bundle, of forage;
— *franche*, (*fenc.*) a "palpable hit;"
— *de fusil*, (*sm. a.*) carbine-bucket;
— *à la housarde*, (*unif.*) light cavalry boot;
— *de lance*, (*sm. a.*) lance-bucket;
— *de manège*, (*unif.*) boot worn by n. c. o. on duty at cavalry school;
— *molle*, (*unif.*) boot worn by French troopers of the *spahis*;
— *de mousqueton*, carbine-bucket;
petite —, (*unif.*) boot worn under trousers by the *gendarmes*;
— *porte-carabine*, v. — *de mousqueton*;
porter, pousser, une —, (*fenc.*) to make a thrust, to thrust;
serrer la —, (*man.*) to close the legs in on the horse.
botté, a., booted; (*hipp.*) balled.

bottelage, m., bundling of straw, hay, etc.
botteler, v. a., to bundle hay, etc.
botteleur, m., workman who puts hay, etc., into bundles.
botter, v. a. and r., to boot; (*hipp.*) to ball.
bottier, m., bootmaker; (*mil.*) regimental cobbler, bootmaker;
maître- —, (*Fr. a.*) service term for the corporals (shoemakers) in charge of the general shoe shops.
bottillon, m., half-boot; (*mil.*) truss of straw, used in loading saddles, carriages, etc., on a railway car;
— *à soufflet*, sort of half-boot (with a flap forming the front of the leg).
bottine, f., (*mil.*) short-legged boot (cavalry, mounted men of artillery, engineers, and train).
bouchage, m., stopping up, plugging; luting.
bouche, f., mouth, orifice, opening, entrance; mouth (of a river); (*sm. a.*, *art.*) muzzle (of a gun, rifle, etc.); (in combination, *art.*) muzzle-loading;
assurer la — *à un cheval*, (*man.*) to ride with a light hand, to lighten the reins;
avoir la — *faite*, (*hipp.*) to have all his teeth (at 5 years);
n'avoir ni — *ni éperon*, (*man.*) to obey neither bit nor spur;
— *chatouilleuse*, (*man.*) oversensitive mouth;
cheval qui a bonne —, (*hipp.*) horse that will eat anything;
— *à droite* (*gauche*), (*art.*) muzzle right (left);
— *dure*, (*man.*) hard mouth;
— *égarée*, (*man.*) mouth tender to the last degree;
— *à feu*, (*art.*) gun, piece of artillery;
— *à feu composée*, (*art.*) built-up gun;
— *à feu cuirassée*, (*art.*) armored gun, i. e., in armored mount;
exécution de la — *à feu*, (*art.*) service of the piece, standing gun drill;
— *à feu lisse*, (*art.*) smoothbore;
à feu rayée, (*art.*) rifled gun;
— *à feu de rechange*, (*art.*) spare gun; (more especially in France, *pl.*) reserve of guns forming part of the armament of a *place de guerre*;
fort en —, (*man.*) hard-mouthed;
— *fraîche*, (*hipp.*) mouth that fills with foam on insertion of bit;
— *-s inutiles*, (*siège*) inhabitants that consume food without adding to the defense;
maladie de —, (*hipp.*) carney;
munitions de —, (*mil.*) provisions, victuals;
sans —, (*man.*) hard-mouthed;
— *sensible*, — *tendre*, (*man.*) tender mouth;
— *-trou*, stop-gap.
boucher, m., butcher.
boucher, v. a., to stop, plug; to close up (as windows, doors, etc.); (*art.*) to bush; (*elec.*) to plug (in resistance box);
— *un cheval*, (*hipp.*) to examine a horse's mouth for age;
— *la lumière*, (*art.*) to serve the vent (obs.).
boucherie, f., butcher shop; (*fig.*) massacre.
bouchon, m., stopper, plug, cork; wisp; wad; tompion; familiar term for *brosse en chiendent*, q. v.; (*art.*) fuse-plug, screw cap of a fuse; fuse-seat of a projectile; (*sm. a.*) cap or plug (as of the Lebel magazine);
— *d'amorce*, (*torp.*, etc.) primer-plug, fuse-plug;
— *-caoutchouc*, (*torp.*, etc.) rubber plug;
— *de charge*, (*art.*) wad (obs.);
— *concutant*, (*art.*) plug containing the time-plunger in certain fuses (German, etc.);
— *détonant*, (*art.*) primer-plug (fuse);
— *disjoncteur*, (*elec.*) disconnecting key or plug;
— *écrou*, (*sm. a.*, etc.) screw-plug;
— *fileté*, (*art.*) threaded fuse-plug; screw plug;
— *fusible*, (*elec.*) fusible or safety plug;
— *de fusil*, (*sm. a.*) tompion (obs.);
— *d'œil*, (*inst.*) eye-cap (of a telescope);

bouchon *porte-amorce,* (*art.*) primer-plug (of a fuse);
——*de queue,* (*art.*) base plug or screw (of a fuse);
——*rugueux,* (*art.*) striker and plug in one (of a fuse);
——*de sûreté,* (*elec.*) fusible plug, safety plug;
——*de vidange,* (*mach.*, etc.) drip-cock;
——*à vis,* screw-plug.

bouchonner, v. a., to rub down a horse with a wisp, with straw.

boucle, f., buckle, clasp; eye, ring; (*sm. a.*) band; (*cord.*) half-hitch, kink, twist; (*top.*) turn, elbow, or bend of a river; (*in pl.*) irons, fetters, shackles;
——*d'amarrage,* (*art.*, etc.) mooring-ring, lashing-ring;
——*-crampon,* buckling staple;
——*enchapée,* buckle at the end of a strap, swinging buckle;
——*d'étoupille,* (*art.*) eye or loop of a friction primer;
——*de hart,* (*siege*) withy knot;
——*de manœuvre,* (*F. m. art.*) maneuvering loop; maneuvering chain eye;
——*ouverte,* open buckle or link;
——*porte-étrivière,* (*harn.*) stirrup-leather buckle;
——*de quai,* mooring-ring;
——*rênoir,* (*harn.*) rein-buckle on pommel of certain packs, saddles, etc.;
——*roulante,* ——*à rouleau,* roller-buckle;
——*de sangle,* (*harn.*) girth-buckle;
sous ——*s,* in irons.

bouclement, m., (*hipp.*) ringing (of a mare).

boucler, v. a. and n., to knot, to tie; to buckle, clasp; to lock up (prisoners); to bulge out (of a wall badly built);
——*une jument,* (*hipp.*) to ring a mare.

bouclerie, f., generic term for buckle and parts of buckles.

boucleteau, m., (*harn.*) strap with buckle at the end (used with *contre-sanglon,* q. v., to distinguish ends of same strap, *boucleteau* being the buckle end); tug;
——*antérieur,* loin-strap tug;
——*postérieur,* hip-strap tug.

boucletoir, m., (*harn.*) tug.

boucleton, m., (*harn.*) near-side neck-strap.

bouclier, m., shield; (*art.*) gun shield; (*fort.*) armor plate or shield, esp. plate through which embrasure of turret is pierced, embrasure plate;
——*cuirassé,* (*art.*) armored shield;
——*de sape,* (*siege*) sap shield;
——*en tôle,* (*art.*) gun shield (esp. mobile and r. f. pieces).

bouder, v. n., to pout;
——*contre les jambes,* (*hipp.*) to use or move the legs sluggishly or unwillingly;
——*sur son avoine,* (*hipp.*) to be off his feed (horse).

boudin, m., spiral or helical spring; wad; flange of a wheel (as *r. r.*); nosing (of a stair); (*r. r.*) buffer; (*artif.*) hose or tube for a *saucisson:* (*mil. min.*) synonym of *saucisson* (latter the better term); (*harn.*) the coil into which a forage or picketrope is rolled; (*cord.*) sort of coil (as, of a rope packed in a chest);
——*à ressort,* (*r. r.*) spring buffer.

boue, f., mud, mire.

bouée, f., buoy;
——*d'amarrage,* mooring-buoy;
——*de sauvetage,* life-buoy;
——*sonnante,* bell-buoy.

boueux, a., muddy, miry.

bouffée, f., puff, whiff;
——*de fumée,* puff of smoke;
——*de vent,* squall.

bouffette, f., (*harn.*) head knot, rosette, of a bridle.

bouge, f., bilge, swell (of a cask, ship); rounding (as of a deck);
——*de moyeu,* swell of the nave of a wheel·
——*de mur,* swell or bulge of a wall;
——*de planche,* warping of a plank.

bougie, f., candle;
——*décimale,* (*elec.*) practical unit of light.

bouillante, f., (*mil. slang*) soup.

bouilleur, m., (*steam*) boiler-tube; heater (of a boiler); one of the tubes of a tubular boiler.

bouillon, m., (*unif.*) epaulet-fringe.

boulange, f., unbolted flour.

boulanger, m., baker.

boulangerie, f., bakery, bakehouse, bake-oven;
——*de campagne,* (*mil.*) field bakery;
——*légère de campagne,* (*Fr. a.*) field bakery for areas inaccessible to regular field bakery.

boule, f., ball; time-ball;
——(*mil. min.*) box trap;
——*du régulateur,* (*steam.*) governor ball;
——*de son,* (*mil. slang*) loaf of bread.

bouleau, m., birch (tree and wood).

boulet, m., (*art.*) shot, solid shot (very little used, having given way to *obus,* in its various meanings); (*hipp.*) pastern-joint, fetlock-joint; (*mil.*) ball and chain (punishment);
——*barré,* bar shot (obs.);
——*cerclé,* (*hipp.*) pastern-joint surrounded by exostoses;
——*à chaîne,* (*art.*) chain shot (obs.);
condamner au ——, (*mil.*) to condemn to drag a ball and chain;
——*couronné,* (*hipp.*) scarred pastern joint, showing traces of falls, i. e., strained; crowned pastern joint;
——*creux à incendier,* (*art.*) incendiary shell;
——*enchaîné,* v. ——*à chaîne;*
——*incendiaire,* (*art.*) incendiary shell;
——*messager,* shell containing a message (obs.);
——*mort,* (*art.*) spent shot;
——*plein,* (*art.*) solid shot;
——*ramé,* (*art.*) bar shot (obs.);
——*rouge,* (*art.*) hot shot (obs.);
——*de rupture,* (*art.*) armor-piercing projectile;
——*torpille,* (*art.*) generic term for a torpedo projectile, i. e., a torpedo fired by an explosive; a (projected) shell to be stopped by ship's side and then to explode after dropping under the ship's bottom;
traîner le ——, (*mil.*) to drag a ball and chain (punishment).

bouleté, a., (*hipp.*) said of a horse whose fetlock joint is crowned, swelled, dislocated;

boulette, f., small ball;
——*d'amorce,* ——*fulminante,* (*artif.*) pellet primer, wax primer.

bouleture, f., (*hipp.*) straightening of, and throwing forward of, bony parts of fetlock joint, causing the horse to stand on his toe.

bouleux, m., (*hipp.*) short, thickset horse.

boulevard, m., bulwark, rampart (obs. as a military term, except fig.); avenue planted with trees.

boulimie, f., voracious, abnormal appetite; bulimy.

bouline, f., (*cord.*) bowline.

bouliner, v. a., (*cord.*) to haul on a bowline.

boulineur, m., camp-robber, -follower.

bouloir, m., larry, lime-rake.

boulois, m., (*mil. min.*) punk (obs.).

boulon, m., bolt;
——*d'assemblage,* through bolt, assembling-bolt; (*art.*) transom bolt;
——*d'attelage,* drag-bolt;
——*d'axe,* bolt serving as an axis of rotation;
——*de bande,* tire-bolt;
——*de blindage,* armor-bolt;
——*à bout percé,* eyebolt;
——*de carène avec écrou,* holding-down bolt and nut;
——*à charnière,* hinge-bolt;
——*de la charnière,* (*art.*) plug-tray hinge-bolt;
——*de choc,* (*r. f.*, *art.*) buffer-bolt;
——*à clavette,* eyebolt, collar-bolt, key bolt;
——*à coulisse,* sliding bolt;
——*à croc,* ——*à crochet,* hook-bolt;
——*de cuirassement,* armor-bolt;
——*échelon,* (*art.*) foot-rest bolt (of a gin);
——*à écrou,* bolt and nut;

boulon *d'embarrage*, (*art.*) maneuvering, embarring bolt;
— *d'entretoise*, (*art.*) transom bolt;
— *-entretoise*, (*art.*) bolt-transom, i. e., bolt that serves as a transom;
— *à ergot*, bolt and spur;
— *d'excentrique*, eccentric bolt;
— *de fermeture*, fastening bolt (of a hasp);
— *de fixation*, fixing, securing bolt;
— *de fondation à ancre*, (*cons.*) anchor bolt;
— *à gorge*, grooved bolt;
— *à goupille*, (*mach.*) pin, bolt;
— *de loquet de console*, (*art.*) latch bolt;
— *de manille*, shackle-bolt;
— *de manœuvre*, (*art.*) maneuvering bolt;
— *à mentonnet*, dog bolt;
— *d'obturateur*, (*F. m. art.*) obturator-bolt;
— *Palliser*, l'alliser (armor) bolt;
— *à patte*, bolt with holdfast attached to head;
— *à quatre pans*, square-headed bolt;
— *de rappel*, adjusting bolt, regulating bolt;
— *à ressort*, spring-bolt, latch;
— *à*, — *de, scellement*, cramping-bolt (iron or metal, to stone or concrete);
— *de sellette*, (*art.*) bolster-bolt, axletree bolt;
— *de serrage*, fixing-pin, set-bolt, packing-bolt;
— *à six, etc., pans*, hexagonal, etc., bolt;
— *taraudé*, threaded bolt, screw-bolt;
— *de tenue*, armor-bolt;
— *à tête bombée*, round-headed bolt;
— *à tête carrée*, square-headed bolt;
— *à tête de champignon*, round-head bolt;
— *à tête coudée*, bolt with elbow head;
— *de timon*, pole-pin, pole-bolt;
— *à tourniquet*, bolt and turnbuckle;
— *à virole*, clinch-bolt;
— *à vis*, screw-bolt.

boulonner, v. a., to bolt, bolt down, pin down.

boulonnet, m., small bolt.

bouquet, m., bunch, cluster;
— *d'arbres*, clump of trees;
— *d'artifices*, (*artif.*) girandole;
— *de paille*, wisp of straw.

bourbe, f., mud, mire.

bourbeux, a., muddy.

bourbier, m., (*top.*) slough, bog, bog wallow.

bourcette, f., horse-mash.

bourdaine, f., black alder (tree and wood).

bourdonnière, f., (*mach.*) step, step-bearing; collar.

bourgade, f., small market town; straggling village.

bourgeois, m., civilian;
en —, in plain clothes, in mufti; in "cits" (U. S. A.).

bourgeron, m., (*unif.*) fatigue coat or jumper of canvas, worn by corporals and men on certain fatigues and drills;
— *blouse*, long canvas blouse, worn by artillery, cavalry, and train;
— *de manœuvre*, working-jacket, jumper.

bourlet, m., v. *bourrelet*.

bournous, m., v. *burnous*.

bourrade, f., blow with butt of a gun; kick, of a gun.

bourrage, m., (*min.*) tamping, of a blast, mine, etc.;
longueur de —, distance from center of charge to outer end of tamping surface, measured in a straight line.

bourrasque, f., squall.

bourre, f., plug; (*harn.*) pad, padding; stuffing, wad, wadding; (*sm. a.*) wax disk (in ammunition); (*fond.*) horsehair, tow, etc., used in molding;
— *de canon*, (*art.*) wad, wadding;
— *de soie*, (*art.*) silk-waste (for cartridge bags);
— *de sûreté*, (*expl.*) safety tamping;
tire-—, (*art.*) worm.

bourreau, m., executioner.

bourrelé, a., flanged.

bourrelerie, f., business, trade, of harness maker.

bourrelet, m., rim; (*sm. a.*) rim forming bottom of bayonet socket; flange or rim of a cartridge case; (*art.*) swell at the muzzle of a gun; centering-band, centering-shoulder; (*fort.*) reinforce of metal whose embrasure of a turret-gun (Cane); (*top.*) fold of the ground, long mound; (*hip*.) coronary band, cutidure; (*r. r.*) flange of a wheel, head of a rail, (*harn.*) pad, cushion; (*in gen.*) any long pad or cushion;
— *-avant*, (*art.*) centering band or shoulder;
— *de centrage*, (*art.*) centering-shoulder;
— *-guide*, guide flange;
— *périoptique*, (*hipp.*) perioplic band or circle;
— *postérieur*, (*r. f. art., sm. a.*) cartridge-head flange or rim;
— *de terre*, (*fort.*) small parapet (as of a shelter trench);
— *en tulipe*, (*art.*) swell of the muzzle.

bourreletage, m., operation of putting on a *bourrelet* (esp., *art.*), on cartridge-cases).

bourrelier, m., harness maker; pack harness maker; (*Fr. art. and train*) workman that keeps harness in repair, saddler.

bourrer, v. a., to pack; to tamp; to stuff, pad, wad; to push or strike with the butt of the rifle. (*man.*) to bolt; (*min.*) to ram, tamp, a mine, charge or blast;
— *sur l'obstacle*, (*man.*) of a horse, to bolt at a hurdle or jump and try to tear the reins out of the rider's hands.

bourretage, m., act of putting in a *bourre*.

bourrique, f., ass; (*hipp.*) poor, miserable horse.

bourriquet, m., small ass; (*tech.*) horse trestle; (*min.*) box or basket for hoisting earth, etc., from bottom of a mine by a winch.

bourroir, m., (*min.*) tamping-rod.

bourse, f., purse; saddlebag; stock exchange; scholarship in a school (military or civil); (*med.*) suspensory bandage; (*hipp., etc.*) testicle-sack;
baril à —, (*art.*) budge-barrel;
demi- —, half-scholarship (schools).

boursier, m., pupil in enjoyment of a scholarship; gambler on the stock exchange.

boursoufflement, m., (*met.*) blister, bubble, air cell; (*chem.*) increase of volume by chemical action.

boursouffler, v. a., to swell; v. r., (*chem., etc.*) to increase in volume by chemical, etc., action.

boursoufflure, f., (*top.*) swell, rise, in the ground.

bousillage, m., loam and straw (adobe, almost); (*fig.*) poor piece of work.

bousiller, v. a., to wall, revet, with loam and straw; (*fig.*) to botch, to do work badly.

bousin, m., outer soft crust (of a stone, of a rock).

boussole, f., (*inst.*) compass; (*elec.*) galvanometer;
— *affolée*, defective, erroneous compass (v. *affolé*);
— *-alidade*, compass fitted with an alidade;
— *d'arpenteur*, surveyor's compass;
— *de batterie*, (*art.*) compass specially designed for use with field batteries (no longer carried);
— *-breloque*, very small compass, used in orienting maps;
— *-éclimètre*, *à éclimètre*, clinometer compass, leveling compass;
— *électrique*, (*elec.*) galvanometer;
— *à main*, hand compass;
— *marine*, (*nav.*) mariner's compass;
— *nivelante*, leveling compass;
— *de poche*, pocket compass;
— *portative*, pocket, hand, compass;
— *à prisme*, prismatic compass;
— *-rapporteur*, compass and protractor, for sketching;
— *à réflexion*, reflecting compass;
— *-roulette*, sort of universal instrument for field sketching;
— *des sinus*, (*elec.*) sine-galvanometer;
— *suspendue*, swinging compass;
— *des tangentes*, (*elec.*) tangent-galvanometer;

boussole *topographique*, surveyor's compass;
— *à viseur*, compass fitted with a sighting tube.
bout, m., end; extremity, tip, nib; butt;
— *à* —, end to end;
de — *en* —, from end to end;
— *pour* —, end for end;
— *d'affût*, (*art.*) trail-plate;
bon —, (*cord.*) hauling part of a warp;
le bon —, the lower layer of leather composing the heel of a shoe;
— *de cordage*, — *de corde*, (*cord.*) rope's end;
— *de crosse*, (*art.*) trail-plate;
— *de crosse-lunette*, (*art.*) trail-plate and lunette;
— *décommis*, (*cord.*) fag end of a rope;
— *d'essieu*, axletree end;
faire mordre le —, (*cord.*) to fasten an end down under a seizing;
— *fileté*, any threaded end (as, *sm. a.*, of a ramrod);
— *de fourreau*, (*sm. a.*) chape of a scabbard;
— *de garant*, (*cord.*) fall of a tackle;
gros —, butt end (of timber);
— *gros du moyeu*, hind end of nave;
— *de lance*, (*sm. a.*) shoe or socket of a lance;
— *libre*, (*cord.*) free end of a rope;
— *de madrier*, block, skid;
— *de masselotte*, (*fond.*) deadhead;
— *du nez*, (*hipp.*) muzzle;
— *petit du moyeu*, fore end of the nave;
— *de pied*, (*art.*) prypole head;
à — *portant*, (*ball.*) point-blank.
boutant, a., used only with *arc*.
boute, f., water-cask, tub.
bouté, a., (*hipp.*) straight-legged;
cheval —, horse whose legs are straight from knee to coronet.
boute-charge, m., (*Fr. a.*) trumpet call, signal for placing loads on horses or in wagons (train, convoys, and expeditions in Africa; used also by artillery and cavalry and sometimes by infantry in the field).
boutée, f., (*fort, pont., etc.*) abutment.
boute-en-train, m., (*hipp.*) stallion turned loose with mares.
boute-feu, m., (*art.*) linstock (obs.).
bouteille, f., bottle; (*mach., etc.*) any elongated vessel;
— *alimentaire*, (*steam*) feed apparatus used before invention of Giffard injector;
— *de Leyde*, (*elec.*) Leyden jar;
— *à poudre*, (*mil. min.*) elongated receptacle for charge;
— *en tôle*, (*mil. min.*) receptacle for charge, 0.3m in diameter.
bouterolle, f., bushing; snap, snap head of a rivet; riveting set; ward (of a lock, key); (*sm. a.*) escutcheon, bushing, recoil bushing (of a revolver); (*sm. a.*) fence; (*sm. a.*) chape of a scabbard;
— *de butée*, (*art.*) prypole bearing (of a pin);
— *de corps de platine*, (*sm. a.*) fence;
— *de détente*, (*sm. a.*) trigger bushing;
— *à rivet*, snap.
boute-selle, m., (*mil.*) boots and saddles (practically obs., Fr. a.)
boutisse, f., (*mas.*) header; (*fort.*) sod laid headerwise, head-sod.
boutoir, m., (*farr.*) butteris, buttrice;
— *à guide*, butteris used in Charlier system of shoeing.
bouton, m., button; stud, lug; knob, handle; (*r. f. art.*) crank-stud (Hotchkiss); (*sm. a.*) thumb-piece; threaded end of barrel; (*fenc.*) button of a foil;
— *d'amorce*, (*art.*) primer-plug (fuse);
— *d'arrêt*, (*cord.*) any knot limiting motion of a loop, slide, etc.;
— *arrêtoir*, (*sm. a.*) stop;
— *de bielle*, (*mach.*) crank-pin;
— *de bosse*, (*cord.*) knot or crown of a stopper;
— *de canon*, (*sm. a.*) breech-pin;
— *à clavette*, (*mach.*) collar, collar-pin;
— *de commutateur*, (*elec.*) switch-plug;
— *de contact*, (*elec.*) contact key or plug;

bouton *coulant*, (*harn., etc.*) sliding button or loop;
— *de culasse*, (*art.*) knob of the cascable; (*sm. a.*) breech-pin; (*r. f. art.*) adjusting-ball (Gatling);
— *du cylindre*, (*sm. a.*) lug joining cylinder and bolthead (French, model 1874);
— *-écrou*, (*sm. a.*) firing-pin nut; thumbpiece (nut to which firing-pin is attached in boltguns);
— *d'étai*, (*cord.*) mouse of a stay;
— *farcineux*, (*hipp.*) farcy-bud;
— *de fiche*, hinge pin;
— *fileté*, (*sm. a.*) threaded (breech) end of a rifle barrel;
— *de fourreau*, (*sm. a.*) bayonet-scabbard tip or button;
— *de manivelle*, (*mach.*) crank-pin;
— *de mire*, (*art.., sm. a.*) front sight;
— *mobile*, (*harn.*) running button;
— *molleté*, milled head;
— *en olive*, knob, handle;
— *-poussoir*, firing-button (Maxim machine gun);
— *de repère*, (*Fr. art.*) v. s. v. *repère;*
— *à ressort*, (*elec. etc.*) spring-button;
— *de sonnerie*, (*elec.*) bell-push, push-button;
— *de tir*, firing-button (Maxim machine gun);
— *de tournevire*, (*cord.*) Turk's head;
— *à vis*, (*r. f. art.*) screw stud, cascable, cascable button (Gatling, Gardner).
boutonner, v. a., to button; (*fenc.*) to touch.
boutonnière, f., buttonhole; (*harn.*) runningloop; . (*unif.*) distinctive mark worn on the collar by officers of the territorial army in France; good-conduct badge of military convicts in French army.
bouvet, m., (*carp.*) grooving-plane.
bouveter, v. a., (*carp.*) to tongue and groove, to groove.
box, m., loose box (stable).
boxe, f., boxing, art of boxing.
boxer, v. n., to box.
boyau, m., (*siege*) boyau, zigzag; connecting trench between parallels or places of arms; (*mil. min.*) small gallery, branch, of a mine; (*in gen.*) pipe, hose; (*hipp.*) belly, flank;
— *de communication*, — *de tranchée*, (*siege*) zigzag, trench between parallels, between places of arms.
brabant, m., iron strap or band;
— *à fourche*, forked holdfast;
— *à patte*, holdfast, brace, clip.
bracelet, m., bracelet, any ring or circular band (esp. of metal); (*sm. a.*) band of a sword-' of a bayonet-, scabbard; (*art.*) loop of a lanyard;
— *explorateur*, (*man.*) Marey's pneumatic bracelet, for registering gaits by compressed air;
— *-pontet*, (*sm. a.*) band and bow of the French infantry rifle (Lebel).
bracon, m., strut.
brague, f., (*nav. and s. c. art.*) breeching, breechtackle;
— *de canon*, breech-tackle;
croc de —, breech-tackle hook.
braguet, m., (*cord.*) heel-rope.
braguette, f., front opening of trousers; (*cord.*) heel-rope.
brai, m., tar, pitch; rosin;
— *gras*, mixture of tar, colophany, and pitch; tar;
— *sec.* pitch.
braie, f., (*fort.*) bray, advanced wall projecting from tower over main entrance of fortress (obs.);
fausse —, (*fort.*) false bray (obs.).
braise, f., (*wood*) coals.
braisine, f., (*fond.*) molding sand mixed with cow dung.
brancard, m., shaft (of a wagon); handbarrow; iron shoe of a drag; (*med.*) litter, stretcher (for wounded); (*art., etc.*) side rail (of a caisson, ammunition wagon, etc.);
— *de caisse*, wagon-body side bar (in wagons having — *de limonière*, q. v.);
cheval de —, off-wheeler;
— *de limonière*, side rail in prolongation of shafts of certain two-wheelers;

brancard *mobile*, shifting shaft.
brancardier, m.. litter-bearer, stretcher-bearer;
—— *d'ambulance*, (*Fr. a.*) litter-bearer belonging to field hospital (*ambulance*);
—— *d'artillerie*, (*Fr. a.*) litter-bearer furnished by musicians of the artillery school;
caporal ——, (*Fr. a.*) corporal in charge of litter-bearer, one per battalion;
—— *de frontière*, (*Fr. a.*) volunteer litter-bearer, to assist medical service and the French Red Cross;
—— *d'infanterie*, (*Fr. a.*) litter-bearer furnished by musicians and reservists.
—— *régimentaire*, regimental litter-bearer;
sergent ——, (*Fr. a.*) sergeant in charge of litter-bearers, one to each infantry regiment.
branche, f., branch; subdivision of a subject; any part of a thing that may be distinguished as a *branch*; branch, bough, of a tree; leg (of a tripod, of a compass); (*harn.*) cheek of a bit; (*farr.*) part of a horseshoe from *mamelle* to extremity or heel; (*fort.*) longest side of a crown work, **of a** horn work;
—— *d'amorce*, (*expl.*) fuse-wire; fuse lead;
—— *ascendante* (—— *descendante*) *de la trajectoire*, (*ball.*) ascending (descending) branch of the trajectory;
—— *d'avaloire*, (*harn.*) bearing strap;
avoir de la ——, (*hipp.*) to have a long neck and shoulders and good withers;
bas de la ——, (*harn.*) lower part of bit-cheek;
—— *d'en bas*, lower arm of a knee (boats);
—— *de coquille*, (*sm. a.*) bar of a sword-guard;
—— *coudée de baïonnette*, (*sm. a.*) bayonet shank;
—— *de crosse*, (*sm. a.*) rear end (toward butt) of trigger-guard plate;
—— *de devant*, (*sm. a.*) fore end of trigger-guard plate;
—— *d'échelle*, side, side rail, of a ladder;
—— *d'épée*, (*sm. a.*) basket-hilt;
—— *d'étrier*, (*harn.*) leg or upright of a stirrup;
—— *de fer*, (*farr.*) quarter of a horseshoe;
—— *de fortification*, (*fort.*) v. —— *d'ouvrage de fortification*;
—— *de garde d'épée*, (*sm. a.*) bow of a sword hilt;
grande ——, (*sm. a., r. f. art.*) long branch, main branch, of the hammer spring;
haut de la ——, (*harn.*) top part of bit-cheek;
—— *de manivelle*, (*art.*) elevating-screw arm;
—— *mobile*, v. *grande* ——;
—— *de mors*, (*harn.*) branch of a bit;
œil de la ——, (*harn.*) eye of a bit-cheek;
—— *d'ouvrage de fortification*, (*fort.*) crest or part of a work more or less normal to the front, (hence) face, etc.;
—— *de percussion*, (*sm. a.*) percussion spring (e. g., French revolver, model 1892);
—— *petite*, (*sm. a.*), *r. f. art.*) short branch of a hammer spring;
—— *de rebondissement*, (*sm. a.*) rebound spring (revolver);
—— *de rivière*, (*top.*) arm of a river;
—— *de sabre*, (*sm. a.*) bow of a sword hilt;
—— *de sous-garde*, (*sm. a.*) trigger-guard plate;
—— *supérieure d'une courbe*, standard or upper part of a knee (boats);
—— *de support de timon*, (*art.*) pole-yoke branch.
brancher, v. a., to tie up to a bough; to hang (as a deserter); to connect (tubes, pipes, etc.); (*mil. min.*) to cause to break out from (e. g., *rameau* from *écoute*, qq. v.).
brandebourg, m., (*unif.*) frogs and loops.
brandevin, m., brandy.
brandevinier, m., (*mil.*) sutler (rare).
brandir, v. a., to brandish (a saber); (*carp*) to stay, strengthen.
branle, m., impulse, impetus; path of a pendulum regulator;
donner, mettre, le ——, to set in motion.
branle-bas, —— *de combat*, (*nav.*) clearing for action;
faire le ——, to clear the decks for action.

branler, v. a. n., (*mil.*) to waver; give away; (*in. gen.*) to shake loose, move stir;
—— *au manche*, to shake loose in the handle (tools).
branloire, f., rock staff (field forge, etc.).
braquer, v. a., to fix a telescope, a field glass, on an object; (*art.*) to lay, traverse, a gun (not used technically);
se —— *sur le mors*, (*man.*) to bore on the bit.
bras, m., arm; brace, tie; handle; link (in many; relations); lever, crank-lever; (*top.*) arm of a river; (*mach.*) spoke of a pulley; (*sm. a.*) carrier-plate; carrier-lever; (*hipp.*) true arm; (*r. f. art.*) cocking-cam (Hotchkiss); (*art.*) leg of a gin;
à ——, by hand;
à —— *en arrière* (*avant*), (*art.*) by hand to the rear (front);
à —— *francs*, (*sm. a.*) offhand, from the shoulder
à —— *d'hommes*, by hand;
—— *d'ancre*, anchor arm;
—— *d'avaloire*, (*harn.*) hip-strap;
—— *d'aviron*, arm or handle of an oar;
—— *de balance*, arm of a balance;
—— *de brancard*, shaft;
—— *-canon*, (*art.*) gun-arm of the Watkin position finder;
—— *de cheval*, (*hipp.*) horse's arm from shoulder to knee;
—— *de chèvre*, (*art.*) leg of a gin;
—— *de courbe*, leg of a knee (boats);
—— *directeur*, (*r. f. art.*) action lever;
double ——, (*mach.*) arm brace;
—— *excentrique*, (*mach.*) excentric rod, side rod;
—— *de force*, bolt acting as a brace to keep a door or shutter closed or open;
—— *inutiles*, useless hands (e. g., in a siege);
—— *de levier*, lever arm;
—— *de limonière*, (*art. etc.*) single shaft;
—— *des machines*, (*mach.*) side rod;
—— *de manivelle*, (*mach.*) crank-lever;
—— *de mer*, (*top.*) inlet, arm of the sea;
—— *mort d'une rivière*, (*top.*) stagnant arm;
—— *de pointâge*, (*art.*) elevating arm (esp. in French naval turrets);
pompe à ——, hand-pump;
—— *de rappel*, (*mach.*) righting-lever, radius-rod, bridle-rod;
—— *de rivière*, (*top.*) arm of a river;
—— *de roue*, spoke;
—— *-support*, (*mach., etc.*) arm, support;
—— *de traverse*, (*fort.*) inclined postern under a traverse.
brasage, m., brazing, operation of brazing; hard soldering.
braser, v. a., to solder, hard solder; to braze, braze on.
brasque, f., (*met.*) brask, luting.
brasquer, v. a. (*met.*) to lute.
brassage, m., (*met.*) stirring, rabbling.
brassard, m., (*unif.*) arm-badge; (*Fr. a.*) insignia of officers of the *service d'état-major*, and of the men of certain *services*, e. g., drivers, telegraph operators, etc.;
—— *blanc à croix rouge*, —— *de la convention de Genève*, (*med.*) Geneva cross, red cross.
brasse, f., fathom (1.66m).
brasser, v. a., (*met.*) to stir, to mix, rabble.
brasseur, m., (*met.*) puddler, rabbler (workman);
—— *mécanique*, mechanical rabbler, puddler.
brassiage, m., fathoming; depth of water (ascertained by fathoming).
brassicourt, m., (*hipp.*) horse bandy-legged by nature.
brassières, f., pl., (*mil.*) knapsack straps, braces.
brasure, f., hard soldering; brazing, brazing on or together; seam, suture.
brave, a., brave, gallant; m., brave soldier;
—— *à trois poils*, true as steel;
vieux ——, brave veteran.
braver, v. a., to brave face, defy, danger, an enemy, etc.
bravoure, f., bravery, courage.

braye, f., v. *broye.*
brayer, m., *(mil.)* belt with shoe or socket for colors; *(cord.)* pulley-tackle.
brayer, v. a., to pitch or tar (a boat); to pay the seams.
brayette, f., v. *braguette.*
brèche, f., *(siege)* breach;
  batterie de ——, (art.) breaching battery;
  battre en ——, to breach;
  faire ——, to breach (a wall, parapet, etc.); *(mil.)* to break through a line;
  mise en ——, breaching;
  —— praticable, practicable breach.
bredindin, m., *(cord.)* garnet, burton; Spanish burton; small stay tackle; whip for raising light weights.
bréhaigne, a., sterile (of animals, *(hipp.)* of a mare).
brelage, m., *(pont.)* lashing, racking, racking down; *(in gen.)* lashing;
  anneau de ——, lashing ring;
  chaîne de ——, (art.) lashing chain, keep-chain;
  —— à garrot, racking;
  —— de la pièce à l'avant-train, (art.) mechanical maneuver, the "carry piece;"
  —— de poutrelles, (pont.) balk-lashing.
breler, v. a., *(cord.)* to woold, to rack; to lash; to secure with slings for purpose of raising; *(pont.)* to lash, rack down;
  —— la pièce à l'avant-train, (art.) to "carry piece" (mechanical maneuver).
breleur, m., lasher, *(esp. pont.)* balk lasher.
breloque *(mil.* drum beat, for "break ranks" or "dismissed;" (also, but not official) dinner call, mess call;
  battre la ——, to beat "dismissed" (mess call, not official).
brequin, m., gimlet, centerbit.
bretaudé, a., *(hipp.)* crop-eared.
bretauder, v. a., *(hipp.)* to crop a horse's ears.
bretelle, f., suspender; sling of a handbarrow, litter, etc.; waist-belt strap; *(unif.)* shoulder strap of a cuirass; *(sm. a.)* sling, rifle-sling;
  —— de cartouchière, (mil.) cartridge-box suspender or sling;
  —— de fusil, (sm. a.) rifle-sling;
  —— de havre-sac, (mil.) knapsack strap;
  —— porte-effets, (mil.) strap formerly used in artillery and train to sling effects;
  —— porte-obus, (art.) shell-strap.
brette, f., *(sm. a.)* long sword (obs.).
bretté, a., toothed, jagged.
brettelement, m., *(cons.)* scratch coat of plaster.
bretteler, v. a., *(cons.)* to scratch (as a coat of plaster).
bretteur, m., duelist, one fond of fighting with the sword.
bretture, f., teeth of tools; notches, scratches, made by them.
breuvage, m., draught; *(hipp.)* drench.
brève, f., *—— s et longues, (sig.)* dots and dashes, shorts and longs.
brevet, certificate; *(mil.)* commission, officer's commission; brevet (U. S. A. and English army);
  avancement à ——, brevet promotion (U. S. A. and English army);
  —— de capacité, certificate of fitness;
  —— d'état-major, (Fr. a.) staff college certificate (field officers take a special examination before a special commission);
  —— d'invention, patent, letters patent;
  —— de nomination, warrant; cadet warrant (U. S. M. A.).
breveté, a., having a certificate, or the certificate of . . . .; patented;
  —— d'état-major, (mil.) having a staff certificate; passed the staff college (English army);
  officier ——, (Fr. a.) officer who has a staff certificate.
brick, m., *(nav.)* brig.
bricole, f., *(harn.)* strap, thong, breast-strap, breast-collar; drag-rope; *(art.)* ricochet fire;

bricole, *banderole de ——,* shoulder strap of man-harness;
  —— de bout d'essieu, (art.) drag-rope;
  —— de canonnier, —— de manœuvre, man-harness.
bricolier, m., off horse.
bricolière, f., shoulder strap of man-harness.
bricoque, f., prypole of a gin (rare).
bride, f., strap (in assemblages, e. g., fork strap of a limber), bridle, stay, loop, band; over-casting; iron bolt or bar on which some other part plays or runs; *(mach.)* flange, thimble, bridle, strap, stay; *(harn.)* bridle;
  à —— abattue, (man., etc.) at full speed;
  —— d'armons, (art.) shifting bolt; pole-bolt;
  —— d'arrêt (de console), (F. m. art., 32 m) plug tray stop;
  —— de la bielle, (mach.) connecting-rod strap or stirrup;
  —— carpienne, (hipp.) check tendon;
  —— de chaîne de brelage, —— d'embrelage, (art.) keep-chain traversing bar;
  —— d'écartement, (art., etc.) spreader or distance piece;
  —— d'épaulette, (unif.) lace-binder, keep-strap, of an epaulet or shoulder knot;
  —— d'étrier d'essieu, (art.) coupling-bar or plate;
  —— d'excentrique, (mach.) eccentric strap;
  —— de fourchette, fork-strap;
  —— de hausse, (art.) rear-sight supporting spring, or loop;
  lâcher la ——, (man.) to give a horse his head;
  ——-licol, (harn.) bridle-halter for off horse of artillery teams;
  —— de manœuvre, (art., etc.) part on which some other part plays or runs;
  —— de noix, (sm. a.) bridle (obs.);
  —— à œillères, (harn.) bridle fitted with blinkers;
  —— à oreilles, (mach.) a flange fitted with or having ears;
  —— de parallélogramme, (mach.) radius-bar;
  —— de poignée de revolver, (sm. a.) revolver butt-ring;
  pousser à toute ——, (man.) to urge a horse at full speed;
  tourner ——, to turn and ride in the opposite direction;
  —— de traille, (pont.) rope fastening trail to cable;
  —— de tuyau, flange-joint, pipe-thimble.
brider, v. a., to overcast; to curb, tie fast; *(harn.)* to bridle; *(cord.)* to seize, frap; to make a cross-seizing; to frap two ropes together; *(mil.)* to blockade (obs.);
  —— des cordages, (cord.) to seize "all parts" together by a rope or lashing.
bridon, m., *(harn.)* snaffle with articulated bit; bridoon;
  en ——, having a snaffle on;
  —— d'abreuvoir, watering-bridle;
  ——-licol, snaffle and halter for artillery teams;
  —— à œillères, snaffle fitted with blinkers;
  scier du ——, (man.) to saw upon the bit, to pull the reins alternately.
bridonner, v. a., *(harn.)* to put on a snaffle or bridoon.
bridure, f., *(cord.)* frapping, seizing.
brigade, f., brigade (2 regiments); squad, of from 4 to 7 men and 1 corporal, of *gendarmerie;* squad, gang, party, of men (no fixed number), as, of engineers, workmen, draughtsmen, etc., working together for a specified end;
  —— d'artillerie, (Fr. a.) brigade of 2 regiments to each corps, except the sixth;
  —— de corps d'armée, (mil.) cavalry brigade attached to corps;
  demi-——, (Fr. a.) Revolutionary name for regiment;
  être de —— avec, faire —— avec, to be brigaded with;
  faire ——, to be brigaded together;
  —— de gendarmerie, 4 to 7 men and a corporal of gendarmerie;
  —— du génie, (Fr. a.) 2 regiments of engineers belonging to the garrison of Paris;

brigade *de ligne*, brigade of (French) cavalry, considered with reference to its duties in the *ligne d'attaque* and *ligne de manœuvre*, qq. v.;
—— *mixte*, brigade composed of all arms of the service;
—— *de sûreté*, brigade of corps cavalry;
—— *topographique du génie*, (*Fr. a.*) section of engineer troops charged with execution of contoured surveys (now *service géographique de l'armée*).

brigadier, m., (*art., cav., and train*) corporal; corporal of gendarmes; corporal in charge of the *garde républicaine*; the commanding general of a brigade (rare); bow oar (boats);
—— *de boulangers*, v. —— *de four*;
—— *b'ancardier*, sort of corporal of litterbearers, one to each group of mounted batteries (n. e.);
—— *de chambrée*, corporal in charge of squad room;
—— *d'escouade*, corporal of cavalry who, in his squad room, looks after the men of his squad;
—— *de four*, (*Fr. a.*) baker, baker in charge of oven;
—— *fourrier*, (*n. e.*) quartermaster-corporal;
—— *fourrier d'ordre*, (*cav.*) quartermaster-corporal who takes the morning report of the regiment to the headquarters of the place and who brings back the orders of the day, etc.;
—— *général*, brigadier-general (not a French grade, but used by French to denote the grade of brigadier-general in United States);
—— *infirmier*, (*n. e.*) a corporal of the hospital corps;
—— *moniteur d'escrime*, assistant fencingmaster;
—— *d'ordinaire*, mess corporal;
—— *de pièce*, (*art.*) corporal in charge of and responsible for the men of his piece;
—— *premier ouvrier armurier, bottier, cordonnier, sellier, tailleur*, (*Fr. a.*) workmen of these various trades, of the grade of corporal, and acting as foreman in the shops in which these trades are carried on;
—— *-sapeur*, (*cav.*) corporal of sappers;
—— *de semaine*, corporal on duty for the week;
aux ——*s de semaine*, call for the assembly of the ——*s de semaine*;
—— *trompette*, corporal of field music.

brigand, m., brigand.

brigandage, m., (*mil.*) pillage, robbery.

brigantine, f., (*nav.*) brigantine.

brigue, f., cabal, party, association for some improper purpose.

brimade, f., hazing.

brimbale, f., handle, lever, of a pump.

brimer, v. a., to haze.

brin, m., small bit, piece; (*cord.*) strand; quality of hemp; yarn; a part (in combination, fold) of a tackle; tackle-fall or free end of a cable on a winch or windlass;
bois de ——, small timber; tree roughly squared;
chêne de ——, v. s. v. *chêne*;
—— *conducteur*, (*mach.*) driving side of a belt;
—— *conduit*, (*mach.*) driven side of a belt;
—— *libre*, (*cord.*) free end of a rope; (*mach.*) driving side of a belt;
—— *menant*, (*mach.*) driving side of a belt;
—— *mené*, (*mach.*) driven side of a belt;
—— *moulé*, (*cord.*) rove-tackle;
premier ——, best part of hemp, hemp of the best quality;
second ——, inferior part of hemp (combings).

bringuebale, f. v. *brimbale*.

brique, f., brick;
—— *blanche*, fire brick;
—— *creuse*, hollow brick;
—— *crue*, unburnt brick;
—— *pilée*, brick dust; brick dust and oil, to remove rust;
—— *réfractaire*, fire brick;
—— *vernissée*, glazed brick.

briquemann, briquemon, m., (*mil. slang*) cavalry saber.

briquet, m., steel (for striking a light); (*sm. a.*) old term for infantry sword.

briquetage, m., brickwork; imitation brickwork.

briqueter, v. a., to imitate brickwork.

briquette, f., brick of patent fuel;
—— *de charbon*, —— *de houille*, patent fuel,

bris, m., breaking open; escape from jail, jail-breaking; breaking smashing; (*art.*) smashing of a projectile against an armor plate;
—— *d'armes*, (*mil.*) v. illful damage or injury to arms.

brisant, m., breaker, wave; breakwater; sand bank, shoal, reef or shelf of rocks under water.

brisant, a., (*expl.*) "brisant," sudden;
poudre ——*e*, 'brisant" powder, sudden in its action.

brise, f., breeze;
—— *de mer*, sea breeze;
—— *de terre*, land breeze.

brisé, a., broken; folding, jointed; (*fort.*) tenailled consisting of alternate salients and reentrants;
tracé ——, (*fort.*) broken, tenailled, trace.

brisé, m., (in singlestick) moulinet (downward in front of the body).

brise-circuit, m., (*elec.*) circuit-breaker; (*torp.*) cut-off.

brise-cou, m., (*man.*) horse-breaker, rough rider.

brise-courant, m., (*elec.*) make-and-break key.

brise-glace, m., ice-fender; cutwater (ice-breaker) of a bridge.

brise-lames, m., breakwater.

brisement, m., breaking, smashing; breaking o sea along the shore.

briser, v. a., to break, shatter, shiver, smash, break down; (*elec.*) to break a circuit, a current;
—— *une résistance*, (*mil.*) to overcome, break down, a resistance;
se —— *sur la trajectoire*, (*art.*) to burst in the air (of a projectile).

brise-vent, m., shelter against wind; windguard.

brisquard, brisque, m., (*mil. slang*) old soldier, with long-service stripes.

brisure, f., any part broken; break, crack; (*fort.*) brisure, break in a parapet to prevent enfilade;
—— *de l'embrasure*, (*fort.*) neck of the embrasure.

broche, f., spike, peg, picket; spike-nail; tent peg; gudgeon; felly-tenon (of a spoke); (*mach., etc.*) spindle; (*art.*) mandrel (gun making); gauge, measuring-point, hoop-gauge; (*elec.*) plug of a resistance box; (*pont.*) pointed bolt (assemblage of trestle); (*sm. a.*) the pin on which the hammer strikes, in *cartouche à* ——; spindle of a revolver cylinder; (*artif.*) rammer, drift;
arme à ——, (*sm. a.*) rifle using *cartouche à* ——, q. v.;
—— *d'assemblage*, riveting-pin;
carrée, (*artif.*) punch; square rammer;
cartouche à ——, (*sm. a.*) cartridge fitted with a projecting pin near base, on which the hammer strikes, pin-cartridge (obs.);
—— *de chargement*, (*art.*) sort of lever used to support breech of French 138^{mm} gun (on siege carriage) during loading;
—— *à étoupilles*, (*artif.*) spindle;
—— *à expansion*, (*art.*) measuring-point, hoop-gauge;
—— *de percussion*, (*sm. a.*) pin or striker of *cartouche à* ——;
—— *de rais*, felly tenon of a spoke;
—— *de réglage*, adjusting-pin;
—— *de sûreté*, (*art.*) safety-pin of a fuse.

brocher, v. a., to stitch sheets together (pamphlets, books, etc.); (*farr.*) to drive nails into a hoof;
—— *un clou*, to prick a horse in shoeing;
—— *en musique*, (*farr.*) to clinch nails at different heights;

brocher *trop à gras*, (*farr.*) to set a nail too close to the quick, or too high;
—— *trop à maigre*, (*farr.*) to engage a nail too near the outer surface of the hoof.
brocheté, a., formed or made upon a spindle.
brochoir, m., (*farr.*) shoeing-hammer, farrier's hammer.
brocheur, m., spike-driver.
brodequin, m., high (lace) shoe;
——-*botte*, combination of shoe and boot.
broderie, f., (*unif.*) insignia of rank or of function.
broiement, m., pounding, crushing, pulverizing, grinding.
brome, m., (*chem.*) bromine.
bromure, m., (*chem.*) bromide.
bronchade, f., (*man.*) stumbling, tripping (of a horse).
broncher, v. n., (*man.*) to stumble, trip, take a false step (horse).
bronolite, f., (*expl.*) bronolite.
bronzage, m., (*art., sm. a.*) browning.
bronze, m., (*met.*) bronze; (*mil. slang*) artillery, (*in pl.*) artillerymen;
——-*acier*, steel-bronze, Uchatius bronze;
——-*aciéré*, steely bronze;
——-*d'aluminium*, aluminium bronze;
——-*à canon*, gun metal;
——-*à cloches*, bell metal;
——-*dépassé*, bronze that has been over-stirred, and that has lost part of its malleability;
——-*dur*, —— *durci*, hardened bronze;
——-*au manganèse*, manganese bronze;
——-*moulu*, bronzing powder;
——-*naturel*, common bronze;
——-*de nickel*, nickel bronze;
——-*phosphoreux*, phosphor bronze;
——-*siliceux*, silicon bronze;
——-*suraffiné*, v. —— *depassé.*
bronzer, v. a., to brown (small arms); to bronze.
broquette, f., tack;
clou ——, headless nail, sprig.
brosse, f., brush; underbrush on the edge of a wood;
——-*à cheval*, horse brush;
——-*en chiendent*, —— *en crin*, v. —— *à cheval*;
——-*double*, blacking brush;
——-*écouvillon*, (*art.*) vent-sponge;
——-*en fil de métal*, wire brush;
——-*à fusil*, v. —— *à graisse*
——-*à graisse*, —— *grasse*, oiling-brush (for arms);
——-*passe-partout*, small dusting brush (e. g., for the breech mechanism of a rifle);
——-*à patience*, polishing brush, button brush;
——-*rude*, stiff brush.
brosseur, m., (*mil.*) officer's servant, soldier servant; "striker" (U. S. A.).
brouette, f., wheelbarrow;
——-*de chargement*, (*art.*) shot barrow;
——-*de déroulement*, —— *à dérouler le fil*, (*teleg.*, etc.) wire drum, unreeling barrow;
——-*à deux roues*, handcart.
brouetter, v. a., to wheel, carry in a barrow.
brouillard, m., fog.
brouiller, v. a., to confuse, to throw into disorder.
brouillon, m., scratch pad; rough draught or copy.
broussaille, f., bramble, underbrush (used also in plural); (*const.*) riprap work;
——-*en pierres sèches*, dry stone riprap.
brout, m., term applied to a boring bit or tool when cutting untrue.
broutage, m., **broutement**, m., (*mach.*) vibration, unsteadiness, state of being out of true, of line (of a boring or cutting instrument).
brouter, v. n., (*mach.*) to vibrate, to cut untrue (of borers, etc.); to browse (of animals).
broyage, m., v. *broiement*; (*cord.*) breaking (manufacture of rope, etc.).
broye, f., (*cord.*) breaking-machine.
broyé, m., (*cord.*) material that has been passed through the breaking-machine.

broyer, v. a., to bruise, crush, pulverize, grind; (*cord.*) to break.
broyeur, m., crusher, (*met.*) crushing-mill;
——-*pulvérisateur*, crusher and pulverizer.
bruine, f., mist, drizzle, Scotch mist.
bruit, m., noise; rumor; rattle, din, disturbance;
——-*de caisse*, (*mil.*) drum music, beat of drums;
——-*de grenouille*, (*hipp.*) churning noise of the sheath (esp. in the trot);
——-*de guerre*, rumor of war.
brûlage, m., burning off dry grass, etc., from fields.
brûlé, a., (*hipp.*) burnt (coat of horse); (*met.*) burnt, overheated.
brûle-amorce, m., igniter, primer.
brûle-pourpoint, à, adv., at the muzzle, close up, at very close quarters.
brûle-queue, m., (*hipp.*) circular cautery (used after docking).
brûler, v. a., to burn, scorch, sear, blast; (*met.*) to overheat, burn (as iron, steel), (*mil.*) to fire;
sans —— *une amorce*, (*mil.*) without firing a shot;
—— *une cartouche*, (*mil.*) to fire a shot;
se —— *la cervelle*, to blow out one's brains;
—— *le pavé*, to go very fast;
—— *les signaux*, —— *les stations*, (r. r.) to rush past signals, stations, etc., without stopping.
brûleur, m., burner;
——-*de Bunsen*, (*chem.*) Bunsen burner;
——-*intermittent*, —— *permanent*, (*mach.*) (Otto) gas engine.
brûlis, m., part of a forest on fire; burning off weeds, etc., from the fields.
brûlot, m., brand; fire-raft, fire-ship, any burning floating body to set fire to and burn a bridge; (*powd.*) half-burnt charcoal.
brumaille, f., thin mist.
brume, f., mist, fog;
cornet de ——, fog horn.
brumeux, a., foggy, misty.
brun, a., brown; m., brown color;
——-*châtain*, (*hipp.*) chestnut;
——-*clair*, (*hipp.*) light-brown;
——-*foncé*, (*hipp.*) dark-brown.
brunir, v. a., to burnish, (also) to brown (arms, etc.).
brunissage, m., burnishing, browning (arms, etc.).
brunissoir, m., burnisher;
—— *à roue*, polishing wheel.
brunissure, f., browning, burnishing (arms, etc.).
brusqué, (*mil.*) sudden, without preparation or notice (i. e., without stopping at, or going through, the usual intermediate steps, as of an attack, a siege, etc.);
attaque ——*e*, (*mil., siege*) sudden attack, assault (no intermediate steps).
brusquer, v. a., (*mil., siege*) to make a sudden and violent effort without stopping at intermediate stages; (*more esp. siege*) to start trenches on counterscarp and work back;
—— *une attaque*, to go straight to the objective by a *coup de main* (e. g., fortress);
—— *un cheval*, (*man.*) to start a horse suddenly;
—— *une place*, —— *un poste*, (*siege*, etc.) to attempt to seize a fort, etc., by assault and without a siege.
brut, a., rough, unwrought, unpolished; gross;
bois ——, timber in the rough, undressed timber;
charge ——*e*, gross weight;
——-*de coulée*, (*fond.*) rough-cast, just as cast (i. e., without finishing or dressing);
ferrure ——*e*, rough-forging;
——-*de fonte*, (*fond.*) just as cast;
——-*de forge*, rough-forged.
brutal, m., (*mil. slang*) cannon.
bruyère, f., heath, broom, heater.
buanderie, f., laundry.
buandier, m., laundryman.
bûche, f., stick of firewood.
bûcher, m., woodhouse.

bûcher, v. a., to dress down a piece of timber; (*mas.*) to remove projections, irregularities from a stone.
bûcheron, m., woodcutter.
bûchilles, f. pl., borings, bore-chips (metal).
budget, m., (*adm.*) budget, yearly estimates of expenses;
— *extraordinaire,* estimates for unusual or special expenses;
— *de la guerre,* estimates of the war department;
— *ordinaire,* regular yearly estimates.
budgétaire, a., belonging to the budget;
dépense —, expense for which an appropriation has been made.
buffle, m., (*mil.*) buff-stick, emery stick.
buffleterie, f., (*mil.*) slings and belts, "trimmings" (U. S. M. A.);
gant de —, leather glove.
bugle, m., (*mus.*) key bugle.
buis, m., box (wood and tree).
bulle, f., bubble; blister, flaw;
— *de niveau,* bubble of a level;
niveau à — *d'air,* (*inst.*) spirit level.
bulletin, m., bulletin, notice, report, document; (*mil.*) report of a battle, of operations, dispatches; (*adm.*) blank, blank form (as, for orders, etc.).
— *administratif,* (*adm.*) any memorandum, etc., having relation to the administration, e. g., of repairs to arms, of arms turned in, etc.;
— *d'avis d'expédition,* (*adm.*) notice of shipment made;
— *de dépôt,* (*adm.*) certificate of delivery to administrative magazines of goods furnished by contractor;
— *de guerre,* v. — *des opérations;*
— *météorologique,* weather report;
— *officiel du ministère de la guerre,* (*Fr. a.* called generally — *officiel*) periodical official publication of war department, in two parts, the first (*partie règlementaire*) containing all laws, circulars, etc., of permanent character; the second (*partie supplémentaire*), containing decisions, etc., of more or less temporary nature;
— *des opérations,* (*mil.*) report of operations, official report or reports of a battle, etc. (sometimes, *bulletin* merely);
— *de tir,* (*art., ball.*) table of practice, reports of practice, of experimental firing, etc.
bure, f., drugget, coarse woolen cloth; (*min.*) shaft, pit-hole.
bureau, m., office; writing table or desk; (*Fr. a.*) subdivision of *cabinet du ministre* (war office), of the general staff, and of the various *directions* of the war department; bureau, i. e., department;
— *de l'administration centrale* (v. *supra,* under meanings, *Fr. a.*);
— *arabe,* (*Fr. a.*) bureau having charge of native affairs in Algerian territory not under normal administration;
— *de compatabilité,* auditor's office, accounting office;
— *de la guerre* (v. *supra,* under meanings, *Fr. a.*);
— *de recrutement,* (*Fr. a.*) recruiting bureau, one in each regional subdivision (except in the departments of the Seine, Seine-et-Oise and Rhône).
burette, f., burette; (*art.*) bottom board of ammunition wagon;
— *d'huile,* oil can, oiler.
burin, m., graver, graving tool, cutter; wedge-driver; (*mach.*) screw-thread cutter; cutter (of a lathe, etc.); (*cord.*) splicing-fid;
— *d'alésoir,* cutter of a boring machine;
— *de serrurier,* locksmith's chisel;
— *de volée,* (*art.*) muzzle-roller (in machanical maneuvers).
burinage, m., engraving, graving; (*met.*) cutting out of cracks from ingots.
buriner, v. a., to grave, engrave; to cut metal with a burin; to wedge, drive in a wedge; (*met.*) to cut out cracks from ingots.

burnous, m., (*unif.*) burnoose, hooded cape, borrowed by the French from the Arabs. (The spelling *burnous* is preferred to *bournous*.)
busc, m., sill, miter-sill of a canal lock; angle between leaves of sluice gate when shut; vertical front edge of a breastplate; (*sm. a.*) edge on back of stock (just in rear of small of the stock);
montant de —, miter post, of a lock-gate.
buse, f., muzzle, nozzle (as of bellows); blast pipe (of a furnace); ventilator; pipe; channel, mill-channel;
— *d'airage,* channel, air-channel;
— *de gabions,* (*pont.*) gabions united end to end by an axial stake, and used in construction of temporary bridges;
— *de soufflet,* tuyere, nozzle, blast-pipe.
busque, m., v. *busc.*
busqué, a., arched; (*hipp.*) Roman-nosed.
but, m., object, aim, object aimed at; (*art., sm. a.*) target; butt;
— *cuirassé,* armored target;
— *en blanc,* (*ball.*) second point of intersection of trajectory with line of sight;
de — *en blanc,* (*ball.*) point-blank;
— *à éclipse,* (*t. p.*) disappearing target;
encadrer le —, (*art.*) to get the target between a short and an over;
— *instantané,* (*art.*) target to be fired on as soon as it (suddenly) appears; surprise target;
— *instantané et à éclipse,* (*art.*) instantaneous and disappearing target, one that shows itself for a moment or two only, disappearing surprise target;
— *mobile,* (*art., t. p.*) moving target;
portée de — *en blanc,* (*ball.*) point-blank range.
bute, f., (*farr.*) butteris, buttrice.
butée, f., v. *buttée.*
buter, v. a. n., to buttress, to buttress up; to strike, strike against; to bear against; (*t. p.*) to hit the target.
butin, m., spoil, booty, pillage.
butiner, v. a., to loot, pillage.
butoir, m., (*farr.*) buttrice, cf. *boutoir;* (for other meanings v. *buttoir*).
butte, f., (*top.*) hill; low, broad hill; (*t. p.*) butts, target-butt, bullet stop;
être en — *à,* to be exposed to.
buttée, f., butting, tapping, striking, of one part on another (in machinery, etc.); measuring-shoulder, fixed point (of calipers, measuring rods, etc.); lug, limiting-motion lug, stop; (*pont.*) bridge abutment; (*mach.*) thrust-bearing; (*r. f. art.*) resistance piece (machine guns); (*sm. a.*) chamber-cone; (*Fr. art.*) frame supporting or abutting the pintle block (24ᶜᵐ s. c. gun);
— *de chargement,* (*harn.*) in the carriage pack of mountain artillery, a hook or leg to keep the load from slipping forward;
— *à collets,* (*mach.*) thrust-bearing; thrust-block;
— *de la culasse mobile,* (*sm. a.*) bearing surface for the front reinforce of the cylinder (Lebel);
— *à disque(s),* — *à rondelle(s),* (*mach.*) thrust-bearing, thrust-ring and disk, thrust-collar;
— *tronconique,* (*sm. a.*) conical surface, catching or holding the neck of the cartridge.
(The spelling *butée* is more common than *buttée,* but the latter is preferable.)
butter, v. a. n., to strike, hit, tap upon; to prop. to support, (a wall, etc.); to buttress up; to earth up, to bank up earth against; (*t. p.*) to hit the mark; (*hipp.*) to kick, strike against, pebbles, inequalities, etc., in the road, (hence) to stumble, trip, come near falling.
buttoir, m., (*mach.*) catch, tappet, driver; dog; collar and set-screw; guide; valve-guard; (*r. r.*) buffer; (*art.*) retracting-stud; seating-pin; stud, projection; stop;

buttoir *d'auget*, (*sm. a.*) tappet or lug of the carrier, heel of the carrier, carrier-stud (Kropatschek, Lebel, etc.);
—— *de descente*, (*sm. a.*) front lug or tappet of the carrier (Kropatschek), by which the front of the carrier is caused to drop;
—— *de fermeture*, (*art.*) closing-stop;
—— *à guide courbe*, (*art.*) circular bar-buffer (Schumann);
—— *d'ouverture*, (*art.*) opening-stop;
—— *de recul*, (*art.*) recoil-buffer;

buttoir *de relèvement*, (*sm. a.*) stud or tappet on the carrier, by which the latter is made to rise (Kropatschek, Lebel);
—— *-repère*, stop, contact with which assures correct position of the part concerned;
—— *à ressort*, (*sm. a., etc.,*) spring-tappet; spring-stud.
(*Butoir*. is seen perhaps more often than *buttoir*.)

buveur, m., —— *d'encre*, (*mil. slang*) quill-driver, clerk.

C.

caban, m., (*unif.*) cape with hood and sleeves (no longer worn, *Fr. a.*).
cabane, f., cabin, hut.
cabestan, m., capstan; crab, winch; (*siege*) fascine-choker;
armer le ——, to rig the capstan;
—— *de carrier*, (*Fr. art.*) capstan or winch used for hauling heavy loads up inclines or along horizontal surfaces;
choquer le ——, to surge the capstan;
cloche de ——, capstan barrel;
dévirer au ——, to unwind the capstan, to come up the capstan;
équiper un ——, to rig a capstan;
—— *pour fascines*, (*siege*) fascine-choker;
garnir le ——, to rig the capstan;
—— *hydraulique*, hydraulic capstan;
—— *multiple*, patent-geared capstan;
—— *à n brins*, capstan equipped with an n-fold tackle;
tenir bon au ——, to avast heaving;
—— *à vapeur*, steam capstan;
virer au ——, to heave away the capstan;
—— *volant*, portable winch, crab.
cabillot, m., toggle;
—— *à griffe*, (*art.*) hand extractor; lanyard toggle and extractor in one (*F. m. art.*).
cabine, f., (*nav.*) stateroom, cabin.
cabinet, m., cabinet; ministers of government;
—— *d'aisances*, water-closet, "rear;"
—— *du ministre*, (*mil.*) office (personnel) of the minister or secretary of war.
câblage, m., (*cord.*) laying up (of strands into rope), twisting up into a cable (wire, strands, etc.);
chemin de ——, ropewalk (proper).
câble, m., (*cord.*) cable, rope; (*elec., etc.*) telegraph cable, electric cable; (*nav.*) cable-length; (*mach.*) driving rope, cable;
—— *en acier*, wire rope (steel);
—— *d'amorce*, (*torp.*) torpedo-cable; fuse-cable;
—— *d'ascension*, captive cord or cable of a balloon;
bander le ——, to haul upon the cable;
—— *chaîne*, chain cable;
—— *de charge*, (*mach., etc.*) any cable, etc., used to support a load;
—— *de chèvre*, (*art.*) gin-fall;
—— *à deux conducteurs*, (*elec.*) two-wire cable, secondary conductor;
—— *à un conducteur*, (*elec.*) one-wire cable, main conductor;
—— *double*, (*elec.*) two-wire cable;
—— *électrique*, (*Fr. eng. park.*), two-wire cable;
—— *encablure*, cable length;
—— *étançonné*, stud chain cable;
—— *d'extraction*, hoisting cable (mines, shafts, etc.);
—— *non étançonné*, close-link chain, crane-chain;
—— *en fer*, (iron) wire rope;
—— *en fil d'acier*, steel wire rope;
filer du ——, to pay off or out the cable;
—— *lâche*, (*cord.*) long-jawed rope, loosely laid up rope or cable;
—— *métallique*, wire rope;
—— *militaire*, (*teleg., etc.*) insulated wire, for military telegraph, etc., for firing batteries;
parer un ——, to clear a cable;
passer le ——, (*cord.*) to untie and pass a knot through a block or pulley;
—— *plat*, flat rope;

câble *de pont volant*, (*pont.*) mooring;
recette de ——, cable landfall;
—— *remorque*, —— *de remorque*, towing line or cable;
—— *de retenue*, captive cable or cord of a balloon;
—— *serré*, (*cord.*) tightly laid up rope or cable, short-jawed rope;
—— *sous-marin*, submarine cable (telegraph, etc.);
—— *souterrain*, underground cable;
—— *télodynamique*, (*mach.*) power-transmitting cable;
—— *à tension renforcée*, (*mach.*) tightened cable, tightened driving-rope;
—— *de traille*, (*pont.*) sheer-line;
—— *de transmission*, (*mach.*) power cable;
—— *volant*, loose cable.
câbleau, m., (*cord.*) mooring-rope, boat rope, painter.
câbler, v. a., (*cord.*) to lay up.
câblot, v. *câbleau*.
cabo, m., (*mil. slang.*) corporal (inf.);
élève ——, lance corporal.
caboche, f., hobnail; nave-nail (for nave-hoops).
cabotage, m., coasting trade;
grand ——, coasting trade from headland to headland;
petit ——, coasting trade close inshore.
caboter, v. n., to coast, to sail along the coast.
caboteur, m., coaster.
cabotier, m., coasting vessel.
cabre, f., sort of gin.
cabrade, f., (*man.*) rearing.
cabrer, v. r., (*man.*) to rear; m., rearing;
faire ——, to cause a horse to rear.
cabriole, f., (*man.*) capriole.
cabrioler, v. n., (*man.*) to execute caprioles, to capriole.
cabriolet, m., cabriolet, cab; truck.
cabron, m., (*mil.*) buff-stick, emery stick.
cache, f., hiding-place.
cache-carte, m., (*mil.*) map-cover (transparent and waterproof, to permit use of maps in wet weather).
cache-ceinture, m., (*art.*) band-guard (used in the transportation of large projectiles).
cache-éperons, m., spur-hood, spur-cover.
cache-lumière, m., (*art.*) vent-cover.
cache-mèche, m., (*artif.*) slow-match case (obs.).
cache-platine, m., (*sm. a.*) lock-cover.
cachet, m., seal;
—— *de service*, (*mil., etc.*) official stamp or seal.
cache-tête, m., (*hipp.*) hood put over the head of a horse that is cast.
cachot, m., cell, prison cell.
cachou, m., catechu, cutch, cashoo.
cacolet, m., (*med.*) cacolet.
cadastral, a., having to do with, or referring to, the register of lands (*cadastre*), cadastral.
cadastre, m., public record or register of realty (value and area), cadaster.
cadastrer, v. a., to survey lands (for the *cadastre*).
cadavre, m., corpse.
cadenas, m., padlock;
—— *de chaîne*, fetter-lock.
cadence, f., (*mil.*) cadence; (*man.*) the precision or accuracy of a horse's gaits;
—— *du pas*, (*mil.*) cadence of the step, of marching;

cadence, *perdre la* —— *du pas,* (*mil.*) to lose step; *tenir la* —— *du pas,* (*mil.*) to keep step.
cadencer, v. a., (*mil.*) to cadence;
—— *une allure,* (*man.*) to regularize, to steady, a gait;
—— *le pas,* (*mil.*) to keep step in marching, to take up a cadenced step
cadet, m., (*mil.*) cadet.
cadette, f., flat paving stone, suitable for a sidewalk.
cadmium, m., cadmium.
cadole, f., latch;
—— *à ressort,* spring latch.
cadran, m., face, dial, plate; dividing-plate; shake (in wood); (*Fr. s. c. art., F. m. art.*) breech-plate;
fermeture à ——, (*F. m. art.*) breech-closure in which the console runs in the grooves of a dial (*cadran*) (antiquated);
—— *indicateur de pointage,* (*art.*) elevation indicator or dial;
—— *solaire,* sun-dial.
cadranures, f. pl., shakes (in wood).
cadre, m., frame, outline, skeleton; border (of a map); (*art.*) bracket-block; (*artif.*) quickmatch rest; (*mil.*) fundamentally, officers (commissioned and noncommissioned) who frame in (*encadrer*) the men; (by extension) all the officers, n. c. o.'s, and corporals of a battery, troop, or company; all the officers of a regiment; (by a further extension) the officers of the staff, of the *intendance, santé* etc., of the various departments (*contrôle, archives,* etc.) connected with the military service; the staff of a military school; the officers on recruiting, remount, etc., service; (in a limited sense, esp. in pl.) the n. c. o.'s of a company; (as applied to units) the elements necessary to the continuous organic existence of the unit, irrespective of the men to put it on an effective basis; a class of officers (e. g., —— *de réserve, infra.*)
   I. Frame (material); II. Military organization.
   I. Frame:
—— -*affût,* (*min.*) stand or frame of a drill;
avec ——*s arc-boutés,* (*pont.*) single-locked;
—— -*cible,* (*ball.*) velocity-frame, frame and wires for determination of muzzle velocities;
—— *coffrant,* (*mil. min.*) casing (of a shaft), sort of box frame;
—— *dévidoir,* reel;
—— *extenseur,* (*phot.*) stretching-frame;
faux ——, (*mil. min.*) temporary frame, somewhat larger than permanent, used in sinking shafts in difficult (crumbling) soil;
—— *de galerie,* (*Fr. art.*) arrangement or frame on caisson chest, to carry the *sacs* of the men (caisson, model 1888);
—— *à glissières,* (*mach.*) slide-frame;
—— -*interrupteur,* (*ball.*) v. —— -*cible;*
—— *de manœuvre,* (*pont.*) frame used in construction of trestle and of pile bridges (for swinging trestle into position and for pile-setting);
méthode des ——*s,* (*pont.*) method of construction of trestle bridges in which ——*s de manœuvre* are used;
—— *de mine*(*s*), (*mil. min.*) shaft-frame;
—— *multiplicateur,* (*elec.*) galvanometer coil;
—— *à oreilles,* (*mil. min.*) top frame of a shaft, the ends or horns (*oreilles*) of which stick out beyond the square;
—— *porte-lunette,* (*mach.*) poppet-head;
—— *à sécher,* (*artif., powd.*) drying-frame, tray board;
—— -*support,* any supporting frame;
—— *des tamis,* (*powd.*) shaking-frame, shaker;
—— *uni,* (*mil. min.*) shaft-frame (i. e., has no projecting horns, (*oreilles*);
—— *à vis,* screw-frame.
   II. Military organization:
—— *actif,* active army (as opposed to reserve), troops of the active army (used in a very general way);

cadre, *alimenter le* ——, *les* ——*s,* to strengthen the *cadres,* add to them;
—— *armé,* the armed or arm-bearing portion of a unit (as opposed to the unarmed or noncombatant);
—— *complémentaire,* supernumerary officers (*Fr. a.,* of infantry);
—— *de conduite,* detachment for conducting recruits to the regiments to which they are assigned (generally, *Fr. a.,* of n. c. o.'s, but sometimes commanded by a lieutenant);
exercices des ——*s,* drill and instruction of officers and n. c. o. (generally of the latter);
—— *fixe,* permanent staff (as of a school);
hors ——, seconded (n. e., U. S. A.); status of an officer of the active list, on the staff, at a military school, either as instructor or student, etc., and so not counted as belonging to a statutory organization;
—— *instructeur,* staff or detail for instruction (drill);
loi des ——*s,* law organizing the army;
manœuvres avec ——*s,* v. *exercices des* ——*s;*
—— *mobile,* the attached, transient, or detailed staff (as of a school);
—— -*officier,* the officers of a unit;
—— *des officiers de réserve,* officers of the reserve;
—— *d'organisation,* the elements of the organic existence of a unit;
—— *de réserve,* (*Fr. a.*) the second section of the *état-major général de l'armée* (i. e., body of general officers), composed of generals and of functionaries of assimilated rank that have reached retiring age or are retired through disability; (in a general sense) officers, etc., of the *réserve,* as contradistinguished from those of the active army;
troupe- ——, skeleton organization.
caducée, m., caduceus, (*Fr. a.*) emblem of the *personnel militaire* of the medical department.
café, m., coffee;
—— *torréfié,* roasted coffee;
—— *vert,* raw, unroasted coffee.
caffût, m., old iron, scrap iron; (*art.*) fragments of projectiles, old guns, etc.
caffûter, v. a., (*art.*) to break up guns, projectiles, etc.; to condemn.
cage, f., cage, casing, housing; well (of a staircase, the case proper); casing of a stairway; (*sm. a.*) cylinder recess of a revolver; (*fort.*) well for the counterpoise of a drawbridge;
—— *du barillet,* (*sm. a.*) recess of a revolver frame, space in which the cylinder revolves;
—— *de culasse,* (*art.*) breech-casing of the *canon à balles;*
—— -*guide,* (*sm. a.*) lock-case (Borchardt pistol);
—— *de monte-charges,* (*fort.*) frame or well of a shell-hoist;
mur de ——, wall of a staircase;
renverser, verser en ——, (*art.*) to turn a gun and its carriage upside down, carriage uppermost (said also of any other vehicle).
cagnard, m., skulker.
cagnarder, v. n., to skulk.
cagneux, a., (*hipp.*) open (as to leg); cross-footed, pigeon-toed (as to foot) (hocks and elbows turned out and toes in).
cahier, m., blank book;
—— *des charges,* (*adm.*) specifications (for contracts, bids, etc.);
—— *des charges générales,* (*adm.*) general specifications (one set of, for each *service,* adopted by minister of war);
—— *des charges spéciales,* (*adm.*) specifications peculiar to each contract made within a service;
—— *de correspondance,* letter-book;
—— *d'enregistrement,* (*mil. adm.*) company (battery, troop) daybook of receipts and expenses;
—— *d'ordinaire,* (*mil. adm.*) mess book (accounts) of a company, etc.;
—— *à souche,* blank book with stubs;

cahier de visite des hôpitaux, (des infirmeries), (med.) record of patients in hospital (infirmary);
— de visite régimentaire, (med.) sick-book.
cahot, m., jolt (of a carriage).
cahotage, m., jolting.
cahoter, v. a. n., to jolt; to be jolted, tossed about.
cahute, f., poor house or building; (fort.) contemptuous name for a small fort, or for a post badly fortified and in bad condition.
caillebottis, m., (nav.) grating.
caillot, m., clot of blood.
caillou, m., pebble, small stone.
cailloutis, m., metaling, broken stone for a road; en —, metaled, macadamized.
caisse, f., chest, box, case; frame; shell (of a block, pulley); shell (of a drum); money chest, (hence) treasury, funds; body of a railway car or carriage; (sm. a.) arm-chest;
— à air, air-tight compartment (e. g., of a life-boat, etc.);
— d'alimentation, feed case (of a machine gun);
— d'amarrage, mooring-buoy;
— d'ambulance, chest for medical supplies;
— d'ancrage, (pont.) makeshift anchor (box filled with stones);
— à archives, (mil.) records-chest;
— d'armes, (sm. a.) small-arm packing case;
— d'artifices, (art.) powder-chest (obs.);
— à bagages, (mil.) regulation army-chest (officers'), for field service;
— à bascule, (mil. r. r.) tilt-box or case (for the transportation of earth);
— blanche, (sm. a.) small-arm ammunition box (unpainted, whence named), carried by the parks;
— blanche de double approvisionnement, (Fr. mountain art.) reserve ammunition box;
bruit de —, (mil.) beat of drum, drum music;
— à cartouches, (sm. a.) small-arm ammunition chest or box;
— de cémentation, (met.) cementation-box;
— à charbon, (art.) coal-box of a field-forge;
— claire, (mus.) snare drum;
— de comptabilité, (mil.) field-chest for papers, money, etc.;
— des dépôts et consignations, (Fr. adm.) treasury for sums due heirs of deceased soldiers, or for sums that for any reason can not be paid out directly from the military chest;
— à dynamite, (mil.) chest for transportation of pétards de cavalerie, by horse artillery or artillery parks;
— à eau, water tank;
— à effets, (mil.) chest for reserve supplies of clothes (in the field);
— à feu, fire box;
— à flambeaux, (art.) chest for transportation of fiambeaux;
— à fusées, (art., etc.) signal-rocket chest;
— à gargousses, (siege) cartridge chest (left in expense magazines of batteries);
grosse —, (mus.) bass-drum;
— à instruments, any instrument box; (art.) case for battery telescope and stand;
— militaire, military (money) chest;
— à munitions, (art.) ammunition chest;
— à obus (de siège), (siege) case for packing of shells, and for transportation of same from parks to batteries; (mil. r. r.) case or chest for transportation of shells and of small articles;
— d'outils, tool-chest;
— de parc, (siege) chest for tools, spare parts, tools for fireworks, etc.;
— de piston, (mach.) piston body;
— à poudre, powder chest;
— de poulie, shell of a block;
— de régiment, regimental (money) chest;
— de réparation, (mil.) chief armorer's tool-box;
— de réserve, (expl.) storing-box (for gun cotton);
— de résistance, (elec.) resistance-box;
— de roue, (nav.) paddle-box;

caisse roulante, (mus.) snare drum;
— à son, (art.) large box filled with bran, for recovery of projectiles;
— de tambour, (mus.) shell of a drum;
— à tasseaux, small-arm packing case;
— de transport, chest for tools, etc., carried by mountain batteries, telegraph sections (mountain), ambulance park, etc.; (expl.) gun-cotton shipping case;
— de trésorier, (mil.) money chest, pay chest;
— de voiture, body of a wagon, railway carriage.
caisson, m., caisson (esp. an artillery caisson); (in gen.) any closed wagon used in the military service, for the transportation of ammunition, tools, supplies; (cons.) caisson;
— d'ambulance, hospital wagon;
— d'appui, (fort.) supporting sheet-iron boxing or framing of a turret;
— de bataillon, (inf.) infantry ammunition supply wagon (now disused, Fr. a.);
desservir un —, to serve ammunition from a caisson;
— à dynamite (mélinite), dynamite (melinite) wagon of the engineer parks;
— des équipages militaires, (Fr. a.) caisson similar to that of artillery, used a long time for transportation of subsistence supplies;
— à fusées, rocket carriage;
— d'infanterie, infantry ammunition supply wagon;
— à munitions, ammunition wagon;
— à poudre, (Fr. eng.) powder-wagon of the engineer park.
cal, m., the joint or knitting of a fractured bone;
— osseux, (hipp.) bone spavin.
calade, f., (man.) slope of a rising manège ground.
calage, m., wedging, chocking, blocking; (inst.) leveling (as a transit); adjustment or setting of various parts of an instrument (or machine) with a view to keying or clamping them in proper position; clamping; operation of fixing, e. g., an axis, in its proper position; adjustment; (art., sm. a.) jamming of breech mechanism;
angle de —, (elec.) lead of the brushes; (steam) angle between eccentric radius and crank;
anneau de —, (art.) key ring;
— à chaud, (mach.) hot forcing, hot pressure;
frette de —, (art.) locking hoop;
— à froid, (mach.) cold pressure, cold forcing.
calaison, f., (nav.) load water-line.
calamite, f., calamite, loadstone.
calandre, m., (steam) the lower tube of a two-tube boiler.
calcaire, a., calcareous; m., limestone.
calcanéum, m., (hipp.) point of hock, os calcis.
calcinage, m., roasting, calcining.
calcination, f., calcination, roasting.
calciner, v. a., to roast, to calcine.
cale, f., key, wedge; chock, block; quoin; prop, beam; (top.) cove, small bay; (mach.) key; (harn.) small pad (as on headgear of pack horse); (nav.) hold (of a vessel); shipbuilding stocks; (mach.) liner;
— d'ajustage, tightening key, adjusting key;
— en bois, wooden thickness piece;
— de construction, shipyard;
grande —, (art., etc.) wheel chock;
grande — de roue, (r. r.) wheel chock;
— longue, (art.) roller chock;
petite —, (art.) gun chock;
— de radoub, graving dock;
— de roue, (art., etc.) wheel chock;
— de rouleau, (art.) roller chock;
— sèche, dry dock.
calé, a., said of a level when the bubble is bisected by origin of graduations.
caleçon, m., drawers.
calepin, m., memorandum book; (sm. a.) greased patch, patch of a rifled bullet.

caler, v. a., to chock, scotch (wheels); to key; to wedge, wedge in; (*nav.*) to draw (so many feet of water); to strike, lower colors; (*inst.*) to level (as a transit); to adjust in a given position, to set, to clamp (an instrument when set up); (*mach.*) to set (e. g., the cranks of pumps coupled up together); (*elec.*) to give the brushes their lead; (*art.*) to jam (a turret, or a gun, by a well-placed shot);
—— *à chaud* (*froid*), (*mach.*) to force, press on hot (cold);
—— *les roues*, to chock, scotch, wheels;
se ——, to be keyed on.

calfait, m., calking iron.

calfat, m., calking, calker, calking iron;
fer de ——, calking iron;
fer de —— *double*, calker's making-iron;
maillet de ——, calking mallet.

calfatage, m., calking.

calfater, v. a., to calk;
maillet à ——, calking mallet.

calfateur, m., calker.

calibrage, m., (*mach.*, *sm a.*, *art.*, etc.) operation of giving a piece its exact form and dimensions; calibration, gauging; verification of form and dimensions by means of gauges, templets, etc.; verification of the caliber of a cylinder esp. of the bore of firearms; (*sm. a.*) in manufacture, the increase in the diameter of a boring, and removal of irregularities left by first boring.

calibre, m., gauge, templet; calipers; wire-gauge; (*art.*, *sm. a.*) caliber; (*art.*) fuse-gauge, shot-gauge;
—— *d'un arbre*, (*mach.*) diameter of a shaft, axle;
compas de ——, calipers;
—— *coulant*, sliding gauge;
—— *à débiter*, model (*sm. a.*, gunstock model);
—— *à entailles*, stand measure, surface length measure; line-standard;
être de ——, to be of proper size, caliber, diameter, etc.;
—— *fort et faible*, maximum and minimum gauge;
gros ——, (*art.*) large caliber (24ᶜᵐ and above);
moyen ——, (*art.*) medium caliber, 90ᵐᵐ to 19ᶜᵐ;
petit ——, (*art.*) small caliber, 37ᵐᵐ to 65ᵐᵐ;
—— *pour pas de vis*, screw-pitch gauge;
—— *type*, caliber gauge.

calibré, a., sized, of proper size and form, fitted to size;
chaîne —— *e*, sprocket chain, pitch-chain.

calibrer, v. a., to take, measure, the size of; to gauge; to measure with calipers, gauge, etc.; to calibrate an instrument; (*art.*, *sm. a.*) to take, measure, the caliber of a gun, to fix the caliber of a firearm and of its projectile according to determined rules;
—— *une bouche à feu*, —— *un projectile*, (*art.*) to gauge a gun, a projectile.

calice, m., (*art.*, *sm. a.*) cup of a fuse, of a primer; (*artif.*) mouth of a rocket-case.

caliorne, f., (*cord.*) purchase-tackle, any large 2- or 3-fold tackle;
—— *de retenue*, retaining tackle.

calme, a., calm; **m.,** calm (no wind);
—— *plat*, dead calm.

calorie, f., calorie (unit of heat);
—— *gramme-degré*, v. *petite* ——;
grande ——, (unit) amount of heat necessary to raise 1 kilogram of water from 0° C. to 1° C.;
—— *-kilogramme*, v. *grande* ——;
—— *négative*, amount of heat taken out, as by a freezing mixture, etc.;
petite ——, (unit) amount of heat necessary to raise one gram of water from 0° C. to 1° C.

calorifère, m., heater, heating apparatus; **a.,** heat-conducting;
—— *à air*, hot-air apparatus;
—— *à eau*, hot-water heating apparatus;
—— *à la vapeur*, steam heater.

calorifuge, a., nonconducting (of heat).

calorimètre, m., calorimeter.

calorimétrie, f., calorimetry.

calot, m., (*unif.*) stable cap; crown of a *képi* or shako.

calottage, m., capping, heading.

calotte, f., cap, knob, head; (*unif.*) stable cap (mounted troops); (*sm. a.*) back piece of a sword hilt, steel cap, of the Schmidt rifle bullet, butt cap of a revolver; (*inst.*) spherical cup or bearing (as of a plane table); (*fort.*) roof of a cupola, skin or lining of a cupola roof, of overhead turret armor; (*art.*) cap or capping of a projectile;
—— *-chéchia*, (*unif.*) fez, sort of fez worn by zouaves;
—— *de campagne*, (*unif.*) sort of Scotch cap;
—— *d'écurie*, (*unif.*) stable cap;
—— *sphérique*, (*fort.*) head or cover (as of a land turret).

calpin, m. v., *calepin*.

calquage, m., act, operation, of tracing (a drawing).

calque, m., tracing, copy, counterdrawing.

calqué, m., the same as *dessin calqué*.

calquer, v. a., to make a tracing (of a drawing), to trace;
papier à ——, tracing paper;
toile à ——, tracing cloth.

calquoir, m., tracing-point or instrument.

calus, m., callosity, blister.

camail, m., (*hipp.*) blanket covering the forehand of a horse (sick horses, horses from remount establishments, while on the road).

camarade, m., comrade; (*mil. slang*) regimental hairdresser;
batteries par ——, v. s. v. *batterie*;
battre par ——, v. s. v. *battre*;
—— *s de chambrée*, (*mil.*) men of same squad-room;
—— *s de combat*, (*mil.*) men that must not separate during drill in dispersed order, nor in battle;
—— *s de lit*, bedmates (during routes).

cambouis, m., oil and dirt, turned black by friction, dust, etc., on the bearing parts, working parts, of machinery, etc.

cambre, m., cambering, bulging.

cambré, a., curved, bent, broken-backed, (of a shoe) blocked; (*hipp.*) bow-legged (fore legs).

cambrement, m., curving, cambering, bulging, bending.

cambrer, v. a., to bend, curve, crook, warp, cast.

cambrion, m., shank, waist piece of a shoe.

cambriser, v. a., (*mil. min.*) to plank the gallery of a mine.

cambrure, f., bend, curve, flexure; arch of the instep; curve, hollow, of the instep of a shoe, of the foot; filling (of a shoe); (*sm. a.*) curve of a sword blade.

came, f. (*mach.*) cam, lifter, wiper, tappet; cog;
à —— *s*, cogged, fitted with cams;
—— *à bander*, cocking-cam (machine guns);
—— *de détente*, trigger-cam, tappet;
—— *fixe*, "island" cam (Maxim r. f. mach. gun);
—— *de manœuvre*, action-cam (Nordenfeldt);
—— *de platine*, lock-cam (Gardner).

camelote, f., poorly done work; trash, poor stuff; worthless merchandise.

camion, m., truck, dray; (*art.*) sling-cart;
—— *à dynamite*, (*expl.*) dynamite truck of the French engineer park.

camionnage, m., transportation of material between railway and storehouse.

camisade, f., (*mil.*) night attack (obs.).

camisard, m., (*mil. slang*) soldier of the *bataillon d'Afrique* (punishment corps).

camouflet, m., (*mil. min.*) camouflet (i. e., a mine producing no exterior crater);
—— *-contrepuits*, camouflet intended to act as a *contrepuits* (i. e., to destroy shafts, etc., of the attack without damage to the defense), and prepared by boring from a gallery;
machine à —— *s*, camouflet-borer (sort of auger).

camp, m., (*mil.*) camp, encampment;
aligner un ——, to lay out a camp;

camp, *asseoir un* ——, to pitch a camp;
  assiette d'un ——, site of a camp;
  —— *baraqué,* —— *de baraques,* hut-camp;
  —— *à cheval (sur un cours d'eau),* camp with a stream running through it;
  —— *décousu,* camp irregularly laid out;
  —— *défensif,* v. —— *retranché;*
  dresser un ——, to lay out a camp;
  établir un ——, to pitch a camp;
  —— *d'exercice,* camp of instruction;
  —— *fixe,* permanent camp;
  garde du ——, camp-, quarter-guard;
  —— *d'instruction,* camp of instruction, of exercise (to-day in France synonymous with *champ de tir,* target-practice camp);
  lever le ——, to strike camp;
  ligne magistrale d'un ——, line of encampment, color-line;
  maladie de ——, camp fever; typhus;
  —— *de manœuvre,* camp of exercise, instruction (almost obs.);
  —— —— *de marche,* temporary camp;
  maréchal de ——, major-general (obs.);
  marquer un ——, to lay out a camp;
  —— *offensif,* the camp of a besieging army;
  —— *de passage,* —— *passager,* temporary camp;
  —— *de rassemblement,* camp in which troops are assembled previous to a campaign (common in United States in 1861 and in 1898);
  —— *retranché,* (formerly) any intrenched camp; a camp under the walls of a fortress and commanded by its guns (Vauban, obs.); (to-day) a military position, strongly fortified and intrenched, and supported by permanent works, behind which position an army may refit, recruit, etc. (In other words, a *camp retranché* means a *place à forts détachés,* an equivalent that is semiofficial in France, though held inaccurate by some writers. Other authorities go so far as to say that the expression *camp retranché* should be banished from the military vocabulary);
  —— *de séjour,* permanent camp;
  service de ——, camp duties;
  —— *de siège,* camp of a besieging army, siege-camp;
  —— *tendu,* —— *tenté,* —— *de tentes,* camp under canvas, regular camp;
  tracer un ——, to lay out a camp;
  —— *volant,* flying, temporary camp.

campagne, f., country; (*mil.*) campaign, service in a campaign; series of military operations; a whole war (rare); (the) field;
  année de ——, v. s. v. *année;*
  artillerie de ——, (*art.*) field artillery;
  avoir, avoir fait (tant de) ——*s,* to have so many years' service;
  avoir fait la —— *de,* to have served in such and such a campaign;
  battre la ——, (*mil.*) to scout; to make incursions in the enemy's country;
  couverture de ——, soldier's blanket;
  décompte des ——*s,* estimation of campaigns in length of service, v. s. v. *année;*
  demi- ——, (*Fr. a.*) half-campaign, allowance of half the time actually served on a man-of-war (peace), or in ports or seacoast forts (war), v. s. v. *année;*
  en ——, in the field;
  entrer en ——, to take the field;
  être en ——, to be in the field;
  faire ——, to take the field;
  faire une ——, to serve through or in a campaign;
  indemnité d'entrée en ——, (*Fr. adm.*) allowance (of money) paid to an officer joining an army in the field, or a mobilized body of troops;
  mettre en ——, to put or bring into the field;
  se mettre en ——, to take the field;
  ouvrir la —— *contre,* to take the field against;
  pièce de ——, (*art.*) field piece;
  en rase ——, in the open field;
  tenir la ——, to hold the country, keep the field;
  tenue de ——, (*unif.*) field dress; marching order for the field.

campêche, m., Campeachy wood, logwood.

campement, m., (*mil.*) encampment; (synonym of) camp; the establishment of a camp; party sent forward to select a cantonment or bivouac; camping-party (U. S. A.);
  chef, commandant, du ——, officer or n. c. o. in command of a *campement;*
  couverture de ——, soldier's blanket;
  effets, matériel, du ——, camp equipage;
  service administratif du ——, (*Fr. adm.*) department that furnishes material for shelter and bedding in camps and cantonments.

camper, v. n., (*mil.*) to camp, encamp;
  —— *en potence,* to camp with flanks thrown back.

camper, m., (*hipp.*) spreading out of the four feet from the normal and regular base of support (manner of standing in which the horse does not keep his feet well under him, but spreads them out front and rear).

camphre, m., camphor;
  —— *-poudre,* (*expl.*) camphorated powder.

camphrer, v. a., (*expl.*) to camphorate.

campylomètre, m., improved curvimeter, instrument to measure lengths of curves on maps.

camus, a., flat-nosed;
  cheval à tête ——*e,* (*hipp.*) hollow-nosed horse.

can, m., edge (as of an armor plate).

canal, m., canal; channel, bed, bed of a river; groove; conduit, tube, drain; (*art.*) rear-sight socket; (*fond.*) drain-metal;
  —— *d'aérage,* air-hole, ventilating flue;
  —— *d'amorce,* (*expl.*) detonator-hole of a gun-cotton disk; (*art.*) vent channel of percussion locks (obs.);
  —— *d'assèchement,* drainage-canal;
  —— *de baguette,* (*sm. a.*) ramrod groove;
  —— *du canon,* (*sm. a.*) barrel groove of the stock;
  —— *de communication,* (*art.*) communicating groove or channel of a fuse;
  —— *de coulage,* (*fond.*) v. —— *de coulée;*
  —— *de coulée,* (*fond.*) runner; (less accurately) gate, ingate;
  —— *de dérivation,* supply channel (of a mill);
  —— *de dessèchement,* v. —— *d'assèchement;*
  —— *d'écoulement,* channel of a river;
  —— *-évent,* (*art.*) flame passage of a fuse;
  —— *de fuite,* waste channel for escape of water after use in a mill;
  —— *fusant,* (*art.*) time-train; time-channel, of a fuse;
  —— *de hausse,* (*art.*) rear-sight socket or slot;
  —— *longitudinal,* (*art.*) powder channel of a fuse;
  —— *de lumière,* (*art.*) vent;
  —— *de navigation,* ship canal;
  —— *du percuteur,* (*sm. a.*) seat of the firing-pin or striker;
  —— *de poulie,* channel of a block, pulley;
  —— *de la tige,* (*mach.,* etc.) piston-hole;
  —— *de visée,* (*art.*) aiming slot or hole (Grüson turret, etc.).

canali, m., (*cons.*) sort of roofing tile, convex upward, used in combination with *tegoles,* q. v.

canalisation, f., the confinement of a river between artificial banks; sewerage; operation of digging canals; (*steam*) connections (generic term); (*elec.*) wiring (with special reference to conduits or channels).

canaliser, v. a., to inclose a river between artificial banks; to render navigable (a stream); to dig canals; (*elec.*) to wire (with special reference to conduits, etc.).

canapsa, m., tool-bag; (*mil.*) knapsack (obs.).

canarder, v. a., (*mil.*) to fire somewhat at random, from under cover.

canardière, f., (*sm. a.*) duck-gun.

candéfaction, f., (*met.*) state of being at a white heat.

candelette, f., (*cord.*) burton, top-burton; fish-tackle.

canevas, m., canvas, sailcloth; (*surv., top.*) network of triangles, triangulation, skeleton of triangles;
  —— *de détail,* triangulation that forms the immediate basis for filling in, whether of details of surface or of leveling;

canevas *de détail de nivellement*, triangulation for filling in differences of level, on which the detail work of leveling is based;
—— *directeur du tir*, (*siege*) skeleton survey connecting observatories, batteries, and natural bench-marks or stations (attack);
—— *d'ensemble*, general framework or skeleton of the whole area to be surveyed, primary triangulation (applies to leveling as well as to areas);
—— *d'ensemble de nivellement*, primary triangulation for the determination of differences of level;
—— *général*, framework or skeleton of triangles amplifying or extending the
—— *d'ensemble*, secondary triangulation;
—— *géodétique*, geodetic triangulation;
—— *de nivellement*, triangulation for the determination of differences of level;
—— *polygonal*, triangulation, system of triangles;
—— *trigonométrique*, triangulation;
—— *trigonométrique de nivellement*, (accurate) determination of differences of level in a survey.
caniveau, m., channel or conduit for shafting, belting, or transmissions in general; (*elec.*) subaqueduct for wires.
canne, f. , cane, stick;
cours de ——, single-stick instruction;
—— *hippométrique*, (*hipp.*) stick for measuring height of horses;
jeu de ——*s*, singlestick;
—— *de tambour-major*, drum-major's baton.
cannelé, a., grooved (as, *mach.*, a roller; *sm. a.*, a bullet).
canneler, v. a., to channel, groove, flute.
cannelle, f., cock, spigot, tap, faucet.
cannelure, f., channel, groove, cannelure, fluting; rib, spline (rare); (*sm. a.*) hollow of a blade;
balle à ——*s*, (*sm. a.*) cannelured bullet.
cannetille, f., (*unif.*) flat braid of twisted gold or silver wire, worn on collar and *képi* by certain military functionaries; (*garde d'artillerie, adjoint du génie, etc.*).
cannette, f., v. *cannelle*.
canon, m., (*art.*) gun, cannon, (sometimes, in singular) all the guns of a position, of an army, or of part of army engaged, (*fig.*) artillery fire; (*sm. a.*) gun-barrel (inclusive of *boîte de culasse*); (*mach., etc.*) any bored cylinder; (*harn.*) beam of a bit; (*hipp.*) canon, canon bone;
—— *en acier*, steel gun;
—— *en acier fondu*, cast-steel gun;
—— *d'alarme*, alarm gun;
—— *à âme amovible*, field gun with removable tube (Oboukoff);
—— *à âme lisse* (*rayée*), smooth-bore (rifled) gun;
—— *d'assaut*, gun used in preparation of an assault;
—— *d'attaque*, gun used in armor-plate proofs;
au ——, by artillery fire;
—— *automatique*, (*r. f. art.*) automatic gun;
—— *à balles*, the French mitrailleuse of 1870-71, now assigned to ditch-flanking;
—— *de batterie*, (*nav.*) broadside gun;
—— *à bêche*, gun whose carriage is fitted with a recoil-checking spade;
—— *de bombardement*, any gun delivering high-angle fire;
—— *de bord*, naval gun;
—— *-bouche*, —— *se chargeant par la bouche*, muzzle-loading gun;
—— *à boulons*, Schultz wire-gun, in which the longitudinal stress is taken up by rods (*boulons*);
—— *brisé*, (*harn.*) jointed beam of a bit;
—— *de campagne*, field gun;
—— *de cavalerie*, horse-artillery gun;
—— *cerclé*, hooped gun;
—— *de chasse*, (*nav.*) bow chaser;
—— *à cheval*, horse-artillery gun;
—— *choke-bored*, (*sm. a.*) choke-bored barrel (sporting);

canon *de combat*, an expression for a siege or position gun used on the battlefield (as distinguished from the same gun used in sieges, etc.);
—— *de côte*, seacoast gun;
coup de ——, gunshot;
—— *court*, shell-gun, howitzer;
—— *cuirassé*, —— *à cuirasse*, gun with shield mounting, gun mounted in a turret;
—— *-culasse*, —— *se chargeant par la culasse*, breech-loading gun;
—— *de débarquement*, (*nav.*) landing gun;
—— *démontable*, gun that takes apart, jointed gun (e. g., some mountain guns), take-down gun;
à deux ——*s*, (*sm. a.*) double-barreled (sporting);
—— *à deux hausses*, gun with two sight-seats;
—— *à dynamite*, Zalinski dynamite gun;
—— *d'embarcation*, (*nav.*) boat gun;
enclouer un ——, to spike a gun;
faire ——, (*min.*) of a charge, to blow out or to detach only a very small amount of material;
faux ——, Quaker gun;
—— *à fils*, —— *à fils d'acier*, wire-gun, wire-wound gun;
—— *à fils à segments*, (Brown) segmental wire-wound gun;
fonderie de ——*s*, gun factory;
—— *en fonte*, cast-iron gun;
—— *de forge*, (*sm. a.*) bar or skelp of which the barrel is made;
—— *de fourrés*, "exploration" gun (i. e., a gun for use in heavily wooded countries, in thickets, during exploration, (Krupp);
—— *fretté*, built-up gun;
—— *de fusil*, (*sm. a.*) gun-barrel;
—— *à gargousse obturatrice*, any gun using a self-sealing cartridge;
—— *à grande longueur*, gun of great length of bore, (hence) high-power gun;
—— *de gros calibre*, heavy gun, (sometimes) high-power gun;
—— *à jaquette*, turret-gun mounted in a recoil-sleeve; wire-wound gun, in which a cast-iron jacket surrounds the wiring;
—— *lance-torpilles*, (*torp.*) torpedo-launching tube, torpedo tube;
—— *lisse*, smooth-bore gun;
—— *à lumière centrale*, gun with axial vent;
—— *à lumière dans le renfort*, gun with radial vent;
—— *-machine*, machine gun;
marcher au ——, (*mil.*) to march to the sound of the guns;
—— *de marine*, naval gun;
—— *mécanique*, generic term for machine and for r. f. guns;
—— *de montagne*, mountain gun;
—— *-obusier*, shell gun, howitzer;
—— *-pendule*, (*ball*), gun-pendulum;
—— *perce-cuirasse*, a. p. gun, high-power gun;
—— *de perforation*, a. p. gun, high-power gun;
—— *à pivot*, swivel gun;
—— *de place*, fortress gun;
—— *pneumatique*, air-gun; Zalinski gun.
—— *pneumatique américain*, —— *pneumatique à la dynamite*, dynamite gun, Zalinski gun;
—— *porte-amarre*, life-saving gun or mortar;
à portée de ——, within cannon range;
—— *de position*, gun of position;
—— *de 8, 10, etc., pouces*, 8", 10", etc., gun;
—— *de quatre, (six, etc.)*, 4-pounder; gun firing a 4-kg. projectile, etc.;
—— *rayé*, rifle, rifled gun; (*sm. a.*) rifled barrel;
—— *à recul supprimé*, —— *sans recul*, nonrecoil gun;
—— *de remise*, term used by Brialmont, descriptive of guns kept in store near emplacement and mounted when wanted;
—— *de retraite*, (*nav.*) stern-chaser;
—— *-revolver*, (*r. f. art.*) Hotchkiss revolving cannon;

canon *rond,* (*hipp.*) a canon having a soft limp tendon;
—— *de rupture,* a. p. gun;
—— *segmentaire,* the Brown segmental wire-gun;
—— *semi-automatique,* (*r. f. art.*) semi-automatic gun;
—— *de siège,* siege gun;
—— *sous-marin,* (*nav.*) submarine gun;
—— *à sphère,* muzzle-pivoting gun (spherical joint);
—— *à tête sphérique,* v. —— *à sphère;*
—— *à tir rapide,* r. f. gun;
—— *-torpille,* torpedo gun;
trou à ——, (*fort., art.*) gun-pit;
—— *tubé,* gun having a lining-tube;
—— *tubé et fretté,* built-up gun;
—— *de torpilleur,* (*nav.*) torpedo-boat gun;
—— *de 4, etc.,* gun firing a 4-kg., etc., projectile;
—— *de 80, etc.,* an 80mm, etc., gun.
canonnade, f., (*art.*) cannonade.
canonnage, m., (*nav.*) the art of practical gunnery.
canonner, v. a., (*art.*) to batter, cannonade;
—— *en plein bois,* to hull a ship.
canonnier, m., (*art.*) cannoneer, artillery-man, gunner; (*sm. a.*) barrel-maker; (*hipp.*, in pl.) the two upper lumbrical muscles;
—— *-conducteur,* (*art.*) driver;
—— *de la pièce humide,* (*mil. slang*) hospital attendant, private of the hospital corps;
—— *sédentaire,* (*Fr. art.*) cannoneer belonging to the *bataillon de* ——s *sédentaires* of Lille, q. v.;
—— *-servant,* (*art.*) cannoneer, member of gun detachment (generally *servant* simply);
—— *vétéran,* (*nav.*) warrant officer or quartermaster who, after serving 4 years on board as a *canonnier,* is sent on board a special ship to perfect his training; seaman-gunner (U. S. N.)
—— *volant,* (*art.*) horse artilleryman (obs.).
canonnière, f., (*nav.*) gunboat; (*fort.*) embrasure (obs.);
—— *cuirassée,* armored gunboat.
canot, m., small boat;
—— *de* ....*avirons,* boat of (so many) oars;
—— *à couple,* double-banked boat;
—— *pliable,* folding, collapsible boat;
—— *porte* (——) *fusée,* rocket-boat;
—— *porte-torpille,* spar-torpedo boat;
—— *de sauvetage,* lifeboat;
—— *à vapeur,* steam launch.
canotage, m., art of managing small boats; (*pont.*), boat drill.
canotier, m., boatman.
cantine, f., small box or chest; (*mil.*) canteen, post exchange (U. S. A.);
—— *d'ambulance,* field-hospital chest (medicines, surgical instruments, and records);
—— *à bagages,* chest for officer's baggage and for material of the regimental staff;
—— *de compatabilité,* (*Fr. a.*) chest for papers, etc. (field bakery);
—— *médicale,* field medical chest;
—— *régimentaire,* regimental chest (no longer regulation, *Fr. a.,* though some are still in service);
—— *vétérinaire,* horse-medicine, etc., chest;
—— *à vivres,* officers' (field) mess-chest.
cantinier, m., (*mil.*) mess-steward; sutler.
cantinière, f., (*Fr. a.*) woman (soldier's wife) in charge of canteen (n. e.);
—— *-vivandière,* (*Fr. a.*) the same as *cantinière* (n. e.).
(In addition to canteen duties, these women are required to keep the mess for n. c. o.'s and men who are not required to live at the company mess.)
canton, m., canton; (*r. r.*) block (in block system).
cantonnement, m., (*mil.*) cantonment;
—— *-abri,* v. s. v. *abri.*
cantonner, v. a. n., (*mil.*) to be or stay in cantonments; to put troops into cantonments, to canton.

cantonnier, m., road-mender, road-worker.
canule, f., nozzle of a syringe; canula.
caoutchouc, m., india rubber, caoutchouc; tire (of a bicycle);
—— *anti-dérapant,* (bicycle) pebble-tread tire;
—— *creux,* single-tube bicycle tire;
—— *durci,* ebonite;
—— *pneumatique,* pneumatic tire (bicycle, etc.);
—— *sulfuré,* vulcanite, vulcanized rubber;
—— *tringlé,* bicycle tire with rod inside;
—— *vulcanisé,* vulcanite.
caoutchouter, v. a., to cover or treat with india-rubber.
cap, m., cape; (*nav.*) head of a vessel;
—— *de maure,* (*hipp.*) horse whose head is much darker than the rest of his body (sometimes spelled *more*);
mettre —— *au large,* (*nav.*) to stand out to sea;
—— *de mouton,* (*cord.*) dead-eye;
—— *de mouton à corde,* (*cord.*) rope-bound dead-eye;
—— *de mouton ferré,* (*cord.*) ironbound dead-eye;
de pied en ——, cap-à-pie.
capacité, f., generic term for a recipient or vessel; (*elec.*) capacity;
—— *inductrice,* (*elec.*) inductive capacity.
caparaçon, m., (*harn.*) caparison.
caparaçonner, v. a., (*harn.*) to caparison.
cape, f., hooded cloak;
—— *de batardeau,* (*fort.*) cover of a batardeau.
capeler, v. a., (*art.*) to slip over (as a jacket over a tube);
—— *un four,* to rig a heating furnace about a gun (e. g., one to receive a lining-tube, etc.).
capelet, m., (*hipp.*) capped hock.
capiston, m., (*mil. slang*) captain;
—— *bêcheur,* v. *capitaine bêcheur.*
capitaine, m., (*mil., nav.*) captain; (*fig.*) great general.
(Except where otherwise indicated, the following terms relate to the French services:)
—— *adjudant-major,* adjutant (v. *adjudant-major*);
—— *d'armes,* (*nav.*) warrant officer, ranking as *adjudant,* infantry instructor on board a man of war;
—— *bêcheur,* (*mil. slang*) officer acting as prosecutor of a court-martial;
—— *commandant,* (*cav.*) the commanding officer (first captain) of a cavalry troop; (*art.*) the first captain of a field battery (this designation, though no longer official, is still retained in current service);
—— *en deuxième,* (*art., cav.*) second captain (rare, usually —— *en second*);
—— *de distributions,* captain who superintends the issue of meat, forage, and other stores as furnished by contractors or by responsible agents (*comptables*);
—— *divisionnaire,* (*inf.*) captain commanding a division (two companies);
—— *écuyer,* instructor of equitation in a military school;
—— *de frégate,* (*nav.*) commander (U. S. N., R. N.);
—— *d'habillement,* officer in charge of, and responsible for all, the *matériel* of a regiment, including arms, commands the *section hors rang,* q. v.;
—— *ingénieur,* officer in charge of all technical material in the Paris fire brigade (*sapeurs-pompiers*);
—— *inspecteur d'armes,* captain of artillery (one or more to each *corps d'armée*), permanently detailed to inspect small arms and ammunition, and to report on same;
—— *instructeur,* (*art., cav.*) officer charged with various *manège* and hippological duties in a regiment, must have passed through the cavalry school of application;
—— *de logement,* captain commanding the *logement,* or party preceding a column on the march, and who reports the expected arrival of the column, etc., to the c. o. of the garrison or *place* through which the column is to pass;

capitaine *au long cours,* in the commercial navy, a master qualified for the foreign trade, i. e., to undertake long voyages; deep-water captain;
—— *-major,* captain who discharges the functions of *major,* in separate battalions (*bataillons formant corps*), in squadrons of the train, and in battalions of foot artillery;
—— *d'ordinaire,* captain that superintends mess issues;
—— *de pavillon,* (*nav.*) flag captain;
—— *de piquet,* captain commanding a *piquet;*
—— *en premier,* (*art., cav.*) first captain;
—— *de recrutement,* recruiting officer;
—— *de remonte,* remount officer;
—— *en second,* (*art., cav.*) second captain;
—— *de semaine,* captain detailed on duty for the week;
—— *de la soupe,* (*mil. slang*) officer who has never been under fire;
—— *de tir,* instructor of small-arm practice and range officer;
—— *-trésorier,* paymaster, officer who has the money responsibility of the regiment, and keeps the records (*archives*) of the regiment;
—— *de vaisseau,* (*nav.*) captain (U. S. N. R. N.);
—— *de visite,* captain detailed every day to inspect hospitals and prisons.

capitainerie, f., —— *générale,* (*mil.*) in Spain, the district or department commanded by a captain-general.

capital, m., capital;
—— *-actions,* capital stock;
—— *-obligations,* liabilities.

capitale, f., capital, metropolis; capital letter; (*fort.*) capital (either the bisector of a salient or reentrant angle, or the perpendicular at the middle point of the exterior side in a front);
—— *de champ de tir,* line down the middle of the range, perpendicular to the butt at its middle point;
—— *droite,* upright capitals, for legends of maps and names of towns;
en ——, (*fort.*) along the capital;
marcher en ——. (*siege*) to advance against a salient from both sides of the capital;
—— *penchée,* inclined capitals for names of villages and small towns, on maps.

capitonnage, m., impression or depression made by balloon cords in balloon.

capitulant, m., (*mil.*) German name for *rengagé,* i. e., soldier who reenlists.

capitulation, f., (*mil.*) capitulation; German name for reenlistment;
—— *par écrit,* written capitulation;
—— *verbale,* capitulation by word of mouth.

capituler, v. n., (*mil.*) to capitulate.

caponnière, f., (*fort.*) caponier;
—— *en as de pique,* caponier shaped like a pike-head;
—— *en capitale,* caponier along the capital;
—— *casematée,* casemated caponier;
—— *à ciel couvert* (*ouvert*), covered (open) caponier;
—— *- coupole,* a cupola performing the duty of a caponier;
—— *double,* double caponier;
—— *enterrée,* sunken caponier;
—— *en fer à cheval,* horseshoe caponier;
—— *à orcilles de chat* (lit. "cat-eared"), a caponier with a projection at each end of its head;
—— *en palissades,* palisade caponier;
—— *pleine,* full caponier;
—— *à redan,* caponier with redan-shaped head;
—— *de revers,* rear caponier;
—— *simple,* simple, single caponier;
tête de ——, caponier head, part farthest from center of work.

caporal, m., corporal (*Fr. a.,* infantry and engineers, is not counted a n. c. o.);
—— *armurier,* assistant of the *chef armurier;*
—— *brancardier,* corporal in charge of the litter-bearers;
—— *de chambrée,* corporal in charge of squad room (property, behavior of men, etc.);

caporal-*clairon,* corporal of the field music (bugles), charged with the instruction of the buglers;
—— *conducteur,* corporal in charge of the wagons, teams, etc., of the regiment;
—— *de consigne,* ranking corporal of a guard, has immediate charge of guardhouse and property;
élève- ——, lance-corporal;
—— *élève-fourrier,* corporal who assists the *sergent-fourrier,* q. v. (*chasseurs à pied*);
—— *d'escouade,* corporal of a squad, looks after it in barracks, sees that it observes regulations, etc.;
—— *-fourrier,* assistant of the *sergent-fourrier,* sort of quartermaster-corporal (n. e.);
—— *de garde,* corporal of the guard;
—— *d'infirmerie,* corporal responsible for the police, condition of the infirmary, watches convalescents, etc.;
—— *moniteur d'escrime,* assistant of the fencing-master;
—— *d'ordinaire,* mess corporal;
—— *de planton aux cuisines,* corporal detailed daily to supervise the police and discipline of the kitchen;
—— *de pose,* corporal of the guard that posts and relieves sentries, instructs them in their duties, etc.;
—— *premier ouvrier cordonnier,* corporal that acts as foreman in the shoe-shops of the regiment;
—— *premier ouvrier tailleur,* corporal foreman of the regimental tailor-shop;
—— *sapeur,* corporal in immediate charge of the *sapeurs ouvriers d'art,* q. v., of the regiment;
—— *secrétaire,* clerk of the major, treasurer, *officier d'habillement, d'armement,* etc.;
—— *de semaine,* assistant of the *sergent de semaine,* q. v., in all the duties of the *service de semaine;*
—— *-tambour,* corporal of field music (drums), instructs the drummers.

capot, m., cover, hood;
—— *de cabestan,* drum, head, of a capstan;
—— *de canon,* (*art.*) gun-cover;
—— *de cheminée,* funnel cowl;
faire ——, (*nav.*) to capsize, overset, turn turtle.

capote, f., (*unif.*) watch-coat, sentry's greatcoat; buffalo coat (U. S. A.); (*med.*) hood (of a litter);
—— *de faction,* —— *de guérite,* sentry's watchcoat; buffalo coat (U. S. A.).

capsule, f., (*sm. a.*) cap, capsule, primer; priming-cap; percussion-cap (*obs.*); (*expl.*) detonator, primer, detonating primer; (*mach., etc.*) external gauge, ring-gauge, external cylindrical gauge;
—— *-amorce,* (*expl.*) detonator, detonating primer;
—— *à balle,* (*t. p.*) gallery practice cartridge;
—— *fulminante,* fulminate cap, or primer.

capter, v. a., to take.

capteur, m., captor.

captif, a., captive; m., captive, (and popularly) captive balloon;
ballon ——, captive balloon.

captivité, f., captivity.

capture, f., (*mil.*) apprehension of a deserter; (*nav.*) prize, captured vessel; (*mil., law*) seizure of contraband goods;
prime de ——, (*mil.*) bounty or reward for the capture of a deserter.

capturer, v. a., to capture, esp. to make a prize, as a vessel.

capuchon, m., (*unif.*) hood; (*in gen.*) hood; (*cord.*) whipping.

capuchonner, v. n., (*man.*) to arch the neck (horse).

capucine, f., (*sm. a.*) band, lower band;
—— *du centre,* middle band;
—— *d'en haut,* upper band;
—— *premiére,* lower band.

carabine, f., (*sm. a.*) carbine;
—— *de cavalerie,* carbine of the French cavalry (except *cuirassiers*);
—— *de cuirassiers,* carbine with india-rubber butt plate, of the French cuirassiers;

carabine	64	**carré**

carabine-*express*, express rifle;
— *à magasin*, magazine carbine;
— *merveilleuse*, carbine with hair rifling;
— *à répétition*, repeating, magazine carbine;
— *de salon*, parlor rifle;
— *à tige*, carbine with a pin (*tige*) screwed in the bottom of the bore, to cause expansion of the bullet on ramming (obs.);
— *de tir*, target practice carbine.
carabinier, m., (*mil.*) carbineer.
caracole, f., (*man.*) caracole;
escalier en —, winding stair (sometimes written *caracol*, m.).
caracoler, v. n., (*man.*) to caracole.
caractère, m., **caractérisation**, f., (*mil.*) in German army, assignment of the next higher grade to an officer without the commission; assignment of next higher grade, by way of favor, to an officer leaving the active service.
caractérisé, a., (*mil.*) in German army, endowed with the next higher grade to the one for which a commission is actually held.
caractéristique, f., (*inst.*) the inclination that will make the air bulb of a level travel one division of the graduation, (hence) the value of one division of the bulb graduation, (*elec.*, *expl.*) characteristic;
— *externe*, — *interne*, (*elec.*) external (internal) characteristic.
carapace, f., (*fort.*) shell or surface, as a whole, of an armored fort; overhead cover in an (otherwise) open turret; cover, hood (of rounded form, over a shaft, etc.).
carapata, m., (*mil. slang*) cavalry recruit.
carboazotine, f., (*expl.*) safety blasting powder.
carbodynamite, f., (*expl.*) carbodynamite.
carbonatage, m., (*expl.*) operation of removing all acid from guncotton by alkalies, etc.
carbonate, m., (*chem.*) carbonate.
carbonater, v. a., (*chem.*) to convert into a carbonate; to saturate with carbonic acid.
carbone, m., carbon;
— *combiné*, (*met.*) combined carbon;
— *fixe*, (*met.*) free carbon, fixed carbon;
— *graphitique*, (*met.*) free carbon;
— *de recuit*, (*met.*) (in the cellular theory of steel) carbide of iron forming the *ciment* (q. v.), cementite;
teneur en —, proportion of carbon (in steel, armor plates, etc.);
— *de trempe*, (*met.*) free carbon held in solution.
carbonisation, f., carbonization; charring (as of foot of telegraph poles).
carboniser, v. a., to carbonize, to char.
carbonite, f., (*expl.*) carbonite.
carburateur, m., carburizer.
carburation, f., (*met.*) carburation.
carbure, m., (*chem.*) carbide;
— *à noyaux fermés*, benzine and its derivatives.
carburé, a., holding carbon in combination (e. g., steel).
carburet, m., (*chem.*) carburet, carbide.
carcas, m., (*met.*) refined cast iron.
carcasse, f., skeleton, frame; (*pont.*) frame of the U. S. A. advanced guard bridge equipage; (*torp.*) shell, case, of a torpedo; (*sm. a.*) frame (of a revolver); (*harn.*) skeleton tree of a saddle; (*artif.*) carcass, fireball;
— *de balles à feu*, (*artif.*) shell of a fire ball;
— *de képi*, (*unif.*) shell or body of a forage cap;
— *de piston*, (*mach.*) piston body;
— *de shako*, (*unif.*) body of a shako.
carcel, f., (*elec.*) carcel (unit of light, about 9½ British candles).
carder, v. a., to card (as cotton waste for guncotton).
cardinal, a., cardinal;
point —, cardinal point.
carène, f., (*nav.*) bottom of a ship; under-water hull.
caresser, v. a., (*man.*) to stroke, to make much of a horse.

caret, m., (*cord.*) rope-maker's reel;
fil de —, rope yarn.
cargot, m., (*mil. slang*) canteen man.
carie, f., rot (in wood); (*hipp.*) caries, ostitis;
— *dure*, (*hipp.*) a form of caries in which the bone, though dead, retains its structure;
— *molle*, (*hipp.*) a form of caries in which the bone disintegrates;
— *des mur (aille)s*, (*mas.*) efflorescence;
— *sèche*, dry rot (in wood).
carié, a., having dry rot (of wood); caried.
carieux, a., carious.
carillon, m., iron rod (applied to a variety of uses, e. g., to brace beams in iron floors);
fer de —, iron in small rods.
carlingue, f., (*nav.*) keelson.
carmin, m., carmine (color).
carnage, m., carnage, massacre.
carne, f., salient angle of a stone, of a table, etc.
carneau, m., flue, fire tube (of a furnace).
carnet, m., notebook, memorandum book; (*top.*) notebook, field notes. (Carnet, Fr. a., may be followed by almost any name descriptive of the sort of notes or accounts kept);
— *des accidents de tir*, (*t. p.*) record of accidents due to defective cartridges;
— *de caisse*, money account book of the council of administration;
— *de campagne*, record, kept by functionaries of the *intendance*, of orders received, incidents, etc., during fall maneuvers and in the field;
— *de comptabilité*, record kept by c. o. 's of administrative units, beginning on first day of mobilization, of all property and money transactions of their respective units;
— *de correspondance*, (*mil. sig.*) record of messages sent and received;
— *des déserteurs*, deserters' descriptive book;
— *médical*, (*in peace*) register of inventories, and of administrative matters in general; (*in war*) register of dead, sick, wounded;
— *de mobilisation*, register of details of mobilization, kept by captains;
— *de pointures*, register of the labels affixed to shoes, clothes, etc., kept in storehouses and magazines;
— *de réglage du tir*, notebook in which the range officer (*lieutenant d'armement*) enters certain required observations on the arms and ammunition in service;
— *de réquisition*, book containing blank requisition forms;
— *à souche*, stub-book;
— *de tir*, (*t. p.*) notebook containing information needed in order to carry on infantry indirect fire at a given place;
— *de tir de bataillon*, (*t. p.*) blank form for results or record, etc., of target practice.
carnier, m., game-bag.
caronade, f., (*nav. art.*) carronade (obs.).
carotte, f., (*mil. slang*) medical inspection.
carotter, v. a., — *le service*, (*mil. slang*) to shirk (vulgar); to dead beat (U. S. M. A.).
carpe, m., (*hipp.*) carpal bones, knee bones.
carre, f., angle, face; (*sm. a.*) face of a sword-blade, bayonet, etc.
carré, a., square;
en —, (so may inches, feet, etc.) square.
carré, m., square part (of anything); (*met.*) iron of square cross-section; (*cord.*) rope-maker's sledge; (*mil.*) square; (*man.*) rectangle (30m × 90m) laid out in the open, for instruction in riding;
— *du bouton*, (*art.*) cascable square.
— *du chien*, (*sm. a.*) tumbler-pin hole (obs.).
faire le —, (in driving) while holding both reins in left, to seize the right rein in right hand, hands 0.10m apart, and both reins equally stretched;
former le —, *se former en* —, (*mil.*) to form square;
grand —, (*man.*) large rectangle (300m × 100m) in the open, for instruction in riding;

carré, *position du* ——, v. *faire le* ——.
carreau, m., flagstone; floor; ground; fender-strake of a boat; window pane; large coarse file; (*expl.*) guncotton disk (parallelopiped in form);
 coucher sur le ——, to sleep on the floor;
 coucher, jeter, quelqu'un sur le ——, to kill or wound (one's adversary, etc.);
 demeurer, rester, sur le ——, to be left dead on the field;
 —— *de vitre*, window pane.
carreau-module, m., square, reference-square on a map.
carrefour, m., crossroads;
 —— *de galerie*, (*min.*, *mil. min.*) intersection of two or more galleries.
carrelet, m., large needle, packing needle; (*sm. a.*) sword of thin triangular blade; (*expl.*) cubical grain.
carrer, v. a., (*mil.*) to form square (obs.).
carrier, m., quarryman.
carrière, f., career; quarry; course; (*man.*) out-of-doors riding school or ground, open *manège*;
 cheval de ——, (*man.*) riding school horse;
 donner —— *à*, (*man.*) to give a horse his head;
 donner (*prendre*) ——, to increase the gallop (obs. as a mil. term);
 galop de ——, (*man.*) open *manège* gallop;
 officier de ——, (*mil.*) regular officer;
 système de ——, (*mil.*) system in which an officer accomplished his service in his own arm, as opposed to *système de passage*, q. v.
carrossable, a., (of roads) practicable for carriages, vehicles, passable by vehicles.
carrossage, m., set (of an axle-arm).
carrossier, m., draught horse, carriage horse;
 grand ——, coach horse;
 petit ——, small carriage horse.
carrousel, m., military festival or pageant by groups of horsemen in costume, who go through various exercises, etc.
carroyage, m., checkering (of cross-section or profile paper);
 copier par ——, to copy by means of squares on cross-section paper.
carrure, f., breadth across the shoulders.
cartahu, m., (*cord.*) whip, girt-line;
 —— *double*, whip upon a whip;
 —— *des projectiles*, (*art.*) shell-whip.
cartayer, v. n., to drive so as to keep a rut between the horses and between the wheels.
carte, f., map, chart; card, pasteboard;
 —— *d'assemblage*, in France, map consolidating the labors of the *brigade topographique*, and used in the orientation, etc., of siege artillery fire;
 —— *chorographique*, general map, map of a country;
 —— *en courbes*, contoured map;
 —— -*croquis*, sketch-map;
 —— *à l'effet*, map representing surface forms;
 —— *d'ensemble de tir*, (*art.*, *fortress*) map showing what guns bear on a given point;
 —— *de l'état-major*, French official map of France, on scale of 1:80,000, begun in 1818 and finished in 1882;
 —— *générale*, general map, ordinary geographical map;
 —— *géographique*, geographical map;
 —— *en hachures*, hachured map;
 —— *hydrographique*, hydrographic chart;
 lecture des ——*s*, map-reading;
 lire une ——, to read a map;
 —— *marine*, marine chart;
 —— *militaire*, military map or plan;
 —— -*minute*, map made directly from survey notes to the scale of the survey, and inked;
 —— *de mobilisation*, map issued when troops are mobilized for war, of the probable field of operations;
 —— *modelée à l'effet*, map showing relief of surface of the ground, by hachures, or by flat tints, suitably graduated, etc.;
 —— *de moulage*, (*artif.*) cartridge-paper, stiff board;
 —— *de l'observation*, (*Fr. art.*) plan used by observer in an *observatoire de tir*;

carte *orographique*, orographic map, mountain map;
 —— *orométrique*, map showing relief by contours, contoured map;
 —— *particulière*, special map;
 —— *plane*, —— *plate*, plain chart;
 pointer la ——, to prick the chart;
 porter un relèvement sur la ——, to lay down a bearing on the chart or map;
 —— *réduite*, Mercator's chart, plain chart;
 —— *routière*, road map;
 —— *topographique*, map of a small area, topographic chart, giving details;
 —— *universelle*, map of the world;
 —— *de visibilité*, map by which visibility of points may be determined with reference to a given point.
cartel, m., (*mil.*) cartel, cartel of exchange;
 —— *d'échange*, cartel of exchange;
 —— *d'extradition*, agreement between neighboring governments to surrender each other's deserters.
cartographie, f., cartography.
carton, m., pasteboard; any pasteboard box or receptacle; porfolio; (*artif.*) paper for pyrotechnic purposes;
 —— *à bretelles*, (*top.*) sketching case, slung around neck;
 —— *Bristol*, bristol board;
 —— *de cartouches*, (*sm. a.*, etc.) cartridge packing case;
 —— *de correction*, (*art.*) correction card;
 —— *à dessin*, sketching-board;
 —— -*planche*, sketching-board or -case (for military sketching, reconnaissance);
 —— -*portefeuille*, sketching-board, case (reconnaissance, etc.).
cartouche, m., ornamental scroll or drawing inclosing the legend of a map; frame or cartouch to receive a card or paper (as on a railway car); (*mach.*, etc.) receiver, cylinder (e. g., *sm. a.*, the receiver of the Giffard liquefied-gas gun); (*artif.*) rocket-case; (*ball.*) recorder of the Le Boulengé chronograph;
 —— *récepteur*, (*ball.*) recorder of the Le Boulengé chronograph.
cartouche, f., cartridge (esp. *sm. a.*, but said of artillery in a few cases); (*expl.*) cartridge, demolition cartridge, blasting cartridge;
 aller aux ——*s*, (*mil.*) to go for cartridges;
 —— -*amorce*, (*expl.*, *torp.*) cartridge serving as primer for a charge, e. g., of dynamite, of frozen dynamite;
 —— *à amorce centrale*, (*sm. a.*) center-fire cartridge;
 —— *à amorce périphérique*, (*sm. a.*) rim-fire cartridge;
 avaler sa ——, (*mil. slang*) to die;
 —— *à balle*, (*sm. a.*) ball-cartridge;
 —— *sans balle*, (*sm. a.*) blank cartridge;
 —— *à balles*, (*art.*) a term long used for case-shot; case-shot and its charge in one (obs.);
 —— *à balles multiples*, (*art.*) round of ammunition for the *canon à balles* (flank defense);
 —— *à blanc*, —— *blanche*, blank cartridge;
 —— *à bourrelet*, flanged, rim-, cartridge;
 —— *sans bourrelet*, rimless cartridge;
 —— *à bourrelet et à gorge*, flanged and crimped cartridge;
 —— *à broche*, pin-cartirdge, pin-fire cartridge;
 —— *s de cartouchière*, cartridges issued for guard and other security purposes;
 —— *à collet*, bottle-shaped cartridge;
 —— *de combat*, (*r. f. art.*) service cartridge;
 —— *combustible*, self-consuming (paper) cartridge;
 —— *complète*, (*sm. a.*, *r. f. art.*) round of fixed ammunition;
 —— *à culot replié*, folded-head cartridge;
 —— *de dynamite*, dynamite petard;
 —— *à eau*, (*expl.*) sort of water jacket surrounding the explosive (safety device);
 —— *s d'économie*, unexpended cartridges;
 —— *emboutie*, (*sm. a.*, *r. f. art.*) solid-head, solid-drawn cartridge;
 étui de ——, cartridge case;
 —— *à étui combustible*, paper cartridge;

cartouche à étui métallique, metallic cartridge;
— à —, d', exercice, (r. f. art., sm. a.) drill cartridge, dummy cartridge;
—s d'exercice, yearly allowance of ammunition;
— explosive, (mil. min.) melinite cartridge used in making forages, q. v.;
fausse —, dummy cartridge;
— à fumée, (art.) (signal) smoke-producing cartridge to represent bursting of shells at target practice;
— à gorge, grooved-head cartridge;
— à gorge et sans bourrelet, crimped rimless cartridge;
— de guerre, regulation service cartridge;
— à inflammation centrale, center-fire cartridge;
— à inflammation périphérique, rim-fire cartridge;
— d'instruction, drill, dummy cartridge;
— métallique, metallic cartridge;
— à mitraille, (r. f. art.) case-shot; (sm. a.) fragmentation cartridge (containing several bullets instead of one, segments, etc.);
—s de mobilisation, (Fr. a.) supply of cartridges kept near assembling point of a corps, to be issued on mobilization;
— obturatrice, self-seating cartridge;
— de pansement, (med.) first-aid packet;
poche à —s, cartridge pouch;
— à poudre, blank cartridge;
—s de sac, v. —s de sûreté;
— spéciale, self-igniting cartridge;
—s de sûreté, cartridge issued to the guard, etc.;
— de tir réduit, (sm. a.) gallery practice cartridge;
— de tube à tir, subcaliber cartridge.
cartoucherie, f., cartridge factory.
cartouchière, f., (sm. a.) cartridge-box; (sometimes) quick-loader;
— automatique, box in which the cartridges are pushed out by a spring to the same point, one after another.
cascade, f., (top.) cascade, waterfall, falls.
cascane, f., (mil. min.) cascan, shaft from terreplein near rampart (obs.).
case, f., cabin, hut; compartment, pigeon-hole; (in blank forms) divisions made by lines cutting columns at right angles; (mil. min.) intersection of several galleries;
— aux charges, (art.) cartridge compartment of an ammunition chest;
— aux obus, shell compartment of an ammunition chest;
— aux projectiles, projectile compartment of an ammunition chest.
casemate, f., (fort.) casemate;
— armée, (art.) casemate supplied with its equipments;
— à canon, gun-casemate (direct fire);
— de contrescarpe, counterscarp casemate, under the glacis, at the salients of the front;
— défensive, generic term for gun-casemate;
— à l'épreuve, bomb-proof casemate;
— d'escarpe, scarp casemate, opposite and facing a ditch;
— à étages, tiered casemate;
— à un (deux) étage(s), single- (double-) tier casemate;
— de flanquement, flanking casemate (sometimes equivalent to caponier);
— d'habitation, casemate-quarters;
— à la Haxo, Haxo casemate;
— logement, casemate-quarters;
— de revers, counterscarp casemate;
— servie, (art.) casemate ready to open fire;
tête de —, head of a casemate, part turned toward enemy;
— à tir indirect, indirect-fire (mortar, generally) casemate.
casemater, v. a., (fort.) to casemate.
caser, v. a., to put away; to put away in a pigeonhole; to pigeonhole.
caserne, f., (mil.) barracks;
— blindée, (s. c. fort.) armored barrack.

casernement, m., (mil.) barracking, quartering in barracks; (Fr. a.) a generic term for all buildings, quarters, etc., used for the service or duties, quarters, and instruction of the men, for administrative departments, etc.;
commission de —, (Fr. a.) board of officers which fixes yearly the distribution and assignment of barracks and of hired quarters, determining the number of men to be sheltered, etc.;
effets de —, barrack-equipage;
masse de —, barrack-fund;
officier de —, (Fr. a.) lieutenant who, under the major, is responsible for the details of the casernement;
remettre un —, to turn over barracks to troops;
remise de —, turning over barracks to troops;
reprendre un —, (Fr. a.) to take over barracks quitted by troops (engineers and intendance);
reprise de —, taking over of barracks quitted by troops.
caserner, v. a., n., (mil.) to quarter, be quartered, in barracks;
faire —, to put into barracks.
casernier, m., (mil.) barrack-keeper, caretaker.
casier, m., set of pigeonholes for papers;
— judiciaire, (Fr. adm.) record kept by each civil court (in pigeonholes, hence name) of all legal proceedings, civil or military, against every person born in the district (arrondissement).
casque, m., (unif.) helmet;
— colonial, cork helmet worn by colonial troops;
— à pointe, German "pickelhaube."
casquette, f., cap; (mil. slang) bugle call "aux champs;"
la —, (in camp) signal of the arrival of the mail.
cassage, m., crushing (as of ore).
cassant, a., (met.) brittle, short;
— à chaud, hot-short, red-short;
— à froid, cold-short.
cassation, f., (mil.) reduction to the ranks; (law) reversal of sentence of lower court;
cour de —, court of appeals; in France, supreme court.
casse, f., (met.) crucible; (mil.) cashiering (obs.).
casseaux, m., pl., (hipp.) clamps used in French mode of castration.
casse-cou, m., (man.) horse-breaker, roughrider.
casser, v. a., to break, snap; (mil.) to reduce to the ranks; (law) to reverse, set aside a decision, a verdict;
— les genopes, to break the stops (of a flag);
— de son rang et remettre, replacer, soldat, to reduce a. n. c. o. to the ranks.
casse-tête, m., sand bag, loaded stick.
cassette, f., box, casing, case.
casseur, m., — de sucre à 4 sous, (mil. slang) member of a compagnie de discipline.
cassine, f., (fort.) fortlet of no importance; town or post not susceptible of defense; small isolated country house that may be defended (Marbot); (met.) limestone flux.
cassure, f., (met.) fracture;
— conchoïde, conchoidal fracture;
— crystalline, crystalline fracture;
— à éclats, splintering fracture;
— à fibres, — fibreuse, fibrous fracture;
— à fines fibres, silky fracture;
— à grosses fibres, fracture of coarse fiber;
— à grain(s), — granulaire, granular fracture;
— inégale, unequal fracture;
— lamelleuse, lamellar fracture;
— à nerf, fibrous fracture.
castor, m., **castorin**, m., (mil. slang) officer that rarely leaves his garrison.
castoriser, v. a., (mil. slang) to like to make a long stay in a pleasant garrison town.
castramétation, f., (mil.) castrametation.
castration, f., (hipp.) castration, gelding; (of a mare) spaying (rare);

castration *par casseaux*, gelding by the clamp method;
— *à testicules couverts (découverts)*, covered (uncovered) operation.
castrer, v. a., (*hipp.*) to geld, cut, castrate; to spay.
catalan, a., Catalan;
méthode à la ——*e*, (*met.*) Catalan process.
catalogue, m., catalogue.
cataplasme, f., (*med.*) poultice.
cataracte, f., (*top.*) waterfall, cataract; (*mach., steam*) cataract, regulator for single-acting engines; (*fort.*) sort of portcullis, herse; (*hydr.*) difference of level between upstream and downstream waters of a bridge.
catégorie, f., category; (*mil.*) class (of soldiers, i. e., with respect to entry into service); class of horses (officers' remounts, and in census of horses and of mules for mobilization).
cathétomètre, m., (*inst.*) cathetometer.
cathode, m., (*elec.*) cathode.
catoptrique, a., (*opt.*) catoptric.
caustique, a., (*chem.*) caustic.
cautère, m., cautery; (*hipp.*) rowel.
cautérisation, f., cauterization.
cautériser, v. a., to cauterize, sear.
caution, f., (*adm.*) bond, security, surety, bail;
—— *personnelle*, bondsman.
cautionnement, m., (*adm.*) sum deposited as security (in contracts, etc.);
—— *matériel*, money or property presented as security;
—— *personnel*, bondsman.
cavalcade, f., cavalcade.
cavalcader, v. n., to go on a cavalcade.
cavale, f., (*hipp.*) mare.
cavalerie, f., cavalry;
—— *d'exploration*, v. —— *indépendante*;
grosse ——, heavy cavalry, (*Fr. a.*, cuirassiers);
—— *indépendante*, cavalry of exploration, covers mobilization and concentration, forms the outer cavalry screen;
—— *légère*, light cavalry, (*Fr. a.*, chasseurs and hussards);
—— *de ligne*, cavalry of the line, (*Fr. a.*, dragoons);
—— *de réserve*, reserve cavalry, to support other cavalry (*ligne, légère*), and to act against broken infantry;
—— *du service d'exploration*, v. —— *indépendante*;
—— *du service de sûreté*, the cavalry of an army corps, performs security service in immediate neighborhood of the corps;
—— *de sûreté*, v. —— *du service de sûreté*.
cavalier, m., rider, horseman; staple (as, for telegraph wire, etc.); (*cav.*) trooper; (*fort.*) cavalier; (*r. r.*) spoil-bank;
—— *de bois*, (*man.*) dumb-jockey;
—— *conducteur*, (*mil.*) driver of the *train des équipages*;
—— *élève-brigadier*, (*Fr. a.*) lance-corporal of spahis;
—— *élève-télégraphiste*, (*Fr. a.*) trooper who takes a course of telegraphy at Saumur;
—— *espagnol*, (*man.*) sort of dumb jockey;
—— *léger*, light cavalry man;
—— *de manège*, trooper charged with care of horses, etc., in certain military schools;
—— *de remonte*, trooper in a remount establishment;
——*s rouges*, (*mil. slang*) spahis;
surfaix-——, (*man.*) sort of dumb-jockey;
—— *de tranchée*, (*fort.*) trench-cavalier (siege, attack).
cave, f., cave; cellar;
—— *à canon*, (*fort.*) sort of covered battery (arch overhead covered by 2^m or 3^m of earth, and behind rampart), for indirect fire;
—— *à mortier*, (*fort.*) mortar casemate, proposed by Carnot.
caveau, m., underground vault.
cavecé, a., (*hipp.*) headed;
—— *de noir*, black-headed horse

caveçon, m., (*man.*) cavesson.
cavée, f., hollow road.
caver, v. a., to dig holes, to hollow, undermine, excavate; (*fenc.*) to uncover, so as to give adversary a chance to hit.
cavesson, m., (*man.*) cavesson.
cavillot, m., v. *chevillot*.
cavin, m., hollow road; defile.
cavité, f., hollow, cavity; hole; (*art.*) pit, pitting, in a bronze gun.
ceaux, m., (*mil.*) abbreviation of *faisceaux*, used in commands only.
cécité, f., blindness;
—— *des couleurs*, color-blindness;
—— *lunatique*, moon-blindness.
céder, v. a., to cede, yield, surrender, resign, give up; give way;
—— *l'avantage du vent*, (*nav.*) to lose the weather gauge.
cèdre, m., cedar (wood and tree).
cédule, f., (*mil. law*) summons of a witness before a court-martial, signed by judge-advocate.
ceindre, v. a., to surround, compass, encompass; to put around; to inclose, fence in; (*mil.*) to surround a position, an army;
—— *une épée*, to buckle on a sword.
ceinture, f., belt, waistband; (*r. r.*) belt railway; (*nav.*) armor belt; (*fort.*) line or belt of works, surrounding a place; (*pont.*) wale or strip on interior of upper ends of the knees of a pontonboat, clamp; (*art.*) rifling band, rotating band, centering band; fillet, ring;
—— *d'appui*, (*art.*) centering-band (sometimes centering shoulder);
——-*arrière*, (*art.*) rifling-band, rotating band or ring;
——-*avant*, (*art.*) centering-band (in some projectiles, centering shoulder);
—— *de blindage à la flottaison*, (*nav.*) water-line armor belt;
—— *de centrage*, (*art.*) centering-band;
chemin de fer de ——, belt railway;
—— *à expansion*, (*art.*) rotating band; rifling band;
—— *directrice*, (*art.*) rotating-band;
—— *de flanelle*, waistband of flannel;
—— *de flottaison*, (*nav.*) water-line armor belt;
—— *forçante*, —— *à*, —— *de*, *forcement*, (*art.*) forcing, rifling, rotating, band;
—— *de natation*, swimming belt;
prolongation de ——, (*pont.*) blocks, blocking (under rowlocks, U. S. bridge equipage);
—— *de sauvetage*, life-belt, life-preserver;
taquet de ——, (*pont.*) cleat supporting *prolongation de* ——.
ceinturer, v. a., (*art.*) to seat a band on a projectile.
ceinturon, m., (*unif.*) waistbelt, swordbelt;
allonge de ——, belt-strap;
plaque de ——, waistbelt plate.
céleustique, f., (*mil.*) pedantic term for military signaling (obs.).
cellule, f., cell; (*mil.*) prison cell, punishment cell (solitary confinement);
—— *composée*, (*met.*) (in the cellular theory of steel) agglomeration of globulites, group of simple cells;
—— *de correction*, (*mil.*) solitary confinement (in the *régiments étrangers* and *bataillons d'Afrique* alone);
—— *de dépôt*, cell in which a military person may be temporarily confined;
—— *simple*, (*mil.*) ordinary solitary confinement (in France); (*met.*) (in the cellular theory of steel) a grain of iron surrounded by cementite.
celluloïd, m., celluloid.
cellulose, f., (*chem.*) cellulose.
cément, m., (*met.*) cementation powder; (*hipp.*) cement (of teeth).
cémentation, f., (*met.*) cementation, case-hardening;
four, fourneau, de ——, cementation furnace.
cémenté, a., (of armor plates) face-hardened (e. g., Harvey plates).
cémenter, v. a., (*met.*) to cement, to caseharden.

cendrage, m., (*fond.*) washing over, ashing over, of molds.
cendre, f., ashes, cinders; (in the arts, etc.) residues of combustion, ash;
— **de plomb,** (*sm. a.*) dust-shot, mustard seed (*cendrée* is more commonly used).
cendrée, f., dross, sweepings; (*sm. a.*) dust-shot, mustard seed.
cendrer, v. a., (*fond.*) to blacken, ash, wash, line, a mold.
cendreux, a., (*met.*) filled with, or having, ash-spots.
cendrier, m., ash-pit, ash-hole (of a furnace, etc.).
cendrure, f., (*met.*) ash, or cinder-spot in iron, etc.
centésimal, a., centesimal.
cent-gardes, m., pl., Napoleon III's bodyguard.
centiare, m., one square meter (very little used).
centigramme, m., centigram.
centilitre, m., centiliter, a hundredth of a liter.
centimètre, m., centimeter.
centime, m., centime;
— **s de baguettes,** (*Fr. a.*) retention of 5 centimes a day on pay of drummers and buglers, as a fund to keep instruments in repair, etc.;
— **d'ordinaire,** (*Fr. a.*) part of corporals' and of soldiers' pay turned into the mess fund;
— **de poche,** (*Fr. a.*) part of pay of corporals and of soldiers given to them in money.
centrage, m., centering, operation of centering, in general; (*top.*) centering; (*art.*) operation of finding axis of gun forging before shaping the outside and boring the inside; centering of a projectile in the bore; (*mach.*) centering (in a lathe);
— **du canon,** (*sm. a.*) operation of placing a centering collar on barrel, before turning.
entralisation, f., centralization; (*adm., Fr. a.*) consolidation and balance of quarterly accounts of funds received and expended;
registre de —, (*adm., Fr. a.*) consolidated account of quarterly receipts and expenditures, showing balance kept by each regiment and establishment counting as a regiment.
centraliser, v. a., to centralize; (*Fr. a., adm.*) to consolidate accounts, returns, etc.; (of an official) to supervise, to give the necessary orders, directions, etc., and to cause all matters to pass through his hands;
— *le service de,* (of an official) to cause all matters relating to a given service to pass through his hands, etc.
centre, m., middle point; (*mil.*) center;
— *d'approvisionnement,* (*mil., etc.*) center of supply;
— *du groupement,* (*ball.*) center of impact;
— *de résistance,* (*mil.*) point or position organized to offer a serious resistance to the enemy.
centrer, v. a., to find the center, to mark the center (as in lathe work); (*top.*) to center; (*art.*) to center a projectile in the bore; (*mach.*) to center work in a lathe, etc. (e. g., a gun, a projectile).
centrier, m., centripète, m., (*mil. slang*) foot soldier.
cerceau, m., hoop.
cercle, m., circle, hoop; tire of a wheel; ring, band; club; (*inst.*) limb; surveying or astronomical instrument for angle measurement; (*man.*) closed curve described by a horse in working, exercising; circling (by a squad); (*mus.*) hoop keeping drumhead down; (*mil.*) formation in circle to listen to publication of orders; circle of officers and n. c. o. receiving orders; subdivision of military territory in Algeria;
— *d'alignement,* (*inst.*) surveyor's transit;
— *annuel,* ring (of a tree);
— *arabe,* district of military territory in Algeria;
— *d'arrêt,* (*mach.*) a circle that limits or arrests the motion of some other piece;
— *azimuthal,* (*inst.*) azimuth circle;

cercle *de butée,* (*s. c. art.*) bearing circle (of a platform);
— *de cible,* (*t. p.*) circle of a target;
— *de commotion,* (*min.*) v. — *de friabilité;*
— *de contact de piston,* (*mach.*) piston-ring;
— *à crémaillère,* (*art., mach., etc.*) ratchet-circle;
— *décrit,* (*nav.*) turning-circle of a ship;
demi- —, (*art.*) calipers to measure exterior diameters;
— *directeur,* (*fort.*) directing circle;
— *de division,* (*mach.*) v. — *primitif;*
— *de la distribution,* (steam) valve-circle of a valve diagram;
— *dynamométrique,* (*expl.*) General Abbot's ring apparatus;
— *d'étoupe,* (*mach.*) junk ring;
— *de friabilité,* (*mil., min., etc.*) base of truncated cone through which the effects of explosion reach the surface of the ground, i. e., the circle whose radius is the crater radius;
— *fusant,* (*art.*) priming circle of a fuse;
— *geodésique,* (*inst.*) surveyor's transit with a single-trunnion telescope;
— *d'investissement,* (*siege*) circle of investment;
— *limite,* (*mach.*) clearance circle (of gearing);
— *militaire,* (*mil.*) military club;
— *noir,* (*t. p.*) bull's-eye;
ordre au —, (*mil.*) orders given to officers and n. c. o.'s. formed in a circle;
— *de plomb pour joint,* lead joint-ring;
— *primitif,* (*mach.*) pitch-circle;
— *proportionnel,* (*mach.*) ratio-circle;
quart de —, quadrant;
— *de recouvrement,* (*steam*) lap-circle of a valve diagram;
— *de réflexion,* (*inst.*) reflecting-circle;
— *répétiteur,* (*inst.*) repeating-circle;
— *de rupture,* (*min.*) circle of rupture;
— *de serrage,* gathering band or collar at the top of a balloon;
— *à taquet,* clasp hoop;
— *du tiroir,* (*steam*) valve-circle of a valve diagram;
— *à vis,* screw-hoop;
— *zénithal,* (*inst.*) zenith circle.
cercler, v. a., to hoop; (*art.*) to hoop a gun.
cercueil, m., coffin.
céréales, f. pl., cereals.
cérémonial, m., ceremonial.
cérémonie, f., ceremony.
cerf-volant, m., kite;
— *porte-amarre,* life-saving kite.
cerne, f., annual ring (of a tree).
cerner, v. a., (*mil.*) to surround an army, a position.
certificat, m., certificate.
(The following terms relate to the French army):
— *d'acceptation,* certificate from colonel or c. o. of recruiting bureau to a young man accepted as a volunteer;
— *administratif,* certificate that a responsible person assumes, or is relieved from, responsibility in the case in hand;
— *d'aptitude,* (in promotion by selection) certificate of fitness given to candidates after successfully passing the required examinations;
— *d'aptitude professionnelle,* certificate of proficiency presented by a young man wishing to serve in certain technical branches (artificers, artillery workmen, engineers, train, etc.);
— *d'aptitude au service,* document given by c. o. of a recruiting bureau to a young man who wishes to enlist and is recognized as fitted for the service;
— *de bien vivre,* certificate of good behavior during absence from regiment, presented by a detachment on rejoining, and furnished by the civil authorities; similar document for a rear-guard during a march in France (more or less obsolete);
— *de bonne conduite,* good character given on expiration of service, or on discharge;

certificat de bonne vie et mœurs, civil document setting forth correct life and morals, to be furnished by every Frenchman who offers himself for enlistment;
—— *de cessation de paiement*, certificate showing up to what date pay has been received, and the stoppages, if any, to be made (final statements, U. S. A.);
—— *de contre-visite*, certificate of medical inspection or examination, given as a check, or serving as a check, on a previous medical inspection (sick leave, discharge from hospital, etc.);
—— *d'entrée,* —— *de rentrée, en solde*, certificate of return to a status of pay (suppressed in 1883);
—— *d'examen*, medical certificate of examination for the retired list in all grades, and for *réforme No. 1* in case of the men;
—— *d'exemption*, certificate of total unfitness for military service;
—— *d'incurabilité*, certificate of chief surgeon of a hospital (on demand of *réforme*, by officers, etc.) of incurable disability not incurred in line of service;
—— *de non-opposition*, certificate that no objection to payment exists, on presentation of a check or order at a *caisse* other than that on which drawn;
—— *d'origine de blessures ou d'infirmités*, attest that wounds, maladies, etc., have been received in line of duty;
—— *de présence sous les drapeaux*, a certificate of presence with the colors, issued under certain conditions, on demand of interested parties;
—— *de vérification*, medical certificate, check on *d'examen*;
—— *de vie*, document vouching for the holder's being alive on a certain date, and required before certain payment can be made;
—— *de visite*, medical certificate given in certain well-defined cases (sick leave, transfer from one arm to another, etc.).

certifier, v. a., to certify;
—— *une copie*, to make "a true copy."

céruse, f., white lead.

cervelle, f., brain; (of earths, soils) cohesiveness.(?)

cessation, f., cessation, stop (of hostilities, payment, etc.)

cesse, v. a., (*in commands*) avast! (heaving); (*art.*) halt!

cession, f., cession (of territory); (*mil. adm.*) issue or grant from one department of the army to another.

chable, f., (*cord.*) stout rope passed through a pulley for raising a weight.

chableau, m., (*cord.*) towline, warp.

chabler, v. a., (*cord.*) to fasten a rope to anything in order to raise it.

chabotte, f., anvil-block.

chabraque, f., v. *schabraque*.

chaînage, m., (*surv.*) chaining; (*cons.*) bracing of walls by anchors, angle-irons, etc.

chaîne, f., chain; chain of mountains; tie, shackle, band; warp (of cloth); anchor, tie-rod (to bind walls together); main chain or cable (of a bridge); (*top.*) chain of triangles (as distinguished from a network); (*mil.*) chain (of skirmishers); the line engaged, the firing line; chain (of sentinels, etc.);
—— *allemande*, open-link chain;
—— *d'amarrage*, mooring-chain, anchoring-chain;
—— *d'ancrage en bois*, wooden anchor-chain, float-chain;
—— *anglaise*, close-link chain;
—— *de l'ancre*, chain cable;
anneau de ——, link;
—— *d'arpenteur*, (*surv.*) surveyor's chain;
—— *d'arrêt*, locking-chain; (*harn.*) breast-, pole-chain;
—— *à articulations*, —— *articulée*, v. —— *de galle;*
—— *d'attache*, stall-chain;
—— *d'attelage*, (*harn.*) breast-chain; (*r. r.*) coupling-chain;

chaîne *d'attelage latéral*, (*harn.*) side horse coupling chain;
—— *à barbotin*, sprocket-chain;
—— *de bout de timon*, (*harn.*) pole-chain, breast-chain;
—— *de bout de trait*, (*harn., esp. art.*) trace-chain;
—— *de brêlage*, keep-chain; lashing-chain (e. g., for chess in chess-wagon);
câble- ——, —— *-câble*, chain cable;
—— *calibrée*, sprocket chain; pitch chain;
—— *de charge*, (*mach., etc.*) any supporting chain;
clef de ——, (*pont.*) suspension chain hook or toggle;
—— *en coins à articulations*, (*mach.*) Clissold belt;
—— *de combat*, (*mil.*) the line engaged, the firing line;
—— *à crochets*, v. —— *à la Vaucanson;*
—— *décamètre*, (*surv.*) surveyor's chain;
—— *de démontage*, (*art.*) sling-chain;
—— *d'échappement*, (*art.*) that part of a locking or drag chain formed into a loop and not in bearing when wheel is on drag-shoe;
—— *d'embrelage*, —— *à embreler*, (*art.*) limber-chain, keep-chain;
—— *d'enrayage*, —— *à enrayer*, (*art.*) drag-chain, locking chain;
—— *d'équipement*, (*art.*) sling-chain;
—— *à étais*, studded chain, stay-link chain;
—— *sans étais*, unstudded chain, crane chain;
—— *étançonnée*, —— *à étançons*, v. —— *à étais;*
—— *sans fin*, endless chain;
—— *fixe*, stationary chain;
—— *de fourrage*, (*mil.*) chain of troops to protect foraging parties;
—— *de frein*, (*art.*) brake-chain;
—— *-galles*, —— *de galle*, pitch-chain, sprocket-chain, flat-link chain;
—— *à godets*, chain pump;
—— *de levage*, (*art., etc.*) hoisting, lifting chain;
—— *de licou*, (*harn.*) halter chain;
—— *à maillons*, link-chain (ordinary chain);
—— *à maillons étroits*, close-link chain;
—— *à larges maillons*, open-link chain;
—— *de manœuvre*, (*art.*) traversing chain; (*in gen.*) maneuvering chain;
—— *métrique*, (*surv.*) surveyor's chain;
—— *mobile*, running chain;
—— *de montagnes*, (*top.*) chain of mountains;
—— *ouverte*, generic term for open-link chain;
—— *de paratonnerre*, chain, conductor, of a lightning-rod;
—— *de pierres*, (*mas.*) course of dressed stones, in a wall of small stones, etc., intended to strengthen it; chain course;
—— *en plâtre*, (*cons.*) plaster bracing for floor sleepers;
—— *de pointage*, —— *de pointage en direction*, (*art.*) traversing chain;
—— *de port*, boom of a harbor;
—— *de postes*, (*mil.*) chain of posts;
—— *de retenue*, keep-chain;
—— *de retraite*, (*harn., art.*) breast-chain, backing chain;
—— *de rive*, (*nav.*) mooring chain, shorefast;
—— *de rochers*, (*top.*) ridge, ledge, of rocks;
—— *à saucissons*, (*siege*) fascine-chain, choking-chain;
—— *de sécurité*, (*r. r.*) safety-coupling chain; (*art.*) safety chain (as of a shell-hoist);
—— *de sûreté*, v. —— *de sécurité;* (*art.*) safety chain (on French ammunition caissons);
—— *de suspension*, (*pont.*) suspension chain (trestle-cap); (*art.*) the part of a sling-chain that bears the weight;
—— *de timon*, (*art., etc.*) pole-chain;
—— *de tirailleurs*, (*mil.*) line of skirmishers;
—— *de touage*, towing chain;
—— *de traction*, hauling, traction, chain;
—— *de transmission*, (*mach.*) power chain, driving chain;
—— *à la*, —— *de*, *Vaucanson*, pitch-chain, sprocket-wheel chain. (In this chain the links are not closed, but each link serves as a clasp to the next.)

chaînée, f., (*surv.*) chain-length.
chaîner, v. a., (*surv.*) to chain, chain off.
chaînette, f., small chain; catenary (curve); (*harn.*) slobbering-chain; (*sm. a.*) (*art.*) stirrup;
— *-arrêt*, keep-chain;
— *d'enrayage*, (*art.*) brake-chain;
former ——, to sag, bend down in the middle;
— *de servante*, (*art.*) pole-prop chain;
— *de susbande*, (*art.*) cap-square key chain;
— *à T*, small chain and toggle.
chaîneur, m., (*surv.*) chainman, (*esp.*) hind chainman;
— *porte-chaîne*, hind chainman.
chaînon, m., link (of a chain); cable-length; (*top.*) short chain of mountains;
— *en bois*, link or element of a chain of logs end to end.
chair, f., flesh; flesh side of leather; (*r. r.*) chair, railway-chair;
— *cannelée*, (*hipp.*) lamellar tissue of the foot.
— *à canon*, (*mil.*) food for powder;
— *feuilletée*, (*hipp.*) v. —— *cannelée*;
— *de pied*, (*hipp.*) envelope of the interior parts of the foot, keratogenous membrane; (*fam.*) flesh of the foot;
— *veloutée*, (*hipp.*) velvety tissue of the foot.
chaise, f., chair, seat; (*cons.*) frame, timber-work; (*r. r.*) rail-chair; (*mach.*) hanger; (*nav.*) armor shelf;
— *à articulation sphérique*, (*mach.*) ball-and-socket hanger;
— *à colonne*, (*mach.*) post hanger;
— *de la cuirasse*, (*nav.*) armor shelf;
— *à nervures*, (*mach.*) ribbed hanger;
— *à rotule*, (*mach.*) adjustable hanger;
— *de transmission et de renvoi*, (*mach.*) hanger.
chaland, m., lighter;
— *à cheval*, lighter for embarking horses (also spelled *chalan*).
chaleur, f., heat;
— *blanche*, (*met.*, *etc.*) white heat;
bouche de ——, heating pipe or flue;
conduit de ——, v. *bouche de* ——;
coup de ——, sunstroke;
— *rouge*, (*met.*) red heat;
— *soudante*, (*met.*) welding heat;
— *spécifique*, specific heat;
— *suante*, (*met.*) welding heat.
châlit, m., bedframe, frame of a bed.
chaloupe, f., launch;
— *canonnière*, (*nav.*) gunboat;
— *pliante*, collapsible boat, folding boat;
— *à vapeur*, steam-launch.
chalumeau, m., blowpipe;
— *à gaz*, gas blowpipe.
chamade, f., (*mil.*) drum or bugle for a parley with the enemy (obs.);
battre la ——, to sound or beat a parley; (*fig.*) to surrender, yield.
chamailler, m., (*mil.*) slight skirmish, brush, of no importance; v. n., to have a slight brush with the enemy.
chambard, m., (*mil. slang*.) at the École Polytechnique, act of smashing the furniture, etc., of new cadets, making hay in a new cadet's room.
chambarder, v. n., (*mil. slang*) at the École Polytechnique, to smash the furniture, etc., of new cadet's room.
chambrage, m., (*art.*) ratio of chamber diameter to caliber; (*mil. min.*) operation of preparing or forming the mine chamber by the use of high explosives (or of hydrochloric acid in calcareous rock); operation of increasing the diameter of a boring (by the same process); the mine-chamber itself, prepared by high explosives.
chambranle, m., frame, casing; door or window casing; (*min.*) frame of a mine gallery;
contre- ——, outer or false casing.

chambre, f., room, chamber; Chamber of Deputies (France); (*art.*, *sm. a.*) chamber (for the charge the projectile); chamber of a revolver; (*art.*) cavity, blowhole, defect, honeycomb (bronze and cast-iron guns);
— *à air*, air-chamber; inner tube of a bicycle tire;
— *ardente*, (*sm. a.*) a circular recess or space in rear of paper cartridge of first breech loader used, to insure expulsion, combustion, of unburnt parts of case;
— *à balles*, (*art.*) bullet-chamber (of a shrapnel);
— *des canons*, (*fort.*) gunroom (of a turret);
— *d'un canot*, stern sheets;
— *de charge*, (*torp.*) explosive chamber (of a torpedo);
— *de chaudière*, boiler room;
— *de chauffe*, fire-room, stoke-hole;
— *claire*, camera lucida;
— *claire mégaloptique*, magnifying camera lucida (Colonel Goulier's);
— *de combustion*, (*steam*) fire-box;
— *de compression*, (*mil. min.*) the (mathematical) volume over which the interior effects of an explosion are manifested; (*mach.*) the (exaggerated) clearance of the Otto gas-engine cylinder;
— *à crasse*, (*sm. a.*) v. —— *ardente*;
— *cylindrique*, (*art.*) case for the bursting charge of a shell;
— *d'éclatement*, (*art.*) bursting space behind a proving-ground target, admitting of the recovery of shell fragments in experimental firing;
— *à feu*, (*art.*) powder-chamber of a fuse;
— *de fonderie*, (*fond.*) defect, in a casting;
— *aux lampes*, (*fort.*) lamp-recess of a powder magazine;
— *aux lumières*, v. —— *aux lampes*;
— *de la machine*, engine-room;
— *de manœuvre*, —— *des manœuvres*, (*fort.*) working or operating room of a turret;
— *de mine*, (*mil. min.*) chamber of a mine;
— *noire*, (*phot.*) camera;
— *à poudre*, (*art.*) powder-chamber (of a gun, of a shell, of a shrapnel, of a fuse);
— *aux poudres*, (*fort.*) powder-room of a magazine;
— *des poudres*, (*min.*) powder-chamber;
— *du projectile*, (*art.*) shot-chamber;
— *de rail*, (*r. r.*) seat of a rail;
— *des régulateurs d'immersion*, (*torp.*) balance or "secret" chamber of the Whitehead torpedo;
— *à*, —— *de*, *sable*, (*art.*) sand inclosure, or inclosed sand-butt (proving ground, for safety and for recovery of projectiles);
— *de tir*, (*fort*) gun-room of a turret;
— *du tiroir*, (*steam*) valve-chest;
— *tournante*, (*met.*) rotary chamber (of a mechanical puddler);
— *à*, —— *de, vapeur*, (*steam*) steam-chest; steam-space of a a boiler.
chambré, a., (*met.*) blistered, honeycombed; (*art.*) chambered (of a gun).
chambrée, f., (*mil.*) number of men in one squad-room, squad.
de ——, in charge of a squad, or barrack-room.
chambrer, v. a. n., to chamber; (*mil.*) to mess and live together in the same room or tent; (*mil. min.*) to make the chamber of a mine, (*art.*, *etc.*) to chamber (a gun);
se ——, (*art.*) to become pitted or honeycombed (of a gun).
chambreux, a. v., *chambré*.
chambrière, f., long whip, lunging whip; cart-prop; (*cord.*) lacing.
chameau, m., camel;
— *à une bosse*, dromedary;
— *à deux bosses*, camel proper;
corps de —*x*, camel corps of the Egyptian army.
chamois, m., chamois leather, shammy.
champ, m., field; edge (i. e., least dimension, as of a plank, brick, etc.); (*inst.*) field, interior space (of a telescope, etc.); (*mil.*) battlefield; (*art.*) field (of fire);

champ, aux ——s (*mil.*), call or battery sounded on the approach of the president or sovereign, of ministers of state, and of general officers toward troops under their command; when a body of troops passes another, or before a post or guard; at the ceremony of relieving guard; and when a soldier sentenced to death arrives at the place of execution;
—— *de bataille*, battlefield;
—— *de bataille défensif*, defensive battlefield, (hence, technically) a strongly intrenched battlefield;
—— *de bataille défensif-offensif*, a battlefield strongly intrenched, but with wide intervals, permitting offensive operations;
—— *de bataille offensif*, offensive battlefield, (hence, technically) one but slightly intrenched;
de ——, on edge, on end, upright (of planks, bricks, etc.); (*mas.*) headers and stretchers (of bricks);
—— *défensif*, (*mil.*) that part of a battlefield intended to be merely held;
—— *d'égal éclairement*, (*opt.*) field of equal illumination;
—— *d'épreuve*, (*art.*) proving-ground;
—— *d'expériences (des torpilles)*, (torpedo) testing ground;
—— *d'exploitation*, (*min.*) working place or room in a mine;
—— *de feu*, (*ball.*) obsolete name for trajectory;
—— *de force*, (*elec.*) field of force;
—— *galvanique*, (*elec.*) magnetic field;
—— *de lumière*, (*art.*) vent-field;
—— *magnétique*, (*elec.*) magnetic field;
—— *de manœuvre*, (*mil.*) drill-ground;
—— *de Mars*, (*mil.*) drill, parade ground; square in Paris, formerly used as a drill, etc., ground;
—— *offensif*, (*mil.*) that part of a battlefield in which the main, i. e., the offensive, effort will be made;
—— *de tir*, (*inf.*) small-arm target range; (*art.*) proving-ground, practice-ground; field of fire (of a gun, of a battery, of a work, of a line, of a position);
—— *tournant*, (*elec.*) rotating field;
—— *de vision*, (*opt.*) field of view (field-glass, telescope, etc.);
—— *visuel*, visual field; space seen or covered by one eye or two eyes (*monoculaire*, *binoculaire*);
—— *de vue*, v. —— *de vision*.

champignon, m., mushroom, (*art.*) mushroom, i. e., head of the De Bange gas-check; (*hipp.*) scirrhous cord; (*elec.*) knot or mushroom on arc-light carbon; (*r. r.*) head of a rail (with reference to shape of cross-section); (*fond.*) sleeker;
rail à double ——, (*r. r.*) double-headed rail;
à tête de ——, mushroom-headed.

chance, f., chance;
——*s de guerre*, (*mil., etc.*) fortune of war.

chanceler, v. n., to totter, stagger; (*mil.*) to waver (of troops).

chancelier, m., chancellor (of an order, etc.).

chancellerie, f., chancellery, administration (of an order, e. g., Legion of Honor);
grande ——, administration of the Legion of Honor.

chancre, m., canker (trees, horses, etc.).

chandelier, m., candlestick; stanchion, upright; crutch, swivel; (*siege*) chandelier, fascine-chandelier (obs.); (*r. f. art.*) fork-pivot, pillar of a pillar-mount (in many r. f. mounts), cross-head (sometimes); (*art.*) firing-support (Russian 15ᶜᵐ mortar);
affût à ——(*r. f. art.*) any fork-pivot mount; pillar-mount;
—— *de blinde*, v. —— *de tranchée*;
à fourche, (*r. f. art.*) fork-pivot;
—— *de mine*, (*mil. min.*) miner's candlestick;
—— *de tranchée*, (*siege*) a wooden framing used in building up cover by means of fascines (obs.).

chandelle, f., candle; prop, shore, stay; (*mil. slang*) musket, sentry;
—— *romaine*, (*artif.*) Roman candle.

chanée, f., (*cons.*) hollow tile, when laid with hollow up.

chanfrein, m., chamfer, bevel; (*hipp.*) forehead;
en ——, beveled;
tailler en ——, to chamfer, to bevel.

chanfreiner, v. a., to chamfer, bevel.

changement, m., change (of direction, of garrison, etc.);
—— *à aiguilles*, (*r. r.*) point-switch; split-switch;
—— *à double aiguille*, (*r. r.*) double switch; double-slip switch;
—— *d'arme*, (*mil.*) transfer from one arm to another;
—— *de corps*, (*mil.*) transfer from one regiment to another;
—— *de dimension*, (*mil. min.*) passage from a gallery to a smaller (in construction);
—— *de direction*, (*mil.*) change of direction (of a column or line of troops, to the right, etc., by the flank, etc.); also, the command for a change of direction (the movement and the command are in some cases equivalent to "column right," etc.; in others, to "(squad) right," etc., U. S. A.);
—— *de front*, (*mil.*) change of front;
—— *de garnison*, (*mil.*) change of garrison;
—— *de main*, (*man.*) change of hands;
—— *de marche*, (*mach.*) reversing gear;
—— *de marée*, turn of the tide;
—— *d'objectif*, (*art.*) change of target, also command for the purpose;
—— *de pente*, (*mil. min.*) change of slope (in constructing galleries, etc.);
—— *de pied*, (*man.*) change of the (leading) foot in galloping;
—— *de pied en l'air au tact*, (*man.*) change of foot (made in the air on contact);
—— *de pied en l'air au temps*, (*man.*) change of foot made during *temps de suspension*, q. v.;
—— *de postes*, (*art.*) change of posts (at a gun);
—— *de sape*, (*siege*) change from one kind of sap to another;
—— *triple*, (*r. r.*) three-throw switch;
—— *à trois voies*, (*r. r.*) v. —— *triple*;
—— *de voie*, (*r. r.*) switch; action of switching; siding;
—— *de voie à rails mobiles*, (*r. r.*) stubswitch.

changer, v. a., to change;
—— *le pas*, (*mil.*) to change step. (For other adjuncts, as —— *de corps*, —— *de main*, —— *de voie*, etc., v. s. v. *changement*.)

chantepleure, f., sort of funnel with a rose at the end; opening in a wall for the passage of water; barrel faucet, or plug; outlet for water; pipe, spout; drain, sort of gutter.

chantier, m., skid, small beam; (*art.*) block, full-block (for mechanical maneuvers); (*nav.*) yard, shipyard, stocks; (*min.*) thirling, chamber of excavation, working-room (of a tunnel); (*in gen.*) any place where men are at work (e. g., working parties, in a siege, etc.); (*r. r.*) embarking track (troops);
—— *de calage*, skid;
—— *de chevalets à fascines*, (*fort., siege*) fascine bench;
—— *de commettage*, (*cord.*) laying-machine;
—— *à commettre*, v. —— *de commettage;*
—— *de construction*, (*nav.*) shipyard, shipbuilding yard; (*in gen.*) construction shop;
demi- ——, (*art.*) half-block;
entrée en ——, (*fig.*) commencement of any operation, work;
entrer en ——, (*fig.*) to begin work, an operation, etc.;
—— *d'exploitation d'une mine*, (*min.*) excavation room of a mine;
—— *de manœuvres*, (*art.*) block (for mechanical maneuvers), beam, or skid;
—— *de marine*, (*nav.*) dockyard;
mettre en ——, to put to work (as, *mil.*, working parties);
mettre un navire sur le ——, (*nav.*) to put a ship upon the stocks.

chantignole, f., bracket, wooden purlin-block, or cleat; (*mas.*) sort of brick of half the thickness of the common brick.

chantournage, m., cutting, operation of cutting in profile; sawing out along a curve; to trace out on wood, marble, etc.

chantourner, v. a., to cut a profile; to saw out along a curve; to trace on marble, wood, etc.

chanvre, m., hemp;
—— *femelle*, female hemp;
—— *mâle*, male hemp;
—— *de Manille*, manila hemp;
—— *porte-graine*, v. —— *mâle*.

chaparder, v. a., (*mil. slang*) to loot, steal, go marauding.

chapardeur, m., (*mil. slang*) marauder.

chape, f., cap (of a compass or magnetic needle); bearing of the knife-edge of a balance; shell (of a block), pulley-frame, roller-frame; tongue (of a buckle); (*fond.*) cope (of a mold); molding clay; (*art.*) frame holding chassis wheels; outer barrel or case for a powder-barrel, chape; (*harn.*) rectangular frame used to connect two straps, leather loop formed by folding end of strap back on itself; (*mach., etc.*) strap (e. g., connecting-rod strap), clasp, clevis, loop, chape, strap and loop or eye, swivel-frame, bearing, axle-bearing; (*fort. mas.*) coating of cement laid on extrados of a vault, of an arch; (*sm. a.*) top-locket (of a scabbard); chape (of a scabbard), holster-cover;

—— *d'articulation*, hinged strap;
—— *à crochet*, shell (of a block) and hook;
—— *de crochet*, the frame supporting a hook;
—— *de croupière*, (*harn.*) crupper loop;
—— *en dos d'âne*, (*fort.*) capping (of a casemate);
—— *d'équipement*, (*art.*) rigging-shell of a gin-pulley;

—— *-piton*, staple or strap with an eye or loop in the middle;
plan de ——, (*cons.*) weathering;
—— *à, —— de, soulèvement*, (*art.*) roller-frame for throwing the top carriage in gear; roller-frame of the trail of the *affût à soulèvement*;
—— *de sûreté*, (*mach.*) safety- or locking-strap.

chapeau, m., hat, cap; cover; gland, top, lid; cap (in constructions in general); (*art.*) fuse-cap, fuse-cover; breech-plug cap; bearing-plate; cap of a gunlift; (*cons.*) hollow tile, when laid with hollow down; girt, intertie; (*mach.*) press-block; gland of a stuffing-box; (*mil. min.*) cap (of a frame); (*pont.*) trestle-cap; (*unif.*) hat worn by generals, staff officers; cocked hat;
—— *en bataille*, (*unif.*) cocked hat worn cross-wise;

—— *de bott à étoupe*, (*mach.*) stuffing-box gland;
—— *de cabestan*, drumhead of a capstan;
—— *de châssis*, (*mil. min.*) cap, upper crosspiece of a frame;
—— *de cheminée*, funnel-cowl, chimney-hood;
—— *chevalet*, (*pont.*) trestle-cap;
—— *à claque*, (*unif.*) cocked hat;
—— *en colonne*, (*unif.*) cocked hat worn length-wise (i. e., in the usual way);
—— *à cornes*, (*unif.*) cocked hat;
—— *à coudes*, (*pont.*) the cap of the Belgian trestle (allows a variation of slope to first bay);
—— *de fusée*, (*art.*) fuse cap or cover;
—— *de la hausse*, (*art.*) cross-bar, deflection-bar, of a rear sight;
—— *de lumière*, (*art.*) vent-cover;
—— *mobile*, v. —— *de la haussée*;
—— *de palée*, cap, capping-piece, string-piece;
—— *de palier*, (*mach.*) plummer-block top or cap, bearing cap;
—— *porte-cran de mire*, (*art.*) traversing head of a rear sight;
—— *-protecteur*, (*art.*) fuse-cover, -cap;
—— *de réglage*, (*art.*) movable cap for setting a fuse;
—— *-sommier*, (*pont.*) capping-piece;
—— *-travon*, v. —— *-sommier*.

chapelet, m., rosary, chaplet; chain-pump; paternoster-pump, noria; raft of casks; (*harn.*) stirrup-leather; pair of stirrup-leathers with stirrups, fastened to pommel of saddle and connected at the top by a sort of leather buckle; connecting-strap or -piece (pack saddle, off saddle of an artillery team); (*expl.*) string, row, or series of cartridges; (*sm. a.*) holster-iron; (*hipp.*) sort of chain or series of splints; apparatus or device to keep a horse from biting, etc., a place that has been dressed;
—— *électrique*, (*expl.*) set of fuses connected so as to produce simultaneous explosion of several charges; (*torp.*) assemblage of torpedoes connected up so as to go off together;
en ——, (*expl.*) in a row, string, necklace;
faire manœuvre en ——, (*art.*) to parbuckle a piece on skids;
monter une pièce en ——, (*art.*) to parbuckle a piece on skids;
—— *de torpilles*, (*torp.*) line or range of mines.

chapelle, f., chapel; (*mach.*) cylinder-case of a blowing-cylinder; valve-box (of a pump, etc.);
—— *de campagne*, (*mil.*) field chapel (portable furniture for religious services).

chaperon, m., cover, hood; coping (of a wall); (*sm. a.*) holster-cap;
—— *du bec*, (*pont.*) capping of the cutwater;
—— *de fonte*, (*sm. a.*) holster-cap, cover;
—— *de la lumière*, (*art.*) vent-cover.

chapiteau, m., head, top, cap; cover, hood; capital (of a column); head-beam; bridge (of a boring-machine); (*art.*) sort of vent-cover made of planks; (*artif.*) head, cone, of a rocket, etc., carcass-heading; (*sm. a.*) top or edge of the locket of certain side-arms;
—— *d'artifice*, —— *de fusée*, (*artif.*) head and cone of a rocket.

chapska, m., (*unif.*) lancer's helmet.

chapuis, m., (*harn.*) wooden saddle-, pack saddle-frame.

char, m., cart (2-wheeled vehicle).

charançon, m., weevil.

charançonner, v. r., to become weevil-eaten.

charbon, m., coal, charcoal; disease of men and of animals, accompanied by gangrenous tumors;
—— *bitumineux*, bituminous coal;
—— *blanc*, (*hipp.*) anasarca;
—— *de bois*, charcoal;
—— *de cornue*, gas-coke;
dépôt de ——, (*nav.*) coaling station;
—— *électrique*, (*elec.*) arc-light carbon;
faire du ——, (*nav.*) to coal;
filtre à ——, charcoal filter;
—— *de forge*, smithy coal;
—— *gros à la main*, screened coal;
—— *de houille*, coke; sea-coal, pit-coal, anthracite coal;
—— *incombustible*, anthracite coal;
—— *luisant*, anthracite coal;
manche à ——, coal chute;
—— *minéral*, hard coal, anthracite;
—— *noir*, black charcoal (for black powders);
pelle à ——, coal shovel;
—— *de pierre*, stone coal, anthracite;
poussier de ——, coal dust;
—— *roux*, red, brown charcoal (for cocoa powders);
soute à ——, (*nav.*) coal bunker;
—— *de terre*, coal, anthracite coal;
—— *de tourbe*, peat coal;
—— *d'usine*, forge coal;
—— *végétal*, charcoal.

charbonnage, m., coal-mine; coal-mining.

charbonné, a., (*hipp.*) marked with black (of a horse's coat).

charbonner, v. a., to reduce to charcoal, to carbonize.

charbonnier, m., fireman; collier; coal-hole; coal-burner.

charbonnure, f., (*hipp.*) black mark, of variable size and distinctness, in a horse's coat.

chardon, m., (*mil.*) creeper, worn on shoe in an assault (obs.).
charge, f., load, burden, freight; order, direction, command, commission; office, place, employment; charge (of a furnace); charge or pressure corresponding to the difference of two piezometric levels; (*adm.*) condition imposed on bidders or contractors, (hence, in pl.) specifications; (*harn.*) pack; (*hipp.*) horse-poultice; (*mil.*) charge, onset, attack, signal for attack; (*art., sm. a.*) charge (of powder, of explosive), bursting charge (of a shell); loading, operation of loading; (*expl., min.*) charge; (*law*) charge, accusation.
    I. Military; II. Technical and Miscellaneous.
    I. Military:
—— -*amorce*, (*art., expl., torp.*) priming-charge; flash charge, dry gun-cotton priming of a torpedo;
—— *d'amorçage*, (*art., etc.*) priming charge; flash charge;
—— *arrière*, —— *à l'arrière*, (*art.*) rear burster;
à —— *arrière*, (*art.*) with rear burster (of a shell);
—— *avant*, —— *à l'avant*, (*art.*) front burster;
à —— *à l'avant*, (*art.*) fitted with a front burster (of a shell);
—— *à la baïonnette*, bayonet charge;
—— *en bataille*, charge in line;
battre la ——, to beat the charge;
—— *en blanc*, blank charge;
—— *de combat*, (*art.*) service charge;
—— *à cylindre*, (*art.*) reduced charge, the powder being confined in an interior cylinder;
—— *d'éclatement*, (*art.*) bursting charge;
—— *en échelons*, (*cav.*) charge in echelons;
—— *entière*, (*art.*) full charge;
—— *d'épreuve*, (*art., sm. a.*) proof charge;
—— *d'exercice*, (*art.*) practice charge, reduced or drill charge;
—— *explosive*, (*art.*) bursting charge;
—— *faible*, (*art., sm. a.*) reduced charge;
—— *à fond de train*, (*cav.*) charge pushed home;
—— *en fourrageurs*, (*cav.*) charge in extended order; charge as foragers;
—— *de guerre*, (*art.*) full charge, full service charge;
—— *en haie*, charge in line;
—— *intérieure*, (*art.*) bursting charge of a shell;
la ——, *la* —— *en fourrageurs*, signal or call for the charge, etc.;
—— *en ligne*, —— *en muraille*, (*cav.*) charge in line;
—— *normale de tir*, (*art.*) usual service charge;
—— *nouvelle*, (*mil. law*) additional charge;
petite ——, (*art.*) reduced charge;
—— *pleine*, v. —— *entière*;
procédé des ——*s successives*, (*mil. min.*) system in which each explosion prepares a chamber for a second charge;
quotient de la ——, (*art.*) ratio of weight of powder to weight of projectile;
—— *réduite*, v. —— *petite*;
—— *de salut*, (*art.*) saluting charge;
—— *de service*, (*art.*) service charge;
sonner la——, to sound the charge;
—— *en deux temps*, (*r. f. art.*) system of loading in which cartridge and projectile are separate;
—— *en un temps*, (*r. f. art.*) system in which cartridge and projectile form one piece;
—— *en temps et mouvements*, (*art., etc.*) loading by detail, by the numbers;
—— *de tir*, (*art.*) gun charge as distinguished from the bursting charge;
—— *usuelle*, (*Fr. art.*) reduced charge for curved fire;
—— *à volonté*, (*sm. a.*) loading at will.
    II. Technical and miscellaneous:
à ——, (*law*) for the prosecution;
à —— *et à décharge*, (*law*) for and against the prisoner;

charge *d'aplatissement*, the load that will just flatten out a set of (Belleville, etc.) springs;
—— *brute*, gross weight;
cahier des ——*s*, specifications for contractors, bidders, etc.;
cheminée de la ——, (*met.*) throat of a blast furnace;
contre- ——, counterpoise;
coussinet de ——, (*harn.*) breast harness pad;
—— *en cueillette*, a general cargo (ship's);
—— *d'épreuve*, test-load;
—— *d'étoupe*, thread of oakum;
être en ——, to be loading (ship);
—— *en grenier*, bulk cargo (ship's);
—— *en hauteur*, edge pressure or load;
ligne de ——, (*nav.*) load water line;
—— *mobile*, live load, rolling load;
pièce à ——, (*law*) any object that has been used in the commission of a crime, or that may serve to support the truth;
—— *entre deux points*, difference of piezometric level of two points;
sans rompre ——, without breaking bulk;
—— *de rupture*, breaking load;
—— *de sécurité*, safe load;
—— *de sûreté*, safe load;
taux de ——, load per unit of section or of area;
témoin à ——, witness for the prosecution;
—— *de traction*, load behind teams.
chargé, a., (*art., sm. a., etc.*) loaded, filled (of a shell); (*mach., etc.*) stressed, loaded;
—— *debout*, receiving stress on end (as a connecting-rod, etc.);
—— *d'encolure*, —— *de ganache*, —— *de tête*, (*hipp.*) heavy-headed.
chargé, m., —— *d'affaires*, chargé d'affaires.
chargement, m., loading (of a wagon, etc.); (*adm.*) record of registry of a letter, etc.; (*art., sm. a.*) loading (operation of) of a gun, etc.; (*art.*) filling (of a shell); (*sm. a.*) filling (of a magazine); all the operations of assembling the parts of a complete cartridge; (*min.*) charging (of a mine);
atelier de ——, (*art.*) filling-room;
à —— *central*, (*art.*) central-loading (said of turrets when ammunition is brought up in center);
—— *coup par coup*, (*sm. a.*) single-loading (said of magazine rifle when magazine is not used);
cylindre de ——, (*art.*) loading-tube;
densité de ——, (*art.*) density of loading;
éléments de ——, (*art.*) elements of loading;
—— *à espace d'air*, (*art.*) air-spaced loading;
planchette de ——, (*art.*) loading-tray;
position de ——, (*art.*) loading position;
—— *en deux, trois, etc., temps*, (*art.*) loading in two, three, etc., motions;
voie de —— *et de déchargement*, (*r. r.*) side-track, siding.
chargeoir, m., (*art.*) loading-ladle (obs.).
charger, v. a., to load (in general); (*art., sm. a.*) to load a gun, a rifle, etc.; to fill a shell; to prepare, fill, primers, fireworks, etc.; (*art.*) to mount a gun upon its carriage; (*min.*) to load a mine; (*met., etc.*) to charge, feed, a furnace; (*mach.*) to load (an engine); (*adm.*) to register a letter; (*mil.*) to charge, attack, the enemy; (*sm. a.*) to fill (magazine);
—— *à ...., —— en ....*, (*art.*) to fill a shell with .... (e. g., melinite);
—— *un accumulateur*, (*elec.*) to charge an accumulator;
—— *d'air comprimé*, (*mach., etc.*) to fill with compressed air;
—— *une capsule*, (*expl., sm. a., etc.*) to fill a primer;
—— *en échelons*, (*mil.*) to charge in echelons;
—— *à espace d'air*, (*art.*) to load with air spacing;
—— *à fond*, —— *à fond de train*, (*cav.*) to charge home;
—— *en fourrageurs*, (*cav.*) to charge in extended order;
—— *une fusée*, (*artif.*) to load, fill, a fuse, rocket;
—— *en guerre*, (*art.*) to fill a shell with its (service) bursting charge;

charger *en ligne*, v. —— *en muraille*;
—— *un magasin*, (*sm. a.*) to fill a magazine;
—— *en muraille*, (*cav.*) to charge in line with no intervals;
—— *en queue*, (*mil.*) to attack in rear;
—— *en deux, trois, etc., temps*, (*art.*) to load in two, three, etc., motions;
—— *en ville*, (*cav. slang*) to go into town.

chargette, f., (*sm. a.*) powder measure (gallery practice).

chargeur, m., (*met.*) lift; (*art.*) cannoneer who loads (esp. r. f. art.); (*sm. a.*) loader, cartridge packet, clip, loading clip;
—— *automatique*, (*sm. a.*) self-loader;
—— *à deux branches*, (*sm. a.*) loader supporting point and base of cartridge;
fusil à ——, (*sm. a.*) rifle to which a loader may be attached, or requiring a loader;
—— *d'introduction*, (*sm. a.*) loading-strip, i. e., one not fixed to, or introduced in, weapon;
lame- ——, (*sm. a.*) loader, loading-strip;
—— *à lame simple*, (*sm. a.*) loader holding only base of cartridge;
—— *à mâchoires*, (*sm. a.*) clip;
—— *à main*, (*sm. a.*) loader, hand-loader, **quick** loader;
—— *mobile*, (*sm. a.*) detachable loader;
—— *rapide*, (*sm. a.*) quick loader;
—— *séparable*, v. —— *mobile*.

chariot, m., chariot, wagon (four-wheeled), trolley, truck, railway carriage, carriage; (*art.*) cradle; trunnion carriage (esp. of r. f. guns); (*sm. a.*) cartridge-carrier; (*mach., etc.*) slide, slide-rest (as of a lathe), saddle, traversing-head, traveler, any sliding tool-holder, any sliding device for holding the work, sliding carriage; (*inst.*) in some instruments, the rotating plate supporting the instrument proper.
I. Military:
—— *d'artillerie*, (*art.*) battery wagon (U. S. art.);
—— *de bagages*, baggage wagon;
—— *de batterie*, (*art.*) battery wagon, ammunition wagon;
—— *à canon*, (*art.*) mortar wagon, heavy sling-cart, truck wagon;
—— *des dérives*, (*art.*) deflection slide;
—— *d'équipages*, baggage wagon;
—— *d'explosifs*, (Fr. *cav.*) explosive wagon (accompanies a division of cavalry for demolition purposes);
—— *de forage*, —— *de forerie*, (*art.*) boring-bench;
—— *forge*, (*art.*) traveling-forge;
—— *fournil*, bakery-cart, kneading cart, of a field bakery;
—— *fourragère*, forage wagon;
—— *de mine*, (mil. min.) miner's truck or car;
—— *mobile*, (*art.*) trunnion-carriage, trunnion slide or cradle (e. g., Hotchkiss 57ᵐᵐ);
—— *à palan*, (*art.*) short traveler or trolley, supporting the projectile by a tackle; tackle-trolley;
—— *de parc*, (Fr. *art., etc.*) siege wagon, park or transport wagon, (also) bridge-train wagon, ammunition supply wagon (is used for a variety of purposes);
—— *à perches*, (Fr. *a.*) pole wagon of a telegraph section;
—— *porte-barre*, (*art.*) rifling-bar rest;
—— *porte-corps*, (*art.*) mortar wagon, heavy truck wagon; block-carriage (R. A.);
—— *porte-munitions*, ammunition supply wagon;
—— *simple*, (*art.*) shot traveler or trolley (the projectile being directly supported by a chain);
—— *télégraphique*, (Fr. *a.*) wagon for transportation of telegraph material, coils of wire, etc.;
—— *de travail*, (Fr. *a.*) construction wagon of a telegraph section;
—— *à vis*, (*ball.*) screw-slide (Marcel Deprez).
II. Technical and miscellaneous:
—— *d'alésoir*, (*mach.*) boring-block;
—— *à commettre*, —— *à couchoir*, (*cord.*) **laying-top**;

chariot *de forage*, —— *de forerie*, (*mach.*) boring-bench;
—— *à glissières*, (*mach.*) slide-rest, sliding support;
le grand ——, the Great Bear, the Dipper;
—— *de mine*, (*min.*) miner's truck or car;
le petit ——, the Little Bear;
—— *porte-fardeau*, truck or traveler of a crane;
—— *porte-fraise*, (*mach.*) reamer holder;
—— *porte-pièce*, (*mach.*) table, plate, slide carrying the work;
—— *de réglage*, (*mach.*) cross-slide of a lathe;
—— *à ridelles*, cart or wagon with a rack, open-side wagon;
—— *de roulement*, expansion roller (metallic trusses, etc.);
—— *à roulettes*, truck wagon;
—— *de tour, lathe-slide*;
—— *de tournage*, (*mach.*) turning-down lathe;
—— *transversal*, (*mach., etc.*) cross-slide.

chariotement, m., motion by or in a chariot or slide (cranes, lathes, etc.); (*mach.*) displacement of a tool, as carried by its slide.

charlemagne, m., (*mil. slang*) saber-bayonet.

charme, m., hornbeam (wood and tree).

charnière, f., hinge; joint, turning-joint;
à ——, hinged, fitted with a hinge;
—— *à cog*, hook-and-eye hinge, gate hinge;
femelle de ——, —— *femelle*, female-joint, axis of a hinge;
mâle de ——, —— *mâle*, male joint, revolving part of a hinge;
—— *à pli*, (*harn.*) joint of a broken bit, mouth joint.

charpe, f., mooring-rope for pontons.

charpente, f., framing, framework, timberwork; carpentry;
—— *en fer*, iron framing, framework;
—— *en fonte*, cast-iron framework;
—— *de mine*, (*min.*) timbering, casing of a mine.

charpenterie, f., carpentry, frame- or timber-work.

charpentier, m., carpenter.

charpie, f., lint.

charretée, f., cart load.

charretier, m., driver (of a cart); a., practicable for carts;
chemin ——, cart road;
voie ——*e*, cart-track, distance between wheels, gauge.

charrette, f., cart, two-wheeled cart;
—— *à bras*, handcart, pushcart; tumbril;
—— *à cartouches*, (*sm. a.*) ammunition cart;
—— *à munitions*, (*sm. a.*) ammunition cart;
—— *porte-obus-torpille*, (*art.*) torpedo-shell cart;
—— *porte-projectile*, (*art.*) shell cart;
—— *de siège* (siege) transport cart;
—— *de tranchée*, (siege) trench cart.

charriage, m., cartage; (*art.*) transportation of guns, etc., by wheeled carriages;
—— *des glaces*, downflow of ice in a stream.

charrier, v. a., to cart, to transport in carts;
to carry down ice (of a river).

charroi, m., teaming, transportation by wagon.

charron, m., wheelwright.

charronnage, m., wheelwright's trade.

charroyer, v. a., to transport in carts.

chartagne, f., intrenchment or retrenchment in a wood.

charte-partie, f., charter party, charter (of a vessel).

chas, m., eye of a needle; mason's plumb-rule.

chasse, f., chase, pursuit (esp. nav.); set, setter, drift, punch; (of drains, etc.) flushing; (of vehicles) the greater or less facility of being moved forward; (of a bicycle) the horizontal distance from the vertical projection of the front axle on the ground, to the steering post prolonged; (*mach.*) play; (*sm. a.*) frame on which sword-blade is shaped; impact of the hammer; (*artif.*) filling, bouncing; (*fort.*) rush of water through a ditch, flooding; (*hipp.*) propulsive power of the hind legs;
—— *d'affe*, (*mil. slang*) chasseur d'*Afrique*;

chasse à *biseau*, bevel setter;
—— –*bloc*, (*art*.) drift (special drift for dismounting breech mechanism of 270ᵐᵐ s. c. mortar;)
—— –*boîte*, driving-bolt;
—— –*caillou*, v. —— –*pierres;*
—— –*calotte*, (*sm. a.*) driver of hard wood, for setting the *calotte* on the tongue of a sword blade;
—— *carrée*, square setter;
—— –*chaîne*, chain-spacer, chain guide;
—— *crochet de pontage*, lashing-hook setter;
—— *demi-ronde*, half-round setter;
donner —— *à un navire*, (*nav.*) to give chase to a ship;
—— *d'eau*, sudden flooding, flushing, (*esp. fort.*) sudden flooding of the ditches caused by very rapid opening of water-gates, so as to sweep away enemy's ditch crossing;
écluse de ——, (*fort.*) flash-gate;
—— –*étui*, (*art.*) shell-driver (e. g., of a Gardner gun, etc.);
—— –*fusée*, (*art.*) fuse-setter;
—— –*goupille*, pin-setter, pin-bolt drift;
lever la ——, (*nav.*) to give up the pursuit;
—— –*marée*, (*mil. slang*) chasseur d'Afrique;
—— *morte*, random shot, miss;
—— –*mouche*, fly net;
—— –*noix*, (*sm. a.*) blunt punch or driver;
—— *à parer*, set hammer;
—— –*pattes*, (*mil. slang*) chasseur à pied;
pièce de ——, (*nav. art.*) bow chaser;
—— –*pierres*, (*r. r.*) cowcatcher (U. S.), wheel guard;
—— –*poignée*, —— –*pommeau*, driver for driving hilt on tongue of sword-blade;
réservoir de ——, flushing-tank, reservoir;
—— –*rivet*, setter, riveting set;
—— *ronde*, round setter;
—— –*roues*, wheel-post, spur-post.

chassepot, m., (*sm. a.*) the French rifle model, 1866.

chasser, v. a., (*mil., etc.*) to drive before one, to drive out of a position; (*tech.*) to drive in, to set; (*man.*) to set forward at an increased gait; (of an anchor) to drag;
—— *les ancres*, to drag anchors;
—— *en avant*, (*man.*) to urge a horse forward;
—— *les fesses sous soi*, (*man.*) to sit well down in the saddle.

chasseur, m., huntsman; (*Fr. a.*) chasseur (cav. and inf.);
—— *s d'Afrique*, (*Fr. a.*) light cavalry for African service;
—— *s alpins*, (*Fr. a.*) light infantry for service in the Alps;
—— *s à cheval*, (*Fr. a.*) mounted chasseurs, light cavalry;
—— *s forestiers*, (*Fr. a.*) rifle corps recruited, on mobilization, from the personnel of the government forestry department;
—— *s à pied*, (*Fr. a.*) rifles, light infantry.

chassis, m., frame; (*art.*) chassis; (*tech., etc.*) saddle, bed; (*mil. min.*) gallery frame; (*r. f. art.*) gun-case (Maxim-Nordenfeldt); (*fond.*) molding-box, flask;
—— *circulaire*, (*art.*) circular chassis (Canet);
—— *à claire voie*, openwork frame;
—— *coffrant*, (*mil. min.*) case, mine case;
—— *de combat*, (*mil. min.*) case of a *rameau de combat*, q. v.;
faux ——, (*mil. min.*) false frame, temporary or movable frame;
—— *de fermeture*, (*art.*) breech-frame;
grand ——, (*art.*) chassis proper, as distinguished from *petit* —— (q. v.), of some French guns;
—— *grillé à dépêches secrètes*, ciphering frame (in cryptography);
—— *hollandais*, (*mil. min.*) v. —— *coffrant;*
—— *inférieur*, (*fond.*) drag, lower flask;
—— *du milieu*, (*fond.*) middle flask;
—— *de mine*, (*mil. min.*) mine frame, gallery frame;
—— *mobile de pieds* (*de tête*), (*med.*) forward (rear) movable part of a litter;
—— *de moulage*, (*fond.*) molding box, flask;
—— *négatif*, (*phot.*) plate-holder (of a camera);
—— *petit*, (*art.*) pintle-frame (French 138ᵐᵐ);
—— *de plate-forme*, (*art.*) traversing carriage;
—— *des platines*, (*r. f. art.*) lock-frame (Gardner gun);
—— *de pont*, (*pont.*) beam, sleeper (e. g., to unite casks, etc.);
—— *porte-scie*, saw-frame;
—— *porte-tourillons*, (*r. f. art.*) frame (Hotchkiss revolving cannon);
—— *positif*, —— –*presse*, (*phot., etc.*) printing-frame;
—— *supérieur*, (*fond.*) cope, top-flask;
—— *à tabatière*, skylight with movable window;
—— *de transport*, (*art.*) sort of truck for transporting carriages, etc., in garrison (somewhat like United States mortar wagon);
—— *vitré*, skylight.

chassoir, m., driver, drift; fuse-mallet.

chat, m., (*art.*) searcher (obs.);
—— *de pont volant*, —— *de potence*, —— *de traille*, ferry-block.

châtaigne, f., (*hipp.*) chestnut.

châtaigner, m., chestnut (wood and tree).

châtain, a., chestnut (color); (*hipp.*) chestnut horse.

château, m., castle, fort, citadel;
—— *d'eau*, (*r. r.*) water tank.

châtier, v. a., to punish.

chatière, f., cat-hole, small opening in roof.

châtiment, m., punishment.

chatouiller, v. a., to tickle; (*man.*) to touch a horse lightly with the spur.

châtrage, m., (*hipp.*) castration, gelding;
—— *des roues*, shortening of fellies, tightening up of fellies and spokes.

châtrer, v. a., (*hipp.*) to geld, cut;
—— *une roue*, to set up the fellies and spokes, to remove part of the fellies.

chatterton, m., a waterproof material composed of tar, rosin, and gutta-percha.

chaud, a., hot; (frequently, *met.*) red-hot;
joindre à ——, (*met.*) to weld;
placer à ——, (*met.*) to shrink or put on hot;
travailler à ——, (*met.*) to work hot.

chaude, f., (*met.*) heat, heating;
—— *blanc de lune*, —— *blanche*, —— *grasse*, welding heat, white heat;
donner la ——, *une*, *deux*, *etc.*, ——s, to heat (for tempering), to heat once, twice, etc.;
—— *grasse*, white heat;
—— *rouge*, red heat;
—— *rouge cerise*, cherry-red heat;
—— *sombre*, dark-red heat;
—— *soudante*, —— *suante*, welding heat.

chaudière, f., copper, kettle; (*steam*) boiler, steam boiler;
—— *alimentaire*, feed boiler;
—— *auxiliaire*, donkey boiler;
batterie de ——s, range or battery of boilers;
—— *à bouilleurs*, French boiler (with under-body heaters);
—— *à brai*, pitch kettle;
—— *à carneaux*, flue boiler; wagon boiler;
—— *chauffée des deux bouts*, double-ended boiler;
chemise de ——, boiler-jacket, case, cleading, lagging, deading;
—— *à circulation rapide*, Field boiler;
—— *cylindrique*, cylindric boiler, barrel boiler;
—— *cylindrique à bouilleurs*, cylindrical boiler with tubes exposed to heat;
—— *cylindrique à foyer intérieur*, Cornish boiler;
—— *double*, double boiler;
—— *à éléments à circulation*, water-tube boiler;
—— *à éléments multitubulaires*, multitubular boiler;
enveloppe de ——, boiler shell;
—— *à flamme directe*, locomotive boiler;
—— *à foyer extérieur* (*intérieur*), externally (internally) fired boiler;
—— *sans foyer*, internally fired boiler;
—— *française*, French boiler;

chaudière, *lit de pose d'une* ——, boiler setting;
— *marine*, marine boiler;
— *à petit cheval*, v. —— *auxiliaire;*
— *à basse (haute, moyenne) pression*, low (high, mean) pressure boiler;
— *de raffinage*, refining boiler (e. g., for saltpeter);
— *rectangulaire*, box boiler;
— *à retour de flamme*, return-flame boiler;
— *sectionnelle*, sectional boiler, water-tube boiler;
— *semi-tubulaire*, boiler in which the lower heater (*bouilleur*) has its place taken by tubes;
support de ——, knee;
— *à*, —— *en*, *tombeau*, wagon boiler; caravan boiler;
— *tubulaire*, —— *à tubes*, —— *tubulée*, (ordinary) tubular boiler (i. e., fire-tube);
— *en* V, Thornycroft, Yarrow, boiler;
— *à vaporisation instantanée*, instantaneous generator.

chaudronnerie, f., coppersmith's trade, (more extensively) plate-working in general;
grosse ——, work in heavy plates or bars; boiler making;
petite ——, work in light plates, etc., done by hand.

chaudronnier, m., coppersmith, plate-worker; boiler maker; (*mil. slang*) cuirassier.

chauffage, m., fuel; firing; heating of barracks and quarters; stoking; heating (in general);
bois de ——, firewood;
— *d'un coussinet*, —— *d'un essieu*, (*mach.*) heating of a bearing;
— *à la vapeur*, steam heating.

chauffe, f., stokehole; furnace; fire-hole of a furnace; (*met.*) fireplace (as of a reverberatory furnace); (*met.*) action of heating;
donner une ——, (*met.*) to heat;
— *du fourneau*, fire-hole of a furnace;
surface de ——, heating surface.

chauffer, v. a., to heat, to warm; to fire up; (*steam*) to get up steam; to keep in fuel, give firing to; (*mil.*) to grow hot (of a combat fam.); (*hipp.*) to grow heated, as a result of injury;
— *blanc*, (*met.*) to bring to a white heat (and similarly of the other heats, v. *chaude*);
— *une forteresse*, —— *une troupe*, (*mil.*) to keep up a brisk, steady fire on a fort, a body of troops;
— *à la vapeur*, to heat by steam.

chaufferie, f., (*met.*) bar-iron furnace; (*steam, etc.*) stokehole, fire-box; range of boilers.

chauffeur, m., fireman, stoker.

chauffure, f., (*met.*) brittleness; burnt, overburnt, iron.

chaufour, m., limekiln.

chausse, f., sock;
—*s*, breeches;
— *d'aisances*, waste-pipe, soil-pipe.

chaussée, f., high road; causeway; (more narrowly) the central part of a road, the carriage way;
— *bombée*, barreled road;
— *de branchage*, road of mattress of boughs, etc., for swampy ground;
— *en déblai*, road formed by cutting;
— *en empierrement*, metaled road;
— *ferrée*, metaled road;
— *à la MacAdam*, —— *macadamisée*, macadamized road;
— *en pavé*, paved highway;
— *en remblai*, road formed by filling in, as a sustaining wall, by embankment;
— *de tronc d'arbres*, corduroy road.

chausser, v. a., to put on (stockings, shoes); (*man.*) to stick the feet home in the stirrups;
— *les éperons aux ennemis*, (*mil.*) to pursue the enemy;
— *les étriers*, (*man.*) to ride with feet home in the stirrups.

chausse-trape, f., (*mil.*) crow's-foot, caltrop.

chaussette, f., sock;
—*s*, (*mil. slang*) gloves.

chausson, m., (*Fr. a.*) kind of buskin or sock, of flannel, etc., worn over stockings, and issued to mounted men, to men in camp, etc.;
—*s*, (*art.*) magazine slippers.

chaussure, f., footwear, foot-covering; shoe of a pile;
— *exploratrice*, (*man.*) Marey pneumatic footbulb (for registering gaits);
— *de mobilisation*, (*Fr. a.*) reserve pair of shoes, kept in company storehouse;
— *de pointe d'arcade*, (*harn.*) sort of shoe or point of a saddle panel, into which the *pointe d'arcade* fits;
— *de repos*, (*Fr. a.*) second pair of shoes (including gaiters).

chauvir, v. n., to prick up the ears (mule, ass).

chaux, f., lime;
— *anhydre*, quicklime;
blanc de ——, whitewash;
— *calcinée*, quicklime;
— *caustique*, quicklime;
crépir à la ——, to whitewash;
donner un blanc de ——, to whitewash;
— *éteinte*, slaked lime;
— *éteinte à sec*, air-slaked lime;
four à ——, limekiln;
— *fusée*, slaked lime;
— *grasse*, fat lime, air lime;
— *hydraulique*, hydraulic lime, water lime;
lait de ——, whitewash;
— *maigre*, poor lime;
— *morte*, dead-burned lime, overburned lime;
peindre à la ——, to whitewash;
pierre à ——, limestone;
— *en poudre*, air-slaked lime;
— *vive*, quicklime.

chavirement, m., (*nav.*) capsizing.

chavirer, v. n., (*nav.*) to capsize.

chéchia, f., (*unif.*) long fez worn by Algerian troops.

chef, m., chief, leader, principal, master, head (*chef* in the military vocabulary denotes either the grade, or the functions or duty, whether permanent or temporary, of the person to whom applied; for examples see *infra*; it is the term usually employed by the men to designate the sergeant-major); (*mil.*) (in many relations) commanding officer, head;
— *armurier*, (*Fr. a.*) chief armorer (regimental);
— *artificier*, (*Fr. art. and inf.*) chief artificer (a n. c. o., v. s. vv. *maréchal* and *sergent*);
— *d'attaque*, (*siege*) engineer officer (captain, generally) in charge of the approaches, etc., in each attack or important part of each attack; is detailed every day;
— *de bande*, (*mil.*) guerrilla chief;
— *de bataillon*, (*inf.*) major;
— *de batterie*, (*art.*) c. o. of a battery;
— *de brigade*, colonel (obs.); chief of a brigade of gendarmerie;
— *de caisson*, n. c. o. in charge of an infantry battalion ammunition caisson;
— *de casemate*, (*art.*) sergeant in charge of a casemate, is also *chef de pièces;*
— *de chambrée*, corporal in charge of a squadroom, or ranking private if there be no corporal;
— *de complot*, ringleader;
— *de convoi*, c. o. of a convoy;
— *de corps*, c. o. of any body of troops having an organic existence or independence, i. e., generally of a battalion or of a regiment; the commanding general of an army corps; in the absence of limitations, *chef de corps* may be always safely translated colonel (of a regiment);
— *de détachement*, c. o. of a detachment;
en ——, in chief;
— *d'équipe*, foreman of a gang, crew, party, shop, etc.; (*art.*) n. c. o. in charge of a mechanical maneuver;
— *d'escadron*, (*art.*) major;

chef *d'escadrons,* (*cav.*) major;
— *d'escouade,* (*inf.*) corporal; squad leader;
— *d'état-major,* chief of staff;
— *d'état-major général de l'armée,* (*Fr. a.*) chief of the general staff (also called — *d'état-major de l'armée*);
— *de fanfare,* (*mil.*) bandmaster (brass band);
— *de file,* file leader;
— *de gare,* (*r. r.*) station-master;
— *du génie,* (*Fr. eng.*) engineer officer in command of a *chefferie,* q. v.; district engineer (U. S. A.);
— *-lieu,* county-town; capital of a *département*; (*mil.*) seat or headquarters of an administrative subdivision of the territory (e. g., — — *de corps d'armée*);
— *de manœuvre,* (*art.*) corporal in charge of the rotation of a turret;
— *mécanicien,* (*Fr. art.*) master machinist (regimental);
— *de musique,* (*mil.*) bandmaster;
— *de nage,* stroke (in a rowboat);
— *d'ordinaire,* (*Fr. a.*) corporal in charge of men's mess;
— *-ouvrier,* (*in gen.*) master-workman; (*Fr. art.*) employee of the artillery, serving as foreman, etc. (identical with *garde d'artillerie*); (*Fr. a.*) general designation for any n. c. o. at the head of a workshop, but more especially the chief armorer;
— *d'ouvriers,* foreman;
— *de patrouille,* c. o. of a patrol;
— *de peloton,* chief of platoon (section, British);
— *de pièce,* (*art.*) gunner (U. S. Art.);
— *de pièces,* (*art.*) sergeant in immediate charge of the guns of a turret;
— *de petit poste,* c. o. of a picket;
— *de poste,* c. o. of a guard;
— *de révolte,* ringleader of a mutiny, riot, etc.;
— *de secteur,* (*siege*) engineer officer in charge of siege operations in each sector of attack (daily detail);
— *de section,* (*inf.*) section leader; (*art.*) chief of platoon (U. S. Art.), section commander (R. A.);
— *de service,* (*mil.*) head of a department or service (e. g., chief medical officer);
— *de tourelle,* (*art.*) c. o. or n. c. o. in charge of the service of a turret;
— *de train,* (*r. r.*) conductor of a train (U. S.);
— *de voiture,* (*art.*) chief of piece (hitched).

chefferie, f., (*Fr. a.*) district under the charge of a *chef du génie,* consisting generally of a main *place* with its dependencies (sometimes — *du génie*).

chemin, m., road, way, path, track; means, course; (*r. r.*) line; (*nav.*) way.
avant- — *couvert,* (*fort.*) advanced covered way;
— *d'accès,* avenue;
— *couvert,* (*fort.*) covered way;
croisée de — *s,* crossroads;
— *direct,* (*nav.*) distance run;
— *d'écoulement,* (*sm. a.*) gas-port (in case of ruptured shell);
— *est,* (*nav.*) easting;
— *est et ouest,* (*nav.*) departure;
faire du — *à l'est, au nord, etc.,* (*nav.*) to make easting, northing, etc.;
— *de fer,* railway, railroad;
— *de fer de ceinture,* belt railway, belt;
— *de fer à chevaux,* horse tramway;
— *de fer électrique,* electric railway;
— *de fer funiculaire,* cable road;
— *de fer à rails plats,* tramroad;
— *de fer stratégique,* military railway, strategic railway;
— *de fer à deux, quatre, etc., voies,* double-track, four-track, etc., road;
— *de fer à double voie,* double-track road;
— *de fer à voie étroite,* — *à voie réduite,* narrow-gauge road;
— *de fer à voie normale,* standard-gauge road;
— *de fer à une seule voie,* — *de fer à voie unique,* single-track road;

chemin *ferré,* metaled road;
grand — , high road;
— *de halage,* towpath;
— *militaire,* military road; (*fort.*) the road between the works and the houses, etc., of the town;
— *national,* high road;
— *nord,* (*nav.*) northing;
— *nord et sud,* (*nav.*) difference of latitude;
— *à ornières,* tramroad;
— *ouest,* (*nav.*) westing;
— *particulier,* private road;
— *passant,* public road, great thoroughfare;
— *des pontonniers,* (*pont.*) space between side rails and ends of chess in military (ponton) bridges;
— *de ronde,* (*fort.*) chemin de ronde; exterior corridor;
— *de roulement,* (*art.*) roller path;
— *rural,* crossroad;
— *sud,* (*nav.*) southing;
— *de terre,* country road; "dirt road" (U. S.);
— *de traverse,* crossroad;
— *vicinal ordinaire,* (*Fr.*) road kept up by a commune;
— *vicinal de grande communication et d'intérêt commun,* (*Fr.*) road kept up by the interested communes.

cheminée, f., chimney; fireplace; hearth; smokestack, funnel; (*sm. a.*) nipple (obs.); (*art.*) flame-passage (of a fuse, of a primer, etc.);
— *en acier,* (*sm. a.*) barrel-cover (e.g., of Belgian Mauser);
— *d'appel,* ventilating shaft, air-shaft (e. g., of a magazine, etc.);
— *de la charge,* v. — *supérieure;*
— *à coulisse,* (*nav.*) telescopic funnel;
étai de — , chimney stay;
papillon mobile d'une — , regulator, damper of a chimney;
— *supérieure,* (*met.*) throat and main body (above bosses) of a blast furnace.

cheminement, m., (*in gen.*) advance along a road; (*siege*) advance of the zigzags toward the place besieged; the zigzags themselves, works executed by besiegers in approaching a place; progress of siege works; (*mil.*) approach or advance of troops to position from which they are to attack; (*top., surv.*) method in which plane-table, etc., is taken from vertex to vertex of the polygon to be surveyed;
— *géodésique,* (*top.*) method of survey consisting of going from point to point, followed when from the nature of the ground triangulation is impossible;
— *tactique,* (*mil.*) procedure and methods of measuring paces, steps, gaits, etc., of the various arms.

cheminer, v. n., to advance, as along a road; (*siege*) to push on, advance, the zigzags in siege operations.

chemise, f., shirt; jacket, case, casing; (*fort.*) synonym of *enceinte*; (*sm. a.*) bullet-jacket, casing, or mantle; (*sm. a.*) barrel-casing or -guard; (*steam*) steam-jacket; (*met.*) outer casing, shell, of a furnace;
— *de batterie,* (*fort.*) epaulment;
— *de canon,* (*sm. a.*) barrel-casing or -guard;
— *de cheminée,* funnel-casing;
— *de chaudière,* (*steam*) boiler-casing, cleading, lagging;
— *de cylindre,* (*steam*) steam-jacket, cylinder jacket;
— *d'eau,* water-jacket (e. g., of Gardner gun);
— *-enveloppe,* (*sm. a.*) barrel-casing;
— *d'escarpement,* (*fort.*) revetment, facing, of a slope;
— *de fortification,* (*fort.*) the *enceinte* of an earthwork; the main work itself (in permanent fortifications) when of little strength;
— *de fourneau,* lining of a furnace chimney;
— *métallique,* (*sm. a.*) metallic bullet-cover;
— *de pierre,* (*mas., etc.*) stone facing;

chemise *de plomb,* (*art.*) lead coating of a projectile (obs.);
— *de rempart,* (*fort.*) rampart wall, revetment wall of a rampart;
— *résistante,* strengthening jacket;
— *du tiroir,* (*steam*) slide-valve casing;
— *de vapeur,* (*steam*) steam-jacket.

chemiser, v. a., (*sm. a.*) to jacket, mantle, a bullet.

chenal, m., channel, fairway (esp. to or from a port).

chêne, m., oak (tree and wood);
— *de brin,* oak sapling;
— *à feuilles vertes,* live oak;
— *à grappes,* British oak (a variety of);
— *des Indes,* teak wood;
— *liège,* cork oak, cork tree;
— *rouvre,* British oak;
— *vert,* evergreen oak, water oak;
— *de vie d'Amérique,* live oak.

cheneau, m., gutter (on a cornice).

chenet, m., andiron, firedog.

chenevis, m., hemp seed.

chènevotte, f., (*cord.*) boon, odds and ends of rope, yarn, etc.; bullen.

cherche, f., (*top. drawing*) "searcher," "seeker," in finding curves of equal shade.

chercheur, m., searcher, finder (of a telescope).

cheval, m., horse; (*tech.*) horsepower; (*steam*) donkey-engine.
 I. Cheval (horse), qualified by a noun or an adjective, etc.; II. Cheval (horse), as the object of, or qualifying, a verb; III. Cheval (horse), qualified by a relative clause; IV. Technical expressions, not hippological.

 I. Cheval, qualified by a noun or an adjective, etc.

à —, (call) to saddle, bridle, and mount, "to horse;"
abatage de — *aux,* casting of horses;
les aplombs du —, the various positions of the legs of the horse with respect to the vertical;
— *anglaisé,* horse with tail nicked and docked;
— *en arbalète,* leader of a spike team;
— *d'armes,* troop horse, service horse;
— *arqué,* knee-sprung horse, horse over in the knees (when due to wear and tear); bandy-legged horse;
arrêt du —, stopping or halting of a horse;
— *arzel,* horse whose off hind foot is white;
— *atteint d'éparvins,* spavined horse;
— *d'attelage,* generic term for a carriage horse, draught horse, harness horse;
— *balzan,* black or bay horse with a white foot, or with all four white;
— *barbe,* barb;
— *de bât,* pack horse;
— *de bataille,* charger;
— *bâti en girafe,* horse of abnormally high croup;
— *bégu,* horse that never loses the mark;
— *belle-face,* white-faced horse;
— *bien culotté,* horse with fine, well-set buttocks;
— *bien embouché,* horse with good mouth, to which the bit applies normally;
— *bien fait de la main en avant,* horse with a fine forehand;
— *bien gigoté,* horse with handsome, strong thighs;
— *bien mis,* well-trained, thoroughly broken horse;
— *bien ouvert,* horse with broad chest and straight legs;
— *bien ouvert du derrière,* v. — *bien culotté;*
— *bien ouvert du devant,* broad-chested horse;
— *bien pris,* well-built horse;
— *bien soudé,* horse with wide, thick joints;
— *bien suivi,* horse whose lines run into one another, offering no marked contrasts;
— *bien traversé,* broad-shouldered horse;
— *bien troussé,* well-made horse;
— *bistourné,* horse castrated by torsion;

cheval *fort en bouche,* — *sans bouche,* hard-mouthed horse;
— *bouleté,* horse with crowned, dislocated fetlock;
— *bouté,* horse whose leg is straight from knee to coronet;
— *de brancard,* shaft horse;
bras de —, horse's arm from shoulder to knee;
— *brassicourt,* bandy-legged horse, horse over in knees, knee-sprung (if congenital);
— *brétaud,* — *bretaudé,* horse with ears cropped halfway up;
— *cagneux du derrière,* horse cross-footed behind, whose hocks are turned out;
— *cagneux du devant,* horse cross-footed in front, with elbows turned out;
— *cambré,* bow-legged horse;
— *campé du derrière,* horse whose hind legs splay out to the rear too much;
— *campé du devant,* horse whose fore legs splay forward, approach too near the vertical from the point of the shoulder;
— *cap de maure,* black-headed horse (rest of the body of another color);
— *de carrière,* race-horse, running-horse; (*mil.*) riding-school horse;
— *de cavalerie légère,* (*Fr. a.*) horse adapted to hussars and to chasseurs;
— *cavecé de maure,* Moor-faced horse (inferior part of head is darker than all the rest of the body);
— *en chair,* horse in condition;
— *chargé d'encolure, de ganache, de tête,* heavy-headed horse;
— *de chasse,* hunter;
— *cillé,* horse marked by white hairs about temples;
— *clabaud,* horse whose ears are horizontal at rest, and shake from down up in motion;
— *claqué,* broken-down horse;
— *clos du derrière,* horse whose hocks are too close together, cow-hocked;
— *à collier,* draught horse;
comité d'achat de — *aux,* (*mil.*) horse-board;
— *cornard,* roarer;
— *cornu,* horse with prominent haunches;
— *couronné,* broken-kneed horse;
course de —, ride;
course de — *aux,* horse-race;
— *de course,* race-horse;
— *de course plate,* running horse (flat race);
— *courtaud,* — *courtaudé,* horse with ears cropped and tail docked;
— *courte-queue,* docked horse;
— *court-jointé,* short-pasterned horse;
couverture de —, horse blanket;
cran de —, ridge in horse's mouth;
— *crapecé,* — *craps,* v. — *bretaud;*
— *crochu,* horse whose hocks are turned in too much, are too close together, cow-hocked horse;
— *cryptorchide,* horse whose testicles have not descended into the scrotum; cryptorchis;
— *débourré,* horse that suddenly loses its fleshy appearance (dealers' trick);
— *décousu,* horse showing lack of proportion between parts, ill-built horse;
— *du dedans,* (*art.*) horse on side toward which wheel or turn is made;
— *déferré,* horse with shoes removed, horse that has cast a shoe;
— *déhanché,* hipshot horse;
— *du dehors,* (*art.*) horse on side opposite to that toward which a wheel or turn is made;
— *demi-belle-face,* semiwhite-faced horse;
— *à demi-châtré,* rig;
— *demi-sang,* half-bred horse;
— *de derrière,* (*art.*) wheel-horse, wheeler;
— *derrière les jambes,* horse that throws himself on his hind quarters;
— *derrière la main,* horse that creeps, that is behind the bit;
— *désajusté,* horse not true in his paces;
— *détaché,* horse that keeps the lead (sporting);
— *de deux cœurs,* obstinate horse;
— *à deux fins,* saddle and draught horse;

cheval à deux mains, saddle and draught horse;
— de devant, (art.) lead horse;
— difficile, stubborn horse, one hard to manage;
dompteur de —, horse-breaker;
— à dos ensellé, saddle-backed horse, sway-backed, hollow-backed;
— doux, quiet, gentle horse;
— droit, sound horse;
— droit sur jambe, short-pasterned horse;
— droit et léger aux aides, sound horse, well broken, and obedient to the aids;
— droit sur les boulets, horse whose pasterns approach the vertical, straight-pasterned horse;
— écloppé, lame horse, unfit for service;
— d'école, trained horse;
— écourté, docked horse;
— efflanqué, thin-flanked horse;
— encastelé, horse with contracted hoof;
— encloué, horse pricked in shoeing;
— ensellé, saddle-, sway-backed, horse;
— entier, stallion;
— entier à main droite (gauche), horse that refuses to turn to the right (left), that disobeys spur or bridle urging to the right (left);
— entier aux deux mains, horse that refuses to turn to the right or left, that is not bridlewise;
entraîneur de — aux, horse-trainer (for racing);
— époinlé, hipshot horse;
— errant, wild horse;
— d'escadron, regimental horse;
— étoffé, stout, stocky horse;
— fait, fully developed horse;
— farcineux, farcied, glandered horse;
faux —, horse that gallops false;
— faux bégu, horse whose external dental cavity persists when it should have disappeared;
fer à —, de —, horseshoe;
fer de — cramponné, horseshoe with calkins;
— ficelle, very slenderly built horse;
— en forme de coin, horse with a fine croup;
— fourbu, foundered horse;
— de fourgon, (mil.) baggage-wagon horse;
— franc du collier, free horse in harness;
— franc de trait, willing puller (draught horse);
— à front plat, mare-faced horse;
— à gauche, near horse;
— gras fondu, horse that has the grease;
— gros d'haleine, thick-winded horse;
— de guerre, charger, cavalry horse;
— de harnais, draught horse, harness horse;
— haut-jointé, long-legged, long-pasterned horse;
— haut-le-pied, spare horse, held horse;
— haut-perché, long-legged horse (legs too long);
— hérissé, rough-coated horse;
heure de —, hour's ride;
homme de —, thorough horseman, lover of horses;
— hongre, gelding;
— hors d'âge, old horse, horse that has lost the mark;
— incertain, restless, fidgety horse;
— d'industrie et de commerce, ordinary draught horse;
— isabelle, clay-bank horse;
— à jarrets oscillants, horse with rotating hocks;
— de ligne, (Fr. cav.) horse adapted to, or fitted for, dragoons;
— long-jointé, long-pasterned horse;
— louvet, clay-bank horse;
— lunatique, wall-eyed horse;
— de luxe, pleasure horse;
— à main, off horse;
— de main, led horse; held horse; off horse; hand horse;
— mal coiffé, lop-eared horse;
— de manège, riding-school horse;
— méchant, vicious horse;

cheval du milieu, (art.) swing horse;
— miroité, dappled-bay horse;
— moineau, horse with ears cropped halfway up;
— monorchide, monorchis, ridgel, ridgeling, rig;
— monté sur des allumettes, horse having slender canon bones;
— moreau, black horse;
— morveux, glandered horse;
— moucheté, flea-bitten horse;
— nu, hairless horse;
— d'obstacles, hurdle-racer, steeplechaser;
— d'officier, officer's horse, charger;
— ombrageux, skittish horse;
ongle de —, horse's hoof;
— d'ordonnance, horse pressed for service;
— oreillard, lop-eared horse (sometimes spelled orillard);
— ouvert du derrière, horse with broad croup;
— ouvert du devant, horse whose fore legs are too wide apart;
— panard, crooked-legged horse, horse that dishes, outbow-footed horse;
— panard du derrière, cow-hocked, close hammed horse, horse outbow-footed behind;
— panard du devant, horse with crooked fore legs, outbow-footed in front;
pantalon de —, riding-trousers, -breeches;
— aux de même parure, well-matched horses;
pas de —, horse's pace;
— de pas, pacer;
pension de — aux, livery stable;
passe- —, horse ferry;
— pesant à la main, hard-mouthed horse, one that bears on the bit;
— peureux, skittish, shying horse;
— de phaéton, small carriage horse;
— pie, piebald horse;
— plat, lank-sided horse;
— pointu du derrière, horse whose croup is narrow behind;
— porteur, (art.) near horse;
— porteur d'outils, (mil.) tool horse;
— poussé de nourriture, overfed horse;
— poussif, broken-winded horse;
— près de terre, short-legged horse;
promenade à —, ride;
— pur sang, — de pur sang, thoroughbred horse;
quartiers de —, quarters of the horse;
— de race, blooded horse; thoroughbred;
— ragot, stocky horse;
— ramassé, compactly built horse; stocky horse;
— ramingue, horse that refuses to obey the spur;
râteau à —, hay-rake;
— refait, made horse;
— aux de réforme, cast horses;
réforme de — aux, casting of horses;
registre de — aux, stud-book;
— de relai, relay, fresh horse;
— de remonte, remount horse;
— de remplacement, spare horse;
— renforcé, strong, thickset horse;
— de renvoi, return horse;
— de réquisition, (mil.) horse pressed into service by requisition;
— de réserve, (Fr. cav.) horse adapted to cuirassiers; (in gen.) spare horse;
— rétif, stubborn, obstinately disobedient horse;
robe de —, horsehide, coat of the horse;
— aux de même robe, horses matched in color;
— rouan, roan horse;
— roulé, stocky horse, compactly built;
— rude, rough, uneasy horse;
— rueur, kicking horse;
— ruiné, broken-down horse;
sabot de —, horse's hoof, coffin;
— sage, steady horse, of good disposition;
— sage au feu, horse steady under fire, that will stand under fire;
— sain et net, horse warranted sound and free from vice;

cheval *de sang,* thoroughbred horse, (better) blooded horse (not necessarily thoroughbred; v. —— *pur sang,* with which —— *de sang* is often taken synonymously);
—— *de selle,* saddle horse;
—— *serré du derrière,* horse whose hind legs are too close together;
—— *serré du devant,* horse whose fore legs are too close together, horse closed in front, knock-kneed;
—— *de somme,* pack horse;
—— *soupe de lait,* cream-colored horse;
—— *souris,* mouse-colored horse;
—— *sous la main,* off or hand horse;
—— *sous lui de derrière,* horse whose hind legs are too much under him;
—— *sous lui du devant,* horse that has his fore legs too much under him, horse over in the knees, (sometimes, even) knee-sprung;
—— *de sous verge,* (art.) off horse;
—— *de steeple chase,* steeplechaser;
—— *de suite,* pack horse;
—— *taillé à coups de hache,* horse the bones of whose croup stand out, rawboned horse;
—— *tartare,* saddle horse, half wild, with some Arab blood;
—— *de tête,* officer's horse;
—— *en tête,* horse with medium-sized white spot on head;
—— *fortement en tête,* horse with large white spot on head;
—— *en tête interrompu,* horse with white spot on head interrupted or broken by a colored spot or place;
—— *légèrement en tête,* horse with a more or less considerable number of white hairs on head;
—— *marqué en tête,* horse with white markings on head;
—— *en tête prolongé,* horse with white marks or spots coming down to chamfer;
—— *tigré,* spotted horse;
—— *de timon,* wheel-horse, wheeler;
—— *tiqueur,* horse given to some trick or habit, as crib-biting, wind-sucking (term ordinarily means crib-biting);
—— *de tirage,* draught horse;
—— *tirant au renard,* horse that pulls or hangs on halter in stable, with a view to breaking it;
—— *tisonné,* branded gray horse, marked with black spots;
tondeur de —— *aux,* horse-clipper;
tondeuse de —— *aux,* horse-clipper;
—— *à tous crins,* horse with flowing mane and tail;
—— *à toutes mains,* horse that may be driven or ridden indifferently;
train de ——, horse's pace, rate;
train de derrière d'un ——, hind quarters of a horse;
—— *de trait,* draught horse;
—— *de gros trait lent,* horse for heaviest draught;
—— *de gros trait rapide,* omnibus horse, express-wagon horse, etc.;
—— *de trait léger,* cab-horse, etc.; (also) light-draught horse for artillery service, engineers, etc.;
—— *trapu,* strong, thickset horse;
travail à ——, (*cav.*) mounted work, exercises, in widest sense (not drill so much as equestrian work pure and simple);
—— *trop enlevé,* horse whose ribs do not come down as low as the elbow;
—— *trop fait en montant,* horse of abnormally high croup;
—— *trop ouvert du derrière,* horse whose hind legs, on the whole, are too wide apart;
—— *de trot,* trotter;
—— *de troupe,* (*mil.*) soldier's horse (subdivided into —— *de selle,* for the cavalry, etc., and —— *de trait* and —— *de trait léger,* for the artillery); troop-horse;
—— *truité,* flea-bitten horse, trout-colored horse;
—— *uni,* horse that gallops smoothly, well-paced horse;

cheval *vert au montoir,* horse hard to mount;
—— *vicieux,* vicious, headstrong horse;
—— *vif,* high-spirited horse;
voiture pour dresser les —— *aux,* horse-break;
—— *de volée,* leader, lead horse;
—— *zain,* whole colored horse (total absence of white hairs).

II. Cheval as the object of or qualifying a verb:
abattre un ——, to cast a horse;
acheminer un ——, to teach a (young) horse to advance in a straight line;
achever un ——, to finish breaking a horse;
ajuster un ——, v. *achever un* ——;
appuyer un ——, to passage a horse, to cause him to move with shoulders and hind quarters on parallel tracks;
arrondir un ——, to break a horse;
asseoir un ——, to train a horse; to throw him well on his haunches;
asseoir un —— *sur ses jambes,* to train a horse to gallop with his croup lower than his shoulders, to execute *airs de manège;*
assouplir un ——, to give a horse a regular and easy motion;
assurer la bouche à un ——, to ride with a light hand;
assurer un ——, to make a horse steady, i. e., train him to execute all movements with accuracy;
attaquer un ——, to spur a horse, to use the spur;
avertir un ——, to urge, encourage, a horse, to gather him;
boucher un ——, to look at a horse's teeth to determine his age;
brusquer un ——, to start a horse suddenly;
conduire un —— *par la figure,* to lead a horse by the head;
couper un ——, to geld a horse;
déferrer un ——, to remove the shoes from a horse;
dépêtrer un ——, to free or disentangle a horse from harness;
donner feu à un ——, to fire a horse;
donner haleine à un ——, to breathe a horse, give him his wind;
donner un point d'appui à un ——, to cause a horse to bear on the hand while urging him forward;
donner du talon à un ——, to spur a horse;
dresser un ——, to train a horse;
enlever un ——, to urge a horse vigorously forward;
entraîner un ——, to train a horse for a race;
envelopper un ——, to grasp or clasp a horse with the legs (of the seat);
être mal à ——, to have a poor seat;
faire la toilette à un ——, to trim a horse;
faire parader un ——, to make a horse show off his paces;
faire prendre le vert à un ——, to put a horse on green food, to put a horse out to grass;
faire quitter le vert à un ——, to take a horse from grass;
ferrer un ——, to shoe a horse;
flatter un ——, to make much of a horse;
franchir un ——, to jump over a horse;
gourmander un ——, to use a horse roughly with the hand, to increase its severity;
lancer un ——, to urge a horse, to set him at speed;
marcher à ——, to ride;
mettre quelqu'un à ——, to teach one how to ride;
mettre un —— *au vert,* to turn a horse out to grass;
mettre un —— *sur les dents,* to override, founder a horse;
monter à ——, to mount a horse, to ride;
monter à —— *sous quelqu'un,* to learn to ride from some one;
monter un ——, to ride;
monter un —— *à nu, à poil,* to ride bareback;
outrer un ——, to override, to founder, a horse;
parcourir à ——, to ride over;
piquer un ——, to prick a horse in shoeing, to drive a nail into the quick; (hence) to lame a horse; to spur a horse;

cheval, *plier un* ——, to break a horse;
porter un —— *d'un talon sur l'autre*, to passage a horse;
pousser un ——, to push a horse, try his mettle;
promener un ——, to lead a horse up and down, to walk him up and down;
promener un —— *par le droit*, to make a horse advance on a straight line;
rabaisser les hanches d'un ——, to throw a horse on his haunches;
rassembler un ——, to gather a horse;
recevoir des ——*aux en pension*, to keep a livery stable;
rechercher un ——, to animate a horse, to stir him up, excite him;
réduire un ——, to break a horse;
refaire un ——, to make a horse;
réformer un ——, to cast a horse;
relayer de ——*aux*, to change horses;
remettre un —— *à sec*, to take a horse from grass and put him on dry food again;
retenir un ——, to curb, restrain a horse;
retirer un —— *du vert*, to take a horse from grass;
rompre l'eau à un ——, to stop, check, a horse in drinking;
rompre un ——, to break a horse;
sauter à ——, to vault into the saddle;
serrer un ——, to keep a horse together, well in hand;
sonner à ——, to sound "to horse;"
soutenir un ——, to keep a horse up, keep him at (e. g., increase of gait);
surmener un ——, to override, to founder, a horse;
tenir bien à ——, to have a good seat;
tenir un —— *en main*, to keep a horse in hand, hold him by the head;
unir un ——, to give a horse regularity of motion;
vider un ——, to rake a horse.

III. Cheval qualified by a relative clause:

—— *qui a de l'action*, stepper;
—— *qui a beaucoup d'action*, high stepper;
—— *qui a bonne bouche*, horse that will eat anything;
—— *qui a de l'école*, trained horse;
—— *qui a les jambes foulées*, foundered horse;
—— *qui a les jambes sèches*, clean-limbed horse;
—— *qui a les jarrets bien vidés*, clean-hocked horse;
—— *qui n'a ni bouche n iéperon*, horse that obeys neither bit nor spur;
—— *qui a tout mis*, horse that has all his teeth, full-grown horse (at least 5 years old);
—— *qui a un effort*, strained, foundered horse;
—— *qui s'arme*, —— *qui s'arme de son mors*, horse that gets the bit in his teeth, that bends down his head so that bridle-cheeks rest on his chest;
—— *qui s'atteint*, horse that overreaches;
—— *qui s'attrape*, horse that overreaches;
—— *qui bat du flanc*, horse that wheezes;
—— *qui se berce*, horse that rocks in trotting;
—— *qui billarde*, horse that paddles;
—— *qui boit dans son blanc*, white-lipped horse;
—— *qui commence à marquer*, young horse that begins to show his age by his teeth;
—— *qui se coupe*, horse that cuts himself (leg in air cuts leg bearing on the ground);
—— *qui se dérobe de dessous l'homme*, horse that slips from under his rider;
—— *qui fait magasin*, horse that keeps part of his food between his teeth and cheeks;
—— *qui fait quartier neuf*, horse that is quarter cast;
—— *qui fauche*, horse that paddles (not so much as —— *qui billarde*);
—— *qui forge*, horse that forges;
—— *qui galope faux*, horse that gallops false;
—— *qui harpe*, horse that bends at the hocks, on starting, backing, turning corners, etc.;
—— *qui s'immobilise*, balky horse;
—— *qui jette sa gourme*, glandered horse;

cheval qui laboure la terre, digger;
—— *qui marque*, horse that still has the mark;
—— *qui ne marque plus*, horse that has lost the mark;
—— *qui se monte en dame*, horse broken to a side-saddle;
—— *qui part bien de la main*, horse that starts as soon as he feels the hand slacken;
—— *qui porte sa tête en beau lieu*, horse that carries his head well;
—— *qui se ramène bien*, horse that carries his head well;
—— *qui rase*, horse that is losing mark of mouth;
—— *qui rase le tapis*, horse that goes too near the ground;
—— *qui renifle sur l'avoine*, horse that is off his feed;
—— *qui tâte le pavé*, tender-footed horse;
—— *qui tire à la main*, horse that bores, that pulls hard, heavy in hand;
—— *qui tire au renard*, horse that pulls or hangs on his stable halter;
—— *qui trousse*, horse that raises his fore legs too much;
—— *qui va large*, horse that goes wide.

IV. Technical expressions, not hippological:

à ——, (gym., etc.) astraddle, on both sides of;
à —— *sur*, (*une rivière*, etc.) (mil., etc.) on both sides (of a river, road, etc.);
—— *de bois*, wooden horse (gymnasium);
—— *de l'adjudant*, (mil. slang) camp-bed of a cell;
—— *électrique*, (elec.) electric horsepower (746 watts);
fer à ——, (fort.) horseshoe;
—— *de force, force de* ——, horsepower;
—— *de frise*, (fort., etc.) cheval-de-frise;
—— -*heure*, horsepower per hour;
—— *nominal*, nominal horsepower;
petit ——, donkey-engine;
—— *de vapeur*, horsepower;
—— -*vapeur*, horsepower; donkey-engine;
—— -*vapeur effectif*, (indiqué, nominal) effective (indicated, nominal) horsepower;
—— *de voltige*, vaulting horse (of a gymnasium).

chevalement, m., prop, stay, shore.
chevaler, v. a. n., to prop up, shore up; (man.) to passage (v. *appuyer*).
chevalet, m., frame; trestle; rack; horse; stand; (siege) fascine-horse, (mil.) arm-rack; (pont.) trestle of a bridge; (mach.) bit of a borer, (art.) rocket-stand;
—— *d'armements*, (mil.) equipment rack;
—— *d'armes*, (mil.) arm-rack (in camp, etc.);
—— *belge*, (pont.) Belgian trestle (cap on tripods);
—— -*bigue*, (pont.) sheer-trestle, two-legged trestle; spar trestle;
—— *Birago*, (pont.) ordinary ponton-bridge trestle;
—— *de bombardement*, (siege) siege-rocket stand;
chapeau de ——, (pont.) trestle cap;
—— *à chapeau mobile*, (pont.) trestle with adjustable cap;
—— *de corps de garde*, (Fr. a.) saw-horse (part of furniture of a guardhouse);
—— *de fusées*, (art.) rocket stand or frame;
—— *de lancement*, (siege) launching trestle of a flying bridge;
—— *de natation*, swimming-stand or trestle (gymnasium);
—— -*palée*, (pont.) two-legged trestle and cap, driven down like piling;
—— *à deux, quatre, etc., pieds*, (pont.) two-, four-legged trestle, etc.;
—— *de pointage*, (sm. a.) aiming-stand, aiming-drill rest;
—— *rapide*, (pont.) trestle that may be set up in 15 to 20 minutes;
—— *support*, (pont.) trestle set up in a ponton boat;
—— -*trépied*, any three-legged trestle; rocket stand.

chevalier, m., knight (of an order); (Fr.) lowest grade of the Legion of Honor;

3877°—17——6

chevalier-*garde*, (*mil.*) chevalier-garde (Russian).
chevaline, a., (*hipp.*) equine (e. g., *bête* ——, *race* ——, *etc.*).
chevauchée, f., (*cav.*) mad, d e s p e r a t e charge.
chevauchement, m., (*tech.*) riding, overlapping; (*cord.*) overriding; (*hipp.*) overriding of heels.
chevaucher, v. a., (*man.*) to go on horseback; (*tech.*) to lap, overlap, to ride; (*hipp.*) to override (of heels); (*cord.*) to ride over;
—— *court* (*long*), (*man.*) to ride with short (long) stirrups.
chevau-léger, m., (*cav.*) sort of light-horseman;
——*s*, light cavalry (suppressed in France in 1830).
chevet, m., bolster; (*sm. a.*) end of small-arm chest; (*pont.*) beam supporting balks of a bridge.
chevêtre, f., (*harn.*) halter; (*cons.*) trimmer, header (of a wooden floor); bracing, brace (of an iron floor).
cheveux, m. pl., hair; (*sm. a.*) hair-rifling.
chevillage, m., bolting, pinning, fastening with pins;
—— *en cuivre*, copper bolting;
—— *double*, double bolting;
—— *en fer*, iron bolting;
—— *simple*, single bolting.
cheville, f., peg, pin; dowel; bolt, spike; key (of a joint); wooden pin or plug; treenail;
—— *arrêt d'excentrique*, (*art.*, *etc.*) eccentric stop;
—— *d'arrêt*, locking-pin;
—— *arrêtoir*, stop-pin, locking-pin; (*art.*) travelling trunnion-bolt;
—— *d'assemblage*, through bolt, shackle bolt, assembling bolt;
atteler en ——, (*harn.*) to hitch a horse in front of a shaft horse;
—— *à barbe*, —— *barbelée*, —— *barbue*, rag-bolt;
—— *bardelée*, spike-nail;
—— *de bois*, —— *en bois*, treenail;
—— *à boucle*, ringbolt, eyebolt;
—— *à bout-perdu*, short-drove bolt;
—— *à charger les mines*, (*min.*) **shooting-needle**, blasting-needle, nail;
chasser une ——, to drive a bolt;
—— *clavetée*, clench-bolt;
—— *à croc*, hook-bolt;
—— *à ergot*, rag-, sprig-, barb-bolt;
—— *en fer*, iron pin;
—— *à fiche*, rag-bolt;
—— *fixe*, steady pin;
—— *à goujon*, common bolt;
—— *à grille*, —— *grillée*, rag-bolt;
—— *hâche*, (*pont.*) crossbar key of the Belgian trestle;
—— *maîtresse*, kingbolt;
—— *de manœuvre*, (*art.*) gun-lift pin;
—— *à mentonnet*, (*art.*) cap-square bolt;
—— *ouvrière*, pivot, pivot-bolt, spindle; mainpin; bolster bolt; (*art.*) pintle, pintle bolt;
—— *ouvrière arrière*, (—— -*avant*, —— *de l'avant*), (*art.*) rear (front) pintle bolt;
—— -*pivot*, (*art.*) central pivot, pivot mounting, pivot-bolt or pillar (esp. of armored mounts);
—— *à repousser*, drift-bolt, starting bolt;
—— *à la romaine*, shifting-pin (of a carriage);
sellette de —— *ouvrière*, pintle-block;
—— *à tête carrée*, square-headed bolt;
—— *à tête de diamant*, diamond-headed **bolt**;
—— *à tête encastrée*, countersunk bolt;
—— *à grosse tête*, saucer-headed bolt;
—— *à tête plate*, flat-headed bolt;
—— *à tête ronde*, round-headed bolt;
—— *de tournage*, (*cord.*, *etc.*) belaying pin;
—— *à tourniquet*, screw-bolt; (*pont.*) rack-stick;
—— *à virole*, clinch-bolt;
—— *à vis*, (*art.*) screw-pillar;
—— *visée*, screw-plug.
cheviller, v. a., to bolt, to bolt together; to peg, to peg a shoe, etc.;
—— *d'un bord à l'autre*, to bolt through;
—— *en cuivre*, to copper fasten;

cheviller *en fer*, to fasten with iron spikes, bolts, etc.;
—— *à travers bois*, to bolt through.
chevillette, f., pin, key, keep-pin;
—— -*arrêt*, keep-pin, stay-pin;
—— *à clavette*, key-pin;
—— -*clef*, pole key (of a wagon); key-pin;
—— -*clef de timon*, (*art.*) fork-strap key;
—— *de crochet cheville ouvrière*, (*art.*) pintle key.
chevillot, m., (*art.*) toggle, belaying pin.
chèvre, f., (*art.*, *etc.*) gin;
—— *ascenseur*, (*art.*) lifting-gin (used in fortresses);
—— *brisée*, field-gin;
—— *en cabestan*, gin rigged as a capstan;
câble de ——, gin-fall;
—— *à déclic*, garrison gin;
dresser la ——, to raise the gin;
entretoise de ——, crossbar of a gin;
équiper la —— *en cabestan*, to rig the gin as a capstan;
équiper la —— *à n brins*, to rig the gin with an n-fold tackle;
—— *à haubans*, gin used as sheers; gin as sheers;
pied de ——, leg of a gin;
—— *de place*, garrison gin;
—— *postiche*, sheers;
—— *de tranchée*, (*siege*) trench- or siege-gin; (*art.*) gun-lift.
chevrette, f., small or light gin; small tripod used in cleaning wagons, etc.; setter;
—— *double*, setter with two uprights;
—— *simple*, setter.
chevron, m., (*cons.*) rafter, couple, jack rafter of a roof; (*mil.*) n. c. o. chevron (not worn in France), service stripe;
—— *d'ancienneté*, (*mil.*) service stripe;
—— -*passant*, (*cons.*) rafter forming support of eaves.
chevronnage, m., frame, framework.
chevrotine, f., (*sm. a.*) buckshot.
chicane, f., (*mil.*) slight skirmish, affair of small importance (to worry the enemy); obstacle, material obstacle of whatever sort, to delay or annoy the enemy, to dispute his possession of the ground; (*steam*, *etc.*) obstacle (e. g., to passage of flame, so as to have a greater heating surface);
—— *du fossé*, (*siege*) defense of the ditch by every possible means;
guerre de ——, (*mil.*) generic term for small operations designed to annoy or delay the enemy;
lutte de ——*s*, (*mil.*) struggle in which the advance of the enemy is opposed at every step by obstacles;
terrain de ——, (*mil.*) ground possession of which is disputed by means of obstacles, etc., (in gen.) disputed ground.
chicaner, v. a., to dispute;
—— *le terrain*, (*mil.*) to dispute the ground.
chicot, m., tree-stump.
chicstrac, m., (*mil. slang*) refuse, offal, etc.;
corvée de ——, (*mil.*) fatigue for cleaning cesspools and the like.
chien, m., dog; (*sm. a.*) hammer (lock guns); cocking piece, cocking-stud (of modern military rifles); (*r. f. art.*) hammer; (*mach.*) dog.;
armer le ——, (*sm. a.*) to cock;
bander le ——, (*sm. a.*, *etc.*) to cock;
—— *à charnière*, hinged dog;
—— *de compagnie*, (*mil. slang*) sergeant-major;
conduire le —— *à l'abattu*, (*sm. a.*) to let down the hammer;
débander le ——, (*sm. a.*) to let down the hammer;
désarmer le ——, v. *débander le* ——;
—— *de guerre*, (*mil.*) dog trained for military purposes (to give alarm in outposts, etc.);
mettre le —— *à l'abattu*, (*sm. a.*) v. *conduire*, *etc.*, *supra*;
un officier ——, (*mil. slang*) a martinet;
—— *de régiment*, (*mil. slang*) an adjutant;
sans ——, (*sm. a.*) hammerless (e. g., shotgun).

chien *vert*, (*mil. slang*) any supply department (e. g., *intendance*); course of law at military schools; law instructors of military schools.

chiendent, m., quitch grass, couch grass.

chiffraison, f., (*inst.*) the figures on a graduated limb, etc.; (hence) graduation.

chiffre, m., number, figure; cipher;
— *arrondi*, round number;
— *carré*, cipher of Blaise de Vigenère (so called from its base being a square);
— *chiffrant*, enciphering key of a cryptogram;
— *déchiffrant*, deciphering key of a cryptogram;
— *à grilles*, cipher based on the use of enciphering frames;
— *indéchiffrable*, — *par excellence*, v. — *carré*.

chiffrer, v. a., to encipher.

chiffreur, m., cipherer.

chinois, m., (*mil. slang*) "cit" (compare *pékin*).

chio, m., (*met*.) tap-hole of a firing forge, for withdrawing scoriæ;
plaque de —, slag-hole plate.

chirurgical, a., surgical.

chirurgie, f., surgery.

chirurgien, m., surgeon (obs. as a military term); — -*major*, (*mil.*) surgeon-major (obs.).

chlorure, m., (*chem.*) chloride.

choc, m., shock, clashing, blow, concussion, collision, jerk; (*mil.*) attack, onset, encounter of two lines or bodies (esp. cavalry); (*art., sm. a.*) impact (of projectiles);
— *direct*, (*ball.*) normal impact;
énergie au —, (*ball.*) striking energy;
— *oblique*, (*ball.*) oblique impact;
troupes de —, (*mil.*) assaulting troops;
vitesse au —, (*ball.*) striking velocity.

choisi, a., picked (horses, men, etc.).

choix, m., choice, selection;
avancement au —, (*mil.*) promotion by selection;
de —, picked (men, horses, etc.).

choke-bored, m., (*sm. a.*) choke-bore shotgun.

chômage, m., (of a mill, etc.) state of being idle, not in operation, out of blast (furnace).

chômer, v. n., (of a mill, engine, etc.) to be idle, not in operation, out of blast (furnace).

choquer, v. a., to shock, strike, collide;
— *au cabestan*, to surge, check at the capstan.

chorographie, f., chorography.

choucarde, f., (*mil. slang*) wheelbarrow.

chouette, chouettard, chouettaud, (*mil. slang*) "crack."

chouffliqueur, m., (*mil. slang*) shoemaker.

chromate, m., (*chem.*) chromate.

chrome, m., chromium.

chromographe, m., chromograph (copying or manifolding apparatus, rude papyrograph).

chrono-électrique, a., chrono-electric.

chronographe, m., (*ball.*) chronograph;
— *à chariot*, tram chronograph;
— *de chute*, falling-weight chronograph;
— *à diapason*, tuning-fork chronograph;
— *électrique*, any electric chronograph;
— *à enregistrement mécanique*, chronograph with clockwork or other mechanical register;
— *à enregistreur*, generic term for any chronograph equipped with a registering device;
— *à pointage*, stop watch for measuring times of flight;
— -*vélocimètre*, time-recording velocimeter.

chronomètre, m., chronometer; (*ball.*) instrument for measuring intervals of time; chronometer of Le Boulengé chronograph; sort of stop watch for measuring times of flight;
— -*étalon*, standard chronometer;
— *à pointage*, (*ball.*) stop watch for times of flight.

chronoscope, (*ball.*) chronoscope.

chronotélémètre, m., (*mil.*) range finder.

chryptorchide, m., (*hipp.*) cryptorchid, (according to some authorities) ridgeling.

chute, f., fall; (*art.*) fall (of a projectile); (*met.*) sinking head, part rejected of an ingot;
angle de —, (*art., ball.*) angle of fall;
— *du bas*, (*met.*) (of an ingot, etc.) lower end rejected;
— *d'eau*, waterfall; head of water;
— *du haut*, (*met.*) (of an ingot, etc.) upper end rejected;
— *du jour*, fall of day;
point de —, (*art., ball.*) point of fall;
— *de potentiel*, (*elec.*) fall of potential;
tomber en — *libre*, to fall freely.

cible, f. (*mil*), target (for small-arm practice);
— *de buste*, head-and-shoulders target (a disappearing target); silhouette of man's bust;
cadre- —, (*ball.*) velocity-frame (in determination of muzzle velocities);
— *cuirassée*, plated target;
— *à éclipse*, disappearing target;
— *fixe*, stationary target;
— *flottante*, floating target;
— *mobile*, moving target;
— *oscillante*, rocking target;
— *parlante*, self-recording target;
— *rectangulaire*, (*Fr. a.*) ordinary small-arm target;
— *à relèvement*, disappearing target;
— *à ressort*, spring-target;
— *à signaux automatiques*, indicator target;
— -*silhouette*, silhouette target, skirmish target (U. S. A.);
tir à la —, target practice;
— *pour tir réduit*, gallery practice target;
tirer à la —, to fire at a target.

cicatrice, f., scar.

cicatriser, v. a. and r., to scar, cicatrize; to skin over, to be cicatrized.

ciel, m., sky; weather; roof (of a mine gallery, of a furnace); (*fort.*) top (of a cupola);
à — *couvert*, cloudy weather;
à — *couvert*, under cover, with overhead cover;
à — *ouvert*, (*art., fort.*) without overhead cover; open;
plancher de —, (*mil. min.*) roof timbers;
planches de —, (*mil. min.*) roof planks of a mine gallery;
— *pommelé*, mackerel sky;
— *serein*, clear sky;
travaux à — *ouvert*, (*min.*) ordinary blasting as distinguished from tunneling operations.

cierge, m., candle; (*sm. a.*) curve separating illuminated from dark portion of interior of barrel, when tested by the method *à la planchette* (v. *dressage à la planchette*).

cigale, f., anchor-shackle.

cigogne, f., (*mach., etc.*) web, crank lever.

cimaise, f., (*sm. a.*) part of guard supporting the knob of the hilt (model 1817).

cimbale, f., v. *cymbale*.

cime, f., (*top.*) crest, ridge, summit, peak.

ciment, m., cement; (*hipp.*) cement (of a tooth); (*met.*) material, as quartz, added to clay in making furnace linings; in the cellular theory of steel, the carbide of iron surrounding a *cellule simple*, q. v.;
— *armé*, cement (blocks, etc.) reinforced or strengthened by iron;
béton de —, concrete;
— *hydraulique*, hydraulic cement, pozzuolana;
— *à prise lente*, slow-setting cement;
— *à prise rapide*, quick-setting cement;
— *prompt*, quick-setting cement;
— *romain*, Roman cement.

cimenter, v. a., to cement.

cimeterre, m., (*sm. a.*) scimitar.

cimetière, m., cemetery, burying ground.

cimier, m., (*unif.*) crest (of a helmet).

cimolée, f., **cimolie,** f., pipeclay.

cinématique, f., kinematics.

cinabre, m., cinnabar, red lead.

cincelle, f., **cincenelle,** f., v. *cinquenelle*.

cinglage, m., (met.) shingling;
 laminoir de ——, shingling-roll;
 marteau de ——, shingling hammer;
 scorie de ——, shingling slag.
cinglard, m., (met.) shingling hammer.
cingler, v. a., to lash (with a whip); (met.) to shingle;
 machine à ——, (met.) squeezer, shingling rolls.
cingleresse, f., (met.) shingling-tongs.
cingleur, m., (met.) squeezer, shingler;
 —— *rotatif*, shingling rolls.
cinquenelle, f., (pont.) sheer-line.
cintrage, m., plate-bending (armor plate, ship plates, etc.).
cintre, m., curb; center-bit; bend, curve (as of an armor plate); (cons.) arch, center of an arch; (art.) bend or curve of a carriage;
 —— *de crosse*, (art.) rounding of the trail;
 —— *en ogive*, (cons.) pointed arch;
 plein ——, full-centered arch, semicircular arch;
 —— *rampant*, rampant arch;
 —— *surbaissé*, surbased, elliptical arch;
 —— *surhaussé*, arch whose rise is more than twice the span;
 —— *surmonté*, surmounted arch.
cintrement, m., sag, cambering.
cintrer, v. a., to arch; to curve; to bend, camber; to sag; to bend to shape (as a plate); to shape, curve to shape; (cons.) to center an arch, to put in the center; (mil.) of a line, to sag back a little in the center;
 machine à ——, plate-bending machine.
cipahi, m., (mil.) Sepoy, Sipahi.
cipaye, m., v. *cipahi*.
cirage, m., shoe-blacking; varnish (for equipments, etc.).
circonférence, f., circumference;
 —— *d'un cordage*, (cord.) size of a rope;
 —— *primitive*, (mach.) pitch-circle.
circonscription, f., (mil., etc.) administrative, judicial, military subdivision of territory;
 —— *d'étapes*, in a *ligne d'étapes*, a subdivision of three or four days' march;
 —— *de fourniture*, (Fr. a.) subdivision of territory for purposes of supply;
 —— *militaire*, any subdivision of territory for military purposes, of no matter what nature.
circonstance, f., circumstance, occasion;
 général de ——, a general made so by circumstances, as distinguished from one by training, by profession; in the United States, a general of volunteers, appointed from civil life, and without previous military experience, (sometimes) "political" general;
 soldat de ——, an emergency or makeshift soldier, as opposed to a regular soldier.
circonvallation, f., (fort.) circumvallation.
circuit, m., circuit; (elec.) circuit;
 —— *de charge*, charging circuit (of a storage battery);
 court ——, short circuit;
 —— *de décharge*, discharging circuit (storage battery);
 —— *dérivé*, shunt, derived circuit;
 fermer un ——, to close a circuit;
 —— *d'inflammation*, (torp., etc.); firing circuit;
 ouvrir un ——, to open a circuit;
 —— *primaire*, primary circuit;
 —— *de sautage*, fuse circuit;
 —— *secondaire*, secondary circuit;
 —— *d'utilisation*, external circuit (as in a light system).
circulaire, a., circular; f., fifth wheel (of a wagon); circular band; (mil.) circular (of war department); (art.) racer, traverse-circle;
 bande ——, fifth wheel (wagon, etc.);
 —— *dentée*, (art.) cogged racer, traversing rack;
 —— *à gorge*, (art., etc.) chain groove;
 —— *lisse*, (art.) ordinary flat racer;
 —— *de pointage*, (art.) traversing rack, aiming circle (in turrets, etc.);
 —— *de roulement*, (art.) roller-circle;
 —— *de tir*, (art.) graduated circle for aiming in direction (cupolas, etc.).

cire, f., wax, sealing wax;
 —— *à cacheter*, sealing wax;
 —— *d'Espagne*, sealing wax;
 —— *à giberne*, (mil., etc.) heel ball.
cirer, v. a., to wax; to varnish; to polish (shoes, leather in general); to make (cloth) waterproof.
cirque, m., circus, amphitheater; (top.) a natural depression of the ground, bowl, especially a more or less circular valley;
 —— *d'effondrement*, sink-hole.
cisaillage, m., shearing, cutting.
cisaillement, m., shear, shearing; shearing stress; guillotining (as of a cartridge).
cisailler, v. a., to shear, cut with shears, to guillotine.
cisailles, f. pl., (mach., etc.) shears, plate- or metal-cutting shears; wire-nippers;
 —— *de cavalerie*, (Fr. cav.) wire-nippers (for destroying telegraph lines, etc.);
 —— *circulaires*, (mach.) circular, rotatory shears;
 —— *cylindriques*, v. —— *circulaires*;
 —— *à ébarber*, (sm. a.) bullet-trimmer (gallery practice);
 —— *à main*, snips.
ciseau, m., chisel, (in pl.) scissors; turner's tool;
 les ——*x*, (man.) the movement of facing the horse's tail, from normal position in saddle;
 en ——*x*, (nav.) wing and wing;
 —— *bédane*, crosscut chisel; parting-tool;
 —— *de calfat*, calking-iron;
 —— *de calfat double*, double-notched calking-iron;
 —— *de calfat simple*, single-edged calking-iron;
 —— *de charpentier*, carpenter's chisel;
 —— *de charron*, cant-chisel;
 —— *à chaud*, rod chisel, hot chisel (for iron when hot);
 —— *de côté*, side-tool of a lathe;
 —— *à ébaucher*, hollowing chisel;
 —— *à froid*, cold chisel;
 —— *à main*, hand-chisel;
 —— *à mastiquer*, cement-calking tool;
 —— *à mortaiser*, mortising chisel;
 —— *à raboter*, smoothing chisel;
 —— *rond*, round tool.
ciselage, m., chasing, graving; action of chasing, etc.; paring, cutting down.
ciseler, v. a., to cut, carve, chase.
ciselet, m., graver.
ciselure, f., graving, cutting, chasing.
citadelle, f., citadel.
citadin, m., (mil.) civilian, "cit" (U. S. A.).
citation, f., (law) summons; (mil.) mention in orders, as for a gallant act; (more esp.) mention in a general order to the army.
cité, f., city, (generally) the oldest part of a city;
 droit de ——, rights of a citizen; freedom of a city.
citer, v. a., to cite, summon (as before a court);
 —— *à l'ordre*, (mil.) to publish, mention, in orders.
citerne, f., cistern;
 —— *du condenseur*, (steam) hot well;
 —— *flottante*, water-boat;
 —— *à vapeur*, steam water-boat; steam harbor fire-engine;
 voiture ——, tank cart.
citerneau, m., reservoir, (esp., fort.) reservoir of filtered water.
citoyen, m., citizen.
citrouillards, m. pl., (mil. slang) dragoons.
civière, f., handbarrow; (art.) shot barrow;
 —— *à bobines*, wire-drum barrow;
 —— *à bombes*, (art.) barrow for spherical projectiles;
 —— *à chaîne(s)*, (art.) chain and tool barrow;
 —— *de chargement*, (art.) shot-bearer, loading barrow;
 —— *à crochet*, (art.) shot-bearer;
 —— *à obus*, (art.) shell barrow;
 —— *à poudre*, (art.) powder barrow;

civière à *projectile*, (*art.*) shot barrow;
— à *toile*, v. — à *poudre*.

civil, a., civilian (i. e., nonmilitary); m., civilian; *en* —, in plain clothes, in mufti, in "cits" (U. S. A.);
état —, (*adm.*) vital statistics, department of vital statistics.

clabaud, a., (*hipp.*) said of a horse with ears horizontal at rest, and shaking from down up in motion.

claie, f., screen; (*fort.*) wickerwork, hurdle, wattle;
passer à la —, to screen;
revêtement en —, (*mil.*) hurdle revetment.

clair, a., clear; written out in full, in plain words (of a cipher message);
— *-de-lune*, moonlight.

claire-voie, f., skylight; interval or space between joists, etc., when too large; opening closed by a grating;
à —, with openwork, checkerwork of open and closed spaces.

clairière, f., (*top.*) clearing in a wood, glade.

clairon, m., (*mil.*) bugle; bugler;
élève- —, lance-bugler (i. e., soldier learning the bugle);
maître- —, chief bugler.

clamauder, v. a., to clamp, secure with a cramp-iron.

clameau, m., clamp, dog, holdfast; cramp-iron, timber-dog;
— à *une face*, cramp-iron with points in same direction;
— à *deux faces*, cramp-iron with directions of points at right angles;
— *plat*, v. — à *une face*;
— à *pointe et à crochet ouvert*, cramp-iron with a point and a hook, respectively, at the ends, timber-dog;
— *simple*, v. — à *une face*;
tracé à —*x*, (*fort.*) trace of covered way in which the crest breaks at right angles to pass traverses.

clameur, f., (*mil.*) seditious cry.

clamp, m., v. *clan*.

clampin, m., (*mil.*) straggler; lame soldier.

clamponier, m., (*hipp.*) long-pasterned horse.

clan, m., sheave-hole (of a pulley, block).

clapet, m., (*mach.*, etc.) clack, valve-clack, clack-valve, flap-valve, pump-valve;
— à *articulations*, hinged valve, flap valve;
— *d'aspiration*, suction valve;
— à *couronne*, bell-shaped valve, cup-valve;
— *de décharge*, delivery valve;
— *double*, double-flap valve;
— *du fond*, foot-valve;
— *inférieur*, v. — *du fond*;
— *de pied*, v. — *du fond*;
— *de pompe*, pump-sucker, clapper;
— *de pompe à air*, blow-valve;
— *de refoulement*, delivery valve;
— *de retenue*, retaining valve; stop valve;
— *de ventilation*, damper.

clapotage, m., rippling, chopping, turbulent motion (of the sea, lakes, etc.), and noise of same.

clapoter, v. n., to ripple, to chop, to make a rippling noise (of the sea, etc.).

clapoteux, a., choppy (of the sea).

claquage, m., (*hipp.*) overstretching, (and hence) breaking down.

claque, m., (*unif.*) cocked hat.

claquement, m., crack, cracking (of a whip); (*steam*, etc.) water-hammer, water-hammering.

claquer, v. a. n., to crack a whip; (*hipp.*) to overstretch, and (hence) to break down.

claquette, f., musical instrument imitating the cracking of a whip.

clarinette, f., clarinet; clarinet player.

classe, f., class, classification; (*mil.*) class (i. e., officers, men, of the first, of the second class, etc.);
— *de mobilisation*, (*Fr. a.*) class to which a man belongs when summoned for service, and depending on number of years of service accomplished;

classe, *première* —, (*Fr. a.*) the upper half of the list of captains and of lieutenants (except infantry captains);
— *de recrutement*, (*Fr. a.*) class to which a man belongs as fixed by drawing lots.

classement, m., class, classifying;
— *de chevaux*, (*Fr. a.*) classification as to fitness for military service, after inspection, by a commission, of all horses and mules 5 years old on January 1 of year in which board operates;
— *de sortie*, graduation, final, standing (schools).

claudication, f., lameness; act of limping.

clausoir, m., (*cons.*) keystone of an arch.

claveau, m., (*cons.*) keystone, arch stone, voussoir;
— *de naissance*, springer.

clavet, m., calking-iron.

clavetage, m., (*mach.*, etc.) keying, system or set of keys;
— *d'assemblage*, keyed connection;
— *de hauteur*, edged keying.

claveter, v. a., (*mach.*, etc.) to key, to gib.

clavette, f., key, pin, peg; driving-wedge, cotter; forelock; (*art.*) cap-square key; (*harn.*) sort of key or toggle for fastening watering or bridoon bit to halter;
— *en arc*, curved key;
— *de bielle*, connecting-rod key;
boulon à —, cotter-pin;
— *de calage*, tightening key, adjusting key;
— *de charge*, draft key;
contre- —, gib;
— *et contre-* —, cotter and gib; gib and key;
— *double*, spring-key, split pin;
— *de dressage*, tightening key;
— *excentrée*, eccentric key, forming wedge;
— *fendue*, split pin, split-cotter pin;
— à *fourche*, forked cotter, split pin;
— *de hauteur*, edge key;
— à *mentonnet*, gib;
— *rasante*, side key, rectangular key;
— *de serrage*, tightening key;
— *de susbande*, (*art.*) cap-square key;
— *de torsion*, torsion key.

clayon, m., (*fort.*) small rod used in gabion making, gabion rod.

clayonnage, m., (*fort.*, etc.) wickerwork, wattling; fence of hurdles and turf; waling.

clayonner, v. a., (*fort.*, etc.) to make the web of, to wattle, a gabion; to wattle.

clef, f., (sometimes spelled *clé*) key (of a lock); key (of a cipher); chock; fid; wedge; cotter; plug (of a cock); spanner, wrench; toggle, hook, at the end of a chain; key of a locking-chain; (*cons.*) keystone; (*art.*) stop (of a carrier-ring); (*cord.*) hitch, knot; (*mil.*) key (of a position);
— *anglaise*, — à *l'anglaise*, screw-wrench, monkey wrench;
— *anglaise à écrous*, monkey wrench;
— *d'assemblage*, assembling key, key;
— à *béquille*, spanner;
— *de calage*, adjusting, tightening, key;
— *de chaîne*, fetter; (*pont.*) toggle or hook;
— *de chaîne d'embrelage*, keep-chain hook;
— *de chaîne d'enrayage*, locking-chain hook;
— *de chaîne de suspension*, (*pont.*) suspension-chain hook;
— *de charpenterie*, (*cons.*) reinforce, reinforcing piece (of a beam);
— *de chiffre*, cipher key;
— *de couplage*, (*art.*) assembling key of German double-action fuse;
— *de décharge*, (*elec.*) discharging key (as of a condenser);
demi- —, (*cord.*) half-hitch;
— *double*, double wrench, double-ended wrench;
— *double à écrous*, double screw-wrench;
à *double* —, said of a cipher in which the alphabet is changed at each word or letter, in which the reference of the symbol changes;
— *double à fourche et à tenons*, (*art.*) screw- and adjusting-wrench;

clef *à douille*, key-wrench, socket-wrench;
— *à écrous*, screw-wrench, nut-wrench;
— *à ergot*, spanner;
— *à fourche*, wrench, fork-wrench, spanner, spanner-wrench, alligator wrench;
— *à*, — *de, fusées*, (*art.*) fuse-wrench;
— *pour fusées à temps*, (*art.*) time-fuse key or wrench;
— *d'interruption*, (*elec.*) make-and-break key;
à — *invariable*, said of a cipher in which the key remains unchanged;
— *d'inversion*, (*elec.*) reversing-key;
— *de manœuvre*, (*art., etc.*) cam-lever;
mot —, key word (of a cipher system);
— *ouverte*, spanner;
— *de palier*, plummer-block key;
— *pendante*, (*cons.*) hanging post;
— *à platine*, (*art.*) lock-wrench (Gardner gun);
— *de position*, (*mach.*) setting key, flat key;
— *de poutre*, (*cons.*) anchor, or support (on the end of a beam);
— *à quatre branches*, combination wrench;
— *de rappel*, adjusting key;
— *de réglage*, (*art.*) fuse-wrench;
— *de robinet*, plug of a cock;
— *de serrage*, wedge-key, tightening key;
— *simple*, simple wrench;
à simple —, said of a cipher in which the key remains unchanged;
— *stratégique*, (*mil.*) strategic key-point;
— *tactique*, (*mil.*) key-point of a position, of a battlefield, etc.;
— *à tétons*, sort of socket-wrench;
— *tourne-à-gauche*, tap-wrench;
— *universelle*, monkey wrench;
à — *variable*, said of a cipher system in which the key changes;
— *à vis*, turn screw, screw-wrench, screw-spanner;
— *de vis écrou*, (*art.*) fuse-wrench;
— *à volant* (*r. r.*) brake key;
— *de voûte* (*cons.*) keystone.

clenche, f., bolt, latch (also *clenchette*).

clepsydre, f., clepsydra;
— *électro-balistique*, (*ball.*) electro-ballistic clepsydra.

clerc, m., clerk (esp. lawyer's); candidate for the ministry;
compter de — *à maître*, (*adm.*) to account for receipts and expenses only, without further responsibility;
régime de — *à maître*, (*adm.*) administration or management by an agent for a principal.

clichage, m., stereotyping.

cliché, m., stereotype plate; (*phot.*) negative;
— *à demi-teintes*, process plate;
— *instantané*, instantaneous negative;
— *négatif*, negative;
— *négatif retourné*, reversed plate;
— *photographique négatif*, full term for negative;
— *posé*, negative;
— *retourné*, reversed plate.

clicher, v. a., to stereotype.

climat, m., on maps, space between two parallels; (hence) climate.

clin, m., (of boats) clinker-work;
bordé à —, clinker-built.

clinomètre, m., (*inst.*) clinometer.

clinquant, m., Dutch leaf, foil; foil of any metal; tinsel.

clique, f., (*mil. slang*) squad of drummers and buglers;
— *pour percer*, ratchet-drill.

cliquet, m., click, pawl, detent.
— *à contrepoids*, gravity pawl;
— *à percer*, ratchet drill;
— *à ressort*, (*art.*) spring latch.

cliquetis, m., (*mil.*) clash (noise) of arms.

clisimètre, m., (*inst.*) clinometer;
— *à collimateur*, collimating clinometer;
— *à perpendicule*, plummet clinometer.

clisse, f., rod or stem of osier; splint; little hurdle, wicker, etc.

cloche, f., bell; any inverted vessel (in manufacturing, etc.); barrel, spindle (e. g., of a capstan); (*chem.*) bell-jar;
— *de brouillard*, fog bell;
— *à*, — *de, plongeur*, diving-bell.

clocher, m., steeple, belfry, bell-tower; v. n., to limp, hobble;
course au —, race across country, steeplechase.

cloison, f., partition, partition wall; (*sm. a., etc.*) partition of a revolver cylinder, of a cartridge box, etc.; (*nav.*) bulkhead (*art., sm. a.*) land;
— *cuirassée*, (*nav.*) armored bulkhead;
— *directrice*, guide blade (of a Fourneyron turbine);
— *étanche*, (*nav.*) water-tight bulkhead;
— *transversale*, (*nav.*) bulkhead;
— *transversale cuirassée*, (*nav.*) armored bulkhead.

cloisonnement, m., (*nav.*) generic term for bulkheads and partitions.

cloque, f., (*sm. a., etc.*) blister-like defect in a cartridge-case, etc.; blister produced by firing.

clos, m., (*top.*) field, cultivated, and inclosed by walls, hedges, etc.; walled inclosure;
— *d'équarrissage*, (*hipp.*) inclosure for holding autopsy on horses, etc.

clôture, f., inclosing fence, wall, hedge, etc.; closure (of debate); close, closing, as of the fiscal year.
donner —, (*sig.*) to close (a station).

clou, m., nail; spike; rivet; stud; (more accurately) a rectangular nail; (*mil. slang*) guard room, cell; bayonet; (*mil. slang in pl.*) "dough boys," "mud crushers;"
— *becquet*, shoe nail, shoe peg;
— *de bouche*, tack;
— *à cheval*, (*farr.*) horseshoe nail;
— *couvre-lumière*, (*art.*) vent-cover (s. c. pieces);
— *de demi-tillac*, sixpenny nail;
— *d'enclouage*, — *à enclouer*, (*art.*) spike;
— *d'enclouage à ressort*, (*art.*) spring-spike, temporary spike;
— *épingle*, brad;
— *de fer à cheval*, — *à ferrer*, v. — *à cheval*;
— *fraisé*, countersunk nail;
— *à glace*, (*farr.*) calkin, frost-nail;
— *de maréchal*, v. — *à cheval*;
— *de route*, — *de rue*, (*hipp.*) pick-up, picked-up nail, etc.; inflammation due to a picked-up nail, etc.;
— *de soufflet*, tack;
— *de tillac*, tenpenny nail;
— *à vis*, screw nail;
— *vissé*, (*farr.*) screw nail.

clouer, v. a., to nail, nail on, spike, spike on.

coaccusé, m., one accused along with others.

coak, m., coke, gas coke.

coalisé, a., in coalition, leagued;
— *s*, m. pl., members of a league, of a coalition.

coaliser, v. r., to form a league, a coalition.

coalition, f., coalition, league; any unlawful combination of functionaries, contractors, workmen, etc.

coaltar, m., coal-tar.

coaltariser, v. a., to cover or paint with coal-tar.

cobalt, m., cobalt.

cocarde, f., (*unif., etc.*); cockade (*art.*) part of cartridge-bag above the choke; (*steam.*) ventilator of a steam engine;
prendre la —, (*mil.*) to become a soldier, to enlist.

cocardier, m., (*mil. slang*) soldier fond of his profession.

coche, f., slit, notch; score;
en —, (*cord.*) chock-a-block, two-block (of blocks).

coche, m., barge, passenger barge.

cochon, m., hog, pig.

coco, m., (*mil. slang*) horse;
la botte à —, "stables."

cocons, m. pl., (*mil. slang*) first-term cadets of the Polytechnique.

code, m., code, code of laws;

code *de justice militaire*, —— *militaire*, (*mil.*) military law;
—— *militaire pénal*, articles of war.

co-efficient, m., coefficient;
—— *balistique*, (*ball.*) ballistic coefficient;
—— *économique*, (*steam, mach., etc.*) efficiency;
—— *de sécurité pour la charge*, (*cons.*) (load) coefficient of safety;
—— *de sécurité contre la rupture*, (*cons.*) (breaking) factor of safety.

cœur, m., heart; heartwood, or heart of a tree;
à ——. to the interior (as of tempering);
courbe en ——, (*mach., etc.*) heart, heart wheel;
—— *de croisement*, (*r. r.*) frog;
en ——, heart-shaped (cam, etc.);
excentrique en ——, (*mach.*) heart wheel; heart-cam;
plaque de ——, (*r. r.*) frog, frog plate;
plateau de ——, (in timber) the diametral slab containing the heart of the tree;
roue en ——, (*mach.*) heart wheel;
—— *de tour*, (*mach.*) heart-shaped driver or carrier.

coffin, m., (*mil. slang*) peculiar sort of desk in use at the Polytechnique (so called from the inventor, General Coffin).

coffrage, m., planking, sheeting, sheathing (as, *min.*, of a gallery); crib, cribwork;
planches de ——, (*min.*) plank lining.

coffre, m., box, chest, trunk; (*min.*) mine-chamber (obs.); (*fort.*) flanking casemate, caponier; (*siege*) epaulment, parapet, of a siege battery;
—— *d'amarrage*, mooring buoy; shore fast;
—— *d'avant-train*, (*art.*) lumber-chest;
—— *à avoine*, (*cav.*) grain-chest;
—— *de contrescarpe*, (*fort.*) counterscarp gallery, flanking gallery in the angle of the ditch; sort of flanking casemate;
—— *défensif*, (*fort.*) caponier;
demi- ——, (*fort.*) German name for single caponier;
—— *d'encaissement*, v. —— *de la voie*;
—— *flanquant*, (*fort.*) flanking casemate or works (in modern works); caponier, especially an infantry caponier in a short ditch; sort of blockhouse for flanking the ditch;
—— *flanquant double*, (*fort.*) double caponier, etc.;
—— *flanquant simple*, (*fort.*) single caponier, etc.;
—— *de flanquement*, (*fort.*) sort of caponier, v. —— *flanquant*;
—— *-fort*, strong-box, money-chest;
—— *de fortification*, (*fort.*) obsolete flanking work in the ditch of a bastion; German name for double caponier;
—— *de mine*, (*mil. min.*) obsolete name for mine-chamber;
—— *à munitions*, (*art., sm. a.*) ammunition chest;
—— *à outils*, tool-chest (traveling-forge, etc.);
—— *à vapeur*, (*steam*) steam-chest;
—— *de la voie*, (*r. r.*) ballast-bed of a railway.

coffrer, v. a., to plank, to line with planks, to sheathe (as a mine gallery, a shaft, etc.).

coffret, m., small chest, box;
—— *d'affût*, (*art.*) trail-chest;
—— *d'avant-train*, (*art.*) limber-box;
—— *de caisson*, (*art.*) caisson- or body-box;
—— *de cartouchière*, (*sm. a.*) v. —— *de giberne*;
—— *d'essieu*, (*art.*) axle-box; (*in gen.*) nave-box of a wheel;
—— *de flèche*, (*art.*) trail-chest;
—— *de giberne*, (*sm. a.*) leather body of a cartridge box;
—— *-marchepied*, (*art.*) small chest under foot-boards;
—— *d'outils*, tool chest;
—— *de rempart*, (*siege, etc.*) ammunition box.

cognée, f., ax;
—— *de bûcheron*, felling-ax.

cogner, v. a., to strike on a nail, bolt, etc., to drive it; to wedge together.

coiffage, m., (*art.*) fuse-cap.

coiffe, f., cap, cover, hood; guard; any strong metal plate covering or joining adjacent parts, or any parts; sling or strap (e. g., on a gin, to take an additional block); inside band of a hat, etc.; (*unif., Fr. a.*) white linen attachment (*manchon*) to which the havelock is fixed (Algiers and Tunis), (in France) white cover, marking the enemy at fall maneuvers; (*art.*) soft nose or cap for a. p. projectiles;
—— *d'armons*, (*art.*) futchel strap;
—— *de culasse*, (*art.*) breech cover;
—— *d'écouvillon*, (*art.*) sponge-cover;
—— *extérieure*, (*unif.*) cap- or headgear-cover;
—— *de fusée*, (*art.*) fuse-cap;
—— *intérieure*, (*unif.*) band of a cap or of other headgear;
—— *de lisoir*, (*art.*) bolster-plate;
—— *de lunette*, (*art.*) pintle-plate;
—— *de manœuvres*, (*unif.*) v. supra, *coiffe* under (*unif., Fr. a.*);
—— *d'obturateur*, (*art.*) obturator guard;
projectile à ——, (*art.*) soft-nosed or capped a. p. shell;
—— *de sassoire*, (*art., etc.*) sweep-bar plate;
—— *de sellette*, (*art.*) bolster-plate, pintle-plate;
—— *de volée*, (*art.*) muzzle-cover.

coiffer, v. a., to cap, fit over, anything;
—— *la chèvre*, (*art.*) to apply or fit a sling to a gin;
—— *la fusée*, (*art.*) to put a cap (*coiffage*) on a fuse.

coiffure, f., generic term for head cover, head-gear.

coin, m., wedge, quoin; corner, angle, solid angle; die; (*hipp.*) corner tooth; (*art.*) chock; the Krupp, or any other similar breech-closing wedge; (in timber) the slab or part of a log lying between the *plateau* and the *plateau de cœur*, qq. v.; (*in pl.*) screw-dies;
—— *d'arrêt*, (*sm. a.*) nose of the cocking-piece, cocking-stud; (*art.*) chock or stop (at the end) of a traverse circle; small recoil-checking wedge;
—— *d'arrimage*, stowing, securing, wedge or chock;
arrondir les ——s, (*man.*) to cut corners in riding-hall;
—— *à canon*, (*art.*) gun-chock (in the transportation of guns);
—— *de collier de guindage*, (*pont.*) balk-stirrup wedge;
—— *cylindro-prismatique*, (*art.*) cylindro-prismatic wedge;
—— *à fendre*, ordinary splitting wedge;
—— *de fermeture*, (*art.*) breech-closing wedge;
—— *-guide*, (*art.*) sledge-chock (transportation of guns);
—— *de guindage*, (*pont.*) side-rail wedge;
—— *à manche*, calking-iron; (*art.*) chock;
—— *de mire*, (*art.*) elevating quoin (obs.);
—— *à poignée*, wedge with handle;
—— *de pointage*, v. —— *de mire*;
—— *prismatique double* (*simple*), (*art.*) double (single, flat) wedge;
—— *de recul*, (*art.*) recoil-checking wedge (siege, etc., guns);
—— *de retour en batterie*, (*art.*) v. —— *de recul*;
—— *roulant*, (*mach.*) rolling wedge;
—— *-sabot*, (*art.*) wheel-chock, drag-shoe;
—— *-sabot élévateur*, (*F. m. art.*) recoil chock.

coincer, v. a., to wedge, to jam; v. r., to jam (as bullets in the magazine).

coite, f., launching-cradle, way;
—— *de vindas*, check of a windlass.

coke, m., coke, gas coke.

col, m., neck, collar; (*unif.*) stock; (*top.*) saddle, pass, col;
—— *de cygne*, (*art.*) "swan neck" (siege carriage); (*mach., etc.*) goose-neck.

colback, m., (*unif.*) colback, busby; (*mil. slang*) raw recruit.

colichemarde, f., (*sm. a.*) fencing sword (long thin blade, very broad near guard).
colique, f., colic, gripes.
colis, m., package, box, bale.
colismarde, f., v. *colichemarde.*
collage, m., pasting, gluing; sizing.
collatéral, a., collateral;
 points ——*aux*, points halfway between the cardinal points.
colle, f., glue, paste;
 —— *forte*, glue;
 —— *de menuisier*, carpenter's, joiner's glue;
 —— *à pierre*, cement;
 —— *de poisson*, isinglass.
collecteur, m., collector; (*elec.*) collector;
 —— *simple*, (*elec.*) collector (as of alternating-current machine);
 —— *de vapeur*, (*steam*) steam drum (Belleville boiler, etc.).
collection, f., collection;
 —— *d'effets* (*Fr. a.*) the collection of tools, objects, etc., necessary to a given service or duty; (more particularly the clothing and necessaries of the men, kept in the company magazines, and subdivided into: —— *No. 1* (*guerre et parade*), clothes, equipments, etc., issued in peace only on mobilization or on occasions of great ceremony; —— *No. 2* (*extérieur*), the next best articles, etc., after No. 1, and used by men when they go into town individually; —— *No. 3* (*instruction*), all articles unfit for active service; in general, all other articles in use in the company;
 —— *d'effets de pansage*, horse-kit, stable-kit;
 —— *de matériel de bivouac*, bivouac outfit.
coller, v. a., to glue, to stick, to adhere; to cake; to fit close; to size paper; to find deficient (a candidate, student), to "find" (U. S. M. A.), to "bilge" (U. S. Naval Academy);
 —— *au bloc*, (*mil. slang*) to send to the guard house;
 —— *à onglet*, to mount on strips (as, maps for binding, an extra sheet for indorsements, etc.).
collerette, f., rim, flange, ring; polar ring or collar of a balloon; (*mach., etc.*) collar; (*expl.*) necklace (of cartridges); (*fort.*) glacis armor, outer protecting ring (of a land turret or cupola) (may be of concrete); (*nav.*) capping ring or flange (of an open turret).
colleron, m., (*harn.*) neck harness or strap, harness collar (supports pole, worn only by wheel horses).
collet, m., collar; shoulder; neck; throat (of an anchor, of a knee); (*mach., etc.*) bearing of an arbor, axle, spindle, etc.; brass, pillow, pillow-bearing, spindle, collar, neck, shank, throat, flange, gland; (*sm. a.*) neck of a cartridge case, shank of the bolthead, neck or narrowing of barrel-cover (German Mauser); (*hipp.*) neck (of a tooth); (*cons.*) end of stair farthest from supporting wall; (*unif.*) collar; (*art.*) neck (of the cascable);
 —— *de l'ancre*, throat of an anchor;
 —— *d'arbre*, (*mach.*) shaft collar;
 —— *d'un aviron*, shaft or loom of an oar;
 —— *à billot*, (*art.*) gun sling;
 —— *de buttée à bourrage*, (*mach.*) packing ring;
 —— *du bouton*, (*art.*) neck of the cascable;
 —— *de buttée*, (*mach.*) thrust ring or collar;
 —— *à capuchon*, (*unif.*) collar and hood;
 —— *d'une courbe*, throat of a knee;
 —— *-criméenne*, (*unif.*, *Fr. a.*) sort of tippet worn by officers (not regulation);
 —— *de cylindre*, (*steam*) cylinder flange;
 —— *à éperon*, neck (of a spur);
 —— *de l'étui*, (*sm. a.*) neck of a cartridge case;
 —— *inférieur*, (*mach.*) step, lower brass;
 —— *de jonction*, (*mach.*), steam, etc.) shoulder, flange, collar;
 —— *de lance*, (*sm. a.*) neck of a lance;
 —— *de manchon*, (*sm. a.*) shank, neck;
 —— *manteau*, (*unif.*) sort of cape worn by men of the *gendarmerie*;
 —— *de mât*, step of a mast;
 —— *de la mortaise du té*, (*mach.*) cross-tail butt;

collet *de selle*, (*harn.*) hollow of the pommel of a saddle;
 —— *supérieur*, (*mach.*) upper brass, collar;
 —— *de la tête mobile*, (*sm. a.*) bolthead shank;
 —— *de tulipe*, (*art.*) neck of the muzzle of a gun;
 —— *de vis*, screw-neck.
collier, m., collar, hoop, ring; heel-piece of a spur; (*inst.*) ring, bearing ring (as of a level in the Y's); (*pont.*) rack collar, suspension-chain ring; (*hipp.*) draught horse;
 —— *d'antitiqueur*, (*hipp.*) collar to prevent cribbiting;
 —— *d'attache*, (*harn.*) hitching collar, picketing collar; pack-mule collar;
 —— *de cheval*, (*harn.*) horse-collar;
cheval de ——, draught horse;
cheval franc de ——, willing, energetic puller;
 —— *de conduite d'excentrique*, (*mach.*) eccentric guide collar;
coup de ——, (of teams) pull on the collar, (hence) great effort;
 —— *d'éperon*, neck or bow of a spur;
 —— *d'excentrique*, (*mach.*) eccentric hoop, belt, strap;
 —— *de force*, (*hipp.*) collar or brake for restraining horses (in veterinary operations, etc.);
 —— *à galets*, (*mach.*) friction-roller collar;
 —— *-guide*, guide-collar;
 —— *de guindage*, (*pont.*) balk collar, balk stirrup;
se jeter dans le ——, (of teams) to jump forward, struggle, work against the collar;
 —— *à martingale*, (*harn.*) collar fitted with breast-straps;
 —— *de pointage*, (*art.*) elevation collar, muzzle collar (as in French 155^{mm} casemate);
 —— *porte-guidon*, (*art.*) sight collar;
 —— *de poussée*, (*mach.*) thrust ring;
 —— *de presse-étoupe*, (*mach., etc.*) stuffing-box collar;
 —— *de pression*, v. —— *de poussée*;
 —— *à la prussienne*, (*pont.*) iron stirrup, in permanent military bridges;
 —— *à vis*, screw-collar.
collimateur, a., collimating;
niveau ——, (*top. inst.*) collimating level.
collimation, f., (*inst.*) collimation.
colline, f., (*top.*) hill.
collodion, m., collodion.
collodine, f., (*expl.*) collodine (sporting powder, of the Volkman type).
colloxyline, f., (*expl.*) colloxylin (synonym of soluble cellulose).
colombage, m., (*carp.*) row of columns or joists in a wall or partition.
colombier, m., dovecote;
 —— *fixe*, stationary dovecote;
 —— *militaire*, military dovecote;
 —— *mobile*, traveling or movable dovecote;
 —— *de pied*, v. —— *fixe*;
 —— *sur piliers*, v. —— *mobile*.
colombin, m., (*mas.*) mixture of plaster of paris and baryta.
colombine, f., pigeon dung.
colombophile, m., pigeon-fancier, -breeder.
colombophilie, f., pigeon-fancying, -breeding.
colon, m., colonist; (*mil. slang*) colonel;
petit ——, (*mil. slang*) lieutenant-colonel.
colonel, m., colonel;
 —— *propriétaire*, personage of high rank, after whom, by way of compliment, a regiment is named;
 —— *de tranchée*, (*siege*) colonel detailed every day for each of the principal attacks and who commands the troops on duty in the trenches.
colonelle, f., the colonel's wife; (*mil.*) the first company in a regiment (obs.).
colonial, a., colonial.
colonie, f., colony.
colonne, f., (*cons.*) column; (*mil.*) column; (in the fullest development a column consists of: 1, troops proper; 2, *trains de combat*; 3, *ambulances*; 4, *trains régimentaires*; 5, *convois*);

colonne   89   **combat**

colonne(s) *accouplées*, (*cons.*) grouped columns;
—— *d'assaut*, (*mil.*) storming party;
—— *d'attaque*, (*mil.*) attacking column;
—— *de bataillon*, (*inf.*) battalion in column of companies in company columns;
—— *de bataillon en masse*, (*inf.*) battalion in close column of companies (at 6 paces);
—— *de bataillons en masse*, (*inf.*) of a regiment, column of battalions in close column of companies (6 paces);
—— *par batteries*, (*art.*) (of a group) column of batteries in line at 14 meters distance;
—— *par brigade en ligne de masses*, (*cav.*) of a division, column of brigades in line of masses;
chapeau en ——, (*unif.*) cocked hat worn in the usual way, horns or peaks "fore and aft";
—— *à cheval*, (*art.*) the teams, carriages, of artillery marching alone;
—— *de combat*, (Fr. a.) so much of the column as consists of *troupes*, *trains de combat*, and *ambulances*:
—— *de compagnie*, (*inf.*) company column (section front);
—— *contre la cavalerie*, (*inf.*) column against cavalry;
—— *à demi-distance*, (*inf.*) column at half distance;
—— *directe*, column right in front;
—— *à distance entière*, column at full distance;
—— *à distance entière par section* (*par compagnie*), column of sections (of companies) at full distance;
—— *à distance de six pas*, (*inf.*) English quarter column;
—— *de*, —— *par*, *divisions*, (Fr. cav.) column of divisions (front of two platoons);
—— *double*, (*inf.*) of a battalion, double column at 6, 24, or more, paces interval, with columns in column of company columns; (*cav.*) double column, each of a half regiment in column of platoons, interval of 12 paces (or more or fewer); (in a brigade or division) two regiments side by side in column of platoons; (also) two or more double columns of regiments, one behind the other;
—— *double ouverte*, (*inf.*) double column with more than 24 paces interval;
—— *double de régiment*, (*inf.*) column of battalions in double column at 24 paces, one behind the other at section distance plus 12 paces;
—— *d'escadrons*, (*cav.*) column of squadrons (troops, U. S. cav.);
—— *d'escadrons à demi-distance* (*à distance entière*), (*cav.*) column of squadrons at half (full) distance;
—— *expéditionnaire*, column on special duty, on detached duty, on a special expedition;
fût de ——, (*cons.*) shaft of a column;
gros de la ——, main body of a column;
ligne de ——*s*, (*art.*) line of batteries in column of platoons (sections, R. A.) at open or closed intervals; (*cav.*) line of squadrons in column of platoons at deploying intervals; (*inf.*) line of companies in company columns (at various intervals);
ligne de ——*s de compagnie*, (*inf.*) line of company columns;
ligne de ——*s en masse*, (*cav.*) line of squadron columns at close intervals;
ligne de ——*s par pièce* (*par section*), (*art.*) line of column of pieces or of platoons (sections, R. A.);
—— *de manœuvre*, formation for moving troops on battle field;
—— *de marche*, marching column; (also) the march formation of a corps, division, etc.;
marcher en ——, to march in column;
—— *de masses*, (*art.*) column of masses (several groups); (*cav.*) column of regiments in mass (brigade, etc., formation);
—— *mobile*, flying column, one acting independently, and composed of all arms of the service.

colonne *de munitions*, ammunition column, to supply ammunition on battle field (this designation holds, no matter what the formation of the wagons); in the German artillery, the same as the French "*section de munitions*";
—— *à nervures en croix*, (*cons.*) cruciform column;
ordre en ——, column formation;
—— *parallèle*, any column formed parallel to a previous line;
—— *de pelotons*, —— *par peloton*, (*inf.*) column of platoons; (*cav.*) column of platoons (squadron, regiment, brigade);
—— *perpendiculaire*, any column formed at right angles to a previous line;
—— *par pièce*, (*art.*) column of pieces (battery, group);
—— *à pied*, (*art.*) the unmounted men of artillery marching alone;
—— *-pivot*, (*r. f. art.*) pillar-mount; balance-pillar; supporting pillar;
profondeur de ——, depth of a column;
—— *de purge*, (*mach.*) drain pipe (of a pump) between pump and accumulator;
—— *par quatre*, (*cav., inf.*) column of fours, squads;
queue de ——, rear of a column;
—— *de régiment*, (*inf.*) regiment with battalions one behind the other at section distance plus 12 paces, each in *colonne de bataillon*;
—— *de rendez-vous*, column formed in a *position d'attente*;
rompre en ——, to break into column;
—— *de route*, route column, cavalry in fours or twos, infantry in 4 ranks (exceptionally, in 6 or 8); (*art.*) column of pieces;
—— *par section*, (*art.*) column of platoons (U. S. Art.), of sections (R. A.) (battery and group formation);
—— *serrée*, (*inf.*) close column; (*art.*) close column of batteries in line, distance of 14 meters; (*cav.*) close column of squadrons at 18 paces;
serrer en ——, to close in mass;
serrer une ——, to close a column;
tête de ——, head of a column;
—— *de tête*, leading column;
tronc, *vif*, *de* ——, v. *fût de* ——;
—— *du vent*, (*nav.*) weather column;
—— *de sous le vent*, (*nav.*) lee column.
colonette, f., short column, post.
colophane, f., colophony.
colta, m., **colthar**, m., coal tar.
coltinage, m., act of carrying a load on the shoulders or back.
coltiner, v. a., to carry a load on the shoulders or back.
coltineur, m., stevedore; (*in gen.*), a man that carries a load on his shoulders or back.
combat, m., (*mil.*) combat, fight; action, affair, engagement (less in importance than a battle);
—— *d'attente*, delaying action;
—— *à contre-bord*, (*nav.*) engagement broadside on;
—— *corps à corps*, hand-to-hand fight;
—— *décisif*, real attack (i. e., flank attack) while *démonstratif* is going on;
—— *démonstratif*, front attack, feint, attack to engage attention of enemy (synonymous now with front attack) (v. —— *de front*, *d'usure*);
—— *encadré*, combat by troops acting in concert, e. g., by two divisions supporting each other, by two corps, etc. (opposite of —— *isolé*);
engager le ——, *un* ——, to engage, to come into action;
éviter le ——, to shun, avoid, battle;
—— *à feu*, fire action;
—— *d'exercice*, sham fight;
formation de ——, attack formation;
fort du ——, thick of the fight;
—— *de front*, front attack; (in modern tactics) the demonstrative attack, to engage attention of enemy while the real attack is going on somewhere else (generally a front attack);
hors de ——, disabled;

combat *isolé*, combat by a single unit, unaided, single-handed, or not in concert with other troops;
ligne de ——, fighting line (in many relations);
—— *de mer*, —— *naval*, (*nav.*) naval engagement;
—— *de nuit*, night attack, engagement;
—— *offensif*, offensive combat (two phases, one demonstrative, the other really offensive, combined afterwards in a single attack pushed home if possible);
ordre de ——, attack formation;
—— *en ordre dispersé*, attack, combat, in open order;
—— *à pied*, (*cav.*) dismounted action;
—— *en pointe*, (*nav.*) engagement bows on;
—— *de près*, combat at close quarters;
—— *rapproché*, v. —— *de près*;
—— *en retraite*, running fight;
rompre le ——, to break off a fight;
—— *simulé*, sham fight;
—— *singulier*, single combat;
—— *à terre*, land action;
—— *en tirailleurs*, action in open order;
train de ——, ammunition wagons and ambulances;
—— *traînant*, combat of which the object is to allow the real attack time to make its dispositions;
—— *d'usure*, (v. —— *démonstratif*, —— *de front*, synonyms) in the preparation for the attack, the back and forth of defensive and offensive fighting, in which the artillery supports the infantry, drives off the adversary, backs up offensive movements, etc.
combattant, m., (*mil.*) combatant; *non*-——, noncombatant.
combattre, v. a., to fight, be engaged; to make war on;
—— *corps à corps*, —— *de près*, to fight hand to hand.
combe, f., (*top.*) valley between two mountains (-*combe*, of English compounds).
combinaison, f., combination; chemical combination; contrivance.
combiné, a., of, or belonging to, two or more allied powers.
combiner, v. a., to combine, join, unite.
comble, m., top, ridge; garret, roof, generic term for roof with all its parts; summit, height, zenith; complement, consummation; a., heaped, full;
—— *en dôme*, domed roof;
—— *à deux égouts*, ridge roof;
—— *à un égout*, shed roof;
—— *à la Mansard*, mansard roof;
—— *en pavillon*, pyramidal or conical roof;
pied ——, (*hipp.*) pumiced foot.
combleau, m., (*cord.*) strong rope used in mechanical maneuvers to lift, drag, and, at a pinch, for hitching horses to ponton wagons.
comblement, m., filling up (as of a ditch, ditch of a besieged place, etc.).
combler, v. a., to fill, fill up; supply, make good; crown, fulfill complete.
comburant, a., (*chem.*) oxidizing, burning; m., oxidizing agent, burning agent.
comburer, v. a., (*chem.*) to burn, oxidize.
combustibilité, f., combustibility.
combustible, a., combustible; m., combustible, fuel;
alimenter en ——, to fuel, keep in fuel;
—— *ancien*, generic term for coal;
—— *d'éclairage*, any lighting material operating by combustion;
—— *minéral*, v. —— *ancien*;
—— *moderne*, generic term for wood and peat.
combustion, f., combustion; (*expl.*) propagation of flame to interior of grain or mass;
—— *lente*, slow burning or combustion;
—— *au libre*, combustion in free air;
—— *spontanée*, spontaneous combustion;
—— *vive*, quick burning, combustion.

comestible, a., eatable, comestible; m. pl., victuals, food.
comité, m., committee, commission, (*esp.*, *mil.*) board of officers;
—— *d'achat de chevaux*, (*Fr. a.*) horse board;
—— *d'administration centrale*, (*Fr. a.*) a war department board of which the minister is president, and whose object is to secure and maintain uniformity of views and of principles in respect of the various branches of the war department, and to coordinate the general measures to be taken in any given case;
—— *consultatif*, advisory board;
—— *consultatif des poudres et salpêtres*, (*Fr. a.*) war department board to consider administrative and technical questions bearing on explosives;
—— *de défense*, (*Fr. a.*) board to consider the defensive system of France (suppressed in 1888, and functions assigned to *conseil supérieur de défense*);
—— *permanent des subsistances*, (*Fr. a.*) consulting and investigating board or committee taking cognizance in the widest sense of all questions bearing on the supply, in war, of armies in the field, of fortified places, and, in general, of all questions relating to resources, both domestic and foreign;
—— *de santé*, (*Fr. a.*) the technical committee of the medical department;
—— *de surveillance des approvisionnements de siège*, (*Fr. a.*) committee having charge of the question of supplies for a fortress (garrison and civilian population);
—— *technique*, (*Fr. a.*) board investigating all technical questions submitted to it by minister of war (one to each arm and to each department of the service).
commandant, m., (*mil.*) commanding officer (troops, department, service, etc.); (colloquially) major; commandant, commander;
—— *d'armée*, the commander in chief of an army;
—— *d'armes*, the c. o. of a garrison, or of a fort, in peace;
—— *d'artillerie de tranchée*, (*siege*) field officer of artillery charged in each attack, or important part of each attack, with the direction and service of the batteries;
—— *des avant-postes*, the c. o. (field officer) of outposts;
—— *de batterie*, c. o. of a battery;
—— *des batteries*, c. o. of a group of batteries;
—— *du campement*, c. o. or officer in charge of a campement, q. v.;
—— *de cercle*, (*Fr. a.*) in Algiers, the chief of administration in each subdivision of military territory;
—— *en chef*, commander in chief (reserved in France to c. o. of a group of armies, or of an army operating independently);
—— *d'étapes*, officer in command of a post on a line of communication, i. e., of a *gîte d'étapes*;
—— *de la force publique d'une division*, c. o. of the gendarmerie of a division;
—— *de gare*, (*Fr. a.*) officer discharging functions of *commissaire de gare*, in a *commandement de gare*, i. e., acts as —— *d'armes* in respect of passing troops, etc.;
—— *de groupe d'armées*, commander in chief; v. —— *en chef*;
officier ——, commanding officer;
—— *par intérim*, temporary commander;
—— *de place*, c. o. of a place (obs. in a technical sense);
—— *de quartier général*, c. o. of a headquarters, officer in charge of a headquarters;
—— *de recrutement*, (*Fr. a.*) officer in charge of a recruiting bureau;
—— *de secteur d'attaque*, (*siege*) in the French army, an officer specially detailed to command a sector of attack on receipt of order of mobilization;
—— *en sous-ordre*, subordinate commander;
—— *supérieur de l'artillerie (du génie)*, (*siege*) the senior artillery (engineer) officer present for duty;

commandant *supérieur de la défense*, (*Fr. a.*) c. o. of a group of forts, charged with preparation of plans for the defense of the same, with questions of supply, etc. (as a rule, is a general officer);
—— *supérieur de la région*, (*Fr. a.*) the c. o. or governor of a *place à camp retranché*;
—— *territorial*, (*Fr. a.*, etc.) c. o. of a district, i. e. general of division, of brigade, etc., under the authority of the corps commander.

commande, f., (*cord.*) lashing; (*mach.*) driver, feed, controlling gear or device; operation, working, control;
de ——, (*mach.*) working, operating, controlling (a motion or mechanism, etc.);
—— *de billot*, (*pont.*) rack-stick lashing;
—— *de brêlage*, (*pont.*) back lashing;
à —— *cinématique*, (*mach.*) kinematically controlled;
à —— *desmodromique*, (*mach.*, etc.) controlled, operated, by belts, straps, or bands;
à —— *directe*, (*mach.*) direct acting;
—— *de guindage*, (*pont.*) side-rail lashing;
—— *de pontage*, (*pont.*) balk lashing;
—— *de poutrelle*, (*pont.*) balk lashing.

commandé, a., (*mil.*) ordered, detailed, told off for duty;
—— *de service*, detailed for duty.

commandement, m., (*mil.*) command (at drill); military authority in general; command, i. e., all that relates to the commanding authority, as functions, duties, etc.; (more especially) the commanding authority, those in command, (and hence, sometimes) the commanding officer, (and even) the supreme commanding authority in the army; command, as applied to a district, to duties; control, as over a line in action; (*fort.*) command (by a position, of a position). (*Commandement* is applied particularly in France to marshals and to corps commanders; it may also mean the territory or troops under one and the same commander, just as we say "command" in more or less the same way);
—— *absolu*, (*fort.*) height of crest of a work above work or territory commanded, i. e., the difference of level of the points, etc., considered;
—— *d'avertissement*, (*drill*) warning command, cautionary command;
avoir le —— *sur*, (*mil.*) to rank, take precedence of; to have military authority over;
bâton de ——, marshal's baton;
centre de ——, central authority;
—— *en chef*, (*mil.*) command exercised in virtue of powers conferred by the executive; (*in gen.*) command in chief;
—— *au clairon*, (*drill*) command given by bugle;
couvrir le ——, to drown the commands (of noise);
—— *défensif*, (*fort.*) command furnishing protection from hostile fire;
—— *d'étapes*, (*mil.*) the personnel of a *gîte d'étapes*, q. v.;
—— *d'exécution*, (*drill*) command of execution;
—— *faible*, (*fort.*) command of less than 6 meters;
—— *de gare*, (*mil.*) board composed of a military c. o. and of a station master, on duty in stations beyond the *stations de transition*, cares for sick and wounded, expedites transports, etc.;
—— *hiérarchique*, (*mil.*) command exercised in one's grade by virtue of one's commission or warrant;
—— *ad latus*, (*mil.*) state of being second in command;
lettre de ——, (*Fr. a.*) letter of appointment to command;
marque de ——, (*mil.*) distinguishing mark of a c. o., e. g., a headquarters flag;
—— *offensif*, (*fort.*) command permitting fire at an enemy or into an enemy's work, state of having crests commanded by crests of work in rear (of a general system of works); command permitting fire to be delivered on an enemy in front;

commandement *préparatoire*, (*drill*) preparatory command;
—— *rasant*, v. —— *faible*;
—— *relatif d'un point sur un plan*, (*fort.*) the vertical height of a point above the plane;
remettre le ——, (*mil.*) to turn over the command;
—— *en second*, v. —— *ad latus*;
—— *au sifflet*, (*drill*) command by whistle;
tourelle de ——, (*nav.*) conning tower;
—— *à la voix*, (*drill*) command by word of mouth.

commander, v. a., (*mil.*) to command, to order; to detail, tell off, to warn (for guard, or any other duty); (*fort.*) to dominate, to command; (*mach.*) to drive, to work, to operate, to control;
se ——, (*mach.*) to be controlled, to gear together, to work together (said of parts that gear or work together);
—— *l'exercice*, (*mil.*) to give the commands at drill, to be in command, in charge, at drill;
—— *la manœuvre*, (*mil.*) v. —— *l'exercice*;
—— *de piquet*, (*mil.*) to detail, so as to be ready at a moment's notice;
—— *le service*, (*mil.*) to keep the roster of regular routine duties, to make details for the regular routine.

commandeur, m., third grade in Legion of Honor; commander (of an order).

comme ça! (in command) steady! so!

commencer, v. a., to begin, commence;
—— *le feu*, (*mil.*) to begin firing, to open the action, to open fire.

commensal, m., mess-mate.

commettage, m., (*cord.*) lay of a rope; rope-laying; laying up of yarn to make a strand;
atelier de ——, ropewalk;
—— *à 140%*, etc., laying in the ratio of 140 meters, etc., of strand to 100 meters, etc., of rope;
—— *au* $^m/_n$, laying so that m meters of strand shall form n meters of rope ($m > n$).

commettre, v. a., to intrust, commit; (*cord.*) to lay, lay up;
—— *à 140%*, (*cord.*) to lay so that 140 meters of strand shall make 100 meters of rope;
—— *au* $^m/_n$, to lay so that m of strand shall form n of rope ($m > n$);
—— *en aussière*, to lay hawser fashion;
—— *deux fois*, to lay cablewise;
—— *de droite à gauche*, to lay up against the sun, cablewise, left-handed;
—— *en garochoir*, to lay up the wrong way;
—— *de gauche à droite*, to lay hawser fashion, with the sun, right-handed;
—— *en grelin*, to lay cablewise;
—— *de main torse*, to lay the wrong way;
—— *au quart*, to lay soft (twisted one-quarter);
—— *en quatre*, to lay up in four strands, shroud fashion;
—— *en queue de rat*, to point a rope;
—— *au* ——, *en tiers*, to lay up hard (twisted one-third);
—— *en toron*, to lay hawserwise;
—— *en trois*, to lay up in three strands.

commis, p. p., (*cord.*) laid, laid up, v. *commettre*; empowered, delegated, appointed.

commis, m., clerk;
—— *aux écritures*, (*mil.*) clerk of the administration; general-service clerk (U. S. A.);
—— *greffier*, (*Fr. mil. law*) clerk in the archives or record offices of courts-martial;
—— *de la guerre*, (*mil.*) subordinate clerk of the war department;
—— *et ouvriers d'administration*, (*Fr. a.*) artisans, mechanics, etc., and clerks, enlisted as such, and organized into 25 sections for the military service (v. *section*).

commissaire, m., commissioner, member of a commission; (*Fr. nav.*) paymaster;
aide- ——, (*Fr. nav.*) assistant paymaster;
—— *adjoint*, (*Fr. nav.*) passed assistant paymaster;
élève- ——, (*Fr. nav.*) cadet paymaster (n. e., U. S. N.);

commissaire *du gouvernement*, (*Fr. mil. law*) field officer or *assimilé*, public prosecutor before courts-martial; representative of the government in military cases appealed; (*in gen.*) any officer (general or other) charged with a special mission;
—— *du gouvernement rapporteur*, (*Fr. mil. law*) public prosecutor of a court-martial, for an army in the field, and for a place invested or besieged;
—— *des guerres*, (*Fr. a.*) commissary and quartermaster (suppressed at the Restoration);
—— *de la marine*, (*Fr. nav.*) paymaster, etc.;
—— *militaire de gare*, (*Fr. a.*) officer member of *commission de gare*, and acting as c. o.;
—— *de 1ère classe (2ème classe)*, (*Fr. nav.*) pay director or inspector;
sous-——, (*Fr. nav.*) assistant paymaster (2 classes of);
—— *de surveillance administrative*, (*mil. r. r.*) functionary in a railroad station who looks after soldiers that have missed their way or lost their transportation;
—— *technique de gare*, (*mil. r. r.*) agent of the railway, member of the *commission de gare*, looks after transport of troops, etc.

commissariat, m., (*mil.*) commissariat (does not exist as such in France); the corps of *commissaires de guerre* (obs.); (*Fr. nav.*) naval pay, quartermaster, etc., corps.

commission, f., order, charge, duty; warrant, authority, mandate; commission, board, committee (e. g., of officers, in many applications); (*Fr. a.*) warrant or *lettre de service*, given to *commissionnés*, q. v. (In the following list will be found the more prominent boards relating to the French military service:)
—— *d'abatage*, board that determines whether a horse shall be cast or not;
—— *d'achat*, purchasing board (horses, etc.);
—— *d'adjudication*, board to settle assignment of contracts;
—— *administrative des hôpitaux mixtes*, board to regulate the administration of civil hospitals used by the military forces of the nation;
—— *d'admission*, board that passes on the admissibility of would-be bidders, and on the nature, etc., of their guaranty or security;
—— *d'aérostation militaire*, board, investigates questions of aerial navigation and reports to head of engineer committee;
—— *d'armes*, promotion board for captains and majors;
—— *de casernement*, board to assign, etc., quarters, etc., in barracks;
—— *centrale des travaux géographiques*, board that assists the minister of war in the coordination, etc., of all topographic work done by the state;
—— *de certificat de bonne conduite*, board of officers to pass on claims of n. c. o. and men to a certificate of good conduct;
—— *de chemin de fer de campagne*, board, in war, charged with railway transportation beyond the base of operations; that is, beyond the *station de transition*, q. v. (uses military railroad forces);
—— *de classement pour l'avancement, la Légion d'honneur et la médaille militaire*, board to recommend candidates for promotion, the Legion of Honor and the military medal;
—— *de classement de chevaux*, biennial board classifying and inspecting all horses and mules 5 years old on January 1 of year in which census is taken;
—— *de classement et de réquisition des chevaux et voitures*, census board for horses and wagons for military service, in case of war;
—— *de classement des sous-officiers*, board drawing up lists of n. c. o., candidates for civil employment;
—— *consultative des subsistances*, mixed board assisting the *intendance*, when called on, in respect of questions of administration, etc., of stores and supplies, in the *places* where these are accumulated;

commission *consultative de télégraphie militaire*, war department board, investigates all questions bearing on military telegraphy;
—— *de défense*, commission of officers charged in peace, in each *place*, with the revision of plans of mobilization or of defense, in accordance with ministerial instructions;
—— *de défense du littoral*, board for the consideration of all questions relating to seacoast defense;
—— *d'enquête*, board of survey, board of inquiry (for cases of officers only);
—— *d'études*, committee to investigate all questions bearing on *transports stratégiques*, under the direction of the *commission supérieure*;
—— *d'études pratiques du tir*, artillery committee having charge of all questions relating to improvements in artillery practice;
—— *d'évaluation des indemnités*, board to settle amounts due for supplies furnished on mobilization of troops, of the army, or on assembling of troops for any purpose;
—— *d'examen*, examining board (schools);
—— *d'examen de denrées refusées en distribution*, board called to settle differences between contractors or issuing officers (*service des subsistances*), on the one hand, and receiving parties, on the other, in case of refusal to accept stores issued or furnished;
—— *d'examen des inventions intéressant les armées de terre et de mer*, a war department board investigating all inventions, of interest to army and to navy, submitted by their authors (board of ordnance and fortification, U. S. A.);
—— *d'expériences de l'artillerie*, gun board, experimental artillery board;
—— *de gare*, board charged in war with transport of stores, troops, issues of stores, care of sick and wounded, in the railway stations assigned to it (has varying designations according to duties);
—— *d'hygiène hippique*, war department board reporting on all hippological questions referred to it by the minister;
—— *de ligne*, name taken by the —— *d'études*, q. v., as soon as the *transport stratégique* begins;
—— *locale de réception*, board of artillery officers for the inspection and test of small-arm cartridges;
—— *militaire supérieure des chemins de fer*, commission of officers and of civilians, under the presidency of the chief of staff of the minister of war, having charge of all general questions of railway service and organization, such as transport of troops, organization of new lines, etc.;
—— *mixte des travaux publics*, board of officers and civilians for the examination and discussion of such projects, relating to the frontier and to the territory around fortified positions, as may concern the national defense and civil and marine interests;
—— *des ordinaires*, mess (men's) council or board;
—— *des prises*, (*nav.*) prize commission;
—— *de réception des cartouches*, board of officers of all arms, one at each cartridge factory, which inspects and tests small-arm ammunition as it is manufactured;
—— *de réception des matières*, board of officers and of one expert to inspect and examine, etc., stores received by the government;
—— *de réforme*, retiring board;
—— *de remonte*, remount board or council;
—— *de réquisition*, board that, on mobilization, examines and classifies horses, mules, etc., of which a census has been made or that satisfy certain conditions, looks after the supply of wagons for transportation, etc.;
—— *de réseau*, board that, in peace, has charge of all questions bearing on the railway service of a given system of lines, in a military point of view, the railway personnel being employed; on mobilization this board is assisted by a *sous*-—— *de réseau*;

commission *spéciale de réforme*, board deciding whether men who present themselves shall be discharged (*réformés*) or not; one to each regional subdivision;
—— *spéciale de réforme extraordinaire*, special board to pass on all cases of old soldiers whose gratuities or pensions expire during the year, with a view to their renewal or continuance;
—— *des substances explosives*, technical war department board, investigates scientific questions referred to it by the minister and furnishes the *comité des poudres et salpêtres* with the means of deciding on questions of manufacture and use of all sorts of explosives;
—— *supérieure de classement*, promotion board for lieutenant-colonels, colonels, and major-generals.

commissionnaire, m., agent, (*esp.*) a buyer or seller for a third person.

commissionné, m., (*Fr. a.*) n. c. o. and corporal or man performing certain special duties (e. g., fencing-masters, drummers, tailors, etc.), allowed to remain with the colors after term of service in the active army is completed.

commissionner, v. a., to give or deliver a commission.

commissure, f., commissure (of the lips).

communal, a., belonging to a commune.

commune, f., (*Fr.*) commune, parish, township.

communicateur, m., (*mach.*) countershaft.

communication, f., communication, transmission; (*mach.*) communication, transmission, as of power, motion; (*expl., etc.*) quick-match, etc., used to communicate explosion to the charge; (*steam*) necking (of a French boiler); (*elec.*) channel, for wires or conductors; (*mil.*), usually in pl.) communications, lines of communication; (*fort.*) communications, i. e., passages, gateways, etc., in a fortress;
boyaux de ——, (*siege*) zigzags;
de ——, in connection, connecting;
en —— *avec*, in gear with;
ligne de ——*s*, (*mil.*) line of communications;
mettre en ——, to bring, put into communication; (*mach.*) to put in gear;
—— *de mouvement*, (*mach.*) transmitting-, connecting-gear;
prendre, recevoir, —— *de*, to take cognizance of;
—— *avec la terre*, (*elec.*) ground circuit, ground, earth.

communiquer, v. a., to communicate; to have intercourse with (after raising of quarantine).

commutateur, m., (*elec., etc.*) commutator; cut-out; interrupter; switch; keyboard;
—— *à bouchon*, plug-switch;
—— *à chevilles*, pin-switch;
—— *conjoncteur*, circuit-closer;
—— *disjoncteur*, contact-breaker, cut-out;
—— *à glissement*, sliding-contact;
—— *de ligne*, (*mil. tel.*) line commutator;
—— *à pédale*, lever switch;
—— *permutateur*, universal switch;
—— *redresseur*, commutator proper (continuous-current machine);
—— *de réduction*, rheostat;
—— *simple*, collector (alternating current);
—— *à touches*, key commutator, keyboard, sort of switch-board.

commutation, f., commutation (of punishment).

compact, a., (*mil.*) close, closed up (of order, of formations);
ordre ——, (*mil.*) compact order.

compacité, f., compactness.

compagnie, f., (*mil.*) company (of infantry, engineers, etc.); (also) (*Fr. a., etc.*) the unit of certain organizations of workmen, artificers, etc. (*Fr. art.*) battery of fortress artillery;
—— *d'artificiers*, (*Fr. art.*) company of artificers or laboratory men (3 in land, 1 in marine artillery);
—— *de base*, (*drill*) the company on which a movement is made;
—— *-cadre*, skeleton company;
—— *de brancardiers*, litter-bearer company;
—— *de cavaliers de remonte*, (*Fr. a.*) company of troopers charged with training, etc., horses for military service in remount establishments, military schools, training schools, etc.;
—— *de chasseurs forestiers*, v. *chasseurs forestiers*.
—— *de conducteurs*, (*F. m. art.*) company of drivers;
—— *de débarquement*, (*nav.*) landing party;
—— *de dépôt*, (*Fr. a.*) depot company (zouaves and *régiment étranger*);
—— *de direction*, (*drill*) guiding, directing, company;
—— *disciplinaire*, —— *de discipline*, (*Fr. a.*) punishment company, composed of incorrigible men, of men guilty of self-mutilation, and of men sentenced by way of punishment (serve in Algiers);
—— *disciplinaire des colonies*, (*Fr. nav.*) punishment company (serve in Senegal, Martinique, etc.); (also called —— *de disciplinaires coloniaux*,
—— *de discipline de la marine*);
—— *divisionnaire du génie*, (*mil.*) engineer company of the advanced guard;
—— *d'élite*, (*Fr. a.*) the grenadier and voltigeur companies (obs.);
—— *d'exploitation*, (*r. r.*) company operating a line;
—— *franche*, company made up of volunteers (from other (regular) regiments) for special or extra-hazardous duties;
—— *de fusiliers de discipline*, v. —— *de discipline;*
—— *d'infirmiers*, hospital company;
—— *de jour*, company on duty for the day, or detailed for the day;
—— *de manœuvre*, company formed for drill, company from a drill point of view;
—— *de marche*, a company made up of men drawn from different companies;
—— *d'ouvriers*, (*F. m. art.*) company of workmen, etc.;
—— *d'ouvriers d'artillerie*, v. s. v. *ouvrier*;
—— *par piquets*, company made up of men from different companies of the same or of different regiments;
—— *de pontonniers*, (*pont.*) ponton company;
—— *de raquetiers*, (*art.*) rocket company;
—— *du train des équipages*, company of the train.

compagnon, m., companion;
—— *d'armes*, (*mil.*) brother officer;
—— *de plat*, (*nav.*) messmate.

comparaître, v. n., to appear before (as, a court-martial).

comparateur, m., (*inst.*) comparator, calibrating or standardizing instrument;
—— *à aiguilles*, point comparator;
—— *à levier*, lever comparator;
—— *à réglette*, comparator having a short scale (*réglette*) at one end instead of a needle and graduated arc.

compartiment, m., compartment (of a box, of a railway carriage, of a ship);
—— *étanche*, (*nav.*) water-tight compartment;
—— *de la machine*, engine-room.

compartimentage, m., subdivision into compartments; packing or stowing into compartments.

comparution, f., (*law*) appearance or presentation of one's self before a court.

compas, m., compasses; dividers; calipers; mariner's compass;
—— *d'artisan*, bow compasses;
branche de ——, compass leg;
—— *à branches*, drawing compasses;
—— *à branches courbes*, calipers;
—— *à doubles branches*, (*art.*) shell calipers;
—— *de calibre*, calipers;
—— *à charnière*, ordinary compasses;
—— *à cheveu*, hair compasses;
—— *à coulisse*, proportional compasses;
—— *courbe*, (*art.*) shell calipers, external and internal calipers;

compas   94   **compte**

compas à *diviser*, dividers;
— *élastique* à *ressort*, spring compasses;
— à *ellipse*, oval compass, trammel; ellipsograph;
— *électrique*, (*elec*.) galvanometer;
— *elliptique*, v. — à *ellipsei*
— *d'épaisseur*, (*art*., etc.) calipers for measuring thickness of walls of guns, gun-barrels, and shells; gauge;
— *d'épaisseur* à *crémaillère*, rack compass;
— *d'épaisseur* à *ressort*, spring gauge or callipers;
— *étalon*, standard compass;
— à *glissière*, sliding compasses;
jambe de —, compass leg;
— *liquide*, liquid compass, fluid compass;
— à *lunettes*, double-bowed compasses;
— *de mer*, mariner's compass;
— *mort*, compass whose needle has lost its life;
— à *ovale*, oval compass.
— à *pointes changeantes*, — à *pointes de rechange*, compass with shifting or movable points;
— à *pointes sèches*, dividers;
— à *pompe*, bow-compasses;
— *de précision*, v. — à *cheveu;*
— *de proportion*, proportional compasses;
— à *pression constante*, (*art*.) verifying compasses (for emplacement of hoops in shrinkage);
quart de —, point of the compass (mariner's, etc.);
— à *quart de cercle*, wing compasses;
— à *quart de cercle* à *crémaillère*, rack compasses;
réciter les pointes du —, to box the compass;
— *de réduction*, v. — *de proportion;*
— *renversé*, hanging compass;
— à *ressort*, spring-divi ders, -compasses;
— *de route*, mariner's compass;
— *sphérique*, v. — à *courbe;*
— à *tire-ligne*, compass with shifting points;
— à *trusquin*, beam compasses;
— à *verge*, beam compass; trammel;
— *vérificateur*, (*art*.) inspecting, verifying, compasses (for projectiles, etc.).

compassage, m., (*sm. a*. operation of setting the borings of a gun-barrel so that their axes shall be in the same straight line.

compassement, m., (*mil. min*.) proportioning trains, matches, etc., so as to produce simultaneous explosion of several charges (obs.); (*in gen*.) measuring with compasses;
— *en dérivation*, (*mil. min*.) shunt system of proportioning trains, etc.;
— *des feux*, (*mil. min*.) same as *compassement;*
— *mixte*, (*mil. min*.) compound system of proportioning trains, etc.;
— *en série*, (*mil. min*.) series system of proportioning trains, etc.

compasser, v. a., (*mil. min., etc*.) to proportion the trains, fuses, etc., so as to produce simultaneous explosion of several charges;
— *les feux*, (*mil. min*.) same as *compasser*.

compasseur, m., (*sm. a*.) workman who verifies the preliminary boring of a barrel;
ouvrier —, same as *compasseur*.

compétence, f., (*mil. law, etc*.) jurisdiction; jurisdiction in general.

complément, m., complement;
de —, necessary to bring up anything to its full status (e. g., to a war footing).

complet, a. and m., complement; full, complete;
grand —, full complement;
petit —, reduced complement;
— *règlementaire de la masse individuelle*, (*Fr. adm*.) maximum amount of *l'avoir* à *la masse*, q. v.

compléter, v. a., to complete, fill up; (*mil*.) to fill up the ranks, to bring up a unit to its full number.

complication, f., (*mil. law*) case in which a court finds a prisoner guilty of a charge different from that on which tried.

complice, m., accomplice.

complicité, f., complicity.

complot, m., plot, conspiracy;
chef de —, ringleader.

composante, f., component (force, stress, etc.).

composé, m., chemical compound; a., complex, compound, combined, composite; component (force, stress, etc.);
— *explosif*, (*expl*.) explosive compound.

composer, v. a., to compose, compound; adjust fashion; (*mil*.) to compound for terms, as of capitulation; to come to terms of surrender.

composite, a., composite, i. e., of iron and of wood.

composition, f., composition; compounding; make-up; ingredients; adjustment, settlement; essay or discussion on a given theme in an examination; (*mil*.) capitulation, terms of capitulation, articles of agreement for a surrender; (*artif*.) composition;
— *d'amorce*, primer composition;
— *d'artifices*, firework composition;
— *éclairante*, (*artif*.) any light-giving composition;
faire bonne —, (*mil*.) to grant favorable terms of surrender;
— *fulminante*, (*art., etc*.) priming, detonating composition;
— *fusante*, (*art*.) fuse composition; rocket composition;
— à *fusées*, v. — *fusante*;
— *incendiaire*, (*artif*.) carcass composition;
— *lente*, (*artif*.) slow-burning composition;
— *neuve*, (*powd*.) " green" charge;
— *porte-retard*, (*art*.) delay-action composition (of a torpedo shell);
se rendre par —, (*mil*.) to capitulate;
— *vive*, (*artif*.) quick-burning composition.

compound, a., compound (of armor); (*elec*.) compound; m., compound armor.

compresse, f., (*med*.) compress; pledget.

compresseur, m., compressor; a., compressive, compressing.

compressible, a., compressible; (of ground) insecure, not affording a good foundation.

compression, f., compression, condensation;
globe de —, (*mil. min*.) overcharged mine, globe of compression.

comprimer, v. a., to compress, as gun-cotton, etc.

comptibilité, f., accountability, responsibility; (*adm*.) papers, returns, accounts, paper work;
— *-argent*, v. — *-deniers*;
— *-deniers*, money accountability, money responsibility (sometimes, — *en deniers*);
— *-matières*, property responsibility (sometimes, — *en matières*);
— *nominative*, (*Fr. a*.) papers, etc., in which officers, n. c. o.'s, *rengagés*, etc., are recorded by name;
— *numérique*, (*Fr. a*.) papers, etc., in which n. c. o.'s and men are recorded by their numbers;
pièce de —, voucher.

comptable, a., responsible, accountable, for money, property, etc.; to be taken into account; m., accountable officer or official, e. g., *officier d'administration;*
officier —, (*adm*.) responsible officer (either by virtue of his duties, e. g., *officier d'habillement*, or by virtue of his position, e. g., *officier d'administration*)
place —, (*Fr. a*.) a principal *place*, under the charge of a *comptable*, and to which are annexed smaller *places*, so far as the same service is concerned, depot of issue and receipt;
— *du trésor*, treasury official (*Fr. a*., paying or disbursing officer).

compte, m., account, return, statement: number, quantity;
— *annuel de gestion*, (*Fr. adm*.) the ledger kept in property accounts by the responsible officer, the ledger so kept being turned in as a return;
— *-s-deniers*, money accounts;
— *élémentaire*, (*Fr. adm*.) account furnished by responsible officers, by councils of administration, and by contractors;

compte *en deniers*, quarterly money account;
— *général*, (*Fr. adm.*) account or return submitted by ministers of the government, equal to the sum of —*s élémentaires;*
— *général de l'état*, (*Fr. adm.*) the sum of the *comptes généraux,*
— *général du matériel*, (*Fr. adm.*) a general return of all the property of the war department, i. e., army in its widest sense;
— *de gestion*, (*Fr. adm.*) annual return of government property (*matériel*);
— *en journées*, (*mil.*) return of number of days of treatment in hospital;
— *matières*, (*Fr. adm.*) all and any sorts of property accounts, receipts, invoices, memorandum receipts, etc.;
— *d'ordonnancement*, (*Fr. adm.*) account submitted by officers authorized to warrant the disbursement of funds;
— *du recrutement*, (*Fr. adm.*) ministerial return on the execution of the recruiting laws, made annually to the Chamber of Deputies;
rendre —, to report;
— *rendu*, report;
— *de revues*, (*Fr. adm.*) comparison of amounts due (money or stores) and amounts actually received.
compte-pas, m., (*inst.*) odometer, pedometer.
compter, v. a., to compute, number, reckon, tell off, count; (*hipp.*) to advance and withdraw the knee continually, owing to a shoe set too close to the quick.
compteur, m., counter, meter; (*mach.*) counter, telltale;
— *à eau*, water meter;
— *électrique*, (*elec.*) electric meter;
— *à gaz*, gas meter;
— *de résistances électriques*, (*elec.*) resistance box;
— *-télémètre*, m., (*art.*) proposed range-finder, to act on velocity of sound.
comptoir, m., factory (in a colony).
cônage, m., (*mach.*) coning.
concasser, v. a., to pound, bruise, beat small; to crush, break, stone.
concassement, m., pounding, crushing; stone-crushing.
concave, a., concave.
concavité, f., concavity.
concéder, v. a., to yield, grant.
concentrateur, m., (*sm. a.*) concentrator, pasteboard tube containing bird shot, so as to secure close grouping of shot.
concentration, f., concentration; (*mil.*) concentration, or period of following mobilization; concentration of troops in general;
— *du feu*, (*mil.*) concentration of fire;
— *du tir*, (*art.*) concentration of fire upon a given target;
zone de —, (*mil.*) zone of concentration (mobilization, etc.).
concentré, a., (*chem.*) concentrated.
concentrer, v. a., (*mil., etc.*) to concentrate.
concentrique, a., concentric.
concession, f., concession; grant of land in a colony.
conciliation, f., (*adm.*) act of causing laws, etc., in apparent disagreement, to agree.
concordance, f., agreement;
— *des charges*, (*art.*) selection of charges so as to obtain an agreement of results.
concours, m., competition; competitive examination; match; (*top.*) meeting point of roads;
— *de conduite de voitures*, (*art.*) artillery drivers' competition;
— *au fusil*, rifle match;
— *de marche*, (*mil.*) long-distance march;
— *de pointage*, (*art.*) aiming competition;
— *au revolver*, revolver match;
— *de tir*, (*art. and inf.*) artillery target competition; rifle match, rifle competition (or any other firearm);
— *de voitures*, (*art.*) v. — *de conduite de voitures.*

concret, a., concrete.
concurrence, f., competition (of contractors, etc., commercial, etc.); (frequently) equality.
concurrent, m., competitor.
concutant, a., striking, coming in contact with;
système —, (*art.*) the time system of a double-action fuse;
système — *-percutant*, (*art.*) double-action system of a double-action fuse.
concuteur, m., (*art.*) striker of a fuse; plunger; plunger of a double-action fuse; time plunger, time-train firing-pin, of a double-action fuse.
condamnation, f., conviction, judgment, sentence;
— *sur procédure sommaire*, summary conviction.
condamné, m., person under sentence; convict; a., unfit for service; condemned, sentenced.
condamner, v. a., to condemn, convict, sentence; to seal up, block up, a window or door.
condensabilité, f., condensability.
condensable, a., condensable.
condensateur, m., (*elec.*) condenser.
condensation, f., (*steam, etc.*) condensation; (*mil.*) close order (rare);
auge de —, condensing cistern;
auget de —, v. *auge de* —;
— *par contact*, surface condensing;
eau de —, waste water, water of condensation;
— *par l'extérieur*, surface condensing;
— *par injection*, condensation by injection;
machine à —, condensing engine;
— *par mélange*, condensation by injection;
— *sans mélange*, surface condensing;
— *à sec*, surface condensing;
— *à surface*, v. — *à sec.*
condenser, v. a., to condense; (*mil.*) to close up (rare for *serrer*);
machine à —, (*steam*) condensing engine.
condenseur, m., (*steam, elec.*) condenser; condensing vessel;
baromètre du —, condenser gauge;
— *à double effet*, double-acting condenser;
— *à injecteur*, — *à injection*, jet condenser;
— *à jet*, jet condenser, injection condenser;
— *à* —, *par, mélange*, v. — *à jet;*
— *à simple effet*, single-acting condenser;
— *à surface*, surface condenser;
— *tubulaire*, surface condenser.
condition, f., condition; terms;
— *s ambiantes*, (*ball.*) atmospheric conditions;
sans —, (*mil.*) unconditional (as, a surrender).
conducteur, m., conductor, guide, manager, leader, superintendent; teamster; conductor of a railway train; conductor of heat, etc.); (*art., eng., train*) driver; (*man.*) leading trooper in riding-hall; (*elec.*) conductor, lead wire, main;
— *d'aller*, (*elec.*) main or direct wire;
canonnier- —, (*art.*) (full expression for) driver;
— *de derrière*, (*art.*) wheel driver;
— *de devant*, (*art.*) lead driver;
double —, (*elec.*) double-wire, two-wire conductor;
— *maître*, (*elec.*) lead or main wire; (*mil. min., etc.*) wire connected with firing machine or detonator; (*in gen.*) a conductor with one interior wire;
— *-mécanicien*, locomotive engineer; driver (British);
— *du milieu*, (*art.*) swing driver;
— *de mise de feu*, (*min., mil. min.*) any means of communicating fire, electric or pyrotechnic;
— *moteur*, (*mach.*) driver;
— *primaire*, (*elec.*) lead, lead wire;
— *de retour*, (*elec.*) return wire;
sapeur- —, (*eng.*) driver, engineer driver;
— *secondaire*, (*elec.*) secondary wire; (*mil. min., etc.*) two-wire conductor (to primers, fuses, etc.).

conducteur, a., (*mach.*) driving, transmitting.
conductibilité, f., (*elec.*) conductibility.
conductivité, f., (*elec.*) conductivity.
conduire, v. a. n., to conduct, lead; to direct, guide, manage; to steer (a ship); pursue, continue; to go to (of a road); to drive (a horse); (*sm. a.*) to let down (a hammer, bolt, etc.);
—— *sur (la bride, etc.)*, (*art.*) to use the (bridle, etc.) in driving;
—— *un cheval étroit (large)*, (*man.*) to cause a horse to describe a small (large) circle (riding school);
—— *un cheval par la figure*, to lead a horse by the head;
—— *le chien à l'abattu*, (*sm. a.*) to let down the hammer, bring it down to the safety notch;
—— *au cran de ....*, (*sm. a.*) to bring or let down to (such a notch);
—— *à la Daumont*, to drive from the saddle, postilion-wise;
—— *à grandes guides*, to drive at full speed; to drive four-in-hand;
—— *en guides*, to drive (reins from seat);
—— *une ligne*, to produce a line;
—— *en main*, to lead along a horse (from the saddle);
—— *en selle*, (*art.*) to drive.
conduit, m., a., conduit, pipe, drain; fair-lead; canal, cut, passage;
—— *d'admission*, (*steam*) steam passage of a cylinder;
—— *d'échappement*, (*steam*) exhaust passage;
—— *en entonnoir*, funnel pipe;
—— *à gaz*, gas pipe;
—— *souterrain*, culvert;
—— *de la vapeur*, steam pipe;
—— *de vent*, blast pipe.
conduit, a., (*mach.*) driven; receiving.
conduite, f., conduct, management, leading, leadership; (*mil.*) behavior (as of troops under fire); (*art.*) driving, art of driving; (*steam, etc.*) generic term for pipe, tube, conduit, etc.;
—— *d'amenée*, (*steam, etc.*) delivery pipe, tube;
cadre de ——, (*mil.*) detail of n. c. o. for taking recruits to their regiments;
certificat de bonne ——, (*mil.*) v. *certificat*;
—— *du feu*, (*inf., art.*) fire direction;
—— *en guides*, (*art., etc.*) driving (reins from seat);
—— *du tir*, (*art.*) fire direction and control;
tuyau de ——, delivery pipe;
—— *de vent*, ventilation shaft;
—— *des voitures*, (*art. and train*) driving, act of driving.
cône, m., cone; (*mach., etc.*) taper hole, speed cone;
ajuster ——, to taper;
—— *d'arrachement*, (*expl.*) the cone detached by a blast;
—— *arrière*, (*torp.*) stern cone of the Whitehead torpedo;
—— *de charge*, (*torp.*) war head (Whitehead torpedo);
—— *de commande*, (*mach.*) driving cone, feed cone;
—— *conique*, (*mach.*) cone pulley;
—— *denté*, (*mach.*) toothed-cone clutch;
—— *de dispersion*, —— *de divergence*, (*art.*) cone of spread;
donner du ——, to taper;
—— *d'éclatement*, (*art.*) cone of dispersion of burst;
—— *élastique*, (*art.*) elastic cone mount;
—— *d'entraînement*, (*art.*) gearing cone;
—— *étagé*, (*mach.*) speed cone, speed pulley;
—— *de forcement*, (*art.*) compression slope;
—— *de friction*, (*mach.*) cone clutch, friction clutch, friction pulley;
—— -*poulie*, —— *de poulies*, (*mach.*) speed cone, speed pulley; set of stepped pulleys;
—— *de raccordement*, (*art.*) junction slope, junction cone, chamber slope;
—— *de suspension*, prolongation of net below the body of the balloon

cône(s) *tronqués*, (*mach.*) speed-cone, -pulley, stepped pulley.
côner, v. a., to cone.
confection, f., making up (of cartridges, articles of clothing, etc.).
confectionné, m., any manufactured article.
confectionner, v. a., to make up, prepare, construct, put together, etc.
confectionneur, m., one who makes up articles.
confédération, f., confederation, confederacy; alliance, league.
confédéré, a., confederate, confederated.
conférence, f., meeting, meeting of officers, etc., (hence) lecture, address; lyceum meeting, paper read before a lyceum (U. S. A.).
confin, m., limit, boundary, border.
confiner, v. a. n., to confine, limit; to border on.
confirmation, f., (*Fr. mil. law*) approval by a *conseil de révision*, q. v., of a verdict referred to it.
confirmer, v. a., to confirm;
—— *un cheval*, (*man.*) to finish, perfect, the training of a horse.
conflit, m., conflict, issue, collision; (*law*) clashing or overlapping of authority;
—— *d'attributions*, conflict between the judicial and the administrative authorities;
—— *de juridiction*, conflict between two authorities of the same order or nature;
tribunal des ——*s*, tribunal to which are referred conflicts of authorities.
confluence, f., (*top.*) confluence.
confluent, m., (*top.*) confluent.
confluer, v. n., (of rivers) to meet, to flow one into the other.
conformation, f., conformation.
confrontation, f., (*mil. law*) interview between the accused and witnesses for the prosecution, or between the accused and co-accused, to settle questions of identity.
congé, m., permit, pass; discharge, dismissal; (*mil.*) any cessation of service, whether temporary or permanent, (hence) leave of absence, furlough (if of more than 30 days), discharge from service; (*nav.*) clearance (papers); (*cons.*) rib, bracing rib, congee;
—— *absolu*, (*mil.*) full discharge, discharge from service;
—— *pour affaires personnelles*, (*Fr. a.*) leave on private business;
—— *pour aller à l'étranger*, (*Fr. a.*) leave with permission to visit foreign countries (to go beyond the sea, U. S. A.);
—— *pour aller faire usage des eaux*, (*Fr. a.*) sick leave to go and take the waters;
avoir son ——, (*mil.*) to have one's discharge;
—— *de convalescence*, (*mil.*) sick leave;
avoir un ——, (*mil.*) to be on leave;
donner ——, to discharge, send away;
donner un ——, (*mil.*) to give or grant a leave of absence, etc.;
être en ——, to be on leave, pass, or furlough;
—— *de l'étui*, (*sm. a.*) rim of the cartridge case;
faire un ——, (*Fr. a.*) to serve a full term of three years;
—— *illimité*, (*mil.*) unlimited, indefinite leave; discharge;
—— *de libération de service*, (*mil.*) complete discharge from service;
—— *limité*, (*mil.*) short leave;
—— *par motifs de famille*, (*mil.*) leave on private affairs;
—— *par motifs de santé*, (*mil.*) sick leave;
—— *No. 1*, (*Fr. a.*) discharge on account of physical disability incurred or developed in service;
—— *No. 2*, (*Fr. a.*) discharge for physical disability not incurred in service;
prendre son ——, (*mil.*) to leave the service;
prendre un ——, (*mil.*) to take a leave;
prolongation de ——, (*mil.*) extension of leave;
—— *de réforme*, (*Fr. a.*, etc.) discharge for physical or any other unfitness for service;

congé à *titre de soutien de famille*, (*Fr. a.*) leave of absence on account of applicant being necessary to the support of his family; lasts until beneficiary passes into the reserve of the active army.

congédier, v. a., to discharge; disband; pay off.

congélation, f., congelation, freezing, congealing; cold storage.

congeler, v. a., to freeze, congeal.

congréage, m., (*cord.*) worming, keckling.

congréer, v. a., (*cord.*) to worm, keckle.

congrès, m., congress.

congrève, m., (*artif.*) Congreve rocket; *fusée à la* ——, Congreve rocket.

conicité, f., state of being coned or tapered.

conification, f., (*sm. a.*) operation of tapering the seat of the bullet.

conique, a., conical, tapering; (*mach.*) taper; *fortement* ——, (*mach., etc.*) quick tapering.

conjoncteur, m., (*elec.*) connector, key; —— *disjoncteur*, make-and-break key; —— *de mise de feu*, (*mil. min., expl., etc.*) firing-key.

conjugaison, f., —— *des hausses*, (*art.*) employment (when range is unknown or observation difficult) of several elevations, differing by 100^m or 200^m, in the same battery, in order to determine the range.

conjugué, a., (*art.*) in pairs (of guns in turrets); (*fort.*) having end faces in same line (so as to secure the defilade of a given point, as in a salient).

conjuration, f., conspiracy.

conjuré, m., conspirator.

connaissance, f., acquaintance, knowledge; —— *des temps*, nautical almanac.

connaissement, m., bill of lading.

connaître, v. a. n., to know; to have authority to pass on a matter, take cognizance of; —— *la bride, les éperons, la jambe*, (*man.*) to be bridle-, spur-, leg-, wise.

connexer, v. a., to join (as two shafts, etc.).

connexion, f., connection; —— *directe*, (*mach.*) direct action; *à* —— *directe*, (*mach.*) direct acting.

conquérant, m., conqueror, victor; a., conquering, victorious.

conquérir, v. a., to conquer, overcome, vanquish.

conquête, f., conquest.

consacrer, v. a., to sanction; —— *une prise*, (*nav.*) to legalize a capture.

consanguin, a., (*hipp.*) inbred.

consanguinité, f., (*hipp.*) in-and-in breeding.

conscience, f., breastplate of a bit.

conscription, f. (*mil.*), conscription; familiar expression for recruiting by means of *appels*, q. v.; —— *des chevaux*, (*Fr. a.*) census of horses and mules fit for military service.

conscrit, m. (*mil.*), conscript; in familiar language, a recruit not yet joined.

conseil, m., board, council; —— *d'administration*, (*mil.*) council of administration, (*Fr. a.*) receives funds, etc., from the state and turns them over to the regiment, or corps counting as a regiment); —— *d'administration central*, (*Fr. a.*) the council of the *portion centrale*, q. v.; —— *d'administration éventuel*, (*Fr. a.*) the council for detached portions of at least 6 companies of infantry or 3 troops of cavalry; —— *de l'amirauté*, (*nav.*) admiralty board; —— *de défense*, (*Fr. a.*) board of officers and of functionaries which assists the governor of a place, and the c. o. of forts during a siege (v. *commission de défense*); —— *de discipline*, (*Fr. a.*) court passing on recommendations to the *compagnies de discipline*, q. v.; —— *d'enquête*, (*Fr. a.*) court of inquiry in general; more specially, a court or board passing on certain specified cases, as the *réforme* of an officer, reduction of certain classes of n. c. o., etc.; —— *de guerre*, (*mil.*) council of war; (*mil. law*) court-martial;

conseil *de guerre prévôtal*, (*mil.*) drumhead court-martial; —— *d'honneur*, (*mil.*) court of honor; *passer devant un* —— *de guerre*, (*mil.*) to be tried by court-martial; —— *de régiment*, (*Fr. a.*) board to pass on requests of n. c. o. to be reenlisted; —— *de révision*, (*Fr. a.*) military court of appeal; recruiting board (see next word); —— *de révision (cantonal)*, (*Fr. a.*) board to examine claims of exemption from military service (name frequently given to the —— *de révision* in view of this special duty); —— *de santé*, medical board; —— *supérieur de la guerre*, (*Fr. a.*) board that assists the war minister, deliberates on questions of mobilization, concentration, supplies, etc.

consenti, a., admitted, admissible.

consentir, v. n., to consent; (*tech.*) to spring, give way (as a mast, etc.).

conservateur, m., —— *des bâtiments*, (*Fr. a.*) one of the principal employees of the war department, a sort of chief clerk for military buildings, their construction, etc.

conservation, f., preservation; —— *des tranches*, James Bernouilli's principle of the "parallelism of sections," in the strength of materials.

conserve, f., (*nav.*) convoy, consort; (*fort.*) counter-guard; (*in pl.*) preserved meats, vegetables, soups; *de* ——, preserved (meats, etc.); *naviguer de* ——, (*nav.*) to sail in company with.

conserver, v. a., to preserve, keep, secure; to keep in pay; to keep sight of; —— *les distances*, (*drill*) to keep distances, marching in column; —— *une position*, (*mil.*) to hold a position; —— *la vitesse*, (*art.*) to hold up well (of a projectile that keeps its velocity well).

consignation, f., (*adm.*) deposit of a sum of money, of an article, in the hands of a public official.

consigne, f., (*mil.*) orders, instructions, e. g., special orders, orders for a particular post or duty, especially sentinel's orders; confinement to barracks, quarters; barrack-stove poker; *caporal de* ——, (*Fr. a.*) corporal responsible for guardhouse material (the senior if there are more than one); —— *à la chambre*, confinement to room as punishment (of a n. c. o.); *forcer la* ——, to force a sentry, to break orders; —— *générale*, general orders for a sentinel or post; —— *à gros grains*, (*mil. slang*) imprisonment in the cells; *lever la* ——, to cancel an order; *manquer à la* ——, to break an order; —— *particulière*, special orders for a sentinel or post; *portier-* ——, (*Fr. a.*) gate-keeper of a fortress (retired n. c. o.); —— *au quartier*, confinement to barracks (n. c. o,'s and men); —— *verbale*, temporary or special order.

consigné, m., (*mil.*) man confined to barracks, not allowed to pass the guard, by way of punishment; *salle des* ——s, (*Fr. a.*) sort of guard room for disorderly or insubordinate soldiers in hospital.

consigner, v. a., to consign (merchandise); to deposit (money); to forbid, to prevent access to or egress from; (*mil.*) to confine to barracks or quarters; to have the men ready in an emergency; to give orders to a sentinel or vedette.

consistant, a., tenacious.

console, f., bracket, corbel, console, wall-fixture; (*art.*) loading-tray, plug-tray, bracket; (*sm. a.*) part of revolver frame into which the barrel is screwed; (*mach.*) traveling or movable plate (for the work); *volet-* ——, (*art.*) revolving plug-tray, tray and shutter in one.

consommation, f., consumption; expenditure;
objets de ——, expendable articles.
consommer, v. a., to use up, expend; finish, accomplish.
conspiration, f., conspiracy, plot; combination (in a bad sense).
conspirer, v. n., to plot, conspire.
constante, f., constant quantity, constant;
—— *magnétique d'un lieu,* magnetic elements of a place.
constatation, f., verification, authentication; statement.
constater, v. a., to verify, prove, establish, declare, authenticate.
constituant, a., (*mil.*) forming an integral part of the organic composition of a troop unit.
constitué, a., (*mil.*) having an organic existence of its own, organic.
constituer, v. a., to constitute, appoint, place, depute; to give into custody; (*mil.*) to form, organize, a body of troops, etc.;
—— *une pension,* to settle or grant a pension;
—— *un prisonnier,* to commit a prisoner;
se —— *prisonnier,* to surrender.
constitutif, a., v. *constitué.*
constitution, f., constitution, make-up;
—— *d'un pays,* nature and description of a region, of a country.
constitutionnel, a., constitutional.
constructeur, m., builder, constructor; designing engineer;
———*mécanicien,* mechanical engineer.
construction, f., construction; building, structure, edifice;
—— *d'art,* (*r. r.*) bridge, culvert, etc.;
—— *en bois,* woodwork;
—— *en fer,* ironwork;
—— *en fer et en bois,* composite work;
—— *en pisé,* (*fort.*) pisa work.
construire, v. a., to build, construct; (*min.*) to drive (a gallery); (*fort.*) to throw up (a work).
consul, m., consul.
consultatif, a., consulting (as committee, board).
contact, m., touch, contact; (*mil.*) contact with the enemy; touch (of elbow); (*elec.*) contact, contact-plate, contact-strip;
condensation par ——, (*steam*) surface condensing;
conserver le ——, (*mil.*) to keep (the enemy) in touch, to keep in touch with (the enemy);
—— *du coude,* (*mil.*) touch of elbows in ranks;
escadron de ——, (*cav.*) contact squadron;
être au ——, (*mil.*) to be in contact (with the enemy);
—— *par frottement,* (*elec.*) friction-contact;
garder le ——, (*mil.*) v. *conserver le* ——;
—— *de glissement,* (*elec.*) sliding-contact;
—— *à mercure,* (*elec.*) mercury contact;
perdre le ——, (*mil.*) to lose touch (with the enemy);
prendre le ——, (*mil.*) to establish contact, touch (with the enemy);
rester en —— *avec,* (*mil.*) to remain in touch (with the enemy);
—— *roulant,* (*elec.*) rolling contact.
contagieux, a., contagious.
contagion, f., contagion.
contenance, f., capacity; contents; (*nav.*) burden.
contenir, v. a., to contain, hold; restrain, hold; check (as an advance, the enemy).
contentieux, —— *administratifs,* (*Fr. adm.*) difficulties arising from antagonism of public and of private interests.
contenu, m., contents.
contigu, a., adjoining, contiguous.
continent, m., mainland, continent.
continental, a., continental;
défense ——*e,* (*fort.*) land defense of a seacoast work.
contingent, m., contingent, quota;
—— *annuel,* (*mil.*) annual contingent that a *classe* must furnish.
contondant, a., acting by shock; (*art.*) racking.

contour, m., outline, circuit; plan; (*top.*) contour; (*unif.*) edging of an epaulet, (also) of a flag tassel.
contournement, m., (*fort.*) crochet (of a traverse).
contourner, v. a., to turn, to go around; to contour, give a contour to; to distort, deform.
contracter, v. a. r., to shrink, contract, shorten; to stipulate, make a contract.
contraction, f., contraction, shrinking;
force de ——, contractile force.
contractuel, a., stipulated by a contract.
contracture, f., diminution.
contradictoire, a., (*law*) after having heard both sides; (*adm.*) after examination of both sides of an account.
contrainte, f., constraint, violence; (*adm.*) order to pay decreed against a person owing public money;
—— *par corps,* compulsory, i. e., forced, attendance of a witness.
contraire, a., contrary, opposite, adverse, hostile.
contrarié, a., with joints broken.
contrat, m., contract.
contravention, f., infraction, infringement, violation;
être en ——, to break regulations, etc.
contre, prep., against, near, close by; (in composition, *contre* carries with it the notion generally of opposition, and frequently of juxtaposition);
à ——, (*cord.*) the wrong way, against the sun.
contre, m., (*fenc.*) counter;
—— *de quarte,* quart-, cart-counter;
—— *de tierce,* tierce-counter;
double- ——, double counter.
contre-accélération, f., (*ball.*) retardation.
contre-à-contre, adv., (*nav.*) side by side.
contre-aiguille, m., (*r. r.*) main rail, line or stock rail.
contre-alizé, a., *vent* ——, anti-trade (wind).
contre-amiral, m., (*nav.*) rear-admiral; rear-admiral's flagship.
contre-appel, m., (*mil.*) check roll-call, check; (*fenc.*) caveating; (*teleg.*) call.
contre-approches, f. pl., (*siege*) counter approaches.
contre-appui, m., *système à* ——, (*art.*) system in which the trail rests on the limber instead of over a pintle-hook.
contre-arquer, v. n., to sag.
contre-assaillir, v. a., (*mil.*) to make a counter assault.
contre-assiéger, v. a., (*siege*) to besiege a besieging force.
contre-attaque, f., (*mil.*) counter attack (by the defensive).
contre-aube, f., strip laid against an *aube.*
contre-balancer, v. a., to counterbalance, counterpoise.
contrebande, f., contraband, contraband goods; smuggling;
—— *de guerre,* (*mil.*) contraband of war.
contrebandelette, f., fork-strap (of a pack saddle).
contrebandier, m., smuggler.
contre-bas, *en* ——, adv., below, downward, on a lower level, lower; from down up;
embrasure en ——, (*fort.*) counter-sloping embrasure;
être en —— *de,* to be lower than, on a lower level than.
contre-batterie, f., (*siege*) counter-battery; (more generally) any battery intended to engage a given hostile battery, and particularly battery engaging hostile batteries during the infantry advance.
contre-battre, v. a., (*art.*) to counter batter, to oppose batteries to those of the enemy.
contre-biais, *à* ——, contrariwise.
contrebord, (*nav.*) *aller à* ——, to run aboard or foul (of each other).
contre-bordée, f., (*nav.*) the other tack.
contre-boutant, m., buttress, counterfort.
contre-bouter, v. a., to prop, to support (as a wall by a buttress).
contre-bride, f. counter flange.
contre-brisure, f., (*fort.*) outer brisure.

contre-buter, v. a., to press up against a resisting or supporting mass; to prop, support.
contre-calquer, v. a., (*draw.*) to take a counter-proof of a counterdrawing.
contre-capsule, f., (*art., etc.*) primer-cover or cap.
contrecarrer, v. a., to oppose directly (the efforts of the enemy, etc.), thwart.
contrechangement, m., —— *de main,* (*man.*) double change of hands.
contrechanger, v. n., —— *de main,* (*man.*), to make a double change of hands.
contre-charge, f., counterpoise.
contre-châssis, m., (*fond.*) top-flask; (*cons.*), outer sash.
contre-chef, m., foreman.
contre-clavette, f., gib. tightening key, fox-wedge.
contre-clé, f., gib.
contre-clef, f., (*mas.*) second stone in the crown of an arch.
contre-cœur, m., back plate of a forge, as of a traveling forge; chimney-back; (*r. r.*) guard-rail, wing-rail.
contre-couder, v. a., to bend the other way.
contre-coup, m., return-stroke, counterstroke; rebound; (*expl.*) reverse stroke or effect sometimes remarked in (high) explosive blasts; (*man.*) sudden start, jump (of a horse); (*hipp.*) heaving (of a broken-winded horse);
faire ——, (*mach., etc.*) to receive, support, a blow (as in riveting).
contre-courant, m., back draught, counter-current.
contre-dégagement, m., (*fenc.*) counter-thrust (i. e., a *dégagement,* q. v., made simultaneously with the adversary's *dégagement*); general expression for *contre de quarte, de tierce,* qq. v.; counter-disengage.
contre-dégager, v. n., (*fenc.*) to make a contre-*dégagement,* to counter-disengage.
contre-dérive, f., (*art.*) negative lateral allowance (on a sight).
contre-digue, f., a dike protecting another; construction protecting a dike.
contre-écrou, m., jam nut, lock nut, keep nut, check nut.
contre-enquête, f., counter-inquiry.
contre-épaulette, f., (*unif.*) epaulet having neither bullion nor fringe; shoulder-knot.
contre-épreuve, f., counter-proof, counter-test, check-test.
contre-escarpe, f., (*fort.*) counterscarp.
contre-essai, m., counter-test.
contre-étai, m., (*cord., etc.*) backstay.
contre-étampe, f., swage, upper tool.
contre-expertise, f., counter-valuation.
contrefaçon, f., counterfeiting, illegitimate imitation or copy, in manufactures;
—— *de brevet,* infringement of a patent.
contrefacteur, m., counterfeiter, forger; infringer.
contrefaire, v. a., to counterfeit, forge; to infringe a patent.
contre-feu, m., back plate (of a hearth); a fire lighted to check another (as in woods or plains), back fire.
contre-fenêtre, f., inside sash.
contre-fiche, f., (*cons.*) strut, brace, stud, shore, stay; any compression piece in construction.
contre-fil, m., the opposite direction;
à ——, backwards.
contre-flanc, m., (*art.*) loading-edge.
contre-forger, v. a., (*farr.*) to shape out a horseshoe.
contrefort, m., buttress, shoulder, pillar, pier, supporting mass; reinforcing piece; (of boots, shoes, etc.) stiffener, reinforce, counter; (*top.*) spur of a mountain range, lesser chain of mountains.
contre-fossé, m., counter-drain; (*fort.*) synonym of *avant-fossé.*
contre-foulement, m., upward motion of water in a pipe or tube.
contre-fruit, m., (*mas.*) inside batter (of a wall).

contre-garde, f., (*fort.*) counterguard.
contre-guide, m., (*mach.*) link.
contre-hacher, v. a., to cross-hatch.
contre-hachure, f., cross-hatching.
contre-hauban, m., (*cord., etc.*) back guy.
contre-haut, *en* ——, on a higher level, upwards; from up down;
être en —— *d'un point,* to be higher, on a higher level than.
contre-heurtoir, m., (*art.*) counter-hurter; garnish plate.
contre-jauger, v. a., to counter-gauge.
contre-jour, m., false light, counter-light.
contre-latte, f., (*cons.*) lathwork at right angles to laths already laid on; counter-lath.
contre-latter, v. a., (*cons.*) to counter-lath.
contre-lattoir, m., lath-holder, clincher.
contre-lisoir, m., (*art.*) bolster-block.
contre-lunette, f., (*art.*) lower pintle-plate.
contre-maître, m., overseer, foreman; (*nav.*) boatswain's mate;
aide, ——, assistant foreman.
contremandement, m., countermand, counter order.
countremander, v. a., to countermand.
contre-manivelle, f., (*mach.*) countercrank, return crank.
contre-manœuvre, f., (*mil.*) counter-maneuver (rare).
contre-marche, f., (*mil.*) countermarch; (*cons.*) riser of a stair.
contre-marcher, v. a., (*mil.*) to counter-march.
contre-marée, f., back or contrary tide, eddy tide.
contre-marque, f., check; (*hipp.*) bishoping.
contre-marquer, v. a., to countermark; (*hipp.*) to bishop a horse.
contre-mine, f., (*mil. min.*) countermine.
contre-miner, v. a., (*mil. min.*) to counter-mine.
contre-mineur, m., (*mil. min.*) counter-miner.
contre-mont, adv., up hill, upwards; against the stream.
contre-mot, m., (*mil.*) the "word" given in answer to the *mot d'ordre,* and hence equivalent to the *mot de ralliement;* counter-parole, parole (U. S. A.).
contre-mouvement, m., countermovement.
contre-mur, m., (*fort., etc.*) countermure, contra-mure.
contre-murer, v. a., (*fort., etc.*) to counter-mure.
contre-ordre, m., (*mil.*) counter order; countermand.
contre-ouverture, f., counter opening.
contre-outil, m., any plate, wedge, key, etc., set against a tool to hold it in place.
contre-paroi, f., (*met.*) casing of a blast furnace.
contre-pas, m., (*mil.*) a half-step taken to recover the step, when lost; man that has lost step.
contre-pédaler, v. a., to back-pedal (bicycle).
contre-pente, f., descent or slope toward one, (*fort.*) toward the interior of the work;
à ——, *en* ——, countersloping;
talus à ——, (*fort.*) countersloping glacis (Carnot's substitute for the counterscarp).
contre-percer, v. a., to pierce from the other side; (*farr.*) to pierce through-holes in a horse-shoe.
contre-perçure, f., (*farr.*) through-hole, hole for shank of nail, in a horseshoe.
contre-peser, v. a., to counterbalance.
contre-pied, m., wrong scent, back-scent.
contre-pierre, f., the support or holder of a jewel-bearing (as in the Schmidt chronograph).
contre-plaque, f., reinforcing plate; anchor-plate, etc.; (*fond.*) back-plate; (*art.*) bottom plate, as of a pintle; (*fort.*) underskin, backing-plate (e. g., on which Grüson armor rests);
—— *de poulie,* (*art.*) outer reinforcing plate of a ginleg (near the head).
contre-platine, f., (*sm. a.*) side plate (opposite lock side).

contre-poids, m., counterpoise, counterweight, counterbalance; balance-weight.
contre-poil, m., wrong way of the hair, of the nap.
contre-poinçon, m., counter punch; upper die, or punch.
contre-pointe, f., (*sm. a.*) back edge of a sword, near the point; (*fenc.*) saber or broadsword fencing or exercise; (*mach.*) back-center, dead center, center, center-support, spindle on fixed poppet of a lathe.
contre-porte, f., double door, folding door; (*fort.*) second gate (as, in a passage).
contre-poupée, f., (*mach.*) foot stock, foot-stock center.
contre-pression, f., (*steam, etc.*) back pressure, counter pressure.
contre-puits, m., (*mil. min.*) defensive mine or shaft, prepared in advance, and intended to annoy enemy without destroying the galleries of the defense;
camouflet- ——, camouflet prepared by boring from a gallery (defense).
contre-queue, f., ——*d'hironde*, (*fort.*) work whose branches splay inward, counter swallow-tail work.
contre-quille, f., (*nav.*) keelson.
contre-raid, m., (*mil.*) counter-raid.
contre-rail, m., (*r. r.*) guard-rail, safety-rail, check-rail.
contre-riposte, f., (*fenc.*) counter-thrust.
contre-rivure, f., burr, rivet-plate.
contre-ronde, f., (*mil.*) check-rounds.
contre-sabord, m., (*art. nav., etc.*) port-lid.
contre-salut, m., (*mil.*) return salute.
contre-sanglon, m., (*harn.*) strap, fixed at one end and free at the other, pierced for the tongue of the buckle; girth strap; any small strap buckling or lying over a larger one, or serving to join adjacent parts of harness.
contrescarpe, f., (*fort.*) counterscarp;
—— *à voûtes en décharge*, counterscarp with relieving arches.
contrescarper, v. a., (*fort.*) to counterscarp.
contre-scel, m., counterseal.
contre-sceller, v. a., to counterseal.
contre-seing, m., counter signature.
contre-sens, m., wrong side or way.
contre-signal, m., (*mil.*) any cautionary signal, e. g., the *contre-mot*, q. v.
contresigne, f., (*mil.*) countersign.
contre-signer, v. a., to countersign.
contre-sortie, f., (*mil.*) counter-sortie, attack directed against a sortie.
contre-taille, f., crosscut.
contre-tailler, v. a., to crosscut.
contre-temps, m., accident, disappointment (*hipp.*) the second expiration or "lift" of broken-winded horse; (*man.*) sudden and unexpected passage from action to inaction.
contre-tenir, v. a., to hold up a rivet.
contre-tige, f., (*art.*) the central counter rod of the Canet system.
contre-timbrage, m., counter stamping.
contre-timbre, m., counter stamp.
contre-timbrer, v. a., to counter stamp.
contre-tirage, m., back-draught, reverse draught.
contre-tirer, v. a., to counter prove;
—— *un plan*, to make a tracing of a plan.
contre-torpilleur, m., (*nav.*) torpedo-catcher.
contre-tranchée, f., (*fort., siege*) counter trench.
contreval, à ——, downstream, down the river.
contrevallation, f., (*siege*) countervallation (obs.).
contrevaller, v. a., (*siege*) to construct a countervallation.
contre-vapeur, f., (*steam*) returning, back, steam; steam cushion;
donner ——, to interpose a steam cushion; to use back steam.
contrevenant, m., offender.
contrevenir, v. a., to infringe, violate, break.

contrevent, m., outside shutter; straining piece, brace; (*met.*) blast plate of a firing forge.
contreventement, m., (*cons.*) lateral bracing (of a roof or of a bridge truss); wind-bracing;
pièce de ——, wind-bracing; brace.
contreventer, v. a., (*cons.*) to brace laterally (as a truss).
contre-verrou, m., (*art., etc.*) bolt-stop; any device controlling or affecting the motions or functions of a bolt.
contre-vis, f., differential screw.
contre-visite, f., second inspection, counter inspection, check inspection.
contre-visiter, v. a., to make a counter or check inspection.
contrevolte, f., (*cav.*) countervolt.
contrevolter, v. n., (*cav.*) to countervolt, to make a countervolt.
contribuer, v. a., to contribute, to be laid under contribution.
contribution, f., contribution, tax;
—— *en argent*, (*mil.*) money contribution;
—— *directe*, (*adm.*) direct tax;
—— *de guerre*, (*mil.*) contribution of war;
—— *indirecte*, (*adm.*) indirect tax;
mettre à ——, (*mil.*) to lay under contribution;
—— *en nature*, contribution in kind.
contrôle, m., register, list; check, as on a result; stamp; verification, inspection (as of sm. a. during manufacture); mark, official mark (as on sm. a. after inspection); (*mil.*) muster roll, roll, roster; (*adm.*) inspection or surveillance of administrative transactions, (hence, *Fr. a.*) *le contrôle*, i. e., functionaries or officers, in general, whose duty it is to watch over and inspect the *administration* or the *gestion*;
—— *de l'administration de l'armée*, (*Fr. a.*) special branch of the war department, safeguarding the interests of the treasury on the one hand and of individuals on the other; verifies for the whole service the observance of ministerial orders, decrees, etc.;
—— *annuel*, (*Fr. a.*) muster roll of officers, men, and horses belonging to each administrative unit, made out each year, or is kept up during the year;
—— *central*, (*Fr. a.*) the *contrôle* exercised by the bureaus of the central administration of the war department, by inspectors-general, and by the personnel of the *corps du contrôle*;
corps du —— *de l'administration de l'armée*, (*Fr. a.*) the personnel of the *contrôle* service;
—— *extérieur*, (*Fr. adm.*) the *contrôle* exercised by the courts and by parliament;
—— *général des armes*, (*Fr. a.*) register of all the arms in the hands of a body of troops;
—— *local*, (*Fr. a.*) the care, surveillance, etc., exercised constantly, and on the spot, by the directors of the various *services* (e. g., artillery, engineers, etc.) over their respective personnels, and by the functionaries of the intendance, in a regiment or other troop unit;
—— *nominatif*, (*mil.*) list of names, roster by name;
—— *des personnes mises à la retraite*, (*mil.*) retired list;
porter sur les ——s, (*mil.*) to put on the rolls;
—— *à posteriori*, (*Fr. a.*) a general expression for the ——s *local, central*, and *extérieur*;
rayer des ——s, (*mil.*) to strike off the rolls;
—— *de service*, (*mil.*) duty roster, roster;
—— *de tir*, (*art. and t. p.*) target-practice reports.
contrôlement, m., (*Fr. a.*) subjection of responsible officers to the *contrôle*.
contrôler, v. a., to check; to verify, inspect, compare; (*Fr. a.*) to exercise the duties of the *contrôle*;
—— *le pointage*, (*art.*) to verify the aim of a gun.

contrôleur, m., inspector, person who makes a test; (*Fr. a.*) member of the *corps du contrôle*;
— *d'armes*, (*Fr. a.*) inspector of small arms, employee of the artillery charged with the reception, sale, and maintenance of small arms in small-arm factories and in the artillery directions.
contumace, m., f., (*law*) contumacy, failure to appear for trial; person who, being accused of a crime, fails to appear for trial; defaulter.
contumax, a. and m. (*law*) contumacious; person guilty of contumacy.
convalescence, f., convalescence;
congé de —, v. *congé*.
convalescent, m., convalescent;
dépôt de —*s*, v. s. v. *dépôt*.
convention, f., convention, meeting; agreement;
— *de Genève*, Geneva convention.
conventionnel, a., conventional;
signe —, conventional sign.
convergence, f., convergence; concentration of effort, fire, etc., on a given point;
— *des feux*, — *du tir*, (*art., etc.*) concentration of fire upon a given target
planchette de —, (*art.*) direction-, converging-plate.
convergent, a., converging, as (*art., sm. a.*) of fire.
converser, v. a., (*drill*) to wheel, to wheel about (a general drill term for a change of direction by a line, in line, or by an element, itself in line, of a column; is limited now to artillery and cavalry, and is not a word of command); (*pont.*) to swing a bridge into position.
conversion, f., (*drill*) wheel (artillery and cavalry, v. *converser*); (*pont.*) conversion (U. S. A.), swinging (British);
construire un pont par —, (*pont.*) to construct a bridge by conversion, swinging;
demi- —, (*drill*) about;
demi-tour de —, (*drill*) about;
— *en marchant*, (*drill*) wheel on a movable pivot;
— *de pied ferme*, (*drill*) wheel on a fixed pivot;
— *à pivot fixe*, (*drill*) wheel on a fixed pivot;
— *à pivot mobile, mouvant*, (*drill*) wheel on a movable pivot;
point de —, (*drill*) wheeling point;
pont par —, (*pont.*) bridge by conversion, swinging;
quart de —, (*drill*) wheel of 90°;
repliement par —, (*pont.*) dismantling by conversion;
replier un pont par —, (*pont.*) to dismantle a bridge by conversion.
convertisseur, m., (*met.*) converter, Bessemer converter.
convexe, a., convex.
convexité, f., convexity.
conviction, f., conviction;
pièce à —, (*mil. law*) statement of charges, etc., against a man.
convocation, f., calling out, convocation; (*mil.*) calling out of reservists, territorialists, etc., for service.
convoi, m., (r. r.) train; (*mil.*) train, convoy; (*nav.*) convoy;
— *administratif*, (*Fr. a.*) subsistence supply train, of four sections, one to corps headquarters and one to each division;
— *d'aller*, (r. r.) down-train;
— *d'approvisionnement*, supply train;
— *d'artillerie*, (*Fr. a.*) ammunition supply train;
— *auxiliaire*, (*Fr. a.*) subsistence supply train formed in each corps region and completed on mobilization by wagons obtained by requisition;
— *de bestiaux*, (r. r.) cattle train;
bon de —, (*Fr. a.*) request for wagon transportation;
— *direct*, (r. r.) express train;
— *par eau*, water transportation;
— *éventuel de réquisition*, (*Fr. a.*) transport trains created as requirements develop, and used along the *ligne d'étapes*;
— *exprès*, (r. r.) express train;
— *de grande* (*petite*) *vitesse*, (r. r.) fast (slow, train;
— *de marchandises*, (r. r.) freight train;
— *mixte*, (r. r.) passenger and freight train;
— *de ravitaillement*, (*mil.*) ammunition supply train;
— *régimentaire*, (*Fr. a.*) an incorrect expression for *train régimentaire*, q. v.;
— *de retour*, (r. r.) up-train, return train;
— *de terre*, (*Fr. a.*) wagon transportation;
— *de voyageurs*, (r. r.) passenger train.
convoyer, v. a., (*nav., mil.*) to convoy.
convoyeur, m., (*mil.*) driver (in a convoy); (*more esp., Fr. a.*) an agent or functionary accompanying a convoy and charged with its loading or packing (cf. head-packer, U. S. A.).
co-ordonnée, f., coordinate;
—*s cartésiennes*, cartesian, rectangular, co-ordinates;
—*s polaires*, polar coordinates;
—*s rectangulaires*, rectangular coordinates.
copeau, m., chip, shaving (of wood).
copie, f., copy;
— *de lettres*, letter-press;
pour — *conforme*, "a true copy."
coque, f., shell; (*steam*) body of a boiler, shell (*nav.*) hull of a ship; (*cord.*) kink in a rope;
faire (*défaire*) *des* —*s*, (*cord.*) to kink (unkink) a rope.
coquillard, m., (*mil. slang*) cuirassier.
coquille, f., shell; footboard (of a wagon); thumb (of a latch); (*mach.*) crosshead, bearing, brass; (*fond.*) chill-mold; (*inst., etc.*) clamp of a ball-and-socket joint; (*sm. a.*) dust-cover (Vetterli); basket-hilt, guard, of a sword;
couler en —*s, sur* —, (*fond.*) to cast chill;
— *en fonte*, (*fond.*) cast-iron mold, chill, chill-mold.
coquiller, v. a., (*fond.*) to chill, to cast in a chill-mold.
cor, m., (*hipp., etc.*) corn; (*mus.*) French horn;
à — *et à cri*, hue and cry;
— *de chasse*, (lit., hunting horn; *Fr. a.*) prize for small-arm target practice.
corbeau, m., bolster, bracket, bracket-piece, corbel; grappling-iron.
corbeille, f., basket; (*fond.*) grate-furnace, used to dry molds; (*siege*) small gabion.
cordage, m., (*cord.*) cordage, rope, rigging; outfit of ropes (as for mechanical maneuvers, etc.);
— *d'ancre*, (*pont., etc.*) anchor rope, mooring rope;
— *blanc*, untarred rope;
— *de caisse*, (*mus.*) drum-cord;
— *commis en aussière*, hawser-laid rope;
— *commis en câble*, cable-laid rope;
— *une fois commis*, hawser-laid rope;
— *commis en grelin*, cable-laid rope;
— *commis en quatre*, four-stranded rope, shroud-laid rope;
— *serré*, short-laid rope;
— *composé*, cable-laid rope;
— *à congréer*, worming;
couronne de —, strap, sling;
— *à crochet*, (*art.*) any rope fitted with a hook for maneuvers, etc.;
— *en fil d'acier* (*de fer*), steel- (iron-) wire rope;
— *sans fin*, grommet;
— *goudronné*, tarred rope;
— *à l'huile lourde*, cordage that has been soaked in fat oil;
— *à la main*, handmade rope, rope made of yarn spun by hand;
— *de manœuvre*, (*art.*) elevating rope, maneuvering rope;
— *mécanique*, rope made of yarn produced by machinery;
— *noir*, tarred rope;
— *du pitte*, coir rope;
— *du premier brin*, rope of the best quality;
— *en quatre*, shroud-laid rope, four-stranded rope;

cordage *en queue de rat*, pointed rope;
— *refait*, twice-laid rope;
— *de sauvetage*, (*mil. min.*) life-saving rope;
— *simple*, hawser-laid rope;
— *de trait*, (*harn.*) trace-rope;
— *en trois*, hawser-laid rope;
— *trop tordu*, rope laid up too hard;
— *de volée*, (*F. m. art.*) chase-rope.

corde, f., cord, twine, string; rope, band; belt; (*art., artif.*) choker, choking-line; (*mil.*) drum-cord; tent rope;
à la ——, (*art., cav.*) tied up to a picket line or rope, on the picket line;
ajuster ensemble deux ——*s*, to bend two ropes together;
amarrer une ——, to belay a rope;
— *à anneaux*, (*gym.*) rings;
— *d'un arc*, (*cons.*) span of an arch;
— *d'arrière*, tow-rope;
— *artificielle*, (*artif.*) quick-match (obs.);
— *d'attache*, (*art., cav.*) lariat, picket rope, picketing rope;
— *d'avant*, head rope;
— *à botillon*, forage-cord;
bout de ——, rope's end;
— *à, de, boyau*, catgut;
— *brassière*, v. *cordeau de sûreté;*
— *de brêlage*, lashing-rope (for loads on wagons);
— *de campement*, (*Fr. cav.*) improvised picket-rope (in a bivouac);
— *de charge*, lash-rope (of a pack outfit);
— *à chevaux*, (*art., cav.*) picket, picketing rope;
— *embarrassée*, foul rope;
— *engagée*, foul rope;
— *d'enrayage*, (*art.*) brake-rope, locking-rope;
épisser une —— *à une autre*, to splice a rope on another;
— *équatoriale*, equatorial rope (of a balloon);
— *à feu*, (*artif.*) fuse, slow match;
— *en fil d'acier* (*de fer*), steel- (iron-) wire rope;
— *de fils métalliques*, wire rope;
filer une ——, to ease away a rope;
— *sans fin*, endless rope, band, belt;
— *à fourrage*, (*mil.*) forage cord;
— *gênée*, foul rope;
— *lâche*, slack rope;
— *lisse*, (*gym.*) climbing rope (without knots);
mollir une ——, to slack off a rope;
moucher une ——, to cut off the fag-end of a rope;
— *à nœuds*, (*gym.*) knotted climbing rope;
ourdir une ——, to warp a rope;
— *à piano*, crucible-steel wire;
— *-poitrail*, —— *de poitrail*, (*mil.*) breast-rope (embarkation of horses in railway cars);
— *de pont*, (*pont.*) chord of a bridge;
— *de remorque*, tow rope;
— *de retenue*, guy rope;
— *-signal*, signal rope (as to a diver);
— *de timbre*, drum cord (under drum);
— *de traction*, (*gym.*) rope for a tug-of-war.

cordeau, m., line, tracing-line, chalk-line; measuring-tape; (*sm. a.*) pull-through; (*sm. a.*) brass wire for testing accuracy of gun barrels (obs.);
au ——, by the string, (hence) straight (cf. our familiar expression, "straight as a string");
— *d'alignement*, (*fort.*) tracing-tape;
aligner au ——, to lay out by line;
— *d'amorce*, (*artif.*) quick match;
— *Bickford*, (*artif.*) Bickford fuse;
— *dérivé*, (*artif.*) secondary or subsidiary fuse spliced on to a main fuse;
— *détonant*, (*artif.*) detonating fuse, (esp.) fuse with melinite core;
exercice au ——, (*mil.*) drill with cord held by men (in place of men), in the instruction of cadres;
— *-lacet*, lacing cord;
— *de mise de feu*, (*artif.*) slow match (e. g., Bickford fuse);
— *de piquet*, (*mil.*) tent-cord (i. e., guy-rope);

cordeau *de pointage*, (*art.*) chalk-line; aiming cord (s. b. mortars);
— *porte-feu*, (*expl., mil. min.*) instantaneous fuse, quick match;
— *principal*, (*artif.*) the fuse to which a —— *dérivé* is spliced, main fuse;
— *-roulette*, (*surv.*, etc.) sort of measuring-tape;
— *de sûreté*, safety cord (with arm loops, for use in swimming);
— *de tirage*, tent-cord proper;
— *à tracer*, (*siege, fort.*, etc.) tracing-cord or tape.

cordeler, v. a., (*cord.*) to twist, twine, into a rope.

cordelière, f., cord and tassels; latch; (*unif.*) upper part of fringe of field officer's epaulets.

cordelette, f., small cord.

cordelle, f., warp, towline;
haler à la ——, to tow.

corder, v. a., to cord, tie with a cord.

corderie, f., ropewalk.

cordite, f., (*expl.*) cordite.

cordon, m., string, twist; band; brow-band of a nave, middle nave-hoop; hoop; row; girdle; ribbon (of an order of knighthood); milling, milled edge; (*cord.*) strand of a rope (esp., hawser-laid, to form part of a cable); (*mil.*) line, cordon, of troops; (*art.*) rotating-band, centering-band; (*fort.*) coping, coping-stone (of a scarp, etc.); line of works, of forts, especially line of frontier works;
— *d'appui*, (*art.*) centering band;
— *de clairon* (*trompette*), trumpet cord;
— *de contrescarpe*, —— *d'escarpe*, (*fort.*) cordon or coping;
— *-fourragère*, (*unif.*) helmet cord (U. S. A.);
— *de pétards*, (*expl.*) necklace of high-explosive cartridges (as for felling trees);
— *de postes*, (*mil.*) chain of posts;
— *de sac*, (*siege*) rope-yarn, etc., for tying up sand bags;
— *sanitaire*, sanitary cordon, quarantine line;
— *tire-feu*, (*art.*) lanyard;
— *tire-feu à olives*, (*art.*) slide-lanyard.

cordonner, v. a., (*cord.*) to twist into a rope.

cordonnet, m., (*art.*) choke-lashing of a cartridge bag.

cordonnier, m., shoemaker.

corindon, m., corundum.

cormier, m., service tree and wood.

cornage, m., (*hipp.*) wheezing, roaring.

cornard, m., (*hipp.*) roarer.

corne, f., horn; hoof; stave of a wagon; (*in pl.*) forked end, crotch jaw, of a lifting-jack; (*unif.*) horn or prong of a cocked hat; (*fort.*) horn (of a hornwork); (*mus.*) horn; (*cord.*) splicing-fid; (*hipp.*) hoof; cornet; (*top.*) sharp mountain summit;
— *d'amorce*, (*mil.*, etc.) powderhorn, priming horn (obs.);
— *de bélier*, (*fort.*) ram's horn, sort of tenaille;
— *de cerf*, hartshorn;
chapeau à ——*s*, (*unif.*) cocked hat;
donner un coup de ——, (*hipp.*) to bleed a horse with the cornet;
maladie de ——, horn distemper;
ouvrage à ——*s*, (*fort.*) hornwork;
— *de rancher*, stud-stave;
taquet à ——, (*cord.*) belaying cleat.

cornemuse, f., (*mus.*) bagpipe.

corner, v. n., to sound, wind, a horn; (*hipp.*) to roar.

cornet, m., trumpet, small horn or trumpet;
— *de brume*, fog horn;
— *dentaire externe*, (*hipp.*) external dental cavity;
— *dentaire interne*, (*hipp.*) pulp cavity;
— *à huile*, small oil-can;
— *à piston*, (*mus.*) cornet-a-piston.

cornette, f., burgee, broad pennant; m., (*cav.*) cornet;
— *de commandement*, (*nav.*) senior officer's pennant.

corneur, m., (*hipp.*) roarer.

corniche, f., cornice.

cornichon   103   **corps**

cornichon, m., (*fond.*) curved part (bottom) of runner (in projectile casting); (*mil. slang*) candidate preparing for St. Cyr, "beast" (U. S. M. A.).
cornier, a., relating to the corner, corner; m., cornel tree and wood.
cornière, f., (*cons., etc.*) angle-iron, corner-iron;
—— *à boudin,* bulb-angle;
—— *de butée,* (*art.*) motion-limiting angle-iron (St. Chamond mountain carriage);
—— *-guide,* T-slide; (*art.*) angle-bar confining top carriage to chassis rails;
—— *de renfort,* stiffening angle-iron.
cornouiller, m., dogberry (wood and tree).
cornu, m., (*hipp.*) horse with hips higher than croup.
cornue, f., (*chem.*) retort; (*met.*) converter (Bessemer).
charbon de ——, coke.
corporal, a., corporal (as, punishment).
corps, m., body; main or principal part of anything (as of a building, of a machine); frame (of a bicycle); casing; pump-barrel; (*steam*) shell (of a boiler); (*cord.*) shell of a block; (*art.*) body of a projectile, of a gun; (*fort., etc.*) beam of a stand of *chevaux de frise;* (*pont.*) body of a ponton-boat; (*mil.*) corps; any body of officers or of officers and men, troop unit (hence, frequently) regiment (see below, *corps de troupe*); special corps, as of engineers, etc.; army corps;
—— *d'affût,* (*Fr. art.*) the trail proper (80ᵐᵐ mountain gun); (*in gen.*) the main body of a gun carriage; block-trail (R. A.);
—— *de l'arbre,* (*mach.*) shaft (exclusive of fittings, bearings, etc.);
—— *d'armée,* (*mil.*) army corps;
—— *d'armée de la marine,* (*Fr. a.*) corps (formed in 1892) of the marine artillery and infantry;
—— *de baguette,* (*sm. a.*) rod proper, as distinguished from other parts;
—— *de bataille,* (*mil.*) main body of an army, e. g., as distinguished from the advance guard; combatant body, as opposed to accompanying noncombatants; (*nav.*) main body, fighting part of a fleet;
—— *de bielle,* (*mach.*) connecting-rod proper (exclusive of crosshead, etc.);
—— *de boîte à l'huile,* (*mach.*) grease-box, upper part of axle-box;
—— *à cadre,* diamond frame (bicycle);
—— *à* ——, (*mil.*) hand-to-hand fight;
—— *de caisson,* (*art.*) body of ammunition wagon;
—— *du canon,* (*Fr. art.*) the tube and the hoops taken together; (*F. m. art.*) the part of the gun into which the tube is inserted;
chef de ——, (*mil.*) colonel (of a regiment); sometimes c. o. of a body of troops;
—— *du contrôle de l'administration de l'armée,* (*Fr. a.*) v. *contrôle;*
—— *de cuirasse,* (*cav.*) breastplate;
—— *de débarquement,* (*mil., nav.*) landing party;
debout au ——, end on;
—— *du délit,* (*mil. law.*) body of material proof establishing a *délit militaire;*
—— *de discipline,* (*mil.*) body of military convicts; disciplinary unit;
—— *droit,* straight (bicycle) frame;
—— *d'élite,* (*mil.*) body of picked troops; crack corps or regiment;
—— *d'épaulette,* (*unif.*) body, field, of an epaulet;
—— *des équipages (de la flotte),* (*nav.*) the seamen of a navy, whole body of blue-jackets;
—— *d'essieu,* axletree bed;
—— *d'état-major,* (*mil.*) staff corps;
—— *d'état-major de l'armée navale,* (*nav.*) the officers of a navy, naval officers;
—— *étranger,* (*in gen.*) foreign substance; (*mil.*) regiment or corps of foreigners in the service of France;
—— *expéditionnaire,* (*mil.*) field force, special column, expedition;

corps, *former* ——, (*Fr. a.*) to have the same administrative status as a —— *de troupe,* e. g., a military establishment, a battalion detached, etc.;
—— *fractionné,* regiment occupying more than one station or garrison;
—— *franc,* (*Fr. a.*) corps of volunteers, under the first Republic, the first Empire, and in 1870–71;
—— *de fusée,* (*art.*) fuse-case, fuse-body, or housing;
—— *de garde,* (*mil.*) guardhouse, guardroom; in the field, the post of the guard;
—— *de garde défensif,* (*fort.*) small work commanding an entrance, or road of approach;
garde du ——, (*cav.*) life guards; life guardsman;
garde du corps;
—— *gras,* grease, unguent;
—— *indigène,* (*Fr. a.*) native troops in French service;
—— *d'inspection de la marine,* (*Fr. navy*) inspectors' corps;
—— *d'investissement,* (*mil., siege*) investing body, besieging army;
se lancer à —— *perdu,* (*mil.*) to go in to win or die;
se lancer à son —— *défendant,* (*mil.*) to go in guardedly, half-heartedly;
—— *de logis,* main building, main part of a building;
—— *de manœuvre,* (*art.*) that part of a battery that maneuvers, as opposed to the wagons, etc., that do not;
—— *de mécanisme,* (*sm. a.*) in the Lebel rifle, the plate on which are mounted the feed and trigger mechanisms;
—— *mixte,* (*mil.*) in the colonies, a troop unit containing both Europeans and natives;
—— *mobile,* (*siege*) mobile troops of a defense, used particularly for the exterior defense;
—— *-mort,* shore-beam, mooring-buoy, mooring-sill, dead man, bollard, any permanent or fixed mooring or anchorage;
—— *de musique,* (*mil.*) regimental band, band;
—— *d'observation,* (*mil.*) army of observation;
—— *d'obus,* (*art.*) main body of a shell (exclusive of fittings, etc.);
—— *d'officiers,* (*mil.*) all the officers of the army, the officers of a regiment, of a garrison;
passer sur le —— *à, de,* (*mil.*) to overthrow, rout;
—— *perdu,* makeshift anchor (not intended to be picked up); mooring weight or body;
périr —— *et biens,* (*nav.*) to go down with all hands;
—— *de piston,* (*mach.*) piston body;
—— *de place,* (*fort.*) central position or fort, main work, body of the place, the *enceinte* proper (parapet and ditch) (more or less limited now to old fortifications);
—— *de platine,* (*sm. a.*) lock plate (as of a revolver);
—— *du poitrail,* (*harn.*) breastplate;
—— *de pompe,* barrel of a pump; pump cylinder; (*mil. slang*) staff of the St. Cyr school and of the cavalry school at Saumur;
—— *de ponton,* (*pont.*) body of a ponton-boat, as distinguished from the *becs,* q. v.;
—— *de poulie,* (*cord.*) shell of a block;
prendre ——, *du* ——, to set, to form or harden into a mass;
—— *de presse,* barrel, cylinder, body, of a hydraulic press;
—— *de rais,* body of a spoke (part not mortised in);
—— *de remblais,* (*r. r., etc.*) embankment;
—— *rugueux,* (*art.*) roughened wire (of a friction primer); plunger of a fuse;
—— *de santé,* (*mil.*) medical department;
—— *savant,* (*mil.*) technical corps, scientific corps;
—— *de siège,* (*mil.*) siege corps, siege army; besieging army;
—— *simple,* (*chem.*) element, elementary body;

corps, de *support,* (*pont.*) support of the superstructure of a bridge (either fixed or floating);
—— *du train des équipages,* (*mil.*) the train, train troops;
—— *du train de dessous,* (*art.*) limber-body, caisson-body;
—— *de troupe,* (*Fr. a.,* etc.) in each arm, the largest body whose command and administration are exercised by one and the same chief; (hence, generally) regiment, unless otherwise indicated; troop unit;
les —— *de troupe,* the line of the army;
—— *volant,* (*mil.*) flying body.
correcteur, m., (*art.*) corrector; compensator (on a sight.)
correction, f., correction; (*expl.*) addition of alcohol, when fumes appear, in manufacture of mercury fulminate.
correctionnel, a., (*law*) correctional; punishable (said of certain tribunals, of misdemeanors, and of the punishment of these misdemeanors).
correspondance, f., correspondence; transfer (omnibus, etc.); (*sig.*) communication; (*mil.*) assimilation of rank;
—— *de grade,* (*mil.*) assimilation of rank;
—— *optique,* (*sig.*) communication by visual telegraphy;
ouvrir la ——, (*sig.*) to open communication;
poste de ——, (*sig.*) transmitting station;
—— *de service,* (*mil.*) official correspondence.
correspondre, v. n., to correspond; (*mil.*) to report to;
se ——, to be exactly opposite one another.
corridor, m., corridor, passage-way; (*fort.,* etc.) communication;
—— *d'assèchement,* (*fort.*) drying or ventilating corridor of a powder magazine;
—— *de circulation,* (*siege*) communication along interior crest (obs., see *couloir de circulation*);
—— *de surveillance,* (*fort.*) exterior corridor.
corriger, v. a., to correct.
corroder, v. a., to corrode; (of rivers) to wash, erode.
corroi, m., claying, puddling; layer or lining of earth or of clay; currying of leather.
corrosion, f., corrosion; wash, erosion (of rivers).
corroyage, m., dressing, currying, of leather; (*met.*) refining, shearing (of steel);
—— *de n,* (*met.*) said when the section of the ingot is at least *n* times as great as that of the gun, on leaving the forge.
corroyé-masse, m., (*met.*) scrap-iron.
corroyer, v. a., to dress leather; to trim, dress down (as a felly); to plan ; (*fond.*) to roll sand; to beat up, pug, puddle with clay; (*met.*) to reheat and weld, to shear, refine, (steel) to weld together, to "shut."
corsage, m., (*hipp.*) chest of a horse.
corsaire, m., pirate; (*nav.*) privateer;
—— *autorisé,* (*nav.*) privateer.
cortège, m., escort, cortege, funeral escort.
corvéable, a., liable to contribution of forced labor.
corvée, f., forced labor, statute labor; (*mil.*) fatigue, fatigue duty, fatigue party; (*in pl.*) fatigue call;
de ——, on fatigue duty;
—— *du bois,* fuel fatigue, and party for this duty;
—— *de chicstrac,* (*mil. slang*) fatigue for cleaning cesspools, etc.;
—— *d'eau,* water fatigue, and party;
—— *extérieure,* any fatigue beyond the limits of camp, of quarters;
—— *de fourrage.* forage party or detail (and duty);
—— *d'hommes,* fatigue detail;
—— *d'installation,* labors, duties, etc., of installation in one's cantonments;
—— *intérieure,* post or garrison fatigue;
—— *de l'ordinaire,* kitchen police, (call, party, and duty;)
—— *de propreté,* police;
—— *de quartier,* police call;
service de ——, fatigue duty;

corvée *supplémentaire,* extra fatigue, by way of punishment.
corvette, f., (*nav.*) sloop of war, corvette;
—— *cuirassée,* armored, protected cruiser
cosaque, m., Cossack;
poste à la ——, (*mil.*) Cossack post, irregular post.
cosaquerie, f., (*mil.*) raid, sudden raid.
cosse, f., (*cord.*) thimble.
costal, a., of the ribs, costal.
cote, f., (*top.*) altitude of a point (in meters and fractions of); signal, level; (*ball.*) the coordinates of a point of impact; reference lines on a target; (*fort.*) reference; (*in gen.*) dimensions marked on a tracing, drawing, map, etc.;
—— *de niveau,* (*top.*) reference to a datum plane, level.
côte, f., rib, edge, fillet; coast, seacoast, shore; (*top.*) rise of a hill;
à ——, edge to edge;
artillerie de ——, (*art.*) seacoast artillery;
—— *asternale,* (*hipp.*) false rib;
—— *basse,* (*hipp.*) rib that comes down below the level of the elbow;
batterie de ——, (*art.*) seacoast battery;
défense des ——*s,* (*art.*) seacoast, coast, defense;
—— *descendue,* v. —— *basse;*
matériel de ——, (*art.*) seacoast material;
à mi- ——, halfway up the hill;
—— *sternale,* (*hipp.*) true rib;
——*s de vache,* (*met.*) slit iron (iron in small rectangular rods);
—— *au vent,* weather shore;
—— *sous le vent,* lee shore.
côté, m., side; party; (*nav.*) side of a ship; (*art.*) side rail of chassis (seacoast guns);
bas ——, footpath, or other way, at a lower level than the main road; side aisle; lean-to, penthouse;
—— *de la chair,* flesh side (of leather);
—— *de champ,* edge (as of a plank, etc.);
—— *du chassis,* (*art.*) side rail of a chassis;
de ——, sidewise;
—— *droit,* off side (of a team, etc.);
—— *de droite,* (*harn.*) off-side neck strap (of breast strap);
—— *extérieur,* (*fort.*) exterior side (of a bastion);
—— (*pont.*) side farthest from bank from which the bridge is built or from bank toward which the bridge is dismantled;
—— *gauche,* near side (of a team, etc.);
—— *de gauche,* (*harn.*) near-side neck strap (of a breast strap);
grand ——, (*art.*) chassis side rail; (*harn.*) longer strap of a bridle;
—— *intérieur,* (*pont.*) side nearest bank from which the bridge is built or toward which the bridge is dismantled;
—— *à la main,* near side (of a team, etc.);
—— *de la moelle,* end grain of wood;
—— *montoir, du montoir,* (*man.*) near side of a horse;
—— *hors montoir, hors du montoir,* (*man.*) off side of a horse;
pas de ——, (*man.,* etc.) side step;
petit ——, (*harn.*) shorter strap of a bridle, (v. *grand* ——);
—— *de sous-verge,* off side of a team;
—— *d'un triangle,* side of a triangle.
coteau, m., (*top.*) hill, slope; hill taken in its length; hillock.
coter, v. a., to put down references, measures, dimensions, etc., on a map, plan, drawing, etc.
côtier, a., coast, coasting; m., coaster, coasting vessel.
cotiser, v. a., to assess, to rate.
coton, m., cotton;
—— *azotique,* (*expl.*) gun-cotton;
—— *fulminant,* (*expl.*) gun-cotton;
—— *au nitrate,* (*expl.*) nitrated gun-cotton;
—— *nitraté (de Faversham),* (*expl.*) tonite;
—— *pyrique,* (*expl.*) pyrocotton.
coton-poudre, m., (*expl.*) gun-cotton;
—— *comprimé,* compressed gun-cotton;
—— *à fusil,* small-arm gun-cotton;
—— *humide,* wet gun-cotton;
—— *sec,* dry gun-cotton.

côtoyer, v. a., to coast; to go alongside of.
côtre, m., (*nav.*) cutter;
— *de la douane*, revenue cutter.
cotret, m., small fagot; (*siege*) fascine.
cotter, v. a., to catch, lock, as a wheel (Switzerland).
cou, m., neck;
— *de cygne*, crane neck, goose neck.
couchage, m., (*mil.*) bedding, all that relates to bedding;
— *auxiliaire*, (*Fr. a.*) bed sacks, etc., used by troops when contractor's supplies are lacking or can not be used;
paille de —, bed straw;
sac de —, bed sack;
service de —, (*Fr. a.*) the supply and maintenance, by contract, of bedding for officers and men in barracks, of watch coats for sentinels, and, if required, of washing for the linen of the men.
couchant, m., the west.
couche, f., layer (as of mortar, color, concrete, etc.); seam (of coal, etc.); coat, coating, stratum, bed; ground-sill; annual ring of a tree; (*art.*) tier (of bullets in a case shot); (*sm. a.*) butt end of stock, part of stock between breech and heel; (*fond.*) molding board;
angle de —, (*sm. a.*) slope of the stock;
— *de fibres invariables*, neutral layer;
— *de fulminate*, (*artif.*) priming;
— *d'impression*, priming coat of paint;
longueur de (la) —, (*sm. a.*) distance from shoulder to base of the breech;
— *neutre*, neutral plane, layer (of stressed beams);
pente de —, (*sm. a.*) v. *angle de* —;
plaque de —, (*sm. a.*) butt-plate;
— *de roue*, dish of a wheel.
couché, a., lying down;
position —e, (*mil.*) prone position.
couche-point, m., welt (on the heel of a shoe or boot).
coucher, v. a., to slope, lay down; v. r., to lie down (of troops); to set (of the sun);
— *en joue*, (*sm. a.*) to aim at;
— *la lance*, (*cav.*) to slope a lance;
— *par terre*, (*mil.*) to overthrow completely.
coucher, m., setting of the sun; furniture of a bed; state of being in a recumbent position;
— *abandonné*, (*hipp.*) position of a horse lying on his side.
couchette, f., bedstead.
couchis, m., lagging; bed of sand (under the pavement of a bridge);
— *-revêtement*, (*steam*) lagging of a boiler, cleaning.
coude, m., elbow; any bend or turn in parts of machinery, etc.; bend; knee, angle; elbow (of a pipe, tube, etc.); (*top.*) bend or loop of a river; (*hipp.*) elbow; (*mach.*) crank, cranked portion of a shaft; (*sm. a.*) bend, neck of a bayonet; (*min.*) turn of a gallery;
— *à* —, (*mil.*) alignment in which each soldier must feel the touch of elbow;
— *de gonflement*, inflating tube (of a balloon);
tact des —s, (*mil.*) touch of elbow;
— *universel*, (*mach.*) Brown's joint.
coudé, a., forming a knee or angle; bent, kneed; (*mach.*) cranked;
axe —, (*mach.*) crank.
coudée, f., arm's length.
cou-de-pied, m., instep.
couder, v. a., to bend, to knee; (*mach.*) to crank.
coudran, m., wheel-grease.
couillon, m., — *d'ancre*, anchor-nut.
coulage, m., (*fond.*) casting, cast, act of casting;
— *à cale*, v. — *à siphon;*
— *en châssis*, casting in flasks or boxes;
— *à chaud*, casting at a higher temperature than is required by the nature of the metal;
— *en*, *sur*, *coquille*, chill casting;
— *à*, — *en*, *creux*, casting hollow;
— *à découvert*, casting in open molds;

coulage *à la descente*, casting from the top;
— *à froid*, casting at a lower temperature than is required by the nature of the metal;
— — *massif*, casting solid;
— *d'un morceau*, casting in one piece;
— *à noyau*, casting hollow, casting on a core, with a spindle;
— *plein*, v. — *massif;*
— *à la remonte*, v. — *en siphon;*
— *en sable*, casting in sand molds;
— *en sable gras*, dry sand casting;
— *en sable vert*, green sand casting;
— *à*, *en*, *siphon*, casting from bottom of mold;
— *en source*, casting from the bottom;
— *en terre*, loam casting;
— *à une seule portée*, casting in one piece, all at once.
coulant, a., movable, sliding; crumbling, caving (of earth, soil, etc.); (*mil. slang*) easy, good-natured; m., slide, runner; (*unif.*) slide, keeper (as on a sword belt, etc.); (*inst.*) slide ring, bearing ring; slide of a telescope;
— *porte-fils*, (*inst.*) cross-hair slide, reticule holder.
coulé, m., (*fenc.*) straight lunge feint.
coulée, f., (*top.*) coulée, draw (U. S.); (*fond.*) spray; tapping; runner, sullage-piece; taphole; cast, casting, quantity cast; channel from furnace to mold;
canal de —, gate, ingate, (more accurately) runner;
— *à chaud (froid)*, v. s. v. *coulage;*
— *à la descente*, tapping into the top of a casting pocket or ingot mold;
— *directe*, tapping directly into ingot mold;
faire la — *à*, to tap;
jet de —, feeding head;
pièce de —, casting;
— *en poche*, tapping into casting ladles;
— *à la remonte*, tapping into the bottom of a casting ladle or ingot mold;
— *en siphon*, v. s. v. *coulage;*
— *en source*, tapping into the bottom of an ingot mold or casting pocket;
trou de —, gate, ingate; tap-hole;
venir de —, to be produced or made in the casting; to be cast in one piece with.
couler, v. a. n., to flow, to leak; (*fond.*) to cast; to tap, tap a furnace; (*nav.*) to sink, to run down, to founder; (*man.*) to pass gradually from one gait into another; to give a horse his head in galloping; (*cord.*) of a knot or lashing, to slip;
— *bas*, (*nav.*) to founder, sink; to sink a vessel by artillery fire; to run down;
— *chaud (froid)*, (*fond.*) to cast at a temperature higher (lower) than is required by the nature of the metal;
— *direct*, (*fond.*) to tap or run directly into an ingot mold;
— *à fond*, (*nav.*) to sink, run down, a vessel; to founder;
— *au galop*, (*man.*) to gallop smooth, to have a smooth and flowing gallop;
— *en poche*, (*fond.*) to tap into a casting pocket.
(For other foundry terms, adjuncts of *couler*, v. s. vv. *coulage*, *coulée*.)
couleur, f., color; (*in pl.*, *nav.*) colors of a vessel;
— *lavée*, faint color;
— *s nationales*, (*Fr.*) the tricolor;
— *de recuit*, (*met.*) annealing, tempering, color.
coulis, m., (*fond.*) molten metal.
coulisse, f., (*mach.*, etc.) slot, groove, channel, guide; slotted link, slot-link; Stephenson link motion; guideway, guide slot; valve-gear; slide, shutter; (*pont.*) trestle-cap mortise;
— *à détente variable*, expansion link;
— *-guide*, guide, guide slot;
— *de herse*, (*pont.*) portcullis slide;
monter à —, (*mach.*) to give a sliding motion to a piece.

coulisseau, m., (art.) way-plank; (mach., etc.) slide-block; slipper-block, motion-block, slipper-slide block; guide-block; (sm. a.) handle-guide;
—— *du té*, (steam) crosshead guide.
coulissement, m., (mach.) sliding motion.
coulomb, m., (elec.) coulomb.
couloir, m., (fort., etc.) passage; (art.) feed-guide, hopper, feed-trough;
—— *d'assainissement*, (fort.) ventilation passage or conduit of a powder magazine;
—— *de chargement*, (art.) feed-trough, feed-hopper;
—— *de circulation*, (siege) communication along rear of platforms;
—— *de surveillance*, (fort.) exterior corridor (Carnot trace).
coup, m., blow, stroke, shock; (art., sm. a.) shot, round; (surv.) sight through an instrument; (mach., etc.) stroke; (fenc.) cut, thrust, stroke.
    I. Artillery, small arms, etc.; II. Fencing and related terms; III. Miscellaneous.
    I. Artillery, small arms, etc.
accuser son ——, (t. p.) to "call" one's shot;
—— *amorti*, (art., sm. a.) spent shot;
—— *d'arme à feu*, shot;
—— *d'arrosoir*, (art.) the sheaf or spray of bullets, etc., from a shrapnel; (inf.) the sheaf of bullets produced by a volley;
—— *d'assurance*, (nav.) gun fired by a ship under her true colors;
—— *d'audace*, bold stroke;
—— *de baguette*, (mil.) drum tap;
—— *à balles*, (sm. a.) round or shot fired with ball cartridge;
—— *bas*, (art.) a shrapnel bursting at a height lower than the type height;
—— *très bas*, (art.) shrapnel that bursts below the target;
—— *à blanc*, shot fired with blank cartridge;
—— *de blanc*, (t. p.) bull's-eye;
—— *bleu*, (art., sm. a.) miss;
—— *de bricole*, (art.) shot intended to rebound laterally and strike after the graze (obs.);
—— *au but*, (art., sm. a.) hit;
—— *de canon*, (art.) cannon shot (wound, report, round);
—— *de canon de détresse*, distress gun;
—— *de canon de diane*, (mil.) morning gun, reveille gun;
—— *de canon à l'eau*, (art.) water-line hit, hit between wind and water;
—— *de canon à fleur d'eau*, (art.) v. —— *de canon à l'eau*.
—— *de canon pour indiquer le midi moyen*, time gun, noon gun;
—— *de canon dans l'œuvre vive*, (art.) shot below water line of a ship;
—— *de canon en plein bois*, (art.) shot into the broadside;
—— *de canon à poudre*, (art.) blank round;
—— *de canon à projectile*, (art.) shot or round with projectile;
—— *de canon de réjouissance*, salute gun;
—— *de canon de retraite*, (mil.) evening gun;
—— *de canon de signal*, (mil.) signal gun;
—— *s de canon à une minute d'intervalle*, minute guns;
—— *en chasse*, (nav.) bow shot;
—— *pour* ——, (art.) shot for shot, of a salute;
—— *court*, (art.) short, shot that falls short, the "short" in ranging;
—— *de crosse*, (sm. a.) kick;
—— *de diane*, (mil.) morning gun, reveille gun;
—— *en direction*, (sm. a., art.) line shot;
—— *de doigt*, (sm. a.) involuntary pull of the trigger, by a sudden jerk of the finger; jerk of finger on trigger;
—— *d'écharpe*, (art.) shot fired in direction making a small angle, as with interior crest, etc.;
—— *d'embrasure*, (art.) embrasure hit;
—— *d'enfilade*, (art.) enfilading shot;
—— *d'enfoncement*, (art.) shot to set spade, in guns fitted with trail or axle spades;

coup *d'épaule*, (sm. a.) jerk or drawing back of shoulder through fear of recoil;
—— *d'éperon*, (nav.) blow with a ship's ram, ram, ramming blow;
—— *d'épreuve*, (art., etc.) proof shot;
—— *d'essai*, (art., etc.) proof shot;
faire un beau ——, to make a fine shot;
faire le —— *de feu*, (mil.) to exchange shots; to fight as a private (said of an officer);
faire le —— *de fusil*, (mil.) v. *faire le* —— *de feu*;
faire le —— *de pistolet*, (cav.) to exchange shots;
faire le —— *de sabre*, to exchange blows; to fight with the saber;
faire partir un ——, to fire a shot;
—— *fait*, (art.) round of fixed ammunition (obs.);
sans —— *férir*, (mil.) without firing a shot, without striking a blow;
—— *de feu*, (art., sm. a.) shot; shot wound; report;
—— *fichant*, (art.) plunging shot (angle of fall greater than 4°);
—— *de flambage*, (art.) scaling or warming shot;
—— *de fond*, shot that tells;
—— *fusant*, (art.) time-fuse round or shot;
—— *de fusil*, (sm. a.) rifle shot, bullet wound, report of a rifle;
—— *de fusil de pavillon*, shot at colors;
—— *haut*, (art.) shrapnel, etc., bursting at a height greater than the type height;
—— *très haut*, shrapnel, etc., bursting at a height more than twice the type height;
—— *à hauteur*, (art.) shrapnel, etc., that bursts at the type height;
—— *heureux*, (art., etc.) lucky shot or hit;
—— *isolé*, (art., etc.) single shot;
—— *de langue*, (mil. mus.) short note or blast on a bugle or trumpet, "toot;"
—— *long*, (art., sm. a.) hang fire; "over;" in ranging, the over or long shot;
—— *de main*, (mil.) sudden attack, surprise, bold stroke;
—— *manqué*, (art., sm. a.) miss;
—— *à mitraille*, (art.) generic term for shrapnel, canister, shot;
—— *d'œil militaire*, (mil.) the ability to take in the military situation at a glance;
—— *de partance*, (nav.) sailing gun;
—— *de pavillon*, (nav.) morning (evening) gun;
—— *perdu*, (art., etc.) random shot, unaimed shot;
—— *de pistolet*, (sm. a.) pistol-shot, -wound, -report;
—— *de plein fouet*, (art.) round fired with full service charge in direct fire; (t. p.) direct hit (not a richochet);
—— *plongeant*, (art.) plunging shot;
—— *pointé*, (art., sm. a.) aimed shot;
—— *rasant*, (art.) shot whose direction is parallel, or nearly so, to ground, at point considered;
—— *raté*, (art., sm. a.) miss-fire;
—— *de réglage*, (art., sm. a.) sighting, ranging, shot;
—— *de réjouissance*, (art.) salute gun or round;
—— *de retraite*, (mil.) evening gun, retreat gun;
—— *réussi*, (art.) hit;
—— *de revers*, (art.) shot or hit in reverse;
—— *de revolver*, (sm. a.) revolver shot, report, etc.;
—— *de semonce*, (nav.) bringing-to shot;
—— *de sifflet*, whistle, signal or call made with a whistle;
—— *de sonde*, (nav.) cast with the lead;
—— *tiré en blanc*, (art.) blank round;
tirer à —— *perdu*, to fire at random;
tirer un ——, to fire a shot;
—— *touché*, (art., sm. a.) hit;
—— *à toute volée*, (art.) random shot, shot fired so as to give the greatest range possible;
—— *de vigueur*, (mil.) decisive blow, stroke, action.

II. Fencing and related terms:

coup, *allonger un* ——, to make a pass or thrust;
—— *d'arrêt,* stopping thrust;
asséner un ——, to deliver a violent blow (fist, club, stick, etc.);
—— *en avant,* front thrust (lance);
——*s en avant,* front cuts and thrusts;
—— *de baïonnette,* bayonet thrust;
—— *de banderole,* diagonal cut (broadsword);
—— *de bout de pointe,* point thrust (singlestick);
—— *de bout de talon,* butt thrust (singlestick);
—— *composé,* thrust or lunge resulting from several preceding motions;
——*s composés,* combination of blows (singlestick);
—— *de côté,* side thrust (lance);
——*s de côté,* side cuts and thrusts (saber);
donner le —— *à l'arme,* in a cutting weapon, so to distribute the weight as to make an effective blow possible, (hence) to balance a cutting weapon;
—— *droit,* straight lunge;
—— *d'épée,* sword thrust;
—— *d'estremaçon,* cutting blow; side or back stroke (saber);
—— *de figure,* blow at the face (singlestick);
—— *de figure à droite (gauche),* cut at right (left) cheek (broadsword);
—— *de flanc,* body blow (singlestick); body or flank cut (broadsword);
—— *fourré,* return cut or thrust (said when each combatant is pierced, etc.); simultaneous attack by the adversaries (singlestick);
—— *de lance,* lance thrust;
—— *lancé,* in bayonet exercise, the release of the piece from the left while still holding it in the right hand;
—— *de manchette,* clever blow or stroke on adversary's wrist;
parer un ——, to ward off, parry, a blow, thrust;
—— *de pied,* kick (French boxing);
—— *de pied bas,* a low kick to the front (French boxing);
—— *de pied brisé (en arrière, en avant, de flanc, de figure),* sort of cow kick (to the front, etc.) (French boxing);
—— *de pied brisé tournant en avant (arrière),* cow kick (front, rear), preceded by a face about (French boxing);
—— *de plat de sabre,* blow with flat of sword;
—— *de poing,* blow with the fist;
—— *de poing droit,* blow from the shoulder (boxing);
—— *de poing de masse,* swinging downward blow or cut, sort of moulinet with the fist (French boxing);
—— *de poing de revers,* blow to the rear (French boxing);
—— *de pointe,* thrust, point; point cut (saber);
porter un ——, to give a blow, stab, cut, etc.;
——*s réunis,* combined movements (saber exercise);
—— *de revers,* back stroke, cut, or thrust;
—— *de sabre,* saber cut, sword cut;
—— *simple,* simple thrust or lunge (proceeds from one preceding motion);
—— *de taille,* cut;
—— *de temps,* time thrust or blow;
—— *de tête,* head cut (saber); blow at head (singlestick);
—— *de ventre,* cut at the belly (saber).

III. Miscellaneous:

amortir un ——, to weaken, deaden, a blow;
—— *arrière, (surv.)* back sight;
—— *d'arrière, (steam, etc.)* back stroke (of a piston, etc.);
—— *d'audace,* bold stroke;
—— *d'avance, (steam, etc.)* forward stroke (of a piston, etc.);
—— *avant, (surv.)* fore sight;
—— *d'avant, (steam, etc.)* forward stroke (of a piston, etc.);
—— *d'aviron,* stroke, pull, with an oar;
—— *de balai, avoir* —— *de balai, (hipp.)* to be hipshot;

coup *de bélier, (steam, etc.)* water-hammering;
—— *de boutoir, (farr.)* wound due to paring the hoof too deep;
—— *de chaleur, (hipp.)* heat stroke;
—— *de collier,* great effort, renewed effort;
—— *d'eau,* (in pumps, etc.) impact of water (on piston face, etc.);
—— *d'épée dans l'eau,* vain effort, wasted effort;
—— *de feu,* overheating due to overeating (carrier pigeons, etc.); (in the arts) heating, (sometimes) burning, overburning, overheating;
—— *de fond, (fig.)* home thrust, blow that tells;
—— *de foret,* tool mark, boring mark;
—— *de foudre,* stroke of lightning;
—— *de fouet,* blow or crack with a whip; *(hipp.)* "lift," or second expiration of a broken-winded horse; *(elec.)* time during which a current on first making is more powerful than it is normally; *(expl.)* "whipping out" of a Bickford or other fuse when improperly set, on the flame reaching the end; dislodgment of a second charge by the explosion of the first (before the second goes off);
—— *de foulard, (hipp.)* insertion of a small sponge, etc., in nostrils that discharge (horse dealer's trick);
grand ——, master stroke, *(fig.)* home thrust;
—— *de hache, (hipp.)* marked depression between withers and neck;
—— *de Jarnac,* stab in the dark; mortal wound; underhand blow;
—— *de lance, (hipp.)* depression at junction of neck and shoulders, at base of neck (seen in Turkish and in Spanish horses);
—— *de mer,* heavy sea, wave;
—— *de mine, (expl.)* blast;
—— *de mine crachant, (expl.)* blast in which a part of the charge is thrown out burning;
—— *mortel,* death blow;
—— *de mouchoir,* v. —— *de foulard;*
—— *de niveau, (surv.)* leveling sight;
—— *d'outil,* tool mark;
parer un ——, *(fig.)* to ward off, parry, a blow;
—— *de pied,* kick;
—— *de pied en vache, (hipp.)* cow kick;
—— *de piston, (mach.)* piston stroke;
—— *de piston double, (mach.)* complete piston stroke;
—— *de piston simple, (mach.)* single piston stroke;
—— *de poing,* blow with the fist; *(expl.)* firing-key, exploder;
—— *qui porte,* blow that tells, telling blow;
porter ——, to hit home;
porter le dernier ——, to deal, give, the finishing blow;
sans porter un ——, without a blow;
prendre ——, to settle, get out of plumb (of walls);
—— *de rame,* stroke, pull (rowing);
—— *de rein, (man.)* sort of swing, throw, of the haunches;
—— *de revers,* misfortune;
—— *de roulis,* lurch, roll (of a ship);
—— *de soleil,* sunstroke;
sous le —— *de,* exposed to, threatened by or with;
—— *de surface, (min., expl.)* surface blast;
—— *de tangage,* pitching (of a ship);
—— *de tonnerre,* clap of thunder;
en venir aux ——*s,* to come to blows;
—— *de vent,* squall, gust of wind.

coupable, a., guilty; m., guilty person, culprit;
se déclarer ——, to plead guilty;
se reconnaître ——, to plead guilty;
se rendre ——, to be guilty of.

coupant, a., cutting, sharp; m., edge, cutting edge.

coupe, f., cut, cutting, felling (of trees); chopping; end grain of wood; cup, basin; *(drawing)* section and elevation;
fausse ——, splice;
—— *horizontale,* horizontal section, plan;

coupe *en long*, —— *longitudinale*, longitudinal section;
—— *transversale*, cross-section; end grain (of a tree);
—— *en travers*, cross-section, transverse section;
—— *verticale*, vertical section.
coupé, a., (*top.*) intersected, accidented;
pan ——, (*fort.*) pan coupé; (*in gen.*) any corner cut off (of a room, building, etc.);
terrain ——, (*top.*) ground full of natural obstacles.
coupé, m., (*fenc.*) disengagement over the adversary's point.
coupe-cercle, m., round punch.
coupe-choux, m., (*mil. slang*) infantry sword, "frog-sticker" (U. S. A.).
coupe-circuit, m., (*elec.*) circuit-breaker, cut-out, safety cut-out.
coupée, f., (*nav.*) gangway of a ship.
coupe-filet, m., (*nav.*) torpedo-net cutter.
coupe-gazon, m., sod-cutter.
coupe-gorge, m., (*mil.*) tight place, from which escape is impossible, death-trap.
coupellation, f., (*met.*) cupellation.
coupelle, f., (*met.*) cupel; (*art.*) gas-check cup, reinforcing cup (of a cartridge case); (*sm. a.*) small copper shovel for gunpowder;
essai à la ——, (*met.*) cupel assay;
—— *à languette*, (*art.*) brass cup and wire of French naval electric primer;
mettre à la ——, (*met.*) to submit to cupellation;
—— *obturatrice*, (*art.*) annular obturator (of French double-action fuse);
rondelle à ——, (*art.*) cup-shaped washer on axle-spindle.
coupeller, v. a., (*met.*) to test or assay by cupellation.
coupe-net, m., cutting pliers or nippers.
coupe-paille, m., straw-chopper, chaff-cutter.
coupe-queue, m., (*hipp.*) docking-knife.
couper, v. a., to cut; cross, intersect; to amputate; to chop, strike; to mix, dilute; (*hipp.*) to geld, cut, castrate; (*mil.*) to cut off (a body of troops, a retreat, etc.), to break through (a line); (*cord.*) to wear, chafe, be chafed; (*steam*) to cut off, shut off, steam;
—— *un cheval*, (*hipp.*) to geld a horse;
—— *un correspondant*, (*sig.*) to break off, interrupt, a correspondent;
—— *une jument*, (*hipp.*) to spay a mare;
—— *de longueur*, to cut to proper length;
ne pas y ——, (*mil. slang*) to be in for extra drill, fatigue, or other punishment;
se ——, (*hipp.*) to cut or strike a leg in bearing by one in motion, to interfere;
—— *en travers*, to cut across;
—— *la vapeur*, (*steam.*) to cut off steam.
couperet, m., chopper.
couperose, f., copperas.
coupe-vent, m., wind guard, windbreak (in bicycling).
couple, m., frame, timber; (mechanical) couple; (*elec.*) couple, element, cell;
—— *de l'arrière*, after-body;
—— *de l'avant*, fore-body;
—— *de cavalerie*, (*expl.*) disposition of charges on opposite sides of a rail, in demolition, so as to produce rotation;
—— *à eau*, (*elec.*) cell of a water-battery;
en ——, (*expl.*) on opposite sides, e. g., of a rail, so as to set up a rotation;
—— *de renversement*, (*art.*) overturning couple;
—— *de soulèvement*, (*art.*) raising or lifting couple.
couplet, m., hinge-joint; (*harn.*) backband; (*sm. a.*) gun whose barrel is formed of two parts screwing together.
couplière, f., (*harn.*) hame-strap.
coupoir, m., cutter, cutting-press, punching-cutter.
coupole, f., cupola; (*fort.*) cupola;
—— *-caponnière*, cupola recommended by Brialmont as the equivalent of a caponier;
—— *à éclipse*, disappearing cupola;
—— *oscillante*, oscillating cupola;

coupole *à pivot*, pivot-mounted cupola;
—— *à tir direct* (*indirect*), direct- (indirect-) fire cupola;
—— *tournante*, revolving cupola;
—— *transportable*, (*art.*) traveling shield mounting, traveling cupola.
coupon, m., coupon; remnant.
coupure, f., (*pont.*) cut or opening in a bridge for passage of floating bodies; (*fort.*) palisade, ditch, etc., behind a breach, across a road, obstacle across a road, etc.; cut, coupure; obstacle; coupure or passage in the glacis; break or crochet in a long face; (*top.*) cutting in a wood;
—— *en retirade*, (*fort.*) intrenchment in a bastion when the exterior is destroyed.
cour, f., court, courtyard; (*law*) court;
—— *de cassation*, (*Fr.*) supreme court, court of appeals;
—— *des comptes*, (*law*) auditing office (public funds and execution of financial laws).
courage, m., courage.
courant, m., current (of water, air, electricity, etc.); stream; course; routine; (*cord.*) running part of a tackle; (*hipp.*) set or lay of a horse's hair;
—— *d'air*, draught;
—— *alterné*, —— *alternatif*, (*elec.*) alternating current;
—— *de charge*, (*elec.*) charging current of an accumulator;
—— *dérivé*, (*elec.*) shunt current;
—— *direct*, (*elec.*) direct current;
—— *d'entrée*, (*elec.*) flood stream;
extra- —— *de fermeture* (*de rupture*), (*elec.*) making- (breaking-) contact current;
fil de ——, thread, axis, of a stream;
—— *de flamme*, flue;
—— *de flot*, flood stream;
fort du ——, main stream;
—— *de Foucault*, (*elec.*) eddy current;
—— *inducteur*, (*elec.*) inducing current;
—— *par induction*, —— *induit*, (*elec.*) induced current;
—— *inverse*, (*elec.*) inverse current;
—— *de jusant*, ebb stream;
lit du ——, main stream, bed of the stream;
—— *de marée*, tidal current;
—— *de polarisation*, (*elec.*) polarization current;
—— *primaire*, (*elec.*) charging current of an accumulator;
—— *redressé*, (*elec.*) unidirectional current;
—— *secondaire*, (*elec.*) secondary current, polarization current;
—— *de sortie*, ebb current;
sous- ——, undertow;
—— *sous-marin*, undertow;
—— *terrestre*, (*elec.*) ground-, earth-current.
courant, a., running, flowing; instant; current; (*cord.*) running.
courantin, m., (*artif.*) line-rocket.
courbage, m., bending, curving.
courbant, a., bent, curved; m., compass timber;
bois ——, compass timber.
courbaton, m., knee in a boat.
courbatu, a., (*hipp.*) foundered, chest foundered.
courbature, f., (*hipp.*) founder, foundering, lameness; (*in gen.*) lameness, as from bicycle riding;
—— *générale*, —— *vieille*, (*hipp.*) chest foundering.
courbaturer, v. a., (*hipp.*) to bring on lameness, to founder.
courbe, f., curve; arc; railway curve; (*top.*) contour; (*pont.*) knee, rib, of a boat; (*mach.*) knee, rib; (*harn.*) fork of a saddle; (*hipp.*) curb (not to be confounded with the "curb" of English writers, the French of which is *jarde*);
—— *ballistique*, (*ball.*) ballistic curve;
—— *en bois*, wooden knee;
—— *du chien*, curve of pursuit;
—— *de chute*, (*Fr. art.*) in the *planchette de tir*, the locus of all the points of fall for given angles of elevation;
—— *de la cloche*, probability curve;
—— *en cœur*, (*mach.*) heart wheel;

courbe *de défilement*, (*ball.*) in small-arm fire, curve used in obtaining elevations for plunging fire;
—— *des densités*, (*ball.*) curve showing grouping of shots about the center of impact, (hence) probability curve;
—— *d'égale teinte*, (*top. drawing*) curve of equal shade;
—— *d'épaulement*, (*ball.*) v. —— *de défilement*;
—— *d'évacuation*, (*steam*) initial exhaust line of an indicator diagram;
—— *d'évitement*, (*r. r.*) side track;
—— *en fer*, iron knee;
—— *grosse*, (*top. drawing*) contour drawn in with heavy stroke, in heavy line;
—— *horizontale*, (*top.*) contour;
—— *hypsométrique*, (*top.*) contour;
—— *de l'indicateur*, (*steam*) indicator diagram, curve;
—— *de justesse*, (*ball.*) curve comparing accuracy of different guns;
—— *de liaison*, strengthening knee;
—— *à longue inflexion*, (*mach.*) Watt's link motion;
montant de ——, (*pont.*, *etc.*) upper part of a knee;
—— *de niveau*, —— *de nivellement*, (*top.*) contour;
—— *des pressions*, (*ball.*) curve of pressures; (*cons.*) thrust line (of an arch);
—— *de sections horizontales*, (*top.*) contour;
——*s de tir*, (*ball.*) graphical table of fire, range table;
—— *de Vaucanson*, (*mach.*) heart-wheel.
courbe, a., bent, curved;
devenir —— *par le tir*, (*art.*) to droop at the muzzle;
tir ——, (*art.*) curved fire.
courbé, a., bent, curved, crooked; secured by knees.
courbement, m., bending, incurvation.
courber, v. a. n., to curve, to bend; to shape, bend to shape; to sag, bow, bend.
courbet, m., (*harn.*) bow of a saddle; pack-saddle tree.
courbette, f., (*man.*) curvet, curvetting;
faire des ——, to curvet.
courbetter, v. a., (*man.*) to curvet.
courbure, f., curvature; curve; bend;
—— *de la crosse*, (*sm. a.*) bend or angle of the stock, inclination of the butt;
—— *double*, double curvature, S - shaped curve;
—— *du sol*, curvature of the surface of the ground;
—— *de la terre*, curvature of the earth.
coureur, m., runner; (*mil.*) mounted scout (obs.);
—— *de fond*, long-distance runner;
—— *de vitesse*, sprinter.
courir, v. a. n., to run; to flow; go over; (*mil.*) to overrun (a country), to raid;
—— *aux armes*, to fly to arms;
—— *la bague*, to ride at the rings;
—— *la bouline*, to run the gauntlet;
—— *à bride abattue*, to ride at full speed;
—— *un cheval*, (*man.*) to run, gallop, race, a horse;
—— *une manœuvre*, (*cord.*) to pass along a rope;
—— *sus*, (*mil.*) to make incursions, raids.
couronne, f., crown; hoop; (*fort.*) crown-work; (*hipp.*) coronet; (*mach.*, *etc.*) rim of a wheel, driving face of a pulley, rim; (*art.*) shoulder of a projectile (rare); (*cord.*) strap;
—— *d'appui*, (*fort.*) supporting crown or ring (of a turret); (*art.*) bearing-ring (of the Krupp fermeture);
—— *d'asservissement*, (*mach.*) controlling or communicating bevel-gear wheel;
—— *-avant*, (*art.*) front centering band;
—— *de balles*, (*art.*) layer of bullets in shrapnel;
—— *barbotin*, (*mach.*) sprocket wheel;
—— *de boulets*, (*mach.*, *etc.*) circle of ball-bearings;
—— *de brêlage*, (*pont.*) balk lashing knot;

couronne *du cabestan*, sprocket wheel of the capstan;
—— *circulaire dentée*, (*art.*) traversing rack;
—— *en cuivre*, (*art.*) rifling band, centering band;
—— *dentée*, (*art.*) traversing rack;
double ——, (*fort.*) crown work of three fronts; double crown work;
—— *d'eau*, (*mach.*) (lit., water crown) a device in turret gear permitting the continuous supply and evacuation of water under pressure;
—— *à empreintes*, (*mach.*, *etc.*) sprocket wheel;
—— *de l'étoupage*, (*steam*) packing ring;
—— *foudroyant*, (*artif.*) pitch ring;
—— *de frottement*, (*art.*, *mach.*) friction ring;
—— *de galets*, (*art.*, *etc.*) live-roller ring;
—— *de guindage*, (*pont.*) rack lashing knot;
ouvrage à ——, (*Fort.*) crown work;
—— *de pieu*, pile-hoop;
—— *de piston*, (*mach.*) junk ring, piston cover, piston ring;
—— *de presse-étoupe*, (*mach.*) stuffing-box gland;
—— *à rochet*, (*mach.*) crown wheel;
—— *de roue*, rim of a wheel;
—— *de roulement*, (*art.*) roller path;
—— *de sphères*, (*fort.*) spherical-roller ring;
—— *de tour*, (*fort.*) casting around turret spindle.
couronné, a., (*hipp.*) broken-kneed;
ouvrage ——, (*fort.*) crown work.
couronnement, m., cap, coping; top or summit (of a road, dike, etc.); top piece; (*siege*) crowning;
—— *du chemin couvert*, (*siege*) crowning of the covered way;
—— *d'un entonnoir de mine*, (*siege*) crowning of a mine crater;
—— *de la gabionnade*, (*siege*), laying, act of laying, fascines on top of the gabion work, of the gabion revetment;
—— *pied à pied*, (*siege*) crowning of the covered way by gradual approach;
—— *de vive force*, (*siege*) crowning of the covered way by open assault;
—— *de voûte*, (*mas.*) crown.
couronner, v. a. r., to cap, crown; (*mil.*) to occupy heights, a position, after driving out the enemy; (*siege*) to crown, to make a lodgment in; (of trees) to begin to wither at the top; (of horses) to spring the knees;
—— *le chemin couvert*, (*siege*) to crown the covered way;
—— *un entonnoir*, (*siege*) to make a lodgment in, to crown, a mine crater;
—— *la gabionnade*, (*siege*) to lay fascines on top of the gabion work.
courre, v. a., to run, race (a horse); to hunt (run) (deer, etc.).
courrier, m., courier, messenger;
—— *de cabinet*, bearer of dispatches, king's messenger;
—— *d'ordonnance*, v. *messager de campagne*;
—— *volant*, shell containing message (obs.).
courroie, f., strap, throng; (*mach.*) belt, band, driving belt; (*harn.*) coupling strap, collar-strap, etc.;
—— *d'agrafe de colleron*, (*art.*, *harn.*) pole-yoke strap;
—— *d'arcade*, (*harn.*) yoke strap; (*cav.*, *etc.*) equipment strap;
—— *-arrêt*, (*harn.*) keep strap;
—— *d'attelle*, hame-strap;
—— *de l'avaloire*, (*harn.*) hip-strap, flank-strap;
—— *de botte*, (*cav.*) bucket strap (carbine);
—— *à boucle*, strap and buckle;
—— *de brêlage*, lashing strap;
—— *à brins croisés*, (*mach.*) crossed belt;
—— *à brins parallèles*, (*mach.*) open belt;
—— *-ceinture*, strap belt;
—— *de ceinture*, (*med.*) belt of a cacolet; (*unif.*) waist strap;
—— *-chaîne*, (*mach.*) Rouiller's belt;
—— *de charge*, (*mil.*) luggage strap; (*art.*) shoe-case strap;
—— *de chargement*, (*harn.*) securing strap (for gun-carriage on pack-mule);
—— *de chasse*, (*mach.*) driving belt;

courroie *croisée*, (*mach.*) crossed belt;
— *croisée au quart*, (*mach.*) quarter-crossed belt;
— *demi-croisée*, (*mach.*) quarter-twist, half-crossed belt;
— *directe*, (*mach.*) open belt;
direction de la ——, (*mach.*) set or direction of a belt (so that it will not run off the pulleys);
— *-dossière*, (*harn.*) back-strap (pack, mountain artillery);
— *double*, (*mach.*) belt of double thickness;
— *de dragonne*, (*cav.*) carbine stay-strap;
— *droite*, (*mach.*) open belt;
— *d'entretoise*, (*harn.*) transom strap (pack, mountain artillery);
— *sans fin*, (*mach.*) endless belt, driving belt;
— *de guindage*, (*cav.*) retaining strap;
— *de lance*, (*cav.*) lance sling;
— *de manteau*, (*cav.*) cloak strap;
— *de mousqueton*, (*art.*) carbine belt;
— *ouverte*, (*mach.*) open belt, band;
— *de paquetage*, (*mil.*) pack strap, equipment strap;
— *de pommeau*, (*cav.*) equipment strap (on pommel);
— *de porte-crosse*, (*cav.*) carbine strap (with carbine in boot);
— *porte-équipement*, (*inf.*) equipment strap (supports waist-belt);
— *porte-traits*, (*harn.*) trace strap;
— *de retraite*, (*harn.*) hold-back strap, backing strap;
— *de sac*, (*inf.*) equipment strap;
— *de sacoche*, (*cav.*) brush-bag strap;
— *simple*, (*mach.*) belt of single thickness;
— *-support*, bearing strap, supporting strap;
— *-support de limonière*, (*harn.*) shaft strap (mountain artillery);
— *de surcharge*, (*harn.*) top-pack strap (mountain artillery);
tambour de ——, (*mach.*) belt pulley;
— *de transmission*, —— *de transmission de mouvement*, (*mach.*) belt, driving belt or band;
— *triple*, (*mach.*) belt of threefold thickness;
— *de troussequin*, (*cav.*, *etc.*) coat strap;
— *trousse-traits*, (*harn.*) trace strap, tucking strap.
cours, m., course, direction, extent, length, turn; current, flow, stream; track, path; course of study, term;
— *d'eau*, stream, water course;
— *facultatif*, elective course (in a college, etc.);
— *inférieur*, lower course of a stream, portion of the course in which there is scarcely any fall;
— *moyen*, middle course of a stream, in which the fall is less than 1:1000;
— *préparatoire*, (*Fr. a.*) preliminary school course for illiterate soldiers;
le —— *préparatoire*, (*Fr. a.*) call or signal for men to turn out for school;
— *supérieur*, upper course of a stream, in which the fall varies from 5:100 or 7:100 to 1:1000.
course, f., course, run; race, journey, trip; career; (*mil.*) incursion, inroad, irruption; (*nav.*) cruise; privateering, commerce-destroying; (*art.*) flight of a projectile, (*mach.*) throw (of a piston), travel (of a valve), throw (of an eccentric, of a crank), feed, length of feed;
aller en ——, (*nav.*) to go privateering;
armer en ——, (*nav.*) to fit out for privateering;
— *de la bague*, —— *de bagues*, (*man.*) riding at rings;
à bas de ——, (*mach.*) at the end of the stroke;
à bout de ——, v. *à bas de* ——;
— *à cheval*, ride;
— *de chevaux*, horse race;
— *au clocher*, steeplechase;
— *descendante du piston*, (*mach.*) down stroke, back stroke of a piston·

course, *directe du piston*, *mach.*) forward stroke of the piston;
— *d'éclipse*, (*art. fort.*) "disappearing" distance (of a disappearing cupola or gun carriage);
en ——, (*nav.*) on a cruise, privateering;
— *par enjeux*, sweepstake;
— *d'essai*, (*nav.*) trial trip, trial run;
— *d'excentrique*, (*mach.*) throw of an eccentric;
faire la ——, (*nav.*) v. *aller en* ——;
faire une ——, to take, go for, a run; (*mil.*) to make an inroad;
— *de fond*, long-distance race or run;
guerre de ——, (*nav.*) privateering, commerce-destroying;
— *des haies*, hurdle race;
haut de ——, (*mach.*) top of the stroke;
longueur de ——, (*mach.*) length of stroke;
— *d'obstacles*, steeplechase, hurdle race;
— *à pied*, foot race;
— *de piston*, (*mach.*) throw, travel, of a piston;
— *plate*, flat race;
— *rétrograde*, (*mach.*) return stroke (piston, etc.);
terrain de ——, race course;
— *de têtes*, (*man.*) riding at heads;
— *du tiroir*, (*mach.*) travel of a slide-valve;
— *au trot*, trotting race;
— *de vélocité*, (*mil. gym.*) dash (never exceeds 150 meters);
— *de vitesse*, dash, sprint.
coursier, m., race, mill race; float board; culvert (of a water-wheel); (*hipp.*) charger, steed; (*nav. art.*) bow gun, bow chaser;
— *de roulement*, (*sm. a.*) butt-piece (of the Borchardt pistol).
coursive, f., (*nav.*) waist, wing passage; gangway plank.
court, a., short; limited; scanty; (*art.*) lower (of two elevations, in ranging); shorter (of the elements of the fork in ranging);
à ——, (to be) short of money, etc.);
— *bâton*, (*sm. a.*) generic term for halfpikes, sword canes, etc.);
tourner ——, to turn short.
court-à-pattes, m., (*mil. slang*) foot artilleryman.
courtaud, a., (*hipp.*) thick-set; docked; m., horse docked and cropped.
courtauder, v. a., (*hipp.*) to dock and crop a horse.
courte-épée, f., (*sm. a.*) generic term for any short-bladed weapon (nearly obs.).
courte-queue, m., (*hipp.*) docked horse.
courtine, f., (*fort.*) curtain; (in field works) parapet connecting salients;
former ——, (*mil.*) of men, to form a screen, to close a gap.
court-jointé, a., (*hipp.*) short-pasterned, short-jointed.
court-monté, a., (*hipp.*) short-coupled.
coussin, m., cushion, pad; driver's leggings; bolster, chock, bed;
— *à air*, air-cushion;
— *de mire*, (*art.*) quoin (obs.).
coussinet, m., pad, cushion; (*harn.*) pad (for straps, etc.); (*mas.*) impost (of a *pied-droit*); bolster, wagon-bolster, iron wedge; (*mach.*) box, journal box; bearing, journal bearing; bush, bushing, brass; pillow, pillow-block, plumber block, bolster; collar; (*r. r.*) chair, rail-chair; (*art.*) trunnion-bed lining or plate, trunnion-bed; bolster, carriage-bolster; (*Fr. cav.*) cushion or pad of rubber or leather for carbine (cuirassiers); (*pont.*) bolster, bolster-block;
— *d'alésoir*, (*art.*) cutter; burnisher (of a boring tool);
— *d'assemblage*, (*r. r.*) joint-chair;
— *de bielle*, (*mach.*) crosshead box or bearing proper; connecting-rod end;
— *à billes*, (*mach.*) ball bearings;
— *-bride*, (*mach.*) connecting-rod strap;
— *de charge*, (*harn.*) breast harness pad;
— *de cheval*, (*harn.*) pillion;
— *de chevalet*, (*pont.*) trestle-bolster;
— *de croisement*, (*r. r.*) crossing chair;

coussinet *de culasse,* (*art.*) traveling bolster (breech);
— *dé de* ——, (r. r.) chair-block;
— *double* ——, (r. r.) double chair;
—— *échauffé,* (*mach.,* etc.) hot box, heated bearing;
——-*éclisse,* (r. r.) fish-chair;
—— *à fileter,* (*mach.*) screw-die;
—— *de glissement,* (r. r.) switch-chair; (*mach.*) guide-, slide-, slipper-, motion-block;
—— *inférieur,* (*mach.*) lower brass;
—— *intermédiaire,* (r. r.) simple, ordinary, chair;
—— *à joint,* (r. r.) joint-chair;
—— *mobile,* (*mach.*) link block;
—— *plantaire,* (*hipp.*) plantar cushion, fleshy frog;
—— *de pointage,* (*art.*) breech-bolster, muzzle bolster for s. b. mortars (obs.);
—— *porte-culasse,* (*art.*) breech; bolster;
—— *porte-volée,* (*art.*) muzzle bolster;
— *rafraîchir un* ——, (*mach.*) to cool a bearing;
—— *de selle,* (*harn.*) pad of a saddle;
—— *supérieur,* (*mach.*) upper brass;
—— *de volée,* (*art.*) muzzle traveling bolster.

coût, m., cost;
— *premier* ——, first cost, purchase price.

coutal, m., (*sm. a.*) saber forming a carbine bayonet (obs.).

couteau, m., knife, cutter; knife-edge of a balance; any knife-edge; feather of carrier pigeon to which message is attached;
—— *d'alésoir,* (*art.*) cutter (of a boring head);
—— *anglais,* (*farr.*) drawing knife;
—— *baïonnette,* (*sm. a.*) knife bayonet;
—— *belge,* (*powd.*) granulating knife (for certain powders);
—— *de chaleur,* (*hipp.*) sweat knife, sweating knife;
—— *à conserve,* can opener, opening knife;
—— *à déchiqueter,* (*expl.*) pulping knife, knife roll, knife plate, knife plate and roll;
—— *à deux manches,* draw knife;
— *grand* ——, (*mil., slang*) cavalry saber;
—— -*poignard,* (*sm. a.*) poniard bayonet (Daudeteau);
—— *râcloir,* scraper.

coutelas, m., (*sm. a.*) cutlass.

coutil, m., ticking, drill.

couture, f., seam; fash, burr; rivet seam or burr;
— *battre à plaie* ——, to defeat utterly;
—— *rabattue,* fell, tent seam.

couvercle, m., cover, lid, cap, top; (*sm. a.*) dust cover; (*mach.*) cylinder cover;
—— *à charnière,* hinged lid;
—— *de chemise à tiroir,* (*mach.*) valve-casing cover;
—— *de cylindre,* (*mach.*) cylinder cover;
—— *de piston,* (*mach.*) piston cover.

couvert, a., covered; sheltered, hid; (of the weather) overcast; (*farr.*) wide (said of the bearing-surface of the shoe); (*mil.*) covered; (*top.*) close, wooded;
— *chemin* ——, (*fort.*) covered way;
— *pays* ——, (*top.*) close, wooded country;
— *temps* ——, cloudy weather;
— *terrain* ——, (*top.*) close terrain.

couvert, m., shelter, cover; (*top.*) overgrown, wooded, close country or terrain;
— *à* —— *de,* sheltered from, under cover.

couverte, f., covering; lintel (of a door or window); (*art.*) steel facing, skin, or layer, of an armor (compound) plate; (*met.*) covering or wrapping (of *fer ballé,* q. v.) for packets in making sheet iron; (*mil. slang*) blanket; (*harn.*) saddlecloth;
— *battre la* ——, (*mil. slang*) to sleep;
—— *de campagne,* —— *de campement,* (*mil.*) soldier's blanket;
— *faire sauter à la* ——, to toss in a blanket.

couverture, f., blanket; rug; cover, covering; (*harn.*) saddle-cloth, saddle-blanket; (*cons.*) roof, roofing proper; (*farr.*) width of the bearing-surface of a horseshoe, measured between borders; (*mil.*) the protection or covering of mobilization by covering troops;
—— *d'attente,* (*harn.*) horse-blanket;

couverture *de campement,* —— *du service du campement,* (*Fr. a.*) field blanket;
—— *de cheval,* (*harn.*) horse-blanket;
—— *des lits militaires,* —— *du service des lits militaires,* (*Fr. a.*) ordinary service or garrison blanket (for men, includes also foot-blanket);
—— *de selle,* (*harn.*) saddle-cloth, blanket;
— *troupes de* ——, (*mil.*) troops covering a mobilization, troops garrisoning frontier forts.

couvrante, f., (*fort.*) line of fire, interior crest.

couvre-amorce, m., (*sm. a.*) primer-cover, primer cup or cap.

couvre-bassinet, m., (*sm. a.*) pan-cover (obs.).

couvre-bouche, m., (*art.*) muzzle-cover.

couvre-bouche-couvre-guidon, (*sm. a.*) muzzle and front-sight cover.

couvre-canon, m., (*sm. a.*) muzzle-cover, muzzle tompion.

couvre-charges, m., (*art.*) compartment cover in ammunition wagons.

couvre-chef, m., generic term for headdress, headdress in general.

couvre-clavette, m., (*mach.*) key-cap.

couvre-colback, m., (*unif.*) oilcloth cover for colback.

couvre-culasse, m., (*art.*) breech-cover.

couvre-engrenages, m., (*mach.*) wheelcase, gearing case.

couvre-étoupe, m., (*mach.*) stuffing-box cover.

couvre-face, m., (*fort.*) face cover; counter-guard.

couvre-feu, m., fire-plate.

couvre-file, m., (*mil.*) proposed shield or plate to protect infantry from small-arm fire.

couvre-fonte, m., (*sm. a.*) holster cap.

couvre-frein, m., (*art.*) brake-cover.

couvre-giberne, m., (*mil.*) cartridge-box cover.

couvre-guidon, m., (*sm. a.*) front-sight cover.

couvre-hausse, m., (*art.*) tangent-sight cover.

couvre-joint, m., lapping piece; butt-plate, -strap;
— *double* ——, double butt-plate.

couvre-lumière, m., (*art.*) vent-cover; (*in gen.*) screen.

couvre-molette, m., (*mach.*) cutter-guard.

couvre-nuque, m., (*unif.*) rear peak (of a helmet), havelock; sapper's neck cover.

couvre-obus, m., (*art.*) lid of shell compartment of an ammunition chest.

couvre-pied, m., foot-blanket.

couvre-platine, m., (*sm. a.*) lock-plate cover (obs.).

couvre-pointe, m., point-guard (as for an auger).

couvre-roues, m., (*mach.*) gear-cover, wheel-cover.

couvre-shako, m., (*unif.*) shako-cover (obs. in Fr. a.).

couvre-tranchant, m., edge protector or cover.

couvreur, m., (*pont.*) chess-layer.

couvre-vis, m., (*art.*) breech-screw cover.

couvrir, v. a., to cover, to screen, to overlay; (*mil.*) to cover (a passage, a town, etc.); to cover, protect (the flag covers the merchandise); (*hipp.*) to cover, serve (a mare);
— *se* ——, (*man.*) to cover, check his steps (of a horse);
—— *son chef de file,* (*mil.*) to cover one's front-rank man.

covolume, m., (*ball., expl.*) covolume, the "specific constant" of Clausius.

coxal, m., (*hipp.*) haunch bone.

coyau, m., (*cons.*) rafter foot, eaves lath.

coyer, m., (*cons.*) arris beam.

crachat, m., (*cons.*) poor, worthless building materials; (*slang*) the star indicating high rank in orders of knighthood.

crachement, m., (of boats) starting of seams; (*art., sm. a.*) escape of gas through defective breech closure; (*steam*) priming, foaming, of a boiler; (*fond.*) running over of a mold; (*elec.*) sparking.

cracher, v. a. n., (*art., sm. a.*) to escape (of gas, through defective breech closure); (*steam*) to prime, foam (of boilers); (*fond.*) to run over (of a mold); (*elec.*) to spark;
—— *les étoupes,* to expel, start, calking from seams of a boat.

craie, f., chalk;
être marqué à la ——, (*mil.*) to have troops billeted on a house;
loger à la ——, (*mil.*) to billet;
maison exempte de la ——, (*mil.*) house exempt from billet;
marquer, tracer, à la ——, (*mil.*) to billet or quarter soldiers.

crampe, f., cramp, staple, cramp-iron, catch.

crampon, m., cramping-iron, staple; catch; cramp; holdfast; spike; creeper; brace; (*farr.*) calk of a horseshoe, calkin;
—— *à l'aragonaise,* (*farr.*) pyramidal calk;
—— *d'assaut,* (*mil.*) creeper, worn on shoes in an assault party;
—— *d'attelage,* (*harn.*) shaft-staple;
—— *de fer à cheval,* (*farr.*) frost-nail, calkin;
—— *à glace,* (*farr.*) calkin, frost-nail;
—— *à oreilles de chat,* (*farr.*) v. —— *à l'aragonaise;*
—— *porte-esse de rechange,* (*art.*) spare linch-pin hook;
—— *à vis,* (*farr.*) screw calkin.

cramponner, v. a., to cramp, fasten with a cramp; (hence, *fig.*) to hold on tight (to a wall, etc.); (*farr.*) to rough-shoe a horse, put on calkins.

cramponnet, m., small cramp, cramp-iron; tack; staple.

cran, m., notch, mark, nick; cog, ear; crank; (*met.*) defect in welding; (*hipp.*) ridge of a horse's mouth; (*art., sm. a.*) sight-notch; (*sm. a.*) notch of the firing mechanism;
à ——*s,* cogged, notched;
—— *de l'abattu,* (*sm. a.*) sear-notch, (in the Lebel) the first notch of the cocking piece receiving the sear after firing;
—— *de l'armé,* (*sm. a.*) cocking-notch, full-cock notch;
—— *d'arrêt,* (*sm. a.*) stop notch;
—— *du bandé,* (*sm. a.*) full-cock notch;
—— *du curseur,* (*art., sm. a.*) slide notch (of a sight);
—— *de démontage,* (*sm. a.*) (in the Lebel) dismounting notch in the ejector-head;
—— *de départ,* (*sm. a.*) full-cock notch;
—— *d'embarrage,* (*art.*) maneuvering, embarring, notch;
—— *de la hausse,* (*art., sm. a.*) rear-sight notch;
—— *de mire,* (*art., sm. a.*) sight notch;
—— *mobile,* (*art., sm. a.*) slide notch;
—— *de repos,* (*sm. a.*) half-cock notch; (*r. f. art.*) catch for operating handle;
—— *de sûreté,* (*sm. a.*) safety notch; (*r. f. art.*) safety catch.

crâne, m., skull.

crapaud, m., toad; (*torp.*) anchoring weight, mushroom anchor; (*hipp.*) thrush (aggravated); (*mil. slang*) diminutive man; purse in which soldiers store up savings;
—— *d'amarrage,* mushroom anchor.

crapaudine, f., (*mach., etc.*) step, bearing, step-bearing; pivot-step; thrust-bearing; socket, pivot-hole; bush, collar, pillow; valve of a discharge or escape pipe; grating over the mouth of a gutter pipe, etc.; (*hipp.*) disease or affection of the anterior face of the coronet, leading to fissures in the coronary band;
—— *femelle,* (*mach.*) spindle-bearing;
—— *à grain mobile,* (*mach.*) independent step bearing;
—— *mâle,* (*mach.*) pivot;
—— *à patin horizontal,* step-bearing;
—— *avec plaque de fixation verticale,* (*mach.*) wall step-bearing.

craquer, v. n., to give way, yield, crack, spring.

crasse, f., (*art., sm. a.*) fouling of the bore; (*met.*) dross, cinders, slag.

cratère, m., (*art., min.*) crater (produced by explosion of a shell or of a mine); (*elec.*) hollow or crater of the arc-light carbon.

craticuler, v. a., (*drawing*) to reduce by squares.

cravache, f., riding whip, horse whip.

cravate, m., Croatian horse; horse soldier (obs.).

cravate, f., cravat, necktie; axle-collar, -strap (of a capstan); (*cord.*) top tackle; (*mil.*) bow tied around the top of staff of colors; (*expl.*) quick-match (fastened to a fuse, etc.);
—— *d'avant-train,* (*art.*) pintle-stay.

cravater, v. a., —— *un drapeau,* (*mil.*) to tie a bow, etc., around the head of the color-staff.

crayon, m., pencil, chalk, crayon; pencil-drawing, rough, draught; sketch, outline; (*elec.*) carbon of an arc lamp;
—— *à dessiner,* drawing pencil;
—— *lithographique,* lithographic chalk;
—— *à n becs,* railroad pencil.

crayonner, v. a., to draw in pencil, to sketch, outline.

créance, f., trust, credit; bill, money owing, debt;
élonger en ——, (*cord.*) to make a guess warp;
lettre de ——, letter of credit; ambassador's credentials.

créat, m., (*man.*) riding-master's assistant.

crèche, f., manager, crib.

crédit, m., credit; (*adm.*) appropriation;
—— *extraordinaire,* emergency appropriation;
—— *législatif,* appropriation by legislation, government (congressional, parliamentary) appropriation;
lettre de ——, letter of credit;
—— *ministériel,* funds turned over by a government minister to his subordinates for the expenses of their departments;
—— *sous-délégué,* (Fr. a.) in the *intendance,* funds turned over by an *intendant* to his *sous-intendants;*
—— *supplémentaire,* additional appropriation, (sometimes) emergency appropriation.

crémaillère, f., (*mach., etc.*) rack, rack-work, rack-gear, rack-bar; ratchet, rack-tooth; (*fort.*) indented line; (*art.*) traversing-rack, elevating arc; (*sm. a.*) circular ratchet (Colt's revolver;)
cric à ——, rack-jack;
en ——, indented;
—— *en arc de cercle,* (*art.*) circular elevating arc;
ligne à ——, (*fort.*) indented line;
—— *de pointage,* (*art.*) elevating arc;
réunir en ——, (*carp.*) to joggle;
trace à ——, (*fort.*) cremaillère trace.

crémone, f., sash-fastener.

créneau, m., (*fort.*) loophole, battlement, embrasure; (*mil.*) interval between units, place of a *chef de section* in the line;
—— *de couronnement,* (*fort.*) battlement;
—— *droit,* (*fort.*) normal loophole;
—— *d'étage,* (*fort.*) loophole through a wall at any point;
—— *horizontal,* (*fort.*) horizontal loophole;
—— *-mâchicoulis,* (*fort.*) machicolated loophole;
—— *de pied,* v. —— *mâchicoulis;*
—— *en sacs à terre,* (*fort.*) sand-bag loophole;
—— *vertical,* (*fort.*) vertical loophole;
—— *de visée,* (*fort.*) aiming peep or slit in armored forts).

créneler, v. a., (*fort.*) to loophole, embattle, crenelate; (*in gen.*) to cog, indent, notch, tooth.

crénelure, f., (*fort.*) battlement; (*in gen.*) indenting, toothing, notching.

créosotage, m., creosoting, impregnation with creosote.

créosote, f., creosote.

créosoter, v. a., to creosote, impregnate with creosote.
crêpe, m., crape, piece of crape;
— *porter le* —— *au bras, au sabre*, (*mil.*) to wear mourning on the arm, on the sword.
crépi, m., (*cons.*) plastering, parget; rough coat of plaster;
— *donner un* —— *à*, to rough cast, rough coat.
crépine, f., strainer, rose (of a pump, over the mouth of a pipe, etc.).
crépir, v. a., (*cons.*) to plaster, parget; to rough cast, rough coat;
— *à la chaux*, to whitewash.
crépitement, m., crepitation;
— *de la fusillade*, (*mil.*) crack of musketry.
crépiter, v. n., (*mil.*) to crack (of musketry).
crésylite, f., (*expl.*) cresylite.
crête, f., (*top.*) crest, ridge, top; (*unif.*) crest of a helmet; (*sm. a.*) thumb piece, comb;
— *d'artillerie*, (*mil.*) the ridge line on which guns take position;
— *du chien*, (*sm. a.*) comb of the hammer;
— *couvrante*, (*mil.*) line or position under the shelter of the crest proper;
— *extérieure*, (*fort.*) exterior crest;
— *de feu*, (*fort.*) interior crest;
— *d'infanterie*, (*mil.*) line or position from which the slopes and approaches can be swept by infantry fire (on the slope next the enemy);
— *intérieure*, (*fort.*) interior crest;
— *de manœuvre*, (*sm. a.*) milled head or thumb piece;
— ——*masque*, (*mil.*) covering crest (for artillery);
— *militaire*, (*mil.*) line or position from which the slopes and approaches can be kept under good observation; military crest;
— *quadrillée*, (*sm. a.*) milled head or comb;
— *à ressauts*, (*fort.*) terraced crest, crest in steps;
— *supérieure*, (*fort.*) superior crest;
— *topographique*, (*mil.*) the ridge line proper, watershed.
creusage, m., digging, excavation; deepening.
creuser, v. a., to dig, excavate, deepen; to dredge; to hollow, to hollow out;
— *machine à* ——, dredging machine;
— *au tour*, (*mach.*) to turn hollow.
creuset, m., (*met.*) crucible, hearth, fireplace of a hand-forge; crucible, throat, of a blast furnace;
— *méthode du petit* ——, crucible method of making steel.
creux, a., hollow, empty.
creux, m., hollow, groove, cavity; (*artif.*) fuse socket, fuse cavity; (*art.*) hollow of a shell; (*fond.*) mold; (*mach.*) depth of a cog, clearance, space between adjacent teeth; (*nav.*) trough of the sea; (*sm. a.*) groove of a sword blade;
— *de la lame*, (*nav.*) trough of the sea.
crevasse, f., crevice, crevasse; gap, chink; rift; (*met.*) flaw in metal; (*hipp.*) rat's tail;
— *capillaire*, (*art.*) hair-crack (in armor);
— *au pied*, (*hipp.*) cracked heel.
crevasser, v. a., to split, crack.
crevé, a., (*cord.*) stranded, divided into strands.
crève-faim, m., (*mil. slang*) man who volunteers as a soldier.
crever, v. a. n., to burst, split, crack, break; (of animals) to die; to work (a horse) to death; (*art.*) to burst (a shell, a gun); (*mil.*) to burst through a line, to break a line; (*min.*) to spring a mine;
— *une embarcation*, to stave in a boat.
cri, m., cry, war whoop; shout; (*mil.*) challenge of a sentry.
criblage, m., screening, sifting.
crible, m., sieve, riddle, screen;
— *à grener*, (*powd.*) corning-sieve;
— *passer au* ——, *par le* ——, to sift, to screen;
— *à pied*, screen, sifter;
— *à poudre*, (*powd.*) powder-sieve;
— *sasseur*, bolter;
— *à secousse*, swing-sieve;
— *à tambours*, (*artif.*) mixing-sieve.

cribler, v. a., to sift, to screen; (*sm. a.*) to riddle;
— *de balles*, (*sm. a.*) to riddle.
cric, m., jack, lifting-jack;
— *d'assemblage*, cramp;
— *de campagne*, (*art.*) field jack;
— *composé*, rack and pinion jack;
— *à crémaillère*, rack-jack;
— *à crochet*, hand-jack;
— *à crochets de suspension*, sort of jack for raising and lowering the cap of the Belgian trestle;
— *double*, double jack;
— *à double engrenage*, double rack-jack;
— *à engrenage*, rack-jack;
— *à main*, hand-jack;
— *à noix*, chain-jack;
— *à double noix*, double-clawed jack;
— *à deux pattes*, double-clawed jack;
— *de siège*, trench jack;
— *simple*, ordinary hand-jack;
— *de tranchée*, v. —— *de siège*;
— *à vis*, screw-jack.
crime, m., crime.
criméenne, f., (*unif.*, *Fr. a.*) hooded collar (not regulation) worn with or without the overcoat.
criminel, m. and a., criminal.
crin, m., (*hipp.*) horsehair (of mane or tail);
— *s blancs*, white mane or tail (dark-coated horse);
— *cheval à tous* ——*s*, horse with flowing mane and tail;
— *s lavés*, washed mane and tail (dark-coated horse);
— *s mélangés*, mane or tail containing white hairs, there being none in the coat.
crinière, f., (*hipp.*) mane; (*unif.*) horsehair plume mane of a helmet;
— *en brosse*, (*hipp.*) hogged mane.
crinoline, f., crinoline; *affût à* ——, (*r. f. art.*) crinoline mount.
crique, f., creek, cove; small harbor or port, or basin, made by excavation; (*art.*) hair-crack (of an armor plate); (*met.*) fissure produced by tempering, small crack, seam, hair-crack (of an ingot); (*sm. a.*, etc.) bit of iron, etc. (fraudulently) used to conceal flaws in a gun barrel, etc.; (*top.*) basin-like depression; (*fort.*) cut, ditch;
— *des fossés*, (*fort.*) inundating channel around a fort, to prevent hostile trenchwork.
criquer, v. a., (*met.*) to crack.
criquet, m., (*hipp.*) pony, small horse.
croc, m., hook, crook; claw of a lifting-jack; catch; boat-hook, grapnel, drag;
— *à cosse*, thimble-hook;
— *à déclic*, pawl-hook;
— *double*, clasp-hook;
— *à douille*, socket-hook;
— *à échappement*, slip-hook;
— *à émerillon*, swivel-hook;
— *à mentonnet*, forelock, hook;
— *de palan*, (*cord.*) tackle-hook;
— *à pointe et à crochet*, boat-hook;
— *s porte-écouvillon*, (*art.*) sponge-hooks;
— *s porte-refouloir*, (*art.*) sponge-hooks;
— *porte-seau*, (*art.*) bucket-hook;
— *en S*, S-shaped hook.
crocher, v. a., to hook; to seize with a boat-hook.
crochet, m., hook, little hook; crook; fire-hook; bend; claw; crochet; clasp; hasp; (*art.*) lanyard-hook; (*mil.*) flank, either refused or thrown forward; (*harn.*) near side curb-hook; (*hipp.*) tusk, canine tooth; (*mach.*) hook tool; (*siege*) return (single-sap), rectangular change of direction in the double sap, (also, *siege*) the place of arms formed by a return;
— *d'anneau d'embrelage*, keep-chain hook;
— *s d'armes*, (*sm. a.*) small-arms hooks;
— *d'arrêt*, (*mach.*) pawl, catch;
— *d'arrêt à ressort*, (*art.*) spring-catch (Gardner gun);
— *d'attelage*, (*harn.*) trace-hook; tug-hook;
— *d'attelage de trait*, (*harn.*) trace-hook;
— *à bascule*, (*sm. a.*) trigger-guard catch;
— *à bombe*, (*art.*) shell-hook (obs.);

crochet *de bout de timon,* (*harn.*) pole-hook;
— *de brague,* (*art.*) breeching-hook;
— *de brêlage,* any lashing hook;
— *de carabine,* (*sm. a.*) carbine swivel;
— *de charge,* pack-hook (of pack-saddle);
— *de chemin,* bend or turn of a road;
— *cheville-ouvrière,* (*art.*) pintle-hook;
clou à ——, tenter-hook;
— *de déchargement,* (*art.*) hook for drawing cartridge;
— *défensif,* (*mil.*) defensive crochet, wing or flank thrown back;
— *double,* double hook;
double —— *à anneau,* double tackle hook and ring;
— *double d'équipement,* double tackle (swivel) hook;
— *double de retraite,* (*art.*) double trail hook;
— *double de réunion de chaîne,* double connecting hook (for chains);
— *élévatoire,* lifting hook;
— *d'enrayage de tir,* (*art.*) locking chain hook;
— *d'épars,* (*art.*) crossbar hook of a field gin;
— *d'épontille,* stanchion hook;
— *d'équipement,* (*mach., etc.*) tackle-hook;
— *à étrangler,* (*artif.*) choking hook;
— *extracteur,* (*r. f. art.*) hook extractor (Gardner gun);
— *fendu,* split hook, double hook;
— *fermé,* closed hook;
— *à fourche,* (*art.*) forked hook, sponge-hook;
— *du glacis,* (*fort.*) passage around traverse on the covered way, crotchet;
— *à incendie,* fire hook;
— *jumeau d'amarrage à double bec,* (*art.*) double mooring-hook (*cabestan de carrier*);
— *de levage,* lifting-hook;
— *de manœuvre,* (*art., sm. a.*) any hook used in mounting or dismounting parts of the mechanism;
— *de mousqueton* (*art., etc.*) snap-hook;
— *à obus,* (*art.*) shell-hook;
— *offensif,* (*mil.*) offensive crochet, wing or flank thrown forward;
— *ouvert,* open hook;
— *à piton,* ring-bolt and hook;
— *à poignées,* (*art.*) shell-hook;
— *de pontage,* (*pont.*) lashing hook;
— *s porte - armements,* (*art.*) equipment hooks;
— *port-écouvillon,* (*art.*) sponge-hook;
— *porte-levier,* (*art.*) trail handspike hook;
— *porte-levier portereau,* (*art.*) trail handspike hook;
— *porte-sabots,* (*art.*) locking-shoe hook;
— *porte-seau,* (*art.*) bucket hook;
— *de prolonge,* (*art.*) prolonge hook;
— *-rênoir,* (*harn.*) pack saddle rain-hook;
— *de renvoi,* (*art.*) guide hook (for lanyard);
— *à ressort,* spring hook, spring latch;
— *de retour,* (*fort.*) return;
— *de retraite,* breast or draught hook in heavy carriages; (*art.*) breast-hook, trail-hook;
— *de réunion de chaîne,* chain hook, connecting hook;
ringard à ——, hooked poker;
— *en S,* S-hook, open link, S-shaped hook;
— *de sape,* (*siege*) return;
— *simple,* single hook;
— *de soulèvement,* lifting hook;
— *support de limonière,* (*art.*) shaft hook (mountain artillery);
— *de tête d'affût,* (*art.*) breast-hook;
— *de tête de trait,* (*harn.*) front trace-hook;
— *tire-cartouches,* (*art.*) cartridge-case extractor;
— *tire-chaîne,* chain lock;
— *tire-gargousse,* (*art.*) cartridge-hook, unloading-hook;
— *de trait,* (*harn.*) trace hook;
— *de tranchée,* (*siege*) return;
— *de traverse,* (*fort.*) passage around a traverse on the covered way, crotchet;
— *de volée,* swingletree hook.
crocheter, v. a., to pick a lock.
crochu, m., (*hipp.*) unciform bone.

crocodilo, m., (*mil. slang*) foreign student at St. Cyr.
croisé, m., (*fenc.*) movement by which the adversary's blade or foil is forced from the upper to the lower line.
croisée, f. window, casement; cross (of an anchor); (*top.*) crossing (of roads); (*inst.*) cross-hairs (telescope sight, etc.);
— *d'ancre,* cross of the anchor;
— *de sabre,* (*sm. a.*) cross or guard of a sword.
croisement, m., (*r. r.*) crossing of two tracks; (*hipp.*) crossing of breeds (outbreeding); (*fenc.*) crossing of swords, foils, etc.; (*mach.*) belt crossing;
— *au quart,* (*mach.*) half quarter twist (of a belt).
croiser, v. a. n., to cross, overlap; to lay or set across, to intersect; (*cord., etc.*) to cross; (*hipp.*) to cross, to breed out; (*nav.*) to cruise; (*fenc.*) to cross swords, foils, etc.;
— *la baïonnette,* (*mil.*) to charge bayonets;
se ——, (*cord.*) to be crossed.
croisette, f., (*sm. a.*) sword or foil whose guard is a simple cross.
croiseur, m., (*nav.*) cruiser;
— *à barbette,* barbette cruiser; second-class cruiser;
— *à batterie,* cruiser with broadside battery;
— *à ceinture cuirassée,* belted cruiser;
— *de combat,* armored cruiser;
— *-corsaire,* commerce destroyer;
— *de course,* commerce destroyer;
— *cuirassé,* armored cruiser;
— *protégé,* protected cruiser;
— *à redoute,* cruiser with a redoubt battery;
— *-torpilleur,* torpedo cruiser;
— *à tourelles barbette,* barbette-turret cruiser.
croisière, f., (*sm. a.*) cross, hilt of a sword-bayonet; (*nav.*) cruise, cruising; (*r. r.*) crossing of two tracks; (*pont.*) spring line, cross or diagonal mooring;
en ——, (*nav.*) on a cruise;
guerre de ——, (*nav.*) privateering, commerce-destroying;
tenir ——, (*nav.*) to continue cruising.
croisillon, m., crossarm, crossbar, crosspiece; boss and arms of a water-wheel; (*art.*) slide of a rear sight; (*mach.*) arm of a wheel; (*cons., etc.*) frame of crossed bars;
— *s articulés,* (*mach.*) cross-axis;
— *-poignée,* double tee handle, cross-handle;
— *s à tourillons,* (*mach.*) journaled arms (in some forms of universal joint).
croissant, m., crescent; (*art.*) scraper (for interior of shell, as far as the eye); (*hipp.*) projection of the *os pedis* in founder;
— *de bride,* (*harn.*) heart of the throat-latch.
croix, f., cross, cross of an order; (*cord.*) kink; (*sm. a.*) cross-hilt of a sword;
en ——, crosswise, crossways;
grand- ——, knight grand cross;
grand' ——, the cross or decoration worn by a *grand-croix,* grand cross;
— *de St. André,* diagonal stays, cross-stays; braces joined diagonally.
Croix-Rouge, f., Geneva Cross, Red Cross;
Société de la ——, Red Cross Society.
croquade, f., rough, hasty sketch; draught.
croquemitaines, m. pl., (*mil. slang*) soldiers sent to the disciplinary companies in Africa on account of self-mutilation.
croquer, v. a., to make a rough sketch or draught.
croquis, m., sketch; plan, outline, rough draught;
cahier de ——, sketch book;
— *d'exécution,* working sketch, sketch of detail;
— *pittoresque,* landscape sketch;
— *topographique,* topographical sketch, plan;
— *de tour d'horizon,* (*top.*) sky-line sketch.

crosse, f., (*sm. a.*) butt, (more strictly) butt-stock; butt of a revolver (includes side plates); (*art.*) trail; (*r. f. art.*) stock, shoulder piece; (*mach.*) crosshead;
— *d'affût*, (*art.*) trail;
bout de — *lunette*, (*art.*) trail plate complete (comprising pintle); the end of the trail;
courbure de la —, (*sm. a.*) bend of the stock;
guide de —, (*F. m. art.*) guide for rear part (*crosse*) of transom of 16ᶜᵐ m. l. gun (obs.);
— *à jour*, (*r. f. art.*) openwork stock (e. g., Hotchkiss);
mettre la — *en l'air*, (*mil.*) to surrender, yield;
— *de pistolet*, (*r. f. art.*) pistol stock.
crotte, f., mud, mire (of the road or street).
crottin, m., (*fond.*) horse dung and clay for molds.
croulement, m., falling in, falling down (of buildings).
crouler, v. n., to sink, fall in, give way.
croulier, a., boggy, swampy.
croulière, f., bog, swamp.
croupade, f., (*man.*) croupade.
croupe, f., (*hipp.*) croup, rump, hind quarters, crupper; (*cons.*) hip-roof; (*top.*) long hill or crest, top or brow of a hill, point where brow of a hill drops down to a hill of lesser height;
— *avalée*, (*hipp.*) croup that slopes too much;
balancer la —, (*man.*) to be unsteady of gait (of a horse);
— *biaisée*, (*cons.*) hip-roof whose end wall is oblique to the façade;
— *coupée*, (*hipp.*) very short croup;
— *droite*, (*cons.*) hip-roof whose end wall is perpendicular to the façade;
en —, (*man.*) behind;
monter en —, (*man.*) to get up behind, to ride behind;
— *de mulet*, (*hipp.*) sharp croup;
porter en —, (*man.*) to ride double;
prendre en —, (*man.*) to ride behind;
— *en pupitre*, v. — *avalée*;
— *tranchante*, v. — *de mulet*.
croupé, a., *cheval bien* —, (*hipp.*) horse with fine hind quarters.
croupiader, v. a., to moor by a stern cable.
croupiat, m., small stern-fast or cable.
croupière, f., (*harn.*) crupper; (of boats, etc.) stern-fast.
croupon, m., crup, crup leather, body leather (of a shoe).
croûte, f., crust; (*font.*) skin, casting skin.
cru, a., raw, crude;
monter à —, (*man.*) to ride bareback.
cruche, f., pitcher, water-pitcher.
crue, f., growth, increase, swelling; rise, flood, freshet;
— *d'arbre*, growth and age of a tree.
crusher, m., (*ball.*) crusher, crusher gauge;
— *à écrasement*, crusher gauge.
cryptogramme, m., cipher, cryptogram.
cryptographe, m., cipherer, cryptographer.
cryptographie, f., cryptography.
cryptographier, v. a., to convert into cipher, to encipher.
crystal, m., crystal.
cryptorchide, m., (*hipp.*) cryptorchid.
cryptorchidie, f., (*hipp.*) cryptorchism.
cryptorchidisme, m., (*hipp.*) cryptorchism.
crystallin, a., crystalline.
cubage, m., cubic measurement; cubic contents.
cubature, f., v. *cubage*.
cube, m., cube; third power; volume or cubical amount (to be excavated, etc.);
— *de déblai*, volume to be excavated;
— *de maçonnerie*, volume of masonry, of masonry work.
cuber, v. a., to measure, compute, the cubic contents, the volume of anything.
cubilot, m., (*met.*) cupola furnace; portable furnace, used in manufacture of projectiles;
— -*creuset*, furnace and crucible in one, or in conjunction.
cubique, a., cubic.

cuboïde, m., (*hipp.*) cuboid, one of the tarsal bones.
cueille, f., (*cord.*) turn of a coil of rope, coil, fake.
cueillie, f., first layer of plaster.
cueillir, to gather; (*cord.*) to coil a rope.
cuiller, f., spoon; ladle; (*artif.*) artificer's ladle;
(*fond.*) casting ladle; (*torp.*) spoon or bar of a torpedo tube;
— *à boulet rouge*, (*art.*) hot-shot carrier (obs.).
— *à brai*, pitch ladle;
— *à canon*, (*art.*) gunner's ladle;
— *de coulée*, (*fond.*) casting ladle;
— *enveloppante*, (*mil. min.*) scooping-bit (of an earth auger);
— *à pompe*, pump borer;
— *porte-boulet*, (*art.*) v. — *à boulet rouge;*
— *de tarière*, auger bit.
cuir, m., skin, leather, hide; fold of the skin (under a saddle badly put on);
— *avivé*, harness leather;
— *cru*, raw hide, green hide;
— *fauve*, fair leather;
— *fort*, sole leather;
— *fossile*, asbestos;
— *de harnais*, harness leather;
— *hongroyé*, tawed leather, Hungary leather, white or alum leather;
— *mégi*, white leather;
— *de montagne*, v. — *fossile;*
— *pique*, interwoven leather;
— *de roussi*, — *Russie*, Russia leather;
— *à semelles*, sole leather;
— *en suif*, tallowed leather;
— *en suif à chair propre*, full-tanned leather, free from tallow on the outside and dressed (trimmed) on the flesh side;
— *en plein suif*, full-tanned leather, tallowed and dressed;
— *tanné*, tanned leather;
— -*toile*, enameled cloth, American leather;
— *verni*, patent leather;
— *vert*, — *en vert*, v. — *cru*.
cuirasse, f., (*cav.*) breastplate; (*fort.*) armor, shield; (*nav.*) ship armor; (*sm. a.*) steel cap of the Rubin bullet;
— *de ceinture*, (*nav.*) belt armor;
— *dédoublée*, (*nav., etc.*) sandwich armor;
— *de derrière*, (*cav.*) back plate;
— *à dos*, (*cav.*) cuirass complete (i. e., both breast and back plates);
— *double*, v. — *à dos;*
— *des murailles*, (*nav.*) side armor;
— *de tourelle*, (*nav.*) turret armor.
cuirassé, a., (*fort., nav.*) armored, plated, armor-plated; (*cav.*) wearing a breastplate; (*sm. a.*) armed with a steel cap or cover (of a bullet);
— *à n cm.*, armored up to *n* cm.
cuirassé, m., (*nav.*) armored vessel; (*esp.*) battle ship;
— *à batterie*, broadside battle ship;
— *à batterie centrale*, armored central-battery ship;
— -*bélier*, armored ram;
— *de croisière*, armored cruiser;
— *à éperon*, battle ship and ram, battle ship with ram bow;
— *d'escadre*, battle ship;
— *d'escadre à batterie*, broadside battle ship;
— *d'escadre à réduit central*, battle ship with central redoubt;
— *d'escadre à tourelles barbette*, barbette-turret battle ship;
— *d'escadre à tourelles fermées*, turreted battle ship (closed turrets);
— *à fort central*, battle ship with central redoubt;
— *de ligne*, battle ship;
— *de premier rang*, v. — *de ligne;*
— *à réduit*, v. — *à fort central;*
— *à réduit central*, v. — *à fort central;*
— *de station*, armored cruiser;
— *à tourelles*, turreted battle ship;
— *à tourelles barbette*, barbette-turret battle ship;
— *à tourelles tournantes*, battle ship with revolving turrets.

cuirassement, m., (*fort.*, *nav.*) armor, armor plating; act of armoring a vessel or fort;
—— *en fer et en acier*, compound armor;
—— *mobile*, (*art.*) generic term for traveling shield mounting;
—— *de tranchée-abri*, (*art.*) transportable cupola.
cuirasser, v. a., (*fort.*, *nav.*) to plate, to armor, a vessel, fort, etc.
cuirassier, m., (*cav.*) cuirassier.
cuire, v. a., to cook; to burn (lime, bricks, etc.).
cuisine, f., kitchen.
cuisinier, m., cook; (*mil.*) head cook, first cook;
aide ——, second cook, assistant cook.
cuissard, m., (*steam*) necking (of a French boiler).
cuisse, f., thigh;
élévation des ——*s*, (*man.*) the raising of the thighs to correct for a fork seat;
—— *de grenouille*, (*hipp.*) froggy thigh;
—— *plate*, (*hipp.*) v. —— *de grenouille*;
rotation des ——*s*, (*man.*) working or rotation of the thighs, to supple the legs.
cuissière, f., (*mil.*) leather leg-guard, worn by drummers.
cuisson, f., cooking; burning (of lime).
cuite, f., burning (of lime, bricks); batch of bricks of one burning.
cuivrage, m., coppering.
cuivre, m., (*met.*) copper; (*mus.*) brass instruments (also in pl.); (*mil.*) copper, brass, fittings of equipments, "brasses" (U. S. M. A.), bright work;
—— *brut*, block copper, coarse copper;
—— *jaune*, brass;
—— *noir*, v. *brut*;
—— *rosette*, purified copper;
—— *vide*, (*sm. a.*, etc.) unfilled primer cap.
cuivrer, v. a., to copper, sheathe, plate with copper.
cuivreux, a., coppery;
fer ——, (*met.*) red-short iron.
culvrot, m., drill-box.
cul, m., breech, tail; tail of a cart; turn, bend, of a spring;
—— *de chaudron*, (*mil. min.*) bottom of a mine crater;
faux ——, (*powd.*) the mass at the bottom of the stamper;
mettre (*une voiture*) *à* ——, to tip (a wagon, etc.);
—— *de poule*, (*hipp.*) croup narrow toward the rear;
—— *de poulie*, (*cord.*) tail of a block;
—— *rouge*, (*mil. slang*) soldier with red trousers.
culasse, f., (*art. sm. a.*) breech, (frequently) breechblock, breech-plug, (in composition) breech-loading; (*nav.*, *pont.*) cross of an anchor (*elec.*) yoke-piece;
boîte de ——, (*sm. a.*) casing, breech casing, receiver;
bouton de la ——, (*sm. a.*) threaded end of a gun barrel; (*F. m. art.*) knob of the cascable;
—— *conique*, (*art.*) coned breech-screw;
—— *d'épreuve*, (*sm. a.*) testing proof breech (for proof of barrels);
—— *fausse*, *fausse* ——, (*sm. a.*) false breech, break off, movable breech of a proof barrel;
—— *à filets concentriques*, (*art.*) concentric groove breech-plug;
—— *mobile*, (*sm. a.*) breechblock, breech-closing mechanism and related parts; (*art.*) movable breechblock of the *canon à balles*;
queue de ——, (*sm. a.*) tang;
revue des ——*s mobiles*, (*mil. slang*) monthly medical inspection;
talon de la ——, (*sm. a.*) tang of the breech.
culasser, v. a., (*art.*) to breech (a gun).
culbute, f., somersault; (*mil.*) overthrow, rout.
culbuter, v. a., to overthrow, defeat with great disorder, rout.
culbuteur, m., (*mach.*) reversing device, tripping mechanism or device, tripper.

cul-de-dé, m., (*unif.*) intersection of threads in the lace of a uniform (similar to the checkerwork on a thimble).
cul-de-lampe, m., (*art.*) cascable; (*cons.*) bracket (e. g., supporting a projecting turret or sentry box).
cul-de-porc, m., (*cord.*) wall knot;
—— *double*, double wall-knot, stopper-knot;
—— *simple*, single wall-knot;
—— *avec tête de mort*, crown knot, crowned wall-knot.
cul-de-poule, m., (*sm. a.*) heel of the butt.
cul-de-sac, m., blind alley.
culée, f., (*pont.*) abutment; bank cut down, etc., for a bridge entrance; in a ponton bridge, the abutment sill complete, i. e., anchorage, shore fastening or "dead man," with chess against the ends of the balks; shore-bay; (*fort.*, etc.) abutment masonry, to take up the lateral thrust of one or more arches;
—— *mobile*, (*pont.*) movable bay, draw, of a bridge.
culer, v. n., (*nav.*) to go astern, make sternway;
nager, *scier*, *à* ——, (in rowing) to back all.
culeron, m., (*harn.*) dock.
culière, f., crupper; channel or gutter stone.
culot, m., bottom, bottom part, lower part (as of a spindle, etc.); small washer for a bolt or pin; (*art.*) base, bottom of a projectile; bottom of a cartridge bag; fragment of the cartridge bag at the bottom of the bore, after firing; primer-cup (in certain cartridge cases); (*sm. a.*) head (of a cartridge case); (*fond.*) refuse metal at the bottom of a crucible; (*powd.*) v. s. v. *cul* (*faux*);
—— *en enclume*, (*sm. a.*) base and anvil of the cartridge case (revolver, model 1892);
—— *expansif*, —— *à expansion*, (*art.*) expanding sabot;
faux ——, (*art.*) part of cartridge bag containing the priming-charge;
—— *rapporté*, (*sm. a.*) the head of a built-up cartridge case, one seated or mounted on the case.
culottage, m., (*art.*) the operation of forming or manufacturing a *culot*.
culotte, f., (*unif.*) riding-breeches; (*mach.*) upright tube, upright of a boiler tube; uptake; (*sm. a.*) butt-cap of a revolver;
vieille —— *de peau*, (*mil. slang*) retired general.
culotter, v. a., (*art.*, etc.) to make a *culot*.
culpabilité, f., guilt;
déclaration de ——, (*law*) verdict of guilty.
culte, m., worship, religion;
service des ——*s*, (*mil.*) chaplain's department.
cultellation, f., (*surv.*) cultellation.
culture, f., cultivation (of fields, etc.); (*in pl.*) cultivated fields;
hautes ——*s*, any crops that grow to a considerable height above ground.
cumul, m., state of filling two or more offices at the same time; of drawing pay for more than one office or employment at the same time.
cunéiforme, a., wedge-shaped, cuneiform; m., (*hipp.*) one of the tarsal bones.
cunette, f., (*fort.*) cunette.
curage, m., dredging, cleaning out (rivers, channels, etc.).
cure-feu, m., poker.
curement, m., v. *curage*.
cure-môle, m., dredger.
cure-pied, m., (*hipp.*) hoof-pick.
curer, v. a., to scour, dredge (a river, harbor); to clean, clear out.
curette, f., cleaner, scraper; (*art.*, *sm. a.*) scraper for cleaning guns, rifles, etc.; (*mil. slang*) cavalry saber:
manier la ——, (*mil. slang*) to do saber exercise;
—— *de mineur*, (*mil. min.*) miner's spoon;
spatule- ——, (*mil.*) cleaner or scraper of a soldier's cleaning kit.

curseur, m., (*in gen.*) slide (as, of a ruler, etc.); index, pointer; (*mach.*) slide piece or block; (*art.*, *sm. a.*) slide of a rear sight; (sometimes) the slide-sight itself; (*F. m. art.*) traversing-head slide;
—— *à double cran de mire*, (*art.*, *etc.*) slide with two sight notches;
—— *à rallonge*, (*sm. a.*) slide fitted with an extension piece, extension-slide;
régler le ——, (*Fr. art.*) to set the deviation scale.
cursière, f., (*art.*) guide-groove.
curvigraphe, m., (*drawing*) curve-tracer, cyclograph; op'someter.
curviligne, a., curved, curvilinear.
curvimètre, m., (*top.*, *drawing*) curvimeter.
custode, f., (*sm. a.*) holster-cap.
cutidure, f., (*hipp.*) a synonym of *bourrelet*, q. v.
cuve, f., tub; pneumatic trough; penstock (of a turbine); (*met.*) the interior of a blast furnace, (more especially) the part above the boshes; (*fort.*) the cylindrical part of a turret, the hollow or cylinder in which a turret is set or works;
à fond de ——, flat-bottomed (as, a ditch);
—— *à huile*, (*met.*) oil-tempering tank.
cuvelage, m., (*min.*, *mil. min.*) tubbing, shaft lining.
cuveler, v. a., v. *cuver*.
cuver, v. a., (*min.*, *etc.*) to tub, to line a shaft.
cuvette, f., (*sm. a.*) mouthpiece of a scabbard; hollow in the face of the bolthead; primer cap or cup; (*torp.*) cup of a circuit-closer; (*inst.*) bulb of a thermometer, reservoir of a barometer; (*fort.*) cunette (rare);
batte de la ——, (*sm. a.*) sputcheon;
—— *d'égouttage*, (*mach.*) drip cup.
cuvier, m., laundry washtub.
cyclable, a., fit for wheeling (of a road, path, etc.).
cyclisme, m., wheeling, bicycling.
cycliste, m., (*Fr. a.*) combatant cyclist, i. e., soldier mounted on a bicycle for fighting purposes.
cyclomètre, m., cyclometer.
cyclone, m., cyclone.
cyclostyle, m., cyclostyle (sort of manifolder).
cylindre, m., cylinder; (*mach.*) roller, roll; arbor, shaft, spindle; (*steam*) steam cylinder; (*art.*) cylinder gauge, shot gauge; (*sm. a.*) bolt, bolt cylinder; revolver cylinder;
—— *admetteur*, (*steam*) high-pressure cylinder;
—— *-agrafe*, (*art.*) clip cylinder (of a fuse);
—— *à basse pression*, (*steam*) low-pressure cylinder;
boîte à ——, (*steam*) cylinder box;
—— *cannelé*, (*powd.*) grooved roller; (*in gen.*) any grooved roller;
—— *de charge*, (*torp.*) charge of a waking torpedo;
—— *à chemise*, (*steam*) jacketed cylinder;
chemise de ——, (*steam*) steam jacket;
—— *sans chemise*, (*steam*) unjacketed cylinder;
—— *à cingler*, —— *cingleur*, (*mach.*) shingling roll;
—— *à la Colleton*, (*pont.*) cylinder ponton;
—— *de composition*, (*artif.*) the composition of rockets, etc.;
couvercle de ——, (*steam*) cylinder cover;
—— *dégrossisseur*, (*mach.*) roughing-down roll;
—— *détendeur*, v. —— *à basse pression;*
—— *de détente*, v. —— *à basse pression;*
—— *durci à la surface*, (*mach.*) chilled roll;
—— *ébaucheur*, (*mach.*) roughing cylinder;
—— *-enveloppe*, (*steam*) steam jacket;
enveloppe de ——, (*steam*) cylinder case, casing;
—— *d'épreuve*, (*art.*) primer gauge;
—— *étireur*, (*mach.*) drawing-down roll; finishing roll;
—— *extracteur*, (*sm. a.*) broken-shell extractor;
—— *finisseur*, (*mach.*) finishing roller;
—— *fixe*, (*powd.*) horizontal carbonizing retort;
fond de ——, (*steam*) cylinder head;
—— *forgeur*, (*mach.*) drawing-down roll;
—— *-frein*, —— *de frein*, (*art.*) recoil-checking cylinder;
—— *à fumée*, (*art.*) smoke-producing cylinder (used in the *écoles à feu*, q. v.);
grand ——, (*mach.*) main roll; (*steam*) v. —— *à basse pression;*
—— *à haute pression*, (*steam*) v. —— *admetteur;*
—— *d'horlogerie*, (*mach.*) clockwork cylinder;
—— *incendiaire*, (*art.*) artifice used to convert ordinary into incendiary shells;
—— *d'introduction*, (*steam*) v. —— *admetteur;*
—— *à laminer*, (*mach.*) roller, roll;
—— *de laminoir*, (*mach.*) flatting-mill cylinder;
—— *-lunette*, (*art.*) shot-, shell gauge;
—— *mobile*, (*powd.*) slip (the French slip is on rollers, and forms the retort itself);
—— *oscillant*, (*steam*) oscillating cylinder;
petit ——, (*steam*) v. —— *admetteur;*
plaque de ——, (*steam*) cylinder plate;
—— *de pompe*, pump-barrel;
—— *porte-papier (de l'indicateur)*, (*steam*) indicator cylinder;
—— *préparateur*, (*mach.*) roughing cylinder;
—— *de pression*, (*mach.*) holding-down cylinder (e. g., as in a planer, to hold down the work);
—— *de rayage*, (*art.*) rifling cylinder;
—— *de réception*, (*art.*) shot gauge, cylinder gauge; (*sm. a.*) cartridge gauge;
—— *de recul*, (*art.*) recoil-checking cylinder;
—— *renversé*, (*steam*) inverted cylinder;
—— *soufflant*, (*mach.*) blowing cylinder;
—— *à tôle*, (*mach.*) plate-roll;
—— *-vanne*, cylindrical sluice gate (turbine);
—— *à vapeur*, (*steam*) steam cylinder;
—— *vérificateur*, (*art.*, *sm. a.*) shot gauge, cartridge gauge, cylinder gauge.
cylindrée, f., cylinder-full (of steam, etc.).
cylindrer, v. a., (*mach.*) to roll.
cylindrique, a., cylindrical.
cylindrographe, m., (*top. and phot.*) topographic cylindrograph.
cylindro-ogival, a., (*art.*) cylindro-ogival.
cymbale, f., (*mus.*) cymbal.
cymbalier, m., cymbal player.

D.

D, (*harn.*) D-ring.
dache, m., (*mil. slang*) hairdresser to the zouaves (a mythical personage);
allez raconter cela à ——, go tell that to the marines.
dactylographe, m., typewriter.
dactylographie, f., typewriting.
dague, f., dirk.
daillot, m., (*cord.*) grommet. v. *andaillot*.
dallage, m., flagging, stone flagging.
dalle, f., flagstone.
daller, v. a., to pave with flagstones, to flag.
dalot, m., (*nav.*) scupper, scupper hole.
daltonien, m., color-blind person.
daltonisme, m., color-blindness.
damage, m., ramming, act of ramming (earth).
damas, m., (*met.*) damask surface, damascening, damask steel; (*sm. a.*) Damascus sword.
damasquinage, m., (*met.*) damasking.

damasquiner, v. a., (*met.*) to damascene, damaskeen; to frost.
damasquinerie, f., (*met.*) art of damascening.
damasquineur, m., (*met.*) damascener.
damasquinure, f., v. *damasquinage.*
damasser, v. a., (*met.*) to give a damask surface to.
dame, f., beetle, rammer; earth rammer; rowlock; (*met.*) dam, dam stone; mass of earth left in natural state between excavations, to mark their depth; (*fort.*) turret of a batardeau;
— *plate,* flat beetle (for use on slopes);
— *ronde,* beetle, rammer, paving beetle.
damer, v. a., to ram, pack (earth); (*fond.*) to ram up.
dameur, m., rammer, packer.
damier, m., checkerboard; register (of names); *exploitation en* —, (of mines) excavation by galleries and pillars.
damoiselle, f., rammer, beetle.
danger, m., danger, peril, hazard, jeopardy.
dangereux, a., dangerous, perilous, hazardous.
dans-œuvre, adv., in the clear (of measurements).
dard, m., dart, javelin; spindle, spike; (*sm. a.*) shoe of a scabbard, chape, crampet;
— *de la tête mobile,* (*sm. a.*) firing-pin guide.
darder, v. a., to dart, throw; hurl, strike; to send down (as the rays of the sun); to launch, cast.
darse, f., wet dock, basin (of a harbor).
darsine, f., small basin, dock (of a harbor).
dasymètre, m., (*mach.*) dasymeter.
dauphin, m., dolphin; (*cons.*) lower mouth of a gutter pipe; (*art.*) handle (on a gun).
davier, m., cramp; machine for drawing joints close; (*nav.*) davit.
dé, m., die; thimble; block; prop; plug; dowel; (*art.*) gauge, cylinder gauge; (*cord.*) bush of a sheave, coak of a block; (*mach.*) crosshead; (*harn.*) D-ring;
— *agrafe,* (*art.*) plunger sleeve catch;
— *briseur,* (*art.*) securing-support;
— *de fermeture,* (*sm. a.*) false or movable breech-plug of a proof barrel;
— *à galets,* roller-bush;
— *en pierre,* stone block;
— *de sangle,* (*harn.*) cincha-ring;
— *de surfaix,* (*harn.*) cincha-ring, spider ring.
débâchage, m., untilting of a cart.
débâcher, v. a., to untilt a cart.
débâclage, m., clearing of a harbor.
débâcle, f., confusion, disorder; breaking up of ice in a river.
débacler, v. a. n., to break up (of ice); to clear a harbor; to unbar a door or window.
déballage, m., unpacking.
déballer, v. a., to unpack.
débandade, f., confusion, disorder, stampede;
à la —, in the greatest confusion, helter-skelter;
feu à la —, (*mil.*) random firing, fire without order or system;
marcher à la —, (*mil.*) to straggle;
s'en aller à la —, (*mil.*) to leave the ranks.
débandement, m., (*mil.*) disbanding, leaving, quitting the ranks.
débander, v. a., to take off a band, a bandage; to relax a spring; (*sm. a.*) to let down the hammer, the bolt, to uncock; (*mil.*) to disband;
se —, to go off, disperse; to grow loose, slack; (*mil.*) to break ranks in disorder.
débarcadère, m., (*r. r.*) station; platform, quay; (of a port) wharf.
débarquement, m., landing, debarkation; disembarkation; (*mil.*) detraining (of troops); landing, descent (of troops on a coast);
affût de —, (*nav. art.*) landing gun-carriage;
canon de —, (*nav. art.*) landing gun;
compagnie de —, (*mil.*) landing party;

débarquement, *corps de* —, v. *compagnie de* —;
troupes de —, (*mil.*) troops for a landing.
débarquer, v. a. n., to land, disembark; to unload, discharge; (*mil., etc.*) to leave a train (of troops etc.).
débarrasser, v. a. r., to clear, disentangle; get rid of; clear off, away; to strip, unload.
débarrer, v. a., to unbar, disembar; to withdraw (a handspike, etc.).
débâter, v. a., to take off a pack saddle.
débauchage, m., (*mil.*) act of enticing a soldier (or sailor) to desert.
débaucher, v. a., (*mil.*) to induce a soldier (or sailor) to desert.
débillardement, m., cantling, cutting diagonally.
débillarder, v. a., to cantle, cut off diagonally.
débit, m., debit; cutting up, sawing, of logs, stone; unit volume of discharge or of flow, (hence, frequently) supply; (*mach.*) feed;
grand —, sawing of a log into planks or slabs, first sawing of a log (generally, a large log);
petit —, quartering, splitting, of a log or piece of wood (for small parts, as helves, spokes, etc.).
débiter, v. a., (of timber, stone) to cut up, to saw up, to subdivide by sawing, etc., with a minimum of waste; to discharge (of streams, channels, sluices, etc.).
débitter, v. a., to unbitt a cable.
déblai, m., (*fort. r. r. etc.*) cutting, excavation, cut; excavated earth; rubbish;
en —, excavated; cut;
remblais et —s, cuts and fills;
route, etc., en —, road, etc., below the level of the adjacent ground.
déblaiement, m., clearing (of the ground), removal of rubbish, dug earth, etc.; removal, sweeping away, of earth, as a result of artillery fire.
déblayé, p. p., *entonnoir* —, (*art., mil. min.*) crater into which none of the earth loosened by the explosion falls back.
déblayer, v. a., to excavate, dig; to remove rubbish or excavated earth; to clear; (*r. r.*) to clear the track; (*mil.*) to clear the ground or position of the enemy; to clear the ground in front of a position;
— *un camp,* (*mil.*) to quit an encampment;
— *le terrain,* (*mil.*) to clear the ground of the enemy;
— *la voie,* (*r. r.*) to clear the track.
déblocus, m., (*mil., nav.*) raising of a blockade.
débloquer, v. a., (*mil., nav.*) to raise the blockade, to relieve a besieged place; (*art.*) to unlock (a safety device).
déboisement, m., clearing of woods, of timber; (*mil.*) clearing away of timber from the front of a position.
déboiser, v. a., to clear of woods, of timber; (*mil.*) to clear or remove timber from the front of a position.
déboîtement, m., dislocation (of a bone); (*mil.*) breaking out of column, leaving, act of leaving, column (either temporarily, as at drill, or permanently, when units separate); (in some movements) stepping, or act of stepping, to the right (or left).
déboîter, v. a., to dislocate (a bone); to separate (as joints of a pipe); (*mil.*) to break out of column, to leave the column; (in certain movements) to step to the right (or left).
débonneter, v. a., v. *décoiffer.*
débordement, m., overflow of a river, flood, freshet; (*mil.*) outflanking, overlapping.
déborder, v. a. n., to overflow, to extend beyond, project; to push off (a boat); (*mil.*) to outflank; outwing, to extend beyond the enemy's flank,
— *les avirons,* to unship the oars.
débosseler, v. a., to take the dents or bruises out of anything.
débosser, v. a., (*cord.*) to unstopper (a cable).

débouchage, m., (*art.*) cutting or setting (and operation) of a fuse;
— *de l'évent*, (*art.*) cutting or setting of a fuse.
débouché, m., debouch; outlet, issue (of a postern, defile, pass, etc.), narrow pass, opening; (*mil.*) exit, issue (from a wood, position, etc.); point of arrival of a column; entrance of troops on battlefield; appearance of the head of a column during a march; (*siege*) breaking out of a sap (from a trench);
— *d'un pont*, waterway of a bridge;
— *de sape*, (*siege*) breaking out of a sap (from a trench).
débouchement, m., outlet, issue, opening.
déboucher, v. a. n., to unstop, uncork; (of a river) to run into, empty itself into; to clear away obstacles, inequalities; to make or cut or punch holes (as rivet holes); (*elec.*) to unplug (as in a resistance box); (*mil.*) to debouch, issue (from a defile, wood, position, etc.); to enter or appear upon the battlefield (of troops); to arrive, appear, at a given point (of columns on the march); (*art.*) to punch or set a fuse; (*siege*) to break out (a sap, parallel, from a trench); (*mil. min.*) to break out from a gallery;
— *l'évent*, (*art.*) to punch, set, a fuse;
faire — *une sape*, (*siege*) to break out a sap;
— *d'une tranchée*, (*siege*) to break from the trench.
débouchoir, m., (*art.*) fuse-punch;
pince-—, v. s. v. *pince*;
— *à vrille*, fuse gimlet.
débouchure, f., mouth of a river; burr; part punched out of a rivet hole.
déboucler, v. a., to unbuckle; (*hipp.*) to unring (a mare);
— *un prisonnier*, to unshackle a prisoner.
débourbage, m., removal of mud.
débourber, v. a., to unmire a carriage, pull it out of the mud; to clean (a pond, cistern, etc.).
débourrage, m., (*hipp.*) first training or breaking-in of a horse; (*mil.*) dressing of recruits into shape; (*mach.*) act of clearing a tool of turnings, etc.; (*mil. min.*, etc.) untamping.
débourrer, v. a., (*hipp.*) to begin to break (a horse); (*mil.*) to get recruits into shape; (*mach.*) to clear a tool of turnings, etc.; (*mil. min.*, etc.) to untamp.
débours, m., money advanced on another's account (rare).
déboursé, m., v. *débours*.
debout, adv., upright, erect, end on, ahead; straight toward (as, a sap from foot to crest of glacis); (*mil.*) command to rise, at drill, etc.;
— *au corps*, end on;
— *les avirons*, up oars;
mer —, head sea;
tracer —, (*siege*) to trace straight toward (as a sap from foot to crest of glacis);
vent —, head wind;
— *au vent*, head to wind.
débraser, v. a., to unsolder.
débrayage, m., (*mach.*) disengaging, uncoupling, throwing out of gear; disengaging gear, release, clutch for throwing out of gear.
débrayer, v. a., (*mach.*) to disconnect, uncouple, throw out of gear, release, ungear.
débrêler, v. a., (*cord.*) to unlash.
débridement, m., (*harn.*) unbridling, removal of bridle.
débrider, v. a., (*harn.*) to unbridle; (hence) to halt;
sans —, without interruption.
débridonner, v. a., (*harn.*) to take off the bridoon.
débris, m., fragments; wreck (of an army, etc.); remains; rubbish.
débrouiller, v. a., to make out, discern;
se —, to get out of a difficulty, to know how to take the initiative.

débroussaillement, m., clearing away of underbrush, bushes (as, from the front of a position).
débroussailler, v. a., to clear away underbrush, etc. (as, from the front of a position).
débrutir, v. a., to clear off the rough, to rough polish.
débusquement, m., (*mil.*) dislodgment, expulsion (v. *débusquer*).
débusquer, v. a. n., (*mil.*) to drive out, dislodge (generally by means of a surprise, from under cover, etc.); to come out (as, from cover, etc.).
décade, f., (*Fr. art.*) set of 10 readings with the Goulier *télomètre*.
décagramme, m., decagram, 10 grams.
décalage, m., unscotching (of a wheel); (*mach.*, etc.) unwedging, unkeying.
décaler, v. a., to unscotch a wheel; (*mach.*, etc.) to remove a wedge, a key.
décalibrement, m., (*art.*) modification of the caliber of a gun: 1, by accident; 2, by transformation of bore (conversion).
décalitre, m., decaliter, 10 liters.
décalquage, m., re-tracing, operation of tracing from a tracing.
décalque, m., counter-tracing (tracing from a tracing); counter-drawing, re-tracing.
décalqué, m., counter-traced drawing, re-tracing.
décalquer, v. a., to counterdraw, to trace from a tracing, (in lithography) to transfer.
décamètre, m., decameter, 10 meters; (*inst.*) 10-meter measuring tape;
— *en ressort d'acier*, 10-meter steel tape;
— *à ruban d'acier*, 10-meter steel tape.
décampement, m., (*mil.*) breaking camp; (*fig.*) making off in haste.
décamper, v. a. n., (*mil.*) to break or strike camp; (*fig.*) to decamp, march off precipitately.
décanter, v. a., to decant.
décapage, m., scouring, cleaning; (*met.*) pickling, dipping, (*elec.*) cleaning of electrical connections.
décapeler, v. a., (*cord.*) to unrig a masthead.
décaper, v. a., to clean, scour (as metal); (*met.*) to pickle, dip; (*elec.*) to clean electrical connections;
— *une fusée*, (*art.*) to uncap a fuse;
— *le sol*, to scrape, level, the surface of the ground (as, for a tent).
décarbonisation, f., (*met.*) decarbonization, decarburation.
décarboniser, v. a., (*met.*) decarbonize, decarburate.
décarburation, f., (*met.*) decarburation.
décarburer, v. a., (*met.*) to decarburize, decarbonize.
décarburisation, f., (*met.*) decarburation, decarbonization.
décatir, v. a., to shrink, to sponge (cloth).
décatissage, m., shrinking, sponging (of cloth).
décentrage, m., (*mach.*, etc.) state of being out of true, out of center; operation of putting out of center; (*art.*) boring a gun on an axis different from that of the exterior surface.
décentralisation, f., (*adm.*) decentralization (as, of accounts, purchases, etc.).
décentraliser, v. a., (*adm.*) to decentralize (accounts, auditing, etc.).
décentration, f., (*mach.*, etc.) act or operation of putting centers, axes, foci, etc., out of true.
décentrer, v. a., (*mach.*, etc.) to put out of true (centers, axes, etc.).
décès, m., death, decease.
déchaînement, m., unchaining.
déchaîner, v. a., to unchain.
déchalement, m., retiring of the tide, leaving banks, etc., visible.
déchaler, v. a., to leave a vessel high and dry.
déchaperonner, v. a., to remove the coping from a wall.
déchapper, v. a., (*fond.*) to take a model out of the mold.

décharge, f., unloading; discharge; release, exoneration; lumber room; (*art., sm. a.*) discharge (of a gun, of a rifle), volley; (*steam*) discharge, exhaust; (*cons.*) safety arch, brace; (*law*) defense;
à ——, (*law*) for the defense, in favor of the prisoner;
arceau en ——, (*cons.*) relieving arch;
en ——, (*cons.*) relieving;
—— *générale*, (*art., sm. a.*) salvo, volley;
—— *juridique*, (*law*) receipt; recognition of acquittal of obligations;
mur en ——, (*cons.*) relieving wall;
pièce de ——, lumber room;
soupape de ——, (*steam, etc.*) eduction valve;
témoin à ——, (*law*) witness for the defense;
tuyau de ——, delivery pipe;
voûte en ——, relieving arch.

déchargé, a., unstressed, unloaded (of keys, shafts, etc.).

déchargement, m., unloading, discharging, tilting; (*art., sm. a.*) unloading of a piece; (*art.*) unloading of a shell, of a shrapnel;
entrer en ——, to break bulk.

déchargeoir, m., (*art.*) sort of shell extractor (*canon à balles*).

décharger, v. a., to unload, empty, tilt; to put down, lay down (a load); to relieve, release, exonerate, discharge, acquit; (*art., sm. a.*) to shoot, fire; to draw a load, to unload; (*mil. min.*) to remove the charge from a mine.

déchasser, v. a., to drive out a nail, bolt, etc.

déchaussement, m., blowing up, undermining (of foundations); baring (of a foundation); (*art.*) to expose, lay bare (a foundation) by artillery fire.

déchausser, v. a., to lay bare, to bare; to undermine (as, a foundation); (*man.*) to take the foot out of the stirrup; (*art.*) to undermine, lay bare, by artillery fire;
—— *un pilotis*, to blow up a pile.

déchéance, f., forfeiture of a right, a franchise, statute of limitation.

déchemisement, m., (*sm. a.*) stripping (of a bullet).

déchet, m., waste, loss; wear and tear; (*fond.*) waste of metal in casting;
—— *d'abatage*, waste or loss in slaughtering cattle;
—— *s de coton*, cotton waste, cop;
—— *de distribution*, (*mil.*) the necessary waste in making issues;
—— *de feu*, (*met.*) waste in the furnace.

déchevêtrer, v. a., (*harn.*) to unhalter.

décheviller, v. a., to unpeg, unpin, remove a pin, etc.

déchiffrement, m., deciphering (of a cryptogram, etc.).

déchiffrer, v. a., to decipher (a cryptogram, a cipher dispatch, etc.).

déchiqueté, a., (*top.*) indented (as a coast line).

déchirer, v. a., to rend, tear, break up; to rip off;
—— *de la toile*, (*mil. slang*) to execute platoon firing.

déchirure, f., rent, tear.

déchouer, v. a., to get a grounded boat afloat.

décigramme, m., decigram, tenth of a gram.

décilitre, m., deciliter, tenth of a liter.

décimation, f., (*mil.*) decimation.

décimer, v. a., (*mil.*) to decimate.

décimètre, m., decimeter, tenth of a meter.

décintrement, m., (*cons.*) removal of the center of an arch.

décintrer, v. a., to remove, strike, the centering of an arch.

décisif, a., decisive;
combat ——, (*mil.*) decisive attack (generally on flank).

décision, f., decision (as of war department, of the chief of the state, etc.); (*Fr. art., cav.*) publication of details, punishment list, etc. (the *rapport* of the infantry), by the c. o. of a regiment to his officers and to batteries and troops.

déclanche, f., (*mach.*) disengaging gear, trip-gear.

déclanchement, m., (*mach., etc.*) tripping, trip-gear; release, releasing-gear; operation of tripping, of releasing, of throwing out of gear.

déclancher, v. a., (*mach., etc.*) to trip, release, throw out of gear, disengage a catch.

déclaration, f., declaration, statement;
—— *de guerre*, (*mil.*) declaration of war;
—— *de quittance*, (*adm.*) memorandum or *état de solde* that the sum mentioned has been paid to its rightful owner.
—— *de versement au trésor*, (*adm.*) statement (not a receipt) that a sum due to state by a *conseil d'administration* or by a military person has been paid.

déclarer, v. a. r., to declare, certify; to find (guilty or not guilty); to break out, set in;
—— *la guerre*, (*mil.*) to declare war.

déclassement, m., (*adm.*) passage or transfer of a person, of an article, from one class to another; (*fort.*) abandonment of forts, etc., for military purposes;
—— *d'une bouche à feu*, (*art.*) rejection of a piece as too defective for regular service, accompanied by retention for drill, target practice.

déclasser, v. a., (*adm.*) to pass, transfer, a person, an article, from one class to another; (*fort.*) to abandon a fort, a work, for military purposes; (*art.*) to put "out of system;"
—— *une bouche à feu*, (*art.*) to reject a piece on account of minor defects, unfitting at for regular service, but not for target practice, etc.

déclencher, v. a., to unlatch.

déclic, m., trigger; click (of a pile driver); catch, pawl; (*mach.*) trip, trip-gear; disengaging lever; (*sm. a.*) hair-trigger;
—— *à double détente*, (*sm. a.*) hair-trigger.

déclinaison, f., declination (of the needle); (*inst.*) declination of an instrument.

déclinatoire, m., (*inst.*) declinator, orienting compass; (*law*) exception to the jurisdiction, request of defendant to be sent before a different jurisdiction.
—— *d'incompétence*, (*law*) formal or official statement of lack of jurisdiction.

décliner, v. a., (*inst.*) to declinate, set on a given point or bearing.

décliquer, v. a., to unclinch.

décliqueter, v. a., to let fall a pawl.

déclisser, v. a., (*r. r.*) ro remove a fishplate.

déclive, a., (*top.*) sloping.

déclivité, f., (*top.*) slope, declivity.

déclouer, v. a., to draw nails.

décoffrage, m., (*mil. min.*) unsheathing (act of).

décoffrer, v. a., (*mil. min.*) to unsheathe.

décoiffage, m., (*art.*) uncapping of a fuse, removal of a fuse cap.

décoiffer, v. a., (*art.*) to remove the cap from a fuse.

décoincer, v. a., to unwedge, remove, knock out wedges.

décoilage, (*art., sm. a.*) starting of a cartridge case from its seat, primary extraction.

décollement, m., ungluing, unpasting; (*carp.*) tenoning; (*met. and in gen.*) separation of adjacent parts (e. g., so as to form a crack in a casting, etc.); (*art., sm. a.*) primary extraction, starting of cartridge case, of an obturator, from its seat.

décoller, v. a. r., to unglue, unpaste; to behead; to come apart; (*art., sm. a.*) to start (a cartridge, an obturator, from its seat).

décolletage, m., (*mach.*) screw cutting.

décolleter, v. a., (*mach.*) to turn, cut, screws;
machine à ——, screw machine;
tour à ——, screw lathe.

décoltharisation, f., scraping off, removal, of a layer of coal-tar (as from a gun).

décolthariser, v. a., to scrape off, remove, a layer of coal-tar (as from a gun).

décombres, f. pl., rubbish, chips.

décommander, v. a., (*mil. etc.*) to countermand an order.

décommettre, v. a., (*cord.*) to unlay a rope.

décommis

décommis, a., (*cord.*) fagged, unlaid.
décomposer, v. a., to decompose, break up into constituent parts or elements; (*mil.*) to go through a movement by detail.
décomposition, f., decomposition, breaking up into elements or parts;
—— *de l'effectif*, (*Fr. a.*) report of a very comprehensive character, made annually to the minister of war, setting forth all circumstances of the service of the regiment or *service* for which made, both of an individual and of a general character;
—— *des forces*, resolution of forces.
décompte, m., deduction, discount, allowance; (*adm.*) balancing of accounts; (*mil. slang*) mortal wound;
faire son —— *à quelqu'un*, to settle one's accounts;
—— *de libération*- (*Fr. a.*) auditing and balancing of accounts between a troop unit and the government, showing over- and under-issues;
recevoir son ——, (*mil. slang*) to be done for, have one's accounts settled.
décompter, v. a., to allow; to deduct, discount.
déconfire, v. a., (*mil.*) to defeat thoroughly.
déconfiture, f., (*mil.*) total defeat.
déconjuguer, v. a., (*mach.*) to throw out of gear.
déconstruire, v. a., to pull down, raze, demolish.
décoration, f., decoration, ornament; decoration (of an order);
—— *étrangère*, foreign decoration;
—— *de juillet*, decoration given to combatants in the revolution of July, 1830;
—— *militaire*, (*mil.*) military decoration (as, Medal of Honor, U. S. A.; Victoria Cross, British);
porter une ——, to wear, to have, a decoration.
décorder, v. a., (*cord.*) to unlay, untwist;
—— *les bouts*, to unlay a rope's end.
décordonnage, m., (*powd.*) operation of removing crust from stampers of powder mill.
décorer, v. a., to decorate, adorn; to confer a decoration upon.
découcher, v. n., (*mil.*) to lie or sleep out of quarters.
découcheur, m., (*mil. slang*) man who is habitually absent without leave.
découdre, v. a., to unsew, to rip off;
se ——, (*man.*) of a horse, to become disunited in his gaits, to have broken, irregular gaits.
découlement, m., flowing, running off, trickling.
découler, v. n., to flow, run off, trickle, drop; proceed from.
découpage, m., (*met.*, etc.) cutting out of designs from marked plates.
découpé, a., (*top.*) indented (as, a coast line).
découper, v. a., to cut out.
découpler, v. a., (*art.*) to uncouple (horses).
découpoir, m., (*mach.*, etc.) punch; cutter, stamp.
découpure, f., piece or part cut out by a punch, etc.
décourber, v. a., to straighten.
découronner, v. a., (*mil.*) to clear a crest or height of the troops occupying it, to dislodge troops from a height.
décours, m., wane, decrease.
décousu, a., (*man.*) irregular, broken (of gaits); (*hipp.*) loose-built, badly put on (as, head on neck, fore quarters and hind quarters).
découvert, a., (*top.*) open, treeless; (*mil.*) unsheltered, exposed, without cover;
à ——, exposed to fire, unmasked.
découvert, m., (*steam*) negative lap.
découverte, f., discovery; (*mil.*) reconnaissance, reconnoitering; (*nav.*) lookout;
aller à la ——, (*mil.*) to explore, to scout;
aller à la —— *de*, (*mil.*) to reconnoiter;
envoyer à la ——, (*mil.*) to explore, to scout;
escadron de ——, (*cav.*) scouting troop, troop sent on exploration, scouting duty (independent cavalry);
patrouille de ——, (*cav.*) scouting, exploration, patrol;

défaut

découverte, *service de* ——, (*cav.*) the information service proper of independent cavalry.
découvrir, v. a., to uncover, lay open, expose; to discover, make out; (*mil.*) to leave unprotected, unguarded, to unmask, expose; (*met.*) in tempering of steel, to lose the blackish scale acquired by heating;
se ——, (*mil.*) to come out from under cover; (*man.*) to undercover, to undercheck.
décrassage, m., cleaning, scraping, scouring; (*art., sm. a.*) removal of fouling.
décrasser, v. a., to clean, scrape, scrape out (as tubes, furnaces); (*art., sm. a.*) to remove fouling.
décret, m., decree;
——*-règlement*, (*mil.*) decree having the force of a regulation.
décrépitation, f., decrepitation.
décrépiter, v. n., to decrepitate.
décrire, v. a., to describe.
décrochage, m., unhooking.
décroche-obturateur, m., (*art.*) obturator releasing-gear.
décrocher, v. a., to unhook, unfasten;
—— *les avant-trains*, (*art.*) to unlimber.
décrochoir, m., (*mach.*) disengaging hook, releasing hook.
décroiser, v. a., to uncross;
—— *les échelons*, (*mil.*) to form direct from oblique echelons.
décrotter, v. a., to clean, rub, brush off, scrape off.
décrottoir, m., scraper.
décrouir, v. a., (*met.*) to anneal.
décrouissage, m., (*met.*) annealing.
décroûtage, m., (*fond.*) dressing down, fettling.
décroûter, v. a., (*fond.*) to dress down, fettle.
décrue, f., decrease, fall of water.
décuirasser, v. a., (*nav.*) to reduce the amount of armor carried by a vessel.
décuirassement, m., (*nav.*) reduction of the amount of armor carried by a vessel.
déculassement, m., (*art., sm. a.*) unbreeching, blowing off of the breech, opening or unscrewing or blowing out of the breech by the action of the powder gases.
déculasser, v. a., (*art., sm. a.*) to unbreech.
dedans, adv., inside;
de ——, in the clear;
mettre un cheval ——, (*man.*) to break a horse.
dédommagement, m., compensation, indemnity, indemnification.
dédommager, v. a., to compensate, indemnify, make good.
dédosser, v. a., to plank timber, saw off the outside.
dédoublement, m., (*fenc.*) a doubling in the opposite direction to, and immediately following upon, the adversary's doubling; (*mil.*) formation of column of twos from column of fours, or of single from double file;
—— *constitutif*, subdivision of a troop unit into two, on mobilization, in time of war.
dédoubler, v. a., to split, to divide into two; (*fenc.*) to execute a *dédoublement*;
—— *un amarrage*, (*cord.*) to cast off the turns, the upper layers (of a mooring);
—— *les files*, (*mil.*) to form single rank;
—— *les rangs*, (*mil.*) to form column of twos from column of fours, or single from double file;
—— *un régiment*, (*mil.*) to form two regiments out of one.
défaire, v. a., to undo, unrip, unpin; to loose; (*cord.*) to untie, unfasten; (*mach.*) to disengage; (*mil.*) to rout, defeat, overthrow;
—— *la croix d'un câble*, (*cord.*) to unkink a cable, clear the hawse;
—— *les tours d'une manœuvre*, (*cord.*) to take out the kinks of a rope;
—— *les tours d'un palan*, (*cord.*) to clear away a fall;
—— *une vis*, to unscrew.
défaite, f., (*mil.*) defeat.
défausser, v. a., to straighten that which is warped, to unwarp.
défaut, m., flaw; defect, blemish, break; deficiency, want, default; fault, imperfection;
condamner par ——, (*law*) to bring in a verdict by default;

défaut *de la cuirasse,* break;
— *de l'épaule,* hollow of the shoulder;
— *d'exécution,* defect of workmanship;
faire ——, *(law)* to fail to appear;
—— *de fonctionnement, (art., sm. a.)* failure of any part of the mechanism to work properly;
—— *de soudure, (met.)* defect in welding, defective welding.

défection, f., defection, disloyalty.

défectueux, a., defective, imperfect.

défendable, a., defensible, susceptible of defense.

défendant, a., defending; *(fort.)* flanking;
à son corps ——, guardedly, timorously.

défendeur, m., *(law)* defendant.

défendre, v. a., to defend, protect, maintain, uphold, support; to shelter, screen; to prohibit, forbid;
se ——, *(man.)* to refuse to obey, to resist, to make a show of resistance (of a horse).

défense, f., defense; protection; prohibition; (of boats) fender; *(law)* defense; *(man.)* resistance, countermove (on the part of the horse); *(mil.)* defense, protection; *(fort.)* defensive work, (sometimes) outwork; (in pl. generally) defensive works, fortifications, (more esp.) outworks; *(siege)* defense;
—— *absolue, (fort.)* work or part of a work depending on its own fire;
—— *s accessoires, (fort.)* accessory defenses, i. e., obstacles (exclusive of ditch);
—— *active, (mil.)* defense involving a return to the offensive;
—— *centrale,* —— *centralisée, (fort.)* system of defense in which no attempt is made to stop the invader at the frontier by forts, but in which reliance is placed on a fortified center outside of the enemy's lines of operations, and serving as a point of support to the national army;
—— *continentale, (s. c. fort.)* land defense, landside defense, of a s. c. work;
—— *s continentales, (s. c. fort.)* land defenses (in a s. c. work, to repel land attack);
—— *cuirassée, (fort., nav.)* armored protection;
—— *directe,* v. —— *absolue;*
—— *extérieure, (siege)* defender's operations beyond the *position avancée,* q. v.;
—— *extérieure active, (siege)* operations carried on by the mobile troops of the defense, beyond the permanent works, against the outer works, positions, etc., of the attack;
—— *fixe, (s. c. defense)* defense by guns and submarine mines;
guerre de ——, defensive warfare, defensive war;
ligne de ——, *(siege, etc.)* line of defense;
—— *à la mine, (siege)* defense by mining, by subterranean warfare;
—— *mobile, (siege)* operations by the mobile troops of the defense; *(s. c. defense)* operations, defense, etc., by torpedo boats, mobile torpedoes, etc.;
—— *mobile de terre, (s. c. defense)* use of land troops (infantry, field artillery, etc.) in seacoast defense;
—— *passive, (mil.)* mere defense, passive defense;
—— *périphérique, (fort.)* system of defense in which it is sought to arrest the enemy on the frontier itself by means of forts;
place de ——, *(fort.)* fortified position ready to sustain a siege;
—— *des places, (siege)* defensive siege operations;
—— *radiale, (fort.)* system of defense in which arresting forts are placed on the lines the enemy must follow in his invasion, or in his attempts to reach the capital;
—— *rapprochée, (siege)* close defense, defensive operations when the enemy is close to the work or position; (in a defensive battle) defense at close quarters;
—— *réciproque,* —— *relative, (fort.)* reciprocal defense, defense in supporting relations (c. g., when a work or part of a work supports another);
—— *sous-marine, (s. c. defense)* submarine defense;
—— *souterraine, (siege)* defense by mining.

défenseur, m., *(law)* counsel for defense; *(mil.)* defending troops, the defender.

défensif, a., defensive, pertaining to the defense.

défensive, f., *(mil.)* defensive;
—— *active,* v. —— *offensive;*
—— *offensive,* the offensive-defensive;
—— *passive,* pure defensive, passive-defensive.

déferler, v. n., (of the sea) to break, to form breakers.

déferré, m., *(farr.)* shoe that has been cast; old horseshoe; *(hipp.)* horse that has cast a shoe.

déferrer, v. a., to take off, remove, irons; *(farr.)* to unshoe a horse;
se ——, *(farr.)* to cast a shoe, to lose a shoe.

déferrure, f., act of removing irons; *(farr.)* act or operation of unshoeing.

défi, m., challenge.

déficit, m., shortage, deficit.

défier, v. a., to challenge, defy;
—— *la chaîne,* to stand clear of a chain.

défilateur, m., *(fort.)* defilading instrument.

défilé, m., *(top., etc.)* defile, strait, narrow path; (hence, *mil.)* any narrow passageway, as a dike, bridge, street, etc.; *(mil.)* march past; *(fort.)* passage around a traverse.

défilement, m., *(fort.)* defilade, defilement; act of defilading;
—— *à 1:n,* defilade against shots falling at a slope of 1:n;
—— *par la crête (les crêtes) et par le terreplein,* v. —— *par la masse couvrante;*
—— *horizontal,* defilade in direction;
indice de ——, v. s. v. *indice;*
—— *par la masse couvrante,* defilade by lowering the surface to be covered (parapet being assumed);
plan de ——, plane of defilade;
—— *par le relief,* direct defilade, defilade in altitude;
—— *par la surface à couvrir,* defilade by raising the parapet to the plane of defilade (surface to be covered being given);
—— *par le terreplein,* defilade by lowering the terreplein;
—— *par le tracé,* v. —— *horizontal;*
—— *par des traverses,* defilade by traverse;
—— *vertical,* v. —— *par le relief.*

défiler, v. a. n., *(fort.)* to defile, defilade; to protect from enfilade; (hence, *in gen.)* to cover, to protect (as, by an embankment, r. r. cut, etc.); *(mil.)* to march past; *(art.)* to file off (as teams); —— *au 1:n, (fort.)* to defilade against trajectories of a slope of 1:n;
—— *la parade, (mil.)* to march past, to pass in review; (*mil. slang*) to die;
se ——, *(mil.)* to get under cover;
—— *en tiroir, (mil.)* to change front, when in column, by the march of divisions through one another from the rear.

déflagration, f., deflagration.

déflexion, f., deflection, deviation.

défoncement, m., staving, staving in, beating in.

défoncer, v. a., to stave, beat, or knock in; to break up, dig up, a road; *(mil.)* to break up the enemy, throw him into disorder.

déformable, a., deformable.

déformation, f., deformation, strain, change of form, set; *(art. sm. a.)* upsetting of a projectile; *(art.)* relative displacement of parts of a gun-carriage (in recoil-checking);
affût à ——, *(art.)* generic term for a gun-carriage in which a relative displacement of the parts takes place in checking the recoil;
—— *élastique,* change of form, strain, within the elastic limit;
—— *permanente,* permanent set.

déformer, v. a., to alter, change, the form of anything;
se ——, to undergo a change of form, to be strained, to acquire a set.

défournement, m., uncharging, drawing (of a charge from a furnace, of a batch of bread from an oven, etc.).

défourner, v. a., to draw (a charge from a furnace, a batch of bread from an oven, etc.).
défourrer, v. a., (*cord*.) to take the service off a rope.
défrettage, m., (*art., etc.*) unhooping, removal of a hoop (as, from a gun).
défretter, v. a., (*art., etc.*) to unhoop (as, a gun).
défrichage, m., clearing up of land.
défrichement, m., v. *défrichage*.
défricher, v. a., to clear up land.
dégagement, m., disengagement, extrication release, liberation; escape (of gas, steam, etc.); recess; clearance, clearing; (*mach*.) education clearance, angle of relief, clearance angle; (*fenc*.) disengaging, disengagement, followed by a straight thrust;
—— *du champ de tir*, (*mil*.) clearing the ground in front of a position so as to have a clear field of fire.
dégager, v. a. r., to clear, disengage, free, extricate, disentangle, loosen, relieve, separate; to escape (as gas, steam, etc.); to give out (as heat, etc.); (*mil*.) to clear the ground (before a position, a work, etc.); (*cord*.) to clear (a fall, a tackle); *mach*.) to uncouple, throw out of gear; (*fenc*.) to disengage;
—— *le champ de tir*, (*mil*.) to clear the field of fire;
—— *l'épée*, (*fenc*.) to disengage;
—— *un palan*, (*cord*.) to clear away a fall;
—— *le pivot*, (*mil*.) at drill, to gain ground on the pivot;
se ——, (*mil*.) to decline a proffered combat;
—— *le terrain*, v. —— *le pivot*;
—— *une troupe*, (*mil*.) to rescue, extricate, a body of troops from a critical or difficult position.
dégainer, v. a., to draw, unsheath (a sword, etc.).
dégarnir, v. a., to strip, remove; to dismantle, unrig; (*mil*.) to remove troops (from the center, the wings); (*harn*.) to unharness, to remove the harness from a horse's back;
—— *les avirons*, to unship the oars;
—— *un cabestan*, to unrig a capstan;
—— *une place*, (*mil*.) to remove a part of its supplies, of its garrison, from a fort;
se ——, (*mil*.) to grow thinner (of lines, ranks, etc.).
dégât, m., damage, havoc, waste, dilapidation; (*mil*.) damage to private property by troops; *commission des* ——*s*, (*mil*.) board to appraise damage done to private property by troops in maneuvers.
dégauchir, v. a., to dress roughly, to rough down (as, timber, stone) (*dégrossir* is the usual word), to smooth, straighten, polish, trim.
dégauchissement, m., smoothing, straightening, rough dressing, trimming down.
dégel, m., thaw, thawing.
dégeler, v. a. n., to thaw, to thaw out.
dégenoper, m., (*cord*.) to unseize, unlash (a temporary seizing).
dégonder, v. a., to unhinge.
dégonflement, m., collapse, subsidence, reduction; removal of gas from a balloon.
dégonfler, v. a., to reduce (a swelling), cause to subside, collapse, discharge the gas from a balloon.
dégorgement, m., opening, cutting out; clearing freeing from obstacles;
—— *d'une embrasure*, —— *du parapet*, (*fort*.) cutting out an embrasure in the parapet.
dégorgeoir, m., (*art*.) priming-wire; (*fond*.) venting-wire;
—— *à fraise*, (*art*.) reaming-wire;
—— *à maillet*, vent punch;
—— *ordinaire*, priming-wire fitted with a wooden handle;
—— *à pointe*, ordinary priming-wire;
—— *repoussoir*, (*s. c. art*.) special priming-wire for seacoast guns;
—— *simple*, priming-wire with the ordinary eye or loop;
—— *à taillant plat*, vent-punch;
—— *à vrille*, vent-gimlet.

dégorger, v. a., to clear; to open; to remove an obstruction (as, from a pipe, etc.); (*mach*.) to clear; (*farr*.) to shape out a horseshoe in the rough;
—— *une embrasure*, (*fort*.) to open, pierce, cut an embrasure;
—— *une fusée*, (*art*.) to bore a fuse;
—— *la lumière*, (*art*.) to prick the vent.
dégourdir, v. a., (*mil*.) to form a raw recruit into a well set-up soldier.
dégoût, m., loathing; (*hipp*.) state of being off his feed;
avoir du ——, (*hipp*.) to be off his feed.
dégradation, f., injuring, injury; damage, wear and tear; injury, damage (to small arms and guns, due to use or to accident); degradation (military punishment);
—— *civique*, (*law*) loss of civil rights as a punishment;
—— *militaire*, (*mil*.) military degradation, ignominious dismissal from the army;
—— *de teintes*, (*drawing*) grading or shading of tints.
dégrader, v. a., to degrade, disgrace; to suffer injury, to undergo deterioration, to suffer wear and tear, be worn, to be damaged (of materials in general, of guns, ammunition, etc.); to damage, dilapidate; (*mil*.) to dismiss from the service with ignominy, after stripping of all dignities and of grade; (*drawing*) to grade (as tints, shades).
dégrafer, v. a., to unhook, unclasp.
dégraissage, m., cleaning, scouring; removal of grease.
dégraissement, m., v. *dégraissage*.
dégraisser, v. a., to clean, to remove grease; (*carp*.) to dress down, take the sharp edge off a corner, an angle; to beard, trim, dress down.
dégras, m., degras, wool-grease;
mettre au ——, to free leather from oil, etc., used in tanning.
dégravoiement, dégravoiment, m., washing away, undermining, laying bare (of running water).
dégravoyer, v. a., to undermine, lay bare, wash away (of running water).
degré, m., degree; grade, gradation, pitch, extent; step, stair (usually in plural);
—— *de hausse*, (*ball*.) degree of elevation.
dégréage, m., (*cord*.) unrigging.
dégréement, m., (*nav*.) loss of rigging during an engagement; (*cord*.) unrigging.
dégréer, v. a., (*cord*.) to unrig.
dégréner, v. a., v. *désengrener*.
dégrossage, m., drawing down (of wire).
dégrosser, v. a., to draw (wire) down.
dégrossir, v. a., to rough-hew, rough down, dress down roughly (of materials); to reduce approximately to shape, trim down;
—— *le bois*, to rough-plane wood, to jack it down;
outil à ——, rough-shaping tool;
—— *une recrue*, (*mil*.) to set up a recruit, to get a recruit into shape.
dégrossissage, m., rough dressing, rough shaping, rough hewing, roughing down; act of giving its approximate shape to anything.
dégrossisseur, m., (*mach*.) rough-dressing tool.
dégrossisseur, a., reducing (in size, of lathes, rollers; used also of process itself).
déguerpir, v. n., to pack off, to hasten with extreme rapidity (somewhat familiar as a military word).
déhanché, a., (*hipp*.) hipshot.
déharnachement, m., unharnessing.
déharnacher, v. a., to unharness.
déhocher, v. a., (*fond*.) to tap or strike the model with a wooden mallet to detach or loosen it from the mold.
dehors, adv., outside of; m., outside, exterior; m. pl., (*fort*.) outworks (between the *corps de place* and the covered way);
de —— *en* ——, from outside to outside (in measurements, etc.).
déjecteur, m., —— *anti-calcaire*, (*steam*) scale-ejector.

déjeter — 124 — **déménager**

déjeter, v. r., (of wood, trees) to cast, to warp; (*met.*) to warp, become distorted, to get out of true (as, in tempering).

déjoindre, v. a., to disjoin; start; part, open, gape.

déjouer, v. a., to baffle, to neutralize, to render null and void.

déjour, m., opening left in the fellies of a new wheel.

déjuger, v. r., (*man.*) to undercover, undercheck.

délabrement, m., dilapidation.

délabrer, v. a., to shatter, batter, ruin, dilapidate.

délacer, v. a., to unlace.

délai, m., delay; time, allowance of time (as, to carry out an order, for a change of station, etc.);
—— *d'arrivée*, (*Fr. a.*) time granted every soldier to reach his destination;
dans le plus bref ——, as soon as possible, with the least possible delay;
—— *de grâce*, (*Fr. a.*) allowance of time for joining, granted every recruit or volunteer that fails to join on the assigned day, before being taken up as an *insoumis*, q. v. (1 month in France, 2 months for Algiers, Tunis, etc.);
—— *de recours*, (*Fr. a.*) period (3 months) within which a regular decision of the minister of war must be appealed to the *conseil d'état*;
—— *de repentir*, (*Fr. a.*) time allowed a man absent without leave to return to duty before being declared a deserter or *insoumis*;
—— *de route*, (*Fr. a.*) time allowance for making a journey;
—— *de tolérance*, (*Fr. a.*) 4 days allowed an officer changing residence, in addition to regular allowance of time for the purpose; (*in gen.*) permission to delay in carrying out an order;
—— *de transport*, (*Fr. a.*) time allowed for transportation by rail, wagon, etc.

délaissement, m., abandonment.

délaisser, v. a., to abandon, give up.

délardement, m., chamfer, chamfering; splay, slope; slope of earthwork.

délarder, v. a., to chamfer, splay; to cut sloping (as earthwork).

délavage, m., dilution of a color by water.

délaver, v. a., to dilute a color.

délayé, a., (of the soil) reduced to a state of mud, soaked, soaking wet.

délayer, v. a., to dilute, temper (lime, clay, etc.); to soak.

délégataire, m., (*in gen.*) deputy; one who has received authority (*délégation*) to act in the name and place of another; (*Fr. mil. adm.*) officer or functionary who receives the *délégation* of the funds of the minister of war.

délégation, f., (*adm.*) warrant or authority bestowed upon or granted to a person;
—— *de la commission supérieure*, (*mil. r. r.*) a board, at general headquarters, in charge of transports beyond the base of operations, in the zone operated by the personnel of the regular railway companies, or otherwise, from that base to the *stations de transition* (q. v.);
—— *de signature*, authority to sign;
signature par ——, authorized signature of one person for another;
—— *de solde*, (*mil.*) assignment of his pay, by an officer or soldier, to third persons.

délestage, m., removal of ballast from a vessel.

délester, v. a., to unballast a vessel.

déliaison, f., (*mas.*) unbound masonry.

délicat, a., delicate, difficult;
entrée —— *e*, difficult entrance (of a port).

délicoter, v. r., (*harn.*) to strip off his halter or collar.

délié, a., thin, slender, fine; m., up-stroke of a letter.

délier, v. a., to untie, unfasten.

délimitation, f., delimitation (of a territory, of a sphere of influence).

délimiter, v. a., to delimit.

délinéation, f., delineation.

déliquescence, f., deliquescence.

déliquescent, a., deliquescent.

délit, m., misdemeanor; (*mas.*) wrong bed of a stone;
le corps du ——, the offense itself;
mettre une pierre en ——, (*mas.*) to lay a stone on the wrong bed.

déliter, v. a., (*mas.*) to lay a stone on its wrong bed; to split a stone along its cleaving grain;
—— *la chaux vive*, to slake quicklime;
se ——, (*mas.*) to cleave, split (as slate, etc.); to exfoliate (by the action of frost).

délivrance, f., delivery, issue (of stores, arms, etc.);
—— *par force*, (*mil.*) rescue of a prisoner.

délivrer, v. a., to deliver, deliver over; to free from; set at liberty, free (a prisoner);
—— *par force*, —— *par violence*, to rescue a prisoner.

délogement, m., removal, departure; (*mil.*) change of quarters (of billets) (obs.).

déloger, v. a. n., (*mil.*) to change quarters (of billets) (obs.); to dislodge the enemy, drive him out of his position (v. *débusquer*); to decamp, move out hastily and in disorder (when surprised).

delta, m., (*top.*) delta;
métal ——, delta metal.

délutage, m., unluting.

déluter, v. a., to unlute.

démaçonner, v. a., to undo, take down, masonry work.

démaigrir, v. a., to thin, reduce (a slab, a log, a piece of timber, wood, etc.).

démaigrissement, m., feather-edge; thinning; place where a stone, etc., has been thinned, reduced in thickness.

démailler, v. a., to unlink, unshackle; to unlace

démaillonner, v. a., to unlink a chain.

démanchement, m., unhafting; taking off a haft or handle; removal of a handle.

démancher, v. a., to remove a haft or handle.

demande, f., demand, request, petition, suit, inquiry, application; (*mil.*) requisition; (*cons.*) proportions required in a piece of timber;
à la —— *de*, so as to fit, take, apply to;
—— *en cassation*, (*mil.*) request of the c. o. of a detachment for the reduction of a n. c. o. to the ranks;
filer à la ——, (*cord.*) to pay off a rope according to the stress.

demander, v. a., to demand, require, call for, want; (*mil.*) to submit a requisition;
—— *du câble*, (*cord.*) to require more rope.

demandeur, m., (*law*) plaintiff.

démanillage, m., unshackling.

démaniller, v. a., to unshackle.

démantèlement, m., (*fort.*) dismantling; blowing up works.

démanteler, v. a., (*fort.*) to dismantle, to raze, demolish.

démarcation, f., demarcation;
ligne de ——, line of demarcation.

démarche, f., gait; walk; step; bearing; measure.

démarquage, m., erasure of a mark (as, the letter, number, etc., from a weapon, etc.).

démarquement, m., fraudulent removal of the mark on anything (as, clothes, etc.).

démarquer, v. n., (*hipp.*) to lose mark of mouth.

démarrage, m., (*cord.*) unmooring; casting off of a rope; untying, undoing (of lashings); (*r. r.*) starting of a train; (*mach.*) starting of an engine; (*art.*) starting, moving off (of teams, carriages).

démarrer, v. a., (*cord.*) to cast off a rope, to clear, to unbend; to cast loose; to unmoor; to unlash; (*mach.*) to start an engine; (*art.*) to start, to cause to start (of teams, carriages) (*r. r.*) to start.

démasquer, v. a., (*mil.*) to unmask a front, a position, a movement, an attack.

démêler, v. a., to disentangle, unravel; to part.

déménagement, m., moving out.

déménager, v. a. n., to move out (of one house into another).

démentir, v. a. r., to give way, bulge (building).
démettre, v. a. r., to dislocate, put out of joint; dismiss; turn out; resign; lay down.
demeure, f., lodging; residence;
à ——, fast; fixed; rigidly attached to;
fixer à ——, to secure permanently.
demeurer, v. n., to remain, last, stand, stay; to live in.
demi-à-droite (-gauche), m., (of foot troops, marching, or from a halt) right (left) oblique; (*art., cav.*) right (left) half-wheel.
demi-appel, m., (*Fr. cav.*) signal for the execution of an order previously given.
demi-arrêt, m., half-stop;
—— *des doigts*, (*man.*) slight check or stoppage (of the horse, by the reins).
demi-bastion, m., (*fort.*) half-bastion.
demi-bataillon, m., (*inf.*) half-battalion, wing (obs.).
demi-bateau, m., (*pont.*) sort of half-ponton boat, used by the French engineers.
demi-batterie, f., (*art.*) half-battery (administrative subdivision).
demi-bois, v. *mi-bois*.
demi-brigade, f., (*mil.*) Revolutionary name for regiment.
demi-caponnière, f., (*fort.*) half-caponier, single caponier.
demi-capucine, f., (*sm. a.*) lower band.
demi-cercle, m., semicircle; (*art.*) semicircular gauge; (*inst.*) half-circle; (*fenc.*) half-circle parry;
—— *gradué*, semicircular protractor.
demi-chantier, m., (*art.*) half-block.
demi-châssis, m., (*fond.*) half of a molding-box.
demi-cheminement, m., (*surv.*) a combination of the method by polar coordinates and of *cheminement*, q. v.
demi-clef, f., (*cord.*) half-hitch;
—— *à capeler*, clove-hitch.
demi-colonne, f., (*mil.*) half-column.
demi-comble, m., (*cons.*) area or garret under the roof of a shed.
demi-compagnie, f., (*inf.*) half-company.
demi-cuirasse, f., (*cav.*) breastplate.
demi-cylindre, m., (*art.*) halfroller.
demi-distance, f., (*mil.*) half-distance.
demi-écrou, m., half-nut.
demi-équipage, m., (*Fr. pont.*) half bridge train; (*Fr. art.*) subdivision of the siege train, so called.
demi-escadron, m., (*cav.*) the French equivalent for troop (U. S.).
demi-essieu, m., half-axletree.
demi-étampe, f., (*mach.*) half-die; one of the parts of a die or swage in two parts.
demi-étape, f., (*mil.*) half a day's march.
demi-fer, m., (*farr.*) tip.
demi-ferme, f., (*cons.*) half-frame, half-truss (of a roof).
demi-fiche, f., half the depth to which a pile is to be driven into the earth.
demi-flasque, f., (*art.*) bracket of a block-trail carriage.
demi-flèche, f., (*art.*) middle rail, one of the two parts of the under body prolongation of the *flèche* (French 80^{mm}).
demi-fourniture, (*auxiliaire de campement*), (*Fr. a.*) bedding of straw, issued for lack of regular bed outfit (*appel des réservistes et territoriaux*).
demi-fourreau, m., (*art.*) axle-plate (French 80^{mm}).
demi-fraisé, a., half-countersunk.
demi-galerie, f., (*mil. min.*) low gallery, 1 meter wide by 1.30-1.5 meters high.
demi-gargousse, (*art.*) half-cartridge (the full charge being subdivided for convenience of loading).
demi-gîte, f., (*art., siege*) half-sleeper of a gun platform.
demi-gorge, f., (*fort.*) demigorge.
demi-guêtre, f., (*unif.*) half-legging, short legging.
demi-haut-fourneau, m., (*met.*) blast furnace whose height is less than 7 or 8 meters.

demi-joue, f., (*cord.*) check of a block.
demi-lune, f., (*mach.*) eccentric catch; (*fort.*) demilune, ravelin;
—— *double*, ravelin with a redoubt;
—— *simple*, simple ravelin.
demi-lunette, f., (*mach.*) collar (as, for a gun in a boring bench).
demi-madrier, m., (*pont.*) half-chess; (*art.*) shifting-plank.
demi-merlon, m., (*fort.*) half-merlon.
demi-moule, f., (*fond.*) half-mold.
demi-moyeu, m., half-nave (in the manufacture of iron wheels).
demi-nœud, m., (*cord.*) overhand knot.
demi-parallèle, f., (*siege*) demiparallel, short parallel covering only a part of the work attacked, and constructed in the zigzag zone, after the second parallel.
demi-place d'armes,, f., (*siege*) v. *demi-parallèle*.
demi-portée, f., (*ball.*) half-range.
demi-portière, f., (*pont.*) raft of two pontons.
demi-ration, f., (*mil.*) half-ration, short allowance;
mettre à ——, to put on allowance.
demi-redoute, f., (*fort.*) half-redoubt, sort of lunette open in rear.
demi-régiment, m., (*Fr. cav.*) half-regiment (two squadrons).
demi-revêtement, m., (*fort.*) half-revetment, scarp revetted up to the natural level of the ground.
demi-rond, m., half round of a wagon body; half round (stick, bar, etc.).
demi-sang, m., (*hipp.*) half-bred horse.
demi-section, f., (*Fr. inf.*) half section (of a company).
demi-signalement, m., (*Fr. a.*) summary description or descriptive list.
demi-solde, f., (*mil.*) half-pay (no longer an official expression); officer on half-pay;
en ——, on half-pay.
démission, f., resignation; (*mil.*) resignation from the service (of officers, and also of *commissionnés*);
donner sa ——, to resign from the service;
donner sa —— *de*, to resign an office or employment.
démissionnaire, m., (*mil.*) officer who has resigned;
officier ——, officer who has resigned.
demi-teinte, f., half-tint.
demi-tore, m., half-round.
demi-tour, m., (*cord.*) half-turn, elbow, kink; (*inf.*) about face (from a halt), to the rear (if marching); (*art.*) reverse (of a single carriage team, section, mule), about (of a platoon, of a cannoneer, driver, mounted); (*cav.*) about (individual), (left, right) about (of a platoon);
—— *à droite*, (*inf.*) about face (from a halt), to the rear (if marching); (*cav.*) an individual right about, right about (of a platoon); (*art.*) right about (of a canoneer, driver, mounted); (*fenc.*) right rear volt (bayonet exercise);
—— *à gauche*, (*art.*) reverse (of a single carriage team, section, mule); left about (of a platoon, of a cannoneer, driver, mounted); (*cav.*) individual about; left about (of a platoon); (*fenc.*) left rear volt (bayonet exercise);
—— *individuel*, (*inf.*) about face (from a halt), to the rear (if marching); (*cav.*) an individual about; (*art.*) reverse (of a single team, carriage, section, mule); about (of a cannoneer, driver, mounted);
—— *par section*, (*art.*) left about (of a platoon);
—— *sur les épaules*, (*man.*) half-circle on the fore hand;
—— *sur les hanches*, (*man.*) half-circle on the rear hand.
(*Demi-tour* is frequently used alone, the context indicating the nature of the movement. It is also the name of the call for the movement.)
demi-volte, f., (*man.*) half-circle, followed by a change of hands.
demi-voûte, f., half vault or arch.

démobilisation 126 **denture**

démobilisation, f., (*mil.*) passage of an army, of a regiment, etc., from war to peace footing.
démobiliser, v. a., (*mil.*) to demobilize, to muster out of war service, to go back to peace footing.
demoiselle, f., earth rammer.
démolir, v. a., to demolish, break up, pull down, raze; (*mil.*) to break up stores, etc., condemned; (more esp., *Fr. a.*) to break up cartridges, projectiles, into constituent elements;
—— *une poudre*, (*powd.*) to lixiviate (damaged) gunpowder;
tir à ——, v. *tir.*
démolition, f., demolition, destruction; (*mil.*) the separation of cartridges, projectiles, etc., into their constituent elements, saving the parts that may serve again; military demolition, destruction;
—— *s*, rubbish.
démonstratif, a., (*mil.*) demonstrative;
combat ——, demonstrative action (frequently synonymous with front attack).
démonstration, f., (*mil.*) demonstration.
démontable, a., that takes apart, to pieces, "take-down;"
affût ——, (*art.*) carriage in parts, take-down carriage;
canon ——, (*art.*) jointed gun, gun that takes apart.
démontage, m., (*art., sm. a.*) dismounting, taking apart, (as, a gun, a rifle, etc.); (the destructive) dismounting of the enemy's guns by artillery fire.
démonté, a., (*man.*) unhorsed, dismounted (against one's will).
démonter, v. a., (*art., sm. a.*) to dismount, take apart, to take down (as, a gun, a rifle); (*art.*) to dismount, i. e., remove guns from a work, a gun from its carriage: to dismount, i. e., disable, the enemy's guns; (*mil.*) to supersede (in command); (*man.*) to unhorse; (*in gen.*) to unship, to take to pieces;
—— *la chèvre*, (*art.*) to strike the gin.
démoralisation, f., (*mil.*) loss of morale, demoralization.
démoucheter, v. a., (*fenc.*) to take the button off a foil.
démoulage, m., (*fond.*) lifting, removal of a pattern, of a casting, from a mold.
démouler, v. a., (*fond.*) to lift, remove a pattern, a casting, from a mold.
démunir, v. a., to strip, deprive, of anything (as, a place of its stores, a fort of its ammunition, etc.).
démuseler, v. a., to unmuzzle.
dendrites, m. pl., (*met., min.*) dendrites.
déni, m., refusal of what is due; (*law*) refusal of a judge to *statuer*; action of denying;
—— *de justice*, refusal of a judge to pass on a case, to do justice in a case, whether by open refusal or by negligence.
denier, m., penny; (in pl., as an administrative term) fund, moneys;
allocation en —— *s*, allowance in money;
comptabilité en —— *s*, money accountability, responsibility;
les —— *s de l'état*, public money, public funds.
dénivelée, f., difference of level of two points.
déniveler, v. a., to put out of level.
dénivellation, f., unevenness; state of being out of level; change, fall of level (as in a stream); (*top.*) difference of level.
dénivellement, m., variation of level; settling, sinking (as of an engine, etc.); state of being out of level.
dénombrement, m., census; (*mil.*) computation (man by man) of numerical strength of troops; collective enumeration of the various categories of troops.
dénombrer, v. a., to take a census.
dénomination, f., denomination; (*mil.*) military title of address.
dénoncer, v. a., to declare, publish, the expiration of a treaty, of an armistice, to denounce a treaty; to report to the authorities, to the law, etc.;
—— *la guerre*, to declare war.
dénonciation, f., denunciation (of a treaty, etc.); report made to the authorities, etc.;

—— *de la guerre*, declaration of war.
dénouer, v. a., (*cord.*) to undo a knot, to untie, loose.
dénoûment, m., (*cord.*) untying, loosening.
dénoyage, m., —— *artificiel*, hydro-pneumatism of a turbine.
denrée, f., article, commodity; any article of food (except meat, generally in pl., esp. as a military term), used for both man and beast;
—— *s fourragères*, forage;
—— *s-vivres*, rations, articles composing the soldier's subsistence.
densimètre, m., (*ball.*) densimeter;
—— *à mercure*, mercury densimeter.
densité, f., density;
—— *absolue*, (*ball.*) absolute density (exclusive of pores and interstices);
—— *de chargement*, (*ball.*) density of loading;
—— *des coups*, (*ball.*) grouping of hits, density of grouping;
—— *de feux*, (*art., sm. a.*) density of fire;
—— *gravimétrique*, (*ball.*) gravimetric density, weight of unit volume, inclusive of air spaces;
—— *de groupement*, (*ball.*) density of grouping, ratio of number of shots to unit area on which they fall;
—— *de la ligne de combat*, (*mil.*) number of men per yard of front;
—— *au mercure*, (*ball.*) real density, specific weight, density by mercury;
—— *des points d'impact*, (*ball.*) grouping of hits, density of hits;
—— *réelle*, v. —— *au mercure*;
—— *de section*, (*ball.*) sectional density;
—— *transversale*, v. —— *de section*.
dent, f., tooth; tooth (of a saw, of a horse, etc.); (*mach.*) tooth, catch, cog, pawl; (*top.*) mountain top of prismatic form;
à ——, cogged, toothed;
—— *d'arrêt*, (*r. f. art.*) cocking-toe (Maxim semi-automatic);
—— *d'en bas*, lower tooth;
—— *en bec de perroquet*, gullet tooth (saw);
—— *caduque*, (*hipp.*) milk tooth;
—— *de cheval*, (*hipp.*) adult tooth;
—— *à cheville*, pin;
—— *de dessous*, lower tooth;
—— *de dessus*, upper tooth;
—— *de devant*, front tooth;
—— *d'engrenage*, (*mach.*) ratchet; gear tooth;
—— *s étagées*, —— *en étages*, (*mach.*) stepped teeth, teeth in steps;
—— *d'en haut*, upper tooth;
—— *d'impulsion*, (*mach.*) impulse tooth;
—— *incisive*, incisor;
—— *de lait*, v. —— *caduque*;
—— *de loup*, (*art., mach.*) pawl, catch;
—— *machelière*, jaw tooth;
—— *mitoyenne*, middle tooth;
—— *molaire*, molar tooth;
—— *œillère*, eye tooth;
pas des —— *s*, (*mach.*) pitch of teeth;
—— *persistante*, (*hipp.*) tooth late in coming, and that never falls out;
pied d'une ——, (*mach.*) shoulder of a tooth or cog;
plein d'une ——, (*mach.*) thickness of a tooth or cog;
—— *de remplacement*, v. —— *de cheval*;
roue à —— *s*, (*mach., etc.*) ratchet wheel;
—— *supplémentaire*, (*mach.*) hunting cog;
—— *de sûreté*, (*sm. a.*) safety lug;
—— *triangulaire*, (*art.*) cutter.
denté, a., (*mach., etc.*) cogged, toothed; fitted with teeth;
roue —— *e*, cogwheel;
—— *en scie*, serrated.
dentelé, a., (*top.*) (of a coast) indented.
dentelé, m., (*hipp.*) denticulated muscle.
denteler, v. a., to tooth, notch, indent.
dentelure, f., indentation.
denter, v. a., to tooth, to cog (as, a wheel).
dentition, f., (*hipp.*) dentition;
—— *de cheval*, second dentition;
—— *de poulain*, first dentition.
denture, f., (*mach.*) the teeth or cogs of a wheel, gearing; pitch of gearing;
—— *à développantes de cercle*, evolute teeth;

denture, à flancs droits, radial flank teeth;
— à fuseaux, pin teeth;
— à deux points, double pin gearing;
— à un seul point, single pin gearing.
dépareiller, v. a., to break, spoil, a set, a pair, etc.
départ, m., departure (of a train, of a column, of a traveler, etc.); hour of departure; start (in a race); (art., sm. a.) discharge (of a firearm);
au ——, (art., sm. a.) ready to fire;
— lancé, flying start (in a race);
mécanisme de ——, (art., sm. a., etc.) firing mechanism;
rive de ——, (pont.) bank from which a bridge is built or is dismantled.
département, m., department, subdivision; department (administrative subdivision of France); department (of government administration);
— de la guerre, war department, war office, Horse Guards.
dépassé, a., (met.) overstirred;
bronze ——, bronze that has been overstirred.
dépasser, v. a., to exceed, overreach, surpass, go beyond; (cord.) to unbend, clear, loosen, pay out to the end, unreeve;
— une manœuvre, un palan, to unreeve a rope, a tackle;
— les rênes, to pass the reins over the horse's head;
— les tours (de câble, de chaîne), to clear (the hawse, a chain).
dépêche, f., dispatch;
se battre à —— compagnon, to fight without giving quarter;
— chiffrée, cipher dispatch;
— télégraphique, telegram.
dépendance, f., dependence; (in plural) outbuildings, dependencies in general (as, of a fort).
dépense, f., expense; expenditure; consumption; discharge, flow (of water); dispensary (of a hospital).
déperdition, f., loss (of heat); escape (of gas).
dépérir, v. n., to fall into ruins; to waste, decay.
dépérissement, m., wear and tear.
dépêtrer, v. a., to extricate, disentangle, free, clear;
— un cheval, to extricate a horse entangled in his harness.
déphosphoration, f., (met.) dephosphorization.
déphosphorer, v. a., (met.) to dephosphorize.
dépister, v. r., (man.) to mistrail (itself of the gallop when the imprint of the rear foot is in front of or behind that of the diagonal anterior).
déplacement, m., displacement, change of position; (mil.) change of station, (hence) journey; (nav.) displacement;
— des aiguilles, (r. r.) working or operation of switches;
— d'assiette, (man.) unsteadiness of seat;
— du tir, (art., sm. a.) shift or transfer of fire to a new target.
déplacer, v. a., to change the position of a thing, to displace; to misplace; (mil.) to change, transfer (troops);
— le tir, (art., sm. a.) to change from one target to another;
se ——, to move, change position, station, etc.
déplâtrage, m., removal of plaster.
déplâtrer, v. a., to remove plaster.
déplier, v. a., to unfold;
— la prolonge, (art.) to undo, get out, the prolonge.
déploiement, m., unfolding, display (of a flag); (mil.) deployment;
— en avant, deployment to the front;
— par le flanc, deployment by the flank;
intervalle de ——, deploying interval;
— stratégique, installation of troops in the probable theater of operations; disposition of troops in the order selected or found proper for marching against the enemy;
— en tirailleurs, deployment as skirmishers.

déplombage, m., (art.) stripping (of the lead coating of a projectile).
déplomber, v. a., (art.) to strip off lead (as, in a lead-coated projectile).
déployer, v. a., to unfold, display; (nav.) to display (colors); (mil.:) to deploy, extend;
— en avant, to deploy to the front;
— par le flanc, to deploy by the flank;
se ——, to deploy, extend;
— en tirailleurs, to deploy as skirmishers.
dépointage, m., (art.) loss of aim (of a gun, due to firing), state of being off the target.
dépolarisant, m., (elec.) depolarizer.
dépolarisation, f., (elec.) depolarization.
dépolariser, v. a., (elec.) to depolarize.
dépoli, a., dead, dull; (of glass) ground.
dépolir, v. a., to give a dead surface to anything, to take the polish off; (of glass) to grind.
dépolissage, m., removal of polish; grinding (of glass).
déportation, f., transportation.
déporté, m., convict (transported for life).
déposant, m., deponent (witness).
déposer, v. a., to deposit, lay down; to settle; to depose, to testify; to lodge a complaint;
— les armes, (mil.) to lay down one's arms, surrender, yield, capitulate.
déposition, f., (law) deposition.
déposter, v. a., (mil.) to drive the enemy from a post, a position.
dépôt, m., storehouse, magazine; place of deposit; depot; sediment, settling; (mil.) depot of a regiment; the part of the regiment at the depot;
— d'approvisionnements, supply depot;
— de batterie, (siege) store of tools, supplies, etc. (for working parties);
— central de l'artillerie, (Fr. art.) establishment having a collection of arms of all nations and times, of models, and a library, workshop, etc.;
— de chevaux malades, (Fr. a.) veterinary establishment in rear of an army of operations for sick and wounded horses;
— de convalescents, (Fr. a.) depot for men discharged from the hospitals, but not yet able to resume active service (may exist in peace; in war, are established on the lines of march and of evacuation);
— de corps de troupe, (mil.) regimental depot;
— d'éclopés, (Fr. a.) sort of hospital for men in need of rest, or but temporarily and not seriously disabled (established on the ligne d'étapes);
— d'élevage, v. —— de transition;
— d'étalons, (in France) government establishment in which stallions are kept for the service of private mares;
— des fortifications, (Fr. a.) establishment having charge of fortifications, maps, telegraphy, etc.; suppressed in 1886 by the distribution of its duties to other branches;
— de garantie, (adm.) earnest money, a bidder's guaranty or bond;
— général de la guerre, (Fr. a.) depot of archives (for campaigns and war in general);
— général des mines, (mil. min., siege) depot of material for mining operations (Fr. a., dependency of the main engineer park);
— intermédiaire, (siege) a depot from 1,000 to 1,500 meters in rear of the batteries, used for tools, etc., during their construction, and afterwards for implements, spare carriages, etc.;
— de munitions, (siege) ammunition depot, shell park;
— de prisonniers de guerre, (mil.) place of establishment where prisoners of war are concentrated and kept;
— de projectiles, (siege) shell, projectile, magazine in a battery; projectile expense magazine;
— de recrutement, (Fr. a.) current expression for bureau de recrutement;
— de remonte, (mil. hipp.) remount establishment;

dépôt *de remonte mobile*, (*mil.*) a remount establishment accompanying an army corps in the field;
—— *secondaire*, (*Fr. a.*) supply depot, between the *station-magasin* and *convoi administratif* (supplied by the *station-magasin*);
—— *de télégraphie*, (*Fr. a.*) telegraph establishment (purchase, manufacture, issue, receipt, etc., of telegraph material);
—— *de tranchée*, (*siege*) trench depot, store of tools, implements, etc., immediately in rear of the zigzags;
—— *de transition*, (*Fr. a.*) a remount establishment taking horses at the age of from 3½ to 4 years and keeping them until 5 years of age before sending them into service.

dépouille, f., spoil, booty; tapering, sloping surface; skin of a dead animal; (*fond.*) taper, delivery.

dépouillement, m., spoiling, spoliation; abstract of accounts.

dépouiller, v. a., to strip, lay bare; spoil, plunder; to make extracts from, an abstract of (an account, report, etc.); (*fond.*) to remove the mold, etc., from a casting.

dépréciation, f., wear and tear, wear, waste; depreciation.

dépréder, v. a., to plunder.

déprédation, f., depredation.

dépression, f., depression;
—— *de l'horizon*, dip of the horizon;
télémètre de ——, (*art.*) depression range finder.

dépressitude, f., vertical distance of a point below a datum plane.

déraillement, m., (*r. r.*) leaving the rails, derailing, derailment.

dérailler, v. n., (*r. r.*) to leave, "jump" the track, to run off the track, to be derailed.

dérailleur, m., (*r. r.*) shifting track, derailing track (i. e., short section of track specially contrived for derailing purposes, and transported and used where needed).

déraper, v. n., to slip, trip (of a bicycle tire); to pull out (of an anchor).

dérasement, m., shaving down, cutting down or off (as, a parapet, a traverse, etc.).

déraser, v. a., to cut off, to remove, to clear away; to shave down; to cut down (as, a parapet).

dérayer, v. a., to unscotch a wheel; to work loose (of spokes).

déréglable, a., (*inst.*) liable to get out of adjustment.

déréglage, m., (*inst.*) putting out of adjustment; state of being out of adjustment; (*in gen.*) state of being out of order.

dérégler, v. a., to put out of order; (*inst.*) to put out of adjustment.

dérênement, m., unreining.

dérêner, v. a., to slacken the reins; to unrein.

dérivation, f., deflection, tapping (of a stream, current); (*art., sm. a.*) distance of point of fall from plane of fire; (more rarely) wind deviation; (*mil. min.*) branching of secondary from a central fuse, communicating fire to the separate branches; (*elec.*) shunt, shunted current, derivation, shunting of a current;
canal de ——, sluice;
en ——, (*elec.*) shuntwise, shunted, on a shunt;
fil de ——, (*elec.*) shunt wire;
—— *de route*, turning (of a road);
voie de ——, branch road.

dérive, f., (*art.*) lateral sight allowance, deflection allowance; possible length of lateral travel of rear-sight crosshead; (*nav.*) leeway, departure of a vessel from her course;
contre- ——, (*art.*) negative lateral allowance;
donner la ——, (*art.*) to make the lateral allowance on a sight;
en ——, adrift (of boats, ships, etc.);
—— *du jour*, (*art.*) lateral allowance deduced from the conditions of the day, and true for that day only.

dérivé, a., derived, secondary; (*elec.*) shunt, shunted, in derivation; (*mil. min., etc.*) branching from (one train or fuse to another).

dériver, v. a. n., to derive water or turn a stream from its natural bed; to be derived or turned from its natural course (water stream); to drift (of boats, vessels); to unrivet, unclinch, to remove a rivet; (*elec.*) to shunt.

dérobade, f., (*man.*) sudden movement by which a horse tries to slip from under his rider.

dérobé, a., secret, private (as, a staircase, door, etc.); (*hipp.*) worn so much that a shoe not be put on (of a hoof).

dérobement, m., (*fenc.*) attack in which the foil passes from the upper to the lower line.

dérober, v. a., to rob, steal; to shelter, screen, conceal; (*fenc.*) to execute a *dérobement*;
—— *une marche*, (*mil.*) to steal a march on the enemy;
se ——, (*man.*) to shrink (as, from a jump); to shun, slip away; (*mil.*) to avoid, shun, a combat;
se —— *de dessous l'homme*, (*man.*) to slip from under the rider.

dérochage, m., (*met.*) pickling.

dérocher, v. a., (*met.*) to pickle; to scour metal;
brosse à ——, scratcher.

déroger, v. n., to modify, change, alter, be contrary to.

dérogation, f., (*mil.*) modification of regulations.

dérouiller, v. a., to remove rust.

dérouilleur, m., (*Fr. a.*) soldier who takes care of the *armement de réserve* in the arsenals.

dérouler, v. a., to unroll, to feed (as, wire from a drum).

dérouleur, m., (*mil. teleg.*) man who unwinds the wire from its drum;
aide- ——, assistant to a *dérouleur*.

déroute, f., (*mil.*) rout, defeat;
en ——, in flight (disorderly);
en pleine ——, completely routed;
mettre en ——, to route.

derrière, prep., adv., behind;
cheval de ——, (*art., etc.*) wheel horse;
train de ——, (*hipp.*) hind quarters.

derrière, m., the hind, hinder, part of anything; tail of a cart; (*mil., in pl.*) the rear of the army, (and more esp.) the trains in rear of the army;
assurer ses ——*s*, (*mil.*) to assure a line of retreat;
—— *du système percutant*, (*Fr. art.*) rear block of the firing mechanism of the *canon à balles*.

désablage, m., (*fond.*) removal of sand from castings.

désabler, v. a., (*fond.*) to remove sand from castings.

désaccoter, v. a., to remove props, shores, stays, etc.

désaccoupler, v. a., (*harn.*) to uncouple.

désaciérer, v. a., to cause (a blade, etc.) to lose the properties of steel.

désaffectation, f., (*mil.*) removal or transfer of a soldier from his arm of service because of physical unfitness to serve therein.

désaffleurer, v. n., to jut out, to project, to form a jog.

désaffourcher, v. a., to unmoor.

désagencer, v. a., (*mach., etc.*) to throw out of gear.

désagrafage, m., unclasping, unhooking.

désagrafer, v. a., to unclasp, unhook; (*mach.*) to release, as a catch.

désagréger, v. a., to knock to pieces, to throw into complete disorder; to disarrange (as, the cartridges in a clip).

désaguerrir, v. a., to make unwarlike.

désaiguilleter, v. a., (*cord.*) to unlash.

désaimantation, f., (*elec.*) demagnetization.

désaimanter, (*elec.*) to demagnetize.

désajusté, a., *cheval* ——, horse not true in his paces.
désajuster, v. a., to derange, disorder, disarrange.
désamarrage, m., (*cord.*) taking off (a mooring) or lashing, undoing of a knot or seizing, unlashing, unmooring.
désamarrer, v. a., (*cord.*) to undo a knot, to unmoor, to unbend, to take off a seizing or lashing, to unlash.
désamorçage, m., (*sm. a.*) unpriming of a cartridge case; (*art.*) unpriming of a gun.
désamorcer, v. a., (*art.*) to unprime a gun; (*sm. a.*) to unprime a cartridge case;
pince à ——, (*sm. a.*) unpriming-pincers;
se ——, (*elec.*) to run down (of a dynamo).
désancrer, v. a., to weigh anchor.
désapprovisionnement, m., (*sm. a.*) removal of cartridges from the magazine.
désapprovisionner, v. a., (*sm. a.*) to remove the cartridges from the magazine of a rifle.
désarborer, v. a., (*nav.*) to strike the colors.
désarçonner, v. a., (*man.*) to unseat; to unsaddle; to unhorse; to throw.
désarmé, a., (*nav.*) out of commission.
désarmement, m., (*mil.*) disarming (of prisoners, etc.); disarmament, i. e., a term descriptive of the (proposed) reduction of the excessive numbers of modern armies on a peace footing; passage of an army from war to peace footing; (*fort.*) turning into arsenals, etc., the armament of a fort; removal of a gun from a battery, a fort; (*nav.*) laying up a ship in ordinary.
désarmer, v. a., to disarm; (*mil.*) to pass from a war to a peace footing; to reduce armaments, armies; (*fort.*) to remove a gun from a fort, a battery; (*sm. a.*) to put the hammer, bolt, at the safety notch, to uncock; (*art.*) to unload a gun; to release the brake (after firing); (*fenc.*) to disarm one's adversary; (*nav.*) to put a ship out of commission;
—— *un frein,* (*art.*) etc.) to release a brake.
désarrimer, v. a., (*nav.*) to shift the stowage; (*mil.*) to strip a balloon of its car and accessories.
désarticuler, v. a., to disjoint.
désassembler, v. a., to take apart or asunder, down, to pieces.
désassiéger, v. a., (*mil.*) to raise a siege.
désaturé, a., unsaturated; (of steam) superheated.
désaxer, v. a., (*mach.*) to get out of axis.
désazotation, f., (*chem.*) denitrification.
désazoter, v. a., (*chem.*) to denitrify.
desceller, v. a., to unseal, unfasten, loosen.
descendant, m., ebb tide.
descendant, a., (*mil.*) coming off, marching off (guard, trenches, etc.);
garde ——*e,* old guard.
descendre, v. a. n., to descend, come, get down; to dismount (from a horse); to flow off, to fall, to ebb; to go down, descend (a river); (*art.*) to dismount (cannoneers from chests); to dismount (a gun); (*mil.*) to make a descent, an irruption;
—— *de cheval,* to dismount;
—— *de garde,* (*mil.*) to come off guard;
—— *la garde,* (*mil.*) to be relieved from guard; (*mil. slang*) to die;
—— *une rivière,* to drop down a river;
—— *la tranchée,* (*siege*) to be relieved from duty in the trenches.
descente, f., fall, descent; slope, coming down, letting down; (*cons.*) cellar stairs; (*mil., nav.*) landing, irruption; (*mil.*) coming off (a duty); (*steam,* etc.) down stroke of a piston;
—— *blindée,* (*siege*) blinded descent (into the ditch);
—— *de cave,* (*cons.*) cellar stairs;
—— *de cheval,* (*man.*) dismounting;
—— *à ciel ouvert,* (*siege*) open descent (into the ditch);
—— *à la double attaque,* (*siege*) descent (into the ditch) made in two stories;
faire une ——, (*mil.*) to make a landing or irruption;

descente *de fossé,* (*siege*) descent into the ditch;
—— *de fossé blindée* (*siege*) descent into the ditch by blinded sap;
—— *de la garde,* (*mil.*) coming off, marching off, guard;
—— *de main,* (*man.*) relaxation of the hand, of the reins;
—— *du piston,* (*steam,* etc.) down stroke of a piston;
—— *souterraine,* (*siege*) descent (into the ditch) by gallery;
—— *de la tranchée,* (*siege*) coming off duty in the trenches;
tuyau de ——, soil-pipe.
description, f., inventory, description;
—— *signalétique,* (*mil.*) descriptive list.
deséchouer, v. a., to get a ship or boat afloat.
désemballage, m., unpacking.
désemballer, v. a., to unpack.
désembarquement, m., disembarking, landing.
désembarquer, v. a., to disembark, land, unship.
désemboîter, v. a., to disjoint, get out of joint, dislocate.
désembrayage, m., (*mach.*) throwing out of gear-uncoupling, disengaging, disengaging-gear.
désembrayer, v. a., (*mach.*) to throw out of gear, to uncouple, to disconnect, to disengage.
désemparer, v. a. n., to quit, to abandon; (*nav.*) to disable (a ship, a torpedo, a turret, etc.).
désencasteler, v. a., (*hipp.*) to correct or remove contraction of the hoof.
désencasteleur, m., (*hipp.*) removing or remedying contraction of the hoof.
désenchaîner, v. a., to unchain, remove, take off, a chain.
désenclancher, v. a., (*mach.*) to throw out of gear, disconnect, uncouple.
désenclouage, m., (*art.*) unspiking.
désenclouer, v. a., to draw a nail; (*art.*) to unspike a gun; (*hipp.*) to draw the nail that pricks a horse's hoof.
désengrenage, m., (*mach.*) uncoupling, disengaging; throwing out of gear.
désengrener, v. a., (*mach.*) to throw out of gear; to ungear; uncouple; to take off (e. g., chain from capstan);
—— *la chaîne,* to take the chain off the capstan;
—— *les roues,* (*mach.*) to throw wheels out of gear or play.
désengreneur, m., chain-guard, clearing-guide.
désenrayage, m., unscotching; (*sm. a.*) act or operation of putting safety device out of gear; (*art.*) act or operation of releasing the brake; (*mach.*) uncoupling, disconnecting.
désenrayer, v. a., to unscotch, unchock; to unlock (a wheel), to take off the drag; (*sm. a.*) to put the safety apparatus out of gear; (*art.*) to release the brake; (*mach.*) to put out of gear, to uncouple, disconnect.
désenrôler, v. a., (*mil.*) to discharge, dismiss, a soldier.
désenasbler, v. a., to get off (a boat aground).
désenseller, v. a., to unsaddle (a rider).
désépauler, v. n., (*sm. a.*) to remove, take down, the piece from the shoulder (after firing).
déséquiper, v. a., to take off equipments, gear, working parts, etc.; (*art.*) to take off equipments;
—— *la chèvre,* (*art.*) to unrig the gin;
—— *la pièce,* (*art.*) to remove the equipments and ammunition necessary to the service of a piece;
se ——, (*art.*) to take off equipments (cannoneers).
désergoter, v. a., (*farr.*) to file a hoof to the quick.
désert, m., desert.
déserter, v. n., to quit, abandon; (*mil.*) to desert;

déserter à *l'ennemi*, to go over to the enemy.
déserteur, m., (*mil.*) deserter;
— *ennemi*, deserter from the enemy.
désertion, f., (*mil.*) desertion;
— à *l'ennemi*, desertion to the enemy;
— à *l'étranger*, desertion involving escape to a foreign country; desertion from troops serving abroad;
— à *l'intérieur*, desertion without quitting the country.
déshérité, a., badly formed, imperfect (as, organs of a horse).
déshydrater, v. a., (*chem.*) to remove water from (a substance).
désignation, f., appointment, nomination, assignment; (*mil.*) detail.
désigner, v. a., to appoint, nominate, assign; (*mil.*) detail, tell off for service or duty.
désincorporation, f., (*mil.*, etc.) disincorporation.
désincorporer, v. a., to disincorporate, disunite.
désincrustant, a., (*steam*) scale destroying.
désincrustation, f., (*steam*) destruction, removal, of scale.
désincruster, v. a., (*steam*) to scale a boiler.
désinfecter, v. a., to disinfect, fumigate.
désinfection, f., disinfection, fumigation.
désobéir, v. a., to disobey.
désobéissance, f., disobedience, contumacy.
désoler, v. a., to waste, lay waste.
désordonné, a., disorderly, unruly, ungovernable.
désordonner, v. a., to revoke orders; to confuse, disorder; to throw into disorder.
désordre, m., disorder, confusion; uproar, riot, havoc.
désorganisation, f., disorganization; (*mil.*) confusion or disorder (of troops, of the enemy, etc.); dismantling (as, of a farmhouse, to keep the enemy from making use of it); destruction, disorder (as, of the enemy's siege works, etc.).
désorganiser, v. a., to disorganize, disorder; (*mil.*) to break up the enemy, etc.; to dismantle (as, a farmhouse, to keep the enemy from using it); to disorganize, destroy (the enemy's siege works, etc.).
désorientation, f., (*top.*, *in gen.*) loss of bearings; (*inst.*) state of being out of station.
désorienter, v. a., (*top.*, *in gen.*) to cause to lose one's bearings; (*inst.*) to get out of station;
se —, to lose one's bearings, one's way.
désoxydant, a., (*chem.*) deoxidizing.
désoxydation, f., (*chem.*) deoxidation.
désoxyder, v. a., (*chem.*) to deoxidize.
désoxygénation, f., (*chem.*) deoxidation.
désoxygéner, v. a., (*chem.*) to deoxidize.
dessangler, v. a., (*harn.*) to ungirth, to loosen the girths.
desséchant, a., drying, dessicative.
desséché, a., dried, seasoned (of wood).
dessèchement, m., draining, drying;
fosse de —, drain;
travaux de —, drainage.
dessécher, v. a., to dry, drain; to season (wood).
dessécheur, m., steam-drier.
dessein, m., design, plan; project; resolution, intention.
desseller, v. a., to unsaddle.
desserrement, m., (*mil.*) loosening or loss of distance in columns on the march.
desserrer, v. a., to slacken, to loosen; to unclamp; (*inst.*) to let the compass card down on the pivot; (*art.*) to unship a brake;
se —, to work loose;
— *une vis*, to loosen a screw.
desserroir, m., screw-driver.
desservir, v. a., to supply; to perform, or attend to, the service of anything;
— *un caisson*, (*art.*, *sm. a.*) to pass out ammunition from a caisson.
dessicatif, a., drying; m., (*Fr. a.*) regulation drying oil.
dessication, f., drying; (of wood) seasoning.
dessin, m., drawing, draught, sketch; design, plan; designing;

dessin *d'après nature*, sketch from life, from nature;
— *d'atelier*, working drawing;
cahier de —, drawing book, sketchbook;
— *calqué*, tracing;
— *coté*, drawing, etc., with dimensions marked on it;
— *de construction*, working drawing;
— *au crayon*, pencil sketch;
— *décalqué*, counter tracing;
— *de*, — *en*, *détail*, detailed drawing sketch, of details;
— *d'ensemble*, drawing of the whole, general plan, general view;
— *d'exécution*, working drawing;
— *linéaire*, linear, mechanical, drawing;
— à *main levée*, free-hand drawing;
— *militaire*, sketch for military purposes;
— *négatif*, negative of a drawing (blue print);
— à *la plume*, pen-and-ink drawing;
— *de projets*, working drawing;
— *topographique*, field sketch;
— *au trait*, line-drawing, drawing in outline.
dessinateur, m., draughtsman, designer.
dessiner, v. a., draw, to sketch, design, delineate, outline;
— *d'après nature*, to sketch, draw from nature, from life;
— à *main levée*, to draw free-hand.
dessoler, v. a., (*hipp.*) to draw the sole of a horse's hoof.
dessoudage, m., (*met.*) unwelding, separation along a welding joint.
dessoudé, a., (*met.*) unwelded, unsoldered, separated along a welding joint;
angle —, (*met.*) in plate rolling, the overlapping corners not welded together.
dessouder, v. a., (*met.*) to unsolder, to separate along a weld;
se —, to become unsoldered, separated along a welding joint.
dessoudure, f., (*met.*) unwelding, separation along a welded joint; break or crack in a welded joint or surface; unsoldering.
dessous, adv., under, underneath.
dessous, m., under part or side; wrong side (of cloth);
avoir le —, (*mil.*) to have the worst of it in a combat;
beau —, (*hipp.*) fine, well-shaped, legs;
— *de gorge* (*d'une bride*), (*harn.*) throat-latch;
— *du vent*, (*nav.*) lee-gauge.
dessus, adv., on, upon, over, above;
sens — *dessous*, upside down, topsy turvy.
dessus, m., upper part, top; right side (of cloth); lid, cover;
avoir le —, (*mil.*) to have the advantage, the upper hand, to get the best of it (in a combat);
beau —, *bon* —, (*hipp.*) good back, fine and good upper line of back;
— *de cou*, (*harn.*) support of the breast-collar; neck or shoulder strap;
— *en fer du caveçon*, (*man.*) cavesson-iron;
— *de nez*, (*harn.*) nose-band;
prendre le —, (*mil.*) to get the upper hand, to have the best of it;
— *de tête*, (*harn.*) crown piece (of a headstall);
— *du vent*, (*nav.*) weather-gauge.
destinataire, m., f., addressee.
destination, f., destination;
à —, (*r. r.*) through (of a train).
destituer, v. a., to dismiss, discharge; deprive of employment; turn out; (*Fr. a.*) to dismiss an officer.
destitution, f., dismissal, discharge, removal, deprivation of employment; deprivation of office, of grade; (*Fr. a.*) dismissal of an officer.
destruction, f., destruction, ruin; (*mil.*) intentional destruction, demolition.
désuni, m., (*man.*) horse that drags his hips, that gallops false, or on the wrong foot;
— *au pas*, *au trot*, irregular in his walk, trot.
désunion, f., disunion, separation.

désunir, v. a., to disunite, disjoin, separate, divide; (*mil.*) to break up the enemy.
détaché, a., detached, loose; (*fort.*) detached; m., (*mil.*) officer or man detached from his regiment or corps, person serving on a detachment.
détache-chaine, m., petard for breaking or unhooking a chain.
détachement, m., (*mil.*) draft, detachment (in France, for transportation purposes, must have at least 6 men); (*Fr. a.*) in a subdivided regiment, any part not the *portion principale* or *centrale* (q. v.); (*mach.*) disengagement;
—— *de corvée*, fatigue party.
détacher, v. a., to untie, unfasten, undo, take off, unchain, unrivet, detach, take down, unbind; (*mil.*) to detach, to draft;
—— *à*, (*mil.*) to detach to or with, detail to or for;
—— *une ruade*, to kick, let fly with the heels (horse).
détail, m., detail; (*mil.*) all matters of detail relating to administration and command; (*top.*) filling-in work of a map, reduction and plotting of details;
officier de ——, (*Fr. a.*) lieutenant assisting the c. o. of a detachment of several companies, in administration and in papers, when there is no *conseil d'administration*;
revue de ——, (*mil.*) inspection of men's knapsacks, kits, etc.
détalinguer, v. a., (*cord.*) to unbend (an anchor cable).
détaper, v. a., to open, remove a bung; (*art.*) to take the tompion out of a gun.
dételage, m., (*harn.*) unharnessing, unhitching.
dételer, v. a., (*harn.*) to unharness, to unhitch.
détendeur, m., (*steam*) regulator, pressure regulator (Belleville boiler).
détendre, v. a., to unbend, slacken, loosen; to relax (a spring); (*sm. a.*) to pull the trigger; (*steam*) to expand, to cut off, to allow to expand;
—— *un camp*, —— *une tente*, to strike a camp, a tent.
détenir, v. a., to hold, detain; to keep in custody; to have legal custody of.
détente, f., detent, tension of a spring; expansion of a gas; (*sm. a.*) trigger; (*r. f. art.*) firing mechanism, firing gear (all parts); (*steam*) expansion; cut-off; expansion valve (and gear); cut-off mechanism;
—— *à anneau*, (*sm. a.*) sort of annular trigger in magazine pistols;
appareil de ——, (*steam*) expansion gear, cut off;
came de ——, (*sm. a.*) cylinder cam;
corps de ——, (*sm. a.*) body of the trigger as distinguished from the finger-piece;
—— *de départ*, (*sm. a.*) in a double trigger, the one that determines the discharge of the piece;
—— *double*, (*sm. a.*) double trigger;
double ——, (*sm. a.*) hair-trigger;
—— *à double bossette*, (*sm. a.*) double-pull trigger;
—— *douce*, (*sm. a.*) light trigger, trigger of easy pull;
dur à la ——, (*sm. a.*) hard on the trigger;
échelonnement de la ——, (*steam*) compounding;
—— *gâchette*, (*sm. a.*) trigger and sear in one;
fausse ——, (*sm. a.*) false trigger;
lâcher la ——, (*sm. a.*) to let go the trigger;
machine à ——, (*steam*) expansion engine;
—— *des membres*, (*man.*) stretching out, stepping out;
pièce de ——, (*sm. a.*) trigger plate, guard plate and trigger; (*steam*) cut-off valve, cut-off plate;
—— *de préparation*, (*sm. a.*) in the —— *double*, that which almost disengages the sear;
préparer la ——, (*sm. a.*) to pull the —— *de préparation*;
sous-garde de la ——, (*sm. a.*) trigger plate and guard;
—— *de sûreté*, (*r. f. art.*) safety sear.

détenteur, m., (*law.*) legal custodian of anything.
détention, f., detention, confinement, imprisonment, keeping in custody; (*mil. law*) confinement to a fortress for not fewer than five nor more than twenty years;
—— *préventive*, (*mil.*) confinement before trial.
détenu, m., (*mil.*) officer or man in confinement; prisoner; prisoner in a fortress;
—— *par mesure disciplinaire*, prisoner suffering a disciplinary confinement;
—— *par mesure préventive*, prisoner held for trial by a court-martial.
détérioration, f., deterioration; wear and tear;
—— *par arrachement*, tear;
—— *par l'usure*, wear.
détériorer, v. a. r., to become defaced; to deteriorate, debase, deface.
détirer, v. a., to draw out, stretch.
détiser, v. a., to rake out a fire; calm, quell, still.
détoner, v. n., (*expl.*) to detonate, explode.
détonant, m., (*expl.*) detonating substance, explosive.
détonateur, m., (*art., expl., mil. min.*) detonator; (*art.*) dry gun-cotton priming of a torpedo-shell; the fuse complete of a torpedo-shell, of certain high-explosive shells;
—— *à double effet*, (*art.*) double-action detonator fuse;
—— *retardé*, (*art.*) delay-action detonator (torpedo-shell).
détonation, f., detonation, report; (*expl.*) detonation;
à ——, detonating;
—— *par influence*, (*expl.*) sympathetic detonation, detonation by influence.
détordre, v. a., (*cord.*) to untwist, unlay.
détors, a., untwisted, unlaid.
détortiller, v. a., to untwist, unravel.
détouper, v. a., to unstop, uncork; take out tow or stuffing.
détour, m., winding, turn, detour, circuit; fold, recess.
détournement, m., embezzlement; (*mil.*) making way with, disposing of, arms, accouterments, articles of equipment, etc.
détourner, v. a., to turn, turn aside; (*mil.*) make way with, dispose of, arms, equipment, etc.
détraqué, a., (*man.*) broken-gaited;
allure —— *e*, broken gait.
détraquer, v. a., (*man.*) to break up, ruin, a horse's gaits;
se —— to lose his goodness of gait, his paces (of a horse).
détrempé, a., diluted, soaking, soaked through;
terrain ——, ground soaking wet, soaked through.
détremper, v. a., to dilute, to soak, weaken; (*met.*) to anneal, take out the temper;
—— *de la chaux*, to slake lime.
détresse, f., distress;
en ——, jammed, stuck (as, a shell-hoist);
canon de ——, (*nav.*) distress gun;
signal de ——, (*nav.*) signal of distress.
détroit, m., strait, narrow sound; (*top.*) defile, pass; (*hipp.*) communication between pelvis and abdomen.
détruire, v. a., to destroy, raze, break up.
deuil, m., mourning;
—— *de famille*, private mourning;
—— *militaire*, official mourning.
deux, a., two;
donner des ——, (*man.*) to clap spurs to a horse;
entre ——, interval, space between;
piquer des ——, v. *donner des* ——.
deux-ponts, m., (*vav.*) two-decker.
devancement, m., —— *d'appel*, (*Fr. a., etc.*) entry into service before the legal date fixed for such entry (navy and colonial troops usually, and in a few other cases).
devancer, v. a., to get the start of, outstrip, precede;

devancer *l'appel*, (*Fr. a., etc.*) to enter the service before the legal date fixed for such entry (navy colonial troops, etc.).

devant, prep., adv., before, head;
vent ——, head wind.

devant, m., front, fore part of anything, fore end; breastplate of a cuirass;
cheval de ——, (*art.*) load horse;
les ——*s*, (*mil.*) the front of an army;
gagner le —— *de*, to forge ahead of, outstrip;
jambe de ——, (*hipp.*) fore leg;
—— *de sabot*, (*hipp.*) toe of a hoof;
—— *du système percutant*, (*Fr. art.*) front block of the firing mechanism of the *canon à balles*.

dévastation, f., devastation.

dévaster, v. a., to devastate, lay waste.

développée, f., development (as, of a spiral, etc.).

développement, m., development (of a curve, etc., of a photograph, of a bicycle); (*fenc.*) spread.

développer, v. a., to develop, to uncoil (a wire from a drum); expand; uncover, unfold, extend;
—— *l'allure*, (*man.*) to increase, develop, the gait.

déverrouiller, v. a., to unbolt.

dévers, a., warped, crooked; inclining, out of plumb (as, a wall).

dévers, m., inclination, warp; (*r. r.*) inclination of the vertical axis of the rail to the interior of the track;
—— *de l'essieu*, set of an axle-arm.

déversé, a., sloping.

déversement, m., inclination, state of being out of level, out of plumb; overflowing (of canals, weirs, etc.); (*ball.*) angular displacement of the axis of a bullet with reference to the base, and vice versa;
—— *latéral*, spreading, lateral yield (as, a railway track).

déverser, v. a. n., to bend, curve, to bend out, to put out of level; to bend, incline, to be out of plumb, out of level; to change the direction of a stream, of waters;
se ——, to warp, to get out of level, out of plumb; to fall into, empty into (of a river, canal, etc.).

déversoir, m., weir; sluice-gate.

dévêtir, v. a., (*fond.*) to open a mold.

déviateur, a., (*ball.*) deviating.

déviation, f., deflection, deviation; compass deviation; (*ball.*) deviation, error;
—— *latérale*, lateral error, lateral deviation;
—— *longitudinale*, longitudinal deviation, error in range;
—— *en portée*, v. —— *longitudinale*.

dévidage, m., winding off, unwinding, uncoiling, (as, from a reel, drum, etc.).

dévidement, m., v. *dévidage*.

dévider, v. a., to unwind, reel off (as, wire from a drum).

dévidoir, m., reel, spindle.

dévier, v. n., to deviate, swerve, turn out.

dévirage, m., (*art.*) unscrewing of the breech-block (esp. an accidental and hurtful unscrewing).

dévirer, v. a., turn back (as, a capstan); to come unscrewed, to work backward (as, a winch); to disengage (as, a screw from its seat), to unscrew; to turn an oar flat; to turn upside down, to turn turtle (of a boat); (*cord.*) to take out the kinks of a rope, to unkink;
—— *au, le, cabestan*, to heave back, turn back, the capstan;
—— *le câble*, to surge the cable.

dévis, m., estimate; specifications;
—— *approximatif*, rough estimate;
—— *estimatif*, estimate, of cost, materials, etc.

dévisser, v. a., to unscrew;
—— *la culasse*, (*art.*) to unscrew the breech-plug.

dévoiement, m., flaring; inclination, slope (as, of a chimney, etc.).

devoir, m., duty;
—— *militaire*, military duty (in general);
rentrer dans le ——, to return to one's duty, allegiance (as, after a mutiny).

dévoyer, v. a., to place obliquely, to put out of true; to bend and turn (as, a smoke pipe, a funnel); to set (a saw).

dextrine, f., (*chem.*) dextrine.

diable, m., (*art.*) devil-carriage, truck-carriage, small truck (*Fr. art.*) used in maneuvers with the gin;
—— *à roulettes*, small truck for transporting the carriage of the revolving cannon.

diagonal, a., diagonal.

diagonal, m., (*hipp.*) legs diagonally opposite;
—— *droit*, off fore and near hind legs;
—— *gauche*, near fore and off hind legs

diagonale, f., diagonal brace, tie;
en ——, diagonally.

diagonalisation, f., (*man.*) training of a horse in the extension of his diagonals.

diagonaliser, v. a., (*man.*) to train a horse to extend his diagonals.

diagramme, m., diagram;
—— *de l'indicateur*, (*steam*) indicator diagram.

diamant, m., diamond; lozenge; point of an earth-auger or jumper; (*fort.*) drop ditch, diamond ditch;
—— *de l'ancre*, crown, throat, lower part of the shank of an anchor;
fossé ——, (*fort.*) diamond ditch, drop ditch;
—— *à rabot*, glazier's diamond.

diamètre, m., diameter;
demi- ——, semidiameter;
—— *d'évolution*, (*nav.*) tactical diameter;
—— *primitif*, (*mach.*) pitch diameter.

diane, f., (*mil.*) reveille (and call for same);
battre, sonner, la ——, to beat, sound, reveille;
coup de ——, morning gun;
coup de canon de ——, morning gun.

diapason, m., tuning-fork; (drawing) scale of hachures, of graded tints, of tones.

diaphragme, m., diaphragm (*mach.*) stop valve, sliding stop-valve;
—— *en œil de chat* (*inst.*) diaphragm with adjustable orifice.

diaspongélatine, f., (*expl.*) diaspongelatine.

diastémométrique, a., (*inst.*) measuring distances by direct reading (i. e., without having to go over the distance to be measured).

diatomique, a., (*chem.*) divalent.

diazobenzol, m., (*chem.*) diazobenzole;
azotate de ——, (*expl.*) fulminating aniline.

dictateur, m., dictator.

dictionnaire, m., dictionary;
—— *chiffré*, cipher dictionary, code book (cryptography).

dièdre, a, diedral; diedral angle.

diète, f., diet; (medical) diet;
—— *absolue*, (*Fr. a.*) abstention from food of any sort;
—— *avec aliments*, (*Fr. a.*) hospital diet comprising two articles of the list of articles prescribed;
—— *lactée*, (*Fr. a.*) milk diet.

différentiel, a., differential;
palan ——, (*cord.*) differential tackle;
vis ——*le*, differential screw.

dignitaire, m., dignitary.

digue, f., dike, dam, bank, embankment;
—— *défensive*, (*fort.*) dike or dam of an inundation reservoir;
—— *de fascinage*, fascine-work dike;
—— *d'inondation*, v. —— *défensive*.

diguer, v. a., to dike, dam.

dilaniateur, a., (*expl.*) brisant, rending.

dilatabilité, f., dilatability.

dilatation, f., dilatation, expansion;
construction de ——, (*cons.*) allowance for expansion in the joints of metallic constructions.

dilater, v. a., to expand, dilate.

dimension, f., dimension, size; dimension (of a unit).

diorrexine, f., (*expl.*) diorrexine.

direct, a., direct, straight; (*elec.*) in series; (*fenc. etc.*) straight to the front.

directeur, m., director, superintendent, manager; (*Fr. a.*) head or chief of a "direction;" field officer of artillery, superintendent of a small-arm factory, or of any establishment for the manufacture and preservation of *matériel*; chief of a *direction d'artillerie*, q. v.; head, chief, or director of an operation, of a service;
—— *d'administration centrale*, (*Fr. adm.*) head of a department of the administration, under the direct orders of the minister;
—— *d'administration régionale*, head of a territorial, etc., subdivision of the administration;
—— *d'artillerie*, (*Fr. art.*) chief or head of a *direction d'artillerie*, q. v.;
—— *des attaques de mine*, (*mil. min.*) the officer in charge of mining operations in a siege;
—— *chef de service*, the head of a department or branch of the administration; (more particularly, *Fr. a.*) the chief of a *direction* (as, artillery, powders and saltpeters, etc.);
—— *des chemins de fer aux armées*, (*Fr. a.*) the officer at the head of the military railway service;
—— *des chemins de fer et des étapes*, (*Fr. a.*) the officer at the head of the *service de l'arrière* for an army acting independently;
—— *des constructions navales*, (*Fr. nav.*) chief of construction corps;
—— *de la défense souterraine*, (*mil. min.*) in a siege, the officer in charge of countermining operations;
—— *des étapes*, (*Fr. a.*) chief of the *service des étapes* of an army, is under the orders of its chief of staff;
—— *général des chemins de fer et des étapes*, (*Fr. a.*) general officer at the head of the *service de l'arrière*, is under the orders of the chief of staff, and acts for several armies under one command and in one theater of war;
—— *du génie*, (*Fr. a.*) the colonel or lieutenant-colonel of engineers in charge of a *direction du génie*;
—— *de la manœuvre*, —— *des manœuvres*, (*mil.*) officer who directs a maneuver or maneuvers and criticises and passes on the results;
——*du parc*, (*siege*) the officer in charge of a park;
—— *de service*, (*Fr. a.*) the administrative head of the *services* of artillery, engineers, *intendance*, and of the medical department, respectively, in armies, army corps, and military governments;
—— *du service de santé*, (*Fr. a.*) the chief medical officer of an army corps and of a military government (in peace); chief medical officer of a corps, division, fort, etc. (in war); (*Fr. nav.*) chief medical officer;
sous—— , (*Fr. art.*) major or captain of artillery on duty at a small-arm factory, etc., in special charge of fabrication;
—— *supérieur des chemins de fer et des étapes*, (*Fr. a.*) officer in charge of the *service de l'arrière* of an army acting alone.

directeur, a., directing, guiding;
flanc —— , (*art.*) driving edge.

direction, f., direction, conduct, management, administration, directorship; bearing; steering-post, steering head (of a bicycle); branch of the administration or district, under the charge of a director; (more esp., *Fr. a.*) branch of the war department, under a military director; there are eight in all, one for each arm, and one each for the *contrôle*, "powders and saltpeters," medical department, and administrative services;
—— *d'arme*, (*Fr. a.*) the *direction* of each arm of the service;
—— *d'artillerie*, (*Fr. a.*) territorial subdivision, comprising arsenals, certain manufacturing and repair establishments, magazines, etc.;
—— *des ballons*, (*Fr. a.*) the *direction* in charge of the whole subject of military ballooning;
—— *à billes*, steering-post on ball-bearings (bicycle);
—— *des chemins de fer de campagne*, (*Fr. a.*) direction or administration of transports beyond the base of operations, i. e., between the transition stations and the army, in the zone worked by military railways proper, and by a military personnel; accompanies the main headquarters of the army (sometimes called *direction militaire*, etc.);
—— *de la courroie*, (*mach.*) v. s. v. *courroie;*
—— *de courant*, set of a current;
de —— , (*mil.*) guiding, of direction (said of a unit at drill);
—— *d'étapes*, (*mil.*) the direction or management of the *étapes* service;
—— *du flot*, set of the flood;
—— *générale des chemins de fer et des étapes*, (*Fr. a.*) administration of the whole subject of transport for armies in the field; is adjoined to the commander in chief;
—— *du génie*, (*Fr. a.*) a group of *chefferies du génie* (q. v.);
—— *du jusant*, set of the ebb;
—— *de la marée*, set of the tide;
—— *montante*, upward direction;
—— *plongeante*, downward direction;
—— *de service*, (*Fr. a.*) the direction of a *service*, as, e. g., of the medical department of an army corps, of the *intendance* of one of the military governments;
—— *de la télégraphie militaire*, (*Fr. a.*) department in charge of the military telegraph; in war, one to each army, and to each corps acting independently (the general subject of the telegraph service for a given theater of war constitutes the *direction générale de la télégraphie*).

(It is clear from the foregoing definitions and examples that *direction*, in a military sense, is a more or less comprehensive term. It means—
1. An administrative subdivision of the war department;
2. An administrative subdivision, either of a territory or of troops in the field;
3. In certain special cases, the administration of affairs to secure a specific end. This idea is necessarily connoted in 1 and 2.
Finally, by extension, it may be taken to denote the personnel of a *direction*.)

directive, f., (*mil.*) private information given by the supreme command to subordinate chiefs, to enable the latter to act in conformity with the general object sought.

directrice, f., directrix; (*fort.*) directrix of an embrasure, of a gun port; (*F. m. art.*) support for the *guide de crosse* (q. v.); (*sm. a.*) bayonet guide-rib (Fr. model 1874); (*inst.*) right line tangent to middle point of an air-level.

diriger, v. a., to direct, manage, govern, conduct;
—— *la ligne de mire*, (*art.*) to lay a gun.

disant, m., *le moins* —— , (*adm.*) the lowest bidder.

disciplinaire, a., (*mil.*) disciplinary;
corps —— , disciplinary unit, punishment corps;
peine —— , disciplinary punishment.

disciplinaire, m., (*mil.*) man belonging to a *compagnie de discipline*.

discipline, f., (*mil.*) discipline; punishment;
compagnie de —— , (*Fr. a.*) v. *compagnie;*
conseil de —— , (*Fr. a.*) v. *conseil;*
corps de —— , (*Fr. a.*) v. *compagnie;*
faute de —— , breach of discipline;
—— *des feux*, —— *du feu*, (*art.*, *inf.*) fire discipline;
garder la —— , to maintain discipline;
maintenir dans la —— , to keep under discipline;
—— *de marche*, attention paid to all details on a march, especially to prevent elongation;
peine de —— , disciplinary punishment;
sans —— , undisciplined;
se relâcher de la —— , to relax in discipline.

discrétion, f., discretion;
se rendre à ——, (*mil.*) to surrender at discretion, unconditionally.

disjoindre, v. a. r., to disjoin, disunite; to come apart or asunder.

disjoncteur, m., (*ball.*) disjunctor; (*elec.*) cut-out,
—— *à bouchon*, (*elec.*) plug.

disjonction, f., separation, disjunction; (*ball.*) disjunction;
hauteur de ——, (*ball.*) height of disjunction.

dislocation, f., dislocation, rupture, dismemberment; (*mil.*) separation of troops, after maneuvers, and also after a war, in order to return to their respective garrisons; (at drill) separate action of the members of a unit, e. g., of a battery, or of a group when going into battery;
marche de ——, v. s. v. *marche*.

disloquer, v. a., to break up, dislocate, disperse.

disparition, f., disappearance; (*mil.*) state of being "missing" after an engagement.

disparu, a., (*mil.*) missing (after an engagement).

disparu, m., (*mil.*) man missing (after an engagement).

dispensaire, m., dispensary.

dispense, f., (*Fr. a.*) reduction of military service from 3 years to 1, in certain specified cases; exemption from military service of Frenchmen living regularly abroad, during their absence.

dispensé, m., (*Fr. a.*) the beneficiary of a *dispense*, q. v.

dispenser, v. a., to scatter, dispense;
—— *du service*, (*mil.*) to exempt from military service.

dispenseur, m., (*inst.*) diffuser (of light).

disperseur, m., (*r. f. art.*) oscillating lever, device for causing lateral dispersion of bullets (machine guns).

dispersion, f., (*ball.*) dispersion (of hits); (*art.*) spread, dispersion (of shrapnel bullets and fragments);
cône de ——, (*art.*) cone of spread, of divergence;
ellipse de ——, (*ball.*) the ellipse containing a given percentage of hits;
—— *en hauteur*, vertical dispersion, dispersion in elevation, over a vertical target;
—— *en largeur*, lateral spread, dispersion;
—— *latérale*, v. —— *en largeur*;
—— *en longueur*, dispersion in range, longitudinal dispersion, dispersion over a horizontal target;
rectangle de ——, (*ball.*) the rectangle containing a given percentage of hits.

disponibilité, f., availability; (*Fr. a.*) (of general officers and of assimilated grades) status of nonemployment, of being unattached; (of the men) status of men allowed to return to their homes before the completion of their full term of service;
en ——, unattached, unemployed; sent home before expiration of active service;
non- ——, status of n. c. o.'s and men who, while belonging to the reserve or to the territorial army, in virtue of their position in public services, do not join the colors on mobilization;
solde de ——, half-pay, pay of the status of *disponibilité*;
traitement de ——, half-pay.

disponible, a., available; (*mil.*) in general, available for service; (*Fr. a.*) belonging to, or on the status of, the *disponibilité*;
non- ——, belonging to the *non-disponibilité*.

disponible, m., (*Fr. a.*) general officer or man on the status of *disponibilité*;
non- ——, n. c. o. or man belonging to the *non-disponibilité*.

disposer, v. a., to set, fix, place in position, prepare, arrange;
—— *une fusée*, (*art.*) to set a fuse.

dispositif, m., arrangement; (*mil.*) arrangement or distribution of troops for the common end; (*mil. min.*) mine chamber;

dispositif *de cibles*, (*t. p.*) arrangement of individual targets to represent infantry, cavalry, etc.;
—— *de combat*, (*mil.*) combat dispositions;
—— *en croix*, (*mil. min.*) cross-shaped mine chamber;
—— *des écoles*, (*Fr. mil. min.*) the typical mine system of the *écoles du génie*;
—— *de marche*, (*mil.*) arrangement of troops on the march, order of march, marching arrangements;
—— *de mines*, (*mil. min.*) generic term for a system of countermines, or for the arrangements made in peace, for hasty destruction of railways, bridges, etc.;
—— *des polygones*, v. —— *des écoles*;
—— *restreint de l' École d'application*, (*mil. min.*) mine system regularly taught at the French Artillery and Engineer School of Application;
—— *de stationnement en sûreté*, (*mil.*) general expression for all the measures taken to guard troops in a halt;
—— *de sûreté*, (*mach.*, etc.) lock, locking device, safety device;
—— *en T* (*mil. min.*) T-shaped mine chamber.

disposition, f., arrangement, disposition; order; situation; (*mach.*) gearing, device; (*mil.*) any disposition or arrangement of troops for any purpose whatever;
à la ——, (*Fr. a.*) state of being in the *disponibilité*, q. v.; (*German army*) status of an officer who has retired with a pension, but who may nevertheless be actively employed in both peace and war;
—— *de combat*, (*art.*) measures taken by a battery, either in park or at any halt, to enable it to open fire as soon as possible (as, a command "prepare for action.");
—— *intérieur d'un ouvrage*, (*fort.*) interior arrangement of a work;
—— *de route*, (*art.*) command for passing from the —— *de combat* to the order of march in respect of the *matériel*;
—— *du terrain*, (*top.*) lie, lay, of the ground.

disputer, v. a., to dispute, contend;
—— *le terrain*, (*mil.*) to dispute a position, the ground;
—— *le vent*, (*nav.*) to struggle for the weather gauge.

disque, m., disk; nave-box plate; (*elec.*) disk armature; (*fort.*) embrasure plug or closer (in armored casemates);
—— *d'arrêt*, (*r. r.*) stopping signal, stopping disk;
—— *d'assemblage*, (*r. f. art.*) assembling plate (Hotchkiss revolving cannon);
—— *d'avant de nuit*, (*r. r.*) headlight of a locomotive;
—— *à composition*, (*art.*) fuse-composition disk, time train;
—— *de culasse*, (*art.*) breech-plate;
—— *denté*, (*mach.*) ratchet washer;
—— *à distance*, (*r. r.*) signal to stop, placed 800 to 1,000 meters ahead of stopping point;
—— *à double couronne*, (*Fr. art.*) double ratchet wheel of the French casemate gin;
—— *étalon à références*, standard reference disk;
—— *de garniture*, (*mach.*) packing disk;
—— *-lanterne*, (*r. r.*) headlight of a locomotive;
—— *mobile*, nave-box;
—— *-molette*, (*mach.*) worm;
—— *obturateur*, (*art.*) obturator plate;
—— *de pointage*, (*art.*) elevating drum;
—— *porte-canons*, (*r. f. art.*) assembling disk (Hotchkiss revolving cannon);
—— *de tampon*, (*r. r.*) buffer-plate.

dissimuler, v. a., (*mil.*) to conceal from enemy's view, to cover.

dissipation, f., (*mil.*) making way with arms, effects.

dissolution, f., (*chem.*) solution (act and liquid); (*mil.*) disbanding of a body of troops.

dissolvant, m., (*phot.*) developer.

dissoudre, v. a., to dissolve.

distance, f., distance; (*art., sm. a.*) range (to target);
— à ——, (*mil.*) within range, at a proper distance;
— à —— de, (*mil.*) at (platoon, company, signaling, etc.) distance;
— *appréciation des* ——s, (*mil.*) estimation of distances;
— *apprécier les* ——s, (*mil.*) to estimate distances;
—— *de but en blanc*, (*ball.*) point-blank range, range corresponding to the smallest elevation of a small-arms sight;
—— *de combat*, (*mil.*) fighting range; (*sm. a.*) range corresponding to smallest graduation of rear sight;
— *demi-* ——, (*mil.*) half distance (drill);
—— *entière*, (*mil.*) full distance, wheeling distance;
— *estimation des* ——s, v. *appréciation des* ——s;
— *estimer les* ——s, v. *apprécier les* ——s;
— *évaluation des* ——s, v. *appréciation des* ——s;
— *évaluer les* ——s, v. *apprécier les* ——s;
—— *explosive*, (*expl.*) striking distance, radius of effect, of explosion;
—— *de flèche*, (*ball.*) distance from the origin of fire to the foot of the ordinate passing through the vertex of the trajectory, measured along the line of sight;
—— *franchissable sous la vapeur*, (*nav.*) coal endurance;
—— *de front*, (*mil.*) v. —— *entière;*
— *grande* ——, (*t. p.*) any range greater than 1,200 meters; long range;
—— *graphique*, actual distance on a map, a chart; scale distance;
—— *hyperfocale*, (*phot.*) hyperfocal distance (i. e., that distance beyond which all objects are in focus at the principal focus of the lens; the focusing of a screen in projections);
— *juger les* ——s, (*mil.*) v. *apprécier les* ——s;
— *manuel des* ——s, (*Fr. a.*) table of distances;
—— *de marche*, (*mil.*) distance between the constituent elements of a column on the march;
— *moyenne* ——, (*t. p.*) any range between 600 and 1,200 meters; mean range;
— *petite* ——, (*t. p.*) any range between 0 and 600 meters; short range;
—— *des rails*, (*r. r.*) gauge;
—— *de réglage*, (*Fr. art.*) the distance used in setting the Deport sight;
—— *serrée*, (*mil.*) close distance (drill);
—— *de signal*, —— *des signaux*, (*mil.*) signaling distance;
—— *de tir*, (*art., sm. a.*) range (i. e., distance from gun to target, to be distinguished from *portée*, q. v.);
—— *zénithale*, zenith angle, angle made by the visual ray from the point considered, with the vertical through the station (in geodetic leveling).

distinctif, a., distinctive, characteristic;
— *marque* ——*ve*, (*mil.*) distinctive, distinguishing, mark (as, for faithful service, for success at target practice, etc.).

distinctif, m., (*mil.*) lace, ornament, any distinguishing mark or badge (of arm, grade, of faithful service, etc.).

distinction, f., distinction; distinction, as a decoration; act of distinguishing.

distribuer, v. a., to distribute; to serve out, hand out; (*mil.*) to issue, to make issues;
—— *la vapeur*, (*steam*) to feed, distribute, steam.

distributeur, m., distributer; (*r. f. art.*) carrier, cartridge carrier, distributer; feed-table (Hotchkiss revolving cannon); (*steam*) steam-regulator (of supply, in some engines); (*mil. teleg.*) soldier who hands out implements, etc., from the *chariot de travail*; (*mach.*) distributer (as, of a turbine);
—— *à coulisse*, (*mach.*) Stephenson's link motion;
—— *d'insertion*, (*elec.*) rheostat;
—— *de potentiel*, (*elec.*) transformer, converter;
—— *de la vapeur*, (*steam*) main slide valve, valve motion.

distribution, f., distribution; serving out; arranging; delivery, issue; (*mil.*) issue (of clothing, rations, etc.; frequently used in plural for issues in general, of no matter what sort of stores and supplies); (*steam*) valve gear, expansion valve, expansion valve and gear; introduction of steam; (*mach., etc.*) the gear or device producing a distribution;
— *appareil de la* ——, (*steam*) valve gear;
— *arbre de* ——, (*steam*) valve stem;
—— *à coulisse*, (*steam*) link-motion valve gear;
—— *par déclic*, (*steam*) Corliss valve gear, drop cut-off;
—— *à détente*, link expansion;
—— *à détente fixe*, valve gear with fixed cut-off;
—— *à détente variable*, valve gear with variable cut-off;
— *envoyer aux* ——s, (*mil.*) to send men to draw, receive, stores, issues, etc.;
— *les* ——s, (*mil.*) call for issues;
—— *à levier*, (*steam*) lever motion;
—— *à main*, (*steam*) hand gear for working the slides;
— *mécanisme de* —— *de la vapeur*, (*steam*) steam valve gear;
— *officier de* ——, (*mil.*) issuing officer;
—— *par robinets*, (*steam*) distribution of steam by means of cocks;
—— *par soupapes*, (*steam*) valve gear, distribution of steam by valves;
—— *par tiroirs*, (*steam*) slide-valve gear;
—— *de la vapeur*, (*steam*) regulated supply of steam to the cylinder, valve setting; (hence) valve gear.

divergence, f., divergence, spread; (*art.*) scattering, spread, of shrapnel bullets;
— *cône de* ——, v. *cône*.

diverger, v. n., to diverge.

diversion, f., (*mil.*) diversion.

diviser, v. a., to divide;
— *machine à* ——, dividing engine, graduator.

diviseur, m., (*mach.*) dividing plate; (in cryptography) enciphering table or abacus;
—— *à double transposition*, enciphering plate involving horizontal and vertical inversions;
—— *à simple transposition*, enciphering plate involving only one inversion of each element.

division, f., division; (*mil.*) division (of an army corps, of cavalry); (*Fr. cav.*) division (i. e., two platoons); (*mach.*) indexing mechanism (of a milling machine);
—— *administrative*, (*Fr. a.*) administrative division of territory (e. g., *région de corps d'armée*);
—— *encadrée*, (*mil.*) division forming part of a corps (in action), acting in concert with other troops on both flanks;
—— *indépendante de cavalerie*, (*cav.*) independent cavalry division, division of exploration, cavalry division operating independently in front of an army;
—— *isolée*, (*mil.*) (infantry) division operating alone, not in concert with other troops;
—— *de malades*, (*Fr. a.*) in a hospital, the sick under the charge of one surgeon;
—— *au petit pied*, (*mil.*) a division (infantry) lacking some of its organic parts;
—— *de la rose des vents*, compass division.

divisionnaire, a., (*mil.*) belonging to a division; divisional;
— *artillerie* ——, (*art.*) divisional artillery.

divisionnaire, m., (*mil.*) division commander (not an official term).

dixième, a., tenth;
— *passer au* ——, (*mil. slang*) to die.

docile, a., docile, gentle, tractable.

docilité, f., docility, gentleness; (*mach.*) ease of control;
—— *au feu*, (*hipp.*) steadiness under fire.

dock, m., (*nav.*) dock;
—— *flottant*, floating dock.

document, m., document;
—— *de comptabilité*, (*adm.*) any property return or voucher.

dodécagone, m., dodecagon.
doigt, m., finger; (*r. f. art.*) cocking-toe (Hotchkiss 47ᵐᵐ r. f.);
— *de mise de feu,* (*torp.*) firing trigger of a (Canet) torpedo tube;
— *de mise en marche,* v. — *de mise de feu.*
doigté, m., (*fenc.*) fingering, momentary displacement of the point by the action of the fingers.
doigter, m., (*man.*) pressure or touch of fingers on the mouth through the reins.
doigtier, m., any mechanism operated by the thumb or finger; (*art.*) thumb stall (obs.).
doler, v. a., to work, chip off, with an adze.
dolman, m., (*unif.*) dolman, jacket (really cavalry, but also worn by field artillery, etc., in the French army).
doloire, f., broadax.
dolomie, f., dolomite, f., dolomite.
domaine, m., domain; (*adm.*) public property, state property;
— *de l'état,* state property, public property (frequently synonymous with —— *privé,* q. v.);
— *militaire,* all forts and their dependencies, and all buildings, grounds, etc., belonging to the state and used for military purposes;
— *privé,* state property held by the state as a private citizen holds property;
— *public,* rivers, canals, roads, etc.;
rentrer au ——, to cover into the treasury, to return to the state.
domanial, a., belonging to the state.
domanialité, f., quality of belonging to the state, to the public domain.
dôme, m., dome; cupola; (*steam*) steam dome; (*met.*) reverberating roof;
— *de machine à vapeur,* (*steam*) steam dome;
— *de prise de vapeur,* (*steam*) steam dome of a locomotive;
— *de vapeur,* (*steam*) steam dome.
domicile, m., (*law*) domicile.
dominer, v. a., to dominate, keep in subjection; (*fort.*) to command, overlook.
dommage, m., damage;
—*s-intérêts,* indemnity due a person for damage done him; damages.
dompter, v. a., to reduce, quell, master, keep under; (*man.*) to break a horse.
dompteur, m., tamer, subduer;
— *de chevaux,* (*man.*) horse breaker.
donjon, m., keep, donjon (obs.).
donnée, f., datum.
donner, v. a. n., to give, to hit, strike; (*mil.*) to be engaged, in action; to attack, charge, fight; (*inst.*) to read to; (*nav.*) to make an entrance into a port;
— *l'assaut,* (*mil.*) to storm, to assault;
— *bataille,* (*mil.*) to engage;
— *au but,* (*art., sm. a.*) to hit the mark;
— *carrière à un cheval,* (*man.*) to give a horse full speed;
— *la chaude,* (*met.*) to heat;
— *du cône,* (*mach., etc.*) to taper;
— *un coup à une manœuvre,* (*cord.*) to haul on a tackle, to give a pull to a tackle;
— *la direction,* (*art.*) to lay a gun for direction;
faire ——, (*mil.*) to send in, to put (troops), into the combat;
— *fond,* (*nav.*) to cast anchor;
— *à froid,* (*met.*) to slacken the fire;
— *la hausse,* (*art.*) to elevate; to give the elevation;
— *du mou,* (*cord.*) to slack a rope;
— *à passer,* to set a boat (flying bridge) so that the current will act on it;
— *la pression,* to turn on, apply the pressure; (*steam*) to put on steam;
— *la remorque à,* to take in tow;
— *du résidu,* (*expl.*) to produce fouling (as, in a gun, etc.).
dormant, a., fixed, immovable; standing, dead; (*cord.*) standing;
eau ——*e,* stagnant water;

dormant, faire ——, (*cord.*) to fix, clinch, fasten; to be seized to;
manœuvres ——*es,* (*cord.*) standing rigging, tackle;
pont ——, fixed bridge.
dormant, m., sleeper, post; transom, fan light (of a door or window); (*cord.*) standing part of a tackle, (in pl.) standing rigging;
— *de bretelles,* (*mil.*) upper part of haversack straps.
dos, m., back (in a great variety of relations); (*cav.*) backplate of a cuirass; (*top.*) ridge;
—— *d'âne,* (*fort., etc.*) ridge formed by the intersection of the glacis planes; (*top.*) shelving ridge, saddle;
en —— *d'âne,* (of roads) crowned, barreled;
à, en, —— *d'âne,* saddle-backed;
avoir le sac au ——, (*mil.*) to be in the service, to be a soldier;
— *de carpe,* (*hipp.*) hogged back;
— *creux,* (*hipp.*) sway-back, saddle-back;
— *double,* (*hipp.*) double back (backbone countersunk);
— *droit,* (*hipp.*) straight back;
— *ensellé,* v. — *creux;*
faire le gros ——, (*man.*) to buck;
monter à ——, (*man.*) to ride bareback;
— *de mulet,* (*hipp.*) slightly convex back; mule back;
— *plongé,* (*hipp.*) dipped back;
— *tranchant,* (*hipp.*) sharp back; razorback.
dosable, a., (*chem.*) in proper, equivalent, proportions.
dosage, m., (*chem., expl.*) proportion of ingredients in gunpowder, etc.; percentage composition; action of weighing the ingredients;
être au ——, (of a powder, explosive) to have ingredients of the proper proportion, or according to specification or requirements.
doser, v. a., (*chem., expl.*) to determine proportions of ingredients, as in gunpowder; to use ingredients in proper proportion.
dosse, f., slab (of a log).
dossier, m., back (of a seat, etc.); headboard, backboard; bundle of papers relating to the same subject; wrapper, office jacket (for papers bearing on one subject).
dossière, f., backplate of a (modern) cuirass; (*harn.*) back or ridge strap for supporting the shafts.
dot, f., marriage portion.
dotation, f., endowment (esp., France, of a public institution like the *Invalides,* the *Légion d'honneur*); grant of money to a successful general;
— *de l'armée,* (*Fr. a.*) fund created for the benefit of soldiers with the colors, and to increase their pensions (ceased to exist in 1869).
douane, f., custom-house;
droits de ——, *frais de* ——, customs dues.
douanier, m., custom-house inspector.
doublage, m., (*nav.*) sheathing;
— *en cuivre,* copper sheathing;
— *en métal,* metal sheathing.
double, a., double, duplicate;
— *détente,* (*sm. a.,*) hair-trigger; double trigger;
à —— *effet,* (*art.*) double-acting, combination (of a fuse).
double, m., duplicate, copy, counterpart; (*cord.*) bight; (*mil. slang*) sergeant-major;
en ——, in duplicate;
— *de manœuvre,* (*cord.*) bight;
—*s de papier,* thicknesses of paper.
doublé, a., (*nav.*) bottomed, lined, sheathed;
— *en cuivre,* copper-bottomed.
doublé, m., (*fenc.*) a feint, followed by a thrust made in the same line as the feint.
double-contre, (*fenc.*), v. s. v. *contre.*
double-décimètre, 20-centimeter rule for taking distances, especially curved, from a map.
double-fond, m., (*nav.*) double bottom, double casing.
double-joint, m., —— *universal,* double ball-and-socket joint.

doublement, m., doubling; (*inf.*) doubling of files, ranks; (*mil.*) act or state of marching on same road with (as, artillery with infantry; (*art.*) passage, as, of a piece by its caisson, and vice versa; in general, passage of one vehicle by another in a road; (*fenc.*) doubling;
—— *d'une balle,* (*t. p.*) the enlargement, by a bullet, of the hole made by a preceding bullet;
renforcement par ——, (*mil.*) reenforcement such that the same extent of line is held by an increased number of men.
double-mètre, 2-meter rule or measure.
double-pas, m., double pace or step, in pacing distances.
doubler, v. a., to double; to fold; to lap; to line (as, a coat); (*nav.*) to sheathe; (*art.*) to pass (as, a caisson its piece), and in general, of two vehicles in a road; (*mil.*) to march on the same road (as, artillery with infantry); (*inf.*) to double (files, ranks); (*man.*) in a riding-hall, to pass at right angles from one track to the opposite, preserving the original direction of motion; (*fenc.*) to pass one's blade or foil from one side to the other of the adversary's blade or foil; to double;
—— *les avirons,* to double-bank the oars;
—— *une balle,* (*t. p.*) to strike in the print or hole of a preceding bullet;
—— *un bataillon,* (*mil.*) to extend a battalion to double its normal front in line (obs.);
—— *un cap,* (*nav.*) to double a cape;
—— *une classe,* to repeat a course at school, etc.; to be turned back (U. S. M. A.);
—— *les files,* (*inf.*) to double the files, i. e., form column of two or four files, according as the original line is in single or double rank;
—— *dans la largeur* (*longueur*), (*man.*) in a riding hall, to pass at right angles from one track to the opposite (crosswise or lengthwise), preserving the original direction of motion;
—— *les manœuvres.* (*cord.*) to reeve double;
—— *les rangs,* (*inf.*) v. —— *les files.*
doubler, m. (*man.*) passage at right angles from one track to the opposite in a riding-hall, preserving the original direction of motion.
doublette, f., (of timber) piece of standard size, standard piece.
double-voie, f., (*r. r.*) double-track line.
doublis, m., (*cons.*) eave, eave course (of a roof).
doublure, f., lining (as, of a uniform); fold, lapping; (*met.*) flaw, defect in irons, consisting of scales that peel off, scale; unsound welding.
doucement, adv., easily, gently; handsomely (as, in paying out a rope, etc.).
douceur, f., softness, gentleness;
en ——! handsomely!
amène en ——! (*cord.*) lower handsomely!
doucin, m., freshet; brackish water.
doucine, f., (*art.*) ornamental molding or swell on ordnance.
doucir, v. a., to grind smooth, as by emery, tripoli, etc.
douille, f., socket, bracket, eye; (*mach., etc.*) box, bushing, collet, bush; (*sm. a.*) part of revolver to which the barrel is screwed; bayonet socket; (*sm. a., r. f. art.*) cartridge case; (*steam*) governor sleeve;
—— *d'amorçage,* (*artif.*) primer seat or socket of demolition petard: (*art.*) primer holder of certain German mortar fuses.
—— *-arrêtoir,* (*art.*) locking socket, of a jointed sponge and rammer staff;
—— *à culot rapporté,* (*r. f. art.*) built-up case;
—— *d'embarrage,* (*art.*) embarring socket;
—— *expansible,* (*mach.*) expansion bushing; shell chuck;
—— *en laiton enroulé,* (*r. f. art.*) cartridge case of wrapped brass;
—— *du régulateur,* (*steam*) governor sleeve;
—— *taraudée,* screw-box.
—— *à tenon,* (*art.*) tenon-socket of a jointed sponge and rammer staff;

douille *à tourillons,* (*art.*) elevating crosshead, crosshead elevating nut;
—— *à vis,* v. —— *taraudée.*
dourine, f., (*hipp.*) (so-called) equine syphilis.
douve, f., stave; (*fort.*) cunette; narrow part of ditch capable of being filled with water.
doux, a., gentle (slope, horse, etc.); fair (weather); soft (iron, steel, etc.);
eau ——*ce,* fresh water; potable water;
—— *à la lime,* easy to file;
lime ——*ce,* smooth file.
douzième, a., twelfth;
—— *s provisoires,* (*Fr. adm.*) permission granted to the government to receive and to spend money for one or more months, as fixed by law, in case the regular appropriations have not been made by the 31st of December; emergency appropriation.
dragage, m., dredging; (*torp.*) dragging for torpedoes.
dragée, f., (*sm. a.*) small shot;
—— *de plomb,* small shot.
dragon, m., (*cav.*) dragoon;
——*s-gardes,* dragoon guards (English).
dragonne, f., (*unif.*) sword-knot; (*harn.*) strap;- (*in gen.*) strap or thong (on certain equipments, etc.).
draguage, m., dredging.
drague, f., dredging machine, dredge; (*siege*) sapper's shovel, sort of sweep or hoe to push aside snow, loose earth;
—— *de mine,* (*mil. min.*) miner's sweep (sort of hoe for loose earth);
—— *de sape,* (*siege*) sort of hoe or sweep for loose earth in sap work.
draguer, v. a., to dredge, drag; (*torp.*) to dredge, drag, for torpedoes;
—— *l'ancre,* to sweep or drag an anchor;
machine à ——, dredger.
dragueur, m., dredger;
bateau ——, dredger, dredging boat;
—— *à godets,* bucket dredger.
draille, f., (*cord.*) girt line;
—— *de tente,* ridge rope.
drain, m., draining ditch; draining pipe.
drainage, m., draining, drainage.
drainer, v. a., to drain.
drap, m., cloth;
—— *de lit,* sheet;
—— *de soldat,* (*mil.*) cloth for private's uniforms;
—— *de sous-officier,* (*mil.*) cloth for n. c. o.'s uniforms;
—— *de troupe,* (*mil.*) cloth for the men's uniforms.
drapeau, m., flag, colors, ensign, standard, stand of colors; (*sm. a.*) thumb-piece of the safety device (Spanish Mauser); (*in pl., mil.*) armies;
appeler sous les ——*x,* to call into active service;
arborer un ——, to hoist a color, a flag;
au ——, "to the color;"
battre au ——, to beat "to the color;"
battre les ——*x,* to troop the colors;
bénédiction des ——*x,* consecration of colors;
bénir les ——*x,* to consecrate the colors;
—— *blanc,* flag of truce, white flag;
garde du ——, color-guard;
hampe de ——, color staff, pike;
—— *jaune,* quarantine flag;
—— *parlementaire,* v. —— *blanc;*
planter un ——, to erect a standard;
porte- ——, color-bearer;
se ranger sous les ——*x,* to espouse the cause of;
se rendre aux ——*x,* to join one's regiment;
—— *signal,* signal flag;
sonner au ——, to sound "to the color;"
sous les ——*x,* with the colors, in active service.
drèche, f., malt; spent malt, brewer's grains.
dressage, m., straightening; setting; (*met.*) dressing, fettling; (*sm. a.*) straightening, setting (of a gun barrel during manufacture); (*man.*) training (of horses); (of carrier pigeons) first year's training;
—— *militaire,* (*mil.*) military training (physical);

dressage à *la planchette*, (*sm. a.*) operation and process of detecting defects of straightness in rifle barrels by the reflection of a straight-edge;
vérificateur de ——, (*sm. a.*) optical apparatus for detecting defects of straightness in gun barrels.
dressé, a., (*mil.*) well set up, drilled, and instructed (of the men in ranks); (*man.*) broken;
—— *à la selle*, (*man.*) broken to the saddle.
dressement, m., (*r. r.*) grading, finishing, of earthworks.
dresser, v. a., to dress, straighten, erect; to set up; to make true; to level; to dress; trim (as, a surface, a slope); to draw up, prepare (a report); to arrange; to instruct, train, form; (*man.*) to train (a horse); (*sm. a.*) to straighten, set (a rifle barrel); (*mil.*) to set up drill, and instruct (a recruit); (*mach.*) to line up (an engine);
—— *d'alignement*, to build by the plumb-line;
—— *à l'arçon*, (*sm. a.*) v. —— *au cordeau;*
—— *une batterie*, (*fort., art.*) to trace or mark out a battery; to equip it with guns;
—— *un camp*, (*mil.*) to lay out a camp;
—— *un canon de fusil*, (*sm. a.*) to set and straighten a rifle barrel;
—— *un cheval*, (*man.*) to break a horse;
—— *la chèvre*, (*art.*) to raise the gin;
—— *au cordeau*, to set by line (as, a gun barrel, etc.);
—— *des embûches*, —— *une embuscade*, (*mil.*) to lay an ambush, an ambuscade;
—— *une machine*, (*mach.*) to adjust, line up, an engine;
—— *de niveau*, to level;
—— *à l'œil*, to set by the eye (a gun barrel, etc.);
—— *une recrue*, (*mil.*) to set up, drill, and instruct a recruit;
—— *une tente*, to pitch a tent;
dresseur, m., (*man.*) horse-trainer; (*sm. a.*) barrel-setter.
drille, m., (*mach.*) drill, boring-bit; borer, piercer; (*mil.*) poor soldier;
—— *à archet*, bow-drill;
—— *à arçon*, breast-borer, breast-brace;
bon ——, (*mil.*) old soldier;
—— *en engrenage*, bevel wheel;
—— *à levier*, ratchet-brace, ratchet-drill;
—— *à rochet*, ratchet-drill, ratchet-brace.
drisse, f., halliard;
—— *du pavillon*, flag halliard;
—— *de signaux*, signal halliard.
droit, m., right, rights; title, claim; toll, duty, fee; share; law;
à qui de ——, (*law*) to those having legal rights in the case;
—— *d'asile*, right of shelter, (*esp., nav.*) on a war vessel;
ayant ——, (*law*) person entitled (to payment, redress, etc.), rightful claimant or owner;
—— *de brevet*, patent rights;
—— *de chancellerie*, registration fee;
—— *commercial*, commercial law;
—— *de discipline*, (*mil.*) disciplinary power;
—— *échelonné*, (*adm.*) differential duty;
—— *d'enquête du pavillon*, (*nav. int. law*) right of a man-of-war to make a commercial vessel show colors;
—— *de faire raisonner*, (*nav. int. law*) right of man-of-war to make a commercial vessel answer questions;
—— *des gens*, (*law*) international law, law of nations;
—— *de grâce*, pardoning power;
—— *de la guerre*, (*mil.*) law of war;
—— *maritime*, —— *maritime international*, international maritime law;
—— *militaire*, military law;
—— *des neutres*, law of neutrals;
—— *de prise*, (*nav. int. law*) right of capture;
—— *public militaire*, (*mil.*) code of military legislation, jurisprudence;
—— *de représailles*, law, right, of reprisals;
—— *de vérification de la nationalité*, (*nav. int. law*) right of verifying nationality of a vessel;
—— *de visite*, (*nav. int. law*) right of search.

droit, a., straight, right; plumb; upright, erect, standing up; (*man.*) straight in line, said of a horse when he stands squarely on his legs, with his head and body in the same direction; (*hipp.*) sound, free from lameness;
cheval ——, (*hipp.*) horse warranted sound and free from lameness;
—— *sur jambe*, (*hipp.*) short-pasterned.
droite, f., right line, straight line; right hand, right side; (*mil.*) right of an army, body of troops, etc.;
—— *d'un levé*, (*top.*) line of direction, bearing laid down on a survey.
droit-jointé, (*hipp.*) straight-pasterned (of a horse); straight (of a pastern).
droiture, f., justice, uprightness;
en ——, in a right line.
drome, f., raft, float;
—— *des embarcations*, (*nav.*) in a shipyard, boats, small boats, collection or assemblage of small boats.
drosse, f., (*nav.*) steam steering chain.
dualine, f., (*expl.*) dualin.
ductile, a., (*met.*) ductile.
ductilité, f., (*met.*) ductility.
duel, m., duel;
—— *d'artillerie*, (*art.*) artillery duel;
—— *d'artillerie décisif*, (*art.*) decisive duel, artillery duel proper (assailants in duel position);
—— *d'artillerie préparatoire*, (*art.*) preparatory duel (assailants in reconnoitering position).
Dumanet, (*mil. slang*) French equivalent of Tommy Atkins.
dune, f., (*top.*) dune, sand dune.
dunette, f., (*nav.*) poop.
duplicata, m., duplicate (of report, document).
dur, a., hard; harsh; stiff; (*fond.*) chilled;
—— *au départ*, (*sm. a.*) hard on the trigger;
—— *à la détente*, (*sm. a.*) hard on the trigger;
—— *à la lime*, hard to file, resisting the file.
durabilité, f., durability.
durable, a., durable, lasting.
durci, a., hardened (as, lead for bullets); (*fort.*) faced with concrete (of a parapet); (*fond.*) chilled;
fonte ——*e*, (*met.*) chilled cast-iron;
—— *à la surface*, (*met.*) case-hardened, surface-hardened.
durcir, v. a., to harden; to stiffen; to make hard, stiff; to indurate; (*fond.*) to chill.
durcissement, m., hardening, stiffening; induration; (*fond.*) chilling.
dure, f., bare ground; bare board.
durée, f., duration; life (of a gun, etc.); (*art.*) time of flight; (*ball.*) time interval;
—— *de combustion*, time of combustion (as, of a fuse, a match, etc.); combustion interval;
—— *d'écoulement*, (*mil.*) time necessary to march past a given point;
—— *du jour*, (*art.*) time of combustion of a fuse, as determined by local atmospheric conditions of the day, in contradistinction from the tabular combustion interval;
—— *de la trajectoire*, —— *du trajet*, (*ball.*) time of flight.
durer, v. n., to last, continue; hold out, endure.
dureté, f., hardness;
—— *d'extraction*, (*sm. a., r. f. art.*) difficulty of extraction (of the cartridge case);
—— *de manœuvre*, stiffness, difficulty of operation (of the parts of a mechanism, etc.).
durillon, m., callosity; (*sm. a.*) irregularity of caliber (due to lack of homogeneity in the metal of the barrel);
—— *au pied*, corn.
dynamagnite, f., (*expl.*) Hercules powder.
dynamique, f., dynamics.
dynamite, f., (*expl.*) dynamite;
—— *asbeste*, dynamite with asbestos for absorbent;
—— *à base active simultanée*, dynamite with explosive base;
—— *blanche*, dynamite with base of natural siliceous earth;

dynamite à *cellulose*, lignin dynamite;
—— *au charbon*, dynamite with charcoal absorbent, carbo-dynamite;
—— *au fulmicoton*, glyoxyline;
—— *-gélatine*, explosive gelatine;
—— *-gomme*, gum dynamite, explosive gelatine, blasting gelatine;
—— *grasse*, dynamite with a relatively nonporous base (as compared with —— *maigre*);
—— *grise*, dynamite with a base of resin, charcoal, and sodium nitrate;
—— à *grisou*, fire-damp dynamite, grisoutite;
—— *-grisoutine*, fire-damp dynamite, grisoutite;
—— *de guerre*, service dynamite;
—— *ligneuse*, lignose;
—— *maigre*, dynamite with a very porous base;
—— *au mica*, mica-powder;
—— *molle*, unfrozen dynamite;
—— *noire*, dynamite with a base of pulverized coke and sand;
—— *-paille*, —— à *paille*, straw dynamite;
—— *paléine*, v. —— *-paille*;
—— *rouge*, dynamite with a base of tripoli;
—— à *la silice*, —— *siliceuse*, kieselguhr dynamite;
—— *au sucre*, dynamite with a base of sugar.
dynamiterie, f., (*expl.*) dynamite factory.
dynamiteur, m., (*mil.*) man charged with handling dynamite.
dynamo, f., (*elec.*) dynamo;
armer une ——, to wind a dynamo;
—— *de charge*, dynamo used to charge accumulators, charging dynamo;

dynamo, *compound*, compound-wound dynamo;
—— à *courants alternatifs*, alternating-current dynamo;
—— à *courants redressés*, unidirectional-current machine or dynamo;
—— *électrique*, dynamo;
—— à *enroulement compound*, compound-wound dynamo;
—— à *excitation indépendante*, separately excited dynamo;
—— à *excitation en série*, series-wound dynamo;
—— *excitée en dérivation*, shunt-wound dynamo;
—— *excitée en série*, series-wound dynamo;
—— *-quantité*, low-tension, low-voltage, dynamo;
—— *-tension*, high-tension, high-voltage, dynamo.
dynamogène, m., (*expl.*) dynamogen.
dynamographe, m., dynamometer, dynamograph.
dynamomètre, m., dynamometer.
—— à *compteur*, registering dynamometer;
—— à *courroies*, belt, strap, dynamometer;
—— *d'écrasement*, (*art.*) crusher-gauge;
—— *enregistreur*, registering dynamometer;
—— à *frein*, brake dynamometer;
—— *de rotation*, transmission dynamometer;
—— *de traction*, traction dynamometer;
—— *de transmission*, v. —— *de rotation*.
dyne, f., dyne.
dyssenterie, f., dysentery.

E.

eau, f., water; river; rain; wash; lake; sea; pond; perspiration; urine; (in pl.) waters, territorial waters; fountain, waterworks; watering place; wake;
—— *d'alimentation*, (*steam*) feed water;
—— *basse* ——, low tide;
—— *blanc d'*——, sheet of water of slight depth (e. g., *fort.*, as an accessory defense);
—— *blanche*, mash of bran and water;
—— à *blanchir*, whitewash;
—— *chute d'*——, waterfall;
—— *de ciel*, rain water;
—— *colonne d'*——, head of water;
—— *de condensation*, (*steam*) water of condensation;
couper l'—— à *un cheval*, to keep a horse from drinking too much, from drinking his fill at one draught;
—— *courante*, running water;
—— *cours d'*——, stream;
—— *entre deux* ——*x*, between wind and water; between the surface of the water and the bottom;
—— *dormante*, stagnant water;
—— *douce*, fresh water (i. e., not salt); soft water;
—— *dure*, hard water;
—— *étale*, slack water, slack tide;
étanche d'——, water-tight;
faire ——, to leak;
faire de l'——, (*nav.*) to take in water;
faire une voie d'——, to spring a leak;
—— *ferrée*, water in which a red-hot iron has been quenched;
fil de l'——, the current;
à *fleur d'*——, awash; between wind and water; on the water line;
——*x folles*, surface water, waters not running in a channel, or running in an undefined channel;
—— *fraîche*, fresh water (i. e., not stale, fit for use);
——*x grasses*, greasy water (kitchen waters containing grease, sold by the men's mess in France);
à *grande* ——, (*chem.*) in an excess of water;

eau, *grandes* ——*x*, high water (in a river);
haute ——, high tide;
hauteur de l'——, height of the water (in a gauge);
imperméable à *l'*——, water-tight;
—— *incrustante*, hard water, water producing incrustation;
indicateur de la hauteur de l'——, *du niveau d'*——, (*steam, etc.*) water gauge;
—— *d'injection*, (*steam*) injection water;
——*x aux jambes*, (*hipp.*) grease;
jet d'——, fountain;
ligne d'——, water line;
—— *maigre*, shallow water;
—— *de mer*, salt water, sea water;
—— *mère*, mother liquor;
mettre un bâtiment à *l'*——, (*nav.*) to launch a vessel;
mettre une pièce hors d'——, (*art.*) to depress the muzzle so as to prevent entrance of rain;
mise à *l'*——, (*nav.*) launching;
—— *montante*, rising tide;
——*x mortes*, neap tide;
mortes ——, young flood (of a river);
—— *navigable*, waterway, navigable channel;
——*x d'un navire*, (*nav.*) wake of a ship;
niveau de l'——, water level (as in a boiler);
niveau des ——*x*, watermark, water-level line;
niveau des hautes ——*x*, high-water mark;
au pain et à *l'*——, on bread and water;
—— *panée*, toast-water;
—— *plate*, standing water (as, in a field);
pleine- ——, part of a stream where one may swim freely, body of the stream (as opposed to an inclosed pool, etc.), hence, a swim in full stream;
faire une pleine ——, to take a swim in full stream;
—— *de pluie*, —— *pluviale*, rain water;
—— *potable*, drinking water;
prendre l'——, to leak;
prendre (tant de) pieds d'——, (*nav.*) to draw (so many) feet;
prise d'——, taking in water, watering;

eau, *au ras de l'*——, flush with the surface of the water;
—— *de riz*, arrack;
rompre l'—— *à un cheval*, v. *couper l'*——, etc., *supra*;
—— *rouillée*, v. —— *ferrée*;
—— *saumâtre*, brackish water; pool of brackish water;
—— *x sauvages*, v. ——*x folles*;
—— *de savon*, soapsuds;
—— *seconde*, lye-water; (*chem.*) dilute, weak, nitric acid;
source d'——, spring;
—— *de source*, spring water;
—— *x stagnantes*, stagnant water, stagnant pool;
station à prendre de l'——, (r. r.) watering station;
—— *x territoriales*, territorial waters;
tête d'——, head of water;
tirant d'——, (*nav.*) draught of water;
—— *tombée*, rainfall, amount of rainfall;
—— *de trempe*, (*met.*) chalybeate water, hardening water;
—— *trouble*, turbid water;
tube d'——, water glass;
vif de l'——, height of the tide;
——*x vives*, spring tides;
voie d'——, leak; waterway; general term for canals, rivers, etc., (in questions of transportation).

eau-de-vie, f., brandy.

eau-forte, f., etching; (*chem.*) nitric acid.

ébalIement, m., play.

ébarbage, m., removal of edges, burrs, seams, surface irregularities, from castings, projectiles, plates, and metal objects in general.

ébarber, v. a., to trim (castings, etc.); to nip burrs; to remove irregularities, etc., from castings, projectiles, etc.

ébarbeuse, f., (*mach.*) trimming machine, finishing machine.

ébarboir, m., trimmer, burr-nipper; parer; scraper.

ébarbure, f., burr, fash, beard.

ébattement, m., balancing, rocking, of a carriage.

ébauchage, m., rough hewing, rough shaping; dressing down, turning down; rough filing; (*sm. a.*) turning of exterior of barrel; (*art.*) removal from projectiles with lead centering band of so much of the envelope as does not correspond to the ribs.

ébauche, f., sketch, rough sketch.

ébauché, m., any material got roughly into shape, e. g., (*met.*) a welding packet.

ébaucher, v. a., to rough-work; to dress down, turn or hammer down, approximately to shape; to block out, give its approximate shape to anything; to sketch, sketch roughly; to outline.

ébaudrage, m., (*cord.*) combing out.

ébaudrer, v. a., (*cord.*) to comb out, to remove poor or weak fiber.

ébavurer, v. a., to trim off (seams, burrs, etc.).

ebbe, f., ebb tide.

ébe, f., v. *ebbe*.

ébène, m., ebony.

ébéniste, m., cabinetmaker.

ébergement, m., projection of a slope.

ébonite, m., ebonite.

éboueur, m., road-scraper.

éboulement, m., landslide; falling in (as, of a mine, etc.); crumbling, falling down (of a wall, etc.); pile of rubbish due to crumbling;
—— *de terre*, landslide.

ébouler, v. n., to slide, to fall in (as, a mine); to crumble, give way, tumble down (walls, embankments, etc.).

éboulis, m., rubbish (fallen in); rubble.

ébousiner, v. a., (*mas.*) to remove the *bousin* (q. v.) from stone, from rock.

ébraisoir, m., (*fond.*) fire shovel, furnace shovel.

ébrancher, v. a., to strip or cut off branches (from trees, bushes, etc.).

ébranchoir, m., billhook.

ébranlement, m., shock, concussion; shaking, disturbance, commotion, disorder; (*mil.*) wavering.

ébranler, v. a. r., to shake, unsettle, disturb, stagger; (*mil.*) to shake, to cause to waver; to move, to be set in motion (as, of troops setting out on a march); to waver, to become broken up.

ébrasement, m., splay, splaying;
mur d'——, wing wall, bay wall (as, of an embrasure).

ébraser, v. a., to splay.

ébrécher, v. a., to notch, hack; break a piece out of, make a hole in.

ébrillade, f., sudden jerk with the bridle.

ébroudage, m., plate wire-drawing.

ébroudeur, m., wire-drawer.

ébroudir, v. a., to draw wire through a plate.

ébrouement, m. (*hipp.*) snort (due to surprise or fear).

ébrouer, v. r., (*hipp.*) to snort, sneeze.

ébullition, f., ebullition.

écacher, v. a., to crush; to beat, squeeze, flat.

écagne, f., (*cord.*) part of a skein or hank that has been divided.

écaille, f., splinter, chip; scale;
——*s*, (*met.*) scale;
——*s, de fer*, forge-scale, hammer slag.

écaillement, m., scaling, scaling off (as, an armor plate under fire).

écailler, v. a. r., to scale, peel off; to chip off; to flake.

écaillon, m., tush or tusk of a horse.

écarlate, f., scarlet.

écarrir, v. a., to fine-bore, to polish.

écart, m., any place more or less remote; (in thermodynamics) difference of temperature; (*man.*) shy, sudden start (aside); (*hipp.*) strain; (*opt.*) dispersion, failure of rays to focus on account of spherical aberration; (*carp.*) scarf, joint, (length of) overlapping (in a joint); (*ball.*) error, deviation; distance from center of impact to point aimed at (of a group); distance from point of impact to point aimed at (of a single shot).

(Except where otherwise indicated, the following terms relate to ballistics:)
—— *absolu*, absolute error, deviation: straight line from point considered to center of impact;
—— *absolu moyen*, mean absolute error;
angle d'—— *horizontal*, angle of lateral jump;
—— *angulaire initial latéral*, angle of lateral jump;
—— *angulaire vertical*, vertical angle of jump;
—— *carré*, (*carp.*) butt and butt;
—— *en direction*, lateral deviation;
—— *double*, (*carp.*) hook and butt, two ends of timber laid over each other;
—— *d'éclatement en hauteur*, difference between real height of point of burst and theoretical mean height;
—— *d'éclatement d'un point suivant la trajectoire*, distance of point of burst from plane normal to mean trajectory and passing through mean point of burst;
—— *d'éclatement en portée*, difference between range on burst and mean theoretical range on burst;
—— *d'épaule*, (*hipp.*) shoulder strain;
faire un ——, (*man.*) to shy;
faux ——, (*hipp.*) slight strain of hip joint;
—— *fondamental*, any error that will serve as a standard of comparison (e. g., the probable error, the mean-square error);
—— *géométrique*, hypotenuse of the right triangle, of which the sides are the mean horizontal and vertical deviations, respectively (sometimes called —— *moyen géométrique* and *moyen* ——);
—— *en hauteur*, vertical error, error in range;
—— *horizontal*, lateral error or deviation;
—— *horizontal moyen*, mean lateral error (arithmetical mean);
—— *en largeur*, v. —— *en direction*;
—— *latéral*, v. —— *en direction*;
—— *en longueur*, error in range;

écart *moyen*, mean (arithmetical) error or deviation;
— *moyen en direction*, mean lateral error;
— *moyen géométrique*, v. — *géométrique*;
— *moyen en hauteur*, mean vertical error, mean error in range;
— *moyen en portée*, mean error in range;
— *moyen quadratique*, the square root of the arithmetical mean of the squares of the observed deviations (employed by General Didion);
moyen ——, a synonym, first of —— *géométrique*, and, second, of —— *moyen quadratique*, qq. v.;
— *en portée*, error in range;
— *probable*, probable error (otherwise defined as the radius of the circle, having the center of impact as center and containing 50 per cent of hits);
— *probable absolu*, v. — *probable*, second definition;
— *probable en direction*, — *probable horizontal*,
— *probable en largeur*, probable lateral error;
— *probable en hauteur*, — *probable en longueur*, — *probable en portée*, — *probable vertical*, probable error in range;
— *probable normal*, the probable error, referred to a plane normal to the trajectory;
— *de réglage*, distance between the target and the center of impact of a group, the practice being as accurate as possible;
— *simple*, (*carp.*) v. — *carré*;
—*s totaux*, the sides of the rectangle, equal each to eight times the corresponding probable error;
— *vertical*, v. — *en hauteur*;
— *vertical moyen*, v. — *moyen en hauteur*.

écartement, m., spacing, distance apart; scattering, separation, putting aside; pitch (of rivets); (*fort.*) spacing (of forts, works, on a line);
— *des essieux*, (*r. r.*) wheel base;
— *des flasques d'affût*, (*art.*) distance apart of cheeks;
— *normal*, (*r. r.*) standard gauge;
— *des rails*, (*r. r.*) gauge;
— *des rivets*, pitch of riveting;
— *des roues*, span or width between wheels;
— *de la voie*, (*r. r.*) gauge;
— *de la voie dans œuvre*, (*r. r.*) gauge in the clear.

écarter, v. a. r., to extend, widen, separate; divert, turn aside; ward off; scatter; dispel;
— *un coup*, to ward off a blow.

écavecade, (*man.*) pull with the cavesson.

échafaud, m., scaffold; (*fort.*) balcony made in the window of a house prepared for defense.

échafaudage, m., scaffolding; erection of stages; (*fig.*) preparations.

échafauder, v. a., to scaffold, to stage.

échancrer, v. a., to channel, groove; to hollow, hollow out; to cut, make hollow, sloping; to eat into (as, the sea into a coast, etc.).

échancrure, f., channel, cut, groove; hollowing, notch, indentation; scallop (as in the edge of a saddlecloth); (*fort.*) cut (in the covered way); (*sm. a.*) cut (in the receiver);
— *du glacis*, (*fort.*) defile crotchet.

échange, m., exchange; (*mil.*) exchange (of prisoners);
cartel d' ——, (*mil.*) cartel of exchange;
libre ——, free trade.

échanger, v. a., to exchange; (*mil.*) to exchange (prisoners).

échantignolle, f., bracket, block, lining; axletree bracket; (*art.*) trail bracket (obs.); cheek-piece of a gin.

échantillon, m., sample, specimen, pattern, templet, model; (*fond.*) loam board, modeling board;
— *de chemise*, (*fond.*) thickness piece;
— *droit*, (*fond.*) deadhead board;
être conforme à l' ——, *être d'* ——, to be according to pattern;
à gros —*s*, of large pattern;
— *de noyau*, (*fond.*) core board;

échantillon-*type*, (*adm.*) model or pattern furnished by a contractor.

échantillonnage, m., gauging, verification by a standard.

échantillonné, p. p., made up into patterns, shapes (of manufactures, e. g., iron, etc.).

échantillonner, v. a., to verify, to gauge, to compare with a standard; to make up according to given patterns, etc.

échappé, p. p., m., fugitive; stray, runaway (horse); (*hipp.*) mongrel horse;
cheval — *de normand*, etc., (*hipp.*) horse of Norman, etc., sire and dam of any other breed;
un — *de normand*, etc., (*hipp.*) horse of Norman, etc., sire and dam of any other breed.

échappée, f., space for carriages to turn in; width of a door or window, width of any opening admitting the light; clear space, of a spiral staircase; (*man.*) prank, trick, fit;
—*d'une vis*, distance between threads of a screw;
— *de vue*, vista.

échappement, m., escape; escape, leak (of steam, air, powder gases, etc.); puff, blast (of steam, etc.); escape pipe; flue (of a reverberatory furnace); space for carriages to turn in; height between floor and ceiling; scapement, escapement (of a clock, etc.); (*steam*) exhaust (period);
— *à ancre*, anchor escapement;
— *à cylindre*, cylinder escapement;
— *à l'épine*, deadbeat escapement;
— *à recul*, recoil escapement;
— *à repos*, v. — *à l'épine*;
tuyau d' ——, escape pipe, blast pipe;
— *à verge*, crown escapement.

échapper, v. a. n., to escape, run off or away, break loose; to avoid; (*fenc.*) to retire or withdraw to avoid a thrust, a lunge;
— *la jambe*, (*fenc.*) to withdraw;
laisser ——, to let off (steam, gas, etc.); (*man.*) to put a horse at full speed.

écharpe, f., sash; sling (as, for the arm); shell or frame of a block or pulley; assemblage of one or more sheaves in one shell; jib of a crane, outrigger; tie; (*cons.*) brace, diagonal brace; reenforcing strap; (*unif.*) sash, officer's sash;
batterie d' ——, (*art.*) oblique-fire battery, one firing obliquely at any target;
battre d' ——, (*art.*) to fire, take obliquely;
changer d' ——, to change sides;
coup de sabre en ——, slanting saber cut;
— *à crochet*, block and hook;
d' ——, obliquely;
— *de distinction*, (*unif.*) sash worn over shoulder, as a badge of duty or of command, (obs. in France);
en ——, askew, skew, slanting, in a sling, slantingly, across the shoulders, obliquely.

écharper, v. a., to pass a sling or tackle around a weight; (*mil.*) to cut to pieces (as, a regiment).

échasse, f., stilt; scaffolding pole.

échaudage, m., whitewash, limewash; operation of whitewashing.

échaude, f., (*met.*) welding heat.

échaudé, m., camp stool, folding seat.

échauffé, a., (of wood) moldy, rotten; (of yarn) rotten.

échauffement, m., overheating; (of wood, timber) defect due to the fermentation of sap; (*mach.*) heating of a bearing, journal, etc.; heating of feed water; (*art., sm. a.*) heating of barrel, of gun, due to firing;
— *de la fourchette*, (*hipp.*) disease of the foot.

échauffer, v. a. r., to heat, to become hot, (of timber) to dry rot; (*man.*) to stir up, excite, irritate (a horse).

échauffourée, f., (*mil., etc.*) affray, scuffle; skirmish; ill-conducted affair.

échauguette, f., (*fort.*) watch tower or box, sort of elevated sentry box, bartizan. (Also spelled *écharguette*, obs.)

échaume, m., thole-pin.
échaux, m. pl., drains, irrigating channels.
échéance, f., term, expiration; (*mil.*) expiration of a *délai*;
  à l'——, when due;
  à longue ——, with a long time to run (of a note, etc.).
échec, m., check, repulse, foil, miscarriage;
  être en ——, to be stopped, checked;
  éprouver un ——, to be defeated, repulsed.
échelette, f., rack of a cart.
échelier, m., peg-ladder;
  —— de grue, jib or arm of a crane.
échelle, f., ladder; scale; ship's ladder;
  —— brisée, folding ladder;
  —— de corde, rope ladder;
  à la courte ——, by mounting on one another's shoulders;
  faire la courte ——, to mount on one another's shoulders;
  —— des dixmes, diagonal scale;
  —— double, (*siege*) scaling-ladder of double width; (*in gen.*) double ladder (i. e., two ladders hinged at the top);
  —— d'eau, water-gauge;
  —— d'escalade, (*siege*) scaling-ladder;
  —— de front, plain scale, uniform scale;
  —— fuyante, vanishing scale;
  —— graduée, graduated scale;
  —— graphique, scale;
  —— des hauteurs, vertical scale;
  —— à incendie, fire-ladder;
  —— des longueurs, horizontal scale;
  —— maréométrique, open-sea tide-gauge;
  —— de meunier, staircase without risers;
  —— mobile, sliding scale;
  —— numérique, representative fraction;
  —— de pente, (*fort.*, *top.*) scale of slope;
  pont ——, m., (*siege*) ladder-bridge for crossing ditches;
  —— des ponts, river-gauge;
  —— de projection, scale of projection;
  —— de projection à l'horizon, scale of reduction to the horizon (stadia work);
  —— proportionnelle, diagonal scale;
  —— à rapporter, v. *—— de réduction*;
  —— de réduction, plotting scale, reducing scale; diagonal scale;
  —— de réduction à l'horizon, scale of reduction to the horizon (stadia work);
  —— de siège, scaling-ladder;
  —— simple, ordinary scale of length;
  —— stadimétrique, stadimeter scale.
échelle-observatoire, f., (*mil.*) field observatory (ladder suitably erected and supported).
échelon, m., round, rung, of a ladder; (*mil.*) echelon (in many applications); grade; (*art.*) synonym of *—— de combat*, q. v.;
  —— de combat, (*Fr. art.*) what is left of the *batterie de combat* after taking out the *batterie de tir*, i. e., battery wagon, forge, 6 caissons, 6 spare teams (hence) first line of wagons of a battery;
  décroiser les ——s, (*mil.*) to form direct from oblique echelons;
  —— direct, (*mil.*) echelons parallel to one another and to the line of battle;
  —— de la hiérarchie, (*mil.*) grade;
  —— oblique, (*mil.*) oblique echelon (to guard against a possible flank attack);
  —— de parc, (*Fr. a.*) section of an army corps ammunition park.
échelonné, a., (*mil.*) echelonned;
  tir ——, v. s. v. *tir*.
échelonnement, m., (*mil.*) echelonning, operation, result, of disposing troops in echelon;
  —— de la détente, (*steam*) compounding;
  —— des hausses, (*art.*) use of increasing elevations, in ranging;
  —— du tir, v. *—— des hausses*.
échelonner, v. a., (*mil.*) to proportion; (*mil.*) to form, draw up, post, dispose, troops in echelon;
  les hausses, (*art.*) in ranging, to give or use increasing elevations for the first salvo.
échenal, m., gutter; (*fond.*) gate.
écheneau, m., v. *échenal*.
échenet, m., v. *échenal*.

échet, m., hank.
écheveau, m., hank, skein (French cotton spinning, 1,000 meters).
échevette, f., skein (French cotton-spinning, 100 meters).
échiffe, m., **échiffre**, m., (*cons.*) partition wall of a stair;
  mur d' ——, partition wall of a stair.
échine, f., spine, spinal column.
échiqueté, a., (*mil.*) checkerwise.
échiquier, m., checkerboard; exchequer;
  en ——, (*fort.*) v. s. v. *quinconce* (*mil.*) disposition of troops checkerwise.
échome, m., thole-pin.
échouage, m., (*nav.*) stranding, running aground.
échouement, m., (*nav.*) v. *échouage*.
échouer, v. n., to strand, to go aground; to strike, to run aground; to cast away; (*fig.*) to fail, miscarry, go wrong; to foil, be foiled.
échoué, a., (*nav.*) stranded;
  à sec., stranded high and dry.
éclair, m., lightning; flash;
  —— de chaleur, heat lightning;
  —— de feu, *—— de flamme*, (*artif.*) fixed rocket filled with bright stars.
éclairage, m., lighting, illumination;
  —— électrique, electric lighting;
  —— au gaz, lighting by gas;
  —— du terrain, (*top. drawing*) illumination of a map.
éclaircie, f., clear spot in cloudy sky; (*top.*) clearing or glade in a wood.
éclaircir, v. a. r., to clear, clear up; brighten, polish; clarify; solve, explain; elucidate, (of the weather) to clear up;
  —— les rangs, (*mil.*) to thin the ranks.
éclairement, m., (*top. drawing*) illumination.
éclairer, v. a. n., to light, give light to; to enlighten; (*mil.*) to inform, keep informed; to reconnoiter, watch; to clear or cover the front of an army by scouting;
  —— au gaz, to light by gas;
  —— une marche, (*mil.*) on a march, to send reconnoitering parties ahead;
  —— le terrain, (*mil.*) to explore, scout over, the ground;
  —— une tranchée, (*siege*) to illuminate a trench.
éclaireur, m., (*mil.*) scout; (*nav.*) scout boat;
  aller en ——s, (*mil.*) to scout, to reconnoiter;
  —— cycliste, bicycle scout, scout mounted on a bicycle;
  —— d'escadre, (*nav.*) scout boat;
  —— d'objectif, (*art.*) an officer or n. c. o. who reconnoiters and reports on the number, position, etc., of the enemy's batteries, for the purpose of regulating the fire of his own side;
  —— de position, (*cav.*) scout posted on high ground near the line of march of a column; lookout;
  —— du terrain, (*cav.*) ground scout;
  ——s volontaires, (*Fr. cav.*) a body (19 troops), authorized but never formed, to consist of volunteers from the reserve or the *disponibilité*, satisfying certain fixed conditions, and required to serve only during maneuvers or on mobilization.
éclat, m., splinter, piece, fragment; crack (in wood); explosion, burst, bursting; brilliancy, clearness; (*art.*) splinter (of a shell, etc.); (*sig.*) flash (of a signaling apparatus);
  action d' ——, (*mil.*) brilliant deed of arms;
  —— de bois, splinter of wood;
  —— de l'image, distinctness, clearness, of the image (telescope, etc.);
  —— d'obus, *—— de projectile*, (*art.*) splinter (of a shell, etc.);
  parc-——s, (*fort.*) splinter-proof;
  —— en retour, (*art.*) splinter falling back into shell crater.
éclatante, f., (*artif.*) rocket with stars.
éclatement, m., burst, bursting; breaking out; explosion; (*mil.*) bursting (of a shell, a gun); (*steam*) bursting, explosion (of a boiler);
  charge d' ——, (*art.*) bursting charge, burster;

éclatement, *gerbe d'——,* (*art.*) sheaf of fragments;
—— *de la guerre,* (*mil.*) outbreak of war;
hauteur d'——, (*art.*) height of burst (of a shell, shrapnel);
hauteur théorique moyenne d'——, (*art.*) height of mean point of burst above the ground;
intervalle d'——, (*art.*) horizontal distance of the point of burst to the target;
point d'——, (*art.*) point of burst;
point moyen d'——, (*art.*) mean point of burst;
portée d'——, (*art.*) horizontal distance of the point of burst from the muzzle of the gun;
portée d'—— du point moyen, portée théorique moyenne d'——, (*art.*) range of the mean point of burst;
projectile d'——, (*art.*) any projectile intended to act by explosion; shell, shrapnel;
retard d'——, (*art.*) horizontal distance a shell goes on ricochet before bursting;
surface d'——, (*art.*) area over which effects of burst are felt.

éclater, v. a. n., to burst, shiver, shatter, split, splinter; to explode, blow up; to flash, crack; (*mil.*) to break out (as, war); (*art.*) to burst (shell, gun);
faire ——, to burst, cause to burst, explode;
—— *au repos,* (*art.*) to burst a shell outside of a gun (as, in a sand pit, to determine fragmentation).

écli, m., splinter of wood.

éclimètre, m., (*inst.*) clinometer;
—— *à limbe fixe,* nonadjustable clinometer;
—— *à limbe mobile,* adjustable clinometer;
—— *non-rectifiable,* v. —— *à limbe fixe;*
—— *-pendule,* pendulum clinometer;
—— *rectifiable,* adjustable clinometer.

éclipse, f., eclipse;
affût à ——, (*art.*) disappearing carriage;
but à ——, v. s. v. *but;*
hauteur d'——, (*art.*) drop (of a disappearing gun)
tourelle à ——, (*fort.*) disappearing turret.

éclipsé, p. p., (*art.*) down (of a disappearing gun)

éclipser, v. r., (*art.*) to disappear (of a gun, a turret).

éclissage, m., fishing, fish-jointing.

éclisse, f., splinter; splint; any junction or assembling plate; (*r. r.*) fish, fish-plate;
—— *à cornières,* angular fish-plate, bracket-joint;
—— *plate,* ordinary, flat, fish-plate;
—— *à ressort,* spring fish-plate.

éclissé, adj., fish-jointed.

éclissement, v., *éclissage.*

éclisser, v. a., to splint (as, a broken arm); to join or assemble by a junction plate; (*r. r.*) to fish.

écloppé, a., lame; m., (*mil.*) any man (not an officer) temporarily unable to march in ranks (including sick, wounded, or indisposed not requiring to go into a hospital).

éclopper, v. a., to lame, to cripple.

écluse, f., sluice; dam; lock (of a canal); flood gate; weir; (*fond.*) gate;
—— *s accolées,* chain of locks;
aire d'——, lock chamber;
chambre d'——, v. *aire d'——;*
—— *de chasse,* (*fort.*) sluice gate for introducing water into a ditch;
déversoir à ——, lock weir;
—— *d'entrée et de fuite,* (*fort.*) sluice gate for filling and emptying a ditch;
heurtoir d'——, lock sill;
lâcher une ——, to open a sluice;
—— *de manœuvre,* (*fort.*) sluice gate, for emptying and filling a ditch;
—— *à marée montante,* tide-gate;
péage d'——, lockage;
porte d'——, lock gate;
—— *de retenue,* (*fort.*) sluice gate, flood gate;
—— *à sas,* canal lock;
seuil d'——, lock sill;
—— *à vannes,* flood gate with openings; sash sluice.

éclusée, f., sluice-full, lock-full, of water; water dammed up in a river (otherwise not navigable).

écluser, v. a., to provide with locks, sluices; to pass a boat through a lock; to lock.

écoinçon, m., (*cons.*) diagonal;
—— *de fenêtre,* (*cons.*) reveal of a window;
—— *de porte,* jamb of a door; corner stone of the chamfering.

école, f., school; (*mil.*) school (i. e., drill); instruction, season of instruction, of practice.
I. General terms; II. French military schools.

I. General terms:
—— *de bataillon,* (*inf.*) school of the battalion (and similarly of the other units and duties, of all arms);
cheval qui a de l'——, trained horse;
—— *des distances,* estimating-distance drill;
—— *des clairons,* (*mil. mus.*) bugle practice;
—— *d'équitation,* (*man.*) riding school;
faire l'—— de, to practice at ....;
—— *à feu,* (*art.*)) target practice, season of target practice, practical instruction in target practice;
—— *à feu à longue portée,* (*art.*) long-range practice;
haute ——, (*man.*) "high school" riding;
—— *d'intonation,* (*mil.*) voice drill, instruction in the giving of commands at drill;
—— *de maréchalerie,* school of farriery;
—— *de navigation,* (*pont.*) boat drill;
—— *de nœuds,* instruction in cordage;
—— *de nuit,* (*art.*) night practice;
pas d'——, (*inf.,* etc.) goose step, balance step;
—— *du peloton de la pièce,* (*art.*) standing-gun drill;
—— *de pointage,* (*t. p.*) aiming drill;
—— *de ponts,* (*pont.*) bridge drill;
—— *régimentaire,* (*mil.*) regimental school;
—— *du soldat,* (*inf.,* etc.) recruit drill, squad drill;
—— *supérieure de guerre,* war academy, staff college;
—— *des tambours,* (*mil. mus.*) drum practice;
—— *de tir,* (*inf.*) target practice, practice firing, season of target practice;
tir d'——, (*inf.*) target practice.

II. French military schools:
—— *d'administration militaire,* for officers, etc., of the *intendance,* at Vincennes;
—— *d'aérostation militaire,* school of instruction in ballooning, at Chalais;
—— *d'application de l'artillerie et du génie,* school of application for officers of artillery and of engineers, at Fontainebleau (—— *de Fontainebleau*)
—— *d'application de cavalerie,* school of application for officers of cavalry, at Saumur (admits some artillery and engineer officers, veterinaries, and n. c. o.'s fitted to receive commissions) (—— *de Saumur*);
—— *d'application de médecine et de pharmacie militaire,* medical school of application, Val-de-Grâce, Paris;
—— *d'application des poudres et salpêtres,* school for the study of military chemistry, for *polytechniciens* who enter the *poudres et salpêtres,* at Paris;
—— *d'application du service de santé,* v. *d'application de médecine et de pharmacie;*
—— *d'artillerie,* one in each army corps and one at the headquarters of the Nineteenth artillery brigade, for instruction of artificers and supply of material for target practice, and for other purposes related to the artillery service;
—— *centrale de pyrotechnie militaire,* v. —— *de pyrotechnie militaire;*
—— *du 1er (2ème) degré,* v. —— *régimentaire;*
—— *de dessin topographique,* at the *dépôt de la guerre,* for the training of draughtsmen for the *service géographique;*
—— *de dressage,* an annex of the cavalry school at Saumur, for instruction in the art of training horses;

école *de la Flèche*, preparatory school for sons of officers killed in action or who died without means (*Prytanée militaire de la Flèche*);
—— *du génie*, regimental school for the instruction of n. c. o.'s for the technical instruction of the regiment, the execution of experiments, and the supply of material for engineer service in the field;
—— *d'instruction aérostatique*, v. —— *d'aérostation militaire*;
—— *de médecine et de pharmacie militaire*, v. —— *du service de santé militaire*;
—— *militaire de l'artillerie et du génie*, for instruction of n. c. o.'s of artillery, of engineers, and of the train, fitted to become officers, at Versailles;
—— *militaire d'infanterie*, for instruction of n. c. o.'s fitted to become officers, at Saint-Maixent (—— *de Saint-Maixent*);
—— *militaire préparatoire*, military preparatory school for the sons of soldiers, of subaltern officers, and of deceased field officers (four for the infantry, and one each for the artillery and the cavalry);
—— *nationale vétérinaire*, trains candidates for the position of army veterinary surgeon;
—— *normale de gymnastique*, prepares fencing and gymnastic masters, for the army, at Joinville-le-pont;
—— *normale de tir*, at Châlons, has for object the training of musketry instructors and of musketry experts, the investigation of questions relating to small arms in general, the quarterly inspection of arms as manufactured, etc., and the preparation of rules for infantry fire;
—— *polytechnique*, the Polytechnic, at Paris, furnishes officers of artillery and of engineers, and of certain civil employments, as the *Ponts et Chaussées*, etc.;
—— *de pyrotechnie militaire*, at Bourges, for the instruction of n. c. o.'s recommended for the grade of *maréchal des logis chef artificier*; is a center for the manufacture and investigation of military fireworks;
—— *régimentaire*, one in each regiment and detached battalion, comprising two courses, the first (*école primaire*) for illiterate soldiers, the second (*cours préparatoire*) for n. c. o.'s and corporals reading for the grade of sub-lieutenant. These courses have been substituted for the *écoles du 1er et du 2ème degré*, respectively;
—— *régimentaire du génie*, name ordinarily given to an —— *du génie*, q. v.;
—— *régimentaire de tir*, one to each regiment of infantry, for theoretical and practical instruction in target practice;
—— *régionale de tir*, so-called regional school, suppressed in 1894, had for object to spread among the regiments a knowledge of the progress achieved in all matters relating to small arms and to small-arm practice;
—— *du service de santé militaire*, preparatory to Val-de-Grâce, for the preliminary instruction of medical cadets, at Lyons (—— *de Lyon*);
—— *de sous-officiers de l'artillerie et du génie*, v.
—— *spéciale militaire*, furnishes officers of infantry, of cavalry, and of marine infantry, at Saint-Cyr (—— *de Saint-Cyr*);
—— *supérieure de guerre*, war academy, staff college, at Paris;
—— *de télégraphie militaire*, gives instructions in military telegraphy to a certain number of functionaries of the military telegraph branch (3 in number);
—— *de topographie*, v. —— *de dessin topographique*;
—— *de tir*, v. —— *normale de tir*, —— *régimentaire de tir*, —— *régionale de tir*;
—— *des travaux de campagne*, for instruction of infantry captains in field works (one to each *école du génie*).

écolletage, m., neck.
écolleter, v. a., to beat out, to hammer out.
écoltage, m., v. *écolletage*.
économe, a., economical; m., steward.

économie, f., economy; (*adm.*) the performance of any operation by the government without the intervention of a contractor (as, purchase in the open market, construction by days' labor);(*Fr. a.*) saving of bread and oats (amounts not drawn); (*in pl.*) savings;
—— *politique*, economics, political economy.
écope, f., scoop, swab, ladle.
écoperche, f., outrigger; scaffolding pole; derrick;
—— *double*, double derrick, counterpoise derrick;
—— *simple*, single derrick.
écorce, f., bark.
écorcer, v. a., to bark, strip off bark.
écorcher, v. a. r., to skin (an animal); to gall, rub; (*fort.*) to injure parapets, crests (by destroying regularity of slopes).
écorchure, f., gall, skin rubbed off;
—— *sous la selle*, (*hipp.*) saddle gall.
écorner, v. a., to chip; break off (a corner, etc.); wear off, make a breach in; (*mil.*) to surprise, take, a convoy at one of its extremities.
écornure, f., splinter, corner broken off, chipping; breaking at the edge.
écot, m., share; stump (of a tree), large branch not wholly stripped of twigs, leaves, etc.
écouane, f., rasp.
écouaner, v. a., to rasp.
écouelle, f., cup, saucer; saucer of a capstan.
écouenne, f., v. *écouane*.
écouer, v. a., (*hipp.*) to dock, crop (a horse's tail).
écoulement, m., flow, flowing off, running; draining, discharge; efflux; (*mil.*) passage, marching past, of troops; (*sig.*) passing on, transmission (of messages, signals);
fossé d'——, drain;
galerie d'——, (*min.*) adit, day level;
—— *à gueule bée*, efflux of a filled tube;
—— *nasal*, (*hipp.*) running at the nose;
niveau d'——, drain level;
tuyau d'——, waste pipe;
voie d'——, outlet, water course.
écouler, v. r., to flow off, drain; (*mil.*) to march past a given point; (*sig.*) to transmit, pass on (signals, messages).
écourtage, m., (*hipp.*) docking.
écourter, v. a., (*hipp.*) to dock a horse's tail.
écoute, f., (*mil. min.*) listener, listening gallery.
écouté, a., clear, distinct; precise;
mouvements ——s, (*man.*) precise, measured movements;
pas ——, (*man.*) regular, uniform step;
trot ——, (*man.*) clear, well-marked trot.
écouter, v. a., to listen;
—— *son cheval*, (*man.*) to follow one's horse, to leave him to himself when he is executing his airs properly;
le pas écoute le(s) talon(s), (*man.*) the step is regular.
écouteur, m., (*mil. min.*) listener; earpiece of a telephone; a., v. *écouteux*.
écouteux, a., (*man.*) skittish;
cheval ——, horse that pricks up his ears; horse slow in starting.
écoutille, f., (*nav.*) hatch, hatchway; (*fort.*) manhole.
écouvette, f., sprinkling brush (of a forge).
écouvillon, m., (*art.*) sponge;
—— *articulé*, jointed sponge;
botte d'——, sponge-head;
bouton d'——, v. *botte d'——*;
—— *brisé*, v. —— *articulé*;
—— *à charnière*, v. —— *articulé*;
crosse d'——, v. *botte d'——*;
—— *cuiller*, gun-ladle;
—— *à graisser*, oiling, greasing sponge;
—— *à hampe mobile*, v. —— *articulé*;
—— *à hampe recourbée*, sponge with bent staff;
—— *-levier*, sponge handspike (Fr. 80mm mountain gun);
mouliner ——, *faire mouliner l'——*, to turn, cant, the sponge at drill;

écouvillon-*refouloir*, sponge and rammer staff; *tête d'* —— v. *boîte d'*——.
écouvillonner, v. a., (*art.*) to sponge, sponge out.
écran, m., screen, fire-screen; (*mil.*) screen of troops, e. g., cavalry screen; (*r. f. art.*) deflector (for empty cases); shield; (*ball.*) screen for direct measurement of ordinates of trajectory; (*t. p.*) target;
—— *manipulateur*, screen or cut-off (as, of a signaling apparatus);
—— *de sûreté*, (*mil.*) cavalry screen.
écrasement, m., crushing, crushing in; collapse, ruin, destruction;
épreuve d' ——, crushing test.
écraser, v. a., to crush, bruise; (*fig.*) to ruin, run over; to overwhelm (as, by artillery fire);
moulin à ——, crushing mill.
écrasite, f., (*expl.*) ecrasite.
écrémoire, f., (*artif.*) rubber, grinder, scraper for the measuring table.
écrêter, v. a. (*art.*) to destroy, cut down, the top of a wall, a parapet, by artillery fire; (*fort.*) to pierce (a wall for loopholes);
—— *une côte*, to lower, cut down, a slope;
—— *une route*, to cut down a road.
écrevisse, f., crab; can hooks; large double hooks for raising heavy weights;
—— *de rempart*, (*cav. slang*) foot soldier, "mud crusher."
écriture, f., writing, handwriting; (*in pl.*) papers, official papers; (*top.*) legends, any written or printed indications on a map;
commis aux ——*s*, clerk;
—— *déliée*, fine handwriting;
——*s extérieures*, (*Fr. a.*) accounts and papers kept by a troop unit, or by an establishment, with the state;
——*s intérieures*, (*Fr. a.*) all accounts, papers, not included under ——*s extérieures*;
—— *moulée*, engrossed handwriting (as, in legends on maps); engrossing hand;
—— *à pleins*, full, heavy, writing.
écrivain, m., writer; clerk; purser.
écrou, m., nut, screw-box, screw-nut, female screw; collar; (*adm.*) prison register; (*art.*) holding-down nut; (*sm. a.*) cocking-nut;
—— *ailé*, wing-nut;
—— *à anses*, journaled screw-box;
—— *d'assemblage*, (*art.*, etc.) assembling nut;
billet d' ——, (*mil. adm.*) order of confinement;
—— *borgne*, (*mach.*, etc.) follower, cap follower;
boulon à ——, coupling bolt;
bouton ——, (*sm. a.*) cocking nut, cocking piece;
—— *à branches*, wing-nut, thumb-nut;
—— *à cannelures*, grooved nut;
—— *à chapeau*, (*art.*) wheel-nut, axle-nut;
—— *de culasse*, (*art.*) base ring, breech bushing;
—— *à douille*, (*art.*) brake-nut;
—— *frein*, (*art.*) locking-nut;
—— *de frein*, (*art.*) brake-nut (on axle of some carriages);
—— *à gorge*, grooved nut;
—— *à molette*, milled nut;
—— *moteur*, feed-nut;
ordre d' ——, (*mil. adm.*) order of corps commander to transfer a prisoner from regimental to civil prison;
—— *à oreilles*, wing-nut, thumb-nut;
—— *-pignon*, (*art.*, etc.) pinion-nut;
—— *à pivot*, v. —— *à tourillons*;
—— *à poignée(s)* set-nut and handle;
—— *de pression*, set-nut, jam nut;
—— *prisonnier*, (*mach.*) stationary nut or female screw;
—— *de rappel*, (*mach.*, etc.) adjusting-nut;
—— *de réglage*, (*mach.*) adjusting-nut;
—— *-régulateur*, regulating, adjusting, nut (as in some electric lamps);
—— *de serrage*, clamping-nut, clamp-nut, set-nut; (*art.*) closing cap screw;
—— *à six pans*, hexagonal nut;
—— *à tenons*, (*art.*) pin-wheel (Hotchkiss revolving cannon);
écrou à tourillons, (*art.*) crosshead elevating nut;
—— *de la vis-culasse*, —— *de vis de culasse*, (*art.*) base-ring, breech bushing;
—— *de la vis de pointage*, (*art.*) elevating-screw box.

écrouer, v. a., to enter (name, cause of confinement, etc.) upon the prison register; to imprison.
écroui, a., (*met.*) cold-hammered, springy.
écrouir, v. a. r., (*met.*) to cold-hammer, to hammer-harden; to harden by cold working, as by rolling, drawing, etc.; to make crystalline from fibrous by hammering; to make brittle by hammering; to become springy.
écrouissage, m., (*met.*) cold hammering, hammer hardening, cold hardening.
écrouissement, m., (*met.*) v. *écrouissage*.
écroulement, m., falling in, or down, giving way (of a building).
écrouler, v. r., to crumble in, to fall in; to give way; to fall into decay.
écu, m., shield; escutcheon.
écuanteur, m., disk of a wheel.
écubier, m., (*nav.*) hawse hole.
écueil, m., rock, reef, sand bank; (*fig.*) danger.
écuelle, f., porringer; socket, capstan socket.
éculer, v. r., to run down at the heel (of a shoe).
écume, f., foam, froth, lather; scum, dross;
—— *de fonte*, (*met.*) kish.
écumer, v. a. n., to froth, foam, lather; to skim.
écumoire, f., (*met.*) skimmer.
écurage, m., scouring, cleaning.
écurer, v. a., to scour, clean.
écureuil, m., (*powd.*) granulating cylinder of wire gauze.
écurie, f., stable (esp., stable for a horse); stabling; (*nav.*) horse boat, horse transport.
écusson, m., escutcheon, scutcheon; (*sm. a.*) trigger plate (obs.);
—— *d'attribut*, (*unif.*) the piece of cloth on which are sewed the insignia, the number of the regiment;
—— *d'épaulette*, (*unif.*) pad of an epaulet;
—— *à numéro*, (*unif.*) shield or piece of cloth with regimental number.
écuyer, m., squire; horseman, rider; good horseman; riding master; equerry;
grand ——, grand equerry.
édenter, v. a., to break the teeth (of a saw, etc.).
édification, f., act of building, of construction.
édifice, m., edifice, building, pile (of buildings).
édifier, v. a., to build.
édit, m., edict.
éduction, f., (*steam*) eduction, exhaustion.
effacer, v. a., to efface, expunge, rub out, strike out, scrape out.
effaçure, f., erasure, obliteration, blotting out.
effectif, a., effective.
effectif, m., (*mil.*) strength, numbers, i. e., the total strength, present and absent; (*in pl.*) numbers;
—— *budgétaire*, budgetary strength, numbers as fixed by the annual budget;
—— *de guerre*, war strength, total number on war footing.
—— *moyen*, mean daily strength of the year elapsed, i. e., the numbers obtained by dividing the total number of days present for duty (men and horses) by 365–6;
—— *de paix*, —— *du pied de paix*, peace strength, footing;
—— *du pied de guerre*, war footing, war strength.
effet, m., effect, execution, result, consequence; (*mach.*, etc.) performance; (*in pl.*, *mil.*) kit, outfit, articles of equipment;
à l' ——, (*top. drawing*) so as to show surface forms;
—— *brisant*, (*expl.*) explosive strength, effect;
—— *calorifique*, calorific effect, duty;

effet(s) *de campement*, (*mil.*) camp equipage;
— *contondant*, (*art.*) racking effect;
à double——, double acting, double action (as, a pump, fuse, etc.);
— *d'éclatement*, (*art.*) bursting effect, performance;
— *d'habillement*, (*mil.*) clothing, kit, outfit;
— *d'une machine*, performance of an engine;
— *s de la main*, (*man.*) motions of the hand (in riding);
— *meurtrier*, destructive effect, killing power (as, of a shrapnel);
— *nominal*, nominal effect, performance of an engine, etc.);
— *s de pansage*, (*art., cav.*) grooming kit;
— *perdu*, (*mach., etc.*) lost work;
— *perforant*, (*art.*) perforating, piercing, effect (of a projectile);
— *s de petite monture*, (*mil.*) kit of small parts, articles, etc., carried in saddlebags;
— *s de pointure*, (*mil.*) marked or stamped articles (of the equipment);
— *s de la première portion*, (*mil.*) clothes, *grand équipement*, q. v., belts, etc.; in general, everything normally drawn from the regular administrative magazines;
— *s de la deuxième portion*, (*mil.*) *petit équipement*, q. v.; articles bought in market, and articles exceptionally drawn from administrative magazines;
— *s règlementaires*, (*mil.*) regulation outfit, "uniform" articles (U. S. M. A.);
— *retardé*, (*art.*) delay action (of a fuse);
à simple ——, (*mach., etc.*) single acting, single action;
— *utile*, (*mach., etc.*) useful effect, duty performance; (*ball., sm. a.*) the number of hits by one man in one minute (generally multiplied by 100).

effilé, *a.*, slender;
cheval ——, fine-, thin-shouldered horse.

efflanqué, p. p., (*hipp.*) thin (result of fatigue or of poor food); lean-flanked.

efflanquer, v. a., (*hipp.*) to reduce a horse by fatigue, poor food, etc.

effleurer, v. a., to graze, glance on, rase; to skim the surface of.

efflorescence, f., (*chem.*) efflorescence.

efflorescent, a., (*chem.*) efflorescent.

effondré, a., (of the soil) plowed deep and manured, broken up thoroughly, (hence) boggy, miry; (of a cask, etc.) knocked in.

effondrer, v. a. r., (of the soil) to plow deep, break up thoroughly, and manure, (hence) to bemire, to cut up, as by teaming, etc.; to knock in (a cask, etc.); to crumble, fall.

effort, m., stress, effort, exertion, strain, force, strength; (*hipp.*) strain;
— *de boulet*, (*hipp.*) strain of the pastern joint;
cheval qui a un ——, (*hipp.*) strained, foundered horse;
— *de glissement*, shearing strain;
— *de reins*, (*hipp.*) strain of back and loins, dorso-lumbar strain;
— *de rupture*, breaking stress;
— *de tendon*, (*hipp.*) tendon strain;
— *de tension*, tensile stress;
— *de traction*, tractive effort.

effraction, f., breaking in, for purpose of stealing.

effriter, s'——, to crumble.

effroi, m., fright, terror.

égal, a., equal; (*man.*) even (of a gait); (*top.*) even (ground).

égalisage, m., (*powd.*) equalizing of grains (manufacture).

égalisation, f., (*mil.*) equalization (of platoons, sections, etc.).

égaliser, v. a., to equalize; (*mil.*) to equalize (platoons, etc.); (*powd.*) to classify, separate (grains, according to size).

égaliseur, m., —— *de potentiel*, (*élec.*) potential regulator.

égalisoir, m., (*powd.*) separator, separating sieve.

égalisures, f., pl., (*powd.*) granulated and sieved powder.

égalité, f., equality; smoothness, uniformity; state of level;
— , *angle d'* —— *d'actions*, (*torp.*) angle made by the Whitehead torpedo with the horizontal when the piston and the pendulum neutralize each other's effects;
position d' —— *d'actions*, (*torp.*) position of the Whitehead torpedo when the piston and the pendulum neutralize each other's effects.

égaré, p. p., lost; (*mil.*) missing (report of casualties after a battle);
bouche ——*e*, (*man.*) spoiled mouth.

égarer, v. a. r., to cause to lose one's way; to lead astray; to lose; to lose one's way;
— *la bouche d'un cheval*, (*man.*) to spoil a horse's mouth.

égarrotter, v. a., (*hipp.*) to wring the withers of a horse.

églander, v. a., (*hipp.*) to cut vives out.

église, f., church; the church, especially the Roman Church. (*Eglise* is strictly a Roman Catholic church, Protestants using *temple* instead.)

égohine, f., keyhole saw.

égorger, v. a., to cut the throat of, to slaughter, slay, kill.

égout, m., drip (water); gutter; drain, sink; eaves; slope or face of a roof.

égouttage, m., draining, dripping, drip.

égoutter, v. a. r., to drain, drip; to dry by dripping;
faire ——, to dry.

égouttoir, m., drainer; drip board; grating for tarred rope.

égratigner, v. a, to scratch.

égratignure, f., scratch; (*art.*) scratch, score (in the bore of a gun).

égravoir, m., awl, sharp-pointed bit.

égrenage, m., shelling, picking; ginning (cotton).

égrènement, m., (*art.*) crack, flaw, in the bore of a (bronze) gun.

égrener, v. a. r., to pick, shell; to gin (cotton); (*met.*) to crack, to fly.

égrisé, m., **égrisée**, f., diamond powder (in the arts).

égrugeoir, m., (*powd.*) rubber, mealer.

égruger, v. a., to pound, bruise; (*powd.*) to meal.

égueulement, m., (*art.*) running, spewing, of the metal (at the muzzle of bronze guns).

égueuler, v. a. r., (*art.*) to spoil, break, the muzzle of a gun; (esp. of bronze guns) to run, spew, at the mouth.

éhanché, a., (*hipp.*) hipshot.

éjecter, v. a., (*sm. a., r. f. art.*) to eject.

éjecteur, m., (*sm. a., r. f. art., mach.*) ejector.

éjection, f., (*sm. a., r. f. art.*) ejection.

élagage, m., lopping or thinning of branches; the branches lopped off.

élaguer, v. a., to lop branches, to trim trees.

élan, m., start, spring, jerk, leap; dash; (*mil.*) dash, spirit of troops;
faire un ——, to make a start, spring, dash;
prendre d' ——, to take a start;
prendre un ——, v. *faire un* ——.

élancé, a., lank, thin, slender;
cheval ——, swayback horse, one shrunk in the flank.

élancer, v. a. r., to rush or shoot forward; to make a dash, rush.

élargir, v. a., to enlarge, widen, extend, spread out; to release, set at liberty (a prisoner).

élargissement, m., widening, extension; discharge, release (of a prisoner); (*art.*) enlargement of the bore;
— *de la tranchée*, (*siege*) widening of the trench (by infantry working parties).

élargissure, f., eking piece, piece let in.

élasticité, f., elasticity;
formule d' ——, Hooke's law;
— *de la puissance*, (*steam*) variation of the power of a steam engine from its nominal power.

élasticimètre, m., instrument to measure the elongation of test bars.

élastique, a., elastic.

électeur, m., elector; voter.

électricité, f., electricity;
— *à basse tension*, low-tension electricity;
— *dynamique*, dynamical electricity;

électricité à faible tension, v. —— *à basse tension;*
—— *à forte tension,* high-tension electricity;
—— *à haute tension,* v. —— *à forte tension;*
—— *statique,* statical electricity.
électro, m., (*elec.*) electro-magnet.
électro-aimant, m., (*elec.*) electro-magnet.
électro-ballistique, a., (*ball.*) electro-ballistic.
électro-chimique, a., (*elec.*) electro-chemical.
électrode, f., (*elec.*) electrode.
électro-dynamique, a., (*elec.*) electro-dynamic.
électro-dynamomètre, (*elec.*) electro-dynamometer.
électro-galvanique, a., (*elec.*) electro-galvanic.
électro-galvanisme, m., (*elec.*) electro-galvanism.
électrolyse, f., (*elec.*) electrolysis.
électrolyser, v. a., (*elec.*) to electrolyze.
électrolyte, m., (*elec.*) electrolyte.
électrolytique, a., (*elec.*) electrolytic.
électro-magnétique, a., (*elec.*) electro-magnetic.
électro-magnétisme, m., (*elec.*) electro-magnetism.
électro-métallurgie, f., (*elec.*) electro-metallurgy.
électromètre, m., (*elec.*) electrometer;
—— *à quadrant,* quadrant electrometer.
électrométrie, f., (*elec.*) electrometry.
électro-mineur, m., (*mil. min.*) electric boring, lighting, and ventilating engine.
électro-moteur, (*elec.*) electromotor, motor.
électro-statique, a., (*elec.*) electrostatic.
électrothermique, a., (*elec.*) electro-thermal.
élégir, v. a., to lessen the thickness of a piece of wood; to make narrower, less thick (wood).
élément, m., element; (*elec.*) cell;
—— *s constructifs,* (*mach.*) machine elements;
—— *étalon,* standard cell;
—— *de pile,* —— *de pile primaire,* voltaic cell;
—— *s simples de machine,* v. —— *s constructifs.*
élette, f., (in a shoe) a side lining of leather.
élevage, m., raising, breeding, (of domestic animals).
élévateur, m., lift, hoist, elevator; (*art.*) arm (of a disappearing carriage); (*sm. a.*) platform, cartridge carrier; feed mechanism, feeding lever, carrier lever;
—— *à deux branches,* (*ma. a.*) double-branched cartridge carrier;
—— *à ressorts,* (*sm. a.*) spring carrier, spring feed mechanism;
—— *à ressorts compensateurs,* (*art.*) shell-hoist (Krupp 42cm).
élévation, f., elevation, altitude, height, rise, rising ground, eminence; increase; (*drawing*) elevation;
—— *antérieure,* front elevation;
—— *de côté,* side elevation;
—— *-coupe,* sectional elevation, section and elevation;
—— *de derrière,* rear elevation;
—— *de devant,* front elevation;
—— *de face,* front elevation;
—— *latérale,* side elevation;
—— *postérieure,* rear elevation;
—— *des tourillons,* (*art.*) height of trunnions above platform.
élève, m., pupil; cadet (at a military, a naval, school); (*mil.*) lance (in composition);
—— *d'administration,* (*Fr. a.*) n. c. o. of the administration after graduation from the school at Vincennes, and before appointment as *officier d'administration;*
ancien —— *de ....,* graduate of ....;
—— *-brigadier,* (*art., cav.*) lance-corporal;
—— *-caporal,* (*inf.*) lance-corporal;
—— *de marine,* (*nav.*) naval cadet;
—— *-martyr,* (*cav. slang*) a private in training for corporal's chevrons;
—— *-sapeur,* (*Fr. cav.*) lance-sapper.
élever, v. a., to raise, hoist; build, erect; exalt, advance; to raise, breed (horses, etc.); to put up, throw up (works, fortifications, etc.).

éleveur, m., breeder, (*esp.*) horse-breeder.
élingage, m., (*art.*) slinging of a piece by a strap or sling; the sling itself.
élingue, f., (*cord. art.*) strap, sling, sling chain;
—— *à barrique,* cask-sling;
—— *en chaîne,* (*art.*) sling chain;
—— *de culasse,* (*art.*) breech-sling;
grande ——, v. —— *de culasse;*
—— *à pattes,* can hooks;
petite ——, (*art.*) muzzle sling;
—— *de volée,* v. *petite* ——.
élinguer, v. a., to sling (a weight, cask, gun, etc.).
élinguet, m., capstan pawl.
élite, f., élite;
d'——, (*mil.*) picked (of men); crack (of a regiment, etc.);
corps d'——, (*mil.*) crack corps, body of picked men.
ellipse, f., ellipse; (*art.*) elliptical frame for packing tools on tool mule;
—— *-enveloppe,* f., (*ball.*) ellipse containing 97.4 per cent of all the points of fall.
ellipsoïde, m., ellipsoid;
—— *de bonne rupture,* (*mil. min.*) the ellipsoid whose semiaxes are the vertical and horizontal radii of good rupture, respectively;
—— *de rupture-limite,* (*mil. min.*) the ellipsoid whose semiaxes are the vertical and horizontal radii of limiting rupture, respectively.
éloignement, m., distance;
—— *des forts,* (*fort.*) distance of works from the enceinte, or main central work or position.
éloigner, v. a. r., to stand off, remove, to withdraw, to go away; (*mil.*) to stand off, beat off the enemy.
élonger, v. a., to come alongside;
—— *une manœuvre,* (*cord.*) to stretch out, run out, a rope.
émail, m., enamel;
—— *d'encadrement,* (*hipp.*) external enamel, enamel sheath (of a tooth).
émargement, m., receipt written on the margin of a bill;
feuille d'——, (*Fr. a.*) monthly pay sheet, for officers, reenlisted n. c. o., and *commissionnés.*
émarger, v. a., to cut down or trim a margin; (*adm., Fr. a.*) to receipt for pay (on the margin), (hence) to draw one's pay.
emballage, m., packing (as, of arms, ammunition, clothes, etc.); spurt (bicycle);
fil d'——, packing thread;
toile d'——, packing cloth, burlap.
emballer, v. a. r., to pack, pack up; (of a turbine) to race; (of a bicycle rider) to spurt; (*man.*) to run away (horse).
emballeur, m., packer; (*man.*) horse liable to run away.
embarcadère, m., wharf, quay; *r. r.*) station (strictly, station of departure, v. *débarcadère.*
embarcation, m., boat, small boat; generic name for rowboats and small sailing craft.
embargo, m., embargo;
lever, mettre, un ——, to take off, put on, an embargo.
embarillage, m., (*art., expl.*) packing, stowing, of powder, etc., in barrels.
embariller, v. a., (*art., expl.*) to pack, stow, in barrels, to barrel (powder, etc.).
embarquement, m., embarking; (*r. r.*) entraining.
embarquer, v. a. r., to embark, go on board; (*r. r.*) to get into a train;
—— *le pas de ....,* (*mil.*) to take up the .... step.
embarras, m., obstruction, obstacle, impediment.
embarrasser, v. a., to obstruct, impede, embarrass.

embarrer, v. a. r., to take a purchase; (*art., etc.*) to embar; (of a horse) to get the leg over the pole, over the bail.

embarrure, f., (of a horse) trick of getting the leg over the pole or bail; (*hipp.*) blemish caused by this trick.

embase, f., (*mach., etc.*) shoulder, base, seat; shoulder of a bolt, etc.; (*sm. a.*) front-sight mass; (*art.*) rimbase;
—— *de garniture*, (*sm. a.*) band-shoulder;
—— *du guidon*, (*art., sm. a.*) front-sight mass;
—— *du percuteur*, (*sm. a.*) circular shoulder or band of the striker;
—— *des tourillons*, (*art.*) rimbase.

embasement, m., basement.

embatage, m., tiring, operation of tiring a wheel; *platforme d'*——, tiring platform.

embatailler, v. a., (*mil.*) to draw up in order of battle.

embataillonner, v. a., (*inf.*) to form into battalions.

embâter, v. a., to pack, put on a pack saddle.

embattre, v. a., to tire a wheel.

embauchage, m., engaging, hiring, of workmen; (*mil.*) the crime of enticing and of helping soldiers to desert to the enemy, of enlisting men to serve against their own country.

embaucher, v. a., to hire workmen; (*mil.*) to entice soldiers to desert to the enemy, men to serve against their own country.

embaucheur, m., labor contractor; (*mil.*) person guilty of *embauchage*, q. v.

embauchoir, m., last; (*art.*) semicylindrical block (for taking impressions of the bore).

emblée, *d'*——, adv., at the first onset, at once.

emboire, v. a., to cover with oil or wax (as, a mold, etc.).

emboîtage, m., (in shoemaking) the seat of the heel.

emboîtement, m., fitting; clamping; jointing; putting or fastening together (as, the several parts of a carriage); locking in or together; (*mil.*) action of swinging into, or taking up, the cadenced step.

emboîter, v. a. r., to mortise, to clamp, to tongue; to engage in, fit in; to join, joint, put together; dovetail;
—— *le pas*, (*mil.*) to lock in marching; to swing into, take up, the cadenced step;
s'—— *à feuillure*, (*carp.*) to fit into a rebate or groove.

emboîture, f., socket; nave lining or box; frame (of a door); joint; insertion of one piece, etc., into another; clamp, collar.

embosser, v. a. r., (*nav.*) to moor, anchor, a vessel ahead and astern; to bring the broadside to bear.

embossure, f., (*cord.*) mooring, lashing, stopper; *faire* ——, to put a spring upon a cable.

embouchement, *harn.*) biting (a horse).

emboucher, v. a. r., to put into the mouth; (*harn.*) to bit a horse (i. e., select a suitable bit), to put the bit in the horse's mouth; (of rivers) to flow, empty into; (*mil.*) to use the loopholes of the defense against the defense, to fire through the defense's loopholes; (*nav.*) to enter.

embouchoir, m., mouthpiece (of a trumpet, etc.); (*sm. a.*) upper band, upper slide;
—— *à quillon*, (*sm. a.*) upper band of the Lebel rifle (carries a quillon).

embouchure, f., mouthpiece (of a trumpet, etc.); mouth (of a river, tube, tunnel, etc.); (*harn.*) mouthpiece (of a bit); (*sm. a.*) pipe (of a pistol holster); (*art.*) muzzle (rare); (*fort.*) exterior part, mouth, of an embrasure, a crenelation.

emboudinure, f., puddening (of an anchor).

embouquement, m., entrance of a narrow pass, channel.

embouquer, v. n., (*nav.*) to enter a narrow pass, channel.

embourber, v. a. r., to get into the mire, to stick in the mud.

embourrer, v. a., to stuff, pad (a saddle, etc.).

embout, m., ferrule; (*sm. a.*) tip (metal) of a leather scabbard.

embouter, v. a., to ferrule; put a ferrule on.

embouti, p. p., headed (as, a cartridge case), beaten out, hollowed out by beating, stamping.

embouti, m., any work, piece, produced by swaging or stamping.

emboutir, v. a., (*farr.*) to cut down a horse's hoof; (*met.*) to chase, swage, beat out, stamp out, head, to form, shape (as, curved armor plates); (*sm. a., r. f. art.*) to form the head of, to head, a cartridge case.

emboutissage, m., (*met.*) operation of *emboutir*; swaging, heading up (as, a cartridge case); beating out; shaping, forming (as, curved armor plates);
—— *au pilon*, heading, forming, under the hammer;
—— *à la presse hydraulique*, forming, etc., by hydraulic pressure.

emboutissoir, m., (*mach.*) plate-bending machine.

embouveter, v. a., (*carp.*) to plow and tongue, to join by plow and tongue.

embranchement, m., branch; branching (as, of pipes, etc.); forks (of a road); (*r. r.*) branch line; (*cons.*) brace.

embrancher, v. a. r., to frame, put together; branch, branch out from.

embraquer, v. a., (*cord.*) to sling; to haul in; to haul in the slack;
—— *le mou*, to haul in the slack.

embrasement, m., fire, conflagration, combustion.

embraser, v. a., to set on fire, kindle.

embrasser, v. a., to clasp, span, encircle; to cling to a horse.

embrassure, f., iron hoop, collar (as, for the top of a pile, of a chimney).

embrasure, f., embrasure (of a door, window); (*fort.*) embrasure;
brisure d'——, (*fort.*) embrasure neck;
—— *en contrepente*, (*fort.*) countersloping embrasure;
coup d'——, (*art.*) embrasure hit;
—— *de coulée*, (*met.*) working side, face of a furnace;
dégorger une ——, (*fort.*) to cut out an embrasure;
—— *directe*, (*fort.*) normal embrasure;
évasement de l'——, (*fort.*) splay of the embrasure;
—— *oblique*, (*fort.*) oblique embrasure;
—— *à ouverture minima*, (*fort.*) minimum port embrasure;
tirer à ——, (*art.*) to fire through embrasures;
—— *de travail*, (*met.*) v. —— *de coulée*;
—— *-tunnel*, (*fort.*) embrasure tunnel (casemate embrasure protected by a masonry screen);
—— *de tuyère*, (*met.*) tuyere arch;
—— *à volets*, (*fort.*) shutter embrasure.

embrayage, m., engaging and disengaging gear; coupling;
—— *à cônes* (*de friction*), cone clutch;
—— *à friction*, friction clutch;
—— *à endentures*, clutch.

embrayer, v. a., (*mach.*) to throw in gear; to engage, connect.

embrelage, m., lashing, lashing down; (*pont.*) rack lashing, operation of racking down, racking down;
anneau d'——, lashing ring;
chaîne d'——, (*art., etc.*) keep-chain.

embreler, v. a., to lash, lash down; (*pont.*) to rack down.

embrèvement, m., (*carp.*) mortise, groove; operation of letting one piece into another cut halfway through.

embrever, v. a., (*carp.*) to mortise; to sink or let in.

embrigadement, m., (*mil.*) brigading.

embrigader, v. a., (*mil.*) to brigade together.

embroché, p. p., (*elec.*) joined up to, in circuit with;
poste ——, (*sig., etc.*) way station, intermediate station, connected up with the main line.

embrocher, v. a., to run through with a sword.

embrouillé, p. p., gloomy, overcast (of the weather).
embrouiller, v. r., to become overcast (of the weather).
embrumé, a., foggy.
embûche, f., ambush, snare;
 donner, tomber, dans une ――, to fall into a snare;
 dresser des ――*s à*, to lie in wait for;
 dresser une ――, to lay a snare, ambush.
embuscade, f., (*mil.*) ambush, ambuscade;
 dresser, tendre, une ――, to prepare an ambuscade;
 en ――, in ambush;
 ―― *de tirailleurs*, (*fort.*) sort of small shelter trench to hold 4 or 5 sharpshooters.
embusquer, v. a. r., (*mil.*) to place in ambush, to set an ambuscade; to lie in ambush.
émerger, v. n., to emerge.
émeri, m., emery, emery paper.
émeriller, v. a., to rub, grind, with emery, emery paper.
émerillon, m., swivel; (*cord.*) loper;
 anneau à ――, ―― *-anneau*, swivel ring;
 croc à ――, ―― *à croc*, swivel hook.
émerisé, a., covered with emery, emery;
 papier ――, emery paper;
 toile ――*e*, emery cloth.
émeriser, v. a., to lay on emery, coat, face with emery.
émérite, a., emeritus.
émersion, f., emersion.
émeute, f., riot, disturbance.
émeutier, m., rioter.
émiettement, m., (*art.*) fragmentation, splintering, of a shell.
émietter, v. r., (*art.*) to break up into fragments (shell on plate, etc.).
éminence, f., eminence, elevation, rising ground, hill; (*fig.*) eminence.
émilite, f., (*expl.*) emilite.
émission, f., emission; (*steam*) emission, educting, exhausting, eduction valve;
 lumière d' ――, eduction port;
 tuyau d' ――, eduction pipe.
emmagasinage, m., warehousing, storing (in arsenals, etc.).
emmagasinement, m., v. *emmagasinage*.
emmagasiner, v. a., to store up (in magazines, arsenals, etc.); to store up (work, etc.).
emmanché, a., helved, fitted with a helve, a handle.
emmanchement, m., hafting, helving;
 à baïonnette, bayonet joint.
emmancher, v. a. r., to fit a helve, a handle, to; to slide in or over; to set about, to begin.
emmanteler, v. a., (*fort.*) to surround a place by a fortified enceinte.
emmarchement, m., (*cons.*) width of a stairway;
 ligne d' ――, middle line of a stairway (in plans).
emménagement, m., moving in; (*in pl., nav.*) accommodations.
emménager, v. a., to move in.
emmenotter, v. a., to handcuff.
emmensite, f., (*expl.*) emmensite.
emmétrer, v. a., to arrange materials so as to facilitate their measurement.
emmiellure, f., (*hipp.*) charge, hoof ointment.
emmortaiser, v. a., to mortise.
emmurer, v. a., to wall in or up; to surround by a wall.
émorfiler, v. a., to take off the wire or rough edge.
émouchette, f., fly net for horses.
émoudre, v. a., to grind, whet, sharpen.
émoulage, m., grinding; (*sm. a.*) grinding of barrel to shape.
émouler, v. a., (*mach.*) to grind.
émoulerie, f., grinding shop.
émouleur, m., grinder (workman); (*sm. a.*) barrel grinder.
émouleuse, f., (*mach.*) grinding machine.
émoulu, p. p., sharp.

émoussé, p. p., dull (of edge tools, cutting and thrusting weapons).
émousser, v. a. r., to blunt, dull, to become dull; to chamfer, bevel; (*fig.*) to deaden, weaken;
 ―― *un bataillon, etc.*, (*mil.*) to cut off the corners of a battalion, etc., in square so as to form and octagon.
empaillage, m., wrapping with straw (as, a swing-bar in a stable, etc.).
empailler, v. a., to wrap with straw (as, gun-carriage wheels, etc.).
empan, m., span (in measurements).
empanacher, v. a. r., to adorn with a plume, (*unif.*) to plume (a helmet).
empannon, m., (*cons.*) jack-rafter; (*art., etc.*) hound, brace;
 ――*s de flèche*, (rear) brace, hind hounds;
 ――*s de timon*, (fore) brace, fore hounds.
empaquetage, m., packing, packing up; (*sm. a.*) assembling of cartridges in packets; packing of cartridges in the magazine or the loader;
 ―― *à 6, 8, etc.*, (*sm. a.*) packing of cartridges in sets or packs of 6, 8, etc.
empaqueter, v. a., to pack, pack up: (*sm. a.*) to pack cartridges in cases, in packages generally.
emparer, v. r., (*mil.*) to take a work, position, etc., by assault or main strength.
empâté, p. p., *cheval* ――, coarse, thickset horse, with hairy extremities.
empâtement, m., (*cons.*) foundation, base, footing, of a wall; base, support, of a crane; (*cord.*) splicing, length of the strands in splicing.
empâter, v. a., to foot, to tenon; to shape the foot or tenon of a wheel-spoke; (*cord.*) to splice, to twist the strands of two ropes together when making a splice; (*cons.*) to give footing to a wall.
empattement, m., foot (of a talus); foot, sole (of a rail).
empature, f., scarf, scarfing, joining together (of two beams, end to end).
empêché, p. p., embarrassed, hindered; (*mil.*) prevented from serving, from taking duty;
 cordage ――, (*cord.*) foul or entangled rope.
empêchement, m., impediment, hindrance; obstruction, obstacle; (*mil.*) any hindrance to performance of a duty;
 en cas d' ――, in case of inability.
empêcher, v. a., to stop, prevent, hinder, oppose, obstruct.
empeigne, f., upper, of a shoe.
empêtrer, v. a. r., to entangle; hamper, embarrass; to hobble (a horse); (of horses) to become entangled in the traces, to get the feet, legs, tangled in the traces, harness.
empierrement, m., ballasting; ballast; metaling (of a road, of a railroad bed);
 en ――, metaled, macadamized.
empierrer, v. a., to ballast a road.
empilage, m., piling (of projectiles, etc.).
empilement, m., piling, stacking; pile (of shot, etc.).
empiler, v. a., to pile up, stack (shot, etc.).
empire, m., empire;
 ―― *de la mer*, (*nav.*) sea power.
emplacement, m., seat, site, emplacement; (*art., fort.*) emplacement of a gun; (*mil.*) station (of troops in peace), position of troops, of a line (on the battlefield); (of field artillery) the actual line or position of the guns; the whole artillery position;
 ―― *de batterie*, ―― *des batteries*, (*art.*) the (field) artillery position;
 ―― *cuirassé*, (*art., fort.*) armored emplacement;
 ―― *de la machine fixe*, engine house.
emplanture, f., (*nav.*) step of a mast.
emplâtre, m., plaster, salve.
empléomanie, f., (*Fr. adm.*) craze or mania to hold office under the government.
emploi, m., charge, employment, place, duty, office; (*adm.*) application of funds to a specific purpose;
 double ――, repetition, useless repetition;
 faire double ―― *avec*, to be a duplicate of, to reproduce, repeat, unnecessarily;

emploi, retrait d'——, placing on the unemployed list;
sans ——, (*adm.*) unexpended (of an appropriation).
emplombage, m., (*sm. a.*) leading of the bore.
emplombé, p. p., (*art.*) lead-coated (projectile); (*sm. a.*) leaded.
emplomber, v. a. r., (*sm. a.*) to lead, to become leaded.
employé, m., employee, clerk;
—— *dans les eaux grasses,* (*mil. slang*) clerk of the commissary department;
—— *militaire,* (*Fr. a.*) a functionary of the military establishment (o. g., *officier d'administration, adjoint du génie, etc.*), regarded as an officer, but having no assimilation of rank.
empoigner, v. a., to grasp, gripe, seize with the hand.
empoise, f., (*mach.*) bearing, brass, journal.
empoisser, v. a., to pitch.
emport, m., —— *d'effets,* (*mil. law*) theft of arms, equipment, etc., by a deserter.
emporter, v. a. r., to carry off; (*man.*) to bolt, run away; (*mil.*) to carry, take, by assault or storm;
—— *d'assaut,* (*mil.*) to take, carry by assault, by storm;
—— *la victoire,* (*mil.*) to conquer, to win the victory.
empreinte, f., impression, imprint, stamp; priming (paint); (*art.*) impression of the bore;
—— *à la gutta-percha,* (*art.*) gutta-percha impression of the bore.
emprisonnement, m., imprisonment;
—— *cellulaire,* solitary confinement.
emprisonner, v. a., to imprison, confine.
enarbrer, v. a., to rivet, fix, a wheel, a pinion, on its shaft.
en-bataille, m., (*art., cav.*) the "into line."
en-batterie, m., (*art.*) the "in battery."
encablure, f., cable-length.
encadenasser, v. a., to padlock.
encadré, p. p., framed; (*mil.*) side by side (with other units, as, a company, a battalion, etc., opposed to *isolé*, q. v.); officered; (of young, inexperienced soldiers) supported (framed in) by experienced men on each side; (*man.*) well under one (of the horse);
combat ——, (*mil.*) combat carried on by troops side by side;
troupes bien ——*es,* (*mil.*) well-officered troops.
encadrement, m., frame, framing; (*mil.*) action of *encadrer*, of giving a troop unit its *cadres*, (hence) officering; position of troops, having other troops on the flanks; (*art.*) in ranging, operation of getting the target between a "short" and an "over," (hence) fork, zone, bracket.
encadrer, v. a., to frame, frame in; (*mil.*) to give a troop unit its cadres, (hence) to officer; to enroll; to place a body of troops between other bodies, (hence) to support; to support, encourage, young soldiers by placing them between veterans, (*inf.*) to inclose (said of corporals on the flanks of their half-sections); (*man.*) to get a horse well under one, under the control of the legs;
—— *le but,* (*art.*) in ranging, to get the target between a "short" and an "over."
encaissage, m., (*sm. a.*) packing of rifles, etc., in arm chests, etc.
encaisse, f., cash in hand.
encaissé, p. p., boxed; (of a road, river, etc.) with steep banks, embanked, between banks; commanded by heights (as, a valley).
encaissé, m., cutting (of a road, railway); part of a river between steep banks.
encaissement, m., boxing, packing in a case or box; receipt and deposit of money, funds; (of roads, etc.) embankment; state of being embanked; bed, base;
faire un chemin par ——, to bed, trench, and gravel a road.

encaisser, v. a., to box, pack in cases or boxes; to receive and deposit money; to embank (a river, etc.); to lay a base or bed for a road.
encampané, a., *canon* ——, (*art.*) a gun whose chamber is a truncated cone (obs.).
encan, m., auction; *vendre, vente, à l'* ——, to sell, sale, by auction.
encapeler, v. a., (*cord.*) to make fast.
encapuchonné, a., (*hipp.*) with head curved in (of a horse).
encapuchonner, v. r., (*hipp.*) to bend, curve, the head down.
encartouchage, (*expl.*) operation of making up into cartridges.
encartoucher, v. a., (*expl.*) to make up into cartridges.
en-cas, m., anything prepared and kept for an emergency;
—— *mobile,* (*mil. r. r., Fr. a.*) railroad trains (two or more) on each line of transport, stored with supplies and ammunition, and kept ready for use in emergencies.
encastelé, a., (*hipp.*) hoof-bound; narrow-heeled.
encasteler, v. r., (*hipp.*) to be, become, hoof-bound or narrow-heeled.
encastelure, f., (*hipp.*) state of being hoof-bound or narrow-heeled.
encastrement, m., action, operation, of fitting, bedding, or letting in, one part into another; any bed, groove, hole, cut, mortise, recess, seat, into which some part or piece is or is to be let or fitted; (*sm. a.*) barrel groove;
—— *d'affût,* (*art.*) trunnion bed;
—— *de garniture,* (*sm. a.*) band shoulder;
—— *ordinaire,* (*art.*) firing-bed;
—— *de plaque de couche,* (*sm. a.*) butt-plate seat;
—— *de la platine,* (*sm. a.*) lock mortise;
—— *à ressort,* spring seat and catch;
—— *de route,* (*art.*) traveling-bed; traveling trunnion bed;
—— *de tir,* (*art.*) firing-bed, firing trunnion bed;
—— *des tourillons,* (*art.*) trunnion bed.
encastrer, v. a., to fit, let in; to set in a groove, etc.; to bed in.
encaustique, f., polishing-wax for leather parts of military equipment.
enceinte, f., fence, inclosure; (*fort.*) enceinte, body of the place;
—— *à fronts accolés,* (*fort.*) continuous enceinte;
—— *intérieure,* (*fort.*) enceinte within the main enceinte;
—— *morcelée,* (*fort.*) discontinuous enceinte;
—— *de sûreté,* (*fort.*) sort of interior line to prevent the enemy from swarming into the place on the heels of the defeated garrisons of the outer line.
encenser, v. n., (*hipp.*) to throw the head up and down.
enchaînement, m., chain, links; series, concatenation.
enchaîner, v. a., to chain, link, connect.
enchaînure, f., chain, chainwork.
enchaper, v. a., to inclose in a shell (as, a pulley); (*expl., etc.*) to case (a powder barrel, a pass box).
enchapure, f., (*harn.*) buckle loop, end loop
enchâsser, v. a., to insert, set in.
enchère, f., bidding, auction;
mettre aux ——*s,* to put up for sale by auction;
vendre à l' ——, *aux* ——*s,* to sell by auction;
vente à l' ——, *aux* ——*s,* sale by auction.
enchevaucher, v. a., (*carp.*) to rabbet.
enchevauchure, f., (*carp.*) rabbeting; (*cons.*) overlapping of roofing slates, tiles, shingles.
enchevêtrer, v. a. r., (*harn.*) to halter; to become tangled in the noose, halter, etc.; to get a foot over the halter; (*cons.*) to connect or join two joists by a trimmer.
enchevêtrure, f., (*hipp.*) halter-cast; excoriation of the pastern; (*cons.*) trimming work around a fireplace; trimmed joists.

enclanchement, m., (*mach.*) throwing into gear, coupling, connecting; locking (as, a breech-plug in its seat).
enclancher, v. a., (*mach.*) to throw into gear, to engage, couple, connect, lock.
enclave, f., projection (as, of a closet into a room); recess (for a lock gate when open); (*law*) enclave, enclavure.
enclavement, m., (*law*) enclavement.
enclaver, v. a., (*cons.*) to bolt, to key; (*mas.*) to lay or embed a stone (among other stones already in position); (*law*) to enclave.
encliquetage, m., (*mach.*) click and ratchet, ratchet-gear, -work.
enclos, m., (*top.*) inclosure, inclosed field or yard.
enclouage, m., (*art.*) spiking (obs.).
enclouer, v. a., (*art.*) to spike a gun (obs.); (*farr.*) to prick (a horse, in shoeing).
enclouure, f., (*farr.*) nail prick.
enclume, f., anvil; (*sm. a., etc.*) anvil (of a cartridge, of a primer);
—— *à potence*, horn anvil;
souche d'——, anvil stock;
table d'——, anvil face.
enclumeau, m., hand anvil, portable anvil.
enclumette, f., (*sm. a.*) anvil (of a primer).
encoche, f., notch; (*mach.*) gab, eye, notch.
encochement, m., notching.
encocher, v. a., to notch.
encognure, f., corner, angle; elbow (as, of a knee, etc.).
encollage, m., sizing, size; gumming.
encoller, v. a., to size; paste;
—— *l'ancre*, to put the arms to the shank (of an anchor).
encollure, f., (*met.*) welding point; shutting point.
encolonnement, m., (*mil.*) act of causing troops to form into column, of arranging troops into columns, for the purpose of setting the whole in march; the regulation of the times when, and the places where, the various bodies must enter the column.
encolure, f., of an anchor, point of junction of shank and arms; (*hipp.*) chest, neck, and shoulders of the horse;
—— *d'ancre*, crown of the anchor;
—— *bien sortie*, neck and shoulders, etc., starting out well from the body;
—— *de cerf*, ewe-neck;
chargé d'——, heavy-headed;
—— *chargée*, thick, heavy, neck and shoulders;
—— *de cygne*, cock-throttled;
—— *fausse*, poor neck and shoulders (junction of neck and shoulders too marked);
—— *fichée dans le thorax*, v. —— *fausse*;
—— *mal sortie*, v. —— *fausse*;
—— *penchante*, lop-neck;
pli de l'——, bend of the neck;
—— *renversée*, v. —— *de cerf*;
—— *rouée*, an encolure with the whole upper line of the neck convex upward;
—— *tombante*, v. —— *penchante*.
encombrement, m., crowding, obstructing, obstruction.
encombrer, v. a. r., to obstruct, to encumber, to become obstructed.
encorbellement, m., (*nav.*) sponson.
encorné, a., *atteinte* ——*e*, (*hipp.*) wound due to overreaching, and penetrating into the hoof;
javart ——, (*hipp.*) swelling under the coronet.
encouture, f., clinkerwork (boats).
encouturé, a., clinker-built.
encrassé, a., foul, dirty; choked (of a grate); (*art., sm. a.*) foul (of the bore).
encrassement, m., (*sm. a., art.*) fouling; (*elec.*) fouling of electric contacts; (*steam*) choking (of a grate).
encrasser, v. a. r., to foul, make, foul, dirty; (*art., sm. a., elec.*) to foul; to become foul; (*steam*) to choke (of a grate).
encre, f., ink;
—— *de Chine*, India ink;
—— *à copier*, copying ink;
—— *dragon*, indelible ink;
—— *d'impression*, printer's ink;

encre, *lithographique*, lithographic ink;
mettre à l'——, to ink (a drawing);
mise à l'——, inking (of a drawing);
passer à l'——, v. *mettre à l'*——.
encrené, a., *fer* ——, (*met.*) iron twice heated and hammered.
encrenée, f., (*met.*) iron twice heated and hammered.
encrêper, v. a. r., to put on crape, to go into mourning.
encreur, a., inking;
rouleau ——, inking-roll.
encrier, m., inkstand.
encrier-porte-plumes, m., fountain pen.
encroué, a., *arbre* ——, tree that has fallen upon and caught in another (standing) tree; a (standing) tree supporting another that has fallen upon it.
encroûter, v. a., (*mas.*) to put on a coat of plaster (as, on a wall).
enculassage, m., (*sm. a.*) operation of assembling the barrel and the breech-casing, of screwing the barrel into the breech-casing.
enculassement, m., v. *enculassage*.
enculasser, v. a., (*sm. a.*) to screw the barrel into the breech-casing.
enculasseur, m., workman who performs the *enculassage*.
encurage, m., the consecutive operations, first, of covering a saddletree with a layer of fine ox tendons (*nervage*), and, second, of stretching canvas (*entoilage*) over the tree thus treated.
encuvement, m., (*top.*) hollow, depression, any basin-like depression.
endent, m., square notch.
endente, f., v. *endentement*.
endenté, p. p., cogged;
roue ——*e*, (*mach., etc.*) cogged wheel.
endentement, m., (*mach., etc.*) catching, cogging, toothing; scarfing.
endenter, v. a. r., to indent; to notch, dovetail; to be toothed, notched in; (*mach., etc.*) to furnish with teeth, to cog.
endenture, f., (*mach.*) toothwork; clutches (of a coupling box).
endiguer, v. a., to dam in or up.
endivisionnement, (*mil.***)** the formation of a division, assemblage of troops into a division.
endivisionner, v. a., (*mil.*) to form or constitute a division, (more narrowly) to join brigades into a division, to form a division from two or more brigades.
endosmose, f., endosmose.
endossement, m., (*adm., etc.*) indorsement.
endosser, v. a., to put on (some special dress, as uniform); (*adm., etc.*) to indorse.
enduire, v. a., to wash over, to coat (with paint, plaster, etc.).
enduit, m., coat, coating (of tar, plaster, etc.); grease, unguent; (*fond.*) luting;
—— *à la céruse*, (*art., etc.*) a wash for large metallic objects, made of white lead, charcoal black, linseed oil, etc.;
—— *chatterton*, v. *chatterton*;
—— *de dépouillement*, (*fond.*) a wash applied to a mold to facilitate removal;
—— *drainé*, (*cons.*) network of small drains in the thickness of a layer of concrete;
—— *de noir*, (*fond.*) black wash, wet blacking.
endurcir, v. a. r., to harden, to become hard.
endurcissement, m., hardening.
endurer, v. a. n., to endure; to slacken, slow up (in rowing);
—— *partout*, (of rowing) to hold water.
énergie, f., energy;
—— *au choc*, (*art.*) striking energy.
énergique, a., energetic; (*fig.*) thorough, complete (e. g., of tempering).
énervé, a., (*cord.*) long-jawed, overstretched, fatigued.
enfaîtage, m., (*cons.*) ridge-tile.
enfaîtement, m., (*cons.*) lead on the ridge of a roof, ridge-lead.
enfaîter, v. a., (*cons.*) to put lead on the ridge of a roof.
enfant, m. and f., child;

enfant(s) *perdus*, (*mil.*) forlorn hope;
— *de troupe*, (*Fr. a.*) soldier's son (so recognized with special reference to admission to the *écoles militaires préparatoires*, q. v.).
enfermé, p. p., shut up, in; closed up;
— *entre les terres*, landlocked.
enfermer, v. a., to shut up, in (as, the enemy in a town).
enferrer, v. a. r., to run through (one's adversary); to spit one's self on one's adversary's blade.
enfilable, a., (*art.*) exposed to enfilade fire.
enfilade, f., suite of rooms with doors on same straight line; (*fort.*) enfilade, enfilading; (*art., sm. a.*) enfilading fire;
batterie d'——, (*art.*) enfilade, enfilading, battery;
coup d'——, (*art.*) enfilading shot;
d'——, raking;
tir d'——, (*art., sm. a.*) enfilade, enfilading, fire.
enfiler, v. a., to thread, thread on; to pierce, run through; (*mach.*) to thread on; (*art., sm. a.*) to enfilade, rake;
— *le câble*, (*cord.*) to heave in the cable.
enflé, p. p., swollen;
— *aux jambes*, (*hipp.*) gourdy.
enfler, v. a. n. r., to swell, swell up, rise; to fill (of a balloon).
enflure, f., swelling.
enfonçage, m., bottoming of casks; (*powd.*) packing of gunpowder in barrels.
enfoncement, m., operation of driving, driving in; sinking; depth (as, of a foundation, of a well); breaking in (of a door); recess (in a wall, etc.); hollow (of the ground);
— *de l'âme*, (*art.*) indentation in the bore.
enfoncer, v. a. n. r., to drive in; to set, fix, sink; to drive (nails, etc.); to bottom (a cask); to break, break open; to sink, plunge; (*mil.*) to rout, beat; to break through (enemy's line);
— *un carré*, (*mil.*) to break up, drive in, a square;
— *les éperons*, (*man.*) to spur vigorously;
— *jusqu'au refus*, to drive (a pile) as far as it will go down;
— *une porte*, to drive in, break in, a door.
enfonçure, f., hollow; hole, depression (as, in a road); bottom (of a cask).
enfourchement, m., (*carp.*) open mortise joint; (*mach.*) fork link;
— *double* (*simple*), (*carp.*) an open mortise joint with a double (single) tenon.
enfourcher, v. a., (*man.*) to straddle (a horse).
enfourchure, f., fork; crotch (of the body);
être sur l'——, (*man.*) to have a fork seat.
enfournement, m., operation of setting bread to bake.
enfourner, v. a., to put into a furnace or oven; to set bread to bake.
enfreindre, v. a., to infringe, violate.
enfuir, v. r., to leak; to run away.
enfumement, m., state of being full of smoke (as, a turret, casemate).
enfumer, v., a. r., to fill with smoke; to blacken, darken, with smoke (as glass); (*mil.*) to smoke out (the defenders of a house, etc.).
engagé, p. p., engaged; (*mil.*) engaged, in action; (*cord.*) fouled, fouled; (*art., sm. a.*) jammed (as, a breech mechanism, cartridges in a magazine, a projectile when it can not be seated by an ordinary effort).
engagé, m., (*mil.*) soldier enlisted; (*Fr. a.*) volunteer;
— *conditionnel d'un an*, one-year volunteer;
— *nouveau* ——, recruit;
— *d'un an*, one-year volunteer;
— *volontaire*, volunteer (i. e., person that has enlisted voluntarily).
engagement, m., engagement, contract; (*mil.*) enlistment, enlisting, contract of enlistment; fight, engagement; (*fenc.*) disengage, disengagement;
— *d'un an*, (*mil.*) one-year enlistment;
changement d'——, (*fenc.*) a fresh disengage, made on the opposite side from one previously made;

engagement, *double* ——, (*fenc.*) two immediately successive disengages;
prix d'——, (*mil.*) enlistment bounty;
— *volontaire*, (*mil.*) voluntary enlistment.
(The foregoing translations of fencing terms, taken from the official *Manuel d'escrime* of the French army, differ from those usually given. *Engagement* is generally translated "engage," and is so defined in the Belgian *Règlement d'escrime*.)
engager, v. a. r., to engage; to hire; to entangle; to insert, be inserted (as, a handspike); to slip or fit over, engage in or over; to advance, penetrate (into a defile, pass); to get foul (of an anchor); (*mil.*) to enlist, be enlisted; to engage, to come into action, to begin an action; (*cord.*) to jam, nip, make fast; (*fenc.*) to engage;
— *un bateau*, to run a boat upon the sand;
— *le combat*, (*mil.*) to bring on the action;
— *l'ennemi*, (*mil.*) to engage, attack, the enemy;
— *l'étrier*, (*man.*) to stick, slip, one's foot in the stirrup;
— *un soldat*, (*mil.*) to enlist a soldier;
s' —— *volontairement*, (*mil.*) to enlist voluntarily.
engaîner, v. a., to sheathe, to put up (a sword).
enganter, v. a., (*nav.*) to overtake, come up with, overhaul.
engeance, f., breed (of horses, etc.).
engerbement, m., piling up (as, of powder barrels, etc.); arrangement of small arms in tiered racks.
engerber, v. a., to pile up, stack up (as, powder barrels, carriages, etc.).
engin, m., engine, gear;
— *s de guerre*, (*mil.*) appliances of war;
— *de mise de feu*, (*artif., etc.*) any firing device; any means of firing a charge, etc. (as, Bickford fuse, etc.);
— *de sauvetage*, any life-saving device (fire escape, life belt, etc.).
engorgé, p. p., choked, jammed (as, a breech, a feed mechanism);
jambes —— *es*, (*hipp.*) swelled legs.
engorgement, m., obstruction; choking, jamming (as, a breech mechanism, etc.); (*hipp.*) swelling in a horse's leg.
engorger, v. a. r., to choke, jam (as, a feed mechanism, etc.); (*artif.*) to prime.
engraissement, m., (*carp.*) driving a tenon forcibly into its mortise;
joindre par ——, to fit, to cause to fit, tightly.
engravement, m., (of small boats, etc.) state of being aground on a sand bank (in a river).
engraver, v. a. r., to run aground on a sand bank (of small boats, etc., in a river).
engrenage, m., (*mach.*) gear, gearing of toothed wheels; train of wheels; cogwheel; (*nav.*) stowing;
à ——, geared;
— *d'angle*, bevel-gear;
— *conico-hélicoïde*, skew bevel-gear;
— *conique*, bevel-gear;
— *à crémaillère*, rack and pinion, rack-gear;
— *cylindrique*, spur-gear;
— *à développante*, involute gear-wheel;
— *à développante de cercle*, gear with epicycloidal teeth;
— *droit*, spur-wheel, spur-gearing;
— *droit hélicoïde*, skew spur-wheel;
— *à épicycloïde*, epicycloidal gear;
— *extérieur à fuseaux*, external pin-tooth gearing;
grande roue d'——, driving-wheel;
harnais d'—— *s*, train of gearing; back-gear;
— *intérieur*, internal gearing;
— *intérieur à fuseaux*, internal pin-tooth gearing;
— *intermédiaire*, intermediate gearing;
ôter l'——, to put, throw, out of gear;
— *plan*, v. —— *cylindrique*;
— *à roues d'angles*, v. —— *conique*;
— *à roues coniques*, v. —— *conique*;
roue d'——, brake-wheel;
— *de roues*, gearing;
série d'—— *s*, change gear;

engrenage à vis, screw-gear;
—— à vis sans fin, worm-gearing.
engrènement, m., feeding with grain; (action of) gearing in with.
engrener, v. a. n. r., to feed with corn, beans, etc.; to prime (a pump); (*mach.*) to engage, catch; throw, put, in gear; to tooth, to gear in with; to be put in gear.
engrenure, f., (*mach.*) action of gearing, gearing in together; interlocking.
enharnacher, v. a., (*harn.*) to harness.
enjaler, v. a., to stock an anchor.
enjambée, f., stride (horse, man).
enjamber, v. a. n., to jut, project, stride over.
enjarreté, a., tied by the foot (horse).
enlacement, m., entangling; lacing; twisting, interweaving.
enlacer, v. a. n., to entwine, twist, interlace, entangle, lock;
—— *des papiers*, to file or string papers.
enlaçure, f., auger hole; bolting, pinning, a tenon into its mortise.
enlevage, m., spurt (bicycle).
enlevé, m., (*fenc.*) a moulinet upward in front of the body (singlestick).
enlèvement, m., removal; (*mil.*) storming, carrying by assault (of a redoubt, position, etc.); surprise (of a post); (*pont.*) dismantling (of a bridge).
enlever, v. a., to raise, to lift, lift up, remove, carry away; heave; (*mil.*) to carry by storm; to carry, sweep, along (as, a regiment to the assault); (*pont.*) to dismantle (a bridge);
—— *et changer une roue*, (*art.*) to take off and change a wheel;
—— *un cheval*, (*man.*) to urge a horse vigorously forward, to urge a horse to go;
—— *par une mine*, (*mil. min.*) to blow up;
—— *un quartier*, (*mil.*) to beat up one's quarters;
—— *un régiment*, (*mil.*) to sweep along a regiment (as, to an assault).
enlever, m., (*man.*) dash, bolt;
—— *de l'avant-main*, rise, lift, of the forehand;
l'—— *au galop*, putting into the gallop.
enlier, v. a., (*mas.*) to bind.
enligner, v. a., to align.
enliouber, v. a., (*carp.*) to scarf.
ennemi, a., hostile.
ennemi, m., enemy; adversary, foe;
—— *figuré* (in maneuvers at drill) enemy marked out by a few men only, indicated enemy, outlined enemy;
—— *marqué*, v. —— *figuré*;
—— *représenté*, (in maneuvers, etc.) the enemy represented by a full force, full enemy;
—— *supposé*, (in maneuvers, etc.) assumed enemy, hypothetical enemy.
enquête, f., inquiry, investigation;
commission d'——, (*mil.*) court of inquiry (officers only);
conseil d'——, (*mil.*) court of inquiry;
faire une ——, to hold an inquiry, make an investigation, to investigate.
enraciner, v. a., to bury down, to sink into the ground (as, of foundations).
enraiement, m., v. *enrayage*.
enraser, v. a., to make level or flush.
enrayage, m., chocking, locking, braking (of a wheel); that part of the nave where the spokes enter the mortises; speech; (*mach.*) stopping, jamming, failure to work (of a mechanism); (*art., sm. a.*) general expression for the failure of a mechanism to work properly, jamming (as, of a magazine feed, of a breech-closing mechanism, etc.); (*sm. a.*) act of putting the safety device into play;
—— *à bloc*, braking a wheel so that it slides without turning;
chaîne d'——, drag chain (artillery, etc., carriages);
patin d'——, drag shoe, drag;
—— *de route*, locking of wheels on the road, the march;
sabot d'——, v. *patin d'*——.

enrayement, m., v. *enrayage*.
enrayer, v. a. r., to lock, skid, chock, or drag (a wheel); to put on the drag, the brake, to brake; to spoke (a wheel); (*mach.*) to stop; to jam, to make fast or motionless; to fail to work; (*art., sm. a.*) to jam, stick, fail to work (of a breech mechanism, feed gear, etc.); (*art.*) to secure (as, a gun in a turret, for firing); (*sm. a.*) to put the safety device into play;
—— *à bloc*, to brake a wheel so that it slides without turning;
chaîne à ——, locking chain, drag chain;
corde à ——, drag-rope;
—— *un pignon*, to lock a pinion.
enrayeur, m., manager or conductor of a pile driver.
enrayoir, m., prop, support, stay; locking chain.
enrayure, f., wheel-drag; (*cons.*) radiating framing of a floor (as, under the end of a hip roof, etc.).
enrégimenté, p. p., (*mil.*) regimented; formed into a regiment.
enrégimenter, v. a., (*mil.*) to regiment, to embody, form into a regiment.
enregistrement, m., registration; registry; entry, enrollment; recording (of experiments, etc.);
certificat d'——, certificate of registry.
enregistrer, v. a., to register, record, enter, set down.
enregistreur, m., (*inst.*) any registering or recording device or mechanism (e. g., the quill of a chronoscope); (*ball.*) the registrar of the Le Boulengé chronograph;
projectile ——, (*ball.*) recording, registering projectile, one provided with a registering device;
—— *renclancheur*, (*ball.*) a registering device working on a make-and-break circuit; a register indicating the successive steps of the phenomenon.
enrênement, m., reining in or up (of horses in harness); fastening or bridle-reins to a dumb-jockey, etc.
enrêner, v. a., to rein in or up; to tie by the reins (of horses in harness); to fasten the bridle-reins to a dumb-jockey, etc.
enrênoire, f., hitching-post.
enrimeur, m., driver of a ringing engine.
enrochage, m., (*powd.*) caking, induration.
enrochement, m., (*cons.*) riprap; (*powd.*) induration, caking (of powder in barrels, in shell, against walls, etc., of shell, after firing).
enrocher, v. a. r., (*cons.*) to riprap; (*powd.*) to harden, cake, become indurated, v. *enrochement*.
enrôlé, m., (*mil.*) soldier or sailor whose name is on the rolls of the army or navy, respectively.
enrôlement, m., (*mil.*) enrolling, enrollment, of a name on rolls of army or navy, (of an enlisted man only) enlistment.
enrôler, v. a. r., (*mil.*) to enroll, enlist.
enrôleur, m., (*mil.*) recruiting agent, crimp (obs.).
enrouiller, v. a. r., to rust; to grow, become, rusty.
enroulage, m., winding (as, of wire, etc.); (*met.*) rolling of bars into spirals (to form cylindrical forgings); a cylindrical forging so formed.
enroulement, m., winding (of a dynamo, of a wire gun, etc.); (*mach.*) the "going on" of a strap on its pulley;
—— *compound*, (*elec.*) compound winding;
—— *en dérivation*, (*elec.*) shunt winding;
—— *en série*, (*elec.*) series winding.
enrouler, v. a. r., to roll, roll up, roll into a spiral; to wind (a dynamo, wire, etc.); (*met.*) to roll a bar into a spiral (to make a cylindrical forging); (*mach.*) to run on (of a belt on its pulley).
enrouleur, m., (*met.*) the mandrel on which a bar is rolled, v. *enroulage*.

ensablement, m., sand bank; graveling, ballasting (of a road); filling of a river, harbor, by sand; (*nav.*) running aground; (*sm. a., etc.*) the sanding of the breech mechanism (in firing tests).

ensabler, v. a. r., to ballast a road; to fill up with sand (river, harbor); (*nav.*) to run aground; (*sm. a., etc.*) to cover with sand; to sand (breech mechanism, etc., in firing tests).

ensabotement, m., action of locking, skidding, etc., a wheel; (*art.*) operation of fixing a sabot on a projectile (obs.).

ensaboter, v. a., to lock, skid, etc., a wheel; (*art.*) to fix a sabot on a projectile (obs.).

ensachement, m., act of putting into a bag; (*art.*) of putting powder into a *sachet* (for metallic ammunition).

ensacher, v. a., to put into sacks or bags; (*art.*) to put powder into a *sachet* (for metallic ammunition).

enseigne, f., sign; (*nav.*) ensign, flag; (*in gen.*) colors, standard (poetically and figuratively);
 — *bâton d'*——, flagstaff (stern flag);
 — *s déployées,* with colors flying;
 — *gaule d'*——, v. *bâton d'*——;
 — *grande* ——, (*nav.*) stern flag, national color;
 — *de poupe,* v. *grande* ——.

enseigne, m., (*Fr. nav.*) ensign (U. S. N.); sub-lieutenant (R. N.);
 — *porte-épée,* (*mil.*) the *porte-épée Fähnrich* of the German army;
 — *de vaisseau,* (*nav.*) ensign (U. S. N.).

ensellé, a., (*hipp.*) saddle-backed; sway-backed;
 — *à dos* ——, saddle-backed.

ensellement, m., (*hipp.*) state or condition of being sway-backed; (*top.*) col, saddle, saddle-back.

ensemble, adv., together, at the same time; m., general appearance, effect, etc., as, of a performance, a maneuver; (*mil., etc.*) simultaneous and correct execution of movements.
 — *avoir de l'*——, (*hipp.*) to be well built, have good proportions;
 — *mettre bien* —— *un cheval,* (*man.*) to gather a horse well on his haunches, to keep a horse well together;
 — *nager* ——, to pull together (rowing).

ensemencer, v. a., to sow (grain, etc.).

entaille, f., notch, indent, cut, nick, jag, kerf, mortise, groove, slash, slot; (*art.*) maneuvering notch.
 — *de loquet de console,* (*art.*) latch pocket;
 — *repère,* guide notch or cut (to set one piece in a given position with reference to another, e. g., notch on fuse graduation).

entailler, v. a., to notch, to score, to scarf; cut away;
 — *en queue d'aronde,* to dovetail.

entaillure, f., v. *entaille*.

entamer, v. a. r., to begin (as operations), to enter upon; to cut, to cut into (as, of tools); to touch (of a file); (*mil.*) to break, to begin to break (a square, a line, etc.); (*man.*) to begin, or start in on, a gait;
 — *une affaire,* to begin, open, an affair (an attack, etc.);
 — *le chemin,* (*man.*) to set off;
 — *un cheval,* (*man.*) to begin the breaking of a horse;
 — *du pied droit (gauche),* (*man.*) to lead with the right (left) foot in galloping;
 — *une troupe,* (*mil.*) to throw a body of troops into disorder.

entalinguer, v. a., v. *étalinguer.*
entalingure, f., v. *étalingure.*
entérinement, m., (*law*) ratification.
entériner, v. a., (*law*) to ratify.
entérite, f., (*hipp.*) inflammation of bowels.
enterré, a., (*art., fort.*) sunken (of a battery).
enterrement, m., burial;
 — *militaire,* (*mil.*) military funeral.
enterrer, v. a., to bury.
en-tête, m., letter head; heading of a column (of logarithms, of sines, etc.);
 — *de facture,* billhead.

entier, a., entire, whole, complete; (*hipp.*) entire, uncastrated; (*man.*) obstinate, self-willed;
 — *cheval* ——, (*hipp.*) stallion;
 — *aux deux mains,* (*man.*) unwilling to turn, etc., either to the right or the left;
 — *à une main,* (*man.*) unwilling to turn, etc., to the right or left (as the case may be).

entier, m., (*hipp.*) stallion, entire.

entoilage, m., mounting (pasting on linen, as a map); act of covering with linen, canvas.

entoiler, v. a., to mount (as, a map on linen); to cover with linen, canvas.

entoiser, v. a., to stack or pile material (for convenience of measuring).

entonnoir, m., funnel; (*mil. min.*) mine-crater; (*art.*) shell-crater; crater-like shot-hole in an armor plate (rare); (*top.*) funnel-like pass or outlet;
 — *apparent,* (*mil. min.*) visible mine-crater;
 — *de chargement,* (*r. f. art.*) feed-hopper;
 — *déblayé,* (*art.*) a shell-crater into which none of the earth falls back;
 — *à douille,* (*sm. a.*) powder-hopper (for reloading cartridges);
 — *en* ——, funnel-shaped;
 — *réel,* (*mil. min.*) real crater, i. e., the whole volume dislodged or loosened by the blast.

entorse, f., sprain, strain;
 — *donner une* ——, to sprain;

entourage, m., case, casing; frame; railing.

entouré, p. p., surrounded, hemmed in;
 — *de terre,* landlocked.

entourer, v. a., to surround, hem in.

entours, m., pl., environs; adjacent parts; approaches.

entraîné, p. p., trained, skilled.

entraînement, m., dragging, dragging along; training, state of training (horses, troops, athletes); (*steam*) priming; (*mach.*) any communicating gear; advance, motion, of the work;
 — *de l'eau non vaporisée,* (*steam*) priming;
 — *pièce d'*——, (*mach.*) connecting-bar, -piece, motion-piece.

entraîner, v. a., to drag, drag along, bring on; to make the pace; to train (a soldier, an athlete, a horse); (*mach.*) to advance (the work);
 — *l'eau,* (*steam*) to prime.

entraîneur, m., horse-trainer; pace-maker (in races); (*mach.*) motion-piece.

entrait, m., (*carp.*) tie-beam, straining beam, collar-beam.

entraver, v. a., to shackle, fetter; to impede, entangle, to hobble (a horse).

entraverser, v. a. r., (*nav.*) to set, to lie, across or athwart.

entraves, f. pl., (*harn.*) hobble; (*fig.*) fetters, shackles, clog, hindrance. (This word is also singular.)

entravon, m., (*harn.*) ankle strap or collar (to keep the hobble from chafing).

entraxe, entr'axe, entre-axe, m., distance between axes, interaxial distance (as between adjacent guns, traverses, embrasures, axles of a railway car, etc.).

entrebâiller, v. a. r., to half open, to be half opened.

entrebât, m., the middle part of a pack-saddle.

entrechoquer, v. a., to engage, interfere, clash, strike against each other.

entrecouper, v. a., to intersect, traverse, crosscut.

entrecroiser, v. a., to intersect, cross each other.

entre-deux, m., intermediate space, belt, zone; interval, (of a shoe) welt, half-sole between outer and inner soles; (*steam*) bridge of a slide-valve;
 — *des lames,* trough of the sea;
 — *des orifices,* (*steam*) bridge of a slide-valve;
 — *passer* ——, to pass between.

entrée, f., entrance, mouth; opening, inlet of a channel, pass, etc.); admission; (*mil.*), (*adm.*) turning in of stores, etc., to a magazine;
— *en campagne*, (*mil.*) taking the field;
— *en galerie*, (*siege, mil. min.*) operations preliminary to gallery work proper;
— *d'ordre*, (*mil. adm.*) paper entry or transfer of stores (in a change of classification, etc.);
prendre une —, (*nav.*) to make an entrance (into a port);
— *réelle*, (*mil. adm.*) actual turning in of stores themselves;
— *en solde*, epoch at which pay begins to be due;
tuyau d'—, (*steam, etc.*) admission pipe.

entre-fer, m., (*elec.*) air-gap.

entre-jambes, m., tread (of a bicycle).

entrelacement, m., weaving, twisting, twining; wattling.

entrelacer, v. a., to twine, twist, interweave; to wattle.

entrelacs, m., (*cord.*) twine; knot;
— *à jour*, open twine.

entrelardement, m., defect in timber, caused by freezing of sap.

entre-ligne, m., space between two written lines; interlineation.

entremêler, v. a., to intermix, intermingle, interweave.

entremisage, m., (*nav.*) system of carlings.

entremise, f., mediation, intervention; medium, agency; small wedge; chock; (*nav.*) carling.

entrepas, m., (*man.*) broken, ambling, gait.

entre-pont, m., (*nav.*) lower deck; 'tween decks.

entreposage, m., bonding (goods, merchandise).

entreposer, v. a., to bond, put in bond; to warehouse, to store.

entreposeur, m., bonded warehouse keeper.

entrepôt, m., intermediate depôt; store house; bonded warehouse or store;
à l'—, *en* —, bonded.

entreprendre, v. a., to undertake; (*adm.*) to contract for.

entrepreneur, m., contractor; builder.

entreprise, f., enterprise; contract; undertaking, attempt; encroachment;
à l'—, by contract;
— *de transport*, (*Fr. a., adm.*) contract for railway transportation of material.

entrer, v. a. n., to enter, go in, let in; to march in; to put in, cause to enter, insert;
— *en campagne*, (*mil.*) to take the field;
— *à force*, to force on;
— *au service*, (*mil.*) to enter the army.

entre-serrer, v. r., (*mach.*) to fit too close, to have insufficient play.

entre-sol, m., intermediate story.

entretailler, v. r., (*hipp.*) to cut, to interfere.

entretaillure, f., (*hipp.*) cut, speedy cut, crepance, crepane.

entretenir, v. a. r., to keep, maintain; keep up, keep in repair, in good order; to maintain, support.

entretien, m., support, maintenance, care, preservation; maintenance in good, in working, order;
— *des troupes*, (*mil.*) subsistence of troops.

entretoise, f., cross-bar, -piece, -beam; stay, rib; (*art.*) transom (of a gun carriage); transom, crossbar (of a gin); (*mach.*) stay; (*cons.*) batten; (*pont.*) girder, floor girder; (*harn.*) cross-bar of a bit.
(The following terms, except where otherwise indicated, relate to artillery carriages:)
— *arrière*, rear transom;
— *avant*, front transom;
— *de bout*, hind transom (of an ammunition wagon);
— *de châssis*, chassis transom;
— *à collet*, axle transom (of a winch);
— *de couche*, middle transom;
— *de crosse*, trail transom, rear transom;
— *de devant*, fore bolster, front transom; crossbar of an ammunition wagon;
— *de flèche*, trail transom;
— *de limonière*, shaft-bar of a limber;
— *-lunette*, — *de lunette*, lunette transom;
— *de mire*, middle transom;
— *-pivot*, pivot transom;
— *de plaque de garde*, axle-guard stay;
— *de pointage*, v. — *de mire*;
— *de repos*, v. — *de mire*;
— *de support*, v. — *de mire*;
— *de tête d'affût*, front transom;
— *de volée*, v. — *-avant*.

entretoisement, m., bracing, counterbracing; staying (of a boiler); (*cons.*) battening; (*art.*) the bracing or stiffening of a gun carriage by a transom.

entretoiser, v. a., to batten; to brace, cross-brace.

entrevoie, f., (*r. r.*) space between tracks (of a double-track line).

entr'ouvert, a., half-open, ajar;
cheval —, (*hipp.*) shoulder-pitched horse, horse sprained in shoulder joint.

entr'ouverture, f., (*hipp.*) shoulder pitch; sprain of the shoulder joint.

entr'ouvrir, v. a. r., to half-open, gape; be or set ajar.

enture, f., peg, pin; pin of a peg-ladder; (*sm. a.*) splice, piece let into a stock (to strengthen or repair it); (*carp.*) scarf joint (is also applied to iron beams or pieces);
— *à double mortaise et faux tenon*, doweled joint (pieces vertical);
— *à endent*, (*carp.*) square notching, square notched scarf joint;
— *en fausse coupe*, ordinary oblique scarf joint (pieces vertical);
— *à tenaille*, (*carp.*) rabbet joint (pieces vertical and assembled laterally);
— *à tenon et mortaise*, (*carp.*) ordinary mortise and tenon in vertical pieces.

envagonner, v. a., (*r. r.*) to put on the train, in the cars.

envahir, v. a., to overrun, break in upon; to reach all the parts of (as, fire, of a burning building; water, of a sinking ship, etc.); to encroach upon (as the sea upon the land); (*mil.*) to invade, overrun.

envahissement, m., overrunning; encroachment; extension to all the parts of; (*mil.*) invasion, overrunning.

envahisseur, m., invader.

envaser, v. a. r., to sink into the mud; to stick in the mud.

enveloppe, f., envelope, cover, covering, casing; letter envelope; (*mach., steam*) jacket, case, cleading; outer shell or casing of a double-shelled cylinder; (*art.*) coating of a projectile, the shell or casing of a shrapnel; (*r. f. art.*) water-jacket; (*fort.*) continued counterguard, sillon; (*mil. min.*) main gallery of a countermine; (*ball.*) the zone between the *noyau*, q. v., and the circle of twice the radius of the latter;
— *de cylindre*, (*steam, etc.*) cylinder jacket;
— *isolante*, (*elec.*) insulating cover (as, of a cable, etc.);
— *de pied*, (*hipp.*) flesh, envelope, of the foot; keratogenous membrane;
— *à pinces*, (*art.*) clip-sleeve, of a fuse;
puissance d'—, (*man.*) the grip or clasp of the thighs (of the seat);
— *de vapeur*, (*steam*) steam-jacket.

enveloppée, f., (*fort.*) sillon.

enveloppement, m., action of enveloping, surrounding, etc.; (*mil.*) an enveloping movement;
— *stratégique*, (*mil.*) strategical envelopment, compelling the enemy to fight.

envelopper, v. a., to envelop, surround, inclose, cover, wrap up; (*steam*) to lag, to clead; (*mil.*) to execute an enveloping movement, to invest, hem in.

envergure, f., stretch of a pair of compasses; (*siege*) outflanking (of the trenches).

envers, m., wrong side;
à l'—, wrong side up, inside out, upside down.

envie 156 **épée**

envie, f., (*mach.*) the cutting edge of a *mèche à entailler* (q. v.).
environner, v. a., to surround, beset, inclose encircle, encompass.
environs, m. pl., environs, surroundings, vicinity.
envoi, m., action of sending; thing sent;
—— *aux compagnies de discipline*, (Fr. a.) punishment for incorrigible men (v. *compagnie de discipline*).
envoiler, v. r., (of steel) to bend in tempering, in hardening.
épais, a., thick, large.
épaisseur, f., thickness.
épaissir, v. a. r., to thicken; to grow thick; (*mil.*) to reenforce a line, especially the firing line; to increase the depth of formation of troops.
épanouir, v. a. r., (*art., sm. a.*) to spread, splay (as the base of a projectile, a gas-check, under the pressure of the powder gases);
—— *un obturateur*, (*art.*) to seat an obturator.
épanouissement, m., (*art., sm. a.*) spread, splay (of a gas check, of the base of a projectile, under the pressure of the powder gases); (*elec.*) pole-piece.
éparer, v. r., (*man.*) to kick, fling out the hind legs (rare).
éparpillement, m., scattering; (*art., sm. a.*) scattering, dispersion (of shots, bullets, etc.); (*mil.*) scattering (of troops).
éparpiller, v. a. r., to scatter, spread; (*art., sm. a.*) to scatter, spread (of bullets, fragments, etc.); (*mil.*) to scatter, spread about in small bodies (of troops).
épars, m., spar, bar; crossbar, small transom; crossbar of a wagon; (*art.*) crossbar (of a gin);
—— *de chèvre*, (*art.*) crossbar of a gin;
—— *de fond*, bottom bar (of a wagon);
—— *de limonière*, (*art.*) shaft-bar of a limber;
—— *montant*, standard of a wagon;
—— *du pavillon*, flagstaff.
éparvin, m., (*hipp.*) spavin; ringbone;
—— *de bœuf*, bone spavin;
—— *calleux*, v. —— *de bœuf*;
—— *osseux*, bone spavin;
—— *sanglant*, blood spavin;
—— *sec*. blind spavin, springhalt.
éparviner, v. n., (*hipp.*) v. *harper*.
épaule, f., shoulder.
(Except where otherwise indicated, the following terms relate to the horse:)
avoir de l'——, (*fenc.*) to fence from the shoulder (defect);
avoir les ——*s chevillées, froides*, to be lame, stiff in the shoulders, to have pegged, cold, shoulders;
—— *de bastion*, (*fort.*) shoulder, shoulder angle, of a bastion;
boiter de l'——, to be lame in the shoulder;
boiterie d'——, shoulder lameness;
—— *chargée*, shoulder whose bony base is too deeply embedded.
—— *chevillée*, a permanently stiff shoulder;
défaut de l'——, hollow of the shoulder (of the human body);
—— *descendue*, shoulder lower than the other;
dislocation de l'——, shoulder slip;
—— *droite*, straight, upright, shoulder;
écart d'——, v. s. v. *écart*;
effort d'——, shoulder strain;
l'—— *en dedans*, (with) the shoulder on the inner track (*travail sur deux pistes*);
être pris des ——*s*, to be stiff, lame, in the shoulder;
être sur les ——*s*, (of a horse) to carry his weight too much forward;
—— *froide*, shoulder stiff on leaving stable;
mettre (un cheval) sur les ——*s*, to throw too much of the weight on the fore quarter;
—— *de mortaise*, (*art.*) windlass-bracket;
—— *de mouton*, (*carp.*) chip-ax;
nœud d'——, (*unif.*) shoulder knot;
patte d'——, (*unif.*) shoulder-strap;
—— *plaquée*, flat, sunken shoulder, restrained or stiff in its movements;

épaule, *sur les* ——*s*, on the forehand;
trotter des ——*s*, to trot heavily, stiffly.
épaulée, f., thrust, or push, with the shoulder;
bâtir par ——*s*, (*mas.*) to build, run up (a wall, etc.), irregularly, unevenly.
épaulement, m., shoulder (as, of a tenon, a spoke); (*fort.*) epaulment, parapet, breastwork, gun bank; (*mach.*) shoulder; (*art.*) shoulder, locking-shoulder; (*sm. a.*) recoil-shoulder, firing-pin guide-ring; (*cons.*) revetment wall;
—— *d'appui*, (*sm. a.*) locking-shoulder, locking-lug shoulder;
—— *de batterie*, (*art., fort.*) gun cover, gun parapet;
—— *de campagne*, (*art., fort.*) field-gun epaulment;
hauteur d'——, (*sm. a.*) height of rifle above ground (when fired);
—— *rapide*, (*art., fort.*) hasty epaulment (one to each gun);
—— *en retour*, (*fort.*) an epaulment refused, thrown back.
épauler, v. a. r., to sprain (the shoulder); to assist, support, back; (*fort.*) to protect by an epaulment; (*mil.*) to cover, protect (as, flank by a marsh or other obstacle); (*sm. a.*) to bring the piece to the shoulder (for firing); (*carp.*) to cut down, pare down, a tenon.
épaucaulier, m., (*mil.*) officer inordinately vain of being such.
épaulette, f., (*unif.*) epaulet, shoulder-knot; (*fig.*) commissioned rank;
arriver à l'——, to win, get, a commission;
bouillons d'——, v. *torsades d'*——;
corps d'——, field of an epaulet;
double ——, (Fr. a.) the grade of captain;
—— *à écailles*, shoulder scale;
écusson de l'——, field of an epaulet;
—— *à frange*, fringe epaulet;
—— *à graine d'épinards*, epaulet with large bullion fringe;
—— *à gros grains*, v. —— *à graine d'épinards*;
obtenir l'——, (*mil.*) to become an officer; to get his shoulder-straps, U. S. A.;
—— *à petites torsades*, v. —— *à frange*;
torsades d'——, bullion of an epaulet.
épaulière, f., sort of man-harness; (*unif.*) shoulder-strap of a cuirass.
épée, f., sword; swordsman; (*fig., mil.*) the profession of arms;
abandon d'——, (*fenc.*) lack of control of the blade, (hence) invitation, apparent opening;
absence d'——, (*fenc.*) state of being uncovered, (hence) invite, invitation;
—— -*baïonnette*, (*sm. a.*) sword-bayonet;
se battre à l'——, to fight with swords;
une bonne ——, a good swordsman;
briser son ——, (*mil.*) to quit the service;
coup d'——, sword thrust, sword wound;
l'—— *dans les reins*, (*mil.*) vigorously (as, of a pursuit);
demander son —— *à quelqu'un*, (*mil.*) to arrest;
—— *à l'estoc*, thrusting sword;
le faible de l'——, (*fenc.*) part of the blade near the point;
le fort de l'——, (*fenc.*) part of the blade near the guard;
gens d'——, military men, soldiers;
lame d'——, sword blade;
la main de l'——, (*man.*) sword hand, right hand;
mettre l'—— *à la main*, to draw;
mettre son —— *au service de l'étranger*, (*mil.*) to take service in a foreign army;
le mi fort de l'——, (*fenc.*) the middle part of the blade;
passer au fil de l'——, to put to the sword;
plat d'——, flat of the sword;
poignée de l'——, sword hilt, gripe;
poser l'——, (*mil.*) to return to a state of peace, to stop warring;
remettre l'——, to put up, sheathe, one's sword;
rendre l'——, to give up one's sword; (*mil.*) to surrender;

épée, *rengainer l'——*, v. *remettre l'——;*
tirer l'——, to draw the sword;
—— *vierge*, unfleshed sword.

éperon, m., spur; ice-fender, stream-fender (of a bridge); starling; buttress; mole, breakwater, jetty; (*top.*) mountain spur; counterfort; (*nav.*) stem, ram; (*art.*) trail-spur, trail spade;
chausser les ——s, to put on one's spurs;
—— *à la chevalière*, (*unif.*) officer's spur, ordinary spur (held on by straps and buckles);
déchausser les ——s, to take off, remove, one's spurs;
donner de l'——, des ——s, (*man.*) to spur, put spurs to;
dur à l'——, (*man.*) not spur-wise;
—— *fixe*, screw-spur;
gagner ses ——s, to win one's spurs;
—— *mobile*, v. —— *à la chevalière;*
—— *d'un pont*, starling, fender, of a bridge;
sensible à l'——, (*man.*) spur-wise.

éperonné, a., spurred.

éperonner, v. a. n., (*man.*) to spur; (*fenc.*) to move the foot to the side (in a manner resembling a spurring motion).

éperonnier, m., spur maker.

épervin, m., v. *éparvin*.

épi, m., beard (of wheat, etc.); pile, mass; (*eng.*) stockade of fascine work, jetty, line of mattress, etc., work (along or across a river or channel); (*hipp.*) cowlick; line formed by change of set of the hair;
—— *de barrage*, (*eng.*) stockade thrown across a river;
—— *de bordage*, (*eng.*) fascine, etc., work, to protect the banks of a river or canal;
en ——, (*mas.*) herringbone-wise;
—— *de pierres enrochées*, (*eng.*) pile of riprap;
—— *du vent*, wind's eye.

épidémie, f., epidemic.

épidémique, a., epidemic.

épiderme, m., skin; (of trees) outer bark.

épier, v. a., to spy, to keep a close watch over.

épierrer, v. a., to clear of stones.

épinard, m., spinach;
graine d'——, (*unif.*) large bullion (of an epaulet),
„ v. s. v., *épaulette*.

épine, f., thorn, spine.

épineux, a., thorny, difficult.

épingler, v. a., (*art.*) to prick the vent of a gun.

épinglette, f., (*cord.*) small marlin spike; (*art.*) priming wire; (*sm. a.*) needle; (*Fr. a.*) ornamental pin, prize for target practice;
—— *à vrille*, (*art.*) v. *dégorgeoir à vrille*.

épisser, v. a., (*cord.*) to splice.

épissoir, m., (*cord.*) marlin spike (iron); splicing fid (wood).

épissure, f., (*cord.*) splice;
—— *de câble*, cable splice;
—— *carrée*, short splice;
—— *courte*, v. —— *carrée;*
—— *en diminuant le toron*, tapering splice;
—— *d'estrop*, v. —— *de ganse;*
—— *de ganse*, eye splice;
—— *en greffe*, horseshoe splice;
—— *longue*, long splice;
—— *à œil*, v. —— *de ganse;*
—— *d'œuillet*, v. —— *de ganse;*
—— *ronde*, sailmaker's splice;
—— *avec le toron entier*, full splice.

épite, f., wedge tenoning; small wedge or pin driven into a treenail to swell it.

épitoir, m., iron pin, for making an *épite*.

épizootie, f., (*hipp.*) epizootic, epizooty.

épointé, a., having lost the point, with the point broken off; (*hipp.*) hipshot; sprained in hip or haunch.

épointer, v. a., to break off the point of anything.

épointure, f., (*hipp.*) state or condition of being hipshot.

éponge, f., sponge; (*farr.*) quarter of a horseshoe; (*hipp.*) hygroma of elbow, shoe boil, tumor on point of elbow, in horses that keep their forelegs under them while lying down.

éponger v. a., to wash or clean with a sponge; to dab off.

épontillage, m., stanchioning, shoring up, propping.

épontille, f., stanchion, shore, prop;
—— *volante*, movable stanchion.

épontiller, v. a., to shore up, to prop, support by a stanchion.

épouser, v. a., to wed; (*fig.*) to take the form of.

époussetage, m., dusting; rubbing down (of a horse); (*powd.*) dusting.

épousseter, v. a., to dust, to rub down (a horse); (*powd.*) to dust.

époussette, f., dusting rag, duster.

épreuve, f., proof, trial; test, chemical test; service test; experiment; (*phot.*) proof;
à l'——, under trial; (in combination) proof;
à l'—— de, proof against; (*mil.*) protected against;
à l'—— de l'eau, waterproof;
à l'—— des éclats, (*fort.*, etc.) splinter proof;
—— *d'admissibilité*, v. s. v. *admissibilité;*
—— *d'admission*, v. s. v. *admission;*
—— *à l'air comprimé*, (*art.*) compressed-air test (cast-iron shells);
—— *d'alcalinité*, (*expl.*) alkali-test;
banc d'——(s), (*sm. a.*) proof-bench, proving-bench;
—— *du bloc*, (*sm. a.*) striking test (for the edge of a sword);
—— *de la boîte*, (*sm. a.*) bending test (sword blade);
butte d'——, (*ball.*) proof butt;
—— *de chaleur*, (*expl.*) stability test, heat test;
champ d'——, (*ball.*, etc.) proving ground;
—— *à chaud*, hot test;
—— *au choc*, —— *de choc*, falling-weight test, percussive test;
—— *de conductibilité*, (*elec.*) conductivity test;
—— *de conservation*, v. *quinquennale;*
—— *de contrôle*, v. *semestrielle;*
d'——, probationary;
—— *décennale*, (*powd.*, Fr. a.) decennial inspection and test of certain powders of old type;
—— *de déformation*, bending test;
—— *de déformation par ployage*, v. *de déformation;*
—— *de densité*, (*expl.*) density test;
—— *de durée*, endurance test;
—— *à l'eau*, waterproof test;
—— *de l'eau*, (*sm. a.*, etc.) water test;
—— *d'écrasement*, crushing test;
—— *d'élasticité*, (*sm. a.*) elastic test (sword blades, bayonets, etc.);
—— *d'élasticité par mandrinage*, (*art.*) mandrel test of elastic strength of gun-hoops;
—— *d'emboutissage*, heading-up test (for plates);
—— *encombrante*, pressure test (by weights resting on object);
—— *d'escarpolette*, axletree test;
—— *d'étanchéité*, (*sm. a.*) test to determine whether cartridge cases are water-tight;
faire l'—— de, to try, test, prove, anything;
faire ses ——s, to pass, stand, a test, to be found satisfactory;
—— *à la flexion*, bending test;
—— *de fonctionnement*, working test;
—— *de forge*, forge test;
—— *à froid*, cold test;
—— *d'humidité*, (*expl.*) humidity test;
—— *d'incinération*, (*expl.*) ash test;
—— *individuelle*, test of each piece, as opposed to a piece taken at random from a lot;
—— *d'isolement*, (*elec.*) insulation test;
—— *de la jante*, (*sm. a.*) striking test (flat of sword on curved block);
mettre à l'——, to put to the test, make a trial of;
—— *à mouton*, v. *d'escarpolette;*
—— *à outrance*, resistance test, proof to the limit of ultimate strength, proof to bursting (as, of guns);
—— *de pliage*, folding test;
—— *de ployage*, bending test;
—— *de poinçonnage*, (*art.*) punch test;
—— *positive*, (*phot.*) proof, photograph;

épreuve *des poudres*, (*art.*, *sm. a.*) proof of powder;
— *de la poussière*, (*sm. a.*, *etc.*) dust test;
— *pratique*, (*mil.*) in promotion examinations, practical test in handling troops, etc.;
prendre à l'——, to take on trial;
— *de pression*, pressure test (as, of a boiler);
— *quinquennale*, (*powd.*, *Fr. a.*) inspection and proof of powder every five years;
— *de rabattement*, folding-back test;
— *de recette*, acceptance test;
— *de réception*, v. — *de recette*;
— *de résistance*, resistance test, proof, trial;
— *roulante*, rolling test;
— *de roulement*, (*art.*) carriage test (working test of any new type carriage);
— *semestrielle*, (*powd.*, *Fr. a.*) half-yearly inspection and test of powder;
— *de solubilité*, (*expl.*) solubility test;
— *de soudure*, test of a welded joint;
subir une ——, to stand, pass, a test;
— *à surcharges*, test by overloading, test of ultimate strength;
temps d'——, probation;
— *théorique*, (*mil.*) in promotion examinations, examination in book work;
— *de tir*, (*art.*) firing-proof, proof-firing of artillery, small-arm, proof-firing in general;
— *de tir à outrance*, (*art.*, *ball.*) proof-firing to test ultimate strength;
— *à la vapeur*, (*art.*) steam test (cast-iron shells;)
— *de vitesse*, (*nav.*) speed trial.
éprouvé, p. p., (*mil.*) sorely tried, hard pressed.
éprouver, v. a., to test, try, prove; to experience; to assay, make trial of; to meet with, go through with.
éprouvette, f., (*met.*) test bar, test piece; (*chem.*) test tube; (*ball.*) éprouvette;
— *à crémaillère*, vertical éprouvette;
mortier-——, (*ball.*) powder éprouvette;
— *à poudre*, v. *mortier*-——;
— *à ressort*, (*ball.*) spring éprouvette for testing sporting powders;
— *à roues dentées*, (*ball.*) pistol éprouvette.
épuisement, m., exhaustion; draining, drainage;
d'——, exhausting;
pompe d'——, (*nav.*) bilge pump;
tuyau d'——, exhaust pipe.
épuiser, v. a. r., to use up, wear out (as, troops by fatigue); to exhaust, drain, empty; to run dry (of springs, streams).
épuisette, f., sort of net (for catching carrier pigeons, etc.).
épurage, m., (*cord.*) cleaning or second combing of hemp.
épurateur, m., (*steam*) purifier; (*expl.*) poacher.
épuration, f., refining, refinement, purifying.
épure, f., drawing, mechanical drawing, geometrical drawing; working plan or drawing; diagram (e. g., Zeuner diagram);
— *des efforts*, stress diagram;
— *au lavis*, wash drawing.
épurer, v. a., to purify, cleanse; (*cord.*) to clean hemp, give the second combing to hemp.
équarré, m., square inscribed in the cross section of a log to be squared.
équarrir, v. a., to square (as, timber, stone); to recut, straighten, grooves, etc.; to ease (doors, windows); to kill and cut up animals (beef, horses, etc.).
équarrissage, m., squaring, squareness; scantling (of timber); skinning and cutting up (of animals);
d'——, square, in the square;
— *au cinquième*, squaring of timber, so that the side of the inscribed square shall be the chord of one-fifth of the circumference;
— *de forêt*, rough hewing (of timber, in situ);
formule d'——, v. s. v. *formule*;
— *à vif*, squaring, hewing, that reaches the heartwood.

équarrissement, m., act of squaring, state of being squared; cutting at right angles;
tailler en ——, to square.
équarrisseur, m., nacker, knacker.
équarrissoir, m., square broach.
équateur, m., equator.
équation, f., equation;
— *personnelle*, personal equation, personal error.
équatorial, a., equatorial.
équerrage, m., bevel, beveling;
— *gras*, obtuse bevel;
— *maigre*, acute bevel.
équerre, f., square, rule, scale, carpenter's rule; elbow (of a water-pipe, etc.); strap, junction-strap; knee (of a boat); (*cons.*) corner brace, corner angle-iron, rectangular angle-iron or brace; (*mach.*) any piece branched at or near 90°; (*pont.*) timber-plate.
à ——, *d'*——, *en* ——, square, at right angles;
à l'——, by the square;
— *d'arpenteur*, surveyor's square;
courbe à ——, square knee;
— *de courbe*, (*pont.*) knee-strap of a ponton boat;
— *à diamètre*, universal or center square;
double ——, T-square; (*cons.*) broad strap;
— *double*, T-square;
double — *d'appui*, reinforcing strap;
— *à double réflexion*, (*inst.*) reflecting square;
— *à épaulement*, a square of which one branch is three times as thick as the other;
fausse ——, (*inst.*) bevel square; (*cons.*, *carp.*) corner, bevel, not a right angle;
à fausse ——, out of square;
— *de fer*, iron knee;
— *à miroirs*, (*inst.*) reflecting square;
— *à mitre*, (*carp.*) miter square;
— -*onglet*, —— *à onglet*, v. —— *à mitre*;
— *ordinaire*, T-square;
— *à patte et à coulisse*, (*cons.*) iron strap with groove and holdfast;
— *à piton et à anneau*, iron strap with eyebolt and ring;
— *de précision*, try-square;
— *à prismes*, (*inst.*) optical square;
— *sauterelle*, bevel square;
— *simple*, try-square;
en T, v. —— *ordinaire*.
équerrer, v. a., to square; to bevel.
équestre, a., equestrian.
équiangle, a., equiangular.
équidistance, f., equidistance; (*top.*) equidistance of contours;
— *graphique*, (*top.*) equidistance, difference of level, to the scale of the map;
— *métrique*, (*top.*) difference of level, equidistance, in meters, actual equidistance;
— *naturelle*, v. —— *métrique*;
— *réduite*, v. —— *graphique*.
équidistant, a., equidistant; (*top.*) equidistant (of contours, etc.).
équignon, m., iron band or plate, on the under side of a wooden axle arm.
équilatère, a., **équilatéral**, a., equilateral.
équilibre, m., equilibrium;
formule d'——, Hooke's law;
— *des remblais et des déblais*, (*eng.*) balancing fills and cuts.
équilibrer, v. a., to equilibrate, equalize;
— *le tiroir*, (*steam*) to regulate a slide valve.
équillette, f., **équinette**, f., spindle of a vane.
équipage, m., equipage, equipment; outfit, plant; dress; horse furniture; carriage and horses; (*mach.*) gear, gearing, set of gear wheels; (*nav.*) crew; (*in pl.*) wagons, vehicles, (*hence*, *mil.*) train, wagon train; carriages, wagons, and material of all sorts transported by an army; baggage of an army; (more esp., *Fr. a.*) regimental train; the complete material of a given unit (as, a battery); the wheeled elements (of a siege or artillery park); a large gun, on its carriage, and equipped with all the material necessary to its service; the collection of tools, implements, etc., necessary to execute a given mechanical maneuver;

équipage *d'armée*, (*pont.*) the bridge train of an army;
— *s de l'armée*, (*mil.*) all the transport of an army;
— *s d'artillerie*, (*Fr. art.*) a generic expression for artillery trains and parks, and ammunition columns (two subdivisions, — *s de campagne*, — *s de siège*);
— *d'artillerie d'une armée*, battery material and supplies of ammunition, and of material in general, i. e., ammunition sections, corps parks, and *grand parc*;
— *d'atelier*, machinery, etc., plant of a shop (arsenal, navy-yard, etc.);
— *d'avant-garde*, (*pont.*) advanced guard bridge equipage;
— *s de campagne*, (*mil.*) the baggage of an army in the field; (*Fr. art.*) a generic expression for army and corps (artillery) trains, mountain equipage, and fortress (mobile defense) equipage;
— *de corps d'armée*, (*art.*) the divisional and corps artilleries, with their respective ammunition sections, the infantry ammunition section, and the artillery park;
— *de corps de troupe*, (*Fr. a.*) regimental train (*train régimentaire* and *train de combat*);
demi— —, v. *de siège;*
— *de division*, v. — *d'avant-garde;*
— *de division de cavalerie indépendante*, two horse batteries;
— *de montagne*, (*Fr. art.*) mountain equipage, i. e., the various batteries and ammunition sections assigned to mountain duty;
— *normal*, v. — *de siège;*
— *de place*, (*Fr. art.*) the mobile batteries used in the defense of a fortress;
— *de pompe*, (*mach.*) pump gear;
— *de pont*, (*pont.*) bridge train, bridge equipage;
— *de pont d'armée*, (*pont.*) bridge train of an army, the sum of the two corps trains;
— *de pont de campagne*, (*pont.*) the full name of the bridge train of an army in the field;
— *de pont de corps d'armée*, (*pont.*) the corps bridge train;
— *de pont de place*, (*Fr. a.*) fortress bridge train (maintained at Besançon, Lyons, Toul, and Versailles);
— *s régimentaires*, (*Fr. a.*) regimental wagons, train;
— *de réquisition*, (*mil.*) horses, wagons, etc., taken from the country on requisition;
— *de réserve*, (*mil.*, *pont.*) reserve bridge train;
— *de siège*, (*siège*) siege train (generally, the artillery siege train is meant); consists in France of 5 *équipages normaux*, each comprising 176 guns and mortars of various calibers (exclusive of r. f. guns) and 7,300 rounds, and subdivided into *demi*— *léger* (80 guns) and *demi*— *lourd* (96 guns);
— *de siège de l'artillerie*, (*siège*, *art.*) artillery siege train, the siege train proper;
— *de siège du génie*, (*siège*, *gen.*) engineer siege train (tools, implements, mining outfit, etc.);
train des — *s*, (*mil.*) baggage train; train;
— *de transport*, (*mil.*) the wheeled elements of a siege park, etc.;
— *de vaisseau*, (*nav.*) ship's company, crew.

équipe, f., gang, set, squad; crew or gang of workmen), working party; train of boats; (*art.*) number or party of men necessary to execute a mechanical maneuver;
chef d'— —, foreman of a gang of workmen.

équipée, f., foolish and reckless undertaking.

équipement, m., equipment, outfit; fitting out; stores, equipage; (*mil.*) accoutrement, equipment, (*Fr. a.*) exclusive of clothes and of arms;
— *de cheval*, horse appointments, horse furniture;
grand — —, (*Fr. a.*) belts, gun- and haversack-slings, haversacks, etc. (no longer an official term);

équipement, *petit* — —, (*Fr. a.*) brushes, cleaning kit, small articles generally (no longer an official term);
— *de la 1ère* (*2ème*) *portion*, v. *effets de la 1ère* (*2ème*) *portion*.

équiper, v. a. r., to equip, fit out; to provide, supply, furnish, stock; to arm, man; (*art.*, *etc.*) to rig (a gin, capstan, etc.); (*art.*) to take equipments; to furnish a piece with all the equipments necessary to its service;
— *à n brins*, (*art.*, *etc.*) to rig or equip with an n-fold tackle;
— *une chèvre*, (*art.*) to rig a gin;
— *une chèvre en cabestan*, (*art.*) to rig a gin as a capstan;
— *une chèvre à haubans*, (*art.*) to rig a gin as shears;
— *une fusée*, (*artif.*) to fix a stick to a rocket.

équipeur, m., (*sm. a.*) finisher (in assembling parts);
— *monteur*, stripper and finisher, armorer who assembles, etc., small arms (performs the duties of *équipeur* and of *monteur*, q. v.).

équitation, f., (*man.*) equitation, horsemanship;
— *diagonale*, a system of horsemanship of which the base is the development of the action of the horse's diagonals;
école d'— —, riding school;
maître d'— —, riding master;
— *militaire*, military riding;
— *rassemblée*, riding, equitation, so called from the necessity of the horse getting his hind legs well under him in the execution of the various airs.

érable, m., maple tree, maple wood.

éraflement, m., (*art.*) scratch, score, in the bore of a gun (produced by bursting of shell in bore).

érafler, v. a., to scratch, graze, score.

éraflure, f., slight scratch.

éraillé, a., (*cord.*) worn out, frayed.

érailler, v. a. r., to rub, gall, fret, chafe, fray.

éraillure, f., fraying, galling, chafing.

érection, f., erection, raising, establishment.

éreinté, p. p., broken-backed; tired out; worn out.

éreinter, v. a. r., to break the back; tire out, knock up; to be tired out, knocked up.

erg, m., erg.

ergot, m., hook, spur; toggle; ergot, sort of fungus in oats and other cereals; (*hipp.*) ergot.

ériger, v. a., to erect, establish, institute; (*fort.*) to throw up (works, parapets, etc.).

erminette, f., carpenter's adz.

érosif, a., erosive (as, of powder gases in bore of gun).

érosion, f., erosion; (*art.*) erosion of the bore.

erre, f., (*nav.*) way, headway, steerageway.

erreur, f., error;
— *de la division*, (*inst.*) error of graduation;
— *de fermeture*, (*surv.*) error of closure, closing error;
— *de graduation*, v. — *de la division;*
— *graphique*, the error on the map, corresponding to the actual error in the field, scale error, the field error reduced to the scale of the map;
— *instrumentale*, (*inst.*) instrumental, index error;
— *de lecture*, (*inst.*) error of reading;
— *de niveau apparent*, (*surv.*) error of apparent level;
— *d'observation*, (*inst.*) error in observing, observation error;
— *de pointage*, (*t. p.*) error of aim;
— *de pointé*, (*inst.*) error of sight;
— *de réfraction*, refraction error;
— *de réglage*, (*t. p.*) distance between point of impact and point aimed at (single shot); distance between center of impact and point aimed at (group of shots);
— *de sphéricité*, (*surv.*) error of sphericity;
— *de visée*, (*t. p.*) error of aim (rifle in a rest).

ers, m., fitches, vetches; tares.

erse, f., iron cringle, thimble; (*cord.*) strap, grommet;

erse *en bitord,* selvage, selvage strap;
—— *de poulie,* block strap.
erseau, m., *(cord.)* small grommet, small block strap;
valet ——, grommet wad.
escabeau, m., stool, settle.
escache, f., *(harn.)* oval bit, scatch.
escadre, f., *(nav.)* squadron, fleet;
—— *de blocus,* blockading squadron;
—— *d évolution,* squadron of evolution;
—— *d'observation,* squadron of observation.
escadrille, f., *(nav.)* small squadron.
escadron, m., *(cav.)* squadron, troop;
—— *actif,* (Fr. *cav.*) of a regiment, any squadron not the *dépôt* squadron;
chef d ——, *(art.)* major of artillery;
chef d ——*s,* (*cav.*) major of cavalry;
—— *de contact,* contact squadron;
—— *de découverte,* contact squadron, exploration or scouting squadron;
demi- ——, the French equivalent of the U. S. cavalry troop (in France the *escadron* has 4 platoons, in the U. S. but 2);
—— *de dépôt,* depot squadron;
—— *de dépôt à pied,* in the German cavalry, the depot (dismounted) squadron;
école d' ——, squadron drill, school of the troop;
—— *de garnison,* v. —— *de dépôt à pied;*
—— *de guerre,* (Fr. *cav.*) v. —— *actif;*
—— *mobile,* (Fr. *cav.*) v. —— *actif;*
—— *par quatre,* troop in column of fours;
—— *du train des équipages,* (Fr. *a.*) unit of the train, corresponding to the battalion, and containing 4 companies.
escadronner, v. a., *(mil.)* to maneuver with cavalry; to maneuver as, or in, squadron.
escalade, f., *(mil.)* escalade, scaling; assault with scaling ladders; *(law)* the passage or crossing of any fence, wall, or inclosure, climbing in over a door, so as to get into a house, garden, etc.;
aller à l'——, to advance to the escalade;
donner l'——, to attempt an escalade;
monter à l'——, v. *aller à l'*——*;*
tenter l'——, v. *donner l'*——*;*
vol avec ——, *(law)* theft preceded or accompanied by escalade, v. *supra.*
escalader, v. a., to scale; *(mil.)* to scale, escalade.
escaladeur, m., *(mil.)* escalader.
escale, f., maritime city of the Mediterranean, and more particularly of the Barbary States; *faire* ——, *(nav.)* to touch (at a port).
escalier, m., stairs, staircase, flight of stairs;
—— *de commande,* *(nav.)* accommodation ladder;
—— *dégagé,* —— *de dégagement,* back stairs;
—— *dérobé,* private stairs;
—— *droit,* straight flight of stairs, flyers;
—— *en échiffre,* —— *échiffré,* staircase resting upon a supporting wall;
—— *en escargot,* spiral stairs;
grand ——, principal staircase;
haut d'——, stairhead;
—— *d'honneur,* grand staircase;
—— *en limace,* —— *en limaçon,* cockle stairs, winding stairs, well-stairs;
—— *à noyau plein,* winding stairs;
—— *dans œuvre,* inner stairs;
—— *hors d'œuvre,* turret stairs;
—— *de pointage,* (s. c. *art.*) gunner's step;
——*-rampe,* (*fort.*) ramp and stairs, ramp up the middle of which runs a flight of stairs;
—— *de service,* v. —— *dégagé;*
—— *suspendu,* geometrical stairway;
—— *tournant,* v. —— *en limace;*
—— *à vis,* v. —— *en limace;*
—— *en vis à jour,* geometrical winding stairs with open newel.
(*Escalier* is, strictly, an interior staircase.)
escamoter, v. a., to conjure;
—— *l'arme,* *(mil.)* to slight, slur, the movements of the manual of arms.
escampette, f., (used only in) *prendre la poudre d'*——, to run away as fast as possible.
escapade, f., escapade; *(man.)* fling, caper, of a horse.

escapouler, v. a., *(met.)* to dress down, reduce.
escarbilles, f. pl., cinders, ashes; fragments of unburnt coal in ashes.
escarbite, f., calker's grease or oil box.
escargot, m., *(art.)* gun worm; *(mil. slang)* man with his tent, in the field.
escarmouche, f., *(mil.)* skirmish;
—— *d'exploration,* skirmish on exploration service (cavalry skirmish);
—— *de route,* running fight.
escarmoucher, v. a., *(mil.)* to skirmish.
escarmoucheur, m., *(mil.)* skirmisher.
escarpe, f., *(fort.)* scarp, escarp;
—— *attachée,* full scarp, attached scarp;
—— *attachée avec voûtes en décharge,* full scarp with relieving arches;
—— *à bahut,* semidetached scarp (admitting of fire over the detached portion);
contre- ——, counterscarp;
—— *demi-détachée,* v. —— *à bahut;*
—— *à demi-revêtement,* half-revetted scarp, scarp revetted halfway up;
—— *détachée,* detached scarp;
—— *pleine,* full scarp;
—— *pleine et attachée,* full scarp;
—— *à terre coulante,* scarp left at the natural slope of the earth.
escarpé, a., steep, bluff.
escarpement, m., steepness (as, of banks, etc.); *(top.)* slope; escarpment of a plateau; steepest slope (of a mountain) *(fort.)* escarpment.
escarper, v. a. r., to cut steep down, to form steep slopes; to be steep; *(fort.)* to scarp.
escarpolette, f., swing;
épreuve d'——, *(art.)* axletree test or proof.
escaume, m., v. *échome.*
escavecçada, f., *(man.)* jerk, pull, with the cavesson.
eschare, f., *(hipp.)* scab.
esclavage, m., slavery.
esclave, m., f., slave.
escompte, m., discount.
escompter, v. a., to discount.
escope, f., scoop;
—— *à main,* f., baler.
escoperche, f., v. *écoperche.*
escopette, f., (sm. *a.*) carbine (obs.).
escopetterie, f., (sm. *a.*) small-arm fire (obs.).
escorte, f., *(mil., etc.)* escort; *(nav.)* convoy;
—— *d'honneur,* *(mil.)* escort of honor (of the President, ministers, generals).
escorter, v. a., *(mil., etc.)* to escort; *(nav.)* to convoy.
escouade, f., *(mil.)* squad;
—— *brisée,* squad composed of men of different regiments.
escoupe, f., curved spade, round shovel.
escourgée, f., cat-o'-nine tails; whip; flogging.
escourgeon, m., winter barley.
escousse, f., *(fam.)* start, running start (for a jump).
escrime, f., fencing; m., *(mil. slang)* clerk, "quill driver";
—— *à la baïonnette,* bayonet exercise;
—— *à la contrepointe,* broadsword exercise;
—— *à,* —— *de, l'épée,* fencing with the sword;
maître d ——, fencing master;
—— *à la pointe,* generic expression for fencing in which the point is used;
—— *au,* —— *du, sabre,* broadsword exercise;
salle d'——, fencing school, fencing hall or room; fencing academy (U. S. M. A.).
escrimer, v. a. r., to fence, to fight with swords; to use a stick as a weapon; *(fig.)* to apply one's self to;
s'—— *des dents, s'*—— *des mâchoires,* to be a valiant trencherman.
escrimeur, m., fencer.
espace, m., space, room, interval;
—— *d'air,* *(art.)* air-space, air-spacing;
—— *battu,* *(art., sm. a.)* zone or area swept by fire;

espace, *chargement à —— d'air*, (*art.*) air-spaced loading;
charger à —— d'air, (*art.*) to load with an air-space;
—— *dangereux*, (*ball.*) dangerous space;
—— *mort*, (*fort.*) dead space; (*steam, mach.*) waste space;
—— *nuisible*, v. —— *mort* (*steam, mach.*).

espacement, m., spacing, distance, interval; (*fort.*) spacing (of forts in a line);
d' ——, asunder, with intervals;
—— *des rivets*, (*tech.*) pitch of riveting.

espacer, v. a., to space, place at a proper distance; to leave an interval between.

espadage, m., (*cord.*) tewing.

espade, f., (*cord.*) tewing-beetle.

espader, v. a., (*cord.*) to tew.

espadon, m., v. *espade*.

espadonnage, m., v. *espadage*.

espadrille, f., canvas shoe with twine sole (of Spanish origin).

espagnolette, f., sash, window, fastening or bolt; (*art.*) limber-chest fastening.

espalet, m., (*sm. a.*) shoulder (of the hammer) (obs.).

espars, m., spar.

espatard, m., (*met.*) anvil and hammer.

espèce, f., sort, kind; (*in pl.*) money;
avoir de l' ——, (*hipp.*) elegant, graceful;
—— *s sonnantes*, coin.

espion, m., (*mil., etc.*) spy;
—— *accidentel*, temporary spy;
—— *double*, spy serving both sides;
—— *fixe*, resident spy;
—— *mobile*, traveling spy;
—— *permanent*, professional spy;
—— *par relais*, transmitting spy;
—— *simple*, spy serving one side only.

espionnage, m., (*mil.*) spying, espionage;
contre- ——, counter-spying, check on spying by the use of special spies, of *espions doubles*.

espionner, v. a., to spy, spy out.

esplanade, f., esplanade; (*fort.*) esplanade, parade.

espolette, f., **espoulette**, f., (*artif.*) quick-match (obs.).

esprit, m., spirit;
—— *de bois*, (*chem.*) wood spirit;
—— *cavalier*, (*cav.*) cavalry spirit, dash;
—— *de corps*, (*mil.*, esprit de corps, regimental feeling);
—— *d'initiative*, (*mil.*) the initiative;
—— *de vin*, alcohol.

esquif, m., skiff.

esquille, f., (*med.*) splinter (of bone).

esquilleux, a., splintery.

esquine, f., (*hipp.*) loins (obs.);
faible (*fort*) *d'* ——, weak (strong) in the back;
sauter d' ——, (*man.*) to arch or curve the back in jumping.

esquisse, f., sketch; outline, rough drawing;
cahier d' —— *s*, sketch book;
—— *au crayon*, pencil sketch.

esquisser, v. a., to sketch, outline, make a rough draught of.

essai, m., trial, proof, test; attempt; (*nav.*) trial trip;
—— *sur la base*, (*nav.*) trial over the measured mile;
—— *à chaud*, hot test;
—— *au choc*, falling-weight test;
—— *de cisaillement*, shearing test;
—— *de durée*, endurance trial, test;
—— *de fabrication*, manufacturing test (i. e., test pieces are subjected to the processes of manufacture and the results noted);
—— *de flexion*, bending test;
—— *à froid*, cold test;
—— *de force*, (*art.*) proof to bursting, to ultimate strength; (*nav.*) full speed trial;
—— *de forgeage*, forge test;
—— *au lapidaire*, emery-wheel test of special steels (quality and kind of steel determined by the nature, etc., of the sparks);
—— *à la mer*, (*nav.*) sea test, trial trip;
—— *de montage*, assembling test (strength, soundness, of mounting);
—— *à outrance*, v. —— *de force*, (*art., nav.*);

essai *du petit poinçon*, falling-weight test (of a projectile; the weight strikes a *poinçon* near the center of the base of the projectile);
—— *sur la place*, (*nav.*) dock trial;
—— *de pliage*, —— *de ployage*, folding test;
—— *à la poudre*, powder test, to determine whether the metal tested is strong enough to resist firing pressures;
—— *de pression*, (*art.*) (hydraulic) pressure test (of a projectile);
—— *avec pression de vapeur*, (*steam*) blowing off;
—— *de réception*, —— *de recette*, acceptance test;
robinet d' ——, (*steam*) pet cock.
—— *de rupture*, breaking test;
—— *de rupture à la traction*, tensile test;
—— *de taraudage*, (*art.*) test (of a projectile), consisting in threading a hole previously cut in the base;
—— *à la traction*, —— *de traction*, tensile test;
—— *de trempe*, temper, annealing, test;
(*Essai* is, strictly speaking, a trial, and *épreuve* is a proof; but this distinction, though sound, is not always observed.)

essaim, m., swarm (as, a swarm of skirmishers; not a technical term).

essayage, m., trying on, as of clothes, effects, etc.

essayer, v. a. n., to try, to attempt; (*tech., etc.*) to test; (*art.*) to prove, test (a gun, etc.).

esse, f., linchpin; forelock; S-shaped link, open link; scale-pan hook; wire gauge; (*harn.*) off-side curb-hook;
—— *à anneau*, (*Fr. art.*) linchpin and ring;
—— *d'essieu*, linchpin;
—— *d'essieu à coiffe*, linchpin with a cap;
—— *de gourmette*, (*harn.*) off-side curb-hook;
—— *-marchepied*, (*Fr. art.*) linchpin with a step (for the convenience of cannoneers in mounting);
—— *à ressort*, spring linchpin.

essence, f., kind, species (of wood, of tree).

esser, v. a., (*tech.*) to gauge wire.

essseret, m., long auger.

essette, f., adz; sort of hammer, with one end of head round, the other sharp;
—— *de calfat*, calker's adz.

essieu, m., axle, axletree; spindle; pin (of a block);
—— *amortisseur*, an axle contrived to ease the shock on starting, on striking obstacles;
—— *d'arrière*, trailing axle (of a locomotive);
—— *chargé en plusieurs points*, (*mach.*) multiple loaded axle;
—— *d'avant*, leading axle (of a locomotive);
corps d' ——, axle body;
—— *coudé*, crank axle;
faux- ——, temporary axle arm (as, a substitute for a broken one);
fourreau d' ——, axle sheath, seat, box;
fusée d' ——, axle spindle;
—— *moteur*, (*r. r., etc.*) driving axle;
—— *porte-roue*, —— *porte-roue de rechange*, (*art.*) spare-wheel axle;
—— *de poulie*, pin of a block;
—— *-ressort*, (*art.*) French name for the axle of the R. A. 12-pounder;
—— *simple*, (*mach.*) simple loaded axle;
—— *simple à fuseaux égaux* (*inégaux*), (*mach.*) simple symmetrical (nonsymmetrical) axle.

essorage, m., seasoning (of wood); drying of wood, after steaming; (*powd.*) drying of gunpowder (before granulation); (*expl.*) wringing, drying by wringing (of gun cotton, during manufacture).

essorer, v. a., to dry (wood, etc.); (*expl.*) to dry by wringing, to wring (materials).

essoreuse, f., (*expl., etc.*) centrifugal drying machine, hydro-extractor.

essorillé, p. p., crop-eared;
cheval ——, (*hipp.*) crop-ear, cropped horse.

essoriller, v. a., to crop the ear; (*fam.*) to cut the hair very short.

essoufflement, m., state of being winded, out of breath.

3877°—17——11

essouffler   162   **étalon**

essouffler, v. a. r., to blow, wind (a horse, a man), to become winded.
essourisser, v. a., (*hipp.*) to cut the cartilage of a horse's nose.
essui, m., drying-room.
essuyer, v. a. r., to wipe, wipe off; to dry, dry off; (*fig.*) to bear, endure, support;
—— *le feu,* (*mil.*) to stand fire without flinching.
est, a., m., east, the east; easterly, eastern.
estacade, f., estacade; boom; barricade of piles; (*fort.*) stockade;
—— *fixe,* row of piles, piling, pile stockade (across a river, etc.);
—— *flottante,* boom, floating boom, chain-boom;
—— *mixte,* a combination of —— *fixe* and —— *flottante* (i. e., piles with logs, etc., chained or secured between; does not interfere with navigation);
—— *de,* —— *en, pilotis,* v. —— *fixe;*
—— *de sûreté,* secondary boom protecting the main one, guard-boom.
estafette, f., estafette; courier, express;
—— *montée,* (*mil.*) mounted orderly; courier (U. S. A.).
estafilade, f., gash, cut.
estafilader, v. a., to gash; to cut, slash, the face.
estampage, m., stamping, forming, shaping.
estampe, f., print, engraving; stamp, die, swage;
—— *ronde,* rounding tool.
estamper, v. a., to stamp, emboss; to swage.
estampille, f., stamp, mark (e. g., on stock of guns); (*Fr. a.*) official stamp or seal caused to be affixed by minister of war on samples, models, etc., sent to regiments.
estampoir, m., v. *étampoir.*
estampure, f., v. *étampure.*
estimation, f., estimate, estimation, valuation.
estime, f., estimation; (*nav.*) dead reckoning;
à l'——, by estimation (e. g., of readings, when the index stops between graduations).
estimer, v. a., to estimate, value; (*nav.*) to get the course by dead reckoning.
estoc, m., stump, stock; (*sm. a.*) thrusting sword (obs.), sword point;
arme d'——, thrusting weapon;
coup d'——, thrust;
couper à blanc ——, to cut down a tree to the root;
d'——, (*fenc.*) with the point;
d'—— *et de taille,* (*fenc.*) cut and thrust;
frapper d'——, to thrust.
estocade, f., long sword (obs.); (*fenc.*) thrust, pass; wound produced by a thrust.
estocader, v. a., (*fenc.*) to thrust, make a pass.
estompe, f., stump (drawing).
estomper, to stump (drawing).
estoquer, v. a., to rivet, to jog, to jump; (*sm. a.*) to jump a gun barrel.
estoquiau, m., catch-bolt, catch; pawl, ratchet.
estoupin, m., wad of oakum.
estrac, m., (*hipp.*) lank, lean, heron-gutted horse.
estrade, f., platform;
battre l'——, (*mil.*) of cavalry, to scout, to patrol, so as to keep in touch with the enemy.
estramaçon, m., (*sm. a.*) long, double-edged sword (obs.); (*fenc.*) cut;
coup d'——, (*fenc.*) cut.
estran, m., (*hydr.*) beach or strand between high and low water marks.
estrapade, f., (*gym.*) "skin the cat"; (*man.*) *saut-de-mouton,* q. v., of a horse who stops, kicks, and tries to throw his rider.
estrapasser, v. a., (*man.*) to tire, knock up, override, overwork a horse.
estrapontin, m., hammock.
estrope, f., (*cord.*) strap, block strap;
—— *d'aviron,* oar grommet (to confine the oars to the thole pins);
—— *en bois,* wooden collar;
—— *à crochet,* (*nav., art.*) shell-sling;
—— *en fer,* iron collar;

estrope *de poulie,* block strap.
estroper, v. a., (*cord.*) to strap, to provide with a strap;
—— *une poulie,* to strap a block.
estropié, p. p., lame, crippled, disabled; m., cripple.
estropier, v. a., to cripple, lame, maim, mutilate, disable.
estuaire, m., (*hydr.*) estuary.
établage, m., stabling; stallage; keep (of a horse, etc.); space between shafts.
étable, f., stable (for animals other than horses).
établer, v. a., to stable.
établi, m., workbench; frame (of a lathe, etc.); (*artif.*) laboratory table;
valet d'——, dog, holdfast.
établir, v. a. r., to establish, fix, set up; to institute, erect; to lay down; to take possession of; to construct (a railway, etc.); (*mil.*) to pitch (a camp);
—— *les bois,* to mark wood for sawing;
—— *une machine,* to construct and set up a machine;
—— *les pierres,* (*mas.*) to mark stone for cutting.
établissement, m., establishment; settlement, settling, placing;
——*s d'études,* (*Fr. art.*) a generic expression for all boards, committees, and commissions in charge of technical questions and investigations;
—— *des marées,* —— *d'un port,* establishment of a port;
—— *pénitentiaire,* any penitentiary establishment;
—— *des quartiers,* (*mil.*) cantoning of troops.
étage, m., floor, story (of a house); row, step, layer; degree, class, quality, rank; (*fort.*) tier (as, of casemonts); (*mach.*) step (of a stepped cone or pulley);
——*s de feu,* (*fort., etc.*) tiers of fire;
premier ——, second story (of a house).
étager, v. a. r., to tier, to be arranged in tiers; (*mil.*) to be drawn up in lines one above the other.
étagère, f., set of shelves.
étai, m., strut, brace, stay, prop; (*cord.*) stay;
—— *de cheminée,* (*nav.*) funnel stay;
pomme d'——, (*cord.*) mouse, of a stay;
—— *en sautoir,* diagonal stay;
—— *de tête de bigue,* shear-head stay.
étaie, f., v. *étai.*
étaiement, m., v. *étayement.*
étaier, v. a., v. *étayer.*
étain, m., (*met.*) tin; pewter;
—— *battu,* tin-foil;
—— *en blocs,* block tin;
—— *commun,* common tin;
—— *en feuilles,* sheet-tin;
—— *en saumons,* v. —— *en blocs.*
étalage, m., display; (*in pl., met.*) boshes.
étale, a., slack (of the tide), steady, still; m., slack water;
cordage ——, (*cord.*) rope that slacks after being paid out;
être en ——, to be slack, at a stand;
—— *de flot,* slack of the flood;
—— *de jusant,* slack of the ebb;
marée ——, *mer* ——, flood tide, highest of the flood;
vent ——, wind of moderate velocity.
étaler, v. a., to display; to stem (the tide, a current).
étaleuse, f., (*mach.*) spreader, spreading machine (in rope manufacture).
étalingue, f., (*cord.*) clinch of a cable; shackling of a chain-cable.
étalinguer, v. a., (*cord.*) to bend, clinch, a cable (to the anchor).
étalingure, f., v. *étalingue.*
étalon, m., standard, legal standard (of weights and measures), gauge, standard gauge; (*elec.*) standard of calibration; (*hipp.*) stallion;
—— *approuvé,* (*hipp.*) in France, a stallion recognized by the government as suited to the improvement of the breed, and the owner of which gets a bounty;

étalon *autorisé*, (*hipp.*) in France, a stallion recognized by the government as suited to maintain (i. e., not to improve) the standard reached (owner gets no bounty);
—— *à coulisse*, sliding caliper scale;
—— *à coulisse et à vernier*, standard sliding calipers;
d'——, standard;
—— *départemental*, (*hipp.*) in France, a recognized stallion (one to each *département*);
—— *d'essai*, (*hipp.*) v. *boute-en-train;*
—— *de haras*, (*hipp.*) a stallion attached to a breeding establishment;
—— *national*, (*hipp.*) in France, a stallion belonging to the state;
—— *primé*, v. —— *approuvé;*
—— *rouleur*, (*hipp.*) traveling stallion, "tramp" stallion.

étalonnage, m., gauging, verification, standardizing, comparison with a standard; stamping (after comparison with a standard); (*elec.*) calibration;
—— *du pas*, (*mil.*) determination of the average length of one's pace or step (distances used in France, 10 meters and 100 meters).

étalonnement, m., v. *étalonnage.*

étalonner, v. a. n., to gauge, standardize, compare with a standard; to stamp (after comparison with a standard); (*elec.*) to calibrate; (*hipp.*) to serve;
—— *le pas*, (*mil.*) to find the average length of one's pace or step.

étamage, m., operation of tinning; tinned work;
—— *galvanique*, galvanizing;
—— *de glace*, silvering (of a looking-glass);
—— *au zinc*, zinking.

étambot, m., (*nav.*) sternpost.

étamer, v. a., to tin; to galvanize;
—— *une glace*, to silver (a looking-glass).

étameur, m., tinner.

étamine, f., bolting-cloth; bunting; a flag proper (as distinguished from the head, tassels, etc.);
—— *de crin*, haircloth.

étampage, m., swaging; operation of stamping, punching, or beating out, of giving a cylindrical or truncated-cone shape to anything.

étampe, f., swage, swage-tool, swage-block, stamp; bottom tool; die, boss, drift, mandrel; (*farr.*) horseshoe punch;
dessous d'une ——, bottom swage, bottom tool, lower die;
dessus d'une ——, top swage, upper tool, upper die, swage hammer;
—— *inférieure*, v. *dessous d'une* ——;
—— *ronde*, rounding tool;
—— *supérieure*, v. *dessus d'une* ——.

étamper, v. a., to swage; to stamp out, beat out; to shape, punch, emboss; to drift, open out; to give a cylindrical or truncated-cone shape to anything; (*farr.*) to punch holes in a horseshoe;
—— *gras*, —— *à gras*, (*farr.*) to punch a horseshoe close to the inner edge;
—— *maigre*, —— *à maigre*, (*farr.*) to punch a horseshoe close to the outer edge.

étampure, f., mouth or splay of a hole in a metal plate; (*farr.*) nail-hole (of a horseshoe; strictly, the square hole for the head of the nail).

étamure, f., material for tinning.

étanche, a., tight (steam-, water-, wind-tight);
—— *à l'*, *d'*, *eau*, *à* —— *d'eau*, water-tight;
faire l'——, to make (water-) tight;
mettre à ——, to stanch; to dry, empty (as, a ditch);
—— *à la*, *de*, *vapeur*, (*steam*) steam-tight.

étanchéité, f., tightness (water, steam, of joints, etc.).

étancher, v. a., to stanch, to stop (a leak), quench, slake, dry up;
—— *un vaisseau*, to free a ship by her pumps;
—— *une voie d'eau*, to stop a leak.

étançon, m., prop, stanchion, pillar, shore, stay, stud.

étançonner, v. a., to prop, to shore up; to timber (a mine), to stay, to underprop.

étang, m., pond.

étape, f., (*mil.*) (originally) subsistence issued to troops on the march, (hence, today) halting place at the end of a day's march; (by a further extension) the distance between two halting places, (and, therefore) the day's march itself, (and in gen.) distance to be covered in one march, (more esp.) the regulation distance expected of troops under normal conditions (from 22 to 25 kilometers);
arriver à l'——, to arrive at a *gîte d'*——*s;* (*in gen.*) to reach a halting place;
brûler l'——, to march through or by a *gîte d'*——*s* without halting;
circonscription d'——*s*, in a *ligne d'*——*s*, a subdivision of three or four days' march;
commandant d'——*s*, c. o. of a *gîte d'*——*s;*
commandement d'——*s*, c. o. and the executive and administrative personnel of a *gîte d'*——*s;*
directeur des ——*s*, v. s. v. *directeur;*
direction d'——*s*, v. s. v. *direction;*
faire une ——, to make a day's march, to make the distance from one *étape* to the next;
gîte d'——*s*, halting place on a *route d'*——*s*, q. v.; the various *gîtes* are about 25 kilometers apart, and each has a post-office, a hospital, a telegraph station, a small garrison, etc.;
gîte principal d'——*s* on a *route d'*——*s* (has a stronger personnel than the ordinary *gîte*, a larger hospital, a *sous-intendance*, q. v.; is established only when the *ligne d'*——*s de route* is prolonged beyond 4 *étapes*);
hôpital d'——, the hospital of a *gîte d'*——*s;*
infirmerie d'——*s*, *de gîte d'*——*s*, infirmary established in a *gîte d'*——*s* that has no hospital.
lieu d'——, halting place;
ligne d'——*s*, *ligne d'*——*s de route*, line of ——*s*, (hence, substantially) line of supply, of communications, line of posts;
ligne d'——*s auxiliaire*, a line of ——*s* parallel to a railway;
route d'——*s*, the road on which are situated the *gîtes d'*——*s*, and which connects an army in the field with the terminus of railway transportation;
service des ——*s*, a general term for the whole service of transportation (railways excepted) and supply, as carried on along the *lignes d'*——*s;*
station tête d'——*s de guerre*, most advanced railroad station on line of communication; station where the *route d'*——*s* begins; (in maneuvers, this is called *station tête d'*——*s de manœuvres*);
tête d'——*s de guerre*, v. *station tête d'*——*s de guerre;*
tête d'——*s de route*, on a *route d'*——*s*, the *gîte d'*——*s* nearest the army, advancing and halting with it; (here the transport service of the army itself comes into play);
troupes d'——*s*, troops garrisoning a *gîte d'*——*s;*
zone des ——*s*, the zone, front, or belt crossed by *routes d'*——*s*, (substantially) the supply front.

état, m., the state, the government; state, condition, establishment, position; (*mil. adm.*) report, list, account, return, muster-roll, inventory, estimate, register;
chef d'——, the head of the state;
—— *civil*, (*law*) position occupied by each individual in the family and in society;
—— *de corps*, (*mil.*) register or list of an arm of the service, of a department, etc.;
—— *du corps du génie*, (*Fr. a.*) annual register of the engineer corps;
—— *de demande*, (*mil.*) requisition;
—— *détaillé de l'assiette du logement*, (for each French garrison town) a report on each military building, giving its design, dimensions, capacity, and use;
—— *de distribution*, (*Fr. a. adm.*) in the *service des lits militaires*, paper or voucher on which bedding is issued;
dresser un ——, to make a return of; to draw up a report;
faire l'—— *de*, to make a return, report, of;
—— *de filiation*, return or list, by name, of officers, n. c. o., and men about to embark;

état *de guerre*, state of war; war footing; (*Fr. a.*) the status of a fort or fortified place, upon the publication of the order of mobilization (assumption of command by the designated governor, limitation of the powers and functions of the civil authority, adoption of general measures for the safety of the place, etc.);
— *à —— libre*, (*met.*) in a free state, native;
— *de lieux*, (*Fr. a. adm.*) report on state of barracks when first occupied by new set of troops (so as to fix responsibility for future damages); (*top.*) in fortification surveying, subdivision of the terrain into squares whose vertices have their differences of level determined. These altitudes are recorded on the map; contours also may be run by interpolation on plan (also applicable to roads, cross and axial profiles being run);
— *de malades*, sick report;
— *mensuel d'existence*, (*Fr. a. adm.*) in the *service des lits militaires*, a report or voucher showing what the contractor is entitled to;
— *militaire du corps de l'artillerie*, (*Fr. a.*) annual register of the corps of artillery;
— *naissant*, (*chem.*) nascent state;
— *nominal*, list of names;
— *nominatif d'affectation*, (*Fr. a.*) list of men sent each year from the active army to the reserve;
— *de paix*, state of peace; peace footing; (*Fr. a.*) the status of a fortress or fortified position under peace conditions, in respect of command, police, relations to civil authority, etc.;
— *de perte*, return of losses in action;
porter sur les —— s, to put on the muster rolls;
rayer des —— s, to strike off the rolls;
— *récapitulatif*, (*fort*., *Fr. a.*) report kept in each *place*, showing what works and guns will reach such and such points of surrounding terrain;
— *de réforme*, (*adm.*) inspection report; (*Fr. a.*) detailed list of property to be submitted for *réforme*;
— *de réintégration*, (*Fr. a. adm.*) in the *service des lits militaires*, list or return of property turned in;
— *de service*, military (official) record, record of military service; detail of men for duty in next 24 hours (e. g., in a siege);
— *de siège*, state of siege; (*Fr. a.*) of a fortress or fortified position, state of siege, resulting in the complete transfer of the civil authority to the military governor;
— *signalétique*, descriptive list;
— *signalétique et des services*, (*Fr. a.*) statement of *état civil*, with descriptive list and record of military service;
— *de situation*, return or report showing the strength, condition, and station or stations of an army, regiment, garrison, etc.;
— *de solde*, (*Fr. a. adm.*) pay sheets, on which *conseil d'administration* draws amounts due for pay (officers and men);
— *de solde des officiers*, (*Fr. a.*) officers' pay sheet (monthly);
— *de solde de la troupe*, (*Fr. a.*) men's pay sheet (twice a month);
— *tampon*, "buffer" state;
— *de tir*, (*t. p.*) target-practice report.

état-major, m., (*mil.*) staff; staff office; (*nav.*) commissioned officers, all the commissioned officers of a navy (of a ship);
— *d'armée*, the staff of an army;
— *de l'armée*, the staff of the army, the general staff;
— *de bataillon* (*de brigade*, etc.), battalion (brigade, etc.) staff;
— *d'artillerie*, (*Fr. a.*) the artillery staff of the various *commandements d'artillerie* (in peace), of army corps, etc., and of the various artillery parks (in war);
brevet d'——, staff certificate, staff college certificate (England);
breveté d'——, having a staff certificate; passed the staff college (England);
chef d'——, chief of staff;

état-major, *chef d'—— de l'armée*, chief of the general staff;
corps d'——, staff corps;
— *enseignant*, academic staff (U. S. M. A.);
— *d'escadre*, (*nav.*) fleet staff;
— *de la flotte*, (*nav.*) the whole body of naval officers;
— *général*, (*Fr. a.*) the staff of an army, synonym of —— *d'armée*; (in popular, not official, language) the great general staff, the general staff proper;
— *général de l'armée*, (*Fr. a.*) all the marshals, 110 generals of division, and 220 generals of brigade;
— *du génie*, (*Fr. a.*) engineer staff, formed in time of war (armies, army corps, and army engineer park);
grand ——, grand —— général, general staff, great general staff (of which the German is the exemplar); (more esp.) the staff of a group of armies, the staff of the general in chief or generalissimo;
officier d'——, staff officer;
officier du service d'——, (*Fr. a.*) an officer of the *service d'——*, q. v.;
— *particulier de l'artillerie*, (*Fr. art.*) special artillery staff, i. e., a body of officers in excess of those serving with artillery troops, from whom are drawn the artillery staffs of various units and the personnel of artillery establishments; includes student officers at Fontainebleau, workmen, inspectors of arms, and battery care-takers;
— *particulier du génie*, (*Fr. eng.*) special engineer staff, i. e., officers in excess of those serving with engineer troops, and from whom are taken the engineer staffs of various units and the personnel of engineer establishments and services; includes student officers at Fontainebleau and various engineer employees;
petit ——, (*Fr. a.*) in a regiment, the n. c. o. and men not belonging to the various units, (substantially) the n. c. o. staff and band in the infantry, the noncommissioned staff and certain musicians in the cavalry and artillery;
— *des places*, fortress staff (abolished in France in 1872);
relation d'——, official staff history;
service d'——, (*Fr. a.*) the general staff proper of the French army, composed of 640 officers (field officers and captains, all *hors cadre*) of all arms, who have received a staff certificate; (these constitute a *service*, not a closed corps, and are assisted by *archivistes* and by *secrétaires des bureaux d'——*, who are included in the personnel);
voyage d'——, staff journey.

étau, m., vise;
— *à agrafe*, bench-vise;
— *à attache*, vise and screw clamp;
— *à griffes*, portable vise (in use by the French field artillery; is held in place by *griffes*);
— *-limeur*, filing vise;
mâchoire d'——, jaw of a vise;
— *à main*, hand vise;
— *à mors parallèles*, parallel vise;
— *à mouvement parallèle*, parallel vise;
— *parallèle*, parallel vise;
— *à pattes*, flanged vise;
— *à pied*, standing vise, tail vise;
— *portatif*, hand vise;
— *à queue*, tail vise;
— *tournant*, swivel vise.

étayement, m., propping, shoring, staying, strutting, supporting.

étayer, v. a., to prop, support, stay, shore up;
s'——, (*man.*) to settle one's self firmly in the saddle (e. g., when urging a horse to change gait).

été, m., summer;
— *de la Saint-Denis, —— de la Saint-Martin*, Indian summer.

éteindre, v. a. r., to extinguish; to cancel, abolish, strike out; to diminish, reduce, bring down; to soften, allay, still, quell; to slake (lime); to temper (iron);
— *le feu*, (*art.*) to silence the (enemy's) fire;

éteindre *les feux*, to draw the fires; (*mil.*) to put out lights and fires;
— *le travail*, (*tech.*) (of metals under stress) to take up or stand so much work or stress.
étendage, m., clothes-lines.
étendard, m., standard; (*art., cav.*) colors, regimental color;
à l' —, to the color;
porte- —, color-bearer.
étenderie, f., v. *étendoir*.
tendoir, m., clothes-lines; any drying place.
étendre, v. a. r., to extend, spread, stretch; draw out, lengthen; expand; lay on or over; (*chem.*) to dilute.
étendue, f., extent, length, stretch (of country); extension, duration, volume.
étésiens, a pl., *vents* —, etesian winds, north and northwest winds (in the Mediterranean), blowing for about 40 days after the rising of the Dog Star.
étêtement, m., (of trees) heading, topping, pollarding.
étêter, v. a., (of trees) to top, pollard; to remove the head (as, of a nail, etc.).
éthal, m., (*chem.*) ethal.
éther, m., (*chem.*) ether.
éthérisation, f., (*chem.*) etherization.
éthériser, v. a., to etherize.
étiage, m., (*hydr.*) low-water mark; lowest point or level of a river for the year;
à l' —, at low water.
étinceler, v. n., to spark.
étincelle, f., spark; scale of hot iron;
— *électrique*, (*elec.*) electric spark.
étiqueter, v. a., to label, to ticket.
étiquette, f., label, ticket.
étirable, a., (*met.*) susceptible of being drawn, etc., out.
étirage, m., (*met.*) drawing, drawing down, out; draw; beating out; stretching out; rolling.
étirer, v. a. r., (*met.*) to draw, draw down, out; to beat, roll, stretch out; to draw (wire); to hammer out;
banc à —, draw bench;
— *en fils*, to wire-draw;
à la presse, to draw down under pressure.
étireur, m., v. s. v. *cylindre*.
etnite, f., (*expl.*) etnite, a variety of asphaline.
étoffe, f., stuff, material;
— *d'acier*, the core, heart, or backing of a cutting tool (i. e., the part supporting the cutting edge).
étoffé, p. p., stuffed, stout, stiff; of or with sufficient material;
cheval —, stout, stocky horse;
fer —, iron forming the backing or heart of a cutting tool.
étoffer, v. a., to thicken, stuff, stuff out, stiffen by the addition of material (as, the walls of a gun barrel, in order to resist vibrations).
étoile, f., star; radial cracks (in timber); the star (inaccurately called cross) of the Legion of Honor; crown or set of long radiating meshes near the top of a ballon net; (*art.*) base or plate (on which the gun or forging rests) of a vertical oil-tempering tank; (*hipp.*) blaze; (*Fr. a., unif.*) the star of a general officer (worn on sleeve, a general of brigade having two, a general of division three); star of a retired officer (distinguishing mark of, worn on collar); (*artif., in pl.*) stars of a signal, etc., rocket; (*fort.*) star fort;
— *d'artifices*, (*artif.*) star or decoration of a rocket;
— *à calibrer*, v. — *mobile;*
— *dentaire*, (*hipp.*) dental star;
mèche d' —, — *de mèche*, (*artif.*) star match;
— *mobile*, (*art.*) star gauge;
— *radicale*, v. — *dentaire.*
étoilé, p. p., starred (as, a crack);
fort —, (*fort.*) star fort.
étoiler, v. a. r., to crack radially, to star; to mark with a star.
étonnement, m., astonishment;
— *du sabot*, (*hipp.*) concussion of foot by a shock or blow.
étoqueresse, f., pawl.

étouffement, m., suffocation;
— *d'un incendie*, stifling of a fire.
étouffer, v. a., to choke;
— *un incendie*, to stifle, put out, a fire.
étouffoir, m., (*powd.*) sheet-iron vessel into which charcoal is put as soon as made (to prevent spontaneous combustion); (*steam*) damper.
étoupage, m., (*mach.*) packing, stuffing.
étoupe, f., tow, packing-tow; oakum; stuffing.
— *blanche*, untarred oakum;
boîte à —, (*mach.*) stuffing-box;
garnir d' —, to pack;
— *goudronnée*, tarred oakum;
— *noire*, v. *goudronnée.*
étoupement, m., (*art.*) packing of ammunition in oakum.
étouper, v. a., (*mach., etc.*) to pack, stuff; to stow with tow or oakum, calk; to pack in oakum.
étoupille, f., (*artif.*) match; (*art.*) primer, friction primer, cannon primer;
— *de Bickford*, (*artif.*) Bickford fuse;
— *de campagne*, German friction primer, adapted to lamellar powders;
— *électrique*, electric primer;
fausse —, dummy primer (for use at drill);
— *à friction*, friction primer;
— *à friction en plume*, (*Fr. navy*) goose-quill primer;
— *fulminante*, friction primer;
— *obturatrice*, vent-sealing, obturating, primer;
— *obturatrice à friction*, (*Fr. art.*) subcaliber friction primer;
— *obturatrice à percussion*, percussion self-sealing primer;
— *obturatrice à percussion centrale*, French navy primer, self-sealing and for axial vents;
— *obturatrice à vis*, an obturating primer, screwing into the vent;
pare- —, friction-primer deflector;
— *à percussion*, — *percutante*, percussion primer;
— *à percussion centrale*, French navy self-sealing axial primer;
— *de sûreté*, (*artif.*) safety fuse (Bickford's).
étoupiller, v. a., (*artif.*) to prime, fit, or supply with quick match.
étoupillon, étoupin, m., (*art.*) vent-plug of white oakum (to keep out moisture) (obs.).
étourneau, a., (*hipp.*) flea-bitten; m., flea-bitten horse;
poil d' —, flea-bitten.
étouteau, m., pawl, click; (*sm. a.*) locking-ring pin (of a bayonet);
— *arrêtoir*, (*sm. a.*) stop-pin, pin (bolthead of Lebel, model 1886).
étranger, a., foreign; m., foreigner, stranger;
à l' —, abroad, in foreign parts;
régiment —, (*Fr. a.*) v. s. v. *légion.*
étrangle, f., (*artif.*) choking-string.
étranglement, m., choking, strangling; contraction; (*art., artif.*) choke; (*sm. a.*) choke (of a shotgun); (*steam*) throttling; wire-drawing; (*hipp.*) neck (of a tooth).
étrangler, v. a., to choke, strangle; to compress, condense; (*art., artif.*) to choke; (*steam*) to throttle; to wire-draw; (*cord.*) to seize;
— *les bastions*, (*mil.*) in making rounds, to pass along the gorge of the bastions, i. e., to shirk the inspection of the bastions;
machine à —, (*artif.*) choking-frame;
— *un saucisson*, (*siege*) to choke and bind a fascine.
étrangleur, m., (*fort.*) fascine choker.
étrangloir, m., (*artif.*) choking-frame;
étranguillon, m., (*hipp.*) strangles.
étraque, f., breadth of a strake (boats).
étrave, f., (*nav.*) stem.
être, m., being, existence;
— *s*, parts, disposition, of a house.
être, v. n., to be, stand, lie; to take part in;
— *en armement*, (*nav.*) to be fitting out;
— *à flot*, to be afloat;
— *à mâts et à cordes*, — *à sec*, (*nav.*) to be under bare poles.

étrécir   166   **event**

étrécir, v. a. r., to narrow, shrink; (*man.*) to narrow; to cause a horse to narrow.
étrécissement, m., shrinkage, shrinking, narrowing.
étreignoir, m., clamp, cramp, jack, hand-screw.
étrein, m., (horse) litter.
étreindre, v. a., to bind, clasp, tie.
étrésillon, m., shore, prop, stretcher, stay, strut; (*cons.*) bridging piece.
étrésillonnement, m., (*cons.*) battening, bracing.
étrésillonner, v. a., to stay, to prop, to strut.
étrier, m., shackle, strap, binding-strap, hoop, fork, support (in many relations), stirrup; (*harn.*) stirrup; (*pont.*) trestle-cap strap; (*inst.*) support, fork (of a needle or index);
—— *d'armon*, (*art.*) hound-strap;
—— *de bât*, (mountain art.) side-rack (pack for supporting cargo that can not be lashed to packing hooks);
coup de l'——, stirrup cup;
croiser les ——*s*, (*man.*) to cross the stirrups;
—— *de derrière (devant)*, rear (front) staple or clasp on a pack saddle;
—— *d'essieu*, (*art.*) axletree hoop or stay;
être ferme sur les ——*s*, (*man.*) to have a sure seat;
à franc ——, at full speed;
perdre les ——*s*, (*man.*) to be unhorsed;
pied de l'—— (*man.*) left foot (of the rider); near fore foot (of the horse);
—— *porte-leviers*, (*art.*) handspike ring;
—— *porte-ranchet*, stake socket;
portereau, —— *de portereau*, (*art.*) carrying-bar strap (of a detachable limber chest);
—— *porte-limon de rechange*, (*art.*) spare-pole ring;
—— *de sellette*, (*art.*) bolster hoop or clip;
—— *de suspension*, stirrup (e. g., of a bridge);
vider les ——*s*, v. *perdre les* ——*s*;
vin de l'——, v. *coup de l'*——.
étrière, f., (*harn.*) stirrup strap (i. e., a strap to secure the stirrups to the saddle when not regularly hanging from the stirrup leathers).
étrille, f., currycomb.
étriller, v. a., to curry (a horse).
étripe, f., *aller à* —— *cheval*, to ride a horse furiously.
étripé, a., (*cord.*) worn, fagged.
étriper, v. r., (*cord.*) to become fagged, to wear out.
étriquer, v. a., to scribe a piece of wood.
étrive, f., (*cord.*) elbow, bend, nip; throat seizing; cross seizing; (*in pl.*) cross turns;
amarrage en ——, (*cord.*) seizing of a shroud or stay close to its deadeye.
étriver, v. a. r. n., (*cord.*) to cross, to seize; to become jammed or nipped; to get across, to form an elbow or bend.
étrivière, f., (*harn.*) stirrup leather; (*in pl.*) lashing, whipping.
étroit, a., narrow, tight, close; strict.
étroitesse, f., narrowness.
étrope, f., v. *estrope*.
étroper, v. a., v. *estroper*.
ette, f., (*mil.*) in commands, abbreviation of *baïonnette*.
étude, f., study, investigation;
à l'——, under investigation, research, experiment;
——*s préparatoires*, preliminary experiments, investigation.
étui, m., case, box, sheath; instrument, etc., case; (*sm. a.*) cartridge case;
—— *de cartouche*, (*sm. a.*) cartridge case;
—— *-chargeur*, (*sm. a.*) loading packet;
—— *à collet renversé*, (*sm. a.*) cartridge case the interior of whose neck is cone-shaped;
—— *combustible*, (*sm. a.*) paper case (obs.);
—— *-crosse*, (*sm. a.*) a pistol holster so contrived as to be used for a stock;
—— *de drapeau*, (*mil.*) standard case, flag case;
—— *à épaulement*, (*sm. a.*) shouldered cartridge case;
—— *à gorge*, (*sm. a.*) grooved head cartridge case;

étui de mathématiques, (*inst.*) box or case of mathematical instruments;
—— *métallique*, (*sm. a.*) metallic cartridge case;
—— *-musette*, (*mil.*) haversack; bag;
—— *porte-avoine*, (*cav.*) grain sack;
—— *porte-boîte à mitraille*, (*art.*) canister box or case (on axle, one on each side, field artillery; on gun-pack, mountain artillery);
—— *porte-charge*, (*art.*) leathern cylindrical case for a single round;
—— *à rebord*, (*sm. a.*) flanged cartridge case;
—— *de revolver*, (*sm. a.*) holster.
étuvage, m., seasoning (of wood); (*fond.*) drying of a mold.
étuve, f., stove, drying-stove, drying-oven; room heated by a stove; hot-air or vapor bath; (*fond.*) foundry stove, drying room (for molds);
—— *à eau*, (*chem.*) water-oven, water drying-oven;
—— *-locomobile à désinfection*, traveling disinfecting plant;
—— *à quinquet*, (*chem.*) drying-stove with lamp;
—— *à vapeur*, steam drying-oven.
étuvement, m., bathing, fomenting.
étuver, v. a., to put anything into a drying-stove, stove, to dry; to bathe, foment; (*fond.*) to dry a mold.
euthymètre, m., (*inst.*) horizontal stadia rod used in combination with the *tachéomètre*, q. v.
évacuable, a., (*med.*) that may be removed, taken to the rear.
évacuation, f., (*mil.*, etc.) evacuation; (*med.*) transfer of a patient from one hospital to another, sending of sick and wounded to the rear; (*steam*, etc.) exhaust, eduction, evacuation; (*mach.*) evacuation pipe;
convoi d'——, (*med.*) hospital train or convoy (land or water) conveying sick and wounded to the rear, from field or corps hospitals;
hôpital d'——, (*med.*) "overflow" hospital;
transport d'——, generic term for any method of transporting the sick and wounded in the field.
évacuer, v. a., (*mil.*, etc.) to evacuate, (more esp.) to send back or to the rear (of troops in disorder, broken up); (*med.*) to transfer a patient from one hospital to another; to send the sick and wounded to the rear, home;
faire ——, to clear (a room, etc.) of people.
évadé, m., escaped prisoner.
évader, v. r., to escape, break out of prison;
faire ——, to help, cause, a prisoner to escape.
évaluation, f., evaluation;
—— *des distances*, (*mil.*) estimation of distances.
évaluer, v. a., to value, estimate, rate;
—— *les distances*, (*mil.*) to estimate distances.
évaporation, f., evaporation.
évaporer, v. a. r., to evaporate.
évasé, p. p., bell-mouthed, splayed, wide at the mouth; cupped;
fossé ——, (*fort.*) a ditch whose counterscarp is not parallel to the scarp;
—— *en pavillon*, bell-shaped, hollowed out like a bell.
évasement, m., splay, spreading; (*art.*) enlargement (of the vent, bore); (*fort.*) splay (of an embrasure); state of being not parallel (as, scarp and counterscarp); (*sm. a.*) splay (of cartridge-case neck).
évaser, v. a. r., to enlarge, increase; to widen (an opening) to splay; (*art.*) to become enlarged.
évasion, f., escape (from jail, prison, confinement).
évènement, m., event;
—— *de force majeure*, act, or fact, beyond one's control; act of God, *vis major*;
—— *napoléonien*, (*mil.*) last phase of a battle, crisis upon which depends victory or defeat.
event, m., air-hole; ventilation passage; air-shaft of a mine or gallery; decomposition of cement, etc., by exposure; (*art.*) punching hole (of a fuse); flash-hole, flame passage (of a fuse); (*art.*, *sm. a.*) small cavity, flaw or honeycomb; (*met.*) longitudinal crack (defect); (*fond.*) escape-hole (for gases), gas-port; (*mil. min.*) sort of defensive mine, destroying the effect of a hostile mine, by acting as a line of least resistance;

event, *déboucher l'*——, (*art.*) to punch, cut, a fuse;
—— *de dégorgement,* (*fond.*) riser;
—— *de prise de feu,* (*art.*) flash-hole;
réglage de l'——, (*art.*) determination of the point where a fuse should be punched or cut;
régler l'——, (*art.*) to determine where a fuse should be punched or cut.

éventail, m., fan; fan light; screen; bat-wing gas burner;
en ——, fan-shaped.

éventer, v. a. n. r., to fan, ventilate, air; to give vent to; to discover, to get wind of anything; to hold or pull a weight from the wall while hoisting; to lose strength, decompose, by exposure (of cement, etc.); (*mil. min.*) to discover and destroy or neutralize a hostile mine; (*hipp.*) to raise, elevate, the nose too much;
—— *l'approche* (*la marche, la présence,* etc.) *de l'ennemi,* to discover the enemy's approach (march, presence, etc.);
—— *une mine,* (*mil. min.*) to discover a mine or the wires or train leading to it;
—— *une veine,* to bleed, let blood.

éventoir, m., (*min.*) adit.

éventrer, v. a., to rip, rip up or open.

éventure, f., (*art., sm. a.*) crack, honeycomb (rare).

éversite, f., (*expl.*) eversite.

évidement, m., hollow, aperture, cut, grooving, recess, cavity, guttering;
—— *de la balle,* (*sm. a.*) hollow in the base of a bullet;
—— *du bloc,* (*r. f. art.*) breechblock mortise;
—— *du fût,* (*sm. a.*) lateral groove of the forestock.

évider, v. a., to cut hollow, to make a hollow, groove, recess, or cavity; to scoop out; to hollow out; to groove; to cut sloping or slanting.

évidure, f., hollowing, sloping out, scooping out.

évier, m., stone channel or gutter;
pierre d'——, sink-stone.

évitage, m., navigable breadth, navigable channel (of a river, canal), fairway.

évitée, f., v. *évitage.*

évitement, m., (*r. r.*) turn-out;
gare d'——, turn-out station;
voie d'——, side track (of a single-track road).

éviter, v. a., to avoid, shun.

évolubilité, f., (*nav.*) quickness, readiness, in making evolutions (i. e., in answering helm).

évoluer, v. a., (*mil., nav.*) to make evolutions;
—— *à la vapeur,* —— *à la voile,* to make evolutions under steam, under sail.

évolution, f., (*mil., nav.*) evolution; (*mach.*) (complete) stroke, as of a piston;
——*s de ligne,* (*inf.*) the drill or evolutions of a line of several battalions, of a regiment, of a brigade in line (obs.); (*cav.*) the evolutions of a brigade or of a division of cavalry (obs.);
——*s sous vapeur,* ——*s sous voiles,* (*nav.*) steam tactics, sail tactics.

examen, m., examination; survey, inquiry, investigation;
—— *d'admission,* entrance examination;
—— *d'avancement,* (*mil.*) promotion examination;
certificat d'——, certificate of having passed an examination;
—— *écrit,* written examination;
—— *oral,* oral examination;
passer un ——, to pass an examination;
—— *pratique,* practical examination;
—— *de sortie,* graduation examination;
subir un ——, to undergo an examination;
—— *de théorie,* examination in the theoretical part of a subject, book work.

examinateur, m., examiner.

excavateur, m., excavator, digging machine.

excavation, f., excavation; excavating; hollowing out; digging a foundation;
machine à ——, excavating machine;

excavation *d'une mine,* crater of a mine.

excaver, v. a., to excavate, hollow, dig out.

excédant, a., m., exceeding, in excess; surplus, excess, overweight; (*in pl., adm.*) excess (over what one is responsible for).

excéder, v. a., to exceed; tire, weary; to knock up (a horse); to wear out.

excentré, a., eccentric; out of center; out of true.

excentricité, f., eccentricity.

excentrique, a., eccentric; m., (*mach.,* etc.) eccentric; eccentric gear, wheel, or motion; eccentric chuck (of a lathe); (*r. r.*) switch;
barre d'——, eccentric rod;
bras d'——, v. *barre d'*——;
bras et collier d'——, eccentric rod and belt;
—— *circulaire,* circular eccentric;
—— *en cœur,* —— *en forme de cœur,* heart-wheel, heart-shaped cam;
—— *à deux têtes,* (*Fr. art.*) elevation gear (field piece);
—— *d', de l', expansion,* (*steam*) expansion eccentric;
—— *de la marche en arrière* (*avant*), back, go astern (forward, go ahead) eccentric;
poulie d'——, eccentric sheave;
roue ——, eccentric tappet, gab;
tirant d'——, v. *barre d'*——;
—— *du tiroir,* slide-valve eccentric;
—— *triangulaire,* triangular eccentric.

excès, m., excess;
—— *sphérique,* spherical excess.

excitateur, a., exciting; (*expl.*) determining; m., (*elec.*) discharger, discharging rod;
liquide ——, (*elec.*) liquid of a plunge battery or cell, exciting fluid, electrolyte.

excitation, f., excitement; (*elec.*) excitation;
—— *compound,* —— *en compound,* —— *composée,* (*elec.*) composite excitation;
—— *en dérivation,* (*elec.*) shunt excitation;
double ——, (*elec.*) shunt and series excitation;
—— *en double circuit,* v. *double* ——;
—— *indépendante,* (*elec.*) separate excitation;
—— *séparée,* v. —— *indépendante;*
—— *en série,* (*elec.*) series excitation.

exclusion, f., (*steam*) cut-off;
—— *du service,* (*mil.*) state of not being allowed to serve (of convicts, etc.).

excoriation, f., sore on foot, or abrasion, produced by shoe.

excroissance, f., excrescence (defect of wood).

excursion, f., excursion, inroad; (*mil.*) raid, incursion; (*mach.*) stroke, travel (as, of a piston).

exécuter, v. a., to execute, perform, accomplish; to go through (as, a maneuver), carry out, fulfill; to put to death;
—— *une bouche à feu,* (*art.*) to serve a gun;
—— *les cheminements,* (*siege*) to advance, push forward, the zigzags;
—— *les feux,* (*art., sm. a.*) to fire; to perform, practice, the firings;
—— *militairement un pays,* to punish a country, to subject it to rigorous treatment by way of punishment.

exécution, f., execution (of a sentence, of a maneuver, etc.); achievement, fulfillment; (*adm.*) synonym of *gestion* (rare); (*mil.*) the call for execution;
—— *de la bouche à feu,* (*art.*) standing-gun drill; service of the piece;
—— *des brèches,* (*art.*) breaching;
—— *des feux,* (*art., sm. a.*) firing, practice in the firings;
—— *d'un judgement,* (*law*) execution of a sentence;
—— *militaire,* military execution; forcing the inhabitants to pay their contribution;
mise à ——, execution;
—— *à mort,* execution of a death sentence;
ordre d'——, death warrant, warrant for execution;
—— *des ordres,* carrying out, execution, of orders;
peloton d'——, (*mil.*) firing party;
—— *des réquisitions,* (*mil.*) execution, accomplishment, of requisitions;
—— *de la sape,* (*siege*) sapping.

exécutoire, a., *(adm., etc.)* final, mandatory.
exemplaire, m., model; copy.
exempt, a., exempt, free, excused from;
—— *de service*, *(mil.)* excused from duty (as, from a roll call, etc.);
—— *du service militaire*, *(mil.)* not required to do military service, free from serving.
exemption, f., exemption, freedom from; *(mil.)* exemption from military service;
—— *d'appel*, *(Fr. a.)* exemption from service during maneuvers (of reservists);
—— *de service*, *(mil.)* excused from duty (as, from a roll call, etc.);
—— *du service*, *(mil.)* exemption from serving in the army.
exercer, v. a., to exercise, train, practice; to carry on; *(mil.)* to drill, instruct;
s'—— au tir, —— *à tirer*, *(art., t. p.)* to carry on, do, target practice.
exercice, m., exercise, practice, functions; *(adm.)* fiscal year; *(mil., frequently in pl.)* drill, drilling, exercises;
à l'——, *(mil.)* at drill;
année d'——, *(adm.)* fiscal year; year of office;
—— *s d'assouplissement*, *(mil.)* setting-up drill, exercises;
—— *de la baïonnette*, etc., *(fenc.)* bayonet, etc., exercise;
—— *à blanc*, *(art.)* practice with blank cartridges;
—— *des bouches à feu*, —— *des canons*, *(art.)* gun drill;
—— *budgétaire*, *(adm.)* fiscal year;
—— *de cadres*, *(mil.)* officers' drill, i.e., a drill carried on without men;
—— *à la cible*, target practice;
—— *clos*, *(adm.)* past, closed, fiscal year;
—— *de combat*, *(mil.)* battle exercises;
—— *correctionnel*, *(mil.)* punishment drill, extra drill by way of punishment;
—— *courant*, *(adm.)* fiscal year current;
—— *détaillé*, *(art., etc.)* drill by the numbers, by detail,
—— *à double action*, one in which enemy is represented by a full body of men;
en ——, in practice;
—— *s d'ensemble*, *(mil.)* drill or exercise in which all the units interested take part;
faire l'——, *(mil.)* to be drilled;
faire faire l'——, *à*, *(mil.)* to drill (a company, a battalion, etc.);
—— *à feu*, firing drill;
—— *de lancement*, *(torp.)* torpedo drill, torpedo-launching drill;
—— *s partiels de mobilisation*, *(mil.)* mobilization drill (packing, issue of rations, etc.);
—— *de nage*, rowing drill, boat drill;
—— *s d'occupation*, *(art.)* an exercise involving the manning, equipping, and serving of a battery, as nearly as possible under war service conditions;
—— *de pointage*, aiming drill;
—— *de punition*, v. —— *correctionnel*;
règlement sur l'—— . . . . , *(mil.)* drill regulations for . . . . ;
—— *du sabre*, *(fenc.)* saber drill, exercise;
—— *par temps*, v. —— *détaillé*;
—— *de tir*, *(art., inf.)* target practice.
exhaussement, m., elevation, raising; *(Fr. art.)* plate fixed to the cheek, to raise the trunnions (obs.); sort of frame or rack to increase the capacity of a *chariot de parc*;
—— *du fond*, shallowing of water;
—— *du rail extérieur*, *(r. r.)* elevation of the outer rail;
—— *à hautes (moyennes) ridelles* *(Fr. art.)* long- (short-) raved rack.
exhausser, v. a., to raise; to run up quickly.
exhaustion, f., exhaustion;
Giffard d'——, *(steam)* ejector.
existant, m., *(mil. adm.)* what is actually in store; stores on hand; cash on hand (used also in pl.).
existence, f., existance; *(mil. adm.)* statement of differences (U. S. A.);
certificat d'——, *(Fr. a.)* certificate of existence, submitted by a soldier in certain cases.

exonération, f., *(mil. adm.)* relief from part or all of a debt incurred to the state by involuntary error, or by malversation of an agent, etc.
exosmose, f., exosmose.
exostose, f., *(hipp.)* exostosis.
expanseur, m., *(steam)* tube-expander.
expansibilité, f., expansibility (as, of steam).
expansif, a., expansive, expanding.
expansion, f., *(mach., etc.)* expansion (of steam, of powder gases, etc.);
machine à ——, expansion engine,
expatriation, f., expatriation.
expatrier, v. a. r., to expatriate; to leave one's country,
expédient, m., expedient, shift.
expédier, v. a., to dispatch; to clear; to send; to expedite, hasten; to send on, off; to draw up.
expéditeur, m., person shipping, shipper or sender of anything (goods, material, etc.); sender (of a letter, message, etc.).
expédition, f., expedition; dispatch, speed; dispatch, sending (of material, effects, etc.); *(adm.)* copy, duplicate (of a paper, voucher, of any document); transmission, dispatch, of papers; *(mil.)* expedition;
avis d'——, *(adm.)* invoice (practically, notice that material has been sent or shipped);
en double, triple, ——, in duplicate, triplicate;
être en ——, *(mil.)* to be on an expedition;
—— *militaire*, military expedition.
expéditionnaire, a., *(mil.)* belonging to, or forming part of, an expedition; m., commissioner; clerk;
commis ——, *(adm.)* dispatch clerk;
corps ——, *(mil.)* expedition, body of troops forming an expedition.
expéditionner, v. a., to make an expedition.
expérience, f., experience; experiment, trial, test;
—— *s d'artillerie*, *(art.)* gunnery trials, proof firing;
—— *à blanc*, test (as, of a torpedo circuit);
—— *s de giration*, *(nav.)* turning trials;
—— *de recette*, acceptance test;
—— *s de tir*, *(art.)* ballistic firing, trials, experiments;
—— *de traction*, tensile test.
expérimental, a., experimental.
expérimentation, f., experimentation.
expérimenté, a., skilled, expert.
expérimenter, v. a., to try, to make tests, experiments on, about, anything; to test.
expert, a., m., expert.
expertise, f., expert investigation;
conseil d'——, board of survey of experts;
faire une ——, to make a survey.
expiration, f., expiration (of a patent), end (of a year, term, etc.); *(mach.)* discharge (of water, steam, etc.).
exploit, m., exploit, achievement, deed.
exploitable, a., fit to be worked.
exploitation, f., working, operating (of a railroad, mine, factory, etc.); business; execution of works; *(min.)* winning;
—— *des bois*, felling of trees;
champ d'——, working place, room;
chantier d'——, *(min.)* chamber of excavation;
chef d'——, manager (of a railroad, etc.);
compagnie d'——, *(mil. r. r.)* a company charged with operating a railway;
—— *par compartiments*, *(min.)* panel working;
en ——, in operation;
—— *par grande taille*, *(min.)* long work;
matériel d'——, working material, stock;
mettre en ——, to put into operation (as, a railway);
—— *par piliers et compartiments*, v. —— *par compartiments*;
—— *par piliers et galeries*, *(min.)* working by post and stall, pillar and breast;
taille d'——, *(min.)* head of excavation, winning headway;
—— *des voies ferrées*, *(r. r.)* railway administration.
exploiter, v. a., to work (as, a railroad, mine, etc.), employ, use, win;
—— *un bois*, to fell a wood;

exploiter à *la poudre*, to blast.
explorateur, m., explorer; secret agent, spy (rare);
a., exploring, inspecting;
appareil ——, (*art.*) apparatus or device for inspecting the bore of a gun.
exploration, f., exploration; (*mil.*) exploration service performed by independent cavalry, the general permanent exploration work carried on at a distance from the main army;
front d'——, (*mil.*) front of exploration, *supra*;
secteur d'——, (*mil.*) sector or area to be explored or scouted over;
service d'——, (*mil.*) v. *supra*;
—— *stratégique,* (*mil.*) exploration service, the purpose of which is to dislocate and paralyze the troops of the enemy before the campaign or particular operation in hand defines itself.
explorer, v. a., to explore; (*mil.*) to explore, scout, the front of an army.
exploser, v. n., (*expl.*) to burst.
exploseur, m., (*elec., expl., etc.*) exploder, firing-key; (*art.*) safety exploder, used in casemates provided with embrasure masks;
—— *magnétique,* dynamo-electric machine.
explosible, a., (*expl.*) explosive.
explosif, a., explosive;
distance——*ve,* (*elec.*) in a nonconducting medium, the distance beyond which a spark will not go; (*expl.*) v. s. v. *distance.*
explosif, m., (*expl.*) explosive, (*esp.*) chemical explosive; (sometimes) powder (see *infra*);
—— *amide,* nitroglycerine and amide powder;
—— *binaire,* an explosive consisting of two substances;
—— *s à la cellulose nitrée,* gun cotton and its variants;
—— *chimique,* a "modern high explosive," a high explosive (chemical compound);
—— *Favier,* Favier explosive, miner's safety explosive (England);
—— *Gotham,* Gotham powder;
—— *instable,* unstable, unsafe, explosive;
—— *mécanique,* a mechanical mixture, a low explosive; generic term for black powder and its variants;
—— *s à la nitroglycérine,* dynamite and its variants;
—— *réglé,* safe, stable, explosive;
—— *simple,* an explosive consisting of but one substance;
—— *stable,* v. —— *réglé;*
—— *de sûreté,* Favier explosive.
explosion, f., explosion, bursting, blast, blowing up;
faire ——, to burst, explode;
—— *de deuxième ordre,* (*expl.*) ordinary explosion, as of gunpowder;
—— *de premier ordre,* (*expl.*) detonation, explosion of the first order;
rayon d'——, radius of effect;
—— *simple,* (*expl.*) explosion proper (not detonation);
—— *sympathique,* (*expl.*) sympathetic detonation, detonation by influence.
exportation, f., exportation.
exporter, v. a., to export.
exposant, m., exhibitor; petitioner; exponent, index.
exposé, m., statement, outline, account.
exposer, v. a. r., to expose; to expose one's self (to danger).
exposition, f., exhibition, exposure; statement, account; aspect.
exprès, m., (*r. r.*) express train;
train ——, express train.
expression, f., expression;
—— *des pentes,* (on maps, plans, etc.) slope.
exprimer, v. a., to express; squeeze; strain; wring out; (of a map) to show, to represent.
expropriation, f., condemnation by right of eminent domain; (*mil.*) condemnation of land for military purposes;
jury d'——, assessment board (to determine value of land).
exproprier, v. a., to take, condemn, by right of eminent domain; to condemn (as, land for military purposes).

expulser, v. a., to expel, drive out; throw out; to exclude, banish.
expulsion, f., expulsion, banishment.
exsudation, f., exudation; (*expl.*) leakage (as, of nitroglycerine from dynamite).
exsuder, v. n., to exude; (*expl.*) to leak (nitroglycerine from dynamite).
extensif, a., extending, expanding.
extension, f., extension; (*tech.*) strain, stress, of extension.
extérieur, a. m., exterior, external, outside; out of doors; the outside; foreign parts; (*mil.*) said of duties beyond a line of outposts, or outside of a place or camp; (*pont.*) said of the side farthest from first bay built or last dismantled;
à l'——, abroad; out of doors; on the outside;
travail ——, (*mil., etc.*) drill, exercises out of doors, in the open country.
exterritorialité, f., (*law*) exterritoriality, extraterritoriality.
extincteur, m., fire engine, fire extinguisher.
extinction, f., extinction; reduction; diminution; quelling, suppression; slacking of lime;
—— *par aspersion,* slaking by sprinkling;
—— *des feux,* (*mil.*) "lights out," "taps";
—— *à grande eau,* slaking by sluicing;
—— *spontanée,* (of lime) air-slaking.
extrac, a., (*hipp.*) slim, slender.
extra-calciné, a., supercalcined.
extra-courant, m., (*elec.*) self-induced current;
—— *direct,* extra-current on breaking;
—— *de fermeture,* extra-current on making;
—— *inverse,* v. —— *de fermeture;*
—— *d'ouverture,* v. —— *direct.*
extracteur, m., (*sm. a., art.*) extractor;
—— *à bascule,* (*sm. a.*) rocking extractor;
griffe de l'——, extractor hook;
—— *avec projection,* extractor that throws out the case;
—— *à vis,* (*art.*) screw extractor for projectiles.
extraction, f., extraction, discharge; quarrying of stone; (*art., r. f. art., sm. a.*) extraction of cartridge case;
—— *à ciel ouvert,* surface quarrying;
—— *en galerie,* quarrying by gallery;
pompe d'——, (*mach.*) discharging pump;
tuyau d'——, (*mach., etc.*) discharge pipe.
extradition, f., (*law*) extradition, surrender of a criminal;
traité d'——, treaty of extradition.
extrados, m., (*cons.*) extrados of an arch.
extradossé, a., (*cons.*) extradosed.
extradosser, v. a., (*cons.*) to make the extrados of an arch;
—— *en arc,* to make an extrados with a curve similar, but not parallel, to that of the intrados;
—— *en chape,* to build up the extrados ridgewise;
—— *de niveau,* to build up the extrados of an arch flat;
—— *parallèlement,* to make the extrados parallel to the intrados.
extra-dynamite, f., (*expl.*) extra dynamite.
extraire, v. a., to extract, draw out; to take a prisoner out of prison.
extrait, m., extract, abstract; certificate (of birth, etc.);
—— *d'actes,* (*adm.*) certified copy of any legal or other document.
extraordinaire, a., extraordinary; m., anything extraordinary;
—— *des guerres,* (*mil.*) contingent fund for extraordinary expenses of an army during war.
extrapontin, m., v. *estrapontin.*
extra-règlementaire, a., (*mil.*) not prescribed by the regulations.
extra-territorial, a., (*law*) exterritorial, extraterritorial.
extraterritorialité, f., (*law*) v. *exterritorialité.*

F.

fabricant, m., manufacturer.
fabrication, f., manufacture, make.
fabrique, f., manufactory, factory; works, mill; manufacture, make;
—— *d'armes*, small-arm factory;
prix de ——, manufacturer's price, cost at the factory.
fabriquer, v. a., to construct, make, manufacture.
façade, f., front, face, façade;
droit de ——, frontage.
face, f., face; surface; aspect; flat (of a sword, etc.); (*fort.*) front (of a work); flank (of a caponier); face (of a bastion); (*hipp.*) face; (*mil.*) face, facing, faced (at drill);
—— *d'appui*, bearing face;
—— *en arrière*, (*mil.*) face, faced, to the rear; to the rear (infantry skirmishers);
—— *d'assise*, seat or surface of any piece, etc., on which piece rests;
—— *en avant*, (*mil.*) face, faced, to the front;
belle ——, v. *belleface*;
—— *carrelée*, (*sm. a.*) checkering of the comb of a hammer;
de ——, in front, front;
—— *à droite (gauche)*, (*mil.*) faced to the right (left); (*fenc.*) right (left) volt, (bayonet exercise;)
en —— *de*, in front of;
faire —— *à*, to face; to oppose (as, an enemy);
faire —— *de tous côtés*, (*mil.*) to face in all directions;
—— *de lame*, (*sm. a.*) flat of the blade;
—— *libre*, (*fort.*) end face (i. e., not attached to or running into any other part of the work);
—— *motrice (du piston)*, (*mach.*) steam side or end of a piston;
—— *de puits de mine*, (*mil. min.*) side of shaft;
—— *résistance (du piston)*, (*mach.*) exhaust side or face of a piston;
—— *de travail*, (*met., etc.*) working face or side.
facette, f., facet; (*art.*) seat (flat surface) of gunner's quadrant.
façon, f., make, form, shape, workmanship; operation; care, attention;
bois à ——, dimension, lumber.
façonnage, m., shaping, forming; (in Fr. sm. a. manufactories) work along a curve, either open or closed.
façonner, v. a., to form, model, mold; to make, work, fashion; to figure, shape, construct; to face; to finish off; to accustom.
factage, m., carriage, porterage, delivery;
compagnie de ——, parcels delivery company.
facteur, m., factor; carrier; agent; railway porter; postman;
—— *parasite*, numerical factor necessary to pass from one system of units to another; converting or reducing factor;
—— *de sécurité*, factor of safety.
factice, a., sham, artificial, mock;
guerre ——, (*mil.*) sham fight; maneuvers.
faction, f., faction; (*mil.*) standing sentry, sentry duty, sentry go;
capote de ——, watch coat; buffalo coat (U. S. A.);
en ——, on sentry, on post;
entrer en ——, to go on sentry or post;
être en ——, to be on sentry duty;
faire ——, to be on sentry, stand sentry, to do sentry go;
mettre en ——, to post a sentry;
monter la ——, to go on sentry or post, to walk post, do sentry go;
prendre la ——, to go on post;
relever la ——, to relieve sentry;
remettre la ——, to come off post;
sortir de ——, to come off post.
factionnaire, m., (*mil.*) sentry, sentinel; soldier detailed for any service;
—— *de la portière*, (*nav.*) gangway sentry;
poser un ——, to post a sentry;
relever un ——, to relieve a sentry.

factum, m., case; statement of a case.
facture, f., invoice, bill;
—— *des comptes en deniers*, (*adm.*) money bill, bill for services rendered, for goods delivered, etc.;
—— *des comptes en matières*, invoice of material shipped, transferred, delivered, etc.;
—— *d'expédition*, invoice of goods shipped;
—— *de livraison*, invoice of goods delivered in person;
—— *à talon*, invoice with stub.
faculté, f., power, ability;
—— *giratoire*, (*nav.*) turning power of a ship.
fagot, m., fagot; (*mil.*) fagot, man falsely mustered as a soldier; (*art.*) packet or bundle of powder in making up cartridges; (*fort.*) sap fagot, fagot; fascine; (*met.*) fagot;
—— *d'allumage*, kindling wood;
âme de ——, core of a fagot, made of the smallest sticks;
—— *ardent*, (*artif.*) tarred fascine, for incendiary purposes;
bois de ——, bavin;
châtrer un ——, to take a few sticks from a fagot;
—— *goudronné*, v. —— *ardent*;
lien de ——, fagot band;
parement de ——, bavin, large stick of a fagot;
petit ——, small fagot;
—— *de sape*, (*fort.*) sap fagot or fascine.
fagotage, m., fagot making; fagot wood.
fagotaille, f., (*fort.*) brushwood revetment.
fagoter, v. a., to make up into fagots; to fagot.
faguette, f., (*siege*) small fascine.
faible, a., weak, feeble, slight, slender;
—— *d'effectif*, weak in numbers (of an army, etc.);
front ——, narrow front (of troops, formations, etc.).
faible, m., (*sm. a.*) part of sword blade near point.
faiblesse, f., weakness.
faiblir, v. n., to weaken, lose strength; to die down (of the wind); (of troops) to lose courage.
faille, f., (geological) fault.
failli, m., bankrupt.
faillir, v. n., to fail, to become bankrupt.
faillite, f., failure, bankruptcy.
faim, f., hunger; famine;
réduire par la ——, (*mil.*) to starve into surrender.
faim-calle, f., v. *faim-valle*.
faim-valle, f., (*hipp.*) hungry-evil (balk caused by hunger, and cured only by eating).
faire, v. a., to make, to do, to construct; (*mil.*) to execute (a maneuver);
—— *abatage*, to raise a weight by means of a fulcrum and lever;
—— *l'appel*, (*mil.*) to call the roll;
—— *des, les, armes*, (*fenc.*) to fence;
—— *assaut*, (*mil.*) to storm, assault;
—— *l'avant-garde*, (*mil.*) to be the advance guard; to do advance-guard work or duty;
—— *du bifteck*, (slang) to become galled from riding;
—— *du bitord*, (*cord.*) to make spun yarn;
—— *du bois*, (*nav.*) to take wood on board;
—— *bonnes garcettes*, (*cord.*) to nip or lash well together;
—— *des bordées*, (*nav.*) to tack;
—— *branle-bas*, (*nav.*) to clear for action;
—— *brèche*, (*fort.*) to breach;
—— *capot*, (*nav.*) to turn upside down, to turn turtle;
—— *le charbon*, (*nav.*) to take in coal;
—— *ciseaux*, (*hipp.*) to move the jaw incessantly;
—— *cône*, (*mach.*) to taper;
—— *le coude*, to make a turn, elbow, or bend;
—— *le coup de fusil*, v. s. v. *coup*;

faire *dormant*, (*cord.*) to make fast (to make standing fast);
—— *eau*, (*nav.*) to leak, be leaky;
—— *de l'eau*, (*nav.*) to take in water;
—— *l'école*, (*mil.*) to drill, instruct;
—— *faction*, (*mil.*) to be on post;
—— *ferme*, to hold out;
—— *feu sur*, (*mil.*) to fire, to open fire, on;
—— *des feux*, to hang out lights as signal of distress;
—— *force de rames*, (*boats*) to pull hard;
—— *les forces*, v. —— *ciseaux;*
ne —— *pas de force*, to put no strain on; to bear no strain;
—— *la garde*, (*mil.*) to mount, to be on, guard;
—— *grenier*, (*hipp.*) to keep bits of food between the molars and the cheeks;
—— *la guerre*, (*mil.*) to make war, to fight;
—— *la guerre de*, (*mil.*) to serve (generally **in past**) in a particular war;
—— *le guet*, to be on the lookout;
—— *long feu*, (*sm. a. art.*) to hang fire;
—— *magasin*, v. —— *grenier;*
—— *la main*, to make the hand skillful;
se —— *la main*, to get one's hand in;
—— *main basse*, to put to the sword;
—— *l'office*, (*mil.*) to perform divine service;
—— *paquet*, (*art.*) to hold well together in the trajectory (of bullets from shrapnel, case, etc.);
—— *partir*, to set off, to discharge (a rocket, mine, shot, etc.);
—— *pavillon*, (*nav.*) to hoist, show, colors;
—— *pavois*, (*nav.*) to dress ship;
—— *la planche*, to float (swimming);
—— *plus loin*, (*art.*) to elevate;
—— *plus près*, (*art.*) to depress;
—— *prise*, (of mortar, cement, etc.) to set;
—— *quarantaine*, —— *la quarantaine*, to perform quarantine;
—— *quartier*, to spare, to give quarter;
—— *des recrues*, (*mil.*) to recruit;
—— *revenir* (*l'acier*), (*met.*) to temper, anneal;
—— *la révérence*, (*man.*) to stumble, trip;
—— *la revue de*, (*mil.*) to review; to inspect (accounts);
—— *la ronde*, (*mil.*) to go the rounds, to make rounds;
—— *rougir*, (*met.*) to heat, heat red-hot;
—— *la route*, (*mil.*) (of a march) to cover or make the distance between two points;
—— *sentinelle*, (*mil.*) to be on sentry go, to do sentry go;
—— *des signaux*, to signalize, make signals;
—— *des soldats*, (*mil.*) to raise troops;
—— *la soupe*, (*mil.*) to make or take a meal;
—— *taire*, (*art.*) to silence (a battery);
—— *son temps*, (*mil.*) to serve one's time;
—— *tête (à)*, to oppose, resist; (*cord.*) to become taut or stiff;
—— *travailler à compression*, etc., to subject to a stress of compression, etc.;
—— *les vivres*, (*nav.*) to put stores on board;
—— *voile*, (*nav.*) to set sail;
—— *voile sur*, (*nav.*) to make sail for.

faisceau, m., bundle, fagot, sheaf; truss, pile; (in pl.) *fasces;* (*sm. a.*) arm rack; stack of arms;
—— *aimanté*, (*elec.*) compound magnet;
—— *d'armes*, (*mil.*) stack of arms; arm rack; stakes for supporting colors;
—— *du canon-revolver*, (*r. f. art.*) assemblage or sheaf of the five barrels of the (Hotchkiss) revolving cannon;
en ——, in a pile, piled up;
former les ——*x*, (*sm. a.*) to make stacks;
—— *de lumière*, —— *lumineux*, pencil of light;
mettre en ——, (*sm. a.*) to stack arms;
rompre les ——*x*, (*sm. a.*) to break stacks;
—— *de trajectoire*, (*ball.*) sheaf of trajectories.

fait, a., settled; grown; full-grown; fit, qualified; steady (said of weather and of wind);
cheval ——, full-grown horse broken in.

fait, m., fact, deed, exploit;
—— *administratif*, administrative act or operation;
—— *d'armes*, (*mil.*) distinguished feat of arms;
de ——, de facto;

fait *disciplinaire*, any act calling for disciplinary control or repression;
—— *de guerre*, act of war; (*mil.*) combat;
—— *guerrier*, presonal feat of arms;
—— *juridique*, any act or operation calling for action by the courts;
haut ——, high crime; (*in pl., mil.*) feats of arms;
voie de ——, violence, assault, blows.

faîtage, m., (*cons.*) ridge; ridge lead; roof timber; ridge board of a roof, ridgepiece;
—— *double*, under ridgepole.

faîte, m., ridge (of a house); top, coping, summit; (*top.*) watershed, summit level; watershed between two thalwegs;
ligne de ——, (*top.*) crest line, line joining the various summits of a chain of heights.

faîtière, a., belonging to a ridge; f., (*cons.*) ridge; ridgepiece; ridge of gutter tile; ridgepole;
lucarne ——, skylight;
tuile ——, ridge tile.

faix, m., burden, load, weight;
prendre son ——, to settle, to take its set.

falaise, f., (*top.*) cliff; shore cliff, bluff; headland.

falcade, f., (*man.*) falcade; quick curvet.

falot, m., (*mil.*) lantern used by patrols, guards, etc.; (*mil. slang*) military cap;
—— *de ronde*, (*mil.*) guardhouse lantern (for making rounds).

falourde, f., fagot of thick sticks.

falque, f., washboard of a boat; (*man.*) falcade.

falquer, v. n., (*man.*) to execute falcades.

falsificateur, m., falsifier, one who adulterates.

falsification, f., adulteration; forgery.

falsifier, v. a., to adulterate; to debase (metal);
—— *les allures d'un cheval*, (*man.*) to spoil a horse's gaits.

famine, f., famine;
prendre par la ——, (*mil.*) to reduce by famine.

fanal, m., lantern; beacon; signal light; light, beacon light; blaze or flame; light-house;
—— *de signaux*, (*mil.*) signal lantern;
—— *sourd*, dark lantern;
—— *de soute*, (*mil.*) magazine lantern;
—— *pour soutes à poudre*, v. —— *de soute.*

faner, v. a., to turn (grass, etc., in order to cure it for hay).

faneur, m., haymaker.

fanfare, f., brass band (cavalry and chasseurs à pied); flourish of trumpets.

fange, f., mud, mire.

fangeux, a., muddy, miry.

fanion, m., pennon, flag; (*mil.*) distinguishing flag, camp color, headquarters flag; flag for target practice;
—— *d'alignement*, flag for dressing line of troops; marker's flag;
—— *d'ambulance*, Red Cross flag;
—— *de commandement*, headquarters flag;
—— -*éventail*, swallow-tailed flag;
—— *en forme de flamme*, swallow-tailed flag;
—— *en forme de pavillon*, rectangular flag;
—— *de lance*, lance pennon;
porte- ——, color bearer;
—— -*signal*, signaling flag;
—— *de tir*, (*t. p.*) red flag used for various purposes at target practice; e. g., danger flag.

fanon, m., pennon; (*hipp.*) fetlock.

fantabosse, } (*mil. slang*) infantryman.
fantasboche, }

fantassin, m., (*mil.*) foot soldier, infantryman.

fantôme, m., ghost;
—— *magnétique*, (*elec.*) magnetic curve, figure.

faquin, m., wooden dummy for various military exercises.

farad, m., (*elec.*) farad.

farcin, m., (*hipp.*) farcy, glanders;
corde de ——, farcy knot, bud, or cord.

farcineux, a., (*hipp.*) farcied, glandered; affected with glanders;
bouton ——, farcy bud.

fardeau, m., load, burden;
   passer le ——, (*art.*) in mechanical maneuvers, to lift (a weight, gun) over (ns, a wall).
farder, v. n., to sink, sink in (of a wall); to run or go too close to (one boat to another in a river).
fardier, m., dray for carrying stone.
fargue, f., washboard of a boat.
farine, f., flour; powder or meal produced by trituration;
   —— *de bois*, (*expl.*) sawdust;
   —— *fossile*, infusorial earth, kieselguhr;
   —— *de montagne*, v., ——*fossile.*
farouche, a., shy; shying; ferocious.
fascicule, m., part, number (of a serial publication); (*Fr. a.*) sheet sewed in the *livret individuel* of a soldier quitting the active army and containing certain documents (*ordre de route* in case of mobilization, and other data).
fascinage, m., brush-revetting (of a river bank); (*fort.*) fascine-making, fascine work; (*field fort.*) generic term for gabion, fascine, and hurdle work.
fascination, f., (*fort.*) fascine revetment.
fascine, f., fagot; bavin; hurdle; (*fort.*, *siege*) fascine; bundle of small branches for fascines;
   —— *d'ancrage*, *à ancre*, anchoring fascine;
   —— *de blindage*, covering fascine;
   —— *de couronnement*, trench fascine;
   —— *à farcir*, filling fascine;
   —— *fondrière*, —— *à fossé(s)*, water fascine, sinking fascine;
   —— *goudronnée*, (*artif.*) pitched, tarred fascine for inflammatory and illuminating purposes;
   —— *de gravier*, fascine filled with gravel (hydraulic engineering);
   —— *de pierre*, fascine filled with broken stone (hydraulic engineering);
   —— *de premier rang*, fascine of the lowest course;
   —— *de retraite*, securing fascine;
   —— *à revêtir*, revetting fascine;
   —— *à tracer*, tracing fascine.
fasin, m., mixture of earth, ashes, and small twigs.
fatigue, f., fatigue, weariness; toil, hardship; fatigue of metals, etc. (i. e., strain, long-continued strain);
   cheval de ——, horse used for the hardest kind of work.
fatigué, a., fatigued; strained; (*cord.*) worn, fagged.
fatiguer, v. a. n., to fatigue, tire, weary; to try, strain; to labor, work, overwork; to be strained (said of springs when weakened, of the parts of a mechanism).
fau, m., beech tree.
faubert, m., (*nav.*) mop, swab.
fauberter, v. a., (*nav.*) to mop, swab.
fauberteur, m., (*nav.*) swabber.
faubourg, m., suburb, outskirt (of a town).
fauchage, m., mowing.
fauchaison, f., mowing time.
fauchant, m., sweeping, oscillating motion (of machine guns).
fauche, f., a mowed crop; mowing time.
faucher, v. a. n., to mow, to reap; to throw the legs sidewise in walking; (*hipp.*) to dish (said also of a lame shoulder); (*mil.*) to sweep by fire, to mow down (as, by skirmishes or by the horizontally oscillating motion of a machine gun).
fauchère, f., (*harn.*) wooden rod; crupper for pack mules.
faucheur, m., mower.
faucille, f., reaping hook, sickle.
fauciller, v. a., to cut with a sickle.
faucillon, m., billhook;
   bois ——, *bois à* ——, brushwood.
fauconneau, m., jib of a crane.
fauconnier, m., *monter à cheval en* ——, (*man.*) to mount a horse on the off side.
fauconnière, f., saddlebag.
faufil, m., basting thread.
faufilage, m., basting.
faufiler, v. a., to baste;
   se ——, to creep in or through.
faufilure, f., basting, act of basting.

faulde, f., charcoal pit;
   charbon en ——, charcoal made in open air for common use (unfit for powder manufacture).
faune, f., fauna.
faussage, m., v. *faussement.*
faussé, a., (*sm. a.*, etc.) sprung' bent;
   —— *à court pli*, sharply bent or sprung;
   —— *à long pli*, sprung over a considerable length.
fausse aiguille, v. s. v. *aiguille.*
fausse(-)alarme, v. s. v. *alarme.*
fausse(-)alerte, v. s. v. *alerte.*
fausse âme de chargement, v. s. v. *âme.*
fausse-axe, f., pattern spindle.
fausse-bande, f., (*harn.*) false saddle-bar.
fausse-boîte, f., —— *de roue*, (Fr. *art.*) nave box on spare-wheel axle, 95^{mm} gun.
fausse(-)braie, f., (*fort.*) low rampart; false bray;
   —— *détachée*, detached, advanced, false bray.
fausse-cheminée, f., sheet-iron stovepipe.
fausse-coupe, f., v. s. v. *coupe.*
fausse-culasse, f., v. s. v. *culasse.*
fausse-équerre, f., v. s. v. *équerre.*
fausse-fusée, f., —— *d'exercice*, v. s. v. *fusée.*
fausse-gourme, f., v. s. v. *gourme.*
fausse-gourmette, f., v. s. v. *gourmette.*
fausse-lance, (*art.*) v. s. v. *lance.*
fausse-manche, f., v. s. v. *manche.*
fausse-marche, f., v. s. v. *marche.*
fausse-martingale, f., v. s. v. *martingale.*
faussement, m., (*sm. a.*) springing of a gun barrel, state of being sprung.
fausse-parois, f., v. s. v. *parois.*
fausse-pièce, f., v. s. v. *pièce.*
fausse-porte, f., v. s. v. *porte.*
fausse-poutrelle, f., v. s. v. *poutrelle.*
fausser, v. a. r., to bend, warp, break; to make or become crooked; to strain, force; to set, get wrong; (*mil.*) to be irregular in line, out of line (of troops that have lost the dress); (*sm. a.*, etc.) to spring, be sprung (of a gun barrel, etc.);
   —— *une lame*, (*sm. a.*) to spring, bend (said of a sword blade, in block test, when it fails to recover its first direction);
   —— *une serrure*, to force or spoil a lock.
fausse-rêne, f., v. s. v. *rêne.*
fausses-bottes, f. pl., v. s. v. *botte.*
fausset, m., faucet; spigot; pin.
faute, f., fault; (*mil.*) breach (of discipline);
   —— *contre la discipline*, (*mil.*) breach of discipline;
   —— *contre l'honneur*, (*mil.*) any breach of military honor.
fauteur, m., promoter, inciter (as, of a mutiny, of a riot, etc.).
fauve, a., drab, fair (of leather);
   bêtes ——*s*, deer.
faux, f., scythe; m., falsehood; forgery.
faux, a., false, artificial; sham, mock; imitation; counterfeit; blank; (*man.*) false, broken (of a gait);
   —— *sse aire*, (*cons.*) coarse material on which the true flooring (*aire*) is laid;
   —— *collier*, collar of a preventer stay;
   —— *sse côte*, (*hipp.*) false rib;
   être en porte-à- ——, v. *porter à* ——;
   faire ——*sse route*, to lose one's way;
   porter à ——, to have an overhang, to project from; to be out of the perpendicular;
   —— *sse position*, double position (in computations);
   —— *sse route*, error in one's reckoning, loss of one's way;
   —— *soldat*, (*mil.*) false muster.
faux-bouton, m., (*art.*) square, chuck square, cascable square.
faux-cadre, m., v. s. v. *cadre.*
faux-cercle, m., (*r. r.*) under hoop.
faux-châssis, m., v. s. v. *châssis.*
faux-côté, m., (*nav.*) list.
faux-coup, m., (*fenc.*) random thrust, miss.
faux-cul, m., v. s. v. *cul.*
faux-entrait, m., (*carp.*) top beam.
faux-étai, m., preventer stay.
faux-fond, m., v. s. v. *fond.*
faux-fourreau, m., v. s. v. *fourreau.*
faux-frais, m. pl., v. s. v. *frais.*

faux-frêt, m., v. s. v. *frêt.*
faux-fuyant, m., bypath.
faux-guindage, m., v. s. v. *guindage.*
faux-jour, m., v. s. v. *jour.*
faux-mantelet, m., v. s. v. *mantelet.*
faux-marqué, m., v. s. v. *marqué.*
faux-plancher, m., v. s. v. *plancher.*
faux-pont, m., v. s. v. *pont.*
faux-quartier, m., v. s. v. *quartier.*
faux-rais, m., v. s. v. *rais.*
faux-ranchet, m., v. s. v. *ranchet.*
faux-rond, m., v s. v. *rond.*
faux-sabord, m., v. s. v. *sabord.*
faux-siège, m., v. s. v. *siège.*
faux-tampon, m., v. s. v. *tampon.*
faux-titre, v. s. v. *titre.*
faux-tranchant, m., v. s. v. *tranchant.*
faveux, a., relating to favus.
favus, m., v. *teigne faveuse.*
fécule, f., fecula.
fédéral, a., federal.
fédéraliser, v. a., to federalize.
fédération, f., federation.
fédéré, a., federated, federal.
feindre, v. a. n., to feign; (*hipp.*) to be very slightly lame, to limp very slightly.
feint, a., sham, feigned, counterfeit, imitation.
feinte, f., feint, sham, pretense; (*hipp.*) slight halting or lameness; (*fenc.*) feint.
feld-maréchal, m., (*mil.*) field marshal.
feldspath, m., feldspath.
fêle, f., iron blowpipe.
fêler, v. a., to crack.
fêlure, f., crack; (*hipp.*) splitting of bone.
femelle, a., female; (*mach., etc.*) female (i. e., having a hollow or bore into which a correlative fits).
femme, —— de l'adjudant, (*mil. slang*) lockup, "jug;"
—— **de régiment,** (*mil. slang*) big (bass) drum.
fenderie, f., cutting, slitting; slitting mill.
fendeur, m., splitter, cleaver.
fendiller, v. r., to crack.
fendilies, f. pl., small cracks (in iron, etc.).
fendre, v. a. r., to split, crack, slit, cut open, burst, cleave, open, rend; to break through; to slot (a screw); (*fenc.*) to lunge out; to thrust the body forward; to open (attack); to thrust out;
—— **l'oreille,** (*mil. slang*) to place on the retired list;
se ——, (*art.*) to step off; to break;
se —— en arrière, (*art.*) to break to the rear (of the man who pulls the lanyard).
fenêtre, f., window (opening and panes); opening, in general; (*steam*) admission port; (*inst.*) hollow of a level; (*sm. a.*) rectangular opening in breech casing of the French model 1874, allowing extractor to enter its seat;
—— **en baie,** bay window;
châssis de ——, window frame;
—— **cintrée,** bow window;
condamner une ——, to block up a window;
fausse ——, sham window;
—— **en saillie,** v. —— *en baie.*
fenil, m., hayloft.
fente, f., crack, chink; slit, slot; hollow, guideway; cut, groove; opening, notch; rent; cranny, crevice, rip, etc.; (*fenc.*) lunge;
—— **de repère,** any adjusting or assembling mark, fiducial slit or mark (e. g., on a rifle barrel, to show when it is properly connected with the breech casing);
—— **de visière,** (*art.*) aiming slit (in turrets).
fenton, m., iron cramp or tie; iron rod used in bracing beams of iron floors; peg-wood; nail rod; small iron rod of square cross section.
fer, m., (*met.*) iron; (*tech., cons.*) tool; tool blade or cutter; cutting instrument or tool; ironwork, shoe of a pile; (*farr.*) horseshoe; (*in pl.*) fetters, chains, irons, bilboes, gyves, shackles; ironwork; iron or steel furnished by the trade under fixed and determined shapes; (*sm. a.*) sword; point, head of a lance; (*fenc.*) foil.
    I. Metallurgy; II. Structural iron; III. Farriery; IV. Miscellaneous.
    I. Metallurgy:
—— **acéré,** steeled iron;

fer, aciéreux, steely iron;
—— **adoucir le ——,** to soften iron;
affinage de ——, refinement of iron;
—— **affiné,** refined pig;
—— **aigre,** brittle iron;
—— **ballé,** iron that has passed once through the rollers, raw iron brought to a white heat and drawn out under the rollers;
battre le —— à froid, to cold-hammer;
—— **battu,** wrought iron;
—— **blanc,** white cast iron;
—— **bocard,** iron made from earthy ore;
—— **au bois,** charcoal iron;
—— **brûlé,** burnt iron;
—— **brut,** crude iron; esp. the bloom when formed into bars under the hammer or the rolls (muck bar);
—— **carbonaté,** spathic iron;
—— **carbonaté, argileux,** spherosiderite;
—— **carbonaté lithoïde,** spherosiderite;
—— **carbonaté spathique,** siderite, spathic iron ore;
—— **de cartilage,** soft iron;
—— **cassant,** short, brittle iron;
—— **cassant à chaud,** hot short, red short, iron;
—— **cassant à chaud noir,** black short iron;
—— **cassant à froid,** cold short iron;
—— **cassant à froid et à chaud,** iron short under all circumstances;
—— **à la catalane,** iron made by the Catalan process;
—— **cémenté,** cement iron, steely iron;
—— **cendreux,** weak, flawy iron;
—— **au charbon de bois,** charcoal pig (iron);
—— **chromaté minéral,** chromate of iron;
corroyer le ——, to weld iron, to forge iron;
—— **coulé,** pig-, cast-, iron;
—— **cru,** crude iron, pig iron;
—— **cylindré,** rolled iron;
cylindrer le ——, to roll iron;
débris de ——, scrap iron;
—— **doux,** soft iron, malleable iron, soft wrought iron;
—— **ductile,** ductile iron;
—— **dur,** fine-grained iron, brittle iron;
—— **écroui,** hammer-hardened iron;
—— **encrené,** iron twice welded;
—— **étiré,** rolled iron;
—— **de ferraille,** scrap iron;
—— **fibreux,** fibrous iron;
—— **fin,** refined iron, puddled iron, charcoal iron;
—— **fin au bois,** charcoal iron;
fonderie de ——, iron foundry;
fondre le ——, to melt, cast, iron;
—— **fondu,** cast iron, ingot iron;
—— **fondu à air chaud,** hot-blast iron;
—— **fondu à air froid,** cold-blast iron;
—— **fondu en coquilles,** case-hardened cast iron;
—— **de fonte,** cast iron;
—— **de forge, —— forgé,** wrought, forged, hammered, iron;
forger le ——, to forge iron; to work iron under the hammer;
—— **fort,** wrought iron neither cold nor hot short;
—— **à la française,** iron reduced by a process almost identical with the Catalan process;
—— **galvanisé,** galvanized iron;
—— **à grain,** granular iron;
—— **gratté,** iron whose outer black crust has been removed by filing or brushing;
—— **gris,** gray iron, gray pig;
—— **à gros grains,** coarse-grained iron;
—— **en gueuse,** pig iron;
—— **homogène,** homogeneous iron;
—— **laminé,** rolled iron, laminated iron;
laminer le ——, to roll iron;
limaille de ——, iron filings;
—— **limé,** iron bastard filed;
lopin de ——, ball, lump, of iron; bloom;
—— **à loupe,** bloom iron;
—— **en loupes,** iron blooms;
—— **de lune,** iron brought to white heat;
—— **magnétique,** magnetic iron;
—— **malléable,** malleable iron;
maquette de ——, half-forged bloom;

fer *marchand*, merchant iron, bar iron;
— *martelé*, hammered iron;
— *à massiau*, bloom iron;
— *mêlé*, cast iron between white and gray;
— *météorique*, meteoric iron;
— *métis*, red short iron;
— *micacé*, micaceous iron ore;
— *de mine limoneuse*, muddy ore;
— *de minerai en roche*, stony ore;
— *de minerai terreux*, earthy ore;
— *mitis*, alloy of iron and aluminium;
— *mou*, soft iron;
— *natif*, native iron;
— *à nerf*, fibrous iron;
nerf de ——, fiber of iron, fibrous texture;
— *nerveux*, fibrous iron;
— *noir*, black cast iron; sheet iron for making tin plate;
— *oligiste*, oligist iron ore;
— *oligiste micacé*, micaceous specular iron;
— *oxyde rouge*, red oxide of iron;
— *oxydulé*, magnetic iron ore, magnetite;
— *pailleux*, weak iron, iron full of flaws;
pièce de ——, piece into which bloom is divided for working;
— *platiné*, rolled iron;
— *puddlé*, puddled iron;
— *raffiné*, refined iron;
riblons de ——, scrap iron;
— *rouverin*, red short iron;
— *en saumon*, pig iron;
scorie de ——, iron dross;
— *soudé*, welded iron, weld iron;
souder le ——, to weld iron;
— *spathique*, spathic iron;
— *spéculaire*, specular iron;
— *tenace*, ductile, tough, iron;
— *tendre*, cold short iron;
— *touché*, "aired" bar (cementation process);
— *en tournure*, iron turnings;
usine de ——, ironworks, smelting works;
— *zingué*, galvanized iron.
II. Structural iron, etc.:
— *d'angle*, angle iron;
— *aplati*, hoop iron;
— *de bandage*, tire iron;
bande de ——, iron strap;
— *en bandes*, hoop iron;
barre de ——, iron bar;
— *en barres*, bar iron;
— *en berger*, bar iron;
— *à biseau*, wedge-shaped iron;
— *en bottes*, bar iron;
— *à boudin*, bulb iron;
— *à C*, C-iron;
— *à clou*, nail iron; nail plate;
— *conique*, ∧ iron (*fer zorès*);
— *à cornière*, angle iron;
— *creux*, v. ——*à boudin*;
— *à croix*, + iron;
— *demi-rond*, half-round iron;
——*s ébauchés*, forgings;
— *ébroudi*, iron wire;
——*s échantillonés*, iron bars, rods, of definite cross section, and according to specifications or model (includes iron wire);
— *à écrou*, screw plate;
— *émaillé*, enameled iron;
équerre de ——, iron knee;
— *façonné*, special iron;
— *fendu*, slit iron, nail rod;
— *feuillard*, strap iron, hoop iron;
— *en feuilles*, sheet iron;
fil de ——, iron wire;
——*s forgés*, iron that has been more or less shaped and prepared for use;
gros ——*s*, iron used in main or heavy work, and but roughly shaped;
— *à l'H*, *en H*, double-T iron, channel iron;
— *en I*, I-iron;
— *en lames*, sheet iron;
— *à la mode*, v. —— *aplati*;
— *à nervure*, flanged iron;
— *ordinaire à plancher*, I-iron;
plaque de ——, iron plate;
— *en plaques*, boiler plate;
— *plat*, hoop iron;
——*s profilés*, v. ——*s spéciaux*;

fer *de roulage*, thick iron wire;
— *roulé*, rolled iron;
— *en ruban*, hoop iron;
——*s spéciaux*, iron of special cross section; i. e., U, T, I, V, L, etc., iron;
— *à té*, T iron;
— *à triple nervure*, double-I beam;
— *en T*, T iron;
— *à double T*, *en T double*, channel iron, H iron, double-T iron;
— *tubulaire*, ∩ iron (*fer zorès*);
— *en*, *à*, ∩, ∩ iron;
— *en V*, V-iron;
— *en* ∩, ∩ iron;
— *zorès*, ∩-, ∧-shaped beam iron.
III. Farriery.
— *anglais*, a shoe without projecting border, and grooved on the lower face;
— *à anneaux*, ringed shoe;
— *à bec*, shoe with prolongation on toe;
— *à bords renversés*, shoe with edges turned down;
— *broché*, horseshoe put on;
— *à caractère*, irregular shoe, for special shape of foot; shaped shoe;
— *charlier*, v. —— *périplantaire*;
— *à charnière*, hinged shoe;
— *à*, *de*, *cheval*, horseshoe;
— *à cheval crampponné*, calked shoe;
— *couvert*, wide shoe;
crampon de ——, calk, calkin;
— *à crampons immédiats*, shoe with calkins, either fixed or movable;
— *à crampons médials*, a removable horseshoe, fitted with calkins;
— *à croissant*, half-shoe;
— *dégagé*, narrow shoe;
— *demi-couvert*, shoe wider and thinner than the (French) regulation shoe;
demi- ——, half-shoe;
— *à demi-pantoufle*, v. —— *à pantoufle modifiée*;
— *de derrière*, hind shoe;
— *de devant*, fore shoe;
— *droit*, off shoe;
— *à éponges réunies*, bar shoe;
— *à éponges tronquées*, v. —— *à lunette*;
— *étampé à gras*, shoe with nail-holes far from exterior border;
— *étampé à maigre*, shoe with nail-holes far from interior border;
— *à évidement* (lit., hollowed shoe), shoe to prevent slipping (as on ice);
faire porter le ——, to apply a shoe hot to the sole;
— *forgé à la mécanique*, machine-forged horseshoe;
— *gauche*, near shoe;
— *géneté*, shoe with heels turned up at right angles inside;
— *à glace*, frost shoe, calked shoe;
— *à gorge* (lit., grooved shoe), shoe to prevent slipping;
— *à lunette*, tip;
— *à mamelles tronquées*, interfering shoe;
— *à maréchal*, horseshoe iron;
mettre un ——, to shoe a horse;
— *à pantoufle*, panton shoe;
— *à pantoufle modifiée*, panton shoe with broad heels turned up inside so as to spread the heels;
— *à patin*, sliding shoe, ice shoe;
— *périplantaire*, a shoe let into, and flush with, the exterior surface of the hoof; countersunk shoe;
— *pinçard*, shoe with wide and thick toe, wide and high clip, and high calks;
— *à pince couverte*, shoe wide in the toe;
— *à pince prolongée*, long-toed shoe;
— *à pince tronquée*, hind shoe with truncated toe;
— *à pinçons*, clip shoe;
— *à planche*, bar shoe;
— *à planche à crampon longitudinal*, bar shoe with the rear edge of the bar turned down as a calkin;
— *à plaque*, shoe with a plate of metal, leather, etc., covering the foot, the sole;

fer *Poret*, a shoe without exterior border (thinner near the heel than near the toe, and has neither rights nor lefts);
—— *porte* ——, horseshoe case;
—— *pour pieds combles*, rocking shoe;
relever un ——, to calk or frost a horseshoe;
—— *de rivière*, broad-plate shoe (covering the entire sole except the frog);
—— *à saillie*, shoe fitted with a projection to prevent slipping;
—— *à la turque*, shoe with unequal branches;
—— *à une branche couverte*, shoe with one broad branch;
—— *à une ou deux éponges couvertes*, shoe with one or two broad heels;
—— *à une éponge tronquée*, shoe with one heel removed.

IV. Miscellaneous:
à —— *et à clou*, solidly, firmly;
battre le ——, to fence;
bois de ——, ironwood;
—— *de botte*, boot-heel plate;
—— *à, de, calfat*, calking iron, calking tool;
—— *à cheval*, double winding staircase; (*fort.*) horseshoe fort;
donner le ——, to set the tool (e. g., a rifling tool, after insertion);
—— *à donner, mettre, le feu*, v. s. v. *feu*;
en —— *à cheval*, crescent-shaped;
—— *de gaffe*, head of a boat hook;
—— *de girouette*, iron spindle of a vane;
gris de ——, iron-gray;
—— *de lance*, spearhead, lancehead, head;
manier le ——, to wear the sword;
—— *à marquer*, brand, branding iron;
mettre aux ——*s, un homme aux* ——*s*, to iron a man;
à mi- ——, (of rails, etc.) halved and lapping;
porter la flamme et le —— *dans*, to lay waste with fire and sword, carry fire and sword into;
——*s de prisonnier*, chains, shackles, fetters, irons;
—— *à repasser*, flat-iron;
—— *à souder*, soldering iron; copper bill;
—— *travaillant*, reaming iron;
—— *à vitrage*, window-pane framing made of iron.

fer-blanc, m., (*met.*) tin, tin-plate;
—— *brillant*, iron covered with pure tin;
—— *terne*, iron covered with one-half tin and one-half lead, or one-third tin and two-thirds lead.

ferblanterie, f., tinwork, tinware.
ferblantier, m., tinner, tin-plate worker.
féret, m., ferret; (*met.*) red hematite iron ore.
férir, v. a., to strike; (used only in) *sans coup* ——, without striking a blow.
ferlage, m., (*nav.*) furling.
ferler, v. a., (*nav.*) to furl.
fermage, m., rent; (*mil.*) rent of grass on slope of works.
fermail, m., clasp.
fermant, a., closing; (fitted) with lock and key;
à jour ——, v. s. v. *jour*.
ferme, a., adv., firm, steady; unshaken; fixed, fast, steadfast, strong, stiff; (in commands) steady! heave! pull together!
achat (*marché, vente*) ——, binding purchase, contract, sale (involving certain delivery);
—— *contre prime*, option;
être —— *à cheval, sur ses étriers*, (*man.*) to have a steady seat;
faire ——, (*mil.*) to make a stand;
marché à prix ——, v. s. v. *marché*;
de pied ——, (*mil.*) at or from a halt; on a fixed pivot;
sauter de —— *à* ——, (*man.*) of a horse, to jump without advancing;
tenir ——, hold one's ground, stand fast;
terre ——, terra firma.
ferme, f., farm, farmhouse; lease; (*cons.*) girder, trussed girder; couple, truss, rib, frame, roof truss; pair of principal rafters; main couple; timber frame; centering frame; largest timber on a floor;
à ——, on lease;

ferme *avec arbalétriers munis de contre-fiches*, roof with single-trussed principals;
—— *cintrée*, arched truss;
—— *à* (*une*) *clef pendante*, king-post truss;
—— *à clefs pendantes*, queen-post truss;
—— *à contre-fiches*, roof truss with trussed principals;
—— *courbe*, polygonal roof;
demi- ——, v. *demi-ferme*;
—— *à deux contre-fiches pour chaque arbalétrier*, roof truss with double-trussed principals;
—— *à deux aiguilles pendantes*, queen-post truss;
—— *droite*, roof truss with principals at right angles; simple truss of rectilinear profile (as distinguished from one of curved profile);
—— *droite à deux pannes*, purlined truss, principals at right angles;
—— *droite à entrait*, tiebeam truss or frame, with principals at right angles;
—— *droite à entrait à 2* (*ou 3*) *pannes*, purlined tiebeam, frame or truss, with principals at right angles;
—— *à écharpes*, scissors truss;
—— *à entrait*, tiebeam truss;
—— *de faîte*, ridgepole bracing or frame;
—— *à jambes de force*, roof truss resting on struts (of which the mansard roof is an example);
maîtresse- ——, main truss;
—— *à la Palladio*, a variety of tiebeam truss (frequently used by Palladio); Palladio truss;
—— *à pans inégaux*, truss of unequal principals;
—— *à la Philibert de Lorme*, curved or arched roof truss;
—— *à plusieurs contre-fiches*, English roof truss;
—— *avec poinçon en dessus*, king-post truss;
—— *avec poinçon en dessous*, inverted king-post truss;
—— *à deux poinçons*, queen-post truss;
—— *à deux poinçons en dessous*, inverted queen-post truss;
—— *de remplage*, middle rafter;
—— *simple à poinçon*, simple king-post truss;
—— *simple avec arbalétriers sans contre-fiches*, roof with simple principals;
—— *à tas de bois*, truss in which the rafter is connected with tiebeam by blocks of wood notched into one another and bolted;
—— *triangulaire*, truss.
fermé, a., shut, closed; landlocked; (*law*) closed (of a sea, bay, etc.); (*fort.*) closed;
—— *à l'air*, air-tight;
—— *à l'eau*, water-tight;
—— *par les glaces*, icebound;
—— *à la gorge*, (*fort.*) with a closed gorge;
ouvrage ——, (*fort.*) closed work;
ouvrage mi- ——, (*fort.*) half-closed work (i. e., one whose gorge has a slight parapet, or palisades, etc.);
—— *entre les terres*, landlocked.
ferme-bouche, m., mouth closer, mouth-piece (as, of an air-supplying machine in mines, to be worn by miners).
ferme-circuit, m., (*elec., torp., etc.*) circuit closer;
—— *à boule*, (*torp.*) a ball - contact circuit closer;
—— *à cône*, (*torp.*) a cone-contact circuit closer.
fermer, v. a. r. n., to close, shut, lock, fasten, stop, clinch, bolt, pin, clasp; to bring two objects into one; to inclose; to close a port (by an embargo); to close (a draw in a bridge); (*elec.*) to close (a circuit); (*surv.*) to close; (*man.*) to complete (a figure);
—— *l'angle*, to place one's self so as to bring two objects in the same line;
—— *la communication*, to shut off;
—— *un cours d'assises*, (*mas.*) to lay the last stone of a course;
—— *un croc*, (*cord.*) to mouse a hook;
—— *la marche*, (*mil.*) to bring up the rear, to march in rear;

fermer *deux objets*, to bring two objects (as masts, etc.) in line;
— *les ports*, to close ports by an embargo;
— *à vis*, to screw down;
— *la voie*, (*r. r.*) to stop travel, to close the line;
— *une voûte*, (*mas.*) to put the keystone in place.

fermeture, f., closing, shutting; closing (of a port, of the gates of a fortress); fastening (as, of a window, etc.); (*art.*, *sm. a.*) fermeture, breech closure; system of breech closure; (*elec.*) closing of a current; (*fort.*) closing (as, of a gorge by some defensive work); (*surv.*) closing of a survey;
— *à barillet*, (*sm. a.*) revolving cylinder closure;
— *à bloc*, (*r. f. art.*, *sm. a.*) block fermeture;
— *à cadran*, (*F. m. art.*) breech fermeture, in which the console, after receiving the plug, moves to the right in the grooves of a plate (*cadran*) (obs.);
— *à charnière*, (*F. m. art.*) fermeture in which the console is fastened to the face of the breech by a hinge (obs.);
— *à clapet*, v. s. v. *arme*;
— *à coin*, (*art.*) wedge fermeture;
— *à coin arrondi*, (*art.*) rounded breech plug;
— *à coin cylindro-prismatique*, (*art.*) cylindro-prismatic wedge fermeture;
— *à coin prismatique*, (*art.*) prismatic wedge (rear face of block plane);
— *à. coin unique*, (*art.*) single-wedge fermeture;
— *à console*, (*art.*) any fermeture in which the breech plug, withdrawn, rests upon a console;
— *à coulisse*, (*F. m. art.*) fermeture in which block moves on a slide fixed to the face of the breech (obs.);
— *de culasse*, (*art.*, *sm. a.*) breech mechanism; breech closure; breech-closing mechanism;
— *à culasse glissante*, v. — *à verrou*;
— *à culasse platine*, (*sm. a.*) self-cocking breech closure;
— *à culasse tombante*, (*sm. a.*) falling block fermeture;
— *à culasse tournante*, (*sm. a.*) revolving or hinged breech mechanism;
— *à cylindre*, (*sm. a.*) bolt fermeture;
dé de —, (*sm. a.*) false breech of a proof barrel;
— *directe*, (*sm. a.*) straight-pull closure;
— *à double coin*, (*art.*) double-wedge fermeture (Kreiner) (obs.);
erreur de —, (*surv.*) error of closing;
— *à l'évacuation*, (*steam*) exhaust cut-off;
— *à filets interrompus*, (*art.*) interrupted screw fermeture;
— *à l'introduction*, (*steam*) cut-off;
— *à mouvement de rotation continu*, (*art.*) fermeture in which the breech is opened or closed by a continuous movement; one-motion breech fermeture;
— *à piston*, (*art.*) piston fermeture (Wahrendorf) (obs.);
— *de ports*, embargo;
— *des portes*, (*fort.*) closing of the gates (at nightfall, etc.);
— *rectiligne*, (*sm. a.*) straight-pull closure;
— *à segments hélicoïdaux*, (*art.*) breech screw with helicoidal slots;
— *à tabatière*, (*sm. a.*) hinged-block closure, with side axis;
— *à verrou*, (*sm. a.*) bolt fermeture; (*art.*) the Wahrendorf piston fermeture;
— *à verrou à mouvement combiné*, (*sm. a.*) turn-down pull fermeture (the usual case of modern military rifles);
— *à verrou à mouvement rectiligne*, (*sm. a.*) straight-pull fermeture (Mannlicher);
— *à vis*, (*art.*) screw fermeture; screw plug;
— *à vis et à charnière*, (*art.*) fermeture in which breech screw, etc., works on a hinge in opening the breech (obs.);
— *à vis à filets hélicoïdaux*, (*art.*) helicoidal breech fermeture;

fermeture *à vis à filets interrompus*, (*art.*) interrupted screw fermeture;
— *à volet*, (*art.*) system in which plug is supported, when breech is open, by a carrier ring (*volet*) instead of a plug tray;
— *à la Wahrendorf*, v. — *à piston*.

fermoir, m., clasp, snap; mortising chisel; (*sm. a.*) cut-off (of the Schmidt rifle);
— *à nez*, — *nez rond*, corner chisel.

ferrage, m., tiring (a wheel); ironing (a criminal); (*farr.*) horseshoeing, act of shoeing; farriery.

ferraille, f., scrap iron; iron (as in *bruit de ferraille*);
fer de —, scrap iron.

ferraillement, m., production of iron bits by wear in gearing, etc.

ferrailler, v. a., to fence clumsily or badly; to fight with the sword; to make a noise by clashing swords together; (*fig.*) to dispute.

ferrailleur, m., bully; junk dealer.

ferrant, a., pertaining to blacksmithing;
maréchal —, farrier, horseshoer.

ferré, p. p., ironbound, iron-shod; hooped with iron; metaled, ballasted (of roads);
bout —, end shod with iron, ferrule;
chemin —, metaled, macadamized, road;
eau —*e*, forge water;
— *à glace*, (*farr.*) roughshod; frosted, frost-nailed; calked;
route —*e*, metaled road;
soulier —, hobnailed shoe;
voie —*e*, railway.

ferrement, m., any iron tool or instrument; (*in pl.*) ironwork (as, of a building);
—*s d'assemblage*, iron straps, connecting straps;
—*s de protection*, guard plates, straps, etc.

ferrer, v. a., to bind, shoe, fit, etc., with iron; to shoe a horse; to put a shoe on a pile; to tire a wheel; to metal a road;
— *à chaud*, (*farr.*) to shoe a horse hot;
— *à éclisses*, to fish;
— *à froid*, (*farr.*) to shoe a horse cold;
— *à glace*, (*farr.*) to rough-shoe a horse, to calk.

ferrerie, f., ironwork; ironmongery.

ferret, m., (*unif.*) metallic pencil of an aiguillette.

ferretier, m., farrier's hammer.

ferrière, f., farrier's pouch.

ferro-aluminium, (*met.*) ferro-aluminium.

ferro-chrome, m., (*met.*) ferro-chrome.

ferro-manganèse, m., (*met.*) ferro-manganese, artificial spiegeleisen.

ferro-nickel, m., (*met.*) nickel steel.

ferronnerie, f., ironworks, iron warehouse; hardware.

ferro-silicium, m., (*met.*) ferro-silicon.

ferro-tungstène, m., (*met.*) ferro-tungsten.

ferrugineux, a., ferruginous; chalybeate.

ferrure, f., ironwork, iron mounting; ironwork in general (carriages, machinery, etc.); (*in pl.*) ironwork; small parts (especially in guns, etc.); (*farr.*) shoeing of a horse; manner of shoeing; (strictly) art of forging horseshoes; 4 shoes and 32 nails;
— *anglaise*, so-called English shoeing, v. *fer anglais*;
— *brute*, rough forging;
— *Charlier*, v. — *périplantaire*;
— *à chaud*, (*farr.*) shoeing hot;
— *de cheval*, (*farr.*) horseshoeing;
— *cornière*, (*cons.*) angle-iron work;
— *courante*, (*Fr. a.*) horseshoes in use;
demi- —, pair of spare shoes, with calks and nails;
— *ébauchée*, dimension iron (from forge);
— *façonnée*, ironwork in a finished state;
— *à froid*, (*farr.*) cold shoeing;
— *à glace*, (*farr.*) rough-shoeing, calking;
— *grattée*, rough-filed iron (crust removed);
— *normale*, (*farr.*) shoeing of a sound and normal foot;
— *orthopédique*, (*farr.*) special shoeing intended to remedy defects of bearing in the foot;
— *pathologique*, (*farr.*) special shoeing for deformed or diseased feet;

ferrure *périplantaire*, (*farr.*) system of shoeing in which the shoe is let into a groove around the hoof (Charlier system);
—— *Poret*, v. *fer Poret;*
—— *porte-armement*, (*art.*) generic term for all keys, bolts, etc., carrying equipments;
—— *de poulie*, iron block strap;
—— *de rechange*, (*Fr. a.*) spare shoes;
—— *de réserve*, (*Fr. a.*) reserve shoes kept in magazine;
—— *sans clous*, nailless shoeing.

fertier, m., farrier's shoeing hammer.

féru, p. p., (*hipp.*) struck.

fervoie, f., (*r.r.*) (proposed term for) railway.

fesses, f. pl., (*hipp.*) buttocks, rump.

fessier, m., (*hipp.*) muscle of the croup, gluteus.

fête, f., feast, fête;
—— *nationale*, in France, the 14th of July, anniversary of the capture of the Bastille;
—— *de Sainte Barbe*, feast of St. Barbara, patron saint of the artillery (4th of December).

feu, m., fire, flame; heat; conflagration; combustion; fireplace, fire irons, hearth, (hence, *fig.*) home; light, light-house; signal light; (*met.*) fireplace, hearth; fire; blast; (*mil.*) fire, firing; command to fire; (*in pl.*) the fires of an army; (*art.*) action; (*hipp.*) fire, firing.

I. Military; II. Miscellaneous.

I. Military:

à l'abri du ——, sheltered, protected, from fire;
—— *ajusté*, accurate fire;
aller au ——, to go under fire, into action;
—— *de l'amorce*, misfire;
—— *allongé*, (*art.*) fire over the heads of one's own people;
arme à ——, *arme à* —— *portative*, (*sm. a.*) firearm;
—— *en arrière*, (*art.*) action rear;
—— *d'artifice*, fireworks;
—— *d'artifice de guerre*, military fireworks;
—— *d'artillerie*, artillery fire;
—— *d'attaque*, (*sm. a.*) the fire delivered by a line on arriving at 350ᵐ of the hostile position (not regulation in France);
—— *en avant*, (*art.*) action front;
balle à ——, (*artif.*) fireball;
baptême de ——, baptism of fire, first exposure to fire;
—— *de bataillon*, fire of a (single) battalion;
—— *par bataillon*, fire by battalion;
—— *de batterie*, battery practice; fire by or from a battery;
—— *de Bengale*, (*artif.*) Bengal light, blue light;
—— *de bilbaude*, —— *de bilboque*, (*inf.*) independent firing (obs.);
—— *bien servi*, well-sustained fire;
—— *à blanc*, v. —— *à poudre;*
bouche à ——, piece of artillery, gun;
—— *à bout portant*, point-blank fire;
—— *de but en blanc*, (*ball.*) point-blank fire;
—— *de canon*, artillery fire;
—— *à cartouches comptées*, (*inf.*) firing of so many rounds, indicated in the command;
cesser le ——, to cease firing;
—— *de chaussée*, (so-called) street firing (obs.);
—— *collectif*, (*inf.*) the fire of a body, as opposed to individual firing;
—— *de colonne*, v. —— *de chaussée;*
—— *à commandement*, (*inf.*) fire by command;
commencer le ——, to begin firing;
conduire le ——, to direct the fire;
conduite du ——, fire direction;
—— *convergent*, converging fire;
—— *couché*, (*inf.*) fire lying down;
—— *de couleur*, colored light (signal);
coup de ——, shot; gunshot wound;
—— *coup par coup*, (*inf.*) firing with the magazine rifle used as a single-loader;
—— *courbe*, (*art.*) curved fire;
à couvert du ——, sheltered or protected from fire;
crête de ——, (*fort.*) interior crest;

feu(x) croisés, cross fire;
—— *à la débandade*, skirmish firing (obs.);
—— *debout*, (*inf.*) fire standing;
densité du ——, volume, intensity, density, of fire;
—— *direct*, (*art.*) direct fire;
discipline du ——, fire discipline;
—— *divergent*, divergent fire;
—— *de division*, (*inf.*) fire by division;
—— *droit*, straight fire from guns laid horizontally with full charge;
—— *d'écharpe*, oblique, slant fire;
—— *échelonné*, echeloned fire, fire by echelon;
école à ——, v. s. v. *école;*
—— *d'enfilade*, (*art.*) enfilade fire;
—— *de l'ennemi*, hostile fire;
entre deux ——*x*, between two fires;
—— *x d'ensemble*, (*inf.*) volley firing;
essuyer un ——, *un coup de* ——, to receive a fire, to be fired at;
—— *x d'étages*, —— *x étages*, tiered fire, fire from various tiers;
éteindre un ——, (*art.*) to silence (hostile) fire;
être en prise du —— *de*, to be exposed to the fire of;
exercice à ——, v. s. v. *exercice;*
extinction des ——*x*, v. s. v. *extinction;*
—— *de face*, front fire;
faire ——, to fire (of a body of troops), to discharge a piece (of a sentry, etc.);
faire —— *en chasse* (*retraite*) *directe*, (*nav.*) to fire straight ahead (astern);
faire le coup de ——, v. s. v. *coup;*
faire faux ——, to miss fire;
faire long ——, to hang fire;
faux ——, misfire;
faux ——*x*, signals made by burning small quantities of powder;
—— *fichant*, (*art.*) any fire giving an angle of fall greater than 4°;
—— *de file*, (*inf.*) file firing;
—— *fixe*, (*mil. sig.*) steady light (i. e., not a flash);
—— *de flanc*, —— *flanquant*, flanking fire;
—— *de front*, v. —— *de face;*
—— *de fusil*, musket shot, rifle shot;
—— *à genou*, (*inf.*) fire kneeling;
—— *grégeois*, Greek fire;
—— *x de guerre*, (*t. p.*) practice at silhouette targets, under service conditions, simulating as nearly as possible the conditions of battle; real (i. e., war) firing;
—— *à guillocher*, (*artif.*) double catharine wheel;
halte au ——, (*art.*) command for the temporary cessation of the fire of a battery;
—— *indirect*, (*art.*) indirect fire;
—— *individuel*, (*inf.*) individual fire, fire at will;
—— *d'infanterie*, small-arm fire;
lance à ——, (*art.*) port fire (obs.);
—— *lent*, (*Fr. art.*) fire at the rate of one round a minute;
ligne de ——, (*fort.*) interior crest;
—— *de ligne*, (*inf.*) line firing, fire from or by a line;
long ——, hang fire;
—— *de masse*, (*inf.*) fire in four ranks (first two kneeling);
mettre à —— *et à sang*, to lay waste with, to put to, fire and sword;
mettre entre deux ——*x*, to bring a cross fire to bear;
mettre le ——, to discharge, fire, a piece;
—— *meurtrier*, deadly fire;
mise de ——, discharge; firing mechanism;
—— *à mitraille* (*à obus*, *à shrapnel*, etc.), (*art.*) case (shell, shrapnel, etc.) fire;
—— *de mousqueterie*, v. —— *d'infanterie;*
ne point donner prise au —— *de*, not to be exposed to the fire of;
—— *nourri*, heavy fire, well directed and thoroughly kept up; (*Fr. art.*) fire at the rate of six rounds a minute;
—— *ordinaire*, (*Fr. art.*) fire at the rate of three rounds a minute;
ouverture du ——, opening, beginning, of fire, of firing;
ouvrir le ——, to open fire;

feu *de peloton,* (*inf.*) platoon firing, fire by platoon;
—— *par pièce,* (*art.*) fire by piece;
—— *de plein fouet,* (*art.*) direct fire with full service charges;
—— *plongeant,* (*art.*) plunging fire;
—— *x de polygone,* target practice, firing, on known range;
pot à ——, (*artif.*) fire pot (obs.);
—— *x à poudre,* blank-cartridge firing;
—— *de poursuite sur un ennemi,* running fire in pursuit;
—— *de, par, rang,* (*inf.*) fire by rank;
—— *de deux rangs,* (*inf.*) file firing;
—— *rapide,* (*inf.*) rapid fire; (*Fr. art.*) fire at the rate of 10 to 12 rounds per minute;
—— *rapide coup par coup,* (*sm. a.*) rapid fire, using the piece as a single loader;
—— *rapide à répétition,* rapid magazine fire (at close quarters);
—— *rasant,* (*art.*) grazing fire;
—— *x réels,* any firing or practice in which projectiles are fired;
—— *règlé,* fire that hits the target;
répartir le ——, (*art.*) to distribute the fire over a battery, to have the battery fire as a whole after getting the range;
répartition du ——, (*art.*) distribution of fire after getting the range;
—— *à répétition,* (*sm. a.*) magazine fire;
—— *en retraite,* (*art.*) prolonge firing (obs.);
—— *de revers,* reverse fire;
—— *roulant,* (*inf.*) running fire; uninterrupted small-arm fire;
—— *de, par, salve,* (*art.*) salvo firing; (*inf.*) volley firing;
—— *de salve à répétition,* (*sm. a.*) magazine volley firing;
—— *par salve de section* (*de batterie*), (*art.*) platoon (battery) salvo firing;
—— *de salve à volonté,* (*inf.*) salvos at captain's discretion;
—— *de signaux,* signal fire;
—— *x de signaux,* signal fireworks;
sous le ——, under fire;
soutenir le ——, to stand up well under fire;
soutenir un ——, to keep up (the) firing;
—— *de, en, tirailleurs,* (*inf.*) skirmish firing;
—— *vertical,* (*art.*) vertical fire (mortars);
—— *vif,* smart, brisk, fire;
voir le ——, to be engaged, to take part in a combat;
voir le —— *pour la première fois,* to be under fire for the first time;
—— *à volonté,* (*inf.*) fire at will; independent firing;
—— *à volonté et à cartouches comptées ou limitées,* fire at will up to so many rounds (indicated in the command).

II. Miscellaneous:
à l'abri du ——, fireproof;
accoutumer un cheval au ——, to make a horse gun wise;
activer le ——, *les* —— *x,* to rouse, urge, the fire (in a furnace);
—— *d'affinerie,* (*met.*) firing forge;
alimenter les —— *x,* to feed the fires;
—— *alternatif,* alternating light (lighthouse);
—— *anglais,* (*hipp.*) friction, rubbing, with a vesicating pomade or liquid;
attiser le ——, to poke, stir, the fire;
au ——! (alarm of) fire;
boîte à, de, ——, fire box;
—— *Catalan,* (*met.*) Catalan forge;
cheval de ——, synonym of *cheval-vapeur* (rare);
coffre à ——, fire chest;
coin du ——, fireside;
—— *x Coston,* Coston lights;
—— *x de côté,* (*nav.*) running lights; side lights;
coup de ——, defect due to overburning;
donner le —— *à un cheval,* (*hipp.*) to fire, sear, a horse;
doré au ——, fire gilt;
durcir au ——, to fire-harden;
—— *x d'éclairage,* lights;
—— *à éclats,* flash light (light-house);
—— *à éclipse,* intermittent light (light-house);

feu *à éclipse de minute en minute,* intermittent light (light-house) with one-minute intervals;
—— *électrique,* electric light-house;
employer le ——, (*hipp.*) to fire;
à l'épreuve du ——, fireproof;
faire la part du ——, to try to save something from a fire;
fer à donner, à mettre, le ——, (*hipp.*) canting-iron;
—— *fixe,* fixed, steady light (light-house);
—— *flottant,* light-ship;
garniture de ——, fire-irons;
—— *grisou,* fire damp;
—— *intermittent,* intermittent light (light-house);
jeter bas les —— *x,* to draw the fires;
—— *de joie,* bonfire;
mettre en ——, (*met.*) to fire, to put in blast;
mettre hors de ——, (*met.*) to put out of blast;
mettre le ——, v. *employer le* ——;
mettre ——, *le* ——, *à,* to set on fire;
mettre bas les —— *x,* v. *jeter bas les* —— *x;*
mettre les —— *x au fond* (*du fourneau*), (*steam*) to bank a fire;
mise en ——, (*met.*) firing, putting in blast;
mise hors de ——, (*met.*) putting out of blast;
—— *de navire,* v. —— *de côté;*
—— *d'oxydation,* oxidizing flame;
—— *de paille,* (*hipp.*) a spirited horse of no bottom;
petit ——, low fire; slow fire;
pompe à ——, fire engine; steam pump;
pousser les —— *x,* to urge, excite, the fires;
—— *de raffinerie,* (*met.*) refining hearth;
—— *de reconnaissance,* light-house of the first order;
—— *de réduction,* reducing flame;
rester sur les —— *x avec vapeur,* (steam) to bank the fires and keep steam up;
—— *de route,* v. —— *de côté;*
tache de ——, (*hipp.*) tan spot on a horse;
—— *tournant,* revolving light (light-house).
feuillard, m., hoop wood; hoop iron;
—— *de fer,* iron hoops, hoop iron.
feuille, f., leaf; sheet (of metal, of pasteboard, paper, etc.); plate; scale, spring, branch or blade of a spring; foil; veneer; list; waybill; (*mil.*) generic term for certain reports or documents;
à ——, with leaves, folding;
—— *antérieure,* (*sm. a.*) fore end of trigger guard;
—— *d'appel,* (*mil.*) muster roll;
—— *s de chou,* (*mil. slang*) infantry gaiters;
—— *de cuivre,* sheet copper;
—— *d'émargement,* (*Fr. a.*) officers' pay list, pay roll;
—— *d'étain,* tin foil;
—— *d'évacuation,* (*Fr. a.*) list or statement accompanying men sent singly or in a body from a military hospital;
—— *de fer,* (*met.*) iron plate;
—— *de journées,* (*Fr. a.*) sheet showing number of days present of officers and men, and of horses, as a basis for pay and allowances, of whatsoever nature;
—— *de journées individuelle,* (*Fr. a.*) a quarterly statement showing transfers, etc., during the quarter, and the deductions from the various money allowances (kept for n. c. o., corporals, and privates of regiments and establishments where the *masse individuelle,* q. v., is kept up);
—— *de journées numérique des chevaux,* (*Fr. a.*) quarterly statement showing daily allowances due in forage and in funds from the *masse d'entretien de harnachement et de ferrage;*
—— *de journées numérique des hommes de troupe,* (*Fr. a.*) quarterly statement showing the daily allowance in money due the n. c. o., corporals, and privates of the unit interested, and the allowances in kind due these men, and also the officers on the strength;
—— *de journées des officiers,* (*Fr. a.*) quarterly statement (of money allowances) of deductions, etc., for officers, and of allowances to certain n. c. o. (loss of clothes, etc.), to *enfants de troupe,* etc.;

feuille *maîtresse* ——, longest blade of a carriage spring;

—— -*minute*, (*top.*) sheet on which *minute* is to be drawn;

—— *du personnel de l'officier en campagne*, (*Fr. a.*) sort of personal report on officer's qualities, kept in addition to *feuillet* (q. v.) and taking the place of the latter on mobilization;

porter sur la ——, to insert, put, in the way-bill;

—— *postérieure*, (*sm. a.*) rear end of trigger guard;

—— *de poudre*, (*powd.*) flake of powder;

—— *de pontet*, (*sm. a.*) guard plate;

—— *de prêt*, (*Fr. a.*) numerical sheet or list drawn up by the captain every five days in order to draw the pay of the men; men's pay list;

—— *de rectification*, (*Fr. a.*) sheet on which an *intendant de corps d'armée* enters his decisions about errors noted by him in the matter of pay;

—— *rectificative*, interpolated sheet announcing corrections, amendments, etc.;

—— *de ressort*, plate, leaf, of a spring;

—— *de route*, (*Fr. a.*) voucher or order delivered to a soldier (officer or man) or to a detachment, traveling on duty, and showing the itinerary to be followed and the allowances due on the journey;

—— *de sauge*, (*hipp.*) point of a seton needle;

—— *de solde*, (*mil.*) pay sheet, pay rolls;

—— *de tôle*, plate of sheet iron;

—— *de vérification*, (*Fr. a.*) memorandum of observations (by *contrôleurs* or *intendants*) made in verifying accounts, inspecting stores, etc.;

—— *volante*, loose sheet (of paper, etc.).

feuillé, m., (*top. drawing*) foliage.

feuillée, f., leaves, branches; (*mil.*) camp hut made of branches of trees; camp latrine (surrounded by brushwood).

feuiller, v. a., to rabbet, groove.

feuilleret, m., rabbet plane.

feuillet, m., sheet (two pages); thin plate; thin stone or slate; paneling plank; saw;

—— *matricule*, v. *registre matricule;*

—— *s mobiles*, loose sheets on which are registered names of all men called to the colors;

—— *du personnel des officiers*, (*Fr. a.*) register kept by the lieutenant-colonel of a regiment (by the c. o. of a separate battalion, or the head of a *service*) of the punishments inflicted on officers, and on which are entered semi-annually notes on their private and professional conduct;

—— *à poing*, handsaw;

—— *de punitions*, (*Fr. a.*) register of punishments (of the men) kept in each company of infantry and in each regiment of artillery and of cavalry; defaulter sheet;

—— *de tir*, (*Fr. a.*) sheet in the *livret individuel*, setting forth the soldier's record at target practice.

feuilleter, v. r., to split into thin plates.

feuillure, f., rabbet, rebate; groove; recess, slit; (*sm. a.*) bolthead rim; space between rear surface of chamber and bottom of cu ette;

à tenon et ——, tongued and grooved.

feuquière, f., pack-saddle bow.

feutrage, m., wad, wadding.

feutre, m., packing, felt; (*harn.*) gall leather, piece of leather under buckles, etc., to prevent galling; safe; (*sm. a.*) leather on gun sling to keep button from coming in direct contact with wood of stock.

feutrer, v. a., to pack, pad (a saddle).

fève, f., bean; (*hipp.*) lampas, lampers·

—— *de cheval*, horse bean;

germe de ——, (*hipp.*) marks of (a horse's) mouth.

féverole, f., bean; horse bean.

nourri de ——*s*, bean-fed.

fiacre, m., cab, four-wheeler.

fibre, f., fiber, thread, string; grain (of wood);

—— *invariable*, v. —— *neutre;*

—— *libérienne*, bark fiber;

—— *ligneuse*, wood fiber;

fibre, *neutre*, neutral fiber.

fibreux, a., fibrous, stringy.

ficeler, v. a., to tie, bind; to wire;

—— *de fil de fer*, to wire.

ficelle, f., pack thread, twine, string; (*cord.*) two *fils de caret* (q. v.) laid up together;

—— *de déclenchement*, releasing cord, valve cord (of a balloon);

—— *à fouet*, whip cord.

fichant, a., v. *feu fichant* and *coup fichant*.

fiche, f., stake, rod; hinge pin, hinge hook; plug; hinge; bolt; rag-bolt; surveyor's pin; descriptive card; distance a pile is or may be driven into the ground; (*mas.*) pointing tool; (*art.*) pointing stake; pin, stake; (*fort.*) plunging character (of views, of fire);

anneau à ——, ringbolt;

cheville à ——, rag bolt;

—— *de commutateur*, (*elec.*) plug;

—— *à dents*, mason's filling trowel;

—— *de diagnostic*, (*Fr. a.*) *med.*) mark or tag put on wounded (red for severely hit, but transportable; white for untransportable);

—— *s de mobilisation*, (*Fr. a.*) tables drawn up in peace, and containing data useful on mobilization;

—— *ordinaire à double nœud et à bouton*, pin hinge;

—— *à plomb*, —— *plombée*, (*surv.*) heavy pin used in cultellation;

—— *de pointage*, (*art.*) aiming stake or pointing rod in mortar practice;

—— *s de transport*, (*Fr. a.*) tables giving places of embarking, etc., on mobilization;

—— *à vase*, butt-hinge.

fichée, f., ramming depth (of a pile, etc.).

ficher, v. a. n., to pin, pitch, drive, thrust, stick, or fasten in; (of lines, when prolonged) to plunge into, to be prolonged into; (*mas.*) to point with mortar, plaster.

fidélité, f., loyalty; allegiance;

serment de ——, oath of allegiance.

fiduciel, a., fiducial.

fièvre, f., fever;

—— *des armées*, —— *militaire*, typhus fever.

fiflot, m., (*mil. slang*) infantry soldier, dough boy, mud crusher.

fifre, m., fife, fifer;

premier ——, fife-major.

figement, m., congealing, coagulation.

figer, v. a. n., to congeal, become congealed; to set, become set; (*met.*) to "freeze."

figuratif, a., figurative, symbolic.

figuratif, m., figured way of representing a thing; any diagram;

—— *de tir*, (*art.*, *sm. a.*) plot of shots at target practice.

figure, f., face, figure; diagram; number; (*fenc.*) position (of the body, arm, or sword);

conduire un cheval par la ——, to lead a horse by the head;

—— *schématique*, theoretical diagram (showing relations of parts).

figuré, p. p., figured, representative; m., (*top.*) shape, appearance; representation (by a diagram, plan, or sketch);

ennemi ——, v. s. v. *ennemi;*

—— *particulier*, (*top.*) conventional sign;

plan ——, representative plan;

—— *du terrain*, (*top.*) representation of the ground (as plowed, woods, etc., by special signs).

figurer, v. a., to figure, represent; (*top.*) to represent.

fil, m., thread; string; fiber; wire; chain; yarn; grain (of wood, etc.); edge (of a cutting weapon or tool); stream; vein, crack;

—— *d'accouplement*, (*expl.*) primer wire (joins two primers), fuse wire;

—— *d'acier*, steel wire;

—— *aérien*, overhead wire;

—— *d'aller*, (*elec.*) lead wire (is the insulated wire when only one of a pair is insulated);

—— *d'amorce*, priming wire;

—— *d'araignée*, spider line;

—— *d'archal*, brass wire;

—— *d'argent*, (*farr.*) a rasp of the file given to a horseshoe at the toe and elsewhere to take off sharp edge;

fil (s) *armés*, (*mil.*) obstacle composed of double iron wire twisted, with caltrops at 0.20ᵐ from one another;
—— *blanc*, (*cord.*) untarred yarn;
bois de ——, wood or timber used with the grain parallel to the length of the work;
—— *de*, *du*, *bois*, grain of wood;
—— *de caret*, (*cord.*) yarn, rope yarn;
—— *conducteur*, (*elec.*) conducting wire, conductor;
—— *à congréer*, (*cord.*) worming;
contre le ——, against the grain;
—— *de*, *du*, *courant*, v. —— *de l'eau;*
—— *couvert*, v. —— *recouvert;*
dans le sens du ——, with the grain;
—— *dérivateur,* —— *de dérivation*, (*elec.*) shunt wire, shunt-circuit wire;
donner du —— *à*, to give an edge to;
donner le —— *à*, to whet;
—— *d'eau*, small gutter;
—— *de l'eau*, (*top.*) axis of the stream; current, stream of water; thalweg;
—— *ébroudi*, annealed wire;
écheveau de ——, skein of yarn;
—— *d'emballage*, pack thread;
—— *d'entrée* (*elec.*) leading-in wire; negative wire (of a primary cell);
—— *de l'épée*, edge of the sword;
étirer en ——, to wire-draw;
—— *de fer*, iron wire, wire;
—— *de fer ébroudi*, —— *de fer recuit*, annealed wire;
—— *fin*, fine, small, wire; fine edge; (*cord.*) fine rope yarn;
à —— *fin*, fine-grained;
—— *fusible*, (*elec.*) safety fuse (wire);
—— *à gargousse*, (*art.*) cartridge twine;
—— *goudronné*, (*cord.*) tarred rope yarn;
à gros ——, coarse-grained;
—— *inducteur*, (*elec.*) primary wire of an induction coil;
—— *induit*, (*elec.*) secondary wire of an induction coil;
—— *de lame*, edge of a blade;
—— *à ligatures*, binding wire;
—— *de ligne*, line wire (telegraph, telephone);
—— *de ligne télégraphique*, telegraph wire;
—— *à la main*, (*cord.*) hand-made thread;
—— *de manœuvre*, (*cord.*) fine rope yarn;
—— *mécanique*, (*cord.*) machine-made thread; mill-spun yarn;
mettre le —— *à*, to put an edge on (as bayonets, on mobilization);
mettre au ——, v. *passer au* ——;
—— *de mise de feu*, (*min.*, etc.) firing wire;
—— *noir*, tarred rope yarn;
—— *nu*, (*elec.*) bare, uninsulated wire;
ôter le ——, to take off the edge;
passer au ——, to put to the sword;
—— *plat*, soft yarn;
—— *à plomb*, plummet-line, plumb-line;
—— *à plomb à précision*, accurate plumb-line used in geodesy;
—— *de poste*, (*elec.*) office wire;
—— *de rail*, (*r. r.*) line, set of tracks;
—— *recouvert*, (*elec.*) insulated wire;
—— *retors*, (*cord.*) twine;
—— *de retour*, (*elec.*) return wire;
—— *rosette*, rose copper wire;
—— *de sautage*, (*expl.*, *min.*) fuse-circuit wire of an exploder or igniter;
—— *de sonnerie*, bell wire;
—— *de sortie*, (*elec.*) positive wire (of a primary cell);
—— *souterrain*, underground wire;
—— *télégraphique* (*téléphonique*), telegraph (telephone) wire;
—— *de terre*, (*elec.*) earth wire;
tirer en ——, to wire-draw;
—— *tors*, (*cord.*) hard yarn;
—— *tranchant*, sharp edge;
—— *tressé*, woven wire.
filage, m., (*cord.*) spinning or manufacture of yarn; the yarn itself;
atelier de ——, spinning shop;
—— *à la main*, hand spinning;
—— *mécanique*, mill spinning.
filament, m., filament;
—— *de charbon*, (*elec.*) filament of an incandescent lamp.
filandreux, a., streaked (of wood), flawy (of stone).

filardeau, m., small spike; straight sapling.
filardeux, a., (of stone) streaky.
filasse, f., (*cord.*) tow; harl (of flax, hemp, etc.).
filature, f., spinning mill; (*cord.*) ropewalk.
file, f., row, file; (*mil.*) file; (*Fr. a.*) front-rank man with his rear file, in line; (in formation by the flank) a set of two, or of four, as the case may be;
—— *des amiraux*, (*nav. tactics*) leading column;
l'autre ——, (*nav. tactics*) second column;
—— *de base*, (*drill*) directing or base file;
chef de —— (*mil.*) front-rank man;
—— *creuse*, (*mil.*) blank file, without rear-rank man;
dédoubler les ——s, v. s. v. *dédoubler;*
—— *doublée*, (*Fr. a.*) set of four men when faced by the flank and set of fours formed;
doubler les ——s, v. s. v. *doubler;*
par —— *à droite* (*gauche*), (*inf.*) column right (left) for column of files (squad, section);
par ——s *à droite* (*gauche*), (*inf.*) (of a company) change of direction by the right (left) flank of subdivisions;
ligne de ——, (*nav. tactics*) single line ahead;
—— *de palissades*, (*fort.*) palisading;
rang des serre- ——, (*mil.*) line of file closers;
reprendre son chef de ——, (*mil.*) to fall in, or place one's self again, behind one's front-rank man;
serre- ——, (*mil.*) file closer;
—— *simple*, (*mil.*) set of two men, when faced by the flank.
filé, m., (*top.*) running of a contour.
filer, v. a. n., to spin; (*cord.*) to pay out, to ease off, slack off, ease away; run out; to lower; (*drill*) to file off, past; (*sm. a.*) (of the sear) to slide over the tumbler without engaging in the notches.
—— *en bande*, (*cord.*) to let go amain;
—— *en belle*, v. —— *en douceur;*
—— *par le bout*, (*nav.*) to slip;
—— *un câble*, —— *une chaîne*, (*nav.*) to slip a cable, a chain;
—— *une courbe*, (*top.*) to run a contour;
—— *un cordage*, (*cord.*) to pay off, ease off, a rope;
—— *à la demande*, (*cord.*) to pay off according to the stress or strain, to lower cheerily;
—— *en douceur*, (*cord.*) to ease away slowly, handsomely, hand over hand, with a turn or two;
—— *les eaux*, (*drawing*) to draw water lines around a coast; to draw current lines;
—— *en garant*, (*cord.*) v. —— *en douceur;*
—— *à réa*, (*cord.*) to ease off at will, as fast as possible;
—— *à retour*, (*cord.*) to ease off with a turn around a cleat;
—— *rondement*, (*cord.*) to pay out, ease off rapidly;
—— *à volonté*, (*cord.*) v. —— *à la demande.*
filet, m., thread; net, netting; network in gen.; network (of a balloon); thread of a screw, screw thread; (*harn.*) snaffle; bridoon, watering bridle; (*art.*) fillet, ring (obs.);
—— *d'abreuvoir*, (*harn.*) watering bridle;
—— *arrondi*, rounded screw thread;
—— *carré*, square screw thread;
—— *de combat*, (*nav.*) splinter netting;
à ——s *contraires*, right and left handed thread;
—— *droit*, v. —— *carré;*
—— *à droite*, right-handed screw thread;
—— *femelle*, female screw;
—— *à fourrage*, hay or forage net;
—— *à gauche*, left-handed thread;
—— *interrompu*, interrupted thread;
—— *de manège*, (*harn.*) riding-school snaffle;
mors de ——, (*harn.*) watering bit;
—— *à panurges*, (*harn.*) overdraw checkrein;
—— *pare-torpilles*, (*nav.*) torpedo net;
—— *porteur*, (balloons) supporting net (of the car;
—— *de rayure*, (*art.*) land;
—— *récepteur*, (*r. f. art.*) cartridge-case net;
—— *rectangulaire*, v. —— *carri;*
—— *releveur*, (*harn.*) checkrein;
—— *renversé*, left-handed thread;
scier du ——, (*man.*) to saw upon the bit;

filet *sectionné*, slotted thread;
— *simple*, single thread;
— *taraudé*, female thread;
— *triangulaire*, triangular thread;
— *de vis*, thread of a screw;
vis à — *s interrompus*, (*art.*) interrupted breech screw.
filetage, m., threading, screw-cutting; threading (set of, as a screw).
fileter, v. a., (*mach.*) to cut a thread on;
— *en repassant*, to cut a screw thread with a chaser, to chase a thread;
— *une vis*, to cut a thread, a screw;
— *à la volée*, to cut screws by hand.
filiation, f., *état de* —, (*mil.*) return, list by name of officers, n. c. o., and men who are to embark on one and the same ship.
filière, f., manrope; ridgerope of an awning or tent; (*mach.*) screw plate; tap; draw, drawplate, drawbench; wire gauge; (*cons.*) purlin;
— *à bois*, screw box (wood screws);
— *brisée*, screw stock; die stock;
— *à coussinets*, v. — *brisée*;
— *double*, screw stock and dies;
— *d'envergure*, (*torp.*) ridgerope of a torpedo netting;
— *hiérarchique*, (*mil.*) passage through interested military channels;
plaque —, drawplate;
— *simple*, screw tap;
tarauder à la —, to tap;
— *à tirer*, v. *plaque* —;
— *à vis*, screw plate.
filigrane, f., (*sm. a.*) brass wire on the gripe of a sword hilt.
filin, m., (*cord.*) rope (hawser-laid, right-handed);
— *en abaca*, manila rope;
— *d'amarrage*, seizing;
— *de bastin*, v. — *en caïre*;
— *blanc*, untarred rope;
— *en caïre*, coir rope;
— *composé de cordons*, cable-laid rope;
— *composé de corons*, hawser-laid rope;
— *d'enrayage*, (*art.*) drag rope, locking rope;
franc —, white, untarred, rope;
— *noir*, tarred rope;
— *premier brin*, rope of the finest hemp;
— *en quatre*, four-stranded rope;
— *en quatre avec mèche*, shroud-laid rope;
— *de transfilage*, lacing;
— *en trois*, three-stranded rope.
filite, f., (*expl.*) Italian name of *balistite*.
filtrage, m., filtering.
filtration, f., filtration, percolation.
filtre, m., filter, percolation, filtering stone;
— *à aspiration*, suction filter;
— *à charbon*, charcoal filter;
— *dégrossisseur*, separator;
— *à eau*, water filter;
— *à pression*, pressure filter.
filtrer, v. a., to filter; strain; (*chem., etc.*) to filter;
pierre à —, filter, filtering stone.
fin, a., fine; fine-bred; sharp; slender.
fin, f., end, conclusion, tail, object, purpose; m., (of coal) coal dust;
cheval à toutes — *s*, horse fit to ride or drive.
finage, m., (*met.*) fining, conversion of gray into white pig.
finance, f., cash; (*in pl.*) money affairs, finances;
la —, bankers, capitalists;
ministre des — *s*, secretary of the treasury; chancellor of the exchequer.
finchelle, f., (*cord.*) towrope for boats.
fine, f., coal dust.
finer, v. a., (*met.*) to fine, to convert gray into white pig.
finerie, f., (*met.*) finery (hearth or furnace).
finesse, f., sharpness of an image (as in photography).
fingard, m., restive horse.
finir, v. a., to finish, end, complete;
— *un cheval*, (*man.*) to complete the training of a horse.
finissage, m., finishing of parts of a mechanism, etc., done by hand or with a file; (*sm. a.*) finishing (of barrel), i. e., filing, adjusting, all fine and small work;

finissage *d'un canon* (*art.*), generic term for work on gun after it has been turned down, bored, and rifled (e. g., seats for accessory parts);
— *d'un canon de fusil*, (*sm. a.*) operation of finishing chamber, brazing of sights, and fitting of receiver;
— *des grains de poudre*, (*powd.*) generic term for glazing, drying, and dusting of powder.
finisseur, a., finishing, final (of various operations of manufacture, as of arms); m., finisher (workman); (*mach.*) finishing roll, finishing tool (general).
finiteur, a., *cercle* —, horizon (*obs.*).
fion, m., finishing stroke, last touch.
fisc, m., public funds; administration of public finances.
fiscal, a., fiscal.
fisolle, f., whipping twine, fox made of a single rope yarn.
fissile, a., fissile.
fissilité, f., liability to fissure, fissibility.
fissure, f., crack, fissure.
fissurer, v. a., to fissure.
fixage, m., fixing in place, securing; (*phot.*) fixing.
fixateur, m., (*phot.*) fixing solution.
fixation, f., fixing, setting; appointing;
état de — *s*, (Fr. a.) return drawn up by the minister of war showing the nature and the minimum quantity of war material to be kept in reserve.
fixe, a., fixed, settled, steady, solid, steadfast; firm, stationary; (*art.*) nonrecoiling; (*mil.*) (in commands) front; m., appointed, regular salary; (*in pl.*) fixed bodies;
machine —, v. *machine à vapeur* —;
machine à vapeur —, stationary engine.
fixe-pieds, m., (bicycle) toe-clip.
fixer, v. a. r., to fix, to fasten, to bed, settle, stick; to wedge; (*phot.*) to fix; (*hipp.*) to place (a horse, for a surgical operation);
— *la portée d'un fusil*, (*sm. a.*) to determine the range;
flache, f., hole in a pavement, chuck hole in a road; crack, flaw, depression, in a stone; wane of a slab (of wood); pool of water in a wood.
flacher, v. a., to blaze a tree.
flacheux, a., waney.
flageoler, v. n., (*man.*) to tremble, shake (horse's legs).
flambage, m., singeing; (*art.*) scaling, warming of a gun;
coup de —, (*art.*) scaling shot, warming charge.
flambeau, m., torch, flambeau;
— *d'artifice*, (Fr. a.) torch or flambeau used by troops on the march, or to light up a spot where work is being done, etc. (known as — *Lamarre*);
demi- —, (Fr. a.) a flambeau that will last seven minutes.
flamber, v. a. n., to sear, to singe; to flame, blaze; to purify, disinfect; to sag; (*art.*) to scale a gun; to warm, to fire a warming shot, a scaling or warming round; to fire a friction primer in an unloaded piece;
— *une bombe*, (*art.*) to mark the instant a fuse is burnt out by means of the burster;
— *un moule*, (*fond.*) to dry a mold;
— *un vaisseau*, (*nav.*) to hoist a ship's particular signal to reprove her commander for inattention to orders.
flamberge, f., rapier; strong and heavy sword;
mettre — *au vent*, to draw (sword) (familiar).
flamme, f., flame; blaze; vane, streamer; pennant, pennon, lance pennon, pendant, banderole; signal flag; (*unif.*) pudding bag of a busby; (*hipp.*) fleam for bleeding horses; (*nav.*) man-of-war's pennant; (*art., sm. a.*) flash of discharge;
bâton de —, (*nav.*) staff of a pennant;
— *de Bengale*, (*artif.*) Bengal light;
drisse de —, (*nav.*) halliard of a pennant;
le fer et la —, fire and sword;
— *de lance*, lance pennon;
— *de la messe*, (*nav.*) church pennant;
— *oxydante*, oxidizing flame;
— *de réduction*, reducing flame.

flammèche, f., flake, spark; (*art.*) burning shred of cartridge blown out of a gun.

flan, m., plate, disk; flanch (for coins); (*sm. a.*, *art.*) the disk of which a cartridge case (primer, friction primer, etc.) is made; the plate of which a bayonet scabbard is made.

flanc, m., flank, side; (*top.*) slope or side on each side of a *ligne de thalweg*, q. v.; (*art.*) edge, side of groove in rifling; (*fort.*) flank (of a bastion, etc.); (*hipp.*) flank; (*mil.*) flank;
— *avoir du* ——, (*hipp.*) to have fine flanks, good well-turned ribs;
— *avoir les* ——s *cousus,* (*hipp.*) to be thin-flanked;
— *bas,* (*fort.*) lower flank, casemate;
— *battre de* ——, (*art.*) to take in flank;
— *battre du* ——, (*hipp.*) to heave;
— *de charge,* —— *de chargement,* (*art.*) loading edge of a groove;
— *conducteur,* (*art.*) v. —— *de forcement;*
— *contre-* ——, (*art.*) loading edge;
— *cordé,* (*hipp.*) stringy, corded flank;
— *de courtine,* (*fort.*) auxiliary flank;
— *couvert,* (*fort.*) retired flank;
— *creux,* (*fort.*) concave flank; (*hipp.*) hollow flank;
— *défensif,* (*mil.*) a flank standing on the defensive; the defensive flank in a cavalry charge;
— *directeur,* (*art.*) v. —— *de forcement;*
— *droit,* (*fort.*) straight flank;
— *par le* —— *droit* (*gauche*), (*drill*) by the right (left) flank;
— *extérieur* (*intérieur*), exterior (interior) flank;
— *faire par le* ——, (*drill*) to face to the right (or left) flank;
— *de forcement,* (*art.*) driving edge of a groove;
— *garde,* (*mil.*) flanker;
— *haut,* (*fort.*) upper flank;
— *levretté,* (*hipp.*) greyhound flank;
— *-ligne,* (*mil.*) detachment on the flank of a main line, intended to deceive enemy as to the real points of support of the wings of the main line;
— *de marche,* (*drill*) directing flank;
— *marche de* ——, (*mil.*) flank march;
— *marche, marcher, par le* ——, (*mil.*) march, to march, by the flank;
— *montrer le* —— *à,* v. *prêter le* —— *à;*
— *mort,* (*mach.*) surface opposed to bearing surface;
— *oblique,* (*fort.*) auxiliary flank;
— *offensif,* (*mil.*) a body of troops holding a position menacing a flank of the enemy; the attacking flank in a cavalry charge;
— *de portage,* (*mach.*) bearing surface;
— *prendre de* ——, v. *battre de* ——;
— *prendre en* ——, (*mil.*) to flank, outflank; to take in flank;
— *prêter le* —— *à,* (*mil.*) to be exposed to attack in flank;
— *des rayures* (*art., sm. a.*) the edge of a land;
— *retiré,* (*mil.*) retired flank;
— *retroussé,* (*hipp.*) tucked up flank;
— *de tir,* (*art.*) driving edge of a groove;
— *vif,* (*mach.*) v. —— *de portage.*

flanconade, f., (*fenc.*) blow, pass, or thrust in the flank.

flanelle, f., flannel;
— *ceinture de* ——, (*med.*) flannel waistband.

flanquant, a., (*fort.*) flanking;
— *angle* ——, flanking angle.

flanquement, m., (*fort., mil.*) flanking; flanking defense; flanking fire;
— *bas,* v. —— *rasant;*
— *batterie de* ——, flanking battery;
— *par les crêtes,* v. —— *par le tracé;*
— *direct,* flanking by fire from loopholes in main work (galleries, relieving arches) as opposed to flanking from special works for the purpose;
— *fichant,* flanking by downward fire (as in the bastioned trace);
— *haut,* v. —— *fichant;*
— *rasant,* flanking from works specially constructed for the purpose (as in the polygonal trace);
— *tir de* ——, v. s. v. *tir;*
— *par le tracé,* flanking from the work itself (as in the bastion system).

flanquer, v. a., (*mil., fort.*) to flank, to attack in flank; to defend, secure, guard, the flank of a line or work of a body of troops;
— *8 et 7,* (*mil. slang*) to give a man a fortnight's confinement or arrest.

flanqueur, m. and a., (*mil.*) flanker, flanking.

flaque, f., puddle, stagnant pool.

flasque, a., weak, poor, feeble; not completely filled (of a balloon)

flasque, m., standard or support (as of the driving axle of a dynamo); (*art.*) bracket, cheek, flask; f., (*art.*) rocking-arm (in some Hotchkiss types);
— *s d'affût,* (*art.*) brackets, cheeks (of a gun carriage);
— *de cabestan,* whelp of the capstan;
— *exhaussement de* ——, (*Fr. art.*) cheek plate to fit the old 24 siege-gun carriage for 138^{mm} siege gun;
— *de levage,* (*Fr. art.*) lifting frame (used to lift a *caisse à obus* from its truck);
— *queue de* ——, (*art.*) rear end of cheek.

flatoir, m., flatter, flatting hammer.

flatter, v. a., to flatter;
— *un cheval,* (*man.*) to make much of, to caress, a horse; to yield to a spirited horse, so as gradually to quiet him;
— *un courant,* to divert gradually a stream from a threatened bank;
— *les vagues,* to break the force of waves (by the opposition of an inclined surface).

fléau, m., flail; door beam, crossbeam (to shut a door); beam, bar (of a balance), scale-beam; (*mach., etc.*) beam; knife-edge; link, joint;
— *de porte,* (*fort.*) crossbeam.

flèche, f., arrow; sag, deflection (of a belt or cable, of a beam); sag, bend, departure from a straight line (in falling-weight tests); compression of a spring; underpole connecting the fore and hind parts of a carriage (4-wheel); perch or pole of a carriage; height of the crowning of a pulley; rise of an arch or vault; spire (of a church); stem (of a tree); sweep (of a horse-power); (*art.*) droop of the muzzle; trail; (*ball.*) maximum ordinate of a trajectory (measured from the *ligne de mire,* q. v.); (*min., etc.*) point (of a drill bar); (*fort.*) fleche, redan; (*surv.*) surveyor's pin;
— *d'affût,* (*art.*) trail;
— *de caisson à munitions,* (*art.*) perch;
— *de courbure,* (*pont.*) dip of a bridge cable;
— *directrice,* (*mach.*) guide bar;
— *à élargir,* (*min., etc.*) widener, widening bit;
— *en* ——, tandem (of a team);
— *fausse* ——, —— *fausse,* (*art.*) false trail (used for limbering up short-trailed, heavy carriages, e. g., that of the French 220^{mm} siege mortar);
— *à flancs,* (*fort.*) flanked fleche;
— *de grue,* crane post;
— *de pont-levis,* pliers; swipe beam, of a drawbridge;
— *prendre sa* ——, (of a beam) to get its permanent sag or set;
— *rallonge de* ——, (*art.*) extension piece of a mountain-gun trail;
— *à tambour,* (*fort.*) fleche with tambour;
— *de la trajectoire,* (*ball.*) maximum ordinate of a trajectory;
— *de triqueballe,* (*art.*) shaft, pole, of the truck wagon.

fléchir, v. a. n., to bend, sag; (*in gen.*) to give way, stagger, waver; (*mil.*) to give way, waver.

fléchissement, m., bending, giving way; (*art.*) droop of the muzzle of a gun (esp. of a very long gun).

flectographe, m., (*art.*) flectograph (instrument to measure the flexions of the points of attachment of the various organs of a gun carriage).

flectomètre, m., flectometer.

flette, f., punt, small ferryboat.

fleur, f., flower; flower (i. e., pick); hair side, grain side (of leather, of a skin);
— à —— de, on the same level with;
— à —— d'eau, awash, flush with surface of water; between wind and water (of a shot);
— —— de chaux, v. farine fossile;
— ——s de soufre, flowers of sulphur;
— —— de troupe, (mil.) picked men.

fleuret, m., square boring bit, rock drill (point); (fenc.) foil;
— —— de mine, (min.) mine-borer, miner's borer;
— —— moucheté, (fenc.) foil.

fleuron, m., any ornamental small brass, etc., disk, as in harness (esp. on the headstall of a bridle); lance-like ornament on top of a grating (iron); finial.

fleuve, m., river (large river emptying directly into the sea).

flexibilité, f., flexibility, pliancy, suppleness.

flexible, a., flexible, pliant; m., (mach.) flexible shaft, for small drills, etc.

flexion, f., flexion, bending; (drill) change of direction (rare); (gym.) (leg, trunk, and arm) exercises;
— —— du corps, (gym.) trunk exercise;
— —— des extrémités, (gym.) arm and leg exercise;
— —— simple, flexion;
— —— de la volée (art.) droop of the chase.

flibuster, v. n., to become a filibuster.
flibusterie, f., filibustering.
flibustier, m., filibuster.
flingol, m., (mil. slang.) musket, rifle.
flipot, m., (carp.) piece let in to conceal a defect.

floche, f., bit of cloth that is unraveling;
— coton-poudre en ——s, (expl.) loose gun cotton, gun cotton in yarn.

flocon, m., flake.
flore, f., flora.
floss, m., floss;
— —— d'acier, —— lamelleux, (met.) white cast iron fit to be converted into steel.

flot, m., tide; flood tide; stream, torrent; wave, billow, waters; (in pl.) the sea;
— à ——, afloat; floating, by floating;
— à ——s, in torrents, in waves;
— —— commun, ordinary flood;
— demi- ——, mi- ——, half flood tide;
— direction du ——, set of the flood;
— étale de ——, slack flood;
— —— de fond, ground swell;
— mettre, remettre, à ——, to set afloat, to set afloat again;
— quart de ——, first quarter of the flood tide;
— trois quarts de ——, last quarter of the flood.

flottable, a., navigable for rafts or loose wood, navigable (of a stream); that will float.
flottage, m., rafting.
flottaison, f., (nav.) water line;
— à la ——, on the water line, between wind and water;
— ligne de ——, water line;
— à la ligne de ——, between wind and water.

flottant, a., floating, buoyant; (fig.) irresolute, wavering, fluctuating.

flotte, f., washer (of a wheel); float (as of a cable); (nav.) fleet, squadron, navy;
— aller de ——, (nav.) to sail in company with;
— —— armée, (nav.) fleet of men-of-war;
— —— à crochet, (art.) drag washer;
— —— cuirassée, (nav.) armored fleet;
— —— de débarquement, (nav.) landing fleet for landing operations.

flottement, m., (mil.) wavering or undulation in the ranks; (fig.) hesitation, irresolution.

flotter, v. a. n., to float; to fly out, to float, to flutter (of a flag); to undulate, waver, wave; to hesitate; (man.) to wabble about (of a horse, e. g., a blind horse); to advance, move aimlessly; (mach.) to wabble, be unsteady (of a piston, shaft); (mil.) to waver (of a line), to lose the dress;
— —— un pavillon, (nav.) to float a cable.

flotteur, m., floater; water gauge, float; float (of a torpedo); float board; cable cask or buoy;
— —— d'alarme, (steam), boiler float, to give warning when water is low.

flottille, f., (nav.) flotilla; squadron.

flou, a., (phot.) running, indistinct;
— image ——e, "oyster," shell marks.

flouet, m., vane, weathercock.
flouette, f., v. flouet.
fluet, a., thin, spare, lank, meager.
fluide, a., m., fluid.
flûte, f., (mus.) flute; (nav.) storeship; open boat; troopship;
— —— à bec, (mus.) generic term for clarinets, oboes, and other similar instruments;
— petite ——, (mus.) piccolo.

fluvial, a., fluvial;
— eau ——e, river water.

flux, m., flow, current, stream, flood, influx; (met.) flux;
— —— et reflux, ebb and flow;
— roue à —— et reflux, tide mill.

focal, a., focal;
— longueur ——e, (opt.) focal length.

foi, f., faith, sincerity; parole;
— en —— de, in witness of;
— être prisonnier sur sa ——, (mil.) to be a prisoner on parole;
— faire —— en justice, to be admitted as a witness;
— ligne de ——, fiducial edge;
— sur sa ——, on parole.

foin, m., hay; standing grass;
— botte de ——, bundle of hay;
— —— comprimé, compressed hay;
— —— délavé, hay that has become too wet after mowing;
— —— échauffé, moldy hay;
— faire les ——s, to make hay;
— grenier de ——, hayloft;
— —— lavé, light-colored hay; hay that is too wet, before cutting;
— meule de ——, haystack;
— —— pressé, compressed hay;
— —— rouillé, rusty hay;
— —— vasé, muddy hay.

foire, f., fair.
foirer, v. n., (cord.) to become slack; to slip, get out of its place.

foisonnement, m., swell, increase, growth, in volume (of dug earth).

foisonner, v. n., (of dug earth) to swell, to increase in volume; (nav.) to swell up (as cork filling, etc., of double bottoms).

foliation, f., feathering, foliation.
folio, m., sheet.
folle-avoine, f., wild oats, volunteer oats.
folies, f. pl., (art.) guns whose axes are untrue.
fonçage, m., sinking (of shafts, piles, wells, etc.); driving;
— —— des pieux, pile driving.

fonceau, m., synonym of bossette (harn.), q. v.
foncement, m., v. fonçage.
foncer, v. a. n., to drive, sink (a well, shaft, pile); to deepen; to bottom a cask; to countersink (a rivet);
— —— sur, to fall upon.

foncier, a., landed; based on land, on real estate.
fonçoir, m., a sledge-hammer with a cutting face.

fonction, f., function, duty (esp. in pl.), performance, execution, working;
— —— des pièces, working of gear, gearing.

fonctionnaire, m., functionary (civil); (Fr. a.) functionary, official (generally limited to functionaries of the intendance militaire).

fonctionnement, m., function, method of working, of operating, behavior; (mach., art., etc.) action (of a machine, of a breech mechanism, etc.).

fonctionner, v. n., to work, to work properly, to be in working order; to work, to operate, to go (of a machine, etc.);
— action de ——, play;
— —— à détente dans des cylindres séparés, (steam) to work like a compound engine;
— —— à introduction directe, (steam) to work as a simple engine.

fonçure, f., facing (of a paddle wheel).

fond   184   **fondure**

fond, m., bottom (of a vessel, of a steam cylinder, of a ship, of the sea, of a river); ground, anchoring ground; soundings; head (of a cask); crown (of a hat); depth; bed; groundwork, foundation, base, background; first coat or coats (of paint); center, heart; back, back part; farthest, remotest, part; floor, flooring; (*hipp. etc.*) endurance, bottom; (*fort.*) sole (of an embrasure);
— à ——, bottomed; thoroughly, fully; tight, home;
— *de l'âme*, (*art., sm. a.*) bottom of the bore;
— *d'argile*, clay bottom (stream, sea);
— *avoir du* ——, (*hipp. etc.*) to have endurance, to be sound-bottomed;
— *bas*——, v. *bas-fond;*
— *bon* ——, good anchoring ground;
— *de bonne tenue*, good bottom (for anchoring);
— *de cale*, (*nav.*) bottom of the hold;
— *charger à* ——, *à* —— *de train*, (*mil.*) to charge home;
— *d'une chaudière*, bottom of a boiler;
— *de* —— *en comble*, from top to bottom;
— *couler à* ——, (*nav.*) to founder, to sink, to go down;
— *de cour*, —— *curé*, bottom neither sandy nor muddy (stream, sea);
— *courant de* ——, undertow;
— *de cuve*, hollow, with rounded angles;
— *de cylindre*, (*mach.*) cylinder bottom;
— *de* ——, (*cons.*) resting on a foundation;
— *à deux* ——*s*, (*nav., etc.*) double-bottomed;
— *donner* ——, to cast anchor;
— *double* ——, double bottom or casing; (*nav.*) double bottom;
— *à dragée*, shot mold;
— *dur*, hard bottom (stream, sea);
— *de l'eau*, bottom (of the water), ground;
— *d'embrasure*, (*fort.*) sole of an embrasure;
— *d'embrasure en contrepente*, (*fort.*) counter-sloping sole;
— *faux-* ——, false or movable bottom (as of an ammunition chest);
— *fossé à* —— *de cuve*, ditch with steep sides;
— *du fossé*, (*fort.*) bottom of the ditch;
— *de galerie*, (*min.*) bottom, sole, of a gallery;
— *d'une galerie*, (*min.*) heading;
— *grand* ——, deep water;
— *de gravier*, gravel bottom (stream, sea);
— *haut-* ——, flat, shoal, shallow water, shoal water;
— *haut*, v. *haut-* ——;
— *lame de* ——, ground wave;
— *de lit*, frame or slats (for the mattress);
— *de mauvaise tenue*, bad bottom (for anchoring);
— *de miroir*, back of a mirror;
— *mou*, soft bottom (stream, sea);
— *à mouillage*, anchorage;
— *mouvant*, shifting bottom;
— *d'un navire*, ship's bottom;
— *petit-* ——, depth of water under the keel;
— *peu de* ——, shoal, shallow water;
— *pierreux*, stony bottom (stream, sea);
— *à* —— *plat*, flat-bottomed;
— *de pré*, grassy bottom (stream, sea);
— *prendre* ——, to cast anchor; to touch at a port;
— *de la rayure*, (*art.*) bottom of the grooves;
— *de sable*, sandy bottom (stream, sea);
— *sale*, foul bottom;
— *de son*, bran-colored bottom (stream, sea);
— *à* —— *de train*, as fast as possible (of a charge, gait of horses, etc.);
— *de vase*, muddy bottom (stream, sea);
— *visser à* ——, to screw tight, to screw home.

fondage, m., (*met.*) melting.
fondamental, a., fundamental; radical;
— *pierre* ——*e*, foundation stone.
fondant, a., melting; m., (*met.*) flux.
fondation, f., foundation, groundwork; bed (as of a machine, engine, etc.);
— *à l'air comprimé*, compressed air foundation;
— *asseoir les* ——*s*, to sink a foundation;
— *boulon de* ——, holding-down bolt;
— *avec empatement*, footing foundation;
— *par encaissement*, cofferdam foundation;
— *par encaissement et cuvette*, cofferdam, used in waters liable to rise, (the rise being kept out by a small dam forming *cuvette*);
— *sur fascines*, fascine work; brushwork foundation;
— *jeter une* ——, to lay a foundation;
— *pierre de* ——, foundation stone;
— *plaque de* ——, sole plate, bedplate;
— *reprise sous œuvre*, underpinning.

fondement, m., foundation, (frequently in pl.) base; (*in pl.*) excavation for a foundation; groundwork; substructure; (*fig.*) ground, reliance, trust;
— *à air comprimé*, compressed-air foundation (plenum process);
— *par caissons*, caisson foundation;
— *par enrochement*, random stone, riprap foundation;
— *au moyen de cours de palplanches*, sheet-piling foundation;
— *à pierres perdues*, v. —— *par enrochement;*
— *sur pieux*, pile, (hence) grillage, foundation;
— *pneumatique*, pneumatic foundation;
— *refaire un* —— *de sous œuvre*, to underpin a foundation;
— *reprendre les* ——*s*, to rebuild a foundation;
— *saper le* ——, to sap a foundation;
— *sur le sol*, direct foundation.

fonder, v. a., to found; to lay a foundation; to establish, institute, set up.

fonderie, f., foundry; founding;
— *de bronze*, brass foundry;
— *de canons*, gun foundry;
— *de fer*, iron foundry.

fondeur, m., founder, melter, smelter, molder;
— *de bronze*, brass founder.

fondis, m., settling, sinking of the foundation (of a building).

fondre, v. a. n. r., to melt; to smelt; to fuse; to cast; to found; to blend; to dissolve; to sink in, give way; to fall away; to disappear suddenly, break, break in; to dart, pounce upon;
— *en moule*, (*fond.*) to cast in a mold;
— *en moule creux*, (*fond.*) to cast in a mold without a core;
— *en série*, (*fond.*) to cast a number of small identical pieces together continuously and at one pouring;
— *d'une seule pièce*, (*fond.*) to cast in one piece;
— *sur*, to throw one's self upon (an enemy, line, etc.).

fondrier, a., *bois* ——, wood heavier than water, either naturally or by immersion (water-logged); m., water-logged raft (of logs).

fondrière, f., (*top.*) quagmire, marsh, slough.

fonds, m., the ground or soil of a field, etc.; a sum of money; fund, funds; public funds; business;
— *de commerce*, stock in trade;
— *commun*, (*Fr. a.*) regimental fund used in certain contingencies (part of *masse d'entretien et d'habillement*, q. v.), fund for expenses common to unit concerned;
— *de concours*, (*Fr. a.*) funds supplied by communes, to assist in execution of military works in which the communes are interested;
— *courants*, current funds;
— *d'économie*, savings;
— *de la guerre*, —— *de la marine*, annual appropriations for the war and the navy departments, respectively;
— *particulier*, (*Fr. a.*) fund (part of *masse d'entretien et d'habillement*, q. v.) used to meet special expenses of each administrative unit;
— *publics*, public funds; stocks;
— *de réserve (de siège)*, (*Fr. a.*) funds placed, on mobilization, at the disposition of the governor of a place, and to be used only in case of investment;
— *du trésor*, public money, treasury funds.

fondu, p. p., melted;
— *acier* ——, cast steel;
— *fer* ——, cast iron.

fondure, f., (*hipp.*) grease, molten grease.

fontaine, f., spring, fountain; cistern cock; (*harn.*) hollow in padding of saddle (over a projection on the back);
— *filtrante*, filtering stone.
fontanelle, f., (*hipp.*) rowel.
fonte, f., (*met., fond.*) melting, founding; casting; pig iron, crude iron, cast iron; (*sm. a.*) pistol holster;
— *d'acier*, cast steel;
— *pour acier*, forge pig;
— *aciéreuse*, steel pig, steely pig;
— *d'affinage*, forge pig;
— *à affiner*, v. — *d'affinage*;
— *à l'air-chaud*, hot-blast pig;
— *à l'air froid*, cold-blast pig;
atelier de —, melting house, smeltery;
— *bigarrée*, v. —, *truitée*;
— *blanche*, white pig; hard white cast iron; No. 5 pig;
— *blanche caverneuse*, porous white cast iron;
— *blanche cristalline*, specular iron, spiegeleisen;
— *blanche lamelleuse*, v. — *blanche cristalline*;
— *blanche grenue*, white semigranular pig;
— *blanche miroitante*, v. — *blanche cristalline*;
— *blanchie*, fine iron or metal;
— *blanchie à la surface durcie*, chilled casting, chilled cast iron;
— *de bois*, charcoal pig;
brut de —, as cast, just as cast, rough cast;
— *à canon*, gun pig;
— *à canon en bronze*, gun metal;
— *à canon en fer*, gun pig;
— *de canons*, gun casting (operation);
canon de —, an obsolete term for bronze gun;
— *caverneuse*, porous pig;
— *au charbon de bois*, charcoal pig;
— *au coke*, coke pig;
— *en creux*, hollow casting;
— *crue*, casting, pig iron, cast iron;
— *dure*, chilled cast iron; cast iron especially refined or incompletely refined, for the special purpose of bringing out strength and tenacity;
en —, (made of) cast iron;
— *épurée*, pig refined by hearth process;
de fer, cast iron;
fer de —, cast iron;
— *à fusion crue*, gray pig;
— *à fusion sèche*, white or mottled pig;
— *de 1ère fusion*, cast iron made directly from ore; all-mine pig;
— *de 2eme fusion*, cast iron, made by a mixture of various grades of all-mine pig, or by a mixture of these with scrap iron;
— *graphiteuse*, black pig; graphitic pig;
— *grise*, gray pig; No. 2 pig;
— *grise claire*, light-gray pig; No. 3 pig;
— *grise foncée*, dark-gray pig;
gueuse de —, pig iron;
jeter en —, to cast;
— *lamellaire*, lamellar pig;
— *limailleuse*, graphite iron; black, dark-gray iron;
— *loupante*, metal too rapidly refined and hard;
— *maculée*, mottled pig;
— *malléable*, malleable cast iron;
— *mêlée*, v. — *maculée*;
— *de minerais*, smelting ore;
— *mitis*, mitis metal;
— *de moulage*, foundry pig;
— *noire*, black pig, dark-gray pig; foundry iron, kishy pig; No. 1 pig;
— *à noyau*, cored work, cored casting; hollow casting;
— *de pistolet*, (*sm. a.*) pistol holster;
— *raffinée*, refined pig;
— *de régénération*, cast iron from a regenerative furnace;
— *en saumon*, pig iron;
— *tendre*, gray pig iron;
— *truitée*, mottled pig; No. 4 pig;
venir de — *avec*, to be cast at the same time as, or in one piece with; to be integral with;
venir en —, to be cast;

fonte, *venu de* —, cast in one piece with, cast solid with; integral, integral with;
— *verte*, an obsolete term for bronze.
fonteux, a., (*met.*) resembling pig iron (bad sense).
fonticule, f., (*hipp.*) rowel.
forage, m., (*mach., etc.*) boring, drilling; (*mil. min.*) drilling of a blast hole; also the hole itself (10cm to 30cm in diameter); bored mine or camouflet (small mine);
banc de —, boring bench;
barre de —, boring bar;
puits de —, driven well;
— *des puits*, driving, boring, of wells.
forain, a., foreign; wide enough for two carriages to pass abreast;
mouillage —, (*nav.*) open anchorage;
rade —*e*, (*nav.*) open roadstead.
forant, m., balk, spar.
forban, m., pirate, corsair.
forbannir, v. a., to banish.
forbannissement, m., banishment, transportation.
forçat, m., convict, galley slave.
force, f., force, power, strength; skill, ability; purchase; violence; spirit; plenty; capacity; supporting power (of a crane, bridge, raft, boat, etc.); (*in pl., mil.*) forces, troops; (*cons.*) strut; buttress;
— *accélératrice*, accelerating power;
— *armée*, (*mil.*) armed force; troops of all arms; (in France) the — *publique*, q. v., together with any other body or organization bearing, or susceptible of bearing, arms, e. g., rifle clubs, fire brigades, etc.;
avoir de —, (*cord.*) to be strong enough;
— *du bois*, heart, strength of the wood;
à — *de bras*, by strength of arm;
— *brisante*, v. — *explosive;*
— *centrifuge*, centrifugal force;
— *centripète*, centripetal force;
— *de cheval*, horsepower;
— *en chevaux*, horsepower;
— *en chevaux effectifs*, effective horsepower;
— *en chevaux estimés*, estimated horsepower;
— *en chevaux indiqués*, indicated horsepower;
— *en chevaux nominaux*, nominal horsepower;
— *en chevaux théoriques*, theoretical horsepower;
— *contre électro-motrice*, (*elec.*) counter electromotive force;
— *du coup*, (*art.*) energy of the blow (from a projectile);
— *des eaux*, force of water (against column of air pressing on it);
— *élastique*, elastic force;
— *électro-motrice*, (*elec.*) electro-motive force;
entrer à —, to force on;
être en —, (*cord.*) to bear; (*mil.*) to be in force;
— *expansive*, force due to expansion (as of gases);
— *explosive*, (*expl.*) bursting power, explosive force;
faire — *sur*, (*cord.*) to bring a strain on;
flux de —, (*elec.*) flux of force;
— — *des hommes*, man-power;
— *d'impulsion*, momentum;
— *intrinsèque*, (*mil.*) intrinsic strength (of a position);
jambe de —, strut;
— *de jet*, projectile force or energy;
ligne de —, (*elec.*) line of force;
— *des machines*, available power of an engine or machine;
maison de —, jail;
— *majeure*, act of God, act of Providence; overwhelming necessity, stress of weather; *vis major*;
manœuvres de —, (*art.*) mechanical maneuvers;
— *militaire d'un navire*, offensive power of a ship;
— *motrice*, motive power;

force (s) *navales*, sea forces;
— *de pénétration*, (*art.*) penetrating, perforating power;
plan des ——*s*, force plan;
prendre de ——, to force, to assault;
prendre une ville de, *à*, —— *ouverte*, (*mil.*) to carry a town by storm;
—— *de projection* (*d'une arme à feu*), (*ball.*) projectile, shooting, power;
—— *publique*, (*mil.*) in France, all the organizations that may be called upon to preserve order at home, and to repel a foreign enemy, (hence) the army, navy, gendarmerie, custom-house personnel, and the police, both rural and urban; (more particularly) the gendarmerie, (and in war) all troops, etc., on provost duty;
à —— *de rames*, by pulling hard;
—— *de support*, bearing power, capacity;
—— *de traction*, traction, tractive, power;
—— *tractive*, v. —— *de traction;*
tube de ——, (*elec.*) tube of force;
—— *vive, vis viva*, kinetic energy;
de vive ——, by main strength.

forcé, a., forced, hard, severe (of weather); sprung, strained (of a mast, etc.); (*mil.*) forced (of a march);
marche ——*e*, (*mil.*) forced march;
travaux ——*s*, penal servitude; hard labor.

forcement, m., (*art., sm. a.*) (act, means of) forcing (projectiles to take the grooves); (*art.*) compression system of taking grooves; difference between maximum diameter of rotating band and diameter of bore measured from bottom of grooves;
à ——, (*mach.*) by forcing (of assemblages, etc.);
—— *par affaissement*, v. —— *par compression;*
ceinture à, de, ——, (*art.*) rifling band;
—— *par compression*, (*art.*) forcing projectiles to grooves by compression;
couronne de ——, (*sm. a.*) rifling band;
—— *par excès de calibre*, (*sm. a.*) forcing obtained by making diameter of bullet greater than that of the bore;
—— *par expansion*, (*art., sm. a.*) forcing by expansion;
—— *par inertie*, v. —— *par compression;*
—— *au maillet*, (*sm. a.*, obs.) system in which bullet was splayed out by striking with a mallet on iron ramrod;
système à ——, (*art.*) compression system.

forcer, v. a., to force, strain; to work; to compel; to break open or through; to wrench, to bend; (*mil.*) to overcome, storm;
—— *sur les avirons*, to pull hard;
—— *le blocus*, to force, run, a blockade;
—— *un cheval*, to override a horse;
—— *la consigne*, v. s. v. *consigne;*
—— *le fer*, (*fenc.*) to engage the adversary's blade strongly;
—— *la main*, —— *à la main*, (*man.*) of a horse, to refuse to obey, to force the bit, to bolt;
—— *une, la, marche*, (*mil.*) to make a forced march;
—— *une passe*, (*nav.*) to force a channel, to run by;
—— *un poste*, etc., (*mil.*) to force or carry a post (a defile, passage, etc.);
—— *de, des, rames*, v. —— *sur les avirons;*
se ——, (*art., sm. a.*) to take the grooves.

forces, f. pl., shears (e. g., tailor's, tinmen's);
faire les ——, (*man.*) to open the mouth wide instead of backing (of a horse that his rider is trying to back).

forcettes, f. pl., small shears; masting shears.

forceur, m., —— *de blocus*, blockade runner.

forcite, f., (*expl.*) forcite;
—— *antigrisouteuse*, fire-damp forcite;
—— *gélatine*, a mixture of nitroglycerin and soluble nitrocellulose.

forer, v. a., to bore, to drill, to pierce, to perforate; (*art.*) to bore (a gun);
banc à ——, boring bench;
instrument à ——, borer;
machine à ——, boring machine.

forerie, f., boring house, shop; act of boring; boring machine; (*art.*) gun-boring machine; gun lathe;
banc de ——, boring bench;
—— *horizontal*, horizontal drilling machine;
—— *verticale*, vertical drilling machine.

foret, m., borer, drill, boring bit; boring tool, instrument; twisted bit;
—— *américain*, twist drill;
—— *anglais*, centrifugal drill;
—— *à arçon*, breast drill;
—— *à bois*, —— *de charpentier*, auger;
—— *à canon*, borer, boring bit;
—— *à cuiller*, shell auger;
—— *à langue d'aspic*, flat drill;
—— *à langue de carpe*, first bit in gun boring, rough borer;
—— *à main*, hand drill;
—— *mi-rond*, half-round drill;
—— *à rochet*, ratchet drill, brace;
—— *russe*, (*art.*) annular borer;
—— *tors*, twist drill;
—— *torse*, twisted bit or borer.

forêt, f., forest, forest land, woodland;
—— *sauvage*, trackless forest; one without paths or roads.

foreur, m., borer (workman).

foreuse, f., (*mach.*) drilling machine.

forfait, m., contract, job;
à ——, by the job; by contract;
acheter (*vendre*) *à* ——, to buy (sell) a job lot;
faire un ——, to contract, make a contract for anything.

forge, f., forge; iron works; smith's forge; blacksmith's shop, smithy; (*art.*) battery forge;
—— *d'affinerie*, (*met.*) bloomery, finery;
—— *à l'anglaise*, (*met.*) rolling mill;
brut de ——, rough-forged;
—— *de campagne*, (*art.*) traveling forge; battery forge; field forge;
fine ——, smith's coal;
—— *fixe*, —— *de garnison*, (*art.*) garrison forge (as opposed to a field forge);
fourgon- ——, (Fr. *cav.*) wagon forge, forming part of the cavalry regimental train (one to each squadron);
—— *de maréchal-ferrant*, farrier's forge;
mener un cheval à la ——, (*farr.*) to take a horse to the blacksmith;
—— *de montagne*, (*art.*) mountain forge;
opération de grosse ——, (*met.*) the working of steel or of iron in large masses;
pièce de ——, forging;
—— *portative*, hand forge, portable forge;
—— *roulante*, (Fr. *a.*) the forge in use by the companies of the train (road forge);
—— *stable*, (*art.*) permanent forge (in a siege park);
venir de ——, to be forged at the same time with the main body or piece (said of small parts, e. g., front-sight mass); to be integral with;
venu de ——, forged on (as opposed to inserted or let in); integral with;
—— *volante*, portable forge.

forgé, p. p., forged; wrought;
—— *à chaud* (*froid*), hot (cold) forged;
fer ——, wrought iron.

forgeable, a., forgeable, malleable.

forgeage, m., (*met.*) forging, operation of forging.

forger, v. a., to forge, to hammer, to weld, to forge on; (*hipp.*) to overreach, (strictly) to forge;
action de ——, (*hipp.*) forging, clicking;
—— *à chaud* (*froid*), (*met.*) to forge hot (cold);
—— *à la mécanique*, (*met.*) to work, shape, forge, by machinery (as opposed to shaping, etc., by hand);
presse à ——, (*met.*) forging press.

forgerie, f., smithing, smithery.

forgeron, m., blacksmith.

forgeur, m., blacksmith;
—— *de canons de fusil*, (*sm. a.*) barrel forger;
cannonier- ——, v. —— *de canons de fusil.*

forgeuse, f., —— *américaine*, (*mach.*) forging machine (so called in France because of origin); Brown & Sharpe forging press.

forgis, m., wire iron (of which wire is to be made).

forjet, m., (*cons.*) projection (out of line, out of plumb).

forjeter, v. a., (*cons.*) to jut out, bulge, project; to be out of line, out of plumb.

forjeture, f., v. *forjet.*

format, m., size (of books, of paper);
—— *in-8*, etc., octavo, etc.

formation, f., formation, forming, organization; (*mil.*) formation, organization; training; (*Fr. a, med.*) medical unit (e. g., *dépôt de convalescents, hôpital d'évacuation*);
—— *d'attaque*, (any) attack formation;
—— *d'attente*, any formation preliminary to a maneuver or operation;
—— *en bataille*, (*art., cav.*) formation in line;
—— *par bataillons en masse*, (*inf.*) line of battle with battalions in column of divisions, and deploying intervals, etc. (30 paces);
—— *en batterie*, (*art.*) formation in battery;
—— *des cadres*, instruction of subaltern grades, with special reference to training them to act as drillmasters for recruits;
—— *en colonne*, formation in column;
—— *en colonne par pièce (section)*, (*art.*) formation in column of pieces (sections, i. e., platoons, U. S. Art.);
—— *de combat*, combat formation;
—— *combinée*, a generic term for formations in echelon, for formations involving the use of offensive and defensive flanks, etc.;
—— *constitutive*, the organic constitution of a troop unit;
—— *de défilé*, march-past, review, formation;
—— *déployée*, any deployed formation;
—— *dérivée*, derived formation (e. g., double column);
—— *doublée*, (*inf.*) formation in which two ranks (either double or single) take the place of one (either double or single) when forming for a march by the flank;
—— *échelonnée*, formation in or by echelon;
—— *éventuelle*, formation not carried out at once, as a single maneuver, but involving a stay or delay between the first and last of it;
—— *par file*, formation by file;
—— *par le flanc*, formation by the flank;
—— *groupée*, dense formation, troops in mass;
—— *en ligne*, formation in line;
—— *en ligne de colonnes*, formation in line of columns;
—— *en ligne déployée*, (*inf.*) formation in line (company, battalion, regiment);
—— *de manœuvre*, drill formation;
—— *de marche*, march or route formation;
—— *en marche*, march formation;
—— *en masse*, formation in mass;
—— *normale*, (so-called) normal formation (e. g., line of columns, mass, etc.);
—— *au parc*, v. —— *de rassemblement; passage, passer, d'une* —— *à une autre*, v. s. vv. *passage, passer;*
—— *ployée*, any ployed formation;
—— *de position d'attente*, v. —— *d'attente;*
—— *préparatoire de combat*, (*art.*) the formation taken by a battery on approaching the emplacement whence it is to open fire, consisting of the separation of the *batterie de tir*, q. v., or of the group of such batteries, from the *échelon*, q. v., or group of *échelons;*
—— *de rassemblement*, assembling, close, formation, e. g., of troops assembling from march formation preparatory to passage to combat formation; (*Fr. art.*) a synonym for the formation of a park (—— *avec matériel*): the assembling of the teams and detachments preparatory to hitching up the guns and wagons (—— *sans matériel*);
—— *de rendez-vous*, an assembling formation by which a c. o. gets his troops well together and in hand in a designated place before undertaking definite operations against a specific objective;
—— *de revue*, formation for inspection;
—— *de route*, route formation, road formation;
—— *sanitaire*, (*Fr. a. med.*) any administrative unit of the medical service;
—— *tactique*, tactical formation; any formation relating to a definite tactical object; any formation enabling a troop unit to make the best possible disposition of its means of offense and of defense;
—— *d'un train* (*r. r.*) making up a train.

forme, f., form, shape, figure, make; shoemaker's last; mold, frame; seat; layer of sand for paving stones; block; (*siege*) stage (in trench work); (*hipp.*) ring bone; bone spavin; (*nav.*) dry dock;

forme, *de* ——, of proper, suitable, form; of a special shape (of tools);
—— *flottante*, (*nav.*) floating dock;
—— *de radoub*, (*nav.*) graving dock;
sape sans ——*s*, (*siege*) sap driven at full size (without subsequent enlargement);
—— *sèche*, (*nav.*) dry dock;
—— *tourmentée*, irregular, twisted form.

former, v. a. r., to form, frame, make, shape, mold, fashion, cut out, compose, plan, train, season; (*mil.*) to set up, drill, and instruct soldiers or sailors and to give them the military spirit; (*mil.*) to form up;
—— *le carré*, (*mil.*) to form square;
—— *les faisceaux*, (*mil.*) to make stacks;
—— *la haie*, (*mil.*) to form troops one rank on each side of the street, or of a cortege;
—— *un siège*, (*siege*) to begin a siege, to begin to open the trenches;
—— *une troupe*, (*mil.*) to draw up a body of troops in a definite formation.

formulaire, m., (*med.*) pharmacopœia.

formule, f., formula; form, model (as, for proceedings); (*med.*) prescription;
—— *d'équarrissage*, in strength of materials, a formula showing relation between the exterior forces, factor of safety, and the dimensions (unknown) of the section considered;
—— *d'élasticité*, —— *d'équilibre*, Hooke's law.

fort,, a., strong; thick; considerable; skillful; high (of the wind); heavy (of the sea, rain); (*mil.*) strong (with reference to numbers), numbering; able to resist; strong (of a position), fortified (of a city, position);
bois ——, thick wood;
colle ——*e*, glue;
—— *de* (25,000 *hommes*), (25,000) strong;
place ——*e*, (*fort.*) stronghold; fortress, fortified position;
—— *de poids*, overweighted, overloaded;
terre ——*e*, heavy ground;
ville ——*e*, fortified city.

fort, m., strongest, thickest, or principal part; main point; midst; heart; depth; brunt, center, middle, etc.; skill; strength; (*fort.*) fort (contains no civilian population), esp. detached fort; redoubt; blockhouse; (*mil.*) thick (of a fight); (*sm. a.*) part of the sword blade near the hilt (cf. *faible*).
(Except where otherwise indicated, the following words relate to fortifications:)
—— *d'arrêt*, stop fort; barrier fort (English name);
—— *d'attaque*, fort lying in an attack, hence one that is attacked;
au ——, *de*, (*in gen.*) in the thick of;
—— *bastionné*, bastioned fort;
—— *à batterie basse*, fort with battery near the ground;
—— *à batterie haute*, fort with elevated battery, as in a cavalier, etc.;
—— *de bois*, quarter grain (of wood);
—— *de campagne*, field work;
—— *circulaire*, circular fort, martello tower;
—— *du combat*, (*mil.*) thick of the fight;
—— *continental*, land fort (not intended for seacoast defense);
—— *de côte*, seacoast work;
—— *côtier*, seacoast fort in whose neighborhood troops may be landed, and which must therefore be provided with ditches and flanking defenses; a seacoast fort exposed to attack by landing parties;
—— *à coupoles*, a fort provided with cupolas or turrets;
—— *d'un courant*, main part, middle, of the stream;
—— *à crête basse*, a fort of low crest, for small-arm fire;
—— *à crête haute*, a fort of high crest, for artillery fire;
—— *à crêtes redoublées*, name of forts having cavaliers or parados, having two interior crests;
—— *à crête unique*, fort having but one interior crest;
—— *cuirassé*, armored fort or work;
—— *demi-bastionné*, semibastioned work;
—— *détaché*, detached work;
—— *à étoile*, —— *étoilé*, star fort;

fort *étudier le —— et le faible d'une position*, (*mil.*) to make a thorough study of a position;
—— *d'interdiction*, v. —— *d'arrêt*;
—— *isolé*, a fort on lines of communication, at the forks of a road; a fort intended to close a defile; substantially the same as —— *d'arrêt*, q. v.;
—— *de la lame*, (*sm. a.*) thick of the blade, part near the hilt;
—— *de liaison*, work between two others, and in a sense (military) joining them;
—— *maritime*, seacoast work;
—— *à massif central*, work having a parados or central mass;
—— *de la mêlée*, (*mil.*) thick of the fight, of the mellay;
—— *à la mer*, fort surrounded by water, or at the end of a pier or jetty;
—— *de mer*, seacoast fort; name given by General Brialmont to works built on sand banks, or on rocks that are either submerged or at water level;
mi- ——, (*sm. a.*) middle of a blade;
—— *d'occupation*, a fort occupying a salient point in a fortified perimeter;
—— *plat*, v. —— *à crête unique;*
—— *de protection*, v. —— *d'occupation;*
—— *-réduit*, name applied to the fort to which belong several batteries annexed;
—— *semi-maritime*, a fort partly land, partly sea, intended to works built on sand banks, to repel attacks from both sea and land;
—— *de surveillance*, "watch fort," on high promontory;
—— *tenaille*, tenailled fort; fort of tenaille trace.
forteresse, f., fortress (has a civilian population);
artillerie de ——, v. s. v. *artillerie;*
guerre de ——, siege warfare;
—— *maritime*, —— *de mer*, (*fort.*) seacoast work or fortress;
—— *mobile*, (*fort.*) depot of all necessary supplies, etc., for the creation of a fortified position when needed (as opposed to permanent forts constructed in peace);
troupes de ——, fortress troops.
fortification, (*fort.*) fortifying, fortification; art of fortification; work;
—— *angulaire*, redan system of fortification; tenaille system of fortification;
——*s d'attaque*, offensive work;
—— *bastionnée*, —— *à bastions*, bastion system of fortification;
—— *de campagne*, field fortification (i. e., deliberate); deliberate intrenchments;
—— *du champ de bataille*, field intrenchments; hasty intrenchments;
—— *collatérale*, flanking work;
—— *continentale*, land work (esp. used in connection with seacoast works, to distinguish land from sea works);
—— *côtière*, seacoast fort;
—— *cuirassée*, armored fortification;
—— *extérieure*, exterior fortification; outworks;
—— *improvisée*, v. —— *du champ de bataille;*
—— *irrégulière*, irregular fortification (no system);
—— *maritime*, coast defense;
—— *mixte*, semipermanent fortification;
—— *moderne*, v. —— *polygonale;*
—— *naturelle*, any natural position susceptible, without labor, of being used as a fortification, e. g., woods, swamps, very steep slopes, etc.;
——*s offensives*, works of attack;
—— *passagère*, temporary, improvised, fortification; v. —— *de campagne;*
—— *permanente*, permanent fortification;
—— *perpendiculaire*, perpendicular system of fortification;
—— *polygonale*, polygonal fortification;
—— *portative*, (*inf.*) generic term applied to proposed incasement in armor of a portion of infantry troops;
—— *de position*, deliberate fieldwork;
—— *provisoire*, provisional fortification;
—— *à revers*, a fortification to resist reverse fire;
—— *semi-permanente*, v. —— *provisoire;*
—— *souterraine*, underground fortification;

—— *tenaillée*, tenailled fortification.
fortifier, v. a., to strengthen, to corroborate; (*fort.*) to fortify;
—— *en dedans*, (*fort.*) to construct the bastions inside the polygon;
—— *en dehors*, (*fort.*) to construct the bastions outside the polygon.
fortin, m., (*fort.*) small fort, fortlet, fieldwork, very small fort or work.
fortis, m., (*expl.*) name (in Belgium) of polynitrocellulose.
fortrait, a., overspent, overworked, knocked up (of a horse).
fortraiture, f., (*hipp.*) sickness due to overworking, overfatigue.
fortuit, a., casual, incidental;
cas ——, casualty, emergency.
fortune, f., fortune;
de ——, jury, makeshift;
—— *de mer*, risks of the sea;
officier de ——, (*mil.*) an officer risen from the ranks;
soldat de ——, (*mil.*) soldier of fortune.
forure, f., bore hole, drilled hole.
fosse, f., ditch; trench, pit;
—— *d'aisance*, cesspool;
basse- ——, dungeon;
—— *à couler*, —— *de coulée*, (*fond.*) casting pit;
—— *de fonderie*, (*fond.*) foundry pit, casting pit;
—— *à mouler*, (*fond.*) molding pit;
—— *aux moules*, v. —— *de fonderie;*
—— *à tuber*, (*art.*) assembling pit, shrinkage pit.
fossé, m., ditch, moat, drain, fosse, trench; (*fort.*) moat, ditch;
—— *d'arrêt de l'eau*, catch drain;
avant- ——, (*fort.*) advanced ditch;
—— *de ceinture*, catch-water drain;
—— *du corps de place*, v. —— *principal;*
dégorger un ——, to empty a ditch;
—— *de dehors*, (*fort.*) ditch of outworks;
descente de ——, v. s. v. *descente;*
descente de —— *blindée*, v. s. v. *descente;*
—— *de dessèchement*, drain;
—— *diamant*, (*fort.*) drop ditch, diamond ditch;
—— *d'écoulement*, ditch, drain;
—— *de l'écrou*, (*F. m. art.*) (in some guns, M. 64/66) that part of the breech nut from which threads have been cut away;
—— *à fond angulaire*, triangular ditch;
—— *à fond de cuve*, a ditch with steep sides;
—— *inondé*, (*fort.*) wet ditch;
—— *à manœuvre d'eau*, (*fort.*) ditch with water maneuver; ditch wet or dry at discretion;
—— *passage de* ——, (*siege*) passage of the ditch;
—— *plein d'eau*, (*fort.*) wet ditch;
—— *principal*, (*fort.*) main ditch;
remblayer un ——, to fill up a ditch;
—— *revêtu*, (*fort.*) revetted ditch;
saigner un ——, to drain a ditch;
—— *sec*, (*fort.*) dry ditch;
—— *à sec*, (*fort.*) dry ditch (i. e., emptied of its water);
—— *de séparation*, intervening ditch;
talus d'escarpe et de contrescarpe du ——, slopes of the ditch.
fossoyage, m., digging, excavating, of ditches; ditching, operation of surrounding by a ditch.
fossoyer, v. a., to ditch, moat, trench; to dig moats or ditches; to surround by a ditch.
fou, a., mad, crazy, insane; (*mach.*) loose, idle (of a pulley);
aiguille folle, (*elec.*) a crazy or permanently unsteady needle;
balance folle, unsteady balance, i. e., one that will not come to rest;
boussole folle, unsteady, irregular compass;
pièce folle, (*art.*) a piece whose bore is not true;
poulie folle, (*mach.*) idle pulley.
fouailler, v. a., to whip; (*art.*) to destroy by artillery fire;
—— *de la queue*, to switch the tail.
foudre, m., f., lightning; thunderbolt; m. pl., (*unif., Fr. a.*) distinctive mark (forked lightning) on collar or arm (staff, *archivistes, télégraphistes*);

foudre *de guerre*, hero; thunderbolt of war; (*in pl.*) guns, artillery.
foudroyante, f., (*artif.*) kind of rocket, making a noise intended to represent thunder.
foudroyer, v. a., to strike with lightning; to destroy, crush, ruin; (*art.*) to keep up an incessant, vigorous fire; to batter down; to crush (used of small arms).
fouet, m., whip, horsewhip, lash, cracker, snapper (of a whip); whipping, flogging, cat; whipcord; thong; (*hipp.*) tip of a horse's tail; (*cord.*) whip, tail;
amarrage à ——, (*cord.*) lashing of a tail block;
—— *d'armes,* cat-o'-nine-tails;
atteindre de plein ——, (*art.*) to make a direct hit;
avoir le ——, to get a flogging;
balle de plein ——, (*i. p.*) direct hit;
claquement d'un ——, crack, cracking of a whip;
claquer d'un ——, to crack a whip;
coup de ——, each blow with a whip; (*expl.*) the whipping out of a Bickford, etc., fuse badly inserted and secured, when the flame reaches the end;
coup de plein ——, (*art.*) direct hit;
—— *à neuf cordes,* cat o'-nine-tails;
feu de plein ——, v. s. v. *feu;*
poulie à ——, tail block;
—— *de rênes,* (*art.*) whip attached to reins;
tir de plein ——, (*art.*) direct fire with full service charge.
fouettement, m., whipping; vibrating; (*art.*) bouncing or jumping of a gun when fired; (*mach.*) thrashing, whipping, whip action (of a connecting rod);
—— *de la volée* (*art.*) thrashing of the muzzle.
fouetter, v. a., to whip, lash, horsewhip; to scourge; to beat (as rain, hail); to be cutting (of wind); (*cord.*) to whip; to lash a tail block; (*art.*) to thrash (of the chase), to sweep with shot; (*hipp.*) to tic the scrotum;
—— *en plein sur,* (*art.*) to fall directly upon (of projectiles).
fougade, v. *fougasse.*
fougasse, f., (*mil.*) fougasse;
—— *à bombes,* shell fougasse;
—— *en déblai,* stone fougasse of the usual type (axis at 45°, and the excavated earth forming a parapet behind);
—— *à feu rasant,* grazing fougasse, stone fougasse with horizontal direction of fire;
—— *mobile,* (*art.*) term fancifully applied to modern shell filled with high explosive;
—— *ordinaire,* powder fougasse (i. e., no stones, etc., to be thrown forward);
—— *pierrier,* stone fougasse;
—— *Piron,* fougasse in which powder is on side of excavation instead of at bottom;
—— *rase,* covered fougasse (stone) (flush with the ground, axis at 60°, and excavated earth removed, so as not to show);
—— *en remblai,* a fougasse constructed on the surface of the ground;
—— *surchargée,* overloaded fougasse.
fougère, f., fern; *en* ——, herringbone-wise.
fougue, f., mettle, spirit (of horses); (*artif.*) rocket without stick, moving horizontally.
fouguette, f., (*artif.*) small war rocket.
fougueux, a., fiery, mettlesome, high-spirited (horse); unruly.
fouille, f., digging, excavating; earth dug out or up; excavation; (*art.*) cavity in (brass) guns, caused by long-continued firing; (*met.*) cavity (defect) in a casting, filled with dross, slag, etc.
fouillement, m., v. *fouille.*
fouiller, v. a. n., to dig, to excavate, to search by digging; to trench; to work a mine; to rake, search; to search, as with a search light; (*hipp.*) to make a rectal examination by hand; (*mil.*) to explore, as a wood or thicket;
—— *des obstacles,* (*mil.*) to reconnoiter obstacles;
—— *le terrain,* (*mil.*) to search, explore, the ground; (*mil. sig.*) to search the terrain, to examine it, by telescope or by search light;
—— *un terrain,* (*mil.*) to cover a bit of ground with projectiles, so as to drive the enemy out.

fouine, f., pitchfork.
fouir, v. a., to dig; to sink a well.
fouissement, m., action of digging.
foulage, m., fulling.
foulant, a., forcing;
pompe ——*e,* force pump.
foule, f., fulling; crowd, great number.
foulé, p. p., galled, sprained; (*hipp.*) foundered;
cheval qui a les jambes ——*es,* foundered horse.
foulée, f., tread (of a staircase); blast (of a bellows); (*man.*) tread, step; contact, instant, during which the horse's foot rests, bears, on the ground;
ligne de ——, line of tread (of a staircase).
fouler, v. a., to gall, hurt, sprain, strain;
—— *aux pieds,* to trample under foot.
fouloir, m., (*art.*) rammer (usually *refouloir,* q. v.).
foulon, m., fuller;
argile, terre, à ——, fuller's earth.
foulonnage, m., fulling.
foulure, f., fulling; sprain, bruise, strain; (*hipp.*) surbating, warble, saddle-gall.
four, m., oven; kiln; stove; hearth; fireplace; bakery, bakehouse, bake oven; (*met.*) furnace;
—— *acide,* (*met.*) oven with acid lining;
—— *basique,* (*met.*) oven with basic lining;
—— *à blanchir la fonte,* (*met.*) finery furnace;
—— *à boulanger,* (*met.*) flat coke oven or furnace (for making coke);
—— *à briques,* brickkiln;
—— *de campagne,* (*mil.*) field oven;
capeler un ——, v. *capeler;*
—— *à carneaux,* (*met.*) furnace in which the flame circulates around and about crucibles (these filling the bed) through tubes, e. g., tube boilers;
—— *de cémentation,* —— *à cémenter,* (*met.*) cementation furnace; case-hardening furnace;
—— *de chauffage,* (*art.*) heating furnace (for shrinking, etc.);
—— *à chaux,* limekiln;
—— *à chaux coulant,* perpetual or draw kiln, running kiln;
—— *à cheminée,* furnace working by natural draft;
—— *à coke,* coke oven;
—— *de construction,* (Fr. a.,) masonry (bake) field oven; bake oven built of material found on the spot;
—— *continu,* perpetual kiln;
—— *coulant,* perpetual kiln; (*met.*) tap furnace;
—— *à cuisson continue* (*discontinue*), perpetual (intermittent) kiln;
—— *à cuve,* (*met.*) generic term for furnaces (e. g., blast) that have no separate fireplace;
—— *démontable,* (Fr. a.) bake oven that takes apart;
—— *dormant à tôle,* (*met.*) plate-heating furnace;
—— *à enceinte permanente,* permanent furnace;
—— *à enceinte temporaire,* temporary furnace;
—— *à foyer intérieur,* furnace with interior hearth;
—— *de fusion,* (*met.*) melting furnace;
—— *à gaz,* gas stove or furnace; (*met.*) gas furnace;
—— *intermittent,* intermittent kiln;
—— *locomobile,* (Fr. a.) traveling bake oven;
—— *à manche,* (*met.*) generic term for furnaces less than 2 meters high;
—— *oscillant,* (*met.*) oscillating hearth (for puddling);
—— *à pain,* (*mil.*) baking oven;
—— *à plâtre,* parget kiln;
—— *à poitrine fermée,* furnace with closed breast;
—— *à poitrine ouverte,* furnace with open breast;
—— *portatif en tôle,* sheet-iron portable field oven;
—— *à puddler,* (*met.*) puddling furnace;
—— *à réchauffer,* (*met.*) heating, reheating, furnace;
—— *à recuire,* (*met.*) annealing furnace;

four à *recuire les tôles*, (met.) plate-annealing furnace;
—— *réverbère*, (met.) reverberatory furnace;
—— *rotatif*, (met.) rotating hearth (for puddling);
—— *roulant*, bake oven on wheels; traveling bakery; (mil.) field oven;
—— à *sole*, (met.) any hearth with a bed;
—— à *sole fixe*, (met.) hearth with fixed bed;
—— à *sole mobile*, (met.) hearth with movable bed;
—— à *sole tournante*, (met.) hearth with rotating bed;
—— *soufflé*, (met.) generic term for blast furnace;
—— *tuyère*, (met.) furnace using forced draught;
—— à *vent*, (met.) draught furnace, blast furnace (generic term).

fourbir, v. a., to furbish, to burnish.

fourbissage, m., "bright work."

fourbisseur, m., sword cutler; furbisher.

fourbu, a., foot-sore (of men); (hipp.) foundered.

fourbure, f., (hipp.) foundering, foot founder, laminitis, fever of the foot.

fourche, f., fork; pitchfork; crutch; sheers; bicycle fork; (inst.) Y, wye (of a leveling instrument);
—— *de l'arçon*, (harn.) saddle fork;
—— *-arrêt*, f., double clip, spring clip;
—— *arrière*, —— *d'arrière*, rear fork (of a bicycle);
—— *d'avant*, front fork (of a bicycle);
—— *de bielle d'excentrique*, (mach.) eccentric rod fork;
—— *de carrosse*, prop for a carriage going uphill;
—— *d'embrayage*, (mach.) belt fork, belt shifter;
—— *d'excentrique*, (mach.) eccentric fork;
piton à ——, forked bolt, crotch;
—— *de rampart*, crotch, crutch;
—— *de sape*, (siege, fort.) sap fork, sap hook.

fourchette, f., small fork; fork; kind of calipers or compass used in plane-table surveying; rest, prop; hind guide or futchel of a carriage; (art.) (in ranging) the distance between an over and a short, between the long and the short shot; fork, bracket; (limber) fork; (hipp.) frog, frush; (mach.) belt guide, belt shifter;
—— *en durée*, (art.) the "time fork," in getting the range by observing the times of burst of successive shells;
—— *échauffée et pourrie*, (hipp.) thrush,
—— *étroite*, (art.) the fork corresponding to one-quarter of a turn of the elevating screw for percussion fire, to one-half turn for time-fuse fire;
—— *-guide*, (mach.) belt shifter;
la grande ——, (art.) in ranging, the fork obtained by increasing the elevation from an observed short;
—— *large*, (art.) the fork whose limits are obtained by one turn of the elevating screw;
—— *de percussion*, (sm. a.) percussion fork;
queue de ——, fork of hind guides of a wagon;
régler le tir à la ——, (art.) to get the range by bracketing;
la ——*-resserrée*, (art.) in ranging, the fork obtained by reducing the elevation from an observed over;
—— *de serrage*, (F. m. art.) tightening or compressing fork (24em seacoast);
tir à la ——, (art.) bracket, ranging fire;
tirer à la ——, (art.) to find, determine, the fork;
—— *du tour*, (art.) v. —— *large*.

fourchon, m., prong (of a fork); fork (of a tree).

fourchu, a., forked, branching off; cloven, furcate;
arbre ——, forked tree;
chemin ——, crossroad.

fourchure, f., fork, splitting; bifurcation.

fourgon, m., (mil., etc.) van; victual-, ammunition-, baggage-, wagon; military wagon;
—— à *baggages*, baggage wagon;

fourgon, *de correspondance*, mail wagon;
—— *-forge*, forge wagon;
—— *de levée de boîtes*, mail-collecting wagon;
—— *régimentaire*, regimental baggage wagon;
—— *régimentaire à vivres*, regimental provision wagon;
—— *du service de santé*, medical baggage wagon;
—— à *vivres*, provision wagon.

fourmilière, f., (hipp.) cavity in a chronic foundered foot (due to *fourbure*).

tournaise, f., furnace.

fourneau, m., furnace; fire hole; hearth; stove; kiln; chamber (of a mine); (mil. min.) chamber, mine chamber (strictly, loaded mine, or the charge itself).

I. Military mining, etc.:
—— *x accolés*, series of mines intended to be sprung at the same moment, conjunct mines;
—— *d'artifice*, (artif.) furnace of a pyrotechnic laboratory (of two kinds: —— *de 1ère espèce*, in which the flame reaches the sides and bottom of the boiler; —— *de 2ème espèce*, in which the flame reaches the bottom only);
—— *bas*, lower mine, countershaft mine;
—— à *charge après bourrage*, chamber prepared and tamped and then loaded, one that may be loaded after tamping;
—— *contre-puits*, upper mine (above galleries), countershaft mine;
—— *de*, à *démolition*, demolition mine, (more particularly) a mine prepared to blow up works when abandoned;
—— *de 1ère, 2ème, espèce*, v. —— *d'artifice*;
—— *x-étages*, —— *x étagés*, mines in tiers, tiered mines;
indice du ——, ratio of the line of least resistance to the crater radius;
—— *isolé*, independent, single, mine; (a synonym of) *fougasse*; (*in pl.*) isolated, unconnected mines;
—— *de mine*, the charge of a military mine; mine chamber;
—— *ordinaire*, common mine; mine producing a two-line crater;
—— *retirade*, v. —— *de tête*;
—— à *boulets rouges*, —— à *rougir les boulets*, (art.) hot-shot furnace (obs.);
—— *sous-chargé*, undercharged mine, a mine whose crater radius is less than the line of least resistance;
—— *supérieur*, (in a tiered system) upper mine (synonym of *contre-puits*, q. v.);
—— *surchargé*, overcharged mine (globe of compression, though this term is also applied to undercharged mines); mine whose crater radius is greater than the line of least resistance;
—— *de tête*, a mine having a conjugate (—— *retirade*) and whose explosion causes its conjugate to work.

II. Metallurgy, etc.:
—— *en activité*, furnace in blast;
—— *d'affinage*, refining forge, refinery, refining furnace;
—— à (*l'*)*air chaud* (*froid*), hot- (cold-) blast furnace;
—— à *l'anthracite*, anthracite blast furnace;
—— *arrêté*, furnace out of blast;
arrêter un ——, to put a furnace out of blast;
—— à *avant creuset*, forehearth blast furnace;
bas ——, bloomery furnace or hearth;
—— *de calcinage*, —— à *calciner*, calcining furnace;
—— *de cémentation*, à *cémenter*, steel-converting furnace, cementation furnace;
—— à *charbon*, charcoal pit;
—— à *charbon de bois*, charcoal oven;
chauffe du ——, fire hole of a furnace;
—— *au coke*, coke furnace;
—— à *creuset*, crucible furnace;
—— à *creuset intérieur*, a furnace in which the hearth is wholly within the shell;
—— *de cuisine*, kitchen range;
—— à *échauffer*, heating, annealing, surface;
—— *en feu*, v. —— *en activité*;

fourneau à *foyer indépendant*, (*met.*) furnace where hearth and bed are separate;
fumer le ——, v. *préparer le* ——;
—— *de fusion*, melting furnace;
—— *générateur*, gas generator;
—— *de grillage*, roasting furnace;
haut- ——, blast furnace; iron works;
—— *à manche*, cupola furnace;
mettre un —— *en activité*, *à feu*, to put a furnace in blast;
mettre un —— *hors feu*, to put a furnace out of blast;
—— *ondulé*, corrugated furnace;
ouvrage de ——, hearth of a smelting furnace;
percer le ——, to tap the furnace;
préparer le ——, to heat a furnace;
—— *à puddler*, puddling furnace;
—— *de raffinage*, refining furnace;
—— *à réchauffer*, reheating furnace;
—— *à recuire*, annealing furnace;
refroidir un ——, v. *arrêter un* ——;
—— *à réverbère*, reverberatory furnace;
—— *rotatoire*, rotary furnace;
—— *à rôtir*, roasting furnace;
—— *à sécher*, drying furnace;
—— *à vases clos*, (*met.*) generic term for furnaces in which the charge is separated from and out of contact with the combustible and its products, closed-vessel furnace;
—— *à vent*, blast furnace.

fournée, f., charge of a furnace; baking (batch) length of time needed to bake and turn out a batch of bread.

fourniment, m., (*mil.*) accouterments, etc., (hence) kit.

fournir, v. a. n., to furnish, provide, stock, supply; to contribute, afford; to complete, make up;
—— *son air*, (*man.*) (of a horse) to execute his paces accurately;
—— *la carrière*, (*man.*) to go over the whole course;
—— *un coup d'épée*, (*fenc.*) to make a good thrust.

fournisseur, m., furnisher, contractor;
—— *de l'armée*, (*mil.*) army contractor;
—— *de navires*, ship chandler.

fourniture, f., furnishing, supplying, providing; (hence) materials; things furnished, supply, provision; furniture (esp. bed furniture);
—— *de bureau*, stationery;
—— *en campagne*, (*mil.*) field stores, equipage;
——*s de couchage auxiliaire*, (*Fr. a.*) bed furniture belonging to the state;
—— *en garnison*, (*mil.*) garrison stores, equipage;
——*s d'infirmerie*, (*Fr. a.*) hospital bedding;
——*s des lits militaires*, (*Fr. a.*) bed furniture belonging to contractors who furnish bedding;
——*s d'officier*, (*Fr. a.*) officers' bedding;
——*s à la ration*, (*Fr. a.*) stores supplied in such form as to be ready for immediate use;
—— *en route*, (*mil.*) travel stores, supplies (as bread, transportation of officers' baggage, billets, etc.);
——*s de soldat*, (*Fr. a.*) bedding.

fourrage, m., forage; all that pertains to forage; fodder; (*mil.*) forage, foraging, foraging party; (*cord.*) serving;
aller au ——, (*mil.*) to go on a foraging expedition;
envoyer au ——, (*mil.*) to send out foraging;
nourrissage au ——, stall feeding;
——*s-racines*, edible roots;
—— *sec*, fodder; hay;
—— *au sec*, foraging for hay, etc.;
—— *vert*, grass, green fodder;
—— *au vert*, foraging for grass or any other green fodder.

fourrager, v. a. n., (*mil.*) to forage; to ravage, pilfer, plunder, pillage;
—— *au sec*, to forage for hay;
—— *au vert*, to forage for grass, or to cut grass.

fourragère, f., tail or forage ladder, forage rack (on the tail of a wagon); railwork for carrying fodder; forage wagon; (*unif.*) patrol jacket; helmet cord, busby cord;
chariot- ——, (*mil.*) forage wagon;
—— *de galerie*, (*art.*) sort of rack or frame attached to the *galerie* (q. v.) of the rear chest of a 120^{mm} caisson; (French caisson, model 1888) device on chest for one bag of oats (65 kilograms).

fourrageur, m., forager; marauder; (*Fr. cav.*) (in extended order) trooper using edge weapons;
attaque, *charge*, *en* ——, (*cav.*) charge in extended order;
rouler en ——, to roll up (as a cloak) and wear crosswise over the shoulder.

fourré, p. p., lined; (*top.*) thickly overgrown, wooded, thick (of undergrowth, bushes, etc.); woody, full of thickets;
bois ——, wood full of thickets and underbrush;
bottes de foin ——*es*, mixed hay (good and bad);
coup ——, (*fenc.*) counterstroke, counterthrust; (*in pl.*) interchanged thrusts and blows;
paix ——*e*, false, patched-up peace;
pays ——, country full of woods, hedges, streams, etc.

fourré, m., thicket, brake.

fourreau, m., (*sm. a.*) scabbard, sheath, case, cover (of a sword, bayonet, etc.); pistol holster; (*harn.*) trace pipe, leather pipe for horse traces; (*hipp.*) sheath; (*mach.*) expansion joint;
—— *de baïonnette*, (*sm. a.*) bayonet scabbard;
—— -*calibre*, (*sm. a.*) standard scabbard, into which saber blade must go without difficulty;
coucher dans son ——, to sleep in one's clothes;
—— *d'essieu*, axle sheath or seat; (*art.*) axle reenforcing plate;
faux- ——, outer case, sheath, or scabbard (to protect the real one);
—— *de hausse*, (*art.*) sight sheath, base, bushing;
mettre au ——, to sheathe (a sword);
pisser dans le ——, (*hipp.*) to have the sheath swollen, or too narrow, or obstructed;
—— *de pistolet*, (*sm. a.*) pistol holster;
remettre, *rentrer*, *dans le* ——, to sheathe (a sword);
—— *de tête*, (*art.*) sponge cap;
tirer du ——, to unsheathe (a sword).

fourrelier, m., scabbard maker.

fourrer, v. a., to fur, to line with wood, to pack, to stuff; (*cord.*) to serve (a rope, etc.);
—— *un cordage*, (*cord.*) to serve;
maillet à ——, (*cord.*) serving mallet;
mailloche à ——, v. *maillet à* ——.

fourrier, m., (*Fr. a.*) a term equivalent to quartermaster (indicates a duty, not a grade);
—— *adjoint*, (*inf.*) a n. c. o. who acts as sick marcher, and performs certain clerical duties;
brigadier- ——, (*art.*, *cav.*) quartermaster-corporal (n. e.), performs the duty of —— *d'ordre*, q. v.;
caporal- ——, (*inf.*) quartermaster-corporal (n. e.);
aux ——*s de distribution*, (call) issue;
maréchal des logis- ——, (*art.*, *cav.*) quartermaster-sergeant;
—— *d'ordre*, (*inf.*) a sergeant detailed every week from each battalion to take the regimental report every morning to headquarters (*au rapport*) in the *service des places*, and to bring back details, orders, etc.;
—— *porte-fanion*, (*inf.*) color-bearer (of a battalion);
—— *secrétaire*, assistant or clerk to the *adjutant de semaine*;
sergent- ——, (*inf.*) quartermaster-sergeant.

fourrière, f., wood yard; pound (animals); horse pound.

fourrure, f., plank lining; fur, furring; (*cord.*) service (of a rope).

fouteau, m., beech tree.

foyer, m., hearth, fire in a hearth; home, one's own country; (*mct.*) hearth; (*mach.*) fire box, furnace, stokehole; fire hole; (*nav.*) furnace; (*elec.*) electric light; (*opt.*) focus;
—— *d'affinage*, (*met.*) refining hearth;
—— *d'affinerie*, (*met.*) finery hearth or furnace;
autel de ——, furnace bridge;
avant- ——, v. *avant-foyer;*
barre du ——, fire bar;
—— *chauffé mécaniquement*, self-feeding furnace;
cul de ——, furnace crown;
—— *de finerie*, v. *d'affinage;*
—— *fumivore*, smoke-burning furnace;
—— *de fusée*, fuse cup;
grand ——, (*met.*) boshes of a blast furnace;
—— *à grille*, (*met.*) grate furnace;
grille du ——, fire grate;
—— *lumineux*, generic term for focus or center of light;
—— *de mine*, focus of a mine;
—— *ondulé*, corrugated furnace;
—— *ordinaire*, plain furnace;
porte du ——, fire door;
—— *principal*, (*opt.*) principal focus;
—— *de raffinerie*, v. —— *d affinage;*
—— *à réchauffer*, reheating hearth;
—— *réel*, (*opt.*) real focus;
rentrer dans ses ——*s*, to go back home (from maneuvers, etc.);
renvoyer dans ses ——*s*, to send home;
—— *supérieur*, (*met.*) main body, belly of a blast furnace;
—— *virtuel*, (*opt.*) virtual focus.

frac, m., (*unif.*) full dress single-breasted tunic (obs., except for the *corps du contrôle*, who, however, never wear it).

fraction, f., fraction;
—— *constituée*, (*mil.*) a complete unit; any part of a regiment or organic body whose constitution is provided for by the law on the organization of the army, i. e., any part having an organic existence;
—— *guide*, (*art.*) in ranging, the fraction of the battery intrusted with keeping the range, i. e., that keeps up ranging fire;
—— *non-guide*, (*art.*) the fraction of a battery that does the real firing against enemy with the range already got.

fractionnement, m., (act of) subdividing, of breaking up into parts.

fractionner, v. a., to subdivide, to break up into parts.

fracture, f., breaking, fracture, rupture;
—— *comminutive*, (*med.*) comminuted fracture;
—— *composée*, (*med.*) compound fracture;
—— *simple*, (*med.*) simple fracture.

fracturer, v. a., to break (a leg, an arm).

fragment, m., fragment;
—— *d'obus*, (*art.*) splinter of a shell.

fragmentation, f., fragmentation; (*art.*) breaking up of shell, etc., on bursting; fracture of shell into fragments.

fraicheur, f., cat's-paw; puff of wind.

fraichir, v. n., to freshen, to blow up fresh (of wind).

frais, m. pl., expenses, cost; (*law*) costs;
—— *d'allèges*, lighterage;
—— *de bureau*, allowance for office expenses;
—— *de justice*, (*mil.*) costs;
—— *de déplacement*, v. —— *de route;*
—— *d'entretien*, cost of maintenance;
faux- ——, extra charges, incidental expenses;
—— *d'inhumation*, burial expenses;
—— *de procès*, trial expenses;
—— *de route*, (*mil.*) travel allowance, mileage;
—— *de service*, (*mil.*) allowance for expenses to officers holding certain commands or fulfilling certain functions;
—— *de traversée*, passage money.

frais, a., fresh, cool; recent; fresh (not salted); fresh (rested); m., coolness; (*nav.*) fresh breeze (of wind);
bouche fraiche, (*hipp.*) foaming, moist, mouth;
vent ——, fresh breeze, hard wind.

fraisage, m., reaming, reaming out; work along a right line perpendicular to axis of rotation of reaming bit (in Fr. sm. a. manufactories);
—— *en bout*, end reaming;
—— *latéral*, side reaming.

fraise, f., (*mach.*) cutter, reamer, milling tool, mill; countersink, rose bit; reaming tool, reaming bit; (*fort.*) fraise, pointed stake;
—— *en bout*, end reamer;
—— *circulaire*, cone bit;
—— *conique*, angular cutter;
—— *de côté*, side reamer;
—— *cylindrique*, milling cutter;
—— *à défoncer*, side-milling cutter, straddle mill;
—— *à dents rapportées*, cutter with inserted teeth;
—— *de forme*, formed milling cutter;
—— *de formes américaines*, formed milling cutter;
—— *hélicoïdale*, spiral mill;
—— *à planer*, cylindrical planing tool;
—— *à queue*, end cutter, end mill, reamer;
—— *à rôder*, rose countersink;
—— *à trancher*, metal-slitting saw.

fraisement, m., starling of a bridge; (*fort.*) state of having fraises.

fraiser, v. a., (*fort.*) to fraise; (*mach.*) to countersink; to ream; to mill; to make true;
—— *un bataillon, etc.*, (*mil.*) to prepare a battalion for cavalry, by bayonets, or pickets, etc. (obs.);
—— *en bout*, (*mach.*) to use an end-reamer, to end-ream;
—— *de côté*, (*mach.*) to use a side-reamer, to side-ream;
—— *un retranchement*, (*fort.*) to install, set up, a row of fraises on the scarp or counterscarp slope;
tête à ——, broaching head.

fraiseuse, f., (*mach.*) milling machine;
—— *universelle*, universal milling machine.

fraisil, m., cinders.

fraisure, f., chamfered edge of a rivet hole; beaded border; curled or fringed border; countersinking; ream, reaming; any cavity produced by reaming; (*fort.*) row of fraises.

franc, m., franc (coin and unit).

franc, a., free; (*man.*) free (of a gait); (*law*) free, open (port, etc.);
—— *d'amble*, (*man.*) pacing naturally;
botte ——*he*, v. s. v. *botte;*
—— *de collier*, (horse) free in draught;
—— *d'eau*, free of water;
de —— *étable*, (*nav.*) end on;
à —— *étrier*, v. s. v. *étrier;*
—— *de molestation de guerre*, free from molestation of war;
pierre ——*he*, (*mas.*) sound stone, free from defects;
pompe ——*he*, dry pump;
port ——, free port;
—— *de port*, carriage free, franked;
ville ——*he*, free city.

franc-bord, m., path along a river or canal bank; (*nav.*) planking of a ship's hull; freeboard.

franc-filin, m., v. s. v. *filin*.

franc-funin, m., v. s. v. *funin*.

franchir, v. a. n., to climb (a hill), to overcome (an obstacle); to clear, leap, or jump over; to cross, traverse, pass over; to exceed; to break through, overcome; to start from its seat, to project, as an (sm. a.) extractor; to freshen (of the wind);
—— *un cheval*, (*man.*) to vault over a horse;
—— *la main*, (*man.*) (of a horse) to get rid of the control of, to pull away from, the hand;
—— *un vaisseau*, (*nav.*) to free a leaky vessel by means of the pumps.

franchise, f., exemption, immunity; freedom, liberty; franking privilege; (*man.*) willingness (of a horse); (*law*) right of asylum;
en ——, duty-free;
lieu de ——, place of refuge, of asylum;
—— *postale*, franking privilege;
—— *télégraphique*, telegraphic franking privilege.

franchissement, m., facility of ingress and egress;
—— *d'extracteur*, (*sm. a.*) slipping of the claw over the flange of the cartridge without extraction of the latter;
—— *du fossé*, (*mit.*) passage, crossing, of the ditch;
gradins de —— (*fort.*) steps in the sides of a parallel or trench for facility of ingress and egress;
francisation, f., register of a French vessel.
franco, adv., post-free, carriage-free.
franc-tireur, m., (*mil.*) franc-tireur, French guerrilla.
frange, f., fringe.
frappe, f., striking face (of a hammer, etc.).
frapper, v. a., to strike, hit, stab; to knock; (*cord.*) to frap, to bend, seize, lash on or to;
—— *d'amende*, to fine;
—— *de contributions*, to impose contributions;
—— *un cordage, une manœuvre*, (*cord.*) to seize or lash a rope in its place;
—— *d'estoc*, (*fenc.*) to thrust;
—— *d'estoc et de taille*, (*fenc.*) to cut and thrust;
—— *un palan*, (*cord.*) to clap on a tackle;
—— *palan sur garant*, (*cord.*) to bend luff on luff;
—— *une poulie*, (*cord.*) to strap a block;
—— *de prison*, to imprison;
—— *de servitudes*, (*fort.*) to lay under *servitudes*, q. v.;
—— *de taille*, (*fenc.*) to strike with the edge.
frater, m., (*mil.*) barber (soldier-barber, familiar).
fraude, f., fraud; smuggling;
faire la ——, to smuggle.
frauder, v. a., to defraud; to smuggle;
—— *les droits*, to smuggle.
frayé, p. p., traveled (of a road), beaten.
frayer, v. a., to fray, rub against; to graze, brush; to open a way, trace or mark out a road;
—— *le chemin à la brèche*, (*mil.*) to take the forlorn hope, to be the first to ascend a breach;
se —— *aux ars*, (*hipp.*) to rub, become galled, in the region of the *ars*, q. v.;
se —— *un chemin*, to break out a path; (*mil.*) to break through the enemy.
frégate, f., (*nav.*) frigate;
—— *cuirassée*, armored frigate;
—— *à vapeur*, steam frigate.
frein, m., bit of a bridle; bridle; curb; rein; check; control, restraint; drag (of a wheel); bicycle brake; (*art.*) recoil check; (*art., mach.*) brake; (*sm. a.*) catch, stop.
   I. Artillery; II. Miscellaneous.
I. Artillery:
—— *d'affût*, (*art.*) carriage brake;
armer un ——, v. s. v. *armer*;
—— *d'arrêt*, firing-position brake (15ᶜᵐ Chatillon and Commentry turret);
—— *à barre d'obturation*, throttling-bar recoil check;
—— *à bêche de crosse*, trail-spade recoil check;
—— *à bêche d'essieu*, (*art.*) axle-spade recoil check;
—— *de bout d'essieu à vis*, screw axle brake;
—— *central*, Elswick compressor;
chaîne de ——, v. s. v. *chaîne*;
—— *circulaire*, circular brake (Canet);
—— *à cônes de friction*, friction cone brake;
—— *à contretige*, counter rod brake;
—— *à contretige centrale*, central counter rod brake (Canet);
—— *à cordes*, rope brake, recoil check;
—— *à cordes Lemoine*, Lemoine recoil check;
—— *à côtes inclinées*, (hydraulic) brake with (variable orifices and) inclined sides;
cylindre de ——, (*art.*) recoil cylinder;
—— *de déformation*, generic expression for a brake whose parts undergo a displacement on recoil, elastic brake;
desserrer un ——, to release a brake;
—— *à disques de friction*, friction disk brake;
—— *à effort constant*, brake with variable orifices and constant resistance;

frein *d'enfoncement*, brake to take the downward thrust on the carriage (Locard);
—— *à étrier*, bow compressor;
—— *à excentrique*, eccentric brake;
—— *à fraisures*, hydraulic brake with grooved cylinder;
—— *de, à friction*, friction brake;
—— *à frottement*, friction recoil check;
—— *à frotteurs*, friction brake (plate compressor);
—— *funiculaire*, the Lemoine brake;
—— *hydraulique*, hydraulic brake;
—— *hydropneumatique*, hydropneumatic brake;
—— *à lame circulaire*, circular compressor brake;
—— *à lames*, compressor brake; Armstrong plate recoil check;
—— *à lames et à arcs*, bow compressor;
—— *à lames centrales*, plate compressor;
—— *à lames pendantes*, Elswick compressor;
—— *Lemoine*, Lemoine brake;
—— *à levier*, lever brake;
levier de ——, brake handle, lever;
—— *à mâchoires*, clamp recoil check;
—— *à manette*, clamp and handle (as in r. f. guns);
mettre le ——, to put on a drag or brake;
—— *de moyeu*, nave brake;
—— *de moyeu à cordes*, funicular nave brake;
—— *à orifices constants et écoulement libre*, hydraulic brake with constant orifice and free discharge;
—— *à orifices fermés par une soupape chargée d'un poids constant* (*variable*), hydraulic brake with closing valve of constant (variable) load;
—— *à orifices variables*, hydraulic brake with variable orifices;
—— *à orifices variables et résistance constante*, hydraulic brake with variable orifices and constant resistance;
ôter le ——, to take off the brake or drag;
—— *à patins*, wheel brake, shoe; shoe brake; lock (for wheels, downhill);
—— *à patins et à cordes*, (*Fr. art.*) the Lemoine rope and shoe brake;
—— *à peigne*, (so-called) comb brake;
—— *de pointage*, locking brake for elevating gear; elevation brake (Fr. 120ᵐᵐ short);
—— *à pompe*, hydraulic brake;
—— *de recul*, recoil brake;
—— *de recul à vis*, screw recoil brake;
régler un ——, to adjust a brake, a recoil check;
—— *à ressort*, spring recoil check;
—— *de roues*, wheel brake;
—— *de route*, road brake; traveling brake;
—— *de route à vis*, screw road brake;
—— *à sabot*, block brake; shoe brake;
sabot de ——, brake block;
—— *de serrage*, tightening brake;
—— *serré*, applied brake;
—— *de soulèvement*, brake to resist overturning (Locard);
—— *à soupape chargée*, loaded valve brake;
—— *de sûreté*, safety brake;
—— *de tir*, recoil brake;
—— *à valve*, Vavasseur hydraulic brake; rotating valve brake.
II. Miscellaneous:
à ——, fitted or equipped with a brake;
—— *d'absorption*, absorbing brake or dynamometer;
—— *à air*, (r. r.) air brake;
—— *atmosphérique*, (r. r.) air brake;
—— *automateur*, —— *automatique*, self-acting brake;
—— *continu*, (r. r.) continuous brake;
—— *de couronne de piston*, (*mach.*) junk ring stop;
—— *demi-intérieur*, a bicycle brake lodged in the steering post;
—— *dynamométrique*, (*mach.*) friction brake;
—— *électrique*, (r. r.) electric brake;
—— *d'entraînement circulaire*, circular bicycle brake;
garde- ——, brakesman, brakeman;

frein *invisible*, a bicycle brake, of which only the handle is visible (lodged in steering post and handlebar);
mettre le ——, (*r. r.*) to put on brakes;
ôter le ——, (*harn.*) to unbit, unbridle; (*r. r., etc.*) to take off brakes;
prendre le —— *aux dents*, (*man.*) to bolt, run away;
retenir par le ——, to curb;
ronger le ——, (of a horse) to champ the bit;
—— *à ruban*, friction strap;
—— *de rupture*, (balloons) a set of five parallel cords, the successive breaking of which lessens the shock on the anchor when it finally bites;
—— *à sabot*, (*r. r.*) shoe brake;
tare du ——, (in a friction brake) the tare or allowance due to leverage;
—— *à vide*, (*r. r.*) vacuum brake;
—— *à vis*, screw brake;
—— *de voiture*, carriage brake;
—— *de volant*, clamp of a handwheel.
frêle, a., weak, fragile, brittle.
frémir, v. a., to tremble, vibrate; to simmer (of water about to boil).
frêne, m., ash (wood and tree).
fréquence, f., (*elec.*) frequency, periodicity.
frère, m., brother;
—— *d'armes*, (*mil.*) brother in arms, comrade;
gros —— *s*, (*mil. slang*) cuirassiers.
fret, m., chartering of a vessel, cost of chartering, sea freight (material and cost of transportation); transportation of freight by sea;
—— *d'aller*, outward freight;
donner un vaisseau à ——, to charter a vessel (broker, owner);
faux ——, dead freight;
prendre un vaisseau à ——, to charter a ship (merchant);
—— *de retour*, return freight, homeward freight;
—— *de sortie*, v. —— *d'aller*.
frètement, m., freighting, chartering (of a vessel).
fréter, v. a., to charter a vessel (strictly of the broker or owner); to freight;
—— *cap et queue*, to charter a ship;
—— *à la cueillette*, to charter for a general cargo;
—— *en grand*, to charter in bulk;
—— *en travers*, v. —— *en grand*.
fréteur, m., charterer of a vessel (strictly, of the owner or broker); freighter, shipper.
frettage, m., hooping, tiring (a wheel); (*art.*) hooping, shrinkage; generic term for process of constructing a built-up gun; hoop;
—— *allongé*, (*art.*) system in which hoops extend over a considerable length of the gun; this sort of hooping itself;
—— *biconique*, (*art.*) system (De Bange) of biconic or double-taper hoops;
—— *conique*, (*art.*) taper hooping;
—— *double*, (*art.*) double row of hoops; double hooping;
—— *en fils*, (*art.*) wire hooping;
fossé de ——, v. *puits de* ——;
—— *mixte*, (*art.*) in certain bronze guns, the assemblage of tube and jacket under pressure;
puits de ——, (*art.*) shrinkage pit.
frette, f., hoop, ring (in general); end band (of a wooden nave); ferrule, hoop ring; band, tire, flange; (*art.*) hoop, coil, (esp. to-day) short hoop.
(Except where otherwise indicated, the following terms relate to artillery:)
—— *à adents*, ribbed hoop, cradle hoop;
—— *agrafes*, f., v. —— *à adents*;
—— *d'appui*, (Hotchkiss) mounting hoop;
—— *d'armons*, hoop of fore guides of limber;
—— *arrière*, rear hoop, breech hoop;
—— *d'attache*, hoop for locking or fixing wires (Schultz wire gun);
—— *avant*, first hoop from muzzle of gun;
—— *biconique*, biconic, double-taper, hoop of De Bange;
—— *de butée*, recoil hoop or band (sort of ring or hoop on chase of the gun, e. g. Fr. 120^{mm}, to reduce the shock in case the gun recoils to its limit in the trunnion sleeve);

frette *de calage*, locking hoop or ring;
—— *à crochet*, hoop and hook;
—— *-culasse*, —— *de culasse*, breech hoop;
—— *-écrou*, f., threaded hoop;
—— *à encastrement*, cradle hoop;
—— *de moyeu*, nave hoop;
—— *à oreilles*, a hoop with longitudinal ribs;
—— *de pied*, (*pont.*) iron hoop around top of trestle leg;
—— *à piton*, hoop and eye (on hitching post);
—— *de pointage*, in certain guns, the hoop or band connecting the breech with the elevating gear;
—— *porte-guidon*, front-sight ring or band;
—— *porte-hausses*, sight ring or band;
—— *porte-tourillons*, trunnion hoop;
—— *de recouvrement*, key ring; locking ring;
—— *de roue*, nave band (of a wheel);
—— *à tenons*, assembling hoop (of some hydraulic brakes;
—— *-tourillons*, —— *à tourillons*, f., (*art.*) trunnion hoop; trunnion band;
—— *vissée*, screw hoop;
—— *de volée*, (*art.*) muzzle hoop, chase hoop.
fretté, a., hooped; (*art.*) built up;
canon ——, (*art.*) built-up gun.
fretter, v. a., to hoop; to flange; (*art.*) to shrink on the hoops.
friabilité, f., friability.
friable, a., friable.
friche, f., waste, uncultivated land;
en ——, fallow.
friction, f., rubbing, friction (synonym of *frottement*, but rarely used).
frictomètre, m., friction gauge.
frigorifère, m., freezing, cooling, apparatus; refrigerator;
—— *à gaz*, any apparatus based on the evaporation of a volatile liquid;
—— *à liquide*, any cooling apparatus using a non-freezing liquid;
—— *à affinité*, ammonia apparatus;
—— *à air*, compressed-air cooling apparatus.
frigorifique, a., refrigerating; refrigerator; *appareil* ——, refrigerator.
frimas, m., hoarfrost, white frost.
fringale, (*hipp.*) abnormal appetite.
fringant, a., frisky, mettlesome.
frise, f., frieze; kind of cloth; (*cons.*) flooring board or strip for *parquet*;
cheval de ——, v. s. v. *cheval*.
friser, v. a., to graze, brush, glance upon, touch lightly, approach;
se ——, (*hipp.*) to brush.
frison, m., can or jug.
fritter, v. a., to frit.
froid, m. and a., cold;
épaules —— *es*, (*hipp.*) stiff shoulders;
—— *des épaules*, (*man.*) sluggish, unwilling to start (in increasing gaits);
forger à ——, (*met.*) to cold-hammer.
froissement, m., crushing, bruising; (*fenc.*) sharp, sliding pressure along adversary's foil.
froisser, v. a., to bruise.
fromage, m., cheese;
—— *d'un creuset*, (*met.*) crucible stand.
froment, m., wheat;
—— *cultivé*, wheat;
faux ——, rye grass.
front, m., front, face, facing; head; (*mil.*) (in commands) front! (*fort., mil.*) front; (*hipp.*) forehead;
—— *d'action*, (*mil.*) battlefront;
—— *d'armée*, (*mil.*) front of an army; front occupied by the various units of the first line of an army;
—— *d'attaque*, (*min.*) driving head; face of the work (of a tunnel); (*fort., mil.*) front of attack; (*fort.*) front that will be or is attacked, front selected for attack by besieging enemy;
attaque de ——, (*mil.*) attack in line;
attaquer de ——, (*mil.*) to make a front attack;
—— *de bandière*, v. s. v. *bandière*;
—— *bastionné*, (*fort.*) bastioned front;
—— *de bataille*, (*mil.*) front of all the troops in the first line;
battre ——, v. s. v. *battre*;
changement de ——, *changer de* ——, v. s. vv. *changement, changer*;
cheval à —— *plat*, v. s. v. *cheval*.

front *de combat*, (*mil.*) combat front of a unit;
— *couvert*, (*mil.*) front covered by obstacles, such as a wood, ravine, etc.;
de —, in front, abreast, simultaneously;
— *découvert*, (*mil.*) front open to attack, unprotected; front to be covered by an *escadron de découverte;*
— *de défense*, (*mil.*) defensive front;
— *d'exploration*, (*mil.*) front or zone in front of army advancing, front covered by *service d'exploration;*
faire —, to face, front;
— *de fortification*, (*fort.*) front, all works on one side of the polygon fortified;
— *de gorge*, (*fort.*) gorge front;
— *hérissé*, (*mil.*) front protected by guns, troops, etc.;
— *de marche*, (*mil.*) march front;
marche de —, (*mil.*) march in line;
— *de mer*, (*fort.*) sea front;
— *d'opérations*, (*mil.*) front of operations (imaginary line joining heads of columns in march or in station);
— *polygonal*, (*fort.*) polygonal front;
repousser le —, to drive back the enemy over the ground of his advance;
— *stratégique*, v. — *d'opérations;*
— *de taille*, (*min.*) working face, mine head, heading;
— *tenaillé*, (*fort.*) tenailled front;
— *de terre*, (*fort.*) the land front of a sea-coast work;
— *de tête*, (*fort.*) principal, main, front (i. e., toward enemy), as opposed to — *de gorge*, q. v.; (in a work forming part of defenses of a place) the front facing the exterior of the intrenched camp;
— *de la têtière*, (*harn.*) brow band.

frontal, m., (*harn.*) brow band; a., front, in front;
plaque —*e*, (*fort.*) front plate (of a cupola);
— *sous-gorge*, (*harn.*) brow band and throat-latch in one continuous piece.

frontail, m., (*harn.*) mourning brow band.

fronteau, m., frontlet; (*harn.*) brow band; mourning brow band;
— *de mire*, (*F. m. art.*) front sight;
— *de serre*, (*nav.*) housing chocks;
support de — *de mire*, (*F. m. art.*) sight patch, sight mass.

frontière, f. and a., frontier, border;
passer la —, to cross the border;
ville —, *ville de la* —, border town.

fronton, m., pediment;
— *de bateau*, (*pont.*) ponton stern.

frottage, m., rubbing; dry rubbing.

frottement, m., friction; rubbing;
— *des feuilles*, (*met.*) scouring of plates, after pickling;
— *de glissement*, sliding friction;
monter à — *dur*, to mount with a press fit;
— *de mouvement*, friction of motion;
passer à —, to fit, or go in, with friction; to go in with a press fit;
pièce de —, (*harn.*) galling leather;
plaque de —, friction plate;
— *de repos*, friction of quiescence;
— *de roulement*, rolling friction.

frotter, v. a., to rub, to grind; dry-rub, chafe.

frotteur, m., (*art.*) friction piece or plate for checking recoil; (*elec.*) rubbing contact.

frottoir, m., brush, rubber; scrubbing-brush.

frottis, m., (*top. drawing*) layer of transparent color (e. g., on slopes).

fruit, m., fruit; (*cons.*) batter of a wall; slope, tapering;
avoir du —, to batter;
— *à contre-pente*, reverse batter, counter-sloping batter;
donner du —, to cause to batter;
— *sec*, a cadet or pupil of a government school who fails to pass the examination (originally a Polytechnique term).

fugitif, a., m., fugitive; refugee (U. S.).

fugue, f., (*mil. slang*) *faire une* —, to run away.

fuir, v. a. n., to flee, to run away; to avoid, shun, elude; to leak, run out (of liquids); to be leaky (of a cask, etc.); to escape (of gases);
— *les talons*, (*man.*) to fear the spur.

fuite, f., flight, running away, avoiding, shunning, evasion; leak, leakage (of liquids); escape (of gas);
— *de gaz*, (*art., sm. a.*) escape of gas through imperfect obturation;
mettre en —, to put to flight;
prendre la —, to take to flight.

fulgurite, f., (*expl.*) fulgurite.

fuligineux, a., smoky, sooty.

fulmibois, m., (*expl.*) nitrated sawdust.

fulmi(-)coton, m., (*expl.*) gun cotton;
— *comprimé*, compressed gun cotton;
— *humide*, wet gun cotton;
— *sec*, dry gun cotton.

fulmi-moteur, m., — *à poudre*, explosive engine.

fulminant, a., (*expl.*) fulminating, detonating;
argent —, fulminate of silver;
composition —*e*, fulminating composition;
mercure —, fulminate of mercury;
poudre —, priming powder.

fulminate, m., (*expl.*) fulminate;
— *de mercure*, — *mercurique*, mercury fulminate.

fulminatine, f., (*expl.*) fulminatine.

fulmination, f., (*expl.*) detonation.

fulminer, v. n., (*expl.*) to fulminate, detonate.

fulmi-paille, f., (*expl.*) nitrated straw.

fulmison, m., (*expl.*) nitrated bran.

fumé, m., smoky, smoked;
verre —, smoked glass.

fumée, f., smoke, vapor;
boîte à —, smoke box;
cylindre à —, (*artif.*) smoke-producing cylinder, to simulate at target practice the smoke of an enemy;
poudre à —, (*powd.*) black (brown) powder; smoke-producing powder, smoky powder;
poudre sans —, (*expl.*) smokeless powder;
tuyau conducteur de la —, smoke pipe.

fumer, v. a. n., to smoke, reek, throw out smoke; to expose to smoke; to smoke; (*hipp.*) to steam;
— *un four*, — *un fourneau*, (*met.*) to dry a furnace (e. g., after repairs);
— *sa pipe*, (*hipp.*) (said of horse afflicted with epilepsy) to take food indolently and chew it some time.

fumeron, m., (*powd.*) half-burnt charcoal.

fumier, m., manure, dung, horse dung; dunghill;
tombereau à —, dung cart.

fumigation, f., fumigation.

fumigatoine, a., fumigation.

fumigère, m., smoke composition; a., smoke generating.

fumivore, m., smoke consumer; a., smoke consuming.

fune, f., ridge rope (of an awning).

funèbre, a., funeral.

funer, v. a., (*cord.*) to rig;
— *un mât*, to fix the standing rigging on the masthead.

funérailles, f., pl., funeral, funeral ceremonies.

funiculaire, a., funicular; f., funicular curve, catenary.

funin, m., (*cord.*) ropes and rigging (of a ship);
franc —, white, untarred, rope; hawser; five-stranded rope.

fusain, m., spindle-tree, prickwood; drawing charcoal; charcoal pencil;
crayon de —, charcoal pencil;
dessein au —, charcoal-pencil sketch.

fusair, m., aluminium balloon (for aerial-current investigation).

fusant, a., (*art.*) time; using time fuse;
composition —*e*, (*artif.*) fuse composition;
fusée —*e*, (*art.*) time fuse;
noyau —, (*artif.*) core (as of a Bickford fuse);
section —*e*, (*art.*) section firing time shell (in ranging, as opposed to *section percutante*, q. v.).

fusé, p. p., air-slacked;
chaux —*e*, air-slacked lime.

fuseau, m., spindle; pin; arbor; leaf of a pinion; segment or gore of a balloon;

fuseau *d'aviron*, grommet to retain an oar in the tholes;
—— *de cabestan*, capstan cleat;
—— *d'essieu*, axle arm.

fusée, f., axle arm or spindle; barrel of a windlass; (*art.*) fuse; (*artif.*) match fuse; rocket; (*mach.*) fusee; (*fenc.*) handle of a foil; (*hipp.*) fusee; splint; fistula.
  I. Artillery; II. Rockets, etc.; III. Miscellaneous.

  I. Artillery:
—— *à action retardée*, delay-action fuse;
—— *à ailettes*, wing fuse (MacEvoy);
—— *d'amorce*, priming tube (obs.);
armé de ——, arming of a fuse;
armer une ——, to arm a fuse;
—— *à bobine*, bobbin fuse (in which a wire is unwound from one bobbin to another and a tongue releases a spring that strikes the primer);
—— *en bois*, wooden fuse;
—— *de bombe*, common shell fuse;
bonnetage des ——*s*, v. *bonnetage*;
—— *borgne*, blind fuse, fuse that fails to work;
—— *à boule*, fuse in which the rotation of the central stem is assured by a ball;
—— *à cadran*, disk fuse, ring fuse;
calice de ——, fuse cap;
canal de ——, bore of a fuse;
—— *à chapeau*, cap fuse;
chapeau de réglage de ——, setting collar;
chapiteau de ——, fuse cone;
—— *chargée*, loaded fuse;
chasse- ——*s*, fuse setter;
coiffe de ——, fuse cap;
coiffer la ——, to cap a fuse;
—— *à concussion*, concussion fuse;
corps de ——, fuse case;
—— *de culot*, base fuse;
décoiffer une ——, to uncap a fuse;
—— *à deux étages*, double time-train fuse;
disposer la —— *à temps*, to set a time fuse;
—— *à double effet*, double-action fuse, combination fuse;
—— *à double effet à chapeau mobile*, combination fuse with rotating cap;
—— *à durée(s)*, time fuse;
—— *à éclatement retardé*, delay-action fuse;
—— *à effet retardé*, delay-action fuse;
—— *électrique*, electric fuse;
—— *à étages*, double-ring fuse;
—— *à étoupilles*, v. —— *d'amorce*;
fausse —— *d'exercice*, drill tube;
—— *à friction*, (hand grenade) friction fuse;
—— *fusante*, time fuse;
godet de ——, fuse cup;
—— *à grenade*, hand-grenade fuse;
—— *à inertie*, fuse on which the primer is free and striker fixed;
—— *à inflammation directe*, any fuse inflamed by the discharge;
lumière de ——, v. *canal de* ——;
—— *mécanique*, mechanical fuse;
—— *métallique*, metallic fuse;
—— *métallique à canal circulaire*, ring fuse;
—— *mixte*, double-action fuse, combination fuse;
—— *à obus*, shell fuse;
œil de ——, fuse hole;
—— *à percussion*, —— *percutante*, percussion fuse;
pince de ——, fuse engine;
—— *à plateau*, disk fuse;
raté de ——, failure of a fuse to act;
—— *à refoulement*, fuse in which the primer is fixed and striker free;
—— *réglée*, time fuse;
régler une ——, to set a fuse;
ressort d'armement de ——, fuse spring;
—— *retardée* delay-action fuse;
ruban de —— fuse band;
—— *à six durées*, a (French) six-channeled time fuse;
—— *de sûreté*, safety fuse;
—— *à temps*, time fuse;
—— *à temps à deux étages*, fuse with double time-train;
—— *de tête*, nose fuse, point fuse;
fusée m/n a fuse whose eye is m^mm in diameter, for a shell whose top is n^mm in diameter;

  II. Rockets, etc.:
fusée, *affût à* ——*s*, v. s. v. *affût*;
âme de ——, bore, hollow, of a rocket;
—— *asphyxiante*, (*mil. min.*) asphyxiating device;
baguette de ——, rocket stick;
—— *à baguette*, rocket and stick, ordinary rocket;
—— *sans baguette*, v. —— *à rotation*;
—— *à ballon*, light ball;
—— *de Bickford*, Bickford fuse;
—— *de bombardement*, siege rocket;
—— *à bombe*, shell rocket;
—— *de campagne*, field rocket;
—— *à carcasse incendiaire*, carcass rocket;
chargement de la ——, driving of a rocket;
chevalet à, de, ——*s*, v. s. v. *chevalet*;
—— *à commotion*, concussion rocket;
—— *a la Congrève*, Congreve rocket;
—— *courante*, line rocket;
dégorger une ——, to empty a rocket;
—— *à double vol*, double rocket;
—— *d'éclair*, flash rocket;
—— *d'éclairage*, —— *éclairante*, light rocket;
—— *éclatante*, star rocket;
engorger une ——, to fill a rocket;
garniture de ——, rocket heading;
—— *de guerre*, war rocket (Congrève);
—— *de guerre à ricochets*, ground rockets;
—— *d'honneur*, signal rocket to open fireworks;
—— *incendiaire*, carcass rocket;
—— *instantanée*, instantaneous fuse (*cordeau porte-feu*); Bickford quick match;
—— *lente*, slow match, Bickford fuse;
—— *lumineuse*, rocket light ball;
—— *à mitraille*, case rocket;
—— *à obus*, shell rocket, shell-headed rocket;
—— *à parachute*, parachute rocket (i. e., to furnish a parachute light);
—— *à plusieurs vols*, compound rocket;
—— *porte-amarre*, life rocket;
—— *porte-feu*, (*mil. min.*) firing rocket for firing mines;
rondelle de ——, separator, separating disk;
—— *à rotation*, rifle rocket;
—— *à serpents*, rockets with swarmers;
—— *de signaux*, —— *de signaux*, signal rocket;
—— *de signaux à pétards*, marron-headed rocket; bounce-headed rocket;
—— *à table*, tourbillon;
tête de ——, rocket head;
—— *torpille*, torpedo rocket;
tube à ——, rocket tube;
—— *vive*, v. —— *instantanée*;
—— *volante*, signal rocket, skyrocket;
—— *volante à étoiles*, star-headed rocket.

  III. Miscellaneous:
—— *d'un arbre*, (*mach.*) axle arm;
—— *d'aviron*, mouse of an oar;
—— *de cabestan*, barrel of a windlass;
—— *d'essieu*, axletree arm;
—— *de treuil*, capstan barrel.

fuséen, m., (*art.*) rocketeer.
—— *à cheval*, mounted rocketeer.
fuselé, a., spindle-shaped; slender; taper.
fuser, v. a. n., to fuse, liquify, melt.
fusibilité, f., fusibility.
fusil, m., steel (for striking a light, for sharpening a knife); (*sm. a.*) musket, rifle;
—— *à aiguille*, needle gun;
armer un ——, to cock a gun;
—— *automatique*, (Maxim) automatic gun;
baguette de ——, cleaning rod;
—— *à bloc*, any rifle whose breech is closed by a block containing part of the breech mechanism (e. g., Remington, Peabody);
bois de ——, rifle stock;
bretelle de ——, rifle sling;
—— *à broche*, any rifle firing the cartridge by a striker (*broche*) (e. g., Lebel, Mauser);
canon de ——, rifle barrel;
—— *se chargeant par la culasse*, breech-loading rifle;
—— *à chargeur*, rifle to which a loader may be attached;
—— *de chasse*, fowling piece;
chien de ——, hammer;
—— *à chien non-apparent*, hammerless shotgun;

fusil *choke-bored*, choke-bore shotgun;
— *coup de* ——, v. s. v. *coup;*
— *crosse de* ——, butt of the stock;
—— *à culasse tombante,* v. *arme à culasse tombante;*
—— *à deux canons,* —— *à deux coups,* double-barreled gun;
—— *électrique,* electric gun. (Pieper) gun fired by electricity; also, in tests, etc., any gun fired by electricity (incorrect term);
—— *d'épreuve,* rifle used to try ammunition;
—— *à frette,* (*ball.*) rifle barrel bedded in an enormous cylinder (*frette*), and intended for the test of ammunition;
—— *à gaz,* —— *à gaz liquéfié,* Giffard gun;
—— *de guerre,* v. —— *d'infanterie;*
le —— *haut,* (position of) advance carbine;
—— *d'honneur,* musket of honor, so called, given as a reward during the Revolution, and before the institution of the Legion of Honor;
—— *d'infanterie,* infantry rifle;
—— *lisse,* smoothbore musket;
—— *à longue portée,* long-range rifle;
—— *à magasin,* magazine rifle;
mettre le —— *en joue,* to point; to aim;
—— *de munition,* v. —— *d'infanterie;*
—— *d'ordonnance,* regulation or service rifle;
—— *-pendule,* (*ball.*) gun pendulum, musket pendulum;
—— *à pêne,* v. *arme à pêne;*
—— *percutant,* —— *à percussion,* percussion musket;
—— *de petit calibre,* small-caliber rifle;
—— *à pierre,* v. —— *à silex;*
pierre de ——, flint, gunflint;
—— *à piston,* v. —— *percutant;*
—— *à pompe,* v. —— *percutant;*
portée de ——, rifle range, musket range;
à portée de ——, within musket shot;
—— *rayé,* rifled musket;
régler un ——, to "fire in" a rifle, to get its "personal equation" for the day;
—— *de rempart,* wall piece;
—— *à répétition,* repeating rifle;
—— *-revolver,* gun with cylindrical breech mechanism, like a revolver (v. *arme revolver*);
—— *à rotation rétrograde,* v. *arme à rotation rétrograde;*
—— *à silex,* flintlock;
—— *à tabatière,* v. *arme à tabatière;*
—— *à tige,* v. *arme à tige;*
—— *à tir continu,* rifle from which all the cartridges in a magazine may be fired without taking the piece from the shoulder;
—— *à tir rapide,* rapid-fire small arm;
—— *à tir réduit,* subcaliber gallery piece;
—— *à tiroir,* v. *arme à tiroir;*
—— *-type,* type gun;

fusil *à un coup,* single-barreled gun; single loader;
—— *à vapeur,* steam gun;
—— *à vent,* air gun;
—— *à verrou,* (any) bolt gun, the modern military rifle.

fusilier, m., (*mil.*) fusilier; any person using a small-arm rifle (as, e. g., in flanking fire);
—— *s de discipline,* v. *compagnie de discipline;*
—— *s-marins,* (*Fr. nav.*) sailors trained in the use of the rifle and used on occasion to form landing parties; armed seamen.

fusillade, f., (*sm. a.*) firing, discharge of musketry, fusillade.

fusiller, v. a. r., to shoot, to execute by shooting (military execution); to shoot one another.

fusillette, f., (*artif.*) very small rocket.

fusion, f., fusion, melting; alliance, combination; (*mil.*) combining of troops (as companies, battalions, into one, on account of small numbers);
—— *des corps de troupe,* (*mil.*) dissolution or disbanding of a regiment and passage of elements into other regiments or bodies;
—— *crue,* (*met.*) fusion producing gray pig;
en ——, in fusion, melted;
entrer en ——, to begin to melt;
point de ——, melting point;
—— *réductive,* (*met.*) melting or fusion proper;
—— *sèche,* (*met.*) fusion producing white pig.

fût, m., cask, barrel; barrel or cylinder of a jack; shaft, trunk (of a column); pole; handle (as of a saw); post (of a crane), crane post; (*sm. a.*) wooden lining of a scabbard; stock of a musket (esp. part under barrel), forearm, forestock, tip stock; (*mech.*) brace; (*harn.*) frame (of a saddle);
—— *de girouette,* vane stock;
—— *de grue,* crane post;
—— *et mèche de vilebrequin,* body and tap of a centerbit, stock and bit;
—— *de rabot,* stock of a plane;
—— *de sabre,* scabbard lining (of wood);
—— *de tambour,* barrel of a drum.

futaie, f., (*top.*) forest or wood of high-stemmed trees;
forêt de haute ——, forest of full-grown (old) trees;
haute ——, forest trees of full growth.

futaille, f., cask, barrel; generic term for casks and barrels.

futée, mastic (of glue and sawdust).

fuyant, a., tapering; receding (of shape, e. g., plate of a turret or cupola);
aller en ——, to approach each other (as the edges of a front sight);
échelle ——*e,* reducing scale, flying scale.

fuyard, m., fugitive, runaway.

G.

gabare, f., (*nav.*) transport, lighter.

gabarer, v. a., to scull a boat.

gabari, m., v. *gabarit.*

gabariage, m., act, action, of modeling.

gabarier, v. a., to mold, model.

gabarit, m., gauge, templet, modeling board; (*siege, fort.*) gabion form;
—— *de chargement,* (r. r.) tunnel gauge (i. e., a frame of the interior profile of the tunnel, to gauge trucks or flat cars loaded *en vrac*);
—— *de gabion,* —— *des gabions,* (*siege, fort.*) gabion form;
—— *de pente,* (*mil. min.*) slope gauge;
—— *des rails,* (r. r.) rail section.

gabet, m., vane; sight of a quadrant.

gabier, m., (*nav.*) picked sailor, first-class man;
—— *de hune,* topman, rifleman in the tops.

gabion, m., (*siege, fort.*) gabion;
—— *d'artillerie,* (*Fr. art.*) gabion 1^m high (wattling) by 0.56^m in exterior diameter;
buse de ——*s,* v. s. v. *buse;*
—— *clayonné,* wicker gabion;

gabion *farci,* sap roller;
—— *du génie,* (*Fr. eng.*) gabion 0.80^m high (wattling) and 0.60^m in exterior diameter;
—— *pliant,* collapsible gabion;
—— *roulant,* sap roller;
—— *de sape,* sap gabion;
—— *de tranchée,* trench roller.

gabionnade, f., (*fort.*) gabion work, gabionade, gabion revetment; intrenchment of gabions.

gabionnage, m., (*fort.*) gabioned parapet; gabion-revetted parapet.

gabionner, v. a., (*siege, fort.*) to revet with gabions, to gabion.

gaburon, m., fish, clamp (for a mast).

gâchage, m., mixing of mortar.

gâche, f., nab, catch, staple; wall hook;
—— *de, du, loquet,* (*art.*) latch catch (latch recess);
—— *de serrure,* lock staple.

gâchée, f., amount of mortar, concrete, mixed at one time.

gâcher, v. a., to mix, stir, temper (mortar, plaster); (*fig.*) to bungle, to botch;

gâcher la chaux, to slake lime;
—— *clair*, to make a thin mortar;
—— *lâche*, to make grouting;
—— *serré*, to temper hard; to make a stiff mortar.

gâchette, f., catch; (*sm. a., r. f. art.*) sear; trigger (rare);
—— *détente*, trigger and sear in one;
peser sur la ——, to pull the trigger;
ressort de ——, sear spring;
ressort ——, trigger spring;
—— *de sûreté*, safety sear;
vis de ——, sear screw.

gâcheux, a., sloppy;
terre —— *se*, swampy ground.

gâchis, m., (*mas.*) mortar (made of sand, lime, and plaster).

gaffe, f., boat hook;
—— *à bateau*, (*pont.*) boat hook;
bâton de ——, pole or shaft of a boat hook
fer de ——, head of a boat hook;
—— *à nacelle*, (*pont.*) mooring boat hook;
—— *à pointe et à croc*, ordinary boat hook
porter ——, (*mil. slang.*) to be on sentry duty, to be on post.

gaffer, v. a., to use a boat hook; to keep or hold a boat by a boat hook.

gage, m., pay, hire, wages;
—— *de bataille*, —— *de combat*, challenge.

gagnage, m., pasture land; pasturage.

gagner, v. a., to get, earn, win, obtain, acquire; to recover, fetch, overhaul, draw; (*mil.*) to reach (a position);
—— *de l'avant*, to gain ground;
—— *une bataille*, to win a battle;
—— *les champs*, to run away, to run off, escape;
—— *l'ennemi*, (*mil.*) to reach or overtake the enemy;
—— *au haut*, —— *le haut*, v. —— *les champs;*
—— *à la main*, (*man.*) to pull, to try to get the bit between the teeth (horse);
—— *la mesure*, (*fenc.*) to get an advantage by following up a graze;
—— *du terrain*, to gain ground;
—— *le vent*, (*nav.*) to get the weather gauge;
—— *la volonté d'un cheval*, (*man.*) to break a horse.

gaïac, m., lignum-vitæ, boxwood.

gaie, f., (familiar term for) horse.

gaillard, m., (*nav.*) forecastle;
—— *d'arrière*, quarter-deck;
—— *d'avant*, forecastle deck.

gaille, f., v. *gaie*.

gailleterie, f., what is left of *gaillette*, q. v., when the large pieces have been removed.

gaillette, f., coal in pieces of at least 1^{dm} cube.

gain, m., gain (as, *mil.*), by transfer, promotion, etc.).

gaine, f., sheath, case; shaft; (*mil.*) sheath, scabbard; (*mil. min.*) bored shaft of a *camouflet;* mining case, shaft; (*fort.*) corridor, circulation passage; (*art.*) priming tube (of a melinite shell);
—— *d'aérage*, (*mil. min.*) ventilation shaft;
—— *de chargement*, (*mil. min.*) loading shaft;
—— *de circulation*, (*fort.*) communicating gallery or passage; ventilating gallery; circular passage around a turret;
—— *de combustion*, (*met.*) flue (of a Cowper stove);
—— *détonante*, (*art.*) (German) elongated mortar-shell fuse;
—— *-enveloppe*, f., (*fort.*) gallery of communication and of sanitation, separating living rooms of casemates from parapet or earthwork;
—— *Laloy*, (*mil. min.*) mining case used when a gallery is opened from a mine crater;
—— *de pavillon*, canvas hem (of a flag);
—— *porte-retard*, (*art.*) in some German fuses, the tube containing the retardation mechanism;
—— *du verrou*, (*fort.*) well or shaft of an embrasure closer (*verrou*).

galbe, m., swell, curved outline (as of a connecting rod; entasis, swelling (of a column); (*harn.*) semicircular iron hoop for saddle bow; (*met.*) mass, body, of a blast furnace.

gale, f., wormhole (in wood); (of animals) mange, scab, itch.

galère, f., gallery; (*pont.*) towrope;
avirons en ——, rest on your oars;
conduire un fardeau à bras en ——, (*art.*) to drag a load (as a truck), the power being applied to handspikes thrust through marlin-spike hitches in the drag ropes;
nœud de, en, ——, (*cord.*) marlin-spike hitch;
passer un levier en ——, to pass a lever or a handspike through a marlin-spike hitch.

galerie, f., gallery, corridor, lobby; conduit, drain; (*min.*) adit, level, drift; (*fort., mil. min.*) gallery, mining gallery; (*steam*) flue (of the French boiler); (*mach.*) footboard; (*med.*) end frame of a litter; (*art.*) frame on ammunition chests (to carry or hold the men's *sacs*);
—— *d'adduction*, feeding conduit;
—— *d'allongement*, (*min.*) main level;
—— *d'avancement*, heading of a tunnel;
—— *de captage*, collecting conduit (for subterranean waters);
—— *casematée*, (*fort.*) casemate gallery;
—— *à ciel ouvert*, open gallery;
—— *de communication*, (*fort.*) communicating, transverse, gallery;
—— *de contrescarpe*, (*fort.*) counterscarp gallery;
—— *crénelée*, (*fort.*) loopholed gallery;
—— *en décharge*, counterarched gallery;
demi ——, (*mil. min.*) low gallery;
demi —— *ordinaire*, (*mil. min.*) gallery 1.3 to 1.5^m high by 1^m wide;
—— *descendante*, descending gallery;
—— *d'écoulement*, drainage, draining, gallery;
—— *d'écoute*, (*mil. min.*) listener, listening gallery;
ensemble des —— *s*, (*min.*) drifts;
entrer en ——, (*mil. min.*) to open a gallery;
—— *-enveloppe*, (*fort.*) counterscarp gallery; envelope gallery; (*mil. min.*) the starting gallery of a system of mines (usually the counterscarp gallery);
—— *d'escarpe*, (*fort.*) scarp gallery;
—— *flanquante*, (*fort.*) flanking gallery;
—— *de fourneau*, vent, draught holes, of a furnace
—— *de fusillade*, (*fort.*) musketry-fire gallery (of a caponier);
grande ——, (*mil. min.*) common gallery; part of an *écoute* nearest the exterior of the place;
grande —— *ordinaire*, (*mil. min.*) gallery 1.85 to 2^m high by 1^m wide;
—— *horizontale*, level gallery;
—— *inclinée*, inclined gallery;
—— *magistrale*, (*fort.*) scarp gallery;
—— *majeure*, (*mil. min.*) great, grand, gallery; part of an *écoute* near its origin;
—— *meurtrière*, (*fort.*) loopholed gallery;
en —— *de mine*, (to construct, constructed) after the manner of a mine gallery);
—— *montante*, ascending gallery;
—— *moyenne*, (*mil. min.*) gallery 2^m high by 2.1^m wide;
œil de ——, entrance to a gallery;
en pente, v. —— *inclinée;*
petite ——, (*min.*) heading, creep hole;
—— *principale*, (*min.*) mother-gate; (*mil. min.*) v. —— *majeure;*
—— *de revers*, (*fort.*) counterscarp gallery; reverse gallery, flanking gallery (ditch);
—— *transversale*, (*mil. min.*) cross or communicating gallery.

galérien, m., galley slave.

galet, m., pebble, bowlder, gravel, shingle; (*art. mach.*) roller, friction roller; live roller; chassis wheel;
—— *arrière (avant)*, (*art.*) rear (front) roller;
—— *à boudin*, flanged roller;
—— *de centrage*, centering roller;
—— *de châssis*, (*art.*) traversing roller, chassis wheel;
couronne de —— *s*, (*art.*) live-roller ring, roller circle;
échouer sur le ——, (of boats) to be stranded.
—— *d'équipement*, (*art.*) rigging roller (small pulley used in rigging the gin;)

galet à *friction*, (*mach.*) friction disk;
—— *de friction*, friction roller;
—— *-guide*, roller guide, chain-roller guide;
—— *de relèvement*, (*art.*) eccentric roller; (*Fr. art.*) rear roller of the *affût à soulèvement*;
—— *à ressort*, spring roller, roller with elastic bearings;
—— *de roulement*, (*art.*) recoil roller; friction roller;
—— *de soulèvement*, v. —— *de relèvement* (*Fr. art.*);
—— *tournant*, swivel roller.

galetage, m., (*powd.*) operation of making press cake;
—— *aux meules*, pressing in an incorporating mill;
—— *à la presse*, operation of making press cake by (hydraulic) pressure.

galeter, v. a., (*powd.*) to make press cake.

galetre, f., thin, flat cake; biscuit, cake; (*powd.*) press cake; (*art.*) obturator pad, gas-check pad; separator, separating disk (of a shrapnel);
—— *comprimée*, (*powd.*) press cake;
—— *de l'obturateur*, (*art.*) gas-check pad;
—— *plastique*, (*art.*) gas-check pad;
—— *de poudre*, (*powd.*) mill cake;
promenade ——, (*St. Cyr slang*) general marching out.

galeux, a., mangy, scabby, scabbed.

gallette, f., v. *gaillette*.

galiote, f., (*nav.*) galliot;
—— *à bombes*, bomb-ketch.

galipot, m., white rosin.

galle, f., oak-apple; (*fond.*) knot, aggregation of vitrified sand (in casting sand, if not sufficiently refractory);
noix de ——, nutgalls.

galoche, f., snatch block, leading block, quarter block, shoe block; belaying cleat; (*Fr. a.*) sort of shoe (wooden sole, leather upper);
—— *de bois*, hollow cleat;
—— *de fer*, hanging clamp;
sabot-——, (*Fr. a.*) the army name of the *galoche*, supra.

galon, m., (*unif.*) gold lace, galloon, trimming. stripe, band (as on collar, sleeve, forage cap, saddle cloth, etc.);
arroser ses ——s, v. s. v. *arroser;*
——*s de chevrons*, (*Fr. a.*) woolen stripes worn by corporals;
—— *de distinction*, distinguishing lace, stripes;
——*s de fonctions*, (*Fr. a.*) distinguishing badge (*fourriers*, buglers, drummers);
——*s de grade*, (*Fr. a.*) grade stripes (n. c. o. and corporals);
——*s d'imbécile*, (*mil. slang*) long-service stripes;
——*-montant*, (*unif.*) on a forage cap, lace running from band to crown;
—— *de première classe*, (*Fr. a.*) stripe worn by a private of the first class;
rendre ses ——*s*, to resign one's rank (n. c. o.); to resign one's chevrons (U. S. A.);
—— *-soutache*, (*Fr. a.*) insignia of *adjudants*, honary distinction worn by reenlisted n. c. o.'s and by officers who do not wear the epaulet;
—— *-tournant*, (*unif.*) on a forage cap, lace around the band.

galonner, v. a., (*unif.*) to lace, to galloon.

galop, m., (*man.*) gallop; canter;
—— *aisé*, easy canter;
aller au ——, to gallop;
—— *allongé*, (*Fr. cav.*) a gallop at the rate of 440ᵐ a minute; full gallop;
au ——, at a gallop; (*call*) gallop;
branle de ——, start of a gallop;
—— *de charge*, (*Fr. cav.*) full-speed gallop; charging gallop;
—— *de chasse*, a gallop at the rate of 600ᵐ a minute;
—— *de contretemps*, a broken gallop (gallop, fore legs; *courbettes*, hind legs);
—— *de course*, a gallop at the rate of 800ᵐ to 900ᵐ a minute; running gallop, run (U. S.);

galop *désuni*, disunited gallop;
—— *dur*, rough gallop;
—— *d'école*, v. —— *de manège;* (according to some authorities) a synonym of *galopade*, last definition;
enlever de ——, run off at a gallop;
—— *à faux*, gallop with the wrong foot leading;
—— *forcé*, overreached gallop;
grand ——, full gallop;
—— *juste*, true gallop;
—— *de manège*, (*Fr. cav.*) hand gallop; a slow gallop of variable rate (according to some authorities, of 350ᵐ a minute), employed in training recruits and young horses;
—— *ordinaire*, (*Fr. cav.*) gallop at the rate of 340ᵐ a minute;
petit ——, hand gallop, canter;
plein ——, v. *grand* ——;
—— *à quatre temps*, a gallop in which four beats are heard, (hence) irregular gallop; (according to some authorities) a synonym of —— *de manège;*
au triple ——, as fast as possible;
—— *à trois temps*, true gallop, normal gallop.

galopade, f., (*man.*) galloping, gallop; space galloped over; an air consisting in a short-gaited, well-marked gallop;
faire une ——, to have a gallop.

galoper, v. n., (*man.*) to gallop; to go at a gallop;
—— *désuni*, to gallop disunited;
—— *à droite (gauche)*, to gallop with the right (left) foot leading;
—— *faux*, to gallop false;
—— *juste*, to gallop true;
—— *près du tapis*, to gallop close to the ground (to raise the feet but slightly);
—— *à quatre temps*, to gallop so that the diagonally opposite feet make two beats instead of one;
—— *sur le bon (mauvais) pied*, to gallop with the right (wrong) foot leading;
—— *sur le pied droit (gauche)*, v. —— *à droite (gauche).*

galopeuse, f., secondhand (of a watch).

galvanique, a., (*elec.*) galvanic;
batterie ——, galvanic battery;
élément ——, galvanic cell;
pile ——, galvanic battery.

galvanisation, f., (*met.*) galvanization;
—— *du fer*, galvanizing of iron.

galvaniser, v. a., (*met.*) to galvanize.

galvanisme, m., (*elec.*) galvanism, voltaic electricity.

galvano, m., electrotype.

galvano-cautère, m., (*hipp.*) electric cautery.

galvanomètre, m., (*elec.*) galvanometer;
—— *absolu*, absolute galvanometer;
—— *apériodique*, aperiodic galvanometer, deadbeat galvanometer;
—— *astatique*, astatic galvanometer;
—— *balistique*, ballistic galvanometer;
—— *différentiel*, differential galvanometer;
—— *d'essai*, testing galvanometer (as for torpedoes);
—— *à fil de cocon*, Nobili's astatic galvanometer;
—— *à inductions rapides*, deadbeat galvanometer;
—— *d'intensité*, current-strength galvanometer;
—— *marin*, marine galvanometer;
—— *à miroir*, reflecting, mirror, galvanometer;
—— *à réflexion*, v. —— *à miroir;*
—— *des sinus*, sine galvanometer;
—— *des tangentes*, tangent galvanometer;
—— *à torsion*, torsion galvanometer.

galvanoplastie, f., electroplating.

galvanoplastique, a., galvanoplastic.

galvanoscope, m., galvanoscope.

gamache, f., (*mil.*) spatterdash (obs., *fausses bottes* being now used).

gamelle, f., cup, bowl, platter; (*mil.*) mess kit, mess, dinner (also breakfast);
—— *de campement*, (*Fr. a.*) camp kettle;

gamelle *de cavalerie.* (*Fr. cav.*) individual mess tin;
— *de cuisine,* (*Fr. a.*) soup copper (as for a whole company);
faire — *avec,* to mess with;
— *individuelle,* (*Fr. a.*) individual mess tin;
— *d'infanterie,* (*Fr. inf.*) individual mess tin;
— *-marmite de peloton,* (*Fr. a.*) mess kettle, issued one to each platoon of cavalry;
— *moulin à café,* (*Fr. a.*) a mess tin, having a coffee mill in place of the usual lid;
— *pour 4 hommes,* (*Fr. a.*) mess tin used during marches, maneuvers, on mobilization, and issued four to each squad of infantry, and one to every four men in the artillery, the train, etc.;
prendre —, to take, eat, dinner.

gamme, f., gamut, scale; (*top. drawing*) scale of hachures, tones, tints;
— *des charges,* (*art.*) scale of charges for different angles of fire (plunging).

ganache, f., (*hipp.*) lower jaw; region of the lower jaw;
chargé de —, (*hipp.*) heavy-headed;
écartement des —*s*, splay of the jaw;
faire la —, (*hipp.*) to burn off or remove hairs on the *ganache* (common horses);
— *ouverte,* wide jaw;
— *serrée,* narrow jaw.

ganglion, m., (*hipp.*) spavin.

gangrène, f., (*med.*) gangrene.

gangue, f., (*met.*) gangue; (*mas.*) mass of mortar (rare).

ganse, f., (*cord.*) small cord; loop, eye, cord or rope doubled on itself; bight, double, or bend with which a knot commences; (*art.*) rope handle (for handling projectiles);
— *d'accouple,* (*harn.*) coupling loop;
— *de corde,* loop, bend, eye of a rope;
— *croisée,* (*harn., etc.*) thong-lace;
— *d'étoupille,* (*art.*) friction-primer loop;
— *de prolonge,* (*art.*) prolonge loop;
— *du trait de harnais,* (*harn.*) trace loop;
trou de —, (*art.*) rope-handle eye.

gansin, m., (*cord.*) nipper.

gant, m., glove;
— *bourré,* boxing, fencing, glove;
— *de boxe,* boxing glove;
— *d'escrime,* fencing glove;
— *pour faire les armes,* v. — *d'escrime*;
— *fourré,* v. — *bourré*;
— *de manœuvre,* (*unif.*) sort of drill mitten (*art.*);
— *de manœuvre (dit moufle),* v. — *de manœuvre*;
— *moufle,* mitten;
— *passe-coude,* gauntlet;
—*s de pied,* (*mil. slang*) wooden shoes.

garage, m., docking of boats; (hence) basin, turnout (as in a canal); (*r. r.*) shifting, shunting, side track;
— *de chargement,* (*r. r.*) loading track;
voie de —, (*r. r.*) side track, siding.

garant, m., guaranty, surety; (*cord.*) running, hauling part of a tackle; fall, tackle fall;
— *de caliorne,* (*cord.*) fall of a winding tackle;
— *de candelette,* (*cord.*) fish-tackle fall;
— *de capon,* (*cord.*) cat-rope or fall;
— *de palan,* (*cord.*) tackle fall;
— *d'une poulie,* (*cord.*) strap of a block.

garant, a., guaranteeing.

garanti, m., person guaranteed, bonded person.

garantie, f., warranty, guaranty, security; money deposited as a guaranty.

garantir, v. a., to warrant, guarantee, make sure; to protect, shield.

garantisseur, m., bondsman.

garcette, f., cat; (*cord.*) gasket, knittle, sennit;
— *de fourrure de câble,* (*cord.*) plat for serving a cable;
lever une —, (*cord.*) to unseize;
— *à œillet,* (*cord.*) eye gasket;
— *de tournevire,* (*cord.*) cable nippers.

garçon, m., boy; youngster.

garde, f., guard, guarding, defense, protection, custody, safe-keeping; guard piece; ward of a lock; (*cord.*) check rope; (*r. r.*) rail guard; (*sm. a.*) guard of a sword; (*fenc.*) guard, position of guard; (*art.*) cylinder containing a cartridge; (*mil.*) guard; guard (i. e., body of troops), the Guards.

I. Military; II. Miscellaneous.

I. Military:

arrière —, rear guard;
— *avancée,* outpost (rare);
avant——, advance guard;
— *du camp,* quarter guard (camp, bivouac, cantonment);
— *de caserne,* quarter guard, barrack or police guard;
changer la —, to change guard;
— *à cheval,* horse guards;
corps de —, guardroom, guardhouse, guard post; guard tent;
de —, on guard;
— *descendante,* old guard, guard coming off;
descendre de —, *descendre la* —, to come off, march off, guard;
— *du drapeau,* color guard;
— *d'écurie,* stable guard;
— *extérieure,* exterior guard;
— *extraordinaire,* special guard, additional men for a guard under particular circumstances;
flanc——, m., flanker; flank guard;
— *folle,* v. — *avancée* (obs.);
— *des généraux,* headquarter guard;
grand —, grand guard;
homme de —, member of the guard;
— *d'honneur,* guard of honor;
hors la —, obsolete term for *aux armes* (turn out the guard);
— *impériale,* imperial guard;
inspection avec la —, turning out with the guard (punishment for privates);
— *intérieure,* interior guard, as opposed to grand guard, outposts and the like;
mettre la — *sous les armes,* to turn out the guard;
— *mobile,* v. — *nationale mobile*;
— *montante,* new guard, guard marching on;
monter la —, to mount guard, the guard; to be on guard over, to be on guard at;
— *nationale,* national guard;
— *nationale mobile,* French militia (organized in 1868 and disbanded in 1871);
— *nationale mobilisée,* French militia of 1870 (disbanded in 1871), composed of bachelors between 25 and 35 years of age who had never served in the army;
— *d'officier,* officer's guard (commanded by an officer);
officier de —, officer on, for, or of a guard;
officier de la —, officer commanding the guard;
parade de (la) —, guard mounting;
— *de Paris,* v. — *républicaine*;
partants de la —, men coming off guard;
petite —, n. c. o.'s guard;
— *à pied,* foot guards;
— *de la place,* guard of a fort, etc., and of its gates;
— *de police,* police guard, barrack guard (camp, quarter guard);
prendre la —, to march on, to march on guard;
— *relevante,* old guard, guard marching off;
relever la —, to relieve the guard;
— *républicaine,* the gendarmerie of the city of Paris, under the immediate orders of the prefect of police;
— *royale,* royal guard;
— *de tranchée,* (*siege*) trench guard, infantry in the first parallel;
— *des voies de communication,* in war, troops guarding communications.

II. Miscellaneous:

bonne —, safe-keeping, custody;
— *de cabestan,* capstan swifter;
en —, in custody; (*fenc.*) on guard;
faire la —, to be on the lookout;

garde *se mettre en* ——, (*fenc.*) to take the position of guard;
mettre sous bonne ——, to commit to custody;
reprendre ——, (*fenc.*) to resume position of guard;
—— *de sabre*, (*sm. a.*) guard of a sword or saber;
sous- ——, (*sm. a.*) trigger guard;
se tenir en ——, (*fenc.*) to be on guard;
se tenir sur ses ——*s*, to be careful, to be on the watch.

garde, m., watchman, guardian, keeper; (*mil.*) guardsman;
—— *d'artillerie*, (*Fr. art.*) artillery storekeeper (on duty in the various establishments of the arm, rank as officers; there are five grades: —— *principal de 1ère, 2ème, classe*; —— *de 1ère, 2ème, 3ème classe*);
cent ——*s*, Napoleon III's bodyguard;
cent- ——*s*, a member of the *cent gardes;*
—— *champêtre*, in France, a country policeman;
—— *chef-artificier*, (*Fr. art.*) a —— *d'artillerie* that has been a n. c. o. of artillery and has taken the course at the school of pyrotechny;
—— *chef-ouvrier*, (*F. art.*) a —— *d'artillerie* coming from the *ouvriers d'état*, q. v.;
—— *à cheval*, (*mil.*) horse guardsman;
—— *comptable*, (*Fr. art.*) a —— *d'artillerie* in charge of material, and sometimes responsible for public funds;
—— *du corps*, (*mil.*) life guardsman; in pl., life guard, bodyguard;
—— *forestier*, forester; v. *chasseurs forestiers;*
—— *du génie*, (*Fr. eng.*) an obsolete name for *adjoint du génie;*
—— *impérial*, (*mil.*) member of the imperial guard;
—— *maritime*, v. *garde-pêche;*
—— *à pied*, (*mil.*) foot guardsman;
—— *principal de 1ère (2ème) classe*, (*Fr. art.*) the two highest grades in the hierarchy of the *gardes d'artillerie*;
—— *royal*, (*mil.*) life guardsman, member of royal guard;
—— *des sceaux*, keeper of the seals.

garde à vous, (*mil.*) attention (call and command).
garde-barrière, m., (*r. r.*) crossing keeper.
garde-boue, m., mud guard (bicycle).
garde-caisses, m., (*art.*) in mountain artillery, the cannoneer that passes out ammunition from the chests.
garde-cendre, m., fender.
garde-chaîne, m., chain guard (bicycle).
garde-chambre, m., (*mil.*) room orderly.
garde-chasse, m., gamekeeper.
garde-chevaux, m., (*art.*) horse holder.
garde-coffre, m., (*art.*) the cannoneer at the limber.
garde-consigne, m., person whose duty it is to see that orders are carried out.
garde-corps, m., manrope; hand rail, side rail, balustrade (of a bridge); rail (of a ship); swifter (of a capstan).
garde-côte(s), m., coast guard, coast guardsman; (*nav.*) coast-guard vessel, coast-defense vessel;
—— *cuirassé*, (*nav.*) armored coast-defense vessel;
—— *à tourelles*, (*nav.*) turret coast-defense vessel.
garde-crotte, m., mud guard, dashboard, splashboard.
garde-drapeau, m., (*mil.*) color guard.
garde-étalon, m., stud keeper.
garde-feu, m., fender.
garde-finances, m., in Italy, the customhouse personnel.
garde-flanc, m., (*mil.*) flank guard.
garde-flans, m., side-pad of a saddle.
garde-fou, m., balustrade; rail; railing, manrope, hand-rail.
garde-fourneau, m., (*met.*) assistant founder.
garde-frein, m., brakeman.
garde-frontières, m., in Russia, the customhouse personnel.
garde-jambe, m., (*harn.*) leg guard.
garde-jour, m., screen.

garde-jupe, m., skirt guard (woman's bicycle).
garde-ligne, m., (*r. r.*) track watchman.
garde-magasin, m., storekeeper.
garde-main, m., (*sm. a.*) hand guard, barrel guard.
garde-malade, m., f., (*med.*) nurse.
garde-manger, m., pantry.
garde-marine, m., (*nav.*) naval cadet, midshipman (obs.).
garde-meuble, m., furniture storehouse, lumber-room.
garde-parc, (*Fr. art.*) battery, storekeeper (*sous chef-artificier*) (cares for material, etc.).
garde-pavé, m., curbstone.
garde-pêche, m., in France, a sort of river and coast policeman, coast guard.
garde-phare, m., light-house keeper.
garde-port, m., harbor watch.
garde-poussière, m., dust guard.
garder, v. a., to guard, keep, protect, watch, defend; (*mil.*) to guard; to defend;
—— *les arrêts*, (*mil.*) to observe one's arrest, to be in arrest;
—— *la prison*, to be in jail;
—— *les rangs*, (*mil.*) to remain, keep, in ranks;
se ——, to protect one's self; (*mil.*) to do guard duty, to perform security service;
—— *le terrain*, (*man.*) to keep on the track (of a horse);
—— *en vue*, to keep sight of.
garde-robe, m., wardrobe.
garde-rôle, m., keeper of the rolls.
garde-roue, splasher (of a carriage); guard stone (at a corner); wheel rail (of a bridge of boats).
garde-sable, (*pont.*) hurter.
garde-salle, m., (*fenc.*) fencing-master's assistant.
garde-scellés, m., keeper of the seals.
garde-soleil, m., sunshade (of a telescope, field glasses, etc.).
garde-temps, m., chronometer.
garde-voie, m., (*r. r.*) track watchman.
garde-vue, m., shade, eye shade, screen.
gardien, m., guardian, keeper, watch;
—— *de batterie*, (*Fr. art.*) battery storekeeper (in land and s. c. forts, and in military establishments, unless the engineers have a special representative; nearest equivalent, U. S. A., ordnance sergeant).
gare, f., place where wagons, boats, carriages, etc., may be kept; small wet dock or basin; part of river where vessels are in security; (*r. r.*) station; platform;
—— *d'arrivée*, (*r. r.*) station of arrival, at end of journey;
—— *de bifurcation*, (*r. r.*) station at the intersection or branching of two lines;
chef de ——, (*r. r.*) station master;
—— *de départ*, (*r. r.*) station at the beginning of a journey;
—— *de dépôts*, (*r. r.*) station at which locomotives are changed;
—— *d'eau*, wharf;
—— *d'évitement*, (*r. r.*) turn-out track, side track, passing station (single-track line); (of canals) turn-out (to allow boats to pass in opposite direction);
—— *d'expédition*, (*r. r.*) sending station, starting station;
—— *frontière*, (*r. r.*) frontier station;
grande ——, principal or main railroad station;
—— *halte*, —— *halte-repas*, (*mil.*) covered shed for men halted for meals on railroad journey;
infirmerie de ——, v. s. v. *infirmerie;*
—— *à marchandises*, (*r. r.*) freight depot, goods station;
—— *de point de départ d'étapes*, (*Fr. a.*) v. —— *de rassemblement;*
—— *de rassemblement*, (*Fr. a.*) railroad station designated for each *région de corps d'armée* in the interior, and toward which are directed troops, materials, supplies, etc., coming from the territorial circumscription of the corps (in war, on mobilization;)
—— *de stationnement*, (*r. r.*) railway station;
—— *tournante*, (*r. r.*) turntable;
—— *terminale*, (*r. r.*) terminus;

gare *de tête*, (r. r.) reversing station;
—— *de triage*, (r. r.) station for making up trains;
—— *aux voyageurs*, (r. r.) passenger station.

gare, (interjection) look out!
—— *au câble*, stand clear of the cable!
—— *dessous*, stand from under!
—— *à vous*, look out for yourself!

garer, v. a., to shift, shunt (railway carriages); to put boats in a basin or dock;
se ——, (of carriages) to pass each other on the road by turning out; (r. r.) to pass upon a side track.

gargouille, f., gargoyle; water pipe (in a gutter); spur ring; (harn.) lower cheek eye of a bridle bit.

gargousse, f., (art.) cartridge;
—— *à blanc*, blank cartridge;
—— *de contrôle*, standard cartridge;
—— *à espaces d'air*, air-spaced cartridge;
—— *d'exercice*, practice cartridge;
—— *liée*, choked cartridge;
—— *nue*, exposed cartridge (i. e., not in pass box);
—— *obturatrice*, self-sealing cartridge;
—— *sans poudre*, dummy cartridge;
—— *de salut*, —— *pour salve d'artillerie*, saluting cartridge;
——-*signal*, signal cartridge;
—— *vide*, cartridge bag.

gargoussier, m., (art.) cartridge bag; pass box;
—— *à fond mobile*, (F. m. art.) pass box with movable bottom for convenience of loading.

garni, m., filling; (cord.) rounding.

garnir, v. a., to fit, furnish, equip with; to stock, provide, supply, adorn; to rig; (cord.) to serve (a rope); (mach.) to pack (a stuffing box); to apply (as a chain to a sprocket wheel); (mil.) to man (a line), occupy a position; (harn.) to put the harness (on a horse's back); (artif.) to fit the head and cone of a rocket;
—— *une amarre*, (cord.) to put on a rounding;
—— *les avirons*, to ship the oars;
—— *une batterie*, (art.) to fit out and equip a battery;
—— *le cabestan*, to rig the capstan;
—— *la chèvre*, (art.) to rig the gin;
—— *une épée*, (sm. a.) to put on the guard;
—— *une fusée*, (artif.) to fix the head and cone of a rocket;
—— *les joints*, (mas.) to point;
—— *la ligne de feu*, (mil.) to man, strengthen, the firing line;
—— *le magasin*, (sm. a.) to fill the magazine;
—— *une place*, (mil.) to supply a fortress (guns, etc., stores, etc.);
—— *un saucisson*, (artif.) to fill up a saucisson where it is too thin;
—— *un terrain*, (art.) to cover an area with falling projectiles.

garnisaire, m., a soldier billeted on a family whose son had failed to respond to the conscription (obs.).

garnison, f., (mil.) garrison;
à ——, garrisoned;
—— *de défense*, complete or full garrison of a fortress or fortified position, i. e., the number of men necessary to make a fortress equal to its task in a campaign; (in a limited sense) the (increased) garrison of an attacked sector of defense;
être en —— *à*, to be stationed at;
former la —— *de*, to garrison;
mettre —— *dans*, to garrison;
—— *normale*, the garrison of an unattacked sector of defense (i. e., the number of men necessary to repel an open assault);
—— *de sûreté*, in a fortress or fortified position, the minimum of troops needed to guard against the surprise or open assault;
tenir —— *à*, to be stationed at or in;
ville de ——, garrison town.

garnissage, m., facing, trimming; (sm. a.) general term for all the minor operations of getting a gun barrel to shape.

garnisseur, m., (sm. a.) finisher.

garniture, f., furnishing, furniture, fitting; mounting; set; trimming; filling; quilting; (*in pl.*) mountings, finishings; wards of a lock; lining, liner (*esp. met.*) lining of a furnace, of a converter; (mach.) bearing box; packing, stuffing; (cord.) serving, rigging (standing and running); (farr.) outer edge or border of a horseshoe (projecting beyond the wall of the foot); (sm. a., *in pl.*) fittings, mountings, of a rifle, of a sword (bands, swivels, etc.);
—— *acide*, (met.) acid lining;
artifices de ——, —— *d'artifices*, (artif.) rocket filling;
—— *d'asbeste*, (mach.) asbestos packing;
—— *autoclave*, —— *automatique*, (mach.) self-acting packing;
—— *basique*, (met.) basic lining;
—— *de botte d'étoupe*, (mach., etc.) packing;
—— *en caoutchouc*, (mach.) rubber packing;
—— *de chanvre*, (mach.) hemp packing;
—— *s de coffre*, (mil.) all the fittings, fixed or movable, of ammunition, arms, equipment, and other chests;
—— *en cuir embouti*, (mach.) cup leather packing;
—— *en étoiles*, (artif.) star heading of a rocket;
—— *d'étoupe*, (mach.) tow packing;
—— *de fer*, iron mountings;
—— *de fusée*, (artif.) rocket heading;
—— *de fusil*, (sm. a.) small parts, mountings (bands, etc.) of a rifle;
—— *godet*, (mach.) cup packing;
—— *à membrane*, (mach.) diaphragm packing;
—— *métallique*, (mach.) metallic packing;
pièce de ——, (mach.) patch piece, packing plate;
—— *de piston*, (mach.) piston packing;
plateau de ——, (mach.) packing plate;
—— *à poche*, (mach.) "bag" pump packing;
—— *de pompe*, pump gear;
presse- ——, (mach.) packing-plate, -block;
presse-étoupe, (mach.) stuffing piece;
—— *de rondelle à coupelle*, (art.) cup washer, filling, or packing;
—— *de tête*, (harn.) head gear, head harness.

garochoir, m., (cord.) rope laid the wrong way;
en ——, laid the wrong way.

garouenne, f., (mach.) hanger.

garrot, m., (pont.) rack stick; (hipp.) withers;
—— *de l'arçon*, (harn.) saddlebow;
blessé au ——, (hipp.) wither-wrung, withergalled;
—— *coupé*, (hipp.) withers that are too short;
fistule de ——, (hipp.) fistulous withers;
mal de ——, v. *fistule de* ——;
—— *noyé*, (hipp.) withers that are too low;
—— *de pansement*, (med.) temporary or make shift tourniquet.

garrotté, a., (hipp.) wither-wrung.

garrotter, v. a., (pont.) to rack down.

garroture, f., (cord.) packing hitch.

gâteau, m., cake; (powd.) caked mass at the bottom of the stampers; (expl.) disk (as of gun cotton).

gatling, m., (r. f. art.) machine gun (as, e. g., "le —— *Nordenfeldt*").

gauche, f., left, left hand, left side; (mil.) left, left wing, left flank; (mach.) tap wrench; a., left; crooked; awkward.

gauchir, v. a. n., to warp, shrink, get out of true; to tilt (a gabion).

gauchissement, m., warping, bending, shrinking, getting out of true.

gaule, f., staff, rod, pole;
—— *d'enseigne*, (nav.) flagstaff (small, as at stern of ship);
—— *de pompe*, (mach.) pump spear.

gaulette, f., wattle, withe.

gavage, m., stuffing (as of pigeons).

gaye, *à* ——, astraddle.

gayet, m., (familiar term for) horse;
maquiller un ——, v. s. v. *maquiller*.

gayette, f., v. *galiette*.

gaz, m., gas;
bec de ——, gas burner;

gaz *de combustion*, gases of combustion;
— *compteur à* ——, gas meter;
— *conduit à* ——, gas pipe;
— *conduit à* —— *principal*, gas main;
— *éclairage au* ——, gas lighting;
— *d'éclairage*, illuminating gas;
— *éclairer au* ——, to illuminate by gas;
— *de houille*, coal gas;
— *imperméable au* ——, gas-tight;
— *réservoir de* ——, gas holder, receiver;
— *robinet de* ——, gas cock;
— *usine à* ——, gas works;
— *tuyau à* ——, gas pipe;
— *tuyau à* —— *principal*, gas main.

gazage, m., (*met.*) reduction by gaseous fuel.

gaze, f., gauze;
— *métallique*, wire gauze;
— *à pansement*, (*med.*) dressing gauze.

gazéifier, v. a., to transform into gas.

gazette, f., gazette;
— *lire la* ——, (*hipp.*) to be off his feed (**familiar**).

gazeux, a., gaseous.
gazier, m., gas fitter.
gazogène, m., gas generator.
gazomètre, m., gasometer, receiver;
— *à contrepoids*, counterpoise receiver;
— *à lunette*, telescopic gas holder;
— *à mouvements libres*, free receiver.

gazon, m., grass; turf; sod;
— *de boutisse*, header;
— *motte de* ——, sod (cut out in squares);
— *de panneresse*, stretcher;
— *plaqué*, facing sod;
— *posé de haut*, head sod;
— *de revêtement*, revetment sod;
— *revêtement en* —— *s*, sod revetment;
— *à talus*, v. —— *posé de haut*.

gazonnage, m., (*fort.*) sod work, sodding, revetting with sod.
gazonnement, m., (*fort.*) v. *gazonnage*.
gazonner, v. a., to turf, to sod, revet with sod.
gazonneur, m., (*fort.*) sod layer.

gélatine, f., gelatin;
— *détonante*, v. —— *explosive;*
— *-dynamite*, (*expl.*) gelatin dynamite, Nobel's nitrogelatine;
— *explosive*, (*expl.*) explosive gelatin; gum dynamite;
— *de guerre*, (*expl.*) explosive gelatin, gum dynamite, blasting gelatin;
— *obus à* ——, (*art.*) shell filled with explosive gelatin.

gélatinisation, f., (*expl.*) gelatinization.
gélatiniser, v. a., (*expl.*) to gelatinize.
gelbite, f., (*expl.*) gelbite.
gelée, f., frost;
— *blanche*, hoar frost.
geler, v. a. n. r., to freeze, become frozen.
gélif, a., cracked, or liable to crack, by frost (stones, trees).
gélignite, f., (*expl.*) gelignite.
gélivité, f., liability to crack, etc., by frost.
gélivure, f., frost-crack (in trees, stones).
gemelle, f., (*cons.*) fish, side piece, cheek.
gendarme, m., (*Fr. a.*) gendarme, private of the gendarmerie; (*in pl.*) the gendarmerie;
— *élève-* ——, "lance" gendarme, lowest enlisted grade of the gendarmerie.

gendarmerie, f., (*Fr. a.*) gendarmerie (civil duties, those of a constabulary; army duties, provost guard, prisoner's guard, direction and management of the *trains régimentaires*, when consolidated); headquarters of a gendarmerie post, the post itself;
— *d'Afrique*, the legion (Nineteenth) of gendarmerie serving in Algiers and Tunis;
— *brigade de* ——, v. s. v. *brigade;*
— *coloniale*, colonial gendarmerie;
— *compagnie de* ——, all the gendarmes of a department;
— *départementale*, the gendarmerie serving in the departments of France;
— *légion de* ——, legion of gendarmerie, so called, consisting of all the companies of one and the same army corps region (four regions have two, and one three legions);
— *maritime*, the gendarmerie serving in the military ports of France.

général, m., (*mil.*) general, general officer;
— *d'armée*, army commander (unofficial, not technical);
— *de brigade*, general commanding a brigade (major-general);
— *brigadier-* ——, the English and American term brigadier-general;
— *du cadre d'activité (de réserve)*, (*Fr. a.*) a general officer on the active (reserve) list of the *état-major général de l'armée;*
— *capitaine-* ——, (Spanish) captain-general;
— *en chef*, general in chief (no longer an official designation in France);
— *de circonstnace*, v. s. v. *circonstance;*
— *colonel-* ——, colonel-general (honorary title, not French);
— *commandant une armée*, (*Fr. a.*) official designation of a —— *de division* designated to command, in case of mobilization, an army of two or more corps;
— *commandant un corps d'armée*, (*Fr. a.*) official designation of a —— *de division* assigned to the command of a corps;
— *commandant un groupe d'armées*, (*Fr. a.*) official designation of a —— *de division*, commanding a group of armies on one and the same frontier, or in one theater of war;
— *directeur*, (*Fr. a.*) the general officer at the head of a *direction* of the war department;
— *de division*, general commanding a division (lieutenant-general);
— *d'exécution*, a term used by Marbot for a general fitted to carry out orders, but not to exercise independent command;
— *-feld-maréchal*, field marshal;
— *gouverneur*, (*Fr. a.*) the commanding general of the *places de Paris, de Lyon*, respectively;
— *-inspecteur*, (*Fr. a.*) general officer specially detailed by the minister of war to inspect troops;
— *-lieutenant*, the equivalent, in services other than the French, of —— *de division;*
— *lieutenant-* ——, lieutenant-general (of services other than the French);
— *-major*, the equivalent, in services other than the French, of —— *de brigade;*
— *major-* ——, the English and American term major-general (to be carefully distinguished from the following word);
— *major* ——, (*Fr. a.*) the chief of staff of all the armies in war (*major* is here a noun and *général* an adjective);
— *de tranchée*, (*siege*) c. o. of the trenches (daily detail, for each attack, of a general of brigade).

généralat, m., (*mil.*) dignity or grade of general; duration of functions of a general officer.
générale, f., (*call*) long roll (drums); alarm (bugles); general's wife.
généralissime, m., (*mil.*) generalissimo, general in chief.

générateur, m., (*mach., etc.*) generator (as steam generator); a., generating;
— *de courant*, (*elec.*) current generator;
— *de gaz*, gas generator;
— *turbo-électrique*, Parson's steam turbine, used to work a dynamo;
— *à, de, vapeur*, steam generator, boiler.

génération, f., generation, production (as of steam, gases, etc.);
génératrice, f., (*mach., etc.*) generatrix, (*elec.*) generating dynamo;
— *de l'âme*, (*art.*) surface of the bore;
— *de tourillon*, (*art.*) upper surface of trunnion.

genêt, m., broom; (*hipp.*) jennet.
genêter, v. a., (*farr.*) to bend horseshoes upward.
genette, f., (*harn.*) Turkish bit; kind of snaffle;
— *à la* ——, (*man.*) with short stirrups.

génie, m., engineers, engineering;
— *adjoint du* ——, v. s. v. *adjoint;*
— *chef du* ——, v. s. v. *chef;*
— *civil*, civil engineering;
— *corps du* ——, (*mil.*) engineer corps;

génie, *maritime*, (*nav.*) naval construction; corps of naval constructors;
— *militaire*, (*mil.*) military engineering; corps of engineers.

genope, f., (*cord.*) stopper, nipper; temporary racking, seizing, or lashing;
casser les —s, to break (the stops of a flag).

genoper, v. a. (*cord.*) to knot; to nip; to seize, rack, stop, frap; to jam, become jammed.

genou, m., knee; knuckled joint; loom of an oar; (*inst.*) ball-and-socket joint; spindle and related parts; joint, supporting joint; (*pont.*) knee, timber, of a ponton.
(Except where otherwise indicated, the following terms relate to the horse:)
à —, (*mil.*) (in commands) kneel.
n'avoir rien au dessous des — —*x*, to have slender cannons with bad tendons;
— *arqué*, knee bent forward;
— *de bœuf*, ox knee;
— *cambré*, knock-knee;
— *cerclé*, knee surrounded by *osselets*, i. e., affected with exostosis;
— *de charnière*, (*mach., etc.*) ball-and-socket joint;
— *à coquilles*, (*inst.*) ball spindle;
— *creux*, sunken knee, hollow or sheep knee;
— *effacé*, v. — *creux*;
monté à —, (*tech.*) elbow-jointed, elbowed;
monter à —, (*tech.*) to mount so as to form an elbow;
— *de mouton*, v. — *creux*;
— *de veau*, thick, bulbous, knee.

genouillère, f., kneecap, kneepiece; kneepiece of a boot; (*harn.*) knee pad; (*art.*) knee pad, kneeling pad (for the cannoneer that discharges the piece); (*mil. min.*) miner's or sapper's kneecap; (*r. r.*) hose between tender and locomotive; (*mach., inst.*) toggle joint; (*fort.*) genouillère, so much of the interior slope as lies between the platform and the inner crest of the embrasure; (in barbette batteries) height of parapet above banquette;
hauteur de —, (*art.*) height of trunnions above the ground or platform (an incorrect expression); (*fort.*) height of interior crest, or of embrasure sole, above the surface of the platform.

gens, m. f. pl., people, men, company; (*nav.*) crew, ship's crew;
— *d'affaires*, business men;
— *d'armes*, v. — *d'épée*;
— *sans aveu*, vagrants;
— *de cheval*, cavalry;
— *de cour*, courtiers;
— *d'église*, clergymen;
— *d'épée*, (*mil.*) military **men, soldiers**;
— *de guerre*, v. — *d'épée*;
— *de lettres*, literary men;
— *de mer*, seafaring men;
nos —, (*mil.*) our side, our people;
— *de pied*, infantry;
— *de quart*; (*nav.*) watch;
— *de robe*, lawyers, the legal fraternity.

géodésie, f., geodesy.

géodésique, a., geodetic.

géographe, m., geographer;
ingénieur —, (*nav.*) hydrographer.

géographie, f., geography;
— *astronomique*, astronomical geography;
— *économique*, statistical geography (treats of the resources of a country);
— *mathématique*, mathematical geography;
— *militaire*, military geography;
— *physique*, physical geography;
— *politique*, political geography.

géôlage, m., (*Fr. a.*) the *service* responsible for the food and bedding of soldiers confined in either civil or military prisons.

géologie, f., geology.

géologique, a., geological.

géomètre, m., geometer, mathematician; (*top.*) surveyor, land surveyor;
— *arpenteur*, man who fills in details of a survey;
— *triangulateur*, triangulator.

géométrie, f., geometry;
— *analytique*, analytical geometry;
— *descriptive*, descriptive geometry;
— *descriptive à deux plans*, ordinary descriptive geometry;
— *descriptive par cotes*, one-plane descriptive geometry (fortification drawing and plans);
— *pratique*, surveying;
— *projective*, projective geometry;
— *du sentiment*, (*top., drawing*) the representation of the surface as approximately as possible to real character (of surface forms not susceptible of rigorous representation).

géométrique, a., geometrical.

gérance, f., management, administration, by an agent.

gérant, a., managing, directing; m., agent, manager, director; (*Fr. a. adm.*) agent of the engineer corps authorized to make small payments in matters of current service.

gerbe, f., sheaf; column of water thrown up by a torpedo; (*art.*) cone of spread, of dispersion, sheaf of bullets, of shrapnel, fragments (*ball.*) pencil or sheaf of trajectories:
— *de dispersion*, (*art.*) cone of dispersion;
— *d'éclatement*, (*art.*) cone of burst, sheaf of fragments (of a shrapnel);
— *fusante*, (*art.*) cone of dispersion, sheaf of bullets of shrapnel on burst;
— *de justesse*, (*ball.*) surface of revolution of a *courbe de justesse* about the axis of abscissas;
noyau de la —, (*ball.*) mean trajectory of a sheaf;
— *des trajectoires*, (*ball.*) pencil or sheaf of trajectories.

gerber, v. a., to pile casks one on another; to make up into sheaves.

gerce, f., crack, shake (in wood); plank cracked in its thickness.

gercer, v. a. r., (of wood) to become shaky, to become full of shakes; to crack, cranny; to crack (in general).

gerçure, f., crack, fissure; shake, flaw (in wood); crevice, chink, cleft; (*art.*) hair crack, hair line (in armor, on the outer surface of guns, etc.) flaw produced in the outer surface of bronze guns by long service;
— *de trempe*, (*met.*) crack produced by tempering.

germe, m., germ, seed; shoot;
— *de fève*, (*hipp.*) mark in a horse's mouth.

gerseau, m., (*cord.*) small grommet or block strap.

geste, m., gesture;
— *d'avertissement*, (*mil.*) warning gesture (to attract attention);
— *d'exécution*, (*mil.*) gesture of execution;
— *préparatoire*, (*mil.*) preparatory gesture.

gestion, f., (*adm.*) administration: management or performance of duties according to definite rules, under the obligation of making returns, (hence) officials or persons in charge of stores, of moneys;
— *administrative*, generic term for the performance of any administrative duty;
— *de clerc à maître*, v. — *directe*;
— *collective*, (*Fr. a.*) the administration of a *conseil d'administration*;
— *directe*, the direct performance of a service, etc., by the government (i. e., without the intermediary of a contractor);
— *manutentionnaire*, the actual handling of stores or of moneys, in the performance of a government operation, in carrying on a *service*.

gestionnaire, m., (*adm.*) functionary or official responsible for the performance (*gestion*) of a service; a., in charge, having charge of.

gibecière, f., game bag.

gibelet, m., gimlet, borer.

giberne, f., (*mil.*) cartridge box (old pattern).

giboulée, f., hail shower.

giffard, m., — *d'exhaustion*, (*mach.*) ejector.

gigots, m., pl., (*hipp.*) hind legs of a horse.
gigotté, a., (*hipp.*) having strong thighs.
gigue, f., —— *de crosse*, (*sm. a.*) swell on a gun stock;
—— *monter en* ——, (*sm. a.*) to put or leave a swell on a gun stock.
gilet, m., waistcoat, undershirt;
—— *d'armes*, (*fenc.*) fencing jacket;
donner un —— *à quelqu'un*, (*fenc.*) to outfence one's opponent, to touch him very frequently;
—— *de fer*, (*mil. slang*) cuirassier.
ginguet, m., pawl, (as of a capstan, etc.).
girande, f., (*artif.*) girande, bouquet; any firework turning upon a wheel; girandole chest; collection of rockets placed in a chest in rows, smallest in front and largest in rear.
girandole, f., (*artif.*) girandole, any firework turning upon a wheel; wheel whose circumference is studded with rockets; any firework which revolves horizontally upon its center.
giratoire, a., gyratory;
faculté ——, (*nav.*) turning power (of a ship).
girel, m., windlass, capstan (a Mediterranean term).
giron, m., handle of an oar; movable handle (as of a crank); (*cons.*) tread of a step; (*mach.*) distance between the bearings of a crank;
—— *droit*, (*cons.*) tread of a straight staircase;
—— *triangulaire*, (*cons.*) tread of a winding staircase.
gironné, a., winding;
marche ——*e*, (*cons.*) step of a winding stair.
girouette, f., vane, wind vane, weathercock;
fer de ——, vane spindle.
gisement, m., bed (e. g., of clay); (*nav.*) bearing.
gîte, m., lodging; vein, seam, bed, deposit (of ore, coal, etc.); (*cons.*) floor sleeper; (*art.*) sleeper, sill, of a gun platform; (*mil.*) halting place (esp. at the end of a march, of a day's march);
—— *de calage* (*art.*) rear sleeper, cross sleeper, of a siege platform;
—— *cintrée*, (*art.*) curved sleeper (of a gun platform);
—— *d'étapes*, v. s. v. *étape*;
lambourde- ——, v. s. v. *lambourde*;
—— *principal d'étapes*, v. s. v. *étape*.
gîter, v. n., to lodge, sleep.
givre, m., rime, hoar frost.
glace, f., ice; plate glass; (*mach.*) slide-valve facing, slide-valve seat;
banc de ——, field of ice;
—— *de cylindre*, (*mach.*) cylinder face;
—— *en dérive*, floating ice; iceberg;
montagne de ——, iceberg;
nappe de ——, sheet of ice;
—— *objectif*, (*opt.*) object glass;
patin à ——, (*art.*) runner.
glacer, v. a., to freeze; to glaze.
glacerie, f., manufacture of ice.
glacier, m., glacier.
glacis, m., slope; (*fort.*) glacis;
—— *battu tout juste*, (*fort.*) glacis so constructed that the plane of the superior slope passes not more than 0.50m above its crest nor more than 1m over its foot;
—— *en contre-pente*, (*fort.*) countersloping glacis;
—— *coupé*, (*fort.*) limited glacis, swept by *brisures* in the interior crest of the main work;
—— *cuirassé*, (*fort.*) armored glacis (in Grüson system and others involving use of metals);
—— *masque*, (*fort.*) glacis mask;
—— *de pied de cheval*, (*hipp.*) exterior surface of the hoof;
—— *simple*, (*fort.*) single glacis;
tôle ——, (*fort.*) glacis plate.
glaçon, m., cake of ice, ice floe, icicle.
glais, m., v. *glas*.
glaise, f., loam; clay, potter's clay;
terre ——, clay.
glaiseux, a., clayey, marly;
terrain ——, stiff, clayey, ground.

glaive, m., sword (poetical);
puissance du ——, power of life and death.
gland, m., acorn; tassel.
glande, f., gland; (*hipp.*) tumefaction of the ganglions;
avoir une —— *dans l'auge*, (*hipp.*) to have tumefied, adherent, ganglions.
glandé, a., (*hipp.*) having tumefied, adherent, ganglions; glandered.
glas, m., passing bell; (*mil.*) salvos of artillery fired at the funeral of a sovereign or of a general of very high rank.
glène, f., (*cord.*) coil of rope;
—— *de cordage*, coil of rope;
filer sa ——, to drop the coil from one's hands;
—— *en galette*, flat coil, Flemish coil;
—— *plate*, flat coil, Flemish coil;
—— *plate en S*, French coil;
gléner, v. a., (*cord.*) to coil a rope, to take a few bights in the hand.
glette, f., litharge.
glissade, f., slip, slide, slipping, sliding.
glissant, a., slippery, sliding;
règle ——*e*, sliding rule.
glissement, m., sliding, slipping; (*mach.*) creeping (of a belt).
glisser, v. n., to slide, to slip, glide; (*mach.*) to slide; (of a belt) to creep.
glisseur, m., (*mach.*) slide block.
glissière, (*mach., steam, etc.*) slide, guide, slide face, slide bar, guide bar, motion bar; (*art.*) crosshead slide; (*F. m. art.*) console slide;
—— *de cylindre à vapeur*, (*steam*) slide face;
—— *à pivot*, (*art.*) pivoting slide (Hotchkiss 57mm field);
—— *du té*, —— *de la traverse* (*steam*) crosshead guide;
—— *de la traverse du piston* (*steam*) piston crosshead guide.
glissoir, m., (*mach.*) slide; sliding block.
glissoire, f., (*mach.*) slide, guide, sliding plate;
—— *expansive*, expansive slide;
tige de ——, motion bar.
globe, m., globe; (*powd.*) globe or shot used in testing fine-grained black powders; proof shot;
—— *de compression*, (*mil. min.*) globe of compression; overcharged mine; (sometimes) undercharged mine; (*art.*) a term applied to a torpedo shell on striking;
—— *d'éprouvette*, v. under main word, (*powd.*);
—— *de feu*, (*artif.*) any spherical firework;
—— *fumant*, (*artif.*) smoke ball thrown into the interior of works (obs.);
—— *modèle*, (*fond.*) shot pattern;
gloire, f., glory; (*artif.*) fixed sun in fireworks;
—— *militaire*, military glory.
glôme, m., (*hipp.*) glome.
glonoïne, f., (*expl.*) medical name for nitroglycerin, glonoin.
glu, f., birdlime;
—— *de lin*, (*artif.*) burnt linseed oil (used in the composition of *artifices Lamurre*);
—— *marine*, marine glue.
gluant, a., viscous, sticky, glutinous.
glucose, f., (*chem.*) glucose.
glucodine, f., (*expl.*) nitrated solution of cane sugar.
glycéride, f., (*chem.*) glycerid.
glycérine, f., (*chem.*) glycerin;
nitro- ——, (*expl.*) nitroglycerin.
glycéronitre, f., (*expl.*) another name for polynitrocellulose.
glycéro-pyroxyline, f., (*expl.*) Clark's nitrogelatin, Clark's explosive.
glyoxyline, f., (*expl.*) glyoxylin.
gobelet, m., cup; (*artif.*) rocket-case.
gobeter, v. a., (*mas.*) to point a wall; to stop.
gobille, f., (*powd.*) triturating ball.
godet, m., cup, bucket; bucket of a mill wheel; (*teleg.*) telegraph-line insulator; (*sm. a. art.*) cup or cap of a fuse, primer cap; (*art.*) ventshell in bronze s. b. mortars (obs.); breechplug recess or hollow (de Reffye); sort of washer guard used with the Lemoine brake;
—— *de fusée*, (*art.*) fuse cup;

godet à *graisse*, —— *graisseur*, (*mach.*) oil cup, grease cup;
—— à *huile*, oil cup;
—— à *suif*, tallow cup;
—— *de support*, (*teleg.*) insulator.
godille, f., (*boats*) stern oar, scull.
godiller, v. n., to scull.
godilleur, m., sculler.
godillot, m., (*Fr. a.*) military shoe (so called from name of manufacturer; no longer worn); (*mil. slang*) recruit.
goëlette, f., schooner.
golfe, m., gulf.
gomme, f., gum;
—— *adragante*, gum tragacanth;
—— *arabique*, gum arabic;
—— *élastique*, India rubber;
—— *explosive*, (*expl.*) gum dynamite, explosive gelatin;
——-*gutte*, gamboge;
——-*laque*, gum lac.
gommelaquer, v. a., to gum lac.
gond, m., hinge; hook, pintle; gudgeon collar;
à ——*s*, hinged;
—— *de fiche*, hinge hook;
hors des ——*s*, off the hinges.
gondile, f., v. *godille*.
gondiller, v. a., v. *godiller*.
gondolage, m., warping (of wood).
gondolement, m., v. *gondolage*.
gondoler, v. n., to warp (as a drawing board).
gondolure, f., warping.
gonflement, m., swelling, distension; inflation of a balloon; quantity of gas necessary to fill a balloon; (*sm. a.*) swell in gun barrel (caused by sand, gravel, etc.).
gonfler, v. a., to swell, swell out, distend; to inflate a balloon.
goniasmomètre, m., (*inst.*) goniometer.
goniographique, a., v. *goniométrique*.
goniomètre, m., (*inst.*) goniometer;
—— *de pointage*, (*art.*) aiming goniometer.
goniométrique, a., angle measuring.
gorge, f., throat, neck, mouth, opening; groove, channel, score, of a pulley; (*tech.*) recess, channel, furrow, groove, hollow; (*top.*) gorge, ravine, pass, strait, defile; (*artif.*) choke; (*mach.*) neck (of a journal); (*sm. a.*) groove, gorge, of a cartridge head; (*fort.*) gorge; splay (of a turret port);
—— *brisée*, (*fort.*) broken gorge;
—— *de carte*, roller of a map;
—— *de culasse*, (*art.*) groove on breech (serves to hold *couvre-culasse*, and is used in mechanical maneuvers);
—— *d'un essieu*, (*mach.*) axle neck, shaft neck;
—— *de l'étui*, (*sm. a.*) neck of a cartridge case;
évasé en ——, splayed out;
—— *de fusée*, (*artif.*) choke of a rocket;
—— *de graissage*, grease channel;
—— *de pigeon*, (*harn.*) sharp, severe bit;
—— *de poulie*, score, groove, of a block, of a pulley;
—— *de rupture*, (*art.*) fuse-plug score;
sous- ——, (*harn.*) throat-latch;
—— *de vis*, furrow of a screw.
gorgé, p. p., (*hipp.*) swelled;
jambes ——*es*, swelled legs.
gorger, v. a., to choke, fill up; (*hipp.*) to swell;
—— *une fusée*, (*artif.*) to fill a rocket with composition.
gorget, m., molding plane.
gosier, m., throat.
goudron, m., tar;
—— *des gaz*, v. —— *de houille*;
—— *de houille*, coal tar;
—— *minéral*, mineral tar, coal tar; asphalt;
puits à ——, tar pit of gas works;
usine à ——, tar works;
—— *végétal*, vegetable tar.
goudronnage, m., tarring;
—— *gras*, (*artif.*) composition for tarred links (20 parts pitch to 1 tallow);
—— *sec*, composition for tarred links (equal parts of pitch and of rosin).
goudronner, v. a., to tar, to pay.
goudronnerie, f., tar works.

goudronneur, m., one who pays, lays on, tar.
gouffre, m., gulf, whirlpool.
gouffré, a., ribbed.
gouge, f., gouge, spoonbit;
—— à *baguette*, half-round bit;
—— à *bec de corbin*, bent gouge;
—— à *canon*, barrel gouge;
—— *double*, double gouge (i. e., of ∼-section);
—— à *ébaucher*, turning gouge, round tool;
—— *triangulaire*, parting tool.
gouger, v. a., to gouge.
goujat, m., armed peasant; mortar boy, mason's apprentice; apprentice in a small-arms factory; (*mil.*) soldier's servant (works for his bread only);
—— *de vaisseau*, (*nav.*) powder monkey.
goujon, m., pin, peg, dowel, dowel pin, gudgeon, clip; center pin, coak; (*art.*) plunger pin (of a fuse);
—— *d'arrêt*, stop pin;
—— *de charnière*, hinge pin;
cheville à ——, rag bolt;
—— *perdu*, peg, dowel;
—— *de poulie*, gudgeon;
—— à *vis*, screw gudgeon;
goujonner, v. a., to dowel, to pin.
goujure, f., channel, score, score of a block, notch; groove.
goulée, f., —— *du jet de fonte*, (*fond.*) channel from furnace to mold.
goulet, m., neck of a bottle; gutter; (*top.*) narrow part of a stream, where the current is very swift; gut, channel, entrance of a port or harbor; (*art.*) fuse hole of a shell (rare);
—— *de fossé*, (*fort.*) cunette.
goulette, goulotte, f., small channel, groove, gullet, water channel.
goum, m., an Arabic word denoting the gathering or reunion, under arms, of all the mounted men of a tribe (used by the French as scouts in Algiers and Tunis).
goumier, m., a member of a *goum* (q. v.), used as a scout, (hence) Arab scout.
goupille, f., pin, peg, forelock; joint pin; key of a bolt, bolt pin; leather peg washer;
—— *d'arrêt*, (*art. etc.*) stop pin, stay pin;
—— -*arrêtoir*, (*art.*) pin stop, stay pin;
—— *double*, split pin;
—— *fendue*, split pin, forelock pin;
—— -*pivot*, pivot pin;
—— à *ressort*, split pin;
—— *de sûreté*, (*art.*) safety pin, fuse safety pin.
goupiller, v. a., to pin, peg, dowel.
goupillon, m., sprinkler for a forge.
goupillonner, v. a., to sprinkle water.
gourbi, m., (*mil.*) any improvised light shelter, shack (as of branches, of wattles);
baraque- ——, v. s. v. *baraque*.
gourbillage, m., countersinking.
gourbiller, v. a., to countersink.
gourde, f., cup, drinking cup.
gourdin, m., cudgel, club, rope's end; cobbing board.
gourdiner, v. a., to cob; to cudgel.
gourmander, v. a., to reprimand harshly;
—— *un cheval*, (*man.*) to check a horse violently with the bit, use him roughly with the hand.
gourme, f., (*hipp.*) strangles;
—— *bénigne*, mild strangles;
fausse ——, bastard strangles, improper expression for relapse of the strangles;
jeter sa ——, to run at the nose;
—— *maligne*, severe strangles.
gourmer, v. a., (*man., harn.*) to curb.
gourmette, f., (*harn.*) curb chain; curb;
fausse ——, lip strap;
mors à ——, curb bit.
gournablage, m., doweling, treenailing.
gournable, m., treenail.
gournabler, v. a., to treenail.
goussant, goussaut, a., thickset; m., thickset horse.
gousset, m., armpit; gusset, gusset stay; bracket, brace; (*carp.*) angle brace; (*harn.*) cushion or pad to protect withers;
—— *d'assemblage*, angle plate.

goutte, f., drop;
— *froide*, (*met*.) blister (in a casting), bead or globule of oxidized metal (in an ingot);
— *sereine*, amaurosis;
— *de suif*, (*elec*.) contact point.

gouttière, f., spout, guttering, gutter; synonym of *gélivure*, q. v.; (*sm. a*.) hollow of a blade, of a bayonet; (*fort*.) reentrant angle of intersection of two planes or faces, e. g., of the glacis; (*med*.) cradle (for a broken leg, etc., in bed);
— *de l'estoc*, (*sm. a*.) groove for forging musket barrels;
— *à fusée*, (*art*.) rocket trough;
— *de glacis*, v. under main word, (*fort*.);
— *des jugulaires*, (*hipp*.) jugular gutter;
— *pare-éclats*, (*fort*.) the gutter of a turret, serving also as a protection against splinters;
— *de tourelle*, v. —— *pare-éclats*.

gouvernail, m., (*nav*.) rudder, helm; (*met*.) porter bar; (of a bicycle) handle bar;
effet de ——, (*nav*.) steerageway.

gouvernant, m., governor, ruler.

gouverneur, m., (*mach*.) governor, regulator.

gouverne, f., guidance; government;
pour sa ——, for one's guidance.

gouvernement, m., government; government house; command, direction, management; (*nav*.) steering; (*mil*.) government, command, of a fortified position; (*mil. slang*, Polytechnique) sword;
— *militaire*, (*Fr. a*.) current name of the circumscriptions under the command of the governors of Paris and of Lyons, respectively.

gouverner, v. a. n., to govern, rule; manage, direct, regulate; to take care of, husband; (*man*.) to rein, rein up; (*nav*.) to steer.

gouverneur, m., governor; (*nav*.) helmsman; (*mil*.) c. o. of a fortified position;
— *désigné*, (*Fr. a*.) officer designated to command a place in case of war (has this title in peace only);
— *général*, governor-general;
— *militaire*, (*Fr. a*.) designation of the commanding generals of the *gouvernements* of Paris and of Lyons, respectively;
— *de place*, (*mil*.) c. o. of a fortified town or place.

gouvion, m., large bolt.

grâce, f., pardon; quarter;
coup de ——, finishing stroke;
demander —— to cry mercy, ask for quarter;
droit de ——, pardoning power;
faire —— *de*, to forgive, remit, pardon.

gracier, v. a., to pardon.

grade, m., grade, rank, degree of rank; step; grade (of a circle, unit of angular measure); (*mil*.) grade;
assimilation de ——, (*mil*.) assimilated rank;
— *effectif*, (*mil*.) real or actual grade (as opposed to brevet);
— *honoraire*, (*mil*.) honorary grade, grade higher than the one normally held; brevet grade;
— *honorifique*, (*mil*.) complimentary grade (bestowed usually upon a foreign sovereign);
le plus élevé en ——, (*mil*.) the senior in grade;
monter en ——, (*mil*.) to be promoted;
primauté de ——, (*mil*.) seniority in one's grade;
— *titre*, (*mil*.) brevet commission or grade.

gradé, m., (*mil*.) the holder of a grade (applied usually to noncommissioned grades);
— *d'encadrement*, n. c. o. of the flank of a unit (who frames in the unit, as it were);
— *en serre-file*, n. c. o. in the line of file closers.

gradient, m., (*top*.) gradient, slope.

gradin, m., step; (*fort*.) step; (*art*.) notch; (*sm. a*) slide notch (of a rear sight);
—*s de franchissement*, (*fort*.) steps to get over the parapet, sortie steps;
—*s de fusillade*, (*fort*.) firing steps of a parapet;
—*s de revers*, (*fort*.) steps in the reverse slope of a trench;

gradin(s) *de sortie*, (*fort*.) sortie steps;
—*s de tranchée*, (*fort*.) steps in a trench.

gradine, f., stone cutter's tooth-edged chisel.

graduation, f., (*inst*.) graduation; (*art*.) time scale (of a fuse);
— *en portées*, (*art*.) range scale;
— *de repérage*, (*Fr. art*.) correction scale of a *réglette de correspondance*.

gradué, p. p., (*inst*.) graduated; (*art*., *sm. a*.) sighted;
tige ——*e*, index rod.

graduer, v. a., to mark with divisions, to graduate.

grain, m., grain, seed; (*in pl*.) grain (as wheat, rye, etc.); fiber; squall; shower; (*met*.) grain (due to fracture); (*mach*.) bush, bushing, pillow, box, necking; (*powd*.) grain of powder; (*art*.) direction sight;
à ——*s*, grained; squally;
— *aiguillé*, (*met*.) grain or fracture of crystalline aspect, with radiating needles;
— *d'amorce*, (*artif*.) primer;
— *blanc*, white squall; (*met*.) steel bar (produced by cementation from poorly refined iron);
— *de crapaudine*, (*mach*.) bush, brass, or bearing of a socket;
— *de décollage*, (*art*.) gear ring;
— *à facettes*, (*met*.) large grain or fracture, with facets;
— *fin*, (*met*., *powd*.) fine grain;
à ——*s fins*, (*met*., *powd*.) fine grained;
gros ——, (*met*., *powd*.) large grain;
à *gros* ——*s*, (*met*., *powd*.) large grained, coarse grained;
— *de lumière*, (*art*.) vent bushing;
mettre (*poser, visser*) *un* —— *de lumière*, (*art*.) to bush a gun;
— *noir*, black squall;
— *d'or*, (*artif*.) golden rain in fireworks;
— *d'orge*, (*sm. a*., *art*.) barleycorn fore sight; (*carp*.) sort of dovetail;
réduction en ——*s*, granulation;
réduire en ——*s*, to granulate;
à ——*s serrés*, close grained;
— *truité*, (*met*.) mottled fracture, with small radiating needles arranged in large stars.
— *de vent*, wind squall.

grainage, m., v. *grenage*.

grainailles, f. pl., — *de fer*, (*mil*.) small shot of iron;

graine, f., seed;
à —— *d'épinards*, with large bullion (of epaulets);
— *de giberne*, synonym of *enfant de troupe;*
— *de Turquie*, Indian corn.

grainer, v. a., v. *grener*.

grainoir, m., v. *grenoir*.

grainure, f., (*met*.) character of fracture.

graissage, m., greasing, grease; lubrication;
à —— *automatique*, self-oiling.

graisse, f., grease, unguent;
— *d'armes*, grease for arms in general;
botte à ——, grease box;
— *à chandelle*, an unguent composed of two parts wax and one of kerosene, gun wax;
coupe à ——, *godet à* ——, tallow cup, grease cup;
robinet à ——, grease cock;
— *verte*, unguent for side arms and the exterior parts of firearms.

graisser, v. a., (*mach*., *etc*.) to grease (as with tallow); to lubricate (as with oil);
— *la lime*, to clog a file.

graisseur, m., (*mach*.) lubricator;
— *automatique*, self-acting lubricator;
godet ——, *à godet*, grease cup;
— *mécanique*, self-acting, self-feeding, lubricator.

grakrut, m., (*expl*.) grakrut.

gramme, m., gramme.

grand, a., great, grand, tall, large, main, principal, chief; m., a dignitary, person of dignity; (*in pl*.) grown-up people;
le —— *air*, the open air;
le —— *de l'eau*, highest point to which the tide has risen during the month;
en ——, life-size;
— *d'Espagne*, Spanish grandee.

grand-cordon, m., grand cordon of an order.
grand-croix, m., v. s. v. *croix.*
grand' croix, f., v. s. v. *croix.*
grand-duc, m., grand duke.
grand-duché, m., grand duchy.
grande(s)-marée(s), f. (pl.) v. s. v. *marée.*
grand' garde, f., v. s. v. *garde.*
grandir, v. a. n., to grow; to increase;
— *un cheval,* (*man.*) to bring a horse's haunches under him, and raise his head and neck;
se —, (*man.*) of the horse, to bring the haunches under, and raise the head and neck.
grand(-)livre, m., v. s. v. *livre.*
grand(-)maître, m., v. s. v. *maître.*
grand' messe, f., v. s. v. *messe.*
grand-océan, m., v. s. v. *océan.*
grand-officier, m., v. s. v. *officier.*
grand' rue, f., v. s. v. *rue.*
grange, f., barn;
aire de —, barn floor.
granit, m., granite.
granulation, f., granulation.
granule, m., granule.
granuler, v. a., to granulate, to grain (as powder, explosives, etc.).
granuleux, a., granulated, grained, ingrained.
graphique, a., graphic; m., diagram;
— *de marche,* (*mil.*) march diagram;
— *de tir,* (*ball.*) graphic table of fire.
graphite, m., black-lead; graphite; plumbago.
graphomètre, m., (*inst.*) graphometer;
— *double,* circumferentor;
— *à lunette,* telescopic graphometer.
graphostatique, f., graphostatics, graphical statics.
grappe, f., bunch, grape; (*sm. a.*) cluster of bullets (joined by a web before separation, after casting); (*art.*) grape (obs.); (*farr., in pl.*) sock of calkin; (*met., in pl.*) cluster of sand and small stones in iron ore; (*hipp., in pl.*) grapes;
— *de balles,* v. under main word, (*sm. a.*);
— *de mitraille,* — *de raisin,* (*art.*) grapeshot (obs.).
grappin, m., grapnel, creeper, scraper; (*cons.*) anchor, tie; (*hipp.*) disease of the foot;
— *d'abordage,* (*nav.*) boarding grapnel;
— *à main,* hand grapnel.
grappiner, v. a., to grapple; to warp by means of grapplings and ropes.
gras, a., fat, unctuous, oily, oiled, greasy; slippery; thick, blunt; (of the weather) thick; (*met.*) wet, drossy, clogging the file; (*fond.*) loamy (of casting sand);
cheval — *fondu,* v. s. v. *cheval;*
mortier —, *trop* —, (*mas.*) mortar with too much lime;
pièce de bois — *se,* (*carp.*) piece of timber stronger than necessary;
temps —, hazy weather;
terrain —, clayey, stiff ground;
terre — *se,* heavy, clayey soil;
vue — *se,* (*hipp.*) dim-lighted, wall-eyed.
gras-fondu, m., gras-fondure, f., (*hipp.*) enteritis; grease.
grasset, m., (*hipp.*) stifle, stifle joint.
graticuler, v. a., v. *craticuler.*
gratification, f., gratuity, allowance, bounty; reward, recompense;
— *de première mise,* (*mil.*) outfit allowance (clothing allowance, clothing money, U. S. A.);
— *de réforme,* (*Fr. a.*) allowance or bounty to n. c. o., corporals, and soldiers discharged by *congé No. 1,* q. v.
grattage, m., action of scratching, scraping; erasure (of writing);
— *à vif,* removal of the outside (as, the weathering from stone).
gratteau, m., scratcher, burnisher.
gratte, f., scraper.
gratte-boësse, f., wire brush.
gratte-boësser, v. a., to clean with a scratch brush.
grattebrossage, m., wire brushing.
grattebrosse, f., wire brush.
grattebrosser, v. a., to brush with a wire brush.
gratter, v. a., to scratch, scrape; to erase (writing);

gratter *à vif,* to scrape off the outside; to remove the weathering of stone.
grattoir, m., scraper; rake; eraser; (*art.*) vent-pick;
— *à canon,* (*art.*) gun scraper;
— *-tube,* tube scraper.
gratuit, a., gratuitous;
à titre —, free, gratis, gratuitous.
grave, f., sandy beach; strand; m., heavy body;
a., serious, important, weighty;
blessure —, serious wound.
graveler, v. a., to gravel, cover with gravel.
graveleux, a., gritty, sandy;
crayon —, gritty pencil.
graver, v. a. r., to grave, engrave, impress, imprint; to be eaten (said of iron from which rust has been removed); (*artif.*) to split, crack (as a rocket case);
— *sur bois,* to cut, engrave, on wood;
— *au burin,* to engrave on copper;
— *en creux,* to cut;
— *à l'eau forte,* to etch.
graveur, m., engraver;
— *en creux,* diesinker.
gravier, m., gravel, grit; gravel bottom.
gravimètre, m., (*inst.*) gravimeter.
gravimétrique, a., gravimetric;
densité —, (*ball.*) gravimetric density.
gravir, v. a., to scramble, climb, crawl up; to climb, ascend.
gravitation, f., gravitation.
gravité, f., gravity, weight;
— *absolue,* absolute gravity;
centre de —, center of gravity;
— *spécifique,* specific gravity.
graviter, v. n., to gravitate.
gravois, m., rubble, rubbish; coarse gravel;
— *spécifique,* specific gravity.
gravure, f., engraving; engraved marks or indications (as on a sight);
— *sur bois,* wood engraving;
— *en creux,* diesinking;
— *à l'eau forte,* etching;
— *à la manière noire,* mezzotint;
— *sur pierre,* stone engraving;
— *en taille douce,* copperplate engraving;
— *en taille dure,* steel engraving.
gré, m., will, good will; what is agreeable to the will;
de — *à* —, (of a contract) voluntarily on both sides.
gréage, m., rigging (action of).
grecque, f., fretwork, fret.
gréement, grément, m., rigging; set of ropes needed for a given purpose;
— *en (fil de) fer,* wire rigging;
— *en filin,* hemp rigging;
— *fixe,* standing rigging.
gréer, v. a., to rig.
greffe, m., record office, clerk's office; (*Fr. a.*) clerical element of the permanent part of a court-martial; archives, records, of a court-martial.
greffer, v. a., to graft; (*mil. min.*) to break out from.
greffier, m., (*Fr. a.*) administrative officer of military law, appointed by president, keeps the *greffe* of a military court or of a *conseil de révision;*
adjudant commis- —, *commis-* —, clerk of a *greffe* (has the rank of *adjudant*).
grêle, a., slender, slim, lank.
grêle, f., hail, hailstorm; (*mil.*) shower (hail) of bullets;
orage de —, hailstorm.
grelet, m., pickax.
grelin, m., (*cord.*) warp, hawser; stream cable (strictly, a large rope composed of three or four *aussières* (hawsers) of four *torons* (strands) each);
— *chaîne,* stream chain;
en —, cable-laid;
— *de remorque,* towrope.
grêlon, m., hailstone.
grelot, m., bell, sleighbell.
grenade, f., (*artif.*) grenade; (*art.*) head (of a shrapnel); powder chamber and walls of the *obus à mitraille;* (*unif.*) shell and flame (distinctive mark); (*Fr. art.*) grenade worn by the best gun-layer of the year in a battery;
— *borgne,* (*artif.*) a fuseless grenade;
— *éclairante,* (*artif.*) light grenade, flare;

grenade *éclairante et incendiaire,* (*artif.*) light and incendiary grenade;
—— *extinctrice,* fire grenade;
—— *de fossé,* (*artif.*) ditch grenade;
—— *à main,* (*artif.*) hand grenade;
—— *à perdreaux,* (*artif.*) disk grenade;
porte- ——, grenade pouch;
—— *de rempart,* (*artif.*) rampart grenade;
—— *roulante,* (*artif.*) wall grenade.
grenadier, m., (*mil.*) grenadier; (*sm. a.*) grenade wall-piece.
grenadière, f., (*mil.*) grenade pouch; (*sm. a.*) middle band; (to-day) lower band;
—— *arme,* (command for) sling carbine;
—— *basse,* (*sm. a.*) lower slide;
mettre l'arme à la ——, (*sm. a.*) to sling a musketoon, carbine;
porter à la ——, (*sm. a.*) to carry slung (as a carbine).
grenadine, f., (*expl.*) grenadine (blasting powder).
grenage, m., (*powd.*) granulation, corning; act or operation of making powder grains.
grenaille, f., refuse grain (wheat, etc.); small shot;
en ——, granulated.
grenailler, v. a., to granulate (metal).
grener, v. a., to grain, to granulate; (*powd.*) to granulate, to corn.
grenette, f., (*powd.*) powder grains remaining in sieve.
grenier, m., granary, loft; garret;
en ——, in bulk; in store;
faire ——, v. s. v. *faire.*
grenoir, m., (*powd.*) corning house, corning machine, granulator, sieve;
—— *anglais à cylindres,* cylinder granulating machine;
—— *à cannelures,* granulating machine with ribbed cylinders.
—— *à cribles,* sieve and shaking frame;
—— *à cylindres,* cylinder granulating machine;
—— *a retour,* granulating drum of Colonel Lefébure's device (so called because the grains too large to pass through the second sieve return automatically to the crusher above);
tonne- ——, granulating barrel.
grenouille, f., frog; bush, coak (of a block); antimony sulphide triturator; (*mil. slang*) pay; mess fund;
bruit de ——, v. s. v. *bruit;*
emporter, manger, la ——, (*mil. slang.*) to be dishonest (said of the holder or keeper of the mess fund).
grenouillé, bushed, coaked (of a block).
grenouillette, f., (*hipp.*) disease of the canal of the tongue.
grenu, a., granular; (*powd.*) corned, granulated, grained;
cassure ——, granular fracture;
grès, m., sandstone; freestone; gritstone;
—— *à aiguiser,* grindstone;
—— *dur,* gritstone;
—— *à meule,* millstone grit.
grésil, m., broken, powdered, window glass; sleet;
grésiller, v. n., to sleet.
gresserie, f., sandstone quarry.
grève, f., strike (of workmen); strand, sandy shore; (*mas.*) mortar sand;
faire ——, to strike.
gréviste, m., striker (workman).
grief, m., grievance, injury, tort.
griffage, m., *queue de* ——, (*met.*) tong-hold.
griffe, f., claw, catch, hook; pawl, prong, clasp; guide; facsimile of a signature, and stamp for making same; claw (of a hammer); chuck square; (*sm. a.*) claw of the extractor; (*pont.*) balk cleat or claw; (*art.*) plug-tray guide rail; clip (as of the Merriam fuse); (*farr.*) toe of a horseshoe;
—— *arrêt,* clip stop;
—— *de chargement,* (*art.*) shell hook;
—— *s de la console,* (*art.*) breech-plug guides;
—— *de contreseing,* facsimile signature;
—— *du grand ressort,* (*sm. a.*) mainspring hook;
—— *de mise de feu,* (*art.*) firing catch or hook;
—— *de noix,* (*sm. a.*) tumbler hook;
—— *à ressort,* spring claw.
griffonner, v. a., to scribble; to make a rough drawing, sketch.

griffonnis, m., pen-and-ink sketch.
grignard, m., plaster of Paris.
gril, m., grate, grating; (*cons.*) grillage.
grillage, m., grating; fire grate; wirework, wire lattice; (*met.*) calcining, roasting of ores; (*cons.*) grillage;
—— *de bois,* timber frame;
—— *de la boîte à feu,* stokehole bars;
—— *en fil métallique,* wirework or lattice;
fourneau de ——, (*met.*) roasting, calcining, furnace.
grille, f., lattice, trellis; grate, grating; railing, iron railing; (in cryptography) a reading frame; (*cons.*) grillage; (*harn.*) bearing (of a stirrup); (*met.*) calcination, roasting, furnace, (*art.*) hotshot grate (obs.); (*artif.*) rocket frame; (*fort.*) gate; defensive fence;
—— *anglaise,* iron grating;
barreau de ——, fire bar;
—— *à circulation d'eau,* water-circulation grate;
—— *en escalier,* step grate;
—— *de fer,* iron grating, gate;
—— *à feu,* fire grate;
—— *de forteresse,* (*fort.*) grating; defensive grating (proposed as a substitute for the scarp and counterscarp);
—— *de foyer,* v. —— *à feu;*
—— *fumivore,* smoke-consuming grate;
—— *à gradins,* step grate;
—— *mobile,* revolving grate;
—— *ouvrante,* iron gate;
surface de ——, grate area;
—— *tournante,* v. —— *mobile.*
grillé, a., barbed; p. p., barred.
griller, v. a., to inclose with iron rails, to rail in, bar in; (*met.*) to roast, to calcine.
grilloir, m., roasting kiln.
grimper, v. a., to climb.
grimpeur, a., climbing;
appareil ——, climbing outfit (for telegraph poles, etc.).
grippement, m., (*mach., etc.*) gripping, wedging, wearing (of parts of a mechanism), friction, abnormal or undue friction.
gripper, v. a., to grip, wedge, bind, grind (of parts of a mechanism).
grippette, f., creeper.
grippure, f., (*mach.*) worn bearing or seat of any moving or movable part.
gris, a., gray; (of the weather) gray, raw; (*fam.*) drunk;
faire ——, *faire un temps* ——, to be raw (of the weather);
fer ——, gray iron;
temps ——, raw, gloomy weather.
gris, m., (*hipp.*) gray (horse and coat);
—— *ardoise,* gray shading on a bluish slate color;
—— *argenté,* silver gray;
—— *cendré,* ash gray;
—— *clair,* light gray;
—— *très clair,* almost white;
—— *d'étourneau,* flea-bitten gray;
—— *de fer,* iron gray;
—— *foncé,* dark gray;
—— *louvet,* deep yellow dun;
—— *moucheté,* flea-bitten gray;
—— *ordinaire,* gray of black and white in equal parts;
—— *pommelé,* dapple gray;
—— *porcelaine,* gray with slate-colored spots;
—— *rouan,* roan;
—— *sale,* yellowish gray;
—— *sanguin,* blood roan;
—— *soupe de lait,* light or silver gray;
—— *truité,* flea-bitten gray;
—— *vineux,* blood roan.
grisard, m., very hard freestone or sandstone.
grison, m., purbeck stone; spy, secret, messenger.
grisonné, a., (*hipp.*) grayish (of the coat), i. e., showing white hairs in a black coat, or on the black parts of the body.
grisou, m., fire damp;
mine à ——, a mine with fire damp in it, or liable to fire damp.
grisoutine, f., (*expl.*) grisoutine.
grisoutite, f., (*expl.*) grisoutite; fire-damp dynamite.
grivier, m., (*popular slang*) soldier of the line.

3877°—17——14

grognard, m., grumbler; (*mil.*) old soldier (originally a soldier of the Old Guard).

gronder, v. n., (of guns) to roar, thunder.

gros, a., large, heavy, thick, bulky, big;
— *se caisse*, (*mus.*) bass drum;
— *se cavalerie*, (*cav.*) heavy cavalry, cuirassiers;
— *d'haleine*, (*hipp.*) broken-winded;
— *se mer*, high, rough, heavy sea;
— *ses pièces*, (*art.*) heavy guns, guns of large caliber;
— *ses réparations*, extensive, substantial repairs;
— *temps*, rough, bad weather (at sea);
— *vent*, high wind.

gros, main part of; thick end or part; big or lump coal; wholesale trade; (*mil.*) main body;
— *d'un arbre*, thick end of a tree, trunk;
— *de l'armée*, (*mil.*) main body of the army;
— *d'avant-garde*, (*mil.*) main body of the advance guard;
avoir du —, (*hipp.*) to be strongly built, i. e., to have big bones and corresponding muscles;
de —, (of timber) square (i. e., so many inches square);
— *et détail*, wholesale and retail;
— *de l'eau*, high water at spring tide;
— *de l'été*, — *de l'hiver*, depth of summer, of winter.

gros-blanc, m., strong cement (whiting and glue).

grosse, f., gross (12 dozen); engrossing hand.

grosseur, f., volume.

grossier, a., gross, coarse, rough, thick.

grossir, v. a. n., to magnify, to increase, enlarge (as a photograph); (of the sea) to rise, swell.

grossissement, m., magnifying; magnifying power (of a lens); enlargement (of a drawing, map, photograph).

groupe, m., group, cluster, (*art.*) group of two, three, or four batteries;
— *alpin*, (*Fr. a.*) group of mountain troops (artillery and infantry) commanded by a *chef de bataillon*;
— *d'armées*, (*Fr. a.*) all the armies operating in one theater of war, and under the supreme command of one general;
— *de complément* (*Fr. art.*) in a regiment, so much of the regiment as is not included in the — *s de service* and *de manœuvre*;
— *franc*, (*Fr. a.*) body of men in advance of the advance guard for the purpose of additional exploration;
— *de guerre*, (*Fr. art.*) a group of three *batteries de guerre*, q. v. (field artillery), of two *batteries de guerre* (horse artillery);
— *de manœuvre*, (*Fr. art.*) a group of three *batteries de manœuvre*, q. v.;
— *de service*, (*Fr. art.*) in a regiment, a group charged with the general police, fatigues, etc.

groupé, p. p., grouped;
formation —*e*, (*mil.*) dense formation.

groupement, m., grouping; (*ball.*) grouping of hits about the center of impact; (*elec.*) joining up, connecting up (of cells);
centre du —, (*ball.*) center of impact;
— *collectif*, (*ball.*) the grouping produced by a body of men;
densité de —, (*ball.*) density of grouping, ratio of number of shots to unit area on which they fall;
— *en dérivation*, (*elec.*) joining up in parallel, in quantity;
— *horizontal*, (*ball.*) grouping of shots on a horizontal target;
— *individuel*, (*ball.*) grouping of shots by a single man;
— *en quantité*, v. — *en dérivation*;
— *en série*, (*elec.*) joining up in series;
— *en surface*, v. — *en dérivation*;
— *en tension*, v. *en série*;
— *vertical*, (*ball.*) grouping of shots on a vertical target.

grouper, v. a., to group, bring together; (*elec.*) to join up;

grouper *en dérivation*, — *en quantité*, etc., (*elec.*) v. s. v. *groupement*.

gruau, m., small crane.

grue, f., crane, traveler;
— *alossée*, wall crane;
— *à air comprimé*, crane worked by compressed air;
— *d'alimentation*, water crane;
amarre de —, cheek of a crane;
arbre de —, crane post;
— *à axe tournante*, revolving crane;
bec de —, jib or arm of a crane;
— *à bras*, hand crane;
— *de chargement*, hoisting crane, loading crane;
— *à chariot*, — *à chariot mobile*, overhead, traveling crane;
— *à deux pivots*, wall crane;
— *à double volée*, double crane;
échelier de —, crane jib or arm;
— *électrique*, electric crane, electrically controlled crane;
— *fixe*, stationary crane;
— *hydraulique*, hydraulic crane;
— *hydraulique à palan*, crane in which a tackle is worked hydraulically;
— *hydraulique à piston*, hydraulic piston crane;
— *à lest*, ballast crane;
— *locomobile*, traveling, running, transportable, crane;
— *à main*, hand crane;
— *mâture*, sheers;
— *mobile*, traveling (transportable) crane;
— *monte-projectiles* (*art.*) shell hoist, shot crane;
— *à pierres*, loading crane (for heavy work);
— *à pivot*, revolving crane, post crane;
— *à pivot fixe*, stationary crane;
— *à pivot mobile*, — *à pivot tournant*, turning crane;
— *pivotante*, v. — *à pivot*;
— *à poinçon*, landing or wharf crane; stationary crane;
poinçon de —, crane post;
— *portative*, portable crane, winch, hand winch;
— *rancher de* —, v. *bec de* —;
— *roulante*, v. — *mobile*;
— *à trois os*, three-legged crane;
— *à vapeur*, steam crane;
volée de —, v. *bec de* —.

grume, f., bark left on felled tree;
bois en —, timber with the bark on, unsquared timber (but otherwise trimmed and cut up into lengths);
en —, with the bark on.

grumeau, m., clot, lump;
se mettre en —, to clod, lump, grow thick, coagulate.

grumeler, v. r., to clod, clot.

grumeleux, a., lumpy, clotted.

grumelure, f. (*met.*) small hollow in a casting, in cast metal, pitting.

grumillon, m., (*met.*) bit of hammer scale.

gué, m., ford;
à —, by fording;
passer à —, *traverser à* —, to ford.

guéable, a., fordable;
non—, unfordable.

guéer, v. a., to ford; to water, wash horses, in a river.

guérilla, m., v. *guerrilla*.

guérillero, m., v. *guerrillero*.

guérir, v. a. n. r., to cure, to be cured;
— *de*, to be cured of (a fever, etc.).

guérison, f., cure.

guérite, f., (*fort.*) small watchtower, lookout, operating or conning tower, aiming dome (of a turret), handing room (of a magazine); (*mil.*) sentry box;
— *de barbanon*, (*fort.*) lookout.
— *blindée*, (*art.*, *fort.*) conning tower or aiming turret (of a gun turret);
capote de —, (*mil.*) sentinel's watch coat; buffalo coat (U. S. A.);
— *de locomotive*, (*r. r.*) cab;
officier de —, (*fam.*) a private;
— *téléphonique*, telephone stand or box.

guerlin, m. v. *grelin*.

guerluchone, f., mason's trowel.
guerre, f., (*mil.*) war, warfare, art of war; (more narrowly) campaign; war department; land service (as opposed to navy);
académie de ——, war school, war college;
aller à la ——, to go to war;
allumer la ——, to kindle war;
art de la ——, art of war;
articles de ——, articles of war;
avoir la ——, to be at war;
bâtiment de ——, (*nav.*) man-of-war;
de bonne ——, according to the recognized laws of war;
bureau de la ——, war office, war department;
—— *de cabinet,* any war not a national war, e. g., the French expedition against Madagascar;
—— *de campagne,* war in the open field, as opposed to sieges;
chances de ——, chances of war, luck of war;
cheval de ——, charger;
—— *civile,* civil war;
conseil de ——, council of war; court-martial;
contrebande de ——, contraband of war;
—— *de côte,* coast war (i. e., descents and related operations);
—— *de course,* (*nav.*) commerce-destroying war;
cri de ——, war cry;
—— *de croisière,* v. —— *de course;*
de ——, warlike, of or belonging to war;
déclaration de ——, declaration of war;
déclarer la ——, to declare war;
—— *défensive,* defensive war;
département de la ——, department of war;
droit de la ——, law of war;
école de ——, (any) war school;
l'école de la ——, the school of war;
école supérieure de ——, v. s. v. *école;*
en ——, at war; under service conditions;
état de ——, state of war; war footing;
—— *étrangère,* foreign war;
—— *d'extermination,* war of extermination;
—— *factice,* sham war, simulated warfare;
faire la ——, to wage war, be at war;
faire bien la ——, to understand the art of war;
faire bonne ——, to wage war with all possible humanity; to act honorably to an adversary;
faire une petite ——, to be engaged in minor operations (as opposed to battles);
foudre de ——, great warrior;
franc de molestation de ——, free from molestation of war;
gens de ——, military men, soldiers;
grande ——, war on a large scale; great operations of war;
homme de ——, warrior, soldier;
honneurs de la ——, honors of war;
—— *intestine,* civil war;
jeu de la ——, war game; kriegspiel;
—— *au large,* —— *du large,* (*nav.*) war on the high seas;
lois de la ——, laws of war;
—— *maritime,* naval war, warfare;
—— *de mer,* v. —— *maritime;*
—— *de mines,* subterranean warfare;
ministère de la ——, ministry of war; war department; war office; horse guards;
ministre de la ——, minister of war; secretary of war; secretary of state for war;
—— *de montagne,* mountain warfare;
—— *nationale,* —— *de nations,* war between nations (as opposed to —— *de cabinet*);
—— *offensive,* offensive warfare;
—— *à outrance,* war of extermination;
pain de ——, v. s. v. *pain;*
—— *de partisans,* small operations, as raiding, foraging, etc.;
peste de ——, (*méd.*) typhus fever;
petite ——, minor operations of war; sham fight;
pied de ——, war footing;
place de ——, fortified town, position;
poudres de ——, service powders;
raison de ——, custom of war;
—— *de rue,* street fighting;
ruse de ——, stratagem;
—— *de secours,* war of alliance;

guerre de siège, siege warfare;
soutenir la ——, to carry on war;
—— *souterraine,* underground, siege operations, attack and defense by mines;
stratagème de ——, war stratagem;
temps de ——, war time;
théâtre de la ——, theater, seat, of war;
usage de la ——, custom of war;
vaisseau de ——, (*nav.*) war vessel;
vaisseau moitié ——, *moitié marchandise,* (*nav.*) letter of marque;
vapeur de ——, (*nav.*) war steamer.
guerrier, a., warlike, material; **m.,** warrior, soldier.
guerrilla, f., (*mil.*) guerrilla;
contre- ——, counter-guerrilla.
guerrillero, m., member of a guerrilla organization.
guerroyer, v. a., (*fam.*) to wage or make war.
guerroyeur, m., (*fam.*) warrior.
guet, m., watch, guard, sentry (obs. as a military word);
avoir l'œil au ——, to be on the lookout;
être au ——, to watch, be on the watch;
faire le ——, to watch, keep watch;
mot de ——, watchword;
—— *de nuit,* patrol, night-watch.
guet-apens, m., ambush.
guêtre, f., gaiter; spatterdash;
demi- ——, half-legging;
—— *jambière,* legging (zouaves and *tirailleurs algériens*);
——*s molletières,* leggings.
guêtré, m., (*mil. slang*) foot soldier; trooper who, for any reason whatever, has to go afoot.
guêtrer, v. a. r., to put on gaiters.
guêtron, m., short gaiter.
guette, f., (*cons.*) inclined post or stake, diagonal strut, prick post.
guetter, v. a., to be on the lookout; to watch; to look out for; to dog.
guetteur, m., lookout, watch, watchman;
—— *des électro-sémaphores,* coast signal man.
guetton, m., diminutive of *guette,* q. v.
gueulard, m., (*met.*) mouth, charging door (of a blast furnace);
cheval ——, (*hipp.*) hard-mouthed horse;
tague de ——, (*met.*) iron band encircling the mouth of a blast furnace.
gueule, f., mouth (of animals); mouth (of a furnace, etc.); doucine, cyma; (*fond.*) gate, runner; (*art.*) mouth of a gun;
—— (-)*bée,* discharge by cascade, or in a sheet, from an upper to a lower basin;
futaille à —— *bée,* a cask with its head knocked in;
—— *de loup,* triangular notch in the end of a beam; a funnel that turns with the wind; (*cord.*) black-wall hitch;
—— *de loup double* (*simple*), (*cord.*) double (single) black-wall hitch;
à —— *de loup,* (*carp.*) tongued and grooved;
marcher à —— *bée,* v. s. v. *marcher;*
produire une belle ——, to go well (of a coke furnace);
—— *de raie,* (*cord.*) catspaw, (also) black-wall hitch.
gueusat, m., v. *gueuset.*
gueuse, f., (*met.*) pig, pig-iron; (*nav.*) pig-iron ballast, kentledge;
—— *de fonte,* (*met.*) pig iron;
—— *-mère,* (*met.*) sow, sow iron.
gueuset, m., (*met.*) small pig of iron.
gueusillon, m., v. *gueuset.*
gueux, a., poor, naked.
gui, m., (*nav.*) boom.
guichet, m., wicket, little door; shutter;
—— *à air froid,* damper;
scie à ——, compass saw, table saw.
guidage, m., guiding (of a tool).
guide, f., (*harn.*) rein;
attelage en ——*s,* (*art.*) harnessing up with reins;
conduire en ——*s,* v. s. v. *conduire;*
conduite en ——*s,* v. s. v. *conduite;*
à grandes ——*s,* with reins; four-in-hand; at full speed.

guide, m., guide, leader, conductor; (*mil.*) guide (at drill), guide (of a march); (*in pl.*) special corps of light infantry or cavalry in some armies; (*cord.*) guy; (*mach.*) guide, guide rod, slide bar, radius bar, crosshead guides; (*art.*) ta'cing of the grooves;
— *à ailettes*, (*art.*) taking the grooves by studs;
— *d'alimentation*, (machine guns) feed guide;
— *à articulations*, (*mach.*) link connections;
— *chaîne*, chain guide;
— *de chargement*, (*art.*) loading plate;
— *par chemise de plomb*, (*art.*) taking the grooves by lead coating (obs.);
— *de crosse*, v. s. v. *crosse*;
— *à droite* (*gauche*), (*drill*) guide right (left);
— *général*, (*drill*) general guide;
— *à gorge*, (*mach.*) slotted, grooved, guide;
— *particulier*, (*mil.*) marker, squad or company guide (as opposed to general guide);
— *de platine*, (machine guns) lock guide;
— *principal*, (*mil.*) principal guide;
— *du projectile*, (*art.*) taking the grooves;
— *du té*, (*mach.*) crosshead guide;
— *à tenon*, v. — *à ailettes*;
— *de tête*, (*mil.*) leading guide;
— *de la tige du piston*, (*mach.*) slide and block; piston-rod guide;
— *de la tige du tiroir*, (*mach.*) valve-rod guide; valve-spindle guide.
— *du tiroir*, (*mach.*) valve bridle.
— *de traverse*, (*mach.*) crosshead guide.

guide-âne, m., (*fam.*) guide, directory.
guide-axe, m., axle guard, clout.
guide-cartouche, m., (*sm. a.*) cartridge guide; (machine guns) feed guide.
guide-courroie, m., (*mach.*) strap guide.
guide-noix, m., (*sm. a.*) safety locking mass.
guide-poche, m., pocket guide.
guider, v. a., to guide, lead, conduct, steer, direct, etc.
guide-rope, m., drag, balloon drag (long rope, fitted with iron points, used to check a balloon when near the earth).
guidon, m., handle-bar (of a bicycle); (*mil.*) small flag (synonym of *fanion*); (*nav.*) burgee; (*art.*, *sm. a.*) front sight;
— *annulaire à pointes*, (*art.*) point-ring sight (a front sight consisting of a ring with four equidistant points projecting toward the center);
— *à la bouche*, (*art.*) fore sight;
— *central*, (*art.*) central, axial, sight;
— *à charnière*, (*art.*) hinged fore sight;
— *de commandement*, (*nav.*) broad pennant;
— *à la culasse*, (*art.*) breech-sight (rare);
— *de la douane*, revenue pennant;
— *fin*, (in aiming) fine sight;
— *à fût*, (*art.*) drop sight;
— *latéral*, (*art.*) trunnion or side sight;
— *lumineux*, (*art.*) illuminated sight, night sight;
— *de mire*, (*art.*) front sight;
— *plein*, (in aiming) full sight;
— *rasé*, (in aiming) a sight along the points of the front and back sights;

guidon *de renvoi*, reference mark;
— *à vis*, (*art.*) screw sight.
guigue, f., gig (small boat).
guillaume, m., rabbet-plane; (*powd.*) first sieve or separator; coarsest sieve;
— *à ébaucher*, jack plane.
guilledin, m., (*hipp.*) gelding, hackney.
guillochage, m., engine-turning, rose-engine design.
guilloche, f., rose-engine tool.
guilloché, a., bushed, coaked (of a block).
guillocher, v. a., (*mach.*) to trace or make rose-engine patterns; to engine-turn;
feu à —, (*artif.*) double catherine wheel;
machine à —, (*mach.*) rose-engine.
guillochis, m., (*mach.*) rose-engine pattern; engine-turning.
guillotine, f., guillotine; cut-off, drop-shutter (as of a camera).
guillotiner, v. a., (*mach.*) to guillotine, to shear off.
guindage, m., hoisting, swaying up; (*pont.*) side rail (including rack stick and lashings), riband; racking down, rack lashing; lashing down of chess;
collier de —, (*pont.*) balk collar, balk stirrup;
commande de —, (*pont.*) rack lashing; side-rail lashing;
faux- —, (*pont.*) false balk, false side rail; the joint connecting two consecutive rafts in bridge by rafts;
poutrelle de —, (*pont.*) side rail.
guindal, m., windlass.
guindant, m., hoist of a flag.
guindas, guindeau, m., windlass.
guinder, v. a., to hoist, raise, sway up; to strain, force; (*pont.*) to rack down, lash down.
guipage, m., (*elec.*) wrapping (of a cable).
guiper, v. a., to wind tape around (as a wire).
guipon, m., tar brush.
guipure, f., (*elec.*) wrapping for cables (made of woven thread).
guirlande, f., (*cord.*) marling down; snaking of a seizing;
faire —, (*art.*) (of traces) to sag, hang in the middle; not to be stretched.
gun-cotton, m., (*expl.*) (in France) compressed gun cotton.
gutta-percha, f., gutta-percha.
gymnase, m., gymnasium; school;
— *musical*, (*Fr. a.*) musical school in Paris, trains musicians for the position of *sous-chef de musique*.
gymnaste, m., gymnastic teacher; gymnast.
gymnastique, a., gymnastic; f., gymnastics; gymnasium;
— *appliquée*, applied gymnastics, involving the employment of gymnastic apparatus;
— *d'assouplissement*, free gymnastics;
cloche —, dumb-bell;
moniteur de —, v. s. v. *moniteur*;
pas —, (*mil.*) v. s. v. *pas*.
gypse, m., plaster of Paris; gypsum.
gyroscope, m., gyroscope.

H.

habillage, m., (*fond.*) operation of surrounding a core spindle with tow and clay.
habillement, m., clothing, dress, clothes (*Fr. a.*, includes shoes and headwear);
capitaine d' —, v. s. v. *capitaine*;
magasin d' —, repair and store shop;
officier d' —, v. s. v. *officier*;
service de l' —, (*Fr. a.*) clothing department (furnishes clothing, camp equipage, equipment, accessories of cloth, linen, etc., and (to the cavalry) horse equipment).
habiller, v. a., to dress, clothe; (*fond.*) to surround a core spindle with tow and clay.
habit, m., garment, coat, clothes; (*unif.*) full-dress frock coat;
—*s bourgeois*, plain clothes; mufti, "cits." (U. S. A.);

habit *croisé*, double-breasted coat;
— *d'ordonnance*, (*unif.*) regimentals, regulation clothes, uniform;
— *-uniforme*, uniform coat, or any other uniform clothing.
habitacle, m., (*nav.*) binnacle.
habitant, m., inhabitant.
habiter, v. a., to inhabit, live in, dwell in.
hache, f., ax;
— *d'abordage*, (*nav.*) boarding ax;
— *d'armes*, (*nav.*) boarding ax;
— *de bûcheron*, felling ax, wood-chopper's ax (*Fr. a.*, carried in infantry and engineer tool wagons);
— *de campement*, (*Fr. a.*) camp ax;
— *de charpentier*, carpenter's ax (*Fr. a.*, carried in infantry and engineer wagons);

hache à *main*, hand ax (*Fr. a.*, part of engineer equipment);
— *de parc*, (*Fr. a.*) generic term for ——*s de bûcheron* and *de charpentier;*
— *-pic*, ax and pick in one;
— *portative du génie*, (*Fr. a.*) almost the same as —— *de bûcheron*, and carried by sappers of engineers;
— *portative à main*, (*Fr. a.*) hand ax issued to the infantry;
— *à tête*, (*Fr. a.*) ax and hammer in one, carried by field artillery.

hache-paille, m., straw cutter, chopper.

hacher, v. a., to chop, hew, hack, etc.; (*gen., mil.*) to cut to pieces; (*top. drawing*) to make hachures, to hachure.

hachereau, m., (*mil.*) field hatchet.

hachette, f., hatchet;
— *de campagne*, (*Fr. a.*) small ax or hatchet used by the field artillery to clear the ground of underbrush;
— *de campement*, (*mil.*) hatchet for splitting firewood in bivouacs, etc.;
— *à marteau*, a hatchet with a hammer head.

hachoir, m., chopping board or knife; chaffcutter.

hachotte, f., lath-chopper.

hachure, f., (*drawing*) hachure, hatching;
contre- ——, cross hatching;
— *horizontale*, hachure parallel to contours.

haha, m., ha-ha; sunken fence; any obstacle in a road or communication; (*fort.*) interruption on the landings of stairs of forts; on ramps, covered by a small bridge or draw easily removed in case of need; small ditch (covered by a draw) in front of a postern gate; small ditch with movable bridge (in a communication or passage).

haï(c)k, m., haik.

haie, f., hedge; (*mil.*) lane formed by two ranks of soldiers facing each other; soldiers in line, or in one or more ranks; formation of several ranks into one; (*fort.*) a hedge serving as, or intended for, an obstacle;
— *artificielle*, (*fort.*) iron-wire hedge (obstacle);
border la ——, v. *former la* ——;
doubler la ——, (*mil.*) to form in two ranks facing each other, v. *former la* ——;
être en ——, v. *former la* ——;
former la ——, (*mil.*) to line a road, etc., with troops on each side; to form a lane by two ranks facing each other; to draw up troops along a road for a funeral or ceremony of passage of some person of high rank;
— *morte*, hedge formed of dead thorns;
se ranger en ——, to form line;
— *sèche*, brushwood edge, supported by wire or by pickets;
— *vive*, quickset hedge.

haire, f., haircloth; sackcloth; (*met.*) cast-iron plate, lining hearth of a finery forge, back plate of a finery forge.

halage, m., hauling, traction, towing;
chemin de ——, towpath;
corde de ——, towline, towrope;
— *à la cordelle*, towing;
ligne de ——, towline, towrope;
— *à vapeur*, steam towage.

hâle, m., (*tech.*) clearance, play (of a bolt).

hale-bas, m., (*cord.*) downhaul (of a flag, etc.).

haleine, f., breath, wind;
cheval gros d' ——, (*hipp.*) thick-winded horses;
donner —— *à un cheval*, (*man.*) to breathe a horse;
mettre en ——, to let (a horse, man, etc.) get his wind;
tenir des troupes en ——, (*mil.*) to keep troops in exercise.

halement, m., hauling, towing; pulling, hauling, knot.

haler, v. a., (*cord.*) to haul, haul in; to pull upon a rope;
— *bas*, to haul down;
— *à la cordelle*, to tow;
— *à joindre*, to haul tight, to haul chock-a-block, two blocks;

haler, *à la main*, to haul in by hand;
— *main sur main*, to haul hand over hand;
— *sur une manœuvre*, to haul upon a rope.

halle, f., large open shed; market.

hallebarde, f., halberd.

hallebardier, m., halberdier.

hallier, m., thicket, brushwood.

halogène, a., (*chem.*) halogenous.

haloxyline, f., (*expl.*) haloxylin.

halte, f., stopping, halt (of troops, etc.); stop, stand, resting place, refreshment during a halt; stopping place, halting place; (call or drill command) halt!
— *à la charge*, (*art.*) cease loading! (command given when a battery under fire is about to quit its emplacement);
faire ——, (*mil.*) to halt, make a stand, stop, stay;
— *au feu*, v. s. v. *feu;*
— *gardée*, (*mil.*) halt with sentinels out, with guards, etc., posted;
grande ——, (*mil.*) main or principal halt; long halt for rest, etc., (on march);
— *horaire*, (*mil.*) horary, hourly, halt (of ten minutes);
— *là!* (*mil.*) sentinel's challenge on approach of people;
— *-repas*, (*mil.*) halt or stop for a meal in a *station halte-repas;* dinner halt on a journey;
— *subite*, dead stop; sudden stop.

haltère, m., (*gym.*) dumb-bell.

hamac, m., hammock;
— *pour le passage des blessés*, (*med.*) stretcher.

hameau, m., hamlet.

hameçon, m., fishhook.

hamée, f., (*art.*) sponge staff, handle of a sponge (*écouvillon*).

hammerless, m., (*sm. a.*) hammerless shotgun.

hampe, f., shank, staff; shaft; main spar (of a spar torpedo);
— *de combat*, (*torp.*) war spar (of a spar torpedo);
— *du drapeau*, flag pole or shaft;
— *d'écouvillon*, (*art.*) sponge staff;
— *d'exercice*, (*torp.*) drill spar (of a spar torpedo);
— *de (la) lance*, (*sm. a.*) shaft;
— *de pavillon*, (*nav.*) ensign staff;
— *de pavillon de beaupré*, (*nav.*) jack staff;
— *de refouloir*, (*art.*) rammer staff;
— *d'une torpille*, (*torp.*) spar of a torpedo.

hampette, f., (of a spar torpedo) the part of a torpedo spar that supports the torpedo at its end; secondary spar.

han, m., heave (of a workman striking a heavy blow).

hanche, f., haunch, hip; leg (as of a trestle, e. g., Belgian three-legged trestle); (*art.*) leg of a gin; (*hipp., etc.*) haunch, hip; (*in pl.*) hind quarters of a horse; (*nav.*) quarter of a ship;
— *bien sortie*, (*hipp.*) well-placed, well-set, haunch;
cheval allant sur, paré sur, les ——*s*, (*man.*) horse that in galloping supports himself on his haunches;
— *de (la) chèvre*, (*mach., etc.*) cheeks of a gin;
— *coulée*, —— *effacée*, (*hipp.*) haunch that is not prominent enough;
— *gagnée*, (*man.*) said of a horse when his rider succeeds in controlling his hind quarters;
mettre un cheval sur les ——*s*, (*man.*) to make a horse gallop so that he supports himself on his hind quarters;
— *noyée*, (*hipp.*) sunken launch;
rabaisser les ——*s*, (*man.*) to force a horse to lower his croup;
traîner les ——*s*, (*man.*) to drag (said of a horse that gallops false, or that changes the foot in galloping).

hanet, m., lashing of a tent or of an awning, (*nav.*) reef-line.

hangar, m., shed, penthouse (as gun shed, boat-house, etc.).

hansard, m., pit saw.

hansière, f., (*cord.*) hawser.
hante, f., whip handle.
happe, f., hasp; cramp iron; bed, axletree bed; semicircular wrapping plate or clout;
— *à anneau*, linch hoop and plate, clout plate and loop at end of axletree arm;
— *à anneau et à piton*, wrapping plate and ring (ring attached by means of a *piton*);
— *à crochet*, wrapping plate or clout terminating in a hook;
— *à virole et à crochet fermé*, iron plate beneath the fore end of a limber-pole.
happelourde, m., (*hipp.*) a weak but fine-looking horse.
happer, v. a., n., to seize suddenly;
— *à la langue*, to stick to the tongue (as clayey stones, etc.).
haquenée, f., (*hipp.*) hack, hackney, nag (gentle horse);
aller la —, (*man.*) to amble.
haquet, m., drag or cart without side rails; (*pont.*) ponton wagon;
— *à bateau*, (*pont.*) ponton wagon;
— *à chevalets*, (*pont.*) trestle wagon;
— *à forge*, (*pont.*) forge wagon of a bridge equipage;
— *à nacelle*, (*pont.*) mooring-boat wagon;
— *à ponton*, (*pont.*) ponton wagon;
— *à poutrelles*, (*pont.*) balk wagon.
haras, m., (*hipp.*) stud, breed, of horses; breeding establishment of horses; (*in pl.*) the administration of breeding establishments;
— *d'amélioration*, v. *de perfectionnement*;
— *demi-sauvage*, establishment in which the animals run free in fine weather;
— *domestique*, regular establishment in which the animals are always under care;
— *d'études*, experimental breeding establishment;
— *de mulet*, mule-breeding establishment;
— *parqué*, a regular breeding establishment or stud;
— *de pépinière*, v. — *de perfectionnement;*
— *de perfectionnement*, stud specially intended to improve the breed of horses;
— *de production*, general breeding establishment;
— *sauvage*, stud in which the stallions and mares are at complete liberty all the time;
— *de souche*, — *de tête*, v. — *de perfectionnement*.
harassé, p. p., wornout, tired, done up;
— *par la route*, wayworn.
harasser, v. a., to harass, jade, knock up, fatigue.
haraux, m., stratagem for carrying off an enemy's horses while picketed or at grass.
harcelage, m., v. *harcèlement*.
harcèlement, m., harassing, galling.
harceler, v. a., to harass (the enemy); to gall.
hardes, f., pl., clothes, apparel, clothing.
hardi, a., bold, intrepid, audacious;
branche — *e*, (*harn.*) bent branch or check of bridle bit.
hardiesse, f., boldness, intrepidity, audacity.
haricot, m., bean.
haridelle, f., thin, poor, and worthless horse.
harnachement, m., harness, saddlery in general; horse furniture; (*cav.*) saddle and bridle and parts;
— *d'officier*, officer's horse furniture;
— *de selle*, saddle and its outfit (all parts);
— *de trait*, draught harness.
harnacher, v. a., to harness.
harnacheur, m., harnesser; harness maker.
harnais, m., armor, accouterments; (*harn.*) horse trappings, harness; (*mach.*) train and gear (of wheels); (*unif.*) any kind of military clothing;
— *d'angle*, (*mach.*) bevel gearing;
— *d'attelage monté*, (*art.*) artillery harness (driving from saddle);
— *d'attelage en guides*, (*art.*) harness of a team driven from seat;
— *de bât*, (*art.*) pack-mule harness;
blanchir sous le —, to turn old (gray) in the service, to grow old in the profession of arms;
— *à bricole*, v. — *à poitrail;*

harnais *à bricole renforcé*, (*Fr. art.*) harness used for very heavy weights (as a gun truck);
— *de circonstance*, (*Fr. a.*) the same as — *de limonière*, but lighter;
— *à collier(s)*, collar harness;
— *de derrière*, (*art.*) wheel harness;
— *de devant*, (*art.*) lead harness;
endosser le —, (*mil.*) to turn soldier, go into the army;
— *d'engrenage*, (*mach.*) set or train of gearing; back gear;
— *de limonière*, shaft harness (two-wheeled vehicle);
— *métallique*, (*mach.*) gearing with all teeth metallic;
— *de milieu*, (*art.*) swing harness;
— *mixte*, (*mach.*) gearing with teeth part wood, part metal;
— *à poitrail*, breast harness;
— *de retraite*, anchoring binder;
— *à traits renforcés*, (*art.*) harness with reenforced traces (for very heavy weights).
harpage, m., (*mas.*) toothing.
harpe, f., (*cons.*) corner iron, cramp iron; (*mas.*) tooth, toothing stone, toothing;
— *de fer*, (*cons.*) cramp iron.
harpeau, m., grapnel, grappling iron; fire grappling.
harpé, a., *cheval bien* —, (*hipp.*) shad-bellied horse.
harper, v. a. n., to grapple, gripe; (*hipp.*) to bend or flex the hock on starting, changing direction, or backing;
— *d'une jambe*, (*hipp.*) to raise one hind leg higher than the other without bending the hock.
harpin, m., (*pont.*) boat hook.
harpon, m., harpoon; ripping saw; staple (of a lock); (*mas.*) cramp iron.
harponner, v. a., to harpoon.
harponneur, m., harpooner.
hart, m., (*fort.*) withe, binder, gad; fagot band;
— *de fascine*, fascine band;
— *de retraite*, anchoring withe (siege, revetment); withe band, for fastening a fascine or sod revetment.
harveyiser, v. a., to treat armor plate by the Harvey process.
hast, m., shaft, pole, staff;
arme d' —, v. s. v. *arme*.
hauban, m., guy rope, guy.
hausse, f., anything serving to raise something else; rise (as in the price of stocks, etc.); hind bolster of a wagon; (*met.*) second course of a reverberatory furnace; (*in pl.*) planks laid on a sluice gate to raise the level of the water; (*art. sm. a.*) rear sight; also the elevation itself; (*art.*) the possible length of travel of the rear sight;
— *améliorée*, (*art.*) the elevation corresponding to the inferior limit of the narrow fork (*fourchette étroite*);
— *anglaise*, v. — *à curseur;*
— *à cadran*, v. — *circulaire;*
chapeau de la —, v. s. v. *chapeau;*
— *à charnière*, (*sm. a.*) leaf sight, flap sight;
— *circulaire*, (*sm. a.*) sight whose notch revolves between two plates on which the graduations are marked;
— *à clapet*, (*sm. a.*) leaf sight, flap sight;
— *de combat*, (*art.*) sight for full service charges;
conjugaison des — *s*, v. s. v. *conjugaison;*
— *de contrôle*, (*art.*) sight used in Germany to train gun layers;
— *à correcteur*, (*s. c. art.*) compensating sight;
— *sans correcteur*, (*s. c. art.*) noncompensating sight;
— *courbe*, (*art.*) curved, bow, sight (used with short guns in high-angle fire);
— *à crémaillère*, (*art.*) rack-and-pinion sight;
— *à curseur*, (*sm. a.*) sliding, leaf sight; ordinary type of sight, with slide;
— *à curseur et à gradins*, (*sm. a.*) sight combining the *curseur* and the *gradins*, step and leaf sight;
— *de curseur à rallonge*, (*sm. a.*) sliding sight;

hausse *du demi-tour de manivelle*, (*art.*) elevation corresponding to one-half turn of the elevating-screw handle;
—— *à dérive*, any sight fitted with a device for making lateral allowance;
donner la ——, (*art.*) to give the elevation;
—— *droite*, (*art.*) straight sight;
—— *de droite* (*gauche*), (*art.*) sight on right (left) side of gun;
—— *en échelle*, (*art.*) ladder sight (Hotchkiss 37ᵐᵐ mount, etc.);
——*s échelonnées*, (*art.*) combined elevations;
—— *d'essai*, (*art.*) trial elevation;
—— *d'exercice*, (*art.*) sight for drill and instruction (using *projectile d'exercice*);
—— *d'expériences*, any special sight used in ballistic firing;
—— *fixe*, (*sm. a.*) fixed sight;
—— *fixe à coulisse*, (*sm. a.*) rear sight and slide;
—— *glissante*, (*sm. a.*) sliding sight;
—— *à gradins*, (*sm. a.*) step sight, the foot or base having steps on which the slide rests;
—— *à gradins et à curseur*, v. —— *à curseur et à gradins*;
grande ——, (*art.*) long-range sight;
—— *instrument*, (*art.*) rear sight (rare);
—— *à indications continues*, any sight giving continuous indications;
—— *à indications continues et à cran de mire unique*, any single-notch sight giving continuous indications;
—— *à indications discontinues*, a sight in which the notch can be adjusted in only a few positions;
—— *à indications discontinues et à cran de mire unique*, sight with a single notch, capable of adjustment at fixed ranges only, at a limited number of ranges;
—— *du jour*, (*art.*) the actual elevation used in the field, instead of the tabular elevation, and true for the day only; the elevation found when the range is obtained, corresponding to the determined range;
—— *à lamettes*, (*sm. a.*) leaf-, folding-sight; sight with leaves, plates, of various lengths, each with a notch, for a particular range;
—— *latérale*, (*art.*) side sight;
—— *à lunette*, telescopic sight;
—— *médiane*, (*art.*) central (radial) sight;
—— *de mire et de tir*, tangent scale ; elevation itself;
—— *à miroir*, (*art.*) reflecting sight;
—— *mobile*, (*sm. a.*) movable sight;
—— *à niveau*, (*art.*) spirit-level sight (also *hausse-niveau*);
—— *de n mètres*, (*sm. a., art.*) the elevation corresponding to a distance of *n* meters;
petite ——, (*art.*) short-range sight;
pied de ——, (*sm. a.*) foot, rear-sight bed;
plaque de ——, (*German field art.*) elevation plate;
—— *pratique*, (*ball.*) distance from sight notch to upper element of gun barrel;
—— *provisoire*, (*art.*) the elevation corresponding to the inferior limit of the wide fork (*fourchette large*);
—— *du quart de tour de manivelle*, (*art.*) the elevation corresponding to a quarter turn of the elevating screw;
—— *à rallonge*, (*sm. a.*) extension sight;
—— *réelle*, (*ball.*) the —— *totale*, q. v., corrected for jump;
—— *réelle théorique*, (*ball.*) ratio of the —— *réelle* to the *longueur de mire*;
tête de la ——, v. *chapeau de la* ——;
—— *théorique*, (*ball.*) ratio of the —— *totale* to the *longueur de mire*;
—— *à tige*, (*art.*) rod sight;
—— *totale*, (*ball.*) distance of the sight notch from a parallel to the gun axis through the point of the fore sight;
—— *totale comparative*, (*ball.*) other name of *hausse théorique*;
—— *à trou(s)*, (*sm. a.*) peep sight.
hausse-col, m., (*unif.*) gorget (obs.); collar; (*harn.*) top of the collar.

hausser, v. a. n., to raise, elevate, lift, increase, augment; to heave, hoist; to rise (of a river); to rise (of prices);
—— *le tir*, (*art.*) (*lit.*, "to elevate the fire"); to give the elevation.

haussière, f., v. *aussière*.

haut, a., high, tall, lofty; upper (of a room, story); upper (of a stream, of a country); erect, elevated; deep (of water); high (of a stream, of the tide, of the seas); high (of prices); loud (of the voice); bright (of colors); distant, remote (of time); chief, principal, important, capital; hoisted, flying (of a signal, a flag);
—— *e administration*, the ministers and higher administrative personnel;
les ——*s Alpes*, the central chain of the Alps (and similarly of other mountain systems);
bâtiment de —— *bord*, (*nav.*) ship of the line;
—— *de bord*, (*nav.*) of high free-board;
la chambre ——*e*, the (English) House of Lords;
le —— *commerce*, the higher branches of the commercial world;
—— *e eau*, high water;
—— *école*, v. s. v. *école*;
la ——*e Égypte*, upper Egypt (and similarly of other countries);
la ——*e finance*, capitalists, bankers;
—— *-fond*, v. s. v. *fond*;
—— *-fourneau*, v. s. v. *fourneau* and *haut-fourneau*;
—— *e futaie*, v. s. v. *futaie*;
les ——*es latitudes*, high latitudes;
marée ——*e*, high tide;
—— *e(s) marée(s)*, spring tide; time of spring tide;
mer ——*e*, high tide; high sea;
—— *e mer*, the high seas;
messe ——*e*, high mass;
—— *e-paye*, v. s. v. *haute-paye*;
le —— *Rhin*, the upper Rhine (and similarly of other rivers);
la ——*e Seine*, the Seine above Paris;
le temps est ——*e*, the clouds are flying high;
tenir la bride ——*e*, (*man.*) to hold a tight, a short rein on;
—— *e trahison*, high treason;
la ville ——*e*, the upper town (as at Quebec).

haut, m., height, elevation, top (of a tower, etc.); upper part of anything; (*top.*) mountain, height; summit, top; upper course of a stream; (*nav., in pl.*) top sides; (*harn.*) top cheek (of a bit);
de bas en ——, *de* —— *en bas*, upward, downward; up-and-down motion;
—— *s et bas*, long strokes (at the pump);
—— *de la branche*, (*harn.*) top cheek of a bit;
—— *du jour*, noon.

haut, adv., up, upward;
—— *arme*, (*sm. a.*) advance carbine! (command);
—— *l'arme*, (*sm. a.*) (position of) advance carbine;
—— *les bras*, (*siege*) commence work! (command in trench work);
—— *cheval* —— *chaussé*, (*hipp.*) a horse with a white stocking;
cheval monté ——, —— *monté*, (*hipp.*) a horse with long, thin legs;
—— *la main*, with ease (cf. our familiar expression, "hands down");
mener un cheval —— *la main*, (*man.*) to ride with a high hand;
—— *le pied*, spare (of horses, teams); (*cav.*) held (of horses, troopers dismounted to fight on foot); (*r. r.*) (of a train) containing no passengers;
—— *pistolet*, (*sm. a.*) raise pistol! (command);
—— *le pistolet*, (*sm. a.*) (position of) raise pistol.

haut-bois, m., (*mus.*) oboe.
hautboïste, m., oboe player.
haut-bord, m., (*nav.*) ship of the line; vessel of high free-board; (*in pl.*) marines.
haute-finance, f., v. s. v. *haut*.
haute-futaie, f., v. s. v. *futaie*.
haute-marée, f., high tide; spring tide.

haute-paye, f., (*Fr. a.*) extra or additional pay given to *commissionnés* and to *rengagés* (—— *d'ancienneté*).

hauteur, f., height, depth, altitude (of a triangle, etc.); (*top.*) height, elevation; altitude, hill; rising ground, ascent; (*cons.*) height at which beams are laid; (*mil.*) number of ranks (of a battalion, etc.);

à la —— *de*, off, abreast of; equal to; in a line with; (*drill*) on a line with;

absolue, height above sea level;

—— *apparente*, apparent altitude;

—— *d'appui*, breast height; (*fort.*) height of parapet above banquette (called *genouillère* in barbette battery);

—— *d'aspiration*, height of suction;

—— *d'assise*, (*mas.*) thickness of a course;

—— *du baromètre*, barometer height;

—— *de batterie*, (*nav.*) height of battery above water; height of axis of guns above level of water, above water line;

—— *de chute*, (*ball.*) fall of a projectile due to gravity;

—— *de la chute d'eau*, head of water;

—— *sous clef*, rise of an arch;

—— *d'une colonne*, depth of a column;

de ——, deep, in depth;

—— *d'une dent*, (*mach.*) depth of a tooth;

—— *de disjonction*, (*ball.*) height of disjunction;

donner la ——, to elevate;

—— *de l'eau*, depth of water; (*mach.*) depth of water in a boiler; water line;

écart en ——, (*ball.*) vertical deviation;

—— *d'éclatement*, (*art.*) height of burst from ground, height of the bursting point (of a shell, etc.);

—— *d'entrepont*, (*nav.*) 'tween-decks; between decks;

—— *d'épaulement*, (*sm. a.*) height above the ground at which the gun is fired;

—— *de genouillère*, v. s. v. *genouillère*;

—— *de jet*, (*ball., etc.*) height of projection;

—— *libre*, height in the clear (of a bridge);

—— *du parapet*, (*fort.*) height of the crest above natural ground;

—— *du pas d'une vis*, (*mech.*) pitch of a screw;

—— *piézométrique*, piezometric level;

—— *de plongée*, (*fort.*) height of interior above exterior crest, rise of the superior slope;

—— *du point d'éclatement*, v. —— *d'éclatement*;

pointage en ——, v. s. v. pointage;

—— *de pointe*, (*mach.*) swing;

la —— *de pointe est de* . . . (*mach.*) to swing . . .;

pointer en ——, v. s. v. *pointer*;

prendre ——, *prendre la* —— *du soleil*, to take the sun;

—— *relative*, height of a mountain, etc., above its base;

—— *théorique moyenne d'éclatement*, (*art.*) height of mean point of burst above ground;

—— *du tiroir*, (*mach.*) breadth of the slide;

—— *type*, f., (*art.*) the height above target at which for a given trajectory the shrapnel must burst to give maximum effects;

—— *vraie*, true altitude.

haut-fond, m., shoal, shallow, bank; shoal water.

haut-fourneau, m., (*met.*) blast furnace; ironworks;

—— à *allure chaude* (*froide*), hot-(cold-) blast furnace.

haut-le-corps, m., (*man.*) jump, bound.

haut-mal, m., epilepsy.

haut-monté, a., long-legged.

haut-officier, m., (*mil.*) field officer, general officer (obs.).

haut-pendu, m., a squall of wind or rain, quickly over.

havre, m., harbor, haven.

havre(-)sac, m., (*mil.*) haversack; knapsack; valise; (*met.*) hammer scale.

havresat, m., (*met.*) hammer scale.

hayon, m., front or tail rack of a wagon.

héberger, v. a. r., to lodge, harbor.

hèche, f., sideboard; rave; rack, or plank forming side of wagon.

hectare, m., 100 ares.

hectogramme, m., 100 grams.

hectographe, m., hectograph.

hectolitre, m., 100 liters.

hectomètre, m., 100 meters.

hectostère, m., 100 stères.

heiduque, m., Hungarian foot soldier.

héler, v. a., to hail, challenge.

hélice, f., helix; screw, propeller screw; spiral grooves (as, around the hilt of a sword, etc.); any spiral; (*cord.*) one complete turn or twist;

—— à *ailes amovibles*, built-up propeller;

—— *amovible*, —— *démontable à la mer*, auxiliary propeller;

—— à *deux, trois, etc.*, branches, ailes, two or three bladed, etc., screw;

en ——, spiral, helical;

escalier en ——, spiral staircase;

—— *s jumelles*, twin screws;

—— à *pas droit* (*gauche*), right- (left-) handed propeller;

—— *propulsive*, screw propeller.

hélicé, a., helicoidal.

hélicoïdal, hélicoïde, a., helicoidal; spiral (frequently).

hélingue, f., v. *élingue*.

héliographie, f., heliography.

héliomètre, m., (*inst.*) heliometer.

hélioscope, m., (*inst.*) helioscope.

héliostat, m., (*inst.*) heliostat.

héliotrope, m., (*inst.*) heliotrope, heliograph.

helloffite, f., (*expl.*) hellhofite.

hématite, f., (*met.*) hematite;

—— *brune*, brown hematite;

—— *jaune argileuse*, yellow hematite;

—— *rouge*, red hematite.

hémiplégie, f. (*hipp.*) paralysis of one side.

hémisphère, m., hemisphere;

—— *austral*, southern hemisphere;

—— *boréal*, northern hemisphere.

hémisphérique, a., hemispherical.

hémorrhagie, f., hemorrhage.

hennir, v. n., (*hipp.*) to neigh, whinny.

hennissement, m., (*hipp.*) neigh, neighing, whinnying.

heptaèdre, m., heptaedron.

heptagone, m., heptagon.

héracline, f., (*expl.*) heraclin (a picric powder).

herbe, f., grass, herb;

mauvaise ——, weed.

herbue, f., (*met.*) clay flux.

herco-tectonique, f., (*fort.*) art of fortifying a camp, places, etc.

hérissé, p. p., bristling, erect, rough, armed at all points;

baril ——, v. *hérisson foudroyant;*

cheval ——, (*hipp.*) horse with a staring coat, rough-coated horse.

hérisser, v. a. n., to erect, bristle, stare; to become rough; (*mas.*) to rough-cast with mortar; (*mil.*) to multiply obstacles, accessory defenses, etc.

hérisson, m., hedgehog; sprocket, sprocket wheel; wheel before felly, etc., is put on; (in road-building) a ground bed of large stones covered with small stuff and rolled; spikes on top of a fence; (*mil.*) chevaux de frise; sort of *chausse-trappe* made of 3 lances (used in Algiers); turning beam, herisson; (*mach.*) spur wheel, spur gear, spur fly wheel;

—— *de côté*, (*mach.*) crown wheel;

en ——, (*mas.*) set up on edge (of bricks, flat stones, on the top of a wall);

—— *foudroyant*, (*mil.*) rampart beam (obs.);

—— *de roue*, speech;

—— *roulant*, (*mil.*) strong door studded with iron spikes, used (obs.) to close a breach; had two wheels;

—— *stable*, (*mil.*) same as —— *roulant*, but without wheels.

hérissonner, v. a., (*mas.*) to rough-cast (a wall, etc.).

hermétique, a., hermetic(al), air-tight, close.

hermétiquement, adv., hermetically.

herminé, a., (*hipp.*) ermined (of the coat).

herminette, f., adz.

herminure, f., (*hipp.*) ermine spot in the coat of a horse.

hernie, f., hernia, rupture.
hernier, m., crow's-foot; (*cord.*) kind of dead-eye through which the lines of an awning pass.
herpe, f., (*mil.*) v. *herse*.
herre, f., v. *haire*.
herse, f., harrow; iron cringle, strap; (*cord.*) strap, grommet; (*fort.*) portcullis; (*mil.*) harrow (obstacle); (*pont.*) substitute for anchor in bridge building;
— *articulée*, balloon anchor, grapnel;
— *d'attrape*, (*mil.*) upturned harrow, used as an obstacle;
—*s de la croupe*, (*cons.*) crossbeams;
— *de forteresse*, (*fort.*) sort of portcullis or grating, to close a passage.
hersillon, m., (*fort.*) plank or beam filled with nails, and used to impede a passage; small portcullis.
hetman, m., Cossack chief.
hêtre, m., beech tree, beech wood.
heure, f., hour; hour's distance; moment, time;
avancer l'—, to set a watch, clock, forward;
— *du bord*, ship's time;
— *de chemin*, an hour's distance;
—(*s*) *de cheval*, an hour's, so many hours', ride;
— *chronométrique*, chronometer time;
— *de Greenwich*, Greenwich time;
— *légale*, time by the public clock;
— *locale*, local time;
mettre une montre à l'—, v. *prendre l'*—;
— *de Paris*, Paris time (reckoned from the meridian of Paris); (*r. r.*) Paris time;
prendre l'—, to set a watch, clock;
—*s de présence*, office hours;
retarder l'—, to set a watch, clock, back.
heurt, m., collision, shock, blow; crown of a bridge or pavement.
heurtequin, m., hurter; shoulder washer;
— *à patte*(*s*), hurter with a body clout, or iron wrapping plate; wheel hurter instead of shoulder washer.
heurter, v. a. n. r., to run into, strike, hit, knock against, collide with; (*mil.*) to come in contact with (the enemy, outposts, etc.).
heurtoir, m., door knocker; sill (of a door); (*mach.*) catch, stop tappet; joint bolt; (*art.*) hurter; retracting stud; stop; joint bolt of a cap square; (*r. r.*) hurter;
— *de plate-forme*, (*art.*) platform hurter.
heuse, f., bucket, spear and bucket of a pump.
hexaèdre, m., hexahedron.
hexagone, a., hexagonal; m., hexagon; (*fort.*) six-bastioned work.
hexatomique, a., (*chem.*) hexad.
hie, f., commander, paving beetle, ram, beetle;
battre à la —, to ram.
hiement, m., ramming, paving.
hier, v. a., to ram, drive in with a beetle or rammer.
hiérarchie, f., hierarchy;
— *militaire*, military hierarchy; series of grades, classification of grades, with subordination of lower to higher.
hiérarchique, a., hierarchic; belonging to the hierarchy;
voie —, (*mil.*) military channels.
hiérarchiser, v. a., (*mil.*) to fix or establish a hierarchy, to fix the sequence of grades.
hinguet, m., v. *linguet*.
hippiatre, m., veterinary surgeon.
hippiatrie, f., (*hipp.*) horse medicines.
hippiatrique, f., (*hipp.*) veterinary art; horse medicine; knowledge of horses.
hippique, a., of horses, concerning horses;
concours —, horse show.
hippodrome, m., hippodrome; race course.
hippo-lasso, m., v. *lasso-dompteur*.
hippolithe, f., (*hipp.*) stone in stomach or intestine of a horse.
hippologie, f., hippology.
hippologue, m., hippologist.
hippomane, m., horse lover; person very fond of riding.
hippomanie, f., great love of horses; (*hipp.*) sort of madness, rage.
hippomètre, m., (*hipp.*) hippometer.
hippo-sandale, f., (*hipp.*) horse sandal, sort of box shoe.
hironde, f., *queue d'*—, dovetail.

hirondelle, f., swallow; iron axletree band.
hisser, v. a., to pull up, sway up, hoist, raise;
— *à bloc*, to haul chock-a-block;
— *en douceur*, to hoist handsomely;
— *main sur main*, to haul hand over hand;
— *le pavillon*, to hoist the colors.
histoire, f., history.
historien, m., historian.
historique, m., historical account;
— *de corps*, — *de régiment*, — *régimentaire*, regimental history.
hiver, m., winter;
cœur de l'—, midwinter;
passer l'—; to pass the winter, remain in winter quarters;
quartier d'—, (*mil.*) winter quarters;
queue de l'—, latter end of winter.
hivernage, m., winter season, wintering place, winter fodder; (*mil.*) winter quarters.
hiverner, v. n., to winter; to go into, remain in, winter quarters.
ho! du navire! (*nav.*) ship ahoy!
hobin, m., (*hipp.*) breed of Scotch horses (natural pacers).
hoche, f., notch, nick, hollow.
hocher, v. a. n., to shake the bit, toss the head; to notch.
hochet, m., kind of spade.
hoirin, m., (*nav.*) anchor buoy.
hollander, m., (*expl.*) pulper, rag engine.
homards, m. pl., (*mil. slang*) Spahis.
homme, m., man;
— *d'affaires*, business man; agent.
— *en affectation spéciale*, (*Fr. a.*) any man belonging to or serving in railroad, post, telegraph, customs, forest, etc., departments;
— *sans aveu*, vagrant;
— *de bois*, (*man.*) dumb jockey;
—*s de cadre*, (*mil.*) n. c. o.'s, corporals, trumpeters, saddlers, etc.;
— *de chambrée*, (*mil.*) room orderly;
— *de cheval*, horseman, rider, lover of horses, thorough horseman; (*cav.*) trooper;
— *de communication*, (*mil.*) man used to preserve communication between the parts of an advanced post;
—*s de complément*, (*mil.*) the men needed to bring a unit up to a war footing;
— *de corvée*, (*mil.*) member of a fatigue party;
—*s à la disposition*, (*Fr. a.*) members of a class with the colors, but not in service, or else released before expiration of their terms (since 1895 not an official term);
— *d'église*, clergyman, priest;
— *encadré*, v. — *de rang*;
— *d'épée*, soldier;
— *d'état*, statesman;
— *fait*, full-grown, (hence) able-bodied, man;
— *de guerre*, soldier, warrior;
— *à la mer*, man overboard;
— *de mer*, (*nav.*) able seaman, thorough sailor;
mettre des —*s sur*, to put men at (anything); to man;
—*s non-disponibles*, (*Fr. a.*) on mobilization, men kept for the public good in government departments and in military establishments, so that they may not be broken up by a sudden loss of their personnel;
— *de paille*, man of straw;
— *de peine*, porter;
— *de pied*, (*inf.*) foot soldier, infantryman;
aux —*s punis*, (call or battery) for prisoners to turn out;
— *de rang*, (*mil.*) a term used in a somewhat specialized way for the soldier thoroughly drilled and disciplined, so as to keep his place under all circumstances; man hard to frighten or demoralize;
— *du 1er (2e) rang*, (*mil.*) front- (rear-) rank man;
— *de recrue*, (*mil.*) recruit;
— *de robe*, lawyer;
— *de troupe*, (*Fr. a.*) man in the ranks (strictly, n. c. o., corporal, and private; in current speech, corporal and private).

homogène, a., homogeneous;
  acier ——, (*met.*) ingot steel;
  fer ——, (*met.*) ingot iron.
homologation, f., (*law*) confirmation, legalization.
homolographe, m., (*inst.*) an instrument enabling one to prick mechanically, by mere sighting, the horizontal position of a leveling rod and to read the altitude of the point directly (Peaucellier and Wagner); homolograph.
homologuer, v. a., (*law*) to confirm, legalize.
hongre, a., (*hipp.*) gelded;
  cheval ——, gelding.
hongrer, v. a., (*hipp.*) to castrate, cut, geld.
hongroyé, a., Hungarian, Hungary;
  cuir ——, Hungary leather.
hongroyer, v. a., to taw leather.
honneur, m., honor; (*in pl.*), honors, military honors;
  affaire d' ——, duel;
  arme d' ——, v. s. v. *arme;*
  champ d' ——, (*mil.*) battlefield;
  croix d' ——, cross of the Legion of Honor;
  dettes d' ——, gambling debts;
  faire —— *à*, to honor (as a draft); (*nav.*) to avoid (as a rock, a shoal);
  ——*s funèbres*, funeral honors;
  fusil d' ——, v. s. v. *fusil;*
  ——*s de la guerre*, honors of war;
  légion d' ——, Legion of Honor;
  ——*s militaires*, military honors;
  mourir au lit d' ——, (*mil.*) to be killed in battle;
  ordre d' ——, Legion of Honor;
  point d' ——, point of honor, pundonor;
  rendre les ——*s*, to give military honors, to pay honors;
  sabre d' ——, v. s. v. *sabre.*
honoraire, a., honorary; m. pl., honorarium.
hôpital, m., hospital;
  —— *ambulant*, a "first aid" hospital on the field of battle;
  —— *annexe*, (*Fr. a.*) a hospital administratively attached to a regular military hospital (established in garrisons lacking full hospital facilities);
  —— *auxiliaire*, (*Fr. a.*) Red Cross hospital (in war);
  —— *auxiliaire de campagne*, (*Fr. a.*) Red Cross field hospital;
  —— *de campagne*, field hospital;
  —— *central*, (*Fr. a.*) the hospital to which an —— *annexe* is attached;
  —— *civil militarisé*, —— *civil mixte*, v. —— *militarisé;*
  —— *à destination spéciale*, (*Fr. a.*) a hospital for infectious and contagious diseases;
  —— *d'eaux minérales*, a hospital established at healing springs;
  —— *d'étapes;* the hospital of a *gîte d'étapes;*
  —— *d'évacuation*, hospital at the rear, receiving sick and wounded from the front and forwarding them to interior of the country, "overflow" hospital;
  fièvre d' ——, typhus fever;
  —— *militaire*, post hospital;
  —— *militarisé*, —— *mixte*, (*Fr. a.*) a civil hospital in which wards are specially assigned to the use of soldiers under the care of military surgeons;
  pourriture d' ——, hospital gangrene;
  —— *temporaire*, a hospital established in besieged towns, or else in open towns, to receive the wounded.
horaire, a., hourly; m., railroad time-table;
  halte ——, (*mil.*) hourly halt, horary halt;
  livre ——, (*r. r.*) time-table.
horizon, m., horizon;
  —— *apparent*, v. —— *visible;*
  —— *artificiel*, artificial horizon;
  —— *astronomique*, true horizon;
  en —— *élevé*, above the horizon;
  à mercure, mercurial horizon;
  —— *physique*, v. —— *visible;*
  —— *rationnel*, v. —— *vrai;*
  réduction à l' ——, reduction to the horizon;
  réduire à l' ——, to reduce to the horizon;
  —— *sensible*, v. —— *visible;*
  —— *visible*, *visuel*, apparent, visible, horizon;
  —— *vrai*, true horizon.

horizontal, a., horizontal.
horizontale, f., (*top.*) contour line, contour.
horloge, f., clock;
  —— *à eau*, —— *d'eau*, clepsydra;
  —— *du loch*, log glass;
  —— *à longitudes*, v. —— *marine;*
  —— *marine*, chronometer;
  monter une ——, to wind up a clock;
  régler une ——, to regulate a clock;
  remonter une ——, v. *monter une* ——;
  —— *de sable*, hourglass;
  —— *solaire*, sundial.
horloger, m., clockmaker.
horlogerie, f., clock-, watch-, making; clock work.
hors, m., prep., adv., outside of;
  —— *de batterie*, (*art.*) from battery;
  —— *cadre*, (*mil.*) (of officers on the active list and employed) not on the rolls of a regular organization; in addition to or in excess of the statutory number; seconded (Brit.) (n. e., U. S. A.);
  —— *de combat*, disabled;
  —— *de cour*, (*law*) out of court;
  —— *d'eau*, not level;
  —— *d'école*, (*man.*) long out of riding school (said of the horse);
  être —— *du montoir*, (*hipp.*) to be lame in front;
  —— *feu*, (*met.*) out of blast, not working;
  —— *la garde*, v. s. v. *garde;*
  —— *d'haleine*, out of breath, out of wind;
  —— *la main*, (*man.*) hard-mouthed;
  mettre ——, (*met.*) to put out of blast;
  mettre —— *d'eau*, (*art.*) to depress a gun and close the vent and muzzle so that water will not lodge in it (an expression nearly obsolete);
  le —— *du montoir*, (*man.*) off side;
  —— *d'œuvre*, —— *œuvre*, v. s. v. *œuvre;*
  —— *de la portée* (*du canon*, etc.), out of range of (a gun, etc.);
  —— *rang*, v. s. vv. *peloton*, *section;*
  —— *service*, (*mil.*) unserviceable;
  —— *tour*, (*Fr. a.*) out of turn (said of promotions by selection not counted among those determined by the law).
hors-d'œuvre, m., v. s. v. *œuvre*.
hors-montoir, m., (*man.*) off side.
hors-œuvre, adv., v. s. v. *œuvre*.
hospice, m., hospital;
  —— *civil militarisé*, —— *civil mixte*, v. *hôpital militarisé*.
hospitalisation, f., admission to and sojourn in a hospital.
hospitalisé, a., admitted to a hospital;
  homme ——, man who has been in hospital.
hospitaliser, v. a., to receive in a hospital.
hostile, a., hostile.
hostilité, f., (*mil.*) hostility, enmity;
  commencer, *ouvrir*, *les* ——*s*, to open hostilities;
  conduire, *poursuivre*, *les* ——*s*, to carry on hostilities;
  suspendre les ——*s*, to suspend hostilities.
hotchkiss, m., (*art.*) Hotchkiss gun (esp. the Hotchkiss revolving cannon).
hôte, m., post; (*mil.*) (in France) a person who is compelled to furnish quarters to a soldier.
hôtel, m., hotel; (*mil.*) official residence (as of a general commanding a *corps d'armée*, of the heads of the various military schools, etc.);
  —— *Dieu*, hospital;
  —— *des Invalides*, pensioners' hospital, esp. the "Invalides," at Paris; Soldiers' Home, U. S. A.;
  —— *du ministère de la guerre*, seat of the war ministry in Paris;
  —— *du quartier général*, v. under main word, (*mil.*);
  —— *de ville*, town hall.
hotte, f., scuttle; hod; basket for carrying earth, etc.; (*chem.*, *met.*) hood, funnel hood.
houe, f., hoe.
houer, v. a., to hoe.
houille, f., coal; pit or sea coal; soft coal, bituminous coal; hard coal;
  —— —— *agglomérée*, artificial fuel, pressed fuel;
  —— —— *bitumineuse*, soft, bituminous coal;

houille *brune*, lignite;
 charbon de ——, coke;
 —— *à coke boursouflé*, v. —— *grasse*;
 —— *collante*, v. —— *grasse*;
 —— *demi-grasse*, coal not quite so pasty as —— *grasse*;
 dépôt de ——, coal deposit; store of coal;
 —— *éclatante*, anthracite coal;
 entrepôt de ——, store of coal;
 —— *de forge*, forge coal;
 —— *grasse*, a term descriptive of any coal that becomes pasty; fat coal; smithy coal;
 —— *grasse à courte flamme*, close-, short-burning, short-flaming, coal;
 —— *grasse à longue flamme*, cannel coal; long-flaming, long-burning, free-burning, smith coal (easily inflammable);
 —— *grasse maréchale*, best forge coal;
 —— *luisante*, anthracite coal;
 —— *maigre*, coal containing but little hydrogen; close-burning coal;
 —— *maréchale*, v. —— *de forge*;
 —— *non collante*, v. —— *maigre*;
 —— *proprement dite*, bituminous coal;
 —— *sans flamme*, close-burning coal;
 —— *sèche*, v. —— *maigre*;
 —— *sèche à longue flamme*, long-flaming, long-burning, free-burning, glance coal;
 —— *tendre*, soft coal;
 —— *terreuse*, v. —— *brune*;
 —— *tout-venante*, rough coal, coal as it comes from the mine, mine run.
houiller, a., containing coal, coaly.
houillère, f., coal mine, colliery.
houillite, f., anthracite coal.
houlan, m., *(cav.)* uhlan.
houle, f., swell (of the sea);
 —— *battue*, cross sea;
 —— *de fond*, ground swell;
 —— *longue*, roller.
houlette, f., trowel; spatula; *(fond.)* casting ladle.
houleux, a., (of the sea) rough, rolling, heaving, swelling.
houppe, f., tuft;
 —— *du menton*, *(hipp.)* rounded protuberance on the lower lip.
hourdage, m., *(mas.)* rubblework, rough casting; *(cons.)* pugging.
hourder, v. a., *(mas.)* to roughcast; *(cons.)* to pug.
hourdis, m., *(mas.)* rubblework; *(cons.)* pugging; hollow brick.
hourra, m., hurrah, cheer, sudden attack with cheers.
housard, m., *(cav.)* hussar (rare).
housarder, v. a., *(cav.)* to make a *housardaille*, q. v.
housardaille, f., *(cav.)* sudden and rapid attack or raid of a small party of cavalry (obs.).
houssage, m., sweepings;
 nitre de ——, niter obtained by sweeping walls.
housse, f., *(harn.)* horse cloth, saddlecloth, housing, housing cloth, shabraque;
 —— *en bottes*, croup housing;
 —— *de cheval*, saddlecloth, horse cloth;
 —— *du collier*, collar pad;
 —— *de pied*, —— *en souliers*, a housing coming below the leg of the rider (obs.);
 —— *traînante*, funeral housing.
housser, v. a., to sweep, dust.
houssine, f., switch.
houssiner, v. a., to switch, beat.
houssoir, m., hair broom; whisk of feathers.
houx, m., holly, holly tree.
hoyau, m., grubbing ax;
 pic ——, grubbing pick.
huche, f., hutch, bin, trough, mill hopper; tub.
hue! (in the United States) geel (in England) haw!
huée, f., shouting, whoop, whooping.
huer, v. a., to hoot.
huilage, m., oiling.
huile, f., oil;
 —— *d'amande*, almond oil;
 —— *animale*, animal oil;
 —— *antoxyde*, rust-preventing oil;

huile *de baleine*, whale oil;
 boîte à ——, oil cup;
 —— *de bras*, *(fam.)* "elbow grease;"
 —— *à brûler*, illuminating oil;
 burette à ——, oil can;
 —— *de cétacés*, v. —— *de baleine*;
 —— *de chanvre*, —— *de chènevis*, hemp-seed oil;
 —— *de coco*, coconut oil;
 —— *de colza*, colza oil, rape oil;
 —— *de coton*, cotton-seed oil;
 —— *cuite*, boiled linseed oil;
 —— *douce*, sweet oil;
 éclairage à l'——, illumination by oil, oil lighting;
 éclairer à l'——, to light by oil;
 —— *essentielle*, essential oil;
 —— *explosive*, *(expl.)* Nobel's name for nitro-glycerin, detonating oil;
 —— *de faîne*, beech-nut oil;
 —— *fixe*, fatty, fixed, oil;
 godet à ——, oil cup;
 —— *de goudron*, coal-tar oil;
 —— *grasse*, fatty, fixed, oil; any drying oil boiled with litharge, oxide of zinc, etc.;
 —— *de lampe*, lamp oil;
 —— *de lin*, linseed oil;
 —— *de lin crue*, raw linseed oil;
 —— *de lin cuite*, boiled linseed oil;
 —— *lourde*, paraffin oil; heavy, thick oil;
 —— *lourde de houille*, oil from coal tar, used as disinfectant;
 —— *minérale*, mineral oil;
 —— *de naphte*, petroleum;
 —— *de navette*, rape-seed oil;
 —— *non-siccative*, nondrying oil;
 —— *d'œuillette*, poppy-seed oil;
 —— *d'olive*, olive oil;
 —— *de palme*, palm oil;
 —— *de pétrole*, petroleum, coal oil;
 —— *de pierre*, petroleum, rock oil;
 —— *de pied de bœuf*, neat's-foot oil;
 —— *de poisson*, fish oil;
 —— *à quinquet*, v. —— *à brûler*;
 —— *de résine*, rosin oil;
 —— *de ricin*, castor oil;
 —— *siccative*, drying oil;
 —— *de térébenthine*, (oil of) turpentine;
 —— *de terre*, v. —— *de pierre*;
 —— *végétale*, vegetable oil;
 —— *de vitriol*, oil of vitriol, sulphuric acid;
 —— *volatile*, volatile oil.
huiler, v. a., to oil, lubricate.
huilerie, f., oil refinery.
huilier, m., oil bottle, oiler, oil can.
huis, m., door (obs.);
 à —— *clos*, behind closed doors.
huisserie, f., door frame, doorcase.
huissier, m., usher;
 —— *d'armes*, sergeant-at-arms;
 —— *appariteur*, *(Fr. a.)* n. c. o. (sergeant) attached to a court-martial, and used to serve papers, look after the rooms, etc.
huit-de-chiffres, m., *(man.)* figure of eight, an air in which the horse traces a figure of eight.
huit-pans, m., octagonal nut.
hulan, m., *(cav.)* uhlan, lancer.
humectation, f., moistening, dampening; *(powd.)* dampening, liquoring.
humecter, v. a., to moisten, dampen; *(powd.)* to liquor.
humide, a., wet, damp;
 temps ——, wet weather.
hune, f., *(nav.)* top (of a mast, of a military mast).
hurasse, m., *(mach.)* hurst.
hussard, m., *(cav.)* hussar;
 ——*s de la mort*, Black Hussars (German army);
 —— *à 4 roues*, *(mil. slang)* soldier of the train or army service corps.
hussarde, f., *à la* ——, by means of rapine, etc.;
 botte à la ——, light cavalry boot;
 selle à la ——, *(harn.)* light cavalry saddle.
hutte, f., shed, hut;
 —— *de bois*, log hut.
hutter, v. a., to make an encampment of huts; to make huts, live in huts, etc. (rare).
hyckori, m., hickory wood and tree.

hydarthrose, f., (*hipp.*) scientific name of windgall.
hydrate, m., (*chem.*) hydrate;
— *de chaux,* calcic hydrate;
— *en pâte,* lime powder paste;
— *en poudre,* lime powder.
hydraté, (*chem.*) hydrated.
hydraulique, a., hydraulic; f., hydraulics; hydraulic gear;
bélier —, hydraulic ram;
chaux —, hydraulic lime;
ciment —, water cement;
machine —, ram, hydraulic engine;
ouvrage —, waterworks;
presse —, hydraulic, hydrostatic press;
puissance —, water power;
roue —, water wheel.
hydrocarboné, a., hydrocarbonic.
hydrocellulose, f., (*expl.*) hydrocellulose;
hydrocarbure, m., (*chem.*) hydrocarbon.;
hydrodynamique, a., hydrodynamic.
hydro(-)électrique, a., hydro-electric.
hydro-extracteur, m., (*expl., etc.*) centrifugal drying machine, hydro-extractor.
hydrofuge, a., waterproof, water-tight (said of cements, etc.).
hydrogène, m., hydrogen.
hydrogéné, a., hydrogenized.
hydrographe, m., hydrographer.
hydrographie, f., hydrography.
hydrographique, a., hydrographic.
hydromécanique, a., hydraulic.

hydromètre, m., (*inst.*) hydrometer.
hydrométricité, f., v. *hygrométricité.*
hydrométrie, f., hydrometry.
hydrométrique, a., hydrometric.
hydropneumatique, a., hydropneumatic.
hydropneumatisme, m., hydropneumatism.
hydrostatique, a., hydrostatic, f., hydrostatics.
hydro-ventilateur, m., hydroventilator.
hygiène, f., hygiene.
hygromètre, m., (*inst.*) hygrometer;
— *à absorption (condensation),* absorbing (condensing, or chemical) hygrometer.
hygrométricité, f., hygrometricity.
hygrométrie, f., hygrometry.
hygrométrique, a., hygrometric.
hygroscope, m., (*inst.*) hygroscope.
hyporyctique, a., (*mil. min.*) having to do with the operations of underground war.
hypoténuse, f., hypotenuse.
hypothèque, f., mortgage.
hypothéquer, v. a., to mortgage.
hypothèse, f., hypothesis;
— *tactique,* (*mil.*) tactical situation (in maneuvers, exercises).
hypsomètre, m., (*inst.*) hypsometer.
hypsométrie, f., hypsometry; determination of heights.
hypsométrique, a., hypsometric.
hystérésis, m., (*elec.*) magnetic lag, hysteresis.

I.

ichnographe, m., machine designer.
ichnographie, f., ichnography.
ichnographique, a., ichnographic.
ichthyocolle, f., isinglas, fish-glue.
iconomètre, m., (*inst.*) instrument for measuring distances on maps, plans.
icosaèdre, m., icosahedron.
identité, f., identity;
carte d'—, (*Fr. a.*) document (including a photograph in plain clothes) to be shown by officers and *assimilés* when claiming special rates on railroads; (*in gen.*) identification card, similar in idea to *plaque d'*—, q. v.;
plaque d'—, (*mil.*) identification plate worn in war, issued on mobilization.
idiostatique, a., (*elec.*) idiostatic.
idoine, a., suitable (obs. word still used in administration);
— *expert,* expert.
if, m., yew tree, yew; triangular framework, to support lamps (*lampions*) in illuminations.
igné, a., igneous.
ignition, f., ignition;
en —, ignited;
entrer en —, to ignite;
mettre en —, to ignite, set on fire;
— *spontanée,* spontaneous ignition.
île, f., island.
illettré, a., illiterate; unable to read or write.
îlot, m., islet; (*top.*) block, clump, or mass (of houses, etc.).
image, f., image, picture, illustration; photograph (positive image);
— *directe,* (*opt.*) direct image;
— *électrique,* (*elec.*) electric image;
— *floue,* (*phot.*) "oyster;"
— *négative,* (*phot.*) negative;
— *positive,* (*phot.*) positive;
— *réelle,* (*opt.*) real image;
— *réfléchie,* (*opt.*) reflected image;
— *renversée,* (*opt.*) inverted image;
— *virtuelle,* (*opt.*) virtual image.
imbiber, v. a. r., to soak, imbibe; to soak into.
imbibition, f., soaking, imbibition.
imboire, v. a., to imbue.
imbrication, f., imbrication.
imbricée, a. (feminine only), imbricated, imbricate;
tuile —, hollow tile.

imbrifuge, a., rain-proof.
immalléable, a., (*met.*) immalleable.
immanœuvrable, a., (*nav.*) unmanageable.
immatriculation, f., (*mil.*) taking up on (rolls of men, horses), mustering in (U. S. A.); (*Fr. a.*) inscription of a name on the *registre matricule* (q. v.);
— *au corps,* (*Fr. a.*) taking up of a name on the regimental rolls (*registre matricule*);
— *au recrutement,* (*Fr. a.*) in each regional subdivision, taking up of a name on the recruiting rolls (*registre matricule*).
immatricule, f., enrollment, registry.
immatriculé, m., (*Fr. a.*) soldier or employee of war department inscribed on rolls of the army.
immatriculer, v. a., to enroll, register; (*mil.*) to put on the muster rolls of the army.
immerger, v. a., to immerse.
immersion, f., immersion.
immeuble, a., real (of property); m., landed property, real estate.
immobilier, a., of lands, landed.
immobilisation, f., (*law*) act of giving a real character to property; (*mil.*) tying down of a force to a particular position, immobilization.
immobiliser, v. a., to fix in position; (*law*) to give property a real character; (*mach.*) to stop; (*mil.*) to immobilize, to fasten or tie down to a given position.
immobilité, f., immobility; (*mil.*) state of being still in ranks (as at command, *fixe!*); (*hipp.*) immobility, inability of the horse to execute (voluntary movements); balkiness.
impact, m., impact;
centre d'—, (*ball.*) center of impact;
point d'—, (*ball.*) point of impact.
impair, a., odd, uneven.
impasse, f., blind alley or lane;
se trouver dans une —, to be in a great difficulty.
impediments, impédiments, m. pl., (*mil.*) baggage (especial baggage not directly needed for the combat).
impénétrable, a., impenetrable, impervious.
impénétrabilité, f., impenetrability.
impératrice, f., empress.
impérial, a., imperial.

impériale, f., top of a tent, of a carriage; imperial (tuft on the under lip).
imperméable, a., waterproof, impervious, impermeable; m., mackintosh (overcoat);
— *à l'air*, air-tight;
— *à l'eau*, water-tight;
— *au gaz*, gas-tight;
— *à la vapeur*, steam-tight.
imperméabilisation, f., waterproofing, impermeabilization.
imperméabiliser, v. a., to make water-proof.
impondérabilité, f., imponderability.
impondérable, a., imponderable.
importable, a., importable.
importateur, m., importer.
importation, f., importation;
droit d'—, tariff, impost;
prime d'—, bounty on importation.
importer, v. a., to import.
imposer, v. a., to inflict, levy, impose (as taxes, etc.).
imposition, f., imposition (of a tax, etc.).
imposte, m., transom (of a door or window); (*mas.*) impost.
impôt, m., impost, tax.
impotent, m., person who, for any physical reason whatever, can not earn a living.
impraticable, a., impracticable; impassable (of roads, etc.); (*fort.*) impracticable (of the breach).
imprégnation, f., impregnation (as of wood).
imprégner, v. a., to impregnate (as wood).
imprenable, a., (*mil.*) impregnable.
impression, f., impression; printing; priming;
— *en taille douce*, copperplate printing.
impressionner, v. a., (*phot.*) to "print."
imprimé, m., public document; blank form; (*in pl.*) printed matter.
imprimer, v. a., to print; to impress; to give a first coat of paint;
— *un mouvement*, to set in motion.
imprimerie, f., printing; art of printing; printing establishment;
— *portative*, (*Fr. a.*) portable press accompanying the headquarters of each army and of each group of armies.
imprimeur, m., printer.
imprimure, f., priming, priming coat.
impropre, a., unclean, unfit;
— *au service*, unfit for service, useless.
impulsion, f., impulse, impetus, shock; (*mil.*) impulse given a firing line by the arrival of reserves;
donner l'— *à*, to set in motion; to give, impart an impulse;
force d'—, propelling power.
imputation, f., deduction of a sum of money; sum charged on the debit side of an account; charge.
imputer, v. a., to place to account; to deduct a sum of money.
imputrescibilité, f., imputrescibility.
imputrescible, a., imputrescible.
inabordable, a., inaccessible, as, (*mil.*) a position.
inaccessible, a., inaccessible.
inactif, a., inactive, unemployed.
inaction, f., inaction; (*man.*) keeping a horse still while flexing the *encolure*.
inaffouillable, a., (of earth, soil), not liable to scour, wash.
inaguerri, a., (*mil.*) unwarlike; not trained or inured to war.
inaliénabilité, f., inalienability.
inaliénable, a., inalienable.
inalliable, a., (*met.*) nonalloying.
inamovilité, f., state of being unremovable.
inamovible, f., not removable.
inattaquable, a., unassailable.
incalcinable, a., incapable of calcination.
incalciné, a., uncalcined.
incandescence, f., incandescence; white heat.
incandescent, a., white hot, welding hot.
incapacité, f., (*law*) lack of power to take a legal step.
incarcérable, a., liable to imprisonment.
incarcération, f., imprisonment.

incarcérer, v. a., to imprison.
incendiaire, a., incendiary;
composition —, v. s. v. *composition;*
cylindre —, v. s. v. *cylindre;*
fusée —, (*artif.*) incendiary rocket;
obus —, (*artif.*) carcass, incendiary shell;
projectile —, (*artif.*) incendiary projectile.
incendie, m., conflagration, fire; eruption (of a volcano); (*mil.*) burning, devastation by fire.
appareil de sauvetage pour les —*s*, fire escape;
crochet à —, fire hook;
échelle à —, fire ladder;
extincteur d'—, fire extinguisher, fire engine;
pompe à —, fire engine;
robinet à —, fire plug;
seau à —, fire bucket.
incendier, v. a., to fire, set on fire, burn, burn down.
incertain, a., uncertain, variable;
cheval —, restless, fidgety horse;
joints —, (*mas.*) irregular joints between stones of different sizes.
incessible, a., that may not be ceded, uncedable.
inceste, m., (*hipp.*) in-and-in breeding.
inchavirable, a., (of a boat) non-capsizable.
incidence, f., incidence;
angle d'—, angle of incidence.
incident, m., incident, accident; a., incident;
—*s de tir*, (*art.*) incidents that delay or arrest the fire.
incinération, f., incineration, burning, reduction to ashes;
— *des morts*, cremation.
incinérer, v. a., to incinerate, reduce to ashes, cremate.
incisif, a., incisive;
dent —*ve*, incisor.
incisive, f., incisor, incisive tooth;
— *adulte*, adult incisors, of second dentition;
— *caduque*, incisor of first dentition;
— *de lait*, v. — *caduque;*
— *de remplacement*, v. — *adulte.*
inclinaison, f., inclination; obliquity; gradient, slope; angle; (*elec.*) dip of the needle; (*nav.*) heeling, angle of heel; (*min.*) hade, underlay;
aiguille d'—, (*elec.*) dipping needle;
— *de l'aiguille aimantée*, (*elec.*) dip of the needle;
boussole d'—, (*elec.*) dipping compass; inclination compass;
— *finale des rayures*, (*art.*) muzzle inclination of the rifling;
— *d'une fusée d'essieu*, set of an axletree arm;
— *magnétique*, (*elec.*) dip;
— *d'un mât*, (*nav.*) rake of a mast;
— *des rayures*, (*art.*) angle of rifling; angle between the tangent to the curve and the axis of the bore.
inclinatoir, m., — *magnétique*, (*elec.*) dipping needle.
incliné, m., inclining dial, incliner.
incliner, v. a. r., to bend, bow, stoop, tilt, slope, incline;
— *une pièce sous l'horizon*, (*art.*) to depress a gun.
inclinomètre, m., (*inst.*) inclinometer.
inclure, v. a., to include.
incombustible, a., incombustible.
incombustibilité, f., incombustibility.
incommodé, p. p., inconvenienced, distressed, shattered;
bâtiment —, (*nav.*) a ship that has lost a mast or otherwise suffered injury;
être — *d'un bras, d'une jambe*, to have lost the use of an arm or leg.
incommodité, f., inconvenience, distress;
signal d'—, (*nav.*) signal of distress.
incompatibilité, f., (*law*) impossibility of holding two positions at one and the same time, e. g., civil and military.
incompatible, a., incompatible; (*law*) that may not be held at one and the same time, by one and the same person.
incompétence, f., (*law*) lack of jurisdiction; incompetency.

incompétent, a., (*law*) having no jurisdiction; incompetent.
incomplet, a., incomplete, blank; m., vacancy, blank; (*in pl., mil.*) difference between statutory and budgetary strengths.
incompressibilité, f., incompressibility.
incompressible, a., incompressible.
inconduite, f., bad behavior, misbehavior.
incongelable, a., nonfreezing.
incongelé, a., unfrozen.
inconnue, f., unknown quantity.
incorporation, f., incorporation; (*mil.*) passage of a class, contingent, etc., under the colors, incorporation; action of receiving young soldiers into the regiments to which they are assigned and of placing them on rolls of same (synonym of *immatriculation*); in recruiting, inscription of names on the recruiting rolls; (*powd.*) incorporation.
incorporer, v. a., to incorporate; (*mil.*) to incorporate (a contingent) recruits in a regiment; (*powd.*) to incorporate.
incorrigibilité, f., incorrigibility; (*mil.*) permanent resistance to the requirements of discipline.
incorrigible, a., incorrigible; (*mil.*) permanently undisciplined.
incrasser, v. a., to thicken.
incrimination, f., incrimination.
incriminer, v. a., to incriminate.
incrochetable, a., that can not be picked (lock).
incrustation, f., incrustation; (*steam*) scale, boiler-scale, fur.
incruster, v. a. r., to incrust; (*steam*) to become covered with scale, to become incrusted, to fur.
incuit, m., underburned limestone.
inculpation, f., (*mil. law*) accusation; suspicion of guilt, of having committed a crime or misdemeanor.
inculpé, m., (*mil. law*) the accused, (in the literal sense of the word, before charges or trial).
inculper, v. a., to accuse, criminate.
inculte, a., uncultivated; wild.
incurable, a., incurable.
incurabilité, f., incurability.
incursion, f., incursion; (*mil.*) inroad, raid, irruption, incursion.
incurvation, f., incurvation.
incurver, v. a., to curve in.
indéchiffrabilité, f., quality of being undecipherable.
indéchiffrable, a., undecipherable.
indéfendable, a., (*mil.*) indefensible.
indélébile, a., indelible;
  encre ——, indelible ink.
indélébilité, f., indelibility.
indemne, a., indemnified;
  rendre ——, to indemnify.
indemnisation, f., indemnification.
indemniser, v. a., to indemnify, make good.
indemnitaire, m., person entitled to an indemnity.
indemnité, f., allowance; indemnity; commutation.
  (Except where otherwise indicated, the following terms relate to the French service:)
  —— *pour changement d'uniforme,* allowance to subaltern officers and *adjudants* on involuntary transfer to a regiment having a different uniform;
  —— *de déplacement,* (*in gen.*) traveling expenses; travel allowance;
  —— *aux enfants de troupe,* a yearly money allowance made to *enfants de troupe* who remain at home;
  —— *d'entrée en campagne,* allowance to officers on being ordered to join an active army (one in the field), or to join a fort when its *garnison de défense* is formed.
  —— *extraordinaire de voyage,* allowance depending on distance and on unusual character of mission or duty on which sent;
  —— *de ferrure,* shoeing allowance to gendarmes serving in the provost guard during the fall maneuvers;

indemnité *pour la fête nationale,* a money allowance to n. c. o., corporals, and privates on the occasion of the *fête nationale,* q. v.;
  —— *fixe,* —— *fixe de transport,* allowance of money (officers) for transportation of baggage from quarters to station, and vice versa;
  —— *de fonctions,* allowance accompanying certain functions or details;
  —— *pour frais de bureau,* (*in gen.*) office expenses;
  —— *pour frais de service,* allowance made to officers holding certain positions to meet the expenses incumbent upon them (e. g., chief of staff, governor of Paris, c. o.'s of military schools, etc.);
  —— *de guerre,* (*in gen.*) war indemnity;
  —— *pour habillement,* (*in gen.*) clothing allowance;
  —— *journalière,* daily subsistence allowance to *isolés* traveling under orders;
  —— *kilométrique,* mileage;
  —— *de logement,* commutation of quarters of certain n. c. o. (*rengagés, commissionnés*), who have families; also to master-workmen, etc., for whom there are no quarters in garrison; (*in gen.*) commutation of quarters;
  —— *de monture,* extra pay (mounted);
  —— *pour pertes de chevaux (d'effets),* indemnity for horses killed in action (for loss of effects captured);
  —— *de première mise de harnachement,* allowance to officers upon passing for the first time from dismounted to mounted status;
  —— *de première mise d'équipement,* outfit allowance made in the case of certain promotions (as to n. c. o., who are appointed *adjudants* or officers);
  —— *en rassemblement,* daily allowance to soldiers of all grades stationed in camps or in certain designated fortified positions; to officers who accompany their troops on certain duties (target practice, preparation or organization of ranges, polygons); to soldiers on exceptional occasions of assemblage of troops where provisions are dear;
  —— *en remplacement de vivres,* commutation of rations (in France in time of peace) in certain designated cases, e. g., when troops can easily buy the victuals;
  —— *de rengagement,* (*in gen.*) reenlistment bounty;
  —— *représentative,* (*in gen.*) commutation;
  —— *représentative de fourrage,* (*in gen.*) forage commutation;
  —— *pour résidence en Algérie (Tunisie),* special allowance to officers and military employees ranking as n. c. o. in garrison in Algiers (Tunis);
  —— *pour résidence dans Paris,* special allowance to officers, employees ranking as officers, or as n. c. o. and to n. c. o. *rengagés* or *commissionés* serving within the enceinte of Paris;
  —— *de responsabilité,* allowance to bonded officers of administration responsible for property or money;
  —— *de route,* (*in gen.*) travel allowance, mileage;
  —— *de séjour,* allowance for residence in certain places under certain conditions;
  —— *spéciale,* allowance substituted for pay and rations in case of reservists, territorialists, etc., for the day of their arrival only;
  —— *de table,* (*in gen.*) mess allowance;
  —— *de transport,* (*in gen.*) transportation allowance; travel allowance.
indépendance, f., independence;
  —— *de l'escarpe et du parapet,* (*fort.*) the principle of the independence of the scarp and parapet.
indéréglable, a., that will not get out of adjustment.
Indes, f. pl., the Indes;
  les Grandes ——, the East Indies.
index, m., index, first finger; needle;
  plaque à ——, index plate.

indicateur, m., indicator, index, index plate; (*steam*) steam engine indicator, gauge; (*r. r.*) guide, travelers' guide;
—— *d'aiguille*, (*r. r.*) switch signal;
—— *de chargement*, (*art.*) mark on rammer to indicate when projectile is home;
—— *de chemise*, (*steam*) jacket gauge;
courbe de l' ——, (*steam*) indicator diagram;
—— *de déclivité*, (*r. r.*) gradient post;
diagramme de l' ——, v. *courbe de l'* ——;
—— *de direction*, v. —— *d'aiguille;* (*elec.*) indication of the direction of a current;
—— *de distance*, (*art.*) range finder;
—— *de marche*, (*elec.*) meter;
—— *de niveau*, floating gauge;
—— *de niveau d'eau*, water gauge, water mark;
—— *de niveau d'eau électrique*, electric water gauge;
—— *de niveau à flotteur*, float gauge;
—— *de phase*, (*elec.*) phase indicator;
—— *de pointage*, (*art.*) index (as of a graduated elevation arc);
—— *de position*, (*art.*) position finder;
—— *de pression*, (*steam*) pressure gauge; steam gauge; water gauge;
—— *de pression de vapeur*, (*steam*) steam-pressure gauge;
—— *de profondeur*, (*torp.*) depth indicator;
—— *de recul*, (*Fr. art.*) sort of stud on the chase of the French 120ᵐᵐ C., to indicate when the gun has recoiled to the limit in trunnion sleeve;
—— *repère de pointage*, v. —— *de pointage;*
—— *-robinet*, (*steam*) gauge cock;
—— *de terre*, (*elec.*) ground indicator;
—— *-tube*, (*steam*) gauge glass; water gauge;
—— *du vide*, (*steam*) vacuum gauge;
—— *de la vitesse*, (*mach.*) speed indicator.
indication, f., indication;
—— *de service*, (*mil. sig.*) conventional signals.
indice, m., index; indication; sign; token; proof; probable evidence;
—— *de défilement*, (*fort*), tangent of the angle made with the horizontal by the line of intersection of the plane of defilade with the plane of the profile of the parapet;
—— *d'hydraulicité*, of hydraulic lime, the ratio between the quantities of lime and of siliceous, etc., matter respectively;
—— *de la pression de vapeur*, (*steam*) steam-pressure gauge;
—— *de protection*, (*fort.*) tangent of the angle made by the *ligne couvrante*, q. v., with the horizontal;
—— *de réfraction*, (*opt.*) index of refraction;
—— *de résistance balistique*, (*ball.*) ballistic coefficient.
indigène, a., m., native; a native;
affaires ——s, v. s. v. *affaire;*
corps ——, v. s. v. *corps*.
indignité, f., unworthiness;
—— *de servir*, (*mil.*) unworthiness to serve.
indigo, m., indigo.
indisciplinable, a., unruly; undisciplinable; ungovernable.
indiscipline, f., (*mil.*) violation and transgression of the laws of discipline; lack of discipline; state of being undisciplined.
indiscipliné, a., (*mil.*) undisciplined; refractory to discipline; m., soldier who refuses to submit to discipline.
indisponibilité, f., v. *non-disponibilité*, s. v. *disponibilité*.
indisponible, a., (*mil.*) not available; m., soldier momentarily not available for service.
in-dix-huit, m., 18mo., eighteen-mo.
indomptable, a., indomitable, untamable, unmanageable.
indompté, a., untamed, wild, ungoverned, unvanquished; (of a horse) unbroken, wild.
in-douze, m., 12mo., duodecimo.
inducteur, a., (*elec.*) inducing, primary; m., induction coil; primary coil; field magnet and coil;
—— *différentiel*, differential induction coil.
induction, f., induction, implication; (*elec.*) induction;

induction *dynamique*, dynamic induction, mutual induction;
—— *électro-dynamique*, electro-dynamic induction;
—— *électrostatique*, electro static induction;
—— *magnétique*, magnetic induction;
—— *magnéto-électrique*, magneto-electric induction;
—— *propre*, self-induction;
self- ——, v. —— *propre;*
statique, v. —— *électrostatique*.
inductomètre, m., (*elec.*) inductometer.
induisant, a., (*elec.*) inducing.
induit, p. p., (*elec.*) induced; secondary.
induit, m., (*elec.*) armature of a dynamo;
—— *à anneau*, ring armature;
—— *à disque*, disk armature;
enroulement d' ——, armature coils;
noyau d' ——, armature core;
—— *à tambour*, drum armature.
indurite, f. (*expl.*) indurite.
industrie, f., industry; skill; trade; arts and manufactures, especially private manufacture as opposed to government;
—— *agricole*, farming, agriculture;
—— *commerciale*, commerce, trade;
—— *d'art*, manufacturing;
—— *manufacturière*, manufacturing;
—— *privée*, private manufacture.
industriel, m., manufacturer; a., industrial, pertaining to the arts and manufactures.
inébranlable, a., unmoved, unshaken; resolute, steady (of troops, etc.).
inégal, a., unequal; (*top.*) rough, uneven.
inégalité, f., inequality, roughness, ruggedness; (*top.*) roughness, unevenness of the ground.
inélastique, a., inelastic.
inertie, f., inertia.
inétirable, a., (*met.*) that may not be drawn down or out.
inexécution, f., lack of execution; (*mil.*) failure to carry out orders.
inexercé, a., untrained, unexercised (as troops).
inexpérimenté, a., not subjected to experiment; untested.
inexploitable, a., that can not be worked or developed.
inexploité, a., undeveloped (of natural resources, etc.).
inexplosible, a., nonexplosive, unexplosive.
inexplosif, a., unexplosive.
inexpugnable, a., (*mil.*) impossible of capture by force of arms; inexpugnable, impregnable.
infamant, a., bringing on disgrace, or degrading; *peine* —— *e*, degrading punishment involving loss of certain rights and privileges.
infanterie, f., (*mil.*) infantry;
—— *à cheval*, mounted infantry;
—— *légère*, light infantry;
—— *de ligne*, infantry of the line;
—— *de marine*, marine infantry; in France, not marines, but infantry (in the strict sense of the word), under the orders of the minister of marine, and serving in the ports of France and in the colonies;
—— *montée*, mounted infantry.
infect, infectant, a., infectious.
infecter, v. a. r., to infect, to become infected.
infectieux, a., infectious.
infection, f., infection.
inférieur, a., inferior; of lower rank, lower; lower (i. e., near the sea).
infester, v. a., to infest; (*mil.*) to overrun, ravage, pillage, plunder, etc.
infidélité, f., disloyalty, dishonesty in service or administration.
infiltration, f., infiltration.
infiltrer, v. a. r., to infiltrate.
infirmatif, a., (*law*) reversing, annulling, rendering null and void.
infirmation, f., (*law*) act of reversing, annulling.
infirmer, v. a., to reverse, invalidate, annul, repeal, revoke; (*law*) to reverse the verdict of a lower court.

infirmerie, f., infirmary; hospital for light diseases or slight wounds;
— *d'étapes,* v. s. v., *étape;*
— *de fort,* (*Fr. a.*) sort of hospital established in each besieged or invested fort;
— *-gare, de gare,* (*Fr. a.*) hospital established during war in railroad station for the treatment of sick and wounded in transit, and of those too sick to continue the journey;
— *de garnison,* (*Fr. a.*) same as —— *régimentaire,* but organized in garrisons composed of detachments of different regiments;
— *de gîtes d'étapes,* v. s. v. *étape;*
— *-hôpital,* (*Fr. a.*) a regimental infirmary receiving the sick for treatment (established in garrison towns not having a regular military hospital, and some distance from regular military or civil hospital);
— *régimentaire,* (*Fr. a.*) regimental infirmary;
— *vétérinaire,* horse infirmary.

infirmier, m., hospital attendant; private of the hospital corps (U. S. A.);
— *commis aux écritures,* (*Fr. a.*) (formerly) hospital clerk (had no medical instruction or duties);
— *d'exploitation,* (*Fr. a.*) hospital attendant who does heavy work (loading ambulances, litters, etc., setting up tents, care of field material, etc.);
— *-major,* (*Fr. a.*) sergeant or corporal in charge of a *division de malades,* q. v. (responsible for ventilation, heating, good order, police, etc.);
— *-mécanicien,* (*Fr. a.*) hospital attendant in charge of the engines of the establishment, hospital engineer;
— *militaire,* (*Fr. a.*) hospital attendant belonging to the *service de santé* (i. e., not detailed from the troops);
— *régimentaire,* (*Fr. a.*) soldier detailed for duty in the *infirmerie régimentaire;*
— *de visite,* (*Fr. a.*) hospital orderly in direct attendance upon the sick (dresses wounds, gives medicine, distributes food, etc.).

infixer, v. a., to fix or set in (rare).
inflammabilité, f., inflammability.
inflammable, a., inflammable.
inflammateur, m., (*elec.*) firing key; (*torp., expl.*), igniter;
— *à mouvement d'horlogerie,* clockwork igniter.
inflammation, f., inflammation (esp. of a powder charge); spread of the flames over the surface of grains;
— *centrale,* (*sm. a.*) center-fire (system);
— *électrique,* (*expl., art., etc.*) firing or discharging by electricity;
— *par percussion,* (*sm. a.*) percussion fire;
— *périphérique,* (*sm. a.*), rim-fire (system);
retard d' ——, (*art.*) hang fire;
— *spontanée,* spontaneous inflammation.
infléchir, v. a., to inflect, to bend, to elbow.
inflexible, a., inflexible, rigid.
inflexion, f., inflexion, (*art.*) whipping, thrashing (of the chase of a gun); jumping of the gun when fired; droop.
infliction, f., infliction.
infliger, v. a., to inflict.
influence, f., influence; (*expl.*) sympathy;
détonation par ——, (*expl.*) sympathetic detonation, detonation by influence.
in-folio, m., folio.
information, f., investigation, inquiry; (*in pl., mil.*) intelligence, news, information in the widest possible sense; (*mil. law*) preliminary investigation of a case (interrogatory of the accused and of witnesses and drafting of a report setting forth the facts, with all the circumstances; corresponds to investigation by the judge-advocate, U. S. A., only much more formal and elaborate); report of an investigation;
service d' ——*s,* (*mil.*) a generic term for all means of transmitting information, orders, intelligence, etc.
informé, m., investigation.

informer, v. a. n., to inform, make known, acquaint; (*law*) to investigate a case, inquire into an accusation or charges.
infortifiable, a., (*fort.*) not susceptible of fortification.
infracteur, m., infringer.
infraction, f., infraction, breach, violation.
infrastructure, f., (*r. r.*) roadbed; (more strictly) masonry work and earthwork; (*pont.*) structure under the roadway of a bridge.
infrayé, a., not opened (of a road).
infusibilité, f., infusibility.
infusible, a., infusible.
infusoires, m. pl., infusoria.
ingénieur, m., engineer; (*mil.*) military engineer; (*Fr. nav.*) naval constructor (two classes of, assimilated to captain and commander, respectively, U. S. N.);
— *-artilleur,* artillery engineer;
— *de campagne,* field engineer;
— *en chef,* chief engineer;
— *des chemins de fer,* railroad engineer;
— *civil,* civil engineer;
— *-consultant,* consulting engineer;
— *-constructeur,* (*nav.*) shipbuilder;
— *des constructions navales,* (*nav.*) naval constructor;
— *du corps du génie maritime,* (*nav.*) naval constructor;
— *de cuirassement,* armor engineer;
— *du génie maritime,* (*nav.*) naval constructor;
— *géographe,* surveyor, topographical engineer;
— *hydrographe,* hydrographer; (*Fr. nav.*) hydrographic engineer;
— *hydrographe de 1re (2me) classe,* (*Fr. nav.*) grades assimilated to captain and to commander, respectively, U. S. N.;
— *-inspecteur,* superintending engineer;
— *de la marine,* —— *maritime,* naval constructor;
— *mécanicien,* mechanical engineer;
— *militaire,* (*mil.*) military engineer (a current, not an official, expression);
— *des mines,* mining engineer;
— *opticien,* maker of optical instruments;
— *de place,* garrison engineer;
— *des ponts et chaussées,* (French) government engineer (for bridges and roads);
— *des poudres et salpêtres,* (French) engineer for manufacture of explosives;
— *de routes,* road engineer;
sous- ——, assistant engineer; (*Fr. nav.*) assistant naval constructor (three classes of);
sous- —— *hydrographe,* (*Fr. nav.*) assistant hydrographer (three classes of);
— *-torpilleur,* torpedo engineer.
ingrédient, m., ingredient;
— *de propreté,* (*Fr. a.*) cleaning material (soap, wax, polish, etc.).
inguéable, a., unfordable.
inhumation, f., burial.
inhumer, v. a., to bury.
ininflammabilité, f., uninflammability.
ininflammable, a., uninflammable.
ininflammation, f., failure to inflame.
initial, a., initial.
point ——, (*mil.*) initial point of a march (inaccurate term);
vitesse ——*e,* (*ball.*) muzzle velocity.
initiative, f., (*mil.*) initiative (bold, offensive spirit);
esprit d' ——, spirit of initiative.
injecter, v. a., to inject.
injecteur, m., (*steam*) injector;
— *aspirant,* suction injector;
— *automoteur,* self-adjusting injector;
— *Giffard,* Giffard injector;
— *universel,* universal injector.
injection, f., impregnation of wood; (*steam*) injection;
eau d' ——, (*steam*) injection water;
levier, robinet, tiroir, tuyau, d' ——, (*steam*) injection lever, cock, slide, pipe.
injonction, f., injunction; formal order, imperative command.
innavigabilité, f., state of being unnavigable, (*nav.*) unseaworthiness.

innavigable, a., (of rivers, etc.) not navigable; (*nav.*) unseaworthy.
in-octavo, m., 8vo, octavo.
inoffensif, a., inoffensive; (of mechanisms) out of gear, out of action.
inondation, f., inundation, overflow; (*fort.*) inundation (accessory defense);
—— *inférieure*, (*fort.*) downstream inundation (below town or fort);
—— *latérale*, —— *moyenne*, (*fort.*) an inundation opposite the town or fort;
—— *supérieure*, (*fort.*) upstream inundation (above town or fort);
tendre une ——, (*fort.*) to prepare an inundation.
inonder, v. a., to inundate, overflow.
in-plano, m., full sheet.
in-quarto, m., 4to, quarto.
inquiet, a., restless;
cheval ——, (*hipp.*) restless, skittish horse.
inquiéter, v. a., (*mil.*) to annoy, harass, impede the enemy (a march, the passage of a river).
insaisissable, a., (*law*) not liable to seizure or levy.
inscriptible, a., inscribable.
inscription, f., inscription; notice (as on a signboard); (*mil. adm.*) registry, enrollment;
livre d' ——, register, record book;
—— *maritime,* in France, registry of all persons living by the exercise of some maritime employment.
inscrire, v. a., to register, inscribe, record; (*mil.*) to take up.
inscrit, m., in France, any person forming part of the *inscription maritime*, q. v.;
—— *définitif*, any man belonging to the crew of a merchant vessel and having 18 months' sea experience (must be at least 18 years old);
—— *hors de service*, a sailor of 50, or any sailor discharged as physically unfit;
—— *provisoire*, any minor that can read and write and that intends to follow the sea.
in-seize, m., 16mo., sixteenmo.
insensibilisateur, a., (*expl.*) desensitizing.
insensibiliser, v. a., (*expl.*) to desensitize.
insensible, a., not sensitive;
bouche ——, (*man.*) hard mouth.
insensitif, a., insensitive.
insérer, v. a. r., to insert, be inserted; (*elec.*) to put in circuit.
insertion, f., insertion;
—— *fixe*, (*hipp.*) origin of a muscle;
—— *mobile*, (*hipp.*) insertion of a muscle.
insigne, m., (found generally in the plural) insignia; marks of rank;
——*s de grade*, (*unif.*) insignia of grade;
——*s de grade des sous-officiers hospitalisés*, (*Fr. a.*) insignia worn by n. c. o. and corporals that have been in hospital;
—— *de pointeur*, (*Fr. art.*) grenade worn by the best gun layer of the battery for the year;
—— *de service*, (*mil.*) any badge or mark indicating that the wearer is on duty; (*Fr. a.*) chinstrap down;
—— *de tir*, (*Fr. a.*) insignia (pins or small hunting horns) worn on the left sleeve by the best shots.
insolation, f., sunstroke, heatstroke; (*phot.*) exposure to the sun.
insoler, v. a., to expose to the sun (as hay, a blue print, etc.).
insolubilité, f., insolubility.
insoluble, a., insoluble.
insoudable, a., (*Met.*) unweldable.
insoumis, (*Fr. a.*) recruit that fails to join within the assigned time.
insoumission, f., (*Fr. a.*) failure to join, of a man that has never served before.
inspecter, v. a., to inspect, survey.
inspecteur, m., inspector; surveyor; (*mil.*) inspector; (*Fr. nav.*) inspector (official of the *commissariat,* q. v., n. e. U. S. N.);
—— *adjoint*, (*Fr. nav.*) assistant inspector (grade);

inspecteur *d'armée*, (*Fr. a.*) general of division, member of the *conseil supérieur de la guerre;*
—— *d'armes*, (*Fr. a.*) in each corps, a captain of artillery (sometimes more than one) charged with the yearly inspection of arms in the hands of the troops of the corps region;
—— *des cartoucheries*, (*Fr. a.*) general of brigade, coming from the artillery, inspector of cartridge factories (suppressed in 1897);
—— *en chef de 1re (2me) classe*, (*Fr. nav.*) grades of the *commissariat* (n. e. U. S. N.);
—— *de la défense (d'un groupe)*, (*Fr. a.*) general officer detailed, in peace, to prepare the defense of a group of forts or of fortifications;
—— *des études*, at the Polytechnique, title borne by a captain (of artillery, marine artillery, or of engineers) in charge of drill, discipline, etc. (nearest equivalent, tactical officer, U. S. M. A.);
—— *des forges*, (*Fr. a.*) steel inspector, government inspector at a gun foundry;
—— *général*, (*Fr. a.*) general officer detailed by the minister of war to inspect a regiment, a *service*, an establishment, etc.; provost-marshal-general; (*Fr. nav.*) senior grade of the engineer, construction, and medical departments, respectively;
—— *général permanent de cavalerie*, (*Fr. cav.*) general officer, permanent cavalry inspector (v. s. v. *inspection*);
—— *général permanent des fabrications de l'artillerie*, (*Fr. art.*) general officer, head of this inspection (v. s. v. *inspection*);
—— *général permanent des travaux de l'artillerie pour l'armement des côtes*, (*Fr. art.*) general officer, member of the technical artillery committee, specially charged with this duty;
—— *de manufactures d'armes*, (*Fr. a.*) inspector of small-arm manufactures, member of *section technique* of the artillery (abolished);
médecin——, (*Fr. a.*) medical inspector;
—— *technique*, (*Fr. a.*) an inspector of the special services.
inspection, f., inspection, survey, examination; (*mil.*) inspection; personnel of an inspecting department; charge, functions of an inspector;
—— *arme*, (*sm. a.*) command for inspection of arms
—— *d'armée*, (*Fr. a.*) inspection made by an *inspecteur d'armée;*
—— *d'armes*, (*Fr. a.*) inspection of arms in the hands of troops (made by *inspecteur d'armes*);
faire l' —— *de*, to inspect, examine;
faire une ——, to make a survey, to inspect;
—— *des forges*, (*mil.*) steel inspection duty;
—— *avec la garde*, (*mil.*) (punishment for privates) parading with the guard;
—— *générale*, (*Fr. a.*) inspection made by an *inspecteur général* (or *inspecteur d'armée*) for the information of the minister of war;
—— *générale de l'artillerie des armées*, (*Fr. a.*) general artillery staff of a group of armies (similar staffs exist for the engineer and the administrative services);
—— *générale permanente de cavalerie*, (*Fr. cav.*) the permanent inspection personnel of the French cavalry;
—— *générale des prévôtés des armées*, (*Fr. a.*) the provost *service* at the headquarters of a group of armies;
passer à l' ——, to undergo inspection;
passer une ——, to make an inspection;
—— *permanente des fabrications de l'artillerie*, (*Fr. art.*) inspection department in charge of all technical and administrative matters relating to the manufacturing establishments of the artillery.
inspiration, f., insufflation.
instabilité, f., instability.
instable, a., unstable;
acier ——, (*met.*) steel that loses its properties;

instable, *combinaison* ——, (*chem.*) unstable compound;
équilibre ——, unstable equilibrium.
installation, f., installation (of guns on board, in a work, of troops in a bivouac, cantonment); (*tech.*) plant.
installer, v. a., to install; to mount (as guns).
instance, f., (*law*) instance;
première ——, first instance.
instantané, a., instantaneous; m., instantaneous photograph;
photographie ——*e*, instantaneous photography.
instituteur, m., teacher, tutor, instructor.
instructeur, m., teacher, instructor; (*mil.*) drill instructor, drillmaster; (*man.*) riding master;
capitaine- ——, (*mil.*) captain supervising drill;
caporal- ——, *sergent-* ——, (*mil.*) drill corporal or sergeant;
officier- ——, v. s. v. *officier;*
—— *de tir*, (*mil.*) musketry instructor.
instruction, f., instruction, education; circular; rules, directions, for anything; (*mil.*) instruction, training, (more narrowly) drill; (*law*) preliminary trial or hearing; inquiry, examination (synonym of *information*); (*Fr. a.*) the instruction or order for a march sent by the general in chief to corps or division commanders;
camp d' ——, v. s. v. *camp;*
—— *sur la conduite de voitures*, (*art.*) driving drill;
—— *des demandes de pension*, (*adm.*) all the documents necessary to obtain a pension;
division d' ——, (*Fr. a.*) a division of the École de Fontainebleau, composed of lieutenants detached for one year from their regiments to be trained in riding and as drillmasters;
—— *équestre*, riding instruction;
—— *judiciaire*, v. *information;*
juge d' ——, (*law*) examining magistrate;
lieutenant d' ——, (*Fr. a.*) at Fontainebleau, a lieutenant belonging to the *division d'* ——, q. v.; at Saumur, a lieutenant of cavalry in training for the duties of *capitaine-instructeur;* a lieutenant of artillery in training for the duties of riding instructor and taking a special course in cavalry tactics;
—— *militaire préparatoire*, in France, military instruction of youths of 17 to 20 years by societies organized for the purpose;
—— *ministérielle*, (*mil.*) circular, war department circular;
peloton d' ——, v. s. v. *peloton;*
—— *des recrues*, recruit drill, etc.;
—— *du tir*, target practice.
instruire, v. a., to instruct, teach, explain; (*mil.*) to train, drill; (*law*) to bring up a case or arrange a case for trial;
—— *un cheval*, (*man.*) to train a horse;
—— *une pension*, (*adm.*) to present in due form an application for a pension.
instrument, m., instrument, musical instrument; tool, engine, implement; (*law*) instrument, deed, contract;
—— *affilé*, v. —— *tranchant;*
—— *à anche*, (*mus.*) reed instrument;
—— *d'arpenteur*, surveying instrument;
—— *à bec*, (*mus.*) instrument fitted with a mouthpiece;
——*s de chauffe*, firing tools;
—— *à cordes*, (*mus.*) stringed instrument;
—— *coupant*, v. —— *tranchant;*
—— *de levés*, surveying instrument;
—— *de mathématique*, mathematical instrument;
—— *à mèche*, (*mus.*) reed instrument;
—— *de percussion*, (*mus.*) percussive instrument;
—— *de pointage*, (*art.*) any aiming instrument or device;
—— *de précision*, mathematical instrument;
—— *à réflexion*, reflecting instrument;
—— *tranchant*, edge tool;
—— *à vent*, (*mus.*) wind instrument;
—— *vérificateur*, —— *de vérification*, (*art.*) inspecting, verifying instrument, gauge;
—— *de visite*, (*art.*) inspecting instrument.

instrumentaire, a., *témoin* ——, (*law*) witness to a document.
instrumenter, v. a., (*law*) to draw up legal papers, documents, deeds, contracts, etc.
instrumentiste, m., (*mil.*) band musician.
insubmersibilité, f., (of boats) state of being unsinkable.
insubmersible, a., (of boats) unsinkable.
insubordination, f., insubordination.
insufflation, f., insufflation; (*met.*) blast.
insuffisance, f., insufficiency, lack;
—— *de numéros*, (*Fr. a.*) in drawing by lot, excess of young men over the numbers contained in the urn.
insultable, a., (*mil.*) that may be attacked.
insulte, f., insult, affront; (*mil.*) sudden attack; coup de main; assault, open assault;
hors d' ——, beyond reach of a coup de main;
mettre hors d' ——, (*fort.*) to secure from a coup de main;
—— *à une sentinelle*, interference with, assault on, a sentinel.
insulter, v. a., (*mil.*) to surprise (a place or position).
insurgé, m., insurgent.
insurgence, f., riot, insurrection.
insurger, v. a. r., to excite, cause to rise; to revolt, mutiny.
insurrection, f., rising, revolt;
—— *à main armée*, armed insurrection.
intact, a., untouched, inviolate, unharmed.
intégrateur, m., integrator.
intégromètre, m., (*inst.*) integrometer.
intelligence, f., intelligence; collusion, understanding with, relations with;
avoir des ——*s*, to have an understanding, relations with;
avoir une double ——, to have an understanding, relations, with both sides;
nouer des ——*s*, to establish relations (as with the enemy, with the besiegers, etc.).
intempéries, f. pl., vicissitudes of the weather.
intenable, a., (*mil.*) untenable.
intendance, f., direction, administration of affairs; (*Fr. a.*) one of the branches (*services*) of the administration of the army.
The *intendance* service comprises (1) the services of pay, military subsistence, clothing, camp equipage, cavalry (horse) equipment, travel and transportation, military bedding, and the approval of all disbursements connected with these services; (2) the approval of the outlays made by the troop units and establishments, the verification of issues in cash and *matériel* made from the military chests or from the magazines of these units or establishments; (3) the approval and the verification of the disbursements of the recruiting service and of the service of military justice; and (4) finally the administration of such personnels as are not attached to troop units and of all detached persons (*isolés*) who draw pay, salary, or gratifications (n. e. U. S. A.; army service corps, very nearly, England);
adjoint à l' ——, lowest grade of the *intendance* (rank of captain);
adjoint à l' —— *du cadre actif*, an *adjoint* on the active or permanent list;
adjoint à l' —— *du cadre auxiliaire*, an *adjoint* that comes into the corps on mobilization (belongs to the *cadre auxiliaire*);
attaché de l' ——, special grade (rank of lieutenant or of sublieutenant) of the *cadre. auxiliaire de l'* —— (is not counted a *fonctionnaire*);
cadre actif de l' ——, the permanent list of the corps;
cadre auxiliaire de l' ——, complementary list of the corps formed on mobilization;
corps de l' —— *militaire*, official designation of the personnel of the *intendance;*
fonctionnaire de l' ——, v. s. v. *fonctionnaire;*
—— *militaire*, full designation of the *intendance;*
officier d'administration du service de l' ——, v. s. v. *officier;*

intendance, *service de l'*——, v. under main word;
stagiaire de l'——, an officer, who, after passing the first competitive examination for entrance into the *intendance,* is taking the course in administration at Paris, preparatory to final entry into the corps.

Intendant, m., manager, director; (*Fr. a.*) intendant, officer of the *intendance,* q. v.;
—— *de l'armée,* the —— *général,* or —— *militaire,* at the headquarters of an army, when formed;
—— *du corps d'armée,* the —— *militaire* at corps headquarters;
—— *général,* the highest grade of the *intendance* (ranks with general of division);
—— *militaire,* second grade of the *intendance* (ranks with general of brigade);
sous- —— *militaire,* subintendant (three classes, ranking with colonel, lieutenant-colonel, and major, respectively).

Intensité, f., intensity;
—— *d'aimantation,* (*elec.*) intensity of magnetization;
—— *du champ,* (*elec.*) intensity of field;
—— *du courant,* (*elec.*) current strength;
—— *du tirage,* (*steam*) intensity of draft.

Intenter, v. a., —— *un procès,* (*law*) to bring, enter, an action against.

inter-ars, m., (*hipp.*) part of chest between the *ars* and the forearm.

intercalaire, a., intercalary.

intercalation, f., intercalation.

intercalement, m., (*mil.*) the putting of an increased number of men on the same line.

intercaler, v. a., to intercalate; (*elec.*) to cut in, to insert, to connect with a circuit;
—— *une courbe horizontale entre deux autres,* to find a point or points having a given reference.

intercardinal, a., intercardinal, as NE., NW., etc.

intercepter, v. a., to intercept; to cut off (as a convoy); to close (as communication); (*art.*) to take up (recoil); (*steam*) to cut off steam.

interception, f., interception; (*art.*) taking up of recoil.

interchangeable, a., interchangeable.

interdiction, f., interdiction, prohibition;
—— *de commerce,* prohibition of trade (with a country at war with the prohibiting country);
—— *des droits,* loss of rights (civil, personal, etc.);
—— *légale,* loss of rights, flowing from a legal condemnation (as to the penitentiary, etc.).

interdire, v. a., to prohibit, interdict.

Intéressé, m., interested party.

interférence, f., interference.

intérieur, a., interior, inside, internal; (of measurements) in the clear; (*mil.*) interior, said of duties within the post or *enceinte,* as opposed to duties without; (*pont.*) near (said of the side next to first bay built, or to last dismantled); m., interior, inside; interior department; home (as opposed to abroad).
à l'——, at home (as opposed to abroad);
ministère de l'——, interior department;
ministre de l'——, secretary of the interior, home secretary;
service ——, (*mil.*) interior economy.

Intérim, m., interim; (*adm.*) the exercise of a function by a person not holding the corresponding grade;
commandement par ——, (*Fr. a.*) command held under special orders of the minister of war by an officer not definitely invested with it.

intérimaire, a., ad interim; m., person performing duties of a position *ad interim.*

intérimat, m., state of "interim."

interjeter, v. a., —— *appel,* (*law*) to appeal from a decision.

interligne, f., space between two printed or written lines.

interligner, v. a., to space lines (in writing).

interlinéer, a. v., to interline.

intermédiaire, a., intermediate, middle; m., medium, agent; (*mach.*) intermediate shaft;
arbre ——, (*mach.*) intermediate, transmitting, shaft;
par l'—— *de,* by means of.

intermittence, f., (*mach., etc.*) intermittence.

intermittent, a., (*mach., etc.*) intermittent.

interne, a., internal; m., (*med.*) house surgeon.

interné, m., (*mil.*) interned man.

internement, m., (*mil.*) state of being interned.

interner, v. a., (*mil.*) to intern.

interpolaire, a., (*elec.*) interpolar.

interpolateur, m., interpolator.

interpolation, f., interpolation.

interpoler, v. a., to interpolate.

interprète, m., interpreter;
—— *judiciaire,* (*mil.*) interpreter of a court-martial;
——*s militaires,* (*Fr. a.*) corps of interpreters serving in Algiers and in Tunis (two classes, ——*s titulaires,* Frenchmen by birth or naturalization; ——*s auxiliaires,* not necessarily Frenchmen);
—— *principal,* highest grade in the ——*s militaires;*
——*s de réserve,* (*Fr. a.*) special corps of interpreters attached in war to the various staffs and departments.

interrogation, f., examination of witnesses.

interrogatoire, m., (*law*) examination of accused parties, interrogatory; (*mil. law*) questions by any member of court, and answers thereto; (*Fr. a.*) preliminary examination of a prisoner by the judge-advocate;
—— *contradictoire,* cross-examination.
—— *préliminaire, et lecture de l'acte d'accusation,* arraignment;
subir un ——, to undergo an examination.

interroger, v. a., to interrogate, question, examine.

interrompre, v. a., to close, to cut off, to interrupt (as communications, steam, an electric current, etc.).

interrupteur, m., (*elec.*) interrupter, contact breaker, circuit breaker;
—— *automatique,* automatic interrupter;
—— *d'excitation,* an excitation interrupter;
—— *d'inertie,* an interrupter acting by inertia;
—— *à manette,* switch;
—— *à mercure,* mercury circuit breaker or interrupter.

interruption, f., interruption, obstacle, obstruction; breaking (of a current); break (in a railway track).

intersection, f., intersection; (*surv.*) foresight; method by observations of a point from the extremities of a base line;
—— *en arrière,* (*surv.*) backsight;
méthode d'——, (*surv.*) ordinary method of plane table surveying; in general, angle taking from the ends of a base line.

intervalle, m., interval; (*mil.*) interval; (*mil. min.*) bay; distance between corresponding faces of consecutive mine frames or between consecutive frames;
à ——*s,* bayed;
—— *de déploiement,* (*mil.*) deploying interval;
—— *d'éclatement,*(*art.*) horizontal distance from the point of burst to the nearest point of the target;
—— *de galerie,* bay of a gallery;
—— *ouvert,* (*mil.*) open, full, interval;
—— *de puits,* bay of a shaft;
ouverture d'——*s,* (*mil.*) opening of intervals;
ouvrir les ——*s,* (*mil.*) to open intervals;
—— *serré,* (*mil.*) close interval, reduced interval;
serrement d'——*s,* (*mil.*) closing of intervals;
serrer les ——*s,* (*mil.*) to close intervals;
—— *des torons,* (*cord.*) jaw.

intervention, f., intervention; (*mil.*) intervention, i. e., the entry into action of troops hitherto unengaged.

interversion, f., inversion; (*mil.*) inversion (at drill).

Intervertir, v. a., to invert.
Intestin, a., civil, in *guerre* ——*e*, civil war.
Intimer, v. a., (*law*) to assert or indicate with authority; to give legal notice.
Intonation, f., intonation;
école d'——, (*mil.*) voice drill in giving commands.
Intrados, m., (*cons.*) intrados.
Introduction, f., introduction; (*steam*) admission of steam;
fermer l'——, (*steam*) to shut off steam;
orifice d'——, (*steam*) admission port.
Invalide, a., invalid, disabled; m., invalid, pensioner; (*in pl.*) the Hôtel des Invalides, at Paris;
—— *externe*, out-pensioner;
hôtel des ——*s*, soldiers' home (U. S. A.); the Invalides at Paris; military or naval hospital for invalids.
Invalidité, f., (*law*) invalidity.
Invasion, f., (*mil.*) invasion;
guerre d'——, war of invasion.
Inventaire, m., inventory, return.
Inverseur, m., (*elec.*) commutator.
Inversion, f., (*mil.*) inversion; formation left in front.
Investigation, f., inquiry, investigation.
Investir, v. a., (*mil. fort.*) to invest, blockade, surround.
Investissement, m., (*siege*) investment, blockade;
corps d'——, investing body or army;
—— *définitif*, final or complete investment, when the line of investment is thoroughly established;
ligne d'——, line of investment;
—— *provisoire*, preliminary investment, to dislodge besieged from his advanced positions and to close his communications with the exterior.
Invincible, a., invincible.
Invitation, f., invitation;
—— *de feuille de route*, (*Fr. a.*) request for transportation.
Invulnérable, a., invulnerable;
poste des ——*s*, a synonym of *mont pagnote*, q. v.
Iode, m., (*chem.*) iodine.
Ions, m. pl., (*elec.*) ions.
Ipéca, *le père* ——, (*mil. slang*) regimental surgeon.
Irréductible, a., irreducible, impregnable.
Irrégulier, a., irregular; (*mil.*) irregular, said of troops not belonging to the regular line (practically obs.);
fortification ——*e*, (*fort.*) fortification that for any reason whatever follows no regular definite system.
Irrigation, f., irrigation.
Irriguer, v. a., to irrigate.
Irruption, f., inroad, irruption.
Isabelle, a., (*hipp.*) yellow dun, clay bank; m., a clay-bank colored horse;
cheval ——, clay-bank horse;
couleur ——, dun.
Ischion, m., ischium, hip bone.
Isobarométrique, a., isobaric.
Isocèle, a., isosceles.
Isochrone, a., isochronous.
Isodynamique, a., isodynamic;
ligne ——, isodynamic line.
Isolant, a., (*elec.*) insulating, nonconducting; m., insulator;
tabouret ——, insulating stool.
Isolateur, m., (*elec.*) insulator, nonconductor;
—— *-arrêt*, terminal insulator;
—— *-cloche*, petticoat insulator;
—— *à double arrêt*, shackle insulator; terminal insulator;

Isolateur *double cloche*, double petticoat insulator;
—— *-éjecteur*, (*sm. a.*) the ejector of the Russian model '91 (so called from its two parts);
—— *à paratonnerre*, lightning rod insulator;
—— *à suspension*, suspended insulator; hook insulator;
—— *-tendeur*, stretching insulator.
Isolation, f., (*elec.*) insulation.
Isolé, p. p., isolated, solitary, unconnected, disconnected; (*elec.*) insulated; (*mil.*) detached, separated from his regiment or garrison (of individuals); serving, acting, or operating alone—i. e., not in connection with other troops (said of troop units); carried on alone—i. e., without the help of other troops (said of a combat undertaken by a *bataillon* ——, etc.); (*sm. a.*) separate (said of a bayonet, to which there is no corresponding rifle); m., (*mil.*) man detached or separated from his regiment or garrison; soldier traveling alone; (*in pl.*) detached soldiers, in number insufficient to constitute a detachment responsible for its own administration (in war, clerks, secretaries, orderlies of officers not attached to troops, are classed as *isolés*);
bataillon, etc., ——, (*mil.*) a battalion, etc., operating alone, v. *encadré*;
corps ——, (*elec.*) insulated body;
dépôts des ——*s*, in the more important French ports, depots or establishments charged with the discipline, administration, and subsistence of men about to embark.
Isolement, m., isolation; (*elec.*) insulation; (*art.*) windage over studs (of a projectile) (obs.);
—— *des contagieux*, (*med.*) isolation of sick affected with contagious diseases;
—— *du projectile*, (*art.*) in studded projectiles, the difference between the length of the stud and the depth of the groove (obs.).
Isoler, v. a., to isolate, detach, cut off, separate; (*elec.*) to insulate.
Isoloir, m., (*elec.*) insulating stool.
Isomère, a., (*chem.*) isomeric.
Isomérie, f., **isomérisme**, m., (*chem.*) isomerism.
Isométrie, f., isometry.
Isométrique, a., isometric.
Isotherme, a., isothermal.
Isotrope, a., isotropic.
Isotropie, f., isotropy.
Issue, f., issue, egress, outlet, passage, vent, end, means, expedient; (*in pl.*) odds and ends, waste arising in the handling and preparation of food (*denrées*);
—— *s de la boucherie*, viscera and extremities of animals;
—— *s de la meunerie*, what is left after grinding flour (bran, etc.);
—— *s des ordinaires*, slops, swill.
Isthme, m., isthmus.
Isthmien, a., isthmian.
canal ——, isthmian canal.
Itague, f., (*cord.*) tie, pendant, runner;
—— *de culasse*, (*nav. art.*) breech runner (obs.);
—— *et palan, palan à* ——, runner and tackle;
—— *de retenue*, preventer rope;
—— *de volée*, (*nav. art.*) muzzle runner (obs.).
Italique, a., italic; f., italics.
Itinéraire, m., itinerary; route of march; (*r. r.*) time-table;
levé d'——, (*top.*) road sketch, road map.
Ivoire, m., ivory; dentine, ivory (of the teeth).
Ivre, a., drunk.
Ivresse, f., drunkenness.
Ivrogne, a., drunken; m., drunkard.
Ixomètre, m., instrument to investigate the viscosity of liquids.

J.

jachère, f., fallow land, ground.
jack, m., jack, union jack.
jaillir, v. n., to spout, gush, gush forth.
jaillissement, m., jet, gushing, spouting.
jaler, v. a., to stock (an anchor).
jaline, f., (*expl.*) jaline (a picrate powder).
jalon, m., stake, pole, picket, directing mark; (*surv.*) picket, sighting rod;
—— *à drapeau*, (*surv.*) flag;
—— *-mire*, (*surv.*) sighting rod;
—— *de repérage*, (*art.*) aiming picket (indirect fire).
jalonnage, m., staking out; marking out (a road, a line of march, etc.).
jalonnement, m., (*surv., mil., etc.*) staking out, marking out a line, a direction, etc.
jalonner, v. a. n., to stake out, to mark out with pins, pickets, men, etc.; (*surv.*) to stake out, (*mil.*) (of markers, guides) to mark, establish; the line, a direction;
—— *une direction*, (*mil.*) to take up points in marching;
—— *une ligne*, (*mil.*) to mark, establish, a line.
jalonneur, m., stake planter; (*mil.*) soldier who marks out the ground or takes up points, marker; man who connects the point with the main body of an advance guard and marks out the direction;
procédé des deux ——*s*, (*art.*) indirect aiming in which two men mark out the line of fire.
jalousie, f., grating; Venetian blind; (*nav.*) crankiness; (*mil.*) anxiety, nervousness (as in regard to one's communications);
à ——, (*mach., tech.*) like a grating, having openings like a grating.
jaloux, a., (*nav.*) crank, easily upset; (*mil.*) exposed to attack; perilous.
jambage, m., jamb; chimney jamb; side post; doorpost; upright of a window or door; side face of Λ-iron; leg (as of a steam-hammer frame); (*mas.*) sort of pier or pilaster in a wall, vertical course (as of dressed stone in a brick wall) to ornament or strengthen the wall or to support the main rafters;
——*s d'un tour*, (*mach.*) wooden base or supports of a lathe.
jambe, f., leg; leg from knee to foot; leg (of a compass); branch of a siphon; (*hipp.*) leg (from thigh to hock); (*in pl., fig.*) faculty or power of walking, of marching; (*sm. a.*) shank or lug of the Werder rifle for opening the breech;
—— *artificielle*, artificial leg;
avoir des ——*s*, (*fenc.*) to be firm and steady on the left foot;
avoir les ——*s près*, (*man.*) of the rider, to have the legs close to the flanks of the horse;
n'avoir point de ——*s*, (*hipp.*) to have the fore legs spoiled, ruined;
—— *bien dirigée*, (*hipp.*) well-set leg;
—— *de bois*, wooden leg;
bouder contre les ——, v. s. v. *bouder;*
—— *boutisse*, (*mas.*) vertical course, sort of built-up pilaster (as of dressed stone in a brick wall) bonded header-wise in the wall;
—— *de chèvre*, cheek, leg, of a gin;
—— *de chien*, (*cord.*) sheepshank;
cinquième —— *du cheval*, the bit, if ridden on, (cf. "to ride a horse with three legs");
—— *de dedans*, (*man.*) near leg (front or hind);
—— *de dehors*, (*man.*) off leg;
échapper la ——, (*fenc.*) to withdraw the leg (from a cut) in broadsword exercise;
—— *d'encoignure*, (*mas.*) corner (of quoins);
—— *étrière*, (*mas.*) intermediate pilaster or pillar in a wall;
—— *de force*, strut; (*nav.*) strong back;
—— *hors du montoir*, v. —— *de dehors;*
marcher sur trois ——*s*, (*hipp.*) to be lame;

jambe, *du montoir*, v. —— *de dedans;*
—— *d'ordon*, supporting frame of the trunnions of a hurst;
os de ——, shank;
—— *partie*, (*hipp.*) a leg with a strained tendon;
retenir la —— *de dedans (de dehors) du cheval*, (*man.*) to change the direction of such or such a leg by the action of the reins;
sensible à la ——, (*man.*) leg-wise;
sortir sur les ——*s d'un autre*, (*mil. slang*) to be confined to barracks or guard room;
—— *sous poutre*, (*mas.*) plate course; block or plate, vertical course (as of dressed stone), supporting a rafter or beam;
à toutes ——*s*, as fast as possible.
jambette, f., counter timber; jamb, upright prop; small strut; clasp-knife; (*man.*) high step, tread (of a horse).
jambière, f., gaiter, legging, (*art.*) leg-guard; (*mach.*) sleeve;
guêtre ——, (*unif. Fr. a.*) a high legging, worn by the zouaves and the *tirailleurs algériens*.
jambon, m., ham;
faire un ——, (*mil. slang*) to break one's rifle.
janissaire, m., janissary.
janite, f., (*expl.*) mixture of large-grained black powder and nitroglycerin.
janquette, f., tongue.
jante, f., felly, felloe (of a wheel, of a fifth wheel); face, rim (of a pulley, fly wheel, etc.).
janter, v. a., to rim a wheel, to put on the fellies.
jantière, f., felly plate.
jaquette, f., (*art.*) jacket (in the French marine artillery, a synonym of *corps de canon*); recoil sleeve;
—— *de renfort*, jacket; reenforce jacket;
—— *tourillons*, trunnion jacket (Brown gun);
—— *de volée*, chase hoop.
jard, m., very coarse gravel.
jarde, f., (*hipp.*) jarde (the curb of English writers).
jardin, m., garden;
—— *d'agrément*, pleasure garden;
—— *potager régimentaire*, (*mil.*) regimental vegetable garden; post garden (U. S. A.);
—— *du tambour*, (*nav.*) sponson.
jardon, m., (*hipp.*) a synonym of *jarde*, q. v.
jarret, m., protuberance, projection; long limb (of a tree) stripped of other branches and twigs; elbow formed by the junction of two pipes; (*harn.*) bow of a bit; (*hipp.*) hock;
——*s clos*, hocks too close together, cow hocks;
corde du ——, cord of the hock;
—— *coudé*, hock in which the tibio-tarsal angle is too much closed;
couper le ——, to hamstring;
creux du ——, hollow of the hock;
——*s crochus*, v. ——*s clos;*
—— *droit*, hock in which the tibio-tarsal angle is too wide open;
—— *empâté*, fleshy, full, or doughy hock;
—— *étranglé*, tied-in or strangled hock;
être sur les ——*s*, to be cow hocked;
—— *étroit*, slender, narrow hock;
—— *gras*, v. —— *empâté;*
se mettre sur les ——*s*, (of a horse) to throw himself upon his haunches;
——*s mous*, v. ——*s vacillants;*
—— *net*, neat, clean hock;
pli du ——, fold of the hock;
pointe du ——, point of the hock;
—— *sec*, lean or dry hock;
tendon du ——, hamstring;
——*s trop ouverts*, hocks too wide open;
——*s vacillants*, rotating hocks;
——*s vidés*, clean hocks.
jarreté, a., (*hipp.*) close-hammed.

jarretière   230   **jeu**

jarretière, f., garter; lower strap of a drum holder (around the leg); (*cord.*, *etc.*) lashing; (*art.*) small rope, sling, or strap used in mechanical maneuvers, transporting horses, etc.;
—— *accessoire d'embarquement*, sling or strap to keep poles of wagons up and to tie wheels together in loading on trains;
—— *double*, (*Fr. art.*) sling or strap 10 meters long, used in mechanical maneuvers;
—— *ferrée*, (*Fr. art.*) sling with iron fittings;
ordre de la ——, Order of the Garter.
—— *simple*, (*Fr. art.*) strap 4 meters long (80ᵐᵐ mountain batteries);

jas, m., anchor stock.

jat, m., anchor stock.

jauge, f., gauge, standard; foot rule; (*nav.*) burden, tonnage (of a vessel);
avoir la ——, *être de* ——, to be up to standard, to be of standard measure;
—— *de douane*, (*nav.*) registered tonnage;
—— *de fascine*, (*siege*) fascine gauge;
—— *à tréfiler*, wire gauge;
—— *de vapeur*, (*steam*) steam gauge;
—— *du vide*, (*steam*) vacuum gauge.

jaugeage, m., gauging, measurement, calibration; (*nav.*) tonnage, measurement of a ship.

jauger, v. a., to gauge, measure; (*nav.*) to measure for tonnage.

jaugeur, m., gauger.

jaune, a., yellow;
—— *de chrôme*, chrome yellow;
—— *d'ocre*, yellow ocher;
—— *paille*, (*met.*) straw yellow.

javert, m., (*hipp.*) quitter;
—— *du bourrelet*, v. *encorné*;
—— *cartilagineux*, cartilaginous quitter;
—— *cutané*, cutaneous quitter;
—— *encorné*, tumor under the wall of the hoof;
—— *de la fourchette*, swelling or tumor of the cushion, furuncle of the frog;
—— *simple*, v. —— *cutané*;
—— *tendineux*, tendinous quitter.

javeau, m., bank of mud, sand, in a river.

javelle, f., swath;
tomber en ——, to fall to pieces (as a cask).

jayet, m., jet; jetlike reflection of a horse's coat.

jeanjean, m., (*mil. slang*) recruit.

jectisses, pl. a., *pierres* ——, (*mas.*) stones that may be laid anywhere by hand;
terres ——, earth that has been removed from one place to another.

jet, m., throw, act of throwing, cast, toss, flinging; motion of any body impelled by a force; stream, jet; jettison; set, suit; (*fond.*) casting, pouring; waxen cylinders used to make runners; (*in pl.*) casting pit; (*artif.*) small rocket on the rim of a wheel to make it turn;
—— *d'acier*, steel casting;
amplitude du ——, (*ball.*) range of projection;
arme de ——, missile weapon;
—— *des bombes*, (*art.*) shell practice (obs.);
—— *de condensation*, (*mach.*) condensing jet;
—— *et contribution*, the loss due to jettison;
—— *de coulée*, (*fond.*) feeding head;
—— *d'eau*, fountain jet; weather rail or strip, water strip, rain strip (of a door or window);
faire le ——, to jettison;
—— *de feu*, (*artif.*) fire sheaf, light sheaf; fixed rocket, whose sparks resemble drops of rain in sunbeam;
—— *de flamme*, line or stream of flame;
—— *de fonte*, (*fond.*) deadhead;
force de ——, impetus, velocity (as of a projectile);
—— *d'un fossé*, the soil or earth thrown out of a ditch;
machine de ——, (*art.*, *sm. a.*) gun, rifle;
—— *à la mer*, jettison;
—— *à moule*, (*fond.*) runner, gate channel;
—— *des obus*, v. —— *des bombes*;
—— *de pelle*, throw of a spade;
—— *de pierre*, stone's throw;

jet, *premier* ——, first cast, first casting; rough sketch; (*steam*) priming;
du premier ——, at the first stroke, or cast;
—— *de sable*, sand blast;
d'un seul ——, at one stroke; (*fond.*) at a single casting;
talon de ——, heel of a runner;
—— *de terre*, earth bank;
—— *de vapeur*, steam jet.

jetage, m., (*hipp.*) discharge (of the nose), running at the nose;
—— *d'un pont*, bridging.

jetar, m., (*mil. slang*) prison.

jetée, f., jetty, pier, mole;
—— *-débarcadère*, jetty and unloading pier in one;
musoir de ——, jetty head.

jeter, v. a., to throw, cast, fling, toss; to lay (as a foundation); to cast (an anchor); (*fond.*) to cast; (*mil.*) to throw (arms, stores, reenforcements) into a town, a fort; to throw out (as a brigade, a skirmish line); (*art.*) to throw (as shells, etc.); (*hipp.*) to run or discharge at the nose;
—— *l'ancre*, to cast anchor;
—— *bas*, —— *à bas*, —— *en bas*, to throw down;
cheval qui jette sa gourme, foundered horse;
—— *l'étoupe*, (*nav.*) to spit, spew, oakum;
—— *les fondements de*, to lay the foundation of anything;
—— *en fonte*, (*fond.*) to cast;
—— *le loch*, (*nav.*) to heave the log;
—— *à la mer*, to throw overboard;
—— *en moule*, (*fond.*) to mold, to form, for the casting;
—— *le plomb*, (*nav.*) to heave the lead;
—— *un pont sur*, (*pont.*) to throw a bridge (over a river), to bridge;
—— *la sonde*, v. —— *le plomb*.

jette-feu, m., movable grate, dumping grate (of a locomotive).

jette-sable, m., sand box (of a locomotive).

jeu, m., play, game; gaming, gambling; set (as of tools); (*mach.*, *etc.*) play, clearance; working; backlash; (*min.*) springing of a mine; (*art.*) clearance of a stud (in studded projectiles (obs.)); (*fenc.*) manner of handling the foil; (*cord.*) slack of a rope;
—— *d'accessoires* (*d'armes*), (*sm. a.*) cleaning, mounting and dismounting outfit or kit, for rifle;
—— *des arbres*, (*mach.*) end play;
avoir du ——, to have play, to work loose, to be loose;
—— *x de bois*, generic term for dominoes, checkers, and other similar games;
—— *du citron*, lemon cutting;
—— *composé*, (*fenc.*) complicated play, i. e., employing all possible combinations to deceive the adversary;
donner du ——, to give play, leave loose;
—— *dur*, (*fenc.*) stiff action;
en ——, in gear, working; at work, in action;
—— *d'engrenage*, (*mach.*) backlash;
—— *de ferrures*, set of iron fittings;
—— *de la guerre*, (*mil.*) war game;
—— *de la guerre navale*, naval war game;
—— *de hasard*, gambling game;
laisser du ——, v. *donner du* ——;
—— *de marques*, stamping, marking outfit; stencils;
mettre en ——, to set in motion, to actuate;
—— *d'une mine*, (*min.*) springing of a mine;
mis en ——, actuated;
—— *perdu*, (*mach.*, *etc.*) lost motion;
—— *des pièces*, (*mach.*, *etc.*) play, working;
—— *du piquet*, —— *du piquet de tente*, tent pegging;
—— *du piston*, travel of piston, back-and-forth motion;
—— *de la pointe de l'épée*, (*fenc.*) play of the point;
—— *de poulies*, block (more than one sheave);
prendre du ——, (*mach.*, *etc.*) to work or become loose;
—— *de roues*, (*mach.*) change wheels, change gear of a lathe;
—— *de signaux*, set of signal flags.

joie, f., joy.
feu de ——, bonfire; feu de joie.

joindre, v. a. n., to join, to assemble, put together, (*mil.*) to effect a junction, to join, i. e., attack, the enemy;
— *à* ——, home, block and block, chock-a-block, two blocks;
— *en about*, to join end to end;
— *à chaud*, (*met.*) to weld;
— *à onglet*, to miter, splice;
— *la piste*, (*man.*) to go close to the wall of the riding hall;
— *en V*, — *en pied de biche*, to double splice.

joint, p. p., joined;
— *bout à bout*, joined butt and butt;
— *à chaud*, (*met.*) welded;
ci-——, herewith, hereunto annexed.

joint, m., joint, coupling, hose coupling, union; (*mach.*) link, joint, articulation; (*carp., mas.*) joint (strictly, surface of junction); in an arch, the lines of intersection of a voussoir with the intrados; (*r. r.*) end space between rails;
— *en about*, (*carp.*) butt joint;
— *anglais*, (*teleg.*) telegraphic joint, Britannia joint;
— *à l'anglaise*, (*mas.*) tuck-joint pointing;
— *appuyé*, (*r. r.*) joint supported on a cross-tie;
— *à articulations en croix*, (*mach.*) Hooke's joint;
— *articulé*, (*mach.*) knuckle joint;
assembler à ——*s plats*, (*carp.*) to make a butt joint;
— *d'assise*, (*mas.*) bed joint, coursing joint;
— *à bague*, (*mach.*) ring joint;
— *à boulet*, (*mach., etc.*) ball and socket joint;
— *bout à bout*, (*carp.*) square joint, butt joint, flush joint, jump joint;
— *en bout*, joint between adjacent rows of paving stones;
— *à bride*, (*mach.*) flange joint;
— *brisé*, (*mach.*) knuckle, universal, joint;
à ——*s brisés*, so as to break joints;
— *brut*, (*mas.*) undressed joint, or surface of junction;
— *à la Cardan*, (*mach.*) Hooke's joint;
— *carré*, v. —— *bout à bout*; (*mas.*) square joint;
cercle pour ——, joint ring;
— *de chaîne*, flat link;
— *ciré*, (*mas.*) in pointing, a flat joint;
— *à cisaillement double*, double-riveted lap joint;
— *à cisaillement simple*, single-riveted lap joint;
— *à clef*, (*mach.*) keyed joint;
— *à col de cygne*, (*mach.*) expanding joint;
— *à collet*, (*mach.*) flange joint;
— *compensateur*, (*steam, mach.*) expansion joint;
à ——*s contrariés*, so as to break joints;
coupé par le ——, jointed;
— *en coupe*, joints converging to a center;
— *de la couronne*, (*hipp.*) coffin joint;
— *creux*, (*mas.*) in pointing, a sloping joint to throw off water;
dégarnir un ——, (*mas.*) to scrape out a joint for pointing;
— *démaigri*, (*mas.*) chamfered joint;
— *dérobé*, hidden joint;
— *à double anneau*, (*mach.*) double ring joint;
— *de douelle*, (*mas.*) inside joint of an arch; soffit joint;
— *éclissé*, (*r. r.*) fish joint;
— *élastique*, (*mach.*) elastic joint;
— *espagnol*, (*teleg.*) bell hanger's joint;
— *par étranglement*, v. —— *espagnol*;
— *de face*, (*mas.*) face joint (of an arch);
faire le ——, (*fond.*) to level off and sprinkle the parting plane with facing sand;
— *en fer*, joint iron;
— *feuillé*, rabbet joint, joint of two stones to one-half thickness;
ficher les ——*s*, (*mas.*) to fill joints with mortar;
— *à franc-bord*, butt joint, flush joint;

joint, *garnir les* ——*s*, (*mas.*) to point;
— *glissant*, (*mach.*) expansion joint;
— *gras*, (*mas.*) obtuse angle joint; joint more open than a right angle;
— *de Hooke*, (*mach.*) Hooke's joint;
— *horizontal*, v. —— *d'assise*;
— *s incertains*, (*mas.*) irregular joints between stones of different sizes;
— *latéral*, (*mas.*) heading joint of an arch;
— *libre*, (*r. r.*) a joint not resting on a tie;
— *de lit*, v. —— *d'assise*;
— *à lunette*, telescope joint;
— *maigre*, (*mas.*) acute angle joint, joint less open than a right angle;
— *mâle et femelle*, pin and socket joint;
— *à manchon*, (*mach.*) sleeve joint;
— *montant*, (*mas.*) vertical joint; heading joint of an arch;
— *à onglet*, diagonal joint;
— *perdu*, (*mas.*) dead joint;
placer plein sur ——, to break joints;
plan de ——, (*fond.*) parting plane;
— *plat*, (*carp.*) butt joint;
— *à plat*, v. —— *à recouvrement*;
plein sur ——, so as to break joints;
pointer les ——*s*, (*mas.*) to point flat.
— *en porte-à-faux*, (*r. r.*) unsupported joint (between cross-ties);
— *des rails*, (*r. r.*) rail joint;
recirer un ——, (*mas.*) to smooth a joint (in pointing);
à ——*s recouverts*, with broken joints;
— *à, de, recouvrement*, (*carp., mas.*) lap joint;
à ——*s recroisés*, so as to break joints;
recroiser les ——*s*, to break joints;
— *en rive*, joint between paving stones of the same row;
— *de rivet(s)*, rivet joint;
— *à rotule*, v. —— *universel*;
— *de rupture*, (*mas.*) joint of rupture; breaking joint (of an arch);
— *saillant*, (*carp.*) rebate joint;
— *sphérique*, (*mach.*) ball, spherical, joint;
— *suspendu*, v. —— *en porte-à-faux*;
— *de tête*, v. —— *de face*;
— *par torsade simple*, (*teleg.*) simple twist joint;
— *de tuyaux*, hose coupling;
— *universel*, (*mach., etc.*) universal, Hooke's, joint;
— *vertical*, v. —— *montant*;
— *vif*, sharp, close, joint;
— *à vis*, (*mach., etc.*) screw joint, screw coupling.

jointe, f., (*hipp.*) pastern; m. pl., (*carp.*) tongued and grooved planking of a partition.

jointé, a., (*hipp.*) (in composition; of the horse, pasterned or jointed; of the pastern, jointed);
bas-——, low-jointed, -pasterned;
court-——, short-jointed, -pasterned;
droit-——, straight-jointed, -pasterned;
haut-——, v. *long*-——;
long-——, long-jointed, -pasterned.

jointer, v. a., (*mas.*) to point.

jointif, a., side by side, joined.

jointoiement, m., (*mas.*) pointing, grouting.

jointoyer, v. a., (*mas.*) to point.

jointure, f., joint, joining, seam; joints, articulation, scarfing;
— *courte*, (*hipp.*) short-jointedness;
— *longue*, (*hipp.*) long-jointedness;
— *en T*, elbow joint (tubes, etc.).

jonc, m., cane, rush; twig, osier;
— *de siège*, (*harn.*) seat twist of a saddletree.

jonction, f., junction, joint, joining, coupling, union, binding; railroad junction; (*mil.*) junction;
— *à bride*, flange, joint;
— *à emboîtement*, socket joint.

jonque, f., junk, Chinese junk.

jonquille, m., jonquil yellow.

jouail, m., v. *jas*.

joue, f., cheek; side, sidepiece (in many applications); cheek (of a horse); (*sm. a.*) cheek hollow in the butt; side of the receiver; (*mach.*) shoulder or flange (to keep a belt from slipping off a pulley, a journal from slipping longitudinally); (*mil.*) command "aim";
—— *de bielle*, (*mach.*) connecting-rod fork;
—— *de la chape*, cheek of a pulley;
coucher en ——, (*sm. a.*) to aim; take aim at;
—— *de crosse*, (*sm. a.*) cheek piece of a rifle stock;
demi-——, v. —— *de vache*;
—— *d'embrasure*, (*fort.*) cheek of an embrasure;
en ——! (command) "aim;"
—— *de galerie*, (*mil., min.*) side of a gallery;
mettre en ——, (*sm. a.*) to aim, take aim at;
mettre quelqu'un en ——, (*sm. a.*) to aim at anybody;
mouvement de ——, (*sm. a.*) act or operation of aiming;
position de ——, (*sm. a.*) position of "aim";
—— *de poulie*, cheek of a block;
—— *de vache*, cheek block.

jouée, f., (*cons.*) reveal; (*carp.*) thickness of wood over a tenon, between the face of the tenon and exterior of piece.

jouer, v. a. n., to play; to gamble; to use, handle; to work (as a pump); to warp, start, spring (of wood); to work loose, to be loose; (*mach., etc.*) to have play, clearance; (*min.*) to be sprung, go off, explode; (*art.*) to go off, be discharged;
—— *des couteaux*, to fight sword in hand;
—— *de l'épinette*, (*man., fam.*) to paw the air while rearing;
faire —— *un fourneau*, (*min.*) to spring a mine;
—— *serré*, to discharge one's part or duty thoroughly and well (e. g., in fercing).

jouet, m., clamp, iron fish-plate; (*harn.*) small chain attached to the mouthpiece of a bit.

joug, m., yoke, slavery, subjection; (*mach.*) crosshead.

jouière, jouillère, f., (*harn.*) cheek strap.

joule, m., (*elec.*) joule.

jour, m., day; light, daylight; opening, interval; space, gap, window; (*mach.*) clear space left between adjacent parts of a mechanism;
à ——, open, worked through, cut through (of open work);
capitaine de —— *aux distributions*, (*Fr. art.*) captain superintending daily issues (group of batteries);
——*s de chauffe*, (*nav.*) steaming days;
——*s courants*, running days (including Sundays and holidays);
—— *de coutume*, opening or window made in a wall, not a party wall;
dans le ——, in the clear (of measurements);
demi-——, feeble light;
—— *droit*, a breast-high opening or window;
être de ——, (*mil.*) to be on duty for 24 hours;
à ——*faillant*, at the close of day;
à ——*failli*, after the fall of day;
se faire ——, to cut or break one's way through;
faux ——, false light, poor light; opening or window made in a partition wall;
——*férié*, holiday, *dies non*;
a ——*fermant*, at the close of day;
—— *de grâce*, (*Fr. a.*) allowance of time for the return to duty of persons guilty of various crimes and misdemeanors before proceeding against them;
grand ——, broad daylight;
—— *d'en haut*, light furnished by a skylight;
lieutenant de ——, (*art.*) battery officer of the day;
—— *de marche*, day's march;
—— *de mer*, (*nav.*) sea day; (*in pl.*) days at sea (number of);
mettre à ——, to bring up to date;
——*s de mobilisation*, (*Fr. a.*) on mobilization the days devoted to the assembling of men, horses, and material (reckoned from 12.01 a. m. to 11.59 p. m.);

jour, *ordre du* ——, (*mil.*) order of the day;
—— *ouvrable*, —— *ouvrier*, working day (exclusive of Sundays and of holidays);
—— *de pain*, (*mil.*) one (or pl., so many) day's (supply) of bread;
——*s de planche*, lay days;
—— *à plomb*, light from directly overhead;
prendre —— *sur*, to look out on, to get light from;
——*s de salle de police*, (*mil.*) (so many) days in the guardhouse;
service de ——, (*mil.*) daily duty;
—— *de servitude*, an opening in a party wall, made by agreement, etc.; ancient light;
—— *de souffrance*, opening or window on a neighbor's property made with his consent (always grated and not opening);
——*s de starie*, lay days;
——*s de surestarie*, days on demurrage;
—— *de tolérance*, v. —— *de souffrance*;
—— *de tranchée*, (*siege*) days of trench work;
—— *à verre dormant*, grated window, not opening, overlooking a neighbor's property;
—— *de vivres*, (*mil.*) one (or pl., so many) day's rations.

journal, m., newspaper; journal, diary; daybook;
—— *du bord*, (*nav.*) ship's log, log book;
—— *de caisse*, (*Fr. a. adm.*) register of moneys received and expended (kept in every regiment or establishment having a *conseil d'administration*);
—— *des distributions*, (*Fr. a. adm.*) journal or register of issues;
—— *de la machine*, (*nav.*) engine-room log;
—— *de marche*, (*mil.*) journal, record, of a march, of march operations;
—— *des marches et opérations*, (*Fr. a.*) record kept in each regiment or larger unit of marches and operations executed by the unit;
—— *Militaire Officiel*, (*Fr. a.*) official military gazette (discontinued as such in 1886, its place being taken officially by the *Bulletin Officiel*, and unofficially by the *Journal Militaire*);
—— *de mobilisation*, (*Fr. a.*) register (kept in each troop-unit, establishment and *service*) of the measures to be taken on mobilization in order to pass to a war footing;
—— *Officiel*, in France, the official daily gazette for the publication of all government documents, debates, etc. (two editions, the complete containing all the debates and related documents, government orders and decisions, etc., and the partial containing the journal proper and the debates); (*nav.*) official log book kept by masters of English vessels for the information of the Board of Trade;
—— *d'opérations*, (*Fr. a.*) register kept by *commandants d'étapes*, q. v.;
registre-——, (*Fr. a.*) daily register of all issues, receipts, expenditures, etc., kept by responsible officers;
—— *de route*, (*Fr. a.*) report of veterinarian on the condition of horses during and after a march (to be made 3 days after end of same);
—— *de siège*, (*siege*) siege journal.

journalier, a., daily; variable; m., (*nav.*) fresh, as opposed to preserved; food;
la guerre est ——, war is fickle;
travail ——, day labor.

journalisme, m., journalism; the press.

journaliste, m., journalist, newspaper man.

journée, f., day (between sunrise and sunset), day's work, journey work; day's wages, day wages; journey, day's journey (travel); (*mil.*) day of battle, (hence) battle itself; (*Fr. a. adm.*) 24 hours, from midnight to midnight, the basis of computing all allowances of no matter what nature;
—— *d'absence*, (*mil.*) day absent (leave, sick, confinement, imprisonment, etc.);
à la ——, by the day;
feuille de ——*s*, v. s. v. *feuille*;
gens de ——, day laborers;
à grandes ——*s*, by forced marches;

journée, *homme de* ——, day laborer;
—— *d'hôpital*, (*mil.*) day in hospital;
—— *de marche*, (*mil.*) distance marched in a day (24 kilometers in France, but this is not fixed), day's march;
ouvrage à la ——, journey work;
à petites ——*s*, by easy marches;
—— *de présence*, (*mil.*) day of presence for duty (present, en route to join, on mission);
—— *de route*, (*mil.*) day en route;
—— *de séjour*, (*Fr. a.*) day of travel passed in a given spot;
—— *de solde*, (*mil.*) day of pay (i. e., day of duty entitling one to pay);
tenir la ——, (*siege*) of a besieged place, to agree to surrender by a given day unless before succored (obs.);
—— *de travail*, working day (10 hours in France).

joyeux, m. pl., (*mil. slang*) men of the *bataillon d'Afrique*.

juché, p. p., perched;
cheval ——, (*hipp.*) a horse whose pastern joint is too far forward; horse up on his toes.

juge, m., judge, tribunal; (*Fr. mil. law*) member of a court-martial or court of revision (not the president);
—— *consulaire*, judge for the trial of commercial cases;
—— *d'instruction*, magistrate, examining magistrate;
—— *de paix*, justice of the peace.

jugé, p. p., *affaire, chose,* ——*e*, a settled matter;
au ——, by judgment, by guess.

jugement, m., (*law*) judgment; trial; sentence, finding, verdict; (*in gen.*) opinion;
appeler d'un ——, to appeal from a verdict;
—— *arbitral*, award;
casser un ——, to quash a verdict;
—— *par contumace*, judgment in contempt;
—— *par défaut*, judgment by default;
faire passer en ——, to bring up for trial, send before a court;
infirmer un ——, v. *casser un* ——;
mettre en ——, to bring to trial;
—— *militaire*, sentence of a court-martial;
mise en ——, arraignment;
passer en ——, to be brought up for trial;
prononcer un ——, to give judgment, to pass sentence;
rendre un ——, to pass judgment, to sentence.

juger, v. a., (*law*) to judge, try, put on trial; to sentence; to give judgment; (*in gen.*) to decide, to act as umpire;
se ——, (*man.*) of a horse, to cover, check, his steps (to plant the hind in the track of the fore feet).

jugulaire, f., (*unif.*) chin strap (of a helmet, forage cap);
—— *à coulisse*, sliding chin strap;
—— *d'écailles*, helmet scale;
fausse ——, dummy, false, chin strap (i. e., not used to hold on the cap or helmet).

jumart, m., jumart, hinny.

jumeau, a., twin; (*mach.*) double (generally with the idea of symmetry).

jumelage, m., (*pont.*) overlapping and lashing together of balks in the same row.

jumeler, v. a., to assemble and join two similar objects; to strengthen by means of cheeks, sidepieces; to fish; (*pont.*) to make a *jumelage*, q. v.

jumelle, f., cheek, sidepiece, fish; field glass, opera glass (used also in plural); paved side of a gutter next to the road; (*artif.*) twin rockets (with one stick); (*in pl.*) in various applications, similar pieces symmetrically disposed (as the bearers of a lathe); cheeks, sidepieces, uprights;
—— *de bau*, clamp of a beam;
contre- ——, paved side of a gutter farthest from the road;
—— *à glaces*, field glass fitted with mirrors in front so as to reflect objects to the right and the left of the observer;

jumelle, *grosse* ——, a field glass magnifying 6 or 7 times;
——*s-longues-vues*, telescopic field glass;
—— -*lorgnon*, small field glass;
lunette- ——, telescope field glass;
——*s de sonnette*, guides of a pile driver;
—— -*télémètre*, (*mil.*) field-glass range finder.

jument, f., (*hipp.*) mare;
couper une ——, to spay;
—— *poulinière*, brood mare;
—— *vide*, unimpregnated mare.

jumenterie, f., (*hipp.*) a breeding establishment for the special purpose of producing stallions (Arab and Anglo-Arab stallions in France).

Jupiter, m., Jupiter;
trait de —— *à clef*, joggled and wedged scarf;
trait de —— *à joints droits*, scarf with parallel joints and tabling;
trait de —— *à joints droits et à clef*, scarf with parallel joints and a table on each side;
trait de —— *simple*, tabled scarf.

juré, m., (*law*) juryman (any Frenchman 30 years of age may be sworn to take part in the trial of any matter, civil or criminal).

juridiction, f., (*law*) jurisdiction;
—— *administrative*, the legal power to settle all differences between the government and private individuals;
—— *civile*, civil jurisdiction;
—— *militaire*, military jurisdiction (comprises, in France, the corps commander, the minister of war, and the council of state).

juridique, a., legal.

juridiquement, adv., legally.

jurisprudence, f., jurisprudence.

jury, m., (*law*) jury (criminal cases only); board or commission charged with special functions (e. g., in an exposition);
—— *d'accusation*, (*law*) grand jury (does not exist in France);
chef du ——, (*law*) foreman of the jury;
dresser la liste du ——, (*law*) to empanel a jury;
—— *d'état départemental*, in France, a jury passing on the claim of young mechanics (*ouvriers d'art*) to be exempted from two years of military service;
—— *d'expropriation*, a jury to condemn land;
—— *d'honneur*, in the German army, a board of officers passing on affairs of honor, with a view to prevent duels;
—— *de jugement*, (*law*) trial jury.

jusant, m., ebb, ebb tide, ebb of the tide;
commencement du ——, first quarter of the ebb;
de ——, on the ebb, during the ebb;
direction du ——, set of the ebb;
étale du ——, slack ebb;
fin du ——, last quarter of the ebb;
—— *et flux*, ebb and flood;
mi- ——, half ebb.

juste, a., just, legitimate; exact; correct, right; accurate (as a gait, a firearm, a marksman, target practice); tight, close-fitting; too tight, too short, too narrow; (*cord.*) tight, close-fitting (as a rope in a block); (*tech.*) true;
cheval ——, a horse of regular and steady gaits;
tir ——, (*art. sm. a.*) accurate fire, practice.

justesse, f., justness, exactness; accuracy; (*tech.*) accuracy of fit, truth; (*ball.*) accuracy, precision;
—— *d'une arme à feu*, (*ball.*) precision of a firearm (according to some authorities, *justesse* should be said only of fire, and *précision* of a firearm);
épreuve de ——, (*ball.*) accuracy trial;
—— *de tir*, (*ball.*) accuracy of fire (according to some authorities, the accuracy of fire obtained under ordinary conditions, as distinguished from *précision*, accuracy under selected conditions);
tir de ——, (*ball.*) accuracy trials to determine the precision of a firearm;

justesse *d'un tir*, (*ball*.) per cent of hits made by a group of men firing together under identical conditions.
justice, f., justice, court of justice, law officers, the law, jurisdiction;
  aller en ——, to go to law;
  appeler en ——, to sue;
  citer en ——, v. *appeler en* ——;
  —— *civile*, civil jurisdictions;
  conseil de ——, (*Fr. nav.*) inferior court-martial;
  —— *criminelle*, criminal jurisdiction;
  déni de ——, v. s. v. *déni;*
  maison de ——, v. s. v. *maison;*
  —— *militaire*, (*mil*.) Judge-Advocate-General's department (U. S. A.); department of military justice;
  rendre la ——, to dispense justice;
justice, *traduire en* ——, to prosecute, bring before a court.
justiciable, a., (*law*) amenable, under the jurisdiction of.
justicier, v. a., to inflict corporal punishment; to execute, put to death.
justification, f., (*adm*.) the justification or supporting of all items in public accounts by vouchers; state of being supported by vouchers;
  —— *s comptables*, vouchers made out and presented by the interested parties;
  —— *s d'ordre*, vouchers prepared by superior authority (e. g., orders) independently of the interested parties.
justifier, v. a., to justify; (*adm*.) to support by vouchers.
jute, m., (*cord*.) jute.

K.

kabyle, m., kabyle.
kadmite, f., (*expl*.) kadmite.
kaïr, m., coir, cocoanut fiber.
kaïre, m., (*cord*.) coir rope.
kaléidoscope, m., kaleidoscope.
kalium, m., (*chem*.) potassium.
kaolin, m., kaolin.
kaolinisation, m., kaolinization.
kaoliniser, v. a., to transform into **kaolin**.
kaoutchouc, m., caoutchouc.
karat, m., carat.
karature, f., alloying; alloy;
  —— *blanche*, alloying (gold) with silver;
  —— *rouge*, alloying (gold) with copper.
kathode, m., (*elec*.) cathode.
kébir, m., (*mil. slang*) commander of a corps; colonel (Arab word).
képi, m., (*unif*.) forage cap.
kéracèle, f., (*hipp*.) exterior swelling of the foot.
kéraphylleux, a., (*hipp*.) keraphyllous;
  tissu ——, keraphyllous tissue of the foot.
kéraphyllocèle, f., (*hipp*.) keraphyllocele (horn tumor on inner face of the wall of the foot).
kérapseude, f., (*hipp*.) cracked and seamy horn tumor in some diseases of the foot.
kératogène, a., (*hipp*.) keratogenous.
kérite, f., kerite.
kiliare, m., 1,000 ares.
kilo, m., commercial abbreviation of *kilogramme*.

kilogramme, m., kilogram.
kilogrammètre, m., kilogrammeter.
kilolitre, m., 1,000 liters.
kilométrage, m., measuring by kilometers; operation of putting down kilometer posts along a road.
kilomètre, m., kilometer.
kilométré, p. p., marked off with kilometer posts;
  route —— *e*, a road marked off with kilometer posts.
kilométrer, v. a., to mark off a road with kilometer posts.
kilométrique, a., kilometric;
  indemnité ——, mileage.
kilométriquement, adv., by kilometers.
kina, m., (*med*.) quinine.
kinétite, f., (*expl*.) kinetite.
kinine, f., (*med*.) quinine.
kiosque, m., small house, kiosk, arbor, belvidere; newspaper stand;
  —— *de la barre*, (*nav*.) wheelhouse;
  —— *du capitaine*, (*nav*.) chart room;
  —— *du gouvernail*, v. —— *de la barre;*
  —— *de la passerelle*, (*nav*.) bridge house;
  —— *de veille*, (*nav*.) pilot house.
knout, m., knout.
kolback, m., (*unif*.) colback.
kolpack, m., v. *kolback*.
kriegsspiel, m., (*mil*.) war game.
kyanisation, f., kyanization.
kyaniser, v. a., to kyanize.

L.

labeur, m., labor;
  terre en ——, land under cultivation.
laboratoire, m., laboratory; (*artif*.) establishment for manufacturing fireworks; (*met*.) bed (of a reverberatory furnace);
  —— *de pyrotechnie*, (*artif*.) fireworks laboratory;
  —— *de la section technique de l'artillerie*, (*Fr. art*.) a chemical laboratory belonging to the technical section and making analyses for certain branches of the army.
labour, m., plowing.
labourage, m., plowing; part of a raft of logs under water.
labourer, v. a., to plow; (of an anchor) to drag; (*art*.) to plow, to graze, to tear up the ground; to make wide gaps in the hostile ranks;
  —— *le fond*, (*nav*.) (of a ship) to graze, to touch;
  —— *le terrain*, (*man*.) to stumble, trip.
labrador, m., v. *labradorite*.
labradorite, m., Labrador feldspar, labradorite.

labyrinthe, m., labyrinth; (*mach*.) sinuous guide; (*met*.) launder.
lac, m., (*top*.) lake;
  —— *salant*, salt lake.
laçage, m., operation, act, of lacing.
lacer, v. a., to lace, attach;
  —— *une manœuvre*, (*cord*.) to belay a rope.
laceret, m., small auger or gimlet.
lacet, m., lace, shoe lace; zigzag; zigzag or turn-out in (or from) a road (to avoid an obstruction or obstacle); winding or zigzag of an ascending road; (*cord*., *etc*.) loop, lacing, knot, thong;
  faire un ——, to wind, zigzag;
  mouvement de ——, (*r. r*.) rocking of an engine; (*art*.) motion due to inequality of pressure in recoil cylinders;
  —— *de route*, winding, turn, in a road.
lâchage, m., in river navigation, the descent of a boat.
lâche, a., loose, not packed; slack, started; cowardly, dastardly, mean; m., coward, poltroon;
  temps ——, soft weather.

lâché, p. p., (*mach.*, *cord.*) started, loosened.
lâcnefer, m., (*met.*) tap bar.
lache-mains, m., riding with hands off the handle bars (bicycle).
lâcher, v. a. r., to let go, release, loosen, let loose, let slip; to slacken, slack off, cast off (a rope); to become slack (of a spring); to allow to escape; (*art.*, *sm. a.*) to discharge; m., release, let go (as of carrier pigeons);
— *une arme à feu*, to discharge a firearm;
— *une bordée*, (*nav.*) to pour a broadside into the enemy;
— *la bride*, (*man.*) to slacken the reins;
— *un coup* (*de fusil*, etc.), to fire (a gun, etc.);
— *la détente*, (*sm. a.*) to pull the trigger;
— *un fusil*, (*sm. a.*) to fire a gun;
— *la gourmette*, (*man.*) to loosen a curb chain;
— *la main*, (*man.*) to slacken the hand;
— *la mesure*, (*fenc.*) to fall back, retreat;
— *pied*, to give way, to yield;
— *un robinet*, to turn a cock.
lâcheté, f., cowardice.
lâcheur, m., raftsman.
lâchure, f., synonym of *éclusée*, q. v.
lacis, m., fold, plait, twist.
lacs, m., cord, snare; (*harn.*) hobbles, fetters, shackles.
lacune, f., gap, breach of continuity.
lacunette, f., (*fort.*) cunette.
ladre, a., leprous; (*hipp.*) said of a horse having so-called leprous spots;
— *dans tel naseau*, having a leprous spot in such a nostril;
— *entre les naseaux*, having a spot between the nostrils;
— *entre et dans les naseaux*, having a spot between and in the nostrils.
ladre, m., (*hipp.*) leprous spots about the natural openings (due to absence of cutaneous pigment); horse having leprous spots;
— *bordé*, bordered spot (surrounded by a hairy zone of darker color);
— *interrompu*, crossed spot;
— *marbré*, marbled spot (sown with black spots);
— *mélangé*, mixed spot (partly covered with hair).
lagan, m., lagan.
lagon, m., (*top.*) salt-water lagoon.
lague, f., (*nav.*) wake (obs.).
laguis, m., (*cord.*) running bowline, bowline knot.
lagune, f., (*top.*) lagoon.
laie, f., lane or road through a forest; (*mas.*) tooth-ax, tooth-axing.
lainage, m., woolens; fleece; teaseling.
laine, f., wool;
— *de fer*, filaments of iron;
— *de scories*, slag wool.
lainer, v. a., to tease (wool, leather, etc.).
lais, m. pl., young trees left standing to reach a full growth; (*hydr.*) bank, land gained from the sea (*lais et relais* is often said for *lais* alone).
laisse, f., string, leash; (*hydr.*) strand or beach between high and low water marks; furrow-like deposits of sand, etc., left by the sea on the shore;
— *de basse mer*, low-water mark;
— *de basse mer d'eau vive*, low-water mark in spring tides;
— *de basse mer de morte eau*, low-water mark in neap tides;
— *de haute mer*, high-water mark.
laisser, v. a. n., to let, allow, permit; to let go; to abandon, give up, depart from, quit; to leave off or out; to omit, pass over; to cede;
— *aller*, (of boats) to have way enough; to let down, lower, a weight (as in mechanical maneuvers); to let go;
— *aller une ancre*, to let go an anchor;
— *aller les avirons*, (of boats) to have way enough; to cease rowing;
— *aller une manœuvre*, (*cord.*) to let go by the run;
— *la bride sur le cou à un cheval*, (*man.*) to let a horse go his own way, find the way;

laisser *courir*, to let go;
— *courir les avirons*, v. —— *aller les avirons*;
— *courir une manœuvre*, to let go; to let run;
— *tomber*, to let fall the oars.
laisser-passer, m., (*mil.*) pass through the lines.
lait, m., milk;
— *de chaux*, whitewash;
dent de ——, colt's tooth; milk tooth.
laitance, f., pulpy, gelatinous fluid exuding from concrete deposited in water.
laiterol, m., (*met.*) floss-hole plate.
laitier, m., (*met.*) slag, iron slag; dross, scoria, clinkers;
— *en fer*, iron slag or cinders.
laiton, m., brass;
— *à cartouches*, cartridge-case brass;
— *en feuilles*, sheet brass;
— *de fonte*, cast brass;
— *rouge*, red brass, bronze.
laize, f., breadth (of cloth); a breadth of sail-cloth, of canvas.
lamanage, m., river pilotage (with special reference to the mouth of a river, entrance of a port, etc.).
lamaneur, m., river pilot, coast pilot (especially for entrances, mouths of rivers, etc.).
lambeau, m., shred, fragment, piece.
lambin, m., straggler, loiterer; a., straggling, loitering.
lambiner, v. n., to straggle, loiter.
lambourde, f., a soft calcareous stone; (*cons.*) sleeper; joist; scantling; wall-plate; (*art.*) skid (mechanical maneuvers); sill or sleeper of a gun platform);
— *d'arrêt*, (*art.*) hurter;
— *de calage*, (*art.*) in a platform, a cross-sleeper notched to secure the system;
— *du fond*, sleeper, ground sill;
— *gîte*, ground sill, floor sill laid on the ground; (*art.*) under sleeper, ground sill of a platform;
— *de recouvrement*, (*art.*) flooring sleeper of a gun platform.
lambrequin, m., decorative fringe of a tent.
lambris, m., wainscot; ceiling; paneling;
— *d'appui*, low wainscoting;
bois de ——, wainscoting;
— *de revêtement*, paneling covering the whole wall;
— *de socle*, skirting board.
lambrissage, m., (act of) ceiling, paneling, wainscoting; the wainscoting, etc., itself.
lambrisser, v. a., to panel, wainscot, ceil.
lame, f., plate, blade, bit (of tools); sea, wave, breaker; slat (of a shutter); (*harn.*) side bar (of a saddle); (*sm. a.*) sword, (hence, fig.) swordsman; (*farr.*) width of a horseshoe nail; (*art.*) friction plate (for checking recoil, obs.); (*sm. a.*) clip, magazine clip;
— *d'alésoir*, (*mach.*) cutter;
avoir de la ——, (*hipp.*) to have fine lines;
— *de baïonnette*, (*sm. a.*) bayonet blade;
bonne ——, good swordsman;
— *brisante*, breaker;
— *à canon*, (*sm. a.*) skelp;
— *-charge*, (*sm. a.*) clip, magazine clip;
— *-chargeur*, (*sm. a.*) magazine clip, clip;
— *courbe*, (*sm. a.*) curved blade;
— *courte*, short sea;
creux de la ——, trough of the sea;
— *de Damas*, (*sm. a.*) Damascus blade;
— *damasquinée*, (*sm. a.*) inlaid blade;
— *damassée*, v. —— *de Damas*;
— *debout*, head sea;
— *de la détente*, (*sm. a.*) trigger blade;
dos de la ——, (*sm. a.*) back of the blade;
— *droite*, (*sm. a.*) straight blade;
— *d'eau*, sheet of underground water; (*steam*) water space (of a boiler, of a condenser);
— *d'eau à séparer*, (*steam*) water leg;
entre-deux des ——*s*, trough of the sea;
entre deux ——, in the trough of the sea;
— *d'estoc*, (*sm. a.*) thrusting blade;
— *évidée*, (*sm. a.*) hollowed blade;
faible de la ——, (*sm. a.*) part of the blade near the point;

lame *de, en, fer,* iron strap, band;
—— *de fond,* ground swell or wave;
fort de la ——, *(sm. a.)* middle, main part of a blade;
—— *de frein, (art.)* compressor plate; recoil checking plate or bar; brake strap;
—— *de gaz,* tongue of gas;
—— *à gouttières,* v. —— *évidée;*
grande ——, *(art.)* compressor plate or bar;
—— *haute,* high sea;
—— *de lance, (sm. a.)* lance head;
—— *longue,* long roller;
—— *à la Montmorency,* v. —— *évidée;*
—— *non-évidée, (sm. a.)* flat blade;
—— *pendante,* v. *grande* ——;
—— *percuteur, (sm. a.)* firing pin (of the Spencer rifle);
plat de la ——, *(sm. a.)* flat of the sword;
—— *plate,* flat, smooth, blade;
—— *pleine,* v. —— *plate;*
—— *de plomb,* sheet lead; strip of lead (between the drums of a column);
—— *de pointage, (art.)* elevation bar;
pointe de la ——, *(sm. a.)* point of the blade;
—— *polaire, (elec.)* the strip of a voltaic cell;
—— *de poudre, (powd.)* pressed cake, milled cake;
—— *de rabot,* plane iron;
—— *de ressort,* spring blade;
—— *de sabre, (sm. a.)* sword blade;
—— *de scie,* saw blade;
—— *de support, (art.)* side lever (elevation gear);
talon de la ——, *(sm. a.)* heel of the blade, part near the hilt;
tranchant de la ——, edge of the blade;
—— *de travers,* cross sea;
la —— *use le fourreau, (hipp.)* said of a nervous, irritable horse;
vieille ——, *(mil. slang)* old fellow, or chum.
lamé, a., laminated; plated (with).
lamellaire, a., lamellar; laminated, lamellate, lamellated.
lamelle, f., scale, plate, small disk; *(expl.)* grain, small disk;
—— *de mire, (art.)* the *visirklappe* of the German field-artillery sight (sort of leaf, fixed on the field-gun sight, by means of which the use of *plaques de hausse,* q. v., may be avoided).
lamellé, a., v. *lamellaire.*
lamelleux, a., v. *lamellaire.*
lamette, f., small leaf or plate; cramp, clasp; hoop or ring, used in wood construction, to meet a special stress; *(sm. a.)* leaf of a sight; *(art.)* end band, end band and hoop (of the splinter bar);
amorce de ——, *(art.)* end-band loop;
grande ——, *(art.)* middle band (of a splinter bar);
—— *de hausse, (sm. a.)* sight leaf;
—— *de palonnier, (art.)* swingletree end band;
—— *de volée, (art.)* splinter-bar end band.
laminage, m., *(met.)* rolling, laminating; *(steam)* throttling, wire-drawing;
—— *de tôle, (met.)* plate rolling.
lamination, f., lamination.
laminé, a., rolled.
laminer, v. a., *(mach.)* to roll.
lamineur, a., rolling;
cylindre ——, *(mach.)* roller.
laminitis, m., *(hipp.)* founder.
laminoir, m., *(mach.)* roller (cylinder); rolls, roll; rolling mill;
—— *à bandages,* tire-rolling mill;
—— *cingleur, (met.)* shingling roller;
—— *circulaire,* bending roll;
cylindre de ——, roller;
—— *dégrossisseur,* roughing roll;
—— *double,* three-high roll or mill;
—— *ébauchoir,* roughing roll;
—— *à fers spéciaux,* roller or mill to turn out special forms;
—— *finisseur,* finishing rolls;
—— *de forme,* roll or mill of special form to turn out special work;
—— *à lingots,* ingot rolls;
—— *à mouvement alternatif,* reversing rolls;
—— *à plaque(s),* plate rolls, rolling mill, plate-rolling mill;
—— *de puddlage,* puddling rolls;

laminoir *à rails,* rail mill;
—— *réversible,* reversing roll;
—— *à tôle(s),* plate rollers, plate-bending machine, rolling mill;
—— *à trois cylindres,* three-high roll.
lampas, m., *(hipp.)* lampas;
brûler le ——, to burn off the excrescence of the palate.
lampe, f., lamp, light;
—— *à arc, (elec.)* arc light;
—— *à are voltaïque,* v. —— *à arc;*
cul de ——, *(art.)* cascable;
—— *de Davy, (min.)* miner's safety lamp;
—— *en dérivation, (elec.)* shunt lamp;
—— *à esprit de vin,* spirit lamp;
—— *à essence,* spirit lamp;
—— *à gaz,* gas lamp;
—— *à gaze métallique,* v. —— *de Davy;*
—— *à incandescence,* —— *incandescente, (elec.)* incandescent light;
—— *à main,* hand lamp;
—— *de mine,* —— *de mineur, (min.)* miner's lamp;
—— *à modérateur,* moderator lamp;
—— *à mur,* wall lamp;
—— *de nuit complète, (elec.)* all-night lamp;
—— *portative à retournement, (mil. min.)* miner's electric reversing lamp;
—— *à régulateur, (elec.)* arc lamp (i. e., with a feed or clutch for carbons);
—— *à souder,* soldering lamp;
—— *de sûreté,* v. —— *de Davy;*
—— *à suspension,* hanging, swinging, lamp;
—— *témoin, (elec.)* indicator lamp.
lampion, m., sort of lamp used in illuminations;
—— *de parapet, (siege)* iron pot full of resin, used in lighting up parapets (obs.).
lampisterie, f., lamp locker.
lançade, f., *(man.)* long kick, kick out.
lançage, m., *(nav.)* launching.
lance, f., rod, staff; upright or picket of a grating or iron fence; nozzle; slice, fire slice; *(sm. a.)* lance, spear; *(artif.)* tube filled with composition, used in fireworks (decorative pieces), to sketch out and vary the outline;
allumelle de ——, v. *fer de* ——;
—— *de chevaux de frise, (fort.)* arm or spear of a *cheval de frise;*
coucher la ——, to slope the lance;
coup de ——, v. s. v. *coup;*
courir une ——, *(man.)* to ride at the rings;
—— *de drapeau,* color staff;
fausse ——, *(nav.)* Quaker gun;
fer de ——, *(sm. a.)* lance head;
—— *à feu, (art.)* portfire (obs.); *(steam)* fire slice;
—— *à feu puant, (mil. min.)* stink pot;
—— *de feu, (artif.)* firework used in the defense of a breach;
—— *à fourche, (mil. tel.)* pole with hook, to put wires on branches of trees, etc.;
hampe de ——, *(sm. a.)* lance shaft or pole;
lame de ——, v. *fer de* ——;
lis de ——, v. *fer de* ——;
main de la ——, *(man.)* right hand, saber hand;
mettre la —— *en arrêt,* to put the lance in rest;
pied de la ——, *(hipp.)* off foot;
—— *à pompe,* pump spear;
port de la ——, v. s. v. *port;*
porter la ——, v. s. v. *porter;*
reposer la ——, v. s. v. *reposer;*
—— *de sonde,* sounding rod.
lancé, m., throw or reach; *(steam)* (in indicator diagrams) the parts of a diagram due to momentum and not to steam pressure.
lancement, m., throw, throwing; *(nav.)* launching; *(torp.)* launching, discharge; *(pont.)* operation of bridging;
—— *au-dessus (au-dessous) de la flottaison, (torp.)* above (under) water discharge (of a torpedo);
batterie de —— *(de torpilles), (s. c. fort.)* torpedo battery;
exercice de ——, *(torp.)* torpedo-launching drill; *(pont.)* bridge drill;
—— *d'un pont, (pont.)* construction of a bridge;
poste de ——, *(de torpilles), (torp.)* torpedo-launching station;

lancement *de torpilles*, (*torp.*) torpedo launching;
tube de ——, (*torp.*) torpedo tube.

lancer, v. a. r., to throw, hurl, dart; to rush upon; to cast, let fly; (*art.*) to throw, fire (a projectile); (*nav.*) to launch (a vessel); (*r. r.*) to start (a train); (*cord.*) to throw (a rope);
—— *un cheval*, —— *un cheval au galop*, (*man.*) to urge a horse into a gallop;
—— *un cordage*, to heave a line;
—— *le plomb de sonde*, to heave the lead;
—— *un pont*, (*pont.*) to throw, construct, a bridge;
—— *une torpille*, to discharge, launch, a torpedo;
—— *une torpille au-dessus* (*au-dessous*) *de la flottaison*, to discharge a torpedo above (under) water.

lance-torpilles, n., (*nav.*) torpedo boat; torpedo tube;
bateau-torpilleur ——, torpedo boat.

lancette, f., lancet.

lancier, m., (*cav.*) lancer, uhlan;
oui les ——*s*, (*fam.*) go tell that to the marines.

lancis, m., (*mas.*) operation of repairing a wall by the insertion of fresh and sound stones; stones used for such repair; jamb stone; sort of toothing, binding a revetment wall to its backing;
—— *de l'écoinçon*, interior jamb stone;
—— *du tableau*, exterior jamb stone.

lande, f., (*top.*) heath, moor.

landier, m., andiron.

langage, m., language, speech;
—— *chiffré*, cipher language, i. e., signs of any and of all sorts;
—— *convenu*, in cryptography, conventional language, using ordinary words with altered meaning.

langue, f., tongue, language; (*tech.*) tongue, wooden quoin; (*hydr.*) tongue, neck of land, sand spit;
aides de la ——, (*man.*) use of the voice in urging a horse, clucking;
appel de ——, (*man.*) cluck;
—— *de balance*, index of a balance;
—— *de bœuf*, (*mil. min.*) push pick;
—— *de carpe*, double-edged armorer's tool;
coup de ——, v. s. v. *coup*;
donner de la ——, (*man.*) to excite, urge, a horse by clucking;
doubler la ——, (*hipp.*) to fold the tongue over or under the bit;
—— *de feu*, tongue of fire;
—— *de sable*, (*hydr.*) sand spit;
—— *serpentine*, (*hipp.*) tongue always in motion;
—— *de terre*, (*hydr.*) tongue of land.

languette, f., tongue, tongue of a shoe, shoulder, thin quoin or wedge; iron plate between the sheaves of a block; pawl, detent; (*mach.*) assembling key, feather; (*r. r.*) tongue rail, slide rail; (*sm. a.*) small steel rod, forming, with others, the *trousse*, q. v., from which a sword blade is made; (*carp.*) feather, tongue; (*mas.*) partition wall of a chimney, of a well;
à ——, tongued;
—— *de balance*, index or pointer of a balance;
—— *de ballon*, valve of a balloon;
à rainure et ——, (*carp.*) tongued and grooved;
—— *rapportée*, (*carp.*) inserted tongue.

lanière, f., lanyard; thong, leather strap; tie; lash; linchpin tie; cloak strap;
—— *du crochet d'attelage*, trace hook, thong;
—— -*goupille*, thong used as a connecting pin;
—— *de la lance*, thong of a lance;
—— *de revolver*, (*sm. a.*) revolver strap.

lanterne, f., lantern, light; skylight; small dome or cupola; lantern, lantern light; pommel, knob; (*fond.*) core barrel; mold spindle (of a loam mold); (*mach.*) lantern wheel, trundle; (*art.*) shot scuttle; pass box; powder, cartridge case, box (obs.); tin case or canister; gun ladle (for loading with loose powder, obs.); (*artif.*) copper ladle;

lanterne *d'ambulance*, (*Fr. med.*) hospital lantern, to indicate position of dressing stations (two lanterns—one red, the other white);
—— *applique*, wall lantern;
—— *à canon*, (*art.*) gun ladle;
—— *de chargement*, (*art.*) projectile tray; shell barrow; nose cap (to protect the projectile when carried from magazine to battery);
—— *à charger*, (*art., etc.*) small cylindrical measure for loading shells, etc.;
—— *de disque*, (*r. r.*) signal disk lantern;
—— *d'éclairage*, ordinary illuminating lantern;
—— *à éclipse*, flash lantern;
—— *d'écurie*, stable lantern;
—— *à gargousses*, (*art.*) pass box;
—— *hampée*, (*art.*) sort of loading scoop for muzzle-loading guns, by which the charge could be introduced without a bag (obs.);
—— *indicatrice*, (*mil.*) a lantern to designate headquarters, ammunition parks, etc.;
—— *à main*, portable lantern;
—— *à mitraille*, (*art.*) sort of tin case or canister filled with bullets (obs.);
—— *de niveau d'eau*, (*steam*) water-gauge lamp;
—— *de noyau*, (*fond.*) core-bar; (if large) core-barrel; core-iron, core-rod;
—— *à réflecteurs*, reflecting lantern;
—— -*signal*, signal lantern;
—— *à signaux*, signal lantern;
—— *sourde*, dark lantern, bull's-eye lantern;
—— *de sûreté*, (*Fr. art.*) magazine safety lantern;
—— *de suspension*, swinging lantern.

lanterneau, m., skylight, lantern, lantern light.

lapidaire, m., grindstone (axis vertical).

laque, m., lacker, lacquer, japan.

laque, f., lac, gum-lac, shellac;
—— *en écailles*, shellac.

laquer, v. a., to lacquer.

lard, m., bacon.

lardage, m., thrumming.

larder, v. a., to thrum, pierce, run through; (*fort., siege*) to picket, fasten down, secure with pickets (as fascines, etc.);
—— *un cheval*, (*man.*) to spur a horse until his sides are sore;
—— *de coups d'épée*, (*fam.*) to run through again and again;
—— *un saucisson*, (*fort., siege*) to picket down a saucisson; to fasten two saucissons together by forcing the end of one into that of another.

larderasse, f., (*cord.*) coarse rope made of junk.

lardoire, f., pile-ferrule, band or hoop around a pile.

lardon, m., (*met.*) bit of iron or steel used to conceal or fill a crack in a forging, etc.; (*sm. a.*) plug, etc., for disguising a defect (in a gun barrel); (*artif.*) swarmer, rocket weighing less than 2 ounces.

large, a., broad, wide; loose;
cheval qui va ——, (*hipp.*) horse that goes wide.

large, m., breadth; (*nav.*) the high sea; sea room, offing;
au ——, (*nav.*) off shore, off; on the high seas; (*mil.*) keep off, stand off (sentinel's cry to warn passers-by to keep off, to avoid a point not to be approached, or to keep to the other side of the road);
au —— *de....*, off of....;
pousser au ——, to push off a boat (as in bridge building).

largeur, f., breadth, width; (*nav.*) beam (of a ship);
—— *des cloisons*, (*art.*) width of the lands;
—— *en couronne*, breadth at the top (of a dike or embankment);
—— *extérieure*, outside width;
—— *en fond*, width at the bottom (of a ditch, etc.);
—— *en gueule*, width at the top (of a ditch, etc.);
—— *intérieure*, inside width;

largeur *du jour*, width (of an opening, as windows, etc.);
— *dans œuvre*, width in the clear, inside width;
— *normale de la voie* (*r. r.*) standard gauge;
— *des rayures*, (*art.*) width of the grooves;
— *de voie*, (*r. r.*) gauge.

largue, m., (*cord.*) slack rope; (*nav.*) sea room, offing; a., (*cord.*) loose, slack; (*mach.*) started.

larguer, v. a., (*cord.*) to let go, to ease, cast off, slack, cast loose, let a rope run;
— *en bande*, to let go amain, let go by the run; to cast loose at once;
— *en retour*, to ease off, to ease off handsomely, by degrees, turn by turn, with a turn or so around a post, etc.;
— *la vapeur*, (*mach.*) to blow off steam.

larme, f., tear; drop;
— *de plomb*, small shot.

larmier, m., (*cons.*) drip, undercut; dripstone; water table; coping of a wall or bridge; (*hipp.*) temple (of the horse).

las, a., tired, weary, fatigued.

lascar, m., (*mil. slang*) bold, devil-may-care fellow.

lasser, v. a., to tire, weary, fatigue, wear out.

lasseret, m., v. *laceret*.

lasset, m., v. *lacet*.

lasso-dompteur, m., (*hipp.*) an arrangement of straps to restrain violent horses (acts like a strait-jacket).

last(e), m., load of 2 tons (2,000 kilograms).

latent, a., latent.

latéral, a., lateral, side; m., (*hipp.*) legs on same side of horse.

latitude, f., latitude;
— *d'arrivée*, latitude ascertained at end of day's work;
— *corrigée*, latitude corrected by observation;
— *estimée*, latitude by dead-reckoning;
— *observée*, latitude by observation, observed latitude.

latrine, f., (*mil.*) latrine, "rear";
— *à fosse fixe*, latrine with fixed receptacle;
— *à fosse mobile*, latrine with movable receptacle;
— *à tinette mobile*, v. — *à fosse mobile*.

latte, f., lath, batten, strip; strip of metal; (*sm. a.*) straight saber for heavy cavalry;
— *jointive*, partition lath;
— *volige*, roof lath.

latter, v. a., to lath, to batten;
— *à claire-voie*, to leave a space between laths;
— *à lattes jointives*, to nail laths close together.

lattis, m., lathwork, laths, latticed work;
— *à hauteur d'appui*, ashlering.

lava, f., (*cav.*) the traditional tactical formation of the Cossack cavalry, more or less similar to the formation *en fourrageurs*.

lavabo, m., washstand.

lavage, f., (*in gen., chem., met.*) washing; (*met.*) buddling; (*expl.*) boiling (as of waste for gun cotton);
— *au crible*, (*met.*) riddling;
cuve à —, (*met.*) tossing tub;
— *à la cuve*, (*met.*) tossing;
— *à grande eau*, sluicing;
position de —, (*art.*) elevation permitting water to run out.

lavasse, f., very heavy shower of rain; water; slop, wash; a very hard stone (silicate of alumina with oxide of iron).

lave, f., flat surface stone in a quarry; lava;
— *fusible*, (*mas.*) artificial mastic.

lavé, p. p., light, light-colored; washed;
bai, etc., — (*hipp.*) light bay, etc.;
couleur — *e*, faint color.

laver, v. a., to wash, wash out, off; to wash a drawing, to lay on a wash; (of rivers, the ocean) to wash; (*chem.*) to treat with water, (*met.*) to buddle ore; (*carp.*) to dress, trim (with a twibill);
— *à l'encre de Chine*, to wash (a drawing) with india ink.

laveur, m., (*expl.*) washer, purifier.

lavis, m., (*drawing*) wash, color; washing, tinting (with india ink, etc.);
dessin au —, wash drawing or sketch;
dessiner au —, to wash, to tint, with india ink, etc.;
épure au —, wash drawing.

lavoir, m., washtub, lavatory; (*sm. a.*) washer, cleaning rod; (*met.*) buddle;
— *mécanique*, rotary washer.

lavure, f., metal sweepings (in workshops).

laye, f., v. *laie*.

layer, v. a., to cut a path or lane in a wood or forest; to mark or blaze trees to be left standing.

layette, f., (*powd.*) powder tray.

layeur, m., forest surveyor; person who lays out paths in woods or who marks trees to be left standing.

layon, m., tailboard (of a wagon).

lazaret, m., lazaretto; lazar, pest-house; (*mil.*) German name for hospital.

lazaro, m., (*mil. slang*) prison, "jug."

lé, m., towing path, towpath; breadth (of cloth).

leçon, f., lesson.

lecteur, m., reader; (*Fr. art.*) man who reads the *réglette*;
— *de cartes*, map cover (transparent and checkered to a given scale, for use in wet weather).

lecture, f., reading; reading of an instrument;
— *de l'acte d'accusation*, (*Fr. mil. law*) reading of the order convening the court, and of the results of the *instruction*, q. v.;
— *de la carte*, — *des cartes*, map reading.

légalisation, f., legalization.

légaliser, v. a., to legalize.

légalité, f., legality.

légation, f., legation.

lège, a., light, crank.

légende, f., legend, inscription (esp. legend of a map).

léger, a., light, slight; feeble, slim; faint; buoyant; (*mil.*) light (of certain kinds of troops and of materials); m. pl., a synonym of — *s ouvrages*, q. v.;
artillerie — *ère*, horse artillery (not a technical term);
avoir la main — *ère*, (*man.*) to have a light and good hand;
cavalerie — *ère*, v. s. v. *cavalerie*;
cavalier —, (*man.*) rider that has a good seat, and so does not fatigue his mount;
cheval —, (*hipp.*) spirited horse in good condition;
— *de devant*, (*man.*) light in front (said of a horse that in action keeps his hind quarters lower than his fore quarters);
— *à l'eau*, buoyant;
infanterie — *ère*, (*mil.*) light infantry;
— *à la main*, (*man.*) easy on the hand;
— *s ouvrages*, (*cons.*) light plasterwork (partitions, moldings, coats, etc.);
troupes — *ères*, (*mil.*) light troops.

légion, f., legion; (*Fr. a.*) unit of organization of the *gendarmerie*, q. v.;
— *départementale*, (*Fr. a.*) departmental legion (of gendarmerie); (formerly, during the first part of the Restoration) official designation of a line regiment;
— *étrangère*, (*Fr. a.*) foreign legion, two regiments, serving in Algiers;
— *de la garde républicaine*, (*Fr. a.*) the republican guard (is formed into a legion);
— *d'Honneur*, Legion of Honor (for the various grades see p. xi);
— *de Paris*, (*Fr. a.*) the gendarmerie of the departments of the Seine and of the Seine-et-Oise.

légionnaire, m., knight of the Legion of Honor; (*mil.*) soldier belonging to a legion.

législation, f., legislation;
— *militaire*, in France, an inaccurate expression frequently used to designate the body of laws applying to the military service.

légume, m., vegetable;
gros — *s*, (*mil. slang*) field officers;
— *s secs*, dried vegetables.

lenticulaire, a., lenticular.
lentille, f., lentil; pendulum bob; (*opt.*) lens, microscope, magnifying lens; (*nav.*) deck light;
—— *bi-concave,* double concave lens;
—— *bi-convexe,* double convex lens;
—— *concave-convexe convergente,* convexo-concave lens;
—— *concave-convexe divergente,* concavo-convex lens;
—— *concavo-plane,* plano-concave lens;
—— *concavo-concave,* double-concave lens;
—— *concavo-convexe,* concavo-convex lens;
—— *convergente,* converging lens;
—— *convexe-concave,* —— *convexo-concave,* convexo-concave lens;
—— *convexe-plane,* plano-convex lens;
—— *convexo-convexe,* double-convex lens;
—— *divergente,* diverging lens;
—— *de pendule,* pendulum bob;
—— *plan-concave,* plano-concave lens;
—— *plan-convexe,* —— *plano-convexe,* plano-convex lens.
lésion, f., wound, lesion.
lessivage, m., lixiviation;
—— *de la poudre,* (*powd.*) extraction of saltpeter from damaged powder;
—— *à la vapeur,* steaming (of wood, to drive out sap).
lessivation, f., lixiviation.
lessive, f., lye; lixiviation.
lessiver, v. a., to extract; to lixiviate.
lest, m., ballast (balloons and ships);
aller en ——, to go in ballast;
—— *en béton,* permanent ballast;
—— *d'eau,* water ballast;
—— *en fer,* iron ballast;
—— *en pierres,* stone ballast;
—— *en sable,* sand ballast;
sur ——, *sur son* ——, in ballast;
—— *volant,* shifting ballast.
lestage, m., ballasting.
leste, a., brisk, lively, smart, crack, ship-shape.
lesté, p. p., ballasted; (*art.*) filled with sand (of a projectile, to bring up its weight when no burster is used).
lester, v. a., to ballast.
lesteur, m., ballast lighter (boat and workman); ballast heaver.
lettre, f., letter; letter of the alphabet; (*in pl.*) letters, literature;
—— *administrative,* official letter, letter from a department;
—— *de change,* bill of exchange;
—— *chargée,* registered letter;
—— *en chiffres,* cipher letter;
—— *circulaire,* circular, circular letter;
—— *collective,* (*Fr. a.*) circular letter addressed by the minister of war to corps commanders, heads of *services,* etc., inviting their action in certain cases made and provided;
—— *de commandement,* (*Fr. a.*) letter of appointment to command (of any unit comprising the three arms, of an army, or of all the armies);
—— *de compagnie,* (*mil.*) designating letter of a company of infantry;
——(*s*) *de créance,* credentials; letter of credit;
—— *de crédit,* letter of credit;
—— *d'envoi,* letter of transmittal;
—— *s de marque,* (*nav.*) letters of mark;
—— *de mer,* (ship's) register;
—— *militaire,* (*mil.*) official letter;
—— *nulle,* nonsignificant letter in a cipher message;
——*s numérales,* letters used in the Roman notation;
—— *particulière,* private letter;
—— *de rappel,* letter of recall;
—— *de rebut,* unclaimed letter at the post-office;
—— *recommandée,* v. —— *chargée;*
—— *de représailles,* (*nav.*) letter of reprisals;
—— *de santé,* (*med.*) health certificate, clean bill of health;

lettre *de série,* (*sm. a.*) distinguishing letter stamped on rifle barrels to indicate place of manufacture;
—— *de service,* (*mil.*) official letter; letter of appointment, informing an officer either of his promotion or of his assignment to certain duties or functions;
——*s de signaux,* (*nav.*) signal letters;
en toutes ——*s,* in full (i. e., of words, without abbreviations; of numbers, in words, not in figures);
—— *de voiture,* waybill.
leurre, m., decoy.
leurrer, v. a., to decoy.
levage, m., raising, hoisting.
levain, m., yeast.
levant, m., the east; the Levant; (in the French ports of the Mediterranean) east wind.
lève, f., (*mach.*) wiper, tappet.
levé, m., (*top., drawing, man.*) v. *lever;* p. p., (*mil.*) called into service.
levée, f., act of raising; removing, removal, removal of funds, stores, etc. (from a depository); dike, bank, embankment, causeway, mole; levee; agitation, rising, of the sea; collection of mail from mail boxes, the mail so collected; rising (of a congress, meeting); harvest; arrest of a prohibition, cancellation of an arrest, of a punishment; collection, (of a duty); (*mach.*) cam; cog; lifter; wiper; throw, course (as of a piston); lift, stroke, length of stroke; (*mil.*) levy, levying, raising, formation, of troops, of a body of troops; soldiers levied; (*man.*) act of raising the lance in ring riding; (*top., drawing*) v. *lever;*
—— *d'arrêt,* (*mil.*) termination of arrest, of confinement;
—— *d'un blocus,* (*mil., nav.*) raising of a blockade;
—— *des boîtes,* collection of mail from letter boxes;
—— *de boucliers,* unsuccessful attack; much ado about nothing;
—— *de camp,* (*mil.*) breaking of camp;
—— *d'écrou,* termination of confinement in a jail, either by expiration of sentence or by transfer of the prisoner;
faire une ——, to embark;
—— *en masse,* (*mil.*) raising of the population for the national defense; general uprising of a nation in self-defense;
—— *de piston,* (*mach.*) piston stroke;
—— *de scellés,* (*law*) removal of seals;
—— *d'un siège,* (*mil.*) raising of a siege;
—— *de terre,* (*fort., etc.*) embankment;
—— *de troupes,* (*mil.*) levying of troops (no longer an official term).
lève-gazon, m., turf-spade.
lever, v. a. r., to lift, raise, raise up, heave; to get, pull, take hold; to remove, clear away, break up, relieve; to recall (a permission, an order); to rise (wind, sun, sea); to raise (a quarantine); to clear up (of the weather); (*mil.*) to raise, levy (troops, armies, a regiment, etc.); to break (camp); to raise (a siege, a blockade); to revoke (an order); to relieve (a guard, a sentry, a post); (*cord.*) to remove, take off (a stopper, a service); (*top.*) to survey, to sketch, to plot, to make a plan;
—— *l'ancre,* (*nav.*) to weigh, hoist, anchor;
—— *les arrêts,* (*mil.*) to release from arrest;
—— *le blocus,* (*mil., nav.*) to raise the blockade;
—— *le camp,* (*mil.*) to break camp;
—— *la consigne,* to revoke orders;
—— *une contribution,* (*mil.*) to levy a contribution;
—— *l'étendard,* to proclaim war;
—— *une fourrure,* (*cord.*) to take off a service;
—— *la garde,* (*mil.*) to relieve guard;
—— *un objet,* (*surv.*) to take a bearing on an object;
—— *le piquet,* to retire precipitately, "to pull up stakes";
—— *un plan,* to draw up, make, a plan, a ground plan; to plot;
—— *un poste,* (*mil., sig.*) to break up a station;

lever *rame*, to lie on the oars;
— *des recrues*, (*mil.*) to raise, procure, recruits;
— *le siège*, (*mil.*) to raise a siege.

lever, m., rising, levee; rising (of the sun, moon, etc.); (*man.*) lifting of the foot; (*top.*, etc.) survey, sketch, topographical sketch; surveying; plan, design (of a house, a machine).
(In France the form *levé* is used by the geographical *service* of the army, that of *lever* being retained by the corps of engineers. Except where otherwise indicated, the following expressions relate to surveying:)

— *par alignement*, plotting, survey, by alignment;
— *à la boussole*, compass survey (any other suitable instrument may be substituted for *boussole*);
au —, (*man.*) off the ground, out of bearing;
— *du canevas*, skeleton survey;
— *du canevas de détail*, filling in survey;
— *du canevas d'ensemble*, skeleton survey of the whole area to be covered;
— *par cheminement*, plotting or survey by successive stations;
— *d'une côte*, survey of a coast;
— *à coup d'œil*, eye-sketch made on the terrain (without instruments);
— *de détail*, details, filling-in work, plotting of details; survey for the purpose of filling in;
— *en détail*, v. — *de détail;*
— *d'ensemble*, general survey; survey of a position (e. g., *mil.*, of a set of works forming a system);
— *expédié*, sketch map, quick sketch, field sketch, topographical survey (1:20000 to 1:10000);
— *de fortification*, fortification surveying;
— *général*, general survey;
— *hydrographique*, hydrographic survey;
— *par intersection*, plotting, determination of points by intersection;
— *irrégulier*, v. — *de reconnaissance;*
— *d'itinéraire*, road sketch, road map, road survey;
— *de mémoire*, sketch, map, made from memory;
— *au mètre*, sketch or survey by measurement (made for very short distances, without angle-measuring instruments, and to a scale from 1:500 to 1:100; angles computed from sides measured);
— *de mine*, mining survey;
— *de nivelé*, contoured survey;
— *par les perspectives*, generic expression for photographic and other similar surveys;
— *photographique*, photographic survey;
— *de plan*, — *des plans*, plot, plotting, of a survey;
— *à la planchette*, plane-table survey;
— *de position*, survey of ground for a position, esp. for a fort;
— *de précision*, accurate instrumental survey (1:500 to 1:10000);
— *par rayonnement*, plotting by polar coordinates;
— *de reconnaissance*, (*mil.*) reconnaissance sketch, map;
— *par recoupement*, plotting by resection;
— *régulier*, accurate instrumental survey (1:10000 to 1:5000);
— *par renseignements*, sketch, survey, made on the spot, from information received, and not by actual sight;
— *souterrain*, mining survey;
— *à la stadia*, stadia survey;
— *topographique*, topographical survey, sketch, sum of the operations leading to a map of the terrain (1:20000 to 1:10000);
— *de triangulation*, geodetic survey, triangulation survey;
— *à vue*, eye sketch or map, made on the spot without instruments (1:20000, 1:40000, 1:50000).

lève rames, m., (command) "oars" (boat drill).

leveur, m., — *de plans*, engineer officer (Marbot).

levier, m., lever, hand spike; crow, crowbar; arm; handle; (*sm. a.*) operating handle; carrier lever; spring catch (for detachable magazines); (*mach.*) arm.

I. Artillery, small arms; II. Machinery, miscellaneous.

I. Artillery, small arms.

— *d'abatage*, (*art.*) long handspike or crowbar;
— *d'arrêt*, (*sm. a.*) cartridge-stop lever; (*art.*) stopping lever, jamming lever;
— *-arrêt de répétition*, (*sm. a.*) cut-off;
— *-arrêtoir*, (*F. m. art.*) lever-stop;
— *d'auget*, (*sm. a.*) carrier lever;
— *à bander*, (*r. f. art.*) cocking lever;
— *de chargement*, (*art.*) projectile-hoist lever;
— *de chèvre*, (*art.*) gin lever;
— *de choc*, (spar torpedo) contact lever;
— *à crampon*, (*art.*) lever with staple to hold it to the cheek of the carriage (obs.);
— *cuvette*, — *à cuvette*, (*art.*) gear, socket, lever;
— *de cylindre*, (*sm. a.*) operating handle;
— *de déclenchement*, (*art.*) disengaging lever; (*torp.*) launching lever;
— *de desserrage*, (*s. c. art.*) disengaging-brake lever;
— *de la détente*, (*torp.*) firing lever;
— *directeur*, (*art.*) traversing handspike;
— *directeur à roulette*, (*art.*) roller-, truck-lever;
écouvillon- —, (mountain art.) rammer-lever (rammer used as a lever in packing and unpacking the piece);
— *d'embarrage*, (*art.*) embarring lever;
— *d'embrayage*, (*art.*) in-gear lever;
— *à encliquetage*, (*s. c. art.*) gearing lever, traversing lever;
— *en fer à rouleaux*, (*art.*) iron handspike (service of French s. c. carriages of old model);
— *de frein*, (*art.*) brake lever; brake-nut lever (Fr. 80ᵐᵐ gun);
— *à, de, galet*, (*art.*) roller handspike (one roller);
— *à galets*, (*art.*) roller handspike (two rollers);
— *de galet(s) arrière*, (*art.*) rear eccentric lever;
— *à griffe*, (*s. c. art.*) claw lever (used when, after firing, the breech screw is hard to turn);
— *inférieur*, (*sm. a.*) carrier lever;
— *à lunette*, (*art.*) carrying bar;
— *de manœuvre*, (*art.*) hand lever, maneuvering lever, any operating lever; traversing handspike, trail handspike; (*sm. a.*) bolt handle, bolt lever; operating handle or lever; (of the Lebel) cut-off, magazine cut-off, locking lever;
— *marin*, (*art.*) handspike;
— *mobile*, v. — *à tenon;*
— *à pied de biche*, pinch bar;
— *de place*, v. — *de siège;*
— *placé en galère*, v. s. v. *galère;*
— *poignée*, (*art.*) lever handle; (*sm. a.*) operating handle;
— *de pointage*, (*art.*) trail handspike;
— *-pontet*, (*sm. a.*) lever trigger-guard (Winchester);
— *porte-patin(s)*, (*art.*) brake beam (Lemoine);
— *-portereau*, (*art.*) lifting bar, carrying bar;
— *de prise d'air*, (*torp.*) admission lever;
— *-rancher*, (*art.*) stake of the *porteur*, q. v.;
— *de réglage*, (*art.*) setting, adjusting, lever;
— *de relèvement*, (*F. s. c. art.*) lifting lever;
— *à ressort*, (*art.*) spring lever;
— *de rouleau*, (*art.*) roller handspike;
— *à roulettes*, (*art.*) roller lever;
— *de serrage*, (*Fr. s. c. art.*) compressing lever;
— *de siège*, (*art.*) common handspike;
— *de soulèvement*, (*art.*) eccentric lever;

levier *supérieur*, (*sm. a.*) follower, carrier proper;
—— *à tenon*, (*F. m. art.*) clutch-lever;
—— *à tenon mobile*, v. —— *à tenon;*
—— *de travers*, (*art.*) traversing handspike.
  II. Machinery, miscellaneous.
—— *d'abatage*, long lever;
—— *amplificateur*, multiplying lever;
—— *d'arbre de relevage*, weigh-shaft lever;
—— *à articulations*, toggle joint;
bras de ——, lever arm;
—— *brisé*, joint lever, bent lever, angle lever, angle beam, bell crank;
—— *à cames*, cam lever;
—— *de changement*, —— *de changement de marche*, reversing lever;
—— *à cliquet*, pawl lever;
—— *composé*, combined lever;
—— *à contrepoids*, counterpoised lever; (*r. r.*) switch lever;
—— *coudé*, v. —— *brisé;*
—— *coudé à angle droit*, right-angled lever;
—— *à croc*, claw handspike;
—— *en croix*, cross lever;
—— *de déclenchement*, disengaging, tripping, releasing, lever;
—— *à déclic*, rachet lever;
—— *de démoulage*, (*fond.*) lifting lever;
—— *de la détente*, expansion lever;
—— *à deux branches*, —— *à deux bras*, crooked, broken, lever; double-armed lever;
—— *de distribution*, starting lever, link lever;
—— *de distribution à main*, hand starting lever;
—— *droit*, straight lever;
—— *d'embrayage*, clutch lever, coupling lever;
—— *d'embrayage et de désembrayage*, gear lever, engaging and disengaging lever, clutch lever;
—— *d'enclanche*, gab lever;
—— *d'enclanchement*, engaging, connecting, lever;
—— *d'encoche*, v. —— *d'enclanche;*
—— *à encliquetage*, ratchet lever;
—— *excentrique*, —— *à excentrique*, (*r. r.*) switch lever;
—— *de fer*, crowbar;
—— *à fourchette*, forked lever;
—— *à fourchette et à déclic*, forked ratchet lever;
—— *de frein*, brake lever;
—— *à griffe*, pinch bar;
—— *d'impulsion*, starting lever;
—— *d'injection*, injection lever;
—— *intermobile*, lever of the first order;
—— *interpuissant*, lever of the third order;
—— *interrésistant*, lever of the second order;
—— *à main*, hand lever;
—— *de manœuvre*, any operating leber, generic expression for levers having special functions; (*r. r.*) switch lever;
—— *de mise en marche, en train*, starting lever;
—— *du modérateur*, v. —— *du régulateur;*
—— *à mouvement brisé*, toggle joint;
—— *oscillant*, rocking lever;
pince de ——, claw of a handspike;
—— *de pompage*, pumping lever of a hydraulic jack;
—— *de pompe*, pump lever;
—— *du premier (deuxième, troisième) genre*, lever of the first (second, third) order;
—— *de réglage*, adjusting, setting, lever;
—— *régulateur*, —— *du régulateur*, regulator lever, standard lever;
—— *de régulation*, rocker of the Creusot-Corliss engine;
—— *de relevage*, v. —— *de changement de marche;*
—— *de renvoi de mouvement du registre de vapeur*, throttle-valve lever;
—— *à rochet*, ratchet lever;
—— *de serrage*, compressing lever; tightening, adjusting, lever; compressor;
—— *simple*, simple lever; rocker arm;
—— *simple à plateau*, simple lever-testing machine;
—— *de soupape*, valve handle, valve lever;
—— *de soupape de sûreté*, safety-valve lever;
—— *taillé en biseau*, beveled handspike;
tourillon de ——, lever journal;
—— *avec talon*, shod lever;
—— *à vis*, screw lever.

lévite, f., frock coat; overcoat, surtout.
lèvre, f., lip, edge; rim, lip (of a crater, of a cartridge case);
  s'armer de la ——, (*man.*) to oppose the bit with the lips (said of a thick-lipped horse);
  —— *de l'entonnoir*, (*mil. min.*) rim of a mine crater;
  —— *pendante*, flap;
  aux ——*s pendantes*, flap-mouthed.
lézarde, f., crack, crevice (in masonry, etc.); (*Fr. a. unif.*) sort of gold (silver) lace used to mark the grade of n. c. o.
lézarder, v. a. r., to crack, become cracked.
liage, m., binding; (*powd.*) mixing or incorporating of ingredients.
liais, m., lias.
liaison, f., binding, tie; joint, joining, connection; (*cons.*) strengthening piece, brace; (*mas.*) bond, pointing mortar; (*mil.*) connection or communication to be established between various officers or between various units and officers (field of battle, march; in a more limited sense, artillery and cavalry); the relation or connection between the guide and the unit, or between the guide of the entire unit and the guides of the subunits;
  agent de ——, (*mil.*) any means by which troops, lines, and echelons may be kept in touch with one another, or by which a c. o. may be kept in communication with the various units under his command;
  —— *anglaise*, (*mas.*) English bond;
  —— *à articulation*, (*mach.*) joint;
  —— *croisée*, (*mas.*) cross bond;
  de ——, (*mil.*) communicating (said of units, of the parts of the ammunition supply line from front to rear);
  —— *diagonale*, (*cons.*) diagonal brace;
  en ——, (*mil.*) in touch; (*mas.*) with broken joints, so as to break joints;
  —— *de joint*, (*mas.*) pointing mortar;
  manquant de ——, badly put together;
  —— *à sec*, (*mas.*) dry masonry.
liaisonner, v. a., to join, bind; (*mas.*) to grout, to point, to bond.
liant, a., tenacious, tough (of metals); elastic (of springs); m., elasticity (of a spring); toughness (of metals); cohesion (of earth, land, etc.); (*man.*) suppleness (of a horse).
liasse, f., bundle, file (of papers); string, tape.
libage, m., (*mas.*) large, rough-dressed stone, rough ashlar.
libellé, m., drawing up, wording (of an order, document).
libeller, v. a., to draw up (a report, an opinion, etc.).
liber, m., bast, liber, inner bark.
libérable, m., (*Fr. a.*) soldier that has completed his time in a given category.
libération, f., deliverance, discharge, acquittance; (*mil.*) liberation or discharge from military service after having accomplished all that the law requires;
  —— *de comptable*, (*adm.*) discharge from administrative responsibility; notice sent to responsible officer (money and property) that his accounts are correct and have been passed, after auditing, by ministerial officers.
libéré, m., (*Fr. a.*) man who has accomplished the military service required by law; soldier set at liberty after serving out his sentence, or after pardon.
libérer, v. a., to liberate, set free, exempt; (*mil.*) to discharge (a soldier); (*adm.*) to free from administrative responsibility, to pass the accounts of.
liberté, f., freedom, liberty; (*mach.*) play;
  —— *du commerce*, free trade;
  —— *du cylindre*, (*mach.*) clearance;
  —— *de garrot*, (*harn.*) hollow of pommel;

liberté *de langue*, (*harn.*) port (of a bit);
— *des mers*, freedom of the seas;
mettre en —, to set at liberty, release, free, discharge;
mise en —, discharge, release;
— *du piston*, (*steam*) clearance;
— *de rognon*, (*harn.*) hollow of the cantle;
— *de troussequin*, v. — *de rognon*.

libraire, m., bookseller;
— *-éditeur*, publisher.

librairie, f., bookseller's shop; book trade.

libre, a., free; at liberty, independent, released; loose; at large; open; (*mach.*) out of gear;
à l'air —, in the open air;
espace —, clear ground; plenty of room;
au —, (*ball.*) in free air (of combustion of powder, etc.).

lice, f., lists, field, arena;
entrer en —, to enter the lists.

licence, f., license, permission.

licenciement, m., (*mil.*) disbanding, mustering out.

licencier, v. a., (*mil.*) to disband, muster out.

licet, m., leave, permission.

licol, m., (*harn.*) halter;
bride- —, v. s. v. *bride*;
bridon- —, v. s. v. *bridon*;
— *d'écurie*, stable halter;
— *de force*, shoeing halter (for refractory horse); collar for restraining horses in certain veterinary operations;
— *de parade*, (in the French cuirassiers and dragoons) bridle halter.

licou, m., (*harn.*) halter; head collar, stable collar;
— *d'abreuvoir*, watering bridle.

lie, f., sediment, dregs.

liège, m., cork; (*harn.*) knee-puff;
à —, corked;
— *fossile*, — *de montagne*, asbestos, fossil cork.

liement, m., (*fenc.*) binding.

lien, m., tie, strap, hoop; band, knot; iron band, strap; hoop for strengthening timber; brace; (*in pl.*) shackles, chains, irons, bonds; (*cord.*) lashing; (*mach.*) link (of a link-motion);
— *de fer*, — *en fer*, iron strap or band;
— *incliné*, tie-rod;
— *pendant*, hanging-tie.

lier, v. a., to bind, tie, fasten, lash; to connect, join, unite; (*fenc.*) to bind the adversary's blade.

lierne, f., rail, stringpiece; floor plank of a boat; (*cons.*) wall-plate (of a roof);
— *de palée*, tie piece for supporting king-post; tie, ribbon (bolted to a row of piles to keep them in place).

lieu, m., place, spot; ground; room; order; locus; turn; (*in pl.*) premises;
— *d'aisances*, latrine, water-closet, "rear;"
— *de franchise*, asylum (refuge);
porter en beau —, (*man.*) to carry the head well.

lieue, f., league;
— *géographique*, marine league;
— *marine*, marine league (3,556 meters);
— *de poste*, postal league (3,898 meters);
— *commune*, (so called) common league (4,444 meters);
— *métrique*, metric league, 4,000 meters.

lieutenance, f., (*mil.*) lieutenancy.

lieutenant, m., (*mil. and nav.*) lieutenant; (*fig.*) assistant;
— *adjoint au trésorier*, (*Fr. a.*) assistant of the *capitaine trésorier*;
— *adjudant-major*, (*Fr. a.*) a lieutenant acting as *adjudant-major*;
— *-amiral*, (*nav.*) vice-admiral (rare, not official);
— *d'armement*, (*Fr. a.*) in the infantry, a lieutenant charged with the repair, care, and accountability of the arms in service (acting ordnance officer, U. S. A.);
— *de bord*, (*nav.*) executive officer;
— *-colonel*, lieutenant-colonel;
— *en deuxième*, second lieutenant;
— *-général*, lieutenant-general;

lieutenant *d'habillement*, (*Fr. a.*) the *officier d'habillement* of separate battalions, of the battalions of fortress artillery and of the squadrons of the train;
— *d'instruction*, v. s. v. *instruction;*
— *en premier*, first lieutenant;
second —, — *en second*, second lieutenant;
sous- —, sublieutenant;
— *trésorier*, (*Fr. a.*) the paymaster of separate battalions, of battalions of foot artillery, and of the train squadrons;
— *de vaisseau*, (*Fr. nav.*) lieutenant.

ligan, v. *lagan*.

ligature, f., ligature, tie, binding; (*art.*) choke of a cartridge bag; (*elec.*) twist joint.

lignard, m., (*mil. slang*) foot soldier of the line, linesman.

ligne, f., line; line (of railway, of telegraph, of cable, etc.); (railway) track; order; path in a forest; twelfth part of an inch; the equator; (*fenc.*) line, division; (*man.*) track (followed by the horse in riding school); (*cord.*) small rope, line (*esp. Fr. art.*) rope 75ᵐ long used with the 80ᵐᵐ gun on mountain duty; (*mil.*) line, line of battle, line of troops; rank; the line of the army (as opposed to irregular or light or special troops); line (as opposed to guards); (*in pl.*) lines (quarters); (*fort.*) line of works.

    I. Artillery, ballistics, fortification;
    II. Military operations, drill, formations;
    III. Technical and miscellaneous.

    I. Artillery, ballistics, fortification.

— *d'âme*, (*art.*) axis of the bore;
— *d'approche*, (*siege*) line of approach, trench;
— *d'attaque*, (*siege*) approach, trench;
— *de l'axe*, (*art., sm. a.*) axis of the piece produced;
— *bastionnée*, (*fort.*) bastioned line;
— *de but*, (*art., sm. a.*) range (line to target);
— *capitale*, (*fort.*) capital;
— *de circonvallation*, (*siege*) line of circumvallation;
— *de combat*, (*siege*) the line of works or of prepared positions of a line of investment; (more esp.) the part of the line of investment composed of the two principal lines of works;
— *de combat de la défense extérieure active*, (*siege*) line held by the mobile troops of the defense in advance of the permanent works;
— *s de communication*, (*siege*) communicating trenches;
— *continue*, (*fort.*) continuous line of works;
— *de contre-approche*, (*siege*) line of counter-approaches;
— *de contrevallation*, (*siege*) line of counter-vallation;
— *du cordon*, (*fort.*) magistral;
— *couvrante*, (*fort.*) in defilading, the line below which a standing man will be safe (drawn from the crest to 1.8ᵐ above the farthest point to be defiladed);
— *à crémaillères*, (*fort.*) indented, cremaillere line;
— *de défense*, (*fort.*) line of defense (bastion trace); main line, as distinguished from a flank adjacent; line of frontier defenses; (*siege*) outer line; line of forts; line between the forts and the main body of the place; (*torp.*) line of torpedoes in a pass or channel;
— *à demi-redoutes*, (*siege, fort.*) a continuous line, with half redoubts at intervals;
— *de départ*, (*ball.*) line of departure, the tangent to the trajectory at its origin;
— *d'explosion*, (*expl.*) line, radius, of explosion;
— *de feu(x)*, (*fort.*) interior crest;
hauteur de la — *de mire*, (*ball.*) line of metal elevation (obs.);
— *s intermédiaires*, (*fort.*) defensive lines joining the salients of the *noyau central;* line of works in the second line;
— *à intervalles*, (*fort.*) a line of works with intervals between; noncontinuous line of works, (*lignes Pidoll, Rogniat*);

ligne *d'investissement, (siege)* line of investment;
—— *à lunettes détachées, (fort.)* line of detached lunettes;
—— *magistrale, (fort.)* magistral;
—— *du milieu, (art.)* line of metal (obs.);
—— *de mire, (ball.)* line through sights, line of sight, sighting line;
—— *de mire artificielle, (ball.)* artificial line of sight, (or simply) line of sight;
—— *de mire fixe, (ball.)* line passing through the fixed notches of the rear sight;
—— *de mire de n mètres, (ball.)* line of sight, using the elevation corresponding to a range of *n* meters;
—— *de mire naturelle, (ball.)* line of sight corresponding to the zero of the rear-sight graduation;
—— *de mire primitive, (ball.)* line of sight parallel to the axis of the piece;
—— *d'ouvrages, (fort.)* line of works;
——*s parallèles, (siege)* parallels;
—— *de pointage, (ball.)* v. —— *de mire;*
prendre la —— *de mire, (ball.)* to look through the sights; to aim, take aim;
prendre une telle —— *de mire, (ball.)* to use such and such a division or graduation of the rear sight;
—— *principale de défense, (fort.)* the line of forts;
—— *de projection, (ball.)* line of projection, the tangent to the trajectory at its origin;
—— *à redans, (fort.)* redan line;
—— *à redans et courtines, (fort.)* line of redans connected by curtains;
—— *à redans et à tenailles, (fort.)* redan and tenailled line;
—— *à redoutes, (fort.)* line of redoubts;
—— *à redoutes détachées, (fort.)* line of detached redoubts;
—— *-repère, (art.)* line joining two bench marks, used for laying guns;
—— *retranchée, (fort.)* retrenched line, fortified line;
—— *de retranchements, (fort.)* line of intrenchments;
—— *de site, (ball.)* line from the center of the muzzle to the point aimed at (taken ordinarily as coincident with the —— *de mire);*
—— *de sol extérieur,* v. —— *de terre;*
—— *tenaillée,* —— *à tenailles, (fort.)* tenailled line;
—— *de terre, (fort.)* ground line, ground level;
—— *de tir, (ball.)* line of fire, or of departure (uncorrected for jump), the axis of the piece produced (at the moment of inflammation of the charge); *(t. p.)* direction of fire at target practice on a range;
—— *de tir de polygone, (art.)* firing line of a proving ground;
—— *de tranchée, (fort.)* line of approach, trench.

II. Military operations, drill, formations.
—— *d'approvisionnement,* line of supply;
—— *d'appui,* tactically, the second line in an attack, supporting the firing line;
armée de ——, line of the army;
—— *d'attaque,* tactically, the line, in its most general sense, that executes the attack, the first line; (more esp.) the attacking line, the first line in a cavalry charge;
avant- ——, v. *avant-ligne;*
—— *de bataille,* line of battle in general; front occupied by the leading units (battle, maneuver, review); (more esp.) troops with their *train de combat; (Fr. inf.)* former name of —— *déployée,* q. v.; *(Fr. cav.)* formation in line; (of a brigade, division), line of regiments in line at intervals of 12 meters; *(Fr. art.)* formation in line or in echelons (of several groups of batteries);
—— *de bataillons, (inf.)* any formation in which battalions are ranged side by side on the same alignment;
—— *de bataillons en masse, (Fr. inf.)* in the regiment, etc., line of battalions in mass;

ligne *par brigade en colonne de masses. (Fr. cav.)* of the division, line of brigades in column of (regimental) masses;
cavalerie de ——, v. s. v. *cavalerie;*
—— *de colonnes,* any formation in which columns are side by side on the same alignment; *(Fr. cav.)* of a regiment, line of squadrons in platoon columns, at deploying intervals; of a brigade or division, line of regiments in line of columns at intervals of 12 meters, plus the front of three platoons; *(Fr. art.)* of a group, or of several groups, line of batteries in column of sections (platoons, U. S. Art.);
—— *de colonnes de compagnie, (inf.)* line of company columns (battalion, regiment, etc.);
—— *de colonnes doubles, (mil.)* line of battalions in double column;
—— *de colonnes par pièce (section), (Fr. art.)* of a group, or of several groups, line of batteries in column of pieces (sections, U. S. Art.) or of sections (platoons, U. S. Art.);
—— *de colonnes serrées, (Fr. art.)* of several groups, line of groups in close column of batteries;
en colonne par brigade en —— *de masses,* v. s. v. *colonne;*
—— *de combat,* in general, line of battle, fighting line; (more esp.) tactically, the firing line, the first line, the firing line, with its supports;
—— *de communication,* line of communication;
conduire la ——, *(nav.)* to head the line;
——*s convergentes,* converging lines;
—— *de convoi,* open space between an enemy and its base;
couper la ——, to cut the enemy's line (also a naval term);
—— *de défense,* in general, any line, of whatsoever nature, behind or on which a defense may be made or prepared; any line held to be defended;
—— *déployée, (Fr. inf.)* formation in line;
—— *de direction,* line of direction;
——*s divergentes,* diverging lines;
——*s doubles,* double lines (said of an army pursuing two objectives);
en ——, abreast; "fall in" (command);
—— *en échelons,* line of echelons;
—— *à échiquier,* checkerwise line;
—— *ennemie,* enemy's line;
entrer en ——, to take position in line, on the line;
—— *d'étapes,* —— *d'étapes auxiliaire,* —— *d'étapes de route,* v. s. v. *étape;*
—— *d'étapes routière,* the same as —— *d'étapes de route;*
être en ——, to be in line of battle;
être sur la même ——, to be of the same rank; to be abreast;
——*s extérieures,* exterior lines;
—— *de feu,* the firing line; *(in gen.)* the line on which the action is going on;
—— *de file, (nav.)* line ahead;
forcer la ——, to hold a position in advance of the general line;
formation en ——, any formation in line;
—— *de front, (mil., nav.)* line abreast;
—— *de front de bandière,* magistral line of a camp;
hors les ——*s,* out of quarters;
infanterie de ——, infantry of the line;
——*s intérieures,* interior lines;
—— *à intervalles,* line with intervals;
—— *d'invasion,* line of invasion;
—— *d'invasion naturelle,* line of invasion leading to the capital;
——*s-manœuvres,* maneuver lines (the sum of the directions followed by an army in order to reach its objective);
—— *de manœuvre,* line of troops; in infantry combat formations, the third line, serving, in the case of large bodies, as an *agent de liaison* between the *ailes décisive* and *démonstrative,* respectively; in cavalry, the second line, maneuvering according to the results achieved by the first line, or —— *d'attaque;* (of operations) line leading to the frontier;

ligne *de manœuvres,* line of operations; line cutting probable lines of invasion (used by Brialmont of a junction road in mountainous country);
— *de marche,* line of march;
marcher en —, to march abreast;
marcher sur la même —, to march abreast; to have or hold the same rank; *(nav.)* to sail in line, abreast;
— *de masses, (Fr. art.)* line of groups in mass; *(Fr. cav.)* of a brigade or division, line of regiments in mass;
se mettre en —, to draw up, form line, fall in; to take position in the line of battle;
—*s multiples,* multiple lines of operations (of an army simultaneously pursuing several objectives);
— *d'opération(s),* line of operation, of operations; (sometimes) line of communication with base; (more esp.) line (of operations) parallel to the frontier;
—*s d'opération extérieures (intérieures),* exterior (interior) lines of operations;
—*s d'opération multiples,* lines when the various columns are separated by impassable obstacles, or are out of supporting distance;
— *d'opération simple,* a line of operations when the various columns are within supporting distance;
— *pleine,* solid line, line without intervals;
se porter sur la —, to take position on the line;
première —, front rank;
de première —, (that part, etc.) first used, first called upon, drawn on, or put into action;
— *principale,* in attack formations, the main line;
ranger sur une —, to form line, deploy into line;
— *de ravitaillement,* line of supplies, along which supplies may be pushed to the front;
refuser la —, to refuse the line;
régiment de —, line regiment;
— *de relèvement, (nav.)* quarter line;
— *de résistance des avant postes,* line of grand guards;
rentrer en —, to resume position in line;
— *de retraite,* line of retreat;
rompre la —, to hold a position in advance or in rear of a general line (defense);
serrer la —, to close the line;
— *simple,* single line of operations, of maneuver;
— *stratégique,* (any) strategic line;
tactique de —*s,* cavalry combat tactics (so called from the lines involved, — *d'attaque,* — *de manœuvre,* qq. v., and the re-serve);
— *de tirailleurs,* line of skirmishers;
traverser la — *ennemie,* v. *couper la* — *ennemie;*
troupes de —, troops of the line.

III. Technical and miscellaneous.

— *à adhérence et à crémaillère, (r. r.)* a line using both ordinary and cogged wheels;
— *adiabatique, (steam)* adiabatic line of an indicator diagram;
— *aérienne,* overhead telegraph (telephone) line;
— *d'alignement,* base line;
— *d'amarrage, (cord.)* seizing or lashing line;
— *d'arbres, (mach.)* line of shafting;
— *atmosphérique, (steam)* atmospheric line (indicator diagram);
— *atmosphérique moyenne, (steam)* mean atmospheric line (indicator diagram);
— *d'attrape, (cord.)* heaving line (as, to bring a hawser on board);
— *de l'avance à l'introduction, (steam)* lead line (of an indicator diagram);
— *de base, (surv.)* base line;
— *basse, (fenc.)* lower line;
— *des basses eaux,* low-water mark;
— *blanche, (cord.)* untarred rope, cord;
— *de campagne, (mil. teleg.)* field telegraph line;

ligne *de ceinture, (r. r.)* belt railway;
— *de chaînette,* catenary;
— *de charge, (nav.)* load water-line;
— *de charge extrême, (nav.)* deep-water line;
— *de chemin de fer (r. r.)* railway, railway line;
— *de collimation, (opt.)* line of collimation;
— *de compression, (steam)* compression line (of an indicator diagram);
— *de la contrepression, (steam)* back-pressure line (of an indicator diagram);
— *courbe,* curved line;
— *à crémaillère, (r. r.)* cog railway;
— *de dedans, (fenc.)* inner line;
— *de dehors, (fenc.)* outer line;
— *de douane,* line of frontier custom-houses;
— *droite,* right line;
— *de droite,* v. — *de dehors;*
— *de direction,* line of direction;
— *d'eau,* level line; *(nav.)* water line;
— *d'eau en charge, (nav.)* load water-line;
à — *d'eau, (nav.)* between wind and water;
— *électrique,* telegraph line;
—*s d'embranchement,* v. —*s secondaires;*
— *d'engagement, (fenc.)* line of offense;
être en —, *(hipp.)* to cover (of the fore and hind legs, both standing and moving); *(fenc.)* to be in position to face each other (of the two adversaries);
— *équinoxiale,* the equator;
— *de faîte, (top)* watershed;
— *ferrée, (r. r.)* railway;
— *ferrée de campagne, (mil. r. r.)* field railway;
— *de flottaison, (nav.)* water line, line of flotation;
— *de flottaison en charge, (nav.)* load water-line;
— *de flottaison à lest, (nav.)* ballast line;
— *de flottaison légère, (nav.)* light water-line;
— *de foi, (inst.)* fiducial edge; sighting line; lubber's line (of a compass); spider line;
—*s de force, (elec.)* lines of force;
— *de foulée, (cons.)* line of tread of a staircase;
— *gauche,* — *de gauche,* v. — *de dedans;*
— *goudronnée, (cord.)* tarred line, rope;
grande —, *(r. r.)* main line;
— *de halage, (pont.)* tow line;
— *haute, (fenc.)* upper line;
hors de —, extraordinary, of marked excellence;
hors de —, out of line, out of the way;
— *d'introduction, (steam)* admission line (of an indicator diagram);
— *de joints, (mas.)* edge (of a stone, brick in position);
— *lège,* v. — *sur lest;*
— *sur lest, (nav.)* light water-line;
— *de loch, (nav.)* log line;
— *méridienne,* meridian, meridian line;
mettre en première —, to place in the first rank;
— *mixte, (teleg.)* a line partly overhead, partly underground;
— *de moindre résistance,* line of least resistance;
— *moyenne des basses eaux,* mean low-water line;
— *de niveau, (top.)* contour, level line;
— *parcourue,* sweep;
— *de partage,* — *de partage des eaux, (top.)* ridge between two intersecting slopes, watershed;
— *à plomb,* plumb line, vertical line;
— *de plus grande pente, (top.)* line of greatest slope;
— *pointillée,* dotted line (as in drawings);
— *des pressions zéro, (steam)* absolute, true zero line (of an indicator diagram);
— *de prime,* v. — *haute;*
— *principale, (r. r.)* main line, trunk line;
— *de quarte,* v. — *de dedans;*
— *rampante, (mil. teleg.)* telegraph line of insulated wire, laid directly on the ground;
— *de repère, (top.)* reference line, bench-mark line;

ligne *de respect*, maritime frontier (as at the three-mile limit);
—— *de rupture*, (*expl.*, *etc.*) line of rupture;
——*s secondaires*, (*r. r.*) branch lines, feeders;
—— *de seconde*, v. —— *basse;*
—— *de sonde*, lead line;
—— *sous-marine*, submarine cable; submerged line;
—— *souterraine*, underground (telegraph, telephone) line;
—— *de sûreté*, life line;
—— *suspendue*, v. —— *aérienne;*
—— *de télégraphie militaire*, military telegraph line;
—— *télégraphique* (*téléphonique*), telegraph (telephone) line;
—— *de terre*, ground line;
—— *terrestre*, land (telegraph) line;
—— *de thalweg*, (*top.*) intersection of the sides of a valley, line of junction of waters;
—— *de tierce*, v. —— *de dehors;*
—— *à tracer*, (*fort.*, *etc.*) tracing cord;
—— *trainante*, (*teleg.*) line laid on surface of ground (telephone also);
—— *de visée*, line of sight, of direction;
—— *visuelle*, visual line;
—— *à voie étroite*, (*r. r.*) narrow-gauge road;
—— *à deux voies*, (*r. r.*) double-track road;
—— *à une voie*, —— *à voie unique*, (*r. r.*) single-track road;
—— *volante*, (*teleg.*) flying or temporary telegraph (telephone) line;
—— *volante de télégraphie militaire*, temporary military telegraph.
ligner, v. a., to mark with a chalk line; to rule (paper, etc.).
lignerolle, f., (*cord.*) small line; twine made by hand, fox.
lignette, f., twine.
lignine, f., (*expl.*) v. *lignose*.
lignite, m., lignite, brown coal.
lignose, f., (*expl.*) lignose (mixture of nitroglycerin and sawdust).
ligue, f., league, confederacy.
liguer, v. a. r., to league, combine, form a league.
limace, f., spiral pump, Archimedes's screw.
limaçon, f., snail;
escalier en ——, winding stairs, spiral staircase.
limaille, f., filings, file dust;
—— *de fonte*, (*met.*) kish.
limande, f., (*carp.*) graving piece; (*cord.*) parceling; (*mil. min.*) sheeting;
—— *de garniture*, (*mach.*) packing gasket.
limander, v. a., (*carp.*) to put on or apply a graving piece; (*cord.*) to parcel.
limation, f., filing, rasping.
limatule, f., small file.
limature, f., file dust.
limbe, m., (*inst.*) limb;
à —— *complet*, having a complete graduated circle;
demi- ——, half limb, half circle.
—— *rectifiable*, adjustable limb;
lime, f., file;
—— *d'Allemagne*, straw-packed file;
—— *à archet*, bow file, riffler;
—— *barboche*, ripsaw file;
—— *-barrette*, small file of very fine cut;
—— *bâtarde*, bastard file;
—— *à bouter*, sharp file;
—— *à bras*, arm file, rubber;
—— *à canon*, barrel file;
—— *carrée*, ——, 4/4, square file;
—— *en carrelet*, saw file;
—— *à charnière*, joint file, round-edge joint file;
—— *à contre-taille*, double-cut file;
coup de ——, file stroke;
—— *à couteau*, hack file;
—— *demi-douce*, second-cut file;
—— *demi-ronde*, half-round file;
—— *à deux tranchants*, double-cutting file;
—— *à dossier*, —— *à dossières*, nicking file;
—— *douce*, smooth file;
empâter la ——, to clog the teeth of a file;
enlever avec la ——, to file off;
—— *d'entrée*, entering file;
—— *fendante*, key file;

lime, *à feuilles de sauge*, riffler;
—— *forte*, rough file;
—— *grosse*, coarse file;
—— *à main*, small file, hand file;
manche de ——, file handle;
—— *mordante*, rasp;
—— *moyenne*, fine-toothed file;
—— *à moyenne taille*, bastard file;
—— *en paille*, v. —— *d'Allemagne;*
—— *au paquet*, packet file;
—— *parallèle*, parallel file;
petite —— *ronde*, rat-tail file;
—— *à pilier*, pillar file;
—— *plate*, flat file;
—— *plate à main*, flat hand-file;
—— *plate pointue*, taper file;
—— *en queue de rat*, rat-tail file;
retailler une ——, to recut a file;
—— *ronde*, round file;
—— *rude*, coarse file;
—— *à scie*(*s*), saw file;
—— *sourde*, dead smooth file;
—— *superfine*, dead smooth file;
taillage des ——*s*, file cutting;
taille d'une ——, cut;
—— *à taille croisée*, —— *à taille double*, double-cut file;
—— *à taille simple*, single-cut file, float;
—— *taillée à deux*, double-cut file;
—— *taillée sans croisement*, single-cut file;
tailler une ——, to cut a file;
—— *tiers-point*, —— *en tiers-point*, three-square file;
trait de ——, file stroke;
—— *triangulaire*, saw file, three-square file;
—— *à trois carnes*, —— *trois quarts*, —— 3/4, v. —— *tiers-point*.
limer, v. a., to file;
machine à ——, filing machine;
—— *à la main*, to file by hand;
—— *en long*, to file lengthwise;
—— *en travers*, to file across.
limeur, m., filer; workman who files bayonets; filing machine, filing vise;
étau ——, filing vise.
limeuse, f., (*mach.*) filing-vise, -machine.
limite, f., limit; bound; boundary; landmark, extremity; (as an adjective, appositionally) limiting;
—— *d'âge*, (*mil.*) retiring age; age limit (minimum and maximum) for entering schools; minimum age to enter certain corps or departments;
atteint par la —— *d'âge*, (*mil.*) having reached retiring age;
—— *élastique*, elastic limit;
—— *du tir*, (*art.*) range;
——*s de tolérance*, (*mach.*, *art.*, *etc.*) limits of tolerance;
—— *de vitesse*, limiting (i. e., maximum) velocity.
limiter, v. a., to limit, bound, confine, restrict.
limitrophe, a., bordering, neighboring, frontier, bounding, limiting.
limon, m., mud, slime; shaft; thill; (*cons.*) stringpiece, string, outer string (of a staircase);
—— *en crémaillère*, *faux* ——, (*cons.*) wall string.
limoner, v. n., (of trees) to become large enough for thills.
limoneux, a., slimy, muddy.
limonier, m., shaft horse, wheeler;
—— *sous-verge*, off-wheeler.
limonière, f., shafts; four-wheeled carriage with shafts;
à ——, (*art.*) system in which the shafts are attached to the limber by a tongue.
brancard de ——, in certain two-wheelers, side bar in prolongation of the shaft;
bras de ——, thill;
à bras de ——, (*art.*) system in which the shafts are rigidly attached to the limber;
—— *à bras brisés*, (*mountain art.*) jointed shafts;
voiture ——, *à* ——, four-wheeled carrriage with shafts.
limonite, f., (*met.*) limonite.

limosinage, limousinage, m., (*mas.*) ashlar work; rough walling.
limousiner, v. a., (*mas.*) to construct ashlar work.
limure, f., filing, polish; filings.
lin, m., flax;
— *huile de* ——, linseed oil.
linceul, m., shroud.
linçoir, m., (*cons.*) trimmer, trimmer-beam, hearth-trimmer.
linge, m., linen; underclothes;
— *à pansement,* (*med.*) dressing linen.
lingot, m., (*met.*) ingot; (*sm. a.*) slug.
lingotière, f., (*met.*) ingot mold;
— *à eau,* water-cooled ingot mold (Creusot).
linguet, m., pawl, catch; (*art.*) catch;
— *-arrêtoir,* (*art.*) pawl, catch, stop;
— *de cabestan,* capstan pawl;
— *couronne des* ——*s,* pawl ring;
— *mettre le* ——, to insert, set, the pawl;
— *ressort du* ——, (*art.*) binding spring (Krupp fermeture);
— *de sûreté,* (*art., sm. a.*) safety-bolt, -catch, -key, -lock, -locking arm.
linteau, m., (*cons.*) lintel; (*fort.*) stringer, ribbon (of a row of palisades).
lippitude, f., (*hipp.*) bleardness, lippitude.
liquation, f., (*met., etc.*) liquation, melting.
liquéfaction, f., liquefaction.
liquéfier, v. a., (*met., etc.*) to melt, to fuse.
liqueur, f., liquor; (*chem.*) solution;
— *acide type,* standard acid solution;
— *alcaline type,* standard alkaline solution;
— *épreuve,* testing solution;
— *témoin,* indicator liquid;
— *de virage,* indicator liquid.
liquidation, f., liquidation; (*adm.*) inspection and verification of accounts (to determine whether the expenditures have been made according to law or not, and whether the balances are correct or not);
— *état de* ——, (*Fr. a.*) statement of expenditures (made out by a *sous-intendant* with respect to the warrants drawn by him);
— *rapport de* ——, (*Fr. a.*) quarterly report of inspection and verification of accounts (made out by an *intendant militaire*);
— *revue de* ——, (*Fr. a.*) inspection and verification of regimental accounts, of the accounts of officers not serving with troops, and of *employés militaires.*
liquide, a., liquid; (of money, property) clear, unincumbered; m., liquid;
— *métal* ——, (*met.*) metal in fusion.
liquider, v. a., to liquidate.
lire, v. a., to read (a map, an instrument);
— *la rose des vents,* to box the compass.
liséré, m., binding, edging (as of the ribbon of an order); (*unif.*) piping.
lisière, f., selvage, list; edge, skirt, border, verge (as of a wood, a field).
lisoir, m., bolster, body bolster (of a vehicle); (*art.*) pintle transom, pintle frame;
— *circulaire,* (*art.*) bedplate;
— *directeur,* (*Fr. art.*) sort of slide or frame (supporting the carriage proper in certain casemate mounts, obs.).
lissage, m., (*sm. a.*) glazing of bullets; (*powd.*) glazing; (*fond.*) black wash;
— *tonne de* ——, (*powd.*) glazing drum.
lisse, f., rail, hand rail; outside or vertical edge of the finished sole (of a shoe); (*hipp.*) v. *liste;*
— *-tablette,* (*nav.*) armor shelf.
lisse, a., smooth; (*art., sm. a.*) smoothbored;
— *canon* ——, *à âme* ——, (*art.*) smoothbore gun.
lissée, f., (*powd.*) glazing.
lisser, v. a., (*powd.*) to glaze; (*fond.*) to black wash;
— *au graphite,* —— *à la plombagine,* (*powd.*) to black lead.
lissoir, m., polisher; (*powd.*) glazing barrel; glazing house; (*mas.*) smoothing tool (in pointing).
liste, f., list, roll; (*hipp.*) list (on horse's face);
— *d'admissibilité,* (*Fr. a.*) list of candidates that have passed the *épreuves d'admissibilité,* and are therefore entitled to compete for admission (schools, etc.);

liste *d'admission,* (*Fr. a.*) list of candidates admitted (schools, etc.);
— *d'ancienneté,* (*mil.*) lineal list;
— *d'appel,* (*mil.*) list of the men in a squad room, roll-call list;
— *d'aptitude,* (*Fr. a.*) list (drawn up by inspectors-general and by corps commanders) of candidates competent to discharge certain functions (as pay, color-bearer, etc.); list of officers qualified for promotion;
— *matricule,* (*Fr. a.*) in each recruiting subdivision, a list of the reservists who have transferred their domicile to the subdivision;
— *noire,* black list;
— *de présentation,* (*Fr. a.*) list of officers recommended for special functions, for admission to the Legion of Honor, and for promotion by selection;
— *rayer de la* ——, to strike off the rolls (of the army, etc.);
— *de recrutement cantonal,* (*Fr. a.*) in each canton, a classification list of the young men drawn for military service (made by the *conseil de révision cantonal;* contains seven classes, i. e., fit for service, exempt, rejected, etc.);
— *des réservistes,* (*Fr. a.*) list kept in each company, squadron, and battery of the reservists belonging to it;
— *des sentinelles,* (*mil.*) guard list;
— *de tirage au sort,* (*Fr. a.*) list of drawings (by lot for military service).
listeau, m., fillet, listel, reglet.
listel, m., fillet, listel; (*art.*) fillet (obs.);
— *de la bouche,* (*art.*) muzzle fillet;
— *de bouton,* (*art.*) knob fillet;
— *de cul-de-lampe,* (*art.*) cascable fillet.
lit, m., bed; bed of a stream, channel; layer; (*art.*) layer of bullets in a shrapnel; (*mil.*) guard bed or platform; (*mas.*) natural bed of a stone; bed joint or face of a stone; upper or lower face of a course; bed (of mortar);
— *d'assise,* (*steam*) seating of a boiler; (*mas.*) bed, bed joint;
— *à, en, armoire,* folding bed;
— *bois de* ——, bedstead;
— *be bord,* cot;
— *de bordages,* (*nav.*) planking;
— *brisé,* camp bed, folding (i. e., jointed) bed;
— *brut,* (*mas.*) undressed face (of a stone);
— *de camp,* folding bed, field bed; (*mil.*) guard room stretcher or bunk;
— *de campement,* camp bed;
— *à canapé,* sofa bedstead;
— *de carrière,* (*mas.*) cleaving grain; natural bed of a stone;
— *d'un courant,* main stream;
— *dur,* (*mas.*) lower face (of a stone);
— *entreprise des* ——*s militaires,* (*Fr. a.*) contract service furnishing beds and bedding to the troops;
— *faire le* —— *de la pierre,* (*mas.*) to dress the face of a stone;
— *faux* ——, (*mas.*) wrong bed;
— *à fusion,* (*met.*) mixing bed;
— *de fusion,* (*met.*) charge;
— *de queuse,* (*met.*) sow channel;
— *des hautes eaux,* high-water bed;
— *majeur,* bed occupied by a river at high water;
— *de la marée,* tideway;
— *mineur,* bed occupied by a river at ordinary level;
— *mourir au* —— *d'honneur,* (*mil.*) to be killed in battle;
— *-muraille à bascule,* (*hipp.*) operating table (can be tipped after fastening the animal to it);
— *de pierre,* stratum or layer of stone;
— *d'une pierre,* (*mas.*) face of a stone;
— *de pont,* platform of a bridge;
— *de pose,* (*steam*) boiler seat;
— *de rivière,* channel, river bed;
— *de sangle,* folding bed;
— *tendre,* upper face of a stone;
— *du vent,* wind's eye.

liteau, m., strip; (*fort.*) riband (along palisading);
— *de frottement,* (*art., etc.*) friction band, strake, or guard;
— *de palissades,* (*fort.*) riband (along or on the top of a palisade).
literie, f., bedding, bed and bedding.
litewka, f., (*unif.*) sort of jacket or blouse.
litharge, f., litharge.
lithoclastite, f., (*expl.*) lithoclastite.
lithofracteur, m., (*expl.*) lithofracteur.
lithographe, m., lithographer;
imprimeur —, lithographer.
lithographie, f., lithography, lithographic establishment.
lithographier, v. a., to lithograph.
litière, f., litter, stretcher; (*hipp.*) litter;
être sur la —, (*hipp.*) to be sick, to be down;
faire la —, (*hipp.*) to bed down horses;
— *sautée,* (*hipp.*) litter that has been turned over with a fork.
litre, m., liter.
littéral, m., (*Fr. a.*) parts of the drill book to be memorized by drill masters; "booking" (U. S. M. A.).
littoral, m. and a., coast, seacoast; littoral, sea frontier.
liure, f., (*cord.*) seizing, lashing.
livarde, f., (*cord.*) rubbing cord.
livet, m., (*nav.*) sheer line.
livraison, f., delivery; number of a periodical;
marché de —, contract to furnish.
livrancier, m., deliverer, shipper, of goods, supplies; contractor for supplies; a., delivering.
livre, m., book;
— *de bord,* (*nav.*) ship's journal;
— *à chiffrer,* cipher book, code;
— *de comptabilité,* (*adm.*) account book;
— *de détail,* (*Fr. a. adm.*) register; book of accounts (now *registre de comptabilité*);
grand —, ledger;
— *-journal,* journal or daybook;
— *de loch,* (*nav.*) log book, log;
— *de notes,* (*top.*) notebook; field book;
— *d'ordinaire,* (*mil.*) mess book;
— *d'ordres,* (*mil.*) orderly book;
— *de signaux,* signal book, signal code.
livre, f., pound; a half kilogram; franc;
— *métrique,* a half kilogram;
— *sterling,* pound sterling.
livrer, v. a., to deliver, abandon, yield, give over or up, hand over, surrender; (*mil.*) to give (battle), to make (an assault);
— *un assaut,* (*mil.*) to make an assault;
— *bataille,* (*mil.*) to give battle;
— *à la circulation,* to open for traffic;
— *un combat,* (*mil.*) to fight, give combat;
se —, (*man.*) (of a horse) to yield to the impulse of the reins, of the legs.
livret, m., notebook; (*mil.*) soldier's pocketbook;
— *d'armement,* (*Fr. a.*) register or record of the arms in the hands of detached units;
— *de bouche à feu,* (*art.*) register for each gun (description and record of firing); record of artillery (U. S. Art.);
— *d'emplacement des troupes,* (*Fr. a.*) roster of troops and stations;
— *d'étapes,* (*Fr. a.*) list of *gîtes d'étapes;*
— *individuel,* (*mil.*) soldier's handbook;
— *d'infirmerie,* (*Fr. a.*) sick book or record kept for each horse;
— *matricule de cheval,* (*Fr. a.*) descriptive book for each horse;
— *matricule d'homme de troupe,* (*Fr. a.*) descriptive book made out by the recruiting bureau for each man it enlists (sent to the unit which the man joins);
— *matricule d'officier,* (*Fr. a.*) officer's record or descriptive book (made out by the *major,* or head of department, and kept by the c. o., by the *trésorier* for staff officers);
— *de munitions,* (*Fr. a.*) register of ammunition and of related material received and expended by the regiment;
— *d'ordinaire,* (*mil.*) mess account book;

livret *de solde,* (*Fr. a. adm.*) register or account of sums paid to troops, establishments, officers not serving with troops, etc.
lixiviation, f., (*chem.*) lixiviation.
local, a., local; m., place, premises; (*mil. and fort.*) generic term for any room, apartment, or premises used in the military service, e. g., barracks, storerooms, magazines, kitchens (does not include gun casemates or any other quarters or rooms in which guns are mounted);
— *disciplinaire,* (*mil.*) guardhouse, prison, cell;
— *de punition,* (*mil.*) any place where punishment is served.
localité, f., locality.
locataire, m., f., tenant.
locatif, a., occupier's …., tenant's ….
location, f., hiring, renting.
locatis, m., (*hipp.*) hired hack; screw.
loch, m., (*nav.*) log;
bateau de —, log ship, log chip;
— *breveté,* v. — *enregistreur;*
— *enregistreur,* patent log;
— *de fond,* ground log;
jeter le —, to heave the log;
ligne de —, log line;
livre de —, log book;
— *ordinaire,* common log;
sablier de —, log glass;
tour de —, log reel.
locher, v. n., to be loose (of a horseshoe).
lochet, m., narrow spade.
locomobile, a., locomobile; f., (*mach.*) portable engine, transportable steam engine.
locomoteur, a., producing locomotion.
locomotif, a., locomotive;
force, puissance, — *ve,* locomotive power;
machine — *ve,* (*r. r.*) locomotive engine.
locomotive, f., (*r. r.*) locomotive (ordinarily called "*machine,*" engine);
— *allumée,* locomotive with steam up, fired locomotive;
— *articulée,* double-ended locomotive;
— *chargée,* v. — *allumée;*
express —, express locomotive;
— *de gare,* shifting engine;
— *à grande vitesse,* express engine;
— *de, en, manœuvre,* v. — *de gare;*
— *à marchandises,* freight engine;
— *de montagne,* mountain locomotive;
— *à roues couplées,* coupled engine;
— *à roues indépendantes, libres,* uncoupled engine;
— *à n roues,* n-wheeled engine;
— *routière,* road engine, traction engine;
— *sans feu,* fireless locomotive;
— *-tender,* — *à tender,* tank engine;
— *de terrassement,* bogie engine;
— *à voyageurs,* passenger engine.
loge, f., lodge, hut, cabin; cell, box; booth.
logement, m., lodging, accommodation, quarters; room, house room; (*mil.*) billet, lodging, quarters; quartering, billeting, of troops; (*Fr. a.*) party sent in advance of a column on the march to look up and prepare quarters and supplies for the column on its arrival at destination; (*art., sm. a., mach., tech.*) bed, groove, seat, act of letting in; (*siege*) lodgment; (*art.*) dent, indentation, in the bore at the seat of the projectile;
assiette du —, v. s. v. *assiette;*
billet de —, (*mil.*) billet; billeting ticket;
— *de boulet,* (*art.*) seat (defect) of the projectile in smoothbore guns (due to pressure of gases on the top of the projectile);
— *du canon,* (*sm. a.*) barrel groove;
— *-caverne,* (*fort.*) subterranean shelter or quarters;
— *chez l'habitant,* (*mil.*) billeting;
— *s et emménagements,* accommodations;
— *militaire,* (*mil.*) quarters, quartering;
— *des mines,* (*mil. min.*) mine lodgment (parallel of the offensive, about 20 to 30 meters from most advanced mines of defense, and intended to be used as a starting point for offensive operations);
— *de l'obturateur, etc.,* (*art.*) seat, recess, of the obturator, etc.;

logement, *officier de* ——, (*Fr. a.*) officer in command of a *logement* (reports to the c. o. of the garrison on which the column is marching, receives instructions, prepares for the arrival of the troops, and reports back to his own c. o.);
—— *du projectile*, (*art.*) seat of the projectile;
—— *du tenon de fermeture*, (*sm. a.*) locking shoulder.

loger, v. a. n. r., to lodge, make a lodgment; to stable horses; to set in, seat (used when the part seated projects from the surrounding surface); (*mil.*) to quarter, billet; to be quartered or billeted; to effect a lodgment;
—— *chez l'habitant*, (*mil.*) to be billeted.

logis, m., house, habitation, dwelling;
corps de ——, main building, detached building;
maréchal des ——, v. s. v. *maréchal*.

logistique, f., (*mil.*) logistics.

loi, f., law, rule, authority;
—— *des cadres*, (*mil.*) law organizing the army;
—— *de la guerre*, law (international) of war; law (necessity) of war;
—— *martiale*, martial law;
——*s militaires*, military laws;
—— *du quart*, (*top. drawing*) principle of spacing hachures at a quarter of their length.

loin, adv., far;
au ——, in the distance;
faire plus ——, (*art.*) to elevate.

lolo, m., (*mil. slang*) *gros* ——*s*, cuirassiers.

long, m., length;
au ——, along, at length;
au —— *et au large*, in length and in breadth;
de ——, lengthwise;
en ——, lengthwise;
en —— *et en large*, backward and forward, up and down;
le —— *de* ...., along ....;
sciage de ——, ripping;
scie de ——, ripsaw;
scier de ——, to rip;
scieur de ——, rip sawyer.

long, a., long; much; slow; great; (*art.*) over;
capitaine au —— *cours*, v. s. v. *capitaine*;
un carré ——, a rectangular parallelogram;
—— *côté*, (*fort.*) long face of a work;
coup ——, (*art.*) an over;
—— *feu*, (*art.*, *sm. a.*) hang fire;
prendre le plus ——, to make a detour;
tir ——, (*art.*) overshooting;
voyage de —— *cours*, v. s. v. *voyage*;
vue ——*ue*, long-sightedness.

long-à-la-guerre, m., (*mil.*) detour to avoid a defile;
faire ——, (*mil.*) to remain in the ranks.

long-bois, m., v. *long-à-la-guerre*.

long-courrier, a., *bâtiment* ——, vessel engaged in the foreign trade.

long-cours, m., deep-water sailor.

longe, f., (*harn.*) tether, strap, thong, lunging rein;
—— *d'accouple*, (*art.*, *harn.*) coupling rein;
—— *bouclée*, (*art.*, *harn.*) coupling rein;
—— *de croupière*, (*art.*, *harn.*) crupper strap, back strap;
—— *d'un fouet*, cracker of a whip;
—— *du garrot*, v. s. v. *liberté*;
plate- ——, v. *plate-longe*;
—— *poitrail*, (*Fr. cav.*) lunge (so carried as to act like a *poitrail*);
prise de ——, (*hipp.*) halter-cast;
—— *du rognon*, v. s. v. *liberté*;
—— *de trait*, (*art.*, *harn.*) trace-tug;
travail à la ——, (*man.*) lunging;
travailler à la ——, (*man.*) to lunge a horse.

longer, v. a., to go, walk, run along (as a wood, river bank); to skirt (as a path).

longeron, m., (*cons.*, *mach.*, *etc.*) stringpiece, sleeper, side piece, sill; (*art.*) side rail;
sous- ——, under rail.

long(-)feu, (*art.*, *sm. a.*) hang fire.

longimétrie, f., longimetry, art of measuring distances.

longis, m., (*cord.*) yarn not yet twisted up.

longitude, f., longitude;
bureau des ——*s*. board of longitude (Naval Observatory, U. S.; Royal Observatory, England);
différence de ——, difference of longitude;
—— *estimée*, estimated longitude;
—— *observée*, observed longitude;
—— *par l'horloge marine*, longitude by chronometer;
prendre les ——, to take the longitude.

long-jointé, a., (*hipp.*) long-pasterned; long-jointed.

long-pan, m., (*cons.*) long side of a roof.

longrine, f., sleeper, stringing plank, string beam, stringer, cap; (*art.*) side rail; platform sleeper (*r. r.*) stringer.

longuerine, f., stringer.

longueur, f., length; slowness; duration;
—— *battue*, (*art.*) distance effectively covered by fire;
—— *de câble*, cable length (120 fathoms);
—— *de construction*, pitch length of a chain;
—— *de couche*, v. s. v. *couche*;
couper de ——, to cut to proper length;
en ——, lengthwise; at great length;
—— *équivalente*, (*ball.*) reduced length;
—— *à la flottaison*, (*nav.*) length at the water line;
—— *géodésique*, (*top.*) length of the sides of a triangle reduced to sea level;
—— *graphique*, (*top.*) length or distance measured on the map, scale length or distance;
—— *à la ligne de flottaison*, v. —— *à la flottaison*;
—— *de la ligne de mire*, (*ball.*) sight radius; radius distance;
metre de ——, to give the proper length to (a piece, etc.);
—— *de mire*, v. —— *de la ligne de mire*;
mise de ——, operation of giving the proper length to (a piece, etc.);
—— *entre perpendiculaires*, (*nav.*) length between perpendiculars;
—— *utile*, the considered length (in a test bar).

longue-vue, f., f., telescope, spyglass;
—— *de batterie*, (*Fr. art.*) battery (field, mountain) telescope;
—— *de campagne*, field telescope.

lopin, m., piece, bit, share, portion; (*met.*) bloom (*loupe*) transformed into a more or less regular parallelopipedon; (*farr.*) bit of iron for horseshoes; (*mas.*) stone in burnt lime;
—— *bourru*, (*farr.*) old shoes and bits of iron;
—— *à coquille*, (*farr.*) so-called shelliron;
—— *à quartiers branlants*, (*farr.*) quarters of horseshoes (strung on a wire);
—— *simple*, (*farr.*) piece of new iron;
—— *à vergette*, (*farr.*) bar of new iron (for *fer Charlier*).

loquet, m., latch, clasp, pawl, catch; (*mach.*) locking pin;
—— *d'arrêt*, (*r. f. art.*) stop-latch;
—— *de console*, (*art.*) plug-tray latch;
—— *double*, (*art.*) double-nosed latch;
—— *de fermeture*, (*r. f. art.*) locking latch;
—— *mobile*, (*art.*) latch (De Bange fermeture);
—— *de pression*, (*art.*) clamping latch (Pedrazzoli sight);
—— *à ressort*, (*art.*) spring latch;
—— *de sûreté*, (*art.*) safety latch.

loqueteau, m., latch, small latch, window latch.

lorgnette, f., opera glass; field glass.

lorgnon, m., eyeglass.

losange, f., lozenge;
en ——, lozenge-shaped.

lot, m., lot;
—— *de compagnie*, (*Fr. a.*) collection of effects kept in each company, battery, and troop (issued on mobilization when the unit passes to a war footing);
—— *de corps*, (*Fr. a.*) reserve supply of effects in excess of those forming the ——*s de compagnie*.

lotir, v. a., to divide into lots; (*met.*) to assay.

lotissage, m., separation into lots; (*met.*) assaying.

lotissement, m., division, separation into lots; arrangement, storing, of effects in groups.
louage, m., hire, hiring, renting;
à ——, for hire;
de ——, hired;
donner à ——, to hire, let;
prendre à ——, to rent;
prix de ——, rent.
louche, f., broach, reamer.
louchet, m., narrow spade; (*fort.*) trenching shovel;
—— à aile, —— à gazon, turf-cutters' shovel;
—— de galerie, (*min.*) push-pick.
louer, v. a. r., to hire, rent, let, hire out.
lougre, m., (*nav.*) lugger.
loup, m., wolf; capital defect in wood; nail drawer; (*sm. a.*) broad-bladed sword; (*met.*) bear; (*cord.*) woolder; (*harn.*) sort of bit for refractory horses;
dent de ——, lever; pawl, ratchet; large nail;
faire un ——, to botch a piece of work;
gueule de ——, v. s. v. *gueule;*
—— de haut fourneau, (*met.*) bear;
—— de mer, old salt:
saut de ——, sunken fence;
—— de selle, (*harn.*) straining, straining leather;
trou de ——, (*fort.*) military pit; (*met.*) overflow pit;
loupe, f., knot, lump, flaw (in wood); biconvex lens, magnifying or reading glass; (*met.*) bloom; (*sm. a.*) lump (of a shotgun);
avalement de la ——, (*met.*) balling, balling up;
avaler la ——, (*met.*) to ball, ball up;
—— d'eau, (*opt.*) water-lens;
—— de fer, (*met.*) bloom;
—— -prisme, (*opt.*) eyepiece.
vu à la ——, magnified.
loupeux, a., knotty (as a tree).
lourd, a., heavy, dull, close; (*art.*) heavy (of shells, guns, etc.);
artillerie ——e, special name given to the heavy artillery (field howitzers) accompanying a field army;
batterie ——e, field-howitzer battery.
loustic, m., (*mil.*) good-humored, pleasant soldier, having the gift of amusing his fellowsoldiers and of keeping them in a good humor (familiar).
louve, f., lewis.
louver, v. a., to sling a stone by means of a lewis.
louvet, a., (*hipp.*) yellow-dun (color), fox-color, clay bank.
louveteau, m., wedge, side piece of a lewis.
louveture, f., (*hipp.*) in the coat mixed blackish spot of indistinct outline.
louvoyage, m., (*nav.*) beating.
louvoyer, v. a., (*nav.*) to tack, to beat.
lover, v. a., (*cord.*) to coil down;
—— avec le soleil, to coil with the sun;
—— à contre, to coil the wrong way;
—— en galette, to make a Flemish coil;
—— en S, to make a French coil.
loyer, m., rent, hire;
—— d'entretien, (*Fr. a.*) rent received by contractor (*service des lits militaires*) for articles furnished, whether in hands of troops or not;
—— d'occupation, (*Fr. a.*) rent received by contractor (*service des lits militaires*) for articles in use.
lubrifacteur, m., (*mach.*) self-acting lubricator.
lubrifaction, f., (*mach.*) lubrication, oiling.
lubrifiage, m., v. *lubrifaction.*
lubrifiant, m., lubricant.
lubrificateur, m., (*mach.*) oil cup, self-acting lubricator; (*sm. a.*) bullet lubricant, lubricant;
—— de vapeur, (*mach.*) impermeator.
lubrification, f., v. *lubrifaction.*
lubrifier, v. a., (*mach.*) to oil, to lubricate;
huile à ——, lubricating oil.
lubrifieur, m., (*mach.*) self-acting lubricator; oil cup.

lucarne, f., dormer window; skylight; garret-, roof-window;
—— à la capucine, dormer window;
—— à œil de bœuf, bull's-eye.
lucigène, m., lamp.
lueur, f., (*art.*) flash of a gun.
lugeon, m., runner (placed under a wheel, thus converting a rolling into a sliding carriage: a Swiss term).
lumière, f., light; opening, window; hole, vent; (*steam*) steam port; (*art.*) vent; fuse hole; (*inst.*) sight, sighting slit; (*drawing*) illumination; (*carp.*) slot, through hole or mortise:
—— d'admission, (*steam*) admission port;
—— de l'alidade, (*inst.*) signt slit of an alidade;
—— à arc, —— à arc voltaïque, (*elec.*) arc light;
boucher la ——, (*art.*) to serve the vent (obs.);
bouchon de ——, (*art.*) vent plug;
canal de ——, (*art.*) vent, vent passage;
—— centrale, (*art.*) axial vent;
champ de ——, (*art.*) vent field;
—— de coussinet, (*mach.*) oil hole of a bearing;
—— de cylindre, (*steam*) cylinder port;
—— de décharge, v. —— d'échappement;
dégorger la ——, (*art.*) to prick the vent;
—— directe, (*top. drawing*) vertical illumination (of a map);
—— d'échappement, (*steam*) exhaust port, eduction port;
—— d'émission, v. —— d'échappement;
—— d'entrée de la vapeur, (*steam*) v. —— *d'admission;*
—— d'évacuation, v. —— d'échappement;
évasement de la ——, (*art.*) enlargement of the vent;
—— d'exhaustion, v. —— d'échappement;
—— de fusée, (*art.*) vent of a fuse;
—— à gaz, gaslight;
grain de ——, (*art.*) vent plug, vent bushing;
—— à incandescence, (*elec.*) incandescent light;
—— d'introduction, v. —— d'admission;
masse de ——, (*art.*) vent bush previously fixed in the mold and not tapped (obs.);
nettoyer la ——, (*art.*) to clear the vent;
—— oblique, (*top. drawing*) oblique illumination of a map;
—— d'obus, (*art.*) fuse hole;
—— de pompe, mouth of a pump;
—— de projectile, (*art.*) eye of a shell, etc.;
—— de rabot, slot of a plane;
—— de sortie, (*steam*) exhaust port;
trou de ——, (*art.*) vent;
—— à vapeur, (*steam*) steam port;
—— verticale, v. —— directe;
—— zénithale, v. —— directe.
lunatique, a., (*hipp.*) moon-eyed;
cécité ——, moon-blindness;
cheval ——, moon-eyed horse;
œil ——, moon-eye.
lune, f., moon, lune;
clair de ——, moonlight;
demi-——, (*fort.*) demilune.
luneton, m., (*fort.*) small lunette (defense of covered way and other works).
lunette, f., (*inst.*) telescope, spyglass; (*in pl.*) spectacles; (*cons.*) lunette (of a vault); (*art.*) lunette, gauge, ring gauge, shot gauge; assembling hoop (French 120mm C.); feed valve (Gardner gun); (*fort.*) lunette; (*mach.*) shellchuck; bearing-support; (*farr.*) tip; (*harn.*) lunette;
—— d'affût, (*art.*) upper pintle plate;
—— d'amarrage, (*art.*) mooring eye of a *cabestan de carrier;*
—— d'aplomb, miners' telescope;
—— d'approche, telescope;
—— d'Arçon, (*fort.*) detached lunette invented by d'Arçon, first professor of fortification at the Polytechnic;
—— astronomique, astronomical telescope;
—— de batterie, (*art.*) battery telescope for observation of hits;
—— à calibre, (*art.*) ring gauge, shot gauge, trunnion gauge;

lunette, *de campagne*, field glass;
—*s de cantonnier*, (*t. p.*) stone-breakers' spectacles, eye-protector, worn by markers in the butts;
—— *de chargement*.(*art.*) loading hole of the Krupp wedge;
—— *à charnière*, (*art.*) synonym of *volet*, carrier ring;
—— *de chevaux*, (*man.*) blinders (used on horses hard to mount);
—— *de cheville-ouvrière*, (*art.*) pintle eye;
contre- ——, (*art.*) lower or bottom pintle plate;
cylindre- ——, (*art.*) cylinder gauge; band gauge;
—— *de démoulage*, (*fond.*) lifting ring;
—— *d'équipement*, (*art.*) rigging eye (of a *cabestan de carrier*);
—— *à facettes*, multiplying glass;
—— *fixe*, (*mach.*) center rest;
—— *de flèche*, (*art.*) pintle eye, lunette;
—— *de longue-vue*, telescope;
—— *de marqueur*, (*t. p.*) marker's spectacles (having a wire mesh instead of glasses);
mettre la —— *au point*, to focus a telescope;
—— *micrométrique*, telescope with stadia attachment, for measuring inaccessible distances;
—— *de nuit*, night glass;
plaque de ——, (*art.*) pintle eye-plate;
—— *polarimétrique*, polariscope;
——*-profil*, (*art.*) profile gauge;
—— *pyrométrique*, pyrometric telescope;
—— *de réception*, (*mil. sig.*) receiving telescope; (*art.*) shot or ring gauge;
—— *de réception grande (petite)*. (*art., etc.*) maximum (minimum) gauge;
—— *de repère*, lower telescope of a theodolite; verifying telescope;

lunette *de serrage*, (*art.*) assembling hoop (of the French *canon à balles*, a ring or hoop connecting the rear and front blocks of the firing mechanism);
—— *stadimétrique*, stadia;
—— *terrestre*, terrestrial telescope;
—— *de vérification*, (*art.*) shot, etc., gauge;
—— *à vis de réglage*, (*art.*) adjustable shot gauge;
—— *à viseur*, sighting telescope (e. g., for a compass, etc.);
—— *de volée*, (*art.*) splinter-bar plate.
lupin, m., lupin.
lusin, m., (*cord.*) houseline, housing.
lustrage, m., glossing, polishing.
lustrer, v. a., to polish, to smooth.
lut, m., luting, cement, lute.
luter, v. a., to cement, to lute.
lutte, f., struggle, fight, contest, strife;
—— *d'artillerie*, (*art.*) artillery duel;
—— *générale de traction*, (*gym.*) tug of war.
lutter, v. a., to struggle, fight; to wrestle; to strive.
lutteur, m., wrestler.
lux, m., (*elec.*) unit of light (the illumination produced by one *bougie décimale*).
luxation, f., dislocation; luxation.
luxe, m., luxury;
armes de ——, sporting arms;
train de ——, (*r. r.*) first-class express.
luxer, v. a., to dislocate, put out of joint.
luzerne, f., lucerne (grass).
lycopode, m., lycopodium.
lyddite, f., (*expl.*) lyddite.
lymphangite, f., (*hipp.*) lymphangitis, inflammation of the lymphatic vessels and ganglions.
lyre, f., lyre; (*unif.*) lyre (insignia of musicians).

M.

macadam, m., macadam, macadam pavement.
macadamisage, m., macadamizing.
macadamisation, f., macadamization.
macadamiser, v. a., to macadamize.
macaron, m., (*unif.*) leather patch (under the button of a belt loop); (*mach.*) valve adjuster (of a locomotive).
macération, f., maceration.
macérer, v. a., to macerate.
mâche, f., horse-mash.
mâchecoulis, m., v. *mâchicoulis*.
mâchefer, m., slag, forge scale, hammer slag, scoria, clinkers.
mâchelière, a., of the jaw; f., grinder; jaw tooth, mill tooth;
dent ——, jaw tooth, grinder; mill tooth.
mâchemoure, f., bread dust.
mâcher, v. a., to chew, grind;
—— *le frein*, (*hipp.*) to champ the bit.
machete, m., matchet, machete.
mâchicoulis, m., (*fort.*) machicoulis, machicolation;
créneau- ——, v. s. v. *créneau*.
machinal, a., mechanical.
machinalement, adv., mechanically.
machine, f., machine, engine; machinery, piece of mechanism, contrivance; bicycle; (*r. r.*) locomotive; (*elec.*) dynamo.
  I. Steam; II. Operating machines; miscellaneous; III. Electricity.

  I. Steam:
  (The full expression for steam engine being *machine à vapeur*, the following compounds are occasionally found with the phrase *à vapeur* following the word "machine," e. g., *machine à vapeur à détente*, instead of *machine à détente*, the shorter being the usual form.)
—— *à action directe*, direct-acting engine;
—— *à admission et à échappement indépendants*, scientific name for engines of the Corliss type;

machine *atmosphérique*, atmospheric engine;
—— *à balancier*, beam engine, lever engine;
—— *à balancier de deuxième ordre*, half-beam engine;
—— *à balancier libre*, grasshopper engine;
—— *à balancier simple*, single-beam engine;
—— *à balancier superposé*, overhead-beam engine;
—— *à balanciers latéraux*, side-lever engine;
—— *à basse pression*, low-pressure engine;
—— *à bielle articulée*, engine with jointed connecting rod;
—— *à bielle directe*, direct-acting engine;
—— *à bielle renversée*, v. —— *à bielle en retour*;
—— *à bielle en retour*, back-acting engine; return connecting-rod engine;
—— *sans bielle*, v. —— *oscillante*;
—— *en clocher*, steeple engine;
—— *à commande directe*, direct-acting engine;
—— *compound*, compound engine;
—— *compound à n cylindres*, n-cylinder compound engine;
—— *compound horizontale à cylindres accolés*, horizontal cross compound engine;
—— *compound horizontale à cylindres en tandem*, horizontal tamden engine;
—— *compound à réservoir intermédiaire*, (intermediate) receiver engine;
—— *à condensation*, condensing engine;
—— *à condensation par surface*, surface condensing engine;
—— *à condensation sans détente*, condensing nonexpanding engine;
—— *sans condensation*, noncondensing engine;
—— *à connexion directe*, direct-acting engine;
—— *à connexion indirecte*, beam engine, indirect-acting engine;
—— *Corliss*, Corliss steam engine;
—— *à cylindre oscillant*, oscillating engine;
—— *à cylindres accouplés*, compound engine;
—— *à cylindres combinés*, intermediate-receiver engine;

machine à *cylindres conjugués égaux*, two-cylinder engine with cranks at right angles;
—— à *n cylindres*, n-cylinder engine;
—— à *cylindres renversés*, inverted, overhead-cylinder engine;
—— à *déclic*, drop cut-off engine (generic term for engines of the Corliss type);
—— *demi-fixe*, semiportable engine;
—— à *détente*, expansion engine;
—— à *détente et à condensation*, condensing expansion engine;
—— à *détente (et) sans condensation*, noncondensing expansion engine;
—— à *détente double*, double-expansion engine;
—— à *détente fixe*, fixed expansion engine;
—— à *détente séparée*, (the strict French name for) compound engine;
—— à *détente simple*, single-expansion engine;
—— à *détente variable*, variable expansion engine;
—— à *détentes échelonnées*, multiple expansion engine;
—— *sans détente*, nonexpanding engine;
—— *sans détente et sans condensation*, noncondensing nonexpansion engine;
—— *diagonale*, inclined engine;
—— à *disque*, disk engine;
—— à *double cylindre*, double-cylinder engine;
—— à *double effet*, double-acting engine;
—— à *double expansion*, double-expansion engine;
—— à *expansion*, expansion engine;
—— *sans expansion*, nonexpanding engine;
—— *fixe*, stationary engine;
—— à *fourreau*, trunk engine;
—— à *grande vitesse*, high-speed engine;
—— à *haute et à basse pression*, compound engine;
—— à *haute pression*, high-pressure engine;
—— *horizontale*, horizontal engine;
—— *horizontale à fourreau*, horizontal trunk engine;
—— *inclinée*, v. —— *diagonale*;
—— *lente*, low-speed engine;
—— à *levier*, lever engine;
—— *locomobile*, portable engine;
—— -*locomotive*, (r. r.) locomotive;
—— *marine*, v. —— *marine*, marine engine;
—— *monocylindrique*, single-cylinder engine;
—— à *mouvement de sonnette*, bell-crank engine;
—— à *moyenne pression*, mean-pressure engine;
—— à *multiple expansion*, multiple-expansion engine;
—— *ordinaire*, double-acting engine;
—— *oscillante*, oscillating engine;
—— *pilon*, —— à *pilon*, engine of steam-hammer type; overhead-cylinder engine, inverted direct-acting engine, inverted vertical-cylinder engine;
—— à *piston annulaire*, annular engine;
—— *portative*, portable engine;
—— à *pression intermédiaire*, medium-pressure engine;
—— à *quadruple détente*, quadruple expansion engine;
—— à *quadruple expansion*, quadruple expansion engine;
—— à *réaction*, reaction engine;
—— à *rotation*, —— *rotative*, rotary engine;
—— à *rotule*, sort of oscillating engine;
—— *semi-fixe*, semiportable engine;
—— à *simple coudé*, single-crank engine;
—— à *simple effet*, atmospheric engine, single-acting engine;
—— à *simple expansion*, single-expansion engine;
—— à *soupape*, Sulzer engine;
—— -*tandem*, tandem engine;
—— à *tiroir*, slide-valve engine;
—— à *triple détente*, triple-expansion engine;
—— à *triple expansion*, triple-expansion engine;
—— à *trois cylindres*, three-cylinder engine;
—— à *un cylindre*, single-cylinder engine;
—— à *vapeur*, steam engine;

machine à *vapeur à piston*, piston steam engine;
—— à *vapeurs combinées*, binary engine;
—— *verticale*, vertical engine;
—— *verticale à action directe*, direct-acting vertical engine;
—— à *vitesse moyenne*, medium-speed engine;
—— à *volant rabattant*, type of engine in which the fly wheel (upper part) turns toward the steam cylinder.

II. Operating machines; miscellaneous:
—— *s accolées*, separate engines, working one and the same shaft;
—— *accouplée*, coupled engine;
—— à *affinité*, ammomia freezing apparatus;
—— à *affûter*, tool-grinding machine;
—— à *aiguiser*, tool-grinder;
—— à *air*, compressed-air freezing machine;
—— à *air chaud*, hot-air engine;
—— à *air comprimé*, compressed-air engine;
—— à *aléser*, fine borer, boring machine, cylinder-boring machine;
—— à *aléser et à mandriner*, chucking machine;
—— *alimentaire*, feeding engine;
—— *alternative*, alternating (blowing) engine;
—— à *arracher les pilotis*, pile drawer;
—— à *arrondir*, finishing engine;
—— *aspirante*, exhausting (ventilating) engine;
—— à *aubes*, paddle engine;
—— *automatique à remandriner*, (sm. a.) automatic resizing machine;
—— *auxiliaire*, donkey engine, auxiliary engine;
—— à *barrette*, (sm. a.) pistol-rifling machine;
—— à *battre les pieux*, pile driver;
—— à *bois*, woodworking machine;
—— à *brique*, brick machine;
—— à *buriner*, paring machine;
—— à *cadre*, diamond-frame bicycle;
—— à *calculer*, calculating machine;
—— à *calibrer et à fraiser*, (sm. a.) gauging and reaming machine;
—— à *calibrer et à sertir*, (sm. a.) gauging and crimping machine;
—— *calorique*, v. —— à *air chaud*;
—— à *camouflets*, (mil. min.) camouflet borer;
—— à *canneler*, fluting machine; obsolete term for —— à *rayer*;
—— *centrifuge*, centrifugal machine;
—— à *cercler*, hooping machine;
—— à *chaînette*, rifling machine (worked by a chain, obs.);
chambre de la ——, engine room;
—— à *charger les flambeaux*, (artif.) packing or loading tool;
—— à *cingler*, (met.) squeezer;
—— à *cintrer*, (plate-) bending machine;
—— à *cintrer les tôles*, plate-bending machine;
—— à *cintrer les tuyaux*, tube-bending machine;
—— à *cinq culots*, (sm. a.) machine for making cartridge cases (obs.);
—— à *cisailler*, shears;
—— à *cisailler à excentrique* (à *levier*), eccentric (lever) shears;
—— à *colonne d'eau*, pressure engine, water-pressure engine, hydraulic engine, hydraulic-pressure engine;
—— à *colonne d'eau à double* (*simple*) *effet*, double- (single-) acting pressure engine;
—— *de commettage*, rope-laying machine;
—— *composée*, compound machine (as distinquished from simple);
—— à *composer*, typesetting machine;
—— à *comprimer*, (sm. a.) bullet compressor;
—— à *comprimer l'air*, air-compressing engine;
conducteur de ——, engine driver, engineer;
—— *s conjuguées*, v. —— *s, accolées*;
constructeur de ——*s*, engine builder, engine designer;
—— à *copier*, copying machine;
—— à *corder*, cording machine;
—— à *coudre*, sewing machine;
—— à *couper*, cutting machine, cutter;
—— *de course*, racing bicycle;

machine à *creuser*, dredger;
— à *curer*, v. —— à *creuser;*
— *de dame*, woman's bicycle;
— à *déblayer*, excavator, excavating machine;
— à *débourber*, dredger;
— à *décolleter*, screw machine;
— *de demi-route*, semiroadster (bicycle);
démonter une ——, to take an engine to pieces;
— *dénivelée*, machine, engine, out of line;
— à *désabler*, foundry rattler;
— à *diviser*, dividing machine;
— à *draguer*, dredger;
dresser une ——, to line up an engine;
— à *dresser les écrous*, nut shaper;
— à *échelles mobiles*, v. —— à *monter;*
— à *écrire*, typewriter;
— *électro-mineur*, (*mil. min.*) excavator;
— à *élever l'eau*, water-raising engine;
— à *emboutir*, chaser, chasing machine;
— à *émouler*, grinding machine;
— à *enfoncer les pieux*, pile driver;
— à *engrenage*, geared engine;
— à *enrouler*, (*art.*) winding machine;
— à *envelopper les balles*, (*sm. a.*) jacket-seating machine;
— *d'épuisement*, pumping engine;
— *d'épuisement à pompe*, pumping engine;
— *d'essai*, testing machine;
— à *essayer les huiles*, oil tester;
— à *essayer les matières premières*, testing machine;
— à *estamper*, swage tool;
— à *étaler*, (*cord.*) spreader;
— à *étamper*, v. —— à *estamper;*
— à *étirer*, drawing frame;
— à *étrangler*, choking frame;
— à *excavation*, excavating machine;
— *d'extraction*, hauling, hoisting, winding, engine;
— *d'extraction à vapeur*, steam hoist, steam hoisting engine;
— à *façonner*, shaper, shaping machine;
— à *façonner dite à reproduction*, v. —— à *façonner;*
faire —— *arrière, en arrière*, to reverse the engine, go astern;
faire —— *avant, en avant*, to go ahead;
— *de fatigue*, a bicycle for heavy, rough work;
— à *faucher*, v. —— *récolter;*
— à *fendre les dents*, (*roues*) tooth- (wheel-) cutting engine;
— à *fendre le fer*, slitters, cutters;
— à *fendre les vis*, screw-slotting machine;
— à *filer*, spinning frame, machine; (*cord.*) rope-spinning machine;
— à *fileter*, screw-cutting lathe;
— à *force centrifuge*, centrifugal machine;
— à *forer*, borer, boring machine;
— à *forer* (*les canons*), (*art.*) gun lathe;
— à *foret fixe*, fixed drill (work moves);
— à *forets rotatifs*, boring machine with rotating drills;
— à *foret tournant*, revolving drill (work fixed);
— à *forger*, forger, forging machine;
— à *fraiser*, milling machine; shaper, shaping machine;
— à *fraiser ordinaire*, plain milling machine;
— à *fraiser suivant gabarit*, copying machine;
— à *fraiser verticale*, vertical spindle milling machine;
— *française*, v. —— à *raboter;*
— *frigorifique*, refrigerating machine;
— *frigorifique à air*, v. —— à *air;*
— à *gaz*, gas engine; freezing machine (using a volatile gas);
— à *goutine*, swage tool;
— à *gouverner*, steam steering gear;
— *de guerre*, (*mil.*) (any) engine of war;
— à *guillocher*, rose engine;
— à *hélice*, screw engine;
— *de hissage*, hoisting engine;
— à *huile minérale*, oil engine;
— *hydraulique*, hydraulic engine;
— *infernale*, (*mil.*) infernal machine, torpedo (obs.);

machine à *jet de sable*, sand blast, sand jet;
— *jumelles*, twin engines;
— à *laminer* (*les canons de fusil*), (*sm. a.*) barrel roll;
— à *levier*, any lever engine or machine, lever-testing machine;
— à *limer*, filing machine;
— à *manomètre*, manometric testing machine;
— à *maquer*, (*met.*) squeezer;
— *marteau-pilon*, steam hammer;
— *marteau à vapeur*, steam hammer;
— à *mâter*, sheers; sheer hulks;
mettre la —— *en arrière*, to reverse the engine;
— à *meuler*, grinding machine;
— *de mise en marche*, starting engine;
— *mobile*, transportable engine;
— à *molettes*, winding engine, whin, whim;
— à *monter*, (*min.*) man-engine, mining engine;
monter une ——, to set up an engine, to put an engine together;
monteur de ——, engine fitter;
— à *mortaiser*, slotting, mortising machine;
— *motrice*, generic term for motor in widest sense; any engine furnishing power, prime mover;
— *motrice à gaz*, gas engine;
— à *mouler*, (*fond.*) molding machine;
— à *moulures*, molding machine;
niveler une ——, v. *dresser une* ——;
— *de n chevaux*, an n-horsepower engine;
— *nourricière*, v. —— *alimentaire;*
— à *onglet*, (*carp.*) miter block;
— *opératrice*, any engine in which power is used to perform some specific work; operating machine;
— *oscillante*, oscillating (blowing) engine;
— *-outil*, machine tool;
outil- ——, v. —— *-outil;*
— à *peigner*, (*cord.*) heckle, hatchel;
— à *peler les pommes de terre*, potato-peeler;
— à *percer*, punching machine, rock drill;
— à *percer à colonne*, column-drilling machine;
— à *percer radiale*, radial drill;
— à *percer les rails*, rail drill;
— à *percer verticale*, vertical drill;
— à *perforatrice*, rock-boring machine;
— à *pétrole*, oil engine;
— *-pilote*, (*r. r.*) pilot-engine;
— à *planer*, planing machine;
— à *plateau*, any machine in which the work is placed on a bed or table (*plateau*); single-lever testing machine;
— à *plier les tôles*, plate-bending machine;
— *pneumatique*, air pump, pneumatic machine; ventilator (as for mines);
— à *poinçonner*, punch, punching machine;
— à *poinçonner à main*, hand punch;
— à *polir*, polishing machine;
— à *poser les grains de lumière*, (*art.*) vent drill (obs.);
pourvu de ——*s*, engined;
— *de précision*, generic term for any machine doing accurate work;
— à *pression*, press; (*sm. a.*) bullet-compressing machine;
— à *pression d'eau*, water-pressure engine;
— à *profiler*, profiling machine;
— à *propulsion hydraulique*, jet propeller engine;
puits de ——, engine-pit, -shaft;
— à *pulvériser et à corroyer les terres*, (*fond.*) loam mill;
— à *quart d'effet*, an engine whose shaft makes two revolutions in one complete cycle (e. g., Otto engine);
— à *raboter*, planing machine, planer;
— à *raboter à outil* (*plateau*) *tournant*, planer in which the tool (the work) turns;
— *raboteuse*, v. —— à *raboter;*
— à *rainure(s)*, grooving machine; a synonym of —— à *barrette*, q. v.;
— à *rayer*, (*art., sm. a.*) rifling machine;
— à *rayer à guide latéral*, (*art.*) rifling machine in which the tracer is guided by a groove in a plate;
— *réceptrice*, driven engine;

machine à *récolter*, mowing machine;
— à *recouper, fraiser et calibrer*, (*sm. a.*) trimming, reaming, and gauging machine (for the mouth of the cartridge case);
— à *rectifier les surfaces*, grinding machine;
— à *rectifier les surfaces planes*, surface-grinding machine;
— à *refouler*, upsetter;
— *de renfort*, auxiliary, assistant engine;
— à *reproduire*, copying machine;
— à *retour à vide*, machine or tool that does work in one direction only;
— à *river*, riveting machine;
— à *romaine*, testing machine on steelyard principle;
— à *rotation*, v. — *rotative;*
— *rotative*, rotary engine; rotary blower;
— *de route*, — *routière*, roadster (bicycle);
— à *scier*, sawing machine, saw;
— à *sertir*, (*art.*) band seater; (*sm. a.*) crimping machine;
— *simple*, simple machine;
— à *sonder*, sounding machine;
— *soufflante*, blowing, blast, engine; fan;
— *soufflante à colonne d'eau*, water-pressure blowing machine;
— *soufflante à injection*, steam-jet blower, apparatus;
— *soufflante à piston*, cylinder blowing engine;
— *soufflante à rotation*, rotary blowing machine;
— *soufflante à tiroirs*, slide-valve blowing engine;
— à *soufflet*, v. — *soufflante;*
— à *tailler les engrenages*, gear-cutting machine;
— à *tailler les vis*, screw-cutting lathe;
— à *tambour*, treadwheel;
— à *tarauder*, tapping machine;
tender de —, (*r. r.*) engine tender;
— à *tenons*, tenoning machine;
— à *tête*, cutter head (of a boring machine);
— *thermique*, heat engine;
— *sans tige*, v. — *soufflante;*
— à *tourillonner*, trunnion lathe;
— à *tourner (le bourrelet)*, (*sm. a.*) flange shaper (reloading tool);
— *de traction*, testing machine;
— à *travailler le bois*, woodworking machine;
— à *triturer*, (*artif.*) triturating apparatus;
— à *vapeur pour le changement de marche*, steam reversing gear;
— à *vapeur de mise en marche*, starting cylinder, starting engine;
— à *vent*, generic term for blowing engines, fans, ventilators.

III. Electricity:
— *alternative*, alternating-current dynamo;
— à *anneau plat*, flat-ring dynamo;
— *compound*, compound-wound dynamo;
— *compound à court shunt*, short-shunt compound-wound dynamo;
— *compound à long shunt*, long-shunt compound-wound dynamo;
— *continue*, continuous-current dynamo;
— à *courants alternatifs*, alternating-current machine;
— à *courants continus*, continuous-current dynamo or machine;
— à *courants redressés*, unidirectional dynamo;
— *dynamo*, dynamo;
— *dynamo-électrique*, dynamo-electric machine; dynamo;
— *dynamo-électrique auto-excitatrice*, self-exciting dynamo;
— *dynamo-électrique biphasée*, two-phase, diphase dynamo;
— *dynamo-électrique bipolaire*, bipolar machine;
— *dynamo-électrique hypercompound*, hypercompound dynamo;
— *dynamo-électrique à excitation dérivée*, shunt dynamo; shunt-wound dynamo;
— *dynamo-électrique à excitation en double circuit*, compound dynamo;
— *dynamo-électrique à excitation indépendante*, separately excited dynamo;

machine *dynamo-électrique à excitation simple*, series dynamo;
— *dynamo-électrique à intensité constante*, constant-current dynamo;
— *dynamo-électrique multipolaire*, multipolar machine;
— *dynamo-électrique à potentiel constant*, constant-potential dynamo;
— à *effets de quantité*, quantity dynamo-electric machine;
— à *effets de tension*, tension dynamo-electric machine;
— *électrique*, (any) electric machine;
— *électrique à cylindre*, cylinder electrical machine;
— *électrique à friction*, friction machine;
— *électrique à frottement*, friction machine;
— *électrique par influence*, induction machine;
— *électrique à plateau*, plate machine;
— *électrodynamique*, electrodynamic machine;
— *électromagnétique*, electromagnetic machine;
— *électrophorique*, electrophoric machine;
— *électrostatique*, electrostatic machine;
— à *excitation composée*, compound dynamo;
— à *excitation dérivée*, shunt-wound dynamo, shunt dynamo;
— à *excitation indépendante*, separately excited dynamo;
— à *excitation simple*, series-wound dynamo, or series dynamo;
— *-excitatrice*, exciter;
— *excitée en dérivation*, shunt dynamo;
— *excitée en série*, series dynamo;
— *génératrice*, generator, generating dynamo;
— *hydro-électrique*, hydro-electric machine;
— *d'induction*, induction machine;
— *magnéto*, magneto-electric machine;
— *-magnéto-électrique*, magneto-electric machine;
— à *la main*, hand machine;
— *multipolaire*, multipolar machine;
— *photo-électrique*, searchlight;
— *polyphasée*, polyphase-current machine;
— *rhéostatique*, rheostatic machine;
— *en série*, series-wound dynamo, series dynamo;
— *triphasée*, three-phase dynamo;
— *unipolaire*, unipolar machine.

machinerie, f., machinery.
machiniste, m., machinist;
— *de locomotive*, (*r. r.*) engine driver, locomotive engineer.
mâchoire, f., jaw; (*mach., etc.*) clamp, jaw of a clamp, fork, clip, dog; (*hipp.*) jaw tooth with the mark; (*sm. a.*) clip;
— *de bielle*, (*mach.*) connecting-rod fork;
chargeur à —, (*sm. a.*) clip;
— *cylindrique*, (*r. f. art.*) sleeve;
— *d'étau*, vise jaw;
— *de frein*, brake jaw, clamp.
macmahon, m., (*mil. slang*) head of Medusa on top of helmet.
maçon, m., mason; bricklayer;
aide-—, hod carrier, mortar mixer;
— *en brique*, briqueteur, bricklayer;
maître —, master mason;
— *en pierre*, — *en pierres de taille*, stone mason;
— *piseur*, pisé-mason;
— *-poseur*, rough mason.
maçonnage, m., mason's work.
maçonner, v. a., to do masonry work; to wall up, block up (a door, a window);
— *en degrés*, to build up stair-wise;
— *de remplage*, to do coffer work;
— *par retraites*, v. — *en degrés.*
maçonnerie, f., masonry, stonework, brickwork; bond;
— *appareillée*, ashlar masonry;
— à *assises réglées*, coursed masonry;
— *de béton*, concrete work or construction;
— *de, en, blocage*, filling in, rubblework;
— *de blocaille*, rubblework;
bloquer la —, in foundations, to place the foundation blocks against the face of the excavation;

maçonnerie *de, en, briques*, brickwork;
— *brute*, rubble masonry;
— *croisée*, cross bond;
— *à, en, échiquier*, reticulated work;
— *de fondement*, foundation work;
— *de fourneau*, — *de foyer*, brickwork of a furnace;
— *grossière*, rubblework;
— *hourdée*, mortar masonry (as distinguished from dry work);
— *hydraulique*, waterproof masonry;
— *en liaison*, bonded work;
— *de libage*, masonry of large rough-dressed stones;
— *maillée*, net masonry;
— *de moellons*, ashlar work;
— *de pierres sèches*, dry work;
— *en pierres de taille*, ashlar masonry (dressed stone);
— *de remplissage*, filling, filling in;
— *réticulée*, reticulated work;
— *de soubassement*, v. — *de fondement;*
— *vive*, ashlar masonry (dressed stone).

macrée, f., bore (in rivers).
macule, f., spot, stain, blemish.
madré, a., spotted.
madrier, m., thick, plank; (*art.*) shifting plank; deck plank; platform plank; (*pont.*) chess;
bout de —, (*art.*) short plank;
demi- —, (*pont.*) half chess;
— *d'équipement*, (*art.*) plank used in rigging the gin;
— *gîte*, (*art.*) ground plank, bottom plank, of a gun platform;
— *libre de crosse*, (*art.*) loose plank under the trail (hence easily renewable);
— *de pétard*, (*mil. min.*) board to which a petard is fixed;
— *de plateforme*, (*art.*) deck plank;
— *de pont*, (*pont.*) chess;
poser des —*s*, (*pont.*) to lay chess.

madrure, f., mark, speckle in wood.
maëstral, m., mistral, northwest wind (Mediterranean).
magasin, m., shop; magazine, storehouse, storeroom; provision, store, stock; (*art. fort.*) magazine; (*sm. a.*) magazine; (*mil. slang*) military school;
— *administratif*, (*Fr. a.*) administrative depot or magazine (as distinguished from a troop-unit magazine); receives raw materials, contractors' materials, and acts as a depot of issue to troops (two classes—*intendance*, including subsistence and quartermaster depots; medical, including hospitals and medical supply stores); quartermaster (subsistence) depot (U. S. A.);
— *administratif central*, (*Fr. a.*) depot of issue for current service charged with the manufacture of effects for one, two, or three army corps;
— *administratif général*, (*Fr. a.*) general service depot containing reserve supplies of cloth and of made-up articles;
— *administratif régional*, (*Fr. a.*) depot of issue of made-up articles for current use (has no workshop);
arme à —, (*sm. a.*) magazine rifle;
— *d'armes*, (*mil.*) armory;
— *d'artillerie*, (*siege*) artillery storehouse;
— *automatique*, (*sm. a.*) automatic magazine; loader;
— *de batterie*, (*fort.*) expense magazine;
— *de bouche*, (*mil.*) subsistence depot or storehouse;
— *à cartouches*, (*fort.*) expense magazine in parapet;
— *-caverne*, (*fort.*) powder magazine excavated in rock;
— *central*, (*sm. a.*) central magazine; (*siege*) general magazine for a group of batteries; (*adm.*) v. — *administratif central;*
— *commun*, (*Fr. a.*) magazine in which administration and regimental stores are kept together; mixed magazine;
— *pour la consommation journalière*, (*siege*) expense magazine in or near the battery it supplies;

— *de corps d'troupe*, (*mil.*) regimental storehouse, magazine;
— *à coton-poudre, etc.*, (*fort.*) gun cotton, etc., magazine;
— *divisionnaire*, (*Fr. a.*) in Algiers, the magazine corresponding to the — *régional* of France;
en —, in stock;
— *fixe*, (*sm. a.*) magazine proper; fixed magazine; (*adm.*) stationary magazine, as distinguished from — *roulant*, q. v.;
— *à fourrages*, forage shed, granary;
garde- —, storekeeper;
— *à gargousses*, (*fort., siege*) powder or cartridge magazine of a battery; expense magazine;
— *général*, v. — *administratif général;*
grand — *à poudre*, (*fort.*) main magazine;
— *de guerre*, (*mil.*) ammunition magazine;
— *horizontal*, (*sm. a.*) horizontal magazine;
— *hors classe*, (*Fr. a.*) an administrative magazine of which a special bond is required;
— *intermédiaire*, (*fort.*) magazine common to several batteries, set up, first, when no convenient *magasin de batterie* can be established; second, when the *magasins de secteur* are too far off;
— *mobile*, (*sm. a.*) loader; quick loader; loading clip, clip; detachable magazine;
— *à munitions*, (*mil.*) ammunition magazine;
— *à munitions central*, (*siege*) general supply magazine for a group of batteries (established between the park and the batteries);
— *à munitions d'infanterie*, (*siege*) cartridge magazine;
— *particulier*, (*mil.*) the storehouse of each administrative unit;
petit — *de batterie*, (*fort.*) expense magazine;
— *-pontet*, (*sm. a.*) magazine and guard of the Spanish Mauser;
— *à poudre*, — *aux poudres*, (*mil.*) powder magazine;
— *à poudre central*, (*siege*) magazine of a group of batteries, about 1,000 meters in rear of this group;
— *à poudre principal*, (*siege*) main magazine, containing the supply of powder for a siege (about 10 kilometers from the outer line of works of the place besieged);
— *régional*, v. — *administratif régional;*
— *à répétition occasionnelle*, (*sm. a.*) detachable magazine;
— *de réserve*, (*mil.*) permanent magazine for reserve stores in large quantities;
— *roulant*, (*mil.*) mobile magazine, generic expression for supply train;
— *de secteur*, (*fort.*) main magazine of a sector of defense;
— *de service*, expense magazine;
station- —, (*mil.*) main supply reserve established in a railway station;
— *du temps de guerre*, (*Fr. a.*) service powder-magazine, i. e., filled in war from the supplies of the — *du temps de paix;*
— *du temps de paix*, (*Fr. a.*) ordinary powder-magazine;
— *tubulaire*, (*sm. a.*) tubular magazine;
— *aux vivres*, (*mil.*) subsistence storehouse or magazine; provision storehouse.

magasinage, m., storage, warehousing.
magasinier, m., storekeeper.
magdaléon, m., roll or stick (of sulphur, etc.).
magistral, a., magistral, principal, base;
galerie —*e*, (*fort.*) principal, main gallery of a work;
ligne —*e*, magistral line of a plan, etc.; (*fort.*) magistral, interior crest.
magistrale, f., (*fort.*) magistral.
magistrat, m., magistrate.
magistrature, f., magistracy, **magistrates.**
magma, m., magma.
magnésie, f., magnesia.
magnésium, m., magnesium.
magnétique, a., magnetic.
magnétiser, v. a., to magnetize.

magnétisme, m, magnetism;
— *actif*, active magnetism;
— *apparent*, apparent magnetism;
— *condensé*, bound magnetism;
— *induit*, induced magnetism;
— *libre*, free magnetism;
— *de position*, magnetism of position;
— *rémanent*, residual magnetism;
— *de rotation*, magnetism of rotation;
— *terrestre*, terrestrial magnetism.

magnéto, f., (*elec.*) magneto, magneto-generator; magneto-electric dynamo machine.

magnéto-électricité, f., magneto-electricity.

magnéto-électrique, a., magneto-electric.

magnétographe, m., magnetograph.

magnétomètre, m., magnetometer.

mahogon, m., mahogany.

maie, f., kneading trough, dipping trough (for tarred rope); screening trough (for saltpeter).

maigre, m., (*hydr.*) shoal water (in a river); (*carp.*) weak spot in a timber; gap in a joint;
cheval chargé de —, thin hack;
prendre le — *de la pierre*, to mark out a stone for dressing.

maigre, a., lean, spare, slight, slender, scant; emaciated, mere skin and bones (of a horse); (of coal) nonbituminous; poor, meager (of lime); too small (of stones, columns, etc.);
angle —, acute angle;
en angle —, acute-angled;
en —, sharply;
estamper —, v. s. v. *estamper*;
houille —, close-burning coal.

mail, m., mallet, maul.

maille, f., mesh; link, chain link; eye or loop of a pin or bolt; radial crack (of trees); towing rope; (*harn.*) elongated eye or loop (*art.*) spring of a fuse drawer; clevis of a gin;
— *des attelles*, (*harn.*) breast-strap link;
— *courte*, short link;
débiter contre (*sur*) —, to saw or cut up timber perpendicularly (parallel) to the medullary grain;
— *double*, double link;
— *à écrou*, screw eye;
— *à étai*, stayed link, studded link;
— *sans étai*, unstayed link;
— *étranglée*, figure-of-eight link;
— *à piton*, link and eye;
— *à poignée*, link and handle;
— *de suspension*, (*pont.*) trestle-cap ring;
— *à talon*, link and stud.

maillé, p. p., reticulated (of masonry);
fer —, iron grating.

maillechort, m., German silver.

mailler, v. a., to grate, to lattice.

mailler, m., pack horse.

maillet, m., mallet;
— *à*, *de*, *calfat*, calking mallet;
— *de charpentier*, carpenter's mallet;
— *chasse-fusée*, (*art.*) fuse mallet;
— *à épisser*, (*cord.*) splicing mallet;
— *à fourrer*, (*cord.*) serving mallet;
— *à frapper*, driving mallet;
gros —, maul;
— *à pointes*, (*powd.*) pointed mallet for hand granulation of large-grain powders.

mailletage, m., sheathing of nails (for submerged woodwork).

mailloche, f., beetle, large mallet, maul; (*cord.*) serving mallet;
— *à fourrer*, (*cord.*) serving mallet.

maillon, m., link, chain link; stitch; small ring; chain length (12 fathoms); (*cord.*) noose, running knot, rolling hitch; slip knot;
— *d'affourche*, mooring swivel;
— *d'assemblage*, end link (of a cable);
— *de chaîne*, cable link;
— *double*, twin link;
— *à émerillon*, swivel link;
— *à étai*, stayed, studded, link;
— *sans étai*, plain link, unstayed link;
— *à étançon*, stayed link;
— *étançonné*, stayed link;
— *étroit*, close link;
— *extrême*, v. — *d'assemblage*;
large —, round link;
— *à renfort*, v. — *d'assemblage*;
— *à tige*, stem link;
— *tournant*, swivel.

maillonner, v. a., to link.

maillot, m., sweater.

main, f., hand; hook; handle; fork; body loop (of a carriage); scoop, hand shovel; waiter, plate; (*fig.*) help, assistance; (*man.*) bridle hand; (*powd.*) copper scraper.
I. Military; II. Manège and hippology; III. Miscellaneous.
I. Military.
— *armée*, armed force;
à — *armée*, with arms in the hand, violently;
aux —*s*, engaged, in action;
avoir dans la —, to have under one's control, in hand;
avoir la troupe en —, to keep troops in hand, have them under control;
— *de chargement*, (*r. f. art.*) feed case;
combat de —, *de* — *à* —, hand-to-hand fight;
coup de —, sudden attack, surprise, bold stroke;
dans la —, under control (as, troops in action);
en être aux —*s*, *être aux* —*s*, to fight, to be engaged;
en venir aux —*s*, to come to blows, to engage or begin the fight;
faire — *basse sur*, to show no mercy; to put all to the sword; to seize everything that falls in one's hands; to pillage;
faire sa —, to plunder, pillage;
— *de fer*, very severe discipline;
la — *dans le rang*, the hand at the side;
mettre aux —*s*, to excite war between;
mettre l'épée à la —, to draw; to fight;
— *à poudre*, (*powd.*) copper powder-ladle;
renvoyer la — *dans le rang*, to drop the hand by the side;
se donner la —, to join hands, to effect a junction.

II. Manège and hippology:
appui de la —, bearing on the hand;
arrière- —, hind quarters;
avant- —, fore quarters, forehand;
avoir l'appui, la bouche, à pleine —, to have a good steady mouth;
avoir la — *dure*, to ride with a heavy hand; to bear on the horse's mouth;
avoir la — *légère*, to have an easy hand;
n'avoir pas de —, to use the hand improperly;
baisser la —, v. *lâcher la* —;
— *basse*, the bridle hand;
battre à la —, to throw the head up and down;
— *de la bride*, bridle hand;
bride en —, well in hand;
changement de —, change of hands;
changer de —, to change hands;
cheval bien fait de la — *en arrière*, horse with fine hind quarters;
cheval bien fait de la — *en avant*, horse with a fine forehand;
cheval de —, led horse;
cheval dans la —, well-trained horse;
cheval à deux —*s*, horse fit to ride and to drive;
cheval hors la —, near horse; hard-mouthed horse;
cheval sous la —, off or hand horse;
cheval à toutes —*s*, horse fit to ride or drive;
cheval à une —, horse fit for a single purpose only;
contre-changement de —, v. *contre-changement*;
contre-changer de —, v. *contre-changer*;
en —, (to take a horse) in hand;
entier à une —, v. s. v. *entier*;
entier aux deux —*s*, v. s. v. *entier*;
— *en arrière*, hind quarters;
en arrière de la —, (lit.) back of the hand; said of a horse that backs and rears;
— *en avant*, fore quarters;

main, *en avant de la* ——, (lit.) forward of the hand, said of a horse that plunges, etc.;
—— *de l'épée*, sword hand;
être à —— *droite* (*gauche*), to work to the right (left) (right (left) foot leading);
être bien dans la ——, to obey the hand readily;
forcer la ——, to pull on the arm; to pull, to try to jerk the reins out of the hand;
franchir la ——, to pull completely out of control of the hand;
gagner à, de, la ——, to get his head (away from the control of the horseman);
—— *de la gaule*, v. —— *de l'épée*;
—— *haute*, the right hand;
hors ——, *hors la* ——, off side, off;
—— *ignorante*, unskilful horseman;
lâcher la —— (*à un cheval*), to ride with a slack hand, to give a horse his head;
—— *de la lance*, v. —— *de l'épée*;
léger à la ——, easy on the hand (of the horse);
marcher à —— *droite* (*gauche*), to turn, gallop, etc., to the right (left), right (left) flank of horse inside (riding school);
mener un cheval en ——, to lead a horse in hand;
mener un cheval haut la ——, to raise the hand (e. g., to keep a horse from stumbling);
mettre un cheval dans la ——, to gather a horse;
mise de ——, feeling the bit (by the hand);
mise en ——, getting a horse in hand;
partir de la ——, to take the gallop well; (substantively) a fine start (gallop);
peser à la ——, to bear upon the hand;
pousser un cheval sur la ——, to make the horse constantly feel the hand of the rider, in advancing;
prendre en ——, to lead (horses);
remise de ——, easing of the hand on the bit;
rendre la —— *à un cheval*, to give a horse his head;
—— *savante*, skillful horseman;
sensible à la ——, v. *léger à la main*;
sentir un cheval dans la ——, to feel that a horse answers to the bit;
sous la ——, near side, near;
soutenir la ——, to pull the bridle, to raise the hand;
tenir la ——, to raise the hand;
tenir la —— *haute*, to ride with a high hand;
tenir la —— *à un cheval*, to guide a horse;
tenir un cheval dans la ——, to keep a horse in hand;
tirer à la ——, to pull (of a horse);
tourner à toutes —— *s*, to take all the gaits easily;
—— *de travail*, iron bar of a trave or travis;
travailler à —— *droite* (*gauche*), v. *marcher à* —— *droite* (*gauche*).

III. Miscellaneous.

à la ——, by hand; at hand;
amarrer bonne ——, (cord.) to lash, secure, without slacking; to keep;
—— *avant* ——, hand over hand;
avoir de la ——, (fenc.) to be skillful in evading parries;
avoir la —— *dure* (*légère*), to make a harsh (mild) use of one's authority;
avoir les armes belles, bien, à la ——, (fenc.) to be a graceful fencer;
côté de la ——, (met.) working side;
—— *-coulante*, hand rail;
—— *-courante*, blotter;
—— *-courante de fer*, goose neck;
de la ——, manual;
—— *à denrées*, sort of grocer's ladle (for sugar, salt, etc.);
la dernière ——, the finishing stroke or touch;
à —— *droite* (*gauche*), on the right (left) hand;
s'entretenir la ——, to keep one's hand in;
—— *de fer*, spider, cramp, iron hanging clamp, iron fair leader;
haut la ——, with ease, "hands down;"
à —— *levée*, free hand;
longueur de la ——, span;
mettre les armes, le fleuret, à la ——, (fenc.) to give the first instruction in fencing;

main, *mettre la dernière* —— *à*, to put the finishing touch to anything;
—— *d'olive*, (harn.) trace-tug loop;
—— *de papier*, quire of paper;
posé à la ——, hand set;
—— *de poulie*, block strap;
prêter la —— *à*, to lend a hand, to help;
scie à deux —— *s*, crosscut saw;
—— *de sergent*, clamp, cramp;
sous la ——, at hand, ready;
—— *sur* ——, hand over hand;
tenir bon à la ——, v. *amarrer bonne* ——;
tenir dans sa ——, to be master of;
tenir la —— *à*, to attend to, see to, a thing;
tour de ——, manual skill or dexterity.

mainotte, f., body loop.

main-d'œuvre, f., labor, manual labor; workmanship, handiwork; cost of labor, wages.

main-forte, f., armed assistance, e. g., the support or assistance of the army in the execution of the laws.

main-levée, f., (law) replevin.

main-militaire, f., the armed force of the State (charged with an execution, an expulsion).

main-mise, f., (law) seizure, detention, arrest, attachment, distress; embargo (ships).

maintenir, v. a., to keep up, keep in good order; to maintain, support, uphold, bear up; (mil.) to keep, hold, in check; to hold;
se ——, (mil.) to hold one's ground.

maintien, m., maintenance (of order, discipline, etc.); keeping in repair.

maire, m., mayor.

mairie, f., municipal headquarters; mayor's office of a French commune; mayoralty.

maïs, m., Indian corn, maize, corn (U. S.).

maison, f., house; home, household; family; (mil.) official household, personal staff;
—— *d'aliénés*, insane asylum;
—— *d'arrêt*, house of detention; (Fr. a.) place of confinement of officers punished for disciplinary reasons; of men going to join *compagnies de discipline*; of soldiers traveling under guard of the gendarmery;
—— *de banque*, bank;
—— *en bois blindé*, (fort.) blockhouse;
—— *de campagne*, (mil. slang) cell;
—— *centrale*, penitentiary;
—— *de commerce*, firm, house;
—— *de correction*, v. —— *de détention*; (Fr. a.) place of confinement of officers undergoing sentence, if not dismissed; of men sentenced for less than one year;
—— *de détention*, house of detention;
—— *de force*, v. —— *de détention*;
—— *de justice*, place of confinement of persons awaiting trial; (Fr. a.) place of confinement of persons undergoing trial and of persons awaiting execution of sentence;
—— *militaire*, in France, the military staff or household of the President; (mil.) military family of a general, the personal guard of a sovereign;
—— *de prêt*, pawnbroker's shop;
—— *de santé*, private asylum;
—— *de ville*, town house.

maisonnage, m., building timber.

maistrance, f., (nav.) warrant officers; dockyard staff.

maître, m., master, chief, head, owner; proprietor; teacher, instructor; foreman; in France, style of lawyers; (nav.) boatswain; (Fr. nav.) second grade of warrant officer (ranks with sergeant-major); (Fr. a.) designation of certain workmen (v. examples *infra*), who rank as *maréchal des logis chef*;
—— *d'académie*, principal of a riding school;
—— *d'armes*, (fenc.) fencing master;
—— *armurier*, (Fr. a.) master armorer (incorrect expression for *chef armurier*);
—— *artificier*, (Fr. a.) foreman of a (artillery) laboratory, master artificer;
avoir les pieds en —— *à danser*, (man.) to turn out the toes too much;
—— *de bâtiment*, master of a merchantman;
—— *de bâton*, (fenc.) single-stick teacher;
—— *bau*, (nav.) midship beam;

maître *câble*, best cable, main cable;
— *calfat*, head calker;
— *de canne*, v. — *de bâton;*
— *canonnier*, (nav.) gunner;
— *canonnier en second*, (nav.) gunner's mate;
— *chargé*, (nav.) warrant officer in charge of stores;
— *charpentier*, (mil., nav.) master carpenter;
— *chevron*, (cons.) principal, main rafter;
— *compagnon*, foreman;
— *constructeur*, (nav.) naval constructor;
contre- ——, foreman; (nav.) boatswain's mate;
— *cordier*, (*Fr. a.*) cordage master;
— *couple*, (nav.) midship frame;
— *d'écluses*, sluice master;
— *école*, schoolmaster; (*hipp.*) steady horse, used in setting the pace in training colts to gallop, or hitched up with a young horse under training; good horse used as a decoy in a sale (a poor horse being substituted on delivery);
— *entrait*, (cons.) tiebeam;
— *d'équipage*, (nav.) boatswain;
— *ès-arts*, master of arts;
— *d'escrime*, (fenc.) fencing master; (*Fr. a.*) regimental fencing master;
— *ès-lois*, jurist;
— *en fait d'armes*, v. — *d'armes;*
— *ferrant*, farrier;
— *forgeron*, (*Fr. a.*) chief blacksmith;
— *garçon*, workman in powder mills;
grand(-) ——, grand master;
— *de hache*, v. — *charpentier,*
— *maçon*, master mason;
— *de manège*, riding master;
— *de manœuvre*, v. — *d'équipage;*
— *maréchal ferrant*, (*Fr. a.*) chief farrier in each mounted regiment (*art., cav.*);
— *mécanicien*, chief engineer;
— *de musique*, (mil.) bandmaster;
— *nageur*, swimming teacher;
— *d'œuvre*, foreman;
— *d'ouvrage*, master workman;
— *ouvrier*, (*Fr. art., eng.*) master workman; (*in gen.*) master workman;
— *pilier*, (cons.) arch pillar;
— *pointeur*, (*Fr. art.*) gunner (seacoast, U. S. A.);
— *de port*, harbor master;
— *de poste*, postmaster;
— *poudrier*, master workman of a powder factory;
premier ——, (*Fr. nav.*) senior warrant officer (ranks with *adjudant*);
— *de quai*, wharfmaster;
— *de quart*, (nav.) boatswain's mate;
quartier- ——, (*Fr. nav.*) petty officer (ranks with a corporal);
se rendre — *du feu*, (art.) to subdue the hostile fire;
second ——, (*Fr. nav.*) grade assimilated to sergeant in the army;
second — *canonnier*, (nav.) gunner's mate;
— *sellier*, (*Fr. a.*) saddler's sergeant;
tambour ——, — *tambour*, (mil.) instructor of drum music;
— *torpilleur*, (nav.) torpedo gunner;
— *voilier*, (nav.) sailmaker.

maîtresse, f., mistress, (in combination) principal, chief;
— *ancre*, best Bower;
— *arche*, principal arch (as of a bridge, etc.);
— *bitte*, pawl bitt of a capstan;
— *ferme*, (cons.) main truss or couple;
— *partie*, (nav.) midship bend, dead flat;
— *pièce*, main timber, main piece;
— *poutre*, (cons.) principal beam, girder;
— *tige de pompe*, main pump-rod;
— *varangue*, (nav.) midship floor;
— *voûte*, main vaulting.

maîtrise, f., mastership of a military order;
grande ——, dignity of grand master.

maîtriser, v. a., to master, control, subdue, overcome, keep under, command.

majesté, f., majesty.

majeur, a., great, greater, important;
force ——*e*, act of God, superior force, circumstances beyond one's control, *vis major.*

major, m., (*Fr. a.*) field officer in immediate charge of the administration and accountability of a regiment (function, not a grade); shortened form of *médecin-major;* (mil. slang) at the Polytechnique, the head of the list, number one (*major* is sometimes used by the French to denote the major of services other than the French);
adjudant- ——, v. s. v. *adjudant;*
aide- ——, v. *aide-major;*
aide- — *général*, (*Fr. a.*) general officer, assistant of the — *général*, assistant chief of staff;
clairon- ——, bugle major;
état- ——, v. *état-major;*
gardien ——, in French ports, a sort of chief of police;
— *de la garnison*, (*Fr. a.*) the executive officer of a garrison, assistant to the *commandant d'armes;* town major;
— *général*, (*Fr. a.*) chief of staff of a group of armies in war (for other cases see under *général;* nearest equivalent, U. S. A., adjutant-general); (*Fr. nav.*) chief of staff;
médecin — *de 1ère (2ème) classe*, (*Fr. a.*) surgeon-major (captain);
— *de place*, town major (obs.);
pilote- ——, v. s. v. *pilote;*
— *de queue*, at the Polytechnique, the last man; the "endth," "endth" man, the "goat," (U. S. M. A.);
ronde- ——, v. s. v. *ronde;*
sergent- ——, (*Fr. a.*) sergeant-major (nearest equivalent, U. S. A., first sergeant);
tambour- ——, drum major;
— *de tranchée*, (siege) field officer of the trenches, assistant to the *général de tranchée* in executive duties, police of the trenches, direction of working parties, etc., (permanent detail, one to each attack);
trompette- ——, trumpet major.

majoration, f., allowance in excess of actual needs, to meet contingencies, e. g.:
— *d'essayage*, clothes, etc., in excess of the number of men to be fitted, available in case clothes issued should not fit.

majorité, f., (*Fr. a.*) function of a *major;* office of a *major.*

makis, m., (Corsican) jungle.

mal, a., bad, poor; wrong; sick, ill.

mal, m., hardship, hard work; harm, hurt, evil; disease, malady;
— *d'âne*, (*hipp.*) scratches;
— *blanc*, white cankerous growth (in the throat of pigeons);
— *de bois*, v. — *de brout;*
— *de brout*, (*hipp.*) disease brought on by brousing on young shoots of trees;
— *caduc*, (*hipp.*) epilepsy;
— *de cerf*, (*hipp.*) tetanus;
— *d'encolure*, (*hipp.*) generic name for contusions, etc., on the upper part of the *encolure;*
— *d' Espagne*, v. — *de feu;*
— *de feu*, (*hipp.*) sleepy staggers;
— *de garrot*, (*hipp.*) fistula of the withers, fistulous withers;
— *de mer*, seasickness;
— *de montagne*, mountain sickness;
— *de nuque*, (*hipp.*) poll evil;
— *du pays*, homesickness;
— *de rognon*, (*hipp.*) saddle gall; inflammation (phlegmon) on the haunches;
— *de taupe*, (*hipp.*) poll evil;
— *de terre*, scurvy;
— *de tête de contagion*, (*hipp.*) anasarca glanders.

malade, a. n., sick; patient;
aux ——*s*, (mil.) sick call;
se faire porter ——, (mil.) to get on the sick-report;
rôle des ——*s*, (mil.) sick-report.

maladie, f., illness, sickness, distemper, complaint;
— *de l'aile*, wing disease (carrier pigeon);
— *des camps*, typhus fever;
— *du coit*, v. *dourine*;
— *naviculaire*, (*hipp.*) navicular disease;
— *de terre*, scurvy;
—*s vieilles de poitrine*, (*hipp.*) chronic affections of lungs or of pleura, or of both.

malandre, f., rotten knot in timber; red or white vein appearing in wood before rotting; (*hipp.*) malanders.

malandreux, a., with dead, rotten, knots.

malastique, (*mil. slang*) dirty, slovenly.

malaxer, v. a., to knead, to work (as, dough, clay); to grind up together under a fluid.

mal-bouché, a., v. *mal-denté*.

mal-denté, a., (*hipp.*) *cheval* —, a horse, the malformation or wear of whose teeth makes it difficult to estimate his age; false-mouthed horse.

mâle, a., male; (of a ship) seaworthy; (of the sea) rough; (*mach.*) male (of screws, teeth, etc., of any part fitting into another).

malebête, f., calker's hammer.

mal-façon, f., poor work, bad workmanship, defect in a piece of work.

malfaisant, a., malignant, unhealthy, noxious; *air* —, malaria.

malgache, a., Malagassy.

maline, f., spring tide;
grande —, equinoctial spring tide.

malingre, a., delicate, weak; m., sickly, weak man.

mallard, m., small grindstone.

malle, f., trunk; mail; (*mil. slang*) prison;
faire (*défaire*) *une* —, to pack (unpack) a trunk.

malléabilité, f., malleability.

malléable, a., malleable.

malléine, f., (*hipp.*) mallein.

malle-poste, f., mail coach; mail steamer.

mallier, m., shaft horse.

mal-rasés, m. pl., (*mil. slang*) sappers.

malsain, a., unhealthy, sickly.

malthe, f., mineral tar.

malversation, f., malversation, embezzlement.

malverser, v. a., to embezzle.

mamelle, f., (*hipp.*, *farr.*) part of horseshoe, of hoof, on each side of the toe; (*harn.*) collar pad; lower part of the saddlebow;
en —, rounded.

mamelon, m., gudgeon; (*farr.*) sort of calkin; (*top.*) hummock; (*farr.*) rounded hill or elevation, pap, mamelon; (*mach.*) swell.

mamelonné, a., (*top.*) rounded (of a hill, height).

mamelouk, mameluk, m., Mameluke.

mancelle, f., (*harn.*) pole chain; thill tug.

manche, m., handle, helve, haft; holder for burning signal lights;
— *d'aviron*, loom of an oar;
— *à sabre*, derogatory term for a soldier;
— *universel*, pad.

manche, f., sleeve; hose, pipe, funnel; neck (of a balloon); channel;
— *d'appendice*, filling tube or sleeve (of a balloon);
— *à cendres*, ash hose;
— *à charbon*, coal chute;
— *de cuir*, leather hose;
— *à eau*, watering hose;
fausse- —, oversleeve; (*St. Cyr slang*) fatigue jacket;
— *flexible*, flexible hose;
— *foulante*, delivery hose;
— *de pompe à incendie*, fire-engine hose;
— *avec raccordement à vis*, screw hose;
— *de toile*, canvas hose;
— *à vent*, wind sail; ventilator.

Manche, *la* —, the English Channel.

manchette, f., (*cord.*) lizard; span, bridle; (*art.*) sleeve, gunner's sleeve;
coup de —, v. s. v. *coup*.

manchon, m., sleeve; cylinder; housing; mouthpiece (as of iron piping); joint ring, collarband; (*art.*) hoop (longer than a *frette*), breech, hoop; jacket; cradle, sleeve; recoil sleeve, muzzle sleeve (of a turret gun); trunnion bushing; dirt guard; (*sm. a.*) collar or sleeve (into which the tubular magazine screws); bolthead; dust cover; firing-pin nut; cocking-piece nut; firing-pin clutch; (*unif.*) distinguishing white cover for headdress at manœuvers; (*mach.*) sleeve, box, clutch, transmission clutch; shaft coupling; coupling box, chuck; sleeve or collar in which an arbor turns; mandrel (of a lathe); flange.

I. Artillery, small arms; II. Machinery.

I. Artillery, small arms;

— *-agrafes*, — *à agrafes*, (*art.*) locking sleeve, locking hoop;
— *arrêt* — *d'arrêt* stop-sleeve or collar;
— *de bouche*, (*art.*) muzzle collar (Longridge wire gun);
— *-enveloppe*, (*sm. a.*) outer case, barrel cove.
— *-glissière*, — *glissière à tourillons*, (*art.*) trunnioned slide;
— *-graisseur*, (*art.*) self-oiling box or sleeve; oiling sponge;
— *de manœuvre*, (*art.*) maneuvering clutch;
— *porte-extracteur*, (*sm. a.*) extractor sleeve;
— *porte-hausse*, (*sm. a.*) rear-sight sleeve (Spanish Mauser);
— *de support de timon*, (*art.*) muff;
— *à T*, (*sm. a.*) T-shaped cocking piece;
— *à, de, tourillons*, (*art.*) trunnion sleeve;
— *vissé*, (*art.*) locking ring.

II. Machinery.

— *d'accouplement*, clutch, coupling box, coupling sleeve; automatic coupling;
— *d'alésoir*, cutter head, boring head;
— *d'arbre à manivelle*, crank-shaft flange;
— *d'arrêt*, stop sleeve or collar;
— *articulé*, jointed clutch;
— *d'assemblage*, assembling collar, coupling flange;
— *d'attache*, securing collar or sleeve;
— *-clavette*, keyed shell clutch;
— *à cônes*, cone coupling;
— *à coquilles*, clamp;
— *cylindrique*, cylinder friction clutch;
— *de débrayage*, clutch coupling;
— *denté*, — *à dents*, claw clutch, claw coupling;
— *à disques cannelés*, friction clutch (sort of multiple cone clutch);
— *à double écrou*, screw coupling box;
— *d'écubier*, hawse pipe;
— *embrayage*, clutch coupling, disengaging coupling; coupling box;
— *d'embrayage et de désembrayage*, clutch coupling box;
— *d'essieu de poulie*, coak of a block;
— *fileté*, screw sleeve;
— *fixe*, fixed or permanent coupling;
— *à frettes*, clamp coupling;
— *à friction*, friction coupling box; friction clutch;
— *à griffe*, claw clutch or coupling; Sharp's coupling;
— *à plateaux*, flanged coupling, face-plate coupling;
— *de pression*, adjustable collar or sleeve (sleeve set by a screw); support;
— *à raccord variable*, adjustable clutch;
— *de refroidissement*, (*met.*) cold-water jacket (of a twyer);
— *de soute à charbon*, coal-bunker pipe;
— *de sûreté*, safety coupling;
— *à tenon*, clutch, tenon clutch;
— *d'une seule pièce*, box coupling, muff coupling;
— *à vis*, screw chuck, screw coupling box.

manchonnage, m., (*art.*) operation of putting on, shrinking on, a hoop (applied more particularly to long hoops).

manchot, m., one-armed, one-handed, man.

mandant, m., person who issues a *mandat*.
mandat, m., mandate, commission, check; *(adm.)* order, request (for transportation, money, rations, etc.);
—— *d'amener, (law)* capias;
—— *d'arrêt, (law)* warrant of arrest;
—— *-carte*, v. —— *-poste;*
—— *de comparution, (law)* summons; subpœna;
—— *de convoi, (Fr. a.)* railroad transportation order;
—— *-dépêche*, v. —— *-télégramme;*
—— *de dépôt*, warrant of commitment;
—— *d'étapes, (Fr. a. adm.)* order for forage and rations (issued to troops on march) to be drawn at une *étape;*
—— *de fournitures, (mil. adm.)* requisition;
—— *d'indemnité de route, (mil. adm.)* order to pay over travel expenses;
lancer un ——, to issue a warrant;
—— *de passage*, order for transportation by sea;
—— *de payement*, an order to pay;
—— *-poste*, —— *de poste*, post-office money order;
—— *-télégramme*, —— *télégraphique*, telegraphic money order;
—— *de vivres, (Fr. a. adm.)* ration request, issued to and presented by troops on the march.
mandataire, m., person to whom a *mandat* is issued.
mandater, v. a., to deliver an order for the payment of, deliver an order for.
mandement, m., order, charge.
mander, v. a., to inform, to order.
mandrin, m., mandrel, drift, punch; shaper, former, shape or form; chuck, plug; coupling; holding-up tool; top and bottom swage for rounding bolts; gauge (for hollow or cylindrical objects); *(expl.)* fuse-hole gauge (gun-cotton disks); *(artif.)* rocket-case former;
—— *d'abatage*, holding-up tool;
—— *à l'anneau*, elastic chuck;
—— *de bourrage, (mil. min.)* tamping rod;
—— *brisé*, elastic chuck, cup chuck;
—— *carré*, square drift;
—— *à chaud*, mandrel, etc. (for working metals hot);
—— *de cochon*, sort of screw chuck;
—— *conique*, taper mandrel;
—— *à disque*, disk chuck;
—— *excentrique*, eccentric chuck;
—— *expansible*, shell chuck;
—— *à expansion*, expanding mandrel;
—— *faible, (expl.)* minimum gauge (must fall of its own weight through the fuse hole of a gun-cotton disk);
—— *à foret*, drill chuck;
—— *de forge*, forge mandrel;
—— *fort, (expl.)* maximum gauge (must not of its own weight fall through the fuse hole of a gun-cotton disk);
—— *à gobelet*, bell chuck, cup chuck;
—— *à griffe(s)*, claw chuck (lathe);
—— *à gueule de loup*, split screw chuck;
—— *méplat*, flat drift;
—— *ordinaire*, plain chuck;
—— *à pointes*, flanch or prong chuck;
—— *porte-foret*, drill chuck;
—— *à raises*, rifled drift;
—— *rond*, round puncheon;
—— *taille*, grooved mandrel;
tour à ——, mandrel lathe;
—— *d'un tour*, face plate; —— *à tourner ovale*, oval chuck;
—— *universel*, concentric chuck, universal chuck;
—— *à vis*, screw chuck.
mandriner, v. a., to form, shape, work, upon a mandrel, etc.; to chuck.
manéage, m., *(nav.)* manual labor of loading and unloading cargo, for which no extra pay is demanded.
manège, m., horsemanship, riding, manege, riding-school equitation, manege riding; training, breaking, of horses; riding school, riding-school building or inclosure; drill hall; *(mach.)* horsepower (i. e., mechanical contrivance), horse whim;
manège cavalier de ——, *(cav.)* trooper attached to a riding school;
cheval de ——, riding-school horse; old, worn-out horse;
—— *à chevaux*, horsepower;
—— *civil*, instruction in riding simply, as distinguished from military riding;
—— *à colliers*, horse mill;
—— *découvert*, ring, outdoor riding school;
—— *de guerre*, an uneven gallop, in which the horse frequently changes the leading foot;
—— *par haut*, high school, training of a horse by the *airs relevés;*
maître de ——, riding master.
manette, f., handle, hand gear; hand lever; spoke of a steering wheel; hook or catch (on the bolster of the French 0.60m siege railroad truck).
manganèse, m., manganese.
manganésifère, a., manganiferous.
mangé, p. p., *(cord.)* galled, worn.
mangeoire, f., manger; crib;
musette- ——, v. s. v. *musette.*
manger, v. a., to eat, feed, mess; to corrode, eat out or away; to wear, to grind; to absorb, destroy, eat up; to undermine;
—— *le chemin*, to advance, proceed, with great speed;
—— *ensemble, (mil.)* to mess together;
—— *de la guerre*, to be present at a combat;
—— *du sable*, to cheat the glass;
se ——, to wear, to be ground down;
—— *un pays*, to lay waste a country.
maniable, a., manageable, tractable; handy, pliable, easy to handle, to work.
maniement, m., knack, trick; touch, handling, management, conduct, use, employment;
—— *des armes, (mil.)* manual of arms;
—— *de la lance, (mil.)* manual of the lance;
—— *du sabre, (mil.)* manual of the sword.
manier, m., touching, handling, feel; *(man.)* action of the feet in the three gaits:
au ——, to, by, the touch.
manier, v. a. n., to touch, handle, treat; to manage, administer, use; to substitute new for old paving stones; *(man.)* to work (of the horse); to obey the rider;
—— *les armes, (mil.)* to execute the manual of arms, to handle arms;
—— *à bout, (cons.)* to repair a roof;
—— *un cheval*, to manage a horse;
difficile à ——, unmanageable;
—— *de ferme à ferme, (man.)* to jump without advancing;
—— *sur les hanches, (man.)* to work uopn the haunches;
—— *sur place*, v. —— *de ferme à ferme;*
—— *près de terre, (man.)* to drag the feet.
manifeste, m., manifesto; (ship's) manifest.
manigaux, m., bellows lever.
manilla, m., *(cord.)* manila hemp.
manillage, m., shackling.
manille, f., shackle; m., *(cord.)* manila rope, manila hemp;
—— *d'affourche*, mooring shackle;
—— *d'ancre*, anchor shackle;
—— *d'arrêt*, stop;
—— *d'assemblage*, assembling shackle;
—— *d'attelage, (r. r.)* screw-coupling loop;
boulon de ——, shackle bolt;
—— *de brague*, breech shackle;
—— *brevetée*, patent shackle;
—— *en chanvre*, —— *en cordage*, hemp rope;
—— *d'étalingure*, anchor shackle;
——*s fausses*, preventer gear;
—— *en fer*, wire rope;
——*s majeures*, heavy ropes;
——*s menues*, light ropes;
——*s principales*, v —— *majeures;*
——*s de supplément*, spare ropes;
—— *traversière*, fish shackle.
maniller, v. a., to shackle (a chain, etc.).
manipulateur, m., manipulator; *(teleg.)* key, telegraph key, sender, transmitter;
—— *automatique*, automatic transmitter;
—— *de court circuit, (elec.)* short-circuiting key;

manipulateur *de décharge*, (*elec.*) discharge key;
—— *à inversion du courant*, (*elec.*) reversing key;
—— *télégraphique*, telegraph key.
manipulation, f., manipulation; (*teleg.*) use of key in sending messages, "sending."
manipuler, v. a., to handle, to work.
manivelle, f., (*mach.*) crank, crank arm, hand crank; hand wheel, winch, handle; (*art.*) plug lever;
arbre à ——, crank shaft;
arbre à —— *composée*, built-up crank shaft;
arbre à deux ——*s*, double crank shaft;
arbre à ——*s rapportées*, built-up crank shaft;
axe à ——, cranked axle;
—— *à bouton*, studded crank handle;
bouton de ——, crank pin;
bras de ——, crank arm; web of a crank shaft; (*art.*) shaft handle;
—— *à charnière*, hinged crank;
—— *de cloche*, bellcrank;
—— *de commande*, operating handwheel;
—— *composée*, double crank; built-up crank;
—— *conduite*, trailing crank;
contre- ——, return crank;
—— *décroche-obturateur*, (*art.*) obturator releasing crank;
—— *à deux branches*, (*art., etc.*) double-branched handle;
—— *à disque*, disk crank, wheel crank;
double ——, double-handled crank;
—— *équilibrée*, balanced crank, disk crank, crank disk, crank plate; (*in pl.*) opposite cranks;
—— *d'équipement*, (*art.*) rigging crank;
—— *de frein*, brake crank;
—— *à main*, hand crank;
—— *menanie*, leading crank;
—— *motrice*, driving crank;
moyeu de ——, crank boss;
œil de la ——, crank eye;
—— *s opposées*, opposite cranks;
—— *ordinaire*, single overhung crank;
—— *à pédale*, pedal crank (bicycle);
plateau- ——, disk crank;
—— *de pointage*, (*art.*) elevating crank, elevating handwheel;
puits à ——, crank pit;
—— *rapportée*, crank handle;
rayon de ——, crank arm;
—— *du régulateur*, throttle lever;
—— *simple*, single crank;
soie de ——, crank pin;
—— *de sûreté*, (*r. f. art.*) safety crank;
—— *du tendeur*, screw coupling lever;
tourillon de ——, crank pin;
—— *à tourteau*, disk crank;
—— *de treuil*, winch handle;
—— *à trois coudes*, three-throw crank;
—— *de la vis de pointage*, (*art.*) elevating screw handle.
manne, f., basket.
mannequin, m., mannikin; vaulting horse; (*t. p.*) dummy target; (*fenc.*) dummy or mannikin (saber and bayonet exercise);
—— -*plongeur*, diving dress.
manneton, m., (*mach.*) crank pin.
mannite, f., (*chem.*) mannite, of sugar of manna.
manœuvre, m., laborer, day laborer, unskilled workman.
manœuvre, f., management, operation, working, handling; (*mil.*) maneuver; any drill or operation or work requiring the cooperation of several soldiers, drill in general (*esp. in pl.*); (more accurately) the application of the evolutions of the drill book to the terrain, having in view the positions and movements of the enemy; (*nav.*) evolution(s) of a ship, of a fleet; (*cord.*) rope, cordage, tackle; (*in pl.*) rigging;
à la ——, at drill;
amarrer une ——, (*cord.*) to make fast a rope;
—— *d'automne*, (*mil.*) fall maneuvers, generic term for all maneuvers on a large scale (e. g., fortress, bridge, alpine, independent cavalry, of armies proper, of corps, divisions, brigades, etc.);
—— *en bande*, (*cord.*) slack rope;
—— *à blanc*, (*r. f. art.*) drill with rapid-fire or machine guns;
—— *s des bouches à feu*, (*art.*) artillery drill;
—— *s avec cadres*, v. s. v. *cadre;*
—— *du canon*, (*art.*) gun drill;
champ de ——, v. s. v. *champ;*
—— *s courantes*, (*cord.*) running rigging;
dépasser une ——, v. s. v. *dépasser;*
donner du mou dans une ——, (*cord.*) to slack a running rope;
—— *dormantes*, (*cord.*) standing rigging;
—— *à double action*, (*mil.*) maneuver or operation in which a problem is carried out by full numbers on both sides; competitive maneuver;
—— *d'eau*, (*fort., mil.*) measures taken to inundate at will the ditches of a work, to flood a road, etc.;
—— *en écrevisse*, (*art.*) cross-lifting;
élonger une ——, (*cord.*) to stretch out a rope;
—— *empêchée*, (*cord.*) foul rope;
—— *avec un ennemi marqué*, (*mil.*) maneuver against an indicated enemy;
—— *avec un ennemi représenté*, (*mil.*) maneuver against an enemy in full force;
fausse ——, wrong movement (as in working a breech mechanism);
—— *s à feu*, (*art.*) firing maneuvers;
—— *s à feu de masses d'artillerie*, (*Fr. art.*) firing maneuvers by artillery masses;
—— *s de force*, (*art.*) mechanical maneuvers;
—— *s de forteresse*, (*mil.*) fortress maneuvers;
grandes —— *s*, (*mil.*) fall maneuvers (especially of armies proper);
—— *avec hypothèse*, (*mil.*) a maneuver in which a theme or problem is worked out;
—— *en ligne(s) extérieure(s) (intérieure(s))*, (*mil.*) maneuver on exterior (interior) lines;
—— *manquée*, poor management, working, handling;
passer une —— *dans une poulie*, (*cord.*) to reeve a block;
—— *s de polygone*, (*mil.*) drill-ground maneuvers;
—— *s de pontage*, —— *de ponts*, (*pont.*) bridge drill, pontoon drill;
—— *en queue de rat*, (*cord.*) pointed rope;
règlement de —— *s*, (*mil.*) drill regulations, drill book;
terrain de —— *s*, (*mil.*) drill ground.
manœuvrer, v. a. n., to work, maneuver; (*mil.*) to maneuver; to drill; to direct the movements of an army, of a fleet, of a regiment, etc.;
faire ——, (*mil.*) to maneuver, drill;
—— *en ligne(s) extérieure(s) (intérieure(s))*, (*mil.*) to operate on exterior (interior) lines;
—— *en troupe*, (*mil.*) to drill in a body (said of recruits, as distinguished from individual drill).
manœuvrier, m., (*mil.*) skillful tactician; a., able to maneuver (i. e., seasoned and instructed); skilled in maneuvering.
manœuvrière, f., *bonne* ——, (*mil.*) of an army, skillful maneuverer.
manomètre, m., manometer, gauge; (*steam*) steam gauge, pressure gauge; (*art.*) pressure gauge;
—— *à air comprimé*, (*steam*) compressed-air gauge;
—— *à air libre*, (*steam*) open air pressure gauge;
—— *crusher*, (*art.*) crusher gauge;
—— *à diaphragme métallique*, diaphragm gauge;
—— *différentiel*, differential gauge;
—— *d'écrasement*, (*art.*) crusher gauge;
—— *enregistreur*, recording gauge;
—— *à mercure*, (*steam*) mercurial gauge; vacuum gauge;
—— *métallique*, (*steam*) metallic steam gauge;
—— *à piston*, (*steam*) steam indicator, piston-pressure gauge;
—— *à ressort*, spring pressure gauge;
—— *à siphon*, siphon pressure gauge;

**manomètre, ** *tube* ——, tube gauge;
—— *du vide,* (*steam*) vacuum gauge.
manométrique, a., pertaining to gauges, manometric.
manquant, m., (*mil.*) absentee from roll call; (*in pl.*) the missing.
manque, m., want, failure, miss; (*man.*) false step, misstep.
manquement, m., fault, failure, want, miss;
—— *à l'appel,* (*mil.*) absence from roll call.
manquer, v. a. n., to fail; to go into bankruptcy; to miss; to be missing; to be short of, to be in want of (water, food, etc.), to fall short of; (*sm. a., art.*) to miss fire, to miss; (*cord.*) to become loose:
—— *à l'appel,* (*mil.*) to be absent from a roll call;
—— *le but,* (*t. p.*) to miss the target;
—— *de fer,* (*farr.*) (of a horseshoe) to be too narrow, too thin;
—— *de monde,* to be shorthanded;
—— *à, de, virer,* (*nav.*) to miss stays.
—— *il ne manque personne,* (*mil.*) all are present, Sir.
mansarde, f., attic, garret; mansard roof; garret window;
fenêtre en ——, garret window;
toit en ——, curb roof, mansard.
manteau, m., cloak; (*in pl.*) leaves of a door; (*unif.*) military cloak; cape (U. S. A.); (*met.*) outer casing of a blast furnace; shell of a mold, cope; (*fort.*) cylindrical part of cupola or turret; metal envelope to exclude gases; (*chem.*) hood;
—— *d'armes,* (*mil.*) conical arm cover, ticking to cover a stand of arms in the field;
—— *de cheminée,* mantelpiece;
faux ——, mantelpiece on brackets;
—— *de fer,* iron bar (supporting the lintel of a chimney);
—— *de guérite,* (*unif.*) sentry's greatcoat;
—— *tente,* v. —— *d'armes;*
—— *tente-abri,* (*mil.*) shelter-tent cloak, or half.
mantelet, m., (*art., fort.*) port shutter; embrasure shutter; mantelet, screen; (*harn.*) harness pad; (*nav.*) deadlight;
—— *brisé,* half port;
faux ——, (*nav.*) false port lid; deadlight;
—— *de sabord,* (*nav.*) port lid;
—— *de sapeur,* (*siege*) screen against musketry fire.
manuel, m., manual, handbook, text-book;
—— *des distances,* (*mil.*) official table of distances.
manufacture, f., manufacture, making, factory, mill;
—— *d'armes,* small-arms factory;
—— *de canon,* gun factory.
manufacturer, v. a., to manufacture.
manufacturier, m., manufacturer.
manutention, f., management, maintenance; administration; overhauling; establishment; (*mil.*) bakery.
manutentionnaire, m., manager.
manutentionner, v. a., (*mil.*) to bake bread for troops.
mappe, f., map, chart (rare).
mappemonde, f., map of the world, chart.
maquette, f., (*sm. a.*) skelp;
—— *double,* double skelp (enough for two barrels);
—— *pour sabre,* skelp for a sword blade;
—— *simple,* single skelp (enough for one barrel).
maquetteur, m., (*sm. a.*) barrel forger, skelp forger.
maquignon, m., horse dealer (frequently a dishonest dealer), horse-jockey, jockey; man who bishops horses.
maquignonnage, m., horse dealing, bishoping, horse-jockeying, jobbing.
maquignonner, v. a., to bishop a horse, to make him up for sale.
maquiller, v. a., to make up;
—— *un gayet,* to paint a horse so as to deceive a purchaser.
maquis, m., v. *makis.*

marabout, m., marabout.
marais, m., (*top.*) swamp, marsh, morass, bog;
—— *salant,* salt, salt pit.
maraudage, m., (*mil.*) marauding.
maraude, f., (*mil.*) marauding, freebooting, plundering;
aller à la ——, *en* ——, to go marauding.
marauder, v. a. n., (*mil.*) to plunder, to pillage, rob.
maraudeur, m., (*mil.*) marauder, freebooter, plunderer.
marbre, m., marble; slab, surface plate; heavy plate (of cast iron, etc., on which tests, etc., may be made); barrel, chain drum; (of painting) marbling;
—— *brèche,* marble composed of angular fragments, united by a calcareous cement;
—— *brocatelle* brocatelle marble;
—— *camelote,* marble of uniform color, susceptible of but slight polish;
—— *chiqueté,* painting to imitate granite;
—— *en contre-passe,* marble sawed out normal to its quarry bed;
—— *coquillier,* shell marble;
—— *dur,* granite;
—— *ébauché,* tooth-chiseled marble;
—— *feint,* (of painting) marbling;
—— *fini,* marble that has received its final finish;
—— *jeté,* painting to imitate porphyry;
—— *dans sa passe,* marble sawed out parallel to its quarry bed;
—— *piqué,* rough-pointed marble;
—— *poudingue,* pudding stone;
—— *statuaire,* statuary marble;
—— *de traçage,* laying-out-work table; tracing-table-block.
marbré, a., (*hipp.*) marbled (of the coat).
marbrière, f., marble quarry.
marbrure, f., marbling (of painting); (*hipp.*) light spot, marble spot (of the coat).
marchand, m., merchant;
—— *de marrons,* (*mil. slang*) officer who looks ill at ease in plain clothes.
marchand, a., commercial, mercantile, trading;
bâtiment ——, merchantman;
capitaine ——, shipmaster;
fer ——, merchant iron;
marine ——*e,* commercial navy;
navire ——, v. *bâtiment* ——;
officier ——, v. *capitaine* ——;
prix ——, market price;
rivière ——*e,* navigable river;
vaisseau ——, v. *bâtiment* ——;
ville ——*e,* commercial city.
marchander, v. a., to ask the price of, to bargain for, to haggle.
marchandise, f., merchandise, goods, freight;
—— *en balles,* piece freight;
——*s chargées en grenier,* merchandise in bulk;
——*s de contrebande,* smuggled goods;
—— *au cubage,* bulk freight;
——*s en forêts,* timber cut for fuel, construction;
——*s légales,* lawful merchandise;
moitié guerre, moitié ——, v. s. v. *guerre;*
——*s de pacotille,* goods, merchandise, of poor quality;
——*s principales,* staple articles;
——*s de traite,* export goods.
marche, f., walk, walking; gait, pace, step; course, conduct, procedure; advance, progress; rate (of a chronometer); behavior (as of an engine, furnace, machine); (*tech.*) pedal, treadle; (*cons.*) stair, step of a stair; (*met.*) fire bridge of a reverberatory furnace; (*mus.*) march; (*nav.*) sailing, speed, headway; (*mil.*) march, day's march; marching; frontier, march;
(Except where otherwise indicated, the following terms relate to the art military.)
—— *accélérée,* march at the rate of about 30 kilometers per day (40 kilometers at the maximum), kept up for several consecutive days;
—— *d'angle,* (*cons.*) longest, or diagonal, winder of a staircase;

marche *d'approche*, (in the infantry combat) the march from the combat position to the enemy's line (1,500 meters); also, march up to about 1,500 meters from the enemy's line, i. e., march to the combat position; any march made by troops about to engage; the approach of the columns of attack in Von Sauer's shortened attack; (in field artillery) march toward the emplacement, estimated from the time when the battery or group begins to take up its emplacement;
arrêter la ——, to impede, hinder a march;
—— *arrière*, march to the rear;
—— *en arrière*, stepping back, retreat, march to the rear; (*mach., etc.*) reversed motion, retrograde motion; (*nav.*) sternway;
—— *d'attaque*, march made in executing an attack;
—— *en avance*, gain of a chronometer;
—— *en avant*, march toward the enemy, toward the front; (*mach., etc.*) direct motion, headway; (*nav.*) headway;
—— *en bataille*, march in line;
bataillon de ——, a battalion made up of men of different battalions (for purposes of transportation, etc.);
battre la ——, to strike up the march;
—— *par bonds*, (*inf.*) advance by rushes, in the combat;
—— *au canon*, march to the sound of the guns;
—— *circulaire*, (*man.*) circling;
—— *en colonne*, march in column;
compagnie de ——, v. *bataillon de* ——, substituting *compagnie* for *bataillon*; (*nav.*) landing party;
concours de ——, v. s. v. *concours*;
contre- ——, countermarch;
—— *à contre-vapeur*, (*mach.*) reversing of an engine;
—— *courbe*, (*cons.*) curved step;
—— *dansante*, (*cons.*) winder;
—— *de demi-angle*, (*cons.*) the winder preceding (or following) the longest winder of a staircase;
—— *de départ*, (*cons.*) curtail step;
—— *dérobée*, stolen march;
dérober une ——, to steal a march on, to conceal a march from;
—— *de dislocation*, separation march, v. *dislocation*;
—— *directe*, march straight to the front, as distinguished from an oblique;
dispositif de ——, order, arrangement, on or for a march;
—— *droite*, (*cons.*) flier;
—— *à droite* (*gauche*) *par 3* (*par 4*), march to the right (left) by threes (fours);
en ——, on the march, marching; (*nav.*) underway;
en —— *à la vapeur* (*voile*), (*nav.*) under steam (sail);
—— *d'exercice*, practice march;
faire une ——, to march, make a march;
fausse(-) ——, feigned march;
fermer la ——, to bring up the rear, to march in rear;
—— *de flanc*, flank march;
—— *forcée*, forced march;
formation de ——, march formation;
—— *de front*, march in line;
—— *de front en avant*, march in line to the front;
—— *de front face en arrière*, march in line to the rear;
front de ——, march front;
—— *funèbre*, (*mus.*) dead march;
gagner une —— *sur*, to steal a march on;
—— *gironnée*, (*cons.*) v. —— *dansante*;
graphique de ——, march diagram;
gêner la ——, to impede, hamper a march;
—— *de guerre*, service march, march under service conditions;
—— *inclinée*, v. —— *rampante*;
jour de ——, *journée de* ——, day's march;
la ——, (*cav.*) call for marching past at a walk;
largeur de ——, breadth of area marched over;

marche *en ligne*, march in line;
—— *-manœuvre*, march maneuver, tactical march, e. g., to form or concentrate for a battle; (more generally) a march having for object the execution of some definite tactical purpose; a march whose object is to oppose superior forces by combined movements; (more rarely) a practice march;
mettre en ——, to set in march; (*mach., etc.*) to start; set afoot; to start the engine;
se mettre en ——, to set off, to march;
—— *militaire*, march;
mise en ——, start; (*mach., etc.*) start, starting mechanism, slide-valve link;
—— *de nuit*, night march;
—— *oblique*, oblique march, obliquing;
—— *oblique individuelle*, oblique march (each man obliques); (*mil. slang*) the assembling of soldiers confined to barracks, for roll call;
—— *oblique par troupe*, oblique march of the whole body;
—— *offensive*, offensive march, i. e., to get into position on battlefield whence to engage enemy; (*cav.*) march preceding the charge proper;
—— *ordinaire*, usual march (usual rate);
ordre de ——, order of march; (*nav.*) order of sailing;
ouvrir la ——, to march in front, to lead the march;
—— *-palier*, (*cons.*) last or top step, landing step;
—— *palière*, (*cons.*) first step of a stair;
—— *en parade*, march past;
—— *parallèle*, v. —— *droite*;
—— *au pas cadencé*, march at attention;
—— *au pas de route*, march at route step;
point initial de ——, point at which each element of a column enters the column;
profondeur de ——, depth of area marched over;
ralentir la ——, (*mach.*) to slow down;
—— *rampante*, (*cons.*) weathered step;
—— *rayonnante*, (*cons.*) winder;
régiment de ——, v. *bataillon de* ——, substituting *régiment* for *bataillon*;
—— *de régiment*, regimental march (music);
renverser la ——, (*mach.*) to reverse an engine;
—— *en retard*, of a chronometer, loss;
—— *de*, *en*, *retraite*, retreat, retrograde march;
—— *rétrograde*, euphemism for retreat; march to the rear, away from the enemy;
—— *de route*, march on ordinary roads;
série de ——, box containing butchery implements, scales, weights (used in slaughtering cattle on the march);
service de ——, transportation service;
sonner la ——, so strike up, sound a march;
—— *stratégique*, any march from the base to the objective or from one objective to another; •
suivre une ——, (*in gen.*) to follow a course;
surface de ——, marching area;
—— *à terre*, (*mil. slang*) foot soldier; dough boy (U. S. A.);
—— *en tiroir*, an order of march in which all the elements take the head of the column in turn, each element marching through the preceding elements to reach the head (whence the name);
tringle de ——, (*cons.*) stair rod;
—— *à vide*, (*mach.*) work of an engine unloaded;
vitesse de ——, rate of march;
—— *à volonté*, march at ease;
zone de ——, march zone;
—— *des zouaves*, (*mil. slang*) soldiers who go to medical inspection are said to execute said march.

marché, m., market, market place, any bargain or contract;
—— *d'abonnement*, (*adm.*) contract or engagement to execute a service during a fixed time for fixed price;
—— *par adjudication publique*, (*adm.*) contract awarded to a public bidder;
—— *clef à la main*, —— *clefs en main*, contract to build and finish a house by a given time;

marché, *au comptant*, cash sale;
—— *sur devis*, (*adm.*) contract (for public works) based on estimates furnished and subject to change;
——*ferme*, binding contract (involving certain delivery);
—— *à forfait*, (*adm.*) contract in which everything, both work and price, is fixed in advance;
—— *de fourniture à long terme*, (*adm.*) contract to supply materials, etc., for a given length of time;
—— *de gré à gré*, (*adm.*) contract established between contracting parties, without the intervention of bids; direct contract;
—— *libre*, option;
—— *de livraison*, (*adm.*) contract to deliver specified articles at fixed time and place;
—— *de lots de fournitures*, (*adm.*) contract to supply certain amounts (*lots*) of supplies;
—— *au mètre*, contract for work at so much a meten;
—— *d'ouvrages*, building contract;
passer un ——, to make a contract, purchase, etc.;
—— *à prime*, v. ——*libre;*
—— *à prix ferme*, (*adm.*) government contract (i. e., involving a fixed price);
—— *à la ration*, (*adm.*) contract to furnish supplies to the troops in rations, i. e., ready for immediate use, during a designated period;
—— *sur série de prix*, (*adm.*) contract in which prices are assigned to all works, in a statement (*bordereau*), and by which a contractor binds himself either to carry out all works that may be exacted of him, without knowing their nature, within a certain interval (6 years maximum), or to execute a given work; (useful in works that must be kept secret, or that can not be fixed in advance; requires sanction of minister of war);
—— *à terme*, time bargain;
—— *à terme ferme*, binding time-bargain;
—— *à terme fictif*, optional time-bargain;
—— *de transformation*, (*adm.*) contract by which the contractor is furnished the raw material (and sometimes all or part of his labor) and in which he agrees to convert this raw material into objects for which he receives a set price, fixed in advance.

marchef, (*mil. slang*) abbreviation of *maréchal des logis chef*.

marchepied, m., step, stepladder, steppingstone; footstep; towpath; landing place; stretcher of a boat; (*art.*) footboard; chassis steps; (*nav.*) footrope;
—— *de chargement*, (*art.*) step (of a siege-gun carriage); (*fort.*) loading steps (on which the men stand to pass ammunition to the gunroom of a turret);
—— *de derrière*, (*art.*) platform board;
—— *de devant*, (*art.*) footboard;
—— *de nage*, stretcher of a boat;
—— *de pointage*, (*s. c. art.*) gunner's platform.

marcher, v. n., to walk, go, travel, ride, run, to proceed, move on, go ahead, progress; to take precedence of; to work, behave (of a machine, engine); (*nav.*) to sail, steam (well or ill); (*mil.*) to march; to go on duty, to take duty, to march on, to take duty by roster or in turn; to go out (as on expeditions), (of an aid-de-camp) to "gallop;"
—— *à l'aise*, (*mil.*) to march at ease, at the route step;
—— *après*, (*mil.*) to follow (on the roster);
—— *en arrière*, (*mil.*) to march to the rear, to retreat; (*mach.*) to be reversed; (*nav.*) to go astern;
—— *en avant*, (*mil.*) to march to the front, to advance; (*mach., nav.*) to go ahead;
—— *avec*, (*cord.*) to walk away with a rope;
—— *à l'aveuglette*, to march hither and thither, at random, without definite purpose;
—— *en bataille*, (*mil.*) to march in line;
—— *au canon*, (*mil.*) to march to the sound of the guns;

marcher *à cheval*, to ride;
—— *au coke*, (*bois, etc.*) (*met.*) to use coke (wood, etc.) (as in a blast furnace);
—— *en colonne*, (*mil.*) to march in column;
—— *à condensation*, (*steam*) to use the condenser;
—— *à contre-vapeur*, (*mach.*) to reverse an engine;
—— *en corps*, (*mil.*) to troop;
—— *à échappement*, (*steam*) to exhaust into the air;
faire ——, to march, set in march; set agoing;
—— *au feu*, (*mil.*) to go under fire;
—— *de front*, (*mil.*) to march abreast;
—— *à gueule bée*, of a mill, to work under a full head (water gate lifted clean out of the water);
—— *à grandes journées*, (*mil.*) to advance by forced marches;
—— *large*, to keep off;
—— *à main droite (gauche)*, (*man.*) to turn, gallop, work, to the right (left);
—— *le pas en arrière*, (*mil.*) to step back;
—— *au pas cadencé*, (*mil.*) to march at attention;
—— *au pas de route*, v. ——*à l'aise;*
—— *en sape*, (*siege*) to sap, to advance by sap;
—— *serré*, (*mil.*) to march in close order;
—— *sous bois*, to march in the woods;
—— *sur deux (une seule, etc.) colonnes*, (*mil.*) of an army, to march in two (one, etc.) columns;
—— *en tiroir*, (*mil.*) in column, to march with the rear elements passing through those in front in order to take the head in turn;
—— *sur trois jambes*, (*hipp.*) to be lame;
—— *à la vapeur (voile)*, (*nav.*) to be under steam (sail);
—— *à vanne trempante*, of a mill, to work under a partial head (water gate not wholly drawn out);
—— *à vide*, (*mach.*) to work unloaded;
—— *à volonté*, v. ——*à l'aise.*

marcheur, m., pedestrian, walker; a., (*mil.*) having good marching powers;
bon (mauvais) ——, *grand* ——, (*nav.*) of a ship, good (bad) sailer.

marcheux, m., (in brickmaking) trough or ditch for kneading clay; workman who kneads brick clay.

marchoir, m., (in brickmaking) place or shop for kneading clay.

mare, f., pool, pond.

maréage, m., (*nav.*) hiring of sailors for an entire voyage.

marécage, m., (*top.*) marsh, bog, fen, swamp.

marécageux, a., (*top.*) swampy, marshy, boggy;
pays ——, marsh land;
terrain ——, marsh or moor land.

maréchal, m., blacksmith; veterinarian; (*mil.*) marshal, field marshal;
aide ——, *aide* ——*ferrant*, (*Fr. a.*) underfarrier (art., cav., train, and engineer drivers);
brigadier maître ——*ferrant*, (*Fr. a.*) farrier ranking as corporal (art., cav., train, and engineer drivers);
——*ferrant*, farrier, veterinary;
—— *d'empire*, (*Fr. a.*) (under the first Empire) a marshal of France;
—— *de France*, (*Fr. a.*) marshal of France;
—— *des logis*, (*Fr. a.*) sergeant (art., cav., train, and engineer drivers);
—— *des logis chef*, (*Fr. a.*) sergeant-major (art., cav., train, and engineer drivers; almost the same as first sergeant, U. S. A.);
—— *des logis chef maître-charpentier*, (*Fr. a.*) master carpenter (bridge troops);
—— *des logis chef maître-cordier*, (*Fr. a.*) cordage master (bridge troops);
—— *des logis chef maître-forgeron*, (*Fr. a.*) master blacksmith (bridge troops);
—— *des logis chef mécanicien*, (*Fr. art.*) sergeant, assistant to the captain in charge of the regimental park;
—— *des logis fourrier*, (*Fr. a.*) quartermaster-sergeant (art., cav., train, and engineer drivers);

maréchal   264   **marionnette**

maréchal *des logis garde-magasin*, (*Fr. a.*) sergeant storekeeper;
—— *des logis de garde*, sergeant of the guard;
—— *des logis maître sellier*, (*Fr. a.*) saddler sergeant;
—— *des logis de peloton*, (*cav.*) sergeant in charge of a platoon, and responsible for its discipline, instruction, etc.;
—— *des logis de pièce*, (*art.*) sergeant in charge of a piece, and responsible for the discipline, training, cleanliness, quarters, etc., of the detachment of his piece;
—— *des logis secrétaire*, (*Fr. a.*) sergeant clerk;
—— *des logis de semaine*, (*Fr. art.*) sergeant on duty for the week (roll calls, stables, etc.);
—— *des logis trompette*, (*Fr. art.*) sergeant of field music;
—— *des logis trompette-major*, (*Fr. cav.*) sergeant trumpet-major;
—— *des logis vaguemestre*, (*Fr. a.*) sergeant mail carrier;
maître —— *ferrant*, (*Fr. a.*) master farrier, sergeant farrier;
premier maître —— *ferrant*, (*Fr. a.*) farrier with the rank of sergeant-major;
—— *vétérinaire*, farrier and veterinarian (obs.).
maréchalat, m., (*mil.*) marshalcy, dignity of marshal.
maréchale, f., wife of a marshal, of a field marshal.
maréchalerie, f., farriery;
atelier de ——, blacksmith's shop, farrier's establishment.
maréchaussée, f., mounted police.
marée, f., tide; ebb, flux and reflux;
âge de la ——, v. *retard de la* ——;
aller avec la ——, to go with the tide;
annuaire des ——*s*, tide tables;
—— *baissante*, falling tide;
—— *basse*, *basse* ——, low water; ebb tide, low tide;
—— *basse de vive-eau*, low springtide;
—— *de basse-eau*, neap tide;
changement de la ——, turn of the tide;
contre- ——, eddy tide;
—— *et contre-* ——, tide and half tide;
demi- —— *baissante* (*montante*), half ebb (flood) tide;
—— *descendante*, falling tide, ebb of the tide;
descendre à la faveur de la ——, to tide it down;
direction de la ——, set of the tide;
échelle de ——, tide gauge;
écluse à —— *montante*, tide gate;
—— *d'équinoxe*, —— *des équinoxes*, equinoctial springtide;
établissement de la ——, establishment of a port;
—— *étale*, slack water;
faire ——, to take advantage of the tide;
goulet à forte ——, tide gate;
grande ——, high flood, high tide; springtide;
grande —— *de basse-eau*, high neap tide;
grande —— *de vive-eau*, high springtide;
grande(s) ——(*s*) *des équinoxes*, high springtide;
grandeur de la ——, rise and fall of, range of, the tide;
—— *haute*, *haute* ——, high water, flood, tide;
lit de ——, tideway;
mi- ——, half tide;
montant de la ——, beginning of the flood;
—— *montante*, flood, rising tide; flood tide;
—— *morte*, neap tide, dead neap; dead water;
—— *de morte(s) eau(x)*, neap tide;
moulin de ——, tide mill;
onde de ——, tidal wave;
petite ——, neap tide;
plein de la ——, high water;
pleine ——, high tide, high water;
port à ——, tidal harbor;
—— *portant au vent*, v. —— *qui porte au vent*;
—— *portant sous le vent*, v. —— *qui porte sous le vent*;
prendre la ——, v. *faire* ——;
—— *de quadratures*, neap tide;
—— *de quartier*, neap tide;

marée *qui porte au vent*, weather tide;
—— *qui porte sous le vent*, lee tide;
ras, raz, de ——, tide gate, flood gate;
retard de la ——, age, retard, of the tide;
—— *de syzygies*, springtide;
—— *totale*, tide range;
de toute ——, at all stages of the tide;
—— *de vive(s) eau(x)*, springtide.
marégraphe, maréographe, m., tide gauge.
maréomètre, m., tide gauge.
marge, f., margin, border.
margelle, f., curb (of a well); (*cons.*) rim or curb of the eye of a dome.
margouillet, m., bull's-eye; wooden thimble or traveler; truck; fair leader.
marguerite, f., (*cord.*) sheepshank; messenger; (*hipp.*) (*in pl.*) first white hairs on the temples of an aging horse;
faire ——, *sur*, (*cord.*) to clap a messenger on.
mariage, m., marriage; (*cord.*) lashing two ropes closely together.
marier, v. a., to marry; (*cord.*) to lash together; (two ropes); (*fort.*) to adapt, adjust, the fortification to the site.
marie-salope, f., scow, mud scow.
marigot, m., (*top.*) small stream; pool.
marin, m., sailor, seaman;
—— *canonnier*, (*nav.*) seaman gunner;
—— *du commerce*, seaman of the merchant marine;
—— *d'eau douce*, fresh-water sailor, landlubber;
—— *de l'État*, a seaman (of the war navy);
—— *fusilier*, v. s. v. *fusilier*;
—— *torpilleur*, (*nav.*) torpedo gunner;
—— *s vétérans*, (*Fr. nav.*) (corps of) marine veterans, so called, for garrison and other duty in French ports.
marin, a., sea, seafaring, naval, marine; nautical; (*fam.*) shipshape;
aiguille ——*e*, compass needle (mariner's);
avoir le pied ——, to have one's sea legs;
carte ——*e*, chart;
chaudière ——*e*, (steam) marine boiler;
colle ——*e*, marine glue;
machine ——*e*, marine engine;
montre ——*e*, chronometer;
nœud ——, (*cord.*) reef knot;
sel ——, common salt;
sous- ——, submarine.
marine, f., navy (esp. war navy); naval or sea service; navy department; sea or maritime affairs; navigation;
artillerie de ——, marine artillery; (in France) seacoast artillery, colonial artillery;
—— *de commerce*, merchant navy; merchant marine;
conseil de la ——, naval board;
de ——, naval;
—— *de l'État*, national navy, war navy;
—— *de guerre*, v. —— *de l'État*;
infanterie de ——, v. s. v. *infanterie*;
—— *marchande*, commercial navy;
—— *militaire*, v. —— *de l'État*;
ministère de la ——, navy department; admiralty;
ministre de la ——, secretary of the navy; first lord of the admiralty;
officier de ——, naval officer;
service de la ——, naval, sea service;
soldat de (la) ——, marine; (in France) a soldier belonging to the marine infantry or artillery;
terme de ——, sea term, nautical term;
troupes de la ——, marines; (in France) the infantry and artillery of the marine;
—— *à vapeur*, steam navy, steam shipping;
—— *à voiles*, sail shipping;
marinier, a., of the tide, marine; m., seaman; (in a derogatory sense) bargeman;
—— *de carton*, —— *d'eau douce*, fresh-water sailor;
officier ——, petty officer;
soldat ——, (*mil.*) soldier familiar with boats.
marionnette, f., leading block; nine-pin block.

maritime, a., maritime, naval;
arsenal ——, dockyard, shipyard, navy-yard;
assurance ——, marine insurance;
code, ——, maritime law or code;
divisions ——*s*, (in France) naval prefectures;
législation ——, naval articles of war; maritime law; naval laws and regulations.
marmenteau, a., *bois* ——, wood of tall trees near a house.
marmite, f., kettle, camp kettle; copper; boiler;
—— *de campement*, camp kettle;
—— *de cuisine*, copper, boiler;
—— *à dégraisser*, (*mach.*) soda kettle.
marne, f., marl, fuller's earth;
—— *à foulon*, fuller's earth.
marner, v. n., (of the sea) to rise above the usual level of waters;
—— *de tant de mètres*, (of the tide) to have a range of so many meters.
marnière, f., clay or marl pit.
maroquin, m., morocco (leather).
marprime, f., sailmaker's awl.
marquage, m., marking, tamping; (*hipp.*) branding (on the hoof in French army); (*sm. a.*) markings, stamps, etc., on firearms, showing place and date of manufacture.
marque, f., mark, sign; (commercially) brand, (hence in pl.) goods; brand (of a criminal); cipher; stamp (mark and tool); government stamp (as on dutiable goods); landmark, seamark; buoy; insignia, sign; proof, evidence, testimony; (*sm. a.*) stamp, mark (of date and of place of manufacture, etc.); (*hipp.*) star on a horse's forehead; brand; mark, affixed in an epidemic, to show that the animal so marked is healthy; bean (the mark on a horse's tooth);
—— *de basse marée*, low-water mark;
—— *de bonne conduite*, (*mil.*) good-conduct badge;
—— *de cheval*, (*hipp.*) brand (on hoof, in France);
——*s distinctives*, (*mil.*) distinctive badges (as, for good service, good conduct, of grades, functions); markings (on material objects);
——*s distinctives de commandement*, (*nav.*) insignia or marks of command (pennants, etc.);
—— *de la douane*, custom-house stamp;
—— *d'eau*, water guage;
—— *de l' État*, official stamp of the state; in England, broad arrow;
—— *de fabrique*, trade-mark;
—— *de feu*, (*hipp.*) red spot;
—— *de guide*, leading mark;
—— *de haute marée*, high-water mark;
——*s d'honneur*, insignia (as, of an order); marks of honor, i. e., compliments exchanged (salutes, etc.);
lettre de ——, letters of marque;
—— *de la mer*, trace, mark, left by the sea;
—— *de service*, (*mil.*) sign of being on duty (in French army, chin-strap down);
—— *de terre*, landmark;
—— *en tête*, (*hipp.*) blaze on the forehead;
—— *du tirant d'eau*, (*nav.*) draught mark.
marqué, p. p., marked;
contre-——, v. *faux* ——;
faux- ——, (*hipp.*) bishoped, countermarked;
—— *de feu*, (*hipp.*) red-spotted, red-colored;
—— *en tête*, (*hipp.*) with a blaze or star on the forehead.
marquer, v. a. n., to mark, mark out; to trace; to indicate, read (of a thermometer, clock, etc.); to appoint, assign, determine, fix; to ch∂ck off; (*fenc.*) to make a clear body hit; (*sm a.*) to stamp; (*hipp.*) to brand; to snow the mark (of the teeth); (*fort.*) to trace out (a work);
—— *le camp*, (*mil.*) to indicate the site of a camp;
cheval qui commence à ——, (*hipp.*) young horse that begins to show signs of age by his teeth;
cheval qui marque, (*hipp.*) horse that still has the mark;

marquer, *cheval qui ne marque plus*, (*hipp.*) horse that has lost the mark, that has razed;
—— *le coin*, (*man.*) to take the corners properly (in riding hall);
—— *le coup*, (*fenc.*) to touch one's adversary slightly;
—— *un coup*, (*fenc.*) to feint, to make a feint thrust;
faire ——, to dip (a flag);
—— *une ligne de sonde*, to mark (the divisions on) a sounding line;
—— *le pas*, (*mil.*) to mark time.
marqueter, v. a., to inlay.
marqueterie, f., marquetry, inlaying, inlaid work.
marqueur, m., marker; (*t. p.*) marker; (*mil. teleg.*) marker, man who marks or puts up indications (on houses, trees) for the working party in rear.
marquise, f., large tent, marquee; (*mil.*) headquarters tent (obs.); awning or cover of an officer's tent; (*artif.*) sort of rocket; (*cons.*) (water) shed over a door or window;
double ——, (*artif.*) fifteen-line rocket;
—— *d'une locomotive*, (r. r.) cab.
marre, f., mattock, pickax.
marron, m., chestnut (fruit); mark, ticket, tag; (*artif.*) bouncing powder; maroon, cracker; cube of cartridge paper filled with corned powder; (*fond.*) mold core; (*Fr. a.*) metal disk showing the hour for making rounds (dropped into receptacles prepared for the purpose, by the officer on guard, on making rounds, and thus serving as a check on the performance of the duty);
—— *d'artifices*, (*artif.*) cardboard boxes filled with gunpowder;
—— *de distribution*, (*Fr. a.*) ticket issued to fuel and light contractors and presented as a verification of their claims;
—— *luisant*, (*artif.*) maroon with star composition on the outside, to give a white light before bursting;
—— *de service*, v. under main word (*Fr. a.*).
marronnier, m., chestnut tree, chestnut wood;
faux ——, horse chestnut;
—— *d'Inde*, horse chestnut.
Mars, m., Mars;
champ de ——, (*mil.*) parade ground, drill ground.
marseillaise, f., French national hymn.
marsouin, m., (*mil. slang*) officer or man of the marine infantry.
marteau, m., hammer, mallet, maul, beetle, clapper, knocker; abbreviation of *marteaupilon;* (*art.*) plunger of a fuse, striker; (*r. f. art.*, *F. m. art.*) hammer, striker;
—— *à air comprimé*, hammer worked by compressed air; cushioned hammer;
—— *à l'allemande*, lift hammer;
—— *à l'anglaise*, nose or frontal helve, helve hammer;
—— *d'arbalète*, Jacob's staff;
—— *d'assiette*, paver's hammer;
—— *d'une attaque*, (*mil.*) principal attack in an action;
—— *à balle*, ball-pane hammer;
—— *à bascule*, tilt hammer;
—— *à bouge*, oliver;
—— *treitelé*, tooth ax;
—— *à briser*, miner's pick (ax);
—— *de calfat*, calking hammer;
—— *à cames*, tappet hammer;
—— *cassant*, mason's double-face hammer; sledge hammer;
—— *à chapelel*, sledge hammer (smith's);
—— *de charpentier*, ordinary carpenter's hammer;
—— *chasse-coin*, hammer used for driving in (or out) wedges (sometimes *chasse-coin* alone is used);
—— *cingleur*, (*met.*) shingling hammer;
—— *de cloche*, bell clapper;
—— *à coups terribles*, a hammer worked by compressed air and capable of delivering 500 blows a minute;
dégrossir au ——, to hammer down, to shape down by the hammer;

marteau à *dents*, claw hammer;
— à *deux mains*, sledge hammer, about sledge;
— à *deux pointes*, miner's hammer;
— à *devant*, sledge;
— à *dresser*, planishing hammer;
durcir au —, to hammer-harden;
— à *emboutir*, chasing hammer;
— d'*établi*, carpenter's hammer;
étendre sous le —, to hammer out;
façonné au —, hammer dressed;
faire au —, *faire à coups de* —, to hammer up; to hammer into shape;
fait au —, hammered;
— *fendu*, shoeing hammer;
— *de forge*, forge hammer; smith's hammer;
— à *frapper devant*, sledge hammer, flogging hammer;
— *frappeur*, farrier's hammer;
— à *friction*, friction hammer;
— à *friction plane*, Schönberg friction hammer;
— *frontal*, v. — à l'*anglaise*;
— *granulé*, bush hammer;
gros — *de forge*, — *de grosse forge*, forge hammer;
— d'*horloge*, clock hammer;
— *hydraulique*, hydraulic hammer;
— à *layer*, woodman's hammer;
levée d'un —, v. *volée d'un* —;
— à *levier*, lever hammer;
— à *main*, hand hammer, uphand hammer;
— *de maître*, fitter's hammer;
— à *manche*, helve hammer, tilt hammer;
— *monte-système*, (*Fr. art.*) a special hammer for use with the *canon à balles*;
— à *mouton*, v. — *-pilon*;
— à *nayer*, sort of sheathing hammer (French bridge train);
ouvrier à —, hammerer, hammerman;
— à *panne*, peen hammer;
— à *panne bombée*, ball peen hammer;
— à *panne fendue*, claw hammer;
— à *panne plate*, flat peen hammer;
— à *panne tranchante*, cross peen hammer;
— *de parage*, — à *parer*, dressing hammer;
— *de paveur*, paver's beetle;
— *percutant*, (*art.*, *sm. a.*) hammer, percussion hammer;
— *de percuteur*, (*art. sm. a.*) percussion hammer;
— à *pied de biche*, claw hammer;
— *-pilon*, steam hammer, vertical hammer, (*met*.) shingling hammer;
— *pilon à cames*, tappet hammer;
— *pilon à double* (*simple*) *action* (*effet*), double- (single-) action steam hammer;
— *pilon à friction*, friction hammer;
— *pilon à vapeur*, steam hammer;
— à *piquer le sel de la chaudière*, scaling hammer;
— à *pointe*, (*art.*) a pointed hammer for examining projectiles;
— à *pointes*, flang;
— *de porte*, door knocker;
— *porte-amorce*, (*art.*) plunger (of a fuse);
— à *pression*, squeezing hammer;
— à *queue*, v. — à *bascule*;
— à *refouler*, cutler's set hammer;
— à *relever*, farrier's hammer;
— à *ressorts*, spring (cushioned) hammer;
— à *river*, — *-rivoir*, riveting hammer;
second —, uphand sledge;
— à *souder*, soldering hammer;
— à *soulèvement*, v. — à l'*allemande*;
— à *soyage*, à *soyer*, creasing hammer;
table d'un —, face of a hammer;
— *taillant*, hack hammer;
— à *tête plate* (*ronde*), flat- (round-) head hammer;
— à *tranche*, — *tranchant*, chipping hammer;
— à *vapeur*, steam hammer;
— *vertical*, v. — *-pilon*;
volée d'un —, lift of a hammer.
martel, m., hammer.
martelage, m., hammering, hammering down, forging;

martelage *au martinet*, tilting.
marteler, v. a., to hammer, to hammer down, to forge, to tilt.
martelet, m., small hammer; (*art.*) percussion hammer.
marteleur, m., hammerman.
martellière, f., sluice fitted with water gates.
martello, m., (*fort.*) martello tower.
martial, a., martial; warlike;
code —, articles of war;
cour —*e*, court-martial;
législation —*e*, v. *code* —;
loi —*e*, martial law.
martiner, v. a., to hammer, forge.
martinet, m., hammer, forge hammer, tilt hammer, sledge hammer, forge; cat o' nine tails;
— à *bascule*, small tilt hammer;
— *de forge*, tilt hammer;
forger au —, to tilt;
— *de grosse forge*, forge hammer;
— à *queue*, v. — à *bascule*.
martingale, f., (*mil.*) cartridge-box flap; Y-shaped frog (for bayonet or n. c. o.'s sword scabbard); (*harn.*) martingale;
fausse —, (*harn.*) half martingale.
mascaret, m., bore.
mash, m., (*hipp.*) mash.
masque, m., mask, screen; any plate or shield of wood or of any other suitable material; (*fenc.*) mask, fencer's mask; (*fort.*) mask, cover, and, in general, any concealing cover; embrasure mask, shutter; (*art.*) gun shield, splinter-proof shield; cover or mask of the gun opening of a shield proper; (*mil. min.*) tamping blocks; retaining planks, plate, or shield, used in driving galleries in crumbling soil;
— d'*éponge*, (*mil. min.*) sponge mask (used in succoring asphyxiated miners);
former —, to screen, to mask;
— *protecteur*, (*art.*) gun shield;
— *de tête*, (*siege*) head parapet (of a sap).
masquer, v. a., to mask, screen; (*mil.*) to mask, conceal (a march, etc.); to cover;
— *en alezan*, to paint a horse so as to deceive a purchaser;
— *une batterie*, (*fort.*) to mask a battery;
— *une place forte*, (*mil.*) to mask a fortified position;
— *aux vues*, (*mil.*) to screen from view.
massacre, m., slaughter, massacre, butchery; (*fig.*) bungling; bungler, botcher.
massacrer, v. a., to slaughter, massacre; (*fig.*) to bungle, botch.
masse, f., mass, lump, heap, body, weight, bulk, cake; sledge hammer, mace, maul, beetle, mallet, rammer; bob, ball, of a steelyard; scrap iron; beds of stone in a quarry; fund; (*mil. adm.*) fund; (*top.*) area; (*art.*, *sm. a.*) sight mass; (*mil.*) mass (of troops); (*Fr. cav.*) line of squadrons in platoon columns at 12-pace intervals; (*Fr. art.*) of a group, line of columns (open or closed intervals); (*elec.*) ground;
— *active*, assets;
avoir à la —, *avoir de* —, *avoir la* —, v. s. v. *avoir*;
— *de batterie*, (*Fr. art.*) battery sledge;
— *de bois*, commander, or large wooden mallet;
— *de campement*, (*Fr. a.*) (so called) camp sledge;
— *carrée*, sledge;
— *de casernement*, (*Fr. a.*) barrack fund (special to the regiments and establishments, designated by the minister, that keep their barracks in repair);
— *de chauffage*, (*Fr. a.*) fuel fund (heating and cooking);
commission de —, (*Fr. a.*) board of officers administering a fund;
— *couvrante*, (*fort.*) protective parapet, epaulment;
— *de culture*, (*top.*) cultivated area;
— à *damer*, rammer;
— *demi-affinée*, (*met.*) white pig iron;

masse *des écoles,* (*Fr. a.*) school and instruction fund (to meet expenses connected with target practice, with instruction in fencing, boxing, field fortification, etc.);

en ——, in a mass, together; (*mil.*) massed, in mass;

—— d'entretien du harnachement et ferrage (*Fr. a.*) harness and shoeing fund (applied to a variety of purposes; harness, shoeing, lighting of stables, etc.);

—— d'entretien et remonte, (*Fr. a.*) a fund peculiar to the *gendarmerie*, used to reimburse n. c. o. and men for loss of horses and of effects;

fer de ——, scrap iron;

—— des fourrages, (*Fr. a.*) forage fund;

—— à frapper et à donner, earth rammer, pile driver;

grande ——, (*met.*) main body of a blast furnace;

—— d'habillement et d'entretien, (*Fr. a.*) clothing fund (applied to all expenses connected with the men's clothing, e. g., repairs, alterations, etc.);

—— de harnachemen, (*Fr. a.*) harness fund (used to meet expenses connected with harnessing of horses and mules, shoeing, treatment of sick, etc.);

—— de haut-fourneau, (*met.*) body of a blast furnace;

—— individuelle, (*Fr. a.*) soldier's fund (peculiar to the *spahis, gendarmerie,* and *sapeurs-pompiers de* Paris);

—— d'infirmerie, (*Fr. a.*) hospital fund (applied to the purchase of food for the sick on special diet, and to other small expenses arising in hospitals);

levée en ——, v. s. v. *levée;*

—— de lumière, (*art.*) bush of a gun (cast in with the piece, obs.);

—— magnétique, (*elec.*) magnetic mass;

—— de mire, (*art., sm. a.*) sight mass, sight patch;

—— noire, slush fund (U. S. A.);

—— ordinaire, (*art.*) recoil chock;

par ——, by the lump;

—— passive, liabilities;

petite —— inférieure, (*met.*) hearth of a blast furnace;

petite —— supérieure, (*met.*) shaft;

principe des ——s conjuguées, (*fort.*) in defilading, the principle of conjugate masses (i. e., of covering the opening between two masses by a third mass, the three being arranged checkerwise and having their end faces in the same plane);

—— de remonte, (*Fr. a.*) remount fund (special to the *spahis*);

—— ressuée, (*met.*) copper reduced by liquation;

—— de secours, (*Fr. a.*) relief fund (peculiar to the *gendarmerie* and *spahis*; used in relieving the most needy);

—— à tranche —— à trancher, track iron.

massé, m., (*met.*) ball, bloom.

masseau, m., v. *massiau.*

masselet, m., (*met.*) small bloom.

masselotte, f., (*fond.*) feeding head, riser; deadhead, sinking head; headmetal; (*mach. etc.*) weight; (*art.*) plunger, plunger sleeve (of a fuse); guard;

—— de baïonnette, (*sm. a.*) portion of socket to which the blade is welded;

—— sphérique, (*art.*) hammer of the Merriam fuse.

masser, v. a., (*mil.*) to mass, dispose in mass; to form mass.

masset, m., (*met.*) synonym of *loupe.*

massette, f., miner's hammer; (*artif.*) rocket stamp;

—— de cantonnier, stonebreaker's hammer.

massiau, m., (*met.*) the *loupe* transformed into a more or less regular parallelopipedon.

massicot, m., litharge, massicot.

massif, m., mass of rocks; clump of trees; mass or main body of any building or part of a building, or of any construction (e. g., of a dam, of an embankment); main or solid part of anything; foundation, substructure; shell of a blast furnace; (in *top. drawing*) any solid block or area (as, a pond, house, lake, etc.); (*artif.*) drift; solid part of a rocket; (*top.*) aggregation, mass of mountains about a central summit; (*fort.*) mass of earth supporting or constituting a parapet.

exploitation, ouvrage, par ——s longs, (*min.*) panel working;

—— d'un puits, (*min.*) pillar of a shaft;

—— de terre, (*fort.*) mass of earth supporting a parapet.

massif, a. massive, solid, heavy.

massivage, m., (*cons.*) tamping, packing, of concrete.

massiver, v. a., to mix and settle mortar concrete by tamping or beating.

massoque, f., (*met.*) bloom, slab; bloom obtained by dividing a *massé* in two; (*in pl.*) pieces of iron composing the blast plate of a Catalan furnace; exterior parts of a Catalan bloom.

massoquette, f., (*met.*) bloom obtained by dividing a *massoque* in two.

massoquin, m., (*met.*) synonym of *massoque.*

mastic, m., cement; putty; mastic;

—— artificiel, artificial mastic (mixture of slate dust and coal tar);

—— d'asphalte, —— asphaltique, asphaltic, mastic;

—— bitumineux, bituminous cement;

—— de ciment, mastic cement;

—— de fer, iron cement;

—— de fonte, iron cement;

—— de limaille de fer, v. *—— de fer;*

—— naturel, v. *—— asphaltique;*

—— des vitriers, glazier's putty.

masticage, m., cementing.

mastigadour, m., (*harn.*) slobbering bit; bit of a watering bridle.

mastiquer, v. a., to putty, cement.

masure, f., old hovel; rickety old house.

mat, a., dead, heavy, unpolished; m., dead surface, dead color.

mât, m., masting.

à ——s et à cordes, (*nav.*) under bare poles;

—— d'artimon, (*nav.*) mizzenmast;

—— de beaupré, (*nav.*) bowsprit;

—— de charge, derrick;

—— de fortune, (*nav.*) jury mast;

grand ——, (*nav.*) mainmast;

—— de misaine, (*nav.*) foremast;

—— de pavillon, flagstaff;

—— de pavillon de beaupré, (*nav.*) jackstaff;

—— sémaphorique, (*r. r.*) signal post;

—— de signaux, signal mast;

—— de tente, tent pole.

matable, a., (*met.*) susceptible of being jumped, upset.

matage, m., (*met.*) jumping, upsetting; beating flat; (hence) closing, tightening of joints;

—— à la poudre, (*art.*) process of hardening bronze guns by a *tir de matage,* q. v.

mâtage, m., masting.

matasiette, f., (*expl.*) a dynamite whose dope consists of ocher, sand, powdered charcoal, and a resinous substance (40 per cent nitroglycerin).

matelas, m., mattress; backing; layer; lining; (*fort.*) backing, armor-plate backing;

—— à air, air cushion;

—— en bois, wood backing for armor plates;

—— élastique, any elastic backing;

—— de gravier, gravel layer (foundation for a road etc.);

—— de sable, (*fort.*) sand backing or layer;

—— de terre, layer of earth;

—— à vapeur, steam cushion.

matelasser, v. a., to pad.

matelassure, f., pad, padding; wadded lining of a breastplate; armor-plate backing.

matelot, m., seaman sailor (as a grade corresponds to private in the army);

—— canonnier, seaman gunner;

matelot *de l'État*, blue jacket, man-of-war's man;
—— *gabier*, topman;
—— *de pont*, ordinary seaman.
matelotage, m., seamanship.
mater, v. a., to render dull, deaden; to flatten down (as the edge of a plate); to clinch; to overlay with iron, copper, etc.; to upset, jump; to set metal cold;
—— *une soudure*, to remove, flatten, a solder mark.
mâter, v. a., to step a mast; to set up on end (as, a prop); to set up shears;
—— *les avirons*, to toss or peak the oars;
—— *un canon*, (art.) to set up a gun on end;
machine à ——, masting shears;
machine à —— *flottante*, sheer hulk;
machine à —— *en ponton*, sheer hulk.
mâtereau, m., small mast; pole; staff.
matérialiser, v. a., to render material (as, a center of rotation).
matériaux, m., pl., materials, stuff; rubbish (materials produced by demolition);
—— *bruts*, raw materials, natural products;
—— *de construction*, building material;
résistance des ——, strength of materials.
matériel, m., material, stores; (mil.) *matériel*;
—— *accessoire de la voie*, (r. r.) small parts, fittings, of a railroad;
—— *d'attache*, (mil.) picketing implements;
chef de ——, storekeeper; (r. r.) traffic manager;
—— *d'exploitation*, (r. r.) rolling stock;
—— *fixe*, (r. r.) fixed plant;
—— *de guerre*, military stores, war material of all land and of any sorts;
—— *de mobilisation*, (Fr. a.) furniture, etc., used in *stations halte-repas* during concentrations in *places fortes*, etc.;
petit —— *de la voie*, (r. r.) fittings, small parts;
—— *de pontage*, (pont.) bridge material, especially material for roadway proper;
—— *des ponts*, bridge material;
—— *roulant*, (r. r.) rolling stock.
matériel, a., material, substantial, gross, rough, coarse, rude.
mathématique, f., mathematics;
boîte, étui, de —— *s*, box of mathematical instruments;
instrument de ——, mathematical instrument.
matière, f., matter, stuff, subject; (artif.) composition; (nav.) deck beam;
—— *s d'argent*, silver bullion;
—— *s d'artifices*, (artif.) firework compositions, materials;
—— *civile*, (law) civil case;
—— *criminelle*, (law) criminal case;
—— *s fines*, fine sand;
—— *isolante*, (elec.) insulating material;
—— *en œuvre*, bullion that has been coined;
—— *s d'or*, gold bullion;
—— *(s) première(s)*, raw material;
—— *primitive*, raw material;
—— *soufflée aux poils*, (hipp.) purulent discharge around the coronet (due to inflammation of the sole).
matir, v. a., to deaden metals; to remove a soldering seam.
matoir, m., stamp, riveting hammer, flat hammer, seating tool, upsetter; chisel; soldering iron, copper bit.
matras, m., (chem.) matrass.
matriçage, m., swaging, operation of forming in a matrix or die.
matrice, f., die, matrix; block; boss, bottom swage; lower tool;
—— *à contre-plaque*, sort of block swage used in shaping shrapnel envelopes;
—— *à lunette*, sort of ring swage used in shaping shrapnel envelopes;
—— *de réfection*, (sm. a.) reloading matrix or die.
matricule, f., roll, list; (mil.) regimental roll, muster roll, name list (men and horses);
feuille ——, (mil.) sort of descriptive list;
livret ——, v. s. v. *livret*;
numéro ——, (mil.) regimental number of a soldier; number of a weapon;

matricule, *registre(-)* —— *des chevaux*, (mil.) register of officers' and of public horses, kept by the *trésorier* of each troop unit;
registre(-) —— *des hommes de troupe (des officiers)*, (mil.) men's (officers') register (containing *État civil*, descriptive list, etc.);
user son ——, (mil. slang) to serve in the army.
matriculer, v. a., (mil. slang) to steal.
mattage, m., overlaying with metal; operation of planing, smoothing metal, of beating down, clinching (as, an edge).
matte, f., (met.) matte;
—— *blanche de cuivre*, white metal, regulus;
—— *de cuivre*, copper matte;
—— *de plomb*, lead matte.
matter, v. a., to overlay with iron, copper, etc.; to smooth and plane metals; to beat down, clinch (as, an edge), (same as *mater*).
mature, f., (nav.) masts, masting; spars; masting sheers; mast house;
—— *flottante*, sheer hulk.
mauvais, a., foul, bad (of weather, wind, bottom, etc.); unsound, poor.
maye, f., v. *maie*.
mazage, m., v. *mazéage*.
mazaro, m., (mil. slang) regimental prison; prison cell; "jug."
mazéage, m., (met.) Styrian refining process, charcoal refining; conversion of gray into white pig.
mazé, p. p., (met.) refined;
fonte —— *e*, (met.) refined pig.
mazée, f., (met.) fine metal, refined metal.
mazéer, v. a., v. *mazer*.
mazelle, f., (met.) pig iron produced by melting cast iron with slag.
mazer, v. a., (met.) to refine.
mazerie, f., (met.) refining hearth.
mazette, f., sorry little horse; (mil. slang) recruit.
méandre, m., (top) meander of a river.
mécanicien, m., mechanician, machinist, engine builder; engineer; mechanical engineer; engine driver, locomotive engineer (U. S.); (Fr. nav.) engineer;
—— *en chef*, (Fr. nav.) grade of the engineer corps (relative rank of major);
—— *conducteur*, engine driver;
ingénieur ——, mechanical engineer;
—— *inspecteur*, (Fr. nav.) grade of the engineer corps (relative rank of colonel);
—— *inspecteur général*, (Fr. nav.) chief of naval engineer corps;
ouvrier ——, machinist;
premier ——, chief engineer;
—— *principal de 1ère (2ème) classe*, (Fr. nav.) grades of the engineer corps (relative rank of captain and of lieutenant, respectively);
second ——, second engineer.
mécanique, f., mechanics, mechanism, machine, machinery, piece of machinery;
fait à la ——, machine-made.
mécanique, a., mechanical; automatic.
mécaniquement, adv., mechanically.
mécanisme, m., mechanism, motion; (art.) (by extension) fuse;
corps de ——, (sm. a.) plate on which the parts of the Lebel repeating mechanism are assembled;
—— *de culasse*, (art., sm. a.) breech, breech-closing, mechanism;
—— *dérivé*, derived mechanism;
—— *de détente*, (sm. a., r. f. art.) trigger and related parts, firing mechanism; (mach.) expansion gear, expansion valve gear, cut-off;
—— *à double réaction*, (art.) generic term for any percussion fuse acting by inertia, inertia fuse;
—— *de fermeture*, *de fermeture de culasse*, v. —— *de culasse*;
—— *à friction*, (Fr. nav.) so-called friction fuse;
—— *de mise en marche, en train*, (mach.) slide valve link motion;
—— *d'obturation*, (art.) obturation gear;
—— *percutant*, (art.) percussion fuse; percussion lock;

mécanisme *de renversement*, —— *de renversement de marche, du mouvement*, (*mach.*) reversing gear;
—— *de répétition*, (*sm. a.*) repeating or magazine mechanism;
—— *de sûreté*, (*art., sm. a.*) safety device.
méchant, a., vicious, poor, worthless;
cheval ——, vicious horse.
mèche, f., wick; heart, middle, center or principal part, core; main piece of a steam capstan; bit, flat bit, borer, drill, auger, gouge, centerbit; tinder; lash (of a whip); screw (of a corkscrew); (*cord.*) core of a shroud-laid rope; (*artif.*) match, slow match; fuse; (*mach.*) link; (*art.*) polisher (of the bore); (*carp.*) main piece (of a construction);
—— *d'amorce*, (*mil. min.*) quick match;
—— *anglaise*, nose bit, shell auger, centerbit;
—— *à l'anglaise*, (*artif.*) quick match;
—— *à baguette*, (*sm. a.*) ramrod groove borer;
—— *Bickford*, (*artif.*) Bickford quick match;
—— *blanche*, (*sm. a.*) fine borer, reamer;
—— *à briquet*, tinder;
—— *de cabestan*, capstain barrel, spindle;
—— *à canon*, (*mil. min.*) port fire; slow match (obs.);
—— *de communication*, v. —— *d'amorce*, (*mil. min.*) train (of gunpowder);
—— *de cordage*, —— *de corde*, (*cord.*) heart or mid strand of a rope;
—— *en coton*, v. —— *à, d', étoupilles;*
—— *-cuiller*, —— *à cuiller*, gouge; spoon bit; nose bit;
découvrir la ——, (*mil. min.*) to discover a mine by countermining, and to remove its fuse;
—— *à dégorger*, (*art.*) priming wire;
—— *demi-ronde*, half-round, borer or bit;
—— *à entailles*, reaming bit, reamer;
—— *d'étoile*, (*artif.*) star match;
—— *à, d', étoupille(s)*, (*artif.*) quick match; (yarn dipped in alcohol and mealed powder);
éventer la ——, v. *découvrir la* ——;
—— *à feu*, v. —— *à canon;*
—— *à forer*, boring bar;
—— *de foret*, boring bit;
—— *de fouet*, whiplash;
—— *de graissage*, waste;
—— *de guerre*, v. —— *à canon;*
—— *de guindeau*, spindle of a windlass shaft;
—— *incendiaire*, (*artif.*) slow match (boiled in saltpeter, then soaked in melted *roche à feu*, q. v.);
—— *instantanée*, v. —— *Bickford;*
—— *de mine*, (*min.*) fuse;
—— *à mouche*, metal centerbit; polishing bit;
—— *ordinaire*, v. —— *à canon;*
—— *à percer*, borer, boring bit;
—— *plate*, flat (metal) borer;
—— *de poutre*, (*carp.*) lower element of a solid-built beam;
—— *de rabot*, plane iron;
—— *ronde*, round wick;
—— *à siphon*, siphon wick;
—— *sous-marine*, (*mil. min.*) water-tight fuse (to use under water);
—— *de sûreté*, (*artif.*) slow match;
—— *de sûreté de Bic'ford*, (*artif.*) Bickford fuse;
—— *de tarière*, auger bit;
—— *à tétine*, v. —— *à mouche;*
—— *à teton*, pin drill or borer;
—— *torse*, twisted bit;
tresse de ——, (*artif.*) large match composed of three ordinary strands twisted together;
—— *à trois pointes*, centerbit;
—— *à trois pointes universelles*, expanding bit;
—— *de vilebrequin*, gouge; boring bit; tap of a centerbit;
—— *de vireveau*, main piece of a windlass;
—— *à vis*, worm bit (screw bit).
mécouvrir, v. r., (*man.*) to overcover, overcheck.
médaille, f., medal;
—— *commémorative*, (*Fr. a.*) campaign medal;
—— *de commissionnaire*, (French) St. Helena medal;

médaille *d'honneur*, medal of honor;
—— *militaire*, in general, war medal; (*Fr. a.*) a medal given to n. c. o. and men, and, exceptionally, to generals;
—— *de Sainte Hélène*, (*Fr. a.*) medal given to soldiers who served between 1792 and 1815;
—— *de sauvetage*, life-saving medal.
médaillé, a., medaled; m., medalist, man who has got a medal.
médailler, v. a., to medal, bestow a medal upon.
médaillon, m., medallion.
médecin, m., physician, doctor; (*mil.*) surgeon;
aide- ——, (*Fr. nav.*) assistant surgeon;
—— *aide-major de 1ère (2ème) classe*, (*Fr. a.*) surgeon-lieutenant (sublieutenant);
—— *auxiliaire*, (*Fr. a.*) assistant or auxiliary surgeon, drawn on mobilization from various forces, and ranking as an *adjudant élève d'administration;* "contract doctor" (U. S. A.);
—— *auxiliaire de 2ème classe*, (*Fr. nav.*) subaltern grade, sort of assistant surgeon;
—— *-chef*, senior surgeon of a military hospital; chief surgeon of any unit;
—— *chef de service*, senior medical officer, as of a group of fortified positions;
—— *chef de tranchée*, (*Fr. a.*) in sieges, the chief surgeon of each attack (daily detail);
—— *en chef*, (*Fr. nav.*) full surgeon, fourth grade of the naval medical department;
—— *civil*, (*Fr. a.*) "contract doctor" (U. S. A.);
—— *-inspecteur*, (*Fr. a.*) second grade of the medical department (ranks with general of brigade);
—— *-inspecteur général*, (*Fr. a.*) surgeon-general, chief of the medical corps (ranks with generals of division);
—— *-major de 1ère (2ème) classe*, (*Fr. a.*) surgeon-major (captain);
—— *militaire*, (army) surgeon, medical officer;
—— *de première (2ème) classe*, (*Fr. nav.*) assistant surgeon (sixth and seventh grades);
—— *principal*, (*Fr. nav.*) surgeon (fifth grade);
—— *principal de 1ère (2ème) classe*, (*Fr. nav.*) surgeon-colonel (lieutenant-colonel);
—— *traitant*, surgeon in charge of a case, attending surgeon;
—— *de tranchée* (*Fr. a.*) in sieges, surgeon on duty in the trenches (permanent detail).
médecine, f., medicine.
médical, a., medical.
médicament, m., medicine, medicament;
coffre de ——*s*, medicine chest.
méditerrané, a., surrounded by land.
méditerranée, f., an interior sea, (*esp.*) the Mediterranean.
méditerranéen, a., Mediterranean; surrounded or inclosed by land.
médullite, f., (*hipp.*) inflammation of the marrow.
méfiant, a., (*mil. slang*) fool soldier.
méganite, f., (*expl.*) meganite.
mégie, f., tawing;
passer en ——, to taw.
mégir, v. a., to taw, dress, leather.
mégisser, v. a., to taw.
mégisserie, f., v. *mégie.*
mégissier, m., tawer.
mehari, m., racing camel (*pl. mehara*), (spelled also *mahari*).
mehariste, m., (*Fr. a.*) rider of a *mehari* (a detachment of these exists for service in the Desert of Sahara).
méjuger, v. r., (*man.*) to overcover, overcheck.
mélange, m., mixture, mixing, mingling; (*hipp.*) mash;
—— *d'eau et de vapeur*, (*steam*) priming;
—— *des fils*, (*teleg.*) crossing of wires (contact);
—— *frigorifique*, freezing mixture;
—— *méthodique*, (*powd.*) blending;
—— *réfrigérant*, freezing mixture.
mélangeoir, m., (*powd.*) mixing barrel.

mélanger, v. a., to mix, mingle, intermix, blend, work up together.
mélanose, f., (*hipp.*) melanosis.
mêlé, p. p., mixed, tangled;
— *cheval* ——, horse tangled in the traces;
— *marchandise* ——*e*, merchandise of uneven quality.
mêlée, f., mêlée, mellay, fight, fray, conflict; (at drill) a confusion, or intermingling of the troops, for the purpose of training them to reform.
mêler, v. a., to mix, intermix, mingle;
— *un cheval*, (*man.*) to confuse a horse;
— *se* ——, to fight, take part in a hand-to-hand fight;
— *une serrure*, to spoil a lock.
mélèze, f., larch (wood and tree).
mélonite, f., (*expl.*) melinite.
mélis, m., strong-threaded canvas, sailcloth;
— *toile de* ——, sailcloth.
melon, m., melon; (*mil. slang*) "plebe" (U. S. M. A.), "snooker" (R. M. A.).
mélote, f., sheep's skin with the wool.
mémarchure, f., (*hipp.*) sprain in a horse's leg.
membrane, f., membrane;
— *phonique*, diaphragm of a telephone, vibrating membrane.
membre, m., member, limb, rib; member (of a board); (*nav.*) framing timber; timber.
membrer, v. a., (*mil. slang*) to drill.
membret, m., (*harn.*) thickness of metal at the end of each branch of a spur.
membrière, f., (*cons.*) element of a frame.
membron, m., orle, orlet.
membrure, f., framing; frame; ribs, timbers; framework (of a ship); frame (of a door); cord measure.
mémoire, m., bill; memorandum; memoir; report accompanying a reconnoissance sketch or survey; report; scientific report;
— *de concours*, —— *couronné*, prize essay;
— *descriptif*, descriptive memoir;
— *militaire*, military report;
— *de proposition*, (*Fr. a.*) report accompanying recommendation for promotion, decoration, etc.
mémorial, m., memorandum book; title of certain periodical publications (e. g., *mémorial de l'artillerie de marine*).
menacer, v. a., to threaten;
— *ruine*, (of a building) to be about to fall.
ménager, v. a., to husband, to use cautiously; to take care of; to spare; to make (i. e., contrive);
— *un cheval*, (*man.*) to spare a horse;
— *le coup*, to strike carefully; (*cord.*) to haul handsomely;
— *ses munitions*, (*mil.*) to economize one's ammunition, etc.;
— *des troupes*, (*mil.*) to avoid fatiguing the troops.
menant, pres. p., (*mach.*) driving.
mené, p. p., (*mach.*) driven.
meneau, m., mullion;
— *faux* ——, false mullion.
mener, v. a., to lead (a person, a horse, troops), conduct, guide; to take, carry, convey; to direct, govern; to drive (a team, a horse); to steer (a boat); to carry on (an attack);
— *une barque*, to steer a boat;
— *battant*, (*mil.*) to drive before one;
— *sur le bon pied*, (*man.*) in the gallop, to lead with the off foot;
— *les bras*, to work hard;
— *un cheval droit*, (*man.*) to guide a horse straight ahead (shoulders and hind quarters on same line);
— *un cheval en avant*, (*man.*) to urge forward an unwilling horse;
— *difficile à* ——, (*man.*) hard to ride;
— *l'ennemi battant*, —— *tambour battant*, (*mil.*) to drive the enemy before one;
— *facile à* ——, (*man.*) easy to ride;
— *de front*, to drive abreast;
— *la guerre*, to make war;
— *les mains*, to fight;
— *rudement*, (*mil.*) to subject to severe loss.

meneur, m., leader, driver; ringleader; (*mil.*) ringleader of a mutiny.
ménille, f., shackle;
— *de brague*, (*nav. art.*) breeching shackle (obs.).
ménisque, m., (*opt.*) meniscus.
menotte, f., little hand; handle (of a winch); (*in pl.*) handcuffs; (*mach.*) drag link; (*art.*) trail handle; (*r. r.*, *mach.*) spring hanger; (in the 0.00ᵐ siege railroad truck) sort of coupling rod;
— *mettre les* ——*s*, to handcuff.
mensole, f., (*cons.*) keystone of an arch;
— *de voûte*, keystone.
mensuration, f., mensuration; land surveying.
menton, m., (*hipp.*, *etc.*) chin.
mentonnet, m., catch, lug, nab, ear, tipper, cleat; notched peg; catch-bolt head; (*mach.*) tappet, cam, wiper; (*r. r.*, *mach.*) flange of a wheel; (*art.*) ear (of a shell, obs.); (*sm. a.*) hammer strut;
— *à* ——, fitted with a catch or cog;
— *de bombe*, (*art.*) ear or lug of a shell;
— *de chargement*, (*art.*) roller or lever support;
— *d'établi*, —— *à patte*, holdfast;
— *de revolver*, (*sm. a.*) hammer strut;
— *de soufflet*, lug of a bellows.
mentonnière, f., (*mil.*) chin strap; (*met.*) muffle plate or support.
menu, m., small fragment, small piece; coal dust, small coal;
— —*s des houillères*, small coal.
menu, a., small, fine, minute; minor, lesser;
— —*s achats*, —— *s approvisionnements*, small stores;
— *charbon*, small coal;
— —*s équipages d'une armée*, (*mil.*) spare horses and mules, etc.;
— *filin*, (*cord.*) small rope, small stuff;
— —*es houilles*, fine coal, coal dust;
— *officier*, (*mil.*) noncommissioned officer;
— *plomb*, small shot, bird shot.
menuis(aill)e, f., twigs, brushwood; (*sm. a.*) dust shot, mustard seed.
menuiser, v. a., to do carpenter's, joiner's work; to chop or split up (wood) fine.
menuiserie, f., joinery, joiner's work;
— *en bâtiments*, —— *en bâtisse*, —— *dormante*, house carpentry;
— *mobile*, cabinetmaking.
menuisier, m., joiner, carpenter.
méphitique, a., mephitic.
mépister, v. r., (*man.*) of the horse, to mistrail.
méplat, m., flattened surface (either plane or of greater radius of curvature than adjacent parts); flattening tool; flat side (as opposed to edge); (*art.*) flat top or nose of a projectile.
méplat, a., flat, flatwise;
— *bois* ——, planks, timber, broader than thick; planks laid flatwise;
— *fer* ——, flat iron;
— *solive* ——*e*, flat-laid joist.
mer, f., sea, tide, ocean, waters; (*nav.*) sea service, the navy (as opposed to *terre*, land service, army);
— *à la* ——, overboard; at sea;
— *armée de* ——, the (war) navy; (more narrowly) a fleet, a squadron;
— *basse* ——, low water; shallow sea;
— *basse* —— *de morte eau*, low-water neap;
— *basse* —— *de vive eau*, low-water spring;
— *battre la* ——, to cruise, to cruise about for a long time in the same latitude;
— *belle* ——, smooth water;
— *la* —— *blanchit*, v. *la* —— *moutonne*;
— *bord de* ——, seaside, seashore;
— *bras de* ——, arm of the sea;
— *brise de* ——, sea breeze;
— —— *clapoteuse*, short sea, choppy sea;
— —— *contraire*, cross sea;
— *coupe de* —— , sea (wave); squall;
— —— *courte*, short sea;
— —— *creuse*, cross sea;
— —— *debout*, head sea;
— —— *démontée*, violent sea;

mer *dure*, rough sea;
 écumeur de ——, pirate;
 en ——, at sea;
 la —— *est étale*, it is slack water;
 la —— *est, fait, planche*, the sea is smooth;
 —— *étale*, slack water;
 étale de la pleine ——, high water, slack of the flood;
 —— *fermée*, closed sea, landlocked sea;
 —— *du fond*, ground swell;
 fortune de ——, perils of the deep;
 gens de ——, seamen, seafaring people (plural of *homme de* ——);
 grande ——, the high seas, the main; (*in pl.*) springtide;
 grosse ——, heavy sea;
 haute ——, high seas;
 —— *haute* ——, high water;
 haute —— *d'équinoxe*, high-water spring;
 haute —— *de morte eau*, high-water neap;
 haute —— *de vive eau*, high-water spring;
 en haute ——, on the high seas;
 homme à la ——, man overboard;
 homme de ——, seaman;
 —— *houleuse*, swell; rough sea;
 —— *d'huile*, absolutely calm sea;
 —— *intérieure*, closed, landlocked sea, inland sea; (more especially) the Mediterranean;
 jour de ——, sea day;
 journées de ——, days at sea;
 lame de la ——, wave, billow;
 —— *libre*, interior sea bordering on several States; open sea;
 —— *longue*, long swell;
 mal de ——, seasickness;
 —— *mâle*, rough sea;
 la —— *est mâle*, the sea is rough;
 —— *de maline*, springtide;
 —— *mate*, hollow sea;
 mettre à ——, *en* ——, to put to sea;
 mettre une embarcation à la ——, to lower a boat;
 molle ——, slack water at the lowest of the tide;
 la —— *monte*, the tide is rising;
 la —— *moutonne*, white caps are showing;
 —— *moutonnée*, white caps;
 niveau des basses ——*s*, low-water mark;
 niveau des hautes ——*s*, high-water mark;
 outre ——, *d'outre* ——, from beyond sea;
 par ——, *par voie de* ——, by sea;
 la —— *a perdu*, the tide has fallen;
 périls de la ——, sea risk, perils of the sea;
 plein de la ——, high water;
 pleine ——, high water, slack of the flood; the high seas; offing;
 en pleine ——, on the high seas;
 —— *pleine*, offing, high water;
 pleine —— *d'équinoxe*, equinoctial high water;
 pleine —— *de morte eau*, high-water neap;
 pleine —— *des syzygies, pleine* —— *de vive eau*, high-water spring;
 port de ——, seaport;
 prendre la ——, to set sail;
 —— *privée*, v. —— *territoriale*;
 la —— *rapporte*, the tide begins to increase (after dead neaps);
 la —— *refoule*, the tide ebbs; the water falls;
 tenir la ——, to keep the seas, continue at sea;
 —— *territoriale*, territorial waters;
 —— *tourmentée*, choppy sea;
 —— *du travers*, cross sea.

mercenaire, m., a., mercenary.

merci, f., mercy, grace;
 à la —— *de*, at one's mercy.

mercure, m., mercury;
 —— *fulminant*, (*expl.*) fulminate of mercury.

mercuriale, f., reprimand, censure; periodical list of prices current (especially such lists published by municipal authority in France);
 faire une ——, to rebuke, to reprimand.

mère, f., mother; (*tech.*) master-tap, hub; (adjectively) principal;
 eau ——, mother liquor.

méridien, m., meridian; midday, meridian; a., meridian;

méridien *géographique*, meridian;
 ligne ——*ne*, meridian line, meridian;
 —— *magnétique*, magnetic meridian;
 premier ——, prime meridian;
 —— *terrestre*, meridian.

méridienne, f., meridian, meridian line;
 —— *magnétique*, magnetic meridian.

meridional, a., southern, meridional.

merlin, m., club; long-handled hammer; splitting ax; (*cord.*) marline (in the French artillery, three yarns laid up together);
 —— *blanc*, (*cord.*) white, untarred marline·
 —— *noir*, (*cord.*) tarred marline.

merliner, v. a., (*cord.*) to marl, marline.

merlon, m., (*fort.*) merlon.

merrain, m., clapboard; staves, stavewood.

merveilleux, a., (*sm. a.*) hair (of rifling).

mésair, m., v. *mézair*.

mess, m., (*mil.*) mess.

messager, m., messenger, carrier;
 —— *de campagne*, (*mil.*) proposed mounted orderly for infantry headquarters;
 —— *d'État*, state messenger; king's messenger;
 pigeon ——, carrier pigeon.

messagerie, f., coach office, booking office; coach, stage coach (used sometimes in plural); (*in pl.*) line of mail steamers; (*r. r.*) freight, parcels.

messe, f., mass;
 —— *basse*, low mass;
 grand ——, *haute* ——, high mass;
 petite ——, v. —— *basse*.

mesurage, m., measuring, measurement; mensuration; surveying.

mesure, f., measure, measuring, measurement; dimension; (*hipp.*) cadence; (*fenc.*) proper distance;
 —— *absolue*, absolute measure;
 —— *d'arpenteur*, surveyor;
 —— *s d'assortiments*, copper powder measures of various sizes;
 —— *à bouts*, end measuring;
 —— *à cheval*, measure by horse's gait;
 —— *comble*, heaped-up measure;
 de ——, according to measure;
 —— *dure*, (*fenc.*) stiff action;
 —— *d'élévation d'un piston*, stroke, lift of a piston;
 entrer en ——, (*fenc.*) to take a step toward one's adversary;
 être à la ——, (*fenc.*) to have the proper distance;
 être hors de ——, (*fenc.*) to be out of distance;
 —— *de force*, dynamometer;
 gagner la ——, (*fenc.*) to advance;
 lâcher la ——, (*fenc.*) to retreat;
 —— *de longueur*, measure of length;
 —— *métrique*, metrical measure;
 —— *à la montre*, measure by timing (as, a horse's gait);
 —— *au pas*, pacing off a distance;
 —— *à poudre*, powder measure;
 —— *rase*, strike measure;
 rompre la ——, (*fenc.*) to get out of reach;
 —— *en ruban*, tape measure;
 serrer la ——, (*fenc.*) to close on one's adversary;
 —— *à traits*, measuring between marks;
 —— *à vue*, measure by the eye.

mesurer, v. a., to measure, measure out, mete, mete out; to proportion; (*fenc.*) to measure swords;
 —— *ses armes*, to fight with anyone;
 —— *à cheval*, to measure by a horse's gait;
 —— *à la corde*, —— *au cordeau*, to trace out;
 —— *un coup*, to aim, adjust, a blow accurately;
 —— *les épées*, in a duel, to measure the swords to see if they are of the same length;
 —— *son épée*, (*fenc.*) to fight;
 —— *par équarrissement*, to take as the volume of anything, the volume of the circumscribing rectangular parallelopipedon;
 —— *à la montre*, to measure by timing, to time;
 —— *au pas*, to pace off (a distance;)
 —— *la terre*, to measure one's length on the ground;

mesurer à la vue, to estimate by the eye;
— avec les, des, yeux, to measure, estimate, by the eye.
mesurette, f., small measure.
métacarpe, m., (hipp.) the metacarpal bones.
métacarpien, a., (hipp.) metacarpal; m., metacarpal bone;
os ——s, metacarpal bones;
—— rudimentaire, small metacarpal bone;
—— rudimentaire externe, outer metacarpal bone;
—— rudimentaire interne, inner metacarpal bone.
métacentre, m., (nav.) metacenter.
métacentrique, a., (nav.) metacentric;
hauteur ——, metacentric height.
métairie, f., farm, farmhouse.
métal, m., metal; alloy; ore; (specially) an alloy of tin, copper, and gun metal (founder's term); metaling of a road;
—— d'Aich, an alloy of copper, zinc, and iron;
—— d'Alger, a sort of imitation silver;
—— anglais, Britannia metal;
—— antifriction, antifriction metal; Babbitt metal;
—— antifriction magnolia, magnolia metal;
—— argentin, v. —— anglais;
—— de Bath, Bath metal;
—— blanc, white metal; antifriction metal;
—— britannique, v. —— anglais;
—— brut, coarse metal;
—— à canon, gun metal (i. e., bronze) (more rarely) cast iron;
—— de cloche, bell metal;
—— à collets, antifriction metal;
—— console, an alloy of copper, zinc, and tin;
—— de Darcet, a fusible alloy of bismuth, tin, and lead;
—— décapé, pickled metal;
—— delta, delta metal (iron and brass);
—— doux, antifriction metal;
—— écumé, kish;
fin ——, fine iron, fine metal;
—— de fonte, cast metal;
—— homogène, homogeneous metal;
—— mazé, fine(d) metal;
—— de miroirs, speculum metal;
—— mixte, compound metal (steel and iron, for compound plates);
—— de Muntz, Muntz' metal (zinc and copper);
—— natif, native metal;
—— de prince, super refined copper;
—— du Prince Robert, Prince's metal;
—— rebelle, refractory metal;
—— Roma, an alloy of platinum, manganese, and cobalt;
—— de Rose, Rose's fusible alloy (bismuth, lead, and tin);
—— sterro, sterro metal (an alloy of copper, zinc, iron, and tin);
—— de Stirling, Stirling metal;
—— utile, the quantity of serviceable or workable metal obtained from an ingot or from a charge, v. mise au mille.
métalline nitroleum, f., (expl.) a dynamite whose dope consists of red lead (with or without plaster of paris).
métallique, a., metallic;
gaze, tissu, toile ——, wire gauze.
métallisage, m., plating with metal.
métalliser, v. a., to cover or plate with metal.
métalloïde, m., metalloid.
métallurgie, f., metallurgy.
métallurgique, a., metallurgical.
métatarse, m., (hipp.) the metatarsal bones.
métatarsien, a., (hipp.) metatarsal; m., metatarsal bone;
os ——, metatarsal bone;
—— rudimentaire externe, outer small metatarsal bone;
—— rudimentaire interne, inner small metatarsal bone.
métayer, m., farmer.
méteil, m., maslin (mixed wheat and rye).
météorisation, f., (hipp.) distension of the belly.

météoriser, v. a., (hipp.) to swell, distend, the belly.
météorisme, m., (hipp.) flatulent colic.
météorologie, f., meteorology.
météorologique, a., meteorological.
méthode, f., method; process;
—— de Bapaume, (siege) the classic method of breaching a wall (by attacking it at one-third its height);
—— à la catalane, (met.) Catalan process;
—— de cheminement, (surv.) method in which the sides and angles of the polygon to be surveyed are measured;
—— des doubles visées, (surv.) in compass surveying, method using back as well as fore sights;
—— des gradins, (siege) method of breaching, by making steps in the wall;
—— de l'ile d'Aix, (siege) method of breaching in which the cut in the wall slopes upward;
—— d'intersection, (surv.) usual method of surveying; base-line surveying;
—— des moindres carrés, method of least squares;
—— de rayonnement, (surv.) survey in which points are fixed by polar coordinates from a central station;
—— de recoupement, (surv.) method by resection;
—— de relèvement, —— des segments capables, (surv.) three-point problem.
métier, m., trade, handicraft; (mach.) loom, frame;
de ——, by profession, by trade.
métis, a., (met.) red-short; (hipp.) crossbred; m., crossbred horse;
deuxième ——, (hipp.) three-quarter bred;
premier ——, (hipp.) the offspring of parents of different race.
métissage, m., (hipp.) crossbreeding.
métrage, m., measuring by a meter stick; length in meters.
mètre, m., meter; meter rule;
—— anglais, yard measure, yard;
—— articulé, folding or jointed rule;
—— à bouts, v. —— étalon à bouts;
—— courant, running meter;
double ——, 2-meter rule or measure;
—— étalon, standard meter;
—— étalon à bouts, end standard meter (for comparing base-measuring rods);
—— étalon à traits, line standard meter (for comparing base-measuring rods);
—— -kilogramme, v. kilogrammètre;
—— pliant, folding rule;
—— en ruban, tape measure;
—— -tonne, meter ton;
—— à traits, v. étalon à traits.
métrer, v. a., to measure by the meter.
métrique, a., metric.
métropole, f., metropolis; mother country.
métropolitain, a., metropolitan; home (as distinguished from colonial);
armée ——, home army.
métropolitaine, f., the capital of a State.
mettre, v. a., to put, put on, put to, place, lay, set, employ, use, commit; to hitch up, harness;
—— à l'abri de, to put any body or thing under shelter (e. g., from fire);
—— à l'ancre, to come to anchor;
—— l'avant-train, (art.) to limber up;
—— les avirons sur le plat, to feather oars;
—— les avirons debout, to toss the oars;
—— la baïonnette au canon, (sm. a.) to fix bayonets;
—— à la bande, (nav.) to careen;
—— au bandé, (sm. a.) to cock;
—— au bas, to conclude a letter with a proper formula of respect;
—— bas, to foal, to litter;
—— bas les armes, (mil.) to lay down one's arms, to surrender;
—— en batterie, (art.) to run or go into battery;
—— bien ensemble, (man.) to throw a horse on his quarters;
—— en bois, (sm. a.) to assemble the parts of a rifle on the stock;

mettre à bord, to ship; to put, carry, etc., on board;
— les canons à la serre, (nav.) to house the guns;
— les canons aux sabords, (nav.) to run out the guns;
— le cap sur...., (nav.) to head for....;
— de champ, to set up on edge (as, a plank, bricks, etc.);
— en chantier, (carp.) to mark out; (nav.) to put upon the stocks; (in gen.) to begin anything;
— sur les chantiers, (nav.) to put upon the stocks;
— un cheval dedans, dans la main, dans les talons, (man.) to make a horse obey the hand, the spurs;
— un cheval au galop (pas, trot) (man.) to put a horse to the gallop (walk, trot);
— les chevaux à la voiture, to hitch up;
— dans le circuit, (elec.) to cut in, put in circuit;
— à contribution, to lay under contribution;
— en communication, to connect;
— au cran de départ, (sm. a.) to cock;
— la croupe au mur, (man.) to cause a horse to passage, croup to wall of riding hall;
— dedans, (carp.) to assemble the pieces (as, of a frame);
— au dehors, (nav.) to get out (a boat);
— en demeure, (law) to order, summon;
— les deux bouts en dedans, (man.) to train, supple, the head and the croup of a horse;
— en disponibilité, (Fr. a.) to put on the unemployed list;
— à l'eau, to launch (a ship);
— à l'effet, (top. drawing) to indicate surface forms on a map, topographical sketch;
— en, par, écrit, to put in writing;
— un embargo, to lay an embargo;
— à l'encre, to ink (a drawing);
— d'équerre, to square;
— un fer à un cheval, to shoe a horse;
— à feu, to start up a furnace;
— à feu et à sang, to burn and kill;
— en feu, to burn; (met.) to put a furnace in blast;
— le feu, to set on fire; to touch off (a mine, etc.);
— le feu à, to set on fire (as, a house);
— les feux au fond des fourneaux, to bank the fires;
— au fil de l'épée, to put to the sword;
— à flot, to float (a stranded ship, etc.);
— le frein, to put on the brake (brakes);
— garnison dans, to garrison (a town, etc.);
— un grain de lumière, (art.) to bush a vent;
— en haie, to pile bricks (in rows) for drying;
— un homme aux fers, to iron, shackle, a man;
— hors, v. — hors feu;
— hors du circuit, (elec.) to cut out, put out of circuit;
— hors de combat, to disable;
— hors feu, (met.) to put a furnace out of blast;
— hors d'insulte, (fort.) to secure from sudden assault;
— hors de service, to destroy, make useless;
— à jour, to bring up to date;
— au large, (nav.) to put out, stand out, to sea;
— au lavis, to wash (color), tint, a drawing;
— les linguets au cabestan, to pawl the capstan;
— en marche, to set out on a march; to set out; to set at work (an engine, etc.);
— à la mer, (nav.) to launch a ship; to shove off (a boat); to stand out to sea;
— en mer, to go to sea;
— en mire, (art.) to lay a gun;
— à mort, to put to death;
— le mousqueton à la botte, (sm. a.) to return, boot, a carbine;
— en mouvement, to start (a machine, etc.); to throw into gear;

mettre un navire à la côte, to run a ship ashore;
— au net, to make a fair copy (drawing, notes, etc.);
— à l'ordre, (mil.) to publish in orders;
— au pain et à l'eau, to put on bread and water;
— pavillon bas, (nav.) to haul down one's flag, to surrender;
— son pavillon sur...., (nav.) to hoist one's flag on....;
— une pièce hors d'eau, (art.) to depress the muzzle so as to keep out water;
— pied à terre, to dismount;
— en place, to organize, prepare;
— sur plage, to beach a boat;
— dans le plein, to hit the bull's-eye;
— au point, (inst.) to focus, adjust; in surveying, to set up an instrument accurately over its station;
— à la presse, (mach.) to set or seat by pressure;
— en pression, (steam) to get up steam;
— sur son raide, sur son fort, (carp.) to put the convex side of a piece uppermost;
— à la réforme, (Fr. a.) to condemn (material); to cast a horse; to discharge from the service (for misconduct or physical unfitness);
— au renard, to hold up, keep up, the monkey of a pile driver;
— à la retraite, (mil.) to retire;
— sur roulettes, (art.) to throw a top carriage in gear (on rollers) for return to battery;
— à sac, to sack (a city);
se — sous les armes, (mil.) to get under arms;
— à sec, to pump out;
— le siège devant, to lay siege to;
— en station, to set up an instrument;
— à la suite, (mil.) to attach an officer as supernumerary;
— en talus, (fort.) to slope;
— à terre, to land, disembark, put ashore;
— en train, (mach.) to throw into gear;
— au vert, to put a horse out to grass.

meuble, m., article of furniture.

meuble, a., movable; (of earth, soil) mellow, easily moved or worked;
biens —s, personal property;
terre —, light soil.

meubler, v. a., to furnish.

meulage, m., grinding;
— à l'eau, wet grinding.

meulard, m., large grindstone.

meularde, f., middle-sized grindstone.

meule, f., grindstone, millstone; mow or stack (of hay, fodder, etc.); pile of wood (in charcoal burning); polishing wheel; (powd.) runner, roller; (fond.) foundation for a mold;
— à aiguiser, grindstone;
— à baïonnette, (sm. a.) grindstone for hollowing sides of bayonet blades;
— de carbonisation, (met.) mass to be fired;
champ de —, thickness of a grindstone;
— de charbon, pile of charcoal;
— courante, upper millstone;
— de dessus, upper millstone;
— à eau, wet grindstone;
— d'émeri, emery wheel, buff wheel, buff;
— de foin, etc., hayrick, etc.;
— gisante, nether millstone;
— à gouttières, (sm. a.) grindstone for hollowing sides of swords, etc.;
— s indépendantes, (powd.) runners not working in pairs;
— inférieure, lower, nether, millstone;
mettre en —, to pile (hay, etc.) in ricks;
— mobile, running millstone;
— montée, grindstone on its stool and bed;
moulins à —s, (powd.) incorporating mill;
— roulante, (powd.) runner, roller;
— supérieure, v. — de dessus;
— suspendue, (powd.) runner that does not touch the bed; hanging runner;
— suspendues et indépendantes, (powd.) uncoupled hanging runners;

meule, *trituration sous les* —— *s,* (*powd.*) incorporation.
meuleau, m., small grindstone.
meulier, a., of mills, for grinding; m., millstone maker;
   carrière —— *ère,* millstone quarry;
   pierre —— *ère,* millstone.
meulière, millstone, grindstone; millstone grit, buhrstone;
   pierre de ——, millstone grit, buhrstone.
meurtrier, a., (*mil.*) deadly (of fire, etc.).
meurtrière, f., (*fort.*) loophole.
meurtrir, v. a., to bruise, to kill, slaughter.
meurtrissure, f., bruise, contusion.
mézair, m., (*man.*) an air consisting of a succession of jumps forward.
mho, m., (*elec.*) mho.
mi-bois, m., (*carp.*) scarf, halving;
   assembler à ——, to scarf.
mica, m., mica.
mi-chemin, à ——, halfway.
mi-côte, à ——, halfway down (up) a slope, a hill.
micro-farad, m., (*elec.*) microfarad.
microhm, m., (*elec.*) microhm.
micromètre, micrometer; micrometer caliper;
   —— *à double image,* double-image micrometer;
   —— *objectif,* heliometer;
   —— *optique,* dioptric micrometer;
   —— *prismatique,* v. —— *à double image;*
   —— *de torsion* (*elec.*) torsion micrometer;
   vis à ——, micrometer screw.
micrométrique, a., micrometric.
micron, m., one one-thousandth of a millimeter.
microphone, m., microphone.
microscope, m., microscope;
   —— *composé,* compound microscope;
   —— *simple,* simple microscope;
   —— *solaire,* solar microscope.
microscopie, f., microscopy.
microscopique, a., microscopic.
midi, m., noon, 12 m.; south.
mi-doux, a., (*met.*) half soft.
mi-dur, a., (*met.*) half hard.
mi-fer, *assembler à* ——, to halve and lap (of iron beams, etc.).
mi-flot, m., half flood.
migraine, f., headache.
mi-jusant, m., half ebb.
milice, f., militia; (*in gen.*) troops, army; war, warfare;
   soldat de ——, militiaman.
milicien, m., militiaman.
milieu, m., middle, middle point; middle part; midst; heart; center; cor.; expedient, medium, means; (*art.*) swing;
   attelage du ——, (*art.*) swing team.
miline, f., (*expl.*) miline.
militaire, a., military; soldier-like;
   architecture ——, art of fortification;
   art ——, art of war;
   chantier ——, (*nav.*) navy-yard;
   exécution ——, v. s. v. *exécution;*
   heure ——, "sharp;"
   honneurs, —— *s,* military honors;
   justice ——, v. s. v. *justice.*
militaire, m., soldier (either officer or man);
   être ——, to be a soldier (by profession);
   —— *isolé,* detached soldier.
militairement, adv., in a military, soldier-like, manner; smartly, energetically; by force of arms;
   occuper ——, to occupy with troops.
militant, a., militant.
militariser, v. a., to render military, to fill with the military spirit.
militarisme, m., militarism, domination or supremacy of military ideas.
mille, a., m., thousand;
   grand ——, (paver's term) 1,122 paving stones;
   mise au ——, v. s. v. *mise.*
mille, m., mile;
   —— *d'Angleterre,* statute mile;
   —— *géographique,* geographical (or nautical) mile; (sometimes) the fifteenth of a degree at the equator (7,420 meters);
   —— *géométrique,* v. —— *mesuré;*

mille *marin,* nautical mile (1,852 meters);
   —— *mesuré,* (*nav.*) measured mile;
   —— *métrique,* 1,000 kilometers.
millésime, m., date (on a coin, gun, etc.).
millet, m., millet.
milliaire, m., millestone;
   borne ——, millestone.
milliampère, m., (*elec.*) milliampere.
milliare, m., milliare, one one-thousandth part of an *arc.*
millier, m., thousand; 1,000 kilograms.
milligramme, m., milligram.
millilitre, m., milliliter, one one-thousand of a liter.
millimètre, m. millimeter.
millistère, m., one one-thousandth of *astère,* of a cubic meter.
mils, m. pl., (*gym.*) Indian clubs.
mi-marée, à ——, at half tide.
mi-mât, à ——, at half-mast, half-staff.
miméographe, m., (Edison) mimeograph.
minahouet, minaouet, m., (*cord.*) serving board.
mine, f., mine (of coal, iron, etc.); ore; (*mil.*) mine; (by extension) the shafts, galleries and chambers of a mined site;
   à la ——, by blasting;
   —— *d'acier,* (*met.*) spathic iron;
   —— *à aiguillette,* (*mil.*) bored mine;
   —— *anglaise,* red lead;
   —— *d'attaque,* (*mil.*) offensive mine;
   —— *automatique,* (*mil.*) mechanical mine;
   —— *à bocarder,* raw, crude ore;
   bourrer une ——, to tamp a mine.
   —— *brute,* raw ore;
   —— *cassante,* dry ore;
   —— *sans cervelle,* (*mil.*) gallery of a mine requiring support on account of the loose nature of the soil;
   chambre de ——, (*mil.*) mine chamber;
   contre ——, (*mil.*) countermine;
   —— *en croix,* (*mil.*) triple mine;
   cylindre de ——, (*expl.*) small disk of gun cotton for use in mines;
   —— *défensive,* (*mil.*) defensive mine (countermine);
   —— *à démolition,* (*mil.*) demolition mine;
   —— *dilatée,* (*mil.*) horizontal mine;
   —— *double,* (*mil.*) double mine;
   —— *douce,* zinc blende, sphalerite;
   école des —— *s,* school of mines;
   —— *égarée,* placer, pocket;
   —— *élevée,* (*mil.*) upright mine;
   entonnoir de ——, (*mil.*) mine crater;
   —— *étagée,* (*mil.*) tiered mine;
   éventer une ——, (*mil.*) to discover a mine and destroy it;
   faire jouer, sauter, une ——, (*mil.*) to spring a mine;
   —— *de fer,* (*met.*) iron ore;
   —— *de fer blanche,* (*met.*) spathic iron;
   —— *de fer en grains,* (*met.*) granular (brown) oxide of iron;
   —— *de fer spéculaire,* (*met.*) specular iron (ore);
   —— *fictive,* (*torp.*) dummy mine;
   —— *s fixes,* branchings, veins;
   —— *forée,* (*mil.*) bored mine, synonym of *forage;*
   fourneau de ——, (*mil.*) mine chamber;
   —— *à fourneau ordinaire,* (*mil.*) ordinary mine;
   galerie de ——, (*mil.*) mine gallery;
   en galerie de ——, (*mil.*) after the fashion of a mine gallery;
   guerre de —— *s,* (*mil.*) subterranean warfare;
   —— *grasse,* (*met.*) ore freed from its gangue;
   —— *inflammable à volonté,* (*mil.*) controlled mine, judgment mine;
   logement des —— *s,* (*mil.*) lodgment of mines, of the mine;
   —— *micacée,* (*met.*) specular iron ore;
   —— *militaire,* (*mil.*) military mine (as distinguished from commercial mining);
   —— *offensive,* v. —— *d'attaque;*
   —— *orange,* red lead;
   —— *à percussion,* (*mil.*) percussion mine;
   —— *de plomb,* black lead; lead ore; black lead pencil;

mine, *poudre de* ——, (*expl.*) blasting powder; *poudre de* —— *anguleuse*, (*expl.*) sharp-grained blasting powder;
— *poudre de* —— *fin grain*, (*artif.*) fine-grain powder used in safety fuses;
— *poudre de* —— *ronde*, (*expl.*) round-grained powder (demolition and blasting);
— *de projection*, (*art.*) Brialmont's name for a torpedo shell; synonym of *savartine*, q. v.;
— *puits de* ——, mine shaft;
— *saucisson de* ——, (*mil.*) firing train;
— *simple*, (*mil.*) single mine;
— *sous-chargée*, (*mil.*) undercharged mine, v. s. v. *fourneau;*
— *sous-marine*, f., (*torp.*) submarine mine;
— *surchargée*, (*mil.*) overcharged mine, v. s. v. *fourneau;*
— *système de* ——*s mixte*, (*mil.*) a combination of the two following systems;
— *système de* ——*s du moment*, (*mil.*) a system of mines established when required (i. e., during war);
— *système de* ——*s permanent*, (*mil.*) a system of mines established in peace;
— *en* T, (*mil.*) double mine;
— *en trèfle*, —— *tréflée*, —— *triple*, (*mil.*) triple mine.
miner, v. a., to mine (as coal); to undermine, sap, hollow, eat away; (fig.) to prey upon; to ruin by degrees; (*mil.*) to mine.
minerai, m., ore;
— *d'alluvion*, alluvial ore;
— *bocardé*, crushed ore;
— *broyer les* ——*s*, to crush ore;
— *brut*, raw ore;
— *essai de* ——*s*, assaying;
— *essayer un* ——, to assay an ore;
— *essayeur de* ——, assayer;
— *de fer*, iron ore;
— *de fer en grains*, granular iron ore;
— *en grains*, granular ore;
— *grillage de* ——*s*, roasting of ore;
— *griller les* ——*s*, to roast ore;
— *hydraté globulaire*, kidney iron ore;
— *limoneux*, limonite;
— *des marais*, bog iron ore;
— ——*s menus*, small ore; ore in fragments;
— *riche*, best work;
— *en roche*, compact ore; ore in masses;
— *terreux*, earthy ore;
— *trier les* ——*s*, to sort ore.
minéral, a., m., mineral.
minéralogie, f., mineralogy.
minéralurgie, f., art of treating minerals so as to increase their usefulness.
mineur, m., miner; (*mil.*) miner;
— *corps de* ——*s*, (*mil.*) corps of sappers and miners.
minier, a., mining, mineral.
minière, f., earth, sand, or stone in which ore is found; (more specially) a surface mine.
ministère, m., ministry; department (U. S.); office, administration;
— *des affaires étrangères*, foreign office; Department of State (U. S.);
— *des finances*, treasury department;
— *de la guerre*, war department, war office, ministry of war, Horse Guards;
— *de l'intérieur*, Interior Department (U. S.); home office;
— *de justice*, department of justice;
— *de la marine*, navy department, admiralty, ministry of marine;
— *public*, (*law*) public prosecutor.
ministre, m., minister, secretary; (in France) a Protestant clergyman; (*mil. slang*) a mule;
— *de l'administration de la guerre*, v. —— *de la guerre;*
— *des affaires étrangères*, secretary of State for foreign affairs; Secretary of State (U. S.);
— *d'État*, cabinet minister;
— *des finances*, secretary of the treasury;
— *de la guerre*, minister of war, secretary of state for war; Secretary of War (U. S.);
— *de l'intérieur*, home secretary; Secretary of the Interior (U. S.);
— *de justice*, attorney-general;

ministre *de la marine*, minister of marine; first lord of the admiralty; Secretary of the Navy (U. S.);
— *plénipotentiaire*, minister plenipotentiary;
— *public*, envoy.
minium, m., red lead;
— *de fer*, iron minium;
— *peindre au* ——, to red-lead, cover with red lead;
— *peinture au* ——, red-lead painting;
— *de plomb*, red lead.
minot, m., (*nav.*) davit, boom.
minuit, m., midnight.
minute, f., minute (of time); rough sketch, note, notes, approximate report, minute; minute glass; (*top.*) field notes; graphical translation, in pencil and to the scale of the survey, of field operations;
— *carte-* ——, v. s. v. *carte;*
— *de degré*, minute of arc,
— *demi-* ——, thirty-second glass;
— *quart de* ——, fifteen-second glass;
— *de temps*, minute of time.
minuter, v. a., to make a rough draft or note or copy; to jot down.
minuterie, f., hour train (of a clock).
mi-ouvert, a., half open.
mi-parti, a., half-and-half.
mi-pente, *à* ——, halfway down or up a slope.
miquelet, m., (*mil.*) soldier of the body-guard of a Spanish captain-general.
mirador, m., (*art.*) observation station (for hits of battery).
mirage, m., mirage.
mire, f., (*art.*, *sm. a.*) front sight, fore sight; (*inst.*) leveling rod, sight vane; (*surv.*) directing mark;
— *angle de* ——, v. s. v. *angle;*
— *bouton de* ——, v. s. v. *bouton;*
— *coin de* ——, v. s. v. *coin;*
— *à coulisse*, (*inst.*) sliding leveling rod;
— *cran de* ——, v. s. v. *cran;*
— *fronteau de* ——, v. s. v. *fronteau;*
— —— *graduée*, graduated rod;
— *ligne de* ——, v. s. v. *ligne;*
— *longueur de* ——, v. s. v. *longueur;*
— *longueur de la ligne de* ——, v. s. v. *longueur;*
— *masse de* ——, v. s. v. *masse;*
— *mettre en* ——, v. s. v. *mettre;*
— *de nivellement*, (*inst.*) leveling rod;
— —— *parlante*, (*inst.*) self-reading rod;
— *point de* ——, (*ball.*) point aimed at; (*surv.*) sighting point;
— *prendre la* ——, (*ball.*) to aim;
— *de repérage*, (*art.*) sighting rod;
— *triangle de* ——, (*ball.*) triangle of sight;
— *à voyant*, (*inst.*) leveling rod;
— *de voyant*, (*inst.*) slide vane of a leveling rod.
mirer, v. a., (*ball.*) to aim (rare).
miroir, m., mirror, looking-glass; blaze (on a tree); scutcheon; cavity (on the face of a dressed stone, caused by chipping too deep);
— *à* ——, (*hipp.*) dappled;
— *ardent*, burning glass;
— *grand* ——, (*inst.*) index glass of sextant;
— *d'octant*, (*inst.*) horizon glass;
— *petit* ——, (*inst.*) horizon glass of sextant;
— *de pointage*, (*sm. a.*) in aiming drill, small mirror or reflector used to verify the aim;
— *réflecteur*, (*sm. a.*) reflector for inspecting rifle barrels.
miroitant, a., specular, shining, flashing.
miroité, (*hipp.*) dappled.
miroiter, v. r., to shine, flash.
miroiture, f., (*hipp.*) dapple; regular round spots (either lighter or darker than the main coat, but always of the same color).
mirouetté, a., (*hipp.*) bright dappled bay.
mise, f., laying, laying down, putting, placing, setting, installation, starting; (*met.*) layer;
— *en accusation*, indictment; (*mil. law*) arraignment;
— *-bas*, bringing forth, foaling, littering;
— *en batterie*, (*art.*) act or operation of going into battery; mounting of guns in their emplacements;

mise *hors de batterie,* (*art.*) act or operation of going from battery;
— *à bord,* (*nav.*) installation on board ship;
— *en brèche,* (*siege*) breaching;
— *en cause,* (*law*) summons;
— *en chantier,* setting to work (as, of a working party);
— *en couleur,* (*sm. a.*) browning of a rifle barrel;
— *en demeure,* (*law*) summons, order, to perform an obligation;
— *en disponibilité,* (*mil.*) placing on the unattached list;
— *à l'eau,* (*nav.*) launching of a ship;
— *à l'effet,* (*top. drawing*) representation of surface forms;
— *à l'encre,* inking of a drawing; reproduction in ink of a map or reconnoissance;
— *en état de défense,* (*mil.*) preparation (of a wood, house, etc.) for defense;
— *en état de siège,* (*mil.*) declaration of a state of siege;
être de —, to be presentable;
— *à feu,* (*met.*) beginning of the work;
— *de feu,* (*art.*) discharge of a piece, act of firing a piece; (*min.*) firing;
— *de feu électrique,* (*art., min.*) electric discharge (of a gun, of a mine);
— *de feu multiple,* (*mil. min.*) simultaneous discharge of a group of mines;
— *de feu à répétition,* (*art.*) system in which, in case of miss fire, the same primer may be tried several times in succession without opening the breech;
— *de feu simple,* (*mil. min.*) discharge of a single mine;
— *en feu,* (*met.*) act or operation of putting in blast;
— *hors de feu,* (*met.*) putting out of blast;
— *de fonds,* outlay;
— *en haie,* piling of bricks for drying;
— *hors,* money advanced;
— *en jeu,* starting, putting into play;
— *en joue,* (*sm. a.*) whole operation of aiming;
— *au jour,* bringing to light;
— *en jugement,* trial, putting on trial;
— *au lavis,* washing, tinting, of a drawing;
— *en liberté,* discharge;
— *en lumière,* (*sig.*) flash;
— *en marche,* starting (of machinery, troops, etc.);
— *au mille,* (*met.*) ratio of the gross weight of an ingot to the weight of the finished piece, or the number of kilograms in an ingot or charge needed to make 1,000 kilograms of useful metal *monnaie de* —, v. s. v. *monnaie;*
— *en mouvement,* v. — *en jeu;*
— *au net,* act of making a fair copy;
— *en non-activité,* (*Fr. a.*) placing on the unemployed list;
— *en œuvre,* setting to work, putting in operation;
— *au pain et à l'eau,* putting on bread and water;
— *en place,* seating (as, (*art.*) a rotating band);
— *en plaque,* (*phot.*) adjustment of camera without a lens;
— *au point,* (*inst.*) adjustment of an instrument; focusing of a telescope, lens, etc.; in surveying, the accurate setting up of an instrument over its station, the final operations of;
— *en station,* q. v.;
— *en possession,* (*law*) giving possession;
première —, (*mil.*) original outfit (e. g., on joining for the first time); allowance for outfit on entry into service;
— *en pression,* (*steam*) getting up steam;
— *à prix,* pricing;
— *à rançon,* putting up for ransom;
— *en réforme,* (*Fr. a.*) condemnation (of material); casting (of a horse); discharge (for misconduct or physical unfitness);
— *à la retraite,* (*mil.*) putting on the retired list;
— *sur roulettes,* (*art.*) operation of throwing a top carriage in gear (said of guns having rollers for return to battery);

mise *en route,* setting out on the march, dispatch of recruits, soldiers, etc., to their destination;
— *en scène,* getting up, get up;
— *en service,* (*mil.*) adoption into service;
— *hors de service,* (*mil.*) destruction (as, of condemned articles);
— *en station,* (*surv.*) operation of setting up an instrument;
— *en subsistance,* (*mil.*) assignment of a man to a company or regiment (not his own) for pay and subsistence;
— *à la suite,* (*mil.*) the attachment of an officer as supernumerary;
— *en train,* making ready, starting; (*mach.*) starting of a machine, engine, starting gear;
— *en vente,* putting up for sale;
— *au vert,* putting (horses) out to grass.
misour, m., (in French Mediterranean ports) the south wind.
mission, f., (*Fr. a.*) mission, special duty given by the minister of war to an officer, and temporarily detaching him from his regiment or corps: (frequently) foreign duty.
mistral, m., northwest wind (in the Mediterranean).
mitaine, f., mitten.
mitigation, f., mitigation.
mitiger, v. a., to mitigate, temper, qualify; *pour,* —, in mitigation.
mitoyen, a., middle, intermediate; party; *cloison* — *ne,* partition;
dent — *ne,* (*hipp.*) middle tooth;
mur —, party wall.
mitoyenne, f., (*hipp.*) middle tooth.
mitoyenneté, f., party right.
mitraillade, f., (*art.*) fire of, firing of, case shot; langrage (obs.).
mitraille, f., scrap iron; soldering alloy composed of iron, copper, and silver; (*art.*) case shot (grape, canister); langrage (obs.); any fragmentation material used for scattering effect;
balle à —, (*art.*) case-shot bullet;
boîte à —, (*art.*) case shot;
charge à, de, —, (*art.*) round of case, canister, grape (obs.);
charger à —, (*art.*) to load with canister, etc.;
fer de —, (*met.*) scrap iron;
grosse —, (*art.*) grape (obs.);
obus à —, (*art.*) shrapnel;
pendante, (*met.*) old brass, copper, etc.;
petite —, (*art.*) canister (obs.);
— *rouge,* (*met.*) copper dross;
tirer à —, (*art.*) to fire case shot.
mitrailler, v. a., (*art.*) to fire *mitraille;* (in a general sense) to sweep or cover a line or position with shrapnel, rapid fire, etc.;
— *à brûle-pourpoint,* to overwhelm at close quarters with shrapnel or any other fragmentation fire.
mitrailleur, m., any person using *mitraille* fire; (*art.*) v. *mitrailleuse.*
mitrailleuse, f., (*r.f. art.*) machine gun; (*Fr. art.*) a synonym of *canon à balles;* (*sm. a.*) sort of magazine pocket pistol;
— *automatique* (any) automatic machine gun (e. g., Colt);
— *à canons fixes,* (type of) machine gun with fixed barrels (e. g., Gardner);
— *Gatling, etc.,* Gatling, etc., gun;
— *à n canons, n-*barreled machine gun;
— *à rotation,* (type of) machine gun with revolving barrels (e. g., Gatling).
mitre, f., miter; large paving stone; (*mach.*) cone, mushroom, or spindle valve; (*cons.*) chimney pot or top, cowl; (*r. r.*) rectangular steam dome of a locomotive;
arc en —, (*cons.*) triangular arch;
boîte de —, miter box;
— *de cheminée,* (*cons.*) chimney shaft; chimney top;
équerre de —, miter square;
fenêtre en —, gable window;
ligne de —, miter line; ridge;
sauterelle de —, miter rule.
mi-vitesse, f., half speed.

mixte, a., mixed; (*tech.*) composite (e. g., wood and metal used together in the construction of one object); (*elec.*) mixed (i. e., quantity and series); (*mil.*) composed of all arms, or o. more than one arm, of mounted and of dismounted elements;
avante-poste ——, v. s. v. *avante-poste;*
compagnie ——, (*Fr. a.*) train company in Algiers and Tunis (has both mounted and dismounted men);
navire ——, (*nav.*) composite (iron and wood) vessel; a vessel using both sails and steam;
régiment ——, (*Fr. a.*) regiment composed of units of active and of territorial armies.

mobile, m., moving body (e. g., *art.*, a projectile), body moved; mover, moving power (*moteur* is generally employed instead); (*fig.*) motive, cause; (*mil.*) a member of the *garde nationale mobile,* q. v.;
premier ——, prime mover; leader, head, chief.

mobile, a., mobile, movable; moving, unfixed, unsteady; transient; variable; (*cord.*) running; f., an abbreviation of *garde nationale mobile;* (*art.*) recoiling; (*sm. a.*) detachable (of a magazine);
cadre ——, v. s. v. *cadre;*
colonne ——, (*mil.*) movable column, (more especially) a column composed of the different arms, operating independently in a disturbed country;
être ——, (*hipp.*) to have the staggers;
garde ——, garde nationale ——, v. s. v. *garde;*
timbre ——, adhesive stamp;
troupes ——s, (*mil.*) troops ready to march.

mobilier, m., furniture.

mobilier, a., personal;
biens ——s, personal property, goods and chattels;
crédit ——, loan on personal security;
richesse ——ère, personal property;
valeurs ——ères, transferable securities.

mobilisable, a. (*mil.*) mobilizable;
constructions ——s, a generic term for huts (*baraques*), bridges, observatories, etc.

mobilisation, f., (*mil.*) mobilization;
carnet de ——, v. s. v. *carnet;*
classe de ——, (*Fr. a.*) class to which men belong when summoned to military service for any reason (e. g., mobilization, inspection, etc.);
journal de ——, (*Fr. a.*) register (in each regiment or service), of measures taken (in peace) or to be taken (in war) in case of mobilization.

mobilisé, m., (*mil.*) a member of the *garde nationale mobilisée,* q. v.

mobiliser, v. a., (*mil.*) to mobilize.

moblot, m., popular name for a member of the *garde mobile.*

modelage, m., modeling.

modèle, m., model; pattern; (*fond.*) pattern; (*art.*) model, mark;
—— de fonte, (*fond.*) casting pattern;
—— du jet, (*fond.*) runner stick, pin;
—— type, pattern (e. g., of uniform).

modelé, m., (*top. drawing*) representation (as, of surface forms, etc.), more specially hill shading, or representation of slopes;
—— à l'effet, (*top. drawing*) representation of the surface forms, of the relief of the terrain;
—— en lumière directe (*oblique*), (*top. drawing*) with relief brought out by direct (oblique) illumination.

modeler, v. a., to model, pattern, shape;
le terrain, —— le terrain à l'effet, (*top. drawing*) to represent surface forms, to bring out the relief.

modeleur, modeler.

modérateur, m., (*mach.*) governor; regulator valve; damper;
—— à force centrifuge, conical pendulum.

modérer, v. a., to moderate, check, restrain, curb, temper, slacken, lessen, abate.

modificateur, m., (*mach.*) engaging and disengaging gear.

modifier, v. a., to modify, change, alter; (*art.*) to transform to correspond to a different model (as, of a gun).

modillon, m., modillion, cantilever: bracket; console.

module, m., modulus (of rupture, elasticity, etc.);
—— d'une rivière, means discharge of a river.

moelle, f., marrow; pith (of trees).

moellon, m., (*mas.*) quarry stone; rough, unhewn stone; building stone; ashlar;
—— d'appareil, ashlar; generic term for stone of dressed face and undressed body;
——s d'assise, stones of the same size used in regular courses;
—— brut, rubble, undressed stone; quarry-faced stone;
—— en coupe, stone laid on edge;
—— épincé, face-hammered stone;
—— gisant, quarry stone, more or less regular in form;
—— gros, riprap stone; ashlar;
—— de lit, stone laid on its natural bed;
maçonnerie de ——, ashlar work;
—— de parement, facing stone;
—— piqué, stone with pointed and drafted face;
—— smillé, hammer-dressed stone;
—— de taille, dressed stone, cut stone;
—— tétué, v. *—— épincé.*

mofette, f., choke damp; black damp.

moie, f., heap of sand.

moignon, m., stump of a tree, of a branch; zinc or lead pipe (*med.*) stump of an amputated leg or arm.

moilon, m., v. *moellon.*

moine, m., monk; (*met.*) blister, flaw; (*mil. min.*) monk, punk (obs.).

moineau, m., sparrow; (*hipp.*) cropped horse; (*fort.*) small bastion or lunette at the middle of the curtain;
cheval ——, v. s. v. *cheval.*

moins-perçu, m., (*adm.*) amount not drawn;
—— en deniers, money not drawn or paid out at a payment;
—— en nature, effects, articles, not drawn.

moins-value, f., (*adm.*) deterioration, loss, decrease in value.

mois, m., month; month's pay or allowance.

moise, f., (*cons.*) tie, couple, brace, binding piece, connecting piece, tiebeam, collar or shoulder on a pump barrel; (*powd.*) frame or crib supporting the stamp of a stamping mill.
—— en écharpe, diagonal tie;
—— horizontale, horizontal tie; wale;
—— inclinée, diagonal tie;
——s-jumelles, ——s jumelées, binding pieces;
—— pendante, hanging tie.

moiser, v. a., (*carp.*) to tie, join timbers together by means of a tie; to notch timber together; to bridge over;
—— par entaille, to notch together.

moisi, p. p., musty.

moisir, v. a., to make musty.

moisissure, f., mustiness.

moisson, f., harvest; harvest time; crop.

moissonner, v. a., to reap.

moissonneuse, f., reaping machine.

moite, a., damp, moist, wet.

moiteur, f., dampness, moisture.

moitié, f., half;
à ——, half, halfway;
—— bois, v. *mi-bois;*
à —— chemin, halfway;
à —— prix, half price.

molaire, a., molar; f., (*hipp.*) grinder, molar, molar tooth;
——s en ciseaux, (*hipp.*) beveled molars;
dent ——e, molar tooth;
—— de lait, (*hipp.*) molar of the first dentition;
—— de remplacement, adult molar, molar of the second dentition.

molasse, f., argillaceous sandstone.

môle, m., mole, pier, jetty; dam; jetty head;
—— de, —— d'un, port, breakwater.

molécule, f., molecule.

molestation, f., molestation;
—— de guerre, molestation of war;
franc de —— de guerre, free from molestation of war.

molester, v. a., to molest.
molets, m., (*top.*) bog, swampy ground.
molet(t)age, m., polishing, milling.
molette, f., any small-toothed wheel; tracing wheel; rowel, spur rowel; rope pulley; (*cord.*) whirl; (*mach.*) cutter, milling wheel; (*hipp.*) windgall; tuft of hair;
—— *d'un baritel*, pulley of a winding machine;
—— *chevillée*, (*hipp.*) articular windgall;
—— *coupante*, (*mach.*) cutting wheel;
—— *à égruger*, rubber, mealer;
—— *à fendre*, (*mach.*) screw slotting cutter;
machine à ——*s*, winding engine, whim;
—— *nerveuse*, v. *simple*;
scie à ——*s*, circular saw;
—— *simple*, (*hipp.*) windgall;
—— *soufflée*, (*hipp.*) through windgall;
—— *tendineuse*, (*hipp.*) tendinous windgall.
molet(t)er, v. a., to polish, to mill.
molière, f., grindstone quarry, millstone.
mollet, m., calf of the leg; small nippers; (*hipp.*) blaze;
avoir du ——, (*hipp.*) to have prominent muscles in the gaskin.
molleté, a., knarled.
molletier, a., *bande* ——*e*, (or simply) *molletière*, (*mil.*) putties.
mollière, f., (*top.*) bog, swamp.
mollir, v. a. n., to slack, slacken, abate; to go down, to lull (of the wind); to slack (of the tide, a current); (*cord.*) to ease, ease off, slack;
—— *un câble*, —— *un cordage*, (*cord.*) to slack, loose, case off or away, a cable, a rope.
molybdène, m., molybdenum.
moment, m., moment, momentum, force, leverage.
monarchie, f., monarchy.
monarque, m., monarch.
monceau, m., heap;
mettre en ——*x*, to put in heaps, heap up.
monde, m., the world; people, crew, men, persons; hands;
fournir ——, *garnir de* ——, *mettre du* ——, to man.
monder, v. a., to hull.
mondrain, m., small mound of earth or of sand.
moniteur, m., (*Fr. a.*) monitor, subinstructor, or teacher (as, of boxing, fencing, etc.);
—— *d'escrime*, (*Fr. a.*) assistant regimental fencing master;
—— *général*, head monitor, senior instructor;
—— *de gymnastique*, instructor of gymnastics.
monitor, m., (*nav.*) monitor;
—— *à réduit central*, monitor with a central redoubt or battery.
monnaie, f., change, money; the exchange;
—— *de compte*, money of account;
fausse ——, counterfeit money;
—— *fiduciaire*, bills, paper;
—— *à force libératoire*, legal tender;
—— *imaginaire*, v. —— *de compte*;
menue ——, the copper (and even the small silver) currency;
—— *de mise*, current money;
papier- ——, paper money.
monnayer, v. a., to coin.
monoatomique, a., (*chem.*) univalent.
monocylindrique, a., (*steam*) single-cylinder.
monolithe, m., monolith.
monomachie, f., single combat.
monophote, m., (*elec.*) constant-current arc lamp.
monoptère, a., monopteral.
monorail, m., uno-rail, single rail.
monorchide, m., (*hipp.*) monorchid.
monorchidisme, m., (*hipp.*) monorchidism.
monson, m., monsoon.
mont, m., mountain, mount, hill;
—— *pagnote*, (*mil.*) any elevated place out of harm's way, whence a battle may be viewed without danger.

montage, m., mounting, fitting, putting together, preparation, assembling; (*mach.*) erecting, setting up; (*elec.*) wiring, connection; (*art.*) (operation of) fitting a projectile with its bands, etc., (hence) fittings of a projectile; (*sm. a.*) assembling the parts of a rifle;
atelier de ——, assembling shop;
—— *en dérivation*, (*elec.*) multiple circuit, multiple arc circuit;
—— *en étoile*, (*elec.*) star grouping;
—— *mixte*, (*elec.*) mixed circuit;
—— *en quantité*, (*elec.*) parallel connection;
—— *en série*, (*elec.*) series connection;
—— *en tension*, (*elec.*) series connection;
—— *en triangle*, (*elec.*) delta connection, mesh connection;
—— *à trois conducteurs, fils*, (*elec.*) three-wire system.
montagne, f., (*top.*) mountain;
chaîne de ——*s*, chain of mountains;
—— *de glace*, iceberg;
pays de ——*s*, mountainous country;
torrent de ——, mountain stream; torrent.
montagneux, a., mountainous, hilly.
montant, m., upright, standard, stanchion, guide-post; upright stone, beam, or bar; side timber of a boat; puncheon; beam; stem of a tree; amount, total, sum total; rise (of the tide); stile, post, doorpost; (*in pl.*) cheeks, side beams; (*top.*) rise; (*harn.*) headstall, cheek piece, cheek strap; (*pont.*) upright of pontoon-boat knee; (*min.*) stile (of a frame); (*mil.*) first for duty (on a roster); (*sm. a.*) crane (of a revolver);
—— *d'appui*, fixed part of a drawbridge;
—— *de branloire*, (*art.*) standard of a bellows frame;
—— *de bride*, (*harn.*) headstall; cheek strap;
—— *de buse*, meeting post of a lock gate;
—— *de châssis*, (*min.*) upright of a gallery frame;
—— *de chevalet*, (*pont., etc.*) trestle leg;
—— *de cloche*, belfry;
—— *de courbe*, (*pont.*) upright (of boat timber);
—— *à crochet*, hooked upright, for hooking cacolet to pack saddle;
—— *de droite (gauche)*, (*harn.*) off (near) side neck strap of a saddle;
—— *de l'eau*, rise of water, of the tide;
—— *d'une échelle*, side piece of a ladder;
en ——, uphill;
—— *d'escalier*, (*cons.*) tread; rise of a staircase;
grand ——, (*harn.*) off side, long head;
grand —— *tropical*, (*mil. slang*) riding breeches;
—— *de la marée*, beginning of the flood, rise of the tide;
—— *de poitrail*, (*harn.*) breast strap proper;
—— *de sabord*, (*nav.*) stile of a gun port;
—— *de semelle*, (*pont.*) rising timber;
—— *de tente*, tent pole; awning stanchion;
—— *de la tétière*, v. —— *de bride*.
montant, a., rising;
la garde ——*e*, (*mil.*) the new guard, guard marching on.
monte, f., (*hipp.*) covering; leaping; covering season; serving, time of serving; (*man.*) system or method of riding, of managing a horse.
monté, p. p., mounted, manned; fitted or equipped with (as, a projectile with its rotating band); (*nav.*) commanded by;
bien ——, well supplied, stocked; (*man.*) well mounted, having a good stable;
bien —— *d'hommes*, well manned;
cheval —— *haut, haut* ——, (*hipp.*) long-legged horse (legs too long);
—— *de*, (*nav.*) carrying (so many) guns; having (so many) men;
être bien (mal) —— *en chevaux*, (*man.*) to have a good (poor) stable;
mal ——, ill supplied, stocked, appointed; (*man.*) badly mounted, having a poor stable;
non- ——, unmounted; (*mil.*) dismounted.
monte-charge(s), m., hoist, lift; (*art.*) shell hoist, shell lift, projectile hoist; winch or windlass of (the French) garrison gin;

monte-charge(s) à *barbotin*, (*art.*) winch, windlass of (the French) garrison gin;
—— à *bennes alternatives*, (*Fr. s. c. art.*) counterpoise or alternating shell hoist;
—— à *commande*, power hoist;
—— *hydraulique*, (*art.*) hydraulic shell lift;
—— à *la main*, hand elevator.
monte-courroie, m., (*mach.*) device for mounting a belt on a pulley.
monte-culasse, m., (*art.*) tool used in mounting and dismounting Hotchkiss revolving cannon.
montée, f., ascent; rise, rising approach, ramp; flight of stairs, stair; height (of a building, of a column, etc.); pitch, height (of an arch); height (of a bridge); gradient (of a road).
monte-escarbilles, m., ash hoist.
monter, v. a. n., to mount, ascend; to go, come, up; to rise; to rise (of the barometer, thermometer); to rise in one's profession; to raise; to rise, come in (of the tide); to furnish, supply, stock, equip, fit out; to ship (as, a propeller); to assemble, erect, fit up, put or fit together, mount, put in place; to horse (give a mount to); to be too high (as, a wall); to ride (a bicycle); (*man.*) to ride; (*art.*) to mount (a gun on its carriage); (*nav.*) to go on board of, embark; to command (a vessel); (*mil.*) to be promoted; (*elec.*) to connect up, to wire; (*hipp.*) to serve, cover, leap;
——à *l'anglaise*, (*man.*) to ride with the English seat, to post;
—— à *l'assaut*, (*mil.*) to storm;
—— *un ballon*, to go up, or be, in a balloon;
—— *un bâtiment*, (*nav.*) to hoist one's flag on, to fly one's flag on (such and such) a vessel (of a flag officer);
—— *une batterie*, (*art.*) to arm and equipa battery; (*elec.*) to set up and connect a battery;
—— *le bivac*, (*mil.*) to prepare a bivouac for the night;
—— à *la brèche*, (*mil.*) to storm a breach;
—— *un canon*, (*art.*) to mount a gun;
—— à *cheval*, (*man.*) to mount a horse; to ride; (hence) to put anything astraddle of something else;
—— *un cheval*, (*man.*) to ride; to train a horse;
—— *un cheval à nu, à poil*, (*man.*) to ride bareback;
—— *un chronomètre*, to wind a chronometer;
—— *en croupe*, (*man.*) to get on behind;
—— *un creuset*, (*met.*) to remelt a crucible of metal;
—— à *cru*, (*man.*) to ride bareback;
—— *en dérivation*, (*elec.*) to mount, connect, in multiple circuit;
difficile (facile) à ——, (*man.*) hard (easy) to mount;
—— à *dos*, (*man.*) to ride bareback;
—— *une épée*, (*sm. a.*) to put the hilt on a sword;
—— *en étoile*, (*elec.*) to make a star grouping;
faire ——, to bring, take, get up (as, guns, etc.);
—— *un fleuve*, to ascend a river;
—— *un fusil*, (*sm. a.*) to stock a rifle;
—— *la garde*, (*mil.*) to mount the guard; to serve a tour of guard duty;
—— *sa garde*, (*mil.*) to serve a tour of guard duty;
—— *en grade*, (*mil.*) to be promoted;
—— *une hache*, to helve an ax;
—— *par haut*, (*man.*) to exercise horse in curvetting, in executing *croupades*;
—— *en lunette*, (*mach.*) to mount in or on a shell chuck;
—— *en marchand de cerises*, (*mil. slang*) to ride badly;
—— *sur mer*, to embark;
—— *du monde*, to turn out the hands, the crew;
—— *une montre*, to wind a watch;
—— *un navire*, (*nav.*) to be on board;
—— *entre les piliers*, (*man.*) to take a turn on a *sauteur* (jumping horse fastened between posts);
—— à *poil nu*, (*man.*) to ride bareback;
—— *en quantité*, (*elec.*) to join up, connect, wire, in parallel;

monter *une rivière*, to ascend a river;
—— *en série*, v. —— *en tension*;
—— *tant d'hommes*, (*nav.*) to have a crew of so many men;
se ——, (*man.*) to be ridden;
—— *en tension*, (*elec.*) to connect up in series;
—— *une tente*, to pitch a tent;
—— *la tranchée*, (*mil.*) to serve a tour of guard in the trenches;
—— *en triangle*, (*elec.*) to make a delta connection;
—— à *trois fils*, (*elec.*) to install, wire, under the three-wire system;
—— *un vaisseau*, (*nav.*) to command a ship; to fly one's flag on (such and such) a ship; to man a ship;
—— *sur un vaisseau*, to embark.
monte-ressort, m., (*sm. a.*) lock cramp, spring hook; spring cramp; spring vise.
monte-système, m., (*Fr. art.*) frame used in mounting the *canon à balles*.
monteur, m., engine fitter; (*sm. a.*) gunstock fitter; (*mil. teleg.*) lineman (puts up the line);
—— *en blanc*, (*sm. a.*) gunstock maker;
—— *-fourbisseur*, (*sm. a.*) scabbard maker.
montgolfière, f., hot-air balloon.
monticule, m., (*top.*) small hill, hillock;
en ——, tumular.
mont-joie, f., pile of stones to mark a road.
montoir, m., (*man.*) mounting block; horse block; near side; bearing of the foot in near stirrup (in mounting;
aisé au ——, easy to mount;
côté du ——, near side;
côté hors du ——, off side;
difficile au ——, hard to mount;
docile, doux, au ——, v. *aisé au* ——;
hors ——, *hors le* ——, off side;
pied du ——, near foot;
pied hors du ——, off foot;
rude au ——, v. *difficile au* ——.
montre, f., watch; show case; sample; (*mil.* muster, pay when mustered (obs.); (*man.*) showing a horse's paces; place where horses are shown, paraded, etc., for sale; (*fenc.*) part of a fencing hall in which the foils are kept;
—— *marine*, chronometer;
—— à *répétition*, repeater;
—— à *repos*, stop watch;
—— à *sonnerie*, striking watch.
montrer, v. a., to show, display; to teach;
—— *un cheval*, to show off a horse;
—— *un feu*, to display a light;
—— *un pavillon*, to display a signal.
monteux, (*top.*) hilly, mountainous.
monture, f., mounting, fitting, setting; workmanship; frame; support; (*man.*) mount, nag, hack, hackney; (*nav.*) arming and fitting a ship of war; (*harn.*) headstall (and related parts); (*sm. a.*) hilt and related parts of a sword, of a sword bayonet; stock plate of a revolver; stock of a rifle;
—— *s d'une bouche à feu*, (*art.*) moldings of a gun;
—— *de bride*, (*harn.*) headstall;
—— *en deux pièces*, (*sm. a.*) rifle stock in two pieces (butt and forearm separate);
—— *d'éperon*, spur chain, leather;
—— *de fusil*, (*sm. a.*) stock;
—— *d'officier*, (*cav.*) officer's mount, charger;
petite ——, (*mil.*) cleaning kit and small articles (spoon, comb, etc.);
—— *de sabre*, (*sm. a.*) hilt;
—— *de scie*, saw handle.
moque, f., (*cord.*) dead-eye, bull's-eye;
poulie à ——, dead-eye, block;
—— à *rouet*, clump block;
—— à *trois trous*, oblong three-holed dead-eye;
—— à *trous*, truck, fair leader.
moraller, v. a., (*hipp.*) to use a barnacle.
morailles, f. pl., (*hipp.*) barnacle.
moraillon, m., bolt nab, catch; hasp;
fermer au ——, to hasp, clasp;
—— *et tourniquet*, hasp and turn-buckle.
moraine, f., moraine; (*mas.*) fillet of mortar in pisé work.
moral, m., spirit, morale of troops (Anglicized into morale).

mordache, f., clamp, vise clamp, vise jaw; tongs, pincers;
—— à chanfrein, vise clamp;
—— d'étau, vise claws, vise clamp;
—— à river, riveting clamp;
—— à souder, soldering pincers.

mordant, m., brass nail or stud; double-pointed brass nail used by saddlers; mordant; (of a file, etc.) cutting power.

mordre, v. a. n., to bite; to take hold of, to catch; to corrode; (cord.) to jam, to nip;
—— au cabestan, to ride (one part over another);
faire —— le bout, (cord.) to push down the end of a rope under a loop, a turn, etc.;
—— l'horizon, to dip; (of the sun);
—— la poussière, to bite the dust, to be killed.

mords, m., jaw (of a clamp); cutting or biting edge of nippers, pincers, etc.

mordu, p. p., (cord.) jammed, nipped, stuck fast.

more, m., Moor;
cheval cap de, cavecé de, ——, v. s. v. cheval;
tête de ——, (cord.) Turk's head knot.

moreau, a., m., (hipp.) shining black; horse of shining black coat; hair nosebag.

moril, m., wire edge.

morfondu, (cord.) relaid rope.

morne, m., (top.) low mountain, small mountain (in the French colonies).

mornet, m., (top.) hill.

mornette, f., ferule, ring.

mors, (harn.) bit; (mach.) jaw, chop;
—— d'Allemagne, sort of bit for refractory horses;
—— à l'anglaise, port-mouthed bit, English bit;
—— à branches droites, v. —— omnibus;
—— de bride, bridle bit, curb bit;
—— de bridon, snaffle bit, watering bit;
—— de bridon-licol, snaffle-and-halter bit;
—— à canon brisé, jointed bit, German bit;
—— à canons tordus en spirales, twisted snaffle bit;
—— de filet, snaffle bit;
—— à gorge de pigeon, v. —— à l'anglaise;
—— à, avec, gourmette, curb bit;
—— à liberté montante, v. —— à l'anglaise;
mâcher son ——, to champ the bit;
—— au miroir, a bit to prevent the protrusion of the tongue;
—— omnibus, straight-branched bit;
ôter le —— à, to unbit;
—— parleur, a bit that the horse must be continually champing, thus announcing to rider that he is feeling it; mouthpiece (as in certain kinds of telephone);
—— à pompe, curve-branched bit;
prendre le —— aux dents, to take the bit in the teeth; to run away;
—— rude, severe bit;
—— à simple canon brisé, bit with a jointed mouthpiece.

mort, f., death;
arrêt de ——, v. sentence de ——;
—— civile, (law) loss of all civil rights;
coup de ——, death blow, shot;
crime de ——, capital crime;
mettre à ——, to put to death;
sentence de ——, sentence of death.

mort, a. m., dead, spent, still, dormant, lifeless, dull, inanimate; dead person, body;
angle ——, v. s. v. angle;
argent ——, money returning no interest; dead capital;
balle —— e, v. s. v. balle;
bois ——, deadwood (either felled trees or trees dead standing);
—— bois, brambles, briers; wood of no value;
bras ——, v. s. v. bras;
—— e charge, (art., sm. a., mil. min.) inadequate charge;
à —— e charge, (nav.) overladen;
corps ——, v. s. v. corps;
eau —— e, stagnant water;
le —— d eau, de l'eau, neap tide;
eaux —— es, stern body of a vessel;
—— e(-)eau, neap tide, season of neap tides; low water (of the tide);

mort(e) marée, neap tide, dead low water;
œuvres —— es, (nav.) upper works, dead works;
papier ——, unstamped paper;
—— e-paye, (mil.) pensioner (obs.);
point ——, (mach.) dead center: (of balloons) any instant of cessation of oscillation;
—— e(-)saison, saison —— e, slack time, dull season;
tour ——, (cord.) single turn of rope around a bitt, etc.

mortaisage, m., mortising.

mortaise, f., mortise; hole; recess; (harn.) eye or eyelet (for passage of straps); (cord.) sheave hole; (fond.) tap hole; (art.) wedge slot (Krupp);
—— en adent, (carp.) indented mortise;
assembler à ——, (carp.) to mortise;
—— du cabestan, capstan-bar holes;
—— de clavette, (mach.) keyway;
—— d'étrivière, (harn.) stirrup leather hole;
—— d'hélice, (nav.) propeller key way;
machine à —— continue, groving machine;
—— d'une poulie, (cord.) sheave hole;
—— de sûreté, (art.) lever-handle recess (hence) safety recess;
—— et tenon, (carp.) mortise and tenon.

mortaiser, v. a., to mortise;
machine à ——, mortising machine.

mortaiseuse, f., mortising machine.

mortalité, f., mortality; death rate.

mort-bois, m., v. s. v. mort, a.

morte-eau, f., v. s. v. mort, a.

mortel, a., mortal, deadly;
coup ——, death blow; (fig.) ruin.

mortelier, m., cement-stone breaker.

morte-paye, f., v. f. v., mort, a.

morte-saison, f., v. s. v., mort, a.

mort-gage, m., (law) mortgage.

mortier, m., mortar; sort of cap or headdress (worn by presiding judges in France), (hence) the presiding judge: furnace clay; cob; adobe; lime pit; (art.) mortar; (mas.) mortar; (powd.) mortar piece; (mach.) socket or cup to support the work;
bain de ——, v. s. v. bain;
—— bâtard, (mas.) mortar composed of good and of poor lime;
bâti à chaux et à ——, solidly built;
—— blanc, (mas.) mortar made of poor lime;
—— de campagne, (art.) field mortar;
cave à ——, (fort.) mortar casemate;
—— à chaux et à sable, (mas.) ordinary mortar;
—— de chaux, (mas.) lime mortar;
—— de ciment, (mas.) cement mortar;
—— clair, (mas.) thin mortar;
—— à la Cochorn, (art.) Cochorn mortar;
—— de compas, compass box;
corroyer le ——, (mas.) to beat up mortar;
—— de côte, (art.) seacoast mortar;
—— dur, (mas.) stiff mortar;
—— d'épreuve, v. —— éprouvette;
—— éprouvette, (ball.) eprouvette, proof mortar, trial mortar;
—— gras, (mas.) fat mortar;
—— hydraulique, (mas.) hydraulic mortar, water cement;
—— liant, (mas.) mortar easy to work;
—— liquide, (mas.) grouting;
—— lisse, (art.) smoothbore mortar;
—— maigre, (mas.) thin mortar (lacking in lime);
—— à main, v. —— à la Cochorn;
manège à ——, (mas.) mortar mill;
—— marin, (nav. art.) naval mortar;
—— de moulin à poudre, (powd.) hollow for the stamp of a powder mill;
—— à plaque, (Fr. art.) mortar and bed cast in one piece (obs.);
—— de plâtre, (mas.) plaster;
à plein ——, (mas.) in a bed of mortar;
—— pouzzolanique, (mas.) pozzuolanic mortar;
—— à prise lente (prompte, rapide), (mas.) slow (quick) setting mortar;
—— rayé (art.) rifled mortar;
—— à semelle, v. —— à plaque;
—— de siège, (art.) siege mortar;
à —— soufflant, (mas.) in an excess of mortar;
—— sphérique, (art.) Schumann mortar;

mortier *de terre*, (*mas.*) a generic expression for clay, mud, etc., beaten up and used for rough constructions, or as fire-wall linings, etc.
mort-subite, f., (the American) buggy.
mortuaire, a., mortuary, funeral;
  drap ——, pall;
  extrait ——, certificate of registry of death.
morve, f., (*hipp.*) glanders;
  atteint de ——, glandered.
morveux, a., (*hipp.*) glandered.
mot, m., word, expression; note, memorandum; (*Fr. a.*) the countersign, parole; the *mot* in the French army is composed of two names; the first, known as the *mot d'ordre*, being the name of some great man or celebrated general, and the second, known as the *mot de ralliement*, being the name of a battle, of a city, or of some civic or military virtue; *mot* alone is frequently synonymous of *mot d'ordre*.
  The *mot d'ordre* is the answer to the challenge of posts, patrols, rounds, and of double sentinels, and is itself answered by the *mot de ralliement*; the *mot de ralliement*, used independently, is the answer to the challenge of single sentinels; both may be translated countersign; v. s. v. *avance*.
  avance au ——, v. *avance à l'ordre, au ralliement*;
  —— *clef*, key word (of a cipher);
  donner le ——, (*mil.*) to communicate the countersign;
  —— *de guet*, watchword (obs.);
  —— *d'ordre*, v. *supra*, in definition; (*mil. sig.*) call word;
  —— *de passe*, password;
  porter le ——, (*mil.*) to take, transmit, the countersign;
  prendre le ——, (*mil.*) to receive the countersign;
  —— *de ralliement*, v. *supra*, in definition.
moteur, m., (*mach.*) motor; (*elec.*) electric motor;
  —— *à air*, pneumatic motor;
  —— *à air carburé*, hot-air engine, combustion occurring in cylinder itself;
  —— *à air chaud*, hot-air engine (cylinder separate from point where heat is developed);
  —— *animé*, animal power, living motor (as, man, horse, ox, etc.);
  —— *asservi*, a servo-motor;
  —— *à courants alternatifs*, (*elec.*) alternating-current motor;
  —— *à courant continu*, (*elec.*) continuous-current motor;
  —— *à courant rotatoire*, (*elec.*) rotating-current motor;
  —— *en dérivation*, (*elec.*) shunt-wound motor;
  —— *électrique*, (*elec.*) electric motor;
  —— *à gaz*, gas engine, v. —— *à air carburé*;
  —— *à hélice*, screw propeller;
  —— *hydraulique*, hydraulic motor;
  —— *à pétrole*, oil engine;
  —— *de seconde main*, any motor in which the energy has been stored up;
  —— *en série*, (*elec.*) series-wound motor;
  —— *thermique*, heat engine;
  —— *thermo-pneumatique*, hot-air engine using compressed air;
  —— *à vapeur*, any steam engine.
moteur, a., moving; motive;
  appareil ——, motor;
  axe ——, driving axle;
  force motrice, motive power, prime mover;
  principe ——, prime mover;
  ressort ——, actuating spring;
  roue motrice, driving wheel.
motif, m., motive, reason, consideration;
  par —— *s de famille, santé, service, etc.*, on account of family, health, etc.
motte, f., clod, peat, turf, sod; (*top.*) hillock; (*powd.*) stone bed or foundation for a stamp mill;
  —— *à brûler*, tan balls, tan turf;
  —— *de gazon*, sod;
  —— *de terre*, clod;
  —— *de tourbe*, clod of turf.

mou, a., soft; (of the wind) light; (*cord.*) slack;
  cordage ——, (*cord.*) slack ropes;
  temps ——, soft, heavy weather.
mou, m., (*cord.*) slack (of a rope, cable, etc.);
  avoir du ——, to be slack;
  embraquer le ——, to take up, or in, the slack;
  donner du ——, to slack off a rope.
moucharabi, moucharaby, m., moucharaby.
mouche, f., fly; small bit of paper; point of an auger, of a bit; tuft of hair on under lip, imperial; spy; (*mach.*) sun and planet wheels; (*fenc.*) button of a foil; (*t. p.*) paster; bull's-eye; (*farr.*) small calkin on the heel of the shoe; (*nav.*) dispatch or advice boat;
  abreuvoir à —— *s*, (*mil. slang*) long, broad wound, made by a cutting weapon;
  —— *bretonne*, a hippoboscid that attacks horses;
  chasser les —— *s*, (*fenc.*) to fence wildly;
  —— *de chien*, v. —— *bretonne*;
  couper à la ——, (*fenc.*) to beat off the adversary's blade by a blow near its point;
  faire ——, (*t. p.*) to make a bull's-eye;
  —— *de fleuret*, (*fenc.*) button;
  riposter à la ——, (*fenc.*) to follow the *couper à la* —— by a thrust made in the line toward which the adversary's blade has been beaten;
  —— *du trépan*, point of an earth auger.
moucher, v. a., (*cord.*) to cut off the fag end of a rope; (*carp.*) to trim smooth the end of a piece of timber.
moucheté, a., speckled, spotted; (*hipp.*) flea-bitten (spots black);
  cheval ——, flea-bitten horse;
  épée —— *e*, sword blunted on point for exercise;
  fleuret ——, foil with button on point;
  —— *-truité*, (*hipp.*) said of a flea-bitten coat, containing red and brown spots, as well as black.
moucheter, v. a., (*cord.*) to mouse a hook; (*sm. a.*, *fenc.*) to put a button on a sword, on a foil; (*fenc.*) to hit or reach the adversary with the button of the foil.
moucheture, f., (*hipp.*) slight and superficial scarification; (of the coat) small, round, black spots on the gray and sometimes on the roan (*aubère*).
mouchoir, m., handkerchief;
  en ——, (*mas.*) obliquely;
  —— *d'instruction*, (*mil.*) handkerchief on which are printed various data of use to the soldier.
mouchure, f., waste piece of wood; (*cord.*) fag end (of a rope, after being cut off).
moudre, v. a., to grind, to mill; (*met.*) to grind ore.
mouette, f., v. *mofette*.
moufle, m., (*met.*, etc.) muffle;
  fourneau à ——, muffle furnace, mill.
moufle, f., mitten; (*cons.*) anchor, tie iron; (*cord.*) tackle, block and fall, (strictly) system of pulleys in same shell, but not necessarily on same axis; a block of more than one sheave;
  —— *de chargement*, (*art.*) loading tackle; (projectile) hoisting tackle;
  —— *fixe*, (*cord.*) standing block, fixed block;
  gant ——, v. s. v. *gant*;
  —— *mobile*, (*cord.*) running block;
  rapprocher les —— *s*, to round in the blocks;
  séparer les —— *s*, to overhaul a tackle;
  —— *de traille*, traveling pulley, trail pulley (trolley).
mouflé, a., (*cord.*) fitted with, operated by, a tackle;
  poulie —— *e*, tackle block;
  presse —— *e, roue* —— *e*, press, wheel, operated by a tackle.
moufler, v. a., (*cons.*) to strengthen a wall by ties.
mouflette, f., seat or socket of a boring bit; (*in pl.*) holders for a soldering iron; (*cord.*) a tackle formed of pulleys of unequal radius.

mouillage, m., wetting; depth; anchoring ground, anchorage; anchoring, act of letting go the anchor;
— au ——, (of an anchor) ready to let go;
être au ——, to ride at anchor, be moored;
droit de ——, harbor dues;
venir au ——, to come to anchor.

mouille, f., deep water along the concave bank of a river.

mouillé, p. p., wet;
périmètre ——, wetted perimeter;
surface ——e, wetted surface.

mouiller, v. a., to wet, soak, steep, moisten; to cast anchor; to let go (anchor); to moor; (torp.) to plant torpedoes;
—— en affourchant, to cast anchor on both sides;
—— une ancre, to let go anchor;
—— une ancre en dedans (en dehors), (pont.) to cast the anchor on the near (far) side of the boat, referred to the bank of departure;
—— par l'arrière, (nav.) to anchor by the stern;
—— en croupière, (nav.) to moor bow and stern;
—— en patte d'oie, (nav.) to moor with (so many) anchors (ahead);
—— un vaisseau, (nav.) to anchor a ship.

mouillette, f., sprinkling brush.

mouillure, f., wetting, sprinkling, moistening.

moulage, m., grinding; (fond.) casting, molding; (mas.) brick moulding; (mach.) millwork; (top.) molding, modeling; (cons.) form given to clay; shaped tile;
—— en argile, loam molding;
atelier de ——, casting shop;
—— en châssis, flask molding;
—— en coquilles, chill molding;
—— creux, hollow casting;
—— à creux perdu, dead molding;
—— à découvert, open sand molding;
—— à la machine, machine molding (brick molding and founding);
—— à la main, hand molding (brick molding and founding);
—— mécanique, v. —— à la machine;
—— au modèle, pattern molding;
—— sur plaque, plate molding;
—— au renversé, turn molding, rolling over, turning over;
—— en sable, sand casting;
—— en sable étuvé, casting, molding, in a stove-dried mold;
—— en sable gras, dry sand molding;
—— en sable grillé, molding in a mold dried inside by an open fire;
—— en sable maigre, green sand molding;
—— en sable vert, v. —— en sable maigre;
—— à sec, (in brickmaking) dry shaping or forming;
—— en terre, loam molding;
—— à la trousse, templet molding.

moule, m., (fond.) mold, cast, form; (artif.) filling mold (rocket); (mas.) brick mold: stonecutter's templet; (fig.) model;
—— en argile, clay mold;
—— à balles, bullet mold;
cendrer les ——s, to blacken, wash, line, mold;
——-chape, exterior or case of a mold;
—— en châssis, flask mold;
——-coquille, plaster mold;
—— à creuset, pit;
—— à creux perdu, dead mold;
—— découvert, open sand mold;
enterrer les ——s, to ram down the molds;
flamber un ——, to dry a mold;
—— à fonte, casting mold;
—— de fusée, (artif.) rocket mold;
—— de godet, (artif.) bottom of a rocket mold;
grand ——, (cons.) in France, large size (of roofing tiles);
—— de gueuse, sow, channel;
jeter en ——, to cast; (cons.) to lay (a concrete foundation);
—— de masselotte, deadhead, sinking head mold;
noircir les ——s, v. cendrer les ——s;
—— à oreilles, (hipp.) bent pincers;

moule perdu, v. —— à creux perdu;
petit ——, (cons.) in France, small size (of roofing tiles);
—— en plâtre, v. ——-coquille;
—— de pot de fusée, (artif.) cylinder former;
—— à renverser, roll-over mold;
retourner un ——, to roll over a mold;
—— en sable gras, dry sand mold;
—— en sable maigre, green sand mold.

moulé, p. p., printed; resembling print; m., print; printed matter; print (done with a pen);
lettre ——e, printed letter; letter in imitation of print.

moulée, f., swarf, grit produced by wet grinding; writing in imitation of print.

mouler, v. a., to mold bricks; (fond.) to cast; to mold;
—— à creux perdu, (fond.) to cast in a dead mold;
fausse planche à ——, (fond.) rolling-over board used in projectile casting;
machine à ——, (fond.) molding machine;
—— une pierre, to trace out, mark out, a stone for cutting;
planche à ——, (fond.) molding board.

moulerie, f., (fond.) art of founding, of casting; molding shop.

mouleur, m., founder, molder; brick molder.

moulin, m., mill;
—— à auges, overshot mill;
barrage de ——, milldam;
—— à bras, handmill;
—— à café, coffee mill; (mil. slang) a mitrailleuse;
canal, courant, de ——, race;
—— à chevaux, horse mill;
—— à cylindre, rolling mill;
—— de discipline, treadmill;
—— à eau, water mill;
—— à écacher, flatting mill;
—— à émoudre, grinding mill;
—— à farine, flour mill;
—— à fil d'archal, wire-drawing mill;
—— à forge, ironworks, forge;
—— à marcher, v. —— de discipline;
—— de marée, tide mill;
meule de ——, millstone;
—— à meules, rolling mill; (powd.) incorporating mill;
—— à minerais, ore mill;
—— à pilon, stamping mill;
—— portatif, portable mill;
—— à pots, overshot mill;
—— à poudre, powder mill;
—— à roue de côté, middle-shot mill;
—— à sable, sand mill;
—— à scie, sawmill;
—— à tirer, wire-drawing mill;
—— à vannes, undershot mill;
—— à vapeur, steam mill;
—— à vent, windmill;
—— à vis, (sm. a.) caliber for polishing screw knobs;
—— à volets, v. à vannes.

mouliner, v. a., (cord.) to slip a cable; to ease of slack when two or three turns have been taken around a windlass, etc.

moulinet, m., reel; windlass (of a gin); turnstile; capstan; (mach.) chuck; (fenc.) moulinet (saber, singlestick); (art.) turn or cant of the sponge and rammer staff (obs.);
—— à bitord, spun-yarn reel;
—— de chèvre, (art.) windlass of a gin;
faire le ——, (art.) to turn the sponge in loading (obs.); (fenc.) to make flourishes with a sword, etc.

moulinure, f., gravelly spot (defect) in stone.

moulure, f., border, molding (ornamental raised work); (art.) gun molding, i. e., rings at muzzle or breech, or both (old guns); (mach.) the form of a cutting tool (to cut special shapes);
—— du bourrelet, (art.) muzzle molding;
—— de culasse, (art.) breech molding;
—— traînée au calibre, (cons.) molding obtained by passing a template over the wet plaster.

moulurer, v. a., to cut or make a molding.

mour, m., (*met.*) muzzle or nose of a twyer.
mourant, *en* ——, tapering very gradually; with a very gradual taper.
mourir, v. n., to die; to go out, be extinguished; to stop (as, a bullet or shell);
—— *au champ, au lit, d'honneur*, (*mil.*) to be killed in battle;
faire ——, to put to death.
mousquet, m., (*sm. a.*) musket;
—— *à pierre*, flintlock;
porter le ——, to be a soldier; to carry a musket;
—— *de rempart*, wall piece.
mousquetade, f., (*sm. a.*) discharge of musketry (obs.); musket shot (obs.).
mousquetaire, m., musketeer (obs.).
mousqueterie, f., (*sm. a.*) musketry; rifle fire;
décharge de ——, volley of musketry.
mousqueton, m., (*sm. a.*) musketoon, carbine; (*Fr. a.*) short, light, artillery carbine;
accrocher le ——, to sling the carbine;
banderole porte- ——, carbine sling;
décrocher le ——, to unsling the carbine;
mettre le —— *à la botte*, to return the carbine;
porte- ——, carbine swivel;
—— *à répétition*, repeating carbine.
mousse, m., (*nav.*) cabin boy; powder monkey; f., foam; froth; moss; a., dull, blunt.
mousson, m., monsoon, time of the monsoons.
moustache, f., mustache, (*in pl., hipp.*) hairs or tufts of hair on the upper lip;
brûler la —— *à quelqu'un*, to fire a pistol in one's face;
vieille ——, (*mil.*) old soldier, veteran.
mouton, m., sheep; monkey of a pile driver; rammer, ram, beetle; punch; (*in pl.*) white-caps (waves); testing weight; (*fond.*) driver (to knock out cores);
—— *à bras*, hand pile driver, three-hand maul; paver's beetle;
cap de ——, dead-eye;
—— *de choc*, falling weight (used in tests);
—— *à emporte-pièce*, (hollow) punch;
enfoncer un pieu au refus de ——, to drive a pile till it will go no farther;
épreuve à ——, proof (of axletrees) by dropping a weight;
faire le saut de ——, (*man.*) to buck;
—— *de haquet à bâteau*, stud stave;
—— *à main*, monkey or driver worked by hand;
—— *de mouleur*, wooden driver;
peau de ——, sheepskin;
—— *de pont roulant*, upright, stanchion of a bridge, formed by four-wheeled carriage;
refus de ——, "home";
saut de ——, (*man.*) buck;
—— *de sonnette*, ram, monkey of a pile driver;
—— *à tige*, drive or monkey working on a vertical guide rod;
—— *à vapeur*, steam pile driver.
moutonné, p. p., (of clouds) white and fleecy; rounded by glacial action; (of the sea) showing whitecaps; (*hipp.*) sheep-nosed.
moutonner, v. n., (of the sea) to show white-caps; (*art.*) to jump, thrash (of guns);
—— *un prisonnier*, to worm one's self into the confidence of a prisoner.
moutonneux, a., (of the sea) showing white-caps; (of clouds) fleece-like.
mouture, f., grinding (into flour).
mouvant, a., moving; unstable; m. pl., quicksand;
force ——*e*, moving power;
sable ——, quicksand; shifting, drifting, sand;
terrain ——, slippery ground.
mouvement, m., movement, motion; impulse; action, animation, life; disturbance, commotion; alterations, changes; activity, commercial activity, traffic, volume of traffic; variations in numbers, in prices, etc.; bell crank; (*mach.*) moving, working, parts; motion, gear; wheelwork; (*man.*) action, freedom of the forehand; (*top.*) undulation of the ground; (*nav., in pl.*) evolutions of a fleet; (*mil.*) movement, evolution maneuver; transfer, issue, of stores; promotion (not an official term); (*r. r.*) schedule;
—— *accéléré*, accelerated motion;

mouvement *acquis*, acquired, full, motion;
—— *alternatif*, reciprocating motion;
—— *angulaire*, angular motion;
—— *apparent*, apparent motion;
—— *d'arrêt*, stop motion, disengaging motion;
—— *en arrière*, backward, retrograde, motion; (*mil.*) retreat;
—— *ascendant*, —— *ascensionnel*, upstroke;
—— *en avant*, forward motion, move, stroke; (*mil.*) advance;
—— *de balancement*, —— *de bascule*, see-saw motion, rocking motion;
—— *à bras*, (*art.*) running up by hand;
——*s carrés*, (*drill*) movements at right angles to one another;
—— *central*, central motion;
—— *circulaire*, rotary motion;
—— *composé*, compound motion (due to several forces);
—— *continu*, (*art.*) in mechanical maneuvers, walking away with a rope or fall; (*in gen.*) continuous motion;
contre- ——, countermovement;
—— *en cœur*, —— *en courbe de Vaucanson*, (*mach.*) heart motion;
—— *croissant*, accelerated motion;
—— *curviligne*, curvilinear motion;
—— *descendant*, downstroke;
—— *direct*, direct motion;
—— *différentiel*, differential motion;
en ——, in motion;
—— *d'ensemble*, (*art.*) in mechanical maneuvers, passing a rope along by hand;
—— *enveloppant*, (*mil.*) enveloping movement; outflanking movement (generally in sight of the enemy);
—— *excentrique*, eccentric motion;
—— *de galop*, pitching (of a locomotive);
—— *giratoire*, giratory motion;
—— *de haut en bas et de bas en haut*, up and down motion;
—— *d'horlogerie*, clockwork;
—— *de l'indicateur*, indicator motion;
—— *de lacet*, rocking of a locomotive;
mettre en ——, to set agoing, to set in motion; to throw into gear;
se mettre en ——, to stir, move;
mise en ——, starting;
—— *moyen*, mean motion;
——*s à la muette*, (*mil.*) silent drill, drill by signals (neither commands nor calls);
—— *d'oblique*, (*mil.*) any oblique movement;
——*s obliques*, (*drill*) oblique movements;
—— *opposé*, countermovement;
—— *oscillatoire*, oscillatory movement;
—— *parallélogrammique*, (*mach.*) link motion;
—— *parallèle*, parallel motion;
—— *perpétuel*, perpetual motion;
—— *de piston*, stroke;
—— *d'un port*, number of vessels entering and clearing;
premier ——, start;
—— *d'une prison*, succession of prisoners (in a jail);
—— *projectile*, projectile motion;
quantité de ——, quantity of motion;
—— *réactionnaire*, countermovement;
—— *rectiligne*, rectilinear motion;
—— *réel*, real motion;
—— *de relâche*, slackening motion;
—— *relatif*, relative motion;
renvoi de ——, (*mach.*) motion;
renvoi de —— *du tiroir*, (*mach.*) valve motion;
—— *retardé*, decreasing motion;
—— *en retour*, return motion, return stroke;
—— *rétrograde*, (*mach.*) back stroke, motion; reverse stroke; (*mil.*) retreat, movement to the rear;
—— *de révolution*, revolution;
—— *rotatif*, —— *de rotation*, rotary movement, rotation;
—— *de roulis*, rocking motion;
—— *simple*, simple motion (a single force acts);
—— *à, de, sonnette*, (*mach.*) bell-crank lever;
—— *spontané*, automatic movement;
—— *stratégique*, (*mil.*) strategic movement; maneuver;
—— *tactique*, (*mil.*) tactical movement;
—— *de tangage*, pitching motion, of a locomotive;

mouvement *de terrain*, (*top.*) undulation, variations, (of the ground);
— *de(s) terre(s)*, slipping, crumbling (of earthwork); conveying, transporting (of earth; (*fort.*, *r. r.*) earthwork, cuts and fills;
— *tournant*, (*mil.*) turning movement, made out of sight of the enemy, and intended to take him in flank or rear (strategic move);
— *de translation*, motion of translation;
— *uniforme*, uniform motion;
— *uniformément accéléré (retardé, varié)*, uniformly accelerated (retarded, varied) motion;
— *de va-et-vient*, alternating, reciprocating, motion;
— *variable accéléré (retardé)*, variable accelerated (retarded) motion;
— *varié*, varied, variable motion;
— *vertical*, vertical motion.

mouver, v. n., — *du fond*. (of a stream) to flow at the bottom more rapidly than usual.

mouvoir, v. a., to move, put in motion, drive, impel;
— *à demeure*, (of a screw) to turn without advancing;
faire —, to put in motion: to move, set agoing;
— *en rond*, to wheel, to move round.

moye, f., (*mas.*) soft part of a stone; cavity, flaw, filled with earthy matter.

moyé, a., (*mas.*) flawy;
pierre — *e*, stone sawed in two; stone freed of its soft parts.

moyen, m., way, means, medium, expedient, power; (*in pl.*) means, i. e., riches; (*law, in pl.*) pleas;
voies et — *s*, ways and means.

moyen, a., m., mean, middling, middle; mean (e. g., arithmetical, etc.);
chiffre —, average, on an average;
temps —, mean time;
terme —, average, on an average.

moyenne, f., mean.

moyer, v. a., (*mas.*) to saw a stone in two; to divide a stone along its quarry bed.

moyeu, m., nave; nave box; (*mach.*) boss;
écoltage du —, neck of the nave;
enrayage du —, speeching, mortise holes of the nave;
— *étoilé*, (*lit.*) "star nave," name of peculiar nave in which the spokes are connected near nave by ribs produced under the hammer (*Forges de Couzon*);
gros bout du —, back, hind end, of the nave;
— *d'hélice*, propeller boss;
— *à, de, manivelle*, (*mach.*) crank boss;
— *de piston*, (*mach.*) piston boss.

mucilage, m., mucilage.

muciline, f., (*art.*) axle grease (used by the French artillery, composed of 10 per cent tallow, 10 per cent talc, 5 per cent lime, with resinous and fatty matters).

muder, v. a., (*nav.*) to jibe.

mue, f., molting.

muer, v. n., to molt.

muet, a., dumb, silent;
armes — *tes*, (*sm. a.*) arms that remain silent, dumb weapons (e. g., in the hands of troops too exhausted, etc., to fire them);
carte — *te*, a map without legends;
tir —, v. s. v. *tir*.

muette, f., (*St. Cyr slang*) drill in which cadets purposely fail to make their rifles ring in order to annoy an unpopular officer.

mufle, a., (of shoes) squared-toed.

mufle, m., nozzle of forge bellows; band or strap of iron on the end of a spring;
— *de flèche*, tongue or nose plate on the end of a perch.

muguet, m., — *jaune*, v. *mal blanc*.

muid, m., hogshead.

mulasse, f., young mule (male or female).

mulasserie, f., mule-breeding establishment.

mulassier, a. m., mule breeding; horse of Poitou;

mulassier, *industrie* — *ère*, mule breeding;
jument — *ère*, (or, *mulassière* simply), mule-breeding mare.

mulcter, v. a., to fine, mulct.

mule, f., mule (female);
— *s traversières*, — *s traversines*, (*hipp.*) cracked heels.

mulet, m., mulo (male); (*mil. slang*) marine artilleryman;
— *d'affût*, (*art.*) carriage mule;
— *de bagages*, baggage mule;
— *de bât*, pack mule;
— *de cacolets*, (*med.*) litter (cacolet) mule;
— *de caisses (à munitions)*, (*art.*) ammunition mule;
— *de convoi*, train mule;
— *de forge*, (*art.*) forge mule;
grand —, mule forger (offspring of the jackass and mare);
— *d'outils*, (*art.*) tool mule (other arms as well);
petit —, mule (offspring of the horse and she ass);
— *de pièce*, (*art.*) gun mule;
— *de roues*, (*art.*) wheel mule;
— *de selle*, saddle mule, riding mule.

muletier, a., passable by mules, practicable for mules; m., muleteer;
route — *e*, road practicable for mules; pack trail.

muleton, m., young mule.

multangulaire, a multangular.

multicycle, m., multicycle.

multiple, a., multiple, multiplex; m., multiple;
télégraphie —, multiplex telegraphy.

multiplicateur, m., multiplier; (*elec.*) coil, multiplying coil (of a galvanometer); multiplier;
cadre —, v. s. v. *cadre*.

multiplication, f., multiplication; gear (of a bicycle).

multiplier, v. a., to multiply.

multipolaire, a., (*elec.*) multipolar.

municipal, a., municipal; m., soldier of the municipal guard; m., municipal officials;
conseil —, town council;
garde —, a member of the *garde* — *e*, q. v.;
garde — *e*, name of the *Garde de Paris* between 1830 and 1848.

munir, v. a., to supply, furnish, especially (*mil.*) to supply arms, ammunition, warlike stores in general.

munition, f., (*mil., in pl.*) ammunition; provisions; stores;
— *s de bouche*, victuals, food, food supplies;
de —, regulation;
— *s de grand parc*, v. — *de parc d'armée*;
— *s de guerre*, ammunition and other military stores in general;
— *s d'instruction*, blank cartridges;
— *s de la ligne de bataille*, ammunition in the hands of the troops in the field;
— *s de mobilisation*, (*Fr. a.*) small-arms ammunition kept in magazines for issue on mobilization;
pain de —, (*mil.*) ration bread;
— *s de parc d'armée*, ammunition of the main depot;
— *s de parc d'artillerie*, intermediate supply of ammunition;
— *s de parc de corps d'armée*, reserve ammunition of an army corps;
remplacement des — *s*, (*mil.*) supplying of ammunition in the field;
— *s de sac*, small-arms ammunition carried by the men in their bags;
— *s de sûreté*, (*Fr. a.*) cartridges issued to troops in garrison, and to accustom the infantrymen to carry the weight.

munitionnaire, m., (*mil.*) commissary (obs.).

mur, m., wall; (*fenc.*) salute, prelude of an assault at arms (consisting of conventional disengagements and parries, to prepare the hand and the legs, and accompanied by salutes to the adversary and to the spectators); (*min.*) mine floor;

mur *en ailes*, wing wall;
— *en l'air*, overhanging wall;
— *d'appui*, breast wall;
— *aveugle*, blind wall;
— *à bahut*, (*fort.*) semidetached wall, so arranged as to permit firing over it;
— *de bataille*, (*met.*) tunnel head (in open-top blast furnaces, a sort of chimney surrounding the charging hole to protect the workmen from the flames);
— *de bâtiment*, house wall;
— *bouclé*, bulged or bulging wall (face separated from the mass);
— *de calage*, (*fort.*) a sort of supporting wall for the back of a gun platform (top flush with the platform);
— *de chemise*, revetment wall;
— *de chute*, sluice lift wall, lift wall;
— *de clôture*, inclosing wall;
— *commun*, v. —— *mitoyen;*
contre- ——, v. *contre-mur;*
— *coupé*, wall with cuts, etc., for girders;
— *corrompu*, wall ready to fall;
— *crénelé*, (*fort.*) crenelated, loop-holed wall; wall prepared for defense;
— *crépi*, rough-cast wall;
— *de culée*, abutment wall of a bridge;
de ——, mural;
— *de*, *en*, *décharge*, relieving wall;
— *déchaussé*, wall decayed or destroyed at its foot;
— *de défense*, v. —— *crénelé;*
— *dégradé*, damaged wall;
— *demi-détaché*, (*fort.*) semidetached wall;
— *détaché*, (*fort.*) detached wall;
— *déversé*, wall leaning over;
— *de dossier*, a wall rising above the roof and forming a sort of backing for the chimney;
— *de douve*, interior wall of a reservoir;
— *d'ébrasement*, splay or wing wall;
— *d'échiffre*, string wall of a stair;
— *d'enceinte*, inclosing, main, wall; walls;
— *enduit*, plastered wall;
— *escarpé*, scarped wall;
— *extérieur*, out wall, exterior wall;
— *de façade*, —— *de face*, front wall;
— *forjeté*, overhanging wall, wall leaning over;
— *de foyer*, fire bridge;
— *de fuite*, sluice wall, wall of a sluiceway;
— *de garde*, parapet, breast wall; guard wall of a foundation;
gratter le ——, (*man.*) in riding hall, to go too close to the wall, to hug the wall;
gros ——, main wall;
— *de hourdage*, rubble wall;
— *intérieur*, interior, partition, wall;
— *latéral*, side wall;
— *de masque*, (*fort.*) mask wall;
— *mitoyen*, common wall, party wall;
morts ——*s*, walls of a melting furnace;
— *sans moyen*, private wall (i. e., not a party wall);
— *orbe*, dead wall;
— *ourdé*, ill-built wall;
pan de ——, panel of a wall;
parer au ——, (*fenc.*) to parry the thrusts (tierce and quarte) of a beginner;
— *de purpaing*, perpend wall;
— *pendant*, wall ready to crumble or fall;
— *en pierre sèche*, dry wall;
— *de pignon*, gable wall; wall with a ridge top;
— *planté*, wall built on piles or on a grillage;
— *plein*, dead wall;
— *portant à faux*, v. —— *en l'air;*
— *de pourtour*, exterior wall;
entre quatre ——*s*, in prison;
— *rampant*, sloping or climbing wall;
— *recoupé*, countersloping wall, battered wall;
— *réfractaire*, fireproof wall; fire wall of a forge;
— *de refend*, partition wall;
— *de remplage*, —— *rempli de blocage*, *de hourdage*, coffer wall;
— *de réservoir*, reservoir wall;
— *en retour*, return wall;

mur *de revers*, back wall;
— *de revêtement*, revetment wall;
— *sans jours*, —— *sans autre construction*, dead wall;
— *du sas*, lift wall;
— *de séparation*, v. —— *de refend;*
— *soufflé*, v. —— *bouclé;*
— *de soutènement*, retaining wall;
— *en surplomb*, overhanging wall;
table de ——, wall plate;
— *en talus*, —— *taluté*, scarp, sloped wall;
— *de terrasse*, supporting wall of a terrace;
— *de tête*, (*fort.*) head or end wall of a casemate;
tirer au ——, (*fenc.*) of a beginner, to practice thrusts in quarte and in tierce;
— *en traverse*, crosswall, (sometimes) dwarf wall;
— *de vent*, blast wall of a furnace.

murage, m., masonry, walling;
— *de chaudière*, boiler setting.

muraille, f., wall; wall of a city, of a fortress; (*nav.*) side of a ship; (*hipp.*) wall crust of a horse's foot;
attaque en ——, (*mil.*) charge of cavalry in line without intervals;
——*s de bois*, (*nav.*) "wooden walls";
——*s de Prusse*, artificial inter bed;
— *de revêtement*, revetment wall;
tirer à la ——, (*fenc.*) v. s. v. *mur.*

muraillement, m., walling (especially of galleries, shafts, etc.);
double ——, (*met.*) exterior walls of a blast furnace.

murailler, v. a., to wall, to wall in; to support by a wall; to line with masonry, (as, a shaft, etc.).

mural, a., mural;
carte ——*e*, (*inst.*) wall map;
cercle ——, (*inst.*) mural circle.

muré, p. p., walled, walled up;
non- ——, unwalled;
ville ——*e*, walled town.

mureau, m., (*met.*) tymp arch.

murer, v. a., to wall, wall up, wall in.

mûrier, m., mulberry tree.

museau, m., snout, muzzle; nose (of a pipe, etc.);
— *x de l'essieu*, (*nav. art.*) axle ends;
— *de la tuyère*, (*met.*) blast nozzle.

musée, m., museum;
— *d'artillerie*, artillery or military museum, collection of arms.

museler, v. a., to muzzle; to gag, silence;
— *un croc*, (*cord.*) to mouse a hook.

muselière, f., muzzle, noseband.

muserolle, f., (*harn.*) noseband.

musette, f., nosebag; pouch; (*mil.*) haversack, bag; (*mus.*) (a poetic name for) bagpipe;
— *de cavalier*, (*mil.*) brush bag;
— *mangeoire*, (*harn.*) nosebag;
— *à*, *de*, *pansage*, —— *à pansement*, (*mil.*) grooming kit (mounted service);
— *de propreté*, (*mil.*) cleaning kit.

musicien, m., musician; (*mil.*) bandsman;
les ——*s*, (*mil.*) the band.

musique, f., music, (*mil.*) band, regimental band; (*tech.*) sediment or settlings (as, in a mortar trough);
chef de ——, (*mil.*) bandmaster;
corps de ——, band; (*mil.*) regimental, etc., band;
faire de la ——, (*mas.*) to mix stone dust with plaster;
— *militaire*, —— *de régiment*, regimental band;
pupitre à ——, music stand;
sous-chef de ——, (*mil.*) principal musician (U. S. A.).

musoir, m., head of a jetty, pier head; mitersill head; wing wall of a sluice.

mutation, f., transfer, change (of function or of office); (*mil.*) transfer from one unit to another;
— *de corps*, (*mil.*) change of regiment, transfer from one regiment to another;
—— -*gain*, (*mil.*) gain by transfer;
—— -*perte*, (*mil.*) loss by transfer;
— *de résidence*, change of residence.

mutilation, f., mutilation; maiming; castrating.
mutilé, m., (*Fr. a.*) soldier of a punishment company, sent to it for self-mutilation;
—— *volontaire*, (*mil.*) man who has mutilated himself to escape military service.
mutiler, v. a., to mutilate, maim, mangle, castrate;
se ——, (*mil.*) to mutilate one's self in order to escape military service.
mutin, a., obstinate, headstrong; (*mil.*) mutinous, seditious; m., (*mil. nav.*) mutineer;
en ——, mutinously;
faire le ——, to be mutinous, seditious.
mutiné, p. p., (*mil.*, *etc.*) mutinous, seditious, rebellious.

mutiner, v. a., to drive into rebellion, revolt;
se ——, (*mil.*) to rebel, mutiny.
mutinerie, f., mutiny, revolt; sedition.
muzin, m., (*hipp.*) cross between the tarpan and the domestic horse.
myélite, f., (*hipp.*) inflammation of the spinal marrow.
myope, a., shortsighted.
myopie, f., shortsightedness.
myriagramme, n., myriagram (10,000 grams).
myriamètre, m., myriameter (10,000 meters).
myriare, m., myriare (1 square kilometer).

N.

nabab, m., nabob.
nable, m., scuttle hole, plug hole of a boat; the plug itself;
bouchon de ——, plug.
nacelle, f., boat, skiff, dingey; car of a balloon; (*pont.*) small pontoon boat; mooring boat (U. S. A.); (*nav.*) anchoring boat;
—— *de manœuvre*, (*pont.*) working boat;
—— *régimentaire*, (*pont.*) (so-called) regimental pontoon boat.
nadir, m., nadir.
nage, f., swimming; rowing, pulling (a strictly naval term); part of a rowlock on which the oar rests; way through water; boat; crew; pile of rafted timber;
à la ——, by swimming;
aller à la ——, to swim; to row;
banc de ——, thwart;
chef de ——, stroke, stroke oar;
coup de ——, stroke of an oar;
donner la ——, to pull the stroke oar; to set the stroke;
être en ——, to be in a sweat;
exercice de ——, rowing drill;
passer à la ——, to swim over;
perdre la ——, to catch a crab;
régler la ——, to set the stroke;
tente de ——, boat's awning;
traverser à la ——, to swim over.
nagée, f., stroke in swimming, distance traveled in a single stroke.
nageoire, f., fin; any inflated vessel used to support a swimmer.
nager, v. n., to swim; to float; to row, to pull (a strictly naval term; is also used transitively in this sense); (*man.*) to dish; to paddle;
—— *à l'anglaise*, v. —— *de long*;
—— *en arrière*, to back water;
—— *à couple*, to pull double-banked;
—— *à caler*, *culer*, to back water;
—— *debout à la lame*, to pull head to sea;
—— *debout au vent*, to pull into the wind;
—— *à deux mains*, —— *avec deux avirons*, —— *des deux mains à la fois*, to scull;
—— *ensemble*, to pull together;
faire —— *un vaisseau*, to take a ship in tow;
—— *à faire abattre*, to pull to leeward;
—— *sur le fer*, to pull against the wind or the stream (e. g., when an anchor fails to hold);
—— *de force*, to pull with a will, hard;
—— *lentement*, to pull a long stroke;
—— *de long*, to pull a long stroke;
—— *plat*, to feather;
—— *à*, *en*, *pointe*, to pull a single stroke;
—— *sec*, to pull without splashing, row dry;
—— *à sec*, to row aground; to touch the ground with the oar; to push against a bank, a wharf, to row dry; (*man.*) to compel a horse to go on three legs (by tying up the lame one);
—— *à terre*, to pull for the shore;
—— *vivement*, to pull a short stroke.
In commands,
nage! give way!

nager, *nage bâbord* (*tribord*)! give way port (starboard);
nage en douceur! easy all!
nage ensemble! give way together; pull together!
nage à faire abattre! pull to leeward!
nage ferme! pull away!
nage de force! stretch out; pull hard!
nage le long! bend to your oars!
nage partout! give way all!
nage qui est paré! pull with oars that are shipped!
nage sec! row dry!
nage tribord et scie bâbord! give way starboard, back port!
nage au vent! pull to windward!
nageur, m., swimmer, oarsman, rower (a strictly naval term);
—— *de l'arrière*, stroke oar;
—— *de l'avant*, bow oar;
maître ——, swimming teacher.
naille, f., clamp, small iron plate;
grande ——, iron plate for covering knot holes in a boat plank;
—— *moyenne et petit*, small iron plate to cover seams.
naissance, f., birth; source, starting point, origin (*cons.*) springing line (of an arch); (*mas.*) tooth; toothing stone; band or strip of plaster;
acte de ——, certificate of birth;
—— *de colonne*, —— *de fût*, (*cons.*) apophyge;
—— *d'enduit*, (*cons.*) plaster dressing (around a window, etc.);
ligne de ——, (*cons.*) springing line;
——*s masculines* (*féminines*), number of boys (girls) born in a given time;
niveau de ——, springing level;
—— *des rayures*, (*sm. a.*, *art.*) origin, starting point, of the rifling;
—— *de voûte*, (*cons.*) springing line of an arch.
naissant, a., rising, springing up; (*chem.*) nascent;
état ——, (*chem.*) nascent state;
rouge ——, nascent red (of bodies subjected to heat);
nantir, v. a., to give security (as, for a debt);
se ——, to secure one's self.
nantissement, m., security (as, for a debt).
naphtagil, m., sort of natural bitumen.
naphtaline, f., (*chem.*) naphthalene.
naphte, m., naphtha;
huile de ——, naphtha.
napoléon, m., 20-franc piece.
nappe, f., sheet; nappe;
—— *aquifère*, water-bearing stratum;
—— *d'eau*, sheet of water (both standing and falling);
—— *d'eau souterraine*, soil water;
—— *de glace*, sheet of ice;
—— *paravent*, (*ball.*) name given to the layers of air immediately around and displaced by a projectile in motion;
—— *de pêne*, bolt nap, catch.

narine, f., nostril;
fausse ——, (*hipp*.) false nostril, a fold of the skin in the nostril.
naseau, m., (*hipp*.) nostril.
nasse, f., (*met*.) small hollow in the floor of a melting furnace.
natation, f., swimming;
ceinture de ——, swimming belt;
école de ——, swimming school;
—— *à sec*, preparatory or preliminary swimming exercises on dry land.
natif, a., (*met*.) native.
nation, f., nation;
pavillon de ——, national flag.
national, a., national; m. pl., natives, fellow-countrymen;
garde ——*e*, national guard;
pavillon ——, national flag.
nationalité, f., nationality, (by extension) nation.
natrium, m., sodium.
natron, natrum, m., soda, mineral soda.
natte, f., mat; plait, plaiting; dunnage mat; (*fond*.) straw band;
—— *d'arrimage*, —— *de fardage*, dunnage mat.
natter, v. a., to plait, to mat; to plat (as, a horse's tail);
—— *le trousseau*, (*fond*.) to cover the spindle with straw.
naturalisation, f., naturalization.
naturaliser, v. a., to naturalize.
nature, f., nature; kind, sort, species;
d'après ——, from life, from nature;
en ——, in kind;
prendre ——, (*met*.) to come to nature;
—— *du terrain*, (*top*.) nature, character, of the ground.
naufrage, m., shipwreck;
faire ——, to be shipwrecked;
faire faire ——, to shipwreck.
naufragé, a., shipwrecked, wrecked.
naufrageur, m., wrecker.
nautique, a., nautical.
nautomètre, m., (*nav*.) log (rare).
naval, a., naval;
architecte ——, naval architect;
architecture ——*e*, naval architecture;
armée ——*e*, fleet, naval forces; the navy;
combat ——, sea fight;
construction ——*e*, naval construction, architecture;
tactique ——*e*, naval tactics.
navet, m., turnip.
navette, f., rape seed; shuttle; pig of lead; netting needle; (*mil*.) relief of men in the trenches;
faire la ——, to go back and forth (like a shuttle); to march to and fro (either aimlessly or by design); to ply between a ship and the shore;
—— *de plomb*, pig of lead;
—— *des travailleurs*, (*fort*.) shift of working parties.
naviculaire, a., navicular;
maladie ——, (*hipp*.) navicular disease;
os ——, (*hipp*.) navicular bone, wheel.
navigabilité, f., navigability; seaworthiness.
navigable, a., navigable; seaworthy (of a ship).
navigateur, m., a., navigator, seafaring.
navigation, f., navigation (art, science); navigation (handling of a ship); voyage, trip (on the seas, a lake, a river, a canal); transport of goods (by river or canal);
—— *aérienne*, aerial navigation, ballooning;
—— *par l'arc de grand cercle*, v. —— *orthodromique*;
—— *astronomique*, navigation by observations;
—— *au cabotage*, coastwise trade, coasting trade;
canal de ——, commercial canal (as distinguished from an irrigation canal);
—— *de cap en cap*, v. —— *au cabotage*;
—— *côtière*, v. —— *au cabotage*;
école de ——, navigation school; (*pont*.) boat drill;
—— *en escadre*, fleet sailing;

navigation *à l'estime*, dead-reckoning;
—— *fluviale*, river navigation;
—— *en haute mer*, —— *hauturière*, navigation on the high seas, foreign trade;
—— *intérieure*, inland navigation (lakes, rivers, and canals);
journal de ——, log book;
—— *au long cours*, v. —— *en haute mer*;
—— *le long des côtes*, v. —— *au cabotage*;
—— *orthodromique*, great-circle sailing;
—— *de plaisance*, yachting;
—— *plane*, plane sailing;
—— *de rivière*, river navigation;
—— *à la sonde*, running by the lead;
—— *à vapeur*, steam navigation;
—— *à voiles*, sail navigation.
naviguer, v. a. n., to navigate; to sail, handle a ship; to steer; to row a boat; to be in such and such a trade; to lead a seafaring life;
—— *à l'aventure*, to go in search of freight; to be a "tramp," to "tramp;"
—— *au cabotage*, to be in the coasting trade;
—— *au commerce*, to be in the merchant navy;
—— *en compagnie*, to sail in consort, to keep together;
—— *de conserve*, v. —— *en compagnie*;
—— *à contre courant*, to sail against the stream;
—— *contre la lame*, to head the sea;
—— *en convoi*, to sail in, under, convoy;
—— *en escadre*, to sail in squadron;
—— *à l'estime*, to go by dead-reckoning;
—— *à l'état*, to serve in the national navy;
—— *au long cours*, to be in the foreign trade, to sail the high seas;
—— *en pleine mer*, to sail on the high seas;
—— *à la sonde*, to run by the lead.
naville, f., name given in France to irrigation channels in Lombardy.
navire, m., ship, vessel; sail, bottom.
(*Bâtiment* is a general term for ship; *vaisseau* is usually a war vessel, and *navire* a non-military ship; *bateau* stands for all small boats, with a special application to steamers.)
—— *abandonné*, derelict;
—— *en acier*, steel ship;
—— *d'agrément*, yacht;
—— *allongé*, lengthened ship;
—— *amiral*, flagship;
—— *annexe*, tender;
—— *armé en course*, privateer;
—— *armé en guerre*, armed vessel (e. g., an armed merchantman);
—— *auxiliaire*, auxiliary steamship;
—— *bagne flottant*, convict ship;
—— *baleinier*, whaler;
—— *de bas-bord*, a ship of low free board;
—— *blindé*, armored ship;
—— *en bois*, wooden ship;
—— *bombarde*, mortar boat;
—— *capable de tenir la mer*, seaworthy vessel;
—— *charbonnier*, collier;
—— *de charge*, cargo, freight, ship;
—— *à citernes*, tank steamer;
—— *à coffre*, well-deck ship;
—— *de combat*, battle ship, ship of the line;
—— *de commerce*, trading, merchant, vessel;
—— *composé*, —— *composite*, composite vessel;
—— *contrebandier*, smuggler;
—— *en course*, privateer;
—— *cuirassé*, armored ship;
—— *désarmé*, ship out of commission, in ordinary;
—— *doublé en cuivre*, copper-bottomed ship;
—— *école*, school ship, training ship;
—— *école d'artillerie*, gunnery ship;
—— *d'émigrants*, emigrant ship;
—— *à éperon*, ram, ship fitted with a ram;
—— *en fer*, iron ship;
—— *forban*, pirate;
—— *frère*, v. —— *jumeau*;
—— *garde-côte(s)*, coast-defense vessel; coast-guard vessel;
—— *de garde*, guard ship;
—— *glacière*, refrigerator ship;
—— *de guerre*, man-of-war;
—— *de guerre armé*, ship in commission;
—— *de haut-bord*, a ship of high free board;
—— *à hélice*, screw steamer;

navire à *hélices jumelles*, twin-screw vessel;
—— *hôpital*, hospital ship;
—— *à hurricane deck*, awning-deck ship;
—— *hydrographe*, surveying vessel;
—— *impropre à la navigation*, unseaworthy ship;
—— *jumeau*, sister ship;
—— *lance-torpilles*, vessel fitted with torpedo tubes;
—— *lège*, light ship;
—— *au long cours*, foreign-trading vessel;
—— *maniable*, handy vessel;
—— *marchand*, v. —— *de commerce*;
—— *marchand armé en guerre*, armed merchant-man;
—— *marin*, —— *de mer*, good sea boat;
—— *mixte*, steam-and-sail vessel;
—— *à n mâts*, n-masted vessel;
—— *neutre*, neutral vessel;
oh! du ——! ship ahoy!
—— *pêcheur*, fishing smack;
—— *pénitencier*, prison ship;
—— *de plaisir*, yacht;
—— *à pont abri*, awning-deck ship;
—— *à pont abri partiel*, partial awning-deck ship;
—— *ponté*, decked vessel;
—— *à pont léger*, shelter-deck vessel;
—— *à pont libre*, flush-deck ship;
—— *à n ponts*, n-deck ship;
—— *à pont ras*, flush-deck ship;
—— *à pont tente*, shade-decked ship;
—— *à porques*, web-frame ship;
—— *porte-faix*, v. —— *de charge*;
—— *à puits*, well-deck ship;
—— *de propriété étrangère*, foreign bottom;
—— *qui a le côté droit (faible, fort)*, wall-sided (crank, stiff) ship;
—— *qui fatigue à la mer*, ship that labors;
—— *qui manque de liaisons*, structurally weak ship;
—— *ramassé*, snug, comfortable, ship;
—— *à roues*, —— *à aubes*, paddle steamer;
—— *à shelter deck*, shelter-deck vessel;
—— *à spar deck*, spar-deck ship;
—— *sous-marin*, submarine vessel;
—— *torpilleur*, v. —— *lance-torpilles*;
—— *à tourelles*, turret ship;
—— *à transport*, transport;
—— *à trois mâts*, (sometimes) full-rigged ship;
—— *à vapeur*, steamer;
—— *à vapeur armé en guerre*, armed steamer;
—— *à voiles*, sailing vessel;
—— *à voiles et à vapeur*, sail-and-steam vessel;
—— *en vue!* sail ho!

naye, f., v. *naille*.

nayer, v. a., to sheathe, to cover a seam.

nébuleux, a., cloudy, overcast.

nécessaire, a. m., necessary, needful; what is needful, necessaries;
—— *aux armements*, (art.) implement and spare parts kit (mountain artillery);
—— *d'armes*, (sm. a.) cleaning kit;
—— *de bouche à feu* (art.) cleaning and quick-repair kit.

nécrose, f., (hipp., etc.) necrosis; (of cereals) smut.

nécroser, v. a., to bring on necrosis.

nef, f., nave; main hall or gallery;
—— *de Condé*, sort of boat in use on the northern rivers of France;
moulin à ——, water mill on a boat.

négatif, a. m., negative; (phot.) negative;
épreuve ——*ve*, (phot.) negative proof.

négligence, f., negligence; (mil.) neglect of duty;
—— *en service*, (mil.) neglect on duty.

négliger, v. a., to neglect;
—— *son corps*, (man.) to sit badly.

négoce, m., trade.

négociable, a., negotiable.

négociant, m., merchant.

négociation, f., negotiation; (mil.) parley.

nègre, m., negro; (St. Cyr slang) number one.

négrier, m., slaver, slave ship; slave dealer.

neige, f., snow;
amas de ——, snowdrift;
charrue à ——, snowplow;
chasse- ——, snowdrift; snowplow of a locomotive;
flocon de ——, snowflake;
monceau de ——, v. *amas de* ——.

neigé, a., (hipp.) snowflaky (of the coat).

neiger, v. n., to snow.

neigeures, f. pl., (hipp.) small white spots, resembling snowflakes, on the coat.

neptune, m., look-chamber (of a canal); (nav.) collection of charts; marine atlas.

nerf, m., sinew; fiber; nerve; (met.) fiber, fibrous texture; (cons.) rib;
—— *des affaires*, money;
—— *de bœuf*, ox tendon;
—— *de fer*, (met.) fibrous texture of iron after being well forged;
—— *de la guerre*, money, sinews of war;
—— *d'une tente*, ridgepole of an awning.

nerférer, v. r., (hipp.) to overreach.

nerf-féru, m., **nerf-férure**, f., (hipp.) over-reach; upper attaint, the contusion or cut produced by overreaching.

nervage, m., (harn.) covering of a saddletree with ox tendons.

nervé, a., ribbed, braced or stiffened by a *nervure* or rib.

nerver, v. a., to line, cover, with sinews or tendons; (harn.) to line or cover a saddletree with ox tendons.

nerveux, a., (met.) fibrous.

nervure, f., rib, spline, stem, web, flange, collar; fillet, molding; piping, edging, cording (of a seam); (mach.) feather; (top.) branch, rib of a mountain chain; (carp.) sort of triangular rabbet or groove; (art.) guide, guide clip; (sm. a.) rifling print (on a bullet); guide rib; rail, undercut rail;
—— *extérieure*, (mach.) keyway;
—— *guide*, (sm. a.) guide rib.

net, a., clean, pure, unmixed; free from defect; net (of weights and of prices);
cheval sain et ——, horse warranted sound and free from vice;
faire ——, to clean a horse's manger;
mettre au ——, v. s. v. *mettre*;
mise au ——, v. s. v. *mise*.

netteté, f., cleanness; (opt.) sharpness of definition.

nettoiement, m., cleaning, cleansing;
—— *des grains*, separation of foreign matter from grain; cleaning, scouring, of grain.

nettoyage, m., cleaning, cleansing, scouring; (met.) pickling;
orifice de ——, mud hole;
robinet de ——, purging cock;
trou de ——, v. *orifice de* ——.

nettoyer, v. a., to clean, scour, cleanse, wipe; (met.) to pickle; (mil.) to enfilade, to sweep a work with shot; to clear a village, etc., of the enemy;
—— *le terrain*, (mil.) to clear the ground of the enemy;
—— *la tranchée*, (siege) to drive out the guard and working parties in the trench and destroy their works, etc.

nettoyeuse, f., cleaning machine.

neuf, a., new;
cheval ——, unused horse, fresh horse.

neutralisation, f., neutralization; proclamation of neutrality.

neutraliser, v. a., to neutralize.

neutralité, f., neutrality;
—— *armée*, armed neutrality;
déclaration de ——, declaration of neutrality;
garder, observer, la ——, to remain neutral.

neutre, a., neutral; m. pl., neutral states;
corps ——, (elec.) neutral body;
droit des ——s, rights of neutrals, law of neutrality;
ligne ——, (elec.) neutral line (of a magnet);
pavillon ——, neutral flag.

névé, m., glacier snow, firn.

névrite, f., (hipp.) hypertrophy of the interstitial walls of the nervous organs.

nez, m., nose; nose (of a bit, etc.); (mach.) etc.) front end (of a piece, etc.), end of an arbor or axis; (pont.) head, how, of a ponton boat; (met.) crust of scoria in front of a twyer pipe; (cons.) lug on a flat tile;
—— *d'arrière*, stern beak (of a boat);

nez *d'avant*, beak (of a boat);
— *d'un bateau*, piece of wood at each end of boat, forming stern and bow, and called also *têtière*;
bout du ——, (*hipp.*) part of upper lip between the nostrils;
—— *de busque*, (*sm. a.*) end or curve of the *busc*, q. v., of a rifle stock;
être sur son ——, (*nav.*) to be down by the head;
porter le —— *en terre*, (*man.*) to bore;
porter le —— *au vent*, (*man.*) (of a horse), to carry the head too high;
—— *de renard*, (*hipp.*) fox nose (presence of rusty coloration around the nose and lips);
—— *à ressort*, (*mach.*) feeding finger.
niche, f., niche; (*mil.*) place of a *chef de section* in the line;
—— *à charbon*, (*nav.*) pocket coal bunker, pocket bunker;
—— *à munitions*, (*fort.*) ammunition recess (in a parapet);
—— *de refuge*, niche in a tunnel.
nichette, f., nest (in a dovecote).
nickel, m., (*met.*) nickel.
nickel-acier, m., nickel steel.
nickelage, m., nickel coating, plating.
nickeler, v. a., to nickel; to treat with nickel (as, steel).
nickélifère, a., nickeliferous.
nickelure, f., v. *nickelage*.
nicol, m., (*opt.*) Nicol prism.
nid, m., nest; berth; place; post;
—— *à bombes*, (*mil.*) shell trap (said of a work or position dangerously and uselessly exposed);
—— *de pie*, (*siege*) sap built by the assailants, after the assault of the breach, in order to crown its summit;
—— *à projectiles*, v. —— *à bombes*.
nielle, m., niello; f., smut (of grain).
nieller, v. a., to adorn with niello; (of grain) to blight, smut.
niellure, f., niello work; smut, blight (of grain).
nille, f., wooden tube or guard around a winch handle.
nippe, f., article of clothing; (*in pl.*) old clothes.
nipper, v. a., to stock, fit out, rig out (familiar).
niquetage, m., (*hipp.*) nicking.
niqueter, v. a., (*hipp.*) to nick a horse's tail.
nitramidine, f., (*expl.*) nitramidin (nitrated paper).
nitrate, m., (*chem.*) nitrate;
—— *de potasse*, saltpeter.
nitration, f., (*chem.*) nitrification.
nitre, m., (*chem.*) niter, saltpeter;
—— *du commerce*, crude saltpeter;
—— *cubique*, Chile saltpeter, cubic niter;
esprit doux de ——, sweet spirit of niter;
fleurs de ——, pulverized saltpeter;
—— *de houssage*, v. s. v. *houssage;*
—— *lunaire*, nitrate of silver.
nitrer, v. a., (*chem.*) to nitrate.
nitrésine, f., (*expl.*) nitrated resin (proposed as an absorbent for liquid explosives).
nitreux, a., (*chem.*) nitrous.
nitrière, f., niter bed; saltpeter bed; saltpeter works;
—— *artificielle*, artificial niter bed or works;
—— *s bergeries*, artificial niter bed in sheep stalls or folds.
nitrification, f., (*chem.*) nitrification.
nitrifier, v. a., (*chem.*) to nitrify.
nitrobellite, f., (*expl.*) nitrobellite.
nitrobenzide, f., **nitrobenzine**, f., (*chem.*) nitrobenzine.
nitrobenzole, f., (*expl.*) nitrobenzole.
nitrocellulose, f., (*expl.*) nitrocellulose.
nitrocolle, f., (*expl.*) nitrated glue.
nitro-coton, m., (*expl.*) gun cotton.
nitro-cumène, m., v. *nitrocumol*.
nitrocumol, m., (*expl.*) nitrocumene, nitrocumol.
nitroforme, m., (*expl.*) nitroform.
nitro-gélatine, f., (*expl.*) nitrogelatin, gelatin dynamite;

nitro-gélatine *ammoniacale*, a nitrogelatin with ammonium nitrate base;
—— *picrique*, a nitrogelatin formed by dissolving nitroglycerin in picric acid.
nitrogène, m., (*chem.*) nitrogen.
nitroglucose, f., (*expl.*) nitroglucose.
nitroglycéride, f., (*expl.*) true scientific name of nitroglycerin.
nitro(-)glycérine, f., (*expl.*) nitroglycerin.
nitroglycol, m., (*expl.*) nitroglycol.
nitro-houille, f., (*expl.*) nitro-coal.
nitroleum, m., (*expl.*) another name for nitroglycerin.
nitroline, f., (*expl.*) nitroline.
nitrolite, f., (*expl.*) nitrolite.
nitrolkrut, m., (*expl.*) nitrolkrut.
nitromagnite, f., (*expl.*) dynamagnite, Hercules powder.
nitromannite, f., (*expl.*) nitromannite.
nitromélasse, f., (*expl.*) nitrated molasses.
nitromètre, m., (*chem.*) nitrometer.
nitronaphthaline, f., (*expl.*) nitronaphthaline.
nitrophénol, m., (*expl.*) nitrophenol.
nitropyline, f., (*expl.*) nitropyline (a Volkman powder).
nitrosaccharose, f., (*expl.*) nitrosaccharose, nitrated sugar.
nitrotoluol, m., (*expl.*) nitrotoluene, nitrotoluol.
niveau, m., level; (*inst.*) level, leveling instrument, spirit level; (*steam*) gauge, water gauge;
—— *à air*, —— *d'air*, (*inst.*) air level, spirit level;
—— *apparent*, apparent level; visual horizon;
au ——, level, horizontal;
au —— *de*, on a level with;
—— *de la basse mer*, low-water level;
—— *des basses eaux*, low-water mark or level;
—— *à boule*, (*inst.*) hand (pocket) level;
—— *à bulle d'air*, v. —— *d'air;*
—— *Burel*, Burel à réservoir d'eau, v. —— *réflecteur;*
—— *à cadran*, (*art.*) sort of gunner's quadrant;
—— *de charpentier*, (*inst.*) carpenter's level;
—— *de chaudière*, (*steam*) water gauge;
—— *collimateur*, —— *à collimateur*, (*inst.*) collimating level;
—— *à collimateur à pendule*, (*inst.*) pendulum level;
de ——, level, horizontal;
de —— *avec*, on a level with;
différence de ——, difference of level;
—— *à l'eau*, (*inst.*) water level;
—— *d'eau*, water-line, -mark, height of water, surface of water; (*inst.*) water level;
—— *d'eau d'une chaudière*, (*steam*) water level in a boiler;
—— *d'eau à fioles*, (*inst.*) water level;
—— *de l'eau*, level of the water, water line, height of water, surface of water;
—— *des eaux*, water-mark, -level, -line;
—— *à équerre*, (*inst.*) mason's level;
établi de ——, on a dead level;
établir de ——, to level;
—— *à fil de plomb*, (*inst.*) plummet level;
—— *à fiole fixe*, —— *à fiole non-réversible*, (*inst.*) a fixed-bulb level;
—— *à fiole indépendante*, —— *à fiole réversible*, (*inst.*) a level with reversible bulb;
—— *à fioles rodées*, a level whose bulb is rubbed true with emery;
—— *flotteur*, float gauge;
—— *des hautes eaux*, high-water mark;
—— *de la haute mer*, high-water level;
indicateur de ——, float, float gauge;
indicateur du —— *d'eau*, (*steam*, etc.) water gauge;
—— *à jambes*, (*inst.*) striding level;
ligne de ——, level line;
—— *à lunette*, (*inst.*) surveyoyor's level;
—— *lyre*, v. —— *à boule;*
—— *de maçon*, (*inst.*) mason's level;
—— *de la mer*, sea level;
—— *de mercure*, mercury level;
mettre au ——, to level;
mettre de ——, to level, bring to a level;

niveau *de pente*, inclined plane; (*inst.*) clinometer, slope level;
— *à pendule*, — *à perpendicule*, — *à plomb*, plummet level, carpenter's level;
plan de —, datum plane, plane of reference;
— *de pointage*, (*art.*) gunner's quadrant;
— *potentiel*, (*elec.*) potential level;
prendre le —, to take the level;
quart de cercle à —, quadrant;
— *réflecteur*, — *à réflexion*, (*inst.*) reflecting level;
— *souterrain*, soil water;
— *sphérique*, (*inst.*) spherical, circular level;
surface de —, level surface;
— *à tube de verre*, (*steam*) glass gauge;
— *à visée directe*, (*inst.*) generic term for hand levels;
voyant du — *à bulle d'air*, (*inst.*) eyepiece of a spirit level;
— *vrai*, true level; true, astronomical, horizon.

nivelée, f., (*top.*) operation of taking or making a level observation.

niveler, v. a., to level, make level, bring to a level; (*surv.*) to level, to run a level;
— *au demi-cercle*, to plumb.

nivelette, f., (*surv.*) boning rod or stick.

niveleur, m., leveler.

nivelle, f., in leveling instruments, a line or directrix serving to show when the line of sight is horizontal; a synonym of *niveau à bulle d'air*, q. v.

nivellement, m., leveling (i. e., operation of making flat or level); (*surv.*) leveling; (*cons.*) determination or establishment of a level line (e. g., for a foundation);
— *barométrique*, (*surv.*) barometric leveling;
— *du canevas général*, (*surv.*) general determination of the differences of level of an area;
— *par cheminement*, (*surv.*) leveling in which the observations are made from different stations by going from station to station;
— *composé*, (*surv.*) operation of getting difference of level between two points when several stations have to be taken;
— *au demi-cercle*, plumbing;
— *de détail*, (*surv.*) filling-in level work;
— *direct*, (*surv.*) leveling by horizontal sights;
— *géodésique*, (*surv.*) geodetic leveling;
— *par l'horizontale*, v. — *direct*;
— *indirect*, (*surv.*) clinometer leveling, leveling by the slope;
mire pour —, (*inst.*) leveling rod;
— *par les pentes*, v. — *indirect*;
— *proprement dit*, v. — *direct*;
— *par rayonnement*, (*surv.*) leveling from a central point or station;
— *simple*, (*surv.*) operation of getting difference of level between two points from one station;
— *topographique*, (*surv.*) operation in which differences of level are got by measuring angles of slope;
— *trigonométrique*, (*surv.*) trigonometric leveling;
— *par visées horizontales*, (*surv.*) Y leveling, leveling by horizontal sights;
— *par visées suivant la pente*, (*surv.*) leveling by sights parallel to the slope.

nobélite, f., (*expl.*) a name for dynamite (from the inventor, A. Nobel).

nœud, m., knot; knot (in wood); node; (*cord.*) knot, tie, bend, hitch; (*nav.*) knot (rate); (*mil.*) key (of a position); (*top.*) plateau from which radiate subsidiary chains;
— *d'agui*, bowline bend;
— *d'agui à élingue*, bowline;
— *d'agui à élingue double*, bowline on a bight;
— *d'ajus*, — *d'ajut*, bend; Carrick bend; granny knot;
— *allemand*, figure-of-eight knot;
— *d'amarrage*, hitch, tie, lashing;
— *d'amarrage en étrive*, seizing;
— *d'amarrage à fouet*, tail block with a rolling hitch, and the end stopped up;

nœud *d'amarrage à plat*, lashing;
— *pour amarrer*, mooring knot;
— *d'amour*, (*artif.*) firework of 3 wheels at the vertices of a triangle;
— *d'ancre*, anchor knot; fishermen's bend;
— *anglais*, fisherman's knot;
— *d'anguille*, timber hitch; running bowline;
— *d'artifice*, (*art.*) choking knot (for cartridge bag);
— *d'artificier*, clove hitch;
— *d'artificier double*, a clove hitch with the free end passed through one of the loops;
— *de bâtelier*, clove hitch; midshipman hitch;
— *de bec d'oiseau*, v. — *à plein poing*;
— *de bois*, timber hitch;
— *de bois et barbouquet*, towing hitch;
— *de bombardier*, bombardier's knot, (German) jar knot;
— *de bosse*, stopper knot;
— *à boucle*, bowknot, slipknot, drawknot;
— *de bouline*, slip clinch;
— *de bouline double*, double slip clinch;
— *de bouline simple*, single slip clinch;
— *de cabestan*, bowline;
— *de capelage*, masthead knot;
— *de chaîne*, chain knot;
— *de chaise*, bowline;
— *de chaise de calfat*, a sort of double bowline;
— *de chaise double*, bowline on a bight; a double bowline (i. e., a double-looped bowline);
— *de chaise simple*, bowline, standing bowline knot;
— *de charrue*, plowman's knot;
— *commun*, overhand knot;
— *de communication*, (*in gen.*) focus or center of communication;
— *coulant*, noose, slip, running knot; rolling hitch;
— *coulant simple*, slipknot, running knot;
— *coulant sur double clef*, timber hitch;
— *courant*, running knot;
— *de croc de palan*, black wall hitch;
— *de croc de palan double*, double black wall hitch;
— *cul-de-porc*, single wall knot;
— *cul-de-porc couronné*, crowned wall knot;
— *cul-de-porc couronné avec une passe double*, double crown knot;
— *cul-de-porc double*, double wall knot;
— *cul-de-porc simple*, single wall knot;
défaire un —, to undo a knot;
demi- —, overhand knot;
— *demi-clef*, half hitch;
— *de demi-clefs*, clove hitch;
desserrer un —, to loosen a knot;
— *double*, double knot; bowline on a bight;
— *double fixe*, rolling hitch;
— *double de tire-veille*, double diamond knot;
— *de drisse*, midshipman hitch;
— *de drisse de bonnette*, studding-sail bend;
— *droit*, reef knot, square knot;
— *droit gansé*, a slip square knot, drawknot, bowknot;
école des —s, cordage drill;
— *d'écoute*, sheet bend, hawser bend;
— *d'écoute double*, double sheet bend;
— *d'écoute simple*, single sheet bend;
— *à élingue*, standing bowline knot;
— *d'épaule*, (*unif.*) shoulder knot;
— *d'épée*, (*unif.*) sword knot;
— *d'épissoir*, marlin-spike kitch;
— *d'étalingure de câble*, cable clinch;
— *d'étalingure de grelin*, fisherman's bend;
faire des —s, to tie knots;
faire un —, to make, tie, a knot;
faire un — *à boucle*, to tie in a bow;
faire un — *serré à*, to tie hard and fast to;
faux —, granny knot;
— *de fil (de) caret*, rope-yarn knot;
filer (tant de) —s, (*nav.*) to go, make, so many knots;
— *de filet*, fisherman's bend;
— *en forme d'un huit*, figure-of-eight knot;
— *de fouet*, rolling hitch;
— *de galère*, marlin-spike hitch; drag-rope knot;

nœud *de galère double*, harness knot or hitch;
— *à, avec le, garrot*, (*pont.*) rack lashing;
— *glissant*, running hitch;
— *de grappin*, fisherman's bend;
— *de greffe*, cut splice;
— *de griffe*, bill hitch;
— *de gueule de loup*, cat's-paw;
— *de gueule de raie*, cat's-paw;
— *du hart*, (*siege*) withy knot of a fascine;
— *de hauban*, shroud knot;
— *de hauban double*, double shroud knot;
— *de hauban simple*, single shroud knot;
— *hongrois*, (*harn.*) cinch knot or lashing; (*unif.*) Austrian knot;
— *de jambe de chien*, sheep shank, dog shank;
— *de jointure*, bend;
— *lâche*, loose knot;
lâcher un —, to loosen a knot;
— *de loch*, (*nav.*) log-line knot;
— *de ma petite sœur*, granny knot;
— *marin*, Carrick bend;
— *d'organeau*, clove hitch with free end seized to standing part; anchor-ring knot;
— *d'orin*, buoy-rope knot;
— *de palan*, tackle hitch (to fasten the hook of a tackle to a cable);
— *de passe*, generic term for any knot requiring the strands to be unlaid; splice;
— *de patte de chat*, black wall hitch;
— *en pattes d'oie*, Magnus hitch and two half hitches (mooring knot);
— *de pêcheur*, fisherman's bend;
— *plat*, reef knot, square knot; Carrick bend;
— *à plein poing*, loop knot; hawser bend;
— *de pontet*, (*sm. a.*) foot, knob, of trigger guard;
— *de poupée*, mooring knot; Magnus hitch;
— *de prolonge*, (*art.*) prolonged knot;
— *de rêne*, (*harn.*) sliding loop;
— *de ride*, Matthew Walker's knot;
— *de ride double*, double Matthew Walker's knot;
— *de ride simple*, single Matthew Walker's knot;
— *de routes*, (*top.*) point of junction of several roads;
— *de saignée*, (*hipp.*) artificer's knot, so called because used in closing a vein after bleeding;
— *serré*, hard knot;
— *simple*, single, simple, overhand knot;
— *s simples*, ordinary, simple, knots;
— *simple bouclé*, drawknot;
— *simple ganse*, draw thumb-knot, slipknot, drag-rope knot;
— *simple de tire-veille*, single diamond knot;
— *de soldat*, granny knot;
— *de tire-veille*, diamond knot;
— *de tisserand*, weaver's knot, sheet bend, single bend;
— *tors*, granny knot;
— *de trésillon*, marlin-spike hitch;
— *de vache*, granny knot; Carrick bend;
— *vicieux*, dead knot in wood;
— *de voleur*, thief knot.

noir, a., black; (of the weather) threatening, dark; (*cord.*) tarred; m., black, mourning; negro, black; (*t. p.*) bull's-eye;
— *d'Allemagne*, German black, Frankfort black;
— *animal*, animal black, bone black;
blé —, buckwheat;
— *de charbon*, coal black;
— *de Chine*, China (India) ink;
— *d'Espagne*, cork black, Spanish black;
fonte — *e*, (*met.*) cast iron;
— *franc*, (*hipp.*) uniform black coat (without reflection);
— *de fumée*, lampblack;
— *de fumée calciné*, burnt lampblack;
— *de houille*, coal black;
— *d'impression*, Frankfort black;
— *d'ivoire*, ivory black;
— *jais*, — *jayet*, (*hipp.*) jet black, with brilliant reflection, glossy black;
— *de lampe*, lampblack;

noir *mal teint*, (*hipp.*) dull black, rusty black;
— *minéral*, mineral black;
— *d'os*, bone black;
pain —, black bread;
— *de pêche*, peach black;
pie —, (*hipp.*) black and white;
— *de platine*, platinum black;
point —, small black cloud announcing a storm;
— *végétal*, vegetable coal.

noircir, v. a. r., to make black; to blacken, darken; to smoke; (of the weather) to grow dark, threatening; to lower, become cloudy; (*fond.*) to black wash.

noircissement, m., blackening; (*fond.*) black washing.

noircissure, f., black spot.

noisette, f., little nut;
casser la —, (*hipp.*) to agitate the lower lip rapidly.

noix, f., nut; plug (of a cock); drum, head (of a capstan); (*cons.*) semicircular groove (as of a window stile); tongue fitting into such a groove; (*sm. a.*) cocking notch; lower part of cocking piece; cocking-piece nose; cocking piece, cocking nut; tumbler; (*r. f. art.*) tumbler; (*mach.*) chain pulley;
— *de cabestan*, head of a capstan;
— *excentrique en forme de cœur*, (*mach.*) heart wheel;
— *de galle*, gall nut;
— *de platine*, (*sm. a.*) tumbler of a gunlock;
— *de robinet*, plug of a cock;
treuil à —, chain winch;
vis de —, (*sm. a.*) tumbler pin.

nolet, m., v. *noue*.

nolis, m., freight (by sea); price of transportation (a Mediterranean term).

nolisateur, m., charterer.

noliser, v. a., to charter a ship.

nolissement, m., chartering of a vessel.

nom, m., name, fame, celebrity;
— *de guerre*, nickname.

nombre, m., number;
— *abstrait*, abstract number;
— *cardinal*, cardinal number;
— *concret*, concrete number;
— *entier*, whole number;
— *impair*, odd number;
— *ordinal*, — *d'ordre*, ordinal number;
— *pair*, even number;
— *premier*, prime number;
— *rond*, round number.

nombrer, v. a., to number; to count, enumerate.

nombreux, a., numerous.

nomenclature, f., nomenclature, list;
prix de —, published, list, price.

nominal, a., nominal (of effect, horsepower, etc.).

nominatif, a., nominative;
état —, list (of names);
liste — *ve*, list of names.

nomination, f., (*mil. etc.*) appointment; nomination; right of nomination;
à la — —, in the appointment, gift, of ....

nommer, v. a., to name; to appoint, to promote.

non-activité, f., (*mil.*) state of being detached from active duty, of being on waiting orders (said of officers whose corps are disbanded, who return from captivity, whose functions are canceled, etc.); (as a punishment) suspension;
mettre en —, to put on waiting orders; to relieve from active duty; (as a punishment) to suspend;
— *par retrait d'emploi*, (*Fr. a.*) suspension not to exceed three years, at the expiration of which period a *conseil d'enquête* declares whether the officer shall be discharged or not;
— *par suspension d'emploi*, (*Fr. a.*) suspension, no substitute being appointed for a year, before the expiration of which the officer may be restored to duty.

non-combattant, m., (*mil.*) noncombatant.

non-conducteur, m., a., (*elec.*) nonconductor.

non-disponible, a., m., v. s. v. *disponible*.

non-exercé   **nu**

non-exercé, a., (*Fr. a.*) green, untrained (said of men that have not had at least three months' active service).
nonius, m., (*inst.*) vernier; nonius.
non-lieu, m., (*law*) quashing of an indictment, *nolle prosequi*;
   ordonnance de ——, (*law*) order to quash a case.
non-valeur, f., deficiency, waste; bad debt, worthless paper; (*mil.*) any person deducted for any reason whatever from the effective fighting strength (e. g., orderlies, man who does not report with his class of the contingent).
noquet, m., (*cons.*) gutter lead; flashing, apron.
nord, m., north, the north, north wind; north point;
   —— *du compas,* compass north, north by compass;
   faire le ——, (*nav.*) to run north;
   —— *magnétique,* magnetic north;
   —— *du monde,* v. —— *vrai*;
   perdre le ——, (*nav.*) to lose the course;
   —— *vrai,* true north.
noria, f., noria, chain of buckets, chain pump.
normal, a., normal, standard;
   bougie ——*e,* (*elec.*) standard candle;
   écartement ——, (*r. r.*) standard gauge;
   ligne ——*e,* norinal line.
normale, f., normal, normal line.
nostalgie, f., nostalgia, homesickness;
   —— *de l'épaulette,* (*mil.*) hankering for the service, longing to be in the service again.
notable, a., m., notable, noteworthy; principal personage (e. g., of a town or of a commune);
   —— *idoine,* (*adm.*) expert, who pronounces in certain cases of differences between contractors and government.
notaire, m., notary.
notarial, a., notarial.
notariat, m., functions, duties, of a notary.
notarié, a., done by a notary, sworn to and subscribed before a notary;
   acte ——, v. s. v. *acte.*
note, f., note; (musical) note; memorandum; mark (of standing in schools); rating of professional standing (as of officers); bill; diplomatic communication;
   —— *des avaries,* average statement;
   —— *ministérielle,* (*mil.*) war department memorandum.
noter, v. a., to note, note down; to take (as the time from a chronometer).
notification, f., notification; notice;
   —— *de blocus générale,* (*ou*) *diplomatique,* notice of blockade sent to governments;
   —— *de blocus locale,* notice of blockade sent to place to be blockaded;
   —— *de blocus spéciale,* notice of blockade given to ships in port, or making the port.
noue, f., pasture land; low-lying meadow; (*cons.*) valley (of a roof); pantile, gutter tile, gutter stone; valley rafter;
   —— *cornière,* (*cons.*) valley of a roof.
nouement, m., knotting.
nouer, v. a., (*cord.*) to tie, hitch, lash, bend;
   —— *des intelligences avec l'ennemi, avec la place,* to establish secret communication with an enemy or with a besieged town.
nouet, m., a synonym of *mastigadour,* q. v.
nouette, f., (*cons.*) hip tile.
noueux, a., knotty, gnarled, full of knots.
noulet, m., (*cons.*) valley of a roof, gutter of a roof;
   —— *chevron, chevron à* ——, valley rafter.
nourri, p. p., fed, sustained; full, thick laid; (of the wind) steady and strong; (of clouds) passing freely; (*art., sm. a.*) steady, well-delivered (of fire).
nourrice, f., nurse; (*mach.*) small auxiliary pump (attached to some hydraulic machines; called also *mère* ——).
nourrir, v. a., to feed, maintain, keep up, supply, cherish, entertain;
   —— *la pièce,* (*met.*) to feed the piece.
nourrissage, m., feeding, fattening;
   —— *des lampes,* (*elec.*) feeding.

nourrisseur, cow keeper;
   tuyau ——, feed pipe.
nourriture, f., food; nourishment; diet; subsistence of men, horses, etc.;
   —— *de cheval,* horse feed.
nouveau, a., new; fresh; inexperienced; raw;
   —— *modèle,* latest, most recent, model;
   —— *elles troupes,* fresh troops.
nouvelle, f., news, intelligence, tidings;
   envoyer aux ——*s,* to send out for intelligence.
novembre 33, (*mil. slang*) officer or n. c. o. who sticks close to regulations.
novice, m., youngster; novice; apprentice; (*nav.*) ordinary seaman.
noyau, m., heart, core, nucleus; knot, group; groundwork, origin; chosen, select, band; (*tech.*) plug (of a cock); stem, shank (of a screw, of a bolt); nave (of a fly wheel); (*cons.*) newel of a staircase (both solid and hollow); (*fond.*) core, former; (*sm. a.*) core (of a jacketed bullet); (*ball.*) central part of a group of hits (containing the better half of the hits); (*powd.*) small irregular lump (becomes spherical in process of manufacture); (*fort.*) central position or fort; the enceinte of a *place à forts détachés*; (*mil. slang*) recruit;
   arbre à ——, (*fond.*) core bar, core spindle;
   boîte à ——, (*fond.*) core box;
   —— *central,* (*fort.*) town surrounded by an enceinte (said of modern system of detached forts);
   —— *en corde,* (*cons.*) a newel with a spiral hand rail;
   enveloppe du ——, (*ball.*) in a sheaf of trajectories, the part surrounding the —— *de la gerbe,* and containing 40 per cent of hits;
   —— *d'escalier,* (*cons.*) newel;
   —— *de fond,* (*cons.*) continuous newel;
   —— *fusant,* (*artif.*) core (as of a Bickford fuse);
   —— *de la gerbe,* (*ball.*) in a sheaf of trajectories, that part (nearest the axis) containing 50 per cent of the hits;
   —— *de moule,* (*fond.*) core of a mold;
   —— *d'un robinet,* plug of a cock;
   sable à ——, (*fond.*) core sand;
   —— *suspendu,* (*cons.*) interrupted newel;
   —— *d'une vis,* stem of a screw.
noyé, p. p., drowned; ruined; clouded; (*tech.*) let in flush; (*nav.*) hull down; (*nav. art.*) too low (of the battery), too near the water;
   —— *dans,* (*mas.*) bedded in.
noyer, m., nut, walnut (wood and tree);
   —— *noir,* black walnut.
noyer, v. a. r., to drown; to flood; to inundate, deluge, swamp; to confuse colors; (*tech.*) to countersink; to let in, to sink in (as one piece into another); (*mas.*) to bed in (as in concrete);
   —— *la chaux,* (*mas.*) to overslack lime;
   —— *un clou,* to countersink a nail;
   —— *les poudres,* (*mil.*) to flood a magazine;
   —— *le rivet,* (*farr.*) to clinch a horseshoe nail;
   —— *la soute aux poudres,* (*nav.*) to flood the magazine; to scuttle a ship;
   —— *la terre* (*un vaisseau*), to sink the land (a ship) below the horizon;
   —— *la tête d'une vis,* to countersink the head of a screw.
noyon, m., v. *noyure.*
noyure, f., countersink (as for the head of a screw), hollow.
nu, a., m., naked, bare; naked or bare part; (*man.*) bareback; without harness; (*art.*) (of cartridges) uncovered, exposed, i. e., not carried in a pass box;
   à ——, naked, bare; (*man.*) bareback;
   cheval ——, horse bought or sold without saddle or bridle;
   à dos ——, (*man.*) barebacked;
   épée ——*e,* (*sm. a.*) bare sword;
   feu ——, (*chem.*) flame in direct contact with the object heated;
   maison ——*e,* unfurnished house;
   métal ——, native metal;
   mettre à ——, to strip, unclothe, lay bare, open;

nu, *monter un cheval à* ——, *à dos* ——, *(man.)* to ride bareback;
—— *du mur,* bare part of a wall;
œil ——, naked eye;
pays ——, *(top.)* bare, treeless country;
——*s en plâtre,* (*mas.*) plaster ribs on a wall to be plastered (to determine the surface);
—— *e propriété,* (*law*) property without usufruct;
—— (-)*propriétaire,* (*law*) the possessor of a ——*e propriété.*

nuage, m., cloud, haze, mist, darkness; (*hipp.*) cloudiness (disease of the eye), slight opalescence of the cornea;
—— *balle de coton,* cumulus;
—— *en bandes,* cirro-stratus;
—— *en bandes gras,* stratus;
——*s bas,* lower clouds;
—— *de chat,* v. —— *queue de vache;*
—— *d'été,* summer cloud, cumulus;
—— *à grain,* v. —— *à pluie;*
——*s hauts,* upper clouds;
—— *moutonné,* fleecy cloud;
—— *moutonneux,* v. —— *pommelé;*
—— *orageux,* thundercloud, nimbus;
—— *à pluie,* rain cloud;
—— *pommelé,* mackerel sky, cirro-cumulus·
—— *queue de vache,* cirrus, mare's tail;
—— *de vent,* —— *venteux,* wind cloud.

nuageux, a., cloudy.

nuaison, f., steady wind; time during which a wind blows steadily.

nuance, f., shade, tint; (*met.*) brand (of steel).

nuancer, v. a., to shade, tint, variegate.

nuée, f., cloud; storm; (*fig.*) host, multitude, flock.

nuer, v. a., to shade (a drawing, etc.).

nuit, f., night;
avoir la —— *franche,* (*mil.*) to have the night off duty; to have a night in bed (U. S. A.);
binocle de ——, night glass;
—— *d'un cheval,* hay, straw, given to a horse during the night;
de ——, by night;
lunette de ——, night glass;
permission de ——, (*mil.*) night pass;
quart de ——, night watch;
—— *de repos,* (*mil.*) night in bed (U. S. A.);
ronde de ——, (*mil.*) night rounds, rounds.

nuitée, f., duration of the night; night's work.

nulle, f., in cryptography, a nonsignificant letter, word, or phrase.

nullité, f., in columns of numbers, ditto marks in blank spaces; (*law*) fatal defect in a document, etc.;
—— *d'un brevet,* voidance of a patent;
à peine de ——, under penalty of disapproval.

numéraire, a., m., numerary; of legal value (money); coined money; specie, cash; (*nav.*) numerary (of a signal); number signal;
—— *de la banque,* reserve of a bank;
envoi de ——, transfer of specie;
—— *fictif,* paper money;
pierre ——, milestone;
—— *réel,* coin;
valeur ——, legal value, legal tender.

numéral, a., numeral;
arithmétique ——*c,* arithmetic proper (as distinguished from algebra);
lettres ——*es,* v. s. v. *lettre.*

numération, f., numeration, action of counting; notation.

numérique, a., numerical.

numéro, m., number (in a series); number of a periodical; number of a vessel, of a regiment, of a battery, etc.; (*art.*) cannoneer, "number;" (*mil.*) number (drawn by lot);
bon ——, (*mil.*) a number exempting the drawer from military service;
—— *du contrôle annuel,* (*Fr. a.*) number of an officer, man, or horse on the *contrôle annuel* (q. v.) of the administrative unit interested;
—— *international,* (*nav.*) official number of a ship in international code;
mauvais ——, (*mil.*) a number requiring military service to be performed;
—— *matricule,* v. s. v., *matricule;*
—— *d'ordre,* serial number;
—— *du répertoire* (*des réservistes et disponibles*), (*Fr. a.*) number of a man who has passed into the reserve of the active army or into the *disponibilité;*
—— *de signaux,* (*nav.*) signal number.

numérotage, m., numbering, telling off.

numéroter, v. a., to number;
se ——, (*mil.*) to call off, to count fours.

nuque, f., nape; (*hipp.*) poll;
couvre ——, v. *couvre-nuque;*
mal de ——, (*hipp.*) poll-evil.

nutation, f., habitual shaking of the head; nutation.

nysébastine, f., (*expl.*) nysebastine (a dynamite with a chlorate-mixture base).

O.

oasis, f., oasis.

obéir, v. n., to obey; to yield; to give way before;
—— *à la barre,* (*nav.*) to answer the helm;
—— *à l'éperon,* —— *à la main, etc.,* (*man.*) to heed the spur, the hand, etc.

obéissance, f., obedience;
jurer —— *à,* to swear allegiance to.

obélisque, m., obelisk.

obier, m., snowball tree, guelder rose.

objectif, a., (*mil.*) *opt.*) objective;
ligne ——*ve,* (*mil.*) line to the objective;
verre ——, (*opt.*) object glass.

objectif, m., (*opt.*) object glass, objective, lens (*mil.*) objective;
—— *achromatique,* achromatic lens;
—— *aplanétique,* (*phot.*) aplanatic lens;
—— *composé,* built-up lens, object glass;
—— *diaphragmé au 1/n,* (*phot.*) lens in which the ratio of diaphragm opening to focal distance is 1:*n;*
—— *double,* doublet lens;
—— *d'émission,* object glass of a field signaling apparatus;
—— *à grand angle,* —— *grand angulaire,* (*phot.*) wide-angle lens;
—— *principal,* (*mil.*) the final objective, the objective proper;

objectif *rectilinéaire,* (*phot.*) rectilinear objective;
—— *rectilinéaire grand angle,* (*phot.*) wide-angle rectilinear lens;
—— *secondaire,* (*mil.*) intermediate objective, one necessary to carry out the —— *principal;*
—— - *simple,* (*opt.*) simple objective;
verre d'——, (*opt.*) objective.

objet, m., object, thing;
——*s de consommation,* expendable articles, stores;
——*s hors de service,* condemned, unserviceable articles;
——*s ouvrés,* prepared articles;
——*s de rechange,* spare parts.

obligataire, m., bondholder.

obligation, f., obligation; bond (e. g., a railroad bond);
—— *au service militaire,* (*mil.*) liability to military service.

obliquangle, a., oblique-angled.

oblique, a., oblique, askew, slanting, slant; indirect; (substantively) an oblique line;
—— *à droite* (*gauche*) (*mil.*) right (left) oblique;
faire une route ——, (*nav.*) to traverse;
feux ——*s,* (*mil.*) oblique fire;

oblique, *marche* ——, (*mil.*) oblique march;
navigation ——, loxodromics;
ordre ——, (*mil.*) oblique order of battle;
port ——, (in France) a port not the residence of a maritime prefect;
rendre ——, to slant;
route ——, (*nav.*) traverse.
obliquement, a., obliquely, slant, askew.
obliquer, v. n., (*mil.*) to incline, oblique.
obliquité, f., obliquity;
—— *du tir*, (*art.*) oblique impact (as on armor plates).
oblitération, f., obliteration; canceling of postage stamps; (*hipp.*) the filling up of the pulp cavity.
oblitérer, v. a., to obliterate; to cancel (a postage stamp).
oblong, a., oblong.
obron, m., v. *auberon*.
obronnière, v. *auberonnière*.
observateur, m., observer (especially, *art.*, of the points of fall).
observation, f., observation (instrumental and other); observance, fulfillment; accomplishment;
armée d'——, (*mil.*) army of observation; an army watching the frontier of a neutral but suspected state;
bâtiment d'——, (*nav.*) lookout boat;
corps d'——, (*mil.*) corps of observation;
—— *des coups*, (*art.*) observation of shots (points of fall);
être en ——, (*mil.*, *nav.*) to be on the lookout;
faire, *prendre*, *des* ——*s*, to make, take, observations;
—— *par miroir*, mirror observation;
rapporter des ——*s*, to lay down, plot, reduce, observations.
observatoire, m., observatory (*mil.*) any position, whether natural or artificial, whence the progress of a battle, etc., may be observed;
(*fort.*) observing station;
—— *blindé*, (*fort.*) armored, protected, observing station;
—— *de campagne*, (*mil.*) field observatory;
—— *cuirassé*, (*fort.*) conning tower, armored observatory;
—— *à éclipse*, (*fort.*) disappearing observatory or conning tower;
—— *phare*, (*fort.*) electric-light observing station;
—— *volant*, (*siege*) mobile observatory (attack).
observer, v. a., to observe; to keep in view, look out, watch; to take an observation, a bearing, an angle, etc.; to fulfill, perform;
—— *les distances*, (*mil.*) to observe, maintain, distances;
—— *les hanches*, *la ligne*, (*man.*) (of the horse) to keep on the line;
—— *le terrain*, v. *garder le terrain*.
obstacle, m., obstacle, incumbrance, bar, hindrance, impediment; (*mil.*) obstacle.
obstruction, f., obstruction;
—— *de l'âme par le projectile*, (*art.*) jamming of a projectile in the bore.
obtempérer, v. n., to submit, to obey;
—— *à un ordre*, to obey an order.
obturant, obturateur, a., (*art.*) obturating sealing.
obturateur, m., (*tech.*) flap, plug, stopper, screen, cut-off, shutter; (*mach.*) generic term for valve, for any device used to close a tube; (*art.*, *sm. a.*) obturator, gas check; (*fort.*) embrasure plug (in turrets and armored forts);
anneau ——, (*art.*) ring obturator;
—— *à boule*, (*art.*) pellet obturator;
—— *à bloc-bascule*, (*sm. a.*) the obturator of the Guèdes rifle;
—— *Broadwell*, (*art.*) Broadwell ring;
—— *en carton*, pasteboard obturator;
—— *à clapet*, (*phot.*) flap shutter;
—— *composite*, (*art.*) Schneider's obturator (Broadwell ring and plastic pad);
—— *conique*, miter;
—— *à contre-poids*, (*fort.*) counterpoise embrasure plug (Gruson turret);
décroche ——, (*art.*) obturator release;
—— *à demeure*, (*sm. a.*) fixed obturator;

obturateur *d'embrasure*, (*fort.*) embrasure plug (of a turret gun);
—— *fixe*, (*art.*) an obturator fixed in position in the gun;
—— *par glissement*, (*mach.*) slide valve;
—— *à guillotine*, (*phot.*, etc.) drop shutter, camera slide;
—— *de lumière*, (*art.*) vent sealer, vent-sealing device;
—— *d'oculaire*, (*inst.*) eyepiece cap;
—— *plastique*, (*art.*) plastic obturator;
—— *à rideau*, (*phot.*) blind shutter;
—— *à secteurs*, (*phot.*) sector shutter;
—— *par soulèvement*, (*mach.*) lift valve.
obturation, f., obturation; (*art.*, *sm. a.*) gas checking, sealing, obturation;
appareil d'——, (*art.*) gas-check mechanism;
—— *automatique*, (*art.*) automatic obturation (due to the special construction of the parts involved);
—— *à boule*, (*art.*) pellet obturation;
—— *par expansion*, (*art.*, *sm. a.*) obturation by expansion of case;
—— *de lumière*, (*art.*) vent sealing; vent-sealing device;
—— *par serrage automatique*, (*art.*) obturation produced by the pressure of the powder gases on the obturating device;
—— *par serrage initial*, (*art.*) obturation produced by the initial tight fit of the obturating device.
obturer, v. a., to close, close up; (*art.*, *sm. a.*) to obturate.
obtus, a., (of an angle) obtuse.
obtusangle, a., obtuse-angled.
obus, m., (*art.*) shell; shell gun (rare);
—— *en acier*, steel shell;
—— *à ailettes*, studded projectile (the studs are set approximately parallel to the rifling, obs.);
—— *allongé*, elongated shell; a synonym of *torpille*; a shell of great interior capacity;
—— *à anneaux*, ring shell;
—— *armé*, live shell;
—— *à balles*, shrapnel;
—— *à balles à chambre arrière* (*postérieure*), shrapnel with rear burster;
—— *à balles à chambre avant*, shrapnel with front burster;
—— *à balles libres*, a shrapnel whose bullets are not packed or set in sulphur or plaster;
—— *de brèche*, battering shell;
—— *brisant*, a shell with high-explosive burster;
—— *canon*, term descriptive of shrapnels whose walls do not burst, and whose bullets are projected as it were from a gun;
—— *à ceinture*, a shell fitted with a rotating band;
—— *à ceinture expansive*, a shell fitted with an expanding rotating band;
—— *à charge arrière*, shell with rear burster;
—— *à charge centrale*, shell with central burster;
—— *chargé*, loaded shell, filled shell;
—— *à chemise*, coated shell;
—— *concentrique*, concentric shell;
—— *à cordons de plomb*, shell with leaden rotating rings;
—— *à couronnes dentées*, sort of segment shell;
—— *à culot rapporté*, shell whose base is assembled to the body;
—— *à culot vissé*, a shell whose base is screwed onto the body;
—— *à diaphragme*, diaphragm shell;
—— *double*, double shell;
—— *à double paroi*, double-walled shell;
—— *éclairant*, light shell, illuminating shell;
éclats d'——, shell splinters;
enveloppe d'—— *à mitraille*, shrapnel body;
—— *à étoile*, star shell;
—— *excentrique*, eccentric shell;
—— *à expansion*, expanding shell (takes the grooves by expansion);
—— *explosif*, torpedo shell;
—— *à explosif*, a high-explosive shell;
—— *fictif*, dummy shell;
—— *en fonte*, cast-iron shell;

obus *en fonte dure*, chilled (cast-iron) shell;
— *fougasse*, thin cast-iron globe filled with explosive compound (obs.); (to-day) a thin-walled shell of high capacity charged with high explosives;
— *à fragmentation*, generic term for shell, with reference to its producing its effects by bursting into fragments;
— *à fragmentation systématique*, a shell in which fragmentation is sought to be made regular;
— *à fusée*, common shell;
— *à fusée à double effet*, a double-action shell (i. e., fitted with a double-action fuse);
— *à fusée percutante*, percussion shell;
— *à fusée à effet retardé*, shell fitted with a delay-action fuse;
— *à fusée à temps*, time shell;
— *à gerbe creuse*, a shell whose fragments are thickest on outside of sheaf;
— *à gerbe étroite, mince*, shell giving a narrow cone of dispersion;
— *à gerbe ouverte*, shell giving a wide cone of dispersion;
— *à gerbe pleine*, a shell whose fragments are uniformly scattered;
— *à grande capacité*, a shell of high interior capacity;
— *incendiaire*, carcass, incendiary shell;
— *lesté*, plugged shell;
— *lesté en explosif*, shell filled with its burster, but not fused;
— *à mélinite, etc.*, melinite, etc., shell.;
— *à méplat*, flat-nosed shell;
— *-mine*, German name of torpedo shell;
— *de mine*, mine shell;
— *à mitraille*, shrapnel;
— *multiple*, multiple shell (so called from its construction, containing a number of small shells, so to say, or grenades);
— *non chargé*, empty, unloaded shell;
— *non crevé*, blind shell;
— *oblong*, oblong shell;
— *ogival*, — *ogivo-cylindrique*, ogival shell;
— *à ogive vissée*, a shell whose ogive is screwed on to the body;
— *ordinaire*, common shell;
— *ordinaire en fonte*, common cast-iron shell;
— *à percussion*, — *percutant*, percussion shell;
— *de perforation*, armor-piercing shell;
— *à pétrole*, petroleum (incendiary) shell;
— *pierrier*, stone fougasse;
— *de rupture*, armor-piercing shell; battering shell;
— *à segments*, segment shell;
— *en un seul morceau*, shell made in one piece;
— *shrapnel*, — *à la shrapnel*, shrapnel shell;
— *à tenons*, studded shell (the studs, *tenons*, are circular, obs.);
— *tête de-mort*, v. — *incendiaire*;
— *à tête plate*, flat-headed shell;
— *-torpille*, torpedo shell.

obuser, v. a., (*art.*) to shell.

obusier, m., (*art.*) howitzer, shell gun;
— *de campagne*, field howitzer;
— *cuirassé*, armored howitzer (armored mount);
— *lisse*, smoothbore howitzer;
— *rayé*, rifled howitzer;
— *à tir rapide*, rapid-fire howitzer.

occasion, f., opportunity; occasion; occurrence; (*mil.*) engagement, fight (obs.).

occident, m., west, the west;
à l' —, westward;
— *d'été (d'hiver)*, point where the sun appears to set on the longest (shortest) day of the year.

occidental, a., western, westerly, west.

occlure, v. a., (*chem., met.*) to occlude.

occlusion, f., (*chem., met.*) occlusion; (steam) cut-off.

occultateur, m., cut-off, screen (as in a searchlight).

occupation, f., occupation; (*mil.*) seizure, occupation, of a country, of a position; (*tech.*) in a joint, the surface that receives the *about*, q. v., bearing surface;
armée (brigade, corps) d' —, (*mil.*) army (brigade, corps) of occupation; (more narrowly) an army occupying a friendly or allied country (as a guard against surprise, invasion, etc.).

occuper, v. a., to occupy; to employ; (*mil.*) to occupy (a position, etc.).

océan, m., ocean; (especially to Frenchmen) the Atlantic;
grand —, the Pacific Ocean.

ocre, f., ocher;
— *brune*, brown ocher;
— *jaune*, yellow ocher, mineral yellow, limonite;
— *rouge*, red ocher (Spanish red, Venetian red).

octant, m., (*inst.*) octant.

octave, f., (*fenc.*) octave parry.

octogone, m., octagon; (*fort.*) an eight-bastioned work.

octroi, m., town dues (municipal tariff); office where such dues are paid; personnel charged with the collection of these dues.

oculaire, a., ocular; m. (*opt.*) eyepiece, eyeglass, ocular;
— *composé*, (*opt.*) built-up ocular;
— *négatif*, (*opt.*) negative ocular;
— *positif*, (*opt.*) positive ocular.

odomètre, m., (*inst.*) odometer.

œil, m., eye; opening, mouth, hole; loop (as of a drag chain); reefing eye; (*mach., tech.*) opening, eye, aperture, hole (in a great variety of applications); (*cord.*) eye, loop; (*harn.*) eye (of a bit, etc.); (*cons.*) any small opening; (*art.*) fuse hole; (*met.*) tap hole;
à l' —, by the eye;
— *d'ancre*, eye of an anchor;
avoir l' — *au guet*, to be on the lookout;
— *de l'arbre de manivelle*, (*mach.*) large eye of a crank;
— *des attelles*, (*harn.*) hames ring;
— *de bœuf*, (*hipp.*) gross eye, bovine eye;
— *-de-bœuf*, bull's-eye window; bull's-eye, in hot countries, small cloud appearing suddenly and announcing a tornado; scar or wound in a tree (produced by a rotten branch); water gall, weather gall; wooden traveler (the plural is *œils-de-bœuf*);
— *de charnière*, hinge-bolt hole;
— *d'une chèvre*, opening for the passage of the ropes;
— *de cochon*, (*hipp.*) small eye, pig eye;
— *de coulée* (*met.*) tap hole;
coup d' —, survey, view, glance, peep;
coup d' — *militaire*, (*mil.*) ability to take in a military situation at a glance, to recognize possibilities of a position, eye for ground;
— *couvert*, v. — *de cochon*;
— *de dôme*, (*cons.*) eye of a dome;
— *d'un étau*, screw hole of a vise;
être sur l' — (man.) to shy;
— *d'un étrier*, (*harn.*) stirrup eye;
— *à la flamande*, (*cord.*) Flemish eye, made eye;
fondre par l' —, (*met.*) to melt without closing the tap hole;
— *de gâchette*, (*sm. a.*) sear hole;
— *de gaine*, tack of a flag;
— *de galerie*, (*min.*) entrance of a gallery;
grand — *de la manivelle*, (*mach.*) large eye of the crank;
— *gras*, v. — *de cochon*;
— *gros*, v. — *de bœuf*;
— *de manivelle*, (*mach.*) small eye of a crank;
hauteur de l' —, height of the eye;
— *d'une grue*, opening for the passage of the ropes, the chains;
— *de marteau*, eye or socket of a hammer;
— *d'une meule*, eye of a millstone;
— *du mors*, (*harn.*) eye of a bit;
— *d'obus*, (*art.*) fuse hole;
— *petit*, v. — *de cochon*;
— *de pie*, (*cord.*) reefing eye; eyelet;

œil *du pivot de la bride*, (*sm. a.*) pivot hole of the tumbler bridle of a gunlock;
— *de pont*, weeper of a bridge;
— *de roue*, nave hole;
— *à toc*, (*art.*) plug-lever cam eye, tappet eye;
— *de la tuyère*, (*met.*) twyer hole;
— *vairon*, (*hipp.*) wall-eye;
— *du vent*, wind's eye;
à vue d'——, by the eye, by sight.

œillard, m., middle-sized grindstone; eye of a millstone; opening for the shaft of a mill; aspirating mouth (of a ventilator);
— *moitié*, (in a ventilator) opening whose diameter is one-half the diameter of the wheel;
— *au quart*, (in a ventilator) opening whose diameter is one-fourth the diameter of the wheel.

œillère, f., canine tooth; eyetooth; (*harn.*) blinder, blinker; blind; eyeflap;
dent ——, eyetooth.

œillet, m., eye, eyehole; eye of a bolt; (*cord.*) eye; eyelet; eyelet hole; thimble;
— *d'amarrage*, (*cord.*) lashing eye;
— *de l'ancre*, anchor-stock eye;
— *artificiel*, (*cord.*) Flemish eye;
bague d'——, grommet of an eyelet;
— *d'un boulon*, eye of a bolt;
—— *coulant*, roller stud (sort of stud pierced by an eyelet containing a small roller);
— *épissé*, (*cord.*) eye-splice;
— *d'estrope*, eye of a block strap;
— *d'étai*, eye of a stay;
— *de fer*, eye of a bolt;
— *à la flamande*, v. —— *artificiel*;
— *de pavillon*, tack of a flag;
— *de poulie*, eye of a block strap;
— *rond*, (*cord.*) half a crown;
— *de voilier*, (*cord.*) reefing eye.

œilleton, m., (*art.*, (*sm. a.*) eyehole, peephole (of a sight); (*inst.*) eyehole of a telescope.

œstre, m., gadfly.

œuf, m., egg;
courbe en ——, (*steam*) slide-valve diagram;
— *électrique*, (*elec.*) electric egg.

œuvre, f., work, piece of work, fabric;
à l'——, at work; (*mil.*) in action;
atelier de grosses ——*s*, (*nav.*) that part of an arsenal where capstans and other heavy pieces are made;
——*s blanches*, tool-maker's cutting tools;
bois d'——, construction timber;
en ——, at work (being worked at);
——*s extérieures*, (*nav.*) outboard works;
——*s intérieures*, (*nav.*) inboard works;
main d'——, v. *main d'œuvre*;
maître des ——*s*, foreman;
mettre à l'——, to put to work;
mettre en ——, to employ (a workman); to utilize; (*mas.*) to set a stone;
— *mortes*, (*nav.*) deadworks, upper works;
— *pisée*, cofferwork, cobwork; earthwork; pisé work;
— *vives*, (*fort.*) generic term for working parts of a fort (guns, emplacements, magazines, etc.); (*nav.*) lower works.

œuvre, m., construction, building; (*met.*) argentiferous lead;
dans ——, inside of the building; inside measurement;
gros ——, (*cons.*) the heaviest walls;
hors ——, exterior measurement;
hors d'——, outside of the main walls; outside measurement; projecting, bulging out;
hors-d'——, a projecting construction; outwork;
se jeter hors d'——, to project, bulge out;
mesure dans (*hors, hors d'*) —— ——, inside (outside) measurement;
mesurer dans (*hors, hors d'*) ——, to take an inside (outside) measurement;
à pied d'——, at hand, in the neighborhood (of the work to be done);
reprendre sous ——, *reprendre en sous*- ——, (*cons.*) to underpin, rebuild from the foundation; to alter the groundwork;
reprises en sous- ——, (*cons.*) repairing, rebuilding, alteration, of foundations;

œuvre, *sous* ——, (*cons.*) underpinning, shoring up, foundation;
en sous- ——, (*cons.*) at the foundation.

offensif, a., (*mil.*) offensive; (*torp.*) ready to act;
armes ——*s*, offensive weapons, weapons of offense;
guerre ——*ve*, offensive war;
retour ——, counterstroke, return stroke;
traité ——, offensive treaty;
traité —— *et défensif*, offensive and defensive treaty.

offensive, f., (*mil.*) offensive;
garder l'——, to remain on the offensive;
prendre l'——, to assume the offensive.

office, m., duty, employment; service; divine service; turn, functions; diplomatic note, paper; bureau;
d' ——, officially, by virtue of one's office; spontaneously, of one's own accord;
—— *divin*, divine service, public worship;
faire l' —— *de*, to perform the service, functions, duty of.

officiant, m., officiating chaplain, clergyman.

officiel, a., official;
bulletin ——, v. s. v. *bulletin*;
journal ——, v. s. v. *journal*.

officier, v. n., to perform divine service, to officiate.

officier, m., officer; official; lowest grade but one in the Legion of Honor; (*mil.*, *nav.*) officer;
— *d' Académie*, in France, person that has received the decoration known as the *palmes académiques*, q. v.;
—— *acheteur*, member of a horse-purchasing board;
—— *en activité* (*de service*), officer on the active list;
—— *adjoint à l'habillement*, (*Fr. a.*) assistant of the *capitaine d'habillement*, q. v.;
—— *d'administration*, (*Fr. a.*) officer in charge of an administrative department, or who assists the head or director of a department in his accounts and papers. Administrative officers serve in three services—the *intendance*, military justice, and hospitals—and enjoy the privileges of officers proper. The grades are —— *d'administration principal*, —— *d'administration de 1re* (*2e*) *classe*, —— *d'administration adjoint de 1re* (*2e*) *classe*, and have no assimilation with military grades proper except as to pay and retirement (from major to second lieutenant);
—— *d'administration greffier*, v. —— *d'administration de la justice militaire*;
—— *d'administration de la justice militaire*, (*Fr. a.*) administrative official of military courts (—— *d'administration greffier*, or *greffier*, q. v.) and of military prisons;
—— *d'administration du services des hôpitaux*, (*Fr. a.*) official in charge of accounts, etc., of sanitary establishments, under the orders of the medical officer;
—— *d'administration du service de l'intendance*, (*Fr. a.*) accountant or administrative official of the *intendance* (serves in the bureaus of the *intendance*, in the subsistence, and in the clothing departments);
—— *amiral*, (*nav.*) flag officer;
—— *d'approvisionnement*, (*Fr. a.*) subsistence officer of a regiment; regimental commissary officer (U. S. A.);
—— *d'armement*, (*Fr. a.*) the assistant of the *capitaine d'habillement* in matters relating to the arms and ammunition of a regiment (almost the same as acting ordnance officer, U. S. A., in some respects);
aux ——*s*, officers' call;
—— *auxiliaire*, (*Fr. nav.*) an officer of the merchant navy serving in the war navy as a complementary or auxiliary officer;
bas ——, noncommissioned officer;
—— *de la batterie*, (*nav.*) officer in charge of the gun deck;
—— *bleu*, auxiliary, supernumerary officer who takes another's turn of duty when sick or absent
——*s du bord*, (*nav.*) ship's officers;

officier *en bourgeois*, v. —— *en civil;*
—— *breveté*, officer with staff certificate; officer who has passed the staff college (English army);
—— *canonnier*, (*nav.*) gunnery lieutenant;
—— *de casernement*, (*Fr. a.*) officer who acts as barrack master (the color bearer);
—— *chargé du détail*, (*nav.*) executive officer;
—— *chargé du matériel d'artillerie*, (*nav.*) gunnery lieutenant;
—— (*chargé*) *des montres*, (*nav.*) navigator, navigating officer;
—— *chef d'attaque*, (*siege*) in each attack, an officer charged with the direction of the work, with the execution of the orders of the engineer in chief;
—— *civil*, functionary or official of the civil service;
—— *en civil*, officer in mufti;
—— *commandant*, commanding officer;
—— *de compagnie*, company officer;
—— *comptable*, accountant officer;
—— *de corps*, —— *de corps de troupe*, regimental officer, line officer;
corps d'——*s*, the whole body of officers (of the army and the navy);
—— *corps de pompe*, (*mil. slang*) officer well up in scientific and technical subjects;
corps des ——*s de santé*, medical staff;
—— *de demi-batterie*, officer in direct charge of a half battery (police, discipline, etc.);
—— *en demi-solde*, officer on half pay;
—— *démissionnaire*, officer who has resigned;
—— *descendant de garde*, officer coming, marching, off guard;
—— *de détail*, v. s. v. *détail*;
—— *du détail général*, (*nav.*) executive officer;
—— *en disponibilité*, officer unattached, on waiting orders;
—— *de distribution*, issuing officer;
—— *-élève, élève-* ——, student officer;
—— *de l'état civil*, registrar, official in charge of vital statistics (duties performed by the *intendance* in the army);
—— *d'état-major*, staff officer;
—— *de l'état-major général*, (*nav.*) staff officer;
—— *de l'état-major de la marine*, (*nav.*) naval officer;
—— *de garde*, officer on guard, or of the guard; (*nav.*) officer of the deck in port;
garde d'——, officer's guard;
—— *général*, general officer;
—— *général de l'armée de mer, de marine*, (*nav.*) flag officer;
grade d'——, officer's rank;
grand- ——, second grade of the Legion of Honor;
—— *de guérite*, (*mil. slang*) private;
—— *d'habillement*, v. *capitaine d'habillement;*
—— *hors cadre*, seconded officer (n. e., U. S. A.)
—— *indigène*, native officer (in native troops);
—— *instructeur*, officer in charge of drill (the "*instructeur*" of the drill book);
—— *d'instruction*, student officer;
—— *de l'Instruction publique*, in France, the senior grade of the *palmes académiques*, the lower of which is —— *d'Académie*.
—— *de jour*, officer of the day, orderly officer of the day;
—— *de justice*, police official;
—— *de la Légion d'honneur*, lowest grade but one in the Legion of Honor;
—— *de logement*, v. s. v. *logement;*
—— *de majorité*, (*nav.*) officer on the admiral's staff;
—— *de manœuvre*, (*nav.*) officer of the deck;
—— *de marine*, —— *de la marine*, naval officer (of the line only);
—— *marinier*, (*nav.*) warrant officer;
—— *-marchand*, merchant captain;
—— *mécanicien*, (*nav.*) engineer officer;
—— *militaire*, combatant, as distinguished from an administrative officer;
l'—— *le moins ancien* (*élevé*), the junior officer;d;
—— *montant de garde*, officer marching on guar
—— *monté*, mounted officer;

officier *d'ordonnance*, orderly officer, aide-de-camp (*nav.*) flag lieutenant;
—— *de paix*, justice of the peace;
—— *parlementaire*, officer bearing or with a flag of truce;
—— *payeur*, paymaster; (*mil. slang*) comrade who treats the company to drink;
—— *de peloton*, officer in direct charge of a platoon (police, discipline, etc.);
l' —— *le plus ancien*, *l'* —— *le plus ancien dans le grade le plus élevé*, the senior officer;
—— *de police*, police official;
—— *de port*, officer in charge of a dockyard, dock master;
—— *porte-drapeau*, color bearer;
poste d' ——, v. *garde d'* ——, *supra;*
premier ——, first mate (commercial navy);
—— *de prise*, (*nav.*) officer in command of a prize;
—— *de quart*, (*nav.*) watch officer;
—— *de recrutement*, recruiting officer;
—— *réformé*, half-pay officer;
réformer un ——, to put an officer on half pay;
—— *de réserve*, (*Fr. a.*) an officer of the reserve of the active army;
—— *en retraite*, —— *retraité*, retired officer;
—— *de ronde*, officer on rounds;
—— *sans troupe*, (*Fr. a.*) officer who has not under his direct orders an organic troop unit; officer holding a staff appointment;
—— *de santé*, health officer; quarantine officer; (*mil.*) medical officer;
—— *en second*, (*nav.*) executive officer;
—— *de semaine*, officer on duty for the week;
—— *en serre-file*, supernumerary officer; file closer;
—— *de service*, officer on duty, on duty for the day; orderly officer;
—— *du service d'état-major*, staff officer;
—— *du service de santé*, (*Fr. a.*) medical officer (includes surgeons, apothecaries, and medical administrative officers);
sous- ——, noncommissioned officer;
sous- —— *de marine*, (*nav.*) petty officer; warrant officer;
—— *stagiaire*, (*Fr. a.*) officer who, having taken the course at the *École supérieure de guerre* and received certificate, is required to serve two years in some staff employment, including service in arm or arms different from his own;
—— *subalterne*, company officer;
—— *subordonné*, subordinate officer;
—— *à la suite*, supernumerary officer, officer attached as supernumerary of his rank;
—— *supérieur*, field officer; (*nav.*) flag officer;
—— *supérieur de jour*, field officer of the day;
—— *supérieur de service*, field officer of the day (on duty);
—— *supérieur de visite*, (*Fr. a.*) (in garrison) field officer who inspects guards;
—— *surnuméraire*, additional officer; officer in excess of the statutory number; supernumerary officer;
table des ——, officers' mess;
—— *territorial*, (*Fr. a.*) officer of the territorial army;
—— *de tir*, musketry instructor; range officer (U. S. A.);
—— *de torpilles*, —— *torpilleur*, (*nav.*) torpedo lieutenant;
—— *de troupe*, regimental officer, line officer (as distinguished from staff officer);
—— *vaguemestre*, officer in command of the *train régimentaire;*
—— *de vaisseau*, (*nav.*) naval officer;
—— *vélocipédiste*, officer mounted on a bicycle, belonging to a bicycle corps.
officieux, a., nonofficial.
offre, f., (contractor's) bid; supply;
—— *et demande*, supply and demand.
ogivage, m., (*art.*) operation of making the ogive.
ogival, a., ogival; (*art.*) ogival.
ogive, f., pointed arch, ogive; (*art.*) ogive, ogival head, of a projectile;
en ——, ogival.

ohm   298   **ordonnancement**

ohm, m., (*elec.*) ohm.
ohmmètre, m., (*elec.*) ohmmeter.
oignon, m., onion; (*hipp.*) swelling of the *os pedis*, bony tumor of the foot.
oing, m., grease, lubricant, wheel grease;
   vieux ——, wheel grease.
oiseau, m., bird; mason's hod;
   à vol d' ——, as the crow flies;
   à vue d' ——, bird's-eye view.
olécrane, m., (*hipp., etc.*) olecranon.
olécranien, a., olecranoid.
oléfiant, a., (*chem.*) olefiant;
   gaz ——, ethylene, olefiant gas.
oléonaphte, m., Russian mineral oil.
oligiste, a., m., (*met.*) oligistic; oligistic iron;
   fer ——, specular iron.
olive, f., olive; (in general) button, knob, olive-shaped button; (*art.*) lanyard button, slide; (*harn.*) olive bit;
   poulie à ——, shoe block;
   —— *de tirage*, gathering or assembling button;
   tire-feu à ——*s*, (*art.*) slide lanyard.
olivier, m., olive tree and wood.
ombrageux, a., (*man.*) shy, skittish;
   cheval ——, skittish horse;
   être ——, to be shy.
ombre, f., shade, shadow;
   —— *brûlée*, burnt umber;
   —— *naturel*, raw umber;
   terre d' ——, umber.
ombrer, v. a., (*drawing*) to shade.
ombromètre, m., rain gauge.
omelette, f., omelet;
   —— *de la pleurésie*, (*hipp.*) clots due to pleurisy.
omis, m., (Fr. a.) young man whose name has not been inscribed on the *tableau de recensement*, q. v.
omission, f., omission;
   —— *de numéros*, (*mil.*) omission of numbers (in drawing by lot).
omnibus, m., omnibus, 'bus.
onde, f., wave; undulation; (*mach.*) (in pl.) surface irregularities caused by defective machining;
   longueur d' ——, wave length;
   —— *lumineuse*, light wave;
   —— *marée*, tidal wave;
   —— *sonore*, sound wave.
ondé, a., undulated, wavy, corrugated;
   bois ——, grained wood.
ondée, f., shower.
ondoyant, a., undulating, undulatory; waving, wavy.
ondoyer, v. n., to undulate, to wave, to rise in billows.
ondulation, f., undulation;
   —— *du sol*, (*top.*) undulation of the ground.
ondulatoire, a., undulatory.
ondulé, a., corrugated;
   fer ——, corrugated iron;
   tôle ——*e*, *v. fer* ——.
onduler, v. n., to undulate, wave, float, play loosely; to flutter (as a flag).
onéreux, a., onerous;
   à titre ——, at one's own cost; paying all charges.
ongle, m., nail;
   —— *de cheval*, (*hipp.*) (horse's) hoof;
   chute de l' ——, (*hipp.*) casting the hoof;
   —*s du poing de la bride*, (*man.*) nails of the left hand.
onglet, m., graver; band or strip on which maps, etc., are pasted for binding; ungula; hollow (in the flat side of a rule); notch, nail-cut (in a lid, in a knife blade); (*carp.*) miter, miter joint; (*min.*) conical cut in the excavation of a mine; (*hipp.*) inflammation of the nictitating membrane; (*art.*) vent shell of a mortar (obs.); (*sm. a.*) fullering (of a blade);
   à ——, mitered;
   assemblage à, *en*, ——, (*carp.*) miter joint;
   assembler à, *en*, ——, (*carp.*) to miter;
   boîte à ——, (*carp.*) miter box;
   coller à ——, to mount on strips (as maps);
   couper à ——, (*carp.*) to miter;
   en ——, mitered;
   —— *avec enfourchement*, (*carp.*) a mortised miter joint;
   équerre à ——, miter square;
   tailler à ——, *v. couper à* ——.

onglette, f., small, flat graver, punch; nail cut for opening anything with the thumb; (*sm. a.*) fullering of a blade.
oolithe, m., oolite.
ope, f., (*cons.*) scaffolding hole left in a wall.
opérateur, m., (*mach.*) operator, motor;
   —— *à eau*, (*mach.*) water motor.
opération, f., operation, performance; working; speculation; (*mil.*) operation;
   base d' ——, (*mil.*) base of operations;
   front d' ——, (*mil.*) front of operations;
   —— *de guerre*, (*mil.*) military operations, operations of war;
   ligne d' ——, (*mil.*) line of operations;
   —— *de nuit*, (*mil.*) night operation;
   théâtre d' ——, (*mil.*) theater of operations.
opercule, (*art.*) diaphragm (separating bullets from powder chamber in certain shrapnel).
opérer, v. a. n., to work, perform, effect; operate; work out, etc.; (*mil.*) to operate (against);
   —— *contre*, to counterwork.
ophicléide, m., (*mus.*) ophicleide.
opposer, v. a., to oppose; (*fenc.*) to parry by opposition, to oppose;
   —— *les épaules aux hanches*, (*man.*) to cause the shoulders to follow the haunches (so as to keep a horse straight).
opposition, f., opposition; attachment or claim on pay; (*fenc.*) motion of the hand by which a thrust is parried;
   être en ——, (*fenc.*) to keep one's point directed against the adversary's breast while covering one's own with the guard;
   faire des ——*s*, (*man.*) (of the rider) to use the hands and legs simultaneously;
   —— *juridique*, *v. saisie-arrêt*.
optique, a., f., optical; optics; (*mil.*) visual signaling.
or, m., gold;
   —— *fulminant*, (*expl.*) fulminate of gold.
orage, m., storm, tempest, thunderstorm; storm of rain, wind;
   —— *magnétique*, magnetic storm.
orageux, a., stormy, tempestuous;
   coup ——, storm.
orbe, a., that which causes contusion;
   coup ——, heavy blow, severe bruise;
   mur ——, (*cons.*) dead wall.
orbevoie, f., blind arcade.
orbières, f. pl., (*harn.*) blinkers.
orbite, f., orbit; socket (of the eye).
ordinaire, a., ordinary, customary; (of functionaries) sitting the whole year.
ordinaire, m., (*mil.*) soldier's mess; mess fund;
   caporal d' ——, mess corporal;
   chef de l' ——, n. c. o. or soldier in charge of mess;
   —— *de la troupe*, (Fr. a.) mess fund (applied also to the lighting of barracks, to the purchase of cleaning materials, etc.).
ordon, m., (*met.*) hammer frame;
   —— *à bascule*, tilt frame.
ordonnance, f., order, regulation, decree, administrative order (in these senses no longer used in the French army); ordering, disposition; (*med.*) prescription; (*law*) decision of a judge; (*adm.*) warrant to pay or disburse; (*mil.*) orderly (sometimes masculine in this sense);
   à l' ——, according to regulations;
   caporal d' ——, orderly corporal;
   cavalier d' ——, (*mil.*) mounted orderly;
   d' ——, (*mil.*) regulation;
   —— *de délégation*, (*adm.*) act by which a minister delegates his powers to an *ordonnateur secondaire*, q. v.;
   en ——, on orderly duty;
   habit d' ——, (*unif.*) uniform (obs.);
   —— *de non-lieu*, v. s. v. *non-lieu*;
   officier d' ——, orderly officer, aide-de-camp;
   —— *de la place*, (*mil.*) standing order, post order (U. S. A.);
   ——*s de police*, police regulations;
   soldat d' ——, orderly.
ordonnancement, m., (*adm.*) written order of payment (at the bottom or across the face of a bill).

ordonnancer, v. a., (*adm.*) to write the order for the payment of anything, to warrant a payment;
—— *une dépense*, to authorize a payment or disbursement.
ordonnateur, m., (*adm.*) person empowered to order disbursements;
—— *secondaire*, (*adm.*) person to whom the power to order disbursements is delegated by the one who holds it in virtue of his office (e. g., *Fr. a.*, the *directeurs* of the various *services*).
ordonnée, f., ordinate.
ordonner, v. a., to order; to regulate, direct, command, enjoin, prescribe;
—— *de*, to make regulations for;
s' ——, (*med.*) to be prescribed.
ordre, m., order, rank, class, arrangement; (*mil., nav.*) order, command; order (formation);
à l' ——, n. c. o.'s call, first sergeant's call (U. S. A.); by order;
à l' —— *de*, by the order of;
—— *d'appel*, (*Fr. a.*) order to young men of a given class to report at the capital of the regional subdivision in order to join;
—— *de l'armée*, general order;
aux ——*s de*, under the orders of; by order of;
avance à l' ——, v. s. vv. *avance* and *mot;*
—— *de bataille*, list of troops composing an army, a corps, etc.; order of precedence of troops; battle array, order of battle;
—— *en bataille*, (*cav.*) order in line; (*art.*) order of formation in line (battery or group);
—— *en bataille à intervalles ouverts* (*serrés*), (*art.*) order in line with open (closed) intervals;
—— *en batterie*, (*art.*) order of formation in battery (battery or group);
—— *du cabinet*, cabinet order;
—— *de chasse*, (*nav.*) order of pursuit;
—— *en colonne*, (*art., cav.*) column formation;
—— *en colonne par pièce* (*section*), (*art.*) formation in column of pieces (sections) (platoons, U. S. Art.);
—— *en colonne serrée*, (*art.*) formation in close column;
—— *de combat*, combat order, disposition for battle;
—— *compact*, compact formation, compact order of battle;
—— *complet*, order that, in addition to the exact description of the duty to be performed, gives some idea of the *ensemble* of the whole operation;
—— *concave*, concave order of battle;
conforme à l' ——, according to orders;
—— *constitutif*, organic order of formation (e. g., of companies in the battalion, of battalions in the regiment);
contre- ——, counter order;
—— *convexe*, convex order of battle;
—— *de convocation*, order convening a court; (*Fr. a.*) order to reservists and territorialists to join for a period of instruction;
copier l' ——, (*mil. slang*) to do fatigue duty;
—— *déployé*, deployed order;
—— *dispersé*, extended order;
—— *à distance entière*, order of formation at full distance;
—— *à distance serrée*, close order;
donner —— *de*, to give an order to (advance, etc.);
donner l' —— *à*, to order, appoint, direct, provide for;
donner un ——, to give an order;
—— *en échelons*, echelon formation;
—— *d'écrou*, order to confine a man;
en ——, in order;
—— *en échiquier*, checkerwise;
être mis, porté, à l' —— *du jour*, to be mentioned in orders;
—— *d'extraction*, (in military prisons) order releasing a prisoner or person acquitted from confinement;
—— *de front*, (*nav.*) line abreast;
—— *général*, general order;
—— *hiérarchique*, order of rank;

ordre *d'informer* (*law*) order to investigate a case;
infraction aux ——*s*, breach of orders;
—— *inverse*, inverted order;
—— *du jour*, general order, one publishing a matter or matters of importance; order of the day;
jusqu'à nouvel ——, until further orders;
—— *de levée d'écrou*, order releasing a man from confinement;
—— *linéaire, en ligne*, order in line;
—— *en ligne de colonnes*, (*art., cav.*) order in line of columns;
—— *en ligne de colonnes* (*par pièce ou par section*), (*art.*) order of formation in line of columns (of pieces or of sections) (platoons, U. S. Art.);
—— *en ligne de file*, (*nav.*) order in line ahead;
—— *en ligne de front*, (*nav.*) order in line abreast;
livre d' ——*s*, orderly book;
—— *de marche*, order of march, march formation; (*nav.*) order of sailing;
—— *en masse*, formation in mass;
mettre à l' ——, to put or publish in orders;
mettre à l' —— *de l'armée, du jour*, to publish in general orders;
—— *mince*, shallow formation (wide front and little depth);
mise à l' ——, publishing in orders;
mise à l' —— *de l'armée, du jour*, publishing in general orders;
—— *de la mise en jugement*, (*law*) order to try a case;
—— *de mise en route*, march order;
—— *de mobilisation*, order to mobilize;
à moins d' ——*s contraires*, in the absence of other orders;
mot d' ——, v. s. v. *mot;*
—— *de mouvement*, march order; order for the movement of troops; transportation order;
—— *de mouvement rapide*, order for railway transportation issued on mobilization or in emergencies;
—— *en muraille*, formation in line without intervals;
—— *naturel*, direct order;
—— *s de nuit*, night orders;
—— *oblique*, oblique order of battle;
par ——, by order, by command;
—— *parallèle*, parallel order (of battle, etc.);
—— *particulier*, special order;
—— *permanent*, standing orders;
—— *perpendiculaire*, order in which one body of troops is perpendicular to another;
—— *ployé*, ployed order;
—— *en potence*, formation or order with one flank refused or thrown forward;
pour ——, to define or give a status;
—— *précis*, order that exactly describes the duties of the detachment, etc., to whom issued;
—— *profond*, deep formation (small front and great depth);
—— *de prompte formation*, formation without any condition of order;
—— *à rangs serrés*, close order;
recevoir l' —— *de*, to be ordered to;
rédaction des ——*s*, preparation of orders;
rédiger un ——, to draw up an order;
registre d' ——*s*, orderly book;
—— *renversé*, inverted order;
renverser l' ——, to invert the order of formation;
—— *de réquisition*, requisition order (i. e., ordering a requisition to be made);
rétablir l' ——, to restore order;
—— *de retraite*, order of retreat;
révoquer un ——, to cancel, revoke, an order;
—— *rostral*, v. —— *convexe;*
—— *de route*, (*Fr. a.*) railway transportation order in the *livret individuel*, q. v., to be used under contingencies specified in advance (e. g., mobilization);
selon les ——*s*, according to orders;
—— *serré*, close order;
—— *de service*, relative order in which duties are performed, precedence of duties;

ordre, *service d'——*, department, personnel, charged with the maintenance of order;
— *en sous-* ——, subordinate, of lower grade, of a subordinate capacity;
— —*supplémentaire*, after-order;
— —*s de tenue*, dress orders, dress regulations;
— —*en tirailleurs*, skirmishing order;
— *transmettre un* ——, to transmit an order;
— *de transport*, order for the railway transportation of stores and of *matériel* in general.

ore, f., (*met.*) twyer plate.

orée, f., skirt, border, of a wood (obs.).

oreillard, a., lop-eared;
— *cheval* ——, lop-eared horse.

oreille, f., ear; (*tech.*, etc.) lip, lug, corner, handle, projection, foot; (*sm. a.*) thumb piece, comb; (*mil. min.*) horn (of a caso); (*top.*) either crest or summit of a double-peaked mountain; (*carp.*) sort of arch (upper part straight in front, bottom rounded); (*mas.*) cut or notch at the end of a window- or door-sill:
— *aller de l'——*, v. *boiter de l'——*;
— *d'ancre*, anchor fluke;
— *d'âne*, kevel, cleat;
— *bec d'——*, point, bill, of a fluke;
— *de boîte de roue*, ear of a nave box;
— *boiter de l'——*, (*hipp.*) to limp at the ear (said of an ear that flops during motion);
— *cassée*, (*hipp.*) broken ear;
— *de chat*, (*farr.*) sort of calkin;
— *de cochon*, (*hipp.*) swine ear; thick, broad ear, falling over;
— *d'un cordage*, (*cord.*) ear of a rope;
— *coucher les* ——*s en arrière*, (*hipp.*) to throw back the ears;
— *coupée*, (*hipp.*) cropped ear;
— *dresser les* ——*s*, (*hipp.*) to prick up the ears;
— *hardie*, (*hipp.*) well set, fine ear;
— *inquiète*, (*hipp.*) restless ear, uncertain ear (always in motion);
— *plaquée*, v. —— *de cochon*;
— *de pointage*, (*art.*) aiming ear or lug (Russian field mortar);
— *de renard*, v. *hardie*;
— *taquet à* ——*s*, belaying cleat.

oreillère, f., (*harn.*) ear ornament.

oreillon, m., (*fort.*) v. *orillon*; (*sm. a.*) languet; (*harn.*) head knot of a bridle, tassel.

organe, m., organ; (*mach.*, etc.) part of a machine, of a mechanism;
— —*s actifs*, the moving, working parts of a machine shop;
— *de déformation*, (*art.*) generic term for hydraulic cylinders, buffers, etc., used in taking up and checking recoil;
— —*s passifs*, the fixed parts of a machine shop (hangers, plummers, etc.);
— *de réglage*, adjustment.

organeau, m., ring; anchor ring;
— *d'ancre*, anchor ring.

organisation, f., organization; (*mil.*) preparation for defense (wood, house, village, etc.);
— *défensive d'un mur, d'une maison*, etc., (*mil.*) preparation of a wall, house, etc., for defense;
— *intérieure*, (*fort.*) interior arrangement (of a fort or work).

organiser, v. a., to organize, arrange, prepare; (*fort.*, *mil.*) to prepare;
— *défensivement*, (*mil.*) to prepare anything for defense (as a village, house, etc.).

orge, f., barley;
— *d'automne*, v. —— *carré*;
— *carré*, winter barley;
— *grain d'*——, v. s. v. *grain*;
— *d'hiver*, v. —— *carré*;
— *mondé*, Scotch barley;
— *perlé*, pearl barley.

orgette, f., (*mach.*) beam or shaft of a mandrel lathe.

orgueil, m., prop. fulcrum of a lever.

oriasite, f., (*expl.*) oriasite.

orient, m., east; the east;
— *l'——d'une carte*, right-hand side of a map;
— —*d'*——*e (d'hiver)*, point where the sun rises in summer (winter);
— *l'extrême* ——, the Far East (China, Japan).

oriental, a., eastern, oriental.

orientation, f., orientation; the getting of one's bearings, of the points of the compass; (*nav.*) trim (of the sails);
— *d'un lever*, putting the north and south line on a map.

orientement, m., proper facing (i. e., situation) of a building; (*nav.*) set of the sails.

orienter, v. a., to orient (one's self, a plane table, etc.); to find out one's position or bearings;
— *un lever*, to draw a north-and-south line on a map;
— *la planchette*, to set up a plane table;
— *les voiles*, (*nav.*) to trim the sails.

orifice, m., hole; orifice; nozzle; mouth, opening; manhole; (*steam*) port;
— *d'admission*, admission port;
— *d'admission (d'introduction) du cylindre*, cylinder port;
— *d'admission du tiroir*, slide-valve port;
— *d'aspiration*, aspirating port or orifice;
— *de charge*, (*art.*) filling hole (of a shell);
— *de décharge*, discharge hole or vent;
— *d'émission*, exhaust port;
— *d'émission du cylindre*, cylinder exhaust port;
— *d'émission du tiroir*, slide-valve exhaust port;
— *à l', d', évacuation*, eduction port, exhaust port;
— *d'évacuation du cylindre*, cylinder exhaust port;
— *de graissage*, (*mach.*) oil hole;
— *d'introduction*, —— *d'introduction de la vapeur*, steam port, admission port;
— *de nettoyage*, —— *de nettoiement*, mud hole;
— *de prise d'eau à la mer*, sea cock.

original, m., (of a document, etc.).

origine, f., origin; (*hipp.*, etc.) origin of a muscle;
— *des coordonnées*, origin of coordinates;
— *d'une ligne*, (*r. r.*) starting point of a railway;
— *des rayures*, (*art.*) origin or beginning of the grooves;
— *du tir*, (*ball.*) point at which the projectile leaves the piece.

orillard, a. v., *oreillard*.

orillon, m., ear; handle; (*fort.*) orillon (obs.);
— *bastion à* ——*s*, (*fort.*) bastion with orillons.

orin, m., (*cord.*) buoy rope;
— *d'ancre*, anchor-buoy rope.

oringuer, v. a., to haul upon a buoy rope to see if the anchor is all right; to raise an anchor by the buoy rope.

oripeau, m., tinsel.

orle, m., lip or rim of a volcano crater.

orme, m., elm (tree and wood).

ormeau, m., young elm.

orne, m., manna ash.

ornière, f., rut (of a wheel); (*r. r.*) rail; groove or cut (for a wheel flange);
— *angulaire*, (*r. r.*) edge rail;
— *en bois*, wooden (tram) rail;
— *chemin à* ——*s*, tram road;
— *chemin à* ——*s de pierre*, stone way;
— *creuse*, tram rail;
— *garder l'*——, to follow or keep in the track of a leading carriage;
— *plate*, —— *plate à rebord*, tram rail;
— *saillante*, v. —— *angulaire*.

orographe, m., (*inst.*) orograph.

orographie, f., orography.

oromètre, m., (*inst.*) orometer, mountain barometer.

orométrique, a., altitude measuring.

orpiment, m., **orpin**, m., orpiment.

ort, a., (of weights) gross.

orthochromatique, a., orthochromatic.

orthographie, f., orthographical projection; elevation; (*fort.*) normal profile or section;
— *externe*, elevation;
— *interne*, section.

ortie, f., nettle; (*hipp.*) rowel, seton;
— *de Chine*, ramie, China grass;

ortie — **outre-passer**

ortie, *pratiquer une* —, (*hipp.*) to rowel.
os, m., bone;
  avoir de l'—, (*hipp.*) to have a thick or big cannon bone;
  —— *crochu,* (*hipp.*) carpel bone, trapezium;
  —— *naviculaire,* (*hipp.*) navicular bone, sesamoid bone;
  —— *de pied,* (*hipp.*) coffin bone;
  —— *suscarpien,* (*hipp.*) v. —— *crochu.*
oscillant, a., oscillating.
oscillation, f., oscillation, vibration, fluctuation;
  —— *double,* double or complete oscillation;
  —— *simple,* single oscillation.
oscillatoire, a., oscillatory, oscillating.
osciller, v. n., to oscillate, vibrate, fluctuate.
osculateur, a., osculatory.
osculation, f., osculation.
oseraie, f., osier bed.
osier, m., osier, wicker; withe, withy.
osmose, f., osmose;
  —— *électrique,* electric osmose.
ossature, f., skeleton, (hence) framework; (*surv.*) skeleton of a survey; network of triangles; (*mil., min.*) skeleton of a system of mines.
osselet, m., (*hipp.*) exostosis (of the knee, fetlock, pastern).
ostéite, f., (*hipp.*) inflammation of bony tissue, ostitis.
ostéomalacie, f., (*hipp.*) softening of the bone caused by lack of lime salts, osteomalacia.
ôtage, m., hostage.
ôter, v. a., to remove, take off, take out;
  —— *l'avant-train,* (*art.*) to unlimber;
  —— *la charge,* (*art.*) to draw a load;
  —— *l'engrenage,* (*mach.*) to throw out of gear.
otite, f., (*hipp.*) inflammation of the mucous membrane of the ear, otitis.
ouate, f., wadding, packing; cotton wool.
ouater, v. a., to wad; to pack.
oubliette, f., dungeon.
oucher, v. a., (*sm. a.*) to mark a barrel with a file or graver; to mark the quantity of metal to be taken off.
ouest, m., west; the west; the west wind;
  —— *du compas,* west by compass.
ouillage, m., addition of the same liquid to a cask, etc. (to make up for loss by evaporation, etc.).
ouiller, v. a., to add the same liquid to a cask, etc. (to make up for loss by evaporation).
ouragan, m., hurricane.
ourdage, m., strong wooden frame, used to support piles when making a wharf.
ourdir, v. a., to warp; to plat straw; (*cord.*) to warp yarn; (*mas.*) to roughcast (a wall); (*fig.*) to plot, contrive;
  —— *un complot,* to hatch a plot.
ourdissage, m., (*cord, etc.*) warping.
ourdissoir, m., warper.
ourler, v. a., to hem.
ourlet, m., hem; (of metals) seam, lap joint.
ours, m., bear; (*mil. slang*) guardhouse, prison.
ourse, f., she bear;
  la grande ——, the Great Bear, the Dipper;
  la petite ——, the Little Dipper.
oursin, m., (*unif.*) bearskin (headdress).
ousseau, m., **ousset,** m., well-room (of a boat).
outil, m., tool, implement, instrument;
  —— *d'ajustage,* adjustable tool (as the cutters, etc., of machine tools);
  —— *de bourrage,* (*min.*) tamping tool;
  —— *de campement,* (*mil.*) camping, bivouac, tools;
  —— *de carrier,* quarry tool; (*mil. min.*) mining tool (in general);
  —— *à centrer,* center punch;
  —— *s de chauffe,* (*steam*) fire tools;
  coffret d'—s, tool chest;
  —— *de côte,* side tool;
  —— *cylindrique,* in machine tools, any tool that is set in a tool holder;
  —— *à dégrossir,* rough-shaping tool, trimming tool;
  —— *de démontage,* dismounting implement;

outil(s) *de destruction,* (*mil.*) demolition tools, cutting tools;
  —— *double,* tool with two cutting edges, double-edged tool;
  —— *à élargir,* widening bit or tool (as in a boring);
  —— *s de ferrage,* farriers' tools;
  —— *à forer,* boring tool;
  —— *de forme,* tool holder and cutter in one;
  —— *à grain d'orge,* round-nosed tool;
  jeu d'—s, set of tools;
  —— -*machine, machine-* ——, machine tool;
  —— *s de machine,* engine tools;
  —— *s de matage,* calking tools;
  —— *de mine,* mining tools;
  nécessaire à ——, tool chest;
  —— *s de parc,* (*mil.*) transported tools;
  —— *à percer,* boring tool;
  —— *s à percer et à tirer,* (*min.*) blasting implements;
  —— *s de pétardement,* (*mil.*) demolition outfit;
  —— *s à pionniers,* (*mil.*) intrenching tools, pioneers' tools.
  —— *à planer,* planing tool or cutter;
  —— *portatif,* (*mil.*) tool carried by soldier (intrenching tool, etc.); portative tool, hand tool;
  porte- ——, tool holder, tool post;
  —— *simple,* tool with one cutting edge, single-edge tool;
  —— *de sondage,* —— *à sonder,* boring tool;
  —— *de terrassier,* (*mil.*) intrenching tool;
  —— *de tour,* —— *à tourner,* lathe tool;
  —— *tranchant,* edge tool, cutting tool;
  —— *s de tranchée,* (*mil.*) intrenching tools;
  —— *s de transport,* (*mil.*) tools carried in wagons;
  voiture d'—s, (*mil.*) tool wagon.
outillage, m., tools, tool outfit; plant; (*mil.*) tool equipment; (*art.*) general term for the buckets, picks, etc., transported on the gun carriage;
  —— *d'un atelier,* plant of a (machine, etc.) shop;
  —— *de combat,* (*mil.*) generic expression for intrenching, etc., tools;
  —— *à distribution,* (*Fr. a.*) outfit of tools used in making issues of stores and supplies;
  —— *de guerre,* (*mil.*) all tools, etc., necessary in war;
  —— *de l'infanterie,* (*mil.*) infantry tool equipment;
  —— *de la machine,* engine tools;
  —— *de pionnier,* (*mil.*) pioneer outfit;
  —— *de pionnier portatif,* (*mil.*) pioneer's hand tools;
  —— *d'un port,* machinery, etc., for the service of a port;
  —— *de tir réduit,* (*t. p.*) outfit of tools, etc., for gallery practice.
outillé, p. p., equipped with tools, shops, magazines, construction and repair shops, etc.
outiller, v. a., to furnish, supply, or stock with tools; to fit out (a shop, factory, etc.).
outilleur, m., tool maker, tool seller.
outrance, f.,
  à ——, to the uttermost, to the limit;
  combat à ——, (*mil.*) combat to the death (no quarter and no retreat on either side);
  éprouvé à ——, test to limit of resistance.
outre, prep., adv., beyond; out of, above; in addition to;
  d'—— *en* ——, through and through;
  d'————-*Manche,* English;
  d'————-*mer,* beyond seas, from beyond the seas;
  d'————-*mont,* on the other (Italian) side of the Alps, ultramontane;
  passer ——, to proceed; to go on, go ahead; to disregard, ignore, a protest or claim.
outre, f., leathern bag or skin.
outré, p. p., strained; spent; jaded; extravagant; unreasonable.
outre-passer, v. a., to go, extend, beyond; to exceed (an order, a limit).

outrer, v. a., to overload, overwork, strain, exaggerate, overwhelm; (*man.*) to override (a horse).

ouvert, a., open; free (of a port); (of a roadstead) open, unsheltered; (of a river) open; (of a circuit) open; (*fort.*) unfortified, without a parapet; open (i. e., without defenses along the gorge); (*r. r.*) clear; (*nav.*) undecked; (*top.*) open (of the terrain);
— à l'——, open, abreast of (a town, etc.);
— *cheval bien* ——, (*hipp.*) horse with broad chest and straight legs;
— à *force* ——*e*, by open force;
— — à *la gorge*, (*fort.*) open-gorged;
— *guerre* ——*e*, open (i. e., unconcealed) war;
— à *livre* ——, at sight;
— *ouvrage* ——, *ouvrage* —— à *la gorge*, (*fort.*) open work;
— *pays* ——, country without fortifications;
— *ville* ——*e*, open, unfortified, unwalled town.

ouverture, f., opening, act of opening, orifice, hole, aperture; crack, chink, rip, vent; width, gap; spout, mouth, unlocking, overture, means, expedient; opening (of Congress, of Parliament); (*pont.*) width of a bay; waterway; (*elec.*) opening of a circuit; (*steam*) port; (*met.*) mouth, charging door, of a furnace; (*r. r.*) opening of a line; (*cons.*) generic term for doors, windows, or any other opening; width of a door or window; span of an arch; (*opt.*) aperture of a lens; (*mil., nav.*) beginning of a campaign, of a war, etc.; (*fort.*) mouth of an embrasure; (*top.*) opening or valley between two hills;
— à *l'*——, bayed;
— — à *l'admission*, (*steam*) cylinder port;
— — *d'un arc*, span of an arch;
— — *d'arçon*, (*harn.*) stuffing hole of a saddle;
— — *d'une baie*, mouth (entrance) of a bay;
— — *de la campagne*, (*mil.*) opening, beginning, of a campaign;
— — *de chargement*, —— à *charger*, (*met.*) charging door;
— — *de la chauffe*, stokehole;
— — *des colonnes*, (*mil., nav.*) interval between columns;
— — *d'un compas*, opening, spread, of a compass;
— — *de la coquille du piston* (*mach.*) crosshead hole;
— — *de coulée*, (*met.*) tap hole;
— — *de décharge*, (*steam*) evacuation port;
— — *extérieure*, (*fort.*) outer mouth of an embrasure;
— —. *du feu*, (*mil., art., siege*) opening or beginning of fire;
— — *d'un fourneau*, (*met.*) charging hole of a furnace;
— — *des hostilités*, (*mil.*) opening of hostilities;
— — *intérieure*, (*fort.*) inner mouth of an embrasure;
— — *d'intervalles*, (*mil.*) opening of intervals;
— —*s de paix*, overtures of peace;
— — *de la pièce*, (*art.*) opening of the breech (act of);
— — *plate*, (*cons.*) opening or window in a cupola;
— — *de pointage*, (*art.*) sighting hole in a turret top;
— — *d'une porte*, etc., (opening) width of a door, etc.;
— — *des portes*, opening of the gates of a fort or fortified town;
— — *d'un puits*, mouth of a well;
— — à *la scorie*, (*met.*) slag hole;
— — *de sortie*, (*steam*) exhaust port;
— — *de la tranchée*, (*siege*) opening of the trenches, beginning of siege operations, establishment of the first parallel;
— — *de travée*, width of a span;
— — *utile de l'objectif*, (*opt.*) working aperture of the objective;
— — *de la vapeur*, (*mach.*) steam port;
— — *de vidange*, (*steam*) mud hole;
— — *d'une voûte*, v. —— *d'un arc*.

ouvrable, a., *heure* ——, working hour;
— *jour* ——, working day.

ouvrage, m., work, piece of work; workmanship labor, performance; (*met.*) hearth of a blast furnace; interior of a blast furnace; (*fort.*) work; (*min.*) method of working.
(Except where otherwise indicated, the following terms relate to fortification.)
— à *l'*——, at work;
— — *annexe*, subsidiary work in the first line;
— — *d'arrêt*, generic term for any work or *place* destined to arrest enemy on the frontier itself;
— —*s d'art*, (*r. r.*) constructions, generic term for bridges, culverts, viaducts, etc.;
— — *d'attaque*, offensive work;
— — *avancé*, advanced work; work outside of the covered way, but flanked from the *corps de place, enceinte;*
— — *blanc*, (*tech.*) tool making;
— *bois d'*——, v. s. v. *bois;*
— — *de campagne*, fieldwork;
— — *casemate*, casemated work;
— — *de champ de bataille*, generic term for any defensive work on field of battle;
— — *collatéral*, collateral work;
— — *par compartiments*, (*min.*) panel working;
— — à *cornes*, hornwork;
— — à *cornes couronné*, a crown hornwork;
— — à *couronne*, —— *couronné*, crownwork;
— — *crénelé*, loopholed, crenelated work;
— — *de défense*, —— *défensif*, defensive work;
— — *dégradé*, work damaged by hostile fire;
— — *demi-revêtu*, half-revetted work;
— — *détaché*, detached work;
— — *donné* à *l'entreprise*, (*adm.*) contract work;
— — *écharpé*, work reached by oblique fire;
— — *enfilé*, enfiladed work;
— — *épaulé*, any work protected by epaulments;
— — *d'escadron*, work holding one squadron (dismounted and using carbines);
— — *éventuel*, (*in gen.*) job; job work;
— — *extérieur*, outwork, work outside the enceinte ditch;
— — *fermé*, closed work; work with closed (fortified) gorge;
— — à *flancs*, work with flanks;
— — *flanquant*, flanking work;
— — *flanqué*, flanked work;
— — *de fonte*, (*fond.*) casting;
— — *en fonte* à *noyau*, (*fond.*) cored work;
— — *de fortification*, fortification;
— — *en gradins*, (*min.*) stoping;
— — *en, par, gradins droits*, (*min.*) underhand stoping;
— — *en, par, gradins renversés*, (*min.*) overhand stoping;
— — *par grandes tailles*, (*min.*) long work;
— *gros* ——*s*, (*cons.*) a generic term for foundation walls, façades, party walls, arches, etc.; (*in gen.*) heavy work;
— — *d'interdiction*, stop fort, barrier fort.
— — *intérieur*, work within the *corps de place*, or main enceinte; interior work; keep;
— — *intermédiaire*, intermediate work; a work in the first line (to support the main permanent works of same, occupied when needed); a work between the line of detached works and the *noyau central;*
— — *isolé*, isolated work;
— — *par massifs longs*, v. —— *par piliers et galeries;*
— *menus* ——*s*, (*cons.*) generic term for chimneys, ceilings, etc.;
— *mettre* à *l'*——, (*in gen.*) to put to work, set to work;
— *se mettre* à *l'*——, (*in gen.*) to go to work, set to work;
— — *mi-fermé*, half-closed work;
— — *noir*, (*tech.*) blacksmith work, black work;
— — *non revêtu*, unrevetted work;
— — *ouvert*, open work;
— — *passager*, temporary work;
— — *permanent*, permanent work;
— — *par piliers et galeries*, (*min.*) working by post and stall, pillar and breast;
— — *par piliers et compartiments*, (*min.*) panelwork;

ouvrage *de position*, a generic term for *têtes de pont*, works in defiles, and in any important and quasi-permanent position near important lines or positions;
—— *de première ligne*, outer work of a fortified position;
—— *principal*, main work;
—— *de protection*, in mountainous countries, a work constructed on a point which, if held by the enemy, might threaten a barrier fort;
——*s publics*, (*in gen.*) public works, public buildings;
—— *réticulé*, (*mas.*) net masonry;
—— *retiré*, retired work;
—— *revêtu*, revetted work;
—— *à seaux*, (*tech.*) scoop wheel;
—— *secondaire*, secondary, auxiliary, work;
—— *de seconde ligne*, intermediate work (between the line of detached works and the *noyau central*);
—— *simple*, a work of simple trace, covering but little ground by its fire;
—— *à la tâche*, (*in gen.*) piecework;
—— *de*, *en*, *terre*, earthwork;
——*s de sujétion*, (*cons.*) a generic term for concealed, difficult, or special work;
—— *en travers*, (*min.*) cross system of mining;
—— *-type de compagnie*, type of work to hold one company;
—— *vu de revers*, work commanded in rear.

ouvragé, a., wrought.

ouvrager, v. a., to work, to adorn;
—— *la mine*, (*met.*) to scum the ore in a furnace.

ouvrant, a., opening;
à jour ——, at daybreak;
à porte(s) ——*e(s)*, at the opening of the gates of a town.

ouvré, p. p., wrought, shaped;
fer ——, iron wrought to shape;
matières ——*es*, made articles.

ouvreau, m., (*met.*) working hole; air vent.

ouvrer, v. a., to work.

ouvrier, m., workman, laborer, artisan, operative, mechanic, journeyman artificer; (*Fr. a.*) regimental workman or artisan;
—— *d'administration*, (*Fr. a.*) special workman (as baker, miller, tinner, saddler, etc.; v. s. v. *commis*);
—— *armurier*, (*mil.*) armorer;
—— *d'art*, skilled workman;
—— *d'artillerie*, (*art.*) artificer;
—— *de batterie*, (*art.*) battery artificer;
—— *en bois*, wood worker;
—— *chauffeur*, stoker, fireman;
chef- ——, v. s. v. *chef*;
——*s des chemins de fer*, (*Fr. a.*) railway operatives (9 sections, called into service on mobilization only);
—— *civil*, (*Fr. a.*) workman hired in civil life;
——*s des corps*, (*Fr. a.*) soldiers having a trade and plying it in the regiment;
——*s d'état*, (*Fr. a.*) in the artillery and engineers, n. c. o.'s ranking as *adjudant* and acting as foremen or chief workmen in arsenals and in the artillery and engineer schools;

ouvrier *à façon*, outworker;
—— *en fer*, worker in iron;
——*s de l'habillement et du campement*, tinners, saddlers, etc. (belong to the ——*s d'administration*);
—— *manœuvre*, journeyman;
—— *de port*, 'longshoreman; dock laborer;
premier ——, (*Fr. a.*) foreman of a regimental workshop;
——*s du service des substances*, (*Fr. a.*) bakers, millers, etc. (belong to the ——*s d'administration*);
—— *vétéran*, v. —— *d'état*.

ouvrier, a., operative, mechanical, working;
cheville ——*e*, any large pin connecting parts of a carriage so as to permit free play; (*art.*) pintle;
jour ——, working day.

ouvrir, v. a., to open, make an opening, break open, rip, push, or pull open, unclose, unlock, unstop, unpeg, unfold; to begin, open (as negotiations), to open (a railway, a shop, school, etc.); to begin digging (a canal, etc.); (*élec.*) to open (a circuit); (*mach.*) to open out (an engine, a cylinder, etc., as for inspection);
—— *un chemin de fer*, (*r. r.*) to open a railway line;
s'—— *un chemin*, (*mil.*, etc.) to cut one's way through;
—— *un circuit*, (*élec.*) to open a circuit;
s'—— *en conversant*, (*mil.*) to lose touch toward the pivot in turning;
—— *l'ennemi*, (*mil.*) to break through the enemy;
—— *le feu*, (*mil.*) to open fire;
—— *la ligne de bataille*, (*nav.*) to open out (the fleet);
—— *un passage*, (*mil.*) to break or pierce through;
—— *un port*, to throw open a port to trade; (*nav.*) to open (draw near to) a port;
—— *ses portes*, (of a town) to surrender;
—— *les rangs*, (*mil.*) to open ranks;
—— *la tranchée*, (*fort.*) to open the trenches, begin the first parallel;
—— *un tunnel*, to drive a tunnel.

ouvroir, m., workshop.

ovale, a., m., oval.

ovalisation, f., (*art.*) enlargement, ovalization of the bore (especially of bronze guns); (*steam*) collapse of a boiler.

ovaliser, v. r., to lose the circular cross section; to become oval; (*steam*) to collapse (of a boiler).

ovation, f., ovation.

ove, m., ovolo, quarter round.

oxomite, f., (*expl.*) oxomite.

oxydable, a., (*chem.*) oxidizable.

oxydation, f., (*chem.*) oxidation.

oxyde, m., (*chem.*) oxide;
—— *salin*, (*met.*) magnetite, magnetic iron ore.

oxyder, v. a., (*chem.*) to oxidize.

oxygène, m., (*chem.*) oxygen;
gaz ——, oxygen gas.

ozocérite, ozokérite, f., ozocerite.

P.

pa, m., heavy drum stroke made with the left hand.

pacage, m., pasture, pasture land.

pacager, v. a. n., to pasture, to graze.

pacification, f., pacification.

pacifier, v. a., to pacify.

packfong, m., (*met.*) German silver (written also *packfung, packfond*).

pacotille, f., goods shipped by a passenger to sell at a venture, (hence) venture; package or parcel of freight; lots or parcels composing the cargo; supplies or stores of poor quality;

pacotille, *de* ——, cheap, poor, made to sell, worthless, shoddy;
marchandises de ——, v. s. v. *marchandise*.

pacotiller, v. a., to trade at a venture, in shoddy goods.

pacotilleur, m., person who trades at a venture, who deals in shoddy goods.

pacquet, m., (*met.*) tempering water, cementation water.

pacte, m., pact, convention, agreement.

pactiser, v. n., to make agreements, conventions,

pagaie, f., paddle;
aller à la ——, to paddle.
pagaie, *en* ——, adv., helter-skelter, at random.
pagayer, v. a., to paddle.
pagayeur, m., paddler.
page, m., page; f., page (of a book);
—— *de garde*, fly leaf (of a book).
pagnote, m., coward, dastard; a., cowardly, dastardly;
mont ——, v. s. v. *mont.*
pagnoterie, f., cowardice.
paiement, m., payment.
paillage, m., (action of) covering, wrapping, with straw.
paillasse, f., straw mattress, bed, bedsack; all masonry raised upon a floor; upper part of a furnace.
paillasson, m., mat; straw mat, door mat.
paille, f., straw, bit of straw; chaff; (*met.*) flaw in metal; crack; scale, blister; (*in pl.*) forge scale, hammer slag;
aller à la ——, (*mil.*) to draw straw;
avoir la —— *au cul*, to be for sale (of horses, and also figuratively); (*mil. slang*) to be discharged;
—— *de blite*, windlass norman;
—— *cariée*, pale straw, of little or no nutritious quality;
un cent, etc., de ——, one hundred, etc., bundles of straw;
—— *de couchage*, (*mil.*) bed straw;
couleur ——, (*met.*) straw color (in tempering);
—— *d'emballage*, packing straw;
s'en aller —— *au cul*, (*mil. slang*) to leave the regiment with a term of imprisonment or confinement unserved;
envoyer les soldats à la ——, (*mil.*) to give the men a rest during drill;
—— *de fer*, (*met.*) hammer scale; (*mil. slang*) the bayonet, cold steel;
—— *fourrageuse*, forage straw;
—— *hachée*, chopped straw;
homme de ——, nonentity; straw man (e. g., a straw bidder);
menue ——, chaff;
—— *rouillée*, mildewed straw;
—— *terrée*, straw beaten flat by rain, that has been soaking wet;
tirer à la courte ——, to draw straws or cuts.
pailler, m., farmyard; heap of straw; loft; straw shed.
pailler, v. a., to cover, wrap with straw.
paillet, m., mat, straw mat; (*cord.*) chafing mat;
—— *de brasseyage*, paunch mat;
—— *de collision*, fender mat;
—— *de portage*, chafing mat;
—— *uni*, sword mat.
paillette, f., spangle; small metal plate or facing; reenforce of leather for a shoe; (*met.*) scale, forge scale.
pailleux, a., made of straw, containing straw; (*met.*) flawy, unsound.
paillon, m., straw; foil; link; solder; spangle; sheet tin, drop or small lump of tin; silvered copper or brass;
—— *de soudure*, leaf solder.
paillonner, v. a., to tin, cover with tin.
paillot(t)e, f., straw hut.
pain, m., bread; loaf of bread, cake; cake (as of asphalt, etc.); pig, lump (of metal);
—— *d'acier*, (*met.*) lump of steel;
au —— *et à l'eau*, on bread and water;
—— *bis*, brown bread;
—— *biscuité*, (*Fr. a.*) soldier's bread cooked longer than usual (1 hour and 10 minutes instead of 50 minutes);
—— *à blanchir*, cake of whiting;
—— *à cacheter*, wafer;
—— *de commission*, (*mil.*) ration bread;
—— *de cire*, wafer; wax patch;
—— *de craie*, lump of chalk;
—— *de crasse*, (*met.*) pig of waste metal;
—— *de cuivre*, (*met.*) copper brick;
—— *de fournisseur*, v. —— *de munition*;
—— *frais*, soft bread;
—— *de guerre*, (*mil.*) field-service bread;
pain *de liquation*, (*met.*) an alloy of copper and of lead (1:3);
—— *de mer*, ship biscuit;
—— *de munition*, (*mil.*) ration bread;
—— *de plomb*, (*met.*) pig of lead;
—— *rassis*, stale bread;
soute au ——, bread room;
—— *tendre*, soft bread;
—— *de vieux oing*, cake of grease.
pain-vin, m., ray grass;
avoir un ——, (*hipp.*) to have the excretory canal of the maxillary gland inflamed by the lodgment of bits of forage.
pair, a., even, equal, similar;
non ——, odd, uneven.
pair, m., peer.
paire, f., pair; brace; couple;
à la ——, in pairs;
—— *de limons*, pair of shafts;
par ——, in a pair;
par ——*s*, in pairs;
—— *de solives armée*, (*cons.*) solid-built beam.
paix, f., peace;
en ——, in peace;
—— *fourrée*, —— *plâtrée*, patched-up peace;
pied de ——, peace footing.
pal, m., impaling stake; (*met.*) large clinker bar or rake.
palade, f., distance traveled over by a boat between successive strokes of the oar; stroke of an oar.
palais, m., palace;
—— *de justice*, law court;
terme de ——, law term.
palan, m., (*cord.*) tackle, purchase.
(A *palan* consists of two *moufles*, q. v., equipped with suitable cords or chains.)
affaler un ——, to overhaul a tackle;
—— *d'appareil*, heavy purchase;
—— *de bout*, heel tackle;
brider un ——, to frap a tackle;
—— *à*, —— *de*, *caliorne*, winding tackle; any large two or three fold tackle;
—— *de candelette*, fish tackle;
—— *à*, *de*, *canon*, gun tackle;
—— *de capon*, cat tackle;
—— *à chaîne*, chain tackle;
—— *de charge*, garnet;
——*s conjugués*, combined tackles (e. g., luff on luff, whip on whip);
—— *à cordages*, rope tackle;
—— *de côté*, side tackle;
courants de ——, running parts;
—— *à croc*, hook tackle;
croc de ——, tackle hook;
—— *du davier*, boat tackle;
dépasser un ——, to unreeve a tackle;
—— *de descente*, lowering tackle;
—— *à deux poulies doubles*, twofold purchase;
—— *à deux poulies quadruples*, fourfold purchase;
—— *à deux poulies simples*, gun-tackle purchase;
—— *à deux poulies triples*, threefold purchase;
—— *à deux poulies à violon*, long tackle;
—— *différentiel*, differential tackle;
—— *de dimanche*, luff tackle;
—— *de direction*, training tackle;
—— *double*, double tackle;
—— *d'embarcation*, boat tackle;
—— *d'étai*, stay tackle;
—— *à fouet*, tail tackle, tail jigger;
—— *sur garant*, luff upon luff;
garant de ——, tackle fall or rope;
—— *de qui*, boom tackle;
—— *grand* ——, main tackle;
—— *à hélice*, tackle in which the lifting chain is actuated by an endless screw, itself worked by a hauling chain;
—— *hydraulique*, (*mach.*) hydraulic tackle;
itague et ——, runner and tackle;
—— *à itague*, runner purchase;
—— *mobile*, luff tackle;
—— *à moufles en bois*, wooden block tackle;
—— *à moufles métalliques*, metallic block tackle;
—— *sur palan*, luff upon luff;
passer un ——, to reeve a tackle;

palan, *passer en* —— *franc,* to reeve a tackle (said of manifold blocks when fall first goes through end sheave);
 passer les garants d'un ——, to reeve a tackle;
 —— *de pied (des bigues),* heel tackle;
 —— *de pointage, (nav. art.)* training tackle;
 —— *portatif,* v. *mobile;*
 —— *de poulie,* purchase block;
 poulie double de ——, long tackle block with two separate axes and two sheaves on each block;
 poulie simple de ——, single-tackle block;
 —— *de recul,* v. —— *de retraite;*
 —— *renversé,* any tackle in which the power is greater than the resistance;
 reprendre un ——, to overhaul a tackle;
 —— *de retenue,* preventer tackle, relieving tackle;
 —— *de retraite,* relieving tackle;
 —— *de revers,* whip upon a whip;
 —— *des sabords,* gun tackle, side tackle;
 —— *simple,* luff tackle;
 —— *de traversière,* v. —— *de candelette;*
 —— *Verlinde,* —— *Verlinde à hélice,* v. —— *à hélice;*
 —— *volant,* jigger tackle; watch tackle.

palançon, m., stake or beam used for building mud walls.

palandeau, m., v. *palardeau.*

palanque, f., *(fort.)* loopholed stockade of round timbers, stockaded fort;
 —— *tournante,* tambour stockade.

palanquer, v. a., *(cord.)* to bouse.

palanquin, m., palanquin; *(cord.)* small tackle.

palardeau, m., plug or stopper for boats; hawse plug.

palastre, m., main plate of a lock; plug or stopper for boats.

palâtre, m., *(sm. a.)* part of guard of sword, shaped like a spade.

pale, f., pile; blade of an oar; dam, sluice gate, flood gate; paddle board; bung; stopper, stopple;
 arbre de ——, paddle shaft;
 —— *articulée,* feathering float;
 —— *d'aviron,* blade of an oar;
 —— *fixe* radial float;
 —— *de rame,* v. —— *d'aviron;*
 roue à ——*s,* paddle wheel.

paléage, m., unloading (a ship) with shovels.

palée, f., row of piles, sheet piling;
 —— *basse,* lower row of trestlework (of a bridge);
 —— *basse de pont,* foundation piling;
 chapeau de ——, stringpiece, cap;
 —— *haute de pont,* trestle;
 —— *de pont,* pier of a pile bridge;
 —— *de rive,* shore pier (of a pile bridge);
 —— *supérieure,* upper trestle, trestle on a trestle or piling.

palefrenier, m., groom, hostler.

palefroi, m., *(hipp.)* palfrey, lady's horse.

paleine, f., *(expl.)* straw dynamite.

paleron, m., shoulder blade, bone (of a horse, etc.).

paletot, m., great coat; overcoat.

palette, f., palette; pallet armature; paddle, float board; copper shovel (used in powder mills and magazines); *(tech.)* drill plate; *(t. p.)* marker (for aiming drill); *(harn.)* horn or projection of the hind forks (of some cavalry saddles); *(teleg.)* ticker;
 à ——, undershot;
 —— *de l'arçon, (harn.)* horn or projection of the hind fork (of some cavalry saddles);
 —— *d'aviron,* blade of an oar;
 —— *de drille, (tech.)* drill plate;
 —— *à forer, (tech.)* drill frame;
 —— -*levier, (teleg.)* ticker;
 —— *mobile,* feathering paddle;
 —— *de rame,* v. —— *d'aviron;*
 roue à ——, undershot wheel.

palier, m., landing (of a staircase, of a shaft); horizontal, level, stretch of a road, of a railroad; *(mach.)* plummer, plummer block, pedestal; hanger; journal box, bearing, shaft bearing.
 (Except where otherwise indicated, the following terms relate to machinery.)

palier *d'appui,* thrust bearing;
 —— *d'arbre,* shaft bearing;
 —— *d'arbre de couche,* main bearing, crank-shaft bearing;
 —— *d'arbre coudé,* —— *d'arbre de couche;*
 —— *d'arbre d'excentrique,* eccentric shaft bearing;
 —— *d'arbre à manivelle,* v. —— *d'arbre de couche;*
 —— *d'arbre de parallélogramme,* parallel motion shaft bearing;
 —— *d'arrivée, (cons.)* top landing;
 —— *articulé,* adjustable pillow block;
 —— *de bouton de manivelle,* crank-pin bearing;
 —— *de buttée,* thrust block;
 —— *à cannelures,* multiple collar bearing;
 —— -*chaise,* hanger;
 chapeau de ——, cap;
 —— *de charge,* lateral bearing;
 —— *à chevalet,* pedestal bearing;
 —— *circulaire, (cons.)* landing of a spiral staircase;
 —— *à collets,* neck bearing;
 —— *à colliers,* collar bearing;
 —— *de communication, (cons.)* communicating landing;
 —— -*console,* —— *à console,* bracket, wall bearing;
 —— -*console à rotule,* adjustable wall bearing;
 coussinet de ——, bearing brass;
 —— *de déchargement, (min.)* pit mouth;
 demi- ——, *(cons.)* intermediate landing;
 —— -*échauffé,* heated bearing;
 —— *à fourchette,* yoke bearing;
 —— *frontal,* front bearing, wall bracket;
 —— *en gaïac,* lignum-vitæ bearing;
 —— *de galerie, (mil. min.)* return landing;
 —— -*graisseur,* plummer block;
 graisseur de ——, oil cup for a bearing;
 —— *à graisseur automatique,* self-lubricating bearing;
 grand ——, main plummer block;
 —— -*guide,* lateral bearing;
 —— *horizontal,* pillow block;
 —— *horizontal articulé,* adjustable pillow block;
 —— *horizontal à trois coussinets,* three-part bearing;
 lumière de ——, oil hole of a bearing;
 marche- ——, v. s. v. *marche;*
 —— *à patin,* broad base bearing;
 —— *sans patin,* narrow base bearing;
 —— *pendant,* hanger;
 —— *de poussée,* thrust block;
 —— *principal,* main bearing, crank-shaft bearing;
 —— *de repos, (cons.)* intermediate landing;
 —— *à rotule,* v. —— *articulé;*
 —— *de support,* pedestal;
 —— *de suspension,* hanger;
 —— *pour tourillons d'appui,* thrust bearing;
 —— *pour tourillons de charge,* lateral bearings;
 —— *à trois coussinets,* v. —— *horizontal à trois coussinets.*

palière, f., top stair.

palification, f., compression, preparation, of soil for foundations by pile work; foundation pile work.

palifier, v. a., to prepare the soil for foundations by piling.

palis, m., pale, paling; fence of stakes; inclosure of palisades; *(fort.)* palisade stake.

palissade, f., hedge; *(fort.)* palisade, stockade; palisade stake;
 —— *caponnière, (fort.)* palisade caponier;
 —— *défensive, (fort.)* palisade, stockade;
 —— *tournante, (fort.)* swing palisade, barrier gate.

palissadement, m., *(fort.)* stockade, stockade work; palisading, palisade work.

palissader, v. a., *(fort.)* to palisade; to pile; to put up a stockade.

palissandre, m., rosewood.

palmage, m., dressing, trimming (as of a mast, etc.).

palme, f., palm; palm tree; blade of an oar; m., hand (measure of length);

palme(s) *académiques,* in France, decoration given by the Minister of Public Instruction for literary or scientific services (consists of a double spray of leaves, silver for an *officer d'Académie,* gold with a rosette for an *officier de l'Instruction publique*);

palmer, v. a., to flatten, beat down; to trim down a yard, a mast.

palmer, m., gauge for metallic plates.

palmier, m., palm (tree and wood).

palombe, f., v. *palonne.*

palonne, f., (*cord.*) in a ropewalk, the rope regulating the motion of the laying-top.

palonneau, m., swingletree.

palonnier, m., swingletree, singletree, whiffletree; horse hitched or harnessed to the singletree;
—— *d'appel,* (*art.*) rear crossbar (of the Lemoine brake);
—— *de frein,* (*art.*) front crossbar (of the Lemoine brake);
—— *mobile,* swing splinter bar or singletree;
lamette de ——, v. s. v. *lamette.*

palot, m., turf cutter's spade.

palper, v. a., to touch, feel with the hand;
—— *l'eau,* to hold water (with the oar).

palplanche, f., sheet pile, sheet piling; pile plank; cofferdam plank.

paltoquet, m., lubber, lout;
en ——, lubberly, loutish.

pâmer, v. r., (*met.*) (of steel) to burn; to lose temper.

pan, m., skirt (of a coat or other garment); face, surface, part, front, side; panel; side of a wall; side or face of a roof; flat part (of a timber, etc.), cant; side of a boltshead; (*sm. a.*) flat surface, cant, on the breech; flat of a sword blade;
—— *de bastion,* (*fort.*) face of a bastion;
—— *de bois,* (*cons.*) partition;
à —— *s de bois,* (*cons.*) timber framed;
—— *de canon,* (*art., sm. a.*) flat (of a gun, of a barrel);
—— *en charpente,* (*cons.*) timber framing;
—— *de comble,* (*cons.*) panel of a roof;
—— *coupé,* (*fort., in gen.*) pan coupé, cant, broken corner;
à —— *s coupés,* cantwise;
—— *creux,* (*sm. a.*) hollow of a bayonet or sword blade;
—— *de cuirasse,* lower border of a cuirass;
—— *extérieur,* (*r. r.*) tread of a wheel;
fronton à —— *s,* (*cons.*) gable under a half hip;
—— *latéral,* (*met.*) side wall of a blast furnace;
—— *long,* (*cons.*) long side of a roof;
—— *de lumière,* (*sm. a.*) flat of a barrel;
—— *de mur,* —— *de muraille,* part, piece, of wall; panel;
—— *de roue,* (*art., etc.*) wheel purchase; the surface on which the purchase is taken;
—— *de la tête d'affût,* (*art.*) head plate;
—— *de voûte,* (*cons.*) sectroid.

panaceau, m., (*artif.*) thin slip of wood or of pasteboard applied to a rocket instead of a stick.

panache, m., (*unif.*) plume;
faire ——, to take a header (from a horse, a bicycle).

panacher, v. a., to plume, decorate with a plume.

panais, m., parsnip.

panard, a., (*hipp.*) turned out (of the feet), outbow footed, cowhocked; m., horse with fore feet turned out;
—— *du boulet,* with pasterns turned out or too open;
cheval ——, horse whose fore feet are turned out;
cheval —— *du derrière,* horse whose hind feet are turned out.

pancarte, f., placard; bill; folder or holder for papers in current use.

panclastite, f., (*expl.*) panclastite.

pandour(e), m., Hungarian soldier; robber, pillager.

paneton, m., sort of basket into which dough is put before baking.

panier, m., basket, hamper; (*sm. a.*) basket hilt;

panier *d'ancrage,* basket anchor, wicker anchor makeshift anchor of basket and stones;
anse de ——, basket handle;
—— *d'armements,* (*art.*) equipment basket;
—— *à charger,* (*met.*) charging basket;
—— *conique,* feed-pipe strainer;
—— *à feu,* (*artif.*) sort of incendiary projectile (obs.);
—— *de guidon,* handle-bar basket (bicycle);
—— *laveur,* (*phot.*) plate washer;
—— *de mine,* —— *de mineur,* (*mil. min.*) miner's basket;
—— *à papier,* wastebasket;
—— *à parapet,* (*fort.*) gabion;
—— *à pierres,* (*art.*) stone basket (service of stone mortars, obs.);
plâtre au ——, (*mas.*) plaster passed through a wicker screen;
—— *de sabre,* (*sm. a.*) basket hilt;
—— *à terre,* miner's basket;
—— *à trous,* perforated shaker (for cleaning tools, etc.).

panifiable, a., susceptible of conversion into bread.

panification, f., conversion of flour into bread.

panifier, v. a., to convert into bread.

panique, a., f., panic;
pris de ——, panic-stricken;
terreur ——, panic.

panne, f., pane (of a hammer); hammer face; face (of a press); (*cons.*) purlin, rib; binding or longitudinal rafter, tiebeam; pantile; (*nav.*) heaving to, bringing to;
—— *de brisis,* (*cons.*) angle purlin, hip purlin (of a mansard roof);
—— *-chevron,* (*cons.*) purlin which supports roofing directly;
—— *courante,* (*nav.*) heaving to under sail;
en ——, (*mil.*) to be exposed to enemy's fire without being able to return it;
être en ——, (*nav.*) to be lying aback;
—— *fendue,* v. —— *à pied de biche.*
laisser en ——, (*mil.*) to leave a body of troops exposed to fire;
—— *de marteau,* pane or face of a hammer;
mettre en ——, (*nav.*) to heave to;
—— *à pied de biche,* hammer claw;
—— *sèche,* (*nav.*) heaving to under bare poles.

panneau, m., panel; pane of glass; snare, trap; (*mas.*) filling; templet (for stone cutting); face of a dressed stone; (*cons.*) space between the stringboard and the sleeper of a staircase; paneling between stringboard and sleeper of a staircase; (*nav.*) hatchway, hatch; hatch cover; deadlight; (*harn.*) pad (of a saddle); sort of pack or flat saddle used in the French artillery (two-horse teams); (*t. p.*) target; skirmish target, silhouette target;
à —— *x,* paneled;
donner dans un ——, to fall into a trap;
—— *d'échiffre,* (*cons.*) paneling of a staircase;
—— *d'éclatement,* (*art.*) bursting panel (sort of target used in the determination of the angle of the cone of dispersion);
—— *d'écoutille,* (*nav.*) hatch;
—— *x flottés,* overlapping panels;
—— *de lit,* (*mas.*) lower bed of a stone (cut);
—— *en planches,* (*art.*) board target;
porte à —— *x,* paneled door;
—— *de sabord* (*nav.*) port lid;
—— *pour le tir,* target;
—— *de selle,* (*harn.*) pad;
—— *de tête,* (*mas.*) head, face, of a stone;
—— *à verre,* —— *à vitre,* window sash;
—— *de verre,* —— *de vitre,* window pane.

panneresse, f., (*mas.*) stretcher (said also of bricks, of sods);
en ——, stretcherwise;
gazon de ——, v. s. v. *gazon.*

panneton, m., key bit; handle of a window fastening; catch.

pannetonné, fitted with a *panneton.*

pannon, m., dogvane.

panoplie, f., panoply.

pansage, m., grooming of a horse, (hence) (*art., cav.*) "stables;"
effets du ——, stable utensils, stable kit

pansage | **paquet**

pansage, *jeu d'effets de* ——, horse kit;
—— *à la main,* rubbing down, hand rubbing; grooming.
panse, f., stomach of a horse; (*harn.*) broadest part of a horse collar.
pansement, m., grooming of a horse; (*med.*) dressing of wounds;
lieu de ——, (*med.*) dressing station;
—— *de la main,* grooming;
paquet individuel de ——, v. s. v. *paquet.*
panser, v. a., to groom a horse; (*med.*) to dress a wound;
—— *de la main,* to groom a horse.
panseur, m., (*Fr. a.*) current name for *infirmier de visite,* q. v.
pantalon, m., trousers, breeches;
—— *de cheval,* riding breeches; (*unif.*) regulation cavalry trowsers;
—— *d'ordonnance,* (*unif.*) regulation trowsers, made of cloth;
—— *de treillis,* (*unif.*) canvas trowsers.
pantélégraphe, m., pantelegraph.
pantéléphone, m., pantelephone.
pantographe, m., pantograph.
pantomètre, m., pantometer.
pantopollite, f., (*expl.*) pantopollite (nitroglycerin and naphthaline).
pantoufle, f., slipper; (*farr.*) panton shoe;
fer à ——, (*farr.*) panton shoe.
panurge, m., (*harn.*) sort of link or loop (to support the reins).
pape, m., the Pope.
paperasse, f., old, worthless paper.
paperasser, v. a., to dabble in papers; to scribble.
paperasserie, f., "red tape."
paperassier, m., scribbler, "paper" man, "red-tape" man.
papier, m., paper; document; bill, paper (in a commercial sense); government securities; (*in pl.*) papers; passport;
—— *albuminé,* albumenized paper;
—— *albuminé brillant,* double albumenized paper;
—— *en bloc,* sketching block, pad;
——s *de bord,* ship's papers;
—— *Bristol,* bristol paper;
—— *brouillard,* blotting paper;
—— *bulle,* scribbling paper, common writing paper;
—— *buvard,* v. —— *brouillard;*
—— *calque,* —— *à calque,* tracing paper;
—— *à calquer,* tracing paper;
—— *-carton,* pasteboard;
—— *à, de, cartouches,* cartridge paper;
—— *à cartouches de(s) fusée(s),* rocket-case paper;
—— *ciré,* oiled (tracing) paper;
—— *Colas,* black print paper (black lines on white ground);
—— *coquille,* letter paper;
—— *couché,* (*phot.*) baryta paper;
—— *court,* a short-term note;
—— *à crayons,* v. —— *Bristol;*
—— *au cyano-fer,* blue-print paper (blue lines on white ground);
—— *à décalquer,* tracing paper; carbon paper;
—— *à dessin,* —— *à dessiner,* drawing paper;
—— *dioptique,* waxed tissue paper;
—— *doré sur tranche,* "gilt-edged" paper (commercial);
——s. *doubles,* (*nav.*) duplicate (hence, probably false) papers;
—— *écolier,* common paper, scribbling paper;
—— *à écrire,* letter paper;
—— *d'emballage,* wrapping paper;
—— *à l'émeri,* —— *émeri,* —— *d'émeri,* —— *émerisé,* emery paper;
——s *et enseignements,* (*nav.*) ship's papers; passport;
—— *enveloppe,* (*sm. a.*) patch, bullet patch;
—— *à envelopper,* envelope, wrapping paper;
—— *d'épreuve,* (*chem.*) test paper; litmus paper;
—— *explosif,* (*esp.*) explosive paper;
—— *au ferro-prussiate,* blue-print paper (white lines on blue ground);
—— *filtre,* —— *à filtrer,* filter paper;
—— *fossile,* asbestos paper;

papier *fulminant,* pyro paper;
—— *à gargousses,* (*art.*) cartridge paper;
—— *gauffré,* goffered paper (ridged);
—— *gélatine,* tracing paper;
—— *au gélatino-bromure,* gelatino-bromide paper;
—— *au gélatino-chlorure,* gelatino-chloride paper;
—— *-glace,* v. —— *-calque;*
—— *-goudron,* —— *goudronné,* tarred paper;
—— *d'indicateur,* (*steam*) indicator card;
—— *ioduré,* iodized paper;
—— *joseph,* filter paper;
—— *à lettres,* letter paper;
—— *libre,* unstamped paper;
—— *long,* a long-term note;
—— *mâché,* papier mâché;
main de ——, quire of paper;
—— *Marion,* blue-print paper (white lines on blue ground);
—— *marqué,* v. —— *timbré;*
—— *-monnaie,* paper money;
—— *de montagne,* asbestos;
—— *mort,* v. —— *libre;*
—— *de musique,* music paper;
—— *négatif,* (*phot.*) negative paper;
—— *nitroglucose,* nitroglucose paper;
—— *d'oiseau,* very thin paper (e. g., for pigeon messages);
—— *parchemin,* —— *parcheminé,* parchment paper, vegetable parchment;
—— *à patrons,* cross-section paper, profile paper;
—— *Pellet,* blue-print paper (white lines on blue ground);
—— *pelure,* tissue paper (carrier-pigeon service);
—— *photographique,* photographic paper;
—— *plombaginé,* black-leaded paper;
—— *à poncer,* tracing paper;
—— *positif pelliculaire,* (*phot.*) transferotype paper;
—— *de poste,* letter paper;
—— *poudre,* (*expl.*) Melland's paper powder;
—— *de quadrillage,* —— *quadrillé,* cross-section paper;
rame de ——, ream of paper;
—— *réactif,* (*chem.*) litmus paper; test paper;
—— *de rebut,* waste paper;
—— *réglé,* ruled paper;
—— *sablé,* sandpaper;
—— *salé,* plain salted paper;
—— *sensible albuminé,* (*phot.*) ready sensitized paper;
—— *de serpente,* tissue paper;
—— *similiplatine,* (*phot.*) similiplatine paper;
—— *de soie,* silver, tissue, paper;
sur le ——, on paper;
—— *de sûreté,* bank-note paper;
—— *témoin,* (*expl.*) standard-tint paper (heat test);
—— *timbré,* stamped paper;
—— *toile,* tracing cloth, linen;
—— *-tournesol,* —— *de tournesol,* (*chem.*) litmus paper;
—— *végétal,* tracing paper;
—— *vélin,* vellum paper;
—— *de verre,* glass paper;
—— *volant,* loose sheet.
papillon, m., butterfly; inset (as of a map); searcher, finder (of a map); insert, extra sheet for indorsements; (*mach.*) butterfly valve, throttle valve, damper, regulator; milling-tool grinder or cutter;
—— *de détente,* (*mach.*) expansion valve;
—— *mobile,* regulator, damper, of a chimney or flue;
—— *registre,* (*mach.*) throttle valve;
soupape à ——, (*mach.*) butterfly valve.
papyrographe, m., papyrograph.
papyrographie, f., papyrography.
paquebot, m., packet, mail steamer;
—— *postal,* —— *poste,* —— *de poste,* mail steamer;
—— *à vapeur,* steam packet;
—— *à voiles,* sailing packet.
paquet, m., packet, package; bundle; (*met.*) pile, fagot;
——s *cachetés,* (*nav.*) sealed orders;

paquet   308   **parc**

paquet *de cartouches*, (*sm. a.*) packet, cartridge packet;
—— *double*, (*sm. a.*) double skelp;
—— *d'eau*, v. *éclusée*;
—— *en* ——, without order; by the run; quick, cheerily;
—— *individuel de pansement*, (*med.*) first-aid packet;
—— *de mer*, green sea;
—— *de mitraille*, v. *grappe de mitraille*;
trempe en ——, (*met.*) casehardening;
tremper en ——, (*met.*) to caseharden.

paquetage, m., pack, packing; (*mil.*) packing (as of equipment); (*sm. a.*) putting into packets (cartridges); (*met.*) piling, fagoting.

paqueter, v. a., to pack, put up in packets.

paraballe, m., (*t. p.*) bullet stop (gallery practice); (*mil.*) defensive armor proposed in Belgium.

parabole, f., parabola.

paraboloïde, m., paraboloid.

paracel, m., sunken reef, coral reef.

parachute, f., parachute; safety stop (of a hoisting apparatus); (*art.*) sort of piston valve (recoil-checking).

paracrotte, m., mud guard.

parade, f., show, pageant; (*mil.*) review, parade, especially parade for guard; (*fenc.*) parry; (*man.*) stopping, halting of a horse (synonym of *arrêt*); place where horses are sold;
à la ——, on parade;
—— *de banderole*, (*fenc.*) parry against a diagonal cut (broadsword);
cheval sûr à la ——, horse easy to stop;
—— *de corps*, (*fenc.*) body parry (broadsword);
défiler la ——, (*mil. slang*) to die;
demi- ——, v. *demi-arrêt*, s. v. *arrêt*;
en ——, (*art.*) (of a gun detachment) facing to the front (from the epaulment, if there is one);
—— *d'exécution*, (*mil.*) parade of troops to witness a military execution;
faire ——, (*nav.*) to dress ship;
faire la ——, (*mil.*) to parade; attend, go to, parade;
—— *de figure à droite* (*gauche*), (*fenc.*) right (left) cheek parry (broadsword);
—— *de flanc*, (*fenc.*) flank parry (broadsword);
—— *de garde*, (*mil.*) guard mounting;
—— *manquée*, (*man.*) refusal of a horse to stop;
—— *de pointe*, (*fenc.*) point-cut parry (broadsword);
—— *de tête*, head parry (broadsword);
—— *de ventre*, v. —— *de banderole*.

parader, v. n., *faire* —— *un cheval*, (*man.*) to put a horse through, make him show off, his paces.

parados, m., (*fort.*) parados;
—— *réduit*, parados that may serve as a keep in case of need.

paradosser, v. a., (*fort.*) to build a parados.

paraffine, f., paraffin.

paraffiner, v. a., (*expl.*, etc.) to paraffin.

paraflanc, m., (*fort.*) flanking traverse, epaulment.

parafoudre, m., (*elec.*) lightning arrester, lightning protector.

parafouille, m., (*cons.*) piling to prevent the washing of foundations.

parage, m., stretch of coast, waters, shores, latitudes, parts; trimming, polishing, of metallic surfaces;
—— *des pilotes*, pilot waters.

paraglace, m., ice fender, ice breaker.

parallaxe, f., parallax.

parallèle, a., parallel; m., parallel (of latitude); parallel rule; f., parallel line; (*fort.*) parallel trench;
avant- ——, v. *avant-parallèle*;
demi- ——, v. *demi-parallèle*;
ouvrir la ——, (*siege*) to begin the construction of the parallel;
règles ——*s*, f., parallel rulers.

parallèle, *système des* ——*s*, (*mil. teleg.*) system of laying military telegraph in which each headquarters is followed by its own line.

parallélépipède, parallélipipède, m., parallelopipedon.

parallélogramme, m., parallelogram; (*mach.*) parallel motion;
—— *articulé*, (*mach.*) parallel motion;
—— *flexible*, v. —— *articulé*;
mouvement du ——, (*mach.*) parallel motion;
tiges de ——, bars, arms, of a parallelogram, (*mach.*) of a parallel motion;
—— *de Watt*, v. —— *articulé*;
—— *de Wheatstone*, (*elec.*) Wheatstone bridge.

paralléliographe, m., parallel rulers.

paramagnétique, a., (*elec.*) paramagnetic.

paramagnétisme, m., (*elec.*) paramagnetism.

paraneige, m., snow fender or guard, snow beam (as on Alpine huts).

parapet, m., parapet; (*fort.*) parapet;
border un ——, v. s. v. *border*;
—— *en clayonnage*, (*fort.*) hurdle parapet;
—— *durci*, (*fort.*) lit., "hardened" parapet, i. e., a parapet faced with beton;
—— *en gabionnade*, (*fort.*) gabion parapet;
—— *de poutres*, timber or log parapet;
—— *en terre*, earthen parapet.

paraplégie, f., (*hipp.*) paralysis of the abdominal members.

parapluie, m., umbrella; hood (umbrella-shaped hood); (*fond.*) splashboard;
—— *de cheminée*, funnel bonnet.

parasoleil, m., lens hood.

parasouffle, m., (*art.*) flash rim.

paratonnerre, m., (*elec.*) lightning rod; lightning arrester;
—— *à air raréfié*, vacuum protector;
—— *à fil préservateur*, fusible lightning arrester;
—— *à lame isolante*, strip fuse, safety strip;
—— *à plaque isolante*, plate lightning arrester;
—— *à plaques*, v. —— *à plaque isolante*;
—— *à pointes*, comb protector; comb lightning arrester;
—— *à stries*, ribbed lightning arrester (the intersections of the ribs on the two plates acting as points);
tige de ——, lightning rod proper.

paravent, m., screen, folding screen;
—— *à feuilles*, folding screen.

parc, m., park, inclosure; (*mil.*) park; emplacement where stores are kept and repairs are made (as during a siege); supplies, material, equipment; train of ammunition, guns, etc., accompanying an army in the field; gun park, wagon park;
—— *aérostatique*, balloon park;
—— *d'aérostiers*, v. —— *aérostatique*;
—— *d'armée*, v. *grand* —— *d'artillerie d'armée* and —— *de bétail*;
—— *d'artillerie de corps d'armée*, (*Fr. a.*) the artillery park of an army corps (four sections), commanded by a lieutenant-colonel;
—— *d'artillerie de siège*, (*siege*) siege (artillery) park;
—— *attelé*, (*Fr. a.*) the wagons of a company of railway troops;
—— *à ballons*, balloon park;
—— *de bétail*, (*Fr. a.*) cattle on the hoof, to last 12 days, subdivided into—
    —— *divisionnaire*, 2 days;
    —— *de corps d'armée*, 4 days;
    —— *d'armée*, 2 days;
    —— *de réserve*, 4 days;
—— *de bois à plate-formes*, (*siege*) timber depot;
—— *de(s) bouches à feu*, gun park;
—— *à boulets*, (*nav.*) shot locker;
—— *à charbon*, coal yard;
chariot de ——, v. s. v. *chariot*;
—— *de corps d'armée*, v. —— *d'artillerie de corps d'armée* and —— *de bétail*;
de ——, after the name of a tool, etc., means of the model adopted for the service;
directeur de ——, (*Fr. a.*) lieutenant-colonel in charge of a park;

parc *divisionnaire*, v. —— *de bétail;*
—— *élémentaire de siège,* (*Fr. a.*) engineer park having enough material for the siege of a fort or of a town of simple enceinte;
entrer au ——, (*art.*) to enter the park;
—— *aux équipages de pont,* bridge-train park;
former le ——, (*art.*) to form park, to go into park;
—— *du génie,* (*siege*) engineer park, containing all the tools and other appliances needed in a siege (attack);
—— *du génie d'armée,* (*Fr. a.*) engineer park of an army;
—— *du génie de corps d'armée,* corps engineer park;
—— *du génie de siège,* engineer siege park;
—— *de gonflement,* (balloons) filling park;
grand ——, a synonym of —— *d'armée;* (*siege*) main park, main artillery park, where most of the material (except ammunition) is kept till wanted (tools, timber, guns, spare carriages, etc.);
grand —— *d'artillerie d'armée,* main park of an army (supplies the corps parks);
—— *à hydrogène comprimé,* (*Fr. a.*) in a balloon park, the wagons and material used in filling the balloons;
—— *de matériel* (*r. r.*) car yard;
—— *aux munitions,* (*siege*) ammunition depot, shell park (includes places for the preparation of ammunition);
—— *ordinaire,* (*Fr. a.*) in a balloon park, a subdivision containing mostly baggage and equipment wagons;
—— *d'outils,* (*siege*) tool depot;
petit ——, (*siege*) repair shops, magazines, near the works, for the service of material in daily use;
—— *de pilotis,* (*cons.*) piling, pile framing;
—— *sur rails,* (*Fr. a.*) the rolling stock of a company of railway troops;
—— *de réserve,* reserve park; reserve telegraph park attached to each section of the second line; four days' supply of beef on the hoof, as a reserve for the army, v. —— *de bétail;*
rompre le ——, (*art.*) to unpark;
section de ——, (*Fr. a.*) subdivision of the corps park, commanded by a captain;
—— *de section,* (*Fr. a.*) telegraph park, attached to each telegraph section of the first line;
—— *de siège,* siege park (artillery, usually);
sortir du ——, (*art.*) to leave the park;
—— *télégraphique,* (*Fr. a.*) telegraph park (accompanies an army in the field);
—— *aux vivres,* supply depot;
—— *de voitures,* wagon park.
parcellaire, a., *cadastre* ——, register of land, cadaster;
plan ——, plan or plot of land subdivisions.
parcelle, f., part, particle; plot or parcel of land.
parchemin, m., parchment;
—— *animal,* parchment;
—— *végétal,* vegetable parchment.
parcourir, v. a., to go, get over, go through; (*fig.*), to survey, overhaul, examine;
—— *à cheval,* to ride over;
—— *les coutures,* to examine the seams of a boat with a chisel;
—— *à pied,* to walk over.
parcours, m., trip, road, way covered; regular trip, regular journey between two places; line (of a public conveyance by sea or land); course (of a stream); track of a hurricane;
—— *libre,* (*r. r.*) pass; (*in gen.*) freedom of way;
—— *d'une manœuvre,* (*cord.*) lead of a rope.
par-dessus, m., greatcoat, overcoat.
paré, p. p., clear (of a rope, an anchor); (*nav.*) cleared for combat; (*fenc.*) parried;
pied ——, (*hipp.*) worn hoof.
pare-balles, m., splinter-proof; bullet shield, for land service (proposed only, and intended to cover, say, front of company); (*t. p.*) marker's shelter;
—— *à rabattement,* (*art.*) folding shield.
pare-boue, m., mud scraper (of a bicycle).

pare-douilles, m., (*r. f. art.*) deflector, cartridge-case deflector.
pare-éclats, m., (*fort.*) splinter-proof; (*art.*) splinter-proof shield.
pare-étincelles, (*r. r.*) spark arrester.
pare-étoupilles, (*art.*) primer deflector.
pareil, m., equal; fellow; match; a., similar, like.
parement, m., curb, curbstone; facing of a building; face (of a wall, of a beam); (*carp.*) outer, visible, face of a construction; (*mas.*) outer face of a stone; (*in pl.*) outer stones of a wall; (*fort.*) face of a cutting; (*unif.*) facings (on the cuff);
—— *d'appui,* (*cons.*) stones supporting a casement;
—— *brut,* (*mas.*) rough, undressed stone in the face of a wall;
—— *de couverture,* (*cons.*) fillet or border of mortar;
—— *droit,* plain face;
—— *s d'un fagot,* outer (large) sticks of a fagot;
faux ——, (*mas.*) facing of a wall;
—— *intérieur,* (*mas.*) back of a wall;
—— *de moellon,* (*mas.*) setting of a stone in position;
—— *de pavé,* uniform arrangement of paving stones;
—— *d'un pavé,* surface of a pavement;
placer, poser, en ——, (*mas.*) to lay stones lengthwise;
—— *postérieur,* v. —— *intérieur;*
—— *de sujétion,* (*mas.*) ornamental facing (stones arranged in regular designs);
—— *de tête,* (*mas.*) the dressing and placing in position of the stones in the head of an isolated wall;
—— *de voûte,* (*cons.*) face of an arch.
parementer, v. a., to make even, smooth (as the face of a wall).
parenchyme, m., pith.
pare-pied, m., fender; fire screen.
parer, m., (*man.*) halt, stop (fore legs raised).
parer, v. a. n., to deck, set off, trim, make neat; to get ready, prepare; to parry, ward off, fend, keep off, defend, guard, screen; (*fenc.*) to parry (*cord.*) to clear, to get clear, to coil; (*tech.*) to flatten, finish (as a rivet head); (*nav.*) to weather, double, clear, go clear of; (*siege*) to trim (gabions, fascines); (*mas.*) to set stones in position;
—— *un abordage,* (*nav.*) to ward off a boarding party, a collision;
—— *une ancre,* to see an anchor all clear for coming to;
—— *une botte,* (*fenc.*) to parry a thrust;
—— *un câble,* to get a cable clear;
—— *un cap,* (*nav.*) to double a cape;
—— *au corps,* (*fenc.*) to remove the body from the line of a thrust, etc.;
—— *un coup,* (*fenc.*) to parry a blow, a cut;
—— *à droite (gauche),* (*fenc.*) to make right (left) parry (bayonet exercise);
—— *sur les hanches,* (*man.*) to bear or rest on the haunches (of a horse in the gallop);
—— *les manœuvres,* (*cord.*) to clear the ropes, the tackle;
—— *le pied d'un cheval,* (*farr.*) to trim down the hoof in order to fit on the shoe; to dress the plantar surface;
—— *un pavillon,* to clear a flag;
—— *de la pointe,* (*fenc.*) to beat off the adversary's blade with the point;
—— *un saucisson,* to trim a fascine;
se ——, to stand by;
—— *en tête,* (*fenc.*) to make the head parry (bayonet exercise);
—— *en tête à droite (gauche),* (*fenc.*) to make right (left) head parry (bayonet exercise).
parère, m., expert advice, counsel (especially in commercial matters).
paresseux, a., idle, lazy; (of the compass) sluggish; said of rotary fireworks that fail to rotate.
pare-torpilles, m., (*nav.*) torpedo net;
filet ——, torpedo net.
parfumer, v. a., to perfume;
—— *des lettres, une maison, un navire,* to fumigate letters, a house, a ship.

parisien, m., lubber, clumsy sailor; "screw," worthless horse; (*mil. slang*) active, cheery, knowing soldier.
parisienne, f., overalls.
parlement, m., Parliament; United States Congress.
parlementage, m., (*mil.*) parleying.
parlementaire, a. m., (*mil.*) bearer of a flag of truce; (*nav.*) cartel ship;
drapeau ——, flag of truce;
vaisseau ——, cartel ship.
parlementer, v. n., to parley; to come to terms.
parler, v. a., to speak, talk;
—— *un navire*, to speak a ship.
parleur, m., transmitter of a telephone; receiver, sounder, of a telegraph; (*mil. teleg.*) transmitter (for testing the line as it is put up).
paroi, f., wall; partition wall (of masonry); face of a wall; shell of a boiler; wall of a tent; (*art.*) wall, side of a gun or of a shell; (*sm. a.*) wall of the barrel; (*min.*) side of a gallery; (*met.*) wall of a furnace; (*hipp.*) crust of the hoof;
—— *de l'âme*, (*art., sm. a.*) surface of the bore;
—— *antérieure*, v. —— *tubulaire*;
—— *de la boîte à feu*, fire plate (of a locomotive);
—— *de la chambre*, (*art., sm. a.*) wall of the chamber;
—— *de chaudière*, (steam) boiler shell;
contre- ——, v. *contre-paroi*;
fausse ——, (*met.*) shell of a furnace, outer casing;
—— *forçante*, (*art.*) driving edge of the rifling;
—— *principale*, (*met.*) inner lining (of a furnace);
—— *de rochers*, (top.) wall of rocks;
—— *tclon*, (*art.*) loading edge of rifling, side or edge opposite to the driving edge;
—— *d'une tranchée*, (*r. r.*) side of a cutting;
—— *tubulaire*, (steam) tube plate (of a boiler).
paroir, m., scraper; (*farr.*) farrier's paring knife; farrier's butteris.
paroisse, f., parish.
parole, f., word; promise; (*mil.*) parole, countersign, word of command;
avoir la ——, to have the floor;
prendre la ——, to take the floor;
sur ——, (*mil.*) on parole (of prisoners).
parotide, f., (*hipp.*) vives, parotid glands;
engorgement, gonflement, des ——, (*hipp.*) vives (i. e., swelling of).
parpaigne, a., *pierre* ——, (*mas.*) through stone, header.
parpaing, m., (*mas.*) through stone, through binder, header; bonder, perpend stone, perpender;
—— *s d'appui*, stones supporting a casement;
—— *d'échiffre*, wall supporting the steps of a staircase;
en ——, laid as a through stone, headerwise;
faire ——, to go through (of a stone).
parquer, v. a., to pen, to put in an inclosure; (*mil.*) to park; (as a neuter verb) to be parked.
parquet, m., inclosure; inlaid floor, flooring; (*adm.*) place where officers of the public ministry hold their meetings (hence) such officers themselves; bench (of a law court), (also) bar; (*Fr. mil. law*) the permanent (as opposed to the detailed) part of a court-martial or court of appeal, consisting of the *commissaire du gouvernement*, q. v., etc.;
—— *à l'anglaise*, (*cons.*) floor of which the strips are all parallel;
—— *à bâtons rompus*, (*cons.*) herringbone floor (square headings);
—— *à boulets*, (*nav.*) shot locker;
—— *à la capucine*, (*cons.*) a variety of herringbone floor (heading joints in same vertical plane, and at 45°, more or less, with edge of boards);
—— *de chambre de chauffe*, (steam) engine platform;
—— *de chargement*, (*art.*) loading platform;
—— *de chauffe*, (steam) stoke-hold platform;

parquet *à compartiments*, (*cons.*) paneled floor;
—— *en feuilles*, (*cons.*) inlaid floor;
—— *de justice*, bar of a court;
—— *de manœuvre*, (steam) engine-room floor;
—— *militaire*, v. *supra*, under main word;
—— *à point de Hongrie*, (*cons.*) miter-laid floor.
parquetage, m., (*cons.*) parquetry.
parrain, m., sponsor; person who christens a vessel; sponsor of a candidate for admission to an order; (*Fr. a.*) person selected, by a soldier about to be shot, to bandage his eyes.
parseinte, f., piece of parceling to cover a seam.
part, f., part; share; hand; interest, concern; side;
—— *des accidents du laminoir*, (*mach.*) breaking pieces or spindles;
de —— *en* ——, through and through, from front to back;
donner ——, to communicate officially (diplomatic term);
faire ——, to inform;
—— *d'un navire*, share;
—— *de prise*, (*mil., nav.*) prize money.
partage, m., partition, division; distribution; share;, portion, lot; summit level;
bief de ——, v. s. v. *bief*;
canal à point de ——, summit canal;
ligne de ——, watershed;
point de ——, summit level; watershed;
à point de ——, with a dividing ridge.
partager, v. a., to share, to have a share in; to divide;
—— *les rênes*, (*man.*) to take a rein in each hand;
se ——, to divide, fork (as a road);
—— *le soleil*, in a duel, to divide the sun equally;
—— *le vent*, (*nav.*) of a vessel unable to get the weather gauge, to maneuver so as to keep the enemy from getting it.
partance, f., (*nav.*) departure;
coup (de canon) de ——, signal gun for departure;
en ——, about to start;
pavillon de ——, blue peter.
partement, m., (*artif.*) small rocket; (*nav.*) difference between the last observed and the actual longitude of a ship.
parti, m., party, side, cause; defense, expedient, means; condition, terms; (*mil.*) party, detachment;
aller en ——, (*mil.*) to go out to harass the enemy;
—— *bleu*, (*mil.*) marauding party (obs.);
envoyer en ——, (*mil.*) to send on detached duty;
prendre ——, (*mil.*) to enlist;
—— *réglé*, (*mil.*) party sent out on legitimate duty (as distinguished from —— *bleu*; obs.).
particularités, f., pl., (*hipp.*) markings of a horse.
particulier, a., particular, special, private; m., private individual; (*mil. slang*) civilian.
partie, f., part, portion; game, match; (*law*) party (to a lawsuit), client; (*man.*) start; (*mus.*) part;
—— *adverse*, (*law*) the other side;
—— *s belligérantes*, (*mil.*) belligerents;
—— *civile*, (*law*) plaintiff;
—— *comparante*, (*law*) party appearing in court;
construire un pont par —— *s*, (*pont.*) to build a bridge by parts;
—— *défaillante*, (*law*) party failing to appear in court;
à, en, —— *double*, by double entry (bookkeeping);
—— *exécutive*, (*adm.*) person or officer that makes a payment, an issue of stores;
faire une —— *de la main*, (*man.*) to slacken the hand for a gallop;
—— *filetée*, screw threads;
—— *forçante* (*art.*) driving edge;
—— *frottante* (*mach.*) rubbing, bearing, part;
—— *d'honneur*, (in games) rubber;

partie *morte, (fort.)* dead space;
—— *nulle,* drawn game; (in races) dead heat;
—— *occupante, (adm.)* tenant;
—— *plaignante, (law)* plaintiff;
—— *(s) prenante(s), (adm.)* person(s) who receive(s) issues (clothing, etc.); public creditors;
prendre à ——, *(law)* to sue;
procéder par petites ——*s,* to proceed, work, advance, etc., by degrees;
—— *publique,* legal designation of the public ministry, as representing the public; solicitor-general;
—— *secrète, (mil.)* the spy service of a staff;
à, en, —— *simple,* by single entry (bookkeeping).

partir, m., *(man.)* start (of a horse).

partir, v. n., to go, leave, set out, proceed, start, spring, to sail (of a ship); to go off (of a shot, a firearm); to be carried away (of a spar, etc.);
—— *bien de la main, (man.)* to take the gallop as soon as the hand slackens;
faire —— *un coup,* to fire a shot;
—— *juste, (man.)* to take the gallop with the leading foot;
—— *de la main, (man.)* to take the gallop easily;
—— *du pied droit (gauche),* to step off with the right (left) foot;
—— *de pied ferme, (man.)* to pass from the walk to the gallop.

partisan, m., partisan; *(mil.)* partisan, guerrilla; (in pl.) raiding party;
guerre de ——*s,* v. s. v. *guerre.*

parure, f., adornment; *(hipp.)* parings of a horse's foot;
chevaux de même ——, well-matched horses, of same size.

parvenu, m., *(mil.)* ranker.

pas, m., pace, step, footstep; stride, gait, progress; precedence; threshold of a door, step before an entrance; strait; obstacle, bad place, in a road, etc.; *(top.)* narrow, difficult, pass in a valley or mountain; *(carp.)* notch (as for the foot of a rafter); *(cons.)* going of a stair; *(cord.)* turn of a rope; *(tech.)* pitch (of a screw, of a gearing); thread of a screw, turn of a screw; *(art., sm. a.)* pitch (of rifling); *(mus.)* march.

I. Military, manège; II. Technical, artillery; III. Miscellaneous.

I. Military, manège

—— *accéléré, (mil.)* quick step, quick time (0.75^m, 128 to the minute); the call for quick time;
aller au ——, to go at a walk;
allongement du ——, *(mil.)* lengthening of the step; *(man.)* walking out;
—— *allongé, (mil.)* stepping out;
allonger le ——, *(mil.)* to step out, lengthen the step; *(man.)* to walk out;
—— *d'armes, (mil.)* post that one undertakes to defend (obs.);
—— *en arrière, (mil.)* back step;
au ——, *(mil.)* at a walk; "step!" (to troops marching in line, when they lose the step); call for walk;
—— *averti, (man.)* regular, well-marked stepping of a horse (sort of gait between a walk and slow trot);
—— *cadencé, (mil.)* cadenced step;
changer de ——, *(mil.)* to change the gait;
changer le ——, *(mil.)* to change step;
—— *de charge, (mil.)* charging pace (0.75^m, 140 to the minute; cadenced, as a rule; may be as fast as possible, according to circumstances);
cheval de ——, pacer;
—— *complet, (man.)* completed step (i. e., all the motions made between two successive and identical positions of a given member);
à —— *comptés, (man.)* gradually;
—— *de conscrit, (man.)* high step with fore legs;
contre- ——, *(mil.)* short step or half step (to recover the step);
—— *de course,* running step, run, v. —— *gymnastique;*

pas *de côté, (mil.)* side step; *(man.)* side step;
—— *d'école, (mil.)* short step, balance step; *(man.)* well-marked walk (sort of gait between a walk and a slow trot);
—— *écouté, (man.)* regular, uniform step;
—— *espagnol, (man.)* an air in which the horse, at a walk, gives all the extension possible to each of his fore legs;
étalonner le ——, *(mil.)* to find the average length of one's step or pace;
faire aller au ——, to walk a horse;
faire les cent ——, *(mil.)* to do sentry duty;
faux ——, *(man.)* stumble;
—— *gymnastique, (mil.)* double time (0.80^m, 170 to the minute); call for same; double step;
hâter le ——, to hasten, to hurry, the march;
longueur du ——, *(mil.)* length of the step;
marcher au ——, to walk, pace, move at a walk;
marquer le ——, *(mil.)* to mark time;
mettre au ——, to walk a horse;
se mettre au ——, *(mil.)* to take step;
—— *ordinaire, (mil.)* marching step, ordinary step or pace;
ouvrir le ——, *(mil.)* to begin marching;
passer d'un —— *à un autre, (mil.)* to change the gait;
raccourcir le ——, *(mil.)* to shorten step;
—— *redoublé, (mil.)* double-quick;
—— *règlementaire, (mil.)* regulation step;
—— *relevé, (man.)* high step, high action;
remettre un cheval au ——, *(man.)* to come down to a walk;
—— *rétrograde,* step backward;
rompre le ——, *(mil.)* to break, or lose, or get out of, step, as in crossing a ponton bridge;
—— *de route, (mil.)* the route step (1 kilometer in 11 to 12 minutes);
—— *saccadé,* irregular pace.

II. Technical, artillery:

à l'anglaise, v. —— *carré;*
au —— *de...., (sm. a., art.)* making one turn in....;
—— *carré,* square thread;
—— *constant, (art., sm. a.)* uniform twist (also of a screw, of a propeller);
—— *de la corde,* mean pitch;
—— *court,* fine thread, low pitch;
—— *des dents, (mach.)* pitch of teeth;
—— *diamétral, (mach.)* diametral pitch;
double ——, double thread (screw);
—— *à droite (gauche),* right- (left-) handed twist;
—— *d'écrou,* thread of a nut;
—— *d'engrenage,* pitch of a gearing;
—— *d'entrée, (art., sm. a.)* initial pitch;
—— *final, (art., sm. a.)* final twist;
—— *du flanc directeur, (art., sm. a.)* twist of the driving edge;
—— *à gaz,* gas thread;
hauteur du —— *d'une vis,* pitch, one complete turn, of a screw;
—— *de l'hélice,* pitch of the propeller;
inclinaison du —— *des rayures, (art., sm. a.)* angle of rifling;
—— *initial, (art., sm. a.)* initial twist;
—— *long,* coarse thread, steep pitch;
longueur du ——, *(art., sm. a.)* distance in which projectile makes one turn;
moyen, v. —— *de la corde;*
—— *multiple,* multiple thread (screw);
—— *progressif, (art., sm. a.)* increasing twist;
—— *raccourci,* short twist;
—— *de(s) rayure(s), (art., sm. a.)* twist of rifling;
—— *d'une roue dentée, (mach.)* pitch of a toothed wheel;
—— *de Sellers,* Sellers's thread;
—— *simple,* single thread (screw);
—— *à la sortie,* final pitch;
—— *uniforme, (art., sm. a.)* uniform twist;
—— *qui va à droite (gauche), (art., sm. a.)* right- (left-) handed twist;
—— *variable, (art.)* varying pitch;
—— *de vis,* furrow of a screw; pitch, thread;
—— *de vis américaine,* United States standard thread;

pas *de vis anglaise,* English standard thread;
— *de Whitworth,* Whitworth thread.

III. Miscellaneous:

— *d'âne,* (*hipp.*) bit to keep horse's mouth open for examination; balling iron; (*harn.*) sharp bit for a horse, sharp bridle bit; (*sm. a.*) basket hilt; (*fort.*) "stepped" ramp for pedestrians and pack animals;
— *en arrière,* (*hipp.*) breeding back;
un — *en arrière* (*avant*), (*fenc.*) in bayonet exercise, advance (retire);
avoir le — *sur,* to have the precedence of;
— *de Calais,* Straits of Dover;
— *de clerc,* blunder, mistake;
— *de coq,* (*htpp.*) springhalt;
double — *en arrière* (*avant*), (*fenc.*) front (rear) pass (bayonet exercise);
. *un* — *à droite* (*gauche*), (*fenc.*) right (left) step (bayonet exercise);
faire un, des, — *en arrière,* to step back;
faux —, stumble, misstep; mistake;
— *géométrique,* geometrical pace of 5 feet;
à grands —, rapidly;
marcher à — *comptés,* to walk slowly and in a dignified manner;
marcher à grands (*petits*) —, (*fenc.*) to advance or retire by long (short) steps;
mauvais —, bad place or obstacle in a road;
mesurage au —, (*top.*) measuring by pacing;
mesurer au —, (*top.*) to measure by pacing;
— *métrique,* pace assumed equal to 1 meter;
outils de toutes sortes de —, tools of all sizes;
— *de porte,* step stone, stone sill of a door;
— *de souris,* (*fort.*) pas de souris, small steps (in permanent works) between the ditch and the covered way.

passade, f., (*fenc.*) thrust, passade; (*man.*) passade.

passage, m., passage, passing (as of a column); transit, crossing; voyage (beyond seas), passage, passage money, fare; road, way, thoroughfare, ferry; arcade, gallery; passage (short corridor); strait; undercut of a vehicle; (*r. r.*) crossing; (*mil.*) crossing, passage (of a stream, defile, ditch, etc.); transfer (as from active army to reserve, from one regiment to another); (*art.*) track or print of a projectile; passing, handling, of ammunition (as from magazine to gun); (*man.*) sort of trot (action well marked and cadenced);

— *balisé,* buoyed channel;
barrer le —, to stop the way, stand in the way;
bateau de —, ferryboat;
— *de bateau,* boat-crossing of a stream; (*top.*) representation of a boat-crossing on map;
— *blindé,* (*siege*) covered or protected passage in batteries and trenches;
— *d'un chemin de fer,* railroad crossing;
— *à contre-bord,* (*nav.*) broadside on;
de —, in transit; temporary;
— *en dessous* (*dessus*), (*r. r.*) street crossing under (over) a railway;
donner — *à,* to give vent to;
— *de la flamme,* (*met.*) fire bridge;
— *d'une formation à une autre,* (*mil.*) passage from one formation to another;
— *de fossé,* (*siege*) passage of the ditch;
se frayer un —, to break one's way through;
gagner son —, to work one's passage;
garder le —, to maintain, keep, the pass or passage;
— *sur la glace,* passage on ice;
grand — *développé,* (*man.*) an air consisting of the *passage* highly developed;
— *à gué,* fording (of a river);
instrument de —*s,* (*astr.*) transit;
lieu de —, thoroughfare;
— *de la ligne,* crossing of the line (equator);
— *de munitions,* (*nav.*) ammunition passage;
— *à la nage,* passage by swimming;
— *à, de, niveau,* (*r. r.*) grade crossing;
point de —, (*top.*) col;
prix de —, passage money;
— *de rivière,* (*mil.*) passage, crossing, of a river;

passage *de roues,* undercut of a vehicle;
— *des sangles,* (*hipp.*) girth (i. e., part of body around which girth passes);
— *par St. Cyr, etc.,* training, education, at St. Cyr, etc.;
— *souterrain,* underground passage;
système de —, (*mil.*) system in which officers serve first in one and then another arm, in the staff, etc., (as opposed to *système de carrière,* q. v.);
— *de vapeur,* (*steam*) steam-passage, -way.

passager, v. a. n., (*man.*) to passage.

passager, m., passenger;
— *de l'avant,* steerage passenger;
— *à la chambre,* cabin passenger;
— *de l'entrepont,* 'tween-decks passenger;
— *de pont,* deck passenger.

passant, m., crosscut saw; (*harn.*) keeper; (*unif.*) shoulder-knot loop or keeper;
— *coulant,* (*harn.*) sliding keeper or loop;
— *d'épaulettes,* (*unif.*) keeper, binder;
— *fixe,* (*harn.*) fixed loop;
— *mobile,* v. — *coulant.*

passation, f., (*adm.*) conclusion or accomplishing of a bargain, contract, etc.

passavant, m., permit (to pass goods through a custom-house).

passe, f., pass, situation, case, state; small channel, fairway, entrance of a harbor; loop, keeper (e. g., a leather loop to hold a file against a wall); (*fenc.*) pass; (*cord.*) end strand, passing through the eye in making a splice; turn (of a mooring); (*harn.*) loop or binder; (*mach., met.*) pass;
— *d'attaque,* (*mach.*) working pass;
— *d'un cordage,* (*cord.*) turn of a rope (over a pulley, in a knot, etc.);
— *d'ébauchage,* (*mach.*) rough-shaping pass;
— *de finissage,* (*mach.*) finishing or shaping pass;
— *de fleuve,* channel;
mesure de —, (*fenc.*) position such that the two blades can cross near the point;
mot de —, (*mil.*) password;
nœud de —, v. s. v. *nœud.*

passe-appareil, m., (*cord.*) whip line.
passe-balle, m., (*sm. a.*) bullet gauge (obs.).
passe-bombe, m., (*art.*) shell gauge (obs.).
passe-boulet, m., (*art.*) shot gauge (obs.).
passe-campane, m., (*hipp.*) capped hock.
passe-canal, m., canal ferry.
passe-cheval, m., horse ferry.
passe-corde, m., (*harn.*) drawing awl.
passe-cordon, m., large needle for thread or string.
passe-coudes, m., gauntlet.
passe-debout, m., permit for passage of goods through a city without payment of *octroi* dues.
passe-droit, m., (*mil.*) injustice done to anyone by promoting a junior over him; overslaughing.
passée, f., passage (of troops); track.
passe-lacet, m., bodkin.
passement, m., lace, trimming;
— *de jambe,* (*gym.*) cut-off (rings);
— *de pied en avant* (*arrière*), step to the front (rear) (French boxing).
passementerie, f., lace, trimmings.
passe-méteil, m., maslin (two-thirds wheat, one-third rye).
passe-parole, m., (*mil.*) word passed from mouth to mouth, from the head to the tail of a column.
passe-montagne, m., sort of headdress (either a knitted woolen shawl wrapped around the head and neck or a cap pulled down over the ears).
passe-partout, m., master key, pass key; large crosscut saw, keyhole saw, pad saw; passport; (*nil. r. r.*) short (0.25m) piece of track, used (in the Decauville system) to connect two tracks end to end;
— *de batterie,* (*Fr. art.*) long crosscut saw.
passe-poil, m., (*unif.*) piping; welting, welt.
passe-port, m., passport; pass; sea letter;
— *de navigation,* sea letter;
visa d'un —, indorsement of a passport;
viser un —, to indorse a passport.

passer, v. a. n., to pass, pass up, hand, go, clear; to cross, go across, take across; to perish; to put on; to exceed; to present one's self; to take, pass, an examination; to sift; (*cord.*) to reeve, put on, run, pass;
—— *à*, (*mil.*) to join (as a battery or regiment);
—— *sur son adversaire*, to close with one's adversary;
—— *un amarrage*, (*cord.*) to pass a lashing;
—— *par les armes*, (*mil.*) to shoot, to be shot, to execute by shooting;
—— *l'arme à gauche*, (*mil. slang*) to die;
—— *à l'arrière*, (*nav.*) to pass astern;
—— *en arrière (avant)*, (*fenc.*) to retire (advance);
—— *par les baguettes*, to run the gauntlet; to compel a man to run the gauntlet;
—— *sous le beaupré*, (*nav.*) to pass under the bows;
—— *un billet à l'ordre de quelqu'un*, to make a check payable to the order of;
—— *au bleu*, —— *jaune, etc.*, (*met.*) to temper up to blue, yellow, etc.;
—— *le câble*, (*cord.*) when a knot or lashing (as, for example, in two ropes lashed together) comes to the blocks, to untie, etc., so that the cable may pass;
—— *capitaine*, —— *général*, (*mil.*) to become captain, general, etc.;
—— *à un conseil de guerre*, (*mil.*) to go before a court-martial;
—— *un contrat*, to draw up, make, a contract;
—— *à contre*, (*cord.*) to reeve the wrong way;
—— *sur le corps à une armée*, (*mil.*) to defeat an army;
—— *par les courroies*, (*cav.*) to run the gauntlet, to cause to run the gauntlet;
—— *par dessus le bord*, (*nav.*) to pass over the rail;
—— *devant*, to have precedence of;
—— *à l'encre*, to ink a drawing;
—— *à l'ennemi*, (*mil.*) to desert to the enemy;
—— *par le fer*, —— *par le fil de l'épée*, (*mil.*) to be destroyed, put to death;
—— *au fil de l'épée*, (*mil.*) to put to the sword;
—— *d'une formation à une autre*, (*mil.*) to pass from one formation to another;
—— *les garants d'un palan*, (*cord.*) to reeve a tackle, fall, etc.;
—— *d'un grade à un autre*, (*mil.*) to be promoted;
—— *par tous les grades*, (*mil.*) to go, pass, through all grades;
—— *à gué*, to ford (a river);
—— *un habit*, to put on a coat;
—— *un homme à un officier*, (*mil.*) to allow an officer to have a man (soldier) as a servant;
—— *l'inspection*, (*mil.*) to inspect;
—— *à la main*, to pass along by hand;
—— *au large*, to go around;
—— *une manœuvre*, (*cord.*) to reeve a rope;
—— *un marché*, to make a contract or purchase;
—— *à la nage*, to swim (a river);
—— *au noir*, (*fond.*) to black wash;
—— *outre*, v. s. v. *outre*;
—— *un palan*, (*cord.*) to reeve a tackle;
—— *en palan franc*, (*cord.*) to reeve a tackle (said of manifold blocks, where fall first goes through end sheave);
—— *sur le pied de guerre (de paix)*, (*mil.*) to pass to a war (peace) footing;
—— *les poudres*, to pass cartridges on or up to gun;
—— *à poupe*, (*nav.*) to pass close astern;
—— *avec la protection de Mme. Berger-Levrault*, (*mil. slang*) to be promoted by seniority;
—— *de la queue à la tête*, (*mil.*) to become the head from the tail of the column;
—— *en revue*, (*mil.*) to review, inspect; to be reviewed, inspected;
—— *la revue de*, (*mil.*) to review (a given regiment, etc.);
—— *une revue*, (*mil.*) to hold a review, an inspection;
—— *une rivière à gué*, to ford a river;
—— *par St. Cyr, etc.*, to go through St. Cyr, etc.;
—— *au tamis*, to sift;
—— *les torons*, (*cord.*) to splice;

passer *dessus un vaisseau*, (*nav.*) to run a vessel down;
—— *au vent*, (*nav.*) to pass to windward;
—— *sous le vent*, (*nav.*) to pass to leeward;
—— *sur le ventre à une armée*, (*mil.*) to defeat an army;
—— *par les verges*, v. —— *par les baguettes;*
—— *la visite de*, to inspect;
—— *une visite*, to make an inspection.

passerelle, f., foot bridge, bridge (of a ship); gang plank;
—— *de commandement*, (*nav.*) captain's bridge;
—— *d'entrée*, (*nav.*) gangway;
—— *flottante*, floating gang plank;
—— *roulante*, (*pont.*) light bridge on axle and wheels for passing very slight streams;
—— *de vigie*, (*nav.*) look-out bridge;
—— *volante*, (*nav.*) connecting bridge.

passeresse, f., (*cord.*) ship line, lace.

passe-rivière, m., sort of raft, used to cross a river rapidly.

passeur, m., ferryman.

passe-violet, m., (*met.*) violet tempering color.

passe-volant, m., (*mil.*) man falsely mustered or enlisted, false muster, false enlistment.

passibilité, f., liability (to trial, punishment, etc.).

passible, a., liable to;
—— *d'un conseil de guerre*, (*mil.*) liable to trial by court-martial;
—— *de la peine de mort*, punishable with death.

passif, a., passive; (*fort.*) organized or prepared for protection merely (as an obstacle, abattis, shelter, etc.);
abri ——, (*fort.*) mere shelter;
commerce ——, excess of imports over exports.

passoire, f., strainer; bird-shot mold.

pastel, m., crayon, pastel; pastel drawing;
au, en, ——, in crayon;
dessiner au ——, to draw in crayon;
peindre en ——, v. *dessiner au* ——;
peintre en ——, pastelist;
peinture au ——, pastel work.

pastèque, f., snatch block.

pastille, f., (*sm. a.*) small round or disk of wax;
—— *fulminante*, v. *boulette d'amorce.*

pastourelle, f., (*mil. slang*) trumpet call for extra drill.

patache, f., tender (of a vessel);
—— *d'avis*, advice boat;
—— *de la douane*, custom-house boat;
—— *de la police*, harbor police boat.

patarasse, f., calking iron, horsing iron;
—— *cannelée*, making iron.

patarasser, v. a., to calk, drive with a calking iron; to horse.

pâte, f., paste; pulp; dough; (*powd.*) mill cake;
—— *de fourneau*, clay, lute, loam, of a furnace;
—— *à grener*, (*powd.*) mill cake;
métal en ——, (*met.*) pasty metal;
—— *de poudre*, v. *galette de poudre;*
—— *de pulvérin*, (*powd.*) priming-cake.

pâté, m., blot of ink; block, mass, of buildings; (*met.*) bundle of pieces of old iron to be welded together; (*fort.*) small isolated work, v. *infra*.
—— *de fortification;* (*cons.*) convex mass of earth, etc., sometimes used as the center of an arch;
construire une voûte sur ——, (*cons.*) to build an arch on an earthen center;
—— *de fortification;* (*fort.*) outwork, or isolated work, especially when surrounded by water or when in inundated country;
—— *de roches*, (*hydr.*) bank of rocks, small reef, bank or small reef, etc.

patelette, f., (*mil.*) flap (of a knapsack, cartridge box, pouch, etc.).

patenôtre, f., endless chain (of a chain pump).

patente, f., bill, patent, diploma; license; bill of health; patent axle;
—— *brute*, foul bill of health;
—— *douteuse*, tainted bill of health;
essieu à ——, patent axle;
—— *de navigation*, (*nav.*) register;

patente *nette*, clean bill of health;
— *de santé*, bill of health;
— *suspecte*, v. — *douteuse*.

pateux, a., muddy; (*met.*, *etc.*) pasty;
chemin —, muddy, greasy, road;
pain —, soggy bread.

patience, f., (*mil.*) button stick;
— *-planchette*, button stick.

patin, m., skate; clump sole; (*in pl.*) magazine slippers; sole, plate; (*farr.*) patten shoe; (*cons.*) sill of a staircase; batten; flange or foot of channel, or other special beams; grillage; foundation sill, sleeper; (*harn.*) part of the stirrup bar riveted to the saddle; (*mach.*) slide block, crosshead slide block; guide-, motion-, slipper-block;
— *d'amarrage*, timberhead, bitt, bollard, kevel head;
— *de calfat*, v. *patarasse;*
— *de charpente*, grating;
— *d'enrayage*, locking shoe;
— *d'entretoise*, (*art.*) batten, or tie, etc., of a transom;
faire — (of wheels) to slide;
— *de faîte*, (*cons.*) ridgepiece;
fer à —, (*farr.*) patten shoe;
— *à neige*, ski;
rail à —, (*r. r.*) foot rail;
— *du rail*, (*r. r.*) rail foot; flange;
— *de traverse*, (*mach.*) guide block.

patinage, m., skating; (*r. r.*) of locomotive wheels, slipping, turning without advancing.

patine, f., (the beneficial) weathering of a stone.

patiner, v. a., to skate; (*r. r.*) of locomotive wheels, to slip, turn without advancing.

paton, m., piece of leather, in a shoe, to preserve the shape of the upper.

pâton, m., cake of dough for baking.

patouille, f., (*met.*) ore separator.

patouillet, m., (*met.*) washing or separating cylinder; shop where separation is carried on;
— *de séparation*, washing cylinder.

patrie, f., native country; the part of the country from which one comes.

patriote, m., patriot.

patriotisme, m., patriotism.

patron, m., patron; master, "boss," captain, skipper; model, sample; templet; stencil; (*art.*) cartridge-bag pattern;
— *d'embarcation*, coxswain;
faire un —, to draw, take, a pattern.

patronne, f., v. *patron*, s. v. (*art.*).

patronner, v. a., to stencil; to draw, take, a pattern.

patronneur, m., pattern maker.

patrouille, f., (*mil.*) patrol; rounds (detachment and duty);
aller en —, to patrol;
— *d'avant-postes*, cavalry patrol at the outposts, security patrol;
— *de cavalerie*, cavalry patrol;
— *de combat*, combat patrol (a cavalry patrol sent out before a battle or when expecting early contact with the enemy);
— *de communication*, intermediate or communicating patrol;
— *de découverte*, scouting or exploration patrol (sent out by independent cavalry);
en —, on patrol, on rounds;
faire —, *faire la*, *une*, —, to make rounds, go the rounds;
— *de flanc*, flank patrol;
grande —, strong patrol;
— *indépendante*, independent patrol;
— *offensive*, combat patrol (i. e., one not avoiding combat);
— *d'officier*, officers' patrol;
— *de police*, night rounds (in garrison);
— *de sûreté*, v. — *d'avant-postes*.

patrouiller, v. n., to churn (of a turbine); (*mil.*) to patrol, to go the rounds, make rounds.

patrouilleur, m., (*mil.*) member of a patrol; man specially trained for scouting service.

patte, f., claw, foot; clamp of a vise; holdfast; tug, draught chain; flap, flange; batten; cringle; nave tenon (of a spoke); fluke of an anchor, claw of a grapnel; (*unif.*) shoulder strap, any band sewed on uniform, as on collar;
— *d'ancre*, fluke of an anchor;
— *d'anspect*, claw of a handspike;
— *d'araignée*, (*mach.*) oil groove, lubricating channel;
— *d'attache*, (*harn.*) cincha ring;
— *d'attelles*, (*harn.*) top of saddle-bow frame;
— *de bois*, batten; cramp iron;
— *en bois*, cramp iron;
— *à boucle*, tug with a buckle;
— *de bouline*, bowline cringle;
— *de chaîne d'enrayage*, (*art.*) skid chain holdfast;
— *de chat*, (*cord.*) midshipman's hitch, marlinspike hitch;
— *de collet*, (*unif.*) collar patch;
— *de cric*, foot of a jack;
— *s de crapaud*, (*mil. slang*) epaulets;
— *à crochet*, tug and hook;
élingue à —*s*, —*s d'élingue*, can hooks;
— *d'épaule*, (*unif.*) shoulder strap, knot;
— *de fer*, cramp iron; click, pawl; (*r. r.*) rail foot;
— *en fer*, click, (*r. r.*) rail foot;
— *-fiche*, cramp, holdfast;
— *-fiche à bois*, cramp iron fastened to wood;
— *-fiche à plâtre*, cramp iron driven through plaster;
— *s à futailles*, v. *élingue à* —*s;*
garnir la — *d'ancre*, to shoe the anchor;
— *de lièvre*, (*r. r.*) wing rail;
— *de loup*, (*carp.*) joggling, dovetailing;
mouiller en — *d'oie*, to moor with three anchors ahead;
— *d'oie*, point of junction of several roads; meeting point of two gutters; (*cord.*) bridle, bridle-mooring; any point where several cords meet; (*mil. min.*) three small branches at the end of a gallery; (*carp.*) junction of three timbers;
— *s d'oie équatoriales*, in a balloon, the cords starting at the equatorial line of knots;
— *à piton*, holdfast with an eye or loop;
— *en plâtre*, cramp iron (for stones);
— *du rais*, scarf tenon, foot of a spoke;
— *à ressort*, (*art.*) spring catch or clip;
— *de tension*, chain adjuster of a bicycle;
— *à tige*, sort of clamp or cramp iron with a stem;
— *de trait*, (*harn.*) trace ring or loop;
— *de volée*, (*art.*) splinter-bar loop.

pâturage, m., pasturage, pasture, pasture land, grazing.

pâture, f., food, feed, pasture, pasture ground;
vaine —, right of the inhabitants of a commune to graze their cattle on one another's land at certain times.

pâturer, v. n., to pasture, graze.

pâtureur, m., (*mil.*) herdsman.

paturon, m., (*hipp.*) pastern;
— *bas-jointé*, low pastern;
— *court-jointé*, short pastern;
— *droit-jointé*, straight pastern;
— *long-jointé*, long pastern;
os de —, pastern bone.

paume, f., palm; (*hipp.*) hand (measure).

paumelle, f., (sailor's) palm; (*cord.*) ropemaker's list; rubbing cord; (*cons.*) hinge, door hinge;
— *s à boules*, door hinge ornamented with acorns;
— *s doubles*, door hinge;
— *s à équerre*, garnet hinge;
— *s à gonds*, eye hinge (for masonry);
— *s à nœud bouché*, hinge with closed knuckle.

paumer, v. r., to be towed by hand.

paumet, m., sailor's palm.

paumoyer, v. a., (*cord.*) to underrun a rope;
se —, to climb hand over hand.

pause, f., suspension, stop, rest; stint, fixed amount of work;
— *d'exercice*, (*mil.*) rest; period, duration, of drill;

pause *de pas*, (*mil.*) said of marching at a walk, as a rest from a more rapid gait.

pavage, m., paving; paving work; pavement; paving materials;
— *en asphalte* — *en bois, etc.*, asphalt, wood, etc., paving.

pavé, m., paving block; pavement, paving; paved road, paved floor; carriage way;
— *d'asphalte*, asphalt pavement;
— *à bain de mortier*, mortar paving;
— *en béton*, concrete pavement;
— *de blocage*, pebble pavement;
— *en bois*, block wood pavement;
— *en brique(s)*, brick pavement;
brûler le —, to ride furiously, go whip and spur;
— *en cailloutis*, pebble pavement;
— *d'échantillon*, standard paving stone;
— *épincé*, face-hammered paving stone;
gros —, street, road, paving stone;
haut du —, wall side; inside of the pavement;
— *de mosaïque*, tessellated pavement;
petit —, paving stone for courtyard, etc.;
— *de pierre*, stone pavement;
— *piqué*, dressed paving stone;
— *d'un pont*, floor of a bridge;
— *à sec*, dry paving;
— *smillé*, hammer-dressed paving stone;
tâter le —, to examine the ground.

pavement, m., act of paving; paving materials.

paver, v. a., to pave;
— *à bain de mortier*, to lay a pavement on a coat of mortar;
— *de carreaux*, to flag;
pierre à — paving stone.

pavillon, m., pavilion, summerhouse, tent; mouth, bell (of a bugle or other instrument); mouthpiece (of a telephone); standard, colors, flag, ensign (strictly a naval term); (*fig.*) navy, sea power; (*mil.*) officers' quarters in barracks;
amener —, *amener son* —, to strike one's colors;
amener le —, to strike, to haul down, the colors;
amener un —, to haul down, strike, colors;
— *amiral*, admiral's flag;
appuyer le —, *appuyer son* —, (of a ship) to fire a gun under its proper colors;
arborer le —, to hoist the colors, to hang out the colors;
— *d'armateur*, house flag;
— *d'assistance*, signal of distress;
— *d'assurance*, signal of peace;
assurer le, *son* —, —, v. *appuyer le* —;
baisser —, to strike sail, to yield;
bâton de —, flagstaff;
battant de —, fly of colors;
battre — *de* ...., to fly the flag of ....;
— *de beaupré*, jack;
— *en berne*, v. — *d'assistance*;
— *blanc*, — *en blanc*, flag of truce;
caisse aux —s, flag chest;
capitaine de —, (*nav.*) flag captain;
— *carré*, square flag;
— *du chef de l' État*, President's flag;
— *de combat*, signal for action;
— *de commandement*, headquarters flag; designating flag (as an admiral's);
le commerce suit le —, trade follows the flag;
— *de conseil*, signal for council of war;
— *consulaire*, consular flag;
coup de —, morning and evening gun;
— *cornette*, burgee;
le — *couvre la marchandise*, the flag covers the goods;
— *de départ*, blue peter;
— *distinctif*, — *de distinction*, distinguishing flag;
— *de la douane*, custom-house flag, revenue flag;
drisse de —, ensign halliard;
— *d'embarcation*, boat flag;
— *étoilé*, the "Stars and Stripes";
faire — *blanc*, to strike colors in token of peace or of surrender;
— *fendu*, swallowtail flag;

pavillon *de garde*, guard flag;
gaule de —, flagstaff;
— *de guerre*, man-of-war's flag, man-of-war's pennant;
— *en guidon*, swallowtail flag;
guindant de —, hoist of colors;
hisser le —, to hoist the colors;
— *d'honneur*, flag of honor;
— *impérial*, imperial standard;
— *jaune*, quarantine flag;
jeu de —s, set of colors or flags;
— *de justice*, punishment flag;
— *de malle-poste*, mail flag;
— *marchand*, merchant flag;
mât de —, flagstaff;
mettre — *bas*, to lower one's flag;
— *à mi-corne*, — *à mi-drisse*, — *à mi-mât*, flag at half mast;
— *du ministre de la marine*, naval secretary's flag;
— *national*, national flag;
— *neutre*, neutral flag;
— *noir*, black flag, pirate flag;
— *de paix*, flag of truce;
parer un —, to clear a flag;
— *parlementaire*, flag of truce;
— *de partance*, blue peter;
— (*de*) *pilote*, pilot flag;
porter — *de* ..., to fly the flag of ....;
porter le — *et la flamme*, to wear the colors of a man-of-war;
— *-poste*, mail flag;
— *de poupe*, stern ensign, main ensign;
— *de quarantaine*, quarantine flag;
— *renversé*, flag with union down (distress);
— *rouge*, powder flag (ship carries explosives);
— *royal*, royal standard;
série de —s, set of flags;
— *de signaux*, signal flag;
— *de société*, house flag;
vaisseau —, flagship (obs.).

pavois, m., (*nav.*) bulwark; weatherboards; flags for dressing ship;
déployer le —, to display colors;
grand —, full dress of ship;
montrer le —, to show colors;
petit —, dressing ship with masthead flags;
rentrer le —, to haul down colors.

pavoisement, m., decoration by means of flags; (*nav.*) dressing of ship.

pavoiser, v. a., to deck with flags; (*nav.*) to dress ship;
— *à l'anglaise*, (*nav.*) to dress ship fore and aft;
— *à la française*, (*nav.*) to dress ship over the yardarms.

paye, f., pay; paymaster; wage;
à la basse —, ordinary pay;
— *entière*, full pay;
haute —, v. *haute-paye*;
jour de —, pay day;
morte- —, (*mil.*) pensioner (obs.);
sans —, unpaid.

payement, paiement, m., payment.

payer, v. a., to pay, pay off;
— *l'équipage*, to pay off the crew.

payeur, m., payer; (*mil.*, *nav.*) paymaster;
— *adjoint*, (*Fr. a.*) assistant paymaster (ranks as captain);
— *général*, (*Fr. a.*) chief paymaster of an army (ranks as *général de brigade*);
— *particulier*, (*Fr. a.*) division paymaster (ranks as major);
— *principal*, (*Fr. a.*) corps paymaster (ranks as colonel).

pays, m., country, countryside, region; part of a country from which one comes, home; climate;
au —, home, at home;
battre le —, (*mil.*) to explore the ground;
— *de chicane*, (*mil.*) intersected, broken country;
— *coupé*, (*mil.*) rough, intersected country, full of natural obstacles;
haut —, mountainous part of a country;
mal du —, homesickness;
— *ouvert*, (*mil.*) open country;
— *plat*, flat country;

pays, *plat*, ——, (*mil.*) open country, as distinguished from fortified places.
paysage, m., landscape.
paysan, m., peasant.
péage, m., toll, tollhouse;
— *barrière de* ——, tollgate, toll bar.
péager, a., *chemin* ——, toll road.
peau, m., hide, skin; (*met.*) skin or crust of an ingot or casting;
— *dont la* —— *adhère aux muscles*, hidebound;
— *de batterie*, batter head of a drum;
— *de caisse*, drumhead;
— *crue*, rawhide;
— *de* ——, leathern;
— *inférieure*, snare head;
— *préparée*, dressed skin or hide;
— *rouge*, redskin, North American Indian;
— *de tambour*, drumhead;
— *verte*, rawhide.
péchard, m., familiar term for roan horse.
pêche, f., fishing.
pêcher, v. a., to fish.
pêcherie, f., fishing grounds.
pêcheur, m., fisherman.
peckin, m., v. *pékin*.
péculat, m., peculation (of public moneys);
— *militaire*, theft of part of booty or of public moneys by accountable persons.
péculateur, m., functionary or official who steals public moneys; peculator.
pédale, f., (*mach.*) treadle, footboard, pedal; pedal of a bicycle;
— *à billes*, pedal on ball bearings (bicycle);
— *à caoutchouc(s)*, rubber pedal (bicycle);
— *dynamométrique*, dynamometric pedal (gives pressure of foot on the pedal of a bicycle);
— *à scie*, rat-trap pedal (bicycle);
— *d'un tour*, lathe pedal.
pédaler, v. n., to pedal (bicycle).
pédalier, m., crank hanger (bicycle).
pédiluve, m., footbath; place where horses' feet are washed.
pédestre, a., pedestrian.
pédomètre, m., pedometer.
pédométrique, a., pedometric.
pé¹on, m., courier on foot, runner.
peignage, m., combing (as of wool, etc.).
peigne, m., comb; (*tech.*) chaser; (*hipp.*) a variety of *crapaudine*, q. v., so called because the hairs stand up like the teeth of a comb (used also in plural); (*cord.*) gill;
— *de déclanchement*, (r. f. art.) trigger comb;
— *à ébaudrer*, (*cord.*) hatchel;
— *à épurer*, (*cord.*) cleaning comb;
— *femelle*, (*tech.*) internal chaser;
— *à fileter*, (*tech.*) chasing tool, chaser;
— *sans fin*, (*cord.*) gill;
— *mâle*, (*tech.*) external chaser.
peigner, v. a., to comb; (*cord.*) to taper (the end of a rope); to hatchel (hemp).
peigneuse, f., (*cord.*) hatcheling machine.
peignon, m., (*cord.*) twist of hemp.
peinchebec, m., pinchbeck.
peindre, v. a., to paint, draw; to coat;
— *à l'aquarelle*, to paint in water colors;
— *à la chaux*, to whitewash;
— *à l'huile*, to paint in oils.
peine, f., penalty, punishment; labor, toil, trouble, fatigue;
— *afflictive et infamante*, punishment of grave crimes, involving loss of liberty, forced labor, death;
— *du boulet*, (*mil.*) ball and chain;
— *disciplinaire*, —— *de discipline*, (*mil.*) disciplinary punishment;
— *frapper d'une*, to impose a penalty;
— *homme de* ——, porter, any man without a trade, earning his living by severe physical labor;
— *infamante*, ignominious punishment (banishment, civil degradation);
— *de mort*, death penalty;
— *sous* —— *de* ...., under penalty of ....;
— *sous* —— *de la mort, de la vie*, under penalty of death.
peintre, m., painter; (*mil. slang*) sweeper.
peinturage, m., house painting.
peinture, f., painting; picture; paint, colors;
— *d'apprêt*, —— *en apprêt*, priming coat;

peinture *à l'aquarelle*, water-color painting;
—— *blanche*, white paint (*Fr. a.*, used for distinctive marks);
—— *à la chaux*, whitewashing;
—— *à la cire*, encaustic painting;
—— *à la colle*, kalsomining;
— *couche de* ——, coat of paint;
—— *à fresque*, fresco work;
—— *grise*, second coat on ironwork;
—— *à l'huile*, oil painting;
—— *au minium de plomb*, red-leading (first coat on ironwork);
—— *noire*, black paint (third and final coat on ironwork);
—— *noire sur minium*, black paint over red lead;
—— *olive*, olive painting (wooden parts of gun carriages);
—— *en pâte*, paste from which paint is made;
—— *à la plombagine*, black leading;
—— *au sable*, covering paint with sand;
—— *silicatée*, silicate painting;
—— *au vernis*, varnishing.
pékin, m., (*mil. slang*) "cit," cf. *chinois*;
—— *de bahut*, (St. Cyr slang) cadet who has finished his studies;
— *en* ——, in plain clothes, in "cits."
pelage, m., (*hipp.*) color, coat.
pelard, a., *bois* —— (or simple *pelard*), barked wood, wood stripped of its bark.
pélardeau, m., (*nav.*) plug (as for a shot-, a hawsehole).
pélastre, pélâtre, m., pan of a shovel.
pelé, a., bare, threadbare; (*top.*) bare, treeless, verdureless.
peler, v. a., to peel, bare, strip; to remove the hair from a skin; to bark; (*fort.*) to cut sods;
— *la terre*, to cut sods.
pèlerine, f., (*unif.*) hooded cape.
pélican, m., iron dog.
pelisse, f., (*unif.*) pelisse, hussar pelisse.
pellage, m., spading.
pellardeau, m., v. *pélardeau*.
pellâtre, m., v. *pélastre*.
pelle, f., shovel, scoop; oar blade; (*met.*) firinghole slide;
—— *d'Arras*, sort of sod cutter, useful in ground that holds together well, but is easy to dig;
—— *d'aviron*, blade of an oar;
—— *bêche*, (*mil.*) spade or shovel in one, with a saw up one side of blade; spade-shovel;
—— *carrée*, spade;
—— *à charbon*, coal shovel;
—— *à, de, feu*, fire shovel;
— *jet de* ——, throw of a shovel;
— *ramasser une* ——, (*fam.*) to get a fall from a bicycle;
— *remuer à la* ——, to shovel, to use the shovel;
—— *ronde*, round shovel;
—— *tranchante*, (*fort.*) trenching shovel.
pellée, f., shovelful.
peller, v. a., to shovel, to use a shovel.
pellerée, f., v. *pellée*.
pelletée, f., v. *pellée*.
pelleteur, m., shoveler, man who uses a shovel, shovel man.
pellicule, f., (*phot.*) sensitive film;
—— *à celluloïd(e)*, celluloid film;
—— *d'huilage*, negative paper.
pelote, f., ball; clew; (*sm. a.*) pellet of small shot; (*hipp.*) blaze, star, on a horse's forehead; (*met.*) sheet copper rolled up and ready for melting;
—— *d'émeri*, emery ball;
—— *à feu*, (*artif.*) light ball;
—— *de glaise*, (*artif.*) wad of clay (rockets).
peloter, v. a., to make up into, wind in, a ball; to clew; (*met.*) to stretch under the hammer; (*sm. a.*) to cluster (of small shot).
peloton, m., ball, lump; group, knot, party; (*mil.*) platoon; (French cavalry) platoon (12 files);
—— *de chasse*, (*mil. slang*) extra drill;
— *défaire un* ——, to unclew;
—— *d'exécution*, (*mil.*) firing party;
— *faire* —— *de punition*, to belong to the punishment squad, to serve punishment;
— *feu de* ——, (*mil.*) platoon firing;

peloton *hors rang*, (*Fr. cav. and art.*) in each regiment, a special group comprising the *petit état-major*, and various workmen, clerks, fencing masters, etc.;
—— *d'instruction*, (*Fr. a.*) platoon of soldiers taking special drill;
—— *de (la) pièce*, v. s. v. *pièce*;
—— *des pièces*, (*art.*) gun detachment (as in a turret);
—— *de punition*, (*mil.*) punishment squad;
—— *des servants*, (*art.*) gun detachment.
pelotonnement, m., (*mil.*) platoon formation in three ranks (rare).
pelotonner, v. a. r., to clew; to gather into knots, lumps; (*mil.*) to gather into small groups (e. g., to resist cavalry).
pelouse, f., lawn; lawn grass.
peloux, m., earth washed down by rains from mountains.
pelucheux, a., shaggy.
pénalité, f., penal law; (in pl.) punishments (legal).
penchant, m., slope (of a hill, earthwork, etc.), declivity, brink.
pencher, v. a. n., to incline, lean; bend; stoop; be out of plumb; (*nav.*) to list, have a list.
pendant, a., hanging; pending;
clef ——*e*, (*mas.*) hanging keystone; v. s. v. *clef*;
mitraille ——*e*, (*met.*) old brass.
pendant, m., (*mil.*) sword-belt sling; frog; (*nav.*) man-o'-war's pennant;
—— *brut*, (*mas.*) long flat stone used especially in arches.
pendentif, m., pendentive; sleeve, neck, of a balloon;
en ——, hanging.
penderoles, f. pl., (*mil.*) trumpet cords.
pendeur, m., (*cord.*) pendant.
pendu, m., (*St. Cyr slang*) instructor.
pendulaire, a., pendular, pendulous.
pendule, m., pendulum; f., clock;
—— *balistique*, (*ball.*) ballistic pendulum (the part that receives the projectile);
canon ——, (*ball.*) gun pendulum (the part that carries the gun);
—— *à canon*, (*ball.*) artillery ballistic pendulum (the whole apparatus);
—— *circulaire*, circular pendulum;
—— *compensateur*, —— *à compensation*, —— *compensé*, compensating pendulum;
—— *composé*, compound pendulum;
—— *conique*, (*mach.*) Watt's governor, conical pendulum;
—— *cycloïdal*, cycloidal pendulum;
—— *électrique*, electric pendulum;
—— *à force centrifuge*, v. —— *conique*;
fusil ——, (*ball.*) musket pendulum (i. e., the part that carries the rifle);
—— *à fusil*, (*ball.*) small-arm ballistic pendulum (the whole apparatus);
lentille de ——, pendulum bob;
poids de ——, pendulum weight;
—— *à réveil*, alarm clock;
—— *à secondes*, seconds pendulum;
—— *simple*, simple pendulum;
tige, *verge*, *de* ——, pendulum rod.
pendulographe, m., (*drawing*) pendulograph.
pêne, f., mop; (*in. pl.*) the small cords hanging from a fly net.
pêne, m., bolt (of a lock); (*sm. a.*) bolt; (*art.*) working bolt, operating bolt;
arme à ——, v. *arme*;
—— *à coulisse*, sliding bolt;
—— *à demi-tour*, v. —— *à ressort*;
—— *dormant*, dead bolt, bolt proper;
—— *à ressort*, latch bolt.
péneau, m., state of being a-cockbill (of an anchor);
en ——, a-cockbill.
pénétration, f., penetration; impregnation (of wood); (*art.*, *sm. a.*) penetration of a projectile.
pénétrer, v. a. n., to penetrate; (*art.*, *sm. a.*) to penetrate (of projectiles).
péniche, f., pinnace.
péninsule, f., peninsula;
la ——, Spain and Portugal.

pénitencier, m., penitentiary;
—— *militaire*, military penitentiary (in France, for soldiers sentenced to more than one year's imprisonment).
pénitentiaire, a., reformatory.
penne, f., mop.
pennon, m., pennon.
penon, m., dogvane.
pension, f., pension, annuity, allowance; food, keep; livery; boarding school, boarding house;
accorder une ——, to bestow a pension;
—— *alimentaire*, board money;
cheval en ——, horse at livery;
—— *pour les chevaux*, livery stable;
envoyer en ——, (horses) to put out at livery;
—— *sur l'État*, government pension;
être de la même, —— to mess together;
mettre en ——, to put (horses) at livery;
—— *proportionnelle*, v. —— (*de retraite*) *proportionnelle*.
recevoir des chevaux en ——, to keep a livery stable;
—— *de réforme*, (*Fr. a.*) (officers') pension, granted for physical or other unfitness;
—— *de retraite*, (*Fr. a.*) retiring pension (officers and men);
—— *de retraite pour ancienneté de services*, (*Fr. a.*) retiring pension, based on length of service (officers and men);
—— (*de retraite*) *proportionnelle*, (*Fr. a.*) retiring pension (men only) granted for length of service between 15 and 25 years;
—— *de retraite par suite de blessures et d'infirmités*, (*Fr. a.*) retiring pension for wounds and disability in line of duty;
—— *pour services éminents*, (*Fr. a.*) good-service pension, with reversion to widows and orphans;
tenir en ——, to keep in livery;
—— *viagère*, life pension.
pensionnaire, m., pensioner.
pensionnat, m., boarding school.
pensionner, v. a., to pension, to grant a pension or allowance.
pentaspaste, m., five-block tackle, for raising heavy weights.
pentagone, a., m., pentagonal; pentagon.
pente, f., slope, inclination; descent, ascent; declivity, acclivity; uphill, downhill; gradient, grade; pitch (of a roof); talus (of a wall); fall (of a stream); face (of a hill); (*sm. a.*) jeg; fall or bend of the stock; (*tech.*) jig; (*carp.*) set or angle of a tool blade or cutter;
—— *aisée*, good gradient;
aller en ——, to incline, slope, shelve;
angle de ——, angle of slope;
—— *ascendante*, ascent, acclivity;
avec ——*s et rampes*, undulating;
avoir de la ——, v. *être en* ——;
—— *de couche*, (*sm. a.*) angle of slope of stock;
—— *courante*, slope of a road (in the direction of its length);
—— *descendante*, descent, declivity;
—— *difficile*, v. —— *raide*;
—— *douce*, good gradient; easy slope;
—— *d'enrayage*, slope on which wheels must be locked or chocked;
être en ——, to incline, be inclined, sloping;
—— *d'un fusil*, (*sm. a.*) fall, bend, of musket stock;
galerie en ——, (*min.*) adit of a mine;
—— *du glacis*, (*fort.*) slope of the glacis;
—— *latérale*, side slope of a road;
—— *limite accessible*, limiting practicable slope;
—— *de mât*, (*nav.*) rake of a mast;
—— *de plâtre*, (*cons.*) layer of plaster, to receive the leads of a gutter;
—— *raide*, steep, bad, slope;
—— *rapide*, steep slope;
—— *d'une rivière*, fall of a river;
—— *de tente*, side walls of a tent; valance of an awning;
—— *du terrain*, (*top.*) tangent of the angle of slope (of the ground); gradient;
terrain en ——, sloping ground;
—— *d'un toit*, pitch of a roof.

penture, f., hinge; hinge strap; iron hangings of a door or window; iron brace; pintle; eye of a clamp; holdfast;
—— *automate,* self-acting hinge;
—— *à charnière,* hinge joint;
—— *à crapaudine,* v. —— *à piton;*
—— *double,* forked hinge;
—— *à équerre,* garnet hinge;
—— *en F,* cross tailed hinge;
—— *flamande,* double hinge strap;
—— *à gond,* hook-and-eye hinge;
—— *longue,* strap hinge;
—— *à piton,* —— *à pivot,* pin-and-socket hinge;
—— *de sabord,* (*nav.*) port hinge;
—— *à talon,* cramp hinge.

péquin, m., v. *pékin.*
péralite, f., (*expl.*) peralite.
péras, pérat, m., patent fuel, briquettes.
perçage, m., boring, punching; (*art., sm. a.*) first operation of boring.
perçant, m., (of a horse) keenness, spirit; (*mach.*) piercer, borer, puncher.
perce, f., size of opening (in screens, sieves, etc.); borer, drill, punch;
—— *à main,* hand punch.
percé, p. p., pierced;
—— *à jour,* openworked.
perce-carte, m., (*mil.*) flag pin, used to mark the position of troops on a map.
perce-cartouche, m., (*art.*) priming wire (obs.).
perce-chaussée, m., teredo navalis.
perce-clous, m., gimlet.
perce-droit, m., straight borer.
percée, f., opening in a wood, vista; opening (of a canal, etc.); (*mil.*) breaking or cutting the enemy's line; (*met.*) tap hole;
faire la ——, (*met.*) to tap;
faire une ——, to break through (as the enemy); to make an opening (door, window) in a wall;
trou de ——, (*met.*) tap hole.
perce-fournaise, m., (*met.*) tapping bar, bott stick.
perce-fusées, m., (*art.*) fuse cutter.
percement, m., piercing, perforation, boring; tunneling, tunnel piercing; piercing of an isthmus; (*mas.*) opening in a wall (for a door or window); (*met.*) tapping; (*min.*) driving of a gallery, sinking of a shaft.
percepteur, m., collector, taxgatherer.
perception, f., collection of taxes; collectorship; (more narrowly, *mil.*) action of receiving stores, supplies, issues, (hence) stores, issues, drawn.
percer, v. a. n., to pierce, bore, drill, punch; to broach, tap; to break through, open; to penetrate, go through; to wet through; to tunnel, drive a tunnel; to stick, lance; to run through (with a sword); (*fort.*) to pierce (loopholes); (*min.*) to drive, sink, shafts, galleries; (*met.*) to tap (a blast furnace); (*mil.*) to break through, pierce (the hostile line); to rise in rank by merit;
—— *un bois,* to make roads through a wood;
—— *une croisée,* to make an opening for a window;
—— *dans deux rues,* (*of a house*) to open on two streets;
—— *un escadron, etc.,* (*mil.*) to break, dash, through a squadron, etc.;
—— *à foret,* to drill;
—— *une forêt,* v. —— *un bois;*
—— *le haut fourneau,* (*met.*) to tap the furnace;
instrument à ——, borer;
machine à ——, punching machine;
—— *un navire,* (*nav.*) to pierce a ship for guns;
—— *outre,* to dig through;
—— *d'outre en outre,* —— *de part en part,* to pierce through and through; to run, shoot, through;
—— *un pays,* to make roads through a country;
—— *une porte,* to open a door (as in a wall);
—— *un puits,* to sink a shaft, well;
—— *une rue,* to open a street;

percer *un tonneau,* to broach a cask;
—— *les tôles,* to punch the plates;
—— *une tranchée,* (*siege*) to break through a sap;
—— *un tunnel,* to drive a tunnel.
percerette, f., gimlet, borer.
percette, f., borer, drill.
perceur, m., borer (workman); (*met.*) tapper;
—— *de culasse,* (*sm. a.*) breech-screw maker;
—— *mécanique,* (*mach.*) punching machine.
perceuse, f., (*mach.*) boring machine.
percevoir, v. a., to collect taxes, revenue; to get, draw (as clothes, supplies, etc.).
perche, f., perch, pole; shaft (of an oar); (*surv.*) measuring rod; (*gym.*) pole, hanging pole, jumping pole, climbing pole;
—— *d'aviron,* shaft of an oar;
—— *à brasser,* (*met.*) stirring pole (for metal in fusion);
—— *double,* (*mil. teleg.*) extension pole (two joints);
—— *à houblon,* (*mil. slang*) lance;
—— *à mesurer,* measuring rod;
—— *oscillante,* (*gym.*) swinging pole;
saut à la ——, (*gym.*) pole vaulting;
sauter à la ——, (*gym.*) to pole-vault;
—— *à sauter,* (*gym.*) vaulting or leaping pole;
—— *de soufflet,* rock staff, bellows lever;
—— *triple,* (*mil. teleg.*) extension pole (three joints).
percheron, a., m., Percheron (horse).
perclus, a., crippled.
perçoir, m., punch, awl; broach, borer, drill; (*met.*) tapping bar;
—— *à couronne,* rose bit;
—— *à couteaux mobiles,* slotting bar, grooving tool with movable cutter;
—— *à levier,* v. —— *à rochet;*
—— *multiplex,* automatic punch;
—— *à rochet,* rack drill; ratchet brace; ratchet drill;
—— *à tête crêpée,* keyway drill.
percolateur, m., percolator, percolating coffeepot.
percussion, f., percussion, shock, blow; hammering, striking, of a timber, to test its soundness; (*art.*) shock (of trunnions in beds, of trail on platform, due to firing); the blow on deck, due to firing; (*art., sm. a.*) percussion;
amorce à ——, (*sm. a.*) percussion cap (obs.); percussion priming;
arme à ——, (*sm. a.*) percussion arm;
—— *centrale,* (*sm. a.*) center fire;
fusil à ——, (*sm. a.*) percussion musket (obs.);
—— *périphérique,* (*sm. a.*) rim fire;
presse à ——, (*mach.*) stamping press;
système à ——, (*art.*) percussion principle;
—— *du tir,* (*art.*) the blow on the carriage, etc., due to firing;
vis de ——, fly, screw, press.
percutant, a., percussive, striking; (*art.*) using percussion shell;
appareil ——, (*art.*) percussion apparatus, device;
chien ——, (*sm. a.*) hammer;
mécanisme ——, v. *appareil* ——;
platine ——*e,* (*art.*) hammer lock;
section ——*e,* (*art.*) in ranging, section using percussion shell, as distinguished from *section fusante;*
système ——, (*art.*) percussion system of double-action fuze;
tir ——, (*art.*) percussion fire, fire with percussion shell.
percuteur, m., (*sm. a.*) striker; (*art.*) striker, hammer; firing pin (of a fuse); percussion firing pin; plunger of a fuse;
—— *libre,* (*sm. a.*) loose striker, striker not actuated by a spring, free to move under a blow (as Springfield, Winchester);
—— *à ressort,* (*sm. a.*) spring striker;
—— *de sûreté,* (*art.*) safety striker.
perdant, m., ebb of the tide;
—— *des marées,* take-off of the tides.
perd-fluide, m., (*elec.*) ground end of a lightning rod.

perdre   319   **personnel**

perdre, v. a. n., to lose, waste, ruin; to drag (of an anchor); to carry away (sails, yards, etc.); to fall in price; to fall (of the tide); to take off (of the tide);
— *une bataille*, (*mil.*) to lose a battle;
— *son chemin*, to lose one's way;
se — *corps et biens*, (*nav.*) to be lost with all hands;
— *fond*, (of an anchor) to drag;
— *le fond*, (*nav.*) to get off soundings;
— *sa marche*, to lose speed;
— *la nage*, to catch a crab (rowing);
se —, (of a river) to fall into;
— *la sonde*, v. — *le fond*;
— *du terrain*, (*man.*) in executing vaults, to lose ground;
— *terre*, (*nav.*) to lose sight of land;
— *de vue*, to lose sight of.

perdu, p. p., lost, ruined; extinct; (*nav.*) carried away (yard, etc.); lost (ship);
ballon —, free balloon;
à corps —, headlong, desperately without regard to consequences;
coup —, random shot;
à coup —, at random;
enfants —*s*, (*mil.*) forlorn hope;
fondation à pierre(s) —*e(s)*, random, riprap, foundation;
ouvrage à pierres —*es*, v. *fondation à pierres* —*es*;
pierre —*e*, (*mas.*) stone completely embedded in mortar or concrete (as in filling work);
puits —, dry well (i. e., whose sandy bottom absorbs the water);
sentinelle —*e*, (*mil.*) advanced sentry;
à tête —*e*, flush, driven flush (as a bolt, a peg).

peréquation, f., equalization (as of promotion, of taxation).

perfection, f., completion, perfection.

perfectionnement, m., improvement.

perfectionner, v. a., to improve.

perforant, a., perforating;
effet —, (*art.*) perforating effect.

perforant, m., (*hipp.*) perforans tendon.

perforateur, a., perforating; m., drill, borer; rock drill;
— *à air comprimé*, pneumatic drill, compressed-air drill;
— *à injection d'eau*, injection drill;
— *à percussion*, jumper; percussion drill;
— *à rotation*, churn drill; rotating drill.

perforation, f., perforation (*esp. art.*) of an armor plate;
— *mécanique*, mechanical boring;
— *à percussion*, jumping;
puissance de —, (*art.*) perforating power (of a gun);
— *à rotation*, churn drilling.

perforatrice, f., borer, drill, rock drill;
— *à diamant*, diamond drill.

perforé, m., (*hipp.*) perforatus tendon.

perforer, v. a., to bore, drill; to perforate.

péricliter, v. n., to be in danger; hence (*mil.*) to run down (in discipline, spirit).

péridot, m., olivin, chrysolite.

péridotite, f., peridotite.

périer, m., (*fond.*) tapping bar.

périgraphe, m., — *instantané*, (*phot.*) radial perspective camera (for photographic surveying).

périgueux, m., glass soap.

péril, m., peril, danger, risk, hazard;
en —, (of a building, a façade) about to fall;
—*s de la mer*, dangers of the sea.

périmètre, m., perimeter;
— *mouillé*, wetted perimeter (of a river).

période, f., period.
(The following terms relate to steam.)
— *d'admission*, admission;
— *de l'avance à l'échappement*, inside lead;
— *de l'avance à l'introduction*, outside lead;
— *de compression*, compression;
— *de contre-pression*, back pressure;
— *de détente*, expansion;
—*s de la distribution de la vapeur*, behavior of steam in a double piston stroke;
— *d'évacuation*, exhaust;

— *d'introduction*, admission;
— *de l'introduction naturelle*, admission, steam port wide open.

périople, f., (*hipp.*) periople.

périoplique, a., (*hipp.*) perioplic;
bourrelet —, v. s. v. *bourrelet*.

périoste, m., (*hipp.*) periosteum.

périostéite, f., (*hipp.*) periostitis.

périostose, f., (*hipp.*) periostosis.

périphérie, f., rim, periphery.

périphérique, a., rim, peripheric.

péripneumonie, f., (*hipp.*) consumption.

périr, v. n., to perish; to decay; to sink, be lost (of a ship);
— *corps et biens*, to perish (crew and cargo); to be completely destroyed.

périssable, a., perishable, subject to decay.

perlite, f., perlite, pearlstone.

permanence, f., permanence;
armée en —, standing army;
en —, permanent, continual.

permanent, a., permanent, lasting;
armée —*e*, standing army;
fortification —*e*, permanent fortification.

perméabilité, f., permeability.

perméable, a., permeable, pervious.

permis, p. p., permissible, admissible; m., written permission, permit;
— *de circulation*, (*r. r.*) pass;
— *d'exportation*, permit to export;
— *de séjour*, permit of residence.

permission, f., leave, permission; (*nav.*) liberty; (*mil.*) pass; leave (*Fr. a.*, not to exceed 30 days);
accorder la —, to grant leave, permission;
— *d'aller à terre*, (*nav.*) shore leave;
avec —, with leave, permission;
avoir — *de 24 heures*, (*mil.slang*) to be on guard;
dépasser une — (*mil.*) to overstay a leave;
en —, (*mil.*) on pass;
— *de soirée*, (*mil.*) night pass;
en — *de soirée*, (*mil.*) on night pass;
— *à titre de sursis*, (*mil.*) delay (as in joining, etc.).

permissionnaire, m., (*mil., nav.*) officer on leave; man on pass, on liberty, on furlough.

permissionner, v. a., to license.

permutateur, m., (*elec.*) plug.

permutation, f., permutation; (*elec.*) switching of a current; (*mil.*) transfer;
— *par convenance personnelle*, (*mil.*) transfer at officer's request; exchange;
— *d'office*, (*mil.*) transfer by order.

permuter, v. a., (*mil.*) to transfer, to exchange.

péroné, m., (*hipp.*) fibula.

perpendiculaire, a., perpendicular; f., (*fort.*) perpendicular (in the bastioned system).

perpendicule, m., plummet (obs.).

perquisition, f., search.

perré, m., dry wall; stone packing, stone pitching; seashore covered with stones or shingle; water wing of a bridge; (*r. r.*) small culvert, drain, filled with broken stone; rubble drain.

perreyé, m., packed with stone; covered or faced with stone (as a dike);
sable —, facing sand (of a dike).

perrier, m., (*met.*) tapping bar; bott stick.

perrière, f., (local name for) quarry; (*met.*) tapping bar.

perron, m., the steps and landing of the main entrance of a building, perron; step (of a waterfall going over a series of steps); (*r. r.*) platform;
— *de chargement*, (*r. r.*) freight platform;
— *à pans*, a perron with cut-off corners.

perruquier, m., barber.

persienne, f., Persian blind.

personnage, m., person, personage.

personne, f., person;
— *civile*, (*law*) the State, a community, corporation, association;
de sa —, in person;
la — *du roi*, the king.

personnel, m., (*mil., etc.*) personnel;
— *ouvrier*, workmen.

perspectif

perspectif, a., perspective;
  plan ——, plane of the picture, prespective plane.
perspective, f., perspective; perspective drawing;
  —— *aérienne,* aerial perspective;
  —— *cavalière,* cavalier-perspective;
  —— *linéaire,* linear perspective;
  —— *spéculative,* science of perspective;
  —— *à vol d'oiseau,* bird's-eye view;
  —— *en vue accidentelle,* oblique perspective;
  —— *en vue de face,* parallel perspective.
perte, f., loss; loss (of a battle, position, town, of men, of a ship); waste; escape (of heat, etc.); ruin;
  —— *censée totale,* (nav.) constructive total loss;
  —— *de charge,* (tech.) loss of pressure; (elec.) waste (as in the circuit);
  —— *de courant,* (elec.) current leakage;
  ——*s graves,* (mil.) serious, heavy, losses;
  ——*s légères,* (mil.) slight losses;
  —— *par les supports,* (elec.) weather contact;
  —— *à la terre,* (elec.) ground leakage;
  —— *totale,* (nav.) total loss;
  à —— *de vue,* as far as the eye can reach.
pertuis, m., hole; ward of a key; hole or opening of a screw plate; waste, drain hole (of a basin, etc.); drain; sluice; opening for boats in a dike or levee; strait (between two islands, between an island and the mainland); narrow channel in, to, or about a harbor; (in France) rapids of the Seine; narrow part (of a river navigable by flushing only, point where the retaining dam is established); (top.) col; (met.) tap hole.
perturbateur, a., perturbating, disturbing.
perturbation, f., perturbation; disturbance;
  —— *de l'aiguille,* perturbation of the needle;
  —— *atmosphérique,* atmospheric disturbance.
pesade, f., (man.) rearing of a horse, pesade;
  —— *de chèvre,* pesade in which the horse does not bend his fore legs under him.
pesage, m., weighing; weighing in.
pesant, m., horse that does not rise easily on his fore legs.
pesant, a., heavy, ponderous; slow, sluggish;
  cheval ——, horse sluggish, heavy, in front;
  cheval —— *à la main,* hard-mouthed horse, horse that bears on the bit;
  grain ——, heavy squall;
  vent ——, heavy wind.
pesanteur, f., weight; gravity, force of gravity;
  —— *spécifique,* specific gravity;
  —— *universelle,* universal gravitation.
pèse-acide, m., (chem.) acidometer, acid densimeter.
pesée, f., weighing, weight; amount weighed at one and the same time; effort made with a lever; pull down on a rope;
  faire une ——, to bear down on a lever, to take a purchase.
pèse-esprit, m., alcoholmeter.
pèse-lettres, m., letter scale.
pèse-liqueur, m., areometer; hydrometer.
pèse-papier, m., paper weight.
peser, v. a. n., to weigh, to weigh in; to bear hard; to press; to be heavy; to lie, hang, heavy on; (cord.) to haul down on a rope; (fig.) to weigh, consider;
  —— *sur un anspect,* to bear down on a handspike;
  bascule à, pour, ——, weighing machine;
  —— *brut,* to weigh gross;
  —— *fort,* to bear hard, heavily, on;
  —— *sur un levier,* to bear down on a lever;
  machine à, pour, ——, weighing machine;
  —— *à la main,* (man.) to pull on the bit;
  —— *sur une manœuvre,* (cord.) to pull or haul on a rope;
  —— *net,* to weigh net.
pèse-sel, m., salinometer.
peson, m., steelyard;
  —— *à contre-poids,* steelyard;
  —— *à ressort,* spring balance.
pesse, f., Norway pine.
peste, f., pestilence, plague.
  —— *de guerre,* typhus fever.

phare

pétard, m., (artif.) petard, cartridge; (r. r.) torpedo; (min.) blast hole;
  —— *d'amorce,* (artif.) the primed petard (in a charge of several);
  —— *de cavalerie,* (mil.) demolition cartridge carried by cavalry;
  —— *de circonstance,* (mil.) any petard or torpedo made on the spot out of any material at hand;
  —— *de dynamite, etc.,* (expl.) dynamite, etc., cartridge;
  —— *pour écoles à feu,* (art.) a petard or cartridge to simulate burst of shell on strike;
  —— *de mines,* (min.) blast hole (made by explosives);
  —— *de poudre,* (artif.) (black) powder petard.
pétardement, m., act, action, of blasting, or of destroying by explosives; (min.) blasting;
  trou de ——, (min.) blast hole.
pétarder, v. a., (mil., min.) to blast, to blow up; to destroy, demolish, by blasting.
pétardeur, m., petard worker or thrower.
pétardier, v. *pétardeur.*
péterolle, f., (artif.) cracker.
pète-sec, m., (mil. slang) martinet.
pétillement, m., crackling; (sm. a.) crackling of musketry.
pétiller, v. n., to crackle (of a fire, of musketry, etc.).
petit, a., small, short, little, insignificant;
  —— *autel,* flue bridge;
  ——*s bois,* v. s. v. *bois;*
  —— *équipement,* v. *équipement;*
  —— *état-major,* v. *état-major;*
  ——*e guerre,* v. s. v. *guerre;*
  ——*e monture,* v. s. v. *monture;*
  —— *pied,* (hipp.) coffin bone.
petit-cheval, m., (mach.) donkey engine;
  —— *alimentaire,* (steam) feed-water donkey engine;
  —— *compresseur,* donkey compressor.
petits-vivres, m. pl., (mil.) a generic term for salt, sugar, coffee, dried vegetables, etc.
pétralite, f., (expl.) petralite.
pétrin, m., kneading trough, pug mill.
pétrir, v. a., to knead (dough, clay); to mold, form, make.
pétrissage, m., **pétrissement,** m., kneading.
pétrofracteur, m., (expl.) petrofracteur.
pétrole, m., petroleum, mineral oil.
pétrolerie, f., (rock-)oil refinery.
pétrolier, m., tank steamer.
peuplade, f., colony; tribe.
peuple, m., people, nation; host, throng, crowd.
peuplier, m., poplar (tree and wood);
  —— *blanc,* white poplar;
  —— *franc,* black poplar;
  —— *de Hollande,* v. —— *grisard;*
  —— *grisard,* gray poplar;
  —— *noir,* v. —— *franc;*
  —— *pyramidal,* Lombardy poplar;
  —— *tremble,* aspen.
peur, f., fear.
peureux, a., timorous;
  cheval ——, skittish, shying, horse.
phalange, f., joint (of a finger, a toe); (hipp.) phalanx;
  deuxième ——, (hipp.) cornet bone, coronary or small pastern bone;
  première ——, (hipp.) great pastern bone, fetterbone;
  troisième ——, (hipp.) coffin bone.
phalangette, f., (hipp.) coffin bone.
phalangien, m., (hipp.) phalanx.
phalangine, f., (hipp.) cornet bone.
phare, m., light-house;
  —— *alternatif,* alternating light (red and white lights without eclipse);
  droits de ——, light dues;
  —— *électrique,* electric light-house; (fort.) electric-light observatory or observing station;
  —— *à éclats,* flash light;
  —— *à feu fixe,* steady light;
  —— *fixe à éclats,* flash light (the flash being preceded or followed by short eclipses);
  —— *flottant,* light-ship;

phare, *gardien de* ——, light-house keeper;
— *intermittent,* intermittent light;
— *tournant,* revolving light (the intensity of the light increases and decreases gradually).

pharillon, m., small light-house.

pharmacie, f., pharmacy, medicine chest, dispensary of a hospital;
— *d'approvisionnement,* (*Fr. a.*) depot of medical supplies; distributing pharmacy;
— *portative,* hospital medicine chest;
— *régionale,* (*Fr. a.*) analyzing, distributing, inspecting pharmacy for each *corps d'armée* not provided with a military hospital.

pharmacien, m., apothecary; (*mil.*) military apothecary;
— *militaire,* (*Fr. a.*) military apothecary; the grades, with their assimilation, are as follows:
— *inspecteur, général de brigade;*
— *principal de 1ère (2ème) classe,* colonel (lieutenant-colonel)
— *major de 1ère (2ème) classe,* major (captain);
— *aide-major de 1ère (2ème) classe,* lieutenant (sublieutenant).

phase, f., phase; (*elec.*) phase;
décalage, différence, de ——, (*elec.*) difference of phase;
retard de ——, (*elec.*) retardation of phase;
— *d'un signal,* (*top.*) the phase or appearance of a target, due to light and shade, to the nature of background, etc.

phénique, a., (*chem.*) phenic.

phénol, m., (*chem.*) phenol.

phénylamine, f., (*chem.*) aniline.

phlébite, f., (*hipp.*) phlebitis.

phonographe, m., phonograph.

phonomètre, m., phonometer.

phonoscope, m., phonoscope.

phonotélémètre, phonotelemeter.

phosphore, m., phosphorus;
— *amorphe,* amorphous, red, phosphorus;
— *-bronze,* (*met.*) phosphor bronze;
— *rouge,* v. —— *amorphe.*

photo-avertisseur, m., (*r. r.*) device telling whether a signal lantern is out or not.

photocollographie, f., photocollography.

photocopie, f., (*phot.*) positive, print.

photo-électrique, a., photo-electric.

photogène, m., photogen.

photographe, m., photographer.

photographie, f., photograph, photography;
— *sur bois,* photography on wood;
— *instantanée,* instantaneous photograph;
— *s vitrifiées,* enamels.

photographier, v. a., to photograph.

photogravure, f., photogravure;
— *directe,* line and half-tone block making;
— *de mercure,* mercurography;
— *sur zinc,* zincography.

photo-incision, f., process of making gelatine plates.

photolithographie, f., photolithography.

photolithographier, v. a., to photolithograph.

photolivre, m., (*phot.*) book camera.

photomécanique, a., photomechanical.

photomètre, m., photometer;
— *de dispersion,* dispersion photometer.

photométrie, f., photometry.

photomicrographie, f., photomicrography.

photophone, m., photophone.

photoretardographe, m., (*ball.*) photoretardograph.

photoscope, m., v. *photo-avertisseur.*

phototélégraphe, m., telephote.

phototélégraphie, f., telephotography.

phototype, m., phototype.

phototypie, f., phototypy.

phototypographie, f., phototypography.

phototypogravure, f., line and half-tone block making.

photozincographie, f., photozincography.

3877°—17——21

phrase, f., phrase;
— *d'armes,* (*fenc.*) a succession of blows given and taken without interruption.

physique, f., physics.

piaffement, m., pawing the ground (of horses).

piaffer, v. a., (of a horse) to paw the ground; (*man.*) to execute the *piaffer,* q. v.

piaffer, m., (*man.*) piaffer (air in which the diagonal bipeds are alternately raised and lowered without advancing or backing);
— *dépité,* a mincing piaffer;
— *espagnol,* Spanish walk (i. e., the piaffer well marked);
grand —— *passagé,* an air consisting of the *passage,* without advancing (the motions being ample, clean cut, and well separated).

piaffeur, a., pawing the ground (of horses); (hence) m., a stately horse (one that paws the ground).

piano, m., *jouer du* ——, (*man., slang*) to have disunited gaits.

pic, m., pickax, pick, bush hammer; (*top.*) very high, steep mountain;
à ——, perpendicular; (*nav.*) a-peak;
côte à ——, bold, bluff, shore;
— *à feu,* (*met.*) poker;
— *à feuille de sauge,* (*min.*) miner's pick for sand and gravel;
— *hoyau,* mattock; grubbing pick;
— *d'infanterie,* (*mil.*) portable pickax;
— *à roc,* miner's pick; rock pick;
— *à roc à tête,* poll pick;
— *à rocher,* v. —— *à roc;*
— *à tête,* poll pick;
— *à tranches,* mattock;
— *à tranche et à pointe,* double pick (i. e., with a point and with a chisel edge).

picoche, m., scaling hammer.

picolet, m., clamp, bolt clamp; holdfast.

picorée, f., plundering, marauding.

picorer, v. a., to plunder, maraud, go plundering.

picoreur, m., plunderer, freebooter, marauder.

picot, m., pointed hammer; splinter left on a stump, etc., not cut completely through; (*min.*) wedge used in the framing of a mine shaft.

picotage, m., (*min.*) wedging of a mine shaft.

picoter, v. a., (*man.*) to prick, spur lightly, to feel with the spur; (*min.*) to wedge the framing of a mine shaft.

picotin, m., peck, feed of oats, feed measure.

picrate, m., (*chem.*) picrate.

picrique, a., (*chem.*) picric;
acide —— (*expl., etc.*) picric acid.

pie, f., magpie; (*hipp.*) piebald, calico, horse;
nid de ——, (*fort.*) lodgment in angle of bastion or of demilune or ravelin.

pie, a., (*hipp.*) piebald; (*pie* may be followed by the name of the color which, with the white, produces the piebald effect);
— *bai,* piebald horse more white than bay;
bai ——, piebald horse more bay than white;
cheval ——, piebald horse.

pièce, f., piece, part (as of a clock, of a machine); fragment, bit; cask, barrel; piece of money, coin; apartment, room; apiece, each, (*adm., etc.*) document, voucher; (*carp.*) piece of wood 6′ x 6″ x 6″ (sort of unit of measure); (*met.*) a shingled bloom; (*mach.*) the work; (*cons.*) half tile, half a roofing slate; (*art.*) gun, piece, cannon; (more rarely) gun and carriage, gun and limber, more especially, in France) subdivision of a field battery on a war footing (the first six *pièces* have charge each of a gun and caisson, the seventh of three caissons, the eighth of the forge, etc., and the ninth of the baggage and supply wagon; the men serving several siege or fortress guns; any group of cannoneers.

I. Artillery, etc.; II. Miscellaneous.

I. Artillery, etc.:
— *sur affût d'aviso,* swivel gun;
— *sur affût marin,* (*nav. art.*) gun on truck carriage (obs.);
— *d'alarme,* alarm gun;
— *à âme lisse,* smoothbore gun;
— *à âme rayée,* rifle;
— *d'arrêt,* (*sm. a.*) stop; cut-off;

pièce d'artillerie, gun;
— armée, a piece ready to be served, i. e., with its equipments at hand (siege and seacoast artillery);
—s d'armes, (sm. a.) parts, spare parts, of arms;
— d'armons, iron band embracing end of fore guide;
— d'avant-train, v. — d'armons;
— de, en, batterie, (nav. art.) broadside gun;
— en batterie, (any) gun in battery;
— de bombardement, any piece for delivering curved or high-angle fire with shells of large bursting charge (as distinguished from — de perforation, q. v.);
— bouche, muzzle-loading gun;
— de campagne, field piece;
— de canon, piece of artillery;
— de chasse, (nav. art.) bow chaser;
chef de —, chief of piece;
— de côte, seacoast gun;
— culasse, breech-loading gun;
— de débarquement, (nav. art.) landing gun;
— de décoration, (artif.) set piece;
demi- —, a half detachment;
descendre une —, to dismount a gun;
— de descente, v. — de débarquement;
— détachée, (fort.) detached work;
— de détente, (sm. a.) trigger piece; guard plate;
— directrice, directing gun;
— à double détente, (sm. a.) hair-trigger plate;
— égalisée, gun with no dispart;
— d'embarcation, (nav.) boat gun, landing gun;
— de fermeture, (sm. a.) breechblock; breech-closer proper;
— feu! number .... fire! (command);
— de flanquement, flanking gun;
— folle, piece not bored true;
— de fortification, (fort.) any part of a fortification;
— fusante, piece firing time shell;
— de gros calibre, gun of large caliber;
—s jumelées, guns in pairs in turrets;
— lisse, smoothbore gun;
— de lumière mobile, removable vent piece;
— de mer, seacoast gun;
monter une —, to mount a gun;
— nette, gun well cast and bored true (obs.);
papillon de —, (Fr. art.) a searcher or locator for locating on the map the common target of a group of batteries;
peloton de —, (Fr. art.) the pièce (as defined under main word);
peloton de la —, gun detachment (siege and seacoast artillery);
— percutante, piece firing percussion shell;
— de perforation, high-power armor-piercing gun for direct fire (as distinguished from — de bombardement, q. v.);
— de petit calibre, gun of small caliber;
— -pivot, — à pivot, — sur pivot, pivot, swivel, gun;
— de position, siege gun;
— de pouce, (sm. a.) small plate let into a stock;
— rayée, rifled piece;
— de rempart, a gun permanently mounted, not intended to be removed from a work or fort; wall-piece;
— de retraite, (nav. art.) stern chaser;
— de salut, saluting piece or gun;
— servie, a gun in action or ready to open fire at once, equipments in place and men at their posts (siege and seacoast artillery);
— de siège, siege gun;
— de sûreté, (sm. a.) safety device;
— de surveillance, (s. c. art.) gun commanding as great a field of fire as possible, or having an all-around fire;
tailler une armée en —s, (mil.) to cut an army to pieces;
— de tourelle, turret gun;
— tournante, swivel gun;
— de travers, (nav. art.) broadside gun;
— type, type gun;
— versée en cage, — en panier, gun upset with its carriage upside down;

pièce, voiture- —, gun and limber;
à vos —s, detachments opposite your pieces! (command);
— à vue directe, any piece laid habitually directly on the target (e. g., field gun);
— de n, n-pounder;
— de n po(uces), an n-inch gun.
II. Miscellaneous:
à la —, by the job, piecework;
— à l'appui, (adm.) voucher;
— d'appui, (cons.) breast rail;
— d'arrêt, (mach.) locking piece;
— de blé, field of wheat;
— d'attelage, (r. r.) coupling;
— de bétail, head of cattle;
— de bois, generic term for beam, plank, etc.;
— brute de forge, (met.) rough forging;
— carrée, (carp.) a sort of square or gauge (for trying rectangular assemblages);
— de ceinture, gunwale of a boat;
— chargée debout, generic term for any piece receiving an end stress or load;
— de charpente, (cons.) any piece of light timber (used in construction of ships, carriages, etc.);
— de comptabilité, (adm.) voucher;
— de construction, timber (as of a ship, etc.);
— de contreventement, (cons.) crossbeam;
— de conviction, document establishing proof or facts of guilt, or of the circumstances leading to arrest;
— de cordage, (cord.) length or piece of rope (120 fathoms);
— de coulée, (fond.) casting;
— croisée, (cons.) crosspiece;
— à culotte, (mach.) breeches piece (breeches pipe);
— de décharge, (adm.) relieving document;
demi- —, half barrel;
— de dépense, (adm.) pay account or voucher, statement of money due;
— à eau, water cask;
— d'eau, ornamental water;
— d'écriture, page written with great care;
emporte- — punch; stamp iron;
enlever la —, to punch, to stamp;
— d'entraînement, (mach.) connecting bar;
— d'estomac, flannel waistband;
fausse- —, (fond.) false bottom;
— frondée, anything slung for raising;
— de frottement, friction piece, plate; (harn.) galling leather, safe;
— du harnais, v. — de frottement, s. v. (harn.);
— de jointure, connecting piece;
— justificative, (adm.) voucher;
—s justificatives comptables, (adm.) vouchers proper;
—s justificatives d'ordre, (adm.) generic term for orders, reports, certificates, etc., drawn up and published by superior authority;
— de liaison, (cons.) tie, etc.;
maîtresses —s, (cons.) main pieces (e. g., sills, sleepers);
— de membrure, frame (of a ship);
— de métal forgé, (met.) forging;
mettre en —s, to rend, tear, to pieces;
mettre une —, to patch, piece;
mettre une — en perce, to broach a cask;
— de paiement, (adm.) pay voucher, statement of moneys due;
— de pont, arc or main piece of a pont volant;
—s de pont, (pont.) cross bracing;
— principale, main piece (e. g., of a windlass);
— de rapport, piece, patch, of inlaid or tesselated work; piece let in to conceal a defect; (fond.) drawback, false core;
— rapportée, piece put on, let in, etc., not forming one body with main piece; (fond.) drawback, false core;
rassemblage de —s, patchwork;
— de recette, (adm.) receipt;
— de rechange, spare part; duplicate;
— de recouvrement, butt strap;

pièce(s) *régulières*, (*adm.*) regular routine papers (e. g., invoices and receipts), regular vouchers;
—— *de remplissage*, liner;
—— *de renfort*, stiffening piece, plate, etc.;
—— *de réserve*, spare part;
—— *de terre*, piece, parcel, of land;
de toutes ——*s*, at all points, complete in all respects;
—— *à travailler*, (*mach.*) the work;
travailler à la ——, *à ses* ——*s*, to do piecework, to be paid by the piece;
—— *en treillis*, (*cons.*) framed beam.

pied, m., foot; foot (of a tree, mast, column, of a mountain, of a perpendicular); foot (12 inches); foot rule; leg (as of a telescope stand, etc.); base, footing, stem, support; (*art.*) leg of a gun lift; pry pole of a gin; (*pont.*) leg of a trestle; (*mach.*) end (as of a piston rod); (*mil.*) footing; (*mil. slang*) recruit.

I. Hippology; II. Miscellaneous.

I. Hippology:

blanchir le ——, (*farr.*) v. s. v. *blanchir;*
—— *de bœuf*, toe-crack;
—— *cagneux*, cross foot, parrot-toed foot;
—— *cerclé*, rammy, circled, foot;
changement de ——, (*man.*) change of the leading foot (in galloping);
changer de ——, (*man.*) to change the leading foot (in galloping);
—— *comble*, full foot, pumiced foot;
—— *dérobé*, broken foot;
—— *de derrière*, hind foot;
—— *de devant*, fore foot;
—— *encastelé*, narrow-heeled, contracted, foot;
faire haut le ——, to run away;
faire —— *neuf*, to get a new hoof;
—— *fourbu*, foundered foot;
galoper sur le bon (*mauvais*) ——, (*man.*) to gallop true (false);
—— *grand*, large foot;
—— *gras*, soft foot;
haut le ——, spare (of horses);
—— *s inégaux*, feet unequal in size;
—— *long en pince*, long-toed foot;
—— *maigre*, dry foot;
—— *du montoir*, (*man.*) near fore foot;
—— *hors du montoir*, (*man.*) off fore foot;
—— *mou*, v. —— *gras;*
—— *de mulet*, mule foot;
—— *panard*, outbowed foot;
petit ——, third phalanx; coffin bone;
—— *petit*, small foot;
—— *pinçard*, foot touching the ground only at the toe;
—— *plat*, *plat* ——, flat foot;
—— *à quartiers resserrés*, foot with contracted quarters;
—— *rampin*, a variety of the —— *pinçard*, q. v.;
—— *sec.* v. —— *maigre;*
—— *à talons bas* (*hauts*), low-heeled (high-heeled) foot;
—— *à talons chevauchés*, foot with overriding heels;
—— *à talons fuyants*, foot with sloping heels;
—— *à talons resserrés*, foot with contracted heels;
—— *de travers*, foot unevenly pared; crooked foot.

II. Miscellaneous:

à ——, dismounted; on foot;
armée sur ——, (*mil.*) standing army;
avoir du ——, (of a wall) to have a batter; (of a ladder) to have a safe inclination;
—— *de banc*, (*mil. slang*) sergeant;
en ——*s de banc*, bench-legged;
—— *-de-biche*, crowbar; pinch bar; claw hammer; ratchet, pawl, catch; small iron lever for drawing nails;
—— *bleu*, (*mil. slang.*) recruit;
—— *à calotte sphérique*, ball-joint stand (for plane tables, etc.);
de —— *en cap*, from head to foot, cap-a-pie;
—— *de chevalet*, (*pont.*) leg of a trestle;
—— *-de-chèvre*, pinch bar; crowbar; (*art.*) pry pole (of a gin);

pied *cornier*, corner (i. e., boundary) tree; (*carp.*) corner piece of a framing;
—— *à coulisse*, calipers, sliding calipers, micrometer gauge;
—— *à coulisse de précision*, vernier calipers;
coup de ——, v. s. v. *coup;*
—— *courant*, running foot;
—— *d'une dent*, shoulder of a tooth or cog;
en ——, full length (as a portrait);
—— *étalonné*, standard foot; standard sliding calipers;
être en ——, (*mil.*, etc.) to be on the establishment; to be on full pay;
faux ——, (*pont.*) false leg of a trestle;
de —— *ferme*, obstinately, without flinching; (*mil.*) at, from, a halt; on a fixed pivot; (*fenc.*) without advancing or retiring;
gens de ——, (*mil.*) foot soldiers;
—— *du glacis*, (*fort.*) foot of the glacis;
—— *de guerre*, (*mil.*) war footing, establishment;
—— *de guidon*, (*sm. a.*) front-sight mass;
—— *de hausse*, —— *de la hausse*, (*sm. a.*) foot or base of rear sight;
lâcher ——, *lâcher le* ——, to give way, to lose ground;
marcher du même ——, to keep step;
—— *marin*, sea legs;
mettre sur ——, (*mil.*) to enlist men, raise troops;
mettre —— *à terre*, to dismount, alight, from a horse; to get out of a carriage; to go ashore;
—— *du mur*, foot, footing, of a wall;
—— *du nuage*, blending or continuity of a cloud with the horizon;
à —— *d'œuvre*, v. s. v. *œuvre*, m.;
—— *de paix*, (*mil.*) peace footing, establishment;
perdre ——, to get out of one's depth;
de plain ——, (rooms) on same floor;
—— *poudreux*, deserter, vagrant; (*mil. slang*) "rounder" (man who deserts from one regiment to another);
prendre ——, to touch bottom (as in a ford, etc.);
—— *à profondeur*, depth gauge;
—— *du rail*, (*r. r.*) rail foot;
réduire un plan au petit ——, to draw a plan, etc., on a small scale;
—— *de rustine*, (*met.*) charging door;
sur ——, on foot; standing (as wheat, oats, etc.); (*mil.*) ready, ready to serve;
—— *du talus*, foot of a slope;
—— *à terre!* (*mil.*) dismount! (call and command);
—— *à terre*, (*mil.*, *cav.*) afoot, dismounted;
—— *de la tige du piston*, (*mach.*) piston-rod end;
—— *du vent*, direction in which the wind blows;
y avoir ——, to be within one's depth.

pied-à-terre, m., "box" or "place" where one stops.

pied-droit, m., (*cons.*) supporting wall, abutment, pier (of an arch, etc.); one of the stones of a pier, of an abutment; jamb; small end of a gutter; sheet of lead over the framing of a dormer window;
—— *de cheminée*, chimney jamb;
—— *extrême*, end or main abutment;
—— *de fenêtre*, window jamb;
—— *intermédiaire*, intermediate pier;
—— *de pont*, bridge pier;
—— *de porte*, door jamb;
—— *de voûte*, pier or support of an arch.

piédestal, m., pedestal; stepping-stone.

pied-livre, m., foot-pound.

piédouche, m., pedestal bracket.

pied-tonneau, m., foot-ton.

piège, m., snare, trap, decoy;
dresser, *tendre*, *un* ——, to set a snare.

pierraille, f., broken stone; rubble; road metal.

pierre, f., stone; flint, rock, gravel;
—— *à adoucir*, slip;
—— *à affiler*, whetstone;
—— *de l'air*, aerolite;
—— *à aiguiser*, whetstone, grindstone;
—— *d'aimant*, loadstone;

pierre *almantée*, magnetic stone;
— *angulaire*, corner stone (i. e., at a corner);
— *d'appareil*, ashlar;
— *appareillée*, hewn stone;
— *arénacée*, sandstone;
— *argileuse*, argillaceous stone;
— *artificielle*, artificial stone; brick;
assise de —, course of stone;
— *d'attente*, toothing: block left to be carved;
— *à bahut*, rounded block of a coping;
— *en bossage*, stone left to be carved;
— *en boutisse*, bonder;
— *à briquet*, flint;
— *à broyer*, grindstone (for colors);
— *à brunir*, red hematite; burnishing stone;
— *brute*, unhewn stone; rubble, ashlar;
— *calcaire*, any calcareous stone;
— *carrée*, squared stone;
— *en chantier*, stone ready or in position for cutting;
— *à chaperon*, v. — *à bahut;*
— *à chaux*, limestone (for making lime);
— *s concassées*, broken stone; road metal;
— *de construction*, building stone;
— *coupée*, triangular paving stone (to finish a course laid diagonally);
de —, of stone, stony;
dé en —, stone block;
délarder une —, to dress a stone with the pick;
— *en délit*, stone laid on breaking grain;
— *délitée*, stone that cleaves easily;
— *demi-dure*, building stone of moderate hardness;
— *dernière*, stone on which is cut a date, etc.;
— *dure*, hard stone, building stone;
— *d'échantillon*, stone cut or dressed to proper size;
— *d'émeri*, emery;
en —, of stone, stony;
— *d'encoignure*, quoin;
— *équarrie*, v. — *carrée;*
— *d'évier*, sink, sink stone;
— *à feu*, firestone, (*sm. a.*) gunflint (obs.);
— *filtrante*, — *à filtre*, filter, filtering, stone;
— *de fond*, sole;
— *fondamentale*, foundation stone;
— *de foudre*, v. — *de l'air;*
— *franche*, stone without defects;
— *de fusil*, v. — *à feu;*
garnir de —, to face with stone;
— *gypseuse*, parget;
— *à l'huile*, oil stone, whetstone;
— *infernale*, lunar caustic;
jet de —, stone's throw;
— *en lit*, stone laid on cleaving grain;
— *lithographique*, lithographic stone;
— *de meule*, — *meulière*, — *de meulière*, millstone;
— *milliaire*, milestone;
— *de mine*, iron ore;
— *nette*, stone squared and dressed;
— *noire*, ampelite;
— *à papier*, (stone) paper weight;
— *de parement*, — *de parure*, facing stone;
— *parpaigne*, bonder, heart bond; perpend, perpend stone;
— *à paver*, paving stone;
— *s perdues*, loose stones, thrown into deep water for a foundation, "riprap;"
piquer une —, to pick a stone;
— *plate*, stone slab;
— *à plâtre*, gypsum;
— *pleine*, sound stone, free from all defects;
— *ponce*, pumice;
— *poreuse*, gravel stone;
— *pourrie*, rotten stone;
première —, corner stone;
— *quarrée*, broad stone;
— *à récurer le pont*, (*nav.*) holystone;
— *réfractaire*, firestone;
— *à remouleur*, grindstone;
— *à repasser*, hone;
— *retournée*, stone squared and parallel on opposite sides;
— *rustiquée*, rusticated stone;

pierre *schisteuse*, schistous stone;
— *scintillante*, silicate;
— *de scorie*, slag stone;
— *s sèches*, dry stone;
— *de sujétion*, any stone that can occupy but one place in a construction, whose position is marked;
— *à taillandier*, grindstone;
— *de taille*, ashlar; cut stone;
taille de la —, stonecutting;
— *taillée*, hewn stone;
tailler une —, to dress, cut, a stone;
tailleur de —, stonecutter;
— *tendre*, soft stone;
— *traversée*, cross-hatched stone;
— *velue*, — *verte*, quarry-faced stone.

pierreau, m., (*mil. slang*) recruit; soldier who has been one year in the corps or regiment.

pierrée, f., stone drain, rubble drain; sort of riprap composed of boxes filled with small stones.

pierrelle, f., stone packing of a drain.

pierreux, a., stony.

pierrier, m., (*art.*) (stone) mortar (obs.);
— *fougasse* —, stone fougasse.

pierrière, f., stone pit.

pierrot, m., (*mil. slang*) soldier who shirks and incurs punishment; recruit.

piétage, m., (*nav.*) draft mark.

piétiner, v. a., to trample under foot.

piéton, m., pedestrian; (*mil.*) foot soldier, infantryman (obs.).

pieu, m., pile, spile; stake, palisade, post;
arracher un —, to draw a pile;
battre un —, to drive, sink, a pile;
— *en béton*, a foundation pile formed by filling a pile hole with concrete;
enfoncer un —, v. *battre un* —;
— *d'essai*, proof pile;
faux —, punch;
— *ferré*, ferruled pile;
ficher un —, v. *battre un* —;
— *de fondement*, foundation pile;
— *de garde*, (*fort.*) guard pile;
lardoire de —, shoe of a pile;
rang de —*x*, row of piles;
récéper un —, to square or saw off the head of a pile;
— *en sable*, a foundation pile formed by filling a pile hole with sand;
— *à vis*, screw pile or stake.

piézomètre, m., piezometer.

pif, m., (*hipp.*) horse whose testicles have not descended (*fam.*).

pigeon, m., pigeon; (*mas.*) trowelful;
clou à —*s*, large hooked nail, tenter hook;
— *jeune d'été*, pigeon hatched between February and July;
— *messager*, carrier pigeon;
— *militaire*, carrier pigeon (military service);
— *voyageur*, carrier pigeon.

pigeonnier, m., — *militaire*, dovecote.

pigeonnage, m., (*mas.*) act of laying on mortar by hand.

pigeonner, v. a., (*mas.*) to lay on mortar by hand (without packing).

pignon, m., gable, gable end (of a house); (*mach.*) pinion; cogwheel; (*art.*) lid prop;
— *d'angle*, (*mach.*) bevel pinion or wheel;
— *conique*, (*mach.*) bevel pinion;
— *cylindrique*, (*mach.*) spur pinion;
— *à double denture*, (*mach.*) double-toothed pinion;
— *à double manivelle*, pinion worked by a double-handled winch;
— *droit*, spur pinion;
— *-écrou*, pinion working an inside screw;
— *-galle*, (*mach.*) sprocket;
— *à manchon d'assemblage*, (*mach.*) clutch pinion;
— *de réglage*, regulating adjusting pinion;
— *de renvoi*, transmitting pinion;
— *de serrage*, tightening pinion;
— *de torsion*, twist wheel.

pilage, m., pounding.

pilastre, m., pilaster; stanchion and thick planking;
— *cornier*, corner pilaster;

pilastre *rampant*, —— *de rampe*, post of a sloping balustrade.

pile, f., pile (of shot, etc.); starling of a bridge; bridge pier; masonry pier; vat; stem or trunk of a tree fit to convert into lumber; (*powd.*) hollow (of a powder mill) in which pilon works; (*elec.*) battery, element, cell; (*hipp.*) horse whose testicles have not descended, ridgeling (*fam.*).

(Except where otherwise indicated, the following terms relate to electricity.)

—— *à auges*, trough battery;
—— *au bichromate de potasse*, bichromate battery;
—— *à charbon*, carbon cell; (*in gen.*) pile of charcoal;
—— *au chlorure d'argent*, silver chloride battery;
—— *de compensation*, compensating battery;
—— *constante*, constant battery;
—— *à couronne de tasses*, crown cell battery;
—— *culée*, abutment pier (of a bridge);
—— *à déchiqueter*, (*expl.*) pulper (gun cotton);
démonter une ——, to dismount a battery;
—— *à densité*, gravity battery;
—— *à deux liquides*, two-fluid battery;
—— *à eau*, water cell;
—— *électrique*, electric battery;
—— *étalon*, standard battery;
—— *fluviale*, (*pont.*) river pier;
—— *galvanique*, galvanic, electric, battery;
—— *à gravité*, gravity battery;
—— *hydro-électrique*, hydro-electric battery;
—— *à immersion*, plunge battery;
—— *d'inflammation*, (*torp.*) firing battery;
—— *intermittente*, v. —— *des parcs du génie;*
—— *de ligne*, line battery;
monter une ——, to assemble and connect up a battery;
—— *à mortier*, (*powd.*) bed of a powder mill;
—— *des parcs (du génie)*, bichromate battery, plunge battery (used as a firing battery in the French army);
—— *percée*, (*pont.*) open pier;
—— *plongeante*, v. —— *des parcs du génie;*
—— *de polarisation*, polarization, storage battery;
—— *d'un pont*, (*pont.*) bridge pier;
—— *primaire*, primary cell;
—— *à renversement*, reversing cell;
—— *réversible*, v. —— *à renversement;*
—— *sèche*, dry battery;
—— *secondaire*, accumulator, storage battery, secondary battery;
—— *terrestre*, earth battery;
—— *thermo-électrique*, thermo-electric pile;
—— *à un liquide*, single-fluid cell;
—— *de Volta*, —— *voltaïque*, voltaic battery.

piler, v. a., to pound, bruise;
—— *menu*, to break up small;
—— *du poivre*, (*mil. slang*) to mark time (said of the rear units of a column); (*man.*) to bump up and down in the saddle, not to sit tight.

pileur, m., pounder, beater.

pilier, m., pillar; standard; pier (of a bridge); post (in a stall); (*man., in pl.*) the posts between which a *sauteur* (q. v.) is hitched;
—— *antérieur*, head post (in a stable); ;
—— *d'assise*, bed pile, foundation pile;
—— *b(o)utant*, (*cons.*) close buttress;
monter entre les ——*s*, (*man.*) to take a turn on the *sauteur;*
—— *postérieur*, heel post (in a stable);
sauter entre les ——*s*, (*man.*) to train a horse to jump without losing or gaining ground (by hitching him between the posts);
—— *de station*, (*tech.*) supporting station (rope transmission).

pillage, m., plunder, pillage, looting;
abandonner, livrer, au ——, to give up to plunder.

pillard, m., pillager, plunderer, looter; a., plundering, pillaging.

piller, v. a., to pillage; to plunder, loot.

pilleur, m., pillager, plunderer, looter.

pilon, m., stamp, pestle, crusher; (*powd.*) rammer; (*met.*) shingling hammer; (*mach.*) hammer;

pilon *atmosphérique*, (*mach.*) compressed-air hammer;
—— *du bocard*, (*met.*) stamp;
forger au ——, (*met.*) to work or forge under the hammer;
marteau- ——, v. s. v. *marteau;*
—— *mécanique*, (*mach.*) power hammer;
—— *à planche de friction*, (*mach.*) friction-roll hammer;
—— *à vapeur*, (*mach.*) steam hammer.

pilon(n)age, m., ramming; (*mach.*) hammering, forging.

pilon(n)er, v. a., to ram (earth, concrete, etc.); (*mach.*) to steam hammer, to hammer.

pilon(n)eur, m., workman who rams, etc.

pilonnier, m., man who works a steam hammer.

pilot, m., pile;
—— *d'ancrage*, anchoring pile;
appointer un ——, to point a pile;
arracher un ——*s*, to draw piles;
battre les ——, to drive piles;
—— *de bordage*, gauge, gauged, pile;
coiffer les ——, to cap piles;
enfoncer les ——, to drive piles;
faux ——, false pile, used for sinking a pile below level of the ram or driver;
ferrer le bout d'un ——, to shoe a pile;
—— *de fondement*, foundation pile;
fretter un ——, to hoop a pile;
—— *de grillage*, foundation pile of a grating;
mettre un —— *en fiche*, to plant, drive, a pile;
pointe de ——, foot of a pile;
—— *de pont*, bridge pile;
—— *à rainure*, sheet pile;
—— *de remplage*, —— *de retenue*, filling pile;
retirer les ——*s*, to draw piles;
sabot de ——, shoe of a pile;
saboter un ——, to shoe a pile;
—— *de support*, foundation pile of a grating;
tête de ——, head, top, of a pile.

pilotage, m., pile driving; piling, pile work; pilotage, pilot office;
bois de ——, pile timber, piling;
droits de ——, pilot fees;
—— *d'entrée*, inward pilotage;
—— *de l'État*, government pilotage;
—— *hauturier*, sea pilotage;
—— *à niveau du terrain*, flush (foundation) piling;
—— *obligatoire*, compulsory pilotage;
—— *avec pieux saillants*, projecting pile work;
—— *de sortie*, outward pilotage.

pilote, m., pilot;
—— *autorisé*, licensed pilot;
—— *de canal*, channel pilot;
—— *côtier*, coasting pilot;
—— *de glace*, ice master;
—— *hauturier*, v.—— *de mer;*
—— *lamaneur*, v. —— *côtier;*
—— *major*, pilot master;
—— *de mer*, sea pilot;
—— *de port*, dock, harbor, pilot;
—— *pratique*, unlicensed pilot;
—— *reçu*, licensed pilot;
—— *de rivière*, river pilot.

piloter, v. a., to pile; to drive piles; to strengthen with piles; to pilot a ship;
—— *le terrain*, to drive piles into the ground (as for a foundation).

pilotin, m., pilot apprentice.

pilotis, m., pile; set of piles, piling;
—— *d'ancre*, anchoring picket;
—— *de bordage*, border pile;
—— *creux*, hollow pile;
—— *de garde*, fender piles;
ouvrage de ——, pile work;
pont de ——, v. s. v. *pont;*
—— *de retenue*, cofferdam piling;
—— *à vis*, screw pile.

pin, m., pine (tree and wood);
—— *blanc*, white pine;
—— *jaune*, yellow pine;
—— *de mâture*, Norway pine;
—— *de poix*, pitch pine;
—— *rouge*, red pine, Norway pine;
—— *à trochets*, v. —— *de poix.*

pinasse, f., pinnace.

pinçard, m., horse that wears his shoe fast at the toe; (*man.*) rider that sits tight, with a firm seat.

pince, f., pinchers; pliers; nippers; clamp screw; clamp; pinch bar, crowbar; toe or wedge (of a lever) or pinch bar; tongs; hold, square, of a handspike; trowsers clip (bicycle); purchase; twibill; (*farr.*) toe (of a horseshoe); (*hipp.*) pincher, foretooth; toe of the foot; (*r. f. art.*) clip, catch (Maxim automatic 25 mm.);
— *à amorçage*, (*expl.*) fuse pliers, cutters, or pinchers;
— *d'assemblage*, riveting flange, shoulder, or rim (French boiler);
— *automatique*, (*Fr. art.*) automatic fuse punch;
avoir de la ——, (*man.*) to sit tight;
— *en bois*, vise chops, vise claws, vise clamps;
— *à clichés*, (*phot.*) plate lifter;
— *à côté*, side-cutting pliers;
— *à coulant*, pin-tongs, sliding tongs;
— *coupante*, cutting pliers;
— *à couper*, cutting pliers;
— *à couper et à sertir*, cutting and crimping pliers;
— *à crochet*, small hooked lever;
— *-débouchoir*, (*art.*) adjustable fuse pinchers, fuse punch;
— *à déclic*, tongs of a pile driver;
— *à désamorcer*, (*sm. a.*) unpriming pliers or pinchers, unpriming machine (reloading tool);
— *à double cran*, cutting and crimping pliers (for Bickford and other fuses);
— *de fer à cheval*, (*farr.*) toe;
— *à fil de fer*, wire pinchers;
— *de forgeron*, smith's tongs;
— *à fusée*, (*art.*) fuse drawer;
— *à goupille*, pin pliers;
— *-levier*, pinch bar;
— *à main plate coupant de côté*, side-cutting flat-nose pliers;
— *de manœuvre*, pinch bar, iron handspike;
— *de mineur*, (*mil. min.*) miner's crowbar;
— *-œilleton*, (*art.*) spring clip (for the peep of a sight);
— *à oreilles*, round-nosed pliers;
— *pantalon*, trowsers clip (bicycle);
petites ——*s*, tweezers;
— *à pied de biche*, claw handspike;
— *plate*, flat-nose(d) pliers;
— *pressante et coupante*, cutting and pressing pliers;
— *de pression*, clamp (as on a sextant, etc.);
— *à ressort*, spring pliers, spring catch; a synonym of ——*-œilleton;*
— *à riper*, crow or pinch bar;
— *à sertir*, crimping pliers;
— *tire-culot*, (*art.*) cartridge-head extractor or pinchers (*canon à balles, etc.*);
— *tire-étoupilles*, (*art.*) primer-case extractor or pinchers;
— *tire-goupille*, pin extractor;
— *de treillageur*, wire pliers (lattice maker's);
— *à trois branches*, (*elec.*) three-legged tongs;
— *à vis*, hand vise.

pincé, p. p., (of a lever) caught, pinched, under a load, weight, etc.

pinceau, m., brush; (*mil. slang*) broom; (of balloons, *in pl.*) the two pencils of cords to which the car is directly attached;
— *à goudron*, —— *à goudronner*, tar brush;
— *à peindre*, paint brush.

pince-balle, m., (*sm. a.*) bullet nippers; (*art.*) hot-shot nippers (obs.).

pince-dur, m., (*mil. slang*) an *adjudant*.

pincer, v. a., to pinch, nip, hold fast, press; screw; (*man.*) feel, press, gently with spur;
— *le vent*, (*nav.*) to go close to the wind.

pincette, f., pinchers, nippers; (*in pl.*) tongs, fire tongs, small nippers, tweezers;
— *ronde*, round pliers, round-nose pliers.

pincebeck, m., pinchbeck.

pinçon, m., (*farr.*) clip.

pinne, f., (*fort.*) spear of *chevaux de frise*.

pinnule, f., (*inst.*) vane, sight vane; sight;

pinnule *à crin*, slotted, slit sight;
— *objective*, object vane;
— *oculaire*, eye vane;
— *à œilletons*, aperture sight.

pioche, f., pickax; pick, mattock; (*slang*) hard study; cramming for examination; (at the *Polytechnique*) serious application to mathematics;
— *aux gazons*, (*fort.*) sod mattock;
— *du génie*, (*Fr. eng.*) a pick of special model, used by engineer troops;
— *à marteau*, stone pick;
— *portative*, (*mil.*) portable pick;
temps de ——, (*slang*) examination time.

plocher, v. a., to dig; (*slang*) to study hard; v. n., (*mach.*) said of a fly wheel, the upper part of which turns from the steam cylinders;
— *son examen*, (*slang*) to cram for an examination;
— *l'x*, (*slang*) to apply one's self to mathematics, to "bone math" (U. S. M. A.).

plocheur, m., digger; man who uses a pick; (*slang*) hard student, grind.

pionnage, m., (*mil.*) pioneering.

pionnier, m., (*mil.*) pioneer.

pioupiou, m., (*slang*) familiar term for an infantryman; "doughboy" (U. S. A.); "Tommy Atkins" (British army).

pipette, f., pipette.

pipo(t), m., (*mil. slang*) the *École Polytechnique;* a cadet of the *Polytechnique*.

pique, f., pike; poker;
— *d'abordage*, (*nav.*) boarding pike.

piqué, p. p., (of wood) worm-eaten; (*steam*) pitted (of a boiler); (*met.*) pitted; (*mas.*) pointed and drafted (of the face of a stone); (substantively) a pointed and drafted stone;
moellon ——, v. s. v. *moellon;*
— *de vers*, worm-eaten.

pique-châsse, m., (*artif.*) awl.

pique-feu, m., poker.

piquer, v. a., to prick, prick through; to interweave; to backstitch; to puncture; to prick off; (*man.*) to spur a horse; (*carp.*) to mark out (work);
— *les absents*, to note absentees;
— *la chaudière*, (*steam.*) to scale a boiler;
— *un cheval*, (*man.*) to spur a horse; (*farr.*) to drive a nail into the quick;
— *la cloche*, v. —— *l'heure;*
— *des deux*, (*man.*) to clap both spurs to a horse;
— *l'heure*, (*nav.*) to strike bells;
— *l'horloge*, v. —— *l'heure;*
— *du nez*, (of a torpedo) to turn out of the course;
— *la mazette*, (*fam.*) to be poorly mounted, to ride a poor horse;
— *une pierre*, (*mas.*) to point a stone;
— *en râpe*, (*sm. a.*) to checker a gunstock;
— *le vent*, (*nav.*) to go close to the wind.

piquet, m., picket (small stake), peg, holdfast, stake; tent peg; (*cav., etc.*) picketing post; (*surv.*) sighting, alignment, stake; (*fort., siege*) tracing picket; (*mil.*) picket, detachment; (more specially) detachment or party holding itself in readiness to turn out at a moment's notice for any duty whatsoever, or for duty other than duty by roster, e. g., for an extraordinary guard; (in the field) so much of the fraction on duty for the day as is not on post, men off post; picket (military punishment, so called, obs.), extra duty (U. S. M. A.);
— *d'ancrage*, anchoring, mooring, picket;
— *arrêt*, (*art.*) stake (for the carriage of the *canon-revolver*);
— *d'attache*, picket pin;
au ——, (*mil.*) on picket; standing in one place as a punishment; (call) prisoners, turn out!
— *de campement*, (*mil.*) tent peg;
commander de ——, (*mil.*) to detail, so as to be ready at a moment's notice;
— *de cour d'assise*, (*Fr. law*) detachment of troops furnished a law court, when requested by latter;

piquet, de ——, (*mil.*) ready to turn out at a moment's notice (as the quarter of the *grand' garde*, the remainder resting);
—— *-directeur*, (*fort.*) directing picket;
être au ——, (*mil.*) to be on picket; to be standing in one place as a punishment;
—— *d'exécution*, (*mil.*) firing party (execution);
faire le ——, to stand still in one place;
—— *ferré*, shod picket;
former en ——, to picket;
—— *fretté*, hooped picket;
—— *de gabion*, (*fort.*) gabion stake;
—— *à gradins*, (*sm. a.*) firing rest;
—— *à larder*, (*siege*) fascine picket;
lever le ——, to decamp, strike camp;
—— *à mentonnet*, (*fort.*) anchoring, fascine, picket;
petit ——, (*mil.*) sharpened stake by way of obstacle (in plural, generally);
planter le ——, to encamp, pitch a tent;
—— *de plateforme*, (*art.*) securing stake;
—— *de repérage*, (*art.*) direction stake, steel pin or peg used in the *repérage* of a piece;
—— *de retraite*, holdfast; (*siege*) fascine picket;
—— *saboté*, shod picket;
—— *de service en cas d'incendie*, (*mil.*) fire picket;
—— *-support*, (*art.*) equipment picket or stake (siege battery);
—— *de tenon de manœuvre*, (*art.*) securing picket;
—— *de tente*, tent peg;
—— *de terre*, (*teleg.*) reel picket;
—— *à tracer*, (*top., fort.*) tracing picket.

piquetage, (*mil.*) planting small pickets as obstacles; (*fort.*) staking out the lines or *tracé* of a work.

piqueter, v. a., to stake out (a line); to fasten by means of pickets; to picket; (*fort.*) to trace; (*carp.*) to mark out (the work);
—— *des fascines*, (*fort.*) to picket or secure fascines;
—— *les gazons*, (*fort.*) to picket down sods.

piqueur, m., overseer; stud groom, horse breaker; outrider;
—— *d'écurie*, head groom.

piqûre, f., prick, pricking; puncture; worm hole; backstitching; (*farr.*) pricking; (*met.*) pitting;
—— *de ver*, worm hole (in wood).

pirate, m., pirate.

piraterie, f., piracy.

pirogue, f., canoe.

piron, m., (*mach.*) pin, pivot; lower gudgeon of a vertical shaft.

pirouette, f., (*man.*) pirouette, circle;
—— *sur les épaules*, circle on the forehead;
—— *sur les hanches*, circle on the hind quarters;
—— *militaire*, circle whose vertical axis passes through center of the saddle;
—— *ordinaire*, v. —— *sur les épaules*;
—— *renversée*, v. —— *sur les hanches*.

pirouetter, v. n., (*man.*) to circle.

pisé, m., pisé; clay, rammed earth; pisé work; (*met.*) lining material (ganister, fire clay) for converters, etc.;
en ——, in, of, beaten earth.

piser, v. a., to do pisé work; to construct *en pisé*.

pissalphalte, m., pissasphaltum, mineral tar.

pissée, f., (*met.*) slag duct or channel.

piste, f., track, trace; (*gym., etc.*) cinder path; (*man.*) longing ring, race track; track of a horse at work;
—— *double*, (*man.*) double trace or trail, marked by both fore and hind quarters at the same time;
marcher d'une ——, (*man.*) to make one trace (hind follow fore quarters);
marcher de deux ——s, (*man.*) to make a double trace (quarters on different but parallel tracks);
—— *simple*, (*man.*) single trace or trail.

pister, v. r., (*man.*) (of the horse) to trail himself (in the gallop, to plant one foot of the rear biped on same cross line as diagonally opposite anterior).

pisteur, m., follower (in a flight of pigeons).

pistolet, m., (*sm. a.*) pistol; (*art.*) pistol grip and appurtenances of a rapid-fire gun; (*min.*) head, point, or borer of an earth auger, of a mine auger, of a boring bar;
apprêter le ——, (*sm. a.*) to cock the pistol;
—— *d'arçon*, (*sm. a.*) horse pistol;
—— *double*, (*sm. a.*) double-barreled pistol (obs.);
faire le coup de ——, (*mil.*) of a trooper, to leave the ranks and challenge the enemy to single combat;
haut le ——, (*sm. a.*) raise pistol (position);
—— *de mine*, (*min.*) borer or drill of an earth auger or of a mine auger;
—— *de mineur*, v. —— *de mine*;
—— *de pétardement*, (*min.*) drilling bar;
—— *de poche*, (*sm. a.*) pocket pistol;
—— *à répétition*, (*sm. a.*) magazine pistol;
replacer le ——, (*sm. a.*) to return pistol;
—— *-revolver*, revolver;
—— *-signaleur*, signal pistol.

pistolier, m., good pistol shot.

piston, m., (*mach.*) piston, sucker, forcer, plug, bucket, etc.; (*sm. a.*) tube, nipple (obs.); follower (of a tubular magazine);
—— *aspirant*, valve piston;
—— *à basse pression*, low-pressure piston;
—— *à brosses*, piston with a brush packing;
cercle de ——, piston ring;
—— *de chargement*, (*r. f. art.*) loading piston (Hotchkiss);
clapet de ——, piston clapper;
clef de ——, piston key;
collerette du grand ——, duplicate piston ring;
—— *compensateur*, valve-balance piston;
coquille du ——, piston crosshead;
corps du ——, piston body;
coup de ——, piston stroke;
couronne de ——, junk ring;
course du ——, length of stroke of piston;
couvercle du ——, piston cover;
—— *du cylindre à vapeur*, steam piston;
—— *déplaceur*, (Stirling's hot-air engine) driving piston;
—— *à disques*, leather-packed piston;
distribution à ——, piston-valve motion;
—— *à double effet*, double-action piston;
—— *à double tige*, piston with two rods;
—— *à étoupage*, packed piston;
—— *à étoupage de chanvre (métallique)*, hemp-packed (metallic) piston;
—— *expansif*, expansive piston;
—— *de fermeture*, (*sm. a.*) piston fermeture;
—— *foré*, v. —— *perforé*;
—— *à fourreau*, trunk piston;
garniture de ——, piston packing;
—— *à garniture de chanvre*, hemp-packed piston;
—— *à garniture métallique*, metallic piston;
grand ——, steam piston;
guide de la tige du ——, piston-rod guide;
—— *à haute pression*, high-pressure piston;
—— *intermédiaire*, intermediate-pressure piston;
jeu de ——, length of stroke of a piston, travel;
levée de ——, piston stroke;
marche de ——, length of piston stroke;
—— *métallique*, metallic piston;
—— *moteur*, main piston;
—— *percé*, —— *percé à jour*, v. —— *perforé*;
—— *perforé*, hollow piston, perforated piston;
plaque de ——, piston packing;
—— *plein*, solid piston;
—— *plongeur*, plunger, ram, ram plunger;
—— *pneumatique*, dashpot of the Corliss engine;
—— *de pompe*, sucker; bucket;
—— —— *de pompe foulante*, forcer;
presse-étoupes de ——, piston stuffing box;
—— *à pression intermédiaire*, intermediate pressure piston;
queue de ——, tailpiece of a piston;

piston à *rainures*, grooved piston;
—— *récepteur*, main piston;
ressort de ——, packing spring;
—— *à simple effet*, single-action piston;
—— *à soupape(s)*, valve piston;
soupape de ——, piston valve;
—— *suédois*, Swedish piston;
T (ê) de (tige de) ——, piston-rod crosshead;
tête du ——, piston crosshead;
—— *-tige*, —— *à tige*, rod piston;
tige de ——, piston rod;
—— *à tige compensatrice*, (*art.*) compensating piston;
—— *tournant*, radial piston or plate (as of an oscillating pump);
—— *à vapeur*, steam piston.

pitaine-crayon, m., (*slang*), orderly who waits on drawing classes (*Ecole Polytechnique*).

pitance, f., (*Fr. a.*) soldier's expression for his share of the ration.

pitchpin, m., Georgia pine, yellow pine.

pite, f., *Agave americana;* coir; m., hemp, cordage, made from the agave.

pithomètre, m., cask gauge.

piton, m., eyebolt, ringbolt, screw ring, ring, eye; (*top.*) peak, top, of a mountain; sharp point or crag or spur; cone- or point-shaped mountain;
—— *d'affût*, (*art.*) eyebolt on a carriage;
—— *d'amarrage*, anchoring eyebolt;
—— *à anneau*, eyebolt, ringbolt;
—— *s et anneaux*, (*sm. a.*) loops and rings (of a scabbard);
—— *à boucle*, ringbolt;
—— *carré*, crutch;
—— *de chargement*, packing eye or ring;
—— *de clameau*, cramp-iron eyebolt;
—— *à croc*, hook bolt;
—— *à crochet*, ring hook;
—— *double*, double eyebolt;
—— *à fourche*, shackle bolt; crutch;
—— *de ganse*, (*art.*) carrying ring and bolt (projectile);
—— *à œil*, eyebolt;
—— *de retour*, (*art.*) guide ring (for a lanyard);
—— *de sellette*, (*art.*) baggage-strap staple;
—— *de tirant*, eye of a cramp;
—— *à vis*, screw eyebolt;
—— *à vis à bois*, wood screw eyebolt.

pitte, f., coir;
cordage en ——, coir rope.

pivot, m., pivot, axis, spindle; pin, stud; taproot of a tree; stud (or pin) of a spring in general; (*mil.*) pivot; (*sm. a.*) bridle stud;
abattre un arbre en ——, to fell a tree through its taproot;
affût à ——, (*art.*) pivot carriage;
—— *d'un arbre*, (*mach.*) journal, gudgeon;
—— *d'articulation*, (*mach.*) shipper dog;
assembler à ——, to pivot together;
—— *avancé*, (*art.*) embrasure pivot;
—— *de boussole*, center pin of a compass;
—— *de cabestan*, spindle of a capstan;
canon à ——, (*art.*) swivel gun;
—— *de compas*, v. —— *de boussole;*
—— *fixe*, (*mil.*) fixed pivot;
fourche- ——, (*art.*) fork pivot;
—— *à fourchette*, (*art.*) fork pivot;
—— *hydraulique*, (*fort., mach., art., etc.*) hydraulic pivot;
—— *de manœuvres*, (*fort.*) a fortified position regarded as a pivot in the operations of an army;
—— *mobile*, movable pivot;
monté sur ——, pivoted;
—— *mouvant*, (*mil.*) movable pivot;
à —— *mouvant*, (*mil.*) on a movable pivot;
—— *de la noix*, (*sm. a.*) tumbler stud;
—— *du ressort de batterie*, (*sm. a.*) hammer-spring stud;
—— *stratégique*, (*fort.*) a name sometimes given to a fortress or fortified position in virtue of its influence on strategical operations;
—— *-support*, (*r. f. art.*) pivoting support.

pivotement, m., (*art.*) pivoting (ease of training or traversing).

pivoter, v. n., to pivot; to fell a tree through its taproot; (*mil. slang*) to be drilled, to be on fatigue;
faire ——, to slue.

placage, m., (*carp.*) veneering; (*fort.*) rammed earth facing; (*cons.*) facing wall (against a backing of rock); (*met.*) plating; (*mas.*) loam mortar.

placaque, m., (*mas.*) loam mortar.

placard, m., placard; (diplomatic) document on parchment and not folded; sort of flat pulley; galley proof; cupboard (*nav.*) scupper leather; (*carp.*) panel.

placarder, v. a., to placard; (*carp.*) to panel.

place, f., space, room, ground; seat (in a train, etc.); spot; stall (in a stable); square (of a town); town; change (commercial); employment, office, post, position, place, standing (in one's class); (*mil.*) fortified town or position; garrison town;
—— *d'appui*, (*mil.*) position of support in military operations; (*fort.*) a position supporting a —— *à camp retranché*, q. v.; a supporting fort;
—— *d'armes*, (*mil.*) parade, parade ground; open space or place of assembly in the center of a town; general parade or assembling ground of a camp or cantonment; alarm post; frontier town serving as a depot of supplies; (*fort.*) place of arms; (*siege*) parallel; (sometimes) approaches;
—— *d'armes rentrante*, (*fort.*) reentrant place of arms;
—— *d'armes retranchée*, (*fort.*) retrenched place of arms;
—— *d'armes saillante*, (*fort.*) salient place of arms;
artillerie de ——, garrison artillery;
—— *s basses*, (*fort.*) the lower part (casemates, etc.) of a bastion flank;
batterie de ——, v. s. v., *batterie;*
—— *à camp retranché*, (*fort.*) position or city defended by a *camp retranché*, q. v.;
—— *à ceinture de forts*, (*fort.*) a position surrounded by a girdle of forts;
—— *de la chaudière*, (steam) boiler room;
—— *comptable*, (*mil.*) depot of issue and of receipt; (*Fr. art.*) *arrondissement* where a *garde d'artillerie* is stationed;
—— *de dépôt*, (*fort.*) generic term for fortifications considered as centers of supply or as supply depots;
—— *avec enceinte et ceinture de forts*, v. —— *à forts détachés;*
—— *d'essieu*, letting up for the axletree;
—— *d'exercice*, (*mil.*) drill ground;
—— *forte*, (*fort.*) fortress; fortified town;
—— *à forts détachés*, (*fort.*) usual name of a modern fortified position (consisting of an interior position, *enceinte* or *noyau*, and an exterior line of forts);
—— *à fossés d'eau*, (*fort.*) fort with wet ditches;
—— *à fossés secs*, (*fort.*) fort with dry ditches;
—— *frontière*, (*fort.*) frontier fort, position, town;
—— *de garnison*, (*mil.*) garrison town;
grandes —— *s*, high offices of the government;
—— *de guerre*, (*fort.*) fortified town; fortified place or position (isolated fort);
homme de ——, a high-placed official;
—— *de la machine*, (*mach.*) engine room;
—— *de manœuvre*, (*fort.*) a name given to fortifications with reference to their influence on operations;
—— *à manœuvres d'eau*, (*fort.*) a fort whose ditches may be filled with or emptied of water at will;
—— *maritime*, a military port;
mettre en ——, to stow away, put into its proper place; to prepare, organize;
mise en ——, putting in place, stowage;
—— *mobile*, v. *forteresse mobile;*
—— *ouverte*, (*fort.*) open town;
—— *ouverte avec forts extérieurs*, (*fort.*) open town defended by exterior forts;
—— *à ouvrages détachés*, (*fort.*) town surrounded by detached works;

place, *petite* ——, (*fort.*) small fortress in a very strong position;
—— *de refuge*, (*fort.*) name given to a fortress regarded as a refuge for a beaten army;
—— *à simple enceinte*, (*fort.*) place defended by an enceinte only (no detached or outer works);
—— *de simple fabrication*, (*Fr. a. adm.*) place in which contractor simply makes the bread, flour being supplied by the administration;
—— *de sûreté*, (*fort.*) place held as a guarantee of the execution of a treaty;
—— *vide*, (in a forest) an open clearing;
volte sur ——, (*man.*) leap (in the horse's tracks);
à vos ——*s*, (*mil.*) posts! (command).

placé, p. p., placed; placed (in a race);
cheval bien ——, horse that holds his head well.

placement, m., placing, setting; investment (money).

placer, m., (*man.*) upright stand or attitude of a horse; normal, proper, position.

placer, v. a., to place, put, set, seat; to sell; to invest; to find a place, a position, for anyone;
—— *de l'argent*, to put money out at interest;
—— *bien un coup*, (*fenc.*) to thrust home;
—— *de champ*, to set up on edge;
—— *un cheval*, (*man.*) to keep a horse in equilibrium;
se —— *à cheval*, (*man.*) to take the correct position on horseback;
—— *un homme à cheval*, (*man.*) to make a rider take the proper seat;
—— *une sentinelle*, (*mil.*) to post a sentry.

plafond, m., (*tech., cons.*) ceiling; top, ceiling, of a caisson; crown (of a fire box, etc.); bottom (of a canal, of a reservoir); under face (of a staircase); (*min.*) roof of a gallery; (*nav.*) floor of a ship; floor grating of a boat;
—— *en augets*, (*cons.*) ceiling with the spaces between joists filled with plastering (trough-shaped on top);
—— *à caissons*, (*cons.*) coffered ceiling;
faux ——, (*cons.*) false ceiling.

plafonnage, m., (*cons.*) ceiling, ceiling work.

plafonner, v. a., (*cons.*) ceil.

plafonneur, m., ceiling plasterer.

plage, f., beach, strand; sandy shore; quarter of the wind;
—— *de galets*, shingly beach;
les quatre ——*s principales*, the cardinal points;
—— *de sable*, sandy sloping coast.

plaidant, a., (*law*) litigant;
avocat ——, barrister.

plaider, v. a. n., (*law*) to plead, to argue a case; to go to law with.

plaideur, m., (*law*) suitor.

plaidoirie, f., (*law*) pleading, plea.

plaidoyer, m., (*law*) speech, argument.

plaie, f., wound, sore;
—— *de marche*, (*mil.*) foot sore, foot blister.

plaignant, m., (*law*) plaintiff.

plain, a., flush (with anything); level. even, plain, m., the high seas;
de ——*-pied*, on the same floor;
le ——*-pied*, rooms, quarters, on the same floor.

plaine, f., (*top.*) plain;
—— *d'eau*, broad stretch of calm waters.

plainte, f., complaint;
—— *en conseil de guerre*, (*mil. law*) charge and specifications;
loger ——, *porter* ——, to lodge a complaint.

plan, a., plane;
angle ——, plane angle;
carte ——*e*, plain chart; Mercator's chart;
——*-concave*, (*opt.*) plano-concave;
——*-convexe*, (*opt.*) plano-convex;
figure ——*e*, plane figure;
navigation ——*e*, plane sailing;
nombre ——, a number containing but two factors;
surface ——*e*, plane surface.

plan, m., plane, plane surface; plan, scheme, project, design; ground (as foreground, background) of a picture; tier (as of barrels); (*top. drawing*) ground plan, ground plot, (more specially) plot, plat, map of a small area of ground, detailed map;
arrière- ——, background;
—— *d'arrimage*, (*nav.*) tier of the hold;
—— *automoteur*, self-acting plane;
—— *cadastral*, (*surv.*) cadastral plat;
—— *de campagne*, (*mil.*) plan of campaign;
—— *de chape*, (*cons.*) v. s. v. *chape;*
—— *de collimation*, (*inst.*) plane of collimation;
—— *de combat*, (*mil.*) plan of battle;
—— *de comparaison*, (*fort.*, etc.) ground plane, plane of reference, reference plane;
—— *de construction*, working drawing;
—— ——*s cotés*, v. *méthode des* ——*s cotés;*
—— *coupe*, sectional plan, plan and section;
—— *des crêtes*, (*fort.*) in defilading, the plane in which the crests of the work or works must lie;
—— *de décharge*, discharge plane;
—— *de défilement*, (*fort.*) plane of defilement, of defilade;
—— *de départ*, (*ball.*) plane of departure;
—— *de détail*, plot or plan showing details;
—— *diamétral*, sectional elevation;
—— *directeur* (*des attaques*), (*siege*) map or chart on which are plotted all the positions, trenches, batteries, etc., in the siege of a place; directing plan;
—— *directeur d'ensemble des attaques*, v. —— *directeur;*
—— *directeur des attaques rapprochées*, (*siege*) directing plan of the attack at close quarters;
—— *de direction*, plane of direction;
—— *d'eau*, the level of the water in a river;
—— *d'écartement*, plane of spread;
—— *d'enfilade*, v. —— *de revers;*
—— *d'engagement*, (*mil.*) c. o.'s plan for getting troops into action;
—— *d'ensemble*, general plan (of a fort, etc.);
—— *d'épreuve*, (*elec.*) proof plane;
être au ——, (*mil. slang*) to be in arrest;
—— *d'exécution*, plan of execution; working drawing, plan;
faire le —— *de*, to survey;
faire l'élévation d'un ——, to construct the elevations of a plan;
faire un ——, to draw, make a plan;
faire lever le —— *de*, to survey;
—— *des fibres invariables*, neutral surface, plane of neutral fibers;
—— *de flexion*, plane of flexion;
—— *de fondation*, foundation plane;
—— *des forces*, force plan;
—— *du foyer*, focal plane;
—— *fuyant*, vanishing plane;
—— *géométral*, ground plan;
—— *de glissement*, (*steam*) slide;
—— *géométrique*, plan;
—— *graphique*, any plan of a town, fort, building, to given scale;
—— *de guerre*, plan of war;
—— *horizontal*, ground plan; horizontal plane;
—— *d'immersion*, (*torp.*) plane of immersion;
—— *d'incidence*, plane of incidence;
—— *incliné*, inclined plane;
—— *de joint*, (*fond.*) parting plane;
levée d'un ——, survey;
lever un ——, to make, draw, get up, a plan;
lever des ——*s*, to survey;
—— *longitudinal*, longitudinal plan;
—— *manœuvre*, engine plane;
méthode des ——*s cotés*, system in which elevations are represented solely by references on the ground plan (as in fortification drawing); "one-plane descriptive" (U. S. M. A.);
—— *de mire*, (*ball.*) plane of sight;
—— *de mobilisation*, (*mil.*) plan of mobilization; (*fort.*) plan of the measures necessary to put a fort in a good state of defense;
—— *de niveau*, datum plane;
—— *parcellaire* (*du cadastre*), v. s. v. *parcellaire;*

plan *perspectif*, perspective plane;
— *premier* ——, foreground;
— *principal*, general plan, principal plan;
— *de projection*, (*ball.*) plane of projection;
— *rampant*, (*fort.*) plane of slope;
— *de réflexion*, plane of reflection;
— *de réfraction*, plane of refraction;
— *relevé*, vertical projection of the roofs and other upper parts of a building; roof plan;
— *relief*, —— *en relief*, (*top.*) relief map, map showing relief of ground;
— *de repère*, datum plane;
— *de revers*, (*fort.*) plane of reverse defilade;
— *de rez*, datum plane;
— *second* ——, middle ground;
— *de site*, (*fort.*) plane of site (in defilading), rampant plane (parallel to and 1.5 meters below the plane of defilade); (*ball.*) plane perpendicular to the plane of fire and containing the *ligne de site;*
— *de site artificiel*, (*fort.*) plane of site on sloping ground;
— *de site naturel*, (*fort.*) plane of site on level ground;
— *de symétrie*, plane of symmetry;
— *du tableau*, directing plane;
— *de terrain*, plot;
— *de tir*, (*ball.*) plane of fire;
— *de tir initial*, (*ball.*) initial plane of fire;
— *de tiroir*, (*mach.*) slide face;
— *topographique*, topographical map or plan;
— *tracer un* ——, to draw, make, a plan;
— *transversal*, cross section;
— *troisième* ——, background;
— *vertical extérieur*, elevation;
— *vertical intérieur*, section, sectional elevation;
— *de visée*, plane of collimation;
— *à vol, à vue, d'oiseau*, bird's-eye view.

planche, f., plank, board; shelf, plate; (*nav.*) gang plank; putting in port; (*farr.*) planch shoe; (*harn.*) foot rest (of a stirrup); (*sm. a.*) leaf of rear sight; (*top.*) plane table;
— *armée de clous*, (*mil.*) spiked board (accessory defense);
— *aux armements*, (*art.*) in a turret, equipment shelf;
— *d'arrimage*, (*nav.*) dunnage plank;
— *à bagages*, (*mil.*) clothing shelf in a squad room;
— *à bascule*, tilting, rocking, plank;
— *de caisson*, (*art.*) board or plank for an ammunition wagon;
— *de ciel*, (*mil. min.*) top plank of the casing of a mine gallery, overhead plank(s), overhead sheeting, roofing plank;
— *circulaire*, (*siege*) gabion bottom;
— *à clous*, v. —— *armée de clous;*
— *de coffrage*, (*mil. min.*) plank lining, sheathing;
— *de côté*, v. —— *porte-fourrage;*
— *côtière*, (*cons.*) (in a roof) outer plank of a gutter;
— *à débarquer*, (*nav.*) gang plank;
— *à dessin*, drawing board;
— *-directrice*, (*fort.*) wooden bottom, for making gabions;
— *à effets*, (*mil.*) equipment, clothing, shelf, in a squad room;
— *d'embarcation*, gang plank;
— *entière*, full-sized plank;
— *s d'entrevoux*, (*cons.*) planks covering spaces between joists;
— *faire* ——, (of the sea) to be calm;
— *faire la* ——, to float, in swimming; (*fig.*) to lead the way; to show how to do anything;
— *fausse* —— *à mouler*, (*fond.*) rolling-over board;
— *de fond*, (*fond.*) molding board, bottom board; (*pont.*) flooring board of a pontoon;
— *de friction*, (*mach.*) friction-roll plank;
— *grande* ——, (*nav.*) main gang plank;
— *de hausse*, (*sm. a.*) sight leaf;
— *jointive*, sheet pile (feathered and grooved);
— *jours de* ——, (*nav.*) lay days;

planche-*marchepied*, footboard of a wagon; (*art.*) footboard;
— *marteau à* ——, (*mach.*) drop hammer controlled by friction or geared rolls (the hammer head is attached to a board);
— *minute*, (*top. drawing*) engraved plate of a minute;
— *mobile*, loose plank; (*sm. a.*) movable leaf, leaf (of a sight);
— *à mouler*, (*fond.*) bottom, molding, board;
— *de mulet*, (*farr.*) planch shoe;
— *à pain*, bread tray; (*Fr. a.*) bread shelf in a squad room;
— *porte-fourrage*, (*art.*) side board for carrying forage (ammunition wagon);
— *de profil*, (*art.*) chamber gauge;
— *de reproduction*, (*sm. a.*) guide plate (for shaping the stock);
— *à rétablissement*, (*gym.*) plank fixed to a wall for exercising men in pulling themselves up;
— *de roulis*, (*nav.*) washboard; leeboard (of a berth);
— *à trousser*, (*fond.*) loam board.

planchéiage, m., planking, boarding.

planchéier, v. a., to board, plank; to cover with planks; to floor; to lath.

plancher, m., (*cons.*) floor, flooring (strictly, a wooden floor); ceiling, covering, roof; framing on which the roofing rests; platform, stage; (*harn.*) stirrup tread; (*r. r.*) engine driver's stand; (*tech.*) horizontal position of a stone;
— *aire de* ——, floor proper, surface of a floor;
— *bois de* ——, flooring;
— *charge de* ——, thickness added to a floor (to bring it up to the level of surrounding floors);
— *de chargement*, (*art., met.*) loading platform;
— *de ciel*, (*mil. min.*) roof planks of a mine gallery;
— *à compartiments*, paneled floor;
— *creux*, hollow floor;
— *d'une darse*, apron;
— *d'une écluse*, bottom of a sluice;
— *faux-* ——, (*cons.*) false floor (to reduce the height of ceiling); sound boarding;
— *de frise*, parquetry floor;
— *d'en haut*, ceiling;
— *hourdé*, rubble floor;
— *hourdis de* ——, filling between joists;
— *du lavoir*, (*met.*) slime pit;
— *ordinaire*, floor made of full-sized planks;
— *perdu*, v. *faux-* ——;
— *plein*, a floor filled in solid (masonry, etc.) between joists;
— *d'un pont*, roadway of a bridge;
— *à remplissage*, v. —— *plein;*
— *roulant*, (*r. r.*) traveling platform;
— *des vaches*, (*fam.*) dry land, terra firma;
— *voûté*, vaulted floor.

plancher, v. n., (*mil. slang*) to be confined in the cells or guardroom.

planchette, f., small board; (*harn.*) lady's stirrup; (*art.*) crossbar, deflection bar (of a sight); (*mil. min.*) trap, box trap; (*teleg. and gen.*) wire board (in a telegraph office); (*inst.*) plane table; (*mil.*) cleaning board (leather equipment);
— *aux armements*, (*art.*) small equipment shelf (siege battery);
— *à astiquer*, (*mil.*) board for cleaning and polishing leather parts of equipment;
— *de batterie*, (*art.*) battery plane table (*Fr. art.*); traversing, pointing, scale; deflection scale;
— *à calotte sphérique*, (*inst.*) plane table with spherical cup or bearing;
— *de chargement*, (*art.*) loading tray;
— *de convergence*, (*inst.*) circumferentor;
— *à courbes d'égal angle de tir*, (*art.*) a scale giving the angle of elevation for a given target and the bearing of the line from gun to target;
— *déclinée*, v. —— *orientée;*
— *des dérives*, (*art.*) deflection leaf, deflection scale (of a sight);
— *à douille*, (*inst.*) plane table with socket bearing or spindle;

planchette *graduée de pointage*, (*art.*) aiming scale, pointing scale;
— *de hausse*, (*sm. a.*) sight leaf;
— *d'inflammation*, (*art.*) little tongue of wood used in some shrapnel or shell to facilitate the combustion of the burster and to prevent its caking;
— *de lever*, (*inst.*) plane table;
— *d'objectif*, (*phot.*) lens board;
— *orientée*, (*inst.*) plane table, oriented set to or on given bearings;
— *photographique*, (*inst.*) photographic plane table;
— *porte-armements*, (*art.*) small equipment tray (of an ammunition chest);
— *de pression*, (*art.*) in an ammunition chest, a strip or plank pressed down on the contents to keep them from moving;
— *à règle à curseur*, (*art.*) a scale giving angles of elevation and the distance apart of the collars on a *règle à curseur*, q. v.;
— *de repérage*, v. —— *de tir*;
— *de soutien*, any block or bit of wood, etc., placed under a tie or strap to assist in bearing the stress;
— *de tir*, (*Fr. art.*) range-laying scale (gives elevation, direction, charge, and time of flight; consists of a map, 1:20000, of the field of fire, suitably mounted, graduated, etc., to furnish the necessary indications);
— *à trous*, (*sm. a.*) pierced plank or tray for holding cartridges.
plançon, m., plank timber.
plane, f., chisel, turning chisel; spokeshave; drawing knife; m., plane (tree);
— *de charron*, spokeshave.
planer, v. a., to smooth, planish; (of a balloon) to soar over (a position, etc.);
machine à ——, jointing machine.
planeuse, f., jointing machine.
planigraphe, m., (*inst.*) planigraph.
planimètre, m., planimeter.
planimétrie, f., mensuration, planimetry; operation of making a plat or plan of a given region.
planitude, f., flatness;
— *des champs*, (*opt.*) flatness of field.
planomètre, m., planometer, surface plate.
plantaire, a., (*hipp.*) plantar.
plantation, f., planting; plantation of trees; (*fort.*) trees, shrubs, etc., used to screen a work.
planté, p. p., erect;
bien ——, well situated, placed;
poil ——, (*hipp.*) staring coat, hair erect, bristling.
planter, v. a., to plant, set, fix, set up; to place, drive in; (*cons.*) to prepare a foundation; (*fort.*) to set (palisades, etc.);
— *des échelles*, to plant or set up ladders;
— *un édifice*, (*cons.*) to begin the construction of a building;
— *l'étendard*, (*mil.*) to set up or raise the standard, the flag (over a captured town, a work carried by assault, etc.);
— *le piquet*, to encamp, take up one's quarters.
planton, m., (*mil.*) orderly;
— *à cheval*, mounted orderly;
— *de conseil de guerre*, orderly of a court-martial;
être de ——, to be on as orderly.
plaque, f., plate; slab; sheet of metal, leaf of metal, star of an order; (*met.*, etc.) back, back plate; (*art.*) face plate; (*nav.*, *fort.*) armor plate; (*sm. a.*) sword-belt plate; part of guard surrounding the hand; (*unif.*) helmet-, cap-plate; (*r. r.*) sole (of a railway chair).
I. Artillery, etc.; II. Miscellaneous.
I. Artillery, etc.:
— *en acier*, steel armor plate;
— *d'appui*, (*art.*) face plate; wheel guard;
— *d'appui de crosse*, (*art.*) trail plate;
— *d'appui mobile*, (*r. f. art.*) spring bar;
— *d'appui d'obturateur*, (*art.*) obturator plate, gas plate;
— *d'armons*, (*art.*) foreguide plate;

plaque *d'arrêt*, (*art.*) stop or keep plate of an ammunition box;
— *de blindage*, armor plate;
— *à boutonnière(s)*, (*unif.*) end of shoulder (cuirass) strap;
— *de butée*, (*art.*) front part or face of the firing pin recess (*canon à balles*);
— *carrée*, (*art.*) trail transom plate;
— *de ceinturon*, (*mil.*) waist-belt plate; sword-belt plate;
— *de coin*, (*art.*) wedge plate of the Krupp fermeture;
— *de combat*, (*nav.*) shot plug;
— (*en*) *compound*, compound armor plate;
— *de contact*, (*torp.*) circuit-closing plate (of a ball-contact circuit closer);
— *de couche*, (*sm. a.*) butt plate;
— *de crosse*, (*art.*) trail plate;
— *de cuirasse*, —— *de cuirassement*, armor plate;
— *de culasse*, (*art.*) breech plate, rear barrel plate (Hotchkiss, etc.);
— *de déclanchement*, (*art.*) firing-pin plate (*canon à balles*, etc.);
— *de dessous*, (*art.*) bottom plate (of a trail); bottom axle plate;
— *de dessus*, (*art.*) top plate (of a trail);
— *directrice*, (*r. f. art.*) breechblock;
— *d'embrasure*, (*fort.*) port plate;
— *d'épaulement*, (*Fr. art.*) upper reenforce of the pry pole of the steel gin;
— *d'épreuve*, (*art.*) proof or test plate;
— *d'extraction*, (*art.*) base plate of Boker cartridge, by which it was extracted;
fausse —— *d'appui*, (*art.*) inserted face plate;
— *en fer*, iron armor plate;
— *en fer et en acier*, compound plate (armor);
— *en fer forgé*, wrought-iron armor plate;
— *en fer laminé*, laminated armor plate;
— *en fonte durcie*, chilled cast-iron armor plate;
— *de front*, (*r. f. art.*) front barrel plate;
— *de fondation*, (*art.*) base plate;
—*s de garde-champêtre*, (*mil. slang*) old sergeant's stripes;
— *de garde*, (*art.*) axle strap; garnish plate;
— *à godet*, (*Fr. art.*) sort of embarring or maneuvering plate (obs.);
— *de hausse*, (*art.*) elevation plate or slide (German artillery);
— *d'identité*, (*mil.*) identification tag or plate;
— *isolante*, (*art.*) curved sheet of metal formerly used with muzzle-loading rifle, surrounding rear of cylindrical part of projectile to prevent ballotting (obs.);
— *de large*, steering rudder (of a torpedo);
— *de lunette*, (*art.*) pintle plate;
— *mixte*, compound armor plate;
mortier à ——, v. s. v. *mortier*;
— *en nickel-acier*, nickel-steel armor plate;
— *à oreilles*, (*art.*) metallic base plate of French *boîte à mitraille*; hinge plate and lugs of a breech-loading gun; (*sm. a.*) repeating-mechanism plate (of the Lebel rifle);
— *de pivot*, (*art.*) pivot plate;
— *de plomb*, (*art.*) leaden vent apron;
— *de pont*, (*nav.*) deck armor plate;
— -*pontet*, (*sm. a.*) trigger and plate;
— *porte-cartouche*, (*r. f. art.*) cartridge plate;
— *de poulie*, (*art.*) pulley-supporting plate (of a gin);
— *de protection*, (*r. f. art.*) deflector;
— *de recouvrement*, (*art.*) face plate; breech cap (Krupp fermeture); (*sm. a.*) frame cap, side plate (of a revolver);
— *de recouvrement d'essieu*, (*art.*) top axle plate;
— *de repérage*, (*art.*) sort of deflection scale (German artillery);
— *de semelle*, (*art.*) swing, bedplate;
— *surcémentée*, face-hardened armor plate;
— *de tête*, (*art.*) top or head plate of a gin;
— *de tête d'affût*, (*art.*) breast transom, front transom;
— *de tir*, (*art.*) face plate;

plaque *de toiture*, roof place of a Grüson turret, etc.;
—— *de tuyère*, boss or back plate of a traveling forge;
—— *de tuyère de contre-cœur*, boss or plate of a field forge;
—— *d'union*, (*pont.*) assembling or joining plate.
  II. Miscellaneous:
—— *d'about*, (*r. r.*) shoe, bedplate;
—— *d'accouplement*, junction plate;
—— *antérieure*, (*met.*) front plate;
—— *d'appui*, locking plate (wagons); (*harn.*) galling leather;
—— *d'arrêt*, (*élec.*) terminal plate;
—— *d'arrière*, back plate;
—— *d'assemblage*, assembling plate or strap; junction plate;
—— *d'assise*, socket of a coiled spring;
—— *d'assise pour rails*, (*r. r.*) bedplate;
—— *d'avant*, front plate;
—— *en bois*, veneer;
—— *de boue*, mud guard, cuttoo plate;
—— *de buttoir*, (*r. r.*) buffer disk;
—— *de chio*, (*met.*) (or simply *chio*) tap-hole plate (of a firing forge);
—— *cisaille*, (*mach.*) shearing or cutting plate;
—— *de contrevent*, (*met.*) blast plate;
—— *de cylindre*, (*mach.*) cylinder cover;
—— *de dame*, (*met.*) dam plate;
—— *diaphragme*, division plate;
—— *disjonctrice*, disjunction plate;
—— *de division*, (*mach.*) division, index, plate;
—— *d'enclume*, anvil face;
—— *entière*, (*phot.*) whole plate;
—— *entre les portes et les barres du fourneau*, dead plate, coking plate;
—— *excitatrice*, (*élec.*) exciting plate, electrode;
—— *de feu*, back of a chimney;
—— *filière*, (*mach.*) draw plate; bar plate;
—— *de fixation*, sole, bedplate, of an engine, a machine;
—— *fixe*, dead plate;
—— *de fond*, (*met.*) bottom plate;
—— *de fondation*, (*mach.*) bedplate; foundation plate; sole plate; base plate;
—— *de foyer*, (*steam*) furnace plate;
—— *de frein*, (*r. r.*) brake block;
—— *de friction*, friction plate;
—— *frottante*, (*mach.*) slide-valve face;
—— *de frottement*, friction plate;
—— *de frottement de l'essieu*, linch plate; axle-tree clout;
—— *de frottement (de sassoire)*, (*art.*) sweep plate;
—— *de garde*, (*r. r.*) axle guard plate;
—— *de jonction*, connecting plate;
—— *à lunettes*, eye or lunette plate (of a *cabestan de carrier*);
—— *de mire*, (*inst.*) vane;
—— *pelliculaire*, (*phot.*) film;
—— *à piton*, plate and eye, plate and loop;
—— *de poulie*, plate separating or inclosing the sheaves of a block;
—— *de recouvrement*, any strengthening plate, reenforce; axle strap or plate; cover plate (in general); lap; lap strip (for a joint, etc.);
—— *de renfort*, reenforcing plate, stiffening plate;
—— *de rustine*, (*met.*) back plate;
—— *sensible*, (*phot.*) sensitive plate;
—— *à soupape*, valve plate (forge bellows);
—— *souple*, (*phot.*) film;
—— *de sûreté*, guard plate;
—— *de terre*, (*élec.*) ground plate;
—— *de tête (des tubes)*, (*steam*) tube plate (boilers);
—— *tournante*, (*r. r.*) turntable;
—— *de travail*, (*met.*) working plate (of a firing forge);
—— *de trou de sel*, mud-hole door (of a boiler);
—— *à tubes*, —— *des tubes*, —— *tubulair*, tube plate (of a boiler);
—— *de tuyère*, (*met.*) twyer plate.

plaqué, a., (*hipp.*) stiff, with little or no parotid channel (said of junction of head to neck).

plaquer, v. a., to veneer, to plate, plate over; to lay down turf or sods; to lay on mortar or plaster;
—— *du bois*, to veneer;
feuille à ——, veneer, veneering;
—— *du gazon*, to sod, cover with sod or turf;
—— *du plâtre*, to lay on a coat of plaster.

plaquette, f., small plate, small disk; (*sm. a.*) stock plate of revolver;
—— *droite*, (*sm. a.*) right stock;
—— *gauche*, (*sm. a.*) left stock;
—— *de monture*, (*sm. a.*) stock plate of a revolver;
—— *de recouvrement*, (*sm. a.*) stock plate of a revolver.

plaquis, m., stone casing.
plasticité, f., plasticity.
plastique, a., plastic.
plastron, m., breastplate, front cuirass; (fencer's) plastron; (*mach.*) drill plate;
—— *de cuirasse*, breastplate;
—— *à manchette*, sleeved plastron.

plastronner, v. n., (*fenc.*) to practice fencing, to practice feints, thrusts, etc. (i. e., against the teacher, who wears his plastron.)

plat, a., flat, level; dead; (of boats) flat-bottomed; m., dish; plate; flat part of anything; flat (of the oar, of the sword); (*met.*) iron in plates or sheets;
à ——, flatwise;
—— *d'aviron*, blade of an oar;
—— *de balance*, scale of a balance;
battre à —— *e couture*, v. s. v. *couture;*
calme ——, dead calm;
camarade de ——, messmate;
carte —— *e*, plain chart;
cheval ——, (*hipp.*) lank-sided horse;
course —— *e*, flat race;
—— *de crosse*, (*sm. a.*) part of stock opposite to the *joue;*
eaux —— *es*, still waters (as distinguished from fountains);
—— *d'épée*, (*sm. a.*) flat of a sword;
—— *extérieur*, (*sm. a.*) lock side of a gunstock (obs.);
faire —— *ensemble*, to mess together;
à fond ——, flat-bottomed (of a boat);
—— *intérieur*, (*sm. a.*) brass-side, interior side (obs.);
—— *de malades*, (*nav.*) sick mess;
matériel de ——, mess outfit;
mettre à ——, to lay flat, flatten the oars;
nœud ——, v. s. v. *nœud;*
pays ——, flat, level country;
souliers —— *s*, heelless shoes;
—— *du sabre*, flat of the sword.

platane, m., plane (tree and wood);
—— *d'Amérique*, buttonwood;
faux ——, sycamore-maple.

plat-bord, m., (*nav.*) gunwale, plank sheer; covering board.

plate, f., small flat-bottomed barge.
platée, f., (*cons.*) entire foundation.
plateau, m., waiter, tray; plate, disk, board, flange; (earth-) rammer; scale pan (of a balance); slab, thick plank; one of the two main beams that may be sawed out of a log; (*mach.*) bed of a machine; bed or table for the work; face plate of a lathe; piece performing the office of valve plate in Farcot's Corliss engine; (*mil. min.*) plate or slab covering the charge of a mine; shield of a fougasse; (*art.*) time scale of a fuse; (*top.*) plateau, table-land, flat-topped hill; (*powd.*) lignum-vitæ granulator;
—— *d'arbre à manivelle*, (*mach.*) crank shaft flange;
—— *d'assemblage*, (*mach.*) coupling flange;
—— *de buttoir*, (*r. r.*) buffer disk;
—— *cannelé*, (*art.*) grooved board for carrying spikes, punches, etc., nailed to side of ammunition wagon;
—— *circulaire*, (*art.*) dust guard (seacoast gun);
—— *de cœur*, the plate containing the heart proper of a log (faces tangent to heart);

plateau *collecteur*, collecting plate;
—— *condenseur*, condensing plate;
—— *conducteur*, (*mach.*) wrist plate, rocking disk, motion disk;
—— *de cylindre*, (*steam*) cylinder cover;
—— *de cylindre inférieur* (*supérieur*), (*steam*) cylinder bottom (top);
—— *de distribution*, (*mach.*) vibrating disk; valve plate in Farcot's Corliss engine;
—— *diviseur*, (*mach.*) index plate;
—— *de foyer*, (*steam*) hearth plate;
—— *de friction*, (*mach.*) friction plate or disk;
grand —— *de pétard*, v. —— *de pétard;*
—— *inférieur du cylindre*, (*steam*) cylinder bottom;
—— *d'une machine*, bedplate;
—— *-manivelle*, (*mach.*) disk crank, crank disk, crank plate;
—— *mobile*, (*mach.*) follower;
—— *à mors*, (*mach.*) jawed chuck;
—— *de mortier à plaque*, (*Fr. a.*) sole or bed of the mortier à plaque;
—— *oscillant*, (*mach.*) wrist plate;
—— *permutateur*, (*elec.*) switchboard;
—— *de pétard*, (*mil.*) board to which a petard is fixed;
petit —— *de pétard*, (*mil.*) thin board laid over charge before fixing petard to its bed;
—— *de piston*, (*steam*) piston top plate;
—— *de pointe*, (*art.*) shoe of a gin;
—— *porte-pièces*, (*mach.*) table (of piercing machine, planer, etc.);
—— *à roulis*, (*nav.*) swinging tray;
—— *supérieur du cylindre*, (*steam*) cylinder top, front cylinder cover;
—— *de tampon*, buffer disk;
—— *du, de, tour*, (*mach.*) faceplate (of a lathe);
—— *tournant*, (*r. r.*) turntable;
—— *à trous*, (*mach.*) index plate;
—— *de la vis de pointage*, (*art.*) elevating-screw bed.

plate-bande, f., (*cons.*) platband, lintel; iron strap (for joining two beams, etc., end to end); iron rod; (*carp.*) casing, furring; (*art.*) cap square; fillet, ring; (*fort.*) flat, circular border (of a turret);
—— *de baie*, (*cons.*) lintel;
—— *de la bouche*, (*art.*) muzzle ring;
—— *de bout d'affût*, (*art.*) trail plate;
—— *de culasse*, (*art.*) base ring;
—— *de pavé*, border stone;
—— *de roulement*, roller path;
—— *à talon*, iron dog;
—— *de tête d'affût*, (*art.*) head plate;
—— *de volée*, (*art.*) muzzle ring.

plate-cheville, f., (*art.*) pintle plate (12cm gun, Gruson turret), flat pintle.

plate-forme, f., platform (in general); mudsill of a bridge; (*art.*) platform; (*carp.*) pole plate; any timber 2 inches (or 4 inches) by 1 foot or more (length variable); (*fort.*) platform (as of a casemated work); (*pont.*) carriage way; (*r. r.*) railroad truck; roadbed; (*steam*) seat of a slide valve;
armer une ——, (*art.*) to mount a gun on its platform;
—— *de balance*, platform of a weigh bridge;
—— *en bois*, (*art.*) wooden platform;
—— *en caillebottis*, grating of a boat;
—— *de chargement*, (*art.*) loading platform; (*met.*) charging platform of a blast furnace;
—— *de circonstance*, (*art.*) extemporized platform;
—— *de circonstance à double pivot démontable*, (*Fr. art.*) sort of double-pivot platform for firing in opposite directions;
—— *circulaire*, (*art.*) traversing platform;
—— *de côte*, (*art.*) seacoast platform;
—— *à diviser*, division plate;
—— *évasée en arrière*, —— *en éventail*, (*art.*) splayed platform;
—— *de fondation*, (*cons.*) sleepers;
—— *mixte*, (*art.*) a platform of concrete and of wood;
—— *de mortier*, (*art.*) mortar platform;
—— *d'obusier*, (*art.*) howitzer platform;
—— *ordinaire*, (*art.*) rectangular platform;
—— *en pierre*, (*art.*) stone platform;
—— *à pivot démontable*, (*Fr. art.*) shifting pivot platform (155mm, 120mm guns);

plate-forme *de place*, (*art.*) garrison platform;
—— *de pointage*, (*art.*) aiming platform (seacoast);
—— *à la prussienne*, (*art.*) transportable platform (for siege guns);
—— *de roulage*, (*min.*) trolley;
—— *roulante*, (*r. r.*) truck; (*art.*) traveling platform, i. e., a railroad truck serving as a gun platform;
—— *de siège*, (*art.*) siege platform;
—— *à tablier continu*, (*art.*) siege platform, decked platform;
—— *à tablier superposés*, (*Fr. art.*) three-decked platform (155mm C.);
—— *en terre*, (*art.*) gun bank;
—— *tournante*, (*r. r.*) turntable;
—— *volante*, (*siege*) literally, flying platform, i. e., three planks (under wheels and trail, respectively).

platelage, m., armor skin, interior skin (armor); (*pont.*) floor of a bridge.

plate-longe, f., (*harn.*) kicking strap; leading rein; hitching strap; lunging rein, sideline; (*Fr. art harn.*) breastband (buckling to the breeching).

platière, f., terrace at the foot of a hill.

platin, m., (*hydr.*) small bank, flat, awash at low tide.

platine, f., plate, lock plate; (*sm. a.*) lock; (*r. f. art.*) lock, action block; m., (*met.*) platinum;
—— *à aiguille*, (*sm. a.*) needle-gun lock;
—— (*en*) *arrière*, (*sm. a.*) lock whose mainspring is behind the tumbler; back-action lock;
—— (*en*) *avant*, (*sm. a.*) lock whose mainspring is in front of the tumbler (e. g., bar lock);
corps de ——, (*sm. a.*) lock plate, lock frame (of a revolver);
—— *à double détente*, (*sm. a.*) hair-trigger lock;
—— *de lumière*, (*art.*) apron, vent apron;
—— *à mèche*, (*sm. a.*) matchlock (obs.);
—— *à percussion*, (*sm. a.*) percussion lock (obs.); (*art.*) percussion lock;
—— *à pierre*, (*sm. a.*) flintlock (obs.);
—— *de rail*, (*r. r.*) bedplate;
—— *rebondissante*, (*sm. a.*) rebounding lock (revolver);
—— *renversée*, (*sm. a.*) back-action lock;
—— *à rouet*, (*sm. a.*) wheel lock (obs.);
—— *à silex*, v. —— *à pierre;*
—— *spongieux*, platinum sponge;
—— *de sûreté*, safety device.

platiner, v. a., (*sm. a.*) to lock a gun; (*met.*) to overlay (copper) with an amalgam of tin and mercury.

platineur, (*sm. a.*) lock maker.

plat-pays, m., (*hydr.*) large sand bank.

plâtrage, m., (*cons.*) plaster work; plastering; lath and plaster.

plâtras, m., rubbish of old walls, plaster, etc., from which saltpeter may be extracted.

plâtre, m., plaster; gypsum; (*in pl., cons.*) plaster work of a house, (more esp.) moldings, cornices, etc.;
—— *aluné*, plaster treated with 2 per cent of alum;
amour du ——, unctuosity (of well-burnt plaster of Paris);
battre le ——, to powder gypsum;
—— *blanc*, white plaster (freed from coals);
carrière à ——, gypsum quarry;
—— *-ciment*, a variety of carbonate of lime, largely mixed with clay and iron oxide;
—— *clair*, thin plaster (excess of water);
—— *cru*, unburnt plaster;
—— *cuit*, calcined plaster;
—— *déposé*, plaster rubbish;
enduit en ——, stucco;
—— *éventé*, weathered plaster;
—— *fin*, white parget;
fleur de ——, flowers of gypsum;
—— *gâché*, gypsum paste;
—— *gras*, healthy plaster;
—— *gris*, gray plaster (not freed from coals);
—— *gros*, unsifted plaster;
—— *luisant*, v. —— *fin;*

plâtre *mouliné*, powdered plaster;
— *noyé*, thin plaster (excess of water);
— *de Paris*, plaster of Paris;
— *(passé) au panier*, coarse-sifted plaster;
— *(passé) au sas*, fine-sifted plaster;
— *à la pelle*, v. *fleur de* —;
pierre à, *de*, —, plaster stone;
— *serré*, stiff plaster (mixed with but little water);
— *au tamis de soie*, finest sifted plaster;
— *vert*, underburnt plaster.

plâtrer, v. a., to plaster; (*fig.*) to patch, piece up.

plâtreux, a., chalky.

plâtrier, m., plasterer.

plâtrière, f., gypsum quarry.

plein, a., full, whole, entire, complete; close-worked; thorough; solid (not hollow); (of animals) gravid; (*art.*) solid (of a projectile);
en — *air*, out of doors;
— *bois*, (*nav.*) part of a ship above the water line;
bois —, close-grained wood; thick forest;
boulet —, (*art.*) solid shot (obs.);
en —*e campagne*, out in the country;
en — *champ*, in the fields;
en —*e déroute*, (*mil.*) completely routed;
— *d'eau*, water-logged;
— *fouet*, v. s. v., *fouet;*
en — *hiver*, in the middle of winter;
jarrets —*s*, (*hipp.*) broad houghs;
jour —, a day of 24 hours;
—*e lune*, full moon;
en —*e marche*, (*mil.*) in full march;
—*e mer*, *mer* —*e*, v. s. v. *mer;*
en —*e*, *mer*, v. s. v., *mer;*
mois —, any month containing 30 or 31 days (as distinguished from *mois cave*, February);
ouvrages en — *bois*, (*carp.*) glued work (i. e., not tenoned or doweled);
—*e pression*, (*steam*, etc.) full pressure;
en —*e retraite*, (*mil.*) in full retreat;
—*e vitesse*, full speed.

plein, m., full or down stroke of a letter; beach between high and low water mark; high water; full, full part, of anything; full cargo;
à —, fully;
absous à pur et à —, honorably acquitted;
avoir son —, to be fully laden (ship);
battre son —, (of the tide) to be at the full;
— *d'un bois*, etc., middle of a wood, etc.;
dans le —, *en* —, in the middle;
— *d'une dent*, (*mach.*) thickness of a tooth or cog;
— *de l'eau*, high water;
— *-sur-joint*, —*s sur joints*, so as to break joints;
— *de la lune*, full of the moon;
— *de la marée*, high water;
— *de la mer*, high water;
mettre dans le —, *mettre en* —, (*t. p.*) to hit the mark;
— *d'un mur*, mass of a wall;
—*s des rayures*, (*art.*, *sm. a.*) lands;
tant — *que vide*, (in measuring areas) including windows; (*mil.*) with intervals.

plein-cintre, m., (*arc en* ——) Roman arch.

pleine-eau, f., v. s. v., *eau*.

plénipotentiaire, m., plenipotentiary;
ministre —, minister plenipotentiary.

plet, m., (*cord.*) fake;
faux —, catch fake.

pleureur, v. s. v. *puits*.

pleuvoir, v. n., to rain; (*fig.*) to fall, come down (as bullets, shells);
— *à flots*, to rain in torrents;
— *par ondées*, to be showery;
— *à seaux*, v. — *à flots;*
— *à verse*, to rain in torrents.

pleyon, m., osier; twig.

pli, m., fold; bend; plait; seam (as in soldering); reentrant angle (of a building); manner of folding a letter; cover, envelope (of a letter, etc.); official paper or document; (*art.*) clip of a case shot; (*cord.*) fake, coil, bight;
— *de câble*, (*harn.*) fake, coil (of a cable or rope).

pli *cacheté*, (*nav.*) sealed orders;
charnière à —, v. s. v. *charnière,*
— *de l'embouchure*, (*harn.*) joint of a bit;
— *de l'encolure*, (*hipp.*) bend of a horse's neck;
enfermer sous le —, v. *mettre sous le* —;
filer un —, (*cord.*) to veer away some of the cable;
— *de manœuvre*, v. — *de câble;*
mettre un cheval dans un beau —, (*man.*) to make a horse arch his neck;
mettre sous (*le*) —, to inclose (letters);
— *du paturon*, (*hipp.*) fold of the pastern;
sous —, herewith, inclosed;
— *de terrain*, (*top.*) inequality, fold, in the ground.

pliable, a., flexible, bending, pliant.

pliage, m., folding;
— *de la poudre de chasse*, putting up sporting powder in rolls.

pliant, a., bending, flexible, pliable, supple, folding; m., folding chair, camp chair, camp stool.

plier, v. a. n., to fold, fold up, do up; to bend, curve, sag, bow; to bring under; to yield, submit, give way; to lose ground; (*mil.*) to give way; (*nav.*) to run away, quit the combat (of a ship);
à —, folding;
— *bagage*, (*mil.*) to decamp; to beat a retreat;
— *à bloc*, to fold double (in plate testing);
se — *à bloc*, to bend over a block (test for various articles);
— *un cheval*, (*man.*) to break a horse; to bend a horse's head down to right or left, so as to give him a flexible neck;
— *le cou d'un cheval*, (*man.*) to give a horse a flexible neck;
faire —, to bend, curve;
faire — *l'ennemi*, (*mil.*) to force the enemy to lose ground, give way;
— *la fortification au terrain*, (*fort.*) to adapt a work to the site;
— *les jarrets*, (*man.*) to work the hocks (i. e., to be supple in the hocks);
— *les reins*, (*man.*) in galloping, to bring the croup down;
— *les tentes*, (*mil.*) to strike camp.

plieuse, f., (*mach.*) folding machine.

plinthe, f., plinth; washboard; skirt, skirting board.

plioir, m., folding knife; paper knife, pliers.

plissage, m., (act of) folding.

plissement, m., corrugation; creasing, plaiting, fulling.

plisser, v. a., to corrugate, crease, plait.
— *du fer*, (*met.*) to corrugate iron.

ploc, m., (*nav.*) sheathing felt.

ploiement, m., (*mil.*) ployment.

ploie-ressort, m., v. *ploye-ressort*.

plomb, m., lead; lead stamp or seal; plumb line, plummet, plumb bob; (*cons.*) roof lead, leads; (*sm. a.*) bird shot;
à —, plumb, upright;
— *aigre*, hard lead;
balle de —, (*sm. a.*) leaden bullet;
— *blanc*, hard lead; white lead;
blanc de —, white lead;
— *blanchi*, lead mixed with tin;
— *brûlé*, lead ash;
cendre de —, (*sm. a.*) mustard-seed shot;
— *de chasse*, (*sm. a.*) bird shot;
de —, leaden;
— *durci*, (*sm. a.*) hardened lead (for bullets)
feuille de —, sheet of lead; sheet lead;
— *en feuille* (*s*), sheet lead;
fil à —, plumb line, rule;
— *filé*, lead wire;
— *fusible*, (*elec.*, *steam*) safety fuse;
garnir le — *de sonde*, (*nav.*) to arm the lead;
grand — *de sonde*, (*nav.*) deep-sea lead;
— *granulé*, (*sm. a.*) shot;
— *en grains*, (*sm. a.*) bird shot;
— *laminé*, sheet lead; rolling lead;
lancer le — *de sonde*, (*nav.*) to heave the lead;
menu —, (*sm. a.*) small shot;

plomb *de mer*, graphite;
mettre à ——, to plumb, mark the perpendicular;
mine de ——, graphite, black lead; lead ore;
minerai de ——, lead ore;
—— *mou*, soft lead;
—— *à niveau*, plumb line, plummet; mason's level;
—— *noir*, lead used in the arts;
—— *d'œuvre*, pig lead; argentiferous lead;
—— *d'œuvre*, leaden goods or wares;
—— *rouge*, red lead;
—— *en rouleau*, rolled lead;
—— *en saumon*, pig lead;
—— *de sonde*, (*nav.*) lead, sounding lead;
—— *de sûreté*, (*mach.*) safety plug, fusible plug;
—— *en tôle*, sheet lead;
—— *de vitrier*, came.

plombage, m., plumber's work; sealing with lead; (*art., sm. a.*) leading.

plombagine, f., black lead, graphite, plumbago; *de* ——, plumbaginous.

plombaginer, v. a., to black lead, to cover with black lead.

plombée, f., trying by plummet; (*mas.*) layer or course five bricks deep (in the brick facing of a concrete or rubble wall).

plomber, v. a., to try, test, with a plummet; to cover with lead; to fill with lead; to affix a leaden stamp; to tamp, pack, the ground; (*sm. a.*) to lead the bore of a gun.

plomberie, f., plumbery; plumber's shop; lead works.

plombier, m., plumber.

plomée, f., (*mas.*) *faire les* ——s, to dress, trim, the face of a stone to the middle.

plongeade, f., (*man.*) plunge (of a horse).

plongeant, a., plunging;
feu ——, (*art.*) plunging fire;
tir ——, (*art.*) high-angle fire;
vue ——*e*, view from above.

plongée, (*fort.*) superior slope; parapet around a turret; sole of an embrasure;
—— *du parapet*, dip; superior slope;
tirer de, par, ——, (*art.*) to fire downward.

plongement, m., dip, plunging; (*r. r.*) pitching of a locomotive; (*nav.*) pitching of a ship.

plongeon, m., duck, diver; (*artif.*) water rocket;
faire le ——, to duck, dive; to duck when a shot is fired; to flinch, recoil, back out.

plonger, v. a. n., to plunge, dive, sink; to dip, immerse; to pitch, plunge (of a ship in a sea); to go downward, take a downward direction; to look down upon; (*art.*) to come down upon (of fire); to fire downward; (*met.*) to quench;
—— *le coup*, (*art.*) to depress a gun;
machine à ——, diving bell.

plongeur, m., plunger, ram; diver; diving apparatus, a., plunging;
cloche à ——, diving bell;
machine de ——, diving bell;
piston ——, plunger.

plot, m., short, thick, piece; (*elec.*) brasses of a resistance box, contact stud.

ployer, v. a. n., to fold, put up, bend, bow; (*fig.*) to force to submit, yield, give way; to curb; to overthrow; to yield, submit, give way; (*mil.*) to ply, form into column; (*carp.*) to separate timbers and pile them up.

ploye-ressort, m., (*sm. a.*) spring iron, spring cramp.

ployeuse, f., plate-bending machine.
ployure, f., fold.
pluie, f., rain; shower; spray;
—— *battante*, pelting rain, pouring rain;
craint la ——, to be kept dry;
—— *de feu*, (*artif.*) rain, shower, or cascade of fire;
—— *menue*, fine rain, drizzle;
—— *d'or*, (*artif.*) golden rain;
petite ——, fine rain, drizzle.
plumail, m., feather, plume, feather brush.
plumasseau, m., feather brush; pledget.
plume, f., feather; pen; quill; (*unif.*) plume of a cocked hat;

plume *bec de* ——, nib;
—— *à dessiner*, drawing pen, right-line pen;
donner des ——*s*, (*hipp.*) to rowel; to put in a seton;
en ——*s*, feathered;
—— *métallique*, metallic pen;
—— *rectrice*, rectrix;
tuyau de ——, barrel of a pen.

plumeau, m., pen tray; dusting brush.

plumet, m., (*unif.*) plume; pompon of feathers;
—— *droit*, (*unif.*) aigrette;
—— *flottant*, (*unif.*) long plume for a helmet, etc.;
—— *de pilote*, (*nav.*) dogvane.

plumitif, m., minute book; minutes;
tenir le ——, to keep the minutes of the proceedings.

plus(-)offrant, m., (*adm.*) highest bidder.
plus-value, f., gain, increase in value.
pluvieux, a., rainy.
pluviomètre, m., rain gauge.
pneu, m., (*fam.*) pneumatic tire (bicycle).
pneumatique, a., pneumatic; f., pneumatics; m., pneumatic tire (bicycle);
—— *antidérapant*, corrugated, pebble-tread, tire;
machine ——, air pump, pneumatic engine;
pompe ——, air pump;
—— *à talons*, flanged (detachable) tire (bicycle);
—— *à triangles*, wired tire (bicycle).

pochade, f., rough, rapid sketch.
poche, f., pocket, pouch, sack;
argent de ——, v. *centimes de* ——;
—— *d'avoine, etc.*, sack of oats, etc.;
centimes de ——, v. s. v. *centime;*
—— *de coulée*, (*met.*) casting ladle, foundry ladle;
—— *à couler*, (*met.*) crane ladle;
—— *à fer*, (*cav.*) (spare) horseshoe bag;
—— *de fonderie*, (*met.*) casting ladle;
—— *à huile*, (*mach.*) oil pan;
—— *de mineur*, hanging compass.

pochette, f., small pocket.
podomètre, m., (*inst.*) podometer; (*farr.*) horseshoe gauge.
podométrique, a., *ferrure* ——, (*farr.*) horseshoe made to measure.
poêle, m., stove; pall (funeral);
—— *à sécher*, drying stove.

poids, m., weight; weight (of a balance, of a clock); load, burden;
—— *absolu*, absolute weight;
—— *d'une bordée*, (*nav. art.*) weight of metal of a broadside;
—— *brut*, gross weight;
—— *de charge*, (*steam*) load of a safety valve;
—— *excédant*, overweight;
—— *mobile*, sliding weight (of a balance);
—— *mort*, dead weight, dead load; dead freight;
—— *net*, net weight;
—— *propre*, v. —— *mort;*
—— *public*, weigh house;
—— *remorqué*, (*r. r.*) train;
—— *spécifique*, specific gravity, weight;
—— *utile*, (*mach.*) effective, useful, weight; (*r. r.*) paying freight.

poignard, m., poniard, dagger;
—— *-baïonnette*, (*sm. a.*) dagger-bayonet;
—— *à deux tranchants*, double-edged dagger.

poignarder, v. a., to stab (with a dagger).
poignée, f., handle, grip, gripe; handful; handle of an oar (*nav.*) steering-wheel spoke; (*sm. a.*) small of the stock; pistol stock; sword hilt or gripe; (*art.*) trail handle; breech-plug handle; stock or handle of a rapid-fire gun;
—— *d'aviron*, handle of an oar;
—— *de (la) baïonnette*, (*sm. a.*) bayonet hilt;
—— *centrale*, (*F. m. art.*) central gripe or handle;
—— *de coffre*, (*art.*) chest handle;
—— *en croix*, cross handle;
—— *de crosse*, (*art.*) trail handle;
—— *de culasse*, (*art.*) breech-screw handle;
—— *d'épée*, (*sm. a.*) sword hilt;
—— *fixe*, (*art.*) block handle;
—— *de flèche*, (*art.*) perch handle;
levier ——, (*art.*) breech-plug lever handle;

poignée-*magasin*, (*sm. a.*) the stock and magazine of a magazine pistol;
— *de manœuvre*, (*art.*) maneuvering, operating handle, working handle;
— *de manœuvre à deux branches*, (*r. f. art.*) double-branch maneuvering handle or crank;
— *de pistolet*, (*r. f. art.*) pistol grip;
— *de pointage*, (*r. f. art.*) stock handle;
ressaut de la —, (*sm. a.*) pistol grip;
— *à ressort*, (*r. f. art.*) spring handle;
— *de revolver*, (*sm. a.*) butt of a revolver;
— *de robinet*, handle of a faucet.

poignet, m., wrist; wristband;
manquer de —*s*, (*hipp.*) to have a small and weak fetlock joint, to be weak on the pins.

poil, m., (*hipp.*) hair (of the body, also the hair of the nostrils, ears); coat;
à —, barebacked;
avoir l'éperon au —, (*man.*) to spur a horse;
d'un beau —, sleek and smooth;
— *composé*, composite coat;
faire le — *à*, to clip (a horse);
— *de laitier*, (*met.*) slag wool, silicate cotton;
à — *long et rude*, shaggy;
— *mixte*, mixed coat;
monter à —, to ride barebacked;
à — *nu*, (*man.*) bareback;
— *piqué*, — *planté*, hair that stands up;
— *de scorie*, (*met.*) slag wool, silicate cotton;
— *simple*, simple coat;
souffler au —, (*hipp.*) to discharge (pus) above the hoof.

poiler,, a., hairy, shagged.

poinçon, m., bodkin, awl; drift, punch; die; coak, dowel; (*cons.*) king or queen post; hanging post, joggle piece, joggle post; (*in pl.*) uprights of the center of an arch; (*farr.*) pritchel; (*man.*) sort of goad (used to urge horses to jump between the posts), *piliers*, q. v.; (*mach.*) vertical arbor or spindle; (*art.*) (Rodman) pressure gauge;
— *de contrôle*, hall mark;
— *à découper*, hollow punch;
— *d'échafaudage*, (*cons.*) scaffolding pole;
— *effilé*, brad awl;
— *à emboutir*, punch;
— *à épisser*, v. — *du voilier*;
— *d'épreuve*, proof mark, as (*sm. a.*) on a barrel after it has stood a firing test;
— *à grain d'orge*, (*mil. min.*, *etc.*) diamond-pointed punch, diamond point tool;
machine à —, (*mach.*) punching machine;
— *à main*, hand punch;
percer au —, to punch;
— *à piquer*, pricking punch;
— *de rebut*, stamp or die to mark rejected articles;
— *de réforme*, stamp or die to mark condemned articles;
— *à river*, rivet set;
— *Rodman*, (*art.*) Rodman's crusher gauge;
— *tire-gargousse*, (*Fr. art.*) cartridge extractor (used when the cartridge sticks to the walls of the chamber);
— *à vis*, (*mach.*) screw punch;
— *du voilier*, marlin spike.

poinçonnage, m., stamping, punching; official stamping (after proof).
poinçonnement, m., punching, stamping.
poinçonner, v. a., to stamp; to punch;
machine à —, stamping, punching, machine.
poinçonneuse, f., punching machine; plate-punching machine.
poindre, v. n., to dawn.
poing, m., fist;
— *de la bride*, (*man.*) the left hand, bridle hand;
coup de —, blow of the fist; (*elec.*) firing key, exploder; (*sm. a.*) pocket pistol.
point, m., point; (geometrical) point, place; ship's position (on the chart); mark; break of day; hole in a strap; stitch; degree, height, extent, pitch, condition, state; (*mach.*) center.

I. Artillery, etc.; II. Miscellaneous.

I. Artillery, etc.:

— *d'alignement*, (*mil.*) point of alignment;

point *d'appui*, (*mil.*) supporting point (as for a flank); point of defense (as a farm, village, ect.); any important defensive point in a line; fixed point; point on which line is to rest (drill);
— *d'arrivée*, (*ball.*) point at which the projectile strikes the ground;
— *d'attaque*, (*mil.*) point of attack;
au —, (*art.*) on the target;
— *du bastion*, (*fort.*) point, salient, of a bastion;
— *à battre*, (*art.*, *sm. a.*) point aimed at; target;
— *de chute*, (*ball.*) point of fall; (strictly) the point at which the projectile meets prolongation of *ligne de mire;*
— *de chute moyen*, (*ball.*) mean point of fall;
— *de concentration*, (*art.*) point on which fire is concentrated in target practice, service, etc.;
— *de conversion*, (*drill*) wheeling point;
— *de la courtine*, (*fort.*) curtain point;
— *culminant*, (*art.*) highest point, vertex, of a trajectory;
— *dangereux*, (*fort.*) (in defilading) points within the dangerous area or zone from which the enemy may observe the position;
— *de défense*, v. — *d'appui;*
— *de départ*, (*ball.*) point of departure; first point in which trajectory cuts horizontal through muzzle;
— *de direction*, (*mil.*) point of direction for troops to march upon;
— *de direction plus à droite* (*gauche*), (*drill*) incline to the right (left);
— *de dislocation*, (*mil.*) point at which troops, after a march, break column to go to their quarters;
— *dominant*, (*mil., fort.*) commanding point;
— *d'éclatement*, (*art.*) bursting point, point of burst, of a shell or shrapnel;
— *d'épaule*, (*fort.*) shoulder point (of a bastion);
— *gamma*, (*slang*) time of final examination (*École Polytechnique*);
— *de la gorge*, (*fort.*) gorge point, or intersection of two curtains;
— *d'impact*, (*ball.*) point of impact;
— *d'impact moyen*, (*ball.*) center of impact;
— *d'impact de la trajectoire moyenne*, (*ball.*) center of impact;
— *initial* (*de marche*), (*mil.*) initial point of a march, (when troops are forming a column of march) the point at which each element must appear at a fixed time to take its assigned place in the column (an inaccurate term);
jusqu'au — m., (*slang*) to a certain point (*École Polytechnique*);
— *de mire*, (*art.*, *sm. a.*) sighting point (on a sight); aiming point; point to aim (on target); target;
— *mort*, (*art.*) point at which a combination fuse must be set to make it act as a percussion fuse;
— *moyen* (*d'impact*), (*ball.*) center of impact;
— *de niveau*, (*ball.*) second point in which the trajectory cuts the horizontal through muzzle;
— *plongeant*, (*fort.*) (in defilade) dominating, commanding point (v. — *dangereux*);
principe des trois —*s*, (*fort.*) v. (s. v. *masse*) *principe des masses conjuguées;*
— *de ralliement*, (*mil.*) rallying point; place of assembly for each squadron, battery, or company in a cantonment;
— *stratégique*, (*mil.*) strategic point;
— *visé*, (*ball.*) point aimed at;
— *de visée*, v. — *visé;*
— *de vue*, v. — *de direction.*

II. Miscellaneous:

— *d'alignement*, any point of alignment or direction, as in running lines;
— *d'amarrage*, anchorage;

point *d'appui*, support, prop, fulcrum; center of motion (as of a balance); (*mas.*) pillar, column, any isolated supporting mass of masonry; (*man.*) center of motion;
—— *d'arrêt*, stop;
—— *arrière*, back stitch;
—— *d'arrivée*, (*nav.*) point sailed to; destination;
—— *d'aspect*, point of view;
—— *de bifurcation*, fork;
—— *broché*, stitch in a flat seam;
——*s cardinaux*, cardinal points;
—— *de chaînette*, chain stitch;
——*s de chaussure*, number of a shoe, (strictly, a division of a shoemaker's rule);
—— *de clef*, crown (point) of an arch;
—— *de concours*, (*opt.*) point of convergence (of rays);
—— *de congélation*, freezing point;
—— *conséquent*, (*elec.*) consequent point;
—— *de contact*, point of contact;
—— *corrigé*, (*nav.*) corrected position;
——*s courants*, (*drawing*) a broken line;
—— *d'une courroie*, hole (of a strap);
—— *culminant*, culminating point; (*top.*) dividing ridge;
—— *debout*, stitch in a round seam;
—— *de départ*, initial point; starting point or post; (*mach.*) the points, lines, or surfaces serving to fix the work on the operating machine; (*nav.*) point sailed from;
—— *de destination*, destination (of a ship);
—— *devant*, darning stitch;
—— *de dispersion*, (*opt.*) point of dispersion (of rays);
—— *d'ébullition*, boiling point;
—— *d'équilibre*, center of gravity;
——*s équinoxiaux*, equinoctial points;
—— *estimé*, (*nav.*) position of the ship by dead reckoning;
faire le, son, ——, (*nav.*) to prick (the position of the ship on the chart or map);
—— *faufilé*, backstitch;
—— *fixe*, fulcrum; (*surv.*) fixed, permanent, point in triangulation;
——*s fixes*, fixed points on a thermometer, etc.;
—— *fixé*, (*surv.*) fixed point, permanent point, in triangulation;
—— *de froid*, freezing point;
—— *de fuite*, vanishing point;
—— *de fusion*, fusing point;
—— *fuyant*, vanishing point;
—— *géodésique*, (*surv.*) triangulation point (i. e., an accurately determined point);
à —— *de Hongrie*, (*carp.*) miter-laid;
—— *d'honneur*, pundonor;
—— *d'incidence*, (*opt.*) point of incidence;
—— *d'intersection*, point of intersection;
—— *de jonction*, meeting point;
—— *du jour*, break of day;
—— *marquant*, (*top.*, etc.) conspicuous point;
—— *mettre au* ——, (*inst.*) to focus;
—— (*du*) *milieu*, (*mach.*) middle point of an arbor or shaft, of a trunnion, etc.;
—— *mort*, (*mach.*) dead point, dead center; (of a balloon) moment or instant of rest (in oscillations);
—— *mort d'une scie*, point of a saw where the teeth stop;
—— *de mouvement*, center of motion, fulcrum;
—— *du navire*, (*nav.*) ship's position on the chart;
—— *de niveau*, extremity of a horizontal line;
—— *noir*, threatening cloud;
—— *observé*, (*nav.*) ship's position by observation;
—— *d'ombre*, brilliant sun spot (in getting meridian by corresponding altitudes of the sun);
—— *de partage*, (*top.*) dividing ridge; watershed;
—— *perdu*, (*surv.*) temporary point (of reference) in triangulation;
—— *piqué*, middle stitch;
—— *principal de triangulation*, (*surv.*) principal station, primary triangulation point;
—— *radieux*, (*opt.*) radiant point;

point *de réflexion*, (*opt.*) point of reflection;
—— *de réfraction*, (*opt.*) point of refraction;
—— *de repère*, bench mark; reference-, datum-point;
—— *de rosée*, dew point;
—— *secondaire de triangulation*, (*surv.*) secondary vertex or point;
—— *de section*, v. —— *d'intersection*;
—— *de serrage*, (*mach.*) point where work is held or receives pressure, etc.;
——*s solstitiaux*, solstitial points;
—— *de soudure*, soldering seam;
—— *de suspension*, point of suspension;
—— *de sustentation*, point of support;
de tout ——, completely;
—— *trigonométrique*, (*surv.*) point determined trigonometrically;
—— *de vaporisation*, evaporation point;
——*s verticaux*, the zenith and nadir;
—— *de vue*, point of view; (*opt.*) focus.
pointage, m., (*art.*) aiming, pointing; art of laying guns; (*nav.*) pricking one's position on a map or chart.
(The following terms relate to artillery:)
arc de ——, elevating arc;
appareil de —— *latéral*, side sights;
—— *en chasse*, (*nav. art.*) laying (a gun) on object in front of bows;
cordeau de ——, chalk line;
—— *direct*, direct aiming;
—— *direct à la hausse*, direct aiming by sights;
—— *en direction*, aiming in direction;
école de ——, aiming drill;
exercices de ——, aiming drill;
—— *au fil à plomb*, laying by plumb;
—— *à la hausse*, laying by sights;
—— *en hauteur*, aiming for elevation;
—— *horizontal*, laying point blank;
—— *indirect*, indirect aiming;
—— *latéral*, lateral aiming; aiming for direction; side aiming, i. e., with side sights;
—— *à la manivelle*, aiming corrected by turns (or fractions of turns) of the elevating-screw hand-wheel;
—— *minutieux*, fine, accurate aiming;
—— *négatif*, depression (act of);
—— *au niveau*, quadrant aiming;
—— *de nuit*, night aiming;
—— *positif*, elevation (act of);
—— *en retraite*, (*nav. art.*) laying a gun on object astern;
—— *vertical*, aiming for elevation;
vis de ——, *volant de* ——, elevating screw.
pointal, m., prop, upright stay; (*art.*) firing support or strut (on some carriages);
—— *de levier*, fulcrum.
pointe, f., point, tapering point; tack, sprig, wire nail; point of a bit, auger, etc.; double-pointed pick; bit, boring bit; point of a blade; gore (of cloth); point of the compass; dawn, break of day; (*top.*, *hydr.*) point of land, point of land at the confluence of two rivers; foreland, promontory; top, peak (of a mountain); (*mach.*) center point of a movable poppet; pivot of a vertical shaft; (*min.*) boring bit; (*fenc.*) thrust, point; (*mil.*) extremity of a wing; point (of an advanced guard); rapid, bold advance beyond or off the line of operations; (*art.*) point of a star-gauge; (*man.*) a synonym of *cabrer*, q. v.;
à la —— *des armes*, by force, by force of arms;
—— *de l'arçon*, (*harn.*) saddletree point;
—— *d'arrière-garde*, (*mil.*) rear-guard point;
—— *d'avant-garde*, (*mil.*) advance-guard point;
—— *basse*, (*hydr.*) low point or spit;
—— *de bastion*, (*fort.*) bastion salient;
—— *conductrice*, center point of a boring bit;
—— *de cœur*, (*r. r.*) tongue of a frog;
——*s collectrices*, (*elec.*) collecting points;
coup de ——, (*fenc.*) saber thrust;
couper la ——, (*fenc.*) to pass one's point over the adversary's blade;
—— *de diamant*, diamond point, glazier's diamond;
à la —— *de l'épée*, v. *à la* —— *des armes*;
être en ——, (*mil.*) to be (march) in front or on the side of the main body of troops;

3877°—17——22

pointe, *extrême* ——, (*mil.*) extreme point of the advance guard;
faire une ——, (*mil.*) to make a movement (generally bold and rapid) beyond one's lines of operations; (*man.*) (of the horse) to leave the circle or track of the ring;
—— *de feu,* (*hipp.*) firing needle or iron;
—— *fixe,* (*mach.*) dead center, dead point, of a lathe;
—— *à graver,* drawing point;
grosse ——, double-pointed pick;
—— *du guidon,* (*sm. a.*) point of the fore sight;
—— *haute,* (*hydr.*) headland;
hauteur de ——, (*mach.*) swing;
—— *du jour,* dawn;
—— *morte,* v. —— *fixe;*
—— *d'officier(s),* (*mil.*) small detachment of officers sent out on exploration service when near the enemy;
parer de la ——, (*fence.*) to ward off the adversary's point;
—— *de Paris,* French nail, wire nail;
—— *de pavé,* fork-shaped junction of two street gutters;
—— *percutante,* (*torp.*) nose;
—— *du pied,* point of the foot;
prendre entre ——*s,* (*mach.*) to swing;
—— *rivée,* clinched point;
tailler en ——, to taper;
terminer en ——, to taper;
—— *de terre,* (*hydr.*) foreland, cape, headland, point;
—— *de toile,* gore;
—— *de tour,* (*mach.*) lathe point;
—— *à tracer,* scribe, drawing point, tracing point;
—— *tranchante,* nicker of a centerbit;
—— *-et-tranche,* pick;
—— *de vent,* light breeze.

pointé, m., (*inst.*) sighting, directing, pointing (of an instrument, as a transit); (*fenc.*) thrust (bayonet exercise).

pointeau, m., punch or stamping tool; center punch;
coup de ——, nick, notch, cut; punch mark.

pointement, m., outcrop; appearance (on the surface, in a flame, etc.); (*art.*) pointing, laying, of guns (rare).

pointer, v. a. n., to prick, pierce, mark, stick; to spear; to check off a list; to mark or check a list of names (present and absent); to point (i. e., direct upon, as a telescope, etc.); (*art.*) to lay or aim a gun; (*pointer* is to lay a gun; *viser* is to dispose the piece so as to see the target through the sights; (*fenc.*) to strike with the point, thrust; (*man.*) to rear; (*mach.*) to mark the center or points of support in lathe work; (*carp.*) to mark out work, transfer dimensions, etc., from a working drawing;
—— *en avant,* (*fenc.*) to thrust (bayonet exercise);
—— *en belle,* (*nav.*) to aim a gun at the hull;
—— *de but en blanc,* (*art.*) to aim point blank;
—— *les canons,* (*art.*) to aim;
—— *la carte,* to prick off in the chart, to prick (the position of a point on) the chart;
—— *en chasse,* (*nav. art.*) to aim forward of the beam;
—— *en direction,* (*art.*) to aim in direction, to give the direction;
—— *à la hausse,* (*art.*) to lay by the sights;
—— *en hauteur,* (*art.*) to give the elevation;
—— *à la manivelle,* (*art.*) to lay a gun (i. e., correct the aim) by the elevating hand-wheel;
—— *au niveau,* (*art.*) to lay by quadrant;
—— *en plein bois,* v. —— *en belle;*
tailler en ——, to sharpen, to sharpen to a point.

point(e)rolle, f., (*min.*) miner's pitching tool; poll picks.

pointeur, m., (*art.*) cannoneer who aims the piece, gunner (U. S. Art.); No. 1 (R. A.);
aide-——, (*Fr. art.*) assistant gunner;
formation des ——*s,* training of gunners;
maître-——, (*Fr. art.*) master gunner.

pointeur-*servant,* cannoneer who assists the gunner, No. 3 (U. S. Art.).

pointier, m., (*cons.*) (vertical) scaffolding pole.
pointillage, m., dotting, stippling.
pointillé, p. p., dotted, stippled; m., dotting, stippling;
au ——, *en* ——, drawn dotted.
pointiller, v. a. to dot, to stipple.
pointu, a., pointed, sharp-peaked, sharp-pointed.
pointu-carré, m., (*mil. slang*) stick-in-the-mud.
pointure, f., size of shoes, etc.; mark, tag, or labe showing size and class of shoes, clothes, etc.).
poire, f., pear; powder horn; ball, weight of a steelyard; bulb;
—— *d'amorce,* priming horn (obs.);
—— *de buttoir,* (r. r.) buffer box;
—— *de caoutchouc,* rubber bulb;
—— *de dégonflement,* (balloon) releasing bulb;
—— *à feu,* (*artif.*) smoke ball;
—— *à poudre,* powder flask (obs.);
roue à ——, tub wheel;
—— *s secrètes,* (*harn.*) pear bit.
poirier, m., pear (tree and wood).
pois, m., pea;
—— *chiche,* chick-pea.
poison, m., poison.
poisser, v. a., to pitch, to wax (a thread).
poisseaux, a., pitchy, pitch-like.
poisson, m., fish.
poitrail, m., (*hipp.*) chest; (*harn.*) breast harness; breast plate, (*cons.*) girder; beam; lintel; breast summer; capping plate; (*tech.*) breastplate (of a drill);
—— *armé,* (*cons.*) trussed girder;
corps de ——, (*harn.*) body of a breast strap;
—— *enfoncé,* (*hipp.*) hollow chest;
—— *de harnais,* (*harn.*) breastplate, breast piece, breast strap;
large de ——, (*hipp.*) broad-chested;
montant de ——, (*harn.*) upright strap of a breast strap; supporting strap.
poitrine, f., breast, chest; (*met.*) breast or front of a furnace;
—— *d'acier,* (*mil. slang*) cuirassier;
—— *d'enclume,* anvil pillar;
—— *de fond,* (*hipp.*) chest in which the ribs come down below the level of the elbow;
—— *près de terre,* —— *de fond;*
—— *de velours,* (*mil. slang*) engineer officer.
poitrinière, f., (*harn.*) breast strap.
poivre, m., pepper;
piler du ——, (*mil. slang*) to mark time.
poix, f., pitch;
—— *bâtarde,* v. —— *noire;*
—— *blanche,* —— *de Bourgogne,* Burgundy pitch;
—— *élastique,* india rubber;
—— *grasse,* v. —— *blanche;*
—— *de Judée,* asphalt;
—— *juive,* asphalt, Jew's pitch;
—— *minérale,* mineral tar;
—— *navale,* v. —— *noire;*
—— *noire,* common black pitch;
—— *résine,* rosin;
—— *sèche,* hard pitch; stone pitch.
polaire, a., polar; (*elec.*) polar; f., pole star;
charbons ——*s,* (*elec.*) electric-light carbons;
étoile ——, pole star;
surface ——, (*elec.*) polar surface.
polarimètre, m., (*inst.*) polarimeter, polariscope.
polarisateur, a., polarizing.
polarisation, f., (*elec.*) (*opt.*) polarization;
courant de ——, (*elec.*) polarization current.
polariscope, m., (*inst.*) polariscope.
polariser, v. a., (*elec., opt.*) to polarize.
polariseur, m., (*opt.*) polarizer
polarité, f., (*elec.*) polarity;
renversement de ——, change of polarity;
renverser la ——, to reverse the polarity.
pôle, m., pole; (*elec.*) pole;
—— *de l'aimant,* magnetic pole;
—— *d'aimantation,* pole of magnetization;
chercheur de ——, pole finder, indicator;
—— *s conséquents,* consequent poles;
indicateur de ——, v. *chercheur de* ——;

pôle *magnétique*, magnetic pole;
— *négatif*, negative pole;
—*s de nom contraire*, opposite poles;
—*s de même nom*, poles of the same name, like poles;
— *d'une pile*, battery pole;
— *positif*, positive pole.

poli, a., polished, bright, shining; m., polish, polishing;
— *des armes*, (*sm. a.*) burnishing.

police, f., police; public order and safety (of a State or town); police regulations, police administration; regulations, rules of conduct and of order of an assembly, of an association; policy (of insurance); (*mil.*) police, public order, etc. (of a post, a garrison);
agent de —, policeman;
— *d'assurance*, insurance policy;
bonnet de —, (*unif.*) forage cap;
— *de chargement*, bill of lading;
commissaire de —, police superintendent;
— *correctionnelle*, police court;
détachement de —, (*mil.*) in a march, troops bringing up rear, so as to pick up stragglers and make them join;
faire la —, to maintain order;
— *flottante*, open policy (of insurance);
garde de —, (*mil.*) quarter, barrack, guard;
garde de — *du camp*, (*mil.*) rear guard in camp;
haute —, police power of a State;
.... *jours de salle de* —, (*mil.*) (so many) days in the guardhouse;
— *judiciaire*, court of law;
— *dans les marches*, (*mil.*) rules of conduct on the march;
— *médicale*, v. — *sanitaire;*
— *militaire*, (*mil.*) military police;
préfet de —, prefect of police (police commissioner);
salle de —, (*mil.*) prisoner's room (guard);
— *sanitaire*, sanitary measures or rules;
— *de sûreté*, measures, etc., to preserve public order;
tribunal de (simple) —, police court.

policer, v. a., to police, establish law and order.

policier, a., police.

poliment, m., polishing, grinding.

poliorcète, m., capturer of towns.

poliorcétique, a., (*fort.*) pertaining to sieges; f., poliorcetics.

polir, v. a., to polish, burnish, to smooth; (*sm. a.*) to fine-bore a gun or gun barrel;
— *une arme à feu*, (*sm. a.*) to burnish;
— *à la ponce*, to pumice.

polissage, m., polishing; (*sm. a.*) fine-boring; polishing (as of a gunstock, etc.);
— *à l'émeri*, emery polishing;
— *au gras*, polishing by emery powder and oil;
— *à sec*, dry polishing (e. g., by means of an emery wheel).

polisseur, m., polisher (workman); (*sm. a.*) fine-borer.

polissoir, m., polisher, burnisher, buff stick; polishing iron or brush; polishing wheel; (*powd.*) glazing barrel; (*sm. a.*) finishing borer;
— *à brosse*, (*mach.*) brush wheel;
— *à l'émeri*, emery stick.

polissure, f., polishing, burnishing;
— *à l'émeri*, grinding with emery.

politique, a., political; m., politician; f., (public) policy, polity.

polochon, m., (*mil. slang*) bolster of a bed.

poltron, m., coward, skulker, poltroon.

poltronnerie, f., cowardice.

polyèdre, m., polyhedron.

polygonal, a., polygonal;
fortification —*e*, (*fort.*) polygonal fortification;
tracé —, (*fort.*) polygonal trace.

polygone, m., polygon; (*art. and eng.*) target range, target-practice ground, proving ground, testing ground; (*sm. a.*) range (generally, *champ. de tir*); (*art.*) field artillery drill ground;
— *d'appui*, polygon of support;

polygone *articulé*, v. — *funiculaire;*
— *du canon*, (*art.*) artillery drill ground;
— *d'essai*, (*art.*) proving ground;
— *extérieur*, (*fort.*) exterior polygon;
— *à fortifier* (*fort.*) polygon the sides of which are to receive the works to be built;
— *funiculaire*, funicular polygon;
— *intérieur*, (*fort.*) interior polygon;
manœuvres de —, (*art.*) drill-ground maneuvers;
méthodes des —*s et traverses*, (*surv.*) system in which the plane table, etc., is taken from vertex to vertex of the polygon to be surveyed;
petit —, (*art.*) place for gun drill, etc., in winter; regimental drill ground;
pièce de —, (*art.*) proving-ground piece (e. g., a piece that has not stood the test of service, or a piece specially designed for proving-ground work, as distinguished from a service piece);
— *de tir*, (*art.*) proving ground, firing ground.

polygoner, v. r., to get out of center, to be out of true.

polygraphe, m., polygraph; copying pad, hectograph.

polygraphie, f., cryptography.

polymètre, m., (*inst.*) an instrument for taking horizontal distances and differences of level.

polynitrocellulose, f., (*expl.*) polynitrocellulose.

polyphasé, a., (*elec.*) polyphase;
courant —, polyphase current;
machine —*e*, polyphase generator.

polyphote, m., (*elec.*) constant potential arc lamp.

polytechnicien, m., cadet, graduate, of the École Polytechnique at Paris.

pomme, f., apple; ball, knob; mouse; acorn; (*nav.*) truck (of a mast); (*mil.*) pommel on drum-major's baton;
— *d'arrosoir*, rose (of a sprinkler);
— *de conduite*, fair-lead truck;
— *d'étai*, mouse of a stay;
— *d'étrier*, (*cord.*) diamond knot;
— *de girouette*, vane truck;
— *gougée*, fair leader;
— *de mât*, truck;
— *de mât de pavillon*, flagstaff truck;
— *de pavillon*, truck of the flagstaff;
— *de racage*, parrel truck;
— *de terre*, potato;
— *de tire-veille*, (*cord.*) double diamond knot.

pommeau, m., (*sm. a.*) pommel (of sword, pistol); (*harn.*) pommel (of a saddle);
— *d'arçon*, (*harn.*) saddle pommel.

pommelé, a., dappled; overcast with dapple clouds; (*hipp.*) dapple (more or less restricted to gray horses by the French);
gris —, (*hipp.*) dapple gray.

pommelle, f., strainer; pipe grating; (*min.*) jack;
— *simple en Té*, cross-tailed hinge.

pommelure, f., (*hipp.*) dapple (limited to gray by the French).

pommette, f., (*sm. a.*) pommel, knob, or cap of a pistol.

pommier, m., apple tree.

pompe, f., pump; syringe; pomp; state; (*St. Cyr slang*) study;
— *à action directe*, direct-acting pump, a pump whose piston is worked directly by the piston of the steam engine;
— *à air*, air pump;
— *à air à double (simple) effet*, double- (single-) action air pump;
— *à air horizontale (verticale)*, horizontal (vertical) air pump;
— *alimentaire*, feed pump, feed-water pump;
— *alimentaire horizontale (verticale)*, horizontal (vertical) feed pump;
— *d'alimentation*, v. — *alimentaire;*
— *d'alimentation à bras*, hand feed pump;
— *d'alimentation des chaudières*, boiler-feed pump;

pompe *d'alimentation à vapeur*, steam feed pump;
allumer la ——, v. *amorcer la* ——;
—— *alternative*, reciprocating pump;
âme de ——, pump bore;
amorcer la ——, to fetch the pump;
appareil de ——, pump gear;
—— *aspirante*, suction pump;
—— *aspirante et élévatoire*, suction and lift pump;
—— *aspirante et foulante*, lift and force pump;
—— *atmosphérique*, atmospheric pump;
—— *auxiliaire*, auxiliary pump;
—— *de l'avant*, head pump;
—— *de bouchain*, bilge pump;
—— *à bras*, hand pump, hand engine;
bras de ——, pump handle;
—— *à bringueballe*, brake pump;
bringueballe de ——, pump handle, pump lever;
—— *de cale*, bilge pump;
—— *centrifuge*, centrifugal pump;
—— *à chapelet*, chain pump;
charger la ——, to fetch the pump;
chopine de ——, lower pump box;
—— *de circulation*, circulating pump;
—— *de circulation à double (simple) effet*, double- (single-) action circulating pump;
clapet de ——, pump valve;
—— *de compression*, compressing pump; air compressor; condensing pump;
—— *à comprimer l'air*, air compressor;
—— *s conjuguées*, pumps in tandem, one above the other;
corps de ——, pump barrel, pump stock; (*St. Cyr slang*) staff of instructors; academic staff (U. S. M. A.);
crépine de ——, pump-kettle, strainer;
croc de ——, pump hook;
curette de ——, pump scraper;
cylindre de ——, v. *corps de* ——;
la —— *se décharge*, the pump has lost water;
dépôt de —— *à incendie*, engine house;
—— *à douches*, shower-bath pump (for horses);
—— *à double effet*, double-action pump;
—— *à double piston*, double-piston pump;
—— *à eau chaud (froide)*, hot- (cold-) water pump;
—— *élévatoire*, lift pump;
engrener la ——, v. *amorcer la* ——;
—— *d'épuisement*, bilge pump; drain pump;
—— *d'extraction de la chaudière*, discharging pump;
—— *à faire le vide*, exhausting pump;
faire prendre la ——, v. *amorcer la* ——;
—— *à feu*, fire pump;
—— *fixée*, fixed pump;
—— *flottante à incendie*, fire tug;
—— *à force centrifuge*, centrifugal pump;
—— *foulante*, force pump;
—— *foulante et aspirante*, lift and force pump;
la —— *est franche*, the pump sucks;
franchir la ——, to clear, free, the pump;
garniture de ——, pump gear;
—— *à glycérine*, (*art*.) glycerine pump (in turret gear);
heuse de ——, pump spear and box;
—— *à huile*, oil pump;
—— *à incendie*, fire engine;
—— *à incendie portative*, portable fire engine;
—— *à incendie à vapeur*, steam fire engine;
—— *à jet*, jet pump;
—— *à jet aspirant*, suction jet pump;
—— *à jet d'eau*, water-jet pump;
—— *à jet de vapeur*, steam-jet pump;
—— *jumelle*, double-cylinder pump;
—— *à main*, hand pump;
manche de ——, pump dale;
manivelle de ——, pump handle;
—— *de mer*, waterspout;
—— *à mercure*, mercury pump;
—— *mixte*, v. —— *aspirante et foulante*;
—— *oscillante*, oscillating pump;
—— *de petit cheval*, donkey pump;
piston de ——, pump bucket;
—— *à piston*, piston pump;
—— *à piston plongeur*, plunger pump;
—— *pneumatique*, air pump (rare).

pompe *portative*, portable pump;
—— *des prêtres*, monk's pump, lift and force pump;
la —— *est prise*, the pump is fetched;
—— *(à propulseur) hélicoïde*, propeller pump, spiral pump;
reprise de ——, pump lift;
—— *à rotation*, —— *rotative*, rotary, centrifugal pump;
—— *royale*, common pump;
—— *à simple effet*, single-acting pump;
sonde de ——, pump gauge;
—— *suspendue*, sinking pump;
temps de ——, (*St. Cyr slang*) examination time;
tige de ——, pump spear, pump rod;
—— *à trois corps*, —— *à trois manivelles*, three-throw pump;
—— *à vapeur*, steam pump;
verge de ——, pump spear;
volant de ——, fly wheel of a pump;
—— *sans volant*, direct-acting pump;
—— *volante*, sinking pump.

pomper, v. a. n., to pump; to suck, suck up; (*sm. a.*) to shake (said of a gun barrel when not properly fitted to stock); (*mil. slang*) to rain; (*St. Cyr*) to study, work.

pomperie, f., pump work.

pompier, m., fireman (fire brigade); (*mil. slang*) dirty, slovenly, soldier;
sapeur- ——, fireman, especially *régiment de sapeurs-* ——*s de la ville de Paris*, (*Fr. a.*) the Paris fire brigade (forms part of the infantry of the army).

pompon, m., (*unif.*) pompon;
vieux ——, (*mil. slang*) old soldier.

ponant, m., the west; (in French Mediterranean ports) the west wind; the Atlantic Ocean.

ponçage, m., rubbing with pumice; pouncing.

ponce, f., pumice stone;
—— *commune*, v. *pierre* ——;
pierre ——, pumice stone.

ponceau, m., culvert, little bridge, small bridge of a single arch (in the military service, generally extemporized);
—— *avec contre-fiches*, strutted bridge;
—— *avec étais*, frame bridge with stays;
—— *sur cadres arc-boutés*, single- (double-) lock bridge;
—— *sur fermes en perches*, single-sling bridge;
—— *de rigole*, (*r. r.*) culvert;
—— *avec traverses*, small bridge of two balks supporting the roadway.

poncel, m., v. *ponceau*.

poncelet, m., (*elec*.) proposed unit of activity, 100 kilogrammeters per second (approximately one kilowatt).

poncer, v. a., to pumice; to prick a drawing and rub it over with powdered charcoal or chalk.

poncette, f., pounce bag.

poncif, m., v. *poncis*.

poncis, m., drawing pounced, pricked, and rubbed over with powdered charcoal or chalk.

ponctuer, v. a., to dot (as lines in a drawing).

poney, m., pony;
—— *double*, cob.

ponghée, m., China silk, pongee.

pont, m., bridge; (*nav.*) deck of a ship (when used alone, the main deck is meant); (*met.*) fire bridge, flame bridge; (*elec.*) multiple circuit, bridge, in electric lighting; bridge of an electric fuse; (*sm. a.*) well cover; dust cover; (*tech.*) junction piece, tie; bridge-like support.

I. Military, naval; II. Bridges in general, miscellaneous.

I. Military, naval:

(Except where otherwise indicated, the following terms are military.)
—— *abri*, (*nav*.) awning deck;
—— *abnormal*, special (pontoon) bridge for special circumstances (v. —— *normal*);
—— *d'argent*, loophole of escape left to a defeated and beaten enemy, lest he should become desperate and reverse the results;

pont *d'assaut*, assault bridge;
—— *avant*, (*nav.*) foredeck;
—— *de bateaux*, bridge of boats;
—— *de bateaux de commerce*, bridge of ordinary boats picked up in or near the river, etc.;
—— *de bateaux d'équipage*, regulation bridge of material carried by bridge train;
—— *par bateaux successifs*, bridge by successive pontoons, (English, by "booming out");
—— *de la batterie*, (*nav.*) gun deck, main deck;
—— *blindé*, (*nav.*) armored (protected) deck;
bordé de ——, (*nav.*) deck planking; deck plating;
—— *de bout*, —— *à bout*, (*nav.*) flush deck;
—— *en buse de gabions*, bridge of gabions attached end to end by a stake (laid parallel to current, with roadway across);
—— *de caisses*, bridge laid on air-tight boxes;
—— *de châssis*, bridge of balks on floating piers;
—— *de chevalet(s)*, trestle bridge;
—— *de chevalet à chapeau mobile*, (*mil.*) bridge on two-legged trestles, with movable transoms;
—— *sur chevalet suspendu*, bridge on suspended trestle;
—— *de circonstance*, temporary, extemporized bridge;
—— *continu*, (*nav.*) flush deck;
—— *par conversion*, bridge by conversion (English, by "swinging");
—— *à la Congreve*, Congreve's carriage bridge;
—— *à coulisse*, sliding bridge, run out when wanted;
—— *coupé*, (*nav.*) deck open amidships;
—— *cuirassé*, (*nav.*) armored deck;
demi- ——, (*nav.*) half-deck;
—— *à deux tabliers*, bridge with two separate and distinct roadways;
—— *dormant*, fixed bridge (in a fort);
—— *de dunette*, (*nav.*) poop deck;
—— *-échelle*, ladder or flying bridge to cross ditches in an assault;
enlever un ——, to take up, dismantle, a (pontoon) bridge;
—— *entier*, (*nav.*) flush deck;
entre- ——, (*nav.*) 'tween decks, height between decks;
équipage de ——, bridge train;
—— *d'équipage*, pontoon bridge;
—— *étagé*, bridge of more than one tier, as trestle on pontoon, trestle on trestle;
faux ——, (*nav.*) orlop deck;
—— *en fer*, (*nav.*) iron deck;
—— *en fermes*, frame bridge;
—— *en fermes composées*, double-lock bridge;
—— *en fermes simples*, single-lock bridge;
—— *à flèche*, small drawbridge before a postern;
—— *de gabions*, gabion bridge;
—— *à, de(s), gaillard(s)*, (*nav.*) upper deck; forecastle deck;
—— *sur gueule*, v. —— *coupé*;
—— *inférieur*, (*nav.*) lower deck;
—— *intermédiaire*, (*nav.*) middle deck;
jeter un —— *sur une rivière*, to throw a bridge over a river;
—— *de manœuvre*, (*nav.*) gun deck;
—— *militaire*, military bridge;
—— *normal*, bridge of standard length (58 yards Austrian), usually occurring type;
—— *d'or*, v. —— *d'argent*;
—— *d'outres*, bridge of inflated skin;
—— *pare-éclats*, (*nav.*) protective deck (engines, boilers, etc.);
—— *par parties*, bridge by parts (English, by "forming up");
—— *de passerelle*, (*nav.*) bridge deck;
—— *de plein-pied*, (*nav.*) flush deck;
—— *poisson*, portable bridge (for crossing the ditch in an assault);
pontée d'un ——, v. *pontée*;
—— *de pontons*, pontoon bridge;
portière d'un ——, raft, or single bay, of a bridge;
—— *par portières*, bridge by rafts;
—— *principal*, (*nav.*) main deck;

pont *de promenade*, (*nav.*) promenade, hurricane, deck;
—— *protecteur*, (*nav.*) protective deck;
—— *provisoire*, temporary bridge;
—— *de radeaux*, raft bridge;
—— *de radeaux d'arbres*, timber-raft bridge;
—— *de radeaux de tonneaux*, cask bridge;
—— *-rails*, "rail bridge," i. e., sort of truck for hoisting, shifting, materials in fortifications, barracks, etc.;
—— *ras*, (*nav.*) flush deck;
replier un ——, to take up, dismantle, a bridge;
—— *de rondins*, spar bridge;
—— *roulant*, rolling bridge; any vehicle run into a ditch, rivulet, etc., and serving as a foundation for a (temporary) bridge; bridge built on such a carriage;
—— *de sortie*, sally bridge;
—— *sous-marin*, (*nav.*) under-water deck;
—— *supérieur*, (*nav.*) upper deck;
—— *à tablier rétréci*, bridge with chess laid down obliquely;
—— *-tente*, (*nav.*) shade deck;
tête de ——, (*fort.*) bridge head; tête de pont;
—— *de tonneaux*, bridge of casks;
—— *de tonneaux à l'anglaise*, cylinder pontoon bridge.
travée de ——, bay of a bridge;
—— *sur traverses en encorbellement*, extemporized cantilever bridge;
—— *de voitures*, bridge over wagons used as supports, bridge resting on vehicles; carriage bridge;
—— *volant*, flying bridge; portable bridge for crossing ditches in an assault; trail bridge; (*n iv.*) bridge;
—— *volant d'équipage*, trail bridge of two pontoon boats.

II. Bridges in general, miscellaneous:
—— *américain*, —— *à la Town*, Town's lattice truss bridge;
—— *américain en dessous*, lattice truss bridge, roadway on top;
—— *américain en dessus*, lattice truss bridge, roadway below;
—— *-aqueduc*, aqueduct bridge;
—— *en arcs métalliques*, iron arched bridge;
—— *à arches*, arched bridge;
armature de ——, truss frame of a bridge;
—— *à armatures*, hanging or truss-framed bridge;
—— *à armatures et contre-fiches*, strut- and truss-framed bridge;
—— *arqué*, v. —— *à arches*;
—— *articulé*, floating drawbridge of a landing stage;
—— *à bascule*, drawbridge; (*mach.*) weighbridge; track scale;
—— *à bascule inférieure*, dropping drawbridge;
—— *à bascule supérieure*, lifting drawbridge;
—— *de bateaux*, bridge of boats;
—— *biais*, skew bridge;
—— *en chaînes*, chain bridge;
—— *sur chaînette*, suspension bridge with roadway resting on cables;
—— *à charbon*, (*elec.*) carbon bridge;
—— *de châssis*, strut-framed bridge;
—— *de la chauffe*, (*met.*) fire bridge;
chaussée d'un ——, roadway of a bridge;
—— *s et chaussées*, bridges and roads; government civil engineers (in France);
—— *sur contre-fiches*, strut-framed bridge;
—— *de cordage(s)*, —— *de cordes*, rope bridge;
—— *en dessous*, bridge below roadway;
—— *en dessus*, bridge above roadway;
—— *dormant*, fixed, permanent, bridge;
—— *à double (triple) voie*, bridge of two (three) roadways;
—— *d'échappement*, (*met.*) flue bridge of a reverberatory furnace;
—— *de fascines*, brushwork causeway;
—— *en fer*, iron bridge;
—— *de fil de fer*, wire bridge;
—— *fixe*, v. —— *dormant*;
—— *flottant*, floating bridge;
—— *en fonte*, cast-iron bridge;
—— *de jambes de force*, strut-framed bridge;
—— *à jambettes*, strut-framed bridge;

pont *de jonc*, brushwood causeway;
 lancer un ——, to throw a bridge across;
 ——*-levis*, v. *pont-levis*;
 lit de ——, platform of a bridge;
 —— *à longerons*, v. —— *à poutres*;
 —— *de maçonnerie*, stone bridge;
 —— *mixte*, bridge with both fixed and floating supports;
 —— *mobile*, any movable bridge;
 —— *oblique*, skew bridge;
 œuil de ——, opening through spandrels of a bridge;
 palée de ——, pier of a pile bridge:
 —— *sur parabole*, ordinary, usual, suspension bridge;
 —— *de pierre*, stone bridge;
 —— *en pierres*, stone bridge;
 —— *à, pour les, piétons*, footbridge;
 —— *de pilotis*, pile bridge;
 plancher d'un ——, roadway of a bridge;
 —— *à plusieurs voies*, bridge of more than one roadway;
 potence d'un —— *volant*, horse of a swing or pendulum bridge;
 —— *à poutres*, timber bridge;
 —— *à poutres armées*, trussed girder bridge;
 —— *à poutres en treillis*, lattice girder bridge;
 —— *roulant*, traveling crane; (*r. r.*) rolling platform;
 —— *de service*, temporary bridge; footbridge;
 —— *stable*, permanent bridge;
 —— *à supports fixes*, bridge on fixed supports;
 —— *à supports flottants*, floating bridge, bridge on floating supports;
 —— *sans supports intermédiaires*, any bridge without intermediate supports;
 sur ——, f. o. b.;
 —— *suspendu*, suspension bridge;
 —— *suspendu à chaînes*, chain bridge;
 —— *suspendu à cintres*, bowstring bridge;
 —— *suspendu à des cintres*, tension bridge; bowstring bridge;
 —— *suspendu à cordages*, rope suspension bridge;
 —— *suspendu en fil de fer*, wire suspension bridge;
 tablier d'un ——, roadway of a bridge;
 —— *tournant*, turn-, swing-, or draw-bridge; (*r.r.*) turntable;
 —— *à la Town*, v. —— *américain*;
 —— *transporteur*, transfer bridge;
 travée de ——, bay of a bridge;
 —— *en treillis*, lattice bridge;
 ——*-tube, en tube,* —— *tubulaire*, tubular bridge;
 —— *volant*, trail bridge; (*r. r.*) sort of gangway for loading material;
 —— *de Wheatstone*, (*elec.*) Wheatstone bridge.
pontage, m., bridge building; bridging; (*pont.*) bridge drill or operations, manner of constructing the bay of a pontoon bridge;
 —— *à bateaux à contrepoids*, bridging with counterpoised boats;
 —— *par bateaux successifs*, bridging by successive pontoons;
 —— *par conversion*, bridging by conversion;
 —— *à grande portée*, bridging with wide intervals (each balk rests on three gunwales, two of one boat and one of the next);
 —— *par parties*, bridging by parts;
 —— *à petites portées*, bridging with reduced intervals (each balk on four gunwales);
 —— *par portières*, bridging by rafts;
 —— *à très grande portée*, bridging with extended intervals (each balk on two gunwales).
ponté, a., (*nav.*) decked;
 non- ——, open (of a boat).
pontée, f., (*pont.*) floating pier (inclusive of the adjacent bay); the material necessary to the construction of one bay; (*nav.*) deck load.
ponter, v. a., to deck a boat; (*pont.*) to bridge; to set up (as a trestle); to lay the balks, chess, side rails; to lay down the roadway of a bridge.
pontet, m., curved ornament (on the bars of a grating); (*pont.*) small bridge; (*harn.*) saddlebow; (*sm. a.*) trigger guard; small arch of a bayonet socket (for the passage of the front sight, obs.); catch securing a bayonet scabbard to the frog; scabbard holder;

pontet, *bracelet-* ——, (*sm. a.*) bayonet scabbard catch;
 —— *de fourreau de baïonnette*, (*sm. a.*) scabbard catch;
 ——*-magasin*, (*sm. a.*) trigger guard and magazine in one;
 —— *de selle*, (*harn.*) bow of a saddle;
 —— *de la sous-garde*, (*sm. a.*) trigger guard.
pontille, f., v. *épontille.*
pont-levis, m., drawbridge; (*man.*) rearing;
 —— *à bascule*, drawbridge worked by a counterpoise lever;
 —— *à bascule en dessous*, v. —— *à tape-cul*;
 —— *à bascule en dessus*, drawbridge with counterpoise above roadway, lifting drawbridge;
 —— *à chaînes*, chain drawbridge;
 —— *à contrepoids*, counterpoise drawbridge;
 —— *à engrenage*, drawbridge worked by gearing;
 —— *à flèche*, drawbridge worked by pliers;
 flèche de ——, pliers;
 —— *à la Poncelet*, lifting bridge with chain counterpoise;
 —— *à spirale*, drawbridge worked by a chain passing over a spiral drum;
 —— *à tape-cul*, drawbridge with counterpoise under roadway, dropping drawbridge;
 tape-cul de ——, counterpoise of a drawbridge;
 —— *à trappe*, lever drawbridge.
ponton, m., pontoon, ponton; hulk;
 —— *d'accostage*, lighter;
 —— *allège*, camel;
 bac à ——, pontoon raft;
 ——*-bigue*, sheer pontoon; crane barge;
 cadre à ——, pontoon frame;
 —— *à canon*, gun hoy;
 —— *charbonnier*, coal hulk;
 —— *a la Colleton*, (*mil.*) cylinder pontoon;
 —— *à creuser*, —— *à curer*, —— *cureur*, dredging boat, dredge;
 —— *cylindrique*, (*mil.*) cylinder pontoon;
 —— *flottant*, pontoon;
 haquet à ——, (*mil.*) pontoon wagon;
 —— *en liège*, (*mil.*) cork pontoon;
 —— *à mâter*, —— *-mâture*, sheer hulk;
 —— *métallique*, (*mil.*) metallic pontoon;
 —— *mouilleur pour torpilles*, torpedo-planting hulk;
 —— *en toile goudronnée*, (*mil.*) canvas pontoon;
 ——*-tonneau*, (*mil.*) cylinder pontoon.
pontonnier, m., (*mil.*) pontonier.
porc, m., hog, pig; (*met.*) slag containing unmelted ore;
 cul-de- ——, (*cord.*) single wall knot;
 cul-de- —— *couronné*, (*cord.*) crown knot;
 cul-de- —— *double*, (*cord.*) double wall knot;
 soie de- ——, bristle;
 tête de ——, (*mil.*) wedge-shaped formation of a battalion.
porée, f., (*fond.*) fine earth for molding.
poreux, a., porous; (*met.*) blistered (of a casting).
porges, m., (*met.*) twyer side (of a Catalan forge).
porphyre, m., porphyry.
porque, f., (*nav.*) rider.
port, m., port, harbor, haven; city around a port; quay; tonnage, burden; postage: carriage; wearing (as of uniform); mode of carrying (as carbine, saber, etc.); (*man.*) carriage (of the head);
 —— *abrité*, landlocked port;
 aller, arriver, à bon ——, to arrive in safety; to come, to get, into harbor;
 —— *d'arme*, (*sm. a.*) (position of) "carry arms;"
 —— *d'armement*, port of registry; (*mil.*) fitting-out port;
 —— *d'armes*, license to shoot or to carry arms; (*sm. a.*) (position of) "carry arms;"
 —— *d'attache*, headquarters port; port of assembly, of concentration, of young men liable to summons for sea service;
 avant- ——, outer harbor;
 avoir ses —— *s francs*, to have the postal franchise;

port à barre, a bar harbor;
  bâtiment du —— *de* ...., ship of .... burden;
  —— *d'un bâtiment*, tonnage, burden of a ship;
  —— *de cabotage*, —— *à, de, caboteurs*, coasting-trade port;
  capitaine de ——, harbor master; port warden;
  —— *de commerce*, commercial port;
  droits de ——, port charges;
  —— *d'échouage*, harbor of refuge;
  —— *à écluse*, sluice port;
  —— *d'embarquement*, port of departure;
  —— *d'entrée*, tidal harbor that may be entered at all times;
  —— *à entrepôt*, bonded port;
  —— *d'état*, v. —— *de guerre;*
  —— *étranger*, foreign port;
  —— *fermé*, landlocked port;
  fermer un ——, to close a port;
  —— *franc*, free port;
  franc de ——, postage paid;
  —— *à la grenadière*, (*sm. a.*) slung over the shoulder;
  —— *de guerre*, military port;
  —— *de la lance*, (*sm. a.*) "carry" of the lance (right hand on staff);
  —— *de lettre*, postage;
  —— *libre*, free, open port;
  —— *marchand*, commercial port;
  —— *à, de, marée*, tidal harbor (may be entered only at high tide);
  —— *de mer*, seaport; outport;
  —— *militaire*, v. —— *de guerre;*
  —— *de navire*, burden of a ship;
  —— *ouvert*, open port;
  —— *payé*, prepaid;
  —— *de pêche*, fishing harbor;
  —— *de pilotage*, pilot harbor;
  prendre ——, to enter port;
  —— *de ravitaillement*, port of call, for supplies, etc.;
  —— *de refuge*, harbor, port, of refuge;
  —— *de relâche*, port of call;
  —— *de rivière*, river harbor; close port;
  —— *à la selle*, (*sm. a.*) fastened to saddle (sabre);
  —— *de toute marée*, v. —— *d'entrée;*
  —— *de l'uniforme*, (*mil.*) wearing of uniform.

portage, m., carriage, land carriage, conveyance, porterage; portage; (*mach.*) bearing; (*cord.*) nip (place where two ropes cross and bear one on the other);
  faire ——, to carry boat from one river to another.

portail, m., portal.

portant, a., bearing;
  à bout ——, v. s. v. *bout;*
  chaîne ——*e*, bearing or supporting chain;
  cheval —— *bas*, horse that hangs his head;
  —— *en faux*, overhanging;
  —— *de fond*, built up from the ground;
  vent ——, wind that carries far (e.g., sounds).

portant, m., (chest) handle; lengthening strap; (*elec.*) keeper of a magnet;
  —— *au vent*, (*hipp.*) horse that carries his head too high.

portatif, a., hand, portable;
  armes ——*ves*, (*sm. a.*) hand firearms;
  outils ——*s*, portable tools.

porte, f., gate; door; doorway; gateway; postern; portal; entrance; flood gate; (*in pl.*) defile; eye for a hook; (*sm. a.*) gate (of a revolver); (*sm. a.*) gate, magazine gate (as in Krag);
  —— *d'agrafe*, eye of a hook;
  —— *d'amont*, crown gate, head gate of a lock;
  —— *d'aval*, tail gate of a lock;
  bateau ——, boat sunk at the entrance of a port to close it;
  —— *battante*, swing door;
  bouton de ——, door knob;
  —— *blindée*, armor-plated door;
  —— *brisée*, jointed door (i. e., one whose upper half folds down on the lower);
  —— *busquée*, lock gate;
  —— *du cendrier*, damper, ash-pit door;
  —— *de charge*, (*met.*) charging door; (*torp.*) loading door, filling gate;
  —— *de chargement*, (*sm. a.*) gate, magazine gate (as in the Krag);

porte *de chauffe*, (*met.*) fire door;
  —— *à claire-voie*, v. —— *à jour;*
  —— *cochère*, carriage entrance;
  —— *de coffre*, lid (as of an ammunition chest);
  —— *coulante*, portcullis;
  —— *à coulisse*, sliding door; trapdoor;
  —— *coupée*, half door;
  —— *-croisée*, French window;
  —— *cuirassée*, v. —— *blindée;*
  —— *de dégagement*, back door;
  —— *dérobée*, back door;
  —— *de derrière*, back door;
  —— *à deux battants*, folding door;
  —— *à deux vantaux*, folding door;
  —— *de devant*, front door;
  —— *d'écluse*, sluice gate;
  —— *d'entrée*, front door;
  enfoncer une ——, to break in a door;
  —— *étanche*, (*nav.*) water-tight door;
  fausse ——, sham, blind, door; secret door; (*fort.*) postern gate or door, sally port;
  —— *feinte*, sham, blind, door;
  à —— *fermante* (*ouvrante*), (*mil.*) at the closing (opening) of the gates;
  fermer ses ——*s*, (*mil.*) (of a town or fort) to refuse to surrender, to resolve to stand a siege;
  —— *de fermeture*, (*art.*) breech door (Hotchkiss revolving cannon);
  —— *à fermeture étanche*, (*nav.*) water-tight door;
  —— *flamande*, a grated door surmounted by ornamental ironwork;
  —— *du foyer*, fire door;
  —— *de galerie* (*min.*) air gate;
  —— *de garniture*, (*mach.*) packing port;
  —— *de glace*, mirror door;
  —— *à jour*, grated door;
  —— *latérale*, charging door of a furnace;
  —— *de mesure*, (*nav.*) measurement capacity;
  —— *de mouille*, v. —— *d'aval;*
  la —— *Ottomane*, Turkey;
  ouvrir ses ——*s*, (*mil.*) (of a town or fort) to capitulate;
  —— *à panneaux*, paneled door;
  —— *perdue*, concealed door;
  petite ——, postern;
  —— *à placard*, vestibule door, door folding back into a recess;
  petite ——, postern;
  —— *pleine*, ordinary nonpaneled door;
  —— *en poids*, (*nav.*) dead-weight capacity;
  pousser une ——, to leave a door ajar;
  —— *principale*, main gate;
  ressort de ——, (*sm. a.*) gate spring;
  —— *de rue*, street door;
  —— *roulante*, rolling door, overhung door;
  —— *de secours*, (*fort.*) sally port;
  —— *de sortie*, exit door;
  la Sublime ——, the Sublime Porte;
  —— *de sûreté*, (*fort.*) defensible gate, i. e., a gate in a communication, and that may be closed and defended;
  —— *de tête*, v. —— *d'amont;*
  —— *tournante*, (*fort.*) balance gate; flood gate;
  —— *en tour creuse* (*ronde*), inner (outer) door (of a round tower);
  —— *de travail*, (*met.*) working hole, door;
  —— *à un vantail*, single door;
  —— *vitrée*, glazed door.

porte-à-faux, m., overhang; state of being out of plumb;
  coudé en ——, (*mach.*) overhung crank;
  en ——, not in bearing, unsupported;
  être en ——, to overhang; to be out of plumb.

porte-aigle, m., (*mil.*) eagle bearer.

porte-aiguille, m., (*art.*) seat or holder of the firing pin (of a fuse); (*sm. a.*) needle holder.

porte-allumettes, m., match box.

porte-amarre, m., life-saving apparatus;
  canon ——, life-saving mortar;
  cerf-volant ——, life-saving kite;
  fusée ——, life-saving rocket.

porte-amorce, m., (*art.*) primer spindle, primer holder, fuse cup; (*sm. a.*) primer holder or cap; (*mil. min.*) primer holder;
  bouchon ——, (*art.*) fuse plug.

porte-appareil, m., —— *de sûreté*, (*sm. a.*) safety-locking mass.

porte-auge, m., hodman.
porte-avaloire, (*harn.*) hip, flank, strap.
porte-avoine, m., *étui* ——, oat bag.
porte-bagages, m., parcel holder (bicycle).
porte-baguette, m., drumstick holder; (*sm. a.* ramrod thimble; ejector-rod frame of a revolver;
  —— *à queue*, (*sm. a.*) tail pipe.
porte-baïonnette, m., (*sm. a.*) bayonet frog.
porte-balai, m., broomstick; (*elec.*) brush holder.
porte-balance, m., balance stem.
porte-bandeau, m., blindfolded person.
porte-barres, m., (*harn.*) pole ring, collar ring.
porte-bêche, m., (*mil.*) spade case.
porte-bobine, m., reel (wire).
porte-boîte, m., —— *à mitraille*, (*art.*) canister holder (of an ammunition chest).
porte-bonnet, m., (*mil.*) cap straps (on a cartridge box).
porte-bottes, m., (*mil. slang*), trooper;
porte-boulet, m., (*art.*) hot shot bearer (obs.).
porte-brancards, m., (*harn.*) shaft strap.
porte-branche, m., (*harn.*) checkrein (in double harness).
porte-broche, m., (*mach.*) tool holder.
porte-cadenas, m., hasp and staple.
porte-canon, m., (*sm. a.*) carbine bucket; (sometimes) muzzle support.
porte-canons, m., (*fort.*) generic word for fort or fortification, as being chiefly arranged to hold and work guns.
porte-carabine, m., (*sm. a.*) carbine swivel.
porte-carte(s), m., map case;
  étui ——, map case.
porte-cartouches, m., (*sm. a.*) loader (serves at same time to carry cartridges on the person).
porte-cartouchière, m., pouch belt.
porte-ceinture, m., (*mach.*) band holder (of a band setting machine).
porte-chaîne, m., (*surv.*) chainman (esp., fore chainman).
porte-chaleur, m., (*hipp.*) heater (for cauteries).
porte-charbon, m., (*elec.*) carbon holder.
porte-charge, m., (*art.*) cartridge compartment of an ammunition chest; ammunition holder or transporter of a turret;
  —— *revolver*, (*art.*) revolving ammunition carrier or turntable (of a turret).
porte-charnière, m., hinge support.
porte-chevalets, m., (*pont.*) trestle carrier.
porte-clapet, m., valve piece (of a pump).
porte-clefs, m., turnkey, key ring.
porte-corps, m., (*art.*) sling wagon, platform carriage, mortar wagon.
porte-coussinet, m., (*mach.*) hanging carriage.
porte-couteau, m., (*mach.*) tool holder, cutter bar.
porte-crapaudine, m., pintle or hollow pillar, supporting a *crapaudine*.
porte-crayon, m., pencil case.
porte-crosse, m., (*r. f. art.*) part of the stock connecting the shoulder piece with breech; (*sm. a.*) carbine holster; (sometimes) support for the stock.
porte-drapeau, m., (*mil.*) color bearer (of infantry).
portée, f., range, reach; length, extent, compass; capacity, power, ability, pitch; range of a light, of a signaling apparatus; distance between telegraph, etc., poles; bearing, i. e., distance, clear horizontal space, between supports; part of a beam, etc., resting on supports; bearing length, bearing surface, span of an arch, radius, reach, of a crane; (*pont.*) span of a bay; (*mach.*) bearing, bearing surface, neck, journal, shoulder; liner; plane bearing surface to fix work in a lathe; turned surface for measurements, bearings, etc.; (*mil. min.*) amplitude, reach of a mine; (*surv.*) chain length; (in geodesy) (length of) 3 (or 4) base-measuring rods taken together; operation of using one set of rods; (*ball.*) distance from the gun to the point of fall; (*art., sm. a.*) range;

portée, *à* —— *de*, within reach, range, of;
  à la —— *de*, within range of;
  —— *d'un arc*, span of an arch;
  —— *de but en blanc*, (*ball.*) point-blank range;
  —— *de canon*, (*art.*) cannon range;
  coulée à une seule ——, (*fond.*) casting all at once;
  couler à une seule ——, (*fond.*) to cast in one piece, all at once;
  courte ——, (*art., etc.*) short range;
  demi- ——, (*art., sm. a.*) half range, half (cannon, etc.) range;
  —— *d'eau*, volume of water discharged by a stream in a given time;
  —— *d'éclatement*, (*art.*) distance of point of burst from muzzle;
  —— *efficace*, (*art., sm. a.*) effective range;
  —— *d'essai*, (*art., sm. a.*) trial range;
  —— *d'un feu*, range of a light;
  —— *de fusil*, etc., (*sm. a.*) rifle, etc., range;
  —— *d'une grue*, radius of a crane jib;
  hors de (la) —— *de*, out of range of;
  longue ——, (*art., sm. a.*) long range;
  —— *lumineuse*, v. —— *d'un feu*;
  —— *maximum*, (*art.*) maximum range;
  —— *moyenne*, (*art.*) mean range;
  petite ——, (*art., sm. a.*) short range;
  la plus grande ——, (*art.*) maximum range;
  —— *de la poudre-type*, (*art., sm. a.*) standard range;
  —— *des poutrelles*, (*pont.*) span of a balk;
  à —— *des signaux*, within signaling distance;
  —— *du son*, range of sound;
  se tenir à ——, (*mil., etc.*) to keep within range, reach;
  se tenir hors de la ——, (*mil., etc.*) to keep out of range;
  —— *théorique moyenne d'éclatement*, (*art.*) distance of mean point of burst from the muzzle;
  —— *totale*, (*art.*) extreme range;
  —— *à toute volée*, (*art.*) greatest range possible, using service charge, and greatest angle of elevation;
  trouver la ——, (*art.*) to find the range;
  —— *visuelle*, visual range;
  —— *de (la) voix*, range of the voice; hailing distance;
  —— *d'une voûte*, span of an arch;
  vraie ——, (*art., sm. a.*) true range;
  —— *de la vue*, seeing distance.
porte-écouvillon, m., (*art.*) sponge holder.
porte-éjecteur, m., (*sm. a.*) ejector holder.
porte-enseigne, m., (*mil.*) ensign bearer.
porte-épée, m., (*sm. a.*) frog of a n. c. o.'s sword belt.
porte-épée-baïonnette, m., (*sm. a.*) bayonet frog.
porte-éperon, m., spur strap.
porte-épingle, m., (*hipp.*) instrument for making a suture after bleeding.
porte-éponge, m., sponge holder.
porte-étendard, m., (*cav.*) color bearer; stirrup socket or bucket.
porte-étriers, m., (*harn.*) a synonym of *étrière*, q. v.
porte-étrivières, m., (*harn.*) stirrup bar;
  boucle ——, stirrup-leather buckle.
porte-faix, m., porter.
porte-fanion, m., (*t. p.*) flagman; (*mil.*) bearer of a battalion color; n. c. o. carrying *fanion* of a general officer.
porte-fer, m., (*cav.*) horseshoe case.
porte-feu, m., (*art.*) flame passage (of a shell, etc.); (*mil. min.*) leader, train, igniter; a., fire transmitting (in composition);
  —— *Bickford*, (*artif.*) Bickford igniter.
porte-feuille, m., portfolio; (*fig.*) functions of a minister of state;
  ministre à ——, minister in charge of a department;
  ministre sans ——, minister without a department; without portfolio.
porte-filière, m., (*mach.*) die holder.
porte-flambeau, m., torchbearer; torch holder.
porte-foret, (*mach.*) drill holder, drillstock.
porte-fouet, m., whip socket.
porte-fourreau, (*mil.*) sword case.

porte-fraise, a., (*mach.*) fraise, or reamer holding, etc.; **m.,** (*mach.*) reamer holder; milling-tool holder;
— *arbre* —, spindle of a milling machine.
porte-fusée, m., (*mil. min.*) fuse holder (in priming boxes, etc.).
porte-fusil, m., (*sm. a.*) device for carrying a gun on a bicycle.
porte-gamelle, m., (*Fr. a.*) gamelle holder for men on guard, on duty.
porte-gargousse, m., (*art.*) pass box.
porte-giberne, (*mil.*) cartridge-box belt.
porte-gouvernail, m., (*pont.*) steering-post of a pontoon boat.
porte-grenade, m., (*mil.*) grenade pouch.
porte-grosse-caisse, m., bass-drum strap.
porte-guide, m., (*harn.*) driving rein (i. e., outer rein in double harness).
porte-guidon, m., (*mil.*) guidon bearer; (*art., sm. a.*) front-sight mass or holder; a., sight-holding.
porte-hache, m., (*mil.*) ax case; also, ax carrier;
— *de campement,* cavalry ax case.
porte-isolateur, m., (*teleg.*) pin (of an insulator).
porte-lame, m., (*mach.*) cutter head, boring head.
porte-lance, m., (*art.*) portfire holder (obs.); (*mil.*) lance bucket, lance strap.
porte-lanterne, m., lantern holder or support.
porte-lettres, m., letter case.
porte-loupe, m., magnifying-glass stand or stem.
porte-lumière, m., heliostat.
porte-lunette, m., (*mach.*) poppet head.
porte-madriers, m., (*pont.*) chess carrier.
porte-mandrin, m., (*mach.*) chuck holder; spindle.
porte-manger, m., meal tin.
porte-manteau, (*mil. and gen.*) valise;saddlebags; portmanteau; clothes hook; (*nav.*) stern boat.
porte-masselotte, m., (*art.*) plunger-sleeve holder.
porte-matrice, m., (*mach.*) matrix bed or holder.
porte-mèche, m., (*mach.*) bit holder.
porte-mire, m., (*surv.*) rodman.
porte-molette, m., (*mach.*) milling-tool rest.
porte-montre, m., watch holder of a bicycle; watch stand in general.
porte-mors, m., (*harn.*) heading rein; bit ring; cheek billet, cheek piece (of a bit);
— *anneau* —, bit ring.
porte-mousqueton, m., (*sm. a.*) carbine hook or swivel; snap hook.
porte-niveau, m., level arm.
porte-objectif, m., (*inst.*) objective holder.
porte-obus, m., (*art.*) shell compartment of an ammunition chest; (*mach.*) shell holder (of a shell-fitting press).
porte-oculaire, m., (*inst.*) eyepiece holder.
porte-œilleton, m., (*inst.*) eyepiece holder.
porte-outil, m., (*mach.*) tool holder; cutter bar; head stock; tool rest; (*art.*) sort of tool frame for transport of tools.
porte-percuteur, m., (*r. f. art.*) U-shaped bolt of the Gardner gun.
porte-pièce, a., m., (*mach.*) work holding or carrying; work holder, table.
porte-pied, m., pedal.
porte-pioche, m., (*mil.*) pickax case.
porte-pistolet, m., (*mil. min.*) bit holder, borer clamp.
porte-plate-longe, a., *maille* —, (*harn.*) link or eye through which the *plate-longe* is passed.
porte-plume, m., penholder.
porte-poignée, m., (*r. f. art.*) lower part of shoulder piece of a rapid-fire gun (part that carries the gripe).
porte-poinçon, m., (*mach.*) punch holder.

porte-pompon, m., (*unif.*) pompon socket.
porte-poutrelles, m., (*pont.*) balk carrier.
porte-poutres, m., v. *porte-poutrelles*.
porter, v. a. n., to support, sustain, bear, carry, transport, take; to expend; to suffer, stand; to bear up, be supported by, rest, lie, lean on; to bear (of a fro en ri er, etc.); to wear (a coat, a sword); to carry (a pistol); to carry, reach (of sound or light); to set (as a current); to price (a bearing on a chart); to measure (as so many feet); (*adm., mil.*) to inscribe, carry (as on a muster roll), hence to recommend (for promotion, a decoration, etc); (*mas*) of a stone, to have such and such dimensions; (*art., sm. a.*) to hit, to range, to carry; (hence, *fig.*) to take effect, do execution; (*mach.*) to come into bearing; (*nav.*) to mount (so many guns); (*cord.*) to nip;
—— *l'arme*, (drill) to "carry arms;"
—— *les armes,* (*mil.*) to be a soldier; to make war;
—— *un arrêt,* to make a decision;
—— *en avant,* (*man.*) to go forward, advance;
— *se* —— *en avant,* (*mil., etc.*) to forge ahead, advance;
—— *la barbe,* to wear a beard;
—— *bas,* (*man.*) to carry the head low;
—— *bateau,* (of a river) to be navigable;
—— *beau,* (*man.*) to carry the head well;
—— *bien son plomb,* (*sm. a.*) to make a small pattern;
—— *une botte,* (*fenc.*) to make a thrust;
—— (*tant de*) *bouches à feu,* (*nav.*) to carry (so many) guns;
—— *le cap à,* (*nav.*) to head for;
—— *la carabine à la grenadière,* (*sm. a.*) to sling carbine;
—— *son cheval,* (*man.*) to keep up, sustain, one's horse;
—— *en compte,* to charge to;
—— *à cru,* to be built or resting on the ground;
—— *en corbellement,* v. —— *à faux;*
—— *un coup,* to deal a blow;
—— *un coup d'épée,* (*fenc.*) to make a thrust;
—— *le deuil (de),* to be in mourning (for);
—— *l'épée,* (*mil.*) to be an officer;
—— *à faux,* to be out of plumb; to overhang; to project (as a balcony, a turret); (*cord.*) to bear badly;
—— *des fers,* to be a slave;
—— *à fond,* to be erected from the ground, to rest on the foundation;
—— *à gauche (droite),* to bear off to the left (or right);
—— *haut,* (*man.*) to make one's horse carry his head high; (of the horse) to carry his head high;
—— *juste,* (*art., sm. a.*) to hit the target;
—— *la lance,* (*mil.*) to "carry lance;"
—— *une loi,* to enact a law;
—— *malade,* (*mil.*) to put on the sick report;
—— *moustache,* to wear a moustache;
—— *le mousquet,* (*mil.*) to be a soldier;
—— *le nez au vent,* v. —— *au vent;*
—— *au nord, etc.,* (*nav.*) to stand to the northward, etc.;
—— *le pavillon de* ...., (*nav.*) to fly (such and such) colors;
—— *à plat,* to lie or bear flat (as saddle bar on back of horse);
—— *pour,* to recommend or propose (for promotion, etc.);
—— *règlement sur,* (*mil.*) to fix regulations for;
—— *un relèvement sur,* to prick or lay off a bearing;
—— *la robe,* to be a magistrate;
—— *en route,* (*nav.*) to lay the course;
—— *le sabre,* (drill) to "carry saber;"
—— *en saillie,* v. —— *à faux;*
—— *en sautoir,* to wear crosswise (as shoulder belts);
— *se* ——, to go; to comport one's self; to declare one's self; to be a candidate, to run, to stand;
— *se* —— *à,* to advance to (an attack);
— *se* —— *en avant,* to press forward;
—— *sur,* to bear against (one part against another); (*nav.*) to steer or stand for;
—— (*tant*) *de tonneaux,* (*nav.*) to carry (so many) tons;
—— *à terre,* (*nav.*) to stand for the shore;

porter *au vent*, (*man.*) (of the horse) to carry the head high and horizontal.

porte-rame, m., rowlock (for either side oar or steering oar).

porte-rancher, m., **porte-ranchet**, m., stake socket.

portereau, m , (*art.*) carrying bar;
levier- ——, v. s. v. *levier*.

porte-rêne, m., (*harn.*) rein billet.

porte-retard, m., (*art.*) delay-action device or mechanism of a fuse.

porte-réticule, m., (*opt.*) tube holding the reticule.

porte-roue, m., —— *de rechange*, (*art.*) spare-wheel axle.

porte-rugueux, m., (*art.*) plunger spindle (of a fuse).

porte-sabre, m., (*mil.*) saber holder (of a saddle, of a bicycle); (*sm. a.*) sliding frog for n. c. o.'s sword belt.

porte-sac, m., (*mil.*) medical-kit bearer.

porte-sacoches, m., (*mil.*) medical kit bearer.

porte-scie, m., (*mil.*) saw case.

porte-selle, m., saddle rack, rest, bar, block, or holder.

porte-servante, m., (*art., etc.*) pole prop hook.

porte-taraud, m., (*mach.*) tap holder, tap wrench.

porte-tarière, m., (*mach.*) brace head, drill stock.

porte-timon, m., (*art.*) pole-prop.

porte-torpille(s), m., (*nav.*) (torpedo) spar; spar-torpedo boat;
bateau ——, *canot* ——, spar-torpedo boat.

porte-to(u)let, m., rowlock.

porte-trait, m., (*harn.*) trace loop.

porte-tuyère, m., (*art.*) ring or collar (of a twyer of a field forge).

porteur, m., carrier, bearer, porter; (*art.*) near horse of a team; (*Fr. a.*) truck of the Decauville transportable railway;
au ——, payable to bearer;
—— *Decauville*, (*Fr. a.*) Decauville transportable railroad;
—— *de dépêches*, bearer of dispatches;
—— *de derrière*, (*art.*) near wheel horse;
—— *de devant*, (*art.*) near lead horse;
—— *d'eau*, irrigation channel;
méthode par ——*s*, (*siege*) in a flying sap, method in which all the gabions are put down together;
—— *de milieu*, (*art.*) near swing horse;
—— *d'outils*, (*mil.*) the pack animal carrying the tools of an infantry company;
—— *de la voie*, (*pont., etc.*) stringpiece, sleeper;
wagonnet- ——, (*Fr. a.*) the truck of the Decauville railway.

porte-valises, m., valise support or holder (bicycle).

porte-vent, m., blast pipe (of a blowing engine); (*met.*) nose pipe, nozzle, tuyere.

porte-verres, m., slide of a signal lantern.

porte-vis, m., (*sm. a.*) side plate of a gun lock.

porte-voix, m., speaking trumpet, speaking tube;
fixe, speaking tube.

portier, m., porter;
——*-consigne*, v. s. v. *consigne*.

portière, f., curtain; coach door; door; flap (as of a leather case); (*pont.*) raft, cut (part of a bridge of boats formed by two or three boats completely covered); movable part or draw of a bridge of boats; (*art.*) old name for *volet*; mantelet; (*fort.*) shutter (of an embrasure);
construire un pont par ——*s*, (*pont.*) to build a bridge by rafts;
—— *en cordages*, (*art.*) rope mantelet;
—— *d'embrasure*, (*fort.*) embrasure shutter or mantelet or blind;

portière *entière*, (*pont.*) a raft of three pontoon boats;
—— *de manœuvre*, (*pont.*) construction raft for bridge building;
—— *de navigation*, (*pont.*) movable part or draw of a bridge of boats;
—— *de pont*, (*pont.*) cut, swing bay or raft of a bridge;
pont par ——*s*, (*pont.*) bridge by rafts;
repliement par ——*s*, (*pont.*) dismantling bridge by rafts, by breaking up into rafts;
replier par ——*s*, (*pont.*) to dismantle a bridge by rafts;
—— *roulante*, sliding door.

portillon, m., small postern.

portion, f., share, part, portion;
—— *active*, v. —— *détachée*;
—— *centrale*, (*Fr. a.*) in a divided regiment, the part containing the depot, or designated to form it in case of mobilization;
—— *circulaire*, (*siege*) circular place of arms;
—— *du contingent*, (*Fr. a.*) subdivision of contingent (when the annual contingent is larger than the appropriations admit, it is divided into two parts; the first (lower numbers on the list) serves three years, the second (higher numbers on the list) serves one year, and is then sent home *en disponibilité*);
—— *détachée*, (*Fr. a.*) the part of a divided unit of troops that is not comprised in the *portion centrale*, q. v. (also called *détachement*);
effets de la 1ère, 2ème, ——, v. s. v. *effets* (sometimes shortened to *première portion, deuxième portion*);
—— *principale*, (*Fr. a.*) in a subdivided regiment, the part directly commanded by the senior officer of the regiment;
—— *séparée*, v. —— *détachée*.

portique, m., porch; (*siege*) covered passage in the trenches; blindage (of a blinded sap); (*gym.*) rectangular frame, board, or platform 3.5^m or 4.5^m above the ground, used in various gymnastic exercises;
—— *avec échelle*, (*gym.*) a *portique* fitted with a ladder.

portugaise, f., (*cord.*) sheer lashing, head lashing;
amarrage en ——, sheer lashing.

posage, m., laying, placing, laying down; setting up, hanging.

pose, f., setting, setting up, planting, laying, laying down (as of a railway, a telegraph line); posture, attitude; stint; set; shift (of workmen); stationing; (*mach.*) seating (as of a rifling band) (*mil.*) posting of a sentry, relief of a guard; relief of a working party; (*phot.*) exposure; (*mas.*) operation of laying stone, bricks;
—— *d'un câble*, cable laying;
caporal de ——, (*mil.*) v. s. v. *caporal*;
—— *de ceinture*, (*art.*) banding of a projectile;
—— *des corps de chaudières*, (*steam*) boiler setting;
—— *descendante*, (*mil.*) relief coming off;
—— *insuffisante*, (*phot.*) under exposure;
—— *d'une ligne télégraphique*, laying down of a telegraph line;
—— *montante*, (*mil.*) relief going on;
—— *de n heures*, relief or shift after *n* hours' work;
—— *de nuit*, (*mil.*) night guard; additional sentries posted after retreat;
—— *d'ouvriers*, shift of workmen;
—— *au pas cadencé*, (*mil.*) march at attention;
—— *au pas de route*, (*mil.*) march at ease;
première, *etc*. ——, (*mil.*) first, etc., relief;
—— *de la première pierre*, laying of the corner stone;
temps de ——, (*phot.*) duration of exposure, exposure;
—— *de la voie*, (*r. r.*) track laying.

posée, f., (*man.*) the instant during which the foot is in bearing.

poser, m., (*man.*) v. *posée*.

poser, v. a. n., to put, place, set, lay, lay down; to put in position, in place; to stand, plant; to post, station; to bear, lie, rest on; to lay down a railway track; to set or put up a telegraph line; (*mil.*) to pitch a camp; to post a sentry; (*mas.*) to lay stone, bricks;
—— *à l'anglaise*, (*cons.*) to lay a floor in parallel strips; (*mas.*) to lay bricks in English bond;
—— *l'arme à terre*, (*drill*) to "ground arms;"
—— *les armes*, (*mil.*) to lay down one's arms; surrender; to make peace, a truce;
—— *les armes bas à terre*, (*mil.*) to lay down one's arms, to surrender;
—— *sur la carne*, to set on edge (as a plank);
—— *de champ*, to lay on edge;
—— *en coupe*, v. —— *en délit;*
—— *à cru*, (*cons.*) to erect a timber framework without a foundation;
—— *en décharge*, to set (a timber) as a prop;
—— *en délit*, (*mas.*) to lay (a stone) contrary to its cleaving grain, on edge;
—— *l'épée*, (*mil.*) to leave the military profession;
—— *en épi*, (*mas.*) to lay (bricks) herringbonewise;
—— *sur le faux lit*, v. —— *en délit;*
—— *un grain de lumière*, v. s. v. *grain;*
—— *en lit*, —— *sur son lit*, (*mas.*) to lay (a stone) on its cleaving grain;
—— *à main*, to set or lay by hand;
—— *en panneresse*, —— *en parement*, to lay (stones, etc.) lengthwise, stretcherwise;
—— *à plat*, to lay flat;
—— *de plat*, (*mas.*) to lay (a stone) on its cleaving grain;
—— *plein sur joint*, (*mas.*) to lay so as to break joints;
—— *la première pierre*, to lay a corner stone;
—— *la quille*, (*nav.*) to lay the keel;
—— *à recouvrement*, to lap over;
—— *à sec*, (*mas.*) to lay down dry, lay stones without mortar;
—— *une sentinelle*, (*mil.*) to post a sentry;
—— *une sonnette*, to hang a bell;
—— *sur la tranche*, v. —— *sur la carne;*
—— *la voie*, (*r. r.*) to lay the track.

poseur, m., builder, fitter; mason, bricklayer; (*r. r.*) plate layer, track layer;
contre- ——, mason's assistant.

positif, m., (*phot.*) positive;
—— *pour projection*, lantern slide;
—— *sur verre*, transparency.

position, f., position, situation, stand, place, posture; (*mil.*) position; (*mil. adm.*) status; (*art.*) the ground occupied by the artillery in a battle (as distinguished from *emplacement*, the position of the guns alone); (*man.*) position on horseback;
—— *d'absence*, (*mil.*) status of absence (*Fr. a.*. leave, sick, confined, under trial, absent without leave);
—— *d'activité*, (*mil.*) status of active service;
—— (*administrative*) *générale*, (*Fr. a. adm.*) generic term for the status of soldiers as a whole (i. e., peace and war footings);
—— (*administrative*) *individuelle*, (*Fr. a. adm.*) administrative status of the individual with respect to pay and allowances (e. g., *activité, non-activité, disponibilité*, etc.);
—— *d'arrêt*, (*mil.*) a position prepared in advance to check or stop the enemy, (more specially) such a position in a mountain pass;
—— *d'attente*, (*mil.*) position in which the enemy is awaited; temporary position, under cover, if possible, assumed before passing to the —— *de combat;* (*art.*) transitory position of field artillery just before going into battery;
—— *avancée*, (*fort.*) advanced positions occupied by the defense to keep enemy from using them for his guns; line of fortified advanced posts of an intrenched camp;
batterie de ——, (*art.*) battery of position (position guns);
—— *de chargement*, (*art.*) loading position of a gun;

position *de combat*, (*mil.*) in general, combat position; (*siege*) intrenched position of a besieging army for the special purpose of repelling sorties; (*fort.*) main line of works of an intrenched camp; (of a turret) firing position;
éclaireur de ——, v. s. v. *éclaireur;*
—— *d'éclipse*, (*art.*) loading position of a disappearing gun;
fausse ——, in computation, double position; (*surv.*) false position;
—— *de fermeture*, (*art.*) position of breech plug when home;
—— *générale*, v. —— (*administrative*) *générale;*
—— *hors rang*, (*Fr. a.*) status of being *hors rang;*
—— *individuelle*, v. —— (*administrative*) *individuelle;*
—— *d'ouverture*, (*art.*) position of breech plug when open;
prendre une ——, (*mil.*) to take up a position;
—— *de présence*, (*Fr. a. adm.*) status of being on duty (present, on mission, en route);
—— *principale* (*de défense*), (*fort.*) the line of forts;
—— *de ralliement*, (*mil.*) rallying position;
—— *de rassemblement*, (*mil.*) point where troops collect after executing march maneuvers, in rear of the positions where they are to take combat formation;
—— *de rendez-vous*, v. —— *de rassemblement;*
—— *de repli*, (*mil.*) position selected in advance to fall back on if repulsed;
—— *de retraite*, (*mil.*) position to fall back on;
—— *de route*, (*art.*) traveling bed; traveling position (siege guns);
—— *en route*, (*Fr. a. adm.*) status of being en route;
—— *de secours*, (*mil.*) succoring or supporting position on line of retreat of an army;
—— *du soldat sous les armes*, (*mil.*) (position of) attention;
—— *de soutien*, (*mil.*) position in rear to which troops may fall back; (*fort.*) second line of forts; (of an intrenched camp) position to render the parts of the —— *de combat* in enemy's hands untenable, to protect retreats, etc.;
—— *en station*, (*Fr. a. adm.*) status of duty (i. e., of being at one's post of duty);
—— *de tir*, (*art.*) firing position (esp. of siege guns); (*fort.*) firing position of a turret;
—— *du tireur*, (*inf.*) firing position of the individual soldier;
—— *zéro*, position of rest; (*art.*) setting of a shrapnel fuse to zero.

possession, f., possession; occupation;
prise de ——, (*mil.*) capture.

poste, f., post house or station; stage (distance of two leagues); post, post-office; mail; (*sm. a.*) swan shot;
bâtiment- ——, mail steamer;
bureau de ——, post office;
chevaux de ——, post horses;
courir la ——, to travel by post;
directeur de la ——, postmaster;
directeur des ——s, postmaster-general;
grande ——, general post-office;
malle- ——, v. *malle-poste;*
—— *militaire*, (*mil.*) field post (office);
petite ——, local, city, post-office;
—— *restante*, general delivery, to be called for;
train- ——, (*r. r.*) mail train.

poste, m., post, station, employment, place; (*mil.*) post, any position or point occupied by troops, or where troops may be quartered; particular duty or station; post of a guard, guardhouse, guard, main guard; the garrison of a post; position taken by a line; (*nav.*) berth; position or post of a ship; station; (*art.*) post of a cannoneer at the gun;
à ——, at home;
à son ——, at one's post, in attendance;
affaire de ——s, (*mil.*) an affair of outposts;
—— *à aiguilles*, (*r. r.*) operator's (switching) tower;
—— *d'alarme*, (*mil.*) rendezvous, alarm post;

poste *d'amarrage,* (*nav.*) mooring berth;
— *d'arrêt,* (*fort.*) small field work recommended for mountain passes;
— *avancé,* (*mil.*) advanced post; post nearest enemy; in a defensive position, points in front of main position (wood, farm, village, etc.) that may be advantageously held by infantry 800m to 1,000m in front;
avant— —, (*mil.*) outpost;
avant— —*à la Cosaque,* (*mil.*) Cossack post;
avant— —*irrégulier,* (*mil.*) Cossack post;
avant— —*régulier,* (*mil.*) ordinary outpost;
— *des blessés,* (*nav.*) cockpit;
— *de campagne,* (*mil.*) field post, such as a church, mill, etc.;
— *central,* (*mil., fig.*) central station;
changer de —*s,* (*art.*) to change posts;
chef de —, (*mil.*) c. o. of a guard, of a post, etc.;
— *de combat,* (*nav.*) quarters, station in action;
— *de correspondance,* (*mil.*) transmitting station, relay, dispatch post, communicating post; transmission or dispatching post (in exploration service from front to rear);
— *à la Cosaque,* (*mil.*) Cossack outpost;
— *détaché,* (*mil.*) detached post; (more esp.) an advanced post more than rifle range in front of the main position;
— *de discipline,* (*mil.*) prisoners' guard; (in cantonments) place where men found guilty of breaches of discipline are confined;
— *embroché,* (*élec.*) telephone station in main line (i. e., not shunted);
enlever un —, (*mil.*) to carry off a post;
— *d'examen,* (*mil.*) examining posts (deserters, etc.);
— *extérieur,* (*mil.*) exterior post of a guard;
faire entrer (*sortir*) *les servants à leurs* —*s,* (*art.*) to take (leave) posts;
à — *fixe,* (*mil.*) permanently posted or stationed;
— —*s,* (*mil.*) outpost operations;
guerre de —, (*mil.*) outpost operations;
— *d'honneur,* (*mil.*) dangerous post or position; post designated to pay honors; post or guard of honor;
— *intérieur,* (*mil.*) interior post of a guard;
— *intermédiaire,* (*mil.*) intermediate post, detached corps placed between others for support;
— *des invulnérables,* (*mil.*) place out of the reach of harm (v. *mont pagnote*);
— *isolé* (*mil.*) detached post;
— *jaloux,* (*mil.*) post of danger;
— *de lancement* (*de torpilles*), (*torp.*) torpedo-launching point or post;
ligne de —*s,* (*mil.*) picket line;
— *des malades,* (*nav.*) sick bay;
— *d'observation,* (*mil.*) advanced post, recommended for observing mountain passes (in mountain warfare) during mobilization; (*torp.*) range station; observation station;
petit —, (*mil.*) picket post; small outpost; small outpost established by day by the cavalry;
petit —, *spécial,* (*mil. teleg.*) communicating station, or central station of lines connecting parts of a large fort, or a central fort with detached works;
— *de police,* (*mil.*) post of the police guard;
— *de pompiers,* fire-engine house;
prendre un —, to take one's station;
— *de quatre hommes,* (*mil.*) special post of four men, Cossack post;
— —*relai,* (*sig.*) transmitting station; (*teleg.*) relay;
relever un —, (*mil.*) to relieve a post;
se rendre à son —, (*mil.*) to go to one's station or duty;
rentrer à son —, (*art.*) to resume one's post;
— *retranché,* (*fort.*) intrenched post;
— *de sauvetage,* life-saving station;
— *secondaire,* (*mil. sig.*) communicating station;
— *de secours,* life-saving station; (*mil. med.*) dressing station;
— *sémaphorique,* (*sig.*) semaphore station;

poste *de signaleurs,* (*mil.*) signal station; outlying station (i. e., one nearest enemy, observing station);
— *de soutien,* (*mil.*) succoring post, for *poste d'arrêt,* recommended for mountain passes during mobilization;
— *des soutiens,* (*mil.*) reserve picket;
— *spécial,* (*mil.*) post at specially important points;
— *de surveillance,* (*fort.*) lookout (as for ditches); defensive lookout, near and for a *fort d'arrêt;*
— *télégraphique,* telegraph station;
— *téléphonique,* telephone station;
— *de veille,* (*torp.*) lookout station, observation station;
— *vigie,* (*mil.* and *gen.*) lookout station;
voiture- —, (*mil. teleg.*) telegraph wagon (for the rapid installation of a telegraph station);
— *volant,* (*mil. teleg.*) flying or mobile telegraph station.

poster, v. a. r., (*mil., etc.*) to post, station, place; to take post, station, stand.

postérieur, a., (*hipp.*) hind; m., (*hipp.*) hind leg.

postiche, a., substituted, false, sham, artificial; (*mil.*) temporarily attached to; doing duty provisionally;
canon —, Quaker gun, dummy gun;
— *caporal,* lance corporal.

postillon, m., postilion; runner of a tent cord.

pot, m., pot; bucket; jug; headpiece (obs.); (*artif.*) pot; rocket head; (*cord.*) pot;
à — —*s,* overshot;
— *d'aigrette,* (*artif.*) brilliant firework so called;
— *à brai,* pitch kettle;
— *à colle,* glue pot;
— *à eau,* (*artif.*) water cartridge;
— *à feu,* (*artif.*) sort of lantern (pot filled with resin, etc.), for illuminating purposes; stink-pot; kind of case or stand from which several rockets may be discharged simultaneously;
— *de fusée,* (*artif.*) rocket cylinder; rocket head;
— *de garniture,* v. — *de fusée;*
— *au noir,* doldrums;
— *de pompe,* lower pump box;
papier au —, v. *papier écolier;*
— *de presse,* (*art., fort.*) pivot socket or cylinder of a turret; pressure cylinder;
roue à —*s,* overshot wheel;
— *à suffoquer,* (*mil.*) stinkpot;
— *en terre,* (*cons.*) pot-shaped hollow brick (as for filling floors);
— *en tête,* leading sapper's helmet (obs.).

potasse, f., (*chem.*) potash;
— *caustique,* caustic potash;
— *nitratée,* saltpeter.

poteau, m., post, pillar; telegraph pole; stake; (*mil. slang*) post; (*in pl.*) framing (as of a winch);
— *d'amarrage,* bollard, mooring post;
— *d'arrêt,* (*teleg.*) terminal pole; (*elec.*) double terminal insulator;
— *de bifurcation,* (*teleg.*) pole at the fork of a telegraph line;
— *de cloison,* (*cons.*) quarter, stud;
— *à contre-fiche,* (*cons.*) braced pole;
— *cornier,* (*teleg.*) corner pole; (*cons.*) corner post;
— *à couples,* (*teleg.*) ∧-(shaped) pole;
— *de croisée,* (*cons.*) jamb, side post of a door or window;
— *de distance,* (in races) distance post; (*r. r.*) section mark;
— *double,* (*teleg.*) double pole;
— *d'écurie,* stable post;
— *d'exécution,* place of execution;
— *de fond,* (*cons.*) stud or post extending the whole depth of a wall;
— *gagnant,* (in races) winning post;
— *de galerie,* (*min.*) stay, strut;
— *guide,* hand post, signpost;
— *d'huisserie,* (*cons.*) window post or stud;
— *indicateur,* signpost

poteau *de jonction*, (*teleg.*) pole or post, as on a railway bridge;
—— *x jumelés*, (*teleg.*) poles bolted together;
—— *à lumière*, light spindle, spar, or buoy;
—— *x de mine*, (*min.*) pit wood;
—— *montant*, (*cons.*) jamb;
—— *de raccordement*, v. —— *de jonction*;
—— *de remorque*, towing post;
—— *de remplage*, (*cons.*) stud, quartering, quarter;
—— *télégraphique*, telegraph pole;
—— *valet*, fastening or securing post (of a sluice gate).

potée, f., putty (of tin); (*fond.*) luting loam;
—— *d'émeri*, emery dust;
—— *d'étain*, tin putty;
—— *de fer*, rust putty;
—— *de montagne*, rotten stone;
—— *de rouille*, v. —— *de fer*.

potelet, m., small strut, post, stanchion.

potence, f., crutch; standard or rule for measuring the height of men and of horses; gallows, gibbet; arm (to hang anything on); pin tongs; saddle post (of a bicycle; (*surv.*) offset staff; (*man.*) ring post; (*mil.*) angle in a line (e. g., formed by a flank with the line); (*cons.*) bracket; crossbeam; (*art.*) shaft or stem (of a loading crane); (*pont.*) pier of a (military) suspension bridge; horse of a flying bridge;
brider la ——, (*man.*) in riding at the rings, to hit the post instead of the ring;
—— *de chargement*, (*art.*) shell crane;
chat de ——, cat of the horse (of a flying bridge);
comble en ——, (*cons.*) shed roof;
—— *à deux liens*, (*cons.*) double bracket;
en ——, (*mil.*) with a flank forming an angle with the main line;
étai à ——, (*cons.*) strut fitted with a bracket;
—— *à forer*, —— *de foret*, boring frame;
—— *d'un pont volant*, horse of a flying bridge;
—— *à un lien*, (*cons.*) single bracket;
sous ——, under standard height;
troupe en ——, (*mil.*) line with a flank forming an angle.

potentiel, m. (*elec.*, *etc.*) potential;
chute de ——, fall of potential;
différence de ——, difference of potential;
—— *d'élasticité*, potential of elasticity;
perte de ——, loss of potential.

potentiomètre, m., (*elec.*) potentiometer.

potentite, f., (*expl.*) potentite.

poterie, f., pottery; (*cons.*) stoneware pipes; hollow bricks (for floorings and arches).

poterne, f., sort of tunnel under a quay (opening on the river from the street); (*fort.*) postern.

potiche, f., hole bored, cut made, in a timber, to examine it.

potin, m., (*met.*) pewter; pinchbeck;
—— *gris*, cock metal;
—— *jaune*, brass.

pouce, m., thumb; inch;
—— *d'eau de fontainier*, —— *de fontainier*, —— *fontainier*, unit of flow; quantity of water passing in a minute through a circular aperture, 27^{mm} in diameter, pierced in a vertical wall, under a head of 15.8^{mm} (about 13 liters a minute);
—— *d'eau métrique*, unit of flow through an aperture pierced in a wall 17^{mm} thick, under a head of 5·^{m} (24 cubic meters in 24 hours).

poucet, m., (*cons.*) wiper, lifter, nipper, cam.

poucettes, f. pl., handcuffs; manacles for the thumb.

poucier, m., thumbstall.

poudre, f., dust, powder; gunpowder; explosive. (Except where otherwise indicated, the following terms relate to explosives:)
—— *aciéreuse*, (*met.*) cementation powder;
—— *amide*, —— *amidée*, amide powder;
—— *à l'ammoniaque*, ammoniakrut;
—— *angulaire*, —— *anguleuse*, angular powder (angular grains);
—— *Atlas*, Atlas powder;
—— *avariée*, damaged powder;
—— *s aux azotates*, nitrate powders;

poudre *azotée*, nitrogen powder (generic term for modern powders);
battre la, ——, (*man.*) to paw the ground;
—— *binaire*, powder composed of two substances (e. g., gunpowder without any sulphur);
—— *blanche*, white German or American powder; Augendre's powder; white powder;
—— *au*, *de*, *bois*, nitrated sawdust;
—— *au bois pyroxylé*, Schultz powder;
—— *brisante*, brisant powder, quick powder; high explosive; any chemical explosive (i. e., no mechanical mixture);
—— *à bronzer*, (in the arts) bronze powder;
—— *brune*, brown powder;
—— *-caillou*, pebble powder;
—— *à canon*, cannon powder;
—— *à canon des pilons*, ordinary powder;
—— *de carabine*, rifle (sporting) powder;
—— *de carrière*, blasting powder;
—— *cémentante*, —— *cémentatoire*, v. —— *aciéreuse*;
—— *au charbon*, gunpowder, so called to distinguish it from new powders;
—— *de charbon*, (*in gen.*) coal dust;
—— *de chasse*, sporting powder; (in France, three kinds: *fine*, *superfine*, *extra-fine*);
—— *chloratée*, chlorate powder;
—— *s aux chlorates*, chlorate powders;
—— *chocolat*, brown powder, cocoa powder;
—— *de commerce extérieur*, powder manufactured for the foreign trade;
—— *à combustion lente*, slow-burning powder;
—— *à combustion rapide*, quick-burning powder;
—— *comprimée*, compressed powder, v. —— *moulée en charges*;
—— *-coton*, —— *de coton*, gun cotton, pyroxylin;
—— *cuite*, powder made by boiling ingredients to a paste;
—— *cylindrique*, cylinder powder;
—— *demi-écrasée*, bruised powder;
—— *de démolition*, blasting powder;
épreuve de ——, proof, test, of powder;
essai de ——, v. *épreuve de* ——;
—— *d'exercice*, (*art.*) blank-cartridge powder;
exploiter à la ——, (*min.*, *etc.*) to blast, blow up;
—— *explosive*, Noble's explosive powder;
—— *extra-fine*, a variety of sporting powder (French);
—— *faible*, low explosive;
—— *à faible fumée*, a powder producing but little smoke;
faire parler la ——, (*mil.*) to open hostilities;
—— *fine*, small-grained powder; a variety of sporting powder (French);
—— *forte*, high explosive;
—— *au fulmicoton*, gun-cotton powder;
—— *fulminante*, fulminating, detonating, powder; priming powder, mixture of mercury fulminate and other bodies; powder in which the chlorate is substituted for the nitrate of potassium;
—— *au fulminate de mercure*, fulminate of mercury powder;
—— *à fumée*, ordinary black (smoke-producing) powder;
—— *avec fumée*, ordinary black powder (to distinguish it from smokeless);
—— *sans fumée*, smokeless powder;
—— *à fusil*, rifle powder, musket powder, small-arms powder;
—— *de fusion*, (*met.*) a sort of flux (niter, sulphur, and sawdust);
—— *gallique*, Horsley's gallnut powder;
—— *géante*, giant powder;
—— *à giboyer*, v. —— *de chasse*;
—— *en grain*, grained powder;
—— *à grains fins*, fine-grain (small-arm) powder;
—— *grenée*, grained powder; a commercial name for —— *pyroxylée*, q. v.;
—— *grise*, gray powder, grakrut;
—— *à gros grains*, (*art.*) pebble powder (England); mammoth powder (United States); (*sm. a.*) large-grain powder;

poudre *de guerre*, service powder;
—— *Hercule*, Hercules powder;
—— *en lamelles*, powder in flakes or disks;
—— *lente*, slow-burning powder;
—— *lissée*, glazed powder;
magasin à ——, v. s. v. *magasin*;
mesurette à ——, powder measure;
—— *de mine*, blasting powder (in France, three kinds: —— *anguleuse*, —— *fin(s) grain(s)*, —— *ronde*);
mordre la ——, (*mil.*) to bite the dust, be killed in action;
—— *moulée en charges*, powder compressed into large cakes (no cartridge bag needed, obs.);
—— *moulée en grain*, molded-grain powder;
moulin à ——, powder mill;
—— *à mousquet*, —— *de mousqueterie*, small-arm powder;
—— *neuve*, new powder;
—— *nitratée*, nitrated powder;
—— *noire*, ordinary black powder;
—— *noire comprimée*, compressed black powder;
—— *normale*, standard powder;
noyer les ——s, to flood the magazine(s);
—— *offensive*, v. —— *brisante*;
—— *papier*, paper powder;
—— *pellet*, pellet powder;
—— *au picrate*, picrate powder;
—— *au picrate d'ammoniaque (de potasse)*, ammonium (potassium) picrate powder;
—— *picratée*, picrate powder;
—— *picrique*, Abel's powder, picric powder;
—— *de plomb*, (*sm. a.*) dust shot, mustard seed;
poire à ——, powder flask (obs.);
—— *prismatique*, prismatic powder;
—— *progressive*, progressive powder;
—— *pyroxylée*, a nitrated cellulose;
—— *radoubée*, powder worked over again;
—— *rapide*, quick-burning powder;
—— *refaite*, v. —— *radoubée*;
—— *en roche*, powder that has caked;
—— *ronde*, round-grain powder;
—— *à rondelles comprimées*, a powder consisting of compressed disks;
——s *et salpêtres*, (Fr. a.) one of the subdivisions of the *administration centrale de la guerre*, dealing with technical questions of explosives and related matters, and having general charge of the manufacture of explosives (government monopoly);
——s *salpêtrées*, generic term for black and for brown powders;
—— *à salut*, saluting powder;
sentir la ——, (*mil.*) to have been frequently the seat of war;
soute aux ——s, powder magazine;
—— *de sûreté*, name given at first by M. Nobel to dynamite;
—— *superfine*, a variety of sporting powder (French);
—— *ternaire*, a powder composed of three bodies or substances (e. g., gunpowder);
—— *à tirer*, v. —— *de chasse*;
—— *aux tonnes et presses*, powder that has not been incorporated (that has passed directly from *tonnes* to *presses*);
—— *tonnerre*, thunder powder;
—— *de traite*, (originally) powder manufactured for the slave trade; (now) export powder for the colonial trade;
—— *à tremper*, (met.) cementation tempering powder;
—— *à trituration réduite*, powder produced by a shortened process of incorporation;
—— *type*, standard type powder, standard powder;
—— *de vente*, powder in the trade;
—— *verte*, new powder; powder not thoroughly dried; a variety of picric powder;
—— *vive*, quick-burning powder;
—— *en vrac*, loose powder;
—— *de Vulcain*, Vulcan powder.
poudrer, v. a., to pounce.
poudrerie, f., (*powd.*) powder mill, powder works.
poudreux, a., dusty.

poudrier, m., hand or workman of a powder mill;
poudrière, f., powder mill; powder magazine (not a military term).
pouf, a., (of stones) weak, crumbling; m., (met.) proper amount of resistance required in a mold to bear the weight of metal in fusion.
pouilleux, a., (of timber, tree) with cankers or ulcers.
poulain, m., colt, foal, filly; sledge (for heavy loads);
faire un ——, (*mil. slang*) to fall from one's horse, to come a cropper (to bite tan bark, U. S. M. A.);
maladie de ——, (*hipp.*) colt evil.
poule, f., hen, fowl; (in racing, etc.) pool, sweepstakes);
acier ——, (*met.*) blistered steel;
—— *d'essai*, (racing) maiden stakes;
—— *des produits*, (racing) futurity stakes.
poulet, m., chicken;
—— *d'Inde*, (*mil. slang*) cavalry horse.
poulevrin, m., (*powd.*) priming powder; pulverized, mealed powder; powder flask (obs.);
—— *vert*, powder previous to granulation;
vieux ——, powder caked from age and damp.
pouliage, m., blocks, system of blocks.
pouliche, f., (*hipp.*) filly.
poulie, f., (*cord., mach.*) pulley, block; sheave, (more rarely) tackle;
—— *d'appareil*, large purchase block; fourfold block;
—— *d'assemblage*, made block;
bande de ——, block strap;
—— *barbotin*, —— *à barbotins*, chain pulley;
—— *-baraquette*, sister block;
——s *à bloc*, block and block;
—— *en bois*, wooden block;
—— *en bois d'assemblage*, made wooden block;
—— *bombée*, (*mach.*) rounded, crowning pulley;
—— *à bras divisés*, (*mach.*) split or divided-rim pulley;
caisse de ——, shell of a block;
—— *caliorne*, —— *de caliorne*, winding tackle block;
—— *de capon*, cat block;
—— *de cartahu*, girt-line block;
—— *à chaîne*, chain pulley;
—— *à chaîne à la Vaucanson*, sprocket (chain) pulley;
—— *à chapeau*, shoulder block;
clan de ——, sheave hole; swallow;
—— *de commande*, (*mach.*) (the pulley working a particular tool or engine or machine) driving pulley;
—— *conductrice*, (*mach.*) driving pulley;
—— *conduite*, (*mach.*) driven pulley;
—— *de conduite*, leading block;
—— *à cônes étagés*, (*mach.*) stepped pulley;
—— *à contre*, shoe block;
—— *à corde*, rope pulley;
—— *à corde métallique*, wire-rope pulley;
—— *à cosse*, block with lashing eye or thimble;
—— *coupée*, snatch block;
—— *coupée et ferrée*, ironbound snatch block;
—— *courante*, running block;
—— *à croc*, hook block;
—— *à crochet*, traversing pulley;
—— *à dé*, bushed block;
—— *à demi-joue*, cheek block;
—— *à dent*, snatch block;
—— *destropée*, —— *détropée*, unstrapped block, block shaken out of its strap;
—— *à deux engoujures*, double-scored block;
—— *à deux estropes*, double-strapped block;
—— *à deux joues*, (*mach.*) double-flanged pulley;
—— *en deux morceaux, pièces*, split pulley;
—— *à deux rangs de rais*, (*mach.*) double-arm pulley;
—— *différentielle*, differential pulley;
—— *différentielle double (simple)*, double (single) differential pulley or tackle;
—— *directrice*, (*mach.*) guide pulley;
—— *disque*, (*mach.*) disk pulley;
—— *dormante*, standing block;

poulie *double*, (*mach.*) double cable pulley; (*cord.*) double block;
— *double à canon*, double gun-tackle block;
— *à double gorge*, double-grooved pulley;
— *double de palan*, long tackle block;
— *double à tourniquet*, iron block with a swivel hook;
— *à émerillon*, swivel block;
— *à empreinte*, chain pulley;
— *enchapée*, frame pulley;
— *enchapée double*, double frame pulley;
— *enchapée simple*, single frame pulley;
engoujure de —, score of a block;
erse de —, block strap;
estrope de —, block seizing or strap;
— *à estrope double*, double-strapped block;
— *à estrope en fer*, iron-strapped block;
— *à estrope interne*, internal (-bound) block;
— *à estrope simple*, single-strapped block;
— *estropée*, strapped block;
— *estropée en fer*, iron-bound block;
— *excentrique*, eccentric sheave;
— *d'expansion*, (*mach.*) expansion drum;
— *en fer*, iron block;
— *ferrée*, iron-bound, -strapped block;
— *fixe*, (*mach.*) fast pulley, fixed pulley; (*cord.*) standing block;
— *fixe et folle*, fixed and loose pulley;
— *folle*, (*mach.*) loose, idle, pulley;
— *en fonte*, cast-iron pulley;
— *à fouet*, tail block;
— *française à hélices*, a synonym of *palan à hélice*, q. v.;
frapper une —, to seize a block;
— *à, de, friction*, (*mach.*) friction pulley;
— *à gorge*, grooved pulley;
gorge de —, score of a block;
— *-guide*, (*mach.*) guide pulley;
— *libre*, (*mach.*) idle pulley;
— *marionnette*, nine-pin block;
— *menante*, (*mach.*) driving pulley or wheel;
— *menée*, (*mach.*) receiving pulley or wheel;
— *mobile*, (*mach.*) movable pulley; loose sheave; (*cord.*) running block;
— *à moque*, dead-eye;
— *motrice*, (*mach.*) main driving pulley;
— *mouflée*, tackle (set of blocks, block and tackle);
— *à mouvement universel*, (*mach.*) pulley having a motion around three rectangular axes;
— *non-estropée*, unstrapped block;
— *œil*, — *àl'œil*, v. — *à cosse*;
— *à olive*, shoe block;
— *d'orientation*, (balloons) direction pulley;
— *ouvrante*, snatch block;
— *de palan*, — *pour palans*, tackle block;
— *de pied de mât*, nine-pin block;
— *plate*, cheek block;
— *pleine*, (*mach.*) solid pulley (as distinguished from one with arms);
— *portante*, bearing pulley;
réa de —, shea 'e;
— *à n réas*, n-sheaved block;
— *de renvoi* (*mach.*, *cord.*) guide pulley;
— *résistante*, (*mach.*) driven pulley;
— *de retour*, leading block;
— *ronde et courte*, clump block;
rouet de —, sheave;
— *à n rouets*, n-fold pulley;
— *à sabot*, v. — *à chapeau*;
— *de secours*, (*mach.*) guide pulley (to keep belts on main pulley;
— *d'une seule pièce*, mortised block;
— *simple*, (*cord.*) single block; (*mach.*) single-cable pulley, one-cable pulley;
— *simple à croc*, single-hook block;
— *simple estropée à œillet*, eye block;
— *simple à fouet*, tail block;
— *support*, — *de support*, (*mach.*) supporting, intermediate, pulley (of a rope transmission);
— *à talon*, shoulder block;
— *de tension*, (*mach.*) tightening pulley;
— *de tête de mât*, masthead block;
— *de touage*, (balloon) towing pulley;
— *à tourniquet*, gin block, monkey block;
— *de transmission*, (*mach.*) communicating, transmitting, pulley; belt pulley;
— *de traversière*, fish-tackle block;

poulie *triple*, threefold block, triple block;
— *Verlinde*, a synonym of *palan à hélice*, q. v.;
— *vierge*, sister block;
— *vierge double*, double sister block;
— *vierge simple*, single sister block;
— *à violon*, fiddle block;
— *-volant*, (*mach.*) driving fly wheel.

poulier, v. a., to work, hoist, draw up by a pulley;

poulierie, f., block shed.

poulieur, m., block maker.

poulin, m., dog shore.

poulinaille, f., (*hipp.*) foal's cramp (displacement of kn 'e pan);

poulinement, m., (*hipp.*) foaling.

pouliner, v. a., to foal.

poulinière, f., (*hipp.*) brood mare;
bonne —, good breeder;
jument —, brood mare.

poupe, f., (*nav.*) stern, (*torp.*) rear end of a spar torpedo;
enseigne de —, (*nav.*) stern flag, national color;
passer à — *d'un navire*, (*nav.*) to pass close under the stern.

poupée, f., doll; warping-cone, -end, of a windlass; (*pont.*) timber head, mooring post (of a pontoon boat); (*fond.*) tow winding of a core bar; (*elec.*) binding post; (*mach.*) bar, rod (in many applications); (more esp.) poppet, center, headstock; spindle or shaft and related parts;
contre- —, v. *contre-poupée*;
— *contre-pointe*, v. — *mobile*;
— *de derrière*, (*mach.*) back poppet;
— *de devant*, (*mach.*) fore poppet;
— *à diviseur*, (*mach.*) spiral head, index head, index center;
faire la —, (*fond.*) to wind tow, etc., around the spindle of a core box;
— *fixe*, (*mach.*) headstock, live head, fixed poppet or head (of a lathe);
— *à jour*, (*mach.*) shank mandrel;
— *à lunette*, (*mach.*) cone plate, boring collar;
— *mobile*, poppet proper, tailstock, movable headstock, back poppet; deadhead; sliding poppet;
— *à pointe*, v. — *mobile*;
— *à roder*, (*mach.*) grinding poppet.

pour-cent, m., percentage.

pourette, f., (*hipp.*) grease complicated by grapes.

pourfendre, v. a., to cleave with a sword.

pourparler, m., parley;
en —, parleying in treaty;
entrer en —, to enter into a parley;
être en —, to parley; to be in treaty.

pourri, p. p., rotten, corrupt; damp, wet; m., rottenness, rotten part;
pierre —, rotten stone;
temps —, damp, wet, weather.

pourriture, f., rottenness; (of wood) rot;
— *d'hôpital*, (*med.*) hospital gangrene;
— *humide*, wet rot;
— *sèche*, dry rot;
tomber en — to rot, become rotten.

poursuite, f., pursuit; chase; search; (*law*) proceedings; prosecution; (*mil.*) pursuit; (*man.*) exercise in riding school in which each trooper pursues and tries to touch his adversary;
— *directe*, (*mil.*) pursuit in contact with enemy and behind him;
diriger des —*s*, (*law*.) to institute, take, proceedings;
frais de —, (*law*) costs; court fees;
— *indirecte*, (*mil.*) pursuit without combat alongside of the enemy, and in which contact is never lost;
intenter des —*s*, v. *diriger des* —*s*;
— *stratégique*, (*mil.*) strategic pursuit, so-called, the object of which is to reap the fullest advantage possible from a victory;
— *tactique*, (*mil.*) tactical pursuit, i. e., actual pursuit of a defeated enemy for the purpose of further breaking him up (said esp. of cavalry in pursuit).

poursuivre, v. a., to pursue, chase; to go on, proceed with; to prosecute; (*law*) to prosecute; (*mil.*) to pursue.

pourtour 352 **poutrelle**

pourtour, m., circumference, circuit, length (around, etc.), compass;
— *de la chambre*, (*art.*) interior surface of the chamber of a gun.
pourvoi, m., (*law*) appeal;
— *en cassation*, appeal to the supreme court;
— *en grâce*, appeal for a pardon.
pourvoir, v. a. n., to provide, supply, furnish, stock; to see to, attend to; to appoint (to an office);
se —, to appeal; (*law*.) to appeal a case;
se — *en cassation*, (*law*) to appeal to the supreme court;
se — *en grâce*, (*law*) to appeal for pardon.
pourvoyeur, m., purveyor; (*mil.*) in general, soldier who brings up ammunition, as from the wagons to the line; (*mil.*) cannoneer who serves ammunition to the piece (limber to gun, magazine to gun, etc.);
— *-servant*, (*art.*) cannoneer who serves ammunition.
pousse, f., shoot, growth; dust (commercial term); (*hipp.*) heaves, broken wind.
poussé, p. p. pushed; (of ironwork) filed, i. e., unpolished;
bon —, (of hardware) state of being half polished (between polish and total lack of polish);
cheval — *de nourriture*, overfed horse, pampered horse.
pousse-cailloux, m., (*mil. slang*) foot soldier.
poussée, f., push; pressure, thrust (as of an arch, of an embankment); jerk (as of a machine); (*man.*) spurt, dash;
— *de l'eau*, pressure of water; buoyancy of water;
faire le trait des — *s des voûtes*, (*cons.*) to compute and mark out the thickness necessary to resist the thrust of an arch;
— *de l'hélice*, thrust of the screw;
— *indiquée*, indicated thrust;
— *de la lame*, heave of the sea;
— *normale*, normal thrust;
— *des terres*, thrust of earthwork;
— *de vapeur*, (*steam*) (intermittent) thrust or delivery of pressure from a steam cylinder in a steam vacuum pump;
— *verticale de l'eau*, buoyancy;
— *au vide*, (*cons.*) an unsupported thrust.
pousse-goupille, m., (*sm. a.*) driver, pin drift.
pousser, v. a. n., to push, to push forward, to thrust, to impel; to extend, carry on; to press, press forward; to drive, drive back, along, in, on; to shove, shove off (a boat); to jut, bulge out; (*fenc.*) to thrust; (*hipp.*) to heave, to be broken-winded; (*mil.*) to drive the enemy; (*siege*) to drive, push (a sap, a trench); (*cons.*) to thrust, to have or develop a thrust;
— *une attaque*, (*mil.*) to press an attack;
— *en avant*, to push ahead or on;
— *une botte à quelqu'un*, (*fenc.*) to make a thrust at anyone;
— *à bout*, to press home, carry to a successful issue;
— *une charge*, (*mil.*) to press a charge;
— *un cheval*, (*man.*) to push a horse, try his mettle;
— *un clou*, to drive a nail;
— *un coup d'épée*, (*fenc.*) to make a pass or thrust;
— *une découverte*, (*mil.*) to explore with a view to fixing enemy's position;
— *en dehors*, to bulge out, swell out (as a wall);
— *l'ennemi*, (*mil.*) to drive the enemy;
— *ferme*, to push, thrust hard; to strike, hit hard; to carry on fast;
— *le feu*, to urge a fire; to fire up;
— *les feux au fond de la grille*, to bank the fires;
— *de fond*, to pole a boat;
— *une garde, etc.*, (*mil.*) to push forward an advanced party or guard, etc.;
— *au large*, to push, shove off (a boat, a pontoon);
— *les marches*, (*carp.*) to put the moldings on the front of the stairs;

— *des moulures*, (*cons.*) to form, make moldings;
— *de nourriture*, (*hipp.*) to overfeed;
— *outre*, to carry too far;
— *la planche*, (*nav.*) to put out a gang plank;
— *une reconnaissance*, (*mil.*) to make a reconnaissance suddenly and boldly;
— *la sape*, (*siege*) to advance the sap;
— *une tranchée*, (*siege*) to drive a trench;
— *trop fort*, — *trop loin*, to overdrive;
— *au vide*, (*cons.*) to have an unsupported thrust.
poussier, m., dust (of coal, gunpowder, etc.); (*mas.*) stone chips;
— *de charbon*, coal dust;
— *de foin*, bits of hay (on the floor of a hayloft);
— *de minerai*, slime;
— *sec*, (*powd.*) gunpowder dust, after drying, rolling, and glazing;
— *vert*, (*powd.*) dust produced in granulation.
poussière, f., dust, powder;
battre la —, (*man.*) to paw the ground;
— *de charbon*, coal dust;
faire de la —, to be dusty;
— *de foret*, stone dust;
mordre la —, to bite the dust;
— *de la poudre*, gunpowder dust;
tomber en —, to crumble, crumble to dust.
poussif, a., (*hipp.*) broken-winded; pursy.
poussoir, m., driver, punch; (*sm. a.*) spring button; push button, push piece, pusher (on a bayonet); ejector-rod head (of a revolver);
— *à ressort*, spring push button.
poussolane, f., pozzuolana.
poutrage, m., (*cons.*) framing, timber work.
poutraison, f., (*cons.*) framing of joists, timber work.
poutre, f., (*cons.*) girder, beam; (*pont.*) balk;
— *armée*, trussed beam or girder;
— *armée par fourrures*, fished beam;
— *d'assemblage*, built-up beam;
— *cellulaire*, cellular girder;
— *cintrée*, — *courbée*, bent, curved, beam;
— *cloisonnée*, chambered beam;
— *de culée*, (*pont.*) shore balk, trestle balk;
— *à deux contre-fiches*, double-trussed beam;
— *en double, en double* T, **I**-girder;
— *foudroyante*, (*fort.*) rampart beam;
— *à griffes*, (*pont.*) claw balk;
— *de guindage*, (*pont.*) side rail;
— *horizontale*, (*gym.*) horizontal beam;
— *à latice*, lattice girder;
— *de la machine*, (*mach.*) engine beam;
maîtresse- —, main girder;
— *ordinaire*, common balk; (*pont.*) balk;
— *-plancher*, (*pont.*) trussed beam, in form a solid of equal resistance, with floor fixed in position to cover small streams;
— *pleine*, tubular girder;
— *à plusieurs contre-fiches*, multiple-trussed beam;
porte- —*s*, (*pont.*) balk carrier;
— *quarderonnée*, beam with a rounded corner;
— *-rail*, rail beam (i. e., beam and rail in one— the track of a *pont-rail*, q. v.);
— *ramée*, trussed beam;
— *de rampe*, (*pont.*) balk used in the construction of trestle bridges, to slide the trestle into place;
sous- —, sleeper;
— *à une seule contre-fiche*, simple-trussed beam;
— *en* T, T-girder;
— *en tôle*, plate beam;
— *tournante*, swing bar of a gate;
— *traversière*, straining beam, straining piece;
— *à treillis*, — *en treillis*, lattice girder;
— *à trois contre-fiches*, triple-trussed beam;
— *tubulaire*, tubular girder;
— *en ventre de poisson*, fish-bellied truss.
poutrelle, f., joist, skid, small beam; (*pont.*) balk;
— *articulée*, (*pont.*) jointed balk;

poutrelle, à *charnière(s)* (*pont.*) hinged, jointed, balk;
—— *du cheval de frise*, (*fort.*) barrel of a cheval de frise;
—— *courte*, (*pont.*) short balk;
—— *de culée*, (*pont.*) shore balk;
fausse —— (*de guindage*), (*pont.*) false balk (used in method by rafts);
—— *à galets*, small sluice gate;
—— *à griffes*, (*pont.*) claw balk;
—— *de guindage*, (*pont.*) side rail;
—— *s jumelées*, (*pont.*) fished spars;
—— *longue*, (*pont.*) long balk;
—— *en madriers*, (*pont.*) trussed balk made of chess;
—— *de manœuvres*, (*art.*) skid, parbuckling skid;
—— *en planches*, (*pont.*) trussed balk made of planks;
—— *de rampe*, (*siege*) sap roller skid;
—— *de support*, (*pont.*) balk proper;
—— *de tablier*, (*pont.*) road bearer;
—— *traînante*, (*art.*) small spar lashed under the axle, with one end dragging, as a substitute for a broken or disabled wheel.
pouvoir, m., power; authority, power of attorney; influence;
—— *calorifique*, heating power;
—— *conducteur*, (*elec.*, etc.) conductivity;
—— *disciplinaire*, (*mil.*) disciplinary power;
—— *éclairant*, illuminating power;
—— *évolutif*, (*nav.*) maneuvering power of a ship;
—— *multiplicateur*, multiplying power;
—— *optique*, v. —— *séparateur*;
—— *perforant*, (*art.*) perforating effect (of a projectile);
—— *des pointes*, (*elec.*) action of points;
—— *séparateur*, (*opt.*) defining power of a lens.
pouzzolane, f., pozzuolana;
—— *artificielle*, artificial pozzuolana;
—— *naturelle*, natural pozzuolana;
—— *en pierre*, trass;
—— *en poudre*, pozzuolana powder,
prairie, f., meadow; prairie (U. S.).
prame, f., praam.
praticable, a., practicable; passable (of a road); (*mil.*) practicable (of a breach).
praticien, m., practitioner; practical man.
pratique, f., practice; execution (of a design, etc.); method of execution, custom, habit; experience; customer; (*nav.*) pratique, communication with the shore; (in southern Europe), pilot; (*law, med.*) practice; (*mil. slang*) deadbeat, coffee-cooler (U. S. A.);
entrer en libre ——, (*nav.*) to have permission to enter, to land (no quarantine);
homme de ——, (*law*) jurist; lawyer;
libre ——, (*nav.*) free communication with the shore (no quarantine);
—— *de la mer*, practical knowledge of seamanship;
mettre en ——, to put into execution;
pierre de ——, (*mas.*) undressed stone used in construction.
pratique, a., practical, experienced;
cours ——, practical course;
homme ——, experienced man;
marin, pilote, ——, seaman, pilot, familiar with given waters.
pratiquer, v. a., to practice, put in practice; to tamper with; to practice (medicine, the veterinary art); to make, contrive, drive, run; to work in, introduce, let in, cut (as a wicket in a door, a staircase in the thickness of a wall, etc.); (*nav.*) to have free access;
—— *un chemin*, to cut a road;
—— *une embrasure*, (*fort.*) to cut an embrasure;
—— *la pierre*, (*mas.*) to cut a stone with as little waste as possible;
—— *des rameaux*, (*mil. min.*) to run out branches.
pré, m., meadow;
—— *salé*, salt marsh.
préachat, m., prepayment.
préau, m., prison yard.
préceinte, f., (*nav.*) wale.
précession, f., precession.
précipice, m., precipice.

précipitation, f., precipitation; (*chem.*) precipitation.
précipité, m., (*chem.*) precipitate.
précipiter, v. a., to precipitate, hasten, plunge; (*chem.*) to precipitate;
se ——, to throw one's self; to rush; (*chem.*) to be precipitated.
précis, m., abstract, summary, compendium.
précis, a., precise, exact; (*ball.*) holding well together (said of bullets all of which fall close to the center of impact, though all may be off the target, as distinguished from *réglé*, accurate, i. e., on the target).
précision, f., precision, accuracy; (*ball.*) precision (v. *justesse*);
arme de ——, v. s. v. *arme;*
atelier de ——, v. s. v. *atelier;*
de ——, accurate;
instrumen de ——, v. s. v. *instrument.*
précompte, m., stoppage or deduction of pay.
préfecture, f., office, jurisdiction, functions, term of office, headquarters, residence, city of residence, of a prefect;
—— *maritime*, headquarters of a maritime district;
—— *de police*, functions, office, of the prefect of police;
sous- ——, v. *sous-préfecture.*
préfet, m., prefect;
—— *maritime*, (*Fr. nav.*) maritime prefect (a vice- or rear-admiral);
—— *de police*, police commissioner;
sous- ——, v. *sous-préfet.*
préhender, v. a., to seize.
préjudice, m., harm, damage, injury, loss.
préjugé, m., prejudice; (*law*) preceden..
préjuger, v. a., to prejudge.
préla(r)t, m., tarpaulin;
—— *de crin*, haircloth for the floor of a powder magazine. (The form *prélat* is used by the French artillery.)
prélèvement, m., selection, as of specimens for test; retention of pay, of money, for any specific purpose; installment.
prélever, v. a., to select from, pick out of, a lot; to retain (a sum of money, part of a soldier's pay).
préliminaire, a. m., preliminary;
les —— *s de la paix*, (*mil.*) preliminaries of peace.
premier, a., m., first (in rank, order, time, or place); former, early, next; prime (of numbers); leader, chief; first story of a house (second, U. S.); (*mil.*) first drum, first call;
au ——, on the first floor (second, U. S.);
le —— *de l'an*, New Year's day;
—— *étage*, first floor (second, U. S.);
être reçu le ——, to pass first;
—— *figure*, a (geometrical) figure, not susceptible of subdivision into simpler figures;
les —— *s juges*, judges of the lower courts;
—— *lieutenant, lieutenant en* ——, (*mil.*) first lieutenant;
matières —— *es*, raw materials;
—— *ministre*, prime minister, first minister;
le —— *du mois*, first day of the month;
—— *Paris*, leading article of a newspaper;
—— *pont*, (*nav.*) lower deck.
prémunir, v. a. r., to warn, caution.
prenable, a., (*fort.*) that may be captured.
prenant, a., —— *n ans*, (*hipp.*) rising *n* years;
partie ——*e*, v. s. v. *partie.*
prendre, v. a. n. r., to take, take hold of, lay hold of, seize, obtain; to catch, take, a disease; to arrest, take, a person; to surprise, overtake, catch at; to take quarters, to occupy; to carry, go, turn; to take fire, burn, begin to burn; to congeal; to freeze (of a river, etc.); to set (of mortar, cement); to bite (of an anchor); to take hold (of a screw); (*cord.*) to become entangled; to get foul; (*mil.*) to take, capture (a fort, town, stores, arms, guns, prisoners, etc.); (*art.*) to take (in flank, in reverse, etc.);

prendre *les aides des jambes,* (*man.*) to become legwise;

—— ....*ans,* (*hipp.*) to be rising....years;
—— *les armes,* (*mil.*) to get under arms; to take up arms; to turn out (of a guard);
—— *l'arme à la main,* (*mil.*) to trail arms;
—— *d'assaut,* (*mil.*) to take by storm;
—— *de l'avantage pour monter à cheval,* v. s. v. *avantage;*
—— *chair,* (*hipp.*) to take on flesh after a long sickness;
—— *en charge,* (*adm.*) to take up (property, etc.);
—— *chasse,* (*nav.*) to run away from a pursuing enemy;
—— *la clef des champs,* to run away;
—— *les coins,* (*man.*) to take corners in riding hall;
—— *un cordage à retour,* (*cord.*) to take a turn with a rope;
—— *au corps,* to arrest, make prisoner;
—— *le corps mort,* to make fast to moorings;
—— *la cuirasse,* (*mil.*) to turn soldier;
—— *ses degrés,* to take one's degree;
—— *le deuil,* to go into mourning;
—— *les dents,* (*hipp.*) to begin the second dentition;
—— *les distances,* (*mil.,* etc.) to open out;
—— *à dos,* (*art.*) to take in reverse;
—— *à droite,* (*gauche*), to go to the right (left);
—— *sa droite,* to place one's self on the right of any one;
—— *d'écharpe,* (*art.*) to take obliquely;
—— *d'enfilade,* (*art., etc.*) to enfilade;
—— *son équipage,* (*nav.*) to ship the crew;
faire —— *le feu,* to make the fire burn;
faire —— *le mortier,* (*mas.*) to set the mortar;
—— (*le*) *fait et cause pour,* (*law, gen.*) to side with;
—— *feu,* (*sm. a., etc.*) to go off (as fuses, caps, etc.);
—— *en flanc,* (*mil.*) to take in flank;
—— *fond,* (of an anchor) to bite, to take hold;
—— *fuite,* to scamper off, run away;
—— *le galop,* (*man.*) to break into a gallop;
—— *la garde,* to go on guard;
—— *ses grades,* v. —— *ses degrés;*
—— *la haute mer,* (*nav.*) to get out to sea;
—— *hauteur,* to take an altitude;
—— *les intervalles,* (*mil.*) to extend intervals;
—— *ses jambes à son cou,* v. —— *fuite;*
—— *jusqu'à....de diametre,* (*mach.*) to swing (so many inches, etc.) in diameter;
—— *la lame debout,* to head the sea;
—— *langue,* to obtain intelligence by secret means;
—— *le large,* v. —— *fuite;*
—— *du lest,* to take in ballast (ships and balloons);
—— *la ligne de mire,* (*art., sm. a.*) to aim;
—— *la mire,* (*art., sm. a.*) to take aim;
—— *le mors aux dents,* (*man.*) to take the bit between the teeth; to run away;
—— *mouillage,* (*nav.*) to take up an anchoring ground;
—— *des nivellements,* (*surv.*) to level, to take a line of levels;
—— *des observations,* to take observations;
—— *la parole,* to take the floor;
—— *parti,* (*mil.*) to enlist;
—— *le parti de l'épée,* (*mil.*) to turn soldier;
—— *le pas sur,* to take precedence of;
se —— *les pieds,* to get the feet entangled (e. g., horse, in the traces, etc.);
—— *une position,* (*mil.*) to take up a position; to capture a position;
—— *les rayures,* —— *dans les rayures,* (*art., sm. a.*) to take the grooves;
—— *la remorque,* to be towed, to take the tow;
—— *à la remorque,* to take in tow;
—— *à revers,* (*art., sm. a.*) to take in reverse;
—— *en rouage,* (*art.*) to take a line or mass of matériel in flank; to sweep by enfilade fire;
—— *la semaine,* (*mil.*) to go on duty for the week;
—— *sur le temps,* (*fenc.*) to thrust while the adversary is busy with a movement of his own;
—— *tour,* (*cord.*) to take a turn;
—— *le train,* (*r. r.*) to take the train;

prendre *des troupes,* (*nav.*) to take troops on board;
—— *le vent,* (*nav.*) to get the weather gauge of.

preneur, m., captor; lessee;
—— *à bail,* lessee;
vaisseau ——, (*nav.*) ship that has taken a prize.

préparateur, m., coach (for examinations).

préparatifs, m., preparations;
—— *de guerre,* preparations for war.

préparation, f., preparation;
—— *de l'attaque par l'artillerie,* (*mil.*) artillery preparation for the infantry attack;
—— *des bois,* steeping, impregnation, injection of wood;
—— *mécanique des minerais,* (*met.*) dressing of ores;
—— *du temps de guerre,* (*fort.*) in France, assumption of command by designated c. o., and adoption of all measures to put a fortress on a war footing as soon as mobilization is ordered;
—— *du temps de paix,* (*fort.*) in France, preparation in time of peace of plans of defense by the governor of a fortress, aided by the *commission de défense,* q. v.

préparatoire, a. m., preparatory, preliminary;
commandement ——, (*mil.*) cautionary command;
machine ——, (*met.*) dressing machine;
signal ——, (*mil.*) cautionary signal.

préparer, v. a., to prepare, arrange; (*cord.*) to see all clear;
—— *l'attaque,* (*mil.*) to prepare the (infantry) attack by artillery fire;
machine ——, (*met.*) dressing machine.

prépondérance, f., (*art.*) preponderance.

prépondérant, a., preponderant;
voix ——*e,* casting vote.

préposé, m., officer or person in charge; (*adm.*) agent or representative of a contractor at the place specified in the contract;
—— *de la douane,* custom-house officer.

près, adv., near;
au plus ——, (*nav.*) close to the wind;
faire plus ——, (*art.*) to depress.

presbyte, a., m., far-sighted; far-sighted person.

presbytie, f., long-sightedness.

prescription, f., precept; prescription; (*law*) prescription; expiration by limitation of the right to prosecute, to punish; (*Fr. law*) closing of accounts, etc., in favor of the state if not settled after five years; (*med.*) prescription;
par ——, by the statute of limitation.

prescrire, v. a., to order, command; (*law*) to acquire by prescription; (*med.*) to prescribe.

préséance, f., precedence;
avoir la ——, to take precedence of.

présence, f., presence; (*mil.*) status of being present for duty;
solde de ——, (*mil.*) full pay, active-duty pay.

présent, a., m., present, "here;"
la ——*e,* this letter;
le —— *porteur,* the bearer of this;
les ——*s et les absents,* the present and absent.

présentation, f., presentation; submission (as of a list of candidates for the Legion of Honor, etc.).

présenter, v. a., to present, offer, hold out, bring forward, introduce; to try a piece of work to see if it is right before fixing it in its place; to submit (as a list of candidates for the Legion of Honor);
—— *l'arme* (*les armes*), (*mil.*) to present arms;
—— *la baïonnette,* (*mil.*) to charge bayonets;
—— *la bataille,* (*mil.*) to offer battle;
—— *un cheval,* to show, present, a horse to a buyer, to his rider;
—— *la gaule,* (of a groom) to salute with the whip;
—— *une manœuvre,* (*cord.*) to snatch a rope, put it into a snatch block.

préservateur, a., preservative.

présidence, f., presidency.

président, m., president; (*law*) senior, presiding judge.
presqu'île, f., peninsula.
pressage, m., pressing.
presse, f., press; crowd; the press; (*nav.*) press gang; (*mach.*) press, printing press;
—— *à amorcer*, (*sm. a.*) priming machine, priming press;
—— *à approche rapide*, (*mach.*) quick advance press;
—— *à bras*, hand press;
—— *à coins*, wedge press;
—— *à copier*, copying press;
—— *à cylindre*, (*mach.*) roller press;
—— *à découpoir*, (*mach.*) punch, punching press;
—— *à désamorcer*, (*sm. a.*) unpriming press;
détachement de la ——, (*nav.*) press gang;
—— *à emboutir*, (*mach.*) stamping press;
—— *de l'établi*, bench screw;
—— *à étamper*, v. —— *à emboutir;*
exercer la —— *contre*, (*nav.*) to impress seamen;
—— *à forger*, (*mach.*) forging press;
—— *à fusée*, (*artif.*) rocket press;
—— *à galeter*, (*powd.*) gunpowder press;
—— *hydraulique*, (*mach.*) hydraulic press;
—— *à levier*, (*mach.*) lever press;
—— *lithographique*, lithographic press;
—— *à main*, cramp, screw clamp;
—— *à mandriner*, (*mach.*) mandrel press;
—— *des matelots*, (*nav.*) impressment;
—— *à mater*, (*mach.*) hydraulic press (for seating bands home on projectile);
—— *mécanique*, (*mach.*) engine press, fly press, power press;
mettre à la ——, to seat by pressure (as a rifling band, etc.);
—— *à mouler*, (*fond.*) molding press;
ouvrage fait à la ——, presswork;
—— *de pointage*, (*art.*) elevating (hydraulic) cylinder;
pot de ——, v. s. v. *pot;*
—— *à poudre*, (*powd.*) powder press;
—— *à river*, riveting clamp;
—— *à rotule*, (*mach.*) circular toggle press;
—— *à rouleau*, v. —— *à cylindre;*
—— *à satiner*, (*phot.*) rolling machine;
—— *à satiner à chaud*, (*phot.*) burnishing machine;
—— *à serrer*, v. —— *à main;*
sous ——, in the press (of books);
—— *à tourillons*, (*art.*) trunnion cylinder;
—— *à vis*, (*mach.*) screw press, fly press.
presse-étoupe(s), m., (*mach.*) stuffing box; stuffing-box gland;
chapeau de ——, stuffing-box gland;
collier de ——, collar or hoop of a stuffing box;
—— *à chapeau en écrou*, screw-gland stuffing box;
—— *à contre-chapeau*, counter-gland stuffing box;
couronne de ——, stuffing-box gland;
faire le ——, to pack a stuffing box;
grain de ——, collar or hoop of a stuffing box;
—— *à lanterne*, lantern-brass stuffing box;
serrer un ——, to press down the packing of a stuffing box;
siège de ——, v. *couronne de* ——.
presse-fusée, m., (*art.*) fuse setter (obs.).
presse-garniture, m., (*mach.*) (leather) packing box; stuffing box; collar screwed into cylinder heads and keeping packing in position.
pressement, m., compression.
presse-papier, m., paper weight.
presser, v. a. n., to press, compress, squeeze, jam; to urge, hasten; to crowd, throng; (*man.*) to ride hard;
—— *un cheval*, (*man.*) to urge a horse, keep him from slackening his gait;
—— *le pas*, to hasten on.
presseur, a., m., pressing, pressman.
pression, f., pressure; (*fenc.*) slight pressure of point on point to deflect the adversary's blade;
—— *absolue*, absolute pressure;
—— *actuelle*, actual pressure;
—— *atmosphérique*, atmospheric pressure;
avoir de la ——, to have steam up;
—— *barométrique*, barometric pressure;

pression, *basse* ——, (*steam*) low pressure;
centre de ——, center of pressure;
—— *de chaudière*, boiler pressure;
—— *constante*, constant pressure;
contre- ——, back pressure;
courbe des ——s, curve of pressures;
échelle de ——, scale of pressure;
—— *d'éclatement*, bursting pressure;
—— *effective*, useful, effective, working, pressure;
en ——, (*steam*) with steam up, under pressure, under steam;
épreuve de ——, pressure test;
être sous ——, to have steam up (as a steamer, etc.);
—— *exigée*, necessary pressure;
—— *extérieure*, external pressure;
faire monter la ——, to get up steam;
—— *finale*, final pressure;
haute ——, high pressure;
—— *hydraulique*, hydraulic pressure;
—— *initiale*, initial pressure;
—— *intérieure*, internal pressure;
—— *intermédiaire*, medium pressure;
—— *-limite*, limiting pressure;
mettre en ——, (*steam*) to get up steam;
mise en ——, (*steam*) act, operation, of getting up steam;
moyenne ——, (*steam, gen.*) mean pressure;
—— *normale*, normal pressure;
pleine ——, full pressure;
—— *de régime*, working pressure;
—— *requise*, v. —— *exigée;*
sous ——, with steam up;
tenir de la ——, to keep steam up;
—— *de travail*, working pressure;
—— *de la vapeur*, tension or pressure of steam;
—— *voulue*, v. —— *exigée*.
pressoir, m., rolling or pressing mill; pinch bar.
prestataire, m., person compelled (by statute) to work on the public roads; (*mil.*) soldier (regarded from the point of view of the allowances, etc., drawn).
prestation, f., swearing fidelity, taking an oath; compulsory labor on the public roads; stores, supplies; (*Fr. a.*) generic term for pay and allowances, supplies;
—— *en deniers*, (*mil.*) pay and allowances;
—— *en nature*, (*mil.*) supplies in kind furnished by the government.
prêt, m., action of lending money, loan, sum of money lent; (*Fr. a.*) pay of n. c. o. and privates;
—— *franc*, (*Fr. a.*) pay free from any detention whatever, given to men not living at the mess;
—— *à la grosse*, (*law*) respondentia.
prêt, a., ready, prepared;
se tenir ——, to be ready, in readiness, prepared, etc.
prêter, v. a. n., to lend, give, attribute; to yield, stretch (as cloth);
—— *le côté*, (*nav.*) to prepare for action;
—— *le flanc*, (*mil.*) to expose one's flank;
—— *la main* —— *les mains*, to help, lend a hand;
—— *main-forte*, (*law*) to carry out a verdict;
—— *secours*, to help;
—— *serment*, to take an oath.
prêtre, m., priest (esp. Roman Catholic priest);
bonnet de ——, (*fort.*) outwork with three salients in front (obs.).
preuve, f., proof, trial, evidence, testimony;
commencement de ——, incomplete proof;
—— *s par écrit*, written proof;
faire ses ——*s*, to give proof of one's ability, courage, etc.;
—— *s induites des circonstances*, circumstantial evidence;
—— *littérale*, written proof;
—— *résultant des présomptions*, presumptive evidence;
—— *testimoniale*, proof by witnesses, oral proof.
preux, a., valiant.
prévenir, v. a., to precede, anticipate; to prevent, forestall; to forewarn;

prévenir *un cheval,* (*man.*) to stop a horse when about to change the foot.
prévention, f., (*law*) accusation; (*Fr. mil. law*) confinement of a man until publication of verdict;
—— *mise en* ——, (*law*) commitment for trial.
prévenu, p. p., (*law*) accused; m., prisoner before trial; (*mil.*) person ordered to be tried on charges preferred.
prévision, f., prediction;
—— *du temps,* weather prediction.
prévôt, m., (*mil.*) provost marshal; (*Fr. a.*) c. o. of *gendarmerie* of a *corps d'armée;* (*fenc.*) fencing master;
—— *d'armes,* (*fenc.*) fencing master's assistant;
brigadier —— *d'armes,* (*cav.*) assistant fencing master (corporal);
drague de ——, cat-o'-nine tails;
élève ——, (*mil.*) soldier candidate for fencing master;
—— *d'escrime,* (*mil.*) assistant fencing master;
—— *des étapes,* (*Fr. a.*) c. o. of *gendarmerie* of a *direction d'étapes,* q. v.;
grand ——, (*Fr. a.*) c. o. of *gendarmerie* of an army in the field; provost-marshal-general.
prévôtal, a., (*mil.*) relating to provost duty;
cour ——*e,* (*in gen.*) temporary criminal court sitting without appeal;
service ——, (*mil.*) provost duty;
unité ——*e,* (*mil.*) any unit on provost duty; (in France) the various units of the *gendarmerie* serving with troops in the field.
prévôté, f., (*mil.*) provost-marshal's troops; (in France) the *gendarmerie* attached to the various headquarters and under the orders of a *prévôt;*
—— *militaire,* provost duty, (in France) carried on by the *gendarmerie.*
prévôtalement, adv., by martial law.
primage, m., primage, hat money.
primauté, f., (*mil.*) seniority, "rank";
—— *de grade,* seniority in one's grade.
prime, f., premium; (*mil.*) bounty; any amount allowed for some special service, or to keep up a fund (*masse*); (*fenc.*) prime;
—— *d'assurance,* insurance premium;
—— *de chargement,* primage;
—— *d'engagement,* enlistment bounty;
—— *d'enrôlement,* v. —— *d'engagement;*
—— *de fonctions,* (*Fr. a.*) allowance to bands, bandsmen;
parade de ——, (*fenc.*) prime parry;
—— *de rengagement,* (*mil.*) reenlistment bounty;
—— *de travail,* (*mil.*) extra-duty pay (*Fr. a.*, artillery workmen, regimental *ateliers,* workmen of administration).
primer, v. a. n., to take the lead; to give a premium to; (*steam*) to prime;
—— *la marée,* (*nav.*) to take advantage of the tide.
prince, m., prince; (*mach.*) tilt-hammer beam;
—— *de l'Église,* cardinal; bishop;
—— *royal,* crown prince;
—— *du sang,* prince of the blood royal.
princesse, f., princess;
—— *royale,* princess royal.
—— *du sang,* princess of the blood royal.
principal, a., capital, principal; (*Fr. a.*) chief (used in the designation of the grades of various *services* and departments); m., capital;
somme ——*e,* capital.
principat, m., principality.
principauté, f., principality.
principe, m., principle; (*hipp.*) white marking surrounding the leg, and not rising above the coronet; beginning of white foot;
—— *de l'état initial et de l'état final,* (*chem.*) (in thermo-chemistry) principle of the calorific equivalency of chemical transformations, principle of the initial and final states;
—— *du travail maximum,* (*chem.*) in thermo-chemistry, the principle of maximum work;

principe *des travaux moléculaires,* (*chem.*) in thermo-chemistry, the principle of molecular work.
pris, p. p., taken, caught; fixed, settled (as an hour); made, formed, shaped; congealed, set, frozen; entangled (as a horse in the traces, in the halter);
—— *de calme,* (*nav.*) becalmed;
—— *de chaleur,* (*hipp.*) suffering a pulmonary congestion;
cheval bien ——, (*hipp.*) well-made, well-built, horse;
—— *des épaules,* (*hipp.*) stiff in the shoulders;
—— *de la fumée,* (*hipp.*) affected by staggers;
—— *par les glaces,* frozen (of rivers, ports, etc.); ice-bound, frozen in (of a boat, ship);
non ——, loose;
—— *de la nuit,* overtaken by night;
—— *dans œuvre,* measured in the clear, inside measure;
—— *hors d'œuvre,* outside measure.
prise, f., taking, hold, purchase, gripe, handle; quarrel; (*tech.*) valve, cock; (*mas.*) setting (of mortar); (*cons.*) end of stair seated in wall of case; (*mach.*) hold, bearing; (*mil.*) capture, conquest, taking; way in which fire has reached the enemy, as, *prise d'écharpe,* —— *de flanc,* etc.; (*nav.*) capture; prize;
—— *d'air,* (*tech.*) air passage; draught passage;
—— *d'armes,* (*mil.*) taking arms, getting under arms, for a duty; parading, turning out, under arms; duty performed under arms;
arrêt de ——, v. s. v. *arrêt;*
—— *d'assaut,* (*mil.*) capture by storm, by assault;
avoir —— *sur,* to have a hold on; to have a hold on; (*mach.*) to bear upon, engage with;
beaucoup (*moins*) *de* ——, greater (lesser) object or mark presented;
bonne ——, (*nav.*) lawful prize;
capitaine de ——, (*nav.*) prize master;
—— *en charge,* (*adm.*) taking up (of property, etc.);
code des ——*s,* (*nav.*) prize code;
consacrer une ——, (*nav.*) to legalize a capture;
—— *de corps,* arrest, apprehension;
—— *de courant,* (*elec.*) point at which a current comes in;
droit de ——, (*nav.*) right of capture;
—— *d'eau,* source of supply of water for manufacturing purposes; supply of water for manufacturing purposes; water channel, or gutter, or pipe; (*r. r.*) water tank;
—— *d'eau à la mer,* sea cock;
(*en*) *être aux* ——*s* to clash, come to blows, fight;
être aux ——*s avec,* to struggle, be struggling, with;
être de bonne ——, (*nav.*) to be a lawful prize;
être en —— *avec,* (*mach.*) to have engaged;
être en —— *au feu de,* (*mil.*) to be exposed to the fire of;
être hors de ——, to be out of reach;
faire ——, (of mortar, cement, etc.) to set;
faire une ——, (*nav.*) to make a capture, prize;
—— *de feu,* (*art.*) flame passage (in the head of certain cartridge cases);
lâcher ——, to let go, let go one's hold; (*mil.*) to quit the combat;
—— *lente,* slow setting (of mortar, etc.);
—— *de longe,* (*hipp.*) halter cast;
mortier à —— *lente* (*rapide*), slow- (quick-) setting mortar;
part(*s*) *de* ——, (*nav.*) prize money;
ne point donner —— *au feu,* (*mil.*) not to expose one's self to fire;
—— *de possession,* act of taking possession; (*mil.*) capture (of a town or position);
—— *de service,* (*mil.*) taking up of duty;
station —— *d'eau,* (*r. r.*) water tank;
tribunal de ——, (*nav.*) prize court;
—— *de vapeur,* (*steam*) taking, drawing, delivery, of steam; steam cock, steam valve, steam port; steam dome; mouth of the steam (also "steam pipe") at the boiler;
en venir aux ——*s* v. (*en*) *être aux* ——*s;*

prise *au vent*, (of balloons) exposure to the effects of wind.
prismatique, a., prismatic.
prisme, m., prism;
— *à réflexion totale*, (*opt.*) total reflection prism;
— *de remplissage*, volume of water (of a canal);
— *-télémètre*, (*mil.*) (Souchier) range finder.
prison, f., prison, jail; imprisonment, captivity; (*mil.*) confinement, (punishment of n. c. o., corporal, and men); (*powd.*) collar (of the stamper of a powder mill);
bris de —, breach of prison;
conduire en —, to take to jail;
directeur de —, governor of a jail;
en —, in confinement;
forcer une —, to break jail, break out of prison;
gardien de —, jailer;
mettre en —, to jail;
— *de quartier*, (*mil.*) confinement to barracks;
sortir de —, to come out of jail, be released from confinement.
prisonnier, m., prisoner; (*mil.*) prisoner; (*mach.*) set bolt; stud-bolt, -pin; (adjectively, *mach.*) fixed; (*fond.*) piece (e. g., a band on a projectile) set in position in the mold before casting, and which after casting is in its proper place on the object cast;
avoir (tant de) —*s*, (*mil.*) to lose (so many) prisoners;
— *de cabestan*, drop bolt;
— *de chapeau de presse-étoupes*, (*steam*) stuffing-box gland stud;
se constituer —, (*mil.*) to surrender one's self prisoner;
faire —, to take into custody; (*mil.*) to take prisoners;
fers de —, fetters;
garder —, v. *tenir* —;
— *de guerre*, (*mil.*) prisoner of war;
— *sur parole*, (*mil.*) prisoner on parole;
se rendre —, v. *se constituer* —;
tenir —, to keep in custody.
privation, f., privation;
— *de sortir*, (*mil.*) stoppage of pass.
privé, a., private; m., privy;
acte sous seing —, document not drawn before a public officer;
homme —, private citizen.
prix, m., price, cost; reward; prize;
— *d'achat*, v. — *coûtant*;
— *de l'armée*, v. — *de tir de l'armée*;
— *courant*, current price; price current;
— *coûtant*, first cost, prime cost;
— *d'évaluation*, valuation;
— *fait*, price agreed on;
— *ferme*, fixed and unalterable price;
— *fixe*, fixed price;
— *fort*, full price;
— *de journée*, daily wages;
juste —, moderate price;
— *-limite*, limiting price;
— *-limite débattu*, upset price;
marché à — *fait*, v. *marché à forfait*;
— *de marché*, market price;
mettre la tête d'un homme à — —, to put a price on a man's head;
— *net*, trade price;
— *de nomenclature*, published or list price;
— *de passage*, money, fare;
— *sur place*, v. — *de marché*;
— *de recousse*, salvage money for recapture;
— *de remorquage*, towage;
— *de revient*, net cost;
— *de tir*, (*sm. a.*) target-practice prize;
— *de tir de l'armée*, (*Fr. a.*) prize in target practice;
vendre à non- —, to sell at a loss.
— *de vente*, selling price.
probabilité, f., probability;
— *d'atteindre*, (*ball.*) probability of hitting;
— *du tir*, v. — *d'atteindre*.
probant, a , conclusive in proof.
probation, f., probation, trial;
de —, probational, on probation.
problème, m., problem;
poser un —, to state a problem;

problème, *résoudre un* —, to solve a problem.
procédé, p. p., (*law*) tried;
bien jugé, mal —, (*law*) judgment sound, procedure wrong.
procédé, m., process; proceeding, procedure; conduct, behavior; operation; tip of a billiard cue;
— *Bessemer*, (*met.*) Bessemer process;
— *au charbon*, (*phot.*) carbon process;
— *au collodion humide* (*sec.*), (*phot.*) wet (dry) collodion process;
— *contondant*, (*art.*) racking method;
— *de la double (simple) cuisson*, double (single) calcination (of lime);
— *du double transfert*, (*phot.*) double transfer carbon process;
— *au gélatino-bromure d'argent*, (*phot.*) gelatino-bromid process;
— *négatif*, (*phot.*) blue-print process (white lines on blue ground);
— *perforant*, (*art.*) perforating method;
— *positif*, (*phot.*) blue-print process (blue lines on white ground);
— *Siemens*, (*met.*) Siemens process;
— *par le transparent*, (*ball.*) method of determining the vulnerability of a given target by applying to the plot of the results of practice a transparent figure of the target and noting the hits that fall within it.
procéder, v. n., to proceed, pass on; to arise from, originate in; to act, behave; (*law*) to try;
— *criminellement*, (*law*) to proceed criminally against;
— *militairement*, (*law*) to proceed without observing due forms;
— *par-devant un tribunal*, (*law*) to proceed judicially.
procédure, f., procedure, proceedings; (*law*) legal proceedings;
— *civile*, — *criminelle*, (*law*) civil, criminal, proceedings.
procès, m., (*law*) trial; process, suit;
au —, on trial;
— *civil*, civil action;
— *criminel*, criminal action;
faire le — *à*, to bring to trial;
faire un — *à*, to involve in a lawsuit;
intenter un — *à*, to institute proceedings;
sans forme de —, without trial.
procès-verbal, m., proceedings, report, official report (as of a board);
dresser un —, to draw up a report.
procès-verbaliser, v. n., to draw up a report, make out proceedings.
proclamation, f., proclamation.
proclamer, v. a , to proclaim.
procuration, f , (*law*) power of attorney, proxy.
procureur, m , proxy, authorized agent;
— *général*, solicitor-general, attorney-general.
producteur, a., productive; m., producer;
action — *trice*, efficiency.
produire, v. a., to produce, bring forward, introduce, show;
— *de la vapeur*, (*steam*) to generate steam.
produit, m., product, produce, performance, output.
profane, a., m., lay, profane.
professeur, m., professor.
profession, f., profession, declaration;
de —, by profession.
profil, m., profile, section; gauge; (*art.*) profile gauge; (*fort.*) profile (of a work, etc.); profile, (frame) used in constructing a parapet, etc.;
— *américain*, (*art.*) saw-tooth rifling;
de —, in profile;
— *défensif*, (*fort.*) defensive or protecting profile, one not intended for offensive action;
— *directeur*, (*fort.*) guiding, directing, profile;
— *double*, (*art.*) double gauge (to test the rectilinearity of the generatrices of a projectile);
élever des —*s*, (*fort.*) to set up profiles;
— *extérieur*, (*art.*) exterior gauge for a projectile;

profil *intérieur*, (*art.*) interior gauge for a projectile;
— *en lattes*, (*fort.*) lath profile;
— *en long*, — *longitudinal*, longitudinal section;
— *naturel*, natural section (horizontal and vertical dimensions on same scale);
— *de niveau*, (*top.*) vertical section;
— *primitif*, (*fort.*) profile of the whole works;
procéder par — *s échelonnés*, (*fort.*) to construct a parapet progressively (i. e., when an attack threatens, to throw up a mere shelter, deepening and widening as time permits);
— *de rayure*, (*art.*) right section of the bore;
— *simple*, (*art.*) single gauge;
— *surbaissé*, section in which the vertical is less than the horizontal scale;
— *surhaussé*, section in which the vertical is greater than the horizontal scale;
— *en talus*, (*fort.*) profile of the end face of a parapet (includes end slopes);
— *de terrassement*, (*fort.*) earthwork profile;
— *en travers*, transverse section;
— *triangulaire*, (*fort.*) profile in which the scarp is suppressed and the ditch is reached by direct fire from work;
— *vérificateur*, (*art.*) rifling gauge.

profilé, m., iron, steel, of special cross section (generally in plural).

profilement, m., (*fort.*) profiling.

profiler, v. a., to cut to a special profile; (*fort.*) to set up profiles; to establish profiles, to profile; (*surv.*) to run profiles, to profile.

profond, a., deep, profound; complete, dark; (*mil.*) deep (of a formation).

profondeur, f., depth; (*mil.*) depth (length) of a column; depth of a formation;
de —, deep, in depth;
— *de foyer*, depth of focus;
— *des rayures*, (*art.*) depth of the grooves.

programme, m., programme; bill;
— *de tir*, (*ball.*) programme, order, of experimental firing.

progrès, m., progress, course, improvement, efficiency.

progressif, a., progressive; (*art.*) increasing;
rayure — *ve*, (*art.*) increasing twist.

progressivité, f., (*expl.*) progressiveness (of powders, etc.).

projecteur, m., searchlight;
— *électrique*, searchlight;
— *fixe*, fixed searchlight;
— *mobile*, transportable searchlight.

projectile, a., projectile;
force —, projectile force; impetus.

projectile, m., projectile; (*art.*) projectile;
— *à ailettes*, studded projectile (v. s. v. *obus*);
— *allongé*, elongated projectile;
— *d'amorce*, (*art.*) in tests, a projectile destined to explode others near by, in test of explosibility;
— *arrondi sur l'avant*, round-nosed projectile;
— *à ceinture*, banded projectile;
— *à ceinture de forcement*, projectile fitted with a rifling band;
— *chargé en guerre*, shell filled and fused;
— *à chemise*, jacketed projectile;
— *à chemise de plomb*, lead-jacketed projectile;
— *à coiffe*, (*art.*) soft-nosed shell, capped shell;
— *à concussion*, racking shot;
— *à cordons de plomb*, lead-ribbed projectile;
— *creux*, shell;
— *creux à charge brisante*, shell and high explosive;
— *cuirassé*, Rubin bullet, capped with steel cap, called *cuirasse*;
— *à culot évidé*, v. — *expansif*;
— *cylindrique*, cylindrical projectile;
— *cylindro-ogival*, cylindro-ogival projectile;
— *discoïde*, discoid projectile;

projectile *durci*, chilled shot;
— *éclairant*, light ball;
— *éclatant*, shell;
— *d'éclatement*, generic term for shell;
— *enregistreur*, (*ball.*) registering projectile (fitted with a device registering law of motion);
— *à enveloppe*, v. — *à chemise*;
— *d'exercice*, drill projectile;
— *expansif*, expanding projectile, one taking the grooves by expansion;
— *explosif*, shell filled with high explosive;
faux —, dummy projectile;
— *en fonte dure*, chilled shot;
— *forcé*, projectile fitted with forcing bands;
— *à fragmentation systématique*, shrapnel;
— *fulminant*, percussion shot (rare);
— *incendiaire*, carcass, incendiary projectile;
— *lenticulaire*, lenticular projectile;
— *à main*, hand projectile (e. g., hand grenade);
— *à manteau*, v. — *à chemise*;
marche d'un —, flight of a projectile;
— *massif*, solid shot;
— —*mine*, term descriptive of modern shell of great capacity; torpedo shell;
— *oblong*, elongated projectile;
— *ogival*, ogival projectile;
— *ogivo-cylindrique*, ogivo-cylindrical projectile;
— *perce-cuirasse*, armor-piercing projectile;
— *plat sur l'avant*, flat-nosed projectile;
— *plein*, solid projectile;
— *ramé*, bar shot (obs.);
— *à*, *de*, *rupture*, armor-piercing projectile;
— *sphérique*, round shot;
— *à tenons*, studded projectile (v. s. v. *obus*);
— *à tête plate*, flat-headed projectile;
— —*torpille*, (*art.*) torpedo projectile, torpedo shell;
— *à vent*, a projectile with windage;
— *vérificateur*, studded projectile, with handle, for testing bore and grooves of rifled guns.

projection, f., projection; (*man.*) the instant (in the trot) in which the animal is completely off the ground; (*ball.*) projection (*fond.*) pouring;
— *centrale*, central projection;
— *de côté*, side view;
— *cylindrique*, cylindrical projection;
— *d'eau*, (*steam*) priming;
— *de face*, front view;
— *horizontale*, horizontal projection;
— *isocylindrique*, isocylindric projection;
ligne de —, (*ball.*) axis produced, line of projection;
mouvement de —, projectile motion;
— *orthogonale*, orthogonal projection;
— *orthographique*, orthographic projection;
plan de —, plane of projection;
— *polaire*, polar projection;
— *stéréographique*, stereographic projection;
— *verticale*, vertical projection.

projecture, f., projection; jutting out.

projet, m., project, plan, design, scheme; rough draft;
— *d'acte*, rough draft of a legal paper on unstamped paper;
— *d'attaque et de défense*, (*fort.*) (probable) plan of attack, (and hence) of defense, of a fortress;
faire un — *de machine*, to design a machine;
— *de loi*, bill (in Parliament, Congress);
— *de machine*, machine design;
— *de siège*, (*siege*) siege plan (based on information obtained).

projeter, v. a. r., to project, project on, jut out, stand out; to scheme, plan, contrive.

prolongation, f., prolongation, lengthening; extension, continuation of a certificate, etc.;
— *de ceinture*, v. s. v. *ceinture*;
— *de congé*, (*mil.*) extension of leave.

prolonge, f., (*art.*) prolonge; (*mil.*) forage wagon, heavy general-service wagon; ammunition wagon; picket line;
— *de campement*, (*mil.*) picket line;
déplier les —s, (*art.*) to fix prolonge;
— *double*, (*Fr. art.*) rope 25ᵐ long, used in mechanical maneuvers;
—*ferrée*, (*art.*) prolonge fitted with its (iron) loops;
replier les —s, (*art.*) to unfix prolonge;
— *à ridelles*, rack wagon;
— *simple*, (*Fr. art.*) rope 15ᵐ long, used in mechanical maneuvers.

prolongement, m., prolongation; (*fort.*) prolongation of the faces of a work; (*art.*) extension of a platform;
— *de plate-forme*, (*art.*) extension of the platform to receive the trail;
— *des rayures*, (*F. m. art.*) prolongation of three grooves beyond origin of rifling in certain pieces (for which *plaques isolantes* are necessary);
renforcement par —, (*mil.*) reenforcing a line by extending it, using additional troops (infantry).

prolonger, v. a., to prolong, extend; to lengthen, eke out, stretch out; (*art.*) to secure with a prolonge;
— *une côte*, (*nav.*) to skirt a coast.

promenade, f., walk; drive, road;
— *en bateau*, row;
— *à cheval*, ride;
— *des chevaux*, (*man.*) horse exercise;
— *sur l'eau*, sail;
— *galette*, (*St. Cyr slang*) general marching out;
— *militaire*, (*mil.*) practice march;
— *à pied*, walk;
— *de plaisir*, trip;
— *à la voile*, sail;
— *en voiture*, drive.

promener, v. a. r., to take out, lead; to take a walk (ride, drive, etc.);
— *un cheval*, to walk a horse up and down;
— *un cheval en main*, to lead a horse;
— *un cheval dans la main et les talons*, (*man.*) to control a horse by hand and spur;
— *un cheval entre les deux talons*, to walk a horse straight ahead;
se — *à cheval*, to take a ride.

promontoire, m., promontory, foreland, headland.

promotion, f., elevation of several persons to the same dignity; (*mil.*) promotion, (of military schools) class;
camarade de —, classmate;
faire une —, to make a promotion;
la — *de 1900*, at the Polytechnique, the class that enters in 1900; at St. Cyr, the class graduated in 1900.

promouvoir, v. a., to promote, raise, advance.

promu, p. p., promoted; m., person promoted.

promulgation, f., promulgation (of a law, etc.).

promulguer, v. a., to promulgate (a law, etc.).

prononcé, m., (*mil. law*) verdict of a court-martial.

prononcer, v. a., to pronounce; to decide; to render salient; to delineate strongly, bring out, give prominence to; (*law*) to bring in a verdict; to adjudicate.

pronunciamento, m., (*mil.*) military plot, followed by insurrectionary acts.

propagation, f., propagation (as of sound, light, etc.).

propager, v. r., to be propagated (of light, sound).

propension, f., propensity;
— *aciéreuse*, (*met.*) propensity of pig iron to be converted into steel.

proportion, f., proportion, relation, ratio; (*mach.*) ratio or relation between two wheels gearing together;
à —, in proportion;
à — *de*, in proportion to;
compas de —, proportional compasses.

proposer, v. a., to propose; (*mil.*, *etc.*) to recommend formally for promotion, the Legion of Honor, etc.

proposition, f., formal recommendation for promotion, Legion of Honor, or for any employment, duty, etc.

propre, a., own, fit, proper; adapted, fitted to; clean, neat.

proprement, adv., properly, exactly, neatly, cleanly; (*cord.*) snugly.

propreté, f., neatness, cleanliness;
revue de —, (*mil.*) inspection of arms and clothing (Saturday morning inspection, U. S. A.).

propriétaire, m., owner;
— *foncier*, landed proprietor.

propriété, f., property, distinctive quality, propriety, ownership; landed property;
— *abandonnée*, (*nav.*) derelict;
— *littéraire*, copyright.

propulseur, a. m., propelling; (*mach.*) propeller, propelling apparatus; (*nav.*) screw;
— *à hélice*, (*nav.*) screw propeller.

propulsion, f., (*mach.*, *etc.*) propulsion.

prorogation, f., prolongation, postponement; adjournment (of Congress, etc.);
— *d'enquête*, (*law*) continuation of an investigation;
— *de juridiction*, (*law*) transfer of jurisdiction.

proroger, v. a., to prolong, put off; to adjourn (Congress, etc.).

proscription, f., proscription.

proscrire, v. a., to proscribe.

proscrit, m., outlaw; (political) refugee.

protecteur, m., protector.

protection, f., protection, cover, defense; support; (*fort.*) cover;
— *commerciale*, protection, high tariff;
escadron de —, (*cav.*) covering, protecting, squadron;
— *au moyen du charbon*, (*nav.*) coal protection;
système de —, protective (high-tariff) system;
sous la — *de*, under cover of.

protectionnisme, m., high-tariff system.

protectionniste, m., advocate of a high tariff.

protectorat, m., protectorate.

protégé, p. p., protected;
croiseur —, (*nav.*) protected cruiser.

protège-main(s), m., (*sm. a.*) hand guard (to protect against heating).

protéger, v. a., to protect, defend, shield; (*mil.*) to cover, protect by cover.

protêt, m., protest (of a note, check); protest (of a shipmaster);
— *de mer*, shipmaster's protest.

protocole, m., protocol.

proue, f., prow (of a boat); nose (of a torpedo).

prouesse, f., prowess.

prouy, m., (*cord.*) mooring rope, painter.

provenance, f., origin; production, growth, manufacture (generally in plural);
en — *ou en destination de*, coming from or going to;
lieu de —, place of origin.

provende, f., provender, provisions (*fam.*).

province, f., province.

provision, f., provision, supply, stock, store; (*in pl.*) victuals, food; (*law*) provisional damages (before final decision); (in commerce) commission; money deposited to be drawn on;
— *alimentaire*, (*law*) allowance for support granted by a court; alimony;
—s *de bouche*, victuals;
—s *de guerre*, war supplies.

provisoire, a., temporary, provisional; (*law*) provisional;
fortification —, (*fort.*) temporary fortification;
matière —, perishable material.

provocation, f., challenge; instigation;
— *en duel*, challenge to fight a duel.

provoquer, v. a., to challenge, call out, instigate, promote, incite; (*law*) to initiate, originate, legal action, a suit.

prud'homme, m., expert, umpire, arbitrator.
prytanée, m., —— *militaire (de la Flèche)*, *(Fr. a.)* preparatory military school for the sons of officers without means or of officers that have been killed in action.
pseudo-vitesse, f., *(ball.)* the horizontal component of a projectile's velocity.
psychromètre, m., psychrometer.
public, a., public; m., the public;
le bien ——, the public good;
l'autorité ——*que*, officials charged with the public administration;
charges ——*ques*, taxes;
la chose ——*que*, the State;
droit ——, science of government, constitutional law;
édifices ——*s*, public buildings;
femme ——*que*, prostitute;
force ——*que*, v. s. v. *force;*
maison ——*que*, house of ill-fame;
ministère ——, v. s. v. *ministère;*
personnes ——*ques*, public officials;
puissance ——*que*, the national power;
partie ——*que*, v. s. v. *partie;*
officier ——, public official;
domaine ——, v. s. v. *domaine.*
publication, f., publication;
—— *périodique*, periodical.
publiciste, m., political writer.
puchet, m., *(pont.)* pitch pan, pitch **ladle**.
pucheux, m., v. *puchet.*
puddlage, m., *(met.)* puddling;
—— *chaud*, wet puddling, pig boiling;
—— *au four bouillant*, v. *chaud;*
—— *au gaz*, gas puddling;
—— *gras*, v. —— *chaud;*
—— *humide*, v. —— *chaud;*
—— *maigre*, dry puddling;
—— *mécanique*, mechanical puddling;
—— *sec*, v. —— *maigre.*
puddler, v. a., *(met.)* to puddle;
four à ——, puddling furnace.
puddlerie, f., v. *puddlage.*
puddleur, m., *(met.)* puddler (workman); puddling furnace.
—— *mécanique*, puddling machine;
—— *rotatif*, rotary puddler, furnace; revolving furnace.
pudrolithe, f., *(expl.)* pudrolithe.
pueil, m., *bois en* ——, wood not three years old.
pugilat, m., boxing.
pugiliste, m., boxer, pugilist.
puine, f., brushwood.
puisage, m., drawing water.
puisard, m., waste well; cesspool; drain well, trap, sink; drain pipe embedded in a wall; *(min.)* sump; water sump; *(nav.)* well;
—— *absorbant*, sort of sink or catch basin (used when the slope is too gentle to carry off water).
puiser, v. a., to draw, dip from, imbibe; *(fig.)* to borrow, take from; *(farr.)* to take off, remove, horn with the shank of the nail.
puisette, f., sort of ladle or scoop.
puisoir, m., ladle.
puissance, f., power; force, authority; domination, empire; property; virtue; power (of a microscope); thickness (of a vein); government, sovereign state; effectiveness; *(tech.)* horsepower; *(art.)* effectiveness (as of a gun, shrapnel, etc.); quantity of explosive in a projectile;
—— *balistique*, *(art.)* ballistic power;
—— *en bougies*, *(elec.)* candlepower;
—— *calorifique*, heating power;
—— *centrifuge*, centrifugal force;
—— *de cheval*, horsepower;
—— *en chevaux*, horsepower;
—— *défensive*, *(nav.)* armor;
—— *dilaniatrice*, *(expl.)* rending strength; explosive, tearing, power;
—— *effective*, actual power; actual, effective, horsepower; effective horsepower per second;
élasticité de la ——, quality of a steam engine to furnish power carrying from nominal power;
—— *électrique utile*, *(elec.)* net output;
—— *électrique totale*, *(elec.)* total output;

puissance *d'évaporation*, —— *évaporatoire*, *(steam)* evaporative power (of a boiler);
—— *exécutive*, the executive (in a government);
—— *d'expansion*, expansive power;
—— *garante*, guaranteeing power (government);
hautes ——*s contractantes*, high contracting powers;
—— *hydraulique*, water power;
—— *impulsive*, impulsive, impelling, power;
—— *indiquée*, indicated horsepower per second;
—— *législative*, the legislature;
—— *logistique d'une ligne*, *(r. r.)* circulating capacity of a railway (sixty to seventy trains a day in each direction of a double-track railway);
—— *de la machine*, *(mach.)* power, efficiency, of the machine or engine;
—— *d'une machine en chevaux*, *(mach.)* horsepower of an engine;
—— *maritime*, naval power;
—— *mécanique*, simple machine;
—— *meurtrière*, *(art.)* killing power (as of a shrapnel);
—— *militaire d'un navire*, *(nav.)* offensive qualities of a ship;
—— *motrice*, motive power; *(nav.)* engine (generic term);
—— *nette*, v. —— *effective;*
—— *nominale*, nominal power; nominal horsepower;
—— *offensive*, *(nav.)* armament;
—— *perforante*, —— *de perforation*, *(art.)* perforating, punching, power of a projectile;
—— *réelle*, effective power;
à toute ——, at full power;
—— *de traction*, —— *tractive*, tractive power;
—— *utile*, useful power;
—— *de la vapeur*, steam power;
—— *de vaporisation*, *(steam)* evaporating power.
puits, m., well; water tank; *(min.)* pit, shaft; *(mil.)* military pit; *(mil. min.)* shaft; *(nav.)* well; locker;
—— *absorbant*, drain well;
—— *d'aérage*, air shaft;
—— *à air*, ventilating shaft;
—— *d'appel*, *(min.)* uptake; upcast shaft;
—— *artésien*, artesian well;
—— *à la barre à mine*, (auger-) bored well;
—— *à la Boule*, *(mil. min.)* shaft *à la Boule*, one in which the frames (*châssis coffrants*) are placed one on top of the other;
—— *à bras*, pump well;
—— *à chaîne(s)*, *(nav.)* chain locker;
—— *de chargement*, *(art.)* shell-hoist shaft (of a turret); ammunition shaft;
—— *à coffrage*, *(mil. min.)* lined shaft;
—— *sans coffrage*, *(mil. min.)* unlined shaft;
—— *à combustion*, *(met.)* flue;
—— *de descente*, *(min.)* footway shaft;
—— *à l'eau*, water tank;
—— *d'éclatement*, *(art.)* explosion chamber;
—— *d'épuisement*, *(min.)* sump shaft;
—— *d'extraction*, *(min.)* hoisting shaft;
—— *de forage*, —— *foré*, driven well, drive well tube well;
—— *de frettage*, *(art.)* shrinkage pit;
grand ——, *(mil. min.)* great gallery shaft;
—— *de l'hélice*, *(nav.)* propeller well;
—— *à la hollandaise*, *(mil. min.)* shaft lined with *châssis hollandais;*
—— *instantané*, tube well;
—— *intérieur*, *(min.)* winze;
—— *à manivelle*, *(mach.)* crank pit;
—— *de marée*, tide shaft (sort of tide indicator);
—— *militaire*, *(mil.)* military pit;
—— *de mine*, *(mil. min., etc.)* shaft;
moyen ——, *(mil. min.)* common branch shaft;
percer un ——, to sink a shaft;
—— *perdu*, blind well (a well of sandy bottom, absorbing the water);
petit ——, *(mil. min.)* shaft from which opens a cased branch;
—— *pleureur*, seepage well;
—— *à poulie*, draw well;
—— *à roue*, draw well;

puits *de service,* —— *de travail,* working shaft; (*mil.*) trou-de-loup;
—— *souterrain,* (*min.*) sump;
—— *à tube-flèche,* tube well;
—— *de tunnel,* tunnel shaft.
pulpage, m., pulping.
pulper, v. a., to pulp.
pulsateur, m., (*mach.*) pulsator.
pulsomètre, m., (*mach.*) steam vacuum pump; pulsometer.
pulverin, m., (*powd.*) mealed powder; priming powder;
—— *d'amorce,* priming powder;
toile à ——, (*mil. min.*) linen or calico covered with mealed powder paste;
—— *vert,* powder previous to pulverization;
—— *vieux,* powder caked from damp, age, etc.
pulvérisateur, m., pulverizer; (*mach.*) sprayer of an oil motor.
pulvérisation, f., pulverization.
pulvériser, v. a., to pulverize, reduce to powder; (*powd.*) to meal.
pulvérulence, f., pulverulence.
pulvérulent, a., pulverulent.
punaise, f., thumb tack.
punir, v. a., to punish.
—— *de mort,* to punish by death.
punition, f., punishment;
—— *disciplinaire,* (*mil.*) disciplinary punishment;
lever une ——, to remit a punishment;
livre de ——*s,* (*mil.*) defaulter book; record of previous trials and convictions (U. S. A.);
peloton de ——, (*mil.*) punishment squad.
pupille, m., f., ward.
pureau, m., (*cons.*) tail, uncovered or exposed part of a shingle or tile; distance between two laths.
purgation, f., (*met.*) refining of metals.
purge, f., cleaning, disinfection; (*steam*) blowing through; (*mach.*) blow-through valve;
robinet de ——, (*mach.*) blow-off cock, pet cock.
purgeoir, m., filtering basin or tank.
purger, v. a., to purge, to clean, purify, clear; to strip (bark and sap from wood); (*law*) to purge; (*steam*) to blow through;
—— *une condamnation,* (*law*) to serve out a sentence;
—— *la contumace,* (*law*) to purge one's self of contempt;

purger *la machine,* (*steam*) to blow through;
—— *les métaux,* (*met.*) to refine metals;
—— *la quarantaine,* to lie out the quarantine, to spend the quarantine;
robinet à ——, (*mach.*) drain cock, blow-through cock;
soupape de ——, (*mach.*) drain valve; blow-through valve.
purgeur, m., (*mach.*) blow-off gear; purging gear.
purifier, v. a., to purify, clean, cleanse.
pur(-)sang, m., (*hipp.*) thoroughbred (horse).
putzen, m., (*met.*) unmelted ore sticking to sides of furnace.
puy, m., (*top.*) eminence, peak, puy.
pylône, m., (*mach.*) supporting station or tower (of a rope transmission).
pyr, m., (*elec.*) a *bougie décimale;* pyr (unit of intensity of light, producing one lux).
pyramide, f., pyramid; (*top.*) mountain or elevation more or less pyramidal in shape.
pyrite, f., pyrites.
pyrobalistique, m., pyroballogy, ballistics.
pyrocoton, m., (*expl.*) pyrocotton.
pyroélectricité, f., (*elec.*) pyroelectricity.
pyroglycérine, f., (*expl.*) Sobrero's name for nitroglycerin.
pyrolite, f., (*expl.*) pyrolit(h)e.
pyromètre, m., pyrometer;
—— *à air,* air pyrometer;
—— *à cadran,* dial pyrometer;
—— *électrique,* electric pyrometer;
—— *optique,* optical pyrometer (photometer really);
—— *thermo-électrique,* thermo-electric pyrometer.
pyronitrine, f., (*expl.*) pyronitrine.
pyronôme, m., (*expl.*) pyronome.
pyrotechnie, f., pyrotechny.
pyrotechnique, a., pyrotechnical.
pyrothèque, m., (*elec.*) explodor (induced-current machine).
pyroxylam, m., (*expl.*) pyroxylam, Uchatius' white powder.
pyroxyle, m., (*expl.*) gun cotton, pyroxylin.
pyroxyline, f., (*expl.*) gun cotton; pyroxylin.
pyroxylol, m., (*expl.*) name proposed for nitrocellulose.

Q.

quadrangle, m., quadrangle.
quadrangulaire, a., quadrangular.
quadrant, m., (*inst.*) quadrant; Hadley's quadrant; (*elec.*) henry.
quadrature, f., quadrature; dial work.
quadricycle, m., two bicycles side by side.
quadrilatère, m., quadrilateral; (*mil.*) quadrilateral, (esp.) the quadrilateral of northern Italy;
—— *italien,* (*mil.*) the Italian quadrilateral (Mantua, Peschiera, Verona, Legnano).
quadrillage, m., flagging; checkering (of cross section or profile paper); (*sm. a.*) checkering (on a hammer).
quadrille, f., (*man.*) group of horsemen executing fanciful movements in a carousel.
quadrillé, a., m., checkered; any checkered part of a mechanism, etc.; checker work.
crête ——*e,* (*sm. a.*) checkered cocking piece;
papier ——, cross-section paper;
systèmes ——*s,* (*cons.*) framing by parallel pieces braced by other parallel pieces at right angles;
à tête ——*e,* with checkered head.
quadriller, v. a., to checker; to cover with lines at right angles.
quadruplette, f., quadruplet, (four-seat bicycle).
quadruplex, a., quadruplex;
système ——, quadruplex telegraphy.
quai, m., quay, wharf; (*r. r.*) platform;
—— *à charbon,* coal wharf;

quai *de débarquement,* (*mil.*) point where troops detrain;
—— *d'embarquement,* (*mil.*) point where troops entrain;
emplacement à ——, (*nav.*) quay berth;
être à ——, (*nav.*) to be at the wharf;
—— *flottant,* landing stage;
—— *à houille,* coal wharf;
—— *de rive,* river quay.
qualification, f., qualification; name, style.
qualifier, v. a., to qualify; to call, style.
qualité, f., quality; property; position, title, rank; (*law*) qualification; proposition, motion, of a litigant;
avoir —— *pour,* to have legal or statutory competency or qualification; to be qualified to;
bois de ——, (*cons.*) good timber;
—— *s d'évolution,* —— *s évolutives d'un navire,* (*nav.*) maneuvering qualities of a ship;
—— *s du jugement,* (*law*) the facts of the case;
—— *s militaires,* (*nav.*) offensive qualities of a ship;
—— *s nautiques,* —— *s de navigation,* (*nav.*) seaworthiness of a ship.
quantième, m., day of the month; date;
montre à ——*s,* watch giving the day of the month.
quantité, f., quantity; (*elec.*) quantity, parallel;
—— *d'eau motrice,* quantity of driving water;
en ——, (*elec.*) in quantity, in parallel;
monter en ——, (*elec.*) to join up in parallel, in quantity;

quantité *de mouvement,* quantity of motion, momentum;
—— *de travail,* quantity of work.

quarantaine, f., quarantine; (*cord.*) ratline stuff;
être en ——, to be quarantined;
faire (*la*) ——, to perform quarantine;
lever la ——, to admit to pratique;
—— *d'observation,* light quarantine, observation;
prendre la ——, to begin quarantine;
purger sa ——, to clear one's quarantine, to spend the quarantine.

quarantainier, quarantenier, quarantinier, m., (*cord.*) ratline stuff;
gros ——, twelve-thread ratline stuff;
petit ——, nine-thread ratline stuff.

quarderonner, v. a., to quarter round; to round off.

quart, m., quarter; point of the compass; cup or tin; (*mil.*) soldier's drinking cup; (*nav.*) watch (duty and party).

I. Naval; II. Miscellaneous.

I. Naval:

—— *d'un aire de vent,* quarter point of the compass;
—— *de bâbord,* port watch;
chef de ——, captain of the watch; engineer of the watch;
chef de —— *aux chaudières,* leading stoker;
—— *de commandement,* officers' watch;
demi-——, half point of the compass;
—— *de deux heures,* dog watch;
—— *de diane,* morning watch;
être de ——, to be on watch;
faire bon ——, to keep a good lookout;
faire le ——, to keep the watch;
faire son ——, to stand one's watch;
—— *de huit heures à midi,* forenoon watch;
—— *de huit heures à minuit,* first watch;
—— *du jour,* morning watch; day watch;
libre de ——, watch below;
maître de ——, boatswain's mate;
—— *de mécanicien,* engine-room watch;
—— *de midi à quatre heures,* afternoon watch;
—— *de minuit à quatre heures,* middle watch;
—— *de mouillage,* anchor watch;
—— *de nuit,* night watch;
officier de ——, watch officer;
petit ——, dog watch;
prendre le ——, to go on watch, to take the watch;
régler le ——, to set the watch;
remplacer le ——, to relieve the watch;
rôle de ——, watch bill;
—— *de tribord,* starboard watch;
—— *de vent,* point of the compass.

II. Miscellaneous:

—— *biscuité,* (*mil.*) bread cooked only one-fourth as long as biscuit;
—— *de cercle,* (*inst.*) quadrant; (*art.*) gunner's quadrant;
—— *de cercle en bois,* (*art.*) wooden quadrant;
—— *de cercle à niveau* (*à bulle d'air*), (*art.*) spirit-level quadrant;
—— *de cercle à pendule,* (*art.*) pendulum quadrant;
—— *de cercle à règle,* (*inst.*) common quadrant;
—— *de conversion,* (*mil.*) wheel (of 90°);
—— *en* ——, (*man.*) sort of volt;
être —— *de cheval,* to have a fourth interest in race horse;
loi du ——, v. s. v. *loi;*
—— *de minute,* 15-second sand glass;
—— *de nonante,* quadrant;
—— *à pendule,* (*art.*) pendulum quadrant;
—— *de réflexion,* v. *octant;*
—— *de réserve,* one-quarter of the timber belonging to public establishments left to grow;
—— *de rond,* (*cons.*) quarter round (tool, molding, and strip);
travailler un cheval de —— *en* —— (*man.*) to take a horse three times over each side of the square.

quarte, f., (*fenc.*) quarte; (*hipp.*) quarter crack;

quarte, *contre de* ——, v. s. v. *contre,* m.;
fausse ——, (*fenc.*) feint quarte;
parade de ——, (*fenc.*) quarte parry.

quartier, m., quarter; part, piece; quarters, lodging; quarter, ward, of a town; study room or hall (of a school); sum paid quarterly; quarter (of a year, of the moon, of a shoe, etc.); principal part of a coat; (*farr.*) quarter (of a horseshoe); (*hipp.*) quarter of the hoof; (*inst.*) quadrant; quadrant (of a compass card); (*harn.*) saddle flap; (*mil.*) open garrison town; barracks, quarters (in a town); quarter (i. e., sparing the life of an enemy);
—— *d'assemblée,* (*mil.*) rendezvous, alarm post; place of assembly in general (e. g., to form for a march, a review);
bois de ——, cleft (fire) wood;
—— *de cantonnement,* (*mil.*) cantonments; cantoned troops;
cheval qui fait —— *neuf,* (*hipp.*) horse that is quarter cast;
——*s de cheval,* quarters of a horse;
—— *de campement,* (*mil.*) camp quarters;
demander ——, (*mil.*) to ask quarter;
donner ——, (*mil.*) to give quarter; (*mas., carp.*) to turn a stone or timber over on its next face; (in stereotomy) to represent a stone in position;
doubles ——*s,* (*harn.*) saddle skirts;
en ——, in quarters; (*cord.*) four-stranded;
enlever un ——, (*mil.*) to beat up quarters;
entrer en ——, (*hipp.*) to cut away the quarters of a hoof;
faire ——, (*mil.*) to give quarter;
faire lever les ——*s,* (*mil.*) to force the enemy to break up and abandon his post;
faire —— *neuf,* (*hipp.*) to be quarter cast;
faux ——, (*hipp.*) false quarter; (*harn.*) under flap, false quarter (of a saddle);
—— *de fourrage,* (*mil.*) cantonments where troops have to be scattered for subsistence;
—— *général,* (*mil.*) headquarters (includes the staff proper and all the *services* and departments); officers composing the headquarters; quarters of c. o.;
grand —— *général,* (*mil.*) headquarters of the army, main headquarters, commander in chief's headquarters; officers composing the main headquarters;
——*s d'un habit,* four principal pieces of a coat;
—— *d'hiver,* (*mil.*) winter quarters;
lever les ——*s,* (*mil.*) to break up one's quarters and retire in good order;
—— *de la lune,* quarter of the moon;
—— *-maître,* (*mil.*) quartermaster (obs. in France); (*nav.*) quartermaster;
—— *-maître général,* (*mil.*) quartermaster-general (obs. in France);
petit —— *général,* (*mil.*) private headquarters;
—— *de pierre,* large block of stone; said also of dressed stones, so many to the cubic meter;
—— *de rafraîchissement,* (*mil.*) summer quarters, or quarters where troops may recruit in a campaign ;
—— *de réflexion,* (*inst.*) Hadley's quadrant;
—— *de selle,* (*harn.*) saddle flap, skirt;
—— *de soulier,* quarter, heelpiece, of a shoe;
—— *de terre,* bit of land;
—— *tournant,* (*cons.*) winding (stair) between two flights;
—— *de troupes,* (*mil.*) part of a town where troops are stationed;
—— *du vent,* quarter of the wind;
—— *de ville,* quarter, district, ward;
—— *de vis suspendue,* (*cons.*) part of a staircase joining two apartments not on the same floor;
—— *des vivres,* (*mil.*) place where supplies are stored.

quartz, m., quartz.

quatre, adv., four;
en ——, (*cord.*) four-stranded;
par ——, (*mil.*) right forward, fours right;

quatre, *par la gauche par* ——, (*mil.*) left by fours, (squads, U. S. A.);
se compter ——, (*mil.*) to count fours.
quatre-mâts, m., (*nav.*) four-master.
quatre-quart, m., little square file.
quayage, m., wharfage.
quenouille, f., distaff; (*met.*) sort of bott stick (to regulate the flow of molten metal from a furnace).
quenouillère, f., (*met.*) bott stick.
quenouillette, f., v. *quenouillère*.
quenouillon, m., pledget of oakum used in calking.
querelle, f., quarrel;
chercher ——, to pick a quarrel;
prendre ——, to have a quarrel with.
quereller, v. a., to quarrel.
question, f., question; point; torture;
—— *de cabinet*, vital or cabinet question by which a ministry will stand or fall;
—— *de fait*, matter of fact;
—— *de lieu*, question of place;
—— *de personne*, question of person;
—— *préalable*, the previous question.
quête, f., (*nav.*) rake of the sternpost.
queue, f., tail, end, tip; rear, tail, of a procession; file, string, of people; queue; tail of a letter; billiard cue; train (of a cloak, etc.); pendant, tailpiece; fly of a flag, point of a guidon; tang; shank of a button; tail-race, mill tail; tail of a block; whetstone; (*mas.*) hinder, back, face of a stone; (*hydr.*) end of a tongue of land, of a bank; (*mach.*) end of a piston rod; (*cons.*) the end of a stair seated in the wall of the case; (*cord.*) down haul; (*ball.*) tailings (zone containing 6 per cent of the hits and lying between the *enveloppe* and the circle whose radius is three times that of the *noyau*); (*mil.*) rear, tail (of a column, etc.)
à ——, tailed;
—— *à* ——, in file;
à la ——, behind, in file, in a string;
à la —— *de*, at the heels of; in pursuit;
—— *d'affût*, (*art.*) rear of top carriage;
—— *à l'anglaise*, —— *anglaisée*, (*hipp.*) docked and nicked tail;
—— *d'armée*, (*mil.*) part of an army opposite to the *front de bandière*;
—— *d'armons*, (*art.*) fork or junction of foreguides;
—— *d'aronde*, (*cons.*) dovetail, dovetail scarf; (*fort.*) crown or horn work whose branches approach each other toward the body of the place;
—— *d'aronde couverte*, (*carp.*) lapped dovetail;
—— *d'aronde percée*, (*carp.*) ordinary dovetail;
assemblage à —— *d'aronde*, —— *à* —— *d'hironde*, (*carp.*) dovetailing;
assemblage à —— *percée*, (*carp.*) assemblage with visible joints;
assemblage à —— *perdue*, (*carp.*) assemblage with covered joints;
assembler à —— *d'aronde*, (*cons.*) to dovetail;
attaquer en ——, (*mil.*) to attack the rear of a column;
—— *d'auget*, (*sm. a.*) tail or end of the cartridge carrier;
aviron de ——, steering oar;
—— *en balai*, (*hipp.*) tail ending in a sharp point (i. e., docked, but with the hair long and coming to a point);
—— *d'un banc*, (*hydr.*) end or tail of a bank;
—— *de bataillon*, etc., (*mil.*) rear or tail of a battalion, etc., in column;
—— *bien attachée*, (*hipp.*) well-attached tail (dock clear of buttocks, and in continuation of median line of croup);
bouton à ——, shank but' on;
—— *de camp*, (*mil.*) rear of camp, part opposite to *front de bandière*;
—— *de carpe*, sort of iron cramp;
—— *en catogan*, (*hipp.*) docked tail, with the hairs long on each side;
charger en ——, to load the rear end of anything;
—— *de chat*, cat-o'-nine tails;

queue *de cheval*, (*mil.*) horse tail (so called), used as a standard;
cheval —— *de rat*, (*hipp.*) rat-tailed horse;
—— *de cochon*, auger with spiral point;
—— *collée*, v. *mal attachée*;
—— *d'une colonne*, (*mil.*) tail of a column;
—— *d'un convoi*, etc., (*mil.*) tail of a convoy, etc.;
—— *d'un cordage*, (*cord.*) rat's tail;
faire un cordage en —— *de rat*, (*cord.*) to point at rope;
—— *d'un coup de vent*, tail of a storm;
—— *d'un couteau*, tang of a knife blade;
—— *de (la) culasse*, (*sm. a.*) tang of the breech;
—— *en cul-de-lampe*, (*cons.*) pendant (keystone);
de ——, (*drill*) last, rearmost (in a column);
à la —— *des dépôts*, (*mil.*) (of recruits, etc.) to be allowed to remain behind (to be put at the foot of the list of men to be sent off to the army);
—— *de détente*, (*sm. a.*) finger piece of a trigger, trigger proper;
donner en, *sur*, *la* ——, v. *prendre en* ——;
—— *d'eau*, tail water;
—— *écourtée*, (*hipp.*) bauged tail;
—— *d'empanons*, v. —— *de fourchette*;
en ——, behind, at one's heels;
—— *d'enrayement*, (*r. r.*) brake rod;
épissure en —— *de rat*, (*cord.*) pointed splice;
être à la ——, to be in a file, in a column of file;
—— *d'extracteur*, (*sm. a.*) tang of the extractor;
faire —— *de colonne*, (*mil.*) to be at the tail, to be the tail, of a column;
fausse ——, (*hipp.*) false tail;
—— *de flasque*, (*art.*) rear end of top carriage cheek;
—— *de forgeage*, (*met.*) tong hold (of an ingot);
—— *de fourchette*, fork of hind guides of a wagon;
—— *de la gâchette*, (*sm. a.*) arm or tail;
garnir de ——*s de rat*, (*cord.*) to point ropes;
—— *du glacis*, (*fort.*) foot of the glacis;
—— *d'un grain*, final shower of a squall;
—— *de la Grande Ourse*, star at the end of the Dipper;
—— *de greffage*, (*met.*) tong hold;
—— *d'un haut-fond*, (*hydr.*) tail of a shoal;
—— *de l'hiver*, end of winter;
lettre à, *sur*, *double* ——, letter with its seal on a band passed through the parchment;
lettre scellée sur simple ——, *lettre à simple* ——, letter with its seal on a corner of the parchment;
—— *d'une ligne*, (*nav.*) last ship in a line;
longueur de ——, (*mas.*) length of a stone perpendicular to the face of the wall;
—— *mal attachée*, (*hipp.*) tail badly attached (when dock is not well clear of buttocks);
—— *d'un martinet*, (*mach.*) tail of a tilt hammer;
mettre un —— *de rat*, (*cord.*) to point (a rope);
—— *de morue*, long wedge-shaped plank;
—— *niquetée*, (*hipp.*) nicked tail;
ouvrage à —— *d'aronde*, (*fort.*) work whose faces or branches splay outward;
ouvrage à contre- —— *d'aronde*, (*fort.*) work whose branches run inward, toward the body of the place;
—— *de paon*, (*cons.*) fan-shaped tracery or inlaying;
—— *à patte*, (*carp.*) lapped dovetail;
—— *d'un pavillon*, tip of a flag;
—— *d'un pavillon de la série*, (*nav.*) tack of a signal flag;
—— *perdue*, (*carp.*) mitered dovetail;
—— *de la Petite Ourse*, the pole star;
pièce à ——, (*carp.*) dovetailed piece;
—— *d'une pierre*, (*mas.*) back face of a stone;
porter la —— *en trompe*, (*hipp.*) to carry the tail straight out;
—— *d'une poulie*, tail of a block;
prendre en ——, (*mil.*) to take (a column) in the rear;
—— *à procédé*, tipped billiard cue;

queue *d'une rade*, innermost part of a roadstead;
— *de rat*, reamer, rat-tail file; (*hipp.*) rat's tail (disease); rat('s) tail (hairless or thin-haired tail); (*cord.*) point;
— *recouverte*, — *à recouvrement*, (*carp.*) lapped dovetail;
— *de renard*, firmer chisel;
— *ronde*, trundle tail;
— *de la tige du piston*, (*mach.*) piston-rod tail-piece;
— *à tous crins*, (*hipp.*) switch, full, tail (undocked and hairs full length);
— *de la tranchée*, (*siege*) tail of the trench, point where the attack keeps materials and munitions;
— *en trompe*, (*hipp.*) tail carried high, trumpet-formed tail;
tronçon de —, (*hipp.*) dock of a horse's tail;
— *de troupe*, v. — *d'armée*;
vaisseau de —, (*nav.*) rearmost ship.

queurse, f., v. *queux*.

queux, m., hone, whetstone;
— *à l'huile*, oilstone.

quiboquet, m., mortise gauge.

quille, f., cap, capping piece, runner, string piece; trestle cap; stand, post, support, stem, shank; slater's wedge; curved arm or support; wheelwright's auger; (*sm. a.*) screw tap; (*nav.*) keel;
— *en barre*, bar keel;
— *de bouchain*, bilge keel;
contre —, rising wood of a keel;
— *creuse*, hollow keel; dish keel;
fausse —, false keel;
— *en fer plein*, bar keel;
— *en gouttières*, v. — *creuse*;
— *inférieure*, lower keel;
— *latérale*, bilge keel;
laton d'une —, skeg of a keel;
— *massive*, v. — *en barre*;
— *plate*, flat plate keel;
— *principale*, main keel;
— *de repos*, (of wagons) cart prop;
— *supérieure*, v. — *principale*;
— *de sûreté*, safety keel;
— *à tôles latérales*, plate keel.

quillon, m., (*sm. a.*) quillon (curved part of a sword or bayonet hilt).

quincaille, f., hardware; ironmongery.

quincaillerie, f., hardware; ironmongery.

quinconce, m., quincunx.

quinçonneau, m., toggle.

quinine, f., quinine.

quinquet, m., argand lamp;
— *de gaz*, gas burner.

quinquina, m., cinchona bark.

quintal, m., quintal; hundredweight (112 pounds);
— *métrique*, 100 kilogrammes.

quintan, m., quintain.

quinte, f., (*man.*) dead stand made by a horse; (*fenc.*) quint;
fausse —, (*fenc.*) feint quint;
parade de —, (*fenc.*) quint parry.

quintelage, m., kentledge.

quinteux, a., (*man.*) freakish; irritable; liable to come to a dead stop; balky.

quintuple-mètre, m., (*surv.*) 5-meter rule.

quintuplette, f., quintuplet, (a five-seat bicycle).

quittance, f., receipt.

quittancer, v. a., to receipt.

quitter, v. a., to quit, leave; to give up, renounce; to retire from; to discontinue; to go away, let go, lay down, abandon;
— *le commandement*, (*mil.*) to give up command;
— *l'épée*, (*mil.*) to leave the service;
— *les étriers*, (*man.*) to quit the stirrups (voluntarily or not);
— *le fond*, (*nav.*) to go off soundings;
— *les rangs*, (*mil.*) to step out of ranks.

quitus, m., (*adm.*) final receipt given to a person accountable for public moneys; the closing of an account after settlement.

qui va là? qui-va-là, (*mil.*) (of a sentry) "who goes there?" (sentry's challenge).

qui vive? qui-vive, (*mil.*) challenge of a sentry, of a patrol;
aller au —, to patrol;
arrêter par le —, to challenge;
crier —, to challenge;
être sur le —, to be on the lookout; to keep a sharp lookout;
sur le —, on the lookout.

quoailler, v. n., (*hipp.*) to switch the tail continually.

quote-part, f., quota, portion, contingent, etc.

quotient, m., quotient;
— *de la charge*, (*ball.*) ratio of weight of charge to that of projectile.

quotité, f., share, portion, percentage; quota; rating.

R.

a, m., (*mil.*) stroke on a drum, so as to produce a very short roll.

rabais, m., diminution, shrinkage of value, reduction of price, abatement, deduction; (*adm.*) award of a contract to the bidder offering the greatest reduction from a given price (method specially followed by the French artillery and engineers);
adjudication au —, Dutch auction;
au —, by contract at a reduction from a given price; by Dutch auction.

rabaissement, m., lowering; reduction (in value, of price).

rabaisser, v. a., to lower, abate, diminish, lessen, bring down; to reduce a price, value; (*man.*) to force a horse to use his hind quarters (said of horses that have a tendency to march and work with their shoulders or to raise themselves on their hocks);
— *les hanches d'un cheval*, (*man.*) to throw a horse on his haunches;
se —, (*man.*) to come nearer to the ground (said of a horse that can not keep up a high curvetting).

raban, m., (*cord.*) roband, rope band, earing;
— *de pavillon*, rope band of a flag;
— *de volée*, (*art.*) muzzle lashing.

rabat, m., discount; reduction of price; scriber; (*mach.*) spring beam (of a trip hammer);
rabat de défaut, (*law*) annulment of a verdict by default.

rabat(-)eau, m., splashboard.

rabattement, m., bending back or down, as in metal tests; (in descriptive geometry) revolution into the plane of ....;
— *du défaut*, (*law*) v. s. v. *rabat*;
levier à —, folding lever.

rabattoir, m., dressing tool (for slates).

rabattre, v. a., to lower, abate, diminish, lessen, depress; to beat down or back; to flatten; to bring (cut, force, lay, pull, put, take, turn) down; to turn, turn off; (*met.*) to bend, fold, back or down (in metal testing); to beat out under the hammer; to forge; to jump up; to hammer even; to shape (as the head of a nail); (*fenc.*) to parry, ward off; (*mil.*) to sit down before a place; to deploy; (*man.*) v. — *les courbettes*, *infra*;
— *un clou*, to clinch a nail;
— *un coup*, (*fenc.*) to parry by beating down the adversary's blade;
— *une couture*, to flatten a seam;
— *les coutures*, to horse the seams (of a boat);
— *les courbettes*, (*man.*) to force a curvetting horse to bring down his feet simultaneously;
— *court*, to follow quickly with the next blow (of anvil work);

rabattre *un défaut*, (*law*) to annul a finding of contempt;
se ——, to turn down (as a collar);
se —— *sur*, (*mil.*) to fall back on;
—— *un volant*, (*mach.*) to make the upper part of a fly-wheel turn toward the steam cylinder.

rabattu, p. p., flattened, smoothed;
épée ——*e*, blunted sword (has neither edge nor point).

rabiau, rabio, rabiot, m., (*mil. slang*) what is left of food after all have had their share; profits on forage, food supplies, etc.; remainder of a term of service; term of service in a *compagnie de discipline;* convalescent soldier.

râble, m., rake, stirrer, stirring iron or bar; (*met.*) rabble; (*pont.*) frame;
—— *en bois*, wooden skimmer for metal in fusion;
—— *à crochet*, —— *en fer*, (*met.*) rabble, rabbling tool;
—— *à main*, (*met.*) rabbling tool;
—— *mécanique*, (*met.*) mechanical rabbler.

râbler, v. a., to stir, rake, a fire; (*met., etc.*) to rabble;
—— *le plâtre*, to clear plaster of Paris of coals.

râblure, f., rabbet, channel, grooving.

rabot, m., plane; jack plane; road scraper; hard paving stone; sort of hoe for stirring and mixing mortar; (*met.*) stirrer (used during the washing of ores); (*powd.*) spreader;
—— *à baguette*, (*sm. a.*) ramrod-groove plane;
—— *à canon*, (*sm. a.*) barrel plane;
—— *à chaux*, sort of hoe for mixing mortar;
coin de ——, plane wedge;
—— *à diamant*, diamond pencil;
diamant à ——, glazier's diamond;
—— *d'établi*, bench plane;
fer de ——, plane iron;
—— *à feuilleret*, rabbet plane;
fût de ——, plane stock;
—— *à gorge*, spout plane;
lumière de ——, plane-iron hole;
mèche de ——, v. *fer de* ——;
—— *à mitre*, miter plane;
—— *plat*, smoothing plane;
—— *à rainure*, grooving plane;
—— *à repasser*, v. —— *plat;*
—— *rond*, compass plane.

rabotage, m., planing; (*sm. a.*) smoothing (of the lands).

rabotement, m., planing.

raboter, v. a., to plane, plane down; (*farr.*) to rasp a hoof;
—— *la courbe*, (*ball.*) in constructing a trajectory, to cause the curve to pass as nearly as possible through the observed points;
machine à ——, planing machine;
—— *le mortier*, to beat mortar;
—— *à travers*, to plane across the grain.

raboteur, m., planer (workman).

raboteuse, f., (*mach.*) planing machine;
—— *de côté*, horizontal slotter or planer;
—— *-tour*, vertical lathe;
—— *verticale*, slotting machine.

raboteux, a., rough, uneven (as the ground, a road); knotty, knotted (of wood, planks, etc.).

raboture, f., shaving.

rabougri, a., (of trees) stunted, crooked.

rabougrir, v. a. n., to stunt, to be stunted (of trees).

rabouter, v. a., to join on, end to end (of ironwork); (*mach.*) to ferrule.

raboutissage, m., joining end to end (of ironwork).

rabraquer, v. a., (*cord.*) to haul in anew.

racage, m., (*nav.*) parrel.

racambeau, m., (*nav.*) traveler.

racastillage, m., (*nav.*) repair of the upper works.

racastiller, v. a., (*nav.*) to repair the upper works.

raccommodage, m., repair, repairing.

raccommoder, v. a., to repair, mend, piece, patch, set right, correct.

raccord, m., (*tech.*) joining, junction, union, leveling, smoothing; adjustment, adjusting; joint, flush joint, assembling or connecting joint; coupling (as of tubes); connecting pipe; (*elec.*) connecter;
—— *de rail*, (*r. r.*) rail joint;
—— *à vis*, union (for hose, etc.).

raccordement, m., junction, joining; (*tech.*) coupling (as of tubes, etc.), joint (of wires), connecting pipe; (*art.*) junction cone, chamber slope, compression slope; (*sm. a.*) junction slope or cone between the body and the neck of a cartridge; (*mil.*) establishment of contact between troops that have concentrated for any purpose (esp. for night attacks); (*surv.*) junction of two triangulations; (*r. r.*) junction (of the tracks, of the rails);
cône de ——, (*art.*) junction cone;
courbe de ——, junction curve.

raccorder, v. a., to join; to level, adjust, smooth, unite; to make parts agree; to make flush.

raccourci, m., foreshortening; any short cut; (*r. r.*) short line making a short cut between two points;
en ——, foreshortened.

raccourcir, v. a. n., to shorten, take up, contract, draw up, abridge, foreshorten; to become shorter, shrink;
—— *la bride*, (*man.*) to shorten the reins;
—— *un cheval*, (*man.*) to slacken the gait; to keep a horse down;
—— *un cordage*, (*cord.*) to shorten a rope; to make a sheepshank (on or of);
—— *des demi-voltes*, (*man.*) to make half circles of a smaller radius;
—— *les étriers*, to shorten, take up, stirrups;
—— *le pas*, (*mil.*) to shorten step.

raccourcissement, m., shortening, contraction; foreshortening.

raccroché, a., (*man.*) having a long seat (of a rider); "hunched up" (of a seat on horseback).

raccrocher, v. a., to hook again.

race, f., race; stock; breed (of horses);
cheval de ——, thoroughbred horse;
—— *chevaline*, horse species;
—— *croisante*, (*hipp.*) breed whose blood is introduced;
—— *croisée*, (*hipp.*) breed whose blood is to be improved;
pure ——, (*hipp.*) unmixed breed, pure breed (not to be confused with *pur sang*).

rachat, m., redemption, recovery, commutation;
—— *des chevaux*, (*Fr. a.*) purchase by the state of horses issued to officers;
faculté de ——, privilege of repurchase;
pacte de ——, v. *faculté de* ——;
—— *de pain*, (*mil.*) money equivalent of bread, when not drawn.

rache, f., dregs of pitch and tar; compass trace on wood.

racher, v. a., to scribe.

racheter, v. a., to buy back, redeem, ransom; to commute; to compensate for, make up for;
—— *un jeune homme*, (*mil.*) to buy a young man's exemption from military service;
se ——, (*mil.*) to purchase exemption from military service; to buy one's discharge.

racheux, a., knotty, stringy (of wood).

rachevage, m., finishing (of any piece of work), last touch.

rachever, v. a., to finish, put the last touch to (a piece of work).

racheveur, m., finisher (workman).

rachis, m., (*hipp.*) vertebral column.

rachitique, a., (*hipp.*) rickety.

rachitisme, m., (*hipp.*) rickets.

racinage, m., generic term for edible roots.

racinal, m., (*cons.*) sleeper; sole, sill; ground timber; (*in pl.*) grillage;
—— *du busc*, main sill at the entrance of a sluice;
—— *de comble*, corbel;
—— *d'écurie*, manger post;
—— *de grue*, sole of a crane;
—— *de palée*, sleeper or sill of a grillage;

racinal *de pont*, bolster for a stringpiece.
racine, f., root; (*top.*) foot of a mountain;
—*s fourrages*, edible roots.
raclage, m., thinning of underbrush.
racle, f., strike; scraper; (*hydr.*) natural basin in a river.
racle-curette, f., scraper.
raclement, m., scraping.
racler, v. a., to scrape; to strike a measure; to thin out underbrush.
raclette, f., small scraper.
racleur, m., scraper (workman).
racloir, m., scraper; boiler-tube scraper; road scraper; (*powd.*) scraper, rake.
racloire, f., spokeshave; strike, striker, strickle.
raclure, f., scrapings.
racolage, m., (*mil.*) kidnapping of recruits, crimping; enticing men to enlist (obs.).
racoler, v. a., (*mil.*) to get soldiers by fraud or violence; to kidnap, to crimp, to entice, men to enlist (obs.).
racoleur, m., (*mil.*) kidnapper, crimp (obs.).
radasse, f., swab, mop.
rade, f., (*hydr.*) roads, roadstead;
aller en ——, to leave a port;
être en ——, to be at anchor in the roads;
—— *foraine*, open roadstead;
grande ——, outer roadstead;
mettre en ——, v. *aller en* ——;
petite ——, inner roadstead.
radeau, m., raft; raft, train, of logs; (*pont.*) raft; (*nav.*) floating stage;
—— *de barriques*, (*pont.*) raft of casks;
—— *de bois de construction*, float or raft of timber;
—— *de circonstance*, (*pont.*) temporary raft; (*nav.*) improvised raft (as to escape from a sinking ship);
—— *double*, (*pont.*) raft made up of two sets of logs, end to end; double raft;
—— *de fortune*, v. —— *de circonstance* (*nav.*);
—— *de manœuvre*, (*pont.*) construction raft (used for setting trestles in place);
pont de ——*x*, (*pont.*) bridge of rafts;
—— *de sauvetage*, life raft;
—— *simple*, (*pont.*) raft made up of single logs side by side; single raft;
—— *de tonneaux*, v. —— *de barriques*;
——-*traille*, trail or flying raft; (*pont.*) raft bridge.
rader, v. a., to strike (wheat, corn, etc.); to divide a block of stone with a chisel; (*nav.*) to bring a ship into the roads.
radial, a., radial;
(*machine*) ——*e*, (*mach.*) any radial machine; (more esp.) radial drill.
radiant, a., radiant.
radiation, f., radiation; erasure, striking out (of a word, of an item from a bill); (*mil.*) striking off the rolls (said of horses as well as of men; may be, and is, said when a man or a horse is transferred from one company to another of the same regiment, or from one regiment to another, or has gone into the reserve or the *disponibilité*, or is dead); striking off the rolls of the army for any reason whatever, (hence, sometimes) dismissal.
radical, a., radical; m., (*chem.*) radical;
vice ——, (*law*) fatal defect.
radier, m., (*cons.*) frame of ground timber (for the foundation of a sluice, etc.); floor, platform (of a lock); bed of a sluice; apron (of a basin, dock, between the piles of a bridge, the walls of a lock gate); head race; (more generally) concrete or masonry work (to strengthen a foundation in compressible soil), inverted arch, invert;
arrière-——, downstream apron;
avant-——, upstream apron;
bassin de ——, graving dock, dry dock;
faux ——, apron of a dam;
—— *général*, (in compressible soil) strengthening layer of concrete, etc., supporting the whole foundation of a building;
—— *de pont*, apron between the piles of a bridge.
radieux, a., radiant.
radiophone, m., (*elec.*) radiophone.

radoub, m., (*nav.*) repair, refitting (of a ship); dockyard; (*powd.*) drying, reworking, of powder; reworked powder;
bassin de ——, v. s. v. *bassin*;
cale de ——, (*nav.*) graving slip;
en ——, (*nav.*) in the dockyard, under repair;
forme de ——, (*nav.*) graving dock;
poudre de ——, (*powd.*) reworked powder.
radoubage, m., (*powd.*) remanufacturing, working over again.
radouber, v. a. r., (*nav.*) to refit, repair (a vessel); to be under repair; (*powd.*) to dry (work over) powder.
radoubeur, m., (*nav.*) calker.
radoucir, v. a. r., to soften, become softer; (*met.*) to anneal.
rafale, f., squall, wind squall;
à ——*s*, squally;
par ——*s*, in squalls.
raffermir, v. a., to harden, consolidate, make firm; to strengthen, fasten, confirm.
raffermissement, m., hardening, strengthening.
raffinage, m., refining (of saltpeter, etc.); (*met.*) refining, fining.
raffiner, v. a., to refine (as saltpeter, etc.); (*met.*) to refine, to fine.
raffinerie, f., (*met.*, etc.) refinery.
raf(f)utage, m., putting tools in complete order.
raf(f)uter, v. a., to put tools in complete order.
rafiau, **rafiot**, m., small sail and row boat.
raflouage, m., v. *renflouage*.
raflouement, m., v. *renflouement*.
raflouer, v. a., v. *renflouer*.
rafraîchir, v. a., to cool, freshen, refresh; to renew, renovate, repair, make over, do up, rub up; to rest, cause to rest; (*mas.*) to retrim, re-dress, joints, faces (of a stone); (*mil.*) to throw supplies, reenforcements, etc., into a place; (*cord.*) to cut off fag-ends; (*expl.*) to trim (a Bickford fuse, etc.);
—— *une amarre*, (*cord.*) to freshen a rope, cable, at the nip;
—— *les amarres*, (*cord.*) to make a knot or lashing over again;
—— *une bouche à feu*, (*art.*) to cool a gun, to swab it out;
—— *le bout d'un cordage*, (*cord.*) to cut off the fag-end;
—— *un câble*, v. —— *une amarre*;
—— *un canon* (*art.*) v. —— *une bouche à feu*;
—— *un cordage*, v. —— *les amarres* and —— *le bout d'un cordage*;
—— *un coussinet échauffé*, (*mach.*) to cool a heated bearing, a hot box;
—— *le cuivre*, (*met.*) to alloy copper with (three times its weight of) lead;
fourneau de ——, (*met.*) refining furnace;
—— *la fourrure*, (*cord.*) to put on a fresh service;
—— *un fusil*, (*sm. a.*) to clean a gun;
—— *le mortier*, (*mas.*) to thin mortar;
—— *une place*, (*mil.*) to throw fresh troops and supplies into a besieged town;
—— *une queue*, to trim a horse's tail even;
—— *les rayures*, (*art.*) to rifle over, to reset, to recut the grooves;
—— *une scie*, to sharpen a saw;
—— *les troupes*, (*mil.*) to halt, refresh, and rest troops and horses.
rafraîchissement, m., refreshment, cooling;
——*s*, supply of fresh provisions, fresh provisions;
—— *du cuivre*, (*met.*) alloying of copper with (three times its weight of) lead.
ragage, m., v. *raguage*.
raganage, m., trimming of the lower branches of trees.
ragot, m., shaft hook (of a wagon); breeching hook; a., thickset, stocky;
cheval ——, thickset horse.
ragréage, m., dubbing; (*nav.*) refitting of rigging.
ragréer, v. a., to finish off, work over, dub; to refit, repair, restore, renovate; to remove inequalities; (*nav.*) to refit the rigging;

ragréer *une maison*, to renovate a house.
ragrément, m., finishing, finishing off, dubbing, repairing, renovation, restoration;
—— *simple*, (*mas.*) mere finishing, trimming, of a surface (includes pointing of joints);
—— *à vif*, (*mas.*) finishing, dressing (involves removal of the surface to a depth of 5 or 10^{mm}; said of soft stone; includes pointing of joints).
raguage, m., (*cord.*) chafe, chafing; rubbing, galling.
rague, f., fair-lead, truck.
raguer, v. a., (*cord.*) to chafe, rub, gall.
ragure, f., chafing, chafed part.
raid, m., (*mil.*) raid.
raide, a., stiff, rigid, tight, inflexible; steep (of a slope); swift (of the course of a stream); rapid (of a descent); (*cord.*) taut;
—— *mort*, stone dead;
se tenir ——, to stand stiff upright;
tuer ——, to kill outright, on the spot.
raideur, f., stiffness, rigidity, tightness, inflexibility, toughness; steepness (of a slope); rapidity (of a descent); (*cord.*) stiffness of cordage;
—— *des cordages*, (*cord.*) stiffness of cordage.
raidillon, m., rise, ascent, in a road, little hill.
raidir, v. a. n., to stiffen, to stretch; to tighten; to become stiff; to bear up against; to withstand; to make steep (as a slope); (*cord.*) to haul tight (of a rope, etc.), to haul taut; to set up, to have everything well set up;
—— *en abattant*, (*cord.*) to sway down on;
—— *un étai*, to stiffen a prop;
—— *à la main*, (*cord.*) to haul taut by hand;
—— *les manœuvres*, (*cord.*) to haul in the slack;
se ——, (*man.*) (of a horse) to balk.
raie, f., stroke; stripe; line (of a spectrum); furrow; (*art.*) rifling (rare).
—— *de misère* (*hipp.*) groove or depression in thin horses, between muscles of buttocks and elsewhere;
—— *de mulet*, (*hipp.*) mule ray or stripe.
rail, m., (*r. r.*) rail;
—— *américain*, foot rail;
—— *à base plate*, v. —— *américain*;
—— *central*, central rail;
—— *à champignon*, rail resting on chairs, chair rail;
—— *à champignon avec semelle*, v. —— *américain*;
—— *à champignon symétrique*, double-headed rail;
—— *à champignons inégaux*, rail of unequal heads;
—— *en charpente*, wooden rail of a tramroad;
—— *contre-*——, guard-, wing-, safety-, side-rail;
—— *contre-aiguille(s)*, stock rail, main rail of a switch;
—— *s convergents*, junction rails;
—— *à coussinet*, v. —— *à champignon*;
—— *croiseur*, crossing rail;
—— *denté*, toothed rail;
—— *à deux bourrelets*, double-headed rail;
distances des ——*s dans œuvre*, gauge;
—— *divergent*, wing rail;
—— *à double bourrelet*, double-headed rail;
—— *double T*, v. —— *à champignon symétrique*;
—— *à double champignon*, v. —— *à champignon symétrique*;
—— *en* ∩, v. —— *à pont*;
—— *à éclisse*, fish-plate rail;
—— *s éclissés*, fished rails;
entre-voie des ——*s*, gauge;
—— *extérieur*, exterior (higher) rail of a curve;
—— *extérieur d'un changement de voie*, v. —— *divergent*;
—— *fixe*, v. —— *contre-aiguille*;
—— *en H*, channel rail;
—— *mobile*, switch, point, rail;
—— *ondulé*, fish-bellied rail;
—— *à ornière*, tram rail;
—— *parallèle*, v. —— *à champignon symétrique*;
—— *à patin*, foot rail, single-headed rail;
—— *patte-de-lièvre*, v. —— *divergent*;
—— *plat*, plate rail;
—— *à pont*, bridge rail;

rail *prismatique*, plate rail;
—— *à rebord*, v. —— *saillant*;
—— *saillant*, edge rail;
—— *à simple champignon*, v. —— *américain*;
—— *simple T*, v. —— *américain*;
—— *à surface bombée* (*plane*), flat- (round-) headed rail;
—— *symétrique*, double-headed rail;
—— *à tête d'acier*, capped rail;
—— *à un seul champignon*, v. —— *américain*;
—— *à ventre de poisson*, v. —— *ondulé*;
—— *Vignole*, v. —— *américain*.
raille, f., furnace rake.
railroute, f., (*r. r.*) proposed name for railway.
rain, m., (*top.*) skirt of a forest or wood.
raineau, m., (*cons.*) timber binding piece (for piles).
rainer, v. a., to flute; to groove, to score.
rainette, f., tracing iron or tool; (*hipp.*) foot probe, paring knife;
—— *simple*, v. *rogne-pied*.
rainure, f., groove, slot, rabbet, furrow, opening, channel, guide way, keyway;
—— *d'arrêt*, (*sm. a.*) rim or shoulder (limiting the entry of the cartridge); (*F. m. art.*) seating groove;
—— *à baguette*, (*sm. a.*) ramrod groove;
—— *à coin*, key bed, keyway;
—— *de dégagement*, spiral channel (of a drill or auger or bit);
—— *de départ*, (*sm. a.*) cocking groove;
—— *directrice*, (*r. f. art.*) cam groove;
—— *d'extraction*, (*art.*) extraction groove;
—— *de fût*, v. —— *à baguette*;
—— *hélicoïdale*, (*sm. a.*) helicoidal groove;
—— *à huile*, (*mach.*) oil way;
—— *et languette*, (*carp.*) tongue and groove;
à —— *et languette*, (*carp.*) tongued and grooved;
—— *de poulie*, score of a block;
—— *transversale*, cross groove;
—— *de visée*, (*fort.*) sighting slot or aperture, sighting groove (of a revolving turret).
rais, m., spoke; rowel, of a spur; (*mach.*) spoke or arm of a wheel;
broche de ——, tenon embedded in the nave of a wheel;
—— *creux*, tubular spoke;
—— *évidé*, beveled spoke;
faux- ——, fish (for mending a broken spoke);
patte de ——, v. *broche de* ——;
serre- ——, spoke cramp, setter.
raisin, m., grapes; (*art.*) grape (obs.);
grappe de ——, (*art.*) grapeshot (obs.);
tirer à grappe de ——, (*art.*) to fire grapeshot (obs.).
raison, f., reason, proof, ground; ratio; name (of a firm);
avoir —— *de*, to get the better of;
demander —— *à*, to challenge to a duel;
—— *de guerre*, (*mil.*) military custom, practice, martial law;
mettre des pièces de bois en leur ——, (*carp.*) to place timbers in their respective places;
rendre —— *à*, to fight a duel with;
—— *de service*, (*mil.*) service reasons;
—— *sociale*, firm name.
raisonner, v. n., to argue, reason; (*nav.*) to show papers;
droit de faire ——, v. s. v. *droit*;
faire —— *un navire*, to oblige a ship to come near and show papers; to make a commercial vessel answer questions.
rajuster, v. a., to readjust (instruments, etc.).
rajusteur, m., sealer of weights and measures.
ralentir, v. a. n. r., to slacken, retard; to lessen; to reduce the velocity of; to become slower; (*mach.*) to slow down an engine; (*man.*) to reduce a gait;
—— *un cheval*, (*man.*) to slow up a horse;
—— *la marche*, (*mach.*) to ease the engines to slow down; (*mil.*) to slacken a march.
ralentissement, m., retardation, slackening, loss of velocity; diminution, cooling; (*mach.*) slowing down (of an engine); (*man.*) slowing up, slackening of a gait.

ralingue, f., (*nav.*) boltrope.

ralinguer, v. a., (*nav.*) to sew on the boltropes of a sail.

ralliement, m., (*mil.*) rallying, rally, rallying into groups; rallying of skirmishers; rallying of troops after a charge or battle; (*nav.*) rallying, assembling, of ships;
le ——, (*mil.*) call for a rally;
mot de ——, (*mil.*) v. s. v. *mot;*
point de ——, (*mil.*) rallying point;
signal de ——, (*nav.*) rallying signal for ships to take respective stations;
signe de ——, (*mil.*) rallying sign;
sonner au ——, (*mil.*) to sound the rally.

rallie-papier, m., paper chase.

rallier, v. a., (*mil.*) to rally; to re-form; (*nav.*) to join (a ship); to approach (the land, a port);
—— *le camp*, (*mil.*) to rejoin camp;
—— *le corps*, (*mil.*) to rejoin one's regiment;
—— *un port*, (*nav.*) to approach a port;
—— *la terre*, (*nav.*) to stand for the land;
—— *un vaisseau*, (*nav.*) to rejoin a ship.

rallonge, f., any extension piece, eking piece in general; leaf of a table; (in balloons) rope forming the extension of the net proper; (*sm. a.*) rear-sight extension leaf;
curseur à ——, (*sm. a.*) extension slide of a sight;
—— *de flèche*, (*Fr. art.*) trail extension piece of the 80ᵐᵐ mountain gun;
—— *de trait*, (*harn.*) the rope extension piece of the French artillery trace.

rallongement, m., eking, lengthening; splicing (of timbers).

rallonger, to lengthen; to eke out; to splice (timbers).

rallumer, v. a. r., to rekindle, light again; to break, burst out again.

ramandot, m., (*powd.*) crust of powder (forming in barrels during glazing).

ramas, m., heap.

ramasse, f., sort of sled used in the Alps; (*tech.*) sort of reaming rod or tool (for enlarging a cylindrical hole in wood or metal); (*sm. a.*) steel cleaning rod.

ramassé, p. p., thickset, squat; compact (e. g., an engine); (*hipp.*) short in the joints.

ramasser, v. a., to pick up, take up; to gather; to rake and scrape together; to collect; to drag on a *ramasse*, q. v.; (*cord.*) to take in;
—— *de la boite*, (*mil. slang*) to be locked up;
—— *les fourreaux de baïonnette*, (*mil. slang*) to come up after a battle has been fought;
—— *une pelle*, (*fam.*) to get a tumble from a bicycle.

rame, f., oar, scull, paddle; ream of paper;
—— *à bateau*, (*pont.*) pontoon oar;
canot à ——, rowboat;
à force de —— *s*, by dint of rowing, by hard pulling;
lève —— *s*, oars! unship oars!
lever les —— *s*, to lie or rest on the oars;
à n —— *s*, *n*-oared;
—— *à nacelle*, (*pont.*) mooring boat oar;
tirer à la ——, to pull at the oar.

ramé, p. p., rowed;
boulet ——, (*art.*) bar shot (obs.).

rameau, m., branch, bough; (*mil. min.*) branch or small mine gallery.
(The following terms relate to military mining):
—— *ascendant*, ascending branch;
—— *de combat*, small branch (0.70ᵐ high and 0.60ᵐ wide, requires cases);
—— *descendant*, descending branch;
—— *à double retour d'équerre*, (*mil. min.*) double rectangular branch (i. e., a branch breaking out at right angles and then making a rec tangular change of direction);
—— *écoute*, branch between, and parallel to, listening galleries (*écoutes*);
entrer en ——, to open a branch;
grand ——, great branch (1ᵐ high and 0.80ᵐ wide, requires cases);
—— *grimpant*, climbing or ascending branch;
—— *hollandais*, —— *à la hollandaise*, v. *petit* ——;

rameau, *petit* ——, small branch (0.80ᵐ high and 0.65ᵐ wide, requires cases);
—— *à retour d'équerre*, branch broken out at right angles from a gallery.

ramené, p. p., brought back; m., (*man.*) the position of the horse's head, when brought down and in;
être ——, (*mil.*) to be pursued, forced to retreat, defeated; to be taken, brought, into camp (said of cavalry).

ramener, v. a. r., to bring, take, lead, drive, back; to recall, retrieve; to recover, restore; (*man.*) (of the rider) to lower and bring in the head of a horse (is also said of a bit); (substantively) the act of lowering, etc., the head; (of the horse) to carry the head; (*mil.*) to handle roughly; to take into camp (said of cavalry);
—— *bien*, (*man.*) to carry the head well;
—— *un cheval*, (*man.*) to force a horse to bring his head down and in;
cheval qui se ramène bien, horse that carries his head well;
—— *l'ennemi*, (*mil.*) to defeat and drive back the enemy, to take him into camp;
se ——, (of a horse) to carry the head well.

ramèneret, m., stroke with a chalk line.

ramer, v. a., to row, to pull (the naval expression is *nager*); to paddle.

rameur, m., oarsman (*nageur* in the navy).

ramification, f., ramification.

ramifié, a., ramified; (*hydr.*) having arms (said of a bay).

ramille, f., stick, twig, bavin.

ramingue, f., restive; (*man.*) disobedient to the spur;
cheval ——, v. s. v. *cheval*.

ramollir, v. a., to soften.

ramonage, m., chimney sweeping.

ramoner, v. a., to sweep chimneys.

ramoneur, m., chimney sweep.

rampant, a., crawling, creeping, climbing, rising; (*cons.*, etc.) inclined, sloping; m., slope, sloping part; (*met.*) sloping side of the bed of a reverberatory furnace; flue of a reverberatory furnace (between the crown and the chimney proper);
arc ——, rampant arch;
bandage ——, (*med.*) spiral bandage;
fenêtre —— *e*, (*cons.*) oblique staircase window;
limon ——, (*cons.*) continuous (unbroken) stringpiece;
lunette —— *e*, (*cons.*) sectroid;
marche —— *e*, v. s. v. *marche;*
voûte —— *e*, rampant arch.

rampe, f., slope, ascent, descent, declivity, acclivity; (*cons.*) flight of straight stairs, slope or inclined plane on which the steps are constructed; hand rail, railing (of a staircase); (*fort.*) ramp; (r. r.) slope or gradient, grade; loading ramp;
—— *d'accès*, approach of a bridge;
—— *courbe*, (*cons.*) curving flight of stairs;
—— *d'armement*, (*fort.*) gun ramp of a platform;
—— *de débarquement*, (r. r.) unloading ramp, brough;
—— *de dégagement*, (*sm. a.*) disengagement curve or surface of a breech mechanism;
—— *d'entrée*, slope or bank leading down to a ford;
—— *hélicoïdale*, (*sm. a.*) helicoidal guide surface of a breech mechanism;
—— *improvisée*, temporary ramp;
—— *à longrines en fer*, (r. r.) girder ramp;
—— *mobile*, (*mil. r. r.*) transportable ramp for embarking and debarking;
—— *de sortie*, slope leading from a ford;
—— *voûtée*, arched or bow ramp.

ramper, v. n., to creep, crawl; (*cons.*) to incline according to a given slope.

rampin, a., (*hipp.*) that wears his shoes at the toes; "over" in the feet;
cheval ——, horse over in the foot, horse that stands on his toes;
pied ——, "over" foot, foot bearing on the toe.

rance, f., skid, parbucking skid; a., rancid.

ranche, f., round or peg of a ladder (esp. of a peg ladder), peg or pin of a gin prypole; rack of a crane or pile driver;
— *de haquet, (pont.)* stud stave of a pontoon boat.

rancher, m., peg ladder; stud stave; *(pont.)* iron upright of a pontoon wagon, to keep a boat in (in the fore and hind bolster of the wagon); *(art.)* iron stirrup on the side of an ammunition wagon for spare pole;
— *de grue,* jib, arm, of a crane.

ranchet, m., upright stud (in the framing of a wagon body), stay of a pile driver (pegged to serve as a ladder):
faux —, movable stanchion;
— *de haquet, (pont.)* stud stave;
levier- —, stud iron or stanchion of the Decauville truck.

rancir, v. n., to become rancid.

rançon, m., ransom;
mettre à —, to set a ransom on; to levy contribution on, pillage;
mise à —, ransoming; pillaging.

rançonnement, m., ransoming.

rançonner, v. a., to ransom; to levy contribution on, pillage.

randanite, f., *(expl.)* a variety of infusorial earth or kieselguhr.

rang, m., row, order; rank, station, degree, place, class, precedence; turn; tier; rate; *(man.)* part of the riding hall where the line of troopers or horsemen is drawn up; *(mas.)* course (of bricks, of stones); *(mil.)* rank, line; *(Fr. a.)* rank (a line of men standing side by side and a file of men when marching by the flank, i. e., the designation of rank does not change); *(nav.)* rate of a vessel.
(Except where otherwise indicated, the following terms relate to the art military:)
aligner par — *de taille,* to size (a company, etc.);
au —! *(cav.)* form rank! (command);
— *d'ancienneté,* seniority; lineal list;
par — *d'ancienneté,* according to seniority;
arrivé par les —s, risen from the ranks;
avoir — *après, avant, (in gen.)* to come after, before (in point of precedence);
— *de bataille,* tactical order of troops in line or in column in battle;
— *de bataille des troupes,* precedence of troops;
— *de canons,* tier of guns;
casser de son —, to reduce to the ranks; to cashier;
casser de son — *et renvoyer soldat,* to reduce to the ranks;
compléter les —s, to fill the ranks, bring numbers up to their full complement;
de —, *(in gen.)* in a row, abreast;
dépouiller de son —, to deprive of one's rank;
dernier —, rear rank; *(in gen.)* lowest rank;
— *double, (tech.)* double row of rivets;
— *doublé,* double column of files;
à double (simple, etc.) — *de rivets, (tech.)* double- (single-, etc.) riveted;
doubler le(s) —(s), to form (from line in single or double rank) a column of two or four files by the flank;
éclaircir les —s, to thin the ranks;
— *d'écurie,* in a stable, number of horses tied up to the same rack;
en —, arrayed, drawn up;
enfoncer les —, v. *percer les* —;
entrer dans les —s *d'une armée,* to be incorporated in an army;
être au premier —, *(in gen.)* to stand in the first rank;
être au second —, *(in gen.)* to stand in the second rank;
être sur les —, *(in gen.)* to be a candidate, a competitor;
— *de fascines, (fort.)* course, row, of fascines;
formez vos —s! fall in! (command);
garder les —s, to keep the ranks, to stay in ranks;

rang, grand —, in a stable, the row or line containing the greatest number of, or the finest, horses;
homme de —, man in the ranks;
homme du premier —, front-rank man;
homme du second —, rear-rank man;
hors —, v. s. vv. *peloton, section;*
la main dans le —, the hand at the side;
— *de marche,* place, order, in a march;
mettre au —, *(in gen.)* to set in a row;
mettre au — *de, (in gen.)* to place among;
mettre en —, to draw up in line;
se mettre sur les —s, *(in gen.)* to become a candidate, a competitor;
sur n —s, in *n* ranks;
— *d'ordre,* rank, order of precedence;
ouvrir les —s, to open ranks;
par —s, *(in gen.)* in rows;
peloton hors —, v. s. v. *peloton;*
percer les —, to break through a line;
porter les —s *au complet,* to fill up the ranks, bring up the number to the limit;
premier —, front rank;
rentrer dans le —, (of officers' servants) to take their places in ranks in order to march and fight;
— *de rivets, (tech.)* row of rivets;
rompre les —s, to break off, to break ranks; to be dismissed (drill);
— *de saucissons, (fort.)* course of fascines;
— *et séance, (in gen.)* precedence;
second —, rear rank;
section hors —, v. s. v. *section;*
— *de serre-file,* line of file closers;
serrer les —, to close ranks, to close up;
en —s *serrés,* closed up, in close ranks;
— *simple, (tech.)* single row of rivets;
à son —, *(in gen.)* in one's turn;
sortant des —s, promoted, risen, from the ranks;
— *de sortie,* (of schools) graduation, final standing;
sortir des, du, —(s), to rise from the ranks; to leave, step out of ranks;
— *de taille,* size (of an individual man), order of size, position according to size in a company of infantry;
le — *de taille,* sizing;
par — *de-taille,* according to size;
tenir les —, to keep the ranks, stay in the ranks, keep in the ranks;
— *triple, (tech.)* triple row of rivets;
— *des troupes,* precedence of troops;
sur un —, in single rank;
vaisseau de premier, etc., —, *(nav.)* vessel of the first, etc., rate.

range, f., row of paving stones of the same size;
— *losange,* paving stones set lozengewise.

range, p. p., arranged in a certain order; placed; steady; in good order;
bataille —*e, (mil.)* pitched battle;
— *au-dessous de la forme,* (of the insole of a shoe) neither smaller nor narrower than the bottom of the last.

rangée, f., row, line, course; *(art.)* tier of guns; *(nav.)* line, row, of ships in a dock.

ranger, v. a., to put in a place, to range, to arrange, to put to rights, in order, etc.; to put aside, out of the way; to keep back; to keep out of the way; to subdue, to subject; to rank, number, reckon, place; *(mil.)* to draw up, marshal, troops; *(nav.)* to draw up a fleet;
— *à,* to subject to;
— *en bataille, (mil.)* to draw up a body of men in "battle array," to draw up in the order of battle;
— *un cheval à droite (gauche), (man.)* to range a horse to the right (left);
— *en...., (mil.)* to draw up in....;
— *les épaules, (man.)* to put or range the shoulders on a line;
— *les gazons, (fort.)* to lay sods;
— *les hanches, (man.)* to put or range the hind quarters on a line;
— *à l'honneur, (nav.)* to pass close astern (for orders); (by extension) to pass very close to a buoy, a coast, etc.;

ranger *du monde sur une manœuvre*, (*cord.*) to man a rope;

— *se* ——, to place one's self, to take up a certain position; to leave the way clear, to get out of the way; (of the wind) to veer; (*man.*) of the horse, to take up a proper position; (*mil.*) to fall in, to be drawn up for battle;

— *se* —— *à bord, à quai*, (*nav.*) to come up along side of a ship, of a quay;

— *se* —— *sous les drapeaux de* ...., (*mil.*) to take service under....;

—— *des troupes*, (*mil.*) to draw up troops.

rangette, f., sheet iron for stovepipes.

ranguillon, m., tongue (of a buckle).

râpage, m., rasping.

rapatelle, f., horsehair cloth for sieves.

rapatriement, m., sending home (of soldiers, sailors, citizens, from a foreign country or expedition, etc.).

rapatrier, v. a., to send home, to one's country; to send home (from the colonies, from foreign expeditions, time-expired men, etc.).

râpe, f., rasp, file, grater; (*hipp.*) in plural, malanders with a transverse crack;

—— *à bois*, wood rasp;

—— *à bois à fine* (*grosse, moyenne*) *piqûre*, fine- (coarse-, medium-) cut wood rasp;

—— *à chaud*, rasp for hot iron;

—— *à froid*, rasp for cold iron;

—— *de maréchal*, v. —— *à sabot*;

—— *à sabot*, (*farr.*) farrier's rasp.

râper, v. a., to rasp, to scrape, to grate;

machine à ——, rasping mill.

rapetassage, m., patching.

rapetasser, v. a., to patch, patch up, piece, botch.

rapetisser, v. a. n., to lessen, make shorter, shorten, to become little.

raphé, m., (*hipp.*) median raphe.

rapide, a., rapid, quick; steep (of a slope); m., rapid(s) (of a river);

canon à tir ——, (*art.*) rapid-fire gun;

—— *de marée*, v. s. v. *raz de* ——;

tir ——, (*mil.*) rapid fire.

rapidité, f., rapidity, velocity;

indice de ——, (*phot.*) ratio of working aperture to focal length;

—— *d'un objectif*, (*phot.*) rapidity of a lens;

—— *du tir*, (*art., sm. a.*) rapidity, rate of fire.

rapiécer, v. a., to piece, to put in a piece, to patch.

rapière, f., (*sm. a.*) rapier.

rapine, f., rapine, pillage, plunder, plundering.

rapiner, v. a., to commit rapine; to pillage, plunder, etc.

rapointir, v. a., to repoint (as a broken tool).

rappareiller, v. a., to match, to get a pair, a match;

—— *des chevaux*, to match horses.

rapparier, v. a., v. *rappareiller*.

rappel, m., recall; (*mil.*) battery or call for assembly, i. e., first call; summons to service; withdrawal from command; (*art.*) return to battery; (*nav.*) beating to quarters; (*adm.*) further payment, payment of what remains due;

—— *d'un ambassadeur*, recall of an ambassador;

battre le ——, (*mil.*) to beat the assembly; (*nav.*) to beat to quarters;

—— *pour l'incendie*, fire call;

lettre de ——, letter of recall (of an ambassador);

—— *à l'ordre*, call to order (in deliberative assemblies);

le —— *de pied ferme* (*tambours et clairons réunis*), (*mil.*) the assembly proper, signal to fall in (sounded by the assembled field music);

—— *pneumatique*, (*mach.*) dashpot gear of Frikart-Corliss engine;

—— *de prêt*, v. —— *de solde*;

—— *à la question*, bringing back to the point (in deliberative assemblies); the previous question;

—— *de roulis*, (*nav.*) weather roll;

—— *de solde*, (*mil.*) further payment, restoration of retained pay;

rappel, *le* —— *aux tambours et aux clairons*, (*mil.*) call for the assembly of field music (assembly of trumpeters, U. S. A.).

rappeler, v. a., to recall; to call up (by telegraph); (*cord.*) to grow; (*mil.*) to beat, sound, the assembly;

—— *à* ...., (*mil.*) to sound the call for ....;

—— *à, pour, la garde*, (*mil.*) to sound the call for guard mounting;

—— *aux malades*, (*mil.*) to sound sick call;

—— *à l'ordre*, to call to order (deliberative assemblies);

—— *à la question*, to bring back to the point (deliberative assemblies); to call for the previous question;

—— *au règlement*, to protest against a violation of regulations.

rappointis, m., plastering nail; (*in pl.*) generic term for nails, spikes, hooks, etc.

rapport, m., action of bringing; report (in general); statement, account; return (from property), income, revenue, profit; reimbursement; evidence, testimony; ratio, relation, connection; (*mil.*) orderly hour, orderly call; (regimental) daily report, morning report, guard report; publication of orders, etc.; (*fond.*) false core, drawback; (*mach.*) take-up for wear;

au ——, (*mil.*) at orderly hour;

au —— *de*, on the report of;

avant le ——, (*mil.*) before orderly hour;

—— *d'avaries*, (*nav.*) damage report;

—— *de la charge au projectile*, (*art.*) ratio of the charge to the projectile;

—— *sur la conduite*, (*mil.*) confidential report on conduct;

—— *d'experts*, report of a survey, etc.;

—— *général de la place*, (*mil.*) consolidated guard report;

heure du ——, (*mil.*) orderly hour;

—— *journalier*, (*mil.*) daily (morning) report;

—— *de marée*, volume of water brought by the tide into a canal, river, etc.;

—— *du matin*, (*mil.*) morning report;

—— *de mer*, (*nav.*) shipmaster's protest against responsibility for damage to ship or cargo, or for a collision;

pièce de ——, v. s. v. *pièce*;

salle des ——*s, du* ——, (*mil.*) orderly room;

—— *de santé*, (*mil.*) medical return;

——-*situation, situation-*——, (*mil.*) morning report of a company, troop, or battery; report, return;

—— *du soir*, (*mil.*) evening report;

terres de ——, made ground.

rapporté, a., inlaid, inserted, not in one piece with; (*mach.*) with take-up for wear;

pièce ——*e*, inlaid or inserted piece.

rapporter, v. a. n., to bring back, to add; to report, make a report; to state, show; to join, fit in a piece, insert, inlay, join; to repeal; to yield, to produce, bring in; to repay; (of the tides) to increase; (of the sea) to rise; (*drawing*) to set down, trace, protract, plot a survey; (*nav.*) to bring up, to show (so many fathoms, of a sounding line);

—— *une loi, etc.*, to repeal a law;

—— *une pièce*, to inlay; to insert a piece;

—— *des terres*, to fill in with (transported) earth;

—— *un toron*, (*cord.*) to lay in a strand.

rapporteur, m., reporter; secretary of a board, chairman of a commission, etc.; (*mil.*) recorder of a board; in France, the *major* in a *conseil d'administration*, judge-advocate of a court-martial; (*inst.*) protractor, semicircular protractor;

—— *d'atelier*, (*inst.*) bevel protractor;

—— *du budget*, (*adm.*) official charged with the preparation of the budget;

——*-éclimètre*, (*inst.*) clinometer protractor;

—— *à limbe complet*, (*inst.*) circular protractor;

officier ——, (*mil.*) judge-advocate of a court-martial.

rapprochement, m., bringing, placing, putting together; conjunction; junction.

rapprocher, v. a., to bring nearer or together; (of a telescope) to bring up, bring near.

râpure, f., rasping.
râpuroir, m., copper boiler (for potassium nitrate).
raquette, f., racket; snowshoe; sort of saw for ripping curved pieces; (*artif.*) rocket without a stick;
—— *perceuse,* ratchet drill.
raréfaction, f., rarefaction;
tuyau de ——, air pipe.
raréfier, v. a., to rarefy.
ras, a., close, smooth, cropped close (of hair) flush; bare, open, flat; level with the surface of;
à au, —— *de,* to the level of;
à —— *du sol,* flush, level, with the ground;
au —— *de l'eau,* to the water's edge;
bateau ——, flat-bottomed boat;
bâtiment ——, dismasted ship;
——*e campagne,* open, flat, country (no hills, nor trees, rivers, etc.);
en ——*e campagne,* (*mil.*) in the open, in the field (as distinguished from forts, etc.);
mesure ——*e,* strike measure;
à —— *terre,* along the ground, down to the ground.
ras, m., reef (awash); race; punt; floating stage; flat-bottomed boat;
—— *de carène,* floating stage;
—— *de courant,* race;
—— *de marée,* race, tide gate; bore; eagre;
—— *de métal,* (*art.*) face of the tulip of a gun.
rasage, m., shaving.
rasance, f., —— *de la trajectoire,* (*ball.*) flatness of the trajectory.
rasant, a., (*art.*) sweeping, grazing; (*fort.*) rasant;
feu ——, (*art.*) grazing fire (i. e., going along, or but little above, the object fired at);
flanc ——, (*fort.*) flank sweeping the face of the opposite bastion;
fortification ——*e,* (*fort.*) a fortification having little or no command; rasant, low-built work or fortification;
ligne ——*e, v. flanc* ——;
tir ——, (*art.*) horizontal fire;
vue ——*e,* view of an open, flat country.
rase, f., (*nav.*) a mixture of rosin, sulphur, ground glass, etc., to protect the underwater hull against insects; turpentine oil;
eau de ——, oil of turpentine.
rasé, p. p., shaven; razed to the ground; (*nav.*) razeed, dismasted; (*hipp.*) having lost the mark of mouth.
rasement, m., razing to the ground, demolishing, pulling down; (*hipp.*) loss, disappearance, of the mark of mouth; (*art.*) shearing (of the studs, obs.).
raser, v. a., to shave; to dismantle; to raze; to demolish, overthrow; to pass very near to (e. g., the enemy); to lay flat, level with the ground; (*hipp.*) to raze or lose mark of mouth, to lose the mark; (*art.*) to sweep; to demolish by artillery fire;
—— *la campagne,* (*art.*) to sweep the terrain with shot, shells, etc.;
—— *la côte,* (*nav.*) to keep the land aboard;
faire —— *une tuyère,* (*met.*) to diminish the inclination of a twyer;
—— *un navire,* (*nav.*) to cut down a ship; to dismantle a ship (in battle);
—— *au niveau du sol,* (*art.*) to raze to the ground;
se ——, (*r. r.*) to follow the level of the ground;
—— *le tapis,* (*man.*) (of a horse) to go too near the ground.
rase-tapis, m., (*hipp. and slang*) horse that goes too near the ground; daisy cutter.
rasoir, m., razor;
—— *de sûreté,* safety razor.
rassemblage, m., (*mach.*) connection.
rassemblement, m., gathering, assemblage, crowd; (*mil.*) mustering, collecting, assembling, troops preparatory to forming a marching order; re-forming in normal order of troops (infantry) that have been engaged in the dispersed order; (more esp., *Fr. a.*) formation of a squad in line, of a company in column of sections, of a battalion in double column, at six-pace intervals; command for these movements; assembly of skirmishers; (on a large scale) temporary formation of troops between the order of march and that of battle; (*art.*) forming up of a gun detachment, of a battery; command for these movements; (*fenc.*) return to the upright position from that of guard; assemble, finish;

rassemblement *en arrière,* (*fenc.*) return, etc., in which the right is brought back to the left foot;
—— *en avant,* (*fenc.*) return, etc., in which the left is brought up to the right foot;
formation de ——, (*mil.*) v. s. v. *formation;* (*art.*) general term for the assemblage of the various subunits of a battery into greater units, these forming up the battery;
gare de ——, v. s. v. *gare;*
—— *de la garde,* (*mil.*) assembly of guard details; call for same;
—— *au parc,* (*art.*) formation of a battery at its park;
—— *sans matériel,* v. s. v. *formation.*
rassembler, v. a., to gather, collect, draw, bring together; to put together, to mount; (*mil.*) to collect, assemble (troops, an army, material, etc.); to assemble troops for a march formation; (more esp.) to re-form in normal from dispersed order (v. s. v. *rassemblement);* (*art.*) to form up a gun detachment, a battery; (*man.*) to gather a horse; (*fenc.*) to assemble, finish; to return to an upright position from that of guard; m., (*man.*) position of having hind legs under horse, gathering up (operation of bringing the four feet under base of support);
—— *un cheval,* (*man.*) to gather a horse;
—— *les quatre jambes,* (*man.*) (of the horse) to gather (the legs) for a jump;
se ——, (*mil.*) to form in normal from dispersed order; (*art.*) to form up (detachment, battery);
se —— *en arrière (avant),* (*fenc.*) to finish to the rear (front), v. s. v. *rassemblement;*
se faire ——, (*mil. slang*) to get reprimanded or punished.
rasseoir, v. a., to seat again; (*farr.*) to refasten a horseshoe.
rassiéger, v. a., (*mil.*) to besiege anew.
rassis, p. p., reseated; reset; m., (*farr.*) horseshoe set on again with new nails;
pain ——, stale bread;
terre ——*e,* packed, well-settled earth.
rassurer, v. a., to settle, clear up; to secure, strengthen (as a weak bridge).
rastel, m., (*fort.*) ramp (out of the covered way through the glacis).
rastelle, f., v. rastel.
rat, m., rat; (*Polytechnic slang*) cadet who is late at a formation;
—— *de pont,* (*Polytechnic slang*) cadet whose marks are not high enough to put him into the Ponts et Chaussées;
prendre un ——, (*sm. a.*) to miss fire;
queue de ——, v. s. v. *queue.*
ratafia, m., ratafia.
rataplan, m., "rub-a-dub."
ratapoil, m., unconditional partisan of militarism.
raté, p. p., missed; m. (*art., sm. a.*) misfire; (*min.*) misfire;
—— *de charge,* (*art.*) (of the charge) misfire, failure to go off; (more esp.) misfire caused by pushing a reduced charge too far into the breech (powder chamber).
râteau, m., rake; rack; coal fork; ward (of a lock); fair-lead;
—— *à cheval,* hay rake;
faire son ——, (*mil. slang*) of reservists, to remain over time with the regiment by way of punishment.
râtelage, m., raking.
râtelée, f., rakeful.
râteler, v. a., to rake.
râtelier, m., rack; hayrack; tool rack; (*mil.*) armrack, rack of arms;
—— *d'armes,* (*mil.*) gun rack, armrack;
—— *d'écurie,* hayrack;
—— *d'établi,* tool rack;
remettre les armes au ——, (*fig.*) to sheathe one's sword, to retire from the service.

rater, v. a. n., (of firearms in general) to miss fire; of the man behind the gun, to miss.
ratification, f., ratification (of a treaty, etc.).
ratifier, v. a., to ratify (a treaty, etc.).
ration, f., (*mil., etc.*) ration, allowance, (for horses) feed;
— *à la* ——, on ration, on ration list;
— —— *administrative*, regulation ration;
— —— *collective de chauffage*, (*Fr. a.*) daily allowance of fuel for each stove or range and for each coffee boiler;
— —— *diminuée*, short allowance;
demi- ——, half ration;
à la demi- ——, on half allowance;
— —— *double*, double ration;
— —— *entière*, full ration;
— —— *d'entretien*, ration, feed, sufficient to keep in good order (for horses that do not work);
— —— *d'étape(s)*, route ration; travel ration (U.S. A.);
— —— *d'été*, (*Fr. a.*) summer ration (horses);
être à la ——, (*mil.*) to be on short allowance;
— —— *fixe annuelle de chauffage*, (*Fr. a.*) yearly fuel allowance for libraries, schools, etc.;
— —— *forte (de campagne)*, (*Fr. a.*) increased field ration (issued during active operations);
— —— *de fourrage*, feed, ration, of forage;
— —— *de guerre*, (*Fr. a.*) field ration (horses);
— —— *d'hiver*, (*Fr. a.*) winter ration (horses);
— —— *individuelle de chauffage*, (*Fr. a.*) daily allowance of fuel to each man when no range is available, or to each married noncommissioned officer;
marché à la ——, v. s. v. *marché*;
mettre à la ——, (*mil.*) to put on short allowance, on short rations;
mettre à la demi- ——, (*mil.*) to put on half allowance, half ration;
mettre à la petite ——, (*mil.*) to put on a short allowance;
— —— *normal de campagne*, (*Fr. a.*) normal field ration, issued during *stationnements* of some duration and during any period of war when duty is not excessive;
— —— *normale de guerre*, v. —— *forte*;
— —— *de poêle*, fuel allowance per stove for heating barrack rooms;
— —— *réduite*, short allowance;
— —— *de route*, travel ration (horses);
— —— *de tirage*, (*Fr. a.*) feed or allowance for draft horses;
— —— *de transport*, (*Fr. a.*) travel ration or feed (horses).
rationnaire, m., (*mil.*) man who draws or is entitled to draw rations, not necessarily a combatant.
rationné, a., (*mil.*) in rations, i. e., ready for immediate use (v. *marché à la ration*).
rationnement, m., (*mil., etc.*) short allowance of rations; act of putting on short allowance.
rationner, v. a., (*mil., etc.*) to put on rations; to put on short allowance.
ratissage, m., scraping.
ratisse-caisse, m., v. *ratisse-moule*.
ratisse-moule, m., (*fond.*) raking board, scraper.
ratisser, v. a., to rake, scrape.
ratissette, f., smith's tool scraper.
ratissoire, f., rake, scraper.
ratissure, f., scrapings.
rattachage, m., action of reknotting.
rattachement, m., (*surv.*) operation of connecting a point or station with the known point or points.
rattacher, v. a., to tie again, to fasten, link with; (*surv.*) to connect (a point or station) with the known point or points;
— —— *pour ordre*, (*mil.*) to attach for the purpose of defining a status.
rature, f., erasure, word crossed out.
raturer, v. a., to erase, scratch, cross out, blot out.
ravage, m., ravage(s), ruin, devastation, pillage, plundering.

ravager, v. a., to ravage, spoil, waste, lay waste, pillage.
ravageur, m., pillager, plunderer.
ravalement, m., trimming of branches; (*mas.*) rough plastering, rough casting, rough coat of plaster; a synonym of *ragrément à vif*, q. v.; (*carp.*) hollowing out of a piece, sawing out a curve (so as to bring other parts out in relief); (*cons.*) hollow recess.
ravaler, v. a., to swallow; to trim down branches; (*mas.*) to rough cast, put on a rough coat of plaster; (*carp.*) to hollow out, saw out (so as to bring other parts out in relief).
ravaleur, m., (*mas.*) rough caster.
ravinade, m., (*top.*) ravine; dry bed, or stream whose bed is often dry.
ravine, f., torrent, mountain torrent, bed of a torrent.
ravinement, m., scouring, wash (by torrents, etc.), gullying.
raviner, v. a., to scour, to gully.
ravineux, a., gullied.
ravitaillement, m., revictualing; feeding, supplying (as a machine with fuel and water); (*mil.*) revictualing (a fort, a town); esp., supplying of ammunition on battlefield to troops engaged; replenishing of ammunition after a battle; furnishing of supplies in general, (hence) supplies themselves, supply train, ammunition;
— —— *en munitions*, replenishment of ammunition;
— —— *en vivres*, replenishment of victuals.
ravitailler, v. a., to revictual; (*mil.*) to supply (as ammunition, food, etc.); to resupply a town, a fort.
raviver, v. a., to freshen, revive; to stir (and excite) a fire; to urge, touch up; (*expl.*) to cut the end of a Bickford fuse, so as to expose the powder;
— —— *les rayures*, (*art.*) to freshen, recut, the grooves;
— —— *une surface limée*, to refile a surface.
rayage, m., (*art., sm. a.*) rifling; operation of rifling;
banc de ——, rifling bench;
— —— *à côtes*, rib rifling.
rayé, scratched out, erased; (*mil.*) struck off (the rolls); (*art., sm. a.*) rifled.
rayer, v. a., to streak; to rule; to erase, strike off; (*art., sm. a.*) to rifle (*mil.*) to strike off the rolls (of the army);
— —— *des contrôles*, (*mil.*) to strike off of the muster rolls;
— —— *à droite (gauche)*, (*art., sm. a.*) to rifle to the right (left);
machine à ——, (*art., sm. a.*) rifling machine;
tringle à ——, (*art., sm. a.*) rifling rod.
rayère, f., loophole of a tower; wheel race of a mill.
rayon, m., ray (of light, of heat); ray of a star; radius; spoke of a wheel; shelf (of a bookcase, etc.); (*hipp.*) radius; region of the body; (*nav.*) handle (of a steering wheel);
— —— *d'action*, (*mil.*) radius of the area over which a body of troops, an arm (e. g., artillery), or a fort makes its influence felt; (*nav.*) coal radius; (*torp.*) destructive radius;
— —— *d'action d'un navire*, (*nav.*) radius of action, coal radius;
— —— *d'activité*, (*mil.*) in a siege, the area covered by the operations of the defense;
— —— *astronomique*, (*inst.*) Jacob's staff;
— —— *d'attaque*, (*mil.*) in a siege, the area covered by the operations of the attack;
— —— *de bonne rupture*, (*mil. min.*) radius of useful or effective rupture (i. e., distance within which enough of a gallery would be ruined to make it impassable);
— —— *calorifique*, heat ray;
— —— *chimique*, chemical ray;
— —— *de courbure*, radius of curvature;
— —— *de défense*, (*mil.*) radius of action of a fort or fortified position;
— —— *d'entonnoir*, (*mil. min.*) crater radius;
— —— *d'évolution d'un navire*, (*mil.*) tactical radius;
— —— *d'excentricité*, —— *d'excentrique*, (*mach.*) throw of the eccentric; eccentric radius;

rayon *d'explosion*, (*mil. min.*) radius of explosion, of destructive effect;
—— *extérieur*, (*fort.*) line from the center of a place to the flanked angle of a bastion;
—— *d'une forteresse*, (*mil.*) sphere of action, of influence, of a fortress;
—— *de giration*, radius of gyration;
—— *intérieur*, (*fort.*) line from the center of a place to the center of a bastion;
—— *d'investissement*, (*mil.*) in a siege, radius of investment;
jeter les métaux en ——*s*, (*met.*) to cast metal into bars;
—— *majeur*, (*fort.*) v. —— *extérieur;*
—— *mineur*, (*fort.*) v. —— *intérieur;*
—— *myriamétrique*, (*fort.*, Fr. a.) 10-kilometer radius around the principal French forts (v. *servitude*);
—— *d'opération*, (*in gen.*) radius or sphere of operation; (*nav.*) steaming radius, coal endurance, coal radius;
—— *photogénique*, v. —— *chimique;*
—— *régulateur*, (*mach.*) radius bar; bridle rod;
—— *de roue*, arm of a wheel; paddle arm;
—— *de rupture*, (*expl., min.*) radius of rupture, of destruction;
—— *de rupture-limite*, (*mil. min.*) limiting radius of rupture (beyond which a gallery, e. g., would suffer no harm);
—— *tangent*, bicycle spoke;
tige de ——, radius rod;
—— *visuel*, visual ray; (*top.*) line of direction; (*art.*) line of sight.

rayonnement, m., radiation, radiance, beaming; (*surv.*) method by use of polar coordinates from a central station.

rayonner, v. a. r., to radiate, shine, irradiate; to put up shelves; (*fig.*) to make itself felt at a distance (e. g., an army).

rayure, f., stripe, streak, scratch; action of striking off (as a name from a list); (*cons.*) a synonym of *enrayure*, q. v.; (*carp.*) groove, furrow; (*art., sm. a.*) groove, (hence) rifling.
(The following terms relate to firearms:)
—— *alternative*, v. —— *double;*
—— *en anse de panier*, (*art.*) an obsolete form of rifling whose profile resembled a flattened basket handle, whence the name;
arête de la ——, edge of the groove;
—— *à arêtes rectilignes*, straight rifling;
—— *arrondie*, rounded groove;
—— *capillaire*, hair rifling;
—— *de charge*, loading groove;
—— *à chargement automatique*, shunt rifling;
—— *s à cheveu*(*x*), hair rifling;
—— *s en coin*, grooves whose width decreases from the breech to the muzzle;
—— *à colonnes*, rifling in which the lands are rounded off;
—— *concentrique*, groove whose bottom is concentric with the surface of the bore;
—— *à côtes*, grooves in which the lands are narrower than the grooves; rib rifling;
—— *en crémaillère*, rifling the profile of which resembles the teeth of a saw, saw-tooth rifling;
—— *s cunéiformes*, v. —— *s en coin;*
—— *de décharge*, discharging groove;
—— *double*, alternate rifling;
—— *droite*, straight groove, straight rifling;
—— *à droite*, right-hand rifling;
—— *à étoiles*, rifling of which the lands are broader at the base than at the surface of the bore;
—— *excentrique*, a groove the bottom of which is not concentric with the bore; eccentric groove;
—— *à facettes*, a groove with facets;
—— *à flancs rectilignes*, groove the profile of whose flanks is rectilinear;
flanc des ——*s*, v. s. v. *flanc;*
—— *à fond incliné*, shunt groove;
—— *s en forme de coin*, y. —— *s en coin;*
—— *fuyante*, v. —— *à fond incliné;*
—— *gauche*, left-hand rifling;
—— *en hélice*, helical or spiral twist; groove whose development is a right line;

rayure *hélicoïdale*, helicoidal rifling or twist;
inclinaison de la ——, angle made at any point by the tangent to the groove with the axis of the bore;
—— *s merveilleuses*, v. —— *s à cheveu*(*x*);
—— *multiple*, polygroove rifling;
—— *parabolique*, parabolic rifling, groove whose development is a parabola;
—— *s parallèles*, grooves of uniform width;
paroi forçante des ——*s*, driving edge;
pas des ——*s*, pitch of rifling;
—— *à pas constant* (*progressif*), rifling of uniform (increasing) twist;
pas final de la ——, final twist of rifling;
—— *à pas variable*, rifling of varying twist;
—— *peu profonde*, shallow groove;
prendre les ——*s*, *prendre dans les* ——, to take the grooves;
profil de ——, curve or profile obtained by a plane perpendicular to the axis of a gun;
—— *à profondeur décroissante*, v. —— *à fond incliné;*
—— *s à profondeur progressive*, groove of progressive depth;
—— *progressive*, rifling of increasing twist, progressive groove; (also) groove growing shallower toward the muzzle;
rafraîchir les ——*s*, to recut, put a fresh edge on, the grooves;
—— *rectangulaire*, rectangular groove;
—— *en redans*, v. —— *à étoiles;*
—— *rétrécie*, shortened groove, shunt groove;
—— *à rétrécissement*, v. —— *à fond incliné;*
—— *à rochet*, v. —— *à crémaillère;*
sens des ——*s*, direction of rifling;
sortir des ——*s*, to fail to take the grooves;
—— *en spirale*, spiral groove;
tracé de la ——, development of the groove;
—— *uniforme*, groove of constant depth;
—— *en voûte*, groove with rounded edges or bottom.

raz, m., —— *de marée*, v. s. v. *ras*, m.

razzia, f., (*mil.*) raid.

razzier, v. a., (*mil.*) to raid, make a raid.

réa, ria, m., (*cord.*) sheave, sheave of a block.

réactif, a., reactive; m., (*chem.*) reagent.

réaction, f., reaction; (*man.*) jar or jolt of the rider (produced by the motion of the horse); (*chem.*) reaction;
procédé par ——, (*met.*) open-hearth process.

réadmission, f., readmission; (*mil.*) reentry into service of certain n. c. o.'s as *commissionnés*, q. v.

réagir, v. a., to react; (*chem.*) to react.

réaimanter, v. a., (*elec.*) to remagnetize.

réaléser, v. a., to rebore.

réalgar, m., (*chem.*) realgar.

réapprovisionnement, m., replenishing of supplies, etc.; (*sm. a.*) refilling of a magazine.

réapprovisionner, v. a., to replenish (supplies, etc.); (*sm. a.*) to refill a magazine;
—— *en munitions*, (*mil.*) to replenish with ammunition, to refill the ammunition chests;
—— *en vivres*, (*mil.*) to revictual.

réarmer, v. a., to rearm; (*nav.*) to put a vessel into commission again.

réarpentage, m., resurvey.

réarpenter, v. a., to resurvey.

réassurance, f., reinsurance.

réassurer, v. a., to reinsure.

réatteler, v. a., to put horses to, again.

réattaquer, v. a., to reattack.

rebander, v. a., to bind up again (as a wound); (*sm. a.*) to cock (a piece) again.

rebarbe, f., (*tech.*) beard, fash, burr.

rebataille, f., (*mil.*) renewed battle.

rebâtir, v. a., to rebuild;
—— *par le pied*, to underpin.

rebattage, m., any operation of beating up (as a mattress); (*art.*) hammering (of projectiles, so as to remove inequalities).

rebattre, v. a., to beat up, to beat over (a mattress, etc.); to travel, go, over again;
—— *les boulets*, (*art.*) to hammer a projectile (so as to remove seams, inequalities, etc.);
—— *une place*, (*art.*) to turn the guns upon a position anew;
—— *un tonneau*, to set up the hoops of a cask.

rebelle, a., rebellious, rebel; obstinate; ill-adapted; refractory; m., rebel.
métal ——, (*met.*) refractory metal;
substance ——, refractory substance or material.
rebeller, v. a., to rebel, revolt, rise in arms.
rébellion, f., rebellion, revolt, insurrection.
rebiffer, v. r., (*mil.*) to have a soldier-like bearing; to "brace" (U. S. M. A.), (*fam.*).
reboiser, v. a., to replant trees.
rebond, m., rebound.
rebondir, v. n., to rebound.
rebondissant, a, rebounding;
chien ——, (*sm. a.*) rebounding hammer;
platine ——*e*, (*sm. a.*) rebounding lock.
rebondissment, m., rebound, resilience; (*sm. a.*) rebounding of a hammer;
branche de ——, (*sm. a.*) rebounder.
rebondonner, v. a., to replug (a cask).
rebord, m., rim, shoulder, edge, border, flange; (*art.*) guide rail (of the breech tray);
—— *-arrêtoir*, (*mach.*, etc.) a flange acting as a stop;
—— *de chaudière*, boiler stay, boiler flange;
—— *de roue*, (*r. r.*) tire flange;
—— *saillant*, (*mach.*) tappet, wiper, driver, lifter;
—— *de soupapes*, (*steam*) lapping of the valve.
rebotter, v. r., to put on one's boots again.
reboucher, v. a., to plug, cork up, anew.
reboulonner, v. a., to rebolt.
rebours, m., wrong way, wrong side; twist of grain due to wind; a., reverse, contrary;
à ——, *au* ——, against the grain, backward, the wrong way; reversed;
bois ——, *bois à*, *de*, ——, crossgrained wood;
cheval ——, obstinate horse (horse that backs, kicks, etc., that pays no attention to correction).
rebousse, f., starting bolt.
rebrousse-poil, *à* ——, against the hair; (*fig.*) the wrong way.
rebrousser, v. a. n., to set, or turn, back; to make no impression (of a cutting tool);
—— *chemin*, to retrace one's road; to turn back;
faire ——, to turn up the edge (of a cutting tool).
rebroyer, v. a., to grind over again.
rebut, m., rejection; waste, rubbish, anything rejected, rejected piece; fag end; repulsion;
arme de ——, condemned arm;
bois de ——, refuse timber;
bureau des ——*s*, dead-letter office;
canons de ——, condemned guns;
charbon de ——, refuse coal;
de ——, condemned, waste;
être au ——, to be thrown aside;
mettre au ——, to throw aside, condemn.
rebuté, p. p., condemned, rejected (as stores, articles, etc.); (of horses) worn out by fatigue; (*mil.*) disgusted, discouraged.
rebuter, v. a., to reject, condemn; to reject after testing; to refuse as unserviceable; (*mil.*) to discourage (the enemy, as by repeated repulses);
—— *un cheval*, to make a horse obstinate by overworking him.
recaler, v. a., to smooth, level.
recalescence, f., (*met.*) recalescence.
recalfatage, m., recalking.
recalfater, v. a., to calk over again.
recalibrer, v. a., (*sm. a.*) to resize (a cartridge case).
recaloir, m., smoothing plane.
récapitulatif, a., recapitulative.
état ——, v. s. v. *état*.
recarboniser, v. a., (*met.*) to recarbonize.
recarburation, f., (*met.*) recarburization.
recarburer, v. a., (*met.*) to recarburize, to recarbonize.
recarburisation, f., (*met.*) recarburization.
recarreler, v. a., to repave, pave anew.

recel, m., (*law*) receiving stolen goods.
recèlement, m., concealing; (*law*) concealment of stolen goods; (*mil.*) concealment of deserters;
lieu de ——, "fence."
receler, v. a., to hide, conceal; (*law*) to receive, conceal, stolen goods; (*mil.*) to conceal deserters.
receleur, m., (*law*) concealer of stolen goods, "fence."
recensement, m., census; return; new inventory, fresh examination (as of weights, etc.);
tableaux de ——, (*Fr. a.*) annual list of young men who reached the age of 20 in the year preceding.
recenser, v. a., to take a census; to check off.
recenseur, m., enumerator (of a census).
recepage, m., cutting down.
recepée, f., part of a wood cut down to the ground, clearing.
receper, v. a., to cut down (trees) to the ground, to the stump; to cut off close to the ground, under water (as a pile); to cut off, saw off, the top of a stake, the top of a pile, when driven home; to bring a broken pile, stake, etc., to a required height by fitting on a new piece;
machine à ——, machine for cutting away stakes;
—— *une muraille*, (*mas.*) to cut down a wall to the ground.
récépissé, m., receipt, formal receipt, acknowledgment of receipt;
—— *comptable*, (*Fr. a. adm.*) receipt for stores received (in two parts: one delivered to the issuing party, the other (*talon de récépissé*), kept by the *comptable* or regiment as a voucher);
—— *provisoire*, (*Fr. a. adm.*) receipt for stores received with respect to weight, number of packages, and to external appearance;
talon de ——, v. *comptable*.
réceptacle, m., receptacle; basin;
—— *de la vapeur*, (*steam*) steam chamber.
récepteur, m., receiver; (*mach.*) receiver, steam receiver; working wheel (of the Fourneyron turbine); (telephone) receiver; (*teleg.*) receiver, recorder, register, ink writer, of Morse's telegraphic instrument; indicator;
—— *à aiguille* (*teleg.*) needle receiver;
—— *à cadran*, (*teleg.*) dial indicator;
—— *hydraulique*, (*mach.*, etc.) hydraulic motor;
—— *Morse écrivant*, (*teleg.*) ink writer;
—— *Morse à pointe sèche*, (*teleg.*) Morse dry-point receiver;
—— *à noir fumée* (*teleg.*) carbon (lampblack) recorder;
—— *à vapeur*, (*steam*) steam chamber.
réception, f., receipt (of goods, horses, stores, etc.); reception, acceptance, receiving; (*teleg. sig.*) "acknowledgment;" (*mil.*) (shortened expression for) all measures of verification, etc., followed on the receipt of supplies, guns, ammunition;
accuser —— *de*, to acknowledge receipt of;
—— *à la cour*, levee, drawing-room;
épreuve de ——, acceptance test;
tir de ——, (*art.*, *sm. a.*) acceptance firing;
—— *et vérification des bouches à feu*, (*art.*) reception and proof of guns.
réceptrice, f., (*elec.*) dynamo driven as a motor.
recerclage, m., rehooping (of casks).
recercler, v. a., to new hoop, hoop over again.
recette, f., receipt; recipe; prescription; receipts (money); receipt (of stores, etc.); acceptance;
—— *des armes à feu*, etc., (*sm. a.*) examination of small arms, followed by acceptance on approval;
bois de ——, timber fit for service;
—— *de câble*, landfall of a cable;
épreuve de ——, acceptance test.
recevoir, v. a., to receive (a letter, present, etc., an ambassador); to accept (after examination and test); to admit (to a school); to hold a reception, a levee, a drawing-room; (*mil.*) to receive the enemy; (*mil. sig.*) to receive;
—— *un abatage*, (*mil. slang*) to receive a reprimand; "to be jumped all over" (U. S. A.);

recevoir, aller —— les ordres de ...., to report to .... for orders;
—— *une balle,* to be hit by a bullet;
—— *la bataille, (mil.)* to accept battle;
—— *à bord, (nav.)* to receive on board, to ship;
—— *un coup d'épée,* to get a sword thrust;
—— *un coup de vent au mouillage, (nav.)* to ride out a gale;
—— *dans son grade,* v. —— *un officier;*
—— *une lame, (nav.)* to be struck by a sea;
—— *le mot d'ordre, (mil.)* v. *prendre le* ——, s. v. *mot;* to receive the *mot* from a sentinel, etc.;
—— *un officier,* (Fr. a.) to receive an officer (formality upon his joining after a promotion);
se ——, *(man.)* (of a horse) to land after a jump.

rechange, m., spare part; duplicate (generally in the plural);
de ——, spare;
—— *s d'école,* (*Fr. art.*) spare parts for target practice (*école à feu*);
pièce de ——, duplicate;
roue de ——, (*art.*) spare wheel.

rechargement, m., reloading; remetaling of a road; reballasting of a railway.

recharger, v. r., to reload; to put on a load or burden again, to lade again; to restore the worn parts of a (metal) tool; to remetal a road; (r. r.) to reballast; (*mil.*) to recharge, make a fresh charge; (*art., sm. a.*) to reload;
—— *un essieu,* to line an axletree, to repair it where it is much worn by friction;
—— *un plancher,* to renew the worn parts of a floor;
—— *un prisonnier par un nouvel écrou,* (law) to reconfine a prisoner.

rechasser, v. a., to drive back.

réchaud, m., chafing dish;
—— *de rempart,* (*fort.*) rampart grate.

réchauffage, m., (*met., etc.*) reheating.

réchauffement, m., heating (as feed water).

réchauffer, v. a., to heat, to heat over again;
—— *un cheval,* (*man.*) to warm a horse up to his work;
—— *la machine,* to warm the engine.

réchauffeur, m., (*steam*) feed-water heater; *tube* ——, feed-water heater.

rechausser, v. a., to shoe over again; to bank up the foot of a tree;
—— *un mur,* to line the foot of a wall, underpin it, strengthen it;
—— *une roue,* (*mach.*) to new cog a wheel.

rêche, a., rough or hard to the touch.

recherche, f., search, pursuit, investigation, inquiry, research; repair of a roof, of a pavement;
—— *des déserteurs,* (*mil.*) search for deserters;
—— *du lieu de défaut,* (*elec.*) localization of a fault;
—— *minutieuse,* minute investigation, inquiry.

rechercher, v. a., to search for (into, out); to inquire after; (*chem.*) to test for;
—— *un cheval,* (*man.*) to animate a horse;
—— *la longe,* (*man.*) to make a horse take the full length of the longing rein.

rechevillé, a., rebolted.

rechute, f., relapse (of an illness); (*fort.*) bonnette.

récidive, f., repetition of an offense or crime after condemnation for the same previous offense.

récidiver, v. n., to repeat an offense or crime.

récidiviste, m., old offender.

récif, m., (*hydr.*) reef;
—— *barrière,* barrier reef;
—— *en bordure,* fringing reef;
—— *de côte,* v. —— *en bordure.*

récipiangle, m., recipiangle (obs.).

récipiendaire, m., person received formally into the Legion of Honor, the Academy, into a grade, etc.

récipient, m., recipient; receiver; well, cistern; condenser;
—— *-déjecteur,* (*steam*) settling receiver;
—— *d'eau chaude,* (*steam*) hot well;
—— *à huile,* oil receiver.

réciprocation, f., (*mach.*) reciprocation.

réciprocité, f., (*mach.*) reciprocating motion.

réciproque, a., reciprocal; (*mach.*) reciprocating; reversible.

recirer, v. a., to wax again;
—— *un joint,* (*mas.*) in pointing, to smooth the mortar in a joint.

réciter, v. a., to recite;
—— *la rose des vents,* to box the compass.

réclamateur, m., claimant.

réclamation, f., claim, demand; objection, complaint; protest.

réclamer, v. a., to claim, demand, reclaim; to object, make objection; to protest against;
course à ——, selling race.

reclamper, v. a., to fish a mast, etc., when sprung.

reclasser, v. a., to reclass (a ship).

réclinaison, f., inclination toward the horizon.

reclouer, v. a., to nail on again, to nail up again.

réclusion, f., solitary confinement, imprisonment, in a penitentiary.

récolement, m., examination, verification; verification of an inventory, of the building line of a house;
—— *de bois,* verification of a fall of timber.

récoler, v. a., to examine, verify.

recoller, v. a., to paste, glue, on again.

récolte, f., harvest;
—— *sur pied,* standing crop.

récolter, v. a., to harvest.

recombattre, v. a., to fight anew.

recomblement, m., (*mil. min.*) falling back of earth into a crater.

recombler, v. a., to fill in again.

recommandé, p. p., registered (of a letter).

récompense, f., reward, recompense; compensation, indemnity, amends.

récompenser, v. a., to reward, recompense, compensate, indemnify, repay, make up for.

reconduire, v. a., to take back, to accompany (a visitor) to the door;
se faire ——, to be compelled to beat a hasty retreat;
—— *à la frontière,* to expel (a foreigner) from the country.

reconnaissance, f., recognition; recognition of one government by another; examination, verification, inspection; exploration of a country; acknowledgment of indebtedness, "I O U;" (*nav.*) generic term for buoys, etc.; (*mil.*) reconnaissance, reconnoitering party;
(The following expressions are military:)
aller en ——, to go on a reconnaissance;
—— *de champ de bataille,* one having for its object discovery of enemy's position, etc.;
de ——, reconnoitering;
envoyer en ——, to send on a reconnaissance;
être en ——, to be reconnoitering, on a reconnaissance;
faire la —— *de,* to reconnoiter, examine, verify, inspect;
faire une ——, to reconnoiter, make a reconnaissance;
—— *en force,* reconnaissance in force;
—— *générale,* reconnaissance, exploration, made in time of peace and having for its object the collection of information in general about a probable theater of war or of operations;
—— *journalière,* (*mil.*) daily reconnaissance necessary for safety of camps, posts, etc.;
levé de ——, reconnaissance survey;
—— *offensive,* reconnaissance in force, to compel the enemy to show his hand and to deploy in his fighting position; reconnaissance made by the advance guard;
—— *d'officiers,* reconnaissance by officers (in the exploration service of independent cavalry);
—— *ordinaire,* a reconnaissance having for its object the nature of the ground, the roads, and local resources, and the evident intentions of the enemy;

reconnaissance *spéciale*, reconnaissance in the neighborhood of enemy (as distinguished from —— *générale*); reconnaissance with a special object; reconnaissance having for its object the special reconnoitering of the terrain, with reference to the march of the troops, of the various positions to be occupied either offensively or defensively, of the position and forces of the enemy.

reconnaître, v. a., to recognize; to recognize a government; to acknowledge, proclaim; to reward, recompense, return; to explore, discover, find out; (*mil.*) to reconnoiter, make reconnaissance; to challenge (patrols, rounds, etc.);
—— *un écueil*, (*hydr.*) to take the bearings of a reef;
faire ——, to proclaim;
faire —— *un officier*, (*mil.*) to proclaim an officer;
faire —— *les rondes*, (*mil.*) to challenge the rounds;
se ——, to get one's bearings;
—— *un vaisseau*, (*nav.*) to approach and hail a ship (with a view to learning its nationality, etc.).

reconquérir, v. a., to reconquer.

reconquête, f., reconquest.

reconstitué, p. p., (*mil.*) (of troops) reformed, as after an attack.

reconstituer, v. a., to reconstitute, reorganize.

reconstitution, f., reconstitution, reorganization.

reconstruction, f., rebuilding.

reconstruire, v. a., to rebuild.

record, m., record (sports).

recorder, v. a., to tie up again; (*cord.*) to lay up a rope anew.

recoupe, f., waste, clippings, stone chips; rubble; pollard, first bran; second cutting (of clover, hay, etc.).

recoupement, m., offset, recess in a wall or building; (in range finding) intersection; (*surv.*) resection.

recouper, v. a., to recut; (*fort.*) to trim, to pare; (*surv.*) to fix by intersection.

recoupette, f., sharps, middlings.

recoupon, m., v. *recoupette*.

recourber, v. a., to bend around or down.

recourir, v. a. n., to run again; to resort, have recourse to; apply to, betake one's self;
—— *un câble*, (*cord.*) to underrun a cable, a rope;
—— *les coutures*, (*mil.*) run over the seams (as of a pontoon boat), or to calk them slightly and expeditiously;
faire —— *une manœuvre*, to take a rope to the point to which it should go;
—— (*sur*) *une manœuvre*, v. —— *un câble*;
—— *un prisonnier*, to deliver a prisoner.

recours, m., recourse; resort; resource; refuge; help;
avoir —— *à*, to resort to, to apply to;
—— *en cassation*, (*law*) writ of error;
—— *en grâce*, (*law*) petition for pardon, for mitigation of punishment;
—— *en revision*, (*Fr. a.*) appeal for retrial or rehearing of a case.

recousse, f., (*nav.*) recapture;
droit de ——, salvage on recapture;
prix de ——, salvage money upon recapture.

recouvert, p. p., covered, overcast again; recovered; (*mas.*) not visible (of joints); covered with plaster (as a partition); (*elec.*) insulated (of wire).

recouvrement, m., recovery (of debts, of health, etc.); collection; covering, action of covering; overlap, overlapping, overlapping plate or strip; part of a stone or tile covering a joint; cap (e. g., of a telescope); (*harn.*) cover (of a saddlebag), hood of the American cavalry stirrup; (*mas.*) coat of plaster on a partition; (*steam*) lap of a slide valve; (*sm. a.*) lip of a cartridge carrier;
à ——, by overlapping;
—— *d'avarie particulière*, particular average;

recouvrement, *bande de* ——, butt strap; (*art.*) cheek plate, garnish plate; (*mach.*) flap joint;
—— *du chemin*, (*r. r.*) ballasting;
—— *du côté de l'admission*, (*steam*) steam lap;
—— *du côté de l'émission*, (*steam*) exhaust lap;
—— *en dedans*, (*steam*) inside lap;
—— *en dehors*, (*steam*) outside lap;
être à ——, to lap over, to be lapping over;
—— *à l'évacuation*, (*steam*) inside lap, exhaust lap;
—— *extérieur*, v. —— *en dehors*;
—— *des flancs*, (*mil.*) flanking party;
—— *de flasque*, (*art.*) garnish plate;
—— *intérieur*, v. —— *en dedans*;
—— *à l'introduction*, (*steam*) outside lap, steam lap;
—— *négatif*, (*steam*) negative lap;
plaque de ——, garnish, cover, plate;
poser à ——, to lap over;
—— *positif*, (*steam*) positive lap;
souder à ——, (*met.*) to lap-weld;
soudure à ——, (*met.*) lap-welding;
—— *de talus de flasque*, (*art.*) upper garnish plate on cheeks of a traveling carriage;
—— *du tiroir*, (*steam*) lap of a slide valve;
—— *de tôles*, overlap of plating.

recouvrer, v. a., to recover, regain; get in, levy, collect; (*cord.*) to haul in a rope hanging slack in the water (as into a pontoon boat).

recouvrir, v. a., to cover, cover anew, cover over, conceal, mask; to wash metals; (*adm.*) to cover into the Treasury;
se ——, to become cloudy again.

recrépiment, m., v. *recrépissage*.

recrépir, v. a., to parget, rough-cast again; to rough-coat again; to replaster.

recrépissage, m., pargetting, rough coating again.

recroiser, v. a., to cross again; to weave in and out (as in making gabions); (*hipp.*) to cross again;
—— *les joints*, to break joints.

recroqueviller, v. r., to shrivel (under the influence of heat).

récrouer, v. a., to re-imprison.

récrouir, v. a., (*met.*) to reheat; (*cord.*) to ret.

recru, a., tired out, worn out, weary; jaded, spent.

recru, m., fresh growth, shoots on trunks of trees.

recrue, f., (*mil.*) new levy, recruit, recruits, body of recruits; recruiting;
aller en ——, to go on recruiting service;
dresser la ——, to drill, instruct, recruits;
faire des ——*s*, to recruit, to enlist men;
former la ——, to set up a recruit;
lever des ——*s*, to raise recruits.

recrutement, m., (*mil.*) recruiting, enlisting;
—— *de l'armée de mer*, (*nav.*) recruiting for the navy;
classe de ——, (*Fr. a.*) young men summoned to draw for military service;
service de ——, (*Fr. a.*) recruiting service (in charge of the yearly enrollments and dispatch to regiments).

recruter, v. a., (*mil.*) to recruit;
—— *des soldats*, to beat up recruits, soldiers.

recruteur, m., (*mil.*) recruiter, recruiting officer.

rectangle, a., rectangular; m., rectangle;
—— *enveloppe*, (*ball.*) rectangle inclosing all the shots; (*sm. a.*) packet of revolver cartridges;
—— *en papier*, (*Fr. art.*) paper wrapping for cartridge bags;
—— *de probabilité*, (*ball.*) probable rectangle.

rectangulaire, a., rectangular;
—— *de champ*, (of a beam) set up on edge;
—— *à plat*, (of a beam) laid flat.

rectifiable, a., susceptible of correction; rectifiable; (*inst.*) adjustable;
limbe ——, (*inst.*) adjustable limb.

rectificateur, m., (*art.*) workman who verifies and assures axial direction in gun boring.

rectification, f., straightening (as of a road), correction; rectification of a curve; rectification (of spirits); (*inst.*) adjustment; (*mil.*) rectification of an alignment;
— *de route* (*mil.*) measures taken to prevent two columns from meeting when their roads touch.

rectifier, v. a., to straighten; to rectify, adjust, correct; to rectify (a curve, spirits); (*inst.*) to adjust an instrument; (*mach.*) to grind true;
— *l'alignement*, (*mil.*) to rectify the alignment;
— *le tir*, (*art.*) to correct the range.

rectiligne, a., rectilinear, right-lined;
à mouvement —, (*sm. a.*) straight pull.

recto, m., right-hand page; first page of a leaflet, odd-numbered page.

reçu, p. p., received;
être —, to be admitted (to a school after examination); to be admitted (formality, as to the French Academy);
être — *le premier*, to pass first (as for the Polytechnique);

reçu, m., receipt;
au — *de*, on receipt of;
sur —, after passing receipts.

recueil, m., collection.

recueillir, v. a., to gather, collect; to harvest; (*cord.*) to wind string on a spool;
— *un mur*, (*cons.*) to connect a repaired wall with that above (e. g., a face or party wall).

recueilloir, m., (*cord.*) rope maker's top.

recuire, v. a., to cook over; (*met.*) to anneal, let down; to heat before cooling;
— *au bleu, etc.*, (*met.*) to heat up to a blue, etc., heat;
four, fourneau, à —, (*met.*) annealing furnace;
— *à l'huile flambante*, (*met.*) to blaze off.

recuisson, m., (*met.*) annealing.

recuit, m., (*met.*) annealing;
— *après temps*, (*met.*) annealing proper;
couleur de —, (*met.*) annealing color;
— *au bleu, etc.*, (*met.*) annealing by heating to a blue, etc., heat;
— *à l'huile flambante*, (*met.*) blazing off.

recul, m., recoil in general; (*art., sm. a.*) recoil; kick (of a rifle); (*nav.*) slip of the screw; (*elec.*) fall of the armature of an arc-light regulator;
aller au —, (*art.*) to go "from battery;"
angle de —, (*art.*) angle of recoil;
— *apparent*, (*nav.*) apparent slip of the screw;
— *arrêté*, (*art.*) restrained recoil;
être au —, (*art.*) to be "from battery;"
— *de l'hélice*, (*nav.*) slip of the screw;
— *libre*, (*art.*) free recoil;
mettre au —, (*art.*) to put "from battery;"
mise au —, (*art.*) putting "from battery;"
— *négatif*, (*nav.*) negative slip of the screw;
palan de —, (*cord.*) relieving tackle;
— *réel*, (*nav.*) true slip of the screw;
sans —, (*art.*) nonrecoil;
à — *supprimé*, (*art.*) nonrecoil, nonrecoiling;
supprimer le —, (*art.*) to take up or check recoil;
vent de —, wind shifting through west from south-southwest to north-northwest or north, and back again.

reculade, f., retreat, retrograde movement, recoiling, falling back, slipping back; backing (of a carriage).

reculement, m., backing (of carriages); (*harn.*) breeching; (*cons.*) batter of a wall.

reculer, v. a. n., to back, carry back, move back; to delay, put back; to go backward; to retire, recede, withdraw, give way; to back (of a carriage, of a horse); to back, rein back, a horse, a team; (*mil.*) to retire, fall back; (*art.*) to recoil; (*sm. a.*) to recoil, kick;
faire —, to drive, push, back;
— *les frontières*, to extend a state, to "expand."

reculer, m., (*man.*) backing, act of backing.

reculons, à —, adv., backward, the wrong way; *aller à* —, to go backward; *marcher à* —, v. *aller à* —.

récupérateur, m., (*art.*) recuperator (of a carriage); (*met.*) recuperator;
— *de frein*, (*art.*) recuperator;
— *d'inertie*, (*mach.*) inertia recuperator;
— *à ressort*, (*art.*) spring recuperator.

récurer, v. a., to scour metal, to do bright work.

récusable, a., (*law*) that may be challenged (as a witness).

récusation, f., (*law*) plea in bar of trial; denial of jurisdiction, of the competency of a witness, of an expert; exception.

récuser, v. a., (*law*) to challenge, take exception to.

rédacteur, m., person who draws up a paper, a report, proceedings, etc.; editor of a newspaper.

rédaction, f., preparation of a report, etc.; editing of a newspaper; editorial staff; (*mil.*) drawing up of an order; (*top. drawing*) construction of a map from notes;
travail de —, (*top. drawing*) construction of a map from notes; indoor work.

redan, m., notch; (*cons.*) skew back; recess in a pilaster or vertical chain (to form a bond with the masonry of the main wall); set-off; (*in pl.*) corbel steps; (*fort.*) redan;
batterie à —*s*, (*fort.*) indented battery;
double —, (*fort.*) double redan, i. e., a redan flanking its own faces;
— *à flancs*, v. *double* —;
ligne à —*s*, (*fort.*) indented line;
— *renforcé*, (*fort.*) v. *double* —.

reddition, f., submitting an account for audit; (*mil.*) surrender of a town, of a fortified place.

rédemption, f., redemption, ransom.

redevable, a., m., owing; debtor.

redevance, f., rent; dues, sum payable at fixed intervals.

rédhibition, f., (*law*) action to set aside a contract of sale.

rédhibitoire, a., (*law*) redhibitory, setting aside a contract of sale;
cas —, (*hipp.*) disease or defect that annuls the sale of a horse (is applied to all domestic animals);
vice —, v. *cas* —.

redingote, f., frock coat;
— *croisée*, double-breasted coat;
— *droite*, single-breasted coat;
— *de voyage*, riding coat.

redonner, v. a., to give back; (*mil.*) to charge, fall on, again.

redoublage, m., (*nav.*) resheathing.

redoublé, p. p., doubled, redoubled; increased;
pas —, (*mil.*) double-quick time, step, march.

redoublement, m., increase; (*fenc.*) two (or more) attacks in quick succession from the same lunge, following a parry without *riposte*.

redoubler, v. a., to double, redouble, increase; to lie over again; (*fenc.*) to make a succession of attacks from the same lunge without rising.

redoute, f., (*fort.*) redoubt;
— *batterie*, batterie- —, in semipermanent fortification, redoubt in the intervals between the forts of a line;
— *blindée*, blockhouse;
— *carrée*, square redoubt;
— *casematée*, casemated bomb-proof redoubt;
— *circulaire*, circular redoubt;
— *de compagnie*, redoubt (standard) for one company;
— *couverte*, v. — *blindée*;
— *en croix*, cross redoubt;
demi- —, half redoubt;
— *à mâchicoulis*, redoubt of several stories, the upper projecting beyond the wall (so as to see and defend the bottom);
ligne à demi- —*s*, v. s. v. *ligne*.

redressage, m., straightening, as of a bar, etc.

redresse, f., (*cord.*) relieving rope;
câble de ——, relieving cable.
redressé, p. p., straightened; (*elec.*) unidirected, commuted, redressed;
courant ——, (*elec.*) commuted current.
redressement, m., straightening; making straight; new erection; unwarping; (*fig.*) redress, reparation;
—— *d'un courant,* (*elec.*) commutation of a current.
redresser, v. a., to straighten; to make straight again; to erect, reset, set up again; to rectify, redress, put, set, to right; (make) right; to repair, relieve, remedy, correct, adjust; (*man.*) to train over again;
—— *les armes,* (*sm. a.*) to "recover arms;"
—— *les courants* (*alternatifs*), (*elec.*) to commute currents;
—— *les rayures,* (*art.*) to freshen the grooves by recutting;
se ——, (*drill*) to resume one's original direction.
redresseur, m., —— *de courant,* (*elec.*) commutator.
réducteur, a., (*chem., met.*) reducing (as agent).
réduction, f., reduction; (*chem., met.*) reduction; (*mil. and in gen.*) subjugation; reduction of a position, fortress, etc.;
—— *au centre,* (*surv.*) reduction to the center;
—— *de la course du tiroir,* (*mach.*) linking up;
—— *à l'horizon,* reduction to the horizon;
—— *de pression,* reduction, lowering of pressure;
—— *de la route,* (*nav.*) working up the day's run;
—— *au sommet,* (*surv.*) in leveling, reduction of zenith distance to summit of station.
réduire, v. a., to reduce, change; to subdue, subjugate, subject; to compel, oblige, force; to diminish, abate, cut down; (*chem., met.*) to reduce; (*mil.*) to reduce a place;
—— *l'acier,* (*met.*) to reduce steel to iron;
—— *en cendres,* to reduce to ashes;
—— *un cheval,* (*man.*) to break a horse;
—— *à l'horizon,* to reduce to the horizon;
—— *un plan,* to reduce a plan to a smaller scale;
—— *en poudre,* to grind, reduce, to powder;
—— *en poussière,* to crumble to dust;
—— *la pression,* to reduce, lower, a pressure;
—— *la ration,* (*mil.*) to reduce or cut down the ration;
—— *la route,* (*nav.*) to work up the course and distance;
—— *au silence,* (*art.*) to silence (a battery, etc.);
—— *au sommet,* in leveling, to reduce to the summit;
—— *la vitesse,* to cut down the speed of, to slow down.
réduit, p. p., reduced, diminished;
à nomore ——, (*drill*) with diminished numbers;
carte ——*e,* map on Mercator's projection.
réduit, m., (*fort.*) keep, redoubt; (*nav.*) redoubt;
—— *blockhaus,* (*fort.*) blockhouse in the reentering place of arms;
—— *central,* (*nav.*) central redoubt; central battery, keep, or casemate in certain types of armored ships;
—— *de groupes de batteries,* (*fort.*) a redoubt subserving a group of detached batteries at close intervals;
—— *de demi-lune,* (*fort.*) demilune redoubt;
—— *de place,* (*fort.*) keep of a fort;
—— *de place d'armes,* (*fort.*) redoubt of the place of arms;
—— *de position,* (*mil.*) keep of any position (village, tête de pont), to which one may fall back until succor comes, or in order to prolong defense;
—— *du ravelin,* (*fort.*) ravelin redoubt.
réel, a., real; (*law*) real (as distinguished from personal);
offres ——*les,* cash offer.
réenclenchement, m., (*mach.*) recoupling (of separated parts).
réengagement, m., (*mil.*) reenlistment.
réengager, v. a., (*mil.*) to reenlist.

réfaction, f., doing up again; (commercially) allowance;
—— *de droits,* rebate.
refaire, v. a., to do or make over again, to remake, repair, mend; to begin again; (*mil.*) to rest, refresh (troops);
—— *un cheval,* to make a horse;
se ——, to recruit one's strength; to "pick up."
refait, p. p., done over again; (*cord.*) twice laid;
bois ——, (*cons.*) squared timber, prepared for use and dressed upon all sides;
cheval ——, made horse;
cordage ——, (*cord.*) twice-laid cordage or rope.
réfection, f., repair, reconstruction; (*sm. a.*) preparation of cartridge cases for reloading.
réfectionner, v. a., (*sm. a.*) to prepare cartridge cases for reloading.
réfectoire, m., mess room.
refend, m., splitting; (*carp.*) strip removed from a plank, etc., that is too large; (*cons.*) party wall, partition wall; (*mas.*) joint of a wall;
bois de ——, cleft, hewn, wood; quartered timber; timber sawn lengthwise into several pieces;
mur de ——, (*cons.*) partition wall, bearing wall;
pierre de ——, (*cons.*) corner stone.
refendage, m., (*met.*) slitting of iron.
refendre, v. a., to split, cleave; to saw wood, stone, etc., lengthwise; to quarter timber; to saw stone into slabs.
refente, f., splitting, sawing, lengthwise.
referrer, v. a., (*farr.*) to reshoe; to new-shoe.
refeuillement, m., (*carp.*) rabbet made *in situ.*
refeuiller, v. a., (*carp.*) to make a double-lapping rabbet.
refeuillure, f., (*carp.*) double-lapping rabbet.
reficher, v. a., to sink in anew; (*mas.*) to pick out the joints of a wall (in repairing the pointing).
réfléchir, v. a. n., to reflect, be reflected.
réfléchissement, m., reflection.
réflecteur, m., reflector; (*sm. a.*) reflector for inspecting barrels;
—— *à miroir,* (*sm. a.*) reflector for inspecting barrels;
—— *parabolique,* parabolic reflector.
reflet, m., reflection.
refléter, v. a., to reflect.
réflexion, f., reflection.
instrument à ——, reflecting instrument.
refluer, v. n., flow back, run back (of the tide), to ebb; to flow.
reflux, m., ebb, ebb tide, reflux.
refonçage, m., reheading of a cask.
refondre, v. a., to remelt, melt down; to overhaul thoroughly; to make thorough repairs; to correct, improve, remodel.
refonte, f., remelting, recasting, remolding; thorough repair; overhauling.
reforer, v. a., to bore up, ream out, enlarge.
reforeur, m., borer, workman who bores gun barrels.
réforme, f., reform, reformation; (*Fr. a.*) condemnation of material; casting of horses; enforced retirement of officers and of men (from physical disability or from unworthiness to serve);
chevaux de ——, cast horses;
—— *de chevaux,* casting of horses;
congé de ——, v. s. v. *congé;*
de ——, cast, condemned;
être en ——, (*Fr. a.*) to be on the compulsory retirement list (either from misconduct or from physical disability);
mettre à la ——, v. s. v. *mettre;*
traitement de ——, (*Fr. a.*) reduced pay.
réformé, p. p., (*Fr. a.*) condemned (of material); cast (of horses); on the compulsory retirement list (officers and men);

réformé, *officier* ——, (*Fr. a.*) officer on the compulsory retirement list (physical disability or unworthiness to serve).

reformer, v. a. r., (*mil.*) to re-form, form up again; to be re-formed (as after an assault).

réformer, v. a. r., to reform, rectify, correct, retrench; (*mil.*) to retire for physical disability, for unworthiness to serve; to cast (a horse); to condemn (stores, equipment);

—— *un arrêt, etc.*, (*law*) to modify a decision, etc.;
—— *des chevaux*, (*mil.*) to cast horses;
—— *un officier*, (*Fr. a.*) to put an officer on reduced pay (v. s. v., *réforme*);
—— *un soldat*, (*Fr. a.*) to give a soldier his *congé de réforme*;
—— *des troupes*, (*mil.*) to disband, discharge, soldiers; to reduce numbers.

refortifier, v. a., to refortify.

refouillement, m., (*carp.*) hollow in a timber; (*mas.*) polyhedral hollow in, or hollowing of, a stone.

refouiller, v. a., to carve, chisel, mortise out (as in a timber, a stone).

refoulage, m., (*met.*) jumping up, upsetting.

refoulement, m., stemming (of the tide, of a current); ebbing (of the tide); flowing back (of a current); driving back, compression; forcing; seating or setting (by hammering); backing (of a train, of a locomotice); delivery of a pipe or conduit; (*met.*) jumping up, upsetting; flow of metal; (*art.*) enlargement of the chamber; jumping up, upsetting, of a projectile; ramming down of the charge; (*steam*) compression;

—— *d'un bandage*, jumping or slipping of a tire;
—— *d'une bouche à feu*, (*art.*) enlargement of the chamber;
période de ——, (*mach.*) exhaust period (of a hot-air engine);
tuyau de ——, delivery pipe.

refouler, v. a. n., to stem (the tide, a current); to fall, ebb (of the tide); to flow back, cause to flow back; to back, drive home, or back; (of a pin, etc.) to refuse to go in, to go the wrong way; to start, to drive; to drive out (a pin, bolt, etc.); to upset (a pin, etc.); to deliver (of a pipe, etc.); to force; to seat, set (by hammering); (*art.*) to ram the charge home; to seat a projectile; to upset a projectile; (*met.*) to hammer on all sides so as to unite the particles; to cause to flow; to jump up, upset; (*mil.*) to drive back, repel (outposts, the enemy); (*man.*) to cause a horse to throw his weight back, to bring a horse back;

—— *le canon de fusil*, (*sm. a.*) to jump a gun barrel;
—— *la charge*, (*art.*) to ram down the charge;
—— *une cheville*, to jump a pin;
—— *l'éponge*, (*farr.*) to jump a horseshoe (by hammering on the end of the branch);
—— *un rivet*, to close a rivet;
se ——, (*art.*) of a projectile, to become upset.

refouloir, m., force pump.

refouloir, m., jumper, monkey; ram; (*farr.*) farrier's jumping hammer; (*art.*) rammer; instrument used for jumping up a vent bushing; (*min.*) rammer (for loading mines);

—— *articulé*, (*art., mil. min.*) jointed rammer;
—— *écouvillon*, (*art.*) sponge and rammer staff;
—— *à expansion*, (*art.*) rammer with expanding head, as for taking gutta-percha impressions;
hampe de ——, (*art.*) rammer staff;
—— *hydraulique*, (*art.*) hydraulic rammer; (*min.*) tamping bar; (*met.*) plate on which the bloom or loop is hammered to clear off scoriæ, etc., before going under hammer;
—— *télescopique*, (*art.*) telescopic rammer;
tête de ——, *art.* rammer head.

refournir, v. a., to refurnish;
—— *les colliers*, (*harn.*) to stuff collars.

réfractaire, a., stubborn, rebellious, disobedient; fireproof; refractory; m., (*mil.*) refractory man or recruit that refuses to join his regiment;

argile ——, fire clay;
brique ——, fire brick;
métal ——, stubborn metal;
mine ——, refractory ore.

réfracter, v. a., to refract.

réfracteur, m., (*inst.*) refracting telescope.

réfractif, a., refractive;
puissance ——*ve*, refractive power.

réfraction, f., refraction;
à ——, refracting;
—— *astronomique*, —— *atmosphérique*, atmospheric refraction;
double ——, double refraction;
—— *terrestre*, terrestrial refraction.

réfractomètre, m., (*inst.*) refractometer;
—— *interférentiel*, interference refractometer.

refrain, m., (*mil.*) march or air peculiar to a regiment or corps.

refranchir, v. a., to free, clear; to discharge by pumping.

réfrangibilité, f., refrangibility.

réfrangible, a., refrangible.

réfrigérant, a., cooling; m., water cooler, cooler;
mélange ——, freezing mixture.

réfrigérateur, m., refrigerator.

réfrigération, f., refrigeration.

réfrigérer, v. a., to cool, refrigerate.

réfringence, f., refringency.

réfringent, a., refringent.

refroidir, v. a. r., to cool, chill; to get cold, cool; to become chilled; to refrigerate.

refroidissement, m., cooling, chill, coldness.

refroidisseur, a., cooling; m., cooling vessel.

refuge, m., refuge, shelter; (*fort.*) shelter in Voorduin's work.

refugié, m., refugee.

refugier, v. r., to take refuge, shelter one's self.

refuite, f., (*carp.*) unnecessary depth of a mortise, etc.; auger hole bored too deep;
donner de la ——, to give play.

refus, m., refusal; (*tech.*) refusal (as of a bolt, pile), (more strictly) the distance a pile sinks under a volley of blows;

à ——, home, as far as possible (of a moving, sliding, etc., piece);
—— *absolu*, in pile driving, the final refusal of the pile;
au ——, home, as far as it will go;
battre à ——, to drive (as a pile) until it will sink no farther;
battre jusqu'à —— *de mouton*, to drive a pile until it will go no farther;
en ——, set or driven home;
enfoncer jusqu'à ——, \. *battre à* ——;
être au ——, to be set or driven home;
—— *relatif*, in pile driving, the refusal due to friction.

refuser, v. a. n., to refuse, deny, decline, avoid, resist, withstand, shun, throw back; (of tools) to refuse to work, to fail to cut, etc.; (of a pile) to resist, refuse, any farther (deeper) driving; (*man.*) (of the horse) to refuse obedience; (*mil.*) to refuse;

—— *l'aile*, (*mil.*) to refuse a wing;
—— *la bataille*, (*mil.*) to refuse combat, to decline a battle;
—— *le front*, (*mil.*) to refuse the front;
se —— *des hauteurs dangereuses*, (*fort.*) to put the crest as nearly as possible at right angles to the line of fire from dangerous heights;
—— *le service*, (*mach.*) to get out of order;
—— *une tranchée*, (*siege*) to oblique a trench to the right or left (to cause its axis to make a wider and wider angle with the direction of advance as one advances);
—— *de virer*, (*nav.*) to miss stays.

regagnage, m., (*expl.*) recovery of acids in the manufacture of explosives.

regagner, v. a., to regain, gain back, retake, recover, retrieve; to rejoin, to reach, to meet again;
— *le chemin,* to get back into the road;
— *un corps, (mil.)* to rejoin;
— *le dessus du vent, (nav.)* to regain the weather gauge;
— *ses foyers,* to return home;
— *un ouvrage,* — *un poste, (mil.)* to retake a work, a post;
— *du terrain, (mil.)* to recover lost ground, to drive back the enemy from the ground he has taken;
— *le terrain, (man.)* to bring a horse back to the track, to the ground that he had left;
— *le vent sur,* v. — *le dessus du vent.*

regain, m., second crop of grass; second cut of forage (clover, etc.); amount by which a timber or stone is too large for the place it is to occupy.

régalage, m., v. *régalement.*

régalement, m., leveling (of ground); *(r. r.)* grading, finishing, of earthwork; *(fort., etc.)* distribution, spread, of earth (as in a parapet); *(adm.)* equitable distribution of a tax.

régaler, v. a., to level (the ground, upturned earth, an embankment); *(fort., etc.)* to distribute, spread, earth, soil; *(r. r.)* to spread ballast even.

régaleur, m., *(fort., r. r., etc.)* spreader, leveler (of earth, as in or on a parapet, etc.).

regard, m., look, aspect, survey, glance; supervision, superintendence; cardinal point; *(cons.)* opening, eye, skylight, hole; draught hole; opening in an aqueduct; *(met.)* sight hole; *(fort.)* opening, as in some disappearing rapid-fire turrets; giving access from one part to another;
en —, facing, opposite to.

regarder, v. a. n., to look on or at; to face, front, be opposite to; to point toward; to survey, examine;
— *de l'arrière,* — *de l'avant,* (of a block) to lead astern, forward;
— *de près,* to be nearsighted;
— *dans la volte, (man.)* to have the head turned inward or in the direction of advance.

réga(t)te, f., regatta;
— *à rames,* boat race (oars);
— *à voiles,* yacht race.

regayer, v. a., *(cord.)* to hatchel.

regayoir, m., *(cord.)* hatchel.

regayure, f., *(cord.)* refuse of hemp.

regazonnement, m., resodding.

regazonner, v. a., to resod.

regel, m., freezing anew.

regeler, v. a. n., to freeze again.

régence, f., regency.

régénérateur, m., regenerator (of a hot-air engine); *(met.)* regenator;
appareil —, *(met.)* regenerative furnace.

régénération, f., regeneration; *(met.)* regeneration, regenerative principle.

régent, m., regent.

régie, f., administration, excise;
en —, directly and wholly under government control; (in general) in the hands of trustees.

regimbement, m., kicking (of saddle animals).

regimber, v. n., to kick, to wince (of saddle animals).

régime, m., diet, regimen; rule, system, government, administration, management; régime, flow (of a river or stream); *(art.)* behavior, peculiarities, of a gun (whether due to atmospheric conditions or to the essence of the piece);
l'ancien —, in France, the times preceding the Revolution of 1789;
— *blanc, (hipp.)* diet of straw and of bran or barley mixed with drinking water;
— *de combustion,* control, regulation, of combustion;
de —, normal;
— *de détonation, (expl.)* the limiting state for the propagation of the explosive wave;

régime *en direction, (art.)* peculiarities of a gun as regards deflections;
le nouveau —, in France, the times following the Revolution of 1789;
— *en portée, (art.)* the peculiarities of a gun as regards range;
— *sanitaire, (med.)* sanitary measures;
— *spécial, (med.)* diet in hospital recommended by surgeon.

régiment, m., *(mil.)* regiment;
— *actif,* *(Fr. a.)* infantry regiment of the active army;
— *d'artillerie-pontonniers, (Fr. a.)* bridge-train regiment, formerly part of the artillery (suppressed in 1894);
— *-cadre,* skeleton regiment;
chien de —, v. s. v. *chien;*
demi- —, *(Fr. cav.)* half regiment;
— *de corps, (art.)* regiment of corps artillery;
— *divisionnaire, (art.)* regiment of division artillery;
— *étranger, (Fr. a.)* regiment composed of foreigners;
— *-frontière,* a regiment stationed or serving on the frontier;
historique de —, regimental history;
— *de manœuvre, (Fr. cav.)* drill regiment, i. e., regarded as an organic unit from point of view of drill only;
— *de marche,* a regiment composed of men belonging to different regiments; (more esp.) such a regiment in 1870;
— *mixte, (Fr. a.)* regiment formed of two or more kinds of troops, e. g., dragoons and chasseurs; infantry regiment composed of one active and two territorial battalions;
— *régional, (Fr. a.)* infantry regiment of four battalions of four companies (formed in peace); (more generally) territorial regiment raised in a particular region and intended for service in it;
— *de salade,* paltry regiment;
service intérieur du —, interior economy of a regiment;
situation du —, monthly regimental return;
— *subdivisionnaire, (Fr. a.)* name given to the regiments of the infantry of the line on account of the fact that each belongs to a *subdivision* (q. v.) of the *région de corps d'armée* (q. v.);
— *de tirailleurs algériens, (Fr. a.)* regiment of Algerian rifles;
— *territorial, (Fr. a.)* infantry regiment of the territorial army.

régimentaire, a., regimental;
bibliothèque —, v. s. v. *bibliothèque;*
école —, v. s. v. *école;*
service —, v. s. v. *service;*
train —, v. s. v. *train.*

région, f., region; *(Fr. a. art)* subdivision of the *planchette de tir* bounded by parallels and meridians; *(hipp.)* region (of the horse);
— *basse,* lower course of a river;
— *de corps d'armée, (Fr. a.)* military district of an army corps in time of peace, has eight *subdivisions de région,* one to each infantry regiment of the corps;
— *d'étapes, (Fr. a.)* subdivision or region of the *étapes service;*
— *fortifiée, (fort.)* fortified region, behind whose forts armies may assemble and maneuver, or an army incapable of taking the field may still offer resistance to an enemy;
— *-frontière, (mil.)* frontier region or military subdivision of territory;
— *haute,* upper course of a river;
— *s officielles,* circles of the administration;
— *principale, (fort.)* that part of the enceinte most exposed to fall into power of enemy;
— *secondaire, (fort.)* such part of an enceinte as is less likely to fall into power of enemy;
subdivision de —, *(Fr. a.)* v. — *de corps d'armée;*
— *territoriale, (Fr. a.)* territorial subdivision or region.

régir, v. a., to govern; to manage; to administer;

régir *à l'économie*, (of the government) to administer, manage, carry on works, etc., without the intervention of a contractor.

régisseur, m., manager, administrator.

registre, m., register, blank book, record; (*tech.*) damper, regulator, smoke valve; (*nav.*) ship's register;
—— *de cendrier*, ash-pit damper;
charger un ——, to enter in an account book;
—— *de cheminée*, funnel, chimney, damper or regulator;
—— *de chevaux*, (*hipp.*) stud book;
—— *de comptabilité*, (*mil. adm.*) account book;
—— *comptable*, (*Fr. a. adm.*) register of property, of stores, etc., to be accounted for;
décharger un ——, to credit on the books;
—— *des délibérations*, record of the minutes of a board, etc.;
—— *de détente*, (*steam*) expansion valve;
—— *à écrou(s)*, (*mil.*) register of confinements;
—— *d'effectif*, (*Fr. a. adm.*) register, kept by the paymaster of the regiment, of variations of strength;
—— *des entrées*, (*med.*) register of patients received;
—— *des entrées et sorties*, (*mil. adm.*) register of receipts and issues;
faire (le) ——, to record;
—— *à feuillets mobiles*, register with detachable sheets;
——*-journal*, v. s. v. *journal;*
manœuvre du ——, (*steam*) throttling;
—— *matricule*, v. s. v. *matricule;*
—— *non-comptable*, (*Fr. a. adm.*) current register or account of property, of stores, in hand;
—— *d'ordinaire*, (*mil.*) mess book;
—— *d'ordres*, (*mil.*) orderly book;
—— *de prise de vapeur*, (*steam*) throttle valve;
—— *des punitions*, (*mil.*) defaulter book (British); record of previous trials and convictions (U. S. A.);
—— *de route*, (*Fr. a. adm.*) record of *feuilles de route* and of related payments;
—— *de service*, (*mil.*) daily duty roster;
—— *des sorties*, (*med.*) register of patients discharged;
—— *à souche*, stub book (of any sort);
tenir ——, to keep a register, an account book;
—— *de tir*, (*art., sm. a.*) target record, target-practice record book;
—— *de vapeur*, (*steam*) throttle valve.

réglable, a., adjustable.

réglage, m., adjustment (as of an instrument), regulation; (*art.*) ranging, i. e., all operations necessary to put the center of impact on the target;
—— *de l'arme*, (*sm. a.*) measure taken to insure accuracy of an arm;
—— *automatique*, self-adjustment;
axe de ——, adjusting spindle;
—— *d'un chronomètre*, rating of a chronometer;
coup de ——, (*art.*) sighting shot, ranging shot;
de ——, adjusting, setting;
—— *en direction*, (*art.*) regulation of fire in direction;
distance de ——, v. s. v. *distance;*
—— *de l'évent*, (*art.*) fuse setting (punching, cutting);
—— *à la fourchette*, (*art.*) determination of the range by the "fork;"
—— *d'une fusée*, (*art.*) setting of a fuse;
—— *à la main*, adjustment by hand;
—— *d'une montre*, regulating of a watch;
—— *en portée*, (*art.*) determination of the range, ranging (operation of getting one "short" and one "over");
—— *du tir*, (*art.,²sm. a.*) ranging, all operations necessary to put the center of impact on the target (includes fuse setting in time-shell fire);
transport de ——, (*art.*) v. s. v. *transport;*
—— *au zéro*, (*inst.*) setting at zero.

règle, f., rule, ruler, straightedge; rule (of action, of conduct, etc.); pattern, model; (*surv.*) measuring rod, base-measuring rod; (*mach.*) guide bar, motion bar;
affaire en ——, a duel in which all formalities are observed; (*mil.*) combat in strict accordance with the rules of war;
—— *à anneau carré*, (*art.*) sort of trunnion gauge;
—— *à araser*, straightedge;
attaque en ——, (*mil.*) regular attack (all rules observed);
attaquer une place dans les ——*s*, (*mil.*) to lay regular siege before a place;
—— *à bout*, end standard, end rule;
—— *à calculer*, v. —— *à calculs;*
—— *à calcul(s)*, computing rule or scale;
—— *de charpentier*, measure, measuring rule;
—— *à coulisse*, (*art.*) relaying scale (to give a piece after firing the direction it has before firing);
—— *à curseur*, (*Fr. art.*) direction scale;
—— *à dessiner*, drawing rule, T-square;
—— *dioptrique*, (*art.*) dioptric laying scale;
—— *de direction*, (*art.*) direction scale;
—— *divisée*, scale;
—— *de dressage*, straightedge;
—— *à éclimètre*, (*inst.*) clinometer and rule in one, for use with plane table;
en ——, according to rule, proper, in proper shape, on proper terms;
——*-équerre*, carpenter's square;
—— *de fausse position*, (in computing) rule of double position;
—— *géodésique*, (*surv.*) base-measuring rod;
—— *des glissoires*, (*mach.*) motion-, guide-, bar;
—— *glissante*, sliding rule; computing scale;
—— *graduée à curseur*, (*ball.*) measuring rule of the Le Boulengé chronograph;
—— *du grossissement*, (*top. drawing*) in slopes of 1:2, the rule of separating hachures by a distance equal to their thickness;
—— *à jour*, open rule (for ruling lines);
—— *à lunette*, v. *alidade plongeante;*
—— *de marée*, tide scale;
—— *des mineurs*, (*mil. min.*) the ordinary rule for a common mine, i. e., that the charge=constant × line of least resistance cubed;
—— *montée*, v. —— *pliante;*
—— *de nivellement*, level rule;
—— *à parallèles*, parallel ruler;
—— *pliante*, folding rule; curve-tracing rule;
—— *de pointage*, (*art.*) scale or graduated rule used to lay guns in direction;
——*s de proportion*, proportional rulers;
—— *du quart*, v. s. v. *loi;*
—— *à raser*, striker; strickle;
—— *réduite*, reduced scale;
—— *de rejointoyeur*, (*mas.*) jointing rule;
—— *de retraite*, (*fond.*) contraction rule;
——*s de la route*, (*nav.*) rules of the road;
—— *de surhaussement*, (*r. r.*) gauge for setting the outer rail of a curve;
siège en ——, (*mil.*) regular, formal, siege;
—— *topographique de campagne*, (*surv.*) field sketching instrument;
—— *à trait*, line standard;
—— *de trois*, rule of three.

réglé, p. p., ruled; regular, regulated, steady; (*art.*) accurate (of fire);
appareil ——, (*mas.*) masonry of courses of uniform height;
assises ——*es*, (*mas.*) courses of the same height;
tir ——, (*art.*) accurate fire;
troupes ——*es*, (*mil.*) regular troops;
vent ——, trade wind.

règle-allures, m., (*man.*) gait regulator.

règlement, m., rule, regulation; settlement of an account; (*mil.*) regulations; drill book (sometimes in plural);
—— *d'avarie*, average statement;
—— *des avaries communes* (*simples*), adjustment of general (particular) average;
—— *d'un compas*, adjustment of a compass;
—— *d'exercices*, (*mil.*) drill regulations, drill book;
——*s de la douane*, customs regulations;
—— *sur l'exercice de* ...., (*mil.*) drill book;

règlement *de manœuvres,* v. —— *d'exercices* (also simply *règlement*);
—— *s de police,* police regulations;
—— *du port,* port regulations;
—— *de soupape,* (*mach.*) adjustment of a valve (e. g., safety);
—— *du tir,* (*art.*) regulation of fire.

règlementaire, a., of or pertaining to the regulations; "regulation."

règlementation, f., process of reducing to rule, or regulations.

régler, v. a., to rule, rule lines; to regulate, order, put in order, settle, wind up, adjust, determine, decide;
—— *un chronomètre,* to rate a chronometer;
—— *un compas,* to adjust the compass;
—— *un compte,* to settle a bill;
—— *une fusée,* (*art.*) to cut, or set, or adjust a fuse;
—— *une montre,* to regulate a watch; to set a watch by;
—— *une rivière,* to give a river a constant régime;
—— *le(s) quart(s),* (*nav.*) to set the watch(es);
—— *le service,* (*mil.*) to regulate (i. e., make rules for) any series of duties;
—— *le tir,* (*art.*) to get the range, i. e., to take all the steps necessary to put the center of impact on the target;
—— *le tir à la fourchette,* (*art.*) to get the range by means of the "fork;"
—— *les tiroirs,* (*mach.*) to adjust slide valves;
—— *au zéro,* to set an instrument at zero.

réglet, m., carpenter's rule; (*art.*) small fillet at the muzzle of a gun (obs.);
—— *double,* winding sticks.

réglette, f., reglet; small rule, ruler; (*art.*) graduated rule for correcting, aiming laterally; deflection slide or scale;
—— *d'altitude,* (Fr. *art.*) altitude scale (for determining the distance of a ship from the known altitude of the station, and from the measured micrometric distance of the water line from the horizon);
—— *de conversion,* (*art.*) rule or scale for converting turns of the elevating screw into degrees;
——*correcteur,* — *à correcteur,* (Fr. *art.*) scale for correcting the distance given by a range finder (seacoast practice);
—— *de correspondance,* (Fr. *art.*) scale, rule giving the points at which a fuse should be punched for given elevations (and vice versa);
—— *curseur* (*art.*) the graduated slide of a direction scale;
—— *de direction,* (Fr. *art.*) scale or rule for observing deflections in seacoast practice;
—— *d'un jet de coulée,* (*fond.*) metal filling the runner;
—— *de lecture,* (Fr. *art.*) general name for the —— *s d'altitude* and —— *de mâture;*
—— *de mâture,* (Fr. *art.*) scale for obtaining the distance of a ship from the measured micrometric height of its masts;
—— *porte-fiole,* (*art.*) level-arm of a gunner's quadrant;
—— *de repérage,* (*art.*) direction bar or rule;
—— *de repère,* (Fr. *art.*) strip or bar of wood used in laying guns for direction.

régloir, m., ruler, burnisher.

réglure, f., drawing or ruling of lines; distance between two ruled lines.

régnant, a., ruling; reigning; prevailing; predominating; prevalent;
vent ——, prevailing wind.

règne, m., reign; predominance; kingdom (e. g., animal, vegetable).

régner, v. n., to rule, reign, govern; to prevail; to reach, run along, extend (over).

régnicole, m., (*law.*) native; (by extension) naturalized.

regonflement, m., swelling or rising of waters by some obstruction; refilling of a balloon.

regonfler, v. a. n., to swell, rise, overflow (of rivers, etc.); to refill a balloon.

regorgement, m., overflow, overflowing; flowing back (as through a drain); superfluity.

regorger, v. n., to overflow, run over, flow back; to abound.

regrattage, m., scraping (as of a wall).

regratter, v. a., to scrape, rub down.

regrossir, v. a., to make thicker, thicken.

reguinder, v. a., to hoist again.

régularisation, f., putting in order, regular order;

régularisation, f., putting in order, regular order; setting right; regularization (of accounts, etc.); written confirmation (of an oral order, etc.) (*mil. adm.*) formal approval of mileage and other travel accounts.

régulariser, v. a., to put in order, regular order, regulate, set right; to confirm an oral order; (*mil. adm.*) to approve mileage, etc., accounts.

régulateur, m., regulator; (*mach.*) regulator, governor, pressure regulator; conical windlass; (*elec.*) regulator;
—— *à action directe,* (*elec.*) direct-acting regulator; gravity regulator;
—— *à action pneumatique,* pneumatic governor;
—— *à ailettes,* momentum wheel governor;
—— *à air,* air vessel;
—— *alimentaire,* —— *d'alimentation,* (*mach.*) feed regulator;
—— *à boules,* (*steam.*) ball governor;
—— *conique,* conical pendulum;
—— *de courant,* (*elec.*) current regulator;
—— *à déclic,* (*elec.*) clutch regulator;
—— *en dérivation,* (*elec.*) shunt regulator;
—— *de descente,* safety apparatus of a hoist;
—— *différentiel,* (*elec.*) differential regulator, regulator with differential coils;
—— *excentrique,* eccentric governor;
—— *à force centrifuge,* Watt's governor; pendulum governor;
—— *à friction,* friction governor;
—— *à gaz,* gas governor;
—— *d' injection,* injection-valve cock;
—— *d'immersion,* (*torp.*) immersion or depth regulator;
—— *à jalousie,* (*steam*) regulator valve (for admission of steam);
—— *à main,* hand governor;
—— *monophote,* (*elec.*) monophotal arc-light regulator;
—— *polyphote,* (*elec.*) polyphotal arc-light regulator;
—— *à potentiel courant,* (*elec.*) current-potential regulator;
—— *de pression,* (*torp.*) v. —— *d'immersion;*
—— *de résistance,* (*elec.*) any regulating resistance;
—— *à ressorts,* (*mach.*) spring regulator;
—— *en série,* (*elec.*) series regulator;
—— *de tension,* (*elec.*) potential regulator;
—— *de tirage,* draft regulator.

régulateur, a., regulating;
force —— *rice,* regulating force.

régule, m., (*met.*) regulus, matt.

régulier, a., regular, exact, in order; m., (*mil., in pl.*) regulars, regular soldiers.

réhabilitation, f., rehabilitation; recovery of civil rights.

réhabiliter, v. a. r., to reinstate; to recover civil rights.

rehaussement, m., raising, raising higher; touching up (a drawing).

rehausser, v. a., to raise, raise higher; to touch up (a drawing).

rein, m., kidney; (*in pl.*) loins, back; (of animals, esp. of the horse) lower part of the back, loins (either singular or plural); (*cons.*) rein, side of an arch, spandrel; (strictly) joint making an angle of 30 degrees with the horizon;
avoir du ——, to be strong in the back;
—— *bas,* (*hipp.*) loins not on same horizontal with croup;
—— *du chien,* (*sm. a.*) neck of the hammer (obs.);
à —— *court,* (*hipp.*) short-coupled;
—— *(s) double(s),* (*hipp.*) double loins;
effort de ——, (*hipp.*) strain of the loins;
à —— *long,* (*hipp.*) long-coupled;
—— *mal attaché,* (*hipp.*) loins much below the croup;

rein, *poursuivre l'épée dans les* ――*s*, to pursue hotly;
―― *soudé,* v. ―― *mal attaché;*
tour de ――, v. *effort de* ――;
―― *tranchant,* (*hipp.*) sharp, narrow loins;
―― *de voûte,* (*cons.*) spandrel.

reine, f., queen.

réintégration, f., reinstatement; (*adm.*) return, or turning into, a magazine of stores unconsumed; (*nav.*) reclassing of an (old) ship done over;
―― *des chevaux,* (*Fr. a.*) return of government horses to public use (as, e. g., when an officer entitled to be mounted at public expense turns his mount in).

réintégrer, v. a., to reinstate; (*adm.*) to turn back into (as supplies to a store, cartridges, etc., to a magazine); (*nav.*) to reclass a ship;
―― *un cheval,* (*Fr. a.*) to return a public animal to public use.

réitérateur, a., (*inst.*) repeating.

réitération, f., reiteration; (*inst.*) frequent measurement of a magnitude, so as to reduce the final error;
―― *par tour d'horizon,* (*surv.*) continuous reading around the horizon.

rejaillir, v. n., to gush, gush out, break forth, spout out; to rebound, fly back; to flash; to reflect, be reflected.

rejaillissement, m., gushing, gushing out; spouting forth, out, up; rebounding, flying back; reflection, flashing.

rejet, m., rejection; outthrow; shoot (of a tree); earth thrown out (of a ditch); overflow pipe; trap; (*sm. a., r. f. art.*) throwing out of an empty cartridge case; (*adm.*) disallowance (of accounts, of expenses); (*fond.*) overflow.

rejeter, v. a., to reject, throw, throw away; to refuse, set aside; to wash up, to cast up (as by the sea); to send out shoots (of a tree); (*sm. a., r. f. art.*) to throw out an empty cartridge case.

rejeton, m., shoot (of a tree, a bush).

rejingot, m., (*cons.*) weather strip or shoulder of a window sill.

rejoindre, v. a., to rejoin; (*mil.*) to join, rejoin, one's post, regiment.

rejointoiement, m., (*mas.*) repointing.

rejointoyer, v. a., to repoint.

relâche, m., interruption; intermission, discontinuance, respite, rest. relaxation; f., (*nav.*) call, stay, or sojourn in a port (for repairs, rest, etc.); port of call;
faire une ――, (*nav.*) to call at a port;
――*forcée,* (*nav.*) enforced sojourn in a port (e. g., because wind bound).

relâchement, m., relaxation; rest, refreshment; decline, falling off; moderation of the weather;
―― *de discipline,* (*mil.*) decay, falling off, of discipline.

relâcher, v. a. n., to relax, slacken, loosen, loose, release, enlarge; to set free, at liberty (a prisoner); to let out, let go, yield, remit, abate, diminish; to moderate (cold, etc., of the weather); (*nav.*) to touch (at a port), put in, put into, a port, call at a port; to let a vessel proceed on its voyage;
―― *la discipline,* (*mil.*) to relax discipline;
faire ――, to beat back.

relai, relais, m., relay of horses, relay station; relay, relief, of workmen; working place; distance excavated earth is transported in a barrow; distance to which excavated earth may be thrown with a shovel; reshoveling of earth; earth to be reshoveled; ground left bare by a running stream, by the sea; (*mil.*) method of joining up several charges or petards in line, connected by a detonating fuse; (*fort.*) berm (obs.); (*mil. sig.*) transmitting station; (*teleg.*) relay station; local, relay, battery;
être de ――, to be unemployed;
―― *indicateur de tension,* (*elec.*) potential indicator;
merlon- ――; (*siege*) (in constructing a battery) the natural ground on which the merlon is to rise;

relai, relais, *à la pelle,* distance to which earth may be thrown with a shovel;
pelleter à n ――*s*, to reshovel earth *n* times;
poste- ――, v. s. v. *poste;*
transmission par ――*s*, (*mil. sig.*) transmission by relay, by transmitting stations;
traverse- ――, v. s. v. *traverse.*

relation, f., relation, account; official report.

relaxation, f., release, enlargement, of prisoners.

relaxer, v. a., to release, enlarge.

relayer, v. a., to relieve, relieve each other; to serve as a relief for; to change horses;
―― *de chevaux,* to change, take, fresh horses;
se ――, to alternate in a duty.

relégation, f., banishment without loss of civil or political rights.

relevé, p. p., raised, elevated; (*fig.*) noble, exalted;
air ――, v. s. v. *air.*

relevé, m., act of raising; bearing (of a coast); (*adm.*) abstract, extract, summary (as from an account, a register, etc.); report; (*farr.*) removal, shifting, of a horseshoe;
―― *de centralisation,* (*adm.*) abstract;
―― *de comptabilité,* (*Fr. a. adm.*) monthly account of funds made to the minister of war by *ordonnateurs.*

relève, f., (*mil.*) relief, as of troops, sentry, etc.; change of station (rare);
de ――, relieving.

relèvement, m., act of raising anything; taking up, removing (as torpedoes); picking up (of a cable, of a telegraph line); statement, account, list; (*surv.*) bearing; (*mil.*) relief (of an outpost, a sentry), relieving; (*ball.*) jump, angle of jump;
angle de ――, (*ball.*) angle of jump;
―― *astronomique,* (*surv., nav.*) true bearing;
―― *d'une côte,* bearings, survey, of a coast;
―― *au compas,* compass bearing;
―― *du compas,* setting of the compass;
―― *de la course d'un vaisseau,* (*art.*) vessel tracking;
――*s se croisant,* (*surv.*) cross bearings;
――*s croisés,* (*surv.*) cross bearings;
faire le ―― *de,* to make an abstract or report or statement;
faire un ――, to take the bearings (of a coast, shore, etc.);
―― *magnétique,* magnetic bearing;
―― *magnétique réel,* true magnetic bearing, i. e., corrected;
porter un ―― *sur la carte,* to prick, lay down, a bearing on the map;
prendre un ――, to take a bearing;
―― *du terrain dangereux,* (*fort.*) in defilading, the height of plane of defilade above the point defiladed against;
troupes de ――, relieving troops;
―― *vrai,* true bearing.

relever, v. a. n., to raise, raise up, elevate; to make higher; to take up, out; to take up and repair, lift up or out; to note, register, record; to relieve, replace (in general); to discharge, free, cancel, release (as from an obligation, an oath); to weigh an anchor; (*fig.*) to bring out, make prominent; (*cord.*) to tice up; (*mas.*) to raise a wall, etc.; (*man.*) to raise the feet very high in galloping; (*mil.*) to relieve (a sentry, etc.); to relieve from command; (*surv.*) to take a bearing, to prick a position on a map; (*esp. nav.*) to prick a ship's position on the chart; to bear;
―― *un angle,* (*surv.*) to measure an angle;
―― *un cheval* (*man.*) to raise the forehand by the bridle;
―― *de,* (*mil.*) to report to (for orders, instructions); to be responsible to, be under the immediate orders of;
―― *de faction,* (*mil.*) to relieve from sentry;
―― *un factionnaire,* (*mil.*) to relieve a sentry;
―― *un fer,* (*farr.*) to remove a loose shoe and put it on firmly;
―― *un fossé,* to raise the banks of a ditch;
―― *de garde,* (*mil.*) to relieve from guard;
―― *la garde,* (*mil.*) to relieve the guard;
―― *de maladie,* to rise from a sick bed;

relever *un navire* (*coulé*), to raise a (sunken) ship;
— *les postes*, (*mil.*) to relieve outposts;
— *le quart*, (*nav.*) to relieve, change, the watch;
— *un sabre*, to hook up a sword;
se ——, to rise (from a recumbent position); (*fenc.*) to rise from a lunge;
— *de sentinelle*, (*mil.*) to relieve from post;
— *une sentinelle*, (*mil.*) to relieve a sentinel;
— *une torpille*, (*torp.*) to pick up, raise, a torpedo;
— *les tranchées*, (*mil.*) to relieve the troops in the trenches.

releveur, m., chaser, embosser (tool or man).

reliage, m., new hooping (of a barrel).

relief, m., relief, embossment, embossing, set-off, foil; relief of the ground: projection from the face of a wall; (*fort.*) relief (of a work);
— *de la culasse*, (*art.*) cascable astragal behind base ring, (obs.);
— *faible*, (*fort.*) relief of less than 6m;
— *moyen*, (*fort.*) relief between Cm and 9m;
plan- ——, *plan en* ——, relief map or plan;
— *rasant*, (*fort.*) v. —— *faible*;
— *du sol*, (*top.*) relief of the ground.

relien, m., (*powd.*) coarsely bruised, unsifted, powder (for fireworks);

relier, v. a., to bind; to rehoop or new hoop a cask; to connect (by roads).

religieuse, f., nun.

religieux, a., religious; m., monk.

religion, f., religion;
— *de l'État*, state religion;
surprendre la —— *de*, to impose upon the good faith of; to deceive.

relimer, v. a., to file again, file up, polish up.

remaçonner, v. a., to repair masonry.

remander, v. a., to send again (as an order).

remandrinage, m., remandreling; (*sm. a.*) resizing of cartridge cases.

remandriner, v. a., to remandrel; (*sm. a.*) to resize (cartridge cases).

remanent, a., remanent, remaining.

remaniement, m., repair, mending, working, doing over again; altering; touching, handling again; (*cons.*) repaving, new roofing.

remanier, v. a., to retouch, touch up; to repair, mend, alter, change; to make great alterations in; (*cons.*) to repave, roof anew.

remballer, v. a., to pack up again.

rembarquement, m., reembarkation, reshipping

rembarquer, v. a., to reembark; to go, put, on board again, ship again; to take shipping again; to engage anew.

remblai, m., earth used to make an embankment, to fill a hollow; act of embanking, of filling; embankment, fill;
— *en avant*, embanking from the end, tipping;
— *de côte*, side embanking;
— *par couches*, horizontal filling, embanking, by successive layers;
—*s et déblais*, cuts and fills;
en ——, like an embankment; embankmentwise;
emploi en ——, spreading;
équilibre des —*s et déblais*, balancing fills and cuts;
exploitation par —*s*, (*min.*) system in which ore is taken out from below and the excavations filled.

remblayer, v. a., to make an embankment; to embank; to fill up; to fill an excavation.

remboîtement, m., clamping, putting together again; replacing in a socket; fitting in again.

remboîter, v. a., to reclamp, clamp together again, fit in again; to put together again (as the parts of a gun carriage).

rembourrage, rembourrement, m., stuffing, padding.

rembourrer, v. a., to stuff, to pad.

rembourrure, f., hair for stuffing.

remboursable, a., to be paid back, restored.

remboursement, m., repayment.

rembourser, v. a., to reimburse, pay back.

rembraquer, v. a., v. *embraquer*.

rembrayer, v. a., (*mach.*) to gear in again.

remède, m., remedy.

remédier, v. a., to remedy;
— *à des voies d'eau*, (*nav.*) to stop leaks.

remenée, f., (*cons.*) arch over a door or window.

remener, v. a., to lead, drive, carry, back.

réméré, m., (*law*) privilege of repurchase;
pacte de ——, recognition of the right of repurchase;
vente à ——, sale with the right of repurchase.

remettre, v. a. r., to put back; to restore, give up; to deliver, return; to bring back to health; to set (a broken leg, etc.); to surrender (a prisoner); to pass over, forgive, pardon; to remit; to remit funds; to delay, defer; to commit, confide; to return swords; (*mil.*) to resume, cause to resume, one's original position after a movement;
— *une armée sur pied*, (*mil.*) to raise fresh troops, new forces;
— *l'avant-train*, (*art.*) to limber up;
— *la baïonnette*, (*sm. a.*) to unfix bayonets;
— *un cheval*, (*man.*) to train a horse anew;
— *le commandement* (*d'une armée*, etc.), (*mil.*) to turn over the command to some one else;
— *la consigne*, (*mil.*) to turn over orders (as a sentinel to his relief);
— *une dépêche à destination*, to send a telegram to its address;
— *l'épée dans le fourreau*, (*sm. a.*) to return sword;
— *la garde*, (*mil.*) to turn over the guard;
se —— *en garde*, (*fenc.*) to come back to the position of guard;
— *un navire à flot*, (*nav.*) to get a vessel afloat again;
— *à neuf*, to make as good as new;
— *une pièce en batterie*, (*art.*) to return a piece in battery;
— *le quart*, (*nav.*) to turn over the watch;
— *à la route*, (*nav.*) to resume an (interrupted) course;
— *les sceaux*, (of a minister of state) to give up the seals, resign;
se ——, to recover one's health; (*mil.*) to resume one's original position ("as you were"); to resume one's original formation;
se —— *en route*, to resume one's journey;
— *la semaine*, (*mil.*) to turn over the week (i. e., duty performed by the week);
— *le service*, (*mil.*) to turn over duty to one's relief.

rémige, f., remex (in a carrier pigeon, feather to which the message is attached).

remise, f., return, delivery (of goods, of a letter, etc.); delivery of a medal, of a decoration; remittance (of money); reduction, discount, allowance; surrender, giving up; delay, adjournment; pardon, reduction of a sentence; carriage-, coach-, house; (*mil. adm.*) turning over (of functions, command, orders, property, barracks, quarters, etc.); (*fenc.*) repeated attack (i. e., a thrust following one that has been parried and that is delivered from the same lunge));
— *d'armes*, (*mil.*) laying down of arms (in token of surrender);
— *en batterie*, (*art.*) return to battery;
— *des drapeaux*, (*mil.*) delivery of, presentation of, colors to troops;
faire la ——, (*mil.*) to turn over (troops, material, etc.);
— *de galons*, (*mil.*) resignation of one's chevrons (on a change of regiment or arm, etc.);
— *à l'heure*, setting of a watch;
— *en liberté*, fresh release from confinement;
— *de main*, (*man.*) easing of the hand, of the pressure, on the bit; (*mil.*) counterstroke;
— *à neuf*, making as good as new;
— *de peine*, remittance of a penalty;
— *en possession*, restoration of possession;
voiture de ——, hired carriage.

remiser, v. a., to store, to put up in sheds or storehouses, to keep (under cover); (*fenc.*) to execute a *remise*.

remmancher, v. a., to put on a new handle.

remmoulage, m., v. *renmoulage*.

remmouler, v. a., *renmouler*.

rémolade, f., (*hipp.*) charge for sprains.

remont, m., (*r. r.*) ascent, up grade.

remontage, m., ascent of a river; (*sm. a.*) assembling, remounting (of a firearm dismounted or taken apart); restocking.

remonte, f., ascent of a river; (*mil.*) remount, remounting, remount establishment; remount service; purchase of horses for the army; horses purchased for the army; (*hipp.*) second leap;
— *par abonnement*, (*Fr. a.*) system by which the government shares in the purchase of a horse by an officer (e. g., an officer buys for 2,000 francs and turns in his mount at government price of 1,400 francs; should the officer leave the service he pays back the 1,400 francs and keeps the horse);
capitaine de ——, (*mil.*) remount officer;
cavalier de ——, trooper in a remount establishment;
cheval de ——, second mount; (*mil.*) remount horse;
circonscription de ——, (*Fr. a.*) remount district;
compagnies de (cavaliers de) ——, (*Fr. a.*) companies (8 in number) of troopers for remount establishments, military schools, etc.;
dépôt de ——, (*mil.*) remount establishment;
dépôt de —— *mobile*, (*Fr. a.*) remount establishment accompanying an army in the field;
école de ——, (*mil.*) remount training;
établissement de ——, v. *dépôt de* ——;
service de ——, (*Fr. a.*) the service charged with the purchase of horses, with their care and training until turned over to troops, and with the brood mares intrusted to farmers;
—— *à titre gratuit*, (*Fr. a.*) mount at government expense (subaltern officers);
—— *à titre onéreux*, (*Fr. a.*) mount at one's own expense (general and field officers).

remonter, v. a. n., to go up, to go up again; to rise (of stocks, of the barometer, etc.); to ascend (a river, a hill, a street); to raise, raise in the air; to raise (make higher) a wall, a floor; to equip anew; to assemble, to put together, to mount (anything that has been separated into its parts), to reach, run up, a river (of the tide); (*sm. a.*) to restock a rifle, to assemble a rifle; (*mil.*) to remount (furnish a fresh supply of horses); (*nav.*) to ship a propeller, a rudder;
—— *une arme à feu*, (*sm. a.*) to assemble the parts of a firearm;
—— *un canon*, (*art.*) to put a gun on its carriage again;
—— *sur l'eau*, to rise, come, to the surface of the water;
—— *son écurie*, to buy a lot of new horses;
se ——, to give one's self a fresh mount;
—— *sur le vent*, (*of a rocket*) to come up into the wind.

remontoir, winding key (watch, etc.);
montre à ——, stem-winder.

remordre, v. a., to bite again; (*mil.*) to attack or fight again.

remorquage, m., towing;
câble de ——, towrope;
—— *à couple*, towing alongside;
prix de ——, towage.

remorque, f., towing; tow rope or line;
à la ——, in tow of;
bride de ——, towing bridle;
câble de ——, towrope;
croc de ——, towing hook;
donner la —— *à*, to take in tow;
en ——, v. *à la* ——;
être à la ——, to be in tow;
grelin de ——, v. *câble de* ——;
se mettre à la ——, to get one's self towed;
prendre la ——, to take in tow.

remorqué, m., towed vessel.

remorquer, v. a., to tow; to have, take, in tow.

remorqueur, m., tug, tugboat; towboat; towing vessel;
—— *à aubes*, paddle tug;
bateau- ——, towboat;
—— *à hélice*, screw tug;
—— *à roues*, paddle tug;
—— *à vapeur*, steam tug.

remouiller, v. a., to anchor again.

rémoulade, f., v. *rémolade*.

remoulage, m., v. *renmoulage*.

remouler, v. a., v. *renmouler*.

rémouleur, m., grinder, knife grinder.

remoulin, m., (*hipp.*) old name for blaze or star on a horse's forehead.

remous, m., slack water (of a ship or boat); wash; eddy, back current;
—— *du courant*, rippling of the tide;
—— *de sillage*, eddy or dead water in a ship's wake.

rempailler, v. a., to wrap with straw again; to new bottom.

remparer, v. a. r., to take possession of; to pile earth, stones, etc. (as against a wall); (*mil.*) to cover by a rampart, to fortify, to intrench one's self.

rempart, m., (*fort.*) rampart; (*sm. a.*) rear face of the cut of a breech casing; side of the receiver; rear part or breech piece of a revolver frame;
coffre de ——, (*art.*) expense or portable magazine;
—— *à demi revêtu*, (*fort.*) rampart whose scarp is revetted to the level of the ground;
fusil de ——, v. s. v. *fusil*;
labourer un ——, (*art.*) to plow a rampart with projectiles;
mixte ——, v. *à demi revêtu*;
—— *revêtu*, (*fort.*) rampart whose scarp is revetted.

remplètement, m., (*cons.*) repair of the foot of a wall.

rempléter, v. a., (*cons.*) to repair the foot of a wall.

remplaçant, m., (*mil.*, etc.) substitute.

remplacé, m., (*mil.*, etc.) the principal person who has procured a substitute.

remplacement, m., replacing, substitution; reinvestment (of funds); fresh supply of provisions, stores, etc.; (*mil.*) resupply of stores, etc., whenever exhausted or condemned; fresh supply of men, horses, and ammunition in action; finding a substitute (obs.);
en —— *de*, to replace, in lieu of, vice;
—— *des grains de lumière*, (*art.*) robushing of a vent;
—— *des lignes*, (*mil.*) relief of one line by a fresh one.

remplacer, v. a., to replace; to substitute for; to succeed; to supply; to supply the place of; (*mil.*) to serve as a substitute; to officiate;
se faire ——, to get a substitute; to be replaced;
—— *les hommes*, etc., (*mil.*) to bring up fresh men, etc.

remplage, m., filling up (as of a cask); (*mas.*) rubble; (as for the inside of wall); (*carp.*) lining or filling pieces;
chevron de ——, (*cons.*) intermediate rafter;
ferme de ——, v. s. v. *ferme*;
pièces de ——, (*cons.*) scantlings;
poteau de ——, v. s. v. *poteau*.

rempli, a., filled, swamped (of boats); m., edging, tuck.

remplir, v. a., to fill again; to fill, fill up; to stock, supply; to fulfill, execute; to reimburse; to discharge, perform, the duties of; (of boats) to fill, swamp; (*adm.*) to fill up the blanks (of papers, documents); (*mil.*) to fill up the numbers of a troop unit; (*art.*) to fill a shell;
se —— *par trop plein*, (of a basin, etc.) to fill by the overflow of another basin, etc.

remplissage, m., filling, stuffing, packing, padding, make-weight; any timber filling up an empty space; (*cons.*) lining, filling (of masonry, of timber); (*art.*) filling of a shell; bursting charge; (*nav.*) deadwood;
couple de ——, (*cons.*) filling timber;
maçonnerie de ——, (*mas.*) filling in, rubble; cofferwork;

remplissage, *pièce de* ——, (*cons.*) filling piece.
remporter, v. a., to bear, carry, carry off;
—— *la victoire*, to win the victory.
remué, p. p., moved, shaken, stirred;
terre ——*e*, transported earth.
remuement, remûment, m., commotion, disturbance; stirring;
—— *des terres*, removal of earth.
remuer, v. a., to move, shake, jog; to stir up, agitate;
—— *la gaule*, (*man.*) to swish a riding whip;
—— *les poudres*, (*powd.*) to sun powder;
—— *de la terre*, to remove earth or soil;
—— *la terre*, (*mil.*) to dig, turn over, ground (in sieges, trench work, mines).
renâcler, v. n., (of a horse) to snort.
renard, m., fox; cant hook; timber dog; opening, gap, chink, crack (as in pipes, reservoirs, etc.); (*met.*) bloom, loop; (*cons.*) blind wall; (*mas.*) stone tied to a string by way of plumb line; (*nav.*) traverse board;
faire la guerre en ——, (*mil.*) to be full of stratagems;
mettre au ——, v. s. v. *mettre;*
queue-de- ——, root or vegetable growth in a water pipe; firmer chisel;
tirer au ——, (of a horse) to pull on the halter in the stable and try to break it.
renardière, f. (*met.*) refinery, refining furnace.
renclanchement, m., the coming into play of the various parts of a ballistic instrument on the current being made.
renclouer, v. a., to spike again;
se ——, (of a horse) to pick up a nail again, get a nail in the foot again.
rencontre, f., meeting, adventure, chance, juncture, encounter, chance encounter, clash, collision; duel, meeting; (*mil.*) unforeseen encounter; any encounter or combat;
bataille de ——, (*mil.*) chance, unforeseen, battle;
combat de ——, v. *bataille de* ——;
de ——, second hand;
roue de ——, balance wheel.
rencontrer, v. a., to meet, encounter, fall in with; to light, hit, upon; (*mil.*) to meet, i. e., engage (the enemy);
—— *un navire*, (*nav.*) to meet, speak, a ship; to run foul of a ship.
rendement, m., efficiency (of a gun, machine, engine, etc.); duty; (*art.*) ratio of the weight of explosive to the weight of the shell;
—— *d'une bouche à feu*, (*art.*) efficiency of a gun (ratio of useful to total work);
—— *commercial*, v. —— *industriel;*
—— *économique,* economic efficiency (of a machine), (ratio between amount of steam used and effective power);
—— *effectif,* efficiency (ratio of useful work to the work corresponding to the quantity of heat developed by the combustible agent);
—— *électrique,* (*elec.*) electric efficiency of a dynamo (ratio of the net to the gross output);
—— *industriel,* commercial efficiency (ratio of the net output to the power necessary to drive the machine);
—— *organique,* ratio of useful to indicated work;
—— *thermique,* heat efficiency (ratio of useful work to the work corresponding to the heat absorbed);
—— *total,* total efficiency.
rendez-vous, m., meeting place, rendezvous, rallying point; (*mil.*) place where troops take the *formation de rassemblement.*
rendre, v. a. n. r., to restore, give back; to deliver, surrender, yield; to accord; to bestow, do (as a favor); to convey, transport, take; to produce; to proceed, to go to; to lead to, open upon; to strike, fall into; to be spent, knocked up; to pronounce sentence, find a verdict; to give (honors); (*cord.*) to pay out; to round in or up; to stretch, give; (*mil.*) to surrender, yield; (*man.*) to slacken the hand;

—— *l'appel,* v. s. v. *appel;*
—— *les armes,* (*mil.*) to lay down one's arms, to surrender;
—— *un arrêt,* (*law*) to pronounce a verdict;
—— *son bâtiment,* (*nav.*) to turn over one's ship on going out of commission;
—— *la bride,* (*man.*) to slacken the bridle;
se —— *par capitulation,* (*mil.*) to capitulate;
—— *combat,* (*mil.*) to return an attack, to resist an attack;
—— *par composition,* (*mil.*) to surrender upon terms;
—— *compte,* to give an account;
—— *sous condition,* to surrender upon articles;
—— *les crosses,* (*art.*) to traverse in aiming;
se —— *à discrétion,* (*mil.*) to surrender at discretion;
—— *l'épée,* (*mil.*) to deliver up one's sword, to surrender;
—— *une forteresse,* (*mil.*) to surrender a fortress;
—— *ses galons,* (*mil.*) to resign one's chevrons;
—— *les honneurs,* (*mil.*) to pay, render, or do honors;
se —— *avec les honneurs de la guerre,* (*mil.*) to surrender with the honors of war;
—— *à la liberté,* to set at liberty;
—— *à la main,* (*man.*) (of the horse) to obey the hand;
—— *la main,* (*man.*) to slacken the hand, to give a horse his head;
—— *la parole,* to release from an engagement, from one's word;
se —— *prisonnier,* (*mil., etc.*) to give one's self up as a prisoner;
—— *le quart,* (*nav.*) to turn over the watch;
—— *raison à,* to give satisfaction (as by a duel);
—— *un salut,* (*mil.*) to return a salute.
rendre, m., deserter; repayment, return.
rendu, p. p., returned, restored; worn out, spent, exhausted, wearied; (*cord.*) two-block, chock-a-block (of a tackle); home, properly stretched (of a rope); (*art.*) in its seat (of a cartridge); m., (*mil.*) soldier who has laid down his arms;
compte ——, report.
rendurcir, v. a. r., to harden; to make harder; to become hard.
rêne, f., (*harn.*) rein, bridle rein, bearing rein; (*mach.*) bridle, strap;
abandonner les ——*s,* to let go the reins;
ajuster les ——*s,* to adjust the reins (to take and hold up the reins with the thumb and forefinger of the right hand);
—— *d'appui,* v. —— *opposée;*
——*s de bride,* bridle reins proper;
cinquième ——, (*fam.*) the pommel or the mane (sometimes used by unskillful horsemen);
—— *coulante,* running rein;
—— *directe,* (*man.*) said of rein when right (left) rein is opened to right (left) to bring horse's head to right (left); direct rein;
fausse ——, strap used as a sort of checkrein to force a horse to arch his neck;
——*s de filet,* snaffle reins;
grand côté de ——, v. s. v. *côté;*
lâcher les ——*s,* to loosen, slacken, the rein; to give a horse his head;
—— *opposée,* (*man.*) said of the rein when right (left) rein is borne against right (left) side to bring head to left (right);
partager les ——*s,* to take a rein in each hand;
petit côté de ——, v. s. v. *côté;*
porte- ——, rein billet;
reprendre les ——*s,* to take up the reins again;
séparer les ——*s,* to take a rein in each hand;
sensible aux ——*s,* (*man.*) bridlewise;
sixième ——, (*fam.*) tail of the horse.
rêner, v. a., (*man.*) to rein a horse.
rênette, f., tracing iron; (*farr., hipp.*) paring knife foot probe;
—— *à guide,* v. *boutoir à guide;*
—— *tourne-à-gauche,* saw-set.
rénetter, v. a., (*hipp.*) to groove, probe, a hoof with a paring knife.

renfaîtage, m., (*cons.*) repair of the ridge of a roof.

renfaîter, v. a., (*cons.*) to repair the ridge of a roof.

renfermer, v. a., to shut up, confine; to imprison; to comprise, include; (*mil.*) to shut up in a place; to close in and invest an army, etc.;
— *un cheval*, (*man.*) to have a horse well under the control of the hand and legs.

renflé, p. p., swelled, swollen; inflated; fat; (*nav.*) bluff (of a ship).

renflement, m., swell, increase of diameter, of thickness, in a piece; boss, hub seat of an axle; fur, furring; rising of waters; (*art.*) swell or shoulder of a projectile; (*sm. a.*) cheekpiece of a stock;
— *de bourlet*, — *de bourrelet*, (*art.*) swell of the muzzle of a gun;
— *de centrage*, (*art.*) centering shoulder of a projectile;
— *central*, center boss.

renfler, v. a., to swell, swell out.

renflouage, m., (act of) getting afloat again (a stranded boat, etc.).

renflouement, m., refloating (a stranded boat, etc.).

renflouer, v. a., to get afloat again (a stranded boat, etc.).

renflure, f., inflation of the hollow of the eye in old horses (horse dealer's trick).

renfoncement, m., cavity, hollow; break, dent, recess; (*fort.*) recess, passage around a traverse.

renfoncer, v. a., to sink, pull down; to new head and bottom a cask.

renforçage, m., act of reenforcing; (*phot.*) intensification of the blacks.

renforcé, p. p., a., strong; reenforced; thickset.
canon —, (*art.*) reenforced gun;
cheval —, thickset, strong, horse.

renforcement, m., strengthening, reenforcing; (*mil.*) reenforcement; (*phot.*) intensification;
— *par doublement*, (*mil.*) reenforcement by putting more men on the same line;
— *par prolongement*, (*mil.*) reenforcement by extending the line;
— *par resserrement*, (*mil.*) reenforcement by making men stand closer together.

renforcer, v. a., to strengthen, to reenforce; to increase, augment; (*mil.*) to reenforce;
— *par doublement*, (*mil.*) to reenforce by adding to the men on a given line or position;
— *par prolongement*, (*mil.*) to reenforce by extending a line or position;
— *par resserrement*, (*mil.*) to reenforce by diminishing intervals, and thus making room for more men.

renformir, v. a., (*mas.*) to repair a wall by giving it a thick coat of plaster and replacing missing stones; to dub out.

renformis, m., (*mas.*) repair of damaged parts (by the insertion of fresh stones and addition of a thick coat of plaster); dubbing out.

renfort, m., fresh strength, strengthening; (*mach.* etc.) reinforce, strengthening piece, rib, shoulder; any part added to give strength; (*art.*) reinforce, part of the breech over the powder chamber, breech hoop; (*sm. a.*) reinforce, heel of a carrier lever; front face of the breech casing (in some magazine arms); (*mil.*) reinforcement; support (in attack formation);
— *antérieur*, (*sm. a.*) front reinforce of a bolt;
— *d'armons*, (*art.*) cross bar or piece between fore guides;
— *de chaudière*, (*steam.*) boiler stay;
— *du cylindre*, (*sm. a.*) cylinder reinforce;
de —, strengthening, stiffening;
— *de fût*, (*sm. a.*) swell of a rifle stock;
— *du levier*, (*sm. a.*) base or foot of the operating handle; locking shoulder, locking mass;
— *postérieur*, (*sm. a.*) rear reinforce of a bolt;

renfort, *premier* —, (*art.*) first reinforce;
second —, (*art.*) second reinforce;
— *de tête mobile*, (*sm. a.*) reinforce of the movable head; slotted rib holding the extractor;
— *de tourillons*, (*art.*) trunnion cleat.

rengagé, a., (*mil.*) reenlisted.

rengagé, m., (*mil.*) reenlisted soldier.

rengagement, m., (*mil.*) reenlistment.

rengager, v. a., (*mil.*) to reenlist;
— *le combat*, to renew the combat.

rengaîner, v. a., (*sm. a.*) to sheathe, put up (sword).

rengréner, v. a., (*mach.*) to throw into gear again; to reengage.

reniflard, m., (*mach.*) shifting valve, blow valve.

reniflement, m., sniffing;
— *sur l'avoine*, (*hipp.*) state of being off his feed.

renifler, v. n., to sniff at;
— *sur l'avoine*, (*hipp.*) to be off his feed.

renmoulage, m., (*fond.*) assembling the various parts of a mold.

renmouler, v. a., (*fond.*) to assemble the various parts of a mold.

rênoir, m., (*harn.*) rein toggle (of a mountain artillery pack saddle).

renommer, v. a., to reelect.

renoncer, v. a., to renounce, give up, relinquish, surrender, throw up, disclaim, disown, waive.

renonciataire, m., (*mil.*) man who surrenders his dispensation from military service.

renonciation, f., renunciation, waiver.

renouer, v. a., (*cord.*) to knot again, tie up again; (*fig.*) to resume, renew; to join, put, together.

renouveler, v. a., to renew;
— *les rayures*, (*art.*) to recut the grooves.

renseignement, m., information, intelligence, direction, indication;
—*s directs*, (*mil.*) information obtained, by reconnaissance, from prisoners, spies, by troops themselves, etc.;
—*s indirects*, (*mil.*) information obtained from neutrals on hostile territory, private letters, enemy's newspapers, etc.;
prendre des —*s*, to inform one's self, get information;
service des —*s*, (*mil.*) intelligence department.

renseigner, v. a. r., to inform, direct; to get information.

rente, f., rent, income; public funds;
— *sur l'État*, funds;
— *foncière*, ground rent;
— *viagère*, life annuity.

rentier, m., person of property, bondholder.

renton, m., rabbet, groove, splice.

rentraire, v. a., to darn, patch; to renter.

rentraiture, f., (*harn.*) overcasting.

rentrant, a., reentering, reentrant; (*fort.*) reentrant; m., reentering angle; (*top.*) bend of a river;
angle —, reentrant angle;
nombre —, (*mach.*) the number of teeth of a pinion which an exact divisor of the number of teeth of the wheel geared into.

rentrayage, m., operation of rentering.

rentrayer, v. a., (*harn.*) to stitch buffalo skins.

rentrée, f., return, return home; return to school after the holidays; resumption of functions in general; getting in of crops; receipt of moneys, collection of taxes; (*law*) reopening of the courts; (*mil.*) voluntary surrender (as of a deserter); return of prisoners to their country; return home from maneuvers of men that do not belong to the active army; return of men to barracks.

rentrer, v. a. n., to return, return home, reenter; to fit in together; to come in (of moneys); to take, get, bring, force, pull, haul, heave, in; to unship; (*mil.*) to rejoin, come in (from detached duty, etc.); to dress back (of an individual in the line);
— *l'ancre*, to take the anchor aboard;
— *dans l'armée*, (*mil.*) to go back into the army;

rentrer *les avirons*, to take in boat, the oars;
— *en batterie*, (*art.*) to return to battery;
— *un canon*, (*art.*) to run in a gun;
— *un cordage*, (*cord.*) to haul in a rope;
— *au corps*, (*mil.*) to rejoin, as from detached duty;
— *les couleurs*, (*nav.*) to haul down, strike, colors;
— *dans le, son, devoir*, (*mil.*) to return to duty (after mutiny, etc.);
— *en fonction*, to resume one's duties;
— *dans les foyers*, (*mil.*) to return home (after maneuvers, etc.);
— *au régiment*, v. — *au corps;*
— *en relâche*, (*nav.*) to put back;
rentrez devant! in bows!
rentrez partout! rowed of all! way enough!

rentrure, f., register (of the parts of a map mounted on linen).

renvahir, v. a., to invade anew.

renverse, f., shift, turn (of a current, tide, wind); (*man.*) diagonal motion.

renversé, p. p., reversed, inverted, (of a flag) with union down; thrown down (as a wall);
canon — *en cage*, (*art.*) gun upset, but still hanging onto its carriage;
compas —, hanging compass;
encolure —*e*, (*hipp.*) ewe-neck;
ligne de file —*e*, (*nav.*) line ahead, inverted;
moulage au —, v. s. v. *moulage;*
ordre —, (*nav., mil.*) inverted order.

renversement, m., overturning, inversion, upsetting; disorder, overthrow, destruction; throwing down (as of a wall); turn of the tide, shift of the wind; (*mach.*) reversing (of an engine); (*siege*) destruction (as of the counterscarp); (*gym.*) act of turning over the horizontal bar; (*man.*) sharp reining up of a horse (to make him change the leading foot);
canon à —, (*sm. a.*) drop-hinge barrel (as of the Smith & Wesson revolver);
— *de courant*, (*elec.*) reversal of a current;
descendre par —, (*gym.*) to dismount (from a horizontal bar, etc.) by turning over;
— *de la machine*, (*mas.*) reversing of an engine;
— *de la marche*, (*mach.*) reversing of an engine;
mécanisme de — *du mouvement* (*à la main, à la vapeur*), (*mach.*) (hand, steam) reversing gear;
tiroir de —, (*mach.*) reversing valve;
— *de la vapeur*, v. — *de la machine.*

renverser, v. a., to turn upside down, invert, upset; to reverse; to knock down, overthrow, throw down (as a wall, etc.); to bear, beat, down; to derange, throw or put into disorder; to ruin, subvert; to spill out, to cant; to capsize (a boat); (*mach.*) to reverse an engine; (*mil.*) to defeat, rout, destroy, overthrow;
appareil pour —, (*mach.*) reversing gear;
— *un cheval*, (*man.*) to rein up a horse sharply (in order to make him change the leading foot);
— *un corps de troupe*, (*mil.*) to defeat a body of troops;
— *un corps de troupe sur un autre*, (*mil.*) to drive back one force on another, so as to involve them both in the same rout;
— *la machine*, (*mach.*) to reverse the engine;
— *la marche*, (*mach.*) to reverse (an engine, etc.);
se —, (*gym.*) to "skin the cat;" to turn over (rings, etc.); (*mil.*) to fall back in disorder;
— *une tige*, to reverse a rod;
— *les travaux de l'ennemi*, (*mil.*) to destroy the works of the enemy (as in a siege).

renvoi, m., sending back, return; discharge, dismissal; adjournment, postponement; explanation (as of the signs on a map); reference, note; (*mil.*) discharge of men, of classes, whose term of service has expired; discharge from the active army; (*mach.*) general term for gearing; transmission gear; countershaft; comprehensive term for overhead works;
— *par des arbres*, (*mach.*) shafting;

renvoi de l'armée, (*mil.*) dismissal from the army;
chevaux de —, return horses;
— *de l'excentrique*, (*mach.*) eccentric-rod gear;
— *de mouvement*, (*mach.*) gear, gearing; return motion; transmission gear, reversing gear; the particular shaft from which a particular machine gets its power, special shaft;
— *de mouvement à la main*, (*mach.*) hand gear;
— *de mouvement d'excentrique*, (*mach.*) eccentric (rod) gear;
— *de mouvement de mise en train*, (*mach.*) starting gear, starting-valve gear;
— *de mouvement du registre de vapeur*, (*mach.*) throttle-valve gear;
— *de mouvement de la soupape d'arrêt*, (*mach.*) stop-valve gear;
— *de mouvement du tiroir*, (*mach.*) valve gear, slide-valve gear;
— *de mouvement du tiroir de détente*, (*mach.*) expansion-valve gear;
— *de la 1ère à la 2ème classe*, (*Fr. a.*) reduction from first to second class (punishment of men);
— *de raccordement*, (*mach.*) connecting gear;
— *du son*, reverberation of sound;
— *de, du, tiroir*, (*mach.*) valve gear, valve motion.

renvoyer, v. a., to return, send back, away, off; to decline; to discharge, dismiss, turn off; to delay, put off; to refer; to reverberate; to reflect (heat, light, etc.); (*law*) to acquit; to send a case before such and such a judge;
— *d'accusation*, (*law*) to dismiss a complaint;
— *un accusé*, (*law*) to acquit an accused person;
— *de la plainte*, (*law*) to acquit a prisoner;
— *de la 1ère à la 2ème classe*, (*Fr. a.*) to reduce from the first to the second class (punishment of men).

réorganisation, f., reorganization.

réorganiser, v. a., to reorganize.

repaître, v. a. n., to feed, bait (horses).

répandre, v. a., to spill, pour out, send out, scatter, spread, diffuse;
— *du sang*, to spill, shed, blood;
— *les troupes dans les villages*, (*mil.*) to distribute troops among the villages.

réparation, f., repair, repairs, mending;
— *par les armes*, duel;
— *d'avaries*, damage repairs of a ship;
—*s civiles*, (*law*) exemplary damages;
en —, under repair;
—*s d'entretien*, keeping in repair;
entretenir en —, to keep in repair;
grosses —*s*, (*cons.*) heavy repairs (as of walls, arches, new roofs, etc., made by the owner);
—*s locatives*, v. *menues* —*s;*
menues —*s*, (*cons.*) minor repairs (made by the tenant).

réparer, v. a., to repair; to refit, mend, make amends; to restore, recover, recruit; to redress wrongs; (*tech.*) to remove file or hammer marks.

repartir, v. n., to go off again; to leave again (as a train); (*hipp.*) of a tendon, to become strained again;
faire — *un cheval*, (*man.*) to give a horse his head again; to bring a horse upon the track again.

répartir, v. a., to distribute, divide, divide among; to assign, apportion; to dispense; (*surv.*) to distribute (an error);
— *le feu*, (*art.*) to "distribute" the fire, to make it general for a given battery or position, after the range has been got.

répartiteur, m., (*mil.*) billet master.

répartition, f., division, distribution (of duties, etc.); adjustment (of a loss); (*surv.*) distribution of an error;
— *du feu*, (*art.*) distribution of fire, making it general over a line or battery.

repas, m., meal; feed, feeding (of horses);
le — *des chevaux*, (*mil.*) call to feed horses.

repassage, m., setting, honing, whetting, grinding; ironing (of linen).

repasser, v. a. n., to return; to go, come, back; to cross again (as a river, a mountain); to hone, set, grind, whet, sharpen (a knife, etc.); to work over again; to iron (linen); (*nav.*) to overhaul rigging; (*met.*) to temper over again;
— *le canon de fusil*, (*sm. a.*) to heat the barrel of a rifle and hammer it slightly;
— *un compte*, to verify an account;
— *sur un cuir*, to strop;
— *une manœuvre*, (cord.) to reeve a rope again;
— *des manœuvres*, (*nav.*) to reeve new running rigging;
— *sur la meule*, to grind;
— *sur la pierre*, to whet, hone.

repasseur, m., grinder.

repassoir, m., grinding-stone, grindstone;
— *à crayon*, pencil sharpener.

repérage, m., marking, laying landmarks; picketing (as a line of direction); (*art.*) laying by bench marks, by observation of some fixed and prominent point;
— *en arrière (avant)*, (*art.*) laying by reference to an object in rear (front);
— *en arrière (avant) à la hausse*, (*art.*) laying by sighting on an object in rear (front);
— *en arrière (avant) à la hausse et au niveau*, (*art.*) laying by sight and level on an object in rear (front);
mire de —, (*art.*) sighting rod;
— *de nuit*, (*art.*) marking the position of a gun, so as to fire by night;
point de —, mark, bench mark;
pendre — *sur tel ou tel point*, to make an observation on such and such point;
— *à la règlette et au niveau*, (*art.*) laying a gun by rule and level;
— *de la station*, (*surv.*) fixing position of station by reference to conspicuous objects on a map and thus finding it on terrain;
— *du tir*, (*art.*) correction or adjustment of practice by means of bench marks or any observed points (*point de repère*).

repère, m., mark, bench mark, adjusting mark, reference mark, marks (on a level bulb, on a test bar); extension marks on a telescope; picket marking the height of a cut or of a fill; *bouton de* —, (*Fr. art.*) button or stud on some carriages, used as a guide in laying the gun; *fente de* —, v. s. v. *fente*;
— *fixe*, any fixed mark, (more esp.) fixed mark used in adjusting the Deport range finder;
— *de flottaison*, floating mark used in adjusting the Deport range finder;
— *de nivellement*, (*mil. min.*) bench marks for determining differences of slope in galleries;
point de —, trench mark, datum point; (*Fr. art.*) reference mark on some gun carriages, used in laying);
— *de pointage*, (*Fr. art.*) reference mark (on some gun carriages, used in laying);
règle de —, (*Fr. art.*) aiming scale;
— *sur* —, (*art.*) said of a shell or shrapnel fuse when so set as to cause explosion on striking.

repèrement, m., (*surv.*) reference to a bench mark; recording such references, sketches, etc., as may serve to fix the position of a point or points in case the pegs should be lost.

repérer, v. a., to mark; to establish a bench mark; to mark out a line by stakes; (*surv.*) to refer to bench marks; to connect with bench marks;
— *en arrière (avant)*, (*art.*) to lay a gun by reference to an object in rear (front);
— *en arrière (avant) à la hausse*, (*art.*) to lay a gun by sighting on an object in rear (front);
— *une distance*, to determine a distance (by means of bench marks);
— *à la hausse*, (*art.*) to lay a gun by sighting on some distant object;

repérer *une pièce*, (*art.*) to mark the position of a gun aimed (elevation and direction);
— *à la règlette et au niveau*, (*art.*) to lay a gun by rule and level;
— *le tir*, (*art.*) to correct the fire by reference to observed points or conspicuous marks.

répertoire, m., inventory, table; list;
— *des réservistes et disponibles*, (*Fr. a.*) list kept by the major of each regiment, of the reservists and *disponibles*, arranged in categories.

répéter, v. a., to repeat; to examine, question (students) on a subject of study; to strike (of a watch or clock); (*law*) to claim;
— *un prisonnier*, (*mil.*) to demand the return of a prisoner;
— *les signaux*, (*mil.*, *nav.*) to repeat signals.

répétiteur, m., instructor, tutor; examiner (in important scientific schools); (*nav.*) signal repeating ship.

répétition, f., repetition; private lesson; (*inst.*) multiplication of the measures of a magnitude (so as to reduce the error of the final result); (*sm. a.*) short expression for repeating mechanism; (*lau*) claim;
arme à —, (*sm. a.*) magazine arm;
arrêt de —, (*sm. a.*) cut-off, magazine cut-off;
double —, (*sm. a.*) double feed;
feu à —, (*sm. a.*) repeating fire;
montre à —, repeater;
pendule à —, striking clock;
système à —, (*sm. a.*) repeating system.

repiler, v. a., to pound over again.

repiocher, v. a., to dig over.

repiquage, m., repair of the surface of a road (removal of broken stones and substitution of new ones).

repiquer, v. a., to repair the surface of a road.

replacer, v. a., to put back; to reinvest funds;
— *l'arme*, (*mil.*) to "recover arms" in the fire exercise;
— *le pistolet*, (*mil.*) to return pistol.

replanir, v. a., (*carp.*) to finish (a surface);
— *un plancher*, to plane a floor smooth.

replat, m., (*top.*) flat place, shelf, on a mountain.

replâtrage, m., new plastering.

replâtrer, v. a., to replaster.

repli, m., fold, plait, crease; (*top.*) winding and turn of a river, of a path; sinuosity; (*mil.*) supporting troops (on which, e. g., outposts may fall back); falling back;
— *de matière*, (*met.*) fold on the surface of sheet metal;
— *du terrain*, (*top.*) fold of the ground;
— *s de la tranchée*, (*siege*) turns, or changes of direction, of a trench.

repliement, m., folding; (*pont.*) dismantling or taking up of a pontoon bridge;
— *par bateaux successifs*, (*pont.*) dismantling by successive pontoons (British, by "booming in");
— *par conversion*, (*pont.*) dismantling by conversion (British, by "swinging");
— *par parties*, (*pont.*) dismantling by parts;
— *par portières*, (*pont.*) dismantling by rafts (British, by "breaking up into rafts").

replier, v. a. r., to fold up again; to turn in, back, up; to double upon; to coil up, bend up; (of a stream, etc.) to turn and twist; (*mil.*) to drive, force, back (a line, outposts, etc.); to fall back; retire; to cause to fall back; (*pont.*) to dismantle a pontoon bridge;
— *par bateaux successifs*, (*pont.*) to dismantle by successive pontoons (by "booming in," British);
— *un chevalet*, (*pont.*) to dismantle a trestle;
— *par conversion*, (*pont.*) to dismantle by conversion (by "swinging," British);
— *par parties*, (*pont.*) to dismantle by parts;
— *un pont*, to dismantle a bridge;
— *par portières*, (*pont.*) to dismantle by rafts;

replier, se —— sur soi-même, (man.) to turn suddenly around (horse).
replonger, v. a. n., to plunge again, to dive again.
repolir, v. a., to repolish.
repompement, m., repumping.
repomper, v. a., to repump.
répondre, v. n., to answer; to reply; to answer (a letter, a signal); to take an examination; to correspond to; to lead, go; (mil.) to return a salute;
—— *aux aides,* (man.) to obey (the leg, spur, hand, etc.);
—— *de,* to be responsible for, answer for;
—— *à un signal,* to answer a signal.
réponse, f., answer.
report, m., transfer (as to paper, of observations); transfer (of designs, etc., from one to several other lithographic stones); (commercially) contango; "brought (carried) forward" (in accounts).
reporter, v. a., to transfer, carry, place, carry over; to carry back, reconvey; to retransfer (in lithography);
à ——, to be carried forward (in accounts);
—— *un mortier,* (art.) to cross lift a mortar.
reporter, m., leader of a flight of carrier pigeons.
repos, m., repose, rest, quiet, peace; resting place, seat, (cons.) sort of landing place on stairs; (mil.) rest during drill; (sm. a.) halfcock; (mach.) rest, state of being at rest; (fort.) berm;
arme en ——, (sm. a.) arm at halfcock;
au ——, (sm. a.) halfcocked; (mil.) standing at ease;
cran de ——, v. s. v. *cran;*
—— *du curseur,* slide rest, index rest;
de ——, quiescent;
échappement à ——, dead beat escapement;
en —— still, at rest, quiescent; (mach.) out of gear;
escalier sans ——, (cons.) continuous flight of stairs;
être au ——, (teleg.) to be silent (of the key);
long ——, (mil.) long rest or halt;
mettre au ——, (sm. a.) to set at halfcock;
mettre en ——, to set at rest; (mach.) to put out of gear;
montre à ——, stop watch;
nuit de ——, (mil.) night in bed (U. S. A.);
pièce de ——, (mach.) rest of a lathe;
en place ——! (mil.) place, rest!
valeur à couverture de tout ——, paper fully secured.
reposé, f., reconstruction (as of a railroad).
reposé, p. p., —— *sur l'arme,* (mil.) at "order arms."
repose-pieds, m., coaster (of a bicycle).
reposer, v. a. n. r., to put, place, set, lay again; to rest; to rest upon; to lean, lie, upon; to settle; to relay (as a railroad track, a telegraph line) (mil.) to ground, to order, arms;
—— *l'arme,* (mil.) to order, come to order, arms;
reposez arme! (mil.) order, arms;
—— *la lance,* (man.) to pass the arm through the loop and let the lance swing behind the right arm;
(se) —— sur la main, (man.) (of the horse) to bear on the hand.
reposter, v. a., (mil.) to post again.
repous, m., sort of mortar made of brick dust.
repoussé, m., chased work, repoussé.
repoussement, m., reaction; (sm. a.) recoil, kick.
repousser, v. a. n., to push back; to keep off, keep at a distance; to reject; to beat, force, throw, drive, back; (of a spring) to resist; to drive out (a bolt); to throw out shoots (of trees, etc.); (mil.) to repulse; to repel, beat off, an attack; (sm. a.) to recoil, kick; (art.) to chase, adorn, engrave (a gun);
compas à ——, inside adjusting calipers.
repoussoir, m., drift, drive bolt, starting bolt, punch, pin driver, starting punch; (r. f. art.) shell starter; (artif.) small rammer; (farr.) farrier's drift; (powd.) plow;
—— *à manche,* driving, starting, bolt;

—— *un chemin,* to get back into the road again;
—— *le dessus,* (mil., etc.) to get the better of it; to recover a lost advantage;
—— *haleine,* to take breath, get back one's wind;
—— *le large,* (nav.) to get out on the high seas again;
—— *une manœuvre,* (cord.) to sheepshank, shorten, a rope;
—— *le mouillage,* (nav.) to return to moorings;
—— *un mur,* (mas.) to repair a wall;
—— *un mur (en) sous-œuvre, par dessous œuvre,* (mas.) to repair the foundation of a wall;
—— *un palan,* (cord.) to overhaul, fleet, a tackle;
—— *le pas,* to change pace, step;
—— *son poste,* (mil., etc.) to resume one's station or post;
—— *une prise,* (nav.) to recapture a prize;
—— *les rangs,* (cav.) to form rank;
—— *en sous-œuvre,* (mas.) to rebuild a foundation; to underpin;
—— *terre,* (of a swimmer) to get within one's depth.
représaille, f., reprisals, retaliation (used generally in the plural);
lettres de ——s, letters of marque and reprisal.
représentant, a., m., representative.
représentatif, a., representative, (substantively) representative government.
représentation, f., representation; exhibition;
—— *de l'ennemi,* (mil.) in maneuvers, impersonation of the enemy;
frais de ——, money for keeping up the official state.
représenté, p. p., re-presented; represented;
ennemi ——, v. s. v. *ennemi.*
représenter, v. a., to re-present; to represent; to personate;
—— *l'ennemi,* (mil.) in maneuvers, etc., to personate the enemy.
réprimande, f., reprimand;
—— *du capitaine,* (Fr. a.) reprimand of n. c. o. by a captain (disciplinary punishment);
—— *du colonel,* (Fr. a.) reprimand of officer, of n. c. o., by the colonel (disciplinary punishment);
—— *des généraux,* (Fr. a.) reprimand of officers by a general officer (disciplinary punishment).
réprimander, v. a., to reprimand.
repris, p. p., —— *de justice,* (law) condemned; (as a noun) old offender, jail bird, ticket-of-leave man.
reprise, f., taking again, taking up again; recovery, return, renewal, resumption; mending, patching, of clothes; (in pl., law) claims; (mas.) joint in concrete work; repair of a wall; (tech.) soldering of a cracked pipe or conduit; (mil.) recapture; resumption (of hostilities, etc.); (nav.) recapture, recaptured ship; (fenc.) renewal of an attack (without rising from the lunge); bout; (man.) each part of a riding lesson (for both horse and man); riding squad (under instruction); ride;
aller à ——, (of water) to be raised by two pumps of different level;
—— *du casernement,* (mil.) occupation of barracks by a body of troops;

reprise, en ——, (*man.*) command to close (in riding school) on such a trooper (i. e., so as to form a squad);
—— *de froid,* return of cold weather (after a mild spell);
—— *du manège,* (*man.*) each part of a lesson to horse or to rider (after which a rest is taken); set of troopers or horsemen practicing or exercising together;
—— *en sous-œuvre,* (*mas.*) repair of the foundation of a wall.
repriser, v. a., to mend, patch, clothes.
reproduction, f., reproduction; copy; (*mach.*) shaper, guide, copying guide.
reproduire, v. a., to reproduce; (*mach., etc.*) to copy.
république, f., republic.
répulsif, a., repulsive, repellant.
répulsion, f., repulsion.
requérant, m., (*law.*) petitioner, plaintiff.
requérir, v. a., to request, require, claim, summon, demand, call upon; (*law*) to demand, claim; (*mil.*) to make requisition.
requête, f., request, application, demand; (*law*) petition.
requiper, v., a. to reequip, fit out again, rig out anew.
réquisition, f., requisition, demand, application; (*mil.*) requisition; conscription;
de ——, (*mil.*) by requisition, got by requisition; issued, obtained, on requisition;
cheval de ——, v. s. v. *cheval;*
mettre en ——, (*mil.*) to requisition; to put into requisition;
ordre de ——, v. s. v. *ordre;*
zone de ——, (*mil.*) zone or area to furnish supplies, etc., by requisition.
réquisitionnaire, m., (*mil.*) conscript.
réquisitionner, v. a., (*mil.*) to levy requisitions on, etc.
réquisitoire, m., (*law*) public prosecutor's speech, charge (said also of military courts).
resoif, m., v. *récif.*
rescinder, v. a., to rescind.
rescision, f., rescinding.
rescription, f., check, money order.
réseau, m., net, network; network (of railways, of roads in general, of the lines of travel of a steamship company); (*surv.*) skeleton of a survey;
—— *de chemins de fer,* (*r. r.*) railway system, division; network of railways;
commission de ——, v. s. v. *commission;*
—— *de courants,* (*elec.*) net of currents;
—— *électrique,* (*mil. and in gen.*) network of electric-telegraph lines, system of telegraph lines;
—— *ferré,* v. —— *de chemins de fer;*
—— *de fil de fer,* (*fort.*) wire entanglement;
—— *géodésique,* (*surv.*) continuous network of triangles (not overlapping);
—— *optique,* (*mil.*) network of military visual telegraph stations, of visual signal stations;
—— *de l'observation,* (*mil.*) wiring (network of wires) of a fortified position, for the transmission of intelligence as to the enemy;
—— *des ordres,* (*mil.*) wiring (i. e., network of wires) of a fortified position for the transmission of orders;
—— *x de postes,* (*mil.*) line of posts;
—— *routier,* (*top.*) network of roads; (*mil.*) all the roads in the march front;
—— *de sûreté,* (*mil.*) generic term for outposts, patrols, and all other agents to secure the safety of an army;
—— *télégraphique,* telegraph system;
—— *de triangles,* —— *triangulaire,* —— *trigonométrique,* (*surv.*) system of triangles, triangulation.
réserve, f., reservation, reserve; resist paste; trees left standing in a cut; store; (*mil.*) reserve, (more esp.) reserve on the battlefield, reserve of the firing line; (*nav.*) reserve; (*Fr. art.*) auxiliary supply of ammunition and of spare parts for a mountain battery; (*pont.*) one pontoon wagon and one *chariot de parc,* q. v.; (*law*) protest;

réserve *de l'armée active,* (*Fr. a.*) reserve of the active army;
—— *de l'armée territoriale,* (*Fr. a.*) reserve of the territorial army;
—— *d'avant-postes,* (*mil.*) outpost reserve;
—— *de bataillon,* —— *de brigade, etc.,* (*mil.*) battalion, brigade, etc., reserve;
—— *de batterie,* (*German art.*) reserve for the service of the piece and to replace drivers;
bois de ——, trees left standing in a cut;
cadre de ——, v. s. v. *cadre;*
corps de ——, (*mil.*) reserve, body of troops in reserve;
de ——, in reserve, spare, in store;
en ——, (*mil., etc.*) in reserve; (*nav.*) in ordinary;
garder en ——, (*mil., etc.*) to keep, hold, in reserve;
—— *générale,* (*mil.*) the reserve proper of an army engaged in battle; (*siege*) a body of men held in reserve to face emergencies (in order to avoid drawing on the garrisons of the various sectors); (more esp.) mobile troops charged with all operations outside of the *place;* (*Fr. art., siege*) general reserve of guns, intended to increase the number of guns on the side attacked, and in particular to equip the *batteries de circonstance;*
—— *de guerre,* (*mil.*) material kept in store for war;
lieutenant, etc., de ——, (*Fr. a.*) a lieutenant, etc., of the reserve army;
mettre en ——, (*mil., etc.*) to put in reserve;
officier de ——, (*Fr. a.*) an officer of the reserve army;
—— *de recrutement,* (*mil.*) Ersatz reserve;
——*s de régiment,* (*mil.*) in the attack, battalion in the second line;
—— *de secteur,* (*siege*) mobile troops of a sector of defense;
——*s stratégiques,* (*mil.*) in mountain warfare, troops occupying central positions for action whenever their presence is needed;
——*s tactiques,* (*mil.*) in mountain warfare, the troops that first engage the enemy until backed up by the ——*s stratégiques.*
réserver, v. a., to reserve, keep in store, keep back;
—— *de manœuvre,* (*mil.*) to keep off other duty so as to have drill;
—— *de service,* (*mil.*) to keep off drill, etc., so as to perform some other duty.
réserviste, m., (*mil.*) reservist, man belonging to the reserve.
réservoir, m., reservoir, pond, cistern, tank, receiver, holder, well, receptacle;
—— *alimentaire,* feed tank;
—— *à air,* —— *d'air,* air box;
—— *alimentaire,* —— *d'alimentation,* feed tank; hot well;
—— *d'avant-train,* limber locker;
—— *de chasse (d'eau),* flushing tank;
—— *collecteur à huile,* (*mach.*) oil tank;
—— *à eau,* —— *d'eau,* water reservoir, tank, cistern; (*r. r.*) water tank;
—— *d'eau chaude,* (*steam*) hot well, hot-water well of a steam engine;
—— *à eau salée,* brine tank;
—— *à gaz,* gasometer, gas holder;
—— *à graisse,* axle box, grease box;
—— *à huile,* oil can; oil tank;
—— *intermédiaire,* (*steam*) receiver of a compound engine;
—— *de pompe,* cistern, well, of a pump;
—— *à,* —— *de la, vapeur,* (*steam*) steam chamber, reservoir, dome, space.
résidu, m., residue, residuum, settlement, settling; (*art., sm. a.*) fouling, residue, in the bore of a gun;
—— *de la poudre,* (*powd.*) fouling, beads.
résiduel, a., residual;
charge ——*le,* (*elec.*) residual charge;
magnétisme ——, (*elec.*) residual magnetism.
résignation, f., (*law*) resignation (as of rights).
résigner, v. a., to resign; give up.
résiliation, f., cancellation, annulment of a contract, etc. (by agreement of the parties).

résilier, v. a., to cancel, annul, a contract (by agreement of the parties).

résine, f., resin, rosin;
— *blanche*, white resin;
— *commune*, coarse resin;
— *copal*, gum copal;
— *élastique*, caoutchouc;
— *jaune*, yellow rosin;
— *liquide*, soft resin;
— *solide*, hard resin;
— *de térébenthine*, common black resin;
— *de terre*, mineral resin.

résiner, v. a., to tap a pine for resin; to cover with resin.

résineux, a., resinous.

résistance, f., resistance, strength, strength of materials; opposition, contumacy; endurance; (*elec.*) resistance; (*man.*) opposition (of the horse);
— *de l'air*, (*ball.*) resistance of the atmosphere;
— *apparente*, (*elec.*) apparent resistance;
bobine de ——, (*elec.*) resistance coil;
boîte de ——, (*elec.*) resistance box;
— *à chaud*, (of metals) resistance hot;
— *au cisaillement*, shearing strength;
— *de compensation*, compensating resistance;
— *à la compression*, strength of compression;
— *des cordes*, strength of cordage;
— *critique*, critical resistance;
— *à l'écrasement*, crushing strength;
— *essentielle*, v. — *intérieure*;
— *à l'extension*, tensile strength;
— *extérieure*, (*elec.*) external resistance;
— *à la flexion*, bending strength;
— *à froid*, (of metals) resistance cold;
— *de frottement*, frictional resistance;
— *au glissement*, sliding friction;
— *intérieure*, (*elec.*) internal resistance;
— *d'isolement*, (*elec.*) insulation resistance;
— *des machines*, internal resistances of machines;
— *magnétique*, (*elec.*) magnetic resistance;
— *des matériaux*, strength of materials;
— *des pièces chargées debout*, resistance to buckling;
— *au roulement*, rolling friction;
à —*s séparées*, (*art.*) of wire guns, said when different parts take up the different kinds of strains (tangential and longitudinal);
solide de la moindre ——, solid of least resistance;
— *à la tension*, tensile strength;
— *à la torsion*, torsional strength;
— *de tout un circuit*, (*elec.*) total resistance;
— *à la traction*, tensile strength;
— *vive élastique*, resistance within the elastic limit;
— *vive à la*, *de*, *rupture*, limit of resistance to rupture; breaking stress.

résister, v. n., to resist, offer resistance, withstand, oppose; to abide, bear, endure; to stand, hold out; (*tech.*, to support, to take (as a stress); (*mil.*) to resist; (*man.*) of the horse, to refuse obedience;
capable de ——, proof against.

résolution, f., resolution, solution; (*law*) cancellation (as of a contract) by superior authority.

résonnance, f., resonance; (*sm. a.*) rattling of a ramrod, of a gun, in the manual of arms.

résonnateur, m., resonator.

résonnement, m., resounding; clank (of spurs, saber, etc.), reverberation.

résonner, v. n., to resound, clank (spurs, saber), clang.

résoudre, v. a., to resolve, settle, solve, decide, determine; to reduce, dissolve; (*law*) to cancel (a contract).

respirateur, m., respirator;
— *anglais*, (*mil. min.*) respirator worn on going into a foul mine.

responsabilité, f., responsibility;
dégager la ——, to free from responsibility;
engager la ——, to make responsible for.

responsable, a., responsible, liable, accountable.

ressac, m., surf.

ressangler, v. a., to regirth, to tighten the girth.

ressaut, m., projection, lug, jog, shoulder; break or change of level (as of an interior crest); change of direction; (*art.*, *sm. a.*) junction slope; (*mil. min.*) abrupt change of level in a change of gallery; (*surv.*) change of level in chaining up (or down) hill;
— *d'une façade*, projection of the façade of a building;
faire ——, to project, to stand out of a line or range;
— *de la poignée*, (*sm. a.*) pistol grip;
— *de projectile*, (*art.*) shoulder of a projectile;
— *de terrain*, (*top.*) abrupt fall of the ground.

ressauter, v. a. n., to leap, jump over, to leap back, jump back (as over a ditch); to project, stand out of a line or range.

resseller, v. a., to saddle again.

ressemelage, m., new soling.

ressemeler, v. a., to new sole.

resserrement, m., tightening; (*mil.*) closing up;
— *de l'argent*, tightness of money;
renforcement par ——, v. s. v. *renforcement*.

resserrer, v. a., to tie, tie again, fasten, bind up; to tighten; to confine, restrain, confine more closely, keep in close confinement; to contract, compress, crowd; (*mil.*) to invest (a town), pen up (a garrison); to close up (ranks);
— *son argent*, to hoard one's money;
se ——, to grow colder.

ressif, m., v. *récif*.

resservir, v. n., (*mil.*) to serve again.

ressort, m., elasticity, springiness; (*fig.*) force, energy, activity, incentive; (*law and in gen.*) jurisdiction, sphere, function; (*tech.*) spring.

   I. Artillery, small arms, etc.; II. Springs in general.

   I. Artillery, small arms, etc.:
— *d'armement*, (*art.*) cocking spring; restraining spring of a fuse;
— *arrêt*, ——, *d'arrêt*, (*sm. a.*) spring stop;
— *arrêtoir*, (*sm. a.*) spring stop; cut-off;
— *d'attache*, (*sm. a.*) spring catch;
— *d'auget*, (*sm. a.*) carrier spring;
— *de baguette*, (*sm. a.*) ramrod spring;
— *de batterie*, (*sm. a.*) sear spring;
— *de barrette*, (*sm. a.*) hand spring (of a revolver);
— (*de*) *Belleville*, (*art.*) Belleville spring, spring disk;
— *en bois*, v. —— *de garniture*;
— *de bouche*, (*sm. a.*) band spring;
— *à boudin*, (*r. f. art.*) lock spring;
— *à brassardelle*, (*sm. a.*) upper band spring;
— *de capucine*, (*sin. a.*) lower band spring;
— *de détente*, (*r. f. art.*) mainspring; (*sm. a.*) trigger spring;
— *distributeur*, (*sm. a.*) spring clip hook;
— *d'embouchoir*, (*sm. a.*) upper band spring;
— *de fermenture*, (*sm. a.*) breech-closing spring;
— *gâchette*, (*sm. a.*) spring sear;
— *de gâchette*, (*sm. a.*) sear spring;
— *de garniture*, (*sm. a.*) band spring;
grand ——, (*sm. a.*) mainspring;
— *de grenadière*, (*sm. a.*) lower band spring, v. *grenadière*;
— *de la hausse*, (*sm. a.*) rear-sight spring;
— *de lancement*, (*torp.*) discharging spring (of a torpedo tube);
— *de mise de feu*, (*art.*) firing spring (turret);
— *de noix*, (*sm. a.*) mainspring of a gunlock;
— *de percussion*, (*art.*) percussion of spring;
petit ——, (*sm. a.*) sear spring;
— *à pince(s)*, (*art.*) clip spring; spring catch;

ressort *plat,* (*r. f. art.*) spring buffer or buffer spring;
—— *de poussoir,* (*sm. a.*) push-button spring;
—— *de rappel* (*en batterie*), (*art.*) return spring; counter-recoil spring;
—— *de recul,* (*sm. a., etc.*) recoil spring;
—— *de sûreté,* (*art.*) safety spring (of a fuse);
—— *tire-feu,* (*F. m. art.*) spring on some (old) breech-loading guns, forming part of the safety device of the breech closure.
  II. Springs in general.
à —— (*s*), fitted with a spring (or springs);
acier à ——, (*met.*) spring steel;
—— *antagoniste,* antagonistic spring;
—— *en arc,* bow spring;
—— *atmosphérique,* air buffer, air cushion;
—— *d'attelage,* (*r. r.*) draw spring;
balance à ——, spring balance;
bander un ——, to bend, set, a spring;
boîte à —— (*s*), spring box; (*r. r.*) buffer box;
boudin à ——, spring buffer;
—— *à boudin,* spiral spring;
—— *à boudin à droite et à gauche,* right- and left-hand spiral spring;
—— *à branches multiples,* zigzag spring;
—— *à branches plates,* flat-branched spring;
bride de ——, spring shackle;
—— *à C,* C spring;
candole à ——, spring latch;
—— *en caoutchouc,* india-rubber spring;
charge à ——, spring weight;
—— *de choc,* (*r. r.*) buffer spring;
—— *de choc à feuilles,* (*r. r.*) plate buffer spring;
—— *de choc à spirale,* (*r. r.*) spiral buffer spring;
—— *à clavette,* key spring;
—— *à coches,* notched spring;
compas à ——, spring dividers;
—— *s compensateurs,* restoring springs;
—— *de compression,* compressor spring;
—— *de connexion,* connecting spring; (*elec.*) contact spring;
—— *conique,* coiled or conical spring;
—— *à coupelle,* plate or disk spring;
coussin à ——, spring cushion;
croc, crochet, de ——, spring hook;
détendre un ——, to ease, let go, relax, a spring;
—— *de dérivation,* (*elec.*) insulated spring;
—— *double,* double-branched spring; figure-of-eight spring;
—— *à double lame,* double-branched spring;
—— *elliptique,* elliptic spring;
—— *d'enclanchement,* spring catch;
étrier de ——, *v. bride de* ——;
faire ——, to spring, to fly back; to spring back;
feuille de ——, spring plate;
—— *à feuilles,* plate spring;
—— *à feuilles étagées,* carriage spring;
—— *de flexion,* flexion spring;
—— *-frein,* spring brake;
—— *frotteur,* friction spring (as for making electrical contacts);
grand ——, mainspring;
—— *à griffe,* spring catch;
—— *à hélice,* helical spring;
—— *d'indicateur,* (*steam*) indicator spring;
lame de ——, plate of a spring;
—— *à lame(s),* plate spring;
—— *à lame simple,* single plate spring;
—— *à lames à deux bras,* double-armed plate spring;
—— *à lames étagées,* zigzag spring; carriage spring;
—— *à lames multiples,* v. —— *à lames étagées;*
logement de ——, spring hole; seat, socket, of a spring;
loquet à ——, spring latch;
—— *du manomètre,* (*steam*) gauge spring;
—— *de marteau,* spring of a hammer;
monte- ——, spring hook or cramp;
—— *de montre,* v. —— *à, en, boudin;*
—— *moteur,* actuating spring;
—— *à pince(s),* V spring;
—— *de piston,* (*mach.*) piston spring; packing spring;
—— *plat,* flat spring, plate spring;

ressort *à pompe,* spiral spring;
—— *de rappel,* antagonistic spring;
—— *de réception,* (*elec.*) receiver spring;
—— *de segment de piston,* (*mach.*) piston ring;
—— *spiral,* spiral spring, hairspring;
—— *à, en, spirale,* v. —— *à boudin;*
—— *de soupape,* valve spring;
—— *de soupape de sûreté,* safety-valve spring;
—— *de suspension,* suspension spring (of a railroad car or of a carriage); half-elliptic spring;
—— *de suspension à spirale,* volute spring;
—— *de tambour,* (*r. r.*) buffer spring;
—— *-tampon,* spring buffer;
—— *de tampon,* v. —— *de tambour;*
tampon de choc à ——, (*r. r., etc.*) spring buffer;
tige à ——, spring rod;
tige de ——, spring pin;
—— *de tiroir,* (*mach.*) slide-valve spring;
—— *de torsion,* torsion spring;
—— *tourné en hélice,* conical spring;
—— *à, de, traction,* draft spring, traction spring, tug spring;
—— *de traction à spirale,* volute drawspring;
traverse du ——, spring bar;
—— *en V,* V spring;
verrou à ——, spring bolt;
—— *vibrant,* oscillating spring;
—— *de voiture,* carriage spring;
—— (*en*) *volute,* volute spring.

ressortir, v. n., to go out again; (*law*) to be under the jurisdiction of.
ressoudir, v. a., (*met.*) to reweld; to resolder.
ressoudure, f., (*met.*) rewelding, resoldering.
ressource, f., resource, means;
—— *de l'arrière,* (*mil.*) supplies pushed forward by the *service des étapes;*
—— *s de l'avant,* (*mil.*) resources, supplies, etc., obtained locally;
—— *statistiques,* classified, tabulated, resources (e. g., of a country).
ressuage, m., sweating (as of a wall); soaking (of bread); (*met.*) liquation, eliquation.
ressuer, v. a. n., to sweat again; to sweat (of walls, etc.); to soak (of bread); (*met.*) to separate by liquation; to shingle; to reheat;
faire ——, (*met.*) to eliquate, to separate by liquation; to shingle.
ressuyage, m., drying, airing.
ressuyer, v. a., to dry; to air; to bring about the evaporation of moisture;
—— *la poudre,* (*powd.*) to dry, to air, powder before granulation.
restauration, f., restoration; restoration of a dynasty.
restaurer, to restore, to repair; to restore a dynasty.
reste, m., rest, residue, remainder; (*in pl.*) mortal remains.
rester, v. n., to remain; to be left, stay, continue; (*nav.*) to keep, stand, bear;
—— *à l'ancre,* (*nav.*) to remain at anchor;
—— *de l'arrière,* (*nav.*) to bring up the rear;
—— *en arrière,* to lag behind, to stay behind, to be left behind, to fall behind, hang back;
—— *derrière la main,* (*man.*) (of the horse) to be slow, to be continually slackening the gait;
en —— *à,* to go no farther; to stop at, discontinue, leave off;
—— *sous le feu,* (*mil.*) to remain exposed to the enemy's fire;
—— *à long pic,* (*nav.*) to ride short;
—— *dans la main,* (*man.*) (of the horse) to remain under the control of the bit;
—— *en panne,* (*nav.*) to remain hove to.
restituer, v. a., to return, restore; (*law*) to restore to one's original status.
restitution, f., restitution, restoration; (*law*) restoration to one's original status, rehabilitation; (*surv.*) determination of a point by the intersection of lines of perspective.
résultante, f., resultant.
résumer, v. a., to recapitulate, to sum up.
rétablir, v. a., to reestablish, repair thoroughly, restore; to reinstate;
—— *les affaires,* —— *la bataille,* —— *le combat,* (*mil.*) to regain the upper hand;

rétablir, se ——, (*gym.*) to pull one's self up, to mount on (a horizontal bar, etc.).

rétablissement, m., restoration, reestablishment; thorough repair, doing up; (*gym.*) act of mounting, of seating one's self on the horizontal bar, etc.;

—— *par renversement*, (*gym.*) act of pulling one's self up, of mounting on the bar, etc., by turning over it.

retaille, f., shred, cutting, clipping (as of plates, sheets, etc.); furrow of a millstone.

retailler, v. a., to cut again; to recut a file; (*carp.*, etc.) to cut and fit again;

—— *la lance*, (*mil.*) to shorten the lance in the hand.

rétamer, v. a., to retin.

retard, m., delay, retardation, slowness, loss: lag or lost time of any apparatus; (*in gen.*) the time necessary to cause any apparatus to work, or any operation to proceed; (*mil.*) delay in reporting, in returning to barracks; (*art. artif.*) delay of inflammation; (*artif.*) delaying fuse;
—— *d'aimantation*, (*elec.*) magnetic lag;
—— *de déclanchement*, (*elec.*) interval of time between the breaking of a current through an electromagnet and the flying back of the armature;
—— *de désaimantation*, (*elec.*) lag of demagnetization;
—— *d'éclatement*, (*art.*) horizontal distance a shell goes on ricochet before bursting;
en ——, behind time; backward, slow;
être en —— sur, (of a watch) to be slow in comparison with;
—— *de l'étincelle d'induction*, (*elec.*) lag of an induction spark;
—— *d'inflammation*, (*art.*) delay of inflammation, hang fire;
—— *à l'introduction*, (*steam*) state of being late (of a slide valve);
—— *de la marée*, —— *des marées*, v. s. v. *marée*;
—— *de la phase du courant*, (*elec.*) lag of a current;
—— *de renclanchement*, (*elec.*) interval of time between the making of a current and the springing into place of the armature.

retardataire, a., in arrears, lagging; (*mil.*) slow in joining, slow in reporting, m., person in arrears, delinquent; (*mil.*) recruit slow in joining; man who overstays his pass; person late in reporting for duty.

retardation, f., retardation, negative acceleration.

retardateur, a., retarding.

retardé, p. p., delayed, retarded;
effet ——, (*art.*) delay action (of a fuse).

retardement, m., delay;
jours de ——, lay days;
—— *des marées*, retard of the tide.

retarder, v. a. n., to retard, delay, defer, put off, put back; to lose (as of a watch); to go, be, slow; (of the moon) to wane.

retassement, m., (*met.*) pipe (defect in steel).

retassure, f., (*met.*) pipe (defect in steel).

retenir, v. a., to get hold of again; to withhold, retain; to hold, hold back; to keep back, reserve, deduct, take off; to restrain, curb, check, fold in; to hire, engage, secure; (of horses in harness) to hold back (as going downhill); (*man.*) to curb, hold in, check (a horse); (*cord.*) to guy;
—— *une cause*, (*law*) of a judge, to declare his competency in a given case;
—— *une poutre*, (*cons.*) to cramp down a beam;
—— *prisonnier*, to detain anyone as a prisoner;
se ——, to stop, to hold on; (*man.*) (of the horse) to be unwilling to advance; to prance instead of going forward.

rétention, f., retention; (*law*) reservation; retention of a case by a judge.

retentir, v. a., to resound, to peal; (*art.*) to boom (of guns).

retentissement, m., resounding; (*art.*) (of a gun) pealing, booming.

retenu, p. p., restrained, checked, etc.;
cheval ——, horse unwilling to start;
—— *par les glaces*, ice bound.

retenue, f., deduction, stoppage, stoppage of pay, charge against one's pay; keeping in (punishment in schools); confinement (U. S. M. A.); basin between two sluices; reservoir of water; (*cord.*) guy rope;
câble de ——, balloon rope; (*nav.*) anchor cable;
—— *de câble-chaine*, (*nav.*) cable reliever;
—— *de chasse*, flushing basin or reservoir;
corde de ——, (*cord.*) pendant, tackle guy;
mettre en ——, (of schools) to keep in: to confine (U. S. M. A.);
palan de ——, (*cord.*) relieving tackle;
—— *d'un pont suspendu*, fixture of a suspension bridge;
—— *de solde*, (*mil.*) retention of pay.

réticule, m., (*inst.*) reticule, cross hairs, cross wires.

réticulé, a., reticulated.

rétif, a., (*man.*) obstinate, stubborn, refusing to obey.

retirade, f., (*mil. min.*) mine that bursts into a *fourneau de tête* (q. v.); (*fort.*) retreat, retrenchment, cut (obs.).

retirer, v. a. r., to draw over again; to withdraw, to retire; to shelter, obtain; to draw from; to remove, unship; to leave, go; to contract, shrink; (of waters) to fall, subside; (of the tide) to ebb; (*art., sm. a.*) to withdraw (a charge, a load); to fire off again; (*mil.*) to retire, retreat, draw off; to retire from the service; (*cord.*) to shrink;
—— *sa parole*, to release one's self from a promise;
se —— à la prolonge, (*art.*) to withdraw with prolonges fixed.

rétiveté, rétivité, f., (*hipp.*) obstinacy, refusal to obey.

retombée, f., (*cons.*) courses of voussoirs near the springing line; (more strictly) elbow of a voussoir laid near the springing line.

retordre, v. a., to retwist, to twist again; to twist together (as two or three strands, cords, etc.).

retors, a., twisted; m., (*cord.*) second twist.

rétorsion, f., (*law*) retorsion.

rétorsoir, m., cord wheel; spun-yarn winch.

retouche, f. finishing, dressing, trimming; fitting (as of clothes, i. e., that have to be done over or made to fit); alteration (in any object, to make it serviceable, e. g., articles of uniform).

retoucher, v. a. n., to retouch; to dress, finish, trim (as after casting); (*mach.*) to dress, to finish (after main work has been done).

retour, m., return; home journey, return journey; angle, corner, turn, winding turn, twist; return pipe; (*cord.*) lead; round turn; (*law*) reversion; (*fort.*) flank of a battery; flank or end traverse; return, change of direction, of a sap, trench; (*mil. min.*) return of a mine;
—— *en arrière*, (*hipp.*) breeding back;
—— *en batterie*, (*art.*) return to battery;
bois sur le ——, tree decaying from age;
choc en ——, return stroke;
—— *d'un cordage*, (*cord.*) return lead or part of a rope led through a leading block;
côté en ——, return side;
courant de ——, (*elec.*) return current;
—— *de courant*, eddy of a stream;
de ——, back, homeward, homeward bound;
—— *droit*, v. —— *d'équerre*;
en ——, at right angles; (*mil.*) thrown back (as of a flank, or of a line);
—— *d'équerre*, (*siege, mil., min.*) rectangular change of direction;
en —— d'équerre, at right angles;
être sur le ——, (of trees) to begin to die, to fade;
faire son ——, (*nav.*) to return to the point of departure (of a ship);
fil de ——, (*elec.*) return wire;
filer un cordage à ——, (*cord.*) to pay out a rope with a turn;
—— *d'une manœuvre*, (*cord.*) hauling or running part;
—— *de la marée*, change, turn, of the tide;

retour *de mine*, (*mil. min.*) change of direction at the intersection of two mine galleries;
—— *oblique*, (*siege, mil. min.*) oblique change of direction;
—— *offensif*, (*mil.*) counter attack, return stroke; attempt to recover a position from which just driven; renewed attack (by the offensive);
—— *du piston*, (*mach.*) back stroke, return stroke, of a piston;
poulie de ——, leading block, quarter block;
prendre à ——, (*cord.*) to take a turn;
prendre un ——, (*cord.*) to take a round turn;
—— *rapide*, (*mach.*) quick return motion, quick return;
route de ——, homeward route;
—— *de sape*, (*siege*) change of direction in the sap;
—— *de sape double en sape double*, (*siege*) double sap return from a double sap;
—— *de sape double en sape simple*, (*siege*) single sap return from a double sap;
—— *de sape simple en sape double*, (*siege*) double sap return from a single sap;
—— *de sape simple en sape simple*, (*siege*) single sap return from a single sap;
tenir à ——, (*cord.*) to hold on with a round turn;
—— *en tête*, (*siege*) change of direction of same gallery;
—— *au type*, v. —— *en arrière*;
vivres de ——, unexpended provisions;
voyage de ——, return voyage.

retracer, v. a., to retrace.

retrait, a., *bois* ——, wood warped in seasoning.

retrait, m., repurchase; withdrawal; shrinkage; (*fond.*) shrinkage, contraction, of cast metal on cooling; (*fort.*) v. *retirade;*
à double ——, (*fond.*) with allowance for double shrinkage (said of a model, itself intended to furnish a second model);
—— *d'emploi*, (*mil.*) suspension, removal from an office or duty (punishment);
être en ——, *sur*, to be back from, not in the same surface with;
—— *de la mer*, (gradual) withdrawal of the sea;
mettre en —— *d'emploi*, (*mil.*) to deprive of one's office or place;
—— *de mouture*, shorts.

retraite, f., retreat, withdrawal, retiring, superannuation; place of retreat; (commercially) renewing, counterdraft; draft on the last indorser of a protested note; (*tech., met.*) contraction, shrinkage; (*cons.*) offset, setback; (*harn.*) sort of leading rein; (*fenc.*) retire; (*farr.*) part of nail left in the hoof; (*cord.*) free end (of a gin, windlass, or capstan rope, etc.); tackle fall; (*art.*) recoil; (*mil.*) retreat (before the enemy); retreat (in garrison), call for retreat; pension, retired pay; retired status, (strictly, *Fr. a.*) status of a soldier definitively struck from the rolls of the army with a retiring allowance; retired list; (*fort.*) berm.
(Except where otherwise indicated, the following expressions are military:)
avoir sa ——, to be on pension;
battre en ——, to retreat, to draw off;
se battre en ——, to make a running fight;
battre la ——, to beat retreat;
caisse de ——, superannuation fund;
contrôle nominatif des personnes mises à la ——, retired list;
coup de canon de ——, retreat gun, evening gun;
couper la ——, to cut off the retreat;
de ——, retiring;
demander sa ——, to apply for the retired list;
—— *directe*, direct retreat upon object to be covered;
donner —— *à* (*in gen.*) to shelter, harbor;
donner sa —— *à* to superannuate, pension off;
—— *des eaux*, (*in gen.*) subsidence of water;

retraite, en ——, in retreat; call to retreat; on the retired list, retired; to the rear (as, e. g., guns behind an intrenched line); (of teams) pulling back, so as to pull back (in steep descents); (*cons.*) setback;
état nominatif des personnes mises à la ——, v. *contrôle nominatif des personnes mises à la* ——;
être en ——, (*cons., etc.*) to be set or sit back, to retreat;
être en —— *les uns sur les autres*, (*con., etc.*) to be set back one from the other;
être mis à la ——, to be put on the retired list;
faire ——, (*cons.*) to sit or be back, retreat;
faire une ——, to make a retreat; (*in gen.*) to live in retirement for a time;
—— *latérale*, retreat in a direction perpendicular to the direction of the object to be covered;
lieu de ——, (*in gen.*) place of retreat, hiding place;
mettre à la ——, to pension off, to put on the retired list; to retire;
mise à la ——, retiring;
mise à la —— *d'office (pour les commissionnés)*, (*Fr. a.*) compulsory retirement of *commissionnés;*
obtenir sa ——, to get one's pension, to get on the retired list;
officier en ——, officer on the retired list;
opérer une ——, to effect a retreat;
palan de ——, (*cord.*) training tackle;
—— *parallèle* (*mil.*) retreat parallel to the base;
pension de ——, retiring pension or allowance;
sans pension de ——, unpensioned, without an allowance; wholly retired (U. S. A.);
—— *perpendiculaire*, retreat perpendicular to the base;
pièce de ——, (*nav., art.*) stern chaser;
placer en ——, to cause men and teams to hold back wagons on steep descents;
prendre sa ——, to go on the pensioned list; to retire upon a pension; to go, or be put, on the retired list;
quitter sa ——, to come out of retirement;
sans ——, v. *sans pension de* ——;
sonner la ——, to sound retreat, the retreat;
—— *stratégique*, the retreat of a defeated army after the battle;
—— *tactique*, withdrawal of a body of troops defeated in an assault;
—— *tirée*, (*nav.*) evening-gun fire (in military ports).

retraité, a., (*mil.*) retired, on the retired list, pensioned; m., person on the retired list; (strictly, *Fr. a.*) person who has a *pension de retraite;*
non ——, unpensioned, without an allowance (but still not in service; wholly retired, U. S. A.)
officier ——, retired officer.

retraiter, v. a., (*mil.*) to retire.

retranchement, m., curtailment, abridging; recess; (*fort.*) retrenchment, intrenchment.
(The following terms relate to fortifications:)
—— *de campagne*, field intrenchment;
—— *de champ de bataille*, field intrenchment, shelter trench;
—— *expéditif*, rapid intrenchment (constructed by working parties on both sides, and hence the same on both sides);
—— *d'infanterie*, infantry parapet or intrenchment (in front of main line of works);
—— *intérieur*, interior retrenchment (sort of keep or cavalier);
ligne de ——*s*, line of intrenchments;
—— *ordinaire*, intrenchment of the usual type;
—— *progressif*, progressive intrenchment (i. e., one that may be strengthened, from a mere shelter trench to the dimensions of the rapid intrenchment);
—— *rapide*, rapid intrenchment (a deliberate intrenchment);

retranchement *rapide normal*, normal hasty intrenchment (constructed if time permits);
—— *rapide perfectionné*, intrenchment less elaborate than the usual type (lower parapet, narrower and shallower ditches, etc.);
—— *rapide simplifié*, modification of the usual type (narrower and shallower ditches, lower and thinner parapet, etc.).

retrancher, v. a., to retrench, cut off, deduct, stop the allowance of, strike off, take away; (*mil.*) to intrench;
se ——, (*mil.*) to intrench one's self; to be intrenched.

retravailler, v. a., to work over again; to file; to retouch.

rétrécir, v. a., to narrow, straiten, take in, contract, shrink, limit, confine;
—— *un cheval*, (*man.*) to narrow a horse.

rétrécissement, m., narrowing, shrinking; shrinkage, contraction; narrowness, cramping.

retreindre, v. a., to hammer out, shape under the hammer;
—— *au martelage*, to raise, shape, under the hammer;
—— *au tour*, to shape in a lathe.

retreinte, f., hammering out, shaping under the hammer;
—— *au martelage*, raising; shaping under the hammer;
—— *au tour*, shaping in a lathe.

retremper, v. a., to dip, soak, again; (*met.*) to temper again, anew; (*fig.*) to strengthen, give new vigor to.

rétribuer, v. a., to remunerate, repay; to reward.

rétribution, f., remuneration, salary.

retrier, v. a., to sort over again.

rétrocéder, v. a., (*law*) to cede back.

rétrocession, f., (*law*) retrocession; (*Fr. a.*) retrocession, by sale, of horses issued to officers.

rétrogradation, f., retrogradation; (*Fr. a.*) reduction of a noncommissioned officer or corporal to a lower grade.

rétrograde, a., retrograde;
mouvement ——, retrograde motion; (*mil.*) retrograde movement (not necessarily a retreat).

rétrograder, v. n., (*Fr. a.*) to be reduced to a lower grade (noncommissioned officers and corporals); (*mil.*) to fall back.

retrotter, v. n., to retrot.

retroussé, p. p., tucked up, turned, trussed;
entrait ——, (*cons.*) hammer beam;
flancs ——*s*, (*hipp.*) hollow, tucked up, flanks.

retroussement, m., turning, tying, tucking up.

retrousser, v. a., to turn, tie, tuck, truss up;
—— *la queue*, to carry the tail high (horse).

retroussis, m., (*unif.*) facings (i. e., part turned back and showing the lining);
—— *de botte*, boot top;
bottes à ——, top boots.

retrouver, v. a., to find, find again;
—— *son chemin*, to find one's road;
—— *le fond*, (*nav.*) to get on soundings again.

rétuver, v. a., to foment a wound.

réunion, union, junction, collection; (*mach.*) connection;
—— *à contre-appui*, (*art.*) v. *contre-appui*;
—— *à suspension*, (*art.*) pintle-and-hook connection of the parts of a carriage.

réunir, v. a. r., to join, unite, bring together, annex, collect, muster, assemble, summon; to meet;
—— *à adent*, (*carp.*) to joggle;
—— *en crémaillère*, v. —— *à adent*.

réusinage, m., machining over again.

réusiner, v. a., to machine over again.

revaporisation, f., (*steam*) revaporization.

revaporiser, v. a., (*steam*) to revaporize.

revêche, a., harsh;
fer ——, (*met.*) iron that becomes hard on annealing.

réveil, m., awakening; alarm (of a clock); (*mil.*) reveille, call for reveille;
battre, sonner, le ——, (*mil.*) to beat, sound, reveille.

réveil(le)-matin, m., alarm, alarm clock; (*mil.*) reveille.

réveiller, v. a., to wake, rouse, call;
—— *le compas*, to touch, tap, the compass (so as to quicken the needle).

révélateur, m., (*phot.*) developer.

révélation, f., revelation, disclosure; (*phot.*) development.

révéler, v. a., to reveal, disclose; (*phot.*) to develop, bring out.

revendeur, m., retailer, broker.

revenir, v. n., to come back, return; (*met.*) to recover original qualities by tempering and annealing;
—— *à la charge*, (*mil.*) to renew an attack after a repulse;
faire —— *l'acier*, v. s. v. *faire*.

revenu, m., income; (*adm.*) revenue; (*met.*) annealing.

revenue, f., return; new growth of timber.

réverbération, f., reverberation, reflection.

réverbère, m., reflector; street lamp; (*met.*) dome of a reverberatory furnace;
feu de ——, (*met.*) reverberatory fire;
four à ——, (*met.*) reverberatory furnace;
fourneau à, de, ——, v. *four à* ——;
foyer à ——, v. *four à* ——;
lampe à ——, reflecting lamp.

reverdie, f., spring tide at new and at full moon.

revers, m., back, other side, wrong side, reverse; back stroke, back-handed blow; facing, lapel (of a coat); (*fenc.*) reverse; (*top.*) reverse slope of a hill, etc.; (*sm. a.*) back of a sword blade; (*fort.*) reverse (reverse slope);
à ——, (*mil.*) in reverse;
batterie de ——, (*art.*) reverse battery;
battre à, de, ——, (*mil.*) to take (a work, a line, etc.) in reverse;
bottes à ——, top boots;
—— *de botte*, boot top;
coup de ——, back blow, stroke, back-handed blow; (*fenc.*) reverse;
de ——, (*mil.*) in reverse;
double ——, said of a road with a gutter in the middle;
—— *d'eau*, (*cons.*) sort of pavement surrounding the walls of a house, to protect the foundations against rain water;
—— *du fossé*, (*fort.*) opposite side of the ditch (from the interior of the place);
frapper de ——, to strike a back-handed blow;
—— *d'un habit*, facing of a coat;
—— *de pavé*, side pavement, inclined part of the pavement from the house to the gutter;
prendre à, de, ——, (*mil.*) to take in reverse;
—— *de la tranchée*, (*siege*) reverse of the trenches (side toward the open country); (incorrectly) the exterior side of the parapet.

reverseau, m., (*cons.*) weather-strip on the sill of a window, of a casement.

reversement, m., change, turn, shift (of wind, current, tide, monsoon, etc.).

reverser, v. a., to pour out again; to transfer from one account to another.

réversibilité, f., reversibility; (*elec.*) reversibility of a dynamo.

réversible, a., reversible; (*elec.*) reversible (of a dynamo).

réversion, f., (*hipp.*) breeding back.

revêtement, m., revetment, facing, coating, casing; (*fort.*) revetment; (*steam*) lagging;
(Except where otherwise indicated the following terms relate to fortification:)
—— *en bois*, (*fort., nav.*) wood backing of armor; (*nav.*) wooden sheathing;
—— *en bordages*, (*nav.*) planking;
—— *en boue de route*, mud revetment;
—— *en branchages*, branch revetment;
—— *en briques*, brick revetment;
—— *en charpente*, plank revetment;
—— *de chaudière*, (*steam*) boiler lagging;
—— *en chiendent*, pisa revetment;
—— *en claies*, hurdle revetment;
—— *en clayonnage*, wicker revetment;
—— *de contrescarpe*, counterscarp revetment;
—— *du cylindre*, (*steam*) cylinder lagging;
—— *en décharge*, counter-arched revetment;

revêtement, demi- ——, half revetment;
—— *détaché*, detached revetment;
—— *d'escarpe*, scarp revetment;
—— *extérieur*, (nav.) outer skin of a ship;
—— *en fascines*, fascine revetment;
—— *en gabions*, gabion revetment;
—— *en gazons*, sod revetment;
—— *en gazons par boutisses et panneresses*, sod revetment laid header- and stretcher- wise;
—— *en gazons posés de plat*, facing sod revetment;
—— *d'huisserie*, (cons.) jamb lining;
—— *intérieur*, (nav.) inner skin of a ship;
lambris de ——, v. s. v. *lambris*;
—— *en maçonnerie*, masonry revetment;
—— *en madriers*, v. —— *en charpente*;
—— *mixte*, revetment of gabions and fascines;
—— *mort*, fascine revetment (supports no thrust);
mur de ——, v. s. v. *mur*;
—— *en nattes*, straw-band revetment;
—— *d'un navire*, (nav.) skin of a ship;
—— *à parement incliné*, inclined revetment;
—— *à parement vertical*, vertical revetment;
—— *en pierres*, stone revetment;
—— *en pierres sèches*, revetment of stone laid dry, without mortar;
—— *en pisé*, pisa revetment;
—— *plein*, full revetment;
—— *en sacs à terre*, sand-bag revetment;
—— *en saucissons*, fascine revetment;
—— *terrassé*, v. *mort*;
—— *en tôles*, (nav.) plating;
—— *en torchis*, mud and chopped-straw revetment.

revêtir, v. a., to clothe; to face, cover; to put on, buckle on; to vest; to assume; to fit, equip, with; to confer upon; (cons., fort.) to revet; to wainscot;
—— *un acte de . . .* , to draw up a paper (in due form); to put (a signature) to a paper;
—— *de signatures en rond*, to sign round a robin.

revêtissement, m., v. *revêtement*.

revidage, m., reboring, enlargement, of a hole.

revider, v. a., to rebore, enlarge, a hole.

revient, m., *prix de* ——, net cost.

revirement, m., turn of the tide; transfer; adjustment of accounts on the books; (adm.) exchange of stores.

réviseur, m., examiner, reviser.

révision, f., view, inspection; medical examination; (sm. a.) revision, reexamination, of small arms after proof;
—— *des chaudières*, (steam) boiler testing;
conseil de ——, v. s. v. *conseil*;
se pourvoir en ——, (mil. law) to lodge an appeal.

révocation, f., revocation; (Fr. a.) revocation of warrant (chefs de musique and commissionnés); cancellation of commission (officers with the reserve and of the territorial army).

revolin, m., eddy wind; back or reverse current.

révolte, f., revolt, mutiny, rebellion;
en ——, in revolt, rebellion;
exciter, porter, à la ——, to excite, stir up, revolt.

révolté, m., rebel.

révolter, v. a. r., to revolt; to rebel; to mutiny; to rise up in arms against.

révolu, a., accomplished:
avoir n ans ——*s*, to have completed one's nth year.

révolution, f., revolution, turn; revolution, i, e., rebellion; turn of a rope around a windlass, capstan;
la —— *de février*, (French) revolution of 1848;
la ——, *la grande* ——, (French) revolution of 1789;
—— *de juillet*, (French) revolution of 1830;
la première ——, (French) revolution of 1789;
la —— *du 4 septembre*, (Parisian) revolution of September 4, 1870.

revolver, m., (sm. a.) revolver;

revolver, *à double mouvement*, double-action revolver;
—— *à feu continu*, self-cocking revolver;
—— *à mouvement continu*, self-cocking revolver;
—— *à mouvement intermittent*, single-action revolver;
—— *à mouvement simple*, single-action revolver;
—— *à n coups*, n-chambered revolver;
—— *d'officier*, officer's revolver;
pistolet- ——, navy revolver (France).

révoquer, v. a., to revoke, recall; (mil.) to discharge; (Fr. a.) to withdraw one's employment (v. *révocation*).

révoyeur, m., canal dredge (boat).

revue, f., examination, survey, inspection; (mil.) review, muster, inspection; also "review and inspection," U. S. A.;
—— *administrative*, (Fr. a.) review made by the administration, as distinct from a purely military inspection;
—— *de détail*, (Fr. a.) general kit or equipment inspection, examination of account books, etc. (general's inspection; may take place in quarters);
—— *d'effectif*, (Fr. a.) general inspection (made by the *intendance*, or else by an official of the *contrôle*, for the purpose of verifying numbers); muster (U. S. A.);
—— *d'ensemble*, (mil.) general inspection;
faire la —— *de*, to review, survey, examine, make an inspection;
—— *de ferrure*, (mil. slang) said of a horse that plunges and kicks out (shows his shoes);
grande ——, (mil.) grand, ceremonious, review; field day;
—— *d'honneur*, (Fr. a.) review held at a general inspection;
—— *d'inspection*, (mil.) kit inspection;
—— *d'installation*, (Fr. a.) review held by the general commanding a corps, on taking command of it;
—— *de liquidation*, (Fr. a. adm.) quarterly comparison of regimental accounts, in order to balance and settle them;
—— *de linge et chaussures*, v. —— *de petit équipement*;
passer en, la, une, ——, (mil.) to review;
passer la —— *de*, (mil.) to hold a review; to review a particular regiment or corps, etc.; to be passed in review by, to be reviewed by; to muster, to be mustered;
—— *de petit équipement*, (mil.) kit inspection;
—— *de propreté*, (mil.) inspection (of clothes, arms, etc.); Saturday morning inspection (U. S. A.);
—— *de santé*, medical inspection.

rez, m., level.

rez, prep., close to, on the level with;
—— *de mur*, interior measurement, in the clear;
—— (-)*pied*, even, level, with the ground;
—— (-)*terre*, v. —— *pied*;
—— *tronc*, close to the trunk.

rez-de-chaussée, m., ground level; street level; ground floor;
à ——, level with the ground;
au ——, on the ground floor.

rez-mur, m., (cons.) inner face of a wall; clear distance between walls;
avoir tant de pieds de ——, (of a beam) to be so long in the clear.

rez-terre, m., even surface of the ground.

rhabillage, m., repair, mending, patching up, so as to render serviceable; (more esp. sm. a.) repair of damaged small arms.

rhabiller, v. a., to repair; to dress again.

rhabilleur, m., mender, patcher;
—— *ajusteur*, (sm. a.) workman who mends and fits.

rhéostat, m., (elec.) rheostat;
—— *à clef*, plug rheostat;
—— *d'excitation*, motor rheostat;
—— *à liquide*, hydrorheostat;
—— *à manette*, sliding rheostat;
—— *de réglage*, regulating rheostat.

rhexite, f., (expl.) rhexite.

rhumb, m., v. *rumb*.

ria, m., v. *réa.*
riaule, f., miner's instrument, sort of hook fitted with a handle.
riblon, m., scrap iron;
   fer de ——, scrap iron.
ribordage, m., (*nav.*) damage caused by a collision;
   droit de ——, collision damages.
ricin, m., castor-oil plant;
   huile de ——, castor oil.
ricochable, a., (*art.*) exposed to richochet; that will cause a projectile to ricochet.
ricocher, v. a. n., (*art.*) to ricochet; to fire at the long faces of a work.
ricochet, m., (*art., sm. a.*) ricochet;
   à ——, ricochet, ricocheting;
   balle mise par ——, (*t. p.*) ricochet hit;
   batterie à ——, (*art.*) ricochet battery;
   battre à ——, (*art.*) to use cicochet fire;
   faire des ——*s*, (*art.*) to cause a shot to ricochet;
   feu à ——, (*art.*) ricochet fire;
   —— *mou,* ——*plongeant,* (*art.*) plunging ricochet;
   —— *raide,* (*art.*) flat ricochet;
   —— *tendu,* v. —— *raide;*
   tir à ——, (*art.*) ricochet fire.
ride, f., wrinkle; fold; ripple (on the surface of water); (*cord.*) lanyard;
   nœud de ——, (*cord.*) Matthew Walker's knot;
   —— *à palan,* (*cord.*) lanyard serving as tackle to a stay.
ridé, p. p., ribbed, corrugated, fluted.
rideau, m., curtain, screen; mound or elevation raised above a road, a canal, supporting wall of such a mound; screen of trees, of bushes, wind screen; blower (of a fireplace); (*pont.*) the rods, chains, etc., supporting the floor of a suspension bridge; (*phot.*) shield (of a plate holder); (*mil. top.*) undulation of the ground (behind which troops may be concealed); (*mil.*) screen of troops (esp. cavalry); screen of skirmishers;
   —— *de cavalerie,* (*mil.*) screen of cavalry;
   —— *défensif,* (*fort.*) a line of works barring hostile advance; (more esp.) fortified region on or near the frontier;
   —— *protecteur,* (*mil.*) generic term for all troops acting as a screen in front of an army (advance guard, etc.);
   —— *de tente,* curtain of an awning.
ridelle, f., cart rack, rave; (*in pl.*) oaken saplings, reserved for wheelwright's work; (*art.*) upper rail (of a wagon frame); upper rail or spar at the top of an artillery wagon;
   chariot à ——*s,* v. s. v. *chariot.*
rider, v. a., to ripple; to corrugate; (*cord.*) to set up, tighten, haul taut.
ridoir, m., (*cord.*) rigging screw, stretcher;
   —— *à vis,* stretching screw, regulating screw.
riflage, m., operation of using a *rifloir.*
riflard, m., large double-handled plane; jack plane; mason's parting tool; toothed chisel; coarse, heavy, metal file.
rifler, v. a., to file; to use a jack plane; to furrow a grindstone.
rifloir, m., rifle; crooked file: file bent at end for working in the bends of different pieces of metal or wood.
rigidité, f., rigidity, stiffness;
   —— *de la trajectoire,* (*ball.*) rigidity of the trajectory.
rigodon, m., (*t. p.*) call or battery announcing a bull's-eye.
rigole, f., drain, culvert, gutter; small ditch; trench, furrow; small foundation trench; small stream; (*mach.*) oil groove; (*fond.*) runner, channel for melted pig iron; (*sm. a.*) gas port; (*fort.*) trench in which first layer of fascines is set to one-half their diameter; trench for sod revetments;
   —— *de châssis de place,* (*art.*) groove for trail-truck
   —— *d'échappement,* (*sm. a.*) gas port;
   —— *d'écoulement,* culvert;
   —— *latérale,* side drain of a road;
   —— *du milieu,* main drain.

rigueur, f., rigor; severity of the weather;
   arrêts de ——, v. s. v. *arrêt.*
rime, f., rhyme; stroke (in rowing).
ringard, m., (*met., etc.*) poker, fire iron; tapping bar; porter bar;
   barre de ——, picker bar;
   —— *à crochet,* hook, fire hook;
   —— *à lance,* slice, fire slice.
ringarder, v. a., (*met., etc.*) to stir, poke, the fire.
ringot, m., (*cord.*) small grommet, becket.
ripage, m., (*mas.*) smoothing of a stone, removal of plaster, with a scraper; (*cord.*) slipping of a rope; shifting of weights in a horizontal direction, as (*r. r.*) sliding of cross-ties along the ground or on the track, either by hand or by crowbars;
   —— *de la voie,* (*r. r.*) side shifting of the track.
ripe, f., (*mas.*) scraper; drag.
ripement, m., boiling of the sea (due to the meeting of submarine currents); (*cord.*) slipping of one rope on another.
riper, v. a. n., to shift a weight (in a horizontal direction); to boil (of the sea); (*mas.*) to scrape or make smooth (with a drag); (*nav.*) to shift (of cargo); (*cord.*) to slip, surge;
   pince à ——, v. s. v. *pince;*
   —— *sur,* (*cord.*) to wear, chafe, against;
   —— *la voie,* (*r. r.*) to shift the track sidewise.
riposte, f., (*fenc.*) parry and thrust, ripost, return thrust; (*man.*) start, jump (of a horse on feeling the spur).
(The following terms relate to fencing:)
   —— *composée,* compound ripost (i. e., the result of a combination of movements);
   contre- ——, counter thrust (attack following a parried ripost);
   —— *droite,* a ripost in the same line as the parry;
   —— *d'opposition,* high ripost (hand elevated on account of the adversary's *opposition;*
   —— *simple,* simple, straight, ripost;
   —— *du tac au tac,* a thrust or blow following immediately on a parry;
   —— *à temps d'arrêt,* ripost preceded by a pause.
riposter, v. a., (*fenc.*) to parry and thrust; (*mil.* and *art.*) to fire back; to return fire;
   —— *au feu,* (*art.*) to return an enemy's fire;
   —— *à la mouche,* v. s. v. *mouche.*
riquette, f., (*met.*) scrap iron; clipping from skelp when there is more metal than is required;
   fer de ——, scrap iron.
ris, m., (*nav.*) reef.
risban, m., (*fort.*) risban, (artificial) bank or flat piece of ground on which a battery may be constructed for the defense of a harbor;
   —— *armé* port battery.
risberme, f., risberm; pile and stone work to protect a foundation against washing; space between the sheet piling and the body of a cofferdam; (*fort.*) risberm; fascine work at the foot of an embankment.
risée, f., squall, gust (of wind).
riser, v. a., (*nav.*) to reef.
risque, m., risk;
   —— *de capture,* risk of capture;
   —— *s de guerre,* risks of war;
   —— *s de mer,* perils of the deep; dangers of the sea;
   à ses ——*s et périls,* at one's own risk.
risse, f., (*cord.*) gripe, lashing, frapping, rope.
risser, v. a., (*nav.*) to lash to the deck.
risson, m., four-clawed grapnel.
rivage, m., river bank; shore (of the sea, of a lake); (*tech.*) riveting.
rive, f., bank, shore, border (of a stream, of a lake); towpath; skirt, edge, border (of a wood); (*farr.*) edge, border (of a horseshoe);
   —— *amie,* (*mil.*) our side, our bank, of the river;
   —— *d'arrivée,* (*pont.*) bank to which a bridge is built or toward which it is dismantled;
   —— *de départ,* (*pont.*) bank from which a bridge is built or from which it is dismantled;
   deuxième, ——, v. —— *d'arrivée;*
   —— *droite,* right bank (of a stream);

rive *ennemie*, (*mil.*) the enemy's side, the other side of a stream;
— *externe* (*interne*), (*farr.*) outer (inner) border of a horseshoe;
— *d'un four*, border of a furnace;
— *gauche*, left bank (of a stream);
première —, v. — *de départ;*
seconde —, v. — *d'arrivée.*

rivé p. p., bent over (of a sword point, etc.).

rivement, m., clinching (of a nail).

river, v. a., to clinch; to clinch a nail; to rivet;
mordache à —, v. s. v. *mordache;*
presse à —, v. s. v. *presse;*
— *sur virole*, to clinch on a ring.
(For other compounds, see *rivetage* and *rivure*.)

riverain, m., person living on the bank of a river or lake; owner of property situated along a road, a forest; (*in pl.*) inhabitants of both sides of a street.

riverain, a., riparian; bordering on a forest; owning property situated along a forest, road, or street;
propriétaires —*s*, (near) neighbors;
propriétés —*es*, neighboring property.

rivet, m., rivet, clinch; (*farr.*) clinch of a horseshoe nail; border of a horseshoe; (*cons.*) edge of a roof at the gable;
— *d'assemblage*, rivet connecting two or corresponding parts;
— *bouterollé*, snap-head rivet;
chasse- —, riveting set;
— *de clou d'un fer à cheval*, (*farr.*) horseshoe clinch;
écartement des —*s*, pitch of rivets;
enfoncer la tête d'un —, to countersink a rivet;
— *à entretoise*, stay bolt;
— *fileté*, threaded bolt;
foncer un —, v. *enfoncer la tête d'un* —;
— *fraisé*, countersunk rivet;
fraiser la tête d'un —, v. *enfoncer la tête d'un* —;
marteau à —, riveting hammer;
— *noyé*, countersunk rivet;
noyer un —, (*farr.*) to clinch a nail;
rang de —*s*, row of rivets;
refouler un —, v. s. v. *refouler;*
— *à tête affleurée*, flush-head rivet;
— *à tête conique*, conical-head rivet;
— *à tête fraisée*, flush, countersunk, rivet;
— *à tête perdue*, countersunk, flush, rivet;
— *à tête ronde*, round-head rivet;
trou de —, rivet hole.

rivetage, m., rivet work, riveting;
— *des abouts*, butt riveting;
— *à la bouterolle*, snap-head rivet work;
— *en chaîne*, chain riveting;
— *à chaud*, riveting hot;
— *à couvre-joint double* (*simple*), double (single) lap-joint riveting;
— *double*, — *à double rang*, double riveting;
— *en échiquier*, zigzag, staggered, snake riveting;
— *fait à la machine*, machine riveting;
— *fait à la main*, hand riveting;
— *des joints longitudinaux*, edge riveting;
— *à la main*, hand riveting;
— *mécanique*, machine riveting;
— *quadruple*, — *à quadruple rang*, quadruple riveting;
— *en quinconce*, zigzag riveting;
— *à un rang*, single riveting;
— *simple*, single riveting;
— *à tête conique*, conical riveting;
— *à tête fraisée*, countersunk riveting;
— *triple*, — *à trois rangs*, triple riveting;
— *vis-à-vis*, chain riveting;
— *en zigzag*, zigzag riveting.

riveter, v. a., to rivet.

rivet-paillette, m., (*Fr. art.*) primer reenforce (De Reffye cartridge).

riveur, m., riveter.

riveuse, (*mach.*) riveting machine;
— *à vapeur*, steam riveting machine.

rivière, f., river; (*top.*) narrow valley containing a few feeble springs (rare);
fausse —, (*top.*) temporary or overflow channel of a river;

rivière, *la* — *de Gênes*, the Riviera;
— *haute*, full river;
— *marchande*, navigable river.

rivoir, m., riveting hammer; rivet set.

rivure, f., hinge pin; pin-connected joint; (action of) riveting; the head of a rivet (shaped *in situ*); riveted joint(s), rivet joint; clinched end of a rivet;
— *bombée*, snap-head riveting;
— *en carré*, chain or square riveting;
— *à chaîne(s)*, chain riveting (butt riveting, butt-joint riveting);
— *à chaînes double*, double chain riveting (double butt riveting);
— *à chaînes simple*, single chain riveting (single butt riveting);
— *conique*, conical, hammered joint, riveting;
— *convergente*, group riveting;
— *convergente à n rangs*, n-group riveting;
— *à deux rangs*, double riveting;
— *double*, double riveting;
— *double à recouvrement*, double lap riveting;
— *en échiquier*, zigzag, snake, staggered, riveting;
— *de force*, special riveting, riveting of special strength;
— *fraisée*, countersunk riveting;
— *en goutte de suif*, round-head riveting;
— *de joints*, edge riveting;
— *plate*, flat riveting, flush riveting;
— *à plat-joint*, butt-joint riveting;
— *à pointe de diamant*, conical riveting;
— *à recouvrement*, lap (joint) riveting;
— *à recouvrement double* (*simple*) double (single) lap (joint) riveting;
— *rectangulaire*, square or chain riveting;
— *simple*, single riveting;
— *simple à recouvrement*, single lap (joint) riveting;
— *à trois rangs*, treble riveting;
— *zigzag*, — *en zigzag*, staggered, zigzag, riveting;

rixe, f., quarrel, disturbance; combat, scuffle.

riz, m., rice.

rizière, f., rice field, rice plantation.

riz-pain-sel, m., (*mil. slang*) anyone connected with the commissary department.

robe, f., dress, robe (of a judge, etc.); the magistracy; (*hipp.*) (coat of a horse); name for the whole of the hairy covering of a horse, tail and mane included;
— *binaire*, coat of two elements;
— *de cheval*, coat of a horse;
cheveaux de même —, horses matched in color;
— *composée*, composite coat (more than one color);
— *de deux poils*, mixed coat (two colors, e. g., gray);
— *d'un seul poil*, simple coat (only one color);
— *simple*, simple coat (single color);
— *ternaire*, coat of three elements;
— *de trois poils*, mixed coat (three elements).

robinet, m., cock; tap; plug of a cock; (*art.*) connecting tube (of a combination fuse); (*steam*) Corliss valve;
— *d'admission*, steam cock, admission cock;
— *à air*, air cock;
— *alimentaire*, — *d'alimentation*, feed cock;
— *angulaire*, angle cock;
— *d'arrêt*, stopcock;
— *d'arrosage*, water-service cock;
— *d'aspiration*, suction cock;
— *automoteur*, self-acting cock;
— *pour boire à même*, bib cock;
boisseau de —, shell of a cock;
— *à boisseau*, ordinary faucet, plug cock;
— *de boîte à graisse*, grease cock;
clef de —, plug of a cock;
— *de chaudière*, gauge cock;
— *à clef creuse*, hollow plug cock;
— *de communication*, communicating cock;
— *condenseur*, condenser cock;
— *de cylindre*, drain cock, pet cock;
— *conique*, conical cock;

robinet *de décharge*, delivery cock;
— *de détente*, expansion cock;
— *distributeur*, regulator cock, distributing cock;
— *de distribution*, v. — *distributeur;*
— *à eau*, water cock;
— *d'eau*, gauge cock;
— *à eau salée*, brine cock;
— *d'échappement*, exhaust cock;
— *de l'écume*, scum cock;
— *d'entrée de vapeur*, steam regulator;
— *équilibré*, balanced cock;
— *d'épreuve*, gauge cock;
— *d'essai*, gauge cock, try cock;
— *d'essai de pompe*, pet cock;
— *s étagés*, gauge cocks (one above the other at different levels of water);
— *d'évacuation*, exhaust cock;
— *d'extinction des feux*, steam cock for putting out fires;
— *d'extraction*, blow-off cock;
— *d'extraction par le fond*, bottom blow-off cock;
— *d'extraction à la surface*, brine cock, scum cock, surface blow-off cock;
— *à feu*, fire plug;
— *flotteur*, float gauge;
— *de godet*, tallow cock;
— *de graissage*, — *à graisse*, — *-graisseur*, grease cock, tallow cock, lubricating cock;
— *à huile*, oil cock;
— *à incendie*, fire plug;
— *d'incendie*, steam fire cock;
— *indicateur*, gauge cock, indicator cock;
— *d'injection*, injection cock;
— *d'injection à la cale*, bilge injection cock;
— *d'injection par mélange*, jet injection cock;
— *d'introduction*, steam cock;
— *jauge*, gauge cock;
— *jauge de niveau d'eau*, water-gauge cock;
— *jauge supérieur*, top gauge cock;
— *de manomètre*, gauge; steam-gauge cock;
— *de mise en train*, starting cock;
— *modérateur*, regulator tap; throttle; ball cock;
— *de niveau de l'eau*, gauge cock;
noix de —, plug of a cock;
— *à n canaux (eaux, faces, fins, orifices, ouvertures, tubulures, voies)*, n-way cock;
— *ordinaire*, stopcock;
— *papillon*, winged tap; butterfly cock;
— *paraglace*, ice cock;
poignée de —, handle of a cock;
— *de prise de vapeur*, steam cock;
— *de prise d'eau à la mer*, sea cock;
— *de purge*, drain cock, air cock, clearing cock; pet cock of a pump;
— *purgeur*, drain cock;
— *purgeur de cylindre*, cylinder drain cock;
— *réchauffeur*, heating cock;
— *de regard*, gauge cock;
— *de remplissage*, filling cock (as, *art.*, of a recoil cylinder);
— *réparateur*, relief cock;
— *de retenue*, stopcock;
— *à rotation*, rotary cock;
— *de sel*, sediment cock;
— *souffleur*, blast cock;
— *à soupape*, valve faucet, disk plug valve;
— *de sûreté*, safety cock, shut-off cock, stopcock, safety water plug or faucet; test cock;
— *valve*, gate valve;
— *à valve*, rotating valve;
— *vanne*, gate valve; sluice gate, plug, cock, faucet; feed-valve or cock;
— *à vapeur*, steam cock;
— *à vapeur extincteur*, steam fire-extinguishing cock;
— *vérificateur*, pet cock;
— *de vidange*, blow-off cock, drain cock, blowdown cock;
— *de vidange de l'eau*, water-outlet cock;
— *à vis*, screw faucet.
robinetterie, f., cocks and valves.
robre, m., oak, English oak.
roburite, f., (*expl.*) roburite.
roc, m., rock, mass of rock;
de —, rocky.
rocaille, f., rockwork.
rocailleux, a., flinty, stony.

rocantin, m., v. *roquentin.*
roche, f., rock, rocks, mass of rock; hardest stone of a quarry; calcareous, semicompact stone, containing shell prints; lump (of coal); crude borax; (*in pl.*) overburnt or vitrified bricks; (*met.*) horse;
banc de —, (*hydr.*) reef;
— *couverte à la marée*, (*hydr.*) half-tide rock;
— *à feu*, (*artif.*) Valenciennes composition, firestone;
— *à fleur d'eau*, (*hydr.*) rock on a level with water;
fond de —, (*hydr.*) rocky bottom;
— *noyée*, (*hydr.*) sunken rock;
— *qui découvre à mi-marée*, (*hydr.*) half-tide rock;
— *à sel*, rock salt;
— *à souder*, soldering stone;
— *sous l'eau*, (*hydr.*) sunken rock.
rocher, m., high, steep, rock; (*hydr.*) rock (above water), small islet;
banc de —, (*hydr.*) reef;
— *qui découvre*, (*hydr.*) rock above water.
rocher, v. a., to sprinkle with borax.
rochet, m., (*mach.*) rack, ratchet, ratchet wheel; ratch;
— *d'encliquetage*, ratchet wheel;
rayure à —, v. s. v. *rayure;*
roue à —, ratchet wheel, click.
rochette, f., (*artif.*) war rocket (obs.).
rocheux, a., rocky.
rochoir, v. a., solder box.
rodage, m., grinding;
— *à l'émeri*, grinding of a lens (with emery).
roder, v. a., to grind, to rub together; to polish (as bird shot);
— *à l'émeri*, to grind with emery.
rôder, v. n., to work smoothly (of a machine, a mechanism); to have play;
— *sur son ancre*, (*nav.*) to veer at anchor;
— *sur un pivot*, to turn on a pivot.
rodet, m., horizontal mill wheel.
rodoir, m., polisher; grinder;
— *à l'émeri*, emery stick.
rogatoire, a., *commission* —, (*law*) requisition by one judge on another.
rogne, f., scab, mange, inveterate itch; mildew on wood.
rogne-pied, m., (*farr.*) farrier's butteris, rasp.
rogner, v. a., to pare, scrape, trim; to cut off;
— *le pied d'un cheval*, (*farr.*) to pare a hoof.
rogneux, a., scabby, mangy, itchy.
rognon, m., kidney; kidney stone; (*harn.*) cantle (of a saddle);
bande de —, (*horn*) cantle plate;
blessure aux —*s*, (*hipp.*) navel gall;
— *s de fer noyés dans l'asphalte*, (*cons.*) iron concrete;
mine en —*s*, (*met.*) kidney ore.
rognure, f., shred, clipping, paring.
roi, m., king.
rôle, m., sheet (of writing, on both sides); list, catalogue; (*mil.*) roll, rolls, roster, muster roll (obs. in France; *contrôle nominatif* is used instead); (*law*) calendar (of cases to be tried); (*mach.*) roller; (*nav.*) bill;
— *de combat*, (*nav.*) quarter bill;
— *d'équipage*, (*nav.*) list, muster roll, of the ship's company;
— *d'incendie*, (*nav.*) fire-station bill;
porter sur le —, (*mil.*) to put or carry on the rolls;
— *de poste*, (*nav.*) station bill;
— *de quart*, (*nav.*) watch bill;
— *des soldats*, (*mil.*) muster bill;
à tour de —, in turn; (*mil.*) by roster (of duty).
romaillet, m., v. *rombaillet.*
romain, a., Roman; (*mil.*) infantryman (rare).
romaine, f., steelyard; lever (of a weighbridge); sliding-weight testing machine; rope-testing machine; roman (lettering for maps);
— *droite*, upright roman (for villages on maps);

romaine à *leviers multiples*, compound-lever testing machine;
—— *oscillante*, steelyard;
—— *penchée*, inclined roman (for hamlets on maps).

rombaillet, m., graving piece, dutchman.

romite, f., (*expl.*) romite.

rompre, v. a. r. n., to break, break in pieces, asunder; to burst through; to destroy (as roads, bridges, etc.); to break the effect of; to interrupt, discontinue, prevent, avert, break off, divert, keep off; to fail in an obligation; to fatigue; to train, accustom to, become accustomed to; (*opt.*) to refract; (*fenc.*) to retire (said also of boxing); (*cord.*) to part; (*mil.*) to break an enemy's line; to reduce, diminish, the front (as of a column for a march); to pass from line to column; to break into column; to lose one's formation;
—— *une armée*, (*mil.*) to break up or disband an army;
—— *les arrêts*, (*mil.*) to break one's arrest;
—— *un bataillon*, etc., (*mil.*) to break through a battalion, etc.;
—— *le camp*, (*mil.*) to break up camp;
—— *un carré*, (*mil.*) to break a square; to reform into column from square;
—— *charge*, to break bulk;
—— *un chemin*, to break up a road;
—— *un cheval*, (*man.*) to break a horse;
—— *un cheval à une allure*, (*man.*) to train a horse to a gait;
—— *en colonne*, (*mil.*) to break into column;
—— *le combat*, (*mil.*) to break off a fight;
—— *le cou d'un cheval* (*man.*) to make the head of a horse flexible;
—— *un coup*, to weaken or deaden a blow;
—— *un cours d'eau*, to divert a stream; to turn a stream off;
—— *un courant*, (*elec.*) to break a current; to open a circuit;
—— *à droit*, (*mil.*) to break into column to the right;
—— *l'eau, à un cheval*, to interrupt a horse while drinking, to keep him from drinking his fill;
—— *les faisceaux*, (*mil.*) to break stacks;
—— *à la fatigue*, to inure to fatigue;
—— *le fil*, v. —— *un cours d'eau*;
—— *à gauche*, (*mil.*) to break into column to the left;
—— *la glace*, to break the ice;
—— *un gué*, to destroy a ford;
—— *à la guerre*, (*mil.*) to inure to war;
—— *la mesure*, (*fenc.*) to retire while parrying;
—— *la mesure à son adversaire*, (*fenc.*) to keep one's adversary from delivering his thrust;
—— *le pas*, (*mil.*) to break step, to take route step (as in going over bridge);
—— *un pont*, to destroy a bridge;
—— *la prison*, to break jail;
—— *les rangs*, (*mil.*) to break ranks, to be dismissed (as after drill);
—— *le salpêtre*, to pulverize saltpeter;
—— *la semelle*, (*fenc.*) to retire a foot's length;
—— *une troupe*, (*mil.*) to drive in a body of troops, throw it into disorder.

rompu, p. p., broken; trained; (*nav.*) carried away; (*mil.*) defeated, broken up;
chemin ——, broken road;
—— *à la fatigue, guerre*, etc., inured to fatigue, war, etc.;
train ——, (*man.*) broken gait, resembling the *aubin* and the *traquenard*, qq. v.

ronce, f., bramble, brier;
—— *artificielle*, (*mil., fort.*) obstacle or entanglement composed of spiral or barbed wire, so as to form artificial thorns.

rond, a., round, circular, cylindrical, rounded off full, even; round (of a number); stiff (of the wind);
lettre ——*e*, round hand.

rond, m., washer; round; ring; circle; wheel; traversing plate; circular sweep, fifth-wheel, of a wagon; (*met.*) iron of round cross section; (*pl.*) bar iron; (*man.*) volt;
—— *d'arrière-train*, upper part of a fifth-wheel;

rond *d'avant-train*, lower part of a fifth-wheel;
couper le ——, (*man.*) to cut a volt;
—— *de cuir*, leather washer; leather round for a chair; (hence, *fam.*) person who never leaves his desk, paper man, chair warmer;
demi- ——, v. *demi-rond;*
—— *de dessous de train*, v. —— *d'avant-train;*
—— *de dessus de train*, v. —— *d'arrière-train;*
—— *d'eau*, circular reservoir or basin of water;
en ——, round, in a ring;
faux- ——, (*met.*) bend or curve due to irregularity of heating or of cooling (as in a gun barrel or other steel tube, or in any other piece that should be cylindrical);
—— *de papier*, (*sm. a.*) paster, target practice;
quart de ——, quarter round;
—— *de roue*, tire;
tourner en ——, to turn round;
travailler un cheval en ——, (*man.*) to work a horse in a ring.

ronde, f., round; round hand; (*mil.*) rounds (duty and party); (*man.*) round, volt;
à la ——, round, around;
chemin de ——, (*fort.*) chemin de ronde;
—— *de commandant d'armes*, (*mil.*) rounds made by the c. o.;
contre- ——, (*mil.*) check rounds;
en ——, (*mil.*) on rounds;
être de ——, (*mil.*) to be on one's rounds;
faire la ——, (*mil.*) to make rounds;
—— *-major*, (*mil.*) round made by a field officer or by the major of the garrison; grand rounds;
—— *de nuit*, (*mil.*) night rounds;
—— *d'officier*, (*mil.*) officer's rounds;
officier de ——, (*mil.*) officer of the rounds;
—— *de sous-officier*, (*mil.*) n. c. o.'s rounds;
—— *supérieure*, (*mil.*) grand rounds.

rondeau, m., (*art.*) astragal, round molding on a gun.

rondelle, f., any (small) round of wood or metal; washer, packing washer, scutcheon, disk, wad; round-nosed chisel; (*powd.*) disk, pellet; (*art.*) face plate, obturator plate; vent-bush plate; small washer for the points of a star gauge; (*mach.*) dial; (*met.*) disk of (purified) copper;
—— *-agrafe*, (*art.*) primer spindle catch;
—— *annulaire*, ring washer;
—— *d'arrêt*, (steam) screw of safety-valve lever;
—— *-arrêtoir(e)*, stop; (*sm. a.*) cut-off; (*art.*) wooden base plug or bottom of a canister;
—— *Belleville*, Belleville disk or spring;
—— *de bout*, (*art.*) axle-end washer;
—— *de bout à crochet*, hook washer;
—— *de bout d'essieu*, linchpin washer;
—— *de bout à piton*, eyebolt washer;
—— *de butée*, (*mach.*) thrust collar, supporting ring or collar (as of a spiral spring);
—— *de carton*, pasteboard disk;
—— *en carton-pâte*, (*art.*) pasteboard obturator;
—— *de collet d'essieu*, shoulder washer;
—— *à coupelle*, (*art.*) cup washer (of an axle);
—— *à crochet*, (*art.*) drag hook; hook washer;
—— *à crochet à oreille*, (*art.*) drag hook;
—— *de cuir*, leather disk or washer;
—— *de cuir embouti*, cup leather washer;
—— *à cuvette*, elastic cup washer (as for armor bolts);
—— *de démoulage*, (*fond.*) ring serving to ease the removal of a pattern (in projectile casting, France);
—— *élastique*, elastic washer;
—— *d'épaulement*, (*art.*) shoulder washer (o- axle);
—— *d'ensabotage*, (*art.*) sabot disk of tin (for spherical projectiles, obs.);
—— *entre deux*, v. *fausse* ——;
—— *d'étoupage*, (*mach.*) packing washer;
—— *d'essieu*, drag washer plate;
fausse ——, (*mach.*) intermediate disk (of slitting rollers);
—— *de flèche*, (*art.*) flat ring, lining the opening for the head of the perch, cut in the bolster and hind axletree;
—— *fusible*, (*mach.*) fusible plug, safety plug;
—— *de garniture*, (*mach.*) packing washer;

rondelle *d'obturateur*, (*F. m. art.*) gas-check plate;
— *obturatrice*, (*mach.*) packing washer;
— *ouverte*, semicircular washer;
— *de papier*, (*t. p.*) paster;
— *en papier*, (*artif., etc.*) paper round, patch;
— *porte-couronne d'appui d'obturateur*, (*F. m. art.*) bearing plate of an obturator collar;
— *porte-obturateur*, (*F. m. art.*) gas-check plate;
— *de poudre*, (*art.*) powder ring (of a double-action fuse); (*powd.*) powder pellet;
— *de pression*, buffer-spring disk;
— *de refroidissement*, (*fond.*) cooling ring (used to protect the band of a projectile in projectile casting, v. *prisonnier*);
— *de réglage*, (*art.*) adjusting disk (of a sub-caliber tube);
— -*ressort*, (*art.*) spring safety disk (of a fuse);
— *de rivure*, rivet plate, washer;
— *de serrage*, (*art.*) tightening or filling disk;
— *de sûreté*, v. — -*ressort;*
— *en talus*, beveled washer (for bolts);
— *de transmission*, (*art.*) powder ring (communicating fire to the time train of a double-action fuse);
— *de vérification*, (*art.*) flat iron ring used as a gauge for shot molds.

rondement, adv., smartly, quickly.

rondeur, f., roundness, circumference.

rondin, m., round log or spar (of small dimensions); round wood, pole, billet; (*fort.*) short palisade;
chemin de —*s*, corduroy road;
— *d'une palanque*, (*fort.*) filling-in timber of a stockade;
plancher en —*s*, corduroy floor or roadway;
— *de voie*, (*r. r.*) round tie.

roud-point, m., point of intersection of several streets, circuit.

ronflement, m., snoring; peal, boom;
— *de canons*, roar of guns;
— *de la mer*, roar of the sea.

ronfler, v. n., to snore; to roar, rumble, peal; to boom (of guns); to snort (of a horse).

rongé, p. p., eaten, worm-eaten, corroded; pitted (of iron);
— *de vers*, worm-eaten.

ronger, v. a., to gnaw; to corrode, eat away, waste;
— *le frein*, to champ the bit.

roquentin, m., (*mil.*) in France, old soldier allowed a residence in citadels, forts, etc.

roquet, m., (*artif.*) war rocket.

roquette, f., v. *roquet*.

rosbif, m., roast beef;
— *de mouton*, leg of mutton (rare).

rose, f., rose; rose color; (*art.*) wind card, wind diagram;
— *de l'acier*, (*met.*) irisated spot in steel;
bois de —, rosewood;
— *de la carte*, fleur-de-lis of the north point (on a map);
— *de compas*, compass card, compass rose;
— *de correction*, (*art.*) wind card, showing correction or allowances for deviation and ranges;
— *d'orientation*, v. *d'orientement;*
— *d'orientement*, (*top. drawing*) the rose or "compass" of a plan or map;
réciter la — *des vents*, to box the compass;
— *des vents*, v. — *de compas*.

roseau, m., reed;
étoupille de —, (*art.*) reed friction primer.

rosée, f., dew;
point de —, dew-point.

rosette, f., rosette; red ink; red chalk; burr, rivet plate, disk, washer, scutcheon; spur rowel; clout nail; speck (defect) in wood, etc.; (*met.*) cake of (purified) copper; (*cord.*) rose lashing;
— *de battant*, (*sm. a.*) fore-swivel stud;
— *à boucle et à anneau*, rivet plate, with loop and ring attached;
cuivre de —, v. s. v. *cuivre;*
— *crochet*, (*art.*) hook and plate;
— -*piton de chaîne d'enrayage*, (ordinary wagons) lock-chain plate;
— *en talus*, beveled washer.

rossard, m., (carrier pigeons) poor, worthless, pigeon.

rosse, f., (*man.*) worthless horse; jade.

rossignol, m., nightingale; picklock; quarryman's fork; (*carp.*) mortise wedge; (*mach.*) air valve, snifting valve; (*hipp.*) artificial fistula (ignorantly supposed to help the breathing of a broken-winded horse).

roster, v. a., (*cord.*) to woold.

rosture, f., (*cord.*) woolding;
faire une —, to woold.

rotang, m., v. *rotin*.

rotatif, a., rotatory, rotating;
machine —*ve*, rotary engine;
outil —, rotating tool.

rotation, f., rotation; (*man.*) action of turning on the spot;
crémaillère à —, (*mach.*) revolving rack.

rotative, f., (*mach.*) rotary engine.

rotatoire, a., rotatory.

roter, v. n., — *sur l'avoine*, (*hipp.*) to be off his feed;
— *sur la besogne*, (*hipp.*) to be lazy, unwilling to work.

rôtie, f., (*cons.*) running up of a (party) wall at one-half the thickness of the lower part.

rotin, m., rattan.

rôtir, v. a., to roast; (*met.*) to roast ore.

rôtissage, m., (*met.*) roasting, calcining.

rotoide, m., solid of revolution.

rotule, f., kneepan; whirlbone; (*mach.*) small roller or wheel;
s'articuler à —, v. s. v. *articuler*.

rot-vert, m., (*hipp., fam.*) cryptorchid, cryptorchis, (popularly) ridgeling, rig.

rouable, m., rake, fire rake;
— *à vase*, mud rake.

rouage, m., (*mach.*) wheelwork; wheels; train wheels; machinery, mechanism;
batterie en —, v. s. v. *batterie;*
bois de —, wheel timber;
— *à chaîne*, (*mach.*) chain wheelwork;
— *à courroie*, (*mach.*) strapped wheelwork;
débit du —, (*mach.*) travel of wheelwork in unit time;
prendre en —, v. s. v. *prendre*.

rouan, a., m., (*hipp.*) roan; roan horse;
— *cap de more*, roan with black mane and tail;
cheval —, roan horse;
— *clair*, light roan (white predominates);
— *foncé*, dark roan (brownish red predominates);
— *ordinaire*, ordinary roan (red, white, and black in substantially equal quantities);
robe —*ne*, roan coat;
— *vineux*, red (blood, strawberry, wine) roan (red in predominance).

rouanne, f., marking tool, branding iron;
— *à marquer*, timber marker;
— *de pompe*, pump borer.

rouanné, a., (*hipp.*) said of a gray coat with chestnut hairs.

rouanner, v. a., to mark, brand; to bore;
— *une pompe*, to enlarge the bore or channel of a pump.

rouannette, f., small marking tool, scraper, pointer.

roue, f., wheel.

I. Machinery; II. Water wheels; III. Vehicles and miscellaneous.

I. Machinery:
— *à action directe* (*indirecte*), direct (indirect) acting wheels;
— *à aile*, fly wheel;
— *à alluchons*, cogwheel;
— *d'angle*, bevel wheel, bevel gear; miter wheel
— *sur l'arbre*, wheel and axle;
— *s d'assortiment*, interchangeable gear;
— *barbotin*, sprocket wheel;
calage d'une —, keying, wedging, of a wheel;
— *calée*, wheel keyed or wedged on;
caler une —, to wedge, to key, on a wheel;
— *à cames*, cam wheel, wiper wheel;

roue *de carrière*, quarry capstan;
— *à chaîne(s)*, chain wheel, chain pulley;
— *de champ*, crown wheel, face wheel;
changement de —*s*, change wheels;
— *à chevrons*, wheel with V-shaped teeth;
— *en cœur*, heart wheel;
— *à coins*, wedge friction wheel;
— *de commande*, driving wheel;
— *conductrice*, driving wheel, main wheel;
— *conduite*, driven wheel;
— *conique*, v. — *d'angle*;
— *conique menante*, main cone driving wheel (of a steam windlass);
— *à corde*, cable pulley;
— *à couronne*, v. — *de champ*;
— *de courroie*, belt pulley;
— *à couteaux*, rotary cutter (wood working);
— *creuse*, internal wheel;
— *cylindrique*, spur wheel;
— *cylindrique à dents hélicoïdales*, cylindrical spiral gear;
décalage d'une —, unkeying of a wheel;
— *décalée*, loose, idle, wheel;
décaler une —, to unkey, release, a wheel;
— *à déclic*, ratchet wheel;
dent de, d'une, —, cog;
— *dentée*, toothed or cogged wheel; gear wheel;
— *dentée à action directe (indirecte)*, direct- (indirect-) acting toothed gearing;
— *dentée d'angle*, v. — *d'angle*;
— *dentée à couronne*, v. — *de champ*;
— *dentée à couronne intérieure*, crown wheel with internal teeth;
— *dentée cylindrique*, spur wheel;
— *dentée à dents de bois*, mortise wheel;
— *dentée droite*, v. — *dentée cylindrique*;
— *dentée épicycloïdale*, v. — *épicycloïdale*;
— *dentée en étages*, v. — *en étages*;
— *dentée intérieure*, v. — *intérieure*;
— *dentée à rochet*, v. — *à déclic*;
— *à dents*, cogwheel;
— *à dents de bois*, mortise wheel;
— *à dents de côtés*, v. — *de champ*;
— *à dents droites*, straight-toothed wheel;
— *à dents étagées*, v. — *en étages*;
— *à denture extérieure (intérieure)*, spur (internal) wheel;
— *à dents inclinées, obliques*, oblique-toothed wheel;
— *sans denture*, friction wheel;
désengrener les —*s*, to ungear the wheels;
— *détachée de l'essieu*, loose, idle, wheel;
— *de distribution*, starting wheel;
— *droite*, spur wheel;
— *double à filets symétriques*, v. — *à chevrons*;
— *à emboîtement*, double-flanged wheel;
— *à empreintes*, sprocket wheel;
— *d'encliquetage*, ratchet wheel;
— *endentée*, cog wheel;
— *d'engrenage*, gear wheel;
engrenage de —*s*, gearing;
— *d'engrenage conique*, miter wheel, bevel wheel;
— *d'engrenage intérieur*, annular wheel, internal wheel;
— *d'engrenage à vis*, worm wheel;
— *engrenant extérieurement (intérieurement)*, — *à engrènement extérieur (intérieur)*, spur (annular) wheel;
— *épicycloïdale*, epicycloidal wheel;
— *en étages*, stepped gearing, Hooke's gearing;
— *étoilée*, star wheel;
— *excentrique*, eccentric wheel;
— *à filets coniques*, spiral bevel-gear wheel;
— *à filets hélicoïdaux*, spiral-gear wheel;
— *folle*, loose, idle, wheel;
— *à fourchette*, forked wheel;
— *à friction*, friction wheel;
— *à friction conique*, friction wheel for inclined axes;
— *à friction cylindrique*, friction wheel for parallel axes;
— *à frottement*, friction wheel or gearing;
— *à frottement à action directe (indirecte)*, direct- (indirect-) acting friction wheel, friction gearing;

roue *sans frottement*, Hooke's gearing;
— *à gorge*, double-flanged wheel;
grande — *d'engrenage*, driving wheel;
— *(à, en,) hélice*, worm, worm wheel, screw wheel;
— *hélicoïdale*, spiral-gear wheel;
— *hyperbolique*, skew wheel, skew-bevel wheel;
— *indépendante*, loose wheel;
— *intérieure*, internal wheel, annular wheel;
— *intermédiaire*, intermediate wheel (or gearing);
— *à lanterne*, lantern wheel;
— *libre*, v. — *indépendante*;
— *à main*, handwheel;
— *de manœuvre*, operating wheel, working wheel;
— *menante*, driving wheel;
— *menée*, driven wheel;
— *motrice*, driving wheel, main wheel;
petite —, scroll wheel;
— *à pignon*, wheel and pinion;
— *plane*, plane gear wheel;
planétaire, planet gear wheel;
— *pneumatique*, ventilator;
— *à rainures*, grooved wheel;
rechausser une —, to recog a wheel;
— *s non réciproques*, single wheels;
— *de renvoi*, reversing wheel;
— *à rochet*, ratchet wheel;
— *satellite*, sun-and-planet wheel;
— *solaire et planétaire*, sun-and-planet wheel;
— *striée*, worm wheel;
— *subordonnée*, driven wheel;
— *s symétriques*, wheels belonging to the same set or gear;
tour de —, (mach.) reciprocation;
— *d'un tour*, lathe wheel;
— *travailleuse*, driving wheel;
— *vis*, screw wheel;
— *à vis sans fin*, worm wheel;
— *volante*, fly wheel.

II. Water wheels:
arbre de —, paddle shaft;
aube de —, paddle;
— *à aubes*, paddle wheel, mill wheel;
— *à aubes articulées*, feathering paddle wheel;
— *à aubes courbes*, paddle wheel with curved float board;
— *à aubes fixes*, radial paddle wheel;
— *à aubes en porte-à-faux*, overhanging paddle wheel;
— *à augets*, bucket wheel;
— *à augets de côté*, breast water wheel, bucket wheel;
— *à augets de dessus, en dessus*, overshot wheel;
— *à axe vertical*, turbine;
— *à baquets*, bucket wheel;
caisse de —, paddle box;
chemin de la —, wheel race;
— *à choc*, v. — *à palettes planes*;
— *de côté*, middle-shot wheel; breast wheel;
— *de côté classique*, standard or type water wheel;
couverture de —, v. *caisse de* —;
— *par derrière*, high breast wheel;
— *en dessous*, undershot wheel;
— *en dessus*, overshot wheel;
— *à eau*, water wheel;
— *élévatoire*, water (raising) wheel;
— *à godets*, scoop wheel, Persian wheel;
— *hélicoïde*, Archimedean screw;
— *hydraulique*, hydraulic wheel, water wheel;
— *hydraulique à augets*, bucket wheel;
— *hydraulique de côté*, breast wheel, middle-shot wheel;
— *hydraulique par derrière*, high breast wheel;
— *hydraulique en dessous (dessus)*, under- (over-) shot wheel;
— *hydraulique à moulin*, mill wheel;
— *hydraulique à palettes*, flash wheel;
— *hydraulique persane (persique)*, scoop wheel, Persian wheel;
— *mobile*, working wheel (of the Fourneyron turbine);

roue, *moyeu de* —— *à aubes,* paddle-wheel boss;
—— *mue en dessous,* undershot wheel;
—— *mue en dessus,* overshot wheel;
—— *ordinaire,* ordinary paddle wheel;
—— *à pales,* v. —— *à aubes;*
pale de ——, paddle;
—— *à palettes,* flash wheel;
—— *à palettes planes,* undershot wheel;
—— *pendante,* stream wheel (no dam);
—— *à piston,* chain pump;
—— *à poire,* tub wheel;
—— *de poitrine,* breast wheel;
—— *en porte-à-faux,* overhanging water wheel;
—— *à pots,* overshot wheel;
—— *de puits,* water wheel;
rayon de ——, paddle arm;
—— *à réaction,* reaction wheel;
—— *à seaux,* v. —— *à godets;*
tambour de ——, v. *caisse de* ——;
tourteau de —— *à aubes,* v. *moyeu de* —— *à aubes;*
—— *turbine,* turbine;
—— *à tympan,* v. —— *à godets.*

III. Vehicles and miscellaneous:
—— *s accouplées,* coupled wheels (of a locomotive);
—— *arrière,* hind wheel; hind wheel of a bicycle;
—— *d'arrière,* hind wheel; stern wheel of a river boat; trailing wheel (as of a locomotive);
—— *de l'arrière-train,* hind wheel of any four-wheeled vehicle;
—— *d'avant,* leading wheel (as of a locomotive); fore wheel of any four-wheeled vehicle;
—— *d'avant-train,* (art.) limber wheel;
—— *s de l'avant-train,* fore wheels of any four-wheeled vehicle;
—— *avec rais,* spoke wheel;
bandage de ——, tire;
bateau à ——*s,* (nav.) side-wheeler;
—— *de bois,* wooden truck (wheel);
boîte de ——, nave box;
—— *de câble,* (cord.) cable coil;
calage d'une ——, scotching of a wheel;
—— *calée,* scotched wheel;
caler une ——, to scotch a wheel;
—— *de canon,* hour wheel;
changer une ——, (art.) to change wheels;
châtrer une ——, v. s. v. *châtrer;*
—— *de chaussée,* minute wheel;
—— *à chevilles,* quarrier's treadmill;
—— *de commande,* v. —— *conductrice;*
—— *commandée,* (locomotive) driver;
—— *de commettage,* (cord.) laying wheel;
—— *de compte,* ——*-comptcur,* register;
—— *conductrice,* driving wheel (of a locomotive);
—— *de cordage,* coil of rope;
décaler une ——, to unscotch a wheel;
—— *dentée,* sprocket wheel of a bicycle;
—— *de derrière,* v. —— *d'arrière;*
—— *de devant,* v. —— *d'avant;*
—— *directrice,* fore, steering, wheel of a bicycle;
—— *à disque,* solid, plate, wheel;
—— *à échappement,* escape wheel;
embattre une ——, to tire a wheel;
—— *à émeri,* emery wheel;
en ——, wheel-shaped;
enrayage d'une ——, spoking of a wheel;
—— *enrayée à bloc,* wheel skidded tight;
faire la ——, to wheel about; (gym.) to turn handsprings sideways;
faire des pans de ——, to make or take a wheel purchase;
fenderie des ——*s,* wheel cutting;
—— *ferrée,* tired wheel;
ferrer une ——, to tire a wheel;
feu de ——, wheel fire;
—— *à feu,* (artif.) catharine wheel;
—— *du gouvernail,* (nav.) (steering) wheel;
—— *pour gouverner à la main (à la vapeur),* (nav.) hand (steam) steering wheel;
—— *hérissée,* a wheel before its fellies are put on, speech;
jante de ——, rim of a wheel;
machine à fendre les ——*s,* wheel cutter;

roue à manche, spoke wheel;
—— *s du milieu,* driving wheels (of a locomotive);
—— *de minute,* minute wheel;
—— *mixte,* composite wheel of wood and iron;
—— *s montées,* pair of wheels;
—— *motrice,* driving wheel (of a locomotive); hind wheel, driving wheel (of a bicycle);
—— *à moyeu préparé,* wheel with a metal nave;
—— *s non réciproques,* single wheels;
passage de ——*s,* v. *passage;*
—— *pénitentiaire anglaise,* v. —— *à tambour;*
—— *pleine,* v. —— *à disque;*
—— *à poignée,* spoke wheel, handwheel;
ponton à ——, dredging machine;
—— *portante,* trailing wheel;
—— *de poulie,* sheave of a block;
pousser à la ——, to push on the wheels (in order to relieve the horses);
—— *à rais,* spoke wheel;
rayon de ——, wheel arm;
—— *de rechange,* spare wheel;
—— *de rencontre,* balance wheel;
—— *à ressorts,* a wheel in which double blades of steel take the place of the spokes; spring wheel;
sabot de ——, drag;
—— *à sabots,* wheel in which the spokes are joined to the fellies by shoes of steel;
—— *s de support,* trailing wheels of a locomotive;
—— *à tambour,* treadmill;
—— *à voussoirs,* wheel with a metal nave.
rouelle, f., cut, slice, of any round article; (hipp.) rowel; circular seton.
rouer, v. a., (cord.) to coil down a rope; to ret; (fig.) to harass, wear out;
—— *à contre,* (cord.) to coil against the sun;
—— *à tour,* (cord.) to coil with the sun.
rouet, m., spinning wheel; sheave, sheave of a block; curb of a well; (mach.) gear wheel of a mill;
—— *en bronze,* brass sheave;
—— *à chaîne,* chain sheave;
—— *à cylindre,* multiplying sheave;
—— *à dé,* bushed sheave;
—— *de gaïac,* lignum-vitæ sheave;
lumière d'un ——, oil hole of a sheave;
—— *de poulie,* sheave of a block;
—— *de serrure,* scutcheon of a lock.
rouette, f., osier, twig.
rouge, a., m., red, red hot; red pigment, color; red heat;
—— *blanc,* (met.) white heat; welding heat;
bois ——, wood beginning to decay; fir wood, red deal;
boulet ——, (art.) red-hot shot (obs.);
—— *cerise,* (met.) cherry-red;
—— *cerise clair,* (met.) bright cherry-red;
chauffé au ——, red hot;
cuivre ——, (met.) red copper;
fer ——, red-hot iron;
—— *au minium,* engineer's cement;
—— *naissant,* (met.) nascent red;
petit ——, (met.) light red;
—— *rose,* (met.) rose-red;
—— *suant,* v. —— *blanc;*
—— *vif,* (met.) bright red.
rougeur, f., redness; (in pl.) reddish spots (announcing decay of wood).
rougir, v. a. n., to redden; to grow red; (met.) to heat red hot;
—— *à,* —— *en, blanc,* (met.) to bring to welding heat;
faire ——, (met.) to heat red hot.
rouillage, m., rusting (esp. of iron and steel in the open air).
rouille, f., rust;
—— *du cuivre,* verdigris.
rouiller, v. a. r., to rust;
se ——, to become rusty.
rouilleux, a., rusty.
rouillure, f., rustiness (of metals, and of oats, etc.).
rouir, v. a. (cord.) to ret.
rouissage, m., (cord.) retting.
roulage, m., rolling; ease of rolling; cartage, wheel transportation; transfer (of baggage); transfer office; transfer company; bicycling;
cercle de ——, flesh hoop of a drum;

roulage, *par le* ——, by land transportation;
petit ——, carrier;
vergette de ——, v. *cercle de* ——;
voiture de ——, transfer, baggage, express, wagon.

roulant, a., rolling, easy, (of roads) easy running, (of vehicles) easy for teams;
armée ——*e*, (*slang*) chain gang; (*mil.*) stragglers;
boulet ——, (*art.*) shot without its sabot (of spherical shot, obs.);
chemin ——, road easy for carriages;
feu ——, (*mil.*) rolling fire, running fire;
fonds ——, funds for current expenses;
matériel ——, (*r. r.*) rolling stock.

roule, m., log (for sawing into planks, etc.).

roulé, a., shaky (of wood).

rouleau, m., roll, roller; cylinder; log (for sawing up, etc.); road roller; sheave, drum; (*mach.*) roll, cylinder, friction roller; (*art.*) roller (in mechanical maneuvers); roller of a top carriage; (*siege*) sap roller; (*cord.*) coil; (*mas.*) course, ring, of voussoirs in a thick arch;
—— *alimentaire*, (*mach.*) feed roller;
—— *de bandes*, (*med.*) field dressing;
—— *compresseur*, (*mach.*, etc.) press roll; road roller;
—— *de culasse*, (*art.*) breech roller;
—— *délivreur*, (*mach.*) delivery roll or roller;
—— *à dépiquer*, thrashing roll;
—— *de dilatation*, (*cons.*) expansion rollers;
—— *-écraseur*, road roller;
—— *étireur*, (*mach.*) drawing roller;
grand ——, (*art.*) long roller;
—— *-guide*, (*mach.*) roller, guide roller, belt guide;
—— *de manœuvre*, (*art.*) roller;
petit ——, (*art.*) short roller;
—— *de pression*, (*mach.*) press roll;
—— *de secours*, (*med.*) roll containing articles of first aid to asphyxiated, drowned, etc., persons;
—— *-support*, (*mach.*) supporting roller (of a cable transmission);
—— *-tendeur*, (*mach.*) tension roller;
—— *de tension*, (*mach.*) tightening pulley, tension roller;
—— *à vapeur*, steam roller.

roulement, m., roll, rolling; ease of rolling (bicycle, etc.); circulation of money; rotation of office; (*mus.*) roll of a drum, ruffle; (*mil.*) alternation of duty by roster; transfer (as of troops from one point to another); (*met.*) working season, continuance of blast;
—— *des effets*, (*adm.*) movement or shifting of stores in magazines, so that longest in shall be used first;
faire un ——, to roll;
fonds de ——, funds for current expenses;
—— *de fonds*, rapid circulation of money;
—— *du haut fourneau*, (*met.*) working season of a furnace.

rouler, v. a. n., to roll; to wind, roll, up; to "wheel" (bicycle); to tumble; to turn, run, on wheels; to roll (of sounds); to rotate in the performance of a duty; (*mil.*) to be on the roster; to take duty, service, by roster; (*nav.*) to roll;
—— *à cheval*, (*man.*) to wobble in the saddle, to have absolutely no seat at all;
—— *ensemble*, (*mil.*) to take duty by turns on the same roster (said of officers of the same grade);
—— *entre eux*, v. —— *ensemble;*
faire ——, to trundle;
—— *panne sur panne*, (*nav.*) to roll gunwale under;
—— *sur une roue*, to wheel.

roulette, f., roller, runner, caster, truck; friction roller; small wooden wheel; cycloid; (*art.*) chassis wheel; truck wheel;
—— *-arrière*, (*art.*) rear-truck wheel;
—— *-avant*, (*art.*) front-truck wheel;
—— *de châssis*, (*art.*) truck wheel;
—— *de friction*, (*mach.*, etc.) friction roller.

rouleur, m., cask roller; workman who wheels a barrow; (*nav.*) ship that rolls heavily;

rouleur, *méthode par* ——*s*, (*siege*) in the flying sap, method in which the gabions are put down successively, advancing them beyond those already down;
ouvrier ——, traveling journeyman.

roulier, m., wagoner, carter.

roulis, a., *bois* ——, shaky wood.

roulis, m., (*nav.*) rolling of a ship;
coup de ——, heavy lurch.

roulon, m., round, rundle, rung, of a wagon, stable rack, etc.

roulure, f., (of wood) shake.

roussin, m., (*hipp.*) stocky, more or less undersized stallion, of common breed.

rouster, v. a., (*cord.*) to woold.

rousture, f., (*cord.*) woolding.

rousturer, v. a., (*cord.*) to woold.

route, f., road through a forest; road, route, track, path, way; direction; (*nav.*) course; (*mil.*) route, order for a march, march;
—— *apparente*, (*nav.*) steered course;
barrer une ——, to stop a road;
barrer la —— *à quelqu'un*, to stop, check (as troops, etc.);
—— *à barrière*, turnpike;
—— *battue*, beaten road, path;
—— *bombée*, convex, barreled, road;
—— *en caillou*, metaled road, broken-stone road;
—— *à carrosser*, carriage road;
—— *chaussée*, high road;
chaussée de ——, road led;
circulation sur une ——, traffic on a road;
clôture de ——, road fence;
clou de ——, v. s. v. *clou;*
—— *combinée*, (*nav.*) compound course;
—— *au compas*, (*nav.*) compass course;
compas de ——, (*nav.*) steering compass;
construction des ——*s*, road making;
construire une ——, to make a road;
—— *corrigée*, (*nav.*) true course;
—— *corrigée de la dérive*, (*nav.*) course corrected for leeway;
—— *corrigée de la déviation*, (*nav.*) course corrected for deviation;
—— *creuse*, hollow road;
—— *dégradée*, broken road;
—— *départementale*, country road; (in France) road kept up by a *département* and having a number in the *département* only;
détérioration d'une ——, wear and tear of a road;
détérioration par arrachement d'une ——, tear of a road;
direction des ——*s*, line of draft;
donner la ——, (*nav.*) to set the course;
—— *par eau*, water journey;
—— *empierrée*, stone-ballasted road;
en ——, on the way, road, march;
—— *encaissée*, road with high banks;
encaissement de ——, base, bed, of a road;
—— *enfoncée*, road full of chucks;
—— *estimée*, (*nav.*) dead-reckoning;
—— *d'étapes*, v. s. v. *étape;*
faire ——, (*nav.*) to sail, navigate;
faire —— *à*, (*nav.*) to stand for;
faire la ——, to/cover the distance between two points;
faire fausse ——, to take a wrong road, a wrong way; to lose one's way; (*rav.*) to take the wrong course;
faire —— *à pied*, to foot it;
—— *fatiguée*, heavy road;
—— *de fer*, railway;
—— *ferrée*, metaled road; broken-stone road;
feuille de ——, v. s. v. *feuille;*
—— *à flanc de coteau*, hillside road;
frais de ——, travel expenses or allowance;
grand', grande, ——, high road;
—— *impériale*, national road;
indemnité de ——, v. s. v. *indemnité;*
—— *inégale*, rough, rugged, road;
ingénieur des ——*s*, road engineer;
—— *dans l'intérieur*, (*mil.*) march from one point to another at home;
itinéraire de ——, (*mil.*) road report;
—— *kilométrée*, road with kilometers marked off;

route, *livrer une* —— *à la circulation,* to open a road to the public;
—— *à la macadam,* —— *macadamisée,* macadam road;
—— *magnétique,* (*nav.*) compass course;
—— *par mer,* sea journey;
se mettre en ——, to march, go, off; to set out;
mettre le cap en ——, (*nav.*) to head for;
—— *militaire,* military road;
—— *muletière,* horse road;
—— *nationale,* high road; (in France) national highway built and kept up by budget of public works, and numbered in series for the whole of France;
—— *de niveau,* road on a level with the surrounding country;
nœud de ——*s,* v. s. v. *nœud;*
—— *ordinaire,* common road, "dirt road" (U. S.);
ouvrage d'art de ——, road masonry, stonework;
ouvrir une ——, to open a road;
—— *parcourue,* (*nav.*) course made good;
—— *de partance,* (*nav.*) outward voyage;
—— *pavée,* paved road;
—— *à péage* turnpike;
position de ——, (*art.*) traveling position of a gun;
—— *à profil partie en déblai, partie en remblai,* road half excavation, half embankment;
—— *publique,* high road, government road;
—— *raboteuse,* rough, rugged, road;
remblai de ——, road embankment;
se remettre en ——, to leave, set out, start, again;
—— *de retour,* homeward voyage;
—— *royale,* king's road, national road, parliamentary road;
—— *stratégique,* v. —— *militaire;*
tenir une ——, (*nav.*) to steer a particular course;
—— *en terrain naturel,* surface road, country road;
—— *de terre,* land road;
tracé d'une ——, line, direction, of a road;
tracer une ——, to lay out, line out, a road;
travaux de ——*s,* road works;
—— *de traverse,* crossroad;
(à-) *vau-de-* ——, in disorder, in confusion;
—— *vicinale,* parish road;
—— *vraie,* (*nav.*) true course.
routier, m., guide (rare); road book; (*nav.*) sailing directions;
vieux ——, old soldier, old stager.
routier, a., road, of a road;
carte ——*e,* road map;
locomotive ——*e,* traction engine.
routine, f., routine;
par ——, by rote.
routiner, v. a., to teach anyone to do anything by rote.
routinier, a., m., person who does everything by routine; routine, "red tape."
routoir, m., (*cord.*) steeping tank.
rouverin, a., (*met.*) hot short, red short; (used only in) *fer* ——, red short iron.
rouvieux, m., (*hipp.*) mange on the upper part of the neck; a., mangy.
rouvre, m., English oak.
rouvrir, v. a., to reopen.
roux, a., reddish;
—— *alezan,* (*hipp.*) red bay;
vents ——, —— *vents,* cold April winds.
roux-vieux, m., an incorrect form of *rouvieux,* q. v.
royal, a., royal; m., (*pl.*) the royal troops;
altesse ——*e,* royal highness (title);
bastion ——, (*fort.*) large bastion;
chemin ——, national road;
famille ——*e,* children and grandchildren of the king (reigning or dead);
maison ——*e,* all the princes and princesses of the blood;
parapet ——, (*fort.*) parapet of the *enceinte;*
route ——*e,* national road;
ville ——*e,* capital; Paris.
royale, f., tuft (on the under lip).
royaliste, a., m., royalist.

royaume, m., kingdom.
royauté, f., royalty.
ru, m., channel, bed, of a rivulet.
ruade, f., (of the horse) kick with one foot or both feet (always accompanied by an elevation of the hind quarters);
allonger une ——, to kick out, lash out, with the hoof;
détacher une ——, v. *allonger une* ——;
lancer une ——, v. *allonger une* ——.
ruban, m., ribbon; ribbon (of an order); binding hoop; (*sm. a.*) riband or strip (for the manufacture of gun barrels);
—— *d'acier,* (*inst.*) steel tape, steel measuring tape;
canon à ——, (*sm. a.*) twisted barrel;
fer à ——, hoop iron;
—— *de frein,* (*tech.*) brake strap;
—— *-mesure,* measuring tape.
rubanner, v. a., (*sm. a.*) to twist into a barrel.
rubican, a., (*hipp.*) rubican (having white hairs on a black, bay, or chestnut coat); m., rubican coat, rubican horse.
rubrique, f., red ocher, chalk; imprint of a book; title, head.
rude, a., rough, rugged, unpolished; coarse, hard, rough hewn, uneven, harsh; severe; troublesome, violent, fatiguing;
—— *adversaire,* redoubtable adversary;
brosse ——, stiff brush;
chemin ——, rough, rugged, road;
cheval ——, rough, hard-set, uneasy, horse;
coup ——, heavy, severe, blow;
épreuve ——, severe trial;
froid ——, bitter cold;
main ——, (*man.*) hard, heavy, hand;
train ——, (*hipp.*) uneasy, rough, action.
rudération, f., paving of small stones or pebbles; (*mas.*) rough coat of plaster on a wall.
rudesse, f., roughness, harshness.
rudoiement, rudoyement, m., (*man.*) rough handling of a horse.
rudoyer, v. a., (*man.*) to spur, whip, a horse violently; to maltreat a horse.
rue, f., street; in a quarry, space from which stone has been removed; (*mil.*) street of a camp;
—— *borgne,* by-street;
—— *de camp,* (*mil.*) camp street;
clou de ——, v. s. v. *clou;*
—— *écartée,* v. —— *borgne;*
grand' ——, high street of a provincial town; (*mil.*) street perpendicular to the rows of huts or tents, general parade (of a camp);
grande ——, high street of a provincial town; (*nav.*) waist;
guerre de ——, (*mil.*) street fighting;
—— *militaire,* v. —— *de rempart;*
—— *passante,* thoroughfare;
petite ——, (*mil.*) street parallel to huts or tents, company street (of a camp);
en pleine ——, in the middle of the street;
—— *du rempart,* (*fort.*) in a fortified town, natural ground or space between the nearest houses and the foot of the rampart slope;
—— *de traverse,* cross street.
ruelle, f., small street; (*mil.*) street of, or in, a camp.
ruellée, f., (*cons.*) joint or covering of plaster, or of mortar, along the intersection of a roof with a vertical wall.
ruer, v. n., (of a horse, etc.) to kick up, throw the heels in the air;
—— *à la botte,* to kick at the rider as he mounts, or when he closes his leg;
—— *en vache,* to kick like a cow.
rueur, m., a kicking horse;
cheval ——, kicking horse, kicker.
rugination, f., (*hipp.*) bone scraping.
rugine, f., (*hipp.*) rugine, bone rasp.
ruginer, v. a., (*hipp.*) to scrape or rasp the bones.
rugosité, f., rugosity; rough place, corrugation, seam, wrinkle.
rugueux, a., rough; m, (*art.*) roughened wire (of a friction primer); striker (of a fuse).

ruine, f., fall or destruction of a building; ruins; ruin, overthrow, destruction; (*hipp.*) breaking down of a horse;
  battre en ——, (*art.*) to batter down;
  en ——, ruinous, in ruins;
  tomber en ——, to become ruinous, fall to decay, go to waste, etc.
ruiné, p. p., useless, spoiled, broken down, ruined, downfallen;
  bouche ——*e*, (*man.*) spoiled mouth;
  chemin ——, destroyed, broken up, road;
  cheval ——, (*hipp.*) spoiled, broken-down horse;
  jambes ——*es*, (*hipp.*) sprung legs.
ruiner, v. a. r., to ruin, destroy, spoil, lay waste, overthrow; to baffle, subvert; to decay, fall to decay; (*art.*) to batter; (*hipp.*) to break down, spoil, a horse; (*carp.*) to groove.
ruineux, a., ruinous, decayed, in decay; ready, threatening, to fall; falling to ruin, to decay.
ruinure, f., (*carp.*) groove, housing (of a beam).
ruisseau, m., stream, brook, rivulet, rill, gutter, kennel.
ruisson, m., small channel to drain a swamp.
rumb, m., r(h)umb;
  —— *de vent,* point of the compass.
rupture, f., rupture, breaking, breaking off, bursting, destruction, fracture; rupture, breaking off (of relations); annulment (of a contract, etc.); (*med.*) hernia; (*elec.*) break (of a current); (*mach.*) breakdown of machinery; (*mil.*) breaking from line into column; reduction of the front of a column; (*expl.*) destruction, demolition, by explosives;
  boulet de ——, (*art.*) armor-piercing projectile;
  charge de ——, (*tech.*) breaking load; (*art.*) a breaking charge used in incendiary projectiles (to keep the enemy from getting hold of them);

rupture *double*, (*expl.*) destruction of a rail by explosives on both sides;
  —— *double en couple,* (*expl.*) destruction of a rail by explosives so placed as to form a torque;
  —— *des galeries* (*mil. min.*) destruction of gallerie;
  —— *de mouvement,* (*mil.*) breaking from one formation to another;
  obus de ——, (*art.*) armor-piercing shell;
  —— *d'une porte,* breaking in of a door;
  —— *par quatre,* (*mil.*) breaking into column of fours;
  rayon de ——, v. s. v. *rayon;*
  rayon de bonne ——, v. s. v. *rayon;*
  rayon de —— *limite,* v. s. v. *rayon;*
  —— *des routes,* (*mil.*) destruction of roads;
  —— *simple,* (*expl.*) destruction of a rail by explosives placed on one side only;
  solide de bonne ——, v. s. v. *solide;*
  solide de —— *limite,* v. s. v. *solide;*
  —— *stratégique,* (*mil.*) separation of forces of an army into two distinct bodies; breaking of enemy's center;
  —— *par subdivisions,* (*mil.*) break into column of subdivisions.
ruse, f., stratagem, ruse, artifice, art;
  —— *de guerre,* (*mil.*) stratagem.
rusticité, m., primitiveness; (of machines, gun carriages, etc.) simplicity (with a connotation of rudeness of design).
rustine, f., (*met.*) back plate of a forge hearth; back of a blast furnace.
rustiquage, f., (*mas.*) rustic work (both of stone and of plaster).
rustique, a., rustic; primitive, simple (e. g., of machines, etc.); twisted (of the grain of wood); (*mas.*) rustic;
  bois ——, wood of twisted or curling grain;
  genre, ouvrage ——, (*mas.*) rustic work.
rustique, m., mason's hammer point.
rustiquer, v. a., (*mas.*) to rusticate; to cover (a wall) with a rustic coat of plaster; to hatch or nig the surface of a stone.
rutilant, a., (*chem., expl.*) rutilant.

S.

S, any S shaped cramping iron or plate, spare couple, S-shaped link;
  en ——, S-shaped, sigmoid;
  —— *de la gourmetie,* (*harn.*) curb hook.
sabatte, f., anchor shoe;
  —— *patin,* sole for heel of sheers.
sabaye, f., boat rope, painter, mooring rope, stern rope.
sable, m., sand, gravel, ballast; (*nav.*) glass, hour-glass;
  —— *arène,* sand formed *in situ* by the spontaneous decomposition of rock;
  —— *argileux,* clayey sand;
  —— *asphaltique,* bituminous sand;
  bain de ——, (*cham.*) sand bath;
  banc de ——, (*hydr.*) sand bank;
  —— *battu,* binding sand;
  boîte à ——, (*cons.*) sand box (used in striking centers);
  ——*s boulants,* quicksand;
  —— *bouillant,* quicksand;
  carrière de ——, sand pit;
  —— *de carrière,* quarry sand;
  —— *de cave,* v. —— *terrain;*
  de ——, sandy;
  —— *doux,* silver sand;
  —— *d'étuve,* —— *étuvé,* (*fond.*) burned sand;
  —— *femelle,* v. —— *mâle;*
  —— *de fer,* (*artif.*) iron filings used in fireworks;
  —— *fin,* fine, sharp sand (no grain larger than 1^{mm} in diameter); (*fond.*) facing sand;
  —— *fin de moulage,* (*fond.*) facing sand;
  fond de ——, sandy bottom;
  —— *de fondeur,* v. —— *de moulage;*
  forme de ——, bed of sand;
  —— *fossile,* fossil sand (sand transported by water);
  —— *de fouille,* —— *fouillé,* pit sand;

sable *gras,* (*fond.*) dry sand;
  ——-*gravier,* coarse, angular sand;
  —— *grillé,* (*fond.*) dried or roasted sand;
  —— *gros,* coarse sand (diameter of grains 2^{mm} or 3^{mm});
  lit de ——, sand bed;
  —— *maigre,* (*fond.*) green sand;
  —— *mâle,* dark-colored sand (so called to distinguish it from —— *femelle,* lighter-colored sand in the same bed);
  manger du ——, (*nav.*) to cheat the glass, turn it over too soon;
  —— *de mer,* sea sand;
  mer de ——, sand flood;
  —— *de mine,* pit sand;
  —— *mobile,* v. ——*s mouvant;*
  —— *de moulage,* (*fond.*) molding sand;
  —— *à mouler,* —— *à moules,* v. —— *de moulage;*
  ——(*s*) *mouvant*(*s*), shifting sand(s);
  —— *à noyaux,* (*fond.*) core sand;
  —— *recuit,* v. —— *d'étuve;*
  —— *de rivière,* river sand;
  —— *de sablonnière,* pit sand;
  sac à ——, (*siege*) sandbag;
  —— *sec,* (*fond.*) dry sand; parting sand;
  —— *à souder,* (*met.*) welding sand;
  —— *terrain,* —— *de terre,* pit sand;
  —— *vasard,* muddy sand;
  —— *vert,* v. —— *maigre.*
sablé, p. p., sanded; graveled;
  allée ——*e,* graveled walk;
  fontaine ——*e,* sand filter.
sabler, v. a., to sand, gravel, cover with sand or gravel; to sandbag.
sablerie, f., (*fond.*) workshop for preparing sand molds.
sableur, m., (*fond.*) sand molder.
sableux, a., sandy.

sablier, m., hourglass; sand box, glass; sanddigger; sand box of a locomotive;
— *de demi-heure*, — *de demi-minute*, etc., half-hour, half-minute, etc., glass;
— *d'une heure*, hourglass;
— *de loch*, (*nav.*) log glass.

sablière, f., sand pit; (*cons.*) wall plate; ground timber, sole, or plate; (*r. r.*) ground plate; (*fort.*) metallic ring surmounting the cylindrical part of a turret;
— *basse*, (*cons.*) sill;
— *haute*, (*cons.*) girt, intertie.

sablon, m., very fine sand;
écurer avec du —, to scour with sand;
récurer avec du —, v. *écurer avec du* —.

sablonner, v. a., to scour with sand; (*met.*) to scatter, sprinkle, with sand.

sablonneux, a., sandy, gritty.

sablonnière, f., sand pit, gravel pit; (*fond.*) (molding-) sand-box.

sabord, m., (*nav.*) port, porthole; gun port; (*fort.*) gun port; embrasure;
— *d'aérage*, (*nav.*) air port;
— *d'arrière*, — *de l'arrière*, (*nav.*) stern port;
— *de l'avant*, (*nav.*) bow (loading) port;
— *des avirons*, (*nav.*) row port;
barre de —, (*nav.*) port bar;
— *de belle*, v. — *de charge;*
— *brisé*, (*nav.*) half port;
cadre de —, (*nav.*) port frame;
— *de casemate*, (*fort.*) casemate, cupola, port, or embrasure;
— *de charge*, (*nav.*) cargo, loading, port;
— *de dégagement*, (*nav.*) bulwark port;
— *d'échelle*, v. — *d'entrée;*
— *d'entrée*, (*nav.*) gangway port;
faire le —, (*art.*) to fire a (hostile) shot through the embrasure;
faux —, (*nav.*) sham port; deadlight;
fermer les —s, (*nav.*) to close the ports;
fermeture de —, v. *barre de* —;
— *hublot*, (*nav.*) scuttle port;
— *à lester*, (*nav.*) ballast port;
mantelet de —, (*nav., fort.*) port lid;
— *de nage*, v. — *des avirons;*
— *de pavois*, v. — *de dégagement;*
— *de tourelle*, (*nav., fort.*) turret port, embrasure.

sabordement, m., (*nav.*) scuttling.

saborder, v. a., to scuttle (a ship).

sabot, m., wooden shoe; drag, drag shoe, skid; shoulder block; (*hipp.*) hoof; (*tech.*) shoe (of a post, pile, etc.); (*cons.*) rafter shoe; curved part of the stringboard of a staircase; (*cord.*) laying top; (*mach.*) clamp, pump piston; (*r. r.*) chair; (*art.*) sabot of a projectile; recoil-checking shoe; socket of a gin shoe; (*sm. a.*) sabot of a bullet; ferrule of a lance;
— *d'ancre*, shoe of an anchor.
— *de l'arbre*, cog, catch, of a wheel;
armer d'un —, to shoe;
— *de bigues*, step for the heel of a sheers;
— *en bois*, (*art.*) wooden bottom of a case shot;
— *de boîte à balles*, (*art.*) case-shot bottom;
— *à boulet*, (*art.*) round-shot bottom (obs.);
— *cassant*, (*hipp.*) brittle hoof;
— *-chaussure*, armature;
— *de cheval*, (*hipp.*) hoof, coffin;
devant du —, (*hipp.*) toe of a horse's hoof;
— *dérobé*, (*hipp.*) hoof so worn that no shoe can be put on;
— *d'enrayage*, (*art., etc.*) shoe, brake shoe, drag shoe, chock;
— *d'enrayage de campagne*, (*art.*) field-gun drag shoe;
— *d'enrayage de siège*, (*art.*) siege-gun drag shoe;
— *à enrayer*, drag, shoe, drag shoe; locking skid for a wheel;
— *d'épars*, (*art.*) the slotted shoe or support of the first cross bar of the French garrison gin;
étonnement du —, (*hipp.*) concussion of foot by shock or blow;
— *fendu*, cloven hoof;
— *de frein*, (*mach.*) brake block;

sabot-*galoche*, (*Fr. a.*) barrack shoe (wooden sole, leather upper);
— *à glace*, ice shoe, for going down hill covered with ice; creeper;
— -*guide*, (*sm. a.*) sabot of a tubular bullet;
— *de lance*, (*sm. a.*) shoe of a lance;
— *d'obus*, (*art.*) shell bottom;
— -*patin*, (*art.*) wheel runner (to be used on snow);
— *de pied*, iron shoe plate (for a trestle leg, etc.);
— *de pilotis*, pile shoe;
— *de pompe*, pump piston;
— *à rainures*, (*art.*) grooved sabot;
— *de traîneau*, sledge shoe.

sabotage, m., (*r. r.*) operation of fixing the chairs on the ties; cut or notch for the foot of a foot rail.

saboter, v. a., to put on a sabot or shoe; to shoe (a pile, etc.); to scamp, skimp, work.

sabotte, f., — *de chevalet*, (*pont.*) collar or shoe (to prevent a trestle leg from sinking too deep).

sabre, m., (*sm. a.*) sword, saber; (*mas.*) mason's filling trowel;
— *d'abordage*, (*nav.*) cutlass;
— -*baïonnette*, sword bayonet;
— -*baïonnette isolé*, sword bayonet to which there is no corresponding rifle;
— -*briquet*, short infantry sword (obs.);
— *au clair*, sword bared;
coup de —, sword cut, sword wound;
coup de plat de —, blow with the flat of a sword;
— *droit*, straight sword, saber;
— *d'escrime*, fencing saber;
— *de grosse cavalerie*, cut-and-thrust sword;
haut le —! (*command*) raise saber!
— *d'honneur*, sword of honor (officers' reward for gallantry, given during the Revolution, before the institution of the Legion of Honor);
— *main*! (*command*) draw saber!
— *à la main*, (*call*) draw saber;
manche à —, soldier (regarded as a mere killing machine; term of contempt);
— -*Montmorency*, light cavalry saber;
— -*poignard*, sort of dirk;
port du —, position of carry saber;
porter le —, to carry saber;
présenter le —, to present saber;
remettre le —, to return saber;
reposer le —, to order saber;
— *traînant*, cavalry saber;
traîneur de —, blustering soldier (term of contempt);
— -*yatagan*, curved sword bayonet.

sabrer, v. a., (*mil.*) to saber, to cut down with the saber; (*fig.*) to hurry over, to patch up (work) (cf. our expression to "soldier" or "soger");
— *en aveugle*, (*mil.*) to saber recklessly, without doing much;
— *et pointer*, to cut and thrust.

sabretache, f., (*mil.*) sabretache.

sabre-tout, m., fire-eater.

sabreur, m., (*mil. slang*) brave soldier, without great knowledge of his profession; bloodthirsty soldier; (*fig.*) slovenly workman (cf. our "soldier" or "soger").

sac, m., sack, bag, pouch; sack (of a town); (*mil.*) knapsack, haversack; (*min.*) pocket;
— -*abri*, (*mil.*) shelter-tent half;
— *d'ambulance*, (*med.*) sort of medical field kit;
— *à avoine*, (*mil.*) oat sack for field equipment;
— *de batterie*, (*art.*) grain sack;
— *à battre la poudre*, (*powd.*) calfskin for pulverizing powder;
— *à blanchissage*, wash bag;
— *à blé*, corn sack;
— *de bourrage*, (*mil. min.*) sand bag used for tamping;
bretelles de —, (*mil.*) slings of a knapsack;
— *à cartouches*, (*mil.*) cartridge bag for carrying cartridges;
— *à charge(s)*, (*art.*) cartridge bag;
— *de couchage*, bed sack, sleeping sack;
course en —, sack race;

sac, cul de ——, blind alley;
dans le ——! (*nav.*) turn in!
—— de dépêches, mail bag;
—— à, de, distribution, (*Fr. a.*) sack used in making issues of supplies; issue bag or sack;
—— à éponges, (*art.*) sponge sack;
—— à étoupilles, (*art.*) primer pouch;
faire le ——, (*mil.*) to pack a knapsack;
—— à feu, v. —— foudroyant;
—— foudroyant, (*mil.*) bag containing 2 or 3 pounds of powder, used in the defense of a breach (obs.);
—— à fourrage, (*harn.*) nosebag;
—— de guidon, bag on handle bars of a bicycle;
—— à laine, (*fort.*) woolsack, for making parapet;
—— à malices, (*mil. slang*) soldier's bag for brushes, thread, needles, etc.;
mettre à ——, to sack, plunder, pillage (as, e. g., a captured city);
se mettre dans le ——, (*nav.*) to turn in;
—— de mitraille, (*art.*) langrel, langridge, bag (obs.);
—— à munitions, (*art.*) cartridge pouch;
—— natatoire, swimming bag (for cavalry cross ing rivers);
—— de nuit, traveling bag;
—— à paille, straw sack, for bolster of bed (bedding);
porte- ——, (*med.*) man who carries the —— d'ambulance;
—— porte-armements, (*art.*) implement bag;
—— à poudre, (*in gen.*) powder bag; v. —— foudroyant; (*artif.*) bag holding the bouncing of some fireworks;
poitrail de ——, (*mil.*) breast strap of a knapsack;
radeau- ——, (*mil.*) sack filled with straw (for crossing streams);
—— à sable, (*siege*) sand bag;
—— à tente, (*mil.*) tent sack (for shelter tent);
—— tente-abri, (*mil.*) shelter-tent half;
—— à terre, v. —— à sable;
——s à terre! (*command*) unsling knapsacks!
toile à ——, sackcloth, bagging;
vivres du ——, v. s. v. vivre.
saccade, f., jerk, start, shake, jolt; (*man.*) saccade, jerk (on the bit); sudden jerk with the reins;
donner une ——, to jerk, shake, jolt, give a shake to;
par ——s, by jerks.
saccadé, p. p., respiration ——e, (*hipp.*) respiration of a broken-winded horse.
saccader, v. a., (*man.*) to jerk a horse.
saccagement, m., sacking, pillaging; sack, plunder, pillage.
saccager, v. a., to sack, plunder, pillage, ransack; (*fam.*) to throw into confusion.
sachet, m., small sack, bag; oat bag (hung around a horse's neck; (*art.*) cartridge bag (empty);
—— de cartouche, (*art.*) cartridge bag;
—— d'étamine, (*art.*) empty cartridge bag of bolting cloth;
—— à, de, gargousse, (*art.*) (empty) cartridge bag;
—— de mitraille, (*art.*) grapeshot (obs.);
—— de pansement, (*med.*) first-aid packet
—— plein, (*art.*) cartridge;
—— à vivres, (*Fr. a.*) provision bag, containing the petits vivres de 4 jours.
sacoche, f. (*harn.*) saddle pocket, saddlebags;
—— d'ambulance, (*med.*) the cavalry sac d'ambulance (French army);
—— à charbon, (*art.*) coal sack of a field forge;
—— de maréchal-ferrant, farrier's tool bag;
—— à outils, (bicycle) tool bag.
sacome, m., (*cons.*) molding, profile of a molding.
sacre, m., coronation, consecration.
sacré, a., sacred, consecrate, holy; cursed; (*hipp.*) having to do with the os sacrum.
sacrer, v. a. n., to crown; to consecrate; to curse and swear.
sacrum, m., (*hipp.*) sacrum.
safran, m., crocus.

sagaie, sagaïe, f., assegai.
sage, a., wise; quiet, gentle; (*hipp.*) kind, steady, of good disposition;
cheval ——, quiet horse.
sagène, m., sachine (Russian measure of length, 2.13ᵐ).
saie, f., hard brush.
saignée, f., bleeding, bloodletting; small of the arm; trench, drain; draining; long narrow notch; cut, groove; (*mil. min.*) opening in a powder hose for the communication of fire;
—— baveuse, (*hipp.*) bleeding in which blood comes out in a sheet;
—— blanche, (*hipp.*) unsuccessful bleeding (no blood flows);
—— d'irrigation, irrigation channel.
saigner, v. a. n., to bleed; to be open, green (of a wound); to drain (a ditch, a swamp, etc.); to divert (a stream); to cut or open a dam; (*mil.*) to drain a stream (so as to prevent an inundation); to cut the dam (of an inundation);
—— un fossé, (*fort.*) to drain a ditch;
—— une gargousse, (*art.*) to reduce the charge of a cartridge;
—— une inondation, (*mil.*) to drain flooded ground;
—— du nez, (*art.*) (of the muzzle) to dip, drop, at the moment of fire; to spew, run at the mouth;
—— une rivière, to turn the course of a part of a river.
saillant, a., salient, projecting, prominent, protuberant; (*fig.*) striking;
angle ——, (*fort.*, etc.) salient angle;
assise ——e, (*mas.*) projecting course.
saillant, m., (*fort.*) salient, salient angle.
saillie, f., projection; projection on or from a building; projecting piece, jutting out; lug, flange, set-off; ledge; rabbet; (*r. r.*) check of (a chair); flange of a wheel; (*hipp.*) leaping, covering;
à ——s, flush;
—— -arrêtoir, (*sm. a.*) projection on the cylinder of a revolver (French, model 1892), acting in connection with the cylinder stop;
(s')avancer en ——, to project, protrude, stick out;
carte de ——, (*hipp.*) certificate of service (by French Government stallions);
construire en ——, to cause to jut out;
—— des embases, (*art.*) projection of the rim bases;
en ——, projecting, jutting out;
en —— sur, projecting from;
être en ——, v. (s')avancer en ——;
faire ——, v. (s')avancer en ——;
faire une —— dans, to rabbet;
fondation en ——, (*cons.*) projecting foundation;
formant —— (sur), flanged (on), jutting out (from);
mettre en ——, to stick out, up; to bring out;
porter en ——, to project;
—— -repère, (*art.*) origin on some fuse graduations;
—— du sol, (*top.*) undulation of the ground.
saillir, v. a. n., to gush out, protrude; to stand out, project, jut out; (*hipp.*) to cover, serve, leap, horse; (*cord.*) to haul upon a hawser.
sain, a., sound, healthy, clear; (*art.*) sound (of armor plate), i. e., no cracks within 2.5 calibers of the point of impact;
cheval —— et net, v. s. v. cheval;
côte ——e, clear coast;
mouillage ——, safe anchorage;
plaque ——e, (*art.*) plate (armor) without any cracks (even though it may have a shot hole through it).
saindoux, m., hog's lard.
sainfoin, m., sainfoin.
saint-cyrien, m., cadet of St. Cyr; theoretically educated officer.
sainte-barbe, f., gun room; powder magazine; patron saint of artillerymen (written with capitals in this sense; v. s. vv. barbe and fête);

sainte-barbe, *jour de la* ——, artillerymen's festival. (v. s. v. *fête*).
saisi, m., (*law*) debtor distrained upon;
  tiers ——, third person distrained upon.
saisi, p. p., (*law*) seized of; charged with.
saisie, f., (*law*) seizure, distraint; (*nav.*) capture or seizure of a neutral vessel;
  —— *-arrêt,* (*law*) garnishee; seizure by creditor of goods of debtor in hands of third person;
  —— *-brandon,* (*law*) siezure of crops and fruits;
  —— *-exécution,* (*law*) seizure of furniture; execution, distraint;
  —— *-gagerie,* (*law*) distraint;
  —— *-revendication,* (*law*) seizure under prior claim.
saisine, f., (*law*) seizin; (*nav.*) gripe.
saisir, v. a., to seize, capture, catch, lay hold of, take hold of; to grip; to put in the possession of, to submit (a document); (*law*) to seize; (*cord.*) to lash, secure; to seize; to fasten with a seizing; to nip;
  —— *ensemble,* (*cord.*) to lash all up together;
  —— *un tribunal de,* (*law*) to lay a case before a court.
saisissable, a., (*law*) subject to levy or seizure.
saisissant, m., (*law*) seizing or levying person.
saisissement, m., (*law*) seizure; (*fenc.*) seizure of the adversary's blade;
  —— *de froid,* chill.
saison, f., season.
salade, f., salad; (*hipp.*) mess of wine and of bread for a horse;
  régiment de ——, (*mil.*) regiment of men taken from different regiments; (hence) poor regiment;
  troupes de ——, (*mil.*) poor troops.
salaire, m., pay, wages, salary, hire, reward.
salaison, f., salting (of meat, etc.); salt provisions (*in pl.* usually).
salant, a., salt;
  marais ——, salt marsh;
  puits ——, salt well.
salarié, p. p., m., paid, salaried; salaried person.
salarier, v. a., to pay, fee, hire.
sale, a., dirty; (*nav.*) foul (ship's bottom); foul (of a coast).
salé, m., salt pork.
saler, v. a., to salt (meat, etc.);
  —— *un navire,* to salt a ship;
  —— *et sécher au soleil,* to jerk (meat).
salière, f., saltcellar; (*hipp.*) hollow (over the eye).
salin, a., saline; m., salt marsh; potash (crude, obtained by evaporation).
saline, f., salt fish; salt works, pan, pit.
salinomètre, m., salinometer.
salir, v. a., to soil, make dirty.
salite, f., (*expl.*) salite.
salle, f., hall, room (large); ward of a hospital;
  —— *d'armes,* fencing hall; (*sm. a.*) armory;
  —— *d'artifice(s),* (*artif.*) fireworks laboratory;
  —— *d'attente,* waiting room;
  —— *d'audience,* audience hall;
  —— *de bains,* bathroom;
  —— *de billiard,* billiard room;
  —— *de chauffe,* (steam) fire room;
  —— *aux compositions,* (*powd.*) mixing house, mixing room or shop;
  —— *des consignés,* in military hospitals, prisoners' ward;
  —— *des convalescents,* convalescent ward;
  —— *de correction,* fencing hall; (*sm. a.*) v. —— *de discipline;*
  —— *de dépôt,* lockup;
  —— *de désinfection,* disinfection ward of a veterinary hospital;
  —— *de discipline,* (*mil.*) guardroom (for prisoners), confinement in the guardhouse;
  —— *d'école,* schoolroom;
  —— *d'épreuve,* (*sm. a.*) room in which gun barrels are loaded for proof;
  —— *d'escrime,* fencing hall;
  être de ——, (*mil.*) to be on guard at the palace;

salle *de gabarits,* (*nav.*) mold loft;
  —— *des gardes,* guardroom of a palace;
  —— *d'hippiatrique,* room containing models, skeletons, used in teaching hippology;
  —— *d'honneur,* (*Fr. a.*) kind of regimental hall, containing patriotic busts, etc., lists of officers and men killed in action, and used for meetings of officers and for various regimental councils, etc.;
  —— *d'hôpital,* ward of a hospital;
  —— *d'humidité,* (*powd.*) damping room for testing powder barrels;
  . . . . *jours de* —— *de police,* (*mil.*) (so many) days in the guardhouse;
  —— *de lecture,* (*Fr. a.*) men's reading room;
  —— *des malades,* v. —— *d'hôpital;*
  —— *de manœuvre,* (*mil.*) drill hall;
  —— *à manger,* mess room;
  —— *de musique,* band practice room;
  —— *de police,* (*mil.*) prisoners' room, guardroom, confinement in the guardhouse;
  prévôt de ——, (*fenc.*) under fencing teacher;
  —— *des, du, rapport(s),* (*mil.*) orderly room;
  —— *de réception,* v. —— *d'audience;*
  —— *de recette,* (*sm. a.*) room for receipt and examination of arms and component parts;
  —— *de révision,* (*sm. a.*) room for examining gun barrels after proof;
  —— *du trône,* throne room;
  —— *de visite,* reception room (of a military hospital).
salon, m., drawing room, parlor; gallery, exposition; (*nav.*) saloon;
  —— *des courses,* betting stand.
salpétrage, m., formation of saltpeter in artificial beds.
salpétration, f., efflorescence on buildings.
salpêtre, m., saltpeter, niter, potassium nitrate; (by extension, poetical) gunpowder;
  —— *brut,* crude, commercial saltpeter;
  —— *du Chili,* Chile saltpeter;
  —— *de conversion,* saltpeter made from sodium nitrate;
  —— *de deux (trois, etc.) cuites,* potassium nitrate refined by second (third, etc.) boiling;
  —— *de deux eaux,* v. —— *de deux cuites;*
  —— *de houssage,* potassium nitrate obtained by efflorescence;
  —— *naturel,* crude, natural, saltpeter;
  —— *raffiné,* refined potassium nitrate;
  —— *de roche,* saltpeter after the fourth boiling;
  —— *terreux,* efflorescence on damp walls.
salpêtrer, v. a., to cover (a walk, etc.) with a mixture of saltpeter and earth; to bring out saltpeter.
salpêtrière, f., niter works.
saluer, v. n. a., to bow to; to take off one's hat to; to greet; to greet by letter; (*mil., nav.*) to salute, (of persons and of guns); to fire a salute, to exchange salutes; (of colors) to droop; to cheer;
  —— *du canon,* (*mil.*) to fire a salute with guns;
  —— *de . . . . coups de canons,* (*mil.*) to salute with (so many) guns;
  —— *du drapeau,* (*mil.*) to droop the colors;
  —— *de l'épée,* (*mil.*) to salute with the sword;
  —— *de la main,* (*mil.*) to salute with the hand;
  la mer salue la terre, the ship fires the first salute when making a port;
  —— *de la mousqueterie,* to salute with musketry;
  —— *avec le, du, pavillon,* to dip the flag;
  —— *par un vival,* to cheer;
  —— *de la voix,* v. —— *par un vivat.*
salure, f., degree of saltness (of the ocean, etc.).
salut, m., safety, salvation, welfare; greeting, bow; salute; (*mil. nav.*) salute (person and guns);
  —— *des armes,* (*mil.*) military salute; (*fenc.*) salute with the foils;
  —— *du canon,* (*mil., nav.*) salute with guns;
  —— *de canot à canot,* (*nav.*) boat salute;
  contre- ——, (*mil.*) return salute;
  donner un ——, (*mil., nav.*) to fire a salute;
  —— *au drapeau,* (*mil.*) salute of the colors;
  —— *du drapeau,* (*mil.*) salute by the colors;

salut, *échanger des* ——*s*, (*mil.*, *nav.*) to exchange salutes;
—— *de l' épée*, (*mil.*) salute with the sword;
—— *de l' État*, commonwealth;
faire un ——, (*mil.*, *nav.*) to fire, give, a salute;
faire un —— *de* ....*coups*, (*mil.*, *nav.*) to fire a salute of (so many) guns;
—— *de mer*, (*nav.*) salute by a ship to another, or to a fleet;
—— *militaire*, (*mil.*) military salute;
——*s militaires*, (*mil.*) military honors, compliments;
—— *public*, the public welfare;
rendre un ——, (*mil.*, *nav.*) to return a salute;
rendre le —— *coup pour coup*, (*mil.*, *nav.*) to return a salute gun for gun;
—— *du sabre*, (*mil.*) salute with a sword.

salve, f., (*art.*) salvo; (*sm. a.*) volley;
—— *d'artillerie*, (*art.*) salvo; salute by a salvo (honors paid in French army in certain cases);
—— *basse*, (*art.*) salvo whose average point of burst is below the type height (of burst);
—— *d'essai*, (*sm. a.*) sighting volley;
faire une ——, to fire a salvo, volley;
feu(x) de ——, volley firing, etc.;
—— *haute*, (*art.*) a salvo whose average point of burst is above the type height (of burst);
—— *à hauteur*, (*art.*) a salvo at the proper height of burst;
—— *percutante*, (*art.*) a shrapnel salvo that strikes before bursting;
—— *très haute*, (*art.*) a salvo whose point of burst is more than twice as high as the type height;
tirer une ——, v. *faire une* ——.

sancir, v. n., (*nav.*) to founder, sink;
—— *sur ses amarres*, v. —— *à l'ancre;*
—— *à l'ancre*, to founder at anchor;
—— *à la mer*, to founder at sea.

sandale, f., sandal; (*harn.*) lady's stirrup; (*fenc.*) fencing shoe, sandal; (*gym.*) gymnasium shoe; (*in pl.*) magazine slippers.

sandaraque, f., pounce, sandarac;
poudre de ——, pounce.

sandwich, a., sandwich (armor, etc.).

sang, m., blood; race, family; the blood royal; (*hipp.*) blood; full blood;
avoir du ——, (*hipp.*) to be blooded, part blooded;
se battre au premier ——, to fight until blood is drawn and no longer;
cheval pur ——, full-bred horse;
de ——, (*hipp.*) blooded;
demi ——, (*hipp.*) half bred;
demi- ——, (*hipp.*) half-bred horse;
deuxième ——, v. *demi-* ——;
faire couler le ——, to cause a war;
jour de ——, bloody day;
mettre à feu et à ——, to put to fire and sword;
mettre une ville à ——, to put a town to the sword;
près du ——, (*hipp.*) nearly thoroughbred;
pur ——, (*hipp.*) thoroughbred;
pur(-) ——, (*hipp.*) thoroughbred horse;
répandre du ——, to shed blood;
tout en ——, covered with blood;
trois-quarts de ——, (*hipp.*) three-quarters bred horse.

sang-froid, m., coolness, calmness;
de ——, in cold blood.

sanglade, f., cut, lash, with a whip.

sanglant, a., bloody (fight, defeat, war, etc.).

sangle, f., strap, band; paunch, mat; (*harn.*) girth, saddle girth, cincha; neck harness, neck strap; (*mach.*) ring;
—— *à cartouches*, (*sm. a.*) cartridge-packet strap;
—— *de chanvre*, hemp strap;
—— *croisée*, web strap;
—— *dossière*, back strap of a litter;
—— *d'embarcation*, boat gripe;
—— *faite à la main*, hand mat;
—— *au métier*, sword mat;
passage des ——*s*, (*hipp.*) girth;
—— *porte-caisse*, (*art.*) chest strap;
première ——, (*harn.*) under girth;
seconde ——, (*harn.*) middle girth;
—— *de selle*, (*harn.*) saddle girth;

sangle, *sous-ventrière*, belly strap of a litter.

sangler, v. a., to bind; to put a strap on; (*harn.*) to girth; to strap; to cinch; to cinch up;
—— *de coups de fouet*, to lash, horsewhip;
—— *un coup*, to give a lash, a blow.

sanglon, m., (*harn.*) small girth, strap.

sanglot, m., v. *sanglon*.

sanguine, f., red hematite; red chalk;
—— *à brunir*, polishing stone (for metals);
—— *à crayon*, red (drawing) chalk;
crayon en ——, red chalk (for drawing).

sanitaire, a., sanitary;
cordon ——, sanitary line of troops; line of troops on quarantine duty.

sans-culotte, m., Jacobin, sans-culotte.

santé, f., health; medical service; quarantine service;
agent de la ——, quarantine officer;
appeler ——, to make quarantine signal;
bateau de ——, quarantine boat;
billet de ——, bill of health;
bureau de ——, board of health;
canot de la ——, v. *bateau de* ——;
pour cause de ——, on medical certificate;
corps de ——, (*mil.*, *nav.*) medical department;
demander la ——, v. *appeler* ——;
garde de ——, quarantine official;
maison de ——, private asylum;
par motif de ——, v. *pour cause de* ——;
officier de ——, v. *s. v. officier;*
patente de ——, bill of health;
service de ——, medical staff, corps or department; (also) duties, functions, of the medical staff; quarantine service.

sape, f., undermining (of a wall, a tower, etc.); hole made in the foot of a wall to bring it down; (*siege*) sap, sapping.
(The following terms relate to sieges:)
—— *accidentelle*, generic term for saps belonging to no regular system, but adjusted to circumstances; improvised sap (as in going across a rock that can not be dug);
—— *ancienne*, Turkish sap;
blinde de ——, sap mask or blind;
—— *blindée*, blinded sap;
bouclier de ——, sap shield;
—— *à, en, crémaillère*, indented sap, sap with crotchets all on one side;
crochet de ——, sap hook;
—— *debout*, direct double sap;
—— *à découvert*, open sap;
—— *demi-double*, half-double sap;
—— *demi-pleine*, same as single sap, except that a *gabion farci* is not used;
—— *dérobée*, sap hastily executed (when enemy's fire slacks, or he does not keep a good watch);
—— *double*, double sap;
—— *double à terre roulante*, double sap (head parapet pushed forward);
—— *double traversée*, sap with double traverses;
—— *double à traverses en crémaillère*, double sap with indented traverses;
—— *entière*, full sap;
exécuter une ——, to drive a sap;
fagot de ——, sap fagot, fascine;
faire déboucher une ——, to break out a sap;
—— *forée*, bored sap (i. e., a boring is made parallel to the surface of the ground, loaded, and blown up);
—— *sans formes*, sap driven at full size (and therefore requiring no subsequent enlargement); (in case of oblique fire, when one of the two forms of the simple regular sap may be suppressed);
fourche de ——, sap fork;
—— *en gabions*, gabion sap;
—— *en galeries de mine*, underground sap (driven under ground);
marcher en ——, to sap, to drive a sap;
masque de ——, head parapet of a sap;
—— *sans masque*, sap without covering parapet at its head;
—— *ouverte*, open sap;
—— *sans parapet*, deep sap without parapet or *masque de tête*;

sape *pied à pied*, single or full sap, progressive sap;
— *pied à pied en sacs à terre*, single sandbag sap;
— *pleine*, full sap;
— *pleine double à traverses tournantes*, double sap with cube traverses;
portion circulaire de —, "circular portion" (of the sap);
pousser une —, to push a sap;
— *profonde*, v. — *sans parapet*;
retour de —, change of direction of a sap;
— *en sacs à terre*, sand-bag sap;
— *simple*, single sap;
— *simple à terre roulante*, single sap (head parapet pushed ahead);
— *simple traversée*, — *à traverses*, single traversed sap;
— *simple en zigzag*, simple zigzag sap;
— *simultanée*, simultaneous sap, all dug at the same time;
— *sinueuse*, v. — *tournante*;
— *successive*, v. — *pied à pied*;
— *à tambour*, serpentine sap, traversed sap;
— *à terre roulante*, sap in which the excavated earth at the head of the trench is gradually pushed forward;
— *à terre roulante double (simple)*, double (single) sap with the sap head pushed on;
tête de —, sap head;
— *tournante*, serpentine sap;
travaux de —, trench work;
— *à traverses*, traversed sap;
— *à traverses tournantes*, sap with cube traverses;
— *volante*, flying sap, flying trench work;
— *volante dérobée*, hastily executed flying trench work (enemy napping, or his fire kept down);
— *volante en gabions et sacs à terre*, gabion and sand bag flying sap, executed in earth too hard to be dug;
— *en zigzag à angles arrondis*, serpentine sap.

sapement, (*siege*) sap work, trench work.

saper, v. a., to undermine; (*siege*) to sap.

sapeur, m., (*mil.*) sapper, pioneer;
— *-aérostier*, (*Fr. a.*) balloon man (one company in each of the first four engineer regiments);
brigadier —, v. s. v. *brigadier*;
caporal —, v. s. v. *caporal*;
— *de chemin de fer*, (*Fr. a.*) soldier of the engineer railway regiment;
— *conducteur*, v. s. v. *conducteur*;
élève —, v. s. v. *élève*;
— *du génie*, engineer sapper;
— *-mineur*, (*Fr. a.*) engineer soldier proper (sappor);
— *-ouvrier*, (*Fr. a.*) private of the engineer railway regiment;
— *s ouvriers d'art*, (*Fr. inf.*) intelligent men, skilled in the use of tools, employed in the manipulation of ammunition, in the care of target-practice material, etc.; in the care and repair of barracks, of ranges, gymnasiums, etc.; in the construction of field works, etc.;
— *-pionnier*, pioneer;
— *-pompier*, v. s. v. *pompier*;
— *-s-pompiers de la ville de Paris*, v. s. v. *pompier*;
— *porte-hache*, (*Fr. a.*) the classic sapper of the infantry (suppressed in 1871);
— *s porteurs d'outils*, v. *soldats porteurs d'outils*;
— *de régiment*, (*Fr. a.*) a regimental (e. g., infantry) sapper, as distinguished from one belonging to the engineers;
— *-télégraphiste*, (*Fr. cav.*) telegraph operator;
— *de tête*, leading sapper.

saphène, f., (*hipp.*) saphena (vein).

sapin, m., fir (wood and tree; in commerce, generic term for resinous woods);
ais de —, deal;
— *blanc*, silver fir;
bois de —, deal;
faux —, pitch pine;
— *rouge*, red fir, spruce; pitch pine.

sapine, f., fir joist, plank; scaffolding timber or pole; lift scaffold; sort of hoisting or winding engine.

sarbacane, f., pea shooter; speaking trumpet.

sarche, f., frame of a sieve.

sarcler, v. a., to weed.

sarcloir, m., hoe.

sardine, f., sardine; (*in pl., mil. slang*) stripes on the sleeves of a tunic;
— *s blanches*, gendarme's stripes.

sarrasin, m., buckwheat; (*met.*) waste, sweepings;
blé —, buckwheat.

sarrasine, f., (*fort.*) portcullis, sarrasin.

sarrau, m., smock-frock, blouse, laboratory frock, wagoner's frock, veterinary's frock, surgeon's frock when operating.

sas, m., sieve, bolt; lock-chamber;
— *à air*, air lock (of the plenum process);
— *délié*, fine sieve;
— *fin*, v. — *délié*;
gros —, coarse sieve;
— *mobile*, cradle, rocker;
passer au —, to sift, to bolt.

sasse, f., (*pont.*) water scoop.

sassement, m., sifting.

sasser, v. a., to sift, bolt, winnow.

sasset, m., small sieve.

sassoire, f., sweep bar, sway bar, riding bed or bolster.

satellite, m., satellite;
veine —, (*hipp.*) vein nearest the arteries.

satiner, v. a., to press paper, to satin;
— *à chaud*, to hot press;
— *à froid*, to cold press.

satisfaction, f., satisfaction;
donner —, to give satisfaction (as by a duel).

satisfaire, v. a., to satisfy;
— *à la conscription*, (*mil.*) to draw lots for military service.

saturation, f., (*chem., elec., steam, etc.*) saturation.

saturer, v. a., (*chem., etc.*) to saturate.

saturomètre, m., (*steam*) salt gauge, salmometer.

saucier, m., shoe for the foot of a stanchion to hold it in place; saucer of a capstan;
— *couronne des linguets*, pawl rim of a capstan.

saucisse, f., sausage; (*fort.*) saucisson; (*artif.*) powder hose; bag containing powder, or powder itself.

saucisson, m., sausage; (*artif.*) saucisson; powder hose; (*fort.*) long fascine;
— *de pétards*, (*expl.*) string or necklace of petards;
— *de retraite*, (*siege*) saucisson to which fascine-work revetment is anchored; anchoring fascine;
— *volant*, (*artif.*) long petard choked in the middle.

sauf-conduit, m., (*mil.*) safe conduct, pass.

sauge, f., sage;
feuille de —, (*min.*) miner's pickax; (*hipp.*) point of a seton needle.

saule, m., willow (tree and wood);
feuille de —, feather-edge file;
— *pleureur*, weeping willow.

saumache, a., v. — *saumâtre*.

saumâtre, a., brackish.

saumon, m., salmon; charge of coal in a coke furnace; (*fond.*) feeding head, sinking head; sullage piece; (*met.*) pig of metal; (*in pl., nav.*) kentledge;
— *de fer*, kentledge;
— *de fonte*, (*met.*) casting pig, iron pig;
fonte en —, (*met.*) pig iron;
plomb en —, (*met.*) pig lead.

saumure, f., brine.

saupoudrer, v. a., to powder, dust, sprinkle with;
(*fond.*) to blacken the mold.

saupoudreur, m., (*fond.*) workman who blackens mold.

saure, a., sorrel.

saut, m., jump, leap, vault; jerk; waterfall; (*mil.*) military spring, bound; (*hipp.*) leaping, covering;

saut *avec armes*, (*mil. gym.*) jump armed;
— *avec armes et bagages*, (*mil. gym.*) jump armed and equipped;
— *d'un cordage*, (*cord.*) surge, jerk, on a rope;
— *dedans et dehors*, (*man.*) in-and-out jump;
— *dessus et au-delà*, (*man.*) jump over and beyond;
donner un — *à la gourmette*, to shorten a curb chain;
faire des —*s*, (*man.*) to curvet;
franchir de plein —, to clear at a jump (as a ditch);
— *de haut en bas*, (*man.*) descending leap;
— *en hauteur*, high jump;
— *en largeur*, broad jump;
— *de loup*, (*fort.*, etc.) ha-ha;
— *de mouton*, (*man.*) buck, buck jump; sometimes, a rise of the fore and rear quarters alternately;
— *d'obstacles*, (*man.*) jumping of obstacles;
pas et le —, (*man.*) an air consisting of a shortened gallop, a curvet, and a capriole;
— *à la perche*, (*gym.*) pole vaulting;
— *de pie*, (*man.*) short jump (as before breaking into gallop);
— *au pilier*, (*man.*) jump on a *sauteur*, q. v.;
— *de pied ferme*, (*gym.*) standing jump; (*man.*) standing jump;
— *précédé d'une course*, (*gym.*) running jump;
— *en profondeur*, (*gym.*) downward jump;
— *rampant*, (*man.*) longitudinal leap or jump;
— *de volée*, (*man.*) flying jump.

sautage, m., exploding; act of blowing up; blasting.

saute, f., — *de vent*, sudden shifting, veering, change of wind.

sauter, v. a. n., to jump, jump over; to leap, leap over; to skip; to spring, to spring at; to jump (i. e., omit, pass over); to burst, explode; (of the wind) to veer, shift; (*fenc.*) to raise the foot too high, and hence describe a curve instead of a straight line; (*hipp.*) to leap, cover, serve;
— *à l'abordage*, (*nav.*) to board;
— *le bat-flanc*, (*mil.*) to "run it" out of barracks;
— *à cheval*, (*man.*) to vault into the saddle;
— *en croupe*, (*man.*) to jump up behind;
— *entre les piliers*, (*man.*) to jump between the posts (v. *pilier* and *sauteur*);
faire —, to spring, disengage, unhook; (*expl.*, etc.) to blow up;
se faire —, to blow one's self up; (*nav.*) to blow up the ship, one's ship;
— *de ferme à ferme*, v. s. v. *ferme*, adv.;
— *d'un grade à un autre*, (*mil.*) to jump from one grade to another (without passing through the intermediate grade or grades);
— *en liberté*, (*man.*) to jump free (of a horse, not tied between the posts);
— *à la perche*, (*gym.*) to pole vault;
— *en selle*, v. — *à cheval*.

sauterelle, f., bevel, bevel square; swing bale catch;
— *à mitre*, — *à onglet*, miter rule, bevel rule.

sauteur, m., (*man.*) bucking horse, pillar horse, of riding schools (horse tied between posts and trained to buck and jump);
grand —, high jumper (horse);
— *en liberté*, leaper at liberty (i. e., not tied between posts);
— *dans*, *entre*, *les piliers*, leaper between posts.

sautoir, m., St. Andrew's cross; (*gym.*) jumping bar, hurdle;
en —, crosswise; (*cord.*) with a cross seizing, with cross turns;
étai en —, diagonal stay;
porter en —, to wear crosswise (as shoulder belts);
porter un ordre en —, to wear an order upon a ribbon around the neck.

sauvage, a., wild, savage, barbarous; (*in pl.*, as a noun) savages (uncivilized people);
eaux —*s*, v. *eaux folles*.

sauvegarde, safeguard, safe-keeping; custody; protection; (*nav.*) manrope; (*mil.*) party or detachment to guard a place against pillage;
— *d'aviron*, oar lanyard;
—*s du gouvernail*, (*nav.*) rudder chains, pendants.

sauvegarder, v. a., to safeguard; to protect.

sauvement, m., salvage.

sauver, v. a., to save (a man, a ship, etc.), rescue, deliver, preserve, keep;
se —, to escape, run away, flee, make one's escape; to take refuge, to refugee;
se — *à cheval*, to escape on horseback;
se — *de prison*, to break jail, prison; to break out of prison;
sauve qui peut! everyone for himself!

sauvetage, m., salvage, recovery of anything wrecked; rescue (of men fallen overboard, of persons in danger of drowning);
appareil de —, life-preserver;
appareil de — *pour incendies*, fire escape;
bouée de —, life buoy;
canot de —, lifeboat;
ceinture de —, life belt, life-preserver;
droits de —, salvage money;
ligne de —, life-preserving line;
médaille de —, life-saving medal;
poste de —, life-saving station;
prix de —, salvage, salvage money;
société de —, salvage association.

sauveter, v. a., to rescue, save, from a wreck.

sauveteur, m., salvor; a., life-saving.

savane, f., (*top.*) savannah;
— *noyée*, a savannah on the seacoast.

savartine, f., (*mil.*) sort of fougasse to throw barrels of powder (fused) into the enemy.

savate, f., old shoe; sole (as of sheers); (*gym.*) savate, French system of boxing; (*mil.*) sort of punishment given a soldier by his comrades (really a severe spanking); (*nav.*) anchor shoe.

savon, m., soap, soaping;
— *blanc*, castile soap;
— *de cuivre*, v. — *métallique*.
— *dur*, hard soap;
eau de —, soapsuds;
— *jaune de résine*, yellow soap, turpentine soap;
— *métallique*, (*art.*) mixture of soap and sulphate of copper (used to diminish friction);
— *mou*, soft soap;
— *-ponce*, sand soap;
— *vert*, green, black, soap; soft soap.

savonnage, m., soaping, washing with soap.

savonner, v. a., to wash with soap.

saxhorn, m., (*mus.*) saxhorn.

saxifragine, f., (*expl.*) saxifragine.

saxon, m., (*artif.*) turning staff.

saxophone, m., (*mus.*) saxophone.

scalène, a., scalene.

scaphandre, m., cork, swimming, jacket; diving dress;
casque de —, diver's helmet;
pompe de —, diving pump.

scaphandrier, m., diver.

scaphoïde, a., m., scaphoid; scaphoid bone.

scarificateur, m., scarifier, scarificator.

scarification, f., scarification.

scarifier, v. a., to scarify.

sceau, m., seal; impression of a seal; (*in pl.*) state seal; functions of the chancellor;
— *de l'Etat*, the state seal;
garde des —*x*, lord chancellor.

scellé, m., seal (on doors, locks, etc., put on by authority); sealed document;
apposer le(s) —(*s*), to seal up;
bris de —*s*, breaking of seals (misdemeanor);
lever le(s) —(*s*), to remove the seals.

scellement, m., (*mas.*) cramp, cramping, sealing, bedding; the end cramped or sealed in.

sceller, v. a., to seal, put on a seal; to seal up; (*fig.*) to ratify; (*mas.*) to bed, fix, cramp in (with mortar, melted lead, etc.);
— *hermétiquement*, to seal hermetically;
— *au plomb*, (*mas.*) to cramp or fix with melted lead.

scène, f., scene; the stage.

schabraque, f., (*unif.*) shabraque.

schako   414   **scorie**

schako, m., (*unif.*) shako.
schapska, m., (*mil.*) uhlan headdress, lancer's helmet.
schelling, m., shilling.
schéma, schème, m., diagram, idea, plan, construction.
schématiquement, adv., diagrammatically.
schiste, m., schist.
schisteux, a., schistous.
schlague, f., (*mil.*) flogging (obs.);
   donner une ――, to flog;
   recevoir une ――, to be flogged.
schlamm, m., (*met.*) slime.
schrapnel, m., v. *shrapnel*.
sciable, a., that may be sawed, fit for sawing.
sciage, m., sawing (operation of);
   bois de ――, sawn timber; timber fit for sawing;
   ―― *en bois de feu*, (act of) sawing wood into lengths merely;
   ―― *en bois d'œuvre*, (act of) sawing wood to shape;
   ―― *à bras*, hand sawing;
   ―― *en grume*, timber sawing;
   ―― *de long*, rip sawing;
   ―― *mécanique*, mechanical sawing;
   ―― *en travers*, sawing against the grain.
sciagraphie, f., v. *sciographie*.
sciasse, f., (*cord.*) stretching rope; rope with several strings or straps attached.
sciatère, f., sun dial.
scie, f., saw; web;
   ―― *allemande*, frame saw;
   ―― *alternative*, alternating saw;
   ―― *anglaise*, scroll saw; jig saw;
   ―― *à araser*, tenon saw;
   ―― *à arc*, bow saw;
   ―― *en archet*, v. ―― *à arc*;
   arçon de ――, saw frame;
   ―― *articulée*, jointed or folding saw; articulated saw;
   bran de ――, sawdust;
   ―― *à bras*, pit saw;
   ―― *à chantourner*, sweep saw; compass saw; bow saw; fret saw;
   ―― *à charnière*, hinge saw;
   ―― *de charpentier*, carpenter's saw;
   ―― *à châssis*, frame saw;
   châssis de ――, saw frame;
   ―― *à chaud*, metal saw, to cut metal hot;
   ―― *circulaire*, circular saw; buzz saw;
   ―― *circulaire au tour*, bench saw;
   ―― *à contourner*, scroll saw; buhl saw;
   coup de ――, saw, cut;
   ――*-couteau*, ―― *à couteau*, fret saw; keyhole saw;
   ――*cylindrique*, cylinder saw; drum saw;
   ―― *à débiter*, v. ―― *de charpentier*;
   débiter a la ――, to saw up;
   dent de ――, saw tooth;
   ―― *à dents*, (specifically) a toothed stone saw;
   ―― *à dents en bec de perroquet*, gullet saw; briertooth saw;
   ―― *à dents crochues*, v. ―― *à dents en bec de perroquet*;
   ―― *sans dents*, stone saw;
   ―― *à dents fines*, fine-toothed saw;
   ―― *à deux mains*, crosscut saw;
   donner du chemin à une ――, to set a saw;
   donner la voie à une ――, v. *donner du chemin à une* ――;
   ―― *à dos*, back saw;
   ―― *à dossière*, v. ―― *à dos*;
   ―― *droite*, straight saw;
   ―― *à échancrer*, v. ―― *à chantourner*;
   ―― *égoïne* (*égoïne*), pruning saw; small handsaw; keyhole saw;
   ―― *d'entrée*, keyhole saw; fret saw;
   ―― *à évider*, v. ―― *à chantourner*;
   fer de ――, two men working a ripsaw;
   ―― *sans fin*, band saw; belt saw;
   ―― *fine*, sash saw;
   ―― *à froid*, metal saw, to cut metal cold;
   fût de ――, v. *châssis de* ――;
   ―― *grande*, ripsaw;
   ―― *à grosses dents*, coarse saw;
   ―― *à grume*, timber saw;
   ―― *à guichet*, keyhole saw, compass saw;
   ―― *horizontale*, horizontal saw;

scie, *lame de* ――, saw blade;
   ―― *à lame sans fin*, v. ―― *à rubans*;
   ―― *de long*, long saw, ripsaw;
   ―― *à main*, handsaw;
   ―― *à manche*, pad saw;
   ―― *à manche d'égohine* (*d'égoïne*), handsaw;
   ―― *mécanique*, power saw, bench saw;
   ―― *à métaux*, hack saw;
   ―― *à molettes*, circular saw;
   ―― *montée*, ―― *à monture*, frame saw, buck saw;
   monture de ――, saw handle;
   ―― *à mouvement alternatif*, v. ―― *alternative*;
   ―― *à passe-port*, keyhole saw;
   ―― *passe-partout*, crosscut saw;
   ―― *à pierre*, stonecutter's saw;
   ―― *à pédale*, pedal saw;
   ―― *à placage*, veneering saw;
   ―― *à planche fâcheuse*, slab saw;
   ―― *pliante*, folding saw;
   ―― *à poignée*, handsaw;
   porte-――, saw pad;
   ―― *à refendre*, ripsaw, pit saw;
   ―― *ronde*, v. ―― *circulaire*;
   ―― *à ruban*(*s*), band saw, belt saw, ribbon saw;
   ―― *de scieur de long*, ripsaw, pit saw;
   ―― *à taillader*, v. ―― *fine*;
   ―― *à tenon*, tenon saw, sash saw;
   ―― *à tournefond*, v. ―― *à chantourner*;
   ―― *tournante*, ―― *à tourner*, compass saw;
   trait de ――, kerf of a saw;
   ―― *de travers*, crosscut saw;
   ―― *à tronçonner*, felling saw;
   ―― *verticale*, straight saw;
   ―― *ventrue*, crosscut, felling, saw;
   voie d'une ――, set of a saw;
   ―― *à voleur*, keyhole saw.
science, f., science;
   ―― *de la guerre*, science of war;
   ――*s militaires*, military arts and sciences.
scier, v. a. n., to saw, saw off; to reap, cut down; (of boats) to back, back water; to hold water, back oars;
   ―― *avec les avirons*, to hold, back, water;
   ―― *bâbord*, to back with the port oars;
   ―― *du bridon*, (*man.*) to saw the mouth; to pull on the reins alternately;
   ―― *à culer*, to back water, back astern; to back;
   ―― *contre le fil*, to saw against the grain, across the grain;
   ―― *du filet*, v. ―― *du bridon*;
   ―― *de long*, to saw (wood) with the grain; to rip;
   ―― *partout*, to back all;
   ―― *en travers*, v. ―― *contre le fil*.
scierie, f., sawmill; sawing machine; saw, power saw;
   ―― *alternative*, reciprocating saw;
   ―― *à chantourner*, sweep saw, scroll saw;
   ―― *à chariot*, sawmill in which the work advances on a cradle;
   ―― *circulaire*, circular sawmill;
   ―― *à cylindres*, roller sawmill;
   ―― *en grume*, log sawmill;
   ―― *à lame circulaire*, circular sawmill;
   ―― *à lame sans fin*, band saw (mill);
   ―― *mécanique*, sawmill;
   ―― *à mouvement alternatif*, v. ―― *alternative*;
   ―― *à mouvement continu*, continuous-motion saw;
   ―― *à placage*, veneering saw;
   ―― *à plusieurs lames*, gang saw.
scieur, m., sawyer, reaper;
   ―― *au dessous*, pit man;
   ―― *au dessus*, top sawyer;
   fosse ―― *de long*, saw pit;
   ―― *de long*, sawyer.
sciographie, f., sciagraphy.
sciographique, a., sciagraphic.
sciot(t)e, f., stonecutter's handsaw.
sciure, f., sawdust;
   ―― *de bois*, (wood) sawdust;
   ―― *de marbre*, marble sawdust.
scorbut, m., scurvy.
scorie, f., (*met.*) slag, scoria, dross;
   ―― *crue*, slag poor in oxide of iron, tap cinder;

scorie *douce*, slag rich in oxide of iron;
—— *de fer*, iron slag;
—— *de forge*, hammer slag;
—— *pauvre*, v. —— *crue;*
—— *riche*, v. —— *douce.*
scorification, f., (met.) scorification.
scorifier, v. a., (met.) to scorify.
scrutin, m., ballot, election;
—— *de liste,* ballot count.
scrutin, m., ballot, election;
—— *par assis et levé,* voting or election in which the voters rise and remain standing till counted;
—— *cadveré,* secret ballot;
—— *découvert,* open ballot;
—— *direct,* direct ballot;
—— *individuel,* ballot containing but one name;
—— *de liste,* blanket ballot;
—— *secret,* secret ballot;
urne de ——, balloting urn;
voter au ——, to cast a secret vote.
sculptable, a., that may be sculptured.
sculpter, v. a., to carve, sculpture.
sculpteur, m., sculptor, carver.
sculpture, f., sculpture, carving; (art.) ornamental work on guns (obs.).
séance, f., meeting (of committees, boards, etc.); meeting, session (as of a court);
lever la ——, to close the meeting;
ouvrir la ——, to open the meeting;
—— *publique,* open session of a court;
—— *tenante,* before rising; on the spot;
—— *de tir,* firing, "shoot."
seau, m., bucket, pail; bucketful;
—— *d'abreuvoir,* watering bucket;
—— *d'affût,* (art.) sponge bucket;
—— *en bois,* etc., wooden, etc., bucket;
—— *à cendres,* ash bucket;
—— *à charbon,* coal scuttle;
—— *d'eau,* water bucket;
—— *d'écurie,* stable bucket;
—— *de forge,* forge bucket;
—— *à goudron,* tar bucket;
—— *à graisse,* slush bucket;
—— *à incendie,* fire bucket;
—— *de manœuvre,* v. —— *d'affût;*
—— *à puiser,* well bucket, draw bucket;
—— *à suif,* slush bucket;
—— *en toile,* canvas bucket; bucket for watering horses in cars.
sébastine, f., (expl.) sebastine.
sébile, f., (wooden) bowl.
sec, a., dry, thin, lean, spare, hard; dried (of fruit, vegetables, etc.); waterless; without springs (as a buffer); (of some metals) brittle; m., dryness; dry food, manger food (for horses);
à ——, high and dry; dry, drained, exhausted; dry (as a ditch); (gym.) on dry land (of preparatory or preliminary swimming exercises); (nav.) under bare poles;
banc ——, (hydr.) bank uncovered at ebb tide;
bois ——, seasoned wood;
cheval qui à les jambes ——*hes,* clean-limbed horse;
courir à ——, (nav.) to scud under bare poles;
être à ——, (boats) to be high and dry;
faire ——, (weather) to be dry;
fossé ——, *fossé à* ——, v. s. v. *fossé;*
fourrage ——, v. s. v. *fourrage;*
fruit ——, v. s. v. *fruit;*
muraille de pierres ——*hes,* dry wall;
nager ——, to row dry;
nager à ——, to row ground, to touch the ground with the oars;
orage ——, rainless storm;
à pied ——, dry shod;
pierres ——*hes,* v. s. v. *pierre;*
remettre un cheval à ——, to take a horse from grass and put him on dry food again;
son ——, coarse bran.
sécante, f., secant.
sécateur, m., cutter (sort of shears).
sécession, f., secession;
guerre de la ——, war of secession (United States, 1861–1865).
sécessioniste, m., secessionist; Confederate (United States, 1861–1865).
séchage, m., drying (as of powder); "stoving;" seasoning (of wood);

séchage *à l'air,* air drying;
—— *de la vapeur,* drying of steam.
séchager, v. a., to squilgee and swab dry.
sèche, f., (hydr.) sand, rock, left dry at low water.
séché, p. p., (slang) found deficient in examination; found (U. S. M. A.); (military schools) punished.
séchée, f., (mil. slang) punishment (schools).
sécher, v. a. n., to dry; to dry up, parch; to grow dry; to season; to drain (as a ditch); (met.) to heat;
—— *à l'air,* to air-dry;
cadre à ——, (artif.) drying frame;
—— *un élève,* (slang) to find deficient; to find (U. S. M. A.).
table à ——, (powd.) drying tray.
sécheresse, f., dryness, barrenness; drought.
sécherie, f., drying room or place; (powd.) drying room, stove.
sécheur, a., m., (appareil) ——, —— *de vapeur,* (steam) drying pipe, steam trap.
séchoir, m., drying room, place, or floor; (powd.) drying room;
—— *à air chaud,* hot-air drying room;
—— *à l'air,* air drying room.
second, a., m., second, other; inferior; second, assistant; second floor (third floor in American and English houses); (nav.) first mate, first officer (commercial navy);
au ——, on the second (third) floor;
capitaine en ——, (mil.) second captain;
commandant en ——, (mil.) second in command;
eau ——*e,* v. s. v. *eau;*
en ——, second, in the second order;
—— *entrait,* (cons.) collar beam;
lieutenant en ——, (mil.) second lieutenant;
—— *maître,* (nav.) boatswain's mate;
—— *pont,* (nav.) upper deck;
—— *de quart,* (nav.) mate of the watch;
—— *rang,* second rate;
vaisseau ——, (nav.) tender, consort.
secondaire, a., secondary, accessory; (elec.) secondary.
seconde, f., second (of arc, angle, time, etc.); (fenc.) seconde (parry).
seconder, v. a., to second, assist; to back, support, promote.
secouement, m., shaking, jolting, shake.
secouer, v. a., to shake, shake off, cast off; to get rid of; (mil.) to shake off an attack, an enemy;
—— *le joug,* to shake off the yoke.
secoueur, m., (met.) mallet (to break up molds after casting).
secourable, a., capable, susceptible, of relief.
secourir, v. a., to succor, relieve, help, rescue, aid;
—— *un cheval,* (man.) to use the aids when the gait slackens.
secours, m., succor, rescue, help, relief, aid, assistance; men who help on capstan bars; (Fr. a. adm.) allowance by minister of war of funds to the *masse d'habillement et d'entretien;* (mil.) relief of a besieged place; troops sent to relieve a besieged place;
—— *éventuel,* (Fr. a. adm.) occasional money help from the government to worthy soldiers (or their families) who have no claim to a pension;
—— *permanent,* (Fr. a. adm.) permanent sum from the government to worthy soldiers or their families who have no claim to a pension (considered as compensation for title almost acquired to a *pension de retraite*);
porte de ——, v. s. v. *porte.*
secouru, p. p., succored;
—— *ou non secouru,* "rescue or no rescue" (obs.).
secousse, f., shock, shake, jolt, jerk, concussion, shaking; check; attack, blow, stroke, toss, tossing; (mil.) shock of attack;
—— *de la bride,* (man.) v. *saccade.*
secquière, f., path or road through a forest.
secret, a., secret, private, inward; m., secret; concealment, close confinement; secret drawer (of a cabinet, etc.); place from which a fire ship is fired by the captain;

secret, au ——, in close confinement, "incomunicado;"
escalier ——, private staircase;
fonds ——*s*, secret-service money.
secrétaire, m., secretary; writing table;
—— *d'ambassade, etc.*, secretary of an embassy, etc.;
—— *d' État*, secretary of state;
—— *d'état-major*, (*Fr. a.*) staff clerk (general service clerk, U. S. A.);
——*s d'état-major et de recrutement*, v. s. v. *section;*
—— *du recrutement*, (*Fr. a.*) recruiting clerk.
secrétaire-archiviste, m., (*Fr. a.*) staff clerk (record clerk).
secrétariat, f., functions, duties, term of office, place of office, of a secretary.
secteur, m., sector; (*fort.*) sector (of defense); (*siege*) sector of attack; (*mach.*) slot link; double bar link;
—— *d'attaque*, (*siege*) sector of attack;
—— *conducteur*, (*art.*) guide piece of a star gauge;
—— *à crémaillère de l'appareil de pointage*, (*art.*) toothed elevating arc;
—— *de défense*, (*fort.*) sector of defense;
—— *sans défense*, (*fort.*) dead sector, undefended line in front of a salient;
—— *denté*, (*mach.*) toothed, cogged, arc; (*art.*) segmental rack;
—— *à dents*, v. —— *denté;*
division en ——*s*, (*mil.*) the subdivision of a zone of defense or of attack, or of a wood or village to be defended, into sections;
—— *d'exploration*, (*mil.*) zone, or area, or sector, covered by exploration;
—— *d'explosion*, (*ball.*) sector of explosion;
—— *sans feu*, v. *sans défense;*
—— *fileté*, (*art.*) threaded sector of a breech plug;
—— *glissant*, (*mach.*) slide sweep (of an oscillating engine);
—— *de gouvernail*, (*nav.*) steering quadrant;
—— *gradué*, graduated sector;
—— *du levier de changement de marche*, (*mach.*) reversing-gear sector (of a locomotive);
—— *lisse*, (*art.*) smooth sector, (blank), of a breech plug;
—— *plein*, (*art.*) v. —— *fileté;*
—— *privé de feux*, (*fort.*) dead space or sector; sector not covered by fire;
—— *de réglage* (*du frein*), (*art.*) adjusting arc (of a brake);
—— *de serrage* (*du frein*), (*art.*) compressor arc (of a brake);
—— *vide*, (*art.*) v. —— *lisse.*
section, f., section; profile; (*top., etc.*) profile; (*r. r.*) section; (*Fr. a.*) subdivision of the French general staff; name applied to certain units of non-combatants and workmen employed in the army (v. examples *infra*); shortened term for *section hors rang;* one of the four subdivisions of the *convoi administratif*, q. v.; (*Fr. inf.*) one-fourth of a company; (*art.*) section (two guns), (platoon, U. S. Art.);
—— *active*, (*Fr. a.*) active section (i. e., belonging to the active army); section of the staff in charge of matters touching the active army;
—— *attelée*, (*mountain art.*) section hitched;
—— *de bateaux*, (*pont.*) pontoon section;
—— *du centre*, (*art.*) center section (platoon, U. S. Art.);
chef de ——, (*art., inf.*) chief of section (*art.*, chief of platoon, U. S. A.);
——*s de chasseurs forestiers*, v. s. v. *chasseur* (organized into companies, sections, and detachments, according to numbers available);
——*s de chemin de fer de campagne*, v. —— *s techniques d'ouvriers de chemins de fer;*
—— *de chevalets*, (*pont.*) trestle section;
——*s de commis et ouvriers militaires d'administration*, (*Fr. a.*) sections (25) of general service clerks and skilled workmen assigned to army corps, to Algiers, to Paris, and to Lyons;
—— *en croix*, (*tech.*) cruciform section;

section *de culées*, (*pont.*) shore-bay section;
—— *dangereuse*, (*tech.*) dangerous section;
demi-——, (*Fr. inf.*) half section (two squads);
——*s de deuxième ligne*, (*Fr. mil. teleg.*) sections working in rear of the army and connecting the army with home;
——*s de douaniers*, in France, sections of the custom-house personnel (organized into battalions, companies, and sections, to assist the army, second category; and into companies and sections for fortress warfare, first category);
—— *de droite*, (*art.*) right section (platoon, U. S. Art.);
—— *d'écoulement*, area of aperture for the passage of liquids (as in the piston head of a recoil cylinder);
—— *d'égale résistance*, (*tech.*) section of equal resistance;
—— *d'étapes et de chemin de fer*, (*Fr. a.*) communication and railway section, French military telegraph service;
—— *de forge*, (*pont.*) forge section;
—— *de forteresse*, (*Fr. a.*) telegraph party for connecting *places fortes* with the army in the neighboring territory;
—— *franche*, v. *groupe franc;*
—— *fusante*, v. s. v. *fusant;*
—— *de gauche*, (*art.*) left section (platoon U. S. Art.);
—— -*guide*, (*art.*) the section charged with getting the range, ranging section;
—— *horizontale*, horizontal section; (*top.*) contour;
——*s d'infirmiers militaires*, (*Fr. a.*) sections (25) of hospital orderlies assigned to army corps, etc., and varying in strength according to the needs of the service;
—— *intermédiaire*, (*top.*) section lying between equidistant sections;
—— *longitudinale*, longitudinal section;
—— *mouillé*, (*tech.*) area of waterway;
—— *de munition(s)*, (*mil., art.*) ammunition column;
—— *de munitions d'artillerie*, (*Fr. a.*) artillery ammunition column;
—— *de munitions d'infanterie*, (*Fr. a.*) infantry ammunition column;
—— *de munitions mixte* (*Fr. a.*) artillery and infantry ammunition column of a mountain brigade;
—— *de munitions des parcs*, (*Fr. a.*) reserve ammunition column;
——*s non-guides*, (*art.*) in ranging, sections regulating the height of burst;
—— *normale de la voie*, (*r. r.*) standard gauge;
—— *de parc*, (*Fr. a.*) one of the subdivisions of the general park;
—— *percutante*, v. s. v. *percutant;*
—— -*pivot*, (*art.*) pivot section (platoon, U. S. Art.);
—— *la plus fatiguée*, v. —— *dangereuse;*
—— *portée*, (*mountain art.*) section packed (i. e., guns on pack animals);
—— *de première ligne*, (*Fr.mil. teleg.*) section operating in the zone of operations of the army;
——*s principales*, (*top.*) sections determined by equidistant planes;
—— *de la queue*, (*hipp.*) docking;
—— *hors rang*, (*Fr. inf.*) in each regiment, a special group comprising the *petit état-major*, the band, workmen, clerks, etc.;
—— *de réserve*, (*Fr. mil. teleg.*) reserve section;
rompre par ——*s*, (*mil.*) to break off by sections;
—— *des secrétaires d'état-major et de recrutement*, (*Fr. a.*) sections of clerks for staff and recruiting duty, assigned to army corps and to Paris, and varying in strength according to the needs of the service;
—— *technique*, (*Fr. a.*) experiment or investigating committee attached to each *comité d'arme* or *de service* (a sort of subcommittee of the *comité technique;* it investigates all questions, experimental and other, new inventions, etc., respecting the arm to which it belongs);

section(s) *techniques d'ouvriers de chemins de fer,* (*mil.,* r. r., France) sections of skilled railway operatives, formed in peace, to serve in conjunction with the engineer railway troops, such railways as are not worked by the companies themselves (recruited from the personnel of the companies, and form no part of the army);
— *s de télégraphie* (*militaire*), (*Fr. a.*) telegraph sections (v. —— *de 1ère, 2ème, ligne*);
—— *télégraphique légère,* (*Fr. cav.*) the reunion of the *ateliers,* q. v.;
—— *transversale,* transverse, cross, section;
—— *verticale,* vertical section.
sécurite, f., (*expl.*) securite.
sécurité, f., security;
service de ——, v. s. v. *service.*
sédentaire, a., sedentary; v. s. v. *bataillon;*
garde nationale ——, national guard not intended to go into the field;
troupes ——*s*, (*mil.*) local troops, militia; garrison soldiers (i. e., soldiers that never change garrison).
sédiment, a., sediment, dregs, grounds; (*pl.*) boiler incrustation.
séditieux, a., seditious.
sédition, f., sedition.
segment, m., segment; (*steam*) piston ring, packing ring;
—— *capable* (*d'un angle*), segment in which a given angle may be inscribed;
—— *de garniture,* (*mach.*) segment of metallic packing;
—— *métallique de piston,* (*mach.*) piston ring, packing ring;
obus à ——*s*, v. s. v. *obus;*
—— *s de piston,* (*mach.*) piston rings, piston packing.
ségrairie, f., wood (forest) owned in common (with the state or with other persons).
ségrais, m., (*top.*) wood separated from a larger one.
ségrégation, f., segregation, separation.
seigle, m., rye.
seigneur, m., lord.
seigneurie, f., lordship.
seille, f., wooden pail.
seilleau, seillot, m., v. *seille;*
—— *à graisse,* slush pot.
seime, f., (*hipp.*) sand crack; seam, crack;
—— *en barre,* seam of the bars;
—— *en pince,* toe crack;
—— *quarte,* quarter crack.
sein, m., small landlocked bay (obs.); belly of a sail.
seine, f., seine.
seing, m., signature; sign manual;
blanc- ——, signature in blank;
faux- ——, forged signature;
—— (-) *privé,* signature not made before a public officer;
sous- ——, private deed or document.
séjour, m., stay, abode; visit; (*mil.*) halting day of a march; (*nav.*) stay in a port;
bête de ——, (*hipp.*) sick horse, and so kept in the stable;
camp de ——, (*mil.*) camp for one day or night only;
carte de ——, permit of residence;
—— *facultatif,* (*r. r.*) stop-over privilege.
séjourner, v. n., to abide, remain, tarry, make a sojourn in.
sel, m., salt; (*chem.*) salt;
—— *ammoniac,* sal ammoniac;
—— *commun,* common salt;
dépôt de ——, (*steam*) boiler incrustation;
—— *gemme,* rock salt;
—— *de roche,* v. —— *gemme.*
self-induction, m., (*elec.*) self-induction;
coefficient de ——, (*elec.*) inductance.
sellage, m., saddling.
selle, f., (*harn.*) saddle; (*met.*) slacken;
—— *allemande,* dragoon saddle; heavy cavalry saddle;
—— (*à l'*)*anglaise,* (English) hunting saddle;
—— *à l'anglaise piquée et rembourrée,* somerset;
—— *à bandes mobiles,* saddle with movable side bars;

selle *à bandes à panneaux,* artillery saddle; officer's saddle (padded side bars);
—— *à bandes sèches,* cavalry saddle (unpadded side bars);
—— *à basque,* saddle without a tree;
bois de ——, saddletree;
boute- ——, (*mil.*) "boots and saddles" (call);
—— *de calfat,* calker's calking box;
—— *de cavalerie légère,* hussar saddle; saddle of light cavalry;
chercher le fond de la ——, (*man.*) to try to sit well down in the saddle;
—— *de chasse,* v. —— *anglaise;*
cheval de ——, saddle horse;
cheval de —— *et de trait,* saddle and draft horse;
corps de ——, saddle proper, without accessories;
—— *de côté,* sidesaddle;
—— *course sur route,* road racing saddle (bicycle);
—— *de dame,* v. —— *de côté;*
—— *demi-route,* semiroadster saddle (bicycle);
en ——, on horseback;
être bien en ——, (*man.*) to have a good seat;
—— *de femme,* v. —— *de côté;*
—— *française,* manege saddle;
gagner le fond de la ——, (*man.*) to sit well down in the saddle;
—— *de grosse cavalerie,* v. *allemande;*
—— (à la) *hongroise,* v. —— *de cavalerie légère;*
—— *à la hussarde,* v. —— *de cavalerie légère;*
—— *intermédiaire,* v. *coussinet intermédiaire;*
jeter une —— *sur un cheval,* to saddle in haste;
—— *de joint,* v. *coussinet à joint;*
—— *à lames mobiles,* v. —— *à bandes mobiles;*
—— *de manège,* v. —— *à la française;*
mettre une —— *à,* to saddle;
se mettre en ——, to mount;
—— *à monter,* (riding) saddle;
—— *à piquer,* v. —— *à la française;*
—— *pneumatique,* pneumatic saddle (bicycle);
—— *à la polonaise,* v. —— *à basque;*
—— *qui porte,* well-fitting saddle;
—— *pour rails,* (*r. r.*) rail plate, rail saddle;
—— *rase,* v. —— *anglaise;*
remettre en ——, to put back on the horse (after falling);
renverser d'une ——, to unseat;
—— *de route,* road saddle (bicycle);
—— *de la roue,* nave block;
—— *royale,* road saddle;
sans ——, unsaddled, bareback;
sauter en ——, (*man.*) to vault into the saddle;
—— *sans tenue,* saddle of bad shape;
—— *à tous chevaux,* saddle that will fit any horse;
—— *de voltige,* (*man.*) vaulting or leaping saddle.
seller, v. a., to saddle.
sellerie, f., saddle making; saddlery; saddle room; harness room.
sellette, f., stool; seat; bed; lower bolster of a wagon, axletree bolster; (*harn.*) pad, off-horse, saddle; (*art.*) limber bolster; seat on the leg of some rapid-fire mounts; base plate of an *affût-truc* (q. v.), of a rapid-fire mount; saddle of some rapid-fire mounts; base ring of a seacoast carriage; pedestal; pintle block or plate;
—— *d'avant-train,* (*art.*) pintle, bolster, limber bolster;
boulon de ——, bolster bolt;
—— *de cheville ouvrière,* (*art.*) pintle block or frame;
—— *de derrière,* (*art.*) hind bolster;
—— *de devant,* (*art.*) fore bolster;
étrier de ——, bolster hoop, strap, band;
—— *de flèche,* (*art.*) trail seat;
—— *-fourchette,* (*art.*) fork and bolster;
frette de ——, v. *étrier de* ——;
seye de ——, v. *boulon de* ——;
—— *de sous-verge,* (*art.*) off saddle.
sellier, m., harness maker; saddler, saddle maker.
semaine, f., week; week's work; week's wages;
(*mil.*) the week, i. e., duty by the week;
être de ——, (*mil.*) to be on duty for the week;

semaine, *de grande* ——, (*Fr. art.*) on duty (week) with a group of batteries;
officier de ——, v. s. v. *officier;*
de petite ——, (*Fr. art.*) on duty (week) with a battery, on "battery officer" (U. S. Art.);
prendre la ——, (*mil.*) to begin, enter on, one's duties as *officier de semaine.*

sémaphore, m., semaphore, semaphore station;
poste de ——, semaphore station.

sémaphorique, a., semaphore;
signal ——, semaphore signal.

semaque, f., boat.

seméiotique, f., v. *sémiotique.*

semelle, f., sole (of a shoe); pace, step; sill, ground sill, sleeper; bed, bedplate; floor timber of a boat, of a bridge; chord, stringer, boom, of a bridge; bottom of a furnace; shoe of a capstan; (*cons.*) tiebeam; cap piece (of an arch center); (*nav.*) leeboard; (*mil. min.*) ground sill of a frame; (*pont.*) shoe of a trestle leg; lower part of a pontoon-boat knee; bottom strake of a pontoon boat (at the edge); chafing batten; (*art.*) bed of some carriages; firing support (French 220ᵐᵐ siege mortar); shoe of a gin leg; sill (of a gun lift); racer support of a gun platform; side rail of a truck wagon, of a mortar wagon;
—— *d'affût*, (*art.*) swing bed (obs.);
—— *d'ancre*, anchor shoe;
—— *d'appui de tir*, (*art.*) shoe (for firing support Russian 6-inch field mortar);
—— *de bateau*, (*nav.*) leeboard;
—— *d'ancre*, shoe of an anchor;
—— *d'assemblage*, (*cons.*) ground plate, ground sill;
—— *de châssis*, (*min.*) ground sill of a frame; (*art.*) chassis side rail;
—— *de chevalet*, (*pont.*) trestle-leg shoe;
—— *de comble*, (*cons.*) pole plate;
—— *d'un coussinet de rails*, (*r. r.*) chair foot;
—— *courbe*, (*cons.*) curb plate;
——*s de croisillon*, (*art.*) sills at right angles to one another, on which rests the *sellette de cheville ouvrière*, in some old platforms;
—— *de dérive*, (*nav.*) leeboard;
double- ——, **double-** —— *entre-deux*, middle sole of a shoe;
—— *d'embrèvement*, (*cons.*) bearing surface or shoulder of a silt-and-tongue joint;
—— *d'état*, sill of a stay;
—— *extérieure*, (*pont.*) chafing batten;
—— *inférieure*, lower chord of a bridge; stringer of an arch center;
—— *intérieure*, (*pont.*) bottom board of a boat or pontoon laid lengthwise;
—— *mobile d'une vis de pointage*, (*art.*) swing bed (obs.);
—— *oscillante*, (*mil. r. r.*) rocking support (on some of the trucks of the Decauville system);
—— *de palier*, (*mach.*) plumber-block bottom, or bedplate, or ground plate;
—— *première*, insole, inner sole;
—— *d'un rabot*, face of a plane;
—— *d'un rail*, (*r. r.*) foot of a rail;
reculer, rompre, d'une ——, (*fenc.*) to retire a foot's length;
—— *seconde*, outer sole of a shoe;
—— *supérieure*, top chord of a bridge;
—— *de (d'une) vis de pointage*, (*art.*) elevating-screw bed.

semence, f., seed; fine sprig, sprig nail.

semer, v. a., to sow; (*art.*) to examine a gun;
—— *des chausse-trapes*, (*mil.*) to spread, scatter, caltrops.

semestre, m., half year, six months; one-half year's pay, revenue, duty, rent, etc.; (*mil.*) six months' leave of absence, furlough; soldier on furlough for half a year (obs.);
de ——, half-yearly, semiannual;
congé de ——, (*mil.*) six months' leave of absence;
être de ——, to be in one's half-year's duty;
être en ——, to be on six months' leave of absence;
être hors de ——, to be off duty for six months;
—— *de janvier*, semester beginning January 1;

semestre, *par* ——, half-yearly.

semestre, a., half-yearly, semiannual.

semestrier, m., (*mil.*) soldier absent on six months' furlough.

sémiotique, f., (*mil.*) the art of drilling troops by signal (no commands).

semonce, f., (*nav.*) summons to show colors, to heave to.

semoncer, v. a., (*nav.*) to summon a ship to show colors, to heave to.

sénat, m., senate.

sénateur, m., senator.

senau, m., (*nav.*) sort of brig.

sens, m., sense, meaning, import, signification, acceptation, construction; (*mach., mil.*) direction, side, way, sense;
bon ——, (*mach.*) direction of rolling;
en —— *contraire de*, in a contrary direction;
—— *du courant*, direction of a current;
dans le ——, in the right way, in the right direction;
—— *dessus dessous*, upside down, wrong side up;
—— *devant derrière*, hind part foremost;
en —— *inverse*, in a contrary direction;
en —— *long*, with the grain; (*mach.*) with the direction of rolling;
—— *du laminage*, v. —— *en long*, (*mach.*);
—— *de la marche*, (*r. r.*) direction of the train;
mauvais ——, (*mach.*) direction opposed to the direction of rolling (also at right angles to);
mettre —— *dessus dessous*, to turn upside down;
—— *de rotation*, direction of rotation;
—— *travers*, across the grain; (*met.*) across the direction of rolling;
dans tous les ——, in all directions;
—— *du tir*, (*mil.*) direction of fire.

sensibilisateur, a., sensitizing;
bain ——, (*phot.*) sensitizer.

sensibilisation, f., sensitization.

sensibiliser, v. a., to sensitize, to make sensitive (as a fuse, a photographic plate, etc.).

sensibilité, f., (*expl., etc.*) sensitiveness of an explosive, fuse, primer, of a balance, etc.; (*elec.*) figure of merit of a galvanometer;
—— *au choc*, —— *à la friction*, (*expl.*) sensitiveness to shock, friction;
—— *de la main*, (*man.*) lightness and quickness of hand.

sensible, a., perceptible, sensitive; (*expl., etc.*) sensitive (fuse, plate, thermometer, etc.);
bouche ——, v. s. v. *bouche*;
—— *au choc*, —— *à la friction*, (*expl.*) sensitive to shock, friction;
—— *à l'éperon*, (*man.*) spurwise;
—— *à la jambe*, (*man.*) legwise;
—— *aux mouches*, (*hipp.*) sensitive to flies;
—— *à la rêne*, —— *aux rênes*, (*man.*) bridlewise.

sensitif, a., sensitive;
fusée ——*ive*, (*art.*) sensitive fuse.

sente, f., v. *sentier.*

sentence, f., sentence; (*law*) judgment, (more specially) decree, sentence, of death;
appeler d'une ——, to appeal from a judgment;
confirmer une ——, to confirm a sentence;
prononcer une ——, to sentence, to pass a sentence.

sentencier, v. a., to sentence, to condemn.

senti, p. p., conspicuous, noticeable, marked.

sentier, m., path, footpath, track;
—— *battu*, beaten path;
—— *détourné*, bypath; (*mach.*) footway (in machinery);
—— *muletier*, mule path.

sentiment, m., feeling, sensation; sentiment; estimate, guess, judgment;
au ——, by judgment, by guess'(e. g., in running lines, contours).

sentine, f., well of a boat.

sentinelle, f., (*mil.*) sentinel, sentry;
—— *d'augmentation*, extra sentinel (posted at night);
—— *avancée*, advanced sentinel;
—— *de communication*, warning or communicating sentry;
—— *devant les armes*, sentinel at the guardhouse, No. 1;
——*s doubles*, double sentry;

sentinelle, *être en* ——, to be on post, to be on sentry duty, to stand sentry;
faire ——, v. *être en* ——;
—— *fixe*, stationary sentry;
—— *d'honneur*, a complimentary sentry, e. g., before a general's quarters, etc.;
—— *intermédiaire*, v. *de communication;*
lever une ——, to relieve a sentry;
mettre en ——, to put on sentry;
—— *d'observation*, look out;
—— *perdue*, advanced outpost, sentry, sentry posted in a very dangerous situation;
placer une ——, to post a sentry;
poser une ——, v. *placer une* ——;
—— *du poste*, v. —— *devant les armes;*
relever une ——, v. *lever une* ——;
—— *simple*, single sentinel;
—— *volante*, flying sentinel, running sentry.

sentir, v. a., to feel, be sensible of; (*man.*) to feel a horse;
—— *la barre, le gouvernail,* (*nav.*) to answer the helm;
—— *le fond,* (*nav.*) to be almost touching the bottom (of a ship moored).

sep, m., (*nav.*) belaying cleat.

séparage, m., sorting.

séparateur, m., separator; separating funnel;
—— *d'eau*, separator.

séparation, f., separation; partition (as in an ammunition chest); plate between the sheaves of a block;
—— *de la batterie,* (*art.*) in ranging, the separation of the battery into the *section-guide*, and the *sections non-guides*, qq. v.;
entonnoir de ——, separating funnel.

séparer, v. a., to separate, sort, sort out, part; (*cord.*) to part;
—— *la batterie,* (*art.*) to divide a battery into the *section-guide* and the *sections non-guides*, qq. v.;
—— *les rênes*, v. s. v. *rêne;*
se ——, (of a river) to divide; (*mil.*) (of an army) to break up.

sépé, m., (*sm. a.*) boring bench.

sépia, f., sepia; sepia drawing.

septentrion, m., north, the Little Bear.

septentrional, a., north, northerly.

septime, f., (*fenc.*) septime, half-circle parry.

séquestration, f., (*law*) sequestration; separation of diseased from healthy animals.

séquestre, m., (*law*) sequestration; thing sequestrated; depository, sequester; (*hipp.*) sequestrum (piece of dead bone remaining in its place).

séquestrer, v. a., (*law*) to sequester; to make an illegal arrest.

séranine, f., (*expl.*) seranine.

serein, a., (of the weather) clear and calm.

sereine, f., (*nav.*) quarantine, isolation (of a man with a contagious disease).

serge, f., serge; (*art.*) cartridge flannel.

sergent, m., (*mil.*) sergeant (*Fr. a.*, foot troops); (*tech.*) screw frame, clamp, holdfast, cramping frame;
—— *-artificier*, master artificer, laboratory sergeant;
—— *brancardier*, v. s. v. *brancardier.*
—— *de crottin,* (*mil. slang*) noncommissioned officer of cavalry school at Saumur;
—— *d'encadrement*, second sergeant in the last section of a battalion;
—— (-)*fourrier,* (*Fr. a.*) quartermaster-sergeant of a company;
—— *de garde*, sergeant of the guard;
—— *garde-magasin,* (*Fr. a.*) sergeant in charge of the clothing magazine of the regiment;
—— *d'hiver,* (*mil. slang*) soldier of the first class (allusion to his woolen stripes, humorously supposed to keep him warm in winter);
—— *instructeur*, drill sergeant;
—— *-major*, sergeant-major (nearest equivalent, U. S. A., first sergeant);
—— *-major-artificier*, master artificer;
—— *secrétaire,* (*Fr. a.*) sergeant acting as clerk or secretary;
—— *de section,* (*Fr. a.*) sergeant responsible for the instruction, barrack room, arms, and equipment of his section (cf. line sergeant, U. S. A.);

sergent *de semaine,* (*Fr. a.*) sergeant under the orders of *officier de semaine*, assisting him in the execution and administration of his duties;
—— *-tambour*, sergeant of field music;
—— *de tir,* (*Fr. a.*) sergeant who brings up poor shots and has charge of details;
—— *de ville*, policeman.

série, f., series, set, as of signal flags, etc.; (*elec.*) series;
en ——, (*elec.*) in series;
—— *d'engrenages,* (*mach.*) change gear;
—— *de marche,* (*Fr. a.*) implements for measuring, weighing, etc., provisions (at issue to troops); box containing butchery implements, scales, etc.;
montage en ——, v. s. v. *montage;*
monter en ——, v. s. v. *monter;*
—— *de pavillons*, set of flags;
—— *régimentaire d'outils de boucher,* (*Fr. a.*) butcher's tools;
—— *Z,* (*Fr. a.*) said of or applied to scabbards for which no bayonets are issued, belonging to *infirmiers* neutralized by Geneva convention.

sérieux, a., serious, grave, important.

sérimètre, m., serimeter, instrument for testing silk thread.

seringue, f., syringe, squirt;
—— *pour graisser*, —— *à huile*, lubricating syringe;
—— *à main*, hand fire engine;
—— *à suif*, tallow syringe.

seringuer, v. a., to syringe; to rake a ship fore and aft.

serment, m., oath;
avoir —— *en justice*, to have been sworn by a court;
déclarer sur la foi du ——, to swear;
déférer le —— *à*, to administer an oath to; to swear a person;
—— *d'entrée en charge*, oath of office;
faire —— *de*, to swear to (a thing, a fact);
faire un faux ——, to commit perjury;
faire prêter —— *à*, to put on .... oath;
fausser son ——, to break one's oath;
—— *de fidélité*, oath of allegiance;
formule d'un ——, form of an oath;
prestation de ——, taking an oath, swearing;
prêter ——, to take oath, an oath; to be sworn;
rompre son ——, v. *fausser son* ——;

serpe, f., billhook, pruning hook; (*mach.*) worm.

serpéger, v. a., (*man.*) to take a horse over a sinuous track or path.

serpent, m., serpent, snake; (*mil. slang*) leather belt used as a purse; (*Polytechnique slang*) one of the first fifteen on the list after entrance examination.

serpente, m., silver paper, tissue paper; f., (*cord.*) snaking line;
papier (*de*) ——, tissue paper.

serpenteau, m., young serpent; (*artif.*) serpent, squib; iron hoop hung round with grenades armed with spikes, for the defense of a breach (obs.);
—— *broché,* (*artif.*) swarmer; small rocket driven on a spindle;
—— *à étoile,* (*artif.*) serpent with star composition over the powder;
—— *incendiaire,* (*artif.*) squib or serpent fired from a musket to set buildings on fire (obs.);
—— *à pirouette,* (*artif.*) serpent full of composition, with holes at each end and on each side to give whirling motion.

serpenter, v. a., to meander, wind (river, road); (*man.*) to take a horse over a sinuous path or track; (*cord.*) to snake a rope;
—— *deux cordages,* (*cord.*) to snake two ropes together;
faire —— *un cheval,* (*man.*) to cause a horse to describe a sinuous track.

serpentin, a., serpentine; m., (*met., mach., etc.*) worm, coil;
avoir la langue ——*e,* (*hipp.*) to have a pendulous and restless tongue;
—— *fusant,* (*art.*) time train of a fuse;
—— *rafraîchisseur*, refrigerating coil; worm condenser;

serpentin, à *vapeur,* steam coil.
serpentine, f., (*man.*) sinuous path or track described by riders in column of files.
serpette, f., pruning knife; billhook; (*pont.*) sort of calking tool.
serpeyer, v. a., v. *serpéger.*
serpiller, v. a., to reduce the thickness of a palisade by cutting off the sides.
serpillière, f., sarplar, coarse packing cloth.
serrage, m., fastening, securing, tightening, adjusting (as of a screw); uniting together; tightness of fit (as a cartridge in its seat); (*art.*) tension or strain (of gun metal); shrinkage; (frequently) compression; in obturating devices, the pressure or bearing of the device against the walls of the gun; (*artif.*) crimping of Bickford fuse, etc.; (*mach.*) vertical path or course of a bit (boring machines, etc.); the play or course or travel of a cutting edge (machine tools); locking; (*sm. a.*) longitudinal distance traveled by the cylinder necessary to close the breech, measured from the point where the stop-screw comes in contact with the lateral groove; distance from handle up to handle down after the bolt has been pushed as far forward as possible without turning;
 clef de ——, wedge key;
 de ——, adjusting, tightening (in many combinations);
 —— *diamétral,* (*art.*) shrinkage;
 fossé de ——, (*art.*) shrinkage pit;
 —— *gras,* tight fit (as of nuts, washers, etc.);
 —— *naturel relatif,* (*art.*) the difference per unit of length between the exterior diameter of the tube and the interior diameter of the adjacent hoop;
 —— *pratique absolu,* (*art.*) the actual expansion of a hoop necessary to put it in place;
 puits de ——, v. *fossé de* ——.
serre, f., ship plank; greenhouse; pressing together; press;
 à la ——, v. *en* ——;
 donner une ——, to press together;
 en ——, housed.
serré, p. p., hard, tight, close, fast, compact, close together, shut up in; (of wood, grain) close grained; (*mach.*) of work, hammer, rolls, etc.) as close to the final shape as possible; (*mil.*) close (of formations, etc.); (*man.*) (of a gait) short, well together;
 cheval ——, v. s. v. *cheval;*
 feu ——, (*mil.*) close, well-delivered fire;
 jouer ——, v. s. v. *jouer;*
 ligne ——*e,* (*mil.*) dense line;
 —— *en masse,* (*mil.*) closed in mass.
serre-cou, m., (*hipp.*) compress used in bleeding a horse at the neck.
serre-écrous, m., nut setter.
serre-étoupe, m., (*mach.*) stuffing box.
serre-feu, m., (*met.*) fire screen.
serre-fil(s), m., (*elec.*) binding screw; binding post; (*teleg.*) connector;
 —— *à deux vis,* double connector.
serre-file, m., (*mil.*) file-closer; last soldier of a file (*nav.*) sternmost ship of a column or fleet (whether in line or column);
 en ——, (*mil.*) in the line of file-closers;
 être en ——, (*mil.*) to be in the line of file-closers;
 —— *général,* (*mil.*) general file-closer detailed under special circumstances to bring up rear; gives information to any elements that may have fallen behind; may be an officer;
 rentrer en —— (*s*), to go or fall into the line of file-closers.
serre-frein, m., (*mach.*) brake tightener or adjuster; brake setter (of a bicycle); (*r. r.*) brakeman.
serre-garniture, m., (*mach.*) gland.
serre-joints, m., clamp, holdfast.
serrement, m., squeezing, pressure; (*drill*) closing of intervals; (*min.*) dam or partition to keep out water.
serre-nez, m., (*hipp.*) horse twitch.
serre-papiers, m., paper weight; paper holder or clip; set of pigeonholes, file for papers.

serrer, v. a., to close, close up, press, press close, tighten, squeeze; to crowd, crowd on, gain on, push back; to clasp; to lock, lock up, put away or aside, take in; to pass, go, close to; to tie, arrange, fasten, wedge in, make secure, screw down, shrink, crimp; (*mach.*) to lock; (*met.*) to bloom; (*man.*) to narrow the area over which a horse is moving (as in volts); (*mil.*) to close, close up;
 —— *son adversaire,* v. —— *la botte,* (*fenc.*);
 —— *à bloc,* to set up: to set home; to cause all parts to bear together properly;
 —— *la botte,* (*fenc.*) to thrust home and hard; (*man.*) to close in the legs on the horse;
 —— *les bottes,* (*cav.*) to close in boot to boot;
 —— *la bride,* (*man.*) to pull on the bridle;
 —— *un canon,* (*nav. art.*) to house a gun;
 —— *un cheval,* (*man.*) to keep a horse together, well in hand;
 —— *une colonne,* (*mil.*) to make a column close to a denser formation;
 —— *le compas,* to lift the card off the point (pivot);
 —— *contre,* to push, force, back, on or to;
 —— *la côte,* (*nav.*) to hug the coast;
 —— *la demi-volte,* (*man.*) in a demivolt, to bring the horse accurately back to his starting point;
 —— *à distance,* (*drill*) of the rear rank, etc., to close up to proper distance;
 —— *l'ennemi,* (*mil.*) to push, press, the enemy, press him hard;
 —— *l'éperon,* —— *les éperons,* (*man.*) to put spurs to a horse;
 —— *une fascine,* (*siege*) to choke a fascine;
 —— *les files,* v., —— *les rangs;*
 —— *le frein,* to put on a brake; to brake;
 —— *une garniture,* (*mach.*) to tighten, screw down, a packing;
 —— *les intervalles,* (*mil.*) to close intervals;
 —— *un joint,* to tighten a joint;
 —— *la ligne,* (*mil.*) to close or contract a line of battle; (*nav.*) to close up, form a close line;
 —— *la mesure,* v. —— *la botte,* (*fenc.*);
 —— *un nœud,* (*cord.*) to tighten, to secure a knot or lashing;
 —— *une place,* (*siege*) to annoy, press upon, a place, cut off its communications;
 —— *de près,* (*mil.*) to press the enemy hard;
 —— *les rangs,* (*mil.*) to close up, to close ranks;
 —— *les sangles,* to tighten the girths;
 —— *le saucisson,* (*siege*) to choke a fascine;
 se ——, (*man.*) (of the horse) not to take enough ground;
 —— *sur,* (*mil.*) to close upon (the enemy, the unit in front); (*nav.*) to close upon;
 —— *une tente,* (*mil.*) to furl a tent;
 —— *la terre,* (*nav.*) to hug the coast;
 —— *sur la tête,* (*mil.*) to close up on the head of a unit, column, etc.;
 —— *une ville de près,* (*siege*) to invest a town;
 —— *à vis,* to screw, screw down;
 —— *la volte,* (*man.*) to narrow a volt.
serre-rais, m., spoke setter.
serre-rayons, m., spoke setter of a bicycle.
serre-tête, m., headband, band.
serre-tube, m., (*mach.*) apparatus for riveting locomotive tubes, riveting clamp.
serrière, f., (*met.*) plug of iron, stopple of a furnace.
serrure, f., lock;
 —— *bénarde,* lock with a solid key, i. e., lock that may be turned from both sides;
 —— *à broche,* piped-key lock;
 brouiller une ——, to derange a lock;
 —— *de cabinet,* latch-, spring-, lock;
 —— *cachée,* v. —— *encastelée,*
 —— *à clef femelle,* v. —— *à broche;*
 —— *à clef mâle,* lock with a French key;
 —— *à combinaison(s),* combination lock;
 crocheter une ——, to pick a lock;
 crocheteur de ——, picklock, lock picker;
 —— (*à*) *demi-tour,* spring lock, spring catch;
 —— *à deux pênes,* spring stock lock;
 —— *à deux tours,* lock with a double turn, double lock;
 —— *à double tour,* v. —— *à deux tours;*

serrure *encastelée*, mortise lock;
—— *entaillée*, v. —— *encastelée;*
entrée de ——, scutcheon of a lock;
—— *à fourreau*, v. —— *encastelée;*
—— *à houssette*, v. —— (à) *demi-tour;*
—— *à mortaise*, v. —— *encastelée;*
—— *à palastre*, rim lock;
—— *pendante*, padlock;
—— *à pêne dormant*, stock lock;
—— *à ressort*, spring lock;
—— *secrète*, combination lock; rim lock;
—— *à secret*, v. —— *secrète;*
—— *de sûreté*, safety lock;
—— *tréfilière*, v. —— *à broche;*
—— *à un seul pêne*, bolt lock.

serrurerie, f., locksmith work, business, art, etc.; (*cons.*) ironwork of a building;
grosse ——, (*cons.*) ironwork (of a building).

scrrurier, m., locksmith;
—— *en bâtiments*, person who furnishes or is occupied with construction-iron work;
—— *charron*, mechanic who does the ironwork of vehicles;
—— *mécanicien*, engine builder.

sertir, v. a., to set, seat; to put into a seat (as, *art.*, a rifling band); (*sm. a.*) to crimp a cartridge case;
bague à ——, (*mach.*) crimping-, seating-, band or collar;
machine à ——, v. s. v. *machine.*

sertissage, m., setting, seating; (*sm. a.*) crimping (of a cartridge case about a projectile).

sertissement, m., v. *sertissage.*

sertisseur, m., (*sm. a.*) crimping tool.

sertissure, f., seating; (*sm. a.*) crimping.

servant, m., (*art.*) cannoneer, gunner (R. A.);
—— *de manivelle de la vis de pointage*, handle of the elevating screw;
peloton des ——*s*, v. s. v. *peloton;*
pointeur- ——, v. s. v. *pointeur;*
pourvoyeur- ——, v. s. v. *pourvoyeur.*

servante, f., drag staff; (*art.*) pole prop; prop (for lid of limber chest); (*powd.*) scraper of a cylinder incorporating powder mill.

servi, p. p., served, etc.; (*art.*) ready for action (as a turret, etc.);
feu bien ——, (*mil.*) well-sustained fire.

service, f., service, duty, office; (*mil.*) service (i. e., the army); military duty or work in general; special military duty or work; any duty or class of duties; department, arm, etc., in charge of, or performing, a special or specified duty; department as distinguished from *corps de troupe.*
(Unless otherwise indicated, the following expressions are military:)
abandonner le ——, to leave, give up, the service;
—— *actif*, active service, service with the colors;
en activité de ——, on the active list; in full active service;
——*s administratifs (de l'armée)*, administrative services (correspond to the so-called staff departments, U. S. A.);
—— *d'administration*, administrative service;
——*s de l'administration de l'armée.* (*Fr. a.*) the administration of the artillery, engineers, *intendance*, powders and saltpeters, and medical departments, respectively;
—— *des affaires indigènes*, (*Fr. a.*) administration of the military territory in Algiers;
—— *ambulancier*, hospital service in the field;
—— *armé*, —— *en armes*, service under arms;
—— *des armées en campagne*, field service in general;
armement en ——, v. s. v. *armement;*
—— *de l'armement*, all duties, etc., relating to the armament;
—— *à l'arrière*, service at the rear;
—— *de l'arrière*, in widest extension, the transport forward of ammunition, supplies, and men; rearward, of the sick and wounded; includes the guarding and care of lines of communication, the care of sick and wounded in transit, and of material and men going forward; (*Fr. mil. teleg.*) the telegraph service connecting the army with the base of operations; (*Fr. a. med.*) medical service on the lines of communication.

service *de l'arrière de la zone des armées*, v. —— *de l'arrière* (general definition);
—— *d'arrière-garde*, rear-guard duty or service;
—— *de l'artillerie*, artillery service duties, in the most general sense (similarly of other arms or corps, as —— *du génie*);
au ——, in the service;
——*s auxiliaires*, (*Fr. a.*) (state of war only) comprehensive term for railroad, military telegraph, pay and post, *étapes*, and any other services that may be performed by men not armed (noncombatants);
—— *à l'avant*, service at the front;
—— *de l'avant*, the supply service, etc., carried on by the army itself from the local resources; (*Fr. a.*) medical service at the front, with the troops engaged;
—— *d'avant-poste*, —— *des avant-postes*, outpost duty;
avoir du ——, to have seen service;
avoir .... ans de ——, to have served (so many) years;
le bien du ——, the good of the service;
—— *à bord*, (*nav.*) sea service (U. S. N.); service afloat;
—— *de (la) bouche à feu*, (*art.*) standing-gun drill;
—— *de campagne*, field service;
—— *du campement*, generic term for everything relating to camping;
—— *de cantonnement*, generic term for everything relating to cantonments;
—— *central*, (*Fr. a.*) generic term for all operations, etc., that are taken cognizance of by the minister of war or his direct subordinates;
centraliser le ——, to centralize the service, to concentrate all responsibility, direction, etc.;
charge de ——, (*art.*) normal, service, charge;
—— *des chemins de fer*, (*Fr. a.*) branch of the —— *de l'arrière*, railway service between the army and home in time of war;
commander le ——, to make details for regular routine duties;
commander pour le ——, to detail (for such and such a duty);
—— *au commerce*, (*nav.*) merchant service;
—— *de communication*, communication service;
congé absolu de ——, *congé de libération de* ——, discharge from the service;
contrôle de ——, duty roster;
—— *du contrôle (de l'administration de l'armée),* (*Fr. a.*) v. s. v. *contrôle;*
—— *du corps d'armée*, (*Fr. a.*) the services (departments) forming the link, as regards supply, between the *tête d'étapes de route* and the corps;
—— *de correspondance*, transmission service (as from the front to the rear);
—— *de la correspondance*, communication service (between the various bodies of troops engaged in the same operation);
—— *de corvée*, fatigue duty, extra (fatigue) duty;
—— *courant*, daily service, ordinary routine service; current service;
en —— *courant*, in service, forming part of the system;
—— *de couchage*, v. s. v. *couchage;*
—— *de courte durée*, "short" service;
—— *des cultes*, (*Fr. a.*) department of public worship in the army;
de ——, on duty;
—— *de découverte*, name applied to the special function of the exploration cavalry, of discovering enemy, keeping in contact with him, etc.;
—— *de défense*, (*siege*) defense of a place (fortified place or fort only);
dégagement de ——, discharge from the service;

service de *deuxième ligne,* (*Fr. mil. teleg.*) telegraph service connecting the headquarters of various armies with the main headquarters, and all the armies with home;
—— *de deuxième tour,* duty of the second class;
directeur de ——, v. s. v. *directeur;* head of any department or duty;
—— *de direction,* v. —— *d'administration;*
direction de ——, v. s. v. *direction;*
—— *divin,* (*in gen.*) divine service;
—— *sous les drapeaux,* service with the colors, in active army;
durée de ——, period, length, of service;
—— *des écuries,* stable duty;
entrer au ——, to enter the service;
—— *d'estafette,* messenger, courier, duty;
—— *des étapes,* v. s. v. *étape;*
état de ——, certificate of service;
—— *de l' État,* (*in gen.*) service of the state; (*nav.*) service in the war navy;
—— *d',* de *l', état-major,* staff service (in *Fr. a.,* not to be confounded with *état-major général*);
—— *à l'étranger,* foreign service;
être au ——, to serve;
être de ——, to be on duty;
être en activité de ——, to be on the active list;
être appelé au ——, to be called into service;
être de —— *auprès de,* to be on duty with, to attend upon, one;
—— *éventuel,* subsistence duty in the field;
—— *d'exécution,* v. —— *de gestion;*
exempt du ——, excused from duty;
—— *d'exploitation,* (*r. r.*) operation of a railway;
—— *d'exploration,* the exploration service of independent cavalry;
expulsion du ——, dismissal;
expulsion infamante du ——, drumming out, dishonorable discharge;
—— (*à l'*)*extérieur,* service or duty beyond the enceinte, beyond the line of sentinels; foreign service;
faire le —— *de,* to perform the duties of a given position;
faire son ——, to do one's duty, one's tour of duty;
faire son temps de ——, to finish one's time of service, to serve out one's time;
finir son ——, v. *faire son temps de* ——;
—— *des forges,* (*Fr. a.*) steel-, gun-, inspection duty;
—— *de forteresse,* fortress duty or service;
—— *des fourrages,* forage (supply) service;
—— *de garde,* guard duty;
—— *de garnison,* garrison service, whether open or fortified town; routine garrison duty;
—— *de la gendarmerie aux armées,* (*Fr. a.*) provost service;
——*s généraux de l'armée,* (*Fr. a.*) term describing certain staff and administrative departments having relations with the whole army, as the general staff, the *contrôle* and *administration,* etc.;
—— *géographique de l'armée,* (*Fr. a.*) a department of the French staff charged with astronomical, geodetical, topographical, and cartographic duties;
—— *de gestion,* v. s. v. *gestion;*
—— *de l'habillement,* (*Fr. a.*) army clothing department;
—— *du harnachement de la cavalerie,* (*Fr. a.*) harness supply department;
—— *d'honneur,* honorary service; (e. g., *auprès d'un roi, d'un empereur*);
hors de ——, unserviceable, unfit for service; out of service; retired from the service;
—— *des hôpitaux militaires,* hospital staff;
impropre au ——, unfit for duty or service;
inhabilité au ——, unfitness for the service;
—— *de l'intendance,* v. *intendance;*
—— *intérieur,* interior economy; (*Fr. a.*) branch of the *administration centrale* having certain duties with respect to its material and personnel, to archives, etc.;
—— *à l'intérieur,* service in a fort or place inside of a line of sentinels; home service;

service *des interprètes,* v. *interprète;*
—— *de jour,* daily duty;
—— *de la justice militaire,* law service of the army (judge-advocate's department, U. S. A.);
lettre de ——, v. s. v. *lettre;*
licenciement de ——, discharge from the service;
—— *des lits militaires,* v. —— *de couchage* and *entreprise des lits militaires;*
—— *de marche,* march, transportation, service;
—— *de la marine,* naval, sea, service;
—— *maritime,* v. —— *de la marine;*
marque de ——, v. s. v. *marque;*
—— *médical,* medical service;
—— *de mer,* v. —— *de la marine;*
—— *de messagerie,* (*in gen.*) transport, freight, etc. service;
mettre hors de ——, to wear out;
—— *militaire,* military service or duty;
—— *militaire de la marine,* naval service (proper);
obligation du ——, obligation to serve;
—— *obligatoire,* compulsory service;
—— *optique,* (*mil. sig.*) field or military visual signaling;
—— *de l'ordinaire,* mess duty;
—— *d'ordonnance,* orderly duty;
ordre de ——, v. s. v. *ordre;*
—— *d'ordre,* v. s. v. *ordre;*
—— *de pansage,* stable duty;
——*s particuliers de l'armée,* (*Fr. a.*) term descriptive of staff and other departments relating to particular arms or departments or branches of the army, as military schools, chaplains, Algerian affairs, interpreters, remount, etc.;
—— *permanent,* subsistence duty pertaining to permanent camps and garrison towns;
—— *du personnel,* department in charge of personal affairs ("appointment, commission, and personal" division, A. G. O., U. S. A.);
—— *de*(*s*) *place*(*s*), garrison and fortification service; routine garrison duty;
—— *de planton,* orderly duty;
—— *de police,* police duty;
—— *du port,* (*nav.*) harbor duty;
—— *des poudres et salpêtres,* (*Fr. a.*) the department or duties of the *poudres et salpêtres,* v. s. v. *poudre;*
—— *de première ligne,* (*Fr. mil. teleg.*) telegraph service connecting the main headquarters of the army with headquarters of corps or other designated points; (*Fr. a. med.*) advanced medical service; medical service with troops (includes regimental service, dressing stations, field hospitals);
—— *du premier tour,* duty of the first class;
prendre le ——, to go on duty; to take over duty;
—— *prévôtal,* —— *de la prévôté,* provost duty;
propre au ——, fit for service;
—— *du quartier-général,* headquarters service or duty;
quitter le ——, to leave the service;
raison de ——, service reasons;
—— *de reconnaissance,* reconnaissance duty;
—— *du recrutement,* recruiting service;
—— *régimentaire,* (*med., Fr. a.*) regimental medical service; first attendance, on halts, during marches and combats, (*mil. in gen.*) regimental service;
—— *religieux,* v. —— *divin;*
remettre le ——, to turn over a duty, service, guard, etc.;
—— *de la remonte,* remount service;
—— *des renseignements,* intelligence department; (more esp.) corps cavalry exploration service;
—— *de réparation,* department of repairs;
—— *de réserve,* (*Fr. a.*) service having in view the maintenance in peace of supplies of all kinds in view of mobilization, and keeping forts fully supplied with material for sieges;
se retirer du ——, to withdraw from, leave, the service;
—— *de ronde,* patrol duty; rounds;

service *de santé*, medical department, medical corps, medical service;
—— *de santé de l'arrière*, (*Fr. a. med.*) medical service on the lines of communication;
—— *de santé de l'avant*, (*Fr. a. med.*) medical service at the front, i. e., with the troops engaged;
—— *de santé en campagne*, (*Fr. a. med.*) medical field service (includes —— *de l'arrière* and —— *de l'avant*);
—— *de sécurité en marche*, all measures taken to insure safety on the march (advance guard, flankers, etc.);
—— *de sécurité en station*, all measures taken to insure safety during halts; outpost service;
—— *de semaine*, garrison duty taken by the week; v. s. v. *semaine;*
—— *des signaleurs*, —— *des signaux*, signal service, signaling;
—— *de la solde*, pay department;
soumis au —— *militaire*, liable to military service;
—— *de sûreté*, security service (includes protection of columns on the march, outposts, and reconnaissances);
—— *de sûreté en marche (en station)*, v. —— *de sécurité en marche (station);*
—— *de sûreté de première ligne*, duties of information and of protection as carried on by the corps cavalry;
—— *des subsistances militaires*, subsistence department;
tableau de ——, duty roster;
—— *des télégraphes*, (*in gen.*) telegraph service;
—— *de la télégraphie militaire*, military telegraph service;
—— *télégraphique de la première ligne*, (*Fr. a.*) telegraph duty in the zone of operations of the army;
—— *télégraphique de la seconde ligne*, (*Fr. a.*) telegraph duty in the rear of the army and connecting the army with home;
—— *télégraphique de la troisième ligne*, (*Fr. a.*) telegraph duty in forts, arsenals, etc., at home;
temps de ——, period of service;
tenue de ——, service uniform or dress;
—— *de tir*, target practice, target-practice duty;
—— *topographique*, topographic, surveying, duty;
tour de ——, tour (of duty);
—— *de tranchée*, trench duty;
—— *des transports*, (*Fr. a.*) transport service;
—— *de la trésorerie et des postes*, (*Fr. a.*) pay and postal service in the field;
—— *vétérinaire*, (*Fr. a.*) veterinary service;
—— *des vivres*, subsistence department;
volontaire d'un an, one-year volunteer service.

servir, v. a. n., to serve, be in the service of; to serve the state, be in the administration; (*tech.*) to work (as a stream, a mill); (*mil., nav.*) to serve; to be in the service; to perform one's duty; (*hipp.*) to serve (a mare);
—— *à*, to be used for;
—— *une batterie*, (art.) to serve a battery;
—— *à bord*, (*nav.*) to serve afloat;
—— *un canon*, (*art.*) to serve or work a gun;
—— *dans* (*l'artillerie*, *etc.*), to be, serve in (the artillery, etc.);
—— *de*, to serve for or instead;
—— *une pompe*, to work a pump;
—— *sous*, (*mil.*) to serve under (so-and-so);
—— *sur mer*, (*nav.*) to serve afloat.

servitude, f., servitude; (*law*) easement, servitude; (*mil.*) open ground around an explosive magazine (in which no building may be put up);
bateau de ——, (*nav.*) harbor-service boat;
—— *s de la frontière*, (*Fr. a.*) zones along the frontiers in which no ground may be cleared or roads built without the special permission of the *commission mixte des travaux publics;*
—— *s militaires, militaires*, (*fort.*) zones around a fortification in which all building is forbidden, or is allowed only under certain conditions;
navire de ——, (*nav.*) harbor-service vessel.

servo-moteur, m., (*mach.*) servomotor.
sésamoïde, a. m., (*hipp.*) sesamoid, sesamoid bone;
grand ——, sesamoid bone;
petit ——, small sesamoid bone.
séton, m., (*hipp.*) seton, rowell;
aiguille à ——, seton needle;
—— *anglais, rowel;*
—— *animé*, tape smeared with irritant;
—— *à mèche*, seton needle and tape or other thread;
—— *à rouelle*, rowel;
—— *simple*, seton with unsmeared tape.
seuil, m., sill, ground sill; sole, threshold; gate; entrance; doorstep; (*fort.*) interior edge or crest of an embrasure; (*top.*) ridge closing a valley;
—— *d'écluse*, sill, ground sill of a sluice;
—— *de pont-levis*, (*fort.*) counterscarp beam for drawbridge.
seuillet, m., (*nav.*) lower port sill.
sève, f., sap; pith.
sévère, a., severe, stern, harsh, rigid (as of discipline); austere; strict (as a blockade); close (as an investment).
sévérité, f., severity, strictness, etc.
sextant, m., sextant; arc of 60°;
—— *à boîte*, small box sextant;
grand ——, sextant proper;
—— *de poche*, pocket sextant.
sextuplette, f., sextuplet, six-seat bicycle.
seye, f., (*art.*) axletree bolt, bolster bolt;
—— *à l'écrou*, screw bolt.
shako, m., v. *schako;*
shrapnel, m., (*art.*) shrapnel;
—— *à chambre antérieure*, front-burster shrapnel;
—— *à chambre centrale*, central-burster shrapnel;
—— *à chambre postérieure*, rear-burster shrapnel;
—— *percutant*, (*art.*) shrapnel used as a percussion shell, i. e., fuse unset.
siccatif, a., siccative, drying.
siccité, f., dryness;
évaporer à ——, to evaporate to dryness.
sidérite, m., siderite.
sidérotechnie, f., v. *sidérurgie.*
sidérurgie, f., (*met.*) metallurgy of iron.
siècle, m., century; age, period, world.
siège, m., seat; see of a bishop; box (coachman's seat); (*law*) bench; (*mil.*) siege; (*mach.*) valve seat; (*harn.*) seat of a saddle; (*farr.*) flat part of the upper surface of the English shoe;
—— *accéléré*, (*mil.*) hasty, rapid, siege (an irregular siege, batteries erected at once, no approaches);
affût de ——, (*art.*) siege carriage;
—— *d'affût*, (*art.*) axletree seat;
armée de ——, (*mil.*) siege corps or army;
artillerie de ——, (*art.*) siege artillery;
—— *d'assaut*, (*mil.*) sudden attack (no trenches opened);
batterie de ——, siege battery;
bouche à feu de ——, (*art.*) siege gun;
—— *brusqué*, (*mil.*) siege without approaches;
canon de ——, (*art.*) siege gun;
—— *de clapet*, (*mach.*) valve seat;
corps de ——, (*mil.*) besieging army, siege corps or army;
couvrir un ——, (*mil.*) to cover a siege;
de ——, (*mil.*) besieging;
échelle de ——, (*mil.*) scaling ladder;
équipage de ——, (*art.*) siege train (v. s. v. *équipage)*;
—— *d'essieu*, (*art.*) axle seat;
état de ——, v. s. v. *état;*
faire le, un, ——, (*mil.*) to besiege, carry on a siege;
faux ——, (*harn.*) straining, straining leather, cross webbing; facing of cloth and leather on the arch of a pack saddle;

siège *en forme*, —— *dans les formes*, (*mil.*) regular, formal, siege;
une guerre de —— *s*, (*mil.*) a war characterized by many sieges;
la guerre des —— (*s*), (*mil.*) siege warfare;
—— *irrégulier*, (*mil.*) a generic term for any siege in which no trenches are opened (e. g., assault, blockade, etc.);
lever le —— (*d'une place*), (*mil.*) to raise the siege;
mettre le —— *devant*, (*mil.*) to lay siege to ....; to sit down before a place;
mettre en état de ——, (*mil.*) to declare a state of siege; to put in a state of siege;
parc de ——, v. s. v. *parc;*
pièce de ——, (*art.*) siege piece;
—— *en règle*, (*mil.*) regular, formal, siege;
—— *régulier*, (*mil.*) v. —— *en règle;*
—— *à ressort*, (*r. f. art.*) spring seat;
le Saint- ——, the Holy See, See of Rome;
—— *de soupape*, (*mach.*) valve seat;
soutenir un ——, (*mil.*) to stand, or support, a siege;
supporter un ——, (*mil.*) v. *soutenir un* ——;
—— *de violence*, (*mil.*) v. —— *d'assaut.*

siéger, v. n., (*law*) of a court, to sit; (*in gen.*) to preside over an assembly.

sieur, m., sort of legal title used for private persons, in deeds, legal papers, etc.

sifflement, m., whistling, whizzing; (*art.*) hiss, whistling (of a projectile); (*elec.*) hissing of a voltaic arc.

siffler, v. n., to whistle; (*nav.*) to pipe; (*art.*) to whiz, hiss;
—— *un canot*, (*nav.*) to pipe a boat away;
—— *un commandement*, (*mil.*) to give a command by means of a whistle; (*nav.*) to pipe;
—— *la gaule*, (*man.*) to swish a riding whip.

sifflet, m., whistle; steam whistle, call, pipe; (*carp.*) skew scarf; (*met.*) spot or defect in a fracture; (*art.*) (tin) spot in a bronze piece; (*mil. slang*) gun; (*hipp.*) cut in the toe of a hoof (to prevent bearing in the neighborhood of a toe track); aperture ignorantly made under the tail to help a broken-winded horse;
—— *d'alarme*, alarm whistle;
—— *à anche*, "reed" whistle (vibratory whistle);
—— *automoteur électrique*, electric self-acting whistle;
—— *avertisseur*, bicycle whistle;
—— *de brume*, fog whistle;
—— *de brume à vapeur*, steam fog whistle;
commandement avec le ——, (*mil.*) call;
coup de ——, whistle (note); pipe or call;
couper en ——, (*carp.*) to skew, to bevel;
en ——, (*carp.*) skew, bevel;
—— *d'étain*, (*art.*) tin spot in a bronze piece;
—— *de signal*, signal whistle;
—— *sirène*, bicycle siren;
—— *à vapeur*, steam whistle.

siffleur, a., wheezing, whistling, hissing;
cheval ——, roarer.

signal, m., signal, sign; (*top.*) target (over a bench mark, etc.); (*mil.*) signal;
—— *acoustique*, acoustic signal;
—— *d'aiguille*, (*r. r.*) switch signal, showing which track is to be taken;
—— *d'alarme*, signal of alarm;
—— *d'aperçu*, answering signal; "O. K.;"
—— *d'appel*, "call" on a telegraph or telephone line;
appuyer un —— *par un coup*, (*nav.*) to enforce a signal by a gun;
—— *artificiel*, (*top.*) target;
—— *avancé*, v. —— *à distance;*
—— *d'avertissement de tempête*, storm signal;
ballon de —— *aux*, signal ball;
—— *à bras*, homographic signal;
—— *de brume*, fog signal;
—— *de canon*, signal gun;
—— *de cavalerie*, v. —— *à percussion;*
—— *chiffré*, cipher;
—— *clignotant*, intermittent signal;
—— *à cloche*, bell signal;
code de —— *aux*, signal code;
—— *de combat*, (*mil.*) signal for action;
coup de canon de ——, signal gun;

signal *de départ*, starting signal;
—— *de détresse*, signal of distress;
disposer un ——, to bend on a signal;
distance de ——, *de* —— *aux*, signaling distance;
—— *à distance*, (*r. r.*) distance signal;
à distance de —— *aux*, within signaling distance;
donner le —— *de*, to make signal of or for;
—— *à éclats*, flash signal;
—— *par éclats et intervalles*, flash-and-pause signal;
—— *d'embarcation*, boat signal;
—— *d'exécution*, signal of execution;
faire un ——, to make a signal;
—— *à fanaux*, lantern signal;
feu de —— *aux*, signal light;
—— *fixe*, (*r. r.*) stationary signal;
—— *de fusée*, rocket signal;
fusée de —— *aux*, signal rocket;
—— *homographique*, v. —— *à bras;*
hisser un ——, to hoist a signal;
—— *d'incendie*, fire call;
—— *d'incommodité*, signal of distress;
—— *de jour*, day signal;
lettres de —— *aux*, signal letters;
livre de —— *aux*, signal book;
—— *de marée*, tide signal;
—— *mobile*, movable signal;
—— *pour munitions*, (*mil.*) lantern, flag, showing where ammunition is to be had;
—— *de nuit*, night signal;
—— *de nuit Coston*, Coston light;
—— *numérique*, numerical signal;
numéro de —— *aux*, signal number;
—— *optique*, (*mil.*) any visual signal (pyrotechnic or other);
—— *de partance*, (*nav.*) blue peter; sailing signal;
—— *particulier*, private signal;
—— *à pavillons*, flag signal;
—— *à percussion*, (*mil.*) signal light used by vedettes to give warning (consists of a small petard on a stick, fired by striking);
—— *-perche*, (*top.*) rod target;
—— *phonique*, sound signal;
—— *de pilote*, pilot signal;
à portée de —— *aux*, within signal distance;
—— *de quatre signes*, four-flag signal;
—— *de ralliement*, (*mil.*) rallying signal; (*nav.*) signal to leave off chase;
—— *sémaphorique*, semaphore signal;
sifflet de ——, signal whistle;
—— *à sifflet*, whistle;
—— *simple*, flag signal;
tactique des —— *aux*, (*mil.*) signaling tactics, i. e., employed with troops in field;
—— *de tempête*, storm signal;
—— *du temps*, weather signal;
torche de ——, signal torch;
—— *-trépied*, (*top.*) tripod target;
veiller aux —— *aux*, to look for signals.

signalement, m., description (of a person, as in a passport or other document); description of a horse; (*mil.*) descriptive list (U. S. A.); descriptive list of men, animals, guns (France);
—— *des chevaux*, (*hipp.*) description of horses;
—— *s composés*, (*hipp.*) the —— *s simples*, q. v., with the pedigree and performance or record added;
inscrire sur le contrôle de ——, (*mil.*) to take on the strength;
livret de ——, (*mil.*) descriptive book;
—— *s simples*, (*hipp.*) description of the characteristics, main points, of a horse.

signaler, v. a., to give a description (of a person or thing); to point out, mention; to notify; (*mil., nav.*) to make or send signals; to signalize; (*mil.*) to make out (as the enemy); to report anyone to a superior; to send down a descriptive list (obs. in this sense);
se ——, to distinguish one's self.

signalétique, a., descriptive;
état ——, v. s. v. *état.*

signaleur, m., (*mil., nav.*) signalman.

signaliste, m., v. *signaleur.*

signataire, m., signatory.

signature, f., signature;
 à la ——, (submitted, etc.) for signature.
signe, m., sign; mark, token;
 —— *conventionnel*, (*top.*) conventional sign;
 —— *conventionnel militaire*, (*top.*) conventional military sign;
 —— *de grade*, (*mil.*) insignia of rank;
 —— *particulier*, mark on a man by which he may be identified;
 —— *de ralliement*, (*mil.*) rallying signal used instead of *mot de ralliement;* signal by which friends may recognize one another;
 —— *de service*, (*mil.*) badge of being on duty (chin strap down, in France).
signer, v. a., to sign, subscribe; (in the arts) to mark, stamp;
 —— *du fer*, to stamp iron;
 —— *une paix, un traité, etc.*, to conclude peace, a treaty, etc.
signifier, v. a., to signify; to mean; to make known; (*law*) to notify, give formal information of, serve notice;
 —— *les arrêts*, to notify a person that he is in arrest.
signole, f., reel, winch, on a shaft or axle;
 arbre coudé en ——, shaft, axle, cranked so as to form a winch.
siguette, f., (*harn.*) (*mors à la*) ——, a cavesson the noseband of which has iron teeth.
silence, m., silence;
 réduire au ——, (*art.*) to silence (a hostile battery, etc.).
silex, m., flint;
 fusil à ——, flint gun (obs.);
 platine à ——, flintlock (obs.).
silhouette, f., silhouette; (*t. p.*) silhouette target;
 —— *à bascule*, tilting or dipping target;
 cible- ——, silhouette target;
 —— *à contrepoids*, surprise target worked by a counterpoise;
 —— *ressort*, surprise target worked by a spring;
 —— -*tête*, head target (one consisting of a head only);
 —— *tombante*, falling target (falls when hit).
silicate, m., silicate;
 —— *de potasse*, liquid, soluble, or water glass.
silicatisation, f., operation of covering with water glass.
silice, f., flint, silica, quartz.
silicium, m., silicon.
silico-spiegel, m., (*met.*) silico-spiegel.
sillage, m., (*nav.*) headway, course, wake;
 doubler le ——, to make twice as much way;
 montre à ——, patent log.
siller, v. n., (*nav.*) to cut through the water; to run ahead; to have headway; to steer.
sillomètre, m., (*nav.*) patent log.
sillon, m., furrow; dent, groove, depression, trace, mark, trail, train; (*nav.*) wake, track; (*fort.*) trace, traced line; (*hipp.*) ridge of a horse's mouth.
sillonner, v. a., furrow; to plow, cut, groove; (*nav.*) to cut, plow, through the water.
silo, m., silo.
silotvor, m., (*expl.*) silotvor.
simbleau, m., carpenter's radius line for striking large circles.
similigravure, f., similigravure.
simple, a., simple; single; private; common; bare; mere; one; plain; (*chem.*) elementary;
 corps ——, (*chem.*) element, elementary body;
 à —— *effet*, single-acting;
 en —— (*cord.*) not passing through a pulley (of a rope);
 ligne ——, v. *ordre* ——;
 machine à —— *effet*, single-stroke engine;
 —— *matelot*, common seaman;
 nœud ——, v. *nœud;*
 ordre ——, (*nav.*) order in single line;
 poulie ——, single block;
 sape ——, v. s. v. *sape;*
 —— *soldat*, (*mil.*) private soldier, private;
 soulier à —— *semelle*, single-soled shoe.
simulacre, m., imitation; feint, sham;
 —— *de combat*, (*mil.*) sham battle;
 —— *de siège*, (*mil.*) sham siege.

simultané, a., simultaneous.
simultanéité, f., simultaneousness.
sinapisme, m., mustard plaster.
singe, m., monkey; monkey of a pile driver; windlass (mounted on two trestles); hoist, crab; (*inst.*) pantograph;
 —— *clef*, monkey spanner, monkey wrench.
singleau, m., v. *simbleau*.
sinistre, m., disaster, damage (from fire, wreck), accident; loss or damage in anything insured;
 évaluer le ——, to estimate the damage.
sinueux, a., sinuous.
sinuosité, f., bend, turn, sinuosity, meandering.
sinus, m., sine; (*hipp.*) sinus, cavity;
 —— *verse*, versed sine.
siphon, m., siphon; waterspout; (*fond.*) rising gate;
 coulage à ——, v. s. v. *coulage;*
 mèche à ——, siphon wick;
 —— *recorder*, (*teleg.*) siphon recorder;
 robinet de ——, siphon cock.
sire, m., lord, sire.
sirène, f., siren;
 —— *à air comprimé*, compressed-air siren;
 —— *à vapeur*, steam siren.
siroc(o), m., sirocco.
sirtes, f. pl., v. *syrtes*.
site, m., site;
 angle de ——, v. s. v. *angle;*
 en —— *aquatique*, (*fort.*) on a wet site (wet-ditched);
 en —— *élevé*, (*fort.*) on a dry site (i. e., dry-ditched);
 ligne de ——, v. s. v. *ligne;*
 plan de ——, v. s. v. *plan;* otherwise defined as the vertical plane through the *ligne de* ——.
situation, f., situation, condition, position, state; (*nav.*) bearing; (*adm.* and *mil.*) state return; regimental or other return of personnel; report;
 —— *administrative*, (*Fr. a.*) daily report, morning report (as a basis of pay and allowances);
 —— (*administrative*) *de dizaine*, (*Fr. a.*) report of men present and absent, etc., sent in every ten days by fractions detached from an administrative unit;
 —— *de l'enseignement*, (*Fr. a.*) school report;
 état de ——, v. s. v. *état;*
 —— *des fonds*, (*Fr. a.*) quarterly account current of each regiment (includes the various funds (*masses*) entered upon the *registre de centralisation* after the accounts have been audited);
 fournir une ——, (*mil.*) to make or furnish a report;
 —— *des malades*, (*mil.*) sick report or return;
 —— -*rapport*, v. s. v. *rapport;*
 —— *de tir*, target-practice report, kept on range as practice proceeds.
six, m., six;
 —— *as et as*, (*powd.*) old name for black gunpowder (the proportions of niter, charcoal, and sulphur being as 6:1:1).
six-pans, m., hexagonal nut.
sixte, f., (*fenc.*) sixte (parry).
sleeper, m., (*r. r.*) railroad tie, sleeper.
smala(h), f., the collection of tents of a powerful chief (Arab word).
smille, f., stonecutter's pick.
smiller, v. a., to dress stone with a pick.
soc, m., plowshare; (*mil.*) shoe, socket, of a lance or color staff; (*art.*) trail spade, recoil spade.
société, f., society; company; partnership;
 —— *d'actionnaires*, —— *par actions*, stock company;
 —— *anonyme*, joint-stock company;
 —— *d'assurances*, insurance company;
 —— *en commandite*, limited (liability) company;
 —— *de gymnastes*, —— (*de*) *gymnastique*, athletic club;
 la ——, *la* —— *de Jésus*, the Jesuits;
 —— *de secours aux blessés*, society to aid the wounded in war;

société de tir, rifle, shooting, club.
socle, m., footing, base, bottom, pedestal, plinth, blocking, stand; foot, footing, of a wall; foundation of an engine or machine; (*mil. min.*) bracing or pedestal used in tamping ; (*artif.*) block on which rockets, etc., are driven; (*art.*) pedestal (of some mounts);
—— -*bâti*, foundation or pier (as for a stream engine, etc.).
sœur, f., sister;
—— *hospitalière*, sister on army nursing duty.
soie, f., silk; hog's bristle, any bristle; tang (of a tool, sword blade, etc.); (*hipp.*) toe crack; (*mach.*) journal pin;
bourre de ——, v. s. v. *bourre;*
—— *de manivelle,* (*mach.*) crank pin,, connecting-rod journal;
—— *de la poignée,* (*sm. a.*) tang of a hilt.
soif, f., thirst.
soir, m., evening, night (*mil., r. r.*, from 12 o'clock noon to 12 o'clock midnight).
sol, m., soil; ground; surface of the ground; ground plot; sole, bottom (as of a gallery, tunnel, etc.);
—— *compressible*, compressible soil;
creuser le ——, to break ground;
—— *engerbé*, v. —— *d'une traverse;*
relier au ——, (*elec.*) to ground, earth, make a ground connection;
—— *d'une traverse,* (*fort.*) in earthwork, the surface on which a traverse is to rise.
solaire, a., of the sun, solar;
cadran ——, sun dial.
solandre, f. (*hipp.*) solanders, salanders (generally in plural).
solariser, v. a., (*phot.*) to expose to the sun.
solbatu, a., (*hipp.*) surbated, foot-foundered;
rendre ——, to surbate.
solbature, f., (*hipp.*) foot-founder; surbating; closh.
soldat, m., (*mil.*) soldier (when used alone, frequently equivalent to infantry private); (*in gen.*) any member of the military profession; (*in pl.*) the men; a., soldierlike;
—— *d'artillerie*, artilleryman;
—— *de cavalerie,* cavalryman;
—— *du centre*, v. *centrier;*
—— *à cheval*, trooper, mounted soldier;
—— *conducteur,* (*Fr. a.*) driver of (the regimental baggage train);
de ——, soldierlike;
—— *enfant*, young, inexperienced soldier;
faux ——, falsely mustered soldier;
—— *de fortune*, soldier of fortune;
—— *à gages*, mercenary;
—— *du génie*, engineer soldier;
—— *d'infanterie,* infantryman;
jeune ——, (*Fr. a.*) recruit (so called until he gets through his recruit drill); from an administrative point, member of the last class called into service, exclusive of volunteer enlistments;
—— *de* (*la*) *marine*, v. s. v. *marine;*
—— *musicien*, bandsman;
——-*ordonnance*, soldier servant, "striker" (U. S. A.);
—— *d'ordonnance*, orderly;
—— *à pied*, foot soldier;
—— *de planton*, orderly in waiting;
—— *de plomb*, tin soldier;
—— *porteur d'outils,* (*Fr. a.*) soldier skilled in the use of certain tools (so called from his carrying them in the field);
—— *pourvoyeur*, man who passes ammunition from the caissons, etc.;
—— *de première classe,* (*Fr. a.*) private of the first class (title won by good conduct and efficiency);
—— *de rencontre*, untrained soldier, any man picked up and put into the ranks without regard to fitness;
—— *secrétaire*, clerk;
simple ——, private;
—— -*tender,* (*Fr. inf.*) soldier detailed to look after an officer (dismounted) in the field (carries food and various other articles, looks after the officer if wounded, etc.);
—— *du train*, train soldier;
—— *du train d'artillerie*, train soldier;
vieux ——, old soldier.

soldatesque, a., soldierlike; f., soldiery, (esp.) undisciplined, unbridled soldiery.
solde, f., (*mil.*) pay of an officer, noncommissioned officer, or private; currently limited to officers' (commercially) settlement.
(The following terms are military:)
à la —— *de*, in the pay of;
—— *d'absence*, half pay; leave and (*Fr. a.*) captivity pay, one-half of the —— *de présence;*
(—— *et*) *accessoires de* ——, (pay and) allowances;
—— *d'activité*, active-duty pay;
avoir à sa ——, to have in one's pay;
—— *de captivité*, v. —— *d'absence;*
—— *de congé*, leave pay;
demi- ——, half pay;
à, en, demi- ——, on half pay;
—— *de disponibilité,* (*Fr. a.*) waiting-orders pay (generals and *assimilés*);
—— *entière*, full pay;
entrer en ——, to begin to draw pay;
entrée en ——, v. s. v. *entrée;*
état de ——, pay roll;
—— *de guerre*, war pay;
—— *de marche*, travel pay, travel allowance;
—— *de non-activité,* (*Fr. a.*) unemployed pay (three-fifths for lieutenants and below, one-half for all others, if the officer is unemployed through no fault of his own; two-fifths for all grades, if the officer is unemployed through his own fault);
—— *de paix*, ordinary pay;
—— *de permission*, v. —— *de congé;*
prendre à sa ——, to take into one's pay;
—— *de présence*, full duty pay, full pay;
—— *de réforme*, pay of the status of *réforme*, reduced pay (two-thirds of the minimum retired pay in the case of officers retired for physical disability, one-half in the case of officers retired for disciplinary reasons);
—— *de réserve,* (*Fr. a.*) pay of generals and of *assimilés* who have passed into the *cadre de réserve;*
—— *de retraite*, retired pay;
—— *de route*, allowance on a march.
solder, v. a., to pay, settle, liquidate; (*mil.*) to pay troops; (hence, *in gen.*) to keep, retain, in pay.
sole, f., sole, sill, sleeper; ground plate; bottom of a flat boat; (*hipp.*) sole (of hoof); (*met., etc.*) furnace bottom, hearth; dead plate; bed, sole, of a reverberatory furnace; (*art.*) plate at the lower part of some top carriages of the French marine artillery, serving to keep cheeks together (obs.);
—— *d'affût,* (*art.*) transom of a gun carriage;
—— *de bateau*, flat bottom of a boat;
—— *battue,* (*hipp.*) v. *solbature;*
—— *de bigue*, step for the heel of a small shears;
—— *brûlée,* (*farr.*) burned sole (shoe put on too hot);
—— *charnue,* (*hipp.*) fleshy, sensitive, sole of a horse's foot;
—— *chauffée,* (*farr.*) overheated sole (the hot shoe held on too long);
—— *cornée,* (*hipp.*) horny sole;
—— *d'embrasure,* (*fort.*) sole or bed of an embrasure;
—— *de foyer*, dead plate, coking plate of a furnace;
—— *mobile d'affût,* (*F. m. art.*) movable transom;
—— *de sabord,* (*fort., nav.*) port sill;
—— *de sabot,* (*hipp.*) sole of a horse's foot;
—— *de laquel*, step of a kevel.
soleil, m., sun; (*artif.*) fireworks having jets or rocket as radii of a circle;
comme le ——, with the sun;
contre le ——, against the sun;
(*à*) *contre du* ——, against the sun;
—— *couchant*, sunset;
coucher du ——, sunset;
coup de ——, sunstroke;
entre deux ——s, between suns;
—— *d'eau,* (*artif.*) water wheel for fireworks;
—— *fixe,* (*artif.*) fixed sun;
grand ——, midday;
au grand ——, in the blaze of the sun;

soleil *levant*, sunrise;
lever du ——, sunrise;
marche de ——, sunway; as the sun goes;
—— *montant*, (*artif.*) tourbillon, (two rockets composing a cross);
dans le sens du ——, v. *comme le* ——;
—— *tournant*, (*artif.*) sun revolving vertically.

solénoïde, m., (*elec.*) solenoid;
—— *électro-magnétique*, electro-magnetic helix, solenoid; practical solenoid;
—— *fermé*, closed solenoid.

solidaire, a., integral, in one piece with anything, forming one rigid piece with; (*law*) jointly and separably answerable;
—— *de*, integral, rigid with, forming one rigid piece with.

solidarité, f., (*law*) joint responsibility.

solide, a., solid, stout, secure; sound, firm, strong; massive, stable, consistent, hard; m., solid, solid body, cubical contents;
angle ——, solid angle;
—— *d'une batterie*, (*fort.*) cubical contents of earthwork in parapet;
—— *de bonne rupture*, (*expl.*) v. *ellipsoïde de bonne rupture*;
creuser jusqu'au ——, to dig down to solid ground;
—— *de révolution*, solid of revolution;
—— *de rupture limite*, v. *ellipsoïde de rupture limite;*
terrain ——, solid ground (as for foundations).

solidifier, v. a. r., to solidify, to become hard; to make or render solid, acquire solidity.

solidité, f., solidity, density; soundness, stability; compactness; massiveness; strength; stoutness, steadfastness.

solin, m., (*cons.*) distance, space, between joists; mortar filling of the space between joists; (in a tiled roof) the joint or covering of plaster or cement along the intersection of the roof and a vertical wall; joint filled with mortar;
bande de ——, flashing strip (under a *solin* in mortar).

solipède, a., m., soliped, solid-ungulate.

solivage, m., calculation of the number of beams that may be cut out of a log; (*cons.*) timberwork of a house.

solive, f., (*cons.*) joist;
—— *armée*, trussed joist;
—— *boîteuse*, tail joist;
—— *croisée*, cross joist;
—— *d'enchevêtrure*, trimmer joist;
poser des ——*s*, to joist.

soliveau, m., (*cons.*) flooring joist, boarding joist, small joist.

solivure, f., (*cons.*) all the joists of a building.

sollicitation, f., solicitation.

solliciter, v. a., to solicit, petition;
—— *un cheval*, (*man.*) to urge, animate, a horse.

solliciteur, m., petitioner.

soluble, a., soluble.

solubilité, f., solubility.

solution, f., solution; dissolution; resolution; discharge;
—— *de continuité*, breach of continuity.

solvabilité, f., solvency.

solvable, a., solvent.

somache, a., v. *saumache*.

sombre, a., dark, gloomy; (weather) overcast, cloudy;
faire ——, to be cloudy, dark, overcast, etc.

sombrer, v. n., (*nav.*) to sink, to founder.

sommail, m., (*hydr.*) bank, shoal (in a channel).

sommaire, m., summary, compendium; abstract, abridgment, contents.

sommaire, a., summary, concise, short; (*law*) summary; without trial;
condamnation sur procédure ——, (*law*) summary conviction.

sommation, f., summons, process, challenge; reading of the riot act; (*mil.*) summons (to surrender);
* *faire les trois* ——*s*, to read the riot act.

somme, f., sum, amount; burden, load (of a horse, mule); (*hydr.*) bar at the mouth of a port or of a river;
bête de ——, pack animal, beast of burden;
cheval de ——, pack horse;
haute ——, contingent money, large sum paid out by a shipowner;
pays ——, (*hydr.*) shoal, shallow.

sommeil, m., sleep.

sommer, v. a., to summon (as a position, to surrender); to call upon; to challenge, charge; to sum, add up;
—— *une place*, (*mil.*) to summon a besieged town, etc., to surrender;
—— *quelqu'un de sa parole*, to call upon one to keep his promise.

sommet, m., summit; vertex (of an angle, cone, etc.); top (as of a dam, etc.); ridge; height; apex; (*fig.*) pinnacle, zenith, meridian, crown;
—— *de la trajectoire*, (*ball.*) highest point of a trajectory, summit, vertex;
—— *de la triangulation*, (*surv.*) triangulation vertex.

sommier, m., register, ledger; sumpter-, pack-, horse or mule; mattress; piece of wood supporting a heavy mass; springpiece of a bridge; support for long beams; beam of a balance; lower crossbar of an iron gate or grating; double hoop at the end of a barrel or cask; (*cons.*) breastsummer; end stone of a platband; springer, springing stone of an arch; (*fort.*, *nav.*) iron backing or part of compound armor; (*nav.*) upper sill of a port; (*mach.*) bed of a machine (boiler, etc.);
—— *des barreaux*, v. —— *de grille;*
—— *de crin*, hair mattress;
cul de ——, (*cons.*) skewback;
—— *élastique*, spring mattress;
—— *de grille*, fire-bar support;
—— *de lit*, mattress;
—— *de pont*, stringer of a bridge;
—— *de porte*, wooden architrave of a door.

sommière, f., (*harn.*) packing rope, packing strap, packing cord; (*top.*) open space, glade, in a wood.

sommité, f., summit, top, (extremity) end; head;
les ——*s*, highest grades, authorities;
les ——*s du commandement*, (*mil.*) the highest authorities in the army.

son, m., bran; sound;
—— *discordant*, rattling noise, rattling (in machines);
—— *mouillé*, bran mash (for horses).

sondage, m., boring (of earth, as for foundations, etc.); (*nav.*) sounding; (*med.*) probing (of wounds);
—— *chinois*, —— *à la corde*, boring by rope (boring tool raised by a rope);
—— *à la grande sonde*, (*nav.*) sounding with deep-sea lead;
outil de ——, boring tool;
—— *à tige rigide*, boring with a string of tools.

sonde, f., borer, earth auger, boring tool; (*med.*) probe; (*nav.*) lead, sounding lead, sounding line; soundings; sand or gravel sticking to the lead; (*art.*) vent probe; (*surv.*) numeral expression of the distance of a point below a datum plane;
aller à la ——, (*nav.*) to go by the lead;
barre de ——, sounding rod;
—— *s d'une carte*, soundings marked on a chart;
chercher la ——, to make soundings;
crier la ——, (*nav.*) to call the depth;
—— *à crochet*, (*art.*) vent searcher;
être sur la ——, (*nav.*) to be on soundings;
grand plomb de ——, *grande* ——, (*nav.*) deep-sea lead;
jeter la ——, (*nav.*) to heave the lead;
lance de ——, sounding rod;
ligne de ——, (*nav.*) lead line;
—— *à main*, (*nav.*) hand lead;
navigeur la —— *à la main*, v. *aller à la* ——;
—— *à nonius*, (*art.*) shell gauge (thickness of base);
plomb de ——, (*nav.*) lead;

sonde *de pompe*, sounding rod of a pump;
— *de rebut*, (*art.*) vent probe (the passage of which, in examining vents, insures the condemnation of the vent bushing);
— *à tarière*, earth borer, auger;
tête de —, brace head;
— *de Thompson*, (*nav.*) Thompson's patent lead;
verge de —, sounding rod;
— *à vernier*, v. — *à nonius*.

sonder, v. a., to examine by boring, to make borings (as the ground, for foundations); to examine, search (as a beam, etc., to see if it is sound); (*nav.*) to sound, heave the lead; (*med.*) to probe;
— *un bois*, (*mil.*) to explore a wood;
— *les pompes*, (*nav.*) to sound the pumps;
— *le sol*, — *le terrain*, to make borings (for foundations, etc.).

sondeur, m., workman who makes borings; well-borer; (*nav.*) leadsman; sounding apparatus.

sonnant, a., sounding, ringing;
espèces —*es*, gold and silver money, hard cash;
à l'heure —*e*, on the stroke of ....;
horloge, *montre*, —*e*, striking clock, watch.

sonné, p. p., struck;
deux, etc., heures —*es*, on the stroke of 2, etc.

sonner, v. a. n., to sound; to strike (of a clock), to ring, wind, beat; (*mil.*) to sound (a call);
— *la boute-selle*, (*mil.*) to sound "to horse," "boots and saddles;"
— *la charge, etc.*, (*mil.*) to sound the charge;
— *à cheval*, (*mil.*) to sound "to horse;"
— *la cloche*, (*nav.*) to strike bells;
— *au clarion*, (*mil.*) to make a trumpet signal;
— *le quart*, (*nav.*) to strike the bell at the close of a watch.

sonnerie, f., striking train of a clock; electric bell; (*mil.*) call, signal (made on the trumpet or bugle);
— *d'alarme*, alarm, call, bell;
— *des bons de tabac*, (*mil. slang*) trumpet call for men confined in barracks;
— *à carillon*, electric bell; extension call bell;
— *continue*, v. — *à carillon*;
— *à deux battements*, double-stroke bell;
— *électrique*, electric bell; buzzer;
— *des maladroits*, (*cav. slang*) call for infantry drill;
— *répétiteuse*, v. — *électrique*;
— *à trembleur*, — *trembleuse*, v. — *électrique*;
— *à un battement*, — *à un seul coup*, single-stroke bell.

sonnette, f., bell, hand bell; pile driver, pile-driving engine, ringing engine; monkey, ram, for testing axles; (*expl.*) sort of shaker (in the preparation of composition, priming, etc.);
— *artificielle*, v. — *à déclic*;
déclic de —, trigger, catch of a pile driver;
— *à déclic*, pawl pile driver;
mouton de —, monkey;
— *à poudre*, gunpowder pile driver;
— *à tiraude*, ringing engine, common pile driver;
— *à vapeur*, steam pile driver.

sonneur, m., (*teleg.*) sounder.

sorne, f., (*met.*) adherent slag, dross.

sort, m., chance, lot, condition, destiny, fate;
— *des armes*, fate of arms;
— *de la guerre*, luck of war, chances of war;
tirage au —, (*mil.*) drawing (a number, as in a conscription or for military service);
tirer au —, (*mil.*) to draw (a number, as in a conscription or for military service);
tomber au —, (*mil.*) to draw a number for service; to be drawn for service.

sortie, f., departure, going out, coming out; outlet, exit, egress, issue, mouth; export, exportation; graduation from school; issue of stores from magazine or depôt; (*mil.*) sortie; (*steam*) eduction; (*fort.*) postern; sortie passage; (*nav.*) departure of a ship from a port, roads, etc.;

— *annuelle*, (*mil.*) annual loss of horses;
billet de —, (*nav.*) dock pass;
classement de —, final standing (in schools);
à la — *de*, on departing from, on leaving, etc.;
droits de —, (*nav.*) outward dues;
— *extérieure*, (*mil.*) sortie made while the besiegers are at some distance;
fausse —, (*mil.*) feint sortie;
grande —, (*mil.*) sortie in force;
— *intérieure*, (*mil.*) sortie made when the enemy is at close quarters;
— *en masse*, (*mil.*) general sortie;
— *d'ordre*, (*adm.*) paper issue, transfer from one account to another in the same service;
petite —, (*mil.*) small sortie, intended to carry out some particular and limited object;
pièces de —, (*adm.*) all papers accompanying an issue of stores, etc. (e. g., invoice);
porter en —, (*adm.*) to drop from one's papers;
rang de —, graduation, final standing;
— *réelle*, issue, etc., of material itself;
tuyau de —, delivery, eduction, pipe.

sortir, v. a. n., to go, come, out; to depart, sally forth; to get, carry, lead, take out; to project; to leave, come out of; to deviate, swerve from; to be a graduate of; (*mil.*) to turn out (of a guard); to be paraded; to be taken out (of a flag); to come out (individually in dressing a line); (*art.*) to lead out; (*nav.*) to put to sea;
— *en armes*, (*mil.*) to get under arms; to turn out under arms;
— *le deuxième*, v. — *le premier, le deuxième*;
— *de l'eau*, to rise on the horizon;
— *de la garde*, (*mil.*) to have been the guard;
— *de l'herbe*, to take (a horse) from grass;
— *de son lit*, (of a river) to overflow its bed;
— *le premier, le deuxième, etc.*, to pass 1, 2, etc. (in a school);
— *de prison*, to be set at liberty;
— *de la question*, to miss the point, to swerve from the point;
— *des rangs*, (*mil.*) to step out of ranks; to rise or come from the ranks;
— *d'une règle*, to ignore a rule;
— *de la selle*, (*man.*) to show daylight (under the seat);
— *de la voie*, (*r. r.*) to run off the track.

sosie-mannequin, m., (*mil. slang*) bolster arranged so as to look like a man in bed.

sou, m., five centimes, five-centime piece;
— *de poche*, (*Fr. a.*) v. *centime de* —.

soubarbe, f., v. *sous-barbe*.

soubassement, m., (*cons.*) foundation structure, foundation; base, basement; patten of a column;
— *du dôme*, (*steam*) base of a steam chest.

souberme, m., (*unif.*) freshet; torrent swollen by a freshet.

soubise, m., (*unif.*) braiding on the back of a dolman or tunic.

soubresaut, m., start, jolt, shock, jerk; (*hipp.*) lift or second expiration of a broken-winded horse.

soubresauter, v. n., to start, jolt, etc.

souche, f., stump, stock, stem; stub (of a check, etc., book); water pipe of a basin, (*farr.*) end of a horseshoe nail left in the hoof; (*cons.*) part of a chimney above the roof;
à —, fitted with, or having, a stub;
cahier à —, blank book with a stub;
— *de chandelle*, candle (gas) burner;
— *de cheminée*, part of a chimney above the roof;
— *d'enclume*, anvil stock;
livre à —, stub book, check book;
registre à —, v. s. v. *registre*;
talon de —, line of separation between stubs and checks, etc.; stub, counterfoil.

soucherie, f., woodwork, framing, of a (power) hammer.

souchet, m., rag stone, quarry rubbish.

souchon, m., (*met.*) short, thick, bar of iron.

soudabilité, f., (*met.*) weldability.

soudable, a., (*met.*) weldable.
soudage, m., (*met.*) welding; soldering; brazing.
soudant, a., (*met.*) welding;
— *blanc* ——, welding heat.
soudard, soudart, m., familiar and contemptuous name for an old soldier (connotation of bad habits and coarseness).
soude, f., (*chem.*) soda;
—— *boratée*, borax;
—— *caustique*, caustic soda;
nitrate de ——, —— *nitratée*, Chile saltpeter.
soudé, a., (*of the grain of wood*) well knit; (*met.*) welded;
—— *à bout*, (*met.*) jump-welded;
—— *à chaude portée*, (*met.*) lap-welded;
cheval bien ——, v. s. v. *cheval*;
—— *à joints superposés*, (*met.*) lap-welded;
—— *à recouvrement*, (*met.*) lap-welded.
souder, v. a., (*met.*) to solder; to braze; to weld;
barre à ——; soldering iron;
—— *à bout*, to jump-weld;
—— *à chaud*, to weld;
—— *à chaude portée*, to lap-weld;
—— *à chaude suante*, to weld;
—— *au cuivre*, to braze, to solder, with copper; to hard solder;
—— *à l'étain*, to tin solder; to soft solder;
fer à ——, soldering iron;
—— *à joints superposés*, to lap-weld;
poudre à ——, welding powder;
—— *à recouvrement*, to lap-weld;
sable à ——, welding sand;
—— *à soudure forte* (*tendre*), to hard (soft) solder.
soudoir, m., soldering iron or bit, copper bit;
—— *à, en, marteau*, soldering hammer;
—— *pointu*, pointed copper bit.
soudoyer, v. a., to pay, keep in pay, hire (generally with the idea of treachery).
soudure, f., (*met.*) solder, (sometimes) soft solder; welding; shutting, shutting up; soldering, brazing; suture; soldering seam; welding seam; soldered joint; (*mas.*) thick plaster;
—— *d'argent*, silver solder;
—— *autogène*, autogenous, natural, soldering; burning on;
—— *bout à bout*, jump-, butt-, welding;
—— *de cuivre*, hard solder; brazing copper;
—— *électrique*, electric welding;
—— *d'étain*, soft solder;
fil de ——, brazing wire;
—— *fondante*, soft solder;
—— *forte*, hard solder; (operation of) hard soldering;
—— *grasse*, a solder in which tin predominates;
—— *à huit, etc.*, solder of eight, etc.;
—— *de laiton*, brass solder;
—— *maigre*, a solder weak in tin;
—— *non-autogène*, soldering proper (i. e., use of a solder to form a junction);
—— *au quart*, solder with a fourth of copper;
—— *de(s) plombier(s)*, solder of two parts lead to one of tin;
—— *à recouvrement*, lap-welding;
sans ——, weldless;
—— *tendre*, soft solder; (operation of) soft soldering;
—— *au tiers*, solder with a third of copper.
soufflage, m., (*met., etc.*), blowing, blast; glass blowing; (*nav.*) sheathing, furring, added to a ship to increase its stability;
—— *en planches*, (*steam*) cylinder jacket.
soufflant, a., blowing; m., (*mil. slang*) bugler; wind-jammer (U. S. A.);
cylindre ——, blowing cylinder;
machine ——*e*, v. s. v. *machine*.
soufflante, f., (*mach.*) blowing engine, machine.
souffle, m., blast, blasting, puff; breath, wind, breath of air, concussion of air (as in a mine); (*art.*) blast of a gun; (*hipp.*) wind;
—— *du fourneau*, (*mil. min.*) smoke due to saucisson in gallery of a mine;
—— *d'une mine*, (*mil. min.*) blowing out of a mine, when not sufficiently loaded;

souffle *de vent*, breath of wind.
souffler, v. n., to blow (wind, etc.); to blow, breathe, pant (of a horse, man, etc.); to blow out, extinguish; to inflate, swell out; (*powd.*) to rise in dust; (*min.*) to blow out (without doing work expected); (*nav.*) to put on a sheathing (v. *soufflage*);
—— *un canon*, (*art.*) to scale a gun;
—— *une dépêche*, to fail to deliver a dispatch;
—— *une marche*, (*mil.*) to steal a march;
—— *au poil*, v. s. v. *poil*;
—— *le verre*, to blow glass.
soufflerie, f., (*mach., met., etc.*) blowing machine; bellows, blowing engine, air blast; complete assortment of bellows in a forge; fan;
—— *hydraulique*, water blast;
—— *à piston cylindrique*, blowing cylinder, cylinder blowing engine.
soufflet, m., blowing machine; (pair of) bellows; fan, fanner; any piece or part folding or acting like a bellows; bullet, slap in the face;
—— *de bois*, wooden bellows;
—— *à bras*, hand bellows;
—— *à caisse*, chest bellows;
—— *à chaînette*, chain blower;
—— *à charnière*, v. —— *de bois*;
—— *de cheminée*, ordinary fire bellows;
clou de ——, tack;
—— *cylindrique*, blowing cylinder, cylinder blowing engine;
—— *à deux vents*, double bellows;
—— *à double âme, effet, vent*, v. —— *à deux vents*;
—— *de forge*, forge bellows;
—— *à palettes*, v. —— *à chaînette*;
pièce de ——, blast piece, blast pipe;
—— *à piston*, piston blower;
—— *à poudre de pyrèthre*, bellows for insect powder;
—— *pyramidal*, v. —— *de bois*;
—— *simple*, single bellows.
souffleur, m., blower; (*mach.*) blast, blower; (*hipp.*) roarer, roaring horse; (*mas.*) workman in charge of the transportation and setting of stones;
cheval ——, roaring horse;
—— *de verre*, glass blower.
soufflure, f., (*met.*) blowhole, blister, honeycomb, flaw;
—— *de fonte*, blowhole in a casting, bubble, airhole.
souffrir, v. a., to suffer; to resist (as, *mil.*, a siege);
—— *un assaut*, (*mil.*) to resist an assault;
—— *l'éperon*, (*man.*) to be indifferent to the spur.
soufrage, m., treating with sulphur.
soufre, m., sulphur;
—— *en bâtons*, v. —— *en canons*;
—— *brut*, crude sulphur, obtained in the first process of purification;
—— *en canon(s)*, roll sulphur;
fleur de ——, flowers of sulphur;
—— *en fleurs*, v. *fleur de* ——;
—— *de mine*, native sulphur;
—— *naturel*, natural sulphur;
—— *rouge*, realgar;
—— *sublimé*, v. *fleur de* ——;
—— *végétal*, lycopodium powder;
—— *vierge*, virgin sulphur;
—— *vif*, v. —— *naturel*.
usofrer, v. a., to sulphur, to dip in brimstone.
soufrière, f., sulphur mine, pit.
souillard, m., sinkhole in a stone, sink stone; (*cons.*) strut brace;
—— *de pont*, ice guard, ice fender.
soulager, v. a., to lighten; to relieve; to alleviate, to assist; to lift;
—— *avec un levier*, to pry;
—— *un canon*, (*art.*) to lift a gun from its bed;
—— *un navire*, (*nav.*) to throw part of the cargo overboard (in a storm);
—— *un plancher, une poutre*, etc., to lessen the load of a floor, a beam, etc.;
se —— *sur une jambe*, to stand first on one, then on the other leg (of a tired horse).

soulèvement, m., rising; upheaval; rising, beginning, of a revolt, rebellion, or insurrection; (*met.*) turning over the mass on the hearth;
dernier ——, (*met.*) last operation in refining pig iron;
exercices de ——, (*mountain art.*) dismounting and carrying drill;
—— *des flots,* rising of the waves, the sea;
—— *de la masse,* (*met.*) first operation in refining pig iron;
premier ——, (*met.*) v. —— *de la masse.*
soulever, v. a., to raise; to stir up, rouse; to excite to revolt, rebellion, etc.;
—— *les flots,* to cause the waters, etc., to rise;
—— *la poussière,* to raise the dust.
soulier, m., shoe;
—— *d'ancre,* shoe of an anchor;
—— *de cuir,* sort of boot worn by a horse in the stable;
—— *à double* (*simple*) *semelle,* double- (single-) soled shoe;
——*s à neige,* snowshoes;
—— *de repos,* (*Fr. a.*) barrack shoe (worn with garters out of doors);
—— *de troupe,* soldier's shoe.
soulignement, m., underlining, underscoring.
souligner, v. a., to underline, underscore.
soumettre, v. a. r., to subdue, subjugate; to overcome, master, bring into subjection; to submit, yield, give way, succumb.
soumission, f., subjection, submission, obedience; (*adm.*) bid for a contract); tender;
cachetée, sealed bid, tender;
faire ——, to surrender, yield.
soumissionnaire, m., (*adm.*) bidder, tendering party.
soumissionner, v. a., (*adm.*) to bid, make a bid; to tender for.
soupape, f., plug; bung, stopper; (*mach.*) valve;
à ——, valved, furnished with valves;
—— *d'admission,* throttle valve; admission valve; inlet valve;
—— *d'admission de la vapeur,* steam-admission valve;
—— *à air,* air valve;
—— *d'alarme,* alarm valve; telltale valve;
—— *alimentaire,* —— *d'alimentation,* feed **valve;** check valve;
appareil de ——, valve gear;
—— *d'arrêt,* check, stop, shut-off valve;
—— *d'arrêt glissante,* sliding stop valve;
—— *d'arrivée,* inlet valve;
—— *à articulation,* hinged or flap valve;
—— *articulée automatique fermée au retour,* nonreturn valve;
—— *d'ascension,* top clack; forcing valve;
—— *d'aspiration,* foot valve; suction valve; inlet valve;
—— *atmosphérique,* air valve; internal, atmospheric, safety valve;
—— *automotrice,* self-acting valve;
—— *auxiliaire,* auxiliary valve;
—— *à bascule,* balance valve;
boîte à ——, valve box;
—— *à boules,* ball valve; spherical valve;
—— *à boulet,* ball valve;
—— *en caoutchouc,* india-rubber valve;
—— *en champignon,* mushroom valve;
—— *à champignon renversé,* cup valve;
—— *en chapeau,* cup valve;
—— *chargée,* loaded valve;
—— *à charnière,* clack valve; flap valve;
chemise de ——, valve jacket;
—— *de choc,* buffer valve;
—— *à clapet,* v. —— *à charnière;*
—— *à un clapet,* single clack valve;
—— *de communication,* communicating **valve;**
—— *composée,* compound valve;
—— *de condensateur,* foot valve;
—— *conique,* conical valve; cone valve; **plug valve;** mitered valve;
—— *de contrôle,* test valve;
—— *à coquilles,* shell valve;
—— *de Cornouailles,* Cornish **valve;**
—— *à coulisse,* sluice valve;
—— *creuse,* long slide valve (obs.);
—— *en D, dé,* D-valve;

soupape *de décharge,* delivery valve, discharge valve;
—— *de dégonflement,* (*balloon*) emptying valve, escape valve;
—— *de détente,* cut-off valve, expansion valve;
—— *de détente à papillon,* throttle expansion valve;
—— *à disque,* disk valve;
—— *de distribution,* distribution valve;
—— *dormante,* fixed valve, suction valve;
—— *double,* double-beat, equilibrium, valve;
—— *à double orifice,* double-ported valve;
—— *à double siège,* double-seat valve;
—— *à eau,* water valve;
—— *d'échappement,* exhaust valve, escape valve; by-pass valve;
—— *d'écluse,* sluice valve;
—— *électrique,* (*elec.*) electric plug;
—— *d'émission,* eduction, outlet, discharge, valve;
enveloppe de ——, valve jacket;
—— *d'épreuve,* test valve;
—— *d'équilibre,* —— *équilibrée,* balanced valve;
—— *d'essai,* test valve;
—— *d'évacuation,* exhaust valve;
—— *à, d', expansion,* cut-off valve, expansion valve;
—— *d'expiration,* forcing valve, snifting valve;
—— *externe,* external safety valve;
—— *d'extraction,* blow-off valve;
—— *d'extraction par le fond,* bottom blow-off valve;
—— *d'extraction à la surface,* brine-, scum-, surface-, blow-off valve;
fausse ——, false valve of a *ballon-gazomètre,* i. e., substitute for a valve;
—— *de fond,* foot valve;
—— *en forme de cloche,* bell-shaped **valve;**
—— *en forme de coupe,* cup valve;
—— *à garniture,* piston valve;
—— *à gaz,* gas valve;
—— *glissante,* slide valve;
—— *de gonflement,* (*balloon*) filling valve, inflating valve;
—— *à gorge,* throttle valve;
—— *à grille,* gridiron valve;
—— *à guide,* spindle valve;
guide de ——, valve spindle;
—— *hydraulique,* hydraulic clack;
—— *d'injection,* injection valve;
—— *d'injection par mélange,* jet injection valve;
—— *interne,* internal valve;
—— *d'inspiration,* suction valve;
—— *d'introduction,* throttle valve, inlet valve;
jeu de ——*s,* set of valves;
—— *de lancement,* (*torp.*) firing valve;
—— *à lanterne,* v. —— *à grille;*
—— *à levée,* lift valve;
—— *à levée limitée,* limited lift valve;
—— *à levée rectiligne,* lift valve;
—— *de mise en marche,* starting valve;
—— *de modérateur,* throttle valve;
—— *de mise en train,* starting valve;
—— *à papillon,* butterfly valve;
—— *de pied,* foot valve of a **pump;**
piston à ——, valve piston;
—— *de piston,* piston valve;
—— *de pompe,* pump valve;
porte de ——, valve door;
—— *principale,* main valve;
—— *de prise d'eau à la mer,* sea valve;
—— *de prise de vapeur,* steam valve, admission valve;
—— *de purgation,* —— *de purge,* —— *à purger,* blow-through valve, snifting valve;
—— *de rechargement,* filling valve;
—— *de réduction,* reducing valve;
—— *de refoulement,* delivery, forcing, valve; outlet valve;
—— *de réglage,* throttle valve;
—— *régulatrice,* regulating valve;
—— *reniflante,* snifting valve;
—— *renversée,* reverse valve;
—— *à ressort,* spring valve;
—— *de retenue,* check valve; foot valve of a pump; regulator valve of a balloon;

soupape, à *siège conique*, conical valve;
siège de ——, valve seat;
—— *de sortie*, eduction valve; outlet valve;
—— *à soulèvement*, lift valve;
—— *sous clef*, lockup valve;
—— *sphérique*, ball valve;
—— *de sûreté*, safety valve;
—— *de sûreté de chaudière*, boiler safety valve;
—— *de sûreté à corps sphérique*, spherical safety valve;
—— *de sûreté du cylindre*, priming valve;
—— *de sûreté externe*, external safety valve;
—— *de sûreté à levier*, lever safety valve;
—— *de sûreté à ressort*, spring safety valve; spring-balance valve;
système de ——*s*, set of valves;
—— *de tête*, head valve;
tige de ——, valve stem;
—— *à tiroir*, slide valve;
—— *tournante*, rotating, rotary, turning, valve;
—— *de trop-plein*, overflow valve, escape valve;
—— *tubulaire*, tubular valve;
—— *à vapeur*, steam valve;
—— *verticale*, hanging valve;
—— *de vidange*, bottom blow-off valve;
vis- ——, screw valve.

soupe, f., soup; (*Fr. a.*) dinner; mess call;
—— *au*, *de, lait*, (*hipp.*) cream-colored;
faire la ——, (*mil.*) to take or make a meal;
préparer la ——, to cook a meal.

soupente, f., loft; jutty; braces of a carriage; straps holding a horse in a brake; support of a winch.

souper, m., supper; (*mil.*) evening stables.

soupeser, v. a., to poise (in the hand); to weigh by hand.

soupirail, m., air-hole, vent, vent hole; air-hole in ice; (*art.*) telltale channel (to give warning if A tube split);

souple, a., supple, pliant, flexible, limber, tough.

souplesse, f., suppleness, flexibility.

soupoutre, v. *sous-poutre*.

souquenille, f., stable coat, stable frock.

souquer, v. a., (*cord.*) to haul or heave taut, tighten, heave close, any lashing or seizing;
—— *sur les rames*, to pull away with a will.

source, f., spring (of water), fountain; source of a stream;
—— *chaude*, hot spring;
couler de ——, to flow naturally;
—— *d'eau*, spring;
—— *intermittente*, intermittent spring;
—— *du vent*, point whence the wind blows.

sourd, a., deaf; dark, dull; hollow, rumbling (of a sound); absorbing or stifling sounds (as a partition wall); secret, underhand;
lanterne —— *e*, dark lantern;
lime —— *e*, dead file; dead smooth file;
—— *-muet*, deaf mute.

sourdine, f., (*mus.*) sourdine, damper;
à la ——, quietly, softly.

souricière, f., mouse trap; (*mil. min.*) trap, box trap (obs.).

souris, f., mouse; (*min.*) trap, box trap; (*hipp.*) gristle in a horse's nose; mouse color;
cheval ——, mouse-colored horse;
pas-de- ——, (*fort.*) pas de souris;
—— *porte-feu*, (*mil. min.*) mouse by which a mine is fired.

sous, prep., under, beneath; sub-, deputy;
—— *les armes*, (*mil.*) under arms; fighting, at war;
—— *le canon de*, (*mil.*) under the guns of (i. e., under the protection of);
—— *caution*, (*adm.*) under bond;
cheval —— *lui*, v. s. v. *cheval;* horse whose four legs are too much under him;
cheval —— *la main* (*du cocher*), off horse;
cheval —— *poil gris* (*noir*), (*hipp.*) gray (black) horse;
—— *convoi*, (*nav.*) under convoy;
—— *les drapeaux*, (*mil.*) with the colors; in the service (of);
—— *le feu de*, (*mil.*) exposed to the fire of;
—— *pavillon anglais, etc.*, (*nav.*) flying the English, etc., flag;
—— *peine de mort, de la vie*, under penalty of death;
—— *le vent*, (*nav.*) under the lee.

sous-aide, m., under assistant.
sous-arbalétrier, m., (*cons.*) under rafter of a roof.
sous-arrondissement, m., subdistrict, under *arrondissement*.
sous-bandage, m., felly.
sous-bande, f., (*art.*) trunnion bedplate;
—— *d'essieu*, (*art.*) axletree coupling band.
sous-barbe, f., cross piece of a sluice; piece under the tenon of a hammer helve; (*hipp.*) hinder part of a horse's lower jaw; (*harn.*) chin strap of a halter.
sous-berme, m., v. *souberme*.
sous-bief, m., channel connecting two sluices.
sous-bois, m., underwood, underbrush.
sous-boucle, f., (*harn.*) safe.
sous-brigadier, m., (*cav.*) lance-corporal.
sous-cap, m., under foreman, under overseer.
sous-chargé, a., (*mil. min.*) undercharged.
sous-chef, m., under foreman; assistant;
—— *artificier*, (*Fr. art.*) v. s. v. *artificier*;
—— *d'état-major*, (*Fr. a.*) immediate subordinate of a *chef d'état-major*, having charge of a bureau or service of the staff;
—— *de musique*, (*mil.*) principal musician (U. S. A.);
—— *ouvrier*, under master workman.
sous-chevron, m., (*cons.*) under rafter.
sous-collet, m., last hoop of a cask.
sous-comité, m., subcommittee.
sous-commissaire, m., v. s. v. *commissaire*.
sous-commission, f., assistant committee, sub-committee;
—— *de ligne*, (*Fr. mil. r. r.*) subcommittee of the *commission de ligne*, q. v.
sous-contre heurtoir, m., (*art.*) plate lining the under part of the cheek of a field carriage, opposite to the garnish plate.
souscription, f., subscription; share list, order.
souscrire, v. a. n., to subscribe; to take shares in, put one's name down for.
sous-délégataire, m., under deputy; under *délégataire*, q. v.
sous-directeur, m., v. s. v. *directeur*.
sous-direction, f., subdirection, subordinate direction.
sous-écuyer, m., under riding master.
sous-égalisage, m., (*powd.*) fine separation by sieve, separation of grain from dust by third sieve.
sous-égaliser, v. a., (*powd.*) to fine sieve, to pass grain through the third sieve.
sous-égalisoir, m., (*powd.*) fine sieve, third sieve.
sous-embranchement, m., (*r. r.*) branch of a branch road.
sous-entrepreneur, m., subcontractor.
sous-étage, m., generic term for all undergrowth of a forest.
sous-faîte, m., (*cons.*) under roof timber, ridge board or pole.
sous-fondation, f., under pavement; subpavement.
sous-fondeur, m., caster of small articles.
sous-fréter, v. a., to underlet, underfreight.
sous-futaie, f., undergrowth (of a forest).
sous-garde, f., (*sm. a.*) trigger guard, trigger and plate.
sous-gorge, f., (*sm. a.*) throat, neck, of hammer (obs.); (*harn.*) throatlatch.
sous-gouverneur, m., under governor, lieutenant-governor.
sous-gueule, f., (*harn.*) strap end under a horse's mouth.
sous-horizontal, a., (*mil. min.*) under the center of the charge (of effects).
sous-intendance, f., subintendance; (*Fr. a.*) functions, etc., of a *sous-intendant*.
sous-intendant, m., v. s. v. *intendant*.
sous-lieutenance, f., (*mil.*) sublieutenancy.
sous-lieutenant, m., (*mil.*) sublieutenant.

sous-longeron, m., (*cons.*) under stringer; (*pont.*) bolster (under side rail).
sous-main, m., pad, scratch pad.
sous-marin, a., submarine; submerged;
guerre ——*e*, (*nav.*) submarine warfare;
navigation ——*e*, submarine navigation;
navire ——, submarine boat.
sous-mentonnière, f., (*unif.*) chin strap.
sous-œuvre, m., v. s. v. *œuvre*, m.
sous-off, m., (*mil. slang*) "noncom."
sous-officier, m., n. c. o.; (*Fr. a.*, does not include corporal); (*nav.*) petty officer;
—— *chef de pièce*, n. c. o., chief of piece;
—— *comptable*, n. c. o. having money or property responsibility;
—— *de magasin*, n. c. o., storekeeper;
—— *de tir*, noncommissioned range officer, in charge on range.
sous-ordre, m., subordinate; (*law*) garnishee order; (*mil.*) subaltern officer; any officer under the orders of another (may be a corps commander);
en ——, in a subordinate capacity;
officier ——, (*mil.*) subaltern officer, company officer;
officiers en ——, (*mil.*) subordinate officers in general.
sous-pied, m., trowser strap; spur strap; gaiter strap;
—— *de dragon*, (*cav. slang*) foot soldier.
sous-poutre, f., (*cons.*) under girder, corbel piece, bolster.
sous-préfecture, f., district, functions, term, residence, office, city of residence, of a subprefect.
sous-préfet, m., subprefect.
sous-pression, f., underpressure, pressure from below.
sous-secrétaire, m., undersecretary;
—— *d' État*, assistant or under secretary of state.
sous-seing, m., private document (i. e., not drawn up before a public officer); postal franchise.
soussigné, p. p., undersigned;
je, ——, I, the undersigned.
soussigner, v. a., to sign, undersign.
sous-sol, m., subsoil; underground room or apartment;
—— *de chemin de fer*, (*r. r.*) roadbed.
sous-sombrer, v. n., (*nav.*) to sink to the bottom; to founder.
sous-tendeur, m., (*cons.*) tie-rod (metallic truss).
sous-tendre, v. a., to subtend.
sous-traitant, m., subcontractor.
sous-traité, m., subcontract.
sous-traiter, v. a., to make a subcontract.
sous-unité, f., (*mil.*, etc.) subunit.
sous-ventrière, f., (*harn.*) bellyband, surcingle; saddle girth.
sous-verge, m., (*art.*) off horse;
—— *de derrière*, (*art.*) off wheeler;
—— *de devant*, (*art.*) off leader;
être en ——, (*mil. slang*) to be second in command;
limonier ——, v. s. v. *limonier*;
—— *de milieu*, off swing horse.
sous-vétérinaire, m., underveterinarian.
soutache, (*unif.*) embroidery of braid on a uniform.
soutacher, v. a., (*unif.*) to braid.
soute, f., (*nav.*) storeroom, locker, tank;
—— *aux câbles*, cable locker;
—— *à cartouches*, cartridge locker (rapid-fire guns);
—— *à charbon*, coal bunker;
—— *à charbon de côté*, side coal bunker;
—— *à eau*, water-ballast compartment;
—— *à eau et à charbon*, (*r. r.*) water and coal space of a locomotive tender;
—— *à munitions*, filling room;
—— *à obus*, shell room;
—— *à pain*, bread locker;
—— *à poudre*, ——, *aux poudres*, powder magazine;
—— *à projectiles*, projectile magazine;
—— *aux torpilles*, torpedo room;
—— *vitrée*, filling room.

soutènement, m., (*cons.*) support, prop; timbering and walling (of mines, etc.);
mur de ——, retaining wall.
soutenir, v. a., to support, sustain, bear, hold up (a weight, etc.); to stay, prop; to uphold, maintain, keep from falling, keep up (credit, a government, etc.); to sustain (a reputation); to endure, bear, stand, hold out, resist; to back, second, assist; (*mil.*) to support, resist;
—— *un assaut*, (*mil.*) to support an assault;
—— *un bâtiment*, (*cons.*) to prop up a building;
—— *la chasse*, (*nav.*) to maintain a running fight; to follow an enemy closely;
—— *un cheval*, (*man.*) to hold up a horse (going downhill); to keep up, assist, bear up, a horse; to keep a horse at his gait;
—— *un combat*, (*mil.*, *nav.*) to keep up a fight against superior forces;
—— *la guerre*, (*mil.*) to wage war;
—— *un pas*, (*man.*) to keep up a gait;
—— *un siège*, (*mil.*) to stand a siege, to hold out.
soutenu, p. p., (*man.*) (of gaits) well marked, regular, cadenced.
souterrain, a., subterranean, underground;
guerre ——*e*, v. s. v. *guerre*;
nappe d'eau ——*e*, v. s. v. *nappe*.
souterrain, m., underground room or passage; (*r. r.*) tunnel; (*mil. min.*) gallery of a mine; (*fort.*) bombproof; underground portion of a fort;
—— *de caserne*, (*fort.*) casemate barrack;
passage ——, (*min.*) drift, driftway;
puits ——, tunnel shaft, pit.
soutien, m., support; point of support; abutment; (*man.*) lifting of the foot; (*mil.*) support: (esp. *inf.*) the support, i. e., one of the elements of the *ligne de combat* or *ligne de feu*;
compagnie de ——, (*mil.*) company in support;
escadron de ——, (*cav.*) supporting squadron.
soutier, m., (*nav.*) coal trimmer.
soutirage, m., (*met.*) drawing off; (*elec.*) tapping of electricity.
soutirer, v. a., (*met.*) to draw off; (*elec.*) to tap.
souverain, a., m., sovereign.
souveraineté, f., sovereignty.
soyer, v. a., to crease, seam (tin plate, etc.).
spadelle, f., (*met.*) sort of stirring spade.
spahi, f., (*Fr. a.*) spahi (Algerian cavalry, four regiments; the *cadres* are one-half French and one-half native up to lieutenant, inclusive).
spalme, m., (*nav.*) paying stuff.
spalmer, v. a., (*nav.*) to pay a ship's bottom; to pitch, bream, careen.
spalt, m., (*met.*) spalt.
sparte, m., (*cord.*) esparto grass.
sparterie, f., articles made of esparto grass.
spath, m., spar, Iceland spar;
—— *fluor*, fluorspar;
—— *d'Islande*, Iceland spar.
spatule, f., spatula; any spatula-shaped object, e. g., a handle so shaped;
—— *curette*, v. s. v. *curette*;
—— *à étouper*, (*art.*) implement for packing tow around ammunition.
spécification, f., specification.
spécifique, a., specific;
chaleur ——, specific heat;
pesanteur, *poids*, ——, specific weight;
poids —— *relatif d'un corps*, ratio of the specific weight of a body to that of water at 4° C.;
population ——, population per unit of area.
spectral, a., spectral;
analyse ——*e*, spectrum analysis.
spectre, m., spectrum; solar spectrum;
—— *d'absorption*, absorption spectrum;
—— *continu*, continuous spectrum;
—— *de diffraction*, diffraction spectrum;
—— *discontinu*, discontinuous spectrum;
—— *d'émission*, emission spectrum;
—— *magnétique*, magnetic curve; magnetic figure;
—— *prismatique*, prismatic spectrum;
—— *solaire*, solar spectrum;

spectre, *stellaire*, stellar spectrum.
spectromètre, m., (*inst.*) spectrometer.
spectrométrie, f., spectrometry.
spéculaire, a., specular;
—— *fer* ——, specular iron.
speiss, m., (*met.*) speiss.
sphère, f., sphere;
—— *d'activité*, (*mil. min.*) sphere of effect, action;
—— *de broyage*, (*mil. min.*) sphere immediately around the center of disturbance;
—— *de commotion*, (*mil. min.*) v. —— *d'activité;*
—— *de compression*, (*mil. min.*) sphere of compression;
—— *d'éclatement*, v. —— *de broyage;*
—— *d'explosion*, (*mil. min.*) sphere of effect, of action;
—— *de friabilité*, v. —— *de broyage;*
—— *de rupture*, sphere of rupture;
—— *de séparation*, (*mil. min.*) surface along which the disturbance causes separation from the surrounding material;
—— *de volée*, (*art.*) spherical muzzle pivot of a muzzle-pivoted gun.
sphéricité, f., sphericity, curvature;
—— *de la terre*, curvature of the earth.
sphérique, a., spherical.
sphéroïdal, a., spheroidal;
état —— *de l'eau*, (*steam*) spheroidal state of water.
sphéroïde, m., spheroid.
sphéromètre, m., (*inst.*) spherometer.
sphérosidérite, m., (*met.*) spherosiderite.
spickel, m., (*Polytechnique slang*) dress sword.
spiral, a., spiral;
ressort ——, spiral spring.
spiral, m., spiral, spiral spring;
—— *réglant*, regulating spring of a watch.
spirale, f., spiral line; volution; spiral spring;
aller en ——, to wind;
—— *de chauffage*, heating coil;
en ——, spirally;
enroulement en ——, volution; spiral winding;
enrouler en ——, to wind spirally;
se rouler en ——, to roll up or around in a spiral; to roll round.
spire, f., helix, screw, single turn of a screw or helix; spiral, spiral curve; (*elec.*) single turn, coil, of wire;
tour de ——, single turn of a helix.
spiritueux, a., alcoholic.
spongieux, a., spongy.
squelette, m., skeleton; (*nav.*) frame of a ship.
stabilité, stability, steadiness; durability; (*nav.*) stiffness of a ship.
stablat, m., stable, pen for animals.
stable, a., stable, steady, durable, solid; steadfast, permanent, lasting; fast, fixed, settled.
stabulation, f., (of horses, etc.) continuous living in stables.
stabuler, v. a., to keep in stables, under shelter (horses, etc.).
stadia, f., (*inst.*) stadia rod, stadia;
levé à la ——, stadia survey;
lunette à ——, stadia telescope;
—— *à réticule*, reticulated stadia.
stadimètre, m., (*inst.*) horizontal stadia rod.
stadimétrique, a., of or pertaining to a stadia;
lunette ——, stadia telescope.
stadiomètre, m., (*inst.*) stadiometer.
stage, m., term, probation; (*mil.*) course (at a school); tour of probationary duty; (*Fr. a.*) period of instruction for territorial army;
faire un ——, (*mil., etc.*) to take a course, a term; to serve a tour of probationary duty (e. g., on the staff).
stagiaire, a., under instruction; passing a term; (*mil.*) serving a probationary tour; m., (*mil.*) officer taking a probationary tour (e. g., on the staff, for the purpose of learning staff duty);
—— *du génie*, (*Fr. a.*) n. c. o. attached to staff of engineers and taking a course before becoming *adjoint* (will expire by extinction).

stagiaire, *officier* ——, (*mil.*) probationary officer (v. *supra*).
stagnation, f., stagnation;
—— *du compas*, slowness, torpidity, of the compass.
stalle, f., stall; partition between stalls in a stable.
stand, m., (*t. p.*) shooting stand, shooting gallery; gallery practice range; rifle range (rare).
starie, f., (*nav.*) time allowed for loading or unloading cargo;
jours de ——, lay days.
staticité, f., nonsensitiveness.
station, f., standing, stand, station; stop, stay, stoppage; (*nav.*) cruising ground, station; (*mil.*) post, station; (*r. r.*) railway station; (*hipp.*) attitude, manner of standing (horse); (*surv.*) station;
—— *abandonnée*, (*hipp.*) manner of standing of a horse left to himself, free attitude (body supported on three legs);
—— *-abri*, (*t. p.*) firing point of a shooting stand;
—— *pour l'alimentation de l'eau*, (*r. r.*) watering station;
—— *d'arrivée*, (*r. r.*) point of disembarkation;
brûler une ——, (*r. r.*) to run past a station without stopping;
—— *campée*, (*hipp.*) "stretched-out" attitude (fore and rear bipeds splayed out and apart more than three-fourths of the height);
—— *centrale*, (*elec.*) power house;
—— *charbonnière*, (*nav.*) coaling station, dump;
chef de ——, (*r. r.*) station master;
cuirassé de ——, v. s. v. *cuirassé;*
—— *conjuguée*, (*top.*) auxiliary or secondary station trigonometrically connected with its principal (both of them together being called *conjuguées*);
—— *de la côte*, (*nav.*) home station;
—— *de départ*, (*r. r.*) station of embarkation, of departure;
—— *de deuxième ordre*, (*surv.*) secondary station;
en ——, on a station; (*mil.*) said of troops when in cantonments, etc., or when for any reason they occupy the same place for a considerable time; in garrison;
—— *à l'étranger*, (*nav.*) foreign station;
être en ——, (*nav.*) to be on station;
—— *extrême*, (*r. r.*) terminus;
faire la mise en ——, (*surv.*) to set up an instrument;
—— *forcée*, (*hipp.*) forced attitude or manner of standing (the body is supported by all the legs);
—— *halte-repas*, (*Fr. a.*) railroad station in which meals are furnished soldiers in transit by the *administration militaire;*
—— *haute*, (*hipp.*) free station, but on four legs (attentive attitude);
—— *de jonction*, (*r. r.*) railway junction;
—— *intermédiaire*, (*r. r.*) way station;
—— *libre*, (*hipp.*) attitude of a horse standing free, free attitude;
—— *-magasin*, v. s. v. *magasin;*
mettre en ——, (*surv.*) to set up an instrument;
mise en ——, (*surv.*) operation of setting up an instrument;
—— *de partage*, (*mach.*) distributing station, power house (wire transmission);
—— *de passage*, v. —— *intermédiaire;*
—— *de pilote*, (*nav.*) pilot station;
—— *placée*, (*hipp.*) attitude of the horse standing uniformly on all four legs, the fore and hind bipeds being apart three-fourths the height of the horse;
—— *point de départ d'étapes*, v. *gare de rassemblement;*
—— *de premier ordre*, (*top.*) primary station or triangulation point;
—— *quadrupédale*, v. —— *forcée;*
—— *rassemblée*, (*hipp.*) gathered position, i. e., fore and hind bipeds brought under and apart less than three-fourths the height of the horse;

station *de rassemblement*, v. *gare de rassemblement;*
—— *de rebroussement*, v. —— *de tête;*
—— *régulière*, v. —— *placée;*
—— *de repos*, v. *libre;*
—— *de sauvetage*, life-saving station;
—— *de second ordre*, (*surv.*) secondary station or triangulation point;
—— *de signaux*, signal station; (*nav.*) shore signal station;
—— *télégraphique*, —— *téléphonique*, telegraph, telphone, station;
—— *-terminus*, (*r. r.*) terminus;
—— *de tête*, (*r. r.*) terminal station (either beginning or end of the line);
—— *tête d'étapes de guerre*, v. s. v. *étape;*
—— *tête d'étapes de route*, v. s. v. *étape;*
—— *de transition*, (*Fr. mil. r. r.*) railway station beyond which cease the functions of the *commission de réseau*, and separating the *zone de l'intérieur* from the *zone des opérations en temps de guerre*, qq. v.;
—— *de troisième ordre*, (*surv.*) triangulation station of the third order.

stationnaire, a., stationary; m., foreman of a telegraph office; (*nav.*) guard ship; any cruising ship; *machine* ——, stationary engine.

stationnement, m., state of remaining in the same place; stopping of carriages; stand, standing; (*mil.*) halt (not merely a cessation of marching, but halt as for the night or for any other purpose requiring a considerable time, with connotation of the consequent duties); the station or position itself; stop, bivouac, cantonment;
—— *en feu*, (of a locomotive) stopping with steam up.

stationner, v. n., to stop, stand; (*nav.*) to be on station; (*mil.*) to stand (of troops), i. e., to stand up, as when waiting for the command to march.

statique, f., statics;
—— *graphique*, graphical statics.

statuer, v. n., to decide, settle, rule, pass upon and decide a case.

statut, m., law, regulation, statute; by-law.

stéganographe, m., cryptographer.

stéganographie, f., cryptography.

sténographe, m., stenographer.

sténographie, f., stenography.

steppeur, m., (*man.*) high stepper.

stérage, m., measuring of wood (evaluation in *stères*).

stère, m., cubic meter.

stéréogramme, m., stereogram.

stéréographe, m., stereographer; (*inst.*) instrument for the rapid determination of an area.

stéréographie, f., stereography.

stéréométrie, f., stereometry.

stéréoscope, m., stereoscope.

stéréotomie, f., stereotomy.

stéréotypage, m., stereotyping, stereotype printing.

stéréotype, a., stereotype.

stéréotyper, v. a., to stereotype.

stéréotypie, f., stereotypy.

sthénallatique, a., (*inst.*) giving distances reduced to the horizon; sthenallatic;
instrument, lunette, etc., ——, instrument, telescope, etc., furnishing directly distances reduced to the horizon.

stipendiaire, a., hired, stipendiary.

stipendier, v. a., to hire; to keep or retain in pay (said in a bad sense).

stock, m., stock (supplies); stock (shares); stock (of anvil under tilt-hammer).

stopper, v. a., (*mach.*) to stop (an engine); (*nav.*) (of a ship or torpedo) to stop, halt; to stop a cable, rope;
—— *et marcher en arrière*, to stop and reverse.

stoppeur, m., stopper (of a cable, etc.); stopping mechanism (of an engine, etc.).

stoqueur, m., stoker, fire rake.

store, m., window curtain, blind; spring roller.

strapontin, m., bunk; thin, narrow, mattress; hammock used in hot countries; bracket seat of a carriage; (*mil. r. r.*) hanging seat (in a horse car).

stratagème, m., stratagem;
—— *de guerre*, (*mil.*) stratagem of war.

stratégie, f., (*mil.*) strategy.

stratégique, a., (*mil.*) strategic.

stratégiste, m., (*mil.*) strategist; writer on strategy.

stratocratie, f., military government.

stratographie, f., (*mil.*) description of an army.

stratonomie, f., ensemble of laws governing an army.

stratopédie, f., study of arts, etc., useful in an army.

stribord, m., (*nav.*) starboard (unusual, if not obs., for *tribord*).

striction, f., "choke" of a test bar under tension.

stricture, f., (*fort.*) neck (of an embrasure).

strie, f., stria.

strié, a., fluted, striated, corrugated.

strier, v. a., to corrugate, to ridge.

strille, m., mason's double-pointed hammer.

strosse, m., stone in the sole of a mine; stope; (in a tunnel) bench;
exploiter une mine en ——*s*, to stope.

structure, f., structure (in general), construction; grain of a stone.

stuc, m., stucco.

stucateur, m., workman in stucco, stuccoer.

style, m., style;
—— *diviseur*, dividing-point, tracer.

styler, v. a., to train; to bring up; to accustom; (*man.*) to exercise, train, a horse in certain motions.

stylet, m., stiletto.

suage, m., sweating of wood; blacksmith's swage; creasing tool, creasing hammer; tinsmith's anvil; (*nav.*) paying a ship's bottom; paying stuff.

suager, v. a., to crease (tin, etc.); (*nav.*) to pay a ship's bottom.

suant, a., (*met.*) weldable, welding;
blanc ——, welding heat;
chaleur ——*e*, v. *blanc* ——.

subalpin, a., at the foot of the Alps.

subalterne, a., inferior, subordinate, subaltern; m., subaltern;
officier ——, (*mil.*) subaltern officer, company officer.

subdélégation, f., subdelegation.

subdélégué, m., subdelegate.

subdéléguer, v. a., to subdelegate.

subdivision, f., subdivision;
—— *de région*, (*Fr. a.*) subdivision of territory of *région de corps d'armée*, each of which constitutes a recruiting district for a *régiment subdivisionnaire*, q. v.

subdivisionnaire, a., subdivisional;
régiment ——, v. s. v. *régiment.*

subir, v. a., to suffer, undergo, endure;
—— (*un*) *examen*, to take an examination;
—— *un interrogatoire*, (*law*) to answer the questions of a judge;
—— *son jugement*, (*law*) to serve one's sentence.

subjugation, f., subjugation, conquest.

subjuguer, v. a., to conquer, quell, put down, master, subdue, bring into subjection.

sublimation, f., (*chem.*) sublimation.

sublimé, m., (*chem.*) sublimate.

sublimer, v. a., (*chem.*) to sublimate.

submarin, a., submarine.

submerger, v. a., to submerge, drown, flood, swamp.

submersible, a., submersible.

submersion, f., sinking.

subordination, f., (*mil., etc.*) subordination, order.

subordonné, m., subordinate.

subordonner, v. a., to subordinate; to render or make subordinate.

subrécargue, m., (*nav.*) supercargo.

subside, m., subsidy.

subsistance, f., subsistence; (*mil., in pl.*) provisions and forage;

subsistance, *mettre en* ——, (*mil.*) to assign a man to a unit not his own, for pay and subsistence;
—— *s militaires,* (*mil.*) food, forage, illuminating and heating, supplies, etc.;
mise en ——, (*mil.*) assignment of a man to a unit not his own, for pay and subsistence.

subsistant, m., (*mil.*) man temporarily subsisted and paid in a unit not his own (said of *militaire isolé*).

subsister, v. a., to subsist;
faire —— *une armée,* (*mil.*) to subsist an army.

substance, f., substance; matter;
—— *isolante,* (*elec.*) insulating material;
—— *s rebelles,* (*met.*) refractory ores.

substituer, v. a., to substitute.

substitut, m., substitute; deputy.

substitution, f., substitution; (esp. *Fr. a.*) the substitution of one sort of forage for another.

substruction, f., substruction; underpinning; foundation.

subvention, f., relief; subsidy; grant; supply, aid.

subventionner, v. a., to subsidize; to make a grant, a subsidy.

subversion, f., subversion, ruin, overthrow.

subverter, v. a., to overthrow, ruin, subvert.

succomber, v. n., to fail, yield, succumb, be worsted; to sink beneath; to be overthrown.

succursale, f., branch establishment.

sucre, m., sugar;
casser du —— *à deux sous le mètre cube,* (*mil. slang*) to be in punishment companies, breaking stone.

sud, a., south, southerly; m., the south; south wind;
—— *magnétique,* magnetic south;
—— *vrai,* true south.

sudiste, m., Confederate, 1861-1865.

sud-ouest, a., southwest; m., the southwest; southwest wind, southwester; southwester (head gear).

suée, f., (*hipp.*) sweating, sweat.

suer, v. a., to sweat; to perspire;
—— *blanc,* (*hipp.*) to sweat foam; to sweat foamy; to foam;
—— *le fer,* (*met.*) to weld completely.

sueur, f., sweat.

suie, f., soot;
—— *de bois,* wood soot.

suif, m., tallow; suet; (*art.*) tallow disk; (*nav.*) mixture of soap, tallow, brimstone, etc., for paying a ship's bottom;
boîte à ——, tallow box;
bouilloire à ——, tallow kettle;
coupe à ——, v. *godet à* ——;
cuir en ——, v. s. v. *cuir;*
donner un ——, to grease;
godet à ——, tallow cup; grease cup;
goutte de ——, round-head screw, button-head screw;
mettre en ——, to tallow;
—— *minéral,* a variety of talc.

suiffer, v. a., to tallow; (*nav.*) to pay with *suif,* q. v.

suintement, m., oozing, running, exudation (running, leaking, of seams, boilers, etc.); (*expl.*) exudation, leakage, of acids;
—— *de la fourchette,* (*hipp.*) thrush, running thrush.

suinter, v. n., to leak, filter, drip through; to run out, ooze; (*expl.*) to exude, leak (as acids).

suite, f., retinue, train; sequel, continuance, succession, series, set, consequence;
à la ——, (*mil.*) supernumerary;
à la —— *de l'armée,* (*mil.*) of a general officer, in excess of the statutory number;
à la —— *d'un régiment,* (*mil.*) of an officer, attached to a regiment as supernumerary of his grade;
être mis à la ——, (*mil.*) to be discharged with a pension;
officier à la ——, v. s. v. *officier.*

suivage, m., tallowing.

suivant, m., follower in a flight of pigeons.

suiver, v. a., v. *suiffer.*

suivre, v. a., to follow, pursue, go after;
—— *un cours,* to take a course of study (in a college, etc.);
—— *la piste,* (*man.*) to go close to the wall of a riding hall;
—— *le poing de la bride,* (*man.*) to obey the hand readily.

sujet, a., subject to, amenable, liable; prone, given to, addicted to.

sujet, m., subject, topic, matter; subject (of a king);
bon ——, steady, well-behaved man.

sujétion, f., subjection; inconvenience;
pierre de ——, v. s. v. *pierre.*

sulfatisation, f., impregnation, injection, with sulphate of copper.

sulfatiser, v. a., to inject, impregnate, with sulphate of copper.

sulfoglycérique, a., (*chem., expl.*) glycerolsulphuric.

sulfonitrique, a., (*chem., expl.*) sulphonitric.

sulfurite, (*expl.*) sulphurite.

sultan, m., sultan.

sultane, f., sultana.

super, v. a. n., (of a pump) to suck up the packing; (of tubes, etc.) to be stopped, plugged, up.

superficie, f., superficies; surface, area;
à la ——, superficially;
de ——, superficial, in area.

superfin, a., superfine; (of a file) dead smooth.

supérieur, a., superior; upper (of rivers, of countries, e. g., the upper Rhine, etc.);
cour —— *e,* higher court; supreme court;
officier ——, (*mil.*) field officer;
tribunal ——, v. *cour* ——.

superposer, v. a., to superpose.

superposition, f., superposition; (*sm. a.*) overlapping and welding the edges of a skelp.

superstruction, f., (*cons.*) construction above ground.

superstructure, f., superstructure; superstructure of a bridge; (*r. r.*) superstructure (ballast and rails, track and supports).

suppléance, f., substitution; functions of a substitute.

suppléant, a., m., deputy, substitute, assistant.

suppléer, v. a. n., to supply, make good; to furnish, fill up, supply the place of, do duty for, serve instead.

supplément, m., supplement, addition; additional pay; gratuity, bounty; allowance;
en ——, extra, additional;
officier en ——, (*mil.*) supernumerary officer;
—— *de solde,* extra pay or allowance.

supplémentaire, a., supplementary, additional, extra, supernumerary.

supplice, m., punishment, corporal punishment;
le dernier ——, capital punishment.

supplicier, v. a., to execute, inflict capital punishment on.

support, m., support, prop, stay, rest, pillar, column, standard, stanchion, strut, bearing, block, holder, projecting frame, stand, rack; fulcrum, stand; (*fig.*) assistance, help, protection; (*art.*) headstock; mount (rapid-firing gun); (*pont.*) beam supporting balks, etc.; (*mach.*) plumber block; overhanging arm; bracket; (*elec.*) incandescent-lamp socket;
—— *d'alimentation,* (*r. f. art.*) feed box;
—— *d'arbre porte-fraise,* (*mach.*) overhanging arm (as of a milling machine);
—— -*arrière,* (*art.*) rear support (of a star gauge);
—— -*avant (à T),* (*art.*) front or T rest (of a star gauge);
—— *à baïonnette,* (*elec.*) bayonet-joint socket (for an incandescent lamp);
—— *de balancier,* crosshead bracket (of a steam windlass);
—— *de bicyclette,* bicycle stand;
—— *de chargement,* (*art.*) breech support (French 95mm) for fire under great angles;
—— *à chariot,* (*mach.*) slide, slide rest (of a lathe);

support(s) *de la chaudière,* (*steam*) stay bars; girder stays; stays, brackets, of the boiler; knee of the boiler;
— *—conducteur,* (*art.*) guiding support (of a star gauge);
— *continu,* continuous bearing;
— *cornier,* (*teleg.*) angle bracket;
corps de ——, (*pont.*) generic term for boats, trestles, etc., supporting the roadway;
— *à coulisse,* v. —— *à chariot;*
— *de cylindre,* (*mach.*) cylinder column;
— *de devant du haquet,* (*pont.*) upper crosspiece of a pontoon limber;
— *d'essieu porte-roue,* (*art.*) crossbar of the spare-wheel axle;
— *d'étoile mobile,* (*art.*) star-gauge rest;
— *fixe,* (*mach.*) fixed rest; fixed support (of a lathe); (*pont.*) any fixed support (e. g., a trestle);
— *flottant,* (*pont.*) floating pier, any floating support;
— *à fourche,* any forked support, Y;
— *de fronteau de mire,* (*art.*) fore sight, patch, mass;
— *des glissières,* (*mach.*) slide-bar bracket, guide-bar bracket;
— *de(s) grille(s),* (*mach.*) fire-bar support, bearer;
— *de hampe,* (*art.*) staff rest (of a star gauge);
— *de haquet,* (*pont.*) crossbar of a pontoon wagon;
— *isolant,* (*teleg.*) insulator;
— *libre,* loose support;
— *lunette,* (*mach.*) arbor bearing;
— *mobile,* standard;
— *d'un palier,* (*mach.*) plumber-block bearer;
— *pendant,* (*mach.*) hanger;
— *en pierre,* (*r. r.*) stone sleeper, block;
— *à pivot,* pivoting support;
planche de ——, stand;
plaque de ——, stand;
— *du poids,* weigh bar;
— *de pointage,* (*art.*) elevating-gear support; front-sight support; sight rest;
— *de pont roulant,* (*pont.*) crosspiece of a carriage bridge upon which balks are laid;
— *de porte corps,* center transom;
— *des rails,* (*r. r.*) sleeper;
— *à rainures,* (*mach.*) slotted table (to hold the work);
— *de route,* (*art.*) traveling breech support (French 95mm);
— *de tente,* awning boom;
— *de timon,* pole yoke;
— *au tour,* (*mach.*) lathe rest;
— *de tourillon,* (*art.*) traveling trunnion; trunnion bed;
— *à vis,* (*teleg.*) screw bracket;
— *de vis de pointage,* (*art.*) elevating-screw nut.
supporter, v., to support, prop, stay, bear, bear up; to sustain (as attack, etc.); to endure;
— *le feu,* to be fireproof; to stand, resist, fire (as bricks, etc.).
suppression, f., abolishment, suppression, retrenchment, withdrawal;
— *d'emploi,* suppression of an office, of a place.
suprématie, f., supremacy;
— *navale,* sea power, naval supremacy.
sur, a., sour.
sûr, a., sure; steady; safe (of a port);
avoir le coup d'œil ——, to have an accurate eye;
avoir la jambe ——*e, le pied* ——, (*man.*) to be sure-footed.
suraffiné, a., v. *dépassé.*
suranner, v.a., to be more than one year old (passports, etc.).
sur-arbitre, m., umpire deciding a tie between other umpires.
surbaissé, p. p., elliptic, depressed, surbased.
surbaissement, m., surbasement.
surbaisser, v. a., to make elliptic; to surbase.
surbande, f., (*sm. a.*) distance that the hammer may be drawn beyond full cock; (*art.*) cap square.

surbau, m., (*nav.*) frame, framing piece of a hatchway, head ledge.
surbout, m., any piece of wood turning on a pivot.
surcalciné, a., overburned.
sur-cémenté, p. p., face-hardened (armor).
sur-cémenter, v. a., to face-harden an armor plate.
surcharge, f., extra, additional, load; baggage excess; in races, handicap (weight); top pack (pack mules); top load; word(s) or figure(s) written over another or others; (*mas.*) load of an arch; increased thickness of a coat of plaster; increased height of a wall; (*art.*) overcharge (in testing); (*mach.*) overload (of valves);
— *d'épreuve des chaudières,* (*steam*) boiler testing by increased pressure.
surcharger, v. a., to overload; to overburden, oppress, overcharge, overtask, weigh down; to write in a word, etc., over another; (*mil. min.*) to overload a mine.
surchauffage, m., overheating, superheating (as of steam).
surchauffe, f., (*steam*) superheating.
surchauffer, v. a., to overheat; (*met.*) to burn; (*steam*) to superheat.
surchauffeur, m., (*steam*) superheater;
surface de ——*s,* superheating surface.
surchauffure, f., (*met.*) burnt spot (in iron or steel); (operation of) overheating metal.
sur-col, m., (*harn.*) wither strap.
sur-cou, m., v. *sur-col.*
surcuisson, f., overburning of lime.
surdent, m., (*hipp.*) supernumerary tooth; wolf's tooth, supplementary premolar.
surdi-mutité, f., deaf-muteness.
surdos, m., porter's pad; (*harn.*) back band, back strap.
sureau, m., elder tree.
surécartement, m., (*r. r.*) increase of gauge.
surégalisage, m., (*powd.*) passage of grains through the upper (coarse) separating sieve.
surégaliser, v. a., (*powd.*) to pass grains through the upper (coarse) separating sieve.
surégalisoir, m., (*powd.*) coarse sieve, superior or upper separating sieve.
surélargissement, m., v. *surécartement.*
surélévation, f., (*r. r.*) raising of outer rail; (*cons.*) addition to a building (e. g., a story).
surélevé, p. p., *chemin de fer* ——, elevated railroad.
surélever, v. a., to raise; (*cons.*) to raise (a building, e. g., by adding a story).
surémission, f., overissue (of paper money).
surenveloppe, f., outer envelope.
surépaisseur, m., excess of thickness, increased thickness.
surestarie, f., overtime, demurrage;
indemnité pour ——, demurrage.
surestimation, f., overestimate, overvaluation.
surestimer, v. a., to overestimate, overvalue.
sûreté, f., safety, security, safe-keeping, custody, warranty; (commercial) security;
arrêt de ——, (*sm. a.*) safety bolt, catch;
brigade de ——, (*mil.*) a name for the corps cavalry;
caisse de ——, safe;
dispositif de ——, (*mil.*) measures of safety;
dispositif de stationnement de ——, (*mil.*) measures taken in a *stationnement* for security;
— *hypothécaire,* mortgage;
lampe de ——, safety lamp;
lieu de ——, safe place;
parabole de ——, (*ball.*) envelope of the parabolas having the same point of departure and the same initial velocity;
place de ——, place retained by a state as a guarantee of the execution of a treaty;
platine de ——, (*sm. a.*) safety lock;
réseau de ——, (*mil.*) the network of troops established to secure safety;
robinet de ——, safety cock;
serrure de ——, safety lock;
service de ——, v. s. v. *service;*

sûreté, soupape de ——, safety valve;
troupes de ——, (*mil.*) troops employed in *service de sûreté*;
verrou de ——, safety bolt.

surface, f., surface, space, area;
—— *d'appui,* (*mach.*) working surface;
—— *de chauffe,* (*steam*) heating surface;
—— *de chauffe totale,* (*steam*) total heating surface;
—— *de chauffe tubulaire* (*steam*) tubular heating surface;
—— *de comparaison* (*surv.*) level surface of reference;
—— *de condensation,* (*steam, etc.*) condensing surface;
—— *de contact,* surface of contact;
—— *s de départ,* (*mach.*) surfaces by which the work is held in the operating machine;
—— *d'eau,* water surface;
—— *d'éclatement,* (*art.*) space over which effects of burst are felt;
—— *équipotentielle,* equipotential surface;
—— *de frappe,* striking surface (in falling weight test); striking surface of a hammer, etc.;
—— *de frottement,* rubbing surface;
—— *gauche,* skew surface, left-handed surface;
—— *de grille,* (*steam*) grate area;
—— *inégale,* rough, uneven, surface;
—— *libre de la grille,* (*steam*) open grate area;
—— *de niveau,* level surface;
—— *plantaire,* (*hipp.*) plantar surface, bearing surface of the hoof;
—— *raboteuse, v.* —— *inégale;*
—— *réfrigérante,* condensing, cooling, surface;
—— *refroidissante,* cooling surface;
—— *de roulement d'une roue,* (*r. r.*) tread of a wheel;
—— *de surchauffeur*(s), (*steam*) superheating surface;
—— *du terrain horizontal,* level of the ground;
—— *topographique,* the surface of the ground; any surface whose law of generation is unknown and is pierced by a vertical in only one point;
—— *unie,* even, smooth, surface.

surfaix, m., (*harn.*) surcingle, roller;
avoir du ——, (*hipp.*) to have a deep and well-rounded girth;
—— *-cavalier,* v. s. v. *cavalier;*
—— *de couverture,* blanket surcingle;
—— *-dossière,* gun-pack lashing or strap (mountain artillery).

surhaussement, m., elevation, act of raising; exaggeration of vertical scale (in relief maps, etc.); (*r. r.*) elevation of outer rail.

surhausser, v. a., to raise; to run up.

sur-horizontal, a., (*mil. min.*) over the center of the charge (of effects).

surintendance, f., superintendence, overseeing; overseer's duties.

surintendant, m., superintendent, overseer.

surjalé, a., (of an anchor) fouled.

surjaler, v. a. n., to raise an anchor in order to clear the stock; to become unstocked (of the anchor).

surjet, m., overcasting, whipstitching.

surjeter, v. a., to overcast, whipstitch.

surlier, v. a., (*cord.*) to whip (the end of a rope).

surliure, f., (*cord.*) whipping (of a rope end).

surmenage, m., overworking (of a horse).

surmener, v. a., to overstrain, overwork, override, a horse; (*mil.*) to handle roughly.

surmeule, f., runner or upper millstone.

surmonter, v. a., to surmount; to get over; rise above, rise to the surface of; (*fig.*) to overcome, conquer; to surpass, excel.

surmoule, m., mold made from a casting.

surmouler, v. a., to mold from a casting.

surnager, v. n., to float; to float to the surface.

surnuméraire, a., supernumerary; (*mil.*) supernumerary, additional, as additional lieutenant.

suros, m., (*hipp.*) splint;
—— *cerclé, v.* —— *chevillé;*
—— *s en chapelet,* chain splints;
—— *chevillé,* through, double, or pegged splint;
—— *sur la corne,* ringbone;
—— *double, v.* —— *chevillé;*
—— *s fusés,* spindle-shaped splints;
—— *simple,* single splint;
—— *tendineux,* tendinous splint.

surpasser, v. a., to surpass, exceed, go beyond; to overweight, overflow.

surpaye, f., overpay.

surpayer, v. a., to overpay.

surpied, m., spur strap (over instep).

surplomb, m., overhang, overhanging; state of being out of the perpendicular (said of overhanging rocks, etc.);
en ——, overhanging;
être en ——, to overhang.

surplombement, m., overhanging.

surplomber, v. a. n., to overhang; to hang over; to lean out; to be out of the perpendicular (of a wall).

surplus, m., surplus, overplus, remainder.

surpoids, m., overweight.

surprendre, v. a., to surprise, take by surprise; to detect, intercept; to overreach, catch, overtake; to obtain by fraud; (*mil.*) to surprise;
—— *un cheval,* (*man.*) to use the aids suddenly.

surpression, f., (*ball.*) abnormal pressure.

surprise, f., surprise; (*mil.*) surprise, sudden attack; surprise of a fort, of a town;
—— *stratégique,* (*mil.*) state of affairs caused by a lack of information as to the enemy (e. g., a *bataille de rencontre*);
—— *tactique,* (*mil.*) surprise proper (caused by the improper performance of the security service).

sursaturation, f., supersaturation.

sursaturer, v. a., to supersaturate.

surséance, f., suspension, respite, reprieve.

surseoir, v. a., to suspend, delay, put off, arrest, supersede, reprieve, respite.

surseuil, m., (*cons.*) lintel.

sursis, m., delay, reprieve, respite;
accorder un ——, to reprieve;
—— *d'appel,* (*Fr. a.*) delay or postponement of joining;
—— *d'arrivée,* (*Fr. a.*) delay of joining, as a new regiment or service, in consequence of change of residence;
permission à titre de ——, v. s. v. *permission.*

sursoufflage, m., (*met.*) afterblow (Bessemer process).

surveillance, f., supervision, superintendence, inspection, watch; inspectorship;
—— *administrative,* (*Fr. a.*) inspection (by the *intendance*) of the administration of regiments and of establishments ranking as regiments;
—— *de l'ordinaire,* (*mil.*) inspection of company mess.

surveillant, m., watchman, inspector, overseer, supervisor; keeper of a military jail.

surveiller, v. a. n., to superintend; to inspect; to look after, take care of, overlook, watch over, have an eye upon, preside over.

survente, f., increase in the violence of the wind.

surventer, v. a., to blow hard; to blow a gale; to increase in violence (of the wind).

survider, v. a., to empty partly (vessel or receptacle that is too full).

sus-bande, f., (*art.*) cap-square;
—— *à coulisse,* sliding cap-square (slides into position);
—— *de tire-feu,* (*F. m. art.*) kind of strap holding down the lanyard, and forming part of safety device.

suscription, f., address, superscription, of a letter.

susdit, a., aforesaid.

sus-naseau, m., (*hipp.*) part above the nostrils.

suspect, a., suspected, open to suspicion;
cheval ——, horse suspected of a contagious disease;

suspect, *lieu, pays*, —— (*de contagion*), place, country, thought to be infected.
suspendre, v. a., to hang, hang up, sling; to suspend, stay, stop, arrest, discontinue, intermit (as hostilities); to delay, put off; to suspend from office;
—— *un cheval*, to sling a horse;
—— *le compas*, to swing a compass in gimbals.
suspendu, p. p., suspended; swung; (of a vehicle) on springs, having springs; suspended (from duty, from office);
pont ——, suspension bridge.
suspenseur, a., suspensory; m., sling;
câble ——, supporting cable (of a suspension bridge).
suspension, f., suspension, hanging, swinging; discontinuance, interruption, cessation; suspension (from duty, from office); (in balloons) the set or system of suspending cords;
anneau de ——, v. s. v. *anneau*;
—— *d'armes*, (*mil.*) suspension of arms, of hostilities; truce, armistice;
—— *bifilaire*, bifilar suspension;
—— *à la Cardan*, Cardan's, cardanic, suspension;
chaîne de ——, suspension chain;
—— *des chevaux*, (*hipp.*) swinging of horses in case of certain diseases or fractures;
—— *contrefilaire*, counter-filar suspension;
—— *d'emploi*, (*Fr. a.*) suspension from duty (as on account of passage to *non-activité*);
—— *de grade*, (*Fr. a.*) suspension from grade (*chefs de musique*, who have no assimilated rank as officers);
—— *d'hostilités*, v. —— *d'armes*;
mort par ——, death by hanging;
point de ——, point of suspension;
système à ——, (*art.*) lunette-and-pintle system of limbering up;
temps de ——, (*man.*) time or moment after the execution of the parts of a movement;
—— *unifilaire*, unifilar suspension.
suspensoire, f., (*pont.*) suspension rod (as of a suspension bridge); (*med.*) suspensory, suspensory bandage.
sus-pied, m., v. *sur pied*.
suzerain, a., m., suzerain.
suzeraineté, f., suzerainty.
svelte, a., slender, light, slim, graceful, easy.
switch, m., (*r. r.*) switch.
sycomore, m., sycamore tree and wood;
faux ——, pride of India.
syénite, f., syenite.
symétrie, f., symmetry.
symétrique, m., symmetric.
sympiézomètre, m., sympiesometer.
synchrone, a., synchronous.
synchronique, a., synchronic.
synchroniser, v. a., to synchronize.
synchronisme, m., synchronism.
syndic, m., syndic; magistrate; assignee;
—— (*des gens de mer*), in France, seaman or discharged petty officer having police duties with reference to the maritime domain, fisheries, etc., and local representative in all matters relating to the *inscription maritime*.

syndical, a., *chambre* —— *c*, trades committee;
chambre —— *e ouvrière*, trades-union committee.
syndicat, m , office, duties, of a syndic; trades committee.
syndiquer, v. a., to form a syndicate.
synovial, a., synovial.
synoviale, f., (*hipp.*) synovial membrane;
—— *tendineuse*, synovial sheath.
synovie, f., synovia.
synovite, f., (*hipp.*) synovitis.
syrtes, f. pl., (*hydr.*) shifting sands.
système, m., system, project, scheme, plan; in any material or mechanical arrangement, the ensemble of the essential or characteristic parts;
—— *angulaire*, (*fort.*) —— v. *tenaillé*;
—— *d'artillerie*, (*art.*) comprehensive term for everything connected with the arm, especially of a material character;
—— *asservi* (*mach.*) system controlled by a servo motor;
—— *bastionné*, (*fort.*) bastioned system;
—— *à bielle et cadre*, (*mach.*) slit-bar motion;
—— *de bielles combinées*, (*mach.*) connecting-rod system;
—— *en boucle*, (*elec.*) loop system;
—— *de carrière*, v. s. v. *carrière*;
—— *de changement de marche*, (*mach.*) reversing gear;
—— *concentrique*, (*fort.*) system of fortification in which the works are situated in lines, all the fortifications of the national territory being considered;
—— *concutant*, (*art.*) in a double-action fuse, the arrangement that sets off the time system;
—— *concutant-percutant*, v. s. v. *concutant*;
—— *de contremines*, (*fort.*) system of countermines;
—— *à deux fils*, (*elec.*) two-wire system;
—— *de fortification*, fortification system;
—— *fusant*, (*art.*) time train proper of a double-action fuse;
—— *de lumière à arc*, (*elec.*) arc system;
—— *moderne*, (*fort.*) modern polygonal system;
—— *de passage*, v. s. v. *passage*;
—— *percutant*, v. s. v. *percutant*;
—— *de pointage en direction* (*hauteur*), (*art.*) training (elevating) gear;
—— *polygonal*, (*fort.*) v. —— *moderne*;
—— *rayonnant*, (*fort.*) system of fortification in which the works radiate from the capital, along probable lines of invasion, to frontier;
—— *de roues*, system of wheels;
—— *en série*, (*elec.*) series system;
—— *de soupapes*, (*mach.*) set of valves;
—— *tenaillé*, (*fort.*) tenaille system;
—— *à transmission directe* (*indirecte*), (*elec.*) direct- (indirect-) distribution system;
—— *à trois fils*, (*elec.*) three-wire system;
—— *à vide*, vacuum system.
syzygie, f., syzygy;
marée de ——, spring tide.

T.

T, any T-shaped device, any T-shaped part of a mechanism; toggle; T square; T iron; (*siege*) T-shaped trench (in front of the circular place of arms); (*fort.*) T-shaped place of arms; (*mil. min.*) double mine, T-shaped mine; (*mach.*) crosshead;
—— *d'assemblage*, T-shaped assembling or connecting piece;
axe en —— (*à deux leviers opposés*), (*mach.*) cross axle;
essieu en ——, (*mach.*) cross axle;
fer en ——, T iron;
fer à double ——, H iron;
guides du ——, (*mach.*) crosshead guides;
mine en ——, (*mil. min.*) T mine;
—— *du piston*, (*mach.*) piston crosshead;

T, *poutre en* ——, *en double* ——, v. s. v. *poutre*;
—— *renversé*, (*mach.*) cross-tail;
—— *de tige de piston*, (*mach.*) piston-rod crosshead;
—— *du tiroir*, (*mach.*) slide-valve crosshead.
tabac, m., tobacco;
les ——*s*, government sale of tobacco.
tabatière, f., tobacco box;
fusil à ——, v. s. v. *fusil*.
table, f., table; plate; slab; face of an anvil, of a hammer; (in the arts) worktable; table (of logarithms, etc.); index, register; (*top.*) rock with a flat top; table (of a mountain); (*hipp.*) dental table; (*cons.*) flange of an **I**, of a channel, beam; surface or floor of a truss; table, rectangular area in relief, etc., on a wall; panel surrounded by a border of plaster; bed of concrete supporting construction in compressible soils; (*nav.*) signal book; (*mil.*) officers' mess; (*powd.*) bed, bedstone (of the mill); (*mach.*) bed (e. g., of a planer);

table *d'appareil*, worktable;
—— *aux armements*, (*s. c. art.*) implement shelf;
—— *d'attente*, plate or slab for an inscription;
—— *de batterie*, ——*-battière*, molding trough;
—— *en béton*, concrete bed (for construction in compressible soils);
—— *brisée*, folding table;
—— *de charge*, (*artif.*) drift board;
—— *chiffrante*, cipher table, board, or disk;
—— *de construction*, any sheet, plan, etc., containing full details for constructing a mechanism, machine; working plan, working drawing (e. g., small arms); (*Fr. a.*) tables of specifications for the construction of material;
—— *de correspondance*, (*Fr. art.*) table showing variations in the elements of loading for the same piece, using different projectiles;
—— *à coulisses*, v. —— *à rallonges*;
—— *de cylindre*, (*mach.*) cylinder face;
—— *de défilement*, (*art.*) range table for indirect fire (fixes points from which a hidden target may best be reached by indirect fire);
—— *dentaire*, (*hipp.*) dental table;
—— *à dessiner*, drawing table or board;
—— *à double entrée*, double-entry table;
—— *à écrire*, writing table;
—— *à égruger*, (*artif.*) mixing table; mealing tray;
—— *d'enclume*, face of an anvil;
être à la même ——, to mess together;
—— *évier*, tank table;
faire —— *rase*, to obliterate, begin all over again;
—— *en fonte*, cast-iron face plate;
—— *à forer*, (*mach.*) traveling table (of a boring machine);
—— *à fourrer*, (*cord.*) serving board;
—— *de foyer*, hearth plate; dead plate;
—— *de friction*, (*mach.*) slide face;
—— *de glaciers*, glacier table;
—— *d'harmonie*, sounding board;
—— *des hausses*, (*art.*) table of elevations, drift, and of other elements;
—— *à manger*, dining table;
—— *de manipulation*, (*elec.*) switch board; keyboard;
—— *d'un marteau*, hammer face;
—— *des matières*, table of contents;
—— *d'un moulin à poudre*, (*powd.*) bedder, nether millstone;
—— *à ouvrage*, work bench;
—— *pliante*, v. —— *brisée*;
—— *de pointage*, (*Fr. art.*) bed of the lower carriage of the 120ᵐᵐ C.;
—— *porte-projectile*, (*mach.*) projectile table in band-seating machinery;
—— *pratique de tir*, (*ball.*) a ballistic table (so called because it gives only data indispensable for actual execution of fire);
—— *à rallonges*, extension table;
—— *rapportée de cylindre*, (*mach.*) false face of a cylinder;
—— *à rebord*, v. —— *à égruger*;
—— *à rouler*, trundle plank;
—— *saillante*, (*cons.*) panel projecting from its frame;
—— *à sécher*, drying tray;
—— *de tir*, (*art.*) range table;
—— *de tir graphique*, (*art.*) graphical range table;
—— *de tir numérique*, (*art.*) table giving numerical arguments and functions;
—— *de tir plongeant*, (*art.*) range table for curved fire;
—— *de tiroir*, (*mach.*) valve face, slide face;
—— *tournante*, (*min.*) slime table;
—— *à tourner*, (*mach.*) rest of a boring machine;
—— *de visée*, (*nav.*) sort of range table for firing, a torpedo from a ship.

tableau, m., blackboard, announcement board; picture, painting; description, catalogue, list, table, tabulated statement, schedule; diagram; (*cons.*) reveal;
—— *d'assemblage*, index map;
—— *d'avancement*, (*mil.*) promotion roll or list;
—— *de concordance des charges*, (*art.*) table giving the charges of different powders necessary to produce a given muzzle velocity;
—— *de connexion*, (*elec.*) secondary switch board;
—— *de défilement*, v. *table de défilement*;
—— *de distribution*, (*elec.*) switch board;
—— *à double entrée*, double-entry table;
—— *figuratif de tir*, (*ball.*) plotted diagram of practice;
—— (-)*indicateur*, (*art.*) in a seacoast battery, board containing or showing data useful in target practice;
—— *de marche*, (*mil.*) tabulated statement of a day's march (giving elements of column, time when each is to start in order to reach initial point, destination, etc., *Fr. a.*);
—— *de marche des trains*, (*r. r.*) time-table;
—— *de recensement*, v. s. v. *recensement*;
—— *de recherches pour le tir plongeant*, (*art.*) a table giving for known ranges and angles of fall the muzzle velocity, the remaining horizontal velocity, and probable error in height;
—— *secondaire*, (*elec.*) secondary switch board;
—— *de service*, (*mil.*) duty roster; (*r. r.*) time-table;
—— *du service journalier*, (*mil.*) schedule or roster of daily duties;
—— *de service de tranchée*, (*siege*) roster of trench duty;
—— *de tir*, (*art.*) range table.

tablette, f., tablet; shelf; (*in pl.*) writing tablets, notebook; long rib (e. g., in Schultz gun); (*inst.*) plane table; (*nav.*) armor shelf; (*fort.*) coping; (*med.*) tablet;
—— *d'appui*, (*cons.*) window-, door-, sill;
—— *en bahut*, (*cons.*) coping stone;
—— *de cheminée*, (*cons.*) mantelpiece;
—— *à coulisse*, leaf of a table;
—— *de fenêtre*, (*cons.*) window sill;
—— *d'un mur*, coping of a wall;
—— *de platine*, (*sm. a.*) lock plate;
—— *d'une poutre*, —— *d'une solive*, (*cons.*) beam end (resting on a wall).

tablier, m., apron; apron of a carriage; small canvas bag for nails, spikes, etc.; (*pont.*) floor of a suspension bridge; frame and planking of one bay of a frame bridge; roadway of a pontoon bridge; (*art.*) floor of a platform; platform of a two-wheel truck;
à ——, aproned;
—— *à bavette*, cook's apron (high apron);
charpente de ——, road framing of a suspension bridge;
plate-forme à —— *continu*, *à* ——*s superposés*, v. s. v. *plate-forme*;
—— *de pont*, floor, flooring, of a bridge;
—— *de pont flottant*, brow of a floating bridge;
—— *de pont-levis*, platform, movable part, of a drawbridge;
—— *de sapeur*, (*mil.*) sapper's apron;
—— *de siège*, (*fort.*) flooring of a siege platform.

tabors, m., (*mil.*) ring, circle, of baggage, for defense against cavalry.

tabouret, m., stool; (*man.*) riding stool (sort of stool susceptible of motions similar to those produced by a horse in motion);
—— *électrique*, —— *isolant*, (*elec.*) insulating stand.

tac, m., (*hipp.*) rot (eruptive phlegmasia); (*fenc.*) touch.

tache, f., spot, stain; (*art.*) tin spot in gun metal;
—— *d'étain*, (*art.*) tin spot in bronze guns;
—— *dans l'œil*, (*hipp.*) haw(s).

tâche, f., task, job; piecework;
à la ——, by the job;
être à la ——, to work by the job, by the piece;

tâche, mettre à la ——, to put (men) to work by the task (in sieges, 0.4 mc. per hour);
ouvrage à la ——, piecework, job work;
ouvrier à la ——, jobber;
prendre à ——, to undertake a task, etc.;
travail à la ——, v. *ouvrage à la ——;*
travailler à la ——, to work by the task, by the piece.

tachéografie, f., v. *tachygraphie.*

tachéomètre, m., *(inst.)* tachometer; tachymeter, an altimetric and planimetric surveying instrument, invented by Colonel Goulier, of the French engineers.

tacher, v. a., to spot, stain.

tâcher, v. n., to try, endeavor.

tâcheron, m., man on piecework; job hand.

tacheter, v. a., to spot, mark.

tachomètre, m., *(inst.)* tachometer.

tachygraphe, m., stenographer.

tachygraphie, f., stenography.

tachymètre, m., *(inst.)* tachometer, speed gauge.

tachymétrie, f., tachometry.

taconnage, m., *(art.)* defect in cast-iron guns, caused by a flaw of the mold. (obs.).

tact, m., tact, touch, feeling.
—— des coudes, (mil., inf.) touch of elbow;
garder, conserver, tenir le —— des coudes, to keep the touch.

tacticien, m., *(mil.)* tactician;
—— en chambre, closet tactician, theoretical tactician.

tacticographe, m., *(mil.)* writer on tactics.

tacticographie, f., *(mil.)* graphical representation of maneuvers of troops.

tactique, f., *(mil.)* tactics;
—— de combat, battle tactics;
—— de combat spéciale, the drill of each arm;
—— de combinaison v. —— combinée des trois armes;
—— combinée des trois armes, tactics of the three arms, acting in conjunction;
—— élémentaire, drill (of each arm);
—— d'embarcations, (nav.) boat drill, boat tactics;
—— de l'éperon, (nav.) ram tactics;
—— exécutive, term applied to operations in action for the attainment of a definite and immediate purpose;
—— générale, tactics of the three arms in combination;
grande ——, grand tactics;
—— linéaire, linear tactics;
—— de marche, march tactics;
—— navale, (nav.) naval tactics;
—— navale à vapeur, (nav.) steam tactics;
—— perpendiculaire, perpendicular tactics;
—— des signaux, signaling tactics, i. e., signaling with reference to troops in the field, to use under service conditions;
—— de ravitaillement, supply tactics;
—— de stationnement, measures taken on making a halt (v. *stationnement*).

tafia, m., rum.

taie, f., *(hipp.)* film (over the eye, disease);
—— d'oreiller, pillowcase.

taillade, f., cut, gash, slash.

taillader, v. a., to cut, gash, slash.

taillanderie, f., edge tools; edge-tool trade.

taillandier, m., edge-tool maker.

taillant, m., edge, cutting edge (of a tool).

taille, f., size, stature, height; height of a man, figure, height of a horse; cut, cutting, lopping; stone, etc., cutting; cut of a file; tally; coppice wood; *(sm. a.)* edge of a sword;
—— apparente, cutting (stone) to shape, or according to pattern;
arme de ——, v. s. v. *arme;*
ne pas avoir la ——, (mil.) to be too short for military service;
basse ——, basso-rilievo;
—— bâtarde, bastard cut (of a file);
—— de bois, woodcut;
—— brute, rough dressing (of stone);
cheval entre deux ——s, middle-sized horse;
contre ——, crosscut;
coup de ——, v. s. v. *coup;*

taille *courbe fine (layée rustiquée),* v. *—— plane fine,* etc., substituting *courbe* for *plane;*
—— en cuivre, copper plate;
de ——, (fenc.) cut;
de —— et de pointe, (fenc., sm. a.) cut and thrust;
douce ——, smooth cut (of a file);
—— douce, copper plate;
—— dure, steel engraving;
d'estoc et de ——, cut and thrust;
—— d'épaisseur, squaring of stone;
—— d'exploitation, (min.) chamber of excavation; thirling board; wall face;
exploitation par grande ——, (min.) long work;
fine ——, smooth cut (of a file);
frapper d'estoc et de ——, to cut and thrust;
front de ——, v. s. v. *front;*
grande ——, (min.) long work;
grosse ——, coarse cut (of a file);
—— moulurée, cutting of moldings on stone;
—— moyenne, bastard cut (of a file);
pierre de ——, freestone; cut stone; broad stone;
—— plane fine, (of stones, flat surface) fine pointing and drafting;
—— plane layée, (of stones, flat surface) smooth dressing;
—— plane rustiquée, (of stones, flat surface) rustic cut or dressing;
première ——, first course of cuts (of a file);
rang de ——, v. *rang;*
seconde ——, second course of cuts (of a file).

taillé, p. p., *—— de l'avant, (nav.)* built with a sharp bow.

taille-crayon, m., pencil sharpener.

taille-mer, m., *(nav.)* cutwater.

taille-neige, m., snow plow.

tailler, v. a. n., to cut; to hew (timber); to use the edge (of a sword);
—— en adent, to indent;
—— en biseau, v. *—— en chanfrein;*
—— en chanfrein, to chamfer;
—— un cheval, (hipp.) to cut a horse;
—— cône, to taper;
—— des croupières à l'ennemi, (mil.) to pursue a defeated enemy close and hard;
—— en pièces, (mil.) to hew, cut to pieces, utterly destroy;
—— en pointe, to point;
—— en sifflet, to cut skew, to cut askew;
—— une lime, to cut a file.

taillet, m., blacksmith's cutter or hack iron.

tailleur, m., tailor; hewer; cutter;
—— de pierre, stone cutter or mason.

taillis, m., thicket, underbrush; trees of second growth;
bois ——, copse, underwood;
—— composé, —— sous futaie, trees left standing at a cut in order to reach their growth.

tain, m., tin foil; silvering.

taire, v. a., to silence;
faire —— une batterie, (art.) to silence a battery.

taloche, f., float, plasterer's trowel.

talon, m., heel; heel of a shoe; stub; (in the arts) shoulder, heel; shoulder of an axle; *(hipp.)* heel; *(farr.)* heel of a horseshoe; *(man.)* heel, i. e., spur; *(sm. a.)* heel of a lance, of a color staff; shoulder of a sword blade, of a bayonet blade; ear of a sword belt; cocking toe; *(mach.)* flange; *(harn.)* corner of the beam of a bit; *(art.)* holding-down clip; trail of a gun carriage;
à ——, (of documents, papers, etc.) having a stub;
—— de botte de roue, calkin;
—— du chien, (sm. a.) nose of the cocking piece;
—— de coffre à munitions, (art.) stay of an ammunition chest;
connaître les ——s, (man.) to be spur-wise;
—— de crosse, (sm. a.) heel of the butt;
—— de culasse, (sm. a.) tang;
—— de dedans, (man.) heel on the inside of a riding hall;
—— de dehors, (man.) heel next the wall of a riding hall;

talon, *donner du* —— *à un cheval*, (*man.*) to spur a horse;
donner un coup de ——, to strike, to touch, the ground (boats);
entendre les ——*s*, (*man.*) to be spur-wise;
—— *d'épée*, (*sm. a.*) part of blade nearest hilt;
—— *de l'essieu*, joggle, tag, calkin;
être bien dans les ——*s*, (*man.*) to be spurwise;
—— *de flasque*, (*art.*) trail plate;
gros ——, (*mil. slang*, obs.) cuirassier;
—— *de jet*, (*fond.*) heel of a runner;
—— *de lame*, shoulder of a blade;
—— *de mât*, heel of a mast.
—— *de la noix*, (*sm. a.*) back of the tumbler;
pincer des deux ——*s*, v. *serrer les* ——*s*;
—— *de pont-levis*, rear crossbeam of the platform of a drawbridge;
—— *d'une poulie*, shoulder of a block;
porter un cheval d'un —— *sur l'autre*, (*man.*) to give a horse first one, then the other spur;
promener un cheval dans la main et les ——*s*, *entre les deux* ——*s*, v. s. v. *promener*;
—— *de rebondissement*, (*sm. a.*) rebound shoulder (on the hammer of a revolver);
—— *de récépissé*, v. s. v. *récépissé*;
répondre aux ——*s*, (*man.*) to obey the spur;
—— *rouge*, formerly (under the old régime) a nobleman;
serrer les ——*s*, (*man.*) to give a horse the spurs;
sur les ——*s de*, close on the heels of;
—— *de sûreté*, (*art.*) safety stud.

talonner, v. a., (*mil.*, etc.) to follow, pursue closely; (*man.*) to spur; (*nav.*) to strike aground aft.

talonnier, m., (*mil.*) color-staff shoe (i. e., for pole of).

talpack, m., (*unif.*) busby.

talus, m., (steep) slope; sloping surface; talus; (*fort.*) slope; (*cons.*) batter; (*farr.*) seating out of a horseshoe;
aller en ——, to slope, batter, shelve;
—— *d'amont*, upstream slope of a dam;
—— *d'aval*, downstream slope of a dam;
—— *de banquette*, (*fort.*) banquette slope;
—— *de la batterie*, (*fort.*) epaulment slope of a battery;
—— *à contre-pente*, (*fort.*) countersloping glacis;
—— *en contre-pente*, (*fort.*) substitute for counterscarp (proposed by Carnot, had a very gentle slope);
—— *de contrescarpe*, (*fort.*) counterscarp slope;
à —— *coulant*, at its natural slope (of earth, etc.);
couper en ——, to cut slanting, with a feather edge;
donner du ——, *disposer en* ——, to slope;
en ——, sloping, slanting;
—— *d'escarpe*, (*fort.*) slope of the scarp;
être en ——, v. *aller en* ——;
—— *extérieur*, (*fort.*) exterior slope;
—— *de flasques*, (*art.*) slope of the cheeks;
—— *intérieur*, (*fort.*) interior slope;
mettre en ——, to slope, etc.;
mise en ——, sloping (act of);
—— *des remblais*, embankment slope;
—— *de rempart*, (*fort.*) terre-plein slope (joins terre-plein to natural ground);
—— *des tranchées*, slope of a cutting.

talutage, m., sloping (act of).

taluter, v. a., to slope; to give a batter; to slant.

tambour, m., drum; (*mach.*, etc.) drum, barrel, roll, roller, cylinder, casing, chain drum; treadwheel; (*fort.*) tambour; small traverse to prevent the enfilade of a trench; (*nav.*) paddle box; (*elec.*) drum armature; (*powd.*) granulating drum; (*mil.*) drum, drummer, sound of the drum (usually *tambour* is the performer, *caisse* the instrument); (*mil. slang*) an *élève brigadier fourrier*; (*cons.*) center (of a vaulted arch); drum (cylindrical wall of a dome); drum (of a column, of the core of a spiral staircase); point of junction of several pipes; drum connecting two pipes of unequal diameters;

tambour, *le* —— *appelle*, (*mil.*) the drum is beating; the call is going;
aux ——*s*, (call) assembly of drummers;
baguette de ——, drumstick;
bander un ——, to brace a drum;
—— *de basque*, tambourine;
—— *battant*, by beat of drum, with drums beating; energetically, violently, thoroughly;
battement de ——, v. s. v. *battement*;
batterie de ——, beat of drum; (*mil.*) call or signal by beat of drum;
battre du ——, to beat a drum;
battre le ——, to beat a drum (i. e., give a signal by drumbeat);
bruit de ——, drumming;
caisse de ——, case, frame, barrel, of a drum;
caporal ——, v. s. v. *caporal*;
cercle de ——, flesh hoop of a drum;
—— *de chargement*, (*r. f. art.*) (Gatling) feed gear (earliest type);
chasser un soldat au son du ——, (*mil.*) to drum a soldier out;
collier de ——, drum carriage;
—— *conique*, (*mach.*) cone pulley;
corde de ——, brace of a drum;
—— *à*, *de*, *courroie*, (*mach.*) drum of a belt; belt pulley;
—— *de crible*, (*artif.*) composition sieve;
—— *de cuivre*, kettledrum;
dessus de ——, drumhead;
—— *à émeri*, emery wheel;
faiseur de ——*s*, drum maker;
—— *de frottement*, friction drum;
fût de ——, shell, case, of a drum;
grand cercle de ——, snare hoop of a drum;
gros ——, bass drum;
—— -*guide*, (*Fr. art.*) grooved drum of a seacoast sight (controlling the displacement of the deflection scale and automatically correcting for drift);
—— *d'indicateur*, (steam) indicator barrel;
induit en ——, (*elec.*) drum armature;
—— -*magasin*, coiling drum (of a balloon rope);
—— -*major*, (*mil.*) drum major;
—— -*maître*, (*mil.*) corporal of the drums; corporal of field music (U. S. A.);
—— *militaire*, war drum;
—— *d'oculaire*, eyepiece tube (of a telescope);
—— *en palanque*, —— *en palissades*, (*fort.*) stockade tambour;
peau de ——, drumhead;
—— *de porte*, wind screen (of a door);
—— *d'une poulie*, sheave;
rappel pour les ——*s*, (*mil.*) quarter drum;
—— *de roue*, (*nav.*) paddle box;
roulement de ——, roll (of a drum);
—— *roulant*, orchestra or band drum (longer body than the ordinary drum);
—— *d'un tamis*, (*artif.*) composition sieve; drum of a mixing sieve;
tirant de ——, brace of a drum;
—— *d'un treuil*, barrel of a winch;
trou du fût d'un ——, drum hole;
vergette de ——, drum hoop;
—— *voilé*, muffled drum;
voiler un ——, to muffle a drum;
—— *à voûter*, (*mil. min.*) center for vaulted galleries.

tamis, m., sieve; separating sieve; screen; bolter;
—— *de*, *en*, *crin*, horsehair sieve;
—— *à huile*, oil sieve;
—— *de passage*, sieve, sand sieve;
passer au ——, to sift, screen;
—— *de perce-canon*, (*powd.*) canon-powder sieve;
—— *de perce-mousquet*, (*powd.*) rifle-powder sieve;
—— *de*, *en*, *soie*, (*powd.*) silk sieve;
—— *à tambour*, (*artif.*) mixing sieve;
—— *de toile métallique*, wire-gauze screen.

tamisage, m., separation by sieve or screen; sifting, screening, bolting.

tamiser, v. a. n., to separate by sieve or screen; to screen, to sift, to sift through, to bolt; to pass through a sieve, a screen.

tampon, m., plug, stopper; pad; bung; dowel; any plug for stopping a hole or acting as a wad or stopper; internal gauge, internal cylindrical gauge; lid of a water bucket; (*sm. a., art.*) tompion; (*t. p.*) bullet- (shot-) hole plug, marking disk; (*art.*) part of a vent plug near the surface of the bore; (wooden) wad for the charge of a spherical projectile (obs.); (*r. r., mach., etc.*) buffer; (*artif.*) wad;
—— *atmosphérique*, air buffer;
—— *d'attelage*, (*r. r.*) coupling buffer;
boîte à ——, inking and stamping pad;
—— *de bouche*, (*art.*) muzzle tompion;
—— *de buffle*, (*sm. a.*) wad;
—— *de charge*, (*art.*) gun wad (obs.);
—— *de choc*, (*art.*) Belleville spring, used as safety device in case brakes should break; (*art., r. r.*) buffer;
—— *de choc à ressort*, (*art., r. r.*) spring buffer;
—— *de combat*, (*nav.*) shot plug;
coup de ——, (*r. r.*) collision;
disque de ——, (*r. r.*) buffer disk;
—— *d'écubier*, (*nav.*) hawse plug;
—— *d'essieu*, (*art.*) axle buffer (Russian field mortar);
faux- ——, (*r. r.*) buffer box;
—— *de feutre*, (*sm. a.*) wad;
—— *de fusée*, (*artif.*) fuse disk of a rocket; (*art.*) plunger of a fuse;
—— *à galet*, (*art.*) roller buffer;
—— *hydraulique*, hydraulic buffer;
—— *de lumière*, (*art.*) vent plug;
—— *-masque*, (*sm. a.*) gas guard of the Lebel rifle;
—— *du nable*, bont plug;
—— *à ressort*, (*art.*) spring tompion;
ressort de ——, buffer spring;
—— *avec ressorts en caoutchouc* (*en volute*), india rubber (volute) spring buffer;
—— *à rondelles de caoutchouc*, india-rubber buffer;
—— *du seau d'affût*, (*art.*) water-bucket lid;
—— *sec*, (*r. r.*) springless buffer;
—— *de sûreté*, (*art.*) safety plug (of a fuse);
—— *taraudé*, screw plug; screw buffer;
tête de ——, buffer disk;
tige de ——, buffer guide rod;
trou de ——, (*met.*) hole through which metal flows;
—— *de tube*, (*steam*) tube plug;
—— *à vis*, v. *taraudé*.

tamponnage, m., (action of) plugging; plugging, stopping up.

tamponnement, m., v. *tamponnage*.

tamponner, v. a., to plug up, to stop up; to tamp, to pack down; (*t. p.*) to plug; (*r. r.*) to collide;
—— *une fusée*, (*artif.*) to plug a rocket.

tamponneur, m., (*t. p.*) shot-hole plugger, man who fills holes made by shots.

tan, m., tan bark.

tandem, m., open carriage driven tandem.

tangage, m., pitching motion (of a ship, of floating bridge); overbalancing of a locomotive.

tangente, f., tangent.

tangon, m., (*nav.*) swinging boom.

tanguer, v. a., (*nav.*) to pitch; to be down by the head;
—— *en arrière*, to squat.

tangueur, m., lighterman.

tanin, m., v. *tannin*.

tannage, m., tanning.

tannée, f., tan, waste tan, tan bark.

tanner, v. a., to tan.

tannerie, f., tannery, tanyard.

tannin, m., tannin.

taon, m., gadfly, horsefly.

tape, f., slap; plug, stopper; (*art.*) tompion;
—— *d'écubier*, (*nav.*) hawse plug;
—— *d'étoupe*, (*art.*) vent plug;
—— *à expansion*, (*art.*) expanding tompion.

tape-cu(l), m., drawbeam, counterpoise, of a drawbridge;
aller à ——, (*cav. slang*) to ride without stirrups;

taper, v. a., to slap, hit, strike; to plug up, to stop with a plug; (*drawing*) to sketch freely; (*art.*) to put in a tompion; v. n., (*met.*) (of steel) to form or develop interior cracks;
—— *un canon*, (*art.*) to put in a tompion;
—— *un cheval*, to crimp the mane and tail of a horse.

tapereau, m., kind of petard (obs.).

tapin, m., (*mil., fam.*) a drummer, esp. a poor drummer.

tapis, m., table cover, carpet, rug; (*harn.*) saddle cloth;
cheval qui rase le ——, v. s. v. *cheval*;
raser le ——, v. s. v. *raser*;
—— *de selle*, (*harn.*) saddle cloth.

tapure, f., (*met.*) interior crack or fissure (in large ingots) caused by too rapid heating or cooling.

taque, f., (*met.*) plate, lining plate;
—— *de gueulard*, iron ring around the charging hole of a blast furnace.

taqueret, m., fore plate of a forge hearth.

taquerie, f., (*met.*) stokehole of a reverberatory furnace.

taquet, m., cleat, belaying cleat; stop cleat, hurter; angle block; whelp of a capstan barrel; step of a ladder; kevel; wedge; small picket, peg; (*mach., etc.*) catch, stop, driver, tappet, detent; (*sm. a.*) front part of the trigger guard (shaped to receive the ramrod; locking brace; (*art.*) trunnion cleat;
—— *d'arrêt*, (*art.*) stop cleat; (*sm. a.*) bolt stop, stop;
—— *d'arrêt de console*, (*F. m. art.*) tray catch;
—— *arrêtoir*, (*sm. a.*) bolt stop;
—— *de baguette*, (*sm. a.*) ramrod stop;
—— *de baïonnette*, (*sm. a.*) bayonet-ring stop;
—— *de bout*, stop cleat;
—— *de butée*, standard knee of a windlass; (*r. f. art.*) resistance piece;
—— *de cabestan*, capstan whelp;
—— *de ceinture*, v. s. v. *ceinture*;
—— *de chargement*, (*art.*) loading cleat or stud (on various French carriages);
——*s d'un chariot à canon*, (*art.*) trunnion brackets or cleats;
—— *de châssis*, (*art.*) front (or rear) wedge on the chassis side rail (obs.);
——*s d'un châssis de moulage*, (*fond.*) lips or flanges of a molding flask;
—— *de chèvre*, (*art.*) peg on the pole of a gin;
—— *à cœur*, kevel;
—— *à corne*, belaying cleat;
corne de ——, arm, horn, of a cleat, or kevel;
courbe- —— *de bitte*, standard knee of a windlass;
—— *pour crochet de pontage*, (*pont.*) block through which the lashing hook passes;
—— *à dent*, snatch cleat;
—— *de détente*, tappet;
—— *écrou*, (*sm. a.*) ramrod stop (threaded) (in the front end of the guard);
—— *d'embarrage*, (*art.*) embarring bolt or cleat;
—— *d'excentrique*, (*mach.*) catch tappet, eccentric catch; peg;
—— *de fer*, hanging clamp;
—— *fixe*, stop;
—— *en grain d'orge*, stop cleat;
—— *à gueule*, snatch cleat;
—— *de nage*, flat thole;
oreille de ——, v. *corne de* ——;
—— *à oreilles*, thumb cleat or kevel;
—— *de pente*, (*mil. min.*) slope template;
—— *de pont*, deck cleat;
—— *à ressort*, spring cleat (sort of spring cushion);
—— *de retenue*, (*art., etc.*) stop;
—— *de sûreté*, (*r. f. art.*) safety catch;
—— *simple*, belaying cleat;
—— *de tournage*, belaying cleat;
—— *de tringle*, (*pont.*) kevel block;

tarage, m., allowance for tare; (*inst.*) calibration (as, *art.*, of a crusher);
force de ——, (*art.*) force of calibration, i. e., the force producing the final compression of a crusher cylinder.

taranche, f., lever for the turning screw of a press.
tarare, m., winnowing machine.
taraud, m., tap, screw tap; boxing machine;
—— *compensateur*, compensating tap, expanding tap;
—— *conique*, taper tap;
—— *équarrissoir*, square tap-rimer;
—— *à expansion*, v. —— *compensateur;*
manche de ——, wrench, tap wrench;
—— *-mère*, master tap.
taraudage, m., tapping.
taraudé, p. p., tapped;
boulon ——, screw bolt.
tarauder, v. a., to tap; (less strictly) to cut screws;
—— *à la filière*, to cut screws with a die;
—— *à la volée*, to cut screws by hand.
taraudeur, m., workman who taps; tapper.
taraudeuse, f., (*mach.*) screw-cutting engine or machine.
tard, a., late.
tarder, v. n., to delay, put off, loiter, lag, linger.
tare, f., tare; loss, waste; defect, fault; standard of comparison; (*hipp.*) blemish; (*r. r.*) weight of a railway car;
cheval sans ——, horse without blemish;
—— *dure*, (*hipp.*) osseous blemish (e. g., curb, spavin);
—— *molle*, (*hipp.*) generic term for elastic, non-inflammatory tumors near the articulations;
—— *osseuse*, v. —— *dure.*
taré, p. p., blemished;
cheval ——, blemished horse.
tarer, v. a., to damage, injure; to tare; to compare with a weight taken as standard of comparison; to calibrate.
taret, m., teredo, teredo navalis.
targette, f., sliding-bolt, fastener; sash bolt, window fastening;
—— *à ressort*, spring bolt;
—— *de sûreté*, (*sm. a.*) safety device.
tarière, f., auger; wimble, borer;
—— *anglaise*, screw auger; screw earth auger;
—— *à arcade*, (*mil. min.*) an earth scoop strengthened by a curved band (*arcade*) joining the edges of the scoop;
bout de ——, auger bit;
—— *à clapet*, sludger;
—— *creuse*, twisted auger;
—— *à cuiller*, shell auger;
cuiller de ——, shell of an auger;
—— *à élargir*, (*mil. min.*) enlarging auger or earth borer;
—— *à goujon*, dowel borer, hollow auger;
grande —— (*de forage*), (*mil. min.*) name of Colonel Bussière's earth auger;
—— *en hélice*, gimlet;
—— *à jantière*, felly auger;
mèche de ——, v. *bout de* ——;
—— *de mine*, mine auger;
—— *pontée*, v. —— *à arcade;*
—— *à spirale*, v. —— *en hélice;*
tige de ——, shank of an auger;
—— *torse*, twisted auger; (*mil. min.*) earth borer;
—— *tubulaire*, tubular earth auger;
verge de ——, v. *tige de* ——;
—— *à vis*, screw auger.
tarif, m., tariff; charges; table of prices, price list.
tarifer, v. a., to fix the tariff, the fares.
tarir, v. a. n., to dry up, drain, exhaust; to be exhausted, drained; to go, run, dry.
tarissable, a., that dries up.
tarissement, m., draining, drying up, state of being exhausted.
tarpan, m., (*hipp.*) wild horse of the steppes of Mongolia.
tarse, m., (*hipp.*) the tarsal bones.
tartre, m., (*chem.*) tartar; (*steam*) boiler incrustation, scale;
—— *de(s) chaudière(s)*, (*steam*) boiler scale.
tartrifuge, m., (*steam*) scale preventer.
tas, m., heap, pile; small, hand, anvil; polishing block; swage block; holding-up hammer, dolly; straight row of paving stones in the middle of the road; (*cons.*) building ground; building under construction;
tas droit, straight row of paving stones in the middle of the road;
—— *de foin*, haystack;
mettre en ——, to pile, heap up;
—— *à soyer*, creasing tool;
sur le ——, (*cons.*) on the work (said of stones that are dressed, pieces assembled, etc., at or near the point where they are to be used).
tasseau, m., bracket, small block, cleat, torsel; ledge; batten (as in a packing case); binding strip; holdfast; clamp; hand anvil; (*mach.*) tappet;
—— *de clapet*, (*mach.*) valve guard;
—— *de cuir*, leather cup;
—— *de directrice*, (*art.*) bracket or support for the *directrice* of a chassis (truck wagon);
—— *de l'excentrique*, (*mach.*) eccentric tappet;
—— *longitudinal*, rib;
—— *de marche-pied*, (*art.*) footboard bracket;
—— *de volée*, (*art.*) muzzle bracket or support (truck wagon).
tassement, m., settling (of embankments, buildings, pillars, etc.); amount allowed by excess of height for settling; (*sm. a.*) settlement, settling, of powder in a cartridge case.
tasser, v. a., to pile up, heap up; to ram, beat, down (earth); (*fond.*) to ram up; (*sm. a.*) to shake down, settle, powder in a cartridge case;
faire ——, to cause to settle;
se ——, to settle (of buildings, embankments, etc.).
tâter, v. a., to feel; to try, sound;
cheval qui tâte le pavé, v. s. v. *cheval;*
—— *son cheval*, (*man.*) to try one's horse;
—— *l'ennemi*, (*mil.*) to feel the enemy;
—— *le fond*, (*nav.*) to touch bottom;
—— *le pavé*, (*fig.*) to examine the ground;
—— *le terrain*, (*man.*) of a horse, to step hesitatingly.
tâtonnement, m., groping about, attempt, "feeler"; trial and error.
tatouage, m., tattooing.
tatouer, v. a., to tattoo.
taud, m., **taude**, f., (*nav.*) rain awning.
tauder, v. a., (*nav.*) to spread a rain awning.
taugour, m., short lever for various purposes.
taupe, f., (*hipp.*) poll evil;
—— *de rempart*, (*mil. slang.*) engineer soldier.
taupin, m., student of higher mathematics: candidate for the Polytechnique; (*mil. slang*) officer or soldier of engineers;
—— *carré*, second-year student of higher mathematics;
—— *cube*, third-year student of higher mathematics.
taurocolle, f., very strong glue made from tendons, cartilage, etc.
tautochrone, a., tautochronous;
(*courbe*) ——, tautochrone.
tautochronisme, m., tautochronism.
taux, m., rate, rate of discount, of interest; assessment; selling price; fixed price for sale of stores, rate of interest, etc.; (in strength of materials) load;
—— *de charge*, load per unit of section or area (strength of materials);
—— *de sécurité*, (in strength of materials) safe load per unit of area.
taxation, f., valuation, estimate; (*in pl.*) fees.
taxe, f., tax, taxing, rate, price;
—— *militaire*, (*Fr. a.*) tax or duty paid by men exempt from service in the active army;
—— *de port*, (*nav.*) harbor dues, duties.
taxer, v. a., to rate, fix the price of; to charge, to tax.
tayon, m., a tree left standing after the third cut.
tchako, m., v. *schako.*
té, m., v. s. v. *T.*
technique, a., technical, scientific;
projet ——, project for a railway, canal, etc.
technologie, f., technology.
teck, m., teak, teak wood.
tégole, f., (*cons.*) tegula (sort of roofing tile).
teigne, f., (*hipp.*) thrush.

teillage, m., (*cord.*) breaking, stripping of hemp by hand.
teille, f., (*cord.*) shive, fragment separated by breaking.
teiller, v. a., (*cord.*) to break or strip hemp by hand.
teindre, v. a., to dye, stain, tincture.
teint, m., dye, color;
—— *bon*, *grand*, —— fast color;
—— *mauvais* ——, color that will fade.
teinte, f., tint, tinge, tincture, dye; color, touch;
—— *conventionnelle*, conventional tint;
—— *dégradée*, graded tint;
—— *demi-* ——, half tint;
—— *dure*, ground color or tint;
—— *fondue*, blended tint;
—— *forte* ——, hard tint;
—— *panachée*, variegated tint;
—— *de passage*, (*met.*) intermediate tint (between red and green, in a polarizing telescope) used to determine temperatures;
—— *plate*, flat tint;
—— *vierge*, single, unmixed, tint.
teinter, v. a., to tint, color.
teinture, f., dye, hue, tint, tinge, color, tincture.
tek, m., teak wood.
telas, m., coarse sackcloth for sand bags.
télédynamique, a., transmitting power to a distance;
—— *câble* ——, long-distance power cable.
télégramme, m., telegram.
—— *en souffrance*, telegram delayed for any cause whatever.
télégraphe, m., telegraph; (*nav.*) engine-room telegraph;
—— *aérien*, v. —— *optique*;
—— *à aiguille*, v. —— *à cadran*;
—— *à cadran*, dial telegraph;
—— *de campagne*, (*mil.*) field telegraph;
—— *de chambre de machine*, (*nav.*) engine-room telegraph;
—— *écrivant*, writing telegraph;
—— *électrique*, electric telegraph;
—— *imprimant*, —— *imprimeur*, printing telegraph;
—— *à lettres*, v. —— *à cadran*;
—— *militaire*, military telegraph;
—— *optique*, (*sig.*) visual telegraph;
—— *à sonnette*, bell, alarm, telegraph;
—— *sous-marin*, submarine telegraph;
—— *souterrain*, underground telegraph;
—— *à transmission multiple*, multiplex telegraph.
télégraphie, f., telegraphy;
—— *aérienne*, v. —— *optique*;
—— *électrique*, electric telegraphy;
—— *légère*, (*Fr. a.*) telegraphic, visual, and telephonic communication service performed by cavalry;
—— *militaire*, (*mil.*) military telegraph, field telegraph;
—— *multiplex*, multiplex telegraphy;
—— *optique*, aerial, visual, telegraphy, signaling;
—— *sans fil*, wireless telegraphy;
—— *section de* ——, v. s. v. *section*.
télégraphier, v. a., to telegraph; m., telegraph operator.
télégraphique, a., telegraphic;
—— *poste* ——, v. s. v. *poste*;
—— *section* ——, v. s. v. *section*.
télégraphiste, m., telegraph operator (said esp., Fr. a., of employees or agents forming part of telegraph sections).
téléiconographe, m., (*inst.*) instrument (used in perspective surveying) for taking pictures at a distance.
téléologie, f., telelogy;
—— *optique*, visual signaling; strictly accurate term for visual signaling in general.
téléologue, m., (*inst.*) a (proposed) signaling instrument, dependent on the visibility of colors or of luminous objects.
télémètre, m., (*mil.*) telemeter, range finder;
—— *-chronoscope*, range-finding stop-watch;
—— *de combat*, so-called combat range finder, working by sound, invented by Le Boulengé;

télémètre *à curseur*, range finder fitted with an indicator;
—— *de dépression*, depression range finder;
—— *à indications automatiques*, automatic range finder;
—— *indicateur de position*, position finder;
—— *d'infanterie*, the —— *de combat* adapted to infantry use.
télémétrie, f., (*mil.*) range finding.
télémétriste, m., (*art.*) telemetrist.
télémétrographie, f., telemetrography.
téléobjectif, m., (*phot.*) telephotographic camera, with a divergent lens.
téléologue, m., any instrument or device for conveying sound to a distance.
téléphone, m., telephone;
—— *avertisseur*, a telephone that calls directly;
—— *à charbon*, carbon telephone;
—— *à ficelle*, string telephone;
—— *à pile*, long-distance telephone.
téléphonie, f., telephony.
téléphonique, a., telephonic;
—— *station* ——, telephone station.
téléphonographe, m., telephonograph.
téléphote, m., telephote.
téléphotographe, m., telephotograph.
téléphotographie, f., long-distance photography.
télescope, m., telescope, glass;
—— *appareil à* ——, French telescopic signal apparatus;
—— *catoptrique*, catoptric, reflecting, telescope;
—— *dioptrique*, dioptric, refracting, telescope;
—— *à lentilles*, v. —— *dioptrique*;
—— *à miroirs*, v. —— *catoptrique*;
—— *de nuit*, night glass;
—— *de réflexion*, v. —— *catoptrique*;
—— *de réfraction*, v. —— *dioptrique*.
télescoper, v. a., to telescope.
télodynamique, a., telodynamic, transmitting power to a distance;
—— *câble* ——, power cable.
télomètre, m., (*art.*) telometer (telemeter invented by Colonel Goulier, of the French engineers).
témoignage, m., evidence, testimony, witness; testimonial, token;
—— *appeler en* ——, to call as a witness; to call, take, to witness;
—— *faux* ——, perjury, false witness;
—— *porter* ——, to testify, witness, bear witness to;
—— *rendre* ——, to bear witness to; to testify.
témoigner, v. a., to testify, witness, bear witness, give evidence, depose, attest, convey, show;
—— *de la force*, (*hipp.*) to show strength.
témoin, m., testimony, evidence; witness, deponent; second (in a duel); mark, proof, landmark, any mark, etc., by which a determined line, point, etc., may be preserved; boundary mark, boundary tree; mound or lump of earth left in an excavation in order to measure quantity excavated; forge mark in a piece filed and reduced to proper dimensions; (*fort.*) profile for a parapet; (*cord.*) fag end of a new rope; (*mil. min.*) counterpart of the "monk" (lit at the same time to show when the explosion would take place, obs.);
—— *à charge*, witness for prosecution.
—— *de cordage* (*cord.*) untwisted end of rope to show that the whole is correct;
—— *à décharge*, witness for defense;
—— *faux* ——, false witness; addition of earth to a mound left in an excavation in order to obtain false measurement;
—— *interroger un* ——, to examine a witness;
—— *pratiquer un* ——, to tamper with a witness;
—— *prendre à* ——, to call to witness;
—— *récuser un* ——, to challenge a witness.
température, f., temperature;
—— *absolue*, absolute temperature;
—— *de l'eau bouillante*, boiling point;
—— *de combustion*, temperature of combustion;
—— *de la glace fondante*, freezing point;
—— *d'ignition*, igniting point;

température *d'inflammation*, temperature of inflammation.
tempête, f., (violent) storm, tempest;
—— *magnétique*, magnetic storm;
—— *tournante*, cyclone.
temple, m., temple; (Protestant) church.
temps, m., time, season, term, period, interval of time, delay; the present day; leisure; weather; (*fenc.*) opening for a thrust; (*man.*) interval of time between two successive beats (*battues*); (*mil.*) time (as in the manual of arms); interval of time between the successive guns of a salute; (*art.*) (in loading and firing) motion;
au ——, (*mil.*) as you were!
—— *d'arrêt*, stoppage, delay; (*mil.*) halt (esp. halt during infantry attack); (*man.*) action of the hand to slacken a gait;
beau ——, fine weather;
—— *de brouillard, de brume, brumeux*, foggy, misty, weather;
charge en deux, etc., ——, v. s. v. *charge;*
charger en deux, etc., v. s. v. *charger;*
—— *de chute*, time of fall;
—— *clair*, clear weather;
connaissance des ——, nautical almanac, ephemeris;
coup de ——, squall; (*fenc.*) time thrust;
—— *couvert*, thick weather;
—— *en sus*, after time, after hours;
—— *épais*, hazy weather;
à l'épreuve du ——, weather proof;
—— *fait*, settled weather;
finir son ——, (*mil.*) to serve out one's enlistment;
—— *forcé*, stress of weather;
—— *de galop*, (*man.*) short gallop;
galop à quatre, trois, ——, v. s. v. *galop;*
galoper à quatre ——, v. s. v. *galoper;*
—— *à giboulées*, showery weather;
—— *à, de, grain*, squally weather;
—— *gras*, thick weather;
gros, ——, stormy weather;
—— *de guerre*, war time;
—— *humide*, damp weather;
—— *incertain*, unsettled weather;
—— *de jambe*, (*man.*) application of the leg as an aid;
—— *de langue*, v. *appel de langue;*
—— *maniable*, moderate weather;
mauvais ——, bad, foul, weather;
—— *mort*, (*mach.*) lost time; (*s. c. art.*) time between end of aiming and fire of shot; gunner's time;
—— *nébuleux*, cloudy weather;
—— *noir*, thick weather;
—— *d'orage*, —— *orageux*, stormy weather;
—— *perdu*, (*s. c. art.*) lost time, i. e., time beyond latest measure of range and end of aiming; (*mach.*) back lash;
—— *perdu d'une vis*, lost motion, end play;
petit ——, fine weather, with light winds;
—— *de pluie*, ——, *pluvieux*, rainy weather;
—— *de pose*, (*phot.*) exposure;
—— *de repos*, (*mil.*, etc.) rest;
—— *rude*, rough weather;
—— *sec*, dry weather;
—— *serein*, clear weather;
—— *de service*, (*mil.*) period, length, of service;
—— *de suspension*, v. s. v. *suspension;*
tirer sur le ——, (*fenc.*) to make a time thrust;
—— *de trajet*, (*ball.*) time of flight;
—— *variable*, changeable weather;
—— *vilain*, foul weather.
tenable, a., that may be occupied; (of a road) practicable; (*mil.*) tenable, defensible, capable of defense.
tenace, a., tenacious, adhesive, sticky, cohesive, stiff.
ténacité, f., strength; tenacity, stiffness, adhesion; (*met.*) tenacity.
tenaille, f., pincers, nippers, tongs, pliers, pinchers (generally in pl.); (*fort.*) tenaille;
—— *bastionnée*, (*fort.*) bastioned tenaille;
—— *à chanfrein*, beveled nippers;
—— (*s*) *à contourner*, saw-set;
—— (*s*) *à couper*, cutting pincers;
—— (*s*) *à creuset*, (*met.*) crucible tongs;
double ——, (*fort.*) priest cap;

tenaille *à feu*, fire tongs;
—— *à flancs*, (*fort.*) tenaille with flanks;
—— *de forge*, —— *à forger*, —— *de forgeron*, smith's tongs;
—— *s goulues*, wide jawed tongs (opening wide);
—— *s justes*, narrow-jawed tongs (small opening);
ligne à ——*s*, (*fort.*) tenailled line;
—— (*s*) *à loupes*, (*met.*) bloom tongs;
—— *à main*, (*jarr.*) horseshoe tongs;
—— *renforcée*, v. —— *bastionnée;*
—— *simple*, (*fort.*) tenaille without flanks;
—— *à vis*, hand vise.
tenaillé, a., (*fort.*) tenailled.
tenaillon, m., (*fort.*) tenaillon.
tenant, m., (in *pl.*) lands bordering on an estate;
—— *s et aboutissants*, v. s. v. *aboutissant;*
tout d'un, en un, ——, in one piece; in a ring fence.
tendelet, m., awning, canopy, tilt.
tender, m., (*r. r.*) tender.
tendeur, m., stretcher; wire stretcher; coupling screw; (*mach.*) tightening pulley; (*cons.*) generic term for extension piece;
—— *à vis*, turn-buckle; (*r. r.*) screw coupler; (*art.*) chain stretcher (Canet 27^{cm} seacoast cannon).
tendière, f., (*cons.*) (horizontal) scaffolding pole.
tendineux, a., tendinous.
tendon, m., tendon, sinew; (*hipp.*) back sinew;
—— *blanchi*, (*hipp.*) overstretched tendon;
—— *claqué*, (*hipp.*) snapped tendon;
—— *failli*, (*hipp.*) close tendon (united to knee by too deep a depression);
—— *du jarret*, (*hipp.*) hamstring.
tendre, a., tender, soft, delicate;
pierre ——, soft stone.
tendre, v. a., to bend, stretch, hang, set up; to give, hold out; (*fenc.*) to stretch out the arms stiff;
—— *un camp*, (*mil.*) to pitch a camp;
—— *une embuscade*, (*mil.*) to set an ambuscade;
—— *une inondation*, (*mil.*) to prepare or make an inundation;
—— *le nez*, v. *porter au vent;*
—— *un piège*, to set a trap;
—— *une tente*, to pitch a tent.
tendu, p. p., tight, taut, bent, stiff; strained, stretched; (*ball.*) flat;
trajectoire ——*e*, (*ball.*) flat trajectory;
trop ——, strained, overstretched.
teneur, m., tenor; holder; (*met., chem.*) amount, quantity; (*sm. a.*) magazine holder or fastener;
—— *en carbone*, etc., (*met.*) per cent of carbon, etc. (present in iron or steel);
—— *de livres*, bookkeeper.
tenir, v. a. n., to hold; to possess, occupy, keep; to take up, fill (so much space); to contain; to hold (have room for); to stick, adhere; to stand by; to hold (of the weather); to resist, hold out against, stand; to keep (a shop, etc.); to hold (of an anchor, rope, etc.); (*mil.*) to hold out, hold one's position, resist; to hold a line, a position; to keep the field; (*man.*) to sit;
—— *à*, to be contiguous to;
—— *en alerte*, (*mil.*) to keep the enemy on the lookout, to harass him, to leave him no rest;
—— *bien à cheval*, (*man.*) to sit well, to have a good seat on horseback;
se —— *bien* (*mal*) *sur la trajectoire*, (*art.*) to be steady (unsteady) in its flight (of a projectile);
—— *bon*, (*mil.*) to hold on, fast, out; to stick, stand, fast; to stand one's ground; (*cord.*, etc.) to keep (after hauling in, etc.); to keep tight; to hold fast, make fast;
—— *bon sur un cordage*, (*cord.*) to keep, to hold fast; to avast heaving, hauling; to make fast, belay;
—— *bon le coup*, (*cord.*) to haul and hold;
—— *bon partout*, (*cord.*) to keep or hold everything;

tenir *la bride haute*, (*man*.) to raise the hand (in order to restrain a horse);
—— *campagne*, (*mil*.) to be and stay in the field with one's troops;
—— *la campagne*, (*mil*.) to hold the country;
—— *le chemin de*, to be following the road to;
—— *un cheval*, (*man*.) to support, keep up, a horse in his movements;
—— *un cheval en bride* (*main, talons*), (*man*.) to control a horse by the bridle (hand, spurs);
—— *un cheval dans la main*, (*man*.) to have complete control of a horse;
—— *un cheval sujet*, (*man*.) to hold up a horse short when he shies;
—— *les chevaux au filet*, to keep the snaffle on horses (to prevent their eating);
—— *la corde*, to have the inside track;
—— *court*, to hold up a horse short (in driving);
—— *croisière*, (*nav*.) to be cruising (in such and such waters);
se —— *sur la défensive*, (*mil*.) to be on the defensive;
—— *deux objets l'un par l'autre*, to see or keep two objects (e. g., masts) in one;
—— *en échec*, (*mil*.) to keep the enemy from carrying out his plans, to hold him in check;
—— *en embuscade*, (*mil*.) to lie in wait, in ambush;
—— *ferme*, v. —— *bon*;
se —— *sur ses gardes*, (*mil*.) to remain in one's position without seeking to take the offensive;
—— *garnison*, (*mil*.) to be in garrison (at ....);
—— *une garnison dans*, (*mil*.) to keep a garrison in;
—— *en haleine*, (*mil*.) to keep one's troops in condition to fight;
—— *les hanches*, (*man*.) to make a horse step sideways;
—— *un livre*, to keep a book (bookkeeping, etc.);
—— *la main*, to pull the bridle;
se —— *mal à cheval*, (*man*.) to have a poor seat;
—— *la mer*, (*nav*.) to keep the sea; to be master of the sea;
—— *la plume*, to be a secretary;
—— *de la pression*, (*steam*) to keep up steam;
—— *prison*, to be in jail;
—— *en respect*, (*mil*.) to hold or keep off (the enemy, his fleet, etc.); (of a fort) to dominate the surrounding country;
—— *à retour*, (*cord*.) to haul taut by degrees; to haul and hold;
—— *tête*, (*mil*.) to accept combat; to await and resist the enemy.

tenon, m., tenon, stud; (*sm. a.*) stud, bayonet stud; locking lug; brace quoin; (*art*.) stud of a projectile (obs.);
—— *en about*, end tenon;
—— *d'appui*, (*sm. a.*) recoil shoulder;
—— *de l'ancre*, nut of an anchor;
—— *d'arrêt*, stop, stopper (as, *art*., on a canister);
—— *d'arrêt du chien*, (*sm. a.*) cocking-piece stud;
—— *arrêtoir*, (*sm. a.*) lug, stop lug;
assemblage à —— *et mortaise*, (*carp*.) tenoning;
assembler à —— *et mortaise*, (*carp*.) to tenon; to mortise and tenon;
—— *d'attache*, securing, fixing, tenon;
—— *à came*, (*art., mach*.) cam tenon;
—— *de la capucine*, (*sm. a.*) band stud;
—— *à contre-clavette*, tenon and key;
——*s croisés*, cross tenons (tenon and mortise alternating);
—— *d'embouchoir*, (*sm. a.*) upper-band stud; bayonet stud;
—— *d'épée-baïonnette*, (*sm. a.*) bayonet stud (Lebel);
faux ——, false tenon, i. e., one not in one piece with the parts bound together;
—— *de fermeture*, (*sm. a.*) locking lug;
—— *de fusil*, (*sm. a.*) barrel lug or tenon;
—— *de levier*, —— *de manœuvre*, (*art*.) maneuvering bolt; embarring stud;
—— *passant*, (*carp*.) through tenon;

tenon-*pivot*, pivot stud;
—— *porte-hausse*, (*s. c. art*.) sight seat or socket;
—— *à queue*, — *en queue d'aronde*, (*carp*.) dovetail tenon;
—— *de recul*, (*sm. a.*) recoil shoulder;
—— *à renfort*, (*carp*.) tusk tenon;
—— *de sûreté*, (*sm. a.*) safety stud;
—— *de tête mobile*, (*sm. a.*) locking lug, bolt head tenon.

tenonner, v. a., (*carp*.) to tenon.

tenseur, a., stretching.

tension, f., tension; tension (of gases); tensile stress; (*elec*.) series;
assemblage, assembler, en ——, v. s. vv. *assemblage, assembler*;
—— *aux bornes*, (*elec*.) terminal voltage;
—— *chimique*, (*expl*.) chemical tension;
écrou à ——, tightening or adjusting nut;
montage, monter, en ——, v. s. vv. *montage, monter*;
—— *de régime*, normal tension;
—— *de la trajectoire*, (*ball*.) flatness of the trajectory.

tentative, f., trial, attempt, effort.

tente, f., tent; awning; pavilion; (*med*.) lint;
abattre une ——, to strike a tent;
—— -*abri*, —— *d'abri*, (*mil*.) shelter tent;
chandelier de ——, awning stanchion;
—— *conique*, (*mil*.) bell tent;
détendre une ——, to strike a tent;
dresser les ——*s*, to pitch tents;
—— *d'embarcation*, boat awning;
faîtière de ——, ridgepole of a tent;
fune de ——, awning ridgepole;
—— *goudronnée*, tarpaulin;
grande ——, (*nav*.) main-deck awning;
mât de ——, tent pole;
montant de ——, tent pole;
monter une ——, to raise a tent;
muraille de ——, v. *pente de* ——;
—— *de nage*, boat's awning;
pente de ——, tent wall, side wall;
piquet de ——, tent peg;
plier les ——*s*, to strike tents;
—— *de soldat*, bell tent;
tendre une ——, to pitch a tent;
—— *tortoise*, turtle-back tent;
transfilage d'une ——, awning lacing;
traverse de ——, v. *faîtière de* ——;
—— *de troupe*, soldiers', military, tent.

tentement, m., (*fenc*.) double beat against the adversary's blade.

tenue, f., holding; holding ground (for anchors); sitting (of an assembly); appearance, bearing, carriage, deportment, attitude; discipline; dress; (*man*.) seat; (*mil*.) uniform, dress, bill of dress; (sometimes) outfit; appearance, bearing;
avoir une bonne ——, (*man*.) to have a good seat, to sit well on horseback; (*mil*.) to be well set up, to have a soldier-like bearing;
n'avoir pas de ——, (*man*.) to have no seat; (of a saddle) to be badly shaped; (*mil*.) to have no bearing; to have an awkward, slovenly, carriage; not to be well set up; (of weather) to be unsettled;
—— *bahutée*, (*St. Cyr slang*) smart ("spoony" U. S. M. A.) dress;
bonne ——, good holding ground (anchors); (*man*.) good seat; (*mil*.) smart, soldier-like bearing;
en bonne ——, well appointed;
—— *bourgeoise*, (*mil*.) mufti, plain clothes, "cits";
—— *de campagne*, (*mil*.) service marching order, heavy marching order, field dress;
—— *de corvée*, (*mil*.) fatigue dress;
en ——, (*mil*.) in uniform;
—— *d'été*, (*mil*.) summer dress;
—— *d'exercice*, (*mil*.) drill order of dress;
—— *extérieure*, (*mil*.) dress worn on going out of quarters;
—— *de fantaisie*, (*Fr. a.*) nonregulation clothes of finer cut and cloth than the regulation (forbidden);
—— *de fatigue*, (*mil*.) fatigue, working, clothes;

tenue, *grande* ——, (*mil.*) full dress; review order (British army);
—— *de guerre,* (*mil.*) service dress;
—— *d'instruction,* (*mil.*) drill order (of dress);
—— *du jour,* (*Fr. a.*) the usual dress worn from 1 p. m.;
—— *des livres,* bookkeeping;
—— *du matin,* (*Fr. a.*) dress worn until 1 p. m., by officers and men;
mauvaise ——, bad holding (anchoring) ground; (*man.*) poor seat; (*mil.*) slovenly appearance, unsoldier-like bearing;
—— *de nuit,* (*mil.*) dress prescribed for the night;
—— *d'officier,* (*mil.*) officer's dress, outfit;
—— *d'ordonnance,* (*mil.*) regulation outfit or uniform supplied by government;
—— *de parade,* (*mil.*) full dress; review order (British army);
petite ——, (*mil.*) undress;
de première ——, (*unif.*) full dress, best (uniform, etc.);
rectifier la ——, (*mil.*) to inspect and correct a man's appearance, as for guard, etc.;
—— *de route,* (*mil.*) marching order;
selle sans ——, v. s. v. *selle;*
—— *de service,* (*mil.*) service dress;
—— *de soldat,* (*mil.*) soldier's (private's) dress, outfit;
—— *de sortie,* (*Fr. a.*) dress worn out of barracks;
—— *de sous-officier,* (*mil.*) n. c. o.'s outfit, dress;
—— *de travail,* (*mil.*) fatigue, working, dress;
—— *de ville,* street dress.

tenure, f., holding; hole in a block of slate to receive the wedge.

térébenthine, f., turpentine.

terme, m., term, limit, boundary, bound, termination, end; expression; rent, quarter day;
marché, opération, à ——, v. s. v. *marché.*

terminaison, f., termination, close, end; (*hipp.*) insertion of a muscle.

terminer, v. a., to end, limit, stop, finish, close, close up, get through, conclude, wind up, work off.

termite, f., white ant.

ternaire, a., ternary.

terne, a., tarnished, dull.

ternir, v. a., to tarnish, dull, dim.

terrage, m., (*artif.*) claying.

terragnol, m., horse with a heavy forehand, and hard to throw on his haunches.

terrain, m., ground (with special reference to the purpose for which it is to be used); piece of ground; tract, region, stretch of country; dueling ground; (*man.*) ground, track, covered by a horse; (*mil.*) terrain;
—— *accidenté,* (*top.*) rough country, full of natural cover;
aller sur le ——, to fight a duel;
—— *ameubli,* soft ground, for landing from a jump;
—— *argileux,* clayey ground;
attaquer le —— *au pic,* to work, loosen, the ground with the pick;
—— *des attaques,* (*siege*) ground beyond glacis, on which are situated the saps and mines;
—— *avantageux,* vantage ground;
—— *battu,* (*mil.*) space over which projectiles actually fall; all terrain on which bullets fall;
—— *broussailleux,* (*top.*) ground covered with brushwood;
céder le ——, to give way;
—— *de chicane,* (*mil.*) ground in dispute;
chicaner le ——, (*mil.*) to dispute the ground;
—— *de combat,* (*mil.*) battle, fighting, ground;
—— *compressible,* compressible ground (with respect to foundations);
—— *de construction,* building ground;
—— *coupé,* (*top.*) intersected, broken, country;
—— *coupé d'obstacles,* v. —— *coupé;*
—— *couvert,* (*top.*) close country;

terrain *dangereux,* (*fort.*) all ground held by enemy and within range; dangerous ground; undefiladed space; (more generally) any space over which projectiles fall or may fall;
—— *découvert,* (*top.*) open country;
défendre son ——, to hold out, hold one's ground;
—— *délayé,* v. —— *détrempé;*
—— *de derrière,* background;
—— *détrempé,* soft, muddy, ground;
—— *déversé,* sloping ground;
disputer le ——, v. *chicaner le* ——;
—— *dominant,* (*mil.*) commanding ground;
—— *élevé,* upland;
embrasser, du, son, ——, (*man.*) to go wide; to cover the ground;
—— *entrecoupé,* intersected, broken, ground;
—— *d'exercices,* (*mil.*) maneuvering ground, drill ground;
exproprier le ——, to condemn land (e. g., for military purposes);
—— *ferme,* (*fort.*) solid ground (as distinguished from sand or under-water foundations);
—— *fourré,* v. —— *broussailleux;*
gagner du ——, to gain ground;
garder le, son, ——, (*man.*) to keep on the track (horse or man);
garnir un ——, (*art.*) to cover the terrain thoroughly with projectiles;
—— *glaiseux,* loamy ground;
—— *glissant,* slippery ground;
—— *de gymnase,* gynasium ground, athletic field;
—— *houiller,* coal formation;
—— *inondé,* overflowed ground;
—— *d'instruction,* (*mil.*) drill or maneuvering ground;
—— *lourd,* heavy ground;
—— *de manœuvre(s),* (*mil.*) drill ground;
—— *marécageux,* swampy ground;
—— *militaire,* (*mil.*) the area comprised in the *servitudes militaires,* q. v.;
—— *montagneux,* hilly ground;
—— *mou,* soft ground;
—— *naturel,* natural ground;
observer le, son, ——, v. *garder le, son,* ——;
—— *d'opérations,* (*mil.*) field, scene, of operations;
—— *en pente,* sloping ground;
perdre du ——, to lose ground; (*man.*) to narrow;
—— *pierreux,* stony ground;
pli du ——, (*top.*) fold of the ground;
—— *rapporté,* made ground;
—— *rase,* (*mil.*) space in front of *terrain battu,* within which a man standing will be hit;
reconnaître le ——, (*mil., etc.*) to reconnoiter the ground;
se rendre sur le ——, v. *aller sur le* ——;
repli du ——, v. *pli du* ——;
—— *résistant,* hard ground;
—— *rocheux,* rocky ground;
—— *sablonneux,* sandy ground;
sonder le ——, to feel one's way;
sur le ——, on the ground, in the field;
—— *taluté,* sloping ground;
tâter le ——, to feel one's way, v. s. v. *tâter;*
—— *de transport,* —— *transporté,* v. —— *rapporté;*
—— *uni,* smooth, level, ground;
—— *qui va en descendant (montant),* falling (rising) ground;
—— *vague,* waste land;
—— *varié,* varied ground.

terral, m., land breeze.

terrasse, f., terrace, earthwork, platform; flat roof; foreground;
à ——, terraced;
—— *sur comble,* flat roof;
contre- ——, succession of terraces, one raised above another;
en ——, in the form of a terrace;
former en ——, to terrace;
—— *maçonnée,* perron;
travaux de ——, earthwork.

terrassement, m., earthwork (i. e., the moving, etc., of earth), banking, embankment;

terrassement, *conduite des* ——*s,* (*fort., etc.*) management, progression, of earthwork;
travaux de ——, earthwork.
terrasser, v. a., (*fort., etc.*) to execute earthwork; to fill in, bank up; to bank up a wall, a road, with earthwork; to prepare the ground for paving; to throw down on the ground; to fell, to floor;
se ——, (*mil.*) to intrench one's self.
terrassier, m., digger, excavator, navvy; person who contracts to do earthwork.
terre, f., earth, soil, mold, ground; the earth; potter's clay; land, estate, territory, dominions, country (esp. in pl.); (*nav.*) land, the shore; land as distinguished from sea service; (*mil.*) the army, as distinguished from *mer,* the navy; (*elec.*) earth, ground, connection;
à ——, ashore, on shore; aground, stranded:
aller à ——, —— *à* ——, (*man.*) to passage low;
—— *amendée,* v. —— *ré-animée;*
armée de ——, (*mil.*) army proper, land army;
artillerie de ——, (*art.*) land as distinguished from marine artillery;
—— *basse,* low, flat, land on seacoast;
—— *de beurre,* cape flyaway;
bien avant dans les ——*s,* far from the coast, far inland;
—— *à blé, etc.,* wheat, etc., ground;
—— *à briques,* brick clay;
chemin sous la ——, underground road;
commutateur de ——, (*elec.*) earth switch;
—— *coulante,* (*fort.*) loose (unrammed) earth, crumbling earth; earth taking its natural slope;
à —— *coulante,* at the natural slope;
courir ventre à ——, (of a horse) to go at full speed;
se couvrir de ——, (*siege*) to intrench;
—— *cuite,* terra cotta;
dans les ——*s,* inland, up the country;
être à ——, to be aground, ashore; to lie aground (of boats, ships, etc.);
—— *ferme,* mainland, continent (as distinguished from islands); dry land (as distinguished from waters); (*fort.*) solid ground (as distinguished from foundations in sand or under water);
fil de ——, (*elec.*) ground wire;
à fleur de ——, on a level with the ground; close to the ground;
forces de ——, (*mil.*) land forces;
—— *forte,* stiff earth; stiff, clayey, ground (hard to march through);
—— *à foulon,* fuller's earth;
—— *franche,* vegetable mold;
—— *en friche,* fallow land;
—— *s froides,* damp, clayey, ground;
—— *glaise,* pipeclay, potter's earth, luting;
—— *grasse,* loam;
haute ——, upland;
—— *haute,* high land on the coast, bold coast;
homme de ——, landsman;
—— *à n hommes,* (*fort., siege,* and *in gen.*) earth or earthwork that calls for 1 shovel to (*n*-1) picks; (hence) —— *à un homme,* earth that may be directly removed by the shovel (no pick); —— *à un homme* ½, work requiring 2 shovels and 1 pick; —— *à 2, 3, ... hommes,* work requiring 1 shovel and 1, 2, ... picks; —— *à 3 hommes* ½, work requiring 2 shovels and 5 picks, etc.;
—— *d'infusoires,* (*expl., etc.*) infusorial earth, kieselguhr;
—— *en jachère,* v. —— *en friche;*
—— *du Japon,* catechu;
jeter à ——, to throw, tumble, down;
—— *légère,* light, easily worked, ground;
mangé par la ——, landlocked;
—— *maritime,* seacoast, shore;
mettre à ——, to land, set on shore; to touch at a port;
mettre à la ——, (*elec.*) to ground;
mettre pied à ——, v. s. v. *pied;*
—— *à mouler,* (*fond.*) molding loam;
mouvement de ——, earthquake;
—— *moyenne,* (*fort.*) average earth, with respect to crumbling;
—— *neuve,* ground unplowed for a long time;
noyer la ——, to lose sight of land;

terre *d'ombre,* umber;
ouverture de la ——, (*siege*) breaking ground;
ouvrage de ——, (*fort.*) earthwork;
par ——, by land; on, upon, the ground; (*mil.*) by marching;
—— *pelletée à un relai,* earth reshoveled once;
perdre ——, to go swim, out of one's depth; (*nav.*) to lose sight of land;
—— *à pipe,* pipe clay;
—— *à pisé,* pisa earth;
plaque de ——, (*elec.*) ground plate:
—— *à pot,* —— *à potier,* potter's clay;
poterie de ——, earthenware;
—— *pourrie,* rotten stone;
prendre ——, to land, disembark, make land;
—— *de rapport,* —— *rapportée,* made ground;
—— *rassise,* (*fort.*) settled earth;
—— *ré-animée,* earth from which saltpeter is extracted;
reconnaître la ——, to see how the land lies;
—— *réfractaire,* fire clay;
relier à la ——, (*elec.*) to ground;
remuer de ——, to upturn, dig, move, earth (in construction, etc.); (*siege*) to open trenches, do trench work;
remuer la ——, *les* ——*s,* to dig; (*siege*) to intrench;
retour par la ——, (*elec.*) earth return;
sac à ——, (*siege, fort.*) sand bag;
—— *de Sienne,* sienna;
—— *de Sienne brûlée,* burnt sienna;
travailler —— *à* ——, v. *aller* —— *à* ——;
tremblement de ——, earthquake;
—— *vague,* unclaimed land;
—— *végétale,* v. —— *franche;*
—— *au vent,* weather shore;
vent de ——, land breeze;
—— *sous le vent,* (*nav.*) lee shore;
—— *vierge,* earth in its natural state, i. e., not shoveled; ground that has never been cultivated.
terre-à-terre, m., dull, routine, work; (*man.*) low passage (sort of slow gallop sidewise).
terreau, m., garden mold (decomposed leaves).
terre-plein, m., earthen platform; (*fort.*) terreplein; (in temporary fortification) bottom of the ditch behind the parapet; (in siege works) emplacement of the guns of the batteries of first and of second position;
—— *creusé,* (*fort.*) bottom of trench in rear of the parapet (field works).
terrer, v. a., to earth up; to cover with earth; (*artif.*) to plug with clay, with earth;
se ——, (*mil.*) to cover one's self by earth works; to intrench one's self.
terreur, f., terror;
—— *panique,* v. s. v. *panique.*
terreux, a., earthy.
terrine, f., earthenware pan;
—— *filtre,* earthenware filter (used in the manufacture of mercury fulminate).
terrir, v. n., (of boats) to make a landing.
territoire, m., territory;
—— *civil,* (Algiers and Tonkin) territory administered by civilian functionaries;
—— *maritime,* territorial waters;
—— *militaire,* (Algiers and Tonkin) territory administered by the military;
—— *réservé,* (in France) part of the *zone frontière,* as well as all ground within 10 kilometers of a *place forte,* for which severe restrictions exist as to building, etc.;
—— *spécial,* (*Fr. fort.*) area in which woods may not be cleared.
territorial, a., territorial; (*law*) (of the seas) within the 3-mile limit; (*mil.*) in France, belonging to the territorial army; m., (*Fr. a.*) member of the territorial army;
armée —— *e,* (*Fr. a.*) territorial army;
mer —— *e,* seas within the 3-mile limit.
territoriale, f., (*Fr. a.*) the territorial army.
territorialité, f., territoriality.
territorien, m., (*Fr. a.*) member of the territorial army.
terroir, m., soil, ground.
tertre, m., (*top.*) hillock, hummock, knoll, eminence, rising ground, little hill.

tesson, m., potsherd.
test, m., (*met.*) test; cupel.
testament, m., testament, will;
—— *militaire*, soldier's will (made without the formalities usually considered necessary).
testimonial, a., testimonial;
lettre ——*e*, testimonial of good conduct;
preuve ——*e*, proof by witness.
têt, m., v. *test;*
——*s et blocailles*, broken-up castings.
tétanos, m., (*hipp.*) stag-evil; lockjaw.
tétard, m., tongue, foot, of a pole, shaft; square end going in between the guides (*queue d'armons*); pollard tree.
tête, f., head (of a bolt, rivet, etc.); head (of cattle); (in races) head; front end of anything; steering head of bicycle; face of a stone; facing; outside face; face of an arch; life; head (of a canal, of a wood, etc.); (*mil.*) head (of a column, advance guard of an army, etc.); front, front rank; body of troops (thrown out for any purpose; rare);
à la ——, at the head, in front (as in front of a company);
—— *d'affût*, (*art.*) head plate of a gun carriage;
—— *d'alouette*, (*cord.*) crown knot;
—— *d'alouette double*, (*cord.*) double crown knot;
—— *d'alouette simple*, (*cord.*) single crown knot;
—— *d'assemblage*, (*mach.*) connecting point;
—— *en avant*, ahead;
—— *d'avant-garde*, v. s. v. *avant-garde;*
avoir la —— (*en*) *dedans*, (*man.*) to carry the head turned in;
avoir la —— *au mur*, (*man.*) to carry the head turned out (toward the wall);
avoir quelqu'un en ——, to have some one as a rival, an adversary;
—— *baissée*, headlong, undaunted;
—— *de balancier*, v. s. v. *balancier;*
bataillon de ——, (*mil.*) leading battalion;
—— *de bélier*, double hook;
—— *de bielle*, (*mach.*) crosshead (of the rod); stub end; connecting rod (head or end);
—— *de bielle fermée*, (*mach.*) solid head or end;
—— *de bielle à fourche*, (*mach.*) forked end of connecting rod;
—— *de bielle ouverte*, (*mach.*) open end with a strap and key connection;
—— *bien attachée*, (*hipp.*) well-attached head;
—— *de bordage*, (*nav.*) butt end of a plank;
—— *bouchon*, (*sm. a.*) cylinder head, bolthead;
boulet à deux ——*s*, (*art.*) bar shot (obs.);
—— *busquée*, (*hipp.*) arched head (convex face);
—— *de cabestan*, capstan head;
—— *de camp*, (*mil.*) front of the first line, nearest to the enemy; general parade of a camp;
—— *camuse*, (*hipp.*) snub-nosed head;
—— *carrée*, (*hipp.*) straight head;
casser la ——, to shoot a soldier;
—— *de champignon*, mushroom head (of a bolt, of a rail, etc.);
—— *du chapeau*, (*pont.*) trestle cap end;
—— *de chat*, undressed stone more or ess round in shape;
—— *de cheval*, (*mach.*) in a lathe, the slotted bar for the change wheels;
—— *de chevalement*, (*carp.*) crosspiece supported by two stays or props;
—— *du chien*, (*sm. a.*) hammer head;
—— *d'un clou*, head of a nail;
—— *de colonne*, (*mil.*) head of a column;
compagnie de ——, (*mil.*) leading company;
—— *d'un compas*, head or joint of a pair of compasses;
conserver la —— *directe*, (*mil.*) to keep the eyes to the front, as in marching past;
course de la ——, (*man.*) riding at head (lance, saber, or pistol);
—— *croisée*, (*mach.*) crosshead;
de ——, leading;
de —— *en* ——, from stem to stern;

tête *décousue*, v. —— *mal attachée;*
—— *de défilé*, (*top.*) head of a defile next the enemy;
dessus de ——, v. s. v. *dessus;*
donner une ——, to dive headforemost;
—— *à douille*, (bicycle) steering head;
—— *à droite* (*gauche*), (*mil.*) eyes right (left);
—— *d'écouvillon*, (*art.*) sponge head;
en ——, in front, at the head of;
en —— *et en queue*, front and rear;
—— *d'étapes de guerre*, v. s. v. *étape;*
—— *d'étapes de route*, v. s. v. *étape;*
être en ——, (*hipp.*) to be marked about the head (v. s. v. *cheval*);
faire —— *à*, to cope with, to make head against, to face;
faire —— *de colonne*, (*mil.*) to be at the head of a column, to be the head of column;
faire une ——, (*man.*) to make (i. e., strike, etc.)
e. head in riding at the heads;
—— *fixe*, v. *chantier de commettage;*
fourgon de ——, (*r. r.*) baggage car;
—— *fourchée*, forked head;
—— *à fraiser*, (*mach.*) broach, broaching head;
—— *de gâchette*, (*sm. a.*) nose of the sear;
grosse —— *de bielle*, (*mach.*) crank end of the connecting rod;
homme de ——, bold, resolute, man;
—— *de lanterne*, (*art.*) head of a gun ladle;
—— *de lièvre*, (*hipp.*) hare-faced head (convexity limited to forehead);
—— *s de ligne*, (*r. r.*) termini of a railway;
—— *mal attachée*, (*hipp.*) head whose parotid channel is too pronounced;
—— *de manœuvre*, working handle of a drill or boring instrument;
marcher en ——, to lead the way, march in front;
—— *d'un marteau*, etc., head of a hammer, etc.;
marteau à deux ——*s*, double-faced hammer;
—— *de maure* (*more*), (*cord.*) crown knot; v. *cheval cap de maure;*
—— *mobile*, (*sm. a.*) movable head, movable bolthead; (*art.*) movable head (of the breech plug); (*mil. slang*) musketry instructor; (*cord.*) v. *chariot à commettre;*
—— *de mort*, skull; (*artif.*) carcass, incendiary shell; (*cord.*) single crown knot; (*carp.*) hole in which a dowel is broken off below the surface of the piece;
—— *morte*, caput mortuum;
—— *moutonnée*, (*hipp.*) ram's head (convexity on chamfer);
—— *de mur*, thickness of a wall at its end;
obus —— *de mort*, v. s. v. *obus;*
—— *à oreilles*, winged head (as of a screw);
—— *d'un ouvrage à corne*, —— *à couronne*, (*fort.*) front of a horn, crown, work;
—— *perdue*, countersunk head;
petite —— *de bielle*, (*mach.*) crosshead end of a connecting rod;
piquer une ——, v. *donner une* ——;
—— *à pivot*, steering head (of a bicycle);
—— *plaquée*, (*hipp.*) head plastered on (parotid groove effaced);
—— *de platine*, (*r. f. art.*) lock head;
—— *plongeante*, (*fond.*) deadhead;
—— *d'un pont*, approach to a bridge;
—— *de pont à bascule*, front crossbeam of a draw bridge;
—— *porte-outil*, (*mach.*) tool-holder;
porter la —— *sur l'échafaud*, to be beheaded;
—— *de potence de pompe*, cheeks of a pump which support the brake;
—— *-à-queue*, complete turn of a horse to the rear (as, e. g., in refusing a jump);
—— *de rayage*, (*mach.*) rifling head;
—— *de refouloir*, (*art.*) rammer head;
—— *de régiment*, (*mil.*) head of a regiment;
—— *de rhinocéros*, (*hipp.*) rhinoceros head (with concave chamfer);
—— *à la romaine*, round head of a (large) screw;
—— *ronde*, snap head of a rivet;
—— *de sape*, (*siege*) head of the sap;
—— *de sellette*, (*art.*) round part of the bolster of a limber, through which the pintle hole passes;
—— *à six pans*, hexagonal head;

tête *de tampon*, buffer head;
   (*tant de*) ——*s cassées*, (so many) men killed;
   tenir —— *à*, v. *faire* —— *à*;
   tenir la ——, (of a candidate) to have the majority of votes;
   tenir la —— *d'une attaque*, (*mil.*) to lead an attack (said of leading unit);
   —— *de tige*, (*mach.*) crosshead of the piston rod;
   —— *de tranchée*, (*siege*) head of the trench;
   en —— *du travail*, (*mil. min.*) at the head of the mine (of the work);
   —— *de Turc*, head on which blows may be delivered to drive a rod, etc., down (as in devices for boring for water); (*mil. min.*) adjustable striking head (used on a boring bar in resisting soil);
   —— *de vielle*, (*hipp.*) fiddle-shaped, hurdy-gurdy, head;
   —— *de vis*, screw head;
   —— *de voussoir*, face of a voussoir.
tête-bêche, adv., head to tail (of objects adjacent).
tête-de-pont, m., (*fort.*) bridge head;
   —— *double*, double bridge head (i. e., one on each bank).
tetière, m., (*harn.*) headstall; (*pont.*) thwart;
   —— *de bateau*, bow-(stern-) piece of a boat;
   —— *de soufflet*, lug or piece of wood to which the rock staff of a bellows is attached.
tetine, f., dent on a cuirass by a rifle bullet; (*artif.*) nipple of a rocket mold.
teton, m., any small projection or projecting piece; point, end, stud; (*top.*) small round elevation;
   à ——, fitted with, or having, a point or stud;
   —— *de grain de lumière*, (*art.*) end of the bush that enters the bore and is afterwards removed.
tétragone, m., tetragon, quadrangle.
tétrapolaire, a., (*elec.*) tetrapolar.
tétratomique, a., (*chem.*) tetravalent.
tétraposte, f., combination of four pulleys.
tétu, m., mason's cavel;
   —— *d'arête*, cavel.
tétué, a., worked with a cavel.
teugue, f., (*nav.*) forecastle.
texte, m., text;
   —— *chiffré*, enciphered text;
   —— *clair*, text written out in full, in plain words.
texture, f., (*met.*, etc.) texture.
thalweg, m., thalweg;
   axe du ——, axis of a stream, navigable channel of a stream;
   —— (*d'un cours d'eau*), position of the thread moving with the greatest velocity;
   —— *d'une vallée*, the intersection of the slopes of a valley.
thé, m., tea.
théâtre, m., theater; scene of an event; pile of wood in a dockyard; (*powd.*) platform, stage, on which powder is dried;
   —— *de* (*la*) *guerre*, (*mil.*) seat of war, theater of war;
   —— *d'opérations*, (*mil.*) theater of operations.
thème, m., theme, subject, exercise;
   —— *de manœuvre*, (*mil.*) problem to be worked out on varied ground.
théodolite, m., (*inst.*) theodolite;
   —— *autographe*, photographing theodolite;
   —— *à lunette centrée*, a theodolite with axial telescope;
   —— *à lunette excentrique*, a theodolite with eccentric telescope;
   —— *réitérateur*, a theodolite that repeats the reading, whose mean is the angle sought;
   —— *répétiteur*, a theodolite that multiplies the angle observed, so as to minimize the error;
   —— *télémètre*, sort of universal surveying instrument.
théorie, f., theory; (*Fr. a.*) books, works, maps, etc., whose purchase is authorized for the various schools or periods of theoretical instruction; any theoretical instruction given in quarters on various duties, of no matter what sort, (more esp.) drill book;
   arme de ——, v. s. v. *arme*;

théorie, *faire la* ——, (*mil.*) to give or receive theoretical instruction;
   ——*s et placards*, (*Fr. a.*) generic term for the concrete elements of the *théorie* (books, diagrams, etc.);
   —— *pratique*, (*Fr. a.*) application to the terrain of various written principles (applied particularly to the drills of *cadres*);
   —— *récitative*, (*Fr. a.*) recitations in drill, etc. (n. c. o. and candidates for n. c. o. warrants).
théorique, a., theoretical.
thermique, a., relating to heat;
   machine ——, heat engine.
thermo-avertisseur, m., heat alarm.
thermo-baromètre, m., thermobarometer.
thermo-cautère, m., (*hipp.*) thermocautery.
thermochimie, f., thermochemistry.
thermodynamique, a., f., thermodynamic, thermodynamics.
thermo-électricité, f., thermoelectricity.
thermo-électrique, a., thermoelectric.
thermo-magnétique, a., thermomagnetic.
thermo-magnétisme, m., (*elec.*) thermomagnetism.
thermo-manomètre, m., thermometer gauge.
thermomètre, thermometer;
   —— *à air*, air thermometer;
   —— *à alcool*, spirit thermometer;
   —— *à bulle mouillée*, wet-bulb thermometer;
   —— *à bulle sèche*, dry-bulb thermometer;
   —— *centigrade*, centigrade thermometer;
   —— *différentiel*, (*elec.*) thermopile, thermoelectric pile;
   —— *à esprit de vin*, v. —— *à alcool*;
   —— *étalon*, standard thermometer;
   —— *à gaz*, gas thermometer;
   —— *à liquide*, liquid thermometer;
   —— *à maxima*, maximum thermometer;
   —— *à maxima et minima*, maximum and minimum thermometer; self-registering thermometer;
   —— *à mercure*, mercury thermometer;
   —— *à minima*, minimum thermometer;
   —— *métallique*, metallic thermometer.
thermoscope, m., thermoscope.
thermostat, m., thermostat.
théta X, m., (*fam.*) a second-year cadet of the Polytechnique.
thibaude, f., haircloth; cow-hair cloth.
thornycroft, m., (*nav.*) torpedo boat (Thorneycroft).
thran, m., train oil.
tic, m., (*hipp.*) any stable vice;
   —— *en l'air*, —— *en l'air sans appui*, —— *en l'air sans usure des dents*, wind sucking;
   —— *à l'appui*, trick of taking hold of objects with the teeth;
   —— *à l'appui avec usure des dents*, cribbing;
   —— *à l'appui sans usure des dents*, v. —— *à l'appui*;
   —— *avec appui et usure des dents*, cribbing;
   —— *de l'ours*, weaving;
   —— *rongeur*, habit of eating earth, of crib biting, etc.
tierce, f., (*fenc.*) tierce;
   —— *basse*, second tierce;
   porter une, en, une botte en, ——, to thrust in tierce.
tiers, a., m., third; third part; third person; third party;
   ——*s arbitre*, v. *sur-arbitre*;
   —— *ferme*, (*cord.*) hard laid;
   —— *mou*, (*cord.*) soft laid;
   tierce personne, third person.
tiers-point, m., saw file; three-cornered file; (*cons.*) arch of the tierce.
tiers-poteau, m., (*cons.*) small post for slight partitions.
tige, f., stem, boll, of a tree; stalk; rod, stem, shank; shaft, spindle; entire upper leather of a shoe or laced boot, upper and leg of a boot; (*r. r.*) web of a rail; (*sm. a.*) shank of a firing pin; (*art.*) obturator spindle; breech-plug spindle; (*ball.*) rod of the Le Boulengé chronograph; (*mach.*) spindle, rod, piston rod;
   —— *d'ancre*, anchor shank;

tige, (*arbres à*) *basses* ——*s*, trees whose stems are kept down;
(*arbres à*) *hautes* ——*s*, trees whose stems are allowed to grow;
—— *d'arrêt*, (*r. f. art.*) bolt, stop bolt, stop spindle, firing-pin stop;
—— *d'articulation* (*de la pompe à air*), pump rod and links;
—— *articulée*, jointed rod;
—— *d'assemblage*, (*mach.*) link;
bois de ——, forest trees of full growth;
carabine à ——, v. s. v. *carabine*;
—— *de choc*, (*r. r.*) buffer rod; (of a spar torpedo) contact spindle or stem;
—— *de clef*, shank of a key;
—— *de colonne*, shank of a column;
—— *de communication*, (*mach.*) connecting rod;
—— *conductrice*, (*mach.*) guide rod; steam winch connecting rod;
—— *à courroie*, (*mach.*) strap bar, rod strap;
—— *de déclanchement*, (*art.*) piston (of the Merriam fuse);
—— *directrice*, (*mach.*) guide rod, slide rod, tail rod; slide bar;
—— *d'équerre*, arm of a square;
—— *d'excentrique*, (*mach.*) eccentric rod;
—— *filetée*, threaded rod;
—— *de foret*, (*mach.*) boring bar;
—— *du frein*, (*r. r.*) brake rod;
—— *d'une gâchette de platine*, (*sm. a.*) shank of the sear;
—— *de glissoire*, (*mach.*) motion bar;
—— *graduée*, index rod;
—— *graduée de la hausse*, (*art.*) tangent scale or rod;
—— *de guide*, guide rod;
—— *latérale*, (*mach.*) side rod;
—— *de levage*, (*mach.*, etc.) lifting rod, lift;
—— *de la noix*, (*sm. a.*) tumbler axle;
—— *de manœuvre*, (*mach.*, etc.), working rod (e. g., a power-transmitting rod);
—— *de manœuvre*, —— *de refoulement*, Lemoine brake;
——*s de parallélogramme*, (*mach.*) parallel motion; arm or bar of a parallel motion;
—— *de paratonnerre*, lightning rod;
—— *de pendule*, pendulum rod;
—— *percutrice*, (*sm. a.*) striking pin;
—— *de petit cheval*, donkey-pump rod;
—— *de piston*, (*mach.*) piston rod;
—— *à piton*, rod and eye;
—— *à pompe*, hollow stem upon which a piece (*pompe*) may slide;
—— *de pompe*, (*mach.*) pump rod;
—— *porte-aiguille*, (*sm. a.*) needle bolt, needle holder.
—— *de réglage*, adjusting rod or bar;
—— *de régulateur*, (*mach.*) governor arm or rod;
—— *à, de, ressort*, (*art., etc.*) spring pin;
—— *de serrage*, (*art.*) keep or securing pin;
—— *de sondage*, —— *de sonde*, boring rod;
—— *de soupape*, (*mach.*) valve spindle;
—— *de soupape d'arrêt*, (*mach.*) stove-valve rod;
—— *de suspension*, suspension rod (of a suspension bridge);
—— *de la tête mobile*, (*art.*) obturator spindle;
—— *de tiroir*, (*mach.*) valve rod, slide rod, valve spindle or stem, slide-valve spindle;
—— *de transmission*, (*top*) immersion regulator rod (Whitehead);
—— *de vanne*, m., sluice-valve rod;
—— *à vis du frein*, brake screw.

tigré, a., (*hipp.*) spotted, leopard-spotted;
cheval ——, leopard-spotted horse.

tigrure, f., (*hipp*) leopard spot in coat.

tillac, m., (*nav.*) deck (of a merchantman, of a river boat);
clou de ——, tenpenny nail;
clou de demi- ——, sixpenny nail;
franc ——, flush deck.

tille, f., bast; hemp bark; cooper's adz; (*nav.*) cuddy; fore cabin, stern cabin.

tilleul, m., lime tree and wood.

timbale, f., kettledrum, timbal; metal cup.

timbalier, m., kettledrummer; timbal player.

timbre, m., (striking) bell; snare of a drum; stamp, postage stamp, postmark; shell of a helmet; in construction memoirs, the amount and nature of work to be done, written as a catchword opposite each article of the memoir;
adresser sous le —— *de*, address as above (of official letters);
—— *d'alarme*, bicycle bell;
—— *d'annulation*, cancellation stamp.
—— *d'arrêt* (*fort.*) in a turret, bell that rings when embrasures are 180° away from the dangerous sector;
bureau de ——, stamp office;
——*-cachet*, seal press;
——*-dépêche*, telegram stamp;
—— *de dimension*, stamp varying in cost with the size of the paper to which it is attached;
droit de ——, stamp duty;
—— *électrique*, electric bell;
—— *à empreinte*, embossed stamp.
—— *extraordinaire*, a stamp put on a document that should have been written on stamped paper;
—— *fixe*, fixed stamp (impression);
—— *humide*, colored stamp;
—— *losange à pointes*, cancel;
—— *mobile*, attachable stamp (as distinguished from the stamp on stamped paper;
——*-poste*, postage stamp; (*sm. a.*) cartridge (sporting, fam.);
—— *proportionnel*, proportional stamp (varies with amount involved);
—— *-quittance*, —— *de quittance*, receipt stamp (on receipts for more than 10 francs);
—— *sec*, plain stamp (not colored);
—— *de tambour*, snare (of a drum);
——*-télégramme*, v. ——*-dépêche*.

timbrer, v. a., to stamp; to postmark a letter; to head a document with the date and place, and a summary of its contents.

timon, m., pole (of a wagon, gun carriage); (*nav.*) tiller;
—— *de rechange*, (*art*) spare pole;
support de ——, pole yoke.

timonerie, f., (*nav.*) wheelhouse; the quartermasters of a ship.

timonier, m., (*art.*) wheel horse, wheeler; (*nav.*) helmsman.

timpe, f., tympe.

tin, m., (*nav.*) stock, block, for ship's timbers (keel, floor timbers, in construction); boat chock.

tine, f., water cask; (*powd.*) powder tub; (*min.*) tub, kibble.

tinette, f., small tub; soil tub.

tintage, m., blocking, chocking, of casks, etc.

tintenague, f. v., *toutenague*.

tinter, v. a., (*nav.*) to put upon the stocks; to prop, support, with blocks of wood.

tion, m., (*met.*) tool for cleaning crucibles.

tioul, m., (*fond.*) skimmer used in foundries.

tiquer, v. n., (*hipp.*) to have a stable vice; to suck wind, be a crib biter, eat dirt, etc.;
—— *à l'appui*, to crib proper.

tiqueur, m., (*hipp.*) horse that has a stable vice (crib biter, dirt eater, etc.).

tir, m., (*ball., art., sm. a., f. p.*) firing, fire, shooting, target practice; gunnery; round; set or series of shots; shooting gallery, firing ground, practice ground, range.
  I. Artillery, ballistics; II. Small arms, target practice.
  I. Artillery, ballistics:
allonger le ——, v. s. v. *allonger*;
angle de ——, (*ball.*) angle of fire;
—— *attelé*, firing with horses hitched to the carriage;
—— *au-dessous de l'horizon*, v. —— *négatif*;
—— *autour de l'horizon*, all-around fire;
—— *balistique*, ballistic firing, proving-ground practice, practice for ballistic results;
—— *à balles*, shrapnel firing;
—— *sur ballon*, firing at balloons;
—— *à barbette*, barbette firing (U. S. Art.); overbank fire (R. A.);

tir *de batterie,* battery target practice;
boîte pour le —— *fictif.* (*Fr. art.*) case of iron filings used in the —— *fictif,* q. v.;
—— *de bombardement,* shelling;
—— *à boîtes à balle,* shrapnel fire;
—— *en bombe,* v. —— *à feux verticaux;*
—— *des bouches à feu,* artillery fire;
—— *à boulet,* fire with solid shot;
—— *à boulet rouge,* firing red-hot shot (obs.);
—— *en brèche,* breaching fire;
—— *de bricole,* firing at one portion of a work, and under a small angle, so as to glance and hit some part out of reach of direct fire (obs.);
—— *brisant,* practice or fire with high-explosive shells;
bulletin de ——, v. s. v. *bulletin;*
—— *de but en blanc,* (*ball.*) point-blank firing;
—— *contre but mobile,* practice at a moving target;
—— *contre but immobile,* practice at a fixed target;
—— *sur but visible* (*invisible*), practice at visible (invisible) target;
circulaire de ——, direction-aiming circle in cupolas;
—— *convergent,* converging fire;
—— *courbe,* —— *curviligne,* curved fire;
—— *de démolition,* (*siege*) plunging fire on all masonry to be loosened when the breach can not be made directly;
—— *à démonter,* fire the object of which is to dismount hostile pieces (said esp. of siege);
déplacement du ——, shifting of the aim or practice to a new target;
déplacer le ——, to shift the aim or practice to a new target;
—— *au-dessous de l'horizon,* plunging fire;
—— *direct,* direct fire;
—— *en direction,* fire as respects direction;
—— *à distance fictive,* (of experimental firing, as at armor plates, etc.) fire at reduced distances (using reduced charges, so that the initial shall equal the remaining velocity under service charge and service ranges);
—— *à dos,* normal reverse fire;
—— *d', en, écharpe,* oblique fire (from the front);
—— *échelonné,* fire with combined elevations;
école de ——, target practice; school of practice; target season; period of instruction in firing;
—— *d'efficacité,* fire against the enemy after the range has been found;
—— *d'élévation,* v. —— *élevé;*
—— *élevé,* firing at an object above the plane of piece; firing down at an object (rare);
—— *à embrasure,* fire through an embrasure;
—— *d'enfilade,* enfilade fire;
—— *d'ensemble,* ranging practice; trial shots to determine position of target in zone of dispersion corresponding to angle of elevation used;
—— *d'épreuves,* proof firing in general;
—— *d'essai,* ranging practice, trial shots; experimental firing;
—— *à faible charge,* practice with reduced charges;
—— *à feux verticaux,* vertical fire;
—— *fichant,* downward fire, giving an angle of fall greater than 4°.
—— *fictif,* (*Fr. art.*) dummy firing (i. e., firing of cases filled with filings from guns in service, using full service charges, for the purpose of testing brakes and the strength of carriages and of platform; method followed when the range is limited);
flanc de ——, v. s. v. *flanc;*
—— *de flanquement,* flanking fire;
—— *à forcement,* fire in which the projectile is forced to the grooves;
—— *à la fourchette,* bracket firing to get the range, ranging;
—— *fusant,* time-fuse practice; rocket practice;

tir *fusant de plein fouet,* time-fuse direct fire;
—— *fusant plongeant,* time-fuse plunging fire;
—— *de groupe,* group target practice;
—— *de guerre,* practice with full service charges;
—— *en hauteur,* fire as respects elevation;
—— *incliné,* shot fired under horizon;
—— *indirect,* indirect fire;
—— *d'inspection,* special practice (in Russia) ordered by general officers to test capabilities of c. o. of batteries or of groups of batteries;
—— *lent,* slow fire (one round per minute);
ligne de ——, (*ball.*) line of fire;
—— *en ligne droite,* fire under a small angle with full charge;
—— *masqué,* practice at an invisible or unseen target;
—— *de matage,* firing bronze gun with overcharges before final boring, so as to obtain permanent enlargement of powder chamber (obs.);
—— *à la mer,* seacoast practice;
—— *à mitraille,* case-shot fire; in general, any "fragmentation" fire;
—— *montant,* fire at a target above the plane of the piece;
—— *en mouvement,* fire while the turret is rotating;
—— *négatif,* fire under an angle of depression;
—— *au niveau,* practice with quadrant elevation;
—— *de nuit,* night firing, night practice;
—— *oblique,* oblique firing; practice producing an oblique impact;
—— *à obus,* shell practice, shell fire;
—— *à obus à balles,* shrapnel fire;
—— *optique,* practice in which aiming is done by optical means;
—— *ordinaire,* normal rate of fire (two to three rounds per minute);
origine du ——, (*ball.*) origin of fire (muzzle of the piece);
—— *percutant,* practice with percussion shell;
—— *percutant plongeant,* plunging percussion fire;
plan de ——, (*ball.*) plane of fire;
planchette de ——, v. s. v. *planchette;*
—— *de plein fouet,* direct fire with full charges;
—— *de plein fouet direct,* direct fire against visible targets (full charge, flat trajectory);
—— *de plein fouet indirect,* practice against invisible targets (reduced charges, trajectory more or less curved; Brialmont);
—— *sur plate-forme,* fire from a platform;
—— *plongeant,* plunging fire; mortar fire;
—— *plongeant fusant,* time-shell plunging fire;
—— *à pointage indirect,* indirect fire;
—— *par le pointage intérieur,* fire or practice by indirect aiming;
—— *de polygone,* proving-ground fire; range as distinguished from service practice; (*fig.*) accurate fire;
—— *positif,* fire under angles of elevation;
—— *aux pressions,* practice to determine pressures;
—— *progressif,* practice consisting in covering systematically, by progressive zones, the belt situated between the observed inferior limit (short shot) and the superior observed or estimated limit (time shell in sets of six shots, varying by one turn of elevating-screw handle);
—— *des projectiles,* target practice in general;
—— *rapide,* rapid fire (10 to 12 rounds per minute);
à —— *rapide,* rapid fire, quick firing;
—— *rasant,* grazing fire; fire parallel to surface of ground (angle of fall less than 3° or 4°);
—— *de recette,* —— *de réception,* acceptance firing test;
—— *réduit,* practice with reduced charges; sub-caliber practice;

tir *réel*, fire under service conditions as to range, charge, etc. (e. g., said of ballistic firing as compared with —— *à distance fictive*);
réglage du ——, v. s. v. *réglage;*
—— *de réglage*, ranging practice;
—— *réglé*, v. s. v. *réglé;*
règlement de ——, firing regulations, manual;
régler le ——, v. s. v. *régler;*
régler le —— *à la fourchette*, v. s. v. *régler;*
relever le ——, to fire higher, give more elevation;
—— *repéré*, practice corrected by reference to observed points; corrected practice;
repérer le ——, v. s. v. *repérer;*
report du ——, on a change of target, the application to the new ranging of the data obtained for the first target;
reporter le ——, on a change of target, to apply to the new ranging the data obtained for the first target;
—— *de résistance*, firing test, to determine strength;
—— *à revers*, oblique reverse fire;
—— *à ricochet*, ricochet fire;
—— *en rouage*, enfilade, flank, fire (against material objects);
—— *roulant*, smoothbore mortar fire, using small charges, so that shell may roll over crest (obs.);
—— *de rupture*, armor-piercing fire;
—— *de, par, salve*, salvo;
—— *de siège*, siege practice;
—— *simulé*, artillery-fire game;
—— *surbaissé*, downward fire;
—— *à surcharge*, overcharge firing test; test with overcharge;
table de ——, ballistic table, range table;
—— *tendu*, any fire, giving a flat trajectory;
—— *à toute volée*, fire with maximum charge and at maximum elevation, used in proof only; firing at random;
transport du ——, the range being known to a given target, the transfer of the data to a different target; change from percussion to time fire;
transporter le ——, to transfer from one target to another; to change from percussion to time practice;
tube à ——, subcaliber tube;
—— *vertical*, vertical fire;
—— *aux vitesses initiales*, practice to determine muzzle velocity.

II. Small arms, target practice and miscellaneous:
—— *accéléré*, quick firing;
affût de ——, rifle rest;
—— *d'affût*, fire from a rest or stand;
—— *ajusté*, well-aimed fire or practice;
—— *d'amorce*, primer fire, to test cartridge cases;
—— *d'application, (t. p.)* skirmish practice;
appui de ——, firing rest;
—— *sur appui*, fire from a rest;
—— *des armes à feu*, the whole art of pointing and firing guns, etc., accurately;
—— *des armes portatives*, small-arm fire;
assurer le ——, to make sure of one's aim (as of a firing party);
avoir le —— *juste*, to shoot straight;
—— *à balles*, ball practice;
—— *en belle, (nav.)* fire over the beam; horizontal fire;
—— *à, en, blanc*, fire with blank cartridges;
—— *à bras francs*, offhand practice or firing;
—— *de campagne*, field firing;
capitaine de ——, v. s. v. *capitaine;*
—— *à capsule à balle, (sm. a.)* gallery practice;
carnet de ——, v. s. v. *carnet;*
carnet de —— *de bataillon*, v. s. v. *carnet;*
—— *à cartouches à balles*, case-shot firing;
—— *en chambre*, —— *de chambrée*, gallery practice;
champ de ——, v. s. v. *champ;*
chasse au ——, shooting (sport);
—— *en chasse, (nav.)* fire ahead, bow fire;
—— *en chasse directe, (nav.)* fire straight ahead;
—— *à la cible*, target practice;

tir *collectif*, collective fire, group firing (body of troops);
—— *de combat*, actual practice against an enemy; practice under conditions of active service, i. e., at unknown distances;
—— *de combat collectif, (inf.)* collective service practice;
—— *de combat individuel, (inf.)* individual service practice (to teach use of cover, of natural rests);
—— *de concours*, match, competition, firing; competition practice;
—— *continu*, self-cocking fire of a revolver;
—— *couché, (inf.)* practice or fire lying down;
—— *à couler bas, (nav.)* fire, shot, between wind and water;
—— *coup par coup*, fire, shot by shot; single-shot fire (said of a repeating arm, as distinguished from magazine fire);
distance de ——, distance from the target to the muzzle;
—— *à distance connues*, known distance firing;
—— *à distances inconnues*, practice at unknown distances;
—— *d'école*, v. —— *(individuel) d'instruction;*
—— *d'ensemble, (inf.)* collective firing;
—— *d'étude*, test or experimental firing;
—— *d'examen*, firing or practice to test the efficiency of the men;
—— *d'exercice*, practice firing; instruction firing; firing practice; target practice in general;
exercices de ——, firing drill;
—— *d'expérience*, experimental practice;
feuillet de ——, v. s. v. *feuillet;*
—— *à genoux*, fire kneeling;
—— *de guerre*, practice under service conditions;
—— *individuel, (t. p.)* individual target practice;
—— *individuel d'application, (t. p.)* individual skirmish target practice;
—— *(individuel) d'instruction, (t. p.)* individual practice at known distances (includes *tir préparatoire* and *tir d'instruction*);
insigne de ——, v. s. v. *insigne;*
—— *intermittent*, fire as a single loader of a repeating rifle; single-action fire of a revolver;
—— *jusqu'à la destruction*, v. —— *à outrance;*
—— *de justesse;* v. s. v. *justesse;*
—— *à longue portée*, long-range fire;
—— *muet, (t. p.)* snapping in aiming practice;
officier de ——, v. s. v. *officier;*
—— *à outrance*, fire to the limit of resistance, to test ultimate strength;
—— *perdu*, random shot;
—— *de perfectionnement, (t. p.)* extra practice with surplus cartridges;
—— *en plein bois, (nav.)* shot fired into the hull;
—— *de polygone*, practice on range;
—— *à poudre*, blank practice;
—— *de précision*, practice to determine the precision of an arm;
—— *préparatoire, (t. p.)* preliminary practice;
—— *rapide*, rapid fire;
—— *réduit*, gallery practice;
registre de ——, v. s. v. *registre;*
—— *de réglage, (t. p.)* sighting-shot practice at unknown distances;
règlement de ——, firing regulations;
—— *à répétition*, magazine fire;
—— *en retraite, (nav.)* fire astern;
—— *en retraite directe, (nav.)* fire straight astern;
saison du ——, target-practice season;
—— *de, par, salve*, volley fire, firing;
séance de ——, v. s. v. *séance;*
sergent de ——, v. s. v. *sergent;*
—— *de stand*, range as distinguished from service practice;
—— *sur*, fire at (a given target); fire from (a rest, a platform, etc.);
tube à ——, subcaliber tube;
—— *à visée horizontale*, special case of individual fire which is taken on objects on level with muzzle of rifle;

tir, *vitesse de* ——, rate of fire; number of shots by one man in one minute;
—— *de vitesse*, (*t. p.*) rapid firing (i. e., to train men in rapid firing);
—— *à volonté*, fire at will.

tirage, m., draught, traction, pulling load (of vehicles); towpath, towing; quarrying of stone; impression, issue (of a book, etc.); draught (of a chimney, funnel, etc.); (*phot.*) printing; (*mil.*, etc.) drawing of lots; (*min.*) firing of a mine; (*met.*) drawing of metals;
—— *artificiel*, forced draught;
—— *de la chaudière*, furnace, draught;
—— *de la cheminée*, funnel draught;
cheval de ——, draught horse; tow horse;
—— *d'un coup de mine*, (*min.*) (firing of a) blast;
—— *forcé*, forced draught;
—— *naturel*, natural draught;
—— *à la poudre*, (*min.*) blasting;
—— *au sort*, (*mil.*) drawing of lots for service;
—— *au tube*, (*cord.*) formation of strands by drawing threads through a tube of requisite diameter.

tiraille, f., (*mach.*) connecting rod.

tirailler, v. a. n., to pull, pull about; to fire frequently and aimlessly; to bang away; (*mil.*) to skirmish; to fire as skirmishers.

tiraillerie, banging away; useless, wild, firing (*mil.*) skirmishing.

tirailleur, m., bad shot; (*mil.*) rifleman; skirmisher; sharpshooter;
—— *s algériens*, (*Fr. a.*) four regiments of Algerian rifles (Turcos) recruited from the natives, except fourteen men in each company who are French; half the lieutenants are native, captains and field officers French;
—— *s annamites*, (French colonial troops) one regiment of Annam rifles (organized like the —— *algériens*);
en —— *s*, (*inf.*) as skirmishers (also call and command);
—— *d'infanterie*, infantry skirmishers (*tirailleur* alone generally means skirmisher of infantry);
—— *s sénégalais*, (French colonial troops) one regiment of Senegalese rifles (organized like the —— *algériens*);
—— *s tonkinois*, (French colonial troops) three regiments of Tonkinese rifles (organized like the —— *algériens*).

tirant, m., cramp, cramp iron; rib, holdfast; collar; bolt; axle stay; boot strap; purse string; (*cons., mach., etc.*) beam, floor beam, tiebeam, tie-rod, stay, brace, tensile stay, eccentric rod, stay rod; (*unif.*) keeper of a sword belt; (*mus.*) brace of a drum; (*nav.*) draught of water;
—— *d'arbalète*, (*cons.*) straining beam;
—— *de branloire*, bellows chain;
—— *de caisse*, (*mus.*) brace of a drum;
—— *de cavalier*, dog stay;
—— *de (la) chaudière*, boiler stay;
—— *diagonal*, (*cons.*) diagonal stay;
différence du —— *d'eau*, (*nav.*) trim;
sans différence de ——, (*nav.*) on an even keel;
—— *d'eau*, (*nav.*) draught of water;
—— *d'eau égal*, (*nav.*) even keel;
—— *d'eau en charge*, (*nav.*) load water draft;
—— *d'eau lège*, (*nav.*) light water draught;
—— *d'excentrique*, (*mach.*) eccentric rod;
faux ——, (*cons.*) hammer beam;
—— *de fourreau de baïonnette*, (*sm. a.*) bayonet-scabbard loop;
—— *de frein*, brake rod;
maitre ——, (*cons.*) tiebeam;
marque du —— *d'eau*, v. s. v. *marque*;
—— *du milieu de l'avant-train*, (*art.*) middle futchel;
—— *à vide*, v. —— *d'eau lège*;
—— *de voiture*, starting load;
—— *de volée*, (*art.*) splinter-bar stay.

tiraude, f., pulling rope of a pile driver, etc.; strap around a lever;
sonnette à ——, v. s. v. *sonnette*;
—— *de sonnette*, draw line; hand rope of a pile driver.

tire, f., *tout d'une* ——, at a jerk, all at once, at one time.

tire! tire! pull away! pull ahead!

tiré, m., shooting (sport); drawee (bill of exchange).

tire-au-flanc, m., (*mil. slang*) deadbeat; coffee-cooler (U. S. A.).

tire-balle, m., (*sm. a.*) bullet extractor.

tire-botte(s), m., bootjack; boot hook.

tire-boucher, m. v., *tire-boucler*.

tire-bouchon, m., corkscrew; (*art.*) fuse extractor (of a shell).

tire-boucler, m., mortise chisel.

tire-bourrage, m., packing screw.

tire-bourre, m., (*art.*) worm; (*sm. a.*) wad extractor (obs.); (*mach.*) packing hook.

tire-bouton, m., button hook, strap.

tire-braise, m., (*slang*) infantry soldier

tire-cartouche, m., (*sm. a.*) cartridge extractor.

tire-charge, m., (*mountain art.*) loop (of a cartridge box).

tire-clou, m., nail drawer, nail extractor.

tire devant! tire avant! pull ahead! (boats).

tire-douilles, m. v., *tire-cartouche*.

tire-étoupe, m., packing screw.

tire-étoupilles, m., (*art.*) primer extractor.

tire-feu, m., (*art.*) lanyard;
cordon ——, lanyard;
cordon —— *à olives*, —— *à olives*, slide lanyard.

tire-foin, m., (*art.*) gun worm: wad hook (obs.).

tire-fond, m., eyebolt, long wood screw, sort of screw bolt; screw ring; cooper's turrel; (*r. r.*) screw used to fasten chairs to crossties.

tirefonner, v. a., to fasten or secure by an eyebolt, wood screw, etc.

tire-fusée(s), m., (*art.*) fuse extractor, fuse engine (wooden fuses, obs.).

tire-gargousse, m., (*art.*) cartridge worm.

tire-joint, m., mason's smoothing tool.

tire-laine, m., (*fond.*) implement for withdrawing plugs from molds.

tire-ligne, m., ruler; right-line pen.

tire-obturateur, m., (*art.*) obturator extractor (for removing a Broadwell ring).

tire-obus, m., (*art.*) sort of shell hook.

tire-plomb, m., glazier's vise.

tire-ployer, v. a., to unload, discharge, put in order.

tire-racine, driver, punch.

tirer, v. a. n., to draw, pull, pull on, pull out; to haul, drag; to tow; to tighten, stretch; to import, get, obtain, reap, gather, derive, extract; to tap, draw off; to extricate, deliver; to trace, delineate; to draw upon (a. correspondent); to print, run off (so many copies of a book, etc.); to dig (a canal); to draw (of a chimney, a furnace); (*met.*) to draw (tubes, etc.); (*nav.*) to draw (so many feet); (*mil.*) to trace, throw up (an intrenchment); to draw lots for service; to serve; (of firearms in general) to fire, fire off, discharge; to fire at; to go off, be discharged; (*fenc.*) to fence;
—— *à*, (*art., sm. a.*) to use such and such (a charge); to be fired with such and such a charge;
—— *sur affût*, (*sm. a.*) to fire from a rest;
—— *en l'air*, (*sm. a.*) to fire in the air;
—— *sur appui*, (*sm. a.*) to fire from a rest;
—— *des, les, armes*, (*fenc.*) to fence;
—— *des armes à feu*, to shoot;
—— *dans (hors) les armes*, (*fenc.*) to thrust on the left (right) side of the adversary's blade;
—— *sous (sur) les armes*, (*fenc.*) to thrust under (over) the adversary's arm;
—— *à balle(s)*, (*sm. a.*) to fire ball cartridge;
—— *à barbette*, (*art.*) to fire in barbette;
—— *en belle*, (*nav. art.*) to hull a ship;
—— *bien*, to be a good shot;
—— *à, en, blanc*, to fire blank cartridges;
—— *au blanc*, v. —— *au but*;
—— *à, des, boulets*, (*art.*) to fire projectiles;
—— *à boulets rouges*, (*art.*) to fire red-hot shot (obs.);
—— *à bout portant*, (*ball.*) to fire point-blank;

tirer à bras francs, (sm. a.) to fire from the shoulder; to fire standing (said also of a revolver);
— au but, to fire at a target;
— le canon, etc., (art.) to fire a gun (a mortar, etc.);
— de la cellule, (mil. slang) to be confined in a military cell;
— à charge (faible) pleine, (art.) to fire with (reduced) full charges;
— en chasse, (nav. art.) to fire forward of the beam;
cheval qui tire à la main, v. s. v. cheval;
— à la cible, (t. p.) to perform, carry on, target practice; to fire at a target;
— (à la conscription), (mil.) to draw lots for conscription;
— copie, — une copie, to make a copy;
— au cordeau, to trace by the tape;
— un coup, (sm. a., art.) to fire a shot;
— à coups perdus, (sm. a.) to fire at random;
— à couvert (mil.) to fire from behind a protection;
— à dos, (art.) to fire in reverse (normal fire);
— de l'eau, to suck up, absorb, water;
— d'écharpe, (art.) to fire obliquely (e. g., at a line) from in front;
— à embrasure, (art.) to fire through an embrasure;
— d'enfilade, (art.) to enfilade;
— l'épée, to fence, fight, with swords; (fig.) to take up arms;
— l'épée contre quelqu'un, to fight anyone;
— une estocade, (fenc.) to make a pass or thrust (obs.);
faire — l'épée à quelqu'un, to force one to fight;
— des feintes, (fenc.) to feint;
— dans le fer, (fenc.) to thrust, etc., on the side where the opponent is covered;
— le feu, (art.) to use (such and such) a fire;
— un feu d'artifice, to let off fireworks;
— à fleur d'eau, (nav. art.) to hull (a ship);
— au flanc, (mil. slang) to be a deadbeat, a coffee cooler (U. S. A.);
— à la fourchette, (art.) to use the "fork" to determine the range;
— fusant, (art.) to fire time shell;
— une fusée, to let off a rocket;
— au, le, fusil, to fire a gun (and similarly of other small arms);
— juste, to fire true, shoot true, straight;
— une lettre de change (par première, seconde, troisième), to draw a bill of exchange (in duplicate, in triplicate);
— à la main, (man.) to bore;
— la main plate, (fenc.) to raise the hand, nails up;
— à la mer, (s. c. art.) to fire seaward; (nav.) to stand out to sea;
— à mitraille, (art.) to fire canister, case shot;
— au mur, v. s. v. mur;
— à la muraille, v. s. v. muraille;
— à, des, obus, (art.) to fire shell;
— à outrance, (art.) to fire to the limit of ultimate strength;
— percutant, (art.) to fire percussion shell;
— sur plate-forme, (art.) to fire from a platform;
— en plein bois, v. — à fleur d'eau;
— de plein fouet, (art.) to use direct fire (full charges);
— à plomb, to fire small shot (sport);
poudre à —, sporting powder;
— à poudre, v. — à blanc;
— au premier brin, (cord.) to work hemp so as to get the best quality;
— à projectiles, (art.) to fire with projectiles;
— de, en, quarte, etc., (fenc.) to thrust in quarte, etc.;
— race, (hipp.) to breed a mare;
— au renard, (hipp.) to pull on the halter and try to break it; to fall on haunches so as to break the halter (stable vice);
— en retraite, (nav. art.) to fire abaft the beam;
— à revers, (art.) to fire in reverse (oblique fire);

tirer à ricochet, (art.) to use ricochet fire;
— la ruade, (of a horse) to kick out;
— des, par, salves, (art.) to fire salvos; (sm. a.) to fire volleys;
— à sec, to draw boats up on land;
— à six, (art.) to pull together (all six horses);
— au sort, (mil.) to draw lots for military service;
— sur le temps, (fenc.) to make a time thrust;
— à terre, to haul a boat or ship high and dry;
— à toute volée, (art.) v. tir à toute volée;
— à trois, (art.) said of a team when off horses alone pull;
— au tube, (cord.) to draw yarn through tubes in manufacture of rope;
— la tunique, (unif.) so to pull the tunic as not to show more than the prescribed folds in the back;
— au vent, (hipp.) to hold the head high;
— au vol, to shoot on the wing (sport);
— à vue, to draw upon anyone at sight.
tirerie, f., wire-drawing shop; (mil.) wasteful, useless, small-arm firing.
tire-terre, m., quarryman's spade.
tirette, f., register, shutter, drawer, damper; (sm. a., artif.) string (to open cartridge packets, smoke cylinders, etc.).
tireur, m., wire-drawer; fencer; drawer of a note; (mil., etc.) marksman, "shot," rifleman, sharpshooter;
— d'armes, fencer;
arrêter un —, (fenc.) to meet a stop thrust;
— d'aviron, oarsman;
— d'élite, crack shot;
— de mine, miner;
— de position, (mil.) picked shot (inf.) to fire on enemy's artillery during infantry advance.
tire-veille, f., manrope, tide rope;
— de cabestan, swifter of a capstan;
nœud de —, (cord.) manrope knot.
tiroir, m., drawer; (sm. a.) small flat sliding pin fixing the barrel to the stock (sporting arms); (mach.) slide valve; (mil.) second rank (of troops in three ranks).
(Unless otherwise indicated, the following terms relate to machinery:)
arbre du —, eccentric shaft;
armature du —, valve work(s);
— d'asservissement, (mach.) slide valve of a servo-motor, or of a système asservi;
— auxiliaire, auxiliary starting valve;
avance du —, v. s. v. avance;
— à basse pression, low-pressure slide valve;
boîte à —, slide chest;
boîte de —, slide box;
— à boîte, box slide valve;
— compensé, balanced slide valve;
— à, en, coquille, 3-port slide valve, trebleported slide valve; ordinary slide valve;
course du —, travel of the slide valve;
— cylindrique, distributing cock; piston valve;
— en D, D valve;
— en D court (long), short (long) D valve;
défiler en —, v. s. v. défiler;
déployer en —, (mil.) to form line from close column;
— à détente fixe (variable), valve with fixed (variable) cut-off;
— de détente, expansion valve, cut-off valve;
— de détente à piston, piston expansion valve;
— sans détente, valve without lap or lead;
— à deux orifices, double ported slide valve;
— en disque, Brotherhood valve;
distributeur, distributing slide valve;
— de distribution, main slide valve, distributing valve; (art.) feed trough (e. g., of the Hotchkiss revolving cannon);
distribution par —, distributing-valve gear;
— de distribution équilibré, equilibrium slide valve, balanced valve;
— à dos percé, gridiron valve;
— à double orifice, double-ported slide valve;
— en E, E valve;
— équilibré, equilibrium slide valve, balanced slide valve;

tiroir, équilibrer le ——, to balance, equilibrate, the slide;
—— *d'évacuation*, exhaust valve;
—— *à, d' expansion*, cut-off valve, expansion slide valve;
—— *d'expansion à piston*, piston expansion valve;
—— *sans garniture*, long D valve;
glace de ——, slide-valve face;
glissière du ——, slide bar;
—— *à grille*, gridiron valve;
guide du ——, *v. glissière du* ——;
—— *à haute pression*, high-pressure slide valve;
—— *d'injection*, injection slide valve;
—— *intermédiaire*, intermediate pressure slide valve;
—— *d'introduction*, steam valve;
—— *d'introduction directe*, auxiliary starting valve;
marche, marcher, en ——, *v. s. v. marche, marcher;*
—— *à mouvement rectiligne*, slide valve;
partie creuse du ——, exhaust cavity (of a slide valve);
plan du ——, slide face valve;
—— *à pression intermédiaire*, intermediate pressure slide valve;
—— *à recouvrement*, lapped valve;
règlement du ——, adjustment of the valve, valve setting;
—— *de renversement*, reversing valve;
—— *à rotation*, rotating valve or cock;
—— *rond*, pipe valve, piston valve;
—— *à siège cylindrique*, cylindrical valve;
—— *superposé*, superposed valve (Trezel)
tige de ——, slide rod, valve stem;
—— *à trois orifices*, treble-ported slide valve;
—— *à un orifice*, single-ported valve;
—— *de Watt*, long D valve.
tisane, medical drink or potion, tisane.
tisanerie, dispensary for tisanes in hospitals.
tisard, m., (*met.*) air vent.
tison, m., brand, firebrand.
tisonné, a., (*hipp.*) marked with black spots;
cheval ——, branded gray horse, marked with black spots;
cheval gris ——, iron-gray horse.
tisonner, v. a., to stir, poke, stir up, a fire.
tisonnier, m., poker, smith's poker, fire iron.
tisonnure, (*hipp.*) irregular black spot (as though made with charcoal).
tissage, m., weaving.
tisser, v. a., to weave, plait, twine.
tissu, m., tissue, texture, web, textile fabric;
—— *en bourre de soie*, (*art.*) silk waste cloth (cartridge-bag cloth);
—— *en caoutchouc*, rubber cloth;
—— *de chanvre*, hemp cloth;
—— *-crin*, haircloth;
—— *élastique*, elastic band;
—— *imperméable*, waterproof cloth;
—— *métallique*, wire gauze.
titanite, f., (*expl.*) titanite.
titre, m., title; legend of a map; deed, voucher, title deed; (*in pl.*) securities;
en ——, titular; regular by appointment;
faux ——, false title, deed, or document;
à —— *gracieux*, as a free gift, without cost to the beneficiary;
à —— *gratuit*, at Government expense, free of charge;
à —— *onéreux*, at one's own expense;
à —— *remboursable*, on condition of reimbursement.
titré, a., *liqueur* ——*e*, (*chem.*) normal or standard solution.
titrer, v. a., to give a title;
—— *une liqueur*, (*chem.*) to make a standard solution.
titulaire, m., v. a., holder of a title, position, grade, etc.; holding a position, etc.
toc, m., (*mach.*) driver, stop, catch, tappet, dog; catch ring;
—— *d'excentrique*, eccentric stop, catch;
œil à ——, *v. s. v. œil*.
tocsin, m., tocsin, storm bell, alarm bell.
toile, f., linen, linen cloth; canvas, sailcloth; painting, picture; (of a pulley) web joining the rim with the nave; (*fond.*) lash, seam;

toile *d'amiante*, amianth cloth;
—— (*d'*)*amiantine*, (*art.*) silk used for cartridge bags in France;
—— *à calquer*, tracing linen, cloth;
—— *chinée*, water-tight canvas used by French artillery (*chinée*, dyed);
—— *cirée*, oilcloth;
—— *de crin*, horsehair cloth;
déchirer (de) la ——, *v. s. v. déchirer;*
—— *d'emballage*, packing cloth;
—— *-émeri*, emery cloth;
être sous la ——, (*mil.*) to be under canvas; to be in camp;
—— *sans fin*, (*powd.*) sort of endless belt for carrying the cake to the graining cylinders;
—— *goudronnée*, tarpaulin;
—— *grasse*, tarred canvas;
—— *imperméable*, waterproof cloth;
—— *métallique*, wire gauze;
papier- ——, tracing cloth;
—— *peinte*, water-tight canvas (painted, etc., obs.);
—— *à pierre ponce*, pumice cloth;
—— *à prélart*, thick canvas used for tarpaulins;
—— *-prélat*, canvas wagon cover;
—— *à pulvérin*, (*mil. min.*) linen or cloth covered with mealed powder paste;
—— *de tente*, tent canvas;
—— *vernie*, oilskin;
—— *-verre*, glass cloth;
—— *à voile*, sailcloth;
—— *à voile légère*, duck.
toilette, f., trimming a horse;
faire la —— *à un cheval*, to trim a horse.
toise, f., fathom (6 feet); instrument to measure height of persons; pile of broken stone.
toisé, m., measurement with a 6-foot rod.
toiser, v. a., to measure by the fathom.
toiseur, m., measurer, surveyor.
toison, f., fleece; Order of the Golden Fleece;
la —— *d'or*, (the Order of) the Golden Fleece
toit, m., roof; dwelling, house, home; (*min.*) top, upper, stratum of a mine;
—— *en appentis*, v. —— *en potence;*
—— *en batière*, ridge roof;
—— *cintré*, arched roof;
—— *en croupe*, hip roof;
—— *cylindrique*, barrel or tunnel roof;
—— *en demi-croupe*, half-hip roof;
—— *à deux croupes*, span roof;
—— *à deux égouts*, v. —— *en batière;*
égout de ——, eavesdrop;
—— *à l'impériale*, imperial;
—— *à pignon*, gable roof;
—— *plat*, flat roof;
—— *en pointe*, pointed roof;
—— *en potence*, shed, lean-to;
—— *à un seul égout*, v. —— *en potence.*
toiture, f., roofing;
—— *à coyaux*, roofing with projecting rafter foot;
—— *en plomb*, lead roofing;
—— *en tuiles*, tile roofing.
tôle, f., sheet iron; plate, plating; (in combination with names of metals) sheet;
—— *d'acier*, —— *en acier*, sheet steel;
—— *d'appui*, armor shelf plate; armor-plate backing;
—— *à chaudière*, boiler plate;
—— *de choc*, dash plate;
—— *de ciel*, crown plate;
—— *de cloison*, bulkhead plate;
cylindre à ——, plate rolling mill;
—— *diaphragme*, any partition plate;
—— (*de*) *doublure*, lining plate, backing;
—— *d'enveloppe*, (steam) shell plate;
—— *façonnée*, flanged plate;
—— *de faîtage*, (*cons.*) ridge plate;
—— *en fer*, sheet iron;
—— *pour fer blanc*, lattens, trebles;
—— *fine*, latten, treble;
—— *de fond*, bottom plate;
—— *forte*, slab;
—— *galvanisée*, galvanized iron;
—— *gaufrée*, corrugated iron;
—— *-glacis*, (*fort.*) glacis plate of a turret;
—— *gouffrée*, channel plate;
laminage de ——, plate rolling;

tôle *laminée* rolled plate;
— *martelée*, hammered plate;
— *-membrure*, (*nav.*) plate frame;
— *mince*, v. —— *fine;*
— *modératrice*, v. *de choc;*
— *moyenne*, sheet iron for flues, etc.;
— *ondulée*, corrugated iron plate;
plaque de ——, plate of sheet iron;
— *plastron*, breastplate;
— *pliée*, flanged plate;
— *de recouvrement*, cover plate;
— *de renfort*, stiffening, strengthening, plate;
— *ridée*, corrugated iron;
— *support de la cuirasse*, armor shelf plate, armor-plate backing;
triangle en ——, gusset stay;
— *tubulaire*, (*steam*) tube plate;
— *vernie*, japanned sheet iron;
— *zinguée*, galvanized-iron plate.
tolérance, f., toleration; (in the arts, plural generally) tolerance, remedy, allowance.
tolérer, v. a., to tolerate; (in the arts, etc.) to admit (such and such) a tolerance.
tôlerie, f., sheet-iron works; iron-plate manufacture, factory; framework of sheet iron; sheet iron and boiler plate.
tolet, m., thole; thole pin; crutch, fork;
— *à fourche*, swivel rowlock;
renfort de ——, thole-pin reenforce.
tôletière, f., rowlock reenforce; rowlock;
— *à fourche*, swivel rowlock.
tollet, m., v. *tolet*.
tombac, m., tombac; —— *blanc*, white copper; pinchbeck.
tombe, f., tomb, grave.
tombe-cartouche, m., (*sm. a.*) cartridge ejector.
tomber, v. n., to fall; to fall (off, down, into, from, upon, etc.); to drop off; to meet with, light upon; to flag, abate; to fall (of the wind, tide, night, etc.); (*mil.*) to fall upon (with the idea of surprise, of violence); to fall, surrender (of a city, fort);
— *en arrière*, to fall back; (of a mast) to rake;
— *à la conscription*, (*mil.*) to be drawn for service, to be conscripted;
— *dessus*, to fall upon; to set to;
— *ensemble*, to collapse;
— *au fond*, to settle or sink to the bottom;
laisser ——, to let down the fires (in furnaces);
— *à la mer*, to fall overboard;
— *au sort*, (*mil.*) to be drawn for service (at a conscription or at the yearly drawing).
tombereau, m., tumbrel;
— *à bascule*, (*art.*) sort of tilt cart used for transport of earth, material, etc.;
— *à boue*, mud cart;
— *à bras*, hand cart;
roulage au ——, carting.
tome, m., tome, volume.
ton, m., head or cap of a mast; (*drawing*) tone, tint.
tondage, m., horse clipping.
tondeur, m., clipper;
— *de chevaux*, horse clipper.
tondeuse, f., clippers, horse clippers.
tondre, v. a., to clip, shear, shave; to crop; to clip (a horse).
tonite, f., (*expl.*) tonite.
tonnage, m., (*nav.*) tonnage;
— *brut*, gross tonnage;
droit de ——, tonnage (duty);
— *légal*, net tonnage;
— *net*, net tonnage;
— *réel*, burden, true tonnage, dead-weight tonnage;
— *de registre*, net tonnage;
— *sous le pont*, under-deck (gross) tonnage;
— *total*, gross tonnage.
tonnant, a., thundering;
mélange ——, mixture of oxygen and of hydrogen.
tonne, f., cask, barrel; ton, 1,000 kilograms; (*nav.*) can buoy, nun buoy; tank boat; (*powd.*) tumbling barrel, pulverizing barrel;
— *d'arrimage*, (*nav.*) measurement ton;

tonne *binaire*, (*powd.*) barrel in which a *binaire* (q. v.) is pulverized;
— *de capacité*, dead-weight ton;
— *d'extraction*, (*min.*) corf, corb, used in mines;
— *granulateur de mine*, (*powd.*) special granulator for blasting powder;
— *grenoir*, (*powd.*) granulating barrel;
— *légale*, (*nav.*) register ton;
— *de lissage*, (*powd.*) glazing drum;
— *lissoir*, (*powd.*) glazing drum;
— *mélangeoir*, (*powd.*) pulverizing barrel (blasting powder, etc., *poudre de commerce extérieur*);
— *-mètre*, —— *métrique*, metric ton;
— *-poids*, v. —— *de capacité;*
poudre aux ——s, v. s. v. *poudre;*
— *à poudre*, powder barrel;
— *de pulvérisation*, (*powd.*) rolling barrel;
— *réelle*, (*nav.*) ton of dead-weight carrying capacity;
— *de registre*, register ton;
— *ternaire*, (*powd.*) barrel in which all three ingredients of powder are pulverized together;
— *à triturer*, (*powd.*) triturating barrel.
tonneau, m., cask, tun, barrel; (*nav.*) ton (1 cubic meter, 1,000 kilograms); (*art.*) target (barrel on a pole) for mortar practice;
— *d'arrosage*, watering cart, sprinkler;
— *broyeur*, mixing barrel (brickmaking);
— *mélangeur*, mixing barrel;
— *de mer*, 1 cubic meter, 1,000 kilograms;
— *métrique*, v. —— *de mer;*
pont de ——*r*, cask bridge.
tonnelade, f., (*mil.*) breastwork of casks filled with earth.
tonnelet, m., small cask.
tonnelier, m., cooper.
tonnelle, f., arbor; (*cons.*) wagon-headed vault.
tonnellerie, f., cooper's shed, trade.
tonner, v. n., to thunder; to thunder (of artillery).
tonnerre, m., thunder, thunderbolt; (*fig.*) guns, artillery; (*sm. a.*) breech, powder chamber, seat of the charge; (*art.*) reenforce, part of the breech over the powder chamber;
coup de ——, thunder clap, stroke;
éclat de ——, thunder clap, stroke;
fermer le ——, (*sm. a.*) to close the breech;
ouvrir le ——, (*sm. a.*) to open the breech.
tonte, f., shearing; clipping of horses.
tonture, f., clippings, shearings; (*nav.*) sheer.
toper, v. a., (*mil. slang*) to seize, apprehend.
topo, m., (*mil. slang*) topographic survey, general staff map; staff officer.
topographe, m., topographer.
topographie, f., topography.
topographique, a., topographic;
brigade ——, v. s. v. *brigade*.
topogravure, f., photogravure applied to reproduction of topographical drawings.
topométrie, f., (*top.*) topometry (i. e., the science of accurate topographical measurements).
topoplastique, a., topoplastic.
toque, f., jockey's cap.
toquerie, f., (*met.*) hearth, fireplace, of a forge.
toqueur, f., poker, rake, for cleaning a furnace.
torche, f., torch, flambeau; straw mat laid under anything to prevent injury; coil (of wire, etc.);
— *de signal*, (*mil.*) signal torch.
torché, p. p., wiped;
— *à la diable, mal* ——, badly done (work).
torche-fer, m., wet cloth for rubbing soldering irons.
torche-nez, m., (*hipp.*) horse twitch.
torcher, v. a., to wipe, rub, clean; to build with loam and straw mixed.
torchette, f., twisted osier, twig; (*met.*) twyer scraper.
torchis, m., mud, loam, and straw; mud and chopped straw mixed;
mur de ——, mud wall.

torchon, m., rubber, duster, rags; twist of straw under stones (to protect the edges);
se donner un ——, *se rafraîchir d'un coup de* ——, (*mil. slang*) to fight a duel with swords;
——-*serviette*, rubbing-down cloth (for horses).
tordage, m., twisting, twist.
tord-boyaux, m., rotgut.
tordeuse, f., (wire-) cable-twisting machine.
tord-nez, m., (*hipp.*) horse twist.
tordoir, m., (*met.*) stamping mill, ore crusher.
tord-oreille, m., (*hipp.*) ear twist.
tordre, v. a., (*cord.*, *etc.*) to twist; (*fort.*) to twist (as withes);
—— *en garochoir*, (*cord.*) to twist (a rope the wrong way);
—— *de main torse,* (*cord.*) to twist (a rope the wrong way).
tore, m., torus; (*art.*) small semicircular molding on guns.
tornade, f., tornado.
toruado, m., tornado.
toron, m., (*cord.*) strand (must contain not fewer than six yarns);
—— *commis à droite (gauche),* hawser- (cable-) laid strand;
cordage à ——*s cassés,* stranded rope;
en ——, hawser-laid;
intervalles des ——*s,* v. s. v. *intervalle.*
toronnage, m., (*cord.*) formation of strands from yarn;
—— *par ourdissage,* formation of strands by warping;
—— *au tube,* v. *tirage au* ——.
torpédo, m., (*mil.*) torpedo (rare, usually *torpille,* q. v.);
—— *à demeure,* fixed torpedo;
—— *flottant,* drifting torpedo.
torpille, f., (*mil., nav.*) torpedo; ground, land, torpedo; mine;
—— *à aiguille,* v. —— *divergente;*
—— *automatique,* self-acting torpedo, automatic torpedo, clockwork (land) torpedo; contact mine;
—— *automatique-électrique,* electric-contact mine (firing battery in mooring weight);
—— *automatique à fusée chimique (mécanique),* contact mine with chemical (mechanical) fuse;
—— *automobile,* —— *automotrice,* automobile torpedo; locomotive torpedo;
barrage ——, —— *de barrage,* frame torpedo;
bouée ——, —— *bouée,* buoyant torpedo;
boulet——, v. *boulet-torpille;*
carcasse de ——, mine case;
champ (semé) de ——*s,* mine field;
chapelet de ——*s,* line of mines;
—— *de contact,* contact torpedo;
contre- ——, countermine;
—— *défensive,* defensive torpedo;
—— *dérivante,* drifting torpedo;
—— *dérivante de sillage,* drifting torpedo let go astern;
—— *dérivante de traîne,* drifting torpedo towed astern;
—— *dirigeable,* dirigible torpedo;
—— *divergente,* towing torpedo;
—— *dormante,* ground mine; bottom, ground, torpedo (submarine mine);
—— *dormante électrique,* observation ground mine;
—— *dormante à ferme circuit flottant,* ground mine with floating circuit closer;
—— *dormante de fond,* fixed ground mine;
—— *dormante mouillée,* fixed anchored mine;
—— *à double interruption,* v. —— *électro-automatique;*
—— *de drag(u)age,* creeping mine;
—— *électrique,* —— *électrique à simple interruption,* observation mine;
—— *électro-automatique,* electric-contact mine, electro-automatic mine (controlled torpedo, battery ashore);
—— *d'exercice,* dummy torpedo;
—— *fixe,* submarine mine, stationary torpedo; buoyant mine;
—— *flottante,* —— *flottante en dérive,* floating, drifting, torpedo;
—— *de fond,* ground mine (submarine mine); bottom, ground, torpedo;

torpille, *de fond électrique,* observation ground mine;
—— *mixte,* v. *électro-automatique;*
—— *mobile,* mobile torpedo;
—— *mouillée,* —— *mouillée entre deux eaux,* anchored mine; floating torpedo or mine; buoyant torpedo, buoyant mine;
mouiller une ——, to set or anchor a mine;
—— *à mouvement d'horloge,* clockwork torpedo;
—— *d'observation,* observation torpedo (land torpedo);
obus- ——, (*ari.*) torpedo shell;
—— *offensive,* offensive torpedo;
—— -*poisson,* fish torpedo;
—— *portative,* spar torpedo;
—— *portative d'exercice,* dummy spar torpedo;
—— *portée,* spar torpedo;
—— *de pression,* a land torpedo so arranged as to go off when walked on or disturbed;
—— *pyrotechnique,* pyrotechnic torpedo (i. e., one set off by any contact, friendly or hostile);
rangée de ——*s,* row, line, range, of torpedoes;
relever une ——, to pick up a torpedo;
—— *remorquée,* towing torpedo;
—— *remorquée divergente,* torpedo towed astern by a *patte d'oie;*
—— *de sillage,* torpedo let out by a pursued vessel against pursuer;
—— *simulée,* dummy mine;
—— *de terre,* land torpedo;
—— *terrestre,* land torpedo or mine;
—— *de traîne,* towed torpedo (against pursuing vessel);
—— *vigilante,* contact mine;
—— *vigilante automatique-électrique,* v. —— *automatique-électrique;*
—— *vigilante électro-automatique,* v. —— *électro-automatique.*
torpillé, a., planted with torpedoes;
zone ——*e,* zone or area planted with torpedoes.
torpiller, v. a., (*torp.*) to torpedo; to destroy by torpedoes.
torpilleur, m., (*mil.*) torpedo man; (*nav.*) torpedo boat;
—— *autonome,* a torpedo boat that always travels under its own steam;
—— -*aviso,* torpedo gunboat;
bateau ——, torpedo boat;
bélier ——, torpedo ram;
canonnière- ——, torpedo catcher;
contre- ——, destroyer;
—— *à embarquer,* torpedo launch;
d'escadre, v. —— *de haute mer;*
—— *garde-côte,* coast-defense torpedo boat;
—— *de haute mer,* seagoing torpedo boat;
—— -*vedette,* vedette torpedo boat; torpedo scout (boat).
torque, f., torque;
—— *de fil,* coil of wire.
torréfaction, f., roasting; (*met.*) roasting of ores;
—— *du café,* roasting of coffee.
torréfier, v. a., to roast.
torrent, m., torrent, flood, stream.
torrentiel, a., torrential;
pluie ——*le,* torrential rain.
tors, a., twisted, crooked; (*cord.*) laid; m., (*cord.*) twist, lay, of a rope;
bois ——, wood of twisted grain;
cordage qui a trop de ——, (*cord.*) rope laid up too hard;
faux ——, (*cord.*) reverse twist;
maille ——*e,* twisted chain link;
ôter le ——, (*cord.*) to untwist a rope.
torsade, f., twisted fringe; joint made by twisting two wires together; (*unif.*) bullion of an epaulet;
grosses ——*s,* (*Fr. a.*) large bullion (field officer's epaulets);
petites ——*s,* (*Fr. a.*) small bullion (subaltern officer's epaulets).
torsion, f., torsion; (*cord.*) twist;
balance de ——, torsion balance;
—— *fausse,* (*cord.*) wrong twist;
galvanomètre à, *de* ——, v. s. v. *galvanomètre.*

tortiller, v. a., to twist, wreathe; (*cord.*) to twist yarn into strands;
—— *la croupe,* (*man.*) to give a twisting motion to a horse's croup;
—— *une mortaise,* (*carp.*) to begin a mortise with an auger.

tortillon, m., twist (of paper); (*art.*) choke of a cartridge.

tortoir, m., woolding-, packing-, racking-, stick.

tortuer, v. a., to bend, make crooked.

tortueux, a., tortuous, winding, twisting, meandering.

toste, f., v. *tôte.*

totalisateur, m., (*mach.*) dynamometer, for measuring total amount of work done.

tôte, f., thwart of a boat.

touage, m., warping,; towing;
bitton de ——, towing timber.

touchau(d), m., touch-needle.

touche, f., touch; stroke; finger piece; key (of a typewriter, etc.); point (in gauges); (*mach.*) tracer, guide, roller, contact piece (as in copying machines); (*elec.*) touch (v. s. v. *aimantation*); key, contact point, plate, or strip;
pierre de ——, touchstone;
—— *à ressort,* spring point.

touché, p. p., touched; (*fenc.*) hit, touched; (*met.*) "aired;" m., (*i. p.*) hit.

toucher, v. a. n., to touch; to feel try; to get, receive, draw; to reach to; to touch upon; to touch up (a horse); to drive, drive on; (*fenc.*) to touch, hit; (*nav.*) to touch at (a port, an island, etc.), to touch (a reef, a sand bar, etc.);
—— (*l'aiguille d'*) *un compas,* to touch a compass needle with a magnet;
—— *ses appointements,* to receive one's pay;
—— *barre,* v. s. v. *barre;*
—— *de la gaule,* (*man.*) to tap, touch up a horse with a riding whip;
—— *et parer,* to touch and go;
—— *à un port,* to call at a port;
se ——, (*cord.*) to be two-blocks, chock-a-block;
—— *le vif,* (*man.*) to prick a horse.

toue, f., ferryboat; towage, warping.

touée, f., tow, towage; warp, towrope; towline; length of a towline;
à la ——, in tow;
amarre de ——, towline;
ancre de ——, kedge anchor;
—— *de l'arrière,* stern rope, line;
—— *de l'avant,* bow rope;
élonger une ——, to warp, run out a warp;
haussière de ——, towline warp.

touer, v. a. r., to warp, tow, haul one's self ahead;
grelin, haussière, à ——, towrope;
se —— *avec, sur, une ancre de jet,* to kedge.

toueur, m., towboat; kedge anchor; stream anchor;
bateau ——, towboat.

touffe, f., tuft, wisp; cluster, clump, of trees; thicket.

touffu, a., bushy, thick;
bois ——, thick wood (underbrush).

touiller, v. a., (*powd.*) to beat and stir; to mix with a *touilloir,* q. v.

touilloir, m., (*powd.*) curved stick used for mixing ingredients after being moistened.

toulet, m., v. *tolet.*

touletière, f., v. *toletière.*

touline, f., towline.

toupet, m., (*hipp.*) forelock of a horse.

touple, f., (*mach.*) matching machine; matching head.

tour, m., turn, circuit, circumference, winding; round; reel; trick; trip; sprain; turn (i. e., revolution); turn (i. e., order of succession); (*cord.*) turn, round turn; elbow, kink; (*mach.*) lathe, lathe shop; winch, wheel and axle; revolution, turn; (*mil.*) tour (of guard or other duty); turn to go on; class of duty, roster; (*nav.*) log reel.

I. Cordage; II. Lathe; III. Miscellaneous.

I. Cordage:

—— *d'anguille,* passing or frapping of a tail;

tour *à bitord,* spun-yarn reel;
—— *de bitte,* turn around the bitt;
—— *de bitte au câble,* bitting of the cable;
—— *de câble,* foul hawse, elbow or foul in the hawse;
—— *s croisés,* racking turns;
demi-——, elbow, kink;
—— *s extérieurs,* v. —— *s supérieurs;*
—— *mort,* round turn;
—— *mort avec, et, demi-clef,* round turn and a half hitch;
—— *mort avec deux demi-clefs,* round turn and two half hitches;
—— *simple,* turn around a bitt;
—— *s supérieurs,* riding turns, upper turns.

II. Lathe:
—— *en l'air,* chuck lathe: pit lathe;
—— *à aléser,* boring lathe;
arbre de ——, lathe mandrel, spindle;
—— *à archet,* throw lathe;
au ——, in the lathe;
—— *à banc rompu,* break lathe, gap lathe;
banc de ——, lathe frame, bed;
—— *à barre,* bar lathe;
bâti de ——, lathe frame;
—— *à calibre,* (*fond.*) founder's lathe;
—— *à canons,* (*art.*) gun lathe;
—— *à canons de fusil,* (*sm. a.*) barrel lathe (for outside of barrels);
—— *à chariot,* slide (rest) lathe;
—— *à charioter,* slide lathe;
—— *à charioter à main,* hand slide lathe;
—— *à chariots multiples,* turret lathe;
—— *à contrepointes multiples,* v. —— *à chariots multiples;*
—— *à copier,* copying lathe;
—— *à copier dégrossisseur,* roughing down (copying) lathe;
—— *à copier finisseur,* finishing (copying) lathe;
crémaillère de ——, lathe rack;
creuser au ——, to turn hollow;
—— *à crochet,* metal-turning lathe;
—— *cylindrique,* slide (rest) lathe;
—— *à décolleter,* v. —— -*revolver;* screw-cutting lathe;
—— *à dérouiller,* (*art.*) lathe for scraping old paint from projectiles;
—— *à deux burins,* duplex lathe;
—— *à deux pointes,* double-center lathe;
—— *à double outil,* duplex lathe;
—— *à ellipse,* eccentric lathe;
—— *à emboutir,* chasing lathe, spinning lathe;
—— *à engrenages,* geared lathe;
—— *à étrangler,* (*artif.*) choking lathe or press;
fait au ——, lathe-turned;
—— *à fileter,* screw-cutting lathe;
—— *à fileter automate,* self-acting lathe;
—— *à fileter et à charioter,* engine lathe;
—— *à fosse,* pit lathe;
—— *à goupilles,* pin lathe;
—— *à guillocher,* rose engine;
—— *à habiller,* (*fond.*) hand lathe for mounting a core spindle with tow;
machine de ——, winding machine;
—— *à main,* hand lathe;
—— *à la mécanique,* engine lathe, power lathe;
—— *de moulage,* founder's lathe;
—— *à métaux,* v. —— *à crochet;*
—— *à noyaux,* v. —— *de moulage;*
—— *ordinaire,* center lathe;
—— *-o ovale,* oval lathe;
outils de ——, lathe tools;
—— *parallèle,* slide lathe;
—— *à pas de vis,* screw-cutting lathe;
—— *à pédale,* foot lathe;
pédale d'un ——, lathe treadle;
au pied, v. —— *à pédale;*
—— *à plateau,* face lathe;
du potier, potter's wheel;
pointe de ——, lathe center;
—— *à pointes,* center lathe, double-center lathe turn bench;
—— *à pointes fixes,* lathe with fixed centers;
—— *à polir et à finir,* polishing and finishing lathe;

tour *poupée de* ——, poppet;
—— *presseur*, v. —— *à emboutir;*
—— *de précision*, lathe for turning irregular forms; copying lathe;
—— *à repousser*, v. —— *à emboutir;*
—— *revolver*, any lathe in which the work is fixed and the tool turns;
—— *à roder*, grinding lathe;
—— *à rosettes*, rose engine;
—— *à roues*, wheel lathe;
—— *à support*, slide (rest) lathe;
support de ——, lathe rest;
—— *à tourner les tourillons*, (*art.*) trunnioning machine;
—— *à tourillons*, trunnion lathe;
—— *à trois vitesses différentes*, treble-geared lathe;
—— *à verge*, bar lathe;
vis de ——, lathe screw, leading screw;
—— *à vitesse variable*, variable-speed lathe.

III. Miscellaneous:
ampère ——, (*elec.*) ampere turn, ampere winding;
—— *d'ancienneté*, (*mil.*) seniority;
—— *en arrière* (*avant*), turn ahead (astern);
à —— *de bras*, by main strength;
—— *de bateau*, (*hipp.*) strain of the back and loins;
—— *du chat*, space (about half a foot) between a furnace and the adjacent wall;
croquis de —— *d'horizon*, sky-line sketch;
demi- ——, v. *demi-tour;*
—— *de détachement*, (*mil.*) roster of detached service;
deuxième ——, (*mil.*) duty of the second class (in garrison and in the field, duty inside the line of sentinels, police, and other guards; in sieges, piquet, q. v., and reserve);
—— *d'épée*, (*fenc.*) circling of adversary's blade;
faire le —— *de*, to make the circuit of (as of a position);
faire le —— *du compas*, (of the wind) to swing around the compass;
—— *de l'hélice*, (*nav.*) revolution of the screw;
—— *d'horizon*, sky line;
—— *de main*, manual dexterity, skill;
passer à son —— *de bête*, (*mil. slang*) to be promoted by seniority;
premier ——, (*mil.*) duty of the first class (in garrison, duty outside the line of sentinels, exterior guards; in the field, advance and rear guard, flank guard, outposts, exterior guards; in besieged places, guard, service of the guns, exterior fatigue under arms);
prendre son ——, (*mil.*) to take one's tour of duty;
—— *de reins*, v. —— *de bateau;*
à —— *de rôle*, in succession, in turn;
—— *de service*, (*mil.*) service roster; tour of duty, turn for duty; order in which duty is taken;
à son ——, in one's turn;
troisième ——, (*mil.*) duty of the third class (in garrison, personal or individual duty; in the field, and during sieges, general and interior fatigues, issues).

tour, f., tower; bell tower; (*nav.*) turret; (*fort.*) sort of redoubt in seacoast works (obs.);
—— *casematée*, v. —— *maximilienne;*
—— *de commandement*, (*nav.*) conning tower;
—— *creuse*, (*cons.*) concave face of a cylindrical wall;
—— *à feu*, light house;
—— *marine*, —— *martello*, (*fort.*) martello tower;
—— *maximilienne*, (*fort.*) three-storied casemated tower, with a battery of guns on the top (invented by the Emperor Maximilian);
—— *ronde*, (*cons.*) convex face of a cylindrical wall;
—— *de signaux*, signal tower;
—— *tournante*, (*nav.*) revolving turret;
—— *tournante dans l'axe*, (*nav.*) revolving turret in the plane of the keel;

tour *tournante excentrée*, (*nav.*) revolving turret diagonally placed with reference to a similar turret on the other side of the keel.

tourbe, f., peat, turf;
—— *carbonisée*, peat, peat coal.

tourbeux, a., peaty.

tourbier, a., peaty; m., peat cutter (workman).

tourbière, f., peat bog.

tourbillon, m., whirlwind, whirlpool; vortex, eddy; tornado, (*artif.*) tourbillon;
—— *de courant*, whirlpool;
—— *de feu*, (*artif.*) tourbillon;
—— *gazeux*, (*expl.*) vortex action set up by the gases of explosion (in a gun);
—— *de marée*, tide whirlpool;
—— *de poussière*, whirling cloud of dust
—— *de vent*, whirlwind.

tourbillonner, v. n., to eddy, to whirl, wind.

tourdille, a., *gris* ——, (*hipp.*) dirty gray (yellowish tinge, with reddish patches here and there).

tourelle, f., turret; (*nav.*) turret; (*fort.*) turret (land turret); (*mach.*) turret;
affût de ——, (*art.*) turret carriage;
—— *armée*, (*art.*) a turret with its embrasures turned from the enemy, and with the equipments, etc., ready to be taken by the cannoneers;
—— *à axe fixe*, turret with fixed axis;
—— *à axe mouvant*, turret with rotating axis;
—— *à barbette*, (*nav., fort.*) barbette (or open) turret, one permitting barbette fire;
—— *blindée*, armored turret;
—— *à ciel ouvert*, open turret;
—— *de commandement*, (*nav.*) conning tower;
—— *à coupole*, cupola;
—— *cuirassée*, armored turret;
demi- ——, half turret;
demi- —— *en saillie*, sponson;
—— *double*, (*nav.*) double, double-decked, superposed, turret;
—— *à éclipse*, disappearing turret;
—— *fermée*, (*nav.*) closed turret;
—— *fixe*, nonrevolving turret, fixed turret;
—— *fixe à barbette*, nonrevolving barbette turret;
—— *à manœuvre électrique*, electrically controlled turret;
—— *à manœuvre hydraulique*, hydraulically controlled turret;
—— *mobile*, revolving turret;
—— *mobile fermée*, closed revolving turret;
navire à ——, turret ship;
—— *oscillante*, rocking turret;
—— *oscillante à éclipse*, disappearing rocking turret;
—— *ouverte*, open turret;
pièce de ——, (*art.*) turret gun;
—— *à revolver*, (*mach.*) turning turret;
—— *servie*, (*art.*) turret in complete readiness for action (cannoneers at their posts, circuits working, etc.);
—— *tournante*, revolving turret;
—— *transportable*, transportable turret (accompanies an army in the field;
—— *de veille*, (*nav.*) conning tower.

touret, m., wheel, reel; small ring, (bow-) drill, spun-yarn winch, rope-maker's reel; thole pin; swivel; (*mach.*) lantern wheel, trundle;
—— *à archet*, bow drill, fiddle drill;
—— *de bride*, (*harn.*) bridle ring;
—— *de chaînette*, (*harn.*) curb hook;
—— *à forer*, drill holder;
—— *à percer*, drill holder;
—— *de porte-rênes*, (*harn.*) swivel;
—— *à rochet*, ratchet drill (brace);
—— *de tour*, (*mach.*) wheel of a lathe.

tourie, f., carboy (incased in wickerwork).

tourillon, m., (*mach.*) journal, axis, axle, spindle, pivot, bearing, neck, journal, gudgeon, crank pin; (*art.*) trunnion.
(Except where otherwise indicated, the following terms relate to machinery:)
—— *à n ailes*, an *n*-winged journal;
—— *à ancre*, anchored bearing;
—— *d'ancre*, shank nut of an anchor;

tourillon à *ancre artificielle*, keyed-in bearing;
—— à *anneau*, four-winged bearing surrounded by a conical shell;
—— *d'appui*, thrust bearing, pivot;
—— *d'appui à collets*, collar thrust bearing;
—— *d'arbre*, shaft journal;
—— *d'arbre à manivelle*, crank-shaft journal;
—— *d'arbre oscillant*, rocking-shaft journal;
—— *d'arbre de relevage*, weigh-shaft journal;
bande à ——*s*, v. s. v. *bande*;
bande de ——, (*art.*) trunnion plate;
—— à *boulet*, ball pivot, spherical gudgeon;
—— à *cannelures*, multiple collar thrust bearing;
—— *de charge*, multiple journal;
—— *de charge à collets*, multiple neck journal;
—— à *clavette*, v. —— à *ancre artificielle*;
—— à *collets*, neck journal;
—— *en croix*, a journal with securing cross arms cast on it;
—— *de crosse*, wrist pin;
—— *de cylindre*, cylinder trunnion;
demi- ——, half journal;
—— *demi-cylindrique*, half journal;
—— *d'écrou de vis de pointage*, (*art.*) tumbler gudgeon (pivot of the nut of an elevating screw);
encastrement des ——*s*, (*art.*) trunnion bed;
—— *d'essieu*, axle journal;
—— *d'évacuation*, exhaust trunnion (of an oscillating cylinder);
—— *d'extrémité*, end journal;
faux ——, (*art.*) false trunnion (piece fastening a sight rest to the trunnion);
—— à *fourchette*, fork journal;
frette- ——, (*art.*) trunnion hoop;
—— *frontal*, overhung journal;
—— *-guide*, link journal;
—— *inférieur*, lower gudgeon of an upright shaft;
—— *intermédiaire*, v. —— à *collets*;
—— *d'introduction*, admission trunnion (of an oscillating cylinder);
—— *de levier*, lever journal;
—— *de levier de pompe*, pump-lever gudgeon;
logement des ——*s*, (*art.*) trunnion bed;
—— *de manivelle*, crank pin;
palier à ——*s*, *de* ——, trunnion block;
—— *de pied*, v. —— *d'appui*;
—— à *pivot*, ball pivot;
—— *de pointage*, (*art.*) elevation trunnion in guns mounted in cradles;
—— à *pression longitudinale*, end or thrust journal;
—— à *pression transversale*, lateral journal;
—— *simple*, overhung journal;
—— *de soufflet*, (*art.*) bellows gudgeon of a field forge;
—— *supérieur*, upper gudgeon;
—— à *tête sphérique*, spherical bearing;
tour à ——*s*, trunnion lathe;
—— *de traverse*, crosshead journal.

tourillonner, v. n., (*mach.*) to turn on trunnions.

tourloure, tourlouron, m., (*fam.*) infantry recruit.

tourmente, f., gale, tempest; violent storm, sudden storm accompanied by a high wind.

tourmenter, v. a. r., to torment, injure, hurt; to toss, tumble, jolt; to labor, strain (of a ship); to warp (of wood); (*man.*) to worry, fret (one's horse); (of the horse) to fret, be unquiet, annoy his rider.

tourmenteux, a., stormy.

tournage, m., (*mach.*) operation of working in a lathe; (*nav.*) generic term for bitts, belaying pins, cleats, etc.;
taquet de ——, kevel.

tournailler, v. a., to turn around, about, nearly in same place.

tournant, a., turning, revolving;
champ ——, (*elec.*) rotating-current field;
corps ——, (*mil.*) body of troops executing a turning movement;
défaire, faire entrer, en ——, to screw out, screw in;
mouvement ——, (*mil.*) turning movement;
pont ——, swing bridge.

tournant, m., turn, turning, turning room or place; corner; turning point of a race track; warping post (for boats, in the bend of a stream); turning angle (of a vehicle); vortex, eddy, whirlpool;
(*axe*) —— à *volonté*, revolving axle;
—— *d'eau*, vortex, whirlpool;
être ——, (*mach.*) to swivel.

tournante, f., (*unif.*) rim of an epaulet or shoulder knot; (*artif.*) rocket that turns as it rises.

tourne-à-gauche, m., movable handle, used as a lever in various applications (e. g., for working a boring bar); tap wrench; saw-set.

tourne-cheminée, (*sm. a.*) nipple wrench (obs.).

tournée, f., round, visit; (*law*) circuit; (*mil.*, etc.) tour of inspection; journey made on special duty;
en ——, (*mil.*) on a tour of inspection.

tourne-fil, m., steel for sharpening knives.

tourne-gueule, m., (*art.*) shell scraper, tool for cleaning out the inside of hollow projectiles.

tournelle, f., small tower.

tourner, v. a. n., to turn; to turn round, revolve, turn about, over; to wheel, wheel round; to move around; to wind, twirl; to change, cant; (of the wind) to haul, change; (*mach.*) to turn, shape (on a lathe); to revolve; (*mil.*) to outflank, turn the flank of; (*cord.*) to belay; to make fast (around a post, bitt, etc.);
—— *une armée, l'ennemi*, etc., (*mil.*) to turn an army, the enemy; take it, him, in reverse;
—— *ses armes contre*, (*mil.*) to attack, make war against;
—— *en arrière*, to turn back;
—— *un bois*, etc., to pass around a wood, etc.;
—— *bride*, to turn (i. e., ride) back;
—— *le dos*, to flee, run away;
—— à *droite*, —— à *gauche*, (*drill*) to throw the right (left) shoulder forward (obs.); to execute right (left) turn;
—— *le flanc*, (*mil.*) to outflank, turn the flank;
—— *au gras*, to turn to paste (instead of crystallizing; said of saltpeter);
—— *une manœuvre*, (*cord.*) to belay a rope;
—— *au nord*, etc., (of the wind) to shift to the north, etc.;
—— *un ouvrage*, (*mil.*) to cut the communication between an outwork and a place;
—— *le sable*, —— *le sablier*, to turn the glass;
—— *ses souliers*, to run down one's shoes;
—— *tête*, —— *visage*, to turn upon, to face, the enemy;
—— à *toutes mains*, (*man.*) to change direction at any gait; to turn to the right and left equally well.

tournesol, m., (*chem.*) litmus;
papier de ——, litmus paper.

tournette, yarn reel.

tourneur, m., turner.

tourne-vent, m., funnel that turns with the wind.

tournevire, f., (*cord.*) messenger;
bouton de ——, Turk's head;
roue de ——, messenger wheel.

tournevis, m., screw-driver;
—— à *broche*, v. —— à *trou*;
—— à *clef*, (*Fr. art.*) combination wrench and screw-driver;
—— à *trou*, (*Fr. art.*) a double-ended screw-driver (pierced in the middle for a *broche*).

tourniquet, m., turnstile; window latch (to close a sash or to keep open a shutter); turn-buckle; small capstan; screw jack; roller, swivel; (*med.*, etc.) tourniquet; (*mil.*) sort of cheval de frise (beam filled with sharp nails for the defense of a breach, etc., obs.); (*artif.*) synonym of *soleil tournant*, q. v.;
—— *de coffret*, (*art.*) turn-buckle of an ammunition chest;
—— *hydraulique*, reaction wheel.

tournisse, f., (*carp.*) filling-in piece, stud, filling post.

tournoiement, m., turning round and round, wheeling round; dizziness;
—— *d'eau*, whirlpool.

tournoyer, v. a., to turn round and round; to wheel round; to whirl; to wind (of a road).

tournure, f., turnings.

touron, m., v. *toron*.

tourteau, m., oil cake; boss of a wheel; (*powd.*) runner, runner ball;
—— *d'excentrique*, (*mach.*) eccentric sheave;
—— (*goudronné*), (*artif.*) link made of old rope, mèche à canon, etc., covered with an illuminating composition;
—— *de manivelle*, (*mach.*) crank boss;
—— *oscillant*, (*mach.*) rocking disk, wind plate, motion disk (Corliss engine).

toutenague, f., (*met.*) tutenag.

tout-venant, m., coal just as it comes from the mines; mine run.

toyère, f., eye of an ax or any other tool.

traban, m., halberdier.

trabe, f., anchor stock; pole or shaft of an ensign or color.

trac, m., gait, track, trace (of a horse, mule, etc., obs.).

traçage, m., tracing (as of diagrams on plates to be cut out, etc.).

tracasser, v. r., (of a horse) to fret.

trace, f., trace, step, footstep, mark; track, rut, of a vehicle; outline, sketch, draught; lines marked out, traced out, for a construction; (*fort.*) trace; (*art.*) scoring of the bore; (*fond.*) runner;
—— *de balzane*, (*hipp.*) white mark on the coronet (an incomplete *principe*, q. v.).

tracé, p. p., (of a horse) registered.

tracé, m., drawing, plan, ground plan, draught, trace, outline, sketch; laying out grounds; direction, line, of roads; track; tracing, copy; (*fort.*) *tracé*, or trace;
—— *des ateliers*, (*siege, fort.*) marking out of tasks;
—— *bastionné*, (*fort.*) bastioned trace;
—— *à bastionnets*, (*fort.*) trace in which flanking is secured by small bastions (mountainous country);
—— *brisé*, (*fort.*) tenailled trace;
—— *à caponnières conjuguées*, v. —— *à bastionnets*;
—— *de la chambre*, (*art.*) form of the chamber of a gun;
—— *circulaire*, (*fort.*) circular trace;
—— *à clameaux*, v. s. v. *clameau*;
—— *à crémaillère*, (*fort.*) crémaillère trace, indented trace;
faire le —— *de*, to trace out, sketch out, lay out roads, grounds, works, buildings, etc.;
—— *intérieur*, (*art.*) form of the bore of a gun;
—— *mixte*, (*fort.*) trace belonging to both the polygonal and the bastioned systems;
—— *des ouvrages*, (*fort.*) profiling of a work;
—— *perpendiculaire*, (*fort.*) name given by Montalembert to his tenaille trace;
—— *polygonal*, (*fort.*) polygonal trace;
—— *rectiligne*, (*fort.*) rectilinear trace;
—— *de la section*, profile;
—— *tenaillé*, (*fort.*) tenailled trace;
—— *à traverses*, (*siege*) traversed trace of a double sap;
—— *à traverses tournantes*, (*siege*) cube traverse trace.

tracelet, m., tracing point (of a dividing machine).

tracement, m., laying out, tracing out (of a fort).

tracequin, m., marking gauge, carpenter's gauge, scriber.

tracer, v. a., to sketch, trace, mark out, plot; to make a plan or drawing; to lay out roads; (*fort.*) to lay out, on the ground, the dimensions of a work;
cordeau à ——, v. s. v. *cordeau*;
—— *en creux*, to make a cut or groove;
—— *une route sur la carte*, (*nav.*) to lay down a course on the chart.

traceret, m., scriber, scribing awl; tracing point (of a dividing machine).

traceur, m., —— *de route*, (*art.*) position finder.

traçoir, m., scribing awl; tracing point (dividing machine); routing cutter, tool.

tracteur, m., traction engine.

traction, f., traction, draught, tractive power, tension; thrust of a suspension bridge;
de ——, tractive;
force de ——, v. *pouvoir de* ——;
frais de ——, locomotive power;
—— *en palier*, (of a road locomotive) traction along the surface of the ground;
pouvoir de ——, tractive power;
puissance de ——, v. *pouvoir de* ——;
ressort de ——, (r. r.) drawspring.

traditore, m., v. *batterie-traditore*.

traduire, v. a., to translate; (*law*) to indict, arraign; to transfer from one prison to another;
—— *en justice*, (*law*) to indict, bring before a court;
—— *en prison*, to send to prison;
—— *en, devant, par devant, un conseil de guerre*, (*mil. law*) to bring before a court-martial.

trafic, m., trade; (r. r.) freight as distinguished from passenger business;
—— *intérieur*, inland trade.

trafiquer, v. a. n., to traffic, deal in; to negotiate (a bill of exchange).

trahir, v. a., to betray.

trahison, f., treason, treachery;
haute ——, high treason.

traille, f., trail bridge, trail ferry; flying ferry or bridge;
bac à ——, v. s. v. *bac*;
bride de ——, v. s. v. *bride*;
moufle de ——, v. s. v. *moufle*.

traillon, m., small ferryboat.

train, m., pace, rate of going; way, course, manner; retinue, suite; average speed of a bicycle; (fore, after) body of a vehicle; train of cattle; (*man.*) rate, pace, gait, action; (in races) gait, pace; (r. r.) train; (*mach.*) roll, train, mill train; (*hipp.*) quarters; (*mil.*) train, baggage train.

I. Military; II. Railway; III. Miscellaneous.

I. Military:

—— *d'ambulance*, (*med.*) hospital train;
arrière——, v. *arrière-train*;
—— *d'artillerie*, artillery train (suppressed in 1883, place taken by the *sections de munitions*);
—— *auxiliaire*, (Fr. a.) reserve train (supplies the *convois administratifs* and the ——*s régimentaires*);
avant——, v. *avant-train*;
—— *blindé*, (r. r.) armored train;
—— *de bateaux*, (*pont.*) train of boats;
charger à fond de ——, v. s. v. *charger*;
—— *de combat*, (Fr. a.) the fighting train, i. e., wagons accompanying troops directly and carrying supplies needed on the field of battle;
—— *des équipages* (*militaires*), the train;
—— *des équipages de pont*, (*pont.*) bridge train;
équipages de ——, train equipage; wagon train;
—— *d'évacuation*, (*med.*) hospital train taking the sick and wounded home (either a rail or a wagon train);
—— *facultatif* (*militaire*), v. —— *spécial facultatif*;
—— *du génie*, engineer train;
—— *militaire*, (r. r.) troop train;
—— *de pontons*, (*pont.*) pontoon train;
—— *régimentaire*, (Fr. a.) regimental train, provision and baggage train (all wagons not in the —— *de combat*, of both staff and troops); (*Fr. art.* so) much of a battery as is left after taking out the *batterie de combat* (v. s. v. *batterie*), i. e., forage, provision, and baggage wagons;
—— *sanitaire*, (*med.*) hospital train;
—— *sanitaire improvisé*, (*med.*) any improvised hospital train (e. g., one composed of wagons);

train 463 **trait**

train *sanitaire permanent*, railway hospital train (service **organized in peace**);
—— *spécial extraordinaire*, (*Fr. a., r. r.*) emergency train proper;
—— *spécial (facultatif)*, (*Fr. a., r. r.*) train made up and used under unforeseen circumstances; any train of more than eight cars.

II. Railway:
(For military trains, see under Military.)
—— *accéléré*, v. —— *exprès*;
—— *d'aller*, down train;
—— *de bestiaux*, —— *de bétail*, cattle train;
—— *de ceinture*, belt train;
—— *en correspondance*, connecting train;
—— *courrier*, express;
—— *descendant*, down train;
—— *direct*, express;
—— *éclair*, very fast train;
—— *estafette*, v. —— *direct*;
—— *exprès*, express train;
—— *facultatif*, special train;
—— *de grande vitesse*, fast train;
—— *de grand parcours*, long-distance train;
—— *kilomètre*, train kilometer;
—— *lent*, slow, local, accommodation, train;
—— *de luxe*, train of Pullman cars;
—— *de marchandises*, freight (goods) train;
—— *de matériaux*, construction train, repair train;
—— *mixte*, mixed train;
—— *montant*, up train;
—— *-omnibus*, mixed train (draws cars of all classes); local train;
—— *parcourant toute la ligne*, through train;
—— *de petite vitesse*, slow, accommodation, train;
—— *de plaisir*, excursion train;
—— *-poste*, mail train;
—— *à prix réduits*, parliamentary train (England);
—— *de quatre roues*, bogie;
—— *rapide*, express train;
—— *en retard*, train behind time;
—— *de retour*, return train;
—— *de secours*, rescue train;
—— *spécial*, special train;
—— *de transport*, freight train;
—— *de voiture*, under body of a car;
—— *de voyageurs*, passenger train.

III. Miscellaneous:
arrière- ——, v. *arrière-train*;
avant- ——, v. *avant-train*;
ne pas aller de ——, (*man.*) to be unsteady of gait;
—— *de bateaux*, train of boats;
—— *de bois*, raft or tow of rafts;
bon ——, at a great pace;
—— *de charge*, (*man.*) full gallop;
—— *de cheval*, (*man.*) horse's gait;
—— *de chevaux*, string of horses; team of horses;
—— *de derrière*, hind body of a four-wheel vehicle; (*hipp.*) hind quarters of a horse;
—— *de dessous*, under body of a vehicle;
—— *de dessus*, upper body of a vehicle;
—— *de devant*, fore body of a four-wheel vehicle; (*hipp.*) fore quarters of a horse;
—— *duo*, (*mach.*) two-high rolls, two-high mill;
—— *ébaucheur*, (*mach.*) roughing rolls;
—— *épicycloïdal*, (*mach.*) epicyclic train;
—— *à n équipages*, (*mach.*) train with n pairs of rolls;
—— *finisseur*, (*mach.*) finishing rolls;
à fond de ——, at a great pace, as fast as possible; (of a horse) at top speed;
grand ——, v. *bon* ——;
—— *d'une grue*, crane framing;
—— *de laminoir*, (*mach.*) train of rolls;
—— *de marée*, tidal train;
—— *marchand*, (*mach.*) merchant rolls;
mettre en ——, to throw into gear, set going; (hence, generally) to set on foot;
mise en ——, start, starting, putting into gear; (hence, generally) setting on foot, starting;
—— *de puddlage*, (*mach.*) puddling rolls;
—— *à rails*, (*mach.*) rail mill;

train *de serrage*, (*mach.*) roughing-down rolls;
—— *à six chevaux*, six-horse team;
—— *à tôles*, (*mach.*) plate rollers;
—— *rompu*, v. s. v. *rompu*;
—— *-train*, regular way; humdrum; red-tape;
—— *trio*, (*mach.*) three-high rolls, three-high mill.

trainage, m., sledging, sleighing.

trainant, a., dragging;
combat ——, (*mil.*) running fight;
semelle ——*e*, (*cons.*) bottom sill.

trainard, m., (*mil.*) straggler; (*nav.*) rotator and line of the patent log.

traine, f., state of being drawn or dragged; dragging cable; branches, windfalls, on the edge of a wood; (*cord.*) rope maker's sledge;
à la ——, in tow, towing overboard.

traineau, m., sleigh, sledge; sled, truck; drag; (*r. r.*) railroad truck; (*sm. a.*) shoe of a sword scabbard; (*art.*) sledge for the transportation of heavy guns;
—— *à bascule*, tilting or rocking sledge;
—— *du fourreau de sabre*, (*sm. a.*) shoe;
—— *de montagne*, (*art.*) mountain artillery sledge;
—— *de mine*, (*mil. min.*) miner's sledge;
—— *ordinaire*, ordinary sledge carriage;
—— *à patins*, a sledge on slides;
—— *de poterne*, —— *à rouleaux*, (*art.*) roller sledge (for use in narrow ramps, in posterns, etc.).

trainée, f., trail, train;
—— *de la gerbe*, (*ball.*) tailings;
—— *de poudre*, (*mil.*) powder train;
—— *de poussière*, train of dust.

trainement, m., (*art.*) scoring of the bore, made by the projectile (bronze guns, obs.).

traine-paillasse, m., (*mil. slang*) fourrier in charge of the bedding.

trainer, v. a. n., to drag, draw, haul, trail; to lengthen, protract; to be lengthened; (*mil.*) to loiter, straggle; (*sm. a.*) of a lock, to drag (said when the nose of sear stops on the edge of the full-cock notch);
—— *les hanches*, (*man.*) to have unsettled gaits; to gallop false or disunited;
—— *la jambe*, (*man.*) to drag a leg (to be weak or hurt in the leg).

traineur, m., sleigh driver; (*mil.*) straggler, loiterer;
—— *de sabre*, (*slang*) v. s. v. *sabre*;

trait, p. p., drawn (as wire).

trait, m., draught; tow of boats; stroke (of a pen, etc.), line; sketch, draught; cut, mark; arrow, bolt, dart, shaft; kerf; turn of the scale; marking out (of wood for sawing, of stone for cutting); flash (of light); beam (of the sun); (*harn.*) trace (in *Fr. art.* the trace proper runs from the shoulder to the hind quarters of the horse, being eked out by a *rallonge de* ——, q. v.); (*art., sm. a.*) graduation of a sight;
—— *d'arrière*, (*harn.*) wheel trace;
art du ——, art of stonecutting, of cutting out timber;
—— *d'attelage*, (*harn.*) trace;
attelage ——*s sur* ——*s*, (*harn.*) hitching up with the traces in the prolongation of one another;
—— *biais*, (*cons.*) line cutting another under an oblique angle;
—— *de brêlage*, (*art.*) lashing strap or rope;
—— *à canon*, (*art.*) trace rope (for mechanical maneuvers;
—— *carré*, (*cons.*) line cutting another at right angles;
chaîne de bout de ——, (*harn.*) trace chain;
cheval de ——, v. s. v. *cheval*;
—— *de cheval de selle*, (*harn.*) lariat, picketing rope;
—— *avec clef*, (*carp.*) scarf joint and key;
—— *de compas*, division (line) of a compass card;
—— *corrompu*, (*cons.*) free-hand sketch (for stone-cutting, etc.);
couper de, *le*, ——, to make a model in plaster, etc.;
—— *de derrière*, (*harn.*) wheel trace;

trait, *dessin au* ——, line drawing;
—— *de devant,* (*harn.*) lead trace;
—— *de force,* heavy line, stroke, with a pen;
—— *de foret,* (*sm. a.*) mark left by boring bit in the interior of gun barrel;
franc de ——, (of a horse) free in harness;
ganse de ——, (*harn.*) trace loop;
—— *de Jupiter,* v. s. v. *Jupiter;*
longe de ——, (*art., harn.*) trace tug;
—— *à manœuvre,* (*art.*) sort of trace rope;
—— *de milieu,* (*art.*) swing trace;
—— *de niveau,* level mark;
—— *de paysan,* rope for lashing, drawing, light weights;
petit ——, (*harn.*) trace strap;
pièce de ——, construction model;
rallonge de ——, v. s. v. *rallonge;*
—— *de repère,* reference mark;
—— *de scie,* saw cut;
—— *de la scie,* kerf;
—— *simple,* (*carp.*) simple scarf;
—— (*s*) *sur* ——(*s*), v. *attelage* ——*s sur* ——*s;*
—— *de tirage,* (*harn.*) trace;
—— *d'union,* hyphen;
—— *de vent,* point of the compass.

traite, f., trade (especially with coast of Africa); transport, exportation; journey; banking; draft, bill of exchange;
bâtiment de ——, slave ship;
—— *d'ébène,* —— *des nègres,* —— *des noirs,* the slave trade.

traité, m., treatise; contract, treaty, agreement;
—— *d'extradition,* extradition treaty;
—— *de paix,* treaty of peace.

traitement, m., treatment, usage; pay or salary; (*mil.*) full pay (of officers), esp. pay of military members of the Legion of Honor and of holders of the *médaille militaire;* (*med.*) treatment (*hipp.*) means and methods of treating animals;
—— *de congé,* (*mil.*) retired pay;
—— *de la médaille militaire,* (*Fr. a.*) annuity accompanying a military medal;
—— *de table,* table money.

traiter, v. a. n., to treat, handle, negotiate, be in treaty for; to execute; to come to terms; (*med.*) to treat a patient; (*chem., etc.*) to treat.

traître, m., traitor.

trajectoire, f., (*ball.*) trajectory;
abaissement de la ——, v. s. v. *abaissement;*
—— *de l'air,* —— *dans l'air,* trajectory in air, actual trajectory;
base de la ——, right line from the origin to the point of fall;
—— *courbe,* curved trajectory;
faisceau de ——*s,* sheaf of trajectories;
—— *inférieure,* lowermost trajectory of a sheaf;
maintenir sur la ——, to keep a projectile on its trajectory by rotation;
—— *moyenne,* —— *normale,* mean trajectory;
rasance de la ——, flatness, quality of flatness, of the trajectory;
—— *rasante,* flat trajectory;
rigidité de la ——, rigidity of the trajectory;
—— *supérieure,* upper trajectory of a sheaf;
—— *tendue,* flat trajectory;
tension de la ——, flatness of the trajectory;
—— *du, dans le, vide,* trajectory in vacuo.

trajet, m., trip, course, passage (from shore to shore); voyage, journey; (*ball.*) flight of a projectile;
—— *dans l'âme du canon,* (*ball.*) passage or travel of projectile in bore of gun;
de —— *direct,* (*r. r.*) through;
durée du ——, (*ball.*) time of flight;
faire le —— *de,* to make the trip between;
de —— *à stations,* (*r. r.*) local, accommodation;
temps de ——, (*ball.*) time of flight of projectile.

trame, f., woof (of cloth); (*fig.*) course; progress; plot;
ourdir une ——, to lay a plot.

tramer, v. a., to weave; (*fig.*) to hatch, plot, contrive.

tramontain, a., ultramontane.

tramontane, f., the north star; the north wind (in the Mediterranean).

trancade, f., large block of stone, full of cavities, lying near the surface of the ground.

tranchant, a., sharp cutting, edged; (*tech.*) shearing;
instrument ——, edged tool.

tranchant, m., edge (of a sword, etc.); short expression for cutting angle of a tool;
à ——*s,* edged;
angle ——, v. s. v. *angle;*
à deux ——*s,* double-edged;
faux ——, (*sm. a.*) back edge of a sword;
mettre à ——, to put the edge on a tool;
vrai, (*sm. a.*) edge of a sword.

tranche, f., slice; edges (of a book); edge, small dimension (of anything, as a plank); section; face; chisel, anvil cutter; farrier's paring and cutting tool; (*nav.*) compartment of the hold; (*art.*) face (of the muzzle, of the trunnions, of the eye of a shell, etc.);
—— *de la bouche,* (*art.*) face of the muzzle;
—— *de butée,* (*sm. a.*) breech pin;
—— *cellulaire,* (*nav.*) cellular compartment;
—— *cellulaire du bas,* (*nav.*) double bottom;
—— *à chaud,* hot set;
—— *à chaud et à froid,* hot and cold set;
—— *de la culasse,* (*art.*) breech face;
—— *à froid,* cold set;
—— *à main,* rod chisel;
—— *à manche,* rod chisel;
—— *de marbre,* (*cons.*) plate of marble;
—— *du mouleur,* molder's chisel;
parallélisme des ——*s,* Bernouilli's principle of the parallelism of sections;
—— *de la vis de culasse,* (*art.*) face of the breech plug.

tranché, p. p., *bois* ——, cross-grained, knotty, wood.

tranchée, f., trench, ditch; cut across the grain; (*r. r., etc.*) cut, cutting; (*siege, fort., mil.*) trench, or demiparallel; parapet of sand bags, gabions, on rock or soil hard to dig; tour of trench duty;
—— *-abri,* rifle pit, shelter trench;
—— *-abri ébauchée,* rifle pit, shelter trench (cover kneeling);
—— *-abri normale,* normal shelter trench (cover standing);
—— *-abri renforcée,* v. —— *renforcée;*
cavalier de ——, v. s. v. *cavalier;*
—— *à ciel couvert,* covered trench;
—— *à ciel ouvert,* open trench;
colonel de ——, v. s. v. *colonel;*
—— *de communication,* communication trench (to connect gorge with main parapet, or to allow circulation around various elements of a group);
—— *couvrante,* shelter trench used when troops are going to stay any length of time in one position;
—— *couvrante normale,* trench with thickened parapet, protecting reserves and supports against artillery fire;
—— *couvrante simplifiée,* trench protecting against small-arm fire;
—— *à crochet,* zigzag trench;
de ——, on trench duty;
dépôt de ——, v. s. v. *dépôt;*
descendre la ——, v. s. v. *descendre;*
—— *directe,* straight trench;
—— *double,* double trench;
—— *ébauchée,* v. —— *-abri ébauchée;*
—— *explosive,* trench work established by explosives (*Fr. a.,* experimental only);
garde de ——, v. s. v. *garde;*
général de ——, v. s. v. *général;*
—— *intérieure,* trench in rear of a parapet;
major de ——, v. s. v. *major;*
monter la ——, v. s. v. *monter;*
—— *de mur,* (*cons.*) opening left in a wall for a beam;
nettoyer la ——, v. s. v. *nettoyer;*
—— *normale,* v. —— *-abri normale;*
ouverture de la ——, opening of the trench;
ouvrir la ——, to open the trenches;
—— *perfectionnée,* trench with a step in the rear slope;
profil à —— *intérieure,* profile with interior trench;
queue de la ——, v. s. v. *queue;*

tranchée, *refuser une* ——, *(siege)* to cause the trench to make a wider angle with the direction of advance as one advances;
—— *renforcée*, trench affording protection against artillery fire and holding two ranks of men;
—— *de revers*, reverse trench (protecting against enfilade), traversed trench;
——*s rouges*, *(hipp.)* colic, gripes;
—— *en sape dérobée*, "stolen" trench (v. s. v. *sape*);
—— *en sape à terre roulante*, trench in which the head of the sap is pushed forward as the work advances;
—— *en sape volante*, flying trench;
service de ——, trench duty;
—— *simple*, simple, common, trench; simple trench work;
tant de jours de —— *ouverte*, so many days of siege work, i. e., since trenches were opened;
tête de ——, v. s. v. *tête*;
—— *tournante*, enveloping trench;
travaux de ——, trench work or operations.

tranchefil, m., v. *tranchefile*.

tranchefile, f., bar in a shoe; *(harn.)* curb strap; slobbering chain, curb chain.

tranche-fromage, m., *(mil. slang)* sword, "frog-sticker."

tranche-gazon, m., turf cutter, sod cutter.

tranche-papier, m., paper knife.

trancher, v. a. n., to strike off, amputate; to cut off, across; to decide, determine, settle.

tranchet, m., cutting, paring, knife; cutter, hack iron;
—— *de ciseau*, shank, tail, of a chisel.

tranchis, m., *(cons.)* row of slates, etc., cut so as to fit an angle, a valley, etc.;
faire un ——, to cut shingles, tiles, etc., so as to fit a valley, etc.

tranchoir, m., trencher, platter.

tranchon, m., trench cutter.

tranquille, a., quiet, calm (of the weather, sea); steady (of the compass).

tranquillité, f., tranquillity, calm (of the sea, etc.).

transbordement, m., reloading, transfer of cargo from one ship or car to another; transshipment; *(mil.)* of ammunition supply, transfer from one wagon to another, as opposed to exchange of wagons;
par ——, by breaking bulk.

transborder, v. a., to transfer cargo, transship, unload (as freight from one vehicle to another, from ship to ship, ship to shore, etc.); *(mil.)* to shift ammunition from one wagon to another.

transbordeur, m., traveling platform; *pont* ——, transfer bridge.

transcorporation, f., *(mil.)* transfer from one regiment to another (rare).

transférer, v. a., to transfer, remove, convey, transport; to postpone; to make over;
—— *un prisonnier*, to remove, convey, a prisoner.

transfert, m., transfer.

transfil, m., *(cord.)* lacing.

transfilage, m., *(cord.)* marling, marling hitch; lacing of an awning;
faire un ——, to lace.

transfiler, v. a., *(cord.)* to marl; to serve a rope, etc., with marline; to lace.

transformateur, m., *(elec.)* transformer;
—— *à action instantanée*, ordinary transformer with induction coil;
—— *à action différée*, any accumulator or condenser;
—— *à circuit magnétique fermé*, closed-circuit transformer;
—— *à circuit magnétique ouvert*, open-circuit transformer;
—— *à coquille*, ironclad transformer, shell transformer;
—— *de courant à basse tension en courant à haute tension*, step-up transformer;
—— *de courant à haute tension en courant à basse tension*, step-down transformer;
—— *à courants alternatifs*, alternate-current transformer;
—— *à courants continus*, continuous-current transformer;

transformateur *à enveloppe*, v. —— *à coquille;*
—— *à noyau*, cored transformer;
—— *tournant*, rotary transformer.

transformation, f., transformation; *(art., sm. a.)* conversion;
—— *d'un tour de manivelle*, *(art.)* evaluation of a turn of the elevating screw in degrees and minutes.

transformé, p. p., *(art.)* converted (to a different model).

transformer, v. a., to change, transform; *(art., sm. a.)* to convert to a different system;
—— *un tour de manivelle*, *(art.)* to evaluate a turn of the elevation screw in degrees and minutes.

transfuge, m., *(mil.)* deserter to the enemy; *(fig.)* turncoat.

transgresser, v. a., to transgress, violate; to trespass.

transgresseur, m., transgressor.

transgression, f., transgression.

transiger, v. a. n., to compound with, to come to terms; to agree, enter into articles with.

transir, v. a. n., to chill, freeze, benumb; to be, become, chilled, frozen; to paralyze, subdue, overcome.

transissement, m., chill, freezing.

transit, m., transit, transit trade;
droit de ——, transit duty;
en ——, in transit;
marchandises de ——, transit goods;
port de ——, transit port.

transiter, v. a. n., to pass in transit.

translateur, m., *(teleg.)* translator.

translation, f., translation, transfer (of a body, etc.).

transmetteur, m., *(teleg.)* transmitter (of a telephone, telegraph).

transmettre, v. a., to transmit, convey, deliver, forward, send on; to make over, hand over, transfer; *(mil. sig.)* to send.

transmission, f., transmission (as of power, motion, etc.); *(mil.)* transmission (of orders, etc.), transfer (of command); *(mach.)* gear, transmission gear;
arbre de ——, v. s. v. *arbre;*
—— *à courroie*, *(mach.)* belt transmission;
—— *de force*, transmission of power;
—— *de liaison*, *(mach.)* connecting gear;
—— *de mouvement*, *(mach.)* gear, transmission gear;
—— *souterraine*, *(mach.)* floor or underground transmission;
—— *supérieure*, *(mach.)* overhead transmission.

transmontain, a., ultramontane.

transmutation, f., transmutation; *(mil.)* transfer from one regiment, etc., to another (rare).

transparent, a., m., transparent; transparent paper, transparency.

transpercer, v. a., to pierce, strike, run, shoot, through.

transport, m., transport, conveyance; transfer, assignment, removal, traffic; *(nav.)* transport, transport ship, cargo vessel; *(tech.)* transmission (of energy, etc.); *(mil.)* transport, transportation);
bâtiment de ——, transport, transport ship;
chemin de ——, teaming road;
—— *par chemin de fer*, *(mil.)* rail transportation;
commerce de ——, carrying trade;
—— *de concentration*, *(mil.)* transport of troops by rail for concentration;
—— *-écurie*, *(nav.)* horse ship;
entrepreneur de —— *par eau (terre)*, water (land) carrier;
—— *d'évacuation*, v. s. v. *évacuation;*
frais de ——, carriage;
—— *de la guerre*, any conveyance or transportation on War Department account;
—— *des marchandises*, carrying;
—— *maritime*, —— *par mer*, transportation or conveyance by sea; water transportation;
—— *de mobilisation*, *(mil.)* transport of troops by rail for mobilization;

3877°—17——30

transport, *moyens de* ——, means of transportation;
—— *ordinaire*, (*Fr. a.*) rail transport by ordinary train;
prix de ——, carriage dues;
—— *de ravitaillement*, (*nav.*) storeship; (*mil.*) transportation of supplies;
—— *de réglage* (*en direction, en portée*), (*art.*) transfer of a correction (e. g., wind) from one range to another (in direction, in range);
—— *stratégique*, (*Fr. a.*) the transport by rail of troops in great numbers for concentration;
terrain de ——, alluvial ground;
—— *des terres*, teaming;
—— *du tir*, v. s. v. *tir*;
—— *de troupes*, troopship; (*in gen.*) conveyance, transfer, of troops;
vaisseau de ——, transport ship.

transportation, f., (*law, etc.*) transportation.

transporter, v. a., to transport, convoy, carry, remove, transfer, assign, make over; (*mil.*) to transport troops;
—— *à sec*, to carry over dry (as arms, ammunition, etc., across a stream);
—— *le tir*, v. s. v. *tir*.

transporteur, m., transporter, public carrier, hand car; (*sm. a., r. f. art.*) cartridge carrier, transmitter, feeder, food, feed mechanism.

transsudation, f., transudation.

transsuder, v. r., to transude.

transvasement, m., decanting.

transvaser, v. a., to decant.

transversal, a., transverse, cross;
cloison ——*e*, v. s. v. *cloison*.

transversale, f., any transversal line; (*mil. min.*) transversal, transverse gallery, cross gallery; (*surv.*) traverse;
système de ——, (*mil. tel.*) system of cross telegraph lines (system in which, after each day's march, the positions occupied are connected by lines perpendicular to general direction of march followed); (*mil. min.*) system of transverse or cross galleries.

transverse, a., transverse, cross.

trapan, m., top of a staircase.

trapèze, m., trapezium; trapeze of a balloon (frame from which the car swings); (*gym.*) trapeze.

trappe, f., trapdoor, flap door; register (of a furnace, etc.); trap, pitfall; gin, snare; (*mach.*) manhole; (*mil.*) trou-de-loup.

trapu, a., stubby, squat, stocky;
cheval ——, thickset, strong, horse.

traquenard, m., trap (for noxious animals), (hence *mil.*) a fougasse, or other explosive trap, so to say; (*man.*) racking, ambling, gait; racking, ambling, horse.

traquenarder, v. n., (*man.*) to rack, amble.

traquet, m., mill clapper.

trass, m., trass.

trastravat, m., (*hipp.*) horse with white diagonal marks upon his feet.

travail, m., work, labor, workmanship; piece of work, employment; industry, study; report; (in the arts) strain, fatigue (as of joints, lashings, of metals, etc.); (*hipp.*) trave, brake, for refractory horses; (*nav.*) laboring of a vessel; (*mil.*) exercises, practice, drill, training; (more esp.) trench work; (*mach., etc.*) working (of metal, wood, etc.);
—— *aux abandonnés*, dead waste;
—— *aux d'approche*, (*siege*) trench work, approaches;
—— *avec, en, armes*, (*mil.*) exercises under arms;
—— *sans armes*, (*mil.*) exercises without arms (e. g., facings, various steps, etc.);
—— *aux d'art*, (r. r.) constructive works; constructions in masonry, iron, etc. (as bridges, culverts, embankments, tunnels, etc.);
—— *aux d'attaque*, (*siege*) approaches, siege works;
—— *à bascule*, (*hipp.*) tilting travis;

travail, *travaux de bivouac*, (*mil.*) all measures taken on occupying a bivouac;
—— *en bride*, (*mil. man.*) exercises on a horse bridled;
—— *en bridon*, (*mil. man.*) exercises on a horse snaffled;
—— *du camp*, (*mil.*) all labor necessary to get a camp, bivouac, cantonment, etc., into order as soon as troops arrive;
—— *aux de campagne*, (*fort.*) field works;
campagne de ——*aux*, working season;
—— *en carrière*, v. —— *sur de grandes lignes*;
—— *à chaud*, (*met.*) hot working;
—— *à cheval*, (*mil. man.*) mounted drill (and work generally);
—— *chimique*, (*expl.*) change in composition;
—— *aux à ciel ouvert*, (*min.*) ordinary blasting, as distinguished from tunneling operations;
—— *au cisaillement*, shearing strain;
—— *à la compression*, strain of compression;
conseil de ——*aux publics*, board of public works;
—— *aux de démolition, destruction,* (*mil.*) demolition (as of railroads, bridges, etc., in campaign);
—— *effectif*, useful work;
—— *avec élan*, (*mil. man.*) exercises with a running start;
—— *sans élan*, (*mil. man.*) exercises without a running start;
—— *à l'extension*, strain of extension;
—— *extérieur*, (*surv.*) field work, out-of-door work;
—— *aux extérieurs*, (*mil.*) exterior fatigues;
—— *externe*, external work;
faire chômer les ——*aux*, to throw out of work;
—— *à la flexion*, bending strain;
—— *aux de force*, (*art.*) moving heavy guns;
—— *aux forcés*, compulsory labor, hard labor (punishment); transportation;
—— *aux forcés à perpétuité*, hard labor for life; transportation;
—— *aux forcés à temps*, hard labor for a term of years;
—— *de la forge*, forging;
—— *à froid*, (*met.*) cold working;
—— *aux de fortification de campagne*, (*fort.*) field works;
—— *au galop* (*mil. man.*) exercises at gallop;
—— *sur de grandes lignes*, (*mil. man.*) exercises on an out-of-doors manege; track exercises;
—— *aux de guerre*, (*mil.*) field works, etc.;
—— *indiqué*, indicated work (indicator diagram);
—— *intérieur*, (*surv.*) internal work; indoor work; (in pl., *mil.*) garrison fatigue;
—— *à la journée*, work by the day;
—— *à la longe*, (*man.*) longing;
—— *de la machine*, working power of an engine;
—— *à la main*, hand labor;
maison de ——, workhouse;
—— *maximum*, (*expl.*) the mechanical equivalent of heat multiplied by the quantity of heat disengaged; potential of explosive substances;
—— *mécanique*, work done, mechanical effect, mechanical power; machining; (*expl.*) change in state or in condensation;
—— *aux militaires*, (*mil.*) works, military works; fatigues;
—— *moteur*, (*mach.*) work developed by a motor;
—— *de nuit*, night labor;
—— *perdu*, lost work;
—— *à pied*, (mounted troops) foot drill, etc.; (*man.*) v. —— *à vide*;
—— *de pied ferme*, (*mil. man.*) exercises with a horse standing;
—— *à poteaux*, (*hipp.*) ordinary travis;
—— *préparatoire*, (*mil. man.*) preliminary exercises on horseback;
—— *aux préparatoires*, (r. r.) preliminary survey;
—— *aux publics*, public works; (*mil.*) confinement at hard labor;

travail, *quantité de* ——, quantity of work;
—— *résistant*, work necessary to overcome a resistance;
—— *en sens inverse*, (*mil. man.*) work on a double track (one squad on each, going in opposite directions);
—— *avec la selle*, (*mil. man.*) riding exercises, horse saddled;
—— *aux de siège*, (*siege*) sap work, trench work;
—— *avec le surfaix*, (*mil. man.*) riding exercises, surcingle on;
—— *à la tâche*, work by the job, piecework;
—— *aux de terrasse, de terrassement*, (*fort.*, *r. r.*, *etc.*) earthwork;
—— *à la torsion*, strain of torsion;
—— *au tour*, (*mach.*) lathe work;
—— *de tranchée*, (*siege*) trench work;
—— *aux de tranchée*, (*siege*) trench work or operations;
—— *utile*, (*mach.*) useful work, effective work;
—— *à vide*, (*man.*) exercise, work, of a horse without a rider; (*mach.*) work unloaded.

travaillé, p. p., strained, overstrained;
avoir les jambes ——*es*, (*hipp.*) to be stiff in the legs;
cheval (*trop*) ——, overworked horse.

travailler, v. a. n., to work, labor; to fashion, work up; to execute with care; to open, work open, spring, warp; (of material) to bear a stress, to be strained; to work (i. e., prepare material, as iron, wood, etc.); to get out of plumb (of walls, etc.); (*man.*) to exercise a horse; to overwork a horse; (*nav.*) to labor;
—— *un cheval*, (*man.*) to exercise a horse;
—— *un cheval ferme à ferme*, (*man.*) to work a horse without advancing (as in the piaffer);
—— *un cheval dans la main*, (*man.*) to put a horse through his paces by the hand alone;
—— *un cheval de la main à la main*, (*man.*) to change hands;
—— *un cheval de quart en quart*, —— *aux quatre coins*, v. s. v. *quart;*
—— *au cisaillement, etc.*, to bear a shearing, etc., stress;
—— *ensemble*, to bear equally; to bear an equal strain;
faire ——, to work, warp; to stress;
—— *à n kg. par cm²*, to bear a load of *n* kilograms per square centimeter;
—— *à main droite* (*gauche*), —— v. s. v. *main.*

travailleur, m., laborer, operative; mechanic, artisan; close, hard, student or worker; (*mil.*) member of a working party; any soldier doing nonmilitary work; a soldier who has a trade and is required to exercise it in the army;
—— *d'infanterie*, (*siege*) member of a working party.

travat, m., (*hipp.*) horse with white marks on his feet on the same side.

travée, f., (*cons., pont.*) bay; (*mil. r. r.*) element of a track (ready to lay down; of different lengths);
—— *de culée*, (*pont.*) shore bay;
—— *pour les fusils*, (*sm. a.*) arm rack, improvised (two beams and a plank for the butts);
—— *de jonction*, (*pont.*) connecting bay (in bridge by parts, or by rafts);
ouverture de ——, (*pont.*) span of a bay;
—— *de rive*, (*pont.*) shore bay.

travers, m., breadth; spring line; (*art.*) sling, lashing-rope; (*met.*) transverse crack in a casting, forging, etc., in a firearm, in a sword; (*nav.*) side, broadside; beam;
à ——, across, through;
de ——, crooked, awry, askew, the wrong way;
en ——, crosswise, across;
en —— *à*, athwart;
être en —— *de lame*, to lie in the trough of the sea;
ouvrage en ——, (*min.*) cross system of mining;
par le ——, (*nav.*) on the beam, abeam; athwart; (also) parallel, beam and beam;

travers, *pied de* ——, v. s. v. *pied.*

traversant, m., beam, lever, of a balance.

traverse, f., passage across; crossroad; cross-beam, -bar, -piece; whiffletree; horizontal stovepipe; (*cons., etc.*) stay, brace, cross-framing, girder, ground sill, crosspie e; (*r. r., etc.*) sleeper, crosstie; (*hydr.*) sand, mud, etc., bank at the mouth of a port; (*surv.*) traverse; (*mach.*) crosshead; (*fort.*) traverse; (*sm. a.*) cross of a sword; firing-pin crosspiece; deflection slide; (*art.*) traversing box, head, crosshead; partition board of an ammunition chest; sleeper of a platform.

I. Fortification; II. Miscellaneous.

I. Fortification:

—— -*abri*, v. s. v. *abri;*
—— *d'attaque*, traverse, place of arms, in a dry ditch;
bras de ——, inclined postern under a traverse;
—— *en capitale*, traverse along the capital;
—— *casematée*, casemated traverse;
—— *s conjuguées*, conjugate traverses;
—— *contre un commandement*, traverse to screen a work seen or commanded by the enemy;
—— *à crémaillère*, lock traverse;
—— *creuse*, hollow traverse, bombproof traverse;
—— *de défilement*, defilading traverse;
—— *diagonale*, diagonal traverse (axis oblique to interior crest);
—— *enracinée*, attached traverse; traverse running back to and joining the parados;
—— *isolée*, v. —— *tournante;*
—— -*magasin*, magazine traverse;
—— *perpendiculaire*, traverse perpendicular to the interior crest;
—— *pleine*, solid traverse;
—— -*relai*, in a *batterie rapide*, natural ground left between guns, upon which the earth excavated for the *terre-plein* is thrown, and which serves as a traverse on the completion of the battery;
—— *tournante*, cube traverse;
—— *de tranchée*, part of the original ground left as a traverse in a trench.

II. Miscellaneous:

—— *d'appui*, breast rail;
arbre de ——, crosstree;
—— *à arrêtoirs de poutrelles*, (*pont.*) balk stop cross bar of a pontoon wagon;
—— *à charnière articulée*, (*Fr. art.*) (in 120ᵐᵐ C.) bar securing upper to lower carriage for traveling;
chemin de ——, crossroad;
—— *de chevalet*, (*pont.*) trestle brace;
—— *de cloison*, (*cons.*) crosspiece of a partition;
—— *à crémaillère*, (*art.*) toothed or notched traversing head of a sight;
—— *d'échafaud*, (*cons.*) putlog;
—— *de fenêtre*, (*cons.*) window transom;
—— *flottée*, (*cons.*) crossbar concealed behind a partition;
—— *de frein*, brake bar;
—— *avec glissière*, (*mach.*) crosshead for sliding guides;
grande ——, (*mach.*) main crosshead; (*r. r.*) weigh bar;
—— *du grand piston*, (*mach.*) main crosshead;
—— *avec guides à articulations*, (*mach.*) crosshead for link guides;
—— *de grille*, fire-bar bearing; crosspiece of a grating;
—— *d'imposte*, (*cons.*) transom bar (French window);
—— *inférieure*, (*art.*) lower sleeper (of a platform); (*cons.*) bottom rail, weather rail (of a door, of a window);
—— *de joint*, (*r. r.*) joint sleeper;
—— *de limonière*, crossbar of a pair of shafts;
—— *de linteau*, (*cons.*) head rail;
—— *lisoir*, (*art.*) chock of a slide carriage;
—— *d'une manivelle*, shackle bolt;
—— *mobile à huit pans*, shifting (octagonal) bar;

traverse à *mouvement libre*, (*mach.*) free crosshead;
— *moyenne*, (*cons.*) lock rail of a door or window;
— *de palée*, cap;
— *du piston*, (*mach.*) crosshead;
— *de raccordement*, (*art.*) sleeper used to widen a platform;
— *de reculement*, (*harn.*) outrigger, when three horses go abreast;
— *renversée*, (*mach.*) cross-tail;
— *de ressort*, spring bar;
rue de ——, cross street;
— *à un seul patin*, (*mach.*) single guide crosshead;
— *de soupape*, (*mach.*) valve guard;
— *supérieure*, (*art.*) upper sleeper of a platform; (*cons.*) top rail of a door or window;
— *de tente*, v. s. v. *tente*;
— *de tête*, crosshead of a windlass;
— *de la tige*, (*mach.*) crosshead;
— *de tige de piston*, (*mach.*) piston crosshead;
— *d'un wagon*, (*r. r.*) headstock.

traversée, f., passage (as from one port to another, etc.); short sea journey; (*art.*) perforation (of an armor plate);
—— (*de voie*), (*r. r.*) crossing.

traverser, v. a. n., to cross, cross over; to pass, go, get, travel, through; to penetrate; to thwart, lie across; to cut wood against the grain; (*art.*, *sm. a.*) to perforate; (*mil.*) to break through (enemy's line); (*fort.*) to traverse, put up traverses;
—— *l'ancre*, (*nav.*) to fish the anchor;
—— *la lame*, to head the sea;
—— *à la nage*, to swim over;
—— *une pièce*, (*fort.*) to protect a gun by traverses;
—— *une pierre*, to crosscut a stone;
se ——, (*man.*) to traverse; (*nav.*) to present the broadside;
se —— *des épaules* (*des hanches*), (*man.*) to throw the shoulders (the hind quarters) off the line of advance;
—— *un vaisseau devant un fort*, to bring the broadside to bear against a fort or battery.

traversier, a., cross, crossing, going across;
barque ——*e*, ferryboat;
mules ——*es*, v. s. v. *mule*;
poutre ——*e*, v. s. v. *poutre*;
rue ——*e*, cross street;
vent ——, wind right inshore (Mediterranean term).

traversier, m., ferryboat; crosspiece of a banner.

traversière, f., (*cons.*) straining beam, collar beam; (*pont.*) breast line; synonym of *anguille*, q. v.

traversin, m., bolster of a bed; cross-beam, -piece; transom; beam of a balance; thwart of a boat; heading, crosspiece of a cask, barrel, etc.; head- (tail-) piece of a raft of logs; (*fam.*) infantry soldier;
—— *d'affût*, (*art.*) transom of a gun carriage;
—— *de bateau*, thwart;
—— *du gouvernail*, sweep of the tiller;
—— *de nage*, —— *de pied*, stretcher of a boat.

traversine, f., plank to pass from one vessel to another; (*cons.*) crosspiece of a grillage; (*r. r.*) crossing.

travon, m., (*pont.*) stringpiece, capping piece, of a wooden bridge.

travure, f., (*cons.*) framing (of a floor);
—— *composée*, a framing with intermediate supporting beams;
—— *simple*, framing with no intermediate supporting beams.

trébucher, v. n., to trip, stumble; to have an excess of weight; to outweigh, weigh down.

trébuchet, m., pitfall; gold scales.

tréfilage, m., wire-drawing.

tréfiler, v. a., to wire-draw.

tréfilerie, f., wire-drawing works; wire-drawing machine.

tréfileur, m., wire drawer.

trèfle, m., clover; (*mil. min.*) trefle (mine);
double ——, (*mil. min.*) double trefle.

tréflée, f., (*mil. min.*) trefle.

tréfonds, m., ground under the surface of the ground.

treillage, m., trellis, trellis work.

treille, f., vine arbor.

treillis, m., trellis, latticework, grating; drilling, strong ticking, corduroy, buckram;
—— *de fil de fer*, wire lattice.

trémat, m., (*hydr.*) sand bank (in the bends of the lower Seine).

tremble, m., aspen (tree and wood).

tremblement, m., trembling; shaking (e. g., of a bridge); quivering; (*mil. slang*) fight;
—— *de terre*, earthquake.

trembler, v. a., to tremble, quiver, shake (as a bridge).

trembleur, m., electric bell, buzzer.

trembleuse, f., electric bell.

trémeau, m., (*fort.*) obsolete name for merlon.

trémie, f., hopper; (*met.*) charging funnel or cone of a blast furnace; (*sm. a.*) hopper of a reloading tool; (*cons.*) hearth stead;
bande de ——, (*cons.*) iron trimmer strap.

trémion, m., (*cons.*) trimmer; iron band or brace to support the funnel of a chimney.

trempage, m., soaking.

trempant, a., (*met.*) capable of being tempered.

trempe, f., steeping, dipping, soaking, dampening; (*met.*) hardening, tempering; temper; (*hipp.*) firmness of tissues, muscles.
(The following terms are metallurgical:)
—— *à* ...., ... hardening (e. g., oil hardening);
—— *à l'air*, air hardening;
de bonne ——, finely tempered;
—— *par cémentation*, casehardening;
—— *chimique*, hardening by the introduction of foreign substances;
—— *par compression*, hardening by compression;
—— *en coquille*, v. —— *en paquet*;
donner la ——, to harden, temper;
—— *double*, *double* ——, double tempering;
eau de ——, tempering water;
—— *à l'eau*, water tempering;
—— *glacée*, chilling;
—— *à l'huile*, oil hardening, tempering;
—— *négative*, name proposed by Osmond for temper that does not harden steel;
—— *en paquet*, casehardening;
—— *au plomb*, lead hardening, tempering;
—— *positive*, name proposed by Osmond for temper that hardens steel;
—— *de la surface*, v. —— *en paquet*;
—— *à la volée* ordinary tempering.

trempé, a., (*met.*) (of steel, etc.) hardened, tempered.

tremper, v. a. n., to steep, soak, dip, wet through; (*expl.*) to dip (cotton for gun cotton); (*met.*) to temper; to quench; to harden; (less correctly) to anneal; to take temper.
(The following terms are metallurgical:)
—— *à* ...., to .... temper (as to oil temper);
—— *à l'air*, to air harden;
—— *à blanc*, etc., to temper when white, etc., hot;
—— *par cémentation*, to caseharden;
—— *par compression*, to harden by compression;
—— *en coquille*, v. —— *en paquet*;
—— *à l'eau*, to harden by water, to water harden;
—— *la fonte*, to chill cast iron;
—— *à l'huile*, to oil temper;
—— *en paquet*, to caseharden;
—— *son pied dans l'encre*, (*mil. slang*) to be confined to barracks;
—— *au plomb*, to lead harden;
—— *à la volée*, to temper (ordinary process).

trempeur, m., temperer.

trépan, m., bit, boring bit; centerbit, wimble, drill, stone drill; (*min.*) earth auger, rock drill, trepan; tool for boring the roof of a gallery in order to admit air;
donner un coup de ——, to bore the roof of a gallery for air;
—— *de sondage*, —— *de sonde*, (*mil. min.*) ground auger.

trépaner, v. a., (*min.*) to bore the roof of a gallery for air.
trépas, f., very narrow passage, channel; death.
trépidation, f., jarring, tremor, vibration (as of an engine, screw, etc.).
trépied, m., tripod; trivet; three-legged stand (in general);
—— *à doubles branches*, (*inst.*) split-leg tripod;
—— *à fusées* (*artif.*) rocket stand;
—— *de tir*, (*t. p.*) aiming stand;
—— *à vis calantes*, (*inst.*) tripod fitted with leveling screws.
trépignement, m., stamping.
trépigner, v. a., to stamp, tread, under foot.
trépointe, f., welt (of a shoe);
seconde ——, skiving.
très-fonds, m., v. *tréfonds*.
trésaille, f., upper rail of a wagon.
trésillon, m., Spanish windlass; small block of wood placed between planks stacked to dry; (*cord.*) heaver;
nœud de ——, v. s. v. *nœud*.
trésillonner, v. a., to heave with a Spanish windlass; to space planks stacked to dry; (*cord.*) to use a heaver; to fix a tackle to a rope (by a *trésillon*);
trésor, m., treasure; the Treasury;
—— *de l'Etat*, v. —— *public*;
—— *de guerre*, military chest;
—— *public*, public moneys or funds.
trésorerie, f., treasury; public finances; (*mil.*) pay department;
service de la ——, (*mil.*) pay department.
trésorier, m., treasurer; (*mil.*) paymaster, pay officer;
capitaine ——, v. s. v. *capitaine*;
—— *payeur-général*, (in France) chief paymaster or treasury official of a department.
tressage, m., braid work, braid, braiding, plaiting (e. g., around a Bickford or other fuse); (*elec.*) cable wrapping.
tresse, f., plat, plait, twist, braid, plait; (*cord.*) fox, sennit; (*unif.*) braid; (*mach.*) gasket;
—— *anglaise*, (*cord.*) common sennit;
—— *carrée*, (*cord.*) square sennit;
—— *en chanvre*, (*mach.*) packing gasket;
—— *en fils de bitord*, (*cord.*) French sennit;
—— *en fils de caret*, (*cord.*) common sennit;
—— *de garniture*, (*mach.*) packing gasket; coiling for packing;
—— *de joint*, (*mach.*) packing;
—— *plate*, (*cord.*) flat sennit; (*unif.*) flat braid;
—— *ronde*, (*cord.*) round sennit; (*unif.*) round braid, cord;
—— *à tracer*, (*siege*) tracing cord or tape.
tresser, v. a., (*cord*, *etc.*) to braid, twist, plait, weave, wattle.
tréteau, m., trestle, horse; prop.
treuil, m., winch, windlass, crab; barrel or drum (of a crane); wheel and axle; truck;
—— *à barbotin*, v. —— *à noix*;
—— *des carriers*, quarriers' treadmill;
—— *-chariot*, movable or traveling loading crane for heavy work;
—— *à chevilles*, capstan;
—— *de chèvre*, (*art.*) windlass of a gin;
—— *composé*, (double, etc.) geared winch;
—— *conique*, conical drum;
—— *à deux vitesses*, double-speed winch;
—— *différentiel*, Chinese crane; differential windlass;
—— *à double force*, quarriers' treadmill for lifting very heavy stones;
—— *à double puissance*, double-purchase winch;
—— *à engrenages*, geared winch;
—— *à frein*, a winch fitted with a brake;
garnir le ——, (*art.*) to rig the windlass;
—— *à leviers*, windlass;
—— *à main*, windlass;
—— *à manivelle*, common windlass; hand winch;
—— *de manœuvre*, (*art.*) windlass of a gun carriage;
—— *de mine*, (*mil. min.*) miner's windlass, winch;
—— *à noix*, chain winch;
petit ——, crab winch, single-purchase winch;
—— *à simple engrenage*, single-geared winch;

treuil *à tambour* (*et à câble*), ordinary geared hoisting winch;
—— *à vapeur*, steam winch;
vider le ——, to shift the fall.
trêve, f., (*mil.*) truce, suspension of hostilities;
—— *marchande*, truce during which the hostile nations may trade with each other.
trévire, f., (*cord.*) parbuckle.
trévirer, v. a., (*art.*, etc.) to parbuckle.
tréza(il)ler, v. n., (of wood, etc.) to crack, split.
triage, m., sorting (of ores, of the mail, etc.); in a forest, place where timber is being (or has been) felled;
gare de ——, v. s. v. *gare*.
triangle, m., triangle; carpenter's square; (*nav.*) triangular flag; (*elec.*) delta connection, mesh grouping; (*surv.*) triangle of a triangulation;
—— *acutangle*, acute-angled triangle;
—— *du deuxième ordre* (*surv.*) secondary triangle of a triangulation;
—— *en écoperches*, shears;
—— *équilatéral*, equilateral triangle;
—— *d'erreur*, (*surv.*) triangle of error;
—— *isocèle*, isosceles triangle;
—— *de mire*, v. s. v. *mire*;
montage, monter, en ——, v. s. vv. *montage, monter*;
—— *obtusangle*, obtuse-angled triangle;
—— *plan*, plane triangle;
—— *du premier ordre*, (*surv.*) primary triangle of a triangulation;
—— *rectangle*, right-angled triangle;
—— *scalène*, scalene triangle;
—— *sphérique*, spherical triangle;
—— *en tôle*, (*mach.*) gusset stay;
—— *de troisième ordre*, (*surv.*) triangle of the third order of a triangulation;
—— *de visée*, (*Fr. art.*) sighting triangle used in observatories (temporary expedient).
triangulaire, a., triangular.
triangulation, f., (*surv.*) triangulation;
—— *géodésique*, geodetic triangulation (operations over a large area by refined processes);
—— *graphique*, triangulation, plotted from direct observation (when area to be surveyed may all be contained on plane-table sheet);
point principal de ——, principal vertex, primary vertex;
point secondaire de ——, secondary vertex;
—— *trigonométrique*, triangulation proper.
triatomique, a., (*chem.*) trivalent.
tribord, m., (*nav.*) starboard; starboard watch.
tribordais, m., (*nav.*) member of the starboard watch.
tribu, f., tribe;
se mettre en ——, (*mil. slang*) to start a mess.
tribunal, m., tribunal, court;
—— *de l'amirauté*, admiralty court;
—— *d'appel*, court of appeal;
—— *civil*, civil court;
—— *des conflits*, v. s. v. *conflit*;
—— *criminel*, criminal court;
—— *militaire*, military tribunal, court;
—— *de première instance*, court of first instance;
—— *de prise*, (*nav.*) prize court.
tribune, f., gallery; bench, seat; in the French Chamber of Deputies, tribune (the pulpit-like stand from which a member speaks in addressing the house); (*in pl.*) grand stand;
de la ——, parliamentary;
—— *des journalistes*, reporters' gallery;
monter à la ——, to take the floor.
tribut, m., tribute.
tributaire, a., tributary.
tricoises, f. pl., pinchers, nail pinchers; (*farr.*) shoeing pinchers, farriers' pinchers.
tricolore, a., tricolor.
tricot, m., knitting; jersey.
tricoter, v. a. n., to knit; (*man.*) to move the legs rapidly without gaining much ground.
tricouse, f., cloth or knit slipper.
tricycle, m., tricycle.

tricyclette, f., tricycle.
tride, a., (*man.*) quick, cadenced, and regular (said of a horse's movements); (as a noun) quickness and regularity of gait.
trièdre, a., triedral.
trier, v. a., to sort, sort out, select.
trieur, m., sorter (machine and workman).
trilatéral, a., trilateral.
trilatère, m., triangle; (*fort.*) trilateral work.
trimer, v. n., (*fam.*) to march, walk, one's self tired.
trimestre, m., quarter (three months); quarter's pay or rent;
— *par* ——, quarterly.
trimestriel, a., quarterly.
tringle, m., rod, batten, ribbon; tringle; lath; (*mach., in pl.*) links, straps, connecting links; (*art.*) tringle; (*pont.*) end and side beam of a pontoon boat, on which the ribs rest; (*sm. a.*) swivel bar (of a carbine);
—— *de boisage,* (*mil. min.*) batten;
—— *de calfatage,* calking strip of a pontoon boat;
—— *de connexion,* (r. r.) tie-rod of a switch;
—— *à crochet,* (*sm. a.*) swivel bar (of a carbine);
—— *de crochet de pontage,* (*pont.*) cavil (of a pontoon boat);
—— *à dent,* (*sm. a.*) rifling rod;
—— *d'écartement,* (*cons.*) batten (iron roof truss);
—— *d'entraînement,* (*mach.*) drag link;
—— *de mousqueton,* v. —— *à crochet;*
——*s du parallélogramme,* (*mach.*) connecting links;
—— *à rayer,* (*sm. a.*) rifling rod;
—— *recourbée,* v. —— *à crochet;*
—— *de relevage,* (*mach.*) drag link;
—— *à tamponner les tubes,* (*steam*) tube stopper;
taquet de ——, (*pont.*) cavil block;
—— *de tirage,* traction rod;
—— *de traction,* (*art.*) traction rod.
tringler, v. a., to mark out with a chalk line.
tringlette, f., small rod.
tringlos, tringlôt, m., (*mil. slang*) officer or man of the *train des équipages.*
trinitrophénol, n., (*expl.*) trinitrophenol, picric acid.
trio, m., (*mach.*) three-high roll.
triomphe, m., triumph; (*mil.*) great victory.
triompher, v. n., to triumph (over), to conquer.
triplette, f., three-seat bicycle.
tripode, m., tripod (rare).
tripoli, m., tripoli, rotten stone;
—— *brunâtre,* brownish rotten stone.
tripolir, v. a., to polish with tripoli.
triqueballe, f., (*art.*) truck, sling cart;
—— *de chargement,* hand-sling cart;
—— *à roues de charrette,* casemate truck;
—— *à treuil,* sling cart fitted with a winch;
—— *à vis,* (screw) sling cart.
triquer, v. a., to range timber in a yard.
triste, a., gloomy, dull, dreary; overcast (of weather).
trituration, f., (*powd., etc.*) trituration, grinding, pulverization, incorporation;
—— *binaire,* (*powd.*) trituration or incorporation of two constituents together;
—— *réduite,* (*powd.*) shortened incorporation;
—— *séparée,* (*powd.*) trituration of one ingredient at a time;
—— *ternaire,* (*powd.*) trituration or incorporation of the three ingredients together.
triturer, v. a., (*powd., etc.*) to triturate; to grind; to pulverize charcoal; to incorporate.
troc, m., barter, swapping.
trocart, m., (*med., hipp.*) trocar.
trochanter, m., (*hipp.*) trochanter;
grand ——, greater trochanter;
petit ——, lesser trochanter.
trochantérien, a., trochanterian.
trochéamètre, trochéomètre, m., (*inst.*) odometer.
trois-carres, m., v. *trois-quarte.*
trois-fils, m., (*elec.*) three-wire system.

trois-mâts, m., (*nav.*) three-masted vessel; (*mil. slang*) veteran with three stripes.
trois-ponts, m., (*nav.*) three-decker.
trois-quarte, f., coarse triangular file.
trois-quarts, m., v. *trois-quarte;* (*hipp.*) trocar.
trombe, f., waterspout; (*mach.*) trompe, water blast; (*nav.*) ventilator;
—— *de brume à vapeur,* steam fog-horn;
—— *d'eau,* waterspout; (*mach.*) water blast;
—— *de vent,* whirlwind.
tromblon, m., (*sm. a.*) blunderbuss.
trombone, m., (*mus.*) trombone.
trommel, m., (*min.*) trommel;
—— *classeur,* sizing drum;
—— *débourbeur,* washing drum.
trompe, f., trumpet (obs.); hunting horn; speaking trumpet; fog horn; waterspout; (*mach.*) water blast; (*nav.*) ventilator; (*fond.*) gate; runner; (*artif.*) fire pot, or succession of fire pots one above the other; (*cons.*) pendentive, squinch;
—— *de brume,* fog horn;
—— *de voûte,* keystone of a niche.
tromper, v. a., to deceive; (*fenc.*) to avoid a counter, a parry, etc., by shifting one's point;
—— *un cheval,* (*man.*) to turn a horse suddenly.
trompette, m., f., trumpeter; trumpet;
aux ——*s,* (*mil.*) assembly of trumpeters (call);
coup de ——, blast of a trumpet;
—— *-major,* (*mil.*) trumpet major;
—— *parlante,* speaking trumpet.
trompillon, m., (*cons.*) small pendentive or squinch; (*mach.*) air tube or hole of a water blast;
—— *de voûte,* (*cons.*) quoin, angle-stone, of a squinch.
trompion, m., (*mil. slang*) bugler.
tronc, m., trunk (of a tree, of a column, of the body, etc.);
—— *de cône,* truncated cone;
—— *de la queue d'un cheval,* (*hipp.*) dock of a horse's tail.
tronce, tronche, f., short, thick block of wood;
—— *de câble,* bit of cable.
tronçon, m., piece, end; drum of a column; cut, stump, stock; (*hipp.*) dock of a horse's tail; (r. r.) "block" (of a track);
—— *de ligne télégraphique,* section of telegraph line;
—— *de rebut,* waste end;
—— *de voie,* (r. r.) portion of a line, section of a line.
tronconique, a., truncated.
tronçonner, v. a., to truncate, cut into pieces, lop.
trône, m., throne; royalty, royal government;
discours du ——, speech from the throne; king's, queen's, speech at the opening of parliament.
trônière, f., (*fort.*) embrasure.
tronquer, v. a., to truncate, mutilate, maim.
trop, adv., too, too much;
cheval —— *assis,* horse whose hind legs are too much under him;
cheval —— *ouvert* (*serré*), horse whose legs are too far apart (close together) laterally.
trophée, m., trophy;
—— *d'armes,* collection of arms, colors, etc., arranged upon a shaft, column, arch, etc.
tropical, a., tropical.
tropique, m., tropic; (*in pl.*) the tropics;
maladie des ——*s,* yellow fever.
trop-perçu, m., (*adm.*) overissue, drawing of either money or supplies in excess of what is due.
trop-plein, m., overflow, waste water; overflow pipe; basin to catch an overflow;
soupape de ——, overflow valve;
tuyau de ——, waste pipe.
troquer, v. a., to barter, swap.
trot, m., (*man.*) trot (said of both horse and rider);
aller au ——, to trot;
—— *allongé,* trot out, fast trot;
—— *à l'anglaise,* posting;

trot *assis*, close trot (i. e., no posting);
—— au ——, (*call*) trot!
—— (*bien*) *écouté*, v. s. v. *écouté;*
cheval de ——, trotter;
—— *court*, short trot (i. e., one in which the horse does not raise his hind feet high);
—— *doux*, easy trot;
—— *dur*, rough trot;
—— *égal*, v. —— *court;*
—— *enlevé*, posting;
—— *franc*, v. —— *court;*
grand ——, trot out;
au grand ——, at a full trot;
—— *de manœuvre*, (*Fr. art.*) drill trot (240 meters per minute; no longer regulation);
mettre au ——, to bring to a trot;
se mettre au ——, to begin to trot;
petit ——, slow trot;
au petit ——, at a slow trot;
—— *ralenti*, slow trot;
remettre au ——, to resume the trot;
—— *de route*, (*Fr. art.*) road trot (200 meters per minute; no longer regulation);
—— *sec*, rough trot.

trotter, v. a., (*man.*) to trot (said both of horse and of rider);
—— *à l'anglaise*, to post;
—— *un cheval à gauche* (*droite*), to come down to the saddle (seat) when the left (right) diagonal strikes the ground (usual definition, though it is questioned);
—— *en chien*, to dogtrot;
—— *des épaules*, to trot hard;
—— *du genou*, to lift fore legs while gaining but little ground;
—— *menu*, to gain but little ground in the trot;
—— *sec*, to have a rough trot.

trotte-sec, m., (*mil. slang*) foot soldier; field artilleryman.

trotteur, m., trotter, trotting horse;
—— *d'attelage*, driving trotter;
—— *de selle*, saddle trotter.

trotteuse, f., split-second hand;
aiguille ——, split-second hand.

trottinement, m., (*man.*) jog-, dog-, trot.

trottiner, v. n., (*man.*) to step short in trotting; to go a jog trot; to dogtrot; to jogtrot.

trottoir, m., sidewalk, footpath, footway; (*r. r.*) platform.

trou, m., hole, orifice, aperture, gap, mouth, opening;
—— *abri*, —— *d'abri*, (*t. p.*) rifle pit, shelter pit;
—— *d'amarrage*, lashing hole;
—— *d'amorce*, fuse hole (of a torpedo);
—— *d'arrêt*, (*r. f. art.*) stop hole;
—— *borgne*, hole (e. g., in armor plate) not bored completely through;
—— *de boulet*, shot hole;
—— *à canon*, (*art.*) gun pit;
—— *à charbon*, (*nav.*) coal bunker;
—— *de charge*, (*art.*) filling hole of a shell; loading hole (through Krupp wedge, etc.); (*torp.*) filling hole;
—— *de chargement*, (*art.*) filling hole of a shell;
—— *de chat*, (*nav.*) lubber's hole;
—— *de clavette*, keyway;
—— *de coulage*, —— *de coulée*, v. s. v. *coulée;*
—— *de culasse*, (*art.*) breech-screw hole;
—— *d'écubier*, (*nav.*) hawse hole;
—— *d'entretoise de lunette*, (*art.*) trail eye;
—— *d'esse*, linchpin hole;
—— *d'essicu*, pin hole of a block;
—— *de gâchette*, (*sm. a.*) sear hole;
—— *de ganse*, v. s. v. *ganse;*
—— *de graissage*, (*mach.*) lubricating hole, oil hole;
—— *d'homme*, manhole;
—— *de laitier*, (*met.*) slag hole of a blast furnace, of a cupola;
—— *de loup*, (*mil.*) military pit, trou-de-loup; (*met.*) hole where the superfluous metal of the cast is run;
—— *de lumière*, (*art.*) vent; flame passage (of a cartridge);
—— *de mine*, (*min.*) blast hole;
—— *à nettoyage*, mud hole;

trou *de pétard*, —— *de pétardement*, (*mil. min.*) blast hole;
—— *de la porte*, keyhole;
—— *de prise de feu*, (*art.*) flame passage;
—— *de remplissage*, (*art.*) filling hole of a recoil cylinder;
—— *de rivet*, rivet hole;
—— *de sel*, mud hole;
—— *de serrure*, keyhole;
—— *de soute à charbon*, coal-bunker opening;
—— *de tampon*, drain of a furnace;
—— *de tirailleurs*, (*mil.*) rifle pit;
—— *de vidange*, (*art.*) emptying hole of a (recoil cylinder;
—— *de vis de culasse*, (*art.*) breech-screw hole;
—— *de la vis de gâchette*, (*sm. a.*) sear-screw hole;
—— *de visite*, hand hole of a boiler; mud hole of a boiler.

troubade, troubadour, m., (*mil. slang*) infantry soldier.

trouble, m., disturbance, commotion, broil, disorder, confusion; a., turbid, thick, muddy, troubled; dull, overcast (of weather).

troubler, v. a. r., to trouble, disturb; to become, grow, thick, muddy, turbid; to become foggy, thick, cloudy, overcast (of the weather).

troué, a., full of holes.

trouée, f., gap, opening, in a wood, hedge, wall; (*mil.*) break, opening, in the enemy's lines (made by a charge, by artillery fire, etc.); (*top.*) gap (in a mountain chain).

trouer, v. a., to make a hole or holes in; (*mil.*) to make a hole in the enemy's lines.

troupe, f., troop, band, multitude; (*mil.*) party, regiment, body of troops; the men (n. c. o. and men) as distinguished from the officers; (*in pl.*) troops, soldiers, forces.
(The following terms are military:)
—— *s d'administration*, administrative troops (i. e., of the supply and other administrative departments);
—— *s d'approche*, troops that clear the ground for the real attack behind;
—— *s d'assaut*, assaulting body, storming party (i. e., troops expected to get into the enemy's position), troops making the final rush on a position;
—— *s d'attaque*, the assaulting troops (in a battle); troops detailed for attack;
—— *s de chemin de fer*, railroad troops (engineer);
—— *s à cheval*, mounted troops;
—— *s de choc*, in a battle, troops to make the assault (as distinguished from those that have fought up to the assaulting position and are shaken by their work);
—— *s de combat*, combatants (as distinguished from noncombatant elements);
—— *s de communication*, line-of-communication troops;
corps de ——, v. s. v. *corps;*
—— *s de couverture*, covering troops (troops along a frontier, garrisoning frontier towns, and hence covering mobilization);
—— *s de démonstration*, troops making the feint, as opposed to the real attack;
—— *s de deuxième ligne*, in an investment, the troops (of the attack) serving as a reserve to the —— *s de 1ère ligne*, q. v.;
—— *s d'étapes*, line-of-communication troops, troops garrisoning the *gîtes d'étapes* (two sorts: permanent, and detached from the operating army);
—— *s étrangères*, mercenaries;
—— *s d'exploration*, troops sent ahead to prevent mobilization (v. *exploration stratégique*);
—— *s frontières*, frontier troops;
homme de ——, v. s. v. *homme;*
—— *s d'instruction*, depot troops;
—— *s de ligne*, troops of the line (sometimes regular troops, as opposed to irregular);
—— *s de la marine*, v. s. v. *marine;*
—— *s nationales*, national troops (as distinguished from mercenaries);
—— *s d'occupation*, troops stationed in a defeated country;
officier de ——, v. s. v. *officier;*

troupe(s) *d'opérations*, field troops, or troops belonging to a regular field army (as distinguished from troops assigned to a fortified position; said whenever latter happen to be in theater of operations);
——*s à pied*, foot troops;
——*s de première ligne*, in an investment, the sentinels, outposts, etc., of the attack;
——*s réglées*, instructed, disciplined, troops; regular troops;
——*s de relèvement*, relief, relieving, troops, (in forts);
——*s de secteur*, (*siege*) troops assigned to a sector of the defense of a fortified place (exclusive of garrisons, of works, and of reserve);
——*s de sortie*, in a fort, troops for sortie purposes;
——*s de soutien*, supporting forces;
——*s de sûreté*, troops on security service.
troupeau, m., flock, herd, drove (of animals).
troupier, m., (*mil.*) familiar term for a soldier; old n. c. o. advanced to the grade of officer; a., soldier-like;
vieux ——, old campaigner, old soldier, veteran.
trousse, f., bundle of things tied together; truss (of hay, of fodder in general); case (of small objects, of instruments, of tools, etc.); housewife; tail case of a horse; rope for raising moderate weights; (*met.*) pile of fagot; set of slitting cutters; (*cav.*) roll, valise, carried behind the saddle; (*sm. a.*) packet of cartridges; bundle of steel pieces for the manufacture of a sword blade;
——*à cartouches*, (*mil.*) sack for bringing cartridges up to line of battle;
——*de cartouches*, (*sm. a.*) packet of cartridges;
——*en drap*, bundle of rags or of cloth (cleaning kit);
en ——, behind (on a horse);
——*de forets*, set of drills, boring bits, etc.;
monter en ——, (*man.*) to ride behind.
troussé, p. p., well set, well made;
cheval bien ——, well-made horse.
trousseau, m., outfit of clothes, etc. (taken by a student to college); (*fond.*) spindle of a loam mold;
——*de clef*, bunch of keys.
trousse-étrier(s), m., v. *étrière*.
trousse-pas, m., sod-cutter's spade.
trousse-pied, m., (*hipp.*) leg strap (for refractory horses).
trousse-queue, m., tail case (for a horse);
——*de cuir*, tail leather.
troussequin, m., mortise gauge; (*harn.*) cantle of a saddle, hind peak.
troussequiner, v. a., to mark; to gauge with a mortise gauge.
trousser, v. a. n., to tuck up, turn up, tie up; to dispatch, expedite; (*man.*) to raise the fore legs too high (in the trot and walk);
—— *baggage*, to pack up;
—— *la queue*, to tuck up a horse's tail, put it up in a tail case.
trousse-traits, m., (*harn.*) trace strap, tucking strap.
trouvaille, f., anything found;
droit de ——, salvage.
trouver, v. a., to find; to come across, light upon;
—— *le fond*, to find, get bottom in soundings.
trouveur, m., (*inst.*) finder (of a telescope).
truc, m., truck; (*r. r.*) flat or open car; (*mil. slang*) room; military equipment;
——*à plateau tournant*, (*Fr. art.*) truck with turning platform;
——*-support*, (*r. r.*) under frame of a locomotive.
trucheman, truchement, m., interpreter, dragoman.
truelle, f., (*mas.*) trowel;
—— *brettée*, notched trowel.
truellée, f., trowelful
trueller, v. a., to work with a trowel.
truité, a., (*met.*) mottled; (*hipp.*) trout-spotted, flea-bitten (of the coat); red or reddish patches on a base of white or gray;

truité-*moucheté*, (*hipp.*) trout-spotted (spots reddish brown or almost black).
truiture, f., (*hipp.*) small reddish spots on a gray or white coat.
trullisation, f., (*mas.*) coating, rough dressing, with a trowel.
trumeau, m., (*cons.*) pier between window or door openings; paneling between windows; (*fort.*) synonym of *merlon*.
trune, f., sort of alpine shelter.
trusquin, m., mortise gauge; carpenter's gauge; (*art.*) sort of gauge.
trusquiner, v. a., v. *troussequiner*.
trypographe, m., cyclostyle, using a point instead of a wheel.
tsar, m., the czar.
tsarine, f., the czarina.
tsarowitz, m., the czarowitz.
tschako, m., v. *schako*.
tschapka, m., v. *schapska*.
tubage, m., (*art., etc.*) (action of) tubing, insertion of a tube lining;
fosse de ——, (*art.*) shrinkage pit.
tube, m., tube, pipe; (*art.*) tube; ignition tube (of a shrapnel); (*F. m. art.*) lining tube (usually, but not always, extending about half way down the bore from the breech).
 I. Artillery, smallarms; II. Miscellaneous.
 I. Artillery, small arms:
——*en acier*, (*art.*) forging, rough forging;
——*d'amorcage*, priming tube of a *cylindre à fumée*, q. v.;
——*d'amorce*, (*art., etc.*) priming tube;
——*arrêt de piston*, (*sm. a.*) follower stop;
——*buttoir*, (*r. f. art.*) recoil tube, resistance tube;
——*canon*, (*art.*) subcaliber gun; subcaliber tube;
——*à canons*, (*art.*) (gun) forging;
——*de canon*, (*art.*) tube;
——*carcasse*, (*torp.*) launching frame, launching tube; under-water torpedo tube;
——*en caoutchouc*, (*r. f. art.*) rubber shoulder piece, shoulder buffer;
——*central*, (*art.*) ignition tube of a shrapnel;
——*de charge*, (*art.*) loading tube;
——*de chargement*, (*art.*) loading tube; (*mil. min.*) loading tube for a bored mine; (*nav.*) ammunition-supply tube (turrets);
——*composé*, (*art.*) built-up tube (gun construction);
——*de composition*, (*art.*) fuse; friction tube;
——*coussin*, (*r. f. art.*) shoulder piece of rubber;
——*à cuiller*, (*nav.*) torpedo tube fitted with a spoon;
——*détonant*, (*mil. min.*) detonating tube (for communicating fire to mine);
——*directeur*, (*art.*) rocket tube;
——*éjecteur*, (*r. f. art.*) ejecting tube;
——*enveloppe*, (*sm. a.*) barrel cover;
——*d'étoupille*, (*art.*) friction-primer tube;
——*fixe*, (*torp.*) fixed torpedo tube;
——*fusant*, (*art.*) time train (double-action fuse); (*mil. min.*) communicating tube;
——*à fusées*, (*art.*) rocket tube;
——*guide*, (*sm. a.*) guide sleeve;
——*intérieur*, (*art.*) lining tube;
——*de lancement*, (*torp.*) torpedo tube, torpedo-launching tube;
——*de lancement à l'arrière (l'avant)*, (*torp.*) stern (bow) torpedo tube;
——*de lancement au dessous (au dessus) de la flottaison*, (*torp.*) under- (over-) water torpedo tube;
——*-magasin*, (*sm. a.*) tubular magazine;
——*de mise de feu*, (*art.*) communicating tube (of a shrapnel);
——*à percussion*, —— *percutant*, (*art.*) percussion primer;
——*percutant coudé*, (*art.*) rectangular percussion primer;
——*planchette de chargement*, (*art.*) loading tube (protects the thread of the breech nut);
——*à pointage*, (*torp.*) torpedo tube that may be aimed;

tube à *poudre*, (*art.*) powder tube (of some shrapnel);
—— *de renfort*, (*art.*) breech (reenforce) tube or lining;
—— *simple*, (*art.*) single, simple, tube (gun construction);
—— à *tir*, (*art.*, *sm. a.*) subcaliber tube;
—— *de transformation*, (*sm. a.*) gun barrel in the rough;
—— *de transmission*, (*mil. min.*) leaden tube for communicating fire.
II. Miscellaneous:
—— *acoustique*, speaking tube;
—— à *air*, air tube;
—— *alimentaire*, —— *d'alimentation*, (*mach.*) feed pipe;
—— *allonge*, (*mil. min.*) lengthening piece or joint (of a borer);
—— *d'aspersion*, spraying tube;
—— *atmosphérique*, (*mach.*) air tube; vacuum pipe;
bague à, *de*, ——, v. s. v. *bague;*
banc à ——*s*, tube machine;
bouche- ——, (*steam*) tube stopper;
—— *bouilleur*, (*steam*) boiler tube, water tube;
—— à *bride*, flanged tube;
brosse à ——*s*, tube brush;
—— *capillaire*, capillary tube;
—— *de chaudière*, (*steam*) fire tube; boiler tube;
chaudière à ——*s*, tubular boiler;
—— *collecteur d'eau d'alimentation*, (*steam*) supply pipe (of a Belleville boiler);
—— *collecteur de la vapeur*, (*steam*) steam-collecting tube;
—— *de condenseur*, (*steam*) condenser tube;
—— *sans couture*, seamless tube;
—— *en cuivre, etc.*, copper, etc., tube;
—— *cuvette*, bushing or sleeve (of a bicycle axle);
—— *cylindré*, rolled tube;
—— à *deux branches*, double-branched tube;
—— *directeur*, (*steam*) water circulating tube;
—— *d'eau*, (*steam*) water tube;
—— *d'épreuve*, (*chem.*) test tube;
—— (-)*éprouvette* à *effluve*, (*elec.*) tube for testing the effects of an electric flow;
—— *d'essai*, v. —— *d'épreuve;*
—— *d'étambot*, (*nav.*) stern tube;
—— *étiré*, (solid) drawn tube;
—— -*flèche*, drill or point of a well borer;
—— *foyer*, (*steam*) the flue of a Lancashire or Cornish boiler;
—— à *fumée*, (*steam*) fire tube;
—— *fuyant*, leaky tube;
—— *de graissage*, (*mach.*) lubricating tube;
—— *indicateur*, (*steam*) water-gauge glass;
—— *jaugeur*, (*steam*) boiler gauge;
—— *de niveau*, (*steam*) water gauge (boiler);
—— *de noyage*, flooding pipe;
—— *perforateur*, sort of rock drill;
plaque à ——*s*, v. s. v. *plaque;*
—— (-)*rallonge*, lengthening or extension joint;
—— *de retour d'eau*, (*steam*) return tube;
—— *soudé en écharpe*, lap-welded tube;
—— *soudé au point*, butt-welded tube;
—— *sans soudure*, seamless, solid-drawn, tube;
—— *surchauffeur*, (*steam*) superheating tube;
tampon de ——, (*steam*) tube plug;
—— -*tirant*, (*steam*) stay tube;
—— *en V*, V-shaped tube;
—— *de verre*, (*steam, etc.*) glass gauge;
—— *vis*, tubular screw joint;
—— -*viseur*, (*opt.*) sighting tube.
tuber, v. a., to tube; to fit, line, with a tube; to line a shaft with brick, etc.; (*art.*) to tube a gun; to insert a tube lining.
tubulaire, a., tubular;
chaudière ——, (*steam*) tubular boiler;
tôle ——, (*steam*) tube plate.
tubule, f., tube of very small caliber.
tubulé, a., tubular;
chaudière ——, (*steam*) tubular boiler.
tubulure, f., tubulure, neck, nozzle.
tué, p. p., killed;
—— à *l'ennemi*, (*mil.*) killed in action.
tuer, v. a., to kill.
—— *le feu*, to put out, extinguish, the fire.

tuerie, f., butchery, carnage; slaughterhouse.
tuf, m., tufa, tuff;
pierre de ——, tuff.
tuffeau, m., calcareous tufa, tufaceous limestone.
tuile, f., (*cons.*) tile; piece of marble, bronze, etc., serving as a tile;
—— *d'Altkirch*, v. —— *mécanique;*
—— *arêtière*, ridge tile;
—— *de Bourgogne*, hollow tile (in shape, half a truncated cone);
—— -*chaperon*, coping tile (of a wall);
—— *cornière*, v. —— *creuse;*
—— *creuse*, hollow tile; gutter, ridge, crest, tile;
—— à *crochet*, hook tile;
demi- ——, half tile;
—— *écaille*, ornamental tile (i. e., whose tail is rounded, etc.);
—— à *emboîtement*, v. *mécanique;*
—— *faîtière*, gutter, ridge, crest, tile;
—— *flamande*, pantile;
—— *fronton*, facing tile (for the front point of the intersection of two slopes);
—— *gironnée*, wedge-shaped tile (for use on curved surfaces);
—— *de gouttière*, gutter tile;
—— *de Guienne*, sort of hollow tile;
—— *lucarnière*, front tile (for attics, etc.);
—— *s mécaniques*, tongued-and-grooved tile;
—— *pannetonnée*, tile fitted with a catch (*panneton*) by which it may be wired onto the framing;
—— *plate*, flat tile, plain tile;
—— *plate* à *crochet*, a flat hook tile;
—— *réfractaire*, fire tile;
—— *de rive*, gable tile (at end of gable);
—— *romaine*, normal Roman tile, tegula;
—— *de Tarascon*, semicylindrical tile;
—— *en verre*, glass tile (admits light).
tuileau, m., fragment of a tile.
tuilette, f., small tile.
tulipe, f., tulip; (*art.*) tulip, muzzle swell; the mouth or cup of the central tube of a shrapnel; *bourrelet en* ——, (*art.*) swell of muzzle;
collet de ——, (*art.*) neck of a gun.
tulipier, m., tulip tree and wood.
tumeur, f., tumor; (*hipp.*) warbles;
—— *cornée*, (*hipp.*) horn tumor.
tumulte, m., tumult; street disturbance.
tumultueux, a., tumultuous;
réaction ——*se*, (*expl.*) tumultuous reaction.
tumulus, m., tumulus, barrow.
tunage, m., fascine work, mattress work.
tune, f., bed of fascines covered with gravel, mattress.
tungstène, m., tungsten.
tunique, f., (*unif.*) tunic;
—— *abdominale*, (*med.*) coat of the stomach;
—— *vareuse*, (*unif.*) litewka.
tunnel, m., tunnel;
puits de ——, tunnel shaft, pit.
turbinage, m., action of a turbine; (*expl.*) wringing (of gun cotton).
turbine, f., (*mach.*) turbine; (*expl.*) wringer, acid wringer (gun cotton);
—— *centrifuge*, outward-discharge turbine, outflow turbine;
—— *centripète*, inward or central-discharge turbine, inflow turbine;
—— à *choc*, impulse turbine;
—— *composée*, double turbine (i. e., combining outflow and inflow);
—— *double*, double turbine;
—— *mixte*, v. —— *composée;*
—— *noyée*, immersed turbine;
—— *parallèle*, parallel or downward-flow turbine;
—— à *réaction*, reaction wheel or turbine;
—— *réversible*, reversible turbine;
roue- ——, turbine;
—— à *siphon*, siphon turbine;
—— à *vapeur*, steam turbine.
turbo-moteur, m., (*mach.*) turbine motor; steam turbine.
turcie, f., levee or bank along the river.
turco, m., (*Fr. a.*) Turco (Algerian rifleman).

turlutine, f., (*mil. slang*) campaign ration of pounded biscuit, rice, and bacon.
tutelle, f., guardianship, protection.
tuteur, m., guardian, protector; supporting stick or rod or picket (as, *t. p.*, of a silhouette target).
tuyau, m., tube, pipe, hose; stalk, stem; spout, conduit, nozzle; flue of a chimney; funnel (of a ship, etc.); tip (on a horse, hence), trustworthy source of information;
—— *d'accouplement*, coupling pipe;
—— *acoustique*, speaking tube;
—— *d'admission*, inlet pipe;
—— *d'admission de la vapeur*, steam-admission pipe;
—— *d'aérage*, —— *à air*, air pipe;
—— *aérique*, ventilation pipe;
—— *alimentaire*, —— *d'alimentation*, feed pipe;
—— *alimentaire à colonne d'eau*, standpipe;
—— *d'apport*, steam pipe;
—— *d'arrivée*, delivery pipe; main steam pipe, supply pipe;
—— *d'ascension*, (*mach.*) rising main;
—— *aspirant*, *aspirateur*, *d'aspiration*, suction pipe;
—— *atmosphérique*, vacuum pipe;
—— *auxiliaire*, auxiliary pipe;
—— *à bifurcation*, —— *bifurqué*, breeches pipe;
—— *à bourrelet*, flanged pipe;
branche de ——, branch pipe;
—— *à bride*, flanged pipe;
—— *de cale*, bilge pipe;
—— *en caoutchouc*, rubber hose;
—— *de chaleur*, —— *de chauffage*, hot-air flue; heating pipe;
—— *de cheminée*, smoke-pipe, flue, stack, chimney flue, funnel;
—— *de chemise (de cylindre)*, (cylinder) jacket pipe;
—— *de circulation*, circulating pipe;
—— *collecteur d'eau d'alimentation*, v. s. v. *tube*;
—— *à colonne d'eau*, standpipe;
—— *de communication*, connecting pipe;
—— *de communication des soupapes d'arrêt*, induction pipe;
—— *de communication de la vapeur*, steam pipe;
—— *à compensation*, compensation pipe;
—— *des condenseurs à surface*, condenser tube;
—— *de conduite*, delivery pipe, eduction pipe, main pipe, conduit;
—— *à, en, cou de cygne*, elbow pipe;
—— *coudé*, elbowed pipe, bent pipe, kneed pipe;
—— *courbe d'un quart de cercle*, quarter bend;
—— *courbé*, elbow pipe;
—— *en cuir*, leather hose;
—— *cylindré*, rolled pipe;
—— *de décharge*, discharge pipe, delivery pipe; evacuation, blow-off, pipe;
—— *de décharge du condensateur*, waste, overflow, pipe of a condenser;
—— *de dégagement (de la vapeur)*, waste (steam) pipe, escape pipe, blast pipe;
—— *de dégorgement*, discharge pipe; exhaust pipe;
—— *à demi-cercle*, pipe bent to a semicircle;
—— *de départ*, waste pipe; exhaust pipe;
—— *de descente*, lead, vertical lead;
—— *de distribution*, service, distributing, pipe;
—— *de distribution de la vapeur*, steam-distributing pipe;
—— *diviseur*, separating pipe (Belleville boiler);
—— *à eau*, water pipe;
—— *à eau douce*, fresh-water pipe;
—— *d'échappement*, escape pipe, waste pipe, exhaust pipe;
—— *d'échappement des soupapes de sûreté*, safety-valve pipe;
—— *d'échappement de la vapeur*, steam-escape pipe;
—— *d'écoulement*, water pipe;
—— *éjecteur*, ejector pipe;

tuyau élastique, hose pipe;
—— *élévateur*, ascending pipe (of a pump);
—— *à emboîtement*, socket pipe;
—— *d'embranchement*, branch pipe, breeches pipe;
—— *d'émission*, eduction pipe;
—— *d'entrée*, inlet, supply, pipe;
—— *d'épuisement*, exhaust pipe;
—— *étiré*, (solid) drawn tube;
—— *d'évacuation*, waste pipe; exhaust, delivery, evacuation, pipe;
—— *d'évent*, ventilation pipe;
—— *d'extraction*, blow-down pipe; brine, scum, pipe;
—— *d'extraction par le fond*, blow-out pipe;
—— *d'extraction à la surface*, surface blow-off pipe;
—— *d'extraction de la vapeur*, v. —— *de sortie*;
—— *flambeur*, fire tube;
—— *flexible*, hose;
—— *en forme de S*, S-shaped pipe;
—— *à gaz*, gas pipe, tube;
—— *de graissage*, lubricating pipe;
—— *d'indicateur*, gauge tube; indicator pipe;
—— *indicateur du niveau*, glass gauge;
—— *d'indicateur du niveau d'eau*, gauge glass, water-gauge glass;
—— *d'indicateur de vide*, vacuum-gauge pipe;
—— *d'injection*, injection pipe;
—— *d'injection par mélange*, jet injection pipe;
—— *d'introduction*, admission pipe;
—— *d'introduction de la vapeur*, steam-admission pipe;
—— *de jointure*, —— *de jonction*, connecting pipe;
—— *à manchon*, socket pipe;
—— *de manomètre*, gauge pipe;
—— *de manomètre à vapeur*, steam-gauge pipe;
—— *de montée*, (*fond.*) rising pipe;
—— *du niveau d'eau*, water-gauge column or pipe;
—— *de poêle*, (*mil. slang*) regulation boots;
—— *de pompe à incendie*, fire-engine hose;
—— *porte-fond*, cylinder holding up the bottom of the *distributeur* of a turbine;
—— *principal*, main pipe, main;
—— *de prise de vapeur*, steam (delivery) pipe;
—— *de purge*, —— *à purger*, blow-off, blow-through, pipe; exhaust;
—— *purgeur*, drain pipe;
—— *de raccordement*, filling valve;
—— *de rallonge*, extension piece;
—— *de refoulement*, delivery pipe;
—— *réfrigérant*, cooling pipe;
—— *à rotule*, ball-joint pipe;
—— *secondaire*, lateral pipe;
—— *de sifflet à vapeur*, steam-whistle pipe;
—— *à siphon*, siphon pipe;
—— *de sonde*, (*nav.*) sounding pipe;
—— *de sortie*, eduction, delivery, pipe;
—— *de soufflet*, nozzle of a bellows;
—— *souffleur*, blast pipe;
—— *souffleur (de cheminée)*, (funnel) blast pipe;
—— *de soupape de mise en train*, starting-valve pipe;
—— *de tiroir*, valve pipe;
—— *d'un tourillon*, trunnion pipe;
—— *de trop-plein*, overflow pipe;
—— *à vapeur*, steam pipe;
—— *à vapeur principal*, main steam pipe;
—— *à vent*, windbore;
—— *de vidange*, waste, exhaust, bottom blow-off, escape, discharge, pipe; blow-off pipe; escape, discharging, pipe;
—— *de vide*, vacuum pipe;
—— *de vide de cylindre*, cylinder vacuum pipe.
tuyautage, m., piping, set or system of pipes.
tuyauter, v. a., to give a tip on a race horse.
tuyauterie, f., pipe trade or business; piping; set or system of pipes.
tuyère, f., (*met.*) tue iron, tuyere (strictly, the opening in a blast furnace for the passage of what we call tuyere, but the French *porte-vent*, q. v.);
—— *de forge de campagne*, (*art.*) tue iron of a field forge;

tuyère *plongeante*, (*met.*) inclined twyer.
tympan, m., spandrel; die of a pedestal; panel or filling of wood, inclosed by molding; tympanum, treadmill (power applied internally); (*mach.*) pinion mounted on a shaft.
tympe, f., (*met.*) tymp;
fer de ——, tymp iron.
type, a., standard.
type, m., type (printing); model, type, standard;
canon- ——, (*art.*) type gun;
hausse- ——, (*art.*) standard or type sight;
poudre ——, v. s. v. *poudre;*
projectile- ——, (*art.*) standard or type projectile.

typhon, m., typhoon.
typographe, m., typographer, printer;
imprimeur ——, letterpress printer.
typographie, f., typography.
typolithographie, f., typolithography.
typomètre, m., instrument for taking the measure of men's clothes.
tyran, m., tyrant.
tyrannie, f., tyranny.
tyranniser, v. a., to tyrannize over.
tyre, f., tire.
tzar, m., v. *tsar.*

U.

udomètre, m., rain gauge.
uhlan, m., (*mil.*) uhlan, lancer.
u(h)lanka, f., (*unif.*) uhlan tunic.
ukase, m., ukase.
ulcère, m., ulcer, boil, sore;
—— *du bourrelet de la matrice de l'ongle*, (*hipp.*) scratch between heel and pastern joint.
ulmer, v. a., to surrender shamefully, without opposition.
ultérieur, a., ulterior, farther.
ultimatum, m., ultimatum.
ultramarin, a., ultramarine.
ultramontain, a., ultramontane (esp. beyond the Alps).
ultrapont(a)in, m., person living across the river (said of cities crossed by a stream);
pays ——, the Latin Quarter in Paris.
unanimité, f., unanimity;
à l' ——, unanimously.
une-deux, m., (*fenc.*) feint followed by a disengage, in a different line from the feint.
une-deux-trois, m., (*fenc.*) two feints followed by a disengage.
uni, a., joined to, united, level, smooth, even; uniform; simple, plain, without ornament; (*man.*) united, right (of the gallop);
à l' ——, on a level;
cheval ——, horse that gallops united;
les Etats-Unis, the United States;
galop ——, (*man.*) right gallop.
unification, f., unification;
—— *des soldes*, consolidation of pay.
uniforme, a., uniform, regular, even, unvaried;
habit ——, (*unif.*) uniform.
uniforme, m., (*mil.*) uniform, regimentals; military dress in general;
grand ——, full uniform;
en grand ——, in full uniform;
quitter l' ——, to leave the service;
voyant de l' ——, conspicuousness of the uniform; ease with which it may be made out at a distance.
unilatéral, a., (*adm.*, of contracts) binding on one party only.
union, f., union, junction, combination; adjustment, adjusting; concurrence; the Union, the United States; (*man.*) regularity of a horse's motion; (*tech.*) union, union joint, coupling;
—— *des femmes de France*, society of women whose object is to succor the wounded in time of war;
—— *ouvrière*, labor union;
—— *à vis*, screw joint, screw coupling.
unioniste, m., Union man (United States, 1861-1865).
unipolaire, a., (*elec.*) unipolar.
uniphase, m., (*elec.*) single phase.
unir, v. a. r., to join, unite, piece together, combine, link, connect; to level, smooth, make plane;
—— *à chaud*, (*met.*) to weld;
—— *un cheval*, v. s. v. *cheval.*

unitaire, a., of unit of value, unit.
unité, f., unit; (*mil.*) unit of organization;
—— *absolue*, absolute unit;
——*s accolées*, (*mil.*) units side by side;
—— *administrative*, (*mil.*) the smallest organized unit having a distinct administration (e. g., company, troop, or battery);
—— *d'armement* (*et de munitions*), (*mil.*) state or condition of having but one model of small arm;
—— *de chaleur*, unit of heat;
—— *collective*, (Fr. a., *med.*) the outfit (*matériel*) of a sanitary formation, necessary to its duties for a given time;
—— *de combat*, (*mil.*) fighting unit, combat unit (the largest unit that may be controlled directly by the voice—company, troop, battery);
—— *de commandement*, (*mil.*) any unit having a permanent and recognized head (e. g., company, battalion, brigade, divisional artillery);
—— *dérivée*, derived unit;
—— *fondamentale*, fundamental unit;
—— *de longueur*, unit of length;
—— *de manœuvre*, (*mil.*) drill unit;
—— *de marche*, (*mil.*) march unit (e. g., battalion, squadron, battery; a march unit is defined as a combination of groups such that the space between them equals the elongation each group may take in the interval between consecutive halts);
—— *mécanique*, unit of work;
—— *de mesure*, unit of measure;
—— *de poids*, unit of weight;
—— *pratique*, (*elec.*) practical unit;
—— *prévôtale*, v. s. v. *prévôtal;*
—— *de puissance*, unit of force;
—— *de ravitaillement*, (*mil.*) unit of supply;
—— *stratégique*, (*mil.*) strategic unit (group formed of troops of all arms and of necessary services; e. g., army corps, which is the typical strategic unit);
——*s successives*, (*mil.*) units one behind the other (as distinguished from ——*s accolées*);
—— *tactique*, (*mil.*) tactical unit (i. e., the smallest unit to which may be given a definite mission on the battlefield; e. g., battalion, group of batteries, regiment of cavalry; on a larger scale for cavalry, the division of three brigades with three horse batteries);
—— *de travail*, unit of work.
universitaire, a. m., university; university man;
corps ——, faculty of the university.
université, f., university.
uno-rail, m., uno-rail.
urane, m., sort of flying machine (proposed).
urgence, f., urgency;
cas d' ——, urgency, pressing case;
il y a ——, the case is pressing.
us, m. pl., use, usage, custom, ways, practice (found usually in the following expression:)
—— *et coutumes*, usages and customs.
usable, a., liable to wear out.
usage, m., use, usage, practice, habit; employment; service; (*law*) right of pasture; common of estovers;
de bon ——, very serviceable;
hors d' ——, out of use, obsolete.

usance, f., (commercial) usance (thirty days usually);
— *du bois*, time since a wood was cut;
demi- —, half usance;
double —, double usance;
lettre à une, à deux, —(*s*), bill payable in one, two, months.
usblat, m., isinglass (obs.).
usé, a., worn, worn out; (*cord.*) frayed; m., the worn-out part of anything;
cheval —, overridden horse.
user, v. a. r., to wear, wear out, wear away; to rub, become rubbed; to become worn, frayed, etc.; to use, use up, consume; to waste, spend; (*hipp.*) of a tooth, to lose the enamel of the outer edge;
— *de*, to make use of.
user, m., wear, wearing, service;
être d'un bon —, to wear well;
être d'un mauvais —, to wear badly.
useur, a., wearing, grinding.
usinage, m., machining, machine work; working to shape, as by lathe, etc.;
défaut d' —, defective machining;
— *dégrossisseur*, machining to approximate shape;
— *extérieur* (*intérieur*), machining of the outside (inside) of a gun barrel.
usine, f., works, shops, factory, manufactory, machine shop, mill, forge, power house;
— *centrale*, power house;
— *à cuirassements*, armor-plate factory;
— *fixe*, v. — *centrale;*
— *à gaz*, gas works.

usiner, v. a., to machine.
usineur, a., machining;
établissement —, machining establishment.
usne, m., (*cord.*) towrope.
ustensile, m., utensil, implement;
—*s d'artifices*, (*artif.*) laboratory implements;
—*s de campement*, (*mil.*) camping utensils;
—*s de chauffe*, fire tools;
—*s d'écurie*, stable utensils;
—*s de marqueurs*, (*t. p.*) marker's outfit.
usure, f., wear and tear, waste, wearing out; (*hipp.*) loss by wear of the enamel of a tooth (on its outer edge);
— *et accidents*, wear and tear;
— *annuelle*, (*hipp.*) the rubbing surface of the teeth;
corriger l' —, (*mach.*) to take up the wear;
prêter à —, (*mach.*) to work only to wear (as a fly wheel with too high a velocity);
rattrapage de l' —, (*mach.*) take-up, compensation, for wear;
rattraper l' —, (*mach.*) to take up the wear.
usurpateur, m., usurper.
usurpation, f., usurpation.
usurper, v. a., to usurp.
utile, a., useful, serviceable; profitable; expedient;
effet —, v. s. v. *effet;*
en temps —, within the prescribed, the fixed, time;
travail —, useful work.
utiliser, v. a., to utilize, turn to account; employ to advantage.
utilité, f., utility, usefulness, use, benefit, service, avail; profit, advantage.

V.

V, (*tech.*) parting tool.
vacance, f., vacancy; (*mil.*) vacancy; (*pl.*) holidays, vacation;
les grandes —*s*, the long vacation; summer holidays.
vacant, a., vacant, unoccupied.
vacation, f., sitting; vacancy; recess;
chambre des —*s*, (*law*) chambers.
vache, f., cow; cowhide; leather;
amarrer en —, v. s. v. *amarrer;*
côtes de —, (*met.*) slitted iron;
se coucher en —, (*hipp.*) to lie down with legs bent up under the body;
coup de pied en —, (*hipp.*) a kick forward with the hind leg; cow-kick (of a horse);
— *en grain*, grained leather;
nœud de —, v. s. v. *nœud;*
peau de —, cowhide;
petite —, hide of a young cow;
plancher des —*s*, v. s. v. *plancher;*
ruer en —, (*hipp.*) to make a cow-kick.
vacillant, a., vacillating, unsteady;
jarrets —*s*, v. s. v. *jarret*.
vacillation, f., wavering, vacillation; rocking (of a boat); unsteadiness (of a light, etc.).
vaciller, v. n., to waver, falter, vacillate, hesitate; to be unsteady, to rock.
vadel, m., tar-mop handle.
vadrouille, f., swab.
va-et-vient, m., up-and-down motion, back-and-forth motion; ferryboat; ferry rope, pass rope; (*mach.*) reciprocating motion, reciprocating gear; (*mil.*) circulation from front to rear, and vice versa;
en —, pulling back and forth (of a ferryboat);
mouvement de —, (*mach.*) reciprocating motion; back-and-forth motion; (*mil.*) circulation from front to rear, and vice versa.
vagon, m., (*r. r.*) car.
vagonnet, m., (*r. r.*) truck.
vague, f., wave.
vaguemestre, m., (*Fr. a.*) n. c. o. in charge of the post-office; baggage master (in each army corps, an officer of gendarmerie specially charged with this duty; in each division, the *commandant de la force publique*);
officier —, baggage master.

vaigrage, m., (*nav.*) ceiling; operation of ceiling.
vaigre, f., (*nav.*) ceiling plank, ceiling stuff.
vaillamment, adv., valiantly; (*mil.*) gallantly.
vaillance, f., valor, esp. military valor, gallantry.
vaillant, a., valiant; (*mil.*) gallant.
vain, a., vain, vainglorious;
cheval —, horse easily knocked up;
—*e pâture*, v. s. v. *pâture*.
vaincre, v. a., to conquer, vanish, overcome, subdue, defeat, worst, master, surmount, excel, surpass;
— *un cheval*, to break a horse.
vainqueur, m., conqueror.
vairon, a., wall-eyed;
cheval —, wall-eyed horse.
vaisseau, m., vessel; (*nav.*) vessel, ship (esp. war ship);
— *amiral*, flagship;
— *cuirassé*, armored ship;
— *du désert*, camel;
— *école*, school-ship, training ship; gunnery ship;
— *à éperon*, ram; ship fitted with a ram bow;
— *de l' État*, national vessel;
— *x frères*, sister ships;
— *de garde*, guard ship;
— *de guerre*, war ship;
— *de guerre bien battant*, ship on which the guns are well served;
— *à hélice auxiliaire*, auxiliary ship;
— *hôpital*, hospital ship;
— *de ligne*, ship of the line;
— *marchand*, merchantman;
— *mixte*, composite ship;
— *à n ponts*, n decker;
— *rasé*, hulk;
— *à roues*, paddle steamer;
— (*de*) *transport*, transport;
— *à vapeur*, steamship;
— *à voiles*, sailing ship;
— *de 74*, etc., 74, etc., gun ship (obs.).
vaisselle, f., crockery; (*mil. slang*) decorations.
val, m., (*top.*) small valley, vale; (strictly speaking, bottom of valley, between foot of slopes);

val 477 **vapeur**

val, à —— **de.** down from, at the bottom of;
—— *par monts et par vaux,* over hill and dale.
valable, a., valid.
valdrague, en ——, (*nav.*) in confusion, higgledy-piggledy.
valet, m., valet, footman; door weight, rest, support; dog, claw, bench hook, holdfast; (*nav. art.*) junk wad for guns, wad (obs.);
—— *en anneau,* (*art.*) ring wad;
—— *de l'armée,* (*mil.*) officer's servant;
—— *cylindrique,* (*nav. art.*) full wad (obs.);
—— *d'écurie,* stable groom;
—— *erseau,* (*nav. art.*) grommet wad (obs.);
—— *estrope,* v. —— *erseau;*
—— *d'établi,* dog, holdfast;
—— *à griffes,* vise, clamp;
—— *lubricateur,* (*sm. a.*) lubricator, lubricating wad;
—— *de pied,* footman; (*tech.*) bench vise;
—— *de place,* guide;
—— *plein,* v. —— *cylindrique;*
—— *à vis,* screw holdfast or dog.
valeur, f., valor, bravery, gallantry; value; (in races) purse; (*in pl.*) securities;
—— *absolue,* (*mil.*) positive importance or strength of a position;
—— *en compte,* "value in account" (on a bill of exchange);
—— *le 30 courant,* payable the 30th instant;
——*s fictives,* securities having a prospective value;
—— *militaire,* (*nav.*) fighting qualities (of a ship);
—— *nautique,* (*nav.*) seaworthiness;
—— *papier,* generic expression for bank notes, shares (and other paper representing value);
—— *reçue,* value received;
—— *réelles,* securities having a present value;
—— *vénale,* market value.
valeureux, a., valorous, gallant, brave, valiant.
valide, a., valid; in good health; (*mil.*) fit for service, effective.
valider, v. a., to make valid.
validité, f., validity.
valise, f., valise, portmanteau; mail bag;
—— *de cadre,* bicycle valise;
—— *diplomatique,* embassy mail bag.
vallée, f., (*top.*) valley, vale.
vallon, m., (*top.*) small valley, dell.
vallonné, a., (*top.*) full of, or intersected by, valleys.
vallonnement, m., (*top.*) shallow and short valley-like depression.
valtage, m., (*cord.*) woolding.
valter, v. a., (*cord.*) to woold.
valture, f., (*cord.*) woolding; sheer lashing; rope used to lash two masts together.
valve, f., (*mach.*) valve, clack valve;
—— *d'admission,* throttle valve;
—— *d'alimentation,* feed valve;
—— *d'étranglement,* v. —— *d'admission;*
—— *extérieure,* external valve;
—— *d'introduction,* v. —— *d'admission;*
—— *à poignée,* hand valve;
—— *régulatrice,* adjusting valve;
—— *de sortie,* exit valve;
—— *de sûreté,* safety valve;
—— *tournante,* turning, adjusting, valve.
valvoline, f., valvoline.
van, m., cribble; scuttle; sieve; fan; winnowing fan.
vannage, m., winnowing; ventilation; system of water gates in a dam; strong planking used instead of sheet-piling;
—— *en persienne,* arrangement for supplying water to a breast wheel through prismatic ajutages.
vanne, f., flood gate, water gate; hatch, sluice; sluice valve; sluice gate; valve;
—— *d'air chaud,* (*met.*) hot-blast valve;
—— *d'air froid,* (*met.*) cold-blast valve;
—— *ascendante,* lifting flood gate;
—— *de bassin,* dock gate;
—— *de décharge,* waste-water gate;
—— *dormante,* gate holder;
eaux——*s,* waste water of industrial establishments;
—— *lancière,* v. —— *de travail;*

vanne *plongeante,* plunging flood gate;
—— *en persienne,* a plunging gate for a breast wheel (v. s. v. *vannage*);
—— *de travail,* gate admitting water to the wheel.
vanneau, m., flight feather.
vanner, v. a., to winnow, fan, sift; to put in water gates.
vannerie, f., basket work.
vannette, f., basket for sifting oats; stable basket; small water gate.
vannier, m., basket maker.
vantail, m., leaf of a folding door; window shutter; sluice gate;
—— *tournant,* turning sash.
vantil(l)er, v. a., to stop (or shut out) water with planks.
vapeur, f., steam; vapor, damp; exhalation, fume; m., (*nav.*) steamer;
à ——, steam (in combinations);
admettre la ——, to admit steam;
—— *d'admission,* boiler, admission, steam;
aller à pleine, à toute, ——, to go under a full head of steam;
—— *aqueuse,* wet steam;
—— *auxiliaire,* auxiliary steam;
avoir de la ——, to have steam on or up;
bain de ——, (*chem.*) steam bath;
—— *à basse pression,* low-pressure steam;
bateau à ——, steamboat, steamer;
bateau à —— *à hélice simple,* single-screw steamer;
bateau à —— *à hélices jumelles,* twin-screw steamer;
bâtiment à ——, steamship, steamer;
boîte à ——, steam chest;
chambre de ——, steam room, steam chamber, steam space of a boiler;
chariot à ——, steam wagon;
charrue à ——, steam plow;
chaudière à ——, steam boiler;
—— *de la chaudière,* boiler steam, initial steam;
chauffage à la ——, steam-drying;
chauffer à la ——, to steam-dry;
chemise de ——, steam jacket;
cheval——, v. s. v. *cheval;*
—— *à citerne,* tank steamer;
coffre à ——, steam chest;
—— *combinée,* superheated and wet steam; combined steam;
—— *de commerce,* merchant steamer;
—— *condensée,* condensed steam;
conduit de (la) ——, steam port; steam way;
consommation de ——, consumption of steam;
—— *consumée par cheval indiqué,* steam used per indicated horsepower;
contenant la ——, steam-tight;
contre——, return steam, back steam;
—— *coupée,* cut-off steam;
couper la ——, to cut off steam;
—— *de décharge,* exhaust steam;
décharge de la ——, exhaustion of steam;
dégager la ——, to blow off steam;
—— *désaturée,* overheated steam, superheated steam;
la —— *se détend,* the steam expands;
—— *à détente,* expansion steam;
dôme de ——, steam dome;
dôme de machine à ——, v. *dôme de* ——;
dôme de prise de ——, v. *dôme de* ——;
donner la ——, to turn on steam;
donner de la ——, to steam;
—— *d'eau,* steam, vapor of water;
échappement de la ——, escape puff of steam;
—— *d'échappement,* waste or exhaust steam;
—— *échauffée,* heated steam;
en —— with steam on or up;
étanche à la, de, ——, steam-tight;
—— *étranglée,* throttled steam;
étranglement de la ——, wire-drawing of steam;
étrangler la ——, to throttle steam;
étuve à ——, v. s. v. *étuve;*
—— *d'évacuation,* v. —— *d'échappement;*
frégate à ——, (*nav.*) steam frigate;
fuite de ——, escape of steam;
—— *globuleuse,* wet steam;
—— *de guerre,* (*nav.*) war steamer;
—— *à haute pression,* high-pressure steam;

vapeur à *hélice*, screw steamer;
—— *humide*, wet steam;
imperméable à la ——, v. *contenant la* ——;
—— *interceptée*, v. —— *coupée*;
jeter de la ——, v. *donner de la* ——;
—— *jumeau*, twin steamer (as for channel or river work);
laisser échapper la ——, to blow off steam;
lumière à ——, steam port;
machine à ——, steam engine (v. s. v. *machine*);
machine à *détente de* ——, expansion engine;
marcher à *la* ——, (*nav.*) to steam;
marteau à ——, steam hammer;
matelas de ——, steam cushion;
mettre de la, en, ——, to steam; to put on steam, get up steam;
—— *mixte*, combined, mixed, steam; water and steam;
—— *mouillée*, wet steam;
moulin à ——, steam mill;
navigation à *la* ——, steam navigation;
navire à ——, steamer;
—— *non-saturée*, nonsaturated steam;
obtenir de la ——, to get up steam;
orifice à ——, v. *lumière* à ——;
ouverture de la ——, steam port;
paquebot à ——, steam packet;
—— *perdue*, waste steam;
—— *-pilote*, steam pilot boat;
piston à ——, steam piston;
en pleine ——, at full steam, under full steam;
prise de ——, taking, drawing, delivering, of steam; steam dome; mouth of the steam at the boiler;
produire de la ——, to generate steam;
—— *qui a servi*, v. —— *perdue*;
—— *réchauffée*, superheated steam;
—— *régénérée*, v. —— *réchauffée*;
registre de ——, throttle valve;
régulateur de ——, steam regulator;
remorqueur à ——, steam tug;
réservoir de la ——, steam chamber, reservoir (not to be confounded with steam room);
—— à *roues*, paddle (wheel) steamer;
—— *saturée*, saturated steam;
séchage de la ——, drying of steam;
—— *sèche*, dry steam;
sifflet à ——, steam whistle;
sous ——, under steam;
supprimer la ——, to cut off steam;
—— *surchauffée*, superheated steam;
tenir la ——, to be steam-tight;
à *toute* ——, at full steam, under full steam;
trompette à ——, steam siren;
tuyau à ——, steam pipe;
tuyau d'apport de la ——, steam pipe;
tuyau de communication de la ——, v. *tuyau* à ——;
tuyau de prise de la ——, v. *tuyau* à ——;
—— à *une roue*, stern-wheeler;
—— *vésiculeuse*, wet steam;
voiture à ——, steam carriage;
volume de ——, steam space.
vaporifère, m., steam generator.
vaporisateur, m., vaporizer; (*steam*) system of tubes of a Belleville boiler.
vaporisation, f., vaporization.
puissance de ——, evaporating power.
vaporiser, v. a., to steam, vaporize.
vaquer, v. n., to be unfilled (office); to be vacant, unoccupied (house, room);
—— à, to attend to, apply one's self to.
varangue, f., (*nav.*) floor timber.
varec(h), m., wrack, wreckage, sea-wrack.
varenne, f., waste lands.
vareuse, f., (*nav.*) pilot coat, jumper; (*unif.*) loose jacket for undress duties (fatigue jacket; overalls).
variable, a., m., variable, changeable, unsteady;
au ——, on the change.
variation, f., variation; variation (of the needle, etc.);
—— *du compas*, variation of the compass.
varice, f., varicose vein; varix;
—— *du jarret*, (*hipp.*) blood spavin.
varicocèle, m., varicocele;
—— *du jarret*, (*hipp.*) blood spavin.
varié, p. p., varied;
terrain ——, (*top.*) broken, varied, ground.

varier, v. a. n., to vary, change; (of a wind, current, etc.) to shift; (of the needle) to be unsteady.
variole, f., smallpox.
variomètre, m., (*inst.*) variometer.
variqueux, a., varicose.
varlope, f., large carpenter's plane; (*artif.*) rocket press;
demi- ——, two-handled plane;
—— *onglée*, molding plane, match plane.
varme, f., (*met.*) twyer plate of a fining-forge crucible.
vase, f., mud, ooze (at the bottom of rivers, lakes, the sea, etc.); (*in pl.*) mud flats;
banc de ——, mud bank;
fond de ——, muddy bottom.
vase, m., vessel, vase;
—— *clos*, closed vessel;
—— *de condensation*, condenser (distillation, etc.);
—— *poreux*, (*elec.*) porous cell.
vasé, a., *foin* ——, muddy hay.
vaseline, f., vaseline.
vaseux, a., muddy, slimy, miry.
vasière, f., mud hole; muddy, miry, stretch of country.
vasistas, m., bull's-eye window.
vason, m., piece of clay of which a brick or tile is made.
vasque, f., basin of a fountain.
vau, m., (*top.*) small valley; narrows of the gorge of a valley (obs.);
à —— *l'eau*, with the stream, current;
à —— *de route*, (*mil.*) (of a defeat, rout) in the utmost confusion, disorder.
veau, m., calf; calf leather, calfskin; (*cons.*) curved piece cut out of a plank; curved back or block of an arch center; (*mil. slang*) knapsack, "scranbag."
vecteur, m., vector;
arbre ——, v. s. v. *arbre*;
rayon ——, radius vector.
vedette, f., (*mil.*) cavalry sentinel; vedette, mounted sentry; (*fort.*) sentry box; watch tower (rare); (in letters, etc.) conspicuous place for a name, title, etc.;
—— *double*, (*mil.*) double vedette;
être en ——, (*mil.*) to be on vedette duty;
en ——, (*mil.*) on vedette duty; (in letters, printed pages, etc.) conspicuous; displayed; placed so as to catch the eye;
mettre en ——, to station or post on vedette duty;
—— *simple*, single vedette.
véhicule, m., vehicle, carriage; (*inst.*) vehicle, draw tube, slide.
veille, f., keeping awake; eve, day before; watch, watching; (*nav.*) lookout;
à *la* ——, visible above water (said of a projecting or floating object);
ancre de ——, v. s. v. *ancre*;
en ——, à *la* ——; (of an anchor) a-cockbill;
mettre à *la* ——, to cockbill.
veiller, v. n., to stay or be awake; (*nav. and in gen.*) to be on the watch, on the lookout; (of a buoy) rock, etc., to be (partly) above water, in sight, visible.
veilleuse, f., night light;
en ——, serving as a night light.
veillote, f., small haycock.
veinage, m., grain of wood; graining; (*met.*) veining (defect).
veine, f., vein; streak, vein (defect in wood or stone); underground spring; (*min.*) seam, lode, vein;
barrer la —— à *un cheval*, (*hipp.*) to bar a vein;
—— *de bois*, vein (defect) in wood;
—— *d'eau*, underground spring;
—— *fluide*, stream or jet issuing from a thin-walled orifice;
—— *satellite*, (*hipp.*) vein close to an artery;
—— *sombre*, (*met.*) mark or defect resulting from a flattened or welded blowhole.
veineux, a., streaky, veined.
vélin, m., vellum;
papier ——, v. s. v. *papier*.
vélites, m. pl., (*Fr. a.*) battalions of *chasseurs à pied* formed by Napoleon I to furnish n. c. o. to line regiments.

vélo, m., (familiar abbreviation of *vélocipède*) bicycle.
véloce, m., ordinary bicycle.
véloceman, m., bicyclist.
vélocidrome, m., bicycle track.
vélocimètre, m., (*ball.*) velocimeter;
—— *à chariot*, slide velocimeter;
—— *à cylindre tournant*, rotating cylinder velocimeter;
—— *à diapason électrique*, velocimeter with electrically controlled tuning fork;
—— *à diapason et à enregistreurs électriques*, velocimeter with electrically controlled tuning fork and register;
—— *à diapason mécanique*, ordinary tuning-fork velocimeter;
——*-fusil*, gun velocimeter.
vélocipède, m., velocipede, ordinary;
bateau ——, water bicycle.
vélocipéder, v. n., to "wheel," to bicycle.
vélocipédeur, vélocipédier, m., bicyclist.
vélocipédique, a., relating to the bicycle;
service ——, (*mil.*) bicycle service.
vélocipédiste, m., cyclist, bicyclist; (*mil.*) bicycle dispatch rider.
vélocité, f., speed, rate, swiftness, velocity.
vélodrome, m., bicycle track.
vélographe, m., instrument for measuring the speed of a bicycle, bicycle velocimeter.
velours, m., velvet;
—— *à côtes*, corduroy.
veltage, m., v. *valtage*.
velter, v. a., v. *valter*.
velture, f., v. *valture*.
velu, a., hairy, shaggy;
pierre ——*e*, rough stone.
vendre, v. a., to sell;
—— *un cheval crins et queue*, to sell a horse very dear;
—— *un cheval tout nu*, to sell a horse without saddle and bridle.
vengeance, f., vengeance, revenge.
venger, v. a., to avenge, revenge.
vengeur, a. m., avenging, avenger.
venir, v. n., to come, repair, proceed; arrive; to come on, grow; to come out, off; to happen;
—— *à l'appel*, (*cord.*) to begin to draw, to come into bearing;
—— *à bout de ses ennemis*, to overcome one's enemies;
—— *de fonte*, (*fond.*) to be cast on;
—— *de fonte avec*, (*fond.*) to be cast on at the same time with; to be cast in one piece with;
en —— *aux mains*, to begin to fight; (more esp. *mil.*) to come to close quarters; to fight hand to hand;
—— *à point*, (*met.*) to come to nature.
vent, m., air, wind, gale; division of the compass; (*met.*) blast; (*mach.*) clearance; (*art.*) windage (smooth-bore muzzle-loading guns, obs.); blast (of a projectile); difference between turned part of the shoulder of a projectile and the caliber; (*nav.*) weather gauge;
—— *des ailettes*, (*art.*) clearance of studs (obs.);
aire de ——, v. s. v. *aire*;
—— *alizé*, trade wind;
aller au plus près (*du* ——), v. *pincer le* ——;
aller contre —— *et marée*, to have the wind and tide ahead;
—— *d'amont*, wind from upstream; shore wind, offshore wind, land breeze;
—— *arrière*, fair wind;
au ——, in the wind; to windward, windward;
—— *d'aval*, wind from downstream; sea breeze;
avantage du ——, (*nav.*) weather gauge;
avoir —— *arrière* (*devant*), to have a following (head) wind;
avoir le dessus du ——, (*nav.*) to have the weather gauge;
avoir —— *et marée*, to have fair wind and tide;
avoir le —— *sur un navire*, (*nav.*) to have the weather gauge of a ship;
—— *blanc*, a wind that does not blow up clouds;

vent *bon frais*, strong breeze (8 meters per second);
bord du ——, v. *côté du* ——;
bord de sous le ——, v. *côté de dessous le* ——;
bouffée de ——, capful of wind;
—— *de bout*, head wind;
—— *chaud*, (*met.*) hot blast;
chicaner le ——, (*nav.*) to beat;
conduit de ——, (*met.*) blast pipe;
—— *contraire*, head wind, foul wind;
côté de ——, weather side;
côté de dessous le, sous le ——, lee side;
—— *coulis*, wind from a chink or crack; draught, draughty wind;
coup de ——, gust of wind, squall;
dessus du ——, (*nav.*) weather gauge;
disputer le ——, (*nav.*) to maneuver for the weather gauge;
—— *dominant*, v. —— *régnant;*
donner le ——, (*met.*) to blow; to turn on the blast; (St. Cyr slang) to bully; to "devil" (U. S. M. A.)
—— *droit debout*, wind dead ahead; dead wind;
—— *échauffé*, (*met.*) hot blast;
être au —— *d'un navire*, (*nav.*) to have the weather gauge of a ship;
évité au ——, wind rode;
—— *faible*, light breeze;
—— *fait*, steady wind;
—— *favorable*, fair wind;
—— *follet*, baffling wind;
forcé, gale; forced draught;
—— *frais*, fresh breeze (6 meters per second);
—— *froid*, (*met.*) cold blast;
fusil à ——, v. s. v. *fusil;*
gagner le ——, *le dessus du* ——, (*nav.*) to get the weather gauge;
grain de ——, squall, sudden squall;
au gré du ——, with the wind;
—— *grec*, (in the Mediterranean) northeast wind;
—— *impétueux*, violent wind (15 meters per second);
—— *du large*, sea breeze;
—— *du levant*, levanter;
lit du ——, wind's eye;
manche à ——, wind sail;
mauvais ——, foul wind;
—— *de mer*, sea breeze;
mettre l'épée, flamberge, au ——, to draw (familiar);
moulin à ——, windmill;
—— *de mousson*, monsoon;
pincer le ——, (*nav.*) to sail close to the wind;
—— *d'un piston*, (*mach.*) piston clearance;
en plein ——, in the open air;
porter (*le nez*) *au* ——, (*hipp.*) to carry the head high;
—— *du projectile*, (*art.*) windage;
les quatre ——*s*, the cardinal points;
rallier au, le, ——, v. *pincer le* ——;
—— *régnant*, prevailing wind;
rose des ——*s*, v. s. v. *rose;*
rumb de ——, v. s. v. *rumb;*
serrer le ——, v. *pincer le* ——;
—— *de soldats*, soldiers' wind;
sous le ——, (to) leeward, lee, a-lee;
tenir le ——, v. *pincer le* ——;
—— *des tenons*, (*art.*) clearance of studs (obs.);
—— *de terre*, land breeze;
—— *de travers*, wind abeam;
les trente-deux ——*s*, the compass card.
vente, f., sale; felling of wood; place where timber has just been felled;
contra de ——, bill of sale;
—— *aux enchères*, sale by auction;
poudre de ——, v. s. v. *poudre*.
venté, p. p., driven by the wind;
arbre faux ——, wind-shaken tree;
marée ——*e*, tide banked up by the wind.
venteau, m., sluice gate, water gate; air hole of a bellows.
ventelle, f., valve or opening of a lock gate; sprinkling brush.
venter, v. n., to blow hard or fresh (of the wind).
venteux, a., windy; (*fond.*) (of a casting) blistered;
ciel ——, windy sky.

venteux, *temps* ——, windy weather.
ventilateur, m., ventilator, fan;
—— *à ailes,* —— *à ailettes,* fan, blower;
—— *aspirant,* aspirator, suction ventilator;
—— *à bras,* ventilator worked by hand;
—— *centrifuge,* centrifugal ventilator;
—— *à cloche,* Taylor's, Struve's, ventilator;
—— *à force centrifuge,* centrifugal fan;
—— *rotatif,* rotary blower;
—— *soufflant,* blower.
ventilation, f., ventilation; (*law*) valuation, estimate;
puits de ——, ventilating, air, shaft;
—— *renversée,* reverse ventilation (hot air down, cold air up).
ventiler, v. n., to ventilate, fan; (*law*) to estimate;
machine à ——, ventilating engine.
ventiller, v. n., v. *vantiller.*
ventilion, m., valve closing the air hole of a bellows.
ventouse, f., (*met., cons., etc.*) ventilator, air hole, vent; (*med., hipp.*) cupping glass.
ventre, m., belly; body; swell or curve; (*met.*) belly of a blast furnace;
—— *avalé,* (*hipp.*) pot-belly;
avoir (ne pas avoir) du ——, (*hipp.*) to be too full (thin) in the belly;
—— *de biche,* (*hipp.*) fawn-bellied (color);
—— *d'un chien de fusil,* (*sm. a.*) breast;
courir —— *à terre,* to go at full speed;
être sur le ——, (*art.*) to lie on ground (of a dismantled gun);
faire ——, to bulge out, swell out; to lean, lean over (of a wall);
—— *du haut fourneau,* (*met.*) belly of a blast furnace;
—— *lavé,* v. —— *de biche;*
—— *levretté,* (*hipp.*) greyhound belly;
mal au ——, (*hipp.*) belly fretting;
marcher sur le —— *à,* to make head against;
—— *de la noix,* (*sm. a.*) swell or curve of the tumbler;
passer sur le —— *à,* to make head against;
à —— *de poisson,* fish-bellied;
—— *à terre,* at full speed, with whip and spur;
—— *tombant,* v. —— *avalé;*
—— *de vache,* v. —— *avalé.*
ventrière, f., (*harn.*) girth; sling (for hoisting horses); (*cons.*) palisade riband (about halfway up); post for the planking of a sluiceway.
venu, p. p., —— *de fonte,* (*fond.*) cast on, cast in, one piece with; integral;
mal ——, badly done.
ver, m., worm;
—— *de digues,* teredo navalis;
—— *de mer,* v. —— *de digues;*
piqué aux, des, ——*s,* worm-eaten;
rongé aux, des, ——*s,* worm-eaten;
—— *taret,* —— *tarière,* v. —— *de digues;*
—— *de vaisseaux,* v. —— *de digues.*
verbal, a., spoken, verbal, by word of mouth;
procès- ——, v. *procès-verbal.*
verbaliser, v. n., to make a statement of facts, a formal statement; to make an official report (*procès-verbal*).
verboquet, m., (*cord.*) guy for steadying a weight (in hoisting).
ver-coquin, m., (*hipp.*) staggers.
verdet, m., verdigris;
—— *naturel,* carbonate of copper.
verdict, m., verdict, finding;
prononcer, rendre, un ——, to bring in a verdict, finding.
verdillon, m., bent crowbar, pinch bar.
verdir, v. a. n., to paint green; to become covered with verdigris.
verge, f., rod, wand, switch; handle, shaft, spindle; beam of certain balances; (*hipp.*) penis; (*surv.*) surveyor's rod;
—— *de l'ancre,* anchor shank;
—— *d'arpenteur,* (*inst.*) surveyor's rod;
—— *de calibre,* (*art.*) gauge for the bore;
—— *de charpentier,* foot rule;
—— *du collier,* (*harn.*) fore wale;
—— *à compensation,* compensating rod;
——*s crénelées,* nail rod iron;
étain en ——*s,* bar tin;

verge, *faire passer par les* ——*s,* (*mil.*) to flog through the lines (obs.);
faisceau de ——*s,* bundle of rods;
—— *de fer,* iron rod;
fer en ——, rod iron;
—— *de fusée,* (*artif.*) rocket stick;
—— *de girouette,* vane spindle;
—— *de nettoyage,* (*mach.*) cleaning iron;
—— *d'or,* (*inst.*) Jacob staff, cross staff;
——*s du parallélogramme,* (*mach.*) parallel rods;
passer par les ——*s,* to run the gauntlet;
—— *de pompe,* pump spear.
vergeage, m., measuring with a yardstick.
vergelé, m., a soft calcareous stone.
verger, m., orchard.
verger, v. a., to measure (cloth) with a yardstick.
vergette, f., hoop of a drum; (*in pl.*) brush; bundle of rods;
—— *de roulage,* flesh hoop of a drum.
vergeure, f., wire-mark in paper;
verglas, m., sleet.
vergne, m., alder tree.
vergue, f., (*nav.*) yard;
—— *à* ——, yardarm and yardarm.
vérificateur, m., examiner, inspector; gauge, calipers, sliding calipers; standard;
—— *de l'âme et des rayures,* (*art.*) bore and rifling gauge;
—— *de la butée,* (*art.*) instrument for gauging the seating of a projectile;
calibre- ——, (*art., etc.*) gauge;
cylindre- ——, v. s. v. *cylindre;*
—— *de dressage,* (*sm. a.*) an instrument to test the straightness of barrels;
—— *d'ensemble,* (*sm. a.*) an instrument to verify the angle made by the last element of a gun barrel with the line of sight;
—— *de l'excentricité de la lumière,* (*art.*) fuse-hole gauge (eccentricity, diameter, etc.);
instrument ——, gauge;
—— *de la ligne de mire,* (*art.*) instrument for testing the parallelism of a line of sight;
—— *de l'œil,* (*art.*) fuse-hole gauge;
—— *de rebut,* (*art.*) sort of rifling gauge (for gutta-percha impressions);
—— *de réception,* (*art.*) acceptance gauge, so called (to verify the possibility of a projectile reaching its seat properly; obs., muzzle-loading guns);
—— *du taraudage,* (*art.*) thread gauge (of a fuse hole).
vérification, f., verification, test, examination, inspection; auditing of accounts; test of an instrument to see if it is in adjustment; (*surv., etc.*) check or checking of an observation; (*art.*) verification of dimensions, etc.;
—— *de contrôle,* second or counter verification;
—— *faite,* after examination;
—— *de la ligne de mire,* (*art.*) verification of the parallelism of the line of sight;
—— *d'une ligne télégraphique* (*téléphonique*), examination and test of a line (with reference to insulation and continuity of circuit);
soumis à ——, under examination.
vérifier, v. a., to verify, test, try; to test an instrument; to confirm, inspect, audit, prove, go over; to adjust (as a compass); (*cord.*) to mend (a seizing, etc.);
—— *un chronomètre,* to rate a chronometer.
vérin, m., jackscrew; hand screw, lifting jack;
—— *arrache-pieux,* pile-withdrawing screw;
—— *à bouteille,* screw jack;
—— *à chariot,* powerful screw jack for either vertical or horizontal efforts;
—— *à démonter,* dismounting jack;
—— *hydraulique,* hydraulic jack;
—— *à vis,* v. —— *à bouteille.*
vérine, f., (*nav.*) binnacle lamp; (*cord.*) selvages.
vermiculé, a., vermiculate.
vermiculures, f. pl., vermiculated work.
vermillon, m., a., vermilion.
vermoulu, a., worm-eaten, rotten.
vermoulure, f., wormhole, wormhole dust.

vernier, m., vernier; nonius;
— *direct*, direct vernier;
— *rétrograde*, retrograde vernier.
vernir, v. a., to varnish, glaze, japan; to French polish.
vernis, m., varnish, glaze, japan, polish, French polish;
— *à l'alcool*, spirit varnish;
— *d'ambre*, amber varnish;
— *à l'asphalte*, black japan;
— *clair au goudron*, bright varnish;
— *à la copale*, copal varnish;
— *de copal*, copal varnish;
— *d'ébéniste*, cabinet varnish;
— *à l'esprit de vin*, spirit varnish;
— *à l'essence*, turpentine varnish;
— *à fer*, ironwork black;
— *de goudron*, mixture of asphalt, rosin, and coal-tar oil, for fireworks;
— *à la gomme laque*, gum-lac varnish;
— *gras*, oil varnish;
— *à l'huile*, v. — *gras*;
— *hydrofuge*, waterproof varnish;
— *luisant*, glaze;
— *noir au goudron*, black varnish;
— *de plomb*, lead glaze;
— *siccatif*, drying varnish;
— *spiritueux*, v. — *à l'alcool*;
— *à voitures*, carriage varnish.
vernissage, m., varnishing;
— *intérieur*, (art.) varnishing of the interior of projectiles.
vernisser, v. a., to varnish, glaze, japan.
verre, m., glass; lens; glass case; drinking glass; glassful;
— *anallatiseur*, anallatizing lens;
— *ardent*, burning glass;
carreau de —, pane of glass;
— *de champ*, (inst.) field lens;
— *de couronne*, crown glass;
— *de couleur*, stained glass;
— *en canons*, sheet glass, brood glass, cylinder glass;
cul de —, (hipp.) web or cataract in a horse's eye;
— *en cylindres*, v. — *en canons*;
— *dépoli*, ground glass;
— *dormant*, (cons.) small window in a party wall, not opening, v. s. v. *jour*;
— *enfumé*, smoked glass;
— *à glaces*, plate glass;
— *grossissant*, magnifying glass;
— *-indicateur du niveau d'eau*, (steam) water-gauge glass;
— *en manchons*, v. — *en canons*;
— *mort*, v. — *dormant*;
— *moulé*, pressed glass;
— *objectif*, (inst.) objective;
— *oculaire*, (inst.) eyeglass;
— *d'œil*, (inst.) eyeglass;
papier de —, v. s. v. *papier*;
— *de plomb*, lead glass;
— *soluble*, soluble glass;
sous —, under glass;
— *en tables*, plate glass;
— *à vitres*, window glass.
verrerie, f., glass works; art of glass making; glassware.
verrier, m., glass manufacturer.
verrine, f., (inst.) barometer glass.
verrou, m., bolt, bar, hasp, catch bolt; (fort.) embrasure mask; (sm. a.) bolt; maneuvering piece;
— *d'amarrage*, securing bolt;
arme à —, (sm. a.) bolt gun;
— *d'arrêt*, (r. r.) turnable stop bolt;
— *automoteur*, (art.) spring bolt;
— *de combat*, (art.) firing bolt;
contre-—, v. *contre-verrou*;
— *à contrepoids*, (fort.) counterpoise embrasure mask or shutter (Grüson);
— *-crémaillère*, — *à crémaillère*, (art.) ratchet bolt (sort of safety bolt);
— *d'embrayage*, (art.) coupling bolt (French 95mm);
— *d'enclenchement*, (art.) locking bolt;
— *d'enrayage*, securing bolt;
— *d'exercice*, (art.) drill bolt of a firing mechanism;
fermer au —, to bolt;

verrou, *fermeture a* —, v. s. v. *fermeture*;
— *de fermeture*, (art.) locking bolt, latch;
— *glissant*, sliding bolt;
— *à heurtoir*, (art.) tappet bolt;
— *d'immobilisation*, (art.) clamping bolt (fixes a turret gun in the loading position);
— *à mouvement rectiligne simple*, (sm. a.) straight-pull bolt;
— *à platine*, slip bolt;
porter l'épée en —, to wear the sword horizontal;
— *à ressort*, spring bolt (of a door, etc.); (art.) spring safety bolt;
— *de retenue*, stop bolt;
sous les —*s*, in safe custody, in jail;
— *de sûreté*, (art., torp., etc.) safety bolt; (nav.) turret-securing bolt;
tirer le —, to shoot a bolt.
verrouillement, m., (sm. a.) bolt system, system of bolt locking.
verrouiller, v. a., to bolt;
— *la culasse*, (sm. a.) to lock the breech.
versant, m., slope, watershed; slope, side, of a hill, of a mountain; sloping ground on each side of a river;
— *plongeant*, steep slope (plunging into a valley);
— *rasant*, slope (of a hill) of but slight inclination.
versant, a., cranky, liable to overturn.
verse, f., (of cereals) state of being beaten down by rain, etc.; large charcoal basket; (fond.) pouring;
à —, hard, in torrents, (of rain).
versement, m., payment, deposit (of money); (mil. in pl.) turning over, issue (of arms and of material in general);
— *partiel*, installment.
verser, v. a. n., to pour (out, in, forth); to spill, shed, discharge, empty; to lay, beat down; to be beaten down (as crops by rain, etc.); to overturn, upset, be overturned, upset (of vehicles); to throw, cast, on or upon; to deposit, pay in (funds); to turn over, assign (as horses to a troop); (mil.) to issue material; (fond.) to pour;
— *en cage*, — *en panier*, v. s. v. *cage*;
— *le sang*, to shed blood; to kill.
verso, m., left-hand page; second page; even-numbered page.
vert, a., green; raw, fresh; unwrought; (fig.) resolute, firm;
bois —, v. s. v. *bois*;
cheval — *au montoir*, horse hard to mount;
cuir —, rawhide;
fonte —*e*, (met.) brass, gun metal;
pierre —*e*, stone fresh from the quarry;
ration —*e*, allowance, ration, of green food (for horses);
régime —, green food (for horses);
sable —, v. s. v. *sable*.
vert, m., green color; grass, green food, green forage;
au —, at grass;
— *bois*, (top. drawing) green for woods;
cheval au —, horse on green food;
faire prendre le — *à un cheval*, to keep a horse on green food;
faire quitter le — *à un cheval*, to take up a horse from grass;
— *franc*, (top. drawing) green for slopes, orchards;
mettre un cheval au —, to put a horse out to grass;
mise au —, putting a horse out to grass;
— *pré*, (top. drawing) green for pastures, meadows;
retirer un cheval du —, to take a horse from grass;
— *de terre*, verditer.
vert-de-gris, m., verdigris.
vertenelle, f., staple, bolt staple.
verterelle, f., v. *vertenelle*.
vertevelle, f., v. *vertenelle*.
vertical, a., vertical, upright;
la —*e*, the vertical.
verticalité, f., verticalness, verticality.
vertige, m., dizziness, vertigo; (hipp.) staggers;

vertige *abdominal*, (*hipp.*) stomach staggers;
— *essentiel*, (*hipp.*) staggers, blind staggers, mad staggers.
vertigineux, a., causing vertigo; (*hipp.*) causing staggers.
vertigo, m., (*hipp.*) staggers.
vertiqueux, a., spiral, winding.
vesce, f., vetch, tare.
vésicatoire, m., (*hipp., etc.*) blister.
vésiculaire, a., vesicular;
état —, spheroidal state.
vésicule, f., vesicle; (*met.*) blister (in a casting).
vésiculé, a., blistered (as steel, etc.).
vessie, f., bladder; blister;
relâchement de la —, (*hipp.*) profuse staling.
vessigon, m., (*hipp.*) vessicnon, vessignon, windgall;
— *articulaire*, articular windgall;
— *chevillé*, thorough pin;
— *calcanéen*, calcaneal windgall;
— *cunéen*, cuneiform vessignon;
— *du grasset*, stifle;
— *soufflé*, v. — *chevillé.*
— *tarsien*, vessignon of the tarsal sheath;
— *tendineux*, tendinous windgall.
vessir, v. n., (*met.*) to scatter.
veste, f., jacket, round jacket; (*unif.*) undress or fatigue jacket;
— *d'écurie*, (*unif.*) stable jacket;
— *de service*, (*unif.*) undress, fatigue, jacket.
vestibule, m., vestibule, hall, lobby, entrance.
veston, m., short coat.
vêtement, m., clothing, clothes.
vétéran, m., (*mil.*) veteran; (in schools) student who repeats a course; "turn-back" (U.S.M.A.);
— *de mathématique*, "turn-back" in mathematics.
vétérance, f., state of being a veteran.
vétérinaire, a., m., veterinary; veterinarian; farrier;
aide—, (*Fr. a.*) veterinarian (sublieutenant);
artiste —, v. *médecin* —;
— *en 1er (2d)*, (*Fr. a.*) veterinarian (captain, lieutenant);
médecin —, veterinary surgeon; horse doctor;
— *principal de 1ère (2ème) classe*, (*Fr. a.*) veterinarian (lieutenant-colonel, major).
vêtir, v. a., to put on (clothes, a garment).
viabilité, f., practicability, good condition, of a road.
viaduc, m., viaduct.
viager, a., m., during one's lifetime, during life; life interest, life annuity;
pension —*e*, v. s. v. *pension;*
rente —*e*, v. s. v. *rente.*
viande, f., meat;
— *conservée*, preserved meat;
conserves de —, preserved meats;
— *fraîche*, fresh meat;
— *salée*, salt meat.
vibration, f., vibration; oscillation of a pendulum.
vibratoire, a., vibratory.
vibrer, v. n., to vibrate.
vicaire, m., vicar; curate.
vicairie, f., vicariat, m., office of vicar; curacy.
vice, m., vice, defect, flaw, blemish, imperfection;
— *des bois*, defects in wood;
cheval qui a du —, restless horse;
— *de conformation*, (*hipp.*) defect of build;
— *rédhibitoire*, v. s. v. *rédhibitoire;* (*adm.*) fatal defect (in a contract, etc.).
vice-amiral, m., vice-admiral.
vice-amirauté, f., vice-admirality.
vice-consul, m., vice-consul.
vice-gérant, m., deputy manager.
vice-président, m., vice-president.
vice-reine, f., wife of a viceroy.
vice-roi, m., viceroy.
vice-royal, a., vice-royal.

vice-royauté, f., vice-royalty.
vicié, a., corrupted, tainted, foul;
bois —, spoiled wood, timber.
vicier, v. a., to taint, corrupt; to vitiate, make void.
vicieux, a., vicious, defective, faulty, imperfect;
cheval —, vicious horse;
nature — *se*, vice, viciousness, of a horse.
vicinal, a., *chemin* —, crossroad, parish road.
vicomte, m., viscount.
victoire, f., victory, conquest, triumph;
gagner, remporter, la —, to win the day, the victory.
victorieux, a., victorious, triumphant.
victuaille, f., victuals.
vidage, m., emptying; parapet, mound (formed of the earth dug out of a canal and thrown on each side).
vidange, f., clearing, removing; voidance, draining, emptying, blowing off; thinning out (of a forest); empty part (of a cask, etc.); (*in pl.*) soil, night soil; (*mach.*) sediment (in a boiler);
robinet de —, v. s. v. *robinet;*
trou de —, (*steam*) mud hole (of a boiler);
tuyau de — *de chaudière*, (*steam*) blow-off pipe.
vidanger, v. a., to empty.
vidart, m., (*hipp.*) horse liable to diarrhea.
vide, a., empty; void, vacant; devoid; unimpregnated;
à —, empty, in vacuo; (*mach.*) unloaded (of an engine); (*man.*) riderless;
frapper à —, to strike the anvil instead of the work;
porter à —, to be unsupported;
rangés tant pleins que —*s*, (*mil.*) drawn up with intervals equal to the front of the units.
vide, m., empty space, emptiness, void, gap, hole; interstice (as between the bars of a grate, etc.); vacancy; vacuum; (*top.*) chasm; (*cons.*) opening in a wall; span of an arch; (*mil.*) gap in a line; (*art.*) slot (of a breech screw);
— *d'un bastion*, (*fort.*) hollow of a bastion;
combler un —, to fill up a gap; "to fill a long-felt want;"
dans le —, in vacuo;
espace de —, (*steam*) vacuum space;
faire un —, to make a vacuum;
— *de forure*, bore-hole;
— *d'un fourneau*, (*met.*) interior of a blast furnace (synonym of *cuve*, q. v.);
jauge de —, (*steam*) vacuum gauge;
pousser au —, v. s. v. *pousser;*
rester dans le —, (*man.*) of a horse, to carry himself so as not to feel the bit; to cease to feel the bit;
tirer au —, v. *pousser au* —;
volume de —, v. *espace de* —.
vider, v. a., to empty, clear, void, drain, evacuate; to bore, scoop, scoop out; to decide, end, settle, arrange, adjust; to quit, leave, depart from; (*mach.*) to blow off, out;
— *les arçons*, v. s. v. *arçon;*
— *une chaudière*, to blow off a boiler;
— *un cheval*, (*hipp.*) to rake a horse;
— *une clef*, to bore a key;
— *les escarbilles*, to throw out ashes, etc.;
— *une prison*, etc., to clear a prison, etc.;
— *une querelle*, to have a quarrel out by fighting;
— *des terres*, to level ground;
cheval qui a les jarrets bien vidés, clean-hocked horse.
vie, f., life; duration, life (of a gun, etc.).
vieil, a., old.
vieillir, v. n., to grow old; to become obsolete.
vierge, a., f., virgin;
poulie —, v. s. v. *poulie.*
vieux, a., old, antique, obsolete; (of a horse) over 5 years of age;
— *fer et vieilles fontes*, unserviceable iron stores of every description, scrap iron.
vif, a., alive, living; lively, eager, sharp, brisk, quick; spirited; mettlesome; deep, vivid, bright; rapid;

vif, *allure* ——*ve,* (*man.*) rapid gait;
——*ve arête,* v. s. v. *arête;*
arêtes ——*ves,* (*hipp.*) grease, rat-tails, grapes, mangy tumors, on a horse's leg;
argent ——, quicksilver;
attaque, etc., ——*ve,* (*mil.*) sharp attack, etc.;
bois ——, living tree;
bois à ——*ve arête,* timber whose bark or soft wood is cut off to make it square;
——*ve canonnade,* (*art.*) brisk cannonade;
chaux ——*ve,* quicklime;
cheval ——, high-spirited horse;
curer à —— *bord, à* —— *fond,* to clear (a ditch) to regulation depth and width;
eau ——*ve,* running water;
——*ve eau,* v. s. v. *eau;*
feu ——, bright fire; (*mil.*) brisk fire;
——*fonds,* undisturbed ground;
——*ves fontes, fontes* ——*ves,* (*met.*) metals that in fusion remain liquid a long time;
de ——*ve force,* by main force, strength;
forêt ——*ve,* forest of fine and large trees;
——*ve fusillade,* (*inf.*) brisk fusillade;
haie ——*ve,* v. s. v. *haie;*
œuvres ——*ves,* v. s. v. *œuvre,* f.;
roc ——, bed rock;
roche ——*ve,* natural rock (surface unaltered);
de —— *ve voix,* aloud.

vif, m., living person; flesh; quick, core; the thick of anything; solid; (*mas.*) the hardest part of a stone (interior);
—— *d'un arbre,* inside, stem, of a tree;
—— *d'une colonne,* shaft of a column;
—— *du combat,* (*mil.*) thick of the fight;
donation entre ——*s,* deed of gift;
—— *de l'eau,* v. s. v. *eau;*
—— *d'une pièce,* (*art.*) outer surface of a gun as it comes from the mold (obs.);
—— *de piédestal,* die of a column;
piquer jusqu'au ——, (*farr.*) to prick a horse's hoof;
racler au ——, (*mas.*) to scrape off the outside of a stone;
toucher le —— *à,* (*farr.*) to prick a horse's foot.

vif-argent, m., mercury, quicksilver.

vigie, f., (*nav.*) lookout, lookout man; (*r. r.*) box (as near a switch, etc.); (*hydr.*) lurking rock; shoal water;
en ——, on the lookout;
homme de ——, (*nav.*) lookout man.

vigier, v. n., (*nav.*) to look out, watch, be on the lookout.

vigilance, f., vigilance, watchfulness.

vigilant, a., vigilant, watchful, careful, on one's guard.

vigne, f., vine; wash (for sections and elevations).

vignoble, m., vineyard.

vigorite, f., (*expl.*) vigorite;
—— (*américaine*), vigorite.

vilain, a., vile, bad, wretched;
faire ——, to be bad, nasty weather.

vilebrequin, m., brace; stock and bits; (*mach.*) cranked shaft; (*art.*) vent pricker with gimlet point; (*r. f. art.*) crank, arm;
—— *à rochet,* ratchet brace.

village, m., village;
——*fermé,* village with streets;
——*irrégulier,* village without streets;
——*régulier,* v. —— *fermé.*

ville, f., town; the municipal officers;
corps de ——, municipal officers;
—— *d'eaux,* watering place;
—— *de fabrique,* manufacturing town;
—— *de garnison,* (*mil.*) garrison town;
la grande ——, Paris;
hôtel de ——, town hall;
—— *d'industrie,* v. —— *de fabrique;*
—— *de l'intérieur,* inland town;
maison de ——, v. *hôtel de* ——;
—— *maritime,* seaport town;
—— *ouverte,* (*mil.*) open, unfortified town;
sergent de ——, policeman.

vimaire, f., damage to forests by storms.

vin, m., wine;
esprit de ——, spirits of wine.

vinaigre, m., vinegar.

vindas, m., windlass, capstan, crab, drum;

vindas, *clef de* ——, cheek of a capstan;
linguet de ——, pawl of a capstan;
mèche de ——, barrel of a capstan;
taquet de ——, v. *linguet de* ——.

vineux, a., (*hipp.*) wine-colored (of the coat);
rouan ——, wine-colored roan.

vingtaine, f., score; (*cord.*) lashing (for scaffolds); small rone to keep a load from striking the wall (in hoisting).

violation, f., violation;
—— *d'arrêts,* (*mil.*) breaking one's arrest;
—— *de consigne,* (*mil.*) violation of orders.

violence, f., violence, stress;
faire ——, to coerce, strain, stretch;
—— *du temps,* stress of weather.

violent, a., violent; strong, wild, boisterous, ve, hement, fierce, hard.

violenter, v. a., to force, do violence to.

violer, v. a., to violate (as a parole, quarantine, etc.).

violon, m., violin; cell of a guardhouse.

virage, m., heaving, play, of a windlass, of a capstan; hoisting of weights by a capstan; space necessary for the play of a capstan; (*phot.*) toning;
bain de ——, (*phot.*) toning bath;
liqueur de ——, v. s. v. *liqueur;*
rayon de ——, radius of capstan play.

virebrequin, m., v. *vilebrequin.*

virement, m., act of working a capstan; payment of a debt by the transfer of a debt; adjustment of an account by a transfer on the books; clearing; transfer of funds (to equalize accounts); (*adm.*) transfer (frequently illegal) of funds from one article of the budget to another; (*phot.*) toning;
—— *de bord,* (*nav.*) going about;
—— *d'eau,* turn of the tide;
—— *de fonds,* (*adm.*) transfer of funds from one article of the budget to another.

virer, v. a. n., to turn; to turn, to heave (of a capstan); to clear, to transfer an account; to pay a debt by the transfer of a debt; (*nav.*) to go about; (*phot.*) to tone;
—— *l'ancre,* to weigh anchor;
—— *de bord,* (*nav.*) to go about;
—— *au, le, cabestan,* to heave, pull, etc., at the capstan;
—— *le câble,* to haul in the cable;
—— *à courir,* to heave away;
—— *en douceur,* to heave or pull handsomely;
manquer à ——, (*nav.*) to miss stays;
pare à ——! (*nav.*) ready about;
refuser à ——, v. *manquer à* ——;
—— *rondement,* v. —— *à courir;*
tiens bon ——, avast heaving;
—— *vent arrière,* (*nav.*) to wear ship;
—— *vent avant,* (*nav.*) to tack.

vireur, m., (*mach.*) turning gear or device;
—— *à engrenages,* toothed turning device;
—— *de la machine,* turning gear of the engine;
—— *à la main,* hand turning gear;
—— *à la vapeur,* steam turning gear.

virevaude, f., v. *vire-vire.*

virev(e)au, m., winch, small windlass; chain windlass.

vire-vire, m., eddy, whirlpool.

virevolte, f., (*man.*) turning and winding, quick turning.

virite, f., (*expl.*) virite.

virole, f., ferrule; collar, hoop, sleeve, ring; (*mach.*) tube ferrule; (*steam*) section of a boiler; (*sm. a.*) bayonet ring; locking sleeve; (*art.*) base ring; collar of a tube;
—— *à ailettes,* (*sm. a.*) locking ring of a bayonet;
—— *bague,* (*steam*) tube ferrule;
—— *de baïonnette,* (*sm. a.*) locking ring;
—— *de bout d'essieu,* linch hoop; collar of axle-tree arm;
—— *écrou,* (*art.*) base ring, bushing (Longridge wire gun);
—— *porte-cuirasse,* (*fort.*) armor-supporting ring (of Bussière cupola);
—— *de rivure,* clinching ring;
—— *de timon,* (*art., etc.*) pole hoop.

viroler, v. a., to ferrule.

virolet, m., roller; spun-yarn winch.

virure, f., (*nav.*) strake.

vis, f., screw, especially male screw; screw stud; (cons.) spiral staircase; newel of a winding staircase; (art.) breech screw.

I. Artillery, small arms; II. Miscellaneous.

I. Artillery, small arms:

—— *d'amorce*, (art.) screw fuse plug;
—— *d'arrêt*, (art., sm. a.) stop screw; locking screw;
——*-arrêtoir*, (art.) stop bolt; (sm. a.) stop screw;
—— *de bague*, (sm. a) locking ring screw;
—— *du battant*, (sm. a). swivel screw;
—— *de boîte de culasse*, (sm. a.) breech-casing screw;
—— *borgne*, v. s. v. *borgne;*
——*-bouchon*, (sm. a.) screw plug, bolthead and nut; (art.) breech plug;
—— *de butée*, (art.) central bearing support;
—— *calante*, (art.) support (screw) of the St. Chamond disappearing carriage;
—— *du carré*, v. —— *de noix;*
—— *centrale*, (art.) inner screw (of the elevating screw);
—— *conique*, (art.) conical breech screw (Armstrong);
—— *de contre-platine*, (sm. a.) side-plate screw;
—— *de croisière*, (sm. a.) hilt screw;
——*-culasse*, (art.) breech plug, breech screw; screw plug of Creusot projectile tester;
—— *de culasse*, (art.) breech screw, breech plug; (sm. a.) tang screw;
—— *cylindro-conique*, (art.) (Armstrong's) cylindro-conical breech screw;
—— *à dent-de-loup avec chaîne-galle*, (art.) chain and sprocket elevating screw;
—— *de dessus*, (sm. a.) breech pin, breech screw; *double*, —— *double concentrique*, *double* —— *de pointage*, (art.) double elevating screw;
—— *droite*, (art.) single elevating screw;
——*-écrou*, (art.) fuse plug;
—— *à écrou fixe*, (art.) fixed-nut elevating screw;
—— *à écrou oscillant*, (art.) elevating screw with oscillating nut;
——*-éjecteur*, (sm. a.) ejector, ejector screw;
—— *d'élévation*, (art.) elevating screw (rare);
—— *d'entraînement*, (art.) translating screw;
fermeture à ——, v. s. v. *fermeture;*
—— *de fermeture*, (art.) breech screw;
—— *à filets interrompus*, (art.) interrupted screw; slotted screw; stripped screw;
—— *fixe à écrou mobile*, (art.) fixed elevating screw with movable nut;
——*-fourreau*, (art.) outer screw (of the elevating screw);
——*-frein*, (art.) screw set of the Deport sight;
—— *de frein*, (art.) brake screw; screw clamp;
—— *de frein de hausse*, (art.) set screw of a sight;
—— *de gâchette*, (sm a.) sear screw;
——*-goupille*, (sm. a.) screw pin or bolt;
—— *du grand ressort*, (sm. a.) mainspring screw;
grande ——, (art.) outer screw (of a double elevating screw);
—— *guide*, (art., sm. a.) guide screw;
—— *jumelles*, (art.) twin (elevating) screws;
——*-lumière*, (art.) screw vent (Krupp *fermeture*);
—— *de magasin*, (sm. a.) casing screw;
—— *de magasin antérieure*, (sm. a.) front casing screw;
—— *de magasin postérieure*, (sm. a.) rear casing screw;
—— *de mécanisme*, (sm. a.) breech-mechanism screw;
—— *de noix*, (sm. a.) tumbler screw;
petite ——, (art.) inner screw (of a double elevating screw);
—— *à pignon d'angle à écrou tournant*, (art.) bevel gear elevating screw with rotary nut;
—— *à pivot*, (sm. a.) pivot screw;
—— *de plaque*, (sm. a.) butt-plate screw;
—— *de platine*, (sm. a.) lock-plate screw;

vis *de poignée*, (sm. a.) hind screw of the trigger plate;
—— *de pointage*, (art.) elevating screw;
—— *de pointage double*, (art.) double elevating screw;
—— *de pointage simple*, (art.) single elevating screw;
—— *de pontet*, (sm. a.) guard screw;
—— *porte-amorce*, (art.) screw fuse plug,
—— *de pression*, (art.) set screw of a sight;
—— *de rappel*, (art.) elevating (deflection) screw of a sight;
—— *du ressort de gâchette*, (sm. a.) sear-spring screw;
—— *à secteurs hélicoïdaux*, (art.) breech screw with helicoidal slots;
—— *de serrage*, (art.) locking screw, set screw, clamp screw;
—— *à soulèvement*, (art.) raising screw (of an *affût à soulèvement*);
tendeur à ——, v. s. v. *tendeur;*
—— *de translation*, (art.) translating screw;
—— *de triangle*, (art.) swivel-bar screw.

II. Miscellaneous:

à ——, screw, fitted with a screw;
—— *ailée*, wing screw; thumbscrew; wing nut;
—— *d'ajustage*, adjusting screw;
anneau à ——, *à anneau*, screw ring;
—— *à anse*, handle screw;
appareil à ——, screw gear;
—— *d'appui*, bearing or supporting screw;
arbre à ——, screw arbor;
—— *d'Archimède*, Archimedean screw;
—— *d'arrêt*, set screw; screw top;
—— *d'assemblage*, assembling screw, set screw;
—— *d'attache*, assembling screw;
——*-axe*, screw axis;
bague à ——, screw collar;
—— *à billes*, (mach.) ball-bearing screw (thread represented by ball bearings);
—— *en blanc*, screw blank;
—— *à bois*, wood screw;
—— *à bois en fer*, iron wood-screw;
—— *de*, —— *en*, *bois*, wooden screw;
bouchon à ——, screw cap, screw plug;
—— *butante*, (inst.) tangent screw;
—— *de butée*, stop screw, bearing screw;
—— *butoir*, stop screw;
—— *de calage*, (mach.) clamp screw;
—— *calante*, (inst.) leveling screw; (mach.) centering screw;
—— *de centrage*, (mach.) centering screw;
chaise à ——, revolving chair;
chapeau à ——, screw cap;
—— *à chapeau*, (mach.) screw for assembling pipes and collars;
cheville à ——, screw pin; v. s. v. *cheville;*
—— *concave*, female screw;
—— *conductrice*, (mach.) lead screw;
—— *de conduite*, (mach.) lead screw, guide screw;
—— *de correction*, (inst.) tangent screw;
—— *creuse*, hollow screw;
—— *à crochet*, screw hook;
—— *de décharge*, v. —— *soupape de décharge;*
—— *de dégagement*, releasing screw;
—— *à desserrer*, tripping screw (of a jack);
—— *à deux filets*, double-threaded screw;
—— *différentielle*, (mach.) differential screw;
double ——, —— *double*, double screw; double-threaded screw;
—— *à double(s) filet(s)*, double-threaded screw;
—— *à double pas*, double-threaded screw;
—— *à double pas de Prony*, Hunter's screw; differential screw;
—— *de dressage*, leveling screw;
—— *à droite*, right screw;
——*-écrou*, (elec.) binding post;
—— *à écrou*, screw bolt;
écrou d'une ——, female screw, screw box;
—— *d'enrayage*, set screw, binding screw;
escalier à ——, v. s. v. *escalier;*
—— *extérieure*, male screw;
fausse ——, screw die;
——*-femelle*, female screw; screw box, screw nut;
fermer à ——, to screw up; to screw down;

vis, *filet de* ——, screw thread;
—— *à filets carrés*, square-threaded screw;
—— *à filet(s) contraire(s)*, —— *à filets contrariés*, right and left hand screw; translating screw;
—— *à filet double*, double-threaded screw;
—— *à filet plat*, flat-threaded screw;
—— *à filet rapide*, (*mach.*) quick-motion screw;
—— *à filet rectangulaire*, square-threaded screw;
—— *à filet simple*, single-threaded screw;
—— *à filet triangulaire*, triangular-threaded screw; V-threaded screw;
—— *à n filets*, n-threaded screw;
—— *filetée à droite* (*gauche*), right- (left-) hand screw;
—— *de fin calage*, (*inst.*) fine-adjustment screw;
—— *sans fin*, endless screw, worm, perpetual screw; lead screw;
—— *de fixation*, clamping, fixing screw;
foret à ——, screw drill;
—— *-fraise*, v. —— *à tête fraisée*;
—— *de frottement*, friction screw;
—— *à gauche*, left screw;
gland à ——, screw cap;
—— *globique*, —— *globoïdale*, globoid screw;
—— *en goutte de suif*, round-headed screw;
—— *guide*, (*mach.*) leading screw, guide screw;
—— *hollandaise*, water-raising screw (resembles the Archimedean screw, except that the screw and the envelope are separate);
—— *hydraulique*, water screw;
—— *intérieure*, female screw;
—— *à jour*, v. s. v. *escalier*;
—— *mâle*, male, external screw;
mandrin à ——, v. s. v. *mandrin*;
manchon à ——, v. s. v. *manchon*;
—— *-mère*, (*mach.*) feed screw, leading screw, guide screw;
—— *à métal*, metal screw (to engage in metal);
—— *à micromètre*, —— *micrométrique*, micrometer screw;
—— *à noyau plein*, (*cons.*) winding stairs with solid newel;
—— *noyée*, countersunk screw;
—— *à oreille(s)*, winged screw, thumbscrew, male screw with an eye or loop;
pas de ——, v. s. v. *pas*;
—— *à pas à droite* (*gauche*), right- (left-) handed screw;
—— *à pas simple*, single-threaded screw;
—— *de percussion*, (*mach.*) fly press;
—— *perdue*, countersunk screw;
—— *du piston*, piston screw;
piton à ——, v. s. v. *piton*;
—— *à plusieurs filets*, multiple-threaded screw;
poignée à ——, screw handle;
—— *à pompe*, screw surrounded by a spiral spring;
presse à ——, (*mach.*) screw press, fly press;
—— *presse-étoupe*, screw collar of a stuffing box;
—— *de pression*, binding screw, thumbscrew, press screw, adjusting screw, compressor screw, clamp screw, fixing screw;
raccord de ——, screw coupling sleeve;
—— *de rappel*, adjusting screw, regulating screw, set screw; (*inst.*) tangent screw;
—— *de rectification*, (*inst.*) adjusting screw;
—— *de registre*, damper screw;
—— *de réglage*, (*inst.*) adjusting screw;
—— *régulatrice*, (*mach.*) lead screw;
—— *de ressort*, spring screw;
—— *de ridage*, (*cord.*) rigging screws;
—— *-robinet*, screw faucet;
—— *de serrage*, set screw, clamping screw;
—— *-soupape*, (*mach.*) screw valve;
—— *-soupape de décharge*, discharging screw-valve;
—— *à T*, winged screw;
—— *tangente*, (*inst.*) tangent screw;
tendeur à ——, v. s. v. *tendeur*;
—— *du tendeur*, screw of a screw coupling;
—— *de tension*, turn-buckle;
—— *à tête*, v. —— *en goutte de suif*;
tête de ——, screw head;

vis à tête carrée, square-head screw;
—— *à tête fraisée*, bevel-headed screw for countersinking;
—— *à tête molletée*, (k)nurled-head screw, milled-head screw;
—— *à tête noyée*, countersunk screw;
—— *à tête percée de trous*, capstan screw;
—— *à tête perdue*, countersunk screw;
—— *à tête ronde*, round-head screw;
—— *tirante*, (*inst.*) reticule adjusting screw;
—— *de tour*, (*mach.*) lead screw;
—— *à triple pas*, triple-threaded screw;
—— *à trois filets* (*pas*), triple-threaded screw;
trou d'une ——, screw hole;
—— *à volant*, a screw worked by a hand-wheel.
visa, m., visa, visé, signature;
—— *de conformité*, (*Fr. a.*) "true copy" (on copies);
—— *de contrôle*, (*Fr. a.*) sort of countersign on *feuilles de route*, *lettres de service*, etc.;
—— *pour timbre*, "good for a stamp" (on documents, etc.);
—— *de vérification*, (*Fr. a.*) certificate that certain documents (vouchers of sale, receipts, issues and receipts into magazines) have been inspected.
viscope, f., (*mil. slang*) kepi.
visée, f., aim, design; (*art.*, *sm. a.*) aiming, sighting; (*surv.*) sighting (of an instrument);
créneau de ——, v. s. v. *créneau*;
—— *directe*, (*surv.*) front sight;
—— *inverse*, (*surv.*) back sight;
point de ——, (*art.*, *sm. a.*) point aimed at.
viser, v. a. n. (*art. sm. a.*) to aim, take aim at; to lay (a gun) on; (*adm.*, etc.) to visé, to examine and sign; (*surv.*) to take a sight, an observation, on;
—— *le but*, (*art.*, *sm. a.*) to aim at the target.
viseur, m., (*inst.*, etc.) sighting tube; sighting piece; eyepiece;
tuyau ——, sighting tube.
visière, f., (*unif.*) visor (of a cap, etc.); (*sm. a.*) sight; breech sight; front sight; (*fort.*) embrasure, tunnel (cover or protection of masonry for a casemate embrasure; (small arch in front of a caponier); (*art.*) sight; sort of rearsight on a turret;
—— *de canon*, v. —— *de culasse*;
—— *à charnière*, (*sm. a.*) folding, leaf, sight;
—— *à clapet* (*sm. a.*) folding, leaf, sight;
—— *à coulisse*, (*sm. a.*) sliding sight;
—— *de culasse*, (*art.*) breech sight;
fente de ——, (*art.*) aiming slit (turrets);
—— *fixe*, (*sm. a.*) fixed rear sight;
—— *de fusil*, (*sm. a.*) rear sight;
—— *mobile*, (*sm. a.*) folding, leaf, sight;
—— *de protection*, (*fort.*) protective plate of a barbette turret.
vision, f., sight;
—— *directe*, naked eye;
visir, m., v. *vizir*.
visite, f., visit; call (of a doctor); inspection; custom-house inspection; survey; examination; (*nav.*) search; (*law*) search; (*art.*) inspection of gun and carriage after firing;
—— *annuelle*, (*nav.*) annual inspection, survey;
—— *annuelle des chaudières*, (*nav.*) yearly boiler inspection;
—— *annuelle des machines*, (*nav.*) yearly engine inspection;
—— *des armes, bouches, à feu*, (*art.*, *sm. a.*) examination and inspection of guns and small arms;
banc de ——, testing, verifying, bench;
—— *des casernes*, (*mil.*) inspection of barracks;
—— *de corps*, (*Fr. a.*) formal official visit by officers in a body;
—— *domiciliaire*, (*law*) domiciliary visit;
—— *de la douane*, customs inspection;
droit de ——, (*nav.*) right of search;
faire ——, to inspect;
—— *des lieux*, (*law*) search of the premises;
—— *des ponts*, inspection of bridges;
passer une ——, to make an inspection;
—— *des postes*, (*mil.*) inspection of posts, sentries; rounds;

visite *périodique*, (nav.) periodical survey;
— *de réception*, (art.) examination before reception;
— *sanitaire*, (med.) medical, sanitary, health, inspection;
soumis à la —, liable to search, inspection;
— *spéciale*, (nav.) special survey;
— *d'un vaisseau*, (nav.) examination of ship's cargo or papers.
visiter, v. a., to visit, call upon; to inspect, examine, survey, search, look.
visiteur, m., visitor; custom-house inspector, searcher, tide waiter; (nav.) searching vessel.
vissage, m., screwing, screwing up (down); (art.) screwing in of breech plug;
clef de —, wrench.
visser, v. a., to screw (on, down, up, in), fasten with a screw;
— *à fond*, to screw home, tight; to set up a screw.
visuel, a., visual; m., visible object; (t. p.) mark, target.
vite, a., quick, fleet, speedy, rapid.
vitesse, f., rate, velocity, speed, rapidity, fleetness, celerity;
à la — *de*, at the rate of ....;
— *absolue*, absolute velocity;
— *accélérée*, accelerated velocity; (art.) 9 kilometers per hour (artillery marching alone); (r. r.) 300 kilometers per 24 hours;
— *angulaire*, angular velocity;
— *d'arrivée*, (ball.) striking (arriving) velocity;
— *ascensionnelle*, (ball.) upward component of the velocity;
— *de cheminement*, (ball.) horizontal component of the velocity;
— *au choc*, (ball.) striking velocity;
— *de chute*, (ball.) velocity of descent;
— *de combustion*, (expl.) rate of combustion;
— *critique*, critical velocity;
demi- —, half speed;
épreuve de —, (nav.) speed trial;
— *finale*, (ball.) final, terminal, velocity;
gagner de —, to outstrip; to run away with (as a wagon going downhill);
grande —, full speed; (r. r.) quick dispatch;
grande — *en arrière (avant)*, (nav.) full speed astern (ahead);
à grande —, high speed;
— *d'inflammation*, (expl.) rapidity, rate, of inflammation;
— *initiale*, (ball.) initial velocity, muzzle velocity;
lancer à toute —, to set (a battery, a locomotive, ship, etc.) at full speed;
— *au loch*, (nav.) speed by the log;
— *de marche*, (mil.) rate of marching;
mi- —, half speed;
— *moyenne*, mean velocity;
— *normale*, normal velocity; (mach.) usual working speed; (art.) 8 kilometers per hour (artillery marching alone);
parabole de la —, (ball.) the parabola that would be described by a projectile from a given point if the resistance of the air were to cease at and from that point;
petite —, (r. r.) slow dispatch;
— *du piston*, (mach.) speed of the piston;
pleine —, (mach.) full speed;
— *rapide*, (art.) 10 kilometers per hour (artillery marching alone);
— *de régime*, normal speed; (mach.) speed best suited to each engine and peculiar to it;
— *relative*, relative velocity;
— *requise*, desired, required, velocity;
— *respective*, v. — *relative*;
— *restante*, (ball.) remaining velocity;
— *retardée*, retarded velocity;
— *de rotation*, velocity of rotation;
— *terminale*, (ball.) terminal velocity;
— *de tir*, (art.) rate of fire;
toute —, full speed;
à toute —, at full speed;
de toute sa —, at one's utmost speed;
— *de translation*, (mil.) velocity of translation;
— *uniforme*, uniform speed, velocity.
vitrage, m., glazing, glass windows, window glass, glass partition.

vitrail, m., stained-glass window (in pl. generally).
vitre, f., window glass, pane of glass, window;
carreau de —, *panneau de* —, window pane.
vitré, p. p., glazed; glass; vitreous;
porte —*e*, glass door.
vitrer, v. a., to glaze.
vitrerie, f., glazing; glazier's work, trade.
vitrescibilité, f., vitrifiability.
vitrescible, a., vitrifiable.
vitrier, m., glazier; (in pl., mil. slang) *chasseurs à pied*, rifles.
vitrification, f., vitrifaction.
vitrifiable, a., vitrifiable.
vitrifier, v. a., to vitrify.
vitriol, m., vitriol, copperas;
— *blanc*, white vitriol;
— *bleu*, blue vitriol;
— *de cuivre*, v. — *bleu*;
— *de fer*, v. — *vert*;
huile de —, oil of vitriol;
— *vert*, green vitriol;
— *de zinc*, v. — *blanc*.
vivacité, f., vivacity; quickness, eagerness;
— *de la poudre*, (expl.) quickness of powder;
— *de recul*, (art.) quickness, suddenness, of recoil.
vivandier, n., (mil.) sutler, canteen steward.
vivandière, f., (mil.) vivandière.
vivat, m., cheer, hurrah;
crier —, *pousser des* —*s*, *saluer par un* —, to cheer; to hurrah.
vive, v. *qui vive*.
vives-eaux, f. pl., spring tide (v. s. v. *marée*).
vivre, v. n., to live, subsist; to be maintained, subsisted;
— *sur l'habitant*, — *sur le pays*, (mil.) to live on the country.
vivre, m., food; (in pl. esp. mil.) provisions, stores, victuals; commissary department;
—*s d'administration*, (mil.) stores supplied by the administration as distinguished from — *de réquisition*;
administration des —*s*, (mil.) commissary department;
aller aux —*s*, (mil.) to go and draw rations;
bureau des —*s*, victualing office;
—*s de campagne*, v. *petits* —*s*;
—*s de conserve*, preserved (canned, tinned) rations;
—*s-conserves de viande*, iron ration;
—*s de débarquement*, (mil.) rations available on detraining;
être dans les —*s*, (mil.) to be in the commissary department;
faire des —*s*, to victual;
fourgon à —*s*, (mil.) provision wagon;
—*s-pain*, (mil.) bread part of the ration; bread ration;
petits —*s*, (mil.) groceries and dried vegetables;
—*s de réquisition*, (mil.) supplies, etc., got by requisition;
—*s de réserve*, (mil.) reserve supplies;
—*s du sac*, (mil.) two days' rations carried by the soldier;
—*s secs*, dry provisions (i. e., beans, coffee);
—*s-viande*, (mil.) meat part of ration; meat ration.
vivrier, m., (mil.) commissary of provisions.
vizir, m., vizier.
vogue, f., speed of a rowboat (obs.).
vogue-avant, m., bow oar.
voguer, v. n., (of a boat) to be rowed, pulled, (nav.) to sail; to be under sail; to bear, go;
vogue-avant! give way!
vogueur, m., oarsman (obs.); swimming belt.
voie, f., road, way, roadway, trail, track, trace, path, any line of transportation; conveyance, means of conveyance; load, cartload (as much as may be carried at one time, or in a single trip); track (distance between wheels); set of a saw, kerf of a saw; (r. r.) track, line, gauge.

voie

I. Railroads; II. Miscellaneous.
I. Railroads:
voie *accessoire*, side track;
— *assiette de la* ——, (*mil.*) method of laying a road (v. —— *fixe*, and —— *mobile*);
—— *de campagne*, (*mil.*) field railway;
changement de ——, v. s. v. *changement;*
—— *de chargement et de déchargement*, side track, siding;
chemin de fer à une, etc., ——, v. s. v. *chemin;*
—— *de chemin de fer*, railroad;
—— *de circonstance*, (*mil.*) temporary road;
—— *de classement*, v. —— *de triage;*
communication de ——*s*, junction of tracks;
—— *à crémaillère*, cog track;
croisement de ——, crossing;
—— *en cul de-sac*, side track (for storing spare cars);
—— *de débord*, v. —— *en cul-de-sac;*
—— *de départ*, starting track;
—— *de dérivation*, (*mil.*, etc.) branch road;
—— *descendante*, down track;
à deux ——*s*, double tracked;
—— *diagonale*, crossing;
double ——, double track;
—— *à écartement normal*, standard-gauge road;
—— *d'embarquement*, v. s. v. *chantier (r. r.);*
—— *étroite*, narrow-gauge track;
—— *d'évitement*, side track, siding;
—— *de fer*, —— *ferrée*, railway, track;
—— *ferrée portative*, (*mil.*) transportable railway;
—— *fixe*, (*mil.*) permanent railroad;
—— *de formation de train*, track for making up the train;
—— *de garage*, side track, siding;
—— *à grande (petite) section*, broad (narrow) gauge;
largeur de —— *étroite (normale)*, narrow (broad) gauge;
—— *de manœuvre*, shifting track, side track;
—— *mobile*, (*mil.*) mobile track (laid down before and taken up behind the train);
—— *montante*, up track;
—— *normale* standard-gauge track;
—— *permanente*, superstructure (of a railway, bridge, etc.);
—— *de plus grande (petite) largeur*, broad, narrow, gauge;
porteur de la ——, v. s. v. *porteur;*
pose de la ——, v. s. v. *pose;*
—— *principale*, main track;
—— *de raccordement*, junction line;
—— *réduite*, narrow gauge;
—— *de service*, side track;
à une seule ——, single-tracked;
—— *simple*, single track;
sortir de la ——, to leave the track, run off the rails;
—— *supplémentaire*, side track, siding;
traversée de ——, crossing;
—— *de triage*, shifting track;
tronçon de ——, section of the track;
—— *unique*, single track.
II. Miscellaneous:
—— *d'accès*, approach (in general);
—— *d'amortissement*, v. s. v. *amortissement;*
aveugler une —— *d'eau*, to stop a leak temporarily, in a temporary manner;
avoir (ne pas avoir) la ——, (of vehicles) to have (not to have) the regulation track;
avoir une —— *d'eau*, to leak (of boats, ships);
—— *de baguette*, (*sm. a.*) ramrod groove;
—— *de bois*, load of wood;
—— *en bois*, wooden track, tramway;
boucher une —— *d'eau*, to stop a leak;
—— *de charbon*, sack of coal;
—— *de charbon de terre*, a meter cube of coal;
—— *de charretière*, carriage road, wagon road; cart gauge;
—— *de charrette*, cart gauge;
—— *chaude*, hot trail (animals);
par —— *et chemin*, by any road whatever;
—— *circulaire*, (*art.*) racer, traversing circle;
—— *de coke*, a cubic meter and a half of coke;

voie *de communication*, thoroughfare, road; any road of communication, etc.; any road or path, of whatever nature;
par —— *de concours*, by competitive examination;
—— *à crémaillère*, (*art.*) cogged racer, traverse circle;
—— *en déblai*, road below level of country, sunken road;
—— *dérivée*, (*mil.*) connecting road (when main communication has been cut);
donner de la —— *à une scie*, to set a saw;
——*s de droit*, (*law*) resort to legal methods;
—— *d'eau*, load of water (as much as a man can carry, two bucketfuls, usually); waterway; (*nav.*) leak;
—— *économique*, v. *gestion directe;*
—— *d'écoulement*, water course; outlet;
en —— *de*, in a fair way for;
faire une —— *d'eau*, to spring a leak;
—— *de fait*, act of violence;
——*s de fait*, assault, assault and battery; duel; bad treatment;
—— *à flanc de coteau*, road cut in the side of a hill;
—— *fluviale*, waterway;
—— *fumante*, v. —— *chaude;*
la grande ——, the highroad;
—— *hiérarchique*, (*mil.*) regular channels, military channels;
—— *humide*, (*chem.*) wet process;
—— *latérale*, side road, crossroad;
—— *de manœuvre*, (*mil.*) road parallel to the front;
par —— *de mer*, by sea;
—— *mixte*, (*art.*) traverse circle, partly plane, partly toothed;
——*s et moyens*, (*adm.*) ways and means;
—— *navigable*, navigable stream;
—— *à niveau*, road on a level with the country;
—— *ordinaire*, (wagon) road;
par —— *de*, by means of;
par la —— *de*, via, by way of;
—— *de pierre*, cart load of stone;
—— *de plâtre*, twelve sacks of plaster;
—— *de pont*, (*pont.*) bridge way (distance between side rails);
—— *publique*, public road, highway;
—— *de Paris*, unit of volume, 1.92 cubic meters;
—— *de raccord*, —— *de raccordement*, v. —— *dérivée;*
—— *règlementaire*, (*mil.*) regular channels;
—— *de relevé*, cold trail (animals);
—— *en remblai*, road above level of country; raised road;
—— *de roue*, gauge; track of a wheel; (*art.*) traverse circle, racer;
—— *de roulement*, (*art.*) roller track;
—— *de sable*, wagonload of sand;
—— *d'une scie*, set of a saw; kerf of a saw;
—— *sèche*, (*chem.*) dry process (exposure to flame, etc.);
—— *simple*, (*art.*) plane traverse circle;
—— *de terre*, wagon road; land road, as opposed to railroad;
par —— *de terre*, by land (as distinguished from sea);
—— *en tranchée*, v. —— *en déblai;*
—— *vive*, v. —— *chaude;*
—— *de(s) voiture(s)*, track, width between wheels;

voile, f., (*nav.*) sail; (*fig.*) ship; sail;
à la ——, under sail;
aller à la ——, to sail;
——*s de l'avant*, head sails;
basses ——*s*, courses;
diminuer la ——, to shorten sail;
—— *d'eau*, drag sail, drift sail;
être fin de ——, to be a good sailer;
faire ——, to set sail;
faire de la ——, to set more sail;
faire force de ——*s*, to crowd all sail;
faire petite ——, to go under easy sail;
grande ——, mainsail;
——*s hautes*, upper sails;
mettre à la ——, v. *faire* ——;
navire à ——*s*, sailing vessel;
sous ——, under sail.

voile, m., veil; pretext; (*phot.*) obscurity, dimness.
voilé, p. p., clouded, covered; (*nav.*) rigged.
voilement, m., warping, twisting, getting out of true.
voiler, v. a. r., to veil; to conceal, hide, disguise; to warp, to become distorted (steel, wood); (*phot.*) to obscure, dim; (*nav.*) to bend on the sails of a ship;
—— *une caisse, un tambour,* to muffle a drum.
voilerie, f., (*nav.*) sail loft; sailmaking.
voilier, m., (*nav.*) sailmaker; sailing vessel;
bon (mauvais) ——, good (poor) sailor (also *mauvaise voilière*).
voilure, f., warping, distorting (of wood, of steel in tempering); (*nav.*) sails; sailmaking.
voir, v. a., to see; to examine, oversee, attend to, overlook, command, inspect; to frequent; to look out on; (*fort., etc.*) to dominate;
—— *le feu pour la première fois*, (*mil.*) to get one's baptism of fire.
voirie, f., commission of public ways; collective term for all roads and means of communication, whether land or water; sewer, common sewer; garbage, refuse, offal;
grande ——, commission of public roads, railroads, canals, navigable rivers, etc.;
petite ——, commission for local roads, unnavigable rivers, etc.;
—— *rurale,* commission of public ways in rural districts;
travaux de ——, road labor.
—— *urbaine,* commission of public ways in towns.
voisin, a., m., next to, neighboring, adjacent, bordering on, adjoining, contiguous to; neighbor.
voisinage, m., neighborhood.
voiturage, m., carriage, wagon transportation.
voiture, f., method of transportation; conveyance, transportation; carriage, vehicle; fare, load; (*art.*) carriage and limber; caisson, any artillery vehicle; (*r. r.*) car, coach;
—— *d'administration,* (*mil.*) generic term for forage, baggage, and battery wagons;
—— *d'agrès,* (*r. r.*) tool car;
—— *d'ambulance,* (*mil.*) hospital wagon; ambulance (U. S. A.);
—— *d'arrosage,* sprinkling cart;
—— *d'artillerie,* (*art.*) artillery wagon;
—— *automobile,* motor wagon;
—— *automobile de plaisance,* automobile;
—— *à bagages,* (*mil.*) baggage wagon;
—— *à bateaux,* (*pont.*) folding-boat wagon;
—— *à bobines,* (*mil. tel.*) reel wagon;
—— *en blanc,* (*art.*) carriage without the ironwork;
—— *à bras,* push cart, hand cart;
—— *-buffet,* (*r. r.*) buffet car;
—— *-bureau,* (*mil.*) office wagon (of the staff); (*Fr. a.*) telegraph-station wagon;
—— *-cage,* traveling pigeon cote for armies in the field;
—— *-canon,* (*art.*) gun and limber;
—— *des capucins,* "shank's mare";
—— *cellulaire,* prison van;
—— *de chevaux,* string of horses (for sale, etc.);
—— *de cantinière,* (*mil.*) canteen wagon;
—— *à cartouches,* (*mil.*) small-arm ammunition wagon;
—— *des cordeliers,* v. —— *des capucins;*
—— *à n chevaux,* an n-horse carriage;
—— *-citerne,* tank cart;
—— *de .... classe,* (*r. r.*) car of .... class;
—— *à un (deux, etc.) collier(s),* a one- (two-, etc.) horse carriage;
—— *colombier,* (*mil.*) pigeon cart or wagon;
—— *de commerce,* country carriage;
—— *de compagnie,* (*mil.*) company wagon (for tools, ammunition, sometimes knapsacks);
—— *à contre-appui,* (*art.*) carriage in which the trail (or perch) rests on the axle of the limber;
—— *dérouleuse,* (*mil. teleg.*) reel wagon;

voiture *pour dresser les chevaux,* (*hipp.*) horse break;
—— *d'effets,* (*mil.*) baggage wagon;
—— *d'équipement,* (*mil.*) equipment wagon;
—— *d'explosif,* (*mil.*) high-explosive wagon;
—— *fourgon,* (*mil.*) fuel and water, etc., wagon, balloon service;
—— *à fourrage,* (*mil.*) forage wagon;
—— *à frein,* (*r. r.*) carriage with brake;
—— *à hydrogène,* (*mil.*) hydrogen-gas cart (of a balloon park);
—— *-kilomètre,* (*r. r.*) car kilometer;
lettre de ——, waybill
—— *limonière,* —— *à limonière,* shaft carriage;
—— *à marchandises,* (*r. r.*) freight car (U. S.); goods wagon, goods van (England);
—— *médicale,* (*mil.*) hospital wagon;
—— *médicale régimentaire,* (*Fr. a.*) hospital wagon, ambulance;
—— *mixte,* (*r. r.*) mixed car (several classes);
—— *de munitions,* (*mil.*) ammunition wagon;
—— *d'outils,* (*mil.*) tool wagon;
—— *de parc,* (*mil.*) park wagon (v. *parc*);
—— *particulière,* private carriage;
—— *-pièce,* (*art.*) gun and limber;
—— *-pivot,* (*art.*) pivot carriage in changes of direction;
—— *à n places,* an n-seat carriage;
—— *de place,* cab;
—— *postale,* (*r. r.*) mail car;
—— *-poste (dite levée de boîtes),* (*Fr. a.*) mail wagon;
—— *-poste (télégraphique),* v. s. v. *poste;*
prix de la ——, fare;
—— *publique,* stage, diligence;
—— *régimentaire à bateaux,* v. —— *à bateaux;*
—— *de remise,* livery-stable carriage;
—— *de réquisition,* (*mil.*) wagon got by requisition, i. e., when needed;
—— *-réservoir,* tank cart, water cart;
—— *-restaurant,* (*r. r.*) dining car;
—— *à ridelles,* rack wagon;
—— *-salon,* (*r. r.*) parlor car;
—— *de section,* (*Fr. a.*) implement wagon (tools, surveying instruments, tiring gear, etc.);
—— *de service,* (*r. r.*) tender;
—— *suspendue,* any vehicle on springs;
—— *à suspension,* (*art.*) lunette and pintle carriage;
—— *à tournant complet,* —— *à tournant illimité,* cut under;
—— *à tournant limité,* a carriage of limited turn;
—— *à un train, deux trains,* 2-wheeled, 4-wheeled, carriage;
—— *-traîneau,* (*art.*) sort of sled for heavy weights;
—— *-treuil,* reel carriage for the rope of a captive balloon; balloon cart;
—— *à vapeur,* steam carriage;
—— *de vivres,* (*mil.*) ration wagon;
—— *à volonté,* private carriage;
—— *-volière,* v. —— *-cage;*
—— *à voyageurs,* (*r. r.*) passenger coach.
voiturer, v. a., to convey, transport, carry, cart;
—— *par eau,* to convey by water.
voiturier, m., carrier, wagoner, wagon driver, driver.
voiturin, m., driver and owner of a carriage for hire; the carriage itself;
—— *par eau,* shipper.
voix, f., voice; vote; song (sailors') in hoisting, heaving, etc.;
aller aux ——, to take a vote;
—— *active,* right to elect, the suffrage;
—— *active et passive,* right to elect and be elected;
commander à la ——, (*mil.*) to give commands by word of mouth;
—— *consultative,* right to advise, etc., but not to vote;
—— *délibérative,* vote;
donner sa ——, to sing out;
mettre aux ——, to put to vote;
passer la ——, to pass the word;
porte- ——, speaking trumpet;

voix, à portée de ——, within hailing distance, within hail;
de vive ——, aloud.

vol, m., theft, robbery, stealing, thieving; thing stolen; flight (of birds);
—— à bande franche, flight of pigeons composed of reporters alone;
—— à bande mixte, flight of pigeons composed of reporters and pisteurs, q. v.;
—— de corde, (artif.) line rocket;
—— de grand chemin, highway robbery;
à —— d'oiseau, as the crow flies.

volage, a., fickle, inconstant; (of a boat) top-heavy, crank, cranky; (of the compass) unsteady.

volant, a., flying; loose; detached; traveling; floating; transportable; movable (as sash of French window);
artillerie ——e, v. s. v. artillerie;
cabestan ——, v. s. v. cabestan;
camp ——, v. s. v. camp;
câble ——, v. s. v. câble;
fusée ——e, v. s. v. fusée;
ligne ——e, v. s. v. ligne;
pont ——, v. s. v. pont;
roue ——e, v. s. v. roue;
sape ——e, v. s. v. sape.

volant, m., (mach., etc.) fly wheel, hand-wheel (cons.) sash of a French window;
—— à câbles, cable fly wheel;
—— de changement de marche, reversing wheel;
—— à coulisse, fly wheel with (adjustable) sliding weight;
—— denté, geared fly wheel;
—— de direction, (art.) traversing handwheel;
—— d'embrayage, clutch wheel;
—— à gorge, grooved wheel or pulley;
—— à main, handwheel;
—— à manette, fly wheel, or handwheel with handles;
—— -manivelle, —— à manivelle, handwheel;
—— de mise en train, starting wheel;
—— de manœuvre, generic term for any working or controlling wheel;
—— modérateur, regulating wheel;
petit ——, handwheel;
—— de pointage, (art.) elevating handwheel;
—— de pointage en direction, (art.) traversing handwheel;
—— de pointage en hauteur, (art.) elevating handwheel;
—— -poulie, fly wheel and pulley in one;
—— de pression, air reservoir of a pump; name given to the water in a boiler (and in other cases), as compensating irregularities of combustion, of consumption of steam, etc.;
—— rabattant, fly wheel whose upper part turns toward the steam cylinder;
—— de rappel, adjusting handwheel (as, e. g., of a Maillard testing machine);
—— de réglage, v. —— modérateur;
—— -rêne, bridle wheel (steam steering gear);
tourner au ——, to work the fly wheel (in starting an engine).

volateur, m., sort of flying machine.

volcan, m., volcano.

volcaniser, v. a., v. vulcaniser.

volée, f., flight (of a bird); flight, flock, of birds; (in games of ball) volley; number of blows delivered consecutively with a pile driver; jib, overhang, of a crane; splinter bar; mason's hod; (art.) salvo; (also) single round; chase of a gun; flight of a projectile before touching; (mach.) throw of a piston; rise of a hammer; (cons.) flight of stairs;
à la ——, at random, headlong; flying in the air;
—— de bout de timon, neck yoke;
cheval de ——, v. s. v. cheval;
chevaux de ——, swing horses (of a 6-horse team); lead horses (of a 4-horse team);
coup à toute ——, (art.) random shot (full charge and maximum elevation);
—— (de coups de bâton), shower of blows;
—— de derrière, splinter bar;
—— de devant, v. —— de bout de timon;
entretoise de ——, (art.) breast transom;
—— d'escalier, (cons.) flight of stairs;

volée fixe (de derrière), splinter bar;
flexion de la ——, (art.) droop of the muzzle;
—— de grue, jib, arm, of a crane;
jouer de ——, to volley (a ball);
—— mobile (de derrière), doubletree;
poser une ligne télégraphique à la ——, to set up a telegraph line in haste, on the run;
prendre à la, de, ——, v. jouer de ——;
prendre entre bond et ——, to half volley;
raban de ——, (art.) muzzle lashing;
tarauder à la ——, to cut screws by hand;
tir, tirer, à toute ——, v. s. v. tir;
à toute ——, at full speed, under a full head of steam; (art.) with greatest elevation and service charge; at random, without being particularly aimed;
trempe, tremper, à la ——, v. s. vv. trempe, tremper.

voler, v. a. n., to fly (birds); to fly (go at great speed); to rob, steal, pilfer, plunder.

volerie, f., robbing, robbery, theft, pilfering.

volet, m., damper, shutter, widow shutter (strictly, inside shutter); boat compass; paddle of a water wheel; (art.) top board, cover, of interior divisions of ammunition box or carriage; carrier ring; (fort.) embrasure shutter; (phot.) shield of a plate holder; (sm. a.) gate (as of the Krag-Jörgensen);
—— brise, —— de brisure, folding shutter;
—— à charnière, hinged shutter; (phot.) back of a printing frame;
—— -console, v. s. v. console;
—— à coulisse, sliding shutter;
—— de frein, v. —— de patin;
—— de parement, shutter in one piece;
—— de patin, —— porte-patins, (art.) brake shoe rod (of the Lemoine brake);
—— de plateforme, (pont.) chess.

volette, f., horse net, fly net.

voleur, m., thief, robber; (surv.) kink in a surveyor's chain (familiar);
—— d'étiquettes, (mil. slang) quartermaster (said by the men to steal the labels (étiquettes) on their beds so as to charge for new ones).

volige, f., (cons.) batten; very thin plank; roofing strip (slate roof);
construire en ——, to batten;
latte ——, roofing strip (slate roof).

voligeage, m., (cons.) battens; battening (act of).

voliger, v. a., (cons.) to batten.

volontaire, a., voluntary, willing; obstinate, headstrong (said also of horses); spare (of horses); m., obstinate person; (mil.) volunteer, especially man who enlists voluntarily (without being drawn for service);
en ——, (mil.) as a volunteer;
(engagé) —— d'un an, (mil.) one-year volunteer;
s'engager, entrer, comme ——, (mil.) to volunteer.

volontariat, m., (mil.) act of volunteering, voluntary enlistment; period of service (as a one-year volunteer);
—— d'un an, one-year volunteer enlistment.

volonté, f., will, free will, wish, pleasure;
à ——, at will, at pleasure; (mil.) at will (of fire, etc.);
arme à ——, v. s. v. arme;
homme de bonne ——, (mil.) volunteer, man who volunteers for a daring or dangerous enterprise.

volt, m., (elec.) volt;
—— légal, legal volt.

voltage, m., (elec.) voltage.

voltaïque, a., (elec.) voltaic, galvanic;
pile ——, voltaic battery.

voltaïsme, m., (elec.) voltaism, galvanism.

voltamètre, m., (elec.) voltameter.

volt-ampère, m., (elec.) watt.

volt-coulomb, m., (elec.) joule.

volte, f., (fenc.) volt (turn on the left foot to avoid a thrust); (man.) volt, circular tread (circle described from the track toward the middle of the hall or riding ground); the space (round or square) on which volts are made; (gym.) exercises on the wooden horse;

volte, *demi-* ——, v. *demi-volte;*
demi- —— *renversée*, (*man.*) a *demi-volte* reversed;
—— *de deux pistes*, (*man.*) a volt in which the fore and the hind legs take different tracks;
élargir, embrasser, la ——, (*man.*) to widen, broaden, a volt;
—— *de piste*, (*man.*) a volt in which the hind quarters follow the shoulder;
—— *sur place*, v. s. v. *place;*
—— *renversée*, (*man.*) a volt with the head turned in (the fore feet describe a small, the hind a large, circle); volt in renvers;
serrer la ——, v. s. v. *serrer.*
volte-face, f., (*fenc., man., mil.*) wheel, face about;
—— *à droite* (*gauche*), (*fenc.*) right (left) volt (bayonet exercise, accompanied by a leap to the new rear);
faire ——, to wheel or turn about (e. g., to face a pursuing enemy).

volter, v. n., (*fenc.*) to make a volt.

voltige, f., (*man.*) vaulting, etc., on a horse; (*gym.*) vaulting;
selle de ——, v. s. v. *selle.*

voltigement, m., flutter, flicker; vaulting;
de ——, flying.

voltiger, v. a., to flutter, flicker, wave, fly about; (*gym.*) to vault, practice vaulting (on a horse, on a wooden horse).

voltigeur, m., (*gym.*) vaulter (on a horse); vaulting teacher; vaulting horse; (*mil.*) voltigeur (light-infantry man).

volt-mètre, m., (*elec.*) voltmeter;
—— *avertisseur*, potential indicator.

volubilité, f., ease of turning, of rolling (as of wheels)

volume, m., volume (book); volume, bulk, size;
—— *de vapeur*, v. s. v. *vapeur;*
—— *de vide*, v. s. v. *vide.*

volute, f., volute, scroll.

vomir, v. a., to vomit; to throw up;
—— *les étoupes*, (*nav.*) to start oakum from seans in a heavy sea.

vousseau, m., v. *voussoir.*

voussoir, m., (*cons.*) voussoir, archstone; (*fort.*) voussir of a cupola;
roue à ——, v. s. v. *roue.*

voussoirier, m., machine for shaping foot of spokes.

voussure, f., (*cons.*) rise and bend of an arch; curved top of a bay;
arrière- ——, dome-like arch over a door or window.

voûte, f., (*cons.*) vault, vaulting, arch; (*farr.*) hollow of a hoof, of a horseshoe; (*met.*) roof or crown of a reverberatory furnace;
—— *annulaire*, annular arch;
—— *à, en, anse de panier*, basket-handle arch;
—— *en arc de cercle*, segmental arch;
—— *en arc de cloître*, cloistered arch;
—— *d'arête(s)*, groined arch, groined vault; cross vault;
—— *d'arêtes gothique* (*romaine*), gothic (Roman) ribbed arch;
—— *en berceau*, cylindrical arch;
—— *biaise*, —— *en biais*, skew arch;
—— *à caissons*, caissoned arch;
clef de ——, keystone;
—— *composée*, arch or vault composed of two or three kinds;
—— *conique*, conical arch, squinch;
—— *de construction mixte*, arch of mixed masonry (e. g., brick and stone);
contre- ——, counter arch;
—— *croisée*, cross vault;
—— *en cul de four*, semidome, apsidal dome;
—— *cylindrique*, v., —— *en berceau;*
—— *de, en, décharge*, relieving, discharging, arch;
—— *en décharge à terre coulante*, relieving arch with the earth at its natural slope;
—— (*dite*) *à l'impériale*, cloistered arch with a flat top;
—— *en dôme*, dome;
à double ——, bivaulted;
—— *à double arête*, double cross-vault;

voûte *droite*, right arch (axis perpendicular to face);
—— *en ellipse*, —— *elliptique*, elliptical arch;
en ——, arch-like, vaulted;
—— *à l'épreuve*, (*fort.*) shellproof arch;
—— *extradossée en arc*, arch thicker at springing lines than at vertex (extrados not parallel to intrados);
—— *extradossée parallèlement*, arch of uniform thickness (extrados parallel to intrados);
—— *extradossée en chape*, arch whose estrados consists of two symmetrically disposed plane surfaces;
—— *extradossée de niveau*, arch whose extrados is a horizontal plane tangent to the vertex;
—— *du fer à cheval* (*farr.*) hollow of a horseshoe;
—— *forte*, arch to resist heavy loads, and no or but slight vibration (e. g., road arch);
—— *imparfaite*, diminished arch;
—— *en limaçon*, spiral vault;
—— *légère*, arch to resist very slight loads or only its own weight;
—— *moyenne*, arch to resist moderate loads (as in a building);
naissance de ——, springing line of an arch;
—— *oblique*, skew arch;
—— *en ogive*, ogical or pointed arch;
—— *en plate-bande*, straight arch;
—— *en plein cintre*, semicircular arch, full-centered arch;
—— *primitive*, the arch proper of an extradosed arch;
—— *de remplissage*, small filling arch (as in the valley between adjacent arches);
—— *renversée*, inverted arch;
—— *souterraine*, underground arch;
—— *sphérique*, dome;
—— *surbaissée*, surbased, elliptical, arch;
—— *surbaissée au 1:n*, arch whose rise is $1:n$ the span;
—— *surélevée*, v. —— *surhaussée;*
—— *surhaussée*, surmounted vault;
tête de ——, archhead, face of an arch;
—— *très forte*, arch to resist heavy loads and considerable vibration (e. g., of a railway bridge);
—— *tunnel*, (*fort.*) vault or arch constructed in front of a caponier to protect it from direct or plunging fire; embrasure tunnel;
—— *en vis St. Gilles*, inclined annular arch.

voûté, a., vaulted, arched; round-shouldered.

voûtelette, f., (*cons.*) small vault.

voûter, v. a., (*cons.*) to arch; to vault; (*farr.*) to bend a horseshoe.

voûtin, m., (*cons.*) small arch; counter arch.

voyage, m., voyage, trip, journey; run, tour, travels (account of a voyage); (*mil.*) travel (U. S. A.);
—— *d'aller*, outward trip; voyage out;
—— *d'aller et de retour*, entire trip; voyage, out and in;
aller en ——, to go on a journey, trip, voyage;
cheval de ——, good road horse; horse suitable for a long journey;
—— *de*, voyage to (such and such a place, country);
en ——, on a journey;
en cours de ——, outward bound;
—— *entier*, voyage out and in;
être en ——, to be journeying, traveling;
—— *d'épreuve*, v. *d'essai;*
—— *d'essai*, (*nav.*) trial trip;
—— *d'état-major*, —— *d'études*, (*mil.*) staff-college journey or voyage (for the study of military problems on the terrain);
faire ——, v. *être en* ——;
faire un ——, to take or make a journey, make a trip; to go on an errand;
grand, long, ——, long journey, trip;
—— *au, de, long cours*, long (sea) voyage;
petit ——, short journey;
—— *de retour*, return voyage, voyage home;
revenir de ——, to return from one's travels, journey, trip, etc.;
—— *rond*, v. —— *entier;*
—— *stratégique*, (*mil.*) staff journey (to study military problems on the terrain).

voyager, v. a., to travel, make a trip, journey.
voyageur, m., traveler, passenger;
commis ——, commercial traveler.
voyant, a., conspicuous (of colors);
cheval ——, shortsighted horse.
voyant, m., (of colors) conspicuousness, brightness; (*surv.*) vane, target, leveling rod;
mire du ——, slide vane of a leveling rod;
—— *d'un niveau à bulle d'air*, eyepiece of a spirit level;
plaque du ——, v. *mire du* ——;
—— *de l'uniforme*, (*mil.*) conspicuousness of the uniform, ease with which seen at a distance.
voyer, m., road trustee.
vrac, m., disorder, confusion;
charger en ——, to lade in bulk;
en ——, in bulk; (*art.*) without any regular order (of bullets in a case shot).
vrague, vraque, m., v. *vrac*.
vril, m., (*expl.*) vril.
vrille, f., gimlet, wimble, drill, borer, piercer;
—— *en gouge*, kind of auger.
vriller, v. a. n., to bore with a gimlet, etc.; (*artif.*) to whirl (said of a rocket).
vrillerie, f., generic term for files, punches, gimlets, and other small tools (carpenters', armorers', etc.).
vrillon, m., small auger with a gimlet point.
vu, m., sight, examination, inspection; preamble.
vue, f., sight, view, aim, survey; examination, inspection; aspect; (*cons.*) opening, window, looking upon a neighboring house; (*sm. a.*) sighting notch (of a sight);
à ——, at sight (of a draft); in sight (as, a prisoner);
avoir les —— *s fichantes*, (*fort.*) said when the glacis prolonged passes over the interior crest;
avoir les —— *s rasantes*, (*fort.*) said when the glacis prolonged passes through the interior crest;
—— *de bas en haut*, view upward;
—— *basse*, short-sightedness;
—— *de côté*, side view;
—— *par côté*, v. —— *oblique;*
—— *en coupe*, cross section;

vue *courte*, v. —— *basse;*
—— *dérobée*, (*cons.*) lunette;
—— *en dessus*, v. —— *d'oiseau;*
—— *droite*, (*cons.*) window or opening directly overlooking a neighbor's property;
en ——, in sight;
être en ——, to be in sight;
—— *de face*, front view;
—— *géométrale*, geometrical view or presentation;
—— *grasse*, (*hipp.*) v. s. v. *gras;*
—— *de haut en bas*, view downward;
hors de ——, out of sight;
lever à ——, v. s. v. *lever;*
—— *longue*, long-sightedness;
longue- ——, v. *longue-vue;*
lunette de longue ——, telescope;
—— *morte*, window not to be opened;
—— *oblique*, side window or opening (on a neighbor's property);
à —— *d'œil*, visibly, rapidly;
—— *d'oiseau*, bird's-eye view;
à —— *de pays*, by the lay of the ground; (*fig.*) at a cursory glance;
perdre de ——, to lose sight of;
—— *perspective*, perspective view;
à perte de ——, as far as one can see; out of sight;
plan à —— *d'oiseau*, bird's-eye view;
point de ——, point of sight, eye; point of distance; focus;
à portée de ——, within sight;
à la portée de ——, *de la* ——, in, within, sight;
—— *de servitude, de souffrance*, v. s. v. *jour;*
à la —— *simple*, by the naked eye;
à tant de jours de ——, at so many days' sight (of a draft);
—— *de terre*, horizontal, cellar, window;
—— *à vol d'oiseau*, bird's-eye view.
vulcanisation, f., vulcanization.
vulcaniser, v. a., to vulcanize.
vulgarisateur, m., popularizer.
vulgarisation, f., popularization;
ouvrage de ——, popular work or treatise.
vulgariser, v. a., to popularize.
vulnérable, a., vulnerable.
vulnérabilité, f., vulnerability; (*mil.*) liability to be hit.
vulnéraire, a., good for wounds;
eau ——, herb lotion for wounds.

W.

wagon, f., wagon, car; (*r. r.*) carriage, car; [Except where otherwise indicated, the following are railway cars:]
—— *ambulance*, (*mil.*) hospital wagon;
—— *à bagages*, baggage car (U. S.); luggage van (England);
—— *basculant* —— *à bascule*, tilting, dumping, wagon;
—— *à bestiaux*, cattle car;
—— *-boxe*, horse car;
—— *-buffet*, v. —— *-restaurant;*
—— *-cavalier*, horse car;
—— *-chambre*, car divided into rooms;
—— *-chevalet*, trestle car, i. e., a car with trestles or planks instead of seats;
—— *à chevaux*, horse car (transport of horses);
—— *-citerne*, water cart;
—— *-cuisine*, kitchen car;
—— *debout*, fourth-class car (no seats);
—— *découvert*, open car;
—— *dortoir*, sleeping car;
—— *-écurie*, horse car;
—— *d'ensablement*, sand (ballast) car;
—— *fermé*, closed car;
—— *à fourrages*, (*mil. r. r.*) forage car;
—— *-frein*, brake car;
—— *à glace*, —— *-glacière*, refrigerator car;
—— *-grue*, traveling crane (i. e., mounted on a railway truck);
—— *à houille*, coal car;
—— *-lit(s)*, sleeping car;
—— *-locomotive*, dummy;
—— *de marchandises*, freight car (U. S.); goods wagon (England);
—— *à matériel*, (*mil. r. r.*) baggage car;

wagon-*messageries*, express car;
—— *à obus* (*à essieux fixes*), (*s. c. art.*) shell truck (for one projectile of large caliber);
—— *plat*, flat car;
—— *plate-forme*, (*s. c. art.*) shell truck;
—— *à plate-forme découverte*, truck, flat car;
—— *-poste*, mail car;
—— *-réservoir*, reservoir car;
—— *-restaurant*, buffet-car; dining car;
—— *-salon*, parlor car;
—— *à selles*, (*mil. r. r.*) saddle wagon;
—— *-tabagie*, smoking car.
wagonet, m., trolley; miner's truck; (*r. r.*) hand car, small truck;
—— *à manivelles*, (*r. r.*) hand car;
—— *de manœuvre*, (*r. r.*) hand car, truck;
—— *porte-projectiles*, (*art.*) shell-hoist scoop;
—— *-porteur*, (*mil. r. r.*) truck of the Decauville railway.
wagonnage, m., (*r. r.*) transportation of earth, etc., in construction.
wagonnette, f., wagonette.
wagonnier, m., (*r. r.*) conductor.
warme, f., v. *varme*.
watt, m., (*elec.*) watt.
watt-heure, m., (*elec.*) watt hour.
watt-seconde, m., (*elec.*) watt second.
wattmètre, m., (*elec.*) watt meter.
wolfart, m., v. *wolfram*.
wolfe, m., whirlpool or race.
wolfram, m., (*met.*) tungsten.
wootz, m., (*met.*) wootz, Indian steel.
wurst, m., (*art.*) ammunition wagon of light construction (horse artillery, obs.); (*med.*) **ambulance wagon**.

X.

x, m., (*fam.*) a cadet of the Polytechnique;
— *aller à* ——, to go to the Polytechnique;
— *fort en* ——, strong in mathematics;
— *les* ——, mathematics;
— *piocher l'*——, to study mathematics seriously;
— *tête à* ——, good head for mathematics (cf. *thêta* X).

xanthine, f., (*expl.*) xanthine powder.
xyloïodine, f., (*expl.*) xyloïodine.
xyloïdine, f., (*expl.*) xyloidin, nitrostarch.
xylographie, f., xylography.
xylophone, m., (*mus.*) xylophone.

Y.

yac, m., (*nav.*) union jack; (also) yacht;
—— *anglais,* union jack.
yacht, m., yacht;
—— *à glace,* ice boat;
—— *à vapeur,* steam yacht.
yachteur, m., yachtsman.
yak, m., v. *yac.*
yatagan, m., (*sm. a.*) yataghan.
yataganerie, f., collection of arms.
yénite, f., yenite.
yeuse, f., live oak.

yeux, m., plural of *œil.*
yole, f., (*nav.*) yawl;
—— *de l'amiral,* admiral's galley;
—— *du commandant,* captain's galley or gig;
— *grande* ——, galley.
you, m. and interjection, hurrah! halloo!
—— *de guerre,* war cry.
youyou, m., (*nav.*) dingey, jolly-boat.
youyoutier, m., oarsman in a dingey.
ypéreau, m., v. *ypréau.*
ypréau, m., Dutch elm (Ypres elm).

Z.

zagaie, f., assegai.
zain, a., (*hipp.*) whole colored (i. e., no white hairs); pure black or bay;
— *cheval* ——, horse of one color.
zèbre, m., zebra.
zébré, a., (*hipp.*) zebra-marked.
zébrer, v. a., to mark with zebra-like stripes.
zébrure, f., (*hipp.*) zebra mark.
zèle, m., zeal.
zélé, a., zealous.
zénith, m., zenith, vertical point.
zénithal, a., zenithal.
zéphire, zéphyr, m., zephyr; (*mil. slang*) soldier belonging to a disciplinary company; undisciplined soldier.
zéphyrien, a., (*mil. slang*) relating to a *zéphyr,* q. v.
zériba, f., zareba.
zéro, m., zero, cipher;
—— *absolu,* absolute zero;
—— *des cartes marines,* level of lowest tides observed.
zérotage, m., (*inst.*) determination of the zero point.
zigzag, m., zigzag; lazy tongs; (*siege, in pl.,* generally) zigzags.
zigzaguer, v. a., to zigzag.
zinc, m., zinc;
— *blanc de* ——, white zinc, paint;
— *doublage en* ——, zinc sheathing;
— *fleurs de* ——, flowers of zinc;
—— *en plaques,* —— *à souder,* spelter.
zincage, m., v. *zingage.*
zincographie, f., zincography.
zincographier, v. a., to reproduce by zincography.
zingage, m., covering with zinc; galvanizing.
zinguer, v. a., to cover with zinc; to galvanize.
zinguerie, f., zinc works, zinc trade.
zingueur, m., zinc worker.
zinquer. v. a., v. *zinguer.*
zinquier, zinqueur, m., v. *zingueur.*
zodiacal, a., zodiacal.
zodiaque, m., zodiac.
zonaire, zonal, a., zoned.
zone, f., zone; zone of the earth's surface; region; (*cord.*) space between two ropes (on a drum or cylinder);
—— *d'action d'une trajectoire,* (*ball.*) dangerous zone;
—— *d'action d'une hausse,* (*ball.*) depth in which a given target will surely be hit by mean trajectory corresponding to the elevation considered;
—— *d'activité,* (*fort.*) sphere of action of a fortress;
—— *d'alimentation,* (*mil.*) zone of supply;
—— *des alizés,* trade-wind belt;
—— *des armées,* (*mil.*) in the *service de l'arrière,* q. v., territory comprising all the railways, communications, etc., under the direct control of the commander in chief (v. —— *de l'intérieur*);

zone *de l'arrière,* (*mil.*) the area or ground in rear of an army, with reference to lines of communication, of supply, depots, etc.;
—— *d'artillerie,* (*mil.*) zone efficaciously covered by artillery, but not by infantry, fire (begins at 3,000 m);
—— *d'attaque,* (*mil.*) zone of fire action;
—— *battue,* (*mil.*) zone or belt swept or reached by fire;
—— *des calmes,* doldrums;
—— *de concentration* (*mil.*) strategic zone of concentration;
—— *dangereuse,* (*ball.*) dangerous zone, dangerous space;
—— *défilée,* (*fort.*) defiladed zone, belt, area;
—— *de dispersion,* (*ball.*) zone or belt over which shots are scattered in any given case;
—— *d'échauffement,* (*art.*) heat area around shot holes in armor plate;
—— *des étapes,* v. s. v. *étape;*
—— *frontière,* —— *des frontières,* v. *servitudes de la frontière;*
—— *houillère,* coal belt;
—— *de l'intérieur,* (*Fr. a.*) territory comprising all railways, communications, etc., remaining under the direct orders of the minister of war (this, with the —— *des armées,* constitutes the theater of operations of the *service de l'arrière*);
—— *de manœuvre,* (*mil.*) space necessary for a column to bring all its elements from the point where the march formation is abandoned onto the battlefield; zone sheltered from fire (Maillard);
—— *de marche,* (*mil.*) march zone;
—— *militaire,* (*mil.*) area of terrain of a *place forte;* military reservation, U. S. A.;
—— "*de la mort,*" (*mil.*) the terrain 600 to 200 meters from hostile line;
—— *de mousqueterie,* (*mil.*) zone swept by infantry fire (begins at 1,800 meters in open ground);
—— *myriamétrique,* (*Fr. a.*) the country around a *place forte,* to a radius of 10,000 meters;
—— *d'opérations,* (*mil.*) zone of operations;
—— *de protection,* (*fort.*) area or zone protected by defilade, defiladed zone;
—— *rasée,* (*mil.*) fire-swept zone or area;
—— *de recrutement,* (*mil.*) recruiting district;
—— *de réquisition,* v. s. v. *réquisition;*
—— *de servitudes,* v. *servitudes militaires;*
—— *des tirs efficaces,* (*mil.*) zone or area of effective fire;
—— *torpillée,* (*torp.*) mine field;
—— *des vents alizés,* trade-wind belt;
—— *des vents variables,* zone or belt of variable winds.
zouave, (*Fr. a.*) zouave.

APPENDIX.

THE FRENCH REVOLUTIONARY CALENDAR.

This calendar, established by the First Republic, went into effect by retroaction on September 22, 1792, having been legally adopted on November 24, 1793. It was abolished December 31, 1805. The year consisted of twelve months of thirty days each; five, and in leap years six, supplementary days (*sans-culottides*) were added to bring the number of days up to 365 and to 366 respectively. Each set of four years was called a *franciade*. Each month was divided into periods (*décades*) of ten days each, the week having been abolished; the tenth day was a day of rest. The year beginning September 22, 1792, was the year I (*l'an* I) of the Republic.

The correspondence of the Revolutionary with the Gregorian calendar is as follows:

AUTUMN.

vendémiaire (vintage month), September 22 to October 21.
brumaire (fog month), October 22 to November 20.
frimaire (sleet month), November 21 to December 20.

WINTER.

nivôse (snow month), December 21 to January 19.
pluviôse (rain month), January 20 to February 18.
ventôse (wind month), February 19 to March 20.

spring.

germinal (sprout month), March 21 to April 19.
floréal (flower month), April 20 to May 19.
prairial (pasture month), May 20 to June 18.

SUMMER.

messidor (harvest month), June 19 to July 18.
thermidor (hot month), July 19 to August 17.
fructidor (fruit month), August 18 to September 16.

SANS-CULOTTIDES.

les vertus, September 17.
le génie, September 18.
le travail, September 19.
l'opinion, September 20.
les récompenses, September 21.

A FRENCH-ENGLISH MILITARY TECHNICAL DICTIONARY

SUPPLEMENT

By

CORNÉLIS De WITT WILLCOX
Colonel, United States Army
Professor of Modern Languages, United States
Military Academy

[Edition authorized by license from the author, July 17, 1917]

Copyright, 1917,
By
C. De W. Willcox.

TABLE OF CONTENTS.

	Page.
Preface	499
Military abbreviations	501
Miscellaneous abbreviations	503
Compass indications	503
Staff map abbreviations	504
French weights and measures	506
Relations between degrees, grades, and mils	507
Grammatical abbreviations in use in military telegrams	507
Text	509

PREFACE.

Of the words contained in the following pages, the vast majority were collected by reading official and standard texts. The work as it stands was prepared under extreme pressure. It is hoped that this fact will be borne in mind as shortcomings are discovered.

I am indebted to Gen. Paul Vignal, military attaché, French Embassy, for clearing up a few difficulties; to Maj. F. P. Lahm, Signal Corps, as well as to Marchand's Modern Parisian Slang, for some of the slang terms recorded; to Capt. E. L. Gruber, Field Artillery, and to Capt. M. E. de Jarny, French Embassy, for some artillery, and to Col. Wirt Robinson, Military Academy, for some electrical definitions.

My best thanks are due to Sous-Lieut. Maurice Tabuteau, and to Capt. Ernest Jones, Signal Corps, for valued assistance in aeronautical nomenclature.

C. De W. W.

West Point, N. Y., *October 15, 1917.*

MILITARY ABBREVIATIONS.

A............	attelée (said of a battery).
[A]...........	department telegram.
A. C..........	artillerie de corps; auto-canon.
A. C. 9e......	artillerie de corps du 9ème corps.
A. D..........	artillerie divisionnaire.
A. D. 53e.....	artillerie de la 53ème division.
A. G..........	avant-garde.
A. L..........	artillerie lourde.
A. L. A.......	artillerie lourde de l'armée.
A. L. C. A....	artillerie lourde de corps d'armée.
A. L. G. P....	artillerie lourde à grande puissance.
A. L. V. F....	artillerie lourde sur voie ferrée.
A. M..........	auto-mitrailleuse.
Amb...........	ambulance.
A. P..........	avant-poste.
Appt..........	approvisionnement.
Ar............	à retard.
AR............	arrière.
Arr. G........	arrière-garde.
Art...........	artillerie.
A. T..........	artillerie de tranchée.
AV............	avant.
Az............	Aufschlag-zünder (German; percussion fuse or primer).
Bde...........	brigade.
Bie...........	batterie.
Biv...........	bivouac.
B. O. A.......	boulangerie d'armée.
B. O. C.......	boulangerie de campagne.
B. R..........	bulletin de renseignements.
Branc. C......	groupe de brancardiers de corps.
Branc. D_1..	division de brancardiers 1re division.
Btn...........	bataillon.
Bz............	Brenn-zünder (German; time fuse or primer).
C. A..........	corps d'armée.
Cant..........	cantonnement.
Cap...........	capitaine.
C. AR.........	central arrière.
Cav...........	cavalerie.
C. AV.........	central avant.
C. C..........	corps de cavalerie.
Cdt...........	commandant.
Cel...........	colonel.
Ch............	chevaux.
Cie...........	compagnie.
Cit...........	citation.
C. J. M.......	code de justice militaire.
C. M..........	compagnie de mitrailleuses.
C. O. A.......	commis et ouvriers d'administration.
C. T. R.......	court tir rapide.
C. V. A. D....	convoi administratif d'armée.
C. V. A. X....	convoi auxiliaire.
C. V..........	convoi automobile.
D.............	Désaleux (type of projectile, sharp ogive, and truncated base).
D. A..........	direction de l'arrière.
D. C..........	division de cavalerie.
D. C. A.......	défense contre aéronefs.
D. C. F.......	directeur des chemins de fer.
D. E..........	direction des étapes.
D. E. S.......	direction des étapes et services (obsolete).
D. I..........	division d'infanterie; dépôt intermédiaire.
Div...........	division d'infanterie.
Dp. E.........	dépôt d'éclopés.
D. R..........	division de réserve.
[E]...........	instruction telegram.
E. M..........	état-major.
E. M. A.......	état-major de l'armée.
E. N. E.......	éléments non endivisionnés.
Épont.........	équipage de pont.
Esc...........	escadron.
Esca..........	escadrille d'avions.
F. A..........	fonte aciérée.
F. M..........	fusil mitrailleur.
F_n.........	fusant (types indicated by numerical subscript).
G. A..........	groupe d'armées.
G. A. C.......	groupe d'armées du centre.
G. A. N.......	groupe d'armées du nord.
Gal...........	général.
Gaux..........	généraux.
G. B. C.......	groupe de brancardiers de corps.
G. B. D.......	groupe de brancardiers divisionnaires.
G. C. T. A....	groupe de canevas de tir des armées.
Gen...........	génie.
Gend..........	gendarmerie.
Gî. E.........	gîte d'étapes.
G. M. P.......	groupe motopropulseur.
G. O. E.......	gare origine d'étapes.
G. P. A.......	grand parc d'armée.
G. PA.........	grand parc d'artillerie.
G. Q. G.......	grand quartier général.
G. R..........	gare régulatrice.
Gr............	groupe.
G. Rav........	gare de ravitaillement.
G. V. C.......	garde des voies et communication.
H.............	hommes.
H. O. E.......	hôpital d'évacuation.
Hz............	Haubitze-zünder (German; howitzer fuse).
I.............	instantanée.
I. A..........	instantanée allongée.
Inf...........	infanterie.
Int...........	intendant.
Lieut.........	lieutenant.
M.............	mitrailleuse.

(501)

Mcin..........médecin.
M. W..........minenwerfer (German).
N. T..........nontransformé.
[O]...........official telegram.
O. A. C.......obus en acier à amorçage de culot.
O. A. T.......obus en acier à amorçage de tête.
O. F..........offensif fusant.
O. F. T.......obus en fonte à amorçage de tête.
P.............de position (said of a battery).
P. A..........parc d'artillerie; pour ampliation.
P. A. L.......parc d'artillerie lourde.
P. C..........points de contact.
P. C..........poste de commandement.
P. C. A.......parc de corps d'armée.
P. C. A. D....poste de commandement d'artillerie divisionnaire.
P. C. C. A....poste de commandement de corps d'armée.
P. C. D. I....poste de commandement de division d'infanterie.
P. E..........poste d'écoute.
Pel...........peloton.
P. et C. V....parcs et convois.
P. Gén........parc du génie.
P. gén. A.....parc du génie d'armée.
P. H. R.......peloton hors range.
P. I..........point initial.
P. J..........poste de jonction.
P. O..........poste d'observation; par ordre.
P. Racc.......points de raccordement.
P. S..........poste de secours.
P. T. A.......poste téléphonique artillerie.
P. T. I.......poste téléphonique infanterie.
Q. G..........quartier général.
Q. G. A.......quartier général d'armée.
Ravt..........ravitaillement.
Rgt...........régiment.
R. Q..........ravaitaillement quotidien.
S. A..........service automobile; service auxiliaire.
Sct...........section.
S₁ C. V. A. D...section 1 du convoi administratif.
S. Dist.......section de distribution.
Sgt...........sergent.
S. Hos........section d'hospitalisation.
S. M..........station magasin.
Sm. a.........section de munitions d'artillerie.
Sm. i.........section de munitions d'infanterie.
S. Mitr.......section de mitrailleuses.
S. P..........section de parc de campagne.
S. P. G.......section de parc du génie.
S. R..........sans retard.
S. R. A.......service de renseignements de l'artillerie.
S. Ravt.......section de ravitaillement.
S. Res........section de réserve.
S. R. L.......section de repérage par les lueurs.
S. R. O. T....service des renseignements de l'observation du terrain.
S. R. S.......section de repérage par le son.
S. R. T.......section de repérage par la terre.
S. T. A.......service technique de l'aéronautique.
S. T. C. A....section topographique de corps d'armée.
S. T. C. A....service télégraphique de corps d'armée.
S. T. D. I....section topographique de division d'infanterie.
T.............à tracteur (said of a battery).
T. A..........télégraphie acoustique.
T. C..........train de combat.
T. C..........télégramme collationné.
T. E..........tête d'étapes.
T. E. M.......tête d'étapes de manœuvre.
T. M..........convoi de transport de matériel.
T. M..........télégraphie militaire.
T. O..........télégramme ordinaire.
T. P..........tir de place.
T. P. S.......télégraphie par le sol.
T. S. F.......telegraphie sans fil.
T. R..........train régimentaire.
T. U..........télégramme très urgent.
V. B..........Viven Bessières (name of a grenade) appears in Grenade V. B. or sometimes alone: les V. B., the V. B. grenades.
W.............vivres-viande (on earmark or button of cattle).

MISCELLANEOUS ABBREVIATIONS.

A. Aᵉ	Anglo-arabe.
A. C. T	automobile club de France.
A. G. A	association générale automobile.
A. O. E	avance à l'ouverture d'échappement.
B. H. P	British horse power.
C. F	capitaine de frégate.
chx	chevaux (horsepower).
C. V	capitaine de vaisseau.
É. U	États Unis.
F. A. I	Federation aéronatique internationale.
ffon, ffons	faisant fonction(s).
h	heure.
j	jour.
l. m. r	ligne de moindre résistance.
L. V	lieutenant de Vaisseau.
m	minute.
m^2	mètre(s) carré(s).
Mgr	monsignor.
N. M	nord magnétique.
N. S	Nord Sud (on maps).
P. o	pour ordre.
P. S	Pur Sang.
P. S. ar	Pur sang arabe.
R. F. A	retard à la fermeture de l'admission
R. F. E	retard à la fermeture de l'echappement.
R. O. A	retard à l'ouverture de l'admission.
s	seconde.
$\tfrac{1}{2}$s	demi-sang.
T	on documents, indicates that they are to be stamped (timbré).
T. C. F	Touring Club of France.
T. S. V. P	Tournez s'il vous plait.

COMPASS INDICATIONS.

N	nord.
N. N. E	nord nord est.
N. E	nord est.
E. N. E	est nord est.
E. S. E	est sud est.
E	est.
S. E	sud est.
S. S. E	sud sud est.
S	sud.
S. S. W	sud sud ouest.
S. W	sud ouest.
W. S. W	ouest sud ouest.
W	ouest.
W. N. W	ouest nord ouest.
N. W	nord ouest.
N. N. W	nord nord ouest.

STAFF MAP ABBREVIATIONS.

Abat	Abattoir	Slaughterhouse.
Abb^e	Abbaye	Abbey.
Ab^r	Abreuvoir	Watering trough.
Aig^{le}	Aiguille	Needle.
Anc^n	Ancien	Ancient.
Aq^{uc}	Aqueduc	Aqueduct.
Arb	Arbre	Tree.
Ard^{re}	Ardoisière	Slate quarry.
Ars^l	Arsenal	Arsenal.
Art^n	Artésien	Artesian.
As^e	Asile	Asylum.
A^t	Arrêt	Halt.
Aub^{ge}	Auberge	Inn.
Av^{ue}	Avenue	Avenue.
B	Bois	Wood.
Bac	Bac	Ferry.
Bat	Bateau	Boat.
Bat^{ie}	Batterie	Battery.
B^{che}	Bouche	Mouth.
B^{de}	Borde	Farm.
Bg	Bourg	Large village.
B^{ge}	Barrage	Dam.
B^{ide}	Bastide	Country house; redoubt.
B^{ie}	Bergerie	Sheepfold.
B^{in}	Bassin	Basin.
Bl	Blockhaus / Redoute	Blockhouse.
B^{ment}	Bâtiment	Building.
B^{ne}	Borne	Boundary stone.
B^{on}	Buisson / Buron	Thicket. / Cheese factory.
B^{que}	Baraque	Hut.
Br	Broussaille	Brushwood.
B^{re}	Barrière	Barrier, gate.
Bron	Buron	Cheese factory.
Briq^{ie}	Briqueterie	Brickyard.
C	Cap	Cape.
C^{ne}	Cabane	Cabin.
Cab^{et}	Cabaret	Small country inn.
C^{re}	Calvaire	Calvary.
C^{al}	Canal	Canal.
Carref^r	Carrefour	Crossroads.
Carr^{re}	Carrière	Quarry.
C^{ade}	Cascade	Waterfall.
Cas^{ne}	Caserne	Barracks.
Cath^{le}	Cathédrale	Cathedral.
C^{se}	Cense	Farm.
Ch^{et}	Châlet	Swiss cottage.
Ch^{lle}	Chapelle	Chapel.
Ch^{au}	Château	Castle; country house.
Ch^{ée}	Chaussée	Road.
Ch^{in}	Chemin	Way, road.
Ch^{née}	Cheminée	Chimney.
Cim^{re}	Cimetière	Cemetery.
Cit^{lle}	Citadelle	Citadel.
Cit^e	Citerne	Cistern (well).
Cl^{er}	Clocher	Steeple.
Con	Coron	Miner's house.
C^{tal}	Cortal	Shepherd's hut.
Couv^t	Couvent	Convent.
Crx	Croix	Cross.
Déb^{re}	Débarcadère	Station, platform.
Déf^e	Défilé	Defile.
Dig	Digue	Dam.
Dist^{ie}	Distillerie	Distillery.
D^{ne}	Douane	Customhouse.
Dunes	Dunes	Dunes.
E	Eau	Water.
Ec^{se}	Écluse	Lock (flood gate).
E^{le}	École	School.
Ec^{ie}	Écurie	Stable.
Eg^{se}	Église	Church.
Etab^{nt}	Établissement	Establishment.
Et^g	Étang	Pond.
Fab^e	Fabrique	Factory.
F^{bg}	Faubourg	Suburb.
F^{me}	Ferme	Farm.
Fl	Fleuve	Large river.
F^{rie}	Fonderie	Foundry.
Fond^{re}	Fondrière	Quagmire, slough.
F^{ne}	Fontaine	Fountain, spring.
F^t	Forêt	Forest.
F^{ge}	Forge	Smithy, forge.
F^t	Fort	Fort.
Fortif	Fortification	Fortification.
F^{sé}	Fossé	Ditch.
Fr	Four	Furnace.
Gal^{ie}	Galerie	Gallery.
Gl^{er}	Glacier	Glacier.
Gr^{ge}	Gorge	Pass, gorge.
Gr^d	Grand	Large.
G^{ge}	Grange	Barn.
Gr^{te}	Grotte	Grotto.
Gué	Gué	Ford.
Hab^t	Habert	Cowhouse.
H^e	Halte	Halt.
H^{au}	Hameau	Small village, hamlet.
H^{ie}	Haie	Hedge.
H^t	Haut	Height.
Herm^{ge}	Hermitage	Hermitage.
H^{al}	Hôpital	Hospital.
Hosp	Hospice	Asylum.
H^{el}	Hôtel	Hotel.
Hydr	Hydraulique	Hydraulic.
I	Île	Island.
Irrig	Irrigation	Irrigation.
J^{in}	Jardin	Garden.

(504)

Jée	Jetée	Pier, jetty, mole.
K	Ker	House.
L	Lac	Lake.
Lag	Lagune	Lagoon.
L^{de}	Lande	Heath, moor.
L^r	Lavoir	Wash house.
L^{te}	Lette	Swamp.
Loc^{re}	Locature	Farmhouse.
M	Mas	Farm.
Mag^{in}	Magasin	Shop.
M^{on}	Maison	House.
Manuf^{re}	Manufacture	Manufactory.
M$_t$	Marais	Swamp.
M^{ché}	Marché	Market.
Mare	Mare	Pond.
Mar^{ge}	Marécage	Marsh.
Métr^{ie}	Métairie	Small farm.
Mine	Mine	Mine.
Mt	Mont	Mount.
M^{gne}	Montagne	Mountain.
Mont^t	Monument	Monument.
M^{in}	Moulin	Mill.
Mul^r	Muletier (sentier)	Mule path.
Mur	Mur	Wall.
Nat^{le}	Nationale	National.
Navig	Navigable	Navigable.
Obs^{re}	Observatoire	Observatory.
Orat	Oratoire	Oratory.
Os^e	Oseraie	Osier-bed.
Ouvr	Ouvrages extérieurs	Outworks.
Pal^s	Palais	Palace.
P	Parc (à bestiaux)	Peak, cattle pen.
P	Pic	Peak.
Pap^{ie}	Papeterie	Paper manufactory.
P^{ge}	Passage	Passage.
P^{lle}	Passerelle	Footbridge.
P^{on}	Pavillon	Pavilion.
Pêch^{ie}	Pêcherie	Fishery.
Pép^{re}	Pépinière	Nursery garden.
P$^{t(e)}$	Petit, petite	Small.
Ph	Phare	Lighthouse.
Pil^r	Pilier	Pillar.
Pl^{on}	Plantation	Plantation.
Pl^{au}	Plateau	Table-land.
Plâtr^{re}	Plâtrière	Plaster quarry.
P^{nt}	Point	Point.
Polyg	Polygone	Proving ground, artillery range.
P^{ceau}	Ponceau	Culvert.
P	Pont	Bridge.
P^{ton}	Ponton	Pontoon.
P^t	Port, pont	Harbor; bridge.
P^{te}	Porte	Gate.
P^{te}	Poste	Post.
P^{au}	Poteau	Fingerpost, guidepost.
Poud^{ie}	Poudrerie	Powder works.
Poud^{re}	Poudrière	Powdermill; powder magazine.
Pr	Prairie	Meadow.
P^{son}	Prison	Prison.
P^{ts}	Puits	Well.
Q^r	Quartier	Quarter.
R^{au}	Radeau	Raft.
Raf^{ie}	Raffinerie	Refinery.
Rav	Ravin	Ravine.
Red	Redan	Redan.
Red^e	Redoute	Redoubt.
R^{ge}	Refuge	Shelter.
R^{voir}	Réservoir	Tank.
Retr^{nt}	Retranchement	Intrenchment.
Rig	Rigole	Gully.
R	Rivière	River.
R^{er}	Rocher	Rock.
R^{te}	Route Nationale	National road.
R	Rue	Street.
R^{ne}	Ruine	Ruins.
R^{au}	Ruisseau	Brook.
S^e	Sable	Sand.
Sabl^{re}	Sablière	Sand pit.
Sal	Saline	Salt pit.
Sanat	Sanatorium	Sanatorium.
Sap^{re}	Sapinière	Firwood.
Sc^{ie}	Scierie	Sawmill.
Sém	Sémaphore	Semaphore.
S^{al}	Signal	Signal.
Sémin	Séminaire	Seminary.
Sentier	Sentier	Footpath.
Som^t	Sommet	Summit.
S^{ce}	Source	Spring.
S	Sous	Under.
St^{on}	Station	Station.
St^{ue}	Statue	Statue.
Suc^{ie}	Sucrerie	Sugar mill.
Susp	Pont suspendu	Suspension bridge.
Syn^e	Synagogue	Synagogue.
Tan^{ie}	Tannerie	Tannery.
Télég^e	Télégraphe	Telegraph.
Temp	Temple	Temple.
Th^l	Thermal	Thermal.
Tomb	Tombe / Tombeau	Grave.
T^{nt}	Torrent	Torrent.
T^r	Tour	Tower.
Tram	Tramway	Tramway.
Tr^l	Treuil	Windlass.
T^{ie}	Tuilerie	Tilekiln.
T^{el}	Tunnel	Tunnel.
Us^e	Usine	Works.
Vac^{ie}	Vacherie	Cow house.
V^{ée}	Vallée	Valley.
V^{on}	Vallon	Vale.
Verger	Verger	Orchard.
V^{rie}	Verrerie	Glass works.
V^{duc}	Viaduc	Viaduct.
Vign^e	Vigne, vignoble	Vineyards.
V^x V^{lle}	Vieux, vieille	Old.
V^{ge}	Vierge	Virgin.
V^{age}	Village	Village.

FRENCH WEIGHTS AND MEASURES.

I. Length.

Mètre, 3.281 feet.
Décamètre (10 mètres), 32.81 feet.
Kilomètre (1,000 mètres), 1093.633 yards (roughly, ⅝ mile.
Myriamètre (10,000 mètres), 6.2138 miles.
Décimètre (10th of a mètre), 3.937 inches.
Centimètre (100th of a mètre), 0.3937 inch (roughly, 0″.4).
Millimètre (1,000th of a mètre) 0.03937 inch (roughly, 0″.04).

II. Area.

Are (100 square mètres), 0.0988 rood.
Hectare (10,000 square mètres), 2.471 acres.
Centiare (1 square mètre), 1.196 square yards.

III. Volume.

(A) Solid.

Stère (1 cubic mètre), 1.31 cubic yards *or* 35 cubic feet, 547 cubic inches.
Décastère (10 stères), 13.1 cubic yards.
Décistère (10th of a stère), 3 cubic feet, 918.7 cubic inches.
(Cubic meters are more convenient and are more frequently employed.)

(B) Liquid.

Centilitre (100th of a litre), 0.02 pints.
Décilitre (10th of a litre), 0.21 pints.
Litre (1 cubic décimètre), 1.0567 quarts.
Décalitre (10 litres), 2.6417 gallons.
Hectolitre (100 litres), 26.4177 gallons *or* 2.8377 bushels.
Kilolitre (1 cubic mètre), 264.177 bushels.

IV. Weight.

Gramme (weight of 1 cubic centimètre of water in its state of maximum density, at 39½ Fahr., or 4 degrees centigrade), 15.4325 grains troy.
Décagramme (10 grammes), 6.43 dwt.
Hectogramme (100 grammes), 3.527 oz. avoir., *or* 3.216 oz. troy.
Kilogramme (1,000 grammes), 2.2055 lb. avoir., *or* 2.6803 lb. troy.
Quintal métrique (100 kilogrammes), 220.548 lb.
Millier, tonneau de mer (1,000 kilogrammes), (in land transport), tonne, 19 cwt. 12 oz. 5 dwt.
Décigramme (10th of 1 gramme), 1.5432 grain.
Centigramme (100th of 1 gramme), 0.15432 grain.
Milligramme (1000th of 1 gramme), 0.015532 grain.
(Kilogramme is the only multiple of gramme in general use.)

V. Thermometer.

0° Centigrade, melting ice, 32° Fahrenheit.
100° Centigrade, boiling water, 212° Fahrenheit.
0° Réaumur, melting ice, 32° Fahrenheit.
80° Réaumur, boiling water, 212° Fahrenheit.

RELATIONS BETWEEN DEGREES, GRADES, AND MILS (1,600 TO THE QUADRANT).

1° = 1g 11ˋ 11ˋˋ = 17.78 mils (1,600 to quadrant).
1′ (sexagesimal) = 1.ˋ85 (centesimal) = 0.30 mil.
1g = 0° 54′ = 16 mils.
1ˋ (centesimal) = 32″ (sexagesimal) = 0.16 mil.
1 mil = 3′22″.5 = 6.ˋ25.

The German artillery employs:
Field artillery ...... mils system 1,600.
Heavy artillery: $\frac{1}{16}$ of a degree, or
    The circumference = 5,760.
    The quadrant = 1,440.

GRAMMATICAL ABBREVIATIONS IN USE IN MILITARY TELEGRAMS.

Imdt..........Immédiatement.	Rapdt.........Rapidement.
Infér..........Inférieur.	Ss..............Sous.
Mod..........Modèle.	Sup...........Supérieur.
Ns............Nous.	Ts.............Tous.
Pp............Perpendiculaire.	Vs............Vous.
Qq............Quelque.	

A.

abat, m., slaughtering (of cattle);
centre d' ——, (*mil.*) slaughterhouse; point at which cattle are slaughtered for issue.

abatage, m., (*Fr. art.*) operation of taking the firing position, 75 mm.; letting down;
châssis d'——, (*art.*) firing frame.

abatée, f., (*aero.*) downward movement.

abattre, v. t., to lower- to slaughter (cattle); (*aero.*) to haul down (a balloon);
—— *la pièce*, (*Fr. art.*) to set the 75 mm. in position for firing, by placing it on its brake shoes; to let down the piece.

abonné, m., subscriber; (*mil.*) any person operating or using a *ligne d'abonnement*, q. v.

abonnement, m., subscription;
ligne d' ——, v. s. v. *ligne.*

abri, m., (*mil., etc.*) shelter, cover, dugout;
—— *de combat*, (*art.*) combat shelter (used by the detachment in the *séparation de risques*, q. v.); (*mil.*) splinter-proof shelter;
—— *de commandement*, (*mil.*) commanding officer's shelter;
—— *cuirassé*, (*mil.*) armored shelter (trench warfare);
—— *à l'épreuve*, (*mil.*) proof shelter (proof against medium calibers, and isolated shots from heavy calibers);
—— *de guetteur*, (*mil.*) lookout shelter (of steel, trench warfare);
—— *léger* (*mil.*) splinter-proof shelter;
—— *à munitions*, (*trench art.*) ammunition shelter;
—— *renforcé*, reinforced shelter (proof against small and medium calibers);
—— *de repos*, (*art.*) rest shelter (100-150 m. behind the gun, heavy artillery);
—— *téléphonique*, telephone booth or shelter;
—— *de traitement*, (*mil.*) medical shelter.

abri-caverne, m., (*mil.*) (under modern conditions a shell-proof shelter at least 6 meters under virgin earth) dugout.

abri-filtre, m., (*mil.*) (shelter fitted with filters against gas attack) gas shelter.

acare, m., tick.

accalmie, f., calm, lull;
—— *du tir*, (*mil.*) lull in the firing.

accélérateur, m., (*mach.*) accelerator.

accéléromètre, m., (*aero.*) (an instrument forming part of the Doutre stabilizer, and using speed variation to work the rudders, and keep the air plane horizontal) accelerometer.

accolade, f., embrace; (*Fr. a.*) embrace (bestowed on a man by his general when pinning on the Legion of Honor);
tailler en ——, (*unif.*) of a collar patch, to cut brace wise.

accorder, v. t., (*elec.*) to tune (wireless).

accoster, v. t., to accost;
—— *à*, (*nav.*) to come alongside of.

accoudoir, m., arm (as of a carriage seat).

accoupler, v. t., (*elec.*) to connect up, join.

accroc, m., hitch;
fonctionner sans ——, to work without a hitch, smoothly.

accrochage, m., rim-flange clinch (of an automobile wheel); (*sm. a.*) catch (e. g., of a *fusil-mitrailleur*).

accrocher, v. r., (*mil.*) to secure a footing, a hold, on the ground (as in an attack).

accumulateur, m., (*elec.*) accumulator;
—— *d'allumage*, (*elec.*) ignition storage battery;
—— *sec*, (*elec.*) dry cell storage battery.

accus, m., (*elec.*) abbreviation of *accumulateurs.*

accusation, f., (*mil. law*) charge;
chef d' ——, specification;
mettre en ——, (*law*) to arraign;
mise en ——, (*law*) arraignment.

acétylène, m., (*chem.*) acetylene.

acharné, a., stubborn, desperate;
combat ——, (*mil.*) stubborn fight

achat, m., purchase;
—— *de gré à gré*, (*adm.*) purchase by contract;
—— *sur le marché*, (*adm.*) purchase in open market.

acide, m., acid;
—— *au soufre* (*elec.*) a mixture of pure sulphuric acid and water (density 26, Baumé).

acier, m., (*met.*) steel;
—— *double-chromé*, double chrome steel;
—— *forgé*, forged steel.
—— *mi-dur*, (*met.*) semi-hard steel.

acrobaties, f. pl., (*aero.*) stunts.

acte, m., legal paper, document;
—— *d'appel* (*law*) writ of appeal;
—— *d'autorité*, (*adm.*) executive act or course (whether by original or delegated authority);
—— *de gestion*, (*adm.*) administrative step or course;
—— *de mariage*, wedding certificate;
—— *de naissance*, certificate of birth;
prendre ——*de*, to take legal notice of.

action, f., (*law*) lawsuit;
—— *à cheval*, (*cav*) mounted action;
—— *à distance*, distant effect (as of an explosion);
—— *à pied*, (*cav.*) dismounted action;
—— *publique*, (*law*) generic term for all procedure, including prosecution, leading to the detection and punishment of crime.

actionner, v. t., to start, set in motion.

activité, f., (*mil.*) active list;
rappeler à l' ——, to bring from the retired to the active list.

adipeux, m., (*mil. slang*) veterinary surgeon.

adjoint, m., assistant;
—— *tactique*, (*Fr. a.*) assistant to the chief of aeronautics (in matters of purely military character).

adjudant, m., (*Fr. a.*) adjudant, warrant officer;
—— *d'administration*, (*Fr. a.*) noncommissioned officer of the administrative service;
—— *chef*, senior warrant officer.

adjudication, f., award (of a contract);
mettre en ——, to invite bids;
mise en ——, call for bids.

adjupète, m., (*mil. slang*) an *adjudant.*

administration, f., administration;
—— *active* (*France*) functionaries or other agents (e. g., prefects, mayors, etc.);
—— *contentieuse*, (*France*) the judges and tribunals forming part of the administration;
—— *délibérante*, (*France*) the various councils passing on administrative matters (*conseil d'état, de préfecture*, etc.).

administrateur-directeur, m., manager.

admission, f., (*steam, mach.*) intake, admission, inlet, admission period, suction stroke.

aérien, m., (*wireless*) aerial.

aéro, m., (*aero*) familiar abbreviation of *aéroplane.*

aérodrome, m., (*aero*) aerodrome (open space for exercises).

aérodynamique, f., (*aero.*) aerodynamics.

aérologie, f., (*aero.*) aerology (meteorology of the upper atmosphere).

aéromarin, m., (*aero.*) seaplane.
aéromoteur, m., (*aero.*) airplane motor.
aéronat m., (*aero.*), balloon; (more particularly) dirigible balloon.
aéronaute, m., pilot of a free balloon.
aéronef, f., (*aero.*) any aerial vehicle (airplane or balloon), aircraft;
— *s à stabilité dynamique*, generic term for planes and kites;
— *s à stabilité mi-statique mi-dynamique*, dirigibles;
— *s à stabilité statique*, spherical balloons.
aéro-pare, m., (*aero.*) aeronautic station (supplies, shelter, repairs, inflation).
aéroplane, m., (*aero.*) airplane;
— *biplace*, single seater machine;
— *de bombardement*, bombing plane, machine;
— *de chasse*, fighting plane, fighter;
— *de corps d'armée*, reconnaissance plane;
— *à hélice arrière*, pusher;
— *à hélice avant*, tractor;
— *monoplace*, single seater machine;
— *pousseur*, pusher (rare);
— *de reconnaissance*, reconnaissance machine;
— *tracteur*, tractor;
— *triplace*, three seater machine.
aérostable, m., (*aero.*) a stablizer in which the car, acting as a pendulum, controls the tail of the apparatus.
aérostat, m., (*aero.*) any balloon (dirigible or spherical).
aérostation, f., (*aero.*) (limited to-day to) ballooning.
aero-yacht, m., (*aero.*) flying boat.
affaissement, m., collapsing or sinking (as of a pneumatic tire at point of contact).
affaisser, v. r., to collapse, to sink.
affichage, m., posting (as of a public notice).
affiche, f., poster, public notice.
afficher, v. t., to post.
affût, m., (*art.*) gun carriage;
— *agenouillé*, (*mach. gun*) tripod down;
— *barbette*, barbette carriage;
— *à berceau*, cradle carriage;
— *à berceau traîneau*, cradle slide carriage;
— *de crosse*, (*trench art.*) rear carriage (150 T mortar);
— *dressé*, (*mach. gun*) tripod up;
— *à échantignoles*, bracket carriage;
— *à glissement*, any sliding gun carriage;
grand —, (*art.*) châssis;
petit —, top carriage; (*mach. gun*) trail carriage (may move along trail of a trench mount);
— *de rempart*, (*mach. gun*) trench mount;
— *rigide*, rigid carriage (gun and carriage recoil together);
— *de secours*, spare carriage;
— *de tête*, (*trench art.*) front carriage (150 T mortar).
affût-trépied, m., (*mach. gun*) tripod mount;
— *type omnibus*, universal mount (takes any machine gun).
affût-truc, m., (*art.*) railroad truck gun carriage;
— *à glissement*, (*art.*) slide recoil carriage truck.
agent, m., employee, policeman;
— *de change*, (*com.*) stock-broker (is an *officier public* in France);
— *consulaire*, generic term for consuls; more narrowly, lower grade of consuls, as distinguished from *consul* proper;
— *cycliste*, bicycle police;
— *de liaison*, v. s. v. *liaison*;
— *municipal*, policeman;
— *de police*, policeman;
— *voyer*, (*Fr. adm.*) person charged with the construction, repair, and maintenance of communal roads (*chemins vicinaux*), communal roadmaster.
agitateur, m., stirring rod.
agrafe, f., fastener, belt fastener or clip;
— *d'assemblage*, assembling clip, clasp.

aide-chargeur, m., (*mil.*) assistant loader.
aide-grenadier, m., (*mil.*) man who protects a *lanceur*, q. v., in his duty; grenadier protector.
aide-monteur, m., (*mil.*) assistant lineman (telephone, etc.).
aide-pointeur, m., (*art.*) assistant gunner.
aiguille, f., needle; pointer; index; hand of a watch, clock;
— *foudroyante*, needle or hand of a split-second watch.
aiguiller, v. t., (r. r.) to set a switch.
aile, f., wing; (*auto.*) mudguard; (*aero.*) wing; (*aero.*) propeller blade;
bout d' —, (*aero.*) wing tip;
bras d' —, (*aero.*) wing spar.
aileron, m., (*aero.*) wing tip, small movable wing, fin; balancing flap, rudder.
ailette, f., (*mach.*) flange, cooling flange, (*aero.*) fin;
— *de réchauffage*, (*mach.*) heating flange, fin;
— *de refroidissement*, (*mach.*) cooling fin.
air-essence, m., (*auto.*) mixture of air and gasoline.
aire, f., area;
— *couverte par une hélice*, (*aero.*) developed area of a propeller;
— *du cercle balayé par l'hélice*, (*aero.*) disk area of a propeller.
ajouré, a., open (i. e., open worked).
ajusteur, m., (*mach.*, etc.) fitter (person).
alaire, a., (*aero.*) wing.
alboche, a., (*mil. slang*) having German affinities.
alcool, m., alcohol;
— *absolu*, (*chem.*) pure alcohol, absolute alcohol (no water);
— *carburé*, carburetted alcohol;
— *dénaturé*, denatured alcohol;
— *industriel*, commercial alcohol, for use in the arts.
alerte, f., alarm;
— *au gaz*, (*mil.*) gas alarm, notice of a gas attack.
alerter, v. t., (*mil.*) to warn, alarm, give the alarm.
alésage, m., (*mach.*) bore (i. e., interior diameter, of a cylinder); piston diameter;
— *du moteur*, bore, interior diameter, of a motor cylinder.
alésoir-fraise, m., (*mach. gun*) gas chamber cleaner.
alidade, f., (*inst.*) alidade;
— *à coulisse*, sliding alidade (for steep slopes).
aligner, v. t., to line up;
s' —, (*sport*) to come to the scratch.
alimentaire, m., (*mach. gun*) feed piece.
alimentation, f., (*mach.*) feed;
— *par la charge*, gravity feed;
double —, (*mach. gun*) double feed;
mécanisme d' —, (*mach. gun*) feed mechanism;
— *sous pression*, pressure feed.
alimenter, v. t., (*mach. gun*) to feed (ammunition).
alinéa, m., line (in a book or printed document).
aliter, v. r., to take to one's bed.
aliant, a., (*fam.*) with plenty of "go," spirited.
allant, m., "go," spirit (horses and men);
officier qui a de l' —, an officer with plenty of go.
allée, f., alley;
— *cavalière*, bridle path.
alléger, v. t., to lighten.
allongement, m., (*aero.*) in a dirigible, the ratio of the diameter of the principal section to the main longitudinal axis; aspect ratio;
— *du tir*, (*art.*) extension of the range.
allonger, v. t., to lengthen;
— *le tir*, (*art.*) to extend, increase, the range.
allumage, m., (*mil.*) priming (as of a hand grenade); (*mach.*) ignition;
[The following terms relate to motors:]
avance à l' —, sparking advance;
— *à basse tension*, low tension ignition;
— *à brûleurs*, burner ignition;
couper l' —, to cut off ignition, to switch off;
— *électrique*, electric ignition;

allumage *à haute tension,* high-tension ignition;
— *à haute tension par bobine,* jump-spark ignition;
— *à haute tension par magnéto,* high-tension magneto ignition;
— *à,* — *par, incandescence,* incandescent ignition;
— *jumelé,* double ignition;
— *normal,* ignition without advance or delay;
— *prématuré,* premature firing; preignition;
raté d'—, misfire;
— *par rupture,* make-and-break ignition;
— *à temps de pose et par arrachement,* spark produced at a set time.

allumette, f., match;
— *suédoise,* safety match.

allumeur, m., *(mil., etc.)* primer of a grenade; *(elec.)* commutator, contact breaker.
— *à traction, (mil. min.)* friction igniter.

alternateur, m., *(elec.)* alternator;
— *s en cascade,* alternators in series (secondary of each forms primary of next);
— *à haute fréquence,* high frequency alternator;

altimètre, m., *(aero.)* altimeter, height indicator.

alvéole, *(mach.)* ball-bearing cup.

amane, f., *(cord.)* cable;
— *de milieu,* breast line.

ambulance, f., *(mil. med.)* field hospital;
— *automobile,* motor ambulance.

ambulancier-brancardier, m., *(mil.)* litter bearer.

âme, f., *(art., sm. a.)* bore;
fausse —, *(art.)* (the) slot (of the Canet *culasse à bloc*).

aménagement, m., any preparation of ground, place, quarters, terrain for a given purpose; (in pl. *esp. nav.*) arrangements, accommodations; *(mil.)* preparation, reconstruction of a captured trench;
— *électrique,* electrical fittings, plant.

aménager, v. t., to prepare, get ready (as a railway station for an increase of freight, etc.); *(mil.)* to reconstitute, prepare, a captured trench.

amerrir, v. n., *(aero.)* to alight (on the water).

amerrissage, m., *(aero.)* alighting (on water).

amianté, a., amiantine.

amiralat, m., *(nav.)* grade, dignity, functions, etc., of an admiral.

amiralissime, m., *(nav.)* admiral in chief.

ammonal, m., *(expl.)* ammonal.

amonceler, v. t., to pile up (as, *mil.* sand bags).

amoncellement, m., pile, piling up (as, *mil.,* sand bags).

amorçage, m., *(art., sm. a., min.)* priming;
— *en cascade, (mil. min.)* series priming.

amorce, f., *(art., mil. min., etc.)* primer, priming, pellet;
— *au fulminate,* any fulminate fuse;
— *au pulvérin,* powder fuse.

amorcer, v. t., *(art., sm. a., min.)* to prime; to fuse (a shell);
— *en cascade, (mil. min.)* to prime in series.

amortir, v. t., *(adm.)* to redeem; *(mil.)* to absorb a vacancy; v. t., *(elec.)* to damp.

amortissement, m., *(adm.)* amortization; *(mil.)* absorption of a vacancy; *(elec.)* damping.

amortisseur, m., *(mach., aero., and gen.)* damper, shock absorber; tug, trace absorber;
— *à air,* air-buffer, air-cushion, pneumatic spring;
— *de chute, (art.)* shock absorber (gun carriage);
— *de crosse, (mil.)* shock absorber (D. R. firing stand).
— *de direction,* steering gear ball joint;
— *à friction,* friction disk shock absorber;
— *hydraulique,* hydraulic shock absorber;
— *à liquide,* v. — *hydraulique;*
— *pneumatique,* pneumatic shock absorber;
— *progressif,* progressive shock absorber;

amortisseur *à ressorts,* spring absorber, spring buffer;
— *de transmission,* transmission shock absorber.

amovible, a., movable.

ampérage, m., *(elec.)* amperage.

ampèremètre, m., *(elec.)* ammeter;
— *thermique,* hot-wire ammeter.

ampoule, f., bulb, small bulb.

amputé, m., — *d'un bras, d'une jambe,* having lost an arm, a leg, by amputation.

ancien, a., old; *(fam.)* conservative, old fogy.

ancien, m., *(mil.)* old soldier, veteran (not necessarily in the service); *(fam.)* old fogy, reactionary; (*in. pl.*) upper class at St. Cyr and the Polytechnique;
les —*s de 1870,* the veterans of 1870.

ancienneté, f., *(mil.)* seniority, length of service;
— *majorée, (mil.)* length of service increased by a *majoration,* q. v.;
— *pure, (mil.)* length of service;
à titre d'— *(mil.)* by reason of length of service.

ancrage, m., anchoring;
— *en chapelet, (pont)* series anchoring, (several anchors in one file);
— *en patte d'oie, (pont)* anchoring from both sides), double, triple, anchoring;
— *renforcé, (pont)* reinforced anchoring (two or three anchors for one boat).

ancre, f., anchor;
— *champignon (min.)* mushroom anchor;
— *à jas, (nav.)* stock anchor;
repêcher une —, to pick up, take up, an anchor;
— *de sûreté, (pont)* emergency anchor.

ancrer, v. t., to anchor;
— *en chapelet, (pont)* to anchor in file (several anchors in file);
— *en patte d'oie, (pont)* to anchor on both sides.

anémomètre, m., *(aero.)* device forming part of the Doutre stabilizer, and working the rudders.

angle, m., angle;
— *abordable, (aero.)* angle within which motion is possible;
— *d'attaque, (aero.)* (angle between the surface, e. g., of a wing, and the direction of the air stream) angle of incidence;
— *critique, (aero.)* critical angle, i. e., stalling angle, (practical flyers); angle at which maximum lift is obtained (report No. 9, National Advisory Committee); [technically, angle of attack at which lift coefficient is a maximum; flying definition, angle of attack beyond which it is dangerous to fly (the machine stalls). In translating French texts it is safe to use the practical, flying definition];
— *de dérivation, (aero.)* yaw;
— *de hausse, (art., sm. a.)* angle of elevation;
— *horaire,* hour angle;
— *d'incidence, (aero.)* angle of attack, of incidence;
— *optimum, (aero.)* optimum angle;
— *de tir de réglage, (art.)* corrected angle of elevation (i. e., corrected or adjusted trial angle is used in fire for effect);
— *de traînée, (aero.)* trail angle;
— *de transport, (art.)* angle of shift (between point de surveillance and the target);
— *de la voilure avec l'axe longitudinal, (aero.)* sweep back.

anguiller, m., drain hole (bottom of a boat).

anneau, m., ring; *(cav.)* picketing ring;
— *de butée, (trench art.)* tail ring (supports tail of V. D. bomb); *(mach. gun)* bolt stem collar;
— *de crosse, (art.)* trail ring;
— *de levage, (trench art.)* lifting ring;
— *de réglage, (trench art.)* chamber ring;
— *de soulèvement, (trench art.)* lifting ring;
— *de suspension, (trench art.)* suspension ring.

anneau-guide, m., *(art.)* guide ring.

année, f., year;
— *budgétaire, (adm.)* fiscal year;
— *d'exercice, (adm.)* fiscal year;
l'— *terrible,* (in France) 1870-71, (before 1914).

annonciateur, m., annunciator;

annuaire, m., directory.
annulation, f., annulation, lapse;
tomber en ——, (of funds, credits, appropriations) to lapse, to be canceled.
anse, f., (*mach.*) slot, keyway.
antenne, f., antenna; (*wireless teleg.*) aerial, antenna; (*mil. telephone*) branch line, extension line; (*r. r.*) branch line; (*mil. min.*) listening rod (*service des écoutes*);
—— *correspondante*, answering aerial, corresponding aerial;
—— *coudée*, elbow aerial;
—— *en double cône*, double cone aerial;
—— *d'émission*, (*wireless*) sending aerial;
—— *multifilaire*, many-wire aerial;
officier d' ——, (*mil.*) wireless officer (at a wireless station);
—— *parapluie*, umbrella aerial;
—— *en plan horizontal*, flat-top aerial;
—— *prismatique*, prismatic aerial (wires form a prism);
—— *en pyramide renversé*, funnel aerial;
—— *réceptrice*, (*wireless*) receiving aerial;
—— *en rideau*, fan-shaped aerial;
—— *en T*, T aerial;
—— *de transmission*, (*wireless*) sending aerial;
—— *unifilaire*, single-wire aerial.
antiasphyxiant, a., antiasphyxiating.
antiboche, a., German hater.
anti-congélant, m., nonfreezing agent.
antidérapant, a., (*auto, etc.*) nonskidding; m., any device to prevent skidding.
antifriction, f., white metal.
antimilitarisme, m., antimilitarism.
antimilitariste, a., m., antimilitarist.
antiques, m., pl., (at Polytechnique) graduates.
anti-rouille, a., anti-rust.
antitorpilleur, a., torpedo-resisting (guns, to repel torpedo-boat attack).
à-pic, m., (*hydr.*) cliff.
aplomb, m., perpendicularity;
à l'—— *de*, above, vertically above.
appareil, m., apparatus, device;
—— *d'allumage*, igniting gear;
—— *d'aviation*, (*aero.*) airplane;
—— *de désinfection*, disinfecting apparatus;
—— *d' écoute*, (*mil.*) listening apparatus (in sound liaison);
—— *d'extinction*, fire extinguisher;
—— *de filage d'huile*, (*nav.*) oil sprinkler (storms at sea);
—— *à flammes*, (*mil.*) generic term for flame projectors;
—— *de freinage* (*mach.*) brake gear;
—— *à hélice arrière*, (*aero.*) pusher;
—— *à hélice avant*, (*aero.*) tractor;
—— *à oxylithe*, (*mil.*) apparatus to purify vitiated air;
—— *à prélèvement de gas*, (*mil.*) generic term for gas mask;
—— *récepteur*, (telegraphy, etc.) receiving apparatus;
—— *Schilt*, (*Fr. a.*) flame projector;
—— *Strombos*, (*mil.*) Strombos gas alarm (a mechanical bugle);
—— *de tir masqué* (*art.*) a device giving the least elevation needed to clear a mask or screen (hill, wood, parapet, etc.).
appareillement, m., (*hipp.*) matching of horses, teams, etc.
apparier, v. a., (*hipp.*) to match horses for a team.
appel, m., call; (*mil.*) roll-call;
—— *des consignés*, (*mil.*) check roll-call;
—— *magnétique*, (*telephone*) magneto call;
—— *par magnéto*, v. —— *magnétique*;
ordre d' ——, v. s. v. *ordre*;
revue d' ——, v. s. v. *revue*;
—— *de trompe*, (*auto.*) signal, sounding of the horn.
—— *par vibrateur*, (*telephone*) buzzer;
—— *vibré*, (*elec.*) buzzer.
appendice, m., (*aero.*) neck; tail;
—— *sustenale*, (*aero.*) tail.

appoint, m., small quantity over and above some round or great number, "butt" (slang, U. S. A.)
approche, f., approach;
garde des ——*s*, (*mil.*) in sieges, trench guard.
approvisionnement, m., (*art.*) ammunition supply;
—— *d'armée*, (*art.*) army ammunition supply issued under the orders of the general; army reserve ammunition;
—— *de réserve du corps d'armée*, (*art.*) corps reserve ammunition;
—— *du service courant*, (*art.*) ammunition for daily needs;
—— *de sûreté*, (*art.*) ammunition reserved for use in an attack by the enemy.
appui, m., (*mil.*) support.
appui-coudes, m., (*mil.*) elbow-rest (e. g., the berm, sometimes).
appuyer, m., (*man.*) passage; (*mil.*) to support.
aquablindé, a., (*nav.*) water-line armored.
aquarir, v. n., (*aero.*) to alight (on water).
aquatubulaire, a., (*steam*) water-tube (of boilers).
araignée, f., spider;
—— *support de moteur*, (*aero.*) engine spider support.
arbalète, f., cross-bow; (*mil. slang*) a gun (rifle).
arbalétrier, m., strut.
arbitrer, v. n., (*sport*) to umpire.
arbre, m., (*mach.*) shaft.
(The following terms relate mainly to motor vehicles:)
—— *d'attaque*, main or driving shaft;
—— *à cardan*, shaft of a cardanic transmission;
—— *carré*, square shaft;
—— *de chaîne*, chain shaft (chain transmission);
—— *conduit*, driven shaft;
—— *coudée à n coudes*, —— *à n coudes*, an n-cranked shaft; n-throw shaft;
—— *de dédoublement*, half-time shaft, half-speed shaft;
—— *démultiplié*, reducing shaft;
—— *démultiplié à un demi*, half-time shaft;
—— *de direction*, steering post;
—— *de distribution*, (*elec.*) ignition cam shaft;
—— *d'entraînement*, driving shaft;
—— *à excentriques*, cam shaft;
—— *intermédiaire*, lay shaft;
—— *longitudinal*, v. —— *à cardan;*
—— *manivelle*, main shaft, driving shaft;
—— *de moteur*, main shaft, engine shaft;
—— *porte-hélice*, (*aero.*) propeller shaft;
—— *de roue*, wheel shaft; (*motor*) axle half;
—— *secondaire*, countershaft;
—— *des vitesses*, gear shaft, driven shaft.
arc, m., arc; bow; (*elec.*) arc;
—— *de hausse*, (*art.*) elevation arc.
—— *à haute tension*, (*elec.*) high tension arc.
arche, f., arch;
—— *marinière*, fairway arch (of a bridge over water).
ardent, a., quick, ardent; (*aero.*) ready, answering quickly.
arêtier, m., (*aero.*) edge;
—— *arrière*, trailing edge;
—— *avant*, leading edge.
argenté, a., (*inst.*) silvered.
argument, m., argument (of a table).
armature, f., (*elec.*) housing or casing of a spark plug; plate (of a condenser); (*mil., fig.*) armed strength, armed organization.
arme, f., arme; (*inf.*) the rifle;
—— *approvisionnée* (*sm. a.*) piece, magazine loaded;
—— *appuyée* (*t. p.*) piece on or with a rest;
—— *à bascule* (*sm. a.*) any break-off weapon;
—— *à bras franc*, (*t. p.*) piece off hand;
—— *chargée* (*sm. a.*) piece loaded;
prendre les ——*s* (*mil.*) to turn out under arms (as for a ceremony), to parade under arms;
prise d' ——*s* (*mil.*) any parade, etc., of troops under arms.
armé, a., (*sm. a., r. f. art.*) cocked; m., cocking;
—— *automatique*, (*art.*) auto cocking;
à —— *automatique*, (*art., sm. a.*) self-cocking.

armée, f., army; fleet (used alone, rare);
—— *coloniale*, (*France*), the colonial army;
—— *fédérale*, the Swiss army;
—— *métropolitaine*, (*France*), the army proper (service in France).

armée-cadres, f., an army strong in cadres (staff, regimental or other organizations) but weak in numbers.

armée-école, f., (*Fr. a.*) the army from the point of view of a training school in citizenship, developing lofty notions of duty, besides affording instruction in trades, etc., against the return of the soldier to civil life.

armement, m., equipment, equipping, outfitting; (*art., sm. a.*) cocking.

armurerie, f., (*mil.*) armory, arms factory.

arrache-clous, m., nail catcher, nail finder (chain or wire to scrape off nails, etc., that may have stuck on the tire).

arrache-douille, m., (*art.*) cartridge case extractor.

arraisonnement, m., "speaking" of a vessel, act of *arraisonner*, q. v.

arrêt, m., stop; (*mil.*) arrest;
—— *à la chambre*, (*mil.*) confinement to quarters;
—— *de cartouche*, (*mach. gun*) cartridge stop;
—— *facultif*, (*r. r.*) stop-over privilege;
facultié d' ——, (*r. r.*) stop-over privilege;
faire ——, (*nav.*) to stop, come to anchor.

arrêtoir, m., stop;
—— *de bouchon d'appui*, (*mach. gun*) spring guide cap stop;
—— *à ressort*, spring stop.

arrières, m., pl. back (of a motor car);
—— *galbés*, curved back (double curve);
—— *Roi des Belges*, straight back;
—— *tulipe*, curved back (convex outward).

arrimage, m., (*aero*) packing, stowing, of balloon equipment on board; rigging;
—— *de route*, (*art.*) (of guns) travel stowage.

arrimer, v. t., to pack, secure, stow; (*aero*) to pack, stow, equipment on board a balloon, to rig;

arrimeur, m., (*aero.*) rigger (balloons).

arrivée, f., (*mach.*) admission;
—— *d'essence*, gasoline inlet.

arriver, v. n., to succeed; (*fam.*) to "get there."

arrivisme, m., ambition, resolution, to succeed by no matter what means; corrupt ambition; "the state of soul of some men of unscrupulous ambition" (Gen. Bonnal).

arriviste, m., a man of unscrupulous ambition.

arrondi, a., (of numbers) round.

arrondissement, m., (France) political unit composed of several *cantons*, is the jurisdiction of a *sous-préfet* (is not a person of law), arrondissement.

art, m., art;
l' —— *mili*, (*mil. slang*), *l'art militaire*, the art military.

article, m., article;
—— *s d'argent* (*adm.*) generic term for post-office money order.

articulation, f., joint;
—— *à cardan*, (*mach.*) universal joint, Hooke's joint.

articuler, v. a., (*mil.*) to spread out troops, space them apart, so as to diminish the target.

artifice, m., artifice; (*artif.*) composition, fireworks (*in pl.*);
—— *éclairant*, (*mil.*) flare,
—— *s à signaux* (*mil.*) signaling devices.

artiflot, m., (*mil. slang*) artilleryman.

artillerie, f., (*art.*) artillery;
—— *antitorpilleur*, (*nav. art.*) in the modern (dreadnaught) type of ship, the guns intended to beat off torpedo boats;
—— *automobile*, motor-drawn artillery;
—— *d'accompagnement de l'infanterie*, (artillery that assists infantry in an attack both directly and by taking under fire enemy reinforcements, counterattacks, etc.) accompanying artillery;

artillerie *d'appui*, supporting artillery;
—— *d'armée*, army artillery, (as distinguished from corps and division artillery, the artillery of an army over and above the normal corps and division artilleries, and assisting them when needed);
—— *d'assaut*, (generic term for the guns carried in an assault by tanks) tank guns;
—— *de combat*, combat artillery (i. e., artillery, especially field artillery, taking part in trench conflicts, as distinguished from the heavy artillery farther to the rear);
—— *de contre-batterie*, artillery whose object is to destroy hostile batteries and at any moment to neutralize their fire; counter-battering artillery.
—— *courte*, generic term for short guns, (e. g., howitzers);
—— *de destruction*, artillery whose object is the destruction of obstacles, and of principal points in enemy positions (i. e., batteries, depots, etc.) and demoralization of survivors;
—— *légère*, all guns less than 95 mm. in caliber; (*nav.*) secondary battery;
—— *longue*, generic term for long guns;
—— *à longue portée*, long-range artillery;
—— *lourde*, the heavy artillery of a field army, or of an army in the trenches; heavy artillery in general, (all guns larger than 95mm);
—— *lourde d'armée*, heavy field artillery;
—— *lourde de campagne*, mobile heavy artillery;
—— *lourde courbe*, (generic term for) howitzers;
—— *lourde de destruction*, v. —— *de destruction*;
—— *lourde de position*, heavy position artillery;
—— *lourde à grande puissance*, heavy high-powered artillery (distant targets, special targets);
—— *de renforcement*, subdivision of the heavy artillery of a sector comprising the corps or army heavy batteries not included in the
—— *spéciale*, (*Fr. a.*) special artillery, i. e., naval and seacoast guns and certain siege guns used in trench warfare;
—— *de sûreté*, q. v., certain horsed heavy batteries, and foot batteries forming reserve groups;
—— *de sûreté*, (*Fr. a.*) subdivision of the heavy artillery of a sector, comprising foot artillery with siege matériel, and a number of heavy mobile batteries (corps or army batteries);
—— *à tir courbe*, generic term for howitzers, high-angle guns;
—— *de tranchée*, trench artillery.

as, m., (*mil. slang*) a "no. 1" man conspicuous for duty, service, and deeds of valor; (*aero.*) "ace" (man who has brought down five or more enemy planes).

ascension, f., (*aero.*) trip, journey, ascension;
journal des ——*s*, record of ascensions;
mettre en ——, to send up (balloons);
mise en ——, sending up (balloons).

ascensionner, v. n., (*aero.*) to make an ascension.

askar, *askari* (proper name), (*mil. slang*) Arab mercenary.

aspirant, m., (*mil.*) candidate for the grade of officer.

aspirant-officier, m., (*mil.*) candidate for the grade of officer.

aspiration, f., (*mach.*) intake;
—— f., (*mach.*) suction; admission, aspiration, suction stroke.

assaut, m., assault; match;
—— *de canne*, (*fenc.*) single stick match;
détachement d' ——, (*mil.*) assaulting troops, hand-to-hand troops (Stosstruppen, German).
école d' ——, (*mil.*) assault training; (school or point for training in assault tactics) assault school.
—— *de lutte*, (*sport*) wrestling match;
—— *au pistolet*, (*t. p.*) pistol match;
—— *de sabre*, (*fenc.*) sword match.

assemblage, m., (*aero.*) generic term for parts, pieces, etc., used to assemble or join various parts, especially uprights to string pieces;
godet d' ——, assembling cap.

assembler, v. t., (*mach.*) to couple.
asséner, v. t., to strike;
— *un coup*, to deliver a severe blow (with a club, sword, etc.).
assiette, f., (*nav.*) the correct position or level of a ship afloat; (*cons.*) settling; (*aero.*) horizontality;
prendre son ——, to settle.
prendre son ——, (*cons.*) to settle.
assise, f., depth of a seat; (*in pl. law*) assizes.
associé, a., v. s. v., *canot*; (*mil.*) linked, of battalions, etc. (British army).
assujettir, v. t., to secure, fix firmly.
astigmatiseur, m., (*inst.*) astigmatiser (range-finder).
astigmatisme, m., astigmatism.
astralit, f., (*expl.*) astralite.
atelier, m., workshop, working party;
—— *roulant*, traveling workshop;
—— *de séchage*, (*fond.*) drying room;
—— *de télégraphie*, —— *télégraphique*, (*mil.*) telegraph squad;
—— *téléphonique*, (*mil.*) telephone squad (1 commissioned officer and 5 men);
—— *de transformation de charge*, (*art.*) shelter (near a heavy battery) for the conversion and manipulation of charges.
attache, f., clip, paper fastener;
—— *d'aile*, (*aero.*) wing attachment, wing support.
attache-fils, m., wire coupler; (*aero.*) wiring plate.
attaque, f., (*mil.*, etc.) attack; (*mil. min.*) head, head of the work (mine, shaft, or gallery);
chef d' ——, (*mil.*) in sieges, chief of attack, (field officer of engineers in charge of approaches);
double ——, (*mil. min.*) double attack (two-level construction of a gallery);
—— *de, en, flanc*, attack, in flank;
—— *aux gaz* (*mil.*) gas attack;
—— *par gaz*, (*mil.*) gas attack;
—— *gazeuse*, gas attack;
—— *en queue*, attack in rear;
—— *à revers*, attack from the rear;
—— *par surprise*, surprise attack;
—— *par vagues de gaz*, gas attack.
attelage, m., team;
—— *à deux chevaux*, two horse team.
atteler, v. t., (*mil.*) to serve with, have charge of, certain teams, wagons.
atterrir, v. n., (*aero.*) to land, come down.
atterrissage, m., (*aero.*) landing; (*nav.*) landing;
châssis d' ——, (*aero.*) landing chassis, under carriage;
drap d' ——, (*aero.*) landing sheet (indicates position of landing ground);
rampe d' ——, (*aero.*) landing lights;
terrain d' ——, (*aero.*) landing place, landing ground;
train d' ——, v. *châssis d'* ——.
attrape-tiquets, m., tick trap (pigeon lofts).
attribut, m., (*in pl. mil.*) insignia (on uniform, but not of rank).
aubade, f., morning serenade.
audience, f., audience;
—— *de congé*, farewell audience of an ambassador, of a military attaché.
audion, m., (*wireless*) audion.
auditeur, m., auditor; officer of the (French) council of state.
auget, m., trough; (*art.*) rear casing (trench mortar bed).
auto, f., familiar abbreviation of automobile.
auto-allumage, m., (*mach.*) self ignition (overheated cylinders).
auto-ballon, m., (*aero.*) dirigible.
autobloqué, a., self locked.
autobus, m., autobus.
auto-camion, m., auto truck, motor truck.
auto-canon, m., (*art.*) automobile gun (mounted on an auto).

auto-correcteur, m., (*art.*) automatic corrector; (*mil. aero.*) auto-corrector.
auto-cuiseur, m., self-cooker.
automitrailleuse, f., (*mil. auto*) auto-machine gun.
automobile, m., f., automobile;
—— *aérien*, any aerial machine driven by a motor;
—— *à l'alcool*, any automobile in which the energy comes from alcohol;
—— *blindée*, (*mil.*) armored automobile;
canot ——, v. s. v., *canot*;
—— *de course*, racing car;
—— *cuirassé*, (*mil.*) armored motor car;
—— *à essence*, gasoline motor car;
—— *à explosion*, explosion engine car;
—— *de guerre*, (*mil.*) service or military automobile;
—— *mitrailleuse*, (*mil.*) machine gun on automobile carriage;
—— *de tourisme*, touring car;
—— *à vapeur*, any steam automobile;
vedette ——, v. s. v., *vedette*;
—— *à viandes*, (*mil.*) meat automobile.
automobilisme, m., automobilism (sport, science, etc.).
automotrice, f., (*r. r.*) electric locomotive.
autonautisme, m., motor boating.
auto-rupteur, m., (*elec.*) make and break device (adapted to coils not fitted with contact breakers).
auto-taxi, m., (*fam.*) taximeter.
auto-torpillage, m., (*nav.*) self-destruction by torpedo (case where a torpedo turns back on its own ship).
auto-transformateur, m., (*elec.*) auto transformer.
auto-trembleur, m., a synonym of *vibreur*, q. v.
autoviseur, m., small mirror to observe the existence and nature of spark produced.
auto-yacht, m., (*nav.*) motor-yacht.
auxiliaire, m., (*nav.*) auxiliary boat.
avance, f., advance; (*mach.*, etc.) lead;
—— *à l'allumage*, sparking advance, sparking advance gear;
—— *à l'échappement*, exhaust lead;
—— *à l'ouverture d'échappement*, exhaust lead;
—— *sur titre*, (*com.*) loan on security;
—— *par tour*, (of a propeller) the theoretical advance less the slip.
avancement, m., (*mil.*) promotion;
—— *à l'ancienneté*, promotion by seniority;
—— *à l'ancienneté par sélection*, (German Army) promotion by seniority, after elimination of incompetents;
—— *à l'élection*, (*mil.*) promotion by vote or election (French Revolution);
—— *par régiment*, (*mil.*) regimental promotion;
—— *par sélection*, (*mil.*) promotion by selection;
—— *sur toute l'arme*, (*mil.*) lineal promotion.
avant, m., forward part, forepart;
—— *brise-glaces*, (*nav.*) ice-breaker of a ship.
avenant, m., additional clause (as of a contract or treaty).
aviage, m., (*aero.*) flight (word proposed, but not adopted).
aviateur, m., (*aero.*) aviator, pilot, airman.
aviation, f., (*aero.*) aviation;
—— *côtière*, coast aviation (naval purposes);
—— *hauturière*, high sea aviation service.
avier, v. t., (*aero.*) to fly (word proposed, but not adopted).
avion, m., (*aero.*) airplane;
—— *d'accompagnement de l'infanterie*, (*mil.*) airplane that follows the march of advanced units and reserves (one to each division); contact plane;
—— *d'artillerie*, spotting plane;
—— *de bombardement*, bombing plane;
—— *de bord*, ship's plane (carried aboard);
—— *de chasse*, combat plane, pursuit airplane;
—— *de combat*, battle plane;

avion *de commandement,* (*mil.*) airplane to observe the general progress of the combat, the enemy, the distribution of his forces, and indications of counter attack (one to each corps); corps plane.
—— *côtier,* coast plane (i. e., for naval work along a coast);
—— *estafette,* (*mil.*) messenger airplane (to carry information to division, brigade, and regimental commanders);
—— *à hélice propulsive,* pusher;
—— *à hélice tractive,* tractor;
—— *marin,* seaplane;
—— *de reconnaissance,* scouting plane;
—— *de réglage,* (*art.*) fire observation airplane; spotting plane.
—— *type corps d'armée,* generic term for spotting, photographing, etc., planes.
avion-canon, m., (*aero.*) gun plane.
aviso, m., (*nav.*) dispatch boat;

aviso *à roues,* paddlewheel dispatch boat;
—— *-transport,* dispatch transport.
avocat, m., barrister, counselor-at-law.
avoir, v. t., to have; (*mil. slang*) to beat, crush, "do up."
avoué, m., attorney.
axe, m., axis; (*mach., etc.*) shaft;
—— *de blocage,* (*trench art.*) elevation axle;
—— *brisé,* divided shaft;
—— *coucourant,* continuous shaft;
—— *démultiplié,* reducing axis or shaft;
—— *à galet,* (*mach. gun*) roller;
—— *d'observation,* (*mil.*) line of observation (e. g., from a captive balloon).
—— *de pivotement,* pivot, spindle, (of a vehicle).
—— *de retenue,* (*trench art.*) securing axle.
axe-manchon, m., sleeve and axle.
axer, v. t., to mount on an axis, axially.
azimutal, a., azimuth.

B.

bac, m., (*elec.*) jar, of a storage battery.
baccalauréat, m., in France, the examination concluding the course in secondary education; bachelor's degree.
bâche, f., awning;
—— *de campement,* (*mil. aero.*) ground cloth.
bachot, m., punt; (*slang*) the baccalauréat, q. v.
bafouillage, m., chattering; (*mach.*) chattering (of a motor).
bafouiller, v. n., to chatter; (*mach.*) of a motor, to chatter.
baggage, m., baggage;
—— *technique,* (*fam.*) technical qualifications, fitness;
—— *-s-vivres,* (*mil.*) food supplies in the train.
baguage, m., banding (pigeons).
bague, f., ring, leg-band (pigeons);
—— *d'appui,* (*art.*) washer; (*mach. gun*) recoil spring bushing;
—— *d'armement,* (*trench art.*) cocking ring;
—— *d'attache,* (*art.*) locking sleeve;
—— *à billes,* (*mach.*) ball-bearing ring;
—— *de calage,* (*art.*) locking ring;
—— *d'étranglement,* (*mach.*) air throttling ring, valve, or collar, (in some forms of carburetors);
—— *de graissage,* (*mach.*) oil ring, lubricating ring;
—— *de raccord,* (*mach. gun*) front connector bushing;
—— *de réglage,* (*mach.*) adjusting, feed adjusting, collar (some forms of carburetor); (*trench art.*) chamber ring;
—— *de réglage d'air,* (*mach.*) throttle valve, throttle ring valve.
—— *de sûreté,* (*art.*) safety collar or ring.
bague-écrou, f., (*mach. gun*) assembling nut; rear connector bushing.
baguer, v. t., to band (a pigeon's leg).
baguette, f., rod;
—— *de nettoyage,* (*sm. a.*) cleaning rod.
bai, a., m., (*hipp.*) bay;
—— *cerise clair,* light red bay.
bain, m., bath;
—— *de fixage,* (*phot.*) fixing bath.
bain-douche, m., shower bath.
balance, f., balance;
—— *à socle,* ordinary balance.
balancier, m., (*art.*) roller-arm (in some forms of carriage).
baïonnette, f., (*mil.*) bayonet;
—— *basse,* with bayonets at the charge;
—— *courte de tranchée,* trench bayonet;
monter à ——, to mount, assemble by a bayonet joint.
baïonnette-scie, f., (*sm. a.*) saw bayonet.
baladeur, m., (*mach.*) sliding clutch;
système à double (triple) ——, double (triple) reduction gear set;
système à simple ——, simple reduction gear set.

balai, m., broom; (*slang*) gendarme;
—— *rotatif,* (*elec.*) rotary brush, collector.
balance, f., balance;
—— *à socle,* ordinary balance.
balancier, m., (*art.*) roller-arm (in some form of carriage).
balancine, f., (*aero*) diagonal steel wire, supporting the car of a dirigible.
balayeuse, f., street-sweeping machine;
—— *automobile,* motor driven sweeping machine.
baliste, f., balista.
balle, f., (*sm. a., art.*) bullet; (*com.*) bale; (*slang*) one franc;
—— *aciculaire,* v. —— *à pointe effilée;*
—— *blindée,* (*sm. a.*) perforating bullet, (for use against shields);
—— *de coton,* cotton bale;
—— *directe,* direct bullet (point first; cf. —— *renversée*);
—— *efficace,* (*art.*) of a shrapnel bullet, one whose velocity is dangerous;
—— *qui fait champignon,* (*sm. a.*) soft-nosed bullet;
—— *inefficace,* (*art.*) v. —— *inoffensive;*
—— *inoffensive,* (*art.*) harmless bullet, said of a shrapnel bullet whose velocity is harmless;
—— *perforante,* (*sm. a.*) (German) perforating bullet (attack of blindages);
—— *à pointe effilée,* (*sm. a.*) pointed bullet, S-bullet, D-bullet;
—— *renversée,* (*sm. a.*) inverted bullet, (point to the rear);
—— *S,* (*sm. a.*) (German) S-bullet (*Spitze,* pointed bullet);
—— *S M K,* (*sm. a.*) (German) perforating bullet.
ballon, m., (*aero.*) balloon; (*auto.*) top, carriage top; hood; (*com.*) globe (as, of glass);
abattre un ——, to bring down a balloon;
—— *cerf-volant,* (*aero.*) kite balloon;
—— *cigare,* (*aero.*) cigar-shaped balloon;
—— *de commandement,* (*mil.*) balloon observing the course of a combat and reporting directly to the corps commander (one to each corps, duties same as those of *avion de commandement,* q. v.); corps balloon;
—— *déformable,* flexible balloon (dirigible);
—— *demi-rigide,* semi-rigid balloon (dirigible);
—— *démontable,* (*auto.*) detachable top;
—— *à deux nacelles,* (*aero.*) double car balloon;
—— *divisionnaire,* (*mil.*) balloon attached to a division; reports для artillery fire, transmits signals, etc.; divisional balloon;
—— *indéformable,* (*aero.*) rigid balloon (e. g., Zeppelin);
—— *d'infanterie,* (*mil.*) the divisional balloon during an attack (so-called from its duty);
larguer un ——, (*mil. aero.*) to give a balloon cable, to let up a balloon;
—— *normal,* (*Fr. a.*) normal balloon (560 cub. m., free and captive ascensions);
—— *observateur,* observation balloon;
—— *de place,* (*Fr. a.*) fortress balloon (980 cub. m., 1,600 cub. m., free ascensions);

ballon ramener un ——, to bring down a balloon;
—— de siège, (Fr. a.) siege balloon (750 cub. m., captive ascensions);
—— souple, (aero.) flexible balloon (as distinguished from one that has a frame);
—— sphérique, (aero.) spherical balloon.
ballon-école, m., (aero.) instruction balloon.
ballonet, m., (aero.) air bag;
—— à air, ballonet;
—— d'empennage, stabilizing ballonet.
ballonner, v. n., to bulge.
ballon-sonde, m., exploring balloon to get atmospheric records, ballon-sonde.
ballonnet, m., (aero.) ballonet;
—— compensateur, ballonet.
ballot, m., bale; (mil. slang) heavy, dull man.
baluchon, m., (eng.) dredge-bucket, excavator-bucket or shovel; (mil. min.) miner's bucket, (sometimes spelled balluchon).
bandage, m., band; (auto.) tire, shoe;
—— caoutchouc plein, solid rubber tire;
—— creux, pneumatic tire;
—— plein, solid tire.
bande, f., band, newspaper wrapper (usually a narrow strip of paper in France); (mil.) cartridge belt (for machine gun); (auto.) tire, shoe; (med.) bandage;
—— à alvéole, (mil.) sort of bandoleer;
—— de contrôle (Fr. adm.) internal revenue stamp (U. S.);
—— creuse, (auto.) pneumatic tire;
erreur de ——, v. s. v., erreur;
—— jumelles, (auto.) double tires;
—— à maillons métalliques, (mach. gun) metal belt or band;
—— métallique, (mach. gun) metallic feed strip;
—— de mitrailleuse en toile, (mach. gun) canvas feed belt or band;
—— molletière, (unif.) puttee;
—— pleine, (auto.) solid tire;
—— porte-cartouches, (mach. gun) cartridge belt;
retour de ——, (mach. gun) back feed, reverse feed;
—— de roulement, (auto.) shoe, tread (of a tire);
—— rigide, (mach. gun) feed strip;
—— simple, (auto.) single tire;
—— en toile, (mach. gun) canvas ammunition belt.
bandeau, m., (unif.) band of a forage cap.
bande-chargeur, f., (mil.) cartridge belt (machine gun).
bandé, m., compression (of a spring).
bander, v. t., (med.) to bandage.
banquette, f., seat, carriage seat (no intermediate arms); (art.) gunner's seat;
—— de pointeur, (r. r. art.) gunner's platform;
—— de relais, (siege) relay bank;
—— de tir, (mil.) firing step (in trenches).
baptême, m., baptism;
—— de promotion, (mil.) graduation dinner or festival.
baquet, m., bucket seat (motor car); (aero.) cockpit.
baraque, f., wooden hut.
baraquement, m., (mil.) hut (i. e., small wooden house).
barbaque, m., (mil. slang) poor meat.
barbe, f., beard; a shave.
vieilles ——s, (fam.) old soldiers.
barbotage, m., dipping, splash, (as mach.) of a part in oil).
barbotine, f., sort of paste or cement (will pour).
barda, m., (mil. slang) pack.
bardage, m., moving, loading (with a barrow, bard)
pince de ——, sluing bar.
barillet, m., (mach. gun) feed drum;
doigts de ——, feed drum fingers;
—— à ressort, spring drum.
barographe, m., (aero., inst.) barograph.
baro-thermomètre, m., barometer and thermometer.

baro-thermomètre, enregistreur, (aero.) self-recording baro-thermometer.
barrage, m., line of men forming a "hedge" in a street (e. g., gendarmes); line of men (to stop an entrance, block a road, etc.); (fort.) covering or restraining works, e. g., to allow mobilization, to cover an advance, etc.; (mil.) curtain of fire (artillery, grenades);
—— aérien, (aero.) airplane barrage (constituted by airplanes themselves);
—— d'agents, line of police;
—— de feu, (mil.) line of fires lighted to neutralize a gas attack;
—— fixe, (art.) stationary barrage or curtain fire;
—— mobile, (art.) moving barrage or curtain fire (used to assist attacks by infantry);
—— normal, (art.) normal barrage (battery per target);
—— ouvrage de, (fort.) covering work or fort (e. g., during mobilization);
—— renforcé, (art.) reenforced barrage (two or more batteries on same target);
toile de, —— (mil.) specially prepared cloth to exclude gases from shelters, bombproofs, etc. in a gas attack.
barre, f., bar; tidal bore;
—— d'accouplement, (mach.) drag link; (auto.), tiebar, crossbar (front axle);
—— d' attelage, (r. r.) drawbar;
—— de connexion, (auto.) v. —— d'accouplement;
—— de direction, (mach.) drag link; (motor car);
—— d'enlevage, (trench art.) lifting bar;
—— franche, (nav.) tiller; (mach.) steering bar; (motor car);
—— de portage, (trench art.) carrying bar (240 C. T. mortar);
—— de tirage, (mach.) tension rod, pull bar, (motor car);
—— de tension, v. —— de tirage.
barreau, m., grate bar;
—— de manœuvre, (trench art.) hand spike.
barre-entretoise, f., (trench art.) bar-transom (240 C. T. truck).
barrette, f., bar;
—— de mentonnet, (mach. gun) trigger bar.
barretter, m., (wireless) barretter.
barreur, m., (nav.) helmsman.
barricade, f., (mil. min.) tamping mask.
basane, f., sheepskin; (mil. slang) cavalry.
bascule, f., (art., mach.) rocking piece.
basculeur, m., (mach.) rocking lever, tappet lever.
base, f., base (in many relations);
—— de départ, (top) primary base.
bassaing, m., plank 65 mm. thick, and 22-25 cm. wide.
bât, m., pack, pack saddle;
—— de mitrailleuse, (mach. gun) gun pack;
—— de munitions (mach. gun) ammunition pack.
bat d'Af, (mil. slang) an abbreviation of bataillon d'Afrique, q. v.
bataillon, m., (mil.) battalion;
—— d'Afrique, v.;
—— d'infanterie légère d'Afrique;
—— associé, linked battalion (British Army);
—— formant corps, independent battalion.
bataillonneux, m., (mil.) soldier belonging to a bataillon d'Afrique, q. v.
bateau, m., boat;
—— à aubes, side wheeler;
—— à moteur, motor boat;
—— omnibus, Seine passenger boat;
—— poseur de câbles, (nav.) cable ship;
—— sousmarin, (nav.) submarine;
—— submersible, (nav.) submersible boat.
bateau-glisseur, m., (aero.) hydroplane.
bâti, m., (aero.) supporting frame of an airplane (for weight of pilot, motor, etc.).
bâtiment, m., building, (nav.) ship;
—— amiral, (nav.) flagship;
—— hydrographe, (hydr.) coast-survey ship;
—— ravitailleur, (nav.) tender, storeship;
—— de servitude, (nav.) generic term for harbor boats, such as ice, mail, water, etc., boats.

bâton, m., stick;
— *ferré*, iron-shod stick;
— *porte-géophone*, (*mil. min.*) geophone-rod;
— *de remorque*, (*mach. gun*) traction bar (man traction of machine-gun cart).

batterie, f., (*mil., elec.*) battery;
— *A*, horsed battery;
— *d'accompagnement*, (*mil.*) in general, any battery accompanying troops of a different arm, more especially infantry; accompanying battery;
— *d'allumage*, (*elec.*) ignition battery of a motor;
— *alpine*, (*Fr. art.*) Alpine battery, mountain battery;
— *anti-aérienne*, antiaircraft battery;
— *d'attaque*, battery of the attack, infantry battery;
— *en attente*, battery in readiness;
— *avancée*, advancing battery (moves forward with infantry, if necessary by sections);
— *de barrage normal*, any battery executing normal barrage fire (i. e., has its own separate target);
— *à bouclier*, field battery with shielded guns;
— *de brèche*, breaching battery (opens passages in enemy lines);
— *cadre*, skeleton battery;
— *de combat*, (*Fr. art.*) the whole of a field battery, except the *train régimentaire* (consists of the *batterie de tir*, 4 guns and limbers, 6 caissons, and of the *échelon de combat*, 6 caissons, 1 forage and 1 battery wagon);
— *de contre-attaque*, counter-attack battery (held in readiness to resist a counter attack);
— *côtière*, coast battery;
— *courte*, howitzer battery;
en —, of fire hose, ready;
fausse —, (*mil.*) dummy battery, Quaker battery;
— *fausse*, (*mil.*) dummy battery, Quaker battery;
— *de fusils*, (*mil.*) battery of small arm rifles;
— *longue*, battery of long guns;
— *montée de 75* (*Fr. art.*) 75 mm. battery (4 guns);
— *moyenne*, (*nav.*) secondary battery;
— *neutralisée*, neutralized battery (suspends fire, not necessarily out of action);
— *à pied de position*, (heavy) position battery;
— "*portée*," (*art.*) motor battery;
— *de poursuite*, pursuit battery (usually horse artillery);
— *de préparation*, battery of preparation (fires on infantry);
— *de renforcement*, a battery belonging to the *artillerie de renforcement* q. v.
— *en retraite*, battery covering a retreat;
— *de superposition*, any battery that adds its fire to that of another battery on the same target;
— *en surveillance*, battery in position to open fire immediately (assists — *s d'attaque*);
— *T*, tractor battery;
— *de tir*, v. s. v. — *de combat;* firing battery;
— *à tracteurs*, (*art.*) tractor battery;
— *de tranchée*, (*mil.*) trench battery (of trench artillery);
— *V. B.* (*Fr. a.*) battery of V. B. grenade throwers;
— *volante*, horse battery (not an official expression).

batterie-amorce, f., (*art.*) decoy battery.

battre, v. t., to beat; (*art.*) to batter;
— *par le feu*, (*art.*) to take under fire;
— *son plein*, to be in full blast, to be at its height.

bec, m, burner; tip (in general);
— *brûleur*, burner, tip;
— *à incandescence*, incandescent gas burner.

becquetance, f., (*mil. slang*) food.

bengale, m., (*artif.*) Bengal light.

benzine, f., benzine;
— *de la houille*, benzol.

béquille, f., crutch, sprag; (*aero.*) landing prop or support of an airplane, tail skid;
— *d'atterrissage*, (*aero.*) sort of inverted pyramid to assist a dirigible in landing (Lebaudy type), landing prop;
crosse- — *d'arrière*, (*aero.*) rear support tail skid;
— *à ressort*, (*mach. gun*) spring link.

berceau, m., (*auto*) cradle, false frame (so-called if formed of pieces more or less curved); (*art.*) rocker, of the 75 mm.; cradle, chassis; recoil-sleeve, (240 T. R.);
— *du moteur*, (*aero.*) engine bed;
— *de pointage*, (*art.*) elevating carriage.

berne, f., blanket, bed cover;
en —, (of a flag) at half staff (distress and mourning both, in France; "union down" may be translated *en berne*, but more accurately would be *pavillon renversé*).

Bertha, f., (proper name) (*mil. slang*) heavy shell (42 cm.).

bessonneau, m., (*aero. fam.*) canvas hangar.

bestiau, m., (*mil. slang*) horse.

béton, m., (*cons.*) concrete;
— *armé*, reenforced concrete, ferroconcrete.

bétonner, v. t., (*cons.*) to build of concrete;

betterave, f., beet.

beuglant, m., (*mil. slang*) concert hall, (Saumur, slang).

Bicot, m., (*mil. slang*) the Arabs, Berbers, of North Africa; sometimes the Senegalese.

bicyclette, f., bicycle;
— *automobile*, motorcycle;
— *à cadre rigide*, ordinary bicycle;
— *à pétrole*, motorcycle;
— *pliante*, folding bicycle;
— *à vapeur*, steam bicycle.

bidoche, f., (*mil. slang*) meat.

bielle, f., (*mach.*) connecting rod; (*mach. gun*) cartridge guide cam;
— *d'accouplement*, v. *barre d'accouplement;*
— *de commande*, driving rod;
— *de connexion*, v. *barre de connexion;*
— *de direction*, v. *barre de direction;*
— *à fourche*, forked connecting rod;
pied de —, piston end of a connecting rod;
— *de poussée*, radius rod;
— *de réaction*, reaction rod;
tête de —, crank end of a connecting rod.

biellette, f., (*mach.*) link.

biffin, m., (*mil. slang*) infantryman.

bigorre, f., (*mil. slang*) colonial artillery (French Army).

bilatéral, a., bilateral; (*telephone, etc.*) both sending and receiving.

billard, m., billiards, billiard table; (*mil. slang*) operating table (medical).

bille, f., marble;
— *de sécurité*, (*trench art.*) safety ball.

billet, m., (*mil.*) bill; (*r. r.*) ticket;
— *d'entrée*, (*mil. med.*) admission order;
— *de famille*, (*r. r.*) family ticket;
— *individuel*, (*r. r.*) single ticket;
— *au porteur*, (*com.*) promissory note;
— *de sortie*, (*mil. med.*) discharge order;
— *d'urgence*, (*mil. med.*) emergency order.

biphasé, a., (*elec.*) two-phase.

biplace, m., (*aero*) two seater.

biplan, m., (*aero*) biplane.

bique, f., (*mil. slang*) old horse.

bi-réfringent, a., double refracting.

bissecter, v. t., to bisect.

bistro, m., (*mil. slang*) bar keeper.

bizut, m., (*mil. slang*) a "plebe" at the Polytechnique; (also used by other great schools).

blessé, a., wounded;
(*mil.*) *grand* —, severely wounded man.

bleu, a., blue; m., blue sky; (*mil. slang*) "rookie";
— *horizon*, (*Fr. a.*) horizon blue, (color of field uniform);

bleu, *petit* ——, (*fam.*) telegram (in blue envelope); *tirer dans le* ——, (*mil. fam.*) to fire into the blue, into the air.
bleuté, a., bluish.
blindage, m., (*auto.*) dust casing, dust cover; steel sheeting.
bloc, m., block; in the French Chamber of Deputies a "block" supporting certain radical policies (those of Combes, etc.);
à ——, (of straps) in, up to, the last hole, as far as possible;
—— *d'acier,* (*met.*) rough forging;
—— *d'alimentation,* (*mach. gun*) feed block;
—— *de sécurité,* (*art.*) safety lock.
blocage, m., (*mach.*) jamming, of a mechanism, gearing; locking, setting; (*in gen.*) state of being well set up (said of a whole composed of parts in bearing or in contact with one another).
blockhaus, m., (*nav.*) pilot house.
bloquage, m., v., *blocage.*
bloquer, v. t., to lock, set; (*mach.*) to jam; (*in gen.*) to seat "home"; to set, fix (at or in a given position);
—— *un frein,* —— *ses freins,* (*auto.*) to put on brakes suddenly, to jam one's brake suddenly home.
bloquiste, m., in the French Chamber of Deputies, a member of the bloc, q. v.
bobinage, m., (*elec.*) wiring, coil.
bobine, f., (*elec.*) coil; reel;
—— *dérouleuse,* reel;
—— *de réactance, de réaction,* choke coil;
refaire une ——, to rewind a reel, a coil;
—— *à trembleur,* (*elec.*) make and break coil.
—— *à rupteur,* make and break coil;
—— *de self,* self-induction coil;
—— *de syntonisation,* —— *syntonisatrice,* (**wireless**) tuning coil;
bobine-dévidoir, f., reel.
bobosse, m., (*mil. slang*) German soldier.
boche, m., (*mil. slang*) German.
bocherie, f., (*mil. slang*) German cruelty.
bochie, f., (*fam.*) Germany.
bochisant, m., (*mil. slang*) germanophile.
bochiser, v. t., (*mil. slang*) to spy.
bochisme, m., (*mil. slang*) kultur.
bochonnerie, f., (*mil. slang*) asphyxiating gas.
boggie, m., (*rr., mach.*) bogy.
boire, v. t., to drink;
—— *l'obstacle,* (*auto. fam.*) to swallow (i. e., to be indifferent to) obstacles (said of the pneumatic tire).
bois, m., wood;
—— *d'abris,* (*mil.*) shelter lumber, (fit for the construction of shelter);
—— *armé,* armored wood;
—— *brut,* lumber;
—— *œuvrés,* dimension stuff, worked to sizes and shapes;
sous ——, (*mil.*) in the woods, in a wood.
boisseau, m., (*mach.*) throttle valve.
boîte, f., box; (*mil. slang*) guard house (prison), jug;
—— *d'allumage,* (*mach.*) ignition case;
—— *d'appui,* (*art.*) box, case;
—— *à billes* (*mach.*) ball bearing casing;
—— *de branchement,* (*tel.*) junction box, distribution box;
—— *à came,* (*mach.*) cam case;
—— *de cardan,* (*mach.*) sleeve or casing (in a form of cardan joint, receiving the lugs of a *joint à dés*);
—— *de culasse,* (*mach. gun*) breech casing;
—— *à dépêches,* (*mil.*) dispatch box (carrier pigeon service);
—— *de dérivation,* v. —— *de branchement;*
—— *de différentiel* (*mach.*) differential gear casing;
—— *de direction* (*mach.*) steering gear case;
—— *d'émission,* (*mil.*) sounding case (for mechanical bugles, employed in sound liaison);
—— *des engrenages,* v. —— *des vitesses;*
—— *à essieu,* axlebox;

boîte *à graisse,* (*mach.*) oil cup, grease cup;
—— *d'inflammation,* (*mil. min.*) ignition box, (for simultaneous ignition).
mettre à la ——, (*mil. slang*) to put into the guardhouse, to "jug";
—— *de paperasserie,* (*lit.* box of worthless papers, hence, *fig. mil.*) routinist paper man, red tape man;
—— *de percussion.* (*trench art.*) percussion case;
—— *aux poudres,* (*mil. min.*) case (wooden) of a mine charge;
—— *de prise de courant,* (*elec.*) outlet box, terminal box, junction box;
—— *de réception,* (*mil. min.*) junction box (of a listening device), receiving box;
—— *aux rechanges,* (*art.*) spare-part box:
—— *à réparations,* (*auto.*) repair kit, outfit;
—— *de secours,* (*mil. min.*) first-aid box (cases of asphyxiation);
—— *de topage,* carrying case;
—— *de vitesses,* (*mach.*) speed gear box, gear case.
boîte-bâti, f., (*mach.*) air reservoir of the Daimler motor.
boîtier, m., box, case;
—— *de différentiel,* v. s. v. *boîte.*
bombarde, f., (*Fr. a.*) bombard (trench artillery, four infantry barrels suitably mounted on a single frame to fire grenades).
bombardement, m., (*art.*) bombardment;
—— *d'ensemble,* general bombardment.
bombardier, m., (*mil.*) bombardier; (to-day) bomber, trench mortarman.
bombe, f., bomb; (*art.*) bomb, (i. e., trench artillery projectile);
—— *éclairante,* (*artif.*) light ball, flare;
—— *fumigène,* (*artif.*) smoke bomb;
—— *incendiaire,* (*artif.*) incendiary bomb.
bombonne, f., carboy (used, *mil. min.*, as a mine case);
—— *arrêtoir à ressort,* (*mach. gun*) spring stop pin.
bon, a., good; m., check, order;
—— *absent,* (*mil.*) authorized absence;
—— *d'aliments,* (*mil., med.*) order for food;
—— *de médicaments,* (*mil., med.*) order for drugs;
—— *de pansements,* (*mil., med.*) order for dressings;
—— *de poste,* (*Fr. adm.*) postal note payable to bearer (whose name must appear on note);
—— *du Trésor,* (*Fr. adm.*) sort of Treasury bill or warrant, bearing interest.
bondir, v. n., (*mil.*) to rush, make a rush forward.
bonhomme, m., bolt, catch; (*mil. slang*) soldier (in the trenches, the soldier calls himself *bonhomme* and not *poilu*. The plural is *bonhommes*);
—— *d'arrêt,* (*art. mach.*) stop bolt;
—— *arrêtoir de tête mobile,* (*mach. gun*) bolt-head stop;
—— *de bascule,* (*mach.*) rocking piece, bolt.
bonnet, m., cap; (*aero.*) cowl;
—— *parasouffle,* (*nav.*) (cap covering head, as a protection against the blast of discharge); blast protector.
bonnette, f., (*inst.*) eyeglass shade.
bord, m., edge;
—— *arrière,* (*aero.*) trailing edge;
—— *d'attaque,* (*aero.*) leading edge;
—— *d'entrée,* (*aero.*) leading edge;
moyens du ——, (*auto.*) materials (e. g., repair) carried along;
—— *postérieur,* (*aero.*) trailing edge;
—— *de sortie,* (*aero.*) trailing edge.
bordereau, m., indorsement;
—— *d'envoi,* (*mil.*) indorsement on paper going forward or back.
borne, f., milestone; (*elec.*) terminal.
bossage, m., boss; any extra thickness of a part to make it stronger; spoke boss.
botte, f., boot;
—— *à gland,* (*unif.*) tasselled boot.
bottin, m., directory.
bottine, f., half-boot, gaiter;
—— *à boutons,* button shoe;
—— *élastique,* gaiter.

boucher, m., butcher;
— —s noirs, (mil. slang) the French artillery.
boucherie, f., butchers' shop;
— roulante, (mil.) traveling or field butchers' shop.
bouche-trou, m., (slang) a substitute.
bouchon, m., stopper, plug; (mach.) atomizer (of a spray carburetor);
— allumeur, (mil.) ignition priming;
— amortisseur, (art.) buffer (trench mortar);
— d'appui, (mach. gun) spring guide cap;
— cache-poussière, dust cap;
— de chambre à gaz, (mach. gun) gas chamber cap;
— de culasse, (trench art.) breech-closing plug; (210 C. T. mortar);
— à fermeture rapide, quick closing plug;
— de moyeu, hub cap;
— -percuteur, (mil.) plug-striker (of the D. R. grenade);
— porte-retard, (artif.) delay action plug;
— de prise de courant, (elec.) contact plug;
— de remplissage, (mach. gun and in gen.) filling hole plug;
— soluble, (expl.) soluble plug (safety device);
— de valve, valve cap;
— de vidange, drain tap.
boucle, f., buckle;
— double, double irons (punishment).
boucler, v. t., to buckle, buckle together;
— la boucle, (aero.) to loop the loop;
— son budget, (fam.) to make both ends meet;
— deux, trois, etc., lignes (mil. telephone) to loop two, three, etc., lines together.
bouclier, m., shield;
— articulé, (art.) apron, apron shield;
— coulissant, (art.) apron;
— individuel de parapet, (mil.) trench shield;
— mobile, (art.) apron;
— offensif individuel, (mil.) shield for open fighting;
— à rabattement, (mil.) folding-down trench shield;
— roulant individuel, (mil.) rolling shield;
— en tôle et sable, (mil.) double shield, plates separated by siliceous sand.
boudin, m., wad;
— d'air (auto.) pneumatic tire.
bougeoir, m., candlestick.
bougie, f., candle; (mach.) spark plug;
— d'allumage, spark plug;
— à disrupture, external spark-gap plug.
boule, f., (mach.) fly ball of a governor; (inst.) bulb (as of a thermometer); (mil. slang) issue bread, soldier bread;
— prise de son, (mil. min.) sound ball (for tapping sounds in a mine).
boulon, m., bolt;
— de chaîne, sprocket chain bolt;
— de sécurité, safety bolt.
bourdon, m., pole; (mil. slang) old horse.
bourrelet, m., flange, edge, or bead of a tire cover (pneumatic tire).
bourroir, m., (min.) tamper.
bousin, m., rotten ice.
boutefeu, m., (expl.) (generic term for) igniter (e. g., Bickford fuse).
bouteille, f., bottle;
— à gaz, gas cylinder.
bouton, m., button; (elec.) plug;
— d'angle de site, (art.) angle of site knob (155 CTR);
— d'appel, (tel., etc.) push, pushbutton;
— de commande, working head, adjusting knob;
— de dérivation, (art.) drift angle knob;
— double, stud, as on a gun sling;
— de manœuvre, (sm. a.) operating handle; (of the fusil mitrailleur);
mettre un —, to train to a fine point;
— d'oreille, (mil.) sort of earmark or button (on cattle);
— à pointeau, (mach. gun) regulator bolt arm knob;

bouton à poussoir, push button;
— de tir rapide, (mach. gun) releasing button (releases rapid-fire devices, mod. 1907 T).
boutonnière, f., buttonhole; ribbon on button of an order (worn in the button hole).
bowden, (proper name; aero. slang) flexible shaft.
boxe, f., crib (as for grain).
boyau, m., (mil.) boyau (a boyau is a trench of communication, perpendicular to the front; the tranchée is roughly parallel to the front);
— d'accès, approach, boyau of access;
— de commandement, (gives access to commanding officer's post, and receives communications from the rear) commanding officer's boyau;
— debout, boyau straight ahead (toward the enemy);
— d'évacuation, evacuation boyau;
— principal d'adduction, main boyau to the front;
— principal d'évacuation, main boyau to the rear (must be wide enough to take a stretcher);
— secondaire, small boyau;
— de surveillance, lookout boyau (between two trenches, two centers of resistance, etc.);
bracelet-montre, m., wrist watch.
brancard, m., (med., etc.) litter;
— à bretelles, (mil., med.) sling litter;
— traînant, (med.) travois.
branche, f., (aero.) propeller blade;
— à coulisse, sliding joint (as of a tripod leg);
— d'équipression, (aero.) branch of equal pressure (balloon sleeve).
— de fermeture, (mach. gun) breech tenon support;
— de percussion, (mach. gun) percussion arm.
branchement, m., branching; (artif., erpl. mil. min.) combination of two or more fuses with the principal (cordeau maître);
— multiple, multiple branching (more than one secondary fuse);
— simple, simple branching (only one secondary).
brancher, v. t., (elec.) to plug in, to connect.
brande, f., (top.) heath.
braquage, m., (auto.) change of direction of steering wheels;
angle de —, steering angle (i. e., made by plane of wheels with original position).
braquer, v. t., (aero.) to fix and hold the rudder at a given or desired angle.
bras, m., arm;
— de commande, (mach.) any controlling or actuating arm or lever;
— de pivot, pivot arm of a steering bar (motor car).
braser, v. t., (met.) to solder;
— au cuivre, to copper solder.
brasser, v. t., to stir (liquids, gases).
bretelle, f., suspender;
— de suspension (mil.) belt suspenders, belt straps;
— de tirage, breast strap.
brevet, m., (com.) patent (as of an invention);
demande de —, patent application.
breveter, v. t., to patent.
bricole, f., breast strap, dragrope;
— à bras, dragrope;
bride, f., clip;
— d'agrafage, (trench art.) assembling strap (platform);
— d'embrélage, (pont) lashing strap;
brigade, f., (r. r. adm.) detachment of railway mail service;
— neutre, (mil.) neutral brigade (exercises oversight over both parties in mining maneuvers at full strength);
— topographique, (mil.) topography squad.
brigadier, m., "boss" of a gang of workmen;
— d'équipe, (mil.) head or leader of a detail or squad;
— des haras, (Fr. adm.) sort of upper groom in the government studs;

brigadier *de tir*, (*Fr. art.*) (corporal who notes down data of fire, battery in action); recording corporal (corresponds to instrument sergeant, U. S. Art.).
brillant, m., polish (of metal surfaces, etc.); bright work.
brindille, f., twig.
briscard, m., (*mil. slang*) old soldier.
brisque, f., (*mil. slang*) long service stripes.
bristol, m., card, cardboard.
brisure, f., (*mil.*) jog; break in a parapet, to permit flank fire.
broche, f., pin;
—— *de manœuvre*, (*mach. gun*) maneuvering bar.
brochet, m., pike, (fish); (*mil. slang*) (German) trench mortar bomb, (so called from its shape).
brodequin, m., shoe, (U. S.); boot (British);
—— *de marche* (*unif.*) French infantry shoe;
—— *napolitain*, v. —— *de marche*.
brossage, m., brushing;
—— *de décors*, (*fam.*) bluff, play to the gallery.
brosse, f., brush;
—— *à chaîne*, chain brush;
—— *à dents*, tooth brush; (*mil. slang*) moustache.
brouette, f., barrow, wheelbarrow;
—— *porte-brancards*, (*mil. med.*) litter-barrow, wheeled litter.
brouette-caisson, f., (*trench art.*) ammunition barrow.
brouillage, m., (*elec.*) interference (as in a wireless).
brouiller, v. t., (*elec.*) to interfere (as in wireless).
brousse, f., the "brush," "bosque" (Spanish), *guerre de* ——, (*mil.*) bushwhacking; *prendre, reprendre, la* ——, to take to the bush, to the woods, to go into the "bosque."
brûle-gueule, m., (*slang*) short pipe (tobacco).
brûleur, m., burner, heater;
—— *électrique*, electric burner, heater;
—— *à pétrole*, petroleum burner (of an oil engine).
brume, f., fog;
—— *aérienne*, aerial fog (extends above storm clouds).

brun-mat, a., dull brown (i. e., not shiny).
brut, a., raw, gross;
revenus ——*s*, gross receipts.
Brution, m., (*Fr. a.*) graduate of the Prytanée Militaire.
budget, m., budget;
boucler le ——, to make both ends meet.
buisson, m., (*top*) thicket.
bulletin, m., bulletin;
—— *de correspondance*, (*Fr. a.*) (official or administrative note, sent by one staff to another, or to an officer); staff pad.
buraliste, m., (*Fr. adm.*) collector or receiver of certain taxes.
bureau, m., office; (*Fr. a.*) bureau or division of the staff of an army or corps (four in all); the first bureau has charge of personnel and matériel; the second, of information and political affairs; the third, of operations and movements of troops; the fourth, of transport (exists only in armies and groups of armies);
—— *ambulant*, (*adm.*) railway post office;
—— *central météorologique*, central or national weather bureau (at Paris);
—— *de douane*, custom house;
—— *de police*, police headquarters, police station;
—— *sédentaire*, (*Fr. adm.*) stationary, i. e., usual, post office as distinguished from railway post office.
burin, m., cold chisel.
but, m., (*art.*, *sm. a.*) target;
—— *à battre*, (*mil.*) target;
—— *témoin*, (*art.*) auxiliary target, reference target;
—— *flottant*, (*art., nav.*) floating target.
butée, f., lug; thrust collar;
à ——, "home";
—— *de détente*, (*art.*) trigger stop.
—— *de renversement*, (*art.*) tilting stop (French 75).
butteur, m., (*mach.*) tappet, striker.

C.

cabane, f., (*aero.*) cabane.
cabine, f., cabin;
—— *d'isolement*, (*mil.*) private room in barracks, for private toilet;
—— *téléphonique*, telephone booth.
cabinet, m., cabinet, w. c.;
—— *de service*, (*mil.*) office, orderly room.
câble, m., (*cord.*) cable;
—— *d'amarrage*, (*cord.*) mooring cable;
—— *armé*, (*elec.*) armored cable;
—— *d'armement* (*trench art.*) cocking lanyard;
—— *de campagne*, (*mil. telephone*) field cable;
—— *cuirassé*, (*elec., telephone, etc.*) armored cable;
—— *à deux conducteurs*, (*elec.*) double-conductor wire;
—— *double*, double conductor cable;
—— *léger* (*mil. elec.*) light cable;
—— *de mise de feu*, (*trench art.*) firing cable, lanyard;
—— *sous plomb* (*elec.*) leaded cable, cable sheathed in lead;
—— *de tirage*, tension rope, brake rope.
câblogramme, m., cablegram.
cabosser, v. t., to bruise, knock in.
cabot, m., (*mil. slang*) corporal.
cabotage, m., coasting trade;
—— *français*, coasting trade between French ports, (including those of Algiers);
—— *international*, coasting trade between foreign ports, or between French and foreign ports not so far off as to constitute navigation on the high seas.
cabrage, m., (*aero.*) nose-lift.
cabré, m., (*aero*) tail down, nose-lift.
cabrer, v. r., (*aero.*) of an airplane, to rear.
cache-flamme, m., (*art., mach. gun*) flash concealer, flash cut-out, flash protector.

cache-lueur, m., (*art., mach. gun*) flash cut-out flash protector; flash screen.
cache-nez, m., muffler.
cache-soupapes, m., (*mach.*) valve casing (dust guard).
cachet, m., seal, stamp;
—— *de vérification*, inspection stamp, inspector's stamp.
cachou, m., catechu, cutch;
couleur ——, dark buff.
cadence, t., cadence;
—— *du tir*, (*art.*) cadence of fire.
cadre, m., frame: (*aero*) running gear.
—— *bleu*, (*Fr. a.*) at Saumur, all the instructing staff not included in the —— *noir*, q. v.; (blue uniform);
hors ——, (*mil.*) on detached officers' list, D. O. L. (*U. S. A.*);
loi des ——*s*, (*Fr. a.*) the annual law making appropriations for the support of the army, appropriation act (U. S.), mutiny act, (Great Britain);
—— *noir*, (*Fr. a.*) at Saumur, the instructing equitation staff (so-called from their black uniforms);
rajeunir les ——*s*, (*mil.*) to reduce the age of retirement;
rajeunissement des ——*s*, (*mil.*) reduction of retiring age, (so as to get younger officers in each grade);
—— *stabilisateur*, (*aero.*) suspension frame, (Lebandy, acted as a stabilizer).
—— *de suspension*, (*trench art.*) hanging frame (construction of 340 mortar platform);
cafard, m., cockroach; (*mil. slang*) the blues.
cage, f., cage;
—— *d'entrée*, entrance cage (of a pigeon loft).

cagibi, m., (*mil. slang*) hut.
cagna, f., (*mil. slang*) hut.
cagoule, f., (*mil.*) gas mask.
cahoua, m., (*mil. slang*) coffee.
caillebottis, m., (*mil.*) grating, (flooring of a wet trench).
caïd, m., military chief (Arab).
caisse, f., box, chest;
—— *aux armements*, (*art.*) equipment chest;
—— *à grains*, crib;
—— *de charge*, (*min.*) explosive chamber of a mine;
—— *comptabilité*, (*mil.*) document chest;
—— *à détonateurs*, (*art.*) detonator chest;
—— *à ferrures*, (*mil.*) horseshoe chest;
—— *filtrante*, (*mil.*) sort of filter box, used in shelters, against gas attack;
—— *fonds*, (*mil.*) money chest;
—— *à pétards*, (*art.*) demolition chest.
—— *de treuil*, (*min.*) hoisting basket;
caisse-châssis, f., (*art.*) flange, rim (trench-mortar bed) box frame.
caisson, m., caisson, box;
—— *compartimenté*, (*trench art.*), bedplate (340 mortar).
—— *à munitions* (*art.*) caisson;
—— *observatoire*, (*art.*) observing caisson, (captain's post, battery in action).
—— *à renversement*, (*art.*) tilting caisson (French 75).
calage, m., (*elec.*, *etc.*) setting, adjustment; (*art.*) supporting block, (railway artillery).
calbombe, f., (*mil. slang*) lamp.
calculateur, m., counting machine, mechanical computer.
cale, f., key, wedge; (*nav.*) stocks;
—— *d'écartement*, distance piece, separator;
—— *d'épaisseur*, shim.
—— *de halage*, ways;
—— *de ressort*, spring block.
—— *de retenue*, retaining wedge;
caléfaction, f., calefaction.
calendrier, m., calendar; (*mil. fam.*) hairbrush grenade.
caler, v. n., to jam, stick fast, refuse to move or work; (*elec.*, *etc.*) to set, to adjust.
calfeutrage, m., chinking, operation of chinking.
calfeutrer, v. t., to chink (interstices).
califourchon, à, —— astride.
callu, m., wing disease (pigeons).
calot, m., (*unif.*) top of the crown of a forage cap.
calotte, f., (*mil. min.*) dome of a mine case;
—— *de sûreté*, (*expl.*) safety cap or hood (on certain detonators).
camarade, m., comrade;
—— *de promotion*, (*mil.*) classmate.
camaraderie, f., comradeship;
—— *de combat*, (*mil.*) brotherly feeling existing between officers of all arms serving together.
cambouis, m., coom;
les royaux ——, (*mil. slang*) train soldiers (from the grease on the axles of their wagons).
cambrousard, m., (*mil. slang*) peasant.
cambrure, f., (*aero.*) camber.
came, f., (*mach.*) cam;
—— *d'admission*, admission (valve) cam;
—— *d'allumage*, ignition cam; sparking cam;
—— *à bossage*, bossed cam;
—— *à cran*, notched cam;
—— *de décompression*, relief cam;
—— *de distribution*, inlet cam, outlet cam;
—— *d'échappement*, exhaust (valve) cam.
camion, m., truck;
—— *automobile*, motor truck, motor lorry;
—— *à benne basculante*, tilting truck.
—— *de campement*, (*mil. aero.*) camp equipage truck.
camion-atelier, m., (*mil. auto.*) repair truck.
camion-magasin, m., (*mil. aero.*) general supply truck (balloons).
camionnette, f., light truck.
camisole, f., sort of chemise (woman's);
—— *de force*, straight-jacket.
camouflage, m., (*mil.*) disguise, act of disguising (e. g., a gun by painting it in harlequin colors, burying it in bushes, etc.); disguising, concealment (of a road, trench, etc.); "faking."
camoufler, v. t., (*mil.*) to disguise (as, a gun by painting, burying in bushes); to disguise, conceal (a road, a trench, etc.); to "fake."
camp, m., camp;
—— *de concentration*, (*mil.*) concentration camp;
—— *de repos*, (*mil.*) rest camp.
campé, a., (*aero.*) camped (said of aircraft when stopped at a place that offers no shelter).
campement, m., (*mil.*) camp equipage, camp outfit; (*aero.*) stop (at a place that offers no shelter), balloon bed (sphericals);
—— *collectif*, (*mil.*) portions of camp equipage carried by the men, and forming when combined an outfit for several persons; joint camp equipage.
point de ——, (*aero.*) balloon bed.
camper, v. t. n., to camp; to encamp; (*aero.*) to stop (at a place that affords no shelter);
—— *un ballon*, (*aero.*) to camp a balloon (when camped, the balloon is sheltered from enemy fire, shielded from enemy view, masked against airplanes, and properly secured in position).
canal, m., canal;
—— *de sortie*, (*mach. gun.*) band exit;
—— *de conduite*, (*mach.*) guide (as of a valve stem);
—— *à écluses*, lock canal;
—— *latéral*, canal flanking a shoal, fall, in a river; side canal;
—— *à niveau d'eau*, sea-level canal;
—— *à point de partage*, canal joining different basins, or watersheds.
canalisation, f., piping; (*elec.*) wiring.
canapé-lit, m., sofa-bed.
canard, m., duck; (*mil. slang*) soldier.
canarder, v. t., (*mil. fam.*) to snipe.
canardeur, m., (*mil. fam.*) sniper.
canevas, m., (*top.*) triangulation, network of triangles;
—— *d'ensemble*, (*art.*, *mil.*) general triangulation (for artillery and for general military purposes);
—— *de tir*, (*art.*) survey or triangulation for the direction of artillery fire; artillery map or triangulation.
canevas-guide, m., (*top.*) reference triangulation.
caniveau, m., shallow trench.
canon, m., (*art.*) gun (to-day, long gun giving a flat trajectory);
—— *d'accompagnement*, (*mil.*) gun to accompany and be operated by infantry; infantry gun;
—— *anti-aérien*, (*art.*, *aero.*) anti-aircraft gun;
—— *antitorpilleur*, (*nav. art.*) in the modern (dreadnaught) type of ship, a gun whose special business it is to beat off torpedo-boat attack;
—— *automobile blindé*, gun mounted on an armored automobile;
—— *contre-avions*, (*art. aero.*) antiaircraft gun;
—— *contre-torpilleur*, v. —— *antitorpilleur*;
—— *court*, short gun (to-day, practically synonymous with mortier).
—— *éguculé*, gun worn down at the muzzle;
—— *à encombrement normal*, (*art.*) gun occupying normal space;
—— *à fusil*, v. —— *de lancement*;
—— *s jumelés*, (*nav.*) guns in pairs (turret guns);
—— *de lancement*, (*mil.*) discharging barrel (of the *obus à fusil*);
—— *à long recul*, long recoil gun;
—— *à moteur*, gun on a motor carriage;
—— *à recul sur l'affût*, carriage recoil gun;
—— *de tranchée*, trench gun.
canot, m., canoe;
—— *automobile*, (*nav.*) auto boat;
—— *à moteur*, (*nav.*) motor boat.
cantache, f., (*mil. slang*) canteen.
cantoche, f., (*mil. slang*) canteen.

canton, m., (France) political unit composed, for certain joint administrative purposes, of two or more *communes* (is not a person at law); canton.

cantonnement, m., (*mil.*) cantonment;
— *d'alerte*, a cantonment in which an alarm is to be expected (officers stay with their men, no one undresses); alarm cantonment;
— *de rafraîchissement*, (*mil.*) rest cantonment; *surface de* —, billeting area.

cantonnement-bivouac, m., (*mil.*) cantonment in which the troops are lodged as far as accommodations go, with the remainder bivouacking.

caoua, m., (*mil. slang*) coffee.

caoutchouc, m., caoutchouc;
— *durci*, hard rubber.

capiston, m., (*mil. slang*) the captain.

capitaine, m., (*mil., nav.*) captain;
— *de corvette*, (*nav.*) lieutenant commander (grade U. S. N., R. N.);
— *instructeur de tir*, (*Fr. a.*) musketry officer;
— *de jour*, (*mil.*) captain of the day (in attack by mines, assistant to the engineer officer in charge).

capiton, m., cushioned back (of a seat).

capitonnage, m., upholstery, cushioning.

capitonner, v. t., to upholster, to tuft.

caporal-approvisionneur, m., (*mil.*) corporal in charge of an ammunition supply squad; ammunition corporal.

caporal-vigie, m., (*mil.*) look-out corporal.

capot, m., cowl; (*auto.*) bonnet, hood.

capotage, m., (*aero.*) somersault.

capote, f., top, hood (of a carriage, auto, etc.); (*unif.*) (infantry) overcoat.

capoter, v. n., (*aero.*) to turn turtle; to somersault.

captage, m., piping (of water, from its source).

captation, f., telephone tapping.

capter, v. t., to pipe (water); to tap (a telephone).

capteur, m., (*nav.*) captor;
— *d'ondes*, (*wireless*) wave receiver (said of a wire).

capuchon, m., hood;
— *de valve*, valve guard (of a tire).

caramanage, m., mandrelling.

carburant, m., a generic term for the fuel (gasoline, alcohol, petroleum) used in explosion engines.

carburation, f., carburation.

carburateur, m., (*mach.*) carbureter;
— *à alcool, etc.*, alcohol carbureter (and similarly of other combustibles);
— *automatique*, compensation carbureter;
— *automatique à giclage et à niveau constant*, automatic float feed spray carbureter;
— *à barbotage*, splash carbureter, (obs.);
— *à champignon simple*, simple cone carbureter (no needle valve);
— *à contact*, generic term for carbureters of the surface and splash types;
— *à courants parallèles*, (any carbureter in which the air current is sensibly parallel to the direction of spray) parallel flow carbureter;
— *à courants perpendiculaires*, any carbureter in which air flow is perpendicular to spray flow; perpendicular flow carbure.er;
— *à distribution automatique*, generic term for automatic carbureters;
— *à distribution mécanique*, mechanically operated carbureter;
— *à giclage*, spray carbureter;
— *à léchage*, surface carbureter;
— *à mèches*, wick carbureter;
— *à niveau constant*, float feed carbureter;
— *noyé*, flooded carbureter;
— *à pulvérisation*, spray, atomizing, carbureter;
— *à réglage automatique*, compensating carbureter;
— *à réglage de l'air*, extra air carbureter;

carburateur *à réglage de l'air et de l'essence*, extra air gasoline control carbureter;
— *à réglage de l'essence*, gasoline control carbureter.

carburé, a., carburated.

carcasse, f., (*aero.*) frame (over which are stretched the membranes of the wings or body).

cardan, m., (*mach.*) universal joint;
— *à double articulation*, double joint or coupled cardan shaft;
— *longitudinal*, (*auto.*) longitudinal cardanic transmission, (system of transmission from gear box to rear li e axle by a cardanically controlled longitudinal shaft);
— *transversal*, (*auto.*) cross cardanic transmission, (system in which the wheel shafts are directly controlled by cardanic transmission);
— *à une articulation*, single joint or coupled cardan shaft.

carénage, m., (*nav.*) careening;
forme de —, careening ways.

carène, f., (*aero.*) the lower part of the envelope of a dirigible or of its frame; (by extension) the envelope or frame itself.

carlingage, m., (*aero.*) engine bed.

carlingue, f., (*aero.*) car, nacelle; cock pit.

carnet, m., notebook;
— *à dépêches*, (*mil.*) dispatch book;
— *multicopiste*, (*mil.*) copying notebook;
— *de position*, (*art.*) emplacement record;
— *de tir*, (*art.*) battery firing record;

carré, m., (*mach.*) square shaft.

carreau, m., square (of a squared map);
— *de réduction*, reducing square, (for copying a drawing or map on a smaller scale).

carrefour, m., (*mil.*) intersection of boyaux, trenches.

carrelage, m., tiling (small);
— *à caoutchouc à emboîtement*, interlocking rubber tiling.

carrière, f., (*man.*) instruction in jumping (Saumur).

carrosserie, f., (*auto.*) carriage body, carriage proper;
— *à entrée latérale*, side entrance body.

carotte, f., carrot.

carotter, v. t. —, *le pieu*, (*mil. slang*) to sleep late (Saumur).

carroyé, a., squared, in squares (as, a map).

carte, f., map;
— *aéronautique*, (*aero.*) aeronautical map;
— *d'artillerie*, (*art.*) (artillery) fire map;
— *de batterie*, (*art.*) battery chart;
— *d'identité*, (*mil.*) identification card;
— *à main levée*, freehand topographical sketch or map;
— *quadrillée*, squared map, graticulated map, map in squares;
— *du temps*, weather map.

carte-itinéraire, f., (*r. r.*) guide map.

carter, m., (*mach.*) motor casing, engine casing; crank case;
— *à bain d'huile*, splash-about (crank) case;
petit —, gear, case.

carton, m., pasteboard; pasteboard box;
— *bitumé*, (*cons.*) tarred paper.

cartonneux, a., resembling pasteboard, stiff (like pasteboard).

cartouche, f., (*sm. a., artif.*) cartridge; (in signaling) any device or fireworks discharged by a pistol or gun;
— *éclairante*, (*artif.*) flare;
— *éclairante à parachute*, (*artif.*) parachute flare (releases a star fitted with a parachute);
— *à fausse balle* (*sm. a.*) dummy cartridge;
— *de lancement*, (rifle) grenade cartridge;
— *de salut*, (*art.*) saluting cartridge;
— *signal*, (*artif.*) signal cartridge;
— *à signaux*, v. — *signal*;

cartouche *de stand* (*t. p.*) sort of gallery-practice cartridge (to be used, *Fr. a.*, when a range is not available);
—— *de sûreté*, (so called) riot cartridge (fragmentation bullet).
cartouche-indicatif, f., (*mil.*) indentification cartridge (giving the call letter, signal number, or private signal of an airplane).
cartouche-relais, f., (*trench art.*) relay cartridge (German 18 cm. projectile).
cartouche-signal, f., (*mil.*) signal cartridge.
case, f., hut; compartment;
—— *d'armon*, (*art.*) side-rail box;
—— *à nids*, nest case, nest compartment (pigeons).
casematage, m., (*art., fort.*) casemating, protection.
casier, m., case (box or chest divided into compartments);
—— *à bombes*, (*trench art.*) bomb case;
—— *porte-bombes*, (*trench art.*) shell rack (transportation);
—— *téléphonique*, head set.
caso, m., (*mil. slang*) casoar or plume (worn by St. Cyrians on their shakos).
casoar, m., (*unif.*) plume (worn by cadets of St. Cyr on their shakos).
casque, m., (telephone) head set; (*mil.*) steel helmet (casque Adrian);
—— *colonial*, (*unif., Fr. a.*) the helmet worn by French colonial troops;
—— *à pointe* (*mil. slang*) bomb of the Erhardt trench mortar.
casse-pierre, m., (*min.*) rock crusher (drill).
casser, v. t., (*mil. law*) to disapprove the verdict of a court-martial;
—— *un arrêt*, (*law*) to reverse a verdict and send the case back.
cassis, m., water-bar.
catapulte, f., catapult.
catégorie, f., category;
—— *spéciale*, (*Fr. a.*) special class of list of horses for staff or detached officers.
cause, f., case, cause;
hors de ——, (*mil.*) out of action
caution, f., guarantee; (*adm.*) deposit; guarantee;
se porter —— *pour quelqu'un*, to guarantee a person.
cautionnement, m., (*adm.*) bonding.
cautionner, v. t., (*adm.*) to bond.
cavalerie, f., cavalry;
—— *d'armée*, army cavalry (charged with exploration at a distance);
—— *de corps*, corps cavalry;
—— *divisionnaire*, divisional cavalry.
cavalier, a., (*drawing*) cavalier; m., horseman; staple (nail); (*mil.*) trooper, cavalryman;
—— *colombophile*, (*mil.*) cavalry pigeon man;
projection —— *e*, (*drawing*) cavalier projection;
vue —— *e*, (*drawing*) view in cavalier projection.
cavitation, f., (*aero.*) cavitation.
cécité, f., blindness;
—— *monoculaire*, blindness in one eye.
ceinturage, m., (*mil.*) belting of crossroads in trenches, by a passage.
ceinture, f., belt;
(*pont*) wale (metallic boat);
—— *de roue*, caterpillar chain;
—— *de sûreté*, (*aero., etc.*) safety belt;
ceinturer, v. t., to belt.
—— *un carrefour* (*mil.*) to belt a crossroads by a passage.
ceinturon, m., (*mil.*) waistbelt;
—— *porte-grenades*, grenade belt.
cellule, f., cell; (*aero.*) the "box" formed, e. g., by the upper and lower wings of a biplane as joined by stanchions and braces; the aeroplane minus the engine set; cell.
centrage, m., (*aero.*) question of centers of an airplane.

central, m., (*mil., sig., etc.*) central, central post, central station (e. g., central observing station; no reference to post of commanding officer);
—— *aéro*, central aeronautic station;
—— *à 5, etc., directions*, 5-, etc., wire central;
—— *aérostier*, (*mil. telephone*) balloon central station;
—— *de groupement*, (*mil. telephone*) the *central* of a *groupement*, q. v., of artillery;
—— *d'observation*, (*mil. telephone*) station at terminus of an extension line (original transmitting station for observations);
—— *de réglage*, (*mil. telephone*) transmitting station (junction of branches with main line).
central-arrière, m., (*mil. telephone*) central in rear of the front (artillery system) for the higher command, and balloons.
central-avant, m., (*mil. telephone*) front central (near the front; artillery system).
centre, m., center; in general, central point (in many relations); (*aero., etc.*) aviation center, or establishment, for instruction, research, etc. (may be established on battle field);
—— *de compositions et d'examen oral*, place where examinations are held; examining point;
à —— *s confondus*, (*aero.*) with coinciding centers;
—— *d'examen oral*, place where an oral examination is held;
—— *de fabrication de pain*, (*Fr. a.*) central bakery;
—— *hospitalier*, (*mil. med.*) any permanent or temporary hospital in rear of an army in the field, and in the occupied region;
—— *d'instruction*, (*Fr. a.*) officer's school (behind the lines);
—— *d'observation*, (*aero.*) (captive) balloon center (in sieges);
—— *oral*, v. —— *d'examen oral;*
—— *de renseignements*, (*mil.*) collection station, (of information, etc., respecting the enemy, the terrain, etc.; receives airplane observations for large units in march, and is established on the line of march);
—— *de résistance*, (*mil.*) to-day, the combination of *points d'appui*, q. v.;
[1 *centre*=3 (for example) *points d'appui;*
1 *point d'appui*=4 (for example) *segments actifs*;
1 *segment actif*=line or points actively held.]
centre-école, m., (*aero.*) aviation school.
cercle, m., circle, cask;
—— *à calculer*, circular calculation machine;
—— *de charge*, (*aero.*) ring or circle carrying the car; suspension ring;
—— *de suspension*, v. —— *de charge;*
—— *de visée*, (*art.*) aiming circle.
cercle-entretoise, m., (*art.*) circular transom.
cerf-volant, m., kite;
—— *cellulaire*, box-kite.
cerf-voliste, m., (*aero.*) kite man.
chaîne, f., chain, (*art.*) chain, line, (of markers, of runners, may serve to establish a liaison);
—— *d'agrafage*, securing chain;
—— *de relevage*, (*art.*) lifting chain;
—— *à rouleaux*, roller chain;
—— *semi-rigide*, sprocket chain.
—— *silencieuse*, silent chair;
chaînette, f., small chain;
—— *arrache-clous*, v. *arrache-clous;*
—— *d'épaule*, (*unif.*) shoulder chains.
chaîneur, m., (*mil.*) sand-bag passer.
chaise, f., chair;
—— *à bascule*, rocking chair.
chaleur, f., heat;
craint la ——, keep cool, keep away from boilers (on cases, etc., for shipment).
chalumeau, m., blowpipe;
—— *oxyhydrique*, oxyhydrogen blowpipe, compound blowpipe;
—— *oxyacétilénique*, oxyacetylene blowpipe (used for cutting steel).
chambre, f., chamber, room; space;
—— *à air*, inner tube, air tube (of a tire);

chambre, *de carburation*, (*mach.*) mixing chamber, spraying chamber;
—— *de compensation*, (*com.*) clearing house;
—— *de compression*, (*mach.*) compression space (cylinder of an explosion engine); air-compressing chamber (of an air compressor);
—— *des députés*, Chamber of Deputies (lower house, France);
—— *d'escouade*, (*mil.*) squad room;
—— *d'explosion*, (*mach.*) combustion chamber (of an explosive motor);
—— *du flotteur*, (*mach.*) float chamber;
—— *à gaz*, (*mil.*) gas chamber, (through which all officers and men, suitably masked, must pass in order to test them);
mettre la ——, (*trench art.*) to adjust the chamber (of combustion, V. D. mortar);
—— *de mélange*, (*mach.*) mixing chamber;
—— *aux poudres*, (*mil. min.*) powder chamber;
—— *provisoire*, (*trench art.*) mean chamber (V. D. mortar);
—— *de réchauffage*, (*mach.*) hot-air jacket, hot-air chamber;
—— *de soupapes*, (*mach.*) valve chamber;
—— *de vaporisation*, (*mach.*, *etc.*) vaporizing chamber.

chambrée, f., (*mil.*) squad room.

champ, m., field;
—— *d'atterrissage* (*aero.*) landing ground;
—— *de courses*, (*sport*) race track; racing field or ground;
—— *de pointage en direction*, (*art.*) limits of an aim in azimuth;
—— *de pointage en hauteur*, (*art.*) limits of aim in elevation;
—— *de tir*, (*sm. a.*) rifle range;
—— *de tir en hauteur*, (*art.*) limits of elevation of a gun;
—— *de tir latéral*, (*art.*) limits of fire in direction of a gun.

champignon, m., mushroom; (*mach.*) nozzle or spray atomizer; atomizing cone, or atomizer head, cap, (of a valve);
—— *gicleur*, spraying device, spray;
—— *à rainures*, grooved spray, (in some forms of carbureters);

chancre, m., ulcer.

chandelier, m., candlestick; (*mil. slang*) German trench mortar bomb (so called from its shape).

chandelle, f., candle; (*aero. slang*) cabré, tail down.

chandail, m., (*slang*) sweater.

changement, m., change;
—— *de marche*, (*mach.*) reversing gear;
—— *de vitesse*, (*mach.*) change-speed gear; speed gear;
—— *de vitesse à train baladeur*, (*mach.*) slide-block change gear;
—— *de vitesse à embrayage individuel*, v. —— *de vitesse en prise*;
—— *de vitesse en prise*, (*mach.*) individual clutch change-speed gear;
—— *de vitesse à train epicycloïdal*, (*planétaire*), (*mach.*) planetary change gear.

chantier, m., working place; dump, yard (for the collection of material);
—— *auto-maritime*, auto-boat establishment;
carnet de ——, (*mil. min.*) memorandum of tools (received, turned in, or lost) and of special orders;
—— *de débarquement*, (*mil. r. r.*, *etc.*) generic term for unloading points (of whatsoever nature);
—— *d'équarrissage*, v. *clos d'équarrissage*;
mettre en ——, to set men to work; to lay out work;
mise en ——s, operation of setting men to work; laying out of work.

chape, f., (*mach.*) jaw, yoke, fork;
—— *d'attache*, (*mach. gun*) shield bracket;
bras de —— *d'attache*, shield bracket arm;
—— *de pivotement*, Y of a steering axle.

chapeau m., hat; (*surv.*) polygon of error (the number of sides depends on the number of sights taken);
—— *bras*, cocked hat (obs.);
—— *colonial*, (*unif.*) campaign hat;

chapeau *haut de forme*, high hat, pot hat;
—— *de roue*, wheel cap, hub cap;
—— *de serrage*, (*mach.*) clamp, clamping rinrog collar;
—— *de valve*, valve cap (of a pneumatic tire).

chapelet, m., string of beads;
en ——, (*aero.*) in series (of a train of kites).

chapelle, f., (*mach.*) explosion chamber (forms with the *culasse*, the explosion chamber).

char, m., car, cart;
—— *d'assaut*, (*mil.*) tank.

charbon, m., coal;
—— *collecteur*, (*elec.*) brush.

charbonner, v. t., to char; to deposit carbon.
—— *un navire*, (*nav.*) to coal a ship.

charbonneux, a., carbon;
dépôt ——, (*mach.*) carbon deposit (in motors).

chardon, m., thistle.

charge, f., charge, load; (*phys.*) head (pressure);
—— *allongée*, (*mil. min.*, etc.) elongated charge (cartridges end to end);
—— *amorce*, (*art.*) priming charge;
—— *de chambrage*, (*mil. min.*) chambering charge (v. *chambrage*);
—— *d'eau*, (*phys.*) head of water;
—— *faible*, (*trench art.*) reduced charge (mortar);
—— *fondamentale*, (*art.*) unit charge;
—— *forte*, (*trench art.*) full charge (mortar);
—— *d'inflammation*, (*art.*) igniting charge;
—— *par mètre*, (*aero.*) wing loading;
—— *totale*, (*aero.*, etc.) full load;
perte de ——, (*phys.*) loss of head;
rompre ——, to break bulk;
—— *de transmission*, (*expl.*) communicating charge (from primer to burster);
—— *utile*, useful load.

chargement, m., loading;
—— *mixte*, (*art.*) mixed load (i. e. shrapnel and H. E. shell, in limbers, etc.).

chargeur, m., (*art.*, *etc.*) (gun) loader (person); (*elec.*) charging set (storage battery); (*mach. gun*) magazine.

chargeur-pourvoyeur, m., (*trench art.*) loader and server.

chariot, m., wagon;
—— *d'atterrissage* (*aero.*) landing carriage;
—— *de chargement*, (*art.*) shell truck (for loading);
—— *à munitions*, (*trench art.*) projectile truck;
—— *de parc*, (*mil.*) transport wagon;
—— *à pétards*, (*mil.*, *art.*) demolition wagon;
—— *porte-affût*, (*trench art.*) carriage truck (240 C. T. mortar, transportation);
—— *porte-canon*, (*trench art.*) gun truck (transportation);
—— *porte-charges*, (*trench art.*) cartridge truck;
—— *porte-fournil*, (*mil.*) kneading wagon (carries kneading troughs);
—— *porte-mortier*, (*trench art.*) mortar truck (transportation);
—— *porte-outils*, (*trench art.*) tool truck;
—— *porte-plateforme*, (*trench art.*) platform truck (transportation);
—— *porteur*, r. r. *art.*) roller cradle;
—— *de refoulement*, (*art.*) shot truck and rammer, rammer shot truck;
—— *de roulement*, (*art.*) roller truck.

chariot-fournil, m., (*mil.*) kneading wagon.

Charlotte, f., (proper name) (*mil. slang*) the 75-mm. gun.

charnière, f. hinge; knuckle;
—— *à fiche*, pin hinge.

charogne, f., carrion;

charognard, m., (*fam.*) mean, worthless horse (crowbait).

charpente, f., framing;
—— *métallique*, (*cons.*) generic term for metallic girders, beams.

chasse, f., set or inclination of an axle (motor car); (*artil.*) expelling charge;
en ——, (*aero.*) straight ahead (e. g., of fire).

chassebis, m., (*mil. slang*) *chasseur à pied*.

chasse-clavette, m., key driver.

châssis, m., frame;
[The following terms relate mainly to the automobile;]
—— *à arrière relevé*, elevated rear frame;
—— *d'atterrissage*, (*aero.*) landing chassis, under carriage;
—— *à avant rétréci*, cambered frame;
—— *blindé*, v. —— *cuirassé*;
—— *en bois armé*, armored wood frame;
—— *cintré*, curved frame;
—— *cuirassé*, pressed steel frame (in one piece);
—— *à double rideau*, (*phot.*) plate-holder.
—— *droit*, straight frame;
—— *embouti*, pressed steel frame;
faux ——, false frame;
—— *tournant*, (*mil. min.*) turning frame.
—— *en tubes*, pipe frame, tube frame.
châssis-blindé, m., (*mil.*) supporting frame (roof of covered shelter).
châssis-truc(k), m., (*art.*) railway chassis (the truck forms the chassis).
chatterton, m., tape.
chaudière, f., (*steam*) boiler; steam boiler;
—— *aquatubulaire*, water-tube boiler;
—— *aquatubulaire à flamme directe et à retour de flamme*, return flue steam boiler;
—— *à circulation*, circulating boiler;
—— *à éléments*, generic term for tubular boilers;
—— *multitubulaire*, multitubular boiler;
—— *à tubes d'eau*, water-tube boiler.
chauffage, m., heating; (*slang*) cramming for examination;
—— *central*, furnace heating (houses).
—— *à la vapeur*, steam heating;
chauffage-vapeur, m., steam heat, steam heating.
chauffard, m., (*auto.*, *fam.*); chauffeur.
chauffe-bains, m., hot-water heater.
chauffe-pieds, m., foot warmer.
chauffer, v. t., to heat; (*slang*) to cram for examination;
—— *à la vapeur*, to heat by steam.
chauffeur, m., (*auto.*) chauffeur.
chaussette, f., sock.
—— *s à clous*, (*mil. slang*) soldier shoes;
—— *à lacets*, lace shoe;
—— (*dite*) *de repos*, (*mil.*) barrack shoe.
—— *russe*, (*mil.*) (strip or bandage of tallowed cotton, a substitute for socks) Russian sock;
chavirer, v. n., to capsize;
—— *au canon*, (*nav.*) to be sunk by gunfire.
cheddite, f., (*expl.*) cheddite (chlorate of potassium base, and castor oil).
chef, m., chief;
—— *de four*, (*mil.*) head baker;
—— *de gare*, (*Fr. a.*, *r. r.*) military station master;
—— *du mouvement* (*r. r.*) train dispatcher;
—— *de train*, (*r. r.*) conductor.
chemin, m., road, path;
—— *creux*, hollow road, sunken road;
—— *de fer de campagne*, (*mil.*) field railway;
montrer le —— *de St. Jacques*, (of a horse) to point the toe;
—— *de roulement*, (any) roller way.
cheminée, f., chimney;
—— *d'aération*, (*mil.*) ventilating shaft (dugouts, etc.);
—— *d'étoupille*, (*art.*) primer seat.
cheminement, m., (*mil.*) advance under cover, unobserved (not an official word); (hence) ravine, path through standing crops, and in general, any way affording cover;
—— *à ciel ouvert*, advance without overhead cover;
dérober une portion du ——, (*siege*) to steal an advance (e. g., when enemy fire slackens, thus making simpler earthwork possible);
—— *souterrain*, (in sieges) advance by mining;
chemise, f., shirt; (*adm.*) jacket surrounding a number of papers, wrapper;
—— *de circulation*, (*mach.*) water jacket;
—— *de circulation d'eau*, (*mach.*) water jacket;
—— *d'eau* (*mach.*) water jacket;
—— *de réchauffage*, (*mach.*) heating jacket (alcohol motor);

—— *de roulement*, (*art.*) traverse circle, racer.
chèque, m., check (bank check);
—— *postal*, money order.
chercheur, m., seeker;
—— *de pôles*, (*elec.*) pole detector.
cheval, m., horse;
—— *d'âge*, (*mil.*, *Fr. a.*) horse between 4½ to 8 years of age (horse that has the age required for military service);
—— *attelé seul*, single horse;
—— *de bois*, (*aero.*) carrousel (involuntary sharp turn on surface of ground);
—— *calme*, steady horse;
—— *de carrière*, jumping horse;
—— *coulant*, horse of easy, flowing gait;
—— *déclassé*, horse unfit for the class of work intended;
—— *de dehors*, outdoor horse, cross country horse;
dépôt de —— *aux* (*mil.*) horse depot, field depot for remount purposes;
—— *effectif*, (*phys.*) actual horsepower; 75 kilogrammetres;
—— *impressionnable*, nervous, excitable horse;
—— *aux indiqués*, (*phys.*) indicated horsepower;
—— *jeune*, horse less than 4½ years old;
—— *de légère*, (*Fr. cav.*) light cavalry horse;
—— *mis*, well-trained horse;
—— *mis au bouton*, finely trained horse;
—— *nominal*, (*phys.*) 300 kilogrammetres;
—— *placé*, trained horse;
—— *à poil*, bareback horse, i. e., horse without equipments.
cheval-heure, m., (*phys.*) horsepower-hour.
chevalet, m., trestle; (*mil.*) obstacle of barbed wire (fastened to the ends of a sawhorse, as it were, back and forth, diagonally, etc.; (*inf.*) rifle rest;
—— *lance-fusée*, (*mil.*) rocket trestle, stand;
—— *de latrine*, (*siege*) latrine trestle (seat);
—— *de levage*, v. —— *de manoeuvre*;
—— *de manoeuvre*, (*trench art.*) trench gin (340 mortar);
—— *de pointage*, (*mach. gun*) firing stand, support;
—— *de tir*, (*mil.*) firing stand or trestle for rockets and grenades.
chevalet-bigue, m., (*pont.*) sheer trestle.
chevalet-palée, m., (*pont.*) pile trestle.
chevaucher, v. t., (of signals) to overlap.
cheville, f., (*elec.*) jack;
—— *de fixation* (*trench art.*) securing bolt (240 C. T. carriage truck).
chicane, f., obstacle; (*steam*, *etc.*) baffle, baffle plate;
disposer en ——, to stagger (of objects);
en ——, (so as to offer resistance, as it were, to dispute a passage; hence, in many technological relations) with joints broken, elbowed, staggered;
—— *pour les gaz* (any intentional obstacle to the passage of gases so as to lessen shock, violence of action, etc.), baffle plate.
chicasse, f., (*mil. slang*) mixture of chicory and coffee.
chien, m., dog; (*mach.*) dog, catch; (*mach. gun*) bolt body;
—— *d'armement*, (*art.*) cocking piece (trench mortar);
—— *de quartier*, (*mil. slang*) an *adjudant*;
—— *sanitaire*, (*mil. med.*) sanitary dog (trained to search for wounded on battle fields);
tige de —— (*mach. gun*) bolt stem.
chiffon, m., rag;
—— *d'essuyage*, dust cloth.
chlore, m., (*chem.*) chlorine.
choc, m., knocking of an engine, in machinery;
détachement de ——, (*mil.*) hand-to-hand squad (for assaults).
choix, m., choice, selection;
nommer au ——, *nommer au* —— *hors tour*, (*mil.*) to nominate by selection.
chou, m., cabbage.

chronographe, m., any time-measuring instrument, chronograph;
— à *indications multiples*, (*ball.*) continuous-reading chronograph (e. g., Schultz);
— à *indication unique*, (*ball.*) single-reading chronograph (e. g., Boulangé).
chronoscope, m., stop watch, split-second watch.
chronotachymètre, m., (*auto.*) indicator and record, of times, speeds, starts and stops.
chute, f., fall;
— *de matière active*, (*elec.*) scaling (of an accumulator);
— *de pression* (*steam*, etc.) fall of pressure;
— *en tire-bouchon*, (*aero. fam.*) corkscrew fall;
cible, f., target;
— à *avertissement automatique*, (*t. p.*) self-recording, automatic, target;
vérifier la ——, (*t. p., art.*) to inspect and mark a target.
ciel, m., sky;
— *pur*, clear sky.
cigare, m., cigar.
cigarette, f., cigarette;
de quoi faire une —— the "makings."
ciment, m., (*cons.*) cement, concrete;
— *armé*, reinforced concrete.
cinémomètre, m., a synonym of *tachymètre*, q. v.
cingoli, m., caterpillar chain.
circuit, m., circuit; (*elec.*) circuit;
— *d'allumage*, (*elec.*) ignition circuit or wiring; light circuit;
— *bouclé*, (*mil. telephone*), any line switched in; additional circuit;
— *d'éclairage*, (*elec.*) lighting circuit;
— à *étincelles*, (*elec.*) sparking circuit;
— *fermé*, (*elec.*) closed circuit;
— *ouvert*, (*elec.*) open circuit;
— *primaire*, (*elec.*) primary circuit;
— *secondaire*, (*elec.*) secondary circuit.
circonscription, f., subdivision;
— *sanitaire*, (*Fr. adm.*) subdivision of the territory for quarantine purposes.
circulaire, f., circular; (*art.*) racer;
— *d'appui*, (*trench art.*) circular support, supporting ring.
circulation, f., circulation; traffic;
— *forcée*, (*mach.*) forced circulation (as by a pump);
— *par gravité*, (*mach.*) gravity circulation;
— *par pompe*, pump circulation;
— *par thermo-siphon*, thermo-siphon circulation.
circuler, v. n., to circulate, i. e., go back and forth, pass freely;
droit de ——, right of free passage;
— à *pied*, to go about on foot.
ciseler, v. t., to crop (a wing).
citation, f., citation; (*mil.*) mention in orders.
citer, v. a., (*mil.*) to mention in orders.
citerne, f., cistern; (*nav.*) water boat;
— à *vapeur*, (*nav.*) steam water boat.
citoyen-soldat, m., a citizen who is a soldier (has done real military service).
civière, f., barrow;
— *en corde*, (*mil. min.*) cord barrow;
— à *mine* (*trench art.*) bomb carrier (German);
— *en toile* (*mil. min.*) canvas barrow.
civil, a., civil;
au ——, (*law*) in civil cases.
civlot, m., (*mil. slang*) a civilian, a "cit."
clabot, m., (*mach.*) sliding clutch;
— *de prise directe*, high speed sliding clutch.
clabotage, m., (*mach.*) engagement, clutch;
— *de prise directe*, high speed engagement or clutch.
clairon, m., bugle; bugler;
— à *anche* (*mil.*) reed bugle;
— *mécanique* (*mil.*) mechanical bugle, (works by compressed air);

clairon *périscopique*, (*mil.*) periscopic bugle, (ball projects above parapet).
clapet, m., valve; (*more esp.*) head or cap of a valve;
— *de chambre* à *air*, valve plug (pneumatic tire);
— *électrolytique*, (*elec.*) A. C. electrolytic rectifier.
claquement, m., (*sm. a.*) crack, cracking (of a bullet, sound).
claquer, v. n., (*sm. a.*) (of a bullet) to crack (sound).
clarification, f., clarification (of water, etc.).
clarifier, v. t., to clarify (water).
clarinette, f., (*mus.*) clarinet; (*aero.*) gas cylinder manifold.
classe, f., (*mil.*) class of a conscription of levy;
— *d'instruction*, (*mil.*) a squad or number of men under instruction.
classeur, m., file (for letters, papers, cards, etc.).
clavette, f., key, pin;
— à *anneau de bout d'essieu*, linch hoop pin;
chemin de ——, key way;
— *de frein*, securing pin, key;
— *de sûreté*, (*expl.*) safety pin.
clayonnage, m., hurdle.
clef, f., key, wrench;
— *d'appel*, (*telephone*) calling key;
— à *béquille*, pin wrench, socket wrench;
— *en bout*, socket wrench;
— à *bout carré*, socket wrench;
— *de calibre*, double wrench, double ended spanner;
— à *canon*, (*mach. gun*) gas vent ring spanner;
— *carrée*, cock spanner;
— *de déchargement* (*trench art*) breech wrench;
— *de démontage*, wrench;
— à *douille*, box spanner;
— *de fermeture*, (*trench art*), percussion primer locking wrench;
— à *griffes*, claw wrench;
— à *lanterne*, (*trench art*) box spanner.
— *de manœuvre*, spanner;
— à *molette*, monkey wrench;
— *de montage*, assembling wrench;
— *de moyeu*, wheel cap spanner;
— à *ressort*, (*elec.*) spring switch;
— *de serrage*, spanner;
— *serre-fils*, (*elec.*) double connector;
— *tournevis*, screw driver.
— à *tube*, — *en tubes*, jointed, tubular, spanner (for reaching distant parts).
— *universelle*, combination wrench;
— à *vélo*, bicycle wrench.
clin, m., clinker work;
à ——*s*, (*nav.*) clinker built.
clique, f., (*mil. slang*) field music.
cliquet, m., click; dog;
— *d'arrêt*, pawl.
cliqueter, v. n., (of an engine) to knock (from too much spark lead).
cliquette, f., bolting wire.
cloche, f., bell; cap, top (as of a trench sentry box);
— *de cardan*, v. *boîte de cardan*;
— *de direction*, —— *directrice*, (*aero*) steering gear case;
levier à ——, (*aero.*) gear shift lever.
cloison, f., partition; bulkhead;
— *porque*, (*pont.*) reinforcing bulkhead (metallic pontoon).
cloisonné, a., (*aero.*) cellular.
cloisonnement, m., (*aero.*) division of the interior of a dirigible into compartments; cellular construction.
cloisonner, v. t., to divide into compartments.
clos, a., (*adm.*), of the fiscal year, and hence of all budgetary operations) closed.
clos, m., inclosure;
— *d'équarrissage*, place for killing condemned horses and other animals not intended for food.
clôture, f., (*adm.*) closing of accounts, of the fiscal year.
clou, m., nail;
— *cavalier*, staple.

clouterie, f., nail business, nail works.
coco, m., cocoa.
coefficient, m., coefficient;
— *de sécurité*, factor of safety.
coffre, m., chest;
— *d'arrière-train*, (*mach. gun, art.*) caisson chest;
— *d'avant-train*, (*mach. gun, art.*) limber chest;
— *de batterie*, (*art.*) ammunition chest (heavy artillery).
cogner, v. n., (*mach.*) to knock (of an engine).
cohérer, v. n., (*wireless*) to cohere.
cohéreur, m., (*wireless*), coherer;
— *auto décohérent*, self decohering **coherer;**
— *à grenaille*, granular coherer;
— *à limaille*, filings coherer;
— *à poudre*, powder coherer.
récepteur à ——, coherer;
— *à vide*, vacuum coherer.
cohésion, f., (*wireless*) coherence.
coiffe, f., cap, cover;
— *couvre-nuque*, (*unif.*) havelock.
coiffé, a., (*art.*) capped (of projectiles).
coiffer, v. t., (*art.*) to include the target (in bracketing); to cap (a projectile).
coiffure, f., head dress;
— *"à la chien,"* the "mahomet," q. v., down on forehead, and képi pushed back.
coin, m., wedge;
— *de rattrapage*, take-up wedge.
coincement, m., (*mach., etc.*) jamming.
col, m., collar;
— *rabattu*, (*unif.*) falling collar.
colis, m., package;
— *du corps*, (*Fr. a.*) personal parcel delivery;
— *à domicile* (*mil. slang*) shell [lit. parcel to be delivered].
collant, a., tight (as of trousers, etc.).
collationnement, m., comparison; (*mil. tel., etc.*) repeating (of a despatch).
collationner, v. t., to compare; (*mil. tel., etc.*) to repeat (a despatch).
colle, f., (*slang*) oral examination.
collecteur, m., (*mach.*) water cavity; header, receiver, buspipe (e. g., of a muffler); (*elec.*) collector;
— *d'échappement*, (*mach.*) exhaust manifold;
grand ——, (*telephone*) main line, main.
coller, v. n., (*mach., etc.*) to stick.
collerette, f., fluted collar; (*art.*) cartridge hole (of the diaphragm of a limber, etc., chest).
collet, m., (*mach. gun*) bushing.
collet-manteau, m., *à capuchon*, hooded cloak and collar.
collier, m., collar;
— *d'attache* (*trench art*) securing ring;
— *de blocage* (*trench art*) blocking ring;
— *à crochets* (*trench art*) suspension collar (150 T. mortar);
— *de force*, strait collar, (sort of strait-jacket);
— *de pointage*, (*art.*) trunnion-ring (trench mortar);
— *de pointage en hauteur*, (*trench art*) elevating collar;
— *de portage*, (*trench art.*) carrying ring;
— *de serrage*, clamp, clamping ring;
— - *support de miroir*, (*trench art.*) mirror support;
— *à tourillons*, (*trench art.*) trunnion sleeve (150 T mortar).
collimateur, m., collimator;
— *de repérage*, (*art.*) registering collimator.
collier-écrou, m., (*mach.*) screw collar.
collodionage, m., (*phot.*) collodion varnishing.
colmatage, m., warping (flooding).
colo, m., (*mil. slang*) the colonel.
colombier, m., pigeon loft.
colombophile, m., (*mil.*) pigeon man.
colombophilie, f., (*mil.*) pigeon (messenger) service.

colon, m., colonist; (*mil. slang*) the colonel;
le grand ——, (*mil. slang*) the general.
colonel, m., (*mil.*) colonel;
— *fédéral*, colonel in the Swiss Army.
colonne, f., (*mil., etc.*) column; (in some relations) stem; (*inst.*) spindle;
— *attelée*, (*mil.*) horse-drawn column (guns, equipage, etc.);
— *automobile*, (*mil.*) motor-drawn column (guns, equipage, etc.);
— *de direction*, (*auto.*) steering pillar;
— *double*, (*mil.*) double column (of sections, Fr. a., for a battalion, 2 companies in first line and 2 in second, each in line of sections by fours);
— *en échelon*, (*mil.*) column in echelon;
— *en échiquier*, (*mil.*) column checkerwise;
— *en losange*, (*mil.*) column in lozenge formation;
— *par un*, (*mil.*) column of files;
— *en tête de porc*, v. —— *en trapèze;*
— *en trapèze*, (*mil.*) column in trapezium formation;
— *en triangle*, (*mil.*) column in triangle formation.
combat, m., combat;
— *à cheval*, (*cav.*) mounted action;
— *continental* (*mil.*) land fighting as distinguished from —— *maritime*, q. v.
— *maritime*, sea fighting;
— *à pied* (*cav.*) dismounted action;
— *de rues*, (*mil.*) street fight, street fighting;
— *de rupture*, (*mil.*) first assault, carefully prepared, against the whole enemy position battered by artillery; combat in the interior of a position that has been entered.
combiné, m., (*elec.*) set.
combler, v. t., to fill;
— *une vacance*, (*mil., etc.*) to fill a vacancy.
comité, m., committee, board;
— *consultatif des chemins de fer*, (*r. r. adm.*) board reporting to minister of public works on matters relating to new lines, purchase or function of lines, road traction, etc., advisory railway board, or commission;
— *consultatif d'hygiène publique de France*, (*Fr. adm.*) national board of health (assists minister of the interior);
— *de l'exploitation technique des chemins de fer*, (*r. r. adm.*) board charged with all questions affecting police, safety and use of railways (assists minister of public works).
commandant, m., (*Fr. nav.*) the c. o. of a ship (not necessarily a captain);
— *de frégate*, (*nav.*) commander, (grade, U. S. N., R. N.).
commande, f., (*mach.*) operating gear, controlling mechanism; (*inpl., aero.*) controls;
de ——, (in combinations, frequently) working;
— *à distance*, (*mach., elec.*) distant control;
— *double* (*aero.*) double control;
fil de ——, control wire;
mécanisme de ——, (*mach.*) control mechanism;
mécanisme de —— *électrique à distance*, (*elec.*) distant control electric mechanism;
pont de la ——, (*aero.*) control bridge;
— *s à pont*, (*aero.*) bridge controls (dep.);
— *de roue par carré*, (*mach.*) square axle system of transmission;
— *de roue par chaîne*, (*mach.*) chain transmission;
— *de roue par pignon*, (*mach.*) gear transmission;
— *de vitesses*, (*mach.*) speed gear lever.
commandement, m., (*mil.*) commanding officer (frequently); command, i. e., view over adjacent territory; (law) writ;
le haut ——, (*mil.*) (sometimes) the supreme command;
prendre le ——, (*mil.*) to take over command;
— *au sifflet*, (*mil.*) whistle signal;
— *supérieur*, (*mil.*) higher command (division, corps and army commanders).
commander, v. t., (*mach.*) to control, operate.
commis, m., clerk;
— *aux expéditions*, dispatching clerk;
— *de trésorerie*, (*mil.*) paymaster's clerk.

commissaire, m., member of a commission;
— *militaire*, (*mil.*) military member of a mixed commission; (*sport*) military representative on a sporting, race, or athletic committee;
— *de surveillance administrative*, (*r. r. adm.*) local manager (official in each important station, who looks after the police of the railway, and the local exploitation).

commission, f., board, commission;
— *de l'armée*, committee on military affairs, (House, Chamber of Deputies);
— *du budget*, appropriation committee of the House of Deputies;
— *de contrôle*, (*Fr. a.*) control board;
— *d'enquête*, court of inquiry, investigating board;
— *des finances*, the appropriation committee of the Senate (France);
— *de gare*, (*Fr. a. r. r.*) station commission;
— *d'habillement*, (*mil.*) clothing or uniform board;
— *régulatrice*, (*Fr. a. r. r.*) (commission in charge of the supply of an army by rail, and of troop transportation within the army's zone of action) supply commission;
— *rogatoire*, (*Fr. law*) requisition by one judge or court on another to proceed or act in the case referred;
— *de salubrité* (*Fr. a.*) sanitary board (of a cantonment).

commissionnaire, m., porter, messenger.

commotionné, m., (*mil. med.*) man shocked by proximity to H. E. burst, but not directly wounded or hit.

commune, f., (France) municipality, political unit under a *maire* is a person at law; commune.

communication, f., communication; (*mil.*) trench;
— *en crémaillère*, (*siege*) cremaillere trench;
— *à traverses*, (*siege*) traversed trench;
— *à traverses tournantes*, (*siege*) cube traverse trench.

commutatur, m., (*elec.*) commutator;
— *d'antenne*, (wireless) aerial switch.
— *à plots*, (*elec.*) shifting contact switch.

commutateur-interrupteur, m., (*elec.*) contact-shifter.

compagnie, f., company;
— *automobiliste*, (*mil.*) traction company;
— *cycliste*, (*mil.*) cyclist company;
— *de corps*, (*Fr. eng.*) company of army corps engineers;
— *divisionnaire*, (*Fr. eng.*) company of division engineers (part of corps establishment);
— *du drapeau*, color company;
— *du génie de parc*, (*Fr. eng.*) company of the engineer park;
— *légère de cavalerie*, light traction company (for the supply of cavalry, can transport 25 to 30 tons);
— *lourde*, heavy traction company (can transport 125 tons of rations—supply for a corps);
— *Schilt*, (*Fr. a.*) flame projector company;
— *volante*, (*mil.*) flying company.

compagnie-squelette, f., (*mil.*) skeleton company.

compartiment, m., compartment; (*r. r.*) compartment;
— *à couchette*, (*r. r.*) sleeping compartment.

compartimentage, m., subdivision into compartments; (*mil.*) in a military sense, the division of the terrain into first line, support, redoubts, intermediate trenches; of a center of resistance into active segments, works, etc.

compartimenter, v. t., to divide into compartments; (*mil.*) to divide the terrain in a military sense, v. *compartimentage*.

compas, m., compass; bows, frame, of a carriage hood; (*mach. gun*) crossbar (tripod mount, joins legs);
— *à coulisse*, sliding rule;
— *d'embarcation*, boat compass;
— *à liquide*, liquid compass;
— *maitre à danser*, interior compasses (for inside measurements);
— *porte-crayon*, drawing compasses.

complet, m., suit of clothes (all parts of same cloth).

composer, v. t., to take a written examination.

composition, f., composition; (in a written examination) subject; (*artif.*) composition;
amener à ——, to bring to terms;
— *fusante porte-retard*, (*artif.*) delay composition.

compresseur, m., (*mach.*) grease cup;
— *à air*, — *d'air*, (*mach.*) air-compressing engine;
— *à gaz*, (*mach.*) gas compressor.

compression, f., (*mach.*) compression, compression stroke;
— *volumétrique*, (ratio of volume of cylinder to that of chamber of explosion) volumetric compression.

comprimé, m., compressed tablet.

comptabilité, f., (*adm.*) accountability;
apurer la ——, to audit accounts.

comptable, a. m., (*adm.*) accountable; accountant
— *des deniers publics*, official accountable for public moneys;
— *en deniers*, official having a money responsibility;
— *expéditeur*, (*mil.*) administrative shipping officer;
— *en matières*, official having property responsibility.

comptant, a., cash;
au —— , (*com.*) cash down, for cash.

compte, m., account;
— *d'année*, (*Fr. adm.*) account or report for the year current, made by the Minister of Finance; annual account or report (calendar year);
de —— *à demi*, each paying one-half;
— *s d'exercice*, (*adm.*) accounts of the fiscal year;
— *en deniers*, money account;
— *en matières*, property account.

compte-cartouche, n., (*sm. a.*) cartridge counter, (to tell the soldier how many cartridges he has fired).

compte-fils, m., thread counter.

compte-gouttes, m., (*mach.*) drip feed lubricator; dropping tube.

compte-rendu, m., (*Fr. a.*) brief report (made on the spot, and at the time of the occurrence of the event involved. May be oral).

compte-tours, m., (*mach.*) counter (of revolutions).

compteur, m., counter;
— *de cartouches*, v. *compte-cartouche*;
— *de gaz*, gas meter;
— *kilométrique*, mileage recorder.

concentration, f., concentration; (*art.*) the simultaneous or successive fire of several batteries on the same target or zone.

conciliation, f., (*law*) settling of cases out of court.

concours, m., competition;
— *de consommation*, (*auto.*) consumption test or competition;
— *hippique*, horse show.

concurrence, f., concurrence;
à —— *de*, until such and such conditions are satisfied, until such and such a member is reached.

condamnation, f., condemnation;
— *aux dépens*, (*law*) condemnation in costs.

condamner, v. t., to condemn;
— *aux depens*, (*law*) to be found in costs.
— *à mort* (*law, mil.*) to condemn to death.

condensateur, m., (*elec.*) condenser;
— *à air*, air condenser;
— *à plaques*, plate condenser;
— *à plateaux*, plate condenser;
— *à la résonance*, (*wireless*) tuned **condenser**;
— *à verre*, glass condenser.

conductance, f., (*elec.*) conductance.

conducteur, m., foreman; driver; wire, lead;
—— *en guides,* driver (reins, from seat).
—— *simple,* (*mil. min.*) main wire, (wire to detonator);

conductibilité, f., (*elec.*) conductivity;
—— *unilatérale,* unilateral conductivity.

conduire, v. t., (*mach.*) to control (a valve).

conduite, f., channel (in many relations);
—— *d'aspiration,* (*mach.*) suction pipe, admission pipe;
—— *d'eau,* (*mach.,* etc.) water pipe;
—— *du feu,* (*art.*) conduct of fire.

cône, m., cone;
—— *antérieur,* (*sm. a., art.*) shoulder of a cartridge case;
—— *de commande,* (*mach.*) driving cone;
—— *cuir,* (*mach.*) leather cone clutch;
—— *droit,* right cone; (*mach.*) ordinary cone clutch;
—— *inverse,* (*mach.*) inverted cone clutch.

conflit, m., conflict;
—— *négatif,* (*adm.*) conflict in which both authorities disclaim jurisdiction;
—— *positif,* (*adm.*) conflict in which both authorities claim jurisdiction.

congé, (*mach.*) inside beveling (as at the mouth of a cylinder);
—— *illimité,* (*mil.*) indefinite leave of absence.

congrès, m., in France, the Senate and Chamber of Deputies united, or sitting as a national assembly.

conseil, m., council, counsel;
—— *d'état,* (France) a council composed of *conseillers d'état, maîtres des requêtes,* and *auditeurs* q. v., assisting the Government, i. e., the President, the chambers and the ministers, and acting as a supreme court in all administrative litigation;
faire passer en —— de guerre, (*mil.*) to bring before a court-martial;
—— *de l'ordre de la Légion d'honneur,* council of the Legion of Honor charged with the maintenance of the statutes and regulations of the order, and of its establishments;
—— *sanitaire,* (*Fr. adm.*) board of health;
—— *supérieur des haras,* (*Fr.*) horse breeding board, assists Minister of Agriculture and passes on all questions of breeding, etc.

conseiller, m., councilor;
—— *d'état,* (France) a member of the council of state.

conserver, v. t., to preserve;
—— *sa vitesse,* (*art.*) of a projectile, to hold up well.

consigne, f., (*r. r.*) parcel room.
prendre le terrain en ——, (*mil.*) to "spot" the terrain.

console, f., (*mach. gun*) trail-carriage bracket;
—— *de queue d'affût,* (*art.*) trail extension (280 Schneider).

consolidation, f., consolidation; (*mil.*) consolidation of a position, (i, e., repair, strengthening, equipping, said especially of a captured position).

consolider, v. t., to consolidate;
—— *une position,* (*mil.*) to organize, strengthen, repair, equip, and man a position (said especially of a captured position).

consommable, a., (*mil.*) expendable;
non ——, unexpendable.

consommation, f., consumption;
—— *kilométrique,* consumption per kilometer (oil, water, coal, etc.).

contact, m., (*elec.*) contact, switch;
couper le ——, to switch off;
mettre le ——, to switch on;
—— *platiné,* (*elec.*) platinum point.

contacteur, m., (*trench art.*) contact plate (V. D. mortar).

contentieux, m., (*adm.*) controversy, difficulty of governmental or administrative nature between individuals, corporations, between individuals and the state, etc.

contrainte, f., (*law*) act or document authorizing legal measures in certain cases;
porteur de ——, (*law*) (sort of process or writ server) bailiff.

contrarier, v. t., to oppose;
—— *les joints,* to break joints.

contrat, m., contract;
—— *d'assurances,* (*com.*) insurance policy.

contre-arbre, m., (*mach.*) countershaft.

contre-attaque, f., (*mil.*) counter attack (i. e., attack made upon an enemy attack).

contre-batterie, f., (*art.*) (battery engaging the artillery of an attack, or during the attack), counter-battery.

contre-enveloppe, f., protective cover (pneumatic tire).

contre-épaulette, f., (*unif.*) shoulder knot (worn on one shoulder, with epaulet on the other (French army), to indicate certain grades; on left shoulder by adjudant, sous-lieutenant and major; on right shoulder by lieutenant and chef d'escadron and chef de bataillon).

contre-espion, m., counter-spy.

contre-espionnage, m., counter-spying.

contre-feu, m., counter-fire, back-fire;
—— *de barrage,* (*mil.*) counter-fire to neutralize a gas attack.

contre-pente, f., reverse slope.

contre-plaque, f., cheek plate, guard plate.

contreplaqué, a., -ply (in composition); veneered, laminated;
—— *en trois épaisseurs,* three-ply.

contrepoids, m., (*wireless*) counterpoise, artificial ground.

contre-pointe, f., (*elec.*) second point or end of a jump-spark (end of wire to metallic housing).

contre-préparation, f., (*mil.*) counter preparation (to meet an attack in preparation);
—— *offensive,* (*art.*) violent shelling of enemy trenches in which assaulting troops are forming.

contre-rampe, f., (*art.*) lower ramp (fuse).

contretenir, v. t., (*cord.*) to lower handsomely.

contre-torpilleur, a., torpedo-boat resisting (v. s. v. *artillerie, canon*).

contre-vis, f., check screw, stop screw.

co**ntribuable, m.,** (*adm.*) taxpayer.

contrôle, m., (*adm.*) control, check (in many relations and departments);
—— *administratif,* administrative check on accounts, expenditures, etc.;
—— *judiciaire,* judicial check on accounts, maintained by the *Cour des Comptes*;
—— *législatif,* parliamentary check, applied by auditing on the operations of each fiscal year.

contrôleur, m., (*adm.*) inspector (in many relations, departments, etc.); motorman, conductor, of an electric tramcar;
—— *comptable,* (*r. r. adm.*) inspector of railway accounts;
—— *de pression,* air pressure gauge (pneumatic tire);
—— *de route* (*r. r.*) inspector, ticket taker;
—— *du travail,* (*r. r. adm.*) (official who watches over train operations and the execution of regulations) traffic manager.

convalo, m., (*mil. slang*) a convalescent soldier.

convoi, m., train; train of boats (made up of small trains or rafts);
—— *automobile,* (*mil.*) motor train;
—— *de ravitaillement,* (*mil.*) supply train.

convoyer, v. t., to accompany, go with (pigeons in transport).

convoyeur, m., keeper (accompanies pigeons when transported in baskets).

copain, m., (*mil. slang*) chum; "bunkie," U. S. A.

coque-fuselage, m., (*aero.*) hull (of a flying boat).

coquille, f., shell; (*mach.*) casing, housing;
—— (*typographique*), printer's error.

corbeau, m., crow; (*slang*) Roman Catholic priest.
cordage, m., cordage; hempen rope;
— *en acier*, steel rope;
— *en chanvre*, hempen rope;
— *en coco*, fiber rope;
— *en manille*, manila rope.
corde, f., cord; (*cav., art.*) picket line; (*aero.*) chord;
— *de commande*, (*aero.*) maneuvering rope;
— *de déchirure*, (*aero.*) rip cord;
— *à piano*, piano wire.
cordeau, m., line, cord, string; (*artif.*) fuse, match;
— *dérivé*, (*artif.*) match or fuse spliced on the cordeau maître, q. v.;
— *instantané*, (*artif.*) synonym of — *détonant*, q. v.;
— *maître*, (*artif.*) any primed match or fuse;
— *tire-feu*, (*art.*) lanyard.
cordée, f., string of men (in mountain climbing, of men on one rope).
cordier, m., (*aero.*) rope man; cordage expert (balloons).
cordonnet, m., (*unif.*) hat cord (U. S. A.).
cornarder, v. n., (*mil. slang*) to make a mistake.
corne, f., horn;
— *d'appel*, horn.
corner, v. n., (*auto.*) to sound.
cornet, m., horn;
— *acoustique*, ear trumpet.
cornière, f., angle bar; angle iron;
— *bêche*, (trench art.) corner-anchor (240 C. T. mortar).
corps, m., corps; body;
— *d'armée* (*mil.*) army corps (two or three divisions); under modern conditions may be a group of divisions under one command;
— *indigène*, (*mil.*) native regiment, regiment or unit of native troops;
— *nourricier*, (*mil.*) subsisting unit;
— *et services*, (*mil.*) troops and staff departments (U. S. A.).
correcteur, m., reader (of examination papers); (*art.*) corrector;
— *d'efficacité*, (*art.*) (the — *de réglage* increased by two units, final corrector, corrector for effect;
— *de réglage* (*art.*) ranging corrector, adjustment corrector, (gives smoke of burst low enough to observe errors of range);
— *de site*, (*art.*) site corrector.
correction, f., reading of examination papers;
— *de convergence* (*art.*) convergence correction;
— *de station*, (*art.*) (station correction) offset.
correspondance, f., (*adm.*) mail, post;
— *descendante*, (*mil.*) mail going to the army, army mail (in the field);
— *montante*, mail coming from the army (in the field).
corriger, v. t., to read examination papers.
corvéable, a., liable to fatigue duty, to public labor.
cosse-guide, f., (*cord. aero.*) thimble.
cote, f., hill;
en —, (of motor cars) in hill climbing.
côté, m., side;
petit —, (*unif., clothes in general*) side body.
côtier, a., coast;
service —, (*aero.*) coast aviation service (naval purposes).
coton, m., cotton;
— *caoutchouté*, rubber treated cotton, rubberized cotton.
couche, f., layer;
— *d'éclatement*, (*fort.*) in modern field works, protective layer, to protect shelter underneath against burst.
couché, a., (*mil. t. p.*) lying down, lying, prone.
couchette, f., bunk (especially on board ships).
coucou, m., (*mil. slang*) shell; (*aero. slang*) bus (i. e., plane).
coulage, m., (*cons.*) pouring (of concrete).

coulant, m., (*man.*) ease of seat, adjustment to motion of horse.
couler, v. n., to leak; v. t., (*cons.*) to pour (concrete);
— *au canon*, (*nav.*) to sink, to be sunk by gunfire.
coulis, m., (*cons.*) grouting;
— *clair de ciment pur*, v. *barbotine*.
coulisseau, (*mach.*) m., stuffing box (rare); (*art.*) traversing gear; traversing gear (field gun on its carriage);
— *de pointage en direction*, (*mach. gun*) traversing slide.
coulisseau-guide, m., (*mach.*) guide block.
coulissement, m., (*art.*) (sliding back, recoil of gun on carriage), carriage recoil; traversing (of field gun on its carriage);
— *sur l'essieu*, (*art.*) axle-traverse, traverse.
coulisser, v. t., (*art.*) to traverse (a field gun on its carriage).
couloir, m., (*art., etc.*) hopper;
— *d'alimentation*, (*mach. gun*) feed slot (mod. 1907 T and others); feed guide, feed hopper;
— *de circulation*, (*siege*) passage, communication (as around a traverse);
rallonge de —, (*mach. gun*) feed trough extension.
coup, m., blow; call, signal, blast, (trumpet, etc.);
(*fam.*) nip, drink; (*art., etc.*) hit;
— *bas*, (*art.*) low burst;
— *direct*, (*art., etc.*) direct hit;
être sous le — *de*, to be exposed to the attack of;
faire — *double*, to kill two birds with one stone;
faire d'une pierre deux —*s*, to kill two birds with one stone;
— *de feu à blanc* (*art., sm. a.*) blank shot, blank round;
— *de frein* (*mach.*), action of braking;
— *haut*, (*art.*) high burst;
— *de main*, skill, handiness;
— *d'œil militaire*, (*mil.*) eye for ground;
— *de sifflet*, (*mil.*) whistle signal;
— *de téléphone*, telephone call, communication by telephone;
tenir le —, to keep up the stroke.
coup-de-poing, m., (*mach.*) hand oiler.
coupe, f., cut, cutting;
— *des ballons*, (*aero.*) panel cutting;
— *militaire*, of the hair, short everywhere except in front where a "*mahomet*" is left.
coupé, m., coupé, brougham.
coupelle, f., (*mach.*) recess, cup-shaped recess.
couper, v. t., to cut;
— *la communication*, (*telephone*) to ring off;
— *à l'exercice* (*mil. fam.*) to cut drill.
coupe-vent, m., *avant en* —, (*auto.*) sharp front or forebody.
couplage, m., (*elec.*) coupling up, connecting;
— — *lâche*, (*wireless*) loose coupling;
— *en série*, coupling in series;
— *serré*, (*wireless*) tight coupling.
couple, m., (*phys.*) couple;
maître —, (*aero.*) midship frame;
— *moteur*, torque.
coupler, v. t., (*elec.*) to connect up, join.
coupure, f., cut; (*mil.*) straight line trench;
boîte de —, (*elec. telephone*) terminal box.
courant, m., current;
[The following terms are electric.]
— *d'allumage*, lighting current;
— *à basse tension*, current of low potential;
— *de décharge*, discharging, working, current of a storage battery;
— *à haute fréquence*, high frequency current;
— *à haute tension*, current of high potential;
— *de rupture*, break current;
— *de vérification*, testing current;
— *vibré*, (*elec.*) alternating current.
courbure, f., (*aero.*) camber, wing curve section.
coureur, m., (*auto.*) driver, racer (person); (*mil.*) runner, (messenger); mounted scout;
chaîne de —*s*, (*Fr. a*) relay of runners, chain of;

coureur—*s doublés*, (*Fr. a.*) double runners (for messages);
poste de ——*s*, (*Fr. a*) runners' stations (2 or 3 runners).
courir, v. n., to run;
—— *après les projectiles*, (*mil.*) to follow, run after, one's barrage fire.
couronne, f., crown; (*mach.*) crown, crown wheel;
—— *d'agrafage*, (trench art) locking ring (240 L T).
—— *d'appui et d'agrafage*, (trench art.) base ring (240 CT mortar);
—— *de chaîne*, (*mach.*) sprocket wheel;
—— *de corde*, (cord) strap;
—— *dentée*, (*mach.*) gear wheel;
—— *à denture externe*, (*mach.*) external gear;
—— *à denture interne*, (*mach.*) internal gear;
—— *graduée*, (*mach. gun*) gas-vent ring, pressure regulator;
—— *régulateur d'échappement*, v. s. v., *échappement*:
—— *de roulement*, (art.) roller path;
—— *des satellites*, (*mach.*) intermediate pinions, planetary gear, circle.
couronné, a., crowned;
Grand —— *de Nancy*, range of hills around Nancy.
couronnement, m., (*mil.*) top (of a parapet).
courrier, m., (*adm.*) mail, letters, post;
—— *descendant*, (*mil.*) mail going to the army (in the field);
—— *montant*, (*mil.*) mail coming from the army (in the field);
retour du ——, return mail.
courrier-conducteur, m., (*mil.*) mail carrier.
courroie, f., strap;
—— *de portage*, carrying strap, sling;
—— *de suspension*, (*sm. a.*) revolver strap.
cours, m., (*com.*) market price.
course, f., (*aero.*) trip, excursion;
—— *de but*, (*aero.*) trip whose purpose it is to reach a given point;
—— *de haies*, (*sport.*) hurdle race;
—— *de longueur*, (*aero.*) (trip to test staying powers of the balloon); endurance trip;
—— *de relais*, (*sport.*) relay race.
courtage, m., (*com.*) brokerage.
court-circuit, m., (*elec.*) short circuit.
court-circuiter, v. t., (*elec.*) to short circuit.
courtier, m., broker;
—— *d'assurances*, insurance agent, broker;
—— *s interprètes et conducteurs de navires*, ship brokers;
—— *libre inscrit*, registered merchandise broker;
—— *de marchandises*, merchandise broker;
—— *non inscrit*, unregistered broker.
coussinet, m., (*mach.*) bearing, bushing;
—— *bague*, strap bearing, bushing, or brass;
—— *lisse*, unlined bushing (no antifriction metal); smooth, plain bushing (no ball bearings or rollers);
—— *à rouleaux*, roller bearing.;
—— *sphérique*, spherical ball bearing.
couteau, m., knife; primary (feather);
—— *de monteur*, electrician's knife;
—— *de tranchée*, (*mil.*) trench knife.
couture, f., (*nav.*) seam (of a deck).
couvage, m., sitting (on eggs).
couvée, f., hatching.
couver, v. t., to sit (on eggs), incubate, hatch.
couvercle, m., (*art.*) door (of a limber, caisson, chest);
couvert, a., overcast, cloudy;
temps ——, cloudy weather.
couverture, f., cover; (*art.*) operations of covering (e. g., flanks,) by artillery fire.
—— *de campement*, (*Fr. a.*) field blanket;
grandes ——*s*, wing coverts;
manœuvres de ——, (*mil.*) covering operations or maneuvers by *troupes de couverture*, q. v.
petites ——*s*, lesser coverts (wing).
couvre-appareil, m., cover, hood.

couvre-bouton, m., safe (as of a *bouton double*).
couvre-képi, m., (*unif.*) cap cover.
couvre-percuteur, m., (*art.*) striker cover.
couvre-selle, m., (*harn.*) housing, saddle cover.
couvre-soupape, m., (*mach.*) valve cover, valve hood.
couvrir, v. t., (*pont.*) to deck, to lay chess.
crabe, m., crab;
marcher en ——, (*aero.*) to go sideways.
cran, m., notch; (*mil. slang*) day's confinement;
—— *d'armé*, (*sm. a.*) cocking notch.
crapouillot, m., (*mil. slang*) trench mortar, (diminutive of crapaud, toad; sometimes spelled *crapouillaud*).
craticule, f., graticule.
craticulation, f., graticulation.
craticuler, v. t., to graticulate.
cravate, f., cravat;
—— *de commandeur de la Légion d'honneur*, neck ribbon of this grade of the Legion of Honor.
cravater, v. t., to put on a (neck) tie;
—— *de chanvre*, (*fam.*) to hang;
—— *de crêpe*, (*mil.*) to put the colors in mourning.
crayon, m., pencil; (*elec. gen.*) carbon; (*art.*) distance piece (of a fuse);
—— *anti-buée*, (*mil.*) (pencil of composition to prevent formation of mist on lenses of gas masks); anti-mist pencil;
—— *à copier*, copying pencil;
—— *de couleur*, colored pencil;
—— *des états-majors*, a pencil with a scale on it.
crèche, f., manger; (*mil. slang*) shelter (in trenches).
crédit, m., (*adm.*) appropriation;
—— *s extraordinaires*, urgent appropriation;
—— *s législatifs*, appropriation made by the legislature;
—— *ministériel*, (*Fr. adm.*) allotment of public funds made by a minister to his subordinates;
—— *s supplémentaires*, deficiency appropriation.
crémaillère, f., (*mach.*, etc.) rack;
—— *de pointage*, (trench art.) elevating rack.
créneau, m., (*mil.*) loophole;
—— *découvert*, (notch or groove in top of parapet and not perpendicular to the crest) open firing notch.
crépuscule, m., twilight.
crevaison, f., puncture (pneumatic tire).
crever, v. t. n., to burst, to puncture.
cric, m., rack and pinion jack;
—— *à galet*, roller jack.
criminel, a., m., criminal;
au ——, (*law*) in criminal cases.
crin, m., hair (mane and tail);
—— *végétal*, vegetable fiber.
crise, f., crisis;
—— *ministérielle*, disagreement between responsible ministers and parliament, parliamentary crisis.
crissement, m., grating, grinding noise.
crisser, v. n., to grate, to grind.
critique, f., (*mil.*) review, summary, and criticism of operations after maneuvers or exercises.
crochet, m., hook; rim-flange of an automobile wheel;
—— *d'arrêt de chargeur* (*mach. gun*) magazine catch;
—— *à arrêtoir*, (art.) pawl, catch;
—— *d'assemblage*, (*mil.*) coupling hook (assemblage of corrugated iron plates for shelters);
—— *commutateur*, (*telephone*) hook switch;
—— *de cycliste*, trouser clip;
—— *de portage*, carrying hook, bale hook;
—— *de retenue*, (art.) retaining hook;
—— *de traction*, traction hook.
crochet-lyre, m., (trench art.) lyre-shaped hook (240 L. T.).
crocodile, m., crocodile; (*mil.*) (wheeled apparatus for conveying an elongated charge into wire entanglements of enemy, and then exploding it) crocodile.

croiser, v. t., to cross;
— *les feux*, (*mil.*) to cross fire.
croiseur, m., (*nav.*) cruiser;
— *aérien*, (*aero.*) a dirigible of over 9,000 cubic meters.
croiseur-atelier, m., (*nav. aero.*) mother ship.
croisillon, m., (*aero.*) bracing wire;
— *d'un cardan*, cardan cross-bar.
croissant, m., tread (of a rubber or pneumatic tire).
croix, cross; the Legion of Honor (which is not a cross, but a five-pointed star);
— *dite de consolation*, (*Fr. a.*) Legion of Honor given to officers upon retirement;
gagner la — *de bois*, (*mil. slang*) to be killed;
— *de guerre*, (*Fr. a.*) cross of war (decoration);
la —, (*used alone*) the cross of the Legion of Honor;
— *verte*, (*mil.*) green cross (mark on German gas shells).
Croix-Rouge, f., the Red Cross, Red Cross Society.
croquis, m., sketch;
— *d'objectifs*, (*art.*) large scale plan of the terrain on which targets are found (trench artillery);
— *panoramique*, a detailed perspective sketch on large scale;
— *perspectif*, (*mil.*) perspective (landscape) sketch;
— *planimétrique*, sketch plan.
crosse, f., sort of half C-spring;
— *de queue*, (*aero.*) tail skid;
— *tubulaire*, (*art.*) tubular trail.
crosser, v. t., to strike with the butt of the rifle.
croupe, f., (*hipp.*) croup;
— *en dedans*, (*man.*) (of the horse) croup toward center of hall;
— *en dehors*, (*man.*) (of the horse) croup away from center of hall.
croupion, m., rump.
cruchon, m., jug.
cube, m., cube;
— *aplati*, (*mil.*) obstacle of barbed wire (diagonals and sides of a parallelopiped);
— *d'eau*, (*steam*) volume of water of a boiler.
cuir, m., leather;
— *d'embrayage*, (*mach.*) clutch leather;
les —*s*, (*mil. slang*) cuirassiers;
nourrir les —*s*, to keep leather in good condition by rubbing in grease, oil, etc.
cuirasse, f., casing, i. e., auto frame made in one piece.
cuirassé, m., (*nav.*) armored ship;
— *de course*, armored cruiser.
cuirasser, v. t., (*nav.*) to armor;
— *à n mm.*, to put on n mm. armor.
cuisine, f., kitchen;
— *roulante*, (*mil.*) traveling, rolling, kitchen.
cuisinière, f., cook;
— *militaire*, range for military kitchens.
cuistance, f., (*mil. slang*) food.
cuistancier, m., (*mil. slang*) cook.
cuistot, m. (*mil. slang*) cook (sometimes spelt *cuistau*).
cuivre, m., (*met.*) copper;
— *rouge*, copper.
cuivrerie, f., brass work, brasses.

culasse, f., (*mach.*) top of a cylinder (steam, motor, etc.), cylinder head, cap (more esp.) explosion chamber or space; (*art.*) breech block;
— *à bloc*, hemispherical breech plug (Canet);
— *à coin*, (*art.*) wedge breech closure;
— *à eau*, (*mach.*) water cooled cylinder top;
— *à manœuvre rapide*, (*art.*) quick-motion screw;
— *mobile de rechange*, (*mach. gun*) spare breech-block;
tige de —, (*art.*) breech-screw guide rod (155 C.);
— *à vis* (*art.*) screw breech closure.
culbute, f., (*sm. a., art.*) tumbling (of a projectile).
culbuter, v. n., (*art., sm. a.*) to tumble (of a projectile).
culbuteur, m., (*mach.*) rocking lever, tappet lever; rocker;
arbre de —, rocker shaft.
culot, m., (*elec.*) housing or casing of a spark plug, screw plug.
culotte, f., (*mach.*) Y joining two pipes to one, Y-joint, fork; (*art.*) breech-mass (trench mortar);
— *d'échappement*, (*mach.*) bridge piece of a muffler or silencer;
— *à fond doublé*, (*unif.*) reenforced breeches.
cure-dents, m., tooth-pick; (*mil. slang*) bayonet.
cure-pipe, m., pipe cleaner.
cursoir, m., slide.
cursomètre, m., speedometer.
cuve, f., vat, tank;
— *à lavage*, (*phot.*) washing tray;
— *de niveau constant*, (*mach.*) float chamber.
cuvette, f., small cup or cup-shaped vessel (in many relations); (in some relations) female cone; (*top.*) circular depression in the ground, sink hole;
— *-abreuvoir*, horse trough (in a stable);
— *d'un canal*, wetted cross section of a canal;
— *de friction*, (*mach.*) female friction cone;
— *à photographie*, (*phot.*) photographic tray, developing tray.
cycle, m., cycle; bicycle;
— *moteur complet*, (*mach.*) complete cycle of a motor;
— *à n temps*, n-period cycle.
cycliste, m., cyclist;
bataillon —, (*mil.*) cyclist battalion;
compagnie —, (*mil.*) cyclist company.
cyclomètre, m., cyclometer.
cylindre, m., cylinder;
—*s accolés*, (*mach.*) group of cylinders cast in one piece;
— *à ailettes*, (*mach.*) ribbed cylinder;
— *à eau chaude*, (*mach.*) hot-water tank;
— *à gaz*, (*mil.*) gas cylinder;
— *incendiaire*, (*mil.*) incendiary cylinder.
—*s jumelés*, (*mach.*) twin cylinders (cast in sets of twos);
— *de n millimètres d'alésage*, (*mach.*) an n-millimeter cylinder;
— *pliant*, (*mil. sig.*) collapsible cylinder (black cloth, 15 cm. long at rest 1.15 m. long when pulled out mounted on cable of a balloon and used for signaling);
— *à refroidissement par ailettes*, (*mach.*) flange-cooled cylinder;
—*s séparés*, (*mach.*) cylinders cast separately.
cylindrée, f., cylinder-full.
cylindre-frein, m., (*art.*) cylinder brake.

D.

dactylo, m., (*fam.*) typewriter (person).
dactylographe, m., typewriter (person), typist.
dalle, f., (*cons.*) plate, (of glass, etc.);
— *brute coulée*, rough (glass) plate;
— *moulée*, molded (glass) plate;
— *polie*, polished (glass) plate;
— *quadrillée*, checkered (glass) plate.
dam, m., damage.
dais, m., rigid top, roof, of a (an open) carriage, automobile, etc.

dame, f., lady;
— *employée*, (*Fr. adm.*) title of women employed in the services of posts and telegraphs.
dame-jeanne, f., demijohn.
dé, m., thimble;
— *à coudre*, thimble.
débarquer, v. t., to disembark; to detrain.
débat, m., debate, discussion;
—*s d'un procès*, (*law*) argument of a cause or case.

débattre, v. t., to debate; to argue (*law*) a case.
débit, m., delivery, output, outflow;
—— *en ampères,* (*elec.*) intensity;
—— *de tabac,* (*Fr. adm.*) tobacco shop (is a Government monopoly);
—— *visible,* (*mach.*) sight feed, sight feed lubricator (abbreviation of *graissage à débit visible*).
débloquer, v. t., (*mach. guns, etc.*) to release, unclamp.
déboîter, v. t., (*art.*) (of carriages) to wheel out of column, to get on one side.
débouché, m., opening, outset (as, *mil.* of an attack).
déboucher, v. t., (*art.*) to set a fuse;
—— *à la main,* to set a fuse by hand.
déboucheur, m., (*art.*) fuse setter (cannoneer).
débouchoir, m., (*art.*) fuse-setter;
—— *automatique,* automatic fuse setter.
debout, adv., straight ahead, (as, *siege,* straight toward the enemy).
débraillé, a., (*unif.*) sloppy, illfitting.
débrayage, m., (*mach. etc.*) uncoupling, disengaging;
douille de ——, (*mach. gun*) feed drum clutch;
manette de ——, releasing handle.
débrayer, v. t., (*mach.*) to disengage, to throw out; (*auto.*) to change or throw out the speed gear.
débrouillard, a., resourceful, knowing;
officier ——, wide awake, resourceful, officer.
débrouille, f., v. *Système D.*
décalage, m., (*aero.*) stagger (of the wings); (*elec.*) displacement; [*décalage* as defined in Report No. 9, National Advisory Committee for Aeronautics, should not be confused with the *décalage* translated above.]
décalé, a., staggered;
—— *en arrière* (*aero.*) antistaggered;
—— *en avant,* (*aero.*) staggered.
décalement, m., v. *décalage,* (*aero.*)
décaler, v. t., to set, stagger (in a number of relations).
décalotter, v. t., (*art.*) to uncap (a projectile).
décamètre, m., decameter; (*surv.*) 10-meter tape;
—— *en ruban,* (*surv.*) (ordinary linen) tape.
déceinturage, m., removal of a tire; (*art.*) stripping of a rotating band.
déceinturer, v. t., to strip off, remove, a tire; (*art.*) to strip off the rotating band.
décentrement, m., (*mach. etc.*) eccentricity.
décharge, f., (*adm.*) receipt, release; (*elec.*) discharge;
—— *apériodique,* (*elec.*) dead-beat discharge;
courant de ——, (*elec.*) discharging current;
—— *oscillante,* — *oscillatoire,* (*elec.*) oscillating discharge.
décharger, v. t., to unload;
se ——, (*elec.*) (of a storage battery) to run down.
déchéance, f., (*adm.*) limitation, statute of limitation.
déchet, m., waste;
—— *de pétrole,* waste oil.
déchirure, f., tear; (*aero.*) rip;
corde de ——, v. s. v. *corde;*
panneau de ——, v. s. v. *panneau;*
triangle de ——, (*aero.*) rip triangle;
volet de ——, (*areo.*) ripping panel.
décigrade, m., decigrade (tenth of a degree in the centesimal division of the circumference).
décimètre, m., decimeter;
double ——, double 10-centimeter scale (drawing).
décision, f., decision;
—— *présidentielle,* (in France) executive approval of a ministerial proposition (no special decree being issued).
déclanchement, m., v. *déclenchement.*
déclancher, v. t., v. *déclencher.*
déclenchement, m., unlatching, unloosing; (*mach.*) release (act, and gear for act);
appareil de ——, (*mach.*) releasing gear, release;

déclenchement *d'une attaque,* (*mil.*) loosing, launching of an attack;
—— *d'un tir,* (*art.*) opening of fire (connotation of suddenness).
déclencher, v. t., to unlatch;
—— *une attaque,* (*mil.*) to let go, launch, an attack;
—— *le tir,* (*art.*) to open fire, (connotation of suddenness).
décohération, f., (*wireless*) decoherence.
décohérer, v. n., (*wireless*) to decohere.
décohésion, f., (*wireless*) decoherence; (*aero.*) act of leaving the ground (of an airplane, in starting), getting off.
décollage, m., (*mach.*) starting, releasing.
décollement, m., (*fam., mil.*) act of *décoller,* q. v.
décoller, v. n., (*mach.*) to start; (*aero.*) to leave the ground (aeroplane in starting), to get off; (*fam. mil.*) (of cavalry officers of high grade) to cut loose from their troops, ride out in front to see for one's self, to show some independence.
décombres, f. pl., rubbish (of a demolished building).
décommander, v. t., to countermand.
décompresseur, m., (*mach.*) relief valve;
robinet ——, relief cock.
décompression, f., (*mach.*) compression relief (of a motor cylinder); half-compression device; reduction (as of pressure);
came de ——, relief cam;
robinet de —— relief cock.
décoration, f., decoration;
—— *coloniale,* (*Fr. a.*) decoration for colonial service.
découvert, m., (*adm.*) deficiency, (amount by which receipts fail to come up to budgetary expenditures).
découvrir, v. t., (*pont.*) to take up chess.
décret, m., in France, any *acte* of the President; presidential decree;
—— *en la forme des règlements d'administration publique,* a special administrative decree;
—— *général* v. —— *règlementaire;*
—— *individuel,* v. —— *spécial;*
—— *portant règlement,* (*mil.*) a decree having the force of a regulation, or constituting a regulation;
—— *portant règlement d'administration publique,* a decree having the effect of law in the public administration, (must have been passed upon by the *Conseil d'Etat* in general assembly);
—— *règlementaire,* presidential decree, having the character of law, and treating of general interests;
—— *spécial,* presidential decree, administrative in character, and affecting private or individual interests.
décret-loi, m., decree having the force of law.
décrocher, v. a., to unhook; to disengage, tear loose (e. g., (*mil.*) the enemy);
—— *son troisième galon au choix,* (*mil. slang*) to get one's captaincy by selection (said of a lieutenant);
se ——, (*mil.*) to withdraw (from the combat).
déculasser, v. t. or n., (*mil. slang*) to fall from a horse, or cause to fall from a horse.
dédoublement, m., duplication (e. g., of images in certain field glasses, etc.).
défaut, m., defect, default;
faire ——, (*law*) to default, fail to appear.
défendable, a., defensible.
défendre, v. t., (*law*) to be a defendant.
défense, f., defense; (*law*) defense;
—— *côtière,* (*fort.*) coast defense;
—— *s maritimes,* (*fort.*) works on the sea front;
—— *s terrestres,* (*fort.*) works on the land front.
défilement, m., (*mil.*) defilade;
—— *complet,* complete defilade, flash defilade.
défiler, v. t., to cover;
—— *aux vues,* to keep (a company, etc.), from being seen, to protect from view.

déflecteur, m., (*mach.*) baffle plate.
dégauchir, v. t., (*aero.*) to unwarp, to reverse a warp.
dégauchissement, m., (*aero.*) unwarping, reversal of a warp.
dégommer, v. t., (*mach.*) to remove old lubricant, grease, etc.
dégoupiller, v. t., to unpeg, unpin.
dégraisseur, m., cleaner, scourer, grease remover;
—— *d'eau d'alimentation*, (*steam*) feed water purifier.
dégripper, v. t., (*mach.*) to free, release.
dégrossir, v. t., to rough hew; (*art.*) to set approximately (elevations, deflections, etc.).
—— *les réglages*, (*art.*) to make an approximation to the range.
dégrossissage, m., rough hewing;
—— *du tir*, (*art.*) approximate correction of fire.
déguinder, v. t., (*pont.*) to unlash, unrack.
déjeaugeage, m., (of a boat) lifting (of the bow, in motion) lift forward.
déjeauger, v. n., (of a boat) to lift (at the bow, in motion), to lift forward.
délai, m., delay;
—— *de route*, (*Fr. a.*) travel time, not counted in a leave of absence (officers, etc., far from home)
délégation, f., (*mil.*) allotment.
déléguer, v. t., to allot.
délestage, m., (*aero.*) throwing out of ballast, un ballasting.
délester, v. t., (*aero.*) to throw out ballast, to un-ballast.
délit, m., misdemeanor.
demander, v. t., (*law*) to be a plaintiff.
démarrage, m., (*auto*) starting.
démarrer, v. n., (*auto*) to start.
démettre, v. t., to dislocate;
se —— *l'épaule*, to dislocate one's shoulder.
demeure, f., delay;
mettre en ——, (*law*) to summons.
mise en ——, (*law*) summons.
demi-arbre, m., (*mach.*) half-axle or shaft (of a divided axle).
demi-axe, m., (*art.*) half-axle, semi-axle.
demi-botte, f., (*unif.*) Wellington boot.
demi-flèche, f., (*art.*) trail half (split trail).
demi-fourchette, f., (*art.*) half-bracket.
demi-gorge, f., semi-circular channel, recess, groove, socket.
demi-masse, f., (*mil.*) half-mass.
demi-pincette, f., half-elliptic spring.
demi-repos, m., (*mil.*) half-rest.
demi-sang, m., (*hipp.*) (defined as any horse not English, Arab, or Anglo-Arab thoroughbred) half-bred horse.
démissionner, v. t., to resign.
demi-vitesse, f., (*mach.*) half-speed.
démontable, a., dismountable; m., (*auto., etc.*) pneumatic tire (with respect to the ease with which it can be put on or taken off);
pneu ——, pneumatic tire.
démonte-pneu, m., (*auto.*) tire lever.
démonter, v. t., to dismount, take down;
—— *de son commandement*, (*mil.*) to relieve of one's command.
démonte-soupape, m., (*mach.*) valve extractor.
démultiplicateur, a., (*mach.*) reducing; reduction gearing;
engrenage ——, reducing gear.
démultiplication, f., (*mach.*) reduction ratio.
démultiplier, v. t., (*mach.*) to reduce (as speed, in a given ratio); to gear down.
denier, m., penny; (*in pl.*) funds, money;
——*s publics*, public funds.

dent, f., tooth;
—— *de chien*, pawl, catch.
dépanner, v. r., (*auto., etc.*) to overcome, get out of, "trouble."
déparier, v. t., to cross match a team.
départ, m., departure;
—— *arrêté*, (*sport, auto.*) standing start; start from scratch;
—— *d'assaut*, (*mil.*) start, origin, of an assault;
—— *d'un coup*, (*art., sm. a.*) discharge;
—— *à froid*, (*auto.*) start, without preliminary heating;
—— *lancé*, (*auto.*) flying start;
—— *multiple*, (*mach.*) multiple-feed lubricator (for *graissage à départ multiple*);
—— *au pétrole*, (*auto.*) start by the use of petroleum (i. e., without the recourse to gasoline and similarly of other combustibles) direct start;
zone de —— *d'assaut*, (*mil.*) the zone of trenches from which an assault is made.
département, m., (France) department, political unit composed of several arrondissements, is the jurisdiction of a préfet, (has civil personality and is the bond between the state as a whole and the various centers and interests of the people);
—— *ministériel*, department, ministry.
dépêche, f., dispatch;
porter une ——, *porter des* ——*s, à tout le monde*, (*mil. slang*) to have one's horse run away.
dépens, m., pl., (*law*) costs.
dépense, f., expense, expenditure;
—— *kilométrique*, consumption per kilometer.
dépenté, a., sloped, inclined.
déplacement, m., (*phys., mach., etc.*) displacement.
déplacer, v. t., to displace;
se ——, to travel; (*mil.*) to change station.
dépointage, m., (*art.*) loss of aim, after fire.
dépointer, v. t., (*art.*) to lose aim (after each round); (*mil.*) to disturb the aim of, to lose aim.
déporter, v. t., to remove; (*aero.*) to drift.
déposer, v. t., to submit;
—— *un projet de loi*, to propose, introduce a bill (in Congress, Chamber of Deputies, etc.).
—— *un rapport*, to make or submit a report.
dépôt, m., (*r. r.*) parcel room; (*mil.*) dump.
—— *avancé*, (*mil.*) advanced depot;
—— *de charbon* (*nav.*) coaling station;
—— *de chevaux*, v. s. v. *cheval*;
—— *d'étalons*, (*Fr. adm.*) government stud;
—— *intermédiaire*, service dump;
petit ——, (*Fr. a.*) regimental horse depot (in the field; accompanies regiment in the field);
—— *un projet de loi*, introduction of a bill (Congress, etc.);
—— *d'un rapport*, submission of a report.
dépoter, v. t., to decant; to take out of a pot; (*artif.*) to release.
dépouillement, m., examination, inspection (of reports, etc.); opening of the mail; skinning (of a carcass); working up of observations.
dépouiller, v. t., to open the mail; to skin (a carcass); to examine, go over (as reports, returns); to work up (observations).
dépression, f., (*mach.*) vacuum (in a cylinder).
député, m., deputy (member of the lower house, France); deputy in general.
dérapage, m., (*auto*) skidding, slipping, side slip.
déraper, v. n., (*auto*) to skid, side slip.
dératisation, f., extermination of rats and mice.
dératiser, v. t., to exterminate rats and mice.
déréglage, m., disadjustment; (*art.*) maladjustment (of fire).
dérégler, v. r., to get out of order; (*art.*) to lose aim, to get off the target.
dérivation, f., (*mach.*) small tube tapped on a large one.

dérive, f., (*auto*) slipping backward down hill; leeway; (*aero.*) fin, tail fin; drift; leeway;
angle de ——, (*aero.*) (angle between the axis of an aerial vessel and the direction of its proper speed) drift angle;
échelonner la ——, (*art.*) to distribute, spread, the fire;
—— *horizontale*, (*aero.*) stabilizer;
plan de ——, (*aero.*) stabilizing fin, tail fin.

dériver, v. t.; n., to turn, deflect; (of an auto), to slip (backward down hill); (*aero.*) to drift;
—— *le feu*, (*mil.*) to give deflection.

dermatine, f., packing composition (copal dissolved in turpentine forms the base).

dérobade, f., (*man.*) swerving (as before a jump).

dérobé, m., (*man.*) (of the horse) swerve (before a jump).

dérober, v. r., (*man.*) to swerve.

dérouleur, m., (*mil.*) reel man (telephone, etc.);
aide- ——, assistant reel man.

désactiver, v. t., (*min.*) to disarm a mine.

désactivation, f., (*min.*) disarming of a mine.

désaxage, m., (in *auto construction*) the system in which the cylinder and crankshaft axes do not lie in the same plane.

descendre, v. n., to descend, come down; (*aero.*) to nose down;
—— *à plat*, (*aero.*) to pancake;
—— *de voiture*, (*r. r.*, etc.) to leave, get out of, a carriage, car.

descente, f., descent, fall; (*mil. fort.*) descent into a ditch, shelter, bombproof, etc.;
—— *de main*, (*man.*) (of the horse), fall, dropping, of his quarters;
—— *de voiture*, leaving, getting out of, a carriage, a car.

désengraver, v. t., to run off, push off (a boat that has gone aground).

désenrayage, m., (*r.f. art.*, *sm. a.*) unjamming.

désenrayer, v. t., (*r.f. art.*, *sm. a.*) to unjam.

déséquilibre, m., loss of equilibrium.

déséquilibrer, v. t., to cause loss of equilibrium.

désimmobilisation, f., (*mach.*) unlocking.

désimmobiliser, v. t., (*mach.*) to unlock.

désorienter, v. t., (*inst.*) to unseat.

desserrage, m., loosening (as of a screw, bolt, etc.).

desserrement, m., loosening, release;
—— *des casernes*, (*Fr. a.*) removal from barracks of people having no rights to quarters therein.

desserrer, v. t., to unloose, release;
—— *les casernes*, (*Fr. a.*) to remove from barracks, all persons having no right to quarters therein.

désyntonisation, f., (*wireless*) desyntonization.

désyntoniser, v. t., (*wireless*) to desyntonize.

détachement, m., detachment; party;
—— *de découverte*, (*cav.*) reconnaissance detachment thrown out by the army cavalry;
—— *de liaison et d'observation*, (*mil.*) connecting and observation detachment (between infantry and artillery).

détail, m., (*com.*) retail.

détartrage, m., (*steam*, etc.) scaling, furring, of a boiler, pipe, etc.

détartrer, v. t., (*steam*, etc.) to scale, to fur (a boiler, water pipe, etc.).

détecteur, m., (*elec.*) detector;
—— *d'amplitude*, amplitude detector;
—— *à contact imparfait*, crystal rectifier;
—— *électrolytique*, electrolytic detector;
—— *d'énergie*, generic term for thermal, etc., detector;
—— *d'ondes*, wave detector;
—— *à gaz ionisé*, (Fleming's) vacuum detector;
—— *magnétique*, magnetic detector;
—— *thermique*, thermal detector;
—— *avec tube à soupape*, valve-tube detector.

détendeur, m., (*mach.*) reducing valve.

détendu, a., (*steam*) completely expanded.

détente, f., (*mach.*) the working travel of a piston, power stroke; (*sm. a.*, *mach. gun*) trigger;
—— *adiabatique*, (*phys.*) adiabatic expansion;
—— *à chaleur constante*, v. *adiabatique*;
—— *équilibrée*, expansion in a closed vessel;
—— *fixe*, (*mach. gun*) fixed trigger (for intermittent firing, Mod. 1907 T);
—— *isothermique*, (*phys.*) isothermic expansion;
—— *libre*, free expansion;
—— *mobile*, (*mach. gun*) movable trigger (for automatic firing, Mod. 1907 T).

détention, f., imprisonment;
—— *perpétuelle*, (*law*) life imprisonment.

détonateur, m., (*expl.*, etc.) detonator;
—— *chimique*, chemical detonator;
—— *mécanique*, mechanical detonator;
—— *à percussion*, percussion detonator;

détonation, f., (*expl.*) detonation;
—— *franche*, complete detonation.

détournement, m., (*adm.*) application of funds, stores, etc., to one's own use; (hence) embezzlement, theft.

détourner, v. a., (*adm.*) to take from public stores, to apply funds, stores, to one's own use; (hence) to embezzle, to steal.

détroit, m., (sometimes, when used alone) the English Channel.

deuil, m., mourning;
conduire le ——, to head the mourners;
mettre en ——, to put into mourning;
porter le —— *de*, to be in mourning for.

développement, m., length, development;
n. Km. de ——, n kilometers long.

déverrouillage, m., unbolting; (*mach. gun*) bolt release.

déverrouiller, v. t., to unbolt, (*mach. gun*) to release (the bolt).

déversement, m., (*aero.*) inclination (dip, lift) (of a dirigible);
—— *critique*, critical inclination (beyond which it increases very rapidly).

déviation, f., (*elec.*, etc.) deflection (of a needle).

diable, m., devil;
—— *s bleus*, (*mil. slang*) chasseurs (wear blue, renowned for valor).

dévorer, v. t., (*auto.*, *fam.*) to eat up (distances).

dieu, m., god; (the Deity is written with a capital, Dieu);
les ——*x*, (*Fr. a.*) at Saumur the *cadre noir*, q. v. (so called from their horsemanship);
les demi- ——*x*, (*Fr. a.*) at Saumur, the *cadre bleu*, q. v.

différence, f., difference;
—— *en moins*, reduction, decrease;
—— *en plus*, increase.

différentiel, m., (*mach.*, etc.) differential gear, equalizing gear;
—— *à pignons coniques*, bevel gear differential;
—— *à pignons droits*, spur gear differential.

diffuseur, m., (*mach.*) mixer, mixing cone, spray cone; diffuser; Venturi tube.

diligence, f., diligence;
à la —— *de*, by the action, effort, of;
—— *à vapeur*, steam stage coach.

dîner, m., dinner;
—— *d'adieu*, farewell dinner.

dîner-promotion, m., (*mil.*) class dinner.

direct, a., direct;
mettre en ——, ——, to give direct communication (by telephone).

directeur, m., director;
—— *de l'arrière*, (*Fr. a.*) chief or director of transportation;
—— *de dépôt*, (*Fr. adm.*) manager of government stud;
—— *des étapes et services*, (*Fr. a.*) director of supplies; v. s. v. *direction*;
—— *des forêts*, (*Fr. adm.*) chief forester.

direction, f., (*auto.*, *mach.*) steering gear;
—— *à barre*, bar steering gear, tiller steering gear;
—— *à came conique*, cone cam steering gear;
—— *à crémaillère*, rack and pinion steering gear;
—— *à deux pivots*, two-pivot steering gear;

direction

direction *des étapes et services*, (*Fr. a.*) direction of the line of communications service, (the whole subject of communications and supplies);
— *du feu*, (*art.*) fire direction;
— *à essieu brisé*, v. — *à deux pivots;*
— *irréversible*, irreversible steering gear;
— *à levier*, steering bar;
— *à secteur*, v. — *à vis sans fin;*
— *des services automobiles*, (*Fr. a.*) motor transport department;
tige de —, v. s. v. *tige;*
— *à vis sans fin*, worm and sector steering gear;
— *à volant*, wheel steering gear.
direction-repère, f., (*art.*) registration direction, reference direction.
directive, f., (*mil.*) instructions, general directions (as, for an operation, a combat).
dirigeable, a., dirigible, steerable; m., (*aero.*) dirigible balloon;
— *rigide*, rigid dirigible;
— *semi-rigide*, semirigid dirigible;
— *souple*, nonrigid dirigible.
dirigeable-éclaireur, m., (*aero.*) scout dirigible.
dirigeable-vedette, m., (*aero.*) observing dirigible.
diriger, v. t., (*aero.*) to steer.
dislocation, f., (*mil.*) subdivision of troops (i. e., in assignment of stations).
disette, f., scarcity, dearth.
dispensaire-école, m., (*Fr. a.*) sort of dispensary and clinic.
dispense, f., exemption, waiver;
— *d'âge*, waiver of an age limit.
dispositif, m., arrangement, contrivance;
— *de rupture*, (*elec.*) contact breaker.
disposition, f., disposition; arrangement;
— *s de combat*, (*art.*) (may be translated) "take equipments."
disque, m., disk;
— *d'appui*, (*art*) supporting disk;
— *de direction*, (*aero.*) steering disk;
— *de fermeture*, (*r. r.*) block signal.
disrupteur, m., (*elec.*) interrupter, break.
disrupture, f., (*elec.*) disruptive spark gap.
dissimulation, f., dissimulation;
— *de matière*, (*adm.*) incorrect entry of stores, materials; material wrongly taken up.
dissolution, f., (*auto.*) cement (for repairing tires).
distance, f., distance;
— *d'assaut*, (*mil.*) assaulting distance;
— *de bonne rupture*, v. s. v. *rayon;*
à courte —, (*mil.*) at close quarters;
— *explosive*, (*wireless*) spark gap; sparking distance;
aux petites —*s*, (*mil.*) at close quarters;
— *entre les plans*, (*aero.*) gap;
— *de rupture limite*, v. s. v. *rayon.*
distillateur, m., distiller.
distribe, f., (*mil. slang*) distribution of mail.
distributeur, m., (*elec.*) contact maker, make-and-break key (more especially one for periodic makes and breaks);
— *d'allumage*, (*elec.*) ignition switch or contact (for periodic makes and breaks);
— *compte-gouttes*, (*mach.*) drip feed;
— *de courant*, (*elec.*) current distributing key;
— *à haute tension*, v. — *de secondaire;*
— *mécanique*, mechanically operated contact maker;
— *de primaire*, (*elec.*) a synonym of *allumeur* in monocylindric motors;
— *de secondaire*, (*elec.*) high tension contact maker.
distribution, f., distribution; (*mach.*) feed gear, valve gear, valve mechanism.
doigt, m., finger;
— *de débrayage*, (*art.*) releasing dog, jag. (155 C. S.)
domaine, m., domain;
— *éminent*, (*law*) eminent domain;
— *forestier*, national forests;
grand —, a generic term for fiefs of all kinds, estates, lordships, etc.;
— *militaire*, so much of the public domain as is military;

droit

— national, State property;
petit —, a generic term for waste land, swamps, ponds, etc., houses, shops, heaths, moors, etc., (obs.);
— *privé militaire*, generic term for barracks, quarters, hospitals, arsenals, for forts, etc., no longer in commission, etc.;
— *public*, (generic term for roads, streams, seashore harbors and (sea-) roads; gates, walls, etc., of forts or fortresses, and in general for all property of any sort whatsoever not susceptible of private appropriation or use); State property;
— *public artificiel*, canals, roads, railways;
— *public concédé*, all railways not operated by the State; canals not State property, toll bridges;
— *public fluvial*, streams (navigable and floatable);
— *public maritime*, generic term for seashores, ports, roads, havens, whether natural or artificial (in short, for any part of the sea under national control in fishing or navigation);
— *public militaire*, (generic term for all terrain and works used in the public defense); military reservations;
— *public monumental*, generic term for all monuments, churches, and other construction of general interest;
— *public national*, the same as — *public*, q. v.
— *public naturel*, rivers and sea shores;
— *public non concédé*, railways operated by the State, public canals, etc.;
tomber dans le — *public*, to become public property (e. g., an expired patent).
dommages-intérêts, m., pl., (*law*) damages.
donner, v. n., (*art.*) to fire, open fire.
dos, m., back;
— *d'âne*, water bar; (*top.*) small ridge, hog back.
dosage, m., proportions, (as of air and gasoline in an explosive mixture).
doseur, m., (*mach.*) feed regulator;
— *d'air*, air regulator (of automatic carbureters).
dossier, m., papers;
— *du personnel*, (*Fr. a.*) formerly, personal efficiency reports by the lieutenant-colonel of the regiment or head of a *service;*
— *de tir*, (*art.*) (papers, tables, information relating to an emplacement; comprises *planchette de tir* and *carnet de tir*); artillery record.
doublard, m., (*mil. slang*) sergeant.
double, m., (*mil. slang*) sergeant-major.
doublement, m., duplication (of images in certain range finders).
doubler, v. t., to duplicate; to accompany (as an assistant).
doublure, f., substitute, understudy.
douille, f., sleeve, socket;
— *à baïonnette*, (*sm. a.*) bayonet socket (used also in mechanical relations);
— *obturatrice*, (*art.*) obturating cartridge case;
— *porte-amorce*, (*art.*) primer case socket;
— *de prise de courant* (*elec.*) connection socket;
— *de roulement*, nave box;
— *de retenue*, (*art.*) locking sleeve.
douille-entretoisé, f., (*trench art.*) socket-transom (240 C. T. truck).
draperie, f., drapery; bunting used in decoration;
— *d'enseigne*, (*mil.*) flag cloth, bunting.
drisse, f., (*com.*) lashing in general, any temporary lashing.
droguiste, m., druggist.
droit, m., tax, due, fee; law;
— *administratif*, (*adm.*) body of regulations necessary to the public administration;
— *d'asile*, right of asylum;
— *commun*, (*law*) ordinary law (civil or common), as distinguished from military law;
— *constitutionnel*, (*law*) constitutional law;
— *de contrôle*, (*adm.*) right to check (the administration of subordinates);
— *exécutif*, (*law*) right of seizure, of arrest;
— *international*, (*law*) international law;
— *de passage*, (*r. r.*) right of way;

droit *pénal*, (*law*) penal law;
—— *politique*, v. —— *constitutionnel;*
—— *privé*, (*law*) private law, the municipal law;
—— *public*, (*law*) public law (i. e., international, political, penal, etc.);
—— *de réquisition*, (*mil.*) right of requisition;
—— *de saisie*, (*nav. int. law*) right of seizure, of capture.

droitier, a. m., right handed; right-handed man.

dualité, f., duality;
—— *d'origine*, (*Fr. a.*) duality of origin (of officers, some coming from the great schools, the remainder from the ranks).

duite, f., (*cord.*) warp; rope yarn.

duplicateur, m., duplicator (sort of copying machine).

duplice, f., double alliance (France and Russia).

duralumin, f., (*met.*) an alloy of aluminum, non-oxidizable.

durit, m., india-rubber pipe.

dynamite, f., (*expl.*) dynamite;
—— *à base inerte*, inert base dynamite;
—— *gelée*, frozen dynamite.

dynamode, m., (*expl.*) unit of potential (work divided by 1,000).

E.

eau, f., water;
—— *de boisson*, drinking water;
—— *croupie*, stagnant water;
laver à grande ——, (*chem.*) to wash in an excess of water;
—— *ménagères*, slops, waste water;
—— *x métropolitaines*, home waters;
point d'——, any point for the supply of water; water hole;
—— *de refroidissement*, (*mach.*) cooling water;
retenir de l' ——, (*pont, etc.*) to hold water (rowing);
—— *de son*, bran water.

ébouillanter, v. t., to scald, scald out (a vessel, dish, etc.).

écart, m., (*com.*) margin (as, between receipts and expenses);
—— *probable pratique*, (*art.*) the probable error of the field of battle (1.5 times that of the proving ground), field probable error.

écartement, m., spacing, separation;
—— *des plans*, (*aero.*) gap.

écarteur, m., separator.

échancrure, f., (*art.*) cut, recess, in a breech plug;
—— *de visée*, (*mach. gun*) aiming aperture (shield).

échappement, m., (*steam, etc.*) exhaust, escape, exhaust stroke;
—— *libre*, silencer by-pass, relief valve; exhaust into the air (motors);
régulateur d' ——, (*mach. gun*) pressure regulator, gas vent ring.

échaudoir, m., slaughterhouse.

échelle, f., ladder;
—— *de franchissement*, (*mil.*) trench ladder;
—— *de relevage*, (*trench. art.*) lighting gear.

échelon, m., (*mil.*) echelon, (used alone, frequently means) ammunition column, supply, unit;
—— *de batterie*, (*art.*) battery echelon (ammunition supply);
—— *de combat*, v. s. v. *batterie de combat;*
—— *de compagnie*, (*mil.*) ammunition detachment;
—— *de station-magasin*, (*Fr. a.*) ammunition supply kept in one of the S. M. assigned to a given army;
—— *volant*, (*mil., aero.*) the flying section proper of an *escadrille*.

échelonnement, m., (*mil. and in gen.*) echeloning;
—— *des approvisionnements* (*mil.*) supply depots on the line of communications;
—— *de convergence*, (*art.*) convergence difference;
—— *de répartition*, (*art.*) distribution difference;
—— *total*, (*art.*) deflection difference.

éclairage, m., illumination; (*mil.*) enlightenment (rare).

éclairement, m., intensity of illumination.

éclaireur, m., (*mil., etc.*) scout; boy scout; (more particularly to-day) man who explores a cleaned-out trench, (*Fr. a.*) synonym of *aide-grenadier*, q. v.; (*aero.*) a dirigible of from 6,000 to 7,000 cubic meters; (*auto.*) automobile light;
—— *d'artillerie*, v. *officer orienteur*;
—— *monté*, (*mil.*) mounted scout.

éclat, m., splinter;
—— *en retour*, (*art.*) splinter or fragment that flies back, throw-back.

éclatement, m., bursting (of a tire); (*elec.*) sparking;
—— *prématuré*, (*art.*) premature burst;
—— *en surface*, (*art.*) surface burst (on surface of the ground).

éclater, v. n., to burst;
—— *en surface* (*art.*) to burst on the surface (i. e., of the ground).

éclateur, m., (*elec.*) gap, spark-gap;
—— *à boules*, ball spark-gap;
—— *fixe*, (quenched) spark-gap;
—— *à lame de mica*, earth arrester;
—— *multiple*, multiple spark-gap;
—— *tournant*, rotary spark-gap.

éclopé, a., (*mil.*) footsore.

éclosion, f., hatching out.

école, f., school;
—— *de dressage*, (*man.*) in France, school or establishment furnishing horsemen opportunities to train their horses;
—— *de guerre*, v. —— *supérieure de guerre;*
—— *nationale des haras*, (*Fr. adm., hipp.*) school for the development of hippology and the training of the personnel of the government studs;
—— *de palefreniers*, (*Fr. adm.*) school for grooms (forms part of the *école des haras*);

école, f., (continued)
—— *des signaleurs d'armée*, (*Fr. a.*) signal school (for the training of instructors);
—— *de signalisation et de liaison d'infanterie avec avion et ballon*, (*Fr. a.*) corps signal school (instruction of officers and signal men);
—— *de tir*, (*art.*) target practice, target season.

économat, m., stewardship, treasuryship;
commis d' ——, treasurer's clerk.

économe, m., steward, treasurer.

écoulement, m., outlet, (as of a lake).

écoute, f., (*mil. min.*) listening, listening service; (*wireless*) receiving station;
caisse d' ——, (*mil.*) (so-called) "listening box" of a listening post, (contains the receiving trumpets, and angle measuring device);
chef d' ——, n. c. o. in charge of a listening squad;
lunette d' ——, sound direction detector;
—— *téléphonique*, telephonic listening post.

écouteur, m., (*mil. min.*) listener; listening device or apparatus, mine stethoscope.

écoutoir, m., (*mil. min.*) listening device, mine stethoscope; ear piece of a geophone;
—— *double à ressort*, double listener (two ear pieces).

écouvillon-curette, m., (*art.*) sponge and scraper.

écran, m., (*mach.*) baffle plate;
—— *des vitesses*, (*art.*) velocity disk (155 C. S.).

écran-silhouette, m., (*mil.*) silhouette screen.

écriteau, m., signboard.

écrasement, m., collapsing (of a pneumatic tire at point of contact).

écrêtement, m., (*art.*) (of fire) grazing, (top of a parapet or cover, etc.);
—— *indicateur*, (*siege*) trench sign, signboard.

écrou, m., nut;
—— *amortisseur*, buffer nut;
—— *d'arrêt*, locking nut;
—— *d'attelage*, (*art.*) locking nut;

écrou *de blocage*, locking nut, clamping nut;
— *borgne*, screw cap;
— *à chapeau*, screw cover;
— *fendu*, split nut, safety nut (will not unscrew);
— *de fixation*, holding down nut;
— *presse-cuir*, (*art., mach., etc.*) leather-washer jam nut;
— *de retenue*, check nut;
— *de tension*, tightening nut, adjusting nut, setting nut;
— *à tourillons*, (*in gen.*) trunnion-nut.

écuage, m., v. *écuanteur*.
écuanteur, f., dish of a wheel.
écurie-baraque, f., (*mil.*) hut stable, field stable.
écurie-dock, f., box-stall stable.
effacer, v. t., to rub out;
— *les épaules*, to hold the shoulders square and back.
effectif, m., (*mil.*) strength, total strength;
— *budgétaire*, the strength as appearing in parliamentary estimates;
— *mobilisable*, the men available for purely military duty (the — *réel*, diminished by men on non-military duty, *service auxiliaire*);
— *réel*, all the men on the rolls at a given instant.
effet, m., effect;
— *destructif*, (*mil.*) destructive value;
— *meurtrier*, v. — *destructif*;
— *s publics*, (*com.*) Government stocks and bonds;
— *s à la taille de . . .*, clothes, etc., fitting
effiler, v. t., to taper.
effleuré, m., (*man.*) graze (in a jump).
égailler, v. r., to spread, distribute.
éjecteur, m., (steam) ejector.
élargisseur, m., (*art.*) extension piece, extension (platform).
électrisation, f., (*elec.*) electrification.
électriser, v. t., (*elec.*) to electrify.
électrobus, m., (*auto.*) electric bus.
électrogène, a., (*elec.*) generating.
élément, m., element;
— *s non endivisionnés*, (*mil.*) (in a corps, all the elements not organically parts of the divisions, e. g., battalion of engineers, medical units, etc.) nondivisional elements, corps troops.
élévateur, m., (*mach. gun*) carrier.
élève, m., pupil, student;
— *maréchal*, (*Fr. a.*) lance farrier;
— *sortant*, (*mil.*) cadet about to be graduated.
élève-officer, m., (*mil.*) pupil or student who is to become an officer.
élingue, f., sling;
— *de portage*, sling.
Élysée, m., official residence, in Paris, of the President of the French Republic;
palais de l' —, the Élysée Palace.
émailler, v. t., to enamel.
embâcle, m., ice jam, ice pack.
emballage, m., packing case.
emballement, m., (*mach.*) racing; (*man.*) runaway, running away.
emballer, v. t. r., (*mach.*) to race; (*man.*) to run away, bolt.
embarcation, f., small boat;
— *à moteur à pétrole*, petroleum motor boat;
— *a, au, pétrole*, petroleum motor boat;
— *à vapeur*, steam launch.
embardée, f., (*aero.*) yaw, yawing; (*nav.*) yaw; (*auto.*) front skid;
faire une —, (*auto.*) to skid; (*aero., nav.*) to yaw.
embarder, v. n., (*auto.*) to skid; (*aero., nav.*) to yaw.
embase, f., (*mach.*) flange.
emboché, a., (tainted with *bochisme*, or subjected, whether voluntarily or not, to the influence of *bochisme*) "prussianized."

embocher, v. t., (*mil. slang*) to turn boche.
emboîtement, m., interlocking;
à —, interlocking.
embouchoir, m., (*sm. a.*) upper band;
— *porte-guidon*, (*mach. gun*) front right bracket.
embout, m., ferrule, tip.
embouteillage, m., bottling.
embouteiller, v. t., to bottle.
embrayage, m., (*mach.*) engagement (of a clutch, etc.) clutch;
— *à cône(s)*, cone clutch;
— *à cône(s) direct, droit*, cone clutch (the cone moving toward the motor);
— *à cône(s) inversé*, inverted cone clutch;
— *à disques*, friction disk clutch (multiple disks);
— *à enroulement*, v. —*à spirale*;
— *extensible*, expanding clutch;
— *hydraulique*, hydraulic clutch;
— *métallique*, metal on metal clutch;
— *à plateaux*, plate clutch;
— *à ruban*, band clutch;
— *à segments (extensibles)*, expanding clutch (spreading or extension of segments);
— *à spirale*, spiral clutch, coil clutch.
embrayer, v. t., (*auto.*) to put on, use, such and such a speed.
embrun, m., mist, foggy weather.
embuscade, f., (*mil. slang*) situation or state of being a shirker, a slacker; evasion of military duty or service.
embusqué, m., (*mil. slang*) shirker, slacker.
embusquer, v. r., (*mil. slang*) to shirk, to be a slacker; to evade military duty or service.
émetteur, a., sending; m., (*elec., etc.*) sender, sending apparatus, emission device.
émettre, v. t., (*telegraphy, etc.*) to send.
émission, f., *telegraphy, etc.*) sending; (*elec., etc.*) sending, emission;
— *directe*, (*wireless*) direct coupled transmission; direct coupled transmitting station or set;
— *indirecte (par induction) (wireless)* inductively coupled transmission; inductively coupled transmitting set or station.
emmanchement, m., helving;
— *carré*, (*mach.*) square shank;
— *à tournevis*, (*mach.*) sort of tongue-and-groove clutch.
emmancher, v. t., to fit on a handle;
— *en carré*, (*mach.*) to mount on a square shaft.
empattement, m., (*auto., etc.*) wheel base; (*aero.*) width.
empennage, m., (*aero.*) fin, empennage; fin placed at the stern of a dirigible to kill pitching; stability plane; stabilizer; (*art.*) fin of a projectile;
— *pneumatique*, (*aero.*) pneumatic empennage, stability ballonet.
empenner, v. t., (*aero.*) to equip a dirigible with fins; (*mil.*) to equip bombs, grenades, etc., with fins.
emplanture, f., step (of a mast);
— *des ailes* (*aero.*) wing socket.
emplâtre, m., (*auto.*) repairing strip.
empreinte, f., impression;
— *digitale*, finger print.
encadrer, v. t., (*art.*) to bracket.
encagement, m., imprisonment; (*mil.*) inclusion (of a portion of the front, as by grenade fire).
encager, v. t., to imprison; (*mil.*) to include (a portion of the front, as by grenade fire).
en-cas, m., (*any*) emergency provision;
— *mobile*, (*mil. r. r.*) emergency train (supplies, ammunition, etc., ready to start at a moment's notice).
encasernement, m., (*mil.*) barracking, quartering of troops in barracks.
encaserner, v. t., (*mil.*) to quarter in barracks.
enclanchement, m., (*mach.*) coupling gear or device, (also spelled *enclenchement*).
enclenche, f., (*mach.*) (eccentric) gab, (also spelled *enclanche*).

encoche, f., notch;
— *à baïonnette*, bayonet point.
encombrement, m., (*art*.) cubic space (occupied, e. g., by a mortar in battery, measurement over all).
encrassement, m., fouling.
encrasser, v. a. r., to foul, to become foul.
encuivrage, m., (*art*., *sm. a*.) metal fouling.
endaubage, m., preserved (canned) meat.
endauber, v. t., to preserve meat.
enduire, v. t., to dope.
enduit, m., (*aero*.) dope.
enfonce-portes, m., fireman's pickax.
enfourchement, m., fork (in many techincal relations).
engagement, m., (*aero*.), nose dive;
— *spécial, dit du devancement d'appel (Fr. a.)* (enlistment or entry into service before date on which service is due), anticipated enlistment.
engin, m., gear, device;
— *d'arrêt*, (*aero*.) device for stopping a balloon from dragging when landing (rip-strip, anchor, guide rope and valve);
—*s de lancement*, (*artif*.) projecting apparatus;
— *de sauvetage*, life-saving device.
—*s de tranchée*, (generic term for trench guns, mortars, etc.) trench gear;
engrenage, m., (*mach*.) gear, gearing; mesh;
— *de direction*, steering gear;
— *de distribution*, feed-gear.
enroulement, m., (*elec*.) winding;
— *inducteur*, winding of the primary coil, primary winding;
— *induit*, winding of secondary coil;
— *primaire*, v. — *inducteur*;
— *secondaire*, (*elec*.) winding of the secondary coil, secondary winding.
ensablé, a., sand bound;
port —, sand bound port.
entartrage, m., (*mach*.) scale, furring (of tubes, radiators, etc.).
entartré, a., (*mach*.) scaled, furred (of a boiler, pipe, etc.).
entartrer, v. r., (*mach*.) to scale, (become covered with scale).
entassement, m., (*mil*.) the (hurtful) concentration of men on the battle line, or in action, in an assault.
entente, f., understanding;
l' —, the Entente (France and England).
enterrer, v. t., to bury;
— *les morts*, to bury the dead;
s' —, (*mil*.) to dig in, to dig one's self in.
entoilage, m., (*aero*.) (covering of canvas placed over the fuselage, frames, etc., to diminish air resistance); fabric; wing covering; covered body (air plane).
entoiler, v. t., (*aero*.) to cover fuselage, frames, etc., with canvas.
entonnoir, m., funnel;
— *à grille*, a screened funnel.
entraînement, m., (*mach*.) driving;
— *par carré*, square shaft drive;
— *par pignon*, gear driving.
entraîner, v. t., to drag, carry on; (*mach*.) to drive.
entrée, f., entrance; (*mach., etc.*) inlet, intake, admission, introduction;
les —*s*, (*com*.) receipts, income;
— *par l'arrière* (*auto*.) rear entrance;
— *de caisse*, (*auto*.) body entrance;
— *de dent*, (*mach*.) beveling, chamfer of a gear tooth;
— *latérale* (*auto*.) side entrance.
entrefilet, m., paragraph (in a newspaper).
entre-jambes, m., (*unif*.) crotch of a pair of trousers.
entrer, v. n., to enter;
— *à forcement*, to go on or in with a press fit.
entre-rail, m., (*r. r.*) gauge.

entretoise, f., distance piece; cross brace;
— *avant porte-fourche*, (*mach. gun*) bipod block;
— *médiane*, (*mach. gun*) handle block;
— *porte-volet*, (*mach. gun*) barrel catch block.
énucléation, enucleation.
enveloppe, f., shoe (pneumatic tire); (*mach*.) water-jacket; (*aero*.) gas-bag; (*mach. gun*) barrel casing.
envergure, f., (*aero*.) spread; wing span;
— *relative* (*aero*.) aspect ratio;
— *totale*, (*aero*.) spread (from tip to tip).
envol, m., (*aero*.) start.
envolée, f., (*aero*.) trip, journey, flight.
épanouir, v. t., (*mach*.) to splay.
épaulement₂ m., (*aero*.), ferule.
épée, f., sword;
— *de combat*, dueling sword;
faire de l' —, (*fenc*.) to fence with the sword.
épéiste, m., (*fenc*.) skillful sword fencer, swordsman.
éperon₂ m., spur;
— *à boîte*, box spur;
— *à la cavalière*, jack spurs;
— *à courroies*, ordinary spur;
— *mobile*, detachable spur.
épi, m., (*mil. top*.) small salient, point or tongue; (*mil. r. r*.) spur track;
— *courbe*, (*mil. r. r*.) fan-shaped combination of spur tracks (some railway guns can fire only in the direction of the axis of the truck; to allow them to fire in directions other than that of the main line the *épi courbe* is constructed).
épingle, f., pin;
— *à chapeau*, hat pin; (*mil. slang*) rifle grenade;
— *à cheveux*, hair pin; hair pin curve;
— *de nourrice*, safety pin;
— *de sûreté*, safety pin.
épouvantail, m., scare crow.
épreuve, f., test;
— *de conductibilité* (*elec*.) conductibility test;
— *de fond*, (*man*.) endurance test;
— *d'isolement*, (*elec*.) insulation test;
— *d'obstacles*, (*man*.) jumping test;
— *de roulement*, (*art*.) road test, marching test (of field artillery matériel).
épurateur₂ m., purifier, in general.
équerre₂ f., corner, angle plate;
— *de retenue*, (*trench art*) corner iron (for holding platform, 240 C. T., on wagon).
— *de socle*, trusset.
équilibrage, m., (*mach*.) balancing.
équilibre, m., (*aero*.) balance.
équilibrer, v. t., (*mach*.) to balance; (*aero*.) to stabilize.
équilibreur, m., (*aero*.) stabilizer, stabilizing rudder, stabilizing fin; (*art*.) muzzle counterpoise (155 C. T. R.).
équipe, f., (*mil*.) detachment, squad; (*sport*) team.
— *de liaison*, (*mil*.) connecting squad, liaison squad;
— *de salubrité*, (*mil*.) police squad in barracks.
équitation, f., (*man*.) equitation;
— *extérieure*, out-door riding (as distinguished from hall work).
éraillé, a., (*in gen*.) raveled, frayed.
érosion, f., (*art*.) blast (ground in front of a gun, torn up, burnt, etc., by the blast).
erreur, f., error;
— *de bande*, (*aero*.) error due to vertical magnetic influence of earth combined with that of magneto;
— *quadrantale*, (*aero*.) quadrantal compass error (alternating east and west in the four quadrants);
— *semi-circulaire* (*aero*.) semi-circular error (constant direction error).
escadre, f., (*aero*.) squadron (unit, above *escadrille*).
escadrille, f., (*Fr. a., aero*.) unit of aviation troops or personnel; escadrille;
— *d'artillerie*, (*aero*.) artillery escadrille, spotting squadron;
— *de bombardement*, (*aero*.) bombing group or escadrille;

escadrille *de chasse*, pursuit escadrille;
— *de combat*, battle escadrille;
— *d'infanterie*, infantry escadrille (assists infantry combat);
— *de place*, fortress aero squadron;
— *de protection*, protecting escadrille;

escale, f., *(aero.)* landing;
faire —— *à*, to come down at;
port d' ——, station.

escrime, f., fencing;
—— *au fleuret*, foil fencing.

espolette, f., *(artif.)* delay action plug.

esprit, m., spirit;
—— *de sel (chem.)* hydrochloric acid.

essai, m., trial, test;
— *d'étalonnage, (aero.)* calibration test;
— *de portée, (art.)* range trial, trial shot;
— *au sable, (aero.)* sand test;
— *statique*, v. —— *au sable*.

essence, f., gasoline; petrol;
— *à brûler*, fuel oil;
— *à éclairer*, lamp oil, rectified oil;
— *légère*, light oil;
— *lourde*, heavy oil;
— *minérale*, gasoline;
— *de pétrole*, gasoline, petrol.

essieu, m., axle;
[Except where otherwise indicated, the following terms relate to motor vehicles:]
— *arrière*, rear axle; dead or carrying axle of a motor;
— *avant*, front axle, steering axle of a motor;
— *avant surbaissé*, bowed steering axle, dropped axle, hanging axle;
— *brisé*, divided axle;
— *directeur*, steering axle;
— *fictif*, (*Fr. a., r. r.*) unit of capacity of a truck or flat car (in respect of the transportation of vehicles; the *essieu fictif*, is the space taken up by the limber of the 75mm gun or of its caisson. The matériel of a 75mm battery takes 55 *essieux fictifs*);
— *au milieu de l'* ——, *(art.)* at the center of the traverse;
— *moteur*, driving axle;
— *porteur*, rear axle, dead axle, carrying axle;
— *profilé*, I-beam profile;
— *surbaissé*, dropped axle;
— *tournant*, driving axle, live axle (incorrect name for rear axle of a *voiture à cardan*; the correct term is *pont arrière*, q. v.);
— *tubulaire*, tubular axle.

estoquiau, m., *(mach., etc.)* stop, pin, stud (any pin or bolt limiting the play or motion of a part).

estrade, f., platform;
— *de grenadiers, (mil.)* grenade platform (in trenches).

établissement, m., establishment, institution;
— *de répression, (adm.)* penitentiary.

établir, v. t., to establish;
— *le pont volant, (pont.)* to set a flying bridge;

étaler, v. t., to spread, spread out.

étalon, m., standard;
— *à bout*, end standard.

étape, f., *(mil.)* march;
commandement d' ——s, *(Fr. a.)* c. o. and assistant at railhead;
fournir l'——, to make the trip, the distance;
fournir l'—— *de ... à ...*, to march, go, from ... to...

état, m., state, report, return; error (of a clock, watch);
— *de cession*, report accompanying a transfer, as, of a horse, of property, etc.;
— *D*, (*Fr. a.*) report made on an officer setting forth his qualifications for promotion;
— *K', (Fr. a.)* consolidated report made in Minister of War's office from the *états D*, q. v., in the case of officers serving with troops;
— *K', (Fr. a.)* consolidated report made (in Minister of War's office for detached officers) from the *états D*, q. v.
faire —— *de*, to notice, take into account; to report;

— *naissant, (chem.)* nascent state;
— *natif, (chem.)* native state;
— *récapitulatif, (adm.)* consolidated report;
— *riverain*, state bordering on, marching with.

étatisme, m., doctrine of the State (in contradistinction with the doctrine of the individuality).

état-major, m., *(mil.)* staff;
— *administratif*, administrative as distinguished from purely military staff;
— *central*, headquarters staff.

étayer, v. t., *(mil.)* to stiffen, support, back up (e. g., a front under fire).

étendard, m., *(mil.)* standard (cavalry color);
— *étoilé*, the Stars and Stripes.

étincelle, f., *(elec.)* spark;
— *d'allumage*, igniting spark;
— *amortie*, quenched spark;
— *d'arrachement*, break, or breaking spark;
— *d'extra-courant de rupture*, break spark;
— *de rupture*, break spark;
— *de tension*, jump spark, high tension spark, induction spark.

étiquette, f., ticket, label;
— *d'écurie, (mil.)* stable tag (bearing name, sire, dam, and breed of a horse).

étirer, v. t., *(met.)* to draw;
— *à froid*, to draw cold.

étoffe, f., *(hipp.)* build, (in the sense of strength and of endurance);
— *double caoutchoutée, (aero.)* double rubberized cotton;
— *de droit fil, (aero.)* parallel thread cotton;

étoile, f., *(mach.)* star wheel (small star-shaped wheel for adjusting or setting a screw, rod, stem, etc.); *(mil. in pl.)* the grade or dignity of general officer; (*Fr. a*) star on the ribbon of the *croix de guerre*, (bronze for citation in regimental or brigade orders; silver, for division; gold for corps orders);
— *éclairante, (artif.)* light star, flare;
gagner ses ——s, *(mil.)* to get his stars, i. e., to be promoted general;
— *à parachute, (artif.)* parachute flare;
— *de la télégraphie militaire*, a synonym of cartouche signal, q. v.

étoqueau, m., v. *estoquiau*.
étoquereau, m., v. *estoquiau*.
étoquiau, m., v. *estoquiau*.

étoupe, f., waste.

étoupille, f., *(art.)* tube, (primer).

étranger, a., foreign;
au 1er ——, *(mil.)* in the 1st foreign (regiment).

étranglement, m., (in many relations besides steam) throttling; *(auto.)* narrowing, cambering, (of a motor car frame in front).

étrangleur, m., *(mach.)* throttling device, throttle, throttle valve;
— *d'air*, air throttle.

étrave, f., *(aero.)* stem.

étrier, m., stirrup; yoke; clip.

étrier-éperon, m., *(harn.)* spur and stirrup in one (i. e., stirrup with spur permanently attached).

étui-boîte, m., case.

étuve, f., stove;
— *à désinfection, (med.)* disinfecting stove or cylinder.

évacuation, f., *(mach.)* evacuation, exhaust, exhaust stroke.

évader, v. r., to escape from, break jail; to escape from confinement, imprisonment.

évasion, f., escape from jail, from imprisonment.

évent, m., *(mach. gun)* gas port;
— *des erassés, (mach. gun)* cleaning tap;
— *de prise de gaz (mach. gun)* gas port;
— *de réglage, (art.)* adjusted setting (in a series of four shots, two must burst high, and two low).

éventail, m., fan;
— *de la batterie, (art.)* sheaf of fire;
— *du tir*, v. —— *de la batterie*.

examen, m., examination;
— *d'aptitude physique*, physical examination;
— *d'entrée*, entrance examination;
— *de sortie*, graduation or final examination.

excédent, m., excess;
— *de temps*, (sport, man.) exceeding of time limit.

excentrage, m., state of being, operation of putting, out of center (e. g., in a lathe).

excentrique, m., eccentric;
— *d'orientation*, (mach. gun) setting eccentric.

excès, m., excess;
— *de pouvoir*, excess of authority.

excitateur, m., (elec.) exciter;

exercice, m., exercise;
—*s sur la carte*, (mil.) war game;
l'— *est clos*, (adm.) budgetary accounts are closed;
—*s de combat*, (mil.) combat drill (drill in cases that arise in attack and defense);
— *d'évolution*, (mil.) drill to develop suppleness and rapidity of maneuver; development drill;
—*s d'instruction*, (t. p.) preliminary training;
—*s de mine*, (mil. min.) mining maneuvers;
— *de ravitaillement*, (mil.) maneuver to test or apply a supply system; supply maneuver.

exploitation, f., exploitation; (mil.) (of information) circulation, transmission, despatch to interested parties;
— *tactique du succès*, (mil.) operation of clinching a success, of following it up;
— *troupes d'* —— *tactique du succès*, (mil.) troops detailed to clinch, to follow up, a success.

exploiter, v. t., to exploit; (mil.) (of information) to circulate, transmit, send to interested parties;
— *un succès tactique*, (mil.) to follow up a tactical success.

exploration, f., (mil.) exploration;
— *à petite distance*, (mil.) exploration after contact.

exploseur, m., (elec.) exploder;
— *à magnéto*, dynamo-electric exploder;
— *à pile*, battery exploder.

explosif, m., (expl.) explosive;
— *chimique*, chemical explosive (e. g., the picrates);
— *mécanique*, mechanical explosive (e. g., black powder);
— *mixte*, mixed explosive (e. g., dynamite).

explosion, f., (mach.) explosion; power stroke;
— *par influence* v. —— *sympathique*.

extenseur, m., (aero.) rubber shock absorber.

extérieur, a., foreign, out-of-doors (of riding, sports, etc.);
à l'——, in foreign parts;
équitation ——*e*, (man.) riding out-of-doors, cross-country work;
politique ——*e*, foreign politics.

extra-courant, m., (elec.) extra current.
— *de rupture*, extra current.

F.

face, f., face; headlight (rare); (trench art.) end sill (of a mortar platform, e. g., 240 L. T.);
— *à droite (gauche)*, (mil.) right (left) face (command and movement).

facteur, m., agent, postman;
— *distributeur*, postman, mail carrier, (U. S.).

fagot, m., fagot;
— *d'allumage*, kindling wood.

faim, f., hunger;
rester sur la ——, to be kept hungry, (e. g., of a carrier pigeon, away from his station or home).

faire, v. t., to do, make;
— *du n*, (auto.) to do n kilometers per hour.

faisceau, m., sheaf of light, beam; (art.) sheaf of fire (from the guns of a battery); (pont.) bundle (of ropes);
déplacer le ——, (art.) to transfer the sheaf;
— *divergent*, divergent beam;
fermer le ——, (art.) to close the sheaf;
former le ——, (art.) to form the sheaf (by echeloning elevations);
ouvrir le ——, (art.) to open the sheaf;
— *lumineux*, luminous sheaf, beam;
— *parallèle*, parallel beam;
— *des plans de tir*, (art.) sheaf of fire.

fait, m., fact;
—*s divers*, (in newspapers) minor news, gossip, personal notes.

familiarisé, a., familiarized;
— *avec*, (man.) "wise."

famille, f., family;
— *militaire*, (in military newspapers, heading of a column) personal news, personals.

familier, a., familiar; trusting, fearless, (e. g., carrier pigeons, with respect to their keepers).

fanion, m., flag, color;
— *neutralité* (mil. med.) Red Cross flag of a hospital unit.

farine, f., flour;
— *dure*, winter wheat flour;
— *tendre*, spring wheat flour.

fatigué, a., (art.) worn (of a gun, after much firing).

fauchage, m., (art., mach. gun) sweeping fire;
— *en direction*, sweeping fire;
— *double*, (art.) double sweeping (successive rounds separated by six turns of handwheel);
— *en portée*, searching (in depth);
— *simple*, (art.) sweeping proper (successive rounds separated by three turns of handwheel);
tir de ——, mowing fire.

faucher, v. t., to mow; (art., etc.) to mow, sweep, by fire;
— *double*, (art.) to sweep broad (six turns between rounds);
— *double (simple) par n*, (art.) to sweep *n* turns between rounds;
— *simple*, (art.) to sweep (three turns between rounds).

fausse-cale, m., (mil. slang.) nonmobilizable man.

fausser, v. t., to distort, spring, get out of shape.

fauteuil, f., arm chair; (fig.) chair of a committee (i. e., chairman); chair, i. e., membership, in the French Academy.

fauteuil-lit, m., (r. r.) sleeping chair.

faux, a., false; dummy;
— *malade*, (mil.) malingerer.

faux-châssis, m., (auto.) false frame, under frame.

fédéral, a., (in Europe) Swiss;
armée ——*e*, Swiss Army;
colonel ——, colonel of the Swiss Army.

femelle, a., female; (mach.) female.

fenêtre, f., window, opening;
— *d'éjection*, (mach. gun, etc.) ejection opening;
— *à guillotine*, (cons.) sash, (window);
— *objectif*, (mil.) objective opening (trench periscope);
— *oculaire*, (mil.) eye opening (trench periscope);
— *de visée*, (art.) aperture (of a gun shield).

fente, f., slit;
— *d'observation* (mil.) observing slit.

fer, m., iron;
an de ——, a year in irons (punishment);
— *cornière*, (cons.) angle iron.

fermer, v. t., to close, close down;
— *les gaz*, (auto., aero.) to throttle down.

ferré, a., iron bound, shod; (hipp.) shod.

ferrure, f., ironwork; (in pl.) fittings.

feu, fire;
— *d'atterrissage*, (aero.) landing flare;
conduite du ——, (art.) choice of proper procedure (mécanisme) (by the captain);

feu, *direction du* ——, *(art.)* fire direction (by the major);
—— *de discipline,* *(inf.)* salvo firing (impossible without discipline);
—— *éclairant,* *(artif.)* light composition;
être en ——, to be on fire; *(fig.)* to be up, in revolt, in rebellion;
marcher dans le ——, *(mil.)* of an infantry attack, to follow its own artillery fire *(tir de barrage);*
mettre en —— *(explo., min.)* to explode;
—— *nu,* naked flame;
—— *pilote,* *(artif.)* parachute flare;
—— *x de position,* *(mil.)* position lights (e. g., Bengal lights); *(nav., aero.)* running lights;
prendre sous le ——, *(art.)* to take under fire;
—— *de surprise,* *(inf.)* surprise fire (fire at will under the form of a sudden and intense blast or gust).

feuille, f., leaf;
—— *de calcul,* computation sheet;
——*s de chêne,* (lit. oak leaves) insignia of a general officer, oak-leaf embroidery on his uniform;
—— *de déplacement,* *(mil.)* travel order;
descente en —— *morte,* *(aero.)* dead leaf (fall like that of a dead leaf);
—— *rectificative,* correction sheet;
—— *de retouché,* *(mil. min.)* addition sheet (of tracing cloth) progression sheet (of a map or chart).

feutre, m., felt;
—— *graisseur,* grease pad.

fiacre, m., cab;
—— *automobile,* taxi.

fibre, f., fiber;
—— *de bois,* wood fiber.

fiche, f., card (of a card system); (more generally) memorandum (such as can be written on such a card); *(elec.)* jack, jackknife;
—— *nominative,* sort of identification or registry card (e. g., to be filled out before using a library, etc.);
—— *d'objectifs,* *(mil.)* descriptive card of objectives (noted by aviation);
—— *de renseignements,* *(art.)* information cards;
—— *de sûreté,* *(elec.)* plug switch.

fil, m., wire, thread;
—— *adducteur,* *(elec.)* leading-in wire;
—— *d'allumage,* *(elec.)* igniting wire;
—— *d'Ariane,* *(aero.)* plane indicator (balloon above clouds, indicates the vertical plane in which the balloon is moving at the moment of observation);
—— *arrache-clous,* v. *arrache-clous;*
—— *bi-métallique,* *(elec.)* double conductor wire:
—— *de basse tension,* *(elec.)* low tension wire;
—— *de conversation,* *(mil. telephone)* conversation wire;
au double ——, *(elec. mil.)* equipped with return wire;
—— *émaillé,* *(elec.)* enamel wire;
—— *de fer,* *(mil. slang)* a thin man;
—— *de fer barbelé,* *(mil., etc.)* barbed wire;
—— *de fer ronce,* *(mil.)* barbed wire;
—— *ferme-circuit,* *(mil.)* firing wire;
—— *de haute tension,* *(elec.)* high tension wire;
—— *isolé,* *(elec.)* insulated wire; *(mil.)* tripping wire;
—— *de masse,* *(elec.)* ground wire;
—— *primaire,* *(elec.)* primary coil wire;
—— *ramasse-douilles,* *(mil. aero.)* shell catcher;
—— *de réglage,* *(mil. telephone)* firea djustment wire;
—— *de secours,* *(mil. telephone)* rescue wire (communicates with captive balloon);
—— *spécial,* *(teleg.)* private wire;
—— *de terre,* *(elec.)* ground wire;
—— *volant,* *(elec.)* small (temporary) testing wire.

filet, m., net; *(r. r.)* luggage rack (over seats);
—— *protecteur,* *(nav.)* torpedo net.

filet-nasse, m., *(nav.)* sort of flexible net to pocket a torpedo.

fille, f., girl, daughter;
—— *publique,* prostitute.

filtrage, m., *(fam. mil.)* "sifting" through (as, the edge of a wood).

filtre, m., strainer;
—— *à l'amiante,* asbestos filter;
—— *de compagnie,* *(mil.)* company filter;
—— *mécanique,* mechanical filter;
—— *à (au) sable,* sand filter;
—— *en terre,* earth filter.

filtrer, v. n., *(fam. mil.)* to "sift" (as the edge of a wood).

finesse, f., delicacy;
—— *d'un avion,* *(aero.)* ratio of lift to drift coefficients.

fini, a., finite.

firme, f., *(com.)* firm name.

fixe-vitre, a., m., *(mil.)* glass holder, eyepiece holder, (of a gas mask);
rondelle ——, glass holder of a gas mask.

flambage, m., *(art.)* warming, scaling; *(cons.)* yielding (as, a strut to pressure);
charge de ——, *(art.)* warming charge.

flamber, v. n., to give way, to yield (as a strut, to pressure).

flamme, f., flame; pennon;
—— *de signalisation,* *(sig.)* signaling flag.

flammenwerfer, m., *(mil.)* flame projector (German word and invention).

flanc-garde, m., *(mil.)* flanker;
—— *fixe,* stationary flanker;
—— *mobile,* mobile flanker (marches parallel to the column).

flasque, m., bracket, support; flange; *(mach. gun)* side plate;
—— *de manivelle,* *(mach.)* crank arm;
—— *de moyeu,* boss support.

flèche, f., arrow; *(art.)* split-trail half; *(aero.)* camber;
attelage en ——, spike team;
atteler en ——, to hitch up spikewise;
en ——, *(mil.)* ahead of the line (said, e.g., of a battalion in line, with one company out ahead);
—— *d'orientation,* north and south arrow (on maps), meridian;
se placer en —— to place one's self ahead (e. g., of the line);
—— *télescopique,* *(mach. gun)* telescoping trail.

fléchette, f., *(aero.)* aerial arrow.

fléchois, m., *(Fr. a.)* graduate of the *Prytanée Militaire de la Flèche.*

fleurage, m., bran (of grits, also of potatoes).

fleuret, m., *(fenc.)* foil;
faire du ——, *(fenc.)* to fence with the foil.

flexible, a., flexible; m., *(any)* flexible tube or pipe; flexible tubing manifold.

flingot, m., *(mil. slang)* gun, rifle.

flingue, m., *(mil. slang)* infantry rifle.

flocon, m., flake;
—— *de fumée,* puff of smoke.

flottard, a., *(unif., etc.)* absurdly loose.

flotte, f., *(nav.)* fleet;
—— *de la métropole,* home fleet.

flottement, m., wabbling (as of a wheel).

flotteur, m., *(aero.)* float of a hydroplane; *(mach.)* carbureter float;
—— *en catamaran,* *(aero.)* catamaran float;
—— *à redan,* *(aero.)* stepped float.

foin, m., hay;
—— *pressé,* compressed hay.

folie, f., folly;
la —— *du nombre,* craze or mania for large standing armies.

fond, bottom;
—— *de course,* *(mach.)* bottom of the stroke.

fondé, p. p., (of *fonder,* to establish, found);
—— *de pouvoir,* clothed with authority.

fonds, m., *(top.)* swale.

fonte, f., *(met.)* cast iron;
—— *aciérée,* semisteel;
—— *moulée,* cast iron.

force, f., force;
—— *ascensionnelle,* *(aero.)* the lifting power of a gas;

force   543   **friturer**

force lifting power (of a lighter-than-air); lift (of a heavier-than-air);
—— *descensionnelle*, (*aero.*) descending force (dirigibles);
gain de ——, (*aero.*) wake gain;
—— *portante*, (*aero.*) supporting power (airplane).
foret, m., drill, bit;
—— *de nettoyage*, (*mach. gun*) cleaning bit.
forfaitaire, a., (*adm.*) at a price or figure fixed in advance.
formalisme, m., formalism; (sometimes) red tape.
formaliste, m., slave to form, to red tape.
formation, f., (*mil.*) formation;
—— *d'approche*, (*mil.*) approach formation;
—— *dérivée*, (*mil.*) secondary formation, (formation derived from the standard and usual formations; e. g., *colonne double, en losange,* etc.);
—— *d'évolution*, (*mil.*) drill or maneuver formation.
forme, f., form;
—— *de tête*, (*siege*) sap form.
fort, a., strong;
—— *de*, numbering, strong.
forte-self, f., (*elec.*);
—— *formant amortisseur*, choke coil.
fortif, f., (*mil. slang*) fortification.
fortification, f., (*fort.*) fortification;
—— *de campagne légère*, (*mil.*) light field fortification (done by infantry);
—— *de campagne renforcée*, (*mil.*) heavy field fortification (done by engineers).
fosse, f., grave, pit;
—— *à escarbilles*, ash pit, cinder pit;
—— *de recul*, (*art.*) (in high angle fire) recoil pit.
fossé, m., ditch, trench;
—— *à câbles*, (*telephone*, etc.) cable trench.
fossoyeur, m., grave digger.
fou, a., (*mach.*) idle.
fougasse, f., (*mil.*) fougasse;
—— *à la barre à mine*, bored fougasse;
—— *à la mélinite*, melinite fougasse;
—— *à la poudre*, powder fougasse.
fouillé, a., (*top.*, etc.) buried (as a bay in the shore, etc.).
fouiller, v. t., to search, (*mil.*), a prisoner.
four, m., *furnace*;
—— *à chauffage*, heating furnace;
—— *à incinérer*, incinerator.
fourbi, m., (*mil. slang*) soldier's kit, outfit (anything that may be cleaned and polished).
fourbir, v. t., to polish;
—— *au clair*, to burnish.
fourche, f., bracket; (*trench art.*) fork (of the 150 T. mortar mount); (*mach. gun*) bipod.
fourchette, f., fork; (*mil. slang*) bayonet; (*art.*) bracket (defined as the final bracket reached);
—— *de baladage*, (*mach.*) slide block fork;
—— *de commande*, (*mach.*) clutch shifter;
—— *de débrayage*, (*mach.*) clutch throw-out fork.
fourgon, m., (*mil.*) wagon; (*r.r.*) car, van (British);
—— *à archives*, (*mil.*) document wagon (troops in field);
—— *à bagages*, (*r. r.*) baggage car, luggage van, (British);
—— *funèbre*, (*r. r.*) funeral car;
—— *des pompes funèbres*, hearse;
—— *postal*, (*r. r.*) mail carriage;
—— *de queue*, (*r. r.*) rear car (esp. car in rear of a royal car);
—— *de tête*, (*r. r.*) head car (esp. car ahead of a royal car).
fourgon-poste, m., (*r.r.*) mail car, carriage or van.
fourniture, f., (*adm.*) supply under contract.
fourragère-marchepied, f., (*mil.*) tail rack and step.
fourre-tout, m., (*mil.*) soldier's catchall to carry small articles (proposed instead of the *sac*).
fourrure, f., bush plate, riveting plate; (*mach. gun*) barrel sleeve.

foyer, m., focus;
être au ——, (*opt.*) to be in focus;
—— *d'explosion*, (*expl. min.*) center, focus, of explosion.
frais, m., pl. (*law*) costs;
—— *d'exploitation*, (*r. r.*) running expenses;
—— *de représentation*, (*mil. etc.*) entertainment fund.
franc, a., (of space) clear.
français, a., French;
Le petit Francais, (*mil. slang*) the 75 mm. gun.
franc-fileur, m., (*mil.*) runaway.
franchir, v. t., (*mil.*) to "go over the top," to leave the trenches (for an attack), to cross over.
franchise, f., free carriage (luggage on railways, mail, etc.); (*man.*) willingness, good disposition (of a horse);
—— *de bagage*, (*r. r.*) free carriage of luggage;
en ——, free of charge (luggage, etc.).
franchissement, m., (*mil.*) egress from a trench, "going over the top," crossing over;
échelle de ——, (*mil.*) trench ladder;
gradins de ——, (*mil.*) trench steps.
frangine, f., (*mil. slang*) sister (both natural and religious).
frappe, f., stamp, die;
point de ——, (*mil. min.*) origin of sound (sound detection, point whence blows appear to proceed).
frappeur, m., (*mil. etc.*) driver (of stakes, pickets, etc.).
frein, m., brake; (*aero.*) lock wire;
[The following terms relate mostly to machinery.]
—— *à bande*, v —— *à ruban;*
—— *de blocage*, clutch; brake;
—— *à collier*, collar brake, clip brake;
—— *de débrayage*, clutch release lever;
—— *de différentiel*, differential brake;
—— *dynamo-électrique*, motor-testing brake;
—— *d'écrou*, (any device to keep a nut from unscrewing) stop, lock;
—— *à enroulement*, spiral brake;
—— *extensible*, segmental brake;
—— *extérieur*, external brake, outside brake;
—— *intérieur*, internal brake, inside brake;
—— *sur jante*, rim brake;
—— *à mâchoires*, shoe brake, clamp brake;
—— *à main*, hand brake;
—— *de mécanisme*, gear brake (applied directly to the gear or part of the gear);
—— *à pédale*, foot brake;
—— *à pied*, foot brake, pedal brake;
—— *de roue*, wheel brake, hub brake;
—— *à roulement*, strap brake;
—— *ruban*, band brake, strap brake;
—— *à sabots*, v. —— *à mâchoires;*
—— *de secours*, (*r. r.*) emergency brake;
serrer le ——, to apply the brakes;
—— *à tambour*, (*art.*) drum brake.
freinage, braking, arrest, stoppage; (*mach., etc.*) braking, operation of braking.
freiner, v. a., to brake, put on brakes.
fréquence, f., (*elec.*) frequency.
fréquencemètre, m., (*elec.*) frequency meter.
frère, m., brother;
gros —— *s*, (*Fr. cav., fam.*) the cuirassiers.
frette, f., (*art.*) hoop;
—— *à tenons*, (*art.*) maneuvering collar (155 mm. L.).
friable, a., friable.
frichti, m., (*mil. slang*) meal (German: Frühstuck, breakfast).
fricoteur, m., (*mil. slang*) dead beat, coffee cooler, hospital bird.
frigo, m., (*mil. slang*) refrigerated meat.
frire, v. t., to fry.
friture, f., frying; (*telephone*) frying.
friturer, v. n., (*telephone*) to fry.

Fritz, m., proper name, German; (*mil. slang*) German soldier.
front, m., front;
—— *de mer*, (*fort*) sea front;
—— *de terre*, (*fort*) land front.
frontière, f., frontier;
corps ——, (*mil.*) (corps stationed along the frontier) frontier corps, troops.
frottement, m., friction;
à ——, with a friction fit;
à —— *doux*, with easy friction fit.
frotteur, m., (*elec.*) wiping contact, brush.
fuir, v. n., to leak.
fuite, f., leak; leak of information (as to an enemy listening post).
fuseau, m., spindle;
—— *de déchirure*, (*aero.*) ripping strip or panel, rip strip.
fusée, f., (*artif.*) rocket (any device having a rocket stick); (*art.*) fuse; (*auto.*) stub axle;
—— *chenille*, (*artif.*) "caterpillar" rocket (form of signal produced);
—— *dérivée*, (*mil. min.*) secondary train, fuse, match;
—— *drapeau*, (*artif.*) "flag" rocket (form of signal produced);
—— *éclairante*, (*artif.*) flare;
—— *à friction*, (*art.*) friction fuse;
—— *instantanée*, (*art.*) instantaneous fuse;
—— *instantanée allongée*, (*art.*) super instantaneous fuse;
—— *maîtresse*, (*mil. min.*) principal train, fuse, match;
—— *à parachute*, parachute fuse;
—— *porte-retard*, delay action fuse;
—— *à retard*, (*art.*) delay-action fuse (adjustable);
—— *retardée*, (*art.*) fuse set for delayed action; delayed fuse;

fusée *sans retard*, (*art.*) instantaneous fuse;
—— *sensible*, sensitive fuse;
—— *universelle*, universal fuse.
fusée-détonateur, f., (*art.*) detonating fuse, percussion fuse.
fusée-signal, f., (*sig.*) signal rocket.
fuselage, m., (*aero.*) fuselage.
fuselé, a., (*aero.*) faired, streamlined.
fuser, v. t. n., to fuse; (in motors, of the vaporized gases) to form solid drops; to recombine into drops.
fusible, a., fusible; m., (*elec.*) fusible plug, fuse.
fusiforme, a., spindle-shaped.
fusil, m., gun;
—— *automatique*, (*mil.*) automatic rifle;
—— *lance-cartouche*, (*artif.*) signal rifle;
—— *lance-cartouche éclairante* (*artif.*) gun to fire the *cartouche éclairante à parachute*, q. v.;
—— *lance-grenades*, (*sm. a.*) grenade rifle;
—— *mitrailleuse*, (*mil.*) automatic rifle;
—— *signaleur*, (*mil.*) signal rifle.
fusilier, (*mil.*) fusileer; (*Fr. a.*) (name formerly given to the aide-grenadier, q. v.; present name of the soldier carrying and working the *fusil-mitrailleur*, q. v.) automatic rifleman; (in a narrower sense, to-day) man who explores a cleaned-out trench;
—— *mitrailleur*, (*Fr. a.*) man who works a *fusil-mitrailleur*, automatic rifleman.
fusil-mitrailleur, m., (*sm. a.*, *Fr. a.*) automatic rifle, machine rifle.
fusionner, v. a., to consolidate (as reports, accounts, papers, etc.).
fût, m., barrel;
—— *pétrolier*, oil barrel.
futée, f., mastic.

G.

Gabarit, m., gauge;
—— *de chargement*, (*tr. art.*) depth gauge.
gabillot, m., toggle.
gabion, m., (*mil.*) gabion;
—— *pliant*, folding gabion (hinged cases, fold flat for transportation).
gâche, f., recess;
—— *de chargement*, (*tr. art.*) loading slot;
—— *de fermeture*, (*tr. art.*) locking slot;
—— *d'ouverture*, (*tr. art.*) opening slot.
gaffe, f., boat hook;
haut les ——*s*, (*pont*) up boat hooks.
gaffer, v. t., to pole (a boat).
gaffeur, m., poleman (boats).
gagner, v. t., to reach (a place).
gaine, f., (*sm. a.*) holster.
gaine-relais, f., (*trench art.*) booster.
galère, f., galley.
galerie, f., frame or luggage guard, on top of an omnibus, automobile, etc.;
changement de ——, (*mil. min.*) change of gallery (passage from a gallery to a smaller at same level and in same direction);
—— *de départ*, (*mil. min.*) gallery of departure (one from which a return is made);
—— *de retour*, (*mil. min.*) return gallery.
galet, m., (*mach.*) roll; follower, cam follower;
—— *de bouche*, (*art.*) muzzle roller (75 mm., 155 CTR);
—— *à came* (*art., etc.*) cam roller;
—— *de friction*, friction roller;
—— *de roulement*, (*art.*) roller;
—— *de roulement à ressorts*, (*art.*) spring roller.
galetouse, f., (*mil. slang*) bowl, porringer.
galon, m., (*mil.*) stripe (insignia of grade);
à deux, trois, etc., —— (*mil. slang*) method of describing a staff officer having assimilated rank, (e. g. *vétérinaire à trois* —— a veterinarian having assimilated rank of captain).

gamelle, f., cup, bowl; (*mil. slang*) the German 77 mm. shell.
garage, m., (*auto.*) garage; (*mil.*) turn-out, by-pass, (in a trench), recess, storage recess;
—— *à l'eau*, basin for boats;
—— *à sec*, sort of dry basin for storing boats.
garance, f., madder (red);
pur de ——, (*fam. mil.*) free of madder, (said by *chasseurs* of themselves, in allusion to their blue trousers, as compared with the madder trousers of the French infantry of the line).
garde, f., guard; (*art.*) guard plate (to protect gunfirer against recoil);
—— *entière*, (*sm. a.*) solid basket hilt;
prendre la ——, (*mil.*) to march on guard;
prise de la ——, (*mil.*) marching on guard;
—— *de sûreté* (*mach.*) lock (e. g., of an admission cock).
garde, m., watchman;
—— *des voies et communications*, (*mil.*) track watcher (generally a territorial).
garde-boue, m., mud guard (any vehicle).
garde-chaîne, m., chain guard (in general).
garde-crottin, m., mud guard.
garde-écurie, f., (*mil.*) stable guard.
garde-voie(s), m., (*mil.*) railroad watchman (taken from the territorials).
gardiennage, m., (duty, office of a *gardien*) ward, wardship.
gare, f., (r. r.) station;
—— *destinataire*, station of destination;
—— *d'évacuation*, (*Fr. a.*) (origin of railway transportation from the front, i. e., the *gare de ravitaillement* becomes the *gare d'évacuation*), evacuation station;
—— *d'évitement*, (*mil.*) turn-out, by-pass, in a trench;
mettre en ——, to turn over to a railway for shipment, to ship by rail;
—— *de groupement de bétail*, (*Fr. a.*) cattle station (meat for troops);

gare (*d'*)*origine d'étapes*, (*Fr. a.*) (the name of a *gare de ravitaillement*, when more or less distant from troops, v. *ligne d'étapes*); railhead;
—— *point de répartition*, (*Fr. a.*) distributing station (sick and wounded);
—— *de ravitaillement*, (*Fr. a.*) refilling station;
—— *régulatrice*, (*Fr. a.*) (in a given region, a railway station to which all trains are sent for dispatch to destination) regulating, dispatching, station;
—— *de répartition de station-magasin*, (*Fr. a.*) distributing station (of a *station-magasin*, q. v.);
—— *siège d'infirmerie*, (*Fr. a.*) railway infirmary;
—— *de triage*, (*mil. med.*) sorting station.

garer, v. t., (*in gen.*) to put on one side.

garnir, v. t., (*mil.*) to pack (e. g., a machine gun, etc., on a pack animal).

garnison, f., (*mil.*) garrison;
—— *de sûreté*, (*mil.*) safety troops (to hold position in rear if advance falls back).

gars, m., boy;
les ——, (*mil. fam.*) the "boys," i. e., the men (pronounced GA).

gauche, a., left;
jusqu'à la ——, (*slang*) unto death.

gaucher, a., n., left handed.

gauchir, v. n., (*aero.*) to bank, to warp; v. t., to warp (a wing).

gauchissement, m., (*aero.*) warping;
commande de ——, warping control;
levier de ——, warp control.

gaver, v. t., to staff.

gaz, m., gas;
—— *asphyxiant*, (*mil.*) asphyxiating gas;
—— *brûlés*, exhaust gas, burnt gases;
—— *carburé*, mixture of air and of hydrocarbon vapors;
—— *d'échappement*, exhaust gases;
—— *frais*, fresh mixture;
—— *léger*, any gas whose density is less than unity;
—— *pauvre*, a sort of water gas (made by passing a current of air and water vapor over incandescent coke or anthracite);
—— *suffocant*, v. —— *asphyxiant;*
—— *tonnant*, explosive mixture (explosive engines).

gazéifier, v. t., to gasify.

gazer, v. t., (*mil. slang*) to smoke; to drive an automobile.

général, m., (*mil.*) general;
—— *de jour*, (*mil.*) in sieges, the general of the day, of a sector of attack.

générateur, m., (*mach. etc.*) generator (in many relations);
—— *à vaporisation rapide*, (*steam*) quick-steaming boiler.

genouillère, f., (*mach. gun*) knee joint (tripod leg).

géomètre, m., (*top.*) surveyor;
—— *topographe*, (*mil.*) topographer.

géophone, m., (*siege*) geophone (ground listener, sound detector).

geste, m., (*mil.*) (in signaling) motion;
—— *d'avertissement*, warning signal;
commandements au ——, (arm, sabre, and rifle) signals;
—— *commun*, signal common to all arms;
—— *d'exécution*, signal of execution;
—— *particulier*, signal peculiar to each arm;
—— *préparatoire*, preparatory signal.

giclée, f., spray, spray discharge.

gicler, v. n., (*mach.*) to spray.

gicleur, a., (*mach.*) atomizing; m., nozzle, spraying nozzle; jet;
chambre des ——*s*, spray chamber;
—— *-pulvérisateur*, atomizer, spray nozzle.

gifle, f., blow, buffet; (*sm. a.*) kick.

girouette, f., vane;
—— *d'aviation*, (*aero.*) sort of slope, speed, and ascension indicator.

gisement, m., (*surv.*) bearing.

glace, f., carriage window; ice; wind screen (of a motor car);
—— *argentée*, silvered glass;
—— *biseautée*, beveled glass;
—— *brute coulée*, unpolished, rough glass plate;
—— *franche*, glare ice.
—— *s flottantes*, floating ice;
—— *pare-bise*, wind screen of a carriage;
—— *polie*, plate glass.

glacière, f., ice-making machine; ice box, refrigerator.

glacis, m., (*mach.*) sloping surface.

glissade, f., slip, slide;
—— (*sur l'aile*), (*aero.*) side slip.
—— *sur la queue*, tail dive.

glissière, f., (*art.*) recoil slide.

global, a., total, entire, gross;
déplacement ——, (*nav.*) total, gross displacement:
puissance —— *e en chevaux*, total resources, in horses.

gniole, f., (*mil. slang*) brandy.

gnôle, f., (*mil. slang*) brandy.

gnon, m., (*mil. slang*) contusion.

godasses, f. pl. (*mil. slang*) shoes.

gommage, m., (*mach.*) sticking (from oil).

gomme, f., gum, rubber, india rubber.

gommer, v. t., to gum;
se ——, (*mach.*) to stick, (from oil).

gondole, f., gondola; long pleasure boat.

gondolé, a., (*elec.*) buckled, (of a storage battery plate).

gondoler, v. nr., (*elec. etc.*) to buckle, (storage battery, plate, etc.).

gonflage, m., inflation.

gonflement, m., (*auto.*) inflation of tires; (*aero.*) inflation (of a balloon);
culotte de ——, inflating tube.

gonfler, v. t., (*auto.*) to inflate (a tire).

gonfleur, m., inflater, mechanical inflater.

goniomètre, m., goniometer; (*art.*) battery commander's telescope; dial sight;
—— *-boussole*, compass goniometer;
—— *-miroir*, reflecting sight;
—— *panoramique*, panoramic sight;
—— *périscopique*, trench goniometer.
—— *de siege*, (*art.*) siege dial sight, goniometer.
—— *à viseur*, (*art.*) sighting tube goniometer.

goujon, m., pin;
—— *de jonction*, assembling pin.

goupille, f., pin;
—— *de débrayage*, (*mach.*) clutch spring throwout stud;
—— *fendue*, cotter.
—— *de sécurité*, (*mil.*) safety pin, of a hand grenade;
—— *de sûreté*, (*art., grenades*) safety wire.

gourmette, f., chain burnisher, chain pad.

gousset, m., corner plate, stay plate.

goutte, f., drop;
passer entre les ——*s*, (*mil. fam.*) to pass between the falling projectiles (in an attack under artillery fire).

gouttière, f., spout, gutter; (*trench art.*) trough (of a bomb truck).

gouvernail, m., (*aero.*) rudder;
—— *d'altitude*, elevator;
—— *d'assiette*, (*aero.*) horizontal rudder (dirigible).
—— *de direction*, rudder;
—— *horizontal*, v. —— *d'altitude;*
—— *de profondeur*, v. —— *d'altitude;*
—— *de queue*, tail rudder;
—— *de tête*, head rudder;
—— *vertical*, v. —— *de direction*.

gouvernement, m., the administration, the party in power.

gradé, m., (*mil.*) (often) officer; (usually) noncommissioned officer;

grade *de veille*, noncommissioned officer on watch in a trench.
gradient, m., (in meteorology) gradient.
gradin, m., step;
—— *s appui-coude*, (*mil.*) elbow rest;
—— *s de franchissement*, (*mil.*) trench steps (for egress upon assaulting);
—— *de sortie*, (*mil.*) exit steps of a *boyau*.
gradomètre, m., v. *indicateur de pente*.
gradué, m., graduate (e. g., of West Point, of Annapolis).
grain, m., grain, (*mach.*) bush, bushing;
—— *d'appui*, (*mach. gun*) thrust block;
—— *d'orge*, small turner's gouge.
graissage, m., lubrication;
—— *par barbotage*, (*mach.*) splash lubrication;
—— *par circulation*, (*mach.*) lubrication by circulation; ordinary lubrication; (in some cases) gravity lubrication;
—— *à*, *sous, pression*, (*mach.*) forced lubrication.
graisse, f., grease, lubricant;
—— *anti-rouille*, rust preventer;
—— *de bras*, (*fam.*) elbow grease;
—— *consistante*, stiff lubricant.
graisser, v. t., to oil.
graisseur, m., grease cup, oil cup; oiler (person). [The following terms are mechanical:]
—— *à bague*, ring oiler;
—— *central*, central oil feed;
—— *à compte-gouttes*, drip-feed lubricator;
—— *à coup de poing*, hand-pump lubricator;
—— *à débit visible*, sight-feed lubricator;
—— *à départ multiple*, spray oiler;
—— *distributeur*, lubricating-feed mechanism;
—— *par gravité*, gravity-feed lubricator;
—— *mécanique*, mechanical-feed lubricator, force feed oiler;
—— *à percussion*, hand oiler;
—— *à piston*, v. —— *à coup de poing*;
—— *à pompe*, pump-feed lubricator;
—— *à pression de gaz*, pressure-feed lubricator.
grand'garde, f., (*mil.*) support (of the line of outposts).
grand'route, f., highway.
graphique, m., graph.
gratte-papier, m., (*slang*) pen-pusher, quill-driver.
grattoir, m., scraper;
—— *en demi-lune*, semicircular scraper.
graver, v. t., to grave, engrave;
se ——, (*sm. a.*, *etc.*) to be eaten, corroded (rifle barrel, etc.).
gravillon, m., coarse gravel.
graviter, v. n., (*mil. slang*, *St. Cyr*), to go to the hospital.
grenade, f., (*mil.*) grenade;
—— *automatique*, automatic grenade;
—— *défensive*, defensive grenade;
—— *D. R.*, type of rifle grenade, joined to rifle by a mandrel;
—— *d'exercice*, dummy grenade;
—— *fumigène*, smoke-producing grenade;
—— *fusante*, time grenade;
—— *à fusil*, rifle grenade;
—— *incendiaire*, incendiary grenade;
—— *à main*, hand grenade;
—— *offensive*, offensive (used when attacking, synonym of *pétard d'assaut*);
—— *de parapet*, defensive grenade (with special reference to its use in trenches);
—— *suffocante*, asphyxiating grenade;
—— *à tige*, rifle grenade, stick grenade;
—— *V. B.*, the Vivien Bessières grenade.
grenadier, m., (*mil.*) grenadier; (to-day) bomber, grenade thrower;
—— *d'élite*, (soldier specially skilled and bold in the use of grenade) expert grenadier; (is a formal designation in the French Army);
élève- ——, lance grenadier (man qualifying to become a grenadier thrower);
officier ——, officer on grenade duty, grenadier officer.

grenadier *V B*, V B grenadier (throws the V B grenade).
grenaille, f., (*elec.*) granular carbon.
grès, m., glazed earthenware.
griffe, f., claw; clip, clamp;
—— *à agrafage*, (*r. r. art.*) holding-down clamp;
—— *de chargement*, (*art.*) shell hooks;
—— *d'entraînement*, (*trench art.*) cocking catch.
gril, m., frame (as, for boats hauled out).
grille, f., (*elec.*) grid;
à ——, (*mach.*) said of a gear set having both lateral and longitudinal motion (e. g., in change speed gears).
—— *d'entrée d'eaux*, strainer.
grippage, m., (*mach.*) gripping, jamming (of adjacent running parts, not sufficiently oiled).
grippement, m., v. *grippage*.
gripper, v. t., to grip.
grisouteux, a., (of a coal mine) liable to fire damp.
griveton, m., (*mil. slang*) a private.
grolles, f., pl., (*mil. slang*) shoes.
gros, a., large;
en ——, (*com.*) wholesale;
un —— *noir*, (*mil. slang*) large shell.
groupe, m., (*mach.*, *elec.*) set;
—— *de canevas de tir*, (*mil.*) (group of officers, etc., whose duty it is to prepare the *canevas de tir*, q. v., one at each army headquarters); fire triangulation group;
—— *électrogène*, (*elec.*) generator set;
—— *moteur*, (*mach.*) motor set;
—— *moto-propulseur*, (*mach.*) engine set.
groupement, m., (*art.*) group of heavy batteries (corps and army).
grue, f., (*mach.*) crane;
—— *mobile à flèche*, mobile, transportable, crane.
guérite, f., (*mil.*) sentry box;
—— *métallique*, armored sentry box.
guérite-observatoire, f., (*mil.*) lookout shelter.
guerre, f., (sometimes, alone) Franco-Prussian War, *the* war (before 1914);
—— *de brousse*, v. s. v. *brousse*;
—— *de mouvement*, open warfare, war in the open field, as distinguished from trench warfare;
—— *de positions*, position warfare;
—— *sainte*, holy war (Mussulman);
—— *de taupe*, (*mil. slang*) trench warfare; (*taupe* = mole);
—— *de tranchée*, siege operations; (to-day) trench warfare.
guêtre, f., (*auto.*) sleeve or laced wrapping (for punctured tires).
guetteur, m., (*mil.*) (in the trenches) lookout, man constantly on the watch.
gueule, f., (*art.*, *sm. a.*) muzzle;
—— *d'enfer*, (*hipp.*) very hard mouth.
guidage, m., (*mach.*) valve guide.
guide, m., guide-book;
—— *de bielle*, (*mach. gun*) cartridge guide cam cover.
guide-bêche, m., (*trench art.*) anchor guide (platform).
guide-poche, m., pocket-guide.
guide-ropage, m., (*aero.*), (use of the *guide rope*) dragging.
guide-soupape, m., (*mach.*) valve guide.
guidon, m., (*auto.*) handle bar; (*sm. a.*) foresight;
—— *fin*, (*t. p.*) coarse sight;
—— *gros*, (*t. p.*) coarse sight;
—— *pincé à droite* (*gauche*), (*t. p.*) foresight to right (left).
guignol, m., (*aero.*), king post.
guillemets, m. pl., quotation marks.
guitoune, f., (*mil. slang*) tent, shelter.
gutta, f., guttapercha.
gyroplane, m., (*aero.*) sort of helicopter.

H.

habillage, m., fitting;
— *des roues,* (*auto.*) equipment of wheels (with rubber tires).
hale-à-bord, m., (*nav.*) inhaul.
haleur, m., hauler.
halte-repas, f., (*mil.*) the same as *station* —, q. v.
hangar, m., (*aero.*) hangar, shelter;
— *fixe,* stationary hangar.
— *de toile,* canvas hangar.
— *tournant,* revolving hangar.
haras, m., stud farm.
hauban, m., (*aero.*) wire, bracing wire; guy wire
haubannage, m., (*aero.*) bracing, rigging, stays generic term for the *haubans,* q. v.
hausse, f., (*art., sm. a.*) rear-sight, elevation;
angle de —, v. s. v. *angle;*
arc de —, (*art.*) elevation arc;
— *de combat,* (*sm. a., inf.*) battle sight;
— *goniométrique,* (*art.*) goniometric sight;
— *indépendante,* (*art.*) independent sight;
— *du moment,* (*art.*) elevation of the moment (as fixed by local conditions at the time);
— *optique,* (*sm. a.*) telescopic sight.
hauteur, f., height;
— *couvrante,* (*mil.*) height of cover;
prendre de la —, (*aero.*) to rise.
hauturier, a., on the high seas, high sea;
service —, (*aero.*) high sea aviation service.
hecto, m., hectolitre.
hectométrique, a., hectometric.
hélice, f., (*mach., etc.*) screw, propeller; (*aero.*) screw tractor); propeller (pusher);
— *ascensionnelle,* (*aero.*) lifting propeller (of a *hélicoptère);*
— *en bois,* wood propeller;
— *à deux branches,* two bladed propeller;
— *à droite, (gauche),* right (left) handed propeller;
— *métallique,* metal propeller;
— *à pales souples,* (*aero.*) flexible propeller;
pas de l' —, pitch of the propeller;
— *à pas constant,* true screw propeller;
— *propulsive,* (*aero.*) driving propeller;
— *reversible,* reversible screw, screw with reversible blades;
souffle de l' —, (*aero.*) propeller wash;
— *souple,* v. — *à pales souples.*
— *sustentatrice,* (*aero.*) supporting propeller;
— *tractive* (*aero.*) tractor.
hélicoptère, m., (*aero.*) helicopter.
héliographe, m., (*mil. fig.*) heliograph.
hérisson, m., hedgehog; (*mil.*) (polyhedral obstacle of barbed wire nailed to the ends of three stakes at right angles to one another), wire hedgehog, hedgehog.
hétérodyne, a., (*elec.*) heterodyne.
heure, f., hour; time;
— *légale,* (in France) Greenwich time.
heurt, m., shock, collision;
— *d'armes,* shock of arms.
hiérarchiser, v. t., to arrange in order of seniority, or of rank.
hippodrome, m., (*man*) riding field, ground (e. g., Verrie near Saumur).
hippomobile, a., horse-drawn.
hippomobilisme, m., hippomobilism (horse, as distinguished from motor, traction).

homme, m., man;
— *s 32-40,* (*mil., r. r.*) legend on railway car, meaning that 32 men with, 40 without, their equipment, may be taken on board;
— *de chambre,* (*mil.*) room orderly;
— *de communication* (*mil.*) agent of communication (more especially) runner.
homme-baïonnette, m., (*mil.*) man who explores a cleaned out trench, bayonet-man.
hôpital, m., hospital (curable cases);
— *de l'intérieur,* (*mil. med.*) base hospital;
— *militaire,* (*mil. med.*) military, army, hospital.
hôpital-annexe, m., hospital annex.
horaire, m., (*r. r.*) schedule; (any) schedule, time table.
horlogerie, f., clock work;
mouvement d' —, clock work.
hospice, m., hospital (aged persons, and incurables).
hospitaliser, v. t., (*mil. med.*) to send to a hospital.
hostau, m., (*mil. slang*) hospital.
hosteau, m., (*mil. slang*) hospital.
hotte, f., scuttle;
— *porte-bombes,* (*trench art.*), (fort of) bomb scuttle.
houille, f., coal;
— *blanche,* water power.
houppelande, f., great-coat, long coat (e. g., Cossack coat).
houppette, f., tassel; (*unif.*) tuft (as of a helmet).
hourd, m., (sawyer's) trestle.
houseaux, m. pl., leggings.
housse, f., cover; (*auto.*) spare tire cover.
huile, f., oil;
les —*s,* (*mil. slang*) staff officers;
— *à cylindre,* (*mach.*) cylinder oil;
— *de graissage,* (*mach.*) lubricating oil;
— *lampante,* mineral oil; petroleum;
— *légère,* light oil;
— *lourde,* heavy oil;
— *à mécanisme,* (*mach.*) machine oil;
— *à mouvement,* v. — *à mécanisme.*
huileux, a., oily.
huis, m., door (obs. except in the following expression).
à — *clos,* behind closed doors (courts, etc.);
demander le — *clos,* (courts, etc.) to ask that the court be cleared and closed.
humide, a., humid, damp;
— *à n o/o,* (*expl.*) of gun cotton, containing n o/o of moisture.
hutte, f., hut;
— *en clayonnage,* wattle hut.
hydravion, m., (*aero.*) seaplane.
hydroaéroplane, m., (*aero.*) hydroairplane.
hydro-avion, m., (*aero.*) seaplane.
hydrocarbure, f., (*chem.*) hydrocarbon.
hydroplanant, a., (*aero.*) water floating.
hydroplane, m., (*aero.*) hydroplane.

I.

idée, f., idea;
— *de manœuvre,* (*mil.*) fundamental idea of the *plan d'ensemble,* q. v.
ignifuge, a., fireproof.
ignifuger, v. t., to make fireproof.
îlot, m., islet;
— *de résistance,* (*mil.*) (any point, or portion of a position, prepared to resist capture, to offer resistance, e. g., *segment actif,* q. v.); (isolated) point of resistance.

immatriculation, f., (*auto.*) registering.
immeuble, a., real (of estate); m., real property, real estate, house.
immobilisation, f., (*mach.*) locking.
immobiliser, v. t., (*mach.*) to lock; to hold in place.
impédance, f., (*elec.*) impedance.
impériale, f., roof, top of a carriage.

implanteur, m., (*mil.*) (in laying out trenches, etc.), man who sticks pickets into the ground); picket man.
imposition, f., (*law*) tax.
impôt, m., tax;
— *sur le revenu,* income tax.
incendie, m., fire;
— *de forêt,* forest fire.
incidence, f., incidence;
— *rasante,* (*art.*) small angle of incidence.
incliner, v. n., (*aero.*) to bank, tilt.
incolore, a., colorless.
indemnité, f., (*mil.*) allowance, commutation;
— *de bagages,* baggage allowance;
— *pour cherté de la vie,* (*Fr. a.*) allowance due to increased cost of living;
— *pour cherté de vivres,* (*Fr. a.*) allowance for increased cost of food;
— *de commandement,* (*Fr. a.*) commanding officer's allowance;
— *de résidence dans l' Est* (*Fr. a.*) allowance for service on the eastern frontier of France;
— *représentative de vivres,* commutation of rations.
indéréglable, a., (*inst.*) incapable of getting out of adjustment.
indicateur, m., indicator;
— *de pente,* clinometer; (*aero.*) clinometer, slope indicator;
— *de pôles,* (*elec.*) pole indicator;
— *de vitesse,* speed indicator, tachometer; (*aero.*) speed indicator;
— *de vitesse relative,* (*aero.*) air speed indicator.
indicatif, m., (*mil.*) distinguishing number, letter, or cipher; call letter.
indices, m. pl., (*chem.*) trace (as in an analysis).
indisponible, m., (*hipp.*) horse unfit for duty, but not in hospital (generally used in plural).
inductance, f., (*elec.*) inductance;
bobine d' ——, inductance coil.
induit, m., (*elec.*) armature;
— *tournant,* revolving armature.
inencrassable, a., nonfouling.
infiltration, f., infiltration; (*mil.*) the "sifting" of men, troops, into the enemy position.
infiltrer, v. t., to infiltrate; (*mil.*) to "sift" into an enemy position, through a wood, etc.
infirme, a., m., (*mil.*) permanently disabled.
infirmière, f., (*med.*) nurse.
inflammateur, m., (*art.*) igniting charge; (*mach.*) contact point; plug, sparking plug; igniter.
ingénieur-constructeur, m., engine builder, builder and designer of machinery.

ingérance, f., meddling, unauthorized or improper intervention.
inondation, f., inundation;
tendre une ——, (*mil.*) to establish an inundation.
insoumission, f., (*Fr. a.*) state of being *insoumis,* q. v.;
contrôles de l' ——, (*Fr. a.*) muster rolls, or rolls, of the *insoumis.*
inspectorat, m., inspectorate.
instance, f., instance, solicitation;
être en ——, to be an applicant for;
se mettre en —— *de pension,* to apply for a pension;
tribunal de première ——, (*law*) court of first instance.
institut, m., institute;
l' —— *de France,* the Institute (composed of the five academies, French, Inscriptions and Belles Lettres, Sciences, Fine Arts, Moral and Political Sciences).
instructeur, m., instructor;
— *d'équitation,* (*man.*) riding master.
intérieur, a., m., interior;
dans l' —— *des terres,* inland.
internement, m., seclusion (of carrier pigeons);
panier d' ——, seclusion basket (pigeons).
interner, v. t., to seclude (carrier pigeons).
interpeller, v. t., (*mil.*) to challenge (sentry).
interrupteur, m., (*elec.*) plug, switch, key;
— *à compas,* rotary switch;
— *à deux directions,* two way switch;
— *électrolytique,* electrolytic interrupter;
— *à jet de mercure,* (*wireless*) mercury jet interrupter;
— *à main,* hand contact breaker;
— *du manche à balai,* (*aero.*) control lever switch;
— *à marteau,* (*wireless*) hammer break;
— *à mercure,* (*wireless*) mercury interrupter;
— *de sûreté,* (*elec.*) safety switch;
— *à trembleur,* v. —— *à marteau.*
— *à turbine,* (*wireless*) turbine interrupter.
interrupteur-distributeur, m., (*elec.*);
— *d'allumage,* ignition make and break.
invalo, m., (*mil. slang*) invalid soldier.
inverseur, m., (*elec.*) reversing, inverting gear, current reverser.
irrégularité, f., irregularity;
— *d'écriture,* (*adm.*) doctoring of papers.
irréversible, a., (*mach.*) irreversible.
isobare, f., isobar.
isolant, m., (*elec.*) insulator.
isoler, v. t., to isolate, insulate;
— *au caoutchouc,* etc., (*elec.*) to insulate with rubber, etc.
ivraie, f., chaff, tan.

J.

jack, m., (*elec.*) jack, spring jack.
jack-commutateur, annunciator jack, jack.
jalonnement, m., (*mil.*) (visual) indication (of a position or line by Bengal lights, flags, panels, etc.)
jalonneur, m., (*mil.*) man serving to connect two bodies of troops; marker;
chaîne de ——*s,* chain or string of such men, chain of markers.
jambe, leg;
— *d'appui,* support, leg;
— *de force,* stay; strut.
jante, f., felly, rim;
— *acier,* steel rim;
— *amovible,* detachable felly, rim (motor wheel);
— *bois,* wooden rim;
— *démontable,* take down felly.
jarretière, f., garter; (*elec.*) small (temporary) testing wire.

jeter, v. t., to throw;
— *un coup,* (*sm. a., mil.*) to fire a shot (to fire a shot more or less at random, in the hope of getting a hit).
jeune, a. m., young;
— *d'été,* (pigeon hatched out before July), summer pigeon.
joint, m., joint; (*mach.*) packing, gasket;
— *anglais,* (*cord.*) so-called "English hitch" (two ropes joined by a thumb knot from the end of each around the standing part of the other);
— *à dés,* (a form of) cardan joint (in which lugs, *dés,* engage in suitable slots);
— *métallo-plastique,* (*mach.*) (any joint produced by the interposition of a plastic substance between the compressing metal parts); plastic joint;
— *d'Oldham,* (*mach.*) (Oldham joint) tongue and groove clutch.

jointage, m., joining, attachment.
jointer, v. t., to join together (as, by bolts, rivets, etc.), to make a joint.
jonction, f., fork of roads; *(expl.)* junction (of melinite fuse to its igniter).
Joséphine, f., (proper name) *(mil. slang)* the 75 mm. gun.
joue, f., cheek; cheek or plate of a hub (artillery type).
jouer, v. t. n., *(mus.)* to play.
jour, m., day;
—— *de feu, (art.)* the consumption of ammunition during the period of execution following preparation (not necessarily a daily rate, nor a fixed amount); day's supply;
—— *franc,* a whole day;
officer de ——, (in sieges) officer of the day (of a sector of attack; field or general officer);
—— *de vivres (mil.)* one day's rations.
journal, m., journal;
—— *d'écoute, (mil. min.)* record of observation (by listening);

journal *de mines (mil. min.)* journal of mining operations.
Jules, m., (French proper name) *(fam. Fr. a.)* a. synonym of *baquet de propreté,* q. v.
jumelle, f., field glass; *(mach.)* link, spring link;
—— *de Galilée,* Galilean field glass;
—— *à micromètre, (mil.)* micrometer field glass;
—— *micrométrique, (mil.)* micrometric field glass
—— *à prisme(s),* prismatic field glass.
—— *stéréo-prismatique,(inst.)* stereo-prismatic field glass;
—— *stéréoscopique,* stereoscopic field glass.
—— *télémètre, (art.)* range finding field glass.
jument, f., mare;
—— *d'armes,* troop mare.
jury, m., jury, board, commission;
—— *d'honneur, (mil.)* sort of court of honor.
jus, m., *(mil. slang)* soldier's first breakfast (coffee and bread).
juteux, m., *(mil. slang)* an *adjudant.*

K.

kaki, m., *(unif.)* khaki;
toile ——, khaki-colored cloth.
kébour, m., *(mil. slang) képi,* forage cap.
kif-kif, *(mil. slang),* it is all the same, same thing, (Arab word).
kilomètre, m., *(auto.)* kilometer race, dash;
—— *arrêté,* kilometer race, standing start;

kilomètre *lancé,* kilometer race, flying start.
klaxon, m., *(auto.)* claxon horn.
ksar, m., (Berber word meaning, originally) fortified position, (but now indiscriminately applied by the French to any) town, village (plural *ksour*).

L.

labourer, v. t., (of a bullet) to tear open, lay open (as a shoulder, the leg, etc.).
lacet, m., lace, zigzag; *(mil.)* tripping loop;
mouvement de ——*s, (aero.)* yaw, yawing.
lâche, a., *(unif.)* loose, sloppy.
lâcher, v. t., to release;
—— *tout, (aero., etc.)* to let everything go, to throw off.
lacis, m., network;
—— *des tranchées, (mil.)* network of trenches.
lacrymogège, a., tear-producing; m., *(art.)* tear-producing shell.
ladre, m., disease of pigeons (due to inability to get rid of their milk).
laissez-passer, m., *(adm.)* permit, pass.
lame, f., blade of a propeller;
—— *d'arrêt, (mach. gun)* magazine closing plate;
fine ——, great swordsman;
—— *à fourche, (mil. tel., etc.)* lance head;
—— *maîtresse,* longest leaf (of a spring);
—— *platinée, (elec.)* contact strip;
—— *-ressort à griffe,* spring catch.
lame-tournevis, f. screw driver.
lampant, a., clear.
lampe, f., lamp;
—— *à acétylène,* acetylene lamp;
amorçage d'une ——, preparation of a lamp (oil);
amorcer une ——, to prepare a lamp (oil);
—— *à braser,* brazier's lamp, brazing lamp;
—— *à braser dite suédoise,* special brazing lamp (using petroleum);
—— *électrique,* flash light;
—— *(électrique) de poche,* flash light;
—— *à pétrole,* kerosene lamp;
—— *réchauffeuse,* heater, heating lamp (in some forms of explosion engines);
—— *à souder,* plumber's torch;
—— *thermo-ionique, (elec.)* audion, valve.
lance, f., fireman's nozzle; *(mil.)* rod (of *chevaux de frise*);
—— *d'arrosage,* nozzle of a watering hose;
—— *à fourche, (mil. tel., etc.)* lance.
lancé, m., throw.

lance-bombes, m., *(mil.)* trench mortar, *minenwerfer; (aero.)* bomb-launching device, or gear; launching tube; any bomb- or grenade- throwing device or apparatus.
lance-flammes, m., *(mil.)* flame projector.
lance-grenade, m., *(trench art.)* grenade thrower.
lancement, m., *(auto.)* start, starting; *(aero.)* launching (of bombs); *(pont.)* launching of a bridge; *(mil.)* grenade throwing;
—— *de grenades, (mil.)* grenade throwing;
—— *à la main, (auto.)* hand start.
lancer, v. a., *(aero.)* to launch (as, a bomb); *(fenc.)* in bayonet exercise, to thrust out, reach out, with the right arm; *(pont.)* to launch a bridge; *(auto.)* to start;
—— *une dépêche, un télégramme, sur les câbles,* to put a dispatch, etc., on the wire;
—— *à la main, (auto.)* to start by hand, to crank;
—— *un moteur,* to start a motor.
lanceur, m., thrower; *(mil.)* grenade-thrower, grenadier (of a modern army).
landaulet, m., landaulet (carriage and motor car);
—— *trois-quarts, (auto.)* landaulet of which the roof projects over the front seat (motor car).
lanière, f., razor strop.
lanterne, f., case, casing; lantern;
—— *d'agrandissement,* magnifying projector;
—— *d'arrière,* tail light of a motor car;
—— *d'avant,* headlight of a motor car;
—— *électrique,* flash light;
—— *de projection,* projecting lantern;
—— *de repérage, (art.)* registration lantern (night firing);
—— *à serrure, (mil. min.)* miner's lantern;
—— *de signalisation, (sig.)* signalling lantern;
—— *à tracer, (siege)* tracing lantern;
—— *de tranchée, (mil.)* trench-lantern; (may be used also for signalling, and for marking out directions).
lardon, m., patching lace (punctured tire, etc.).
latrine, f., *(mil.)* latrine;
—— *au "tout à l'égout,"* water flushing latrine.
lava, f., *(mil.)* lava (Cossack formation, sotnia or troop in one rank, with a small group in rear as support).

léchage , m., licking;
— *surface de* ——, *(mach.)* (in radiators) surface exposed to the air.
léger, a., light;
— *la* ——*e*, *(Fr. cav.)* the light cavalry;
— *plus* —— *que l'air*, *(aero.)* lighter-than-air (i. e., balloons).
lentille, f., *(opt.)* lens;
— *à échelons*, Fresnel lens.
lest-eau, m., water ballast.
lettre, f., letter;
— *collective*, (France) circular issued by a minister of the Government
— *de créance et de pouvoirs*, minister's, ambassador's credentials;
— *double*, double weight letter;
— *morte*, dead letter;
— *ordinaire*, any personal letter or note (excludes post cards and registered letters);
— *s de rappel*, letter of recall of an ambassador;
— *s sémaphoriques*, semaphore signal letters;
— *simple*, single-weight letter.
lève-auto, m., motor jack.
levée, f., lift of a valve;
— *d'un corps*, (at a funeral) removal of the body for burial; (in some cases exposure of, for purposes of identification).
lever, m., survey;
— *de bâtiment*, *(surv.)* building survey;
— *des travaux souterrains*, *(surv.)* mine surveying;
— *à vue*, *(top.)* place sketch.
lève-roue, m., jack, carriage jack.
levier, m., lever;
[Except where otherwise indicated, the following terms relate to machinery:]
— *d'accrochage*, *(art.)* securing bolt;
— *d'admission (des gaz)*, throttle lever, gas cut-off, gas lever;
— *d'armement*, *(mach. gun)* cocking lever (mod. 1907-T and others);
— *-arrêtoir*, *(mach. gun)* barrel catch;
— *d'avance à l'allumage*, sparking lever, timing lever;
— *bascule*, rocking lever, rocking arm;
— *de braquage*, steering rod;
— *à chaîne*, *(trench art.)* traversing lever and chain;
— *à charnière*, hinged lever;
— *à cloche*, gear shift lever;
— *de commande* *(aero., etc.)* control lever; *(auto.)* steering arm;
— *de débrayage*, clutch throw-out lever; *(mach. gun)* clutch lever;
— *de déclenchement*, *(mil.)* releasing lever (of some forms of grenades); priming device;
— *de démontage*, *(auto.)* tire lever;
— *démonte-pneu*, *(auto.)* tire lever;
— *de direction*, *(auto.)* steering lever;
— *droit*, crowbar;
— *d'équilibre*, compensating lever, reaction rod;
— *fourchu*, forked lever, forked bell-crank;
— *de frein*, brake lever;
— *de gâchette*, *(mach. gun)* sear lever;
— *de gauchissement*, *(aero.)* warping lever, warp control;
— *de manœuvre*, hand lever;
— *de percuteur*, *(mach. gun)* striker arm;
— *à pince*, pinch bar;
— *de portage*, *(trench art.)* carrying bar;
— *porte-galet*, roller lever;
— *à poussette*, spring lever (i. e., controlled or released by a spring);
— *la prise d'air*, air-admission lever;
— *de réglage*, *(mach. gun)* speed-regulating lever (of the mod. 1907-T);
— *à renvoi*, bell-crank lever;
— *à ressort*, *(art.)* spring lever;
— *à rotule*, rock shaft, rocking lever; steering lever;
— *de rupture*, swinging arm, hammer break;
— *de sûreté*, securing lever, locking lever;
— *de tir et de sûreté*, *(mach. gun)* regulator bolt arm;
— *de vitesses*, *(mach.)* change speed lever.

liaison, f., *(mil.)* liaison. [We have no one word in English that exactly and conveniently translates *liaison* in its military relations. It is recommended that the word be adopted into the language.]
— *acoustique*, liaison by sound signals, sound liaison;
— *agent de* ——, (should be an) officer (sometimes, but exceptionally, a noncommissioned officer) to transmit orders and get information; (by popular usage includes any private or n. c. o. that carries orders or goes on simple errands); connecting file, liaison agent;
— *chef de la* ——, (officer charged with the direction of liaison duties) connecting officer, liaison officer;
— *double* ——, duplicate connection, liaison;
— *escadron de* ——, *(cav.)* connecting squadron;
— *latérale*, lateral liaison, side connections (e. g., with an observation point to the right or left);
— *perpendiculaire*, liaison in depth.
libération, f., *(mil.)* discharge;
— *acheter sa* ——, to buy one's discharge;
— *définitive*, final discharge.
libre, a., (in matters of education) not given or supported by the State (France);
— *élève* ——, pupil in a government school, but not intending to go into government service (France).
licence, f., patent.
lieutenant-aérostier, m., *(aero.)* balloon-lieutenant.
ligne, f., line; *(mil.)* line;
[Except where otherwise indicated, the following terms are military:]
— *d'abonnement*, *(mil. telephone)* line always open (as distinguished from others that are opened only when needed, e. g., commanding general's line);
— *bloquée*, *(r. r.)* blocked line;
— *brisée*, broken line (gives flanking fire);
— *de centres de résistance*, (the first position, the one nearest the enemy) line of centers of resistance;
— *à chevaux*, *(mil., pont.)* picket line;
— *de combat*, *(Fr. a., inf.)* (the first line of a *colonne double*, q. v.) combat line;
— *de correspondance*, *(mil.)* (carrier-pigeon) line of correspondence;
— *de démarcation*, *(Fr. a.)* (the line separating the railways of the *réseau des armées*, from those of the *réseau de l'intérieur*, qq. v.) line of demarcation;
— *de demi-sections (escouades) par quatre (deux, un)*, line of half sections (squads) in columns of fours (twos, single file);
— *directe*, *(mil. telephone)* direct telephone line;
— *enterrée*, *(mil. telephone)* buried line, underground line;
— *d'escouades par un accolées*, *(Fr. a.)* line of squads in single file;
— *d'étapes*, *(Fr. a.)* line of supply from railway station to the troops (exists when station is not sufficiently near the troops);
— *de l'intérieur*, *(r. r.)* railway line not in the theater of operations;
— *libre*, *(r. r.)* clear line;
— *de mire du correcteur*, *(art.)* corrector line of sight;
— *de mire indépendante*, *(art.)* independent line of sight;
— *de navigation*, *(com.)* steamship line;
— *neutre*, *(elec.)* in a magnetic field, the line separating the lines of force of opposite polarity; neutral line;
— *la première* ——, first line of fire trenches;
— *prendre la* —— *de mire*, *(t. p.)* to sight, to look along the line of sight;
— *protégée*, *(mil. telephone)* protected line (lies in a trench or conduit);
— *des réduits*, line of works, redoubts;
— *de renforcement*, *(Fr. a., inf.)* (the second line of a *colonne double*, q. v.) supporting line;
— *du réseau des armées*, *(mil. r. r.)* line in the zone of operations;

ligne *réservée* (*mil. telephone*) line reserved for special purpose;
— *de résistance*, line of trenches of the first line, line of resistance;
— *de résistance des avant-postes*, line of supports;
— *de retenue*, (*pont., etc.*) holdfast;
— *de sections par quatre* (*deux, un*), line of sections in columns of fours (squads, U. S. A.) of twos, of single files;
— *de séparation*, (*inst.*) line of separation (range finder);
— *de soutien*, line of support, of supporting trenches;
— *de surveillance*, outer line nearest the enemy, lookout line;
— *de surveillance des avant-postes*, line of observation.
— *de vol*, (*aero.*) line of flight.
limousine, f., (*auto.*) limousine.
lingue, m., (*mil. slang*) knife.
linguet, m., catch;
— *de détente*, (*mach.*) spring catch;
— *d'entraînement*, (*mach.*) driving pawl;
— *de retenue*, (*mach.*) catch.
liquide, m., liquid;
— *excitateur*, (*elec.*) active or exciting liquid of a cell.
lisser, v. t., to smooth, smooth down.
liste, f., list;
— *générale de classement*, general standing, (schools).
listel, m., (*art.*) fillet (of the mushroom head).
lit, m., bed;
— *à rabattement*, folding bed.
lit-canapé, m., sofa bed;
— *à rabattement*, folding sofa bed.
lit-cantine, m., sort of box bed.
lit-salon, m., (*r. r.*) parlor sleeping-car.
livrable, a., (*adm.*) to be delivered (of goods, etc., under a contract).
livre, m., book;
— *bleu*, Blue Book (and so of other colors).
livret-guide-horaire, m., (*r. r.*) railway guide and time-table.

livret-horaire, m., (*r. r.*) time-table.
lobe, f., lobe.
longeron, m., side piece, stringer; (*aero.*) longeron (of the fuselage), spar (of a wing);
— *d'aile*, (*aero.*) wing spar;
— *de queue*, (*aero.*) tailpiece.
longue-vue, f., telescope;
— *binoculaire* (*monoculaire*) *à prismes*, binocular (monocular) prismatic telescope.
louche, f., ladle.
loup, m., mask; (*mil.*) face mask (of a gas mask.)
loupe, f., magnifying glass;
— *d'horloger*, jeweler's magnifying glass.
lourd, a., heavy;
— *du nez*, (*aero.*) head heavy, heavy at the nose (air plane);
— *de la queue*, (*aero.*) tail heavy (air plane);
plus — *que l'air*, (*aero.*) heavier-than-air (i. e., air planes).
lovage, m., (*cord.*) coil of rope, act of coiling.
lover, v. t., (*cord.*) to coil a rope.
lubréfier, v. t., to lubricate.
lueur, f., gleam; flash;
— *de départ*, (*art.*) flash of discharge.
lumière, f., light;
— *aveuglante* (*auto., etc.*) blinding light.
lunette, f., telescope; (*in pl.*) goggles, spectacles; (*mil.*) lunette;
— *acoustique*, (*mil.*) acoustic telescope (listening apparatus, sound liaison), ear trumpet;
— *aplatie*, (*mil.*) half-redoubt (open in rear);
— *d'aviateur*, (*aero.*) goggles;
— *de batterie*, (*art.*) battery commander's telescope;
— *caoutchouc*, (*mil.*) eyepiece, rubber mounted, in certain gas masks;
— *panoramique*, (*art.*) panoramic sight;
— *de pointage*, (*art.*) telescopic sight;
— *sportives*, automobile spectacles, goggles;
— *de visée*, (*art.*) battery commander's telescope.
Luxembourg, (a palace in Paris; by metonymy) the Senate.
lyddite, f., (*expl.*) lyddite.
lyre, f., lyre; stirrup (of a rowlock).

M.

macavoué, m., (*mil. slang*) shell.
machine, f., bicycle; (*mach.*) machine;
— *à condenseur à surface*, (*steam*) surface condensing engine;
— *à découdre*, (*mil. slang*) machine gun;
— *à éprouver*, testing machine;
— *à explosions*, explosion engine;
— *à piston*, generic term for piston, as distinguished from turbine, engines;
— *simple*, (*r. r.*) locomotive without cars;
— *à turbines*, turbine engine.
mâchoire, f., jaw; (*mach., etc.*) clamp;
— *de blocage*, (*trench art.*) clamp.
madrier, m., thick plank;
— *de retenue* (*trench art.*), retaining timber.
magasin, m., (*expl.*) priming charge seat;
— *central automobile*, (*Fr. a.*) mechanical transport repair shop.
magistrat, m., (*law*) magistrate;
— *instructeur*, committing magistrate;
— *du siège*, (*Fr. law*) judge.
magistrature, f., (*law*) magistracy, the bench;
— *assise*, (*Fr. law*) judges proper, as distinguished from — *debout*;
— *debout*, (*Fr. law*) the *parquet*, court officials.
magnéto, f., (*elec.*) magneto;
— *à avance automatique*, automatic lead magneto;
— *à avance fixe*, fixed lead magneto;
— *à avance variable*, adjustable lead magneto;
— *à basse tension*, low-tension magneto;
— *blindée*, cased magneto;
— *à bougies*, high-tension magneto;

magnéto *à haute tension*, high-tension magneto;
— *à inducteur tournant*, stationary armature magneto;
— *à induit oscillant*, oscillating, rocking armature, magneto;
— *à induit tournant*, revolving armature magneto (usual type);
— *à rupteurs*, low-tension magneto;
— *à rupture*, make-and-break magneto;
— *à volet*, — *à volet(s) tournant(s)* stationary armature magneto.
mahomet, m., (*fam.*) hair of the head left long in front.
main, f., hand;
— *courante*, handrail;
en —, (*man.*) in hand (i. e., led, not ridden);
— *à jumelles*, (*mach.*) link spring hanger or carrier;
se mettre en —*s*, (*mil.*) to get, come, to grips;
mise en —*s*, (*mil.*) getting to grips;
remettre en —, (*mil.*) to collect, get, all one's men, troops, together;
— *de ressort*, (*mach.*) spring hanger; dumb-iron;
— *simple*, (*mach.*) spring hanger; dumb-iron.
main-d'oeuvre, f., labor;
— *militaire*, (*mil.*) enlisted workmen.
mainmorte, f., (*law*) mortmain.
maison, f., house;
— *de compagnie*, (*mil. fam.*) company barracks, (i. e., special or separate barracks, and so of other units).
— *de détention*, prison;
— *de force*, prison;

maison *d'invalides*, (*mil.*) "home," to which an invalid soldier may return;
—— *d'unité*, v. —— *de compagnie*.
maitre, m., master;
—— *canonnier*, (*art.*) master gunner (U. S. Art.);
—— *d'équitation*, (*man.*) riding master;
—— *des requêtes*, (France) (officer of the Council of State), master of requests.
maitrise, f., mastery;
—— *des mers*, mastery of the seas, sea-power.
major, m., (*mil.*) (see main work);
—— *des approches*, (*mil.*) (in sieges) major of approaches (field officer responsible for the guard and police of the approaches, assistant to the *général de jour*);
—— *du bivouac* (*du cantonnement*), administrative officer, (assistant of commanding officer in a bivouac or cantonment);
le gros ——, (*mil. slang*) soldiers' name for the major (as being classically old, heavy and fat);
—— *de promotion*, (*Fr. a.*) number one of the graduating class.
majoration, f., allowance (increase);
—— *d'ancienneté*, (*mil.*) (increase in allowance of time of length of service, in computing length of service for promotion), time allowance;
—— *de service*, (*mil.*) (in promotions, credit for increased length of service, as by doubling time in the field, etc.), service allowance.
majorité, f., majority;
—— *absolue*, majority, true majority (i. e., not a plurality);
—— *relative*, plurality (in an election).
maladie, f., sickness; disease;
—— *d'aile*, wing disease (pigeons).
male, a., male; m., (*mach.*) male hole.
malle, f., trunk;
défaire une ——, to unpack a trunk;
faire une ——, to pack a trunk;
—— *d'osier*, (*auto.*) side basket.
malléination, f., (*hipp.*) malleinization.
malleine, f., (*hipp.*) mallein;
épreuve de ——, mallein test for glanders.
mamelon, m., (*top.*) copple, knoll.
manche, f., sleeve, handle;
—— *à air*, ventilating pipe or sleeve;
—— *à balai*, (*aero.*) control lever; joystick;
—— *de gonflement*, (*aero.*) inflating tube;
—— *à vent*, wind cone.
manchette, f., marginal note.
manchon, m., sleeve; (*aero.*) neck (of a balloon); (*mach.*) locking sleeve, clutch; (*art.*) breech casing, gun sleeve, recoil sleeve;
—— *de chambre à gaz*, (*mach. gun*) gas chamber sleeve;
—— *d'embrayage*, (*mach.*) clutch;
—— *excentré*, (*art.*) eccentric sleeve;
—— *à gorge*, (*mach.*) slotted or grooved sleeve;
—— *guide*, (*mach. gun*) (barrel) housing;
—— *de propulsion*, (*mil.*) cartridge sleeve (in some forms of rifle grenade);
—— *réfrigérant*, (*mach. gun*) cooling jacket;
—— *à tourillons*, (*art.*) (trunnioned) recoil sleeve (240 T. R.);
mandat, m., order; (*law*) summons;
—— *d'amener*, (*law*) summons;
—— *sur la poste*, postal money order;
—— *télégraphique*, telegraphic money order.
maneton, m., (*mach.*) crank pin.
manette, f., handle; clamp screw handle; (*aero. auto.*) throttle;
—— *d'accélérateur*, (*mach.*) spark lever;
—— *d'admission des gaz*, (*mach.*) throttle handle, gas lever;
—— *d'allumage*, (*mach.*) ignition handle or lever;
—— *d'avance à l'allumage*, (*mach.*) sparking advance lever or handle;
—— *d'avance à l'allumage*, (*mach.*) sparking advance lever, timing lever, ignition lever;
—— *de blocage*, (*trench art.*) locking handle;
—— *de carburation*, (*mach.*) gas lever;
—— *des gaz*, (*mach.*) throttle lever;
—— *de serrage*, (*inst.*) clamping lever.

manifestant, m., person who joins in a demonstration, rioter.
manifestation, f., demonstration, riot.
manifester, v. n., (of mobs, etc.) to make a demonstration.
manipulant, m., telegraph operator;
aide- ——, assistant operator.
manipulateur, m., key; (*mil. sig.*) sender, sending apparatus (visual signaling).
manivelle, f., (*mach.*) crank;
—— *antiretour*, (*auto.*) safety starting handle.
—— (*d'*) *arrêt* (*de flèche*), (*mach. gun*) trail extension stop crank;
—— *de direction*, (*art.*) deflection handle;
—— *de lancement*, (*auto.*) starting handle or crank;
—— *de mise en marche*, (*auto.*) starting crank, starting handle.
mano-détendeur, m., hand pressure regulator.
manœuvre, f., maneuver; m., (*mil., etc.*) helper, assistant;
—— *s de couverture*, v. s. v. *couverture*;
idée de ——, v. s. v. *idée*;
—— *de garnison*, garrison maneuvers (i. e., carried on by the forces of a garrison).
manque, m., deficiency (of images, in certain range finders).
manteau, m., (*unif.*) cloak, cape (mounted arms);
—— *en caoutchouc*, slipon;
—— *forestier*, (*top*) screen of woods;
—— *d'ordonnance*, (*unif.*) regulation greatcoat, overcoat;
—— *à pèlerine*, (*unif.*) cape and collar.
marchand, m., merchant;
—— *en détail*, retail merchant;
—— *en demi-gros*, merchant who sells habitually to retailers and to consumers;
—— *en gros*, wholesale merchant.
marchandage, m., bargaining;
—— *politique*, log rolling.
marche, f., march; (*mil. r. r.*) (consolidated) graphical schedule (of the trains operated on a line of transport or communication);
—— *arrière*, (*mach.*) reversing gear, reverse;
—— *avant*, (*mach.*) forward gear;
—— *de flexion*, (*mil.*) flexion marching;
—— *funèbre*, (*mus.*) funeral march;
—— *sans cadence*, (*mil.*) route step.
marché, m., bargain; (*adm.*) contract;
passation d'un ——, (*adm.*) making of a contract;
passer un ——, (*adm.*) to make a contract;
revue d'un ——, (*adm.*) examination of a contract.
marchepied, m., footboard; running board (of a train or other vehicle); stepladder; (*in pl.*) stepping stones.
marcher, v. m., (*mil.*) to turn out;
—— *à*, (of a boat, auto, etc.) to use, (e. g., petroleum, gasoline, etc.);
—— *en garde*, (*mil.*) to march with advance, rear and flank guards;
—— *au moteur*, to go under power (of auxiliary boats).
marchichef, m., (*mil. slang*) abbreviation of *maréchal des logis chef*.
mar'gis, m., (*mil. slang*) abbreviation of *maréchal des logis*.
margouiller, m., (*cord.*) dead-eye.
Marie-Louise, m., (*mil. slang*) recruit.
marine, f., water front (Spanish, *marina*).
mariole, a., (*mil. slang*) wide awake.
marmitage, m., (*mil. slang*) *marmite* fire, (v. *marmite*).
marmite, f., kettle; (*mil. slang*) large shell.
marouflage, m., canvas, linen, etc., lining (glued on); (*aero.*) cloth winding.
maroufler, v. t., to line (with canvas, linen, etc., glued on); (*aero.*) to wind on (cloth).
marquage, m., marking (of pigeons).
marque, f., (*com.*) design, patent;
—— *à froid*, cold punching.

marquer, v. t., to mark;
— *à froid*, to cold punch.
marqueur, m., (*t. p.*) marker, sandrat, (slang, U. S. A.).
marron, m., chestnut;
— *à lueurs*, (*art.*) composition producing flash and smoke in a dummy battery (to deceive aerial observation); flash maroon, smoke ball.
marteau, m., hammer;
— *de chaudronnier*, boiler-maker's hammer;
— *de mise de feu*, (*art.*) striker;
— *de percussion*, (*art.*) striker.
martingale, f., (*unif.*) back strap (of an overcoat, etc.).
masque, m., mask; (*mil.*) gas mask; cover;
— *d'arc-boutement*, (*mil. min.*) tamping shield, tail mask;
— *à rabattement* (*mil.*) drop screen (to conceal an observing slit);
— *respirateur*, (*mil.*) respirator (for use in gas attacks), gas mask;
— *pour tireur couché*, cover lying down;
— *à viseurs*, (*mil.*) gas mask fitted with a visor.
masse, f., (*elec.*) ground;
— *cuivre*, copper mallet;
— *d'équilibrage*, counterpoise, counterweight;
fil à la ——, *fil de* —— (*elec.*) ground wire, wire to ground;
— *manœuvre*, — *de manœuvre*, (*mil.*) maneuver mass (so called; under the direct orders of the supreme command, and to be used where most needed as a combat develops; in certain cases, in trench warfare, may be a mass of mobile artillery);
— *polaire*, (*elec.*) pole piece;
avoir un pôle à la ——, (*elec.*) to be connected to ground;
— *par quatre*, (*Fr. cav.*) line of platoons in column of fours.
masselotte, f., (*trench art.*) breech mass (V. D. mortar);
— *à gorge*, (*trench art.*) grooved plunger;
— *à ressort*, (*art.*) spring plunger.
massif, a., powerful, well-organized, massed.
mât, m., mast; (*aero.*) strut;
— *de charge*, boom;
— *à trépied*, tripod mast.
matérialiser, v. t., to mark out, to indicate, materially.
matériel, m., material;
— *de campagne*, (*mil.*) field outfit, field kit;
— *chimique de guerre*, (*mil.*) all chemicals and chemical substances applied to war purposes;
— *à déformation*, (*art.*) generic term for all modern artillery material, as distinguished from the obsolete types in which the various parts had little or no motion with respect to one another;
— (*dit*) *semi-fixe*, (*art.*) semifixed *materiel* (may be transported either by road or by rail).
matière, f., matter;
— *civile*, (*law*) civil cases;
— *contentieuse*, (*adm., etc.*) matter involving conflict of jurisdiction; administrative conflicts;
— *correctionelle*, (*law*) cases of misdemeanor;
— *criminelle*, (*law*) criminal cases;
— *répressive*, (*law*) preventive measures.
matriçage, f., swaging; parts so formed.
matrice, f., (*adm.*) tax register;
—*s cadastrales*, tax registers of unimproved property;
— *générale*, general tax list or register.
matriciel, a., ledger (i. e., having to do with the ledger, relating to, or borne on or in a ledger).
mécanicien, m., (*aero.*) driver, engineer.
mécanisme, m., mechanism;
— *de culasse à coin horizontal*, (*art.*) the Krupp breech-closing system;
— *de culasse à vis*, (*art.*) the screw breech-closing system;
— *à glissière*, (*mach.*) any slide mechanism;
— *de tir*, v. s. v., *tir*.

mécano, m., (*slang*) mechanician, driver.
mécano-morse, m., (*wireless*) automatic sending key.
mèche, f., wick; (*artif.*) fuse, match;
— *d'allumage*, (*artif.*) lighting fuse;
— *d'amorce*, (*min.*) starting drill;
— *blanche*, (*expl.*) Brickford fuse;
— *enveloppante*, (*mil. min.*) spiral earth auger;
— *de fumeur*, tinder;
— *lente* (*expl.*) slow match;
— *porte-feu*, communicating match;
— *de mineur*, (*artif.*) (a form of commercial) Bickford fuse;
— *à ruban*, v. —— *blanche;*
— *à temps*, (*art.*) time fuse (not an official term).
médaillon, m., insert, inset (small picture in a larger one).
médecin, m., doctor;
— *en chef*, (*mil.*) chief surgeon.
mélange, m., mixture;
— *détonant*, v. —— *tonnant;*
— *tonnant*, (*mach.*) explosive mixture (explosion engine).
mêlée, f., (*mil.*) mellay;
exercices de ——, (exercises to train the soldier for hand-to-hand fighting, to develop a spirit of decision at close quarters, to be astonished at nothing that turns up, to be resourceful), hand-to-hand drill.
membre, m., member;
— *inférieur*, (*med.*) lower limb;
— *supérieur*, (*med.*) upper limb.
menotte, f., shackle (joining side to cross springs of a vehicle).
mentonnet, m., catch, lug; (*mach. gun*) hand sear.
mentonnière, f., (*unif.*) chin strap; (*art.*) maneuvering bracket.
mercuriale, f., average price.
mérite, m., merit;
ordre du —— *agricole*, (*France*) order of agricultural merit (for persons who have distinguished themselves in agricultural science).
messagerie, f., transportation (water and rail), express service.
messe, f., (*mil. slang*) *reprise* or ride of the riding masters at Saumur (used only in the expression *aller à la* ——, to go and watch the riding masters when they ride together).
mesurage, m., measuring;
bureau de ——, office of weights and measures.
mesure, f., (*fenc.*) (the greatest distance at which a man may reach his adversary with the bayonet) reach.
métallisation, f., metallization.
métalliser, v. t., to metallize.
métal, m., metal;
— *de canon*, gun metal;
— *ondulé*, corrugated metal.
métallo-plastique, a., (*mach.*) metallo-plastic (said of a joint bearing by metal on a plastic surface).
météorographe, m., (*aero.*) meteorograph.
méthylène, m., (*chem.*) methylene;
— *de bois*, wood spirit, methyl alcohol.
métier, m., craft;
— *de rond de cuir*, (*fam.*) chairwarming, e. g., routine desk work.
métro, m., (*fam.*) abbreviation of *métropolitain*, the Paris underground.
métropole, f., home, home country; (in France) France, as distinguished from the Colonies;
défense de la ——, (*mil.*) home defense, defense of the mother country.
métropolitain, m., the underground at Paris.
metteur, m., setter;
— *de feu*, (*mil.*) firer (sets off bomb-launching devices).
mettre, v. t., to put;
— *un cheval*, (*man.*) to train a horse;

mettre *au point*, to "tune up";
—— *à poste*, (*art.*) to seat a projectile;
—— *à terre*, (*mil.*) to disbark (troops from a vessel).
microphone, m., (*elec.*) microphone;
—— *à crayons*, pencil-microphone;
—— *à grenaille*, (*elec.*) granular microphone.
microtéléphone, m., (*elec.*) microtelephone.
midi, m., noon; the south of France.
mi-enterré, a., half sunken.
mil, m., millet.
milieu, m., middle, center;
—— *x officiels*, official circles.
mille-pattes, m., pl., (*mil.*, *slang*) infantry.
millième, m., (*art.*) mil;
—— *de l'artilleur*, mil;
—— *géométrique*, angle subtended by an arc of 1 meter, at a radius of 1,000 meters.
mine, f., (*mil.*, *etc.*) mine;
—— *de blocus*, (*nav.*) a synonym of *mine automatique* in view of its use in blockade.
minute, f., minute; notes; (the original) drawing (of a survey).
miroir, m., mirror;
—— *parabolisé*, (*opt.*) parabolic mirror;
—— *de pointage*, (*trench art.*) laying mirror.
mise, f., setting, placing, putting;
double —— *de feu*, (*mil. min.*) duplicate ignition;
double —— *du feu par le procédé des relais*, (*mil. min.*) double relayed ignition;
—— *de feu chimique*, (*min.*, *etc.*) chemical ignition;
—— *de feu mécanique*, (*min.*, *etc.*) mechanical ignition;
—— *en feu*, (*expl.*) firing, discharge;
—— *en marche*, (*auto*) starting handle;
—— *en marche automatique*, (*auto*) self-starter;
—— *à poste*, (*art.*) seating (of a projectile in the chamber);
—— *à terre*, (*elec.*) grounding; (*mil.*) disbarking (of troops from a transport or vessel).
mission, f., mission (*mil.*) duty, object, purpose, task, instructions, mission;
—— *de renseignements*, (*mil. aero.*) observation having for its purpose the determination of objectives;
—— *de surveillance*, (*mil. aero.*) duty of observing a sector, so as to mark batteries in action, troops that may be taken under fire;
—— *de tir*, (*art. aero.*) duty of fire observation.
mitaine, f., wristlet.
mitrailler, v. t., (*mil.*) to use machine-gun fire.
mitrailleur, m., (*mil.*) machine-gun soldier, machine gunner.
mitrailleuse, f., (*mil.*) machine gun;
—— *de chasse* (*aero.*) airplane machine gun;
—— *de montagne*, mountain machine gun;
—— *à moteur*, motor machine gun;
—— *type mixte-type alpin*, (Fr. a.) (so-called) mixed type mountain machine gun, (v. *type*).
mitrailleuse-automobile, f., (*mil.*) machine gun on an automobile carriage; auto machine gun.
modérateur, m., (*mach.*) speed reducing device;
—— *de compression*, (*mach.*) compression relief (device);
—— *de vitesse*, (*mach.*) speed regulator.
moineau, m., sparrow; (*mil. slang*) shell.
molette, f., knurled screw head, nut, etc.
monobloc, m., (*mach.*) solid forged.
monocoque, m., (*aero.*) monocoque.
monocylindre, m., (*auto*) single cylinder auto.
monocylindrique, a. & m., (*mach.*, *etc.*) single cylinder; (*auto*) single cylinder car.
monoplace, m., (*aero.*) single seater.
monoplan, m., (*aero.*) monoplane; a., single planed.
monosac, m., (*mil.*) single sack (as distinguished from *bissac*).
monotéléphone, m., wireless telephone.
monotéléphonie, f., wireless telephony.

montage, m., fitting, (*elec.*) wiring;
—— *en circuit-oscillant*, (*wireless*) condenser wiring;
—— *direct*, (*wireless*), direct coupled wiring;
—— *en dérivation*, (*wireless*) shunt coupled wiring;
—— *par induction*, (*wireless*) inductively coupled wiring.
montagne, f., mountain;
prendre la ——, (*fig.*) to take to the hills, i. e., to become a bandit, to take to the road.
montant, m., (*aero.*) strut.
mont-de-piété, m., pawnbroker's shop.
montée, f., (*aero.*) climb.
monte-plats, m., dumb-waiter.
monter, v. t., to mount; (*aero.*) to nose up;
—— *en balance*, to mount on a balanced test bed;
—— *un ballon*, (*aero.*) to operate, manage, a balloon;
—— *à la cardan*, (*mach.*) to mount on a universal joint;
—— *en chandelle*, (*aero.*) to zuhm;
—— *à rotule*, (*mach.*) to mount, assemble, in, or as, a ball-and-socket joint;
se —— *en dame*, (*man.*) may be ridden under a sidesaddle.
montre, f., watch; show case, show window;
—— *à trotteuse compte-secondes*, split-second stop-watch.
montre-bracelet, f., wrist-watch.
montre-chronographe, f., chronograph-watch (second hand centrally mounted, and travels around the face; for computation of small intervals of time).
monture, f., fitting, mounting;
—— *de fixation*, (*artif.*) rocket-stick fixture.
moraillon, m., hasp, clasp;
—— *à ressort*, spring clasp.
mordant, m., spirit, energy.
mortier, m., (*art.*) mortar;
—— *de tranchée*, (*mil.*) trench mortar.
morve, f., pigeon glanders.
moteur, m., (*mach.*) motor, engine;
—— *à ailettes*, flange cooled motor;
—— *atmosphérique*, air motor;
—— *carré*, any motor in which the piston-stroke and cylinder-bore are of the same length;
—— *à combustion* (*interne*), (internal) combustion motor;
—— *à culasse à eau et à cylindre à ailettes*, water- and air-cooled motor;
—— *désaxé*, motor whose cylinder and crank shaft axes do not intersect;
—— *à deux cylindres opposés*, double-opposed two-cylinder motor;
—— *à double alésage*, double-bore motor, (each cylinder is double, has an annular cylinder in addition to the one of ordinary type);
—— *à essence*, gasoline motor;
—— *en étoile*, radial motor;
—— *à explosion(s)*, explosive motor, engine; internal explosion engine;
—— *fixe*, stationary motor;
—— *à foin*, (*fam.*) horse;
—— *horizontal*, horizontal motor;
—— *d'industrie*, any motor used in the arts, manufacturer's motor, industrial motor;
—— *industriel*, v., —— *d'industrie*;
—— *mécanique*, mechanical motor;
—— *à mélange tonnant*, explosion motor;
—— *monocylindrique*, single cylinder motor;
—— *à n cylindres*, n-cylinder motor;
—— *à n temps*, n-phase cycle motor, n-stroke motor;
—— *polycylindrique*, multicylinder motor;
—— *à refroidissement d'air*, air-cooled motor;
—— *à refroidissement d'eau*, water-cooled motor;
—— *rotatif*, rotary motor;
—— *sans soupapes*, silent sleeve valve engine (Knight);
—— *à soupapes commandées*, a motor whose valves are mechanically operated;
—— *thermique*, any heat engine;
—— *tonnant*, explosive motor or engine;
—— *tournant à droite* (*gauche*), right (left) hand engine.

moteur *en V*, V-motor;
—— *vertical*, vertical motor.
—— *à volant extérieur*, motor with exterior flywheel;
—— *à volant intérieur*, motor with interior flywheel.
motocycle, m., (*auto.*) any automobile vehicle not exceeding 200 kg. in weight (a tricycle).
motocyclette, f., (*auto.*) motorcycle.
motocyclisme, m., (*auto.*) motorcycling.
motocycliste, m., motorcyclist.
motodrome, m., (*auto.*) motor race track.
motopropulseur, m., (*aero., etc.*), (propelling) motor.
motoriste, m., (*auto.*) chauffeur.
mototricycle, m., (*auto.*) tricar.
mou, a., soft; (*aero.*) logy, sluggish.
mouchard, m., spy.
moufle, f., muffle (thick glove).
mouillage, m., anchoring, anchorage; (*mach.*) priming (as of a piston-rod, of a pump);
ligne de ——, (*pont.*) anchoring line, line of anchors.
mouiller, v. t., (*mach.*) to wet, prime (a piston-rod, a pump).

mouilleur, m., —— *de mines*, (*art., nav.*) mine planter.
moulin, m., mill; (*aero. slang*) motor engine.
—— *à café*, (*mil. slang*) machine gun.
moulinet, m., reel; air fan;
—— *Renard*, (*aero.*) apparatus for testing the power of a motor.
moustiquaire, m., mosquito bar, net.
mouvementé, a., (*top.*) broken, accidented;
pays ——, (*top.*) broken country.
moyeu, m., hub;
—— *artillerie*, artillery hub (type of wheel);
—— *à billes*, ball-bearing hub;
—— *patent*, (*auto.*) patent hub (motor).
multiplication, f., gear-ratio.
multiplier, v. t., (*mach.*) to multiply, to gear up to.
munition, f., (*mil.*) in pl., ammunition; supplies, stores;
—— *s de salut*, (*art.*) saluting ammunition.
musée, m., museum;
—— *de l'armée*, the army museum at Paris.
musette, f., bag, pouch;
—— *à cartouches*, (*sm. a.*) cartridge pouch;
—— *à vivres*, (*mil.*) ration pouch.
mutualité, f., mutuality (esp. of insurance);
la —— *militaire*, (*mil.*) army mutual relief or other benevolent association.

N.

nacelle, f., skiff; (*aero.*) car, basket, (dirigibles, sphericals, and sometimes, but seldom, airplanes);
—— *de reconnaissance*, (*aero.*) (in Zeppelins, car that may be let down for observation purposes) observation car.
nappe, f., sheet;
—— *de fils*, sheet of telephone wires.
navet, m., turnip.
navire, m., ship;
—— *aérien*, (*aero.*) airship, balloon;
—— *brise-glace*, (*nav.*) ice breaker;
—— *renfloueur*, (*nav.*) wrecking vessel, wrecker;
—— *de plongée*, (*nav.*) generic term for submarine;
—— *poseur de mines*, (*art., nav.*) mine planter.
navire-gigogne, m., (*nav. aero.*) mother ship, hangar ship.
nervure, f., rib; (*aero.*) rib (of a wing).
—— *de guidage*, (*mach. gun.*) guide-rib.
nettoyage, m., cleaning, cleansing; (*mil.*) cleaning up, "mopping" up, of trenches (comprising: (a) destruction of parties of the enemy that continue the defense; (b) cleaning up properly so called, steps to make sure that no enemy are left in the trenches);
—— *des tranchées*, (*mil.*) neutralization or removal of remanent gases in the trenches, after a gas attack; "mopping up" of a trench (i. e., making sure than no enemy is left);
—— *par le vide*, vacuum cleaning.
nettoyer, v. t., to clean; (*mil.*) (to-day) to clean a trench (of its hostile occupants, make sure that none are left);
—— *une tranchée*, (*mil.*) to "mop up" a trench;
—— *par le vide*, to clean by the vacuum process.
nettoyeur, m., (*mil.*) trench cleaner;
—— *de tranchée*, (*mil.*) trench-"mopper."
neutralité, f., neutrality; (in some relations) free from personal comment.
neutre, a., neuter; neutral; (in some relations) free from personal opinion, unvarnished.
nez, m., nose, lug of a key (*clavette*);
—— *à* ——, at the closest possible quarters.
niche, f., niche, recess;
—— *à charge*, (*trench art.*) powder recess;
—— *à grenades*, (*mil.*) grenade recess (in a trench);
—— *de guetteur*, v. *poste de guetteur*;
—— *de tireur*, (*mil.*) rifleman's pit (recess in a trench);

nid, m., nest;
—— *d'abeilles*, (*mach.*) honeycomb radiator;
—— *d'abeilles acoustique*, "honeycomb" megaphone;
—— *d'artillerie*, artillery "nest;"
—— *de batteries*, (*mil.*) group of batteries, close together;
—— *de bombes*, (*mil.*) bomb catcher;
—— *de mitrailleuses*, (*mil.*) group of machine guns;
—— *d'obus*, (*mil.*) shell catcher;
—— *de résistance*, (*mil.*) focus of resistance.
niflette, f., head cold (pigeons).
niôle, f., (*mil. slang*) brandy (the same as *gniole*, *gnôle*, but a better spelling).
niveau, m., (*inst. and in gen.*)
—— *à bulle*, spirit level;
—— *de hausse*, (*art.*) cross level;
—— *d'inclinaison*, (*art.*) cross level;
—— *de repérage*, (*mil.*) registration clinometer, or level;
—— *de site*, (*art.*) angle of site level;
—— *de tir*, (*art.*) firing level (for setting the angle of site);
—— *à totalisation automatique*, (*art.*) integrating level (155 C. T. R.).
nocif, a., hurtful, noxious.
noeud, m., knot;
—— *double fixe*, double bowline;
—— *d'intensité de courant*, (*elec.*) current node;
—— *de jonction*, generic term for any knot joining two ropes end to end;
—— *de voies ferrées*, (*r. r.*) railway junction.
noir, a., black; m., black (substance, color);
—— *de raffinerie*, bone black.
noix, f., nut; (*mach., etc.*) stud.
nombre, m., quorum;
ne pas être en ——, not to have a quorum;
se trouver en ——, to have a quorum.
nomenclature, f., nomenclature;
—— *H I*, (*Fr. a.*) nomenclature and price list of clothing and camp equipage.
non-écrêtement, m., (*art.*) (of fire) close passage over a mask or cover.
non-entrée, f., *en batterie*, (*art.*) failure of gun to return to battery;
non-observation, f., failure to observe, (*art.*) the fall of a shot.
non-percussion, f., (*ins. a., mach. gun.*) non-percussion.

non-remise, f., nondelivery (of a telegram, etc.).
non-retour, m., (*art., mach. gun*) failure of gun to return after fire.
non-valeur, f., (*mil.*) officer taken up on the rolls of a unit, but not serving with it.
noté, a., noted;
  mal ——, of bad repute.
nourrice, f., nurse; (*mach.*) small gravity tank.
noyau, m., nucleus; (*elec.*) core of an induction coil; core (of an electro magnet);
  —— *de résistance*, (*mil.*) focus, nucleus, of resistance.
noyer, v. t., to flood.

nu, a., naked; (*elec.*) uninsulated.
nuage, m., cloud;
  —— *d'éclatement*, (*art.*) smoke of burst.
nul, a., null, void;
  —— *de droit*, null *ipso facto* (from the inherent conditions).
numéro, m., number;
  avoir le —— *n de sa promotion*, to be graduated nth in one's class;
  —— *matricule*, (*in gen.*) serial number.
  sortir avec le —— *n sur m*, to be graduated nth out of m.
numéroteur, m., numbering-machine, -press, or -stamp.

O.

objectif, m., (*mil.*) objective;
  —— *fugitif*, (*art.*) temporary target (visible but a short time, e. g., column of troops).
observateur, m., (*mil., aero.*) observer;
  —— *aérien*, air observer;
  —— *latéral*, (*art.*) (observer on the side of the line of fire), lateral observer;
  —— *terrestre*, land observer (as distinguished from aerial).
observation, f., observation; (*mil.*) observation; (*art.*) observation of fire, of the fall of shots;
  —— *aérienne*, (*aero.*) aerial observation (by airplanes, balloons, etc.);
  —— *bi-latérale*, (*art.*) (observation of points of fall from two stations on sides); bi-lateral observation;
  —— *latérale*, (*art.*) (observation of points of fall from one side only) single observation, lateral observation;
  —— *terrestre*, (*mil.*) land observation (as distinguished from aërial).
observatoire, m., observatory; (*mil.*) observing point or post;
  —— *d'artillerie*, (*art.*) artillery observing point;
  carnet de l' ——, (*mil.*) record book of observations, kept in each observation point;
  —— *de commandement*, (*mil.*) commanding officer's observing point;
  —— *de réglage*, (*art.*) fire adjustment observation station;
  —— *de réglage et de tir*, (*art.*) fire control observation post;
  —— *de renseignements*, (*mil.*) observation station (for the detection of objectives, and particularly of hostile batteries in action); (*art.*) long-distance observation post;
  —— *terrestre*, (*mil.*) (observation station on surface of ground as distinguished from aërial observation), surface station, ground station.
obstacle, m., obstacle; (*man.*) jump;
  —— *aménagé*, (*mil.*) artificial obstacle;
  —— *double* (*triple*) (*man.*) double (triple) obstacle (jump);
  —— *naturel*, (*mil.*) any natural obstacle.
obturateur, m., obturator, shutter;
  —— *de phare*, (*auto.*) light shutter.
obturer, v. t., to plug, stop up.
obus, m., (*auto.*) valve plug (of a pneumatic tire); (*art.*) shell;
  [The following are artillery terms, except where otherwise indicated.]
  —— *acier monobloc*, (*art.*) solid forged shell;
  —— *à amorçage de culot*, base fuse shell;
  —— *chargé en explosif*, high explosive shell;
  —— *éclairant*, —— *à éclairer* (*artif.*) light shell, star shell;
  —— *explosif*, high explosive shell;
  —— *à fausse ogive* (*art.*) screw nose shell;
  —— *fusant*, time shell;
  —— *à fusée instantanée*, shell equipped with an instantaneous fuse;
  —— *à fusil*, (*mil.*) rifle bomb;
  —— *à gaz*, (*mil.*) (asphyxiating) gas shell;
  —— *lacrymogène*, tear shell (tear-producing);
  —— *ogivé*, (*art.*) nosed shell.

obus *avec retard*, slow shell (shell fused to delay burst on strike, attack of deep shelters, of buried targets);
  —— *sans retard*, quick shell (shell fused to burst on strike, attack of blindages);
  —— *de semi-rupture*, (*art.*) semi-armor-piercing shell;
  —— *spécial*, generic term for shell of special purpose, e. g., asphyxiating, tear-producing;
  suivre ses ——, (*mil.*) of an infantry attack, to follow the shell of its own artillery fire (*tir de barrage*, cf. *marcher dans le feu*);
  —— *universel*, universal shell (one type only, for all purposes).
obusier, m., (*art.*) howitzer;
  —— *pneumatique*, (*Fr. a.*) pneumatic howitzer (trench artillery).
obusite, f., (*mil.*) (nervous disease caused by proximity to explosion of large caliber H. E. shell, without being hit); shell shock.
oculaire, m., (*inst.*) eyepiece, ocular;
  —— *à fils*, cross-hair eyepiece;
  —— *de lecture*, reading eyepiece (range finder);
  —— *de visée*, sighting eyepiece (range finder).
œil, m., oye;
  —— *d'immobilisation*, chain stop.
œillère, f., face piece (range finder).
œilleton, m., eyepiece; pinhole.
officemar, m., (*mil. slang*) officer.
officiael, a., official;
  —— *priorité*, on a telegram, means that it has precedence over all others, except *extrême urgence* (v. *télégramme* and *service*).
officier, m., (*mil.*) officer;
  —— *aérostier*, (*aero.*) balloon officer, officer on duty in a balloon section.
  —— *ancien*, officer of length of service;
  —— *d'antenne*, (*Fr. a.*) (at each wireless receiving station, officer who transmits airplane communications) wireless transmitting officer;
  —— *de batterie*, (*art.*) combatant officer;
  —— *s de compagnie*, (*mil.*) company officers;
  —— *de complément*, (*mil.*) supplementary or auxiliary officer (as distinguished from an officer of the active army);
  —— *débrouillard*, resourceful, knowing officer;
  —— *examinateur*, officer member of an examining board;
  —— *légionnaire*, officer belonging to the Legion of Honor;
  —— *de liaison*, connecting officer (e. g., officer knowing both French and English, and acting as intermediary between the two headquarters); liaison officer;
  —— *de manœuvre*, (*aero.*) operations officer (of a balloon company);
  —— *orienteur*, (*art.*) (officer whose duty it is to orient the fire of his battery) objective officer, orienting officer (establishes the communications of a battery by telephone, observatories, etc.);
  —— *de piquet*, (*mil.*) officer on duty;
  —— *de quart*, (*mil.*) watch officer in a trench;
  —— *radiotélégraphiste*, (*mil.*) wireless officer·

officier *de renseignements*, (*mil.*) intelligence, information, officer;
— *reviseur*, (*Fr. a.*) officer on duty with the revision of the map of France;
— *télégraphiste*, (*mil.*) telegraph officer;
— *téléphoniste*, (*mil.*) telephone officer;
— *de tranchée*, (*siege*) officer of the trenches, in charge of trenches;
— *de tranchées*, trench officer, (with reference to construction and organization of a position);
— *s de troupe*, regimental officers, (as distinguished from brigade and division commanders);
un faisant-fonctions d' ——, an acting-officer;
— *de voitures*, (*mil.*) officer of the train.

officier-chimiste, m., (*Fr. a.*) chemical expert belonging to the ammunition inspection of the army.

officier-élève, m., (*mil.*) an officer who is, or becomes, a student at a military school, or war college.

ogive, f., (*art., etc.*) ogive:
fausse ——, (*art.*) nose.

ohmique, a., (*elec.*) ohmic.

olive, f., olive; button, knob;
— *directrice à roulettes;* (*mil. min.*) roller guide (of a mine rammer).

omnibus, m., omnibus;
— *automobile*, (*auto*) autobus.

on, (*mil.*) in commands, abbreviation of *canon*.

onde, f., wave; (*cons.*) corrugation (sheet steel, etc.).
— *de choc*, (*art.*) the shock wave accompanying a projectile in its flight;
— *s entretenues*, (*elec.*) sustained, undamped waves.

ondulé, a., undulated;
métal ——, corrugated metal.

opercule, m., (*art. and gen.*) cover, diaphragm.

or, m., gold;
— *noir*, (*fam.*) coal (cf. *houille blanche*).

ordonnance, f., ordinance, disposition; (*mil.*) (sometimes) soldier servant;
d' ——, according to pattern, to model.

ordre, m., order;
— *d'adieux*, (*mil.*) farewell order;
— *d'appel*, (*Fr. a.*) order to join, (applied also to reservists and territorials);
— *d'éloges*, (*mil.*) order of congratulation, laudatory order;
— *du mérite agricole*, (France) v. s. v., *mérite*; *service d'* ——, v. s. v. *service*;
— *de stationnement*, (*mil.*) orders for stations (at end of march), for outposts.

oreiller, m., pillow;
taie d' ——, pillow slip, pillowcase.

organiser, v. t., to organize;
— *une tranchée*, (*mil.*) to put a (captured or other) trench in order, to equip a trench.

orifice, m., hole, orifice;
— *d'arrivée*, (*mach.*) admission cock or orifice;
— *de remplissage*, (*art.*) filling hole (oil cylinders);
— *de vidange*, (*mach.*) draining orifice or cock.

origine, f., origin;
— *du tir*, (*mil.*) in grenade throwing, point from which a grenade is thrown.

ornithoptère, m., (*aero.*) ornithopter.

orthoptère, m., (*aero.*) orthopter.

oscillateur, m., (*elec., etc.*) oscillator, vibrator.

oscillation, f., oscillation;
— *s amorties*, (*wireless*) damped oscillations.

oscillogramme, m., (*elec.*) oscillogram.

ossature, f., frame (in many practical relations).

osselet, m., knuckle bone; (*mach. gun*) knuckle.

ossomite, f., (*expl.*) a name for emmensite.

ostau, m., (*mil. slang*) hospital.

ouvrage, m., (*fort.*) work;
— *de barrage*, restraining or covering work (e. g., during mobilization, to cover an advance, etc.).

ovalisation, f., (*mach.*) ovalisation (of a cylinder, etc.).

ovaliser, v. t. r., (*mach.*) to ovalize.

oxylithe, m., substance purifying vitiated air.

P.

pacifisme, m., pacifism.

pacifiste, a., m., pacifist, a peace-at-any-price man; (an *antimilitariste* is opposed to armies and to military service; a *pacifiste*, to war).

pageot, m., (*mil. slang*) hospital bed.

paille, f., straw;
— *de couchage*, (*mil.*) bedding straw;
— *pressée*, compressed straw;
— *en rame*, straw in bundles.

pain, m., bread;
— *de guerre*, (*mil.*) hard bread;
— *de pierre*, (*mil.*) hard bread.

paix, f., peace;
— *blanche*, a peace that settles nothing.

pajot, m., (*mil. slang*) hospital bed.

palais, m., palace; Palais Bourbon, (public building in Paris, by metonymy) the Chamber of Deputies;
— *de l'Elysée*, v. *Élysée*.

palan, m., (*con.*) tackle;
— *à chariot*, (*art.*) loading truck, pulley and tackle.

pale, f., (*aero.*) propeller blade.

pâle, a., pale;
un ——, (*mil. slang*) a sick soldier.

palette, f., pallet, splint; (*mil.*) priming spring (in some forms of rifle grenade); (*mil.*) flap of a knapsack; (*aero.*) propeller blade; (*elec.*) armature (of a magnet);
— *de décompression*, (*mach.*) relief valve tappet;
— *de rupture*, (*mach.*) a synonym of *levier de rupture*, q. v.

palier, m., (*mach.*) thrust block, thrust bearing, thrust collar;
— *à bagues*, ring-thrust bearing;

palier *de butée à bagues*, thrust collar, collar-thrust bearing;
— *de butée à billes*, ball-thrust bearing;
— *de chaîne*, sprocket-shaft bearing;
en ——, on the flat, on the level; (motor car, as distinguished from hill climbing).

palme, f., palm; (*Fr. a.*) bronze palm branch on the ribbon of the *croix de guerre*, granted for a citation in orders to the army at large; when five have been obtained, a silver palm is substituted.

palmer, m., micrometer gauge, micrometer caliper.

palonnier, m., (*aero.*) rudder bar, control lever; foot lever; (*mach.*) compensating bar.

panais, m., parsnip.

panache, m., (*unif.*) plume; (*slang*) bluff, play to the gallery;
faire ——, (*auto.*) to turn completely over.

Paname, (*mil. slang*) Paris.

Panbochie, f., (*mil. slang*) Pan-Germany.

paneterie, f., bread-room; store, supply, of baked bread.

panier, m., basket;
— *à pansement*, (*mil. med.*) (basket to carry dressings) dressing basket;
— *à projectile*, (*art.*) shot basket;
— *de voyage*, (*pigeons*) transfer basket.

panier-repas, m., basket meal.

panne, f., (*aero., auto.*) breakdown, accident, trouble.

panneau, m., panel; (*aero.*) side panel;
— *de déchirure* (*aero.*) ripping panel;
— *à éclipse*, (*mil.*) disappearing panel;
— *fixe*, (*mil.*) fixed signal;

panneau *d'identification,* (*mil.*) identification panel (for signaling position to a balloon or air plane);
— *de jalonnement,* (*Fr. a.*) location panel;
— *persienne,* (*aero.*) (sort of shutter white on one side, neutral on other, operated by a string pull); window shutter panel;
— *de signalisation,* (*sig.*) signaling panel or disk;
— *de visite,* (*aero.*) inspection panel.

pansement, m., (*med.*) dressing;
place de, — (*med.*) dressing station;
station de —, (*med.*) dressing station.

panstéréoscope, (m., *inst.*) panstereoscope (admits, among other things, of the immediate examination of ordinary photographs).

pantalon, m., trousers;
— *basané,* — *à basanes,* (*unif.*) spatterdash trousers;
— *à brayette,* trousers opening in front.

pantoufler, v. n., (*mil. slang*) to resign from the army (Polytechnique and Fontainebleau).

paperasse, f., in pl., useless papers; (hence) red tape.

paperassier, m., paper man, scribbler, slave to red tape.

papier, m., paper;
— *carbone,* carbon paper;
— *d'étain,* tinfoil;
— *à étoupilles,* primer paper.

papillon, m., butterfly; leaflet; (*mach.*) wing nut.

paquebot, m., packet boat; (*mil. slang*) ambulance (wagon).

paquet, m., package;
petit —, (*mil.*) small body, small detachment of, troops.

paquetage, m., (*mil.*) packing; the pack itself;
— *de campagne,* field pack.

paradis, m., paradise; (*St. Cyr slang*) the hospital.

parafeux, m., (*art.*) flareback preventer.

parafoudre, m., (*elec.*) spark arrester; (*mach.*) safety gap.

parage, m., regions (usually in plural).

parallèle, f., (*mil.*) parallel; m., parallel of latitude
— *d'attaque,* (*mil.*) parallel from which an attack develops, or is made;
— *de contact,* (*aero.*) (the parallel at which the net leaves the balloon) contact parallel;
— *de départ,* (*mil.*) trench, parallel, from which an assault is made.

parallélisme, m., parallelism; (*art.*) having the same elevation and deflection allowance (said of two pieces, heavy artillery, in ranging).

parapluie, m., umbrella; (*art.*) hood (part of shield over gun, 155 C. S.).

parc, m., park; (*mil.*) train;
— *aérostatique,* (*mil. aero.*) balloon park or establishment;
— *d'ascensions,* (*aero.*) ascension field;
— *automobile,* (*aero.*) truck train (balloon company);
— *à fourrage* (*mil.*) forage park or corral;
grand — *automobile de réserve,* (*Fr. a.*) mechanical transport base;
— *hippomobile,* (*mil.*) horse-drawn train (balloon company);
— *maritime* (*aero.*) dirigible port;
— *mobile,* (*art.*) v. — *à munitions;*
— *à munitions,* (*art.*) expense magazine (railway artillery);
— *sur rails,* (*Fr. a., r. r., eng.*) traveling park (for railway engineer troops);
— *sur route,* (*Fr. a., r. r., eng.*) road park (engineer park, kept to be sent through a first breach);
— *aux tubes.* (*mil. aero.*) gas-cylinder train (balloon company);

parclose, f., front seat (of an auto, etc.).

pare-boue, m., (*auto.*) mudguard.

pare-brise, m., (*auto. aero.*) wind screen.

pare-bruit, m., (*elec.*) silencer.

pare-à-crotte, m., mudguard (said of a spur so put on as to keep the trousers off the ground).

pare-crotte, m., mud guard.

pare-éclats, m., (*mil.*) splinter proof (in a trench).

pare-grenades, m., (*mil.*) grenade stop.

pare-pluie, m., (*mil.*) rain guard (of a gas mask).

pare-poussière, m., dust guard of a motor car).

parer, v. t., to prepare;
— *le frein,* (*art.*) to fill a recoil cylinder.

parlement, m., Parliament; (in France) the Senate and Chamber of Deputies.

parlementarisme, m., parliamentary custom, etiquette; "senatorial courtesy" (U. S.).

parquet, m., (*law, mil. law*) court officials.

partie, f., part;
— *pleine,* (*art., sm. a.*) land (of a rifled tube or barrel);
prendre à —, to take in hand.

pas, m., step; (*mach.*) pitch of a propeller, of a screw; (in a roller chain) pitch (distance between axes of adjacent rollers);
— *de flexion,* (*mil.*) pace in the *marche de flexion;*
— *de parade,* (*mil.*) the *parademarsch* of the Germans.

passage, m., passage;
— *de lignes,* (*mil.*) passage of or through the assaulting battalions by the reserve battalions (in an attack).

passager, m., passenger (in a ship).

passe-montagne, m., mountain climber's hood (adopted by aviators); knitted woolen helmet.

passer, v. n., (*mil.*) to be promoted; v. t., (*adm.*) to make a contract; (*mil.*) to turn over (as, the command);
— *capitaine,* — *colonel,* to become, to be promoted, captain, colonel;
— *sous la protection de M*me *Ve Berger-Levrault,* (*mil. slang*) to be promoted by seniority (see *annuaire;* this book is published by the house of Berger-Levrault, of which at one time M*me* Berger-Levrault was the head).

passerelle, f., (*mil.*) small portable bridge for crossing trenches.

pastille, f., repairing patch (pneumatic tire);
— *d'allumage,* (*art., expl.*) black powder priming charge;
— *incendiaire,* (*mil.*) incendiary tablet (to set houses, etc., on fire; German);
— *microtéléphonique,* (*elec.*) carbon disk (of a microtelephone).

pâtée, f., pigeon milk.

patin, m., spring seat or plate; (*aero.*) runner, skid, landing skid; (*mach., etc.*) shoe; (*Fr. art.*) brakeshoe of the 75 mm. gun;
— *d'appui,* (*art.*) supporting slide;
— *de glissière,* (*aero.*) axle guide bearing;
— *porte-auto,* trestle (to take the weight of a motor car off the tires);
— *de queue,* (*aero.*) tail skid.

patinage, m., (*mach.*) advance, slipping, skidding; (*aero.*) skid, runner.

patiner, v. n., (*mach.*) to turn, work without advancing; to skid.

patron, m., patron, master; (*mil. slang*) the captain.

patrouille, f., (*mil.*) patrol; (*aero.*) air patrol;
— *de contact,* contact patrol.

patte, f., claw, foot;
— *d'attache,* catch, fastening;
— *d'épaule,* (*unif.*) the shoulder part proper of an epaulet;
— *d'épaulette,* (*unif.*) v. — *d'épaule;*
— *de parement,* (*unif.*) band, patch (forming part of facings);
— *s de raccordement,* mortise and tenon (metal).

pavillon, m., sort of funnel; rigid top of an open carriage, motor car, etc.

pavoiser, v. a., to dress, decorate, a building, a street, etc.

péage, m., toll;
   droits de ——, tolls.
péager, m., toll gate keeper.
peau, f., skin;
   —— *des batteurs d'or,* (*aero.*) gold beater's skin.
pédale, f., pedal; (*in pl., Saumur slang*) the stirrups;
   —— (*d'*)*accélérateur,* (*mach.*) accelerator pedal, (motor car);
   —— *de débrayage,* (*mach.*) clutch pedal, (motor car);
   —— *d'embrayage-débrayage,* (*mach.*) clutch pedal (motor car);
   —— *de frein,* (*mach.*) brake pedal (motor car).
pédiluve, m., *automatique,* (*Fr. a.*) automatic foot washing machine.
peiner, v. n., (of a motor) to labor.
peinture, f., painting;
   —— *laquée,* lacquered paint.
pékin, m., (*mil. slang*) *être pékin d'une colle,* to have finished with an oral examination.
pelisse, f., (*Fr. unif.*) sort of short overcoat, with fur collar (not worn under arms); (*auto.*) chauffeur's fur coat.
pelle, f., (*fam.*) "cropper;"
   bêcher une ——, (*mil. slang, man.*) to fall from, or to be thrown by one's horse, to come a cropper;
   ramasser une ——, v. *bêcher une* ——.
pelle-pioche, f. (*mil.*) spade-pick (engineer equipment).
pelliculaire, a., m., film.
pellicule, f., any film.
pelote, f., ball;
   —— *d'étoffe,* (*mil.*) stuff ball, (used as a dummy, or drill, grenade).
peloton, m., (*mil.*) platoon; name given to a battery of 37 mm. T. R.;
   —— *de pièce,* (*art.*) gun detachment.
pelucheux, a., lint producing.
pelure, f., peeling; (*mil. slang*) overcoat.
pencher, v. n. t., to lean, bend; (*aero.*) to bank, tilt.
pendu, m., (*mil. slang*) instructor (not confined to St. Cyr).
pêne, m., (*mach. gun*) cover latch.
penne, f., arrow feather.
penthière, f., (*Fr. adm.*) stretch of 8 or 10 kilometers along frontier, watched by a customs brigade.
pépère, m., (*mil. slang*) a territorial.
péquenot, m., (*mil. slang*) awkward man.
perce, f., drill, punch;
   mettre en ——, to broach (a cask).
percement, m., cutting, digging, of a canal.
percer, v. t., to pierce, cut through (as a canal).
perchette, f., pole, rod.
perchoir, m., perch.
perco, m., (*mil. slang*) news, information.
percuter, v. a., (*art.*) to strike; v. t., to tap, strike; (*art., sm. a., etc.*) to strike; to arm (a grenade).
perfectionnement, m., (*man.*) final training.
périmer, v. n., (*law*) of a patent, etc., to lapse.
période, f., period;
   en ——, (*mil.*) (of reservists) during the period of service.
périscope, m., (*mil.*) periscope;
   —— *double,* double periscope;
   —— *de précision,* (*mil.*) wide field periscope.
   —— *de tranchée,* (*mil.*) trench periscope.
périscopique, a., periscopic;
   effet ——, periscopic effect (distance between eyepiece and objective).
perlot, m., (*mil. slang*) tobacco.
perme, f., (*mil. slang*) permission.
permis, m., permit;
   —— *de circuler,* (*adm.*) permit, license, to circulate, to go and come;
   —— *de conduire,* (*auto.*) chauffeur's license.

permission, f., (*mil.*) pass;
   —— *de la journée,* (*mil.*) all day pass;
   —— *de la demi-journée,* (*mil.*) half-day pass, afternoon pass.
permutant, m., (*mil.*) party of the second part in a transfer.
persienne, f., shutter; (*aero.*) louver.
perte, f., loss;
   —— *de charge,* drop, loss of pressure;
   —— *de vitesse,* (*aero.*) pancake (of a landing); stall (in the air).
pèse-essence, m., gasoline densimeter.
pétard, m., petard; (*mil. min.*) small bored mine;
   —— *d'assaut,* (*mil.*) offensive grenade.
pétard-amorce, m., (*mil. min.*) priming cartridge.
pète-sec, m., (*mil. slang*) gymnastics.
pétrolage, m., oiling (of mosquito holes, etc.).
pétrole, m., petroleum;
   —— *à brûler,* illuminating oil;
   —— *brut,* crude oil;
   déchets de ——, waste oil;
   —— *lampant,* ordinary petroleum;
   —— *ordinaire* v. —— *lampant;*
   —— *rectifié,* lamp oil, rectified oil.
pétroler, v. a., to oil (pools, etc.) against mosquitoes.
pétrolette, f., (*auto.*) motorcycle (name given in France).
phalange, f., (*mil.*) phalanx.
phare, m., automobile lamp, headlight, pilot light; projector, searchlight (rare);
   —— *à acétylène,* acetylin light;
   —— *automobile,* (*mil.*) automobile searchlight;
   —— *de bicyclette,* bicycle lamp;
   —— *électrique,* electric headlight; (electric) searchlight;
   —— *flottant,* (*nav.*) lightship;
   —— *à incandescence,* incandescent light.
photophore, m., sort of electric reflector; (*nav.*) light buoy.
pic, m., pick;
   à ——, vertical, sheer;
   —— *de parc,* (*mil.*) engineer's pick (engineer equipment);
   —— *portatif,* (*mil.*) portable pick (engineer equipment).
pic-hachette, m., (*mil.*) pick-hatchet (infantry equipment).
picriqué, a., (*expl.*) picrated.
pièce, f., document, piece, voucher; room (in a house); apiece, each; (*art.*) (number of) vehicles (used in the service of a gun);
   —— *d'armement,* (*art.*) cocking piece.
   —— *sans dépointage,* (*field art.*) the modern fieldpiece, in respect of its retention of aim;
   ——*s détachées,* separate parts (as of motor cars, etc.).
   ——*s doublées,* (*art.*) gun and caisson, side by side (caisson on left usually);
   —— *à recul sur l'affût,* (*art.*) carriage-recoil piece;
   —— *de sûreté,* (any) safety device.
pièce-guide, f., (*art.*) directing gun (heavy artillery, called *pièce- directrice* in field artillery).
pied, m., foot; support;
   —— *de bielle,* (*mach.*) piston end of the connecting rod;
   —— *panoramique,* (*phot.*) panoramic stand (of a camera).
   prendre ——, to be set on foot;
   sûr de ——, surefooted;
   sûreté de ——, surefootedness.
pierre, f., stone;
   —— *factice,* (*cons.*) artificial stone;
   —— *magnétique,* lodestone;
   —— *tombale,* gravestone.
piétaille, f., (*mil. slang*) infantry.
pieu, m., stake; (*Saumur slang*) bed;
   —— *aiguisé,* (*mil.*) sharpened stake (obstacle);
   carotter le ——, (*Saumur*) to sleep late.
pigeon, m., pigeon;
   —— *d'arrière-saison,* late pigeon (hatched after July);

pigeon *de mobilisation*, (*mil.*) service pigeon (1.5-8 years old).
poste de ——s, (*mil.*) carrier-pigeon station;
—— *de remplacement* (*mil.*) substitute pigeon, pigeon in training.
—— *voyageur*, (*mil. slang*) shell.
pigeonneau, m., young pigeon, squab.
piger, v. n., (*mil. slang*) to understand.
pignon, (*mach.*) pinion; gear;
—— *d'attaque*, driving pinion;
—— *de chaîne*, sprocket wheel or pinion;
—— *de commande*, (*mach.*, etc.) driving pinion;
—— *conduit*, driven gear;
—— *conique*, bevel gear;
—— *de démultiplication*, reducing gear;
—— *distributeur*, —— *de distribution*, valve gear pinion;
—— *d'entraînement*, v. —— *d'attaque*;
—— *manivelle*, (*mach. gun*) breech mechanism pinion.
—— *de marche arrière*, reversing gear;
——*s de prise continue*, constant mesh gear;
—— *satellite*, intermediate pinion or gear;
—— *tendeur*, chain stretcher;
pile, f., (*elec.*) battery;
—— *au bichromate à deux liquides*, two-liquid bichromate cell;
—— *au bichromate à un seul liquide*, single fluid bichromate cell;
—— *bouteille*, bichromate cell;
—— *à dépolarisant*, depolarizing cell;
—— *d'épreuve*, testing battery;
—— *d'essai*, (*elec.*) testing battery.
—— *à liquide immobilisée*, another name for —— *sèche*, dry battery.
pilier, m., pillar;
—— *de direction*, (*auto.*) steering post.
pilote, m., (*aero.*) pilot;
—— *d'aéronat*, (*aero.*) dirigible pilot;
—— *d'altitude*, altitude pilot (dirigible);
—— *d'aréostat*, (*aero.*) balloon pilot;
—— *aviateur*, (*aero.*) airplane pilot;
—— *d'hydroaéroplane*, (*aero.*) hydroplane pilot.
pilote-aéronaute, m., (*aero.*) balloon pilot (spherical).
piloter, v. t., to pilot;
—— *un cheval*, (*man.*) to ride a horse.
pinard, m., (*mil. slang*) ordinary wine.
pince, f., pliers, pincers;
—— *d'électricien*, universal pliers (cutting, pipe wrench, etc., all in one);
—— *à gaz*, pipe wrench, gas pliers;
—— *à main*, hand pliers;
—— *plate coupante*, flat cutting pliers;
—— *à plomber*, seal press;
—— *universelle*, v. —— *d'électricien*.
pinceau, m., brush;
—— *à graisse*, grease brush.
pinceau-graisseur, m., grease brush.
pincette, f., elliptical spring.
pinçon, m., nipping (of a pneumatic tire).
pingouin, m., (*mil. aero., slang*) beginner's practice machine (has rudimentary wings).
pioche, f., pick;
—— *de parc*, (*mil.*) intrenching pick.
pionnier, m., (*mil.*) pioneer;
—— *de cavalerie*, (*cav.*) cavalry pioneer.
pipe, f., pipe; pipe-shaped top of certain single cylinder motors; pipe-shaped turbine.
piquage, m., (*aero.*) nose-dip.
pique-boyaux, m., (*slang*) fencing;
faire du ——, to fence.
piquer, v. n., (*aero.*) to dip, dive, nose down;
—— *du nez*, to nose dive.
piquet, m., (*mil.*) picket;
—— *d'honneur*, (*mil.*) the troops turned out, the *haie*, q. v., on occasions of ceremony, at the Chamber of Deputies, etc.; guard of honor.
—— *métallique à vis*, (*mil. aero.*) ironscrew anchor (balloons).
—— *de terre*, (*elec.*) ground stake;
—— *à vis*, (*mil.*) screw stake (wire entanglements).

piquetage, m., picketing; (*mil.*) any staking or marking out of lines.
piquet-bêche, m., (*trench art.*) anchor (platform).
pisé, m., pisé;
—— *de mâchefer*, (*cons.*) a sort of rough concrete made of clinkers or slag (3 sacks of clinkers to 1 cubic meter of hydraulic lime).
piste, f., trail; (*mil.*) open passage in a network of trenches;
—— *cavalière*, (*man.*) bridle path; ·
—— *de roulement*, (*mach.*) roller path;
—— *en S*, (*mach. gun*) S-groove;
sortir de la ——, (*man.*) to leave the track.
pistolet, m., (*sm. a.*) pistol; (*mil. slang*) bed pan (hospital slang); (*mil. min.*) bit;
—— *d'armement*, (*art.*) cocking gear;
—— *automatique*, automatic pistol (e. g., Colt, etc.);
—— *à diamants croisés*, (*mil. min.*) cross bit;
—— *éclairant*, (*mil.*) flare pistol;
—— *à fusée éclairante*, v. —— *éclairant*.
piston, m., (*mach.*, etc.) piston; (*slang*) influence, interest, "pull"; (*mil. slang*) the captain; (*slang, Polytechnique*), assistant (cares for apparatus, sort of janitor); (*mach. gun*) magazine follower;
—— *à galet*, —— *porte-galet*, (piston in whose face is mounted a roller) roller piston;
—— *moteur*, (*mach. gun*) driving piston;
—— *de réglage*, (*mach.*) feed regulation piston (of some forms of carburetors);
surface de ——, (*mach.*) piston area.
pistonner, v. n., (*slang*) to work a "pull";
se faire ——, (*mil. slang*) to get on by "pull."
piton, m., eyebolt, ringbolt;
—— *de brêlage*, (*harn.*) lashing ringbolt (machine gun).
pivot, m., pivot; (*mach. gun*, etc.) pintle;
—— *directeur*, —— *de direction*, (*mach.*) steering pivot, knuckle;
—— *de pointage en direction* (*trench art.*) pintle.
place, f., place;
la Place, *la place de Paris* (i. e., Paris as a military position);
—— *annexe*, (*fort.*) fort, position, depending upon a main or central fort;
—— *d'armes*, (*mil.*) (trench or trenches serving for the assembling of troops; may be either existing trenches, or trenches laid across the boyaux for the purpose), assembling trench;
—— *du moment*, (*mil.*) any place fortified during the course of a war (e. g., Torres Vedras, Petersburg, Plevna).
placer, m., (*man.*) position of the horse with respect to the leading foot.
plafond, m., ceiling; (*aero.*) limit of height to which an airplane may rise.
plan, m., (*aero.*) wing, synonym of rudder, stabilizing plane; (*mil.*) plan;
—— *d'action*, (*art.*) table showing objectives to be taken under fire; (*mil. in gen.*) complete report on operations to be undertaken; plan of action;
—— *d'action de l'artillerie*, (*art.*) (plan fixing the organization, employment, and action of the artillery in order to secure the results contemplated by the *plan d'engagement*), plan of artillery action;
—— *d'approche*, (*cav.*) plan of approach (i. e., all measures for collecting and forming cavalry in the preparation of the pursuit);
—— *d'armement*, (*mil.*) equipment plan;
—— *de l'assiette*, (*aero.*) horizontal rudder (dirigible);
—— *de barrage*, (*art.*) plan fixing the rôle of each battery in barrage fire;
—— *central*, (*aero.*) central panel;
—— *de contre-préparation*, (*art.*) plan giving all details of the *tir de contre-préparation*;
—— *débordant*, (*aero.*) overhanging plane;
——*s décalés*, (*aero.*) staggered wings;
—— *de défense*, (*mil.*) plan of defense (of a given position);
—— *de déplacement de l'artillerie*, (*art.*) table showing shifts of positions to be made;

plan *de dérivation*, *(ball.)* plane of drift (passing through axis of piece and point of fall);
— *de dérive*, *(aero.)* vertical fin;
— *de destruction des contrebatteries*, *(art.)* table showing hostile batteries to be engaged and destroyed;
— *directeur*, *(mil.)* map or plan for the direction of an attack, battle map; *(art.)* plan, survey, showing the distribution of targets, emplacements, etc., (heavy artillery); *(art. and mil. in general)* directing map (artillery fire, operations in general); *(mil. min.)* map of operations;
— *d'emploi de l'artillerie*, *(art.)* (a plan prepared at each army corps, and division headquarters, covering all the cases of the use of the guns in the unit concerned), artillery plan;
— *d'engagement*, *(mil.)* plan of engagement of troops;
— *d'engagement de l'infanterie*, *(mil.)* plan of attack, in the broadest sense, of an infantry force;
— *d'ensemble*, *(mil.)* plan for the *combat de rupture*, q. v., and the subsequent exploitation of the success achieved; general plan; *(mil. telephone)* general plan of the telephone system of an army;
— *d'exploitation*, *(mil.)* plan made to clinch a successful attack;
— *fixe*, *(aero.)* stabilizer.
— *incliné*, *(phys., etc.)* inclined plane; *(mach. gun)* ramp;
— *inférieur*, *(aero.)* lower wing;
— *des liaisons*, *(mil.)* (measures to secure good communication between troops, troops and commanding officer, infantry and artillery, etc.) liaison plan;
— *de nettoyage*, *(mil.)* cleaning out plan (trenches);
— *de neutralisation*, *(art.)* (table showing batteries to be neutralized as battle progresses) neutralization plan;
— *d'observation*, *(mil.)* observation plan (forms part of the *plan de défense* of the sector, and contains map of observing points, panorama from each of these points, diagram of telephone communications, and statement of the working of the observation service);
— *d'occupation*, *(mil.)* plan of occupation (of enemy position);
— *d'opération*, *(mil.)* plan of operation(s);
— *porteur*, *(aero.)* supporting plane, wing;
— *de poursuite*, *(cav.)* plan of pursuit;
— *rabattant*, *(aero.)* movable plane;
— *supérieur*, *(aero.)* upper wing;
— *sustentateur*, *(aero.)* sustaining plane;
— *de visée*, *(art.)* plane of sight.

planche, f., plank;
— *à coffrer*, *(mil. min.)* sheathing plank, sheathing;
— *repose-pieds*, foot board of a vehicle.

planchette, f., *(inst.)* plane table;
— *de batterie*, *(art.)* battery map (zinc square on which is pasted so much of general map as concerns battery in question; screwed on a wooden board);
— *dévidoir*, standing reel;
— *-rapporteur*, *(art.)* protractor plane table;
— *à rouleaux*, roller map board;
— *stéréoscopique*, *(inst.)* stereoscopic plate (produces stereoscopic effects by the juxtaposition of two photographs of the same object);
— *de tir*, *(art.)* range card.

planement, m., soaring; *(aero.)* volplaning.
planer, v. n., to soar; *(aero.)* to volplane, to glide.
planétaire, m., *(mach.)* driving pinion (of a bevel gear set).
planeur, m., *(aero.)* the aeroplane minus the engine set; glider.
planton, m., *(mil.)* orderly;
— *de surveillance*, *(mil.)* lookout orderly (assigned to be on the lookout, e. g., at a *terrain d'atterrissage*.

plaque, f., plate;
— *d'allumage*, *(mach.)* ignition plate;
— *alvéolée*, *(elec.)* grid (storage battery);
— *d'appui*, *(trench art.)* backplate (trench mortar bed); bedplate;

plaque *à collerettes*, *(art.)* diaphragm (of a limber, etc., chest);
— *de couche*, *(trench art.)* foundation plate.
— *distributrice d'allumage*, *(mach.)* ignition plate;
— *gondolée*, *(elec.)* buckled plate;
— *grillagée*, v. — *alvéolée;*
— *d'identité*, *(auto.)* number plate;
— *de jonction*, *(mach.)* junction plate;
— *à oxydes rapportés*, *(elec.)* plate of the Faure type;
— *de plomb formée*, *(elec.)* formed plate (of a storage battery, Planté type);
— *de recouvrement*, *(art.)* bedplate (trench mortar);
— *de terre*, *(elec.)* ground plate, earth;
— *tordue*, *(elec.)* buckled plate.

plaque-logement, f., *(trench art.)* seat (for the *tige-support*, q. v.);
plaquette, f., *(r. r.)* poster; *(art.)* collar or slide (to increase angle of fall);
— *d'appui*, (small) faceplate, guard plate.
plateau, m., plate, disk; plate (of a caterpillar tire); *(art.)* (circular) deflection scale;
— *d'accouplement*, *(mach.)* assembling or coupling plate;
— *d'embrayage*, driven plate;
— *d'entrainement*, driving plate.
— *gradué de site*, *(art.)* angle of site scale.
plateau-volant, m., *(mach.)* wheel crank, disk crank.
plateforme, f., platform, *(aero.)* (in dirigibles), a platform (forming the junction between the car and the body of the vessel).
platine, *(sm. a.)* lock;
— *d'armement*, *(trench art.)* lock.
plein, m., *(auto.)* solid tire;
trop ——, *(in gen.)* excess.
plombe, f., *(mil. slang)* hour.
plombé, a., lead sealed.
plongée, f., dive (e. g., airplane dive).
plot, m., *(elec.)* plug, contact plug;
— *de contact*, contact plug.
plume, f., pen; feather;
— *caudale*, tail feather;
être sur trois ——s, (of a carrier pigeon) to have three primaries left;
— *de poche*, fountain pen;
— *à réservoir*, fountain pen;
— *de vol*, primary.
pluralité, f., plurality;
— *d'origine*, *(Fr. a.)* plurality of origin (with reference to the various sources from which officers are drawn, v. s. v. *unité*).
pneu, m., tire;
— *ferré*, studded tire (pneumatic);
— *s jumelés*, *(auto.)* (motor truck), twin tires.
pneu-auto, m., *(auto.)* auto equipped with pneumatic tires.
pneu-cuir, m., leather casing (pneumatic tire).
pneumatique, m., *(auto.)* pneumatic tire.
poche, f., pocket;
— *à air*, *(aero.)* air-pocket;
— *à cartes*, *(mil.)* map case;
— *à chargeurs*, *(sm. a.)* loader pouch;
— *à feux*, *(mil.)* "fire pocket" (said of the place inside one's lines to which the enemy has penetrated, and in which he may be pocketed by fire on his flanks and in his rear);
— *de portefeuille*, inside pocket of a coat;
— *de vapeur*, steam bubble (as in a water jacket).
poêle, m., pall;
tenir les cordons du ——, to be a pallbearer.
poids, m., weight;
— *du cheval*, *(auto.)* weight per horsepower;
— *demi-lourd*, *(auto.)* generic term for autos between —— *lourd* and —— *léger* qq. v.;
— *léger*, *(auto.)* generic term for pleasure autos; light delivery wagon;
— *lourd*, *(auto.)* a generic term for heavy draft autos such as trucks, freight-wagons;
— *total*, *(aero.)* lifting capacity;
— *utile*, useful weight carried;
— *à vide*, dead weight.

poignée, f., handle;
—— *de maintien*, (*mach. gun*) handle;
—— *de portage*, (*trench art.*) carrying handle.
poignée-contact, f., (*mach.*) contact lever, contact handle.
poil, m., hair, coat;
à —— *ras*, smooth-haired;
à —— *rude*, wire-haired;
sans avoir un —— *retourné*, without turning a hair.
poilu, m., (*mil. slang*) French soldier, with connotation of virile qualities (v. *bonhomme*).
poinçon, m., stamp (punched in, as on a gasoline tank);
—— *de l'octroi*, official municipal stamp (on tanks, etc., of motor cars).
point, m., dot; stop (punctuation); (*mil. sig.*) dot (of a code);
—— *d'appui*, (*mil.*) combination of active segments, (hence) work, i. e., intrenched position;
—— *d'attache*, (*aero.*) station of a balloon ("home port," as it were);
deux ——*s*, colon;
—— *de direction*, (*mil.*) (in an attack, point situated behind the last assigned objective) direction point;
—— *d'eau*, (*mil.*) water (well, tank);
—— *d'exclamation*, exclamation point;
—— *d'interrogation*, interrogation point;
mettre au ——, to "tune up";
—— *moyen d'éclatement*, (*art.*) mean point of burst;
—— *de pointage*, (*art.*) aiming point;
—— *de repérage* (*art.*) auxiliary aiming point;
—— *de repère*, (*art.*) registration point;
—— *repéré*, (*art.*) registered point;
—— *suspensifs*, —— *de suspension*, row of dots (punctuation);
—— *et virgule*, semicolon;
—— *de visée*, (*art., sm. a.*) point aimed at;
—— *de vitesse*, (*t. p.*) allowance or marks for rapidity of fire.
pointage, m., (*art.*) aim;
dégrossir le ——, (*art.*) to take approximate aim;
au miroir, (*art.*) laying by reflection;
miroir de ——, (*art.*) laying mirror;
plan de ——, (*art.*) plane of aim;
—— *réciproque*, (*art.*) reciprocal laying.
pointe, m., (*mil.*) dash;
—— *de mineur*, (*mil. min.*) drill.
pointeau, m., (*mach.*) needle valve; needle of a float-feed carbureter;
—— *de réglage*, (*mach.*) needle valve;
—— *à ressort* (*mach.*) spring needle valve;
—— *à vis*, (*mach.*) screw needle (to regulate the spray of an atomizer).
pointer, v. t., (*mach., etc.*) to mark relative positions of adjacent parts; (*art.*) to lay a gun;
—— *au miroir*, (*art.*) to lay by reflection.
pointeur-tireur, m., (*trench art.*) gunner.
poire, f., bulb.
poireau, m., leek;
ordre du ——, (*fam.*) the *ordre du mérite agricole*, q. v.
poirot, m., (*mil. slang*) the commanding officer.
pois, m., pea;
—— *jarras*, vetch (usually spelled *jarosse* or *jarousse*).
police, f., police station; in general, all measures taken to insure public order;
—— *administrative*, (all measures looking to the maintenance of order, the well-being and security of the citizens) administrative police;
corps de ——, (*mil.*) constabulary;
—— *générale*, (*mil.*) the general police, (good order, safety, arrest of spies, etc., carried on by the gendarmerie);
—— *judiciaire*, (measures taken to look up violations of order and law, get proofs, and deliver the accused to the courts) criminal police; (*mil.*) duties of gendarmerie (in searching out crime and criminals, and in assisting the regular courts-martial, by holding special courts);
—— *de roulage*, (*auto.*) road regulations;
—— *de sûreté*, police to maintain public order; measures of public safety.

policier, m., policeman, police agent.
politique, f., politics, policy;
—— *de fermeture*, v. —— *d'interdiction*;
—— *d'interdiction*, policy of exclusion (of foreigners, etc.).
polochon, m., (*mil. slang*) bolster (of a hospital bed).
polyconique, a., polyconic.
polycylindre, m., polycylinder motor car.
polylobé, a., (*aero.*) polylobed.
polyphasé, a., (*elec.*) polyphase.
polyplace, m., (*aero.*) "many seater."
Polytechnicienne, f., (*Fr. a.*) mutual benefit society made up of graduates of the Polytechnique.
pompe, f., pump; (*mil. slang*) shoe; theoretical instruction;
—— *à cylindres indépendants*, double-cylinder pump, duplex pump;
—— *à engrenages*, gear pump;
—— *s funèbres*, funeral;
—— *à gonfler*, bicycle pump;
—— *à graisse*, grease pump (lubricating);
—— *à huile*, oil pump (lubricating);
—— *oscillante à piston*, oscillating piston pump;
—— *à palettes*, vane pump;
—— *à pneu*, tire pump, bicycle pump;
—— *à pression*, pressure pump;
—— *à purin*, farmer's pump, (*purin*, liquid manure).
pondre, v. t., to lay (eggs).
pont, m., bridge; (*auto.*) axle (and related parts; e. g., housing); (*r. r. art.*) girder (transmits weight of gun to two trucks); (*trench art.*) traversing cross piece;
—— *aqueduc*, aqueduct bridge;
—— *arrière*, (*auto.*) rear live axle (of a cardanic transmission);
—— *arrière à cardan transversal*, (*auto.*) rear axle (wheel shafts directly controlled by cardanic transmission);
—— *canal*, aqueduct bridge;
—— *à cinquenelle*, (*pont.*) (bridge held by sheer lines) sheer-line bridge;
—— *de fortune*, (*pont.*) temporary bridge, bridge built on the spot from local resources;
—— *non oscillant*, (*auto.*) dead axle;
—— *oscillant*, (*auto.*) floating live axle;
—— *à péage*, (*pont.*) toll bridge;
—— *de portières*, (*pont.*) bridge of rafts;
—— *rigide*, (*auto.*) (system in which the rear axle and springs form one rigid system) rigid axle;
—— *transbordeur*, (*pont.*) transfer bridge;
—— *volant*, (*r. r.*) any portable bridge for loading and unloading; (*mil.*) small portable bridge for crossing trenches;
—— *volant d'équipage*, (*pont. Fr. a.*) flying bridge (of the bridge train; two supports, each of three pontoons);
ponton-grue, m., floating derrick.
popote, f., (*mil. slang*) officers' field mess.
poquette, f., pigeon pox (pigeons);
port, m., port;
—— *d'atterrissage*, (*aero.*) landing station;
—— *d'attache*, (*aero.*) home station, home port;
—— *double*, (*adm.*) double postage;
—— *d'escale*, (*nav.*) port of call; (*aero.*) stopping place;
—— *fluvial*, river port;
—— *de ravitaillement*, (*Fr. a.*) supply station in a river or canal port;
—— *simple*, (*adm.*) single postage;
—— *à traité*, treaty port (e. g., Chinese).
porte, f., closing plug;
—— *de visite*, (*aero.*) inspection door.
porte-affût, m., (*mach. gun*) tripod support (gun cart).
porte-bagages, m., luggage rack.
porte-bec, m., burner holder (of a lamp).
porte-brancard, m., (*trench art.*) carrying bar sleeve or socket.
porte bretelle, m., (*mach. gun*) sling swivel.
porte-canon, m., (*art.*) gun truck (i. e., transports gun proper).

porte-cartes m., (*aero.*) map holder.
porte-cartouches, m., (*sm. a.*) short cartridge belt or holder.
porte-charge, m., (*sm. a.*) cartridge belt.
porte-contact, m., (*elec.*) contact holder.
porte-dépêche, m., (*mil.*) message tube (carrier pigeon).
porte-détonateur, m., (*tr. art.*) detonator cylinder, tube.
portée, f., reach, stretch; (*mach.*) bearing length (of a trunnion or journal, length of the bearing itself); (*fond.*) core holder, or support; (*telephone, etc.*) span;
—— *aérienne,* air span;
ponter à grande ——, (*pont.*) to bridge with extended bays.
porte-galons, a., (*mil.*) chevron wearing.
porte-grenades, a., (*mil.*) grenade carrying;
ceinturon ——, grenade belt;
panier ——, grenade basket.
porte-lance, m., (*mil.*) nozzle man (of a flame projector);
sapeur ——, nozzle man.
porte-lanterne, m., lamp bracket, holder.
porte-manteaux, m., clothes hook.
porte-matériel, m., (*art.*) carrying bar.
porte-matériels, m., (*art.*) equipment cart.
porte-mines, m., (*torp.*) torpedo planter.
porte-mitrailleuse, m., (*mach. gun*) gun support (of a gun cart).
porte-objet, m., slide (of a projection lantern).
porte-parole, m., spokesman.
porte-phare, m., lamp fork or bracket.
porte-plume, m., penholder;
—— *à réservoir,* fountain pen.
porter, v. t., to carry;
—— *une dépêche,* (*mil. slang*) to have one's horse run away.
porte-relais, m., (*art.*) relay-carrier.
porte-savon, m., soap dish, soap holder.
porte-tourillons, m., (*mach. gun*) trunnion bracket.
porteur, m., bearer, holder; (*mil.*) gabion, sandbag, carrier;
—— *de billet,* (*r. r.*) holder of a ticket.
porte-verre, m., (*inst.*) lens-holder (any rim or border holding a lens or glass).
porte-voix, m., speaking trumpet;
mégaphone, megaphone.
portier-consigne, m., (*Fr. a.*) (formerly, gatekeeper of a fortress, usually a retired n. c. o.); foreman (in certain construction shops or offices); clerk or representative of the engineer department (in *places annexes,* q. v.).
portière, f., door curtain, coach door; (*pont.*) raft;
—— *à coulisse,* sliding door;
—— *double,* double raft;
double —— *double,* two double rafts side by side;
—— *de navigation,* (*mil.*) pontoon raft (part of a bridge);
—— *à rabattement,* folding down door;
—— *de transport,* (*mil.*) pontoon raft, (forms no part of a bridge).
portique, (*siege*) covered portion of a *sape debout,* to protect against enfilade.
port-magasin, m., (*Fr. a.*) depot in a river or canal port.
pose, f., laying;
—— *aérienne,* (*mil. tel., etc.*) laying above ground;
—— *à terre,* (*mil. tel., etc.*) surface laying.
poseur, m., fitter; (*siege*) sandbag layer;
—— *de mines,* (*art., nav.*) mine planter.
position, f., position;
—— *d'attente,* (*art.*) waiting position (**pieces** limbered);
—— *de demi-route,* (*art.*) short travel position (transport to short distances);
—— *de déplacement,* (*art.*) change position, (to deceive the enemy);
fausse ——, (*mil.*) dummy position;

position, *de surveillance,* (*art.*) waiting position (pieces unlimbered, against enemy expected at a certain point); position of expectation;
—— *de vol,* (*aero.*) flying position.
poste, f., post office;
petite, —— (in newspapers) questions and answers, answers to correspondents.
poste, m., post; (*art., sm. a.*) "home" or seat of a projectile or cartridge;
—— *d'armée,* (*mil. teleg.*) (station between lines to the front and corps and division headquarters); army transit station;
—— *d'arrivée,* (*mil. sig., etc.*) receiving station;
—— *avancé,* (*mil. telephone*) advanced telephone station;
—— *de commandement,* (*mil.*) commanding officer's station (especially in trench warfare); command post;
—— *de contrôle,* (*Fr. art.*) checking or comparing station (name given to wireless station of a *groupement;* follows the work of airplanes serving groups and the batteries of a *groupement,* as well as the infantry planes in the zone of the *groupement*);
—— *de départ,* (*mil. sig., etc.*) sending station;
—— *d'eau filtrée,* (*mil.*) drinking fountain in barracks;
—— *d'écoute,* (*mil.*) listening post;
—— *d'écoute téléphonique,* telephonic listening post (has telephonic communication with the rear);
—— *émetteur,* (*tel., etc.*) sending station;
—— *d'entretien,* (*mil.*) repair post (of a bridge; engineers);
—— *à l'épreuve,* (*mil.*) shell-proof post or station;
—— *d'examen,* (*mil.*) examining post (outpost duty);
à —— *fixe,* (*mil.*) stationary;
grand ——, (*mil.*) main, central, listening post;
—— *de guetteur,* (*mil.*) lookout post, sentry post (trenches);
—— *intermédiaire en dérivation* (*embroché*), (*telephone*) shunt (series) way-station;
—— *de jonction,* (*mil. teleg.*) connecting station (between lines to the front and those to the rear);
—— *d'observateur,* (*mil.*) observing-post;
—— *d'observation,* (*mil.*) observing station or post;
—— *optique,* (*mil.*) searchlight station;
—— *de pansement,* (*mil. méd.*) dressing station;
petit ——, (*mil.*) sort of outpost in front of the first trenches; small listening post;
—— *photo-électrique,* (*mil.*) searchlight station (dynamo wagon and projector);
—— *de pigeons,* (*mil.*) carrier pigeon station;
—— *de police,* police station, police headquarters; (*mil.*) (infantry) post guarding a bridge; (*mil. in gen.*) any guard post;
—— *portatif de T.S.F.,* portable wireless set;
—— *de protection,* (*mil.*) (in bridge guarding an upstream) post on the lookout for floating objects (may be downstream occasionally);
—— *du quartier général,* (*mil. teleg.*) headquarters telegraph station;
—— *de raccordement,* (*mil. teleg.*) connecting station (between army lines and army group lines);
—— *de rassemblement,* (*mil.*) alarm post (of a camp);
—— *récepteur,* (*telegraphy, etc.*) receiving station;
—— *de réglage,* (*Fr. art.*) ranging station (name given to wireless receiving station of a battery or group of batteries in communication with an artillery plane);
—— *de secours,* (*mil. méd.*) popularly, first-aid station (v. *refuge de blessés,*) dressing station; (*mil. min.*) rescue station;
—— *de secours régimentaire,* (*mil. méd.*) dressing station (behind regiment);
—— *de surveillance,* (*mil.*) lookout post; (*Fr. art.*) general receiving station (at P. C. of the artillery and of the heavy artillery); makes an effort to take all messages of the corps planes, or at least those on general duty, artillery observation, infantry planes, commanding general's planes;
—— *téléphonique,* (*mil.*) telephone station;
—— *de torpilleurs,* (*art.*) torpedo station, mining casemate;
—— *de transit,* (*mil. tel., etc.*) transmitting station;

poste *de T. S. F. d'aviateur*, (*mil.*) airplane receiving station (wireless).
postillon, m., postillion;
— *à déclenchement*, sliding release.
pot, m., pot;
— *d'échappement*, (*mach.*) muffler, silencer, silence-box (of an explosive engine);
— *d'échappement de vapeur*, (*mach. gun*) steam pot (to conceal steam from the Maxim cooling sleeve);
— *à fumée*, (*artif.*) smoke compartment.
— *Ruggieri*, (*artif.*) Bengal light. (Ruggieri was the great maker of fireworks of the Third Empire.)
pote, m., (*mil. slang*) a good friend.
poteau, m., post; (*mil. slang*) a good friend;
— *frontière*, frontier post;
— *de télégraphe*, etc., telegraph, etc., pole.
potentiel, m., (*expl.*) potential, (quantity of work corresponding to the quantity of heat developed by unit weight).
pou, m., louse;
poudre, f., powder;
— *électrique*, (*expl.*) an explosive containing 33 per cent of nitroglycerin (remaining ingredients unknown);
— *géant*, (*expl.*) giant powder;
— *mica*, (*expl.*) a dynamite of mica base;
— *au picrate* (*expl.*) picrate powder;
— *Vulcan*, (*expl.*) Vulcan powder.
poulet, m., chicken, note;
— *sauté*, (*mil. slang*) note thrown into enemy trenches.
poulie, f., pulley;
— *extensible*, expanding pulley;
— *à joues*, cheek pulley.
pourvoi, m., (*law*) appeal;
— *en cassation*, appeal for reversal of verdict, in error.
pourvoyeur, m., (*mil.*) grenade-carrier (assistant to *lanceur*, q. v.), assures supply of grenades; (*Fr. a.*) feeder, supply man (to a *fusil-mitrailleur*); (in gen.) ammunition server or carrier:
— *en charges*, cartridge server;
— *en projectiles*, projectile server.
poussée, f., (*aero.*) lift; (*mach. gun*) gas-pressure; *centre de* —, center of thrust;
tube de —, (*mach. gun*) pressure reduction tube.
poussette, f., push button, spring button.
poussière, f., dust;
— *d'hommes*, (*mil.*) expression to denote the vague, indistinct appearance of a line of men at a distance.
pousse-pousse, m., jinrikisha; pushcart.
poussoir, m., pusher, push button (in general); (*mach.*) valve lifter, cam follower, plunger, push rod, tappet.
poutre, f., (*cons.*) beam, girder;
— *à âme pleine*, (*cons.*) solid beam, solid web beam;
— *à claire-voie*, (*cons.*) lattice beam;
— *de liaison* (*aero.*) (girder connecting principal cellule, q. v., with the rear of the machine) junction girder;
— *souple*, (*aero.*) flexible beam (dirigibles).
poutre-affût, f., (*art.*) girder-carriage (railroad gun).
poutrelle, f., joist, small beam;
— *de blindage*, (*siege*) blindage beam.
pouvoir, m., power, authority;
— *s publics*, public authority, the law.
pré-allumage, m., premature firing, ignition (motor).
préambule, m., preamble; (*mil.*, etc.) office marks or indications of a telegram.
premier, a., first;
mettre en — *ère*, (*auto.*) to put on first speed.
précision, f., accuracy,
— *en direction*, (*art.*) lateral precision;
— *en portée*, (*art.*) precision in range.

prélèvement, m., exclusion (as, *mil.*) of gases.
prélever, v. t., to exclude (as, *mil.*) gas.
prendre, v. t., to take;
— *sous le feu*, (*art.*) to take under fire.
présentation, f., presentation;
— *du drapeau*, (*mil.*) presentation of colors.
présidence, f., chairmanship of a committee, etc.
érpsident, m., chairman of a committee; speaker of the house, of a parliamentary body;
— *de chambre*, (*Fr. law*) the presiding judge of the chamber or division of a court.
pression, f., (*phys.*) pressure;
chute de —, loss of head, fall of pressure.
prêt, m., (*Fr. a.*) soldier's pay;
— *décadaire*, pay every 10 days.
prêter, v. t., to lend;
— *serment entre les mains de*, to swear before.
prétoire, m., (*mil. law*) court room (as of a court-martial).
prévention, f., preventive confinement;
en — *de conseil de guerre*, (*mil. law*) held by or for a court-martial.
prévision, f., forecast, prediction;
— *locale*, local (weather) forecast;
— *à longue échéance*, distant forecast (weather).
prise, f., purchase, grip;
— *d'air*, (*mach.*) air inlet, air intake;
— *de commandement*, (*mil.*) assumption, taking over of a command;
— *de courant*, (*elec.*) jack;
— *directe*, (*mach.*) direct transmission, direct connection;
en —, (*mach.*) engaging with, gearing with;
en — *directe*, (*mach.*) direct-connected;
— *de jour*, (*mil.*, etc.) opening to the air, as in a covered trench;
— *de possession de commandement*, (*mil.*) assumption of command.
— *de terre* (*elec.*) ground, ground connection;
prisme, m., (*opt.*) prism;
— *bi-réfringent* (*inst.*) double refraction prism;
— *déviateur*, (*inst.*) refracting, deflecting, prism (range finder);
— *d'extrémité droite* (*gauche*) (*inst.*) right (left) end prism (of a range finder);
— *de grossissement*, magnifying prism;
— *de repérage*, (*art.*) registration prism;
— *séparateur*, (*inst.*) separating prism (range finder);
— *à toiture*, blunted triangular prism.
prisme-objectif, m., (*opt.*) objective prism.
proboche, a., (*fam.*) German sympathizer.
procès-verbal, m., minutes.
procureur, m., (*law*) attorney;
— *de la République*, (*France*) solicitor general.
produit, m., product;
— *accessoire*, (*chem.*, etc.) by-product.
profil, m., (*aero.*) fair.
profilage, m., (*aero.*) stream lining.
profilé, a., (*aero.*) stream lined; faired.
profondeur, depth;
— *d'une surface*, (*aero.*) length of the chord (wing).
progresser, v. n., (*mil.*) to advance (trench fighting).
progression, f., (*mil.*) advance (trench fighting).
projecteur, m., searchlight;
— *de flammes*, (*mil.*) flame projector;
— *portatif*, (*mil.*) portable searchlight, hand searchlight; hand set;
— *de signalisation*, (*sig.*) Morse lamp.
projectile, m., (*art.*, *sm. a.*) projectile;
— *asphyxiant*, (*tr. art.*) asphyxiating, gas, projectile;
— *à calotte d'acier*, (*sm. a.*) steel-capped bullet;
— *coiffé*, (*art.*) capped projectile;
— *d'exercice*, (*art.*) drill, dummy projectile;
— *lumineux*, (*art.*) tracer shell or projectile;
— *lacrymogène*, (*art.*) tear-producing projectile, lachrymatory shell;

projectile *porte-message*, (*mil.*) message-shell;
— *suffocant*, (*art.*) suffocating projectile;
— *toxique*, (*art.*) poisonous projectile.
projection, f., projection;
— *de gaz à l'arrière*, (*art.*) flareback.
projet, m., project;
— *de budget*, (*adm.*) estimate.
prolonge, f., (*mil.*) service wagon, escort wagon (U. S. A.);
— *à couvercle*, covered wagon;
— *d'outils*, (*eng.*) tool wagon.
promenade, f., promenade;
— *militaire*, (*mil.*) an easy affair.
promenoir, m., promenade; (in pigeon lofts) space for exercise.
promotion, f., promotion; (*Fr. a.*) class (at a school);
baptême de ——, (*Fr. a.*) sort of festival on the occasion of graduation;
dîner de ——, class dinner.
prononcer, v. t., to pronounce;
— *une attaque* (*contre-attaque*); (*mil.*) to develop an attack (a counterattack).
proposable, a., (*mil.*) liable, or fit, to be proposed for promotion, etc.
propulseur, m., propeller (not necessarily a screw);
— *à hélice*, screw propeller;
— *à roue*, paddle wheel.
protecteur, m., (*auto.*) protecting cover (pneumatic tire; said especially of non-skidding covers).
protection, f., protection;
— *collective*, joint protection of a number of people (e. g., *mil.*, against gas attack).
protocolaire, a., relating to the *protocole*.

protocole, m., protocol;
service du ——, (*Fr. adm.*) a bureau or branch of foreign affairs, dealing with a great variety of subjects, such as ceremonial, precedence, reception of ambassadors; credentials and other questions relating to ambassadors, foreign consuls, etc.; the protocol.
pruneau, m., prune; (*mil. slang*) bullet.
puissance, m., power; (powerful) state, power;
— *ascensionelle*, (*aero.*) lifting power;
— *commerciale*, available power, working power;
— *au frein*, (*mach.*) brake horsepower;
— *lumineuse*, (*opt.*) candlepower;
— *massique*, (*mach.*) power per weight (of a motor, ratio of effective power to weight, power per kilogram).
puits, m., well; (*mil.*) (shaft to shelter a machine gun) machine-gun shaft;
— *à cadres coffrants*. (*mil. min.*) cased shaft;
— *sans coffrage*, (*mil. min.*) unlined shaft.
pulsateur, m., (*mach.*) oil (lubricating) pump, pulsator.
pulvérisateur, m., atomizer, sprayer.
pulvérisation, f., atomizing, spraying.
pulvériser, v. t., to atomize, spray.
punch, m., (*mil., fam.*) sort of reception.
pur, a., clear (of the sky);
— *de garance*, v. s. v., *garance*.
pureté, clearness;
— *du ciel*, clearness of the sky.
purgé, p. p. of *purger*;
— *d'air*, with air exhausted.
pylône, m., (*aero., wireless*) pylone;
se mettre en ——, (*aero.*) to land on the nose and remain tail high.

Q.

quadrilatère, m., quadrilateral;
— *articulé*, (*auto.*) (the) linked quadrilateral (formed by the steering axle, bar and pivot arms of a motor car).
quadrillé, a., squared, in squares (as, a map), graticulated.
quadruplice, f., quadruple alliance.
quai, m., quay;
quai d'Orsay, (quay in Paris, by metonymy) the Ministry of Foreign Affairs.
quartier, m., quarter; (*mil.*) the terrain held by a battalion;
donner ——, to turn over (as, a stone, a squared log so that it lies on the side adjacent to the one on which it has been laying).

quatre, m., four;
à droite (*gauche*) *par* ——, (*mil.*) fours (squads, U. S. A.), right (left);
quatre, *en avant par* ——, (*mil.*) right by fours (squads, U. S. A.).
queue, f., tail; (*aero.*) tail; (*art.*) tail (of a mortar bomb, trench artillery);
faire tête à ——, v. s. v. *tête*;
— *de soupape*, (*mach.*) valve stem, rod.
quinzaine, f., fortnight.
quitte, a., (*adm.*) quit (said of an official whose accounts balance, and are found correct).
quitus, m., (*adm.*) official notification that an account has been examined, found correct, and closed; receipt in full, discharge from responsibility.

R.

rabattement, m., bending, folding down or back;
à ——, folding, folding down, folding over.
rabattre, v. t., to revolve (as, a plane, a mechanism, etc.).
rabiau, m., remainder, surplus; "butt" (U. S. A.); (*mil.*) overtime (to make good time lost, e. g., by absence without leave); leavings (after issue of rations); petty theft of rations.
rabiautage, m., (*mil.*) taking the leavings, the "butt."
rabiauter, v. t., (*mil.*) to take the leavings.
rabiauteur, m., (*mil.*) man who takes the leavings.
rabiot, m., v. *rabiau*.
rabiotage, m., v. *rabiautage*.
rabioter, v. t., *rabiauter*.
rabioteur, m., v. *rabiauteur*.

raccord, m., (any piece used to join or connect other pieces or parts) connection; (*expl.*) junction (strands of melinite fuse end to end);
— *chargeur*, (*elec.*) joint or connection (for tapping a current);
— *en équerre*, elbow, elbow joint (tubes, pipes);
— *en I*, I-joint (tubes, pipes).
raccordement, m., coincidence (of images, in certain range finders).
raccord-relais, m., (*mil. min.*) relay-joint.
race, f., race;
— *chevaline*, (*hipp.*) breed of horses, special strain.
raclette, f., scraper.
raclure, f., scrapings;
— *de brique*, brick dust.
radeau, m., raft;
— *de sacs*, (*mil.*) raft of sacks.

radeau-masque, m., *(siege)* raft-cover (passage of a wet ditch).
radeau-sac, m., *(mil.)* raft float.
radiateur, m., *(mach.)* radiator; (in various relations); *(mach. gun.)* radiator, cooling jacket;
—— *à ailettes,* flanged radiator;
—— *cellulaire,* cellular radiator;
—— *multi-cellulaire,* v. —— *cellulaire;*
—— *nid d'abeilles,* —— *à nid d'abeilles,* honeycomb radiator;
—— *soufflé,* fan radiator;
—— *tubulaire,* tubular radiator.
radio, f., radio (telegraphy), wireless telegraphy.
radio-aérien, a., *(elec.)* radio-aerial.
radio-conducteur, m., *(wireless)* coherer.
radiogonomètre, m., *(wireless)* radiogonometer (for controlling the direction of sending).
radiogramme, m., wireless telegram.
radiotélégramme, m., wireless dispatch.
radiotélégraphie, f., radio telegraphy, wireless telegraphy.
radiotéléphonie, f., *(elec.)* wireless telephony.
rafale, f., gust; *(art.)* sudden sweep of shrapnel (sometimes shell) fire; rafale (may consist of only two shots); *(mach. gun, inf.)* gust of fire; *tir par* ——*s, (art., mach. gun)* fire by gusts, by bursts.
rafia, m., raffia, raffia fiber *(Raphia Ruphia* palm; used for withes in farming; also written *raphia*).
raffinerie, f., sugar refinery.
rafraîchir, v. t., to cool;
—— *les cheveux,* to trim the hair.
raid, m., *(auto.)* trip.
raide, a., *(hipp.)* stiff, sore, as, from overwork.
rail, m., rail;
—— *d'atterissage, (aero.)* landing skid, skid;
cœur de ——*, (r. r.)* frog.
—— *de raccord, (r. r.)* junction rail (switches);
—— *de roulement, (art.)* roller way.
rainure, f., guide slot.
rainure-guide, f., guide slot.
raison, f., reason;
—— *commerciale, (com.)* name, firm name.
rajeunir, v. t. to rejuvenate;
—— *les cadres,* v. s. v. *cadre.*
rajeunissement, m., rejuvenation;
—— *des cadres,* v. s. v. *cadre.*
ralenti, p. p. of *ralentir.*
au ——*, (auto.)* at low speed.
ralentir, v. t. n. r., to slow down.
ralentisseur, m., *(mach.)* speed-reducing device, regulator.
ralingue, f., *(cord.)* bolt rope; *(aero.)* strips of strong sail cloth along the sides of a dirigible, to which the suspending cords are attached.
rallonge, f., *(art.)* extension piece (Schneider Canet, allowing aiming above shield);
—— *mobile de pivot, (mach. gun)* pivot extension, (device to increase range of elevation and depression, mod. 1907 T).
ramasse-miettes, m., *(mil. slang)* litter-bearer.
ramasser, v. t., to pick up;
—— *une pelle, (mil. slang)* to fall, or be thrown, from a horse.
rame, f., oar;
border les ——*s,* to unship the oars;
coucher les ——*s,* to stow the oars (lay them in the bottom of the boat);
dresser les ——*s,* to up oars;
haut les ——*s,* oars! way enough!
rameau, m., branch; *(mil. min.)* branch gallery;
changement de ——*, (mil. min.)* passage to a small branch (from a larger);
—— *à châssis coffrants, (mil. min.)* sheathed branch.
ramer, v. n., to row;
se préparer à ——*,* to ship the oars;
préparez-vous à ——*,* let fall!

rampe, f., slope; ramp; bank of lights; border of the stage;
—— *d'atterrissage, (aero.)* landing lights;
feux de la ——*,* footlights;
—— *de graissage, (mach.)* oil distributor;
—— *lumineuse, (aero.)* row of lights (on an airplane);
—— *de service, (pont., etc.)* gangplank.
rancio, m., *(mil. slang)* meal, same as *soupe,* q. v.
randonnée, f., trip.
rang, m., *(mil.)* file (from the point of view of promotion);
de ——*, (mil.)* belonging to, or serving in the ranks;
à —— *serré, (mil.)* in close order.
ranger, v. t., to put on one side, to get out of the way;
se ——*,* of a vehicle, etc., to get out of the way of a following vehicle.
rapace, m., bird of prey.
râpe, f., "scraper" (said, *auto,* of certain very rough roads).
rappel, m., recall, return;
—— *de soupape, (mach.)* return of a valve (to its seat).
rappeler, v. t., *(mach.)* to bring back, (as to a former or initial position); to cause to bear.
rapport, m., *(Fr. a.)* formal, detailed, written report;
—— *général,* general report consolidating the various ministerial budgets (France);
—— *spécial,* the report made on the budget of each ministry (France, v. s. v., *rapporteur*).
rapporteur, m., (France) the member of the appropriations committee of the Chamber of Deputies, reporting on the budget of each department (ministry) of the Government; in the Senate, the member reporting the general budget (only one is appointed);
—— *général,* the member consolidating the reports of the ministerial budgets into one general report.
ras, a., close, smooth;
au —— *du sol,* along the surface of the ground (and very little above it).
rase-mottes, m., *(aero. slang)* grass-cutting.
rata, m., *(mil. slang)* stew.
raté, m.,*(mach.)* misfire; *(art., sm. a.)* misfire, failure;
—— *de,* failure to, failure of, (e. g., —— *d'éclatement,* failure to burst);
—— *d'allumage, de moteur,* misfire.
ration, f., *(mil.)* allowance (of food, forage, fuel, etc.)
—— *de chauffage,* fuel allowance;
—— *de fer,* emergency ration;
—— *de vert,* allowance of green fodder.
rattrapage, m., *(mach.)* take up, (as of the wear of a mechanism);
—— *de jeu,* take up (of wear).
rattraper, v. t., *(mach.)* to take up (as wear of a mechanism);
rave, f., rape (sort of turnip).
ravitaillement, m., *(auto.)* repair and supply station; *(mil.)* refilling, replenishment;
—— *par échange de voitures (art.)* refilling by exchange of vehicles (loaded for empty);
exercice de ——*, (mil.)* maneuvers to train farmers and merchants of locality in the business of supplying armies from local resources;
—— *par transbordement, (art.)* refilling by transfer (of ammunition from the supply train to empty vehicles).
ravitailleur, m., *(mil.)* man who brings up ammunition; server.
rayer, v. t., to strike off;
—— *des contrôles, (mil.)* to take or remove a name from the army list (not necessarily to dismiss).
razzier, v. a., to make a *razzia,* a raid.
réactance, f., *(elec.)* reactance;
bobine de ——*, (elec.)* choke coil.
réarmer, v. t., *(mil.)* to rearm, to reset, (as a fuse, a firing mechanism).

rebroussement, m., turning back; (in roads) Y.
rebut, m., rejection; *(mach. gun)* breech gauge;
 tomber en ——, (of a letter) to go to the dead-letter office.
recalage, m., *(mach., etc.)* reseating, wedging, keying anew.
recaler, v. t., *(mach.)* to reseat, key, wedge anew;
 —— *un candidat,* *(fam.)* to reject a candidate, to "find" (U. S. M. A.).
récepteur, a., m., *(elec.)* receiving; receiver;
 —— *téléphonique,* *(wireless, etc.)* telephone receiver.
réception, f., reception, "at home"; *(wireless)* receiving, receiving set;
 —— *d'adieux,* *(mil.)* farewell reception;
 —— *directe,* *(wireless)* direct connected, direct coupled, receiving; direct-connected receiving set;
 —— *indirecte (par dérivation)* *(wireless)* shunt-coupled receiving; shunt-coupled receiving set;
 —— *indirecte, (par induction)* *(wireless),* inductively connected, coupled, receiving; inductively connected receiving set.
réceptrice, f., *(elec.)* driven dynamo, motor.
receveur, m., *(adm.)* receiver, collector;
 —— *des Domaines,* *(Fr. adm.)* receiver of public moneys.
recevoir, v. t., to receive (company), to be at home.
rechapage, m., *(auto.)* reshoeing a tire.
réchaud, m., heater;
 —— *à gaz,* gas heater.
réchauffeur, m., heater, feed heater.
récidiviste, m., *(mil.)* repeater, (as, a repeated deserter).
recoupement, m., *(mil. min.)* intersection (in sound detection).
rectificateur, m., *(elec.)* rectifier.
rectificatif, m., correction sheet (as, *mil.,* to regulations, drill books, etc.).
rectification, f., *(elec.)* rectification.
rectifier, v. t., *(elec.)* to rectify.
recul, m., slip of the screw;
 —— *absolu,* the linear slip, true slip;
 —— *sur l'affût,* *(art.)* carriage recoil;
 —— *relatif,* ratio of linear slip to the theoretical advance of the screw.
récupérateur, m., *(art.)* recuperator;
 —— *à air,* air recuperator;
 —— *à ressorts métalliques,* return springs.
rédaction, f., preparation, composition;
 —— *convenue,* *(mil.)* cipher agreed upon; adopted cipher.
redan, m., *(aero.)* step (in under surface of a hydroairplane float).
rédiger, v. t., to draw up, prepare (as, *mil.,*) orders.
redoubler, v. t., to repeat a course of study (after a failure to pass); to become a "turn-back," (U. S. M. A.)
redressement, m., *(aero.)* straightening up, recovery;
 manœuvre du —— *au pied,* *(aero.)* reversal of the rudder (banking).
redresser, v. t., to line up (as an engine); *(aero.)* to straighten up.
redresseur, m., *(elec.)* rectifier.
réducteur, m., *(elec.)* reducer;
 —— *à plots,* contact reducer;
 —— *de sensibilité,* *(expl.)* desensitizer.
referrer, v. a., to re-shoe a horse.
refouloir, m., *(art.)* rammer;
 —— *à chaîne semi-rigide,* *(art.)* chain rammer.
refroidi, a., cooled;
 —— *à l'air,* air-cooled;
 —— *à l'eau,* water-cooled.
refroidissement, m., cooling;
 —— *d'air,* air-cooling;
 —— *d'eau,* water-cooling.

refroidisseur, m., *(mach.)* cooling apparatus, device, a synonym of *radiateur,* q. v.;
 —— *cloisonné,* *(mach.)* sectional cooler.
refuge, m., refuge;
 —— *de blessés,* *(mil. med.)* first aid station (immediately behind the troops; sometimes inaccurately called *poste de secours*).
refus, m., *(man.)* refusal (of a jump; horse).
refuser, v. t., *(man.)* to refuse (a jump; horse).
regagner, v. a., to return to (a place); to rejoin.
regard, m., manhole, inspection hole; *(mach.)* a synonym of *compte-gouttes,* q. v.;
 —— *d'égout,* manhole (in a street).
régime, m., rule, system; normal performance;
 —— *du tir,* *(art.)* management, regulation of fire, (e. g., in a *barrage*);
 zone du —— *régulier,* *(aero.)* stream-line flow.
régiment, m., regiment;
 —— *attelé,* *(art.)* regiment serving horsed pieces;
 —— *de France,* *(Fr. a.)* home regiment (as distinguished from colonial or African);
 —— *à tracteurs,* *(art.)* regiment of tractor artillery;
 —— -*squelette,* skeleton regiment.
registre, m., damper;
 —— *de batterie,* *(art.)* battery firing record.
 —— *de prise de vapeur,* *(steam)* intake.
réglable, a., adjustable.
réglage, m., adjustment; *(mach., aero., etc.)* alignment;
 —— *aérien,* *(art., aero.)* ranging by aerial observation;
 —— *d'un appareil,* *(aero.)* lining up;
 coefficient de ——, *(art.)* 1° the ratio of the ranging to the topographic distance (of an auxiliary target, before opening on the real target); 2° ratio of the ranging distance, corrected for atmospheric conditions, to the topographic distance; range-correction factor;
 distance de ——, *(art.)* observed range;
 —— *de frein,* *(mach.)* brake adjustment (take up of wear);
 —— *des hauteurs d'éclatement,* *(art.)* adjustment, determination of the height of burst;
 —— *horizontal,* *(art.)* horizontal adjustment (range finder);
 —— *de l'incidence,* *(aero.)* wash in, wash out;
 —— *percutant en portée,* *(art.)* ranging by percussion;
 —— *terrestre,* *(art.)* ranging by terrestrial observation (usual case, before advent of aeronautics);
 —— *vertical,* vertical adjustment (range finder).
règle, f., rule;
 —— *à calculs,* slide rule;
 —— *de repérage,* *(trench art.)* registration scale.
règlementer, v. t., *(mil.)* to make regulations for.
régler, v. t., *(mach., aero., etc.)* to align.
règlette, f., small rule; scale;
 —— *but,* *(aero.)* enemy speed bar;
 —— *de contre pente,* *(Fr. art.)* counterslope scale (for fire on reverse slope);
 —— *de dérive,* *(aero.)* drift bar;
 —— *de direction,* *(art.)* direction scale;
 —— *de hausse,* *(art.)* elevation scale;
 —— *de mire,* *(trench art.)* sighting alidade.
régloir, m., *(art.)* fuze setter;
 —— *mécanique des fusées,* *(art.)* mechanical fuze setter.
régularité, f., *(auto.)* excellence in all respects, regularity of performance.
régulateur, m., *(mach.)* governor; *(auto.)* speed regulator;
 —— *d'air,* air regulator (of automatic carbureters);
 —— *d'alimentation,* feed control;
 —— *à boules,* ball governor (not necessarily limited to steam);
 —— *centrifuge,* centrifugal governor, regulator;
 —— *de pression,* reducing valve;
 —— *de vitesse,* speed regulator.

régulation, f., regulation;
—— *sur l'allumage par avance (retard)*, (*mach.*) regulation on advance (delay) ignition;
—— *sur l'avance à l'allumage*, regulation on advance ignition;
—— *sur l'échappement*, regulation on exhaust (preventing exhaust);
—— *par étranglement des gaz*, throttle control or regulation, throttling;
—— *sur le retard à l'allumage*, regulation on delay of ignition.

régulier, a., (*auto.*, etc.) regular and steady performance in all respects.

réintégration, f., reinstatement; (more especially, *mil.*), passage of an officer from duty without, to duty with, troops (from *hors cadre* to *cadres*); (*mil.*) return to one's own lines.

réintégrer, v. t., to reinstate; (more especially, *mil.*) to transfer an officer from the status *hors cadre* to that of duty with troops; to return to; get back to;
—— *les lignes*, (*mil.*) to reach, return to, one's own lines, to such and such lines.

relâche, f., (at theaters) no play, no production.

relais, m., relay;
—— *à âme*, (*art.*) cored relay;
—— *de détonation*, (*mil. min.*) relayed detonation;
—— *d'essence*, (*auto.*) relay of gasoline.

relevé, m., abstract, statement, report (in general); *faire le* —— *de*, to plot (upon a map).

relèvement, m., (*mil.*) collecting (of the wounded, after a battle).

relever, v. t., (*mil.*) to pick up, collect (the wounded, after a battle).

relier, v. t., to bind (books); to join;
se ——, (*mil.*, etc.) to get in touch with.

reliquat, m., (*adm.*) unexpended balance.

remercier, v. t., to discharge (from a position, office, etc.).

remettre, v. t., to remit, hand over, deliver;
—— *une décoration*, to deliver a decoration.

rémige, f., remex, wing feather;
——*s bâtardes*, alula, bastard wing;
——*s primaires*, primaries;
——*s secondaires*, secondaries.

remise, f., delivery;
—— *d'une décoration*, delivery of a decoration.

remonte, f., (*mil.*) remount station;
annexe de ——, (*Fr. a.*) annex of a remount depot.

remontoir, m., (*mil. slang*) soldier of the remount troops.

remorque, f., (*auto.*) trailer;
—— *aux agrès*, (*Fr. a. aero.*) kite equipment trailer;
—— *cuisine*, kitchen trailer;
—— *à havresacs*, (*mil. aero.*) knapsack trailer;
—— *à tubes*, (*Fr. a. aero.*) gas trailer.

remorqueur, m., (*auto.*) tractor.

remous, m., (*aero.*) aerial eddy; rocking;
—— *de l'hélice*, (*aero.*) race.

rendement, m., performance, duty; (of vehicles) ratio of useful to total load;
—— *d'appropriation*, (of a propeller, etc.) performance, duty, under service conditions; e. g., when mounted on an airplane;
—— *de construction*, (of a propeller, etc.) shop duty, fixed-test duty.

rendre, v. t., to turn in (as tolls, etc.).

renflouement, m., (*aero.*) inflation; replenishment (of the gas of a balloon after ascension), partial inflation.

renflouer, v. t., (*aero.*) to inflate;
—— *un ballon*, (*aero.*) to replenish (the gas of a balloon, after each ascension), to inflate partially.

reniflard, m., (*mach.*) relief valve.

rentrée, f., return; return of Congress, deputies, etc., after holidays; (of schools) opening;
—— *d'air*, (*aero.*) penetration of air into balloon.

renversable, a., reversible (as a telescope in its bearings).

renversement, m., upsetting; (*aero.*) overturn on the wing;
à ——, letting down.

renverser, v. t., to upset;
se ——, (*art.*) (of a projectile) to tumble.

renvoi, m., return, dismissal;
—— *à, de, sonnette*, (*mach.*) bell crank, bell-crank transmission.

renvoyer, v. t., to send back;
—— *des fins d'une plainte*, (*law*) to acquit.

repas, m., meal;
—— *administratif*, (*mil.*) meal (of fixed type) given in a railway infirmary;
—— *de corps*, (*mil.*) regimental dinner.

repassage, m., razor stropping, stropping.

repasser, v. t., to strop a razor.

repasseur, m., razor strop.

repêchage, m., (*Fr. a.*) inscription of a name on a *tableau d'avancement*, q. v., in year preceding retirement for age.

repérage, m., (*mach.*) adjustment (of gear, etc.); (*art.*) registering;
—— *à la ficelle*, (*art.*) plumb-line registration;
—— *au canon*, (*art.*) registration fire;
—— *par les lueurs*, (*art.*) laying, determination of the range of a target, by the flash of discharge; flash registration;
—— *par le son*, (*art.*) laying, determination of the range, of a target, by the sound of discharge; sound registration;
section de ——, (*art.*) registration detachment.

repère, f., bench-mark;
—— *de pointage*, (*art.*) point, mark, used in laying; registration point.

repérer, v. t., (*mach.*) to adjust (gear, etc.); (*art.*) to register;
—— *au canon*, (*art.*) to fire for registration;
—— *à la ficelle*, (*art.*) to register by plumb-line.

repéreur-stadia, m., (*art.*) range finder.

répertoire, m., repertory;
—— *de concentration*, (*art.*) table showing for each square of a *plan directeur* the batteries that can play upon it; battery reference table.

répétition, f., (*mus.*) practice.

replier, v. t., (*tel.*) to take up a line.

repointage, m., (*art.*) reaiming.

repointer, v. a., (*art.*) to reaim.

report, m., (*com.*) "brought forward," in an account, bill, statement, etc.

reporter, v. t., (*com.*) to bring forward.

reprise, f., resumption of a play (at a theater); mending, repairing;
—— *d'un moteur*, pick up;
—— *de tir*, (*art.*) period of fire.

repriser, v. t., to mend (clothes).

réquisitionner, v. t., to requisition;
—— *des troupes*, (*mil.*) to call out troops.

réseau, m., network of telegraph and telephone lines; (*mil.*) wire entanglement;
—— *d'armée*, (*mil. telephone*) telephone system of an army;
—— *des armées*, (*Fr. a.*, *r. r.*) all railroads under the orders of the commander in chief (this *réseau* is separated from that of the interior by the *ligne de démarcation*); army system;
—— *d'artillerie*, (*mil. telephone*) artillery telephone system of an army corps (connects the artillery commander with groups, and thence with infantry units supported by the groups);
—— *bas*, (*mil.*) low wire entanglement (30 cm. high);
—— *Brun*, (*Fr. a.*) portable wire entanglements, in coils;
—— *de commandement*, (*mil. telephone*) army corps telephone system (from commanding general down to the battalions; connects infantry units with supporting artillery);

éseau, commission de ——, (Fr. a., r. r.) line commission; (is assisted in war by a *sous-commission*);
—— *de corps d'armée,* (*mil. telephone*) army corps telephone system;
—— *d'ensemble,* (*mil. telephone*) telephone system (artillery);
—— *général de commandement,* (*mil. telephone*) telephone system (forms part of the réseau d'armée); connecting the commanding general with stations as far down as the *postes de commandement* of the various corps; headquarters telephone;
—— *général de tir,* (*mil. telephone*) artillery telephone system (so designed as to permit batteries to use a great number of observation stations without being directly connected to them);
—— *de l'intérieur,* (Fr. a., r. r.) all railways under the authority of the minister of war (v. *ligne de démarcation,* —— *des armées, zone des armées,* and *zone de l'intérieur*);
—— *particulier de l'artillerie* (*mil. telephone*) artillery telephone system (artillery command and units);
—— *au ras du sol,* (*mil.*) ground entanglement (about 1 foot high);
—— *de réglage (général),* (*art.*) network of observation points and connections;
—— *de secours,* (*mil. telephone*) relief system, duplicate system (e. g., the *réseau général* may be used in case a *réseau particulier* should break down);
sous-commission de ——, v. *commission de* ——;
—— *téléphonique,* telephone system;
—— *téléphonique d'observation et de réglage,* (*mil. telephone*) telephone fire observation and control system;
—— *en treillage métallique,* fence-wire entanglement.
réservé, a., (*mil. telephone*) private.
réservoir, m., (*inst.*) bulb (as of a thermometer, etc.);
—— *en charge,* (*mach.*) gravity tank;
—— *d'huile,* (*mach.*) oil cup;
—— *sous pression,* (*mach.*) pressure tank.
résiliable, a., (*adm., law*) revocable, subject to cancellation.
résilier, v. a., (*law, adm.*) to revoke;
résistance, f., resistance;
—— *à l'avancement,* (*aero.*) drag, passive resistance;
——s *passives,* (*aero.*) body resistance, head resistance;
——s *parasites,* (*aero.*) head resistance;
—— *de réglage,* (*elec.*) rheostat resistance;
—— *totale à l'avancement,* (*aero.*) drag.
résonance, f., resonance: (*elec.*) resonance, tuning;
—— *aiguë,* sharp tuning;
en ——, tuned.
ressaisir, v. t., to seize again;
se ——, to pull one's self together (e. g., *mil.,* the enemy, after a surprise).
ressaut, m., jog, shoulder; (*mil. min.*) abrupt change of level (in a gallery).
ressort, m., spring;
—— *d'appui de chargeur,* (*mach. gun*) magazine stop spring;
—— *d'armement,* (*art.*) cocking spring (trench mortar);
—— *amortisseur,* buffer spring; shock absorbing spring;
—— *cantilever,* the —— *à crosse,* upside down;
—— *de chargeur,* (*mach. gun*) magazine spring;
—— *à crosse,* three-quarter elliptic spring, crutch spring;
—— *d'écartement,* (*mach. gun*) spreader;
—— *d'embrayage,* (*mach.*) clutch spring;
—— *de levier arrêtoir de canon,* (*mach. gun*) barrel catch spring;
—— *longitudinal,* side spring of a carriage, of a motor car;
—— *de percussion,* (*mil.*) percussion spring of a grenade;
—— *à portière,* sort of bow spring;

ressort *de rebondissement,* (*art.*) striker return spring;
—— *récupérateur,* (*mach. gun*) recoil spring;
—— *de récupérateur,* (*art.*) recuperator spring;
—— *de récupération du canon,* (*mach. gun*) recoil spring;
—— *de récupération de culasse mobile* (*mach. gun*) main spring;
—— *de sécurité,* (*sm. a., etc.*) safety spring;
—— *transversal,* cross spring, (of a carriage, motor car);
—— *trembleur,* (*elec.*) make and break device or contact, (regarded as a spring).
retard, m., (*steam, mach.*) lag;
—— *à la fermeture de l'admission,* closing lag;
—— *à la fermeture de l'échappement,* exhaust (lag);
—— *à l'ouverture de l'admission,* admission lag.
retardataire, m., (*mil.*) laggard, (man who hangs back in an attack or assault).
retenue, f., confinement;
être en ——, to be confined to quarters.
réticule, m., graticule.
retombée, f., (*mil.*) fall of bullets (e. g., those fired at air planes).
retour, m., return;
—— *d'allumage,* (*mach.*) back-fire;
—— *de flamme,* (*mach.*) back-fire;
—— *de manivelle,* kick, reverse stroke of crank;
—— *du moteur,* (*mach.*) back-fire, kick, reverse stroke;
—— *offensif,* (*mil.*) (offensive return, attack to recover a position taken by the enemy, counter thrust;
—— *par la terre,* (*elec.*) ground return.
retourné, a., (*auto.*) turned end for end (in races)
retourner, v. t., to turn;
—— *une tranchée,* (*mil.*) to turn a (captured) trench against its original occupants (the enemy).
retrait, m., contraction; recess;
—— *à mortier,* (*art.*) mortar recess (in a trench).
retraite, f., (*mil.*) retirement;
—— *anticipée,* (*mil.*) premature retirement, retirement in advance of statutory date.
réunion, f., reunion;
—— *des officiers,* (*mil.*) officers' club.
revanche, f., revenge; (more specially sometimes, and used without qualification), revenge for 1870-1871.
revernir, v. a., to revarnish.
revers, m., (*mil.*) (side of the trench opposite to the parapet) reverse; edge or border of the counterscarp.
revêtement, m., (*mil.*) revetment;
—— *en planches,* plank revetment;
—— *en treillis métallique,* wire revetment.
revirement, m., reversal.
revue, f., (*mil.*) review;
—— *d'adieux,* (*mil.*) farewell review or parade (case of a c. o. leaving the corps, of general going on retired list, etc.);
—— *d'incorporation,* (Fr. a.) inspection of recruits, after joining;
passer la —— *de,* to inspect, examine;
passer une ——, (*mil.*) to inspect.
rhigolène, m., (*chem.*) rhigolene.
riciné, a., treated, coated, with castor oil.
ride, f., wrinkle;
—— *de terrain,* (*top.*) fold of the ground.
rideau, m., curtain;
—— *ondulé,* (*cons.*) corrugated screen.
rigolo, m., (*slang*) revolver.
rince-boches, m., (*mil. slang*) the bayonet.
riverain, a., m., riparian, bordering on (a river, sea, lake, road, forest, etc.);
état ——, bordering state, state bordering on a sea, ocean, etc.
robinet m., plug, faucet;
—— *de décompression,* (*mach.*) relief valve, relief cock;

robinet à *deux branches*, (*mach.*) two-way cock;
— à *entonnoir*, v. —— *de décompression*;
— *des gaz*, (*mach.*) throttle;
— *pétroleur*, (*mach.*) oil cock;
— à *pointeau*, plug cock;
— *réducteur*, reducing valve;
— *de refoulement*, feed cock, delivery cock;
— *de sûreté*, (*expl.*) safety cock;
— *vanne*, (*mach.*) throttle valve.

rodage, m., grinding (as of a valve seat).

roder, v. t., to grind (as a valve seat).

rompre, v. n., (*mil.*) to break up, (as a camp) and set out.

rondelle, f., joint ring, strengthening ring;
— *d'appui*, (*trench art.*) stop washer;
— *de bourrage*, (*art.*) wad (trench mortar);
— *de bout d'essieu*, linch-hoop;
— *de calage*, (*trench art.*) lock washer;
— *écrou*, (*mach.*) screw washer.

rondelle-tendeur, f., spring cap, spring compressor.

rongeur, m., rodent.

roquade, f., (*mil.*) shift of troops back and forth.

Rosalie, f., (proper name: in *mil. slang*) the bayonet.

rose, f., rose, card;
— *de la fréquence des vents*, diagram of prevailing winds.

rosette, f., the rosette or button of the grade of officer (Legion of Honor, Instruction Publique);
— *rouge*, the rosette of the Legion of Honor (grade of officer);
— *violette*, the rosette of *officier de l'instruction publique*.

rotin, m., rattan, supple jack.

rotule, f., (*mach.*) ball and socket joint or articulation; knuckle;
— *de direction*, steering knuckle.

roue, f., wheel;
— *arrière*, stern wheel (river boat);
— *caoutchoutée*, rubber-tired wheel (especially solid tired);
— *de chaîne*, sprocket wheel, chain wheel;
— *s directrices*, (*auto.*) front wheels, steering wheels;
— *élastique*, any elastic wheel (i. e., producing resiliency without using pneumatic tires);
— *motrice*, driving wheel;
— *porteuse*, bearing wheel;
— à *rayons métalliques*, bicycle wheel (i. e., wheel of bicycle type);
— à *ressorts*, elastic wheel (using springs to deaden shock);
— *train de* ——s, train of wheels.

rouet, m., reel, winch.

roue-turbine, f., (*mach.*) turbine, turbine wheel.

rouge, a., m., red;
porter au ——, (*met.*) to heat red hot.

rouget, m., (sort of) tick.

roulage, m., transportation;
épreuve de ——, wheel or transportation test.

rouleau, m., eye (of a carriage spring); roller;
— *concasseur*, rock crusher;
— à *jour*, open-work roll, cylinder, or barrel.

roulement, m., (*mach.*) rolling (i. e., bearing) surface; (roller) bearing;
— *annulaire*, annular ball bearing;
— à *billes*, ball bearing;
point de ——, (*mach.*) bearing, journal bearing.

rouler, v. n. t., to roll;
— *au moteur*, (*aero.*) to taxi.

roulette, f., (*surv.*) (ordinary linen) tape (*décamètre* à *ruban*).

rouleur, m., (*mil.*) gabion-roller (soldier); (*aero.*) ground machine.

roulis, m., (*aero.*) roll.

roulotte, f., (*auto.*) sort of large traveling car.

routière, f., road engine, traction engine.

Royal-Cambouis, m., (*mil. slang*) soldier of the *trains des équipages*.

ruban, m., (*mach.*) strap, band;
— *chatterton*, strip of Chatterton, q. v.;
— *copiant*, copying ribbon;
— *hectographe*, hectograph ribbon;
— *rouge*, red ribbon; (*fig.*) the Legion of Honor.

rue, f., street;
— *Royale*, (street in Paris; by metonymy) the Ministry of the Navy;
— *Saint-Dominique*, (a street in Paris; by metonymy) the general staff of the army.

ruelle, f., small street; (*mil.*) company street.

rugueux, m., (*art.*) striker, percussion pin.

ruissellement, m., trickling;
eaux de ——, trickling waters.

rupteur, m., (*elec.*) make-and-break mechanism, key (producing the break spark);
— *distributeur*, make-and-break key (with special reference to periodic makes and breaks).

rupture, f., rupture, burst;
cercle de ——, (*expl.*) circle of destructive effect.

rustique, a., (of mechanical devices, weapons, etc.), not complicated.

S.

sabot, m., block;
— *d'arrêt*, scotch.

sabot-agrafe, m., (*r. r. art.*) rail clip.

sabotage, m., sabotage, (i. e., scamp work, scamping; malicious injury done to the establishment by an employee; intentional garbling of copy by a printer.)

saboter, v. t. (*r. r.*) to notch (a tie, for the foot of the rail).

sabre, m., (*sm. a.*) sword;
— *au poing*, sword in hand, with drawn swords.

sac, m., sack, bag; (*mil.*) knapsack;
— à *allumeurs*, (*art.*) primer bag;
— à *couchage*, sleeping bag;
— à *dos*, (*mil.*) regulation pack;
— *d'insubmersibilité*, (*pont.*) water-tight bag (metallic ponton).

saccadé, a., irregular, jerky.

sachet, m., small bag, pouch.

sac-poncho, m., (*mil.*) poncho (so contrived as to serve as a sack when rolled up).

Saint-cyrienne, f., (*Fr. a.*) mutual benefit society made up of graduates of St. Cyr.

Sainte-Barbe, (proper name) St. Barbara;
célébration de la ——, (*art.*) feast on December 4 to St. Barbara, the patron saint of the artillery;
célébrer la ——, (*art.*) to celebrate St. Barbara, patron saint of the artillery, on December 4.

Saint-maixentaise, f., (*Fr. a.*) mutual benefit society made up of graduates of St. Maixent.

saisir, v. t., (*law*) to refer;
— *un parquet d'une affaire*, (*mil. law*) to refer a case to a *parquet*, i. e., to a court-martial.

salle, f., hall, large room;
— *d'astiquage*, (*mil.*) in barracks, room for cleaning rifles, equipment, etc.;
— *des machines*, machine-shop;
— *de visite*, (*med.*) surgeon's office.

salve, f., (*art.*) salvo: (either rounds fired together, or a series of shots fired one by one);
— *de contrôle*, (*art.*) ranging salvo.

sandow, (proper name; *aero slang*) rubber shock absorber.

santé, f., quarantine (station, personnel, function).

sape, f., (*mil.*) sap, (strictly speaking, *sape* is a procedure, a method of digging);
— *d'attaque*, offensive sap;

sape à *deux formes*, in trenchwork, sap driven in two stages (deepened by a second party behind the first);
— à *une forme*, in trench work, full sized sap;
— *russe*, vaulted unlined gallery (in trench work);
— *volante en sacs à terre*, (*siege*) flying sand bag sap.
sapeur, m., (*mil.*) sapper;
— *colombophile*, (*mil.*) pigeon sapper, pigeon man of the engineer corps.
sapeur-aérostier, m., (*mil.*) soldier belonging to balloon park.
sapeur-bombardier, m., (*Fr. a.*) man assigned to trench artillery duty.
sapeur-cycliste, m., (*Fr. a.*) sapper cyclist.
sapinière, f., (*top.*) pine wood.
satellite, m., (*mach.*) satellite, middle, or intermediate gear or pinion.
satisfaction, f., satisfaction;
lettre de ——, (*mil.*) official letter, as from secretary or minister of war expressing commendation;
témoignage de ——, (*mil.*) expression of commendation, etc., by the authorities.
sauce, f., sauce;
— *poivrade*, Tabasco diet (at one time unlawfully in vogue at West Point).
saucisse, f., sausage; (*mil. slang*) kite balloon.
saucisson, m., sausage; (*mil. slang*) aerial torpedo.
Saumuroise, f., (*Fr. a*) mutual benefit society made up of graduates of Saumur.
saut, m., jump;
— *d'obstacles*, (*man.*) hurdling, jumping.
sauveteur, m., rescuer.
scander, v. t., to measure.
scénario, m., scenario; (*mil.*) plot or plan (whereby the hostile infantry is induced to man its first line trenches when these are to be taken under a *tir de concentration*).
schlitte, f., wood-sledge (for conveying lumber down a mountain side; used in the Vosges);
chemin de ——, continuous track or trestle on which the schlitte slides.
scie, f., saw;
— *égohine*, rip saw;
— à *tenon*, wood saw.
scier, v. t., to saw;
— *la bouche*, (*man.*) of the rider, to saw on the mouth.
séance, f., session;
— à *cheval*, (*man.*) ride, work on horseback.
seau, m., bucket;
— à *charbon*, (*mil. slang*) projectile of the German *Erdmorser*, (buried trench mortar).
sec, a., dry;
être à ——, to be dry, run dry (of reservoirs, etc.);
mélanger à ——, to mix dry (as sand and cement);
piquer une sèche, (*mil. slang*) to know nothing of a subject when questioned, to "fess," (U.S.M.A.)
sèche, f., (*hydr.*) flat (left exposed at low tide).
sécher, v. t., to dry;
— *une manœuvre*, (*mil. slang*) to "cut" a drill.
sécheresse, f., anemia (pigeons).
secrétaire, m., secretary, recorder.
secteur, m., (*mach.*) sector, quadrant, notched quadrant; (*mil.*) sector (the combination of several *centres de résistance* under one command, held usually by a division). [1 *secteur*=3 *centres de résistance*, 1 *centre de résistance*=4 *points d'appui*, 1 *point d'appui*=3 *segments actifs*, 1 *segment actif*= portion of line actively held.] (Numbers are illustrative merely.)
— *d'agrafage*, (*trench art.*) clamping sector;
— *d'appui*, (*trench art.*) base sector (240 L. T.);
— à *barrière*, v. —— à *gorge*;
— à *crans*, (*mach.*) notched quadrant;
— à *gorge*, (*mach.*) lever guide plate;
— *de hausse*, (*art.*) elevation scale (155 C. T. R.);
— *de pointage*, (*mil.*) aiming sector, elevation sector, (D. R. firing stand);
— *de pointage en hauteur*, (*trench art.*) elevation sector;
— *postal*, (*Fr. a.*) postal sector (armies in the field);
— *public d'éclairage d'électricité*, (*elec.*) public electric light station or plant;
— *de site*, (*art.*) angle of site scale;
— *de tir*, (*art., etc.*) sector to be covered by the fire of a piece, sector of fire.
section, f., section;
— *active*, (*Fr. a.*) (of a staff) the section in charge of duties relating to the active army; active section;
— *du courrier*, (*Fr. a.*) in a staff, mail section, (receives and forwards papers);
— *d'hospitalisation* (*mil. med.*) hospital work section;
— *de manœuvre*, (*mach. gun*) maneuver section (firing section and echelon);
— *de parc*, (*Fr. a.*) motor transport column;
— *sanitaire*, (*mil.*) motor sanitary detachment;
— *territoriale*, (*Fr. a.*) (of a staff) territorial section (in charge of territorial army matters);
— *de tir*, (*mach. gun*) firing section;
— *topographique*, (*Fr. a.*) map section (one to each corps and division);
— *de transport*, (*mil.*) motor transport section;
— *de type alpin*, (*Fr. a., mach. gun*) alpine type (pack mules);
— *de type mixte*, (*Fr. a.*) *mach. gun*) section of mixed type (pack and wheel transport).
segment, m., segment; joint (of a caterpillar tire); (*mach.*) piston ring;
— *actif*, (*mil.*) line of trenches actively held;
— *passif*, (*mil.*) portion of a position, held by the flanking fire of the —— *actif*, (it may even be in front of the —— *actif*);
— *de piston*, (*mach.*) piston segment, piston ring;
— *de serrage*, (*trench art.*) locking segment;
— *suédois*, (*mach.*) metallic piston segment or ring;
tiercer les ——*s*, (*mach.*) to set piston rings at 120°.
séismophone, m., (*mil.*) seismophone (listening device, mining).
sel, m., salt;
— *grimpant*, (*elec.*) any salt depositing on binding posts, etc.
self, f., (*elec.*) inductance coil.
self-induction, f., (*elec.*) self-induction, inductance, inductance coil.
selle, f., (*harn.*) saddle;
— *d'armes*, (*mil.*) service saddle;
— *anglaise*, flat or park saddle.
sellette, f., (*trench art.*) bed plate.
semelle, f., (*harn.*) tread of a stirrup; (*auto.*) tread; (*mach. gun*) trail shoe;
— *antidérapante*, nonskidding tread;
— à *ergot*, (*mach. gun*) (sort of, spike shoes tripod leg).
semer, v. t., (*torp.*) to plant torpedoes.
sénateur, m., senator;
— *inamovible*, life senator.
sens, m., (*in gen.*) way, direction;
— *inverse des aiguilles d'une montre*, contra clockwise direction;
— *de la marche des aiguilles d'une montre*, clockwise direction.
sentir, v. t., to feel;
se —— *les coudes*, (*mil.*) (to feel touch of elbow) to have a feeling of collective strength, of power.
séparation, f., separation;
— *des risques*, (*mil.*) separation of risks (i. e., in artillery positions, batteries, the scattering of the gun detachment, so that one hit shall not disable the entire detachment), scattering.
septain, m., (*aero.*) seven-stranded rope (balloons)
sergent, m., (*mil.*) sergeant;
— *arrimeur*, (*mil. aero.*) rigging sergeant (balloons).
série, f., series;
— *régimentaire*, (*Fr. a.*) (in some cases) regimental outfit (as of butchers' tools).

seringue, f., syringe; (*mil. slang*) gun;
— *à graisse*, lubricating syringe.

serment, m., oath;
prêter —— *entre les mains de*, to take an oath before (an officer, etc.) to be sworn by (an officer, etc.).

serpent, m., serpent; (*aero.*) thick rope (used in balloon maneuvers).

serrage, m., (*mach.*) setting (as of a brake); tightening;
—— *à coulisse*, slipping adjustment (said of two bands, one slipping on the other, like the chin strap of a forage cap);
—— *à refus*, (*mach.*) setting, tightening "home" (of a screw, nut, etc.).

serrer, v. t., (*mil.*) to close up;
—— *un frein à bloc*, (*mach.*) to brake home all at once;
—— *à refus*, (*mach.*) to set "home" a nut, screw, etc.

serre-rail(s), m., (*r. r.*) rail plate.

servante, f., (*art.*) carriage prop (155 C.).

service, m., service; duty;
—— *ambulant*, (*adm.*) railway-mail service;
—— *armé*, (*mil.*) the military service in or for a government, as distinguished from the non-military service;
—— *auxiliaire*, (*mil.*) nonmilitary duty (performed by soldiers);
—— *en campagne*, (*mil.*) tactical problems in the field (Saumur);
—— *de chancellerie*, (*mil.*) (in general staff, etc.) paper work;
—— *commandé*, (*mil.*) duty covered by orders;
—— *à court* (*long*) *terme*, (*mil.*) short (long) service;
—— *d'écoute*, (*mil.*) listening service (for airplanes calling stations without previous agreement);
être pris par le ——, (*mil.*) to be on duty;
—— *d'exploitation*, (*r. r., etc.*) operation, service of operation;
—— *extrême urgence*, (*mil.*) on a telegram, means that it has precedence over all others;
—— *d'honneur*, (*mil.*) the duty performed by troops paraded to receive a king or dignitary;
—— *de jour*, (*nuit*), (*r. r.*) day (night) trains;
—— *d'ordre*, (*mil., etc.*) duty of maintaining public order; the troops, police, turned out to maintain order;
—— *de santé*, (*mil.*) sanitary service;
—— *de surveillance*, (*mil.*) lookout duty;
—— *des vérifications*, (*art.*) ammunition-inspection department.

servir, v. n., to be in use.

servo-moteur, m., (*mach.*) servo-motor.

sevrage, m., weaning.

sevrer, v. t., to wean.

shunter, v. t., (*elec.*) to shunt.

sidi, m., (*mil. slang*) the Arabs and Berbers of North Africa.

siège, m., seat;
—— *social*, (*com.*) main office.

sifflet, m., whistle;
en ——, (*in gen., cons.*) skew, bevel-wise.

signal, m., signal; (*artif.*) signal (i. e., device of a rocket, flag, caterpillar, star, etc.);
—— *d'arrêt*, (*r. r.*) stopping signal;
—— *aux d'avertissement*, warning signals, apparatus to produce same;
—— *aux Morse*, (*mil. sig.*) Morse alphabet;
—— *par panneaux*, (*mil.*) panel signal;
—— *parasite*, (*wireless*) interfering message, or signal;
—— *de reconnaissance*, (*mil.*) answering signal, signal of reception (e. g., from a battery to an airplane);
—— *de séparation*, signal made between ciphered messages.

signalisation, f., (*mil.*) signaling, signalizing;
—— *par artifices*, signaling by rockets, by fireworks;
—— *à bras*, arm signaling;
—— *optique*, visual signaling;
—— *par panneaux*, panel signaling.

signe, m., sign;
—— *du temps*, weather indication or sign.

silencieux, m., (*mach.*) muffler, silencer;

singe, m., monkey; (*mil. slang*) canned meat, tinned meat.

sinistré, m., victim of a *sinistre*.

sismographe, m., (*inst.*) seismograph (also spelled *séismographe*).

sismomicrophone, m., (*mil. min.*) long distance listener (device), mine stethoscope.

sismotéthoscope, m., (*mil.*) listening device (*écouteur*); mine stethoscope.

site, m., site;
secteur de ——, (*art.*) site angle sector (scale);
tambour de ——, (*art.*) site angle drum (scale).

sitogonomètre, m., (*art.*) sitogoniometer (for measuring angles of site).

sitomètre, m., (*art.*) sitometer (for measuring angles of site).

situation, f., situation; (*mil.*) report;
—— *de prise d'armes*, (*mil.*) (return of troops, setting out on a march, a duty) field return.

situation-rapport, f., (*mil.*) return;
—— *des cinq jours*, (*Fr. a.*) five-day return.

situer, v. t., to locate.

ski, m., ski.

skieur, m., ski-runner.

soir, m., late afternoon; evening.

sol, m., soil;
—— *ferré*, hard soil (e. g., surface of a hard road).

soldat, m., soldier;
petits ——*s*, (*Fr. a.*) "little soldiers" as a term of endearment;
—— *de rang*, (*lit.* soldier in the ranks) soldier unable to fight usefully when not under the direct orders of his officers, soldier accustomed to fight in the ranks only, routine soldier.

soldat-citoyen, m., (*mil.*) militia man (in bad sense).

soldat-musicien, m., (*mil.*) bandsman.

solde, f., (*com.*) balance of accounts;
—— *de campagne*, (*Fr. nav.*) campaign pay.

soleil, m., sun;
faire le grand ——, (*mil. slang*) to have one's horse turn a somersault.

solidaire, a., solidary;
—— *de*, (*mach., etc.*) integral with.

sortie, f., exit, departure; (*mach.*) outlet, discharge (pipe);
les ——*s*, (*com.*) expenses, outgo.

souder, v. t., to solder;
—— *à l'autogène*, to burn on, to solder autogenously.

soufflage, m., blowing, blast;
—— *d'étincelles*, (*elec.*) blow out (magnetic).

souffle, m., wind, breath;
à bout de ——, out of wind, winded.

souffler, v. n., (*mach.*) to blow.

soufflet, m., (*slang*) gore or gusset (let into a coat to increase its girth).

soulager, v. t., (*mach.*) to release, ease (a valve).

soulier, m., shoe;
—— *bas*, low shoe;
—— *à lacets*, lace shoe.

soupape, f., (*mach.*) valve;
—— *d'air additionnel*, extra-air valve (auto.);
—— *annulaire*, annular valve;
—— *de "campement,"* (*aero.*) station valve (admission of air, dirigible moored at a *campement*);
—— *de chargement*, charging valve;
—— *commandée*, mechanically controlled valve;
—— *à levée variable*, variable lift valve;
—— *de prise d'air*, air inlet valve.

sourdine, f., (*wireless*) damper.

sous-commissaire, m., (*mil.*) member of a *sous commission*, q. v.

sous-groupement, m., (*Fr. art.*) subgroup (v. *groupement*).
sous-marin, m., (*nav.*) submarine;
— *coupe-filet*, wire-cutting submarine;
— *défensif*, coast-defense submarine;
— *offensif*, sea-going submarine.
sous-produit, m., by-product.
sous-quartier, m., (*mil.*) the terrain held by a company.
sous-secteur, m., (*mil.*) sub-sector, fraction of a *secteur*, q. v., held by a brigade or a regiment.
sphérique, m., (*aero.*) spherical balloon.
spire, f., (*elec.*) turn (of a winding);
demi —, half turn of a helix, of a spiral.
spruce, m., spruce (tree and wood).
stabilisateur, m., (*aero.*) elevator (improper name), stabilizer; (*expl.*) stabilizer (e. g., ether, acetone);
— *fixe*, fixed tail plane.
stabilisation, f., (*art.*) waiting situation (before action on either offensive or defensive); (*mil.*) (state or condition of) passivity, as opposed to combat.
stabilité, f., stability;
— *de direction*, v. — *de route*;
— *en hauteur*, (*aero.*) vertical stability;
— *latérale*, (*aero.*) lateral stability;
— *longitudinale*, (*aero.*) longitudinal stability;
— *propre*, (*aero.*) inherent, natural stability;
— *de route*, (*aero.*) directional stability.
stand, m., (*t. p.*) firing point (of a rifle range).
station, f., station;
— *d'aérostat*, (*aero.*) (captive) balloon station, (*sièges*);
— *balnéaire*, watering place, seaside resort, springs;
— *de monte*, (*hipp.*) stand;
— *radiotélégraphique*, (*wireless*) wireless station;
— *réceptrice*, (*mil.*) receiving station;
— *sanitaire*, quarantine station;
— *de télégraphie sans fil légère*, wireless cart;
— *tête d'étapes de guerre*, (*mil., r. r.*) railhead;
— *volante*, traveling or mobile station (e. g., of wireless telegraphy).
stationnement, m., stopping, standing;
point de —, (*aero.*) balloon station.
statoscope, m., (*aero.*) statoscope.
statut, m., by-law, rule or regulation of a corporation, of a commercial company.
stauffer, m., (*mach.*) compression grease cup.
stéréo-jumelles, f., pl., stereoscopic field glass.
stéthoscope, m., (*med.*) stethoscope; (*mil.*) (listening device, mining) stethoscope.
— *à air*, (*mil. min.*) air stethoscope;
— *à eau*, (*mil. min.*) water stethoscope (may be a canteen on the ground).
stock, m., stock;
— *roulant*, (*r. r.*) rolling stock.
stoppage, m., stopping (of an engine, ship, torpedo, etc.).
stopper, v. t., to stop (trains, ships, machinery).
strabisme, m., strabismus.
strapontin, m., folding seat, camp chair.

stylographe, m., stylograph.
submersible, m., (*nav.*) submersible boat.
subsistance, f., subsistence;
prendre en —, (*mil.*) to subsist.
sulfatation, f., (*elec.*) sulphation (of a storage battery).
sulfater, v. r., (*elec.*) to sulphate (storage battery).
support, m., support;
— *pivotant*, (*mach. gun*) top carriage (tripod mount);
— *de pointage et d'armement*, (*art.*) aiming and cocking bracket.
surboche, m., (*slang*) a double-dyed boche.
surépaisseur, f., extra thickness.
sûreté, f., an abbreviation of *la police de sûreté*;
— *générale*, public safety; measures for public safety; personnel responsible for public safety.
surface, f., surface;
— *d'appui*, (*trench art.*), (vertical) bearing surface (of a platform, e. g., 240 L. T.);
— *portante*, (*aero.*) bearing surface, wing;
— *de refroidissement*, (*mach.*) cooling surface (e. g., of a radiator);
—*s verticales*, (*aero.*) keel plane area.
sursaturé, a., supersaturated.
surveillance, f., (*mil., aero.*) observation of a sector;
point de —, (*art.*) (point on which a *batterie en surveillance*, q. v., is laid) point under observation;
position de —, v. s. v. *position*.
surveillant, m., watchman;
— *militaire*, (*mil.*) military watchman; i. e., sentry (as at a palace).
survol, m., (*aero.*) flight.
survoler, v. t., (*aero.*) to fly over (as a zone).
sus-bande, f., (*art., mach. gun*) cap square;
— *à charnière*, hinged cap square;
— *à rotation*, rotating cap square.
suspension, f., hanging lamp; (*aero.*) method of suspending the car (balloon and dirigible), suspension; (*auto.*, etc.) springs, spring mounting (of a vehicle);
— *captive*, the suspension of a captive balloon;
cercle de —, hoop;
— *libre*, (the suspension of a free balloon) free suspension;
— *en triangle*, (*aero.*) triangular suspension (dirigibles, indeformable).
suspensoir, m., (*med.*) suspensory.
suspente, f., (*aero.*) suspending cords (spherical balloon and dirigible), suspension wire.
sustentateur, m., (*aero.*) the same as *planeur*, q. v.
syntonie, f., (*wireless*) syntonism.
syntonisation, f., (*wireless*) tuning, syntonization;
bobine de —, tuning coil.
syntoniser, v. t., (*wireless*) to tune, syntonize.
système D., m., (*mil. fam.*) *système de la débrouille*, i. e., system whose object is to make the trenches, lines, etc., as comfortable as possible by showing resourcefulness under difficult conditions.

T.

tabac, m., tobacco;
mauvais —, (*mil. slang*) bad business, bad affair, "bad medicine."
table, f., table; table (e. g., of logarithms);
— *de chargement*, (*art.*) loading table (155 C. T. R.);
— *pratique de tir*, (*art.*) service range table;
— *de tir à charges constantes*, (*art.*) single charge range table;
— *de tir complète*, (*art.*) ballistic range table (gives all ballistic elements);
— *de tir à double (simple) entrée*, (*art.*) double (single) entry range table.

tableau, m., dashboard (of a motor car);
— *annonciateur*, (*telephone*) annunciator;
— *annonciateur à n directions*, n-drop annunciator;
— *de concours*, (*Fr. a.*) list of competing candidates for Legion of Honor, etc.;
— *à n directions*, (*telephone*) an n-line switchboard;
— *monocorde*, (*telephone*) plug switch annunciator;
— *de parafoudres*, (*telephone*) lightning arrester;
— *à réglettes*, (*telephone*) zinc-strip annunciator.
tabler, v. n., to count upon.

tablier, m., dashboard (of a motor car, of vehicles in general).
tache, f., spot;
 faire la ——— d'huile, (of an army) to spread like an oil spot.
tachymètre, m., revolution counter, tachometer.
tacot, m., *(mil. slang)* any old motor car.
taille, f., cut; *(mach.)* cutting;
 en ———, (unif.) without overcoats;
 —— par fuseaux, (aero.) cutting out by spindles (the envelope);
 —— par panneaux, (aero.) cutting out by panels (the envelope).
tailleur, m., tailor; *(mil. aero.)* fabric expert (balloons).
talon, m., flange, edge, or bead of a tire cover (pneumatic tire).
tambour, m., drum; *(mach. gun)* drum, feed drum, tray;
 —— des directions, (art.) direction drum (scale);
 deflection drum;
 —— de frein, (mach.) brake drum;
 —— de friction, (mach.) friction drum;
 —— des hausses, (art.) elevation drum (scale);
 jugement de ———, (mil.) drumhead court-martial.
 —— moteur, (mach., etc.) driving drum;
 —— de site, (inst.) angle of site micrometer.
tampon, m., plug; *(mil. slang)* orderly; *(med.)* compress;
 —— d'allumage, (elec.) sparking plug;
 —— de presse-étoupes, (mach.) stuffing-box plug;
 —— de rupture, (elec.) sparking plug;
 —— de visite, (mach.) inspection-hole plug.
tangage, m., *(aero.)* pitching.
tangent, a., tangent; *(aero.)* sluggish, logy (in school slang, *tangent* is applied to a student who has almost made the passing mark; more generally, to a person who just misses getting a favor, a decoration, etc.); m., *(aero.)* stalling.
tape-cul, m. (in the following expression):
 faire du ———, (slang) at Saumur, to trot without stirrups.
tapiar, m., *(mil. slang)* drummer.
tapir, f., *(mil. slang.)* topography.
taquet, m., cleat;
 —— de butée, stop;
 —— à glissières, slip cleat, guide cleat, adjustable cleat.
tasseau, m., cleat;
 —— de retenue, stop cleat.
taube, m., *(aero.)* Taube, German airplane.
taupe, f., mole; *(mil. slang)* German soldier;
 guerre de ———, (mil. slang) trench warfare.
taupinière, f., mole hill; *(mil.)* sniper's post:
 —— blindée, (mil.) armored sniper's post.
tauriau, m., (provincial for *taureau,* bull); hence,
 les terribles ———x, (mil. slang) the territorials (connotation of sarcasm).
taxamètre, m., *(auto.)* a variant of *taximètre.*
taxauto, m., *(auto.)* taximeter automobile.
taxe, f., tax;
 —— militaire, (adm.) exemption tax (tax on men exempt or excused from the military service, composed of a head tax which is the same for all, and of two other elements varying with the circumstances of the person and of his family).
taxer, v. t., to fix prices.
taxi, m., *(fam.)* abbreviation of *taximètre,* taxi; *(aero. slang)* bus (i. e., airplane).
taxi-auto, m., *(auto.)* taxicab.
taximètre, m., *(auto.)* taximeter.
taxomètre, m., a variant of *taximètre.*
télégéophone, m., *(siege)* long distance geophone.

télégramme, m., telegram;
 —— d'arrivée, telegram received;
 —— de départ, telegram sent;
 —— officiel, (mil.) official telegram (sent by officers or officials on government or army account);
 —— de service, (mil.) department telegram (sent by personnel of a service, relates to the service);
 —— de transit, telegram in transit (through a transmitting station).
télégramme-mandat, m., telegraphic money order.
télégraphie, f., telegraphy;
 —— sans fil, wireless telegraphy;
 —— sans fil dirigée, controlled wireless, directive wireless.
 —— par le sol, ground telegraphy.
télémécanique, f., *(phys.)* (science of long-distance power transmission) telemechanics.
télémétreur, m., *(mil.)* range taker (man).
téléphane, m., automobile lamp.
téléphone, m., telephone; telephone call;
 appel de ———, telephone call;
 —— de campagne, (mil.) field telephone;
 coup de ———, v. s. N. *coup;*
 —— haut parleur, speaking telephone;
 tribune de ———, telephone booth.
téléphoner, v. t., to telephone; *(mil. slang)* surreptitiously to draw off wine from a cask with a rubber tube.
téléphonie, f., telephony;
 —— sans fil, wireless telephony.
téléphotographie, f., telephotography.
téléscopique, a., telescopic; telescoping.
temps, m., weather, time; *(mach.)* cycle;
 à quatre ———, four cycle (engine).
tendelet, m., small leather strap.
tendeur, m., turnbuckle; jockey pulley; tightener; *(mach.)* stretcher; wire stay;
 —— de chaîne, (mach.) chain adjuster, chain tightener;
 —— de pont, (mach.) reaction rod or lever (to relieve the springs, auto).
tendon, m., tendon;
 —— chauffé, (hipp.) heated tendon.
tendre, v. t., to reach;
 —— la main, (mil.) to join hands with.
tenir, v. n., *(mil.)* to hold out (as, a fort, a line attacked, an army, a nation).
tenon, m., tenon;
 —— d'accrochage, catch;
 —— d'armement, (art.) cocking lug;
 —— d'assemblage, (mach. gun) feed piece assembly stud;
 —— de fermeture, (mach. gun) locking lug;
 —— de manœuvre, (mach. gun) operating lug.
tensiomètre, m., dynamometer, (for determining wind-velocities).
tension, f., traction.
tente, f., tent;
 —— de discipline, (mil.) prisoners' tent;
 —— à enveloppe double (simple), (mil.) double (single) fly tent.
tenue, f., *(mil.)* dress;
 —— d'hiver, (unif.) winter dress;
 —— de route, (mil.) marching uniform, i. e., field.
terme, m., term, limit;
 à ———, time (in commercial transactions, etc.).
terrain, m., ground, soil; *(mil.)* terrain;
 —— coulant, shifting soil;
 —— dénudé, (top.) bare ground;
 —— de pingouin, (aero.) ground school;
 —— de rouleur, v. *—— de pengouin.*
terre, f., earth;
 mettre à ———, (elec.) to ground;
 mise à ———, (elec.) grounding;
 —— végétale, loam.
terre-plein, m., platform or surface of a wharf, railway quay.
territoriale, f., short for *armée ———.*

Tesla, proper name;
—— *primaire* (*secondaire*), (*wireless*) primary (secondary) of a high-frequency transformer.
têtard, m., (*mil. slang*) stubborn horse.
tête, f., head;
—— *à chape*, (*mach.*) crosshead and strap;
—— *à chape de la fourche*, (*mach. gun*) bipod head;
—— *à crochet*, hook end (e. g., of a rod, or tie);
—— *de cylindre* (*mach.*) cylinder head (especially when the head is cast in one piece with the cylinders, *culasse*, q. v., being used for a built-up head);
faire —— *à queue*, (of a horse) to turn suddenly; (aero.) to turn head to tail (unintentional).
—— *de ligne*, (*mil.*, *r. r.*) railhead;
—— *à œil*, eye-end (e. g., of a rod or tie);
—— *simple*, (*mach.*) crosshead.
tétine, f., (*tech.*) nipple.
teuf-teuf, m., (*fam.*) motorcycle.
thermique, a., thermal; m., (*elec.*) hot-wire ammeter.
thermomètre-fronde, m., (*inst.*) whirled thermometer.
thermo-siphon, m., (*mach.*) thermo-siphon.
théorie, f., (*Fr. a.*) the "book."
ticket garde-place, m., (*Fr. r. r.*) (a ticket reserving a seat in advance on long-distance trains, first and second class) seat reservation.
tige, f., rod;
—— *d'amortissement de rentrée*, (*art.*) throttling rod;
—— *de direction*, (*auto.*) steering rod, steering gear rod;
—— *poussoir*, tappet rod;
—— *de renvoi*, (*mach.*) communicating rod, transmission rod;
—— *de soupape*, (*mach.*) valve stem;
—— *support*, (*trench art.*) supporting arm (V. D. mortar).
timbre, m., stamp;
—— *en caoutchouc*, rubber stamp.
timon, m., pole;
—— *de remorque*, (*trench art.*) traction pole (various vehicles).
timonerie, f., (*auto.*) (generic term for all the bars, rods, etc., used in controlling the brake gear) brake gear controls.
tique, f., tick (parasite).
tiquet, m., tick (parasite).
tir, m., (*art.*, *sm. a.*) fire;
—— *ajusté*, aimed fire;
allonger le ——, (*art.*) to increase the range;
—— *d'amélioration*, (*art.*) shot, round, practice, to correct previous practice; application practice;
—— *d'anéantissement*, (*art.*) annihilation fire (against personnel);
—— *d'arrosage*, (*art.*) sprinkling fire (to cover an area by lines of successive parallel bursts);
—— *d'assouplissement*, (*t. p.*) preliminary practice;
—— *sur ballon*, (*art.*, *sm. a.*) fire, practice, at a balloon;
—— *de barrage*, (*art.*, *grenades*) curtain fire; barrage;
—— *de barrage défensif*, (*art.*) (curtain fire against an attack) defensive barrage;
—— *de barrage offensif*, (*art.*) (curtain fire preceding an attack on the enemy) offensive barrage;
—— *bas*, (*art.*) short practice; short;
—— *bloqué*, (*mach. gun*) fire, piece locked in direction;
—— *à la bouteille*, (*Fr. a.*) (fire employing compressed air in tubes, trench artillery), pneumatic fire;
cadence du ——, v. s. v. *cadence*;
chambre de ——, (*mil.*) instruction pit (grenade instruction);
—— *en chambre*, (*art.*) simulated target practice; (*sm. a.*) gallery practice, subcaliber practice;
—— *de concentration*, (*art.*) the fire of several batteries on the same target or targets;

tir de contre-préparation, (*art.*) hostile fire to prevent, annoy, derange, preparation for an attack;
—— *de contrôle*, (*art.*) sighting round, ranging shot or practice;
—— *contrôlé*, (*art.*) in zone fire, practice, after the range has been got, checked by aerial observation;
—— *débloqué*, (*mach. gun*) fire, piece unlocked in direction;
—— *de défense*, (*art.*) defensive fire (to prevent hostile infantry from penetrating first line trenches);
—— *à défilement*, (*art.*) fire from a defiladed position;
—— *à démolir*, (*art.*) demolition fire;
—— *de destruction*, (*art.*) fire against enemy works, against his approaches;
—— *par deux*, (*art.*) two rounds, (and of other numbers);
—— *à distance réduite*, (*t. p.*) gallery practice;
—— *d'efficacité*, (*art.*) fire for effect, real fire;
—— *encadrant*, *d'encadrement*, (*art.*) bracketing fire;
—— *d'ennui*, (*mil.*) annoying fire, (i. e., whose object is merely to disturb);
enrayage de ——, (*art.*, *mach. gun*) jamming;
—— *en éventail*, (*mach. gun*) sweeping fire;
—— *s éventuels*, (*art.*) (generic term for kinds of fire not susceptible of rigorous classification with respect to tactical purposes) casual fire, occasional fire;
—— *avec fauchage* (*mach. gun*) sweeping fire;
—— *fauchant*, (*art.*) sweeping fire;
—— *fauchard*, (*art.*) sweeping fire;
—— *de fonctionnement*, (*mach. gun*) gun-testing practice;
—— *fouillant*, (*art.*) searching fire;
—— *de groupement*, (*t. p.*) (practice the object of which is to produce a small pattern) holding practice, grouping practice;
—— *de harcèlement*, (*art.*) annoying, harassing, fire (e. g., on the trenches used by reliefs, supplies, is a *tir éventuel*, q. v.);
—— *sur hausse unique*, (*art.*) (in a *tir d'efficacité*) a gust of two or four rounds per piece (all using the same elevation);
—— *haut*, (*art.*) over practice; over;
—— *à hauteur*, (*art.*) adjusted fire (the range has been got);
—— *d'instruction*, (*mach. gun*) instruction practice;
—— *d'interdiction*, (*art.*) fire to prevent the enemy from reaching a given point (e. g., a battery destroyed in order to repair it; hence) isolating fire, preventive fire;
mécanisme du ——, (*art.*) fundamentals of firing, of target practice; procedure, method, of fire;
—— *dans la mêlée*, (*t. p.*) (practice of firing in a mellay); snap shooting;
—— *en mitrailleuse*, (*mil.*) machine gun fire (from a *fusil-mitrailleur*);
—— *à obus explosif*, (*art.*) high explosive fire;
—— *pardessus les troupes*, (*art.*) fire over the heads of one's own troops;
—— *de perturbation*, (*art.*) disturbing fire, (any fire to annoy the enemy, e. g., night firing);
—— *par pièce*, (*art.*) fire by piece;
—— *à pointage direct* (*indirect*), (*art.*) direct (indirect) fire;
—— *à la pompe*, (*Fr. a.*) (fire using a bicycle pump to compress air, trench artillery); compressed air fire;
position de ——, (*art.*) firing position (battery or gun);
—— *précis*, (*art.*) accurate fire, (closest possible ranging on fixed targets, and accurate plotting—*repérage*—of the terrain for mobile objectives);
—— *de précision*, (*art.*) accurate fire (to destroy the hostile artillery); demolition fire; (*t. p.*) practice to develop accuracy;
préparation du ——, (*art.*) preparation of fire (e. g., orientation, measure of angle of site, corrector, and range);
—— *progressif*, (*art.*) (fire searching the ground in depth); progressive fire;

tir *de progression*, (*art.*) (fire during a progression) progression fire;
— *par rafales*, (*art.*) rafale fire, gust fire;
— *par rafales échelonnées*, (*art.*) gust fire at increasing elevations;
— *rapide pointé*, (*art.*) rapid aimed fire;
— *de record*, (*art.*) record firing, record practice, regular practice;
— *réel*, (*mil.*) at maneuvers, practice with ball cartridges;
régime du —, v. s. v. *régime*;
— *de réglage*, (*art.*) fire for adjustment;
— *de représailles*, (*art.*) reprisal fire (e. g., on enemy first line trenches, if he himself is shelling first line trenches, is a *tir éventuel*, q. v.);
reprise de, v. s. v. *reprise*;
— *de résistance*, (*art.*) endurance firing (test);
— *de riposte*, (*art.*) return fire, answering fire;
— *roulant*, (*mil.*) rolling, (i. e., uninterrupted fire);
— *par salves échelonnées*, (*art.*) fire by difference salvos;
— *sans fauchage*, (*mach. gun*) fire without sweeping;
— *scolaire*, (*tp.*) rifle practice in schools·
section de —, (*inf.*) firing section.
— *systématique*, (*art.*) (zone fire, using previous ranging, and not susceptible of accurate check by aerial observation) systematic fire;
— *tendu*, (*art.*), flat trajectory fire;
transport du —, (*art.*) shift of fire (to a different target);
transporter le —, (*art.*) to shift the fire (to a different target);
— *d'usure*, (*art., mach. gun*, etc.) wearing-down fire, (at irregular intervals, at unexpected times, so as to wear out the enemy); attrition fire;
— *de vitesse*, (*mach. gun*) rapid fire;
— *sur zones*, (*art.*) zone fire, (to cover the assigned zone with projectiles).
tirage, m., impression (of a book, edition, etc.).
tirailleur, m., (*mil.*) skirmisher (includes, to-day, *grenadiers* and *mitrailleurs*).
tirant, m., tie, brace;
— *de sonnette*, (*mach.*) bell crank arm.
tire-boche, m., (*mil. slang*) the bayonet.
tire-douilles, m., (*mach. gun*) cartridge case extractor.
tire-fond, m., lag screw.
tirer, v. a., to print, to run off (so many copies of a book, etc.);
— *d'après la carte*, (*art.*) to fire by the map.
— *sur ballon*, (*art., sm. a.*) to fire at a balloon;
— *une bordée* (*mil. slang*) to absent one's self without leave;
— *son congé*, (*mil. slang*) to serve one's time;
— *au(x) flanc(s)*, (*mil. slang*) to be a slacker, a shrinker.
tireur, m., (*art.*) gun firer; (*mil.*) firer, marksman;
— *d'élite embusqué*, (*mil.*) sniper;
— *au(x) flanc(s)*, (*mil. slang*) shirker, slacker.
tireveille, f., (*aero., etc.*) tiller-rope.
titre, m., title;
au — *du n* (*ème*) *régiment*, (*mil.*) carried on the rolls of the nth regiment;
à — *précaire*, (*adm.*) revocable (of a license, etc.).
titulaire, m., holder;
être — *de*, to hold or have (as a medal, a decoration).
toile, f., linen;
— *de barrage*, v. s. v. *barrage;*
— *caoutchoutée*, (*auto., aero.*, etc.) repairing tape, strip; rubberized linen;
— *émerisée*, emery cloth;
— *kaki*, khaki cloth.
toilette, f., washstand.
tôle, f., sheet iron, sheet steel; plating;
— *emboutie*, pressed steel (or iron);
— *de support*, (*aero.*) engine spider support.
tolite, f., (*expl.*) tolite.
tombant, m., (*unif.*) fringe of an epaulet.

tomber, v. n., to fall;
— *en tête à queue*, (of a horse) to turn a somersault.
tonneau, m., cask, barrel; tonneau (of a motor car); (*aero.*) horizontal spin.
tonne-kilomètre, f., kilometer-ton.
top, m., voice-signal (made to indicate an instant of time, as in reading chronometers, or in comparing watches).
topette, f., phial.
topinambour, m., Jerusalem artichoke.
topométric, f., (*top. surv.*) the determination of points (topography=forms and objects).
torpédo-fougasse, m., torpedo acting as a fougasse.
torpillage, m., (*nav.*) torpedoing.
torpille, f., (*mil. nav.*) torpedo;
— *sèche*, (*mil.*) land mine.
torsade, f., twist joint.
— *espagnole*, American twist joint.
tortue, f., turtle, tortoise; (*mil. slang*) grenade.
totalisateur, m., (*auto.*) any device or instrument giving distance, run, time spent, etc.
toto, m., (*mil. slang*) trench vermin.
toubib, m., (*mil. slang*) surgeon (*médecin-major;* Algerian word).
touchau, m., (*elec.*) contact, contact piece, plug, or device.
touche, f., (*t. p.*) hit; (*elec.*) contact; (*fenc.*) touch in a fencing match;
— *d'extracteur*, (*art.*) extractor cam;
zéro —, (*fenc.*) untouched (in a fencing match).
touché, m., (*art.*) hit.
toucheur, m., cattle drover.
toupet, m., forelock; (*fam.*) audacity, "cheek," impudence;
avoir du —, to be "cheeky," to be B. J. (U. S. M. A.)
tour, m., (*mach.*) revolution; (*auto*) lap (of races);
— *d'horizon*, complete turn of the horizon;
hors —, (*mil., etc.*) out of turn;
— *de service*, (*mil. eng.*) in sieges, roster of duty; (three classes: 1°, attack, app oaches, saps and mines; 2°, interior service and guards; parks, maintenance of communication; 3° rest).
tourelle, f., (*aero.*) machine gun racer.
tourillon, m., (*art.*) trunnion; (*mach.*) crosshead pin, wrist pin;
— *central*, (*art.*) breechplug spindle (155 C. T. R.).
tourillonnement, m., (*art.*) (state or act of turning on trunnions) trunnioning.
tournant, m., turn, bend, of a road; (*mil.*) turn (around a traverse, etc.); (*aero.*) turning.
tournée, f., round, visit;
— *d'inspection*, (*mil.*) inspecting trip or tour.
tourner, v. t., n., to turn;
— *de n tours*, (*mach.*) make n revolutions.
tourteau, m., oilcake.
traçage, m., (*mil. fort.*) tracing out, staking out, of trenches, works.
tracé, m., design;
— *en ligne brisée*, (*mil.*) broken line trace (gives flanking fire);
— *à redans*, (*mil.*) redan trace;
— *d'un train*, (*r. r., mil.*) graphical schedule of a train (hours of arrival and of departure at various points on its line).
tracteur, m., (*auto*) tractor; (*aero.*) tractor; (*mach. gun.*) cartridge tractor;
à —*s*, using tractors, tractor (of guns, batteries, supply, etc.);
— *remorque*, touring tractor.
traction, f., traction;
— *animale*, animal traction;
— *à bras*, (*mil.*) hand traction;
— *à bras d'hommes*, man traction;
double —, (*r. r.*) double **traction** (two locomotives);

traction **hippomobile,** horse traction;
— *mécanique,* power traction;
triple ——, (*r. r.*) triple traction (three locomotives).
train, m., train (in many relations); (*aero.*) train (of kites);
— *d'atterrissage,* (*aero.*) landing chassis, undercarriage;
— *baladeur,* (*mach.*) sliding gear;
— *bis,* (*r. r.*) second section of a train;
— *à couloir,* (*r. r.*) corridor train;
— *électrique,* train of trolley or other electric vehicles;
— *express,* (*r. r.*) express train;
— *facultatif,* (*r. r.*) occasional train;
former un ——, (*r. r.*) to make up a train;
— *de fourrage,* (*mil.*) forage train;
— *de jour,* (*r. r.*) day train;
— *de nuit,* (*r. r.*) night train;
— *planétaire,* (*mach.*) planetary gear or motion, epicyclic gear, equalizing gear;
— *de ravitaillement,* (*mil.*) supply train;
— *régulier,* (*r. r.*) schedule train (every day);
— *sanitaire improvisé,* (*mil., med.*) emergency train, improvised train;
— *sanitaire de première ligne,* (*mil., med.*) hospital train, sanitary organization at the front.
train-ambulance, m., v. *train sanitaire de première ligne.*
traineau, m., (*art.*) slide or cradle, gun slide.
trainée, f., (*aero.*) drift, trail;
angle de ——, trail angle;
— *de feu,* trail of fire.
traine-pattes, m., (*mil. slang*) ration wagon.
trainglot, m., v. *tringlot.*
train-parc-cantonnement, m., (*Fr. a., r. r. eng.*) railway cantonment (railway engineer troops) railway quarters.
train-poste, m., (*r. r.*) mail train, mail.
trait, m., (*mil. sig.*) dash (of a code).
trajectoire, f., (*art., sm. a.*) trajectory; (*art.*) travel of projectile in the bore (rare); passage of projectile through armor plate (rare); (*aero.*) flight path.
tranche, f., (*drawing*) section; (*mach. gun.*) gas-face (of the driving piston);
— *horizontale,* horizontal section;
— *verticale,* vertical section.
tranchée, f., (*mil., etc.*) trench; (strictly speaking, *tranchée* is a procedure, a particular method of digging);
— *d'avant-postes,* outpost trench (in front of first line trench);
— *s en bretelles,* (*mil.*) (lit. suspender trenches) splaying trenches;
— *de circulation,* passage, circulation trench;
— *de combat,* synonym of — *de tir,* q. v.;
— *de dégorgement,* (siege) draining trench;
— *de départ,* trench from which an attack starts, departure trench;
— *double,* v. — *de doublement;*
— *de doublement,* circulation trench, (trench parallel to — *de tir,* and some 30-40 m. behind it); cover trench;
élément de ——, (a short trench out in front of and parallel to the first line trench, and communicating with it by a ditch or trench); trench element;
fausse ——, dummy trench;
— *forée,* bored trench (bored, then blasted out); blasted trench;
— *à gabions,* flying sap;
guerre de ——, trench warfare;
lacis des ——*s,* network of trenches;
— *en peigne,* (*mil.*) comb-like trench (due to the existence of embrasures);
— *de première ligne,* first line trench;
— *à projectiles,* (trench art.) projectile trench (open magazine);
— *de repli,* refuge trench;
retourner une ——, to turn a captured trench against its original occupants; to reverse a trench;

tranchée *sous bois,* trench in a wood;
— *de soutien,* support trench (150-200 m. behind first line trench);
— *de support,* bombing trench;
— *de surveillance,* lookout trench (in advance of the *tranchée de tir*); if there is no lookout trench the lookout is stationed in the first-line trench, for that reason sometimes called —— *de surveillance;* inaccurate);
— *de tir,* fire trench, first-line trench;
— *de tir fermée,* closed fire trench (i. e., closed on itself so as to form a sort of inclosed work);
— *pour tireur couché,* lying trench;
— *pour tireur debout,* standing trench;
— *pour tireur à genou,* kneeling trench.
transbordement, m., (*r. r.*) change of cars; (*in gen.*) transfer (from one vehicle to another).
transfert, m., transfer.
transformateur, m., transformer; (*wireless*) jigger;
— *à haute fréquence,* high, frequency transformer, oscillation transformer, air-core transformer;
— *de vitesse,* (*mach. auto.*) gear shift, gearshift lever.
transformation, f., (*elec., etc.*) transformation;
rapport de ——, (*elec.*) ratio of transformation.
transmission, f., transmission; (*mach.*) shaft;
— *par chaines* (*mach.*) chain transmission;
— *flexible,* flexible shaft.
transport, m., (*nav.*) troop ship;
— *automobile,* (*mil.*) transportation by motor car;
— *en commun,* public transportation of passengers;
— *à dos d'animaux,* pack transportation;
— *à dos d'homme,* carrier transportation;
gros ——, heavy transportation, transportation of heavy baggage, stores, etc.;
— *sur roues,* wagon transportation, wheeled transportation.
transporteur, m., (*art.*) crane.
transvasement, m., (*mach.*) gas return (intentional before closing of admission valve).
transversale, f., (*mil.*) a synonym of *parallèle.*
travail, work, labor;
— *aux forcés à perpétuité,* (*law*) hard labor for life;
— *aux forcés à temps,* (*law*) hard labor for a limited time;
— *militaire,* (*mil.*) mounted drill (Saumur);
— *sur deux pistes,* (*man.*) two track work;
— *sur une piste,* (*man.*) single track work.
travailler, v. t., to work;
— *sur deux pistes,* (*man.*) to work on two tracks;
— *sur une piste,* (*man.*) to work on one track.
travers, m., breadth;
de ——, (*sm. a.*) of a bullet, keyhole wise (mark on target).
traverse, f., crosspiece, crossbar; (*r. r.*) sleeper;
— *de fourche,* (*mach. gun.*) fork transom;
— *à pivot,* (*mach. gun.*) top carriage support (pack);
— *tournante,* (*fort.*) cube traverse, island traverse.
traverse-blockhaus, m., (*mil.*) sort of defensive traverse.
traversée, f., (*mil.*) crossing, passage, of a river.
traversée-jonction, f., (*r. r.*) crossing (two tracks);
— *double,* double crossing (switch to either track in both directions);
— *simple,* single crossing (crossover possible from one direction only).
trébucher, v. n., to trip.
treillage, m., trellis;
— *de cage à lapin,* chicken wire netting;
— *métallique,* wire fencing.

3877°—17——37

trembleur, m., (*elec.*) make-and-break, contact breaker; buzzer, trembler; (*wireless*) ticker;
—— *magnétique*, electric buzzer (usually and better called *vibreur*);
—— *mécanique*, mechanical (or true) trembler.
treuil, m., winch;
—— *de rappel*, (*aero.*) lowering winch (of a train of kites).
tri, m., sorting (*adm.*, of the mail).
triage, m., sorting.
triangulation, f., triangulation; (*telephone*) tripling of lines.
tribunal, m., court;
—— *prévôtal*, (*mil.*) provost marshal's court.
tricar, m., (*auto.*) generic term for three-wheeled vehicles.
tricot, m., sweater.
tricycle, m., tricycle;
—— *à pétrole*, motor tricycle;
—— *à vapeur*, steam tricycle.
trier, v. t., to sort.
tringle, m., rod; f., rod, stem; (*mach. gun.*) feed bar;
—— *d'armement*, (*art.*) cocking link;
—— *de manœuvre*, (*r. r.*) switch bar, switch rod;
—— *de mise de feu*, (*art.*) firing stem;
—— *de tirage*, (*mach.*) tension link.
tringlot, m., (*mil. slang*) soldier or officer of the *train des équipages*.
trinitrotoluène, m., (*expl.*) trinitrotoluol, T. N. T.
trinitrotoluol, m., (*expl.*) trinitrotoluol.
triphasé, a., (*elec.*) three-phase.
triplace, m., (*aero.*) three-seater.
triplan, m., (*aero.*) triplane.
triplice, f., *la* ——, the Triple Alliance.
tromblon, m., (*sm. a.*) blunderbuss; (*mil.*) (cylinder fitted to the muzzle of a rifle, to take the rifle grenade) grenade sleeve, discharger;
—— *V. B.*, the "discharger" of the V. B. grenade.
trombionnier, m., (*Fr. a.*) grenadier (works the *tromblon*, q. v.).
trommelfeuer, m., drum fire (German word).
trompe, f., (*auto.*) horn.
trompette, f., trumpet; m., trumpeter;
—— *d'appel*, (*elec.*) buzzer.
trotteuse, f., split second hand;
—— *compte-secondes*, split second hand.
trou, m., hole; (*mil. slang*) "soft snap" (U. S.);
—— *d'air*, (*aero.*) air pocket;
—— *de graissage*, v. —— *graisseur;*
—— *graisseur*, (*mach.*) grease, lubricating, hole;
—— *de percussion*, (*art.*) percussion vent;
—— *purgeur*, (*art. mach.*) drain hole;
—— *de sentinelles*, sentinel pit;
—— *de tirailleurs* (*mil.*) individual rifle pits.
troufion, m., (*mil. slang*) private.
troupe, f., troop; (*mil.*) troops;
—— *s allégées*, (*mil.*) infantry, a part of whose pack or equipment is carried in the wagons;
—— *s coloniales*, (*Fr. a.*) colonial troops (belonging to the *armée coloniale*);
—— *s métropolitaines*, (*Fr. a.*) home troops (belonging to the *armée métropolitaine*);
—— *s non-allégées*, (*mil.*) infantry carrying the full pack;
—— *s passagères*, troops in transit.
trousse, f., (*mil.*) medical pouch; (*sm. a.*) bandolier;
—— *coupante*, well case, or casing (for use in loose soils);
—— *d'outils*, tool kit, set of tools;
—— *de soldat* (*mil.*) housewife.
trousseau, m., outfit of clothes.
truc-boggie, m., bogy truck.
trucage, m., special swindle, consisting of the manufacture of antiques; (hence, more generally) "faking"; (*mil.*), quaker work, dummy work.

truquage, m., v. *trucage.*
truquer, v. t., to manufacture "antiques"; (more generally) to "fake."
truck, m., truck (borrowed by the French).
tube, m., tube; (*art.*) (the) gun (itself as distinguished from the carriage);
—— *allonge*, m., v. *tube-rallonge;*
—— *d'admission*, (*mach.*) inlet pipe;
—— *d'amorçage*, (*art.*) priming tube;
—— *de brêlage*, v. —— *de pontage;*
—— *de canalisation*, (*mach.*) pipe;
—— *chargeoir*, (*mil. min.*) loading tube (bored mine);
—— *collecteur*, (*mach.*) collecting tube (of a radiator);
—— *de direction*, (*auto.*) steering column, steering pillar;
—— *d'échappement*, (*mach.*) exhaust pipe;
—— *d'échappement de vapeur*, (*mach. gun*) steam escape pipe;
—— *épontille*, (*pont.*) tube stanchion;
—— *d'évacuation*, (*mach.*) exhaust tube;
—— *de gaz comprimé*, gas cylinder;
—— *lance-bombes*, (*aero.*) launching tube;
—— *lance-fusée*, (*artif.*) rocket trough;
—— *de lancement*, (*Fr. art.*) barrel of a pneumatic howitzer;
—— *à limaille*, (*wireless*) filings-tube coherer;
—— *de niveau*, glass gauge;
—— *de Pitot*, Pitot tube;
—— *plongeur*, siphon (-bottle) tube.
—— *de pontage*, (*pont.*) lashing tube (metallic ponton);
—— *porte-amorce*, (*art.*) primer tube;
—— *porte-dépêche*, (*mil.*) message tube (carrier pigeon);
—— *porte-diaphragme*, (*inst.*) diaphragm slide;
—— *porte-lentille*, (*inst.*) lens holder, slide;
—— *porte-miroir*, (*inst.*) mirror holder, slide;
—— *de poussée*, (*mach. gun*) pressure reduction tube;
—— *de réglage*, (*inst.*) focusing slide;
—— *solaire*, (*inst.*) heliostat;
—— *de sortie*, (*mach.*) outlet tube or pipe;
système à recul de ——, (*art.*) tube-recoil system, independent recoil system;
—— *trompette*, (*mach.*) casing (surrounding certain parts of the rear axle, auto.);
—— *vis*, m., tubular screw joint.
tubulure, f., tubulure;
—— *d'admission*, (*mach.*) admission pipe, joint, neck;
—— *d'échappement*, (*mach.*) escape pipe, joint, neck.
tue-boches, m., (*mil. slang*) the bayonet.
tunique-blouse, f., (*unif.*) (coat having properties of each of elements indicated) loose-fitting tunic.
turban, m., (*unif.*) crown of forage cap.
turbine, f., (*mach.*) turbine;
—— *de basse pression*, low-pressure turbine;
—— *de croisière* (*pour vitesse réduite*), cruising turbine;
—— *à gaz*, gas turbine;
—— *de marche arrière*, reversing turbine.
turbo-alternateur, m., (*elec.*) turbo-alternator.
turbo-interrupteur, m., (*elec.*) turbo-interrupter, turbine interrupter.
turbo-pompe, f., turbine pump.
turbo-ventilateur, m., turbine blowing engine.
tuyau, m., tube, pipe;
—— *à ailettes*, (*mach.*) flanged tube or pipe (as of a radiator);
—— *d'arrivée*, (*mach.*) admission pipe, inlet;
—— *de cheminée*, chimney pot;
—— *en fonte*, cast-iron pipe;
—— *flexible* (*de circulation d'eau*) hose;
—— *de graissage*, lubricating tube;
—— *en grès*, earthenware pipe;
—— *de poêle*, (*fam., trench art.*) "stove pipe", mortar bomb (trench mortar).

tuyauterie   579   **viande**

tuyauterie, f., piping;
—— *d'aspiration*, (*mach.*) inlet piping;
—— *d'échappement*, (*mach.*) exhaust piping.
type, m., type;
—— *alpin*, (*Fr. a.*) (of machine-gun companies) mountain type; (pack transport);
—— *colonial*, (*Fr. a.*) colonial type (of organizations);

type *de France*, (*Fr. a.*) metropolitan type (of organizations, as distinguished from colonial);
—— *hors de France*, (*Fr. a.*) foreign service type (of organizations), colonial type;
—— *mixte*, (*Fr. a.*) combined type (pack and wheel transport);
—— *mixte-type alpin*, (*Fr. a.*) combined alpine or mountain type.

U.

ulan, m., (*mil.*) uhlan (usually spelled *uhlan*).
unité, f., unit, unity; (*mil.*) troop unit;
—— *s d'assaut*, m., (*mil.*) troops told off for assault;
—— *s d'attaque*, (*mil.*) troops told off for attack;
—— *d'origine*, unity of origin, with special reference to (*Fr. a.*) the effort made to get all officers from one and same source;

unité *réservée*, (*mil.*) reserve body of troops, to follow up an attack.
urinoir, m., urinal.
usure, f., wear;
lutte d'——, (*mil.*) (term applied to the war 1914- to describe its wearing-out character) wearing-down conflict, attrition contest.

V.

v, (letter);
—— *longitudinal*, (*aero.*) longitudinal diedral angle;
—— *transversal*, (*aero.*) diedral angle.
vagon-lit, m., (*r. r.*) sleeping car.
vague, f., wave; (*mil.*) wave of gas (gas attack); (*mil.*) line of assaulting troops, wave;
—— *d'assaut*, (*mil.*) assaulting wave;
—— *de nettoyage*, (*mil.*) wave following the assaulting wave, and charged with the duty of "mopping up" the captured trenches; wave or line of trench cleaners, trench "moppers."
vaincre, v. t., to conquer, overcome;
—— *le poids mort*, (*mach.*) to overcome dead centers.
valise, f., valise, bag;
—— *diplomatique*, embassy dispatch bag.
valve, f., (*mach.*) valve;
fausse ——, false valve (used in inserting an air chamber, pneumatic tire).
valvoline, f., (*mach.*) a generic term for cylinder oils.
vanne, f., valve;
—— *d'étranglement*, (*mach.*) (sort of) throttling valve.
vannerie, f., side baskets (of a motor car).
vapeur, m., (*nav.*) steamer;
—— *câblier*, cable ship;
—— *à turbine*, turbine steamer.
vapeur, f., steam;
—— *surchauffée*, superheated steam.
vareuse, f., (*unif.*) blouse;
—— *à capuchon*, (*unif.*) cyclist's hooded jacket (*Fr. a.*).
variole, f., smallpox; pigeon pox (pigeons).
vase, m., vessel;
—— *à niveau constant*, (*mach.*) float chamber.
vedette, f., *vedette*, (*aero.*) French military dirigible cubing 2,000–4,000 meters (obs.).
véhicule, m., vehicle;
—— *automoteur*, (*auto.*) automobile;
—— *à boggie moteur*, (*auto.*) motor truck;
—— *hippomobile*, horse-drawn vehicle;
—— *de transport*, generic term for omnibuses, autobuses, etc.
vélo, m., (*fam.*) bicycling.
vélocimètre, m., (any device to measure speed); speedometer.
venir, v. n., to come;
—— *d'une seule pièce avec*, to be in one piece with, to be one with, integral with.
vent, m., wind;
—— *dominant*, prevailing wind.

ventilateur, m., (*mach.*) fan;
—— *hélicoïde*, spiral fan;
—— *en nids d'abeilles*, (*auto.*) honeycomb radiator;
—— *de remplissage*, (*aero.*) filling sleeve, air intake;
—— *à tubes*, tubular radiator.
ventilation, f., ventilation;
—— *par refoulement*, ventilation by driving out noxious gases.
ventre, m., belly;
faire ——, (*mil. fam.*) to belly out, to bulge (of poor revetments).
vérificateur, m., gauge;
—— *de chute*, (*ball.*) drop gauge.
vérification, (*elec.*) testing, test.
vérifier, v. a., to verify, examine; (*elec.*) to test;
—— *la cible*, v. s. v. *cible*.
vérin, m., screwjack;
—— *à galet*, (*art.*) roller jack.
vernis, m., varnish;
—— *à ballon*, (*aero.*) balloon varnish.
verre, m., glass;
—— *armé*, (*cons.*) wire glass;
—— *protecteur*, (protective) screen (projecting lantern).
verrou, m., bolt;
—— *d'agrafage*, (trench art.) holding-down bolt;
—— *d'assemblage*, (*mach. gun*) assembling bolt;
—— *de crémaillère*, (*art.*) locking rack bolt;
—— *de culasse*, (*art.*) locking bolt (280 Schneider);
—— *de déclenchement*, (*mach.*) releasing, disengaging bolt;
—— *d'entraînement*, (*mach.*) feather, dog, key;
—— *de fermeture*, (*mach. gun*) latch;
—— *de mise de feu*, (*art.*) firing bolt (280 Schneider);
—— *à ressort*, (*art.*) spring firing bolt;
—— *de sécurité*, (*art.*) safety bolt;
—— *du volet*, (*art.*) primer bushing fork.
verrouillage, m., locking device, block system; (*mach. gun*) bolt engagement;
—— *de route*, (*art.*) road lock (120 C. S., secures carriage to axle).
Versaillaise, f., (*Fr. a.*) mutual benefit society composed of graduates of Versailles.
versé, a., (*com.*) paid up.
verser, v. t., to turn in; to overturn; (*com.*) to subscribe, pay in funds; (*mil.*) to transfer, assign (as, so many men to such a battalion).
veston, m., sack coat; (*unif.*) short overcoat.
viande, f., meat;
—— *abattue*, (*mil.*) dressed meat (as distinguished from meat on the hoof);
—— *de conserve*, preserved meat;
—— *fumée*, smoked meat;
—— *marinée*, pickled meat;
—— *sur pied*, meat on the hoof.

vibrateur, m., (*elec.*) vibrator.
vibreur, m., (*elec.*) magnetic make-and-break key, buzzer; contact breaker.
vide, a., empty; unloaded;
— à ——, (*r. r., mach., etc.*) unloaded.
vide-Boches, m., (*mil. slang*) the bayonet.
vieux, a., old;
— le ——, (*mil. slang*) the captain [cf. U. S. A. old man=c. o.].
vif, a., (of an angle) sharp.
ville, f., town;
— *forte*, (*mil.*) fortified town.
Vincennoise, f., (*Fr. a.*) mutual benefit society made up of graduates of Vincennes.
violon, m., fiddle; (*mil. slang*) bedpan.
virage, m., (*aero., auto.*) turn;
— *incliné*, (*aero.*) banking;
— *incliné à droite* (*gauche*), (*auto.*) right (left) bank;
— *de route*, turn of the road.
virer, v. n., (*aero., auto.*) to turn.
virgule, f., comma.
vis, f., screw;
— *-axe*, (*trench art.*) screw axle;
— *-bouchon*, (*trench art.*) screw plug;
— *de blocage*, set screw;
— *de compensation*, (*mach.*) compensating screw, take-up screw;
— *à crémaillère*, (*art.*) elevating screw;
— *-frein*, locking screw, set screw;
— *goujon*, f., screw bolt;
— *de jonction*, assembling screw;
— *à pas rapide*, (*mach.*) quick-motion screw;
— *platinée*, (*élec.*) platinum point, platinum screw;
— *-pointeau*, (*art.*) set screw;
— *de pression*, set screw;
— *de pointage en direction*, (*art.*) horizontal aiming screw; direction screw;
— *de purge*, (*mach.*) drain cock;
— *de toc*, (*mach.*) limiting screw, stop screw.
viseur, m., sight hole; window, inspection opening, (*mil.*) sight;
— *optique*, telescopic sight;
— *périscopique*, periscopic sight.
visite-conférence, f., (*Fr. a.*) visit of officers of the garrison of Paris to *Musée de l'armée*, preceded by lecture on use of museum in instructing the men of the garrison.
vis-pivot, f., pivot screw.
vis-pointeau, f., male center
visser, v. t., to screw;
— *à force*, to screw home.
vitesse, f., speed;
— *commerciale*, (*r. r.*) average speed (includes stops, etc.);
— *conserver sa* ——, (*art.*) (of a projectile) to hold up well;
— *critique*, (*aero.*) stall;
— *de déplacement*, rate of travel;
— *limite*, (*aero.*) terminal speed;
— *de marche*, (*r. r.*) actual running speed;
— *de montée*, (*r. r.*) climbing speed;
— *perdre la* ——, (*aero.*) to stall;
— *perte de* ——, v. s. v. *perte*;
— *par rapport à l'air*, (*aero.*) air speed;
— *par rapport au sol*, (*aero.*) ground speed;
— *de régime*, (*aero.*) (speed of a balloon, when the upward force is balanced by the resistance of the air) normal speed, working speed.
vitre, f., windowglass;
— *anti-buée*, (*mil.*) glass of antimist composition for gas masks.
vivre, f., food, supplies (usually in pl.);
— *-s de chemin de fer*, (*mil. r. r.*) travel ration.
voie, f., road; (*r. r.*) track;
— *de chargement*, (*trench art.*) loading track;
— *de débord*, (*r. r.*) track alongside a road;
— *encaissée*, road in a cut;
— *normale*, (*r. r.*) standard gauge;
— *de tir*, (*r. r. art.*) firing track.

voilure, f., (*aero.*) aero foil.
voirie, f., roads, communications;
— *grande* ——, generic term for highroads (national and departmental navigable and floatable rivers, canals, and railways);
— *petite* ——, communal roads;
— *urbaine*, generic term for streets.
voiture, f., carriage, vehicle;
— *-affût*, (*trench art.*) carriage truck (transportation); (*r. r. art.*) carriage truck;
— *aux agrès*, (*Fr. a., aero.*) kite equipment truck; (*in gen.*) equipment truck;
— *d'ambulance*, (*med.*) ambulance (i. e., vehicle);
— *aux armements*, (*trench art.*) equipment truck;
— *attelée à quatre*, etc., four, etc., horse carriage;
— *automobile*, motor car; motor truck;
— *à bagages*, (*mil. aero.*) baggage truck;
— *à ballon démontable*, (*auto.*) car with removable hood;
— *-berceau*, (*art.*) cradle truck (transportation);
— *à blessés*, (*mil.*) wagon for wounded;
— *bureau-téléphonique*, (*Fr. a., aero.*) telephone truck;
— *-caisson*, (*trench art.*) caisson-truck (transportation);
— *canon*, (*art.*) gun-truck (transportation);
— *capitonnée*, padded van, furniture van;
— *à capote*, (*auto.*) auto with top;
— *à cardan*, (*auto.*) carriage driven by cardanic transmission;
— *-cartes*, f., (*mil.*) map-wagon (carries the material of a topographic or map section);
— *à chaîne(s)*, (*auto.*) chain-driven vehicle;
— *de charge*, freight wagon;
— *de chemin de fer*, (*r. r.*) railway carriage, car;
— *à cheval*, *à chevaux*, any horse-drawn vehicle;
— *-colombier*, f., (*mil.*) pigeon wagon;
— *avec compartiments-couchettes*, (*r. r.*) compartment sleeping car;
— *à couloir*, (*r. r.*) corridor carriage;
— *de course*, (*auto.*) racing car;
— *couverte*, (*auto.*) covered, closed car;
— *à dais*, (*auto.*) half-closed car;
— *demi-fermée*, v., —— *à dais*;
— *découverte*, (*auto.*) open car;
— *de déménagement*, moving wagon, furniture wagon;
— *directe*, (*r. r.*) through carriage, car;
— *d'eau*, (administrative term for any) passenger boat;
— *d'éclairage*, (*mil.*) dynamo wagon (German);
— *électrique*, electromobile;
— *électrogène*, (*mil. wireless*) dynamo wagon;
— *à fauteuil-lit*, (*r. r.*) parlor sleeping car;
— *filtrante*, (*mil.*) filter wagon or cart (for troops in the field);
— *à foyer*, (*mil.*) the cook wagon of a traveling kitchen;
— *hippomobile*, horse-drawn vehicle;
— *à intercirculation*, (*r. r.*) vestibuled carriage;
— *légère à munitions d'infanterie*, (*Fr. a.*) two-horse ammunition wagon;
— *à lits-salons*, (*r. r.*) parlor sleeping car;
— *de livraison*, delivery wagon;
— *avec lits-toilette* (*r. r.*) sleeping car;
— *mixte*, motor car (transmitting motor energy directly, and using electric energy for starts and hill climbing);
— *-mortier* (*trench art.*) mortar-truck (transportation);
— *à moteur animé*, any vehicle drawn by an animal;
— *à munitions*, (*mil.*) ammunition wagon;
— *d'occasion*, (*adm.*) any hired land vehicle;
— *pétroléo-électrique*, (*auto.*) any motor car using hydrocarbon fuel to develop electric energy;
— *photographique*, (*Fr. a., aero.*) photograph truck;
— *-pièce*, f., (*r. r. art.*) gun truck;
— *plateforme*, (*art.*) platform truck (transportation);
— *de poids lourd*, (*auto.*) truck, van;
— *porte-affût*, (*trench art.*) carriage truck (transportation);

voiture *porte-bombes*, (*trench art.*) bomb-truck (transportation);
— *porte-canon*, (*art.*) gun truck (i. e., for transporting heavier calibers);
— *porte-mitrailleuse*, (*mil.*) machine-gun cart;
— *porte-mortier*, (*trench art.*) mortar-truck (transportation);
— *porte-plateforme*, (*trench art.*) platform-truck transportation);
— *publique*, any vehicle for hire (whether rail, water, or ordinary road);
— *-pylône*, f., (*mil.*) pylone wagon (German: for elevating a searchlight);
— *radiotélégraphique*, (*wireless*, *mil.*) radiotelegraph wagon;
— *sanitaire*, (*mil. med.*) generic term for vehicles of the sanitary service;
— *-sellette*, (*trench art.*) bolster truck (transportation);
— *à service régulier*, any public land vehicle (other than rail) observing a regular schedule;
— *stérilisatrice*, sterilizing cart (for drinking water);
— *téléphonique-bureau*, (*mil. aero.*) telephone truck;
— *tender*, (*Fr. a.*, *aero.*) tender (camp equipage);
— *de terre*, administrative term for any public land vehicle other than rail;
— *de tourisme*, (*auto.*) touring car;
— *à transmission électrique*, (*auto.*) motor car all of whose motor energy is converted into electric energy;
— *treuil*, (*mil. aero.*) winch truck ;
— *à tubes*, (*Fr. a.*, *aero.*) tube wagon, gas wagon;
— *à vapeur*, (*auto.*) steam car;
— *à viandes*, (*mil.*) meat wagon;
— *de ville*, (*auto.*) town car (motor car for town use as distinguished from touring);
— *à vivres*, (*mil. aero.*) ration truck;
— *à volonté*, (*adm.*) generic term for cabs.

voiture-atelier, f., (*auto.*, *mil.*) repair truck.

voiture-cantine, f., (*mil.*) canteen wagon.

voiture-citerne, f., (*mil.*) (drinking) water cart.

voiture-cuisine, f., (*mil.*) traveling kitchen.

voiture-observatoire, f., (*art.*) wheeled observatory.

voiture-remorque, f., (*auto.*) trailer.

voiture-tube, f., (*aero.*, *mil.*) gas wagon.

voiturette, f., small carriage; (*mach. gun*) gun and ammunition cart (mod. 1907, T); (*mil.*) small carriage serving as limber for the 37-mm. T. R.
— *porte-bombes*, (*art.*) shell cart;
— *porte-mitrailleuse*, (*mach. gun*) gun cart (transportation);
— *porte-mortier*, (*art.*) (trench) mortar cart;
— *porte-munitions*, (*mach. gun*) ammunition cart.
— *remorque*, f., (*auto.*) trailer.

voix, f., voice, vote;
— *prépondérante*, deciding vote.

vol, m., (*aero.*, etc.) flight;
— *cabré*, (*aero.*) cabré, tail down;
— *piqué*, (*aero.*) sharp dive;
— *plané*, (*aero.*) volplane, glide.
— *ramé*, bird flight.

volant, m., handwheel; (*mach.*) flywheel; (*auto. aero.*) steering wheel; (*mil.*) intermediate dump;
— *de blocage*, (trench art.) locking handwheel (150 T. mortar);
— *extérieur*, (*mach.*) exterior flywheel;
— *intérieur*, (*mach.*) internal flywheel (name sometimes given to a crank disk);
— *de pointage en hauteur*, (*art.*) elevating handwheel;
— *de réglage*, (*trench art.*) chamber regulator (V. D. mortar);
— *ventilateur*, (*mach.*) flywheel fan.

volateur, a., (*aero.*) flying.

volatile, m., bird.

volée, f., flight;
à la —, at speed, by the run.

voler, v. n., (*sm. a.*) to sail (said of some bullets); (*aero.*) to make a flight.

volet, m., (*mach.*) shutter (sort of valve); (*elec.*) oscillating sleeve; (*r. r.*, *art.*) (hinged) bracket (side-support in firing position);
— *d'attelage*, (*art.*) assembling plate;
— *de carburateur*, (*mach.*) butterfly valve;
— *de culasse*, (*trench art.*) breech closing plate (150 T. mortar);
— *de déchirure*, (*aero.*) ripping strip or panel;
— *guide cartouche*, (*mach. gun*) cartridge guide;
— *de magnéto*, (*elec.*) rotating sleeve, make and break;
— *de profondeur*, (*aero.*) elevator;
— *stabilisateur*, (*aero.*) stabilizer;
— *de tir*, (*r. r. art.*) firing bracket (side supports in firing position of certain heavy pieces).
— *tournant*, (*elec.*) revolving sleeve (of a stationary armature magneto).

voltigeur, m., (*Fr. a.*) (to-day, an all-around member of a grenadier squad, assists wherever needed, must also know how to throw a grenade) rifleman.

volume, m., volume;
— *du feu*, (*art.*, *inf.*, etc.) volume of fire.

voyant, a., (of a uniform) conspicuous.

voyer, m., roadmaster;
agent —, roadmaster.

vrille, f., (*aero.*) tail spin.

vulcanisateur, m., (apparatus to produce local vulcanization in repairing pneumatic tires); vulcanizer.

vulcanisation, f., vulcanization.

vulcaniser, v. t., to vulcanize.

W.

wagon, m., wagon; (*r. r.*) car, carriage;
— *ambulant*, (*r. r.*) mail carriage;
— *aménagé*, (*mil.*) car prepared to take on men;
— *cellulaire*, (*Fr. adm.*, *r. r.*) prison van;
— *complet*, (*r. r.*) carload, full carload;
— *à couloir* (*r. r.*) corridor car;
— *à munitions*, (*r. r. art.*) shell car;
— *ordinaire*, (*mil.*) ordinary car;
— *plombé*, (*r. r.*) sealed car;

wagon *royal*, (*r. r.*) royal carriage.
— *vide*, (*r. r.*) idler, empty.

wagonnet, m., truck;
— *de chargement*, (*art.*) shell truck.

wagon-poste, m., (*r. r.*) mail carriage.

wagon-salon, m., (*r. r.*) parlor car, drawing-room car.

westrumite, f., coal-tar preparation for laying road dust.

Z.

zèbre, m., zebra; (*mil. slang*) horse.

zef, m., (*mil. slang*) short for *zéphyr*, q. v.

zéro, m., zero;
erreur du —, (*inst.*) index error.

zig, m., (*mil. slang*) good, shrewd, fellow.

zigomar, m., (*mil. slang*) cavalry saber.

zigouiller, v. t., (*mil. slang*) to kill, to strangle.

zigue, m., (*mil. slang*) v. zig.

zigzaguement, m., zigzagging (*mil.*) of trenches.

zinc, m., zinc; (*aero.*, *slang*) bus (i. e., airplane).

zone, f., zone; (*mil.*) zone, area, region;
—— *de l'armée,* (*mil.*) army zone (practically theater of operations, including supply system).
—— *de l'avant,* (*mil.*) zone of the front, of activities;
—— *de friction,* (*mil.*) zone of activity (i. e., of actual contact between hostile troops); No Man's Land.
—— *frontière,* frontier customs zone (20,000 m. wide; France);
—— *objectif,* (*art.*) the zone taken under fire in concentration fire;

zone *de protection,* (*mil., fort.*) (that part of a defiladed zone in which the trajectory is above the target behind the cover) protected zone;
—— *rasée,* (*mil.*) (for a target of given height, terrain over which the trajectory will not rise higher than the target) danger zone;
—— *de résistance,* (*mil.*) (zone surrounding a *secteur,* q. v., and composed of the *ligne de surveillance, première ligne,* and *ligne de soutien,* qq. v.); zone of resistance.

zouzou, m., (*mil. slang*) zouave.

zyeuter, v. t., (*mil. slang*) to see (*yeux,* eyes).

O. J. S.

www.ingramcontent.com/pod-product-compliance
Lightning Source LLC
Chambersburg PA
CBHW020629230426
43665CB00008B/100